Fourth Edition

Handbook of Parametric and Nonparametric Statistical Procedures

Fourth Edition

Handbook of Parametric and Nonparametric Statistical Procedures

David J. Sheskin

Chapman & Hall/CRC
Taylor & Francis Group
Boca Raton London New York

Chapman & Hall/CRC is an imprint of the
Taylor & Francis Group, an informa business

Chapman & Hall/CRC
Taylor & Francis Group
6000 Broken Sound Parkway NW, Suite 300
Boca Raton, FL 33487-2742

International Standard Book Number-10: 1-58488-814-8 (Hardcover)
International Standard Book Number-13: 978-1-58488-814-7 (Hardcover)

Library of Congress Cataloging-in-Publication Data

Sheskin, David.
    Handbook of parametric and nonparametric statistical procedures / David J. Sheskin. -- 4th ed.
      p. cm.
    Includes bibliographical references and index.
    ISBN-13: 978-1-58488-814-7 (alk. paper)
    ISBN-10: 1-58488-814-8 (alk. paper)
    1. Mathematical statistics--Handbooks, manuals, etc. I. Title.

QA276.25.S54 2007
519.5--dc22
                                            2006029391

Visit the Taylor & Francis Web site at
http://www.taylorandfrancis.com

and the CRC Press Web site at
http://www.crcpress.com

To Vicki and Emily

&

Topspin, Buffy, and Belly Button
my loyal writing companions over the years

Preface

Like the first three editions, the fourth edition of the **Handbook of Parametric and Nonparametric Statistical Procedures** is designed to provide researchers, teachers, and students with a comprehensive reference in the areas of parametric and nonparametric statistics. The addition of material not included in the first three editions (most notably multivariate analyses and material on medical statistics) makes the Handbook unparalleled in terms of its coverage of the field of statistics. Rather than being directed at a limited audience, the Handbook is intended for individuals who are involved in a broad spectrum of academic disciplines encompassing the fields of mathematics/statistics, the social, biological, and environmental sciences, business, and education. My original purpose in writing the Handbook was to provide consumers of statistics with a reference book which I (as well as colleagues and students I had spoken with over the years) had always wanted, yet could never find. To be more specific, my primary goal was to produce a comprehensive reference which covered a scope of material that extended far beyond that which was covered in any single source. It was essential the book be applications oriented, yet at the same time that it address relevant theoretical and practical issues which are of concern to the sophisticated researcher. In addition, I wanted to write a book which is accessible to people who have little or no knowledge of statistics, as well as those who are well versed in the subject. I believe I have achieved these goals, and on the basis of this I believe that the **Handbook of Parametric and Nonparametric Statistical Procedures** will continue to serve as an invaluable resource for people in multiple academic disciplines who conduct research, are involved in teaching, or are presently in the process of learning statistics.

I am not aware of any applications-oriented book which provides in-depth coverage of as many statistical procedures (specifically, more than 160) as the number that are covered in the **Handbook of Parametric and Nonparametric Statistical Procedures**. Inspection of the **Table of Contents** and **Index** should confirm the scope of material covered in the book. A unique feature of the Handbook, which distinguishes it from other reference books on statistics, is that it provides the reader with a practical guide which emphasizes application over theory. Although the book will be of practical value to statistically sophisticated individuals who are involved in research, it is also accessible to those who lack the theoretical and/or mathematical background required for understanding the material documented in more conventional statistics reference books. Since a major goal of the book is to serve as a practical guide, emphasis is placed on decision making with respect to which test is most appropriate to employ in evaluating a specific design. Within the framework of being user-friendly, clear computational guidelines, accompanied by easy-to-understand examples, are provided for all procedures.

One should not, however, get the impression that the **Handbook of Parametric and Nonparametric Statistical Procedures** is little more than a cookbook. In point of fact, the design of the Handbook is such that within the framework of each of the statistical procedures which are covered, in addition to the basic guidelines for decision making and computation, substantial in-depth discussion is devoted to a broad spectrum of practical and theoretical issues, many of which are not discussed in conventional statistics books. Inclusion of the latter material ensures that the Handbook will serve as an invaluable resource for those who are sophisticated as well as unsophisticated in statistics.

It should be noted that although a major goal of this book is to provide the reader with clear, easy to follow guidelines for conducting statistical analyses, it is essential to keep in mind that the statistical procedures contained within it are essentially little more than algorithms which have been derived for evaluating a set of data under certain conditions. A statistical procedure, in and of itself, is incapable of making judgements with respect to the adequacy of the methodology underlying an experiment and/or the reliability of data being evaluated. If either of the latter is compromised (as a result of a faulty experimental design, sloppy methodology, use of inappropriate and/or unreliable measuring instruments, and/or salient violation of assumptions underlying a specific statistical procedure), for all practical purposes, the result of an analysis will be worthless. Consequently, it cannot be emphasized too strongly, that a prerequisite for the intelligent and responsible use of statistics is that one has a reasonable understanding of the conceptual basis behind an analysis, as well as a realization that the use of a statistical procedure merely represents the final stage in a sequential process involved in conducting research. Those stages which precede the statistical analysis are comprised of whatever it is a researcher does within the framework of designing and executing a study. Ignorance or sloppiness on the part of a researcher with respect to the latter can render any subsequent statistical analysis meaningless.

Although a number of different hypothesis testing models are discussed in this Handbook, the primary model that is emphasized is the **classical hypothesis testing model** (i.e., the **null hypothesis significance testing model**). Although the latter model (which has been subjected to criticism by proponents of alternative approaches to hypothesis testing) has its limitations, the author believes the present scope of scientific knowledge would not be greater than it is now if up to this point in time any of the alternative hypothesis testing models had been used in its place. Throughout the book, in employing the classical hypothesis testing model, the author emphasizes its judicious use, as well as the importance of conducting replication studies whenever research evaluating a hypothesis is equivocal.

In order to facilitate its usage, most of the procedures contained in the Handbook are organized within a standardized format. Specifically, for most of the procedures the following information is provided:

I. Hypothesis Evaluated with Test and Relevant Background Information The first part of this section provides a general statement of the hypothesis evaluated with the test. This is followed by relevant background information on the test such as the following: a) Information regarding the experimental design for which the test is appropriate; b) Any assumptions underlying the test which, if violated, would compromise its reliability; and c) General information on other statistical procedures that are related to the test.

II. Example This section presents a description of an experiment, with an accompanying data set (or in some instances two experiments utilizing the same data set), for which the test will be employed. All examples (with the exception of those employed for multivariate analyses) employ **small sample sizes**, as well as **integer data** consisting of **small numbers**, in order to facilitate the reader's ability to follow the computational procedures to be described in Section IV.

III. Null versus Alternative Hypotheses This section contains both a **symbolic** and **verbal** description of the statistical hypotheses evaluated with the test (i.e., the **null hypothesis** versus the **alternative hypothesis**). It also states the form the data will assume when the null hypothesis is supported, as opposed to when one or more of the possible alternative hypotheses are supported.

IV. Test Computations This section contains a step-by-step description of the procedure for computing the test statistic(s). The computational guidelines are clearly outlined in reference to the data for the example(s) presented in Section II. In the case of multivariate analyses, the statistical software package *SPSS* is employed to evaluate the data.

V. Interpretation of the Test Results This section describes the protocol for evaluating the computed test statistic(s). Specifically: a) It provides clear guidelines for employing the appropriate table(s) of critical values to analyze the test statistic(s); b) Guidelines are provided delineating the relationship between the tabled critical values and when a researcher should retain the null hypothesis, as opposed to when the researcher can conclude that one or more of the possible alternative hypotheses are supported; c) The computed test statistic(s) is (are) interpreted in reference to the example(s) presented in Section II; d) In instances where a parametric and nonparametric test can be used to evaluate the same set of data, the results obtained using both procedures are compared with one another, and the relative power of both tests is discussed in this section and/or in Section VI; and e) In the case of multivariate analyses, detailed guidelines are provided for interpreting the *SPSS* output displayed for an analysis.

VI. Additional Analytical Procedures for the Test and/or Related Tests Since many of the tests described in the Handbook have additional analytical procedures associated with them, such procedures are described in this section. Many of these procedures are commonly employed, while others are used and/or discussed less frequently. Many of the analytical procedures covered in Section VI are not discussed (or, if so, only briefly) in other books. Some representative topics which are covered in Section VI are planned versus unplanned comparison procedures, measures of association for inferential statistical tests, computation of confidence intervals, and computation of power. In addition to the aforementioned material, for many of the tests there is additional discussion of other statistical procedures which are directly related to the test under discussion. In instances where two or more tests produce equivalent results, examples are provided that clearly demonstrate the equivalency of the procedures.

VII. Additional Discussion of the Test Section VII discusses theoretical concepts and issues, as well as practical and procedural issues which are relevant to a specific test. In some instances where a subject is accorded brief coverage in the initial material presented on the test, the reader is alerted to the fact that the subject is discussed in greater depth in Section VII. Many of the topics covered in this section are accorded little or no discussion in other books. Among the topics covered in Section VII is additional discussion of the relationship between a specific test and other tests that are related to it. Section VII also provides bibliographic information on less commonly employed alternative procedures that can be used to evaluate the same design for which the test under discussion is used.

VIII. Additional Examples Illustrating the Use of the Test This section provides descriptions of one or more additional experiments for which a specific test is applicable. For the most part, these examples employ the **same data set** as that in the original example(s) presented in Section II for the test. By virtue of using standardized data for most of the examples, the information for a test contained in Section IV (**Test computations**) and Section V (**Interpretation of the test results**) will be applicable to most of the additional examples. Because of this, the reader is able to focus on common design elements in various experiments which indicate that a given test is appropriate for use with a specific type of design.

IX. Addendum At the conclusion of the discussion of a number of tests an **Addendum** has been included which describes one or more related tests that are not discussed in Section VI. As an example, the **Addendum** of the **between-subjects factorial analysis of variance** contains an overview and computational guidelines for the **factorial analysis of variance for a mixed design**, the **analysis of variance for a Latin-square design**, and the **within-subjects factorial analysis of variance**.

References This section provides the reader with a comprehensive listing of primary and secondary source material for each test.

Endnotes At the conclusion of most tests, a detailed endnotes section contains additional useful information which further clarifies or expands upon material discussed in the main text.

In addition to the **Introduction** and a chapter on **Matrix algebra** (which is new to this edition), the **Handbook of Parametric and Nonparametric Statistical Procedures** contains 40 chapters, each of which documents a specific descriptive or inferential statistical procedure/ test. The **Introduction** provides the reader with a comprehensive overview of descriptive statistics and experimental design, since prior familiarity with the latter material facilitates one's ability to use the book efficiently. Following the **Introduction**, the reader is provided with guidelines and decision tables for selecting the appropriate statistical test for evaluating a specific experimental design. Readers should take note of the fact that the term **test** is employed generically for all procedures described in the book (i.e., both inferential tests and measures of correlation/association).

Approximately 500 pages of new material have been added to the fourth edition. The latter includes material added to Chapters/Tests 1–32 (which comprised the third edition) and Chapters/ Tests 33–40 which are new to this edition. A summary of subject matter added to the fourth edition is listed below.

Introduction: Discussion of the **weighted mean**; Expanded coverage of skewness and kurtosis; Discussion of **longitudinal, cross-sectional**, and **cohort-sequential designs**; Overview of **circular statistics**. One new example and seven figures have been added to the **Introduction**.

Test 2: The single-sample *t* test: Computation of sample size for a test of specified power; Additional discussion of confidence intervals and their value in the decision making process. One new figure has been added to this chapter.

Test 4: The single-sample test for evaluating population skewness: Relationship between the computed test statistic and the value for skewness printed out by *SPSS* and other computer software.

Test 5: The single-sample test for evaluating population kurtosis: Description of the **Jarque–Bera test of normality (Test 5b)**; Relationship between the computed test statistic and the value for kurtosis printed out by *SPSS* and other computer software.

Test 6: The Wilcoxon signed-ranks test: Computation of a confidence interval for a population median.

Test 8: Chi-square goodness-of-fit test: Alternative/exact methods for computing a confidence interval for a proportion/binomially distributed variable; Discussion of survey sampling and computation of confidence intervals for surveys.

Test 9: The binomial sign test for a single sample: Computation of sample size for the **z test for two independent proportions** for test of specified power; Computation of a confidence interval for a Poisson parameter; **Test for comparing two Poisson counts (Test 9c)**; Discussion of the **exponential distribution** and its relationship with the Poisson distribution. Eight new examples and one figure have been added to this chapter.

Test 10: The single-sample runs test (and other tests of randomness): Additional discussion of alternative tests of randomness.

Test 11: The *t* test for two independent samples: Alternative methodology for computation of sample size for a test of specified power; **Extreme Studentized deviate test for identifying outliers (Test 11g)**; Discussion of **clinical trials** and **tests of equivalence (The Westlake–Schuirmann test of equivalence of two independent treatments (Test 11h)** and **Tryon's test of equivalence of two independent treatments (Test 11i))**. Four new examples and nine figures have been added to this chapter.

Test 12: The Mann–Whitney *U* test: Computation of a confidence interval for a difference between the medians of two independent populations; Discussion of **probability of superiority** measure of effect size; Discussion of **survival analysis** (including the **Kaplan–Meier estimate (Test 12d)**); General discussion of **censored data**, and discussion of the following procedures for evaluating **censored data** in a design involving two independent samples: **Permutation test based on the median, Gehan's test for censored data (Test 12e)**, and the **log-rank test (Test 12f))** (and equivalency of the **log-rank test** and the **Mantel–Haenszel test of association (Test 16l-c))**. Six new examples and two figures have been added to this chapter.

Test 16: The chi-square test for *r* × *c* tables: Procedure for computing sample size in reference to the power of the *z* **test for two population proportions**; Correction for continuity for a confidence interval for difference between two proportions; Combining the results of multiple 2 × 2 contingency tables through use of **Test 16l: Mantel–Haenszel analysis (Test 16l-a: Test of homogeneity of multiple odds ratios for Mantel–Haenszel analysis, Test 16l-b: Summary odds ratio for Mantel–Haenszel analysis; Mantel–Haenszel test of association, Test 16l-c)**; Test of equivalence for two independent proportions (**The Westlake–Schuirmann test of equivalence of two independent proportions (Test 16m)**); Discussion of the **log-likelihood ratio (Test 16n)**; Discussion of analysis of multidimensional contingency tables with **log-linear analysis (Test 16o)** through use of *SPSS*. Two new examples and four figures have been added to this chapter.

Test 17: The *t* test for two dependent samples: Alternative methodology for computation of sample size for a test of specified power; **The Westlake–Schuirmann test of equivalence of two dependent treatments (Test 17f) and Tryon's test of equivalence of two dependent treatments (Test 17g)**. One new example and two figures have been added to this chapter.

Test 18: The Wilcoxon matched-pairs signed-ranks test: Computation of a confidence interval for a median difference between two dependent populations; Discussion of **probability of superiority** measure of effect size.

Test 20: The McNemar test: Computation of a confidence interval for the McNemar test; Computation of a confidence interval for the odds ratio for the McNemar test; The effect of disregarding matching in a design configured for the McNemar test; **The Westlake–Schuirmann test of equivalence of two dependent proportions (Test 20b)**. Three new examples and four figures has been added to this chapter.

Test 21: The single-factor between-subjects analysis of variance: Discussion of the **Šidák–Bonferroni correction** and **Fisher–Hayter test** for multiple comparisons; **The Levene test for homogeneity of variance (Test 21g)** and **The Brown–Forsythe test for homogeneity of variance (Test 21h)**; Detailed discussion of **trend analysis**; Discussion of the **general linear model**; Three figures have been added to this chapter.

Test 22: The Kruskal–Wallis one-way analysis of variance by ranks: Additional discussion of multiple comparison procedures.

Test 27: The between-subjects factorial analysis of variance: Analysis of a **crossover design** with a **factorial analysis of variance for a mixed design**, and alternative methods of evaluating a **crossover design**; Discussion of **screening designs**; **Analysis of variance for a Latin square design (Test 27j)**; discussion of analysis of higher-order factorial designs. Four new examples and 11 figures have been added to this chapter.

Test 28: The Pearson product-moment correlation coefficient: Detailed discussion of **regression diagnostics** including the **Durbin–Watson test (Test 28h)**; Ecological correlation; Discussion of **cross-lagged panel** and **regression-discontinuity designs**; Expanded discussion of meta-analysis. One new example and nine figures have been added to this chapter; The material on **multiple regression** in this chapter has been revised and expanded (to include analysis of data with *SPSS*), and comprises a separate chapter (**Test 33: Multiple regression**); The material on **factor analysis** in this chapter has also been revised and expanded (to include analysis of data with *SPSS*), and now comprises a separate chapter (**Test 40: Principal components analysis and factor analysis**).

A new section on **multivariate statistical analysis** encompassing nine chapters has been added to the fourth edition. This section, which describes the analysis of multivariate data with computer software (specifically, *SPSS*) is comprised of the following chapters: **Matrix algebra**; **Test 33: Multiple regression**; **Test 34: Hotelling's T^2** (including **the single-sample Hotelling's T^2 (Test 34a)** and **the use of the single-sample Hotelling's T^2 to evaluate a dependent samples design (Test 34b)**); **Test 35: Multivariate analysis of variance**; **Test 36: Multivariate analysis of covariance**; **Test 37: Discriminant function analysis**; **Test 38: Canonical correlation**; **Test 39: Logistic regression**; and **Test 40: Principal components analysis and factor analysis**.

The **Handbook of Parametric and Nonparametric Statistical Procedures** can be used as a reference book or it can be employed as a textbook in undergraduate and graduate courses which are designed to cover a broad spectrum of parametric and/or nonparametric statistical procedures.

The author would like to express his gratitude to a number of people who helped make this book a reality. First, I would like to thank Tim Pletscher of CRC Press for his confidence in and support of the first edition of the Handbook. Special thanks are due to Bob Stern, who, in his role as editor at Chapman and Hall/CRC, is responsible for the subsequent editions of the book. Thanks also to Helena Redshaw at CRC for overseeing production of the manuscript. I am also indebted to Glena Ames who did an excellent job preparing the copy-ready manuscript for the first two editions of the book. Finally, I must express my appreciation to my wife Vicki and daughter Emily, who over the years have both endured and tolerated the difficulties associated with a project of this magnitude.

David Sheskin

Table of Contents
with Summary of Topics

Inferential Statistical Tests Employed with Two Dependent Samples (and Related Measures of Association/Correlation) 741

Test 17: The t Test for Two Dependent Samples 743

Inferential Statistical Tests Employed with Two or More Independent Samples (and Related Measures of Association/Correlation) **865**

Introduction

The intent of this **Introduction** is to provide the reader with a general overview of basic terminology, concepts, and methods employed within the areas of descriptive statistics and experimental design. To be more specific, the following topics will be covered: a) Computational procedures for measures of central tendency, variability, skewness, and kurtosis; b) Visual methods for displaying data; c) The normal distribution; d) Hypothesis testing; e) Experimental design; and f) Basic principles of probability. Within the context of the latter discussions, the reader is presented with the necessary information for both understanding and using the statistical procedures which are described in this book. Following the **Introduction** is an outline of all the procedures that are covered, as well as decision tables to aid the reader in selecting the appropriate statistical procedure.

Descriptive versus Inferential Statistics

The term **statistics** is derived from Latin and Italian terms which respectively mean "status" and "state arithmetic" (i.e., the present conditions within a state or nation). In a more formal sense, **statistics** is a field within mathematics that involves the summary and analysis of data. The field of statistics can be divided into two general areas, **descriptive statistics** and **inferential statistics**.

Descriptive statistics is a branch of statistics in which data are only used for descriptive purposes and are not employed to make predictions. Thus, descriptive statistics consists of methods and procedures for presenting and summarizing data. The procedures most commonly employed in descriptive statistics are the use of tables and graphs, and the computation of measures of central tendency and variability. **Measures of association or correlation**, which are covered in this book, are also categorized by most sources as descriptive statistical procedures, insofar as they serve to describe the relationship between two or more variables. A **variable** is any property of an object or an organism with respect to which there is variation – i.e., not every object or organism is the same with respect to that property. Examples of variables are color, weight, height, gender, intelligence, etc.

Inferential statistics employs data in order to draw inferences (i.e., derive conclusions) or make predictions. Typically, in inferential statistics sample data are employed to draw inferences about one or more populations from which the samples have been derived. Whereas a **population** consists of the sum total of subjects or objects that share something in common with one another, a **sample** is a set of subjects or objects which have been derived from a population. For a sample to be useful in drawing inferences about the larger population from which it was drawn, it must be representative of the population. Thus, typically (although there are exceptions), the ideal sample to employ in research is a **random sample**. A random sample must adhere to the following criteria: a) Each subject or object in the population has an equal likelihood of being selected as a member of the sample; b) The selection of each subject/object is independent of the selection of all other subjects/objects in the population; and c) For a specified sample size, every possible sample that can be derived from the population has an equal likelihood of occurring.

In point of fact, it would be highly unusual to find an experiment that employed a truly random sample. Pragmatic and/or ethical factors make it literally impossible in most instances to obtain random samples for research. Insofar as a sample is not random, it will limit the degree to which a researcher will be able to generalize one's results. Put simply, one can only generalize to objects or subjects that are similar to the sample employed. (A more detailed discussion of the general subject of sampling is provided later in this **Introduction**.)

Statistic versus Parameter

A **statistic** refers to a characteristic of a sample, such as the average score (also known as the **mean**). A **parameter**, on the other hand, refers to a characteristic of a population (such as the average of a whole population). A statistic can be employed for either descriptive or inferential purposes. An example of using a statistic for descriptive purposes is obtaining the mean of a group (which represents a sample) in order to summarize the average performance of the group. On the other hand, if we use the mean of a group to estimate the mean of a larger population the group is supposed to represent, the statistic (i.e., the group mean) is being employed for inferential purposes. The most basic statistics that are employed for both descriptive and inferential purposes are **measures of central tendency** (of which the mean is an example) and **measures of variability**.

In inferential statistics the computed value of a statistic (e.g., a sample mean) is employed to make inferences about a parameter in the population from which the sample was derived (e.g., the population mean). The inferential statistical procedures described in this book all employ data derived from one or more samples in order to draw inferences or make predictions with respect to the larger population(s) from which the sample(s) was/were drawn.

Sampling error is the discrepancy between the value of a statistic and the parameter it estimates. Due to sampling error, the value of a statistic will usually not be identical to the parameter it is employed to estimate. The larger the sample size the less the influence of sampling error, and consequently the closer one can expect the value of a statistic to be to the actual value of a parameter.

When data from a sample are employed to estimate a population parameter, any statistic derived from the sample should be **unbiased**. Although sampling error will be associated with an **unbiased statistic**, an unbiased statistic provides the most accurate estimate of a population parameter. A **biased statistic**, on the other hand, does not provide as accurate an estimate of that parameter as an unbiased statistic, and consequently a biased statistic will be associated with a greater sampling error. Stated in a more formal way, an **unbiased statistic** (also referred to as an **unbiased estimator**) is one whose **expected value** is equal to the parameter it is employed to estimate. The **expected value** of a statistic is based on the premise that an infinite number of samples of equal size are derived from the relevant population, and for each sample the value of the statistic is computed. The average of all the values computed for the statistic will represent the **expected value** of that statistic. The latter distribution of average values for the statistic is more formally referred to as a **sampling distribution** (which is a concept discussed in greater depth later in the book). The subject of bias in statistics will be discussed later in reference to the **mean** (which is the most commonly employed measure of central tendency), and the **variance** (which is the most commonly employed measure of variability).

Levels of Measurement

Typically, information which is quantified in research for purposes of analysis is categorized with respect to the level of measurement the data represent. Different levels of measurement contain different amounts of information with respect to whatever the data are measuring. A data classification system developed by Stevens (1946), which is commonly employed within the framework of many scientific disciplines, will be presented in this section.

Statisticians generally conceptualize data as fitting within one of the following four measurement categories: **nominal data** (also known as **categorical data**), **ordinal data** (also know as **rank-order data**), **interval data**, and **ratio data**. As one moves from the lowest level of measurement, nominal data, to the highest level, ratio data, the amount of information provided by the numbers increases, as well as the meaningful mathematical operations that can be performed on those numbers. Each of the levels of measurement will now be discussed in more detail.

a) **Nominal/categorical level measurement** In nominal/categorical measurement numbers are employed merely to identify mutually exclusive categories, but cannot be manipulated in a mathematically meaningful manner. As an example, a person's social security number represents nominal measurement since it is used purely for purposes of identification and cannot be meaningfully manipulated in a mathematical sense (i.e., adding, subtracting, etc. the social security numbers of people does not yield anything of tangible value).

b) **Ordinal/rank-order level measurement** In an ordinal scale, the numbers represent rank-orders, and do not give any information regarding the differences between adjacent ranks. For example, the order of finish in a horse race represents an ordinal scale. If in a race Horse A beats Horse B in a photo finish, and Horse B beats Horse C by twenty lengths, the respective order of finish of the three horses reveals nothing about the fact that the distance between the first and second place horses was minimal, while the difference between second and third place horses was substantial.

c) **Interval level measurement** An interval scale not only considers the relative order of the measures involved (as is the case with an ordinal scale) but, in addition, is characterized by the fact that throughout the length of the scale equal differences between measurements correspond to equal differences in the amount of the attribute being measured. What this translates into is that if IQ is conceptualized as an interval scale, the one point difference between a person who has an IQ of 100 and someone who has an IQ of 101 should be equivalent to the one point difference between a person who has an IQ of 140 and someone with an IQ of 141. In actuality some psychologists might argue this point, suggesting that a greater increase in intelligence is required to jump from an IQ of 140 to 141 than to jump from an IQ of 100 to 101. If, in fact, the latter is true, a one point difference does not reflect the same magnitude of difference across the full range of the IQ scale. Although in practice IQ and most other human characteristics measured by psychological tests (such as anxiety, introversion, self esteem, etc.) are treated as interval scales, many researchers would argue that they are more appropriately categorized as ordinal scales. Such an argument would be based on the fact that such measures do not really meet the requirements of an interval scale, because it cannot be demonstrated that equal numerical differences at different points on the scale are comparable.

It should also be noted that unlike ratio scales, which will be discussed next, interval scales do not have a true zero point. If interval scales have a zero score that can be assigned to a person or object, it is assumed to be arbitrary. Thus, in the case of IQ we can ask the question of whether or not there is truly an IQ which is so low that it literally represents zero IQ. In reality, you probably can only say a person who is dead has a zero IQ! In point of fact, someone who has obtained an IQ of zero on an IQ test has been assigned that score because his performance on the test was extremely poor. The zero IQ designation does not necessarily mean the person could not answer any of the test questions (or, to go further, that the individual possesses none of the requisite skills or knowledge for intelligence). The developers of the test just decided to select a certain minimum score on the test and designate it as the zero IQ point.

d) **Ratio level measurement** As is the case with interval level measurement, ratio level measurement is also characterized by the fact that throughout the length of the scale, equal differences between measurements correspond to equal differences in the amount of the attribute being measured. However, ratio level measurement is also characterized by the fact that it has a true zero point. Because of the latter, with ratio measurement one is able to make meaningful ratio statements with regard to the attribute/variable being measured. To illustrate these points, most physical measures such as weight, height, blood glucose level, as well as measures of certain behaviors such as the number of times a person coughs or the number of times a child cries, represent ratio scales. For all of the aforementioned measures there is a true zero point (i.e., zero weight, zero height, zero blood glucose, zero coughs, zero episodes of crying), and for each of these measures one is able to make meaningful ratio statements (such as Ann weighs twice as much as Joan, Bill is one-half the height of Steve, Phil's blood glucose is 10 times Sam's, Mary coughs five times as often as Pete, and Billy cries three times as much as Heather).

Continuous versus Discrete Variables

When measures are obtained on people or objects, in most instances we assume there will be variability. Since we assume variability, not everyone or everything will have the same score on whatever it is that is being measured. For this reason, when something is measured it is commonly referred to as a **variable**. As noted above, variables can be categorized with respect to the level of measurement they represent. In contrast to a variable, a **constant** is a number which never exhibits variation. Examples of constants are the mathematical constants *pi* and *e* (which are respectively 3.14159... and 2.71828...), the number of days in a week (which will always be 7), the number of days in the month of April (which will always be 30,), etc.

A variable can be categorized with respect to whether it is **continuous** or **discrete**. A **continuous variable** can assume any value within the range of scores that define the limits of that variable. A **discrete variable**, on the other hand, can only assume a limited number of values. To illustrate, temperature (which can assume both integer and fractional/decimal values within a given range) is a **continuous variable**. Theoretically, there are an infinite number of possible temperature values, and the number of temperature values we can measure is limited only by the precision of the instrument we are employing to obtain the measurements. On the other hand, the face value of a die is a **discrete variable**, since it can only assume the integer values 1 through 6.

Measures of Central Tendency

Earlier in the **Introduction** it was noted that the most commonly employed statistics are measures of central tendency and measures of variability. This section will describe five measures of central tendency: the **mode**, the **median**, the **mean**, the **geometric mean**, and the **harmonic mean**.

The mode The **mode** is the most frequently occurring score in a distribution of scores. A mode that is derived for a sample is a statistic, whereas the mode of a population is a parameter. In the following distribution of scores the mode is 5, since it occurs two times, whereas all other scores occur only once: 0, 1, 2, 5, 5, 8, 10. If more than one score occurs with the highest frequency, it is possible to have two or more modes in a distribution. Thus, in the distribution 0, 1, 2, 5, 6, 8, 10, all of the scores represent the mode, since each score occurs one time. A distribution with more than one mode is referred to as a **multimodal distribution**. If it happens that two scores both occur with the highest frequency, the distribution would be described as a **bimodal** distribution, which represents one type of multimodal distribution. The distribution 0, 5, 5, 8, 9, 9, 12 is bimodal, since the scores 5 and 9 both occur two times and all other scores appear once.

The most common situation in which the mode is employed as a descriptive measure is within the context of a **frequency distribution**. A **frequency distribution** is a table which summarizes a set of data in a tabular format, listing the frequency of each score adjacent to that score. Table I.1 is a frequency distribution for **Distribution A** noted below, which is comprised of *n* = 20 scores. (A more detailed discussion of **Distribution A** can be found later in the **Introduction** in the discussion of **visual methods for displaying data**.) It should be noted that Column 1 of Table I.1 (i.e., the column at the left with the notation *X* at the top) only lists those scores in **Distribution A** which fall within the range of values 22 – 96 that have a frequency of occurrence greater than zero. Although all of the scores within the range of values 22 – 96 could have been listed in Column 1 (i.e., including all of the scores with a frequency of zero), the latter would increase the size of the table substantially, and in the process make it more difficult to interpret. Consequently, it is more efficient to just list those scores which occur at least once, since it is allows for a succinct summary of the data — the latter being a major reason why a frequency distribution is employed.

Distribution A: *22, 55, 60, 61, 61, 62, 62, 63, 63, 67, 71, 71, 72, 72, 72, 74, 74, 76, 82, 96*

Table I.1   Frequency Distribution of Distribution A

X	f
96	1
82	1
76	1
74	2
72	3
71	2
67	1
63	2
62	2
61	2
60	1
55	1
22	1
	n = 20

In addition to presenting data in a tabular format, a researcher can also summarize data within the format of graph. Indeed, it is recommended that researchers obtain a plot of their data prior to conducting any sort of formal statistical analysis. The reason for this is that a body of data can have certain characteristics which may be important in determining the most appropriate method of analysis. Often such characteristics will not be apparent to a researcher purely on the basis of cursory visual inspection — especially if there is a large amount of data and/or one is relatively inexperienced in dealing with data. A commonly employed method for visually presenting data is to construct a **frequency polygon**, which is a graph of a frequency distribution. Figure I.1 is a frequency polygon of **Distribution A**.

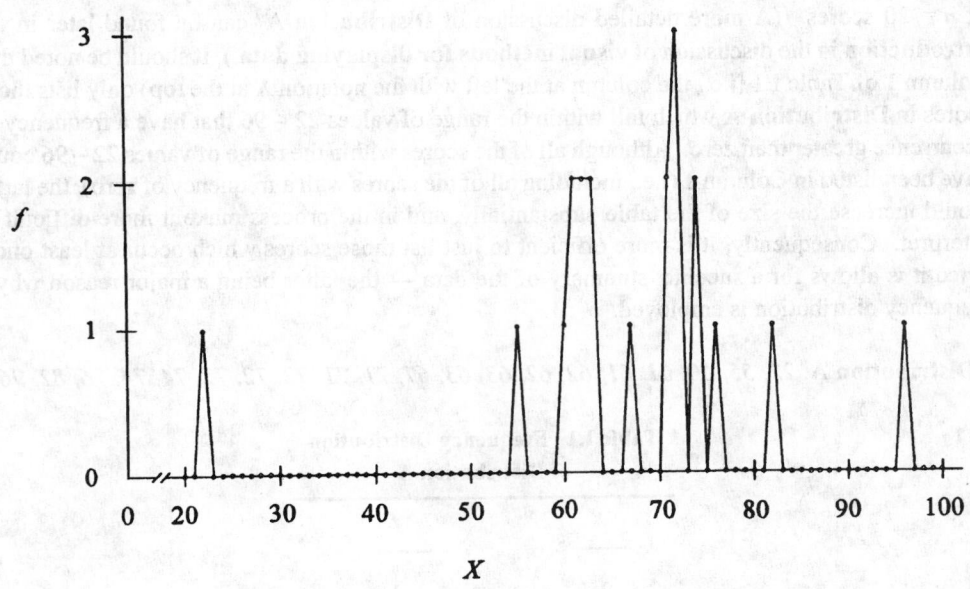

Figure I.1 Frequency Polygon of Distribution A

Note that a frequency polygon is comprised of two axes, a horizontal axis and a vertical axis. The X-axis or horizontal axis (which is referred to as the **abscissa**) is employed to record the range of possible scores on a variable. (The element —/ /— on the left side of the X-axis of Figure I.1 is employed when a researcher only wants to begin recording scores on the abscissa which fall at some point above 0, and not list any scores in between 0 and that point.) The Y-axis or vertical axis (which is referred to as the **ordinate**) is employed to represent the frequency (f) with which each of the scores noted on the X-axis occurs in the sample or population. In order to provide some degree of standardization in graphing data, many sources recommend that the length of the Y-axis be approximately three-quarters the length of the X-axis.

Inspection of Figure I.1 reveals that a frequency polygon is a series of lines which connect a set of points. One point is employed for each of the scores that comprise the range of scores in the distribution. The point which represents any score in the distribution that occurs one or more times will fall directly above that score at a height corresponding to the frequency for that score recorded on the Y-axis. When the frequency polygon descends to and/or moves along the X-axis, it indicates a frequency of zero for those scores on the X-axis. The highest point on a frequency polygon will always fall directly above the score which corresponds to the mode of the distribution (which in the case of **Distribution A** is 72). (In the case of a multimodal distribution the frequency polygon will have multiple high points.) A more detailed discussion of the use of tables and graphs for descriptive purposes as well as a discussion of **exploratory data analysis** (which is an alternative methodology for scrutinizing data) will be presented later in the **Introduction** in the section on the **visual display of data**.

The median The **median** is the middle score in a distribution. If there is an **odd number** of scores in a distribution, in order to determine the median the following protocol should be employed: Divide the total number of scores by 2 and add .5 to the result of the division. The obtained value indicates the **ordinal position** of the score which represents the median of the distribution (note, however, that this value does not represent the median). Thus, if we have a distribution consisting of five scores (e.g., 6, 8, 9, 13, 16), we divide the number of scores in the distribution by two, and add .5 to the result of the division. Thus, $(5/2) + .5 = 3$. The obtained value of 3 indicates that if the five scores are arranged ordinally (i.e., from lowest to highest), the median is the 3rd highest (or 3rd lowest) score in the distribution. With respect to the distribution 6, 8, 9, 13, 16, the value of the median will equal 9, since 9 is the score in the third ordinal position.

If there is an **even number** of scores in a distribution, there will be two middle scores. The median is the average of the two middle scores. To determine the **ordinal positions** of the two middle scores, divide the total number of scores in the distribution by 2. The number value obtained by that division and the number value that is one above it represent the ordinal positions of the two middle scores. To illustrate, assume we have a distribution consisting of the following six scores: 6, 8, 9, 12, 13, 16. To determine the median, we initially divide 6 by 2 which equals 3. Thus, if we arrange the scores ordinally, the 3rd and 4th scores (since $3 + 1 = 4$) are the middle scores. The average of these scores, which are, respectively, 9 and 12, is the median (which will be represented by the notation M). Thus, $M = (9 + 12)/2 = 10.5$. Note once again that in this example (as was the case in the previous example involving an odd number of scores) the initial values computed (3 and 4) do not themselves represent the median, but instead represent the **ordinal positions** of the scores used to compute the median. As was the case with the mode, a median value derived for a sample is a statistic, whereas the median of a whole population is a parameter.

The mean The **mean** (also referred to as the **arithmetic mean**), which is the most commonly employed measure of central tendency, is the average score in a distribution. Typically, when the mean is used as a measure of central tendency, it is employed with interval or ratio level data. Within the framework of the discussion to follow, the notation n will represent the number of subjects or objects in a sample, and the notation N will represent the total number of subjects or objects in the population from which the sample is derived.

Equation I.1 is employed to compute the mean of a sample. Σ, which is the upper case Greek letter **sigma**, is a summation sign. The notation ΣX indicates that the set of n scores in the sample/distribution should be summed.

$$\bar{X} = \frac{\Sigma X}{n}$$ **(Equation I.1)**

Sometimes Equation I.1 is written in the following more complex but equivalent form containing subscripts and superscripts: $\bar{X} = \sum_{i=1}^{n} X_i / n$. In the latter equation, the notation $\sum_{i=1}^{n} X_i$ indicates that beginning with the first score, scores 1 through n (i.e., all the scores) are to be summed. X_i represents the score of the ith subject or object.

Equation I.1 will now be applied to the following distribution of five scores: 6, 8, 9, 13, 16. Since $n = 5$ and $\Sigma X = 52$, $\bar{X} = \Sigma X/n = 52/5 = 10.4$.

Whereas Equation I.1 describes how one can compute the mean of a sample, Equation I.2 describes how one can compute the mean of a population. The simplified version without subscripts is to the right of the first = sign, and the subscripted version of the equation is to the right of the second = sign. The mean of a population is represented by the notation μ, which is the lower case Greek letter **mu**. In practice, it would be highly unusual to have occasion to compute the mean of a population. Indeed, a great deal of analysis in inferential statistics involves employing the mean of a sample to estimate a population mean.

$$\mu = \frac{\Sigma X}{N} = \frac{\sum_{i=1}^{N} X_i}{N} \qquad \textbf{(Equation I.2)}$$

Where: n = The number of scores in the distribution

X_j = The ith score in a distribution comprised of n scores

Note that in the numerator of Equation I.2 all N scores in the population are summed, as opposed to just summing n scores when the value of $\bar{X}$ is computed. The sample mean $\bar{X}$ provides an **unbiased estimate** of the population mean μ, which indicates that if one has a distribution of n scores, $\bar{X}$ provides the best possible estimate of the true value of μ. Later in the book (specifically, under the discussion of the **single-sample** z **test (Test 1)**) it will be noted that the mean of the **sampling distribution** of means (which represents the **expected value** of the statistic represented by the mean) will equal to the value of the population mean. (A **sampling distribution of means** is a frequency distribution of sample means derived from the same population, in which the same number of scores is employed for each sample.) Recollect that earlier in the **Introduction** it was noted that an **unbiased statistic** is one with an **expected value** that is equal to the parameter it is employed to estimate. This applies to the sample mean, since its expected value is equal to the population mean.

The weighted mean   There may be occasions when a researcher has mean values which have been computed for two or more separate samples, yet one wants to compute an overall mean based on all the sample means. In such a case the appropriate value computed for the overall mean is referred to as a **weighted mean**. To illustrate the latter, consider the following situation: A researcher has access to the following average IQ scores of students in a specific school district: $\bar{X}_1 = 100$ for $n_1 = 100$ students who attend School 1; $\bar{X}_2 = 106$ for $n_2 = 200$ subjects who attend School 2; and $\bar{X}_3 = 115$ for $n_3 = 300$ subjects who attend School 3. Assume that the researcher wants to determine the average IQ for the total $N = n_1 + n_2 + n_3 = 600$ students in the district. If the researcher has access to the scores of all 600 students, the latter value can easily be computed through use of Equation I.1 (i.e., sum all 600 scores and divide the latter sum by 600). However, if one only has access to the mean score of each school, the latter computation will not be possible. It would be incorrect to determine the average IQ for all of the students in the district by computing the average of the average IQ scores computed for the three schools — in other words, it would be incorrect to compute the overall mean as follows: $(100 + 106 + 115)/3$ = 107. Note that the latter computation weighs the contribution of each of the three schools equally, and would not be justified unless there were an equal number of students enrolled in each school. In our example, however, the enrollment for School 1 is one-half of that for School 2 and one-third of that for School 3, while the enrollment for School B is two-thirds of that for School 3. Because of the unequal number of students in each school, the most accurate method for computing an overall mean for the school district would be one which weighs each of the three school means by the number of students in each sample. If there are k schools/groups, with n_g subjects per school/group, Equation I.3 can be employed to compute a **weighted mean** (represented by the notation $\bar{X}_W$). Note that when the latter equation is employed for the example under discussion, the computed value $\bar{X}_W = 109.5$ is larger than the previously computed value of 107. The larger value computed for the weighted mean derives from the fact that Equation I.3 allocates greater weight to the mean of School 2 relative to the mean of School 1, as well as greater weight to the mean of School 3 relative to the means of both Schools 1 and 2. Use of Equation I.3 insures that in combining sample means to compute an overall mean, the latter value will be proportionally influenced by the size of the samples used to compute each of the k sample means. Additionally, the value computed for the weighted mean will be identical to the mean value that would be computed if Equation I.1 had been employed to compute an overall mean using the actual scores of the 600 students.[1]

$$X_W = \frac{(n_1)(\bar{X}_1) + (n_2)(\bar{X}_2) + \,..... \, + (n_k)(\bar{X}_k)}{n_1 + n_2 + \,..... \, + n_k}$$

(Equation I.3)

$$X_W = \frac{(n_1)(\bar{X}_1) + (n_2)(\bar{X}_2) + (n_3)(\bar{X}_3)}{n_1 + n_2 + n_3} = \frac{(100)(100) + (200)(106) + (300)(115)}{100 + 200 + 300} = 109.5$$

The geometric mean The **geometric mean** is a measure of central tendency which is primarily employed within the context of certain types of analysis in business and economics. It is most commonly used as an average of **index numbers**, ratios, and percent changes over time. (An **index number** is a metric of the degree to which a variable changes over time. It is calculated by determining the ratio of the current value of the variable to a previous value. The most commonly employed index number is a **price index**, which is used to contrast prices from one period of time to another.) The geometric mean (*GM*) of a distribution is the nth root of the product of the *n* scores in the distribution. Equation I.4 is employed to compute the geometric mean of a distribution.

$$GM = \sqrt[n]{X_1, X_2, \, ... \, , X_n}$$

(Equation I.4)

To illustrate the above equation, the geometric mean of the five values 2, 5, 15, 20 and 30 is $GM = \sqrt[5]{(2)(5)(15)(20)(30)} = 6.18$.

Only positive numbers should be employed in computing the geometric mean, since one or more zero values will render *GM* = 0, and negative numbers will render the equation insoluble (when there are an odd number of negative values) or meaningless (when there are an even number of negative values). Specifically, let us assume we wish to compute the geometric mean for the four values –2, –2, –2, –2. Employing the equation noted above, $GM = \sqrt[4]{(-2)(-2)(-2)(-2)} = 2$. Obviously latter value doesn't make sense, since logically the geometric mean should have a minus sign — specifically, *GM* should be equal to –2. When all of the values in a distribution are equivalent, the geometric mean and **arithmetic mean** will be equal to one another. In all other instances, the value of the geometric mean will be less than the arithmetic mean. Note that in the above example in which *GM* = 6.18, the value computed for the arithmetic mean is $\bar{X} = 14.4$, which is greater than the geometric mean.

Before the introduction of hand calculators, a computationally simpler method for computing the geometric mean utilized **logarithms** (which are discussed in Endnote 14). Specifically, the following equation can also be employed to compute the geometric mean: $\log(GM) = (\Sigma \log X)/n$. The latter equation indicates that the logarithm of the geometric mean is equivalent to the arithmetic mean of the logarithm of the values of the scores in the distribution. The antilogarithm of log (*GM*) will represent the geometric mean.

Chou (1989, pp. 107–110) notes that when a distribution of numbers takes the form of a **geometric series** or a **logarithmically distributed series** (which is positively skewed), the geometric mean is a more suitable measure of central tendency than the arithmetic mean. (The concept of **skewness** (which involves a disproportionate number of high or low scores) is discussed later in the **Introduction**.) A **geometric series** is a sequence of numbers in which the ratio of any term to the preceding term is the same value. As an example, the series 2, 4, 8, 16, 32, 64, ... represents a geometric series in which each subsequent term is twice the value of the preceding term. A **logarithmic series** (also referred to as a **power series**) is one in which successive terms are the result of a constant (*a*) multiplied by successive integer powers of a variable (x) (e.g., ax, ax^2, ax^3, ..., ax^n). Thus, if $x = 3$, the series $a3$, $a3^2$, $a3^3$, $a3^4$, ..., $a3^n$ represents an example of a logarithmic series.

A major consequence of employing the geometric mean in lieu of the arithmetic mean is that the presence of extreme values (which is often the case with a geometric series) will have less of an impact on the value of the geometric mean. Although it would not generally be employed as a measure of central tendency for a symmetrical distribution, the geometric mean can reduce the impact of skewness when it is employed as a measure of central tendency for a non-symmetrical distribution.

The harmonic mean   Another measure of central tendency is the **harmonic mean**. The harmonic mean is determined by computing the reciprocal of each of the n scores in a distribution. (The **reciprocal** of a number is the value computed when the number is divided into 1— i.e., the reciprocal of $X = 1/X$.) The mean/average value of the n reciprocals is then computed. The reciprocal of the latter mean represents the harmonic mean, which is computed with Equation I.5.

$$\bar{n}_h = \frac{n}{\sum_{i=1}^{n}\left[\dfrac{1}{X_i}\right]}$$ **(Equation I.5)**

Where: n = The number of scores in the distribution
   X_j = The ith score in a distribution comprised of n scores

To illustrate the above equation, the harmonic mean of the five values 2, 5, 15, 20 and 30 is computed below.

$$\bar{n}_h = \frac{5}{(1/2) + (1/5) + (1/15) + (1/20) + (1/30)} = \frac{5}{.85} = 5.88$$

Chou (1989, p. 111) notes that for any distribution in which there is variability among the scores and in which no score is equal to zero, the harmonic mean will always be smaller than both the arithmetic mean (which in the case of the above distribution is $\bar{X} = 14.4$) and the geometric mean (which is $GM = 6.18$). This is the case, since the harmonic mean is least influenced by extreme scores in a distribution. Chou (1989) provides a good discussion of the circumstances when it is prudent to employ the harmonic mean as a measure of central tendency. He notes that the harmonic mean is recommended when scores are expressed inversely to what is required in the desired measure of central tendency. Examples of such circumstances are certain conditions where a measure of central tendency is desired for time rates and/or prices. Further discussion of the harmonic mean can be found in Section VI of the **t test for two independent samples (Test 11)** and Chou (1989, pp. 110–113).

Measures of Variability

In this section a number of measures of variability will be discussed. Primary emphasis, however, will be given to the **standard deviation** and the **variance**, which are the most commonly employed measures of variability.

a) **The range**   The **range** is the difference between the highest and lowest scores in a distribution. Thus in the distribution 2, 3, 5, 6, 7, 12, the range is the difference between 12 (the highest score) and 2 (the lowest score). Thus: Range = $12 - 2 = 10$. Some sources add one to the obtained value, and would thus say that the Range = 11. Although the **range** is employed on occasion for descriptive purposes, it is of little use in inferential statistics.

b) **Quantiles, percentiles, quartiles, and deciles** A **quantile** is a measure that divides a distribution into equidistant percentage points. Examples of quantiles are **percentiles, quartiles**, and **deciles**. **Percentiles** divide a distribution into blocks comprised of one percentage point (or blocks that comprise a proportion equal to .01 of the distribution).[2] A specific percentile value corresponds to the point in a distribution at which a given percentage of scores falls at or below. Thus, if an IQ test score of 115 falls at the 84th percentile, it means 84% of the population has an IQ of 115 or less. The term **percentile rank** is also employed to mean the same thing as a percentile — in other words, we can say that an IQ score of 115 has a percentile rank of 84%.

Deciles divide a distribution into blocks comprised of ten percentage points (or blocks that comprise a proportion equal to .10 of the distribution). A distribution can be divided into ten deciles, the upper limits of which are defined by the 10th percentile, 20th percentile, ..., 90th percentile, and 100th percentile. Thus, a score that corresponds to the 10th percentile falls at the upper limit of the first decile of the distribution. A score that corresponds to the 20th percentile falls at the upper limit of the second decile of the distribution, and so on. The **interdecile range** is the difference between the scores at the 90th percentile (the upper limit of the ninth decile) and the 10th percentile.

Quartiles divide a distribution into blocks comprised of 25 percentage points (or blocks that comprise a proportion equal to .25 of the distribution). A distribution can be divided into four **quartiles**, the upper limits of which are defined by the 25th percentile, 50th percentile (which corresponds to the median of the distribution), 75th percentile, and 100th percentile. Thus, a score that corresponds to the 25th percentile falls at the upper limit of the first quartile of the distribution. A score that corresponds to the 50th percentile falls at the upper limit of the second quartile of the distribution, and so on. The **interquartile range** is the difference between the scores at the 75th percentile (which is the upper limit of the third quartile) and the 25th percentile.

Infrequently, the interdecile or interquartile ranges may be employed to represent variability. An example of a situation where a researcher might elect to employ either of these measures to represent variability would be when one wishes to omit a few extreme scores in a distribution. Such extreme scores are referred to as **outliers**. Specifically, an **outlier** is a score in a set of data which is so extreme that, by all appearances, it is not representative of the population from which the sample is ostensibly derived. Since the presence of outliers can dramatically affect variability (as well as the value of the sample mean), their presence may lead a researcher to believe that the variability of a distribution might best be expressed through use of the interdecile or interquartile range (as well as the fact that when outliers are present, the sample median is more likely than the mean to be a representative measure of central tendency). Further discussion of outliers can be found later in the **Introduction**, as well as in Section VI of the *t* **test for two independent samples**.

c) **The variance and the standard deviation** The most commonly employed measures of variability in both inferential and descriptive statistics are the **variance** and **standard deviation**. These two measures are directly related to one another, since the standard deviation is the square root of the variance (and thus the variance is the square of the standard deviation). As is the case with the mean, the standard deviation and variance are generally only employed with interval or ratio level data.

The formal definition of the **variance** is that it is the **mean of the squared difference scores** (which are also referred to as **deviation scores**). This definition implies that in order to compute the variance of a distribution one must subtract the mean of the distribution from each score, square each of the difference scores, sum the squared difference scores, and divide the latter value by the number of scores in the distribution. The logic of this definition is reflected in the **definitional equations** which will be presented later in this section for both the variance and standard deviation.

A **definitional equation** for a statistic (or parameter) contains the specific mathematical operations that are described in the definition of that statistic (or parameter). On the other hand, a **computational equation** for the same statistic (or parameter) does not clearly reflect the definition of that statistic (or parameter). A computational equation, however, facilitates computation of the statistic (or parameter), since it is computationally less involved than the definitional equation. In this book, in instances where a definitional and computational equation are available for computing a test statistic, the computational equation will generally be employed to facilitate calculations. It should be noted that because that they do not directly reflect the meaning of the concept they measure, some sources omit and/or discourage the use of computational equations. Although the latter philosophy can be understood if one always has access to a computer, in instances where one is required to conduct computations by hand or with the aid of a calculator, the use of computational equations can make an analysis less burdensome.

The following notation will be used in the book with respect to the values of the variance and standard deviation.

σ^2 (where σ is the lower case Greek letter **sigma**) will represent the **variance of a population**.

s^2 will represent the **variance of a sample**, when the variance is employed for **descriptive purposes**. s^2 will be a **biased estimate of the population variance** σ^2, and because of this s^2 will generally underestimate the true value of σ^2.

$\tilde{s}^2$ will represent the **variance of a sample**, when the variance is employed for **inferential purposes**. $\tilde{s}^2$ will be an **unbiased estimate of the population variance** σ^2.

σ will represent the **standard deviation of a population**.

s will represent the **standard deviation of a sample**, when the standard deviation is employed for **descriptive purposes**. s will be a **biased estimate of the population standard deviation** σ and because of this s will generally underestimate the true value of σ.

$\tilde{s}$ will represent the **standard deviation of a sample**, when the standard deviation is employed for **inferential purposes**. $\tilde{s}$ will be an **unbiased estimate of the population standard deviation** σ.[3]

Equations I.6–I.11 are employed to compute the values σ^2, s^2, $\tilde{s}^2$, σ, s, and $\tilde{s}$. Note that in each case, two equivalent methods are presented for computing the statistic or parameter in question. The formula to the left is the definitional equation, whereas the formula to the right is the computational equation.

$$\sigma^2 = \frac{\Sigma(X - \mu)^2}{N} = \frac{\Sigma X^2 - \frac{(\Sigma X)^2}{N}}{N} \qquad \textbf{(Equation I.6)}$$

$$s^2 = \frac{\Sigma(X - \bar{X})^2}{n} = \frac{\Sigma X^2 - \frac{(\Sigma X)^2}{n}}{n} \qquad \textbf{(Equation I.7)}$$

$$\tilde{s}^2 = \frac{\Sigma(X - \bar{X})^2}{n - 1} = \frac{\Sigma X^2 - \frac{(\Sigma X)^2}{n}}{n - 1} \qquad \textbf{(Equation I.8)}$$

$$\sigma = \sqrt{\frac{\Sigma(X - \mu)^2}{N}} = \sqrt{\frac{\Sigma X^2 - \frac{(\Sigma X)^2}{N}}{N}} \qquad \textbf{(Equation I.9)}$$

$$s = \sqrt{\frac{\Sigma(X - \bar{X})^2}{n}} = \sqrt{\frac{\Sigma X^2 - \frac{(\Sigma X)^2}{n}}{n}} \qquad \textbf{(Equation I.10)}$$

$$\tilde{s} = \sqrt{\frac{\Sigma(X - \bar{X})^2}{n - 1}} = \sqrt{\frac{\Sigma X^2 - \frac{(\Sigma X)^2}{n}}{n - 1}} \qquad \textbf{(Equation I.11)}$$

The reader should take note of the following with respect to the notation employed in Equations I.6–I.11: a) The notation ΣX^2 represents the **sum of the X^2 scores**. It indicates that each of the n X scores (i.e., each of the n scores which comprise the distribution) is squared, and the sum of the $n X^2$ scores is obtained; b) The notation $(\Sigma X)^2$ represents the **sum of the X scores squared**. It indicates that a sum is obtained for the n X scores, and the latter sum is then squared; c) The notation $\Sigma(X - \bar{X})^2$ indicates that the mean of the distribution is subtracted from each of the n scores, each of the n difference scores is squared, and the sum of the n squared difference scores is obtained.

When the variance or standard deviation of a sample is computed within the framework of an inferential statistical test, one always wants an unbiased estimate of the population variance or the population standard deviation. Thus, the computational form of Equation I.8 will be employed throughout this book when a sample variance is used to estimate a population variance, and the computational form of Equation I.11 will be employed when a sample standard deviation is used to estimate a population standard deviation.

The reader should take note of the fact that some sources employ subscripted versions of the above equations. Thus, the computational form of Equation I.8 is often written as:

$$\tilde{s}^2 = \frac{\sum_{i=1}^{n} X_i^2 - \frac{\left(\sum_{i=1}^{n} X_i\right)^2}{n}}{n - 1}$$

Although the subscripted version will not be employed for computing the values of $\tilde{s}^2$ and $\tilde{s}$, subscripted versions for some statistics may be used later in the book in order to clarify the mathematical operations involved in computing a statistic.

As noted previously, for the same set of data the value of $\tilde{s}^2$ will always be larger than the value of s^2. This can be illustrated with a distribution consisting of the five scores: 6, 8, 9, 13, 16. The following values are substituted in Equations I.7 and I.8: $\Sigma X = 52$, $\Sigma X^2 = 606$, $n = 5$.

$$s^2 = \frac{606 - \frac{(52)^2}{5}}{5} = 13.04$$

$$\tilde{s}^2 = \frac{606 - \frac{(52)^2}{5}}{5 - 1} = 16.3$$

Since the standard deviation is the square root of the variance, we can quickly determine that $s = \sqrt{s^2} = \sqrt{13.04} = 3.61$ and $\tilde{s} = \sqrt{\tilde{s}^2} = \sqrt{16.3} = 4.04$. Note that $\tilde{s}^2 > s^2$ and $\tilde{s} > s$.[4]

Table I.2 summarizes the computation of the unbiased estimate of the population variance $(\tilde{s}^2)$, employing both the definitional and computational equations. Note that in the two versions for computing $\tilde{s}^2$ listed for Equation I.8, the numerator values $\Sigma X^2 - [(\Sigma X)^2/n]$ and $\Sigma(X - \bar{X})^2$ are equivalent. Thus, in Table I.2, the sum of the values of the last column $\Sigma(X - \bar{X})^2 = 65.20$, equals $\Sigma X^2 - [(\Sigma X)^2/n] = 606 - [(52)^2/5] = 65.20$.

Table I.2 Computation of Estimated Population Variance

X	X^2	$\bar{X}$	$(X - \bar{X})$	$(X - \bar{X})^2$
6	36	10.4	$(6 - 10.4) = -4.4$	$(-4.4)^2 = 19.36$
8	64	10.4	$(8 - 10.4) = -2.4$	$(-2.4)^2 = 5.76$
9	81	10.4	$(9 - 10.4) = -1.4$	$(-1.4)^2 = 1.96$
13	169	10.4	$(13 - 10.4) = 2.6$	$(2.6)^2 = 6.76$
16	256	10.4	$(16 - 10.4) = 5.6$	$(5.6)^2 = 31.36$
$\Sigma X = 52$	$\Sigma X^2 = 606$		$\Sigma(X - \bar{X}) = 0$	$\Sigma(X - \bar{X})^2 = 65.20$

$$\tilde{s}^2 = \frac{\Sigma X^2 - \dfrac{(\Sigma X)^2}{n}}{n - 1} = \frac{606 - \dfrac{(52)^2}{5}}{5 - 1} = \frac{65.20}{5 - 1} = 16.3$$

$$\tilde{s}^2 = \frac{\Sigma(X - \bar{X})^2}{n - 1} = \frac{65.20}{5 - 1} = 16.3$$

The reader should take note of the following with respect to the standard deviation and variance:

1) The value of a standard deviation or variance can never be a negative number. If a negative number is ever obtained for either value, it indicates a mistake has been made in the calculations. The only time the value of a standard deviation or variance will not be a positive number is when its value equals zero. The only instance in which the value of both the standard deviation and variance of a distribution will equal zero is when all of the scores in the distribution are identical to one another.

2) As the value of the sample size (n) increases, the difference between the values of s^2 and $\tilde{s}^2$ will decrease. In the same respect, as the value of n increases, the difference between the values of s and $\tilde{s}$ will decrease. Thus, the biased estimate of the variance (or standard deviation) will be more likely to underestimate the true value of the population variance (or standard deviation) with small sample sizes than it will with large sample sizes.

3) The numerator of any of the equations employed to compute a variance or a standard deviation is often referred to as the **sum of squares**. Thus in the example in this section, the value of the sum of squares is 65.2, since $\Sigma X^2 - [(\Sigma X)^2/n] = 606 - [(52)^2/5] = 65.2$. The denominators of both Equation I.8 and Equation I.11 are often referred to as the **degrees of freedom** (a concept that is discussed later in the book within the framework of the **single-sample t test (Test 2)**). Based on what has been said with respect to the sum of squares and the degrees of freedom, the variance is sometimes defined as the sum of squares divided by the degrees of freedom.

Prior to closing the discussion of the standard deviation and variance, a number of other characteristics of statistics will be noted. Specifically, the concepts of **efficiency**, **sufficiency**, and **consistency** will be discussed. (A more in depth discussion of these concepts can be found in Hays and Winkler (1971)).

The concept of **efficiency** is closely related to the issue of whether or not a statistic is biased. An **efficient statistic** is one that provides a more accurate estimate of a parameter relative to an alternative statistic that can also be employed to estimate the same parameter. An example of this is the relative degree to which the mean and median accurately estimate the mean of a symmetrical population distribution. (In a symmetrical population distribution the values of the mean and median will always be identical.) Although both the sample mean and sample median represent unbiased estimators of the mean of a symmetrical population distribution, the sample mean will generally be a more efficient estimator than the sample median, since in most instances the sample mean will have a smaller **standard error** associated with it. The **standard error** (which is discussed in detail under the discussion of the **single-sample z test**) is a measure of variability. Put simply, for a given sample size, the lower value of the standard error of the mean relative to the standard error of the median, the less variability there will be among a set of sample means relative to the amount of variability among an equal number of sample medians. In other words, values computed for sample means will cluster more closely around the true population mean than will values computed for sample medians. In the same respect, in most instances the sample variance and sample standard deviation (when computed with Equations I.8 and I.11) will also be efficient statistics, since they will have lower standard errors than alternative measures of variability.

The mean, variance, and standard deviation also represent **sufficient estimators** of the parameters they estimate. A **sufficient estimator** is one that employs all of the information in the sample to estimate the relevant parameter. The mode and median are not sufficient estimators, since the mode only employs the most frequently occurring score in a sample, while the median only employs the middle score(s). With regard to variability, the range and interquartile range values are not sufficient estimators, since they only employ specific scores in a sample.

Finally, the mean, variance, and standard deviation also represent **consistent estimators**. A **consistent estimator** is characterized by the fact that as the sample size employed to compute the statistic increases, the likelihood of accurately estimating the relevant parameter also increases. The latter is true for all three of the aforementioned statistics.

d) **The coefficient of variation** An alternative, although infrequently employed measure of variability, is the **coefficient of variation**. Since the values of the standard deviation and variance are a direct function of the magnitude of the scores in a sample/population, it can sometimes be useful to express variability in reference to the size of the mean of a distribution. By doing the latter, one can compare the values of the standard deviations and variances of distributions that have dramatically different means and/or employ different units of measurement. The coefficient of variation (represented by the notation CV) allows one to do this. The coefficient of variation is computed with Equation I.12.

$$CV = \frac{\tilde{s}}{\overline{X}} \qquad \text{(Equation I.12)}$$

The following should be noted with respect to Equation I.12: a) When the values of σ and μ are known, they can be employed in place of $\tilde{s}$ and $\overline{X}$; and b) Sometimes the value computed for CV is multiplied by 100 in order to express it as a percentage.

Note that the coefficient of variation is nothing more than a ratio of the value of the standard deviation relative to the value of the mean. The larger the value of CV computed for a variable, the greater the degree of variability there is on that variable. Unlike the standard deviation and variance, the numerical value represented by CV is not in the units that are employed to measure the variable for which it is computed.

To illustrate the latter, let us assume that we wish to compare the variability of income between two countries which employ dramatically different values of currency. The mean monthly income in **Country A** is $\bar{X}_A$ = 40 jaspars, with a standard deviation of $\tilde{s}_A$ = 10 jaspars. The mean monthly income in **Country B** is $\bar{X}_B$ = 2000 rocs, with a standard deviation of $\tilde{s}_B$ = 100 rocs. Note that the mean and standard deviation for each country is expressed in the unit of currency employed in that country. When we employ Equation I.12, we compute that the coefficient of variations for the two countries are CV_A = 10/40 = .25 and $_{C.B.}$ = 100/2000 = .05. The latter CV values are just simple ratios, and are not numbers based on the scale for the unit of currency employed in a given country. In other words, CV_A = .25 is not .25 jaspars, but is simply the ratio .25. In the same respect CV_B = .05 is not .05 rocs, but is simply the ratio .05. Consequently, by dividing the larger value CV_A = .25 by the smaller value CV_B = .05 we can determine that there is five times more variability in income in **Country A** than there is in **Country B** (i.e., CV_A/CV_B = .25/.05 = 5). If we express our result as a percentage, we can say that there is 5 × 100% = 500% more variability in income in **Country A** than there is in **Country B**. If, on the other hand, we had divided $\tilde{s}_B$ = 100 rocs by $\tilde{s}_A$ = 10 jaspars (i.e., $\tilde{s}_B/\tilde{s}_A$ = 100/10 = 10), we would have erroneously concluded that there is ten times (or 10 × 100 = 1000%) more variability in income in **Country B** than in **Country A**. The reason why the latter method results in a misleading conclusion is that, unlike the coefficient of variation, it fails to take into account the different units of currency employed in the two countries.

Measures of Skewness and Kurtosis

In addition to the mean and variance, there are two other measures that can provide useful descriptive information about a distribution. These two measures, **skewness** and **kurtosis**, represent the **third** and **fourth moments** of a distribution. Hays and Winker (1979, p. 161) note that the term **moment** is employed to represent to the expected values of different powers of a random variable. Equations I.13 and I.14 respectively represent the general equation for a moment. In Equation I.13, v_i (where v represents the lower case Greek letter **nu**) represents the population parameter for the ith moment about the mean, whereas in Equation I.14, m_i represents the sample statistic for the ith moment about the mean.

$$v_i = \frac{\Sigma(X - \mu)^i}{N}$$ **(Equation I.13)**

$$m_i = \frac{\Sigma(X - \bar{x})^i}{n}$$ **(Equation I.14)**

With respect to a sample, the **first moment about the mean** (m_1) is represented by Equation I.15. The **second moment about the mean** (m_2, which is the sample variance) is represented by Equation I.16. The **third moment about the mean** (m_3, which as noted above represents **skewness** and is also referred to as **symmetry**) is represented by Equation I.17. The **fourth moment about the mean** (m_4, which as noted above represents **kurtosis**) is represented by Equation I.18.

$$m_1 = \frac{\Sigma(X - \bar{X})}{n} = 0$$ **(Equation I.15)**

$$m_2 = \frac{\Sigma(X - \bar{X})^2}{n} \qquad \text{(Equation I.16)}$$

$$m_3 = \frac{\Sigma(X - \bar{X})^3}{n} \qquad \text{(Equation I.17)}$$

$$m_4 = \frac{\Sigma(X - \bar{X})^4}{n} \qquad \text{(Equation I.18)}$$

Although skewness and kurtosis are not employed for descriptive purposes as frequently as the mean and variance, they can provide useful information. Skewness and kurtosis are sometimes employed within the context of determining the **goodness-of-fit** of data in reference to a specific type of distribution — most commonly the normal distribution. Tests of goodness-of-fit are discussed under the **single-sample test for evaluating population skewness (Test 4)**, the **single-sample test for evaluating population kurtosis (Test 5)**, the **Kolmogorov–Smirnov goodness-of-fit test for a single sample (Test 7)** and the **chi-square goodness-of-fit test (Test 8)**.

Skewness Skewness is a measure reflecting the degree to which a distribution is asymmetrical. A symmetrical distribution will result in two identical mirror images when it is split down the middle. The bell shaped or **normal distribution**, which will be discussed later in the **Introduction**, is the best known example of a symmetrical distribution. When a distribution is not symmetrical, a disproportionate number of scores will fall either to the left or right of the middle of the distribution. Figure I.2 visually depicts three frequency distributions (within the format of **frequency polygons**), only one of which, **Distribution B**, is symmetrical. **Distributions C** and **D** are asymmetrical — to be more specific, the latter two distributions are skewed.

When all the lines connecting the points in a frequency polygon are "smoothed over," the resulting frequency distribution assumes the appearance of the three distributions depicted in Figure I.2. A **theoretical frequency distribution** (or as it is sometimes called, a **theoretical probability distribution**), which any of the distributions in Figure I.2 could represent, is a graph of the frequencies for a population distribution. (The distributions in Figure I.2 can also represent a frequency distribution which can be constructed through use of a specific mathematical equation (such as Equation I.36, the equation for the normal distribution) which is discussed later in the **Introduction**.) As noted earlier in this section, the X-axis represents the range of possible scores on a variable in the population, while the Y-axis represents the frequency with which each of the scores occurs (or sometimes the proportion/probability of occurrence for the scores is recorded on the Y-axis — thus the use of the term **theoretical probability distribution**). It should be noted that it is more precise to refer to the values recorded on the ordinate of a theoretical probability distribution as **density** values, and because of the latter the term **probability density function** is often used to describe a theoretical probability distribution. Further clarification of the concept of density and probability density functions can be found later in this section as well as in Section IX (the **Addendum**) of the **binomial sign test for a single sample (Test 9)** under the discussion of **Bayesian analysis of a continuous variable**.[5]

Returning to Figure I.2, **Distribution B** is a **unimodal symmetrical distribution**. Although it is possible to have a symmetrical distribution that is multimodal (i.e., a distribution that has more than one mode), within the framework of the discussion to follow it will be assumed that all of the distributions discussed are unimodal. Note that the number of scores in the left and right tail of **Distribution B** are identical. The **tail** of a distribution refers to the upper/right and lower/left extremes of the distribution. When one tail is heavier than another tail it means that a greater proportion of the scores fall in that tail. In **Distribution B** the two tails are equally weighted.

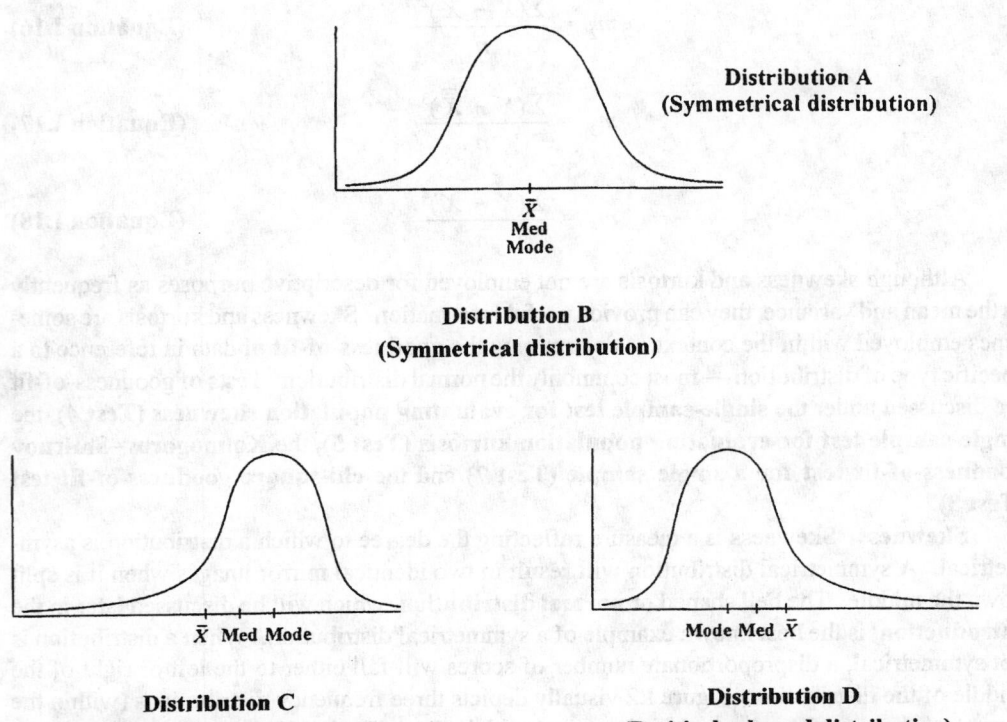

Figure I.2   Symmetrical and Asymmetrical Distributions

Turning to the other two distributions, we can state that **Distribution C** is **negatively skewed** (or as it is sometimes called, **skewed to the left**), while **Distribution D** is **positively skewed** (or as it is sometimes called, **skewed to the right**). Note that in **Distribution C** the bulk of the scores fall in the right end of the distribution. This is the case, since the "hump" or upper part of the distribution falls to the right. The tail or lower end of the distribution is on the left side (thus the term **skewed to the left**). **Distribution D**, on the other hand, is **positively skewed**, since the bulk of the scores fall in the left end of the distribution. This is the case, since the "hump" or upper part of the distribution falls to the left. The tail or lower end of the distribution is on the right (thus the term **skewed to the right**). It should be pointed out that **Distributions C** and **D** represent extreme examples of skewed distributions. Thus, distributions can be characterized by skewness, yet not have the imbalance between the left and right tails depicted for **Distributions C** and **D**.

As a general rule, based on whether a distribution is symmetrical, skewed negatively, or skewed positively, one can make a determination with respect to the relative magnitude of the three measures of central tendency discussed earlier in the **Introduction**. In a perfectly symmetrical unimodal distribution the mean, median, and mode will always be the same value. In a skewed distribution the mean, median, and mode will not be the same value. Typically (although there are exceptions), in a negatively skewed distribution, the mean is the lowest value followed by the median and then the mode, which is the highest value. The reverse is the case in a positively skewed distribution, where the mean is the highest value followed by the median, with the mode being the lowest value. The easiest way to remember the arrangement of the three measures of central tendency in a skewed distribution is that they are arranged alphabetically moving in from the tail of the distribution to the highest point in the distribution.

Since a measure of central tendency is supposed to reflect the most representative score for the distribution (although the word "tendency" implies that it may not be limited to a single value), the specific measure of central tendency that is employed for descriptive or inferential purposes should be a function of the shape of a distribution. In the case of a unimodal distribution that is perfectly symmetrical, the mean (which will always be the same value as the median and mode) will be the best measure of central tendency to use, since it employs the most information. When a distribution is skewed, it is often preferable to employ the median as a measure of central tendency in lieu of the mean. Other circumstances where it may be more desirable to employ the median rather than the mean as a measure of central tendency are discussed in Section VI of the *t* **test for two independent samples** under the discussion of **outliers and data transformation**.

A simple method of estimating skewness for a sample is to compute the value *sk*, which represents the **Pearsonian coefficient of skewness** (developed in the 1890s by the English statistician Karl Pearson). Equation I.19 is employed to compute the value of *sk*, which for the distribution summarized in Table I.2 is computed to be *sk* = 1.04. The notation *M* in Equation I.19 represents the median of the sample.

$$sk = \frac{3(\bar{X} - M)}{\tilde{s}} = \frac{3(10.4 - 9)}{4.04} = 1.04 \qquad \textbf{(Equation I.19)}$$

The value of *sk* will fall within the range −3 to +3, with a value of 0 associated with a perfectly symmetrical distribution. Note that when $\bar{X} > M$, *sk* will be a positive value, and the larger the value of *sk*, the greater the degree of positive skew. When $\bar{X} < M$, *sk* will be a negative value, and the larger the **absolute value** of *sk*, the greater the degree of negative skew.[6] Note that when $\bar{X} = M$, which will be true if a distribution is symmetrical, *sk* = 0.[7]

To illustrate the above, consider the following three distributions **E**, **F**, and **G**, each of which is comprised of 10 scores. **Distribution E** is symmetrical, **Distribution F** is negatively skewed, and **Distribution G** is positively skewed. The value of *sk* is computed for each distribution.

1) **Distribution E**: 0, 0, 0, 5, 5, 5, 5, 10, 10, 10

The following sample statistics can be computed for **Distribution E**: $\bar{X}_E = 5$; $M_E = 5$; $\tilde{s}_E = 4.08$; $sk_E = [3(5 - 5)]/4.08 = 0$. The value $sk_E = 0$ indicates that **Distribution E** is **symmetrical**. Consistent with the fact that it is symmetrical is that the values of the mean and median are equal. In addition, since the scores are distributed evenly throughout the distribution, both tails are identical in appearance.

2) **Distribution F**: 0, 1, 1, 9, 9, 10, 10, 10, 10, 10

The following sample statistics can be computed for **Distribution F**: $\bar{X}_F = 7$; $M_F = 9.5$; $\tilde{s}_F = 4.40$; $sk_F = [3(7 - 9.5)]/4.40 = -1.70$. The negative value $sk_F = -1.70$ indicates that **Distribution F** is **negatively skewed**. Consistent with the fact that it is negatively skewed is that the value of the mean is less than the value of the median. In addition, the majority of the scores (i.e., the hump) fall in the right/upper end of the distribution. The lower end of the distribution is the tail on the left side.

3) **Distribution G**: 0, 0, 0, 0, 0 , 1, 1, 9, 9, 10

The following sample statistics can be computed for **Distribution G**: $\bar{X}_G = 3$; $M_G = .5$; $\tilde{s}_G = 4.40$; $sk_G = [3(3 - .5)]/4.40 = 1.70$. The positive value $sk_G = 1.70$ indicates that **Distribution G** is **positively skewed**. Consistent with the fact that it is positively skewed is that the value of the mean is greater than the value of the median. In addition, the majority of the scores (i.e., the hump) fall in the left/lower end of the distribution. The upper end of the distribution is the tail on the right side.

The most precise measure of skewness employs the exact value of the third moment about the mean, designated earlier as m_3. Cohen (1996) and Zar (1999) note that the unbiased estimate of the population parameter estimated by m_3 can be computed with either Equation I.20 (which is the definitional equation) or Equation I.21 (which is a computational equation).

$$m_3 = \frac{n\Sigma(X - \bar{X})^3}{(n - 1)(n - 2)} \qquad \textbf{(Equation I.20)}$$

$$m_3 = \frac{n\Sigma X^3 - 3\Sigma X\Sigma X^2 + \dfrac{2(\Sigma X)^3}{n}}{(n - 1)(n - 2)} \qquad \textbf{(Equation I.21)}$$

Note that in Equation I.20, the notation $\Sigma(X - \bar{X})^3$ indicates that the mean is subtracted from each of the n scores in the distribution, each difference score is cubed, and the n cubed difference scores are summed. The notation ΣX^3 in Equation I.21 indicates each of the n scores is cubed, and the n cubed scores are summed. The notation $(\Sigma X)^3$ in Equation I.21 indicates that the n scores are summed, and the resulting value is cubed. Note that the minimum sample size required to compute skewness is $n = 3$, since any lower value will result in a zero in the denominators of Equations I.20 and I.21, rendering them insoluble.

Since the value computed for m_3 is in cubed units, the unitless statistic g_1, which is an estimate of the population parameter γ_1 (where γ represents the lower case Greek letter **gamma**), is commonly employed to express skewness. The value of g_1 (which is often referred to as a **skewness coefficient**) is computed with Equation I.22. Readers should take note of the fact that most computer software (e.g., *SPSS*) prints out the value of g_1 to represent the skewness of a distribution.

$$g_1 = \frac{m_3}{\tilde{s}^3} \qquad \textbf{(Equation I.22)}$$

When a distribution is symmetrical (about the mean), the value of g_1 will equal 0. When the value of g_1 is significantly above 0, a distribution will be positively skewed, and when it is significantly below 0, a distribution will be negatively skewed. Although the normal distribution is symmetrical (with $g_1 = 0$), as noted earlier, not all symmetrical distributions are normal. In other words, although a normal distribution will have a skewness coefficient of $g_1 = 0$, not all distributions with the latter skewness coefficient are normal. Examples of nonnormal distributions that are symmetrical are the t distribution and the binomial distribution, when $\pi_1 = .5$ (the meaning of the notation $\pi_1 = .5$ is explained in Section I of the **binomial sign test for a single sample**).

Zar (1999) notes that a population parameter designated $\sqrt{\beta_1}$ (where β represents the lower case Greek letter **beta**) is employed by some sources (e.g., D'Agostino (1970, 1986) and D'Agostino *et al.* (1990)) to represent skewness. Equation I.23 is used to compute $\sqrt{b_1}$, which is the sample statistic employed to estimate the value of $\sqrt{\beta_1}$.

$$\sqrt{b_1} = \frac{(n - 2)g_1}{\sqrt{n(n - 1)}} \qquad \textbf{(Equation I.23)}$$

When a distribution is symmetrical, the value of $\sqrt{b_1}$ will equal 0. When the value of $\sqrt{b_1}$ is significantly above 0, a distribution will be positively skewed, and when it is significantly below 0, a distribution will be negatively skewed. The method for determining whether a g_1 and/or $\sqrt{b_1}$ value deviates significantly from 0 is described under the **single-sample test for evaluating population skewness**. The results of the latter test, along with the results of the **single-sample test for evaluating population kurtosis**, are used in the **D'Agostino–Pearson test of normality (Test 5a)** and the **Jarque–Bera test of normality (Test 5b)**, both of which are employed to assess goodness-of-fit for normality (i.e., whether sample data are likely to have been derived from a normal distribution). When a distribution is normal, both g_1 and $\sqrt{b_1}$ will equal 0.

Table I.3 Computation of Skewness for Distribution E

X	X^2	X^3	$\bar{X}$	$(X - \bar{X})$	$(X - \bar{X})^2$	$(X - \bar{X})^3$
0	0	0	5	−5	25	−125
0	0	0	5	−5	25	−125
0	0	0	5	−5	25	−125
5	25	125	5	0	0	0
5	25	125	5	0	0	0
5	25	125	5	0	0	0
5	25	125	5	0	0	0
10	100	1000	5	5	25	0
10	100	1000	5	5	25	125
10	100	1000	5	5	25	125

Sums: $\Sigma X = 50$, $\Sigma X^2 = 400$, $\Sigma X^3 = 3500$, $\Sigma(X - \bar{X}) = 0$, $\Sigma(X - \bar{X})^2 = 150$, $\Sigma(X - \bar{X})^3 = 0$

$$\bar{X}_E = \frac{\Sigma X}{n} = \frac{50}{10} = 5 \qquad \tilde{s}_E = \sqrt{\frac{\Sigma X^2 - \frac{(\Sigma X)^2}{n}}{n - 1}} = \sqrt{\frac{400 - \frac{(50)^2}{10}}{10 - 1}} = 4.08$$

$$m_{3_E} = \frac{n\Sigma(X - \bar{X})^3}{(n - 1)(n - 2)} = \frac{(10)(0)}{(10 - 1)(10 - 2)} = 0$$

$$m_{3_E} = \frac{n\Sigma X^3 - 3\Sigma X\Sigma X^2 + \frac{2(\Sigma X)^3}{n}}{(n - 1)(n - 2)} = \frac{(10)(3500) - (3)(50)(400) + \frac{(2)(50)^3}{10}}{(10 - 1)(10 - 2)} = 0$$

$$g_{1_E} = \frac{m_{3_E}}{\tilde{s}_E^3} = \frac{0}{(4.08)^3} = 0 \qquad \sqrt{b_{1_E}} = \frac{(n - 2)g_{1_E}}{\sqrt{n(n - 1)}} = \frac{(10 - 2)(0)}{\sqrt{(10(10 - 1)}} = 0$$

At this point employing Equations I.20/I.21, I.22, and I.23, the values of m_3, g_1, and $\sqrt{b_1}$ will be computed for **Distributions E, F**, and **G** discussed earlier in this section. Tables I.3–I.5 summarize the computations, with the following resulting values: $m_{3_E} = 0$, $m_{3_F} = -86.67$, $m_{3_G} = 86.67$, $g_{1_E} = 0$, $g_{1_F} = -1.02$, $g_{1_G} = 1.02$, and $\sqrt{b_{1_E}} = 0$, $\sqrt{b_{1_F}} = -.86$, $\sqrt{b_{1_G}} = .86$.

The use of $\sqrt{\beta_1}$ as a measure of skewness will be further utilized in the chapter on the **single-sample test for evaluating population skewness** when the value of $\sqrt{\beta_1}$ is employed to evaluate a hypothesis about the skewness of an underlying population distribution. The measure of skewness readers are most likely to encounter in other sources (as well as in most computer

Table I.4   Computation of Skewness for Distribution F

X	X^2	X^3	$\bar{X}$	$(X - \bar{X})$	$(X - \bar{X})^2$	$(X - \bar{X})^3$
0	0	0	7	−7	49	−343
1	1	1	7	−6	36	−216
1	1	1	7	−6	36	−216
9	81	729	7	2	4	8
9	81	729	7	2	4	8
10	100	1000	7	3	9	27
10	100	1000	7	3	9	27
10	100	1000	7	3	9	27
10	100	1000	7	3	9	27
10	100	1000	7	3	9	27

Sums: $\Sigma X = 70$, $\Sigma X^2 = 664$, $\Sigma X^3 = 6460$, $\Sigma(X - \bar{X}) = 0$, $\Sigma(X - \bar{X})^2 = 174$, $\Sigma(X - \bar{X})^3 = -624$

$$\bar{X}_F = \frac{\Sigma X}{n} = \frac{70}{10} = 7 \qquad \tilde{s}_F = \sqrt{\frac{\Sigma X^2 - \frac{(\Sigma X)^2}{n}}{n - 1}} = \sqrt{\frac{664 - \frac{(70)^2}{10}}{10 - 1}} = 4.40$$

$$m_{3_F} = \frac{n\Sigma(X - \bar{X})^3}{(n - 1)(n - 2)} = \frac{(10)(-624)}{(10 - 1)(10 - 2)} = -86.67$$

$$m_{3_F} = \frac{n\Sigma X^3 - 3\Sigma X\Sigma X^2 + \frac{2(\Sigma X)^3}{n}}{(n - 1)(n - 2)} = \frac{(10)(6460) - (3)(70)(664) + \frac{(2)(70)^3}{10}}{(10 - 1)(10 - 2)} = -86.67$$

$$g_{1_F} = \frac{m_{3_F}}{\tilde{s}_F^3} = \frac{-86.67}{(4.40)^3} = -1.02 \qquad \sqrt{b_{1_F}} = \frac{(n - 2)g_{1_F}}{\sqrt{n(n - 1)}} = \frac{(10 - 2)(-1.02)}{\sqrt{(10)(10 - 1)}} = -.86$$

software packages) will be the value of g_1 computed with Equation I.22, which represents a **standardized unitless/dimensionless measure of skewness** (which translates into the fact that you can add, subtract, divide, or multiply all values in the distribution by a constant and not change the shape of the distribution).

Readers will find that many sources state that Equation I.24 is the equation for computing the population skewness. The latter equation, in fact, also represents a standardized unitless/dimensionless measure of skewness

$$Skewness = \frac{\Sigma(X - \mu)^3}{N\sigma^3} \qquad \textbf{(Equation I.24)}$$

Depending upon the source, Equation I.25, I.26, or I.27 may be listed as the equation for estimating the skewness of a population from sample data. Most sources state that the estimate provided by Equation I.25 will be biased (it underestimates the value of the population skewness) for small sample sizes, and thus Equation I.26 (which, in fact, is also biased) and Equation I.27 (cited in Cohen (2001, p. 82)) may be employed to correct for bias. Of the three equations, Equation I.27 is the least biased, and will result in the largest value for skewness.

The value for skewness is computed below for **Distribution G** with all three equations. In the calculations conducted with respect to the latter distribution, the following values are employed in Equations I.25–I.27: a) $\Sigma(X - \bar{X})^3 = 624$ (the latter value is obtained by subtracting the mean of the distribution (which is $\bar{X} = 3$) from each of the $n = 10$ scores in the distribution (yielding the 10 difference scores −3, −3, −3, −3, −3, −2, −2, 6, 6, 7), cubing each of the 10 difference scores (yielding the 10 cubed values $-3^3 = -27$, $-3^3 = -27$, $-3^3 = -27$, $-3^3 = -27$, -3^3

Table I.5   Computation of Skewness for Distribution G

X	X^2	X^3	$\bar{X}$	$(X - \bar{X})$	$(X - \bar{X})^2$	$(X - \bar{X})^3$
0	0	0	3	−3	9	−27
0	0	0	3	−3	9	−27
0	0	0	3	−3	9	−27
0	0	0	3	−3	9	−27
0	0	0	3	−3	9	−27
1	1	1	3	−2	4	−8
1	1	1	3	−2	4	−8
9	81	729	3	6	36	216
9	81	729	3	6	36	216
10	100	1000	3	7	49	343

Sums: $\Sigma X = 30$, $\Sigma X^2 = 264$, $\Sigma X^3 = 2460$, $\Sigma(X - \bar{X}) = 0$, $\Sigma(X - \bar{X})^2 = 174$, $\Sigma(X - \bar{X})^3 = 624$

$$\bar{X}_G = \frac{\Sigma X}{n} = \frac{30}{10} = 3 \qquad \tilde{s}_G = \sqrt{\frac{\Sigma X^2 - \frac{(\Sigma X)^2}{n}}{n - 1}} = \sqrt{\frac{264 - \frac{(30)^2}{10}}{10 - 1}} = 4.40$$

$$m_{3_G} = \frac{n\Sigma(X - \bar{X})^3}{(n - 1)(n - 2)} = \frac{(10)(624)}{(10 - 1)(10 - 2)} = 86.67$$

$$m_{3_G} = \frac{n\Sigma X^3 - 3\Sigma X\Sigma X^2 + \frac{2(\Sigma X)^3}{n}}{(n - 1)(n - 2)} = \frac{(10)(2460) - (3)(30)(264) + \frac{(2)(30)^3}{10}}{(10 - 1)(10 - 2)} = 86.67$$

$$g_{1_G} = \frac{m_{3_G}}{\tilde{s}_G^3} = \frac{86.67}{(4.40)^3} = 1.02 \qquad \sqrt{b_{1_G}} = \frac{(n - 2)g_{1_G}}{\sqrt{n(n - 1)}} = \frac{(10 - 2)(1.02)}{\sqrt{(10)(10 - 1)}} = .86$$

$= -27$, $-2^3 = -8$, $-2^3 = -8$, $6^3 = 216$, $6^3 = 216$, $7^3 = 343$), and obtaining the sum of the 10 cubed difference scores which is 624); b) $\tilde{s} = 4.40$. Note that the value 1.017 computed with Equation I.27 is closest to the value $g_1 = 1.02$ computed with Equation I.22. The value computed for skewness by *SPSS* and *Minitab* is 1.02. Readers should take note of the fact that Equations I.25 and I.26 will tend to underestimate population skewness unless an extremely large sample size is employed.

$$\text{Skewness} = \frac{\Sigma(X - \mu)^3}{n\tilde{s}^3} = \frac{624}{(10)(4.40)^3} = .732 \qquad \textbf{(Equation I.25)}$$

$$\text{Skewness} = \frac{\Sigma(X - \bar{X})^3}{(n - 1)\tilde{s}^3} = \frac{624}{(10 - 1)(4.40)^3} = .814 \qquad \textbf{(Equation I.26)}$$

$$\textbf{(Equation I.27)}$$

$$\text{Skewness} = \left(\frac{n}{n - 2}\right)\left(\frac{\Sigma(X - \bar{X})^3}{(n - 1)\tilde{s}^3}\right) = \left(\frac{10}{10 - 2}\right)\left(\frac{624}{(10 - 1)(4.40)^3}\right) = 1.017$$

Kurtosis   According to D'Agostino *et al.* (1990), the word **kurtosis** means **curvature**. Kurtosis is generally defined as a measure reflecting the degree to which a distribution is **peaked**. To be more specific, kurtosis provides information regarding the height of a distribution relative

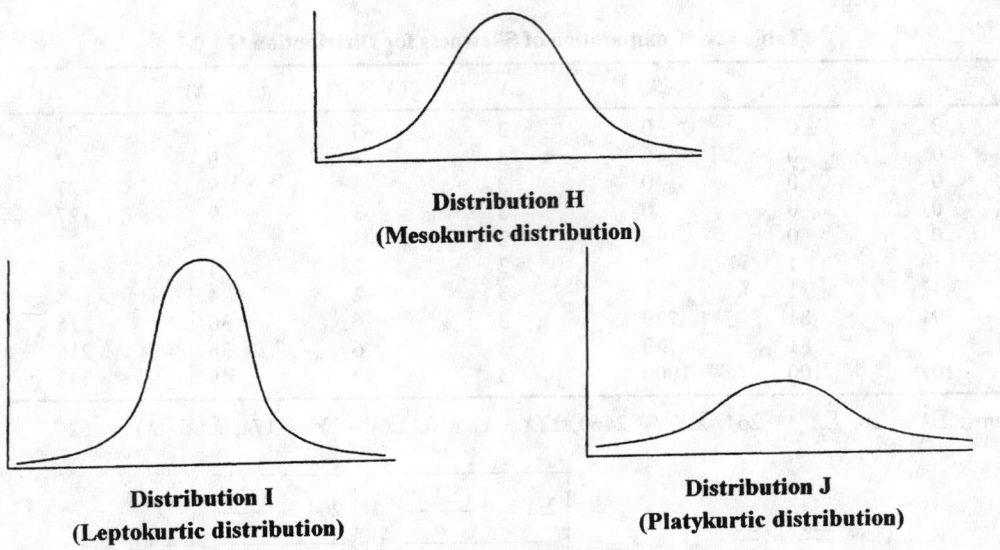

Figure I.3   Representative Types of Kurtosis

to the value of its standard deviation. The most common reason for measuring kurtosis is to determine whether or not data are derived from a normally distributed population. Kurtosis is often described within the framework of the following three general categories, all of which are depicted by representative frequency distributions in Figure I.3: **mesokurtic**, **leptokurtic**, and **platykurtic**.

A **mesokurtic** distribution, which has a degree of peakedness that is considered moderate, is represented by a **normal distribution** (i.e., the classic bell-shaped curve), which is depicted in Figure I.3. All normal distributions are mesokurtic, and the weight/thickness of the tails of a normal distribution is in between the weight/thickness of the tails of distributions that are leptokurtic or platykurtic. In Figure I.3, **Distribution H** best approximates a **mesokurtic** distribution.

A **leptokurtic** distribution is characterized by a high degree of peakedness. The scores in a leptokurtic distribution tend to be clustered much more closely around the mean than they are in either a mesokurtic or platykurtic distribution. Because of the latter, the value of the standard deviation for a leptokurtic distribution will be smaller than the standard deviation for the latter two distributions (if we assume the range of scores in all three distributions is approximately the same). The tails of a leptokurtic distribution are heavier/thicker than the tails of a mesokurtic distribution. In Figure I.3, **Distribution I** best approximates a **leptokurtic** distribution.

A **platykurtic** distribution is characterized by a low degree of peakedness. The scores in a playtykurtic distribution tend to be spread out more from the mean than they are in either a mesokurtic or leptokurtic distribution. Because of the latter, the value of the standard deviation for a platykurtic distribution will be larger than the standard deviation for the latter two distributions (if we assume the range of scores in all three distributions is approximately the same). The tails of a platykurtic distribution are lighter/thinner than the tails of a mesokurtic distribution. In Figure I.3, **Distribution J** best approximates a **platykurtic** distribution.

Moors (1986) defines kurtosis as the degree of dispersion between the points marked off on the abscissa (X-axis) that correspond to $\mu \pm \sigma$. Thus, with respect to the three types of distributions, we can make the statement that the range of values on the abscissa that fall between the population mean (μ) and one standard deviation above and below the mean will be greatest for a platykurtic distribution and smallest for a leptokurtic distribution, with a mesokurtic distribution being in the middle. As will be noted later in the **Introduction**, in the case of a normal distribution(which, as noted earlier, will always be mesokurtic), approximately 68% of the scores will always fall between the mean and one standard deviation above and below the mean.

One crude way of estimating kurtosis is that if the standard deviation of a unimodal symmetrical distribution is approximately one-sixth the value of the range of the distribution, the distribution is mesokurtic. In the case of a leptokurtic distribution, the standard deviation will be substantially less than one-sixth of the range, while in the case of a platykurtic distribution the standard deviation will be substantially greater than one-sixth of the range. To illustrate, let us assume that the range of values on an IQ test administered to a large sample is 90 points (e.g., the IQ scores fall in the range 55 to 145). If the standard deviation for the sample equals 15, the distribution would be mesokurtic (since $15/90 = 1/6$). If the standard deviation equals 5, the distribution would be leptokurtic (since $5/90 = 1/18$, which is substantially less than $1/6$.). If the standard deviation equals 30, the distribution would be platykurtic (since $30/90 = 1/3$, which is substantially greater than $1/6$).

A number of alternative measures for kurtosis have been developed, including one developed by Moors (1988) and described in Zar (1999). The latter measure computes kurtosis by employing specific quantile values in the distribution. The most precise method for estimating the kurtosis of a population can be computed through use of Equation I.28 (which is a definitional equation) or Equation I.29 (which is a computational equation) which compute a statistic to be designated k_4. The reader should take note of the fact that k_4 computed with Equations I.28/I.29 does not in actuality represent the value of the sample fourth moment about the mean – i.e., it is not an estimate of m_4 (which is defined by Equation I.18). The latter is the case because although k_4 can assume a negative value, m_4 cannot (since raising a number to the fourth power, which is the case in Equation I.18, will always yield a positive number).[8]

(Equation I.28)

$$k_4 = \frac{[[\Sigma(X - \bar{X})^4(n)(n + 1)]/(n - 1)] - 3[\Sigma(X - \bar{X})^2]^2}{(n - 2)(n - 3)}$$

(Equation I.29)

$$k_4 = \frac{(n^3 + n^2)\Sigma X^4 - 4(n^2 + n)\Sigma X^3\Sigma X - 3(n^2 - n)(\Sigma X^2)^2 + 12n\Sigma X^2(\Sigma X)^2 - 6(\Sigma X)^4}{n(n - 1)(n - 2)(n - 3)}$$

Note that in Equation I.28, the notation $\Sigma(X - \bar{X})^4$ indicates that the mean is subtracted from each of the n scores in the distribution, each difference score is raised to the fourth power, and the n difference scores raised to the fourth power are summed. The notation ΣX^4 in Equation I.29 indicates each of the n scores is raised to the fourth power, and the n resulting values are summed. The notation $(\Sigma X)^4$ in Equation I.29 indicates that the n scores are summed, and the resulting value is raised to the fourth power. Note that the minimum sample size required to compute kurtosis is $n = 4$, since any lower value will result in a zero in the denominators of Equations I.28 and I.29, rendering them insoluble.

Since the value computed for k_4 is in units of the fourth power, the unitless statistic g_2, which is an estimate of the population parameter γ_2, is commonly employed to express kurtosis. The value of g_2 is computed with Equation I.30. When a distribution is mesokurtic the value of g_2 will equal 0. When the value of g_2 is significantly above 0, a distribution will be leptokurtic, and when it is significantly below 0, a distribution will be platykurtic. Readers should take note of the fact that most computer software (e.g., *SPSS*) prints out the value of g_2 to represent the kurtosis of a distribution.

$$g_2 = \frac{k_4}{\hat{s}^4} \qquad \text{(Equation I.30)}$$

Zar (1999) notes that a population parameter designated β_2 is employed by some sources (e.g., Anscombe and Glynn (1983), D'Agostino (1986), and D'Agostino et al. (1990)) to represent kurtosis. Equation I.31 is used to compute b_2, which is the sample statistic employed to estimate the value of β_2. Some sources refer to g_2 as a **kurtosis coefficient**, while other sources employ b_2 within the latter context.

$$b_2 = \frac{(n - 2)(n - 3)g_2}{(n + 1)(n - 1)} + \frac{3(n - 1)}{n + 1} \qquad \text{(Equation I.31)}$$

When a distribution is mesokurtic, the value of b_2 will equal $[3(n-1)]/(n+1)$. Inspection of the latter equation reveals that as the value of the sample size increases, the value of b_2 approaches 3. When the value computed for b_2 is significantly below $[3(n-1)]/(n+1)$, a distribution will be platykurtic. When the value computed for b_2 is significantly greater than $[3(n-1)]/(n+1)$, a distribution will be leptokurtic. The method for determining whether a g_2 and/or b_2 value is **statistically significant** is described under the **single-sample test for evaluating population kurtosis** (the concept of **statistical significance** is discussed later in the Introduction). The results of the latter test, along with the results of the **single-sample test for evaluating population skewness**, are used in the **D'Agostino–Pearson test of normality** and the **Jarque–Bera test of normality**, both of which are employed to assess goodness-of-fit for normality. As noted earlier, a normal distribution will always be mesokurtic, with $g_2 = 0$ and $b_2 = 3$. It should be noted, however, that although a normal distribution will have a kurtosis coefficient of $g_2 = 0$ or $b_2 = 3$, not all distributions with the latter kurtosis coefficient are normal.

At this point employing Equations I.28/I.29, I.30 and I.31, the values of k_4, g_2, and b_2 will be computed for two distributions to be designated **H** and **I**. The data for **Distributions H** and **I** are designed (within the framework of a small sample size with $n = 20$) to approximate a **leptokurtic distribution** and **platykurtic distribution** respectively. Tables I.6 and I.7 summarize the computations, with the following resulting values: $k_{4_H} = 307.170$, $k_{4_I} = -1181.963$, $g_{2_H} = 3.596$, $g_{2_I} = -.939$, and $b_{2_H} = 5.472$, $b_{2_I} = 1.994$.

The use of g_2 and b_2 as measures of kurtosis will be further utilized in the chapter on the **single-sample test for evaluating population kurtosis** when the value of g_2 is employed to evaluate a hypothesis about the kurtosis of an underlying population distribution. The measure of kurtosis readers are most likely to encounter in other sources (as well as in most computer software packages) will be a value that is comparable to the value computed for g_2 with Equation I.30, which represents a **standardized unitless/dimension less measure of kurtosis** (which translates into the fact that you can add, subtract, divide, or multiply all values in the distribution by a constant and not change the shape of the distribution).

Readers will find that many sources state that Equation I.32 is the equation for computing the population kurtosis. The latter equation, in fact, also represents a standardized unitless/dimensionless measure of kurtosis

$$\text{Kurtosis} = \frac{\Sigma(X - \mu)^4}{N\sigma^4} \qquad \text{(Equation I.32)}$$

Depending upon the source, Equation I.33, I.34, or I.35 may be listed as the equation for estimating the kurtosis of a population from sample data. Most sources state that the estimate provided by Equation I.33 will be biased (it underestimates the value of the population kurtosis)

Table I.6 Computation of Kurtosis for Distribution H

X	X^2	X^3	X^4	$\bar{X}$	$(X - \bar{X})$	$(X - \bar{X})^2$	$(X - \bar{X})^4$
2	4	8	16	10	−8	64	4096
7	49	343	2401	10	−3	9	81
8	64	512	4096	10	−2	4	16
8	64	512	4096	10	−2	4	16
8	64	512	4096	10	−2	4	16
9	81	729	6561	10	−1	1	1
9	81	729	6561	10	−1	1	1
9	81	729	6561	10	−1	1	1
10	100	1000	10000	10	0	0	0
10	100	1000	10000	10	0	0	0
10	100	1000	10000	10	0	0	0
10	100	1000	10000	10	0	0	0
11	121	1331	14641	10	1	1	1
11	121	1331	14641	10	1	1	1
11	121	1331	14641	10	1	1	1
12	144	1728	20736	10	2	4	16
12	144	1728	20736	10	2	4	16
12	144	1728	20736	10	2	4	16
13	169	2197	28561	10	3	9	81
18	324	5832	104976	10	8	64	4096

Sums: $\Sigma X = 200$, $\Sigma X^2 = 2176$, $\Sigma X^3 = 25280$, $\Sigma X^4 = 314056$
$\Sigma(X - \bar{X}) = 0$, $\Sigma(X - \bar{X})^2 = 176$, $\Sigma(X - \bar{X})^4 = 8456$

$$\bar{X}_H = \frac{\Sigma X}{n} = \frac{200}{20} = 10 \qquad \tilde{s}_H = \sqrt{\frac{\Sigma X^2 - \frac{(\Sigma X)^2}{n}}{n - 1}} = \sqrt{\frac{2176 - \frac{(200)^2}{20}}{20 - 1}} = 3.04$$

$$k_{4_H} = \frac{[[\Sigma(X - \bar{X})^4(n)(n + 1)]/(n - 1)] - 3[\Sigma(X - \bar{X})^2]^2}{(n - 2)(n - 3)}$$

$$= \frac{[[(8456)(20)(20 + 1)]/(20 - 1)] - 3(176)^2}{(20 - 2)(20 - 3)} = 307.170$$

$$k_{4_H} = \frac{(n^3 + n^2)\Sigma X^4 - 4(n^2 + n)\Sigma X^3\Sigma X - 3(n^2 - n)(\Sigma X^2)^2 + 12n\Sigma X^2(\Sigma X)^2 - 6(\Sigma X)^4}{n(n - 1)(n - 2)(n - 3)}$$

$$= [[(20)^3 + (20)^2](314056) - 4[(20)^2 + 20](25280)(200) - 3[(20)^2 - 20](2176)^2$$

$$+ 12(20)(2176)(200)^2 - 6(200)^4] / [(20)(20 - 1)(20 - 2)(20 - 3)] = 307.170$$

$$g_{2_H} = \frac{k_{4_H}}{\tilde{s}_H^4} = \frac{307.170}{(3.04)^4} = 3.596$$

$$b_{2_H} = \frac{(n - 2)(n - 3)g_{2_H}}{(n + 1)(n - 1)} + \frac{3(n - 1)}{(n + 1)} = \frac{(20 - 2)(20 - 3)(3.596)}{(20 + 1)(20 - 1)} + \frac{3(20 - 1)}{(20 + 1)} = 5.472$$

Table I.7 Computation of Kurtosis for Distribution I

X	X^2	X^3	X^4	$\bar{X}$	$(X - \bar{X})$	$(X - \bar{X})^2$	$(X - \bar{X})^4$
0	0	0	0	10	-10	100	10000
1	1	1	1	10	-9	81	6561
3	9	27	81	10	-7	49	2401
3	9	27	81	10	-7	49	2401
5	25	125	625	10	-5	25	625
5	25	125	625	10	-5	25	625
8	64	512	4096	10	-2	4	16
8	64	512	4096	10	-2	4	16
10	100	1000	10000	10	0	0	0
10	100	1000	10000	10	0	0	0
10	100	1000	10000	10	0	0	0
10	100	1000	10000	10	0	0	0
12	144	1728	20736	10	2	4	16
12	144	1728	20736	10	2	4	16
15	225	3375	50625	10	5	25	625
15	225	3375	50625	10	5	25	625
17	289	4913	83521	10	7	49	2401
17	289	4913	83521	10	7	49	2401
19	361	6859	130321	10	9	81	6561
20	400	8000	160000	10	10	10	10000

Sums: $\Sigma X = 200$, $\Sigma X^2 = 2674$, $\Sigma X^3 = 40220$, $\Sigma X^4 = 649690$

$\Sigma(X - \bar{X}) = 0$, $\Sigma(X - \bar{X})^2 = 674$, $\Sigma(X - \bar{X})^4 = 45290$

$$\bar{X}_I = \frac{\Sigma X}{n} = \frac{200}{20} = 10 \qquad \tilde{s}_I = \sqrt{\frac{\Sigma X^2 - \dfrac{(\Sigma X)^2}{n}}{n - 1}} = \sqrt{\frac{2674 - \dfrac{(200)^2}{20}}{20 - 1}} = 5.96$$

$$k_{4_I} = \frac{[[\Sigma(X - \bar{X})^4(n)(n + 1)]/(n - 1)] - 3[\Sigma(X - \bar{X})^2]^2}{(n - 2)(n - 3)}$$

$$= \frac{[[(45290)(20)(20 + 1)]/(20 - 1)] - 3(674)^2}{(20 - 2)(20 - 3)} = -1181.963$$

$$k_{4_I} = \frac{(n^3 + n^2)\Sigma X^4 - 4(n^2 + n)\Sigma X^3\Sigma X - 3(n^2 - n)(\Sigma X^2)^2 + 12n\Sigma X^2(\Sigma X)^2 - 6(\Sigma X)^4}{n(n - 1)(n - 2)(n - 3)}$$

$$= [[(20)^3 + (20)^2](649690) - 4[(20)^2 + 20](40220)(200) - 3[(20)^2 - 20](2674)^2$$

$$+ 12(20)(2674)(200)^2 - 6(200)^4] / [(20)(20 - 1)(20 - 2)(20 - 3)] = -1181.963$$

$$g_{2_I} = \frac{k_{4_I}}{\tilde{s}_I^4} = \frac{-1181.963}{(5.96)^4} = -.939$$

$$b_{2_I} = \frac{(n - 2)(n - 3)g_{2_I}}{(n + 1)(n - 1)} + \frac{3(n - 1)}{(n + 1)} = \frac{(20 - 2)(20 - 3)(-.939)}{(20 + 1)(20 - 1)} + \frac{3(20 - 1)}{(20 + 1)} = 1.994$$

for small sample sizes, and thus Equation I.34 (which, in fact, is also biased for small sample sizes) and Equation I.35 (cited in Cohen (2001, p. 84)) may be employed. Of the three equations, Equation I.35 is the least biased, and will result in a largest value for kurtosis.

The value for kurtosis is computed below for **Distribution H** with all three equations. In the calculations conducted with respect to the latter distribution, the following values are employed in Equations I.33–I.35: a) $\Sigma(X - \bar{X})^4 = 8456$ (the latter value is obtained by subtracting the mean of the distribution (which is $\bar{X} = 10$) from each of the $n = 20$ scores in the distribution, raising each of the 20 difference scores (which are $-8, -3, -2, -2, -2, -1, -1, -1,$ $0, 0, 0, 0, 1, 1, 1, 2, 2, 2, 3, 8$) to the fourth power (the values of which are $-8^4 = 4096, -3^4 = 81,$ $-2^4 = 16, -2^4 = 16, -2^4 = 16, -1^4 = 1, -1^4 = 1, -1^4 = 1, 0^4 = 0, 0^4 = 0, 0^4 = 0, 0^4 = 0, 1^4 = 1, 1^4$ $= 1, 1^4 = 1, 2^4 = 16, 2^4 = 16, 2^4 = 16, 3^4 = 81, 8^4 = 4096$), and obtaining the sum of the 20 difference scores raised to the fourth power which is 8456; b) $\tilde{s} = 3.04$. Note that the value 3.613 computed with Equation I.35 is closest to the value $g_2 = 3.596$ computed with Equation I.30. The value computed for kurtosis by *SASS* and *Minitab* is 3.58.

Tabachnick and Fidell (1996, p. 2001, p. 73–74) note that most computer software packages employ the value 0 to designate mesokurtosis. In the event Equations I.33 or I.34 (both of which, as noted, underestimate the population kurtosis) are employed to compute kurtosis, one would have to subtract 3 from the value computed for kurtosis with the latter two equations in order to have a value analogous to (but most likely smaller than) the kurtosis value printed by computer software. Since the value computed with Equation I.35 has the latter subtraction built into it (specifically, the second part of the equation involving the term preceded by the 3), the final value computed with Equation I.35 should be close to the kurtosis value printed by computer software (which, in fact, is the case, since the value 3.613 computed with Equation I.35 is quite close to 3.58).

$$Kurtosis = \frac{\Sigma(X - \bar{X})^4}{n\tilde{s}^4} = \frac{8456}{(20)(3.04)^4} = 4.950 \qquad \textbf{(Equation I.33)}$$

$$Kurtosis = \frac{\Sigma(X - \bar{X})^4}{(n-1)\tilde{s}^4} = \frac{8456}{(20-1)(3.04)^4} = 5.211 \qquad \textbf{(Equation I.34)}$$

(Equation I.35)

$$Kurtosis = \left(\frac{n(n+1)}{(n-2)(n-3)}\right)\left(\frac{\Sigma(X - \bar{X})^4}{(n-1)\tilde{s}^4}\right) - 3\left(\frac{(n-1)(n-1)}{(n-2)(n-3)}\right)$$

$$= \left(\frac{(20)(20+1)}{(20-2)(20-3)}\right)\left(\frac{8456}{(20-1)(3.04)^4}\right) - 3\left(\frac{(20-1)(20-1)}{(20-2)(20-3)}\right)$$

$$Kurtosis = 7.152 - 3.539 = 3.613$$

Visual Methods for Displaying Data

Tables and graphs Earlier in the **Introduction** the importance of employing tables and/or graphs in describing data was discussed briefly. In the latter discussion it was noted that prior to conducting a formal analysis on a set of data with an inferential statistical test, a researcher should examine the data through use of some sort of systematic visual analysis. To be more specific, in any sort of statistical analysis, be it descriptive or inferential, the researcher should evaluate the data with respect to such things as the shape of a distribution, the most prudent measure of central tendency to employ, the most appropriate criterion to employ for assessing variability, whether or not any **outliers** are present, etc. Although it is not the intent of this section to provide exhaustive coverage of the use of tables and graphs in summarizing data, a number of different methods of visual presentation which have been employed by statisticians for many years will be presented. (An excellent source on the use of graphs in summarizing data is *The visual display of data* by John Tufte (1983).) In addition, the subject of **exploratory data analysis (EDA)** will be discussed. **Exploratory data analysis** (which was developed by the statistician John Tukey in 1977) is an alternative approach which many statisticians believe allows a researcher to summarize data in a more meaningful way — more specifically, **EDA** is able to reveal information that might not be obvious through use of the more conventional methods of graphic and tabular display.

Commonly employed tables and graphs   Distribution A below is the same set of data presented earlier in the **Introduction** (which is summarized by the **frequency distribution** in Table I.1 and the **frequency polygon** depicted in Figure I.1). It will be employed to illustrate the procedures to be described in this section. It will be assumed that the scores in **Distribution A** represent a **discrete variable** which represents ratio level of measurement. (Earlier in the **Introduction** it was noted that a **discrete variable** can assume only a limited number of values.) With respect to the variable of interest, we will assume that a score can only assume an integer value. The mode, median, and mean of **Distribution A** are respectively 72, 69, and 66.8 (with $\bar{X} = (\Sigma X = 1336)/20 = 66.8$). The values computed for the **range**, the **second quartile** (i.e., the **25th percentile**), the **third quartile** (i.e., the **75th percentile**), and $\tilde{s}^2$ are respectively 74, 61.5, 73, and 14.00. (The computation of the values for the 25th and 75th percentiles are demonstrated in the latter part of this section within the context of the discussion of boxplots.)

Distribution A: *22, 55, 60, 61, 61, 62, 62, 63, 63, 67, 71, 71, 72, 72, 72, 74, 74, 76, 82, 96*

Table I.8 summarizes the data for **Distribution A**. The five columns which comprise Table I.8 can be broken down into the following types of tables (each comprised of two of the five columns) which are commonly employed in summarizing data: **Columns 1** and **2** of Table I.8 taken together constitute a **grouped frequency distribution**; **Columns 1** and **4** taken together constitute a **relative frequency distribution**; **Columns 1** and **3** taken together constitute a **cumulative frequency distribution**; **Columns 1** and **5** taken together constitute a **relative cumulative frequency distribution**.

At this point a variety of methods for summarizing data within a tabular and/or graphic format will be described. With respect to the use of graphs in summarizing data, Tufte (1983) notes that the following guidelines should be adhered to (which can also be applied in the construction of tables): a) A graph should only be employed if it provides a reader with a better understanding of the data; b) A graph should reflect the truth about a set of data in as clear and simple a manner as possible; c) A graph should be clearly labeled, such that the reader can discern the meaning of any relevant information displayed within it.

The grouped frequency distribution   Earlier in the **Introduction** (under the discussion of the **mode**) the data for **Distribution A** were summarized through use of Table I.1 and Figure I.1. Referring back to Table I.1, note that it is comprised of two columns. The left column contains each of the scores within the range of scores between 22 – 96 which occurs at least one time in the distribution. The right column contains the frequency for each of the scores recorded in the left column. At this point it should be emphasized that tables (as well as a graphs) are employed to summarize data as succinctly as possible. To be more specific, to make data more

Table I.8 Summary Table for Distribution A

Column 1	Column 2	Column 3	Column 4	Column 5
	Frequency	Cumulative	Proportion	Cumulative
X	(*f*)	frequency		proportion
90-99	1	20	1/20 = .05	1
80-89	1	19	1/20 = .05	.95
70-79	8	18	8/20 = .40	.90
60-69	8	10	8/20 = .40	.50
50-59	1	2	1/20 = .05	.10
40-49	0	1	0/20 = .00	.05
30-39	0	1	0/20 = .00	.05
20-29	1	1	1/20 = .05	.05
	n = 20		Sum = 1	

intelligible, a table or graph should organize it a way such that the relevant structural characteristics of the data are clearly delineated. Under certain conditions (e.g., when a broad range of values characterizes the scores the relevant variable may assume) it may be prudent to employ a **grouped frequency distribution**.

A **grouped frequency distribution** is a frequency distribution in which the scores have been grouped into **class intervals**. A **class interval** is a set of scores which contains two or more scores that fall within the range of scores for the relevant variable. A basic principle which should be adhered to in constructing a grouped frequency distribution is that it should summarize the information contained in a set of data without any loss of relevant information. As noted earlier, **Columns 1** and **2** of Table I.8 taken together constitute a **grouped frequency distribution**. In the latter distribution eight class intervals are employed, with each of the intervals containing 10 scores. The eight class intervals are listed in **Column 1** of the table. The determination with respect to the optimal number of class intervals to employ should be based on the principle that too few class intervals may obscure the underlying structure of the data, while too many class intervals may defeat the purpose of grouping the data by failing to summarize it succinctly.

Although there is no consensual rule of thumb for determining the optimal number of class intervals, some sources recommend that the number of intervals be approximately equal to the square root of the total number of scores in a distribution (i.e., $\sqrt{n}$). The use of eight class intervals in Table I.8 (which, since $n = 20$, is more than $\sqrt{20}$ = 4.47, which suggests the use of four or five class intervals) illustrates that a researcher has considerable latitude in determining how many class intervals best communicate the structure of a set of data. Two final points to be made with respect to a grouped frequency distribution are: a) As is the case in Table I.8, each of the class intervals should contain an equal number of scores; b) When a set of data is summarized within the format of a grouped frequency distribution and all the original n scores which comprise the distribution are not available, the mean of the distribution is computed as follows: 1) Multiply the **midpoint** of each class interval by the frequency for that class interval. The **midpoint of a class interval** is computed by dividing the sum of the **lowest** and **upper values** which define the class interval by 2. Thus, the respective midpoints of the eight class intervals in Table I.8 are 24.5, 34.5, 44.5, 54.5, 64.5, 74.5, 84.5, 94.5. When the latter values are multiplied by their respective frequencies, the following values are obtained: $(24.5)(1) = 24.5$; $(34.5)(0) = 0$; $(44.5)(0) = 0$; $(54.5)(1) = 54.5$; $(64.5)(8) = 516$; $(74.5)(8) = 596$; $(84.5)(1) = 84.5$; $(94.5)(1) = 94.5$. The latter products sum to 1370. When 1370 is divided by $n = 20$, the value 68.5 is computed for the mean of the distribution. Note that although the value 68.5 is close to, it is not identical to 66.8, the actual value of the mean computed earlier when the exact values of all 20 scores in the distribution are taken into account. There will generally be a slight discrepancy between a mean computed from a grouped frequency distribution when compared with the actual value of the mean, by virtue of the fact that the computations for the grouped frequency distribution employ the midpoints of class intervals in lieu of the exact scores. In point of fact, when data are only available within the format of a grouped frequency distribution, it will not be possible to determine the exact value of any statistic/parameter of a distribution — in other words, one will only be able to approximate values for the mean, median, mode, and variance, as well, as for skewness, and kurtosis. This is the case, since in order to compute the exact value of a statistic/parameter, it is necessary to know the value of each of the n scores which comprise the distribution.

Earlier in the **Introduction**, it was noted that a **frequency polygon** is a graph of a frequency distribution, and that Figure I.1 represents a frequency polygon of **Distribution A**. A frequency polygon can also be drawn for a grouped frequency distribution. In the latter case, a point is employed to represent the frequency of each class interval. The point representing a class interval is placed directly above the value of the midpoint of that interval (with the latter value being recorded on the X-axis). Figure I.4 represents a frequency polygon for the grouped frequency distribution of **Distribution A**. Note that the values 14.5 and 104.5, which are, respectively, the midpoints of the class intervals $10 - 19$ and $100 - 109$ (which are the class intervals adjacent to the lowest and highest class intervals recorded in Table I.8), are recorded on the X-axis of Figure I.4. The values 14.5 and 104.5 are included in the grouped frequency polygon in order that both ends of the polygon touch the X-axis.

At the beginning of this discussion it was noted that the variable in **Distribution A** is assumed to be a discrete variable which can only assume integer values. However, it is often the case that a researcher is required to construct a frequency distribution and/or polygon for a **continuous variable**. (Earlier in the **Introduction** it was noted that a **continuous variable** can assume any value within the range of scores that define the limits of that variable.) In the latter instance a distinction is made between the **apparent (or stated) limits** of a score (or interval) versus the **real (or actual) limits** of a score (or interval). The **apparent (or stated) limits** of a score or interval are the values employed in the first column of a frequency distribution or grouped frequency distribution. Thus in the case of Table I.8, the **apparent limits** for each of the intervals are the values listed in **Column 1**. If instead of employing a grouped frequency distribution, a simple ungrouped frequency distribution had been employed (with each of the scores between 22 and 96 listed in **Column 1** of the table), the latter values would represent the apparent limits. In the case of a continuous variable it is assumed that the **real limits** of an integer score extend one-half a unit below the **lower apparent limit** and one-half a unit above the **upper apparent limit**. Thus, in the case of the score 22, the lower real limit is $22 - .5 = 21.5$, and the upper real limit $22 + .5 = 22.5$. In the case of the class interval $20 - 29$, the lower real limit of the class interval is $20 - .5 = 19.5$ (i.e., .5 is subtracted from the lowest value in that class interval), and the upper real limit of the class interval is $29 + .5 = 29.5$ (i.e., .5 is added to the highest value in that class interval). The **width** of the latter class interval is assumed to be the difference between the upper real limit and the lower real limit — i.e., $29.5 - 19.5 = 10$ (which corresponds to the number of integer values between the apparent limits of that class interval).

Two additional points to take note of concerning real limits are the following: a) In the event a score is equivalent to a value which corresponds to both the lower real limit and upper real limit of adjacent class intervals (e.g., such as the value 29.5 in Table I.8, which is the upper real limit of the class interval $20 - 29$ and the lower real limit of the class interval $30 - 39$), one can employ a rule such as the following to assign that score to one of the two class intervals: The value is placed in the higher of the two class intervals if the first digit of the lower real limit of

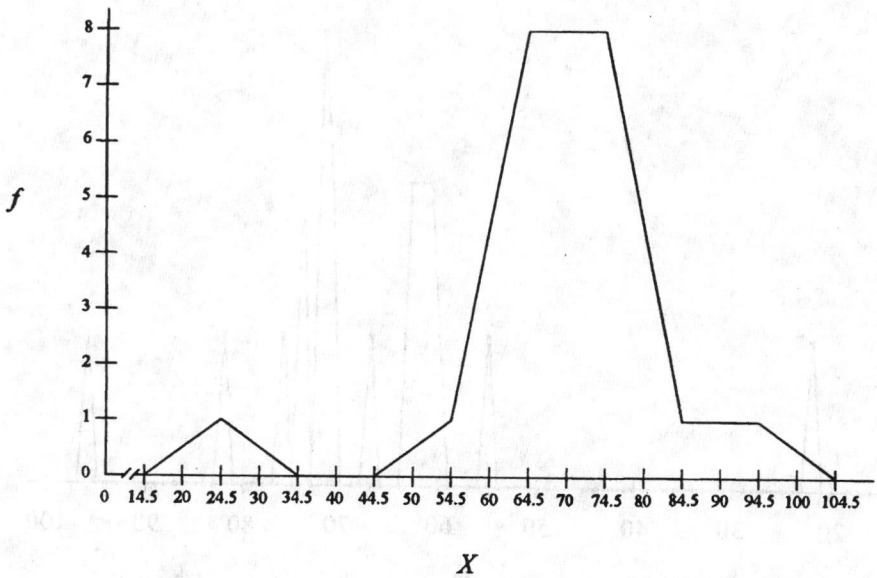

Figure I.4   Grouped Frequency Polygon of Distribution A

that class interval is even, and the value is placed in the lower of the two class intervals if the first digit of the lower real limit of that class interval is odd. Thus, the score 29.5 would be assigned to the class interval 30 − 40 (the real limits of which are 29.5 − 39.5), since the first digit of the lower real limit of that interval is 2 (which is an even number). Note that the real limits of the other class interval under consideration (20 − 29) are 19.5 − 29.5, and that the first digit of the lower real limit is 1 (which is an odd number); b) If the apparent limits of scores are expressed in decimal values, the one-half unit rule noted above for determining real limits is applied in reference to the relevant decimal unit of measurement. Specifically, if the apparent limits of a class interval are 20.0 − 20.5, the real limits would be 20.00 − .05 = 19.95 and 20.50 + .05 = 20.55; if the apparent limits of a class interval are 20.05 − 20.55, the real limits would be 20.050 − .005 = 20.045 and 20.550 + .005 = 20.555, etc. The practical application of the real limits is illustrated later in this section within the protocol for drawing a histogram of a grouped frequency distribution for a continuous variable (Figure I.8).

The relative frequency distribution   **Columns 1** and **4** of Table I.8 taken together constitute a **relative frequency distribution** (more specifically, it can be referred to as a **relative grouped frequency distribution**, since it is based on data grouped in class intervals). Both a **relative frequency distribution** (which is a table) and a **relative frequency polygon** (which is a graph of a relative frequency distribution) employ proportions or percentages instead of frequencies. In a relative frequency distribution, the proportion or percentage of cases for each score (or in the case of a grouped frequency distribution, the proportion or percentage of cases in each class interval) are recorded in the column adjacent to the score (or class interval). In a **relative frequency polygon** (which Figure I.5 represents), the *Y*-axis is employed to represent the proportion or percentage of subjects who obtain a given score, instead of the actual number of subjects who obtain that score (which is the case with a frequency polygon). The use of **relative frequencies** (which proportions or percentages represent) allows a researcher to compare two or more distributions with unequal sample sizes — something the researcher could not easily do if frequencies were represented of the *Y*-axis.

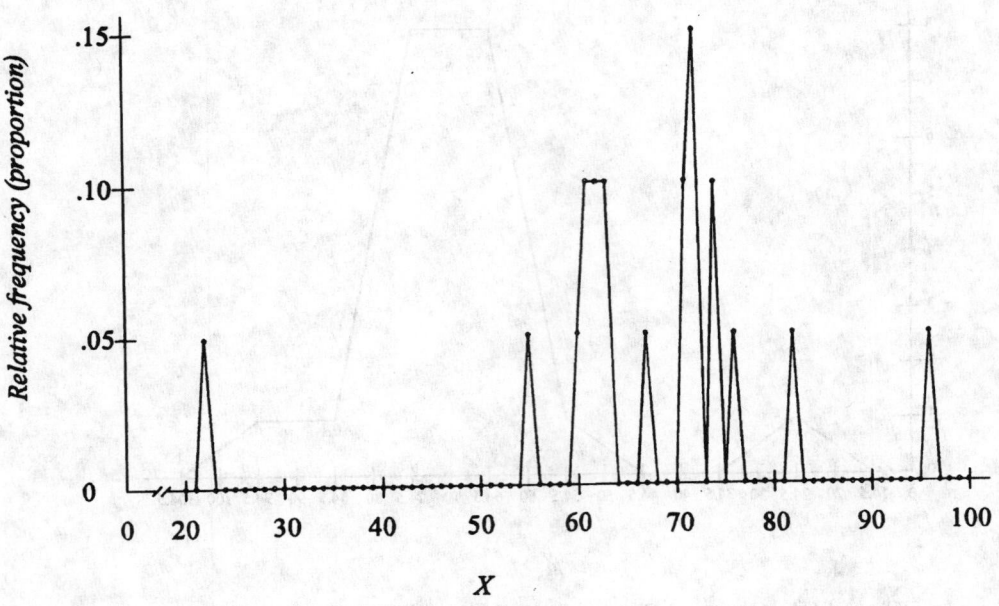

Figure I.5   Relative Frequency Polygon of Distribution A

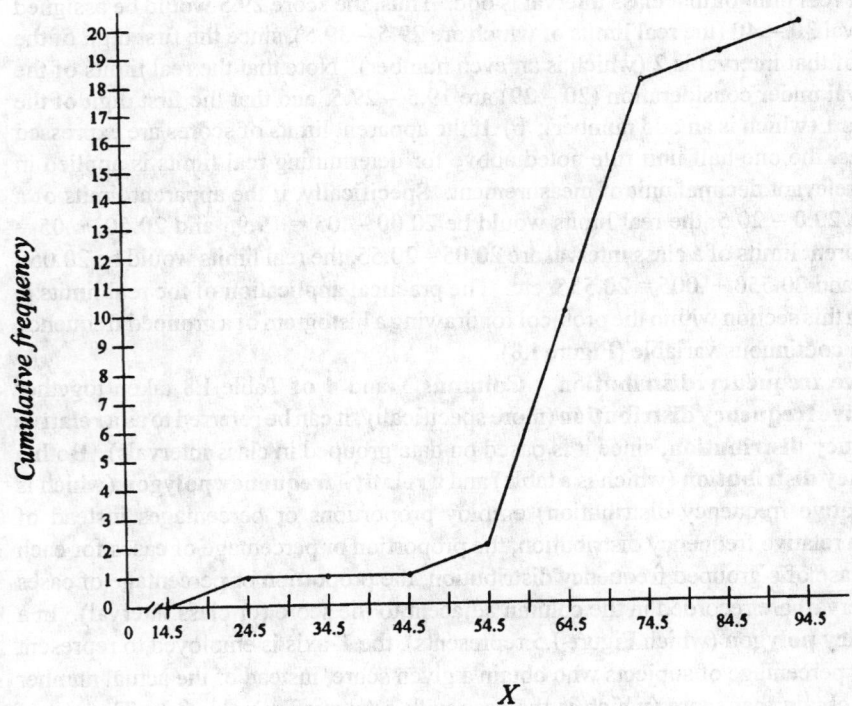

Figure I.6   Cumulative Grouped Frequency Polygon of Distribution A

Note that in Figure I.5 the relative frequencies are recorded for all scores within the range 20 – 100. Any score that occurs three times (i.e., the score of 72) has a relative frequency of 3/20 = .15. Any score that occurs two times (i.e., the scores of 61, 62, 63, 71, and 74) has a relative frequency of 2/20 = .10. Any score that occurs one time (i.e., the scores of 22, 55, 60, 67, 76, 82, and 96) has a relative frequency of 1/20 = .05. Any score in the range of scores 20 – 100 that does not occur has a relative frequency of 0/20 = 0. Although not demonstrated in this section, it is possible to construct a graph of a **relative grouped frequency distribution** – i.e., a **relative grouped frequency polygon**. In such a case, the relative frequency for each class interval is recorded above the midpoint of that class interval at the appropriate height on the Y-axis.

The cumulative frequency distribution   Columns 1 and 3 of Table I.8 taken together constitute a **cumulative frequency distribution** (more specifically, it can be referred to a **cumulative grouped frequency distribution**, since it is based on data grouped in class intervals). In a **cumulative frequency distribution** the frequency recorded for each score represents the frequency of that score plus the frequencies of all scores which are less than that score. In the case of a grouped frequency distribution, the frequency recorded for each class interval represents the frequency of that class interval plus the frequencies for all class intervals which fall below that class interval. Scores are arranged ordinally, with the lowest score/class interval at the bottom of the distribution, and the highest score/class interval at the top of the distribution. The cumulative frequency for the lowest score/class interval will simply be the frequency for that score/class interval, since there are no scores/class intervals below it. On the other hand, the cumulative frequency for the highest score/class interval will always equal n, the total number of scores in the distribution.

Figure I.6 is a **cumulative frequency polygon** of the grouped frequency distribution summarized by **Columns 1** and **3** of Table I.8 (more specifically, it can be referred to as a **cumulative grouped frequency polygon**, since it is based on data grouped in class intervals). Note that in Figure I.6 the cumulative frequency for each class interval is recorded above the midpoint of the class interval. The S shaped curve which describes the shape of a cumulative frequency polygon is commonly referred to as an **ogive**.

The relative cumulative frequency distribution   Columns 1 and 5 of Table I.8 taken together constitute a **relative cumulative frequency distribution** (more specifically, it can be referred to as a **relative cumulative grouped frequency distribution**, since it is based on data grouped in class intervals). In a **relative cumulative frequency distribution**, cumulative proportions or percentages are employed for each score (or class interval) in lieu of cumulative frequencies. In the case of a relative cumulative grouped frequency distribution, the cumulative proportion or percentage is recorded for each class interval. The cumulative proportion (or cumulative percentage) for the lowest score/class interval) will simply be the cumulative proportion (cumulative percentage) for that score/class interval, since there are no scores/class intervals below it. On the other hand, the cumulative proportion (cumulative percentage) for the highest score/class interval will always equal 1 (or 100%).

Additional graphing techniques   This section will describe the following alternative graphing techniques: a) Bar graph; b) Histogram.

Bar graph   Bar graphs are employed when nominal or ordinal data/information are recorded on the X-axis, or when interval or ratio data representing a **discrete variable** are recorded on the X-axis. If we assume that the variable being measured in **Distribution A** is a discrete variable (to be more specific, if we assume that in **Distribution A** scores can only be integer values), we can employ a bar graph such as Figure I.7 to summarize the frequency distribution of the data. In point of fact, Figure I.7, provides the same information presented in

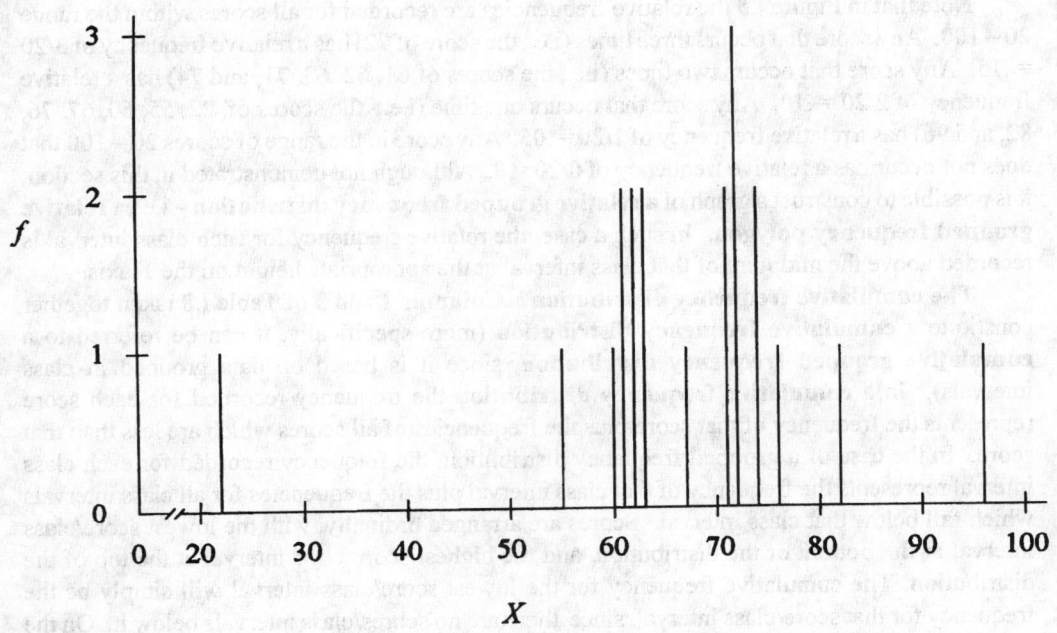

Figure I.7   Bar Graph of Frequency Distribution of Distribution A

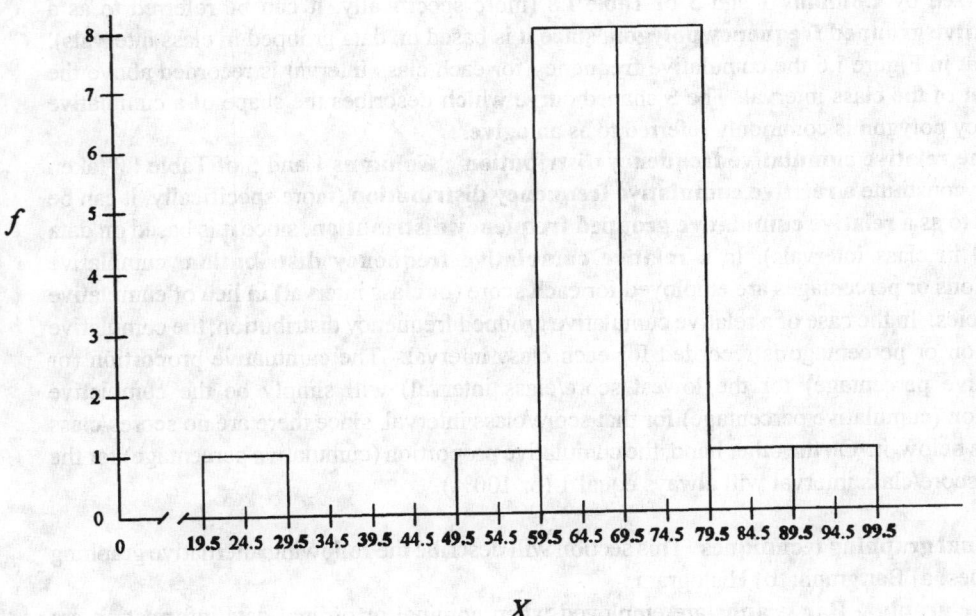

Figure I.8   Histogram of Grouped Frequency Distribution of Distribution A

Figure I.1. Note that in Figure I.7 the top of each of the vertical bars/lines corresponds to each of the points which are connected with one another in the frequency polygon depicted in Figure I.1. (As is the case with the frequency polygon, any score which has a frequency of zero falls on the X-axis, and thus no bar is employed for such scores in Figure I.7.) Note that in a bar graph, since each of the vertical bars is assumed to represent a discrete entity (or value), the bars are not contiguous (i.e., do not touch), although all of the bars are parallel to one another.[9]

Although not demonstrated in this section, it is possible to draw a bar graph for the grouped frequency distribution. In such a case, separate bars are drawn for each class interval noted in Table I.8. The width of each bar will correspond to the width of the **apparent limit** listed in **Column 1** for the class interval. As is the case in Figure I.7, if a discrete variable is involved the bars will not be contiguous with one another.

Histogram A **histogram** is a **bar graph** which is employed to reflect frequencies for **continuous data** In a histogram, since each of the vertical bars represents elements which comprise a continuous variable, the vertical bars are contiguous with one another (i.e., touch one another). If for the moment we assume that a continuous variable is employed in **Distribution A**, Figure I.8 can be employed as a histogram (also referred to as a **frequency histogram**) of the grouped frequency distribution summarized in **Columns 1** and **2** of Table I.8. Note that in Figure I.8, the width of each of the bars corresponds to the **real limits** of the relevant class interval, and the midpoint of each bar is directly above the midpoint of the relevant class interval. As noted above, the bars for adjacent class intervals are contiguous with one another.

Exploratory data analysis As noted earlier, **exploratory data analysis (EDA)** is a general approach for visually examining data that was introduced to the statistical community in 1977 by the statistician John Tukey. Smith and Prentice (1993) note that Tukey (1977) coined the term **exploratory data analysis** to simultaneously represent a philosophy with respect to examining data, as well as a set of tools which were intended to be used for that purpose. The intent of this section is to familiarize the reader with the basic principles underlying **EDA**, as well as to describe the **stem-and-leaf display** and **boxplot**, two of the procedures introduced by Tukey (1977) for visually examining data. Tukey (1977) believed that the latter two procedures were more effective than conventional methods of visual display with respect to their ability to reveal relevant information about a set of data.

The basic philosophy underlying **EDA** is that prior to initiating an analysis a researcher should scrutinize every piece of datum. Only after doing the latter can the researcher make an intelligent decision with respect to what the most prudent strategy will be for analyzing the data. Smith and Prentice (1993, p. 353) note that **EDA** endorses the use of measures which are relatively insensitive to data contamination (e.g., the presence of outliers), and which do not have associated with them strong assumptions regarding the shape of the underlying population distribution — such measures are often referred to as **resistant indicators**. Thus, values such as the **median** (which is commonly employed within the framework of nonparametric statistical analysis) and the **interquartile range** are commonly employed in **EDA**, respectively as metrics of central tendency and variability (in lieu of the **mean** and **standard deviation**). Although the philosophy underlying **EDA** is more concordant with that upon which nonparametric procedures and **robust statistical procedures** are based, Smith and Prentice (1993, p. 356) note that both nonparametric and robust statistical procedures are primarily employed within the framework of statistical inference, whereas **EDA** is employed in order to explore and describe data. A **robust statistical test/procedure** is one which is not overly dependent on critical assumptions regarding an underlying population distribution. Because of the latter, under certain conditions (e.g., when an underlying population distribution deviates substantially from normality) a robust statistical procedure can provide a researcher with more reliable information than an inferential statistical procedure which is not robust (i.e., a nonrobust inferential statistical procedure may yield unreliable results when one or more critical assumption about the underlying population distribution are violated). Further clarification of **robust statistical procedures** can be found in Section VII of the *t* **test for two independent samples** under the discussion of **outliers**, as well as in Section IX (the **Addendum**) of the **Mann–Whitney** *U* **test**. Recommended sources which discuss **resistant indicators** are Grissom and Kim (2005, pp. 16–19) and Wilcox (1987, 1996, 1997, 2001, 2003).

      Stem-and-leaf display A **stem-and-leaf display** summarizes the information contained in a distribution (or **batch**, which Tukey (1977) routinely used as a synonym for a distribution). To be more specific, a stem-and-leaf display simultaneously summarizes the information contained in a frequency polygon, a frequency histogram, and a cumulative frequency distribution, and in doing so it displays all of the original data values. Tukey (1977) believed that the stem-and-leaf display retained all of the original scores in a format that was easier to interpret than the format of more conventional tables and graphs. Among other things, a stem-and-leaf display allows a researcher to easily determine the mode and median of a distribution.

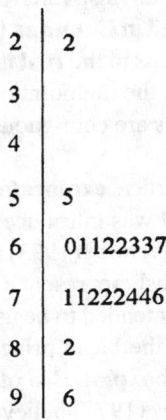

2	2
3	
4	
5	5
6	01122337
7	11222446
8	2
9	6

Table I.9 Stem-and-Leaf Display of Distribution A

      Table I.9 represents a stem-and-leaf display of **Distribution A**. Note that in Table I.9 a vertical line (sometimes referred to as the **vertical axis**) separates two sets of numbers. The numbers to the left of the vertical line represent the **stems** in a stem-and-leaf display, while the numbers to the right of the vertical line represent the **leaves**. In the case of a two digit integer number, the stems represent the first digit of each score (the first digit of a two digit integer number is often referred to as the **tens'**, **leading**, **most significant** or **base digit**). The second digits of a two digit integer number in a stem-and-leaf display will represent a **leaf** (the second digit of a two digit integer number is often referred to as the **units'**, **trailing** or less **significant digit**). Note that in Table I.9, for a given stem value the leaves are the second digit of all scores which have that stem value. Specifically, the score that falls between 20 and 29 (22) has the stem 2 (which is the first digit). The single leaf value (2) recorded to the right of the stem value 2 corresponds to the second digit of the number that falls between 20 and 29. The score that falls between 50 and 59 (55) has the stem 5 (which is the first digit). The single leaf value (5) recorded to the right of the stem value 5 corresponds to the second digit of the number that falls between 50 and 59. All of the scores that fall between 60 and 69 (one 60, two 61s, two 62s, two 63s, and one 67) have the stem 6 (which is the first digit). The eight leaf values (0, 1, 1, 2, 2, 3, 3, and 7) recorded to the right of the stem value of 6 correspond to the second digit of the numbers that fall between 60 and 69. All of the scores that fall between 70 and 79 (two 71s, three 72s, two 74s and one 76) have the stem 7 (which is the first digit). The eight leaf values (1, 1, 2, 2, 2, 4, 4 and 6) recorded to the right of the stem value of 7 correspond to the second digit of the numbers that fall between 70 and 79. The score that falls between 80 and 89 (82) has the stem 8 (which is the first digit). The single leaf value (2) recorded to the right of the stem value 8 corresponds to the second digit of the number that falls between 80 and 89. The score that falls between 90 and 99 (96) has the stem 9 (which is the first digit). The single leaf value (6) recorded to the right of the stem value 9 corresponds to the second digit of the number that falls between 90 and 99. Note that the stem values 3 (which represents any scores that fall between 30 and 39) and 4 (which represents any scores that fall between 40 and 49) do not have any leaf values, since no scores fall in the aforementioned intervals.

Although it is not necessary in the case of Table I.9, it may be desirable to modify a stem-and-leaf display in order to more accurately communicate the underlying structure of a set of data. Specifically, stem values can be employed within the same context of a class interval. As an example, assume we have a bell shaped/normal distribution which consists of 50 scores that fall between the values 50 – 59. In such a case it would not be very useful to construct a stem-and-leaf display comprised of a single row/line which consisted of a single stem value (specifically, the value 5) followed by the 50 leaf values. Instead, a researcher could effectively communicate the underlying structure of the data by employing five rows with a stem value of 5. Each row for a given stem value would only be employed for two leaf values. Thus, any leaf value of 0 or 1 would be recorded in the first row for the stem value 5 (i.e., scores of 50 and 51), leaf values of 2 or 3 would be recorded in the second row for the stem value 5 (i.e., scores of 52 and 53), etc. The latter is illustrated below in Table I.10 for 50 hypothetical scores which fall within the range 50 – 59. (The distribution in Table I.10 will be designated as **Distribution J**.) The number of leaf values that may fall in a given row of a stem-and-leaf display is referred to as the **line width** of the row. Tukey (1977) used typographic symbols or letters such as those noted in Table I.10 following a given stem value to differentiate the leaf intervals within that stem value (i.e., for a given stem value, an asterisk (*) is employed to designate the interval for the leaf values 0 and 1, the letter t is employed to designate the interval for the leaf values 2 and 3, etc.). In contrast to the class intervals in a frequency distribution, the size of a class interval (i.e., the line width) in a stem-and leaf display can only be 2, 5, or some value that is a power of 10 (i.e., $10^1 = 10$, $10^2 = 100$, etc.). The reason for the latter is to insure that each class interval has the same line width (i.e., contains the same number of scores).

5*	00111
5t	222233333
5f	44444444444455555555555555
5s	6667777
5•	89999

Table I.10   Stem-and-Leaf Display Involving Class Intervals for Distribution J

0000011111	5*	00111
2222233333	5t	222233333
4444455555	5f	444444444444555555555555
6666677777	5s	6667777
8888899999	5•	89999

Table I.11   Stem-and-Leaf Display Contrasting Distributions J and K

A stem-and-leaf display can be an effective tool for contrasting two different distributions with one another. The latter is illustrated in Table I.11. Note that the stem values are written along the vertical axis. Two vertical lines are drawn, one to the right of the stem values and one to the left of the stem values. The leaf values for **Distribution J** are recorded to the right of the vertical line which is situated to the right of the stem values. The leaf values for **Distribution K** are recorded to the left of the vertical line which is situated to the left of the stem values. Such an arrangement facilitates a researcher's ability to visually examine whether or not there appears to be major differences between the two distributions. In our example, Table I.11 reveals that **Distribution J** appears to be normal, while **Distribution K** appears to be uniform (i.e., the scores are evenly distributed throughout the range of scores/class intervals in the distribution).

It should be noted that it will not always be the case that the stem values in a stem-and-leaf display will be represented by the tens' value of a score. For instance, if the scores which constitute a set of data fell within the range 100 – 999, the hundreds' digit values would represent the stems of a stem-and-leaf display, while the leaves would be represented by the tens' digit values. In such a case, the units' digit values would be omitted from the table. To illustrate, if the scores in such a distribution which fell between 100 and 199 were 150, 167, 184, and 193, the stem value used to represent the latter four scores would be 1 and the four leaf values would be 5, 6, 8, and 9 (which, respectively, correspond to the first digit or tens' value for 50, 60, 80, and 90). The unit values of 0, 7, 4, and 3 (i.e., that last digit, respectively, of the four scores) would be omitted from the table. As a further example of constructing a stem-and-leaf display for data which involve values other than two digit integer numbers, assume a distribution of data consists of scores that fall within the range .1 – .99. In such a case, the stem values would be the first digit and the leaf values would be the second digit. Thus, if the scores .112, .123, .156, .178 occurred, the stem for the latter values would be 1 and the respective leaf values would be 1, 2, 5, and 7. Once again, the last digit for each score (i.e., the values 2, 3, 6, and 8) would be omitted from the table. Of course, if a researcher deemed it appropriate, in both of the examples cited in this paragraph the use of class intervals could be employed. In other words, as was done in Tables I.10/I.11, one could employ more than one line for a specific stem value. As an example, in the case where scores fall within the range .1 – .99, two rows could be employed for a given stem value, with one row employed for the leaf values 0, 1, 2, 3, 4, and the second row employed for the leaf values 5, 6, 7, 8, 9.

Boxplot   It was noted earlier that proponents of **EDA** endorse the use of descriptive measures which are relatively insensitive to data contamination, such as the **median** (in lieu of the **mean**) and the **interquartile range** (in lieu of the **standard deviation**). A **boxplot** (also referred to as a **box-and-whisker plot**) represents a method of visual display developed by Tukey (1977) which, among other things, provides a succinct summary of a distribution while displaying the **median** and elements called **hinges** (which for all practical purposes correspond to the **25th and 75th percentiles**, which are the boundaries that define the **interquartile range**). A number of elements displayed in a boxplot such as the median and **hingespread** (which is defined later in this section) are commonly described as **resistant indicators**, since their values are independent of the values of any **outliers** present in the data. The latter would not be the case if the mean and variance were, respectively, employed as measures of central tendency and variability. The value of the mean can be greatly influenced by the presence of one or more outliers, and the presence of outliers can dramatically increase the value of the variance. Because of the latter, the mean and variance are not considered to be resistant indicators. Figure I.10 is a boxplot of **Distribution A**.

The following values and/or elements documented in the boxplot depicted in Figure I.10 will now be computed and/or explained: a) **Median**: b) **Hinges**; c) **Hingespread**; d) **Fences, outliers**, and **severe outliers**; e) **Adjacent values**; f) **Whiskers**. The latter values or elements can be determined through use of a stem-and-leaf display, or alternatively by ordinally arranging all of the scores in the distribution as is done in Figure I.9.

Quartile	First	Second	Third	Fourth
	22 55 60 61 61	62 62 63 63 67	71 71 72 72 72	74 74 76 82 96
	↓	↓	↓	

Lower hinge = 61.5   **Median = 69**   **Upper hinge = 73**

Figure I.9   Determination of Hinges for Boxplot of Distribution A

a) The first value we need to compute in order to construct a boxplot for **Distribution A** is the **median**. Earlier in the **Introduction** it was noted that the median is the middle score in the distribution (which corresponds the 50th percentile), and that when the total number of scores is an even number (which is the case with **Distribution A**, since $n = 20$), there will be two middle values. In such a case, the median is the average of the two middle scores. Employing the above rule with **Distribution A**, the two middle scores will be those in the 10th and 11th ordinal position, which are 67 and 71. The average of the latter two scores is $(67 + 71)/2 = \mathbf{69}$, which is the value of the median.

At this point a general protocol will be presented for computing the ordinal position of a score which corresponds to any percentile value, and the score in the distribution which corresponds to that percentile value. The equation $k = np$ is employed, where n represents the total number of scores in the distribution, p represents the percentile expressed as a proportion, and k represents an ordinal position value.

1) If the value computed for k with the equation $k = np$ is **an integer number**, the ordinal position of the score with the desired percentile is the average of the scores in the kth and $(k + 1)$th ordinal positions. Employing this protocol to determine the median/50th percentile of **Distribution A**, we obtain $k = (20)(.5) = 10$. Since $k = 10$ is an integer value, the median will be the average of the scores in the $k = 10$th and the $(k + 1) = 11$th ordinal positions. This is consistent with the result obtained earlier when the values 67 (which is in the 10th ordinal position) and 71 (which is in the 11th ordinal position) were averaged to compute the value **69** for the median of the distribution.

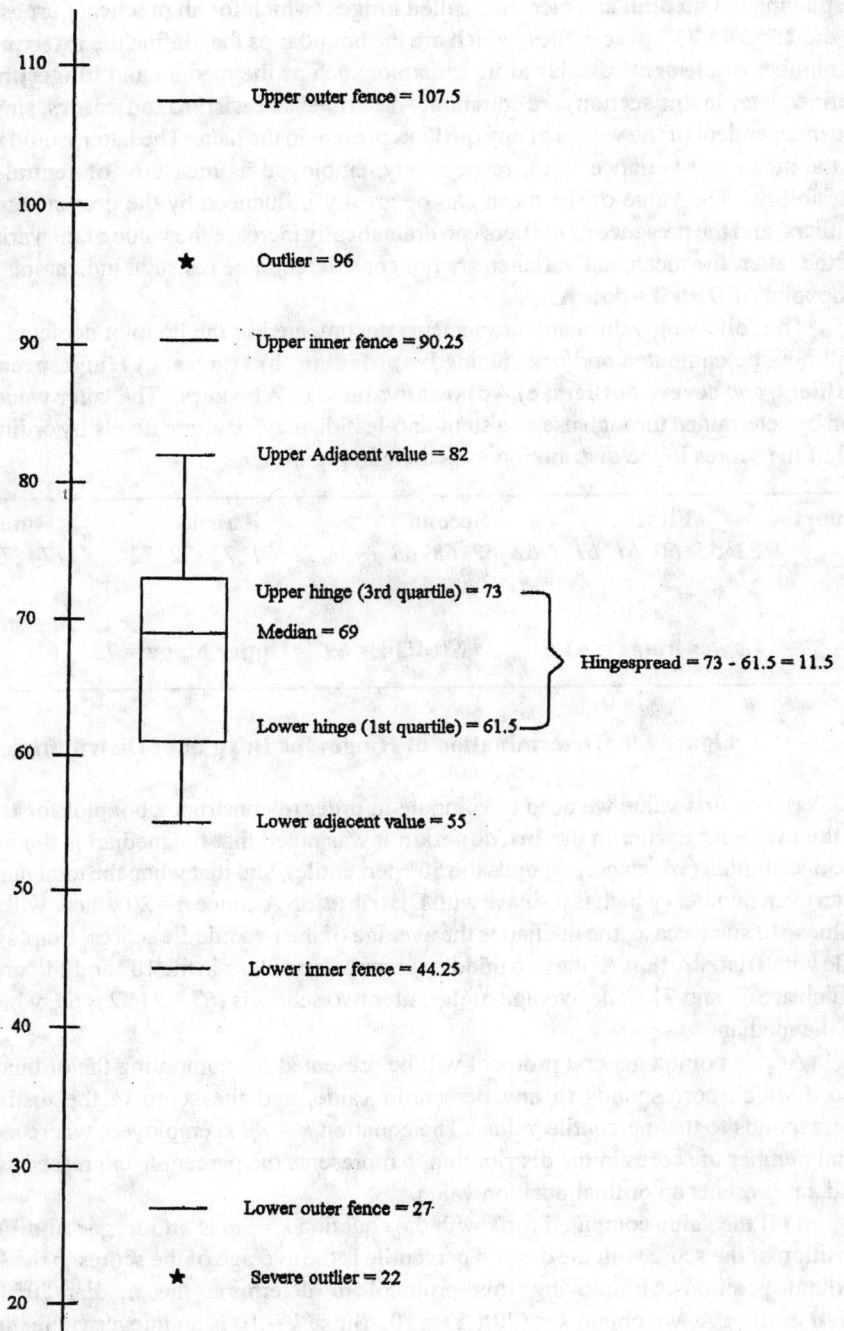

Figure I.10   Boxplot of Distribution A

Employing the above protocol to determine the 90th percentile of **Distribution A**, we obtain $k = (20)(.9) = 18$. Since $k = 18$ is an integer value, the 90th percentile will be the average of the scores in the $k = 18^{th}$ and the $(k + 1) = 19^{th}$ ordinal positions, which are respectively 76 and 82. Thus, the value of the score at the 90th percentile in **Distribution A** is $(76 + 82)/2 = 79$.

2) If the value computed for k with the equation $k = np$ is **not an integer number**, the **ordinal position** of the score with the desired percentile is one unit above the **integer portion** of the value computed for k. Thus, if we wanted to determine the 47th percentile of **Distribution A**, we determine that $k = (20)(.47) = 9.4$. Since the integer portion of $k = 9$, the score in the $k + 1 = 10^{th}$ ordinal position represents the score at the 47th percentile. In the case of **Distribution A** the score in the 10th ordinal position is 67.

It should be noted that the designation of 67 as the score at the 47th percentile is predicated on the assumption that the variable being measured is a discrete variable which can only assume integer values. If, on the other hand, the variable in question is assumed to be a continuous variable, interpolation can be employed to compute a slightly different value for the score at the 47th percentile. (The term **interpolate** means to compute an intermediate value which falls between two specified values.) Specifically, we can determine that the score at the 40th percentile will be 63, since it is the average of the scores in the 8th and 9th ordinal positions (i.e., since $k = (20)(.40) = 8$, we compute $(63 + 63)/2 = 63$). We already know that 69 represents the score at the 50th percentile. Since from the 40th percentile, the 47th percentile is seven tenths of the way up to the 50th percentile, we multiply .7 by 6 (which is the difference between the values 69 and 63 which respectively represent the 50th and 40th percentiles) and obtain 4.2. When the latter value is added to 63 (the value of the 40th percentile) it yields $63 + 4.2 = 67.2$. The latter value is employed to represent the score at the 47th percentile.[10]

b) The **hinges** in Figure I.10 represent the median of the scores above the 50th percentile, and the median of the scores below the 50th percentile (i.e., the points halfway between the median of the distribution and the most extreme score at each end of the distribution). When the sample size is reasonably large and there are relatively few tied scores, the hinge values will correspond closely with the **25th percentile** and the **75th percentile** of a distribution. Because of the latter, many sources define the hinges as the scores at the 25th and 75th percentiles. Figure I.9 illustrates the determination of the hinges, employing the definition of a hinge as the median of the scores above and below the 50^{the} percentile of the distribution. The value 61.5 computed for the **lower hinge** is in between the values 61 and 62, which separate the first and second quartiles of the distribution. 61.5 is the average of 61 and 62 (i.e., $(61 + 62)/2 = 61.5$). The value 73 computed for the **upper hinge** is in between the values 72 and 74, which separate the third and fourth quartiles of the distribution. 73 is the average of 72 and 74 (i.e., $(72 + 74)/2 = 73$). Note that if the equation $k = np$ is employed to compute the scores at the 25th and 75th percentiles, the following values are obtained: a) Since $(20)(.25) = 5$ yields an integer number, the score at the 25th percentile will be the average of the scores in the 5th and 6th (since $(k = 5) + 1 = 6$) ordinal positions — i.e., $(61 + 62)/2 = \mathbf{61.5}$; b) Since $(20)(.75) = 15$ yields an integer number, the score at the 75th percentile will be the average of the scores in the 15th and 16th (since $(k = 15) + 1 = 16$) ordinal position – i.e., $(72 + 74)/2 = \mathbf{73}$.

c) The **hingespread** (also referred to as **H-spread** and **fourth-spread**) is the difference between the **upper hinge** and the **lower hinge**. Thus, **Hingespread** $= 73 - 61.5 = \mathbf{11.5}$. Within the framework of a boxplot, the hingespread is employed as the measure of dispersion (i.e., variability).

d) Tukey (1977) stipulated that any value which falls more than one and one-half hingespreads outside a hinge should be classified as an **outlier**. The term **inner fence** is employed to designate the point in the distribution that falls one and one-half hingespreads outside a hinge (i.e., above an upper hinge or below a lower hinge).[11] Since the value of the hingespread is 11.5, $(1.5)(11.5) = 17.25$. The **upper inner fence** will be the value computed when 17.25 is added to the value of the upper hinge. Thus, $73 + 17.25 = **90.25**$. The **lower inner fence** will be the value computed when 17.25 is subtracted from the value of the lower hinge. Thus, $61.5 - 17.25 = **44.25**$. Any value which is more than three hingespreads outside a hinge is classified by some sources as a **severe outlier**. The term **outer fence** is employed to designate the point in the distribution that falls three hingespreads outside a hinge (i.e., above an upper hinge or below a lower hinge). The **upper outer fence** will be the value computed when $(3)(11.5) = 34.5$ is added to the value of the upper hinge. Thus, $73 + 34.5 = **107.5**$. The **lower outer fence** will be the value computed when $(3)(11.5) = 34.5$ is subtracted from the value of the lower hinge. Thus, $61.5 - 34.5 = **27**$. Note that only two scores can be classified in the outlier or severe outlier category. The score **96** is classified as an **outlier** since it falls beyond one and one-half hingespreads from the **upper hinge**. Although the score **96** falls above the **upper inner fence** of **90.25**, it does not qualify as a **severe outlier**, since it falls below the **upper outer fence** of **107.5**. The score **22** is classified as a **severe outlier** since it falls below the **lower outer fence** of **27**.[12]

e) Those values in the distribution that are closest to the **inner fences**, but which fall inside the **inner fences**, are referred to as **adjacent values** (also referred to as **extreme values**). In other words, the two adjacent values are the value closest to being one and one-half hingespread above the **upper hinge**, but still below the **upper inner fence**, and the value closest to being one and one-half hingespreads below the **lower hinge**, but still above the **lower inner fence**. Put more simply, the adjacent values are the most extreme values in the distribution in both directions which do not qualify as being outliers.[13] In the case of **Distribution A**, the score which is closest to but not above the **upper inner fence** is **82**. Thus, the latter score is designated as the **upper adjacent value**. The score which is closest to but not below the **lower inner fence** is **55**. Thus, the latter score is designated as the **lower adjacent value**.

f) The term **whisker** is employed to refer to each of the vertical lines which extend from the center of each end of the box in the boxplot (i.e., from the upper and lower hinges) to each of the adjacent values. Thus, the whisker at the top of the boxplot extends from the **upper hinge** which designates the score **73** to the **upper adjacent value** which designates the score **82**. The whisker at the bottom of the boxplot extends from the **lower hinge** which designates the score **61.5** to the **lower adjacent value** which designates the score **55**.

Inspection of the boxplot in Figure I.10 reveals the following information regarding **Distribution A**: a) The fact that the median is not in the center of the box element (i.e. the area between the upper and lower hinge) of the boxplot indicates that the distribution is not symmetric; b) The fact that the upper whisker of the boxplot is somewhat longer than the lower whisker indicates that **Distribution A** is negatively skewed — i.e., there are a disproportionate number of high scores in the distribution. The latter is further supported by the fact that the value of the mean 66.8 is less than the median of 69, which is less than the mode, which equals 72.

As in the case of a stem-and-leaf display, by arranging the boxplots of two distributions parallel to one another, one can compare the two distributions. It should be noted that although descriptions of boxplots (as well as the computation of relevant values) may not be consistent across all sources, boxplots derived through use of protocols stated in various sources will generally be quite similar to one another. In the final analysis, slight differences between boxplots based on different sources will be unimportant, insofar as a boxplot is employed as an exploratory device for developing an general picture regarding the structure of a set of data. A more detailed discussion of boxplots and the general subject of **exploratory data analysis** can be found in Mosteller and Tukey (1977), Tukey (1977), Hoaglin, Mosteller, and Tukey (1983), and Smith and Prentice (1993). Cleveland (1985) is also an excellent source on the use of graphs in summarizing data.

The Normal Distribution

When an inferential statistical test is employed with one or more samples to draw inferences about one or more populations, such a test may make certain assumptions about the shape of an underlying population distribution. The most commonly encountered assumption in this regard is that a distribution is **normal**. The **normal distribution** is also referred to as the **Gaussian distribution**, since the German mathematician Karl Friedrich Gauss discussed it in 1809. Zar (1999, p. 65), however, notes that the normal distribution was actually first identified by the French mathematician Abraham de Moivre in 1733 and mentioned as well in 1774 by another French mathematician, Pierre Simon, Marquis de Laplace.

When viewed from a visual perspective, the **normal distribution** (which as noted earlier is often referred to as the **bell-shaped curve**) is a graph of a frequency distribution which can be described mathematically and observed empirically (insofar as many variables in the real world appear to be distributed normally). The shape of the normal distribution is such that the closer a score is to the mean, the more frequently it occurs. As scores deviate more and more from the mean (i.e., become higher or lower), the more extreme the score the lower the frequency with which that score occurs. As noted earlier, a normal distribution will always be symmetrical (with $\gamma_1 = g_1 = 0$ and $\sqrt{\beta_1} = \sqrt{b_1} = 0$) and mesokurtic (with $\gamma_2 = g_2 = 0$ and $\beta_2 = b_2 = 3$).

Any normal distribution can be converted into what is referred to as the **standard normal distribution** by assigning it a mean value of 0 (i.e., $\mu = 0$) and a standard deviation of 1 (i.e., $\sigma = 1$). The **standard normal distribution**, which is represented in Figure I.11, is employed more frequently in inferential statistics than any other theoretical probability distribution. The use of the term theoretical probability distribution in this context is based on the fact it is known that in the standard normal distribution (or, for that matter, any normal distribution) a certain proportion of cases will always fall within specific areas of the curve. As a result of this, if one knows how far removed a score is from the mean of the distribution, one can specify the proportion of cases which obtain that score, as well as the likelihood of randomly selecting a subject or object with that score. Figure I.11 will be discussed in greater detail later in this section.

In a graph such as Figure I.11, the range of values a variable may assume are recorded on the X-axis. In the case of Figure I.11, the scores are represented in the form of z scores/standard deviation scores (which are explained in this section). The values on the Y-axis can be viewed as representing the frequency of occurrence for each of the scores in the population (thus the notation f). As noted earlier, sometimes proportions, probabilities or density values are recorded on the Y-axis. Further clarification of density values is provided below.

Equation I.36 is the general equation for the normal distribution.

$$Y = \frac{1}{\sigma\sqrt{2\pi}}e^{-(X-\mu)^2/2\sigma^2}$$

(Equation I.36)

In point of fact, Equation I.36 and Figure I.11 respectively represent the equation and graph for the **probability density function** of the normal distribution. In mathematics the term **function** is commonly employed to summarize the relationship between two variables such as X and Y. Such an equation summarizes the operations which when performed on the X variable will yield one or more specific values for the Y variable. In the case of a theoretical distribution, such as the normal distribution, a **density function** describes the relationship between a variable and the densities (which for the moment we will assume are probabilities) associated with the range of values the variable may assume. When a density function is represented in a graphic format, the range of values the variable may assume are recorded on the abscissa (X-axis), and the density values (which will range from 0 to 1) are recorded on the ordinate (Y-axis). Through use of Equation I.36 and/or Figure I.11 one can determine the proportion of cases/area in a normal distribution that falls between any two points on the abscissa. Those familiar with calculus will recognize that if we have two points a and b on the abscissa representing two values of a continuous variable, and we integrate the equation of the density function over the interval a to b, we will be able to derive the area/proportion of the curve which falls between points a and b. As will be noted shortly, in lieu of having to employ calculus to compute the proportion that falls between any two points, in the case of the normal distribution the appropriate values can be determined through use of **Table A1 (The Table of the Normal Distribution)** in the **Appendix**. It should be noted that although the densities which are recorded on the ordinate of the graph of a density function are often depicted as being equivalent to probabilities, strictly speaking a density is not the same thing as a probability. Within the context of a graph such as Figure I.11, a density is best viewed as the height of the curve which corresponds to a specific value of the variable which is recorded on the abscissa. The area between any two points on the abscissa however, is generally expressed in the form of a probability or a proportion. Further clarification of a probability density function and the general concept of probability can be found in Section IX (the **Addendum**) of the **binomial sign test for a single sample** under the discussion of Bayesian analysis of a continuous variable.

In Equation I.36 the symbols μ and σ represent the mean and standard deviation of a normal distribution. For any normal distribution where the values of μ and σ are known, a value of Y (which represents the height of the distribution at a given point on the abscissa) can be computed simply by substituting a specified value of X in Equation I.36. Note that in the case of the standard normal distribution, where $\mu = 0$ and $\sigma = 1$, Equation I.36 becomes Equation I.37.[14]

$$Y = \frac{1}{\sqrt{2\pi}} e^{-X^2/2} \qquad \text{(Equation I.37)}$$

The reader should take note of the fact that the normal distribution is a family of distributions which is comprised of all possible values of μ and σ that can be substituted in Equation I.36. Although the values of μ and σ for a normal distribution may vary, as noted earlier, all normal distributions are mesokurtic.

For any variable that is normally distributed, regardless of the values of the population mean and standard deviation, the distance of a score from the mean in standard deviation units can be computed with Equation I.38. The z value computed with Equation I.38 (which is often referred to as a **standard** or **standard deviation score**) is a measure in standard deviation units of how far a score is from the mean. In instances where the population standard deviation is not known (which is often the case in actual research) the estimated population standard deviation $\tilde{s}$ computed with Equation I.11 is employed in lieu of σ in the denominator of Equation I.38 (i.e., $z = (X - \mu) / \tilde{s}$).

$$z = \frac{X - \mu}{\sigma} \qquad \text{(Equation I.38)}$$

Where: X is a specific score
          μ is the value of the population mean
          σ is the value of the population standard deviation

When Equation I.38 is employed, any score that is above the mean will yield a positive z value, and any score that is below the mean will yield a negative z value. Any score that is equal to the mean will yield a z value of zero.

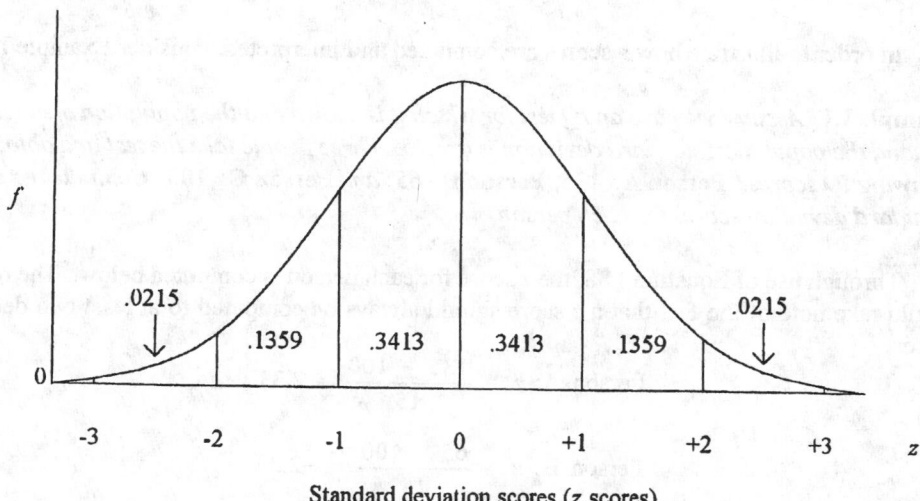

Standard deviation scores (*z* scores)

Figure I.11 The Standard Normal Distribution

At this point we shall return to Figure I.11 which will be examined in greater detail. Inspection of the latter figure reveals that in the standard normal distribution a fixed proportion/percentage of cases will always fall between specific points on the curve. Specifically, .3413 of the cases (which expressed as a percentage is 34.13%) will fall between the mean and one standard deviation above the mean (i.e., between the values $z = 0$ to $z = +1$), as well as between the mean and one standard deviation below the mean (i.e., between the values $z = 0$ to $z = -1$) .1359 (or 13.59%) of the cases will fall between one and two standard deviation units above the mean (i.e., between the values $z = +1$ and $z = +2$), as well as between one and two standard deviation units below the mean (i.e., between the values $z = -1$ and $z = -2$). (One standard deviation unit is equal to the value of σ.) .0215 (or 2.15%) of the cases will fall between two and three standard deviation units above the mean (i.e., between the values $z = +2$ and $z = +3$), as well as between two and three standard deviation units below the mean (i.e., between the values $z = -2$ and $z = -3$).

Note that since the normal distribution is symmetrical, the proportion/percentage of cases to the right of the mean will be equivalent to the proportion/percentage of cases to the left of the mean. If all of the proportions/percentages which fall within three standard deviations of the mean (i.e., between the points $z = +3$ and $z = -3$) are summed, the value .9974 (or 99.74%) is obtained (i.e., 0215 + .1359 + .3413 + .3413 + .1359 + .0215 = .9974). The latter value indicates that in a normally distribution population, .9974 is the proportion of the population which will obtain a score that falls within three standard deviations from the mean. One half of those cases (i.e., .9974/2 = .4987 or 49.87%) will fall above the mean (i.e., between $z = 0$ and $z = +3$), and one half will fall below the mean (i.e., between $z = 0$ and $z = -3$). (Obviously some scores will fall exactly at the mean. Half of the latter scores are generally included in the 49.87% which fall above the mean, and the other half in the 49.87% which fall below the mean.) Only $1 - .9974$ =.0026 (or .26%) of the population will obtain a score which is more or less than three standard deviations units from the mean. Specifically, .0026/2 = .0013 (or .13%) will obtain a score that is greater than three standard deviation units above the mean (i.e., above the score $z = +3$), and .0013 will obtain a score that is less than three standard deviation units below the mean (i.e., below the score $z = -3$). As will be noted shortly, all of the aforementioned normal distribution proportions/percentages can also be obtained from **Table A1**.

In order to illustrate how z scores are computed and interpreted, consider Example I.1.

Example I.1 *Assume we have an IQ test for which it is known that the population mean is μ = 100 and the population standard deviation is σ = 15. Three people take the test and obtain the following IQ scores:* **Person A:** *135;* **Person B:** *65; and* **Person C:** *100. Compute a z score (standard deviation score) for each person.*

Through use of Equation I.38, the z score for each person is computed below. The reader should take note of the fact that a z score should always be computed to at least two decimal places.

$$\text{Person A: } z = \frac{135 - 100}{15} = 2.33$$

$$\text{Person B: } z = \frac{65 - 100}{15} = -2.33$$

$$\text{Person C: } z = \frac{100 - 100}{15} = 0$$

Person A obtains an IQ score that is 2.33 standard deviation units above the mean, Person B obtains an IQ score that is 2.33 standard deviation units below the mean, and Person C obtains an IQ score at the mean. If we wanted to determine the likelihood (i.e., the probability) of selecting a person (as well as the proportion of people) who obtains a specific score in a normal distribution, **Table A1** can provide this information. Although **Table A1** is comprised of four columns, for the analysis to be discussed in this section we will only be interested in the first three columns.

Column 1 in **Table A1** lists z scores which range in value from 0 to an **absolute value** of 4. The use of the term **absolute value of 4** is based on the fact that since the normal distribution is symmetrical, anything we say with respect to the probability or the proportion of cases associated with a positive z score will also apply to the corresponding negative z score. Note that positive z scores will always fall to the right of the mean (often referred to as the **right tail** of the distribution), thus indicating that the score is above the mean. Negative z scores, on the other hand, will always fall to the left of the mean (often referred to as the **left tail** of the distribution), thus indicating that the score is below the mean.[15]

Column 2 in **Table A1** lists the proportion of cases (which can also be interpreted as probability values) that falls between the mean of the distribution and the z score that appears in a specific row.

Column 3 in **Table A1** lists the proportion of cases that falls beyond the z score in that row. More specifically, the proportion listed in Column 3 is evaluated in relation to the tail of the distribution in which the score appears. Thus, if a z score is positive, the value in Column 3 will represent the proportion of cases that falls above that z score, whereas if the z score is negative, the value in Column 3 will represent the proportion of cases that falls below that z score.[16]

Table A1 will now be employed in reference to the IQ scores of Person A and Person B. For both subjects the computed absolute value of z associated with their IQ score is z = 2.33. For z = 2.33, the tabled values in Columns 2 and 3, are respectively, .4901 and .0099. The value in Column 2 indicates that the proportion of the population that obtains a z score between the mean and z = 2.33 is .4901 (which expressed as a percentage is 49.01%), and the proportion of the population which obtains a z score between the mean and z = –2.33 is .4901. We can make comparable statements with respect to the IQ values associated with these z scores. Thus, we can

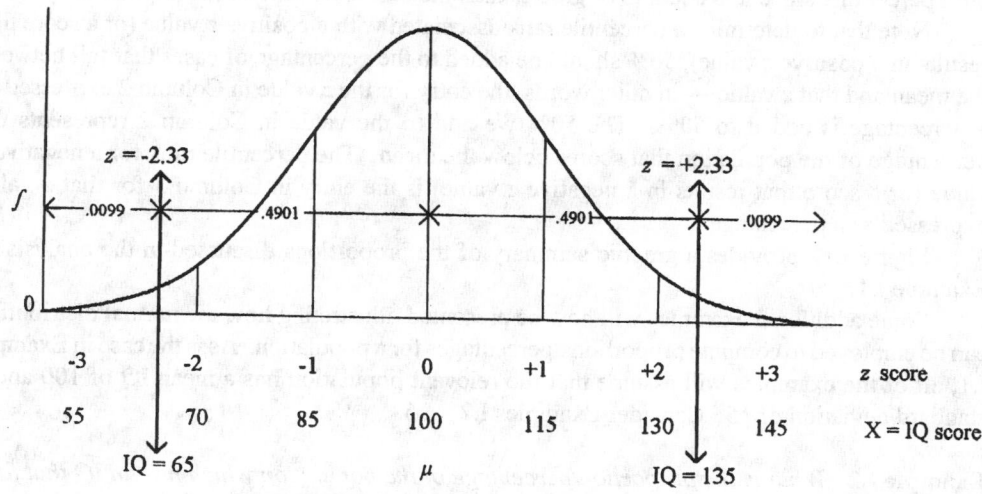

Standard deviation scores (z scores)/IQ scores

Figure I.12 Summary of Example I.1

say that the proportion of the population which obtains an IQ score between 100 and 135 is .4901, and the proportion of the population which obtains an IQ score between 65 and 100 is .4901. Since the normal distribution is symmetrical, .5 (or 50%) represents the proportion of cases that falls both above and below the mean. Thus, we can determine that .5 + .4901 = .9901 (or 99.01%) is the proportion of people with an IQ of 135 or less, as well as the proportion of people who have an IQ of 65 or greater. We can state that a person who has an IQ of 135 has a score that falls at approximately the 99th percentile, since it is equal to or greater than the scores of 99% of the population. On the other hand, a person who has an IQ of 65 has a score that falls at the 1st percentile, since it is equal to or greater than the scores of only approximately 1% of the population.

The value in Column 3 indicates that the proportion of the population which obtains a score of $z = 2.33$ or greater (and thus, in reference to Person A, an IQ of 135 or greater) is .0099 (which is .99%). In the same respect, the proportion of the population that obtains a score of $z = -2.33$ or less (and thus, in reference to Person B, an IQ of 65 or less) is .0099.

If one interprets the values in Columns 2 and 3 as probability values instead of proportions, we can state that if one randomly selects a person from the population, the probability of selecting someone with an IQ of 135 or greater will be approximately 1%. In the same respect, the probability of selecting someone with an IQ of 65 or less will also be approximately 1%.

In the case of Person C, whose IQ score of 100 results in the standard deviation score $z = 0$, inspection of **Table A1** reveals that the values in Columns 2 and 3 associated with $z = 0$ are, respectively, .0000 and .5000. The latter values indicate the following: a) The proportion of the population that obtains an IQ between the mean value of 100 and 100 is zero; b) The proportion of the population that obtains an IQ of 100 or greater is .5 (which is equivalent to 50%), and that the proportion of the population which obtains an IQ of 100 or less is .5. Thus, if we randomly select a person from the population, the probability of selecting someone with an IQ equal to or greater than 100 will be .5, and the probability of selecting someone with an IQ equal to or less than 100 will be .5. We can also state that the score of a person who has an IQ of 100 falls at the 50th percentile, since it is equal to or greater than the scores of 50% of the population.

Note that to determine a percentile rank associated with a positive *z* value (or a score that results in a positive *z* value), 50% should be added to the percentage of cases that fall between the mean and that *z* value — in other words, the entry for the *z* value in Column 2 expressed as a percentage is added to 50%. The 50% we add to the value in Column 2 represents the percentage of the population that scores below the mean. The percentile rank for a negative *z* value (or a score that results in a negative *z* value) is the entry in Column 3 for that *z* value expressed as a percentage.

Figure I.12 provides a graphic summary of the proportions discussed in the analysis of Example I.1.

Some additional examples will now be presented illustrating how the normal distribution can be employed to compute proportions/percentages for a population. As is the case in Example I.1, all of the examples will assume that the relevant population has a mean IQ of 100 and a standard deviation of 15. Consider Examples I.2 – I.5.

Example I.2 *What is the proportion/percentage of the population which has an IQ that falls between* 110 *and* 135?

Example I.3 *What is the proportion/percentage of the population which has an IQ that falls between* 65 *and* 75?

Example I.4 *What is the proportion/percentage of the population which has an IQ that falls between* 75 *and* 110?

Example I.5 *What is the proportion/percentage of the population which has an IQ that is greater than* 110 *or less than* 75?

Figures I.13 and I.14 visually summarize the information necessary to compute the proportions/percentages for Examples I.2 – I.5.

Example I.2 asks for the proportion of cases between two IQ scores, both of which fall above the population mean. In order to compute the latter, it is necessary to compute the *z* value associated with each of the two IQ scores stipulated in the example. Specifically, we must compute *z* values for the IQ scores of 110 and 135. Within the framework of Example I.1, Equation I.38 was employed to compute the value $z = 2.33$ for an IQ of 135. The latter equation is employed below to compute the value $z = .67$ for an IQ of 110.

$$z = \frac{110 - 100}{15} = .67$$

Employing **Table A1**, it was previously determined that a proportion equal to .4901 of the population (which is equal to 49.01%) will have an IQ score between the mean of 100 and an IQ of 135 (which is 2.33 standard deviation units above the mean (i.e., $z = 2.33$)). Employing Column 2 of **Table A1**, we can determine that the proportion/percentage of the population which has an IQ between the mean and a standard deviation score of $z = .67$ (which in the case of a positive *z* value corresponds to an IQ score of 110) is .2486 (or 24.86%). The latter information is visually summarized in Figure I.13.

Note that in Figure I.13, Area A represents the proportion/percentage of the normal distribution that falls between the mean and 2.33 standard deviation units above the mean, while

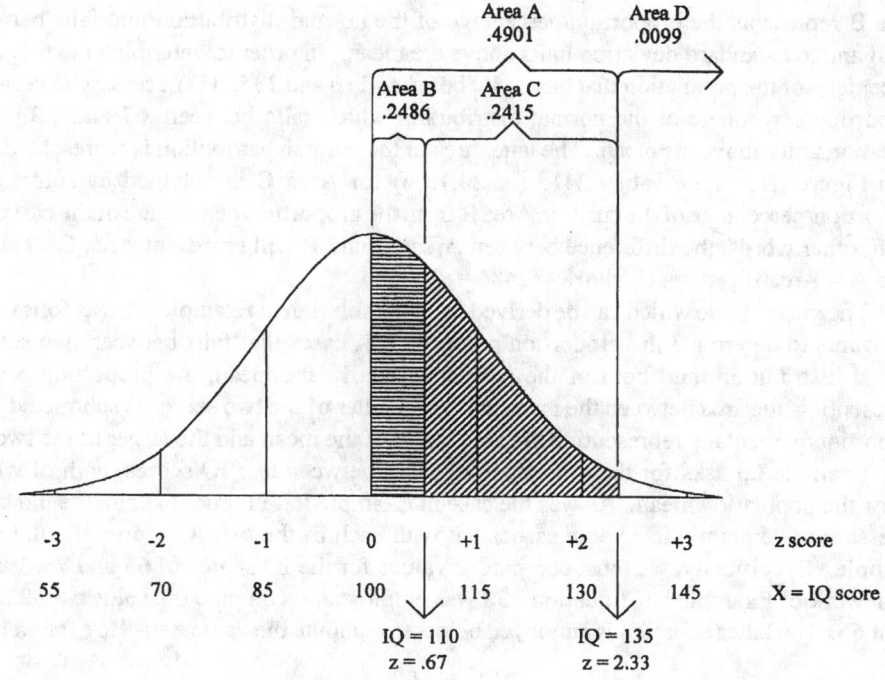

Figure I.13 Computation of Relevant Proportions/Percentages Above the Mean

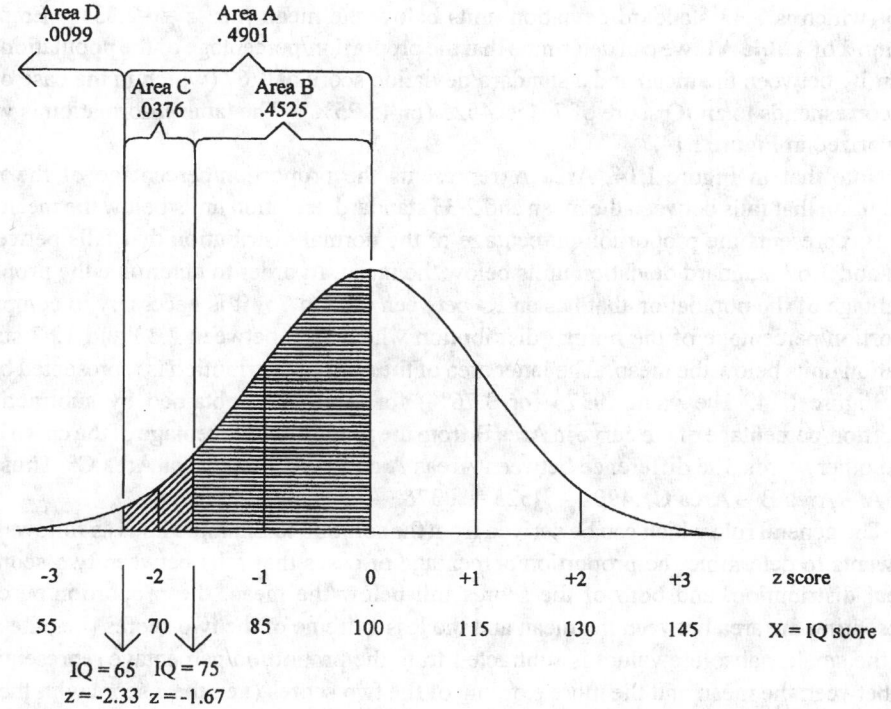

Figure I.14 Computation of Relevant Proportions/Percentages Below the Mean

Area B represents the proportion/percentage of the normal distribution that falls between the mean and .67 standard deviation units above the mean. In order to determine the proportion/ percentage of the population that has an IQ between 110 and 135, it is necessary to compute the proportion/percentage of the normal distribution which falls between .67 and 2.33 standard deviation units above the mean. The latter area of the normal distribution is represented by Area C in Figure I.13. The value .2415 (or 24.15%) for Area C is obtained by subtracting the proportion/percentage of the curve in Area B from the proportion/percentage of the curve in Area A. In other words, the difference between Areas A and B will represent Area C. Thus, since Area A − Area B = Area C, .4901 − .2486 = .2415.

The general rule which can be derived from the solution to Example I.2 is as follows. When one wants to determine the proportion/percentage of cases that falls between two scores in a normal distribution, and both of the scores fall above the mean, the proportion/percentage representing the area between the mean and the smaller of the two scores is subtracted from the proportion/percentage representing the area between the mean and the larger of the two scores.

Example I.3 asks for the proportion of cases between two IQ scores, both of which fall below the population mean. As was the case in Example I.2, in order to compute the latter it is necessary to compute the z value associated with each of the two IQ scores stipulated in the example. Specifically, we must compute z values for the IQ scores of 65 and 75. Within the framework of Example I.1, Equation I.38 was employed to compute the value $z = -2.33$ for an IQ of 65. The latter equation is employed below to compute the value $z = -1.67$ for an IQ of 75.

$$z = \frac{75 - 100}{15} = -1.67$$

Employing **Table A1**, it was previously determined that a proportion equal to .4901 of the population (which is equal to 49.01%) will have an IQ score between the mean of 100 and an IQ of 65 (which is 2.33 standard deviation units below the mean (i.e., $z = -2.33$)). Employing Column 2 of **Table A1**, we can determine that the proportion/percentage of the population which has an IQ between the mean and a standard deviation score of 1.67 (which in the case of $z = -1.67$ corresponds to an IQ score of 75) is .4525 (or 45.25%). The latter information is visually summarized in Figure I.14.

Note that in Figure I.14, Area A represents the proportion/percentage of the normal distribution that falls between the mean and 2.33 standard deviation units below the mean, while Area B represents the proportion/percentage of the normal distribution that falls between the mean and 1.67 standard deviation units below the mean. In order to determine the proportion/ percentage of the population that has an IQ between 65 and 75, it is necessary to compute the proportion/percentage of the normal distribution which falls between 2.33 and 1.67 standard deviation units below the mean. The latter area of the normal distribution is represented by Area C in Figure I.14. The value .0376 (or 3.76%) for Area C is obtained by subtracting the proportion/percentage of the curve in Area B from the proportion/percentage of the curve in Area A. In other words, the difference between Areas A and B will represent Area C. Thus, since Area A − Area B = Area C, .4901 − .4525 = .0376.

The general rule which can be derived from the solution to Example I.3 is as follows. When one wants to determine the proportion/percentage of cases that falls between two scores in a normal distribution, and both of the scores fall below the mean, the proportion/percentage representing the area between the mean and the less extreme of the two scores (i.e., the z score with the smaller absolute value) is subtracted from the proportion/percentage representing the area between the mean and the more extreme of the two scores (i.e., the z score with the larger absolute value).

Based on the values computed for Examples I.2 and I.3 (which are summarized in Figures I.13 and I.14) we have enough information to answer the questions asked in Examples I.4 and I.5. Example I.4 asks what proportion/percentage of the population has an IQ which falls between 75 and 110. In Example I.2 it was determined that .2486 (or 24.86%) of the population has an IQ between 100 and 110, and in Example I.3 it was determined that .4525 (or 45.25%) of the population has an IQ between 100 and 75. The value .2486 is represented by Area B in Figure I.13, and the value .4525 is represented by Area B in Figure I.14. To determine the proportion/percentage of cases that obtain an IQ between 75 and 110 we merely added up the values for Area B in both figures. Thus, .2486 + .4525 = .7011 (or 70.11%). Consequently we can say that .7011 (or 70.11%) of the population obtains an IQ between 75 and 110.

The general rule which can be derived from the solution to Example I.4 is as follows. When one wants to determine the proportion/percentage of cases that falls between two scores in a normal distribution, and one of the scores falls above the mean and the other score falls below the mean, the following protocol is employed. The proportion/percentage representing the area between the mean and score above the mean is added to the proportion/percentage representing the area between the mean and the score below the mean.

Example I.5 asks what proportion/percentage of the population has an IQ which falls above 110 or below 75. In Example I.2 it was determined that .2486 (or 24.86%) of the population has an IQ between 100 and 110. Since a proportion equal to .5 (or 50%) of the cases fall above the mean, it logically follows that .5 − .2486 = .2514 (or 25.14%) of the case will have an IQ above 110. The latter value can be obtained in Figure I.13 by adding up the proportions designated for Areas C and D (i.e., .2415 + .0099 = .2514). Note that the value .2514 corresponds the proportion recorded in Column 3 of **Table A1** for the value $z = .67$. This is the case since the values in Column 3 of **Table A1** represent the proportion of cases which are more extreme than the z score listed in the specified row (in the case of a positive z score it is the proportion of cases above that z score).

In Example I.3 it was determined that .4525 (or 45.25%) of the population has an IQ between 100 and 75. Since a proportion equal to .5 (or 50%) of the cases fall below the mean, it logically follows that .5 − .4525 = .0475 (or 4.75%) of the case will have an IQ below 75. The latter value can be obtained in Figure I.14 by adding up the proportions designated for Areas C and D (i.e., .0376 + .0099 = .0475). Note that the value .0475 corresponds the proportion recorded in Column 3 of **Table A1** for the value $z = 1.67$. This is the case since, as previously noted, the values in Column 3 of **Table A1** represent the proportion of cases which are more extreme than the z score listed in the specified row (in the case of a negative z score, it is the proportion of cases below that z score).

To determine the proportion/percentage of cases that obtain an IQ above 110 or below 75, we merely added up the values .2514 and .0475. Thus, .2514 + .0475 = .2989 (or 29.89%). Consequently we can say that .2989 (or 29.89%) of the population obtains an IQ that is greater than 110 or less than 75.

The general rule which can be derived from the solution to Example I.5 is as follows. When one wants to determine the proportion/percentage of cases that falls above one score that is above the mean or below another score that is below the mean, the following protocol is employed. The proportion/percentage representing the area above the score which is above the mean is added to the proportion/percentage representing the area below the score which is below the mean.

Examples I.6 – I.8 will be employed to illustrate the use of Equation I.39, which is derived through algebraic transposition of the terms in Equation I.38. Whereas Equation I.38 assumes that a subject's score (i.e., the value X) is known and allows for the computation of a z score, Equation I.39 does the reverse by computing a subject's score through use of a known z value.

$$X = \mu \pm z\sigma \qquad \text{(Equation I.39)}$$

Example I.6 *What is the IQ of a person whose score falls at the 75th percentile?*

Example I.7 *What is the IQ of a person whose score falls at the 4.75th percentile?*

Example I.8 *What are the IQ scores which define the middle 98% of the population distribution?*

With respect to Example I.6, in order to determine the IQ score that falls at the 75th percentile, we must initially determine the z score which corresponds to the point on the normal distribution that represents the 75th percentile. Once the latter value has been determined, it can be substituted in Equation I.39, along with the values $\mu = 100$ (the population mean) and $\sigma = 15$ (the population standard deviation). When all three values are employed to solve the latter equation, the value computed for X will represent the IQ score that falls at the 75th percentile. In point of fact, the IQ score 110, which is employed in Example I.2 and visually represented in Figure I.13, represents the IQ score at the 75th percentile. The computation of the IQ score 110 for the 75th percentile will now be described.

In order to answer the question posed in Example I.6, it is necessary to determine the z value in **Table A1** for which the proportion .25 is recorded in Column 2. If no proportion in Column 2 is equal exactly to .25, the z value associated with the proportion which is closest to .25 is employed. The proportion .25 (which corresponds to 25%) is used since a score at the 75th percentile is equal to or greater than 75% of the scores in the distribution. The area of the normal distribution it encompasses is the 50% of the population which has an IQ below the mean (which corresponds to a proportion of .5), as well as an additional 25% of the population which scores directly above the mean (which corresponds to a proportion of .25, which will be recorded in Column 2 of **Table A1**). In other words, $50\% + 25\% = 75\%$.

Employing **Table A1**, the z value with a proportion in Column 2 which is closest to .25 is $z = .67$. Note that since there is no z value exactly equal to .25, the two z values with proportions which are closest to being either above and below .25 are $z = .67$, for which the proportion is .2486, and $z = .68$, for which the proportion is .2517. Of the latter two proportions, .2486 is closer to .25. As previously noted in discussing Example I.2, the proportion .2486 for $z = .67$ indicates that 24.86% of the population has an IQ between the mean and a z value of .67. When the latter z value is substituted in Equation I.39, the value $X = 110.5$ is computed. When 110.5 is rounded off to the nearest integer, the IQ value 110 is obtained. Thus, we can conclude that an IQ of 110 falls at the 75th percentile. The combination of Area B and the area below the mean in Figure I.13 visually represents the area which contains scores that fall below the 75th percentile. (It should be noted that some researchers might prefer to employ the value $z = .68$ in Equation I.39 in the computation of the 75th percentile, since it provides the closest estimate that is equal to or greater than the 75th percentile.)

$$X = 100 + (.67)(15) = 110.5$$

The general rule which can be derived from the solution to Example I.6 is as follows. When one wants to determine the score that falls at a specific percentile (which will be designated as P) and the value of P is above the 50th percentile (i.e., the mean), the following protocol is employed. The proportion in Column 2 of **Table A1** which is equal to the difference $(P - .5)$ is identified. The latter value is designated as Q. Thus, $Q = P - .5$. If there is no value equal to Q, the value closest to Q is employed. The z value associated with the proportion Q in Column

2 is substituted in Equation I.39, along with the values of μ and σ. The resulting value of X represents the score at the Pth percentile.

With respect to Example I.7, in order to determine the IQ score that falls at the 4.75th percentile, we must initially determine the z score which corresponds to the point on the normal distribution that represents the 4.75th percentile. Once the latter value has been determined, it can be substituted in Equation I.39, along with the values μ = 100 (the population mean) and σ = 15 (the population standard deviation). When all three values are employed to solve the latter equation, the value computed for X will represent the IQ score that falls at the 4.75th percentile. In point of fact, the IQ score 75, which is employed in Example I.3 and visually represented in Figure I.14, represents the IQ score at the 4.75th percentile. The computation of the IQ score 75 for the 4.75th percentile will now be described.

In order to answer the question posed in Example I.7, it is necessary to determine the z value in **Table A1** for which the proportion .0475 is recorded in Column 3. (The value in Column 2 that is equal to .5 − .0475 = .4525 can also be employed, since it is associated with the same z value.). If no proportion in Column 3 is equal exactly to .0475, the z value associated with the proportion which is closest to .0475 is employed. The proportion .0475 (which corresponds to 4.75%) is used since a score at the 4.75th percentile is equal to or greater than 4.75% of the scores in the distribution. The area of the normal distribution it encompasses is the 4.75% of the population which has an IQ that falls between the end of the left tail of the distribution and the point demarcating the lower 4.75% of the scores in the distribution (all of which are below the mean, since any percentile less than 50% is below the mean).

Employing **Table A1**, the z value with a proportion in Column 3 which equals .0475 is z = 1.67. Since the percentile in question is less than 50%, a negative sign is employed for the z value. Thus, $z = -1.67$. The proportion .0475 for $z = -1.67$ indicates that 4.75% of the population has an IQ between the end of the left tail of the distribution and a z value of −1.67. When the latter z value is substituted in Equation I.39, the value $X = 74.95$ is computed. When 74.95 is rounded off to the nearest integer, the IQ value 75 is obtained. Thus, we can conclude that an IQ of 75 falls at the 4.75th percentile. The sum of Areas C and D in Figure I.14 visually represents the area which contains scores that fall below the 4.75th percentile

$$X = 100 + (-1.67)(15) = 74.95$$

The general rule which can be derived from the solution to Example I.7 is as follows. When one wants to determine the score that falls at a specific percentile (which will be designated as P) and the value of P is below the 50th percentile (i.e., the mean), the following protocol is employed. The z value is identified which corresponds to the proportion in Column 3 of **Table A1** which is equal to the value of P. If there is no value equal to P, the value closest to P is employed. The z value associated with the proportion P in Column 3 is assigned a negative sign and is then substituted in Equation I.39, along with the values of μ and σ. The resulting value of X represents the score at the Pth percentile.

Example I.8 asks for the IQ scores which define the middle 98% of the distribution. Since 98% divided by 2 equals 49%, the middle 98% of the distribution will be comprised of the 49% of the distribution directly above the mean (i.e., to the right of the mean), as well as the 49% of the distribution directly below the mean (i.e., to the left of the mean). In point of fact, the middle 98% of the normal distribution is depicted in Figure I.12. Specifically, in the latter figure it is the area between the standard deviation scores $z = +2.33$ and $z = -2.33$. Note that in Figure I.12, 49.01% (which corresponds to a proportion of .4901) of the cases fall between the mean and each of the aforementioned z values.

What has been noted above indicates that in order to determine the IQ scores which define the middle 98% of the distribution, we must initially determine the z score for which the proportion listed in Column 2 of **Table A1** is equal to .49 (which is 49% expressed as a proportion). If no proportion in Column 2 is equal exactly to .49, the z value associated with the proportion which is closest to .49 is employed. Equation I.39 is then employed to solve for X for both the positive and negative value of the z score which has a proportion equal to .49. As noted above, the z value with the proportion in Column 2 that is closest to .49 is $z = 2.33$. Along with the values $\mu = 100$ (the population mean) and $\sigma = 15$ (the population standard deviation), the values $z = +2.33$ and $z = -2.33$ are substituted in Equation I.39 below to compute the IQ values 65.05 and 134.95, which when rounded off equal 65 and 135. The latter two values represent the IQ scores which are the limits of the middle 98% of the distribution. In other words, we can say that 98% of the population has an IQ that falls between 65 and 135. This can be stated symbolically as follows: $65 \leq IQ \leq 135$.

$$X = 100 + (2.33)(15) = 134.95$$
$$X = 100 + (-2.33)(15) = 65.05$$

The general rule which can be derived from the solution to Example I.8 is as follows. When one wants to determine the scores that define the middle $P\%$ of a normal distribution, the following protocol is employed. The z value is identified which corresponds to the proportion in Column 2 of **Table A1** which is equal to the value $P/2$ (when $P\%$ is expressed as a proportion). If there is no value equal to $P/2$, the value closest to $P/2$ is employed. Both the positive and negative forms of the z value associated with the proportion $P/2$ in Column 2 are substituted in Equation I.39, along with the values of μ and σ. The resulting values of X represent the limits of the middle $P\%$ of the distribution. The value of X derived for the negative z value represents the lower limit, while the value of X derived for the positive z value represents the upper limit. Thus, the middle $P\%$ of the distribution is any score equal to or greater than the computed lower limit as well as equal to or less than the computed upper limit.

Hypothesis Testing

In inferential statistics sample data are primarily employed in two ways to draw inferences about one or more populations. The two methodologies employed in inferential statistics are **hypothesis testing** and **estimation of population parameters**. This section will discuss hypothesis testing. To be more specific, the material to be presented in this section represents a general approach to hypothesis testing which is commonly referred to as the **classical hypothesis testing model**. The term **classical hypothesis testing model** (also referred to as the **null hypothesis significance testing model (NHST)**) is employed to represent a model which resulted from a blending of the views of the British statistician Sir Ronald Fisher (1925, 1955, 1956) with those of two of his contemporaries, Jerzy Neyman and Egon Pearson (Neyman (1950), Neyman and Pearson (1928), Pearson (1962)). In the next section the evolution of the **classical hypothesis testing model** will be discussed in greater detail. In the latter discussion it will be noted that there were major disagreements between Fisher and Neyman–Pearson regarding hypothesis testing, and, in point of fact (as Gigerenzer (1993, p. 324) notes), the **classical hypothesis testing model** is the end result of other people (specifically the authors of textbooks on inferential statistics during the 1950s and 1960s) integrating the disparate views of Fisher and Neyman–Pearson into a coherent model. Alternative approaches to hypothesis testing will also be discussed in the next section, as well as at other points in the book. The reader should keep in mind that there are those who reject the **classical hypothesis testing model** or argue that it is not always the most appropriate one to employ. Regardless of which hypothesis testing model one favors, a researcher should always be open to employing an alternative model if it provides one with a higher likelihood of discovering the truth. Additionally, within the framework of employing a hypothesis testing model, one should never adopt a mechanical adherence to a set of rules which are viewed as immutable. Put simply, if used judiciously the **classical hypothesis testing model** can provide a researcher with a tremendous amount of useful information. On the other hand, if those who employ it lack a conceptual understanding of the model and/or use it in an inflexible mechanical manner, they may employ it inappropriately and thus arrive at erroneous conclusions.

The most basic concept in the **classical hypothesis testing model** is a **hypothesis**. Within the framework of inferential statistics, a **hypothesis** can be defined as a prediction about a single population or about the relationship between two or more populations. **Hypothesis testing** is a procedure in which sample data are employed to evaluate a hypothesis. In using the term hypothesis, some sources make a distinction between a **research hypothesis** and **statistical hypotheses**.

A **research hypothesis** is a general statement of what a researcher predicts. Two examples of a research hypothesis are: a) The average IQ of all males is some value other than 100; and b) Clinically depressed patients who take an antidepressant for six months will be less depressed than clinically depressed patients who take a placebo for six months.

In order to evaluate a research hypothesis, it is restated within the framework of two **statistical hypotheses**. Through use of a symbolic format, the statistical hypotheses summarize the research hypothesis with reference to the population parameter or parameters under study. The two statistical hypotheses are the **null hypothesis**, which is represented by the notation H_0, and the **alternative hypothesis**, which is represented by the notation H_1.

The **null hypothesis** is a statement of **no effect** or **no difference**. Since the statement of the research hypothesis generally predicts the presence of an effect or a difference with respect to whatever it is that is being studied, the null hypothesis will generally be a hypothesis the researcher expects to be rejected. The **alternative hypothesis**, on the other hand, represents a statistical statement indicating the **presence of an effect or a difference**. Since the research hypothesis typically predicts an effect or difference, the researcher generally expects the alternative hypothesis to be supported.[17]

The null and alternative hypotheses will now be discussed in reference to the two research hypotheses noted earlier. Within the framework of the first research hypothesis which was presented, we will assume that a study is conducted in which an IQ score is obtained for each of n males who have been randomly selected from a population comprised of N males. The null and alternative hypotheses can be stated as follows: H_0: $\mu = 100$ and H_1: $\mu \neq 100$. The null hypothesis states that the mean (IQ score) of the population the sample represents equals 100. The alternative hypothesis states that the mean of the population the sample represents does not equal 100. The absence of an effect will be indicated by the fact that the sample mean is equal to or reasonably close to 100. If such an outcome is obtained, a researcher can be reasonably confident that the sample has come from a population with a mean value of 100. The presence of an effect, on the other hand, will be indicated by the fact that the sample mean is significantly above or below the value 100. Thus, if the sample mean is substantially larger or smaller than 100, the researcher can conclude there is a high likelihood the population mean is some value other than 100, and thus reject the null hypothesis.

As stated above, the alternative hypothesis is **nondirectional**. A **nondirectional** (also referred to as a **two-tailed**) **alternative hypothesis** does not make a prediction in a specific direction. The alternative hypothesis H_1: $\mu \neq 100$ just states that the population mean will not equal 100, but it does not predict whether it will be less than or greater than 100. If, however, a researcher wants to make a prediction with respect to direction, the alternative hypothesis can be stated **directionally**. Thus, with respect to the above example, either of the following two **directional** (also referred to as **one-tailed**) **alternative hypotheses** can be employed: H_1: $\mu > 100$ or H_1: $\mu < 100$.

The alternative hypothesis H_1: $\mu > 100$ states the mean of the population the sample represents is some value greater than 100. If the directional alternative hypothesis H_1: $\mu > 100$ is employed, the null hypothesis can only be rejected if the data indicate that the population mean is some value above 100. The null hypothesis cannot, however, be rejected if the data indicate the population mean is some value below 100.

The alternative hypothesis H_1: $\mu < 100$ states the mean of the population the sample represents is some value less than 100. If the directional alternative hypothesis H_1: $\mu < 100$ is employed, the null hypothesis can only be rejected if the data indicate the population mean is some value below 100. The null hypothesis cannot, however, be rejected if the data indicate the population mean is some value above 100. The reader should take note of the fact that although there are three possible alternative hypotheses that one can employ (one that is nondirectional and two that are directional), **the researcher must select only one of the alternative hypotheses**.

Researchers are not in agreement with respect to the conditions under which one should employ a nondirectional or a directional alternative hypothesis. Some researchers take the position that a nondirectional alternative hypothesis should always be employed, regardless of one's prior expectations about the outcome of an experiment. Other researchers believe that a nondirectional alternative hypothesis should only be employed when one has no prior expectations about the outcome of an experiment (i.e., no expectation with respect to the direction of an effect or difference). These same researchers believe that if one does have a definite expectation about the direction of an effect or difference, a directional alternative hypothesis should be employed. One advantage of employing a directional alternative hypothesis is that in order to reject the null hypothesis, a directional alternative hypothesis does not require there be as large an effect or difference in the sample data as will be the case if a nondirectional alternative hypothesis is employed.

The second of the research hypotheses discussed earlier in this section predicted that an antidepressant will be more effective than a placebo in treating depression. Let us assume that in order to evaluate this research hypothesis, a study is conducted which involves two groups of clinically depressed patients. One group, which will represent Sample 1, is comprised of n_1 patients, and the other group, which will represent Sample 2, is comprised of n_2 patients. The subjects in Sample 1 take an antidepressant for six months, and the subjects in Sample 2 take a placebo during the same period of time. After six months have elapsed, each subject is assigned a score with respect to his or her level of depression.

The null and alternative hypotheses can be stated as follows: H_0: $\mu_1 = \mu_2$ and H_1: $\mu_1 \neq \mu_2$. The null hypothesis states that the mean (depression score) of the population Sample 1 represents equals the mean of the population Sample 2 represents. The alternative hypothesis (which is stated **nondirectionally**) states that the mean of the population Sample 1 represents does not equal the mean of the population Sample 2 represents. In this instance the two populations are a population comprised of N_1 clinically depressed people who take an antidepressant for six months versus a population comprised of N_2 clinically depressed people who take a placebo for six months. The absence of an effect or difference will be indicated by the fact that the two sample means are exactly the same value or close to being equal. If such an outcome is obtained, a researcher can be reasonably confident that the samples do not represent two different populations.[18] The presence of an effect, on the other hand, will be indicated if a significant difference is observed between the two sample means. Thus, we can reject the null hypothesis if the mean of Sample 1 is significantly larger than the mean of Sample 2, or the mean of Sample 1 is significantly smaller than the mean of Sample 2.

As is the case with the first research hypothesis discussed earlier, the alternative hypothesis can also be stated directionally. Thus, either of the following two directional alternative hypotheses can be employed: H_1: $\mu_1 > \mu_2$ or H_1: $\mu_1 < \mu_2$.

The alternative hypothesis H_1: $\mu_1 > \mu_2$ states the mean of the population Sample 1 represents is greater than the mean of the population Sample 2 represents. If the directional alternative hypothesis H_1: $\mu_1 > \mu_2$ is employed, the null hypothesis can only be rejected if the data indicate the mean of Sample 1 is significantly greater than the mean of Sample 2. The null hypothesis cannot, however, be rejected if the mean of Sample 1 is significantly less than the mean of Sample 2.

The alternative hypothesis H_1: $\mu_1 < \mu_2$ states the mean of the population Sample 1 represents is less than the mean of the population Sample 2 represents. If the directional alternative hypothesis H_1: $\mu_1 < \mu_2$ is employed, the null hypothesis can only be rejected if the data indicate the mean of Sample 1 is significantly less than the mean of Sample 2. The null hypothesis cannot, however, be rejected if the mean of Sample 1 is significantly greater than the mean of Sample 2.

Upon collecting the data for a study, the next step in the hypothesis testing procedure is to evaluate the data through use of the appropriate inferential statistical test. An inferential statistical test yields a **test statistic**. The latter value is interpreted by employing special tables which contain information documenting the expected distribution of the test statistic. More specifically, such tables contain extreme values of the test statistic (referred to as **critical values**) that are highly unlikely to occur if the null hypothesis is true. Such tables allow a researcher to determine whether or not the result of a study is **statistically significant**.

The term **statistical significance** implies that one is determining whether or not an obtained difference in an experiment is likely to be due to chance or due to the presence of a genuine experimental effect. To clarify this, think of a roulette wheel on which there are 38 possible numbers that may occur on any roll of the wheel. Suppose we spin a wheel 38,000 times. On the basis of chance each number should occur 1/38th of the time, and thus each value should occur 1000 times (i.e., 38000 ÷ 38 = 1000). Suppose the number 32 occurs 998 times in 38,000 spins of the wheel. Since this value is close to the expected value of 1000, it is highly unlikely that the wheel is biased against the number 32, and is thus not a fair wheel (at least in reference to the number 32). This is because 998 is extremely close to 1000, and a difference of 2 outcomes isn't unlikely on the basis of the random occurrence of events (i.e., chance). On the other hand, if the number 32 only occurs 380 times in 38,000 trials (i.e., 1/100th of the time), since 380 is well below the expected value of 1000, this strongly suggests that the wheel is biased against the number 32 (and is thus probably biased in favor of one or more of the other numbers). Although it is theoretically possible that a specific number can occur 380 times in 38,000 trials on a roulette wheel, the likelihood of such an occurrence is extremely remote. Consequently, because of the latter a casino would probably conclude it was in their best interests to view the wheel in question as defective and thus replace it.

When evaluating the results of an experiment, one employs a logical process similar to that involved in the above situation with the roulette wheel. The decision on whether to retain or reject the null hypothesis is based on contrasting the observed outcome of an experiment with the outcome one can expect if, in fact, the null hypothesis is true. This decision is made by using the appropriate **inferential statistical test**. An **inferential statistical test** is essentially an equation which describes a set of mathematical operations to be performed on the data obtained in a study. The end result of conducting such a test is a final value which is designated as the **test statistic**. A test statistic is evaluated in reference to a **sampling distribution**, which is a theoretical probability distribution of all the possible values the test statistic can assume if one were to conduct an infinite number of studies employing a sample size equal to that used in the study. The probabilities for a sampling distribution are based on the assumption that each of the samples is randomly drawn from the population it represents.

When evaluating the study involving the use of a drug versus a placebo in treating depression, the researcher is asking if the difference between the scores of the two groups is due to chance, or if instead is due to some nonchance factor (which in a well controlled study will be the different treatments to which the groups are exposed). The larger the difference between the average scores of the two groups (just like the larger the difference between the observed and expected occurrence of a number on a roulette wheel), the less likely the difference is due to chance factors, and the more likely it is due to the experimental treatments. Thus, by declaring a difference statistically significant, the researcher is saying that based on an analysis of the sampling distribution of the test statistic, it is highly unlikely that a difference equal to or greater than that which was observed in the study could have occurred as result of chance. In view of this, the most logical decision is to conclude that the difference is due to the experimental treatments, and thus reject the null hypothesis.

Scientific convention has established that in order to declare a difference statistically significant, there can be no more than a 5% likelihood that the difference is due to chance. If a researcher believes that 5% is too high a value, one may elect to employ a 1%, or an even lower minimum likelihood, before one will be willing to conclude that a difference is significant. The notation $p > .05$ is employed to indicate the result of an experiment is not significant. The latter notation indicates there is a greater than 5% likelihood that an observed difference or effect could be due to chance. On the other hand, the notation $p < .05$ indicates the outcome of a study is significant at the .05 level, which means there is less than a 5% likelihood that an obtained difference or effect can be due to chance.[19] The notation $p < .01$ indicates a significant result at the .01 level (i.e., there is less than a 1% likelihood the difference is due to chance).

When the normal distribution is employed for inferential statistical analysis, four tabled **critical values** are commonly employed. These values are summarized in Table I.12.

Table I.12   Tabled Critical Two-Tailed and One-Tailed .05 and .01 z Values

	$z_{.05}$	$z_{.01}$
Two-tailed values	1.96	2.58
One-tailed values	1.65	2.33

The value $z = 1.96$ is referred to as the tabled critical two-tailed .05 z value. This value is employed since the total proportion of cases in the normal distribution that falls above $z = +1.96$ or below $z = -1.96$ is .05. This can be confirmed by examining Column 3 of **Table A1** with respect to the value $z = 1.96$. The value of .025 in Column 3 indicates the proportion of cases in the right tail of the curve which falls above $z = +1.96$ is .025, and the proportion of cases in the left tail of the curve which falls below $z = -1.96$ is .025. If the two .025 values are added, the resulting proportion is .05. Note that this is a two-tailed critical value, since the proportion .05 is based on adding the extreme 2.5% of the cases from the two tails of the distribution.

The value $z = 2.58$ is referred to as the tabled critical two-tailed .01 z value. This value is employed since the total proportion of cases in the normal distribution that falls above $z = +2.58$ or below $z = -2.58$ is .01. This can be confirmed by examining Column 3 of **Table A1** with respect to the value $z = 2.58$. The value of .0049 (which rounded off equals .005) in Column 3 indicates the proportion of cases in the right tail of the curve which falls above $z = +2.58$ is .0049, and the proportion of cases in the left tail of the curve which falls below $z = -2.58$ is .0049. If the two .0049 values are added, the resulting proportion is .0098, which rounded off equals .01. Note that this is a two-tailed critical value, since the proportion .01 is based on adding the extreme .5% of the cases from the two tails of the distribution.

The value $z = 1.65$ is referred to as the tabled critical one-tailed .05 z value. This value is employed since the proportion of cases in the normal distribution that falls above $z = +1.65$ or below $z = -1.65$ in each tail of the distribution is .05. This can be confirmed by examining Column 3 of **Table A1** with respect to the value $z = 1.65$. The value of .0495 (which rounded off equals .05) in Column 3 indicates the proportion of cases in the right tail of the curve which falls above $z = +1.65$ is .0495, and the proportion of cases in the left tail of the curve which falls below $z = -1.65$ is .0495. Note that this is a one-tailed critical value, since the proportion .05 is based on the extreme 5% of the cases in one tail of the distribution.[20]

The value $z = 2.33$ is referred to as the tabled critical one-tailed .01 z value. This value is employed since the proportion of cases in the normal distribution that falls above $z = +2.33$ or below $z = -2.33$ in each tail of the distribution is .01. This can be confirmed by examining Column 3 of **Table A1** with respect to the value $z = 2.33$. The value of .0099 (which rounded off equals .01) in Column 3 indicates the proportion of cases in the right tail of the curve which falls above $z = +2.33$ is .0099, and the proportion of cases in the left tail of the curve which falls below $z = -2.33$ is .0099. Note that this is a one-tailed critical value, since the proportion .01 is based on the extreme 1% of the cases in one tail of the distribution.

Although in practice it is not scrupulously adhered to, the conventional hypothesis testing model employed in inferential statistics assumes that prior to conducting a study a researcher stipulates whether a directional or nondirectional alternative hypothesis will be employed, as well as at what level of significance the null hypothesis will be evaluated. The probability value which identifies the level of significance is represented by the notation α, which is the lower case Greek letter **alpha**. Throughout the book the latter value will be referred to as the **prespecified alpha value** (or **prespecified level of significance**), since it will be assumed that the value was specified prior to the data collection phase of a study.

Type I errors, Type II errors and power in hypothesis testing Within the framework of hypothesis testing, it is possible for a researcher to commit two types of errors. These errors are referred to as a **Type I error** and a **Type II error**.

A **Type I error** is when a true null hypothesis is rejected (i.e., one concludes that a false alternative hypothesis is true). The likelihood of committing a Type I error is specified by the **alpha level** a researcher employs in evaluating an experiment. The more concerned a researcher is with committing a Type I error, the lower the value of alpha the researcher should employ. Thus, the likelihood of committing a Type I error if $\alpha = .01$ is 1%, as compared with a 5% likelihood if $\alpha = .05$.

A **Type II error** is when a false null hypothesis is retained (i.e., one concludes that a true alternative hypothesis is false). The likelihood of committing a Type II error is represented by β, which (as noted earlier) is the lower case Greek letter **beta**. The likelihood of rejecting a false null hypothesis represents the **power** of a statistical test. The **power** of a test is determined by subtracting the value of beta from 1 (i.e., $Power = 1 - \beta$). The likelihood of committing a Type II error is inversely related to the likelihood of committing a Type I error. In other words, as the likelihood of committing one type of error decreases, the likelihood of committing the other type of error increases. Thus, with respect to the alternative hypothesis one employs, there is a higher likelihood of committing a Type II error when alpha is set equal to .01 than when it is set equal to .05. The likelihood of committing a Type II error is also inversely related to the power of a statistical test. In other words, as the likelihood of committing a Type II error decreases, the power of the test increases. Consequently, the higher the alpha value (i.e., the higher the likelihood of committing a Type I error), the more powerful the test. Two other ways to increase the power of a test (and thereby decrease the likelihood of committing a Type II error) are: a) Increasing the sample size employed in a study; and b) Minimizing the amount of variability in a set of data that is attributable to extraneous factors.

The relationship between **Type I error rate**, **Type II error rate** and **power** can be summarized as follows: a) The higher the value of **alpha** the greater the likelihood of committing a **Type I error**. As the value of **alpha** increases, the likelihood of committing a **Type II error** decreases and the **power** of the statistical test increases. In other words, the higher the value of **alpha** the easier it will be for a researcher to reject the null hypothesis. To put it another way, the higher the value of **alpha** the more likely it is that the alternative hypothesis will be supported; b)The lower the value of **alpha** the lower the likelihood of committing a **Type I error**. As the value of **alpha** decreases, the likelihood of committing a **Type II error** increases and the **power** of the statistical test decreases. In other words, the lower the value of **alpha** the more difficult it will be for a researcher to reject the null hypothesis. To put it another way, the lower the value of **alpha** the less likely it is that the alternative hypothesis will be supported.

Table I.13 summarizes the decision making process within the framework of the **classical hypothesis testing model**. In the latter table the **True State of Nature** refers to what is, in fact, the truth regarding the actual status of the null hypothesis (i.e., H_0). The p value represents the probability associated with a specific decision.

Table I.13   Summary of Decision Making Process in Hypothesis Testing

	True State of Nature	
Decision	H_0 **True**	H_0 **False**
Reject H_0	Type I Error $p = \alpha$	Correct Decision $p = 1 - \beta =$ Power
Fail to reject H_0	Correct Decision $p = 1 - \alpha$	Type II error $p = \beta$

Before continuing the reader should take note of the fact that in the **Preface** it was noted that all of the examples employed in the book involve the use of **small sample sizes**. The consequences of the latter are **inflated sampling error** as well as **low power** (which is associated with a **high Type II error rate**) for any inferential statistical test conducted. It should be emphasized that the use of small sample sizes is purely for illustrative purposes in order to facilitate the reader's ability to follow the computational procedures described in the book. In practice, the sample size employed in research should be substantially larger than those employed in the examples in this book. In the final analysis, determination of sample size should be based on what a researcher considers to be an acceptable compromise with respect to the Type I versus Type II error rates associated with a study. As will be noted later, over the years researchers have been criticized (by sources such as Cohen (1962, 1977) and Hunter and Schmidt (1990, 2004)) for conducting studies with inadequate power (which can often be attributed to small sample sizes), thereby making it difficult to reject the null hypothesis, when, in fact, the latter is false.

However, it will also be noted (later in this chapter as well as in the discussion of the **minimum-effect hypothesis testing model** in Section IX (the **Addendum**) of the **Pearson product-moment correlation coefficient (Test 28)**) that as sample size increases, the likelihood the null hypothesis will be rejected approaches 100%.

Example I.9 will be employed to illustrate the application of the above described concepts on hypothesis testing in reference standard deviation (z) scores discussed in the previous section. In point of fact, the analysis of Example I.9 illustrates the use of the **single-sample z test** (discussed in the next chapter) when $n = 1$.

Example I.9 *Assume a researcher wants to determine whether or not it is likely that an individual who has been randomly selected from a population could be a member of a normally distributed population which has a mean IQ of 100 (i.e., $\mu = 100$) and a standard deviation of 15 (i.e., $\sigma = 15$). Thus, employing a sample size of $n = 1$, the researcher evaluates the null hypothesis H_0: $\mu = 100$. Determine what the IQ of the person would have to be in order to reject the null hypothesis at the .05 and .01 levels of significance within the framework of both a one- and two-tailed analysis.*

The results of the analysis of Example I.9 are summarized below.

1: One-tailed (directional) .05 (5%) alpha/significance level employed
H_0: $\mu = 100$
H_1: $\mu < 100$ *or* H_1: $\mu > 100$

a) The likelihood of selecting a person with an IQ of 75.25 or less from a normally distributed population with a mean of 100 and a standard deviation of 15 is .05 (or 5%). The latter is the case since $X = \mu \pm \sigma z = 100 + (15)(-1.65) = 100 - 24.75 = 75.25$. Note that the value $z = 1.65$ delineates the extreme 5% of scores in each tail of the normal distribution. If the directional alternative hypothesis H_1: $\mu < 100$ is employed, the null hypothesis H_0: $\mu = 100$ can be rejected at the .05 level if the person's IQ is equal to or less than 75.25. If the null hypothesis is rejected, there is still a 5% likelihood the researcher's decision to reject the null hypothesis represents a Type I error. In other words, there is still a 5% likelihood the null hypothesis is, in fact, true. Or to put it another way, there is still a 5% likelihood that a person with an IQ of 75.25 or less is a member of a population which has a mean of 100 and a standard deviation of 15. Figure I.15a depicts this example visually.

b) The likelihood of selecting a person with an IQ of 124.75 or greater from a normally distributed population with a mean of 100 and a standard deviation of 15 is .05 (or 5%). The latter is the case since $X = \mu \pm \sigma z = 100 + (15)(1.65) = 100 + 24.75 = 124.75$. Note that the value $z = 1.65$ delineates the extreme 5% of scores in each tail of the normal distribution. If the directional alternative hypothesis H_1: $\mu > 100$ is employed, the null hypothesis H_0: $\mu = 100$ can be rejected at the .05 level if the person's IQ is equal to or greater than 124.75. If the null hypothesis is rejected, there is still a 5% likelihood the researcher's decision to reject the null hypothesis represents a Type I error. In other words, there is still a 5% likelihood the null hypothesis is, in fact, true. Or to put it another way, there is still a 5% likelihood that a person with an IQ of 124.75 or greater is a member of a population which has a mean of 100 and a standard deviation of 15. Figure I.15b depicts this example visually.

2: One-tailed (directional) .01 (1%) alpha/significance level employed
H_0: $\mu = 100$
H_1: $\mu < 100$ *or* H_1: $\mu > 100$

a) The likelihood of selecting a person with an IQ of 65.05 or less from a normally distributed population with a mean of 100 and a standard deviation of 15 is .01 (or 1%). The latter is the case since $X = \mu \pm \sigma z = 100 + (15)(-2.33) = 100 - 34.95 = 65.05$. Note that the value $z = 2.33$ delineates the extreme 1% of scores in each tail of the normal distribution. If the directional alternative hypothesis H_1: $\mu < 100$ is employed, the null hypothesis H_0: $\mu = 100$ can be rejected at the .01 level if the person's IQ is equal to or less than 65.05. If the null hypothesis is rejected, there is still a 1% likelihood the researcher's decision to reject that null hypothesis represents a Type I error. In other words, there is still a 1% likelihood the null hypothesis is, in fact, true. Or to put it another way, there is still a 1% likelihood that a person with an IQ of 65.05 or less is a member of a population which has a mean of 100 and a standard deviation of 15. Figure I.15c depicts this example visually.

b) The likelihood of selecting a person with an IQ of 134.95 or greater from a normally distributed population with a mean of 100 and a standard deviation of 15 is .01 (or 1%). The latter is the case since $X = \mu \pm \sigma z = 100 + (15)(2.33) = 100 + 34.95 = 134.95$. Note that the value $z = 2.33$ delineates the extreme 1% of scores in each tail of the normal distribution. If the directional alternative hypothesis H_1: $\mu > 100$ is employed, the null hypothesis H_0: $\mu = 100$ can be rejected at the .01 level if the person's IQ is equal to or greater than 134.95. If the null hypothesis is rejected, there is still a 1% likelihood the researcher's decision to reject the null hypothesis represents a Type I error. In other words, there is still a 1% likelihood the null hypothesis is, in fact, true. Or to put it another way, there is still a 1% likelihood that a person with an IQ of 134.95 or greater is a member of a population which has a mean of 100 and a standard deviation of 15. Figure I.15d depicts this example visually.

3: Two-tailed (nondirectional) .05 (5%) alpha/significance level employed
H_0: $\mu = 100$
H_1: $\mu \neq 100$

The likelihood of selecting a person with an IQ of 70.6 or less is .025 (or 2.5%), and the likelihood of selecting a subject with an IQ of 129.4 or greater is .025 (or 2.5%). The latter is the case since $X = \mu \pm \sigma z = 100 \pm (15)(1.96) = 100 \pm 29.4$, and $100 - 29.4 = 70.6$ and $100 + 29.4 = 129.4$. Note that value $z = 1.96$ delineates the extreme 2.5% of scores in each tail of the normal distribution. The likelihood of selecting a person with an IQ equal to or less than 70.6 or equal to or greater than 129.4 is $.025 + .025 = .05$ (or 5%). If the nondirectional alternative hypothesis H_1: $\mu \neq 100$ is employed, the null hypothesis H_0: $\mu = 100$ can be rejected at the .05 level if the person's IQ is equal to or less than 70.6 or is equal to or greater than 129.4. If the null hypothesis is rejected, there is still a 5% likelihood the researcher's decision to reject the null hypothesis represents a Type I error. In other words, there is still a 5% likelihood the null hypothesis is, in fact, true. Or to put it another way, there is still a 5% likelihood that a person with an IQ equal to or less than 70.6 or equal to or greater than 129.4 is a member of a population which has a mean of 100 and a standard deviation of 15. Figure I.15e depicts this example visually.

4: Two-tailed (nondirectional) .01 (1%) alpha/significance level employed
H_0: $\mu = 100$
H_1: $\mu \neq 100$

The likelihood of selecting a person with an IQ of 61.3 or less is .005 (or .5%), and the likelihood of selecting a subject with an IQ of 138.7 or greater is .005 (or .5%). The latter is the

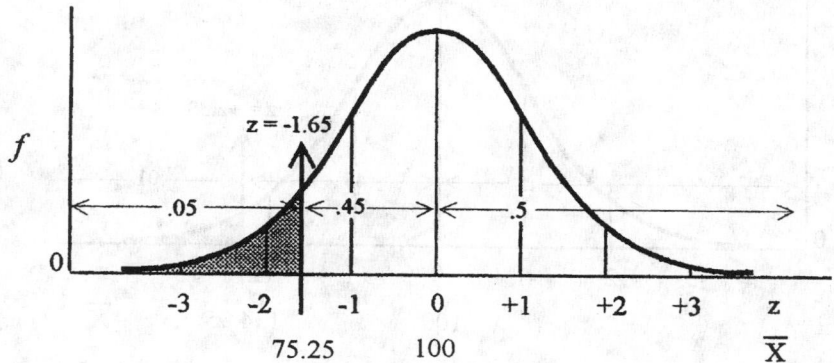

Figure 1.15a   Distribution of Critical One-Tailed .05 z Value for H_1: $\mu < 100$

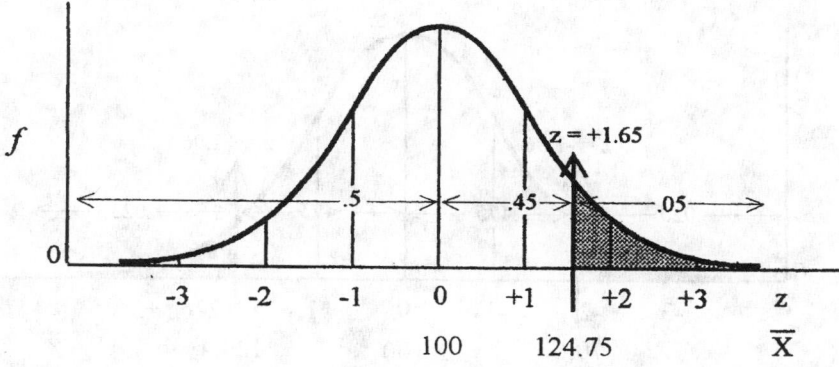

Figure I.15b   Distribution of Critical One-Tailed .05 z Value for H_1: $\mu > 100$

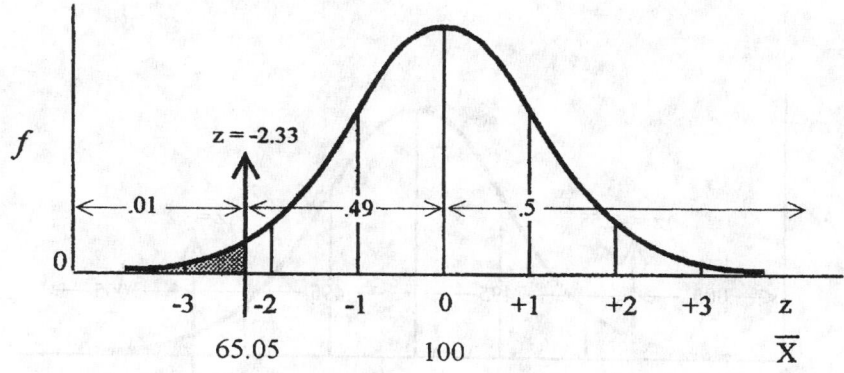

Figure I.15c   Distribution of Critical One-Tailed .01 z Value for H_1: $\mu < 100$

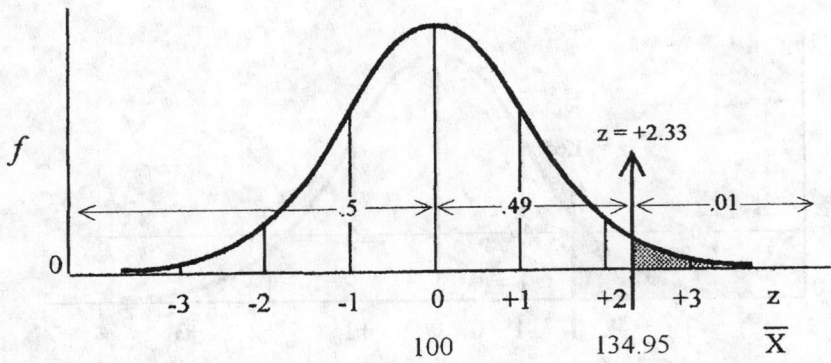

Figure I.15d   Distribution of Critical One-Tailed .01 z Value for H_1: $\mu > 100$

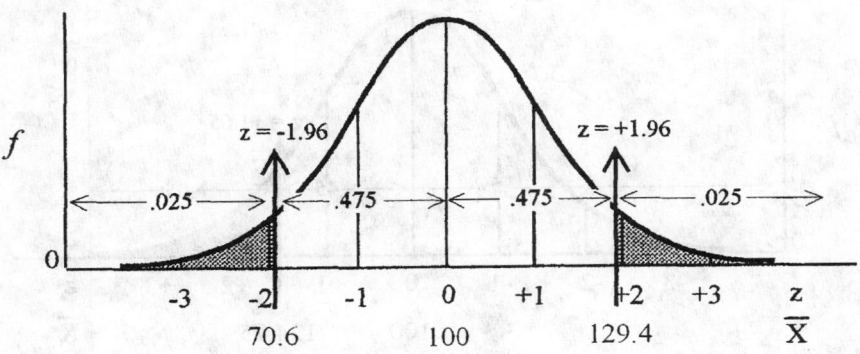

Figure I.15e   Distribution of Critical Two-Tailed .05 z Value for H_1: $\mu \neq 100$

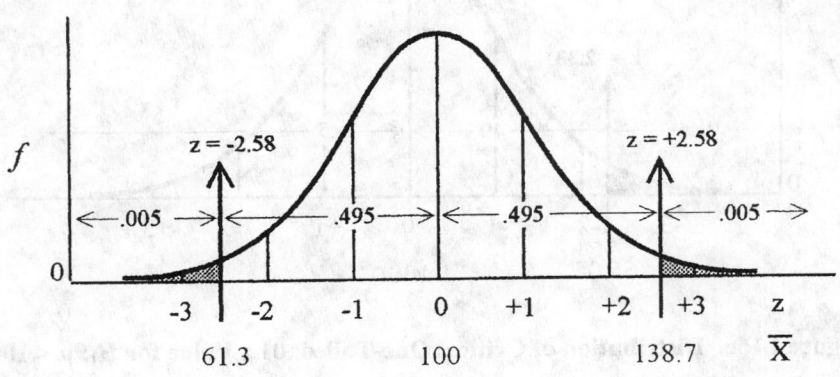

Figure I.15f   Distribution of Critical Two-Tailed .01 z Value for H_1: $\mu \neq 100$

case since $X = \mu \pm \sigma z = 100 \pm (15)(2.58) = 100 \pm 38.7$, and $100 - 38.7 = 61.3$ and $100 + 38.7 = 138.7$. Note that the value $z = 2.58$ delineates the extreme .5% of scores in each tail of the normal distribution. The likelihood of selecting a person with an IQ equal to or less than 61.3 or equal to or greater than 138.7 is $.005 + .005 = .01$ (or 1%). If the nondirectional alternative hypothesis $H_1: \mu \neq 100$ is employed, the null hypothesis $H_0: \mu = 100$ can be rejected at the .01 level if the person's IQ is equal to or less than 61.3 or is equal to or greater than 138.7. If the null hypothesis is rejected, there is still a 1% likelihood the researcher's decision to reject the null hypothesis represents a Type I error. In other words, there is still a 1% likelihood the null hypothesis is, in fact, true. Or to put it another way, there is still a 1% likelihood that a person with an IQ equal to or less than 61.3 or equal to or greater than 138.7 is a member of a population which has a mean of 100 and a standard deviation of 15. Figure I.15f depicts this example visually.

Statistical significance versus practical significance When the term **significance** is employed within the context of scientific research, it is instructive to make a distinction between **statistical significance** and **practical significance**. Statistical significance only implies that the outcome of a study is highly unlikely to have occurred as a result of chance. It does not necessarily suggest that any difference or effect detected in a set of data is of any practical value. As an example, assume that the Scholastic Aptitude Test (SAT) scores of two school districts which employ different teaching methods are contrasted. Assume that the teaching method of each school district is based on specially designed classrooms. The results of the study indicate that the SAT average in School District A is one point higher than the SAT average in School District B, and this difference is statistically significant at the .01 level. Common sense suggests it would be illogical for School District B to invest the requisite time and money in order to redesign its physical environment for the purpose of increasing the average SAT score in the district by one point. Thus, in this example, even though the obtained difference is statistically significant, in the final analysis it is of little or no practical significance. The general issue of statistical versus practical significance is discussed in more detail in Section VI of the *t* **test for two independent samples**, and under the discussion of **meta-analysis and related topics** in Section IX (the **Addendum**) of the **Pearson product-moment correlation coefficient**.

Before proceeding to the next topic, a final comment is in order on the general subject of tests of statistical significance. It cannot be emphasized too strongly that a test of significance merely represents the final stage in a sequential process involved in conducting research. Those stages which precede the test of significance consist of whatever it is a researcher does within the framework of designing and executing a study. If at some point there is a flaw in the design and/or execution of a study, it will have a direct impact on the reliability of the results. Under such circumstances, any data that are evaluated will, for all practical purposes, be of little or no value. The cliche "garbage in, garbage out" is apropos here, in that if imperfect data are evaluated with a test of significance, the resulting analysis will be unreliable. Ultimately, a test of significance is little more than an algorithm which has been derived for evaluating a set of data under certain conditions. The test, in and of itself, is incapable of making a judgement with respect to the reliability and overall quality of the data. To go further, the probability values employed in evaluating a test of significance are derived from theoretical probability distributions. In order for any probability derived from such a distribution to be reliable, one or more assumptions underlying the distribution must not have been violated. In the final analysis, a probability value derived from a test of significance should probably be viewed at best as an estimate of the actual probability associated with a result, and at worst (when data are derived in a poorly designed and/or methodologically weak study) as a woefully inaccurate metric that has little or no relevance to the problem being evaluated.

Although the hypothesis testing model described in this section is based on conducting a single study in order to evaluate a research hypothesis, throughout the book the author emphasizes the importance of replication in research. Aside from the fact that in any given study sampling error may result in a researcher reaching the wrong conclusion, there is also the fact that inferential statistical tests make certain assumptions, many of which the researcher can never be sure have been met. Since the accuracy of the probability values in tables of critical values for test statistics are contingent upon the validity of the assumptions underlying the test, if any of the assumptions have been violated, the accuracy of the tables can be compromised, thus increasing the likelihood the decision a researcher makes will represent either a Type I or Type II error. In view of the aforementioned factors which can compromise the outcome of a single experiment, the most effective way of determining the truth with regard to a particular hypothesis (especially if practical decisions are to be made on the basis of the results of research) is to conduct multiple studies which evaluate the same hypothesis. When multiple studies yield consistent results, one is less likely to be challenged that the correct decision has been made with respect to the hypothesis under study. A general discussion of statistical methods which can be employed to aid in the interpretation of the results of multiple studies that evaluate same general hypothesis can be found under the discussion of **meta-analysis and related topics** in Section IX (the **Addendum**) of the **Pearson product-moment correlation coefficient**.

A History and Critique of the Classical Hypothesis Testing Model

The intent of this section is to provide a brief history of the individuals and events which were responsible for the development of the **classical hypothesis testing model**. The statistical techniques which are most commonly employed in inferential and descriptive statistics were developed by a small group of Englishmen during the latter part of the nineteenth century through the middle of the twentieth century. Readers who are interested in a more comprehensive overview of the individuals to be discussed in this section, as well as biographical information on other notable individuals in the history of statistics should consult sources such as Cowles (1989, 2001), Johnson and Kotz (1997), Kline (2004), Stigler (1999), and Tankard (1984).

During the late 1800s Sir Francis Galton (1822-1911) is credited with introducing the important statistical concepts of **correlation** and **regression**. (Correlation is discussed briefly in the latter part of this **Introduction**, and both correlation and regression are described in detail in the chapter on the **Pearson product-moment correlation coefficient**.) Galton referred to the science of statistics as **biometrics**, since he viewed statistics as a discipline which employed mathematics to study issues in both the biological and social sciences. Another Englishman Karl Pearson (1857-1936) subsequently developed the mathematical procedure for computing a correlation coefficient (which is described in the chapter on the **Pearson product-moment correlation coefficient**). Along with Sir Ronald Fisher, Pearson is probably viewed as having made the greatest contributions to what today is considered the basis of modern statistics. In addition to his work on correlation, Pearson also discovered the chi-square distribution (a theoretical probability distribution) and the **chi-square goodness-of-fit test (Test 8)**. Throughout his professional life Pearson was a source of controversy, much of it resulting from an enthusiasm which he shared with Galton for eugenics (which is the use of selective breeding to enhance the characteristics of the human race). During the latter part of his professional career Pearson was appointed as the first Galton Professor of Eugenics at University College in London (the chair was bequeathed as a gift to the university by Galton). In 1901 Pearson cofounded the influential statistical journal *Biometrika*, in part to provide himself with an outlet to publish some of his more controversial research. Another source of controversy surrounding Pearson revolved around his acrimonious personal and professional differences with Sir Ronald Fisher.

The major contribution of William S. Gosset (1876-1937), who briefly studied under Pearson, was his discovery of the t distribution (a theoretical probability distribution), and his development of the **t-test** (discussed under the **single-sample t test** and the **t tests for two independent and dependent samples (Tests 11 and 17))**. Lehman (1993, p. 1242) states that the modern theory of hypothesis testing was initiated in 1908 with Gosset's development of the **t-test**. Johnson and Kotz (1997, p. 328) note that Gosset was the individual who first introduced the concept of the alternative hypothesis in hypothesis testing. For many years Gosset was employed at the Guiness Brewery in Dublin where he employed the scientific method to evaluate beer making. Throughout his professional career he published under the name *Student*, since the Guiness Brewery wished for his identity to remain anonymous.

Although the three men discussed up to this point had a major impact on the development of modern statistics, those individuals who were most responsible for the development of what was eventually to become known as the **classical hypothesis testing model** were the British statistician Sir Ronald Fisher (1890-1962) and two of his contemporaries Egon Pearson (1895-1980) (who was the son of Karl Pearson) and Jerzy Neyman (1894-1981). As noted earlier, the **classical hypothesis testing model** (i.e., the methodology and concepts discussed in the previous section) is the end result of blending a hypothesis testing model developed by Fisher (1925, 1955, 1956) with an alternative model developed by Neyman and Egon Pearson (Neyman (1950), Neyman and Pearson (1928), Pearson (1962)).

Sir Ronald Fisher (who was also a proponent of eugenics) is credited with developing more key concepts in the field of inferential statistics than any statistician of the modern era. Many of Fisher's ideas were developed during the 14 years he worked at an experimental agricultural station in Hertfordshire, England called Rothamsted. Like Gosset, most of Fisher's discoveries grew out of his need to find solutions to practical problems he was confronted with at work. Fisher's most important contributions were his development of the analysis of variance (described in detail later in the book) and his ideas on the subjects of experimental design and hypothesis testing. In the area of hypothesis testing, Fisher introduced the concepts of the null hypothesis and significance levels. In 1931 and again in 1936 Fisher was a visiting professor at Iowa State University, and during his tenure in the United States his ideas had a major impact on the thinking of, among others, George Snedecor and E. F. Lindquist, two prominent American statisticians. The latter two individuals subsequently published statistics textbooks which were largely responsible for introducing Fisher's ideas on hypothesis testing and analysis of variance to the statistical community in the United States.

By all accounts Fisher was a difficult person to get along with. Throughout their professional careers Fisher and Karl Pearson were bitter adversaries. In his position of Galton Professor of Eugenics at University College in London, Karl Pearson was head of the Galton Laboratory. Among those who worked for or studied under Pearson were his son Egon and Jerzy Neyman, a Polish statistician. When Karl Pearson retired in 1933 Fisher succeeded him as Galton Professor of Eugenics. Since Fisher did not get along with Karl Pearson or his associates, in order to placate the retiring Pearson the school established a separate department of statistics, and appointed Egon Pearson as its head. It was Egon Pearson's collaboration with Jerzy Neyman which resulted in what came to be known as the **Neyman–Pearson hypotheses testing model** (to which Neyman was the major contributor). The latter model became a source of controversy, since it challenged Fisher's views on hypothesis testing.

The last part of this section will focus on the fundamental differences between the **Fisher** and **Neyman–Pearson** models of hypothesis testing. (More comprehensive discussions of the Fisher/Neyman–Pearson controversy can be found in Cowles (1989, 2001), Gigerenzer (1993), Gigerenzer and Murray (1987), and Lehman (1993).) The beginnings of the **classical hypothesis testing model** can be traced to Fisher, who in 1925 introduced **null hypothesis testing** (also known as **significance testing**) (Lehman (1993, p. 1243), Gigerenzer (1993, p. 315)). Fisher (1925) stated that a null hypothesis should be evaluated in the following manner: Employing sample data a researcher determines whether or not the information contained in the sample data deviates enough from the value stated in the null hypothesis to render the null hypothesis implausible. Fisher did not see any need to have an alternative hypothesis. It was Fisher's contention that the purpose of a statistical test was to determine whether or not there was sufficient evidence to reject the null hypothesis. Within the latter context he introduced the .05 level of significance as the standard level for rejecting a null hypothesis, and suggested the .01 level as a more stringent alternative. This latter convention for assessing statistical significance was to become an integral part of the **classical hypothesis testing model**, in spite of the fact that Fisher revised his viewpoint on this matter in the 1950s (Gigerenzer (1993, p. 316) and Lehman (1993), p. 1248)). Fisher's later position was that researchers should publish the exact level of significance computed for a set of data. In other words, if the likelihood computed for obtaining a specific result for a set of data if the null hypothesis is true is .02, instead of stating that the result is significant at the .05 level (i.e., $p < .05$), the researcher should just report the exact probability value (i.e., just state the value $p = .02$).

Gigerenzer (1993, p. 319) notes that in 1925 Fisher stated that a null hypothesis can be disproved but "never proved or established" (Fisher (1925, p. 16)). At another point Fisher (1925, p. 13) stated that "experimenters ... are prepared to ignore all [nonsignificant] results." Gigerenzer (1993, p. 319) suggests that within the scientific community there were many who interpreted the latter statements to mean that a nonsignficant result was worthless, and thus not worthy of publication — an interpretation which resulted in editors of academic journals rejecting virtually all studies which failed to report a significant result. In 1955 Fisher changed his perspective with respect to what the status of the null hypothesis should be if an experiment yielded a nonsignificant result. At this latter point in time Fisher (1955 p. 73) implied that a nonsignificant result could not establish the truth of a null hypothesis, but would merely make it more likely that the null hypothesis was true. Thus Fisher (1955) stated that acceptance of a null hypothesis did not indicate that it was proven and thus should be adopted. In his latest thinking Fisher (1955) took the position that any conclusions reached within the framework of hypothesis testing should be tentative and subject to reevaluation based on the outcome of future research. Fisher thus rejected the use of a mechanized set of rules which obligated the researcher to make a final decision regarding the status of the null hypothesis based on the analysis of a single experiment

It was Fisher's contention that if a null hypothesis was rejected it did not mean that the researcher should adopt an alternative hypothesis (since, as noted previously, Fisher did not believe that an alternative hypothesis should be employed) . Fisher viewed the probability value computed for an experiment entirely within the context of that particular study, and did not view it as having any relevance to potential future replications of the experiment. In contrast, Neyman (Neyman (1950), Neyman and Pearson (1928)) believed that a probability value computed for a single experiment should be viewed within the context of potential future replications of the experiment. (According to Gigerenzer (1993, p. 317) Egon Pearson took a position somewhere in between that of Fisher and Neyman on this issue.) Fisher did not speak of or acknowledge Type I or Type II error rates (since he viewed everything within the context of a single experiment), while, as will be noted shortly, Neyman and Pearson did. In the final analysis, the two main elements Fisher presented which were ultimately integrated into the **classical hypothesis testing model** were: a) The statement of a null hypothesis by a researcher; and b) The convention of employing the probability values .05 and .01 as the minimum standards for declaring statistical significance.

In the late 1920s and early 1930s Neyman and Pearson put forth the argument that a researcher must not only state a null hypothesis, but should also specify one or more alternative hypotheses against which the null hypothesis could be evaluated. Within this context they introduced the concepts of a **Type I error** (i.e., **alpha value**), a **Type II error** (i.e., **beta value**) and the **power** of a statistical test. Fisher, however, refused to integrate the latter concepts into his hypothesis testing model. Neyman and Pearson also rejected the use of a standard level of significance such as the .05 or .01 levels, since they felt it was essential that a researcher achieve a balance between the Type I and Type II error rates. Neyman and Pearson took the position that a researcher should stipulate the specific level of significance one has decided to employ prior to conducting a study, and use that value as the criterion for retaining or rejecting the null hypothesis. On the other hand, as noted earlier, Fisher's (1955) final position on the latter was it was not necessary to stipulate the level of significance beforehand, and that if the result of an experiment was deemed significant, the researcher should report the exact probability value computed for the outcome.

The final hypothesis testing model which resulted from integrating Fisher's views on hypothesis testing with those of Neyman and Pearson constitutes the **classical hypothesis testing model** (i.e., the model described in the previous section on hypothesis testing). As noted in the introductory material on hypothesis testing, the individuals largely responsible for this hybridization of Fisher and Neyman–Pearson were authors of textbooks on inferential statistics during the 1950s and 1960s. Critics of the **classical hypothesis testing model** view the blending of the Fisher and Neyman–Pearson models as an unfortunate and ill-conceived hybridization of two incompatible viewpoints. Some sources (e.g., Gigerenzer (1993) and Gigerenzer and Murray (1987)) argue that the **classical hypothesis testing model** has institutionalized inferential statistics to the extent that researchers have come to view statistical analysis in a dogmatic and mechanized way. By the latter it is meant that the model instructs researchers to employ a standardized protocol which involves an inflexible set of decision making guidelines for evaluating research, and in doing so it neglects to teach researchers to think intelligently about data analysis. Gigerenzer (1993, p. 321) notes that neither Fisher (1955) nor Neyman–Pearson would have endorsed the dogmatic and mechanized approach which characterizes the **classical hypothesis testing model**.

Among others, Gigerenzer (1993), Harlow *et al.* (1997), Hunter and Schmidt (1990, 2004), Kline (2004), Meehl (1967), Rozeboom (1960), Smithson (2003), Thompson (1993, 1999, 2002) and Wilkinson *et al.* (1999) propose that researchers employ alternatives to the **classical hypothesis testing model**, and that, regardless of which alternative one elects to employ, a researcher should not be dogmatic and inflexible in evaluating data. Put simply, researchers should pick and choose from all available methodologies, and select those which are most useful in addressing the problem at hand. Instead of focusing on probability values and viewing them as the sole criterion for drawing conclusions, researchers should employ other criteria in evaluating data. Among the alternatives recommended by Gigerenzer (1993, p. 332) for both evaluating hypotheses and constructing theories are visual analysis of data and analysis of effect size (which is alluded to throughout this book). Gigerenzer (1993, p. 335) concludes his critique of the **classical hypothesis testing model** by stating that, "Statistical reasoning is an art and so demands both mathematical knowledge and informed judgement. When it is mechanized, as with the institutionalized hybrid logic, it becomes ritual, not reasoning." (The term hybrid logic refers to the use of the **classical hypothesis testing model**.) Further discussion of criticism of the **classical hypothesis testing model** can be found in Section IX (the **Addendum**) of the **Pearson product-moment correlation coefficient**. under the discussion of **meta-analysis and related topics**.

With respect to what has been noted above, Anderson (2001) and Everitt (2001) state that one reason why some people are critical of the **classical hypothesis testing model** is that the probability value employed to summarize the result of a test of significance is poorly understood. As noted earlier, the latter probability level refers to the alpha level employed by a researcher (or, in some cases, a published probability value represents the exact probability associated with the outcome of a specific study). Everitt (2001, p. 4) cites a study by Oakes (1986) in which one or more of 70 academic psychologists (who ostensibly were trained in the use of the **classical hypothesis testing model**) stated that one or more of the following statements are true if the result of an analysis comparing the means of two independent groups with a *t* **test for two independent samples** is significant at the .01 level (the percentage of academicians endorsing each statement is noted in parentheses): a) You have absolutely disproved the null hypothesis that there is no difference between the population means (1%); b) You have found the probability of the null hypothesis being true (35.7%); c) You have absolutely proved your experimental/alternative hypothesis (5.7%); d) You can deduce the probability of the experimental hypothesis/alternative hypothesis being true (65.7%); e) You know, if you decided to reject the null hypothesis, the probability that you are making the wrong decision (85.7%); f) You have a reliable experiment in the sense that if, hypothetically, the experiment were repeated a great number of times, you would obtain a significant result on 99% of occasions (60%). Only 4.3% of the academicians endorsed the following correct interpretation of the probability value: If the null hypothesis is true, the probability of obtaining the data (or data that represent a more extreme departure from the null hypothesis) is represented by the .01 probability value. Anderson (2001, p. 49) notes that one might also be tempted to erroneously conclude that the value $(1 - p)$ (which in this case equals $1 - .01 = .99$) is the probability that the null hypothesis is false. In point of fact, the latter statement is not justified. In order to determine the likelihood that the null hypothesis is false one would be required to have access to a sampling distribution which is based on the assumption that the null hypothesis is false. Such a sampling distribution would reflect the presence of an effect size, the magnitude of which would be specified by the researcher (for further clarification see Anderson (2002, p. 44)).[21]

The position of this book is that although the criticisms directed at the **classical hypothesis testing model** certainly have some degree of validity, when used intelligently the model is extremely useful, and to date it has been extremely productive in generating scientific knowledge. Put simply, if those who employ the model lack a clear conceptual understanding of it and/or use it in an inflexible mechanical manner, it can be employed inappropriately and lead to erroneous conclusions.[22] On the other hand, when used judiciously it can be useful and productive. This writer would certainly agree that there are situations where the **classical hypothesis testing model** may not always be most appropriate for evaluating data. In such instances one should be amenable to employing an alternative model if it offers one a higher likelihood of discovering the truth. In the final analysis, however, it is the opinion of the writer that the present scope of scientific knowledge would not be greater than it is now if during the past 100 years any of the alternative hypothesis testing models which have been suggested had been used in place of the **classical hypothesis testing model**.

Among the alternative models which are available for hypothesis testing are two that are discussed later in the book. One alternative is the **minimum-effect hypothesis testing model**, which is described in Section IX (the **Addendum**) of the **Pearson product-moment correlation coefficient**. The crux of the argument put forth by proponents of the **minimum-effect hypothesis testing model** against the **classical hypothesis testing model** is that, in reality, the null hypothesis is always false. Specifically, various sources note that the null hypothesis is a **point hypothesis**, in that it stipulates a precise value — namely zero — for the difference between the experimental conditions. Thus any difference, no matter how negligible, will provide sufficient grounds for rejecting the null hypothesis. It has been pointed out by numerous researchers that the actual difference between two experimental conditions is probably never exactly equal to zero. Although admittedly a difference may be close to zero, if our measuring instrument is sufficiently sensitive and we carry our measurements out to many decimal places, we will probably never record a difference which is exactly equal to zero. And if the latter is true, it means that the null hypothesis will always be false. If, in fact, the null hypothesis is always false, it logically follows that it is not possible to commit a Type I error (which is rejecting a true null hypothesis).

The **minimum-effect hypothesis testing model** employs a null hypothesis which stipulates a value below which any effect present in the data would be viewed as trivial, and above which would be meaningful. As an example, if one were comparing the IQ scores of two groups, the null hypothesis might stipulate a difference between 0 and 5 points, while the alternative hypothesis would stipulate a difference greater than five points. In such a case, any difference of five points or less would result in retaining the null hypothesis, since a difference within that range would be considered trivial (i.e., of no practical or theoretical value). A difference of more than five points would lead to rejection of the null hypothesis, since a difference equal to or greater than five points would be considered meaningful. Note that the null hypothesis in the **minimum-effect hypothesis testing model** stipulates a range of values, whereas in the **classical hypothesis testing model** the null hypothesis stipulates a specific value.

A philosophy similar to that underlying the **minimum-effect hypothesis testing model** is also associated with the use of **tests of equivalence**. In contrast to the **classical hypothesis testing model** (where the alternative hypothesis states that there is a difference between treatments), in a **test of equivalence** the alternative hypothesis states that the treatments are, in fact, equivalent. Conversely, in a **test of equivalence** the null hypothesis states that a difference exists between the treatments. Since it is not mathematically feasible to establish an alternative hypothesis which states exact equality (i.e., a difference of zero) between experimental conditions, when one conducts a **test of equivalence**, prior to conducting a study a researcher stipulates a value which reflects a maximum difference that will be tolerated between two treatments in order that the researcher might conclude that the treatments are equivalent to one another. Any difference equal to or less than the stipulated value would be viewed so small as to be inconsequential. Thus, in a **test of equivalence** if a difference which is equal to or less than the value stipulated by the researcher is detected, the null hypothesis (stating that a difference does exist) can be rejected, and the alternative hypothesis (which stipulates equivalence) can be accepted. **Tests of equivalence**, which are illustrated in a number of chapters of the book, are discussed in detail in Section VII of the *t* **test for two independent samples**.

Another approach to hypothesis testing is the **Bayesian hypothesis testing model** (which Lehman (1993) notes was rejected by Fisher, Neyman, and Egon Pearson). The latter model derives from the work of the Reverend Thomas Bayes (1702-1761), an eighteenth century English clergyman who stated a general rule for computing **conditional probabilities** referred to as **Bayes' theorem**. A **conditional probability** is the probability that an event will occur, given the fact that it is already known that another event has occurred. (**Bayes' theorem** and the concept of **conditional probability** are discussed in greater detail later in the **Introduction**.) The probability value computed within the framework of the **classical hypothesis testing model** is the conditional probability of obtaining a specific outcome in an experiment, given that the null hypothesis is true. In the **Bayesian hypothesis testing model** the probability value computed is the conditional probability that the null hypothesis is true, given the specific outcome in an experiment (see Endnote 21 for further clarification). Although semantically the conditional probabilities computed for the two models sound similar, they are not the same. A detailed discussion of the **Bayesian hypothesis testing model** can be found in Section IX (the **Addendum**) of the **binomial sign test for a single sample**.

Estimation in Inferential Statistics

In addition to hypothesis testing, inferential statistics can also be employed for estimating the value of one or more population parameters. Within this framework there are two types of estimation. **Point estimation** (which is the less commonly employed of the two methods) involves estimating the value of a parameter from the computed value of a statistic. The more commonly employed method of estimation is **interval estimation** (commonly employed within the context of the **classical hypothesis testing model** as well as the **Bayesian hypothesis testing model**), which involves computing a range of values which a researcher can state with a high degree of confidence contains the true value of a population parameter. One such commonly computed range of values is referred to as a **confidence interval** (which was introduced by Jerzy Neyman (1941)). Oakes (1986) notes that whereas a **significance test** provides information with respect to what a population parameter **is not**, a **confidence interval** provides information with respect to what a population parameter **is**. The most commonly computed confidence intervals are the 95% and 99% intervals.

To illustrate the latter, a 95% confidence interval identifies a range of values a researcher can be 95% confident contains the true value of a population parameter (e.g., a population mean). Stated in probabilistic terms, the researcher can state there is a probability/likelihood of .95 that the confidence interval contains the true value of the population parameter. When the result of an inferential statistical test is statistically significant at the .05 level, the 95% confidence interval will not include the hypothesized value of the parameter stipulated in the null hypothesis. Another example of a confidence interval is a range of values a researcher can be confident to a specified degree contains the true difference between two population parameters. Thus, a 99% confidence interval for the difference between two means stipulates the range of values a researcher can be 99% confident contains the true difference between the means of the two populations. When the result of an inferential statistical test is statistically significant at the .01 level, the 99% confidence interval will not include the hypothesized value of the difference between the two population means stipulated in the null hypothesis

During the past 20 years critics of the **classical hypothesis testing model** (e.g., Altman *et al.* (2002), Harlow *et al.* (1997), Hunter and Schmidt (1990, 2004), Kline (2004), Meehl (1967), Rozeboom (1960), Smithson (2003), Thompson (1993, 1999, 2002)) and Wilkinson *et al.* (1999)) have argued that researchers should utilize confidence intervals in decision making, as well as the fact that all published research should include confidence intervals for any relevant statistics. In some instances, those who vigorously advocate the use of confidence intervals argue that the test of statistical significance should no longer be employed for hypothesis testing, and that instead decisions should be made on the basis of values computed for confidence intervals. The latter issue is examined more closely in the discussion of confidence intervals in Section VII of the **single-sample *t* test**.

Another measure which is often estimated within the framework of an experiment is **effect size** (also referred to as **magnitude of treatment effect**). A commonly employed measure of **effect size** is a value which represents the proportion or percentage of variability on a **dependent variable** that can be attributed to variation on the **independent variable** (the terms **dependent variable** and **independent variable** are defined in the next section). Throughout this book the concept of **effect size** is discussed, and numerous measures of **effect size** are presented. At the present time researchers are not in agreement with regard to the role that measures of **effect size** should be accorded in summarizing the results of research. As noted in the previous section, an

increasing number of individuals have become highly critical of the **classical hypothesis testing model** because of its dependence on the concept of statistical significance. These individuals have argued that a measure of **effect size** computed for a study is more meaningful than whether or not an inferential statistical test yields a statistically significant result. The controversy surrounding **effect size** versus statistical significance is discussed in detail under the discussion of **meta-analysis and related topics** in Section IX (the **Addendum**) of the **Pearson product-moment correlation coefficient**. In addition, it is also addressed within the framework of the discussion of measures of **effect size** throughout the book (e.g., in Section VI of both the *t* **test for two independent samples** and the **single-factor between-subjects analysis of variance (Test 21))**.

Relevant Concepts, Issues, and Terminology in Conducting Research

General overview of research methods The three most common strategies employed in conducting research in the natural and social sciences are: a) The **observational method**; b) The **experimental method**; c) The **correlational method**.

a) The **observational method** (also referred to as the **case study, clinical** and **anecdotal method**) involves accumulating information about the phenomenon under study (e.g., the behavior/activity of organisms, inanimate objects, events, etc.) by observing the phenomenon in the real world. Because of its emphasis on observation in the natural environment, this method is sometimes referred to as **naturalistic observation**. Since it is informal and subjective in nature, the observational method is often criticized as being unscientific. It is depicted as a methodology which sacrifices **precision** for **relevance**. Specifically, it opts for **relevance**, which means that it studies real world phenomena, as opposed to studying them under artificial conditions that are often created within the context of laboratory experimentation. The observational method lacks **precision**, which is comprised of the following two elements: **quantification** and **control**. Specifically, more often that not, the observational method does not translate the information it accumulates into quantitative data that can be subjected to statistical analysis. In addition, it does not introduce adequate control into the situations it observes, and because of the latter it cannot clearly identify cause and effect with respect to the phenomenon under study.

b) The **experimental method** (commonly referred to as the **scientific method**) accumulates information by conducting controlled experiments. It is also referred to as the **bivariate method**, since in its simplest form it involves two variables, the **independent variable** and the **dependent variable** (which will be clarified later in this discussion). The experimental method is formal and objective in nature, and depends on the statistical analysis of data to draw conclusions regarding the phenomenon under study. The experimental method is depicted as sacrificing **relevance** for **precision**. Specifically, it opts to sacrifice the reality of everyday life (**relevance**) and instead studies organisms/events in controlled settings (such as a laboratory) in order that it might determine whether or not there is a cause-effect relationship between two variables. The experimental method is often criticized by proponents of the observational method on the grounds that it studies behavior/events in an artificial setting, and information obtained in such an environment may not conform to what actually occurs in the real world. Because of the latter, critics of the experimental method claim that it has poor **external validity** (the **external validity** of an experiment refers to the degree to which its results can be generalized beyond the setting in which it is conducted).

c) The **correlational method** is employed to determine whether or not two or more variables are statistically related to one another. If, in fact, two variables are statistically related to one another, the correlational method allows a researcher to predict at above chance a score on one variable through use of the score on the second variable. For example, the correlational method could be employed to predict how a person will do on a job on the basis of a person's score on a personality test. Although the correlational method provides a blend of both relevance and precision, it lacks the degree of relevance associated with the observational method, as well as the fact that it lacks the degree of precision associated with the experimental method. With respect to the latter, although the correlational method is characterized by **quantification**, it lacks the **control** associated with the experimental method. Data that are evaluated through use of the correlational method can be obtained in either real life or controlled experimental environments. When the correlational method is employed with more than two variables, it is commonly discussed within the context of the **multivariate method** of research (which is discussed in the latter section of the book).

Hypothetical constructs and operational definitions   In simplest terms, a **hypothetical construct** (often just referred to as a **construct**) can be viewed as any theoretical concept. The term hypothetical construct is employed in various disciplines to represent something which is hypothesized to exist, although strictly speaking, whatever it is that one is referring to can never be directly observed. As a result of not being directly observable, a hypothetical construct must be measured indirectly.

To illustrate, intelligence is an example of a hypothetical construct, in that strictly speaking we never directly observe intelligence. We only observe behavior (such as how a person does on an IQ test) which suggests that a person is intelligent to some degree. Perhaps some day we will be able to represent intelligence as a physical reality — in other words, a specific structure or chemical in the brain, which by virtue of its size or concentration will actually represent a person's exact level of intelligence. Until such time, however, intelligence is just a convenient term psychologists employ to reflect individual differences with respect to certain behavior. In spite of the fact that a hypothetical construct such as intelligence can never be directly observed, it is something that scientists such as psychologists often want to measure. The latter can also be expressed by saying that scientists want to **empirically translate** or **operationalize** the hypothetical construct of intelligence. The term **operational definition** (which derives from the term **operationalize**) represents the specific operations/measures which are employed by a researcher to measure a hypothetical construct. Put simply, an **operational definition** is a way of measuring a **hypothetical construct**.

The most common way of measuring the construct of intelligence is the performance of a person on an IQ test. Thus, an IQ test score is an operational definition of the concept of intelligence. Yet there are many other ways one might measure intelligence — for example, how well one does in school or how well one does in solving certain types of problems. In the final analysis, since there is no perfect and direct way of measuring intelligence, any measure of it can be challenged. This will also be the case with respect to any other hypothetical construct which one elects to investigate. Other examples of hypothetical constructs which are frequently the focus of research are traits (such as extroversion, anxiety, conscientiousness, neuroticism, etc.), emotional states (such as anxiety, depression, etc.), a person's status with respect to such things as health or social class, the status of the economy, etc. In addition, conditions which a person (as well as other organisms and inanimate objects) can be exposed to in the environment (such as frustration, boredom, stress, etc.) can also represent hypothetical constructs. Like intelligence, each of the latter environmental conditions is never directly observable. We only observe circumstances and/or behavior which suggests the presence or absence of such conditions to varying degrees.

As noted above, one or more ways can be devised to measure a hypothetical construct. As another example, a trait such as neuroticism, which represents a hypothetical construct, can be measured with pencil and paper personality tests, projective tests, physiological measures, peer or expert ratings, or through direct observation of specific target behaviors which are considered critical indicators for the presence of that trait. In the same respect, one or more ways can be devised to create certain conditions in the environments such as frustration, boredom, stress, etc. To illustrate, the following are a variety of mechanisms a researcher might employ to represent the construct of environmental stress: a) Exposure to a noxious stimulus such as loud noise; b) Threat of physical assault; c) Presenting a person with a task which he or she does not have time to complete. As is the case with intelligence, any of the aforementioned measures or conditions (i.e., operational definitions) representing the constructs of neuroticism and stress may be challenged by someone else who argues that the selected measure or condition is not the optimal way of evaluating the construct in question. Nevertheless, in spite of the fact that one can challenge virtually any operational definition a researcher employs to represent a hypothetical construct, in order to conduct research it is necessary to operationally define hypothetical constructs. Thus, regardless of whether a researcher employs the observational, experimental, or correlational methods of research, one will ultimately have to operationalize one or more constructs that are being evaluated.

Relevant terminology for the experimental method Although the terminology presented in this section will be illustrated through the use of research involving human subjects, it should be emphasized that the basic definitions to be discussed apply to all varieties of experiments (i.e., experiments involving nonhuman subjects, inanimate objects, events, etc.). The typical experiment evaluates one or more hypotheses. Within the context of conducting an experiment the most elementary **hypothesis** evaluated is a formal prediction about the relationship between two variables — specifically, an **independent variable** and a **dependent variable**. As an example of such a hypothesis, consider the statement *frustration causes aggression*. The latter statement is a prediction regarding the relationship between the two variables *frustration* and *aggression*. In order to test the latter hypothesis, let us assume that an experimenter designs a study involving two groups of subjects who are told they will be given an intelligence test. One group (Group 1 — which represents the **experimental group**) is frustrated by being given a very difficult test on which it is impossible for a person to achieve a high score. The other group (Group 2 —which represents the **control group**) is not frustrated, since it is given a very easy test. Note that the group which receives the main treatment (which in this case is the group that is frustrated) is commonly referred to as the **experimental group**. The group which does not receive the main treatment (which in this case is the group that is not frustrated) is commonly referred to as the **control group** or **comparison group**. Upon completing the test, all subjects are asked to play a video game in which a subject has the option of killing figures depicted on a computer screen. The number of figures a subject kills is employed as the measure (i.e., operational definition) of aggression in the study. The experimenter predicts that subjects in the experimental group, by virtue of being frustrated, will record more kills than subjects in the control group. Note that the manner in which the researcher defines frustration can certainly be challenged — specifically, most people can probably think of what they consider a better way to frustrate subjects than by giving them a difficult test. In the same respect one can also challenge the criterion the experimenter employs to measure aggression. However, before being overly critical of the experiment, one must realize that ethical and pragmatic considerations often limit the options available to a researcher in designing an experiment (especially one involving variables such as frustration and aggression).

In the above described experiment there are two **variables**, an **independent variable** (referred to in some sources as an **exogenous variable**) and a **dependent variable** (referred to in some sources as an **endogenous variable**). The **independent variable** is the **experimental conditions** or **treatments** — in other words, whatever it is which distinguishes the groups from one another. In the experiment under discussion the independent variable is the frustration factor, since one group was frustrated and the other group was not. The number of groups represents the number of **levels** of the independent variable. Since in our experiment there are two groups, the independent variable is comprised of two levels. If a third group had been included (which could have been given a moderately difficult test, resulting in moderate frustration), the independent variable would have had three levels. The **dependent variable** in an experiment is the measure that is hypothesized to depend on the level of the independent variable to which a subject is exposed. Thus, in the experiment under discussion, the dependent variable is aggression. If the hypothesis is correct, the scores of subjects on the dependent variable will be a function of which level of the independent variable they served (i.e., of which group they were a member). Thus, it is predicted that subjects in the frustration group (Group 1) will record more kills than subjects in the no frustration group (Group 2).

The simplest way to determine the independent variable in an experiment is to ask oneself on what basis the groups are distinguished from one another. Whatever the distinguishing feature is between the groups represents the independent variable. As noted above, the distinguishing feature between the groups in the experiment under discussion is the degree of frustration to which subjects are exposed, and thus frustration represents the independent variable. In the anti-depressant study discussed earlier in this **Introduction**, the independent variable was whether or not a subject received an antidepressant drug or the placebo (since the latter was the distin-guishing feature between the two groups in the aforementioned study).

The simplest way to determine the dependent variable in an experiment is to ask oneself what the scores represent that the subjects produce at the conclusion of the experiment (i.e., the scores which are employed to compare the groups with one another). As noted above, the level of aggression in the two groups is contrasted with one another, and thus aggression represents the dependent variable. In the antidepressant study discussed earlier, the dependent variable was the depression scores of subjects at the conclusion of the study (since the latter represents the data that were employed to compare the two groups). Although it is possible to have more than one independent and/or dependent variable in an experiment, in this discussion we will only concern ourselves with experiments in which there is a single independent variable and one dependent variable.

Within the framework of the experimental method a distinction is commonly made between a **true experiment** and a **natural experiment** (which is also referred to as an **ex post facto study**). This distinction is predicated on the fact that in a **true experiment** the following applies: a) In a **true experiment** subjects are **randomly assigned** to a group, which means that each subject has an equal likelihood of being assigned to any of the groups employed in the experiment. Random assignment is assumed to result in groups which are likely to be equivalent to one another. It should be noted, however, that although random assignment does not guarantee equivalency of groups, it optimizes the likelihood of achieving that goal. For example, if an experiment employs 100 subjects, half of whom are men and half of whom are women, random assignment of subjects to groups will most likely result in approximately an equal number of men and women in both groups (as well as be likely to make the groups comparable with respect to other relevant demographic variables such as age, socioeconomic status, etc.); b) The independent variable in a **true experiment** is **manipulated** by the experimenter. The frustration-aggression study under discussion illustrates an example of an experiment which employs a **manipulated independent variable**, since in that study the experimenter manipulates each of the groups. In other words, the level of frustration created within each group is determined/manipulated by the experimenter.

In a **natural experiment** random assignment of subjects to groups is impossible, since the independent variable is not manipulated by the experimenter, but instead is some **preexisting**

subject characteristic (such as gender, race, etc.). A **nonmanipulated independent variable** in a natural experiment is often referred to as a **subject variable**, **attribute variable**, or **organismic variable** (since the differentiating feature between groups is some preexisting attribute of the subjects/organisms employed in the experiment). As an example, if we compare the overall health of two groups, smokers and nonsmokers, the independent variable in such a study is whether or not a person smokes, which is something that is determined by nature prior to the experiment. The dependent variable in such an experiment will be a measure of the subjects' health. (Readers should be aware of the fact that some sources limit the use of the term independent variable to a manipulated independent variable, and do not classify a nonmanipulated independent variable as an actual independent variable.)

The advantage of a **true experiment** over a **natural experiment** is that the **true experiment** allows a researcher to exercise much greater control over the experimental situation. Since in the **true experiment** the experimenter randomly assigns subjects to groups, it is assumed that the groups are equivalent to one another, and as a result of this any differences between the groups with respect to the dependent variable can be directly attributed to the manipulated independent variable. The end result of the latter is that the **true experiment** allows a researcher to draw conclusions with regard to cause and effect.

The **natural experiment**, on the other hand, does not allow one to draw conclusions with regard to cause and effect. Essentially the type of information which results from a **natural experiment** is **correlational** in nature. Such experiments can only tell a researcher that a statistical association exists between the independent and dependent variables. The reason why **natural experiments** do not allow a researcher to draw conclusions with regard to cause and effect is that such experiments do not control for the potential effects of **confounding variables** (also known as **extraneous variables**). (Sometimes the term **artifact** is employed for any aspect of an experimental design which may go unnoticed and inadvertently produce a confound.) A **confounding variable** is any variable which systematically varies with the different levels of the independent variable. To illustrate, assume that in a study comparing the overall health of smokers and nonsmokers, unbeknownst to the researcher all of the smokers in the study are people who have high stress jobs and all the nonsmokers are people with low stress jobs. If the outcome of such a study indicates that smokers are in poorer health than nonsmokers, the researcher will have no way of knowing whether the inferior health of smokers is due to smoking and/or job stress, or even to some other confounding variable of which he is unaware. In the case of the **true experiment**, on the other hand, confounding is much less likely as a result of randomly assigning subjects to the experimental conditions.

If an experiment is well controlled so as to rule out any confounding variables, the experiment is said to have **internal validity**. In other words, if a significant difference is found between the groups with respect to the dependent variable in a well designed **true experiment**, the experimenter can have a high degree of confidence that most likely the difference is attributable to the independent variable, and is not the result of some confounding variable.

Earlier in the **Introduction** it was noted that at the conclusion of an experiment the scores of subjects in the different groups are compared with one another through use of an inferential statistical test. In reference to the frustration-aggression experiment alluded to earlier, the purpose of such a test would be to determine whether or not there is a statistically significant difference between the aggression scores of the two groups. To put it another way, by employing a statistical test the experimenter will be able to say whether or not subjects exposed to different levels of the independent variable exhibited a difference with respect to their scores on the dependent variable. In the event a statistically significant difference is obtained between the two groups, the experimenter can conclude that it is highly unlikely that the difference is due to chance. In view of the latter, the most logical decision would be to conclude that the difference in aggression scores between the two groups was most likely due to whether or not subjects were frustrated (i.e., the independent variable).

Before closing this discussion of the experimental method, it should be noted that it is possible to have more than one independent variable in an experiment. Experimental designs which involve more than one independent variable are referred to as **factorial designs**. In such experiments, the number of independent variables will correspond to the number of factors in the experiment, and each independent variable/factor will be comprised of two or more levels. It is also possible to have more than two dependent variables in an experiment. Typically, experiments involving two or more dependent variables (and often simultaneously two or more independent variables) are evaluated with **multivariate statistical procedures**, some of which are discussed later in the book.

Correlational Research   The discussion of correlational research in this section will be limited to the simplest type of correlation — specifically, **bivariate correlation**, which is the use of correlation to measure the degree of association between two variables. It should be noted that correlational procedures can also be employed with more than two variables. The latter type of correlation is commonly discussed within the framework of **multivariate analysis** (e.g., **multiple regression (Test 33)**).

In the simplest type of correlational study, scores on two measures/variables are available for a group of subjects. The major goal of correlational research is to determine the degree of **association** between two variables, or to put it another way, the extent to which a subject's score on one variable can be predicted if one knows the subject's score on the second variable. Usually the variable that is employed to predict scores on a second variable is designated as the X variable, and is referred to as the **predictor variable**. The variable which is predicted from the X variable is usually designated as the Y variable, and is referred to as the **criterion variable**. (Although the terms **independent** and **dependent variable** are typically limited to the variables employed in studies based on the experimental method, some sources refer to the **predictor variable** in correlational research as the **independent variable** and the **criterion variable** as the **dependent variable**.)

The most commonly employed correlational measure is the **Pearson product–moment correlation coefficient (Test 28)**, which is represented by the notation r. The latter value is computed by employing the scores of subjects in an algebraic equation. The value computed for r (i.e., the correlation coefficient) can fall anywhere within the range -1 to $+1$. Thus, the value of r can never be less than -1 (i.e., r cannot equal -1.2, -50, etc.) or be greater than $+1$ (i.e., r cannot equal 1.2, 50, etc.). The **absolute value** of r (represented with the notation $|r|$) indicates the **strength** of the relationship between the two variables. Recollect that the **absolute value** of a number is the value of the number irrespective of the sign. Thus, in the case of the two correlation coefficients $r = +1$ and $r = -1$, the absolute value of both coefficients is 1. As the absolute value of r approaches 1, the degree of relationship (to be more precise, the degree of **linear relationship**) between the variables becomes stronger, achieving the maximum when $|r| = 1$ (i.e., when r equals either $+1$ or -1). The closer the absolute value of r is to 1, the more accurately a researcher will be able to predict a subject's score on one variable from the subject's score on the other variable. The closer the absolute value of r is to 0, the weaker the linear relationship between the two variables. As the absolute value of r approaches 0, the degree of accuracy with which a researcher can predict a subject's score on one variable from the other variable decreases, until finally, when $r = 0$ there is no predictive relationship between the two variables. To state it another way, when $r = 0$ the use of the correlation coefficient to predict a subject's Y score from the subject's X score (or vice versa) will not be any more accurate than a prediction that is based on some random process (i.e., a prediction that is based purely on chance).

The **sign** of *r* indicates the **direction** or **nature** of the relationship that exists between the two variables. A positive sign indicates a **direct** relationship, whereas a negative sign indicates an **indirect** (or **inverse**) relationship. When there is a **direct relationship**, subjects who have a high score on one variable will have a high score on the other variable, and subjects who have a low score on one variable will have a low score on the other variable. The closer a positive value of *r* is to +1, the stronger the direct relationship between the two variables, whereas the closer a positive value of *r* is to 0, the weaker the direct relationship between the variables. Some general guidelines which can be employed are that *r* values between +.70 to +1 represent examples of a **strong direct relationship**; *r* values between +.30 to +.69 represent examples of a **moderate direct relationship**; and *r* values between +.01 to +.29 represents examples of a **weak direct relationship**. When *r* is close to +1, most subjects who have a high score on one variable will have a comparably high score on the second variable, and most subjects who have a low score on one variable will have a comparably low score on the second variable. As the value of *r* approaches 0, the consistency of the general pattern described by a positive correlation deteriorates, until finally, when *r* = 0 there will be no consistent pattern that allows one to predict at above chance a subject's score on one variable if one knows the subject's score on the other variable.

When there is an **indirect/inverse** relationship, subjects who have a high score on one variable will have a low score on the other variable, and vice versa. The closer a negative value of *r* is to –1, the stronger the indirect relationship between the two variables, whereas the closer a negative value of *r* is to 0, the weaker the indirect relationship between the variables. Thus, using the guidelines noted above, *r* values between –.70 to –1 represent examples of a **strong indirect relationship**; *r* values between –.30 to –.69 represent examples of a **moderate indirect relationship**; and *r* values between –.01 to –.29 represents examples of a **weak indirect relationship**. When *r* is close to –1, most subjects who have a high score on one variable will have a comparably low score on the second variable (i.e., as extreme a score in the opposite direction), and most subjects who have a low score on one variable will have a comparably high score on the second variable. As the value of *r* approaches 0, the consistency of the general pattern described by a negative correlation deteriorates, until finally, when *r* = 0 there will be no consistent pattern that allows one to predict at above chance a subject's score on one variable if one knows the subject's score on the other variable.

A common error made in interpreting correlation is to assume that a positive correlation is more meaningful or stronger than a negative correlation. However, as noted earlier, the strength of a correlation is a function of its absolute value. To illustrate this point, let us assume that the correlation between the scores on a psychological test, which will be labeled *Test A*, and how effective a person will be at a specific job is *r* = +.56, and the correlation between another psychological test, *Test B*, and how effective the person will be on the job is *r* = –.78. Of the two tests, *Test B* will predict with a higher degree of accuracy whether or not a person will be effective on the job. This is the case since the absolute value of the correlation coefficient for *Test B* is .78, which is higher than the absolute value for *Test A*, which is .56. In the case of *Test B*, the strong negative correlation indicates that the lower a person's score on the test, the higher his or her job performance, and vice versa.

It should be emphasized that **correlational information does not allow a researcher to draw conclusions with regard to cause and effect**. Thus, although a substantial correlation may be observed between two variables *X* and *Y*, on the basis of the latter one cannot conclude that *X* causes *Y* or that *Y* causes *X*. Although it is possible that *X* causes *Y* or that *Y* causes *X*, one or more other variables which have not been taken into account may be involved in causation for either variable. As an example, if a strong positive correlation (e.g., +.80) is obtained between how much a person smokes (which will represent the *X* variable) and how many health problems a person has (which will represent the *Y* variable), one cannot conclude on the basis of the strong positive correlation that smoking causes health problems. Although smoking may cause health problems or vice versa (i.e., health problems may cause a person to smoke), the correlational method does not provide enough control to allow one to draw such conclusions. As noted in the discussion of the natural experiment, the latter is true since a correlational analysis does not allow a researcher to rule out the potential role of confounding variables. For example, as noted in the discussion of the natural experiment, let us assume that people who smoke have high stress jobs and that is why they smoke. Their health problems may be caused by the high stress they experience at work, and are not related to whether or not they smoke. The strong correlation between smoking and health problems masks the true cause of health problems, which is having a high stress job (which represents a confounding variable that is not taken into account in the correlational analysis). Consequently, a correlation coefficient only indicates the degree of statistical relationship between two variables, and does not allow a researcher to conclude that one variable is the cause of the other. It should be noted, however, that the cause-effect relationship between smoking and health problems has been well documented based on a large number of studies. Many of these studies employ sophisticated experimental designs which allow researchers to draw conclusions that go well beyond those which can be reached on the basis of the simple correlational type of study described in this section. More detailed discussion of the issue of correlation and causation can be found in the chapters on the **Pearson product-moment correlation coefficient** and **multiple regression**, as well as in many sources which address the general subject of experimental design (e.g., Shadish *et al.* (2002), Trochim (2005)).

In the discussion of the experimental method it was emphasized that **natural experiments** only provide correlational information. In other words, a significant result in a **natural experiment** only indicates that a significant statistical relationship/association is obtained between the independent variable and the dependent variable, and it does not indicate that scores of subjects on the dependent variable are caused by the independent variable. In point of fact, correlational measures can be employed to indicate the degree of association between an independent variable and a dependent variable in both a **natural experiment** and a **true experiment**. When correlational measures are employed within the latter context, they are commonly referred to as measures of **effect size**. Measures of **effect size** are described throughout the book within the context of the discussion of various inferential statistical tests.

It should be emphasized that measures of correlation in and of themselves are not inferential statistical tests, but are, instead, descriptive measures, which as noted above, only indicate the degree to which two or more variables are related to one another. In actuality, there are a large number of correlational measures (many of which are described in detail in this book) that have been developed which are appropriate for use with different kinds of data. Although (as is the case with the **Pearson product–moment correlation coefficient** (i.e., *r*)) the range of values for many correlational measures is between −1 and +1, the latter values do not always define the range of possible values for a correlation coefficient.

Experimental Design

Inferential statistical tests can be employed to evaluate data that are generated from a broad variety of experimental designs. This section will provide an overview of the general subject of experimental design, and introduce design related terminology which will be employed throughout the book. An **experimental design** is a specific plan which is employed to investigate a research problem. Winer (1971) draws an analogy between an experimental design and the design an architect creates for a building. In the latter situation the prospective owner of a building asks an architect to draw up a set of plans which will meet his requirements with respect to both cost and utility, as well as the requirements of the individuals who will be using the building. Two or more architects will often submit different plans for the same building, with each architect believing that his are best suited to accomplish the goals stipulated by the prospective owner. In the same respect, two or more experimental designs may be considered for investigating the same research problem. As is the case with the plans of different architects, two or more research designs will have unique assets and liabilities associated with them.

Among the criteria researchers employ in deciding among alternative experimental designs are the following: a) The relative cost of the designs — specifically, relative cost is assessed with respect to the number of subjects required in order to conduct a study, as well as the amount of time and money a study will entail; b) Whether or not a design will yield results which are reliable, and the extent to which the results have internal and external validity; c) The practicality of implementing a study employing a specific design. With regard to the latter, it is often the case that although one design may be best suited to provide the answer to a researcher's question, it is not possible to use that design because of practical and/or ethical limitations. For example, in order to demonstrate unequivocally that a specific germ is the cause of a serious and incurable disease, it would be necessary to conduct a study in which one group of human subjects is infected with the disease. Obviously, such a study would be not be sanctioned in a society that respected the rights and freedom of its members. Consequently, alternative research designs would have be considered for determining whether or not the germ is the cause of the disease.

This section will provide an overview of three general categories of experimental design which were described by Campbell and Stanley (1963), and have since been employed by many sources that discuss the general subject of experimental design. Campbell and Stanley (1963) identified the following three design categories: a) **Pre-experimental designs** (also referred to as **faulty experimental** designs and **nonexperimental designs**): b) **Quasi-experimental designs**; c) **True experimental designs**. The basic difference between **true experimental designs** versus **pre-** and **quasi-experimental designs** is that the internal validity of the latter two designs is compromised by the fact that subjects are not randomly assigned to experimental conditions and/or there is a lack of a control group(s).[23] Since the latter two factors do not compromise the internal validity of **true experimental designs**, such designs are able to isolate cause and effect with respect to the relationship between an independent and dependent variable. The three aforementioned design categories will now be described in greater detail. Readers who require a more detailed exposition of the subject matter on experimental design to be discussed in the section are referred to sources such as Cook and Campbell (1979) and Shadish *et al.* (2002).

Pre-experimental designs Designs which sources categorize as **pre-experimental designs** lack **internal validity**. In other words, such designs do not allow a researcher to conclude whether or not an independent variable/treatment influences scores on some dependent variable/ response measure. Although in most instances this is because a **pre-experimental design** lacks a control group, one **pre-experimental design** will be described which lacks internal validity because subjects are not randomly assigned to the experimental conditions. Campbell and Stanley (1963) and Cook and Campbell (1979) note that it is necessary to control for the potential effects of the following variables in order to insure that a research design has **internal validity**: a) **history**; b) **maturation**; c) **instrumentation**; d) **statistical regression**; e) **selection**; f) **mortality**.

History can be defined as events other than the independent variable which occur during the period of time that elapses between a pretest and a posttest on a dependent variable. To illustrate, assume that 20 clinically depressed subjects are given a pretest to measure their level of depression. Following the pretest the subjects are given antidepressant medication, which they take for six months. After the six months have elapsed a posttest is administered reevaluating the subjects' level of depression. Let us assume that a significant decrease in depression is observed between the pretest and posttest. A researcher might be tempted to conclude from such a study that the antidepressant was responsible for the observed decrease in depression. However, the decrease in depression could be due to some other variable which was simultaneously present during the period of time between the pretest and the posttest. For example, it is possible that all of the subjects were in psychotherapy during the six months they were taking medication. If the latter were true, the decrease in depression could have been due to the psychotherapy (which represents a confound in the form of a historical variable) and not the medication. As a general rule, the longer the period of time that elapses between a pretest and a posttest, the greater the likelihood that a historical variable may influence scores on the dependent variable.

Maturation refers to internal changes which occur within an organism (i.e., biological and psychological changes) with the passage of time. Examples of maturational variables are changes associated with the physical maturation of an organism, an organism developing hunger or thirst with the passage of time, an organism becoming fatigued or bored with the passage of time, and an organism becoming experienced at or acclimated to environmental conditions with the passage of time. To illustrate a maturational variable, assume that 50 one-year-old children who have not exhibited much if any inclination to walk are given a pretest to measure the latter ability. Following the pretest the children are put in a physical therapy program for six months. At the end of the six months a posttest is administered in which the children are reevaluated with respect to their walking ability. Let us assume that a significant improvement in walking is observed in the children when the posttest scores are compared with the pretest scores. A researcher might be tempted to conclude from such a study that the physical therapy was responsible for the improvement in walking. In point of fact, the improvement could have been due to physical maturation, and that with or without the physical therapy the children would have exhibited a significant increase in walking ability.

Instrumentation refers to changes that occur with respect to the measurement of the dependent/response variable over time. To illustrate, a study is conducted to assess the effect of exposure to loud music on blood pressure. A pretest measure of blood pressure is obtained for a group of subjects prior to them being exposed to one half hour of rock music, after which their blood pressure is reevaluated. However, unbeknownst to the researcher, the machine employed to measure blood pressure malfunctions during the posttest phase of the study, yielding spuriously high readings. The researcher erroneously concludes that the rock music was responsible for the increase in subjects' blood pressure, when in fact it was the result of instrumentation error. As another example of instrumentation error, assume a study is conducted in which judges are required to rate subjects with respect to the degree to which they exhibit competitive behavior before and after exposure to one hour of severe heat (which is hypothesized to affect competitiveness). Let us also assume that as the study progresses the judges became bored and/or fatigued, and as a result of the latter their ratings become increasingly inaccurate. Any differences obtained with respect to the judges' ratings on the pretest versus the posttest could be attributed to the unreliable ratings (which represent instrumentation error) and not the experimental treatment (i.e., the heat).

Two other factors related to instrumentation which can compromise the reliability of a study are **ceiling** and **floor effects**. Both of the latter can occur if the measure employed in evaluating the underlying construct represented by the dependent variable is not able to adequately cover the full range of values which a subject can obtain on that variable. To illustrate, a **ceiling effect** would be present if the highest possible score a subject can obtain on a measure of anxiety employed by a researcher, in reality, does not reflect the actual level of anxiety of that subject relative to other subjects who have been assigned the identical maximum score. On the other hand, a **floor effect** would be present if the lowest possible score a subject can obtain, in reality, does not reflect the actual level of anxiety of that subject relative to other subjects who have also been assigned the identical minimum score. Among other things, both ceiling and floor effects will result in underestimating the degree of variability on a dependent variable (since in the case of a ceiling effect there will be additional undetectable variability among those who have the highest possible score, and in the case of a floor effect there will be additional undetectable variability among those who have the lowest possible score).

Statistical regression refers to the fact that a subject yielding an extreme score (i.e., very high or very low) on a response measure will tend to yield a value that is closer to the mean if a second score is obtained on the response measure. As an example, a person may feel terrific on a given day and because of the latter perform extremely well on a specific task (or conversely, a person may feel lousy on a given day and thus perform poorly on a task). When retested, such an individual would be expected to yield a score which regresses toward the mean — in other words, the subject's retest score will be closer to his or her typical (i.e., average) score on the response measure. The regression phenomenon results from the fact that two or more scores for a subject will never be perfectly correlated with one another. The lack of a perfect correlation can be attributed to chance and other uncontrolled for variables which are inevitably associated with measurement. To examine how statistical regression can compromise the internal validity of an experiment, assume that a group of subjects is administered a pretest on a test of anxiety, after which they participate in an exercise program which is hypothesized to facilitate relaxation. Following the exercise program subjects are reevaluated for anxiety (i.e., a posttest is administered). If, in fact, a disproportionate number of subjects are uncharacteristically anxious on the day of the pretest and these subjects obtain substantially lower anxiety scores on the posttest, the latter could just as easily be due to statistical regression as opposed to the exercise program. In other words, it would be expected that if the exercise program had no effect on anxiety, because of statistical regression subjects who obtained uncharacteristically high scores on the pretest would be likely to yield scores that were more representative of their typical level of anxiety on the posttest (i.e., lower scores).

Selection (or **selection bias**) refers to the fact that the method a researcher employs in assigning subjects to groups can bias the outcome of a study. The most common situation resulting in selection bias is when subjects are not randomly assigned to groups. As an example, a study involving two groups is conducted to assess the efficacy of a new antidepressant drug. Subjects in the experimental group are given the antidepressant for six months, while subjects in the control group are given a placebo. In point of fact, subjects in the experimental group are comprised of patients who are being treated at a private clinic that does not generally accept for treatment severely depressed individuals. On the other hand, subjects in the control group are all derived from a clinic at a public hospital which treats many severely depressed patients. Unaware of the difference between the populations treated by the two facilities, the researcher conducting the study assumes that the two groups are comparable with respect to their level of depression, and consequently does not obtain pretest scores for depression. Following the administration of the experimental treatments the researcher finds that the mean depression score for the experimental group is significantly less than the mean score of the control group. The researcher might be tempted to conclude that the antidepressant was responsible for the obtained difference when, in fact, the difference could be due to selection bias. In other words, if the drug does not have an effect on depression, subjects who receive it will probably still yield lower depression scores than subjects who receive the placebo. The latter is the case, since there is an extremely high likelihood that at the beginning of the study the experimental group was less depressed than the placebo group.

Mortality (or **subject mortality**) refers to the differential loss of subjects in one or more groups that are participating in an experiment. During the course of an experiment one or more subjects may die, become ill, move to a different geographical locale, or just elect to no longer participate in the study. When a disparate pattern of subject mortality characterizes the attrition in two or more initially comparable groups, the internal validity of a study can be compromised. To illustrate, a study involving two groups (to which the subjects are randomly assigned) is conducted to assess the efficacy of a new antidepressant drug. It turns out, however, that one-quarter of the subjects in the experimental group (i.e., the group that receives the drug) drop out of the study prior to its completion (primarily due to the fact that they cannot tolerate the side effects of the medication), while only one percent of the control group (i.e., the placebo group) eliminate themselves from the study. Unbeknownst to the researcher, the dropouts in the experimental group happen to be the most depressed subjects in that group. As a result of the latter, at the conclusion of the study the average depression score obtained for the experimental group is significantly lower than the average obtained for the control group. Rather than indicating the efficacy of the antidepressant, the difference between the groups may be attributable to the fact that posttest average of the experimental group is based on the scores of only a portion of the original subjects who comprised that group — specifically, those members of the group who were characterized by the lowest levels of depression. When contrasted with the posttest mean of the control group (which as a result of taking the placebo would only be expected to decrease minimally from its pretest value), the experimental group should yield a lower average due to selective subject attrition.

Campbell and Stanley (1963) and Cook and Campbell (1979) describe the following three types of **pre-experimental designs**, all of which lack internal validity: a) **One-shot case study**; b) **One-group pretest-posttest design**; c) **Nonequivalent posttest-only design** (originally called **static-group comparison** by Campbell and Stanley (1963)).

The **one-shot case study** is also referred to as the **one-group after-only design** and the **single-group one-observation design**. In this design a treatment is administered to a group of subjects, and following the treatment the performance/response of the subjects on some dependent variable is measured. The internal validity of this design is severely compromised since: a) It lacks a control group; and b) It fails to formally obtain pretest scores for subjects which could, at least, be compared with their posttest scores. The **one-shot case study** is summarized in Figure I.16.

Time 1	**Time 2**
Treatment	Response measure

Figure I.16   One-Shot Case Study

As an example of a **one-shot case study**, let us assume that an ostensibly therapeutic surgical procedure is performed on a group of arthritis patients at Time 1 and a year later (Time 2) the researcher measures the severity of symptoms reported by the patients. If at Time 2 the researcher finds that most patients report only minimal symptoms, he might be tempted to conclude that the latter was the result of the surgical procedure. However, the absence of both a control group and a pretest make it impossible for the researcher to determine whether or not the minimal symptoms reported by subjects at Time 2 were due to the treatment or, instead, due to one or more of the extraneous variables noted earlier in this section. For example, it is possible that during the year which elapsed between Times 1 and 2 the patients shared some other common therapeutic experience of which the researcher was unaware — specifically, they might have all engaged in an exercise program which was designed to minimize symptoms.

Consequently, the latter experience, and not the surgery, could be the reason why at Time 2 patients only reported minimal discomfort. In this instance, history would represent an extraneous variable for which the researcher did not control.

The **one-group pretest-posttest design** (which is also referred to as the **one-group before-after design**) is an improvement over the **one-shot case study** in that it includes a pretest. Nevertheless, its internal validity is still compromised since it lacks a control group. The **one-group pretest-posttest design** is summarized in Figure I.17.

Time 1	Time 2	Time 3
Pretest on response measure	Treatment	Posttest on response measure

Figure I.17   One-Group Pretest-Posttest Design

In order to illustrate a **one-group pretest-posttest design**, a pretest will be added to the arthritis study described for a **one-shot case study**. Thus, during Time 1 a pretest to measure arthritis symptoms is administered, the surgical treatment is introduced at Time 2, and a posttest for symptoms is administered at Time 3. If such a design were employed, a researcher would not be justified in concluding that the surgical treatment was responsible for the difference if a decrease in symptoms was found between the pretest and the posttest. Because a control group was not included in the study, the researcher cannot rule out the possible impact of extraneous variables. The historical variable employed in the discussion of the **one-shot case study** (i.e., the exercise program) could still be responsible for a decrease in symptoms observed in Time 3 relative to Time 1. Note that in the discussion of the **one-shot case study** it is assumed that patients have symptoms at the beginning of the study, yet no formal attempt is made to measure them. On the other hand, in the **one-group pretest-posttest design** symptoms are formally quantified by virtue of administering a pretest. Consequently the numerical difference between the pretest and posttest scores obtained in the latter design might be viewed by some as providing at least some minimal evidence to suggest that the surgical treatment is effective. In point of fact, in spite of its limitations, the **one-group pretest-posttest design** is occasionally employed in research — most notably in situations where for practical and/or ethical reasons it is impossible to obtain a comparable control group. In such a case a researcher may feel that, in spite of its limitations, the **one-group pretest-posttest design** can shed some light on the hypothesis under study. In other words, if the alternative is to not conduct a study, the **one-group pretest-posttest design** may be viewed as representing the lesser of the two evils. Additional discussion of the **one-group pretest-posttest design** can be found in Section VII of the *t* **test for two dependent samples (Test 17)**, as well as in the chapters on other inferential statistical procedures which are employed for comparing two or more dependent samples.

	Time 1	Time 2
Experimental group	Treatment	Response measure
Control group	--------	Response measure

Figure I.18 Nonequivalent Posttest-Only Design (Nonrandom assignment of subjects to groups)

The **nonequivalent posttest-only design** attempts to rectify the shortcomings of the two previously described **pre-experimental designs** by including a control group. In the **nonequivalent posttest-only design** one group of subjects (the experimental group) is administered the experimental treatment at Time 1, whereas a control group is not administered the treatment at Time 1. At Time 2 both groups are evaluated with respect to the dependent variable. The internal validity of the design, however, is compromised by the fact that subjects are not randomly assigned to groups. The **nonequivalent posttest-only design** is summarized in Figure I.18.

A **nonequivalent posttest-only design** will be used to evaluate the efficacy of the surgical procedure for arthritis discussed above for the other **pre-experimental designs**. Two groups of subjects who have arthritis are employed in the **nonequivalent posttest-only design**. One group of subjects (the experimental group/surgery group) are patients of Dr. A, while the other groups of subjects (the control group/no surgery group) are patients of Dr. B. One year after surgery is performed on the patients in the experimental group, the symptoms of the two groups are compared with one another. Let us assume that at Time 2 Dr. A's patients exhibit significantly fewer symptoms of arthritis than Dr. B's patients. Since the subjects were not randomly assigned to the groups, it is possible that the surgery was not responsible for the obtained difference at Time 2. Instead, the difference could be due to some uncontrolled for extraneous variable. For example, it is possible that the patients Dr. A treats are younger than those treated by Dr. B, and by virtue of the latter Dr. A's patients have less severe forms of arthritis than Dr. B's patients. If the latter is true, it would not be unexpected that Dr. B's patients would have more symptoms at Time 2 than Dr. A's patients. The lower number of symptoms for Dr. A's patients may not reflect the success of the surgery but, instead, may be indicative of the fact that they had fewer symptoms to begin with. If instead of employing a **nonequivalent posttest-only design**, the study had been modified such that subjects were randomly assigned to one of the two groups, the efficacy of the surgery could have been assessed without compromising internal validity. As is the case with the **one-group pretest-posttest design**, the **nonequivalent posttest-only design** is occasionally employed in research — most notably in situations where for practical and/or ethical reasons it is impossible to randomly assign subjects to groups. Once again, in such a case a researcher may feel that, in spite of its limitations, the latter design can shed some light on the hypothesis under study, and that it represents the lesser of two evils when contrasted with the alternative of not conducting any study. If a researcher elects to employ an inferential statistical test to compare scores on the dependent variable of the experimental and control groups in a **nonequivalent posttest-only design**, the appropriate test to use will be one of the procedures described in the book for comparing two or more independent samples (e.g., the *t* **test for two independent samples**).

Quasi-experimental designs   Since they do not rule out the possible influence of all extraneous variables, **quasi-experimental designs** are also subject to confounding, and thus may lack internal validity. **Quasi-experimental designs**, however, rule out more extraneous variables than **pre-experimental designs**. Because of the latter, **quasi-experimental designs** are preferable to employ when practical and/or ethical considerations do not permit a researcher to evaluate a hypothesis through use of a **true experimental design**. In most instances, lack of random assignment of subjects is responsible for compromising the internal validity of a **quasi-experimental design**.

The following three **quasi-experimental designs** will be described: a) **Nonequivalent control group design**; b) **Separate-sample pretest-posttest design**; c) **Time-series designs**.(**interrupted time-series design** and **multiple time-series design**.

The **nonequivalent control group design** is a modification of the **nonequivalent posttest-only design** to include a pretest measure on the dependent variable. In the **nonequivalent control group design** one group of subjects (the experimental group) is given a pretest on the dependent variable at Time 1, administered the experimental treatment at Time 2, and is reevaluated with respect to the dependent variable at Time 3. As is the case with the experimental group, the control group is administered a pretest and posttest on the dependent variable at Times 1 and 3, but it is not administered any treatment at Time 2. The internal validity of the **nonequivalent**

control group design, however, is still compromised by the fact that subjects are not randomly assigned to groups. The **nonequivalent control group design** is summarized in Figure I.19.

	Time 1	**Time 2**	**Time 3**
Experimental group	Pretreatment response measure	Treatment	Posttreatment response measure
Control group	Pretreatment response measure	---------	Posttreatment response measure

Figure I.19 Nonequivalent Control Group Design (Nonrandom assignment of subjects to groups)

A **nonequivalent control group design** will be described within the context of evaluating the efficacy of the surgical procedure for arthritis discussed previously. The design will be identical to that described for the **nonequivalent posttest-only design**, except for the fact that at Time 1 a pretest measure of symptoms is obtained for both the experimental and control groups. At Time 2 the experimental group has the surgical treatment, while no treatment is administered to the control group. At Time 3 a posttest measure of symptoms is obtained from both groups. As is the case in the discussion of the **nonequivalent posttest-only design**, we will assume that Dr. A's patients comprise the experimental group, and Dr. B's patients comprise the control group. Since, however, once again subjects were not randomly assigned to the groups, if at Time 3 fewer symptoms of arthritis are observed in the experimental group, it could be due to an extraneous variable for which the experimenter did not control. However, if the extraneous variable is that Dr. A's patients are younger than Dr. B's patients, and by virtue of the latter they have fewer symptoms at the beginning of the study, this difference will be identified with the pretest administered during Time 1 of the **nonequivalent control-group design**. The latter illustrates that the **nonequivalent control group design** is more likely to identify potentially confounding variables than is the **nonequivalent posttest-only design**. Unfortunately, due to the lack of random assignment the **nonequivalent control-group design** cannot rule out the potential influence of all extraneous variables. Christensen (2000) and Cook and Campbell (1979) provide excellent discussions of the **nonequivalent control group design** in which they describes how the possible role of extraneous variables can be assessed, through an analysis of the configuration of the pretest and posttest scores of the two groups. The **nonequivalent control group design**, which is frequently employed in research, is more desirable to use than any of the previously discussed **pre-experimental designs** when practical and/or ethical issues make it impossible to employ a **true experimental design** (which always employs random assignment). If a researcher elects to employ an inferential statistical test to evaluate a **nonequivalent control group design**, the analytical procedures recommended for the analysis of the **pretest-posttest control group design** (discussed later in this section under **true experimental designs**) would be used. The latter procedures are discussed in Section VII of the *t* **test for two dependent samples** and Section IX (the **Addendum**) of the **between-subjects factorial analysis of variance (Test 27)**.

The **separate-sample pretest-posttest design** is employed on occasion when circumstances do not allow a researcher to have access to both an experimental and control group throughout the duration of a study. In this design, at Time 1 a pretest measure on a dependent variable is obtained on a random sample of subjects who are derived from the population of interest. The latter group of subjects will represent the control group. At Time 2 an experimental treatment is administered to the whole population. At Time 3 a posttest measure on the dependent variable is obtained from a new group of randomly selected subjects from the population. This latter group will represent the experimental group. The **separate-sample pretest-posttest design** is summarized in Figure I.20.

	Time 1	**Time 2**	**Time 3**
Control group	Pretreatment response measure	Treatment	
Experimental group		Treatment	Posttreatment response measure

Figure I.20   Separate-Sample Pretest-Posttest Design

The **separate-sample pretest-posttest design** is most commonly employed in survey research where it may not be possible to use the same sample before and after some treatment is administered to a population. To illustrate this design, assume that at Time 1 a market researcher solicits the attitude of a random sample of subjects (the control group) derived from a population of consumers about Product X. At Time 2 the experimental treatment (which is an advertising campaign that attempts to convince the population to buy Product X) is administered. At Time 3 the market researcher solicits the attitude toward Product X from a new random sample of subjects (the experimental group) derived from the same population. The difference in consumer attitude between the control group's pretest score and the experimental group's posttest score is employed to assess the effectiveness of the advertising campaign. Campbell and Stanley (1963) note that lack of control for extraneous historical variables is most likely to compromise the internal validity of the **separate-sample pretest-posttest design**. Thus in the example under discussion, negative publicity associated with a competing product (rather than the advertising campaign for Product X) could be the reason subjects express a more positive attitude about the Product X at Time 3 than at Time 1. History, however, is not the only factor which can compromise the internal validity of the **separate-sample pretest-posttest design**. If a researcher elects to employ an inferential statistical test to contrast the control group pretest score with the experimental group posttest score, the appropriate test to use would be one of the procedures described in the book for comparing two or more independent samples (e.g., the *t* **test for two independent samples**). (If matched subjects (discussed later in this section) are employed, an inferential procedure comparing two or more dependent samples would be used.)

In **time-series designs** multiple measurements are obtained for one or more groups on a dependent variable before and after an experimental treatment. An **interrupted time-series design** involves a single group of subjects, whereas in a **multiple time-series design** two or more groups are employed. As a general rule, if more than one group of subjects is employed in a **time-series design**, each of the groups represents a distinct population which can be distinguished from the populations that comprise the other groups on the basis of some defining characteristic (e.g., each group can be comprised of residents who live in different geographical locales during specific time periods). Since an **interrupted time-series design** only involves a single group, the absence of a control group makes it more difficult for a researcher to rule out the potential effects of one or more extraneous variables (e.g., history, maturation, etc.) on the dependent variable. The inclusion of one or more additional groups in a **multiple time-series design** puts the researcher in a better position to rule out the possible impact of extraneous variables. The **interrupted time-series design** and a **multiple time-series design** involving two groups are summarized in Figure I.21. In the latter figure, Time 1 represents the first of $(n-1)$ pretest measures on the dependent variable; Time n represents the time at which the treatment is administered, and Times $(n+1)$ through m represent $(m-n)$ posttest measures on the dependent variable, with Time m representing the final posttest measure. Note that when $n = 2$ and $m = 3$, the **interrupted time-series design** becomes the **one-group pretest-posttest design**. When there are two groups, and $n = 2$ and $m = 3$, the **multiple time-series design** becomes the **nonequivalent control-group design** (since, typically, **multiple time-series designs** do not employ random assignment).

In order to illustrate both the **interrupted** and **multiple time-series designs** let us assume that a researcher wants to determine whether or not a reduction of the speed limit from 65 mph to 55 mph on interstate highways reduces the number of fatal accidents. Let us also assume that the researcher discovers that such a change in the speed limit was implemented in the state of Connecticut effective January 1, 1998. Employing an **interrupted time-series design**, the researcher determines the number of fatal accidents on Connecticut interstate highways during the following time periods: a) Each of the five years which precede the new law going into effect (i.e., 1993, 1994, 1995, 1996, 1997); b) Each of the first five years the law is in effect (i.e., 1998, 1999, 2000, 2001, 2002). This design is summarized in Figure I.22. Note that the summary indicates there are five pretreatment measures of the dependent variable (i.e., the number of fatal accidents on Connecticut interstate highways in 1993, 1994, 1995, 1996, and 1997), the treatment (which is the speed limit reduction law becoming effective January 1, 1998), and the five posttreatment measures of the dependent variable (i.e., the number of fatal accidents on Connecticut interstate highways in 1998, 1999, 2000, 2001, and 2002) .

Interrupted Time-Series Design

Times 1 to $(n-1)$	Time n	Time $(n + 1)$ to m
Pretreatment response measures	Treatment	Posttreatment response measures

Multiple Time-Series Design

	Times 1 to $(n-1)$	Time n	Time $(n + 1)$ to m
Experimental group	Pretreatment response measures	Treatment	Posttreatment response measures
Control group	Pretreatment response measures	---------	Posttreatment response measures

Figure I.21   Time-Series Designs

Times 1-5	Time 6	Times 7-10
R-1993 R-1994 R-1995 R-1996 R-1997	Treatment	R-1998 R-1999 R-2000 R-2001 R-2002

Figure I.22   Interrupted Time-Series Design

If, in fact, the law is effective in reducing the number of fatal accidents, it would be expected that a graphical and numerical analysis of the data will indicate fewer accidents involving fatalities during the time period 1998-2002 versus the period 1993-1997. The main limitation of the above described **interrupted time-series design** is that is does not control for the possibility that some other extraneous variable (such as history) might be confounded with the time periods involved before and after the law going into effect. For example, if the weather was more inclement (e.g., rain, snow, fog, etc.) during the period 1993-1997 than the period 1998-2002, the latter might account for a greater number of accidents involving fatalities between 1993-1997. The researcher would also have to rule out other factors, such as that superior safety features in automobiles (e.g., superior body integrity, better passenger restraint systems, etc.) were integrated into automotive designs in 1998, and could thus account for the reduced accident rate involving fatalities from 1998 on.

If a **multiple time-series design** is employed to evaluate the same hypothesis, one or more control groups are included in the study. For example, let us assume the state of Rhode Island is employed as a control group for the following reasons: a) The Rhode Island interstate highway speed limit is 65 mph throughout the time period 1993-2002 — i.e., no law reducing the 65mph speed limit is passed in Rhode Island within the time period covered by the study; b) Rhode Island is viewed as similar to Connecticut due to its close geographical proximity and its demographic compatibility. The **multiple time-series design** utilizing the two states is summarized in Figure I.23.

Let us assume that employing the above design, a decrease in fatal accidents is observed in Connecticut during the period 1998-2002 relative to the period 1993-1997. In Rhode Island, however, the accident rate is the same during both time periods. Under such circumstances, a researcher would be able to argue more persuasively that the decline in fatal accidents in Connecticut was a direct result of the speed limit reduction law going into effect January 1, 1998. Although still imperfect (since subjects are not randomly assigned to the two groups), the **multiple time-series design** allows the researcher to rule out some of the alternative extraneous variables noted earlier in reference to the **interrupted time-series design**. Specifically, since it is reasonable to presume that the two adjoining states experienced similar weather conditions during the relevant time periods, the weather factor is unlikely to account for the differential fatal accident rates. The automobile safety issue can also be ruled out, since the same types of automobiles would be assumed to be operative in both states. The researcher, however, is still not able to entirely rule out the possible influence of extraneous variables. It should be noted that the external validity of the above described study can be challenged, insofar as one can question whether or not its results can be generalized to other states which are demographically and/or geographically distinct from Connecticut and Rhode Island. (The term **external validity** is sometimes employed to refer to the extent to which a study can be generalized beyond the sample(s) (or in this case, population(s)) employed in a study.) One way to address the latter criticism is to include in the study one or more additional groups representing other states. Such groups can represent states in which the speed limit is reduced from 65 mph to 55 mph (i.e., additional experimental groups representing geographically and/or demographically distinct states) and/or states in which the 65 mph speed limit is not modified (which can represent additional control groups). In spite of their limitations, **time-series designs** are frequently employed in the social sciences (e.g., economics, political science) as well as in business to evaluate "real world" problems that are not amenable to being evaluated through use of the experimental method (i.e., which cannot be evaluated with **true experimental designs**). Further discussion of **time-series** designs can be found in the description of the inferential statistical tests for evaluating two or more dependent samples (e.g., **Cochran Q-test (Test 26)**). An excellent discussion of **time-series designs** can be found in Shadish *et al.* (2002).

	Times 1-5				**Time 6**	**Times 7-10**				
Connecticut	R-1993	R-1994	R-1995	R-1996	R-1997 Treatment	R-1998	R-1999	R-2000	R-2001	R-2002
Rhode Island	R-1993	R-1994	R-1995	R-1996	R-1997 --------	R-1998	R-1999	R-2000	R-2001	R-2002

Figure I.23 Multiple Time-Series Design

Although not categorized under the rubric of **time-series designs**, two other experimental designs which can be employed to evaluate a phenomenon over time are **longitudinal** and **cross-sectional designs** — both of which are most commonly employed to evaluate whether or not changes occur with respect to some developmental process over time. In a **longitudinal design** (also referred to as a **panel design**) a single group of subjects is evaluated over repeated time intervals in order to determine whether or not changes occur with respect to some characteristic of the subjects. A **cross-sectional design** evaluates representative groups of subjects — with each of the groups (which are often referred to as **cohorts**) representing individuals at different age levels — with respect to some characteristic.

The two above noted designs will be illustrated with respect to evaluating the evolution of intelligence over a person's lifetime. If a researcher wanted to employ a **longitudinal design** to evaluate whether or not intelligence improves and/or deteriorates as one ages, she could employ a sample of subjects and evaluate them with respect to intelligence over a prolonged period of time. Specifically, each of the subjects could be evaluated with a standardized test of intelligence at specific time periods (e.g., at the ages of 5, 10, 15, 20, 40, 60, and 80). Two factors which might deter one from conducting the latter type of study are: a) The substantial cost involved in conducting a study over a 75 year period; and b) A **longitudinal study** can easily extend beyond a researcher's professional career and/or life span, and consequently the latter individual might never get to publish (or, for that matter, get credit for) the study during one's lifetime. The variable most often cited as potentially compromising the internal validity of a **longitudinal study** is **subject mortality**. Specifically, the number of subjects lost will increase with the passage of time, and a nonrandom pattern of subject mortality can compromise the internal validity of a study.

If one elected to employ a **cross-sectional design** to assess developmental changes in intelligence across one's life span, multiple samples of subjects representing different ages could simultaneously be evaluated with respect to intelligence. Thus, a researcher could contrast the intelligence of seven groups of subjects, with each group/cohort being comprised of individual's who are the following ages: 5, 10, 15, 20, 40, 60, and 80. The greatest limitation of a **cross-sectional design** is that the equivalence of the groups cannot be insured. To be more specific, the variable most often cited as potentially compromising the internal validity of a **cross-sectional study** is **history**. More specifically, the greater the age discrepancy between any two groups/age cohorts, the greater will be the difference with respect to the social and physical environmental experiences to which they have been exposed during their lifetime. Consequently, any differences in intelligence detected between the different age groups could be confounded by historical factors.

It is interesting to note that the results of experiments employing **longitudinal** versus **cross-sectional designs** investigating the same phenomenon may be inconsistent with one another. To illustrate, Christensen (2000, pp. 58–59) cites Baltes *et al.* (1977) discussion of studies regarding the evolution of intelligence over a person's lifetime. The latter authors note that **cross-sectional** research suggests that intelligence increases up until the age of 30 and then declines as one gets older. **Longitudinal studies**, on the other hand, suggest an increase in intelligence until about the age of thirty, after which there is either no change or an additional slight increase. The inconsistency between the two types of studies is explained on the basis of what is referred to as the **age-cohort effect**. The latter reflects the fact that by virtue of being the same age as all of the other subjects involved in the study, those who participate in a **longitudinal study** are more likely to share with their fellow subjects similar environmental experiences over the duration of their lifetime than subjects who participate in a **cross-sectional study**. Subjects who are involved in a **cross-sectional study** (who may represent different generations) are much less likely share common environmental experiences with subjects in other age groups because of different environmental events which are associated with specific time periods.

Within the framework of **true experimental designs** (which are described in the next section), a **longitudinal design** can be viewed as a **dependent samples design**. In the final analysis, however, due to the likelihood of nonrandom subject mortality, plus the fact that the independent variable of age is not directly manipulated by the experimenter, a **longitudinal design** does not conform to the requirements of a **true experimental design**. Within the context of **true experimental designs**, a **cross-sectional design** can be viewed as an **independent samples design**. Yet, in reality, a **cross-sectional design** does not conform to the requirements of a **true experimental design**, since it does not insure equivalency of the groups.

A compromise between a **longitudinal** and **cross-sectional design** is a **cohort-sequential design**. In the latter design two or more groups/cohorts of subjects whose ages overlap with one another are longitudinally evaluated with respect to a characteristic of interest over a specific period of time. As an example, assume that a researcher wishes to evaluate whether or not changes in intelligence occur between the ages of 5 and 20. A study is initiated during a given year (e.g., 2005) with a sample of five year old subjects, and each of the subjects is administered a standardized intelligence test at the ages of 5 (the year 2005 — the year in which the study commences), 10 (the year 2010), 15 (the year 2015), and 20 (the year 2020). A second sample of five year old subjects is introduced into the study ten years later (e.g., 2015), and each of these subjects is administered the standardized intelligence test at the ages of 5 (the year 2015 — the first year in the study for these subjects), 10 (the year 2020), 15 (the year 2025), and 20 (the year 2030). Note that the study allows the researcher to evaluate two cohorts of subjects longitudinally, yet each of the cohorts can be viewed as representing a different generation since they are born at different points in time. If similar patterns of intellectual change are detected in both samples, it would suggest that changes in intellect are more likely to be related to one's age than to one's generation. On the other hand, if there is a disparity between the patterns of intellectual change in the two samples, it would suggest that generational/historical factors rather than age may be more likely to impact intelligence. The **cohort-sequential design** can be evaluated within the context of a **factorial design**, which is discussed in the chapter on the **between-subjects factorial analysis of variance (Test 27)**. More specifically, the **cohort-sequential design** can be conceptualized as a **mixed factorial design** involving two independent variables — with the four age levels representing one independent variable (involving repeated measures on each subject) and the two cohorts representing a second independent variable (involving two independent groups). Analysis of a **mixed factorial design** is described in Section IX (the Addendum) of the **between-subjects factorial analysis of variance**.

True experimental designs Designs which are categorized as **true experimental designs** are characterized by random assignment of subjects to groups and the inclusion of one or more adequate control groups. The latter conditions optimize control of extraneous variables, and thereby maximize the likelihood of a study having internal validity. The following four **true experimental designs** will be described in this section: a) **Independent samples design**; b) **Dependent samples design**; c) **Pretest-posttest control group design**; d) **Factorial designs**.

An **independent samples design** is also known as an **independent-groups design**, **between-subjects design**, **between-groups design**, **between-subjects after-only research design**, and **randomized-groups design**. In an **independent samples design** each of n different subjects is randomly assigned to one of k experimental groups. In the most elementary **independent samples design** there are $k = 2$ groups, with one group representing the experimental group and the other the control group. Inspection of the latter design, which is summarized in Figure I.24, reveals that it is identical to the **nonequivalent posttest-only design** (described earlier in this section under **pre-experimental designs**), except for the fact that in the

independent samples design subjects are randomly assigned to the two groups. The latter modification of the **nonequivalent posttest-only design** allows a researcher to control for the potential effects of extraneous variables.

	Time 1	**Time 2**
Experimental group	Treatment	Response measure
Control group	---------	Response measure

Figure I.24   Independent Samples Design (Random assignment of subjects to different groups)

The **independent samples design** can be illustrated through use of the arthritis study employed to illustrate the **nonequivalent posttest-only design**, with the modification that subjects are randomly assigned to the two groups. Extensive discussion (including examples and analytical procedures) of the **independent samples design** can be found in the chapters on the *t* **test for two independent samples**, the **single-factor between-subjects analysis of variance**, and other tests involving two or more independent samples.

A **dependent samples design** is also known as a **within-subjects design, within-subjects after-only research design, repeated measures design, treatment-by-subjects design, correlated samples design, matched-subjects design, paired-sample design**, and **randomized-blocks design**. In a **dependent samples design** each of *n* subjects serves in each of *k* experimental conditions. A **dependent samples design** can also involve the use of **matched subjects**. Within the latter context it is commonly referred to as a **matched-subjects design**. In such a design each subject is paired with one or more other subjects who are similar with respect to one or more characteristics that are highly correlated with the dependent variable. The general subject of matching is discussed in detail in Section VII of the *t* **test for two dependent samples**. A **dependent samples designs** (as well as a **matched-subjects design**) is sometimes categorized as a **randomized-blocks design**, since the latter term refers to a design which employs homogeneous blocks of subjects (which matched subjects represent). When a **dependent samples design** is conceptualized as a **randomized-blocks design**, it is because within each block the same subject is matched with himself by virtue of serving under all of the experimental conditions. A **dependent samples design** involving two experimental conditions is summarized in Figure I.25.

	Time 1	**Time 2**
Experimental condition	Treatment	Response measure
Control condition	---------	Response measure

Figure I.25   Dependent Samples Design (All subjects serve in both conditions)

As an example illustrating the **dependent samples design**, assume a researcher wants to evaluate the efficacy of a drug on the symptoms of arthritis. Fifty subjects are evaluated for a six-month period while taking the drug (the experimental condition), and for a six-month period while not taking the drug (the control condition, during which time subjects are administered a placebo). Half of the subjects are initially evaluated in the experimental condition after which they are evaluated in the control condition, while the other half of the subjects are initially evaluated in the control condition after which they are evaluated in the experimental condition. This latter procedure, which is known as **counterbalancing**, is discussed in Section VII of the *t* **test for two dependent samples**. The data are evaluated by comparing the mean scores of subjects with respect to symptoms in the drug/experimental versus placebo/control conditions.

Extensive discussion (including a description of analytical procedures and additional examples) of the **dependent samples design** can be found in the chapters on the *t* **test for two dependent samples**, the **single-factor within-subjects analysis of variance (Test 24)**, and other tests involving two or more dependent samples.

The **pretest-posttest control group design** (also referred to in some sources as the **before-after design**) is identical to the **nonequivalent control group design** (described earlier in this section under **quasi-experimental designs**), except for the fact that subjects are randomly assigned to the two groups. The latter modification of the **nonequivalent control group design** allows a researcher to control for the potential effects of extraneous variables. Note that the **pretest-posttest control group design** is also an improvement over the **one-group pretest-posttest design** (discussed earlier in this section under **pre-experimental designs**), insofar as the latter design lacks a control group. The **pretest-posttest control group design** is summarized in Figure I.26.

The **pretest-posttest control group design** can be illustrated through use of the arthritis study employed to illustrate the **nonequivalent control group design**, with the modification that subjects are randomly assigned to the two groups. The analysis of the **pretest-posttest control group design** is discussed in Section VII of the *t* **test for two dependent samples**.

	Time 1	Time 2	Time 3
Experimental group	Pretreatment response measure	Treatment	Posttreatment response measure
Control group	Pretreatment response measure	---------	Posttreatment response measure

Figure I.26　Pretest-Posttest Control Group Design
(Random assignment of subjects to different groups)

Factorial designs All of the designs described up to this point in this section attempt to assess the effect of a single independent variable on a dependent variable. A **factorial design** is employed to simultaneously evaluate the effect of two or more independent variables on a dependent variable. Each of the independent variables is referred to as a **factor**. Each of the factors has two or more levels, which refer to the number of groups/experimental conditions which comprise that independent variable. If a **factorial design** is not employed to assess the effect of multiple independent variables on a dependent variable, separate experiments must be conducted to evaluate the effect of each of the independent variables. One major advantage of a **factorial design** is that it allows the same set of hypotheses to be evaluated at a comparable level of power by using only a fraction of the subjects that would be required if separate experiments were conducted to evaluate the relevant hypotheses for each of the independent variables. Another advantage of a **factorial design** is that it permits a researcher to evaluate whether or not there is an **interaction** between two or more independent variables — the latter being something which cannot be determined if only one independent variable is employed in a study. An **interaction** is present in a set of data when the performance of subjects on one independent variable is not consistent across all the levels of another independent variable. Extensive discussion (including a description of analytical procedures and examples) of **factorial designs** can be found in the chapter on the **between-subjects factorial analysis of variance**. Among the **factorial designs** discussed in the latter chapter are the **between-subjects factorial design**, the **mixed factorial design**, the **within-subjects factorial design**, and the **Latin square design**.

Single-subject designs An experimental design which attempts to determine the effect of an independent variable/treatment on a single organism is referred to as a **single-subjects design**. Since **single-subject designs** evaluate performance over two or more distinct time periods, such designs can be conceptualized within the framework of a **time-series design**. There is, however, a lack of agreement among researchers with respect to whether **single-subject designs** are best classified as **quasi** versus **true experimental designs**. In the final analysis, the appropriate classification for such a design will be predicated on its ability to rule out the potential effects of extraneous variables on the dependent variable. The ability of a **single-subject design** to achieve the latter may vary considerably depending upon the composition of a specific design.

Single-subject designs are most frequently employed in clinical settings in order to assess the efficacy of a specific treatment on a subject. These designs are often employed in the field of clinical psychology within the framework of behavior modification research. The latter type of research (which derives from the work of the American behavioral psychologist B. F. Skinner) assesses the effect of one or more environmental manipulations (such as reward and punishment) on inappropriate behavior.

Hersen and Barlow (1976) note that researchers are often reluctant to employ **single-subject designs** because of the following limitations associated with them: a) A researcher will be limited by the fact that the results of a study based on a **single-subject design** may not be able to be generalized to other subjects; b) **Single-subject designs** may be susceptible to confounding and/or **order effects**. An **order effect** is where an observed change on a dependent variable is a direct result of the order of presentation of the experimental treatments, rather than being due to the independent variable manipulated by the experimenter. **Order effects** are discussed in greater detail in the chapters on inferential statistical procedure for two or more dependent samples; c) **Single-subject designs** are problematic to interpret when a researcher wishes to simultaneously assess the effect of two or more independent variables on a dependent variable.

In spite of their limitations, Dukes (1965) notes that it may be prudent to employ a **single-subject design** under the following circumstances: a) When the issue of generalizing beyond the subject employed in a study is not of major concern; b) When the dependent variable being evaluated has a low frequency of occurrence in the underlying population (and thus it is impractical or impossible to evaluate the treatment with a large group of subjects); c) When the dependent variable being evaluated is characterized by low intersubject variability. Hersen and Barlow (1976) also note that research in clinical settings is often characterized by the fact that numerous extraneous variables are present which can interact with the treatment variable. Such interactions may be responsible for results which suggest that a treatment is ineffective with a group of subjects. Yet within the group, individual subjects may respond to the treatment, and consequently a **single-subject design** employed with individual subjects may yield positive results.

In order to illustrate a **single-subject design**, consider the following example. A six year old child has temper tantrums which disrupt his first-grade class. Every time the child has a temper tantrum the teacher removes him from the classroom. The school psychologist hypothesizes that the child has temper tantrums in order to get the reward of attention, which removal from the classroom represents. In order to evaluate the hypothesis, the psychologist conducts the following study employing an **ABAB single-subject design**. The letters **A** and **B** represent four distinct time periods which comprise the study, with the letter **A** indicating that no treatment is in effect and the letter **B** indicating the treatment is in effect. Since in an **ABAB design** Time1 is designated **A**, no treatment is in effect. Specifically, every time the child has a temper tantrum he is removed from the classroom. The measure of the subject's behavior during Time 1 is intended to provide the researcher with an initial indicator (referred to as a **baseline measure**) of how often the child exhibits the behavior of interest (commonly referred to as the **target behavior**). This baseline value for temper tantrums will be employed later when it will be compared to the number of temper tantrums emitted by the subject during the times when the treatment is in effect. Since Time 2 is designated **B**, the treatment is administered. Specifically, during Time 2 every time the child has a temper tantrum he is ignored by the teacher, and thus remains in the classroom. If the treatment is effective, a decrease in temper tantrums should occur during Time 2. After the period of time allotted for Time 2 elapses, Time 3 is initiated. During this time the **A** condition is reintroduced. Specifically, once again the child is removed from the classroom any time he has a temper tantrum. If, in fact, the treatment is effective, it is expected that the frequency of temper tantrums will increase during Time 3 and most likely return to the baseline level recorded during Time 1. It should be noted, however, that one problem associated with the **ABAB design** is that if during Time 2 the treatment is "too" effective, during Time 3, the subject may not regress back to the baseline level of behavior. Assuming the latter does not occur, once the time allotted for Time 3 has elapsed, Time 4 is initiated. During this final time period the treatment (i.e., the **B** condition) is reintroduced. If the results of the study indicate a high rate of temper tantrums during Times 1 and 3 (i.e., when no treatment was in effect) and a low rate during Times 2 and 4 (i.e., when the treatment was in effect), the researcher will have a strong case for arguing that the treatment was responsible for reducing the number of temper tantrums. The use of the fourth time period further reduces the likelihood that instead of the treatment, some uncontrolled for extraneous variable(s) (e.g., maturation or history) might have been responsible for the decline in temper tantrums. It should be noted that depending upon the circumstances surrounding a study (i.e., the practical and ethical considerations associated with applying a treatment to a single subject), an **ABAB** design may be truncated to consist of only two or three time periods (in which case it respectively becomes an **AB** or **ABA** design). The reader should keep in mind that although the **ABAB** design has been employed to illustrate a **single-subject design**, it represents only one of a number of such designs that can be employed in research. Since a comprehensive discussion of **single-subject designs** is beyond the scope of this book, the interested reader can find more detailed discussions of the topic in sources such as Christensen (2000), Hersen and Barlow (1976), and Sheskin (1984).

Sampling Methodologies

A great deal of research is based on the use of nonrandom samples (also referred to as **nonprobability samples**). In nonrandom or nonprobability sampling the probability of an object/subject being selected cannot be computed. For example, a sample of subjects who are volunteers or subjects who have been selected purely on the basis of their availability do not constitute a random sample Such samples are commonly referred to as **convenience samples**. Another example of a convenience sample would be those members of a population who return a questionnaire a researcher has mailed to them.

Survey research commonly employs procedures other than simple random sampling, which is the purest form of what is often referred to as the **probability method** of sampling. Some other sampling procedures which are categorized under the probability method of sampling will now be described. Some sources (e.g., Bechtold and Johnson (1989)) refer to the sampling procedures to be described as examples of **restricted random sampling**.

One alternative to simple random sampling is **systematic sampling**. In the latter type of sampling a list of the members who comprise a population is available, and every n^{th} person is selected to participate in a survey. In systematic sampling it is critical that there is no preexisting pattern built into the list, which might bias the composition of the sample derived through this methodology (e.g., an alphabetized list may increase or decrease the likelihood of specific individuals being selected). A systematic sample is not a truly random sample by virtue of the fact that by employing the rule that every n^{th} person be selected a limitation is imposed on the number of different samples that can be selected. Randomizing the names in an alphabetized list prior to the selection process, and then selecting every n^{th} person optimizes the likelihood that the final sample is unbiased, and for all practical purposes is commensurate with simple random sampling.

Another alternative to simple random sampling is **cluster sampling**. It is often the case that the members of a population can be broken down into preexisting groups or **clusters** (such as towns, blocks, classrooms of children, etc.). A researcher selects a random sample of the clusters, and data are then obtained from all the members of the selected clusters. The reliability of cluster sampling is predicated on whether or not the people who constitute the selected clusters are representative of the overall population (and, of course, a function of the response rate within each cluster). An example of cluster sampling would be a market researcher dividing a city into blocks (all of which are assumed to be comparable), randomly selecting a limited number of the blocks, and obtaining opinions from all the people who live in the selected blocks.

Stratified sampling is a methodology employed in survey research which allows a researcher to focus on specific subpopulations embedded within the larger population. In stratified random sampling a population is divided into homogeneous subgroups referred to as **strata**. Random subgroups are then selected from each of the strata. With respect to the latter, one of the following two procedures is employed: a) The number of subjects selected from each stratum corresponds to the proportion of that stratum in the overall population; b) An equal number of subjects are selected from each stratum, and their responses are weighted with respect to the proportion of that stratum in the overall population.

When properly implemented, stratified sampling can provide a researcher with accurate information on those members of the population who comprise the various strata (although, in theory, simple random sampling should accomplish the same goal). As an example of stratified sampling, assume a population is comprised of the following four distinct ethnic groups: Caucasian, Black, Asian, Hispanic. Fifty percent of the population is Caucasian, 20% Black, 20% Asian, and 10% Hispanic. For a sample comprised of n subjects, 50% of the sample is selected from the Caucasian subpopulation, 20% from the Black subpopulation, 20% from the Asian subpopulation, and 10% from the Hispanic subpopulation.

Note that the goal of **stratified sampling** is to identify strata that are similar/homogenous within themselves. Thus, there is a minimum of within-strata variability, but there is a high degree of between-strata variability (i.e., the different strata are dissimilar from one another). On the other hand, in **cluster sampling** the goal is to produce similar clusters, with each cluster containing the full spectrum of the population. Thus, in cluster sampling there is high within-clusters variability but minimal between-clusters variability.

It should be noted that the actual methodology employed in survey sampling may involve additional procedures that go beyond that which has been described above. Nevertheless, regardless of what sampling procedure is employed, the goal of survey research is to minimize **sampling error** (i.e. minimize the difference between a sample statistic and the population parameter it estimates). When conducting surveys which involve a large population the discrepancy between an estimated population proportion obtained from a sample of size n and the actual proportion in the underlying population rarely exceeds $1/\sqrt{n}$. To illustrate, if a sample of $n = 1000$ subjects is obtained from a large population (e.g., ten million people) and it is determined that 55% of the sample endorses a particular candidate, we can be almost certain (approximately, 95% confident) that the range of values 55% ± 3.16% (since $1/\sqrt{1000} = .0316$, which expressed as a percent is 3.16%) contains the true proportion of the population who support the candidate — the latter result is commonly expressed as 55% with a margin of error of ± 3.16% or ± 3.16 percentage points. (Equation 8.6, described in Section VI of the **chi-square goodness-of-fit test**, is employed to compute the exact value of the above noted range of values, which is referred to as the **confidence interval** for a population proportion.) In point of fact, most reputable surveys employ sample sizes of $n \geq 1000$, and in actuality sampling error only decreases minimally as sample size is increased beyond 1000. Further discussion of analysis of survey data can be found in Section VI of the **chi-square goodness-of-fit test** under the section on confidence intervals and in Endnote 10 of the latter test. Among others, Folz (1996), O'Sullivan *et al.* (2002), and Scheaffer *et al.* (1996) provide more detailed discussions of survey sampling.

Basic Principles of Probability

This section will summarize and illustrate a number of basic principles which are commonly employed in computing probabilities.[24] The logic underlying some of the statistical procedures described in the book is based on one or more of the principles described in this section.

Elementary rules for computing probabilities The first rule to be presented for computing probabilities is commonly referred to as the **addition rule**. The latter rule states that if two events A and B are **mutually exclusive** (i.e., if the occurrence of one event precludes the occurrence of the other event), the probability that either event A or event B will occur is the sum of the probabilities for each event. The **addition rule** is summarized with Equation I.40.

$$P(A \text{ or } B) = P(A) + P(B) \qquad \textbf{(Equation I.40)}$$

To illustrate the application of the **addition rule** with two mutually exclusive events, consider the following example. If a card is randomly selected from a deck of playing cards, what is the probability of obtaining either a **Red** card (for which $p = \frac{1}{2}$) or a **Black** card (for which $p = \frac{1}{2}$)? Employing Equation I.40, the probability value $p = 1$ is computed below.

$$P(\text{Red or Black}) = P(\text{Red}) + P(\text{Black}) = \frac{1}{2} + \frac{1}{2} = 1$$

The **union** of two events A and B occurs if either A occurs, B occurs, or A and B occur simultaneously. The simultaneous occurrence of A and B is commonly referred to as the **intersection** of A and B. If the two events A and B are not mutually exclusive (i.e., they intersect/can occur simultaneously), the probability for the union of events A and B is computed with Equation I.41. Note that the symbol $\cup$ is employed to represent the **union** of events.

$$P(A \text{ or } B) = P(A \cup B) = P(A) + P(B) - P(A \text{ and } B) \qquad \textbf{(Equation I.41)}$$

Since the symbol $\cap$ represents the **intersection** of events, Equation I.41 can also be written as follows.

$$P(A \text{ or } B) = P(A \cup B) = P(A) + P(B) - P(A \cap B).$$

To illustrate the application of the **addition rule** with two events which are not mutually exclusive, consider the following example. If a card is randomly selected from a deck of playing cards, what is the probability of obtaining either a **Red** card or a **King**? Employing Equation I.41 the probability 28/52 is computed below.

$$P(\text{Red or King}) = P(\text{Red} \cup \text{King}) = P(\text{Red}) + P(\text{King}) - P(\text{Red} \cap \text{King})$$

$$P(\text{Red or King}) = P(\text{Red} \cup \text{King}) = 26/52 + 4/52 - 2/52 = 28/52$$

The above computations indicate that 26 of the 52 cards in a deck are **Red** (thus, P (Red) = 26/52), and that 4 of the cards are a **King** (thus, P (King) = 4/52). Two of the Kings (the **King of Diamonds** and **King of Hearts**), however, are both a **Red** card and a **King**. When the probability of obtaining a card that is both **Red** and a **King** (2/52) (in other words, the intersection of the latter two events, which is represented with the notation Red ∩ King) is subtracted from the sum of the values P (Red) and P (King), what remains is the probability of selecting a **Red** card or a **King**. Thus, the value 28 in the numerator of the computed probability 28/52 indicates that the 26 red cards (which include the **King of Diamonds** and **King of Hearts**) plus the 2 black Kings (the **King of Spades** and **King of Clubs**) constitute the outcomes which meet the requirement of being a **Red** card or a **King**.[25]

Figure I.27 provides a visual summary of the relationships between two events A and B described in this section. The visual relationships depicted in Figure I.27 are referred to as **Venn diagrams**.[26] A typical Venn diagram employs a circle to represent all possible outcomes of a specific event. Thus, in each part of Figure I.27, all outcomes identified as A are contained within the circle labeled A, and all outcomes identified as B are contained within the circle labeled B. Figure I.27(a) describes the two mutually exclusive events A and B, and as noted above, since A and B do not intersect, P(A ∩ B) = 0. Figure I.27(b) describes two nonmutually exclusive events A and B, where the two circles represents the union of A and B. Figure I.27(c) represents A and its **complement** $\bar{A}$. The **complement** of A is comprised of all possible outcomes other than those contained in A. Figure I.27(d) describes the intersection of the events A and B, which is represented by the area in the center designated by the notation A ∩ B.

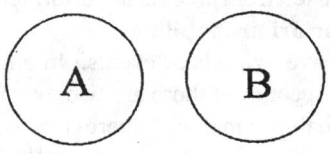

(a) Two mutually exclusive events A and B

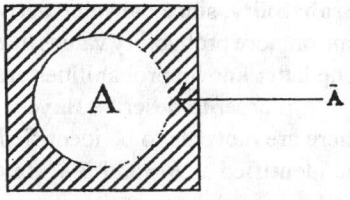

(c) A and Ā, where Ā is complement of A

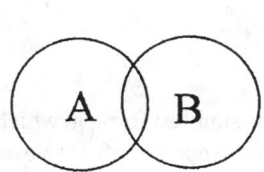

(b) Union of two nonmutually exclusive events A and B

A ∪ B

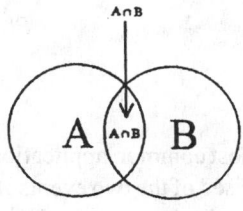

(d) Intersection of events A and B

A ∩ B

Figure I.27   Venn Diagrams

Another commonly employed rule for computing probabilities is the **multiplication rule**. In order to state the latter rule, the concept of **conditional probability** must be employed. The **conditional probability** of an event is the probability the event will occur, given the fact that another event has occurred. To be more specific, the conditional probability of an event A, given the fact that another event B has already occurred (which is represented with the notation $P(A/B)$), can be determined through use of Equation I.42. Note that both the notations $P(A \cap B)$ and $P(AB)$ can be used to represent the intersection of events A and B.

$$P(A/B) = \frac{P(A \cap B)}{P(B)} = \frac{P(AB)}{P(B)} \qquad \textbf{(Equation I.42)}$$

By transposing the terms in Equation I.42 it is also the case that $P(AB) = P(B)P(A/B)$. The latter is often referred to as the **multiplication rule**, which states the probability that two events A and B will occur simultaneously is equal to the probability of B times the conditional probability $P(A/B)$. Since, it is also the case that $P(B/A) = P(AB)/P(A)$, it is also true that $P(AB) = P(A)P(B/A)$. In other words, we can also say the probability that two events A and B will occur simultaneously is equal to the probability of A times the conditional probability $P(B/A)$.

Bayes' theorem   **Bayes' theorem** is an equation which allows for the computation of conditional probabilities.[27] As noted earlier, the **conditional probability** of an event is the probability the event will occur, given the fact that another event has occurred. Thus, the conditional probability of an event A, given the fact the another event B has already occurred (which is represented with the notation $P(A/B)$), can be determined through use of Equation I.42. The latter conditional probability is commonly referred to as a **posterior** or **a posteriori probability**, since the value derived for the conditional probability is computed through use of one or more probability values that are already known or estimated from preexisting information. The latter known probabilities are referred to as **prior** or **a priori probabilities**.

In order to describe **Bayes' theorem**, assume that we have two sets of events. In one set there are n events to be identified as $A_1, A_2, \dots, A_n$. In the second set there are two events to be identified as $B+$ and $B-$. **Bayes' theorem** states the probability that A_j (where $1 \leq j \leq n$) will occur, given that it is known that $B+$ has occurred, is determined with Equation I.43. (To determine the likelihood that A_j will occur, given that it is known that $B-$ has occurred (i.e. $P(A_j/B-)$), the value $B-$ is employed in the numerator and denominator of Equation I.43 in place of $B+$.)

$$P(A_j/B+) = \frac{P(B+/A_j)P(A_j)}{\sum\limits_{i=1}^{n} P(B+/A_i)P(A_i)} \qquad \textbf{(Equation I.43)}$$

The most common application of Equation I.43 involves its simplest form, in which the first set is comprised of the two events A_1 and A_2, and the second set is comprised of the two events $B+$ and $B-$. In the latter case, the equation for **Bayes' theorem** in reference to event A_1 (i.e., $P(A_1/B+)$), becomes Equation I.44. Equation I.45 is employed to compute the conditional probability $P(A_2/B+)$.

$$P(A_1/B+) = \frac{P(B+/A_1)P(A_1)}{P(B+/A_1)P(A_1) + P(B+/A_2)P(A_2)} \qquad \textbf{(Equation I.44)}$$

$$P(A_2/B+) = \frac{P(B+/A_2)P(A_2)}{P(B+/A_1)P(A_1) + P(B+/A_2)P(A_2)} \qquad \textbf{(Equation I.45)}$$

Equation I.46 (from which Equation I.44 can be algebraically derived — see Hays and Winkler (1971, pp. 84–85)) is another way of expressing the relationship described by Equation I.44.

$$P(A_1/B+) = \frac{P(A_1 \cap B+)}{P(B+)} \qquad \textbf{(Equation I.46)}$$

To illustrate the use of **Bayes' theorem**, let us assume that we wish to compute the probability of a person being a male (a male will be represented by the notation A_1) versus a female (a female will be represented with the notation A_2), given the fact that we know the person has a specific disease (a person having the disease will be represented by the notation $B+$, whereas a person not having the disease will be represented by the notation $B-$). In our example it will be assumed that for the population in question we know the following probabilities: a) $P(\text{Male}) = P(A_1) = .40$; b) $P(\text{Female}) = P(A_2) = .60$. In other words, the probability of a member of the population being a male is .40, and the probability of a person being a female is .60. In addition, we know the following: a) $P(B+/A_1) = .01$ (which indicates, the probability a person has the disease, given the fact the person is a male, is .01); b) $P(B+/A_2) = .25$ (which indicates, the probability a person has the disease, given the fact the person is a female, is .25). Employing Equations I.44 and I.45, we respectively determine the following: a) The probability a person is a male, given the fact the person has the disease, is .026; b) The probability a person is a female, given the fact the person has the disease, is .974. Thus, if we know that someone has the disease, it is extremely likely the person is a female.

$$P(A_1/B+) = \frac{(.01)(.40)}{(.01)(.40) + (.25)(.60)} = .026$$

$$P(A_2/B+) = \frac{(.25)(.60)}{(.01)(.40) + (.25)(.60)} = .974$$

A more detailed discussion of **Bayes' theorem** and its application can be found in Section IX (the **Addendum**) of the **binomial sign test for a single sample**.

Counting rules The next group of principles to be discussed are often referred to as **counting rules**, since they allow one to compute/count the number of outcomes/events that can occur within the context of a specific situation.

Rule 1: Assume that in a series comprised of n independent trials, on each trial any one of k_i (where i represents the ith trial) mutually exclusive outcomes can occur. We will let k_1 represents the number of possible outcomes on Trial 1, k_2 the number of possible outcomes on trial 2, ... k_i the number of possible outcomes on the ith trial, ... , and k_n the number of possible outcomes on the nth trial. If the number of different sequences that can result within a series of n trials is designated by the letter M, the value of M is computed as follows: $M = (k_1)(k_2) \ldots (k_i) \ldots (k_n)$.

Table I.14 36 Possible Sequences Employing Rule 1

Heads, 1, Jack	Heads, 2, Jack	Heads, 3, Jack	Heads, 4, Jack	Heads, 5, Jack	Heads, 6, Jack
Heads, 1, Queen	Heads, 2, Queen	Heads, 3, Queen	Heads, 4, Queen	Heads, 5, Queen	Heads, 6, Queen
Heads, 1, King	Heads, 2, King	Heads, 3, King	Heads, 4, King	Heads, 5, King	Heads, 6, King
Tails, 1, Jack	Tails, 2, Jack	Tails, 3, Jack	Tails, 4, Jack	Tails, 5, Jack	Tails, 6, Jack
Tails, 1, Queen	Tails, 2, Queen	Tails, 3, Queen	Tails, 4, Queen	Tails, 5, Queen	Tails, 6, Queen
Tails, 1, King	Tails, 2, King	Tails, 3, King	Tails, 4, King	Tails, 5, King	Tails, 6, King;

To illustrate *Rule 1*, assume that a three trial series is conducted employing a coin, a single six-sided die, and a set of three playing cards comprised of the **Jack**, **King**, and **Queen** of Hearts. On Trial 1 the coin is flipped, on Trial 2 the die is rolled, and on Trial 3 a card is randomly selected from the set of three cards. In our example the number of trials is $n = 3$. The value of $k_1 = 2$, since there are two possible outcomes for the coin (**Head** versus **Tails**). The value of $k_2 = 6$, since there are six possible outcomes for the die (**1, 2, 3, 4, 5, 6**). The value of $k_3 = 3$, since there are three possible outcomes for the playing cards (**Jack, Queen, King**). Employing *Rule 1*, the number of different sequences that can be obtained in the three trial series is $M = (k_1)(k_2)(k_3) = (2)(6)(3) = 36$. The 36 possible sequences that can be obtained are summarized in Table I.14.

A special case of *Rule 1* is when the number of mutually exclusive outcomes that can occur on each trial is a fixed value (i.e., $k_1 = k_2 = ... = k_i = ... = k_n$). When the latter is true, we can employ the notation k to represent the number of possible outcomes on each trial, and use the following equation to compute the value of M: $M = k^n$

To illustrate this special case of *Rule 1*, assume that a fair coin is tossed on three trials. In such a case, there are $n = 3$ trials, and on each trial there are $k = 2$ mutually exclusive outcomes that are possible (**Heads** and **Tails**). Employing the equation $M = k^n$, there are $2^3 = 8$ possible sequences. Specifically, the 8 sequences involving **Heads** and **Tails** that can occur are: **HHH, HHT, HTH, HTT, TTT, TTH, THT, THH**.

As a second example, assume that a single six-sided die is rolled on 4 trials. In such a case, there are $n = 4$ trials, and on each trial there are $k = 6$ mutually exclusive outcomes that are possible (**1, 2, 3, 4, 5, 6**). Employing the equation $M = k^n$, there are $6^4 = 1296$ possible sequences. In other words, within the context of a four trial series there are 1296 possible orderings involving the face values 1 through 6 (e.g., **1111**, ..., **1234**, ..., **1235**, ..., **6543**, ..., **6666**).

Rule 2: Assume that we have k distinct objects. The number of ways that the k objects may be arranged in different order is $k!$.

Before examining this rule in greater detail the reader should take note of the following: a) An ordered arrangement is referred to as a **permutation**; b) The notation $k!$ represents k **factorial**, which is computed as follows: $k! = (k)(k - 1) ... (1)$. (A more detailed discussion of $k!$ can be found in Endnote 13 of the **chi-square goodness-of-fit test**.)

To illustrate *Rule 2*, assume there are three different colored flags (**Red, Blue** and **Yellow**) lined up from left to right in front of a building. How many ways can the three flags be arranged? In this example we have $k = 3$ distinct objects (i.e., the three flags). Thus, there are $k! = (3)(2)(1) = 6$ ways (or **permutations**) in which the three objects/flags can be ordered. Specifically: 1): **Red, Blue, Yellow**; 2) **Red, Yellow, Blue**; 3) **Blue, Red, Yellow**; 4) **Blue, Yellow, Red**; 5) **Yellow, Red, Blue**; 6) **Yellow, Blue, Red**.

Rule 2 can be employed to answer the question of how many different seating arrangements there are for a class comprised of 20 students. If one imagines that instead of arranging three flags in different orders, one is arranging the 20 students in different orders, it is easy to see that the number of possible arrangements equals $k!$ Thus, the number of possible arrangements is $k! = 20! = (20(19) \ldots (1) = (2.432902008)(10)^{18}$, which is an extraordinarily large number.

Rule 3: Assume that we have n distinct objects. If we wish to select and arrange x objects from our pool of n objects (where $x < n$), the number of possible ways that we can do this (which we will designate as M) is computed with Equation I.47.

$$M = \frac{n!}{(n - x)!} \qquad \textbf{(Equation I.47)}$$

The value computed with Equation I.47 can also be represented by the notation P_x^n, which represents the **number of permutations of n things taken x at a time**. To illustrate **Rule 3**, assume that a teacher has four children in her class but only has two seats in the room. How many possible ways are there for the teacher to select two of the four students and arrange them in the two seats? In our example there are a total of $n = 4$ objects/students and we wish to select and arrange $x = 2$ of the objects/students. Thus, employing Equation I.47 we compute that there are $M = 4! / (4 - 2)! = 12$ arrangements possible. To confirm the latter, we will identify the four students as **A, B, C**, and **D**, and assign two of them to seats with the first student listed in each sequence assigned to **Seat 1** and the second student to **Seat 2**. The 12 possible arrangements are as follows: **AB, BA, AC, CA, AD, DA, BC, CB, BD, DB, CD, DC**.

As another example to illustrate **Rule 3**, consider the following. The judges of a beauty contest have selected four semifinalists. However, the judges will only be awarding a first and second place prize. How many possible ways are there for the judges to select a first and second place finisher among the four semifinalists? This example is identical to the previous one, since there are $n = 4$ objects/contestants which the judges must select and arrange in $x = 2$ positions. Thus, we are once again required to compute the number of permutations of 4 things taken 2 at a time. Consequently, there are 12 possible arrangements that can be employed for two of the four semifinalists who are awarded the first and second prize.

Rule 4: Assume that we have n distinct objects. If we wish to select x objects from our pool of n objects (where $x < n$), but are not concerned with the order in which they are arranged, the number of ways that we can do this (which we will designate as M) is computed with Equation I.48.

$$M = \binom{n}{x} = \frac{n!}{x! \, (n - x)!} \qquad \textbf{(Equation I.48)}$$

Equation I.48 is employed to compute the **number of combinations of n things taken x at a time**, which can be written with the notation $\binom{n}{x}$. Note that in computing combinations (unlike in computing permutations), one is not interested in the ordering of the elements within an arrangement.

To illustrate **Rule 4**, assume that a teacher has four children in her class and must select two students to represent the class on the student council. How many ways can the teacher select two students from the class of four students? In this example there are $n = 4$ objects/students, and the teacher must select $x = 2$ of them. The teacher is not concerned with placing them in order (such as would be the case if one student was to be designated the head representative and the other the assistant representative). Since order is of no concern, through use of Equation I.48 the value $M = 6$ is computed.

$$M = \binom{4}{2} = \frac{4!}{2! \, (4 - 2)!} = 6$$

Thus, there are six ways two students can be selected. Specifically, if we designate the four students **A, B, C,** and **D**, the six ways are as follows: **A and B, A and C, A and D, B and C, B and D, C and D**.

As another example to illustrate *Rule 4*, consider the following. The judges of a beauty contest have selected four semifinalists. They will now select two finalists, but will not designate either selectee as a first or second place winner. How many possible ways are there for the judges to select two finalists among the four semifinalists, without specifying any order of finish? This example is identical to the previous one, since there are $n = 4$ objects/contestants, and the judges must select $x = 2$ of them. Thus, as in the previous example, there are 6 ways the judges can select the two finalists. Note that in both of the examples which are employed to illustrate *Rule 4*, the computed value of M is half the number obtained for the examples used to illustrate *Rule 3* (which also employed the values $n = 4$ and $x = 2$). The reason for this is that *Rule 3* takes the order of an arrangement into account. Thus, if the order of the elements is considered for the six arrangements identified with *Rule 4*, each of them can be broken down into two ordered arrangements/permutations. As an example, the **A and B** arrangement computed for *Rule 4* with Equation I.48 can be broken down into the two permutations **AB** and **BA**.

A commonly asked question that *Rule 4* can be employed to answer is the number of different hands which are possible in a game of cards. Specifically, in determining the number of different hands that are possible in a game of poker, one is asking how many ways $x = 5$ cards can be selected from a deck of $n = 52$ cards. *Rule 4* is employed, since one is not concerned with the order in which the cards are dealt. Thus, the number of possible hands is computed with Equation I.48 as follows: $\begin{pmatrix} 52 \\ 5 \end{pmatrix} = \dfrac{52!}{5!\ 47!} = 2{,}598{,}960$.

Rule 5: Assume that we have N distinct objects which we wish to divide into k mutually exclusive subsets, with n_1 objects in subset 1, n_2 objects in subset 2, ... n_i objects in subset i, ..., and n_k objects in subset k. The number of possible ways (which we will designate M) that the N objects can be divided into k subsets is computed with Equation I.49.

$$M = \frac{N!}{n_1!\ n_2!\ \dots\ n_i!\ \dots\ n_k!} \qquad \textbf{(Equation I.49)}$$

To illustrate *Rule 5*, assume that the owner of a restaurant has five applicants **A, B, C, D,** and **E**, and wants to select two applicants to be cooks, two to be bakers, and one to be the manager. Assuming all of the applicants are equally qualified for each of the positions, how many different ways can the five applicants be divided into two cooks, two bakers, and one manager? To be more precise, the question being asked is how many different ways can the $N = 5$ applicants be divided into $k = 3$ subsets – specifically, a subset of $n_1 = 2$ cooks, a subset of $n_2 = 2$ bakers, and a subset of $n_3 = 1$ manager. Employing Equation I.49 the value $M = 30$ is computed.

$$M = \frac{5!}{2!\ 2!\ 1!} = 30$$

The 30 ways that the applicants can be assigned to the three subsets are summarized in Table I.15. In the latter table, the first two letters listed represent the applicants who are given the job of cook, the second two letters represent the applicants who are given the job of baker, and the last letter represents the applicant who is given the job of manager.

As another example to illustrate *Rule 5*, consider the following. Assume that a teacher has 10 children in her class and wants to select two children to be student council representatives, three children to be representatives on the school athletic council, and one child to be the class

Table I.15   Summary of 30 Ways Applicants Can be Assigned to Three Subsets

Arrangement	Cook	Baker	Manager
1	AB	CD	E
2	AB	CE	D
3	AB	DE	C
4	AC	DE	B
5	AC	BE	D
6	AC	DB	E
7	AD	BE	C
8	AD	BC	E
9	AD	EC	B
10	AE	BD	C
11	AE	BC	D
12	AE	DC	B
13	BC	AE	D
14	BC	AD	E
15	BC	DE	A
16	BD	AE	C
17	BD	AC	E
18	BD	CE	A
19	BE	AD	C
20	BE	AC	D
21	BE	CD	A
22	CD	AB	E
23	CD	AE	B
24	CD	EB	A
25	CE	AB	D
26	CE	DA	B
27	CE	BD	A
28	DE	AB	C
29	DE	BC	A
30	DE	CA	B

library representative. How many different ways can the 10 students be divided into two student council representatives, three athletic council representatives, and one library representative? To be more precise, the question being asked is how many different ways can the $N = 10$ students be divided into $k = 4$ subsets. Specifically, the four subsets are a subset of $n_1 = 2$ student council representatives, a subset of $n_2 = 3$ athletic council representatives, a subset of $n_3 = 1$ library representative, and a subset of $n_4 = 4$ students who are not selected to be any kind of representative.

Employing Equation I.49 the value $M = 12,600$ is computed.

$$M = \frac{10!}{2!\ 3!\ 1!\ 4!} = 12,600$$

In point of fact, **Rule 5** is a general rule that can be employed for any value of k, and when $k = 2$ subsets Equation I.49 reduces to Equation I.48, the equation for **Rule 4**. Consequently, **Rule**

4 is a special case of *Rule 5*. In the examples employed to illustrate *Rule 4* (in which a teacher must select two out of four children or the judges of a beauty content must select two out of four contestants) there are two subsets. Specifically, in each example there is the subset of $n_1 = 2$ individuals who are selected and the subset of $n_2 = 2$ individuals who are not selected.

Parametric versus Nonparametric Inferential Statistical Tests

The inferential statistical procedures discussed in this book have been categorized as being **parametric** versus **nonparametric tests**. Some sources distinguish between parametric and nonparametric tests on the basis that **parametric tests** make specific assumptions with regard to one or more of the population parameters that characterize the underlying distribution(s) for which the test is employed. These same sources describe **nonparametric tests** (also referred to in some sources as **distribution-free** or **assumption-free tests**) as making no such assumptions about population parameters. In truth, nonparametric tests are really not assumption free, and, in view of this, Marascuilo and McSweeney (1977) suggest that it might be more appropriate to employ the term **"assumption freer"** rather than nonparametric in relation to such tests.

The distinction employed in this book for categorizing a procedure as a parametric versus a nonparametric test is primarily based on the **level of measurement** represented by the data that are being analyzed. As a general rule, inferential statistical tests which evaluate **categorical/ nominal data** and **ordinal/rank-order data** are categorized as **nonparametric tests**, while tests that evaluate **interval** or **ratio data** are categorized as **parametric tests**. Although the appropriateness of employing level of measurement as a criterion in this context has been debated, in most instances (although there are exceptions) its usage provides a reasonably simple and straightforward schema for categorization which facilitates the decision-making process for selecting an appropriate statistical test.

There is general agreement among most researchers that as long as there is no reason to believe that one or more of the assumptions of a parametric test have been violated, when the level of measurement for a set of data is interval or ratio, the data should be evaluated with the appropriate parametric test. However, if one or more of the assumptions of a parametric test are saliently violated, some (but not all) sources believe it may be prudent to transform the data into a format which makes it compatible for analysis with the appropriate nonparametric test.[28] Related to this is that even though parametric tests generally provide a more powerful test of an alternative hypothesis than their nonparametric analogs, the power advantage of a parametric test may be negated if one or more of its assumptions are violated.

The reluctance among some sources to transform interval/ratio data into an ordinal/rank-order or categorical/nominal format for the purpose of analyzing it with a nonparametric test is based on the fact that interval/ratio data contain more information than either of the latter two forms of data.[29] Because of their reluctance to sacrifice information, these sources take the position that even when there is reason to believe that one or more of the assumptions of a parametric test have been violated, it is still more prudent to employ the appropriate parametric test. Such sources argue that most parametric statistical tests are **robust**. A **robust test** is one which can still provide reasonably reliable information with regard to the underlying population, even if certain of the assumptions underlying the test are violated. Generally, when a parametric test is employed under the latter conditions, certain adjustments are made in evaluating the test statistic in order to improve its reliability.

In the final analysis, in most instances, the debate concerning whether a researcher should employ a parametric or nonparametric test for a specific experimental design turns out to be of little consequence. The reason for this is that most of the time a parametric test and its nonparametric analog are employed to evaluate the same set of data, they lead to identical or similar conclusions. This latter observation is demonstrated throughout this book with numerous examples. In those instances where the two types of test yield conflicting results, the truth can best be determined by conducting multiple experiments which evaluate the hypothesis under study.[30] A detailed discussion of statistical methods which can be employed for pooling the results of multiple studies that evaluate the same general hypothesis can be found under the discussion of **meta-analysis and related topics** in Section IX (the **Addendum**) of the **Pearson product-moment correlation coefficient**.

Univariate versus Bivariate versus Multivariate Statistical Procedures

The term **univariate statistical analysis** is generally employed for descriptive and inferential statistical procedures which evaluate a **single variable** (e.g., a hypothesis about a population mean, a population variance, etc.). The term **bivariate statistical analysis** is generally employed for procedures which allow a researcher to investigate the relationship between **two variables** — specifically, an independent/predictor variable and a dependent/ criterion variable (although some sources limit the use of the term bivariate analysis to simple correlational analysis involving pairs of observations (i.e., scores on two variables) for a sample of subjects).

Grimm and Yarnold (1995, p. 4) note that although researchers are not in complete agreement with respect to the use of the term **multivariate statistics**, it is generally employed for procedures which simultaneously evaluate **multiple** (i.e., two or more) **independent/ predictor variables** and **multiple** (i.e., two or more) **dependent/criterion variables**. Some sources, however, reserve the use of the term **multivariate** to primarily identify procedures which involve two or more dependent variables. It should be noted, however, that **discriminant function analysis (Test 37)** and **logistic regression (Test 39)** (both of which involve one dependent variable) are among the procedures which are commonly identified as multivariate.

Multivariate statistical procedures (which are discussed in detail in Chapters 33–40) afford researchers with a number of advantages over **univariate** and **bivariate procedures**. Specifically, among others, Harlow (2005, Ch. 1) notes the following: a) Because multivariate procedures are able to simultaneously assess the role of multiple variables, they represent a more realistic methodology for evaluating real world phenomena and theoretical models, both of which are typically complex and involve multiple variables; b) Because they evaluate multiple variables, multivariate procedures can minimize the amount of unexplained variability which results from random error; c) Because it takes into account interrelationships/intercorrelations between variables, multivariate statistics are better able to rule out the role of extraneous variables; d) Multivariate procedures allow a researcher to control the overall Type I error rate, which otherwise would be inflated if multiple bivariate analyses were conducted on the same set of variables; and e) Within the context of multivariate analysis, a researcher can conduct an overall analysis involving all of the variables (a macro-analysis) as well as more specific analyses (micro-analyses) assessing the role of the individual variables.

Some disadvantages associated with the use of multivariate procedures are: a) The mathematical operations involved in implementing multivariate procedures are considerably more complex than those required for univariate and bivariate procedures, and, for the most part, multivariate procedures are impractical to conduct without the availability of appropriate computer software; b) Interpreting multivariate data is more complex by virtue of the fact that a researcher is simultaneously examining multiple variables; c) Multivariate analysis requires that certain assumptions (e.g., normality, homogeneity of variance, linearity, etc.), most of which are also associated with parametric bivariate procedures, not be violated; and d) Multivariate analysis requires the use of larger sample sizes.

Selection of the Appropriate Statistical Procedure

The **Handbook of Parametric and Nonparametric Statistical Procedures** is intended to be a comprehensive resource on inferential statistical tests and measures of correlation/association.[31] The section that follows the **Introduction** presents an outline of the statistical procedures covered in the book. Following the outline the reader is provided with guidelines and accompanying decision tables to facilitate the selection of the appropriate statistical procedure for a specific experimental design.

References

Altman, D. G., Machin, D., Bryant, T. N., & Gardner, M. J. (2002). **Statistics with confidence: Confidence intervals with statistical guidelines** (2nd ed.). London: British Medical Journal Books.

Anderson, N. N. (2001). **Empirical direction in design and analysis.** Mahwah, N.J.: Lawrence Erlbaum Associates.

Anscombe, F. J. & Glynn, W. W. (1983). Distributions of the kurtosis statistic. **Biometrika**, 70, 227–234.

Arlinghaus, S. L. & Griffith, D. A. (Eds.) (1996). **Practical handbook of spatial statistics.** Boca Raton, FL: CRC Press.

Baltes, P. B., Reese, H. W. & Nesselroade, J. R. (1977). **Life-span developmental psychology: Introduction to research.** Monterey: CA: Wadsworth Publishing Company.

Batschelet, E. (1981). **Circular statistics in biology.** New York: Academic Press.

Bechtold, B. & Johnson, R. (1989). **Statistics for business and economics.** Boston: PWS-Kent.

Bowley, A. L. (1920). **Elements of statistics** (4th ed.). New York: Charles Scriberner's Sons.

Campbell, D. T. & Stanley, J. C. (1963). **Experimental and quasi-experimental designs for research.** Skokie, IL: Rand–McNally.

Christensen, L.B. (2000). **Experimental methodology** (8th ed.). Boston: Allyn & Bacon.

Cleveland, W. S. (1985). **The elements of graphing data.** Monterey, CA: Wadsworth Advanced Books & Software.

Cohen, B. (1996). **Explaining psychological statistics.** Pacific Grove, CA: Brooks/Cole Publishing Company.

Cohen, B. (2001). Explaining psychological statistics (2nd ed.). New York: John Wiley & Sons.

Cohen, J. (1962). The statistical power of abnormal-social psychological research: A review. **Journal of Abnormal and Social Psychology**, 65, 145–153.

Cohen, J. (1977). **Statistical power analysis for the behavioral sciences.** New York: Academic Press.

Cook, T. D. & Campbell. D. T. (1979). **Quasi-experimentation: Design and analysis issues for field settings.** Boston: Houghton-Mifflin Company.

Cornell University, Office of Statistical Consulting, **StatNews**, 54, November 5, 2002.

Cowles, M. (1989). **Statistics in psychology: An historical perspective.** Hillsdale, NJ: Lawrence Erlbaum Associates, Publishers

Cowles, M. (2001). Statistics in psychology: An historical perspective (2nd ed.). Mahwah, NJ: Lawrence Erlbaum Associates, Publishers

D'Agostino, R. B. (1970). Transformation to normality of the null distribution of g_1. **Biometrika**, 57, 679–681.

D'Agostino, R. B. (1986). Tests for the normal distribution In D'Agostino, R. B. and Stephens, M. A. (Eds.), **Goodness-of-fit techniques** (pp. 367–419). New York: Marcel Dekker.

D'Agostino, R. B., Belanger, A. & D'Agostino Jr., R. B. (1990). A suggestion for using powerful and informative tests of normality. **American Statistician**, 44, 316–321.

Dukes, W. F. (1965). $N = 1$. **Psychological Bulletin**, 1965, 64, 74–79.

Everitt, B. S. (2001). **Statistics for psychologists: An intermediate course**. Mahwah, N.J.: Lawrence Erlbaum Associates.

Fisher, N. I. (1993). **Statistical analysis of circular data**. Cambridge, UK: Cambridge University Press.

Fisher, N. I., Lewis, T., Embleton, B. J. (1987). **Statistical analysis of spherical data**. Cambridge, UK: Cambridge University Press.

Fisher, R. A. (1925). **Statistical methods for research workers**. Edinburgh: Oliver & Boyd.

Fisher, R. A. (1955). Statistical methods and scientific induction. **Journal of the Royal Statistical Society**, Series B, 17, 69–78.

Fisher, R. A. (1956). **Statistical methods and scientific inference**. Edinburgh: Oliver & Boyd.

Folz, D. H. (1996). **Survey research for public administration**. Thousand Oaks, CA: Sage Publications.

Freund, J. E. (1984). **Modern elementary statistics** (6th ed.). Englewood Cliffs, NJ: Prentice–Hall, Inc.

Gigerenzer, G. (1993). The superego, the ego, and the id in statistical reasoning. In Keren, G. & Lewis, C. (Eds.), **A handbook for data analysis in the behavioral sciences: Methodological issues** (pp. 311–339). Hillsdale, NJ: Lawrence Erlbaum Associates, Publishers.

Gigerenzer, G. & Murray, D. J. (1987). **Cognition as intuitive statistics**. Hillsdale, N.J.: Lawrence Erlbaum Associates, Publishers.

Grimm, L. G. & Yarnold, P. R. (1995). Introduction to multivariate statistics. In Grimm, L. G. & Arnold, P. R. (Eds.) (pp. 1–18). **Reading and understanding multivariate statistics**. Washington, D.C.: American Psychological Association.

Grissom, R. J. & Kim, J. J. (2005). **Effect sizes for research: A broad practical approach**. Mahwah, NJ: Lawrence Erlbaum Associates Publishers.

Gurland, J. & Tripathi, R. C. (1971). A simple approximation for unbiased estimation of the standard deviation. **American Statistician**, 25(4), 30–32.

Hair, J. F., Anderson, R. E., Tatham, R. L., & Black, W. C. (1998). **Multivariate data analysis** (5th ed.). Upper Saddle River, NJ: Prentice Hall, Inc.

Harlow, L. L. (2005). **The essence of multivariate thinking**. Mahwah, NJ: Lawrence Erlbaum Associates.

Harlow, L. L., Muliak, S. A. & Steiger, J. H. (Eds) (1997). **What if there were no significance tests**. Mahwah: NH: Lawrence Erlbaum Associates.

Hays, W. L. & Winkler, R. L (1971). Statistics: Probability, inference, and decision. New York: Holt, Rinehart & Winston.

Hersen, M. & Barlow, D. H. (1976). **Single-case experimental designs: Strategies for studying behavior change**. New York: Pergamon Press.

Hoaglin, D.C., Mosteller, F. & Tukey, J. W. (Eds.) (1983). **Understanding robust and exploratory data analysis**. New York: John Wiley & Sons.

Hoffman, P. (1998). **The man who loved only numbers**. New York: Hyperion.

Hogg, R. V. & Tanis, E. A. (1997). **Probability and statistical inferences** (5th ed.). Saddle River, N.J.: Prentice Hall.

Howell, D. C. (1997). **Statistical methods for psychology** (4th ed.). Belmont, CA: Duxbury Press.

Hunter, J. E. & Schmidt, F. L. (1990). **Methods of meta-analysis: Correcting error and bias in research findings** (1st ed.). Newbury Park, CA: Sage Publications.

Hunter, J. E. & Schmidt, F. L. (2004). **Methods of meta-analysis - Correcting error and bias in research findings** (2nd ed.). Thousand Oaks, CA: Sage Publications.

Jammalamadaka, S. R. & SenGuputa, A. (2001). **Topics in circular statistics**. River Edge, NJ: World Scientific Publishing Company.

Johnson, N. L. & Kotz, S. (Eds.) (1997). **Leading personalities in statistical sciences**. New York: John Wiley & Sons, Inc.

Johnson, N. L. & Kotz, S. (Eds.) (1997). **Leading personalities in statistical sciences**. New York: John Wiley & Sons, Inc.

Kline, R. B. (2004). **Beyond significance testing: Reforming data analysis methods in behavioral research**. Washington, D. C.: American Psychological Association.

Lehman, E. L. (1993). The Fisher, Neyman-Pearson theories of testing hypotheses: One theory or two? **Journal of the American Statistical Association**, 88, 1242–1249.

Marascuilo, L. A. & McSweeney, M. (1977). **Nonparametric and distribution-free methods for the social sciences**. Monterey, CA: Brooks/Cole Publishing Company.

Mardia, K. V. (1972). **Statistical of directional data**. New York: Academic Press.

McElory, E. E. (1979). **Applied business statistics**. San Francisco: Holden-Day, Inc.

Meehl, P. E. (1967). Theory testing in psychology and physics: A methodological paradox. **Philosophy of Science**, 34, 103–115.

Moore, M. (Ed.) (2001). **Spatial statistics: Methodological aspects and applications**. New York: Springer.

Moors, J. J. (1986). The meaning of kurtosis: Darlington revisited. **American Statistician**, 40, 283–284.

Moors, J. J. (1988). A quantile alternative for kurtosis. **Statistician**, 37, 25–52.

Mosteller, F. & Tukey, J. W. (1977). **Data analysis and regression: A second course in statistics**. Reading, MA: Addison-Wesley.

Neyman, J. (1941). Fiducial argument and the theory of confidence intervals. **Biometrika**, 32, 128–150.

Neyman, J. (1950). **First course in probability and statistics**. New York: Henry Holt & Company.

Neyman, J. & Pearson, E. S. (1928). On the use and interpretation of certain test criteria. **Biometrika**, 20, 175–240.

Oakes, M. (1986). Statistical inference: A commentary for the behavioral and social sciences. Chichester: John Wiley & Sons.

O'Sullivan, E, Rassel, G., & Berner, M. (2002). **Research methods for public administration** (4th ed.). Boston: Longman.

Pearson, E. S. (1962). Some thoughts on statistical inference. **Annals of Mathematical Statistics**, 3, 394–403.

Ripley, B. D. (2004). **Spatial statistics**. Hoboken, NJ: Wiley-Interscience.

Rosner, B. (2000). **Fundamentals of biostatistics** (5th ed.). Pacific Grove, CA: Duxbury Press.

Rozeboom, W. W. (1960). The fallacy of the null hypothesis significance test. **Psychological Bulletin**, 57, 416–428.

Scheaffer, R. L., Mendenhall, W., & Ott, L. (1996). **Elementary survey samping** (5th ed.). Belmont, CA: Duxbury Press.

Shadish, W. R., Cook, T. D. & Campbell, D. T. (2002). **Experimental and quasi-experimental designs for generalized causal inference**. Boston: Houghton Mifflin Company.

Sheskin, D. J. (1984). **Statistical tests and experimental design: A guidebook**. New York: Gardner Press.

Smith, A. F. & Prentice, D. A. (1993). Exploratory data analysis. In Keren, G. & Lewis, C. (Eds.), **A handbook for data analysis in the behavioral sciences: Statistical issues** (pp. 349–390). Hillsdale, NJ: Lawrence Erlbaum Associates, Publishers.

Smithson, M. (2003). **Confidence intervals**. Thousand Oaks, CA: Sage Publications.

Stevens, J. (2002). **Applied multivariate statistics for the social sciences** (4th ed.). Mahwah, NJ: Lawrence Erlbaum Associates, Publishers.

Stevens, S.S. (1946). On the theory and scales of measurement. **Science**, 103, 677–680.

Stigler, S. M. (1999). **Statistics on the table: The history of statistical concepts and methods**. Cambridge, MA. Harvard University Press.

Tabachnick, B. G. & Fidell, L. S. (2001). **Using multivariate statistics** (4th ed.). Boston: Allyn & Bacon.

Tankard, J. Jr. (1984). **The statistical pioneers**. Cambridge, MA: Schenkman.

Thompson, B. (1993). Statistical significance testing in contemporary practice: Some proposed alternatives with comments from journal editors (Special issue). **Journal of Special Education**, 61 (4).

Thompson, B. (1999). Journal editorial policies regarding statistical significance tests: Heat is to fire as p is to importance. **Educational Psychology Review**, 11, 157–169.

Thompson, B. (2002). What future quantitative social science research could look like: Confidence intervals for effect sizes. **Educational Researcher**, 31, 25–32.

Tolman, H. (1971). A simple method for obtaining unbiased estimates of population standard deviations. **American Statistician**, 25(1), 60.

Trochim, W. M. (2005). **Research methods: The concise knowledge base**. Cincinnati: Atomic Dog Publishing Company.

Tufte, E. R. (1983). **The visual display of quantitative information**. Chesire, CT: Graphics Press.

Tukey, J.W. (1977). **Exploratory data analysis**. Reading, MA: Addison-Wesley.

Upton, G J. & Fingleton, B. (1985). **Spatial data analysis by example: Point pattern and quantitative data.** Volume 1. New York: John Wiley & Sons.

Upton, G J. & Fingleton, B. (1989). **Spatial data analysis by example: Categorical and directional data.** Volume 2. New York: John Wiley & Sons.

van Belle, G. (2002). **Statistical rules of thumb**. New York: John Wiley & Sons.

Watson, G. S. (1983). **Statistical in spheres**. New York: John Wiley & Sons.

Wilcox, R. R. (1987). **New statistical procedures for the social sciences**. Hillsdale, NJ: Erlbaum.

Wilcox, R. R. (1996). **Statistics for the social sciences**. San Diego, CA: Academic Press.

Wilcox, R. R. (1997). **Introduction to robust estimation and hypothesis testing**. San Diego, CA: Academic Press.

Wilcox, R. R., (2001). **Fundamentals of modern statistical methods: Substantially increasing power and accuracy**. New York: Springer.

Wilcox, R. R., (2003). **Applying contemporary statistical techniques**. San Diego, CA: Academic Press.

Winer, B. J. (1971). **Statistical principles in experimental design** (2nd ed.). New York: McGraw–Hill Publishing Company.

Zar, J. H. (1999). **Biostatistical analysis** (4th ed.). Upper Saddle River, NJ: Prentice Hall.

Endnotes

1. There may be situations where a researcher may need to compute an overall mean for a set of k sample means, yet believes that some of the mean values should be accorded more

importance than others (possibly because he believes they are more representative of the overall population mean, or perhaps because has reason to believe that some of the means are less likely to have been influenced by potentially contaminating variables). In the latter situation a researcher may be of the opinion that the weight assigned to a specific mean value should not be proportional to the sample size employed to compute it. Under the latter circumstances Equation I.50 is an alternative equation which can be employed to compute the weighted mean. In the latter equation each mean value is assigned a weight (represented by the notation w_g) reflecting its contribution in computing the overall mean. The sum of all the weights must equal 1 (i.e. $\Sigma w_g = w_1 + w_2 + ... + w_k = 1$). Obviously, in such a situation the assignment of weights to the mean values is subjective and could be subject to challenge.

$$\bar{X}_W = (w_1)(\bar{X}_1) + (w_2)(\bar{X}_2) + ... + (w_k)(\bar{X}_k)$$ **(Equation I.50)**

To illustrate the use of Equation I.50, assume the values $\bar{X}_1 = 100$, $\bar{X}_2 = 106$, and $\bar{X}_3 = 115$ are computed for three sample means, and the researcher elects to assign a weight of .6 to the mean of Sample 1 (since he considers it the most representative of the three samples), a weight of .3 to the mean of Sample 2, and a weight of .1 to the mean of Sample 3 (which he considers the least representative of the samples). Employing Equation I.50 the weighted mean is computed as follows: $\bar{X}_W = (.6)(100) + (.3)(106) + (.1)(115) = 60 + 31.8 + 11.5 = 103.3$. It should be noted that if the weight assigned to each mean is, in fact, directly proportional to the number of subjects in the overall sample employed to compute that mean, then $w_g = n_g/N$. When the latter is true, Equation I.50 becomes identical to Equation I.3. Thus, in the example discussed earlier where $\bar{X}_1 = 100$ and $n_1 = 100$, $\bar{X}_2 = 106$ and $n_2 = 200$, and $\bar{X}_3 = 115$ and $n_3 = 300$, the weighted mean is computed to be: $(100/600)(100) + (200/600)(106) + (300/600)(115) = 109.5$. Note that in the latter computations the weights (w_g) .167, .333, and .5 are respectively employed for the means of Samples 1, 2, and 3.

2. A percentage is converted into a proportion by moving the decimal point two places to the left (e.g. 87% is equivalent to .87). Conversely, a proportion is converted into a percentage by moving the decimal point two places to the right (e.g., .87 is equivalent to 87%).

3. Strictly speaking, $\hat{s}$ is not an unbiased estimate of σ, although it is usually employed as such. In point of fact, $\hat{s}$ slightly underestimates σ, especially when the value of n is small. (The latter type of statistic is said to be **negatively biased**.) Zar (1999) notes that although corrections for bias in estimating σ have been developed by Gurland and Tripathi (1971) and Tolman (1971), they are rarely employed, since they generally have no practical impact on the outcome of an analysis. Kline (2004, p. 26) notes that Equation I.51 (which approaches the value of $\hat{s}$ as the value of n increases) can provide a numerical approximation of an unbiased estimate of σ (designated as $\hat{s}$)

$$\hat{s} = \left[1 + \frac{1}{4(N - 1)}\right]\tilde{s}$$ **(Equation I.51)**

4. The inequality sign > means **greater than**. Some other inequality signs used throughout the book are <, which means **less than**; ≥, which means **greater than or equal to**; and ≤, which means **less than or equal to**.

5. A full discussion of the concept of probability is presented in Section IX (the **Addendum**) of the **binomial sign test for a single sample** under the discussion of **Bayesian hypothesis testing**.

6. The **absolute value** is the magnitude of a number irrespective of the sign. The notation $|x|$ is commonly employed to represent the absolute value of a number designated as x. For example, $|-1| = 1$.

7. McElroy (1979) describes the use of the equation *skewness* $= (\bar{X} - M)/\tilde{s}$ as an alternative approximate measure of skewness. McElroy (1979) and Zar (1999) describe the following measure of skewness, referred to as the **Bowley coefficient of skewness** (Bowley (1920)), which employs quartiles of the distribution (where Q_i represents the ith quartile): *Skewness* $= (Q_3 + Q_1 - 2Q_2)/(Q_3 - Q_1)$. The latter index yields values in the range -1 for a maximally negatively skewed distribution to $+1$ for a maximally positively skewed distribution.

8. a) The author is indebted to Vladimir Britikov for clarifying the latter point in a personal communication; b) Karl Wuensch (2005, online publication) notes that low kurtosis is often confused with high variance. The latter author demonstrates that distributions which have identical values for kurtosis can have different variances, and that distributions with the same variance can have different values for kurtosis.

9. Although the term **bar graph** can be employed in reference to Figure I.7, the vertical elements perpendicular to the X-axis are probably best described as lines rather than bars (since the term bar connotes a width for a vertical element beyond that of a simple line (e.g., ▌ as opposed to |)). The use of noncontiguous bars exhibiting width beyond that of a simple line is most likely to be employed when a nominal/categorical variable is represented on the X-axis. For example, a bar graph can be employed to present the frequency of people in each of four ethnic groups which comprise a population. The identities of each of the four ethnic groups would be recorded on equidistant points along the X-axis. Above each ethnic group a vertical bar would be constructed to a height which reflected the frequency of that group in the population. Thus, a researcher has the option of employing vertical bars, as opposed to vertical lines, when the information recorded along the X-axis represents a discrete and/or categorical variable.

10. a) It should be noted that sources are not consistent with respect to the protocol employed for computing percentile values. In most cases any differences obtained between the various methods which might be employed to identify a score at a specific percentile will be of little or no practical consequence. The methodology employed in the main text of this book for computing percentiles is employed, among others, by Rosner (2000). It should be noted that in the case of **Distribution A**, if the underlying variable upon which the distribution is assumed to be based is discrete, with scores only allowed to assume an integer value, some sources might elect to employ the scores in the 11th and 19th ordinal positions respectively as the scores at the 50th and 90th percentiles. The use of the scores in the 11th and 19th ordinal positions (i.e., 71 and 82) would be based on the fact that they would represent the scores closest to but above the 50th and 90th percentiles (as computed in the main part of the text). On the other hand, some sources might simply designate the score at the $k = (20)(.5) = 10^{th}$ ordinal position as the median (i.e. the score of 67), and the score at the $k = (20)(.9) = 18^{th}$ ordinal position as the score at the 90th percentile (i.e., the score of 76) (since $10/20 = .5$ and

18/20 = .9); b) An alternative more complicated methodology for computing percentiles is described in other sources. Although this methodology can be employed with data that are not in the form of a grouped frequency distribution, it is most commonly employed with the latter type of distributions. It should be noted that when the methodology to be described below is employed with ungrouped data (which is the case in the computation of the values employed in constructing the boxplot in Figure I.9) it will yield slightly different values than those obtained for the Figure I.9 boxplot. The alternative method for computing a score that corresponds to a specific percentile value for a distribution is summarized by Equation I.52 below.

(Equation I.52)

$$P = LL + \left(\frac{np - CF}{f_i} \right)(w)$$

Where: LL = Real lower limit of the class interval which contains the score at the Pth percentile
n = Total number of scores in the distribution
p = Desired percentile value expressed as a proportion
CF = Cumulative frequency of scores in the interval which is directly below the interval containing the score with the desired percentile
f_i = Frequency of scores in the interval which contains the score with the desired percentile
w = width of the class interval which contains the score with the desired percentile.

To illustrate the above equation, we will assume that the data for **Distribution A** are presented in the form of the grouped frequency distribution in Table I.8, and that we wish to compute the score at the 47th percentile. Note that from Column 5 of Table I.8 we can determine that the score at the 47th percentile falls in the class interval 60-69, since it has a cumulative proportion of .50, and the class interval 50-59, which is directly below it, has a cumulative proportion of .10. The computations are noted below, with the value 68.75 being derived for the score at the 47th percentile. Note the discrepancy between the value computed below for the 47th percentile (68.75) and the value computed earlier for the 47th percentile (64.6) when each of the n = 20 scores listed in Table I.1 was employed in the computation. This discrepancy, in part, reflects the fact that the value of any statistic which is based on information recorded in a grouped frequency distribution will not be as accurate as one which takes into account the precise value of every score in the distribution.

It should be noted that when the above equation is employed to compute the score at the 47th percentile for the ungrouped set of data for **Distribution A**, it yields the value 66.9, which is not the same value computed in the main text of the book. Specifically, P = 66.5 + [(20)(.47) − 9]/1](1) = 66.9.

Equation I.53 can be employed to compute the percentile rank for a specific score (designated by the notation X) in a grouped frequency distribution. The latter equation is illustrated employing the score 68.75 computed above, to demonstrate that the latter score falls at the 47th percentile.

(Equation I.53)

$$X = [100] \left[\frac{CF + \left(\frac{X - LL}{w} \right)(f_1)}{n} \right]$$

$$X = [100] \left[\frac{2 + \left(\frac{68.75 - 59.5}{10} \right)(8)}{20} \right] = .47$$

11. Alternative criteria for classifying an outlier are discussed in Section VII of the *t* **test for two independent samples** under the discussion of **outliers**. In addition, protocols for dealing with outliers are also discussed.

12. If the variable being measured is a discrete variable which can only assume an integer value, the values computed in this section would be expressed as integer values. Values above the median are rounded off to the nearest integer value above the last integer digit. In other words, the value 107.5 (or, for that matter, the value 107 with any decimal addition) for the upper outer fence is rounded off to 108. Values below the median are rounded off to the integer value designated in the last integer digit. In other words, if the value 27.5 (or, for that matter, the value 27 with any decimal addition) were derived for the lower outer fence, it is rounded off to 27.

13. Not all sources employ the one and one-half hingespread and three hingespread criteria in, respectively, labeling outliers and severe outliers. For example, Hogg and Tanis (1997, p. 28) consider any score more than one and one-half hingespreads from a hinge as a **suspected outlier**, and any score more than three hingespreads from a hinge an **outlier**. Other sources (e.g., Cohen (2001, p. 78) and Howell (1997, p. 56)) define an outlier as any score that falls beyond an **adjacent value**.

14. The symbol π in Equations I.36 and I.37 represents the mathematical constant **pi** (which equals 3.14159...). The numerical value of π represents the ratio of the circumference of a circle to its diameter. The value e in Equations I.36 and I.37 equals 2.71828.... Like π, e is a fundamental mathematical constant. Specifically, e is the base of the natural system of logarithms, which will be clarified shortly. Both π and e represent what are referred to as **irrational numbers**. An **irrational number** has a decimal notation that goes on forever without a repeating pattern of digits. In contrast, a **rational number** (derived from the word ratio) is either an integer or a fraction (which is the ratio between whole/integer numbers), which when expressed as a decimal always terminates at some point or assumes a repetitive pattern. Examples of rational numbers are $1/4 = .25$ which has a terminating decimal, or $1/3 = .33333...$, which is characterized by an endless repeating pattern of digits (Hoffman, 1998).

    A **logarithm** is the value of an exponent which indicates the power that a number, which is referred to as a **base value**, must be raised in order to yield a specific number value. Typically, e and the number 10 are employed as base values for logarithms. Logarithms that employ e as the base value are referred to as **natural** (or **Naperian**) logarithms, while logarithms that employ 10 as the base value are referred to as **common** (or

Briggsian) logarithms. The notation $\log_e 20 = 2.9957$ or $\ln 20 = 2.9957$ indicates that the value of e (which is the base value of the logarithm) must be raised to the 2.9957^{th} power in order to result in the value 20. Note that if the notation log is employed when e is the base value, the subscript e must be indicated, whereas if the notation ln is employed, it is assumed that e is the base value of the logarithm. If the notation log 20 is employed without the e subscript, it is assumed that the base of the logarithm is 10. Thus, $\log 20 = 1.3010$, indicates that 10 must be raised to the 1.3010^{th} power to equal 20. Logarithms are employed later in the book within the framework of the operations involved in a number of statistical procedures. The **antilogarithm** (or **inverse logarithm**) of a number X is the number whose logarithm is X. Thus in the most recent example, the antilogarithm of 1.3010 is 20.

15. Previously, the term **tail** was defined as the lower or upper extremes of a distribution. Although the latter definition is correct, I am taking some liberty here by employing the term **tail** in this context to refer more generally to the left or right half of the distribution.

16. Although the values in **Column 4** of **Table A1** will not be employed in our example, a brief explanation of what they represent follows. In the case of the standard normal distribution, when a value of X is substituted in Equations I.36 or I.37, the value of X will correspond to a z score. When a z value is employed to represent X, the value of Y computed with Equation I.36/I.37 will correspond to the value recorded for the **ordinate** in **Column 4** of **Table A1**. The value of the ordinate represents the height of the normal curve for that z value. To illustrate, if the value $z = 0$ is employed to represent X, Equation I.36/I.37 reduces to $Y = 1/\sqrt{2\pi}$, which equals $Y = 1/\sqrt{(2)(3.1416)} = .3989$. The resulting value .3989 is the value recorded in **Column 4** of **Table A1** for the ordinate that corresponds to the z score $z = 0$.

17. Kline (2004, pp. 36–37) notes that the null hypothesis is a **point hypothesis** in that it **specifies one numerical value** for a population parameter. He makes a distinction between a null hypothesis which represents a **nil hypothesis** and a null hypothesis which represents a **non-nil hypothesis**.

   A **nil hypothesis** is a null hypothesis which specifies an **absence of an experimental effect**. More specifically, it is a null hypothesis in which the value **zero** is specified for a population parameter or hypothesized for the difference between two or more population parameters. Examples of a **nil null hypothesis** are: a) H_0: $\rho = 0$ (The latter null hypothesis states that the population correlation equals zero. The notation ρ represents the lower case Greek letter **rho**, which is employed to represent a population correlation); b) H_0: $\mu_d = 0$ (which is an alternative way of stating the null hypotheses discussed in Chapters 1 and 2, which states there is no difference between the mean of a population a sample comes from and the value stipulated for the population mean in the null hypothesis (μ_d represents the hypothesized value of the difference)); and c) H_0: $\mu_1 = \mu_2$, which can also be written as H_0: $\mu_1 - \mu_2 = 0$. (The latter null hypothesis states there is no difference between the means of two populations.)

   A **non-nil hypothesis** is a null hypothesis which specifies the **presence of an experimental effect**. More specifically, it is a null hypothesis in which **some value other than zero** is specified for a population parameter or hypothesized for the difference between two or more population parameters Examples of a **non-nil null hypothesis** are: a) H_0: $\rho = .5$. (The latter null hypothesis states the population correlation equals .5.); b)

H_0: $\mu_d = 5$ (which is an alternative way of stating the null hypotheses discussed in Chapters 1 and 2, which in this case states there is a difference of 5 units between the mean of a population a sample comes from and the value stipulated for the population mean in the null hypothesis); and c) H_0: $\mu_1 - \mu_2 = 5$. (The latter null hypothesis states the mean of Population 1 is 5 units greater than the mean of Population 2.)

For the most part, the examples employed in this book (as well as most books) use a **nil null hypothesis,** which is employed on the assumption that prior to conducting a study a researcher does not know whether any effect/difference is present with respect to the parameter(s) under study. When, however, prior to conducting a study a researcher is aware of or strongly suspects a specific effect size/difference with respect to the parameter(s) under study, a **non-nil null hypothesis** may be deemed more appropriate to employ.

Kline (2004, p. 37–38) also notes that the alternative hypothesis is a **range hypothesis,** since it specifies a **range of values** for the population parameter(s) being evaluated. Two examples illustrative of the latter are H_1: $p > 0$ (which hypothesizes that the value of a population correlation is greater than 0), and H_1: $\mu_1 - \mu_2 > 10$ (which hypothesizes that the mean of Population 1 is more than 10 units greater than the mean of Population 2).

18. In actuality, the values of the sample means do not have to be identical to support the null hypothesis. Due to **sampling error** (which, as noted earlier in the **Introduction,** is a discrepancy between the value of a statistic and the parameter it estimates, even when two samples come from the same population) the value of the two sample means will usually not be identical. The larger the sample size employed in a study, the less the influence of sampling error, and consequently the closer one can expect two sample means to be to one another if, in fact, they do represent the same population. With small sample sizes, however, a large difference between sample means is not unusual, even when the samples come from the same population, and because of this a large difference may not be grounds for rejecting the null hypothesis.

19. a) Some sources employ the notation $p \leq .05$, indicating a probability of equal to or less than .05. The latter notation will not be used unless the computed value of a test statistic is the exact value of the tabled critical value; b) Most sources employ the term **critical value** to identify an extreme value of a test statistic which is extremely unlikely to occur if the null hypothesis is true. Anderson (2001), however, suggests that it would be better to refer to such values as **criterial values,** since in the final analysis they are little more than arbitrary values which most members of the scientific community have agreed to employ as criteria within the framework of hypothesis testing. However, in the final analysis, such values are not critical in any objective sense, since other researchers may elect to employ alternative criteria which they consider more suitable for reaching a correct decision in evaluating a hypothesis.

20. Inspection of Column 3 in **Table A1** reveals that the proportion for $z = 1.64$ is .0505. This latter value is the same distance from the proportion .05 as the value .0495 derived for $z = 1.65$. If **Table A1** documented proportions to five decimal places, it would turn out that $z = 1.65$ yields a value that is slightly closer to .05 than does $z = 1.64$. Some books, however, do employ $z = 1.64$ as the tabled critical one-tailed .05 z value.

21. The probability value obtained within the framework of the **classical hypothesis testing model** can be interpreted as a **conditional probability.** A **conditional probability** (which

is discussed in more detail later in the **Introduction** in the section on **basic principles of probability**) is a probability which is contingent upon certain conditions having been met. More specifically, when a null hypothesis is evaluated within the framework of the **classical hypothesis testing model**, the probability value associated with the result of an experiment represents the following **conditional probability**: p **(Outcome of experiment /H_0 true)**, which represents the probability of obtaining the outcome for an experiment, given the fact that the null hypothesis is true. Unfortunately, researchers often erroneously interpret the probability value associated with the result of an experiment to represent the **conditional probability** p **(H_0 true /Outcome of experiment)**, which represents the probability the null hypothesis is true, given the result obtained for an experiment.

To illustrate the aforementioned conditional probabilities, $p < .05$ indicates that the likelihood of the result obtained for an experiment is less than 5% if, in fact, the null hypothesis is true. It does not indicate that the likelihood of the null hypothesis being true is less than 5% given the result obtained for an experiment, and consequently it also does not indicate that the likelihood of the alternative hypothesis being true is greater than 95%, given the result obtained for an experiment. When the result of an experiment is not significant it is generally assumed that $p > .05$. Employing the conditional probabilities noted above, $p > .05$ indicates that the likelihood of the result obtained for an experiment is greater than 5% if, in fact, the null hypothesis is true. It does not indicate that the likelihood of the null hypothesis being true is greater than 5% given the result obtained for an experiment, and consequently it also does not indicate that the likelihood of the alternative hypothesis being true is less than 95%, given the result obtained for an experiment.

22.   The author is in agreement with the following statement by Wainer (1999, p. 212): "To be perfectly honest, I am a little at loss to understand fully the vehemence and vindictiveness that have recently greeted NHT (null hypothesis significance testing). These criticisms seem to focus primarily on the misuse of NHT. The focus on the technique rather than on those who misuse it seems to be misplaced."

23.   a) **Random assignment** is a major prerequisite for insuring the **internal validity** of a study. In contrast, the use of a **random sample** is intended to provide a researcher with a representative cross section of a population, and by virtue of the latter allows the researcher to generalize one's results to the whole population. Although, in practice, random samples are rarely employed in experiments, random assignment is a prerequisite of a **true experimental design**; b) The **true experiment** described earlier in the **Introduction** represents the simplest type of **true experimental design**.

24.   a) Among others, Cowles (1989), Larsen and Marx (1985) and Zar (1999) note that although addressed to some extent earlier, the formal study of the concept of probability can be traced back to two French mathematicians Blaise Pascal (1623-1662) and Pierre de Fermat (1601-1665); b) A more detailed discussion of the rules to be presented in this section can be found in most books that specialize in the subject of probability. The format of the discussion to follow is based, in part, on an excellent presentation of the topic in Hays and Winkler (1971, Ch. 2).

25.   a) If A and B are mutually exclusive, $P(A \cap B) = 0$; b) In the case of three events A, B, or C which are not mutually exclusive, the probability that an event is A or B or C is computed with the equation: $P(A \text{ or } B \text{ or } C) = P(A) + P(B) + P(C) - P(A \text{ and } B) - P(A \text{ and } C) - P(B \text{ and } C) + P(A \text{ and } B \text{ and } C)$.

26. Venn diagrams are named after the British logician James Venn (1834-1923). Hays and Winkler (1971, p. 8) note that such diagrams are also referred to as Euler diagrams, after the Swiss mathematician Lennard Euler (1707-1783).

27. Larsen and Marx (1985) note that although in 1812 the French mathematician Pierre Simon, Marquis de Laplace (1749-1827) made the first explicit statement of Bayes' theorem, the theorem was initially presented in 1763 by Bayes in a posthumously published paper. Cowles (1989, p. 73), however, notes there is a lack of agreement among historians with regard to whether the ideas in Bayes' paper were entirely his own work, or also contain contributions by his friend Richard Price, who submitted the paper for publication two years after Bayes' death.

28. The accuracy of parametric tests is most likely to be compromised when the assumption of normality of the underlying distribution(s) being evaluated is violated. Among others, Wilcox (2003) contends that researchers should employ **computer-intensive procedures**, which are discussed in Section IX (the Addendum) of the **Mann–Whitney U test (Test 12)**, as an alternative to many of the procedures described in this book when there is reason to believe that one or more of the assumptions underlying a parametric or nonparametric test have been saliently violated.

29. Since interval and ratio data are viewed the same within the decision making process with respect to test selection, the expression interval/ratio will be used throughout the book to indicate that either type of data is appropriate for use with a specific test.

30. Use of **computer intensive procedures** referred to in Endnote 28 can often help clarify inconsistent results between parametric versus nonparametric tests employed with the same data when there is reason to believe that one or more assumptions underlying one or more of the tests in question have been saliently violated.

31. One area of statistical analysis which will not be covered in this book is **circular statistics**. The term **circular statistics** (referred to as **angular statistics** or **radial statistics** in some sources) is commonly employed for methods (as well as the associated theory) that are relevant in the analysis of **circular data** — i.e., data based on a **circular scale of measurement**. In the discussion of levels of measurement in the **Introduction**, it is noted that an interval scale of measurement does not have a true zero point. A **circular scale of measurement** is a special case of an interval scale, which not only lacks a true zero point, but is also arbitrary with respect to the magnitude of the values which constitute the scale (i.e., the designation of what constitutes high versus low values on such a scale is arbitrary). Another characteristic of a circular scale is that the maximum and minimum values on the scale intersect/coincide. A common example of a circular scale is the measurement of direction with a compass, which is represented in Figure I.28. The latter scale divides a circle into 360 equal intervals which are referred to as degrees (represented by the notation °). The designation of 0° (which is equivalent to 360°) as north and 180° as south is arbitrary. Note that a direction of 10° is much closer to 350° than it is to 40°, and thus, in the final analysis in an absolute sense it is meaningless to say that 10° is less than (or for that matter greater than) 350°. Zar (1999, p. 592) notes that although circular data are typically measured in angles, such data can also be measured in **radians** (which are discussed in Section VII of the *t* **test for two independent samples** under the discussion of outliers and data transformation). Zar (1999, p. 594) also notes that on occasion angular

data (i.e., data expressed in an angular format) may be arranged on other than a circular cycle — for example, data could be arranged on a semi-circular cycle consisting of 180 °.

Other common examples of circular scales are the time of day, designated with the values 00:00 to 24:00 (which can also be written as 0000 to 2400), which are employed in military time (as indicated in Figure I.28), or the months of the year, if designated by the values 0 through 12 (as indicated in Figure I.28). In the case of time, 12 midnight can be viewed as either the end of one day or beginning of the next day. With respect to military time, if 12 midnight is viewed as the beginning of a day (which is the case in digital watches) it is designated as 00:00, whereas if it is viewed as the end of the day it is designated as 24:00. Note that since the upper and lower extremes of the time scale intersect, 1 AM (01:00 in military time) is closer to 11 PM (23:00 in military time) than it is to 4 AM (04:00 in military time). In the case of months of the year, since the scale is circular, with January designated with the value 0 or 12, February (designated with value 1) is closer to December (designated with the value 11) than it is to May (designated with the value 4).

Visual displays of circular data can be plotted by placing dots on the circumference of a circle (such as that depicted in Figure I.28) to indicate the location of each observation. When a sample includes multiple cases of specific observations, histograms or other types of visual displays can be constructed which employ lines that emanate from the circumference or center of the circle that are proportional to observed frequencies.

Circular variables such as time of day, month of the year, day of the year, etc. can be converted into an angular value (to be designated with the notation $a°$) through use of Equation I.54. In the latter equation, k represents the number of units in a full cycle of the circular scale, and X represents the unit value on the cycle one wishes to convert into an angular value. The latter equation is employed below to convert 6:12 PM (18:12 military time) into 271.8 °. Equation I.54 is also employed to convert the month of October into 270 °. Note that the number 9 is employed to designate October, since January, which represents the beginning of the cycle is designated as 0 (As noted above, since the scale is circular, January can also be designated with the number 12.). In the final illustration of Equation I.54, the date January 15th (which is the 15th day of a 365 day year — i.e., a non leap year) is converted into 14.79 °. If a year is conceptualized on a circular scale, the first and last day of the year intersect, and consequently the latter day can be designated with the number 0 as well as the number 365.

$$a° = \frac{(360°)(X)}{k} \qquad \textbf{(Equation I.54)}$$

$$a° = \frac{(360°)(18.12)}{24} = 271.8°$$

$$a° = \frac{(360°)(9)}{12} = 270°$$

$$a° = \frac{(360°)(15)}{365} = 14.79°$$

Circular statistics are most commonly employed in disciplines such as biology (e.g., tracking directional movements of migrating animals), meteorology (e.g., directional measurement of wind), geology (e.g., directional measurement of faults in rock), epidemiology (e.g., time related measurement of hospital data), and economics (e.g., time

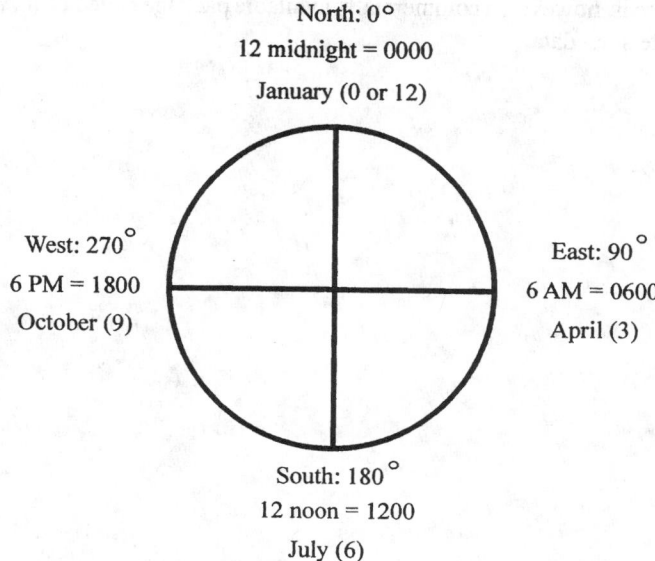

North: 0°

12 midnight = 0000

January (0 or 12)

West: 270°

6 PM = 1800

October (9)

East: 90°

6 AM = 0600

April (3)

South: 180°

12 noon = 1200

July (6)

Figure I.28   Circular scales for Compass Directions, Time, and Months

related measurement of economic data). Because the most commonly described statistical procedures (i.e., those described throughout this book) assume that data being evaluated are arranged on a linear scale, such procedures should not be employed to evaluate circular data.

To illustrate the latter, assume that a population of birds is distributed in a circular field and the location of three randomly selected birds from the center of the field are $10°$, $60°$, and $350°$ — all of which are in a northerly direction. If the mean direction for the latter three birds is computed the usual way, the value $(10° + 60° + 350°)/3 = 140°$ is obtained. Note, however, that the average value $140°$ falls in a southerly direction, and consequently our result does not make sense. Because of the latter, an alternative methodology is required for computing a mean for circular data. In point of fact, depending upon the statistic being computed, the value of each datum for circular data is determined on the basis of its angle and/or distance from a prespecified point on a circle. For the example under discussion, the location of each bird would be determined on the basis of its angle in degrees from some starting direction (e.g., true north). Through use of trigonometric functions — specifically, the sine and cosine of each angle — the appropriate mean value can be computed.

Two areas of analysis related to circular statistics are **spherical statistics** (which is the three dimensional analog of circular statistics) and **spatial statistics** (which, can include both circular and spherical statistics, since it is the application of statistical theory and methods to data that are referenced with respect to spatial coordinates). A full discussion of circular statistics, as well as spherical and spatial statistics, is beyond the scope of this book. Recommended sources on circular statistics are Batschelet (1981), Fisher (1993), Jammalamadaka and SenGupta (2001), Mardia (1972) and Zar (1999, Ch. 26 & 27). Recommended sources on spherical and/or spatial statistics are Arlinghaus and Griffith (1996), Batschelet (1981), Fisher *et al.* (1987), Mardia (1972), Moore (2001), Ripley (2004), Upton and Fingleton (1985, 1989), and Watson (1983). It should be noted that at the present time major statistical software packages do not support the analysis of circular data. There is, however, a commercially available package called *Oriana* which is designed to evaluate such data.

Outline of Inferential Statistical Tests and Measures of Correlation/Association

(One independent variable and one dependent variable is assumed
for all tests unless otherwise indicated)

I. **Inferential statistical tests employed with a single sample**
    A. Inferential statistical tests employed with interval/ratio data
        1. Inferential statistical tests employed with interval/ratio data for evaluating a hypothesis about the mean of a single population
            Test 1: The Single-Sample z Test
            Test 2: The Single-Sample t Test
            Test 34a: The Single-Sample Hotelling's T^2 (Multivariate procedure involving multiple variables)
        2. Inferential statistical tests employed with interval/ratio data for evaluating a hypothesis about a population parameter/characteristic other than a mean
            Test 3: The Single-Sample Chi-Square Test for a Population Variance
            Test 4: The Single-Sample Test for Evaluating Population Skewness
            Test 5: The Single-Sample Test for Evaluating Population Kurtosis
            Test 5a: The D'Agostino–Pearson Test of Normality
            Test 5b: The Jarque–Bera Test of Normality
            Test 10g: The Mean Square Successive Difference Test (for serial randomness)
            Test 11f: Median Absolute Deviation Test for Identifying Outliers
            Test 11g: Extreme Studentized deviate test for Identifying Outliers
    B. Inferential statistical tests employed with ordinal/rank-order data
        1. Inferential statistical tests employed with ordinal/rank-order data for evaluating a hypothesis about the median of a single population, or the distribution of data in a single population
            Test 6: The Wilcoxon Signed-Ranks Test
            Test 7: The Kolmogorov–Smirnov Goodness-of-fit Test for a Single Sample
            Test 7a: The Lilliefors Test for Normality
            Test 9b: The Single-Sample Test for the Median
            Test 12d: The Kaplan–Meier Estimate
    C. Inferential statistical tests employed with categorical/nominal data
        1. Inferential statistical tests employed with categorical/nominal data for evaluating a hypothesis about the distribution of data in a single population
            Test 8: The Chi-Square Goodness-of-Fit Test
            Test 9: The Binomial Sign Test for a Single Sample
            Test 9a: The z Test for a Population Proportion
            Test 10: The Single-Sample Runs Test
            Test 10a: The Runs Test for Serial Randomness
            Test 10b: The Frequency Test (for Randomness)
            Test 10c: The Gap Test (for Randomness)
            Test 10d: The Poker Test (for Randomness)
            Test 10e: The Maximum Test (for Randomness)
            Test 10f: The Coupon Collector's Test (for Randomness)
            Test 16b: The Chi-Square Test of Independence
            Test 16n: The log-likelihood ratio
            Test 16o: Log-linear analysis

II. Inferential statistical tests employed with two independent samples
   A. Inferential statistical tests employed with interval/ratio data
      1. Inferential statistical tests employed with interval/ratio data for evaluating a hypothesis about the means of two independent populations
         Test 11: The *t* Test for Two Independent Samples
         Test 11e: The *z* Test for Two Independent Samples
         Test 11h: The Westlake–Schuirmann Test of Equivalence of Two Independent Treatments
         Test 11i: Tryon's test of Equivalence of Two Independent Treatments
         Test 21: The Single-Factor Between-Subjects Analysis of Variance
         Test 21l: The Single-Factor Between-Subjects Analysis of Covariance[1]
         Test 34: Hotelling's T^2 (Multivariate procedure involving one independent variable and multiple dependent variables)
      2. Inferential statistical tests employed with interval/ratio data for evaluating a hypothesis about variability in two independent populations
         Test 11a: Hartley's F_{max} Test for Homogeneity of Variance/*F* Test for Two Population Variances
         Test 21g: The Levene test for Homogeneity of Variance
         Test 21h: The Brown–Forsythe test for Homogeneity of Variance
   B. Inferential statistical tests employed with ordinal/rank-order data
      1. Inferential statistical tests employed with ordinal/rank-order data for evaluating a hypothesis about the medians, or some other characteristic (other than variability) of two independent populations
         Test 12: The Mann–Whitney *U* Test
         Test 12a: The Randomization Test for Two Independent Samples
         Test 12e: Gehan's test for censored data
         Test 12f: The Log-Rank Test
         Test 13: The Kolmogorov–Smirnov Test for Two Independent Samples
         Test 16e: The Median Test for Independent Samples
         Test 23: The van der Waerden Normal-Scores Test for *k* Independent Samples
      2. Inferential statistical tests employed with ordinal/rank-order data for evaluating a hypothesis about variability of two independent populations
         Test 14: The Siegel–Tukey Test for Equal Variability
         Test 15: The Moses Test for Equal Variability
   C. Inferential statistical tests employed with categorical/nominal data
      1. Inferential statistical tests employed with categorical/nominal data for evaluating a hypothesis about the distribution of data in two independent populations
         Test 9c: Test for Comparing Two Poisson Counts
         Test 16a: The Chi-Square Test for Homogeneity
         Test 16c: The Fisher Exact Test
         Test 16d: The *z* Test for Two Independent Proportions
         Test 16m: The Westlake–Schuirmann Test of Equivalence of Two Independent Proportions
         Test 16n: The log-likelihood ratio
         Test 16o: Log-linear analysis

III. Inferential statistical tests employed with two dependent samples

   A. Inferential statistical tests employed with interval/ratio data

        1. Inferential statistical tests employed with interval/ratio data for evaluating a hypothesis about the means of two dependent populations

            Test 17: The *t* Test for Two Dependent Samples

            Test 17d: Sandler's *A* Test

            Test 17e: The *z* Test for Two Dependent Samples

            Test 17f: The Westlake–Schuirmann Test of Equivalence of Two Dependent Treatments

            Test 17g: Tryon's test of Equivalence of Two Dependent Treatments

            Test 24: The Single-Factor Within-Subjects Analysis of Variance

            Test 34b: The Use of the Single-Sample Hotelling's T^2 to Evaluate a Dependent Samples Design (Multivariate procedure involving one independent variable, but treating one dependent variable as multiple dependent variables)

        2. Inferential statistical tests employed with interval/ratio data for evaluating a hypothesis about variability in two dependent populations

            Test 17a: The *t* Test for Homogeneity of Variance for Two Dependent Samples

   B. Inferential statistical tests employed with ordinal/rank-order data

        1. Inferential statistical tests employed with ordinal/rank-order data for evaluating a hypothesis about the ordering of data in two dependent populations

            Test 18: The Wilcoxon Matched-Pairs Signed-Ranks Test

            Test 19: The Binomial Sign Test for Two Dependent Samples

   C. Inferential statistical tests employed with categorical/nominal data

        1. Inferential statistical tests employed with categorical/nominal data for evaluating a hypothesis about the distribution of data in two dependent populations

            Test 20: The McNemar Test

            Test 20a: The Gart Test for Order Effects

            Test 20b: The Westlake–Schuirmann test of Equivalence of Two dependent proportions

            Test 20c: The Bowker Test of Internal Symmetry

            Test 20d: The Stuart–Maxwell Test of Marginal Homogeneity

IV. Inferential statistical tests employed with two or more independent samples

   A. Inferential statistical tests employed with interval/ratio data

        1. Inferential statistical tests employed with interval/ratio data for evaluating a hypothesis about the means of two or more independent populations which involve one independent variable/factor

            Test 21: The Single-Factor Between-Subjects Analysis of Variance

                Test 21a: Multiple *t* Tests/Fisher's LSD Test

                Test 21b: The Bonferroni–Dunn test

                Test 21c: Tukey's HSD Test

                Test 21d: The Newman–Keuls Test

                Test 21e: The Scheffé Test

                Test 21f: The Dunnett Test

            Test 21l: The Single-Factor Between-Subjects Analysis of Covariance[1]

            Test 35: Multivariate Analysis of Variance (Multivariate procedure involving one independent variable and multiple dependent variables)

Test 36: Multivariate Analysis of Covariance (Multivariate procedure involving one independent variable and multiple dependent variables)

2. Inferential statistical tests employed with interval/ratio data for evaluating a hypothesis about variability in two or more independent populations

Test 11a: Hartley's F_{max} Test for Homogeneity of Variance/F Test for Two Population Variances

Test 21g: The Levene test for Homogeneity of Variance

Test 21h: The Brown–Forsythe test for Homogeneity of Variance

B. Inferential statistical tests employed with ordinal/rank-order data

1. Inferential statistical tests employed with ordinal/rank data for evaluating a hypothesis about the medians, or some other characteristic of two or more independent populations

Test 16e: The Median Test for Independent Samples

Test 22: The Kruskal–Wallis One-Way Analysis of Variance by Ranks

Test 22a: The Jonckheere–Terpstra Test for Ordered Alternatives

Test 23: The van der Waerden Normal-Scores Test for k Independent Samples

C. Inferential statistical tests employed with categorical/nominal data

1. Inferential statistical tests employed with categorical/nominal data for evaluating a hypothesis about the distribution of data in two or more independent populations

Test 16a: The Chi-Square Test for Homogeneity

Test 16n: The log-likelihood ratio

Test 16o: Log-linear analysis

V. **Inferential statistical tests employed with two or more dependent samples**

A. Inferential statistical tests employed with interval/ratio data

1. Inferential statistical tests employed with interval/ratio data for evaluating a hypothesis about the means of two or more dependent populations which involve one independent variable/factor.

Test 24: The Single-Factor Within-Subjects Analysis of Variance

Test 24a: Multiple t Tests/Fisher's LSD Test

Test 24b: The Bonferroni–Dunn Test

Test 24c: Tukey's HSD Test

Test 24d: The Newman–Keuls Test

Test 24e: The Scheffé Test

Test 24f: The Dunnett Test

Test 27j: Analysis of Variance for a Latin Square Design

Test 34b: The Use of the Single-Sample Hotelling's T^2 to Evaluate a Dependent Sample Design (Multivariate procedure involving one independent variable, but treating one dependent variable as multiple dependent variables)

B. Inferential statistical tests employed with ordinal/rank-order data

1. Inferential statistical tests employed with ordinal/rank-order data for evaluating a hypothesis about the medians of two or more dependent populations

Test 25: The Friedman Two-Way Analysis of Variance by Ranks

Test 25a: The Page Test for Ordered Alternatives

C. Inferential statistical tests employed with categorical/nominal data

1. Inferential statistical tests employed with categorical/nominal data for evaluating a hypothesis about the distribution of data in two or more dependent populations

Test 26: The Cochran Q Test

VI. **Inferential statistical tests employed with factorial designs** (evaluates two or more independent variables and one dependent variable)
  A. Inferential statistical tests employed with interval/ratio data
     1. Inferential statistical tests employed with interval/ratio data for evaluating a hypothesis about the means of two or more populations in a design involving two independent variables/factors
        Test 27: The Between-Subjects Factorial Analysis of Variance
           Test 27a: Multiple *t* Tests/Fisher's LSD Test
           Test 27b: The Bonferroni–Dunn Test
           Test 27c: Tukey's HSD Test
           Test 27d: The Newman–Keuls Test
           Test 27e: The Scheffé Test
           Test 27f: The Dunnett Test
        Test 27i: The Factorial Analysis of Variance for a Mixed Design
        Test 27k: The Within-Subjects Factorial Analysis of Variance

VII. **Measures of correlation/association** (One predictor/independent variable and one criterion/dependent variable is assumed unless otherwise indicated)
  A. Measures of correlation/association employed with interval/ratio data
     1. Bivariate measures
        Test 28: The Pearson Product–Moment Correlation Coefficient (and tests for evaluating various hypotheses concerning the value of one or more product-moment correlation coefficients, regression coefficients, or other characteristics of data for which a correlation coefficient is computed **(Tests 28a–28h)**)
     2. More than two samples/sets of scores
        Test 24i: The Intraclass Correlation Coefficient
     3. Multivariate measures
        Test 33: The Multiple Correlation Coefficient (One or more predictor/independent variables and one criterion/dependent variable)
        Test 33a: The Semipartial Correlation Coefficient (One or more predictor/independent variables and one criterion/dependent variable)
        Test 33b: The Partial Correlation Coefficient (One or more predictor/independent variables and one criterion/dependent variable)
        Test 38: Canonical Correlation (Multiple predictor/independent variables and multiple criterion/dependent variables)
  B. Measures of correlation/association employed with ordinal/rank order data
     1. Bivariate measures/Two sets of ranks
        Test 29: Spearman's Rank-Order Correlation Coefficient (and the test for evaluating the significance of Spearman's rank-order correlation coefficient **(Test 29a)**)
        Test 30: Kendall's Tau (and the test for evaluating the significance of Kendall's tau **(Test 30a)**)
        Test 32: Goodman and Kruskal's Gamma (and the test for evaluating the significance of gamma **(Test 32a)**)
     2. Ordinal measure of association for three or more samples/sets of ranks
        Test 31: Kendall's Coefficient of Concordance (and the test for evaluating the significance of the coefficient of concordance **(Test 31a)**)

C. Measures of correlation/association employed with categorical/nominal data

        Test 16f: The Contingency Coefficient

        Test 16g: The Phi Coefficient

        Test 16h: Cramér's Phi Coefficient

        Test 16i: Yule's Q

        Test 16j: The Odds Ratio (and test of significance for an odds ratio (**Test 16j-a)**

        Test 16k: Cohen's Kappa

        Test 28p: Binomial Effect Size Display

D. Other bivariate measures of correlation/association (and effect size) employed when interval/ratio data are used or implied for at least one variable

        Test 28i: The Point-Biserial Correlation Coefficient (and the test for evaluating the significance of a point-biserial correlation coefficient (**Test 28i-a)**)

        Test 28j: The Biserial Correlation Coefficient (and the test for evaluating the significance of a biserial correlation coefficient (**Test 28j-a**))

        Test 28k: The Tetrachoric Correlation Coefficient (and the test for evaluating the significance of a tetrachoric correlation coefficient (**Test 28k-a**))

        Tests 11c/17c/21i/24g/27g: Omega Squared

        Test 11d/21j: Eta Squared

        Test 11b/17b: Cohen's d Index (and Test 2a for one variable)

        Test 21k/24h/27h: Cohen's f Index

VIII. Additional procedures

A. Procedures for evaluating m independent 2×2 contingency tables

        Heterogeneity chi-square analysis (see Section VI of the **chi-square test for $r \times c$ tables**)

        Test 16l: Mantel–Haenszel analysis (Test 16l-a: Test of homogeneity of odds ratios for Mantel–Haenszel analysis, Test 16l-b: Summary odds ratio for Mantel–Haenszel analysis, Test 16l-c: Mantel–Haenszel test of association)

B. Computer intensive procedures (Although **Tests 12b** and **12c** below are discussed within the context of a procedure for ordinal data involving two independent populations, the latter procedures can be employed with designs involving an analysis of one or more independent or dependent population parameters. **Tests 12b** and **12c** typically involve interval/ratio data which in the case of the bootstrap are ultimately rank-ordered.)

        Test 12b: The Bootstrap

        Test 12c: The Jackknife

C. Procedure to detect autocorrelation

        Test 28h: Durbin–Watson Test (for detecting correlation of adjacent residuals)

D. Meta-analytic procedures

        Test 28l: Procedure for Comparing k Studies With Respect to Significance Level

        Test 28m: The Stouffer Procedure for Obtaining a Combined Significance Level for k Studies

        Test 28n: Procedure for Comparing k Studies With Respect to Effect Size

        Test 28o: Procedure for Obtaining a Combined Effect Size for k Studies

E. Additional multivariate procedures

Test 37: Discriminant Function Analysis (Multiple interval/ratio predictor/independent variables and one categorical criterion/dependent variable)

Test 39: Logistic Regression (One or more interval/ratio and/or categorical predictor/independent variables and one categorical criterion/dependent variable)

Test 40: Principal Components Analysis and Factor Analysis (Procedure for reducing multiple interval/ratio variables (with no distinction made between independent and dependent variables) to a limited number of dimensions)

[1]One or more covariates are also employed.

Guidelines and Decision Tables for Selecting the Appropriate Statistical Procedure

Tables I.16–I.19 are designed to facilitate the selection of the appropriate statistical test. Tables I.16–I.19 list the major inferential statistical procedures described in the book, based on the level of measurement the data being evaluated represent. Specifically, Table I.16 lists inferential statistical tests employed with interval/ratio data, Table I.17 lists inferential statistical tests employed with ordinal/rank-order data,[1] and Table I.18 lists inferential statistical tests employed with categorical/nominal data. Table I.19 lists the measures of correlation/association that are described in the book. Using the aforementioned tables, the following guidelines should be employed in selecting the appropriate statistical test.

1. Determine if the analysis involves computing a correlation coefficient/measure of association and, if it does, go to Table I.19. The selection of the appropriate measure in Table I.19 is based on the level of measurement represented by each of the variables for which the measure of correlation/association is computed.

2. If the analysis does not involve computing a measure of correlation/association, it will be assumed that the data will be evaluated through use of an inferential statistical test. To select the appropriate inferential statistical test, the following protocol should be employed.

   a) State the general hypothesis that is being evaluated.
   b) Determine if the study involves a single sample or more than one sample.
   c) If the study involves a single sample, the appropriate test will be one of the tests for a single sample in Tables I.16, I.17, or I.18. In order to determine which table to employ, determine the level of measurement represented by the data that are being evaluated. If the level of measurement is interval/ratio, Table I.16 is employed. If the level of measurement is ordinal/rank-order, Table I.17 is appropriate. If the level of measurement is categorical/nominal, Table I.18 is utilized.
   d) If there is more than one sample, determine how many samples/treatments there are and whether they are independent or dependent. Determine the level of measurement represented by the data that are being evaluated (which represents the dependent variable in the study).

      1) If the level of measurement is interval/ratio, go to Table I.16. Identify the test or tests which are appropriate for that level of measurement with respect to the number and type of samples employed in the study.
      2) If the level of measurement is ordinal/rank-order, go to Table I.17. Identify the test or tests which are appropriate for that level of measurement with respect to the number and type of samples employed in the study.
      3) If the level of measurement is categorical/nominal, go to Table I.18. Identify the test or tests which are appropriate for that level of measurement with respect to the number and type of samples employed in the study.

---

[1] In the case of the following four tests listed in Table I.17, the dependent variable will be interval/ratio data which are converted into a format in which the resulting scores are rank-ordered: The **Wilcoxon signed-ranks test (Test 6)**, the **bootstrap (Test 12b)**, the **Moses test for equal variability (Test 15)**, and the **Wilcoxon matched-pairs signed-ranks test (Test 18)**.

Table I.16   Decision Table for Inferential Statistical Tests Employed with Interval/Ratio Data
(One independent variable and one dependent variable is assumed for all tests unless otherwise indicated)

Number of samples		Hypothesis evaluated	Test
One independent variable			
Single sample		Hypothesis about a population mean	The single-sample z test (Test 1) (σ known) The single-sample t test (Test 2) (σ unknown) The single-sample Hotelling's T^2 (Multivariate procedure involving multiple variables) (Test 34a)
		Hypothesis about a population parameter/characteristic other than the mean	The single-sample chi-square test for a population variance (Test 3) The single-sample test for evaluating population skewness (Test 4) The single-sample test for evaluating population kurtosis (Test 5) The mean square successive difference test (for serial randomness) (Test 10g) The D'Agostino–Pearson test of normality (Test 5a) The Jarque–Bera test of normality (Test 5b) Median absolute deviation test for identifying outliers (Test 11f) Extreme Studentized deviate test for identifying outliers (Test 11g)
Two samples	Two independent samples	Hypothesis about difference between two independent population means	The t test for two independent samples (Test 11) The z test for two independent samples (Test 11e) Westlake–Schuirmann test of equivalence of two independent treatments (Test 11h) Tryon's test of equivalence of two independent treatments (Test 11i) The single-factor between-subjects analysis of (Test 21) The single-factor between-subjects analysis of covariance (Test 21l)[1] Hotelling's T^2 (Multivariate procedure involving one independent variable and multiple dependent variables) (Test 34)
		Hypothesis about two independent population variances	Hartley's F_{max} test for homogeneity of variance/ F test for two population variances (Test 11a) The Levene test for homogeneity of variance (Test 21g) The Brown–Forsythe test for homogeneity of variance (Test 21h)

Table I.16   Decision Table for Inferential Statistical Tests Employed with Interval/Ratio Data
(continued)

Number of samples		Hypothesis evaluated	Test
One independent variable			
Two samples	Two dependent samples	Hypothesis about difference between two dependent population means	The t test for two dependent samples (Test 17) Sandler's A test (Test 17d) The z test for two dependent samples (Test 17e) Westlake–Schuirmann test of equivalence of two dependent treatments (Test 17f) Tryon's test of equivalence of two dependent treatments (Test 17g) The single-factor within-subjects analysis of variance (Test 24) The use of the single-sample Hotelling's T^2 to evaluate a dependent samples design (Multivariate procedure involving one independent variable, but treating one dependent variable as multiple dependent variables) (Test 34b)
		Hypothesis about two dependent population variances	The t test for homogeneity of variance for two dependent samples (Test 17a)
Two or more samples	Two or more independent samples	Hypothesis about difference between two or more independent population means	The single-factor between-subjects analysis of variance (Test 21) The single-factor between-subjects analysis of covariance (Test 21l)[1] Multivariate analysis of variance (Multivariate procedure involving one independent variable and multiple dependent variables) (Test 35) Multivariate analysis of covariance (Multivariate procedure involving one independent variable and multiple dependent variables) (Test 36)[1]
		Hypothesis about two or more independent population variances	Hartley's F_{max} test for homogeneity of variance/ F test for two population variances (Test 11a) The Levene test for homogeneity of variance (Test 21g) The Brown–Forsythe test for homogeneity of variance (Test 21h)
	Two or more dependent samples	Hypothesis about difference between two or more dependent population means	The single-factor within-subjects analysis of variance (Test 24) Analysis of variance for a Latin square design (Test 27j) The use of the single-sample Hotelling's T^2 to evaluate a dependent samples design (Multivariate procedure involving one independent variable, but treating one dependent variable as multiple dependent variables) (Test 34b)

Table I.16　Decision Table for Inferential Statistical Tests Employed with Interval/Ratio Data
(continued)

Number of samples — One independent variable		Hypothesis evaluated	Test
Two samples	Two or more dependent samples	Hypothesis about difference between two dependent population variances	See discussion of sphericity assumption under the single-factor within-subjects analysis of variance (Test 24)
Two independent variables		Hypothesis about difference between two or more population means	The between-subjects factorial analysis of variance (Test 27) The factorial analysis of variance for a mixed design (Test 27i) The within-subjects factorial analysis of variance (Test 27k)

[1]One or more covariates are also employed.

**Table I.17 Decision Table for Inferential Statistical Tests Employed
with Ordinal/Rank-Order Data**

Number of samples		Hypothesis evaluated	Test
Single sample		Hypothesis about a population median or the distribution of data in a single population	The Wilcoxon signed-ranks test (Test 6) [1] The Kolmogorov–Smirnov goodness-of-fit test for a single sample (Test 7) The Lilliefors test for normality (Test 7a) The single-sample test for the median (Test 9b) The Kaplan–Meier estimate (Test 12d)
Two samples	Two independent samples	Hypothesis about two independent population medians, or some other characteristic (other than variability) of two independent populations	The Mann–Whitney U test (Test 12) The randomization test for two independent samples (Test12a) The bootstrap (Test 12b) [1,2] Gehan's test for censored data (Test 12e) The log-rank test (Test 12f) The Kolmogorov–Smirnov test for two independent samples (Test 13) The median test for independent samples (Test 16e) The van der Waerden normal-scores test for k independent samples (Test 23)
		Hypothesis about variability in two independent populations	The Siegel–Tukey test for equal variability (Test 14) The Moses test for equal variability (Test 15) [1]
	Two dependent samples	Hypothesis about the ordering of data in two dependent populations	The Wilcoxon matched-pairs signed-ranks test (Test 18) [1] The binomial sign test for two dependent samples (Test 19)
Two or more samples	Two or more independent samples	Hypothesis about two or more independent population medians, or some other characteristic of two or more independent populations	The Kruskal–Wallis one-way analysis of variance by ranks (Test 22) The Jonckheere–Terpstra test for ordered alternatives (Test 22a) The van der Waerden normal-scores test for k independent samples (Test 23) The median test for independent samples (Test 16e)
	Two or more dependent samples	Hypothesis about two or more dependent population medians	The Friedman two-way analysis of variance by ranks (Test 25) The Page test for ordered alternatives (Test 25a)

[1] The dependent variable for this test will be interval/ratio data which are converted into a format in which the resulting scores are rank-ordered.

[2] Although the bootstrap is discussed within the context of a procedure for ordinal data involving two independent populations, it can also be employed with designs involving an analysis of one or more independent or dependent population parameters.

**Table I.18   Decision Table for Inferential Statistical Tests Employed
with Categorical/Nominal Data**

Number of samples		Hypothesis evaluated	Test
Single sample		Hypothesis about distribution of data in a single population	The chi-square goodness-of-fit test (Test 8) The binomial sign test for a single sample (Test 9) The z test for a population proportion (Test 9a) The single-sample runs test (Test 10) The runs test for serial randomness (Test 10a) The frequency test (for randomness) (Test 10b) The gap test (for randomness) (Test 10c) The poker test (for randomness) (Test 10d) The maximum test (for randomness) (Test 10e) The coupon collector's test (for randomness) (Test 10f) The chi-square test of independence (Test 16b) The log-likelihood ratio (Test 16n) Log-linear analysis (Test 16o)
Two samples	Two independent samples	Hypothesis about distribution of data in two independent populations	Tests for comparing two Poisson counts (Test 9c) The chi-square test for homogeneity (Test 16a) The Fisher exact test (Test 16c) The z test for two independent proportions (Test 16d) Westlake–Schuirmann test of equivalence of two independent proportions (Test 16m) Mantel–Haenszel analysis (Test 16l) (pooling data for multiple 2×2 contingency tables) The log-likelihood ratio (Test 16n) Log-linear analysis (Test 16o)
Two samples	Two dependent samples	Hypothesis about distribution of data in two dependent populations	The McNemar test (Test 20) The Gart test for order effects (Test 20a) Westlake–Schuirmann test of equivalence of two dependent proportions (Test 20b) The Bowker test of internal symmetry (Test 20c) The Stuart–Maxwell test of marginal homogeneity (Test 20d)
Two or more samples	Two or more independent samples	Hypothesis about distribution of data in two or more independent populations	The chi-square test for homogeneity (Test 16a) The log-likelihood ratio (Test 16n) Log-linear analysis (Test 16o)
Two or more samples	Two or more dependent samples	Hypothesis about distribution of data in two or more dependent populations	The Cochran Q test (Test 26)

Table I.19 Decision Table for Measures of Correlation/Association
(One predictor/independent variable and one criterion/dependent variable
is assumed unless otherwise indicated)

Level of measurement		Test
Interval/ratio data	Bivariate	The Pearson product-moment correlation coefficient (Test 28)[1]
	More than two samples/sets of scores	Intraclass correlation coefficient (Test 24i)
	Multivariate	The multiple correlation coefficient (One or more predictor/independent variables and one criterion/dependent variable) (Test 33) The semipartial correlation coefficient (One or more predictor/independent variables and one criterion/dependent variable) (Test 33a) The partial correlation coefficient (One or more predictor/independent variables and one criterion/dependent variable) (Test 33b) Canonical correlation (Multiple predictor/independent variables and multiple criterion/dependent variables) (Test 38)
Ordinal/rank order data	Bivariate/two sets of ranks	Spearman's rank-order correlation coefficient (Test 29) Kendall's tau (Test 30) Goodman and Kruskal's gamma (for ordered contingency tables) (Test 32)
	More than two samples/sets of ranks	Kendall's coefficient of concordance (Test 31)
Categorical/ nominal data	Two dichotomous variables [2]	The contingency coefficient (Test 16f) The phi coefficient (Test 16g) Yule's Q (Test 16i) The odds ratio (Test 16j) Cohen's kappa (Test 16k) Binomial effect size display (Test 28p)
	Two nondichotomous variables	The contingency coefficient (Test 16f) Cramér's phi coefficient (Test 16h) The odds ratio (Test 16j) Cohen's kappa (Test 16k)
Other bivariate correlational measures for which interval ratio/ data are employed or implied for at least one of the variables		Omega squared (One variable, interval/ratio data; second variable, two or more nominal levels) (Tests 11c/17c/21i/24g/27g) Eta squared (One variable, interval/ratio data; second variable, two or more nominal levels) (Test 11d (two nominal levels); Test 21j) Cohen's d index (Test 11b/17b) (One variable, interval/ratio data; second variable, two nominal levels) (with Test 2a for one variable) Cohen's f index (One variable, interval/ratio data; second variable, two or more nominal levels) (Test21k/24h/27h) The point-biserial correlation coefficient (One variable, interval/ratio data; second variable represented by dichotomous categories) (Test 28i) The biserial correlation coefficient (One variable, interval/ratio data; second variable, an interval/ratio variable expressed in form of dichotomous categories) (Test 28j) The tetrachoric correlation coefficient (Two interval/ratio variables, both of which are expressed in the form of dichotomous categories) (Test 28k)

Table I.19   Decision Table for Measures of Correlation/Association
(continued)

Level of measurement	Test
Additional multivariate correlational/predictive measures involving interval/ratio and/or categorical data/variables	Discriminant function analysis (Multiple interval/ratio predictor/independent variables and one categorical criterion/dependent variable) (Test 37) Logistic regression (One or more interval/ratio and/or categorical predictor/independent variables and one categorical criterion/dependent variable) (Test 39) Principal components analysis and factor analysis (Multiple intercorrelated interval/ratio variables (with no distinction made between independent and dependent variables) reduced to a limited number of dimensions)

[1] The Durbin–Watson test (Test 28h) can be employed to determine whether or not adjacent residuals are significantly correlated.

[2] A dichotomous variable is comprised of two mutually exclusive categories.

Inferential Statistical Tests
Employed with a Single Sample

Test 1: **The Single-Sample z Test**

Test 2: **The Single-Sample t Test**

Test 3: **The Single-Sample Chi-Square Test for a Population Variance**

Test 4: **The Single-Sample Test for Evaluating Population Skewness**

Test 5: **The Single-Sample Test for Evaluating Population Kurtosis**

Test 6: **The Wilcoxon Signed-Ranks Test**

Test 7: **The Kolmogorov-Smirnov Goodness-of-Fit Test for a Single Sample**

Test 8: **The Chi-Square Goodness-of-Fit Test**

Test 9: **The Binomial Sign Test for a Single Sample**

Test 10: **The Single-Sample Runs Test (and Other Tests of Randomness)**

Test 1

The Single-Sample z Test
(Parametric Test Employed with Interval/Ratio Data)

I. Hypothesis Evaluated with Test and Relevant Background Information

Hypothesis evaluated with test Does a sample of n subjects (or objects) come from a population in which the mean (μ) equals a specified value?

Relevant background information on test The **single-sample z test** is employed in a hypothesis testing situation involving a single sample in order to determine whether or not a sample with a mean of $\bar{X}$ is derived from a population with a mean of μ. If the result of the **single-sample z test** yields a significant difference, the researcher can conclude there is a high likelihood the sample is derived from a population with a mean value other than μ. The test statistic for the **single-sample z test** is based on the normal distribution. A general discussion of the normal distribution can be found in the **Introduction**.

The **single-sample z test** is used with interval/ratio data. The test should only be employed if the value of the population standard deviation (σ) is known. In the event the value of σ is unknown, the data should be evaluated with the **single-sample t test (Test 2)**. The reader should take note of the fact that some sources argue that even when one knows the value of σ, if the sample size is very small, the **single-sample t test** provides a more accurate estimate of the underlying sampling distribution for the data. Sources that take the latter position are not in agreement with respect to the minimum sample size above which it is acceptable to employ the **single-sample z test** (although it is usually $n \geq 25$).

The **single-sample z test** is based on the following assumptions: a) The sample has been randomly selected from the population it represents; and b) The distribution of data in the underlying population the sample represents is normal. (The importance of this assumption decreases as the size of the sample increases.) If either of the aforementioned assumptions is saliently violated, the reliability of the z test statistic may be compromised.

II. Example

Example 1.1. *Thirty subjects take a test of visual-motor coordination for which the value of the population mean is $\mu = 8$, and the value of the population standard deviation is $\sigma = 2$. If the average score of the sample of 30 subjects equals 7.4 (i.e., $\bar{X} = 7.4$), can one conclude that the sample, in fact, came from a population in which the mean is $\mu = 8$?*

III. Null versus Alternative Hypotheses

Null hypothesis $H_0: \mu = 8$

(The mean of the population the sample represents equals 8.)

Alternative hypothesis $H_1: \mu \neq 8$

(The mean of the population the sample represents does not equal 8. This is a **nondirectional alternative hypothesis**, and it is evaluated with a **two-tailed test**. In order to be supported, the absolute value of z must be equal to or greater than the tabled critical two-tailed z value at the prespecified level of significance. Thus, either a significant positive z value or a significant negative z value will provide support for this alternative hypothesis.)

or

$$H_1: \mu > 8$$

(The mean of the population the sample represents is greater than 8. This is a **directional alternative hypothesis**, and it is evaluated with a **one-tailed test**. It will only be supported if the sign of z is positive, and the absolute value of z is equal to or greater than the tabled critical one-tailed z value at the prespecified level of significance.)

or

$$H_1: \mu < 8$$

(The mean of the population the sample represents is less than 8. This is a **directional alternative hypothesis**, and it is evaluated with a **one-tailed test**. It will only be supported if the sign of z is negative, and the absolute value of z is equal to or greater than the tabled critical one-tailed z value at the prespecified level of significance.)

   Note: Only one of the above noted alternative hypotheses is employed. If the alternative hypothesis the researcher selects is supported, the null hypothesis is rejected.

IV. Test Computations

Assume that the following values represent the scores of the sample of $n = 30$ subjects who take the test of visual-motor coordination in Example 1.1: 9, 10, 6, 4, 8, 11, 10, 5, 5, 6, 13, 12, 4, 4, 3, 9, 12, 5, 6, 6, 8, 9, 8, 5, 7, 9, 10, 9, 5, 4.

   Since X_i can be employed to represent the score of the ith subject, by adding all thirty scores we obtain: $\Sigma X_i = \Sigma X = 222$.

   Equation 1.1 is used to compute the mean of the sample.

$$\bar{X} = \frac{\Sigma X}{n} \qquad \qquad \textbf{(Equation 1.1)}$$

   Employing Equation 1.1, we confirm that the mean of the sample is $\bar{X} = 7.4$, the value stated in Example 1.1.

$$\bar{X} = \frac{222}{30} = 7.4$$

   Before the test statistic can be computed, it is necessary to compute a value that is referred to as the **standard error of the population mean**. This value, which is represented by the notation $\sigma_{\bar{X}}$, is computed with Equation 1.2. A full explanation of what $\sigma_{\bar{X}}$ represents can be found in Section VII.

$$\sigma_{\bar{X}} = \frac{\sigma}{\sqrt{n}}$$ **(Equation 1.2)**

Substituting the values $\sigma = 2$ and $n = 30$ in Equation 1.2, the value of $\sigma_{\bar{X}} = .36$ is computed.

$$\sigma_{\bar{X}} = \frac{2}{\sqrt{30}} = .36$$

It should be noted that $\sigma_{\bar{X}}$ can never be a negative value. If a negative value is obtained for $\sigma_{\bar{X}}$, it indicates a computational error has been made.

Equation 1.3 is employed to compute the value of z, which is the test statistic for the **single-sample z test**. Note that in Equation 1.3, the value that represents μ is the value $\mu = 8$ which is stated in the null hypothesis.

$$z = \frac{\bar{X} - \mu}{\sigma_{\bar{X}}}$$ **(Equation 1.3)**

Employing Equation 1.3, the value $z = -1.67$ is computed for Example 1.1.

$$z = \frac{7.4 - 8}{.36} = -1.67$$

Note that Equation 1.3 will always yield a positive z value when the sample mean is greater than the hypothesized value of μ. The value of z will always be negative when the sample mean is less than the hypothesized value of μ. When the sample mean is equal to the hypothesized value of μ, z will equal zero.

V. Interpretation of the Test Results

The obtained value $z = -1.67$ is evaluated with **Table A1 (Table of the Normal Distribution)** in the **Appendix**. Table 1.1 summarizes the tabled critical two-tailed and one-tailed .05 and .01 z values listed in **Table A1**.

Table 1.1   Tabled Critical Two-Tailed and One-Tailed .05 and .01 z Values

	$z_{.05}$	$z_{.01}$
Two-tailed values	1.96	2.58
One-tailed values	1.65	2.33

The following guidelines are employed in evaluating the null hypothesis for the **single-sample z test**.

a) If the alternative hypothesis employed is nondirectional, the null hypothesis can be rejected if the obtained absolute value of z is equal to or greater than the tabled critical two-tailed value at the prespecified level of significance.

b) If the alternative hypothesis employed is directional and predicts a population mean larger than the value stated in the null hypothesis, the null hypothesis can be rejected if the sign of z is positive, and the value of z is equal to or greater than the tabled critical one-tailed value at the prespecified level of significance.

c) If the alternative hypothesis employed is directional and predicts a population mean smaller than the value stated in the null hypothesis, the null hypothesis can be rejected if the sign of z is negative, and the absolute value of z is equal to or greater than the tabled critical one-tailed value at the prespecified level of significance.

Employing the above guidelines, we can only reject the null hypothesis if the directional alternative hypothesis H_1: $\mu < 8$ is employed, and the null hypothesis can only be rejected at the .05 level. This is the case, since the obtained value of z is a negative number, and the absolute value of z is greater than the tabled critical one-tailed .05 value $z_{.05} = 1.65$.[1] The alternative hypothesis H_1: $\mu < 8$ is not supported at the .01 level, since the absolute value $z = 1.67$ is not greater than the tabled critical one-tailed .01 value $z_{.01} = 2.33$.

The nondirectional alternative hypothesis H_1: $\mu \neq 8$ is not supported, since the obtained absolute value $z = 1.67$ is less than the tabled critical two-tailed .05 value $z_{.05} = 1.96$.

The directional alternative hypothesis H_1: $\mu > 8$ is not supported, since the obtained value $z = -1.67$ is a negative number. In order for the alternative hypothesis H_1: $\mu > 8$ to be supported, the computed value of z must be a positive number (as well as the fact that it must be equal to or greater than the tabled critical one-tailed value at the prespecified level of significance).

A summary of the analysis of Example 1.1 with the **single-sample z test** follows: With respect to the test of visual-motor coordination, we can conclude there is a high likelihood the sample of 30 subjects comes from a population with a mean value other than 8 only if we employ the directional alternative hypothesis H_1: $\mu < 8$, and prespecify as our level of significance $\alpha = .05$. This result can be summarized as follows: $z = -1.67$, $p < .05$.

A more in-depth discussion of the interpretation of the z value computed with the **single-sample z test** is contained in Section VII.

VI. Additional Analytical Procedures for the Single-Sample z Test and/or Related Tests

Procedures are available for computing **power** and **confidence intervals** for the **single-sample z test**. These procedures are discussed in Section VI of the **single-sample t test** (which employs the same protocol for such computations as does the **single-sample z test**).

VII. Additional Discussion of the Single-Sample z Test

1. **The interpretation of a negative z value** The actual range of scores on the abscissa (i.e., the X-axis) of the standard normal distribution is $-\infty \leq z \leq +\infty$. The guidelines outlined in Section V for interpreting negative z values are intended to provide the reader with the simplest and least confusing protocol for interpreting such values. In terms of the actual distribution of z values, it should be noted that although the tabled critical z values listed in Table 1.1 are positive numbers, they are also applicable to interpreting negative z values. Since the critical values recorded in Table 1.1 represent absolute values, the corresponding negative z values are listed in Table 1.2.

Table 1.2   Tabled Critical Two-Tailed and One-Tailed .05 and .01 Negative z Values

	$z_{.05}$	$z_{.01}$
Two-tailed values	−1.96	−2.58
One-tailed values	−1.65	−2.33

Within the framework of the values noted in Table 1.2, if one employs the directional (one-tailed) alternative hypothesis H_1: $\mu < 8$, in order to reject the null hypothesis, the obtained value of z must be a negative number that is **equal to or less than the prespecified tabled critical value**. Thus, to be significant at the .05 level, the obtained z value would have to be equal to or less than $z = -1.65$. The reader should take note of the fact that any negative number which has an absolute value greater than 1.65 is less than −1.65. In the same respect, in order for the alternative hypothesis H_1: $\mu < 8$ to be supported at the .01 level, the obtained z value would have to be equal to or less than $z = -2.33$, since any negative number which has an absolute value greater than 2.33 is less than −2.33. The important thing for the reader to understand is that when one is dealing with a negative number, the larger the absolute value of the negative number, the lower the value of that number.

2. The standard error of the population mean and graphical representation of the results of the single-sample z test   The intent of this section is to provide further clarification of what the z value computed with the **single-sample z test** represents. In order to do this, it is necessary to understand what is represented by the **standard error of the population mean** ($\sigma_{\bar{X}}$), which is the denominator of Equation 1.3. The standard error of the population mean represents a standard deviation of a **sampling distribution of means**. As noted in the **Introduction**, a **sampling distribution of means** is a frequency distribution of sample means derived from the same population, in which the same sample size is employed for each sample. Although such a sampling distribution is theoretical and is based on an infinite number of samples, it is possible to construct an empirical sampling distribution which is based on a smaller number of sample means. In order to construct such a sampling distribution of means, a random sample consisting of n subjects is drawn from a population of N subjects. Upon doing this, the mean of the sample of n subjects is computed. Once again, employing the whole population of N subjects, a second random sample consisting of n subjects is selected, and the mean of that sample is computed. This process is repeated over and over again. At whatever point one decides to terminate the process, a large number of sample means will have been computed, each of which is based on a sample size of n subjects. The frequency distribution of these sample means (which will be distributed normally) is known as a **sampling distribution of means**. The mean of a sampling distribution (represented by the notation $\mu_{\bar{X}}$) that is based on an infinite number of sample means will be the same value as the population mean (μ). As the number of sample means used to construct a sampling distribution increases, the greater the likelihood the computed value of the mean of the sampling distribution equals the value of μ. The standard deviation of a sampling distribution (i.e., the standard deviation of all of the sample means), is the **standard error of the population mean** ($\sigma_{\bar{X}}$), which in many sources is referred to as the **standard error of the mean**.[2]

The z value computed with Equation 1.3 represents the number of standard deviation units (based on the value of $\sigma_{\bar{X}}$) that the sample mean deviates from the hypothesized population mean. Thus in Example 1.1, the value $\sigma_{\bar{X}} = .36$ represents the standard deviation of a sampling distribution of means in which in each sample $n = 30$. The obtained value $z = -1.67$ indicates that $\bar{X} = 7.4$ (the sample mean) is 1.67 sampling distribution standard deviation units below the hypothesized population mean $\mu = 8$ (which as noted earlier has the same value as $\mu_{\bar{X}}$). The difference is statistically significant, since a sample mean of 7.4 obtained with 30 subjects is a relatively unlikely occurrence in a sampling distribution which has a mean of 8 and a standard deviation of .36. If we make the assumption that the distribution of means in the sampling distribution is normal, use of the **single-sample z test** will lead to the conclusion that if, in fact, the true value of the population mean is 8, the likelihood of obtaining a sample mean equal to or less than 7.4 is less than .05 (to be exact, it equals .0475).

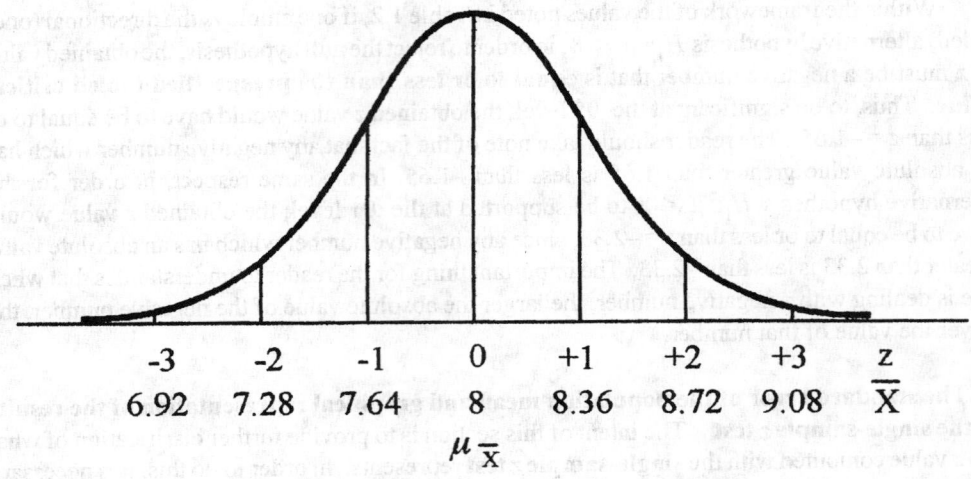

| -3 | -2 | -1 | 0 | +1 | +2 | +3 | z |
| 6.92 | 7.28 | 7.64 | 8 | 8.36 | 8.72 | 9.08 | $\overline{X}$ |

$$\mu_{\overline{X}}$$

Figure 1.1   Sampling Distribution for Example 1.1

Figure 1.1 provides a visual description of the sampling distribution of means for Example 1.1. In Figure 1.1, the numbers 6.92, 7.28, ... , 9.08 along the abscissa identify the values a sample mean will assume if it is 1, 2, and 3 standard deviation (sd) units below and above the mean of the sampling distribution. Since the value of one standard deviation unit for the sampling distribution under discussion is equal to $\sigma_{\overline{X}} = .36$, the value 8.36, which is one standard deviation unit above the mean of the distribution, is obtained simply by adding the value $\sigma_{\overline{X}} = .36$ to $\mu_{\overline{X}} = 8$. The value 8.72, which is two standard deviations above the mean, is obtained by adding two times the value of $\sigma_{\overline{X}}$ to $\mu_{\overline{X}} = 8$, and so on.

Since almost 100% of the cases in a normal distribution fall within three standard deviation units above or below the mean, if a researcher has a sample of 30 subjects which is derived from a population in which $\mu = 8$ and $\sigma = 2$, he can be almost 100% sure that the interval 6.92 and 9.08 (which are the values that correspond to the −3 sd and +3 sd points in Figure 1.1) contains the mean of the distribution.[3] A mean value outside of this range is highly unlikely to occur, and if a more extreme mean value does occur, it is reasonable to conclude that the sample is derived from a population which had a mean value other than 8.

Earlier in the discussion of a sampling distribution it was noted it is assumed that the sample means are normally distributed. The basis for this statement is a general principle in mathematics known as the **central limit theorem**. The central limit theorem states that in a population with a mean value of μ and a standard deviation of σ, the following will be true with respect to the sampling distribution of means: a) The sampling distribution will have a mean value equal to μ and a standard deviation equal to $\sigma_{\overline{X}} = \sigma/\sqrt{n}$; and b) The sampling distribution approaches being normal as the size (n) of each of the samples employed in generating the sampling distribution increases, and as the total number of means used to generate the sampling distribution increases. Although the underlying population each of the samples is derived from does not in itself have to be normal, the more it approximates normality the lower the value of n required for the sampling distribution to be normal. In addition, the more the underlying population each of the samples is derived from approaches normality, the fewer sample means will be required before the sampling distribution becomes normal.

Based on what has been said with respect to the standard error of the population mean, one can determine the value a sample mean will have to be equal to or more extreme than in order

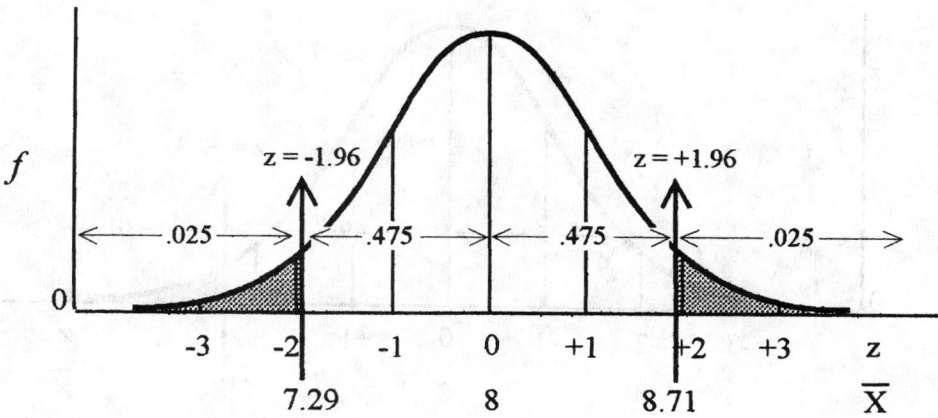

Figure 1.2a   Distribution of Critical Two-Tailed .05 z Value for H_1: $\mu \neq 8$

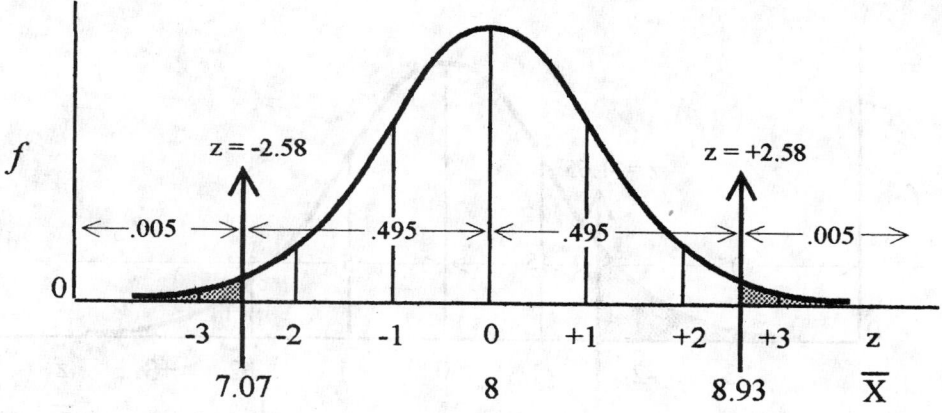

Figure 1.2b   Distribution of Critical Two-Tailed .01 z Value for H_1: $\mu \neq 8$

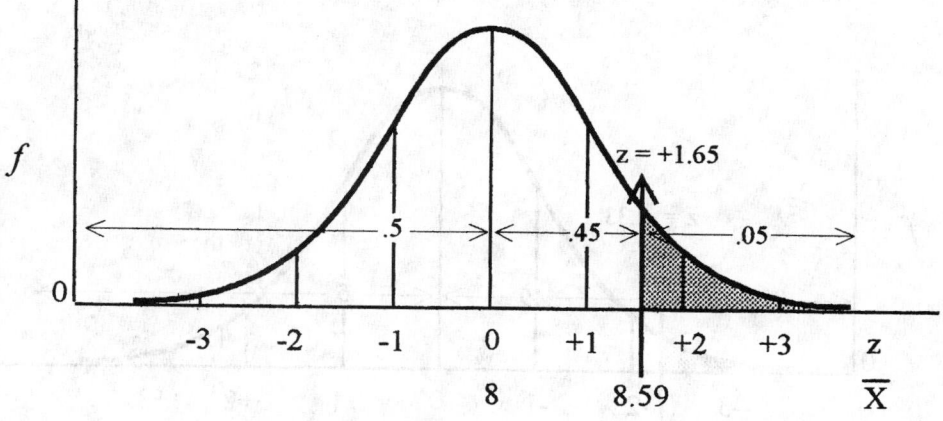

Figure 1.2c   Distribution of Critical One-Tailed .05 z Value for H_1: $\mu > 8$

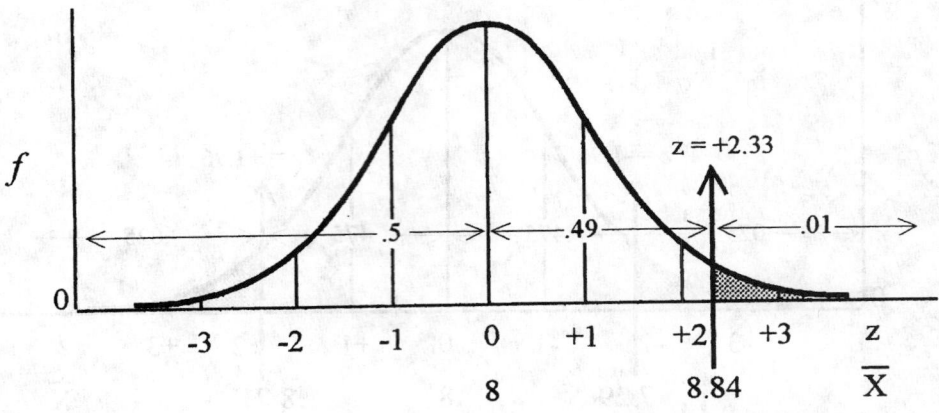

Figure 1.2d   Distribution of Critical One-Tailed .01 z Value for H_1: $\mu > 8$

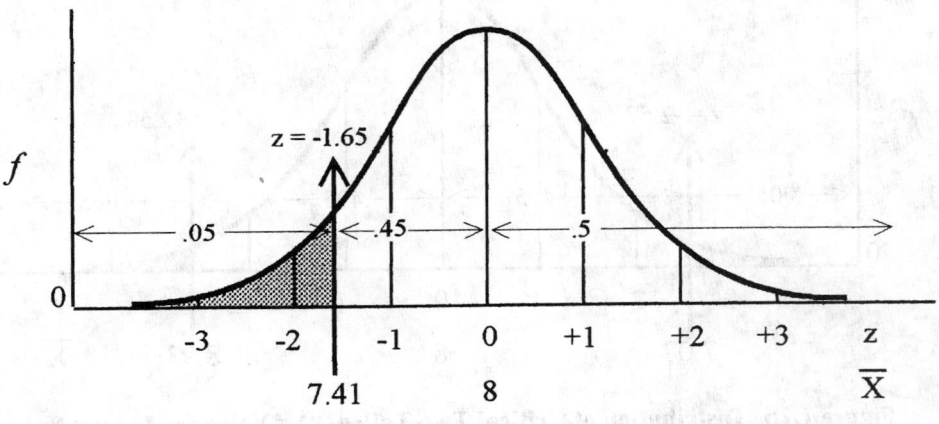

Figure 1.2e   Distribution of Critical One-Tailed .01 z Value for H_1: $\mu < 8$

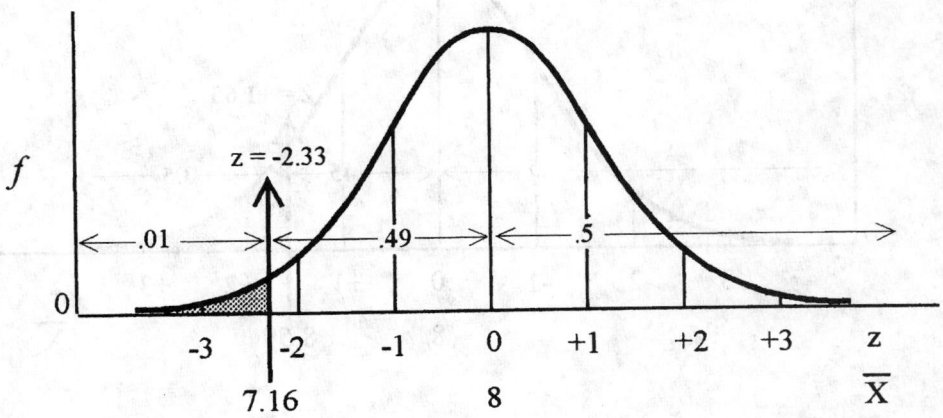

Figure 1.2f   Distribution of Critical One-Tailed .01 z Value for H_1: $\mu < 8$

to reject a null hypothesis at a prespecified level of significance. Figure 1.2 depicts these values for Example 1.1 in reference to the tabled critical one- and two-tailed .05 and .01 z values. Note that in each of the graphs, the value recorded directly below the tabled critical z value for the relevant level of significance is the value the sample mean will have to be equal to or more extreme than in order to reject the null hypothesis H_0: $\mu = 8$.

The values that the sample mean must be equal to or more extreme than in Figure 1.2 are computed with Equation 1.4, which is the result of algebraically transposing the terms in Equation 1.3 in order to solve for the value of $\bar{X}$.

$$\bar{X} = \mu + z\,\sigma_{\bar{X}} \qquad\qquad \textbf{(Equation 1.4)}$$

The value employed to represent z in Equation 1.4 is the relevant tabled critical z value at the prespecified level of significance. By multiplying the latter value by $\sigma_{\bar{X}}$ and adding and subtracting the product from the value of the population mean, one is able to compute the upper limit the sample mean must be equal to or greater than and the lower limit the sample mean must be equal to or less than in order for a result to be significant. This is illustrated below for the case depicted in Figure 1.2a, which describes the upper and lower limits for the sample mean when the nondirectional alternative hypothesis H_1: $\mu \neq 8$ is employed, with $\alpha = .05$.

$$\bar{X} = 8 \pm (1.96)(.36) = 8 \pm .71$$

Since $8 + .71 = 8.71$ and $8 - .71 = 7.29$, in order to be significant at the .05 level, a sample mean will have to be equal to or greater than 8.71 or equal to or less than 7.29. This result can be summarized as follows: $7.29 \geq \bar{X} \geq 8.71$.

A summary of the results depicted in Figure 1.2 follows:

Figure 1.2a: If the nondirectional alternative hypothesis H_1: $\mu \neq 8$ is employed, with $\alpha = .05$, in order to reject the null hypothesis, the obtained value of the sample mean will have to be equal to or greater than 8.71 or equal to or less than 7.29.

Figure 1.2b: If the nondirectional alternative hypothesis H_1: $\mu \neq 8$ is employed, with $\alpha = .01$, in order to reject the null hypothesis, the obtained value of the sample mean will have to be equal to or greater than 8.93 or equal to or less than 7.07.

Figure 1.2c: If the directional alternative hypothesis H_1: $\mu > 8$ is employed, with $\alpha = .05$, in order to reject the null hypothesis, the obtained value of the sample mean will have to be equal to or greater than 8.59.

Figure 1.2d: If the directional alternative hypothesis H_1: $\mu > 8$ is employed, with $\alpha = .01$, in order to reject the null hypothesis, the obtained value of the sample mean will have to be equal to or greater than 8.84.

Figure 1.2e: If the directional alternative hypothesis H_1: $\mu < 8$ is employed, with $\alpha = .05$, in order to reject the null hypothesis, the obtained value of the sample mean will have to be equal to or less than 7.41.

Figure 1.2f: If the directional alternative hypothesis H_1: $\mu < 8$ is employed, with $\alpha = .01$, in order to reject the null hypothesis, the obtained value of the sample mean will have to be equal to or less than 7.16.

Note that with respect to a specific alternative hypothesis in the above examples, the lower the value of alpha, the larger the value computed for an upper limit and the lower the value computed for a lower limit. Additionally, if the value of alpha is fixed, the computed value for an upper limit will be higher and the computed value for a lower limit will be lower when a nondirectional alternative hypothesis is employed, as opposed to when a directional alternative hypothesis is used.

3. Additional examples illustrating the interpretation of a computed z value   To further clarify the interpretation of z values, Table 1.3 lists three additional z values which could have been obtained for Example 1.1 if a different set of data had been employed. Table 1.3 notes the decisions that would be made with reference to the three possible alternative hypotheses a researcher could employ on the basis of each of these z values. The table assumes that H_0: $\mu = 8$.

Table 1.3   Decision Table for z Values

Obtained z value	Alternative hypothesis	Decision
1.75	H_1: $\mu \neq 8$	The null hypothesis cannot be rejected, since the obtained value $z = 1.75$ is less than the tabled critical two-tailed .05 and .01 values $z_{.05} = 1.96$ and $z_{.01} = 2.58$.
	H_1: $\mu > 8$	The null hypothesis can be rejected at the .05 level of significance, since the obtained value $z = 1.75$ is a positive number which is greater than the tabled critical one-tailed .05 value $z_{.05} = 1.65$. The null hypothesis cannot be rejected at the .01 level, since it is less than the tabled critical one-tailed .01 value $z_{.01} = 2.33$.
	H_1: $\mu < 8$	The null hypothesis cannot be rejected, since the obtained value $z = 1.75$ is a positive number.
−2.75	H_1: $\mu \neq 8$	The null hypothesis can be rejected at both the .05 and .01 levels of significance, since the obtained absolute value $z = 2.75$ is greater than the tabled critical two-tailed .05 and .01 values $z_{.05} = 1.96$ and $z_{.01} = 2.58$.
	H_1: $\mu > 8$	The null hypothesis cannot be rejected, since the obtained value $z = -2.75$ is a negative number.
	H_1: $\mu < 8$	The null hypothesis can be rejected at both the .05 and .01 levels of significance, since the obtained value $z = -2.75$ is a negative number and the absolute value $z = 2.75$ is greater than the tabled critical one-tailed .05 and .01 values $z_{.05} = 1.65$ and $z_{.01} = 2.33$.
.75	H_1: $\mu \neq 8$	The null hypothesis cannot be rejected, since the obtained value $z = .75$ is less than the tabled critical two-tailed .05 and .01 values $z_{.05} = 1.96$ and $z_{.01} = 2.58$.
	H_1: $\mu > 8$	The null hypothesis cannot be rejected, since the obtained value $z = .75$ is less than the tabled critical one-tailed .05 and .01 values $z_{.05} = 1.65$ and $z_{.01} = 2.33$.
	H_1: $\mu < 8$	The null hypothesis cannot be rejected, since the obtained value $z = .75$ is a positive number.

4. The z test for a population proportion   Another test which employs the normal distribution to analyze data derived from a single sample is the **z test for a population proportion (Test 9a)**. Equation 9.6, the equation for computing the test statistic for the **z test for a population proportion**, is a special case of Equation 1.3. The use of Equation 9.6 is reserved for evaluating a set of scores for a binomially distributed variable (for which the values of μ and σ can be determined). The **z test for a population proportion** is discussed after a full discussion of the binomial distribution (which can be found under the **binomial sign test for a single sample (Test 9)**).

VIII. Additional Examples Illustrating the Use of the Single-Sample *z* Test

Five additional examples which can be evaluated with the **single-sample *z* test** are presented in this section. Since Examples 1.2–1.4 employ the same population parameters and data set used in Example 1.1, they yield the identical result. Note that Examples 1.2 and 1.3 employ objects in lieu of subjects. Examples 1.5 and 1.6 illustrate the application of the **single-sample *z* test** when the size of the sample is $n = 1$. As noted in the **Introduction**, Example I.9 also illustrates the use of the **single-sample *z* test** when $n = 1$.

Example 1.2. *The Brite battery company manufactures batteries which are programmed by a computer to have an average life span of 8 months and a standard deviation of 2 months. If the average life of a random sample of 30 Brite batteries purchased from 30 different stores is 7.4 months, are the data consistent with the mean value parameter programmed into the computer?*

Example 1.3. *The Smooth Road cement company stores large quantities of its cement in 30 storage tanks. A state law says that the machine which fills the tanks must be calibrated so as not to deviate substantially from a mean load of 8 tons. It is known that the standard deviation of the loads delivered by the machine is 2 tons. An inspector visits the storage facility and determines that the mean number of tons in the 30 storage tanks is 7.4 tons. Does this conform to the requirements of the state law?*

Example 1.4. *A study involving 30 subjects is conducted in order to determine the subjects' ability to accurately judge weight. In the study subjects are required (by adding or subtracting sand) to adjust the weight of a cylinder, referred to as the variable stimulus, until it is judged equal in weight to a standard comparison cylinder whose weight is fixed. The weight of the standard comparison stimulus is 8 ounces. Prior research has indicated that the standard deviation of subjects' judgements in such a task is 2 ounces. Prior to testing the subjects, the experimenter decides she will conclude that a kinesthetic illusion occurs if the mean of the subjects' judgements differs significantly from the value of the standard stimulus. If the average weight assigned by the 30 subjects to the variable stimulus is 7.4 ounces, can the experimenter conclude that a kinesthetic illusion has occurred?*

Example 1.5. *A meteorologist determines that during the current year there were 80 major storms recorded in the Western Hemisphere. He claims that 80 storms represent a significantly greater number than the annual average. Based on data that have been accumulated over the past 100 years, it is known that on the average there have been 70 major storms in the Western Hemisphere, and the standard deviation is 2. (We will assume the distribution for the number of storms per year is normal.) Do 80 storms represent a significant deviation from the mean value of 70?*

Example 1.5 illustrates a problem which would be evaluated with the **single-sample *z* test** in which the value of *n* is equal to one. In the example, the sample size of one represents the single year during which there were 80 storms. When the value of $n = 1$, Equation 1.3 becomes Equation 1.5 (which is the same as Equation I.38 in the **Introduction**).

$$z = \frac{X - \mu}{\sigma} \qquad \textbf{(Equation 1.5)}$$

Equation 1.5, the equation for converting a raw score into a standard deviation (z) score, allows one to determine the likelihood of a specific score occurring within a normally distributed population. Thus, within the context of Example 1.5, Equation 1.5 will allow the meteorologist to determine the likelihood that a score of 80 will occur in a normally distributed population in which $\mu = 70$ and $\sigma = 2$. The analysis assumes that within the total population there are $N = 100$ scores (where each score represents the number of storms in a given year during the 100-year period). Since the frequency of storms during one year is being compared to the population mean, the value of $n = 1$. Note that when $n = 1$, the value of the sample mean ($\bar{X}$) in Equation 1.3 reduces to the value X in Equation 1.5. Additionally, since $n = 1$, the value of $\sigma_{\bar{X}}$ in Equation 1.3 becomes σ in Equation 1.5, since $\sigma_{\bar{X}} = \sigma/\sqrt{n} = \sigma/\sqrt{1} = \sigma$.

Employing Equations 1.3/1.5 with the data for Example 1.5, the value $z = 5$ is computed.

$$z = \frac{80 - 70}{2} = 5$$

The null hypothesis employed for the above analysis is H_0: $\mu = 70$. Example 1.5 implies that either the directional alternative hypothesis H_1: $\mu > 70$ or the nondirectional alternative hypothesis H_1: $\mu \neq 70$ can be employed. Regardless of which of these alternative hypotheses is employed, since the computed value $z = 5$ is greater than all of the tabled critical values in Table 1.1, the null hypothesis can be rejected at both the .05 and .01 levels. Thus, the meteorologist can conclude that a significantly greater number of storms were recorded during the current year than the mean value recorded for the past 100 years. The directional alternative hypothesis H_1: $\mu < 70$ is not supported, since in order to support the latter alternative hypothesis, the computed value of z must be a negative number (which will only be the case if the number of storms observed during the year is less than the population mean of $\mu = 70$).

Example 1.6. *A physician assesses the level of a metabolite in a 40-year-old male patient's blood. Assume that the average level of the metabolite in adult males is 70 milligrams (per 100 milliliters), with a standard deviation of 2 milligrams (per 100 milliliters). If the patient has a blood reading of 80, will his metabolite level be viewed as abnormal?*

As is the case in Example 1.5, Example 1.6 also employs a sample size of $n = 1$. Since this example uses the same data as Example 1.5, the computed value $z = 5$ is obtained, thus allowing the physician to reject the null hypothesis. The value $z = 5$ indicates that the patient has a blood reading that is five standard deviation units above the population mean. Since the proportion of cases in the normal distribution associated with a z score of 5 or greater is so small, there is no value listed in **Table A1** for $z = 5$.

Reference

Freund, J. E. (1984). **Modern elementary statistics** (6th ed.). Englewood Cliffs, NJ: Prentice Hall, Inc.

Endnotes

1. The exact probability value recorded for $z = 1.67$ in Column 3 of **Table A1** is .0475 (which is equivalent to 4.75%). This indicates that the proportion of cases which falls above the value $z = 1.67$ is .0475, and the proportion of cases which falls below the value $z = -1.67$ is .0475. Since this indicates that in the left tail of the distribution there is less than a 5% chance

of obtaining a *z* value equal to or less than $z = -1.67$, we can reject the null hypothesis at the .05 level if we employ the nondirectional alternative hypothesis H_1: $\mu < 8$, with $\alpha = .05$.

2. Equation 1.2 is employed to compute the standard error of the population mean when the size of the underlying population is infinite. In practice, it is employed when the size of the underlying population is large and the size of the sample is believed to constitute less than 5% of the population. However, among others, Freund (1984) notes that in a finite population, if the size of a sample constitutes more than 5% of the population, a correction factor is introduced into Equation 1.2. The computation of the standard error of the mean with the **finite population correction factor** is noted below:

$$\sigma_{\bar{X}} = \frac{\sigma}{\sqrt{n}} \sqrt{\frac{N - n}{N - 1}}$$

Where: *N* represents the total number of subjects/objects that comprise the population.

The finite population corrected equation will result in a smaller value for $\sigma_{\bar{X}}$. This is the case, since as the proportion of a population represented by a sample increases, the less variability there will be among the means which comprise the sampling distribution, and thus the smaller the expected difference between the sample mean obtained for a set of data and the value of the population mean. Thus when $n = N$, employing the finite corrected equation, the value of $\sigma_{\bar{X}}$ will always equal zero. This is the case since when $n = N$, the sample mean and population mean will always be the same value, and thus no error is involved in estimating the value of $\mu_{\bar{X}}$. Since it is usually assumed that the size of a sample is less than 5% of the population it represents, Equation 1.2 is the only equation listed in most sources for computing the value of $\sigma_{\bar{X}}$.

3. Inspection of **Table A1** reveals the exact percentage of cases in a normal distribution which falls within three standard deviations above or below the mean is 99.74%.

Test 2

The Single-Sample *t* Test
(Parametric Test Employed with Interval/Ratio Data)

I. Hypothesis Evaluated with Test and Relevant Background Information

Hypothesis evaluated with test Does a sample of *n* subjects (or objects) come from a population in which the mean (μ) equals a specified value?

Relevant background information on test The **single-sample *t* test** is one of a number of inferential statistical tests that are based on the *t* distribution. Like the normal distribution, the *t* distribution is a bell-shaped, continuous, symmetrical distribution, which to the statistically unsophisticated eye is almost indistinguishable from the normal distribution. *t*, which is the computed test statistic for the **single-sample *t* test**, represents a standard deviation score, and is interpreted in the same manner as the *z* value computed for the **single-sample *z* test**. The only difference between a *z* value and a *t* value is that for a given standard deviation score, the proportion of cases that falls between the mean and the standard deviation score will be a function of which of the two distributions is employed. Except when $n = \infty$, for a given standard deviation score, a larger proportion of cases falls between the mean of the normal distribution and the standard deviation score than the proportion of cases that falls between the mean and that same standard deviation score in the *t* distribution. In point of fact, there are actually an infinite number of *t* distributions — each distribution being based on the number of subjects/objects in the sample. As the size of the sample increases, the proportions (and consequently the critical values) in a *t* distribution approach the proportions (and critical values) in the normal distribution, and, in fact, when $n = \infty$, the normal and *t* distributions are identical. A more detailed discussion (as well as visual illustration) of the *t* distribution can be found in Section VII.

The **single-sample *t* test** is employed in a hypothesis testing situation involving a single sample in order to determine whether or not a sample with a mean of $\overline{X}$ is derived from a population with a mean of μ. If the result of the **single-sample *t* test** yields a significant difference, the researcher can conclude there is a high likelihood the sample is derived from a population with a mean value other than μ.

The **single-sample *t* test** is used with interval/ratio data. The test is employed when a researcher does not know the value of the population standard deviation (σ), and therefore must estimate it by computing the sample standard deviation ($\hat{s}$). As is noted in the discussion of the **single-sample *z* test (Test 1)**, some sources argue that even if one knows the value of σ, when the sample size is very small (generally less than 25), the **single-sample *t* test** provides a more accurate estimate of the underlying sampling distribution for the data.

The following two assumptions, which are noted for the **single-sample *z* test**, also apply to the **single-sample *t* test**: a) The sample has been randomly selected from the population it represents; and b) The distribution of data in the underlying population the sample represents is normal. If either of the aforementioned assumptions is saliently violated, the reliability of the *t* test statistic may be compromised.

II. Example

Example 2.1 *A physician states that the average number of times he sees each of his patients during the year is five. In order to evaluate the validity of this statement, he randomly selects ten of his patients and determines the number of office visits each of them made during the past year. He obtains the following values for the ten patients in his sample: 9, 10, 8, 4, 8, 3, 0, 10, 15, 9. Do the data support his contention that the average number of times he sees a patient is five?*

III. Null versus Alternative Hypotheses

Null hypothesis H_0: $\mu = 5$

(The mean of the population the sample represents equals 5.)

Alternative hypothesis H_1: $\mu \neq 5$

(The mean of the population the sample represents does not equal 5. This is a **nondirectional alternative hypothesis**, and it is evaluated with a **two-tailed test**. In order to be supported, the absolute value of t must be equal to or greater than the tabled critical two-tailed t value at the prespecified level of significance. Thus, either a significant positive t value or a significant negative t value will provide support for this alternative hypothesis.)

or

$$H_1: \mu > 5$$

(The mean of the population the sample represents is greater than 5. This is a **directional alternative hypothesis**, and it is evaluated with a **one-tailed test**. It will only be supported if the sign of t is positive, and the absolute value of t is equal to or greater than the tabled critical one-tailed t value at the prespecified level of significance.)

or

$$H_1: \mu < 5$$

(The mean of the population the sample represents is less than 5. This is a **directional alternative hypothesis**, and it is evaluated with a **one-tailed test**. It will only be supported if the sign of t is negative, and the absolute value of t is equal to or greater than the tabled critical one-tailed t value at the prespecified level of significance.)

   Note: Only one of the above noted alternative hypotheses is employed. If the alternative hypothesis the researcher selects is supported, the null hypothesis is rejected.

IV. Test Computations

Table 2.1 summarizes the number of visits recorded for the $n = 10$ subjects in Example 2.1. In order to compute the test statistic for the **single-sample t test,** it is necessary to determine the mean of the sample and obtain an unbiased estimate of the population standard deviation.
   Employing Equation 1.1 (which is the same as Equation I.1 in the **Introduction**), the mean of the sample is computed to be $\bar{X} = 7.6$.

$$\bar{X} = \frac{\Sigma X}{n} = \frac{76}{10} = 7.6$$

Table 2.1 Summary of Data for Example 2.1

Patient	X	X^2
1	9	81
2	10	100
3	8	64
4	4	16
5	8	64
6	3	9
7	0	0
8	10	100
9	15	225
10	9	81
	$\Sigma X = 76$	$\Sigma X^2 = 740$

Equation 2.1 (which is the same as Equation I.11 in the **Introduction**) is employed to compute $\tilde{s}$, which represents an unbiased estimate of the value of the population standard deviation.

$$\tilde{s} = \sqrt{\frac{\Sigma X^2 - \frac{(\Sigma X)^2}{n}}{n-1}} \qquad \textbf{(Equation 2.1)}$$

Using Equation 2.1, the value $\tilde{s} = 4.25$ is computed.

$$\tilde{s} = \sqrt{\frac{740 - \frac{(76)^2}{10}}{10-1}} = 4.25$$

As is the case with the **single-sample z test**, computation of the test statistic for the **single-sample t test** requires that the value of the **standard error of the population mean** be computed. $s_{\bar{X}}$, which represents an estimate of $\sigma_{\bar{X}}$, is computed with Equation 2.2. Note that $s_{\bar{X}}$, which is referred to as the **estimated standard error of the population mean** (commonly referred to as the **standard error**), is based on the value of $\tilde{s}$ computed with Equation 2.1. A discussion of the theoretical meaning of the standard error of the population mean can be found in Section VII of the **single-sample z test**. Further discussion of $s_{\bar{X}}$ (which represents a standard deviation of a sampling distribution of means and is interpreted in the same manner as $\sigma_{\bar{X}}$) can be found in Section V.[1]

$$s_{\bar{X}} = \frac{\tilde{s}}{\sqrt{n}} \qquad \textbf{(Equation 2.2)}$$

Employing Equation 2.2, the value $s_{\bar{X}} = 1.34$ is computed.

$$s_{\bar{X}} = \frac{4.25}{\sqrt{10}} = 1.34$$

It should be noted that neither $\tilde{s}$ or $s_{\bar{X}}$ can ever be a negative value. If a negative value is obtained for either $\tilde{s}$ or $s_{\bar{X}}$, it indicates a computational error has been made.

Equation 2.3 is the test statistic for the **single-sample t test**.[2]

$$t = \frac{\bar{X} - \mu}{s_{\bar{X}}} \qquad \text{(Equation 2.3)}$$

Inspection of Equation 2.3 reveals that it is similar in structure to Equation 1.3, the equation for the **single-sample z test**. The only differences between the two equations are that: a) Equation 2.3 employs the t distribution as opposed to the z distribution; and b) The value of the standard error of the population mean is estimated in Equation 2.3 from the value of $\tilde{s}$.[3]

Employing Equation 2.3, the value $t = 1.94$ is computed. Note that in Equation 2.3, the value that represents μ is the value $\mu = 5$ which is stated in the null hypothesis.

$$t = \frac{7.6 - 5}{1.34} = 1.94$$

V. Interpretation of the Test Results

Since, like a z value, a t value represents a standard deviation score, except for the fact that a different distribution is employed, it is interpreted in the same manner. The obtained value $t = 1.94$ is evaluated with **Table A2 (Table of Student's t Distribution)** in the **Appendix**.[4] In **Table A2** the critical t values are listed in relation to the proportion of cases (which are recorded at the top of each column in the row labeled p) which falls below a specified t score in the t distribution, and the number of **degrees of freedom** for the sampling distribution that is being evaluated (which are recorded in the left hand column of each row). Equation 2.4 is employed to compute the degrees of freedom for the **single-sample t test**. A full explanation of the meaning of the degrees of freedom can be found in Section VII.

$$df = n - 1 \qquad \text{(Equation 2.4)}$$

Employing Equation 2.4, we compute that $df = 10 - 1 = 9$. Thus, the tabled critical t values which are employed in evaluating the results of Example 2.1 are the values recorded in the cells of **Table A2** that fall in the row for $df = 9$ and the columns with probabilities/proportions which correspond to the one- and two-tailed .05 and .01 values. These critical t values are summarized in Table 2.2.

Table 2.2 Tabled Critical Two-Tailed and One-Tailed .05 and .01 t Values

	$t_{.05}$	$t_{.01}$
Two-tailed values	2.26	3.25
One-tailed values	1.83	2.82

Note that the tabled critical two-tailed value $t_{.05} = 2.26$ is the value in the row $df = 9$ and the column $p = .975$, since $t_{.05} = 2.26$ is the standard deviation score above which (as well as below which in the case of $t = -2.26$) a proportion equivalent to .025 of the cases in the distribution falls. The tabled critical two-tailed value $t_{.01} = 3.25$ is the value in the row $df = 9$ and the column $p = .995$, since $t_{.01} = 3.25$ is the standard deviation score above which (as well as below which in the case of $t = -3.25$) a proportion equivalent to .005 of the cases in the distribution falls. The tabled critical one-tailed value $t_{.05} = 1.83$ is the value in the row $df = 9$ and the column $p = .95$, since $t_{.05} = 1.83$ is the standard deviation score above which (as well as below which in the case of $t = -1.83$) a proportion equivalent to .05 of the cases in the distribution falls. The tabled critical one-tailed value $t_{.01} = 2.82$ is the value in the row $df = 9$ and the column $p = .99$, since $t_{.01} = 2.82$ is the standard deviation score above which (as well as below which in the case of $t = -2.82$) a proportion equivalent to .01 of the cases in the distribution falls.

The following guidelines are employed in evaluating the null hypothesis for the **single-sample _t_ test**.

a) If the alternative hypothesis employed is nondirectional, the null hypothesis can be rejected if the obtained absolute value of t is equal to or greater than the tabled critical two-tailed value at the prespecified level of significance.

b) If the alternative hypothesis employed is directional and predicts a population mean larger than the value stated in the null hypothesis, the null hypothesis can be rejected if the sign of t is positive and the value of t is equal to or greater than the tabled critical one-tailed value at the prespecified level of significance.

c) If the alternative hypothesis employed is directional and predicts a population mean smaller than the value stated in the null hypothesis, the null hypothesis can be rejected if the sign of t is negative and the absolute value of t is equal to or greater than the tabled critical one-tailed value at the prespecified level of significance.

Employing the above guidelines, we can only reject the null hypothesis if the directional alternative hypothesis $H_1: \mu > 5$ is employed, and the null hypothesis can only be rejected at the .05 level. This is the case since the obtained value of t is a positive number which is greater than the tabled critical one-tailed .05 value $t_{.05} = 1.83$. Note that the alternative hypothesis $H_1: \mu > 5$ is not supported at the .01 level, since the obtained value $t = 1.94$ is less than the tabled critical one-tailed .01 value $t_{.01} = 2.82$.

The nondirectional alternative hypothesis $H_1: \mu \neq 5$ is not supported since the obtained value $t = 1.94$ is less than the tabled critical two-tailed .05 value $t_{.05} = 2.26$.

The directional alternative hypothesis $H_1: \mu < 5$ is not supported since the obtained value $t = 1.94$ is a positive number. In order for the alternative hypothesis $H_1: \mu < 5$ to be supported, the computed value of t must be a negative number (as well as the fact that the absolute value of t must be equal to or greater than the tabled critical one-tailed value at the prespecified level of significance).

In Section IV it is noted that the **estimated standard error of the population mean** ($s_{\bar{X}}$) computed for the **single-sample _t_ test** represents a standard deviation of a sampling distribution of means. The use of the t distribution for Example 2.1 is based on the fact that when the population standard deviation is unknown, the latter distribution provides a better approximation of the underlying sampling distribution than does the normal distribution. Figure 2.1 depicts the sampling distribution employed for Example 2.1. This sampling distribution is interpreted in the same manner as the sampling distribution for the **single-sample _z_ test** which is depicted in Figure 1.1.

Inspection of the sampling distribution depicted in Figure 2.1 reveals that the obtained value $t = 1.94$ falls to the right of the tabled critical one-tailed value $t_{.05} = 1.83$. At this point it should be noted that $t_{.05} = 1.83$ is greater than $z_{.05} = 1.65$, the tabled critical one-tailed .05 value employed with the normal distribution. Both of these values demarcate the upper 5% of their respective distributions. If one elects to employ the **single-sample _z_ test**, as opposed to the **single-sample _t_ test**, for an analysis in which the value of σ is unknown, it will inflate the likelihood of committing a Type I error. This is the case since, except when the sample size is very large (in which case the corresponding values of t and z are identical), a tabled critical z

value at a prespecified level of significance will always be smaller than the corresponding tabled critical *t* value at that same level of significance. Thus, in the case of Example 2.1, if we employ the tabled critical one-tailed value $z_{.05}$ = 1.65, the likelihood of committing a Type I error will be greater than .05. This can be confirmed by inspection of **Table A2** which indicates that for df = 9, the proportion of cases in the *t* distribution that falls at or above the value of t = 1.65 is greater than .05 (i.e., a *t* score of 1.65 falls below the 95th percentile of the distribution).[5]

A summary of the analysis of Example 2.1 with the **single-sample *t* test** follows: With respect to the average number of times the doctor sees a patient, we can conclude there is a high likelihood the sample of 10 subjects comes from a population with a mean value other than 5 only if we employ the directional alternative hypothesis H_1: $\mu > 5$, and prespecify as our level of significance α = .05. This result can be summarized as follows: $t(9) = 1.94, p < .05$. (The degrees of freedom employed in the analysis are noted in parentheses after the *t*.)

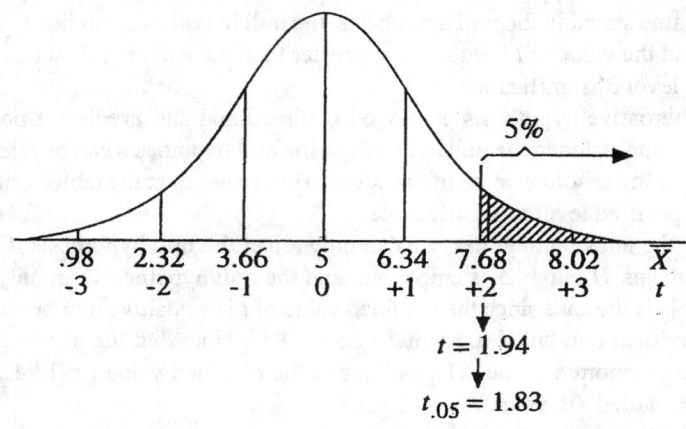

Figure 2.1   Sampling Distribution for Example 2.1

VI. Additional Analytical Procedures for the Single-Sample *t* Test and/or Related Tests

1. Determination of the power of the single-sample *t* test and the single-sample *z* test, and the application of Test 2a: Cohen's *d* index   The power of either the **single-sample *z* test** or the **single-sample *t* test** will represent the probability of the test identifying a difference between the value for the population mean stipulated in the null hypothesis and a specific value which represents the true value of the mean of the population represented by the experimental sample. In order to compute the power of a test, it is necessary for the researcher to stipulate the latter value which will be identified with the notation μ_1. In practice, a researcher can compute the power of a test for any value of μ_1.

The power of the test will be a function of the difference between the value of μ stated in the null hypothesis and the value of μ_1. The test's power will increase as the absolute value of the difference between the values μ and μ_1 increases. This is the case, since if the sample is derived from a population with a mean value which is substantially above or below the value of μ stated in the null hypothesis, it is likely that this would be reflected in the value of the sample mean ($\bar{X}$). Obviously, the more the value of $\bar{X}$ deviates from the hypothesized value of μ, the greater the absolute value of the numerators (i.e., $\bar{X} - \mu$) in Equations 1.3 and 2.3 (which are, respectively, the equations for the **single-sample z test** and the **single-sample t test**). Assuming that the value of the denominator is held constant, the larger the absolute value of the numerator the larger the absolute value of the computed test statistic (i.e., z or t). The larger the latter value, the more likely it is the researcher will be able to reject the null hypothesis (assuming the obtained difference is in the direction predicted in the alternative hypothesis), and consequently the more powerful the test.

Since the obtained value of the test statistic is also a function of the denominator of Equations 1.3 and 2.3 (i.e., the actual or estimated standard error of the population mean), the latter value also influences the power of the test. Specifically, as the value of the denominator decreases, the computed absolute value of the test statistic increases. It happens to be the case that the value of the standard error of the mean is a function of the population standard deviation (which is estimated in the case of the t test) and the sample size. Inspection of Equations 1.3 and 2.3 reveals that the standard error of the mean will decrease if the value of the standard deviation is decreased and the sample size is increased. Thus, by employing an accurate estimate of the population standard deviation (more specifically, in the case of the t test, one that is not spuriously inflated due to sampling error) and a large sample size, one can minimize the value of the denominator in Equations 1.3 and 2.3, and consequently maximize the absolute value of the test statistic. As a result, one can increase the likelihood that the null hypothesis will be rejected, which increases the power of the test.

The power of a statistical test can be represented both mathematically and graphically. Figures 2.2 and 2.3 illustrate the concept of power and its relationship to the Type I and Type II error rates. Both figures contain two distributions which represent the sampling distributions of means for two populations.[6] In each figure, the sampling distribution on the left represents a population with the mean μ (i.e., the value stated in the null hypothesis). The sampling distribution on the right represents a population with the mean μ_1, which we will assume is the true value of the mean of the population from which the experimental sample is derived. Figures 2.2 and 2.3 both assume a fixed value for the sample size upon which the sampling distributions are based, and that each of the underlying populations represented by the sampling distributions has the same standard deviation. Figure 2.2 represents a case in which there is a large difference between the values of μ_1 and μ, whereas Figure 2.3 represents a case in which there is a small difference between the two values. When expressed in standard deviation units, the magnitude of the absolute value of the difference between μ_1 and μ is referred to as the **effect size**. Thus, in Figure 2.2 a **large effect size** is present, whereas Figure 2.3 depicts a **small effect size**.

The reader should note the following with respect to Figures 2.2 and 2.3.

a) The closer the values μ and μ_1 are to one another, the more the sampling distributions of the two populations overlap.

b) In the case of a one-tailed analysis, the value of alpha (α) is represented by area (///) in the distribution on the left. Recollect that α represents the likelihood of committing a Type I error (i.e., rejecting a true null hypothesis). Numerically, α represents the proportion of the left distribution which comprises area (///). In the case of a two-tailed analysis, the proportion of the left distribution represented by area (///) will be equal to $\alpha/2$.

c) The value of beta (β) is represented by area ($\equiv$) in the distribution on the right. β represents the likelihood of committing a Type II error (i.e., retaining a false null hypothesis). Numerically, β represents the proportion of the right distribution which comprises area ($\equiv$).

d) The power of the test is represented by area (\\\) in the right distribution. Note that this is the area in the right distribution which falls to the right of the area delineating β. Numerically, the power of the test represents the proportion of the right distribution that comprises area (\\\). The power of the test can also be represented by subtracting the value of β from 1 (i.e., **Power**

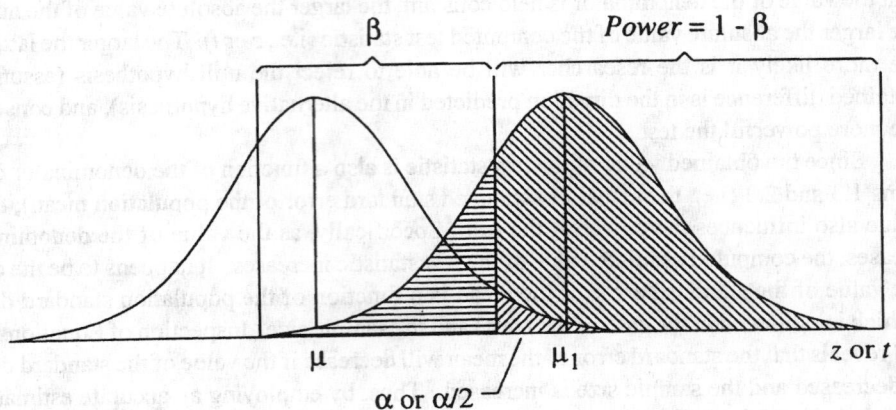

Figure 2.2   Sampling Distributions Employed in Determining the Power of a Test Involving a Large Effect Size

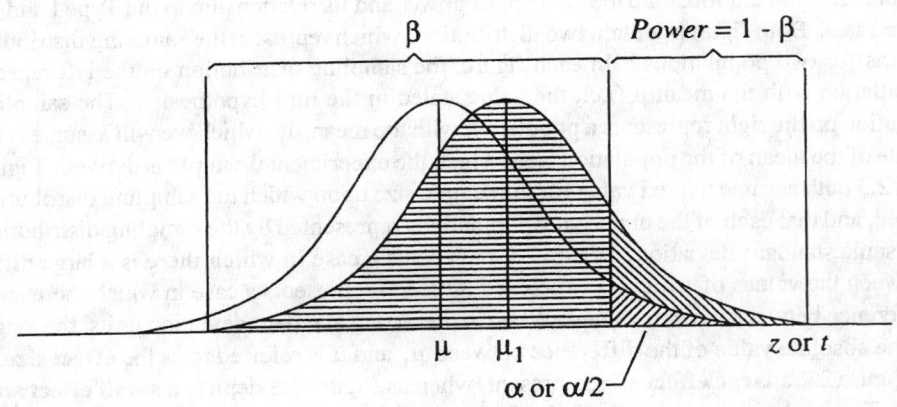

Figure 2.3   Sampling Distributions Employed in Determining the Power of a Test Involving a Small Effect Size

$= 1 - \beta$). Note that the area in the left distribution which represents α overlaps the area in the right distribution representing the power of the test.

e) In order to increase the value of α, one must move the boundary in the left distribution which delineates α to the left. By doing the latter, one will decrease the value of β, since the area in the right distribution that corresponds to β will decrease. By increasing the value of α, one also increases the area in the right distribution which represents the power of the test. This illustrates the fact that if one increases the likelihood of committing a Type I error (α), one decreases the likelihood of committing a Type II error (β), and at the same time increases the power of the test $(1 - \beta)$. In the same respect, to decrease the value of α one must move the boundary in the left distribution which delineates alpha to the right. By doing the latter, one will increase the value of β, since the area in the right distribution that corresponds to β will increase. By decreasing the value of α, one also decreases the area in the right distribution which represents the power of the test. This illustrates the fact that if one decreases the likelihood of committing a Type I error, the likelihood of committing a Type II error increases, and at the same time the power of the test decreases.

f) It should also be noted that at a given level of significance, a one-tailed test will be more powerful than a two-tailed test. This is the case since with a one-tailed test the point which delineates α will be farther to the left in the left distribution than will be the case with a two-tailed test. As an example, when $\alpha = .05$, the tabled critical one-tailed value for z is $z_{.05} = 1.65$, whereas the tabled critical two-tailed value is $z_{.05} = 1.96$. Since both of these critical values are in the left distribution, the former value will be farther to the left, thus expanding the area in the right distribution which represents the power of the test.

Two methods will now be demonstrated that can be used to determine the power of either the **single-sample *t* test** or the **single-sample *z* test**. The first method (which is more time consuming) reveals all of the logical operations involved in computing the power of a test. The second method, which employs **Table A3 (Power Curves for Student's *t* Distribution)** in the **Appendix**, requires fewer computations. It should be emphasized that whenever possible, a power analysis should be conducted prior to the data collection phase of a study. By computing power beforehand, one is able to design a study with a sample size which is large enough to detect the specific effect size predicted by the researcher.

Method 1 for computing the power of the single-sample *t* test and the single-sample *z* test
The first method will initially be demonstrated with reference to the **single-sample *z* test**. The reason for employing the latter test is that it will allow us to use **Table A1 (Table of the Normal Distribution)** in the **Appendix**, which lists probabilities for all z values between 0 and 4. Detailed tables of the t distribution that list probabilities for all t values within this range are generally not available.

Let us assume that in our example we are employing the same null hypothesis which is employed in Example 2.1 (i.e., H_0: $\mu = 5$). It will be assumed that the researcher wishes to evaluate the power of the **single-sample *z* test** in reference to the alternative hypothesis H_1: $\mu_1 = 6$. Note that in conducting a power analysis, the alternative hypothesis states that the population mean is a specific value which is different from the value stated in H_0. In conducting the power analysis, it will be assumed that the null hypothesis will be evaluated with a two-tailed test, with $\alpha = .05$. For purposes of illustration, it will also be assumed the researcher evaluates the null hypothesis employing a sample size of $n = 121$. In addition, we will assume that the value of the population standard deviation is known to be $\sigma = 4.25$.

Employing Equation 1.2, the value of the standard error of the population mean is computed. Thus: $\sigma_{\bar{X}} = 4.25/\sqrt{121} = .39$.

Figure 2.4, which depicts the analysis graphically, is comprised of two overlapping normal distributions. Each distribution is a sampling distribution of population means. The distribution on the left, which is the sampling distribution of means of a population with a mean of 5, will be referred to as Distribution A. The distribution on the right, which is the sampling distribution of means of a population with a mean of 6, will be referred to as Distribution B. We have already determined above that $\sigma_{\bar{X}} = .39$, and we will assume that this value represents the standard deviation of each of the sampling distributions. The area (///) delineates the proportion of Distribution A that corresponds to the value $\alpha/2$, which equals .025. This is the case, since $\alpha = .05$ and a two-tailed test is being employed. In such an instance, the proportion of the curve comprising the critical area in each of the tails of Distribution A will be $.05/2 = .025$. Area ($\equiv$) delineates the proportion of Distribution B which corresponds to the probability of committing a Type II error (β). Area (\\\\) delineates the proportion of Distribution B that represents the power of the test.

The procedure for computing the proportions in Figure 2.4 will now be described. The first step in computing the power of the test requires one to determine how far above the value $\mu = 5$ the sample mean will have to be in order to reject the null hypothesis. Equation 1.3 is employed to determine this minimum required difference. By algebraically transposing the terms in Equation 1.3 we can determine that $\bar{X} - \mu = (z_{.05})(\sigma_{\bar{X}})$. Thus, by substituting the values $z_{.05} = 1.96$ (which is the tabled critical two-tailed .05 z value) and $\sigma_{\bar{X}} = .39$ in the latter equation we can compute that the minimum required difference is $\bar{X} - \mu = (1.96)(.39) = .76$.

Thus, any sample mean .76 units above or below the value $\mu = 5$ will allow the researcher to reject the null hypothesis at the .05 level (if a two-tailed analysis is employed). With respect to evaluating the power of the test in reference to the alternative hypothesis H_1: $\mu_1 = 6$, the researcher is only concerned with a mean value above 5 (which will fall in the right tail of Distribution A).[7] Thus, a mean value of $\bar{X} = 5.76$ or greater will allow the researcher to reject the null hypothesis (since $\mu + (z_{.05})(\sigma_{\bar{X}}) = 5 + .76 = 5.76$).

The next step in the analysis requires that the area in Distribution B that falls between the mean $\mu_1 = 6$ and the value 5.76 be computed. This is accomplished by employing Equation 1.3 and substituting 5.76 to represent the value of $\bar{X}$ and $\mu_1 = 6$ to represent the value of μ.

$$z = \frac{\bar{X} - \mu}{\sigma_{\bar{X}}} = \frac{5.76 - 6}{.39} = -.62$$

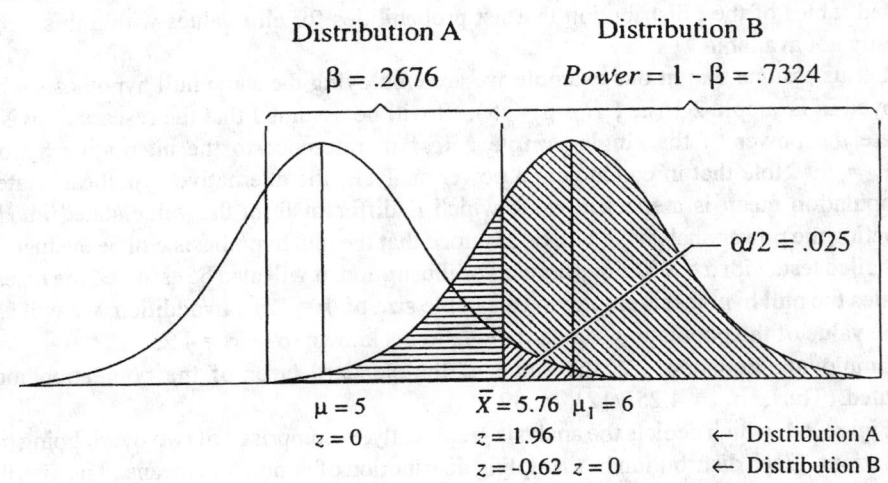

Distribution A

$\beta = .2676$

Distribution B

$Power = 1 - \beta = .7324$

$\alpha/2 = .025$

$\mu = 5$	$\bar{X} = 5.76$ $\mu_1 = 6$	
$z = 0$	$z = 1.96$	$\leftarrow$ Distribution A
	$z = -0.62$ $z = 0$	$\leftarrow$ Distribution B

Figure 2.4 Visual Representation of Power when H_0: $\mu = 5$
and H_1: $\mu = 6$ for $n = 121$

Utilizing **Table A1**, we determine that the proportion of Distribution B which lies between $\mu_1 = 6$ and 5.76 (i.e., between the mean and a z score of $-.62$ or $+.62$) is .2324. Since the value 5.76 is below the mean of Distribution B, if .5 (which is the proportion of Distribution B that falls above the mean $\mu_1 = 6$) is added to .2324, the resulting value of .7324 will represent the power of the test. This latter value is represented by area (\\\\) in Figure 2.4. The likelihood of committing a Type II error (i.e., β) is represented by area ($\equiv$). The proportion of Distribution B that constitutes this latter area is determined by subtracting the value .7324 from 1. Thus: $\beta = 1 - .7324 = .2676$. Based on the results of the power analysis we can state that if the alternative hypothesis H_1: $\mu_1 = 6$ is true, the likelihood that the null hypothesis will be rejected is .7324, and, at the same time, there is a .2676 likelihood that it will be retained. If the researcher considered the computed value for the power too low (which we are assuming is determined prior to implementing the study), she can increase the power of the test by employing a larger sample size.

If the value of σ is not known and has to be estimated from the sample data, the power analysis will be based on the t distribution instead of the normal distribution. In such a case the identical protocol described above for computing power is employed, except for the fact that a tabled critical t value is used in place of the tabled critical z value. Unless the sample size is extremely large, the tabled critical t value will be larger than the tabled critical z value used for the same data. As a result of this, the power of the test computed for the t distribution will be lower than the value computed for the normal distribution.

In the case of the example under discussion, if the t distribution is employed one would use the tabled critical two-tailed .05 t value for $df = n - 1 = 121 - 1 = 120$. From **Table A2** it can be determined that this value is $t_{.05} = 1.98$. Using the latter value and the value $s_{\bar{X}} = .39$, it can be determined that .77 is the minimum required difference in order to achieve significance. When .77 is added to the value $\mu = 5$, it indicates that a sample mean of 5.77 or greater (as well as 4.23 or lower) will allow the researcher to reject the null hypothesis. The t value required to complete the power calculations is determined by utilizing Equation 2.3 and substituting 5.77 to represent the value of $\bar{X}$ and $\mu_1 = 6$ to represent the value of μ. The calculations are noted below.

$$(t_{.05})(s_{\bar{X}}) = \bar{X} - \mu$$

$$(1.98)(.39) = .77$$

$$t = \frac{5.77 - 6}{.39} = -.59$$

Detailed tables of the t distribution indicate that for $df = 120$ the proportion of cases between the mean and a t score of $-.59$ or $+.59$ is approximately .22.[8] The power of the test is derived by adding .5 to the latter value. Thus, Power $= .22 + .5 = .72$, which is slightly lower than the value .7324 obtained for the normal distribution.

It was previously noted that the size of the sample employed in a study is directly related to the power of a statistical test. Thus, in the example under discussion, if instead of using a sample size of $n = 121$, we employ a sample size of $n = 10$, the power of both the **single-sample z test** and the **single-sample t test** will be considerably less than the computed values .7324 and .72. In point of fact, when $n = 10$ the power of the **single-sample z test** equals .1112. The dramatic decrease in power for the small sample size can be understood by determining the minimum amount by which $\bar{X}$ and μ must differ from one another in order to reject the null hypothesis. Since we are still employing the normal distribution, when $n = 10$ the same tabled critical two-tailed value $z_{.05} = 1.96$ is used. However, the value of $\sigma_{\bar{X}}$ is increased substantially, since $\sigma_{\bar{X}} = 4.25/\sqrt{10} = 1.34$. Employing the values $z_{.05} = 1.96$ and $\sigma_{\bar{X}} = 1.34$, we can compute that the minimum required difference in order to reject H_0 when $n = 10$ is 2.63. Specifically: $\bar{X} - \mu = (1.96)(1.34) = 2.63$.

A sample mean that is 2.63 units above $\mu = 5$ is equal to 7.63. The latter value will fall farther to the right in the right tail of Distribution B than the value 5.76 which is computed when $n = 121$. Substituting the values $\bar{X} = 7.63$ and $\mu = 6$ in Equation 1.3, we determine that $\bar{X} = 7.63$ is 1.22 standard deviation units above the mean of Distribution B.

$$z = \frac{7.63 - 6}{1.34} = 1.22$$

If one examines Figure 2.5, which depicts the analysis graphically, it can be seen that in Distribution B the value $z = 1.22$ lies to the right of the mean of the distribution. Thus, when $n = 10$ the power of the **single-sample z test** will be represented by the proportion of Distribution B that comprises area (\\\). Employing the table for the normal distribution, it can be determined that the proportion of the curve to the right of $z = 1.22$ is .1112, which represents the power of the test. On the basis of this, we can determine that the likelihood of committing a Type II error (represented by area ($\equiv$)) will be $\beta = 1 - .1112 = .8888$, which is substantially greater than the value .2676 obtained when $n = 121$. Note that area (\\\) in Distribution B is much smaller than the corresponding area depicted in Figure 2.4 (when $n = 121$). By virtue of area (\\\) being smaller, the proportion of Distribution B in Figure 2.5 representing area ($\equiv$) is substantially larger than the corresponding proportion/area in Figure 2.4.

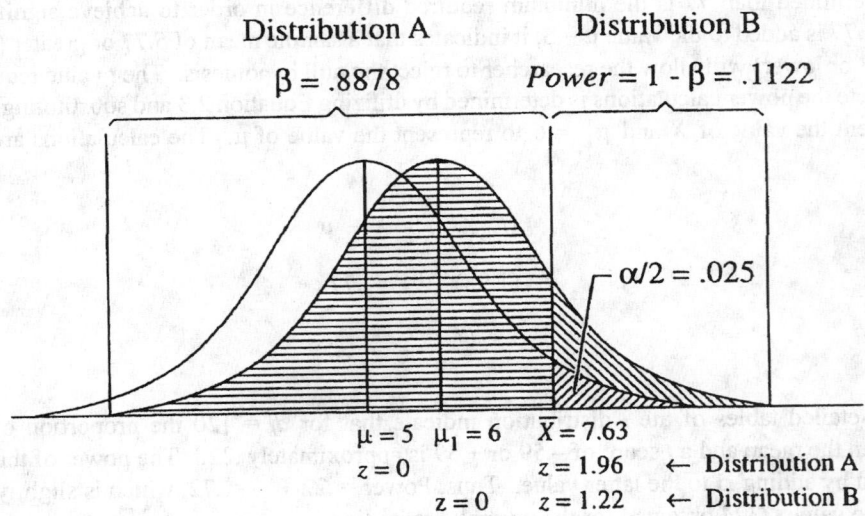

Figure 2.5 Visual Representation of Power when H_0: $\mu = 5$ and H_1: $\mu = 6$ for $n = 10$

The t distribution will now be applied to the above problem. Let us assume that $n = 10$ and the value of σ is unknown. The latter value, however, is estimated from the sample data to be $\tilde{s} = 4.25$. The aforementioned values $n = 10$ and $\tilde{s} = 4.25$ correspond to those employed in Example 2.1. If one wants to compute the power of the **single-sample t test** for Example 2.1

with reference to the alternative hypothesis H_1: $\mu_1 = 6$, the same protocol as described above for the normal distribution is employed, except for the fact that the value $t_{.05} = 2.26$ (which is the tabled critical two-tailed .05 t value for $df = 9$) is used in the analysis. Thus:

$$t_{.05}\, s_{\bar{X}} = \bar{X} - \mu$$

$$(2.26)(1.34) = 3.03$$

$$t = \frac{8.03 - 6}{1.34} = 1.51$$

The use of the value $\bar{X} = 8.03$ in the above t test equation is predicated on the fact that a mean of 8.03 is 3.03 units above the value $\mu = 5$ stated in the null hypothesis. Detailed tables of the t distribution indicate that for $df = 9$, the proportion of cases that falls above a t score of 1.51 is approximately .085 (which corresponds to area (\\\\) in Figure 2.5 if the latter represented the t distribution).[9] The value .085 represents the power of the test. The likelihood of committing a Type II error (which corresponds to area ($\equiv$)) is $\beta = 1 - .085 = .915$.

A comparison of the values obtained for the power of the **single-sample z test** and the **single-sample t test** for the two sample sizes employed in the discussion of power (i.e., for $n = 121$ and $n = 10$), reveals that when the values of both n and the standard deviation are fixed, the **single-sample z test** provides a more powerful test of an alternative hypothesis than does the **single-sample t test** (keeping in mind, however, that the use of the **single-sample z test** is justified only if the value of σ is known).

Test 2a: Cohen's d index (Method 2 for computing the power of the single-sample t test and the single-sample z test) It was noted previously that when the magnitude of the absolute value of the difference between μ_1 and μ is expressed in standard deviation units, the resulting value is referred to as the **effect size**. The computation of effect size, represented by the notation d, can be summarized by Equation 2.5. Throughout the book, the d statistic computed with the Equation 2.5 will be referred to as **Cohen's d index**, since it was first employed as a measure of effect size by Jacob Cohen (1977, 1988).

$$d = \frac{|\mu_1 - \mu|}{\sigma} \qquad \textbf{(Equation 2.5)}$$

In the above equation, in the case of the **single-sample z test** the value of σ will be known, whereas in the case of the **single-sample t test** the latter value will have to be estimated (either from the sample data or from prior research). Cohen (1977; 1988, pp. 24–27) has proposed the following (admittedly arbitrary) d values as criteria for identifying the magnitude of an effect size: a) A **small effect size** is one that is greater than .2 but not more than .5 standard deviation units; b) A **medium effect size** is one that is greater than .5 but not more than .8 standard deviation units; c) A **large effect size** is greater than .8 standard deviation units.

Note that in Equation 2.5 the effect size is based on population parameters and does not take into account the size of the sample. Since the power of a test is also a function of sample size, it is necessary to convert the value of d into a measure that takes into account both the population parameters and the sample size. This measure, represented by the notation φ (which is a symbol for the lower case Greek letter **phi**), is referred to as the **noncentrality parameter**. The value of φ is computed with Equation 2.6.[10]

$$\varphi = d\sqrt{n} \qquad\qquad \textbf{(Equation 2.6)}$$

If the value of φ is computed for a specific sample size, the power of both the **single-sample z test** and the **single-sample t test** can be determined by using **Table A3** in the **Appendix**, which consists of four sets of power curves in which the value of φ is plotted in reference to the power of a test. Each set of power curves is based on a different level of significance. Specifically: **Table A3-A** is employed for either a two-tailed analysis with $\alpha = .01$ or a one-tailed analysis with $\alpha = .005$. Table **A3-B** is employed for either a two-tailed analysis with $\alpha = .02$ or a one-tailed analysis with $\alpha = .01$. Table **A3-C** is employed for either a two-tailed analysis with $\alpha = .05$ or a one-tailed analysis with $\alpha = .025$. Table **A3-D** is employed for either a two-tailed analysis with $\alpha = .10$ or a one-tailed analysis with $\alpha = .05$. Note that each set of power curves is comprised of either eight or ten curves, each of which represents a different degrees of freedom value. When the degrees of freedom computed for an experiment do not equal one of the *df* values represented by the curves, the researcher must interpolate to approximate the power of the test. Regardless of the sample size, the curve for $df = \infty$ should always be used in determining the power of the **single-sample z test**. The latter curve is also used for the **single-sample t test** for large sample sizes.

The protocol for employing the curves in **Table A3** is as follows: a) Compute the value of φ; b) Upon locating φ on the horizontal axis of the appropriate set of curves, draw a line that is perpendicular to the axis which intersects the curve that represents the appropriate *df* value; and c) At the point the line intersects the curve, drop a perpendicular to the vertical axis on which power values are noted. The point at which the latter line intersects the vertical axis will indicate the power of the test.

The noncentrality parameter will now be employed to compute the power of the **single-sample z test** and the **single-sample t test** using the same data employed to demonstrate Method 1. Thus, the power analysis will assume the null hypothesis will be evaluated with a two-tailed test, with $\alpha = .05$. In addition:

$$H_0: \mu = 5 \qquad H_1: \mu_1 = 6 \qquad \sigma = \tilde{s} = 4.25 \qquad n = 121$$

Employing Equation 2.5, the value $d = .235$ is computed.

$$d = \frac{6 - 5}{4.25} = .235$$

Note that using Cohen's (1977, 1988) criteria for effect size, the value $d = .235$ indicates we are attempting to detect a small effect size.

For $n = 121$, Equation 2.6 is employed to calculate the value $\delta = 2.59$.

$$\varphi = (.235)\sqrt{121} = 2.59$$

Employing the power curve for $df = \infty$ in Table **A3-C**, the power of the **single-sample z test** is determined to be approximately .73. Since there is no curve for $df = 120$, the power of the **single-sample t test** is based on a curve that falls between the $df = \infty$ and $df = 24$ power curves. Through interpolation, the power of the **single-sample t test** is determined to be approximately .72. Note that these are the same values that are computed with Method 1.

For the same example, with $n = 10$, $\varphi = (.235)\sqrt{10} = .743$. Employing the power curve for $df = \infty$, the power of the **single-sample z test** is determined to be approximately .11. Since $df = 9$, using the $df = 6$ and $df = 12$ power curves as reference points, we determine that the power of the **single-sample *t* test** is approximately .085. These values are consistent with those computed with Method 1.

It should be emphasized again that, whenever possible, prior to the data collection phase of a study a researcher should stipulate the minimum effect size she is attempting to detect. The smaller the effect size, the larger the sample size which will be required in order to have a test of sufficient power that will allow one to reject a false null hypothesis. As long as a researcher knows or is able to estimate (from the sample data) the population standard deviation, by employing trial and error one can substitute various values of *n* in Equation 2.6 until the computed value of φ corresponds to the desired value for the power of the test. Power tables developed by Cohen (1977, 1988) are commonly employed within the framework of the present discussion as a quick means of determining the minimum sample size necessary to achieve a specific level of power in reference to a specific effect size.

Among others, Rosner (1995, pp. 219–224; 2000, pp. 238–241) and Zar (1999, pp. 105 –108) note that Equation 2.7 provides an alternative way of estimating the sample size (*n*) required in order to conduct the **single-sample *t* test** which has a power of a specified value.

(Equation 2.7)

$$n = \frac{\sigma^2 (z_{\alpha/2} + z_\beta)^2}{\delta^2}$$

Where: δ represents $|\mu_1 - \mu|$, the numerator of Equation 2.5. Although it will not be employed in illustrating the use of Equation 2.7, many sources often employ the value $(\overline{X} - \mu)$ (i.e., the obtained difference between the sample mean and the hypothesized population mean) to represent the value of δ. $(\overline{X} - \mu)$ is commonly employed to represent the value of δ, when after obtaining a nonsignificant result a researcher wants to determine if the power of the test conducted had been set at a specific value, how large a sample size would have been required in order for the **single-sample *t* test** to identify the observed difference between the sample mean and the hypothesized population mean to be significant.

$z_{\alpha/2}$ represents the relevant tabled critical **two-tailed** *z* value in **Table A1**. It is important to note that $z_{\alpha/2}$ is employed in Equation 2.7 when the test of the alternative hypothesis is **nondirectional**. If, on the other hand, a **directional** alternative hypothesis is evaluated, z_α is employed in place of $z_{\alpha/2}$ — i.e., the relevant tabled critical **one-tailed** *z* value in **Table A1** should be employed.

z_β is the *z* value listed in **Table A1** for which a proportion (probability) is recorded that identifies the Type II error rate (i.e., β) employed in the analysis. The protocol to employ to determine the value of z_β is as follows: a) If the **power** stipulated for the test (i.e., the value (1 − β)), is **greater than .5**, z_β is the *z* value for which the proportion recorded in **Column 3** of **Table A1** is equal to the value **stipulated for β**. (It is also the *z* value for which the proportion recorded in **Column 2** of **Table A1** is equal to the value (**Power** − .5).) When the **power** stipulated for the test is **greater than .5**, the sign of z_β will always be **positive**; b) If the **power** stipulated for the test is **less than .5**, z_β is the *z* value for which the proportion recorded in **Column 3** of **Table A1** is equal to the **power** of the test (i.e., the value (1 − β)). (It is also the *z* value for which the proportion recorded in **Column 2** of **Table A1** is equal to the value (β − .5).) When the **power** stipulated for the test is **less than .5**, the sign of z_β

will always be **negative**; c) If the **power** stipulated for the test is .5, $z_\beta = 0$. (In **Table A1** the value $\beta = .5000$ is listed in **Column 3**, and the value .0000 in **Column 2** represents the difference between β and .5 (i.e., $\beta - .5 = 0$, with **Power** $= 1 - \beta = .5$).)

The reader should take note of the following with regard to Equation 2.7: If the value of σ^2 is not known (as will often be the case), $\tilde{s}^2$ should be employed in Equation 2.7 to represent σ^2. Theoretically, when $\tilde{s}^2$ is employed in Equation 2.7 instead of σ^2, the values $t_{\alpha/2}$ and t_β should be employed in the equation in place of $z_{\alpha/2}$ and z_β, and both of the values $t_{\alpha/2}$ and t_β will be larger than their respective analogs $z_{\alpha/2}$ and z_β. However, if one does not have access to exact probabilities for the t distribution, it will be difficult to estimate the appropriate value for t_β. Because of the latter, many sources employ $\tilde{s}^2$ in Equation 2.7 along with the relevant values for $z_{\alpha/2}$ and z_β. As a general rule, the value of n computed under the latter circumstances will underestimate the required sample size (unless the appropriate value for n is greater than 200). The underestimation of the sample size when $\tilde{s}^2$ is employed with $z_{\alpha/2}$ and z_β in Equation 2.7 results from the fact that the normal distribution allows for a more powerful test of an alternative hypothesis than the t distribution. Consequently, if the t distribution is the appropriate distribution to employ, it will require a slightly larger sample size to detect a difference than an analysis based on the normal distribution will indicate.

Equation 2.7 will be employed to demonstrate the computation of an appropriate sample size for the examples illustrating power computations discussed earlier in this section. Specifically, the sample size will be computed for the examples for which the power values .7324 and .085 were computed. It will be assumed that the researcher plans to evaluate the data with a **single-sample t test**, and that a two-tailed analysis will be conducted with alpha set at .05. The value of $\delta = 1$ will be employed in Equation 2.7, since in the previous power calculations in this section the null and alternative hypotheses which were stipulated were H_0: $\mu = 5$ and H_1: $\mu_1 = 6$, with the latter indicating the researcher's intent to determine the necessary power to detect a difference of 1 unit (i.e., $\delta = |\mu_1 - \mu| = 1$).

Initially, we will compute the sample size when the researcher wants the power to equal .7324. In conducting the latter analysis, the following values are employed in Equation 2.7 to compute the number of subjects that will be required: $\tilde{s}^2 = (4.25)^2 = 18.06$, $\delta = 1$, $z_{\alpha/2} = 1.96$ (If a one-tailed analysis is conducted $z_{\alpha/2} = 1.65$ should be employed.), $z_\beta = .62$. The rationale for employing the value $z_\beta = .62$ is as follows: Since **Power** $= 1 - \beta = .7324$, $\beta = .2676$. Employing the guidelines for using Equation 2.7 documented above to compute a sample size when **Power** $> .5$, in **Table A1** $z = .62$ is the z value for which the proportion (probability) recorded in **Column 3** (.2676) is closest to (in fact, equal to) the value $\beta = .2676$. (It is also the case that for $z = .62$, if .5 is added to the value recorded in **Column 2** (i.e., .2324 + .5 = .7324), the resulting value is closest to (in fact, equal to) the value .7324 stipulated for the power of the test (and thus, **Power** $- .5 = .7324 - .5 = .2324$).) Since the power stipulated for the test is greater than .5, the sign of z_β will be positive.

When the appropriate values are substituted in Equation 2.7, the value $n = 120.21$ is computed. Thus, $n = 121$ (which is the next greatest integer value than the integer element of the value computed for n) is the required sample size for the power of the **single-sample t test** to be .7324 in order for it to detect a 1 unit difference. Note that the value $n = 121$ is the value employed earlier in this section when the value .7324 was computed for the power of the test.[11]

$$n = \frac{18.06(1.96 + .62)^2}{(1)^2} = 120.21$$

We will now compute the required sample size when the researcher wants the power of the test to equal .085 (which was the value obtained for power when the *t* distribution was employed). In conducting the latter analysis, the following values are employed in Equation 2.7 to compute the number of subjects that will be required: $\tilde{s}^2 = (4.25)^2 = 18.06, \delta = 1, z_{\alpha/2} = 1.96$ (If a one-tailed analysis is conducted $z_{\alpha/2} = 1.65$ should be employed.), $z_\beta = -1.37$. The rationale for employing the value $z_\beta = -1.37$ is as follows: Since **Power** $= 1 - \beta = .085$, $\beta = .915$. Employing the guidelines for using Equation 2.7 documented earlier to compute a sample size when **Power** $< .5$, in **Table A1** $z = 1.37$ is the z value for which the proportion (probability) recorded in **Column 3** (.0853) is closest to the value **Power** $= .085$. (It is also the case that for $z = 1.37$, if .5 is added to the value recorded in **Column 2** (i.e., $.4147 + .5 = .9147$), the resulting value is closest to the value .915 stipulated for the value of β (and thus, $\beta - .5 = .9147 - .5 = .4147$).) Since the power stipulated for the test is less than .5, the sign of z_β will be negative. As noted above, the value $\delta = 1$ is employed, since in the previous power calculations in this section the null and alternative hypotheses that were stipulated were H_0: $\mu = 5$ and H_1: $\mu_1 = 6$, with the latter indicating the researcher's intent to determine the power to detect a difference of 1 unit.

When the appropriate values are substituted in Equation 2.7, the value $n = 6.29$ is computed, indicating that in order for the power of the **single-sample *t* test** to be .085, the researcher must have at least 7 subjects (which is the next integer value that is greater than the value computed for n). As noted earlier, if the exact values of $t_{\alpha/2}$ and t_β associated with the relevant probabilities were employed in place of $z_{\alpha/2}$ and z_β in Equation 2.7, the value computed for n would be increased (since with small sample sizes the z **test** allows for a more powerful test than the *t* **test**), and thus be close to or equal to the value $n = 10$ which was employed earlier for the example under discussion.

$$n = \frac{18.06(1.96 + -1.37)^2}{(1)^2} = 6.29$$

In closing the discussion of power, it should be noted that if a researcher employs a large enough sample size, a significant difference can be obtained almost 100% of the time. Over the years various researchers who have criticized the **classical hypothesis testing model** have pointed out that the value of a sample mean is rarely if ever equal to the value of μ stated in the null hypothesis — in other words, a null hypothesis is rarely if ever true. Obviously a researcher must discern whether or not a statistically significant difference which reflects a minimal effect size is of any practical or theoretical significance. In instances where it is not, for all practical purposes, if one rejects the null hypothesis under such circumstances, one is committing a Type I error. Criticisms that have been directed toward the **classical hypothesis testing model** (i.e., the model which rejects the null hypothesis of zero difference when a result is statistically significant) are addressed in the **Introduction**, in Section IX (the **Addendum**) of the **binomial sign test for a single sample (Test 9)**, and in the discussion of **meta-analysis and related topics**, which can be found in Section IX (the **Addendum**) of the **Pearson product-moment correlation coefficient (Test 28)**.

2. Computation of a confidence interval for the mean of the population represented by a sample The hypothesis testing procedure described for the **single-sample *t* test** and the **single-sample *z* test** merely allows the researcher to determine whether or not it is reasonable to conclude that the mean of a population is equal to a specific value. **Interval estimation** (which is discussed briefly in the **Introduction**) is another methodology used in inferential statistics that involves computing a range of values which a researcher can state with a high degree of confidence contains the true value of the population parameter. One such commonly computed range of values is referred to as a **confidence interval**. Thus, a confidence interval for the mean is a range of values which a researcher can state with a high degree of confidence contains the true value of the population mean. Smithson (2003, p. 10) notes that "an **unbiased confidence interval** is one whose probability of including any value other than the parameter's true value is less than or equal to $100(1 - \alpha)\%$."

When the value of the population standard deviation is unknown, computation of a confidence interval for a single sample involving interval/ratio data utilizes the t distribution. The following confidence intervals are most commonly computed: a) The **95% confidence interval**: In the case of a 95% confidence interval, a researcher can be 95% confident that the computed interval contains the true value of the population mean. Expressed in probabilistic terms, the researcher can state there is a probability/likelihood of .95 that the confidence interval contains the true value of the population mean. When the result of the **single-sample t test** is statistically significant at the .05 level, the 95% confidence interval will not include the value stipulated in the null hypothesis; b) The **99% confidence interval**: In the case of a 99% confidence interval, a researcher can be 99% confident that the computed interval contains the true value of the population mean. Expressed in probabilistic terms, the researcher can state there is a probability/likelihood of .99 that the confidence interval contains the true value of the population mean. When the result of the **single-sample t test** is statistically significant at the .01 level, the 99% confidence interval will not include the value stipulated in the null hypothesis.

Equation 2.8 is the general equation for computing a confidence interval for a population mean.

$$CI_{(1 - \alpha)} = \bar{X} \pm (t_{\alpha/2})(s_{\bar{X}}) \qquad \textbf{(Equation 2.8)}$$

Where: $t_{\alpha/2}$ represents the tabled critical two-tailed value in the t distribution, for $df = n - 1$, below which a proportion (percentage) equal to $[1 - (\alpha/2)]$ of the cases falls. If the proportion (percentage) of the distribution that falls within the confidence interval is subtracted from 1 (100%), it will equal the value of α.

Equation 2.9 is employed to compute the 95% confidence interval, which will be represented by the notation $CI_{.95}$ (since .95 is equivalent to 95%).

$$CI_{.95} = \bar{X} \pm (t_{.05})(s_{\bar{X}}) \qquad \textbf{(Equation 2.9)}$$

In Equation 2.9, $t_{.05}$ represents the tabled critical two-tailed .05 t value for $df = n - 1$. By employing the latter critical t value, one will be able to identify the range of values within the sampling distribution that defines the middle 95% of the distribution. Only 5% of the scores in the sampling distribution will fall outside that range. Specifically, 2.5% of the scores will fall above the upper limit of the range, and 2.5% of the scores will fall below the lower limit of the range. In the case of the 95% confidence interval, if we obtain an infinite number of samples of size n from the population, 95% of the confidence intervals computed with Equation 2.9 will contain the population mean while only 5% (i.e., $100\% - 95\% = 5\%$) of the confidence intervals will not contain the population mean. Note that in computing a confidence interval it is assumed that the sample mean ($\bar{X}$) is a random variable (since it is free to vary for each sample), while the value of the population mean (μ) is fixed. Through use of the sample mean as a reference point, the 95% confidence interval provides the researcher with a 95% likelihood of including the true value of μ. Unlike the value of the population mean, which as noted earlier is a constant value, the values which define a given confidence interval will vary from sample to sample. The latter is the case since the value of a confidence interval is a function of the value of the sample mean, which itself varies from sample to sample.

Equation 2.10 is employed to compute the 99% confidence interval, which will be represented by the notation $CI_{.99}$ (since .99 is equivalent to 99%).

$$CI_{.99} = \bar{X} \pm (t_{.01})(s_{\bar{X}}) \qquad \textbf{(Equation 2.10)}$$

Note that the only difference between Equation 2.10 and Equation 2.9 is that in Equation 2.10 the critical value $t_{.01}$ is employed. The latter value represents the tabled critical two-tailed .01 value for $df = n - 1$. By using the two-tailed $t_{.01}$ value, one will be able to identify the range of values within the sampling distribution that defines the middle 99% of the distribution. Only 1% of the scores in the sampling distribution will fall outside that range. Specifically, .5% of the scores will fall above the upper limit of the range, and .5% of the scores will fall below the lower limit of the range. In the case of the 99% confidence interval, if we obtain an infinite number of samples of size n from the population, 99% of the confidence intervals computed with Equation 2.10 will contain the population mean while only 1% (i.e., 100% − 99% = 1%) of the confidence intervals will not contain the population mean.

The values $\bar{X} = 7.6$, $s_{\bar{X}} = 1.34$, and $t_{.05} = 2.26$ will now be substituted in Equation 2.9 to compute the 95% confidence interval for the mean of the population employed in Example 2.1.

$$CI_{.95} = 7.6 \pm (2.26)(1.34) = 7.6 \pm 3.03$$

The above result can be summarized as follows: $4.57 \le \mu \le 10.63$. The notation $4.57 \le \mu \le 10.63$ indicates that the value of μ is greater than or equal to 4.57 and less than or equal to 10.63. This result tells us that if a mean of $\bar{X} = 7.6$ is computed for a sample size of $n = 10$, we can be 95% confident (or the probability is .95) that the interval 4.57 to 10.63 contains the true value of the population mean. Thus, with respect to Example 2.1, the physician can be 95% confident that the range of values 4.57 to 10.63 includes the actual value for the average number of visits per patient.[12]

Equation 2.10 will now be employed to compute the 99% confidence interval for the population mean in Example 2.1.

$$CI_{.99} = 7.6 \pm (3.25)(1.34) = 7.6 \pm 4.36$$

The above result can be summarized as follows: $3.24 \le \mu \le 11.96$. This result tells us that if a mean of $\bar{X} = 7.6$ is computed for a sample size of $n = 10$, we can be 99% confident (or the probability is .99) that the interval 3.24 to 11.96 contains the true value of the population mean. Thus, with respect to Example 2.1, the physician can be 99% confident that the range of values 3.24 to 11.96 includes the actual value for the average number of visits per patient.

Note that the range of values which defines the 99% confidence interval is larger than the range of values that defines the 95% confidence interval. This will always be the case, since it is only logical that by stipulating a larger range of values one will be able to have a higher degree of confidence that the interval includes the true value of the population mean. It is also the case that the larger the sample size employed in computing a confidence interval, the smaller the range of values which will define the confidence interval. Figures 2.6 and 2.7 provide a graphical summary of the computation of the 95% and 99% confidence intervals.

Note that in Figure 2.6 the following is true: a) 47.5% of the scores in the sampling distribution fall between the sample mean $\bar{X} = 7.6$ and 4.57, the lower limit of $CI_{.95}$; and b) 47.5% of the scores in the sampling distribution fall between the sample mean $\bar{X} = 7.6$ and

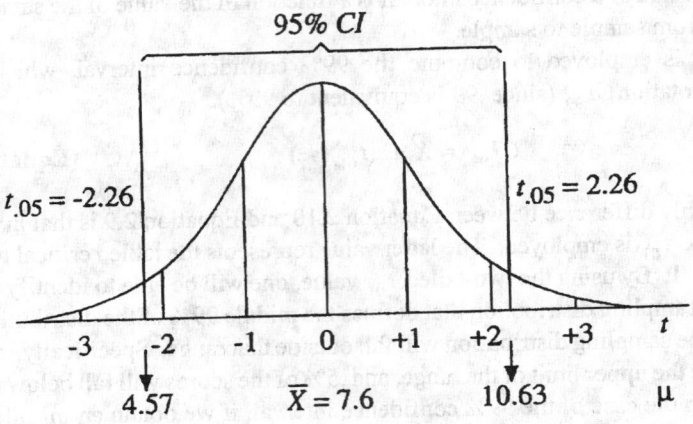

Figure 2.6   Graphical Representation of 95% Confidence Interval for Example 2.1

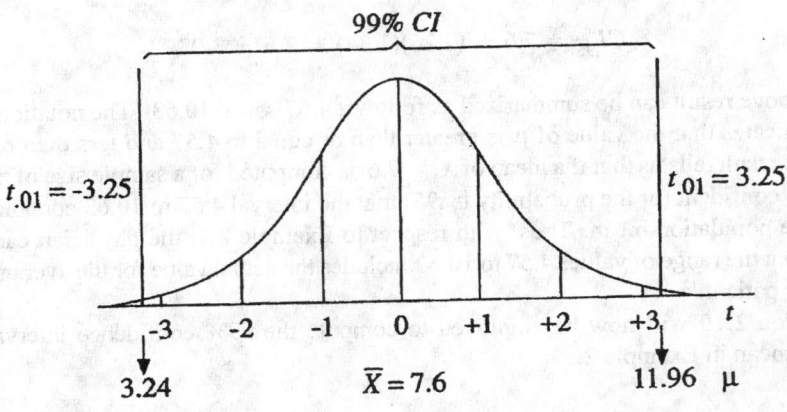

Figure 2.7   Graphical Representation of 99% Confidence Interval for Example 2.1

10.63, the upper limit of CI_{95}. Thus, the area of the curve between the scores 4.57 and 10.63 represents the middle 95% of the sampling distribution. Two and one-half percent of the scores in the distribution fall below 4.57 and 2.5% of the scores fall above 10.63. The scores which are below 4.57 or greater than 10.63 comprise the extreme 5% of the sampling distribution.

In Figure 2.7 the following is true: a) 49.5% of the scores in the sampling distribution fall between the sample mean $\overline{X} = 7.6$ and 3.24, the lower limit of CI_{99}; and b) 49.5% of the scores in the sampling distribution fall between the sample mean $\overline{X} = 7.6$ and 11.96, the upper limit of CI_{99}. Thus, the area of the curve between the scores 3.24 and 11.96 represents the middle 99% of the sampling distribution. One-half of one percent of the scores in the distribution fall below 3.24 and .5% of the scores fall above 11.96. The scores which are below 3.24 or greater than 11.96 comprise the extreme 1% of the sampling distribution.

The reader should take note of the fact that in Figures 2.6 and 2.7 the sample mean is employed to represent the mean of the sampling distribution. This is in contrast to the sampling distribution depicted in Figure 2.1, where the hypothesized value of the population mean is employed to represent the mean of the sampling distribution. The reason for using different means for the two sampling distributions is that the sampling distribution depicted in Figure 2.1 is used to determine the likelihood of a sample mean deviating from the hypothesized value of the population mean, while the sampling distribution depicted in Figures 2.6 and 2.7 reflects the fact that one is engaged in interval estimation, and is thus employing the sample mean to predict the true value of the population mean

Although, as noted previously, CI_{95} and CI_{99} are the most commonly computed confidence intervals, it is possible to calculate a confidence interval at any level of confidence. Thus, if one wanted to compute the 90% confidence interval, the equation $CI_{90} = \overline{X} \pm (t_{.10})(s_{\overline{X}})$ is employed. In the latter equation $t_{.10}$ represents the tabled critical one-tailed .05 t value (which is also the tabled critical two-tailed .10 t value), since the latter value (which for $df = 9$ is $t = 1.83$) establishes the boundaries for the middle 90% of the sampling distribution. Only 10% of the scores in the sampling distribution will fall outside that range (5% of the scores will fall above the upper limit of the range, and 5% of the scores will fall below the lower limit of the range). Thus, for Example 2.1:

$$CI_{90} = 7.6 \pm (1.83)(1.34) = 7.6 \pm 2.45 \quad \text{Thus: } 5.15 \leq \mu \leq 10.05$$

Employing the same logic, the 98% confidence interval can be computed by using the tabled critical one-tailed .01 value $t_{.01} = 2.82$ in the confidence interval equation. The latter t value is employed, since it establishes the boundaries for the middle 98% of the sampling distribution. Thus:

$$CI_{98} = 7.6 \pm (2.82)(1.34) = 7.6 \pm 3.78 \quad \text{Thus: } 3.82 \leq \mu \leq 11.38$$

It should be noted that in order to accurately compute a confidence interval, one must have access to tables of the t distribution which provide the appropriate tabled value for the confidence interval in question. Since most published tables of the t distribution provide only the tabled critical one- and two-tailed .05 and .01 values, they only allow for accurate computation of the following confidence intervals: CI_{90} , CI_{95} , CI_{98} , CI_{99}. **Table A2** is more detailed than most tables of the t distribution, and thus it allows for accurate computation of a greater number of confidence intervals than those noted above. In instances where an exact t value is not tabled, interpolation can be used to estimate that value.

Although the computation of confidence intervals is not described in the discussion of the **single-sample z test**, when the value of the population standard deviation is known, the normal distribution (as opposed to the t distribution) is employed to compute a confidence interval. If a researcher knows the value of the population standard deviation, Equation 2.11 is the general equation for computing a confidence interval.

$$CI_{(1-\alpha)} = \overline{X} \pm (z_{\alpha/2})(\sigma_{\overline{X}}) \qquad \textbf{(Equation 2.11)}$$

Where: $z_{\alpha/2}$ represents the tabled critical two-tailed value in the normal distribution below which a proportion (percentage) equal to $[1 - (\alpha/2)]$ of the cases falls. If the proportion (percentage) of the distribution that falls within the confidence interval is subtracted from 1 (100%), it will equal the value of α.

Note that the basic difference between Equation 2.11 and Equation 2.8 is that Equation 2.11 employs a tabled critical z value instead of the corresponding t value for the same percentile. Additionally, since use of Equation 2.11 assumes that the value of σ is known, the actual value of the standard error of the population mean can be computed. Thus, $\sigma_{\bar{X}}$ is used in place of the estimated value $s_{\bar{X}}$ employed in Equation 2.8.

Generalizing from Equation 2.11, Equation 2.12 is employed to compute the 95% confidence interval for the mean of a population when the normal distribution is used.

$$CI_{.95} = \bar{X} \pm (z_{.05})(\sigma_{\bar{X}}) \qquad \textbf{(Equation 2.12)}$$

If Equation 2.12 is employed with Example 2.1, the only value which will be different from those in Equation 2.9 is the tabled critical two-tailed value $z_{.05} = 1.96$, which is used in place of $t_{.05} = 2.26$. Since we are assuming that $\sigma = \hat{s}$, the values of $s_{\bar{X}}$ and $\sigma_{\bar{X}}$ are equivalent. Thus, $\sigma_{\bar{X}} = \sigma/\sqrt{n} = 4.25/\sqrt{10} = 1.34$.

Equation 2.12 will now be utilized to compute the 95% confidence interval for Example 2.1.

$$CI_{.95} = 7.6 \pm (1.96)(1.34) = 7.6 \pm 2.63$$

The above result can be summarized as: $4.97 \le \mu \le 10.23$. Thus, by using the normal distribution to compute the 95% confidence interval, the physician can be 95% confident (or the probability is .95) that the range of values 4.97 to 10.23 includes the actual value for the average number of visits per patient. Note that when the normal distribution is employed with the data for Example 2.1, the range of values which defines the 95% confidence interval is smaller than the range of the values that is computed with the t distribution. This will always be the case, since, at the same level of confidence, a tabled z value will always be smaller than the corresponding tabled t value.[13] As a result of this, the product resulting from multiplying the tabled value by the standard error of the population mean will be smaller when the normal distribution is employed.

If the value of σ is known, Equation 2.13 (as opposed to Equation 2.10) is used to calculate the 99% confidence interval for the mean of a population.

$$CI_{.99} = \bar{X} \pm (z_{.01})(\sigma_{\bar{X}}) \qquad \textbf{(Equation 2.13)}$$

Note that in contrast to Equation 2.10, Equation 2.13 employs the tabled critical two-tailed value $z_{.01} = 2.58$ instead of the corresponding value $t_{.01} = 3.25$. Equation 2.13 will now be used to compute the 99% confidence interval for Example 2.1.

$$CI_{.99} = 7.6 \pm (2.58)(1.34) = 7.6 \pm 3.46$$

The above result can be summarized as: $4.14 \le \mu \le 11.06$. Thus, by using the normal distribution to compute the 99% confidence interval, the physician can be 99% confident (or the probability is .99) that the range of values 4.14 to 11.06 includes the actual value for the average number of visits per patient. Note once again that the range of values obtained with Equation 2.13 (which utilizes the normal distribution) for the 99% confidence interval is smaller than the range of the values which is obtained with Equation 2.10 (which utilizes the t distribution).

Figures 2.8 and 2.9 provide a graphical summary of the computation of the 95% and 99% confidence intervals with the normal distribution.

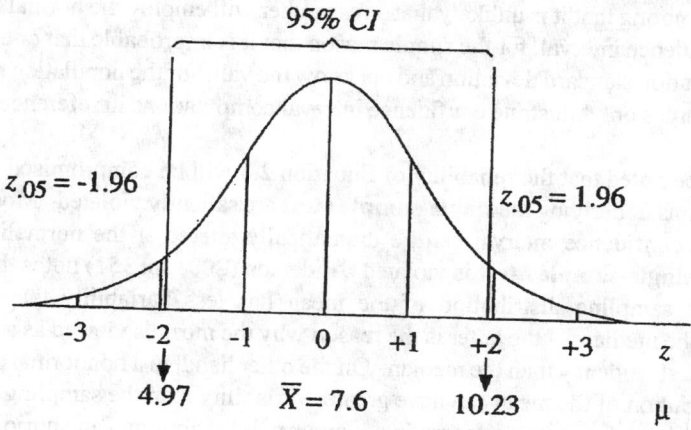

Figure 2.8 Graphical Representation of 95% Confidence Interval for Example 2.1 Through Use of the Normal Distribution

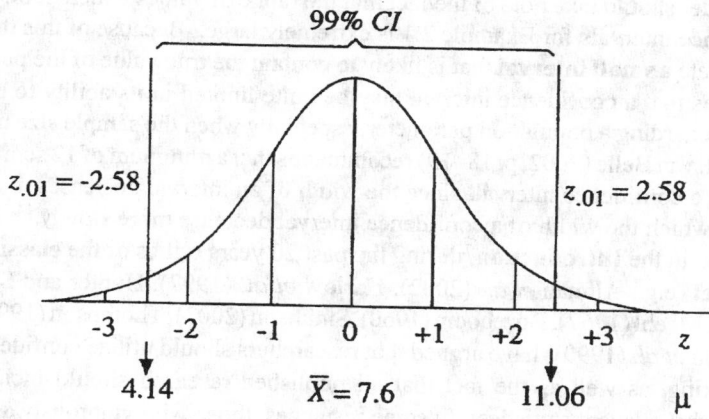

Figure 2.9 Graphical Representation of 99% Confidence Interval for Example 2.1 Through Use of the Normal Distribution

It should be noted that even if the value of σ is known, some researchers would challenge the use of the normal distribution in the above example. The rationale for such a challenge is that, if as a result of employing a **single-sample z test** it is determined that a significant difference exists between the hypothesized value of the population mean and $\overline{X}$, one can question the logic of employing the normal distribution in the computation of a confidence interval. This is the case, since, if we conclude that our sample is derived from a population with a different mean value than the hypothesized population mean, it is also possible that the population standard deviation is different than the value of σ we have employed in the analysis. In such an instance, the best strategy would probably be to employ the sample data to estimate the population standard deviation. Thus, Equation 2.1 would be used to compute $\hat{s}$, and consequently Equations 2.8 and 2.9 (employing the *t* distribution) would be used to compute the 95% and 99% confidence intervals.

It is worth noting that it is unlikely that a researcher will employ the normal distribution to compute a confidence interval, for the simple reason that it is improbable that one will know the value of a population standard deviation and not know the value of the population mean. For this reason most sources only illustrate confidence interval computations in reference to the t distribution.

It should be noted that the reliability of Equation 2.8 will be compromised if one or more of the assumptions underlying the **single-sample t test** are saliently violated. Most specifically, the width of a confidence interval can be dramatically altered if the normality assumption underlying the **single-sample t test** is violated. Anderson (2001, p. 351) notes that in a normal distribution the sampling distribution of the mean has less variability than the sampling distribution of the median — the latter is the reason why the mean is viewed as a more **efficient** measure of central tendency than the median. On the other hand, in a nonnormal distribution the sampling distribution of the mean can have greater variability than the sampling distribution of the median. One or more extreme scores in a nonnormal distribution can spuriously inflate the value of the standard error of the mean, and the latter can result in a drastic increase in the range of values which defines a confidence interval. In addition to affecting confidence intervals, violation of the normality assumption underlying the **single-sample t test** can result in a substantial loss of statistical power in evaluating the alternative hypothesis.

The reader should take note of the fact that the range of values which defines the 95% and 99% confidence intervals for Example 2.1 is extremely large. Because of this the researcher is unable to isolate a **small interval** that is likely to contain the true value of the population mean. This illustrates that a confidence interval may be quite limited in its ability to provide precise information regarding a population parameter, especially when the sample size upon which it is based is small. van Belle (2002, p. 18–19) recommends that a minimum of 12 scores be employed in computing a confidence interval, since the width of an interval decreases most rapidly until $n = 12$, after which the width of a confidence interval decrease more slowly.[14]

As noted in the **Introduction**, during the past 20 years critics of the **classical hypothesis testing model** (e.g., Altman *et al.* (2002), Harlow *et al.* (1997), Hunter and Schmidt (2004), Kline (2004), Meehl (1967), Rozeboom (1960), Smithson (2003), Thompson (1993, 1999, 2002), and Wilkinson *et al.* (1999)) have argued that researchers should utilize confidence intervals in decision making, as well as the fact that all published research should include confidence intervals for any relevant statistics. In some instances, those who vigorously advocate the use of confidence intervals argue that the test of statistical significance should no longer be employed for hypothesis testing, and that instead decision making should be based in large part or entirely on the values computed for confidence intervals.

To be more specific, critics like Smithson (2003, p. 12) note that whereas tests of significance focus on a single parameter value in a null hypothesis, a confidence interval provides an entire range of values of a parameter which cannot be rejected. In addition, he notes that the major advantage of using confidence intervals in lieu of tests of significance is that confidence intervals are much better suited to facilitate comparisons between replications of a study, and in the process allow researchers to reach a conclusion with regard to the status of a research hypothesis.

To illustrate, let us assume that the four sets of data summarized in Table 2.3 are derived from the same population in which four different studies were conducted to evaluate the null hypothesis H_0: $\mu = 5.2$. Careful inspection of Table 2.3 should reveal that Study 2 (in which the sample size is $n = 20$) involves one additional replication of the set of $n = 10$ scores employed for Study 1; Study 3 (in which the sample size is $n = 30$) involves two additional replications of the set of $n = 10$ scores employed for Study 1; and Study 4 (in which the sample size is $n = 40$) involves three additional replications of the set of $n = 10$ scores employed for

Study 1. Since all of the studies involve one or more replications of the same set of data, they yield same mean value $\bar{X} = 6.4$. Note, however, that as the sample size increases the values computed for $\tilde{s}$ and $s_{\bar{X}}$ decrease, resulting in more a powerful test of the alternative hypothesis (which in all four cases is H_1: $\mu \neq 5.2$). The latter is reflected in the fact that as the sample size increases there is a corresponding increase in the value computed for t.

We will assume that prior to evaluating the results of each of the four studies with a **single-sample t test**, the researcher decided that in order to reject the null hypothesis a result would have to achieve significance at the .05 level. The tabled critical two-tailed .05 t values are listed in Table 2.3 for each of the sample sizes evaluated. Inspection of the table reveals three out of the four studies yielded nonsignificant results, and that only Study 4 achieved significance at the .05 level (since $t(39) = 2.09 > t_{.05} = 2.02$). Evaluating the overall results obtained with the four tests of significance, the researcher decides since three-quarters of the studies failed to achieve significance the null hypothesis should be retained — i.e., there was insufficient evidence to indicate that the population mean is some value other than 5.2.

Table 2.3 Hypothetical Replications of Experiments Evaluating
H_0: $\mu = 5.2$ versus H_1: $\mu \neq 5.2$

Data for Study 1
9, 10, 8, 4, 8, 3, 0, 10, 10, 2

| $n = 10$ | $\bar{X} = 6.4$ | $\tilde{s} = 3.78$ | $s_{\bar{X}} = 1.195$ | $t = 1.00$; $p > .05$ | $t_{.05} = 2.262$ |

$CI_{.95} = 6.4 \pm (2.262)(1.195) = 6.4 \pm 2.70$

$CI_{.95}$: $3.70 \leq \mu \leq 9.10$

Data for Study 2
9, 10, 8, 4, 8, 3, 0, 10, 10, 2, 9, 10, 8, 4, 8, 3, 0, 10, 10, 2

| $n = 20$ | $\bar{X} = 6.4$ | $\tilde{s} = 3.68$ | $s_{\bar{X}} = .823$ | $t = 1.45$; $p > .05$ | $t_{.05} = 2.093$ |

$CI_{.95} = 6.4 \pm (2.093)(.823) = 6.4 \pm 1.72$

$CI_{.95}$: $4.68 \leq \mu \leq 8.12$

Data for Study 3
9, 10, 8, 4, 8, 3, 0, 10, 10, 2, 9, 10, 8, 4, 8, 3, 0, 10, 10, 2, 9, 10, 8, 4, 8, 3, 0, 10, 10, 2

| $n = 30$ | $\bar{X} = 6.4$ | $\tilde{s} = 3.64$ | $s_{\bar{X}} = .665$ | $t = 1.80$; $p > .05$ | $t_{.05} = 2.045$ |

$CI_{.95} = 6.4 \pm (2.045)(.665) = 6.4 \pm 1.36$

$CI_{.95}$: $5.04 \leq \mu \leq 7.76$

Data for Study 4
9, 10, 8, 4, 8, 3, 0, 10, 10, 2, 9, 10, 8, 4, 8, 3, 0, 10, 10, 2, 9, 10, 8, 4, 8, 3, 0, 10, 10, 2, 9, 10, 8, 4, 8, 3, 0, 10, 10, 2

| $n = 40$ | $\bar{X} = 6.4$ | $\tilde{s} = 3.63$ | $s_{\bar{X}} = .574$ | $t = 2.09$; $p < .05$ | $t_{.05} = 2.020$ |

$CI_{.95} = 6.4 \pm (2.020)(.574) = 6.4 \pm 1.16$

$CI_{.95}$: $5.24 \leq \mu \leq 7.56$

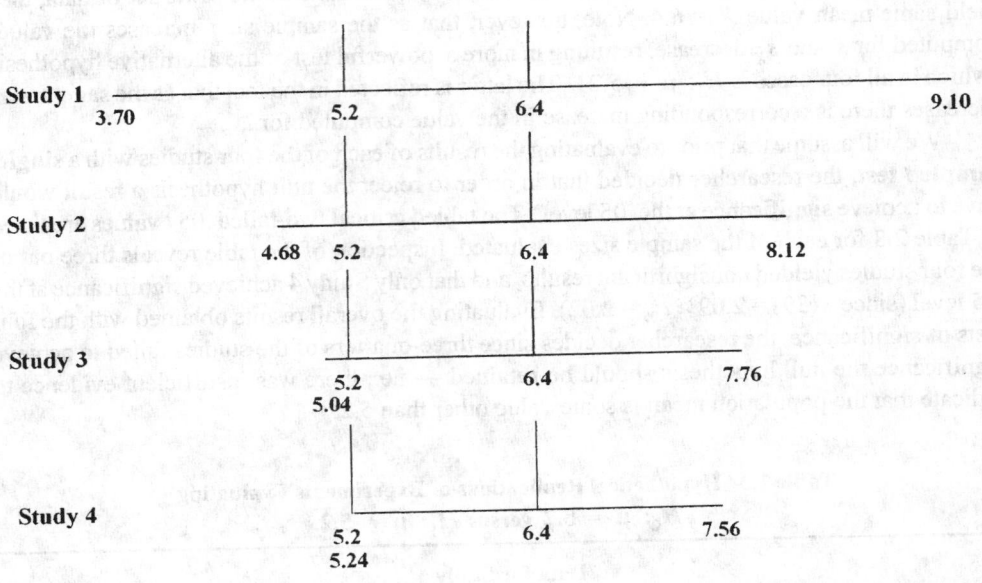

Figure 2.10   95% Confidence Intervals for Studies Summarized in Table 2.3

Now let us consider Figure 2.10 which depicts the 95% confidences intervals computed for the four studies. Note that the four confidence intervals overlap considerably, but more importantly the greater part of all four confidence intervals are clearly to the right of the value 5.2 hypothesized in the null hypothesis. The latter would certainly seem to suggest that the four samples come from a population with a mean value greater than 5.2, and it would appear that it is only because of the low power of the *t* **tests** (due to the small sample sizes) that the null hypothesis cannot be rejected in Studies 1, 2, and 3. (The reader should take note of the fact that if, in fact, the null hypothesis was false and four studies sampling subjects from the same population were actually conducted, there would obviously be more variation between studies with respect to the scores depicted in Figure 2.10. If the latter were the case it would undoubtedly result in some degree of variability among the four sample means. Nevertheless, if the null hypothesis was false and, in fact, the population mean was closer to 6.4 than 5.2, a similar pattern for the confidence intervals (albeit perhaps not as salient) would be obtained as that depicted in Figure 2.10.)

The above example illustrates Hunter and Schmidt's (2004, p. 14) contention that the following two factors make confidence intervals more informative than tests of significance: a) Confidence intervals are centered on the observed value (in this case the sample means) rather than on the value hypothesized in the null hypothesis; and b) Through examination of their width, confidence intervals provide a researcher with a visual picture of the degree of uncertainty in studies which employ small samples. Smithson (2003, p. 13) echoes the same philosophy in stating that "Significance tests can seduce researchers into ignoring statistical power, whereas confidence intervals make power salient by displaying the width (or imprecision) of the interval estimates."

In the final analysis, confidence intervals can be of great value to a researcher, and whenever possible should be reported. Nevertheless, researchers who understand the limitations of tests of significance and are aware of such issues as power and effect sizes (which will be discussed throughout this book) are less likely to misinterpret a test of significance, as well as less likely to design studies which are inadequately powered with respect to detecting an effect size which would be deemed suitable to support an alternative hypothesis. Additionally, researchers who employ tests of significance can employ **meta-analysis** (discussed in Section IX (the **Addendum**) of the **Pearson product-moment correlation coefficient**) to clarify the results of multiple studies evaluating the same hypothesis.

VII. Additional Discussion of the Single-Sample *t* Test

Degrees of freedom   The concept of **degrees of freedom**, which is frequently encountered in statistical analysis, represents the number of values in a set of data that are free to vary after certain restrictions have been placed upon the data. The concept of degrees of freedom will be illustrated through use of the following example.

Assume that one is trying to construct a set consisting of three scores which are derived from a single sample, and it is known that the mean of the sample is $\bar{X} = 5$. Under these conditions, two of the three scores that comprise the set can assume a variety of values, as long as the sum of the two scores does not exceed 15. This is the case since, if $\bar{X} = 5$ and $n = 3$, it is required that $\Sigma X = 15$. Thus, some representative values that two of the three scores may assume are: 1 and 4, 0 and 6, 8 and 6, and 1 and 1. Note that in all four cases the sum of the two scores is less than 15. The value of the third score in all four instances is predetermined based on the values of the other two scores. Thus, if we know that two of the three scores which comprise a set are 1 and 4, the third score must equal 10, since it is the only value that will yield $\Sigma X = 15$.

The rationale for employing $df = n - 1$ in computing the degrees of freedom for the **single-sample *t* test** can be understood on the basis of the above discussion. Specifically, once the size of a sample is set and the mean assumes a specific value, only $n - 1$ scores will be free to vary. In the case of the **single-sample *t* test** (as well as a variety of other inferential statistical tests) degrees of freedom are a function of sample size. However, this is not always the case. For example, when evaluating categorical data through use of the chi-square distribution (discussed later in the book), degrees of freedom are a function of the number of categories involved in the

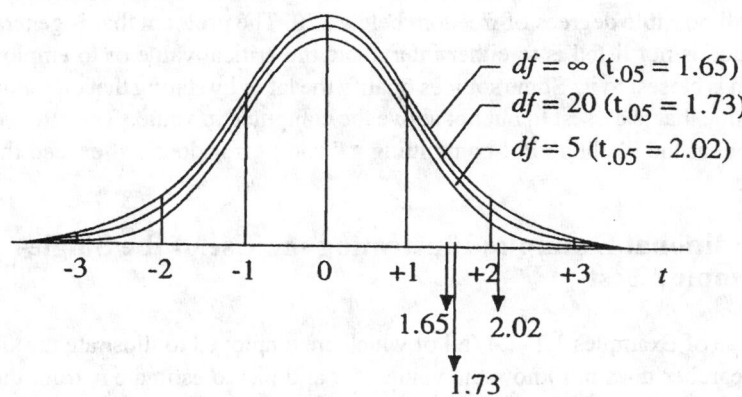

$df = \infty$ ($t_{.05} = 1.65$)
$df = 20$ ($t_{.05} = 1.73$)
$df = 5$ ($t_{.05} = 2.02$)

Figure 2.11   Representative *t* Distributions for Different *df* Values

analysis rather than the size of the sample. In the case of the **single-sample t test**, as sample size increases, the degrees of freedom increase.

Inspection of **Table A2** reveals that as degrees of freedom increase, the lower the value of the tabled critical value which the computed absolute value of t must be equal to or greater than in order to reject the null hypothesis at a given level of significance. Once again, this is not always the case. For instance, when employing the chi-square distribution there is a direct relationship between degrees of freedom and the magnitude of the tabled critical value. In other words, as the number of degrees of freedom increases, the larger the magnitude of the tabled critical chi-square value at a given level of significance.

As noted in Section I, in actuality a separate t distribution exists for each sample size, and consequently for each degrees of freedom value. Figure 2.11 depicts the t distribution for three different degrees of freedom values. Note that the t distribution (which is always symmetrical) closely resembles the normal distribution. As is noted in Section I, except when $n = \infty$, for a given standard deviation score, a smaller proportion of cases will fall between the mean of the t distribution and that standard deviation score than the proportion of cases which fall between the mean and that same standard deviation score in the normal distribution. As sample size (and thus degrees of freedom) increases, the shape of the t distribution becomes increasingly similar in appearance to the normal distribution, and, in fact, when $n = \infty$ becomes identical to it. As a result of this, when the sample size employed in a study is large (usually $n \geq 200$), for all practical purposes, the tabled critical values for the normal and t distributions are identical.[15]

Inspection of the three t distributions depicted in Figure 2.11 reveals that the lower the degrees of freedom, the larger the absolute value of t which will be required in order to reject a null hypothesis at a given level of significance. As an example, the distance between the mean and the one-tailed .05 critical t value (which corresponds to the 95th percentile of a given curve) is greatest for the $df = 5$ distribution (where $t_{.05} = 2.02$) and smallest for the $df = \infty$ distribution (where $t_{.05} = 1.65$).

Most tables of the t distribution list selected degrees of freedom values ranging from 1 through 120, and then list a final row of values for $df = \infty$. The latter set of values is identical to those in the normal distribution, since when $df = \infty$ the two distributions are identical. As a general rule, for sample sizes substantially above 121 (which correspond to $df = 120$), the critical values for $df = \infty$ can be employed. Tables of the t distribution do not include tabled critical values for all possible degrees of freedom below 120. The protocol that is generally used if the exact df value is not listed is to either interpolate the critical value or to employ the tabled df value which is closest to it. Some sources qualify the latter by stating that one should employ the tabled df value that is closest to but not above the computed df value. The intent of this strategy is to insure that the likelihood of committing a Type I error does not exceed the prespecified alpha value.

VIII. Additional Examples Illustrating the Use of the Single-Sample t Test

If, in the case of Examples 1.1–1.4 (all of which are employed to illustrate the **single-sample z test**), a researcher does not know the value of σ and has to estimate it from the sample data, the **single-sample t test** is the appropriate test to use. The 30 scores noted in Section IV of the **single-sample z test** in reference to Example 1.1 can be employed to compute the estimated population standard deviation. Utilizing the 30 scores, the following values are computed: $\Sigma X = 222$, $\bar{X} = 7.4$, and $\Sigma X^2 = 1866$. Equations 2.1–2.3 can now be employed to conduct the necessary calculations for the **single-sample t test**. The null hypothesis that is evaluated is H_0: $\mu = 8$.

$$\tilde{s} = \sqrt{\dfrac{1866 - \dfrac{(222)^2}{30}}{30 - 1}} = 2.77 \qquad s_{\bar{X}} = \dfrac{2.77}{\sqrt{30}} = .506$$

$$t = \dfrac{7.4 - 8}{.506} = -1.19$$

Using the t distribution for Examples 1.1–1.4, the degrees of freedom that are employed are $df = 30 - 1 = 29$. For $df = 29$, the tabled critical two-tailed .05 and .01 values are $t_{.05} = 2.05$ and $t_{.01} = 2.76$, and the tabled critical one-tailed .05 and .01 values are $t_{.05} = 1.70$ and $t_{.01} = 2.46$. Since the computed absolute value $t = 1.19$ is less than all of the aforementioned critical values, the null hypothesis cannot be rejected. Note that when the **single-sample z test** is used to evaluate the same set of data, the directional alternative hypothesis $H_1: \mu < 8$ is supported at the .05 level. The discrepancy between the results of the two tests can be attributed to the fact that the estimated population standard deviation $\tilde{s} = 2.77$ employed for the **single-sample t test** is larger than the value $\sigma = 2$ used when the data are evaluated with the **single-sample z test**.

The **single-sample t test** cannot be employed for Examples 1.5 and 1.6 (which are also used to illustrate the **single-sample z test**), since, when $n = 1$, the estimated value of a population standard deviation will be indeterminate (since at least two scores are required to estimate a population standard deviation). This is confirmed by the fact that no tabled critical t values are listed for zero degrees of freedom (which is the case (and the result obtained with Equation 2.4) if $n = 1$).

Example 2.2, which is based on the same data set as Example 2.1, is an additional example that can be evaluated with the **single-sample t test**.

Example 2.2   *The Sugar Snack candy company claims that each package of candy it sells contains bonus coupons (which a consumer can use toward future purchases), and that the average number of coupons per package is five. Responding to a complaint by a consumer who says the company is shortchanging people on coupons, a consumer advocate purchases 10 bags of candy from a variety of stores. The advocate counts the number of coupons in each bag and obtains the following values: 9, 10, 8, 4, 8, 3, 0, 10, 15, 9. Do the data support the claim of the complainant?*

Since the data for Example 2.2 are identical to that employed for Example 2.1, analysis with the **single-sample t test** yields the value $z = 1.94$. The value $z = 1.94$, which is consistent with the directional alternative hypothesis $H_1: \mu > 5$, is totally unexpected in view of the nature of the consumer's complaint. If anything, the researcher evaluating the data is most likely to employ either the directional alternative hypothesis $H_1: \mu < 5$ or the nondirectional alternative hypothesis $H_1: \mu \neq 5$ (neither of which are supported at the .05 level). Thus, even though the directional alternative hypothesis $H_1: \mu > 5$ is supported at the .05 level, the latter alternative hypothesis would not have been stipulated prior to collecting the data.

References

Altman, D. G., Machin, D., Bryant, T. N., & Gardner, M. J. (2002). **Statistics with confidence: Confidence intervals with statistical guidelines** (2nd ed). London: British Medical Journal Books.

Anderson, N. N. (2001). **Empirical direction in design and analysis**. Mahwah, NJ: Lawrence Erlbaum Associates.

Cohen, J. (1977). **Statistical power analysis for the behavioral sciences**. New York: Academic Press.

Cohen, J. (1988). **Statistical power analysis for the behavioral sciences** (2nd ed.). Hillsdale, NJ: Erlbaum.

Guenther, W. C. (1965). **Concepts of statistical inference**. New York: McGraw–Hill Book Company.

Harlow, L. L., Muliak, S. A. & Steiger, J. H. (Eds) (1997). **What if there were no significance tests**. Mahwah: NH: Lawrence Erlbaum Associates.

Howell, D. C. (2002). **Statistical methods for psychology** (5th ed.). Pacific Grove, CA: Duxbury.

Hunter, J. E. & Schmidt, F. L. (1990). **Methods of meta-analysis: Correcting error and bias in research findings** (1st ed.). Newbury Park, CA: Sage Publications.

Hunter, J. E. & Schmidt, F. L. (2004). **Methods of meta-analysis: Correcting error and bias in research findings** (2nd ed.). Thousand Oaks, CA: Sage Publications.

Kline, R. B. (2004). **Beyond significance testing: Reforming data analysis methods in behavioral research**. Washington, D. C.: American Psychological Association.

Meehl, P. E. (1967). Theory testing in psychology and physics: A methodological paradox. **Philosophy of Science**, 34, 103–115.

Rosner, B. (1995). **Fundamentals of biostatistics** (4th ed.). Belmont, CA: Duxbury Press.

Rosner, B. (2000). **Fundamentals of biostatistics** (5th ed.). Pacific Grove, CA: Duxbury Press.

Rozeboom, W. W. (1960). The fallacy of the null hypothesis significance test. **Psychological Bulletin**, 57, 416–428.

Smithson, M. (2003). **Confidence intervals**. Thousand Oaks, CA: Sage Publications.

Thompson, B. (1993). Statistical significance testing in contemporary practice: Some proposed alternatives with comments from journal editors (Special issue). **Journal of Special Education**, 61 (4).

Thompson, B. (1999). Journal editorial policies regarding statistical significance tests: Heat is to fire as *p* is to importance. **Educational Psychology Review**, 11, 157–169.

Thompson, B. (2002). What future quantitative social science research could look like: Confidence intervals for effect sizes. **Educational Researcher**, 31, 25–32.

van Belle, G. (2002). **Statistical rules of thumb**. New York: John Wiley & Sons.

Wilkinson, L. & APA Task Force of Statistical Inference (1999). Statistical methods in psychology journals: Guidelines and explanations. **American Psychologist**, 54, 594–604.

Endnotes

1. van Belle (2002, pp. 26–28) notes that for small samples (i.e., $n \leq 15$) Equation 2.14 will generally provide a good estimate of the **standard error of the mean**. In the case of Example 2.1, since $Range = 15 - 0 = 15$, the latter equation yields the value $s_{\bar{X}} = 1.5$.

$$\tilde{s}_{\bar{X}} = \frac{Range}{n} = \frac{15}{10} = 1.5 \qquad \textbf{(Equation 2.14)}$$

2. In order to be solvable, Equation 2.3 requires that there be variability in the sample. If all of the subjects in a sample have the same score, the computed value of $\tilde{s}$ will equal zero. When $\tilde{s} = 0$, the value of $s_{\bar{X}}$ will always equal zero. When $s_{\bar{X}} = 0$, Equation 2.3 becomes unsolvable, thus making it impossible to compute a value for t. It is also the case that when

the sample size is $n = 1$, Equation 2.1 becomes unsolvable, thus making it impossible to employ Equation 2.3 to solve for t.

3. In the event that σ is known (and the researcher is confident that the latter value is, in fact, the standard deviation of the population in question) and $n < 25$, and the researcher elects to employ the **single-sample *t* test**, the value of σ should be used in computing the test statistic. Given the fact that the value of σ is known, it would be foolish to employ $\tilde{s}$ as an estimate of it.

4. The t distribution was derived by William Gosset (1876-1937), a British statistician who published under the pseudonym of Student.

5. It is worth noting that if the value of the population standard deviation in Example 2.1 is known to be $\sigma = 4.25$, the data can be evaluated with the **single-sample *z* test**. When employed it yields the value $z = 1.94$, which is identical to the value obtained with Equation 2.3. Specifically, since $\sigma = \tilde{s} = 4.25$, $\sigma_{\bar{x}} = s_{\bar{x}} = 4.25/\sqrt{10} = 1.34$. Employing Equation 1.3 yields $z = (7.6 - 5)/1.34 = 1.94$. As is the case for the **single-sample *t* test**, the latter value only supports the directional alternative hypothesis H_1: $u > 5$ at the .05 level. This is the case since $z = 1.94$ is greater than the tabled critical one-tailed value $z_{.05} = 1.65$ in **Table A1**. The value $z = 1.94$, which is less than the tabled critical two-tailed value $z_{.05} = 1.96$, falls just short of supporting the nondirectional alternative hypothesis H_1: $\mu \neq 5$ at the .05 level.

6. A sampling distribution of means for the t distribution when employed in the context of the **single-sample *t* test** is interpreted in the same manner as the sampling distribution of means for the **single-sample *z* test** as depicted in Figure 1.1.

7. In the event the researcher is evaluating the power of the test in reference to a value of μ_1 that is less than $\mu = 5$, Distribution B will overlap the left tail of Distribution A.

8. It is really not possible to determine this value with great accuracy by interpolating the entries in **Table A2**.

9. If in this example the table for the normal distribution is used to estimate the power of the **single-sample *t* test**, it can be determined that the proportion of cases which falls above a z value of 1.51 is .0655. Although the value .0655 is close to .085, it slightly underestimates the power of the test.

10. a) In the previous edition of this book the notation δ (which is the lower case Greek letter **delta**) was employed to represent the **noncentrality parameter**; b) In Chapter 16 the alternative symbol ϕ is employed to represent the lower case Greek letter **phi**, which is used in the latter chapter to represent a nonparametric measure of association; c) Throughout this book a number of different theoretical probability distributions (e.g., the t distribution, the chi-square (χ^2) distribution, and the F distribution) are employed for conducting inferential statistical tests, as well as in computing additional information such as the power of a test and confidence intervals for a specific parameter. The t distribution (for which **Table A2** displays probability values and corresponding t values) employed in this chapter and throughout this book is sometimes referred to as the **central *t* distribution**. The latter distribution is based on the assumption that a null hypothesis is correct. There is, however,

a **noncentral *t* distribution** which is not based on the latter assumption. Among others, Kline (2004, p. 35) and Smithson (2003) note that the **noncentral *t* distribution** is actually a family of distributions (of which the **central *t* distribution** represents a special case), and that the shape of a specific **noncentral *t* distribution** will be a function of the value of an additional parameter called the **noncentrality parameter**. The latter parameter (which in the case of the **central *t* distribution** equals zero) essentially reflects the degree of deviation from the null hypothesis. Kline (2004, p. 35) notes that as the value of the noncentrality parameter for the **noncentral *t* distribution** becomes increasingly positive, the shape of the *t* distribution becomes increasingly nonsymmetrical and positively skewed. Although noncentral distributions (such as the **noncentral *t* distribution**) can be very valuable for computing such things as power and confidence intervals, they are difficult or impractical to use without the aid of a computer or specialized tables (which are generally not available). However, with the introduction of high speed computers, statisticians are increasingly taking advantage of the benefits noncentral distributions can provide in analyzing data; d) The value of φ can be computed directly through use of the following equation: $\varphi = (\mu_1 - \mu)/(\sigma/\sqrt{n})$. Note that the equation expresses effect size in standard deviation units of the sampling distribution; e) Further discussion of **Cohen's *d* index** (as well as the sample analogue referred to as the ***g* index**) can be found in Endnote 15 of the *t* **test for two independent samples**, and in Section IX (the **Appendix**) of the **Pearson product-moment correlation coefficient** under the discussion of **meta-analysis and related topics**. The *g* **index** for the result of the **single-sample *t* test** can be computed, when in Equation 2.5 $\overline{X}_1$ is employed in lieu of μ_1 and $\tilde{s}$ is employed in lieu of σ.

11. van Belle (2002, pp. 31–33) notes that Equation 2.15 can be employed to estimate the sample size required in order to conduct a **single-sample *t* test** if a nondirectional analysis is conducted with $\alpha = .05$. The value of k in the numerator of the latter equation will depend on the desired power of the test, and for the power values .50, .80, .90, .95, and .975 the values to employ for k are respectively 4, 8, 11, 13, and 16. Equation 2.15 will be demonstrated in reference to computing the necessary sample size for the power of the test to equal .80, if as previously the researcher employs the null hypothesis H_0: $\mu = 5$ versus the alternative hypothesis H_1: $\mu_1 = 6$. If the power of the test is to equal .80, the value, $k = 8$ is employed in the numerator of Equation 2.15. Equation 2.5 is employed to compute the value $d = .235$ (which was computed earlier by dividing the one point difference stipulated by the null versus alternative hypotheses by the value $\tilde{s} = .235$, which represents the best estimate of σ). When the appropriate values are substituted in Equation 2.15, the sample size estimate for the power of the test to equal .80 is $n = 144.86$ which rounded off equals 145.

$$n = \frac{k}{d^2} = \frac{8}{(.235)^2} = 144.86 \qquad \text{(Equation 2.15)}$$

12. A perusal of statistics texts published over the years will reveal that some texts (including the first two editions of this book) would employ the following statement with respect to the 95% confidence interval for Example 2.1: *The physician can be 95% confident that the actual value for the average number of visits per patient (i.e., the population mean) falls within the range of values 4.57 to 10.63.* Technically, the latter statement is incorrect. The subtle difference in language between the latter statement and the statement, *the physician can be 95% confident that the range of values 4.57 to 10.63 includes the actual value for the average number of visits per patient* can be understood if one considers the difference between how probabilities are established within the framework of the **classical hypothesis**

testing model versus the **Bayesian hypothesis test model**. The latter distinction is clarified in Section IX (the Addendum) of the **binomial sign test for a single sample** under the discussion of **Bayesian statistics**. In that discussion it notes that in the **classical hypothesis testing model** an unknown population parameter is viewed as having a fixed value, whereas in **Bayesian hypothesis testing** an unknown population parameter is viewed as a random variable. In **Bayesian statistics** an interval can be computed which is analogous to the confidence interval discussed in this chapter, and in the case of Example 2.1 the Bayesian interval, which is sometimes referred to as a **credible interval**, would allow a researcher to state there is a 95% probability the population parameter falls within that interval. The difference in the language employed in defining a **confidence interval** versus a **credible interval** can be explained on the basis of whether the unknown population parameter is conceptualized as having a fixed value or being a random variable. Further discussion of the language employed in defining a confidence interval can be found in Howell (2002, p. 208) and Kline (2004, pp. 29–30).

13. As noted in Section V, the only exception to this will be when the sample size is extremely large, in which case the normal and *t* distributions are identical. Under such conditions the appropriate values for *z* and *t* employed in the confidence interval equation will be identical.

14. The reader should take note of the fact that although most computed confidence intervals are symmetrical (as is the case in this chapter, where the identical value is added to and subtracted from the sample mean), among others, Kline (2004, p. 27) and Smithson (2003, pp. 5–9) note that a confidence interval is not always symmetrical. Examples of non-symmetrical confidence intervals occur within the framework of employing noncentral distributions, **computer-intensive procedures** such as the **bootstrap** (discussed in Section IX (the Addendum) of the **Mann–Whitney *U* test (Test 12)**), and using **Fisher's *z* transformation** in computing a confidence interval for a correlation coefficient (which is discussed in Section VI of the **Pearson product-moment correlation coefficient (Test 28))**.

15. a) With respect to the fact that a larger proportion of cases falls between the mean of the normal distribution and a given standard deviation score than the proportion of cases which falls between the mean and that same standard deviation score in the *t* distribution, Howell (2002, p. 184) notes the following: The latter can be explained by the fact that the sampling distribution of the estimated population variance (i.e., $\tilde{s}^2$) is positively skewed (especially for small sample sizes). Because of the latter it is more likely that a value computed for $\tilde{s}^2$ will underestimate the true value of the population variance (i.e., σ^2) than overestimate it. The end result of the sampling distribution of $\tilde{s}^2$ being positively skewed is that the *t* value computed for a set of data will be larger than the *z* value which would be computed for the same set of data if, in fact, the value of σ^2 was known and equal to the computed value of $\tilde{s}^2$; b) A *t* distribution is **leptokurtic**, and the degree of leptokurtosis decreases as the sample size increases. Since when $n = \infty$ the *t* distribution is identical to the normal distribution, in the latter instance the *t* distribution will be **mesokurtic**.

Test 3
The Single-Sample Chi-Square Test
for a Population Variance
(Parametric Test Employed with Interval/Ratio Data)

I. Hypothesis Evaluated with Test and Relevant Background Information

Hypothesis evaluated with test Does a sample of n subjects (or objects) come from a population in which the variance (σ^2) equals a specified value?

Relevant background information on test The **single-sample chi-square test for a population variance** is employed in a hypothesis testing situation involving a single sample in order to determine whether or not a sample with an estimated population variance of $\hat{s}^2$ is derived from a population with a variance of σ^2. If the result of the test is significant, the researcher can conclude there is a high likelihood the sample is derived from a population in which the variance is some value other than σ^2. The **single-sample chi-square test for a population variance** is based on the chi-square distribution. The test statistic is represented by the notation χ^2 (where χ represents the lower case Greek letter **chi**).

The **single-sample chi-square test for a population variance** is used with interval/ratio level data and is based on the following assumptions: a) The distribution of data in the underlying population from which the sample is derived is normal; and b) The sample has been randomly selected from the population it represents. If either of the assumptions is saliently violated, the reliability of the test statistic may be compromised.[1]

The chi-square distribution is a continuous asymmetrical theoretical probability distribution. A chi-square value must fall within the range $0 \leq \chi^2 \leq \infty$, and thus (unlike values for the normal and t distributions) can never be a negative number. As is the case with the t distribution, there

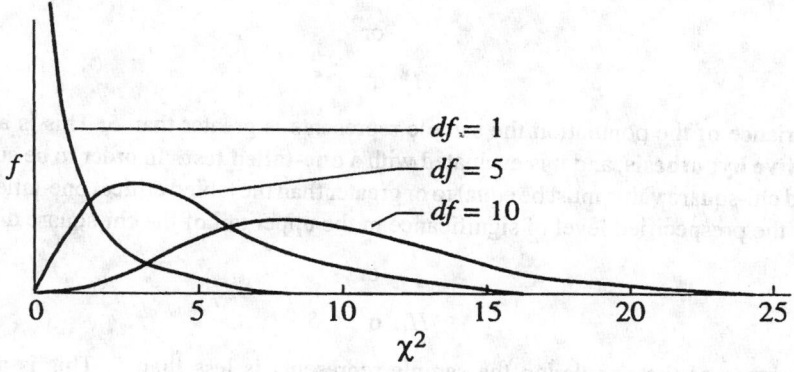

Figure 3.1 Representative Chi-Square Distributions for Different df Values

are an infinite number of chi-square distributions — each distribution being a function of the number of degrees of freedom employed in an analysis. Figure 3.1 depicts the chi-square distribution for three different degrees of freedom values. Inspection of the three distributions reveals: a) The lower the degrees of freedom, the more positively skewed the distribution (i.e., the larger the proportion of scores at the lower end of the distribution); and b) The greater the degrees of freedom, the more symmetrical the distribution. A thorough discussion of the chi-square distribution can be found in Section V.

II. Example

Example 3.1 *The literature published by a company that manufactures hearing aid batteries claims that a certain model battery has an average life of 7 hours ($\mu = 7$) and a variance of 5 hours ($\sigma^2 = 5$). A customer who uses the hearing aid battery believes that the value stated in the literature for the variance is too low. In order to test his hypothesis the customer records the following times (in hours) for ten batteries he purchases during the month of September: 5, 6, 4, 3, 11, 12, 9, 13, 6, 8. Do the data indicate that the variance for battery time is some value other than 5?*

III. Null versus Alternative Hypotheses

Null hypothesis $\qquad\qquad\qquad H_0: \sigma^2 = 5$

(The variance of the population the sample represents equals 5.)

Alternative hypothesis $\qquad\qquad H_1: \sigma^2 \neq 5$

(The variance of the population the sample represents does not equal 5. This is a **nondirectional alternative hypothesis**, and it is evaluated with a **two-tailed test**. In order to be supported, the obtained chi-square value must be equal to or greater than the tabled critical two-tailed chi-square value at the prespecified level of significance in the upper tail of the chi-square distribution, or equal to or less than the tabled critical two-tailed chi-square value at the prespecified level of significance in the lower tail of the chi-square distribution. A full explanation of the protocol for interpreting chi-square values within the framework of the **single-sample chi-square test for a population variance** can be found in Section V.)

or

$$H_1: \sigma^2 > 5$$

(The variance of the population the sample represents is greater than 5. This is a **directional alternative hypothesis**, and it is evaluated with a **one-tailed test**. In order to be supported, the obtained chi-square value must be equal to or greater than the tabled critical one-tailed chi-square value at the prespecified level of significance in the upper tail of the chi-square distribution.)

or

$$H_1: \sigma^2 < 5$$

(The variance of the population the sample represents is less than 5. This is a **directional alternative hypothesis**, and it is evaluated with a **one-tailed test**. In order to be supported, the obtained chi-square value must be equal to or less than the tabled critical one-tailed chi-square value at the prespecified level of significance in the lower tail of the chi-square distribution.)

Note: Only one of the above noted alternative hypotheses is employed. If the alternative hypothesis the researcher selects is supported, the null hypothesis is rejected.[2]

IV. Test Computations

Table 3.1 summarizes the data for Example 3.1.

Table 3.1 Summary of Data for Example 3.1

Battery	X	X^2
1	5	25
2	6	36
3	4	16
4	3	9
5	11	121
6	12	144
7	9	81
8	13	169
9	6	36
10	8	64
	$\Sigma X = 77$	$\Sigma X^2 = 701$

In order to compute the test statistic for the **single-sample chi-square test for a population variance**, it is necessary to use the sample data to calculate an unbiased estimate of the population variance. $\tilde{s}^2$, the unbiased estimate of σ^2, is computed with Equation 3.1 (which is the same as Equation I.8 in the **Introduction**).

$$\tilde{s}^2 = \frac{\Sigma X^2 - \dfrac{(\Sigma X)^2}{n}}{n - 1} \qquad \textbf{(Equation 3.1)}$$

Employing Equation 3.1, the value $\tilde{s}^2 = 12.01$ is computed.

$$\tilde{s}^2 = \frac{701 - \dfrac{(77)^2}{10}}{10 - 1} = 12.01$$

Equation 3.2 is the test statistic for the **single-sample chi-square test for a population variance**.

$$\chi^2 = \frac{(n - 1)\,\tilde{s}^2}{\sigma^2} \qquad \textbf{(Equation 3.2)}$$

Employing the values $n = 10$, $\tilde{s}^2 = 12.01$, and $\sigma^2 = 5$ (which is the hypothesized value of σ^2 stated in the null hypothesis), Equation 3.2 is employed to compute the value $\chi^2 = 21.62$.

$$\chi^2 = \frac{(10 - 1)(12.01)}{5} = 21.62$$

V. Interpretation of the Test Results

The computed value $\chi^2 = 21.62$ is evaluated with **Table A4 (Table of the Chi-Square Distribution)** in the **Appendix**. In **Table A4**, chi-square values are listed in relation to the proportion of cases (which are recorded at the top of each column) that falls below a tabled χ^2 value in the sampling distribution, and the number of **degrees of freedom** (which are recorded in the left-hand column of each row).[3] Equation 3.3 is employed to compute the degrees of freedom for the **single-sample chi-square test for a population variance**.

$$df = n - 1 \qquad \text{(Equation 3.3)}$$

Employing Equation 3.3, we compute that $df = 10 - 1 = 9$. For $df = 9$, the tabled critical two-tailed .05 values are $\chi^2_{.025} = 2.70$ and $\chi^2_{.975} = 19.02$. These values are the tabled critical two-tailed .05 values, since the proportion of the distribution that falls between $\chi^2 = 0$ and $\chi^2_{.025} = 2.70$ is .025, and the proportion of the distribution that falls above $\chi^2_{.975} = 19.02$ is .025. Thus, the extreme 5% of the distribution is comprised of chi-square values that fall below $\chi^2_{.025} = 2.70$ and above $\chi^2_{.975} = 19.02$. In the same respect, the tabled critical two-tailed .01 values are $\chi^2_{.005} = 1.73$ and $\chi^2_{.995} = 23.59$. These values are the tabled critical two-tailed .01 values, since the proportion of the distribution that falls between $\chi^2 = 0$ and $\chi^2_{.005} = 1.73$ is .005, and the proportion of the distribution that falls above $\chi^2_{.995} = 23.59$ is .005. Thus, the extreme 1% of the distribution is comprised of chi-square values that fall below $\chi^2_{.005} = 1.73$ and above $\chi^2_{.995} = 23.59$.

For $df = 9$, the tabled critical one-tailed .05 values are $\chi^2_{.05} = 3.33$ and $\chi^2_{.95} = 16.92$. These values are the tabled critical one-tailed .05 values, since the proportion of the distribution that falls between $\chi^2 = 0$ and $\chi^2_{.05} = 3.33$ is .05, and the proportion of the distribution that falls above $\chi^2_{.95} = 16.92$ is .05. In the same respect, the tabled critical one-tailed .01 values are $\chi^2_{.01} = 2.09$ and $\chi^2_{.99} = 21.67$. These values are the tabled critical one-tailed .01 values, since the proportion of the distribution that falls between $\chi^2 = 0$ and $\chi^2_{.01} = 2.09$ is .01, and the proportion of the distribution that falls above $\chi^2_{.99} = 21.67$ is .01.

Figures 3.2 and 3.3 depict the tabled critical .05 and .01 values for the chi-square sampling distribution when $df = 9$. The mean of a chi-square sampling distribution will always equal the degrees of freedom for the distribution. Thus, $\mu_{\chi^2} = df = n - 1$. The standard deviation of the sampling distribution will always be $\sigma_{\chi^2} = \sqrt{2df}$. Consequently, the variance will be $\sigma^2_{\chi^2} = 2df$. In the case of Example 3.1 where $df = 9$, $\mu_{\chi^2} = 9$ and $\sigma^2_{\chi^2} = 18$.

The following guidelines are employed in evaluating the null hypothesis for the **single-sample chi-square test for a population variance**.

a) If the alternative hypothesis employed is nondirectional, the null hypothesis can be rejected if the obtained chi-square value is equal to or greater than the tabled critical two-tailed chi-square value at the prespecified level of significance in the upper tail of the chi-square distribution, or equal to or less than the tabled critical two-tailed chi-square value at the prespecified level of significance in the lower tail of the chi-square distribution.[4]

b) If the alternative hypothesis employed is directional and predicts a population variance larger than the value stated in the null hypothesis, the null hypothesis can only be rejected if the obtained chi-square value is equal to or greater than the tabled critical one-tailed chi-square value at the prespecified level of significance in the upper tail of the chi-square distribution.

c) If the alternative hypothesis employed is directional and predicts a population variance smaller than the value stated in the null hypothesis, the null hypothesis can only be rejected if the obtained chi-square value is equal to or less than the tabled critical one-tailed chi-square value

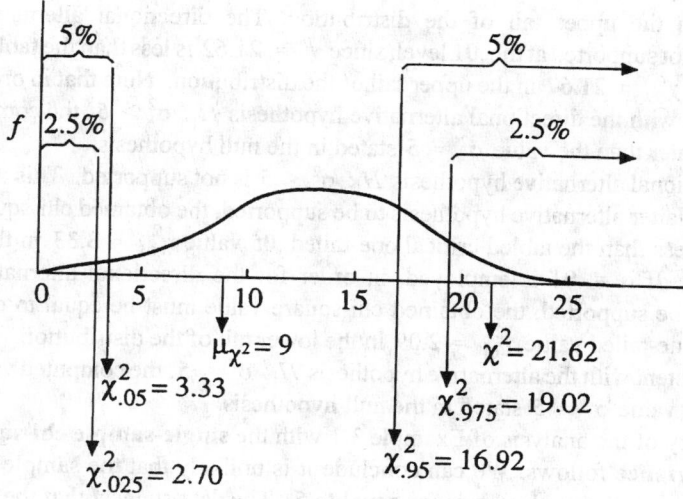

Figure 3.2   Tabled Critical Two-Tailed and One-Tailed .05 χ^2 Values for $df = 9$

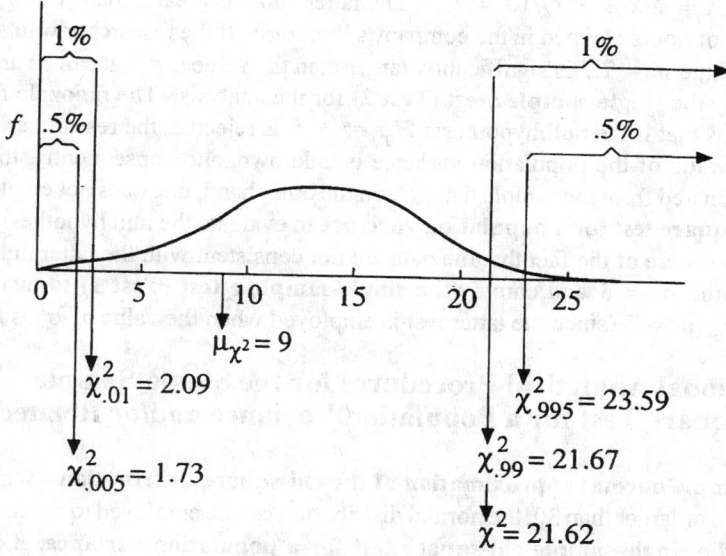

Figure 3.3   Tabled Critical Two-Tailed and One-Tailed .01 χ^2 Values for $df = 9$

at the prespecified level of significance in the lower tail of the chi-square distribution.

Employing the above guidelines, we can conclude the following.

The nondirectional alternative hypothesis H_1: $\sigma^2 \neq 5$ is supported at the .05 level. This is the case since the obtained value $\chi^2 = 21.62$ is greater than the tabled critical two-tailed .05 value $\chi^2_{.975} = 19.02$ in the upper tail of the distribution. The nondirectional alternative hypothesis H_1: $\sigma^2 \neq 5$ is not supported at the .01 level, since $\chi^2 = 21.62$ is less than the tabled critical two-tailed .01 value $\chi^2_{.995} = 23.59$ in the upper tail of the distribution.

The directional alternative hypothesis H_1: $\sigma^2 > 5$ is supported at the .05 level. This is the case since the obtained value $\chi^2 = 21.62$ is greater than the tabled critical one-tailed .05 value

$\chi^2_{.95}$ = 16.92 in the upper tail of the distribution. The directional alternative hypothesis H_1: $\sigma^2 > 5$ is not supported at the .01 level, since χ^2 = 21.62 is less than the tabled critical one-tailed .01 value $\chi^2_{.99}$ = 21.67 in the upper tail of the distribution. Note that in order for the data to be consistent with the directional alternative hypothesis H_1: $\sigma^2 > 5$, the computed value of $\tilde{s}^2$ must be greater than the value σ^2 = 5 stated in the null hypothesis.

The directional alternative hypothesis H_1: $\sigma^2 < 5$ is not supported. This is the case since in order for the latter alternative hypothesis to be supported, the obtained chi-square value must be equal to or less than the tabled critical one-tailed .05 value $\chi^2_{.05}$ = 3.33 in the lower tail of the distribution. If α = .01 is employed, in order for the directional alternative hypothesis H_1: $\sigma^2 < 5$ to be supported, the obtained chi-square value must be equal to or less than the tabled critical one-tailed value $\chi^2_{.01}$ = 2.09 in the lower tail of the distribution. In order for the data to be consistent with the alternative hypothesis H_1: $\sigma^2 < 5$, the computed value of $\tilde{s}^2$ must be less than the value σ^2 = 5 stated in the null hypothesis.

A summary of the analysis of Example 3.1 with the **single-sample chi-square test for a population variance** follows: We can conclude it is unlikely that the sample of 10 batteries comes from a population with a variance equal to 5. The data suggest that the variance of the population is greater than 5. This result can be summarized as follows: χ^2 (9) = 21.62, $p < .05$.

Although it is not required in order to evaluate the null hypothesis H_0: σ^2 = 5, through use of Equations I.1/1.1 it can be determined that the mean number of hours the 10 batteries functioned is $\bar{X} = \Sigma X/n$ = 77/10 = 7.7. The latter value is greater than μ = 7, which is the mean number of hours claimed in the company's literature. If the researcher wants to determine whether the value $\bar{X}$ = 7.7 is significantly larger than the value $\mu = 7$, it can be argued that one should employ the **single-sample t test (Test 2)** for the analysis. The rationale for employing the latter test is that if the null hypothesis H_0: σ^2 = 5 is rejected, the researcher is concluding that the true value of the population variance is unknown, and consequently, the latter value should be estimated from the sample data. If, on the other hand, one does not employ the **single-sample chi-square test for a population variance** to evaluate the null hypothesis H_0: σ^2 = 5 (and is thus unaware of the fact that the data are not consistent with the latter null hypothesis), one will assume σ^2 = 5 and employ the **single-sample z test (Test 1)** to evaluate the null hypothesis H_0: μ = 7 (since the latter test is employed when the value of σ^2 is known).[5]

VI. Additional Analytical Procedures for the Single-Sample Chi-Square Test for a Population Variance and/or Related Tests

1. Large sample normal approximation of the chi-square distribution   When the sample size is equal to or larger than 30, the normal distribution can be employed to approximate the test statistic for the **single-sample chi-square test for a population variance**. Equation 3.4 is employed to compute the normal approximation.

$$z = \frac{\tilde{s} - \sigma}{\dfrac{\sigma}{\sqrt{2n}}}$$ **(Equation 3.4)**

To illustrate the use of Equation 3.4, let us assume that in Example 3.1 the computed value $\tilde{s}^2$ = 12.01 is based on a sample size of $n = 30$. Employing Equation 3.2, the value χ^2 = 69.66 is computed.

$$\chi^2 = \frac{(30 - 1)(12.01)}{5} = 69.66$$

Employing **Table A4**, for $df = 30 - 1 = 29$, the tabled critical two-tailed .05 and .01 values in the upper tail of the chi-square distribution are $\chi^2_{.975} = 45.72$ and $\chi^2_{.995} = 52.34$, and the tabled critical one-tailed .05 and .01 values in the upper tail of the distribution are $\chi^2_{.95} = 42.56$ and $\chi^2_{.99} = 49.59$. Since the obtained value $\chi^2 = 69.66$ is greater than all of the aforementioned critical values, the nondirectional alternative hypothesis $H_1: \sigma^2 \neq 5$ and the directional alternative hypothesis $H_1: \sigma^2 > 5$ are supported at both the .05 and .01 levels.

Equation 3.4 will now be employed to evaluate the data for Example 3.1, if $n = 30$. Since $\sigma^2 = 5$ and $\tilde{s}^2 = 12.01$, it follows that $\sigma = 2.24$ and $\tilde{s} = 3.47$. These values are substituted in Equation 3.4.

$$z = \frac{3.47 - 2.24}{\dfrac{2.24}{\sqrt{(2)(30)}}} = 4.24$$

Employing **Table A1 (Table of the Normal Distribution)** in the **Appendix**, we determine that the tabled critical two-tailed .05 and .01 values are $z_{.05} = 1.96$ and $z_{.01} = 2.58$, and the tabled critical one-tailed .05 and .01 values are $z_{.05} = 1.65$ and $z_{.01} = 2.33$. Since the obtained value $z = 4.24$ is greater than all of the aforementioned critical values, the nondirectional alternative hypothesis $H_1: \sigma^2 \neq 5$ and the directional alternative hypothesis $H_1: \sigma^2 > 5$ are supported at both the .05 or .01 levels. The conclusions derived through use of the normal approximation are identical to those reached with Equation 3.2.

If Equation 3.4 is employed with Example 3.1 for $n = 10$, it results in the value $z = 2.46$. Specifically: $z = [3.47 - 2.24] / [2.24/\sqrt{(2)(10)}] = 2.46$. Since the latter value is greater than the tabled critical two-tailed value $z_{.05} = 1.96$ and the tabled critical one-tailed values $z_{.05} = 1.65$ and $z_{.01} = 2.33$, the nondirectional alternative hypothesis $H_1: \sigma^2 \neq 5$ is supported at the .05 level and the directional alternative hypothesis $H_1: \sigma^2 > 5$ is supported at both the .05 and .01 levels. Recollect that when Equation 3.2 is employed with Example 3.1, the nondirectional alternative hypothesis $H_1: \sigma^2 \neq 5$ and the directional alternative hypothesis $H_1: \sigma^2 > 5$ are both supported, but only at the .05 level. Thus, when the normal approximation is employed with a small sample size, it appears to inflate the likelihood of committing a Type I error (since the normal approximation supports the nondirectional alternative hypothesis $H_1: \sigma^2 > 5$ at the .01 level).

2. Computation of a confidence interval for the variance of a population represented by a sample[6]

An equation for computing a confidence interval for the variance of a population (as well as the population standard deviation) can be derived by algebraically transposing the terms in Equation 3.2. As is noted in the discussion of the **single-sample t test**, a confidence interval is a range of values with respect to which a researcher can be confident to a specified degree contains the true value of a population parameter. Stated in probabilistic terms, the researcher can state that there is a specified probability/likelihood the confidence interval contains the true value of the population parameter. Equation 3.5 is the general equation for computing a confidence interval for a population variance.

$$\frac{(n - 1)\, \tilde{s}^2}{\chi^2_{[1 - (\alpha/2)]}} \leq \sigma^2 \leq \frac{(n - 1)\, \tilde{s}^2}{\chi^2_{(\alpha/2)}} \qquad \textbf{(Equation 3.5)}$$

Where: $\chi^2_{(\alpha/2)}$ is the tabled critical two-tailed value in the chi-square distribution below which a proportion (percentage) equal to $[1 - (\alpha/2)]$ of the cases falls. If the proportion (percentage) of the distribution that falls within the confidence interval is subtracted from 1 (100%), it will equal the value of α.

Equations 3.6 and 3.7 are employed to compute the 95% and 99% confidence intervals for a population variance. The critical values employed in Equation 3.6 demarcate the middle 95% of the chi-square distribution, while the critical values employed in Equation 3.7 demarcate the middle 99% of the distribution.

$$\frac{(n - 1)\, \tilde{s}^2}{\chi^2_{.975}} \le \sigma^2 \le \frac{(n - 1)\, \tilde{s}^2}{\chi^2_{.025}} \qquad \textbf{(Equation 3.6)}$$

$$\frac{(n - 1)\, \tilde{s}^2}{\chi^2_{.995}} \le \sigma^2 \le \frac{(n - 1)\, \tilde{s}^2}{\chi^2_{.005}} \qquad \textbf{(Equation 3.7)}$$

Using the data for Example 3.1, Equation 3.6 is employed to compute the 95% confidence interval for the population variance.

$$\frac{(10 - 1)\,(12.01)}{19.02} \le \sigma^2 \le \frac{(10 - 1)\,(12.01)}{2.70}$$

$$5.68 \le \sigma^2 \le 40.03$$

Thus, we can be 95% confident (or the probability is .95) that the interval 5.68 to 40.03 contains the true value of the population variance. By taking the square root of the latter values, we can determine the 95% confidence interval for the population standard deviation. Thus, 2.38 $\le \sigma \le 6.33$. In other words, we can be 95% confident (or the probability is .95) that the interval 2.38 to 6.33 contains the true value of the population standard deviation.

Equation 3.7 is employed below to compute the 99% confidence interval.

$$\frac{(10 - 1)\,(12.01)}{23.59} \le \sigma^2 \le \frac{(10 - 1)\,(12.01)}{1.73}$$

$$4.58 \le \sigma^2 \le 62.48$$

Thus, we can be 99% confident (or the probability is .99) that the interval 4.58 to 62.48 contains the true value of the population variance. By taking the square root of the latter values, we can determine the 99% confidence interval for the population standard deviation. Thus, 2.14 $\le \sigma \le 7.90$. In other words, we can be 99% confident (or the probability is .99) that the interval 2.14 to 7.90 contains the true value of the population standard deviation. Note that (as is the case for confidence intervals for a population mean) a larger range of values defines the 99% confidence interval for a population variance than the 95% confidence interval.

When $n \ge 30$, the normal distribution can be employed to approximate the confidence interval for a population standard deviation. Equation 3.8 is the general equation for computing a confidence interval using the normal approximation.

$$\frac{\tilde{s}}{1 + \dfrac{z_{(\alpha/2)}}{\sqrt{2n}}} \le \sigma \le \frac{\tilde{s}}{1 - \dfrac{z_{(\alpha/2)}}{\sqrt{2n}}} \qquad \textbf{(Equation 3.8)}$$

Where: $z_{(\alpha/2)}$ is the tabled critical two-tailed value in the normal distribution below which a proportion (percentage) equal to $[1 - (\alpha/2)]$ of the cases falls. If the proportion (percentage) of the distribution that falls within the confidence interval is subtracted from 1 (100%), it will equal the value of α.

Equations 3.9 and 3.10 employ the normal approximation to compute the 95% and 99% confidence intervals for a population standard deviation. The values $z_{.05}$ and $z_{.01}$ used in the latter equations are the tabled critical two-tailed .05 and .01 values $z_{.05} = 1.96$ and $z_{.01} = 2.58$. By squaring a value obtained for the confidence interval for a population standard deviation, one can determine the confidence interval for the population variance.

$$\frac{\tilde{s}}{1 + \dfrac{z_{.05}}{\sqrt{2n}}} \leq \sigma \leq \frac{\tilde{s}}{1 - \dfrac{z_{.05}}{\sqrt{2n}}} \qquad \textbf{(Equation 3.9)}$$

$$\frac{\tilde{s}}{1 + \dfrac{z_{.01}}{\sqrt{2n}}} \leq \sigma \leq \frac{\tilde{s}}{1 - \dfrac{z_{.01}}{\sqrt{2n}}} \qquad \textbf{(Equation 3.10)}$$

For purposes of illustration, let us assume that $n = 30$ in Example 3.1. Using $n = 30$ and $\tilde{s}^2 = 12.01$ for Example 3.1, Equation 3.9 is employed to compute the 95% confidence interval for the population standard deviation and variance.

$$\frac{3.47}{1 + \dfrac{1.96}{\sqrt{(2)(30)}}} \leq \sigma \leq \frac{3.47}{1 - \dfrac{1.96}{\sqrt{(2)(30)}}}$$

$$2.77 \leq \sigma \leq 4.65$$

$$7.67 \leq \sigma^2 \leq 21.62$$

Using $n = 30$ and $\tilde{s}^2 = 12.01$ for Example 3.1, Equation 3.10 is employed to compute the 99% confidence interval for the population standard deviation and variance.

$$\frac{3.47}{1 + \dfrac{2.58}{\sqrt{(2)(30)}}} \leq \sigma \leq \frac{3.47}{1 - \dfrac{2.58}{\sqrt{(2)(30)}}}$$

$$2.61 \leq \sigma \leq 5.18$$

$$6.81 \leq \sigma^2 \leq 26.83$$

If Equations 3.6 and 3.7 are employed with Example 3.1 for $n = 30$ and $\tilde{s}^2 = 12.01$, they yield values close to those obtained with Equations 3.9 and 3.10. As the size of the sample increases, the values which define a confidence interval based on the use of the normal versus chi-square distributions converge, and for large sample sizes the two distributions yield the same values.

Equation 3.6 is employed to compute the 95% confidence interval. Note that for $df = 29$, the tabled critical values $\chi^2_{.025} = 16.05$ and $\chi^2_{.975} = 45.72$ are employed in Equation 3.6.

$$\frac{(30 - 1)\,(12.01)}{45.72} \le \sigma^2 \le \frac{(30 - 1)\,(12.01)}{16.05}$$

$$7.62 \le \sigma^2 \le 21.70$$

$$2.76 \le \sigma \le 4.66$$

Equation 3.7 is employed to compute the 99% confidence interval. Note that for $df = 29$, the tabled critical values $\chi^2_{.005} = 13.21$ and $\chi^2_{.995} = 52.34$ are employed in Equation 3.7.

$$\frac{(30 - 1)\,(12.01)}{52.34} \le \sigma^2 \le \frac{(30 - 1)\,(12.01)}{13.21}$$

$$6.65 \le \sigma^2 \le 26.37$$

$$2.58 \le \sigma \le 5.13$$

In closing the discussion of confidence intervals, the reader should take note of the fact that the range of values which defines the 95% and 99% confidence intervals for Example 3.1 is extremely large. Because of this the researcher is unable to isolate a **small interval** which is likely to contain the true value of the population variance/standard deviation. This illustrates that a confidence interval may be quite limited in its ability to provide precise information regarding a population parameter, especially when the sample size upon which it is based is small.[7] The reader should also take note of the fact that the reliability of Equations 3.5 and 3.8 will be compromised if one or more of the assumptions of the **single-sample chi-square test for a population variance** is saliently violated.

3. Sources for computing the power of the single-sample chi-square test for a population variance   The protocol for computing the power of the **single-sample chi-square test for a population variance** is described in Guenther (1965) (who provides computational guidelines and graphs for quick power computations) and Zar (1999).

VII. Additional Discussion of the Single-Sample Chi-Square Test for a Population Variance

No additional material will be discussed in this section.

VIII. Additional Examples Illustrating the Use of the Single-Sample Chi-Square Test for a Population Variance

With the exception of Examples 1.5 and 1.6, the **single-sample chi-square test for a population variance** can be employed to test a hypothesis about a population variance (or standard deviation) with any of the examples that are employed to illustrate the **single-sample z test** (Examples 1.1–1.4) and the **single-sample t test** (Examples 2.1 and 2.2).

Examples 3.2–3.5 are four additional examples which can be evaluated with the **single-sample chi-square test for a population variance**. Since these examples employ the same population parameters and sample data used in Example 3.1, they yield the same result. The reader should take note of the fact that although in Examples 3.4 and 3.5 the value $\sigma = 2.24$ is given for the population standard deviation, when the latter value is squared, it results in $\sigma^2 = 5$.

A different set of data is employed in Example 3.6, which is the last example presented in this section.

Example 3.2 *A manufacturer of a machine that makes ball bearings claims that the variance of the diameter of the ball bearings produced by the machine is 5 millimeters. A company that has purchased the machine measures the diameter of a random sample of ten ball bearings. The computed estimated population variance of the diameters of ten ball bearings is 12.01 millimeters. Is the obtained value $\tilde{s}^2 = 12.01$ consistent with the null hypothesis H_0: $\sigma^2 = 5$?*

Example 3.3 *A meteorologist develops a theory which predicts a variance of five degrees Celsius for the temperature recorded at noon each day in a canyon situated within a mountain range 100 miles east of the South Pole. Over a ten-day period the following temperatures are recorded: 5, 6, 4, 3, 11, 12, 9, 13, 6, 8. Do the data support the meteorologist's theory?*

Example 3.4 *A chemical company claims that the standard deviation for the number of tons of waste it discards annually is $\sigma = 2.24$. Assume that during a ten-year period the following annual values are recorded for the number of tons of waste discarded: 5, 6, 4, 3, 11, 12, 9, 13, 6, 8. Are the data consistent with the company's claim?*

Example 3.5 *During more than three decades of teaching, a college professor determines that the mean and standard deviation on a chemistry final examination are respectively 7 and 2.24. During the fall semester of a year in which he employs a new teaching method, ten students who take the final examination obtain the following scores: 5, 6, 4, 3, 11, 12, 9, 13, 6, 8. Do the data suggest that the new teaching method results in an increase in variability in performance?*

Example 3.6 *The Sugar Snack candy company claims that each package of candy it sells contains bonus coupons (which a consumer can use toward future purchases), and that the standard deviation for the number of coupons per package is 5. A consumer purchases 10 bags of candy from a variety of stores, and counts the number of coupons in each bag obtaining the following values: 9, 10, 8, 4, 8, 3, 0, 10, 15, 9. Are the data consistent with the company's claim for the product?*

The data for Example 3.6 are identical to those employed for Example 2.1, which is used to illustrate the **single-sample *t* test**. In the computations for the latter test, through use of Equation 2.1, it is determined that the estimated population standard deviation for the data is $\tilde{s} = 4.25$. Since the population variance is the square of the standard deviation, we can determine that $\sigma^2 = (5)^2 = 25$ and $\tilde{s}^2 = (4.25)^2 = 18.06$. Since the company claims that the population standard deviation is 5, the null hypothesis for Example 3.6 can be stated as either H_0: $\sigma = 5$ or H_0: $\sigma^2 = 25$. The nondirectional and directional alternative hypotheses (stated in reference to H_0: $\sigma = 5$) which can be employed for Example 3.6 are as follows: H_1: $\sigma \neq 5$; H_1: $\sigma > 5$; and H_1: $\sigma < 5$. When Equation 3.2 is employed to evaluate the null hypothesis H_0: $\sigma = 5$, the value $\chi^2 = 6.50$ is computed.

$$\chi^2 = \frac{(10 - 1)(18.06)}{25} = 6.50$$

Since $n = 10$, the tabled critical values for $df = 9$ are employed to evaluate the computed value $\chi^2 = 6.50$. Since the latter value is less than the tabled critical two-tailed .05 value $\chi^2_{.975} = 19.02$ in the upper tail of the chi-square distribution and greater than the tabled critical two-tailed .05 value $\chi^2_{.025} = 2.70$ in the lower tail of the distribution, the nondirectional alternative hypothesis H_1: $\sigma \neq 5$ is not supported. Since $\chi^2 = 6.50$ is less than the tabled critical one-tailed .05 value $\chi^2_{.95} = 16.92$ in the upper tail of the distribution, the directional alternative hypothesis H_1: $\sigma > 5$ is not supported. Since $\chi^2 = 6.50$ is greater than the tabled critical one-tailed .05 value $\chi^2_{.05} = 3.33$ in the lower tail of the distribution, the directional alternative hypothesis H_1: $\sigma < 5$ is not supported. Thus, regardless of which alternative hypothesis one employs, the data do not contradict the company's statement that the population standard deviation is $\sigma = 5$.

References

Crisman, R. (1975). Shortest confidence interval for the standard deviation of a normal distribution. **Journal of Undergraduate Mathematics**, 7, 57.

Freund, J. E. (1984). **Modern elementary statistics** (6th ed.). Englewood Cliffs, NJ: Prentice Hall.

Guenther, W. C. (1965). **Concepts of statistical inference**. New York: McGraw–Hill Book Company.

Hogg, R. V. & Tanis, E. A. (1988). **Probability and statistical inference** (3rd ed.). New York: Macmillan Publishing Company.

Howell, D. C. (2002). **Statistical methods for psychology** (5th ed.). Pacific Grove, CA.: Duxbury.

Smithson, M. (2003). **Confidence intervals**. Thousand Oaks, CA: Sage Publications.

Zar, J. H. (1999). **Biostatistical analysis** (4th ed.). Upper Saddle River, NJ: Prentice Hall.

Endnotes

1. Most sources note that violation of the normality assumption is much more serious for the **single-sample chi-square test for a population variance** than it is for tests concerning the mean of a single sample (i.e., the **single-sample z test** and the **single-sample t test**). Especially in the case of small sample sizes, violation of the normality assumption can severely compromise the accuracy of the tabled values employed to evaluate the chi-square statistic.

2. One can also state the null and alternative hypotheses in reference to the population standard deviation (which is the square root of the population variance). Since in Example 3.1 $\sigma = \sqrt{\sigma^2} = \sqrt{5} = 2.24$, one can state the null hypothesis and nondirectional and directional alternative hypotheses as follows: H_0: $\sigma = 2.24$; H_1: $\sigma \neq 2.24$; H_1: $\sigma > 2.24$; and H_1: $\sigma < 2.24$.

3. The use of the chi-square distribution in evaluating the variance is based on the fact that for any value of n, the sampling distribution of $\tilde{s}^2$ has a direct linear relationship to the chi-square distribution for $df = n - 1$. As is the case for the chi-square distribution, the sampling distribution of $\tilde{s}^2$ is positively skewed. Although the average of the sampling distribution for $\tilde{s}^2$ will equal σ^2, because of the positive skew of the distribution, a value of $\tilde{s}^2$ is more likely to underestimate rather than overestimate the value of σ^2. For further discussion of this latter the reader is referred to Howell (2002, pp 184–186).

4. When the chi-square distribution is employed within the framework of the **single-sample chi-square test for a population variance**, it is common practice to employ critical values

derived from both tails of the distribution. However, when the chi-square distribution is used with other statistical tests, as a general rule only critical values in the upper/right tail of the distribution are employed. Examples of chi-square tests which focus on the upper/right tail of the distribution are **the chi-square goodness-of-fit test (Test 8)** and **the chi-square test for $r \times c$ tables (Test 16)**.

5. In Section I of the **single-sample t test** it is noted that some sources argue when the sample size is very small (generally less than 25) the latter test should be employed in evaluating a null hypothesis about a population mean, even if one knows the value of σ.

6. Although the procedure described in this section for computing a confidence interval for a population variance is the one that is most commonly described in statistics books, it does not result in the shortest possible confidence interval which can be computed. Hogg and Tanis (1988) describe a method (based on Crisman (1975)) requiring more advanced mathematical procedures that allows one to compute the shortest possible confidence interval for a population variance. For large sample sizes the difference between the latter method and the method described in this section will be trivial.

7. When data are available from multiple studies, a useful method for evaluating a confidence interval for a population mean is described at the end of the discussion of confidence intervals in Section VI of the **single-sample t test**. This latter method can also be employed in evaluating a confidence interval for a population variance or standard deviation.

Test 4
The Single-Sample Test for Evaluating Population Skewness
(Parametric Test Employed with Interval/Ratio Data)

I. Hypothesis Evaluated with Test and Relevant Background Information

Hypothesis evaluated with test Does a sample of n subjects (or objects) come from a population distribution that is symmetrical (i.e., not skewed)?

Relevant background information on test Prior to reading this section the reader should review the discussion of **skewness** in the **Introduction**. As noted in the **Introduction**, skewness is a measure reflecting the degree to which a distribution is asymmetrical. From a statistical perspective, the skewness of a distribution represents the **third moment about the mean** (m_3), which is represented by Equation 4.1 (which is identical to Equation I.17).

$$m_3 = \frac{\Sigma(X - \bar{X})^3}{n}$$

(Equation 4.1)

Skewness can be employed as one criterion to aid in determining **goodness-of-fit** of data with respect to a normal distribution. Various sources (e.g., Anscombe and Glynn (1983), D'Agostino (1986), D'Agostino and Stephens (1986), and D'Agostino *et al.* (1990)) note that the **single-sample test for evaluating population skewness** and the **single-sample test for evaluating population kurtosis (Test 5)**, when employed together, provide an excellent test for evaluating the hypothesis of goodness-of-fit for normality. The results obtained for the latter two tests are employed within the framework of the following two tests (both of which are described in Section VI of the **single-sample test for evaluating population kurtosis**) for assessing goodness-of-fit for normality: The **D'Agostino–Pearson test of normality (Test 5a)** and the **Jarque–Bera test of normality (Test 5b)**.

In the **Introduction** it was noted that since the value computed for m_3 is in cubed units, it is often converted into the unitless statistic g_1. The latter, which is an estimate of the population parameter γ_1 (where γ represents the lower case Greek letter **gamma**), is commonly employed to express skewness. When a distribution is symmetrical (about the mean), the value of g_1 (which is often referred to as a **skewness coefficient**) will equal 0. When the value of g_1 is significantly above 0, a distribution will be positively skewed, and when it is significantly below 0, a distribution will be negatively skewed. Although the normal distribution is symmetrical (with $g_1 = 0$), as noted earlier, not all symmetrical distributions are normal. In other words, although a normal distribution will have a skewness coefficient of $g_1 = 0$, not all distributions with the latter skewness coefficient are normal. Examples of nonnormal distributions that are symmetrical are the t distribution and the binomial distribution, when $\pi_1 = .5$. (The meaning of the notation $\pi_1 = .5$ is explained in Section I of the **binomial sign test for a single sample (Test 9)**.)

It was also noted in the **Introduction** that some sources (e.g., D'Agostino (1970, 1986) and D'Agostino *et al.* (1990)) convert the value of g_1 into the statistic $\sqrt{b_1}$. The latter is an estimate of a population parameter designated $\sqrt{\beta_1}$ (where β represents the lower case Greek letter **beta**), which is also employed to represent skewness. When a distribution is symmetrical (such as in the case of a normal distribution), the value of $\sqrt{b_1}$ will equal 0. When the value of $\sqrt{b_1}$ is significantly above 0, a distribution will be positively skewed, and when it is significantly below 0, a distribution will be negatively skewed.

The **single-sample test for evaluating population skewness** is the procedure for determining whether or not a g_1 and/or $\sqrt{b_1}$ value deviate significantly from 0. The normal distribution is employed to provide an approximation of the exact sampling distribution for the statistics g_1 and $\sqrt{b_1}$. Thus, the test statistic computed for the **single-sample test for evaluating population skewness** is a z value.[1]

II. Example

Example 4.1 *A researcher wishes to evaluate the data in three samples (comprised of 10 scores per sample) for skewness. Specifically, the researcher wants to determine whether or not the samples meet the criteria for symmetry as opposed to positive versus negative skewness. The three samples will be designated* **Sample E, Sample F,** *and* **Sample G.** *The researcher has reason to believe that* **Sample E** *is derived from a symmetrical population distribution,* **Sample F** *from a negatively skewed population distribution, and* **Sample G** *from a positively skewed population distribution. The data for the three distributions are presented below.*

> **Distribution E:** 0, 0, 0, 5, 5, 5, 5, 10, 10, 10
> **Distribution F:** 0, 1, 1, 9, 9, 10, 10, 10, 10, 10
> **Distribution G:** 0, 0, 0, 0, 0 , 1, 1, 9, 9, 10

Are the data consistent with what the researcher believes to be true regarding the underlying population distributions?

III. Null versus Alternative Hypotheses

Null hypothesis　　　　　　　　H_0: $\gamma_1 = 0$ or H_0: $\sqrt{\beta_1} = 0$

(The underlying population distribution the sample represents is **symmetrical** — in which case the population parameters γ_1 and $\sqrt{\beta_1}$ are equal to 0.)

Alternative hypothesis　　　　　H_1: $\gamma_1 \neq 0$ or H_1: $\sqrt{\beta_1} \neq 0$

(The underlying population distribution the sample represents is **not symmetrical** — in which case the population parameters γ_1 and $\sqrt{\beta_1}$ are not equal to 0. This is a **nondirectional alternative hypothesis**, and it is evaluated with a **two-tailed test**. In order to be supported, the absolute value of z must be equal to or greater than the tabled critical two-tailed value at the prespecified level of significance. Thus, either a significant positive z value or a significant negative z value will provide support for this alternative hypothesis.)

or

$$H_1\text{: } \gamma_1 > 0 \text{ or } H_1\text{: } \sqrt{\beta_1} > 0$$

(The underlying population distribution the sample represents is **positively skewed** — in which case the population parameters γ_1 and $\sqrt{\beta_1}$ are greater than 0. This is a **directional alternative hypothesis**, and it is evaluated with a **one-tailed test**. It will only be supported if the sign of z

is positive, and the absolute value of z is equal to or greater than the tabled critical one-tailed value at the prespecified level of significance.)

or

$$H_1: \gamma_1 < 0 \text{ or } H_1: \sqrt{\beta_1} < 0$$

(The underlying population distribution the sample represents is **negatively skewed** — in which case the population parameters γ_1 and $\sqrt{\beta_1}$ are less than 0. This is a **directional alternative hypothesis**, and it is evaluated with a **one-tailed test**. It will only be supported if the sign of z is negative, and the absolute value of z is equal to or greater than the tabled critical one-tailed value at the prespecified level of significance.)

Note: Only one of the above noted alternative hypotheses is employed. If the alternative hypothesis the researcher selects is supported, the null hypothesis is rejected.

IV. Test Computations

The three distributions presented in Example 4.1 are identical to **Distributions E, F**, and **G** employed in the **Introduction** to demonstrate the computation of the values m_3, g_1, and $\sqrt{b_1}$. Employing Equations I.20/I.21, I.22, and I.23, the following values were previously computed for m_3, g_1, and $\sqrt{b_1}$ for **Distributions E, F**, and **G**: $m_{3_E} = 0$, $m_{3_F} = -86.67$, $m_{3_G} = 86.67$, $g_{1_E} = 0$, $g_{1_F} = -1.02$, $g_{1_G} = 1.02$, and $\sqrt{b_{1_E}} = 0$, $\sqrt{b_{1_F}} = -.86$, $\sqrt{b_{1_G}} = .86$ (the computation of the latter values is summarized in Tables I.3–I.5).

Equations 4.2–4.8 (which are presented in Zar (1999, pp. 115–116)) summarize the steps that are involved in computing the test statistic (which, as is noted above, is a z value) for the **single-sample test for evaluating population skewness**. Zar (1999) states that when $n > 9$, Equation 4.8 provides a good approximation of the exact probabilities for the sampling distribution of g_1 (which is employed to compute the value of $\sqrt{b_1}$ that is used in Equation 4.2). Note that in Equations 4.6 and 4.8, the notation *ln* represents the natural logarithm of a number (which is defined in Endnote 14 in the **Introduction**).

$$A = \sqrt{b_1} \sqrt{\frac{(n + 1)(n + 3)}{6(n - 2)}} \qquad \textbf{(Equation 4.2)}$$

$$B = \frac{3(n^2 + 27n - 70)(n + 1)(n + 3)}{(n - 2)(n + 5)(n + 7)(n + 9)} \qquad \textbf{(Equation 4.3)}$$

$$C = \sqrt{2(B - 1)} - 1 \qquad \textbf{(Equation 4.4)}$$

$$D = \sqrt{C} \qquad \textbf{(Equation 4.5)}$$

$$E = \frac{1}{\sqrt{\ln D}} \qquad \textbf{(Equation 4.6)}$$

$$F = \frac{A}{\sqrt{\frac{2}{C - 1}}} \qquad \textbf{(Equation 4.7)}$$

$$z = E \ln(F + \sqrt{F^2 + 1})$$ **(Equation 4.8)**

Employing Equations 4.2–4.8, the values $z_E = 0$, $z_F = -1.53$, and $z_G = 1.53$ are computed for **Distributions E, F**, and **G**.

Distribution E

$$A = 0 \sqrt{\frac{(10 + 1)(10 + 3)}{6(10 - 2)}} = 0$$

$$B = \frac{3[(10)^2 + 27(10) - 70](10 + 1)(10 + 3)}{(10 - 2)(10 + 5)(10 + 7)(10 + 9)} = 3.32$$

$$C = \sqrt{2(3.32 - 1)} - 1 = 1.15$$

$$D = \sqrt{1.15} = 1.07$$

$$E = \frac{1}{\sqrt{\ln 1.07}} = 3.84$$

$$F = \frac{0}{\sqrt{\dfrac{2}{1.15 - 1}}} = 0$$

$$z_E = 3.84 \ln\left[0 + \sqrt{(0)^2 + 1}\right] = 0$$

Distribution F

$$A = -.86 \sqrt{\frac{(10 + 1)(10 + 3)}{6(10 - 2)}} = -1.48$$

$$B = \frac{3[(10)^2 + 27(10) - 70](10 + 1)(10 + 3)}{(10 - 2)(10 + 5)(10 + 7)(10 + 9)} = 3.32$$

$$C = \sqrt{2(3.32 - 1)} - 1 = 1.15$$

$$D = \sqrt{1.15} = 1.07$$

$$E = \frac{1}{\sqrt{\ln 1.07}} = 3.84$$

$$F = \frac{-1.48}{\sqrt{\dfrac{2}{1.15 - 1}}} = -.41$$

$$z_F = 3.84 \ln\left[-.41 + \sqrt{(-.41)^2 + 1}\right] = -1.53$$

Distribution G

$$A = .86\sqrt{\frac{(10 + 1)(10 + 3)}{6(10 - 2)}} = 1.48$$

$$B = \frac{3[(10)^2 + 27(10) - 70](10 + 1)(10 + 3)}{(10 - 2)(10 + 5)(10 + 7)(10 + 9)} = 3.32$$

$$C = \sqrt{2(3.32 - 1)} - 1 = 1.15$$

$$D = \sqrt{1.15} = 1.07$$

$$E = \frac{1}{\sqrt{\ln 1.07}} = 3.84$$

$$F = \frac{1.48}{\sqrt{\dfrac{2}{1.15 - 1}}} = .41$$

$$z_G = 3.84 \ln\left[.41 + \sqrt{(.41)^2 + 1}\right] = 1.53$$

V. Interpretation of the Test Results

The obtained values $z_E = 0$, $z_F = -1.53$, and $z_G = 1.53$ are evaluated with **Table A1 (Table of the Normal Distribution)** in the **Appendix**. In **Table A1** the tabled critical two-tailed .05 and .01 values are $z_{.05} = 1.96$ and $z_{.01} = 2.58$, and the tabled critical one-tailed .05 and .01 values are $z_{.05} = 1.65$ and $z_{.01} = 2.33$. Since the computed **absolute values** $z_E = 0$, $z_F = 1.53$, and $z_G = 1.53$ are all less than the tabled critical two-tailed value $z_{.05} = 1.96$ and the tabled critical one-tailed value $z_{.05} = 1.65$, the null hypothesis cannot be rejected, regardless of which alternative hypothesis is employed.

The computation of the value $z_E = 0$ for **Distribution E** is consistent with the fact that the latter distribution is employed to represent a symmetrical distribution. Thus, the nondirectional alternative hypothesis H_1: $\gamma_1 \neq 0$ (or H_1: $\sqrt{\beta_1} \neq 0$) is not supported. Whenever a distribution has perfect symmetry, g_1 (as well as $\sqrt{b_1}$), will equal 0, and consequently the value computed for z will also equal 0.

Although not statistically significant, the data for **Distribution F** are consistent with the directional alternative hypothesis H_1: $\gamma_1 < 0$ (or H_1: $\sqrt{\beta_1} < 0$). Similarly, although not statistically significant, the data for **Distribution G** are consistent with the directional alternative hypothesis H_1: $\gamma_1 > 0$ (or H_1: $\sqrt{\beta_1} > 0$). Note that because $g_{1_F} = -1.02$ and $\sqrt{b_{1_F}} = -.86$ are negative numbers, a negative z value is obtained for **Distribution F** (which is hypothesized to represent a negatively skewed distribution). In the same respect, since $g_{1_G} = 1.02$ and $\sqrt{b_{1_G}} = .86$ are positive numbers, a positive z value is obtained for **Distribution G** (which is hypothesized to represent a positively skewed distribution). Whenever a distribution is negatively skewed, the computed value of z will be a negative number, and whenever a distribution is positively skewed, the computed value of z will be a positive number.

The fact that the values $z_F = -1.53$ and $z_G = 1.53$ are not statistically significant (although they are not that far removed from the tabled critical one-tailed .05 value $z_{.05} = 1.65$) can probably be attributed to the small number of subjects (i.e., $n = 10$) in each of the distributions. A small sample size severely reduces the power of a statistical test, thus making it more difficult to obtain a statistically significant result (i.e., in this case, a significant deviation from symmetry). It should be emphasized that in most instances when a researcher has reason to evaluate a distribution with regard to skewness, a substantially larger sample size would be employed than $n = 10$ in Example 4.1.

Section VII discusses tables which document the exact sampling distribution for the g_1 statistic, and contrasts the results obtained with the latter tables with the results obtained in this section.[2] In closing this section, it should be emphasized that a nonsignificant result for the **single-sample test for evaluating population skewness**, in and of itself, provides insufficient evidence for establishing normality of a distribution.

VI. Additional Analytical Procedures for the Single-Sample Test for Evaluating Population Skewness and/or Related Tests

No additional procedures will be described in this section.

VII. Additional Discussion of the Single-Sample Test for Evaluating Population Skewness

1. Exact tables for the single-sample test for evaluating population skewness Zar (1999) has derived exact tables for the absolute value of the g_1 statistic for sample sizes in the range $9 \le n \le 1000$. By employing exact tables, one can avoid the tedious computations for the **single-sample test for evaluating population skewness** (which employs the normal distribution to approximate the exact sampling distribution) described in Section IV. In Zar's (1999) tables, the tabled critical two-tailed .05 and .01 values for g_1 are $g_{1_{.05}} = 1.359$ and $g_{1_{.01}} = 1.846$, and the tabled critical one-tailed .05 and .01 values are $g_{1_{.05}} = 1.125$ and $g_{1_{.01}} = 1.643$. In order to reject the null hypothesis, the computed absolute value of g_1 must be equal to or greater than the tabled critical value (and if a directional alternative hypothesis is evaluated, the sign of g_1 must be in the predicted direction). The probabilities derived for Example 4.1 (through use of Equation 4.8) are extremely close to the exact probabilities listed in Zar (1999). The probability for **Distribution E** is identical to Zar's (1999) exact probability. With respect to **Distributions F** and **G**, the probabilities listed in Zar's (1999) tables for the values $g_{1_F} = -1.02$ and $g_{1_G} = 1.02$ are very close to the tabled probabilities in **Table A1** for the computed values $z_F = -1.53$ and $z_G = 1.53$. Note that the absolute value $g_1 = 1.02$ just falls short of being significant at the .05 level if a one-tailed analysis is conducted. In the case of Example 4.1, the same conclusions regarding the null hypothesis will be reached, regardless of whether or not one employs the normal approximation or Zar's (1999) tables.

2. Note on a nonparametric test for evaluating skewness Zar (1999, pp. 119–120) describes a **nonparametric procedure** for evaluating skewness/symmetry around the median of a distribution (as opposed to the mean). The latter test is based on the **Wilcoxon signed-ranks test (Test 6)**, which is a nonparametric procedure described in a subsequent chapter.

VIII. Additional Examples Illustrating the Use of the Single-Sample Test for Evaluating Population Skewness

No additional examples will be presented in this section.

References

Conover, W. J. (1980). **Practical nonparametric statistics** (2nd ed.). New York: John Wiley & Sons.

Conover, W. J. (1999). **Practical nonparametric statistics** (3rd ed.). New York: John Wiley & Sons.

D'Agostino, R. B. (1970). Transformation to normality of the null distribution of g_1. **Biometrika**, 57, 679–681.

D'Agostino, R. B. (1986). Tests for the normal distribution In D'Agostino, R. B. &Stephens, M. A. (Eds.), **Goodness-of-fit techniques** (pp. 367–419). New York: Marcel Dekker.

D'Agostino, R. B. & Stephens, M. A. (Eds.) (1986). **Goodness-of-fit techniques**. New York: Marcel Dekker.

D'Agostino, R. B., Belanger, A. & D'Agostino Jr., R. B. (1990). A suggestion for using powerful and informative tests of normality. **American Statistician**, 44, 316–321.

Siegel, S. & Castellan, N. J., Jr. (1988). **Nonparametric statistics for the behavioral sciences** (2nd ed.). New York: McGraw–Hill Book Company.

Sprent, P. (1993). **Applied nonparametric statistical methods** (2nd ed.). London: Chapman & Hall.

Tabachnick, B. G. & Fidell, L. S. (2001). **Using multivariate statistics** (4th ed.). Boston: Allyn & Bacon.

Zar, J. H. (1999). **Biostatistical analysis** (4th ed.). Upper Saddle River, NJ: Prentice Hall.

Endnotes

1.   The reader should take note of the fact that the test for evaluating population skewness described in this chapter is a large sample approximation. In point of fact, sources are not in agreement with respect to what equation provides for the best test of the hypothesis of whether or not g_1 (which represents the value for skewness printed out in most computer packages (e.g., *SPSS*)) and/or $\sqrt{b_1}$ deviates significantly from 0. The general format of alternative equations for evaluating skewness which are employed in other sources (e.g., statistical software packages such as *SPSS, SAS, S-Plus*) involves the computation of a z value by dividing the skewness coefficient by the estimated population standard error (Equation 4.9). That a different z value can result from use of one or more of these alternative equations derives from the fact that sources are not in agreement with respect to what statistic provides the best estimate of the population standard error (*SE*).

As an example, Tabachnick and Fidell (2001, p. 73) note that the value of the standard error can be approximated with the equation $SE = \sqrt{6/n}$, and that the value of z can be computed with Equation 4.9.

$$z = \frac{g_1}{SE} \qquad \textbf{(Equation 4.9)}$$

Readers should take note of the fact that an *SPSS* analysis of the examples evaluated in this chapter may yield different z values (see Endnote 2) than those obtained in Section IV. In the final analysis, use of exact tables of the underlying sampling distribution (in Zar

(1999, pp. 119–120)) allows for the optimal analysis of the hypothesis of whether or not a g_1 and/or $\sqrt{b_1}$ value deviates significantly from 0.

2.   Analysis of the data with *SPSS* yields almost identical results. Specifically: a) For **Distribution E**, *SPSS* computes the values *skewness* = g_1 = 0 and *SE* = .687. When the latter values are substituted in Equation 4.9, the value z = 0/.687 = 0 is obtained; b) For **Distribution F**, *SPSS* computes the values *skewness* = g_1 = −1.02 and *SE* = .687. When the latter values are substituted in Equation 4.9, the value z = −1.02/.687 = −1.48 is obtained; and c) For **Distribution G**, *SPSS* computes the values *skewness* = g_1 = 1.02 and *SE* = .687. When the latter values are substituted in Equation 4.9, the value z = 1.02/.687 = 1.48 is obtained.

Since the computed **absolute values** z_E = 0, z_F = 1.48, and z_G = 1.48 are all less than the tabled critical two-tailed value $z_{.05}$ = 1.96 and the tabled critical one-tailed value $z_{.05}$ = 1.65, the null hypothesis cannot be rejected, regardless of which alternative hypothesis is employed — in other words, the researcher is not able to conclude that the data are skewed in any of the underlying populations represented by the samples.

Test 5

The Single-Sample Test for Evaluating Population Kurtosis
(Parametric Test Employed with Interval/Ratio Data)

I. Hypothesis Evaluated with Test and Relevant Background Information

Hypothesis evaluated with test Does a sample of n subjects (or objects) come from a population distribution that is mesokurtic?

Relevant background information on test Prior to reading this section the reader should review the discussion of **kurtosis** in the **Introduction**. As noted in the **Introduction**, kurtosis is a measure reflecting the degree of peakedness of a distribution. From a statistical perspective, the kurtosis of a distribution represents the **fourth moment about the mean** (m_4), which is represented by Equation 5.1 (which is identical to Equation I.18).

$$m_4 = \frac{\Sigma(X - \bar{X})^4}{n}$$

(Equation 5.1)

Kurtosis can be employed as one criterion to aid in determining **goodness-of-fit** of data with respect to a normal distribution. Various sources (e.g., Anscombe and Glynn (1983), D'Agostino (1986), D'Agostino and Stephens (1986), and D'Agostino *et al.* (1990)) note that the **single-sample test for evaluating population kurtosis** and the **single-sample test for evaluating population skewness (Test 4)**, when employed together, provide an excellent test for evaluating the hypothesis of goodness-of-fit for normality. The results obtained for the latter two tests are employed within the framework of the following two tests (both of which are described in Section VI) for assessing goodness-of-fit for normality: The **D'Agostino–Pearson test of normality (Test 5a)** and the **Jarque–Bera test of normality (Test 5b)**.

In the **Introduction** it was noted that since the value k_4 (which is employed to estimate m_4) computed with Equations I.28/I.29 is in units of the fourth power, it is often converted into the unitless statistic g_2. The latter, which is an estimate of the population parameter γ_2 (where γ represents the lower case Greek letter **gamma**), is commonly employed to express kurtosis. When a distribution is mesokurtic, the value of g_2 will equal 0. When the value of g_2 is significantly above 0, a distribution will be leptokurtic, and when it is significantly below 0, a distribution will be platykurtic. The **single-sample test for evaluating population kurtosis** is a test of mesokurtic normality — in other words, whether or not a distribution is mesokurtic.

It was also noted in the **Introduction** that some sources (e.g., Anscombe and Glynn (1983), D'Agostino (1986), and D'Agostino *et al.* (1990)) convert the value of g_2 into the statistic b_2. The latter is an estimate of a population parameter designated β_2 (where β represents the lower case Greek letter **beta**), which is also employed to represent kurtosis. Some sources refer to g_2 as a **kurtosis coefficient**, while other sources employ b_2 within the latter context. When a distribution is mesokurtic, the value of b_2 will equal $[3(n-1)]/(n+1)$. Inspection of the latter equation reveals that as the value of the sample size increases, the value of b_2 approaches 3. When the value computed for b_2 is significantly below $[3(n-1)]/(n+1)$, a distribution will be platykurtic. When the value computed for b_2 is significantly greater than $[3(n-1)]/(n+1)$, a distribution will be leptokurtic.

The **single-sample test for evaluating population kurtosis** is the procedure for determining whether a g_2 and/or b_2 value deviate significantly from the expected value for a mesokurtic distribution. As noted earlier, any normal distribution will always be mesokurtic, with the following expected computed values: $g_2 = 0$ and $b_2 = 3$. It should be noted, however, that although a normal distribution will have a kurtosis coefficient of $g_2 = 0$ or $b_2 = 3$, not all distributions with the latter kurtosis coefficient are normal.

The normal distribution is employed to provide an approximation of the exact sampling distribution for the statistics g_2 and b_2. Thus, the test statistic computed for the **single-sample test for evaluating population kurtosis** is a z value.[1]

II. Example

Example 5.1 *A researcher wishes to evaluate the data in two samples (comprised of 20 scores per sample) for kurtosis. The two samples will be designated* **Sample H** *and* **Sample I**. *The researcher has reason to believe that* **Sample H** *is derived from a leptokurtic population distribution, while* **Sample I** *is derived from a platykurtic population distribution. The data for the two distributions are presented below.*

>   **Distribution H**: 2, 7, 8, 8, 8, 9, 9, 9, 10, 10, 10, 10, 11, 11, 11, 12, 12, 12, 13, 18
>   **Distribution I**: 0, 1, 3, 3, 5, 5, 8, 8, 10, 10, 10, 10, 12, 12, 15, 15, 17, 17, 19, 20

Are the data consistent with what the researcher believes to be true regarding the underlying population distributions?

III. Null versus Alternative Hypotheses

Null hypothesis H_0: $\gamma_2 = 0$ or H_0: $\beta_2 = 3$

(The underlying population distribution the sample represents is **mesokurtic** — in which case the population parameter γ_2 is equal to 0, and the population parameter β_2 is equal to 3.)

Alternative hypothesis H_1: $\gamma_2 \neq 0$ or H_1: $\beta_2 \neq 3$

(The underlying population distribution the sample represents is **not mesokurtic** — in which case the population parameter γ_2 is not equal to 0, and the population parameter β_2 is not equal to 3. This is a **nondirectional alternative hypothesis**, and it is evaluated with a **two-tailed test**. In order to be supported, the absolute value of z must be equal to or greater than the tabled critical two-tailed value at the prespecified level of significance.)

or

H_1: $\gamma_2 > 0$ or H_1: $\beta_2 > 3$

(The underlying population distribution the sample represents is **leptokurtic** — in which case the population parameter γ_2 is greater than 0, and the population parameter β_2 is greater than 3. This is a **directional alternative hypothesis**, and it is evaluated with a **one-tailed test**. It will only be supported if $g_2 > 0$, and the absolute value of z is equal to or greater than the tabled critical one-tailed value at the prespecified level of significance.)

or

$$H_1: \gamma_2 < 0 \text{ or } H_1: \beta_2 < 3$$

(The underlying population distribution the sample represents is **platykurtic** — in which case the population parameter γ_2 is less than 0, and the population parameter β_2 is less than 3. This is a **directional alternative hypothesis**, and it is evaluated with a **one-tailed test**. It will only be supported if $g_2 < 0$, and the absolute value of z is equal to or greater than the tabled critical one-tailed value at the prespecified level of significance.)

Note: Only one of the above noted alternative hypotheses is employed. If the alternative hypothesis the researcher selects is supported, the null hypothesis is rejected.

IV. Test Computations

The two distributions presented in Example 5.1 are identical to **Distributions H** and **I** employed in the **Introduction** to demonstrate the computation of the values k_4, g_2, and b_2. Employing Equations I.28/I.29, I.30 and I.31, the following values were previously computed for k_4, g_2, and b_2 for **Distributions H** and **I**: $k_{4_H} = 307.170$, $k_{4_I} = -1181.963$, $g_{2_H} = 3.596$, $g_{2_I} = -.939$, and $b_{2_H} = 5.472$, $b_{2_I} = 1.994$ (the computation of the latter values is summarized in Tables I.6–I.7).

Equations 5.2–5.7 (which are presented in Zar (1999, pp. 116–118)) summarize the steps that are involved in computing the test statistic (which, as is noted above, is a z value) for the **single-sample test for evaluating population kurtosis**. Zar (1999) states that when $n \geq 20$, Equation 5.7 provides a good approximation of the exact probabilities for the sampling distribution of g_2.[2]

$$G = \frac{24n(n - 2)(n - 3)}{(n + 1)^2(n + 3)(n + 5)} \qquad \textbf{(Equation 5.2)}$$

$$H = \frac{(n - 2)(n - 3)(g_2)}{(n + 1)(n - 1)\sqrt{G}} \qquad \textbf{(Equation 5.3)}$$

$$J = \frac{6(n^2 - 5n + 2)}{(n + 7)(n + 9)} \sqrt{\frac{6(n + 3)(n + 5)}{n(n - 2)(n - 3)}} \qquad \textbf{(Equation 5.4)}$$

$$K = 6 + \frac{8}{J}\left[\frac{2}{J} + \sqrt{1 + \frac{4}{J^2}}\right] \qquad \textbf{(Equation 5.5)}$$

$$L = \frac{1 - \dfrac{2}{K}}{1 + H\sqrt{\dfrac{2}{K - 4}}} \qquad \textbf{(Equation 5.6)}$$

$$z = \frac{1 - \dfrac{2}{9K} - \sqrt[3]{L}}{\sqrt{\dfrac{2}{9K}}}$$ **(Equation 5.7)**

Employing Equations 5.2–5.7, the values $z_H = 2.40$, and $z_I = -1.13$ are computed for Distributions H and I.

Distribution H

$$G = \frac{(24)(20)(20 - 2)(20 - 3)}{(20 + 1)^2(20 + 3)(20 + 5)} = .579$$

$$H = \frac{(20 - 2)(20 - 3)(3.596)}{(20 + 1)(20 - 1)\sqrt{.579}} = 3.624$$

$$J = \frac{6[(20)^2 - 5(20) + 2]}{(20 + 7)(20 + 9)}\sqrt{\frac{6(20 + 3)(20 + 5)}{20(20 - 2)(20 - 3)}} = 1.737$$

$$K = 6 + \frac{8}{1.737}\left[\frac{2}{1.737} + \sqrt{1 + \frac{4}{(1.737)^2}}\right] = 18.325$$

$$L = \frac{1 - \dfrac{2}{18.325}}{1 + (3.624)\sqrt{\dfrac{2}{18.325 - 4}}} = .379$$

$$z_H = \frac{1 - \dfrac{2}{(9)(18.325)} - \sqrt[3]{.379}}{\sqrt{\dfrac{2}{(9)(18.325)}}} = 2.40$$

Distribution I

$$G = \frac{(24)(20)(20 - 2)(20 - 3)}{(20 + 1)^2(20 + 3)(20 + 5)} = .579$$

$$H = \frac{(20 - 2)(20 - 3)(-.939)}{(20 + 1)(20 - 1)\sqrt{.579}} = -.946$$

$$J = \frac{6[(20)^2 - 5(20) + 2]}{(20 + 7)(20 + 9)} \sqrt{\frac{6(20 + 3)(20 + 5)}{20(20 - 2)(20 - 3)}} = 1.737$$

$$K = 6 + \frac{8}{1.737}\left[\frac{2}{1.737} + \sqrt{1 + \frac{4}{(1.737)^2}}\right] = 18.325$$

$$L = \frac{1 - \dfrac{2}{18.325}}{1 + (-.946)\sqrt{\dfrac{2}{18.325 - 4}}} = 1.3779$$

$$z_I = \frac{1 - \dfrac{2}{(9)(18.325)} - \sqrt[3]{1.3779}}{\sqrt{\dfrac{2}{(9)(18.325)}}} = -1.13$$

V. Interpretation of the Test Results

The obtained values $z_H = 2.40$ and $z_I = -1.13$ are evaluated with **Table A1 (Table of the Normal Distribution)** in the **Appendix**. In **Table A1** the tabled critical two-tailed .05 and .01 values are $z_{.05} = 1.96$ and $z_{.01} = 2.58$, and the tabled critical one-tailed .05 and .01 values are $z_{.05} = 1.65$ and $z_{.01} = 2.33$.

With respect to **Distribution H**, since the obtained value $z_H = 2.40$ is greater than the tabled critical two-tailed value $z_{.05} = 1.96$, the nondirectional alternative hypothesis $H_1: \gamma_2 \neq 0$ (or $H_1: \beta_2 \neq 3$) is supported at the .05 level. However, it is not supported at the .01 level, since $z_H = 2.40$ is less than the tabled critical two-tailed value $z_{.01} = 2.58$. The directional alternative hypothesis $H_1: \gamma_2 > 0$ (or $H_1: \beta_2 > 3$) is supported at both the .05 and .01 levels, since the obtained value $g_{2_H} = 3.596$ is a positive number, and the obtained value $z_H = 2.40$ is greater than the tabled critical one-tailed values $z_{.05} = 1.65$ and $z_{.01} = 2.33$.

The directional alternative hypothesis $H_1: \gamma_2 < 0$ (or $H_1: \beta_2 < 3$) is not supported, since in order for the latter alternative hypothesis to be supported, the computed value of g_2 must be a negative number.

With respect to **Distribution H**, the result of the **single-sample test for evaluating population kurtosis** allows one to conclude there is a high likelihood the sample is derived from a population which is **leptokurtic**.

In the case of **Distribution I**, since the obtained absolute value $z_I = 1.13$ is less than the tabled critical two-tailed value $z_{.05} = 1.96$ and the tabled critical one-tailed value $z_{.05} = 1.65$, the null hypothesis cannot be rejected regardless of which alternative hypothesis is employed. Although the value $g_{2_I} = -.939$ computed for **Distribution I** is a negative number, its absolute value is not large enough to warrant the conclusion that the sample is derived from a population that is **platykurtic**. It should be emphasized that in most instances when a researcher has reason to evaluate a distribution with regard to kurtosis, a substantially larger sample size would be employed than $n = 20$ in Example 5.1.

Section VII discusses tables which document the exact sampling distribution for the g_2 statistic, and contrasts the results obtained with the latter tables with the results obtained in this section.[3] In closing this section, it should be emphasized that a nonsignificant result for the **single-sample test for evaluating population kurtosis**, in and of itself, provides insufficient evidence for establishing normality of a distribution.

VI. Additional Analytical Procedures for the Single-Sample Test for Evaluating Population Kurtosis and/or Related Tests

1. Test 5a: The D'Agostino–Pearson test of normality   Researchers would not consider the result of the **single-sample test for evaluating population kurtosis**, in and of itself, as sufficient evidence for establishing goodness-of-fit for normality. In this section a procedure (to be referred to as the **D'Agostino–Pearson test of normality**) will be described which employs the computed value of g_2 and the value computed for g_1 (which is a measure of skewness discussed in the **Introduction** and in the **single-sample test for evaluating population skewness**) to evaluate whether or not a set of data is derived from a normal distribution. The **D'Agostino–Pearson test of normality** was developed by D'Agostino and Pearson (1973), who state that the test is the most effective procedure for assessing goodness-of-fit for a normal distribution. D'Agostino and Pearson (1973) and Zar (1999) claim the **D'Agostino–Pearson test of normality** is more effective for assessing goodness-of-fit than the more commonly employed **Kolmogorov–Smirnov goodness-of-fit test for a single sample (Test 7)** and the **chi-square goodness-of-fit test (Test 8)** (both of which are described in subsequent chapters). The null and alternative hypotheses which are evaluated with the **D'Agostino–Pearson test of normality** are as follows.

Null hypothesis H_0: The sample is derived from a normally distributed population.

Alternative hypothesis H_1: The sample is not derived from a normally distributed population. This is a **nondirectional alternative hypothesis**.

The test statistic for the **D'Agostino–Pearson test of normality** is computed with Equation 5.8.

$$\chi^2 = z_{g_1}^2 + z_{g_2}^2$$

(Equation 5.8)

The values $z_{g_1}^2$ and $z_{g_2}^2$ are respectively the square of the z values computed with Equation 4.8 and Equation 5.7. Equation 4.8 is employed to compute the test statistic z_{g_1} (which is based on the value computed for $\sqrt{b_1}$) for the **single-sample test for evaluating population skewness**, and Equation 5.7 is employed to compute the test statistic z_{g_2} for the **single-sample test for evaluating population kurtosis**. The computed value of χ^2 is evaluated with **Table A4 (Table of the Chi-Square Distribution)** in the **Appendix**. The degrees of freedom employed in the analysis will always be $df = 2$. The tabled critical .05 and .01 chi-square values in **Table A4** for $df = 2$ are $\chi_{.05}^2 = 5.99$ and $\chi_{.01}^2 = 9.21$.[4] If the computed value of chi-square is equal to or greater than either of the aforementioned values, the null hypothesis can be rejected at the appropriate level of significance. If the null hypothesis is rejected, a researcher can conclude that a set of data does not fit a normal distribution. Through examination of the results of the **single-sample test for evaluating population skewness** and the **single-sample test for evaluating population kurtosis**, the researcher can determine if rejection of the null hypothesis is the result of a lack of symmetry and/or a departure from mesokurtosis.

It happens to be the case that **Distribution I** is a symmetrical distribution, and thus $g_{1_I} = 0$ and $\sqrt{b_{1_I}} = 0$. Consequently, the value computed with Equation 4.7 will be $z_{g_1} = 0$. When the latter value, along with the value $z_{g_2} = -1.13$ computed for **Distribution I** with Equation 5.7, is substituted in Equation 5.8, the value $\chi^2 = 1.28$ is obtained.

$$\chi^2 = (0)^2 + (-1.13)^2 = 1.28$$

Since the value $\chi^2 = 1.28$ is less than the tabled critical value $\chi^2_{.05} = 5.99$, we are not able to reject the null hypothesis. Thus, we cannot conclude that the population distribution for **Distribution I** is nonnormal.

It happens to be the case that **Distribution H** is also a symmetrical distribution, and thus $g_{1_H} = 0$ and $\sqrt{b_{1_H}} = 0$. Consequently, the value computed with Equation 4.7 will be $z_{g_1} = 0$. When the latter value, along with the value $z_{g_2} = 2.40$ computed for **Distribution H** with Equation 5.7, is substituted in Equation 5.8, the value $\chi^2 = 5.76$ is obtained (the latter value resulting entirely from the square of the significant value $z_{g_2} = 2.40$, which indicated that **Distribution H** is leptokurtic).

$$\chi^2 = (0)^2 + (2.40)^2 = 5.76$$

Since the value $\chi^2 = 5.76$ is less (although minimally) than the tabled critical value $\chi^2_{.05} = 5.99$, we are not able to reject the null hypothesis. However, analysis of the data in Section IV with the **single-sample test for evaluating population kurtosis** suggests that **Distribution H** is clearly leptokurtic, and that in itself might be sufficient grounds for some researchers to conclude it is unlikely that the underlying population is normal. Thus, in spite of the nonsignificant result with the **D'Agostino–Pearson test of normality**, some researchers might view it prudent to reject the null hypothesis purely on the basis of the outcome of the **single-sample test for evaluating population kurtosis**.

2. Test 5b: The Jarque–Bera test of normality   An alternative test of normality (commonly employed in the field of econometrics) has been developed by Jarque and Bera (1980, 1981, 1987). The test statistic for the latter test is computed with Equation 5.9. Note that the **Jarque–Bera test of normality** employs the values g_1 and b_2 respectively to represent the skewness and kurtosis of a distribution. The chi-square value computed for the **Jarque–Bera test of normality** is always evaluated with two degrees of freedom (i.e., $df = 2$).

$$\chi^2 = n \left[\frac{g_1^{\,2}}{6} + \frac{(b_2^{\,2} - 3)}{24} \right] \qquad \textbf{(Equation 5.9)}$$

Where: g_1 represents the skewness of the distribution
b_2 represents the kurtosis of the distribution

For **Distribution H**, the value $\chi^2 = 5.09$ is computed below. As was the case when the **D'Agostino–Pearson test of normality** was employed to evaluate the same distribution, the latter value is not significant since it is less than the tabled critical value $\chi^2_{.05} = 5.99$. For **Distribution I** the value $\chi^2 = .84$ is computed below. As was the case when the **D'Agostino–Pearson test of normality** was employed to evaluate the same distribution, the latter value is not significant since it is less than the tabled critical value $\chi^2_{.05} = 5.99$. Note that the chi-square values computed for the **Jarque–Bera test of normality** are slightly smaller than those computed for the **D'Agostino–Pearson test of normality**, which indicates that the **Jarque–Bera test of normality** provides a more conservative test of the normality hypothesis.

Distribution H

$$\chi^2 = 20\left[\frac{0^2}{6} + \frac{[(5.472) - 3)]^2}{24}\right] = 5.09$$

Distribution I

$$\chi^2 = 20\left[\frac{0^2}{6} + \frac{[(1.994) - 3)]^2}{24}\right] = .84$$

VII. Additional Discussion of the Single-Sample Test for Evaluating Population Kurtosis

1. Exact tables for the single-sample test for evaluating population kurtosis Zar (1999) has derived exact tables for the absolute value of the g_2 statistic for sample sizes in the range $20 \leq n \leq 1000$. By employing exact tables, one can avoid the tedious computations for the **single-sample test for evaluating population kurtosis** (which employs the normal distribution to approximate the exact sampling distribution) described in Section IV. In Zar's (1999) tables, the tabled critical two-tailed .05 and .01 values for g_2 are $g_{2_{.05}} = 2.486$ and $g_{2_{.01}} = 4.121$, and the tabled critical one-tailed .05 and .01 values are $g_{2_{.05}} = 1.850$ and $g_{2_{.01}} = 3.385$. In order to reject the null hypothesis, the computed absolute value of g_2 must be equal to or greater than the tabled critical value (and if a directional alternative hypothesis is evaluated, the sign of g_2 must be in the predicted direction). The probabilities derived for Example 5.1 (through use of Equation 5.7) are very close to the exact probabilities listed in Zar (1999). In other words, the probabilities listed in Zar's (1999) tables for the absolute values $g_{2_H} = 3.596$ and $g_{2_I} = .939$ are almost the same as the tabled probabilities in **Table A1** for the computed values $z_H = 2.40$ and $z_I = -1.13$. In the case of Example 5.1, the same conclusions regarding the null hypothesis will be reached, regardless of whether one employs the normal approximation or Zar's (1999) tables.

2. Additional comments on tests of normality D'Agostino (1986), D'Agostino *et al.* (1990) and Zar (1999) state that the **D'Agostino–Pearson test of normality** provides for a more powerful test of the normality hypothesis than does either the **Kolmogorov-Smirnov goodness-of-fit test for a single sample** or the **chi-square goodness-of-fit test**. D'Agostino *et al.* (1990) state that because of their lack of power, the latter two tests should not be employed for assessing normality. Other sources, however, take a more favorable attitude towards the **Kolmogorov-Smirnov goodness-of-fit test for a single sample** and the **chi-square goodness-of-fit test** as tests of goodness-of-fit for normality (e.g., Conover (1980, 1999), Daniel (1990), Hollander and Wolfe (1999), Marascuilo and McSweeney (1977), Siegel and Castellan (1988), and Sprent (1993)). Thode (2002) has reviewed over forty procedures for assessing goodness-of-fit, and concludes that the **Shapiro-Wilk test for normality** (Shapiro and Wilk (1965, 1968)) and test statistics derived from moments are generally the best methods for assessing normality.

VIII. Additional Examples Illustrating the Use of the Single-Sample Test for Evaluating Population Kurtosis

No additional examples will be presented in this section.

References

Anscombe, F. J. & Glynn, W. W. (1983). Distributions of the kurtosis statistic. **Biometrika**, 70, 227–234.

Conover, W. J. (1980). **Practical nonparametric statistics** (2nd ed.). New York: John Wiley & Sons.

Conover, W. J. (1999). **Practical nonparametric statistics** (3rd ed.). New York: John Wiley & Sons.

Daniel, W. W. (1990). **Applied nonparametric statistics** (2nd ed.). Boston: PWS–Kent Publishing Company.

D'Agostino, R. B. (1986). Tests for the normal distribution In D'Agostino, R. B. & Stephens, M. A. (Eds.), **Goodness-of-fit techniques** (pp. 367–419). New York: Marcel Dekker.

D'Agostino, R. B. & Pearson, E. S. (1973). Tests of departure from normality. Empirical results for the distribution of b_2 and $\sqrt{b_1}$. **Biometrika**, 60, 613–622.

D'Agostino, R. B. & Stephens, M. A. (Eds.) (1986). **Goodness-of-fit techniques**. New York: Marcel Dekker.

D'Agostino, R. B., Belanger, A. & D'Agostino Jr., R. B. (1990). A suggestion for using powerful and informative tests of normality. **American Statistician**, 44, 316–321.

Jarque, C. M. & Bera, A. K. (1980). Efficient tests for normality, homoscedasticity, and serial independence of regression residuals. **Economic Letters**, 6, 255–259.

Jarque, C. M. & Bera, A. K. (1981). Efficient tests for normality, homoscedasticity, and serial independence of regression residuals: Monte Carlo evidence. **Economic Letters**, 7, 313–318.

Jarque, C. M. & Bera, A. K. (1987). A test of normality of observations and regression residuals. **International Statistical Review**, 55, 163–172.

Hollander, M. & Wolfe, D.A. (1999). **Nonparametric statistical methods**. New York: John Wiley & Sons.

Marascuilo, L. A. & McSweeney, M. (1977). **Nonparametric and distribution-free method for the social sciences**. Monterey, CA: Brooks/Cole Publishing Company.

Siegel, S. & Castellan, N. J., Jr. (1988). **Nonparametric statistics for the behavioral sciences** (2nd ed.). New York: McGraw–Hill Book Company.

Sprent, P. (1993). **Applied nonparametric statistical methods** (2nd ed.). London: Chapman & Hall

Tabachnick, B. G. & Fidell, L. S. (2001). **Using multivariate statistics** (4th ed.). Boston: Allyn & Bacon.

Thode, H. C. (2002). **Testing for normality**. New York: Marcel Dekker.

Zar, J. H. (1999). **Biostatistical analysis** (4th ed.). Upper Saddle River, NJ: Prentice Hall.

Endnotes

1.   The reader should take note of the fact that the test for evaluating population kurtosis described in this chapter is a large sample approximation. In point of fact, sources are not in agreement with respect to what equation provides for the best test of the hypothesis of

whether or not g_2 (which represents the value for kurtosis printed out in most computer packages (e.g., *SPSS*)) and/or b_2 deviates significantly from the expected value for a mesokurtic distribution. The general format of alternative equations for evaluating kurtosis which are employed in other sources (e.g., statistical software packages such as *SPSS*, *SAS*, *S-Plus*) involves the computation of a z value by dividing the kurtosis coefficient (represented by g_2) by the estimated population standard error (Equation 5.10). That a different z value can result from use of one or more of these alternative equations derives from the fact that sources are not in agreement with respect to what statistic provides the best estimate of the population standard error (*SE*).

As an example, Tabachnick and Fidell (2001, p. 73–74) note that the value of the standard error can be approximated with the equation $SE = \sqrt{24/n}$, and that the value of z can be computed with Equation 5.10.

$$z = \frac{g_2}{SE}$$ **(Equation 5.10)**

Readers should take note of the fact that an *SPSS* analysis of the examples evaluated in this chapter may yield different z values (see Endnote 3) than those obtained in Section IV. In the final analysis, use of exact tables of the underlying sampling distribution (in Zar (1999, pp. 121–123)) allows for the optimal analysis of the hypothesis of whether or not a g_2 and/or b_2 value deviates significantly from the expected value for a mesokurtic distribution.

2. In the second and third editions of the book the absolute value of g_2 was employed in Equation 5.3 to compute the value of H. However, further scrutiny of the derivation of the latter equation in D'Agostino et al. (1990) and D'Agostino and Stephens (1986) indicates that the sign of g_2 must be taken into account. More specifically, the latter sources indicate that the value of G computed with Equation 5.2 represents the variance of b_2, and that H represents a standardized value of b_2, which can assume either a positive or negative value. If the absolute value of g_2 is employed in Equation 5.3, the value of z computed with Equation 5.7 can only be positive, which the above noted sources demonstrate will not always be the case.

3. Analysis of the data with *SPSS* yields the following results. Specifically: a) For **Distribution H**, *SPSS* computes the values *kurtosis* = g_2 = 3.58 and SE = .992. When the latter values are substituted in Equation 5.10, the value $z = 3.58/.992 = 3.61$ is obtained (which is substantially greater than the z value obtained with Equation 5.7). Since $z = 3.61$ is greater than the tabled critical two-tailed value $z_{.01}$ = 2.58 and the tabled critical one-tailed value $z_{.01}$ = 2.33, the null hypothesis can be rejected if the nondirectional alternative hypothesis or the directional alternative hypothesis stipulating leptokurtosis is employed — in other words, the researcher can conclude there is a high likelihood the sample is derived from a distribution that is **leptokurtic**; b) For **Distribution I**, *SPSS* computes the values *kurtosis* = g_2 = -.939 and SE = .992. When the latter values are substituted in Equation 5.10, the value $z = -.939/.992 = -.95$ is obtained. Since the absolute value z = .95 is less than the tabled critical two-tailed value $z_{.05}$ = 1.96 and the tabled critical one-tailed value $z_{.05}$ = 1.65, the null hypothesis cannot be rejected regardless of which alternative hypothesis is employed — in other words, there is insufficient evidence to indicate that the sample is derived from a distribution that is not **mesokurtic**.

4. When **Table A4** is employed to evaluate a chi-square value computed for the **D'Agostino–Pearson test of normality**, the following protocol is employed. The tabled critical values for $df = 2$ are derived from the right tail of the distribution. Thus, the tabled critical .05 chi-square value (to be designated $\chi^2_{.05}$) will be the tabled chi-square value at the 95th percentile. In the same respect, the tabled critical .01 chi-square value (to be designated $\chi^2_{.01}$) will be the tabled chi-square value at the 99th percentile. For further clarification of interpretation of the critical values in **Table A4**, the reader should consult Section V of the **single-sample chi-square test for a population variance (Test 3)**.

Test 6

The Wilcoxon Signed-Ranks Test
(Nonparametric Test Employed with Ordinal Data)

I. Hypothesis Evaluated with Test and Relevant Background Information

Hypothesis evaluated with test Does a sample of n subjects (or objects) come from a population in which the median (θ) equals a specified value?

Relevant background information on test The **Wilcoxon signed-ranks test** (Wilcoxon (1945, 1949)) is a nonparametric procedure employed in a hypothesis testing situation involving a single sample in order to determine whether or not a sample is derived from a population in which the median (θ) is equal to a specified value. (The population median will be represented by the notation θ, which is the lower case Greek letter **theta**.) If the **Wilcoxon signed-ranks test** yields a significant result, the researcher can conclude there is a high likelihood the sample is derived from a population with a median value other than θ.

The **Wilcoxon signed-ranks test** is based on the following assumptions:[1] a) The sample has been randomly selected from the population it represents; b) The original scores obtained for each of the subjects/objects are in the format of interval/ratio data; and c) The underlying population distribution is symmetrical. When there is reason to believe that the latter assumption is violated, Daniel (1990), among others, recommends that the **binomial sign test for a single sample (Test 9)** be employed in place of the **Wilcoxon signed-ranks test**.[2] Proponents of nonparametric tests recommend that the **Wilcoxon signed-ranks test** be employed in place of the **single-sample t test (Test 2)** when there is reason to believe that the normality assumption of the latter test has been saliently violated.[3] It should be noted that all of the other tests in this text which rank data (with the exception of the **Wilcoxon matched-pairs signed-ranks test (Test 18)** and the **Moses test for equal variability (Test 15)**) rank the original interval/ratio scores of subjects. The **Wilcoxon signed-ranks test**, however, does not rank subjects' original interval/ratio scores, but instead ranks difference scores — specifically, the obtained difference between each subject's score and the hypothesized value of the population median. For this reason, some sources categorize the **Wilcoxon signed-ranks test** as a test of interval/ratio data. Most sources, however (including this book), categorize the test as one involving ordinal data, because a ranking procedure is part of the test protocol.

II. Example

Example 6.1 *A physician states that the median number of times he sees each of his patients during the year is five. In order to evaluate the validity of this statement, he randomly selects ten of his patients and determines the number of office visits each of them made during the past year. He obtains the following values for the ten patients in his sample: 9, 10, 8, 4, 8, 3, 0, 10, 15, 9. Do the data support his contention that the median number of times he sees a patient is five?*

III. Null versus Alternative Hypotheses

Null hypothesis $H_0: \theta = 5$

(The median of the population the sample represents equals 5. With respect to the sample data, this translates into the sum of the ranks of the positive difference scores being equal to the sum of the ranks of the negative difference scores (i.e., $\Sigma R+ = \Sigma R-$).)

Alternative hypothesis $H_1: \theta \neq 5$

(The median of the population the sample represents does not equal 5. With respect to the sample data, this translates into the sum of the ranks of the positive difference scores not being equal to the sum of the ranks of the negative difference scores (i.e., $\Sigma R+ \neq \Sigma R-$). This is a **non-directional alternative hypothesis** and it is evaluated with a **two-tailed test**.)

or

$$H_1: \theta > 5$$

(The median of the population the sample represents is some value greater than 5. With respect to the sample data, this translates into the sum of the ranks of the positive difference scores being greater than the sum of the ranks of the negative difference scores (i.e., $\Sigma R+ > \Sigma R-$). This is a **directional alternative hypothesis** and it is evaluated with a **one-tailed test**.)

or

$$H_1: \theta < 5$$

(The median of the population the sample represents is some value less than 5. With respect to the sample data, this translates into the sum of the ranks of the positive difference scores being less than the sum of the ranks of the negative difference scores (i.e., $\Sigma R+ < \Sigma R-$). This is a **directional alternative hypothesis** and it is evaluated with a **one-tailed test**.)

     Note: Only one of the above noted alternative hypotheses is employed. If the alternative hypothesis the researcher selects is supported, the null hypothesis is rejected.

IV. Test Computations

The data for Example 6.1 are summarized in Table 6.1. The scores of the 10 subjects are recorded in Column 2 of Table 6.1. In Column 3, a D score is computed for each subject. This score, which is referred to as a **difference score**, is the difference between a subject's score and the hypothesized value of the population median, $\theta = 5$ (i.e., $D = X - \theta$). Column 4 contains the ranks of the difference scores. In ranking the difference scores for the **Wilcoxon signed-ranks test**, the following guidelines are employed:

     a) The **absolute values** of the difference scores ($|D|$) are ranked. (i.e., the sign of a difference score is not taken into account).

     b) Any difference score that equals zero is not ranked. This translates into eliminating from the analysis any subject who yields a difference score of zero.

     c) In ranking the absolute values of the difference scores, the following protocol should be employed: Assign a rank of 1 to the difference score with the lowest absolute value, a rank of 2 to the difference score with the second lowest absolute value, and so on until the highest rank is assigned to the difference score with the highest absolute value. When there are tied scores

Table 6.1 Data for Example 6.1

Subject	X	$D = X - \theta$	Rank of $|D|$	Signed rank of $|D|$
1	9	4	5.5	5.5
2	10	5	8	8
3	8	3	3.5	3.5
4	4	−1	1	−1
5	8	3	3.5	3.5
6	3	−2	2	−2
7	0	−5	8	−8
8	10	5	8	8
9	15	10	10	10
10	9	4	5.5	5.5

$$\Sigma R+ = 44$$
$$\Sigma R- = 11$$

present, the average of the ranks involved is assigned to all difference scores tied for a given rank. Because of this fact, when there are tied scores for either the lowest or highest difference scores, the rank assigned to the lowest difference score will be some value greater than 1, and the rank assigned to the highest difference score will be some value less than n. To further clarify how ties are handled, examine Table 6.2 which lists the **difference scores** of the 10 subjects. In the table, the difference scores (based on their absolute values) are arranged ordinally, after which they are ranked employing the protocol described above.

Table 6.2 Ranking Procedure for Wilcoxon Signed-Ranks Test

Subject number	4	6	3	5	1	10	2	7	8	9
Subject's difference score	−1	−2	3	3	4	4	5	−5	5	10
Absolute value of difference score	1	2	3	3	4	4	5	5	5	10
Rank of $|D|$	1	2	3.5	3.5	5.5	5.5	8	8	8	10

The difference score of Subject 4 has the lowest absolute value (i.e., 1), and because of this it is assigned a rank of 1. The next lowest absolute value for a difference score (2) is that of Subject 6, and thus it is assigned a rank of 2.[4] The difference score of 3 (which is obtained for both Subjects 3 and 5) is the score that corresponds to the third rank-order. Since, however, there are two instances of this difference score, it will also use up the position reserved for the fourth rank-order (i.e., 3 and 4 are the two ranks that would be employed if, in fact, these two subjects did not have the identical difference score). Instead of arbitrarily assigning one of the subjects with a difference score of 3 a rank-order of 3 and the other subject a rank-order of 4, we compute the average of the two ranks that are involved (i.e., $(3 + 4)/2 = 3.5$), and assign that value as the rank-order for the difference scores of both subjects. The next rank-order in the sequence of the 10 rank-orders is 5. Once again, however, two subjects (Subjects 1 and 10) are tied for the difference score in the fifth ordinal position (which happens to involve a difference score of 4). Since, if not equal to one another, these two difference scores would involve the fifth and sixth ranks, we compute the average of these two ranks (i.e., $(5 + 6)/2 = 5.5$), and assign that value as the rank for the difference scores of Subjects 1 and 10. With respect to the next difference score (5), there is a three-way tie involving Subjects 2, 7, and 8 (keeping in mind that the absolute value of the difference score for Subject 7 is 5). The average of the three ranks which would be involved if the subjects had obtained different difference scores is computed (i.e., $(7 + 8 + 9)/3 = 8$), and that average value is assigned to the difference scores of Subjects 2, 7, and 8. Since the remaining difference score of 10 (obtained by Subject 9) is the highest difference score, it is assigned the highest rank which equals 10.

It should be emphasized that in the **Wilcoxon signed-ranks test** it is essential that a rank of 1 be assigned to the difference score with the lowest absolute value, and that the highest rank be assigned to the difference score with the highest absolute value. In most other tests that involve ranking, the ranking procedure can be reversed (i.e., the same test statistic will be obtained if one assigns a rank of 1 to the highest score and the highest rank to the lowest score). However, if one reverses the ranking procedure in conducting the **Wilcoxon signed-ranks test**, it will invalidate the results of the test.

d) After ranking the absolute values of the difference scores, the sign of each difference score is placed in front of its rank. The signed ranks of the difference scores are listed in Column 5 of Table 6.1.

The sum of the ranks that have a positive sign (i.e., $\Sigma R+ = 44$) and the sum of the ranks that have a negative sign (i.e., $\Sigma R- = 11$) are recorded at the bottom of the Column 5 in Table 6.1. Equation 6.1 allows one to check the accuracy of these values. If the relationship indicated by Equation 6.1 is not obtained, it indicates an error has been made in the calculations. In Equation 6.1, n represents the number of signed ranks (i.e., the number of difference scores that are ranked).

$$\Sigma R+ \; + \; \Sigma R- \; = \; \frac{n(n \; + \; 1)}{2} \qquad \textbf{(Equation 6.1)}$$

Employing the values $\Sigma R+ = 44$ and $\Sigma R- = 11$ in Equation 6.1, we confirm that the relationship described by the equation is true.

$$44 \; + \; 11 \; = \; \frac{(10)(11)}{2} \; = \; 55$$

It is important to note that in the event one or more subjects obtain a difference score of zero, such scores are not employed in the analysis. In such a case, the value of n in Equation 6.1 will only represent the number of scores that have been assigned ranks. Example 6.2 in Section VIII illustrates the use of the **Wilcoxon signed-ranks test** with data in which difference scores of zero are present.

V. Interpretation of the Test Results

As noted in Section III, if the sample is derived from a population with a median value equal to the hypothesized value of the population median (i.e., the null hypothesis is true), the values of $\Sigma R+$ and $\Sigma R-$ will be equal to one another. When $\Sigma R+$ and $\Sigma R-$ are equivalent, both of these values will equal $[n(n+1)]/4$, which in the case of Example 6.1 will be $[(10)(11)]/4 = 27.5$. This latter value is commonly referred to as the **expected value** of the **Wilcoxon T statistic**.

If the value of $\Sigma R+$ is significantly greater than the value of $\Sigma R-$, it indicates there is a high likelihood the sample is derived from a population with a median value which is larger than the hypothesized value of the population median. On the other hand, if $\Sigma R-$ is significantly greater than $\Sigma R+$, it indicates there is a high likelihood the sample is derived from a population with a median value that is less than the hypothesized value of the population median. The fact that $\Sigma R+ = 44$ is greater than $\Sigma R- = 11$ indicates that the data are consistent with the directional alternative hypothesis $H_1: \theta > 5$. The question is, however, whether the difference is significant — i.e., whether it is large enough to conclude that it is unlikely to be the result of chance.

The absolute value of the **smaller** of the two values $\Sigma R+$ versus $\Sigma R-$ is designated as the **Wilcoxon T test statistic**. Since $\Sigma R- = 11$ is smaller than $\Sigma R+ = 44$, $T = 11$. The T value is interpreted by employing **Table A5 (Table of Critical T Values for Wilcoxon's Signed-Ranks and Matched-Pairs Signed-Ranks Tests)** in the **Appendix**. **Table A5** lists the critical two-tailed and one-tailed .05 and .01 T values in relation to the number of signed ranks in a set of data. In order to be significant, the obtained value of T must be **equal to or less than** the tabled critical T value at the prespecified level of significance.[5] Table 6.3 summarizes the tabled critical two-tailed and one-tailed .05 and .01 Wilcoxon T values for $n = 10$ signed ranks.

Table 6.3 Tabled Critical Wilcoxon T Values for $n = 10$ Signed Ranks

	$T_{.05}$	$T_{.01}$
Two-tailed values	8	3
One-tailed values	10	5

Since the null hypothesis can only be rejected if the computed value $T = 11$ is equal to or less than the tabled critical value at the prespecified level of significance, we can conclude the following.

In order for the nondirectional alternative hypothesis H_1: $\theta \neq 5$ to be supported, it is irrelevant whether $\Sigma R+ > \Sigma R-$ or $\Sigma R- > \Sigma R+$. In order for the result to be significant, the computed value of T must be equal to or less than the tabled critical two-tailed value at the prespecified level of significance. Since the computed value $T = 11$ is greater than the tabled critical two-tailed .05 value $T_{.05} = 8$, the nondirectional alternative hypothesis H_1: $\theta \neq 5$ is not supported at the .05 level. It is also not supported at the .01 level, since $T = 11$ is greater than the tabled critical two-tailed .01 value $T_{.01} = 3$.

In order for the directional alternative hypothesis H_1: $\theta > 5$ to be supported, $\Sigma R+$ must be greater than $\Sigma R-$. Since $\Sigma R+ > \Sigma R-$, the data are consistent with the directional alternative hypothesis H_1: $\theta > 5$. In order for the result to be significant, the computed value of T must be equal to or less than the tabled critical one-tailed value at the prespecified level of significance. Since the computed value $T = 11$ is greater than the tabled critical one-tailed .05 value $T_{.05} = 10$, the directional alternative hypothesis is not supported at the .05 level. It is also not supported at the .01 level, since $T = 11$ is greater than the tabled critical one-tailed .01 value $T_{.01} = 5$.

In order for the directional alternative hypothesis H_1: $\theta < 5$ to be supported, the following two conditions must be met: a) $\Sigma R-$ must be greater than $\Sigma R+$; and b) the computed value of T must be equal to or less than the tabled critical one-tailed value at the prespecified level of significance. Since the first of these conditions is not met, the directional alternative hypothesis H_1: $\theta < 5$ is not supported.

A summary of the analysis of Example 6.1 with the **Wilcoxon signed-ranks test** follows: With respect to the median number of times the doctor sees a patient, we can conclude that the data do not indicate that the sample of 10 subjects comes from a population with a median value other than 5.

Except for the fact that the mean rather than the median is employed as the population parameter stated in the null and alternative hypotheses, Example 2.1 is identical to Example 6.1 (i.e., the two examples employ the same set of data). Since Example 2.1 states the null hypothesis with reference to the population mean, it is evaluated with the **single-sample t test**. At this point we will compare the results of the two tests. When the same data are evaluated with the **single-sample t test**, the null hypothesis can be rejected when the directional alternative hypothesis H_1: $\mu > 5$ is employed, but only at the .05 level. With reference to the latter alternative hypothesis, the obtained t value exceeds the tabled critical $t_{.05}$ value by a comfortable margin. When the **single-sample t test** is employed, the nondirectional alternative hypothesis H_1: $\mu \neq 5$ is not supported at the .05 level.

When Example 6.1 is evaluated with the **Wilcoxon signed-ranks test**, the null hypothesis cannot be rejected regardless of which alternative hypothesis is employed. However, when the directional alternative hypothesis H_1: $\theta > 5$ is employed, the **Wilcoxon signed-ranks test** falls just short of being significant at the .05 level. Directly related to this is the fact that in some sources the tabled critical values published for the Wilcoxon test statistic are not identical to the values listed in **Table A5**. These differences are the result of rounding off protocol. The critical T values in **Table A5** listed for a given level of significance are associated with the probability that is closest to but not greater than the value of alpha. In some instances a T value listed in an alternative table may be one point higher than the value listed in **Table A5**, thus making it easier to reject the null hypothesis. Although these alternative critical values are actually closer to the value of alpha than the values listed in **Table A5**, the probability associated with a tabled critical T value in the alternative table is, in fact, larger than the value of alpha. With reference to Example 6.1, the exact probability associated with $T = 10$ is .0420 (i.e., this represents the likelihood of obtaining a T value of 10 or less). The probability associated with $T = 11$, which is the critical value of T listed in the alternative table, is .0527. Although the latter probability is closer to $\alpha = .05$ than is .0420, it falls above 05. Thus, if one employs the alternative table that contains the tabled critical one-tailed .05 value $T_{.05} = 11$, the alternative hypothesis H_1: $\theta > 5$ is supported at the .05 level. Obviously in a case such as this where the likelihood of obtaining a value equal to or less than the computed value of T is just sightly above .05, it would seem prudent to conduct further studies in order to clarify the status of the alternative hypothesis H_1: $\theta > 5$.

In the case of Examples 6.1 and 2.1, the results of the **Wilcoxon signed-ranks test** and **single-sample t test** are fairly consistent for the same set of data. Support for the analogous alternative hypotheses H_1: $\mu > 5$ and H_1: $\theta > 5$ is either clearly indicated (in the case of the *t* **test**) or falls just short of significance (in the case of the **Wilcoxon test**). The slight discrepancy between the two tests reflects the fact that, as a general rule, nonparametric tests are not as powerful as their parametric analogs. In the case of the two tests under consideration, the lower power of the **Wilcoxon signed-ranks test** can be attributed to the loss of information which results from expressing interval/ratio data in a rank-order format (specifically, rank-ordering the difference scores). As noted earlier, when two or more inferential statistical tests are applied to the same set of data and yield contradictory results, it is prudent to replicate the study. In the final analysis, replication is the most powerful tool a researcher has at his disposal for determining the status of a null hypothesis.

VI. Additional Analytical Procedures for the Wilcoxon Signed-Ranks Test and/or Related Tests

1. The normal approximation of the Wilcoxon T statistic for large sample sizes   If the sample size employed in a study is relatively large, the normal distribution can be used to approximate the Wilcoxon T statistic. Although sources do not agree on the value of the sample size that justifies employing the normal approximation of the Wilcoxon distribution, they generally state it should be used for sample sizes larger than those documented in the Wilcoxon table contained within the source. Equation 6.2 provides the normal approximation for Wilcoxon T. In the equation T represents the computed value of Wilcoxon T, which for Example 6.1 is $T = 11$. n, as noted previously, represents the number of signed ranks. Thus, in our example, $n = 10$. Note that in the numerator of Equation 6.2, the term $[n(n + 1)]/4$ represents the expected value of T (often summarized with the symbol T_E), which is defined in Section V. The denominator of Equation 6.2 represents the expected standard deviation of the sampling distribution of the T statistic.

$$z = \frac{T - \dfrac{n(n+1)}{4}}{\sqrt{\dfrac{n(n+1)(2n+1)}{24}}} \qquad \textbf{(Equation 6.2)}$$

Although Example 6.1 involves only ten signed ranks (a value most sources would view as too small to use with the normal approximation), it will be employed to illustrate Equation 6.2. The reader will see that in spite of employing Equation 6.2 with a small sample size, it will yield essentially the same result as that obtained when the exact table of the Wilcoxon distribution is employed. When the values $T = 11$ and $n = 10$ are substituted in Equation 6.2, the value $z = -1.68$ is computed.

$$z = \frac{11 - \dfrac{(10)(11)}{4}}{\sqrt{\dfrac{(10)(11)(21)}{24}}} = -1.68$$

The obtained value $z = -1.68$ is evaluated with **Table A1 (Table of the Normal Distribution)** in the **Appendix**. In **Table A1** the tabled critical two-tailed .05 and .01 values are $z_{.05} = 1.96$ and $z_{.01} = 2.58$, and the tabled critical one-tailed .05 and .01 values are $z_{.05} = 1.65$ and $z_{.01} = 2.33$.

Since the smaller of the two values $\Sigma R+$ versus $\Sigma R-$ is selected to represent T, the value of z computed with Equation 6.2 will always be a negative number (unless $\Sigma R+ = \Sigma R-$, in which case z will equal zero). This is the case since, by selecting the smaller value, T will always be less than the expected value T_E. As a result of this, the following guidelines are employed when evaluating the null hypothesis.

a) If a nondirectional alternative hypothesis is employed, the null hypothesis can be rejected if the obtained absolute value of z is equal to or greater than the tabled critical two-tailed value at the prespecified level of significance.

b) When a directional alternative hypothesis is employed, one of the two possible directional alternative hypotheses will be supported if the obtained absolute value of z is equal to or greater than the tabled critical one-tailed value at the prespecified level of significance. Which alternative hypothesis is supported depends on the prediction regarding which of the two values $\Sigma R+$ versus $\Sigma R-$ is larger. The null hypothesis can only be rejected if the directional alternative hypothesis that is consistent with the data is supported.

Employing the above guidelines, when the normal approximation is used with Example 6.1, the following conclusions can be reached.

The nondirectional alternative hypothesis $H_1: \theta \neq 5$ is not supported. This is the case since the computed absolute value $z = 1.68$ is less than the tabled critical two-tailed .05 value $z_{.05} = 1.96$. This decision is consistent with the decision that is reached when the exact table of the Wilcoxon distribution is employed to evaluate the nondirectional alternative hypothesis $H_1: \theta \neq 5$.

The directional alternative hypothesis $H_1: \theta > 5$ is supported at the .05 level. This is the case since the data are consistent with the latter alternative hypothesis (i.e., $\Sigma R+ > \Sigma R-$), and the computed absolute value $z = 1.68$ is greater than the tabled critical one-tailed .05 value

$z_{.05} = 1.65$. The directional alternative hypothesis H_1: $\theta > 5$ is not supported at the .01 level, since the absolute value $z = 1.68$ is less than the tabled critical one-tailed .01 value $z_{.01} = 2.33$. When the exact table of the Wilcoxon distribution is employed, the directional alternative hypothesis H_1: $\theta > 5$ is not supported at the .05 level. However, it was noted that if an alternative table of Wilcoxon critical values is employed, the alternative hypothesis H_1: $\theta > 5$ is supported at the .05 level.

The directional alternative hypothesis H_1: $\theta < 5$ is not supported, since the data are not consistent with the latter alternative hypothesis (which requires that $\Sigma R- > \Sigma R+$).

In closing the discussion of the normal approximation, it should be noted that, in actuality, either $\Sigma R+$ or $\Sigma R-$ can be employed to represent the value of T in Equation 6.2. Either value will yield the same absolute value for z. The smaller of the two values will always yield a negative z value, and the larger of the two values will always yield a positive z value (which in this instance will be $z = 1.68$ if $\Sigma R+ = 44$ is employed in Equation 6.2 to represent T). In evaluating a nondirectional alternative hypothesis, the sign of z is irrelevant. In the case of a directional alternative hypothesis, one must determine whether the data are consistent with the alternative hypothesis that is stipulated. If the data are consistent, one then determines whether or not the absolute value of z is equal to or greater than the tabled critical one-tailed value at the pre-specified level of significance.

2. The correction for continuity for the normal approximation of the Wilcoxon signed-ranks test Although it is not described in most sources, Marascuilo and McSweeney (1977) employ a correction factor known as the **correction for continuity** for the normal approximation of the Wilcoxon test statistic. The **correction for continuity** is recommended by some sources for use with a number of nonparametric tests which employ a **continuous distribution** (such as the normal distribution) to estimate a **discrete distribution** (such as in this instance the Wilcoxon distribution). As noted in the **Introduction**, in a continuous distribution there are an infinite number of values a variable may assume, whereas in a discrete distribution the number of possible values a variable may assume is limited in number. The correction for continuity is based on the premise that if a continuous distribution is employed to estimate a discrete distribution, such an approximation will inflate the Type I error rate. By employing the correction for continuity, the Type I error rate is ostensibly adjusted to be more compatible with the prespecified alpha value designated by the researcher. When the correction for continuity is applied to a normal approximation of an underlying discrete distribution, it results in a slight reduction in the absolute value computed for z. In the case of the normal approximation of the Wilcoxon test statistic, the correction for continuity requires that .5 be subtracted from the absolute value of the numerator of Equation 6.2. Thus, Equation 6.3 represents the continuity-corrected normal approximation of the Wilcoxon test statistic.

$$z = \frac{\left| T - \frac{n(n+1)}{4} \right| - .5}{\sqrt{\frac{n(n+1)(2n+1)}{24}}}$$ **(Equation 6.3)**

If the correction for continuity is employed with Example 6.1, the value of the numerator of Equation 6.3 is 16, in contrast to the absolute value of 16.5 computed with Equation 6.2. Employing Equation 6.3, the continuity-corrected value $z = 1.63$ is computed. Note that as a result of the absolute value conversion, the numerator of Equation 6.3 will always be a positive number, thus yielding a positive z value.

$$z = \frac{\left| 11 - \frac{(10)(11)}{4} \right| - .5}{\sqrt{\frac{(10)(11)(21)}{24}}} = 1.63$$

Since the absolute value $z = 1.63$ is less than the tabled critical one-tailed .05 value $z_{.05} = 1.65$, the directional hypothesis H_1: $\theta > 5$ is not supported. Note that since the obtained absolute value $z = 1.63$ is slightly below the tabled critical one-tailed value $z_{.05} = 1.65$, it is just short of being significant (in contrast to the continuity-uncorrected absolute value $z = 1.68$ computed with Equation 6.2, which barely achieves significance at the .05 level). The result obtained with $z = 1.63$ is consistent with that obtained employing the exact table of the Wilcoxon distribution. In a case such as this, additional research should be conducted to clarify the status of the null hypothesis, since the issue of whether or not to reject it depends on whether or not one employs the correction for continuity.

3. Tie correction for the normal approximation of the Wilcoxon test statistic Equation 6.4 is an adjusted version of Equation 6.2 that is recommended in some sources (e.g., Daniel (1990) and Marascuilo and McSweeney (1977)) when tied difference scores are present in the data. The tie correction results in a slight increase in the absolute value of z. Unless there are a substantial number of ties, the difference between the values of z computed with Equations 6.2 and 6.4 will be minimal.

$$z = \frac{T - \frac{n(n + 1)}{4}}{\sqrt{\frac{n(n + 1)(2n + 1)}{24} - \frac{\Sigma t^3 - \Sigma t}{48}}} \qquad \textbf{(Equation 6.4)}$$

Table 6.4 illustrates the application of the tie correction with Example 6.1.

Table 6.4 Correction for Ties with Normal Approximation

Subject	Rank	t	t^3
4	1		
6	2		
3	3.5		
		2	8
5	3.5		
1	5.5		
		2	8
10	5.5		
2	8		
7	8		
		3	27
8	8		
9	10		
		$\Sigma t = 7$	$\Sigma t^3 = 43$

In the data for Example 6.1 there are three sets of tied ranks: Set 1 involves two subjects (Subjects 3 and 5); Set 2 involves two subjects (Subjects 1 and 10); and Set 3 involves three subjects (Subjects 2, 7, and 8). The number of subjects involved in each set of tied ranks represents the values of t in the third column of Table 6.4. The three t values are cubed in the last column of the table, after which the values Σt and Σt^3 are computed. The appropriate values are now substituted in Equation 6.4.[6]

$$z = \frac{11 - \dfrac{(10)(11)}{4}}{\sqrt{\dfrac{(10)(11)(21)}{24} - \dfrac{43 - 7}{48}}} = -1.69$$

The absolute value $z = 1.69$ is slightly larger than the absolute value $z = 1.68$ obtained without the tie correction. The difference between the two methods is trivial, and in this instance, regardless of which alternative hypothesis is employed, the decision the researcher makes with respect to the null hypothesis is not affected.[7]

Conover (1980, 1999) and Daniel (1990) discuss and/or cite sources on the subject of alternative ways of handling tied difference scores. Conover (1980, 1999) also notes that in some instances, retaining and ranking zero difference scores may actually provide a more powerful test of an alternative hypothesis than the more conventional method employed in this book (which eliminates zero difference scores from the data).

4. Computation of a confidence interval for a population median   Prior to reading this section the reader should review the discussion of the computation of a confidence interval for a population mean in Section VI of the **single-sample *t* test**. Equations 6.5 and 6.6 (also found in Glass and Hopkins (1996), Snedecor and Cochran (1980), and Zar (1999)) can be employed to compute a large sample normal approximation for a confidence interval for a population median.[8]

$$U = \frac{n + 1}{2} + \frac{z_{\alpha/2} \sqrt{n}}{2}$$ **(Equation 6.5)**

$$L = n - U + 1$$ **(Equation 6.6)**

Where: $z_{\alpha/2}$ represents the tabled critical two-tailed value in the normal distribution below which a proportion (percentage) equal to $[1 - (\alpha/2)]$ of the cases falls. If the proportion (percentage) of the distribution that falls within the confidence interval is subtracted from 1 (100%), it will equal the value of α.

U will represent the **ordinal position** of the score which represents the upper limit of the $[1 - (\alpha/2)]^{th}$ confidence interval. Any fractional value is rounded off to the next highest integer number.

L will represent the **ordinal position** of the score which represents the lower limit of the $[1 - (\alpha/2)]^{th}$ confidence interval.

The data for the sample employed in Example 6.1 will be used to demonstrate the use of Equations 6.5 and 6.5 to compute the 95% confidence interval. The scores of the ten subjects which comprise the sample are arranged below from lowest to highest moving from left to right.

0, 3, 4, 8, 8, 9, 9, 10, 10, 15

The values $n = 10$ and $z_{.05} = 1.96$ (which represents the tabled critical two-tailed .05 z value in **Table A1**) are employed in Equation 6.5 to compute the upper boundary of the 95% confidence interval. Substituting $n = 10$ and $z_{.05} = 1.96$ in Equation 6.5 yields the value $U = 8.60$, which rounded off to the next highest integer value is set equal to 9. When $U = 9$ is substituted in Equation 6.6 it yields the value $L = 2$. The result indicates that the score in the **ninth ordinal position** (which is a score of **10** in the above distribution) represents the upper limit of the confidence interval, while the score in the **second ordinal position** (which is the score of **3** in the above distribution) represents the lower limit of the confidence interval.

$$U \geq \frac{10 + 1}{2} + \frac{1.96 \sqrt{10}}{2} = 5.5 + 3.10 = 8.60$$

$$L = 10 - 9 + 1 = 2$$

The above result can be summarized as follows: $3 \leq \theta \leq 10$. The notation $3 \leq \theta \leq 10$ indicates that the value of the population median θ is greater than or equal to 3 and less than or equal to 10. Thus we can be 95% confident (or the probability is .95) that the interval 3 to 10 contains the true value of the population median. With respect to Example 6.1, the physician can be 95% confident that the range of values 3 to 10 includes the actual value for the median number of visits per patient. Note that the range of values $3 \leq \theta \leq 10$ which define the limits of the confidence interval for the population median are close (but not identical) to the range of values $4.57 \leq \mu \leq 10.63$ computed in Section VI of the **single-sample** t **test** for the 95% confidence interval for the population mean for the same set of data.

The confidence interval based on the normal distribution computed in this section is in actuality an approximation of an exact confidence interval which can be computed through use of the **binomial distribution** (which is discussed in detail in the chapter on the **binomial sign test for a single sample (Test 9)**). The methodology for computing a confidence interval for the population median employing the binomial distribution (which employs the probabilities in **Table A6 (Table of the Binomial Distribution, Individual Probabilities)** in the **Appendix**) is described in Higgins (2004), Marascuilo and McSweeney (1977), and Zar (1999). Conover (1980, 1999) and Daniel (1990) describe alternative procedures for computing the large sample normal approximation of the confidence interval.

VII. Additional Discussion of the Wilcoxon Signed-Ranks Test

1. Power-efficiency of the Wilcoxon signed-ranks test and the concept of asymptotic relative efficiency Power-efficiency (also referred to as **relative efficiency**) is a statistic which is employed to indicate the power of two tests relative to one another. It is most commonly used in comparing the power of a nonparametric test with its parametric analog. As an example, assume we wish to determine the relative power of the **Wilcoxon signed-ranks test** (designated as Test A) and the **single-sample** t **test** (designated as Test B). Assume that both tests employ the same alpha level with respect to the null hypothesis being evaluated.

For a fixed power value, the statistic PE_{AB} will represent the **power-efficiency of Test A relative to Test B**. The value of PE_{AB} is computed with Equation 6.7.

$$PE_{AB} = \frac{n_B}{n_A}$$

(Equation 6.7)

Where: n_A is the number of subjects required for Test A and n_B is the number of subjects required for Test B, when each test is required to evaluate an alternative hypothesis at the same power

Thus, if the **single-sample *t* test** requires 95 subjects to evaluate an alternative hypothesis at a power of .80, and the **Wilcoxon signed-ranks test** requires 100 subjects to evaluate the analogous alternative hypothesis at a power of .80, the value of PE_{AB} = 95/100 = .95. From this result it can be determined that if 100 subjects are employed to evaluate an alternative hypothesis with the **single-sample *t* test**, in order to achieve the same level of power for evaluating the analogous alternative hypothesis with the **Wilcoxon signed-ranks test**, it is necessary to employ (1/.95)(100) = 105 subjects.

Conover (1980, 1999) notes that the value computed for **relative efficiency** will be a function of the alpha and beta values a researcher employs in analyzing the data for a study. Pitman (1948) demonstrated that the value of relative efficiency computed for all possible choices of alpha and beta approaches a limiting value as n_A approaches infinity. Pitman referred to this limiting value as the **asymptotic relative efficiency** of the two tests (which is often represented by the acronym **ARE**, and is also referred to as the **Pitman efficiency**). Since asymptotic relative efficiency is a limiting value that is based on a large sample size, it may not be an accurate metric of efficiency when the sample size employed in a study is relatively small. However, in spite of the latter, for some nonparametric tests, the value computed for asymptotic relative efficiency is achieved with a relatively small sample size. The asymptotic relative efficiency of a test is of practical value, in that if a researcher is selecting among two or more nonparametric tests as an alternative to a parametric test, the nonparametric test with the highest asymptotic relative efficiency will allow for the most powerful test of the alternative hypothesis.

Marascuilo and McSweeney (1977, p. 87) present a table of asymptotic relative efficiency values for a variety of nonparametric tests. In the latter table, asymptotic relative efficiency values are listed in reference to underlying population distributions with different shapes. In the case of the **Wilcoxon signed-ranks test**, its asymptotic relative efficiency is .955 (when contrasted with the **single-sample *t* test**) when the underlying population distribution is normal. For population distributions which are not normal, the asymptotic relative efficiency of the **Wilcoxon signed-ranks test** is generally equal to or greater than 1. It is interesting to note that when the population distribution is normal, the asymptotic relative efficiency of most nonparametric tests will be less than 1. However, when the underlying population is not normal, it is not uncommon for a nonparametric test to have an asymptotic relative efficiency greater than 1. As a general rule, proponents of nonparametric tests take the position that when a researcher has reason to believe that the normality assumption of the **single-sample *t* test** has been saliently violated, the **Wilcoxon signed-ranks test** provides a powerful test of the comparable alternative hypothesis.

2. Note on symmetric population concerning hypotheses regarding median and mean

Conover (1980, 1999) and Daniel (1990) note that if, in fact, the population from which the sample is derived is symmetrical, the conclusions one draws with regard to the population median are also true with respect to the population mean (since in a symmetrical population the values of the mean and median will be identical). This amounts to saying that if in Example 6.1 we retain (or reject) H_0: θ = 5, we are also reaching the same conclusion with respect to the null hypothesis H_0: μ = 5. There is, however, no guarantee that the results obtained with the **Wilcoxon signed-ranks test** will be entirely consistent with the results derived when the **single-sample *t* test** is employed to evaluate the same set of data.

VIII. Additional Examples Illustrating the Use of the Wilcoxon Signed-Ranks Test

With the exception of Examples 1.5 and 1.6, the **Wilcoxon signed-ranks test** can be employed to evaluate a hypothesis about a population median with any of the examples that are employed to illustrate the **single-sample z test (Test 1)** and the **single-sample t test**. As noted in Section I, unless the normality assumption of the aforementioned tests is saliently violated, most researchers would employ a parametric test in lieu of a nonparametric alternative. If the **Wilcoxon signed-ranks test** is employed for Examples 1.1–1.4 and Example 2.2 (which has the same data as Examples 2.1 and 6.1), difference scores are obtained by subtracting the hypothesized value of the population median from each score in the sample. All difference scores are then ranked and evaluated in accordance with the ranking protocol described in Section IV.

Example 6.2 (which is a restatement of Example 6.1 with a different set of data) illustrates the use of the **Wilcoxon signed-ranks test** with the presence of zero difference scores.

Example 6.2 *A physician states that the median number of times he sees each of his patients during the year is five. In order to evaluate the validity of this statement he randomly selects 13 of his patients and determines the number of office visits each of them made during the past year. He obtains the following values for the 13 patients in his sample: 5, 9, 10, 8, 4, 8, 5, 3, 0, 10, 15, 9, 5. Do the data support his contention that the median number of times he sees a patient is five?*

Examination of the data for Example 6.2 reveals that three of the 13 patients visited the doctor five times during the year. Since each of these three scores is equal to the hypothesized value of the population median, they will all produce difference scores of zero. In employing the ranking protocol for the **Wilcoxon signed-ranks test**, all three of these scores will be eliminated from the data analysis. Upon elimination of the three scores, the following ten scores remain: 9, 10, 8, 4, 8, 3, 0, 10, 15, 9. Since the ten remaining scores are identical to the ten scores employed in Example 6.1, the result for Example 6.2 will be identical to that for Example 6.1.

If, on the other hand, the **single-sample t test** is employed to evaluate Example 6.2, all 13 scores are included in the calculations resulting in the value $t = 1.87$, which is less than the value $t = 1.94$ obtained for Example 2.1 (which employs the 10 scores used in Example 6.1).[9, 10] The point to be made here is that by not employing the zero difference scores, the same T value is computed for the **Wilcoxon signed-ranks test** for both Examples 6.1 and 6.2. Yet in the case of the **single-sample t test**, which employs all 13 scores for the analysis of Example 6.2, the computed value of t for the latter example is not the same as the computed value of t for Example 2.1 (which employs the same data as Example 6.1). Thus, the presence of zero difference scores may serve to increase the likelihood of a discrepancy between the results obtained with the **Wilcoxon signed-ranks test** and the **single-sample t test**.

Example 6.3 *A college English instructor reads in an educational journal that the median number of times a student is absent from a class which meets for fifty minutes three times a week during a 15 week semester is $\theta = 5$. During the fall semester she keeps a record of the number of times each of the 10 students in her writing class is absent. She obtains the following values: 9, 10, 8, 4, 8, 3, 0, 10, 15, 9. Do the data suggest that the class is representative of a population that has a median of 5 absences?*

Since Example 6.3 employs the same data as Example 6.1, it yields the identical result.

References

Conover, W. J. (1980). **Practical nonparametric statistics** (2nd ed.). New York: John Wiley & Sons.

Conover, W. J. (1999). **Practical nonparametric statistics** (3rd ed.). New York: John Wiley & Sons.

Daniel, W. W. (1990). **Applied nonparametric statistics** (2nd ed.). Boston: PWS–Kent Publishing Company.

Glass, G. V. & Hopkins, K. D. (1996). **Statistical methods in education and psychology** (3rd ed.) Boston: Allyn & Bacon.

Higgins, J. J. (2004). **An introduction to modern nonparametric statistics**. Belmont, CA: Duxbury Press.

Hollander, M. & Wolfe, D.A. (1999). **Nonparametric statistical methods**. New York: John Wiley & Sons.

Marascuilo, L. A. & McSweeney, M. (1977). **Nonparametric and distribution-free method for the social sciences**. Monterey, CA: Brooks/Cole Publishing Company.

Pitman, E. J. G. (1948). **Lecture notes on nonparametric statistical inference**. Columbia University.

Snedecor, G. W. & Cochran, W. G. (1980). **Statistical methods** (8th ed.) Ames, IA: Iowa State University Press.

Wilcoxon, F. (1945). Individual comparisons by ranking methods. **Biometrics**, 1, 80–83.

Wilcoxon, F. (1949). **Some rapid approximate statistical procedures**. Stamford, CT: Stamford Research Laboratories, American Cyanamid Corporation.

Zar, J. H. (1999). **Biostatistical analysis** (4th ed.). Upper Saddle River, NJ: Prentice Hall.

Endnotes

1. Some sources note that one assumption of the **Wilcoxon signed-ranks test** is that the variable being measured is based on a continuous distribution. In practice, however, this assumption is often not adhered to.

2. The **binomial sign test for a single sample** is employed with data that are in the form of a dichotomous variable (i.e., a variable represented by two categories). Each subject's score is assigned to one of the following two categories: **Above the value of the hypothesized population median** versus **Below the value of the hypothesized population median**. The test allows a researcher to compute the probability of obtaining the proportion of subjects in each of the two categories, as well as more extreme distributions with respect to the two categories.

3. The **Wilcoxon signed-ranks test** can also be employed in place of the **single-sample z test** where the value of σ is known, but when the normality assumption of the latter test is saliently violated.

4. It is just coincidental in this example that the absolute value of some of the difference scores corresponds to the value of the rank assigned to that difference score.

5. The reader should take note of the fact that no critical values are recorded in **Table A5** for very small sample sizes. In the event a sample size is employed for which a critical value is not listed at a given level of significance, the null hypothesis cannot be evaluated at that

level of significance. This is the case since with small sample sizes the distribution of ranks will not allow one to generate probabilities equal to or less than the specified alpha value.

6. The term $(\Sigma t^3 - \Sigma t)$ in Equation 6.4 can also be written as $\sum_{i=1}^{s}(t_i^3 - t_i)$. The latter notation indicates the following: a) For each set of ties, the number of ties in the set is subtracted from the cube of the number of ties in that set; and b) the sum of all the values computed in a) is obtained. Thus, in the example under discussion (in which there are $s = 3$ sets of ties):

$$\sum_{i=1}^{s}(t_i^3 - t_i) = [(2)^3 - 2] + [(2)^3 - 2] + [(3)^3 - 3] = 36$$

The above computed value of 36 is the same as the corresponding value $(\Sigma t^3 - t)$ $= 43 - 7 = 36$ computed in the denominator of Equation 6.4 through use of Table 6.4.

7. A correction for continuity can be used in conjunction with the tie correction by subtracting .5 from the absolute value computed for the numerator of Equation 6.4. Use of the correction for continuity will reduce the tie-corrected absolute value of z.

8. Sources are not in agreement with respect to the minimum sample size for which the latter equations should be employed.

9. If the **single-sample t test** is employed with the 13 scores listed for Example 6.2, $\Sigma X = 91$, $\bar{X} = 91/13 = 7$, and $\Sigma X^2 = 815$. Thus, $\hat{s} = \sqrt{815 - [[(91)^2/13]/(13-1)]}$ $= 3.85$, $s_{\bar{X}} = 3.85/\sqrt{13} = 1.07$, and $t = (7-5)/1.07 = 1.87$.

10. Even though the obtained value $t = 1.87$ is smaller than the value $t = 1.94$ obtained for Example 2.1, it is still significant at the .05 level if one employs the directional alternative hypothesis H_1: $\mu > 5$. This is the case, since for $df = n - 1 = 12$ the tabled critical one-tailed .05 value is $t_{.05} = 1.78$, and $t = 1.87$ exceeds the latter tabled critical value.

Test 7

The Kolmogorov–Smirnov Goodness-of-Fit Test for a Single Sample
(Nonparametric Test Employed with Ordinal Data)

I. Hypothesis Evaluated with Test and Relevant Background Information

Hypothesis evaluated with test Does the distribution of n scores that comprise a sample conform to a specific theoretical or empirical population (or probability) distribution?

Relevant background information on test The **Kolmogorov–Smirnov goodness-of-fit test for a single sample** was developed by Kolmogorov (1933). Daniel (1990) notes that because of the similarity between Kolmogorov's test and a goodness-of-fit test for two independent samples developed by Smirnov (1939) (the **Kolmogorov–Smirnov test for two independent samples (Test 13))**, the test to be discussed is generally referred to as the **Kolmogorov–Smirnov goodness-of-fit test for a single sample**.

The **Kolmogorov–Smirnov goodness-of-fit test for a single sample** is one of a number of goodness-of-fit tests discussed in this book. Goodness-of-fit tests are employed to determine whether or not the distribution of scores in a sample conforms to the distribution of scores in a specific theoretical or empirical population (or probability) distribution. Goodness-of-fit tests are somewhat unique when contrasted with other types of inferential statistical tests, in that when conducting a goodness-of-fit test a researcher often wants or expects to retain the null hypothesis. In other words, the researcher wants to demonstrate that a sample is derived from a distribution of a specific type (e.g., a normal distribution). On the other hand, in employing most other inferential tests, a researcher wants or expects to reject the null hypothesis — i.e., the researcher wants or expects to demonstrate that one or more samples do not come from a specific population or from the same population. It should be noted that the alternative hypothesis for a goodness-of-fit test generally does not stipulate an alternative distribution which would become the most likely distribution for the data if the null hypothesis is rejected.

Unlike the **chi-square goodness-of-fit test (Test 8)**, which is discussed in the next chapter, the **Kolmogorov–Smirnov goodness-of-fit test for a single sample** is designed to be employed with a **continuous** variable. (A **continuous variable** is characterized by the fact that a given score can assume any value within the range of values which defines the limits of that variable.) The **chi-square goodness-of-fit test**, on the other hand, is designed to be employed with nominal/categorical data involving a **discrete** variable. (A **discrete variable** is characterized by the fact that there are a limited number of values which any score for the variable can assume.) Further clarification of the distinction between discrete and continuous variables can be found in the **Introduction**.

The **Kolmogorov–Smirnov goodness-of-fit test for a single sample** is categorized as a test of ordinal data because it requires that a **cumulative frequency distribution** be constructed (which requires that scores be arranged in order of magnitude). In the **Introduction** it was noted that in a **cumulative frequency distribution**, the cumulative frequency for a given score represents the frequency of a score plus the frequencies of all scores which are less than that score. Scores are arranged ordinally, with the lowest score at the bottom of the distribution, and the highest score at the top of the distribution. The cumulative frequency for the lowest score will simply be the frequency for that score, since there are no scores below it. On the other hand, the cumulative frequency for the highest score will always equal n, the total number of scores in the distribution. In some instances a cumulative frequency distribution may present cumulative proportions (which can also be expressed as probabilities) or cumulative percentages in lieu of and/or in addition to cumulative frequencies. A cumulative proportion or percentage for a given score represents the proportion or percentage of scores which are equal to or less than that score. (Cumulative frequencies and proportions are summarized in Columns 3 and 5 of Table I.8.) When the term cumulative probability is employed, it means the likelihood of obtaining a given score or any score below it (which is numerically equivalent to the cumulative proportion for that score). Table 7.1 represents a cumulative frequency distribution for a distribution comprised of $n = 20$ scores. Each of the scores which occur in the distribution are listed in the first column. Note that in the third column of Table 7.1, the cumulative frequency values are obtained by adding to the frequency of a score in a given row the frequencies of all scores that fall below it. A cumulative proportion for a score is obtained by dividing the cumulative frequency of the score by n. A cumulative proportion is converted into a cumulative percentage by moving the decimal point for the proportion two places to the right.

Table 7.1 Cumulative Frequency Distribution

X	Frequency (f)	Cumulative frequency	Cumulative proportion	Cumulative percentage
15	3	20	20/20 = 1	100%
14	2	17	17/20 = .85	85%
13	2	15	15/20 = .75	75%
12	0	13	13/20 = .65	65%
11	4	13	13/20 = .65	65%
10	2	9	9/20 = .45	45%
9	1	7	7/20 = .35	35%
8	0	6	6/20 = .30	30%
7	4	6	6/20 = .30	30%
6	2	2	2/20 = .10	10%
	$n = 20$			

In the example to be presented for the **Kolmogorov–Smirnov goodness-of-fit test for a single sample**, a cumulative frequency distribution will be constructed. However, the table containing the cumulative frequency distribution of the test data (Table 7.2) will list the scores in reverse order from that listed in Table 7.1 (i.e., in Table 7.2 the lowest score will be at the top and the highest score at the bottom). This alternative way of arranging the cumulative frequencies is commonly employed to summarize the data analysis for **Kolmogorov–Smirnov goodness-of-fit test for a single sample**.

II. Example

Example 7.1 *A researcher conducts a study to evaluate whether or not the distribution of the length of time it takes migraine patients to respond to a 100 mg. dose of an intravenously administered drug is normal, with a mean response time of 90 seconds and a standard deviation of 35 seconds (i.e., $\mu = 90$ and $\sigma = 35$). The amount of time (in seconds) which elapses between the administration of the drug and cessation of a headache for 30 migraine patients is recorded* below. The 30 scores are arranged ordinally (i.e., from fastest response time to slowest response time).

21, 32, 38, 40, 48, 55, 63, 66, 70, 75, 80, 84, 86, 90, 90, 93, 95, 98, 100, 105, 106, 108, 115, 118, 126, 128, 130, 142, 145, 155

Do the data conform to a normal distributions with the specified parameters?

III. Null versus Alternative Hypotheses

Prior to reading the null and alternative hypotheses to be presented in this section, the reader should take note of the following: a) The protocol for the **Kolmogorov–Smirnov goodness-of-fit test for a single sample** requires that a cumulative probability distribution be constructed for both the sample distribution and the hypothesized population distribution. The test statistic is defined by the point which represents the greatest vertical distance at any point between the two cumulative probability distributions; and b) Within the framework of the null and alternative hypotheses, the notation $F(X)$ represents the population distribution from which the sample distribution is derived, while the notation $F_o(X)$ represents the hypothesized theoretical or empirical distribution with respect to which the sample distribution is being evaluated for goodness-of-fit. Alternatively, $F(X)$ can be conceptualized as representing the cumulative probability distribution for the population from which the sample is derived, and $F_o(X)$ as the cumulative probability distribution for the hypothesized population.

Null hypothesis H_0: $F(X) = F_o(X)$ for all values of X

(The distribution of data in the sample is consistent with the hypothesized theoretical population distribution. In terms of the parameters stipulated in Example 7.1, the null hypothesis is stating that the sample data are derived from a normal distribution, with $\mu = 90$ and $\sigma = 35$. Another way of stating the null hypothesis is as follows: At no point is the greatest vertical distance between the sample cumulative probability distribution (which is assumed to be the best estimate of the cumulative probability distribution of the population from which the sample is derived) and the hypothesized cumulative probability distribution larger than what would be expected by chance, if the sample is derived from the hypothesized distribution.)

Alternative hypothesis H_1: $F(X) \neq F_o(X)$ for at least one value of X

(The distribution of data in the sample is inconsistent with the hypothesized theoretical population distribution. In terms of the parameters stipulated in Example 7.1, the null hypothesis is stating that the sample data are not derived from a normal distribution, with $\mu = 90$ and $\sigma = 35$. An alternative way of stating this alternative hypothesis is as follows: There is at least one point where the greatest vertical distance between the sample cumulative probability distribution (which is assumed to be the best estimate of the cumulative probability distribution of the population from which the sample is derived) and the hypothesized cumulative probability distribution is larger than what would be expected by chance, if the sample is derived from the hypothesized distribution. At the point of maximum deviation separating the two distributions, the cumulative probability for the sample distribution is either significantly greater or less than the cumulative probability for the hypothesized distribution. This is a **nondirectional alternative hypothesis** and it is evaluated with a **two-tailed test**.)

or

$$H_1: F(X) > F_o(X) \text{ for at least one value of } X$$

(The distribution of data in the sample is inconsistent with the hypothesized theoretical population distribution. In terms of the parameters stipulated in Example 7.1, the null hypothesis is stating that the sample data are not derived from a normal distribution, with $\mu = 90$ and $\sigma = 35$. The latter is the case, since there is at least one point at which the vertical distance between the sample cumulative probability distribution (which is assumed to be the best estimate of the cumulative probability distribution of the population from which the sample is derived) and the hypothesized cumulative probability distribution is larger than what would be expected by chance, if the sample is derived from the hypothesized distribution. At the point of maximum deviation separating the two distributions, the cumulative probability for the sample distribution is significantly greater than the cumulative probability for the hypothesized distribution. This is a **directional alternative hypothesis** and it is evaluated with a **one-tailed test**.)

or

$$H_1: F(X) < F_o(X) \text{ for at least one value of } X$$

(The distribution of data in the sample is inconsistent with the hypothesized theoretical population distribution. In terms of the parameters stipulated in Example 7.1, the null hypothesis is stating that the sample data are not derived from a normal distribution, with $\mu = 90$ and $\sigma = 35$. The latter is the case, since there is at least one point at which the vertical distance between the sample cumulative probability distribution (which is assumed to be the best estimate of the cumulative probability distribution of the population from which the sample is derived) and the hypothesized cumulative probability distribution is larger than what would be expected by chance, if the sample is derived from the hypothesized distribution. At the point of maximum deviation separating the two distributions, the cumulative probability for the sample distribution is significantly less than the cumulative probability for the hypothesized distribution. This is a **directional alternative hypothesis** and it is evaluated with a **one-tailed test**.)

Note: Only one of the above noted alternative hypotheses is employed. If the alternative hypothesis the researcher selects is supported, the null hypothesis is rejected.

IV. Test Computations

As noted in Sections I and III, the test protocol for the **Kolmogorov–Smirnov goodness-of-fit test for a single sample** requires that the cumulative probability distribution for the sample data be contrasted with the cumulative probability distribution for the hypothesized population. Table 7.2 summarizes the steps which are involved in conducting the analysis. The values represented in the columns of Table 7.2 are summarized below.

The values of the response time scores of the 30 subjects in the sample (i.e., the X scores) are recorded in **Column A**. There are 29 rows corresponding to each of the scores (with two subjects having obtained the identical score of 90).

Each value in **Column B** is the z score (i.e., standard deviation score) which results when the X score in a given row is substituted in the equation $z = (X - \mu)/\sigma$ (which is Equation I.38), where $\mu = 90$ and $\sigma = 35$. Thus, for each row, the equation $z = (X - 90)/35$ is employed. To illustrate, in the case of **Row 1**, where $X = 21$, the value $z = (21 - 90)/35 = -1.97$ is computed. In the case of the last row, where $X = 155$, the value $z = (155 - 90)/35 = 1.86$ is computed. Note that a negative z value will be obtained for any X score below the mean, and a positive z value for any X score above the mean.

Table 7.2 Calculation of Test Statistic for Kolmogorov–Smirnov Goodness-of-Fit Test for a Single Sample

A (X)	B (z)	C (p)	D $(F_o(X_i)=p\pm.50)$	E $S(X_i)$	F $|S(X_i)-F_o(X_i)|$	G $|S(X_{i-1})-F_o(X_i)|$
21	−1.97	.4756	.0244	1/30 = .0333	.0089	$|0-.0244|$ = .0244
32	−1.66	.4515	.0485	2/30 = .0667	.0182	$|.0333-.0485|$ = .0152
38	−1.49	.4319	.0681	3/30 = .1000	.0319	$|.0667-.0681|$ = .0014
40	−1.43	.4236	.0764	4/30 = .1333	.0569 = M	$|.1000-.0764|$ = .0236
48	−1.20	.3849	.1151	5/30 = .1667	.0516	$|.1333-.1151|$ = .0182
55	−1.00	.3413	.1587	6/30 = .2000	.0413	$|.1667-.1587|$ = .0080
63	−.77	.2794	.2206	7/30 = .2333	.0127	$|.2000-.2206|$ = .0206
66	−.69	.2549	.2451	8/30 = .2667	.0216	$|.2333-.2451|$ = .0118
70	−.57	.2157	.2843	9/30 = .3000	.0157	$|.2667-.2843|$ = .0176
75	−.43	.1664	.3336	10/30 = .3333	.0003	$|.3000-.3336|$ = .0336
80	−.29	.1141	.3859	11/30 = .3667	.0192	$|.3333-.3859|$ = .0526
84	−.17	.0675	.4325	12/30 = .4000	.0325	$|.3667-.4325|$ = .0658
86	−.11	.0438	.4562	13/30 = .4333	.0229	$|.4000-.4562|$ = .0562
90	.00	.0000	.5000	15/30 = .5000	.0000	$|.4333-.5000|$ = .0667 = M'
93	.09	.0359	.5359	16/30 = .5333	.0026	$|.5000-.5359|$ = .0359
95	.14	.0557	.5557	17/30 = .5667	.0110	$|.5333-.5557|$ = .0224
98	.23	.0901	.5901	18/30 = .6000	.0099	$|.5667-.5901|$ = .0234
100	.29	.1141	.6141	19/30 = .6333	.0192	$|.6000-.6141|$ = .0141
105	.43	.1664	.6664	20/30 = .6667	.0003	$|.6333-.6664|$ = .0331
106	.46	.1772	.6772	21/30 = .7000	.0228	$|.6667-.6772|$ = .0105
108	.51	.1950	.6950	22/30 = .7333	.0383	$|.7000-.6950|$ = .0050
115	.71	.2611	.7611	23/30 = .7667	.0056	$|.7333-.7611|$ = .0278
118	.80	.2881	.7881	24/30 = .8000	.0119	$|.7667-.7881|$ = .0214
126	1.03	.3485	.8485	25/30 = .8333	.0152	$|.8000-.8485|$ = .0485
128	1.09	.3621	.8621	26/30 = .8667	.0046	$|.8333-.8621|$ = .0288
130	1.14	.3729	.8729	27/30 = .9000	.0271	$|.8667-.8729|$ = .0062
142	1.48	.4306	.9306	28/30 = .9333	.0027	$|.9000-.9306|$ = .0306
145	1.57	.4418	.9418	29/30 = .9667	.0249	$|.9333-.9418|$ = .0085
155	1.86	.4686	.9686	30/30 = 1.0000	.0314	$|.9667-.9686|$ = .0019

Each value in **Column C** represents the proportion of cases in the normal distribution that falls between the population mean and the z score computed for the X score in a given row (i.e., the z value computed in **Column B**). To illustrate, in the case of **Row 1**, where $X = 21$ and $z = -1.97$, the proportion .4756 in **Column C** (which is the entry for $z = 1.97$ in **Column 2** of **Table A1 (Table of the Normal Distribution)** in the **Appendix**) is the proportion of cases in the normal distribution between the mean and a z score of −1.97. In the case of the last row, where $X = 155$ and $z = 1.86$, the proportion .4686 in **Column C** is the proportion of cases in the normal distribution between the mean and a z score of 1.86.

Each value in **Column D** represents the cumulative proportion for a given X score (and its associated z score) in the hypothesized theoretical distribution (i.e., in a normal distribution with $\mu = 90$ and $\sigma = 35$). To put it another way, if the decimal point is moved two places to the right, the value in **Column D** represents the percentile rank of a given X score in the hypothesized theoretical distribution. For any X score for which a negative z score is computed, the proportion in **Column D** can be obtained by subtracting from .5000 the proportion in **Column C** for that score (it will also correspond to the proportion for that z score in **Column 3** of **Table A1**). For any X score for which a positive z score is computed, the proportion in **Column D** can be obtained by adding .5000 to the proportion in **Column C** for that score (i.e., it will correspond to the sum of .5000 and the proportion for that z score in **Column 2** of **Table A1**). To illustrate, in the case of **Row 1**, where $X = 21$ and $z = -1.97$, the proportion .0244 is equal to .5000 − .4756

= .0244. In the case of the last row, where $X = 155$ and $z = 1.86$, the proportion .9686 is equal to .5000 + .4686 = .9686. The values in **Column D** are commonly represented by the notation $F_o(X_i)$, where the subscript i represents the ith score/row in Table 7.2.

Each value in **Column E** represents the cumulative proportion for a given X score (and its associated z score) in the sample distribution. To illustrate, in the case of **Row 1**, where $X = 21$, its cumulative proportion is its cumulative frequency (1) divided by the total number of scores in the sample ($n = 30$). Thus, $1/30 = .0333$. In the case of the score $X = 100$, its cumulative proportion in the sample distribution is 19 (i.e., a score of 100 is equal to or greater than 19 of the 30 scores). Thus, its cumulative proportion is $19/30 = .6333$. In the case of the score $X = 155$ in the last row, its cumulative frequency in the sample distribution is 30 (since it is the highest score). Thus, its cumulative proportion is $30/30 = 1$. The values in **Column E** are commonly represented by the notation $S(X_i)$, where the subscript i represents the ith score/row in Table 7.2.

Each value in **Column F** is the absolute value of the difference between the proportions in **Column E** and **Column D** — in other words, the difference between the proportions in the sample distribution and the hypothesized population distribution. Thus, $F_i = |E_i - D_i|$ or $F_i = |S(X_i) - F_o(X_i)|$. To illustrate, in the case of **Row 1**, where $D_i = F_o(X_i) = .0244$ and $E_i = S(X_i) = .0333$, we compute the value $F_i = .0089$ as follows.

$$F_i = |E_i - D_i| = |S(X_i) - F_o(X_i)| = |.0333 - .0244| = .0089$$

In the case of the row where $X = 100$, and where $D_i = F_o(X_i) = .6141$ and $E_i = S(X_i) = .6333$, we compute the value $F = .0192$ as follows:

$$F_i = |E_i - D_i| = |S(X_i) - F_o(X_i)| = |.6333 - .6141| = .0192$$

In the case of the last row where $X = 155$, and where $D_i = F_o(X_i) = .9686$ and $E_i = S(X_i) = 1$, we compute the value $F_i = .0314$ as follows:

$$F_i = |E_i - D_i| = |S(X_i) - F_o(X_i)| = |1 - .9686| = .0314$$

As noted in Section III, the test statistic for the **Kolmogorov–Smirnov goodness-of-fit test for a single sample** is defined by the greatest vertical distance at any point between the two cumulative probability distributions. The largest absolute value obtained in **Column F** will represent that value. In Table 7.2 the largest absolute value is .0569, which is designated as the test statistic (represented by the notation M).

Each value in **Column G** is the absolute value of the difference between the proportion in **Column D** for a given row (i.e., $F_o(X_i)$) and the proportion in **Column E** for the preceding row (i.e., $S(X_{i-1})$). In other words, $G_i = |E_{i-1} - D_i|$ or $G_i = |S(X_{i-1}) - F_o(X_i)|$. To illustrate, in the case of **Row 1**, where $D_i = F_o(X_i) = .0244$ and $E_i = S(X_{i-1}) = 0$, we compute the value $G_i = .0244$ as noted below. Note that the value 0 is employed to represent the initial value of $E_{i-1} = S(X_{i-1})$, since that is the value which .0333 is added to in order to get the entry .0333 in **Row 1** of **Column E**.

$$G_i = |E_{i-1} - D_i| = |S(X_{i-1}) - F_o(X_i)| = |0 - .0244| = .0244$$

In the case of the row where $X = 100$, and where $D_i = F_o(X_i) = .6141$ and $E_{i-1} = S(X_{i-1}) = .6000$, we compute the value $G_i = .0141$ as follows:

$$G_i = |E_{i-1} - D_i| = |S(X_{i-1}) - F_o(X_i)| = |.6000 - .6141| = .0141$$

In the case of the last row where $X = 155$, and where $D_i = F_o(X_i) = .9686$ and $E_{i-1} = S(X_{i-1}) = .9667$, we compute the value $G_i = .0019$ as follows:

$$G_i = |E_{i-1} - D_i| = |S(X_{i-1}) - F_o(X_i)| = |.9667 - .9686| = .0019$$

As noted above, the test statistic for the **Kolmogorov–Smirnov goodness-of-fit test for a single sample** is defined by the greatest vertical distance at any point between the two cumulative probability distributions. However, when that value is determined mathematically through use of the value in **Column F**, it is still possible that the largest vertical distance may occur **at some point between one of the scores in the sample distribution**. Since it is assumed that the variable being evaluated is continuous, if there is a larger vertical distance for some score other than those in the sample, the latter score should represent the test statistic, instead of the M value recorded in **Column F**. The method for determining whether there is a larger vertical distance than the maximum value recorded in **Column F** is to compute the values in **Column G** of Table 7.2. If the largest value computed in **Column G** (designated M') is larger than the M value computed in **Column F**, then M' is employed to represent the test statistic. In Table 7.2, the largest value is .0667, and thus $M' = .0667$ becomes our test statistic.[1]

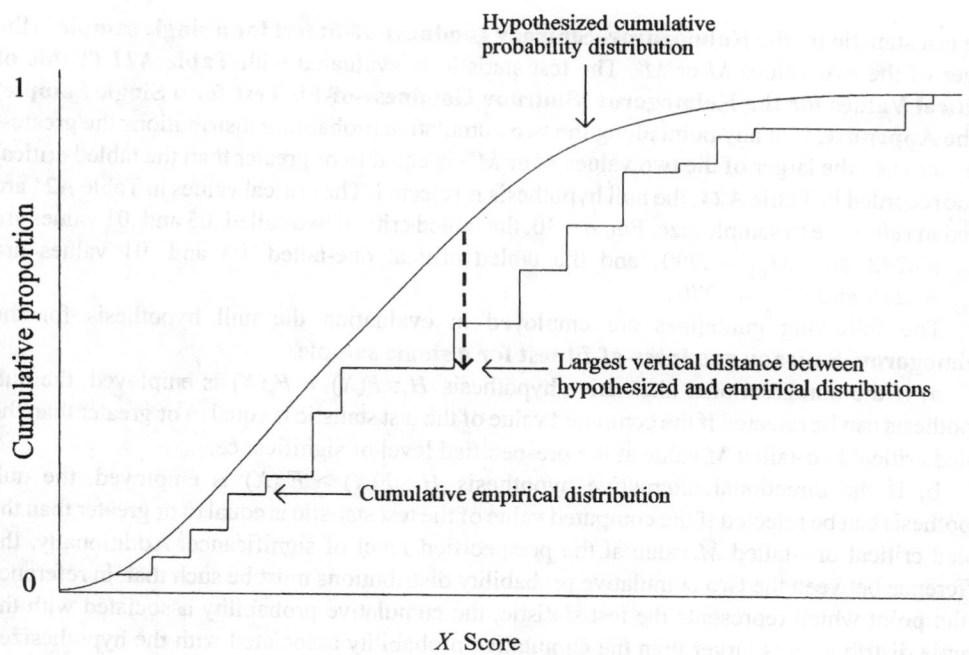

Figure 7.1 Representative Graph Employed for Kolmogorov–Smirnov Analysis Contrasting Hypothesized versus Empirical Cumulative Probability Distributions

An alternative method for determining the **Kolmogorov–Smirnov** test statistic is to construct a graph such as the one presented in Figure 7.1. Before continuing, it should be made clear that the information depicted in Figure 7.1 is not based on the data for Example 7.1 (i.e. the values recorded in Table 7.2 were not employed in constructing the graph). When a graph such as Figure 7.1 is employed for a **Kolmogorov–Smirnov** analysis, it will contain the following two distributions: a) A hypothesized cumulative probability distribution (i.e., a curve which will be based on the relevant values computed for **Column D**); and b) A cumulative empirical distribution (a curve that will be based on the cumulative probabilities for the sample distribution — i.e., the relevant values computed for **Column E**). The largest vertical distance separating the curves for the two distributions is identified, and is employed to represent the **Kolmogorov–Smirnov** test statistic. Such a graph can be employed to determine the values of both M and M' (which for Example 7.1 are respectively computed in **Columns F** and **G**). The graphical method is described in greater detail in Conover (1980, 1999), Daniel (1990), and Sprent (1993). If a graph such as Figure 7.1 were constructed for Example 7.1, most of the points on the cumulative empirical distribution would fall above the curve of the hypothesized cumulative probability distribution (with the value $M = .0569$ representing the vertical distance of one of those points above the latter curve). However, some of the points, including the one resulting in the value $M' = .0667$, would fall below the curve of the hypothesized cumulative probability distribution.

V. Interpretation of the Test Results

The test statistic for the **Kolmogorov–Smirnov goodness-of-fit test for a single sample** is the larger of the two values M or M'. The test statistic is evaluated with **Table A21 (Table of Critical Values for the Kolmogorov–Smirnov Goodness-of-Fit Test for a Single Sample)** in the **Appendix**. If at any point along the two cumulative probability distributions the greatest distance (i.e., the larger of the two values M or M') is equal to or greater than the tabled critical value recorded in **Table A21**, the null hypothesis is rejected. The critical values in Table A21 are listed in reference to sample size. For $n = 30$, the tabled critical two-tailed .05 and .01 values are $M_{.05} = .242$ and $M_{.01} = .290$, and the tabled critical one-tailed .05 and .01 values are $M_{.05} = .218$ and $M_{.01} = .270$.

The following guidelines are employed in evaluating the null hypothesis for the **Kolmogorov–Smirnov goodness-of-fit test for a single sample**.

a) If the nondirectional alternative hypothesis H_1: $F(X) \neq F_o(X)$ is employed, the null hypothesis can be rejected if the computed value of the test statistic is equal to or greater than the tabled critical two-tailed M value at the prespecified level of significance.

b) If the directional alternative hypothesis H_1: $F(X) > F_o(X)$ is employed, the null hypothesis can be rejected if the computed value of the test statistic is equal to or greater than the tabled critical one-tailed M value at the prespecified level of significance. Additionally, the difference between the two cumulative probability distributions must be such that, in reference to the point which represents the test statistic, the cumulative probability associated with the sample distribution is larger than the cumulative probability associated with the hypothesized population distribution. In other words, if, instead of computing an absolute value in **Columns F** and **G** of Table 7.2, we retain the sign of the difference, then a positive sign is required for the directional alternative hypothesis H_1: $F(X) > F_o(X)$ to be supported. Thus, if M is the largest vertical distance, $S(X_i) > F_o(X_i)$, and if M' is the largest vertical distance, $S(X_{i-1}) > F_o(X_i)$.

c) If the directional alternative hypothesis H_1: $F(X) < F_o(X)$ is employed, the null hypothesis can be rejected if the larger of the two values M versus M' is equal to or greater than the tabled critical one-tailed M value at the prespecified level of significance. Additionally, the difference between the two cumulative probability distributions must be such that in reference to the point which represents the test statistic, the cumulative probability associated with the sample distribution is smaller than the cumulative probability associated with the hypothesized population distribution. In other words, if, instead of computing an absolute value in **Columns F** and **G** of Table 7.2, we retain the sign of the difference, then a negative sign is required for the directional alternative hypothesis H_1: $F(X) < F_o(X)$ to be supported. Thus, if M is the largest vertical distance, $S(X_i) < F_o(X_i)$, and if M' is the largest vertical distance, $S(X_{i-1}) < F_o(X_i)$.

The above guidelines will now be employed in reference to the computed test statistic $M' = .0667$.

a) If the nondirectional alternative hypothesis H_1: $F(X) \neq F_o(X)$ is employed, the null hypothesis cannot be rejected, since $M' = .0667$ is less than the tabled critical two-tailed values $M_{.05} = .242$ and $M_{.01} = .290$.

b) If the directional alternative hypothesis H_1: $F(X) < F_o(X)$ is employed, the null hypothesis cannot be rejected since $M' = .0667$ is less than the tabled critical one-tailed values $M_{.05} = .218$ and $M_{.01} = .270$. This is the case in spite of the fact that the test statistic is consistent with the latter alternative hypothesis (i.e., since $[F_o(X_i) = .5000] > [S(X_{i-1}) = .4333]$, if the sign is taken into account, the computed value of M' is a negative value: $M' = S(X_{i-1}) - F_o(X_i) = .4333 - .5000 = -.0667$).

c) If the directional alternative hypothesis H_1: $F(X) > F_o(X)$ is employed, the null hypothesis cannot be rejected, since for the latter alternative hypothesis to be supported the cumulative proportion for the sample distribution must be larger than the cumulative proportion for the hypothesized population distribution.

A summary of the analysis of Example 7.1 with the **Kolmogorov–Smirnov goodness-of-fit test for a single sample** follows: The data are consistent with the null hypothesis that the sample is derived from a normally distributed population, with $\mu = 90$ and $\sigma = 35$.

VI. Additional Analytical Procedures for the Kolmogorov–Smirnov Goodness-of-Fit Test for a Single Sample and/or Related Tests

1. Computing a confidence interval for the Kolmogorov–Smirnov goodness-of-fit test for a single sample Daniel (1990) describes how to construct a confidence interval for the cumulative distribution for the sample proportions.[2] The confidence interval computed for the **Kolmogorov–Smirnov test** statistic is comprised of two sets of limits — a set of **upper limits** and a set of **lower limits**. The reference points for determining the latter values are the $S(X_i)$ scores in **Column E** of Table 7.2. Equation 7.1 is the general equation for computing the limits which define a confidence interval at any point along the cumulative probability distribution for the sample.

$$CI_{(1-\alpha)} = S(X_i) \pm (M_{\alpha/2}) \qquad \textbf{(Equation 7.1)}$$

Where: $M_{\alpha/2}$ represents the tabled critical two-tailed M value for a given value of n, below which a proportion (percentage) equal to $[1 - (\alpha/2)]$ of the cases falls. If the proportion (percentage) of the distribution that falls within the confidence interval is subtracted from 1 (100%), it will equal the value of α.

The **upper limits** for the confidence interval are computed by adding the relevant critical value to each of the values of $S(X_i)$ in **Column E** of Table 7.2. If any of the resulting values is greater than 1, the upper limit for that $S(X_i)$ value is set equal to 1, since a probability cannot be greater than 1. In the case of Example 7.1, 29 upper limit values will be computed, each value corresponding to one of the 29 $S(X_i)$ values recorded in the rows of Table 7.2.

The **lower limits** for the confidence interval are computed by subtracting the relevant critical value from each of the values of $S(X_i)$ in **Column E** of Table 7.2. If any of the resulting values is less than 0, the lower limit for that $S(X_i)$ value is set equal to 0, since a probability cannot be less than 0. In the case of Example 7.1, 29 lower limit values will be computed, each value corresponding to one of the 29 $S(X_i)$ values recorded in the rows of Table 7.2.

The above methodology will now be described in reference to Example 7.1. Let us assume we wish to compute a 95% confidence interval for the cumulative probability distribution of the population from which the sample is derived. Since we are interested in the 95% confidence interval, the value that will be employed for $M_{\alpha/2}$ in Equation 7.1 will be the tabled critical two-tailed .05 M value, which as previously noted is $M_{.05} = .242$. Thus, we will add to and subtract .242 from each of the $S(X_i)$ values in **Column E** of Table 7.2.

To illustrate, the first $S(X_i)$ value (associated with the score of $X = 21$) is .0333. When .242 is added to the latter value we obtain .2753, which is the upper limit for that point on the cumulative probability distribution. When .242 is subtracted from .0333 we obtain the value −.2087. Since the latter value is less than zero, we set the lower limit at that point equal to zero.

In the case of the score of $X = 90$, the value of $S(X_i)$ is .5000. When .242 is added to the latter value we obtain .7420, which is the upper limit for that point on the cumulative probability distribution. When .242 is subtracted from .5000 we obtain .2580, which is the lower limit for that point on the cumulative probability distribution.

In the case of the score of $X = 155$, the value of $S(X_i)$ is 1. When .242 is added to the latter value we obtain 1.242. Since the latter value is greater than 1, we set the upper limit at that point equal to 1. When .242 is subtracted from 1 we obtain .7580, which is the lower limit for that point on the cumulative probability distribution.

As noted earlier, the above described procedure is employed for all 29 points on the cumulative probability distribution for the sample. The resulting set of upper and lower limits defines the confidence interval.

2. The power of the Kolmogorov–Smirnov goodness-of-fit test for a single sample Books which discuss the **Kolmogorov–Smirnov goodness-of-fit test for a single sample** do not describe specific procedures for computing the power of the test. Conover (1980, 1999), Daniel (1990), and Hollander and Wolfe (1999) cite sources that discuss the power of the test and/or describe procedures for determining power. Daniel (1990) and Zar (1999) note that when the **Kolmogorov–Smirnov goodness-of-fit test for a single sample** is employed with grouped data (i.e., scores are categorized in class intervals instead of evaluating each score separately), the test becomes overly conservative (i.e., the power of the test is reduced). Khamis (1990, 2000) has developed a correction factor for the **Kolmogorov–Smirnov test statistic** which can increase the power of the test (also see Harter *et al.* (1984) and Zar (1999)). Zar (1999) states that the **Kolmogorov–Smirnov test** is more powerful than the **chi-square goodness-of-fit test** under the following conditions: a) When the sample size is small; and b) When the expected frequencies for the **chi-square test** are small. Zar, however, notes that since both the **Kolmogorov–Smirnov test** and the **chi-square goodness-of-fit test** are deficient in power, neither test is optimal for assessing normality. Thode (2002) has reviewed over forty procedures for assessing goodness-of-fit, and concludes that the **Shapiro-Wilk test for normality** (Shapiro and Wilk (1965, 1968)) and test statistics derived from moments are generally the best methods for assessing normality.

3. Test 7a: The Lilliefors test for normality Massey (1951) notes that when the population parameters (e.g., μ and σ) are not known beforehand, but are instead estimated from the sample data, the result yielded by the **Kolmogorov–Smirnov goodness-of-fit test for a single sample**

tends to be overly conservative (i.e., the statistical power of the test is less than its power when the values of the parameters are known). Various sources (Conover (1980, 1999) and Daniel (1990)) describe Lilliefors (1967, 1969, 1973) extension of the **Kolmogorov–Smirnov goodness-of-fit test for a single sample** to circumstances in which the values of the population parameters for a variety of distributions (e.g., normal, exponential, gamma) are not known, and thus have to be estimated from the sample data.[3] The procedure to be described here, which is designed to assess goodness-of-fit for a normal distribution when one or both of the population parameters μ and σ are unknown, is referred to as the **Lilliefors test for normality**. The test procedure for the **Lilliefors test for normality** is identical to that described for the **Kolmogorov–Smirnov goodness-of-fit test for a single sample**, except for the following: a) The values of the sample mean $(\bar{X})$ and estimated population standard deviation $(\tilde{s})$ are employed to represent the mean and standard deviation of the hypothesized population distribution. Thus, the values of $\bar{X}$ and $\tilde{s}$ are employed to compute the z values in **Column B** of Table 7.2; and b) Instead of obtaining the critical values from **Table A21**, the values documented in **Table A22 (Table of Critical Values for the Lilliefors Test for Normality)** in the **Appendix** are employed.[4] As is the case with employing the critical values in **Table A21**, in order to reject the null hypothesis when the test statistic is based on the **Lilliefors test for normality**, the computed value for M or M' (i.e., whichever of the two is larger) must be equal to or greater than the tabled critical value in **Table A22** at the prespecified level of significance. The values recorded in **Table A22** are only applicable when both the values of μ and σ are unknown, and must be estimated from the sample data.

Table 7.3 reevaluates the data for Example 7.1 employing the values for the sample mean $(\bar{X} = 90.07)$ and estimated population standard deviation $(\tilde{s} = 34.79)$ in place of the values $\mu = 90$ and $\sigma = 35$ employed for the **Kolmogorov–Smirnov goodness-of-fit test for a single sample**. The values $\bar{X} = 90.07$ and $\tilde{s} = 34.79$ were computed by employing Equations I.1 and I.11 with the 30 scores in Example 7.1.

In Table 7.3 the computed values for M and M' in **Columns F** and **G** are $M = .0594$ and $M' = .0667$. Since the values $\bar{X} = 90.07$ and $\tilde{s} = 34.79$ are quite close to the values $\mu = 90$ and $\sigma = 35$ employed for the **Kolmogorov–Smirnov goodness-of-fit test for a single sample**, it is not surprising that the values in the rows of Table 7.3 are quite close and, in some cases identical, to the values in the rows of Table 7.2. The values $M = .0594$ and $M' = .0667$ obtained in Table 7.3 are either very close or identical to the values $M = .0569$ and $M' = .0667$ obtained in Table 7.2 for the **Kolmogorov–Smirnov goodness-of-fit test for a single sample**. Since $M' = .0667$ is larger than $M = .0594$, $M' = .0667$ will represent the test statistic for the **Lilliefors test for normality**.

As is the case with **Table A21**, the critical values listed in **Table A22** are listed in reference to sample size. Lilliefors' (1967) table only contains two-tailed .40, .30, .20, .10, and .02 values, and one-tailed .20, .15, .10, .05, and .01 values. Daniel (1990) employs more detailed tables for the **Lilliefors test** statistic developed by Mason and Bell (1986). The latter tables have slightly different critical values than those listed in **Table A22**. Mason and Bell's (1986) tables also have additional critical values for when μ is unknown and σ is known, and for when μ is known and σ is unknown.

Employing **Table A22** for $n = 30$, it can be seen that the tabled critical one-tailed .05 and .01 values (which correspond to the two-tailed .10 and .02 critical values) are $M_{.05} = .161$ and $M_{.01} = .187$. Since $M' = .0667$ is less than both of the aforementioned critical values, the null hypothesis of normality cannot be rejected. From the magnitude of the critical values, it is

Table 7.3 Calculation of Test Statistic for the Lilliefors Test for Normality

A (X)	B (z)	C (p)	D ($F_o(X_i) = p \pm .50$)	E $S(X_i)$	F $\lvert S(X_i) - F_o(X_i)\rvert$	G $\lvert S(X_{i-1}) - F_o(X_i)\rvert$
21	−1.99	.4761	.0239	1/30 = .0333	.0094	$\lvert 0-.0239\rvert = .0239$
32	−1.67	.4525	.0475	2/30 = .0667	.0192	$\lvert .0333-.0475\rvert = .0142$
38	−1.50	.4332	.0668	3/30 = .1000	.0332	$\lvert .0667-.0668\rvert = .0001$
40	−1.44	.4251	.0749	4/30 = .1333	.0594 = M	$\lvert .1000-.0749\rvert = .0251$
48	−1.21	.3869	.1131	5/30 = .1667	.0536	$\lvert .1333-.1131\rvert = .0202$
55	−1.01	.3438	.1562	6/30 = .2000	.0438	$\lvert .1667-.1562\rvert = .0105$
63	−.78	.2823	.2177	7/30 = .2333	.0156	$\lvert .2000-.2177\rvert = .0177$
66	−.69	.2549	.2451	8/30 = .2667	.0216	$\lvert .2333-.2451\rvert = .0118$
70	−.58	.2190	.2810	9/30 = .3000	.0190	$\lvert .2667-.2810\rvert = .0143$
75	−.43	.1664	.3336	10/30 = .3333	.0003	$\lvert .3000-.3336\rvert = .0336$
80	−.29	.1141	.3859	11/30 = .3667	.0192	$\lvert .3333-.3859\rvert = .0526$
84	−.17	.0675	.4325	12/30 = .4000	.0325	$\lvert .3667-.4325\rvert = .0658$
86	−.12	.0478	.4522	13/30 = .4333	.0189	$\lvert .4000-.4522\rvert = .0522$
90	.00	.0000	.5000	15/30 = .5000	.0000	$\lvert .4333-.5000\rvert = .0667$ =M'
93	.08	.0319	.5319	16/30 = .5333	.0014	$\lvert .5000-.5319\rvert = .0319$
95	.14	.0557	.5557	17/30 = .5667	.0110	$\lvert .5333-.5557\rvert = .0224$
98	.23	.0901	.5901	18/30 = .6000	.0099	$\lvert .5667-.5901\rvert = .0234$
100	.29	.1141	.6141	19/30 = .6333	.0192	$\lvert .6000-.6141\rvert = .0141$
105	.43	.1664	.6664	20/30 = .6667	.0003	$\lvert .6333-.6664\rvert = .0331$
106	.46	.1772	.6772	21/30 = .7000	.0228	$\lvert .6667-.6772\rvert = .0105$
108	.52	.1985	.6985	22/30 = .7333	.0348	$\lvert .7000-.6985\rvert = .0015$
115	.72	.2642	.7642	23/30 = .7667	.0025	$\lvert .7333-.7642\rvert = .0309$
118	.80	.2881	.7881	24/30 = .8000	.0119	$\lvert .7667-.7881\rvert = .0214$
126	1.03	.3485	.8485	25/30 = .8333	.0152	$\lvert .8000-.8485\rvert = .0485$
128	1.09	.3621	.8621	26/30 = .8667	.0046	$\lvert .8333-.8621\rvert = .0288$
130	1.15	.3749	.8749	27/30 = .9000	.0251	$\lvert .8667-.8749\rvert = .0082$
142	1.49	.4319	.9319	28/30 = .9333	.0014	$\lvert .9000-.9319\rvert = .0319$
145	1.58	.4429	.9429	29/30 = .9667	.0238	$\lvert .9333-.9429\rvert = .0096$
155	1.87	.4693	.9693	30/30 = 1.0000	.0307	$\lvert .9667-.9693\rvert = .0026$

obvious that if more detailed tables were available listing the two-tailed .05 and .01 critical values, the latter values would be greater than $M' = .0667$, and thus the result would not be significant. Since the test statistic is interpreted in the same way as the **Kolmogorov–Smirnov test** statistic, the conclusion drawn from the **Lilliefors test for normality** is identical to that reached with the **Kolmogorov–Smirnov test**. Thus, the null hypothesis of normality is retained.

VII. Additional Discussion of the Kolmogorov–Smirnov Goodness-of-Fit Test for a Single Sample

1. Effect of sample size on the result of a goodness-of-fit test Conover (1980, 1999) notes that if a researcher employs a large enough sample size, almost any goodness-of-fit test will result in rejection of the null hypothesis. In view of the latter, Conover (1980, 1999) states that in order to conclude on the basis of a goodness-of-fit test that data conform to a specific distribution, the data should be reasonably close to the specifications of the distribution. Thus, in some cases where a large sample size is involved, a researcher may end up rejecting the null hypothesis of goodness-of-fit for a hypothesized distribution, yet in spite of the latter, if the sample data are reasonably close to the hypothesized distribution, one can probably operate on the assumption that the sample data provide an adequate fit for the hypothesized distribution.

2. The Kolmogorov–Smirnov goodness-of-fit test for a single sample versus the chi-square goodness-of-fit test and alternative goodness-of-fit tests Daniel (1990) discusses the relative merits of employing the **Kolmogorov–Smirnov goodness-of-fit test for a single sample** for assessing goodness-of-fit versus the **chi-square goodness-of-fit test**. In his discussion Daniel (1990) notes the following: a) Whereas the **Kolmogorov–Smirnov test** is designed for use with continuous data, the **chi-square goodness-of-fit test** is designed to be used with discrete data; b) The **Kolmogorov–Smirnov test** is able to evaluate a one-tailed hypothesis regarding goodness-of-fit, while the **chi-square test** is not suited for such an analysis; c) Whereas the **Kolmogorov–Smirnov test** allows for the computation of a confidence interval for the cumulative population distribution the sample represents, the **chi-square test** does not; d) Since the **chi-square test** groups data into categories/class intervals, it does not use as much information as the **Kolmogorov–Smirnov test**, which generally evaluates each score separately; and e) Whereas the **chi-square test** provides an approximation of an exact sampling distribution (the multinomial distribution), the **Kolmogorov–Smirnov test** employs an exact sampling distribution.

Although the **Kolmogorov–Smirnov goodness-of-fit test for a single sample** and the **chi-square goodness-of-fit test** are the among the most commonly employed (as well as discussed) tests for goodness-of-fit, a number of other goodness-of-fit tests have been developed, including the following which are described in this book: The **D'Agostino–Pearson test of normality (Test 5a)** and the **Jarque–Bera test of normality (Test 5b)** (both of which employ the results obtained for the **single sample test for evaluating population skewness (Test 4)** and the **single sample test for evaluating population kurtosis (Test 5)**). Among other goodness-of-fit tests that are described and/or discussed in nonparametric statistics books are **David's empty cell test** (David (1950)), the **Cramér–von Mises goodness-of-fit test** (attributed to Cramér (1928), von Mises (1931), and Smirnov (1936)), and the **Shapiro–Wilk test for normality** (which is described in Conover (1980, 1999)). Stevens (p. 264, 2002) notes that Wilk, Shapiro, and Chen (1968) in a Monte Carlo (e.g., simulation) study found tests utilizing measures of skewness and kurtosis (e.g., the **single-sample test for evaluating population skewness** and the **single-sample test for evaluating population kurtosis**) and the **Shapiro–Wilk test of normality** provide for the most powerful test of an alternative hypothesis stipulating violation of normality. D'Agostino and Stephens (1986), Daniel (1990) and Thode (2002) contain comprehensive discussions of some or all of the above noted goodness-of-fit procedures. Discussion of goodness-of-fit tests for randomness can be found in Section VII of the **single-sample runs tests (Test 10)**, and under **autocorrelation** (which is a procedure that can also be employed for assessing goodness-of-fit for randomness) in Section VII of the **Pearson product–moment correlation coefficient**.

VIII. Additional Example Illustrating the Use of the Kolmogorov–Smirnov Goodness-of-Fit Test for a Single Sample

Example 7.2 *The results of an intelligence test administered to* 30 *students are evaluated with respect to goodness-of-fit for a normal distribution with the following parameters:* $\mu = 90$ *and* $\sigma = 35$. *The IQ scores of the* 30 *students are noted below.*

21, 32, 38, 40, 48, 55, 63, 66, 70, 75, 80, 84, 86, 90, 90, 93, 95, 98, 100, 105, 106, 108, 115, 118, 126, 128, 130, 142, 145, 155

Do the data conform to a normal distribution with the specified parameters?

Since Example 7.2 employs the same data as Example 7.1, it yields the identical result.

References

Conover, W. J. (1980). **Practical nonparametric statistics** (2nd ed.). New York: John Wiley & Sons.

Conover, W. J. (1999). **Practical nonparametric statistics** (3rd ed.). New York: John Wiley & Sons.

Cramér, H. (1928). On the composition of elementary errors. **Skandinavisk Aktaurietidskrift**, 11, 13–74, 141–180.

D'Agostino, R. B. & Stephens, M. A. (Eds.) (1986). **Goodness-of-fit techniques**. New York: Marcel Dekker.

Daniel, W. W. (1990). **Applied nonparametric statistics** (2nd ed.). Boston: PWS–Kent Publishing Company.

David, F. N. (1950). Two combinatorial tests of whether a sample has come from a given population. **Biometrika**, 37, 97–110.

Harter, H. L., Khamis, H. J., & Lamb, R. E. (1984). Modified Kolmogorov–Smirnov tests for goodness-of-fit. **Communic. Statist. — Simula. Computa.**, 13, 293–323.

Hollander, M. &Wolfe, D.A. (1999). **Nonparametric statistical methods**. New York: John Wiley & Sons.

Khamis, H. J. (1990). The δ corrected Kolmogorov–Smirnov test for goodness-of-fit. **Journal of Statistical Plan. Infer.**, 24, 317–355.

Khamis, H. J. (2000). The two-stage delta-corrected Kolmogorov–Smirnov test. **Journal of Applied Statistics**, 27, 439–450.

Kolmogorov, A. N. (1933). Sulla determinazione empirica di una legge di distribuzione. **Giorn dell'Inst. Ital. degli. Att.**, 4, 89–91.

Lilliefors, H. W. (1967). On the Kolmogorov–Smirnov test for normality with mean and variance unknown. **Journal of the American Statistical Association**, 62, 399–402.

Lilliefors, H. W. (1969). On the Kolmogorov–Smirnov test for the exponential distribution with mean unknown. **Journal of the American Statistical Association**, 64, 387–389.

Lilliefors, H. W. (1973). The Kolmogorov–Smirnov and other distance tests for the gamma distribution and for the extreme-value distribution when parameters must be estimated. Department of Statistics, George Washington University, unpublished manuscript.

Marascuilo, L. A. & McSweeney, M. (1977). **Nonparametric and distribution-free methods for the social sciences.** Monterey, CA: Brooks/Cole Publishing Company.

Mason, A. L. & Bell, C. B. (1986). New Lilliefors and Srinivasan Tables with applications. **Communic. Statis. — Simul.**, 15(2), 457–459.

Massey, F. J., Jr. (1951). The Kolmogorov–Smirnov test for goodness-of-fit. **Journal of the American Statistical Association**, 46, 68–78.

Miller, I. & Miller, M. (1999). **John E. Freund's mathematical statistics** (6th ed.). Upper Saddle River, NJ: Prentice Hall.

Miller, L. H. (1956). Table of percentage points of Kolmogorov statistics. **Journal of the American Statistical Association**, 51, pp. 111–121.

Shapiro, S. S. & Wilk, M. B. (1965). An analysis of variance test for normality (complete samples). **Biometrika**, 52, 591–611.

Shapiro, S. S. & Wilk, M. B. (1968). Approximations for the null distribution of the W statistic. **Technometrics**, 10, 861–866.

Siegel, S. & Castellan, N. J., Jr. (1988). **Nonparametric statistics for the behavioral sciences** (2nd ed.). New York: McGraw–Hill Book Company.

Smirnov, N. V. (1936). Sur la distribution de W^2 (criterium de M. R. v. Mises). **Comptes Rendus** (Paris), 202, 449–452.

Smirnov, N. V. (1939). Estimate of deviation between empirical distribution functions in two independent samples (Russian), **Bull Moscow Univ.**, 2, 3–16.

Sprent, P. (1993). **Applied nonparametric statistical methods** (2nd ed.). London: Chapman & Hall.

Stevens, J.P. (2002). **Applied multivariate statistics for the social sciences** (4th ed.). Mahwah, NJ: Lawrence Erlbaum Associates.

Thode, H. C. (2002). **Testing for normality**. New York: Marcel Dekker.

von Mises, R. (1931). **Wahrscheinlichkeitsrechnung und ihre anwendung in derstatistik and theoretishen Physik.** Leipzig: Deuticke.

Wilk, H.B., Shapiro, S.S. & Chen, H.J. (1965). A comparative study of various tests of normality. **Journal of the American Statistical Association**, 63, 1343–1372.

Zar, J. H. (1999). **Biostatistical analysis** (4th ed.). Upper Saddle River, NJ: Prentice Hall.

Endnotes

1. a) Marascuilo and McSweeney (1977) employ a modified protocol which can result in a larger absolute value for M in **Column F** or M' in **Column G** than the one obtained in Table 7.2. The latter protocol employs a separate row in the table for each instance in which the same score occurs more than once in the sample data. If the latter protocol were employed in Table 7.2, there would be two rows in the table for the score of 90 (which is the only score that occurs more than once). The first 90 would be recorded in **Column A** in a row that has a cumulative proportion in **Column E** equal to 14/30 = .4667. The second 90 would be recorded in the following row in **Column A** with a cumulative proportion in **Column E** equal to 15/30 = .5000. In the case of Example 7.1, the outcome of the analysis would not be affected if the aforementioned protocol is employed. In some instances, however, it can result in a different/larger M or M' value. The protocol employed by Marascuilo and McSweeney (1977) is employed by sources who argue that when there are ties present in the data (i.e., a score occurs more than once), the protocol described in this chapter (which is used in most sources) results in an overly conservative test (i.e., makes it more difficult to reject a false null hypothesis); b) It is not necessary to compute the values in **Column G** if a discrete variable is being evaluated. Conover (1980, 1999) and Daniel (1990) discuss the use of the **Kolmogorov–Smirnov goodness-of-fit test for a single sample** with discrete data. Studies cited in the latter sources indicate that when the **Kolmogorov–Smirnov test** is employed with discrete data, it yields an overly conservative result (i.e., the power of the test is reduced).

2. A general discussion of confidence intervals can be found in Section VI of the **single sample** *t* **test (Test 2)**.

3. The **gamma** and **exponential distributions** are continuous probability distributions. The **exponential distribution** is discussed in detail in Section IX (the **Addendum**) of the **binomial sign test for a single sample (Test 9)**.

4. **Table A22** is only appropriate for assessing goodness-of-fit for a normal distribution. Lilliefors (1969, 1973) has developed tables for other distributions (e.g., the exponential and gamma distributions).

Test 8
The Chi-Square Goodness-of-Fit Test
(Nonparametric Test Employed with Categorical/Nominal Data)

I. Hypothesis Evaluated with Test and Relevant Background Information

Hypothesis evaluated with test In the underlying population represented by a sample are the observed cell frequencies different from the expected cell frequencies?

Relevant background information on test The **chi-square goodness-of-fit test**, also referred to as the **chi-square test for a single sample**, is employed in a hypothesis testing situation involving a single sample. Based on some preexisting characteristic or measure of performance, each of n observations (subjects/objects) which is randomly selected from a population consisting of N observations (subjects/objects) is assigned to one of k mutually exclusive categories.[1] The data are summarized in the form of a table consisting of k cells, each cell representing one of the k categories. Table 8.1 summarizes the general model for the **chi-square goodness-of-fit test**. In Table 8.1, C_i represents the ith cell/category and O_i represents the number of observations in the ith cell. The number of observations recorded in each cell of the table is referred to as the **observed frequency** of a cell.

Table 8.1   General Model for Chi-Square Goodness-of-Fit Test

		Total number of observations
Cell/Category	C_1 C_2 $\cdots$ C_i $\cdots$ C_k	
Observed frequency	O_1 O_2 $\cdots$ O_i $\cdots$ O_k	n

The experimental hypothesis evaluated with the **chi-square goodness-of-fit test** is whether or not there is a difference between the **observed frequencies** of the k cells and their **expected frequencies** (also referred to as the **theoretical frequencies**). The expected frequency of a cell is determined through the use of probability theory or is based on some preexisting empirical information about the variable under study. If the result of the **chi-square goodness-of-fit test** is significant, the researcher can conclude that in the underlying population represented by the sample there is a high likelihood the observed frequency for at least one of the k cells is not equal to the expected frequency of the cell. It should be noted that, in actuality, the test statistic for the **chi-square goodness-of-fit test** provides an approximation of a binomially distributed variable (when $k = 2$) and a multinomially distributed variable (when $k > 2$). The larger the value of n, the more accurate the chi-square approximation of the **binomial** and **multinomial distributions**.[2]

The **chi-square goodness-of-fit test** is based on the following assumptions: a) Categorical/nominal data are employed in the analysis. This assumption reflects the fact that the test data should represent frequencies for k mutually exclusive categories; b) The data evaluated consist of a random sample of n independent observations. This assumption reflects the fact that each observation can only be represented once in the data; and c) The expected frequency of each cell is 5 or greater. When this assumption is violated, it is recommended that if $k = 2$, the **binomial sign test for a single sample (Test 9)** be employed to evaluate the data. When the expected frequency of one or more cells is less than 5 and $k > 2$, the multinomial distribution should be employed to evaluate the data. The reader should be aware of the fact that sources are not in agreement with respect to the minimum acceptable value for an expected frequency. Many sources employ criteria suggested by Cochran (1952), who stated that none of the expected frequencies should be less than 1 and no more than 20% of the expected frequencies should be less than 5. However, many sources suggest the latter criteria may be overly conservative. In the event that a researcher believes one or more expected cell frequencies are too small, two or more cells can be combined with one another to increase the values of the expected frequencies. The latter procedure is demonstrated and discussed in Section VI.

Zar (1999, p. 470) provides an interesting discussion on the issue of the lowest acceptable value for an expected frequency. Within the framework of his discussion, Zar (1999) cites studies indicating that when the **chi-square goodness-of-fit test** is employed to evaluate a hypothesis regarding a **uniform distribution**, the test is extremely **robust**. A **robust test** is one which still provides reliable information, in spite of the fact that one or more of its assumptions have been violated. A **uniform distribution** (also referred to as a **rectangular distribution**) is one in which each of the possible values a variable can assume has an equal likelihood of occurring. In the case of an analysis involving the **chi-square goodness-of-fit test**, a distribution is uniform if each of the cells has the same expected frequency.

II. Examples

Two examples will be employed to illustrate the use of the **chi-square goodness-of-fit test**. Since both examples employ identical data, they will result in the same conclusions with respect to the null hypothesis.

Example 8.1 *A die is rolled* 120 *times in order to determine whether or not it is fair (unbiased). The value* **1** *appears* 20 *times, the value* **2** *appears* 14 *times, the value* **3** *appears* 18 *times, the value* **4** *appears* 17 *times, the value* **5** *appears* 22 *times, and the value* **6** *appears* 29 *times. Do the data suggest the die is biased?*

Example 8.2 *A librarian wishes to determine if it is equally likely that a person will take a book out of the library each of the six days of the week the library is open (assume the library is closed on Sundays). She records the number of books signed out of the library during one week and obtains the following frequencies: Monday,* 20; *Tuesday,* 14; *Wednesday,* 18; *Thursday,* 17; *Friday,* 22; *and Saturday,* 29. *Assume no person is permitted to take out more than one book during the week. Do the data indicate there is a difference with respect to the number of books taken out on different days of the week?*

III. Null versus Alternative Hypotheses

In the statement of the null and alternative hypotheses, the lower case Greek letter **omicron** (o) is employed to represent the observed frequency of a cell in the underlying population, and the lower case Greek letter **epsilon** (ε) is employed to represent the expected frequency of the cell in the population. Thus, o_i and ε_i, respectively, represent the observed and expected frequency of the ith cell in the underlying population. With respect to the observed and expected frequencies for the sample data, the notation O_i is employed to represent the observed frequency of the ith cell, and E_i the expected frequency of the ith cell.

Null hypothesis H_0: $o_i = \varepsilon_i$ for all cells.

(In the underlying population the sample represents, for each of the k cells, the observed frequency of a cell is equal to the expected frequency of the cell. With respect to the sample data this leads to the prediction that for all k cells $O_i = E_i$.)

Alternative hypothesis H_i: $o_i \neq \varepsilon_i$ for at least one cell.

(In the underlying population the sample represents, for at least one of the k cells the observed frequency of a cell is not equal to the expected frequency of the cell. With respect to the sample data this leads to the prediction that for at least one cell $O_i \neq E_i$. The reader should take note of the fact that the alternative hypothesis does not state that in order to reject the null hypothesis there must be a discrepancy between the observed and expected frequencies of all k cells. Rejection of the null hypothesis can be the result of a discrepancy between the observed and expected frequencies for one cell, two cells, ..., $(k - 1)$ cells, or all k cells. As a general rule, sources always state the alternative hypothesis for the **chi-square goodness-of-fit test** nondirectionally. Although the latter protocol will be adhered to in this book, in actuality it is possible to state the alternative hypothesis directionally. The issue of the directionality of an alternative hypothesis is discussed in Section VII.)

IV. Test Computations

Table 8.2 summarizes the data and computations for Examples 8.1 and 8.2.

Table 8.2   Chi-square Summary Table for Examples 8.1 and 8.2

Cell	O_i	E_i	$(O_i - E_i)$	$(O_i - E_i)^2$	$\dfrac{(O_i - E_i)^2}{E_i}$
1/Monday	20	20	0	0	0
2/Tuesday	14	20	−6	36	1.8
3/Wednesday	18	20	−2	4	.2
4/Thursday	17	20	−3	9	.45
5/Friday	22	20	2	4	.2
6/Saturday	29	20	9	81	4.05
	$\Sigma O_i = 120$	$\Sigma E_i = 120$	$\Sigma(O_i - E_i) = 0$		$\chi^2 = 6.7$

In Table 8.2, the observed frequency of each cell (O_i) is listed in Column 2, and the expected frequency of each cell (E_i) is listed in Column 3. The computations for the **chi-square goodness-of-fit test** require that the observed and expected cell frequencies be compared with one another. In order to determine the expected frequency of a cell one must either: a) Employ the appropriate theoretical probability for the test model; or b) Employ a probability which is based on existing empirical data.

In Examples 8.1 and 8.2, computation of the expected cell frequencies is based on the theoretical probabilities for the test model.[3] Specifically, if the die employed in Example 8.1 is fair, it is equally likely that in a given trial any one of the six face values will appear. Thus, it follows that each of the six face values should occur one-sixth of the time. The probability associated with each of the possible outcomes (represented by the notation π, which is the lower case Greek letter **pi**) can be computed as follows: $\pi = r/k$ (**where:** r represents the number of outcomes which will allow an observation to be placed in a specific category, and k represents the total number of possible outcomes in any trial). Since, in each trial only one face value will result in an observation being assigned to any one of the six categories, the value of the numerator for each of the six categories will equal $r = 1$. Since in each trial there are six possible outcomes, the value of the denominator for each of the six categories will equal $k = 6$. Thus, for each category, $\pi_i = 1/6$.[4] Note that the sum of the k probabilities must equal 1, since if the value 1/6 is added six times it sums to 1 (i.e., $\sum_{i=1}^{k} \pi_i = 1$).

The same logic employed for Example 8.1 can be applied to Example 8.2. If it is equally likely a person will take a book out of the library on any one of the six days of the week the library is open, it is logical to predict that on each day of the week one-sixth of the books will be taken out. Consequently, the value 1/6 will represent the expected probability for each of the six cells in Example 8.2. The expected frequency of each cell in Examples 8.1 and 8.2 is computed by multiplying the total number of observations by the probability associated with the cell. Equation 8.1 summarizes the computation of an expected frequency.

$$E_i = n\,\pi_i \qquad \text{(Equation 8.1)}$$

Where: n represents the total number of observations

π_i represents the probability that an observation will fall within the ith cell

Since in both Example 8.1 and 8.2 the total number of observations is $n = 120$, the expected frequency for each cell can be computed as follows: $E_i = (120)(1/6) = 20$.

Upon determining the expected cell frequencies, Equation 8.2 is employed to compute the test statistic for the **chi-square goodness-of-fit test**.

$$\chi^2 = \sum_{i=1}^{k}\left[\frac{(O_i - E_i)^2}{E_i}\right] \qquad \text{(Equation 8.2)}$$

The operations described by Equation 8.2 are as follows: a) The expected frequency of each cell is subtracted from its observed frequency. This is summarized in Column 4 of Table 8.2; b) For each cell, the difference between the observed and expected frequency is squared. This is summarized in Column 5 of Table 8.2; c) For each cell, the squared difference between the observed and expected frequency is divided by the expected frequency of the cell. This is summarized in Column 6 of Table 8.2; and d) The value of chi-square is computed by summing all of the values in Column 6. For both Examples 8.1 and 8.2, Equation 8.2 yields the value $\chi^2 = 6.7$.

Note that in Table 8.2 the sums of the observed and expected frequencies are identical. This must always be the case, and any time these sums are not equivalent it indicates a computational error has been made.[5] It is also required that the sum of the differences between the observed and expected frequencies equals zero (i.e., $\Sigma(O_i - E_i) = 0$). Any time the latter value does not equal zero, it indicates an error has been made. Since all of the $(O_i - E_i)$ values are squared in Column 5, the sum of Column 6, which represents the value of χ^2, must always be a positive number. If a negative value is obtained for chi-square, it indicates an error has been made. The only time χ^2 will equal zero is when $O_i = E_i$ for all k cells.

V. Interpretation of the Test Results

The obtained value $\chi^2 = 6.7$ is evaluated with **Table A4 (Table of the Chi-Square Distribution)** in the **Appendix**. A general overview of the chi-square distribution and guidelines for interpreting the values in **Table A4** can be found in Sections I and V of the **single-sample chi-square test for a population variance (Test 3)**.

The degrees of freedom are employed in evaluating the results of the **chi-square goodness-of-fit test** are computed with Equation 8.3.[6]

$$df = k - 1 \qquad \textbf{(Equation 8.3)}$$

When **Table A4** is employed to evaluate a chi-square value computed for the **chi-square goodness-of-fit test**, the following protocol is employed. The tabled critical values for the **chi-square goodness-of-fit test** are always derived from the right tail of the distribution. Thus, the tabled critical .05 chi-square value (to be designated $\chi^2_{.05}$) will be the tabled chi-square value at the 95th percentile. In the same respect, the tabled critical .01 chi-square value (to be designated $\chi^2_{.01}$) will be the tabled chi-square value at the 99th percentile. The general rule is that the tabled critical chi-square value for a given level of alpha will be the tabled chi-square value at the percentile which corresponds to the value of $(1 - \alpha)$. In order to reject the null hypothesis, the obtained value of chi-square must be equal to or greater than the tabled critical value at the prespecified level of significance. The aforementioned guidelines for determining tabled critical chi-square values are employed when the alternative hypothesis is stated nondirectionally (which, as noted earlier, is usually the case). The determination of tabled critical chi-square values in reference to a directional alternative hypothesis is discussed in Section VII.

Applying the guidelines for a nondirectional analysis to Examples 8.1 and 8.2, the degrees of freedom are computed to be $df = 6 - 1 = 5$. The tabled critical .05 chi-square value for $df = 5$ is $\chi^2_{.05} = 11.07$, which, as noted above, is the tabled chi-square value at the 95th percentile. The tabled critical .01 chi-square value for $df = 5$ is $\chi^2_{.01} = 15.09$, which, as noted above, is the tabled chi-square value at the 99th percentile. Since the computed value $\chi^2 = 6.7$ is less than $\chi^2_{.05} = 11.07$, the null hypothesis cannot be rejected at the .05 level. This result can be summarized as follows: $\chi^2 (5) = 6.7$, $p > .05$. Although there are some deviations between the observed and expected frequencies in Table 8.2, the result of the **chi-square goodness-of-fit test** indicates there is a reasonably high likelihood the deviations in the sample data can be attributed to chance.

A summary of the analysis of Examples 8.1 and 8.2 with the **chi-square goodness-of-fit test** follows: a) In Example 8.1 the data do not suggest that the die is biased; and b) In Example 8.2 the data do not suggest that there is any difference with respect to the number of books taken out of the library on different days of the week.

VI. Additional Analytical Procedures for the Chi-Square Goodness-of-Fit Test and/or Related Tests

1. Comparisons involving individual cells when $k > 2$ Within the framework of the **chi-square goodness-of-fit test** it is possible to compare individual cells with one another. To illustrate this, assume that we wish to address the following questions in reference to Examples 8.1 and 8.2.

a) In Example 8.1, is the observed frequency of 29 for the face value 6 higher than the combined observed frequency of the other five face values? Note that this is not the same thing as asking whether or not the face value 6 is more likely to occur when compared individually with any of the other five face values. In order to answer the latter question, the observed frequency for the face value 6 must be contrasted with the observed frequency for the specific face value in which one is interested.

b) In Example 8.2, is the observed frequency of 29 books for Saturday higher than the combined observed frequency of the other five days of the week? Note that this is not the same thing as asking whether or not a person is more likely to take a book out of the library on Saturday when compared individually with any one of the other five days of the week. In order to answer the latter question, the observed frequency for Saturday must be contrasted with the observed frequency of the specific day of the week in which one is interested.

In order to answer the question of whether **6/Saturday** occurs a disproportionate amount of the time, the observed frequency for **6/Saturday** must be contrasted with the combined observed frequencies of the other five face values/days of the week. In order to do this, the original six-cell chi-square table is collapsed into a two-cell table, with one cell representing **6/Saturday** (Cell 1) and the other cell representing **1, 2, 3, 4, 5/M, T, W, Th, F** (Cell 2). The expected frequency of Cell 1 remains $\pi_1 = 1/6$, since if we are dealing with a random process, there is still a one in six chance that in any trial the face value 6 will occur, or that a person will take a book out of the library on Saturday. Thus: $E_1 = (120)(1/6) = 20$. The expected frequency of Cell 2 is computed as follows: $E_2 = (120)(5/6) = 100$. Note that the probability $\pi_2 = 5/6$ for Cell 2 is the sum of the probabilities of the other five cells. In other words, if it is randomly determined what face value appears on the die or on what day of the week a person takes a book out of the library, there is a five in six chance a face value other than 6 will appear on any roll of the die, and a five in six chance a book is taken out of the library on a day of the week other than Saturday. Table 8.3 summarizes the data for the problem under discussion.

Table 8.3   Chi-Square Summary Table When $\pi_1 = 1/6$ and $\pi_2 = 5/6$

Cell	O_i	E_i	$(O_i - E_i)$	$(O_i - E_i)^2$	$\dfrac{(O_i - E_i)^2}{E_i}$
6/Saturday	29	20	9	81	4.05
1,2,3,4,5/M,T,W,Th,F	91	100	-9	81	.81
	$\Sigma O_i = 120$	$\Sigma E_i = 120$	$\Sigma(O_i - E_i) = 0$		$\chi^2 = 4.86$

Since there are $k = 2$ cells, $df = 2 - 1 = 1$. Employing **Table A4** for $df = 1$, $\chi^2_{.05} = 3.84$ and $\chi^2_{.01} = 6.63$. Since the obtained value $\chi^2 = 4.86$ is larger than $\chi^2_{.05} = 3.84$, the null hypothesis can be rejected at the .05 level (i.e., $\chi^2 (1) = 4.86$, $p < .05$). The null hypothesis cannot, however, be rejected at the .01 level, since $\chi^2 = 4.86 < \chi^2_{.01} = 6.63$. Note that by stating the problem in reference to one face value or one day of the week, the researcher is able to reject the null hypothesis at the .05 level. Recollect that the analysis in Section V does not allow the researcher to reject the null hypothesis.[7]

If the original null hypothesis a researcher intends to study deals with the frequency of **Cell 6/Saturday** versus the other five cells, the researcher is not obliged to defend the analysis described above. However, let us assume that the original null hypothesis under study is the one stipulated in Section III. Let us also assume that upon evaluating the data, the null hypothesis cannot be rejected. Because of this the researcher then decides to reconceptualize the problem as summarized in Table 8.3. To go even further, the researcher can extend the type of analysis depicted in Table 8.3 to all six cells (i.e., compare the observed frequency of each of the cells with the combined observed frequency of the other five cells — e.g., **Cell 1/Monday** versus **Cells 2, 3, 4, 5, 6/T, W, Th, F, S**, as well as **Cell 2/Tuesday** versus **Cell 1, 3, 4, 5, 6/M, W, Th,**

F, S, and so on for the other three cells). If $\alpha = .05$ is employed for each of the six comparisons, the overall likelihood of committing at least one Type I error within the set of six comparisons will be substantially above .05 (to be exact, it will equal $1 - (1 - .05)^6 = .26$). If within the set of six comparisons the researcher does not want more than a 5% chance of committing a Type I error, it is required that the alpha level employed for each comparison be adjusted. Specifically, by employing a probability of $.05/6 = .0083$ per comparison, the researcher will insure that the overall Type I error rate will not exceed 5%. It should be noted, however, that by employing a smaller alpha level per comparison, the researcher is reducing the power associated with each comparison. A detailed discussion of the protocol for adjusting the alpha level when conducting multiple comparisons can be found in Section VI of the **single-factor between-subjects analysis of variance (Test 21)**.

It should also be noted that a researcher can reduce the number of degrees of freedom employed in a chi-square analysis by reconfiguring a table comprised of three or more cells into a table comprised of fewer cells. Reduction of the degrees of freedom will increase the likelihood of rejecting the null hypothesis, since the lower the value of the degrees of freedom, the lower the tabled critical chi-square value at a given level of significance. By employing the latter strategy, a researcher may be able convert a table with three or more cells which does not yield a significant result into a smaller table that does yield a significant result. Obviously, it would be inappropriate to employ such a strategy if its sole purpose is to milk a significant result out of a set of data. Any significant results obtained within the latter context have to be viewed with extreme caution, and should be replicated prior to being submitted for publication.

It is also possible to conduct other comparisons in addition to the ones noted above. For example one can compare the observed frequencies for face values/days of the week **1, 2, 3/M, T, W** with the observed frequencies for **4, 5, 6/Th, F, S**. In such an instance there again will be two cells, with a probability of $\pi_i = 1/2$ for each cell (since $\pi_i = 3/6 = 1/2$). A researcher can also break down the original six cell table into three cells — e.g., **1, 2/M, T** versus **3, 4/W, Th** versus **5, 6/F, S**. In this instance, the probability for each cell will equal $\pi_i = 1/3$ (since $\pi_i = 2/6 = 1/3$).

Another type of comparison which can be conducted is to contrast just two of the original six cells with one another. Specifically, let us assume we want to compare **Cell 1/Monday** with **Cell 2/Tuesday**. Table 8.4 is employed to summarize the data for such a comparison.

Table 8.4   Chi-Square Summary Table for Comparison

Cell	O_i	E_i	$(O_i - E_i)$	$(O_i - E_i)^2$	$\dfrac{(O_i - E_i)^2}{E_i}$
1/Monday	20	17	3	9	.53
2/Tuesday	14	17	3	9	.53
	$\Sigma O_i = 34$	$\Sigma E_i = 34$	$\Sigma(O_i - E_i) = 0$		$\chi^2 = 1.06$

Note that in the above example, since we employ only two cells, the probability for each cell will be $\pi_i = 1/2$. The expected frequency of each cell is obtained by multiplying $\pi_i = 1/2$ by the total number of observations in the two cells (which equals 34). As noted previously, in conducting a comparison such as the one above, a critical issue the researcher must address is what value of alpha to employ in evaluating the null hypothesis. If $\alpha = .05$ is used, for $df = 1$, $\chi^2_{.05} = 3.84$. The null hypothesis cannot be rejected, since the obtained value $\chi^2 = 1.06$ $< \chi^2_{.05} = 3.84$.

A major point which has been emphasized throughout the discussion in this section is that, depending upon how one initially conceptualizes a problem, there will generally be a number of different ways in which a set of data can be analyzed. Furthermore, after analyzing the full set of data, additional comparisons involving two or more categories can be conducted. The various types of comparisons which one can conduct can either be planned or unplanned.The term **planned comparison** is employed throughout the book to refer to a comparison that is planned prior to the data collection phase of a study. In contrast, an **unplanned comparison** is one which a researcher decides to conduct after the experimental data have been collected and scrutinized. A problem associated with unplanned comparisons is that in a large body of data there are a potentially large number of comparisons which can be conducted. Consequently, a researcher can conduct many comparisons until one or more of them yield a significant result. The latter strategy can thus be employed to milk significant results out of a large body of data. It was noted earlier in this section that the larger the number of comparisons one conducts, the greater the likelihood that any significant result obtained for a given comparison will be a Type I error (as opposed to a genuine difference which can be reliably replicated).

Whenever possible comparisons should be planned, and most sources take the position that when a researcher plans a limited number of comparisons before the data collection phase of a study, one is not obliged to control the overall Type I error rate. However, when comparisons are not planned, most sources believe some adjustment of the Type I error rate should be made in order to avoid inflating it excessively. As noted earlier in this section, one way of achieving the latter is to divide the maximum overall Type I error rate one is willing to tolerate by the total number of comparisons one conducts. The resulting probability value will represent the alpha level employed in evaluating each of the comparisons. A comprehensive discussion of the subject of comparisons (which is also germane to the issue of alternative ways of conceptualizing a set of data) can be found in Section VI of the **single-factor between-subjects analysis of variance**.

In closing the discussion of comparisons for the **chi-square goodness-of-fit test**, it should be noted that some sources present alternative comparison procedures which may yield results that are not in total agreement with those obtained in this section. In instances where different methodologies yield substantially different results (which will usually not be the case), a replication study evaluating the same hypothesis is in order. As noted throughout the book, replication is the most effective way to demonstrate the validity of a hypothesis. Obviously, the use of large sample sizes in both original and replication studies further increases the likelihood of obtaining reliable results. An alternative approach for conducting comparisons is presented in the next section.

2. The analysis of standardized residuals An alternative procedure for conducting comparisons (developed by Haberman (1973) and cited in sources such as Siegel and Castellan (1988)) involves the computation of **standardized residuals**. Within the context of a chi-square analysis, the term **residual** is employed to represent the absolute difference between the expected and observed cell frequencies. A **standardized residual** is a residual expressed within the format of a **standard deviation score** (i.e., a z score). By computing standardized residuals, one is able to determine which cells are the major contributors to a significant chi-square value. Equation 8.4 is employed to compute a standardized residual (z_{res_i}) for each cell in a chi-square table.

$$z_{res_i} = \frac{(O_i - E_i)}{\sqrt{E_i}}$$ **(Equation 8.4)**

A value computed for a residual (which is interpreted as a normally distributed variable) is evaluated with **Table A1 (Table of the Normal Distribution)** in the **Appendix**. Any residual with an absolute value that is equal to or greater than the tabled critical two-tailed .05 value $z_{.05} = 1.96$ is significant at the .05 level. Any residual with an absolute value that is equal to or greater than the tabled critical two-tailed .01 value $z_{.01} = 2.58$ is significant at the .01 level. Any cell in a chi-square table which has a significant residual makes a significant contribution to the obtained chi-square value. For any cell that has a significant residual, one can conclude that the observed frequency of the cell differs significantly from its expected frequency. The sign of the standardized residual indicates whether the observed frequency of the cell is above (+) or below (−) the expected frequency. The sum of the squared residuals for all k cells will equal the obtained value of chi-square. Although the result of the chi-square analysis for Examples 8.1 and 8.2 is not significant, the standardized residuals for the chi-square table are computed and summarized in Table 8.5.

Table 8.5 Analysis of Residuals for Examples 8.1 and 8.2

Cell	O_i	E_i	$(O_i - E_i)$	$z_{res_i} = \dfrac{(O_i - E_i)}{\sqrt{E_i}}$	$z_{res_i}^2 = \left[\dfrac{(O_i - E_i)}{\sqrt{E_i}}\right]^2$
1/Monday	20	20	0	0	0
2/Tuesday	14	20	−6	−1.34	1.80
3/Wednesday	18	20	−2	−.45	.20
4/Thursday	17	20	−3	−.67	.45
5/Friday	22	20	2	.45	.20
6/Saturday	29	20	9	2.01	4.05

$\Sigma O_i = 120 \quad \Sigma E_i = 120 \quad \Sigma(O_i - E_i) = 0 \qquad\qquad \Sigma z_{res_i}^2 = \chi^2 = 6.7$

Note that in Column 5 of Table 8.5 the only cell with a standardized residual with an absolute value above 1.96 is **Cell 6/Saturday**. Thus, one can conclude that the observed frequency of **Cell 6/Saturday** is significantly above its expected frequency and, as such, the cell would be viewed as a major contributor in obtaining a significant chi-square value (if, in fact, the computed chi-square value had been significant). It should be noted that this result is consistent with the first comparison which was conducted in the previous section, since the latter comparison indicates that the observed frequency of 29 for **Cell 6/Saturday** deviates significantly from its expected frequency, when the cell is contrasted with the combined frequencies of the other five cells.

3. The correction for continuity for the chi-square goodness-of-fit test Although it is not generally discussed in reference to the **chi-square goodness-of-fit test**, a **correction for continuity** (which is discussed in Section VI of the **Wilcoxon signed-ranks test (Test 6)**) can be applied to Equation 8.2. The basis for employing the correction for continuity with the **chi-square goodness-of-fit-test** is that the test employs a continuous distribution to approximate a discrete distribution (specifically, the binomial or multinomial distributions). The correction for continuity is based on the premise that if a continuous distribution is employed to estimate a discrete distribution, such an approximation will inflate the Type I error rate. By employing the correction for continuity the Type I error rate is ostensibly adjusted to be more compatible with the prespecified alpha value designated by the researcher. Equation 8.5 is the continuity-corrected chi-square equation for the **chi-square goodness-of-fit test**.

$$\chi^2 = \sum_{i=1}^{k} \left[\frac{(|O_i - E_i| - .5)^2}{E_i} \right] \qquad \textbf{(Equation 8.5)}$$

Note that by subtracting .5 from the absolute value of the difference between each set of observed and expected frequencies, the chi-square value derived with Equation 8.5 will be lower than the value computed with Equation 8.2. The magnitude of the correction for continuity will be inversely related to the size of the sample. The correction for continuity for the **chi-square goodness-of-fit test** is only employed when there are $k = 2$ cells. This latter application of the correction is discussed under the **z test for a population proportion**. The use of the correction for continuity with other designs which employ the chi-square statistic is discussed under the **chi-square test for $r \times c$ tables (Test 16)**.

4. Computation of a confidence interval for the chi-square goodness-of-fit test /confidence interval for a population proportion The procedures to be described in this section allow one to compute a confidence interval for the proportion of cases in the underlying population that falls within any cell in a one-dimensional chi-square table.[8] The true population proportion for a cell will be represented by the notation π_i. The analyses to be described will assume that if $k > 2$, the original chi-square table is converted into a table consisting of $k = 2$ cells.

Equation 8.6 provides a **large sample approximation** of the confidence interval for a binomially distributed variable (which applies to the **chi-square goodness-of-fit test** model when $k = 2$). The latter equation represents a commonly employed procedure for computing a confidence interval for a population proportion for a specific cell (i.e., the proportion of cases in that cell in the underlying population), when there are $k = 2$ cells.

$$\left[p_1 - z_{\alpha/2} \sqrt{\frac{p_1 p_2}{n}} \right] \leq \pi_1 \leq \left[p_1 + z_{\alpha/2} \sqrt{\frac{p_1 p_2}{n}} \right] \qquad \textbf{(Equation 8.6)}$$

Where: p_1 represents the proportion of observations in Cell 1. In the analysis under discussion, Cell 1 will represent the single cell whose observed frequency is being compared with the combined observed frequencies of the remaining $(k - 1)$ cells. The value of p_1 is computed by dividing the number of observations in Cell 1 (which will be represented by the notation x) by n (which represents the total number of observations). Thus, $p_1 = x/n$.

$p_2 = 1 - p_1$ The value p_2 represents the proportion of observations in Cell 2. In the analysis under discussion, Cell 2 will represent the combined frequencies of the other $(k - 1)$ cells. p_2 can be computed by dividing the number of observations which are not in Cell 1 by the total number of observations. Thus, $p_2 = (n - x)/n$.

$z_{\alpha/2}$ represents the tabled critical value in the normal distribution below which a proportion (percentage) equal to $[1 - (\alpha/2)]$ of the cases falls. If the proportion (percentage) of the distribution that falls within the confidence interval is subtracted from 1 (100%), it will equal the value of α.

The large sample approximation computed with Equation 8.6 is most accurate when the value of π_1 falls between .3 and .7. Since Equation 8.6 becomes inaccurate if the value of π_1 is close to 0 or 1, when the latter occurs Fleiss *et al.* (2003, pp. 28–29) recommend employing Equations 8.8 and 8.9 for computing the lower and upper limits of a confidence interval. Another alternative to be discussed is computing the exact binomial confidence interval through use of Equations 8.10 and 8.11.

Initially, the use of Equation 8.6 will be demonstrated. If one wants to determine the 95% confidence interval, the tabled critical two-tailed .05 value $z_{.05} = 1.96$ is employed in Equation 8.6. The tabled critical two-tailed .01 value $z_{.01} = 2.58$ is employed to compute the 99% confidence interval. The value $\sqrt{(p_1 p_2)/n}$ in Equation 8.6 represents the **estimated standard error of the population proportion** (which can be represented with the notation s_p). The latter value is an estimated standard deviation of a sampling distribution of a proportion.[9]

If (as is done in Table 8.3) the data for Examples 8.1 and 8.2 are expressed in a format consisting of two cells, Equation 8.6 can be employed to compute a confidence interval for each of the six cells. Thus, if we wish to compute a confidence interval for the **Cell 6/Saturday**, we can determine that $p_1 = x/n = 29/120 = .242$ and $p_2 = (n-x)/n = (120 - 29)/120 = 91/120 = .758$. Substituting the latter values and the value $z_{.05} = 1.96$ in Equation 8.6, the 95% confidence interval is computed below.

$$.242 - (1.96) \sqrt{\frac{(.242)(.758)}{120}} \leq \pi_1 \leq .242 + (1.96) \sqrt{\frac{(.242)(.758)}{120}}$$

$$\pi_1 = .242 \pm .077$$

$$.165 \leq \pi_1 \leq .319$$

Thus, the researcher can be 95% confident that the interval .165 to .319 contains the true proportion of cases in the underlying population which falls in **Cell 6/Saturday**. Stated in probabilistic terms, there is a probability/likelihood of .95 that the interval .165 to .319 contains the true value of the population proportion for **Cell 6/Saturday**.

The 99% confidence interval, which has a larger range, is computed below by employing $z_{.01} = 2.58$ in Equation 8.6.

$$.242 - (2.58) \sqrt{\frac{(.242)(.758)}{120}} \leq \pi_1 \leq .242 + (2.58) \sqrt{\frac{(.242)(.758)}{120}}$$

$$\pi_1 = .242 \pm .101$$

$$.141 \leq \pi_1 \leq .343$$

Thus, the researcher can be 99% confident that the interval .141 to .343 contains the true proportion of cases in the underlying population which falls in **Cell 6/Saturday**. Stated in probabilistic terms, there is a probability/likelihood of .99 that the interval .141 to .343 contains the true value of the population proportion for **Cell 6/Saturday**.[10]

Some sources (Fleiss et al. (2003, pp. 28–29) and Wallis and Roberts (1956)) employ a **correction for continuity** in computing the confidence interval. In the latter case, the width of the computed confidence interval is wider than the width of the confidence interval computed with Equation 8.6. The continuity corrected confidence interval is computed with Equation 8.7.

$$\left[p_1 - z_{\alpha/2} \sqrt{\frac{p_1 p_2}{n}} - \frac{1}{2n} \right] \leq \pi_1 \leq \left[p_1 + z_{\alpha/2} \sqrt{\frac{p_1 p_2}{n}} + \frac{1}{2n} \right] \qquad \textbf{(Equation 8.7)}$$

Since $1/2n = 1/[(2)(120)] = .004$, the latter value is subtracted from the lower limit of the confidence interval computed with Equation 8.6 and added to the upper limit of the confidence interval computed with the latter equation. Thus, in the case of the 95% confidence interval, the continuity corrected interval is $.161 \leq \pi_1 \leq .323$, which as noted earlier is slightly larger than the confidence interval computed with Equation 8.6.

It was noted earlier that when the value of π_1 is close to 0 or 1, Equations 8.8 and 8.9 (which employ a correction for continuity) allow for more accurate computation of the lower (P_{LL}) and upper (P_{UL}) limits of a confidence interval. The latter values are computed below for the values $p_1 = .242$, $p_2 = .758$, and $n = 120$.

(Equation 8.8)

$$P_{LL} = \frac{(2np_1 + z_{(\alpha/2)}^2 - 1) - z_{(\alpha/2)}\sqrt{z_{(\alpha/2)}^2 - [2 + (1/n)] + 4p_1(np_2 + 1)}}{2(n + z_{(\alpha/2)}^2)}$$

$$P_{LL} = \frac{[[(2)(120)(.242)] + (1.96)^2 - 1] - 1.96\sqrt{(1.96)^2 - [2 + (1/120)] + [(4)(.242)][(120)(.758) + 1]}}{2[120 + (1.96)^2]} = .171$$

(Equation 8.9)

$$P_{UL} = \frac{(2np_1 + z_{(\alpha/2)}^2 + 1) + z_{(\alpha/2)}\sqrt{z_{(\alpha/2)}^2 + [2 - (1/n)] + 4p_1(np_2 - 1)}}{2(n + z_{(\alpha/2)}^2)}$$

$$P_{UL} = \frac{[[(2)(120)(.242)] + (1.96)^2 + 1] + 1.96\sqrt{(1.96)^2 + [2 - (1/120)] + [(4)(.242)][(120)(.758) - 1]}}{2[120 + (1.96)^2]} = .322$$

Thus, the 95% confidence interval is $.171 \leq \pi_1 \leq .322$, which is quite close to the values computed employing Equations 8.6 and 8.7.[11]

In Section I it was noted that when $k = 2$, the test statistic for the **chi-square goodness-of-fit test** provides an approximation of a binomially distributed variable. Thus, the confidence intervals computed in this section are estimates of a confidence interval for a binomially distributed variable. Fleiss *et al.* (2003, pp. 25–26) and Zar (1999, pp. 527–529) note that Equations 8.10 and 8.11 allow one to compute the exact confidence interval for a binomially distributed variable. With respect to the analysis discussed in this section, in the latter equations the value x represents the number of observations in Cell 1 (which in the case of the analysis under discussion will be **Cell 6/Saturday**), and thus, $x = 29$. Equations 8.10 and 8.11 also employ notation which indicates the use of F values that are derived from the F distribution, which is most commonly employed for analysis of variance procedures discussed later in the book. (A full discussion of the F distribution can be found in Section VI of the *t* **test for two independent samples (Test 11)** and in the chapter on the **single-factor between-subjects analysis of variance**.) It is because of a relationship between the F distribution and the binomial distribution that critical F values can employed in computing the upper and lower limits for a confidence interval for a proportion.

Tables of critical values for the F distribution are presented in **Table A10 (Table of the F Distribution) in the Appendix**. In **Table A10** critical values are listed in reference to the number of degrees of freedom associated with the numerator and the denominator of a test statistic commonly referred to as an **F ratio** (i.e., df_{num} and df_{den}). Thus, a critical F value has two degrees of freedom values, one for the numerator and one for the denominator of the F ratio. In the case of Equation 8.10, the numerator degrees of freedom is the first value indicated by the subscript below the F, $2(n - x + 1)$ which equals $2(120 - 29 + 1) = 184$, and the denominator degrees of freedom is the second value indicated by the subscript $2x$ which equals $(2)(29) = 58$. In the case of Equation 8.11, the numerator degrees of freedom is the first value indicated by the subscript below the F, $2(x + 1)$ which equals $2(29 + 1) = 60$, and the denominator degrees of freedom is the second value indicated by the subscript $2(n - x)$ which equals $2(120 - 29) = 182$. In computing the 95% confidence interval, the table on the page of critical F values for $F_{.975}$ is employed. (If one were to compute the 99% confidence interval the table of critical F values on the page for $F_{.995}$ would be employed.) Since exact degrees of freedom values corresponding to those employed for the analysis under discussion (i.e., the values 184, 58 in Equation 8.10 and the values 60, 182 in Equation 8.11) are not listed in the table, interpolation has been employed to estimate the appropriate critical values. (Computer software is available which provides exact critical F values for any degrees of freedom values.)

Employing Equations 8.10 (which computes the **lower limit of the confidence interval**) and 8.11 (which computes the **upper limit of the confidence interval**), the 95% confidence interval is $.169 \leq \pi_1 \leq .328$ (the values of which are quite close to the values computed earlier using the alternative methodologies noted in this section). The reader can see that in the case of the analysis under discussion, all of the methodologies described in this section for computing a confidence interval for a proportion yield values which are extremely close to one another.

$$P_{LL} = \frac{x}{x + (n - x + 1)F_{2(n - x + 1),\ 2x}} \qquad \textbf{(Equation 8.10)}$$

$$P_{LL} = \frac{29}{29 + (120 - 29 + 1)(1.55)} = .169$$

$$P_{UL} = \frac{(x + 1)F_{2(x + 1),\ 2(n - x)}}{(n - x) + (x + 1)F_{2(x + 1),\ 2(n - x)}} \qquad \textbf{(Equation 8.11)}$$

$$P_{UL} = \frac{(29 + 1)(1.48)}{(120 - 29) + (29 + 1)(1.48)} = .328$$

The above described procedures for computing a confidence interval for the proportion of cases in the underlying population which falls within any cell in a one-dimensional chi-square table can be repeated for the other five cells in Examples 8.1 and 8.2. In each instance the observed frequency of a cell is evaluated in relation to the combined observed frequencies of the remaining five cells.

5. Brief discussion of the z test for a population proportion (Test 9a) and the single-sample test for the median (Test 9b) In Section I it is noted that when $k = 2$ and the value of n is large, the **chi-square goodness-of-fit test** provides a good approximation of the binomial distribution. Under the discussion of the **binomial sign test for a single sample**, two tests are described which yield equivalent results to those obtained with the **chi-square goodness-of-fit test** when $k = 2$. The two tests are the z **test for a population proportion** and the **single-sample test for the median**. In the latter test, the two cells of the chi-square table are comprised of scores which fall above the median of a specific distribution and scores that fall below the median of the distribution. A full discussion of these tests can be found in Section VII of the **binomial sign test for a single sample**.

6. Application of the chi-square goodness-of-fit test for assessing goodness-of-fit for a theoretical population distribution In analyzing data there are situations when a researcher may want to determine whether or not a distribution of sample data conforms to a specific theoretical population (or probability) distribution. As is the case with the **Kolmogorov-Smirnov goodness-of-fit test for a single sample (Test 7)** and the **Lilliefors test for normality (Test 7a)**, the **chi-square goodness-of-fit test** can also be employed for this purpose. Although the **Kolmogorov–Smirnov** and **Lilliefors tests** are designed to be employed with a **continuous** variable and the **chi-square test** is designed to be employed with a **discrete** variable, the latter test is sometimes employed to assess goodness-of-fit for a continuous variable. The most common application of the **chi-square goodness-of-fit test** with a continuous variable is in assessing goodness-of-fit for a normal distribution, when the population mean and standard deviation have to be estimated from the sample data. Although the **Kolmogorov-Smirnov test** (which stipulates specific values for the population mean and standard deviation) and the **Lilliefors test for normality** (which, like the **chi-square test**, estimates the population mean and standard deviation from the sample data) are better suited for the latter purpose, the **chi-square test** is often used since it requires less computation (which in itself is not sufficient justification for employing a test).

When the **chi-square goodness-of-fit test** is employed to assess goodness-of-fit for a theoretical distribution, Equation 8.3 (i.e., $df = k - 1$) is not appropriate for computing the degrees of freedom. In determining whether or not a distribution of sample data conforms to a specific theoretical distribution (such as the normal, binomial, or Poisson distributions, all of which will be or have been discussed at some point in the book), it may be necessary to estimate one or more population parameters prior to computing the expected frequency of each cell. In such a case, Equation 8.12 is employed to compute the degrees of freedom for the analysis.

$$df = k - 1 - w \qquad \textbf{(Equation 8.12)}$$

Where: w represents the number of parameters that must be estimated

In actuality, $df = k - 1 - w$ is the generic equation for computing the degrees of freedom for the **chi-square goodness-of-fit test**. Equation 8.3 ($df = k - 1$), which has been used in the examples discussed up to this point, represents the form the equation $df = k - 1 - w$ assumes when $w = 0$.

Example 8.3 will be employed to demonstrate the use of the **chi-square goodness-of-fit test** in assessing goodness-of-fit for a normal distribution. In point of fact, Example 8.3 is almost identical to Example 7.1, which is employed in evaluating the same hypothesis with both the **Kolmogorov-Smirnov goodness-of-fit test for a single sample** and **Lilliefors test for normality**. However, the text of Example 8.3 states that the mean and estimated standard deviation of the population are estimated from the sample data (whereas the latter values are stipulated in Example 7.1). The values $\bar{X} = 90.07$ and $\tilde{s} = 34.79$ (which are also employed for the **Lilliefors test**) noted in Example 8.3 were computed by employing Equations I.1 and I.11 with the 30 scores in the sample.

Example 8.3 *A researcher conducts a study to evaluate whether or not the distribution of the length of time it takes migraine patients to respond to a 100 mg. dose of an intravenously administered drug is normal. The amount of time (in seconds) that elapses between the administration of the drug and cessation of a headache for 30 migraine patients is recorded below. The 30 scores are arranged ordinally (i.e., from fastest response time to slowest response time).*

21, 32, 38, 40, 48, 55, 63, 66, 70, 75, 80, 84, 86, 90, 90, 93, 95, 98, 100, 105, 106, 108, 115, 118, 126, 128, 130, 142, 145, 155

The mean and standard deviation of the population are estimated from the sample data to be $\bar{X}$ = 90.07 and $\tilde{s}$ = 34.79. Do the data conform to a normal distribution?

In order to employ the **chi-square goodness-of-fit test**, the researcher must first estimate the values of the population mean (μ) and standard deviation (σ) by computing the values $\bar{X}$ and $\tilde{s}$ from the sample data. Since the latter requires the estimation of two population parameters (i.e., μ and σ), the appropriate degrees of freedom to employ for the analysis will be $df = k - 1 - 2$. Because the value of k represents the number of cells which are employed in the analysis, there must be a minimum of four cells. The latter is true, since if k is less than four, the value of df will be less than 1 (which is impossible). Each of the cells in the chi-square table will represent a **class interval**. A class interval is a limited range of values in which scores in a frequency distribution are grouped. As is the case with previous applications of the **chi-square goodness-of-fit test**, the expected frequency for each cell/class interval is computed and contrasted with its observed frequency.

The null and alternative hypotheses that are evaluated with the **chi-square goodness-of-fit test** in reference to Example 8.3 can be stated either in the form presented in Section III, or as follows.

Null hypothesis H_0: The sample is derived from a normally distributed population.

Alternative hypothesis H_1: The sample is not derived from a normally distributed population. This is a **nondirectional alternative hypothesis**.

The analysis of Example 8.3 with the **chi-square goodness-of-fit test** is summarized in Tables 8.6 and 8.7. In Table 8.7 each of the $n = 30$ scores has been assigned to one of ten cells/categories. The ten cells, which are summarized in Table 8.6, correspond to the ten **deciles** of the normal distribution. In the **Introduction** it is noted that a **decile** divides a distribution into blocks comprised of ten percentage points (or blocks which comprise a proportion equal to .10 of the distribution). The z scores that correspond to the limits of the ten deciles in a normal distribution were determined through use of **Table A1** in the **Appendix**. Thus, the value $z = -1.28$ corresponds to the upper limit of the 10th percentile, since the entry in **Column 3** of **Table A1** for $z = -1.28$ is .1033 (which is the closest value to .1000, which is 10% when expressed as a percentage). Given $\bar{X}$ = 90.07 and $\tilde{s}$ = 34.79, we can compute that the value X = 45.47 corresponds to $z = -1.28$ by employing the equation $X = \bar{X} + (z)(\tilde{s})$. The latter equation is the algebraic transposition of Equation I.38 ($z = (X - \mu)/\sigma$), when $\bar{X}$ is employed in place of μ and $\tilde{s}$ is employed in place of σ. Thus, if we multiply the value $z = -1.28$ by $\tilde{s}$ = 34.79 and add $\bar{X}$ = 90.07 to the product, we obtain: X = 90.07 + (−1.28)(34.79) = 45.54. The latter value indicates that any score less than 45.54 falls in the first decile.

The value $z = -.84$ corresponds to the upper limit of the 20th percentile, since the entry in **Column 3** of **Table A1** for $z = -.84$ is .2033 (which is the closest value to .2000, which is 20% when expressed as a percentage). Thus, the second decile will be represented by scores which fall above the proportion .10 (or the 10% point) up to the proportion .20 (or the 20% point). When the value $z = -.84$ is substituted in the equation $X = \bar{X} + (z)(\tilde{s})$, we obtain X = 90.07 + (−.84)(34.79) = 60.85. The value X = 60.85 is the upper limit of the 20th decile. Thus, any score that is greater than 45.54 but equal to or less than 60.85 falls in the second decile. To complete Table 8.6, the procedure which has been described for the first and second deciles was employed to determine the limits for the eight remaining deciles.

Table 8.6 Class Intervals for Chi-Square Analysis of Example 8.3

Cell/Class interval/Decile	Limits for z values	Limits for X values
1st decile (0 to .10)	$-1.28 \geq z$	$45.54 \geq X$
2nd decile (> .10 to .20)	$-.84 \geq z > -1.28$	$45.54 < X \leq 60.85$
3rd decile (> .20 to .30)	$-.52 \geq z > -.84$	$60.85 < X \leq 71.98$
4th decile (> .30 to .40)	$-.25 \geq z > -.52$	$71.98 < X \leq 81.37$
5th decile (> .40 to .50)	$0 \geq z > -.25$	$81.37 < X \leq 90.07^*$
6th decile (> .50 to .60)	$.25 \geq z \geq 0$	$90.07 \leq X \leq 98.77^*$
7th decile (> .60 to .70)	$.52 \geq z > .25$	$98.77 < X \leq 108.16$
8th decile (> .70 to .80)	$.84 \geq z > .52$	$108.16 < X \leq 119.29$
9th decile (> .80 to .90)	$1.28 \geq z > .84$	$119.29 < X \leq 134.60$
10th decile (> .90 to 1)	$1.28 < z$	$134.60 < X$

*As a general rule, if two or more scores are equal to the value of $\overline{X}$, one-half of the scores are assigned to the 5th decile and one-half to the 6th decile. If only one score equals $\overline{X}$, it can be randomly assigned to either the 5th or 6th decile.

Table 8.7 Chi-Square Summary Table for Example 8.3

Cell/Class interval/ Decile	O_i	E_i	$(O_i - E_i)$	$(O_i - E_i)^2$	$\dfrac{(O_i - E_i)^2}{E_i}$
1st decile	4	3	1	1	.33
2nd decile	2	3	-1	1	.33
3rd decile	3	3	0	0	.00
4th decile	2	3	-1	1	.33
5th decile	2	3	-1	1	.33
6th decile	5	3	2	4	1.33
7th decile	4	3	1	1	.33
8th decile	2	3	-1	1	.33
9th decile	3	3	0	0	.00
10th decile	3	3	0	0	.00
	$\Sigma O_i = 30$	$\Sigma E_i = 30$	$\Sigma(O_i - E_i) = 0$		$\chi^2 = 3.31$

Employing Equation 8.1, an expected frequency of 3 is computed for each of the cells in Table 8.7 (which is the chi-square summary table) by multiplying the sample size $n = 30$ by .1 (i.e., $E_i = (30)(.1) = 3$). The value .1 is employed to represent π_i in the latter equation, since each cell represents an area that corresponds to 10% of a normal distribution. Thus, the likelihood of an observation falling in any of the cells/deciles is .1.

Employing Equation 8.2, the value $\chi^2 = 3.31$ is computed for Example 8.3. Since there are $k = 10$ cells and $w = 2$ parameters which are estimated, the degrees of freedom for the analysis are $df = 10 - 1 - 2 = 7$. Employing **Table A4**, we determine that for $df = 7$ the tabled critical .05 and .01 values are $\chi^2_{.05} = 14.07$ and $\chi^2_{.01} = 18.48$. Since the computed value $\chi^2 = 3.31$ is less than both of the aforementioned values, the null hypothesis cannot be rejected. Thus, the analysis does not indicate the data deviate significantly from a normal distribution. This is consistent with the conclusion which was reached when the same set of data was evaluated with the **Kolmogorov-Smirnov goodness-of-fit test for a single sample** (which employed the population parameters $\mu = 90$ and $\sigma = 35$, which are almost identical to the estimated values $\overline{X} = 90.07$ and $\tilde{s} = 34.79$) and the **Lilliefors test for normality** (which, like the **chi-square test**, employed the values $\overline{X} = 90.07$ and $\tilde{s} = 34.79$).

It should be noted that because of the small sample size employed in the study, the expected frequency of 3 for all of the cells is less than the minimum value recommended for the **chi-square goodness-of-fit test** by many sources. The values of the expected frequencies could be increased by employing fewer cells in the chi-square table. In other words, one could employ quartile blocks (yielding four cells), or blocks consisting of 20% of the cases per block (yielding five blocks), etc. Daniel (1990) notes that the outcome of the **chi-square goodness-of-fit test** is affected by the number of cells which are employed in the analysis, and cites studies (e.g., Dahiya and Gurland (1973)) that address this issue. Further discussion of the application of the **chi-square goodness-of-fit test** with a continuous variable can be found in Conover (1980, 1999), Daniel (1990), and Siegel and Castellan (1988). Conover (1980, 1999) and Daniel (1990) describe the test protocol when the format of data is a frequency distribution which reflects the number of scores in each of k class intervals. The latter format is most likely to be employed if there are a large number of scores, and the researcher elects to group scores in class intervals (such as 20 scores falling within the range 1–10, 15 scores falling within the range 11–20, etc.), since it provides a succinct way of summarizing the data. In other instances where the original data are grouped in class intervals, a researcher may not have access to the exact value of each score, but have only a frequency distribution that categorizes each score within one of k class intervals.

7. Sources for computing the power of the chi-square goodness-of-fit test Cohen (1977, 1988) has developed a statistic called the **w index** which can be employed to compute the power of the **chi-square goodness-of-fit test**. The value w is an **effect size** index reflecting the difference between expected and observed frequencies. The concept of effect size is discussed in Section VI of the **single-sample t test (Test 2)**. It is discussed in greater detail in Section VI of the **t test for two independent samples**, and in Section IX (the **Addendum**) of the **Pearson product-moment correlation coefficient (Test 28)** under the discussion of **meta-analysis and related topics**.

The equation for the **w index** is $w = \sqrt{\Sigma[(P_{alt} - P_{null})^2/P_{null}]}$. The latter equation indicates the following: a) For each of the cells in the chi-square table, the proportion of cases hypothesized in the null hypothesis is subtracted from the proportion of cases hypothesized in the alternative hypothesis; b) The obtained difference in each cell is squared, and then divided by the proportion hypothesized in the null hypothesis for that cell; c) All of the values obtained for the cells in part b) are summed; and d) w represents the square root of the sum obtained in part c).

Cohen (1977; 1988, Ch. 7) has derived tables which allow a researcher to determine, through use of the **w index**, the appropriate sample size to employ if one wants to test a hypothesis about the difference between observed and expected frequencies in a chi-square table at a specified level of power. Cohen (1977; 1988, pp. 224–226) has proposed the following (admittedly arbitrary) w values as criteria for identifying the magnitude of an effect size: a) A **small effect size** is one that is greater than .1 but not more than .3; b) A **medium effect size** is one that is greater than .3 but not more than .5; and c) A **large effect size** is greater than .5.

8. Heterogeneity chi-square analysis Assume that a researcher conducts m independent studies (where $m \geq 2$) which evaluate the same goodness-of-fit hypothesis, and that none of the studies yields a statistically significant result. However, visual inspection of the data suggests a consistent pattern of differences for the observed frequencies of the k categories employed in each of the m studies. The researcher suspects that because of the relatively small sample sizes employed in the studies, the absence of significant results is largely due to a lack of statistical power. In order to increase the power of the analysis, the researcher wants to combine the data for the m studies into one table, and evaluate the latter table with the **chi-square goodness-of-fit test**. Zar (1999, pp. 471–473) notes that the procedure for determining whether or not a researcher is justified in pooling data under such conditions is referred to as **heterogeneity chi-square analysis** (also referred to as **interaction chi-square analysis** or **homogeneity chi-square analysis**).

The null and alternative hypotheses which are evaluated with a **heterogeneity chi-square analysis** are as follows.

Null hypothesis H_0: The m samples are derived from the same population (i.e., population homogeneity).

Alternative hypothesis H_1: At least two of the m samples are not derived from the same population (population heterogeneity).

Example 8.4 will be employed to illustrate the **heterogeneity chi-square analysis**.

Example 8.4 *A researcher evaluates a hypothesis that in the lakes of a specific geographical region the number of fish representing three different species are equally distributed. Over a period of a year four separate studies are conducted, and each study is evaluated with a chi-square goodness-of-fit test. Although none of the studies yields a significant result (which if present would allow the researcher to conclude that the species of fish are not equally distributed), visual inspection of the data suggests that* **Species 3** *is more prevalent than either* **Species 1** *or* **2**, *and that of the three species,* **Species 1** *is the least prevalent. Because he suspects that the nonsignificant results for the chi-square analyses may be due to a lack of statistical power, the researcher would like to combine the data for the four studies, and analyze the pooled data. Is the researcher justified in pooling the data?*

Table 8.8 summarizes the analysis of the data for Example 8.4. **Part A** of Table 8.8 presents the chi-square goodness-of-fit analysis for each of the four individual studies. Column 2 for each of the studies contains the observed species frequency for that study. In each study the expected frequency for any of the $k = 3$ cells is one-third of the total number of observations (i.e., $E_i = (n)(1/3)$, since the latter implies that the species are equally distributed).

The following protocol is employed in the **heterogeneity chi-square analysis**: a) A chi-square value is computed for each of the individual studies. (Although it is not the case in our example, Zar (1999) notes that if the number of cells per study is $k = 2$, the correction for continuity is not used in analyzing the individual tables.); b) The sum of the m chi-square values obtained in a) for the individual studies is computed. The latter value itself represents a chi-square value, and will be designated χ^2_{sum}. In addition, the sum of the degrees of freedom for the m studies is computed. The latter degrees of freedom value, which will be designated df_{sum}, is obtained by summing the value $df = k - 1$ m times; c) The data for the m studies are combined into one table, and the chi-square value, which will be designated χ^2_{pooled}, is computed for the pooled data. The degrees of freedom for the table with the pooled data, which will be designated df_{pooled}, is equal to $df = k - 1$. (Zar (1999) notes that if there are $k = 2$ cells, the correction for continuity is not used in analyzing the table with the pooled data); d) The **heterogeneity chi-square analysis** is based on the premise that if the m samples are in fact homogeneous, the sum of the m individual chi-square values (χ^2_{sum}) should be approximately the same value as the chi-

Table 8.8 Heterogeneity Chi-Square Analysis for Example 8.4

A. Chi-square analysis of four individual studies

Study 1

Cell/Species	O_i	E_i	$(O_i - E_i)$	$(O_i - E_i)^2$	$\dfrac{(O_i - E_i)^2}{E_i}$
1	10	15	−5	25	1.67
2	15	15	0	0	0
3	20	15	5	25	1.67
	$\Sigma O_i = 45$	$\Sigma E_i = 45$	$\Sigma(O_i - E_i) = 0$		$\chi_1^2 = 3.33$

Study 2

Cell/Species	O_i	E_i	$(O_i - E_i)$	$(O_i - E_i)^2$	$\dfrac{(O_i - E_i)^2}{E_i}$
1	13	20	−7	49	2.45
2	21	20	1	1	.05
3	26	20	6	36	1.80
	$\Sigma O_i = 60$	$\Sigma E_i = 60$	$\Sigma(O_i - E_i) = 0$		$\chi_2^2 = 4.30$

Study 3

Cell/Species	O_i	E_i	$(O_i - E_i)$	$(O_i - E_i)^2$	$\dfrac{(O_i - E_i)^2}{E_i}$
1	19	25	−6	36	1.44
2	22	25	−3	9	.36
3	34	25	9	81	3.24
	$\Sigma O_i = 75$	$\Sigma E_i = 75$	$\Sigma(O_i - E_i) = 0$		$\chi_3^2 = 5.04$

Study 4

Cell/Species	O_i	E_i	$(O_i - E_i)$	$(O_i - E_i)^2$	$\dfrac{(O_i - E_i)^2}{E_i}$
1	12	20	−8	65	3.20
2	22	20	2	4	.20
3	26	20	6	36	1.80
	$\Sigma O_i = 60$	$\Sigma E_i = 60$	$\Sigma(O_i - E_i) = 0$		$\chi_4^2 = 5.20$

Sum of chi-square values for four studies = χ_{sum}^2 = 3.33 + 4.30 + 5.04 + 5.20 = 17.87

B. Chi-square analysis of pooled data

Pooled data for $m = 4$ studies

Cell/Species	O_i	E_i	$(O_i - E_i)$	$(O_i - E_i)^2$	$\dfrac{(O_i - E_i)^2}{E_i}$
1	54	80	−26	676	8.45
2	80	80	0	0	0
3	106	80	26	676	8.45
	$\Sigma O_i = 240$	$\Sigma E_i = 240$	$\Sigma(O_i - E_i) = 0$		$\chi_{pooled}^2 = 16.90$

C. Heterogeneity of chi-square analysis

Heterogeneity chi-square = Sum of chi-square values for four studies − Pooled chi-square value

$$\chi_{het}^2 = (\chi_{sum}^2 = 17.87) - (\chi_{pooled}^2 = 16.90) = .97$$

square value computed for the pooled data (χ^2_{pooled}). In order to determine the latter, the absolute value of the difference between the sum of the m chi-square values (obtained in b)) and the pooled chi-square value (obtained in c)) is computed. The obtained difference, which is itself a chi-square value, is the **heterogeneity chi-square value**, which will be designated χ^2_{het}. Thus, $\chi^2_{het} = |\chi^2_{sum} - \chi^2_{pooled}|$. The null hypothesis will be rejected when there is a large difference between the values of χ^2_{sum} and χ^2_{pooled}. The value χ^2_{het}, which represents the test statistic, is evaluated with a degrees of freedom value which is the sum of the degrees of freedom for the m individual studies (df_{sum}) less the degrees of freedom obtained for the table with the pooled data (df_{pooled}). Thus, $df_{het} = df_{sum} - df_{pooled}$. In order to reject the null hypothesis, the value χ^2_{het} must be equal to or greater than the tabled critical value at the prespecified level of significance for df_{het}; and e) If the null hypothesis is rejected the data cannot be pooled. If, however, the null hypothesis is retained, the data can be pooled, and the computed value for χ^2_{pooled} is employed to evaluate the goodness-of fit hypothesis. Zar (1999) notes, however, that if there are $k = 2$ cells (i.e., $df = 1$), the table for the pooled data should be reevaluated employing the correction for continuity, and the continuity-corrected χ^2_{pooled} value (which will be a little lower than the original χ^2_{pooled} value) should be employed to evaluate the goodness-of-fit hypothesis.

The computed chi-square values for the four studies in Table 8.8 are $\chi^2_1 = 3.33$, $\chi^2_2 = 4.30$, $\chi^2_3 = 5.04$, and $\chi^2_4 = 5.20$. Since, in each of the four chi-square tables, there are $k = 3$ cells, $df = k - 1 = 2$ for each table. Thus, the total number of degrees of freedom employed for the four studies is $df_{sum} = (4)(2) = 8$ (i.e., the number of studies (4) multiplied by the number of degrees of freedom per study (2)). Since there are $k = 3$ cells in the table for the pooled data, the degrees of freedom for the latter table is $df_{pooled} = k - 1 = 2$. By summing the chi-square values for the four studies, we compute the value $\chi^2_{sum} = 3.33 + 4.30 + 5.04 + 5.20 = 17.87$. Since, in **Part B** of Table 8.8, we compute $\chi^2_{pooled} = 16.90$, the value for heterogeneity chi-square (computed in **Part C** of Table 8.8) is $\chi^2_{het} = |(\chi^2_{sum} = 17.87) - (\chi^2_{pooled} = 16.90)| = .97$. The degrees of freedom employed to evaluate the latter chi-square value are $df_{het} = (df_{sum} = 8) - (df_{pooled} = 2) = 6$. The tabled critical .05 and .01 chi-square values in **Table A4** for $df = 6$ are $\chi^2_{.05} = 12.59$ and $\chi^2_{.01} = 16.81$. Since the computed value $\chi^2_{het} = .97$ is less than $\chi^2_{.05} = 12.59$, the null hypothesis is retained. In other words we can conclude the four samples are homogeneous (i.e., come from the same population), and thus we can justify pooling the data into a single table.

As noted earlier, in **Part B** of Table 8.8 the value $\chi^2 = 16.90$ is computed for the pooled data. Since there are $k = 3$ cells in the chi-square table for the pooled data, $df = k - 1 = 2$. The tabled critical .05 and .01 values in **Table A4** for $df = 2$ are $\chi^2_{.05} = 5.99$ and $\chi^2_{.01} = 9.21$. Since the value $\chi^2_{pooled} = 16.90$ is larger than both of the aforementioned critical values, the goodness-of-fit null hypothesis for the pooled data can be rejected at both the .05 and .01 levels. In other words, with respect to the pooled data, we can conclude that in the case of at least one of the cells/species there is a difference between its observed and expected frequency. Without conducting additional comparisons, it appears that, as the researcher suspected, the observed frequency for Cell/Species 1 is significantly below its expected frequency, while the observed frequency for Cell/Species 3 is significantly above its expected frequency. Although it does not apply to our example, as noted earlier, Zar (1999) (who provides a comprehensive discussion of the **heterogeneity chi-square analysis**) states that if the number of cells in the table for the pooled data is $k = 2$, the latter table should be reevaluated employing the correction for continuity.

It should be emphasized that a researcher should employ common sense in applying the heterogeneity chi-square analysis described in this section. In point of fact, there may be occasions when even though the computed value of χ^2_{het} is not significant, in spite of the latter it would not be recommended that the researcher pool the data from two or more smaller tables. To be more specific, one should not pool data from two or more tables employing small sample sizes (which when evaluated individually fail to yield a significant chi-square value) in order to obtain a significant pooled chi-square value, when there is an obvious inconsistency in the cell proportions for two or more of the tables. In other words, when the data from m tables are pooled, the proportion of cases in the cells of each of the m tables should be approximately the same. Everitt (1977, 1992), Fleiss (1981) and Fleiss *et al*. (2003, Ch. 10) recommend alternative procedures for pooling the data from multiple chi-square tables. One of the alternative procedures documented in Everitt (1977, pp. 25–27; 1992) is described in the discussion of **meta-analysis** (as well as discussed in more detail in Section VI of the **chi-square test for $r \times c$ tables**). The latter alternative procedure (which is referred to as the **Stouffer procedure for obtaining a combined significance level (p value) for k studies (Test 28o)**) is described in Section IX (the **Addendum**) of the **Pearson product-moment correlation coefficient**.

VII. Additional Discussion of the Chi-Square Goodness-of-Fit Test

1. Directionality of the chi-square goodness-of-fit test In Section III it is noted that most sources state the alternative hypothesis for the **chi-square goodness-of-fit test** nondirectionally, but in actuality it is possible to state the alternative hypothesis directionally. This is most obvious when there are $k = 2$ cells and the expected probability associated with each cell is $\pi_i = 1/2$. Under the latter conditions a researcher can make two directional predictions, either one of which can represent the alternative hypothesis. Specifically, the following can be predicted with respect to the sample data: a) The observed frequency of Cell 1 will be significantly higher than the observed frequency of Cell 2 (which translates into the observed frequency of Cell 1 being higher than its expected frequency, and the observed frequency of Cell 2 being lower than its expected frequency); and b) The observed frequency of Cell 2 will be significantly higher than the observed frequency of Cell 1 (which translates into the observed frequency of Cell 2 being higher than its expected frequency, and the observed frequency of Cell 1 being lower than its expected frequency).

If a researcher wants to evaluate either of the aforementioned directional alternative hypotheses at the .05 level, the appropriate critical value to employ is the tabled chi-square value (for $df = 1$) at the .10 level of significance. The latter value is represented by the tabled chi-square value at the 90th percentile (which demarcates the extreme 10% in the right tail of the chi-square distribution). This latter critical value will be designated as $\chi^2_{.10}$ throughout this discussion. The rationale for employing $\chi^2_{.10}$ in evaluating the directional alternative hypothesis at the .05 level is as follows. When $k = 2$ and the alternative hypothesis is stated nondirectionally, if a computed chi-square value is equal to or greater than $\chi^2_{.05}$ (for $df = 1$), the researcher can reject the null hypothesis if the data are consistent with either of the outcomes associated with the two possible directional alternative hypotheses. If that same tabled critical chi-square value is employed to evaluate one of the two possible directional alternative hypotheses, a directional alternative hypothesis would be evaluated not at the .05 level but at one-half that value — in other words at the .025 level. Thus, if one wants to employ $\alpha = .05$ and states the alternative hypothesis directionally, the alpha level for the directional alternative hypothesis should be .05 multiplied by the number of possible directional alternative hypotheses (which in this instance equals 2). By employing the tabled critical value for $\chi^2_{.10}$, which is a lower value than $\chi^2_{.05}$, an alpha level (and Type I error rate) of .05 is established for the specific one-tailed alternative hypothesis which one employs. The area of the chi-square distribution that corresponds to the alpha level for the latter directional alternative hypothesis will be one-half of the area which comprises the extreme 10% of the right tail of the distribution. The area that corresponds to the alpha level for the other directional alternative hypothesis will be the remaining 5% of the extreme 10% in the right tail of the distribution.

If we turn our attention to Examples 8.1 and 8.2, in both examples there are, in fact, 720 possible directional predictions the researcher can make![12] The latter value is determined as follows: $k! = 6! = (6)(5)(4)(3)(2)(1) = 720$.[13] In other words, a researcher can predict any one of 720 ordinal configurations with respect to the observed frequencies of the six cells. As an example, the researcher might predict that 1/**Monday** will have the highest observed frequency, followed in order by 2/**Tuesday**, 3/**Wednesday**, 4/**Thursday**, 5/**Friday**, and 6/**Saturday**, and only be willing to reject the null hypothesis if the data are consistent with this specific ordering of the observed cell frequencies.

Later in this section it will be explained that when $k = 6$, it is not possible to evaluate any one of the 720 directional alternative hypotheses at either the .05 or .01 level of significance. Indeed, under these conditions the highest value of alpha that can be employed to evaluate a directional alternative hypothesis is approximately .001. In point of fact, when $k = 6$, the tabled critical $\chi^2_{.001}$ value is approximately 2.7, which happens to be the tabled chi-square value at the 28th percentile. This is the case, since if the prespecified alpha value which is employed to evaluate a nondirectional alternative hypothesis (which in this case we will assume is .001, since it is the highest value that will work with $k = 6$) is multiplied by the number of possible directional alternative hypotheses (720) we obtain: $(.001)(720) = .72$. The value .72 demarcates the extreme 72% in the right tail of the chi-square distribution when $df = 5$. Thus, it corresponds to the 28th percentile of the distribution, since $(1 - .72) = .28$. Consequently, in order to reject the null hypothesis with reference to one of the 720 possible directional alternative hypotheses, both of the following conditions will have to be met: a) The obtained value of chi-square will have to be equal to or greater than the tabled chi-square value at the 28th percentile; and b) The data will have to be consistent with the directional alternative hypothesis which is employed. In other words, the ordinal relationship between the observed frequencies of the six cells should be in the exact order stated in the directional alternative hypothesis.

As noted previously, none of the 720 possible directional alternative hypotheses can be evaluated at either the .05 or .01 levels. This is the case, since if either .05 or .01 is multiplied by 720 the resulting product exceeds unity (1) — i.e., $(.05)(720) = 36$ and $(.01)(720) = 7.2$. Since both 36 and 7.2 exceed 1, they cannot be used as probability values. Thus, it is impossible to evaluate any of the directional alternative hypotheses at either the .05 or .01 level. In fact, the largest alpha level at which any of the directional alternative hypotheses can be evaluated is .001388, since $1/720 = .001388$ (i.e., the value .001388 is the maximum number which when multiplied by 720 falls short of 1. Specifically, $(.001388)(720) = .99936$).

On initial inspection it might appear that by employing $\chi^2_{.72}$ to evaluate one of the directional alternative hypotheses, a researcher is employing an inflated alpha level.[14] But as just noted, in this instance the alpha level for a directional alternative hypothesis is, in fact, .001. Of course a researcher may elect to employ a larger critical chi-square value, and if one elects to do so, the actual alpha level for a directional alternative hypothesis will be even lower than .001. For example, if the tabled critical value $\chi^2_{.05} = 11.07$ (which is the tabled value for $df = 5$ at the 95th percentile that is employed in evaluating a nondirectional alternative hypothesis) is employed to evaluate one of the 720 possible directional alternative hypotheses, the actual alpha level which one will be using in evaluating a directional alternative hypothesis will be $.05/720 = .00007$. In such a case there is obviously a minuscule likelihood of committing a Type I error in reference to the directional alternative hypothesis. Yet at the same time, the power of the analysis with respect to that alternative hypothesis will be minimal (thus resulting in a high likelihood of committing a Type II error).

It is also possible in Examples 8.1 and 8.2 to state an alternative hypothesis which predicts two or more, but less than 720, of the possible ordinal configurations with respect to the observed frequencies of the $k = 6$ cells. In other words, a directional alternative hypothesis might state that the null hypothesis can only be rejected if the magnitude of the observed cell frequencies in descending order is either **Cell 1, Cell 2, Cell 3, Cell 4, Cell 5, Cell 6** or **Cell 2, Cell 1, Cell 3, Cell 4, Cell 5, Cell 6**. In such a case, to evaluate the null hypothesis at the .001 level with respect to an alternative hypothesis involving two of the 720 possible ordinal configurations, the tabled critical chi-square value at the 64ᵗʰ percentile is employed (i.e., $\chi^2_{.36}$, since the area of the distribution involved is the extreme 36% $((1 - .64) = .36)$ which falls in the right tail). The general procedure for computing the percentile rank in the chi-square distribution in determining the critical value when evaluating one or more configurations is as follows: a) Divide the total number of possible configurations by the number of acceptable configurations stated in the directional alternative hypothesis; b) Multiply the result of the division by the prespecified alpha level; and c) Subtract the value obtained in part b) from 1.

Applying this protocol to a directional alternative hypothesis in which only 2 out of 720 configurations are acceptable, we derive: a) $720/2 = 360$; b) $(360)(.001) = .36$; and c) $(1 - .36) = .64$. The resulting value of .64 (which can be converted to 64%) corresponds to the percentile rank in the chi-square distribution to employ in determining the critical value. The value .36 obtained in part b) represents the overall proportion of the right tail of the distribution which contains a proportion of the distribution equivalent to .001 that represents the rejection zone for the directional alternative hypothesis under study.

It should be noted that since $2/720 = .0028$, an alternative hypothesis involving 2 out of 720 possible configurations can be evaluated at a level above .001. In point of fact, such an alternative hypothesis can be evaluated at any level equal to or less than .0028. Thus if one elects to employ $\alpha = .002$, using the protocol described above, $(720/2)(.002) = .72$, and $(1 - .72) = .28$. The latter result indicates that $\chi^2_{.72} = 2.7$ is once again employed as the critical value. This is the case, since the latter value represents the tabled value at 28ᵗʰ percentile for $df = 5$.

In closing this discussion, the reader should take note of the fact that all of the critical values for the **chi-square goodness-of-fit test** are derived from the right tail of the chi-square distribution. In point of fact, with the exception of the **single-sample chi-square test for a population variance** (in which case critical values are derived from both tails of the distribution), all of the tests in the book which employ the chi-square distribution only use critical values from the right tail of the distribution.

2. Additional goodness-of-fit tests

In addition to the **chi-square goodness-of-fit test**, the **Kolmogorov–Smirnov goodness-of-fit test for a single sample**, and **Lilliefors test for normality** (both of which are described in the previous chapter),[15] there are a number of other tests that have been developed for evaluating goodness-of-fit, including the following which are described earlier in the book: The **D'Agostino–Pearson test of normality (Test 5a)** and the **Jarque–Bera test of normality (Test 5b)** (both of which employ results obtained for the **single sample test for evaluating population skewness (Test 4)** and the **single sample test for evaluating population kurtosis (Test 5)**). Most alternative goodness-of-fit tests (including the **Kolmogorov–Smirnov** and **Lilliefors** tests) evaluate scores which are assigned to ordered categories.[16] Among other goodness-of-fit tests that have been developed for evaluating ordered categorical data are **David's empty cell test** (David (1950)), the **Cramér–von Mises goodness-of-fit test** (attributed to Cramér (1928), von Mises (1931), and Smirnov (1936)), and the **Shapiro-Wilk test for normality** (Shapiro and Wilk (1965, 1968)) (which is described in

Conover (1980, 1999)). Zar (1999) notes that since both the **Kolmogorov–Smirnov test** and the **chi-square goodness-of-fit test** are deficient in power, neither test is optimal for assessing normality. Thode (2002) has reviewed over forty procedures for assessing goodness-of-fit, and concludes that the **Shapiro-Wilk test** and test statistics derived from moments (i.e., tests which utilized measures of skewness and kurtosis) are generally the best methods for assessing normality. D'Agostino and Stephens (1986), Daniel (1990), and Thode (2002) contain comprehensive discussions of alternative goodness-of-fit procedures. Discussion of goodness-of-fit tests for randomness can be found in Section VII of the **single-sample runs tests (Test 10)**, and under **autocorrelation** (which a procedure that can also be employed for assessing goodness-of-fit for randomness) that can be found in Section VII of the **Pearson product–moment correlation coefficient**.

VIII. Additional Examples Illustrating the Use of the Chi-Square Goodness-of-Fit Test

Three additional examples which can be evaluated with the **chi-square goodness-of-fit test** are presented in this section. Example 8.5 employs the same data set as Examples 8.1 and 8.2, and thus yields the same results. Examples 8.6 and 8.7 illustrate the application of the **chi-square goodness-of-fit test** to data in which the expected frequencies are based on existing empirical information or theoretical conjecture rather than on expected/theoretical probabilities.

Example 8.5 *The owner of the Big Wheel Speedway, a stock car racetrack, asks a researcher to determine whether or not there is any bias associated with the lane to which a car is assigned at the beginning of a race. Specifically, the owner wishes to determine if there is an equal likelihood of winning a race associated with each of the six lanes of the track. The researcher examines the results of 120 races and determines the following number of first place finishes for the six lanes: Lane 1 – 20; Lane 2 – 14; Lane 3 – 18; Lane 4 – 17; Lane 5 – 22; Lane 6 – 29.*

Example 8.6 *A country in which four ethnic groups make up the population establishes affirmative action guidelines for medical school admissions. The country has one medical school, and it is mandated that each new class of medical students proportionally represents the four ethnic groups which comprise the country's population. The four ethnic groups that make up the population and the proportion of people in each ethnic group are: Balzacs (.4), Crosacs (.25), Murads (.3), and Isads (.05).*[17] *The number of students from each ethnic group admitted into the medical school class for the new year are: Balzacs (300), Crosacs (220), Murads (400), and Isads (80). Is there a significant discrepancy between the proportions mandated in the affirmative action guidelines and the actual proportion of the four ethnic groups in the new medical school class?*

Table 8.9   Chi-Square Summary Table for Example 8.6

Cell	O_i	E_i	$(O_i - E_i)$	$(O_i - E_i)^2$	$\dfrac{(O_i - E_i)^2}{E_i}$
Balzacs	300	400	−100	10000	25
Crosacs	220	250	−30	900	3.6
Murads	400	300	100	10000	33.3
Isads	80	50	30	900	18
	$\Sigma O_i = 1000$	$\Sigma E_i = 1000$	$\Sigma(O_i - E_i) = 0$		$\chi^2 = 79.9$

Except for the fact that empirical data are used as a basis for determining the expected cell frequencies, this example is evaluated in the same manner as Examples 8.1 and 8.2. There are $k = 4$ cells — each cell representing one of the four mutually exclusive ethnic groups. The observed frequencies are the number of students from each of the four ethnic groups out of the total of 1000 students who are admitted to the medical school (i.e., $n = 300 + 220 + 400 + 80$ $= 1000$). The expected frequencies are computed based upon the proportion of each ethnic group in the population. Each of these values is obtained by multiplying the total number of medical school admissions (1000) by the proportion of a specific ethnic group in the population. Thus in the case of the **Balzacs**, employing Equation 8.1 the expected frequency is computed as follows: $E_1 = (1000)(.4) = 400$. In the same respect, the expected frequencies for the other three ethnic groups are: **Crosacs**: $E_2 = (1000)(.25) = 250$; **Murads**: $E_3 = (1000)(.3) = 300$; and **Isads**: $E_4 = (1000)(.05) = 50$. Table 8.9 summarizes the observed and expected frequencies and the resulting values employed in the computation of the chi-square value through use of Equation 8.2.

Employing **Table A4**, we determine that for $df = 4 - 1 = 3$ the tabled critical .05 and .01 values are $\chi^2_{.05} = 7.81$ and $\chi^2_{.01} = 11.34$. (A nondirectional alternative hypothesis is assumed.) Since the computed value $\chi^2 = 79.9$ is greater than both of the aforementioned critical values, the null hypothesis can be rejected at both the .05 or .01 levels. Based on the chi-square analysis it can be concluded that the medical school admissions data do not adhere to the proportions mandated in the affirmative action guidelines. Inspection of Table 8.9 suggests that the significant difference is primarily due to the presence of too many **Murads** and **Isads** and too few **Balzacs**. This observation can be confirmed by employing the appropriate comparison procedure described in Section VI.

Example 8.7 *A physician who specializes in genetic diseases develops a theory which predicts that two-thirds of the people who develop a disease called cyclomeiosis will be males. She randomly selects 300 people who are afflicted with cyclomeiosis and observes that 140 of them are females. Is the physician's theory supported?*

In Example 8.7 there are two cells, each representing one gender. Since the expected frequencies are computed on the basis of the probabilities hypothesized in the physician's theory, two-thirds of the sample are expected to be males and the remaining one-third females. Thus, the respective expected frequencies for males and females are determined as follows: **Males**: $E_1 = (300)(2/3) = 200$; **Females**: $E_2 = (300)(1/3) = 100$. Since 140 females are observed with the disease, the remaining 160 people who have the disease must be males. Table 8.10 summarizes the observed and expected frequencies and the resulting values that are employed in the computation of the chi-square value with Equation 8.2.

Table 8.10 Chi-Square Summary Table for Example 8.7

Cell	O_i	E_i	$(O_i - E_i)$	$(O_i - E_i)^2$	$\dfrac{(O_i - E_i)^2}{E_i}$
Males	160	200	−40	1600	8
Females	140	100	40	1600	16
	$\Sigma O_i = 300$	$\Sigma E_i = 300$	$\Sigma(O_i - E_i) = 0$		$\chi^2 = 24$

Employing **Table A4**, we determine that for $df = 2 - 1 = 1$ the tabled critical .05 and .01 values are $\chi^2_{.05} = 3.84$ and $\chi^2_{.01} = 6.63$.[18] Since the computed value $\chi^2 = 24$ is greater than both of the aforementioned critical values, the null hypothesis can be rejected at both the .05 or .01

levels. Based on the chi-square analysis, it can be concluded that the observed distribution of males and females for the disease is not consistent with the doctor's theory.

References

Cochran, W. G. (1952). The chi-square goodness-of-fit test. **Annals of Mathematical Statistics**, 23, 315–345.

Cohen, J. (1977). **Statistical power analysis for the behavioral sciences**. New York: Academic Press.

Cohen, J. (1988). **Statistical power analysis for the behavioral sciences** (2nd ed.). Hillsdale, NJ: Lawrence Erlbaum Associates, Publishers.

Conover, W. J. (1980). **Practical nonparametric statistics** (2nd ed.). New York: John Wiley & Sons.

Conover, W. J. (1999). **Practical nonparametric statistics** (3rd ed.). New York: John Wiley & Sons.

Cramér, H. (1928). On the composition of elementary errors. **Skandinavisk Aktaurietidskrift**, 11, 13–74, 141–180.

D'Agostino, R. B. & Stephens, M. A. (Eds.) (1986). **Goodness-of-fit techniques**. New York: Marcel Dekker.

Dahiya, R. C. & Gurland, J. (1973). How many classes in the Pearson chi-square test? **Journal of the American Statistical Association**, 68, 707–712.

Daniel, W. W. (1990). **Applied nonparametric statistics** (2nd ed.). Boston: PWS–Kent Publishing Company.

David, F. N. (1950). Two combinatorial tests of whether a sample has come from a given population. **Biometrika**, 37, 97–110.

Everitt, B. S. (1977). **The analysis of contingency tables**. New York: Chapman & Hall.

Everitt, B. S. (1992). **The analysis of contingency tables** (2nd ed.). New York: Chapman & Hall.

Feller, W. (1968). **An introduction to probability theory and its applications (Volume I)** (3rd ed.). New York: John Wiley & Sons.

Fleiss, J. L. (1981). **Statistical methods for rates and proportions** (2nd ed.). New York: John Wiley & Sons.

Fleiss, J. L., Levin, B. & Paik, M. C. (2003). **Statistical methods for rates and proportions** (3rd ed.). New York: John Wiley & Sons.

Folz, D. H. (1996). **Survey research for public administration**. Thousand Oaks, CA: Sage Publications.

Glass, G. V. & Hopkins, K. D. (1996). **Statistical methods in education and psychology** (3rd ed.) Boston: Allyn & Bacon.

Ghosh, B. K. (1979). A comparison of some approximate confidence intervals for the binomial parameter. **Journal of the American Statistical Association**, 74, 894–900.

Haberman, S. J. (1973). The analysis of residuals in cross-classified tables. **Biometrics**, 29, 205–220.

Howell, D. C. (1992). **Statistical methods for psychology** (3rd ed.). Boston: PWS–Kent Publishing Company.

Keppel, G. & Saufley, W. H., Jr. (1992). **Introduction to design and analysis: A student's handbook** (2nd ed.). New York: W. H. Freeman & Company.

Kolmogorov, A. N. (1933). Sulla determinazione empirica di una legge di distribuzione. **Giorn dell'Inst. Ital. degli. Att.**, 4, 89–91.

Lilliefors, H. W. (1967). On the Kolmogorov–Smirnov test for normality with mean and variance unknown. **Journal of the American Statistical Association**, 62, 399–402.

Marascuilo, L. A. & McSweeney, M. (1977). **Nonparametric and distribution-free methods for the social sciences.** Monterey, CA: Brooks/Cole Publishing Company.

Miller, I. & Miller, M. (1999). **John E. Freund's mathematical statistics** (6th ed.). Upper Saddle River, NJ: Prentice Hall.

O'Sullivan, E, Rassel, G., & Berner, M. (2002). **Research methods for public administration** (4th ed.). Boston: Longman.

Rosner, B. (1995). **Fundamentals of biostatistics** (4th ed.). Belmont, CA: Duxbury Press.

Scheaffer, R. L., Mendenhall, W., & Ott, L. (1996). **Elementary survey sampling** (5th ed.). Belmont, CA: Duxbury Press.

Shapiro, S. S. & Wilk, M. B. (1965). An analysis of variance test for normality (complete samples). **Biometrika**, 52, 591-611.

Shapiro, S. S. & Wilk, M. B. (1968). Approximations for the null distribution of the W statistic. **Technometrics**, 10, 861–866.

Siegel, S. & Castellan, N. J., Jr. (1988). **Nonparametric statistics for the behavioral sciences** (2nd ed.). New York: McGraw–Hill Book Company.

Smirnov, N. V. (1936). Sur la distribution de W^2 (criterium de M. R. v. Mises). **Comptes Rendus** (Paris), 202, 449–452.

Smithson, M. (2003). **Confidence intervals**. Thousand Oaks, CA: Sage Publications.

Stirling, J. (1730). **Methodus differentialis**.

Thode, H. C. (2002). **Testing for normality**. New York: Marcel Dekker.

von Mises, R. (1931). **Wahrscheinlichkeitsrechnung und ihre anwendung in derstatistik and theoretishen Physik.** Leipzig: Deuticke.

Wallis, W. A. & Roberts, H. V. (1956). **Statistics: A new approach**. Glencoe, IL: Free Press.

Zar, J. H. (1999). **Biostatistical analysis** (4th ed.). Upper Saddle River, NJ: Prentice Hall.

Endnotes

1. Categories are **mutually exclusive** if assignment to one of the k categories precludes a subject/object from being assigned to any one of the remaining $(k-1)$ categories.

2. The reason why the exact probabilities associated with the binomial and multinomial distributions are generally not computed is because, except when the value of n is very small, an excessive amount of computation is involved. The **binomial distribution** is discussed under the **binomial sign test for a single sample**, and the **multinomial distribution** is discussed in Section IX (the **Addendum**) of the latter test.

3. Example 8.6 in Section VIII illustrates an example in which the expected frequencies are based on prior empirical information.

4. It is possible for the value of the numerator of a probability ratio to be some value other than 1. For instance, if one is evaluating the number of odd versus even numbers which appear on n rolls of a die, in each trial there are $k = 2$ categories. Three face values (1, 3, 5) will result in an observation being categorized as an odd number, and three face values (2, 4, 6) will result in an observation being categorized as an even number. Thus, the probability associated with each of the two categories will be $3/6 = 1/2$. It is also possible for each of the categories to have different probabilities. Thus, if one is evaluating the relative occurrence of the face values 1 and 2 versus the face values 3, 4, 5, and 6, the

probability associated with the former category will be 2/6 = 1/3 (since two outcomes fall within the category 1 and 2), while the probability associated with the latter will be 4/6 = 2/3 (since four outcomes fall within the category 3, 4, 5, and 6). Examples 8.6 and 8.7 in Section VIII illustrate examples where the probabilities for two or more categories are not equal to one another.

5.   When decimal values are involved, there may be a minimal difference between the sums of the expected and observed frequencies due to rounding off error.

6.   There are some instances when Equation 8.3 should be modified to compute the degrees of freedom for the **chi-square goodness-of-fit test**. The modified degrees of freedom equation is discussed in Section VI, within the framework of employing the **chi-square goodness-of-fit test** to assess goodness-of-fit for a normal distribution.

7.   Sometimes when one or more cells in a set of data have an expected frequency of less than five, by combining cells (as is done in this analysis) a researcher can reconfigure the data so that the expected frequency of all the resulting cells is greater than five. Although this is one way of dealing with the violation of the assumption concerning the minimum acceptable value for an expected cell frequency, the null hypothesis evaluated with the reconfigured data will not be identical to the null hypothesis stipulated in Section III.

8.   In a one-dimensional chi-square table, subjects/objects are assigned to categories which reflect their status on a single variable. In a two-dimensional table, two variables are involved in the categorization of subjects/objects. As an example, if each of n subjects is assigned to a category based on one's gender and whether one is married or single, a two-dimensional table can be constructed involving the following four cells: **Male-Married**; **Female-Married**; **Male-Not married**; **Female-Not married**. Note that people assigned to a given cell fall into one of the two categories on each of the two dimensions/variables (which are gender and marital status). Analysis of two-dimensional tables is discussed under the **chi-square test for $r \times c$ tables**. In Section VII of the latter test, tables with more than two dimensions (commonly referred to as **multidimensional contingency tables**) are also discussed.

9.   a) Daniel (1990) notes that Equation 8.6 will only yield reliable results when $n\pi_1$ and $n(1 - \pi_1)$ are both greater than 5. It is assumed that the researcher estimates the value of π_1 prior to collecting the data. The researcher bases the latter value either on probability theory or preexisting empirical information. Generally speaking, if the value of n is large, the value of p_1 should provide a reasonable approximation of the value of π_1 for calculating the values $n\pi_1$ and $n(1 - \pi_1)$; b) Although not commonly employed, the **finite population correction factor** stipulated in Endnote 2 of the **single-sample z test (Test 1)** should be used in computing the **estimated standard error of the population proportion** when the size of the underlying population is large and the size of the sample is believed to constitute less than 5% of the population. Under the latter circumstances, the standard error $\sqrt{(p_1 p_2)/n}$ is multiplied by $\sqrt{(N - n)/(N - 1)}$ (which represents the correction factor, where N represents the total number of people that comprise the population). The finite population corrected equation will result in a smaller value for the standard error of the population proportion employed in Equation 8.6.

10. a) Smithson (2003, p. 11) notes that in the case of computing a confidence interval for a proportion which is not an extreme value, for a specific confidence interval level, the width

of the confidence interval will reduce in width by about one-half if the sample size is quadrupled; b) Confidence intervals for population proportions (typically, the 95% confidence interval) are commonly employed in expressing the results of surveys and/or political polls. As an example (employing the proportions $p_1 = 29/120 = .242$ and $p_2 = 91/120 = .758$ obtained for the example under discussion), a pollster might state that in an exit poll for an election (in which we will assume the votes of 120 people were obtained) the proportion (percentage) of votes received by **Candidate A** was .242 (24.2%) versus .758 (75.8%) for **Candidate B**, with a margin of error of ± .101 (which will would most likely be stated as a margin of error of ± 10.1% or ± 10.1 percentage points). Since the latter margin of error is based on the 99% confidence interval, the pollster might also state we can be 99% confident that for a given candidate the sample proportion ± 10.1% contains the actual proportion of people in the population who voted for that candidate. It should be noted that a margin of error of ± 10.1% would be considered too large for a political poll, and because of the latter most polls employ a sample size substantially larger than $n = 120$. The use of a larger sample size in the example under discussion would result in a smaller value for the standard error, and consequently reduce the width of the confidence interval; b) One might intuitively expect that the larger a population, the larger the size of the sample required in a survey in order to achieve a specified level of confidence and/or precision. However, the latter is not the case as long as a population is relatively large. To be more specific, in a large population the proportion of the population sampled will not have an impact on the precision of an estimate or the level of confidence. In point of fact, it is the size of the sample (i.e., n) rather than the proportion of the population the sample represents (i.e., n/N) which determines the precision of the estimate (i.e., the margin of error value) and degree of confidence. Equation 8.13 can be employed to determine the sample size required in order to attain a certain level of confidence and/or precision in a survey. (The latter equation (also discussed in Folz (1996, p. 50) is derived from Equation 8.6, where the width of a confidence interval (represented by w in Equation 8.13) is computed with the term $[\sqrt{(p_1 p_2)/n}\,](z_{\alpha/2}).)$

$$\sqrt{n} = \frac{z_{\alpha/2}\sqrt{p_1 p_2}}{w} \qquad \text{(Equation 8.13)}$$

In order to demonstrate the use of Equation 8.13, assume in a survey in which the sample proportions are $p_1 = .242$ and $p_2 = .758$, we want to determine the required sample size such that we can be 95% confident that the margin of error associated with our estimated population proportions is ± 2%. Substituting the appropriate values in Equation 8.13, we determine the required sample size to be 1761.72 (which rounded off equals 1762).

$$\sqrt{n} = \frac{(1.96)\sqrt{(.242)(.758)}}{.02} = 41.97$$

$$n = (\sqrt{n})^2 = (41.97)^2 = 1761.72$$

It should be noted that the smaller the difference between the values of p_1 versus p_2, the larger the sample size which will be required. For example, a substantially larger sample size will be required at a given level of confidence and precision when $p_1 = p_2 = .50$ than when $p_1 = .90$ and $p_2 = .10$. To illustrate, for $p_1 = p_2 = .50$, if our level of confidence is set equal at 95% with a margin of error of ± 2%, Equation 8.13 yields the value $n = 2401$ (since $\sqrt{n} = [1.96\sqrt{(.5)(.5)}\,]/.02 = 49$ and $n = (49)^2 = 2401$). For $p_1 = .90$ and $p_2 = .10$, the latter equation yields the value $n = 864.36$ (since $\sqrt{n} = [1.96\sqrt{(.9)(.1)}\,]/.02 = 29.4$ and $n = (29.4)^2 = 864.36$); c) Folz (1996, pp. 47–52) notes when it is assumed that the underlying population is comprised of less than 100,000 people, a smaller sample size can be employed than the one computed with Equation 8.13. In such a case an adjusted sample size can be determined which, depending upon the degree of precision one requires, can be anywhere from 50% or less than the size of the actual population. O'Sullivan *et al.* (2002, p. 154) note that an adjusted sample size should be computed through use of the **finite population correction factor** noted in Endnote 9. Examples of tables with adjusted sample sizes for smaller populations can be found in Folz (1996, pp. 52–53); d) The reader should take note of the fact that any numerical value computed in reference to survey data will be of little or no value unless the appropriate sampling procedure has been employed in conducting the survey. Sampling procedures are discussed in the **Introduction** in the section labeled **sampling methodologies**. Among others, Folz (1996), O'Sullivan *et al.* (2002), and Schaeffer *et al.* (1996) provide more detailed discussions of survey sampling.

11. Glass and Hopkins (1996, pp. 325–327) recommend the use of a method proposed by Ghosh (1979) for computing the lower and upper limits of a confidence interval — specifically the use of Equations 8.14 and 8.15. Levin (2005, personal communication) notes that unlike Equations 8.8 and 8.9, Equations 8.14 and 8.15 do not employ a correction for continuity. Because of the latter, any discrepancy between limits of a confidence interval computed with Equations 8.14 and 8.15 versus Equations 8.8 and 8.9 will increase as the sample size decreases. Note that when Equations 8.14 and 8.15 are employed below for the values $p_1 = .242$, $p_2 = .758$, and $n = 120$, they yield a result almost identical to that obtained with Equations 8.8 and 8.9.

(Equation 8.14)

$$P_{LL} = \frac{n}{n + z_{(\alpha/2)}^2}\left[p_1 + \frac{z_{(\alpha/2)}^2}{2n} - z_{(\alpha/2)}\sqrt{\frac{p_1 p_2}{n} + \frac{z_{(\alpha/2)}^2}{4n^2}} \right]$$

$$P_{LL} = \frac{120}{120 + (1.96)^2}\left[.242 + \frac{(1.96)^2}{(2)(120)} - 1.96\sqrt{\frac{(.242)(.758)}{120} + \frac{(1.96)^2}{(4)(120)^2}} \right] = .174$$

(Equation 8.15)

$$P_{UL} = \frac{n}{n + z_{(\alpha/2)}^2}\left[p_1 + \frac{z_{(\alpha/2)}^2}{2n} + z_{(\alpha/2)}\sqrt{\frac{p_1 p_2}{n} + \frac{z_{(\alpha/2)}^2}{4n^2}} \right]$$

$$P_{UL} = \frac{120}{120 + (1.96)^2}\left[.242 + \frac{(1.96)^2}{(2)(120)} + 1.96\sqrt{\frac{(.242)(.758)}{120} + \frac{(1.96)^2}{(4)(120)^2}} \right] = .326$$

Note that the 95% confidence interval $.174 \le \pi_1 \le .326$ is quite close to the values computed with Equations 8.8 and 8.9 (and not that far removed from the values obtained with Equations 8.6 and 8.7). Ghosh's (1979) methodology is one of a number of alternative methodologies which have been proposed for computing the confidence interval for a population proportion. In most instances, any differences for the values of a confidence interval computed with the various methodologies described in this section will be minimal, and, in the final analysis, be of no practical consequence.

12. Since, when $k = 2$, it is possible to state two directional alternative hypotheses, some sources refer to an analysis of such a nondirectional alternative hypothesis as a two-tailed test. Using the same logic, when $k > 2$ one can conceptualize a test of a nondirectional alternative hypothesis as a **multi-tailed test** (since, when $k > 2$, it is possible to have more than two directional alternative hypotheses). It should be pointed out that since the **chi-square goodness-of-fit test** only utilizes the right tail of the distribution, it is questionable to use the terms two-tailed or multi-tailed in reference to the analysis of a nondirectional alternative hypothesis.

13. $k!$ is referred to as k **factorial**. The notation indicates that the integer number preceding the ! is multiplied by all integer values below it. Thus, $k! = (k)(k-1) \ldots (1)$. By definition 0! is set equal to 1. A method of computing an approximate value for $n!$ was developed by James Stirling (1730). (The letter n is more commonly employed as the notation to represent the number for which a factorial value is computed — i.e., the notation $n!$) **Stirling's approximation** (described in Feller (1968), Miller and Miller (1999) and Zar (1999)) is $n! = \sqrt{2n\pi}\,(n/e)^n$ which can also be written as $n! = \sqrt{2\pi}\,(n^{n+.5})(e^{-n})$. As noted in Endnote 14 in the **Introduction**, the value e in the Stirling equation is the base of the natural system of logarithms. e, which equals 2.71828...., is an **irrational number** (i.e., a number that has a decimal notation which goes on forever without a repeating pattern of digits).

14. The subscript .72 in the notation $\chi^2_{.72}$ represents the .72 level of significance. The value .72 is based on the fact that the extreme 72% of the right tail of the chi-square distribution is employed in evaluating the directional alternative hypothesis. The value $\chi^2_{.72}$ falls at the 28th percentile of the distribution.

15. A general discussion of differences between the latter two tests can be found in Section VII of the **Kolmogorov–Smirnov goodness-of-fit test for a single sample**.

16. When categories are ordered, there is a direct (or inverse) relationship between the magnitude of the score of a subject on the variable being measured and the ordinal position of the category to which that score has been assigned. An example of ordered categories which can be employed with the **chi-square goodness-of-fit test** are the following four categories that can be used to indicate the magnitude of a person's IQ: **Cell 1 – 1st quartile; Cell 2 – 2nd quartile; Cell 3 – 3rd quartile; and Cell 4 – 4th quartile**. The aforementioned categories can be employed if one wants to determine whether or not, within a sample, an equal number of subjects are observed in each of the four quartiles. Note that in Examples 8.1 and 8.2, the fact that an observation is assigned to Cell 6 is not indicative of a higher level of performance than or superior quality to an observation assigned to Cell 1 (or vice versa). However, in the IQ example there is a direct relationship between the number used to identify each cell and the magnitude of IQ scores for subjects who have been assigned to that cell.

17. Note that the sum of the proportions must equal 1.

18. Even though a nondirectional alternative hypothesis will be assumed, this example illustrates a case in which some researchers might view it more prudent to employ a directional alternative hypothesis.

Test 9

The Binomial Sign Test for a Single Sample
(Nonparametric Test Employed with Categorical/Nominal Data)

I. Hypothesis Evaluated with Test and Relevant Background Information

Hypothesis evaluated with test In an underlying population comprised of two categories that is represented by a sample, is the proportion of observations in one of the two categories equal to a specific value?

Relevant background information on test The **binomial sign test for a single sample** is based on the **binomial distribution**, which is one of a number of **discrete probability distributions** discussed in this chapter. A **discrete probability distribution** is a distribution in which the values a variable may assume are finite (as opposed to a **continuous probability distribution** in which a variable may assume an infinite number of values). The basic assumption underlying the binomial distribution is that each of n independent observations (i.e., the outcome for any given observation is not influenced by the outcome for any other observation) is randomly selected from a population, and that each observation can be classified in one of $k = 2$ mutually exclusive categories. Within a binomially distributed population, the likelihood an observation will fall in Category 1 will equal π_1, and the likelihood an observation will fall in Category 2 will equal π_2. Since it is required that $\pi_1 + \pi_2 = 1$, it follows that $\pi_2 = 1 - \pi_1$.[1] The mean (μ, which is also referred to as the **expected value**) and standard deviation (σ) of a binomially distributed variable are computed with Equations 9.1 and 9.2.[2]

$$\mu = n\pi_1 \qquad \textbf{(Equation 9.1)}$$

$$\sigma = \sqrt{n\pi_1\pi_2} \qquad \textbf{(Equation 9.2)}$$

When $\pi_1 = \pi_2 = .5$, the binomial distribution is symmetrical. When $\pi_1 < .5$, the distribution is positively skewed, with the degree of positive skew increasing as the value of π_1 approaches zero. When $\pi_1 > .5$, the distribution is negatively skewed, with the degree of negative skew increasing as the value of π_1 approaches one. The sampling distribution of a binomially distributed variable can be approximated by the normal distribution. The closer the value of π_1 is to .5 and the larger the value of n, the better the normal approximation. Because of the **central limit theorem** (which is discussed in Section VII of the **single-sample z test (Test 1)**), even if the value of n is small and/or the value of π_1 is close to either 0 or 1, the normal distribution still provides a reasonably good approximation of the sampling distribution for a binomially distributed variable.

The **binomial sign test for a single sample** employs the binomial distribution to determine the likelihood that x or more (or x or less) of n observations which comprise a sample will fall in one of two categories (to be designated as Category 1), if in the underlying population the true proportion of observations in Category 1 equals π_1. When there are $k = 2$ categories, the hypothesis evaluated with the **binomial sign test for a single sample** is identical to that evaluated with the **chi-square goodness-of-fit test (Test 8)**. Since the two tests evaluate the same hypothesis, the hypothesis for the **binomial sign test for a single sample** can also be stated as follows: In the underlying population represented by a sample, are the observed frequencies for the two categories different from their expected frequencies? As noted in Section I of the **chi-square goodness-of-fit test**, the **binomial sign test for a single sample** is generally employed for small sample sizes, since when the value of n is large the computation of exact binomial probabilities becomes prohibitive without access to specialized tables or the appropriate computer software.

II. Examples

Two examples will be employed to illustrate the use of the **binomial sign test for a single sample**. Since both examples employ identical data, they will result in the same conclusions with respect to the null hypothesis.

Example 9.1 *An experiment is conducted to determine whether or not a coin is biased. The coin is flipped ten times resulting in eight heads and two tails. Do the results indicate the coin is biased?*

Example 9.2 *Ten women are asked to judge which of two brands of perfume has a more fragrant odor. Eight of the women select* Perfume A *and two of the women select* Perfume B. *Is there a significant difference with respect to preference for the perfumes?*

III. Null versus Alternative Hypotheses

Null hypothesis $H_0: \pi_1 = .5$

(In the underlying population the sample represents, the true proportion of observations in Category 1 equals .5.)

Alternative hypothesis $H_1: \pi_1 \neq .5$

(In the underlying population the sample represents, the true proportion of observations in Category 1 is not equal to .5. This is a **nondirectional alternative hypothesis**, and it is evaluated with a **two-tailed test**. In order to be supported, the observed proportion of observations in Category 1 in the sample data (which will be represented with the notation p_1) can be either significantly larger than the hypothesized population proportion $\pi_1 = .5$ or significantly smaller than $\pi_1 = .5$.)[3]

or

$$H_1: \pi_1 > .5$$

(In the underlying population the sample represents, the true proportion of observations in Category 1 is greater than .5. This is a **directional alternative hypothesis**, and it is evaluated with a **one-tailed test**. In order to be supported, the observed proportion of observations in Category 1 in the sample data must be significantly larger than the hypothesized population proportion $\pi_1 = .5$.)

or

$$H_1: \pi_1 < .5$$

(In the underlying population the sample represents, the true proportion of observations in Category 1 is less than .5. This is a **directional alternative hypothesis**, and it is evaluated with a **one-tailed test**. In order to be supported, the observed proportion of observations in Category 1 in the sample data must be significantly smaller than the hypothesized population proportion $\pi_1 = .5$.)

Note: Only one of the above noted alternative hypotheses is employed. If the alternative hypothesis the researcher selects is supported, the null hypothesis is rejected.

IV. Test Computations

In Example 9.1 the null and alternative hypotheses reflect the fact that if it is assumed a fair coin is employed, the probability of obtaining **Heads** on any trial will equal .5 (which is equivalent to ½). Thus, the expected/theoretical probability for **Heads** is represented by $\pi_1 = .5$. If the coin is fair, the probability of obtaining **Tails** on any trial will also equal .5, and thus the expected/theoretical probability for **Tails** is represented by $\pi_2 = .5$. Note that $\pi_1 + \pi_2 = 1$. In Example 9.2, it is assumed that if there is no difference with regard to preference for the two brands of perfume, the likelihood of a woman selecting **Perfume A** will equal $\pi_1 = .5$, and the likelihood of selecting **Perfume B** will equal $\pi_2 = .5$. In both Examples 9.1 and 9.2 the question being asked is as follows: If $n = 10$ and $\pi_1 = \pi_2 = .5$, what is the probability of 8 or more observations in one of the two categories?[4] Table 9.1 summarizes the outcome of Examples 9.1 and 9.2. The notation x is employed to represent the number of observations in Category 1 and the notation $(n - x)$ is employed to represent the number of observations in Category 2.

Table 9.1 Model for Binomial Sign Test for a Single Sample for Examples 9.1 and 9.2

Category		
1 (Heads/Perfume A)	2 (Tails/Perfume B)	Total
$x = 8$	$n - x = 10 - 8 = 2$	$n = 10$

In Examples 9.1 and 9.2, the proportion of observations in **Category/Cell 1** is $p_1 = 8/10 = .8$ (i.e., $p_1 = n_1/n$, where n_1 is the number of observations in **Category 1**), and the proportion of observations in **Category/Cell 2** is $p_2 = 2/10 = .2$ (i.e., $p_2 = n_2/n$, where n_2 is the number of observations in **Category 2**). Equation 9.3 can be employed to compute the probability that exactly x out of a total of n observations will fall in one of the two categories.

$$P(x) = \binom{n}{x} (\pi_1)^x (\pi_2)^{(n - x)} \qquad \textbf{(Equation 9.3)}$$

The term $\binom{n}{x}$ in Equation 9.3 is referred to as the **binomial coefficient** and is computed with Equation 9.4 (which is the same as Equation I.48 in the **Introduction**). $\binom{n}{x}$ is more generally referred to as the **number of combinations of n things taken x at a time**.[5]

$$\binom{n}{x} = \frac{n!}{x! \, (n - x)!} \qquad \textbf{(Equation 9.4)}$$

In the case of Examples 9.1 and 9.2, the binomial coefficient will be $\binom{10}{8}$, which is the combination of 10 things taken 8 at a time. In the combination expression, $n = 10$ represents the total number of coin tosses/women and $x = 8$ represents the observed frequency for Category 1(**Heads/Perfume A**). When the latter value (which equals $\binom{10}{8} = \frac{10!}{8! \, 2!} = 45$) is multiplied by $(.5)^8 \, (.5)^2$, it yields the probability of obtaining exactly 8 **Heads/Perfume A** if there are 10 observations. The probability of 8 observations in 10 trials will be represented by the notation $P(8/10)$. The value $P(8/10) = .0439$ is computed below.

$$P(8/10) = \binom{10}{8}(.5)^8 \, (.5)^2 = (45)(.5)^8(.5)^2 = .0439$$

Since the computation of binomial probabilities can be quite tedious, such probabilities are more commonly derived through the use of tables. By employing **Table A6 (Table of the Binomial Distribution, Individual Probabilities)** in the **Appendix**, the value .0439 can be obtained without any computations. The probability value .0439 is identified by employing the section of the **Table A6** for $n = 10$. Within this section, the value .0439 is the entry in the cell that is the intersection of the row $x = 8$ and the column $\pi = .5$ (where $\pi_1 = .5$ is employed to represent the value of π).

The probability .0439, however, does not provide enough information to allow one to evaluate the null hypothesis. The actual probability which is required is the likelihood of obtaining a value that is equal to or more extreme than the number of observations in Category 1. Thus, in the case of Examples 9.1 and 9.2, one must determine the probability of obtaining a frequency of 8 or greater for Category 1. In other words, we want to determine the likelihood of obtaining 8, 9, or 10 **Heads/Perfume A** if the total number of observations is $n = 10$. Since we have already determined that the probability of obtaining 8 **Heads/Perfume A** is .0439, we must now determine the probability associated with the values 9 and 10. Although each of these probabilities can be computed with Equation 9.3, it is quicker to use **Table A6**. Employing the table, we determine that for $\pi = .5$ and $n = 10$, the probability of obtaining exactly $x = 9$ observations in Category 1 is $P(9/10) = .0098$, and the probability of obtaining exactly $x = 10$ observations is $P(10/10) = .0010$. The sum of the three probabilities $P(8/10)$, $P(9/10)$, and $P(10/10)$ represents the likelihood of obtaining 8 or more **Heads/Perfume A** in 10 observations. Thus: $P(8, 9, \text{ or } 10/10) = .0439 + .0098 + .0010 = .0547$. Equation 9.5 summarizes the computation of a cumulative probability such as that represented by $P(8, 9, \text{ or } 10/10) = .0547$.[6]

$$P(\geq x) = \sum_{r=x}^{n} \binom{n}{x} (\pi_1)^x \, (\pi_2)^{(n-x)} \qquad \textbf{(Equation 9.5)}$$

Where:　$\sum_{r=x}^{n}$ indicates that probability values should be summed beginning with the designated value of x up through the value n

An even more efficient way of obtaining the probability $P(8, 9, \text{ or } 10/10) = .0547$ is to employ **Table A7 (Table of the Binomial Distribution, Cumulative Probabilities)** in the **Appendix**. When employing **Table A7** we again find the section for $n = 10$, and locate the cell which is the intersection of the row $x = 8$ and the column $\pi = .5$. The entry .0547 in that cell represents the probability of 8 or more (i.e., 8, 9, or 10) **Heads/Perfume A**, if there is a total of $n = 10$ observations. Thus, the entry in any cell of **Table A7** represents (for the appropriate value of π_1) the probability of obtaining a number of observations that is equal to or greater than the value of x in the left margin of the row in which the cell is located.

Table A7 can be used to determine the likelihood of x being equal to or less than a specific value. In such a case, the cumulative probability associated with the value of $(x + 1)$ is subtracted from 1. To illustrate this, let us assume that in Examples 9.1 and 9.2 we want to determine the probability of obtaining 2 or less observations in one of the two categories (which applies to Category 2). In such an instance the value $x = 2$ is employed, and thus, $x + 1 = 3$. The cumulative probability associated with $x = 3$ (for $\pi = \pi_1 = .5$) is .9453. If the latter value is subtracted from 1 it yields $1 - .9453 = .0547$, which represents the likelihood of obtaining 2 or less observations in a cell when $n = 10$.[7] The value .0547 can also be obtained from **Table A6** by adding up the probabilities for the values $x = 2$ (.0439), $x = 1$ (.0098), and $x = 0$ (.0010).

It should be noted that none of the values listed for π in **Tables A6** and **A7** exceeds .5. To employ the tables when the value for π_1 stated in the null hypothesis is greater than .5, the following protocol is employed: a) Use the value of π_2 (i.e., $\pi_2 = 1 - \pi_1$) to represent the value of π; and b) Each of the values of x is subtracted from the value of n, and the resulting values are employed to represent x in using the table for the analysis. To illustrate, let us assume that $n = 10$, $\pi_1 = .7$, and $x = 9$, and that we wish to determine the probability there are 9 or more observations in one of the categories. Employing the above guidelines, the tabled value to use for π is $\pi_2 = .3$ (since $1 - .7 = .3$). Since each value of x is subtracted from n, the values $x = 9$ and $x = 10$ are respectively converted into $x = 1$ and $x = 0$. In **Table A6** (for $\pi = .3$) the probabilities associated with $x = 1$ and $x = 0$ will respectively represent those probabilities associated with $x = 9$ and $x = 10$. The sum of the tabled probabilities for $x = 1$ and $x = 0$ represents the likelihood there will be 9 or more observations in one of the categories. From **Table A6** we determine that for $\pi = .3$, $P(1/10) = .1211$ and $P(0/10) = .0282$. Thus, $P(0 \text{ or } 1/10) = .1211 + .0282 = .1493$ (which also represents $P(9 \text{ or } 10/10)$ when $\pi_1 = .7$). The value .1493 can also be obtained from **Table A7** by subtracting the tabled probability value for $(x + 1)$ from 1 (make sure that in computing $(x + 1)$, the value of x which results from subtracting the original value of x from n is employed). Thus, if the converted value of $x = 1$, then $x + 1 = 2$. The tabled value in **Table A7** for $x = 2$ and $\pi = .3$ is .8507. When the latter value is subtracted from 1, it yields .1493.[8]

V. Interpretation of the Test Results

When the **binomial sign test for a single sample** is applied to Examples 9.1 and 9.2, it provides a probabilistic answer to the question of whether or not $p_1 = 8/10 = .8$ (i.e., the observed proportion of cases for Category 1) deviates significantly from the value $\pi_1 = .5$ stated in the null hypothesis.[9] The following guidelines are employed in evaluating the null hypothesis.[10]

a) If a nondirectional alternative hypothesis is employed, the null hypothesis can be rejected if the probability of obtaining a value equal to or more extreme than x is equal to or less than $\alpha/2$ (where α represents the prespecified value of α). If the proportion $p_1 = x/n$ is greater than π_1, a value that is more extreme than x will be any value which is greater than the observed value of x, whereas if the proportion $p_1 = x/n$ is less than π_1, a value that is more extreme than x will be any value which is less than the observed value of x.

b) If a directional alternative hypothesis is employed that predicts the underlying population proportion is above a specified value, to reject the null hypothesis both of the following conditions must be met: 1) The proportion of cases observed in Category 1 (p_1) must be greater than the value of π_1 stipulated in the null hypothesis; and 2) The probability of obtaining a value equal to or greater than x is equal to or less than the prespecified value of α.

c) If a directional alternative hypothesis is employed that predicts the underlying population proportion is below a specified value, to reject the null hypothesis both of the following conditions must be met: 1) The proportion of cases observed in Category 1 (p_1) must be less than the value of π_1 stipulated in the null hypothesis; and 2) The probability of obtaining a value equal to or less than x is equal to or less than the prespecified value of α.

Applying the above guidelines to the results of the analysis of Examples 9.1 and 9.2, we can conclude the following.

If $\alpha = .05$, the nondirectional alternative hypothesis H_1: $\pi_1 \neq .5$ is not supported, since the obtained probability .0547 is greater than $\alpha/2 = .05/2 = .025$. In the same respect, if $\alpha = .01$, the nondirectional alternative hypothesis H_1: $\pi_1 \neq .5$ is not supported, since the obtained probability .0547 is greater than $\alpha/2 = .01/2 = .005$.

If $\alpha = .05$, the directional alternative hypothesis H_1: $\pi_1 > .5$ is not supported, since the obtained probability .0547 is greater (albeit barely) than $\alpha = .05$. In the same respect, if $\alpha = .01$, the directional alternative hypothesis H_1: $\pi_1 > .5$ is not supported, since the obtained probability .0547 is greater than $\alpha = .01$.

The directional alternative hypothesis H_1: $\pi_1 < .5$ is not supported, since $p_1 = .8$ is larger than the value $\pi_1 = .5$ predicted in the null hypothesis. If the alternative hypothesis H_1: $\pi_1 < .5$ is employed and the sample data are consistent with it, in order to be supported the obtained probability must be equal to or less than the prespecified value of alpha.

To summarize, the results of the analysis of Examples 9.1 and 9.2 do not allow a researcher to conclude that the true population proportion is some value other than .5. In view of this, in Example 9.1 the data do not allow one to conclude that the coin is biased. In Example 9.2, the data do not allow one to conclude that women exhibit a preference for one of the two brands of perfume.

It should be noted that if the proportion $p_1 = .8$ had been obtained with a larger sample size, the null hypothesis could be rejected. To illustrate, if for $\pi_1 = .5$, $n = 15$, $x = 12$ and thus $p_1 = .8$, the likelihood of obtaining 12 or more observations in one of the two categories is .0176. The latter value is significant at the .05 level if the directional alternative hypothesis H_1: $\pi_1 > .5$ is employed, since it is less than the value $\alpha = .05$. It is also significant at the .05 level if the nondirectional alternative hypothesis H_1: $\pi_1 \neq .5$ is employed, since .0176 is less than $\alpha/2 = .05/2 = .025$.

VI. Additional Analytical Procedures for the Binomial Sign Test for a Single Sample and/or Related Tests

1. Test 9a: The z test for a population proportion When the size of the sample is large the test statistic for the **binomial sign test for a single sample** can be approximated with the chi-square distribution — specifically, through use of the **chi-square goodness-of-fit test**. An alternative and equivalent approximation can be obtained by using the normal distribution. When the latter distribution is employed to approximate the test statistic for the **binomial sign test for a single sample**, the test is referred to as the z **test for a population proportion**. The null and alternative hypotheses employed for the z **test for a population proportion** are identical to those employed for the **binomial sign test for a single sample**.

Although sources are not in agreement with respect to the minimum acceptable sample size for use with the z **test for a population proportion**, there is general agreement that the closer the value π_1 (or π_2) is to either 0 or 1 (i.e., the further removed it is from .5), the larger the sample size required for an accurate normal approximation. Among those sources which make recommendations with respect to the minimum acceptable sample size (regardless of the value of π_1) are Freund (1984) and Marascuilo and McSweeney (1977), who state that the values of both $n\pi_1$ and $n\pi_2$ should be greater than 5. Daniel (1990) states that n should be at least equal to 12. Siegel and Castellan (1988), on the other hand, note that when π_1 is close to .5, the test can be employed when $n > 25$, but when π_1 is close to 1 or 0, the value $n\pi_1\pi_2$ should be greater than 9. In view of the different criteria stipulated in various sources, one should employ common sense in interpreting results for the normal approximation based on small sample sizes, especially when the values of π_1 or π_2 are close to 0 or 1. Since when the sample size is small the normal approximation tends to inflate the Type I error rate, the latter can be adjusted by conducting a more conservative test (i.e., employ a lower alpha level). A more practical alternative, however, is to use a test statistic that is corrected for continuity. As will be demonstrated in the discussion to follow, when the correction for continuity is employed for the **z test for a population proportion** the test statistic will generally provide an excellent approximation of the binomial distribution, even when the size of the sample is small and/or the values of π_1 and π_2 are far removed from .5.

Examples 9.3–9.5 will be employed to illustrate the use of the **z test for a population proportion**. Since the three examples use identical data they will result in the same conclusions with respect to the null hypothesis. It will also be demonstrated that when the **chi-square goodness-of-fit test** is applied to Examples 9.3–9.5 it yields equivalent results.

Example 9.3 *An experiment is conducted to determine if a coin is biased. The coin is flipped 200 times resulting in 96 heads and 104 tails. Do the results indicate that the coin is biased?*

Example 9.4 *Although a senator supports a bill which favors a woman's right to have an abortion, she realizes her vote could influence whether or not the people in her state endorse her bid for reelection. In view of this the senator decides she will not vote in favor of the bill unless at least 50% of her constituents support a woman's right to have an abortion. A random survey of 200 voters in her district reveals that 96 people are in favor of abortion. Will the senator support the bill?*

Example 9.5 *In order to determine whether or not a subject exhibits extrasensory ability, a researcher employs a list of 200 binary digits (specifically, the values 0 and 1) which have been randomly generated by a computer. The researcher conducts an experiment in which one of his assistants concentrates on each of the digits in the order it appears on the list. While the assistant does this, the subject, who is in another room, attempts to guess the value of the number for each of the 200 trials. The subject correctly guesses 96 of the 200 digits. Does the subject exhibit evidence of extrasensory ability?*

As is the case for Examples 9.1 and 9.2, Examples 9.3–9.5 are evaluating the hypothesis of whether or not the true population proportion is .5. Thus, the null hypothesis and the non-directional alternative hypothesis are: H_0: π_1 = .5 versus H_1: $\pi_1 \neq .5$.[11]

The test statistic for the **z test for a population proportion** is computed with Equation 9.6.

$$z = \frac{p_1 - \pi_1}{\sqrt{\dfrac{\pi_1 \pi_2}{n}}}$$ **(Equation 9.6)**

The denominator of Equation 9.6 ($\sqrt{\pi_1 \pi_2 / n}$), which is the standard deviation of the sampling distribution of a proportion, is commonly referred to as the **standard error of the proportion** (which is represented by the notation s_p).

For Examples 9.3–9.5, based on the null hypothesis, we know that: $\pi_1 = .5$ and $\pi_2 = 1 - \pi_1$ = .5. From the information which has been provided, we can compute the following values:

$p_1 = 96/200 = .48$; $p_2 = (200 - 96)/200 = 104/200 = .52$. When the relevant values are substituted in Equation 9.6, the value $z = -.57$ is computed.

$$z = \frac{.48 - .50}{\sqrt{\frac{(.5)(.5)}{200}}} = -.57$$

Equation 9.7 is an alternative form of Equation 9.6 that will yield the identical z value.

$$z = \frac{x - n\pi_1}{\sqrt{n\pi_1\pi_2}} \qquad \text{(Equation 9.7)}$$

In Section I it is noted that the mean and standard deviation of a binomially distributed variable are respectively $\mu = n\pi_1$ and $\sigma = \sqrt{n\pi_1\pi_2}$. These values represent the mean and standard deviation of the underlying sampling distribution. In the numerator of Equation 9.7, the value $\mu = n\pi_1$ represents the expected number of observations in Category 1 if, in fact, the population proportion is equal to $\pi_1 = .5$ (i.e., the value stipulated in the null hypothesis). Thus, for the examples under discussion, $\mu = (200)(.5) = 100$. Note that the latter expected value is subtracted from the number of observations in Category 1. The denominator of Equation 9.7 is the standard deviation of a binomially distributed variable. Thus, in the case of Examples 9.3–9.5, $\sigma = \sqrt{(100)(.5)(.5)} = 7.07$. Employing Equation 9.7, the value $z = -.57$ (which is identical to the value computed with Equation 9.6) is obtained.

$$z = \frac{96 - (200)(.5)}{\sqrt{(200)(.5)(.5)}} = -.57$$

The obtained value $z = -.57$ is evaluated with **Table A1 (Table of the Normal Distribution)** in the **Appendix**. In **Table A1** the tabled critical two-tailed .05 and .01 values are $z_{.05} = 1.96$ and $z_{.01} = 2.58$, and the tabled critical one-tailed .05 and .01 values are $z_{.05} = 1.65$ and $z_{.01} = 2.33$.

The following guidelines are employed in evaluating the null hypothesis.

a) If the alternative hypothesis employed is nondirectional, the null hypothesis can be rejected if the obtained absolute value of z is equal to or greater than the tabled critical two-tailed value at the prespecified level of significance.

b) If the alternative hypothesis employed is directional and predicts a population proportion larger than the value stated in the null hypothesis, the null hypothesis can be rejected if the sign of z is positive and the value of z is equal to or greater than the tabled critical one-tailed value at the prespecified level of significance.

c) If the alternative hypothesis employed is directional and predicts a population proportion smaller than the value stated in the null hypothesis, the null hypothesis can be rejected if the sign of z is negative and the absolute value of z is equal to or greater than the tabled critical one-tailed value at the prespecified level of significance.

Using the above guidelines, the null hypothesis cannot be rejected regardless of which of the three possible alternative hypotheses is employed. The nondirectional alternative hypothesis $H_1: \pi_1 \neq .5$ is not supported, since the absolute value $z = .57$ is less than the tabled critical two-tailed value $z_{.05} = 1.96$. The directional alternative hypothesis $H_1: \pi_1 > .5$ is not supported, since to be supported the sign of z must be positive. The directional alternative hypothesis $H_1: \pi_1 < .5$ is not supported, since, although the sign of z is negative as predicted, the absolute value $z = .57$ is less than the tabled critical one-tailed value $z_{.05} = 1.65$.

It was noted previously that when there are $k=2$ categories, the **chi-square goodness-of-fit test** also provides a large sample approximation of the test statistic for the **binomial sign test for a single sample**. In point of fact, the large sample approximation based on the **chi-square goodness-of-fit test** will yield a result which is equivalent to that obtained with the z **test for a population proportion**, and the relationship between the computed chi-square value and the obtained z value for the same set of data will always be $\chi^2 = z^2$. Table 9.2 summarizes the results of the analysis of Examples 9.3–9.5 with the **chi-square goodness-of-fit test**, which, as noted earlier, evaluates the same hypothesis as the **binomial sign test for a single sample**. The null hypothesis and nondirectional alternative hypothesis when used in reference to the **chi-square goodness-of-fit-test** can also be stated employing the following format: H_0: $o_i = \varepsilon_i$ for both cells versus H_1: $o_i \neq \varepsilon_i$ for both cells.

Table 9.2   Chi-Square Summary Table for Examples 9.3–9.5

Cell	O_i	E_i	$(O_i - E_i)$	$(O_i - E_i)^2$	$\dfrac{(O_i - E_i)^2}{E_i}$
Heads/Pro-Abortion/ Correct Guesses	96	100	−4	16	.16
Tails/Anti-Abortion/ Incorrect Guesses	104	100	4	16	.16
	$\Sigma O_i = 200$	$\Sigma E_i = 200$	$\Sigma(O_i - E_i) = 0$		$\chi^2 = .32$

In Table 9.2. the expected frequency of each cell is computed by multiplying the hypothesized population proportion for the cell by $n = 200$ (i.e., employing Equation 8.1, $E_i = (n)(\pi_i) = (200)(.5) = 100$). Since $k = 2$, the degrees of freedom employed for the chi-square analysis are $df = k - 1 = 2$. The value $\chi^2 = .32$ (which is obtained with Equation 8.2) is evaluated with **Table A4 (Table of the Chi-Square distribution)** in the **Appendix**. For $df = 1$, the tabled critical .05 and .01 chi-square values are $\chi^2_{.05} = 3.84$ and $\chi^2_{.01} = 6.63$. Since the obtained value $\chi^2 = .32$ is less than $\chi^2_{.05} = 3.84$, the null hypothesis cannot be rejected if the nondirectional alternative hypothesis H_1: $\pi_1 \neq .5$ is employed. If the directional alternative hypothesis H_1: $\pi_1 < .5$ is employed it is not supported, since $\chi^2 = .32$ is less than the tabled critical one-tailed .05 value $\chi^2_{.05} = 2.71$ (which corresponds to the chi-square value at the 90th percentile).

As noted previously, if the z value obtained with Equations 9.6 and 9.7 is squared, it will always equal the chi-square value computed for the same data. Thus, in the current example where $z = -.57$ and $\chi^2 = .32$: $(-.57)^2 = .32$. (The minimal discrepancy is the result of rounding off error.) It is also the case that the square of a tabled critical z value at a given level of significance will equal the tabled critical chi-square value at the corresponding level of significance. This is confirmed for the tabled critical two-tailed z and χ^2 values at the .05 and .01 levels of significance: $(z_{.05} = 1.96)^2 = (\chi^2_{.05} = 3.84)$ and $(z_{.01} = 2.58)^2 = (\chi^2_{.01} = 6.63)$.

To summarize, the results of the analysis for Examples 9.3–9.5 do not allow one to conclude that the true population proportion is some value other than .5. In view of this, in Example 9.3 the data do not allow one to conclude that the coin is biased. In Example 9.4 the data do not allow the senator to conclude that the true proportion of the population which favors abortion is some value other than .5. In Example 9.5 the data do not allow one to conclude that the subject exhibited extrasensory abilities.

Before continuing an additional comment is in order on Example 9.5. With respect to the latter example, the strategy a subject employs in generating responses during the $n = 200$ trials will be a function of what the subject believes with regard to the distribution of digits in the 200 digit series. Specifically, a subject could be told either of the following prior to generating his answers.

a) The subject is told that on a given trial either of the $k = 2$ digit values 0 and 1 can be the correct answer, and that it is not necessary for each digit to occur an equal number of times during the $n = 200$ trials. In other words, it is not required that each of the digits 0 and 1 will be the correct answer on 100 of the 200 trials. When this latter assumption has been made (as, in fact, is the case in Example 9.5), the analysis of the data as described through use of Equations 9.6 and 9.7 provides the optimal normal approximation for the result of the experiment. When the aforementioned scenario depicted in Example 9.5 is described within the context of a deck of cards, it is often stated that it represents an **open deck** testing situation. In other words, the experiment described in Example 9.5 could have been conducted employing a deck containing 200 cards, with each card having a digit of either 0 or 1 representing its face value. In the **open deck** testing situation, it would not be required that each of the two digit values 0 or 1 be the face value of exactly 100 of the 200 cards. It should be noted that if such a deck was randomly generated, it would be expected that the frequencies of each of the $k = 2$ digit values would be reasonably close to 100. However, in a randomly generated deck it is theoretically possible (although not likely) that one of the two digit values will occur at an extremely high or low frequency. (The concept of randomness is discussed in detail within the context of the **single-sample runs test (Test 10)**).

b) The subject is told that the list of $n = 200$ digits will contain an equal number of digits representing each of the $k = 2$ digit values 0 and 1. In other words, there will be 100 trials in which the correct answer is 0 and 100 trials in which the correct answer is 1. When this latter assumption has been made (which is not the case in Example 9.5), the analysis of the data as described through use of Equations 9.6 and 9.7 is not entirely accurate. Under such conditions the value of the denominator of Equation 9.7 (which represents the variance of the sampling distribution) is slightly larger than the value computed with $\sqrt{n\pi_1\pi_2}$. As the value of n increases, the value of the variance approaches the value computed with the denominator of Equation 9.7. The practical consequence of a larger variance is that the computed absolute value of z will be slightly less than the z value computed with Equation 9.6/9/7. When the aforementioned scenario is described within the context of a deck of cards, it is often described as representing a **closed deck** testing situation. In other words, if the experiment described in Example 9.5 employed a deck of cards, in the **closed deck** testing situation it would be required that each of the two digit values 0 and 1 be the face value of exactly 100 of the 200 cards. The protocol for computing the value of the variance to employ in Equation 9.7 for the closed deck testing situation is described in Burdick and Kelly (1977, pp. 89–90). If a closed deck were assumed for the experiment described by Example 9.5, the difference between the absolute value $z = .57$ obtained earlier and the Burdick and Kelly (1977) adjusted value would be minimal.

It was noted previously that when the z **test for a population proportion** is employed with small sample sizes, it tends to inflate the likelihood of committing a Type I error. This is illustrated below with the data for Examples 9.1 and 9.2 which are evaluated with Equation 9.6.

$$z = \frac{.8 - .5}{\sqrt{\dfrac{(.5)(.5)}{10}}} = 1.90$$

Since the obtained value $z = 1.90$ is greater than the tabled critical one-tailed value $z_{.05} = 1.65$, the directional alternative hypothesis H_1: $\pi_1 > .5$ is supported at the .05 level. (Note that since $z^2 = \chi^2$, the value $\chi^2 = (1.90)^2 = 3.61$ would be obtained if the data are evaluated with the **chi-square goodness-of-fit test**. Since the value $\chi^2 = 3.61$ is greater than the tabled critical one-tailed .05 chi-square value $\chi^2_{.05} = 2.71$, the directional alternative hypothesis H_1: $\pi_1 > .5$ is supported at the .05 level.) When the **binomial sign test for a single sample** is employed to evaluate the latter alternative hypothesis for the same data, the result falls just short of being significant at the .05 level. The nondirectional alternative hypothesis H_1: $\pi_1 \neq .5$, which is not even close to being supported with the **binomial sign test for a single sample**, falls just short of being supported at the .05 level when the **z test for a population proportion** is employed (since the tabled critical two-tailed .05 value is $z_{.05} = 1.96$). When the conclusions reached with respect to Example 9.1 and 9.2 employing the **binomial sign test for a single sample** and the **z test for a population proportion** are compared with one another, it can be seen that the **z test for a population proportion** is the less conservative of the two tests (i.e., it is more likely to reject the null hypothesis).

The correction for continuity for z test for a population proportion   It is noted in the discussions of the **Wilcoxon signed-ranks test (Test 6)** and the **chi-square goodness-of-fit test** that many sources recommend a **correction for continuity** be employed when a continuous distribution is employed to estimate a discrete probability distribution. Most sources recommend the latter correction when the normal distribution is used to approximate the binomial distribution, since the correction will adjust the Type I error rate (which will generally be inflated when the normal approximation is employed with small sample sizes). Equations 9.8 and 9.9 are, respectively, the continuity-corrected versions of Equations 9.6 and 9.7. Each of the continuity-corrected equations is applied to the data for Examples 9.3–9.5.

$$z = \frac{|p_1 - \pi_1| - \dfrac{1}{2n}}{\sqrt{\dfrac{\pi_1 \pi_2}{n}}} = \frac{|.48 - .5| - \dfrac{1}{(2)(200)}}{\sqrt{\dfrac{(.5)(.5)}{200}}} = -.49 \quad \textbf{(Equation 9.8)}$$

$$z = \frac{|x - n\pi_1| - .5}{\sqrt{n\pi_1 \pi_2}} = \frac{|96 - 100| - .5}{\sqrt{(200)(.5)(.5)}} = -.49 \quad \textbf{(Equation 9.9)}$$

As is the case when Equations 9.6 and 9.7 are employed to compute the absolute value $z = .57$, the absolute value $z = .49$ computed with Equations 9.6 and 9.7 is less than the tabled critical two-tailed .05 value $z_{.05} = 1.96$ and the tabled critical one-tailed .05 value $z_{.05} = 1.65$. Thus, regardless of which of the possible alternative hypotheses one employs, the null hypothesis cannot be rejected. Note that the continuity-corrected absolute value $z = .49$ is less than the absolute value $z = .57$ obtained with Equations 9.6 and 9.7. Since a continuity-corrected equation will always result in a lower absolute value for z, it will provide a more conservative test of the null hypothesis. The smaller the sample size, the greater the difference between the values computed with the continuity-corrected and uncorrected equations.

Equation 8.5, which as noted previously is the continuity-corrected equation for the **chi-square goodness-of-fit test**, can also be employed with the same data and will yield an equivalent result to that obtained with Equations 9.8 and 9.9. When employing Equation 8.5 there are two cells, and each cell has an expected frequency of 100. The observed frequencies of the two cells are 96 and 104. Thus for each cell, $(|O_i - E_i| - .5) = 3.5$. Thus:

$$\chi^2 = \sum_{i=1}^{k} \left[\frac{(|O_i - E_i| - .5)^2}{E_i} \right] = \frac{(3.5)^2}{100} + \frac{(3.5)^2}{100} = .245$$

Note that $(.49)^2 = .245$ (once again the slight discrepancy is due to rounding off error). As is the case with $\chi^2 = .32$ (the uncorrected chi-square value computed in Table 9.2), the continuity-corrected value $\chi^2 = .245$ is not significant, since it is less than the tabled critical .05 value $\chi^2_{.05} = 3.84$ (in reference to the nondirectional alternative hypothesis).

Although in the case of Examples 9.3–9.5 the use of the correction for continuity does not alter the decision one can make with respect to the null hypothesis, this will not always be the case. To illustrate this, Equation 9.8 is employed below to compute the continuity-corrected value of z for Examples 9.1 and 9.2.

$$z = \frac{|.8 - .5| - \dfrac{1}{(2)(10)}}{\sqrt{\dfrac{(.5)(.5)}{10}}} = 1.58$$

Since the continuity-corrected value $z = 1.58$ is less than the tabled critical one-tailed value $z_{.05} = 1.65$, the directional alternative hypothesis H_1: $\pi_1 > .5$ is not supported. (Note that since $z^2 = \chi^2$, the value $\chi^2 = (1.58)^2 = 2.50$ would be obtained if the data are evaluated with the **chi-square goodness-of-fit test**. Since the value $\chi^2 = 2.50$ is less than the tabled critical one-tailed .05 chi-square value $\chi^2_{.05} = 2.71$, the directional alternative hypothesis H_1: $\pi_1 > .5$ is not supported at the .05 level.) This is consistent with the result obtained when the **binomial sign test for a single sample** is employed. Recollect that when the data are evaluated with Equation 9.6 (i.e., without the continuity correction) the directional alternative hypothesis H_1: $\pi_1 > .5$ is supported. Thus, it appears that in this instance the continuity correction yields a result which is more consistent with the result based on the exact binomial probability.

Sources are not in agreement with respect to whether or not a correction for continuity should be employed. Zar (1999) cites a study by Ramsey and Ramsey (1988) which found that the results with the correction for continuity are overly conservative (i.e., more likely to retain the null hypothesis when it should be rejected).

Computation of a confidence interval for the z test for a population proportion Equation 8.6, which is described in the discussion of the **chi-square goodness-of-fit test**, can also be employed for computing a confidence interval for the z test for a population proportion. The confidence interval computed with Equation 8.6 is obtained by adding and subtracting the product of the appropriate critical z value and the standard error of the proportion to and from the observed probability for the sample data. The reader should take note of the fact that the latter confidence interval represents a **large sample approximation of a confidence interval for a binomially distributed variable**. Equation 8.6 is employed below to compute the 95% confidence interval for Examples 9.3–9.5 for Category 1.

$$p_1 - z_{(\alpha/2)} \sqrt{\frac{p_1 p_2}{n}} \leq \pi_1 \leq p_1 + z_{(\alpha/2)} \sqrt{\frac{p_1 p_2}{n}}$$

$$.48 - (1.96) \sqrt{\frac{(.48)(.52)}{200}} \leq \pi_1 \leq .48 + (1.96) \sqrt{\frac{(.48)(.52)}{200}}$$

$$\pi_1 = .48 \pm .069$$

$$.411 \leq \pi_1 \leq .549$$

Thus, the researcher can be 95% confident (or the probability is .95) that the interval .411 to .549 contains the true proportion of cases in the underlying population in Category 1. The confidence interval for the population proportion for Category 2 (i.e., π_2) can be obtained by adding and subtracting the value .069 to and from the value $p_2 = .52$. Thus, $.451 \leq \pi_2 \geq .589$.

Additional discussion of as well as alternative procedures for computing a confidence interval for a binomially distributed variable can be found in Section VI of the **chi-square goodness-of-fit test**.

Procedure for computing sample size in reference to the power of the z test for a population proportion Fleiss *et al.* (2003, p. 32) note that Equation 9.10 can be employed to estimate the sample size (n) required in order to conduct a z **test for a population proportion** which has a power of a specified value.

(Equation 9.10)

$$n = \left[\frac{z_{\alpha/2} \sqrt{\pi_1(1 - \pi_1)} + z_{\beta} \sqrt{p_1(1 - p_1)}}{\delta} \right]^2$$

Where: δ represents the minimum effect size (in proportion units) the researcher wants to be able to identify. Most sources employ the value ($p_1 - \pi_1$) (i.e., the difference between the hypothesized population proportion and the observed sample proportion) to represent the value of δ. ($p_1 - \pi_1$) is commonly employed to represent the value of δ, when after obtaining a nonsignificant result a researcher wants to determine if the power of the test conducted had been set at a specific value, how large a sample size would have been required in order for the z **test for a population proportion** to identify the observed difference between the two proportions (i.e., ($p_1 - \pi_1$)) to be significant.

$z_{\alpha/2}$ represents the relevant tabled critical two-tailed z value in **Table A1** employed in the analysis.

z_{β} is the z value listed in **Table A1** for which a proportion (probability) is recorded that identifies the Type II error rate (i.e., β) employed in the analysis. The protocol to employ to determine the value of z_{β} is as follows: a) If the **power** stipulated for the test (i.e., the value $(1 - \beta)$), is **greater than .5**, z_{β} is the z value for which the proportion recorded in **Column 3** of **Table A1** is equal to the value **stipulated for β**. (It is also the z value for which the proportion recorded in **Column 2** of **Table A1** is equal to the value (**Power** – .5).) When the **power** stipulated for the test is **greater than .5**, the sign of z_{β} will always be **positive**; b) If the **power** stipulated for the test is **less than .5**, z_{β} is the z value for which the proportion recorded in **Column 3** of **Table A1** is equal to the **power** of the test (i.e., the value $(1 - \beta)$). (It is also the z value for which the proportion recorded in **Column 2** of **Table A1** is equal to the value (β – .5).) When the **power** stipulated for the test is **less than .5**, the sign of z_{β}

will always be **negative**; c) If the **power** stipulated for the test is .5, $z_\beta = 0$. (In **Table A1** the value $\beta = .5000$ is listed in **Column 3**, and the value .0000 in **Column 2** represents the difference between β and.5 (i.e., $\beta - .5 = 0$, with **Power** $= 1 - \beta = .5$.)

Let us assume that in Examples 9.3–9.5 the researcher wanted the power of the *z* **test for a population proportion** to equal .9. In addition to the values $\pi_1 = .5$, $p_1 = .48$, and $z_{\alpha/2} = 1.96$ (If a one-tailed analysis is conducted $z_{\alpha/2} = 1.65$ should be employed.), the value $z_\beta = 1.28$ is employed in Equation 9.10. The rationale for employing the value $z_\beta = 1.28$ is as follows: Since **Power** $= 1 - \beta = .90$, $\beta = .10$. Employing the guidelines for using Equation 9.10 documented above to compute a sample size when **Power** $> .5$, in **Table A1**, $z = 1.28$ is the *z* value for which the proportion (probability) recorded in **Column 3** (.1003) is closest to the value $\beta = .10$. (It is also the case that for $z = 1.28$, if .5 is added to the value recorded in **Column 2** (i.e., .3997 + .5 = .8997), the resulting value is closest to the value .90 stipulated for the power of the test (and thus, **Power** $- .5 = .8997 - .5 = .3997$).) Since the power stipulated for the test is greater than .5, the sign of z_β will be a positive. The observed difference ($p_1 - \pi_1$) = (.48 − .5) = −.02 will be employed to represent the value of δ.

When the appropriate values are substituted in Equation 9.10, the value $n = 6556.85$ is computed, indicating that in order for the power of the *z* **test for a population proportion** to be .90, the researcher must have at least 6557 subjects (which is the next integer value that is greater than the value computed for *n*). Note that the value $n = 6557$ is considerably more than the value $n = 200$ employed in Examples 9.3–9.5.

$$n = \left[\frac{1.96\sqrt{(.5)(.5)} + 1.28\sqrt{(.48)(.52)}}{.48 - .50} \right]^2 = 6556.85$$

It should be obvious that the power of the actual analysis employing 200 subjects was, in fact, quite low. In actuality, when $n = 200$, the power of the test is slightly below .10. The latter can be confirmed by employing the value −1.28 in Equation 9.10 for z_β. The rationale for employing the value $z_\beta = -1.28$ is as follows: Let us assume we wish to compute the sample size if the power of the test is .10. Thus, if **Power** $= 1 - \beta = .10$, $\beta = .90$. Employing the guidelines for using Equation 9.10 documented earlier to compute a sample size when **Power** $< .5$, in **Table A1**, $z = 1.28$ is the *z* value for which the proportion (probability) recorded in **Column 3** (.1003) is closest to the value **Power** $= .10$. (It is also the case that for $z = 1.28$, if .5 is added to the value recorded in **Column 2** (i.e., .3997 + .5 = .8997), the resulting value is closest to the value .90 stipulated for the value of β(and thus, $\beta - .5. = .8997 - .5 = .3997$).) Since the power stipulated for the test is less than .5, the sign of z_β will be negative. When $z_\beta = -1.28$ is substituted in Equation 9.10 the value $n = 289.87$ is computed. Thus, if 290 subjects had been employed (which is 90 more than the actual number employed in Examples 9.3–9.5) the power of the test would equal .10.

$$n = \left[\frac{1.96\sqrt{(.5)(.5)} + (-1.28)\sqrt{(.48)(.52)}}{.48 - .50} \right]^2 = 289.87$$

Fleiss *et al.* (2003, p. 32) state that because it does not employ a correction for continuity, Equation 9.10 may underestimate the required sample size, and that Equation 9.11 may provide a more accurate estimate of the latter value (where *n'* represents the value computed for *n* with Equation 9.10). For the problem under discussion, use of Equation 9.11 would indicate the sample size should be increased from approximately 6557 to 6607 in order to insure that the power of the test is .90. Further discussion of evaluating the power of the z **test for two population proportions**, as well as alternative methodologies, can be found in Fleiss *et al.* (2003, pp. 29–37), Rosner (2000, pp. 253–255) and Zar (1999, Chapter 24).

$$n = \frac{n'}{4}\left(1 + \sqrt{1 + \frac{2}{n'(|p_1 - \pi_1|)}}\right)^2 \qquad \textbf{(Equation 9.11)}$$

$$= \frac{6556.85}{4}\left(1 + \sqrt{1 + \frac{2}{(6556.85)(|.48 - .5|)}}\right)^2 = 6606.76$$

Additional comments on computation of the power of the binomial sign test for a single sample Cohen (1977, 1988) has developed a statistic called the g **index** which can be employed to compute the power of the **binomial sign test for a single sample** when H_0: $\pi_i = .5$ is evaluated. The g **index** represents the distance in units of proportion from the value .5. The equation Cohen (1977, 1988) employs for the g **index** is $g = P - .5$ when H_1 is directional and $g = |P - .5|$ when H_1 is nondirectional (where P represents the hypothesized value of the population proportion stated in the alternative hypothesis — in employing the g **index** it is assumed the researcher has stated a specific value in the alternative hypothesis as an alternative to the value stipulated in the null hypothesis).

Cohen (1977; 1988, Ch. 5) has derived tables that allow a researcher, through use of the g **index**, to determine the appropriate sample size to employ if one wants to test a hypothesis about the distance of a proportion from the value .5 at a specified level of power. Cohen (1977; 1988, pp. 147–150) has proposed the following (admittedly arbitrary) g values as criteria for identifying the magnitude of an effect size: a) A **small effect size** is one that is greater than .05 but not more than .15; b) A **medium effect size** is one that is greater than .15 but not more than .25; and c) A **large effect size** is greater than .25.

2. Extension of the z test for a population proportion to evaluate the performance of m subjects on n trials on a binomially distributed variable Example 9.6 illustrates a case in which each of m subjects is evaluated for a total of n trials on a binomially distributed variable. The example represents an extension of the analysis used for Example 9.5 to a design involving m subjects. The methodology employed in analyzing the data is basically an extension of the **single-sample z test** to an analysis of a population proportion that is based on a binomially distributed variable.

Example 9.6 *In order to determine whether or not a group of 10 people exhibit extrasensory ability, a researcher employs as test stimuli a list of 200 binary digits (specifically, the values 0 and 1) which have been randomly generated by a computer. The researcher conducts an experiment in which one of his assistants concentrates on each of the digits in the order it appears on the list. While the assistant does this, each of the 10 subjects, all of whom are in separate rooms, attempts to guess the value of the number for each of 200 trials. The number of correct guesses in 200 trials for each of the subjects follows: 102, 104, 100, 98, 96, 80, 110, 120, 102, 128. Does the group as a whole exhibit evidence of extrasensory ability?*

Equation 9.12 is employed to evaluate Example 9.6.

$$z = \frac{\bar{X} - \mu}{\frac{\sigma}{\sqrt{m}}}$$ **(Equation 9.12)**

Where: m represents the number of subjects in the sample

$\mu = m\pi_1$

$\sigma = \sqrt{n\pi_1\pi_2}$

The basic structure of Equation 9.12 is the same as that of Equation 1.3 ($z = (\bar{X} - \mu)/\sigma_{\bar{X}}$), which is the equation for the **single-sample** z **test**. In the numerator of Equation 1.3, the hypothesized population mean is subtracted from the sample mean. Equation 9.12 employs the analogous values — employing the sample mean ($\bar{X}$) and $\mu = m\pi_1$ to represent the hypothesized population mean. The denominator of Equation 1.3 represents the standard deviation of a sampling distribution that is based on a sample size of n for what is assumed to be a normally distributed variable. The denominator of Equation 9.12 represents the standard deviation of a sampling distribution of a binomially distributed variable which is based on a sample size of m. In both equations the denominator can be summarized as follows: $\sigma/\sqrt{\text{number of subjects}}$. It should also be noted that when the number of subjects is $m = 1$, Equation 9.12 reduces to Equation 9.7.[12]

Note that in Example 9.6, each of the $m = 10$ subjects is tested for $n = 200$ trials. On each trial it is assumed that a subject has a likelihood of $\pi_1 = .5$ of being correct and a likelihood of $\pi_2 = .5$ of being incorrect. Since we are dealing with a binomially distributed variable, the expected number of correct responses for each subject, as well as the expected average number of correct responses for the group of $m = 10$ subjects, is $\mu = m\pi_1$. As previously noted, the standard deviation of the sampling distribution for a single subject is defined by $\sigma = \sqrt{n\pi_1\pi_2}$. Since there are m subjects, the standard deviation of the sampling distribution for m subjects will be $\sigma/\sqrt{m}$, which is the denominator of Equation 9.12.

The null and alternative hypotheses for Example 9.6 are identical to those employed for Example 9.5. The only difference is that, whereas in Example 9.5 H_0 and H_1 are stated in reference to the population of scores for a single subject, in Example 9.6 they are stated in reference to the scores of a population of subjects which is represented by the m subjects in the sample. In Example 9.6, the mean number of correct guesses by the 10 subjects is computed with Equation 1.1: $\bar{X} = 1040/10 = 104$. The values $n = 200$, $\pi_1 = .5$, $\pi_2 = .5$ are identical to those employed in Example 9.5. Thus, as is the case for Example 9.5, $\mu = (200)(.5) = 100$ and $\sigma = \sqrt{(200)(.5)(.5)} = 7.07$. When the appropriate values are substituted in Equation 9.12, the value $z = 1.79$ is computed.

$$z = \frac{104 - 100}{\frac{7.07}{\sqrt{10}}} = 1.79$$

The obtained value $z = 1.79$ is evaluated with **Table A1**. The nondirectional alternative hypothesis H_1: $\pi_1 \neq .5$ is not supported, since the value $z = 1.79$ is less than the tabled critical two-tailed value $z_{.05} = 1.96$. The directional alternative hypothesis H_1: $\pi_1 > .5$ is supported at the .05 level, since the obtained value $z = 1.79$ is a positive number that is greater than the tabled critical one-tailed value $z_{.05} = 1.65$. The latter alternative hypothesis is not supported at the .01 level, since $z = 1.79$ is less than the tabled critical one-tailed value $z_{.01} = 2.33$. The directional alternative hypothesis H_1: $\pi_1 < .5$ is not supported, since to be supported the sign of z must be negative.

The above analysis allows one to conclude that the group as a whole scores significantly above chance. The latter result can be interpreted as evidence of extrasensory perception if one is able to rule out alternative sensory and cognitive explanations of information transmission. The reader should be aware of the fact that in spite of the conclusions with regard to the group, it is entirely conceivable that the performance of one or more of the subjects in the group is not statistically significant. Inspection of the data reveals that the performance of the subject who obtains a score of 100 is at chance expectancy. Additionally, the scores of some of the other subjects (e.g., 102, 104, 96, 98) are well within chance expectancy.[13] It should also be noted that what was said previously regarding the strategy a subject employs in generating responses in Example 9.5 is also applicable to Example 9.6.

It is instructive to note that in the case of Example 9.6, if for some reason one is unwilling to assume the variable under study is binomially distributed with a standard deviation of $\sigma = 7.07$, the population standard deviation must be estimated from the sample data. Under the latter conditions the **single-sample t test (Test 2)** is the appropriate test to employ, and the following null hypothesis is evaluated: H_0: $\mu = 100$. If each of the 10 scores in Example 9.6 is squared and the squared scores are summed, they yield the value $\Sigma X^2 = 109,728$. Employing Equation 2.1, the estimated population standard deviation is computed to be $\hat{s} = 13.2$. Substituting the latter value, along with $\bar{X} = 104$, $\mu = 100$, and $n = 10$ in Equation 2.3 yields the value $t = .96$. Since for $df = 9$, $t = .96$ falls far short of the tabled critical two-tail .05 value $t_{.05} = 2.26$ and the tabled critical one-tail .05 value $t_{.05} = 1.83$, the null hypothesis is retained. This result is the opposite of that reached when Equation 9.12 is employed with the same data. The difference between the two tests can be attributed to the fact that in the case of the **single-sample t test** the estimated population standard deviation $\hat{s} = 13.2$ is almost twice the value of $\sigma = 7.07$ computed for a binomially distributed variable.

3. Test 9b: The single-sample test for the median

There are occasions when the **binomial sign test for a single sample** is employed to evaluate a hypothesis regarding a population median. Specifically, the test may be used to determine the likelihood of observing a specified number of scores above versus below the median of a distribution. When the **binomial sign test for a single sample** is used within this context it is often referred to as the **single-sample test for the median**.[14] This application of the **binomial sign test for a single sample** will be illustrated with Example 9.7. Since Example 9.7 assumes $\pi_1 = \pi_2 = .5$ and has the same binomial coefficient obtained for Examples 9.1 and 9.2, it yields the same result as the latter examples.

Example 9.7 *Assume that the median blood cholesterol level for a healthy 30-year-old male is 200 mg/100 ml. Blood cholesterol readings are obtained for a group consisting of eleven 30-year-old men who have had a heart attack within the last month. The blood cholesterol scores of the eleven men are: 230, 167, 250, 345, 442, 190, 200, 248, 289, 262, 301. Can one conclude that the median cholesterol level of the population represented by the sample (i.e., recent male heart attack victims) is some value other than 200?*

Since the median identifies the 50th percentile of a distribution, if the population median is in fact equal to 200, one would expect one-half of the sample to have a blood cholesterol reading above 200 (i.e., $p_1 = .5$), and one-half of the sample to have a reading below 200 (i.e., $p_2 = .5$). Although the null hypothesis and the nondirectional alternative hypothesis for Example 9.7 can be stated using the format H_0: $\pi_1 = .5$ versus H_1: $\pi_1 \neq .5$, they can also be stated as follows: H_0: $\theta = 200$ versus H_1: $\theta \neq 200$. Employing the latter format, the null hypothesis states that the median of the population the sample represents equals 200, and the alternative hypothesis states that the median of the population the sample represents does not equal 200.

When the **binomial sign test for a single sample** is employed to test a hypothesis about a population median, one must determine the number of cases which fall above versus below the hypothesized population median. Any score that is equal to the median is eliminated from the analysis. Employing this protocol, the score of the man who has a blood cholesterol of 200 is dropped from the data, leaving 10 scores, 8 of which are above the hypothesized median value, and 2 of which are below it. Thus, as is the case in Examples 9.1 and 9.2, we want to determine the likelihood of obtaining 8 or more observations in one category (i.e., **above the median**) if there are a total of 10 observations. It was previously determined that the latter probability is equal to .0537. As noted earlier, this result does not support the nondirectional alternative hypothesis H_1: $\pi_1 \neq .5$. It just falls short of supporting the directional alternative hypothesis H_1: $\pi_1 > .5$ (which in the case of Example 9.7 can also be stated as H_1: $\theta > 200$).

The data for Example 9.6 can also be evaluated within the framework of the **single-sample test for the median**. Specifically, if we assume a binomially distributed variable for which $\pi_1 = \pi_2 = .5$ and $\mu = 100$, the population median will also equal 100. In Example 9.6, 6 out of the 10 subjects score above the hypothesized median value $\theta = 100$, 3 score below it, and one subject obtains a score of 100. After the latter score is dropped from the analysis, 9 scores remain. Thus, we want to determine the likelihood that there will be 6 or more observations in one category (i.e., **above the hypothesized median**) if there are a total of 9 observations. Using **Table A7**, we determine that the latter probability equals .2539. Since the value .2539 is greater than the required two-tailed .05 probability $\alpha/2 = .025$, as well as the one-tailed probability $\alpha = .05$, the null hypothesis cannot be rejected regardless of which alternative hypothesis is employed. This is in stark contrast to the decision reached when the data are evaluated with Equation 9.12. Since the latter equation employs more information (i.e., the interval/ratio scores of each subject are employed to compute the sample mean), it provides a more powerful test of an alternative hypothesis than does the **single-sample test for the median** (which conceptualizes scores as categorical data).

The **Wilcoxon signed-ranks test** also provides a more powerful test of an alternative hypothesis concerning a population median than does the **binomial sign test for a single sample/single-sample test for the median**.[15] This will be demonstrated by employing the **Wilcoxon signed-ranks test** to evaluate the null hypothesis H_0: $\theta = 200$ for Example 9.7. Table 9.3 summarizes the analysis.

Table 9.3   Data for Example 9.7

Subject	X	$D = X - \theta$	Rank of $|D|$	Signed rank of $|D|$
1	230	30	2	2
2	167	-33	3	-3
3	250	50	5	5
4	345	145	9	9
5	442	242	10	10
6	190	-10	1	-1
7	200	0	—	—
8	248	48	4	4
9	289	89	7	7
10	262	62	6	6
11	301	101	8	8

$$\Sigma R+ = 51$$
$$\Sigma R- = 4$$

The computed Wilcoxon statistic is $T = 4$, since $\Sigma R- = 4$ (the smaller of the two values $\Sigma R-$ versus $\Sigma R+$) is employed to represent the test statistic. The value $T = 4$ is evaluated with **Table A5 (Table of Critical T Values for Wilcoxon's Signed-Ranks and Matched-Pairs Signed-Ranks Test)** in the **Appendix**. Employing **Table A5**, we determine that for $n = 10$ signed ranks, the tabled critical two-tailed .05 and .01 values are $T_{.05} = 8$ and $T_{.01} = 3$, and the tabled critical one-tailed .05 and .01 values are $T_{.05} = 10$ and $T_{.01} = 5$. Since the null hypothesis can only be rejected if the computed value $T = 4$ is equal to or less than the tabled critical value at the prespecified level of significance, we can conclude the following.

The nondirectional alternative hypothesis $H_1 : \theta \neq 200$ is supported at the .05 level, since $T = 4$ is less than the tabled critical two-tailed value $T_{.05} = 8$. It is not supported at the .01 level, since $T = 4$ is greater than the tabled critical two-tailed value $T_{.01} = 3$.

The directional alternative hypothesis H_1: $\theta > 200$ is supported at both the .05 and .01 levels since: a) The data are consistent with the directional alternative hypothesis H_1: $\theta > 200$. In other words, the fact that $\Sigma R+ > \Sigma R-$ is consistent with the directional alternative hypothesis H_1: $\theta > 200$; and b) The obtained value $T = 4$ is less than the tabled critical one-tailed values $T_{.05} = 10$ and $T_{.01} = 5$.

The directional alternative hypothesis H_1: $\theta < 200$ is not supported, since it is not consistent with the data. For the latter alternative hypothesis to be supported, $\Sigma R-$ must be greater than $\Sigma R+$.

Thus, if Example 9.7 is evaluated with the **Wilcoxon signed-ranks test** the nondirectional alternative hypothesis H_1: $\theta \neq 200$ is supported at the .05 level, and the directional alternative hypothesis H_1: $\theta > 200$ is supported at both the .05 and .01 levels. When the same data are evaluated with the **binomial sign test for a single sample/single-sample test for the median**, none of the alternative hypotheses are supported (although the directional alternative hypothesis H_1: $\pi_1 > .5$ falls just short of being significant at the .05 level). From the preceding it should be apparent that the **Wilcoxon signed-ranks test** (which employs a greater amount of information) provides a more powerful test of an alternative hypothesis than does the **binomial sign test for a single sample/single-sample test for the median**.

Examination of Example 6.1 (which is identical to Example 2.1) allows us to contrast the power of the **binomial sign test for a single sample/single-sample test for the median** with the power of both the **single-sample t test** and the **Wilcoxon signed-ranks test**. When the latter problem is evaluated with the **single-sample t test**, the null hypothesis H_0: $\mu = 5$ cannot be rejected if a nondirectional alternative hypothesis is employed. However, the null hypothesis can be rejected at the .05 level if the directional alternative hypothesis H_1: $\mu > 5$ is employed. With reference to the latter alternative hypothesis, the obtained t value is greater than the tabled critical $t_{.05}$ value by a comfortable margin. When Example 6.1 is evaluated with the **Wilcoxon signed-ranks test**, the null hypothesis H_0: $\theta = 5$ cannot be rejected if a nondirectional alternative hypothesis is employed. When the directional alternative hypothesis H_1: $\theta > 5$ is employed, the analysis just falls short of being significant at the .05 level. As noted in the discussion of the **Wilcoxon signed-ranks test**, the different conclusions derived from the two tests illustrate the fact that when applied to the same data, the **single-sample t test** provides a more powerful test of an alternative hypothesis than the **Wilcoxon signed-ranks test**.

If Example 6.1 is evaluated with the **binomial sign test for a single sample/single-sample test for the median**, it would be expected that it would be the least powerful of the three tests. In Example 6.1, 7 of the 10 scores fall above the hypothesized population median $\theta = 5$ and 3 scores fall below it. Thus, using the binomial distribution, we want to determine the likelihood of obtaining 7 or more observations in one category (i.e., **above the median**) if there are a total of 10 observations. Employing either **Table A6** or **A7**, we can determine that for $\pi_1 = \pi_2 = .5$ and $n = 10$, the likelihood of 7 or more observations in one category is .1719. Since the latter value is well above the required two-tailed .05 value $\alpha/2 = .025$ and the required one-tailed .05 value $\alpha = .05$, the directional alternative hypothesis H_1: $\theta > 5$ is not supported. Thus, when compared with the **Wilcoxon signed-ranks test**, which falls just short of significance, the **binomial sign test for a single sample** does not even come close to being significant.

The above noted differences between the **single-sample t test**, the **Wilcoxon signed-ranks test**, and the **binomial sign test for a single sample** illustrate that when the original data are in an interval/ratio format, the most powerful test of an alternative hypothesis is provided by the **single-sample t test** and the least powerful by the **binomial sign test for a single sample**. As noted in the **Introduction**, most researchers would not be inclined to employ a nonparametric test with interval/ratio data unless one had reason to believe that one or more of the assumptions of the appropriate parametric test were saliently violated. In the same respect, unless there is reason to believe that the underlying population distribution is not symmetrical, it is more logical to employ the **Wilcoxon signed-ranks test** as opposed to the **binomial sign test for a single sample** to evaluate Example 6.1.

VII. Additional Discussion of the Binomial Sign Test for a Single Sample

1. Evaluating goodness-of-fit for a binomial distribution There may be occasions when a researcher wants to evaluate the hypothesis that a set of data is derived from a binomially distributed population. Example 9.8 will be employed to demonstrate how the latter hypothesis can be evaluated with the **chi-square goodness-of-fit test**.

Example 9.8 *An animal biologist states the probability is .25 that while in captivity a female of a species of Patagonian fox will give birth to an albino pup. Records are obtained on 100 litters each comprised of six pups (which is the modal pup size for the species) from zoos throughout the world. In 14 of the litters there were 0 albino pups, in 30 litters there was 1 albino pup, in 35 litters there were 2 albino pups, in 18 litters there were 3 albino pups, in 2 litters there were 4 albino pups, in 1 litter there were 5 albino pups, and in 0 litters there were 6 albino pups. Is there a high likelihood the data represent a binomially distributed population with $\pi_1 = .25$?*

The null and alternative hypotheses that are evaluated with the **chi-square goodness-of-fit test** in reference to Example 9.8 can either be stated in the form they are presented in Section III of the latter test (i.e., H_0: $o_i = \varepsilon_i$ for all cells; H_i: $o_i \neq \varepsilon_i$ for at least one cell), or as follows.

Null hypothesis H_0: The sample is derived from a binomially distributed population, with $\pi_1 = .25$.

Alternative hypothesis H_1: The sample is not derived from a binomially distributed population, with $\pi_1 = .25$. This is a **nondirectional alternative hypothesis**.

The analysis of Example 9.8 with the **chi-square goodness-of-fit test** is summarized in Table 9.4. The latter table is comprised of $k = 7$ cells/categories, with each cell corresponding to the number of albino pups in a litter. The second column of Table 9.4 contains the observed frequencies for albino pups. The expected frequency for each cell was obtained by employing Equation 8.1. Specifically, the value 100, which represents the total number of litters/observations, is multiplied by the appropriate binomial probability in **Table A6** for a given value of x when $n = 6$ (the number of pups in a litter) and $\pi_1 = .25$. The latter binomial probabilities are:

$x = 0$ ($p = .1780$); $x = 1$ ($p = .3560$); $x = 2$ ($p = .2966$); $x = 3$ ($p = .1318$); $x = 4$ ($p = .0330$); $x = 5$ ($p = .0044$); $x = 6$ ($p = .0002$). Thus, if the data are binomially distributed with $\pi_1 = .25$, the following probabilities are associated with the number of albino pups in a litter comprised of 6 pups: 0 albino pups: $p = .1780$; 1 albino pup: $p = .3560$; 2 albino pups: $p = .2966$; 3 albino pups: $p = .1318$; 4 albino pups: $p = .0330$; 5 albino pups: $p = .0044$; and 6 albino pups: $p = .0002$. The expected frequencies in Column 3 of Table 9.4 are the result of multiplying each of the aforementioned binomial probabilities by 100. To illustrate the computation of an expected frequency, the value 17.80 is obtained for Row 1 (0 albino pups) as follows: $E_i = (100)(.1780) = 17.80$.

Table 9.4 Chi-Square Summary Table for Example 9.8

Cell	O_i	E_i	$(O_i - E_i)$	$(O_i - E_i)^2$	$\dfrac{(O_i - E_i)^2}{E_i}$
0	14	17.80	−3.80	14.44	.81
1	30	35.60	−5.60	31.36	.88
2	35	29.66	5.34	28.52	.96
3	18	13.18	4.82	23.23	1.76
4	2	3.30	−1.30	1.69	.51
5	1	.44	.56	.31	.71
6	0	.02	−.02	.0004	.02
	$\Sigma O_i = 100$	$\Sigma E_i = 100$	$\Sigma(O_i - E_i) = 0$		$\chi^2 = 5.65$

Employing Equation 8.2, the value $\chi^2 = 5.65$ is computed for Example 9.8. Since there are $k = 7$ cells and $w = 0$ parameters which are estimated, employing Equation 8.12, the degrees of freedom for the analysis are $df = 7 - 1 - 0 = 6$. Employing Table A4, we determine that for $df = 6$ the tabled critical .05 and .01 values are $\chi^2_{05} = 12.59$ and $\chi^2_{01} = 16.81$. Since the computed value $\chi^2 = 5.65$ is less than both of the aforementioned values, the null hypothesis cannot be rejected. Thus, the analysis does not indicate the data deviate significantly from a binomial distribution.

It should be noted that if, instead of stipulating the value $\pi_1 = .25$, the population proportion had been estimated from the sample data, the value of df is reduced by 1 (i.e., $df = 7 - 1 - 1 = 5$). The latter is the case, since an additional degree of freedom must be subtracted for the parameter which is estimated. In Example 9.8 the proportion of albino pups is $p_1 = n_1/N = 167/600 = .278$ (where n_1 is the total number of albino pups, and N is the total number of pups in the 100 litters ($N = (6)(100) = 600$)). The value $n_1 = 167$ is computed as follows: a) In each row of Table 9.4, multiply the value for the number of albino pups in Column 1 by the observed frequency for that number of albino pups in Column 2; and b) Sum the seven products obtained in a). The latter sum will equal the total number of albino pups, and if that value is divided by the total number of pups (N), it yields the value p_1. The latter value will represent the best estimate of the population proportion π_1.

It is important to note that if the value .278 is employed to represent π_1, the expected frequencies will be different from those recorded in Column 3 of Table 9.4. Since the value $\pi_1 = .278$ is not listed in Table A6, the latter table cannot be used to determine the binomial probabilities to employ in computing the expected frequencies. Consequently we would have to employ Equation 9.3 to compute the appropriate binomial probabilities for the values of x (i.e., 0, 1, 2, 3, 4, 5, and 6) when $n = 6$, $\pi_1 = .278$ and $\pi_2 = 1 - .278 = .722$, and then use the resulting binomial probabilities to compute the expected frequencies (once again by multiplying each probability by 100).

VIII. Additional Example Illustrating the Use of the Binomial Sign Test for a Single Sample

Example 9.9 employs the **binomial sign test for a single sample** in a case where the value of π_1 stated in the null hypothesis is close to 1. It also illustrates that the continuity-corrected version of the **z test for a population proportion** can provide an excellent approximation of the binomial distribution, even if the value of π_1 is far removed from .5.

Example 9.9 *A biologist has a theory that 90% of the people who develop a rare disease are males and only 10% are females. Of 10 people he identifies who have the disease, 7 are males and 3 are females. Do the data support the biologist's theory?*

Since the information given indicates that we are dealing with a binomially distributed variable with π_1 = .9 and π_2 = .1, the data can be evaluated with the **binomial sign test for a single sample**. Based on the biologist's theory, the null hypothesis and the nondirectional alternative hypothesis are as follows: H_0: π_1 = .9 versus H_1: π_1 ≠ .9. The data consist of the number of observations in the two categories **males** versus **females**. The respective proportion of observations in the two categories are: p_1 = 7/10 = .7 and p_2 = 3/10 = .3. Thus, given that π_1 = .9, we want to determine the likelihood of 7 or fewer observations in one category if there are a total of 10 observations. Note that since p_1 = .7 is less than π_1 = .9, a value which is more extreme than x = 7 will be any value that is less than 7.

Since π = π_1 = .9 is not listed in either **Table A6** or **A7**, we employ for π the probability value listed for π_2, which in the case of Example 9.9 is π_2 = .1. The probability of x being equal to or less than 7 (i.e., the number of observations in Category 1) if π_1 = .9, will be equivalent to the probability of x being equal to or greater than 3 (which is the number of observations in Category 2) if π_2 = .1. From Table **A7** it can be determined that the latter probability (which is in the cell which is the intersection of the row x = 3 and the column π = .1) is .0702. The same value can be obtained from **Table A6** by adding the probabilities for all values of x equal to or greater than 3 (for π = .1). Thus, the probability of 7 or fewer males if there is a total of 10 observations is .0702.[16] Since the latter value is greater than the two-tailed .05 value $\alpha/2$ = .025 and the one-tailed .05 value α = .05, neither the nondirectional alternative hypothesis H_1: π_1 ≠ .9 nor the directional alternative hypothesis which is consistent with the data (H_1: π_1 < .9) is supported. In other words, p_1 = .7 (the observed the proportion of males) is not significantly below the hypothesized value π_1 = .9. In the same respect p_2 = .3, the observed proportion of females, is not significantly above the expected value of π_2 = .1.

When the **z test for a population proportion** is employed to evaluate Example 9.9, Equation 9.6 (which does not employ the correction for continuity) yields the following result: z = (.7 − .9)/$\sqrt{[(.9)(.1)]/10}$ = −2.11. Equation 9.8 (the continuity-corrected equation) has the identical denominator as Equation 9.6, but the numerator is reduced by 1/[(2)(10)] = .05, thus yielding the value z = −1.58. Since the absolute value z = 2.11 is greater than the tabled critical two-tailed .05 value $z_{.05}$ = 1.96 and the tabled critical one-tailed .05 value $z_{.05}$ = 1.65, without the correction for continuity both the nondirectional alternative hypothesis H_1: π_1 ≠ .9 and the directional alternative hypothesis H_1: π_1 < .9 are supported at the .05 level. When the correction for continuity is employed, the obtained absolute value z = 1.58 is less than both of the aforementioned tabled critical values and, because of this, regardless of which alternative hypothesis is employed, the null hypothesis cannot be rejected. The latter conclusion is consistent with the result obtained when the exact binomial probabilities are employed. Thus, even in a case where the value of π_1 is far removed from .5, the continuity-corrected equation for the **z test for**

a population proportion appears to provide an excellent estimate of the exact binomial probability.

IX. Addendum

This Addendum is divided into two sections. The first section describes five discrete probability distributions, some of which are related to the **binomial distribution**, and the **exponential distribution**, which is a continuous distribution. The second part of the Addendum discusses the concept of **conditional probability** and **Bayes' theorem** (both of which are briefly discussed in the **Introduction**), and then employs the latter topics within the context of presenting a general overview of the **Bayesian hypothesis testing model**. The **Bayesian hypothesis testing model** is an alternative to the **classical hypothesis testing model** (which is employed throughout this book, as well as in most conventional statistics books). As will be noted later in the **Addendum**, in contrast to the **classical hypothesis testing model**, the **Bayesian model** computes probability values not only on the basis of sample data, but also requires that a researcher estimates a prior probability for one or more events and/or hypotheses being evaluated, basing the latter on available preexisting information.

1. Discussion of additional discrete probability distributions and the exponential distribution   In this section the following probability distributions will be described: a) The **multinomial distribution**; b) The **negative binomial distribution**; c) The **hypergeometric distribution**; d) The **Poisson distribution**; e) The **exponential distribution** (which, as noted above, is a continuous distribution that is closely related to the **Poisson distribution**); and f) The **matching distribution**.

a. The multinomial distribution   Earlier in this chapter it was noted that: a) The binomial distribution is a special case of the **multinomial distribution**; b) The **multinomial distribution** is an extension of the binomial model to two or more categories; and c) The **multinomial distribution** is the exact probability distribution that the **chi-square goodness-of-fit test** is employed to estimate (for two or more categories).

The conditions which describe the model for the multinomial distribution are that on each of n independent trials there are k possible outcomes. The probability on any trial that an outcome will fall in the i th category is π_i. Thus, the probabilities for the k categories are π_1, π_2, π_3, ... , π_k. As is the case with the binomial distribution, sampling with replacement (which is defined in Endnote 1) is assumed for the multinomial distribution.

Equation 9.13 is a multinomial generalization of Equation 9.3 (the binomial equation for computing the probability for a specific value of x) which can be employed when there are two or more categories. Equation 9.13 computes the probability that in n independent trials, n_1 outcomes will fall in Category 1, n_2 outcomes will fall in Category 2, n_3 outcomes will fall in Category 3, ... , and n_k outcomes will fall in Category k. The term $n! / (n!_1\, n!_2\, n!_3 \ldots n!_k)$ on the right side of Equation 9.13 is referred to as the **multinomial coefficient**. The multinomial analog of the binomial expansion $(\pi_1 + \pi_2)^n$ (discussed in Endnote 6) is the multinomial expansion described by the general equation $(\pi_1 + \pi_2 + \pi_3 + \ldots + \pi_k)^n$.

(Equation 9.13)

$$P(n_1, n_2, n_3, \ldots, n_k) = \frac{n!}{n!_1\, n!_2\, n!_3 \ldots n!_k}(\pi_1^{n_1})(\pi_2^{n_2})(\pi_3^{n_3}) \ldots (\pi_k^{n_k})$$

When $k = 2$, Equation 9.13 reduces to Equation 9.3. Thus, using Equation 9.13 we can compute the binomial probability .0439 for $x = 8$ computed for Examples 9.1/9.2 in Section IV.

$$P(n_1 = 8, \; n_2 = 2) = \frac{n!}{n_1! \, n_2!}(\pi_i)^{n_1}(\pi_2)^{n_2} = \frac{10!}{8! \, 2!}(.5)^8(.5)^2 = .0439$$

Examples 9.10, 9.11, and 9.12 will be employed to illustrate the application of Equation 9.13 to compute a multinomial probability.

Example 9.10 *An automobile dealer receives a delivery of ten cars. The company that manufactures the cars only delivers cars of the following three colors: silver, red, and blue. Assume the likelihood of a car being silver, red, or blue is as follows: Silver (.2), Red (.3), and Blue (.5). What is the probability that of the ten cars delivered, five will be silver, four will be red, and one will be blue?*

In Example 9.10 there are $n = 10$ observations/trials which correspond to the total of ten cars, and there are the following three categories: Category 1 – Silver cars; Category 2 – Red cars; Category 3 – Blue cars. Thus, $\pi_1 = .2$, $\pi_2 = .3$, and $\pi_3 = .5$. Since we are asking what the probability is that there will be five silver cars, four red cars, and one blue car, we can stipulate the values $n_1 = 5$, $n_2 = 4$, and $n_3 = 1$. Substituting the appropriate values in Equation 9.13, we determine that the probability of the delivery being comprised of five silver cars, four red cars, and one blue car is .0016.

$$P(n_1 = 5, \; n_2 = 4, \; n_3 = 1) = \frac{10!}{(5!)(4!)(1!)}(.2)^5(.3)^4(.5)^1 = .0016$$

Example 9.11 *Assume that during an official time at bat a major league baseball player has a .65 chance of making out, a .18 chance of hitting a single, a .07 chance of hitting a double, a .01 chance of hitting a triple, and a .09 chance of hitting a home run. If, during a game the player has six official at bats, what is the likelihood that he will make out in all six at bats?*

In Example 9.11 there are $n = 6$ observations/trials which correspond to the six at bats, and there are the following five categories: Category 1 – Out; Category 2 – Single; Category 3 – Double; Category 4 – Triple; Category 5 – Home run. Thus, $\pi_1 = .65$, $\pi_2 = .18$, $\pi_3 = .07$, $\pi_4 = .01$, $\pi_5 = .09$. Since we are asking what the probability is that there will be six outs, zero singles, zero doubles, zero triples, and zero home runs, we can stipulate the values $n_1 = 6$, $n_2 = 0$, $n_3 = 0$, $n_4 = 0$, and $n_5 = 0$. Substituting the appropriate values in Equation 9.13, we determine that the probability of the player making out six times is .075.

$$P(n_1 = 6, \; n_2 = 0, \; n_3 = 0, \; n_4 = 0, \; n_5 = 0) = \frac{6!}{(6!)(0!)(0!)(0!)(0!)}(.65)^6(.18)^0(.07)^0(.01)^0(.09)^0 = .075$$

Example 9.12 *A bird watcher spends the day searching for a particular species of bird whose beak can be either red or white and whose tail can be either long or short. Assume that the likelihood of a bird having a red beak and long tail is .10, the likelihood of a bird having a red beak and a short tail is .30, the likelihood of a bird having a white beak and long tail is .40, and* *the likelihood of a bird having a white beak and short tail is .20. If the bird watcher spots three birds of the species in question, what is the probability of observing three birds which conform to the characteristics noted in Table 9.5?*

Table 9.5 Data for Example 9.12

		Tail size		Row Sums
		Long tail	Short tail	
Beak color	Red	0	1	1
	White	1	1	2
Column Sums		1	2	Total 3

In Example 9.12 there are $n = 3$ observations/trials that correspond to the total of three birds, and there are the following four categories: **Category 1** – Red beak/Long tail; **Category 2** – Red beak/Short tail; **Category 3** – White beak/Long tail; **Category 4** – White beak/Short tail. Thus, $\pi_1 = .1$, $\pi_2 = .3$, $\pi_3 = .4$, and $\pi_4 = .2$. Since we are asking what the probability is there will be zero birds with a Red beak/Long tail, one bird with a Red beak/Short tail, one bird with a White beak/Long tail, and one bird with a White beak/Short tail, we can stipulate the values $n_1 = 0$, $n_2 = 1$, $n_3 = 1$, and $n_4 = 1$. Substituting the appropriate values in Equation 9.13 we determine that the probability of the bird watcher sighting the three birds noted in Table 9.5 is .144.

$$P(n_1 = 0, \ n_2 = 1, \ n_3 = 1, \ n_4 = 1) = \frac{3!}{(0!)(1!)(1!)(1!)}(.1)^0(.3)^1(.4)^1(.2)^1 = .144$$

In point of fact, if the bird watcher spots three birds, there are 20 possible configurations of beak color and tail length (which we assume are independent of one another) for which we can compute a multinomial probability. The 20 configurations are summarized in Table 9.6 along with their probabilities. Note that the probability value associated with each of the configurations is quite low, and that the sum of the probability values for all of the configurations adds up to 1. If analogous tables were constructed for Examples 9.10 and 9.11 (in which there are, respectively, ten and six observations), the number of possible configurations would be substantially larger than the 20 configurations for Example 9.12 (which has only three observations).

The values recorded in Columns 2–5 of Table 9.6 represent the number of birds which possess the beak and tail characteristic noted at the top of a column. The configuration represented in Table 9.5, for which the probability .144 is computed, corresponds to **Configuration 19** in Table 9.6.

b. The negative binomial distribution   The **negative binomial distribution** is another discrete probability distribution that can be employed within the context of evaluating certain experimental situations. As is the case with the binomial distribution, the model for the negative binomial distribution assumes the following: a) In a set of n independent trials there are only two possible outcomes; and b) Sampling with replacement. If we identify the two outcomes as Category 1 and Category 2, the negative binomial distribution allows us to determine the probability that exactly n trials will be required to obtain x observations in Category 1. If π_1 represents the likelihood an observation will fall in Category 1, and π_2 represents the likelihood an observation will fall in Category 2, the probability that exactly n trials will be required to obtain x observations in Category 1 is computed with Equation 9.14.

Table 9.6   Color/Tail Configuration for $n = 3$ Birds

Configuration	Red beak/ Long tail	Red beak/ Short tail	White beak/ Long tail	White beak/ Short tail	Multinomial probability
1	3	0	0	0	.001
2	0	3	0	0	.027
3	0	0	3	0	.064
4	0	0	0	3	.008
5	2	1	0	0	.009
6	0	0	2	1	.096
7	2	0	1	0	.012
8	0	2	0	1	.054
9	1	2	0	0	.027
10	0	0	1	2	.048
11	1	0	2	0	.048
12	0	1	0	2	.036
13	0	2	1	0	.108
14	2	0	0	1	.006
15	0	1	2	0	.144
16	1	0	0	2	.012
17	1	1	1	0	.072
18	1	1	0	1	.036
19	0	1	1	1	.144
20	1	0	1	1	.048

Sum = 1.000

$$P(x) = \binom{n-1}{x-1} \pi_1^x \pi_2^{n-x}$$ **(Equation 9.14)**

Although there are special tables prepared by Williamson and Bretherton (1963) for obtaining negative binomial probabilities, the latter values can also be determined through use of tables for the binomial distribution. Miller and Miller (1999) note the probabilities in **Table A6** can be employed to determine a negative binomial probability computed with Equation 9.14, by multiplying the individual probability associated with x in **Table A6** by (x/n). Guenther (1968) notes the negative binomial probability that n or fewer trials will be required to obtain x observations in Category 1 is equivalent to the cumulative binomial probability (in **Table A7**) for x or more observations in n trials. Thus, in **Table A7**, for the appropriate value of π_1, one would find the cumulative probability associated with x for a given value of n.

Examples 9.13 and 9.14 will be employed to illustrate the application of Equation 9.14 to compute a negative binomial probability.

Example 9.13   *The likelihood on any trial that a copy machine will print an acceptable copy is only .25. What is the probability exactly 12 copies will have to be printed before five acceptable copies are printed by the machine? What is the probability that the fifth acceptable copy will be printed on or before the twelfth trial? What is the probability that the fifth acceptable copy will be printed after the twelfth trial?*

Based on the information provided in Example 9.13, we can stipulate the following values we will employ in Equation 9.14: $n = 12$, $x = 5$, $\pi_1 = .25$, and $\pi_2 = .75$. Substituting the appropriate values in Equation 9.14, we compute the probability $p = .04317$. The latter value can also be obtained from **Table A6** by doing the following: a) Go to the section for $n = 12$; b) Find the probability in the cell that is the intersection of the row $x = 5$ and the column $\pi = .25$ (i.e., obtain the probability of five observations in 12 trials) — the latter value is .1032; and c) Multiply the probability obtained in b) (i.e., .1032) by (x/n). The resulting value will represent the likelihood of requiring exactly 12 trials to print five acceptable copies. Thus, $(.1032)(5/12) = .043$.

$$ P(r) = \binom{12 - 1}{5 - 1}(.25)^5(.75)^7 = \binom{11}{4}(.25)^5(.75)^7 = .04317 $$

Example 9.13 also asks for the probability that the fifth acceptable copy will be printed on or before the twelfth trial, and the probability that the fifth acceptable copy will be printed after the twelfth trial. As noted earlier, the probability the xth acceptable copy will be printed on or before the nth trial will correspond to the cumulative binomial probability for $n = 12$, $x = 5$, and $\pi_1 = .25$ in **Table A7** (which contains the probabilities computed with Equation 9.5). In the latter table the appropriate cumulative probability is .1576, which, in the case of Example 9.13, represents the likelihood that the fifth acceptable copy will be printed on or before the twelfth trial. We can also compute the probability that the fifth acceptable copy will be printed on or before the twelfth trial by employing Equation 9.14 with all values of n between five (which is the fewest possible trials in which five acceptable copies can be printed) and twelve, and summing the individual probabilities. When the values five through twelve are substituted for x in Equation 9.14, the following probability values are obtained which sum to .1576 (the minimal difference between the sum of the listed values and .1576 is due to rounding off error): $x = 5$ ($p = .00098$); $x = 6$ ($p = .003675$); $x = 7$ ($p = .00827$); $x = 8$ ($p = .01447$); $x = 9$ ($p = .02171$); $x = 10$ ($p = .02930$); $x = 11$ ($p = .03663$); $x = 12$ ($p = .04317$).

The answer to the last question posed in Example 9.13 (the likelihood that the fifth acceptable copy will be printed after the twelfth trial) is obtained simply by subtracting the value .1576 from 1. Thus, $1 - .1576 = .8424$ is the likelihood the fifth acceptable copy will be printed after the twelfth trial.

Equation 9.15 can be employed to compute the expected value (μ) for a negative binomially distributed variable. μ represents the expected number of trials to obtain x outcomes in Category 1. The standard deviation of a negative binomially distributed variable is computed with Equation 9.16. In the case of Example 9.13, the values $\mu = 20$ and $\sigma = 7.75$ are computed. Thus, if we conducted an infinite number of experiments with the copier, and in each experiment we printed copies until five were acceptable, the average value for n (i.e., the average number of trials required to obtain five acceptable copies) will be 20, and the standard deviation of the sampling distribution will equal 7.75.

$$ \mu = \frac{x}{\pi_1} = \frac{5}{.25} = 20 \qquad \textbf{(Equation 9.15)} $$

$$ \sigma = \frac{\sqrt{x\pi_2}}{\pi_1} = \frac{\sqrt{(5)(.75)}}{(.25)} = 7.75 \qquad \textbf{(Equation 9.16)} $$

Example 9.14 *The likelihood a basketball player will put the ball in the basket each time he shoots is .25. What is the probability that the player will have to take exactly 12 shots before making five baskets? What is the probability that the fifth successful shot will be made on or before the twelfth shot? What is the probability that the fifth successful shot will be made after the twelfth shot?*

Since the data for Example 9.14 are identical to that employed for Example 9.13, it yields the same probabilities. Thus, the probability of requiring exactly 12 shots to make five baskets is .043. The probability that the fifth successful shot will occur on or before the twelfth shot is .1576. The probability that the fifth successful shot will occur after the twelfth shot is .8424.

c. The hypergeometric distribution   The model for the **hypergeometric distribution** is similar to the model for the binomial distribution except for one critical difference — the latter being that in the hypergeometric model **sampling without replacement** (which is defined in Endnote 1) is assumed. In the hypergeometric model, in a set of n trials there are two possible outcomes (to be designated Category 1 versus Category 2), and because sampling without replacement is assumed, the outcome on each trial will be dependent on the outcomes of previous trials. The latter will be the case, since if there are two categories the probability of obtaining an outcome in a given category will change from trial to trial, and on any trial the value of the probabilities will be a function of the number of potential observations in each category that are still available to be selected.

Equation 9.17 is employed to compute a hypergeometric probability. The equation assumes there is a population comprised of N objects, each of which falls into one of two categories. In the population there are N_1 objects in Category 1, and N_2 objects in Category 2. Let us assume that we want to select a sample of n objects from the population employing sampling without replacement. We select x objects from Category 1, and $(n-x)$ objects from Category 2. Equation 9.17 allows us to compute the probability that we will select exactly x objects from Category 1 and $(n-x)$ objects from Category 2.

$$P(x) = \frac{\binom{N_1}{x}\binom{N_2}{n-x}}{\binom{N}{n}} \qquad \textbf{(Equation 9.17)}$$

Examples 9.15 and 9.16 will be employed to illustrate the computation of a hypergeometric probability with Equation 9.17.

Example 9.15 *What is the probability of selecting two boys and one girl from a class of nine students which consists of five boys and four girls?*

Example 9.16 *A researcher predicts that people who suffer from migraine headaches who take 500 milligrams of vitamin E daily are more likely to show a remission of symptoms than patients who don't take the vitamin. Nine patients with a history of migraines participate in a study. Five of the patients take 500 milligrams of vitamin E daily for a period of six months, while the other four patients, who comprise a control group, do not take a vitamin E supplement. At the conclusion of the study, two patients in the vitamin E group exhibit a remission of symptoms, while one person in the control group exhibits a remission of symptoms. What is the probability of this outcome?*

In both of the above examples we are dealing with a population that is comprised of $N = 9$ students/patients. There are $N_1 = 5$ students/patients in Category 1, and $N_2 = 4$ students/patients in Category 2. We let $x = 2$ represent the two boys/patients in the vitamin E group who exhibit symptom remission. Thus, $n - x = 3 - 2 = 1$ represents the one girl/patient in the control group who exhibits symptom remission. Employing Equation 9.17, the value .4762 is computed below for the probability of selecting two boys and one girl, or two patients in the experimental group exhibiting remission and one patient in control group exhibiting remission.

$$P(x = 2) = \frac{\binom{5}{2}\binom{4}{1}}{\binom{9}{3}} = .4762$$

In the case of Example 9.16, we want to determine whether there is a significant difference between the response of the vitamin E group and the control group. The null hypothesis is that there is no difference between the two groups. In order to evaluate the null hypothesis, we must compute the chance likelihood/probability of obtaining an outcome equal to or more extreme than the outcome observed in the study. In order to determine the latter, we must compute the hyper-geometric probabilities for all possible outcomes involving the value $n = 3$. Specifically, the following four outcomes can be obtained in which three out of a total of nine patients exhibit a remission of symptoms: a) All three patients who exhibit remission are in the vitamin E group; b) Two of the three patients who exhibit remission are in the vitamin E group, and the remaining patient is in the control group; c) Two of the three patients who exhibit remission are in the control group, and the remaining patient is in the vitamin E group; and d) All three patients who exhibit remission are in the control group. Note that Outcome b) noted above corresponds to the observed outcome in Example 9.16, and that the sum of the probabilities for the four outcomes is equal to 1 (due to rounding off error, the four probabilities sum to .9999). The hypergeometric probabilities for all four possible outcomes when $n = 3$ are noted below.

$$\text{a) } P(x = 3) = \frac{\binom{5}{3}\binom{4}{0}}{\binom{9}{3}} = .1190 \qquad \text{b) } P(x = 2) = \frac{\binom{5}{2}\binom{4}{1}}{\binom{9}{3}} = .4762$$

$$\text{c) } P(x = 1) = \frac{\binom{5}{1}\binom{4}{2}}{\binom{9}{3}} = .3571 \qquad \text{d) } P(x = 0) = \frac{\binom{5}{0}\binom{4}{3}}{\binom{9}{3}} = .0476$$

If a directional alternative hypothesis is employed (i.e., a one-tailed analysis is conducted), the null hypothesis can be rejected if the following conditions are met: a) The obtained difference is in the predicted direction; and b) The probability of obtaining a value equal to or more extreme than $x = 2$ is equal to or less than the value of α (which we will assume is .05). The data are consistent with the directional alternative hypothesis which predicts a better response by the vitamin E group. However, Outcome a) is in the same direction and more extreme than Outcome b). Thus, if we add the probabilities for Outcomes **a** and **b**, we obtain $.1190 + .4762 = .5952$. Since the latter value is greater than $\alpha = .05$, the null hypothesis cannot be rejected. Thus, we cannot conclude that vitamin E had a therapeutic effect.

If a nondirectional alternative hypothesis is employed (i.e., a two-tailed analysis is con-ducted), the null hypothesis can be rejected if the probability of obtaining a value equal to or more extreme than $x = 2$ is equal to or less than the prespecified level of significance. In the case of a two-tailed analysis, however, we take into account more extreme outcomes in either direction. In actuality, all of the outcomes are more extreme than Outcome b). The latter is the case, since in Outcomes a), c), and d) the proportion of subjects in the group which exhibits remission for two or more subjects is higher than the proportion $2/5 = .40$ for the vitamin E group in Outcome b). Because all of the other outcomes are more extreme than the observed outcome, the null hypothesis cannot be rejected. Use of the hypergeometric distribution in hypothesis testing is discussed in greater detail within the framework of the **Fisher exact test (Test 16c)** (which is discussed in Section VI of the **chi-square test for $r \times c$ tables (Test 16)**).

Equation 9.18 can be employed to compute the expected value (μ) of a hypergeometrically distributed random variable (μ is generally only computed for Category 1). In the case of Examples 9.15/9.16, the latter value is computed to be $\mu = 1.67$. The value $\mu = 1.67$ represents the expected number of outcomes in Category 1, when $n = 3$. If $N_2 = 4$ is employed in Equation 9.18 in place of N_1, the value $\mu = 1.33$ is computed, which represents the expected number of outcomes in Category 2 when $n = 3$. Note that when $\mu = 1.67$ is subtracted from 3, the resulting value is $\mu = 1.33$.

$$\mu = \frac{nN_1}{N} = \frac{(3)(5)}{9} = 1.67 \qquad \textbf{(Equation 9.18)}$$

Equation 9.19 can be employed to compute the expected value of the standard deviation (σ) of a hypergeometrically distributed random variable (σ is generally only computed for Category 1). In the case of Examples 9.15/9.16, the latter value is computed to be $\sigma = .745$. The identical value will be obtained for σ if N_2 is employed in Equation 9.19 in place of N_1.

(Equation 9.19)

$$\sigma = \sqrt{n\left(\frac{N_1}{N}\right)\left(1 - \frac{N_1}{N}\right)\left(\frac{N-n}{N-1}\right)} = \sqrt{3\left(\frac{5}{9}\right)\left(1 - \frac{5}{9}\right)\left(\frac{9-3}{9-1}\right)} = .745$$

When the value of N is very large relative to the value of n, the binomial distribution provides an excellent approximation of the hypergeometric distribution. This is the case, since, under the latter conditions, the differences between the sampling without replacement model and sampling with replacement model are minimized. To further clarify the relationship between the binomial and hypergeometric distributions, in Equation 9.18 the element N_1/N represents the proportion of cases in Category 1 in a dichotomous distribution. Thus, if we represent the element N_1/N with the notation π_1 (since it represents the same thing π_1 is employed to represent for the binomial distribution), an alternative way of writing Equation 9.18 is $\mu = n\pi_1$, which is the same as Equation 9.1, the equation for computing the expected value of a binomially distributed variable.

Now let us turn our attention to Equation 9.19. Once again we can employ π_1 to represent the element N_1/N. The element $1 - (N_1/N)$ represents the proportion of cases in Category 2, and thus we can represent it with the notation π_2 (because $\pi_2 = 1 - \pi_1$), since it represents the same thing π_2 represents for the binomial distribution. If the value of n is very small relative to the value of N, the element $(N - n)/(N - 1)$ will approach the value 1. If the latter is true, Equation 9.19 can be written as $\sigma = \sqrt{n\pi_1\pi_2}$, which is identical to Equation 9.2, the equation for computing the standard deviation of a binomially distributed variable.

Daniel and Terrell (1995) state that, as a general rule, in order to get a reasonable binomial approximation for a hypergeometrically distributed variable, the value of N should be at least ten times as large as the value of n. To illustrate the binomial approximation, let us consider the following values for a hypergeometrically distributed variable: $N = 40$, $N_1 = 10$, $N_2 = 30$, $n = 3$, $x = 1$. When the hypergeometric probability for $x = 1$ is computed below, we obtain the value $p = .4403$.

$$P(x = 1) = \frac{\binom{10}{1}\binom{30}{2}}{\binom{40}{3}} = .4403$$

Employing **Table A6**, we can determined the binomial probability for $x = 1$ when $n = 3$, $\pi_1 = .25$ (since $N_1/N = 10/40 = .25$), and $\pi_2 = .75$ (since $N_2/N = 30/40 = .75$). The latter value, which is also computed below with Equation 9.3, is .4219 (which is very close to the exact hypergeometric probability of .4403 computed above).

$$P(x = 1) = \binom{n}{x}(\pi_1)^x(\pi_2)^{(n-x)} = \binom{3}{1}(.25)(.75)^2 = .4219$$

d. The Poisson distribution With the exception of the binomial distribution, the **Poisson distribution** is probably the most commonly employed discrete probability distribution. The latter distribution is named after the French mathematician Simeon Denis Poisson, who described it in the 1830s (although Zar (1999) notes it was described previously by another mathematician, Abraham de Moivre, in 1718). The Poisson distribution (which Pagano and Gauvreau (1993) note is sometimes referred to as the **distribution of rare events**) is most commonly employed in evaluating a distribution of random events that have a low probability of occurring. The Poisson distribution is employed most frequently in situations where there is an interest in the number of times a particular event occurs within a specified period of time or within a specific physical environment. Feller (1968) and/or Guenther (1968) cite the following as examples of random events whose behavior is consistent with the Poisson distribution: a) The number of automobile accidents per month in a large city; b) The number of meteorites which land in areas of fixed size in a desert; c) The number of typographical errors per page in a manuscript; d) The number of telephone calls a person receives in a 24-hour period; e) The number of defective products manufactured daily on an assembly line; f) The number of atoms per second that disintegrate from a radioactive substance; and g) The number of bombs which hit specified blocks of equal area in London during World War II.

The model for the Poisson distribution assumes that within a given time period or within a given area, the likelihood of a random event occurring (with the occurrence of events being independent of one another) is very low. Consequently, if we have an infinite number of time periods or an infinite block of areas, the likelihood of more than one event occurring within any time period or area is very small (although, theoretically, there is no limit on the number of events which can occur within a specific time period/area).

A Poisson distribution has what is referred to as a **parameter** (also referred to as a **rate parameter**), that is represented by the notation λ (which is the lower case Greek letter **lambda**). The parameter λ (the value of which must be greater than zero) is the average number of events which occur over a given period of time or within a specified area of space. λ also happens to be the variance of a Poisson distributed variable. Thus, Equations 9.20 and 9.21 define the mean/expected value (μ) and variance of a Poisson distribution

$$\mu = \lambda \qquad\qquad \textbf{(Equation 9.20)}$$

$$\sigma^2 = \lambda \qquad\qquad \textbf{(Equation 9.21)}$$

It can be seen from inspection of Equations 9.20 and 9.21 that in a Poisson distribution $\mu = \sigma^2$, and because of the latter it is often assumed that any distribution where $\mu = \sigma^2$ is likely to be Poisson. Equation 9.22 is the general equation for computing a probability for the Poisson distribution. The latter equation computes the probability of x events occurring in a given period of time or within a specified area of space. Since x is a measure of a discrete random variable, any value obtained for x will have to be an integer number.

$$P(X = x) = \frac{e^{-\mu}\mu^x}{x!} \qquad \textbf{(Equation 9.22)}$$

Example 9.17 will be employed to illustrate the computation of a Poisson probability with Equation 9.22.

Example 9.17 *The traffic bureau of a Midwestern city claims on average two accidents occur per day, and that the frequency distribution of accidents conforms to a Poisson distribution. If the latter is true, what are the probabilities for the following numbers of accidents occurring per day: 0, 1, 2, 3, 4, 5, 6, more than 7?*

Given that the average equals two, we can say that $\lambda = \mu = \sigma^2 = 2$. Substituting the values $\mu = 2$ and $e = 2.71828$ in Equation 9.22, the probabilities are computed below for the values of x noted in Example 9.17.[17]

$$P(x = 0) = \frac{e^{-\mu}\mu^x}{x!} = \frac{(2.71828)^{-2}(2)^0}{0!} = .1353$$

This result tells us the likelihood of zero traffic accidents occurring is .1353.

$$P(x = 1) = \frac{e^{-\mu}\mu^x}{x!} = \frac{(2.71828)^{-2}(2)^1}{1!} = .2707$$

This result tells us the likelihood of one traffic accident occurring is .2707.

$$P(x = 2) = \frac{e^{-\mu}\mu^x}{x!} = \frac{(2.71828)^{-2}(2)^2}{2!} = .2707$$

This result tells us the likelihood of two traffic accidents occurring is .2707.

$$P(x = 3) = \frac{e^{-\mu}\mu^x}{x!} = \frac{(2.71828)^{-2}(2)^3}{3!} = .1804$$

This result tells us the likelihood of three traffic accidents occurring is .1804.

$$P(x = 4) = \frac{e^{-\mu}\mu^x}{x!} = \frac{(2.71828)^{-2}(2)^4}{4!} = .0902$$

This result tells us the likelihood of four traffic accidents occurring is .0902.

$$P(x = 5) = \frac{e^{-\mu}\mu^x}{x!} = \frac{(2.71828)^{-2}(2)^5}{5!} = .0361$$

This result tells us the likelihood of five traffic accidents occurring is .0361.

$$P(x = 6) = \frac{e^{-\mu}\mu^x}{x!} = \frac{(2.71828)^{-2}(2)^6}{6!} = .0120$$

This result tells us that the likelihood of six traffic accidents occurring is .0120.

The likelihood of seven or more accidents occurring is the sum of the probabilities for zero through six accidents (which is .9954) subtracted from 1. Thus, the likelihood of seven or more accidents occurring is $1 - .9954 = .0046$.

Under certain conditions the Poisson distribution can be employed to approximate the binomial distribution. When the latter is done it can facilitate the often tedious computations that are involved in determining binomial probabilities. The optimal conditions for approximating the binomial distribution with the Poisson distribution are when n is large and the value of π_1 is very small. Under the latter conditions the value of π_2 will be very close to 1, since $1 - \pi_1 = \pi_2$. In such a case if we set the value of π_2 equal to 1, Equation 9.2 (the equation for computing the standard deviation of a binomially distributed variable) reduces to $\sigma = \sqrt{n\pi_1\pi_2} = \sqrt{n\pi_1}$. If the latter is true, $\sigma^2 = n\pi_1$. Since the expected value of a binomially distributed variable is $\mu = n\pi_1$, under these conditions μ and σ^2 are identical, which is the case with a variable that conforms to a Poisson distribution. Thus, both μ and σ^2 may be represented by the parameter λ.

Consequently, we can say that when n is large and π_1 is very small, the relationship noted below is true (the notation $\approx$ means approximately). The left side of the relationship is Equation 9.3 (the equation for computing the likelihood of a specific value of x for a binomially distributed variable), and the right side of the relationship is Equation 9.22 (the equation for computing the likelihood of a specific value of x for a variable that has a Poisson distribution).

$$\binom{n}{x}(\pi_1)^x (\pi_2)^{(n - x)} \approx \frac{e^{-\mu}\mu^x}{x!}$$

To illustrate the Poisson approximation of the binomial distribution, consider Example 9.18.

Example 9.18 *Assume there is a .03 probability of a specific microorganism growing in a culture. What is the likelihood that in a batch of 200 cultures five of the cultures will contain the microorganism?*

Employing Equation 9.3, we compute the binomial probability for $n = 200$, $x = 5$, $\pi_1 = .03$, and $\pi_2 = .97$. The obtained value is $p = .1622$, which is the likelihood that in a batch of 200 cultures five will contain the microorganism.

$$P(x = 5) = \binom{n}{x}(\pi_1)^x (\pi_2)^{(n - x)} = \binom{200}{5}(.03)^5(.97)^{195} = .1622$$

The Poisson estimate of the binomial probability is determined as follows. First, employing Equation 9.1 we compute the expected value μ, which is $\mu = (200)(.03) = 6$. We then employ the latter value along with $x = 5$ in Equation 9.22, and compute the probability .1606, which is the likelihood that in a batch of 200 cultures five will contain the microorganism. Note that the value $p = .1606$ is very close to the exact binomial probability of $p = .1622$.

$$\mu = \lambda = n\pi_1 = (200)(.03) = 6$$

$$P(x = 5) = \frac{e^{-\mu}\mu^x}{x!} = \frac{(2.71828)^{-6}(6)^5}{5!} = .1606$$

Sources are not in agreement with respect to what values of n and π_1 are appropriate for employing the Poisson approximation of the binomial distribution. Hogg and Tanis (1997) recommend the approximation if $n \geq 20$ and $\pi_1 \leq .05$, or if $n \geq 100$ and $\pi_1 \leq .10$. Daniel and Terrell (1995) concur with the latter, and state the approximation is usually very good when $n \geq 100$ and $n\pi_1 \leq 10$. Rosner (1995, 2000) states the more conservative criterion that $n \geq 100$ and $\pi_1 \leq .01$.

Computation of a confidence interval for a Poisson parameter Zar (1995, p. 575) notes that Equations 9.23 and 9.24 can respectively be employed to compute the lower limit (LL) and the upper limit (UL) of a confidence interval for a **Poisson parameter** (the latter value representing both the mean and the variance of a distribution).

$$LL = \frac{\chi^2_{(\alpha/2), df_{LL}}}{2} \qquad \text{(Equation 9.23)}$$

$$UL = \frac{\chi^2_{[1-(\alpha/2)], df_{UL}}}{2} \qquad \text{(Equation 9.24)}$$

Where: $\chi^2_{\alpha/2}$ represents the tabled critical value in the χ^2 distribution for $df_{LL} = 2x$ below which a proportion (percentage) equal to $\alpha/2$ of the cases falls. If the proportion (percentage) of the distribution that falls within the confidence interval is subtracted from 1 (100%), it will equal the value of α.

$\chi^2_{[1-(\alpha/2)]}$ represents the tabled critical value in the χ^2 distribution for $df_{UL} = 2(x+1)$ below which a proportion (percentage) equal to $[1 - (\alpha/2)]$ of the cases falls.

The application of Equations 9.23 and 9.24 in computing a 95% confidence interval is demonstrated below for Example 9.19.

Example 9.19 *Assume that the Poisson distribution describes the frequencies of a specific microorganism in seawater. Nine of the microorganism are detected in a one liter sample of seawater sent to a laboratory. Compute a 95% confidence interval for the Poisson parameter.*

Since the value of x in Example 9.19 is 9, we can calculate that $df_{LL} = 2x = 2(9) = 18$ and $df_{UL} = 2(x+1) = 2(9+1) = 20$. Since if the proportion (percentage) of the distribution which falls within the confidence interval is subtracted from 1 (100%), it will equal the value of α, then $\alpha = 100\% - 95\% = 5\%$. Employing **Table A4** in the **Appendix** to compute the lower limit for $df_{LL} = 2x = 2(9) = 18$, we must determine the critical value for $\chi^2_{\alpha/2} = \chi^2_{(.05/2)} = \chi^2_{.025}$. The latter value in **Table A4** is $\chi^2_{.025} = 8.23$. In computing the upper limit, for $df_{UL} = 20$, we must determine the critical value for $\chi^2_{[1-(\alpha/2)]} = \chi^2_{[1-(.05/2)]} = \chi^2_{.975}$. The latter value in **Table A4** is $\chi^2_{.975} = 34.17$. Through use of Equations 9.23 and 9.24, we can now determine that the lower and upper limits of the 95% confidence interval are $LL = 8.23/2 = 4.115$ and $UL = 34.17/2 = 17.085$. The latter result indicates the researcher can be 95% confident the range of values 4.115 to 17.085 includes the actual value of the Poisson parameter in the underlying population. Rosner (1995, p. 180) notes that (as is the case for Example 9.19) a Poisson confidence interval will not be symmetric about the value of x, unless the latter value is very large. Rosner (1995, p. 653) also provides a table containing the 90%, 95%, 98%, 99%, and 99.8% Poisson confidence intervals for most values of $x \leq 50$.

Test 9c Test for comparing two Poisson counts Assume that a researcher has two counts X_1 and X_2, each of which is known to have been obtained from a Poisson distribution, and wants to determine whether or not the mean values of the Poisson populations they were derived from are equal. Procedures for evaluating the aforementioned question attributed to Anscombe (1948), Best (1975), Detre and White (1970), and Przyborowski, J. and Wilenski, H. (1940) are described in Parker (1979, pp. 34–35) and Zar (1999; pp. 582–583). Equation 9.25, which employs the normal distribution, can be employed to provide a good approximation to evaluate the difference between two counts obtained from Poisson distributions. The null hypothesis evaluated with the latter equation is H_0: $\mu_1 = \mu_2$ (The mean of the Poisson population Count 1 is derived from equals the mean of the Poisson population Count 2 is derived from.). The nondirectional and directional alternative hypotheses evaluated are respectively H_1: $\mu_1 \neq \mu_2$ (The mean of the Poisson population Count 1 is derived from does not equal the mean of the Poisson population Count 2 is derived from.); H_1: $\mu_1 > \mu_2$ (The mean of the Poisson population Count 1 is derived from is greater than the mean of the Poisson population Count 2 is derived from.); H_1: $\mu_1 < \mu_2$ (The mean of the Poisson population Count 1 is derived from is less than the mean of the Poisson population Count 2 is derived from.). Equation 9.25 is employed below to evaluate Example 9.20.

Example 9.20 *Assume that the Poisson distribution describes the frequencies of a specific microorganism in seawater. Nine of the microorganisms are detected in a one liter sample of seawater sent to a laboratory. The following week 20 of the microorganisms are detected in another one liter sample of seawater sent to the same laboratory. Can the laboratory conclude that the two samples of microorganisms were derived from the same population based on the count data provided?*

$$z = \frac{X_1 - X_2}{\sqrt{X_1 + X_2}} \qquad \text{(Equation 9.25)}$$

When Equation 9.25 is employed to evaluate the data from Example 9.20, the value $z = -2.04$ is obtained.

$$z = \frac{9 - 20}{\sqrt{9 + 20}} = -2.04$$

The same guidelines employed in evaluating the z **test for a population proportion (Test 9a)** described earlier in this chapter are employed to evaluate the z value obtained with Equation 9.25. Specifically: a) If the nondirectional alternative hypothesis H_1: $\mu_1 \neq \mu_2$ is employed, the null hypothesis can be rejected if the obtained absolute value of z is equal to or greater than the tabled critical two-tailed value at the prespecified level of significance; b) If the directional alternative hypothesis H_1: $\mu_1 > \mu_2$ is employed, the null hypothesis can be rejected if the sign of z is positive and the value of z is equal to or greater than the tabled critical one-tailed value at the prespecified level of significance; c) If the directional alternative hypothesis H_1: $\mu_1 < \mu_2$ is employed, the null hypothesis can be rejected if the sign of z is negative and the absolute value of z is equal to or greater than the tabled critical one-tailed value at the prespecified level of significance.

Using the above guidelines, if the nondirectional alternative hypothesis H_1: $\mu_1 \neq \mu_2$ is employed the null hypothesis can be rejected at the .05 level, since the absolute value $z = 2.04$ is greater than the tabled critical two-tailed value $z_{.05} = 1.96$, but it cannot be rejected at the .01 level since $z = 2.04$ is less than the tabled critical two-tailed value $z_{.01} = 2.58$. If the directional alternative hypothesis H_1: $\mu_1 < \mu_2$ is employed the null hypothesis can also be rejected, but again only at the .05 level. The latter is the case since the sign of z is negative and the absolute value $z = 2.04$ is greater than the tabled critical one-tailed value $z_{.05} = 1.65$, but less than the tabled critical one-tailed value $z_{.01} = 2.33$. The null hypothesis cannot be rejected if the directional alternative hypothesis H_1: $\mu_1 > \mu_2$ is employed, since the sign of z is negative. Based on the outcome of the study the researcher can conclude it is unlikely that the two samples of microorganisms came from the same Poisson population.

If the term $\pm. 5$ is omitted from the numerator of Equation 9.26, it is equivalent to Equation 9.25. The .5 in the numerator of Equation 9.26 is a correction for continuity, which yields a slightly smaller absolute z value. (The notation $\pm .5$ indicates that .5 should be added to the difference obtained for the terms in the brackets if the latter has a negative value, and subtracted from the difference if it yields a positive value). Note that if the correction for continuity is employed the absolute value $z = 1.86$ is obtained, which is less than the tabled critical two-tailed value $z_{.05} = 1.96$. Under the latter conditions, the null hypothesis cannot be rejected.

(Equation 9.26)

$$z = \frac{\left(X_1 - \dfrac{(X_1 + X_2)}{2}\right) \pm .5}{\sqrt{\dfrac{X_1 + X_2}{4}}} = \frac{\left(9 - \dfrac{(9 + 20)}{2}\right) - .5}{\sqrt{\dfrac{9 + 20}{4}}} = \frac{-5}{2.69} = -1.86$$

Equation 9.27 (attributed to Anscombe (1948)) is an alternative equation for evaluating the hypothesis under discussion. The z value obtained with Equation 9.27 will be close to (but most likely not identical) that obtained with Equation 9.25. Note that the value $z = -2.05$ obtained with Equation 9.27 leads to the identical conclusions that were reached when the data were evaluated with Equation 9.25.

(Equation 9.27)

$$z = \sqrt{2X_1 + .75} - \sqrt{2X_2 + .75} = \sqrt{(2)(9) + .75} - \sqrt{(2)(20) + .75} = -2.05$$

Evaluating goodness-of-fit for a Poisson distribution   There may be occasions when a researcher wants to evaluate the hypothesis that a set of data is derived from a Poisson distribution. Example 9.21, which is an extension of Example 9.17, will be employed to demonstrate how the latter hypothesis can be evaluated with the **chi-square goodness-of-fit test**.

Example 9.21   *The traffic bureau of a Midwestern city determines that the average number of accidents per day is 2. During a 300-day period the following number of accidents are recorded per day: a) 30 days there are 0 accidents; b) 90 days there is 1 accident; c) 89 days there are 2 accidents; d) 53 days there are 3 accidents; e) 30 days there are 4 accidents; f) 6 days there are 5 accidents; g) 2 days there are 6 accidents; and h) 7 or more accidents do not occur on any day during the 300-day period. Is the distribution of the data consistent with a Poisson distribution?*

The null and alternative hypotheses which are evaluated with the **chi-square goodness -of-fit test** in reference to Example 9.21 can either be stated in the form as presented in Section III of the latter test (i.e., H_0: $o_i = \varepsilon_i$ for all cells; H_i: $o_i \neq \varepsilon_i$ for at least one cell), or as follows.

Null hypothesis H_0: The sample is derived from a population with a Poisson distribution.

Alternative hypothesis H_1: The sample is not derived from a population with a Poisson distribution. This is a **nondirectional alternative hypothesis**.

The analysis of Example 9.21 with the **chi-square goodness-of-fit test** is summarized in Table 9.7. The latter table is comprised of $k = 7$ cells/categories, with each cell corresponding to a given number of accidents per day. The second column of the table contains the observed frequencies for the specified number of accidents. The expected frequency in Column 3 for each cell was obtained by employing Equation 8.1. Specifically the value 300, which represents the total number of days, is multiplied by the appropriate Poisson probability for a given number of accidents (which was previously computed in Example 9.17). The latter Poisson probabilities are as follows: $x = 0$ ($p = .1353$); $x = 1$ ($p = .2707$); $x = 2$ ($p = .2707$); $x = 3$ ($p = .1804$); $x = 4$ ($p = .0902$); $x = 5$ ($p = .0361$), and $x \geq 6$ ($p = .0120 + .0046 = .0166$). To illustrate the computation of an expected frequency, the value 40.59 is obtained for Row 1 (0 accidents) as follows: $E_i = (300)(.1353) = 40.59$.

Employing Equation 8.2, the value $\chi^2 = 8.72$ is computed for Example 9.21. Since there are $k = 7$ cells and no parameters are estimated (i.e., $w = 0$), employing Equation 8.12, the degrees of freedom for the analysis are $df = 7 - 1 - 0 = 6$. Employing **Table A4**, we determine that for $df = 6$ the tabled critical .05 and .01 values are $\chi^2_{.05} = 12.59$ and $\chi^2_{.01} = 16.81$. Since the computed value $\chi^2 = 8.72$ is less than both of the aforementioned values, the null hypothesis cannot be rejected. Thus, the analysis does not indicate that the data deviate significantly from a Poisson distribution.

Table 9.7   Chi-Square Summary Table for Example 9.21 (Poisson Analysis)

Cell/ Number of accidents	O_i	E_i	$(O_i - E_i)$	$(O_i - E_i)^2$	$\dfrac{(O_i - E_i)^2}{E_i}$
0	30	40.59	−10.59	112.15	2.76
1	90	81.21	8.79	77.26	.95
2	89	81.21	7.79	60.68	.74
3	53	54.12	−1.12	1.25	.02
4	30	27.06	2.94	8.64	.32
5	6	10.83	−4.83	23.33	2.15
6 or more	2	4.98	−2.98	8.88	1.78
	$\Sigma O_i = 300$	$\Sigma E_i = 300$	$\Sigma(O_i - E_i) = 0$		$\chi^2 = 8.72$

It should be noted that if, instead of stipulating the value $\mu = 2$ to represent the mean number of accidents, we computed the mean number of accidents from the sample data, the value of df is reduced by 1 (i.e., $df = 7 - 1 - 1 = 5$). The latter is the case since an additional degree of freedom must be subtracted for any parameter that is estimated. In point of fact, in Example 9.21 the sample data yield an average number of accidents equal to $\bar{X} = 1.96$. The latter value is obtained as follows: a) In each row of Table 9.7, multiply the value for the number of accidents in Column 1 by the observed frequency for that number of accidents in Column 2 (multiply 6 by 2 in the last row, since on the two days recorded there were six accidents); and b) Sum the eight products obtained in a). The latter sum will equal the total number of accidents (which comes out to 589), and if that value is divided by the total number of days (300) it yields the value $\bar{X} = 1.96$. The latter value will represent the best estimate of the population mean. Since the latter value is almost equal to $\mu = 2$ employed in the analysis, it would not result in a different conclusion.

It was noted earlier that when n is large and the value of π_1 is very small, the Poisson distribution provides a good approximation for the binomial distribution. In point of fact, the latter conditions apply to Example 9.21. If Example 9.21 is conceptualized within the framework of a binomial model, $n = 300$ and $\pi_1 = .0067$. Within the binomial model there are 300 trials, and on each trial there are two possible outcomes: **accident** versus **no accident**. The value of π_1 can be computed through use of Equation 9.1 as follows: a) Transpose the terms in the latter equation to solve for π_1; b) Since we know the mean of the distribution is $\mu = 2$ and $n = 300$, we obtain $\pi_1 = \mu/n = 2/300 = .0067$.

In point of fact, if the **chi-square goodness-of-fit test** is employed to evaluate the data for Example 9.21 for goodness-of-fit for a binomial distribution, the null hypothesis (which states that the data are derived from a binomial distribution) is supported. Table 9.8 summarizes the latter analysis. The following binomial probabilities are multiplied by 300 to get the expected frequency for each row of the table: $x = 0$ ($p = .1331$); $x = 1$ ($p = .2693$); $x = 2$ ($p = .2722$); $x = 3$ ($p = .1821$); $x = 4$ ($p = .0904$); $x = 5$ ($p = .0369$); and $x \geq 6$ ($p = .0161$). The latter probabilities were obtained through use of Equation 9.3 using the values $n = 300$, $\pi_1 = .0067$, $\pi_2 = .9933$, and the designated value of x. Note that the binomial probabilities are almost identical to the Poisson probabilities for the corresponding value of x.

Employing Equation 8.2, the value $\chi^2 = 8.52$ is computed (which is almost identical to $\chi^2 = 8.72$ computed earlier for goodness-of-fit for a Poisson distribution). Since $df = 7 - 1 = 6$, employing **Table A4** we determine that the tabled critical .05 and .01 values are $\chi^2_{.05} = 12.59$ and $\chi^2_{.01} = 16.81$. Since the computed value $\chi^2 = 8.52$ is less than both of the aforementioned values, the null hypothesis cannot be rejected. Thus, the analysis does not indicate the data deviate significantly from a binomial distribution. The above example illustrates the fact that there are circumstances when a Poisson distribution and binomial distribution will be so similar to one another, that a goodness-of-fit test will not be able to clearly discriminate between the two distributions.

Table 9.8 Chi-Square Summary Table for Example 9.21 (Binomial Analysis)

Cell/ Number of Accidents	O_i	E_i	$(O_i - E_i)$	$(O_i - E_i)^2$	$\dfrac{(O_i - E_i)^2}{E_i}$
0	30	39.93	−9.93	98.60	2.47
1	90	80.79	9.21	84.82	1.05
2	89	81.66	7.34	53.88	.66
3	53	54.63	−1.63	2.66	.05
4	30	27.12	2.88	8.29	.31
5	6	11.07	−5.07	25.70	2.32
6 or more	2	4.83	−2.83	8.01	1.66
	$\Sigma O_i = 300$	$\Sigma E_i = 300$	$\Sigma(O_i - E_i) = 0$		$\chi^2 = 8.52$

e. The exponential distribution The **exponential distribution** is a continuous probability distribution that is closely related to the Poisson distribution. Whereas the Poisson distribution counts the frequency of events which occur in a certain time interval, the exponential distribution measures the time intervals between events. Canavos and Miller (1995, p. 247) note that if you have a Poisson process in which independent occurrences take place at a constant average rate of λ occurrences per unit of time, then the length of time between successive occurrences can be shown to be a continuous random variable that has an exponential distribution with an average length equal to $1/\lambda$. As is the case with the Poisson distribution, the exponential distribution has a **rate parameter** (λ), which is basically the same as that employed for the Poisson distribution. Thus, λ represents the average rate with which random occurrences take place per time unit — e.g., deaths per week, failures per hour, arrivals per minute. Whereas the Poisson distribution is employed to model the number of arrivals or occurrences of some event per interval of time, the exponential distribution is employed to model the time between successive arrivals or occurrences. Among the phenomena the exponential distribution can be employed to model are queueing problems such as the average waiting time for customers in line to be served (e.g., time between arrivals at a supermarket checkout, bank teller's station, toll booth, etc.), time intervals between phone calls, random failures such as time between deaths of a member of a species, time intervals for certain types of random natural deaths among humans (e.g., deaths among very old or extremely ill people), time intervals for certain types of unusual random deaths among humans (e.g., people getting struck by falling objects or getting killed by wild animals), and time intervals between failures of mechanical equipment. Note that although the rate of the aforementioned events may vary depending upon the time of day, month, etc., use of an exponential model assumes a constant rate during the specified time period. The exponential distribution can also be used to model the expected life span of organisms as well as the lifetime of electrical and mechanical components such as batteries, bearings and computer chips (Specific examples are presented in sources such as Canavos and Miller (1995, p. 247), Pitman (2003, pp. 281–283), Selvin (1995, p. 415), and Sinich (1996, p.290).)

The probability density function of the exponential distribution is given by $f(t) = \lambda e^{-\lambda t}$ where both λ and t must be greater than zero. The exact shape of an exponential distribution will depend on the values of λ and t in the probability density function equation. Probabilities are computed by determining the area under the curve between two values of t. Since an exponential distribution is skewed to the right (i.e., positively skewed), the majority of values assumed by an exponentially distributed random variable t will fall below the mean. Figure 9.1 represents a graph of an exponential density function with $\lambda = .5$

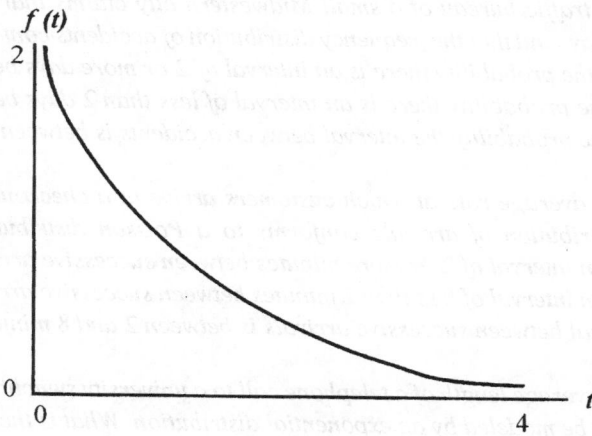

Figure 9.1 Exponential Distribution with $\lambda = .5$

To illustrate an example of a variable modeled by the exponential distribution, assume that time between successive occurrences of accidents at a busy intersection has a rate parameter of $\lambda = 3$ accidents per day. The use of the exponential model assumes that the value of rate parameter is constant, and does not vary according to the day of the week or the time of year. It should also be noted that although in the above example the unit of time employed is a single day, a rate parameter can be expressed in terms of any time unit such as a week (e.g., $\lambda = 3$

accidents per week), an hour (e.g., $\lambda = 3$ accidents per hour), or even an interval such as $\lambda = 3$ accidents every 15 minutes.

The mean/expected value and standard deviation of an exponential distribution can be computed with Equations 9.28 and 9.29. Note that since $\mu = 1/\lambda$, $\lambda = 1/\mu$. Readers should take note of the fact that in an exponential distribution, the rate parameter λ represents the average rate of occurrence of an event over a time period, while μ (the average of the distribution) represents the average length of time between events or the average survival time. Additionally, it should be emphasized that the mean and standard deviation of an exponential distribution are identical, and that the value of λ is equal to the mean of the Poisson distribution which could be employed to model the frequency of events that occur during the specified time unit.

$$\mu = \frac{1}{\lambda} \qquad \text{(Equation 9.28)}$$

$$\sigma = \frac{1}{\lambda} \qquad \text{(Equation 9.29)}$$

Since $\lambda = 3$ accidents per day in the illustration above, Equations 9.23 and 9.24 yield the value $\mu = \sigma = 1/\lambda = 1/3 = .33$. Thus we can state that the average length of time between accidents is $1/\lambda = 1/3$ — more specifically, $1/3^{rd}$ of a day, which equals $(1/3)(24 \text{ hours}) = 8$ hours. In addition to the mean and standard deviation, the median of an exponential distribution can be computed as follows: $Mdn = \ln 2/\lambda$, which in this instance yields $Mdn = (\ln 2)/3 = .231$ days, which is $(.231)(24 \text{ hours}) = 5.544$ hours or 5 hours and 33 seconds between accidents. The mode of an exponential distribution will always be zero.

Examples 9.22–9.27 represent situations in which an exponential model might be employed, and for each of the examples exponential probabilities are easily computed through use of the exponential key (e^x) found on most calculators.

Example 9.22   *The traffic bureau of a small Midwestern city claims that an average of 2 accidents occur per day, and that the frequency distribution of accidents conforms to a Poisson distribution. What is the probability there is an interval of 2 or more days between successive accidents? What is the probability there is an interval of less than 2 days between successive accidents? What is the probability the interval between accidents is between 2 and 8 days?*

Example 9.23   *The average rate at which customers arrive at a checkout counter is 2 per minute, and the distribution of arrivals conforms to a Poisson distribution. What is the probability there is an interval of 2 or more minutes between successive arrivals? What is the probability there is an interval of less than 2 minutes between successive arrivals? What is the probability the interval between successive arrivals is between 2 and 8 minutes?*

Example 9.24   *The average length of a telephone call to a university switchboard is 2 minutes and time lengths can be modeled by an exponential distribution. What is the probability a call will last 2 or more minutes? What is the probability a call will last less than 2 minutes? What is the probability a call will last between 2 and 8 minutes?*

Example 9.25   *The average life span of an engine is 2 years and failures can be modeled by an exponential distribution. What is the probability the engine will survive at least 2 years? What is the probability the engine will survive less than 2 years? What is the probability the engine will survive between 2 and 8 years?*

Example 9.26 *Assume that the average lifetime of a particular insect species is 2 hours and that the lifetime distribution is exponential. A researcher wants to answer the following questions: a) What is the probability a member of the species will live at least 2 hours? b) What is the probability a member of the species will live less than 2 hours? c) What is the probability a member of the species will survive between 2 and 8 hours?*

The questions posed in Examples 9.22–9.26 can all be answered through use of the exponential distribution. In all five of the examples the following three questions are asked and answered as follows.

1) *What is the probability there is an interval of t units of time (e.g., days, minutes) or more between events (where t = 2) or what is the probability an event, object or organism will survive beyond t units of time (e.g., minutes, years, hours)?* The probability there is an interval of t time units or more between events or that an event, object or organism will survive more than t time units is *computed as follows:* $P(x \geq t) = e^{-\lambda t}$.

2) *What is the probability there is an interval of less than t units of time between events (where t = 2) or what is the probability an event, object or organism will survive less than t units of time?* The probability there is an interval less than t time units between events or that an event, object or organism will survive less than t time units is computed as follows: $P(x < t) = 1 - P(x \geq t) = 1 - e^{-\lambda t}$.

3) *What is the probability the interval between events is between s and t units of time (where s = 2 and t = 8) or what is the probability that an event, object or organism will survive between s and t units of time?* The probability there is an interval between s and t time units between events or that an event, object or organism will survive between s and t time units is $P(s \leq x \leq t) = P(x \geq s) - P(x \geq t) = e^{-\lambda s} - e^{-\lambda t}$. The identical value for $P(s \leq x \leq t)$ can also be computed as follows: $P(x < t) - P(x < s)$.

In the case of Examples 9.22–9.23, the average number of events (i.e., $\mu = 2$ accidents in the case of Example 9.22 and $\mu = 2$ arrivals in the case of Example 9.23) represents the mean of a Poisson distribution. Since in a Poisson distribution the values of μ and λ are equal, $\lambda = 2$. Since the time interval of time between events can be modeled with an exponential distribution, the rate parameter for the latter distribution will be the value $\lambda = 2$ specified for the Poisson distribution, and the mean of the exponential distribution will be $\mu = 1/\lambda = \frac{1}{2} = .5$ The questions posed in Examples 9.22 and 9.23 can be answered as follows: 1) Since $t = 2$ and $P(x \geq t) = e^{-\lambda t}$, the probability there is an interval of 2 or more time units (days in the case of Example 9.22 and minutes in the case of Example 9.23) between events (i.e., accidents in the case of Example 9.22 and arrivals in the case of Example 9.23) is $P(x \geq 2) = e^{(-2)(2)} = .018$; 2) Since $P(x < t) = 1 - P(x \geq t) = 1 - e^{-\lambda t}$, the probability there is an interval of less than 2 time units between events is $P(x < 2) = 1 - e^{-\lambda t} = 1 - e^{(-2)(2)} = 1 - .018 = .982$; 3) Since $P(s \leq x \leq t) = P(x \geq s) - P(x \geq t)$, the probability the time interval between successive events will be between $s = 2$ and $t = 8$ time units is $P(2 \leq x \leq 8) = P(x \geq 2) - P(x \geq 8) = e^{(-2)(2)} - e^{(-2)(8)} = .018 - .00000013 = .017999887$.

In the case of Examples 9.24–9.26, the average survival time of 2 time units (e.g., $\mu = 2$ minutes in Example 9.24, $\mu = 2$ years in Example 9.25, and $\mu = 2$ hours in Example 9.26) represents the mean of an exponential distribution. Since $\mu = 2$, the value of λ can be computed to be $\lambda = 1/\mu = \frac{1}{2} = .5$. The latter value represents a rate of .5 calls per minute in the case of Example 9.24, .5 failures per year in the case of Example 9.25, and .5 deaths per hour in the case of Example 9.26. (Selvin (1995, p. 410) notes that within the context of survival studies such as Examples 9.25 and 9.26, the rate parameter is sometimes referred to as the **hazard rate**.) The questions posed in Examples 9.24–9.26 can be answered as follows: 1) Since $t = 2$ and $P(x \geq t) = e^{-\lambda t}$, the probability a phone call will last at least 2 minutes or that a machine engine will survive at least 2 years or that a member of the species will live at least 2 hours is $P(x \geq 2) =$

$e^{(-.5)(2)} = .368$; 2) Since $P(x < t) = 1 - P(x \geq t) = 1 - e^{-\lambda t}$, the probability a phone call will last less than 2 minutes or that a machine engine will survive less than 2 years or that a member of the species will live less than 2 hours is $P(x < 2) = 1 - e^{-\lambda t} = 1 - e^{(-.5)(2)} = 1 - .368 = .632$; 3) Since $P(s \leq x \leq t) = P(x \geq s) - P(x \geq t)$, the probability that the length of a phone call or the life of the machine or the life of a member of the species will be between $s = 2$ and $t = 8$ time units (i.e., minutes, years, hours) is $P(2 \leq x \leq 8) = P(x \geq 2) - P(x \geq 8) = e^{(-.5)(2)} - e^{(-.5)(8)} = .368 - .018 = .35$.

Example 9.27 *In the case of Example 9.26, if a member of the species has survived for 2 hours what is the probability it will die within the next minute?*

Pitman (1993, p. 281) states that since the failure rate remains constant at all times, given the fact that the organism has survived up to 2 hours, the chance of dying within the next minute is computed as follows: $(\lambda)(1/60) = (.5)(1/60) = .0083$. Note that $\lambda = .5$ represents the rate of dying per hour, and the value 1/60 is used since it is the proportion of an hour stipulated in the question.

Selvin (1995, pp. 410–411) notes that for certain situations an exponential distribution can provide an effective and simple model for survival analysis (which is discussed in greater detail in Section IX (the **Addendum**) of the **Mann–Whitney *U* test (Test 12)**). If the survival rate of a whole population is plotted, an exponential survival function begins with a probability of 1 and proceeds to decrease as time progresses. Selvin emphasizes the fact that although an exponential function does not accurately describe human mortality over the course of a life span, it can serve as an accurate descriptor for specific situations or over short time periods. An important property of an exponential function is that it is **memoryless**, which means that the probability of failure or death is independent of the amount of time that has elapsed — i.e., given that a person or object has survived until time *s*, the likelihood of surviving to a further time *t* is the same as surviving to time *t* in the first place. If the latter is applied to human survival, a person would not age, and thus a person's likelihood of dying would be no greater when the individual was 80 than when he or she was 30. Pitman (1993, p. 278) notes that although things such as atoms and electrical components possess the property of memoryless (and can thus be modeled by the exponential distribution), they are exceptional in this respect, since most objects or life forms age, and by virtue of the latter cannot have their life spans modeled exponentially.

f. The matching distribution The **matching distribution** is a discrete probability distribution which can be employed to evaluate certain experimental situations. In order to describe the model for the matching distribution, which is based on **sampling without replacement**, let us assume that we have two identical decks of cards with n_1 cards in Deck 1 and n_2 cards in Deck 2, with $n_1 = n_2$. If we conduct an experiment that is comprised of *n* trials, and on each trial we randomly select one card from Deck 1 and one card from Deck 2, the probability of obtaining *x* matches between cards in the two decks is defined by Equation 9.30. Note that the terms enclosed in the brackets of Equation 9.30 constitute a series (which is a sequence of numbers that are added to and/or subtracted from one another).

(Equation 9.30)

$$P(x) = \frac{1}{x!}\left[\frac{1}{0!} - \frac{1}{1!} + \frac{1}{2!} - \frac{1}{3!} \ldots \pm \frac{1}{(n-x)!}\right]$$

Example 9.28 will be used to illustrate how Equation 9.27 is employed to compute probabilities for the matching distribution.

Example 9.28 *A subject claims that he has extrasensory ability. To test the subject the following five playing cards are randomly arranged face down on a table: Ace of spades; King of spades; Queen of spades; Jack of spades; Ten of spades. The subject is given a set of five cards with identical face values as those on the table, and told to place each of the cards he is holding on top of the corresponding card on the table. Is the subject's performance statistically significant at the .05 level if he matches two of the five cards correctly?*

In this experiment there are $n = 5$ trials, and we want to determine the probability of obtaining two or more matches. The null hypothesis to be evaluated is that the subject will perform within chance expectation (or to say it another way, the performance of the subject will not suggest extrasensory ability). The alternative hypothesis that will be evaluated is the directional/one-tailed alternative hypothesis which states the subject will perform at an above chance level (or to say it another way, the performance of the subject suggests extrasensory ability).

Equation 9.30 is employed below to compute the probability of obtaining $x = 0$, $x = 1$, $x = 3$, and $x = 5$ matches. Note that a subject cannot obtain 4 matches without obtaining 5 matches, since if there are $(n - 1)$ matches there must be n matches. The probabilities for all values of x sum to 1 (there is a slight discrepancy due to rounding off error).

$$P(x = 0) = \frac{1}{0!}\left[\frac{1}{0!} - \frac{1}{1!} + \frac{1}{2!} - \frac{1}{3!} + \frac{1}{4!} - \frac{1}{5!}\right] = .3664$$

This result tells us the likelihood of a subject obtaining 0 matches is .3664.

$$P(x = 1) = \frac{1}{1!}\left[\frac{1}{0!} - \frac{1}{1!} + \frac{1}{2!} - \frac{1}{3!} + \frac{1}{4!}\right] = .3747$$

This result tell us the likelihood of a subject obtaining 1 match is .3747.

$$P(x = 2) = \frac{1}{2!}\left[\frac{1}{0!} - \frac{1}{1!} + \frac{1}{2!} - \frac{1}{3!}\right] = .1665$$

This result tells us the likelihood of a subject obtaining 2 matches is .1665.

$$P(x = 3) = \frac{1}{3!}\left[\frac{1}{0!} - \frac{1}{1!} + \frac{1}{2!}\right] = .0835$$

This result tells us the likelihood of a subject obtaining 3 matches is .0835.

$$P(x = 4) = \frac{1}{4!}\left[\frac{1}{0!} - \frac{1}{1!}\right] = 0$$

This result tells us the likelihood of a subject obtaining 4 matches is 0, since there cannot be 4 matches without 5 matches.

$$P(x = 5) = \frac{1}{5!}\left[\frac{1}{0!}\right] = .0083$$

This result tells us the likelihood of a subject obtaining 5 matches is .0083.

To evaluate the subject's score of $x = 2$ matches, we have to determine the probability of obtaining two or more matches (i.e., $p(x \geq 2)$ when $n = 5$. The latter value is computed by adding the probabilities for 2, 3, and 5 matches computed above, since all of those values are equal to or greater than a score of 2 matches. Thus, $.1665 + .0835 + .0083 = .2583$. Since the obtained value $p = .2583$ is greater than the value $\alpha = .05$, we retain the null hypothesis. Thus, we cannot conclude the subject exhibits evidence of extrasensory ability.

It turns out that the probabilities computed for the matching distribution are quite close to the probabilities which will result if the problem under discussion is reconceptualized within the framework of the **sampling with replacement model**. If the latter model is used, the appropriate distribution to employ to compute probabilities for the number of matches is the binomial distribution. To illustrate the use of the sampling with replacement model, let us assume that after the five test cards are placed face down on the table, the subject is told to randomly select one card from his own identical deck of five cards, and see if it matches the first card which is face down on the table. The subject then puts the card he selected from his own five card deck back into his deck, and randomly selects a second card and sees if it matches the second card that is face down on the table. The subject then puts the card he selected from his own five card deck back into his deck and continues the same process until he has attempted to randomly match a card from his complete five card deck with each of the five cards which are face down on the table. As described, this variant of the experiment involves $n = 5$ trials, and on any given trial there is a one in five chance of the subject being correct. Thus, we are dealing with a binomially distributed variable, where $n = 5$, $\pi_1 = 1/5 = .2$, and $\pi_2 = 4/5 = .8$. To determine the probability of obtaining 0, 1, 2, 3, 4, or 5 matches, we employ the section in **Table A6** for $n = 5$ and $\pi_1 = .2$. (Note that when the sampling with replacement model is employed, it is possible to obtain 4 matches.) The binomial probabilities obtained from **Table A6** are as follows: $x = 0$ ($p = .3277$); $x = 1$ ($p = .4096$); $x = 2$ ($p = .2048$); $x = 3$ ($p = .0512$); $x = 4$ ($p = .0064$); $x = 5$ ($p = .0003$). Note that the latter values are reasonably close to the probabilities for the matching distribution which were previously computed for the original problem.

Larsen and Marx (1985, pp. 244–245; p. 275) note that with respect to the matching distribution, regardless of the value of n, both the expected number of matches and the expected variance for the number of matches will always equal 1. A more detailed discussion of the matching distribution can be found in Feller (1968).

2. Conditional probability, Bayes' theorem, and Bayesian statistics and hypothesis testing

This chapter has focused on the computation of probabilities associated with the value of a variable that is known to be derived from a discrete probability distribution. In discussing inferential statistical procedures, the use of such probabilities has been described within the context of what is commonly referred to as the **classical approach** to statistical inference. Within the framework of the classical approach to statistical inference, the only evidence employed in the decision making process are data which are derived from a sample. An alternative approach, known as **Bayesian statistics**, utilizes not only sample data to compute probability values that are employed within the framework of statistical inference, but also allows a researcher to utilize preexisting information regarding probabilities which is available from other sources. This latter approach to statistical inference is named after the Reverend Thomas Bayes (1702-1761), an eighteenth century English Clergyman who in 1763 (in a posthumously published paper) stated a general rule for computing **conditional probabilities**.

Conditional probability   As noted in the **Introduction**, a **conditional probability** is the probability that an event will occur, given the fact it is already known that another event has occurred. Specifically, the conditional probability that an event, to be designated A, will occur, given the fact it is already known that another event, to be designated *B* has already occurred, is symbolically expressed as follows: $P(A/B)$. The latter conditional probability is commonly referred to as a **posterior** or *a posteriori* **probability**, since the value derived for the conditional probability is computed through use of one or more probability values which are already known or estimated from preexisting information. The latter known probabilities are referred to as **prior** or *a priori* **probabilities**.

Bayes' theorem   **Bayes' theorem** is an equation which allows for the computation of conditional probabilities. In order to describe **Bayes' theorem**, assume that we have two sets of events. In one set there are *n* events to be identified as $A_1, A_2, ..., A_n$. In the second set there are two events to be identified as *B+* and *B–*. **Bayes' theorem** states the probability that A_j (where $1 \le j \le n$) will occur, given that it is known that *B+* has occurred, is determined with Equation 9.31 (which is the same as Equation I.43 in the **Introduction**). To determine the likelihood that A_j will occur, given that it is known that *B–* has occurred (i.e., $P(A_j/B-)$), the value *B–* is employed in the numerator and denominator of Equation 9.31 in place of *B+*.

$$P(A_j/B+) = \frac{P(B+/A_j)P(A_j)}{\sum_{i=1}^{n} P(B+/A_i)P(A_i)}$$ **(Equation 9.31)**

The most common application of Equation 9.31 involves its simplest form, in which the first set is comprised of the two events A_1 and A_2, and the second set is comprised of the two events *B+* and *B–*. In the latter case, the equation for **Bayes' theorem** in reference to event A_1 (i.e., $P(A_1/B+)$), becomes Equation 9.32 (which is the same as Equation I.44 in the **Introduction**). Equation 9.33 (which is the same as Equation I.45 in the **Introduction**) is employed to compute the conditional probability $P(A_2/B+)$.

$$P(A_1/B+) = \frac{P(B+/A_1)P(A_1)}{P(B+/A_1)P(A_1) + P(B+/A_2)P(A_2)}$$ **(Equation 9.32)**

$$P(A_2/B+) = \frac{P(B+/A_2)P(A_2)}{P(B+/A_1)P(A_1) + P(B+/A_2)P(A_2)}$$ **(Equation 9.33)**

Equation 9.34 (from which Equation 9.32 can be algebraically derived and which is the same as Equation I.42 in the **Introduction** — see Hays and Winkler (1971, pp. 84–85)) is another way of expressing the relationship described by Equation 9.32.

$$P(A_1/B+) = \frac{P(A_1 \cap B+)}{P(B+)}$$ **(Equation 9.34)**

Examples 9.29 and 9.30 will be employed to illustrate the application of **Bayes' theorem**.

Example 9.29   *Assume that 1% of the people in a population contract a specific disease. If a blood test is administered to an individual who has the disease, there is a 90% likelihood the test will yield a positive result — i.e., indicate that the person is sick. If the blood test is administered to an individual who does not have the disease, there is a 98% likelihood the test will yield a* *negative result — i.e., indicate that the person is healthy. What is the probability that a person is sick, given the fact that the person's blood test yields a positive result?*

Example 9.30 *Assume that 2% of the people in a population are guilty of committing a specific crime. If a polygraph test is administered to an individual who is guilty of the crime, there is a 99% likelihood the test will yield a positive result — i.e., indicate that the person is guilty. If the test is administered to an individual who is innocent, there is only a 75% likelihood the test will yield a negative result — i.e., indicate that the person is innocent. What is the probability that a person is guilty, given the fact that the person's polygraph test yields a positive result?*

At this point it should be noted that within the context of the test protocol described in Examples 9.29/9.30, the term **positive** is employed to identify a person who has a specific condition which a diagnostician is trying to identify. In the case of Example 9.29, the diagnostician is attempting to identify individuals who are **sick**. In the case of Example 9.30, the diagnostician is attempting to identify individuals who are **guilty**. Thus the term **positive** is employed to refer to individuals who are **sick** (in Example 9.29) and **guilty** (in Example 9.30). Since the term **negative** is employed to refer to individuals who represent the opposite conditions, individuals who are **healthy** (in Example 9.29) and **innocent** (in Example 9.30) represent **negatives**.

Within the framework of our discussion, there are four diagnostic decisions that are possible with respect to an individual. Two of the decisions will be correct and two will be incorrect. Specifically, the four decisions are as follows: a) A **true positive** is an individual who is diagnosed as having the condition in question and, in fact, has the condition. Thus in Example 9.29, a person the blood test identifies as sick who, in fact, is sick will represent a **true positive**. In Example 9.30, a person the polygraph test identifies as guilty who, in fact, is guilty will represent a **true positive**; b) A **true negative** is an individual who is diagnosed as not having the condition in question and, in fact, does not have the condition. Thus, in Example 9.29, a person the blood test identifies as healthy who, in fact, is healthy will represent a **true negative**. In Example 9.30, a person the polygraph test identifies as innocent who, in fact, is innocent will represent a **true negative**; c) A **false positive** is an individual who is diagnosed as having the condition in question and, in fact, does not have the condition. Thus in Example 9.29, a person the blood test identifies as sick who, in fact, is healthy will represent a **false positive**. In Example 9.30, a person the polygraph test identifies as guilty who, in fact, is innocent will represent a **false positive**; d) A **false negative** is an individual who is diagnosed as not having the condition in question and, in fact, does have the condition. Thus in Example 9.29, a person the blood test identifies as healthy who, in fact, is sick will represent a **false negative**. In Example 9.30, a person the polygraph test identifies as innocent who, in fact, is guilty will represent a **false negative**.

Note that the labels employed for the above described four decision categories reflect the following. The terms **positive** and **negative** refer to the diagnostic decision indicated by the test, whereas the terms **true** and **false** refer to whether or not the test result is **correct (true)** or **incorrect (false)**. Thus, a **true positive** and a **true negative** represent **correct/true** diagnostic decisions, whereas a **false positive** and a **false negative** represent **incorrect/false** diagnostic decisions.

Within the framework of medical diagnosis, the probability that a person tests positive for a disease, given the person, in fact, has the disease is commonly referred to as the **sensitivity** of the test. The latter can be represented by the **true positive** rate for the test. The probability that a person tests negative for a disease, given the person, in fact, does not have the disease is commonly referred to as the **specificity** of the test. The latter can be represented by the **true**

negative rate for the test. Among others, Rosner (1995, 2000) notes that in order for a diagnostic test to be an effective indicator of the presence of a disease/condition, it is important that the test be high in both sensitivity (i.e., a high rate of **true positives**) and specificity (i.e., a high rate of **true negatives**).This criterion appears to be met in the case of Example 9.29, where the **sensitivity/true positive rate** is 90% and the **specificity/true negative rate** is 98%. However, in the case of Example 9.30, although the **sensitivity/true positive rate** of 99% is high, the **specificity/true negative rate** of 75% would be viewed as less than ideal by many diagnosticians.

In the case of Example 9.29, A_1 will be employed to represent a person who is sick and A_2 to represent a person who is healthy. $B +$ will be employed to represent a positive blood test — i.e., the result of the test indicates the person is sick. $B-$ will be employed to represent a negative blood test — i.e., the result of the test indicates the person is healthy. In the case of Example 9.30, A_1 will be employed to represent a person who is guilty and A_2 to represent a person who is innocent. $B +$ will be employed to represent a positive result on the polygraph test — i.e., the result of the test indicates the person is guilty. $B-$ will be employed to represent a negative result on the polygraph test — i.e., the result of the test indicates the person is innocent. Based on the information that has been provided for Examples 9.29 and 9.30, we can say the following.

Example 9.29

$$P(Sick) = P(A_1) = .01 \quad P(Healthy) = P(A_2) = 1 - P(A_1) = 1 - .01 = .99$$

Example 9.30

$$P(Guilty) = P(A_1) = .02 \quad P(Innocent) = P(A_2) = 1 - P(A_1) = 1 - .02 = .98$$

The above noted probability values represent **prior probabilities**, since they are known prior to computing the probabilities asked for in each of the examples. The values $P(A_1) = .01$ (for Example 9.29) and $P(A_1) = .02$ (for Example 9.30) can also be referred to as **baserates**. The **baserate** of an event is the proportion of times the event occurs within a population (or the probability that the event will occur). Within the context of our examples, the baserate of an event corresponds to the proportion of positives in the population. Thus, in Example 9.29 the baserate for a person being sick is 1%, and in Example 9.30 the baserate for a person being guilty is 2%.[18] By multiplying the baserate of an event by the total number of events/people in a population, one can compute the number of events/people in the population which represent that event.

In Examples 9.29 and 9.30 we have also been provided with the following conditional probabilities (which also represent **prior probabilities**).

Example 9.29

$$P(B+/A_1) = P(Positive\ blood\ test/Sick) = .90$$
$$P(B-/A_1) = P(Negative\ blood\ test/Sick) = 1 - .90 = .10$$
$$P(B-/A_2) = P(Negative\ blood\ test/Healthy) = .98$$
$$P(B+/A_2) = P(Positive\ blood\ test/Healthy) = 1 - .98 = .02$$

Example 9.30

$$P(B+/A_1) = P(Positive\ polygraph\ test/Guilty) = .99$$
$$P(B-/A_1) = P(Negative\ polygraph\ test/Guilty) = 1 - .99 = .01$$
$$P(B-/A_2) = P(Negative\ polygraph\ test/Innocent) = .75$$
$$P(B+/A_2) = P(Positive\ polygraph\ test/Innocent) = 1 - .75 = .25$$

The above noted conditional probabilities indicate the following: a) In the case of Example 9.29, $P(B+/A_1) = .90$ is the probability of obtaining a positive blood test, given the fact that a person is sick. In the case of Example 9.30, $P(B+/A_1) = .99$ is the probability of obtaining a positive polygraph test, given the fact that a person is guilty. The values noted for $P(B+/A_1)$ represent the **true positive rate** or **sensitivity** of the tests; b) In the case of Example 9.29, $P(B-/A_1) = .10$ is the probability of obtaining a negative blood test, given the fact that a person is sick. In the case of Example 9.30, $P(B-/A_1) = .01$ is the probability of obtaining a negative polygraph test, given the fact that a person is guilty. The values noted for $P(B-/A_1)$ represent the **false negative rate** of the tests; c) In the case of Example 9.29, $P(B-/A_2) = .98$ is the probability of obtaining a negative blood test, given the fact that a person is healthy. In the case of Example 9.30, $P(B-/A_2) = .75$ is the probability of obtaining a negative polygraph test, given the fact that a person is innocent. The values noted for $P(B-/A_2)$ represent the **true negative rate** or **specificity** of the tests; d) In the case of Example 9.29, $P(B+/A_2) = .02$ is the probability of obtaining a positive blood test, given the fact that a person is healthy. In the case of Example 9.30, $P(B+/A_2) = .25$ is the probability of obtaining a positive polygraph test, given the fact that a person is innocent. The values noted for $P(B+/A_2)$ represent the **false positive rate** of the tests.

In Examples 9.29 and 9.30 we are asked to compute the conditional probability $P(A_1/B+)$. In the case of Example 9.29, the latter conditional probability represents the probability of a person being sick, given the fact that the blood test result for the person is positive. In Example 9.30, $P(A_1/B+)$ represents the probability of a person being guilty, given the fact that the polygraph test result for the person is positive. The conditional probability which will be computed for Examples 9.29 and 9.30 is commonly referred to as a **posterior probability** (where the term **posterior** means *coming after in time*), since one or more **prior probabilities** are necessary in order to compute it.

We are now ready to employ **Bayes' theorem** with Examples 9.29 and 9.30. Initially we will illustrate its use with Example 9.29. In the case of Example 9.29 we will begin by employing the theorem (i.e., Equation 9.31/9.32) to compute the value $P(A_1/B+)$. We will also use **Bayes' Theorem** to compute the following additional conditional probabilities: $P(A_2/B+)$, $P(A_2/B-)$, $P(A_1/B-)$.

Employing Equation 9.31/9.32, the value $P(A_1/B+) = .3125$ is computed.

$$P(A_1/B+) = \frac{P(B+/A_1)P(A_1)}{P(B+/A_1)P(A_1) + P(B+/A_2)P(A_2)} = \frac{(.90)(.01)}{(.90)(.01) + (.02)(.99)} = .3125$$

The value .3125 computed for $P(A_1/B+)$ is the probability of a person being sick if the person's blood test is positive. The latter value indicates that if 10,000 people test positive for the disease, 3125 (i.e., $(10,000)(.3125) = 3125$) people will, in fact, have the disease. Various sources (e.g., Pagano and Gauvreau (1993, p. 126)) note that $P(A_1/B+)$ is commonly referred to as the **predictive value of a positive test**.

Using Equation 9.31/9.33, we can compute the value $P(A_2/B+)$, which we determine is $P(A_2/B+) = .6875$. Note that since $(1 - P(A_1/B+)) = P(A_2/B+)$, $(1 - .3125) = .6875$.

$$P(A_2/B+) = \frac{P(B+/A_2)P(A_2)}{P(B+/A_1)P(A_1) + P(B+/A_2)P(A_2)} = \frac{(.02)(.99)}{(.90)(.01) + (.02)(.99)} = .6875$$

The value .6875 computed for $P(A_2/B+)$ is the probability of a person being healthy if the person's blood test is positive. The latter value indicates that if 10,000 people test positive for the disease, 6875 (i.e., $(10,000)(.6875) = 6875$) of the people will, in fact, not have the disease.

Using Equation 9.31, we can compute the value $P(A_2/B-)$, which we determine is $P(A_2/B-)$ $= .999$.

$$P(A_2/B-) = \frac{P(B-/A_2)P(A_2)}{P(B-/A_1)P(A_1) + P(B-/A_2)P(A_2)} = \frac{(.98)(.99)}{(.10)(.01) + (.98)(.99)} = .9990$$

The value .9990 computed for $P(A_2/B-)$ is the probability of a person being healthy if the person's blood test is negative. The latter value indicates that if 10,000 people test negative for the disease, 9990 (i.e., $(10,000)(.9990) = 9990$) of the people will, in fact, not have the disease. The value $P(A_2/B-)$ is commonly referred to as the **predictive value of a negative test**.

Using Equation 9.31, we can compute the value $P(A_1/B-)$, which we determine is $P(A_1/B-)$ $= .001$. Note that since $(1 - P(A_2/B-)) = P(A_1/B-)$, $(1 - .9990) = .0010$.

$$P(A_1/B-) = \frac{P(B-/A_1)P(A_1)}{P(B-/A_1)P(A_1) + P(B-/A_2)P(A_2)} = \frac{(.10)(.01)}{(.10)(.01) + (.98)(.99)} = .0010$$

The value .0010 computed for $P(A_1/B-)$ is the probability of a person being sick if the person's blood test is negative. The latter value indicates that if 10,000 people test negative for the disease, only 10 (i.e., $(10,000)(.0010) = 10$) of the people will, in fact, have the disease.

Note that if we randomly select a person from the population prior to computing the conditional probability $P(A_1/B+) = .3125$, there is only a .01 likelihood that the person who is selected will be sick (where .01 is the baserate $P(A_1) = .01$). By employing the diagnostic test, since $P(A_1/B+) = .3125$, we have increased the likelihood of correctly identifying a sick person by $.3125/.01 = 31.25$. In other words, by using the diagnostic test the probability that a subject is sick is 31.25 times greater than the probability of randomly selecting a sick person from the population. The down side of the analysis is that if a person tests positive for the disease, the likelihood is $P(A_2/B+) = .6875$ (which is $1 - .3125 = .6875$) that the person does not, in fact, have the disease. The severity of the error associated with concluding on the basis of a positive test result that a healthy person is sick, will be a function of the practical consequences of such a decision. Thus, if such a person is treated with a drug which has no long term harmful effects, the error (which constitutes a false positive) will not be very serious. On the other hand, if the treatment employed with the person is more extreme and potentially damaging (e.g., major surgery), the consequences of a person being diagnosed a false positive will be more severe. It should be understood that in medicine a positive test result, in and of itself, may not be viewed as sufficient evidence to initiate treatment. Additional tests may be conducted to further confirm or disconfirm the result of the original diagnostic test. A Bayesian analysis can be employed with any additional diagnostic tests which are conducted, and by virtue of the latter the error rates associated with the final diagnosis can be minimized.

Table 9.9 can be employed to describe the proportion of cases in each of the four diagnostic categories in Example 9.29. In the latter table the baserate values $P(A_1) = .01$ and $P(A_2) = .99$ are recorded in the last column labeled **Sums**. The product obtained within each of the four cells of the table is computed by multiplying the baserate for the row in which a cell appears by the appropriate prior conditional probability. Thus in the upper left cell, the value .009 is obtained by multiplying $P(A_1) = .01$ by $P(B+/A_1) = .90$. In the upper right cell, the value .001 is obtained by multiplying $P(A_1) = .01$ by $P(B-/A_1) = .10$. In the lower left cell, the value .0198 is obtained by multiplying $P(A_2) = .99$ by $P(B+/A_2) = .02$. In the lower right cell, the value .9702 is obtained by multiplying $P(A_2) = .99$ by $P(B-/A_2) = .98$. From Table 9.9 we can confirm the fact that 68.75% of the positives on the diagnostic test represent false positives. The value 68.75% can be obtained by dividing the proportion in the lower left cell (.0198, which represents false positives) by the sum of the proportions in Column 1 (.0288, which represents all of the positives identified by the test). Thus, $.0198/.0288 = .6875$ (which expressed as a percentage is 68.75%), tells us that 68.75% of the positives on the diagnostic test represent false positives.[19]

Table 9.9   Summary of Data for Example 9.29

		Test result		
		Sick (+)	Healthy (−)	Sums
True state of nature	Sick (+)	True positives (.01)(.90) = .009	False negatives (.01)(.10) = .001	.01
	Healthy (−)	False positives (.99)(.02) = .0198	True negatives (.99)(.98) = .9702	.99
Sums		.0288	.9712	1.00

In Example 9.30 the diagnostic test employed is a polygraph (i.e., lie detector test). Many psychologists are highly critical of the polygraph, since they argue the test yields a disproportionately large number of false positives. In Example 9.30 the false positive rate is $P(B+/A_2) = .25$.[20] We will now use **Bayes' Theorem** to compute the following conditional probabilities for Example 9.30: $P(A_1/B+)$, $P(A_2/B+)$, $P(A_2/B-)$, $P(A_1/B-)$.

Employing Equation 9.31/9.32, the value $P(A_1/B+) = .0748$ is computed.

$$P(A_1/B+) = \frac{P(B+/A_1)P(A_1)}{P(B+/A_1)P(A_1) + P(B+/A_2)P(A_2)} = \frac{(.99)(.02)}{(.99)(.02) + (.25)(.98)} = .0748$$

The value .0748 computed for $P(A_1/B+)$ is the probability of a person being guilty if the person's polygraph test result is positive. The latter value indicates that if 10,000 people test positive, only 748 (i.e., $(10,000)(.0748) = 748$) of the people will, in fact, be guilty.

Using Equation 9.31/9.33, we can compute the value $P(A_2/B+)$, which we determine is $P(A_2/B+) = .9252$. Note that since $(1 - P(A_1/B+)) = P(A_2/B+)$, $(1 - .0748) = .9252$.

$$P(A_2/B+) = \frac{P(B+/A_2)P(A_2)}{P(B+/A_1)P(A_1) + P(B+/A_2)P(A_2)} = \frac{(.25)(.98)}{(.99)(.02) + (.25)(.98)} = .9252$$

The value .9252 computed for $P(A_2/B+)$ is the probability of a person being innocent if the person's polygraph test result is positive. The latter value indicates that if 10,000 people test positive, 9252 (i.e., $(10,000)(.9252) = 9252$) of the people will, in fact, be innocent.

Using Equation 9.31, we can compute the value $P(A_2/B-)$, which we determine is $P(A_2/B-) = .9997$.

$$P(A_2/B-) = \frac{P(B-/A_2)P(A_2)}{P(B-/A_1)P(A_1) + P(B-/A_2)P(A_2)} = \frac{(.75)(.98)}{(.01)(.02) + (.75)(.98)} = .9997$$

The value .9997 computed for $P(A_2/B-)$ is the probability of a person being innocent if the person's polygraph test result is negative. The latter value indicates that if 10,000 people test negative, 9997 (i.e., (10,000)(.9997) = 9997) of the people will, in fact, be innocent.

Using Equation 9.31, we can compute the value $P(A_1/B-)$, which we determine is $P(A_1/B-) = .0003$. Note that since $(1 - P(A_2/B-)) = P(A_1/B-)$, $(1 - .9997) = .0003$.

$$P(A_1/B-) = \frac{P(B-/A_1)P(A_1)}{P(B-/A_1)P(A_1) + P(B-/A_2)P(A_2)} = \frac{(.01)(.02)}{(.01)(.02) + (.75)(.98)} = .0003$$

The value .0003 computed for $P(A_1/B-)$ is the probability of a person being guilty if the person's test is negative. The latter value indicates that if 10,000 people test negative, only 3 (i.e., (10,000)(.0003) = 3) of the people will, in fact, be guilty.

Note that if we randomly select a person from the population prior to computing the conditional probability $P(A_1/B+) = .0748$, there is only a .02 likelihood that the person who is selected will be guilty (where .02 is the baserate $P(A_1) = .02$). By employing the polygraph test, since $P(A_1/B+) = .0748$, we have increased the likelihood of correctly identifying a guilty person by .0748/.02 = 3.74. In other words, by using the polygraph test, the probability that a subject is guilty is 3.74 times greater than the probability of randomly selecting a guilty person from the population. The down side of the analysis is that if a person tests positive for being guilty, the likelihood is $P(A_2/B+) = .9252$ (which is $1 - .0748 = .9252$) that the person is, in fact, innocent. Note that the benefit derived from the use of the polygraph test in Example 9.30 is substantially less than that associated with the diagnostic test employed in Example 9.29 (in which case there is a 31.25 times greater likelihood of identifying a positive subject). Note also that the down side associated with the polygraph test is substantially greater than the down side associated with the diagnostic test employed in Example 9.29 — specifically, contrast the values $P(A_2/B+) = .9252$ in Example 9.30 versus .6875 in Example 9.29.

Table 9.10 can be employed to describe the proportion of cases in Example 9.30 in each of the four diagnostic categories. In the latter table the baserate values $P(A_1) = .02$ and $P(A_2) = .98$ are recorded in the last column labeled **Sums**. As noted for Table 9.9, the product obtained within each of the four cells of the table is computed by multiplying the baserate for the row in which a cell appears by the appropriate prior conditional probability. Thus in the upper left cell, the value .0198 is obtained by multiplying $P(A_1) = .02$ by $P(B+/A_1) = .99$. In the upper right cell, the value .0002 is obtained by multiplying $P(A_1) = .02$ by $P(B-/A_1) = .01$. In the lower left cell, the value .245 is obtained by multiplying $P(A_2) = .98$ by $P(B+/A_2) = .25$. In the lower right cell, the value .735 is obtained by multiplying $P(A_2) = .98$ by $P(B-/A_2) = .75$. From Table 9.10 we can confirm the fact that 92.52% of the positives on the polygraph test represent false positives. The value 92.52% can be obtained by dividing the proportion in the lower left cell (.245, which represents false positives) by the sum of the proportions in Column 1 (.2648, which represents all of the positives identified by the test). Thus, .245/.2648 = .9252 (which expressed as a percentage is 92.52%), tells us that 92.52% of the positives on the polygraph test represent false positives. Given the magnitude of the latter value, if within the field of forensics the result of a polygraph test is viewed as valid, the decisions reached through use of such a test would result in identifying an extremely large number of innocent people as guilty.

Table 9.10 Summary of Data for Example 9.30

		Test result		Sums
		Guilty (+)	Innocent (–)	
True state of nature	Guilty (+)	True positives (.02)(.99) = .0198	False negatives (.02)(.01) = .0002	.02
	Innocent (–)	False positives (.98)(.25) = .245	True negatives (.98)(.75) = .735	.98
Sums		.2648	.7352	1.00

   Examples 9.29 and 9.30 illustrate a phenomenon which is commonly referred to as the **low baserate problem** — a diagnostic test which is employed to identify a low baserate event will yield a disproportionately large number of false positives. Example 9.31 represents an even more dramatic case of the **low baserate problem**, in which a researcher attempts to determine whether or not a subject's performance in an experiment is at a level which suggests a subject has extrasensory perception (ESP). A number of issues will be addressed in Example 9.31 which were not discussed in reference to Examples 9.29 and 9.30. For one, the phenomenon under study is controversial, in that there are many members of the scientific community who do not believe there presently is or ever has been any credible evidence to support the existence of extrasensory perception. Consequently, if a researcher elects to employ **Bayes' theorem** to evaluate the relevant hypothesis, a major problem he will have is what criterion to use to estimate the baserate of extrasensory perception (within the context of the experimental situation described in Example 9.31). Note that the application of **Bayes' theorem** becomes much more problematical if the researcher has to estimate the baserate of the target behavior/event, as opposed to actually knowing its exact value. In instances where reliable empirical information regarding the value of the baserate is unavailable, the only option available to the researcher is to estimate the baserate. In such an case, one researcher's estimate may not be the same or even close to another researcher's estimate. With respect to extrasensory perception, there are those who would estimate the baserate to be zero, while there are others who would conjecture some value above zero. With respect to the latter, based on the author's knowledge of research in the area of extrasensory perception, one might conjecture a baserate for extrasensory perception as low as .0000001 to perhaps a maximum baserate of .01.[21] For purposes of illustration, in Example 9.31 I have employed the value .001 as the baserate.

Example 9.31 *In order to determine whether or not a person possesses extrasensory perception (ESP), the following standardized test protocol is employed for each subject. The stimuli employed in the experiment are a list of binary digits (specifically, the values 0 and 1) which have been randomly generated by a computer. For each subject who is tested, the experimenter concentrates on each of the digits in the series of numbers in the order it appears. While the experimenter does this, the subject, who is in another room, attempts to guess the value of the numbers in the order they appear. At the conclusion of the experiment a subject's performance is evaluated with respect to whether or not it is statistically significant. The alpha value employed for determining statistical significance is $\alpha = .05$. The number of trials employed in testing each subject is such that the power of the statistical analysis will equal .99. We will assume that the estimated baserate for ESP in the population is .001.*

Within the framework of Example 9.31, the following four decisions are possible with respect to a subject: a) A **true positive** is an individual who is identified by the test as having ESP and, in fact, has ESP; b) A **true negative** is an individual who is identified as not having ESP and, in fact, does not have ESP; c) A **false positive** is an individual who is identified as having ESP and, in fact, does not have ESP; d) A **false negative** is an individual who is identified as not having ESP and, in fact, does have ESP.

Based on the information which has been provided for Example 9.31, we can say the following.

$$P(ESP) = P(A_1) = .001 \quad P(No\ ESP) = P(A_2) = 1 - P(A_1) = 1 - .001 = .999$$
$$P(B+/A_1) = P(Positive\ test\ for\ ESP/Has\ ESP) = .99$$
$$P(B-/A_1) = P(Negative\ test\ for\ ESP/Has\ ESP) = 1 - .99 = .01$$
$$P(B-/A_2) = P(Negative\ test\ for\ ESP/Does\ not\ have\ ESP) = .95$$
$$P(B+/A_2) = P(Positive\ test\ for\ ESP/Does\ not\ have\ ESP) = .05$$

If we view Example 9.31 from the perspective of the **classical hypothesis testing model** (which is described in the **Introduction**), we can state that the **null hypothesis** is that a subject does not have ESP, whereas the **alternative hypothesis** is that a subject does have ESP. Within such a framework we can say the following with regard to the conditional probabilities noted above: a) $P(B+/A_2) = .05$ corresponds to the value of **alpha**, which represents the likelihood of committing a Type I error. In the case of an individual subject, it represents the likelihood of concluding from a subject's test performance that the subject has ESP, given that the subject, in fact, does not have ESP; b) The value $P(B-/A_2) = .95$ (i.e., $1 - .05 = .95$) represents the likelihood of making a correct decision with regard to the null hypothesis. In the case of an individual subject, it represents the likelihood of concluding from a subject's test performance that the subject does not have ESP, given that the subject, in fact, does not have ESP; c) The value $P(B+/A_1) = .99$ represents the **power** of the test. In the case of an individual subject, it represents the likelihood of concluding from a subject's test performance that the subject has ESP, given that the subject does, in fact, have ESP; d) The value $P(B-/A_1) = .01$ represents the value of **beta** (β), which is the likelihood of committing a Type II error. In the case of an individual subject, the value $P(B-/A_1) = .01$ represents the likelihood of concluding from a subject's test performance that the subject does not have ESP, given that the subject, in fact, does have ESP.[22]

The following conditional probabilities are computed below for Example 9.31: $P(A_1/B+)$, $P(A_2/B+)$, $P(A_2/B-)$, $P(A_1/B-)$.

Employing Equation 9.31/9.32, the value $P(A_1/B+) = .01943$ is computed.

$$P(A_1/B+) = \frac{P(B+/A_1)P(A_1)}{P(B+/A_1)P(A_1) + P(B+/A_2)P(A_2)} = \frac{(.99)(.001)}{(.99)(.001) + (.05)(.999)} = .01943$$

The value .01943 computed for $P(A_1/B+)$ is the probability of a person having ESP if the person's test performance is positive (i.e., significant at the .05 level). The latter value indicates that if 100,000 people test positive for ESP, only 1943 (i.e., $(100,000)(.01943) = 1943$) of the people will, in fact, have ESP.

Using Equation 9.31/9.33, we can compute the value $P(A_2/B+)$, which we determine is $P(A_2/B+) = .98057$.

$$P(A_2/B+) = \frac{P(B+/A_2)P(A_2)}{P(B+/A_1)P(A_1) + P(B+/A_2)P(A_2)} = \frac{(.05)(.999)}{(.99)(.001) + (.05)(.999)} = .98057$$

The value .98057 computed for $P(A_2/B+)$ is the probability of a person not having ESP if the person's test performance is positive (i.e., significant at the .05 level). The latter value indicates that if 100,000 people test positive for ESP, 98,057 (i.e., (100,000)(.98057) = 98,057) of the people will, in fact, not have ESP.

Using Equation 9.31, we can compute the value $P(A_2/B-)$, which we determine is $P(A_2/B-) = .99999$.

$$P(A_2/B-) = \frac{P(B-/A_2)P(A_2)}{P(B-/A_1)P(A_1) + P(B-/A_2)P(A_2)} = \frac{(.95)(.999)}{(.01)(.001) + (.95)(.999)} = .99999$$

The value .9999 computed for $P(A_2/B-)$ is the probability of a person not having ESP if the person's test performance is negative (i.e., not significant at the .05 level). The latter value indicates that if 100,000 people test negative for ESP, 99,999 (i.e., (100,000)(.99999) = 99,999) of the people will, in fact, not have ESP.

Using Equation 9.31, we can compute the value $P(A_1/B-)$, which we determine is $P(A_1/B-) = .00001$.

$$P(A_1/B-) = \frac{P(B-/A_1)P(A_1)}{P(B-/A_1)P(A_1) + P(B-/A_2)P(A_2)} = \frac{(.01)(.001)}{(.01)(.001) + (.95)(.999)} = .00001$$

The value .00001 computed for $P(A_1/B-)$ is the probability of a person having ESP if the person's test performance is negative (i.e., not significant at the .05 level). The latter value indicates that if 100,000 people test negative for ESP, only 1 (i.e., (100,000)(.00001) = 1) person will, in fact, have ESP.

Note that if we randomly select a person from the population prior to computing the conditional probability $P(A_1/B+) = .01943$, there is only a .001 likelihood that the person who is selected will have ESP (where .001 is the baserate $P(A_1)$ = .001). By employing the diagnostic test (i.e., conducting the experiment), since $P(A_1/B+) = .01943$, we have only increased the likelihood of correctly identifying a person who possesses ESP by .01943/.01 = 1.943. In other words, by using the test performance of a subject, the probability that the subject has ESP is 1.943 times greater than the probability of randomly selecting a person who has ESP from the population. As noted earlier, if a person tests positive for ESP, the likelihood is $P(A_2/B+) = .98057$ (which is $1 - .01943 = .98057$) that the person does not, in fact, have ESP. Note that the benefit derived from the use of the test in Example 9.31 is substantially less than that associated with the diagnostic test employed in Example 9.29 (in which case there is a 31.25 times greater likelihood of identifying a positive subject), as well as the polygraph test employed in Example 9.30 (in which case there is a 3.75 times greater likelihood of identifying a positive subject).

Table 9.11 can be employed to describe the proportion of cases in each of the four decision categories in Example 9.31. In the latter table the baserate values $P(A_1)$ = .001 and $P(A_2)$ = .999 are recorded in the last column labeled **Sums**. As noted for Tables 9.9 and 9.10, the product obtained within each of the four cells of the table is computed by multiplying the baserate for the row in which a cell appears by the appropriate prior conditional probability.

Table 9.11   Summary of Data for Example 9.31

		Test result		
		ESP (+)	No ESP (–)	**Sums**
True state of	ESP (+)	True positives (.001)(.99) = .00099	False negatives (.001)(.01) = .00001	.001
nature	No ESP (–)	False positives (.999)(.05) = .04995	True negatives (.999)(.95) = .94905	.999
Sums		.05094	.94905	1.00

Thus in the upper left cell, the value .00099 is obtained by multiplying $P(A_1)$ = .001 by $P(B+/A_1)$ = .99. In the upper right cell, the value .00001 is obtained by multiplying $P(A_1)$ = .001 by $P(B-/A_1)$ = .01. In the lower left cell, the value .04995 is obtained by multiplying $P(A_2)$ = .999 by $P(B+/A_2)$ = .05. In the lower right cell, the value .94905 is obtained by multiplying $P(A_2)$ = .999 by $P(B-/A_2)$ = .95. From Table 9.11 we can confirm the fact that 98.057% of the positives on the ESP test represent false positives. The value 98.057% can be obtained by dividing the proportion in the lower left cell (.04995, which represents false positives) by the sum of the proportions in Column 1 (.05094, which represents all of the positives identified by the test). Thus, .04995/.05094 = .98057 (which expressed as a percentage in 98.057%), tells us that 98.057% of the positives on the ESP test represent false positives. Given the magnitude of the latter value, if a researcher identifies a person as having ESP, the likelihood that the person actually has it is minimal. It should be noted, however, that realistically, a researcher would not conclude a subject has ESP purely on the basis of one test result. Nevertheless, the point to be made here is that in a laboratory test of ESP in which the statistical power of the test is extremely high, a subject for whom a significant result is obtained in all likelihood does not have ESP (i.e., the subject represents an example of a Type I error). Thus, even though the alpha value of .05 employed in the study might suggest there is only a .05 likelihood that a subject with a significant result may, in fact, be a Type I error, the actual likelihood of that subject being a Type I error is .98057, which is substantially greater than .05. Detailed discussions of the impact of baserates on the error rates of diagnostic tests can be found in Meehl and Rosen (1955) and Wiggins (1973).

In Examples 9.29–9.31 a baserate is estimated for only two categories (i.e., A_1 and A_2). By employing Equation 9.31, **Bayes' theorem** can be used to compute conditional probabilities for more than two categories. Example 9.32 illustrates its application in computing conditional probabilities for three diagnostic categories. Example 9.32 also represents another instance in which a researcher has to estimate the baserate of an event, as opposed to actually knowing its precise value.

Example 9.32 *Based on his interpretation of the empirical studies published in the medical literature, a physician employs the following probabilities with respect to whether or not there is some form of pathology present in the prostate gland of a 65-year-old male. The likelihood the man has a normal prostate is. .98, the likelihood he has some type of malignant pathology is .019, and the likelihood there is some type of nonmalignant pathology is .001.*[23] *Since the three aforementioned values represent baserates for mutually exclusive conditions, they can be* *summarized as follows.*

$$P(Normal) = P(A_1) = .98$$
$$P(Malignant\ pathology) = P(A_2) = .019$$
$$P\ (Nonmalignant\ pathology) = P(A_3) = .001$$

Assume that a 65-year-old man tests positive on a new diagnostic test that is designed to identify prostate pathology. Based on his knowledge of the limited medical literature which is available on the test, the physician estimates the prior conditional probabilities noted below.

$$P(B+/A_1) = P(Positive\ test/Normal) = .01$$
$$P(B+/A_2) = P(Positive\ test/Malignant\ pathology) = .99$$
$$P(B+/A_3) = P(Positive\ test/Nonmalignant\ pathology) = .98$$

Employ the above noted prior probabilities to compute the following posterior conditional probabilities: a) The probability a 65-year-old man will have a normal prostate, given the fact his test result is positive; b) The probability a 65-year-old man has a malignant pathology of the prostate, given the fact that his test result is positive; c) The probability a 65-year-old man has a nonmalignant pathology of the prostate, given the fact that his test result is positive

Note that in Example 9.32, the prior probabilities $P(A_1)$, $P(A_2)$, and $P(A_3)$ are subjective estimates made by the physician based on his interpretation of the medical literature, and there is no guarantee that another physician evaluating the same literature (or perhaps even a different body of literature) would come up with identical values. As noted earlier in this section, the application of **Bayes' theorem** becomes more problematical if the researcher has to estimate the probability of a target behavior/event, as opposed to actually knowing its precise value. This latter fact should be kept in mind with reference to Example 9.32 (as well as with Examples 9.33 and 9.34, which are presented later in this section). Put simply, when the prior probabilities employed by a researcher are based on subjective estimates, the confidence one can have in the results of a Bayesian analysis will be a direct function of the degree to which one believes the researcher has accurately estimated the probabilities in question.

The prior conditional probabilities noted in Example 9.32 indicate the following: a) $P(B+/A_1) = .01$ is the probability of obtaining a positive test, given the fact a man has a normal prostate. This value represents the false positive rate for the test; b) $P(B+/A_2) = .99$ is the probability of obtaining a positive test, given the fact a man has a malignant prostate pathology; c) $P(B+/A_3) = .98$ is the probability of obtaining a positive test, given the fact a man has a nonmalignant prostate pathology.

In Example 9.32 we are asked to compute the following conditional probabilities: a) $P(A_1/B+)$, which is the likelihood of a 65-year-old man having a normal prostate, given the fact his test result is positive; b) $P(A_2/B+)$, which is the likelihood of a 65-year-old man having a malignant pathology of the prostate, given the fact his test result is positive; c) $P(A_3/B+)$, which is the likelihood of a 65-year-old man having a nonmalignant pathology of the prostate, given the fact his test result is positive.

Employing Equation 9.31 the above noted posterior conditional probabilities are computed.

$$P(A_1/B+) = .331 \qquad P(A_2/B+) = .636 \qquad P(A_3/B+) = .033$$

Note that when the three conditional probabilities $P(A_1/B+) = .331$, $P(A_2/B+) = .636$, and $P(A_3/B+) = .033$ are summed, they yield the value 1.

$$P(A_1/B+) = \frac{P(B+/A_1)P(A_1)}{P(B+/A_1)P(A_1) + P(B+/A_2)P(A_2) + P(B+/A_3)P(A_3)}$$

$$= \frac{(.01)(.98)}{(.01)(.98) + (.99)(.019) + (.98)(.001)} = .331$$

$$P(A_2/B+) = \frac{P(B+/A_2)P(A_2)}{P(B+/A_1)P(A_1) + P(B+/A_2)P(A_2) + P(B+/A_3)P(A_3)}$$

$$= \frac{(.99)(.019)}{(.01)(.98) + (.99)(.019) + (.98)(.001)} = .636$$

$$P(A_3/B+) = \frac{P(B+/A_3)P(A_3)}{P(B+/A_1)P(A_1) + P(B+/A_2)P(A_2) + P(B+/A_3)P(A_3)}$$

$$= \frac{(.98)(.001)}{(.01)(.98) + (.99)(.019) + (.98)(.001)} = .033$$

The value .331 computed for $P(A_1/B+)$ is the probability of a man having a normal prostate if his test result is positive. The latter value indicates that if 1000 men test positive, 331 (i.e., $(1000)(.331) = 331$) of them will have a normal prostate. The value .636 computed for $P(A_2/B+)$ is the probability of a man having a malignant pathology of the prostate if his test result is positive. The latter value indicates that if 1000 men test positive, 636 (i.e., $(1000)(.636) = 636$) of them will have a malignant pathology of the prostate. The value .033 computed for $P(A_3/B+)$ is the probability of a man having a nonmalignant pathology of the prostate if his test result is positive. The latter value indicates that if 1000 men test positive, only 33 (i.e., $(1000)(.033) = 33$) of them will have a nonmalignant pathology of the prostate. The results of the above analysis indicate that if a man obtains a positive test result, there is a one in three chance the result represents a false positive (i.e., $p = .331$). There is about a two in three chance the man has a malignant pathology (i.e., $p = .636$). Note that if we randomly select a male from the population of 65-year-old males prior to computing the conditional probability $P(A_2/B+) = .636$, there is a .019 likelihood that the man who is selected will have a malignant pathology of the prostate (where .019 is the baserate $P(A_2) = .019$). By employing the diagnostic test, since $P(A_2/B+) = .636$, we have increased the likelihood of correctly identifying a malignant case by $.636/.019 = 33.47$. In other words, by using the diagnostic test, the probability that a 65-year-old man has a malignant pathology of the prostate is 33.47 times greater than the probability of randomly selecting a man with a malignant pathology of the prostate from the population of 65-year-old males.

Although it is unlikely that a man has a nonmalignant pathology of the prostate, given the fact that his test result is positive (i.e., $p = .033$), the use of the test substantially increases the likelihood of identifying such an individual when contrasted with randomly selecting such a person from the population of 65-year-old males. If we randomly select a male from the population of 65-year-old males prior to computing the conditional probability $P(A_3/B+) = .033$, there is a .001 likelihood that the man who is selected will have a nonmalignant pathology of the prostate (where .001 is the baserate $P(A_3) = .001$). By employing the diagnostic test, since $P(A_3/B+) = .033$, we have increased the likelihood of correctly identifying a man with a nonmalignant pathology by $.033/.001 = 33$. In other words, by using the diagnostic test, the probability that a 65-year-old man has a nonmalignant pathology of the prostate is 33 times greater than the probability of randomly selecting a man with a nonmalignant pathology of the prostate from the population of 65-year-old males.

Various sources (Hays and Winkler (1971), Winkler (1972), and Winkler (1993)) note that **Bayes' theorem** can be employed in a sequential manner, with each application of the theorem utilizing the results of the previous analysis in conjunction with additional probabilistic information involving a new variable. Example 9.33 will be employed to illustrate the sequential application of **Bayes' theorem**.

Example 9.33   *Upon returning from a trip to southeast Asia, a patient visits a specialist in tropical medicine complaining of loss of appetite, fatigue, and a variety of additional physical symptoms. Upon conducting a preliminary physical examination in his office and reviewing the patient's medical history, the physician estimates there is a .20 likelihood that the patient has contracted leprosodiasis, an insidious parasitic disease which is indigenous to southeast Asia.*

a) The doctor initially decides to employ a blood test to screen the patient for the disease. The medical literature states that the blood test yields a positive result for 75% of the people who have leprosodiasis, but also yields a positive result for 15% of the people who do not have the disease. Compute the conditional probability that the patient has leprosodiasis given the fact that his blood test is positive.

b) Additional diagnostic information on leprosodiasis can be obtained through use of a liver biopsy. Following the determination of the probability computed in part a), the doctor sends a liver biopsy to the laboratory for analysis. The medical literature states that the liver biopsy yields a positive result for 85% of the people who have leprosodiasis, but also yields a positive result for 18% of the people who do not have the disease. Assume that the biopsy comes back positive. Compute the revised conditional probability that the patient has leprosodiasis given the positive result on the biopsy.[24]

The prior probabilities for Example 9.33 are summarized below. Note that in Example 9.33, instead of representing the baserate of the disease, the values $P(A_1)$ and $P(A_2)$ represent the physician's subjective estimate of the likelihood the patient has the disease ($P(A_1)$) versus the likelihood the patient does not have the disease ($P(A_2)$). With respect to the notation that will be employed for the analysis, the blood test will represent **diagnostic test 1** and the liver biopsy will represent **diagnostic test 2**. Thus, the notation B_1^+ indicates a positive blood test, the notation B_1^- indicates a negative blood test, the notation B_2^+ indicates a positive liver biopsy, and the notation B_2^- indicates a negative liver biopsy.

$$P(Leprosodiasis) = P(A_1) = .20$$
$$P(Does\ not\ have\ leprosodiasis) = P(A_2) = 1 - P(A_1) = 1 - .20 = .80$$
$$P(B_1^+/A_1) = P(Positive\ blood\ test/Leprosodiasis) = .75$$
$$P(B_1^-/A_1) = P(Negative\ blood\ test/Leprosodiasis) = 1 - .75 = .25$$
$$P(B_1^+/A_2) = P(Positive\ blood\ test/No\ leprosodiasis) = .15$$
$$P(B_1^-/A_2) = P(Negative\ blood\ test/No\ leprosodiasis) = 1 - .15 = .85$$

When the above probabilities are employed in Equation 9.31/9.32, the value $P(A_1/B_1^+) = .556$ is computed. The latter value represents the likelihood the patient has leprosodiasis if his blood test is positive. Since $P(A_2/B_1^+) = 1 - P(A_1/B_1^+)$, $P(A_2/B_1^+) = 1 - .556 = .444$ (which can also be computed through use of Equation 9.31), the latter value is the likelihood the patient does not have leprosodiasis if his blood test is positive. Thus, by employing the result of the blood test, the physician can now say there is a .556 likelihood the patient has leprosodiasis (as opposed to his original estimate of .20).

$$P(A_1/B_1^+) = \frac{P(B_1^+/A_1)P(A_1)}{P(B_1^+/A_1)P(A_1) + P(B_1^+/A_2)P(A_2)}$$

$$= \frac{(.75)(.20)}{(.75)(.20) + (.15)(.80)} = .556$$

A second analysis using **Bayes' theorem** will now be conducted which takes into account both the positive blood test and the positive liver biopsy. The prior probabilities which will be employed in the second application of **Bayes' theorem** involve the prior probabilities associated with the liver biopsy, and the posterior probabilities which reflect the revised probabilities derived from the positive blood test. The probabilities that will be employed in this stage of the analysis are summarized below. Note that in the analysis to be conducted, the values $P(A_1)$ and $P(A_2)$ will represent the revised estimates of the likelihood the patient has leprosodiasis (which is now $P(A_1/B_1^+) = .556$) versus the revised estimate of the likelihood the patient does not have leprosodiasis (which is now $P(A_2/B_1^+) = .444$). In other words, the values $P(A_1/B_1^+) = .556$ and $P(A_2/B_1^+) = .444$) computed in the initial analysis will be employed in Equation 9.31/9.32 to represent $P(A_1)$ and $P(A_2)$.

$$P(Leprosodiasis) = P(A_1) = P(A_1/B_1^+) = .556$$
$$P(Does\ not\ have\ leprosodiasis) = P(A_2) = 1 - P(A_1/B_1^+) = 1 - .556 = .444$$
$$P(B_2^+/A_1) = P(Positive\ liver\ biopsy/Leprosodiasis) = .85$$
$$P(B_2^-/A_1) = P(Negative\ liver\ biopsy/Leprosodiasis) = 1 - .85 = .15$$
$$P(B_2^+/A_2) = P(Positive\ liver\ biopsy/No\ leprosodiasis) = .18$$
$$P(B_2^-/A_2) = P(Negative\ liver\ biopsy/No\ leprosodiasis) = 1 - .18 = .82$$

Equation 9.31/9.32 is employed below to compute the revised likelihood that the patient has leprosodiasis. The notation $P(A_1/B_2^+, B_1^+)$ represents the latter likelihood. To be more specific, the computed value $P(A_1/B_2^+, B_1^+) = .855$ represents the likelihood the patient has leprosodiasis if his liver biopsy is positive, subsequent to our previously determining his blood test is also positive.

$$P(A_1/B_2^+, B_1^+) = \frac{P(B_2^+/A_1)P(A_1/B_1^+)}{P(B_2^+/A_1)P(A_1/B_1^+) + P(B_2^+/A_2)P(A_2/B_1^+)}$$

$$= \frac{(.85)(.556)}{(.85)(.556) + (.18)(.444)} = .855$$

Thus, by taking into account both the blood test and the result of the liver biopsy, the physician can now say there is a .855 likelihood the patient has leprosodiasis (as opposed to his original estimate of .20 and the subsequent estimate of .556, which only took into account the result of the blood test). $P(A_2/B_2^+, B_1^+) = 1 - .855 = .145$ is the revised likelihood the patient does not have leprosodiasis.

It should be noted that if the physician initially employed the positive result for the liver biopsy, the likelihood of the patient having leprosodiasis will increase from his original estimate of .20 to .541. Note that in the second stage of the analysis, $P(A_1/B_2^+) = .541$ is employed to represent $P(A_1)$. Since $P(A_2) = P(A_2/B_2^+)$, $P(A_2/B_2^+) = 1 - P(A_1/B_2^+) = 1 - .541 = .459$. The result of the second stage of the analysis yields the revised probability .855, which is the same value obtained previously for the results of the blood test and liver biopsy, when the blood test was evaluated first.

$$P(A_1/B_2^+) = \frac{P(B_2^+/A_1)P(A_1)}{P(B_2^+/A_1)P(A_1) + P(B_2^+/A_2)P(A_2)}$$

$$= \frac{(.85)(.20)}{(.85)(.20) + (.18)(.80)} = .541$$

$$P(A_1/B_1^+, B_2^+) = \frac{P(B_1^+/A_1)P(A_1/B_2^+)}{P(B_1^+/A_1)P(A_1/B_2^+) + P(B_1^+/A_2)P(A_2/B_2^+)}$$

$$= \frac{(.75)(.541)}{(.75)(.541) + (.15)(.459)} = .855$$

We can also determine the likelihood that the patient has leprosodiasis if either of the following occur: a) The blood test yields a negative result, but the liver biopsy is positive; b) The blood test yields a positive result, but the liver biopsy is negative; c) Both the blood test and liver biopsy are negative. Since the same logic is applied in evaluating each of the three aforementioned scenarios, only one of them will be illustrated – specifically, the analysis of a negative blood test and a positive liver biopsy. (It will be assumed that the blood test is administered first, and that after a revised probability is computed, a second revised probability is computed which takes into account the positive liver biopsy.)

Initially, we modify Equation 9.31 to allow us to compute a revised probability of the patient having leprosodiasis if the result of his blood test is negative (i.e., we employ B_1^- in Equation 9.31 in place of B_1^+).

$$P(A_1/B_1^-) = \frac{P(B_1^-/A_1)P(A_1)}{P(B_1^-/A_1)P(A_1) + P(B_1^-/A_2)P(A_2)}$$

$$= \frac{(.25)(.20)}{(.25)(.20) + (.85)(.80)} = .068$$

Thus, based on the negative blood test, the physician can state that the likelihood of the patient having leprosodiasis has been reduced from his original estimate of .20 to .068. $P(A_2/B_1^-) = 1 - .068 = .932$ is the revised likelihood the patient does not have leprosodiasis.

In the second stage of the analysis, **Bayes' theorem** is employed below to evaluate the positive liver biopsy.

$$P(A_1/B_2^+, B_1^-) = \frac{P(B_2^+/A_1)P(A_1/B_1^-)}{P(B_2^+/A_1)P(A_1/B_1^-) + P(B_2^+/A_2)P(A_2/B_1^-)}$$

$$= \frac{(.85)(.068)}{(.85)(.068) + (.18)(.932)} = .256$$

Thus, by virtue of considering the result of the liver biopsy, the revised likelihood of the patient having leprosodiasis is increased to .256. $P(A_2/B_2^+, B_1^-) = 1 - .256 = .744$ is the revised likelihood the patient does not have leprosodiasis. In conclusion, it would appear that if the patient yields a negative blood test and a positive liver biopsy, the physician's original probability of the patient having leprosodiasis is only minimally modified. Specifically, the likelihood has only increased from .20 to .256.

Bayesian hypothesis testing Up to this point, the discussion of Bayesian analysis has been limited to decision making with regard to a single subject. In point of fact, it is far more common in statistics to evaluate one or more hypotheses about a population parameter than it is to arrive at a decision about an individual subject. Thus the intent of this section is to introduce the reader to the use of Bayesian analysis in formal hypothesis testing.

During the past 20 years there has been increasing criticism of the **classical hypothesis testing model**. (Representative sources on this debate are Cohen (1994), Falk and Greenbaum (1995), Gigerenzer (1993), Hagen (1997), Harlow *et al.* (1997), Krueger (2001), Meehl (1978), Morrison and Henkel (1970), and Serlin and Lapsley (1985, 1993).) Among other things, critics of the latter model recommend that it be replaced by (or, minimally, that it be employed in conjunction with) an alternative hypothesis testing model. For a number of years the use of Bayesian analysis for hypothesis testing has been suggested as an alternative. Long before the current debate on the relative merits of the classical versus the Bayesian approach to hypothesis testings, sources such as Edwards, Lindman, and Savage (1963) and Phillips (1973), stated the case for the Bayesian approach. Those who advocate the use of Bayesian analysis in formal hypothesis testing are critical of, among other things, the classical hypothesis testing model because of the fact that the latter model focuses too much attention on the null hypothesis. Critics contend that the classical hypothesis testing model is biased in favor of rejecting the null hypothesis — a hypothesis which stipulates for a population parameter an exact value which, in all likelihood, will never be the actual value of the parameter. In the case of a null hypothesis which stipulates a zero difference between two or more population parameters (the latter type of null hypothesis is evaluated in subsequent chapters), critics of the classical hypothesis testing model argue that, in reality, the difference between two population parameters will probably never equal exactly zero. Thus, within the framework of the classical hypothesis testing model, critics contend it will always be the case that the null hypothesis is false, yet rejection of a null hypothesis in and of itself will provide little information regarding the actual value of a parameter. The rationale for stating that a null hypothesis will always be false is based on the fact that even if the value of a parameter (or the difference between two parameters) is close to the value stipulated in the null hypothesis, if in actuality it is not precisely that value, with a large enough sample size the null hypothesis will always be rejected at the .05 level or less. The latter is directly attributable to the fact that as the sample size employed in a study increases, the power of any statistical analysis conducted will also increase. As the power of a statistical test increases, the likelihood of detecting a difference between a sample statistic and the value of a parameter stated in the null hypothesis (regardless of how minimal the difference may be) also increases.

In contrast to the classical approach, Bayesian statisticians prefer to simultaneously evaluate multiple alternative hypotheses, taking into account prior probabilities (which are ignored in the classical model) for each hypothesis. By evaluating prior probabilities in conjunction with empirical data, relative likelihoods can be computed for the alternative hypotheses. (As will be noted later, relative likelihoods can also be computed within the framework of the classical model.) Each of the alternative hypotheses employed in the Bayesian model stipulates a specific value or range of values for a population parameter, or stipulates a value or range of values for a difference between population parameters. Rather than using the traditional .05 and .01 significance levels, Bayesian statisticians prefer to summarize the result of an analysis by reporting intervals within which it is likely a parameter falls (as well as documenting the relative likelihoods for the various hypotheses evaluated). Lindley (1957) has demonstrated that it is possible for a null hypothesis to be rejected at the .05 level within the framework of the classical model, yet at the same time when evaluated through use of the Bayesian model that same null hypothesis can have a posterior probability as high as .95. This latter paradox is a direct function of the different approaches taken by the Bayesian and classical models in stipulating the experimental hypotheses. A more detailed discussion of the Bayesian critique of the classical hypothesis testing model can be found in Phillips (1973, pp. 333–334). Further discussion of the limitations of the classical hypothesis testing model, as well as a description of another alternative approach to hypothesis testing (the **minimum-effect hypothesis testing model**), can be found in Section IX (the **Addendum**) of the **Pearson product-moment correlation coefficient (Test 28)**.

The reader should take note of the fact that the use of Bayesian analysis for hypothesis testing is a complex subject, and the presentation in this section will be limited in scope. Those who require a more extensive discussion of the subject should consult the sources cited in this section. Example 9.34 will be employed to illustrate the use of **Bayes' theorem** in hypothesis testing.

Example 9.34　*A soft drink company initiates an ad campaign which claims that its fruit flavored soda **Nectaberry** is made with a new formula which gives it a "fruitier" taste than the top ranked fruit flavored soda **Fruizzle**. In order to evaluate the latter claim, a researcher asks each of 14 subjects to drink a four-ounce cup of **Nectaberry** and a four-ounce cup of **Fruizzle**, and indicate which of the two drinks has a "fruitier" taste. After sampling the first drink, each subject rinses her mouth out with distilled water and waits one minute before sampling the second drink. To control for the potential influence of the order of presentation of the drinks, half the subjects drink the **Nectaberry** first followed by **Fruizzle**, while the other half sample the drinks in the reverse order.[25] Nine out of the 14 subjects state that **Nectaberry** has a fruitier taste than **Fruizzle**. Does the latter result support the company's claim that **Nectaberry** is fruitier than **Fruizzle**?*

Throughout this book the classical hypothesis testing model is employed to evaluate an exact null hypothesis against an inexact alternative hypothesis. We will initially employ the **binomial sign test for a single sample** to conduct an analysis of Example 9.34 within the framework of the classical model prior to evaluating it through use of **Bayes' theorem**. In our analysis, the value π_1 will represent the true proportion of subjects in the population who select *Nectaberry* as the fruitier soda. Thus, we will employ H_0: $\pi_1 = .5$ as the null hypothesis. This represents an exact null hypothesis, since it specifies an exact value for π_1. Either H_1: $\pi_1 \neq .5$ or H_1: $\pi_1 > .5$ can be employed as the alternative hypothesis (with the former being nondirectional and the latter being directional). The two aforementioned alternative hypotheses represent inexact alternative hypotheses, since they do not stipulate an exact value which π_1 must equal in order for either alternative hypothesis to be supported. The null hypothesis H_0: $\pi_1 = .5$ states that the proportion of times *Nectaberry* will be selected as the fruitier soda is .5 (or 50%), and consequently it indicates that a subject is equally likely to select either of the two sodas as the fruitier beverage. The use of the nondirectional alternative hypothesis H_1: $\pi_1 \neq .5$ indicates that the proportion of subjects who identify *Nectaberry* as the fruitier soda is some value other than .5. The use of the directional alternative hypothesis H_1: $\pi_1 > .5$ indicates that the proportion of subjects who identify *Nectaberry* as the fruitier soda is some value greater than .5. As is always the case with the classical hypothesis testing model, any inferences which are derived from the analysis will be based on an evaluation of the empirical data in reference to the relevant sampling distribution (which in this case is the binomial distribution).

When the **binomial sign test for a single sample** is employed to evaluate Example 9.34, the null hypothesis cannot be rejected. The latter is the case, since employing **Table A7** (or Equation 9.5) we determine that when $n = 14$ and $\pi_1 = .5$, the likelihood of x being equal to or greater than 9 is .2120. Since the obtained probability .2120 is greater than $\alpha/2 = .05/2 = .025$, the nondirectional alternative hypothesis H_1: $\pi_1 \neq .5$ is not supported at the .05 level. Since .2120 is greater than $\alpha = .05$, the directional alternative hypothesis H_1: $\pi_1 > .5$ is also not supported at the .05 level. Thus, employing the classical hypothesis testing model, the researcher cannot conclude that *Nectaberry* is perceived as fruitier than *Fruizzle*.

Before conducting a Bayesian analysis of Example 9.34, it is worth noting that within the framework of the classical hypothesis testing model a researcher can derive what is commonly referred to as a **likelihood function**. In point of fact, within the classical hypothesis testing model, the **likelihood function** is represented by the relevant sampling distribution (Winkler, (1972, p. 369)). Through use of the **likelihood function**, a **maximum likelihood estimator** can be identified, which represents the most likely value for the population parameter under study. A **maximum likelihood estimator** in the classical hypothesis testing model will always be completely dependent on the information obtained from the sample data. Underlying the derivation of a **likelihood function** is the **principle of maximum likelihood**, which in the classical model assumes that the best estimate of a population parameter will always be based on the value obtained for the relevant sample statistic. In the case of Example 9.34, the parameter the researcher is attempting to estimate is the proportion π_1 (i.e., the actual proportion of people in the population who select *Nectaberry* as the fruitier soda). The sample data for the experiment indicate that the proportion of subjects who select *Nectaberry* is $p_1 = 9/14 = .6429$. If we base the **maximum likelihood estimator** entirely on the information contained in the sample data, it logically follows that the value .6429 will be our best estimate for the value of π_1. To go further, if we consider the possibility that π_1 can be any proportion in the range 0 to 1, we can describe the likelihood function through use of a table or graph. Table 9.12 illustrates how the likelihood function for Example 9.34 can be summarized in a tabular format. Eleven possible values which π_1 can assume (out of the infinite number of possible values between 0 and 1) are recorded in the first column of the table. Each value recorded in Column 2 is the likelihood that 9 out of 14 subjects will select Category 1 (i.e., *Nectaberry* in our example) if the actual value of π_1 is equal to the π_1 value recorded for that row. The latter probabilities can be computed with Equation 9.3 or through use of **Table A6**. (The protocol for determining a probability value in **Table A6** when the value of π_1 is greater than .5 is described at the end of Section IV.) Note that the π_1 value with the highest probability in Column 2 will always be that which corresponds to the p_1 value computed for the sample data (i.e., in the case of Example 9.34, the value $\pi_1 = .6429$).

The information contained in Table 9.12 can also be presented in the form of a graph, which is referred to as a **likelihood curve**. The latter curve is a plot of the likelihoods/ probabilities in

Column 2 (which are recorded on the ordinate/Y-axis) in reference to the values π_1 may assume (which are recorded on the abscissa/X-axis). The probabilities recorded in Column 2 of Table 9.12 (as well as on the ordinate of the graph of the likelihood function) are conditional probabilities, since the likelihood associated with the values of the parameter represents the probability of obtaining the value computed for the sample statistic, given that the true value of the parameter is the value specified in Column 1 of the table (or on the abscissa of a graph).

Table 9.12   Table of Likelihood Function for Example 9.34

π_1	$P(x = 9/ n = 14)$
0	0
.1	.0000
.2	.0003
.3	.0066
.4	.0408
.5	.1222
.6	.2066
.6429	.2181
.7	.1963
.8	.0860
.9	.0078
1	0

Prior to illustrating the use of Bayesian analysis in hypothesis testing (specifically, with Example 9.34), a general discussion of the concept of probability is in order. Among others, Chou (1989) and Phillips (1973) note that there are a number of different ways of conceptualizing probability.[26] From a purely mathematical perspective, the probability of an event is a real number which falls between the values 0 to 1 which reflects the likelihood the event will occur. To further clarify the mathematical definition of probability, we will define a **sample space** as the complete set of all possible events which can occur. Consequently, if we have a sample space in which there are m possible events, the probability for each of the m events must fall between 0 and 1. In addition, the sum of the probabilities for all of the events which comprise the sample space must equal 1. The aforementioned mathematical definition of probability is limited by the fact that it does not provide a way which will allow one to arrive at accurate probabilities for most events that occur in the real world. Because of the latter, the following three approaches are employed for conceptualizing probabilities in the real world: a) The **classical approach**; b) The **relative-frequency approach**; c) The **personalistic approach**.

The **classical approach** for determining probabilities derives from the analysis of games of chance, and is predicated on the idea that if we have a sample space comprised of m events (e.g., the two possible outcomes associated with a coin flip, the six possible outcomes associated with the roll of a die, etc.), and each of the events is deemed equally likely to occur, then the probability of any event within the sample space can be computed by the equation $1/m$. More generally, we can state that the probability of an event is obtained by dividing the number of possible ways the event can occur by the total number of possible events. The accuracy of this approach for defining probability will depend on the validity of its underlying assumption — i.e., that each of the m events is equally likely to occur. Thus, if we assume we have a fair coin, the classical approach establishes the likelihood of obtaining either a **Heads** or **Tails** as 1/2. In the same respect, if we have a fair die, the likelihood of obtaining any of the integer numbers between 1 and 6 is 1/6. If, however, a coin or die is not fair, the probabilities computed through use of the **classical approach** will not be correct. We typically evaluate whether or not we are justified in employing the **classical approach** for establishing probabilities for real world events through use of the **relative-frequency approach**, which is the second approach to be described for establishing probabilities.

The **relative-frequency approach** is predicated on the principle that the only way to establish valid probabilities is through use of repetitive sampling. Specifically, by repetitively studying the events in question, the observed frequencies for the events can be employed as the basis for establishing probabilities for all of the events which comprise the sample space. Thus, if we wish to establish a probability that a flip of a coin will produce a **Heads** versus a **Tails**, we can toss a coin 100 times and use the observed proportion for each of the two outcomes to represent the probability for that outcome. As an example, if we obtain 52 **Heads** and 48 **Tails** on 100 flips of a coin, using the **relative-frequency approach**, we would say that the probability of a **Heads** is 52/100 and the probability of a **Tails** is 48/100. In the eighteenth century the British mathematician James Bernoulli proved the **law of large numbers**, which states that the relative frequency of an event will get closer and closer to its theoretical probability (as defined by the **classical approach**) as the number of trials employed to evaluate the event approaches infinity. Thus, with respect to the coin tossing situation, the **law of large numbers** states that the more times we flip a fair coin, the closer the proportion obtained for both **Heads** and **Tails** will be to the expected/theoretical value of .5. Note, however, that if the coin is not fair, or the person or circumstances involved in flipping the coin are responsible for making one outcome more likely than the other, the probability derived through use of the **relative-frequency approach** will not equal the theoretical probability value derived from the **classical approach**. Phillips (1973, p. 22) notes that one problem with employing the **relative-frequency approach** for establishing probabilities is that the relative frequencies for past events may not be the same as the relative frequencies of future events. In other words, the conditions under which events are evaluated may change, thus invalidating the accuracy of earlier probabilities which were once applicable. Another limitation of the **relative-frequency approach** is that it cannot be applied to unique events. For instance, the latter approach does not allow us to establish the probability that a major earthquake will occur tomorrow in California or that a specific horse will win the Kentucky Derby. To determine probabilities for the latter types of events one must employ the **personalistic approach**, which is the third approach for establishing probabilities.

The **classical approach** for establishing probabilities is objective, since it is logically derived from a set of assumptions. The **relative-frequency approach** is also objective, since the probabilities it generates are purely a function of empirical data. The **personalistic approach**, on the other hand, is subjective, since it employs one's personal opinion as the basis for establishing probabilities. Thus, in all likelihood, the probability you would establish for a major earthquake occurring tomorrow in California will not be the same as the probability someone else would establish for that same event. Each person establishing a probability for the event will base his or her determination on whatever information he or she has available. Establishing probabilities for a horse race illustrates another example of the use of the **personalistic approach**. Let us assume there are five horses entered in a race, and you are asked to establish the odds (which is another way of expressing probabilities) for the race. If you are knowledgeable about horse racing and have followed the horses entered in the race, it is likely that you will assign each horse a probability based on whatever knowledge is available to you (e.g., such things as the horse's past performance, the horse's current state of health, the jockey assigned to ride the horse, etc.). Yet, someone else asked to handicap the same race may establish odds which are quite different from yours. If a person who was totally ignorant of horse racing was asked to establish odds, he might employ a logical strategy such as assigning each of the horses an equal likelihood of winning (i.e., each horse would be assigned a likelihood of 1/5, and thus the odds against each horse winning the race would be 4 to 1). (The concept of odds is explained in detail later in this section.) The latter strategy is predicated on the fact that the less information one has available, the more difficult it will become to make a probabilistic distinction between any of the horses. The condition of assigning an equal prior probability to all events in a sample space, which is often referred to as a **diffuse prior state**, is discussed in more detail later in this section.

Critical to the **personalistic approach** for establishing probabilities is the question of how one can go about establishing a subjective probability for an event. Chou (1989, pp. 187–187; p. 922) describes an ingenious method called the **lottery technique** developed by L. J. Savage (1962) which can be employed to establish subjective probabilities. In order to demonstrate the **lottery technique** let us assume that we want to come up with a subject's best estimate of the probability that an event, which we will label X, will occur. As an example, the *Event X* can be that a major earthquake will occur tomorrow in California or that a specific horse will win the Kentucky Derby. Within the framework of hypothesis testing, if two hypotheses are being evaluated, the *Event X* can be that Hypothesis A is true and Hypothesis B is false. In order to establish a subject's prior probabilities, the subject is presented with the two lotteries outlined below. The subject is told that she must purchase a ticket for one of the lotteries, and that she should select the lottery which is most attractive to her — i.e., the one which she believes offers her the highest likelihood of winning.

Lottery 1

When you purchase a ticket for **Lottery 1** you have a .5 probability of being a winner and a .5 probability of being a loser. If you purchase a winning ticket you win $10,000. If you purchase a losing ticket you win nothing.

Lottery 2

If you elect to play **Lottery 2**, you win $10,000 if *Event X* occurs and you win nothing if it does not occur.

Depending upon which of the two lotteries the subject elects to play, we can draw any one of the three conclusions summarized below with regard to the value of the prior probability the subject has assigned for the occurrence of *Event X*.

a) If the subject believes that the exact likelihood of *Event X* occurring is equal to .5 she will declare that it does not matter which of the two lotteries she plays — i.e., she states that you can give her a ticket for either Lottery 1 or Lottery 2, since she has no preference for either one. By doing the latter, the subject is indicating that her perception is that both lotteries offer her the same likelihood of winning — specifically, .5, which is the likelihood stipulated for Lottery 1.

b) If the subject elects to play Lottery 1, it indicates that she believes the likelihood of *Event X* occurring is less than .5, since if she believed it is greater than .5 she would have elected to play Lottery 2. To determine the precise probability value below .5 the subject will assign as the likelihood of *Event X* occurring, we would again ask her to play the identical two lotteries, with the only difference being that this time the probability associated with a winning ticket in Lottery 1 will be some value less than .5. Thus, we could tell the subject that if she buys a ticket for Lottery 1 she has only a .4 probability of being a winner (and thus a .6 probability of being a loser). Under the latter conditions, if the subject still elects to play Lottery 1 it will indicate that she believes the likelihood of *Event X* occurring is less than .4, since if she believed it is greater than .4 she would have elected to play Lottery 2. We would have the subject continue to play the two lotteries until the probability value associated with a winning ticket results in her stating that she is indifferent with respect to which of the two lotteries she plays. Each time she plays the two lotteries, she would be told that the probability of her selecting a winning ticket in Lottery 1 is less than the previous time she played (i.e., after $p = .4$, the value of p can be reduced to .35, and if she still elects to play Lottery 1 it can be reduced the next time she plays to .3, and so on). Thus, by progressively lowering the likelihood of the subject winning Lottery 1, we can zero in on the point at which the likelihood of winning Lottery 2 is viewed as equivalent to the likelihood of winning Lottery 1, and by virtue of the latter identify the prior probability the subject will assign to Event X.

c) If the subject elects to play Lottery 2, it indicates that she believes the likelihood of *Event X* occurring is greater than .5, since if she believed it is less than .5 she would have elected to play Lottery 1. To determine the precise probability value above .5 the subject will assign as the likelihood of *Event X* occurring, we would again ask her to play the identical two lotteries, with the only difference being that this time the probability associated with a winning ticket in Lottery 1 will be some value greater than .5. Thus, we could tell the subject that if she buys a ticket for Lottery 1 she has a .6 probability of being a winner (and thus a .4 probability of being a loser). Under the latter conditions, if the subject still elects to play Lottery 2 it will indicate that she believes the likelihood of *Event X* occurring is greater than .6, since if she believed it is less than .6 she would have elected to play Lottery 1. We would have the subject continue to play the two lotteries until the probability value associated with a winning ticket results in her stating that she is indifferent with respect to which of the two lotteries she plays. Each time she plays the two lotteries, she would be told that the probability of her selecting a winning ticket in Lottery 1 is greater than the previous time she played (i.e., after $p = .6$, the value of p can be increased to .65, and if she still elects to play Lottery 2 it can be increased the next time she plays to .7, and so on). Thus, by progressively increasing the likelihood of the subject winning Lottery 1, we can zero in on the point at which the likelihood of winning Lottery 2 is viewed as equivalent to the likelihood of winning Lottery 1, and by virtue of the latter identify the prior probability the subject will assign to Event X.

At this point it should be noted that whereas the classical hypothesis testing model is based on the use of the **relative-frequency approach** for establishing probabilities, Bayesian hypothesis testing combines the **relative-frequency approach** with the **personalistic approach**. The Bayesian approach to hypothesis testing differs from the classical hypothesis testing model in a number of different ways. It was noted earlier in this section, that, as a general rule, in the classical hypothesis testing model an exact null hypothesis is contrasted with an inexact alternative hypothesis. On the other hand, in Bayesian analysis a researcher computes probabilities for two or more competing hypotheses, each of which stipulates an exact value (or range of values) for the population parameter under investigation. In addition, the two approaches to hypothesis testing derive different conditional probability values. As noted in Endnote 21 in the **Introduction**, in classical hypothesis testing, a conditional probability (which is based on the relevant sampling distribution) is computed (or, more commonly, derived from tables) which indicates the likelihood of obtaining the sample data in an experiment, given the fact that the null hypothesis is true. In Bayesian hypothesis testing, conditional probabilities are computed which reflect likelihoods that two or more hypotheses are true, given the sample data obtained in an experiment.

Chou (1989, p. 911) notes that another difference between Bayesian and classical hypothesis testing is that in classical hypothesis testing an unknown population parameter is viewed as a fixed/constant quantity, whereas in Bayesian hypothesis testing an unknown population parameter is viewed as a random variable. In order to clarify the latter, a third difference between the two hypothesis testing models must be stipulated — specifically, as noted earlier, the use of subjective/personalistic probabilities within the framework of Bayesian analysis. The latter types of probabilities are rejected by the classical hypothesis testing model, which, as noted earlier, employs the **relative-frequency** view of probability. Specifically, the classical model views an unknown parameter as having a fixed value, and only employs objective evidence (specifically, sample data) to draw conclusions regarding the value of the parameter. Bayesian hypothesis testing operates on the assumption that a parameter can assume any of a number of values, and begins by estimating a prior probability for each of the hypothesized values which are under consideration for the true value of the parameter. The results of empirical studies are then employed in conjunction with the prior probabilities to compute revised probabilities (referred to as **posterior probabilities**) for each of the hypotheses under consideration.

Before proceeding further, it is worth noting a number of circumstances cited by Chou (1989, p. 1038) in which the classical hypothesis testing model may be preferable to employing a Bayesian approach.[27]

a) When there is a large amount of sample data available and the cost of obtaining the data is minimal, the classical hypothesis testing model may be preferable to the Bayesian approach.

b) When the circumstances or the underlying distribution make it extremely difficult or impossible to establish prior probabilities, the classical hypothesis testing model may be preferable to the Bayesian approach.

c) When the consequences associated with the decision making process cannot be easily quantified, the classical hypothesis testing model may be preferable to the Bayesian approach. In point of fact, a major component of Bayesian statistics is **Bayesian decision theory** (which is discussed in detail in books which specialize in Bayesian analysis). Bayesian decision theory involves quantifying the consequences of the decisions which are made on the basis of inferences that derive from the use of **Bayes' theorem**. Bayesian decision theory utilizes procedures which contrast the benefits that result from making correct decisions with any losses which might result from making incorrect decisions. Not surprisingly, the use of Bayesian statistics is more common in disciplines such as business and economics, which routinely employ measures such as profits and losses within the framework of statistical analysis.

d) When a researcher is unable to establish prior probabilities that discriminate between the competing hypotheses, there is nothing to be gained by employing Bayesian statistics in lieu of the classical hypothesis testing model. The inability to establish prior probabilities is most likely to occur when a researcher lacks prior information regarding any of the hypotheses under consideration. Under the latter circumstances the researcher will not have any preconceived biases favoring one hypothesis over another. When the researcher cannot stipulate prior probabilities which discriminate between the competing hypotheses, all of the hypotheses will be assigned the same prior probability (and the sum of all the prior probabilities will equal 1). This latter condition in which all hypotheses are assigned an identical prior probability is commonly referred to as a **diffuse prior state**. An example of a **diffuse prior state** would be if two hypotheses are under consideration, and the researcher assigns a prior probability of .5 to each of them. When all of the hypotheses under consideration are assigned the same prior probability, the posterior probabilities computed with **Bayes' theorem** will only be a function of the sample data. Under such conditions the Bayesian and classical hypothesis testing models yield equivalent results.

Phillips (1973, p. 88) notes that rarely if ever does a researcher conduct a study in which he has no prior opinion regarding its outcome. In other words, more often than not, a researcher has biases in favor of or against one or more hypotheses which are being evaluated in a study. The philosophy of Bayesian statistics is to make such prior opinions public by formally incorporating them in the statistical analysis. Understandably, before a sufficient amount of data has been collected, the prior probabilities established by two or more researchers may be quite different from one another. However, when through use of **Bayes' theorem** the results of empirical research are combined with a researcher's prior probabilities, the revised probabilities (i.e., the **posterior probabilities**) computed by various researchers will be in closer agreement with one another than were their original prior probabilities. The more experimental data that are evaluated, the more similar will become the revised probability values computed by two or more researchers.

At this point in the discussion a Bayesian analysis will be employed to evaluate the relative merits of two hypotheses. It is often the case that when a Bayesian analysis is employed to evaluate two hypotheses, the hypothesis with the higher prior probability is designated as the null hypothesis and hypothesis with the lower prior probability is designated as the alternative hypothesis. Let us assume that in employing a Bayesian analysis with Example 9.34, the researcher is skeptical of the claim that *Nectaberry* is the fruitier soda. Because of the latter, as well as due to some limited information which is available regarding the two sodas, he establishes a prior probability of .7 for the hypothesis that the true value of π_1 in the population is .5. This will represent the null hypothesis. Thus, H_0: $\pi_1 = .5$. He elects to employ H_1: $\pi_1 = .6$ as the alternative hypothesis, and assigns the latter hypothesis a prior probability of .3. In other words, he is stating that there is a .3 likelihood the true value of π_1 in the population is .6. Note that the sum of the prior probabilities for all of the hypotheses under consideration must equal one.

Table 9.13 First Stage of Bayesian Analysis of Example 9.34 Involving Two Hypotheses

(1)	(2)	(3)	(4)	(5)
Hypothesized value of parameter π_1	*Prior probability* $P(A_j)$	*Likelihood* $P(B/A_j)$	*Joint probability =Likelihood x Prior probability* $P(B/A_j)P(A_j)$	*Posterior probability* $\dfrac{P(B/A_j)P(A_j)}{\Sigma[P(B/A_i)P(A_i)]}$
.5	.7	.1222	(.1222)(.7) = .08554	.08554/.14752 = .57985
.6	.3	.2066	(.2066)(.3) = .06198	.06198/.14752 = .42015
	$\Sigma = 1$		$\Sigma = .14752$	$\Sigma = 1$

Table 9.13 summarizes the results of a Bayesian analysis which evaluates the relative merits of the hypothesized π_1 values .5 and .6. Winkler (1972, p. 393) notes that, as a general rule, Bayesian statistical analysis is more concerned with problems of decision making than with problems of inference (which is the major concern of the classical hypothesis testing model). Consequently, the Bayesian statistician is primarily concerned with computing posterior probabilities for all of the competing hypotheses (as is done in Table 9.13) in order to determine which of them is the most tenable. It should be emphasized that any advantage resulting from the use of Bayesian statistics in hypothesis testing (when contrasted with the classical hypothesis testing model) will be a direct function of the accuracy of the prior probabilities estimated by the researcher. The validity of a Bayesian analysis will be compromised if a researcher does not have

sufficient information to allow him to come up with reasonable estimates for prior probabilities and/or the researcher is just poor at estimating probabilities in instances when one has adequate information to do so.

The logic employed in the Bayesian analysis can be best understood through an explanation of what the values in the five columns in Table 9.13 represent

Column 1 lists each of the values the researcher wishes to consider as the actual value of π_1 in the underlying population. Realistically, a researcher should include in Column 1 a row for any probability value which he believes has even a minimal likelihood of being the true value of π_1. In practice, a researcher will most likely not include every possible value, since there can literally be an infinite number of possible values one might consider. For example, we could include in Table 9.13 a row for every proportion that is equal to or greater than .5 and less than or equal to .8. If we employ proportional values which are carried out to three decimal places (e.g., .500, .501, ..., .799, .800) we would have to include 301 probability values in Column 1 of Table 9.13. In the next section, a Bayesian analysis will be employed with Example 9.34 in which four hypotheses are taken into consideration.

Column 2 lists the prior probabilities the researcher has assigned to each of the hypothesized π_1 values in Column 1. As noted earlier, the prior probability values recorded in Column 2 are subjective in nature, and have to be based on any preexisting information the researcher has which would predispose him to view one or more of the hypothesized values in Column 1 as more tenable than one or more of the other values. As noted earlier, the sum of the prior probabilities recorded in Column 2 must equal 1, since we are assuming that the analysis includes every hypothesis the researcher would view worthy of consideration. The notation $P(A_j)$ is employed in Table 9.13 to represent the prior probability assigned to a given π_1 value.

Column 3 contains the binomial probabilities/likelihoods computed for the sample data in reference to the π_1 values listed in Column 1. To be more specific, each value in Column 3 represents the likelihood that 9 out of the 14 subjects will select Category 1 (i.e., *Nectaberry* soda) if the value of π_1 employed in Equation 9.3 is equal to the π_1 value listed for a given row. As noted earlier, these probability values can also be determined through use of **Table A6**. In point of fact, the likelihoods listed in Column 3 are identical to the probabilities recorded for the π_1 values .5 and .6 in Column 2 of Table 9.12. The notation $P(B/A_j)$ is employed to represent the likelihood of obtaining the sample data, given that the true value of π_1 is the value designated for a given row. Note that I have employed the notation B in lieu of $B+$ (which is employed in the previous examples illustrating diagnostic decisions for a single subject) to represent the outcome of the experiment. In the case of Example 9.34, B represents the sample data, which are the outcome $x = 9$ out of the $n = 14$ subjects selecting Category 1.

Column 4 lists the joint probabilities, which for each π_1 value represent the numerator of Equation 9.31 (i.e., the equation for **Bayes' theorem**) when the latter equation is employed to compute the conditional probability $P(A_j/B)$. The joint probability in each row (represented by the notation $P(B/A_j)P(A_j)$) is obtained by multiplying the likelihood $P(B/A_j)$ by the prior probability $P(A_j)$ recorded for that row. Note that the sum of the joint probabilities in Column 4 will represent the denominator of **Bayes' theorem** which will be employed in computing each of the posterior probabilities that are recorded in Column 5.

Column 5 lists for each row the posterior probability $P(A_j/B)$ computed with Equation 9.31 (i.e., it is obtained by dividing the joint probabilities recorded in Column 4 for a given row/π_1 value by the sum of the joint probabilities in Column 4). The posterior probability for each row is the likelihood that the true value of the population proportion is equal to the π_1 value recorded for that row, given the fact that 9 of the 14 subjects selected Category 1 (i.e., *Nectaberry* soda). Note that the sum of the posterior probabilities in Column 5 must equal 1.

The posterior probabilities computed in Column 5 (which are a function of both the researcher's prior probabilities and the information contained in the sample data) are the researcher's revised probabilities for the two hypotheses being evaluated. Thus, by taking into account the prior probability of .7 and the observed data, the likelihood of the null hypothesis H_0: $\pi_1 = .5$ has been reduced from its original prior probability of .7 to .57985. On the other hand, the likelihood of the alternative hypothesis H_1: $\pi_1 = .6$ has been increased from its original prior probability of .3 to .42015.

Let us assume the researcher conducts a second experiment evaluating the same hypotheses. In the latter experiment 10 of 14 subjects select *Nectaberry* as the fruitier soda, while 4 of the subjects select *Fruizzle* as the fruitier soda. Once again, we can employ **Bayes' theorem** to evaluate the data, but this time instead of employing the prior probabilities which are listed in Column 2 of Table 9.13, we will use as our prior probabilities the posterior probabilities which were computed in Column 5 of Table 9.13. The reason for this is as follows. Recollect that the prior probabilities the researcher originally employed were based on whatever information he had regarding the tenability of each of the two hypotheses prior to conducting the first experiment. As a result of evaluating the outcome of the first experiment, the researcher has additional information he can use to establish a new prior probability for each hypothesis. The posterior probability computed for each of the two hypotheses will represent the new prior probability for that hypothesis, since it incorporates the researcher's original prior probabilities with the results of the first experiment.

Table 9.14 Second Stage of Bayesian Analysis of Example 9.34 Involving Two Hypotheses

(1)	(2)	(3)	(4)	(5)
Hypothesized value of parameter π_1	Prior probability $P(A_j)$	Likelihood $P(B/A_j)$	Joint probability=Likelihood x Prior probability $P(B/A_j)P(A_j)$	Posterior probability $\dfrac{P(B/A_j)P(A_j)}{\Sigma[P(B/A_i)P(A_i)]}$
.5	.57985	.0611	(.0611)(.57985) = .03543	.03543/.10051 = .35250
.6	.42015	.1549	(1549)(.42015) = .06508	.06508/.10051 = .64750
	$\Sigma = 1$		$\Sigma = .10051$	$\Sigma = 1$

The results of employing **Bayes' theorem** to evaluate the second experiment are summarized in Table 9.14. Note that Column 1 (which lists the hypothesized values of π_1) is the only column in Table 9.14 which contains values that are identical to those in the corresponding column of Table 9.13. As previously noted, the values of the prior probabilities in Column 2 of Table 9.14 are the posterior probabilities computed for each of the hypotheses based on the results of the first experiment. Note that, as is the case in Table 9.13, the sum of the prior probabilities in Column 2 must equal 1 (since we are assuming that every hypothesis under consideration is included in the analysis). The likelihood values in Column 3 of Table 9.14 are the binomial probabilities computed for the π_1 values listed in Column 1. In the case of the second experiment, each value in Column 3 represents the likelihood that $x = 10$ out of the $n =$

14 subjects will select Category 1 (i.e., *Nectaberry* soda) if the value of π_1 employed in Equation 9.3 is equal to the π_1 value listed for a given row. The joint and posterior probability values in Columns 4 and 5 are determined in the identical manner they were determined in Table 9.13. Note that, as is the case in Table 9.13, the sum of the joint probabilities in Column 4 represents the denominator of **Bayes' theorem** when it is employed to compute each of the posterior probabilities recorded in Column 5. The posterior probabilities in Column 5 represent the new revised probabilities for the two hypotheses. These new revised probabilities are a function of both: a) The researcher's original prior probabilities and the outcome of the first experiment (both of which are documented in Table 9.13); and b) The outcome of the second experiment. Note that, as is the case in Table 9.13, the sum of the posterior probabilities in Column 5 will always equal 1.

Based on the results of the second experiment, the likelihood of the null hypothesis H_0: π_1 = .5 has been reduced from the .57985 (which is computed in Table 9.13) to .35250. Note that the latter value is substantially less than the original prior probability of .7 for the null hypothesis. On the other hand, the likelihood of the alternative hypothesis H_1: π_1 = .6, which had an original prior probability of .3, has been increased from .42015 to .64750. By employing **Bayes' theorem** sequentially with two sets of empirical data, the researcher has substantially modified his perceptions with regard to the tenability of the two hypotheses, and it would seem that at this point in time the most tenable hypothesis (H_1: π_1 = .6) is consistent with the claim that *Nectaberry* is the fruitier soda (since it stipulates a π_1 value above .5).

It should be noted that at this point neither of the two hypotheses has a posterior probability equal to or greater than .95 (which within the framework of the classical hypothesis testing model would be commensurate with achieving significance at the .05 level). If a researcher requires that before he accepts a hypothesis it must have a posterior probability of at least .95 (or some higher value such as .99), additional studies can be conducted until the analysis of the data yields the desired posterior probability value for the hypothesis.

Table 9.15   Composite Bayesian Analysis of Example 9.34 Involving Two Hypotheses

(1)	(2)	(3)	(4)	(5)
Hypothesized value of parameter π_1	Prior probability $P(A_j)$	Likelihood $P(B/A_j)$	Joint probability=Likelihood x Prior probability $P(B/A_j)P(A_j)$	Posterior probability $\dfrac{P(B/A_j)P(A_j)}{\Sigma[P(B/A_i)P(A_i)]}$
.5	.7	.02573	(.02573)(.7) = .01801	.01801/.05111 = .35238
.6	.3	.11033	(.11033)(.3) = .03310	.03310/.05111 = .64762
	Σ = 1		Σ = .05111	Σ = 1

Since the 28 subjects evaluated in the two experiments are independent of one another, the same posterior probabilities computed in Column 5 of Table 9.14 will also be obtained if **Bayes' theorem** is employed using the original prior probabilities in Column 1 of Table 9.13 and the likelihood for the combined results of the two experiments. The latter likelihood is the probability of x = 19 (i.e., 9 + 10 = 19) of the n = 28 (i.e., 14 + 14 = 28) subjects selecting *Nectaberry* as the fruitier soda. The result of the latter analysis is summarized in Table 9.15. (The minimal differences between the posterior probabilities in Column 5 of Tables 9.14 and 9.15 are due to rounding off error.) The likelihoods in Column 3 of Table 9.15 are the binomial probabilities for $x = 19$ and $n = 28$ for the two π_1 values specified in Column 1. Equation 9.3 was employed to determine each of the likelihoods in Column 3. For example, in the case of $\pi_1 = .5$, the probability .02573 in Column 3 is computed as follows: (Note: In Equation 9.3, $1 - \pi_1 = \pi_2$.)

$$P(19/28) = \binom{n}{x} (\pi_1)^x (1 - \pi_1)^{(n - x)} = \binom{28}{19}(.5)^{19}(.5)^9 = .02573$$

Although we have obtained the same result by employing **Bayes' theorem** with the combined results of the two experiments, the reader should keep in mind that the researcher might not have conducted a second study if the results of the first experiment strongly indicated that subjects did not perceive *Nectaberry* to be the fruitier soda (i.e., if substantially less than 50% of the subjects in the first study selected *Nectaberry* as the fruitier soda). Thus, often a researcher might conduct one or more studies, and following each study, based on the computed revised posterior probabilities, he will make a decision with respect to whether or not additional studies are required before committing himself with respect to which of the hypotheses is most tenable. At the point the final study is evaluated, the posterior probabilities obtained for the latter study will equal the posterior probabilities computed for the combined results of the multiple studies (assuming one is dealing with independent events, as is the case in a binomial analysis).

It should be noted that if the classical hypothesis testing model is employed to evaluate the result of the second experiment, the result is not significant. Specifically, when the **binomial sign test for a single sample** is employed (through use of Equation 9.5 or **Table A7**) we determine that when $n = 14$ and $\pi_1 = .5$, the likelihood of x being equal to or greater than 10 is .0898. Since the obtained probability .0898 is greater than $\alpha/2 = .05/2 = .025$, the nondirectional alternative hypothesis H_1: $\pi_1 \neq .5$ is not supported at the .05 level. Since .0898 is greater than $\alpha = .05$, the directional alternative hypothesis H_1: $\pi_1 > .5$ is also not supported at the .05 level. Thus, employing the classical hypothesis testing model, the researcher cannot conclude that *Nectaberry* is perceived as fruitier than *Fruizzle*.

When the classical hypothesis testing model is employed to evaluate the combined result of the two experiments, the nondirectional alternative is not supported but the directional alternative hypothesis is supported. Specifically, when the **binomial sign test for a single sample** is employed, through use of Equation 9.5 (**Table A7** does not list probabilities for values of n above 15), we determine that when $n = 28$ and $\pi_1 = .5$, the likelihood of x being equal to or greater than 19 is .0436. Since the obtained probability .0436 is greater than $\alpha/2 = .05/2 = .025$, the nondirectional alternative hypothesis H_1: $\pi_1 \neq .5$ is not supported at the .05 level. Since .0436 is less than $\alpha = .05$, the directional alternative hypothesis H_1: $\pi_1 > .5$ is supported at the .05 level. Thus, within the framework of the classical hypothesis testing model, the use of a directional alternative hypothesis allows the researcher to conclude that *Nectaberry* is perceived as fruitier than *Fruizzle*.

Within the framework of conducting a Bayesian analysis, by employing the information in Tables 9.13–9.15, it is possible to compute what are commonly referred to as **odds ratios**. Pitman (1993, p. 6) notes that just about any ratio of probabilities (or odds, which are another way of expressing probabilities) can be called an **odds ratio**.

Prior to discussing the computation of **odds ratios**, it will be useful to clarify the concept of **odds**. Odds indicate how much more likely it is that an event will occur as opposed to it not occurring. The odds that an event (to be designated A) will occur are computed by dividing the probability the event will occur by the probability the event will not occur. The latter is summarized by Equation 9.35. In the opposite respect, the odds against an event occurring are computed by dividing the probability the event will not occur by the probability the event will occur. The latter is summarized by Equation 9.36. It is also the case that the probability an event will occur can be determined by dividing the odds the event will occur by (1 + *the odds the event will occur*). The latter relationship is summarized by Equation 9.37 (Phillips (1973), p. 24).

$$Odds(A) = \frac{P(A \ will \ occur)}{P(A \ will \ not \ occur)} = \frac{P(A)}{1 - P(A)} \qquad \textbf{(Equation 9.35)}$$

$$Odds(Against \ A) = \frac{P(A \ will \ not \ occur)}{P(A \ will \ occur)} = \frac{1 - P(A)}{P(A)} \qquad \textbf{(Equation 9.36)}$$

$$Probability(A) = \frac{Odds(A)}{1 + Odds(A)} \qquad \textbf{(Equation 9.37)}$$

The following should be noted with respect to **odds**: a) The computed value for odds can fall anywhere in the range 0 to infinity; b) When the value computed for odds is greater than 1, it indicates that the probability of an event occurring is greater than one-half ($\frac{1}{2}$). Thus, the larger the odds of an event occurring, the higher the probability the event will occur; c) When the value computed for odds is less than 1, it indicates that the probability of an event occurring is less than one-half ($\frac{1}{2}$). Thus, the smaller the odds of an event occurring, the lower the probability the event will occur. The lowest value that can be computed for odds is 0, which indicates that the probability the event will occur is 0; d) When the value computed for odds equals 1, it indicates that the probability of the event occurring is one half ($\frac{1}{2}$) — i.e., there is a 50–50 chance of the event occurring; e) More often than not, when odds are published in the media they are the odds that an event will not occur rather than the odds that the event will occur. As an example, if the bookmakers' odds of a horse winning the Kentucky Derby are 4 to 1, it indicates that the bookmakers are saying there is only a .2 probability the horse will win the derby, and a .8 probability the horse will not win. Consequently, the odds of the horse not winning the derby are computed as follows: *P(Will not win)/P(Will win)* = .8/.2 = 4, which is generally expressed as odds of 4 to 1 (often written as 4:1). If in the opinion of the bookmakers the likelihood that a horse will win the derby is 2 out of 5, then the odds are computed by dividing the likelihood the horse will not win (3/5) by the likelihood the horse will win (2/5), which is 3/2 (or 1.5). The latter odds are expressed as 1.5 to 1 (or 1.5:1) or more commonly as 3 to 2 (or 3:2); f) If the probability of an event is greater than .5, the odds **against** the event will be less than 1, and if the probability of an event is less than .5, the odds **against** the event will be greater than 1. If the probability of an event is .5, the odds against the event will equal 1.[28]

Within the framework of the Bayesian analysis of Example 9.34 the following three **odds ratios** can be computed: a) The **prior odds ratio**, which will be represented by the notation OR_{Prior}; b) The **likelihood ratio**, which Chou (1989, p. 937) notes is also referred to as the **information odds ratio**. The notation *LR* will be employed to represent the **likelihood ratio**; c) The **posterior odds ratio**, which will be represented by the notation $OR_{Posterior}$.

The **prior odds ratio** (OR_{Prior}) is the ratio of the prior probabilities for the two hypotheses under consideration (where H_x and H_y represent the two hypotheses). Equation 9.38 (which employs the values in Column 2 of the Bayesian analysis summary table — i.e., Tables 9.13–9.15) summarizes the computation of the **prior odds ratio**.

$$OR_{prior} = \frac{P(A_x)}{P(A_y)} = \frac{P(H_x)}{P(H_y)} \qquad \text{(Equation 9.38)}$$

If the hypothesis in the numerator of Equation 9.38 is assigned a higher prior probability than the hypothesis in the denominator, OR_{Prior} will be greater than 1. If the hypothesis in the numerator of the equation is assigned a lower prior probability than the hypothesis in the denominator, OR_{Prior} will be less than 1. If the two hypotheses are assigned equal prior probabilities, OR_{Prior} will equal 1. At this point, by using the original prior probabilities recorded in Table 9.13, we will compute a **prior odds ratio** for Example 9.34. If we employ the null hypothesis H_0: $\pi_1 = .5$ in the numerator and the alternative hypothesis H_1: $\pi_1 = .6$ in the denominator, the value $OR_{Prior} = 2.333$ is computed.

$$OR_{Prior} = \frac{P(H_0)}{P(H_1)} = \frac{.7}{.3} = 2.333$$

The value $OR_{Prior} = 2.333$ indicates that the prior probability assigned to the null hypothesis is 2.333 times larger than the prior probability assigned to the alternative hypothesis. Note that we can also reverse the values in the numerator and denominator and compute the value $OR_{Prior} = .3/.7 = .429$, which indicates that the prior probability assigned to the alternative hypothesis is .429 times as large as the prior probability assigned to the null hypothesis.

Let us examine the **prior odds** ratio in more detail by employing Equations 9.35 and 9.36 to compute the **prior odds in favor of** and **against** the null and alternative hypotheses.

$$Odds(\text{In favor of } H_0) = \frac{P(H_0)}{1 - P(H_0)} = \frac{.7}{1 - .7} = 2.333$$

$$Odds(\text{Against } H_0) = \frac{1 - P(H_0)}{P(H_0)} = \frac{1 - .7}{.7} = .429$$

$$Odds(\text{In favor of } H_1) = \frac{P(H_1)}{1 - P(H_1)} = \frac{.3}{1 - .3} = .429$$

$$Odds(\text{Against } H_1) = \frac{1 - P(H_1)}{P(H_1)} = \frac{1 - .3}{.3} = 2.333$$

The value $Odds(\text{In favor of } H_0) = 2.333$ indicates that the researcher is stating (through use of his prior probabilities) that the null hypothesis is 2.333 times more likely to be correct than incorrect (or to put it another way, H_0 is 2.333 times more likely to be correct than H_1). The fact that the odds in favor of the null hypothesis are greater than 1 is consistent with the fact that the null hypothesis has a prior probability greater than .5. In the same respect, the fact that the odds against the null hypothesis are less than 1 (i.e., $Odds \,(\text{Against } H_0) = .429$) is also consistent with the fact that the null hypothesis has a prior probability greater than .5. The value $Odds(\text{In favor of } H_1) = .429$ indicates that the researcher is stating (through use of his prior probabilities) that the alternative hypothesis is .429 times as likely to be correct than incorrect (or to put it another way, H_1 is .429 times as likely to be correct than H_0). The fact that the odds in favor of the alternative hypothesis are less than 1 is consistent with the fact that the alternative hypothesis has a prior probability less than .5. In the same respect, the fact that the odds against the alternative hypothesis are greater than 1 (i.e., *Odds (Against H_1)* = 2.333) is also consistent with the fact that the alternative hypothesis has a prior probability less than .5.

Note that the value OR_{Prior} = 2.333 computed for the **prior odds ratio** is equal to the odds in favor of the null hypothesis (which are identical to the odds against the alternative hypothesis). In addition, the second value computed, OR_{Prior} = .429, is equal to the odds in favor of the alternative hypothesis (which are identical to the odds against the null hypothesis). Thus, the value computed for the **prior odds ratio** will always equal the odds in favor of the hypothesis stipulated in the numerator of the **prior odds ratio** equation (as well as being equal to the odds against the hypothesis stipulated in the denominator of the equation).

Although sources which discuss the **prior odds ratio** (e.g., Chou (1989), Hays and Winkler (1971), and Winkler (1972)) conceptualize it within the context of Equation 9.38, one can also construct a **prior odds ratio** (to be designated $OR_{Prior\ odds}$) based on Equation 9.39.

$$OR_{Prior\ odds} = \frac{Prior\ odds\ in\ favor\ of\ H_x}{Prior\ odds\ in\ favor\ of\ H_y} \qquad \textbf{(Equation 9.39)}$$

Employing Equation 9.39, we can compute $OR_{Prior\ odds}$ = 2.333/.429 = 5.438, which indicates that the odds in favor of the null hypothesis are 5.438 times greater than the odds in favor of the alternative hypothesis. In the same respect, by reversing the values in the numerator and denominator of Equation 9.39, we can compute $OR_{Prior\ odds}$ = .429/2.333 = .184, which indicates that the odds in favor of the alternative hypothesis are .184 times as large as the odds in favor of the null hypothesis.

The information contained in a Bayesian analysis summary table (i.e., Tables 9.13–9.15) also allows a researcher to compute a **likelihood ratio** (*LR*). The latter value, which is based on the conditional probabilities in Column 3 of the Bayesian summary table, is computed with Equation 9.40. Note that the **likelihood ratio** of Hypothesis *x* to Hypothesis *y* is the likelihood of obtaining the observed outcome for the sample data if Hypothesis *x* is true divided by the likelihood of obtaining the observed outcome for the sample data if Hypothesis *y* is true. If the value computed for *LR* is greater than 1, the observed data are more likely to have been obtained if Hypothesis *x* is true than if Hypothesis *y* is true. If the value computed for *LR* is less than 1, the observed data are more likely to have been obtained if Hypothesis *y* is true than if Hypothesis *x* is true. If *LR* = 1, the observed data are equally likely to have occurred if either of the two hypotheses is true.

$$LR = \frac{P(B/A_x)}{P(B/A_y)} = \frac{P(B/H_x)}{P(B/H_y)} \qquad \textbf{(Equation 9.40)}$$

Employing Equation 9.40 with the information contained in Column 3 of Table 9.13, the value *LR* = .591 is computed.

$$LR = \frac{P(B/A_x)}{P(B/A_y)} = \frac{P(B/H_x)}{P(B/H_y)} = \frac{.1222}{.2066} = .591$$

The value *LR* = .591 indicates that the outcome of 9 out of 14 subjects identifying *Nectaberry* as the fruitier soda is .591 times as likely if H_0: π_1 = .5 is true than if H_1: π_1 = .6 is true. By reversing the values in the numerator and denominator of Equation 9.40, we can also compute the value *LR* = .2066/.1222 = 1.691, which indicates that the outcome of 9 out of 14 subjects identifying *Nectaberry* as the fruitier soda is 1.691 times more likely if H_1: π_1 = .6 is true than if H_0: π_1 = .5 is true.

Through use of the information contained in Column 5 of a Bayesian analysis summary table, Equation 9.41 can be employed to compute a **posterior odds ratio**. If the value computed for $OR_{Posterior}$ is greater than 1, the hypothesis in the numerator of the **posterior odds ratio** is more likely to be true than the hypothesis in the denominator. If the value computed for $OR_{Posterior}$ is less than 1, the hypothesis in the numerator is less likely to be true than the hypothesis in the denominator. If the value computed for $OR_{Posterior}$ is equal to 1, the two hypotheses are equally likely to be true.

(Equation 9.41)

$$OR_{Posterior} = \frac{Posterior\ probability\ computed\ for\ Hypothesis\ x}{Posterior\ probability\ computed\ for\ Hypothesis\ y} = \frac{P(A_x/B)}{P(A_y/B)} = \frac{P(H_x/B)}{P(H_y/B)}$$

Employing Equation 9.41 with the information contained in Column 5 of Table 9.13, the value $OR_{Posterior}$ = 1.380 is computed.

$$OR_{Posterior} = \frac{P(A_x/B)}{P(A_y/B)} = \frac{P(H_x/B)}{P(H_y/B)} = \frac{.57985}{.42015} = 1.380$$

The value $OR_{Posterior}$ = 1.380 indicates that the probability of the null hypothesis being true is 1.380 times greater than the probability of the alternative hypothesis being true. By reversing the values in the numerator and denominator of Equation 9.41, we can also compute $OR_{Posterior}$ = .42015/.57985 = .725, which indicates that the probability of the alternative hypothesis being true is .725 times as large as the probability of the null hypothesis being true. Using Equation 9.35, we can determine it is also the case that the posterior odds in favor of the null hypothesis are .57985/.42015 = 1.380 (note that the denominator is 1 − .57985 = .42015), and that the posterior odds in favor of the alternative hypothesis are .42015/.57985 = .725 (note that the denominator is 1 − .42015 = .57985). Consequently, we can say that since 1.380/.725 = 1.903, the posterior odds in favor of the null hypothesis are 1.903 times larger than the posterior odds in favor of the alternative hypothesis (compared with the prior odds in favor of the null hypothesis being 5.438 times greater than the prior odds in favor of the alternative hypothesis).

Following the second experiment, we can compute a revised **posterior odds ratio** based on the values in Column 5 of Tables 9.14/9.15. Employing the values in Table 9.14, we obtain $OR_{Posterior}$ = .35250/.64750 = .544, which indicates that following the second experiment, the probability of the null hypothesis being true is only .544 times as large as the probability of the alternative hypothesis being true (i.e., H_0 is less likely than H_1, since $OR_{Posterior}$ is less than 1). We can also compute $OR_{Posterior}$ = .64750/.35250 = 1.837, which indicates that following the second experiment, the probability of the alternative hypothesis being true is 1.837 times greater than the probability of the null hypothesis being true (i.e., H_1 is more likely than H_0, since $OR_{Posterior}$ is greater than 1). Since 1.837/.544 = 3.377, we can say that the revised posterior odds in favor of the alternative hypothesis are 3.377 times greater than the revised posterior odds in favor of the null hypothesis (compared with the fact that the prior odds in favor of the null hypothesis were 5.438 times greater than the prior odds in favor of the alternative hypothesis when the original odds based on the prior probabilities were contrasted).

The value computed for the **posterior odds ratio** will always equal the product of the **prior odds ratio** and the **likelihood ratio** — i.e., $OR_{Posterior} = (OR_{Prior})(LR)$. This can be confirmed for the data in Table 9.13 (in reference to the **odds ratios** which were computed when H_0 was represented in the numerator of the relevant equations): $[OR_{Posterior} = 1.380] = [(OR_{Prior} = 2.333)(LR = .591)]$.

Table 9.16 Bayesian Analysis of Example 9.34 Involving Four Hypotheses

(1)	(2)	(3)	(4)	(5)
Hypothesized value of parameter π_1	*Prior probability* $P(A_j)$	*Likelihood* $P(B/A_j)$	*Joint probability=Likelihood x Prior probability* $P(B/A_j)P(A_j)$	*Posterior probability* $\dfrac{P(B/A_j)P(A_j)}{\Sigma[P(B/A_i)P(A_i)]}$
.5	.5	.1222	$(.1222)(.5) = .06110$	$.06110/.15683 = .38959$
.6	.3	.2066	$(.2066)(.3) = .06198$	$.06198/.15683 = .39520$
.7	.15	.1963	$(1963)(.15) = .02945$	$.02945/.15683 = .18778$
.8	.05	.0860	$(.0860)(.05) = .00430$	$.00430/.15683 = .02742$
	$\Sigma = 1$		$\Sigma = .15683$	$\Sigma = 1$

Bayesian analysis can also be employed to evaluate the relative merits of more than two hypotheses. For example, in the case of Example 9.34 we can compute the likelihood that π_1 is any of the values recorded in Column 1 of Table 9.12. In point of fact, we will employ **Bayes' theorem** to consider the following four possible values for π_1: .5, .6, .7, .8. Table 9.16 summarizes the results of a Bayesian analysis which evaluates the relative merits of the four aforementioned π_1 values.

As was the case in illustrating an analysis involving two hypotheses, the following should be noted with respect to Table 9.16: a) Column 1 lists each of the values the researcher wishes to consider as the actual value of π_1 in the underlying population; b) Column 2 lists the prior probabilities the researcher has assigned to each of the hypothesized π_1 values in Column 1. As noted earlier, the sum of the prior probabilities recorded in Column 2 must equal 1, since we are assuming that our analysis includes every hypothesis under consideration; c) As noted earlier, it is often the case that if only two hypotheses are contrasted (in which case there will be only two rows in Column 1 containing π_1 values), the hypothesis which is assigned the higher prior probability is designated as the null hypothesis, and the hypothesis which is assigned the lower prior probability is designated as the alternative hypothesis. In the example under discussion the researcher is simultaneously contrasting the following four hypotheses: H_1: $\pi_1 = .5$; H_2: $\pi_1 = .6$; H_3: $\pi_1 = .7$; H_4: $\pi_1 = .8$. Although it is not necessary to do so, one can conceptualize the hypothesis H_1: $\pi_1 = .5$ as the null hypothesis, since it has been assigned the highest prior probability, whereas the other three hypotheses can all be viewed as alternative hypotheses; d) Column 3 contains the binomial probabilities/likelihoods computed for the sample data in reference to the π_1 values listed in Column 1. To be more specific, each value in Column

3 represents the likelihood that 9 out of the 14 subjects will select Category 1 (i.e., *Nectaberry* soda) if the value of π_1 employed in Equation 9.3 is equal to the π_1 value listed for a given row. As noted earlier, these probability values can also be determined through use of **Table A6**. In point of fact, the likelihoods listed in Column 3 are identical to the probabilities recorded for the π_1 values .5, .6, .7, and .8 in Column 2 of Table 9.12; e) Column 4 lists the joint probabilities, which for each π_1 value represent the numerator of Equation 9.31 (i.e., the equation for **Bayes' theorem**) when the latter equation is employed to compute the conditional probability $P(A_j/B)$. Note that the sum of the joint probabilities in Column 4 will represent the denominator of **Bayes' theorem** which will be employed in computing each of the posterior probabilities that are recorded in Column 5; f) Column 5 lists for each row the posterior probability $P(A_j/B)$ computed with Equation 9.31. The posterior probability for each row is the likelihood that the true value of the population proportion is equal to the π_1 value recorded for that row, given the fact that 9 of the 14 subjects selected Category 1 (i.e., *Nectaberry* soda). As noted earlier, the sum of the posterior probabilities in Column 5 must equal 1.

The posterior probabilities computed in Column 5 (which are a function of both the researcher's prior probabilities and the information contained in the sample data) are the researcher's revised probabilities for the four hypotheses being evaluated. Thus, by taking into account the prior probability of .5 and the observed data, the likelihood of the hypothesis H_1: π_1 = .5 has been reduced from its original prior probability of .5 to .38959. On the other hand the likelihood of the hypothesis H_2: π_1 = .6 has been increased from its original prior probability of .3 to .39520, and, in fact, the latter hypothesis is now the most tenable of the four hypotheses. The probability that H_3: π_1 = .7 has also increased from its original prior probability of .15 to .18778. The probability that H_4: π_1 = .8 has been reduced from its original prior probability of .05 to .02742.

We will once again assume that the researcher conducts a second experiment evaluating the same hypotheses. In the latter experiment 10 of 14 subjects select *Nectaberry* as the fruitier soda, while 4 of the subjects select *Fruizzle* as the fruitier soda. We can employ **Bayes' theorem** to evaluate the data, but this time instead of employing the prior probabilities which are listed in Column 2 of Table 9.16, we will use as our prior probabilities the posterior probabilities which were computed in Column 5 of Table 9.16. The rationale for using the latter values as the prior probabilities is that these values provide the researcher with the best estimate of the likelihood for each of the four hypotheses prior to conducting the second experiment.

The results of employing **Bayes' theorem** to evaluate the second experiment are summarized in Table 9.17. Note that Column 1 (which lists the hypothesized values of π_1) is the only column in Table 9.17 which contains values that are identical to those in the corresponding column in Table 9.16. As previously noted, the values of the prior probabilities in Column 2 of Table 9.17 are the posterior probabilities computed for each of the hypotheses based on the results of the first experiment. Note that, as is the case in Table 9.16, the sum of the prior probabilities in Column 2 must equal 1 (since we are assuming that every hypothesis under consideration is included in the analysis). The likelihood values in Column 3 of Table 9.17 are the binomial probabilities computed for the π_1 values listed in Column 1. In the case of the second experiment, each value in Column 3 represents the likelihood that $x = 10$ out of the $n = 14$ subjects will select Category 1 (i.e., *Nectaberry* soda) if the value of π_1 employed in Equation 9.3 is equal to the π_1 value listed for a given row. The joint and posterior probability values in Columns 4 and 5 are determined in the identical manner they were determined in Table 9.16. Note that, as is the case in Table 9.16, the sum of the joint probabilities in Column 4 represents the denominator of **Bayes' theorem** when it is employed to compute each of the posterior probabilities recorded in Column 5. The posterior probabilities in Column 5 represent the latest revised probabilities for each of the four hypotheses. These new revised probabilities are a function of both: a) The researcher's original prior probabilities and the outcome of the first experiment (both of which are documented in Table 9.16); and b) The outcome of the second experiment. Note that, as is the case in Table 9.16, the sum of the posterior probabilities in Column 5 will always equal 1.

Table 9.17   Second Stage of Bayesian Analysis of Example 9.34 Involving Four Hypotheses

(1)	(2)	(3)	(4)	(5)
Hypothesized value of parameter π_1	Prior probability $P(A_j)$	Likelihood $P(B/A_j)$	Joint probability=Likelihood x Prior probability $P(B/A_j)P(A_j)$	Posterior probability $\dfrac{P(B/A_j)P(A_j)}{\Sigma[P(B/A_i)P(A_i)]}$
.5	.38959	.0611	(.0611)(.38959) = .02380	.02380/.13274 = .17930
.6	.39520	.1549	(1549)(.39520) = .06122	.06122/.13274 = .46120
.7	.18778	.2290	(.2290)(.18778) = .04300	.04300/.13274 = .32394
.8	.02742	.1720	(.1720)(.02742) = .00472	.00472/.13274 = .03556
	$\Sigma = 1$		$\Sigma = .13274$	$\Sigma = 1$

Based on the results of the second experiment, the likelihood of the hypothesis H_1: $\pi_1 = .5$ has been reduced from the .3859 (which is computed in Table 9.16) to .17930. Note that the latter value is substantially less than the original prior probability of .5 for that hypothesis. On the other hand, the likelihood of the hypothesis H_2: $\pi_1 = .6$, which had an original prior probability of .3, has been increased from .39520 to .46120. As was the case after evaluating the results of the first experiment, this hypothesis still appears to be the most tenable of the four hypotheses. However, the probability that H_3: $\pi_1 = .7$ has increased substantially — specifically, from .18778 to .32394, which is in marked contrast to its original prior probability of .15. The probability that H_4: $\pi_1 = .8$ is now .03556, and thus still appears to be the least tenable of the four hypotheses.

By employing **Bayes' theorem** sequentially with two sets of empirical data, the researcher has substantially modified his perceptions with regard to the tenability of the four hypotheses. At this point in time, the two most tenable hypotheses are H_1: $\pi_1 = .6$ and H_2: $\pi_1 = .7$, both of which are consistent with the claim that *Nectaberry* is the fruitier soda (since they stipulate a π_1 value above .5). Based on his analysis, the researcher can state there is a .82070 likelihood that the true value of π_1 is greater than .5, since [$P(H_2$: $\pi_1 = .6) = .46120$] + [$P(H_3$: $\pi_1 = .7)$ = .32394] + [$P(H_4$: $\pi_1 = .8) = .03556] = .82070$.

Since the total of 28 subjects evaluated in the two experiments are independent of one another, the same posterior probabilities computed in Column 5 of Table 9.17 will also be obtained if **Bayes' theorem** is employed using the original prior probabilities in Column 1 of Table 9.16 and the likelihood for the combined results of the two experiments. The latter likelihood is the probability of $x = 19$ (i.e., $9 + 10 = 19$) of the $n = 28$ (i.e., $14 + 14 = 28$) subjects selecting *Nectaberry* as the fruitier soda. The result of the latter analysis is summarized in Table 9.18 (The minimal differences between the posterior probabilities in Column 5 of Tables 9.17 and 9.18 are due to rounding off error.) The likelihoods in Column 3 of Table 9.18 are the binomial probabilities for $x = 19$ and $n = 28$ for each of the four values of π_1 specified in Column 1. Equation 9.3 was employed to determine each of the likelihoods in Column 3.

Table 9.18 Composite Bayesian Analysis of Example 9.34 Involving Four Hypotheses

(1)	(2)	(3)	(4)	(5)
Hypothesized value of parameter π_1	Prior probability $P(A_j)$	Likelihood $P(B/A_j)$	Joint probability=Likelihood x Prior probability $P(B/A_j)P(A_j)$	Posterior probability $\dfrac{P(B/A_j)P(A_j)}{\Sigma[P(B/A_i)P(A_i)]}$
.5	.5	.02573	$(.02573)(.5) = .02187$	$.02187/.07177 = .17932$
.6	.3	.11033	$(.11033)(.3) = .03310$	$.03310/.07177 = .46120$
.7	.15	.15497	$(.15497)(.15) = .02325$	$.02325/.07177 = .32395$
.8	.05	.05096	$(.05096)(.05) = .00255$	$.00255/.07177 = .03553$
	$\Sigma = 1$		$\Sigma = .07177$	$\Sigma = 1$

Odds and **odds ratios** can also be computed for the data in Tables 9.16–9.18. Note, however, that in this instance we are dealing with four hypotheses instead of two (which was the case for the data in Tables 9.13–9.15). Employing Equation 9.35 with the original prior probabilities in Tables 9.16/9.18, we can compute the **prior odds in favor of** and **against** the four hypotheses.

$$\text{Prior odds(In favor of } H_1) = \frac{P(H_1)}{1 - P(H_1)} = \frac{.5}{1 - .5} = 1$$

$$\text{Prior odds(Against } H_1) = \frac{1 - P(H_1)}{P(H_1)} = \frac{1 - .5}{.5} = 1$$

$$\text{Prior odds(In favor of } H_2) = \frac{P(H_2)}{1 - P(H_2)} = \frac{.3}{1 - .3} = .429$$

$$\text{Prior odds(Against } H_2) = \frac{1 - P(H_2)}{P(H_2)} = \frac{1 - .3}{.3} = 2.333$$

$$Prior\ odds(In\ favor\ of\ H_3) = \frac{P(H_3)}{1 - P(H_3)} = \frac{.15}{1 - .15} = .176$$

$$Prior\ odds(Against\ H_3) = \frac{1 - P(H_3)}{P(H_3)} = \frac{1 - .15}{.15} = 5.667$$

$$Prior\ odds(In\ favor\ of\ H_4) = \frac{P(H_4)}{1 - P(H_4)} = \frac{.05}{1 - .05} = .053$$

$$Prior\ odds(Against\ H_4) = \frac{1 - P(H_4)}{P(H_4)} = \frac{1 - .05}{.05} = 19$$

Note that, except for the hypothesis H_1: $\pi_1 = .5$ (which has a prior probability of .5), the values of the prior *odds in favor* of the other three hypotheses are less than 1 (which will always be the case when the prior probability for a hypothesis is less than .5). In the same respect, except for the hypothesis H_1: $\pi_1 = .5$, the values for the prior *odds against* the other three hypotheses are greater than 1 (which will always be the case when the prior probability for a hypothesis is less than .5).

Employing Equations 9.35 and 9.36 with the posterior probabilities in Tables 9.17/9.18, we can compute the final **posterior odds in favor of** and **against** the four hypotheses.

$$Posterior\ odds(In\ favor\ of\ H_1) = \frac{P(H_1)}{1 - P(H_1)} = \frac{.17930}{1 - .17930} = .218$$

$$Posterior\ odds(Against\ H_1) = \frac{1 - P(H_1)}{P(H_1)} = \frac{1 - .17930}{.17930} = 4.577$$

$$Posterior\ odds(In\ favor\ of\ H_2) = \frac{P(H_2)}{1 - P(H_2)} = \frac{.46120}{1 - .46120} = .856$$

$$Posterior\ odds(Against\ H_2) = \frac{1 - P(H_2)}{P(H_2)} = \frac{1 - .46120}{.46120} = 1.168$$

$$Posterior\ odds(In\ favor\ of\ H_3) = \frac{P(H_3)}{1 - P(H_3)} = \frac{.32394}{1 - .32394} = .479$$

$$Posteior\ odds(Against\ H_3) = \frac{1 - P(H_3)}{P(H_3)} = \frac{1 - .32394}{.32394} = 2.087$$

$$Posterior\ odds(In\ favor\ of\ H_4) = \frac{P(H_4)}{1 - P(H_4)} = \frac{.03556}{1 - .03556} = .037$$

$$Posterior\ odds(Against\ H_4) = \frac{1 - P(H_4)}{P(H_4)} = \frac{1 - .03556}{.0355} = 27.121$$

Note that the values of the *posterior odds in favor* of all four hypotheses are less than 1 (which will always be the case when the probability for a hypothesis is less than .5). In the same respect, the values for the *posterior odds against* for all four hypotheses are greater than 1 (which will always be the case when the probability for a hypothesis is less than .5).

Any of the **odds ratios** (i.e., OR_{Prior}, LR, $OR_{Posterior}$) discussed previously can be computed in reference to any of the four hypotheses evaluated in this section. The specific **odds ratio** that is computed can contrast the probability for one of the four hypotheses with the combined probabilities for the other three hypotheses, or it can contrast the probability for one of the four hypotheses with the probability for one of the other hypotheses (or even with the combined probability for two or more of the other hypotheses).

To illustrate, the **prior odds ratio** (based on the original prior probabilities) for H_1: $\pi_1 = .5$ versus the other three hypotheses is computed below through use of Equation 9.38.

$$OR_{Prior} = \frac{P(H_1)}{P(H_{2/3/4})} = \frac{.5}{.5} = 1$$

In this instance, the value $OR_{Prior} = 1$ indicates that the prior probability assigned to Hypothesis 1 is equal to the combined prior probabilities assigned to Hypotheses 2, 3, and 4. Note that the value $OR_{Prior} = 1$ is equal to the prior odds in favor of Hypothesis 1 (which are identical to the prior odds against Hypothesis 1).

The **prior odds ratio** (based on the original prior probabilities) for H_2: $\pi_1 = .6$ versus the other three hypotheses is computed below.

$$OR_{Prior} = \frac{P(H_2)}{P(H_{1/3/4})} = \frac{.3}{.7} = .429$$

In this instance, the value $OR_{Prior} = .429$ indicates that the prior probability assigned to Hypothesis 2 is .429 times as large as the combined prior probabilities assigned to Hypotheses 1, 3, and 4. Note that the value $OR_{Prior} = .429$ is equal to the prior odds in favor of Hypothesis 2 (which are identical to the prior odds against the other three hypotheses (i.e., *Prior odds (Against H_1, H_2, H_3)* $= (1 - .7) / .7 = .429$).

We can also compute a prior odds ratio contrasting Hypotheses 1 and 2. The latter **prior odds ratio** is computed below.

$$OR_{Prior} = \frac{P(H_1)}{P(H_2)} = \frac{.5}{.3} = 1.667$$

The value $OR_{Prior} = 1.667$ indicates that the prior probability assigned to Hypothesis 1 is 1.667 times larger than the prior probability assigned to Hypotheses 2. Note, however, that in this instance (unlike in the case of the previous **prior odds ratios** computed), the value of the **prior odds ratio** does not equal the prior odds for or against either Hypothesis 1 or Hypothesis

2 (since the latter will only be the case when the sum of the probabilities which are involved in computing an **odds ratio** equals 1 — i.e., when the probabilities for all four of the hypotheses are included in the analysis).

Employing the information in Tables 9.17 or 9.18, the following **likelihood ratios** and **posterior odds ratios** are computed below through use of Equations 9.40 and 9.41: a) Hypothesis 1 versus the combined probabilities for Hypotheses 2, 3, and 4; b) Hypothesis 1 versus Hypothesis 2; c) Hypothesis 2 versus the combined probability for Hypotheses 1, 3, and 4. Although not computed below, the reader should take note of the fact that one can also compute **likelihood ratios** and **posterior odds ratios** in which Hypotheses 3 and 4 are represented in the numerator of the **odds ratio** equations. The **likelihood ratios** computed below (which are based on the $n = 28$ subjects in the two experiments) employ the probabilities in Column 3 of Table 9.18. Analogous **likelihood ratios** can be computed if one employs the probabilities in Column 3 of Tables 9.16 and 9.17 (which are **likelihood ratios** for the $n = 14$ subjects in each of the two experiments). The **posterior odds ratios** computed below employ the values in Column 5 of Table 9.17 (which, except for rounding off error, are identical to the values in Column 5 of Table 9.18).

Odds Ratios with Hypothesis 1 in the Numerator

$$LR = \frac{P(B/A_1)}{P(B/A_{2/3/4})} = \frac{.02573}{.97427} = .026$$

The value $LR = .026$ indicates that the outcome of 19 out of 28 subjects identifying *Nectaberry* as the fruitier soda is .026 times as likely if H_1: $\pi_1 = .5$ is true than if any of the other three hypotheses are true. (Note that the value .97427 in the denominator of the equation is $1 - .02573 = .97427$, which is the combined probability for the other three hypotheses.)

$$LR = \frac{P(B/A_1)}{P(B/A_2)} = \frac{.02573}{.11033} = .233$$

The value $LR = .233$ indicates that the outcome of 19 out of 28 subjects identifying *Nectaberry* as the fruitier soda is .233 times as likely if H_1: $\pi_1 = .5$ is true than if H_2: $\pi_1 = .6$ is true.

$$OR_{Posterior} = \frac{P(A_1/B)}{P(A_{2/3/4}/B)} = \frac{P(H_1/B)}{P(H_{2/3/4}/B)} = \frac{.17930}{.82070} = .218$$

The value $OR_{Posterior} = .218$ indicates that the final posterior probability of H_1: $\pi_1 = .5$ being true is .218 times as large as the combined final posterior probability for any of the other three hypotheses being true. (Note that the value .82070 in the denominator of the equation is $1 - .17930 = .82070$, which is the combined final posterior probability for the other three hypotheses.)

$$OR_{Posterior} = \frac{P(A_1/B)}{P(A_2/B)} = \frac{P(H_1/B)}{P(H_2/B)} = \frac{.17930}{.46120} = .389$$

The value $OR_{Posterior} = .389$ indicates that the final posterior probability of H_1: $\pi_1 = .5$ being true is .389 times as large as the final posterior probability of H_2: $\pi_1 = .6$ being true.

Odds Ratios with Hypothesis 2 in the Numerator

$$LR = \frac{P(B/A_2)}{P(B/A_{1/3/4})} = \frac{.11033}{.88967} = .124$$

The value $LR = .124$ indicates that the outcome of 19 out of 28 subjects identifying *Nectaberry* as the fruitier soda is .124 times as likely if H_2: $\pi_1 = .6$ is true than if any of the other three hypotheses are true. (Note that the value .88967 in the denominator of the equation is $1 - .11033 = .88967$, which is the combined probability for the other three hypotheses.)

$$LR = \frac{P(B/A_2)}{P(B/A_1)} = \frac{.11033}{.02573} = 4.288$$

The value $LR = 4.288$ indicates that the outcome of 19 out of 28 subjects identifying *Nectaberry* as the fruitier soda is 4.288 times more likely if H_2: $\pi_1 = .6$ is true than if H_1: $\pi_1 = .5$ is true.

$$OR_{Posterior} = \frac{P(A_2/B)}{P(A_{1/3/4}/B)} = \frac{P(H_2/B)}{P(H_{1/3/4}/B)} = \frac{.46120}{.53880} = .856$$

The value $OR_{Posterior} = .856$ indicates that the final posterior probability of H_2: $\pi_1 = .6$ being true is .856 times as large as the combined final posterior probability for any of the other three hypotheses being true. (Note that the value .53880 in the denominator of the equation is $1 - .46120 = .53880$, which is the combined posterior probability for the other three hypotheses.)

$$OR_{Posterior} = \frac{P(A_2/B)}{P(A_1/B)} = \frac{P(H_2/B)}{P(H_1/B)} = \frac{.46120}{.17930} = 2.572$$

The value $OR_{Posterior} = 2.572$ indicates that the final posterior probability of H_2: $\pi_1 = .6$ being true is 2.572 times larger than the final posterior probability of H_1: $\pi_1 = .5$ being true.

In the introductory remarks concerning the use of Bayesian analysis in hypothesis testing, it was noted that the greater the amount of empirical data collected the closer in agreement will be the values multiple researchers compute for the posterior probabilities for two or more competing hypotheses. To illustrate this point, let us assume that two researchers are considering four hypotheses, but are initially in disagreement with respect to the prior probabilities they employ. In order to determine which hypothesis is most tenable successive studies are conducted, and after each study **Bayes' theorem** is employed by each researcher to compute a revised posterior probability for each of the four hypotheses under consideration. As the number of studies evaluated increases, the posterior probability computed by each researcher for the most tenable of the hypotheses will approach 1. (Keep in mind that the most tenable hypothesis will be the one which stipulates a value for the population parameter which is closest to the actual value of the parameter in the underlying population.) In addition, each researcher's posterior probabilities for all of the remaining hypotheses will approach 0. It should be emphasized, however, that if a posterior probability of 1 is obtained for one of the hypotheses, it does not necessarily mean that the parameter stipulated in that hypothesis is the actual value of the parameter in the population. This latter point will now be clarified.

Let us assume that in Example 9.34 the researcher is considering the following four hypotheses: H_1: π_1 = .5; H_2: π_1 = .6; H_3: π_1 = .7; H_4: π_1 = .8. We will also assume that in actuality the true value of π_1 in the population is π_1 = .9. Note that the latter value is not included among those which are stipulated in the four hypotheses under consideration. Now, let us assume that we conduct a large number of studies, and that in each study we determine the proportion of subjects who select *Nectaberry* as the fruitier soda. Assuming that the subjects employed in the studies are randomly selected from the population, as the number of studies increases the posterior probability computed for the hypothesis H_4: π_1 = .8 will approach 1, while the posterior probabilities computed for the other three hypotheses will approach 0. The latter will occur, since regardless of the prior probability values a researcher stipulated for the four hypotheses, if the empirical outcomes of the studies are representative of what is true in the underlying population, the overwhelming proportion of subjects in most of the studies should select *Nectaberry* as the fruitier soda. In other words, in each of the studies it is more likely that the proportion of subjects who select *Nectaberry* as the fruitier soda will equal .8 (which is the value closest in proximity to the true value of the population parameter π_1 = .9) than it is that it will equal .5, .6, or .7 (which are further removed than .8 from the true value π_1 = .9). By virtue of the latter, each successive application of **Bayes' theorem** will most likely result in a higher posterior probability for the hypothesis H_4: π_1 = .8, and a lower posterior probability for each of the three hypotheses H_1: π_1 = .5, H_2: π_1 = .6, and H_3: π_1 = .7.

Suppose, however, that in the example under discussion H_5: π_1 = .9 had been included among the hypotheses under consideration. If the latter were the case, at some point in analyzing successive studies, the posterior probability computed for the hypothesis H_5: π_1 = .9 would equal 1, and at the same time the posterior probability computed for the hypothesis H_4: π_1 = .8 (along with the other three hypotheses) would equal 0. The fact that under such circumstances the posterior probability computed for the hypothesis H_4: π_1 = .8 will equal 0, would be attributable to the fact that it did not correspond to the actual value of π_1 in the underlying population, while at the same time one of the other hypotheses under consideration (H_5: π_1 = .9) did correspond to the actual value. Yet when the hypothesis H_5: π_1 = .9 was not included among the hypotheses taken into consideration, a posterior probability of 1 would eventually be computed for the hypothesis H_4: π_1 = .8, since, as noted earlier, of all the hypotheses considered, it stipulates the value which is closest to .9. Consequently, even if H_5: π_1 = .9 had been included among the hypotheses under consideration, there is no way the researcher can be 100% sure that π_1 = .9 is the correct value for the population parameter. Thus, if the posterior probability computed for a hypothesis equals 1, it does not necessarily mean that the hypothesis is true. In point of fact, it only means that of all the hypotheses under consideration, it is the hypothesis which has the highest likelihood of being correct. In conclusion, the point to be made here is that although it is possible for a researcher to disprove a hypothesis (by virtue of deriving a low posterior probability), he can never prove a hypothesis with 100% certainty.

Bayesian analysis of a continuous variable The Bayesian analysis employed for Example 9.34 is an oversimplification of how some researchers would conceptualize the analysis. In reality, rather than just state hypotheses which stipulate the four probability values .5, .6, .7, and .8 for the population parameter π_1, a researcher might prefer to employ hypotheses which stipulate intermediate probability values. In other words, a researcher might elect to evaluate the hypothesis that the true value of π_1 falls within the interval .5 and .8. Consequently, the following two hypotheses might be evaluated: H_1: .5 ≤ π_1 ≤ .8 versus H_2: *Not H_1*. It would be required that the parameter π_1 be conceptualized as a continuous random variable rather than as a discrete random variable in order for an analysis to include all values which fall within the interval .5 to .8. In Bayesian hypothesis testing it is not uncommon to conceptualize the variable under study as a continuous rather than as a discrete variable.

When the variable being evaluated is conceptualized as continuous rather than discrete, the number of values the variable may assume becomes infinite. To illustrate this, let us assume a researcher is interested in all proportional values that fall within the interval .5 to .8. The only thing which limits the number of values he can identify within that interval is the precision of measurement he elects to employ. If a researcher elects to measure proportional values to three decimal places there will be 301 values which fall within the interval .5 to .8 (i.e., .500, .501, ... , .799, .800). Yet if he elects to employ four decimal places, 3001 values can be stipulated within that interval (i.e., .5000, .5001, ... , .7999, .8000). Each additional decimal place the researcher employs will increase by tenfold the number of proportional values which fall within the interval .5 to .8. It logically follows that as the number of decimal places employed increases, the number of proportional values which can be stipulated approaches infinity. Under the latter circumstances (i.e., when there are an infinite number of possible values a variable can assume), the theoretical likelihood of any specific value occurring becomes so minimal that for all practical purposes it equals zero. Because of the latter, it is meaningless to discuss a probability associated with a specific value of a continuous variable. Instead, probabilities are discussed only for values of the variable that fall within a specified interval (e.g., between .5 and .8). When probabilities are discussed within the latter context they are based on a **density function** (often referred to as a **probability density function**).

Before proceeding further, at this point it will be useful to review material discussed in the **Introduction** of the book concerning a **probability density function**. In mathematics a **function** is commonly employed to summarize the relationship between two variables such as X and Y. Specifically, if for each value X can assume there are one or more corresponding values for Y, we can state that Y is a function of X. The general form of the equation which summarizes the functional relationship between X and Y is $Y = f(X)$. The latter equation summarizes the operations which when performed on the X variable will yield one or more specific values for the Y variable. A **density function** (which can be summarized both mathematically in the form of an equation, as well as visually as a graph of the equation) describes the relationship between a variable (e.g., a population proportion as in Example 9.34, a population mean, etc.) and the densities (which for the moment we will assume are probabilities) associated with the range of values the variable may assume. When a density function is represented in a graphic format, the range of values the variable may assume are recorded on the abscissa (X-axis) and the density values (which will range from 0 to 1) are recorded on the ordinate (Y-axis). Equation I.36, the equation for the normal distribution, and its corresponding graph, Figure I.11, represent a density function. Both the equation and graph allow one to determine the proportion of cases/area under the curve which falls between any two points on the abscissa. Those familiar with calculus will recognize that if we have two points a and b on the abscissa representing two values of a continuous variable (which in Figure I.11 is a normally distributed variable), and we integrate the equation of the density function over the interval a to b, we will be able to derive the area/proportion of the curve which falls between points a and b. In lieu of having to employ calculus to compute the proportion that falls between any two points, in the case of the normal distribution the appropriate values can be determined through use of **Table A1**. It should be noted that although the densities which are recorded on the ordinate of the graph of a density function are often depicted as being equivalent to probabilities, strictly speaking a density is not the same thing as a probability. Within the context of a graph such as Figure I.11, a density is best viewed as the height of the curve which corresponds to a specific value of the variable which is recorded on the abscissa. The area between any two points on the abscissa however, is generally expressed in the form of a probability or a proportion.

A general outline of the procedure employed for Bayesian hypothesis testing when the variable under study is conceptualized as a continuous variable will now be described.

a) Initially a researcher must construct a prior distribution for the variable being evaluated. In order to do the latter for Example 9.34, prior to collecting the data the researcher would be required to stipulate a range of values which he believes includes any value for which there is a minimal likelihood is the actual value of π_1. He would then construct a graph depicting the prior probability distribution for the variable. Phillips (1973, p. 108–109) notes that such a graph can be constructed by doing the following. Record on the abscissa the range of values one is considering for the actual value of the variable in the underlying population. In the case of Example 9.34, let us assume that we are open to the possibility that any value which falls within the range .1 to .9 can be the actual value of π_1. If the latter were the case, the researcher would mark off equally spaced points on the abscissa representing proportional values (such as .1, .2, ... , .8, .9). On the ordinate, probability values would be recorded which will be used to indicate the researcher's prior probabilities for the values of the variable recorded on the abscissa. The probability values on the ordinate will range from 0 (which will be placed at the origin of the graph) to 1 (which will be at the top of the ordinate). Once the appropriate values are recorded on the axes, the researcher would insert a dot directly above the value on the abscissa which he believes is most likely to be the actual value of the variable in the population. The latter dot will be aligned with a height on the ordinate which corresponds to the prior probability for that value of the variable. The aforementioned dot will be the highest point on the graph of the prior probability distribution. The researcher would then select other values the variable might assume, and place a dot above each value at a height which corresponds to the researcher's prior probability for a given value. After a reasonable number of dots have been drawn, all of the dots are connected. The resulting continuous line will represent the prior distribution (more formally called the **prior density function**) for the variable.

b) The second step in the analysis requires that the researcher collect empirical data. The empirical data along with the information contained in the prior distribution will be employed in **Bayes' theorem** to construct a posterior probability distribution (which in all likelihood will be different from the prior probability distribution). The process of acquiring empirical data and employing it in **Bayes' theorem** can be repeated over and over again. Each time the process is repeated a revised posterior distribution is derived. As illustrated earlier when **Bayes' theorem** was employed with Example 9.34, the most recently computed posterior probability distribution is used to represent the prior probability distribution each time an analysis involving a new set of empirical data is conducted.

The above described procedure is essentially the same as that described for evaluating a discrete variable. The major difference is that in the case of a continuous variable the prior probabilities and subsequent posterior probabilities are based on the density function for that variable, and because of the latter different computational procedures are required to compute posterior probabilities. Bayesian analysis of a continuous variable is made considerably easier by the fact that graphs are available which depict various distributions/density functions which a researcher might want to use to represent a prior distribution. Sources which contain such graphs (e.g., Phillips, 1973) also contain additional information which allows the researcher, among other things, to derive the equation for the density function of a prior or posterior distribution. The equation for a mathematical function will always contain constant numerical values, which are commonly referred to as its parameters. If the general form of the equation for the function is known and the values of the parameters are stipulated, the equation for the function can easily be derived. For example, the general form for the equation of a straight line (which represents a function describing the relationship between two variables X and Y) is $Y = aX + b$. The values a and b in the latter equation are the parameters of the function. Thus, if we are told that we have a function which describes a straight line, and that its parameters are $a = 6$ and $b = 1$, the equation for the straight line is $Y = 6X + 1$. The latter line can easily be graphed by selecting any two values for X and solving for Y. Thus if we substitute the value $X = 0$ in the equation, it yields the value $Y = 1$, and if we substitute $X = 2$ in the equation, it yields the value $Y = 13$. A dot is then employed to represent each of the two points $(0, 1)$ and $(2, 13)$, and the two resulting dots are connected. The line which results from connecting the two dots represents the straight line described by the equation $Y = 6X + 1$. (The procedure for drawing the graph of a straight line is demonstrated in Section V of the **Pearson product-moment correlation coefficient**.) As is the case with a straight line, the equation for a density function will also have a general equation, and the latter equation will contain one or more constants/parameters. If the general form of the equation for the density function is known and the values of the parameters are stipulated, the equation for the density function can easily be derived.

The three most common distributions which are employed to represent a prior distribution within the framework of a Bayesian analysis of a continuous variable are a **rectangular/uniform distribution** (in which case all values of the variable are equally likely to occur, and thus are assigned equal prior probabilities), the **normal distribution**, and the **beta distribution**. At this point we will concentrate on the **beta distribution**, since it is the most appropriate distribution to employ with respect to prior probabilities when a hypothesis evaluating a population proportion is evaluated (as is the case in Example 9.34). Equation 9.42 is the general equation for a beta distribution.

$$Y = \frac{(v + w - 1)!}{(v - 1)!\,(w - 1)!} X^{(v - 1)} (1 - X)^{(w - 1)} \qquad \textbf{(Equation 9.42)}$$

In Equation 9.42 the values v and w represent the parameters of a beta density function. X represents the variable being evaluated. Since X must represent a proportion, a value of X must fall within the range 0 to 1. The range of prior probability values which are considered for X are recorded on the abscissa of the graph of a beta density function. Y represents the computed density value. The range of density values Y may assume are recorded on the ordinate of the graph of a beta density function. Y can be computed for any value of X by substituting X along with the values of the parameters v and w in Equation 9.42. Unlike the normal distribution, which is symmetrical, the shape of a beta distribution/density function can be symmetrical or asymmetrical. Depending on the values of its parameters, a beta distribution/density function can assume any of a variety of shapes including a uniform distribution (which is the case when $v = 1$ and $w = 1$, since when the latter is true Equation 9.42 reduces to the straight line $Y = 1$), a U-shaped distribution (which results when both v and w fall between 0 and 1), a normal distribution, or a skewed distribution. Phillips (1973, p. 126) notes that when the values of v and w are equal, the resulting distribution is symmetrical, and when they are unequal the distribution is asymmetrical. The larger the values of v and w, the more peaked the distribution. The more disparate the values of v and w, the more skewed the distribution.

When the beta distribution is employed in a Bayesian analysis, a researcher must visually inspect a large number of such distributions and select the most appropriate one for approximating the prior probabilities he has stipulated. Selection of the appropriate beta distribution is accomplished through inspection of a collection of beta distributions which can be found in sources such as Phillips (1973). Once the appropriate beta distribution is identified, the values of the parameters v and w will be known, since they will be provided along with the graph of each distribution. By employing the parameters of a beta distribution in equations (which will not be presented here) which have been derived by Bayesian statisticians, a researcher can compute the mean, mode, and standard deviation of that distribution.[29] After compiling empirical data, the researcher can use the appropriate equations to compute a revised mean, mode, and standard deviation for the posterior distribution. At that point the mean, mode, and standard deviation values computed for the prior and posterior distributions can be compared with one another. The analysis also allows the researcher to compute a **credible interval** for both the prior and the posterior distributions, and the credible intervals for the two distributions can be compared with one another. A **credible interval** is similar to but not identical to a **confidence interval**. Whereas a **confidence interval** is a probability statement about an interval, a **credible interval** is a probability statement about a population parameter.

Earlier in the discussion of Bayesian hypothesis testing it was noted that one difference between the Bayesian and classical hypothesis testing models is that in classical hypothesis testing an unknown population parameter is viewed as having a fixed value, whereas in Bayesian hypothesis testing an unknown population parameter is viewed as a random variable. Consequently, in the classical hypothesis testing model a confidence interval may vary from sample to sample, but the population parameter will always remain a fixed value. It is because of the difference in conceptualizing a population parameter in the two hypothesis testing models that a confidence interval and a credible interval are defined differently. A *C* **percent credible interval** allows a researcher to state there is a C% probability the population parameter falls within that interval. On the other hand, a *C* **percent confidence interval** allows a researcher to state there is a C% probability the interval contains the true value of the population parameter. Thus, in the case of Example 9.34, if the 95% credible interval for a posterior probability distribution is .5 − .8, the researcher can state there is a 95% probability the true population proportion falls within the interval .5 − .8. On the other hand, if the 95% confidence interval for the population proportion (which would be computed within the framework of the classical hypothesis testing model) is .5 − .8, the researcher can state there is a 95% probability that the interval .5 − .8 contains the true value of the population proportion. The use of the beta distribution in evaluating a hypothesis about a proportion will not be developed beyond what has been described in this section. As noted earlier, Bayesian hypothesis testing is a complex subject and the intent of this discussion has been to introduce the reader to the basic principles underlying this alternative approach to hypothesis testing. For a comprehensive discussion of how the beta distribution (as well as other distributions) is employed in evaluating a hypothesis concerning a proportion, the reader should consult sources such as Chou (1989), Phillips (1973), and Winkler (1972). For the mathematically sophisticated reader, Gelman *et al.* (2004) is an excellent reference on the general subject of Bayesian data analysis.

References

Alcock, J. E. (1981). **Parapsychology: Science or magic? A psychological perspective.** Oxford: Pergamon Press.

Anscombe, F. J. (1948). The transformation of Poisson, binomial, and negative binomial data. **Biometrika**, 35, 246–254.

Berry, D. A. (1996). **Statistics: A Bayesian perspective.** Belmont, CA: Duxbury Press.

Best, D. J. (1975). The difference between two Poisson expectations. **Australian Journal of Statistics**, 17, 29–33.

Broughton, R. S. (1991). **Parapsychology: The controversial science**. New York: Ballantine Books.

Burdick, D. S. & Kelly, E. F. (1977). Statistical methods in parapsychological research. In Wolman, B. B. (Ed.), **Handbook of parapsychology** (pp. 81–129). New York: Van Nostr& Reinhold.

Canavos, G. C. & Miller, D. M. (1995). **Modern business statistics**. Belmont, CA: Duxbury.

Chou, Y. (1989). **Statistical analysis for business and economics**. New York: Elsevier.

Christensen, R. (1990). **Log-linear models**. New York: Springer-Verlag.

Christensen, R. (1997). **Log-linear models and logistic regression** (2nd ed.). New York: Springer–Verlag.

Cohen, J. (1977). **Statistical power analysis for the behavioral sciences**. New York: Academic Press.

Cohen, J. (1988). **Statistical power analysis for the behavioral sciences** (2nd ed.). Hillsdale, NJ: Lawrence Erlbaum Associates, Publishers.

Cohen, J. (1994). The earth is round ($p < .05$). **American Psychologist**, 49, 997–1003.

Cowles, M. (1989). **Statistics in psychology: An historical perspective**. Hillsdale, NJ: Lawrence Erlbaum Associates, Publishers.

Daniel, W. W. (1990). **Applied nonparametric statistics** (2nd ed.). Boston: PWS–Kent Publishing Company.

Daniel, W. W. & Terrell, J. C. (1995). **Business statistics for management and economics** (7th ed.). Boston: Houghton Mifflin Company.

Detre J. and White, C. (1970). The comparison of two Poisson distributed observations. **Biometrics**, 26, 851–854.

Edwards W., Lindman H., & Savage. L. J. (1963). Bayesian statistical inference for psychological research. **Psychological Review**, 70, 193–242.

Falk, R. W. & Greenbaum, C. W. (1995). Significance tests die hard. **Theory and Psychology**, 5, 75–98.

Feller, W. (1968). **An introduction to probability theory and its applications (Volume I)** (3rd ed.). New York: John Wiley & Sons.

Freund, J. E. (1984). **Modern elementary statistics** (6th ed.). Englewood Cliffs, NJ: Prentice Hall.

Gelman, A., Carlin, J. B., Stern, H. S., & Rubin, D. R. (2004). **Bayesian data analysis** (2nd ed.). Boca Raton, FL: Chapman & Hall.

Gigerenzer, G. (1993). The superego, the ego and the id in statistical reasoning. In Keren, G. & Lewis, C. (Eds.), **A handbook for data analysis in the behavioral sciences: Methodological issues** (pp. 311–339). Hillsdale, NJ: Lawrence Erlbaum Associates, Publishers.

Guenther, W. C. (1968). **Concepts of probability**. New York: McGraw–Hill Book Company.

Hagen, R. L. (1997). In praise of the null hypothesis statistical test. **American Psychologist**, 52, 15–24.

Hansel, C. E. M. (1989). **The search for psychic power: ESP and parapsychology revisited**. Buffalo: Prometheus Books.

Harlow, L. L., Mulaik, S. A., & Steiger, J. H. (Eds.) (1997). **What if there were no significance tests?** Mahwah, NJ: Lawrence Erlbaum Associates, Publishers.

Hays, W. L. & Winker, R. L. (1971). **Statistics: Probability, inference, and decision**. New York: Holt, Rinehart, & Winston.

Hines, T. (2002). **Pseudoscience and the paranormal: A critical examination of the evidence** (2nd ed.) Buffalo: Prometheus Books.

Hogg, R. V. & Tanis, E. A. (1997). **Probability and statistical inference** (5th ed.). Upper Saddle River, NJ: Prentice Hall.

Irwin, H. J. (1999). **An introduction of parapsychology** (3rd ed.). Jefferson, NC, McFarland & Company.

Krueger, J. (2001). Null hypothesis significance testing. **American Psychologist**, 56, 16–26.

Larsen, R. J. & Marx, M. L. (1985). **An introduction to probability and its applications**. Englewood Cliffs, NJ: Prentice Hall.

Lindley, D. V. (1957). A statistical paradox. **Biometrika**, 44, 187–192.

Marascuilo, L. A. & McSweeney, M. (1977). **Nonparametric and distribution-free methods for the social sciences**. Monterey, CA: Brooks/Cole Publishing Company.

Meehl, P. & Rosen, A. (1955). Antecedent probability and the efficiency of psychometric signs, patterns, or cutting scores. **Psychological Bulletin**, 52, 194–216.

Meehl, P. (1978). Theoretical risks and tabular asterisks: Sir Karl, Sir Ronald, and the slow progress of soft psychology. **Journal of Consulting and Clinical Psychology**, 46, 806–834.

Miller, I. & Miller, M. (1999). **John E. Freund's mathematical statistics** (6th ed.). Upper Saddle River, NJ: Prentice Hall.

Morrison, D. E. & Henkel, R. E. (Eds.) (1970). **The significance test controversy: A reader**. Chicago: Aldine.

Pagano, M. & Gauvreau, K. (1993). **Principles of biostatistics**. Belmont, CA: Duxbury Press.

Parker, R. E. (1979). **Introductory statistics for biology** (2nd ed.). Cambridge: Cambridge University Press.

Phillips, L. D. (1973). **Bayesian statistics for social scientists**. New York: Thomas Y. Crowell Company.

Pitman, J. (1993). **Probability**. New York: Springer–Verlag.

Przyborowski, J. and Wilenski, H. (1940). Homogeneity of results in testing samples from Poisson series. **Biometrika**, 31, 313–323.

Ramsey, P. H. & Ramsey, P. P. (1988). Evaluating the normal approximation to the binomial test. **Journal of Educational Statistics**, 13, 173–182.

Rao, P. V. (1998). **Statistical research methods in the life sciences**. Pacific Grove, CA: Duxbury Press.

Rosner, B. (1995). **Fundamentals of biostatistics** (4th ed.). Belmont, CA: Duxbury Press.

Rosner, B. (2000). **Fundamentals of biostatistics** (5th ed.). Pacific Grove, CA: Duxbury Press.

Savage, L. J. (1962). **The foundations of statistical inference**. London: Methuen.

Selvin, S. (1995). **Practical biostatistical methods**. Belmont, CA: Duxbury.

Serlin, R. A. & Lapsley, D. K. (1985). Rationality in psychological research: The good-enough principle. **American Psychologist**, 40, 73–83.

Serlin, R. A. & Lapsley, D. K. (1993). Rational appraisal of psychological research and the good-enough principle. In Keren, G. & Lewis, C. (Eds.), **A handbook for data analysis in the behavioral sciences: Methodological issues** (pp. 199–228). Hillsdale, NJ: Lawrence Erlbaum Associates, Publishers.

Siegel, S. & Castellan, N. J., Jr. (1988). **Nonparametric statistics for the behavioral sciences** (2nd ed.). New York: McGraw–Hill Book Company.

Sinich, T. (1996). **Business statistics by example**. Upper Saddle River, NJ: Prentice Hall.

van Belle, G. (2002). **Statistical rules of thumb**. New York: John Wiley & Sons.

Wiggins, J. (1973). **Personality and prediction: Principles of personality assessment**. Reading, MA: Addison–Wesley Publishing Company.

Williamson, E. & Bretherton, M. (1963). **Tables of the negative binomial probability distribution**. New York: John Wiley & Sons.

Winker, R. L. (1972). **Introduction to Bayesian inference and decision**. New York: Holt, Rinehart, & Winston.

Winkler, R. L. (1993). Bayesian statistics: An overview . In Keren, G. & Lewis, C. (Eds.), **A handbook for data analysis in the behavioral sciences: Statistical issues** (pp. 201–232). Hillsdale, NJ: Lawrence Erlbaum Associates, Publishers.

Zar, J. H. (1999). **Biostatistical analysis** (4th ed.). Upper Saddle River, NJ: Prentice Hall.

Endnotes

1. a) The binomial distribution is based on a process developed by the Swiss mathematician James Bernoulli (1654-1705). Each of the trials in an experiment involving a binomially distributed variable is often referred to as a **Bernoulli trial**. The conditions for Bernoulli trials are met when, in a set of repeated independent trials, on each trial there are only two possible outcomes, and the probability for each of the outcomes remains unchanged on every trial; b) The binomial model assumes **sampling with replacement**. To understand the latter term, imagine an urn which contains a large number of red balls and white balls. In each of n trials one ball is randomly selected from the urn. In the **sampling with replacement model**, after a ball is selected it is put back in the urn, thus insuring that the probability of drawing a red or white ball will remain the same on every trial. On the other hand, in the **sampling without replacement model**, the ball that is selected is not put back in the urn after each trial. Because of the latter, in the sampling without replacement model the probability of selecting a red ball versus a white ball will change from trial to trial, and on any trial the value of the probabilities will be a function of the number of balls of each color which remain in the urn. The binomial model assumes sampling with replacement, since on each trial the likelihood an observation will fall in Category 1 will always equal π_1, and the likelihood an observation will fall in Category 2 will always equal π_2. The classic situation for which the binomial model is employed is the process of flipping a fair coin. In the coin flipping situation, on each trial the likelihood of obtaining **Heads** is $\pi_1 = .5$, and the likelihood of obtaining **Tails** is $\pi_2 = .5$. The process of flipping a coin can be viewed within the framework of sampling with replacement, since it can be conceptualized as selecting from a large urn that is filled with the same number of **Heads** and **Tails** on every trial. In other words, it's as if after each trial the alternative which was selected on that trial is thrown back into the urn, so that the likelihood of obtaining **Heads** or **Tails** will remain unchanged from trial to trial. In Section IX (the **Addendum**) the **hypergeometric distribution** (another discrete probability distribution) is described, which is based upon the sampling without replacement model; c) The binomial distribution is actually a special case of the **multinomial distribution**. In the latter distribution, each of n independent observations can be classified in one of k mutually exclusive categories, where k can be any integer value equal to or greater than two. The multinomial distribution is described in detail in Section IX (the **Addendum**).

2. The reader should take note of the fact that many sources employ the notations p and q to represent the population proportions π_1 and π_2. Because of this, the equations for the mean and standard deviation of a binomially distributed variable can be written as follows: $\mu = np$ and $\sigma = \sqrt{npq}$. The use of the symbols π_1 and π_2 in this chapter for the population proportions is predicated on the fact that throughout this book Greek letters are employed to represent population parameters.

3. Using the format employed for stating the null hypothesis and the nondirectional alternative hypothesis for the **chi-square goodness-of-fit test**, H_0 and H_1 can also be stated as follows for the **binomial sign test for a single sample**: H_0: $o_i = \varepsilon_i$ for both cells; H_1: $o_i \neq \varepsilon_i$ for both cells. Thus, the null hypothesis states that in the underlying population the sample represents, for both cells/categories the observed frequency of a cell is equal to the expected frequency of the cell. The alternative hypothesis states that in the underlying population the sample represents, for both cells/categories the observed frequency of a cell is not equal to the expected frequency of the cell.

4. The question can also be stated as follows: If $n = 10$ and $\pi_1 = \pi_2 = .5$, what is the probability of two or less observations in one of the two categories? When $\pi_1 = \pi_2 = .5$, the probability of two or less observations in Category 2 will equal the probability of eight or more observations in Category 1 (or vice versa). When, however, $\pi_1 \neq \pi_2$, the probability of two or less observations in Category 2 will not equal the probability of eight or more observations in Category 1.

5. The number of **combinations of n things taken x at a time** represents the number of different ways that n objects can be arranged (or selected) x at a time without regard to order. For instance, if one wants to determine the number of ways that 3 objects (which will be designated A, B, and C) can be arranged (or selected) 2 at a time without regard to order, the following 3 outcomes are possible: 1) An A and a B (which can result from either the sequence AB or BA); 2) An A and a C (which can result from either the sequence AC or CA); or 3) A B and a C (which can result from the sequence BC or CB). Thus, there are 3 combinations of ABC taken 2 at a time. This is confirmed below through use of Equation 9.4.

$$\binom{3}{2} = \frac{3!}{2! \, (3 - 2)!} = \frac{3!}{2! \, 1!} = 3$$

To extend the concept of combinations to a coin tossing situation, let us assume that one wants to determine the exact number of ways 2 **Heads** can be obtained if a coin is tossed 3 times. If a fair coin is tossed 3 times, any one of the following 8 sequences of **Heads** (H) and **Tails** (T) is equally likely to occur: $HHH, HHT, THH, HTH, THT, HTT, TTH, TTT$. Of the 8 possible sequences, only the following 3 sequences involve 2 **Heads** and 1 **Tails**: HHT, THH, HTH. The latter 3 arrangements represent the combination of 3 things taken 2 at a time. This can be confirmed by $\binom{3}{2} = \frac{3!}{2! \, 1!} = 3$.

When the order of the arrangements is of interest, one can compute the number of **permutations of n things taken x at a time**. The latter is represented by the notation P_x^n, where $P_x^n = n!/(n - x)!$ (which is the same as Equation I.47 in the **Introduction**). Thus, if one is interested in the order when the events A, B, and C are taken/selected 2 at a time, the following number of permutations are computed: $P_2^3 = 3!/(3 - 2)! = 3!/1! = 6$. The 6 possible arrangements taking order into account are AB, BA, AC, CA, BC, and CB.

Although based on the definition of a permutation which has been presented above, one might conclude that the three combinations HHT, THH, HTH take order into account, and thus represent permutations, they can be conceptualized as combinations if one views the binomial model as follows: Within each of the three combinations HHT, THH, HTH, the 2 **Heads** are **distinguishable** from one another, insofar as each of the **Heads** can be assigned a subscript to designate it as distinct from the other **Heads**. To illustrate, within the combination HHT, H_1 and H_2 can be employed to distinguish the two **Heads** from one another. If we assume that the two **Heads** are randomly selected from an urn, and one **Heads** is assigned the label H_1, and the other **Heads** is assigned the label H_2, the following two permutations are possible: H_1H_2T, H_2H_1T. Thus, the arrangement HHT is a combination which summarizes the two distinct permutations H_1H_2T and H_2H_1T. Based on what has been said, it follows that the following 6 permutations comprise the 3 combinations HHT, THH, and HTH: H_1H_2T, H_2H_1T, TH_1H_2, TH_2H_1, H_1TH_2, H_2TH_1. In point of fact, many sources (e.g., Marascuilo and McSweeney (1977, pp. 12–13)) describe the value computed for the binomial coefficient as a permutation, since it can be viewed as representing a value based on two sets of different but identical objects.

A more detailed discussion of combinations, permutations, and the general subject of counting within the framework of computing probabilities can be found in the **Introduction**.

6. The application of Equation 9.3 to every possible value of x (i.e., in the case of Examples 9.1 and 9.2, the integer values 0 through 10) will yield a probability for every value of x. The sum of these probabilities will always equal 1. The algebraic expression which summarizes the summation of the probability values for all possible values of x is referred to as the **binomial expansion** (summarized by Equation 9.5) which is equivalent to the general equation $(\pi_1 + \pi_2)^n$ (or $(p + q)^n$ when p and q are employed in lieu of π_1 and π_2). Thus:

$$\sum_{r=x}^{n} \binom{n}{x} (\pi_1)^x (\pi_2)^{(n-x)} = (\pi_1 + \pi_2)^n$$

To illustrate, if $n = 3$, $\pi_1 = .5$, and $\pi_2 = .5$, the binomial expansion is as follows.

$$(\pi_1 + \pi_2)^3 = (\pi_1)^3 + 3(\pi_1)^2(\pi_2) + 3(\pi_1)(\pi_2)^2 + (\pi_2)^3$$

$$= (.5)^3 + 3(.5)^2(.5) + 3(.5)(.5)^2 + (.5)^3$$

Each of the four terms which comprise the above noted binomial expansion can be computed with Equation 9.3 as noted below. The computed probabilities .125, .375, .375, and .125 are respectively the likelihood of obtaining 3, 2, 1, and 0 outcomes of π_1 if $n = 3$ and $\pi_1 = .5$.

$$\textbf{Term 1 } (P(3/3)) = (\pi_1)^3 = \binom{3}{3}(.5)^3(.5)^0 = .125$$

$$\textbf{Term 2 } (P(2/3)) = 3(\pi_1)^2(\pi_2) = \binom{3}{2}(.5)^2(.5) = .375$$

$$\textbf{Term 3 } (P(1/3)) = 3(\pi_1)(\pi_2)^2 = \binom{3}{1}(.5)(.5)^2 = .375$$

$$\textbf{Term 4 } (P(0/3)) = (\pi_2)^3 = \binom{3}{0}(.5)^0(.5)^3 = .125$$

7.   If $\pi = .5$ and one wants to determine the likelihood of x being equal to or less than a specific value, one can employ the cumulative probability listed for the value $(n-x)$. Thus, if $x = 2$, the cumulative probability for $x = 8$ (which is .0547) is employed since $n - x = 10 - 2 = 8$. The value .0547 indicates the likelihood of obtaining 2 or less observations in a cell. This procedure can only be used when $\pi_1 = .5$, since, when the latter is true, the binomial distribution is symmetrical.

8.   If in using **Tables A6** and **A7** the value of π_2 is employed to represent π in place of π_1, and the number of observations in Category 2 is employed to represent the value of x instead of the number of observations in Category 1, then the following are true: a) In **Table A6** (if all values of π within the range from 0 to 1 are listed) the probability associated with the cell that is the intersection of the values $\pi = \pi_2$ and x (where x represents the number of observations in Category 2) will be equivalent to the probability associated with the cell that is the intersection of $\pi = \pi_1$ and x (where x represents the number of observations in Category 1); and b) In **Table A7** (if all values of π within the range from 0 to 1 are listed) for $\pi = \pi_2$, the probability of obtaining x or fewer observations (where x represents the number of observations in Category 2) will be equivalent to for $\pi = \pi_1$, the probability of obtaining x or more observations (where x represents the number of observations in Category 1). Thus if $\pi_1 = .7$, $\pi_2 = .3$, and $n = 10$ and there are 9 observations in Category 1 and 1 observation in Category 2, the following are true: a) The probability in **Table A6** for $\pi = \pi_2 = .3$ and $x = 1$ will be equivalent to the probability for $\pi = \pi_1 = .7$ and $x = 9$; and b) In **Table A7** if $\pi = \pi_2 = .3$, the probability of obtaining 1 or fewer observations will be equivalent to the probability of obtaining 9 or more observations if $\pi = \pi_1 = .7$.
      A modified protocol for employing **Table A7** when a more extreme value is defined as any value which is larger than the smaller of the two observed frequencies, or smaller than the larger of the two observed frequencies is described in Section VIII in reference to Example 9.9.

9.   It will also answer at the same time whether $p_2 = 2/10 = .2$, the observed proportion of cases for Category 2, deviates significantly from $\pi_2 = .5$.

10.   Since, like the normal distribution, the binomial distribution is a two-tailed distribution, the same basic protocol is employed in interpreting nondirectional (i.e., two-tailed) and directional (one-tailed) probabilities. Thus, in interpreting binomial probabilities one can conceptualize a distribution which is similar in shape to the normal distribution, and substitute the appropriate binomial probabilities in the distribution. (In the interest of accuracy, one should take note of the fact that the greater the discrepancy between the values of π_1 and π_2, the more dissimilar in shape the binomial becomes from the normal distribution.)

11.   In Example 9.4 many researchers might prefer to employ the directional alternative hypothesis H_1: $\pi_1 < .5$, since the senator will only change her vote if the observed proportion in the sample is less than .5. In the same respect, in Example 9.5 one might employ the directional alternative hypothesis H_1: $\pi_1 > .5$, since most people would only interpret above chance performance as indicative of extrasensory perception.

12.   Equation 9.7 can also be expressed in the form $z = (X - \mu)/\sigma$. Note that the latter equation is identical to Equation I.38, the equation for computing a standard deviation score for a

normally distributed variable. The difference between Equations 9.7 and I.38 is that Equation 9.7 computes a normal approximation for a binomially distributed variable, whereas Equation I.38 computes an exact value for a normally distributed variable.

13. The reader may be interested in knowing that in extrasensory perception (ESP) research, evidence of ESP is not necessarily limited to above chance performance. A person who consistently scores significantly below chance or only does so under certain conditions (such as being tested by an extremely skeptical and/or hostile experimenter) may also be used to support the existence of ESP. Thus, in Example 9.6, the subject who obtains a score of 80 (which is significantly below the expected value $\mu = 100$) represents someone whose poor performance (referred to as **psi missing**) might be used to suggest the presence of extrasensory processes.

14. When the **binomial sign test for a single sample** is employed to evaluate a hypothesis regarding a population median, it is categorized by some sources as a test of ordinal data (rather than as a test of categorical/nominal data), since when data are categorized with respect to the median it implies ordering of the data within two categories (i.e., **above the median** versus **below the median**).

15. a) In the discussion of the **Wilcoxon signed-ranks test**, it is noted that the latter test is not recommended if there is reason to believe the underlying population distribution is asymmetrical. Thus, if there is reason to believe that blood cholesterol levels are not distributed symmetrically in the population, the **binomial sign test for a single sample** would be recommended in lieu of the **Wilcoxon signed-ranks test**; b) Marascuilo and Mc-Sweeney (1977) note that the **asymptotic relative efficiency** (discussed in Section VII of the **Wilcoxon signed-ranks test**) of the **binomial sign test for a single sample** is generally lower than that of the **Wilcoxon signed-ranks test**. If the underlying population distribution is normal, the asymptotic relative efficiency of the binomial sign test is .637, in contrast to an asymptotic relative efficiency of .955 for the **Wilcoxon signed-ranks test** (with both asymptotic relative efficiencies being in reference to the **single-sample *t* test**). When the underlying population distribution is not normal, in most cases the asymptotic relative efficiency of the **Wilcoxon signed-ranks test** will be higher than the analogous value for the **binomial sign test**.

16. The reader should take note of the fact that the protocol in using **Table A7** to interpret a π_2 value that is less than .5 in reference to the value $\pi_2 = .1$ is different than the one described in the last paragraph of Section IV. The reason for this is that in Example 9.9 we are interested in (for $\pi_2 = .1$) the probability that the number of observations in Category 2 (**females**) are equal to or greater than 3 (which equals the probability that the number of observations in Category 1 (**males**) are equal to or less than 7 for $\pi_1 = .9$). The protocol presented in the last paragraph of Section IV in reference to the value $\pi_2 = .3$ describes the use of **Table A7** to determine the probability that the number of observations in Category 2 are equal to or less than 1 (which equals the probability that the number of observations in Category 1 are equal to or greater than 9 for $\pi_1 = .7$). Note that in Example 9.9 a more extreme score is defined as one which is larger than the lower of the two observed frequencies or smaller than the larger of the two observed frequencies. On the other hand, in the example in the last paragraph of Section IV a more extreme score is defined as one that is smaller than the lower of the two observed frequencies or larger than

the higher of the two observed frequencies. The criterion for defining what constitutes an extreme score is directly related to the alternative hypothesis the researcher employs. If the alternative hypothesis is nondirectional, an extreme score can fall both above or below an observed frequency, whereas if a directional alternative hypothesis is employed, a more extreme score can only be in the direction indicated by the alternative hypothesis.

17. Endnote 14 in the **Introduction** states that the value e is the base of the natural system of logarithms. e, which equals 2.71828... , is an **irrational number** (i.e., a number that has a decimal notation that goes on forever without a repeating pattern of digits).

18. If instead of employing **sick** and **guilty** as the designated events, we specified **healthy** and **innocent** as the designated events, the values $P(A_2) = .99$ (for Example 9.29) and $P(A_2) = .98$ (For Example 9.30) represent the **baserates** of the events in question.

19. The proportion .0288 (i.e., the sum of the proportions in the first column) is commonly referred to as the **selection ratio**, since it represents the proportion of people in the population who are identified as positive. The term **selection ratio** is commonly employed in situations when individuals whose test result is positive are selected for something (e.g., a treatment, a job, etc.). The proportion .9712 (i.e., the sum of the proportions in the second column) is ($1 - selection\ ratio$), since it represents the proportion of people who are not selected.

20. Another type of test which is commonly employed in assessing honesty (typically, within the framework of screening people for employment) is the **integrity test** (also referred to as an **honesty test**). Like the polygraph, the latter type of test (which is a pencil and paper questionnaire) is subject to criticism for the large number of false positives it yields. The author has seen false positive rates for integrity tests that are close to .50 (i.e., 50%). It should be noted, however, that as a general rule the only consequence of being a false positive on an integrity test is that a person is not hired for a job. The seriousness of the latter error would be viewed as minimal when contrasted with a false positive on a polygraph, if a positive polygraph result is construed as evidence of one being guilty of a serious crime (although, in reality, the use of polygraph evidence in the courts is severely restricted).

21. Alcock (1981), Hansel (1989), and Hines (2002) are representative of sources which would argue that the baserate for ESP is zero. On the other hand, Broughton (1991) and Irwin (1999) are representative of sources which would argue for a baserate above zero.

22. In order to provide an even more dramatic example of the **low baserate problem**, I have taken a number of liberties in conceptualizing the analysis of Example 9.31. To begin with, a null and alternative hypothesis would not be stated for a Bayesian analysis, and, consequently, the terms Type I and Type II error rates are not employed within the framework of such an analysis. As will be noted later in this section, in a Bayesian analysis, based on preexisting information (i.e., prior probabilities), posterior probabilities are computed for two or more specific hypotheses, neither of which is designated as the null hypothesis. It should also be emphasized that within the framework of Example 9.31, I will assume that any subject who performs at a statistically significant level, performs at exactly the .05 level. The reason for this is that if the probability associated with a subject's level of performance is less than .05, the posterior probabilities computed for that subject will not

correspond to those computed for the example. Although my conceptualization of Example 9.31 is a bit unrealistic, it can nevertheless be employed to illustrate the pitfalls involved in the evaluation of a low baserate event.

23. The values employed for Example 9.32 are hypothetical probabilities which are not actually based on empirical data in the current medical literature.

24. The reader should take note of the fact that, in actuality, there is no such disease as leprosodiasis, and the probability values employed in Example 9.33 are fictional.

25. A comprehensive discussion of the subject of controlling for order effects (i.e., controlling for the order of presentation of the experimental conditions) can be found in Section VII of the *t* **test for two dependent samples (Test 17)**.

26. Other sources, such as Cowles (1989), describe alternative ways of conceptualizing probability.

27. It should be noted that the Bayesian hypothesis testing model was rejected by Ronald Fisher, Jerzy Neyman, and Egon Pearson, the three men upon whose ideas the classical hypothesis testing model is based.

28. An excellent discussion of odds can be found in Christensen (1990, 1997). Berry (1996, pp. 116–119) and Pitman (1993, pp. 6–9) provide good discussions on the use of odds in betting.

29. Since the mean, mode, and standard deviation of a beta distribution are a function of the parameters, the latter values can be computed with Equations 9.43, 9.44 and 9.45.

$$\mu = \frac{v}{v + w} \qquad \textbf{(Equation 9.43)}$$

$$Mode = \frac{v - 1}{v + w - 2} \qquad \textbf{(Equation 9.44)}$$

$$\sigma = \sqrt{\frac{vw}{(v + w)^2(v + w + 1)}} \qquad \textbf{(Equation 9.45)}$$

Test 10

The Single-Sample Runs Test
(and Other Tests of Randomness)
(Nonparametric Test Employed with Categorical/Nominal Data)

I. Hypothesis Evaluated with Test and Relevant Background Information

Hypothesis evaluated with test Is the distribution of a series of binary events in a population random?

Relevant background information on test By definition a **random series** is one for which no algorithm (i.e., set of rules) can be generated that will allow one to predict at above chance which of the k possible alternatives will occur on a given trial.[1] The **single-sample runs test** is one of a number of statistical procedures that have been developed for evaluating whether or not the distribution of a series of N numbers is **random**. The test evaluates the number of **runs** in a series in which on each trial the outcome must be one of $k = 2$ alternatives. Within the series, one of the alternatives occurs on n_1 trials and the other alternative occurs on n_2 trials. Thus, $n_1 + n_2 = N$. A **run** is a sequence within a series in which one of the k alternatives occurs on consecutive trials. On the trial prior to the first trial of a run (with the exception of Trial 1 in the series) and the trial following the last trial of a run (with the exception of the Nth trial in the series), the alternative that occurs will be different from the alternative which occurs during each of the trials of the run. The minimum length of a run is one trial, and the maximum length of a run is equal to N, the total number of trials in the series. To illustrate the computation of the length of a run, consider the three series noted in Figure 10.1. Each series is comprised of $N = 10$ trials. On each trial a coin is flipped and the outcome of Heads (H) or Tails (T) is recorded.

Trial	1	2	3	4	5	6	7	8	9	10
Series A:	H	H	T	H	H	T	T	T	H	T
Series B:	T	H	T	H	T	H	T	H	T	H
Series C:	H	H	H	H	H	H	H	H	H	H

Figure 10.1 Illustration of Runs

In **Series A** and **Series B** there are $n_1 = 5$ Heads and $n_2 = 5$ Tails. In **Series C** there are $n_1 = 10$ Heads and $n_2 = 0$ Tails.

In **Series A** there are six runs. Run 1 consists of Trials 1 and 2 (which are Heads). Run 2 consists of Trial 3 (which is Tails). Run 3 consists of Trials 4 and 5 (which are Heads). Run 4 consists of Trials 6–8 (which are Tails). Run 5 consists of Trial 9 (which is Heads). Run 6 consists of Trial 10 (which is Tails). This can be summarized visually by underlining all of the runs as noted below. Note that all the runs are comprised of sequences involving the same alternative. Thus: <u>H H</u> <u>T</u> <u>H H</u> <u>T T T</u> <u>H</u> <u>T</u>.

In **Series B** there are 10 runs. Each of the trials constitutes a separate run, since on each trial a different alternative occurs. Note that on each trial the alternative for that trial is preceded by and followed by a different alternative. Thus: **T H T H T H T H T H**. As noted in the definition of a run, Trial 1 cannot be preceded by a different alternative, since it is the first trial, and Trial 10 cannot be followed by a different alternative, since it is the last trial.

In **Series C** there is one run. This is the case, since the same alternative occurs on each trial. Thus: **H H H H H H H H H H**.

Intuitively, one would expect that of the three series, **Series A** is most likely to conform to the definition of a random series. This is the case, since it is highly unlikely a random series will exhibit a discernible pattern that will allow one to predict at above chance which of the alternatives will occur on a given trial. **Series B** and **C**, on the other hand, are characterized by patterns that will probably bias the guess of someone who is attempting to predict what the outcome will be if there is an eleventh trial. It is logical to expect that the strength of such a bias will be a direct function of the length of any series exhibiting a consistent pattern.[2]

The test statistic for the **single-sample runs test** is based on the assumption that the number of runs in a random series will be expected to fall within a certain range of values. Thus, if for a given series the number of runs is less than some minimum value or greater than some maximum value, it is likely that the series is not random. The determination of the minimum allowable number of runs and maximum allowable number of runs in a series of N trials takes into account the number of runs, as well as the frequency of occurrence of each of the two alternatives within the series.

It should be noted that although the **single-sample runs test** is most commonly employed with a binomially distributed variable for which $\pi_1 = \pi_2 = .5$ (as is the case for a coin toss), it is not required that the values π_1 and π_2 equal .5 in the underlying population. It is important to remember that the runs test does not evaluate a hypothesis regarding the values of π_1 and π_2 in the underlying population, nor does it make any assumption with regard to the latter values. The test statistic for the runs test is a function of the proportion of times each of the alternatives occurs in the sample data/series (i.e., $p_1 = n_1/N$ and $p_2 = n_2/N$). If the observed proportion for each of the alternatives is inconsistent with its actual proportion in the underlying population, the **single-sample runs test** is not designed to detect such a difference. Example 10.7 in Section VIII illustrates a situation in which the **single-sample runs test** is employed when it is known that $\pi_1 \neq \pi_2 \neq .5$.

II. Example

Example 10.1 *In a test of extrasensory ability, a coin is flipped 20 times by an experimenter. Prior to each flip of the coin the subject is required to guess whether it will come up Heads or Tails. After each trial the subject is told whether his guess is correct or incorrect. The actual outcomes for the 20 coin flips are listed below.*

H H H T T T H H T T H T H T H T T T H H

To rule out the possibility that the subject gained extraneous information as a result of a nonrandom pattern in the above series, the experimenter decides to evaluate it with respect to randomness. Does an analysis of the series suggest that it is nonrandom?

III. Null versus Alternative Hypotheses

Null hypothesis H_0: The events in the underlying population represented by the sample series are distributed randomly.

Alternative hypothesis H_1: The events in the underlying population represented by the sample series are distributed nonrandomly. (This is a **nondirectional alternative hypothesis** and it is evaluated with a **two-tailed test**.)

or

H_1: The events in the underlying population represented by the sample series are distributed nonrandomly due to too few runs. (This is a **directional alternative hypothesis** and it is evaluated with a **one-tailed test**.)

or

H_1: The events in the underlying population represented by the sample series are distributed nonrandomly due to too many runs. (This is a **directional alternative hypothesis** and it is evaluated with a **one-tailed test**.)

    Note: Only one of the above noted alternative hypotheses is employed. If the alternative hypothesis the researcher selects is supported, the null hypothesis is rejected.

IV. Test Computations

In order to compute the test statistic for the **single-sample runs test**, one must determine the number of times each of the two alternatives appears in the series and the number of runs in the series. Thus, we determine that the series described in Example 10.1 is comprised of $n_1 = 10$ Heads and $n_2 = 10$ Tails. Note that $n_1 + n_2 = N = 20$. We also determine there are $r = 11$ runs, which represents the test statistic for the **single-sample runs test**. Specifically, as one moves from Trial 1 to Trial 20: **H H H** (Run 1); **T T T** (Run 2); **H H** (Run 3); **T T** (Run 4); **H** (Run 5); **T** (Run 6); **H** (Run 7); **T** (Run 8); **H** (Run 9); **T T T** (Run 10); and **H H** (Run 11). This can also be represented visually by underlining each of the runs. Thus:

<u>**H H H**</u> <u>**T T T**</u> <u>**H H**</u> <u>**T T**</u> <u>**H**</u> <u>**T**</u> <u>**H**</u> <u>**T**</u> <u>**H**</u> <u>**T T T**</u> <u>**H H**</u>

V. Interpretation of the Test Results

The computed value $r = 11$ is interpreted by employing **Table A8 (Table of Critical Values for the Single-Sample Runs Test)** in the **Appendix**. The critical values listed in **Table A8** only allow the null hypothesis to be evaluated at the .05 level if a two-tailed/nondirectional alternative hypothesis is employed, and at the .025 level if a one-tailed/directional alternative hypothesis is employed. No critical values are recorded in **Table A8** for the **single-sample runs test** for very small sample sizes, since the levels of significance employed in the table cannot be achieved for sample sizes below a specific minimum value. More extensive tables for the **single-sample runs test** which provide critical values for other levels of significance can be found in Swed and Eisenhart (1943) and Beyer (1968).[3]

    Note that in **Table A8** the critical r values are listed in reference to the values of n_1 and n_2, which represent the frequencies which each of the alternatives occurs in the series. Since in Example 10.1, $n_1 = 10$ and $n_2 = 10$, we locate the cell in **Table A8** that is the intersection of these two values. In the appropriate cell, the upper value identifies the **lower limit** for the value of r, whereas the lower value identifies the **upper limit** for the value of r. The following guidelines are employed in reference to the latter values.

a) If the nondirectional alternative hypothesis is employed, to reject the null hypothesis the obtained value of r must be equal to or greater than the tabled critical upper limit at the prespecified level of significance, or be equal to or less than the tabled critical lower limit at the prespecified level of significance.

b) If the directional alternative hypothesis predicting too few runs is employed, to reject the null hypothesis the obtained value of r must be equal to or less than the tabled critical lower limit at the prespecified level of significance.

c) If the directional alternative hypothesis predicting too many runs is employed, to reject the null hypothesis the obtained value of r must be equal to or greater than the tabled critical upper limit at the prespecified level of significance.[4]

Employing **Table A8**, we determine that for $n_1 = n_2 = 10$, the tabled critical lower and upper critical r values are $r = 6$ and $r = 16$. Thus, if the nondirectional alternative hypothesis is employed (with $\alpha = .05$), the obtained value of r will be significant if it is equal to or less than 6 or equal to or greater than 16. In other words, it will be significant if there are either 16 or more runs or 6 or less runs in the data. Since $r = 11$ falls inside this range, the nondirectional alternative hypothesis is not supported.

If the directional alternative hypothesis predicting too few runs is employed (with $\alpha = .025$), the obtained value of r will only be significant if it is equal to or less than 6. In other words, it will only be significant if there are 6 or less runs in the data. Since $r = 11$ is greater than 6, the directional alternative hypothesis predicting too few runs is not supported.

If the directional alternative hypothesis predicting too many runs is employed (with $\alpha = .025$), the obtained value of r will only be significant if it is equal to or greater than 16. In other words, it will only be significant if there are 16 or more runs in the data. Since $r = 11$ is less than 16, the directional alternative hypothesis predicting too many runs is not supported.

In point of fact, it turns out that if more detailed tables are employed neither the nondirectional or either of the directional alternative hypotheses are supported at either the .05 or .01 levels. Thus, the analysis indicates that regardless of which alternative hypothesis one employs, the null hypothesis cannot be rejected. Consequently, the data do not allow the researcher to conclude that the series is nonrandom.

VI. Additional Analytical Procedures for the Single-Sample Runs Test and/or Related Tests

1. The normal approximation of the single-sample runs test for large sample sizes The normal distribution can be employed with a large sample size/series to approximate the exact distribution of the **single-sample runs test**. The large sample approximation is generally employed for sample sizes larger than those documented in **Table A8**. Equation 10.1 is employed for the normal approximation of the **single-sample runs test**.

$$z = \frac{r - u_r}{\sigma_r} = \frac{r - \left[\dfrac{2n_1 n_2}{n_1 + n_2} + 1\right]}{\sqrt{\dfrac{2n_1 n_2(2n_1 n_2 - n_1 - n_2)}{(n_1 + n_2)^2(n_1 + n_2 - 1)}}} \qquad \textbf{(Equation 10.1)}$$

In the numerator of the above equation, the term $[(2n_1 n_2)/(n_1 + n_2)] + 1$ represents the mean of the sampling distribution of runs in a random series in which there are N observations. The latter value may be summarized with the notation u_r. In other words, given $n_1 = 10$ and $n_2 = 10$, if in fact the distribution is random, the best estimate of the number of runs one can expect to observe is $\mu_r = 11$. The denominator in Equation 10.1 represents the expected standard deviation of the sampling distribution for the normal approximation of the test statistic. The latter value is summarized by the notation σ_r.

Employing Equation 10.1 with the data for Example 10.1, the value $z = 0$ is computed.

$$z = \frac{11 - \left[\dfrac{(2)(10)(10)}{10 + 10} + 1\right]}{\sqrt{\dfrac{(2)(10)(10)[(2)(10)(10) - 10 - 10]}{(10 + 10)^2(10 + 10 - 1)}}} = \frac{0}{2.18} = 0$$

Since $\mu_r = 11$ and $\sigma_r = 2.18$, the result of the above analysis can also be summarized as follows: $z = (11 - 11)/2.18 = 0$.

The obtained value $z = 0$ is evaluated with **Table A1 (Table of the Normal Distribution)** in the **Appendix**. To be significant, the obtained absolute value of z must be equal to or greater than the tabled critical value at the prespecified level of significance. The tabled critical two-tailed .05 and .01 values are $z_{.05} = 1.96$ and $z_{.01} = 2.58$, and the tabled critical one-tailed .05 and .01 values are $z_{.05} = 1.65$ and $z_{.01} = 2.33$. The following guidelines are employed in evaluating the null hypothesis.

a) If the nondirectional alternative hypothesis is employed, to reject the null hypothesis the obtained absolute value of z must be equal to or greater than the tabled critical two-tailed value at the prespecified level of significance. In Example 10.1 the nondirectional alternative hypothesis is not supported, since the obtained value $z = 0$ is less than both of the aforementioned tabled critical two-tailed values.

b) If the directional alternative hypothesis predicting too few runs is employed, to reject the null hypothesis the following must be true: 1) The obtained value of z must be a negative number; and 2) The absolute value of z must be equal to or greater than the tabled critical one-tailed value at the prespecified level of significance. In Example 10.1 the directional alternative hypothesis predicting too few runs is not supported, since the obtained value $z = 0$ is not a negative number (as well as the fact that it is less than the tabled critical one-tailed values $z_{.05} = 1.65$ and $z_{.01} = 2.33$).

c) If the directional alternative hypothesis predicting too many runs is employed, to reject the null hypothesis the following must be true: 1) The obtained value of z must be a positive number; and 2) The absolute value of z must be equal to or greater than the tabled critical one-tailed value at the prespecified level of significance. In Example 10.1 the directional alternative hypothesis predicting too many runs is not supported, since the obtained value $z = 0$ is not a positive number (as well as the fact that it is less than the tabled critical one-tailed values $z_{.05} = 1.65$ and $z_{.01} = 2.33$).

Thus, when the normal approximation is employed, as is the case when the critical values in **Table A8** are used, the null hypothesis cannot be rejected regardless of which alternative hypothesis is employed. Consequently, we cannot conclude that the series is not random.

2. The correction for continuity for the normal approximation of the single-sample runs test

Although it is not described by most sources, Siegel and Castellan (1988) recommend that a correction for continuity be employed for the normal approximation of the **single-sample runs**

test. Equation 10.2, which is the continuity-corrected equation, will always yield a smaller absolute z value than the value derived with Equation 10.1.[5]

$$z = \frac{|r - \mu_r| - .5}{\sigma_r} \qquad \textbf{(Equation 10.2)}$$

Employing Equation 10.2 with the data for Example 10.1, the value $z = -.23$ is computed.

$$z = \frac{|11 - 11| - .5}{2.18} = -.23$$

Since the absolute value $z = .23$ is lower than the tabled critical two-tailed value $z_{.05} = 1.96$ and the tabled critical one-tailed value $z_{.05} = 1.65$, the null hypothesis cannot be rejected, regardless of which alternative hypothesis is employed (which is also the case when the correction for continuity is not employed). Thus, we cannot conclude that the series is not random.

3. Extension of the runs test to data with more than two categories Wallis and Roberts (1956) and Zar (1999) note that Equations 10.3 and 10.4 can be employed for the **single-sample runs test** when the data fall into more than two categories. Equation 10.3 computes the mean of the sampling distribution (i.e., μ_r, the expected number of runs) and Equation 10.4 computes the expected standard deviation of the sampling distribution (σ_r). When there are two categories, Equation 10.3 is equivalent to the term on the right side of the numerator of Equation 10.1 (i.e., μ_r), and Equation 10.4 is equivalent to the denominator of Equation 10.1 (i.e., σ_r). The values computed for μ_r and σ_r are substituted in the normal approximation equation for the **single-sample runs test** — i.e., Equation 10.1 (the continuity-corrected version Equation 10.2 may also be employed). The computed value of z is interpreted in the same manner as when there are two categories.

$$\mu_r = \frac{N(N + 1) - \Sigma n_i^2}{N} \qquad \textbf{(Equation 10.3)}$$

$$\sigma_r = \sqrt{\frac{\Sigma n_i^2 \left[\Sigma n_i^2 + N(N + 1) \right] - 2N\Sigma n_i^3 - N^3}{N^2(N - 1)}} \qquad \textbf{(Equation 10.4)}$$

With respect to the notation employed in Equations 10.3 and 10.4, note that there will be k categories with the following number of trials/observations for each category: n_1, n_2, n_3, ..., n_k. Since n_i represents the number of trials/observations for the ith category, $N = \Sigma n_i$. The notation Σn_i^2 and Σn_i^3, respectively, indicate that the number of observations in each category are squared and cubed, and the latter values are summed.

To illustrate the **single-sample runs test** when there are more than two categories, assume we have a three-sided die which on each trial can come up as face value **A**, **B**, or **C**. Assume the pattern of results noted below is obtained.

$$\underline{A} \; \underline{A} \; \underline{B} \; \underline{C} \; \underline{C} \; \underline{B} \; \underline{A} \; \underline{A} \; \underline{B} \; \underline{B} \; \underline{C} \; \underline{C} \; \underline{A} \; \underline{C} \; \underline{B} \; \underline{A} \; \underline{A} \; \underline{B} \; \underline{B} \; \underline{B}$$

Each of the runs (which are comprised of one or more consecutive identical outcomes) is underlined, yielding a total of 12 runs. If we let n_1 represent the number of trials in which **A** appears, n_2 represent the number of trials in which **B** appears, and n_3 represent the number of trials in which **C** appears, then $n_1 = 7$, $n_2 = 8$, $n_3 = 5$, and $N = 20$. We can compute that $\Sigma n_i^2 = 7^2 + 8^2 + 5^2 = 138$ and $\Sigma n_i^3 = 7^3 + 8^3 + 5^3 = 980$. Substituting the appropriate values in Equations 10.3, 10.4, and 10.1, we compute the value $z = -1.06$.

$$\mu_r = \frac{20(20 + 1) - 138}{20} = 14.1$$

$$\sigma_r = \sqrt{\frac{138[138 + 20(20 + 1)] - (2)(20)(980) - (20)^3}{(20)^2(20 - 1)}} = 1.98$$

$$z = \frac{r - \mu_r}{\sigma_r} = \frac{12 - 14.1}{1.98} = -1.06$$

The correction for continuity can be employed by subtracting .5 from the absolute value in the numerator of Equation 10.1. When the correction for continuity is employed, $z = (|12 - 14.1| - .5)/1.98 = -.81$. Since the absolute values $z = 1.06$ and $z = .81$ are lower than the tabled critical two-tailed value $z_{.05} = 1.96$ and the tabled critical one-tailed value $z_{.05} = 1.65$, the null hypothesis cannot be rejected, regardless of which alternative hypothesis is employed. Thus, we cannot conclude that the series is not random. The negative sign for z just indicates that the observed number of runs was less than the expected value.

The reader should note that **Table A8** cannot be employed to evaluate the results of the analysis described in this section, since the latter table is only designed for use with two categories. Zar (1999) notes that O'Brien (1976) and O'Brien and Dyck (1985) have developed a more powerful version of the runs test which can be employed when there are more than two categories. Other tests of randomness that can be employed when there are more than two categories are discussed in Section IX (the **Addendum**).

4. Test 10a: The runs test for serial randomness A variant of the **single-sample runs test** described in this section is the **runs test for serial randomness** (also referred to as the **up-down runs test**). The use of the term **serial** refers to the analysis of the sequence of events in a series. Attributed to Wallis and Moore (1941), the **runs test for serial randomness** (which is also discussed in Schmidt and Taylor (1970) and Zar (1999)) is employed when the data being evaluated are in a quantitative rather than a categorical format. In such a case a researcher might want to determine if the shifts in the direction (i.e., up or down) of a sequence of scores is in a random order. Within the framework of this test, each shift in direction represents the beginning of a new run. The total number of runs is the total number of directional shifts in a set of data. To illustrate, consider the following set of ten scores:

$$+ \ - + + \ - \ - \ - \ - \ +$$
$$2, \ 3, \ 1, \ 6, \ 7, \ 4, \ 3, \ 2, \ 1, \ 7,$$

Note that a plus sign (+) is recorded at the upper right of a score if the score that follows it is larger, and a minus sign (–) is recorded if the score that follows it is smaller. Runs are determined as they were with the **single-sample runs test**. Thus, each string of plus signs or minus signs constitutes a run. In the above example there are five runs. The first two runs are comprised of a plus and a minus sign (+ –), followed by a run comprised of two plus signs (+ +), followed by a run comprised of four minus signs ($----$), followed by a run comprised of a plus sign (+). Note that the total number of plus and minus signs is one less than the total number of scores.

The null hypothesis evaluated by the **runs test for serial randomness** is that in a set of data the distribution of successive changes in direction (i.e., runs) is random. The nondirectional alternative hypothesis is that in a set of data the distribution of successive changes in direction (runs) is not random. The alternative hypothesis can also be stated directionally. Specifically, one can predict a nonrandom pattern involving an excessive number of shifts in direction (resulting in a higher than expected number of runs), or a nonrandom pattern involving very few shifts in direction (resulting in a lower than expected number of runs).

Zar (1999) has prepared a table of exact probabilities for the **runs test for serial randomness** when $N < 50$. However, for large sample sizes (generally 50 or more trials/observations), Equation 10.7 can be employed to evaluate the results of the test. Note that although the latter equation has the same structure as the normal approximation equation for the **single-sample runs test**, in the case of the **runs test for serial randomness** Equations 10.5 and 10.6 are employed to compute the values of the expected number of runs (μ_r) and the expected standard deviation (σ_r). The computed value of z is interpreted in the same manner as it is for the **single-sample runs test**. As is the case with the latter test, it will require either a very large or very small number of runs to reject the null hypothesis, and thus conclude that the data indicate a lack of randomness.

$$\mu_r = \frac{2N - 1}{3} \qquad \textbf{(Equation 10.5)}$$

$$\sigma_r = \sqrt{\frac{16N - 29}{90}} \qquad \textbf{(Equation 10.6)}$$

$$z = \frac{r - u_r}{\sigma_r} \qquad \textbf{(Equation 10.7)}$$

Example 10.2 will be employed to illustrate the **runs test for serial randomness**. Although the sample size in Example 10.2 is less than the value recommended for the normal approximation, it will be employed to demonstrate the test. Since it provides for a more conservative test, one has the option of employing the correction for continuity (through use of Equation 10.2) to lower the likelihood of committing a Type I error.

Example 10.2 *A quality control study is conducted on a machine that pours milk into containers. The amount of milk (in liters) dispensed by the machine into 21 consecutive containers follows:* 1.90, 1.99, 2.00, 1.78, 1.77, 1.76, 1.98, 1.90, 1.65, 1.76, 2.01, 1.78, 1.99, 1,76, 1.94, 1.78, 1.67, 1.87, 1.91, 1.91, 1.89. *Are the successive increments and decrements in the amount of milk dispensed random?*

The sequence of up-down shifts is summarized below.

$$\underline{+\ +}\ \ \underline{-\ -\ -}\ \ \underline{+}\ \ \underline{-\ -}\ \ \underline{+\ +}\ \ \underline{=}\ \ \underline{+}\ \ \underline{=}\ \ \underline{+}\ \ \underline{-\ -}\ \ \underline{+\ +}\ \ \underline{0}\ \ \underline{=}$$

The following 20 symbols are recorded above (one less than the total number of observations): 9 pluses, indicating an increase from one measurement to the next; 10 minuses, indicating a decrease from one measurement to the next; and one zero indicating no change (for the two values of 1.91). When one or more zeroes are present in the data, the number of runs are determined if a zero is counted as a plus, as well as if a zero is counted as a minus. If the zero is counted as a plus, the total number of runs will equal 12. This is the case since prior to the zero there are 11 runs. If the zero is counted as a plus it extends the 11th run (which will now consist of three pluses instead of two pluses), and the last minus constitutes the 12th run. If the zero is counted as a minus there will still be 12 runs, since there are 11 runs up to the zero, and if the zero is viewed as a minus it joins with the last minus to comprise the 12th run, which will now consist of two minuses. In some cases if a zero is present, a different total will be obtained for the number of runs, depending upon whether the zero is viewed as a plus or minus. When the latter is true, a test statistic is obtained for each run value, and a decision is made based on both values. If more than one zero is present in the data the analysis can get quite tedious, since one has to consider all possible combinations of counting any zero as a plus or minus. In such a case, it would probably be advisable to employ the **single-sample runs test** (see Example 10.5) or some alternative test of randomness (some of which are discussed in Section IX (the **Addendum**)). The data for Example 10.2 will now be evaluated employing Equations 10.5–10.7. For our example, $N = 21$, which represents the total number of observations, and $r = 12$, which represents the number of runs.

$$\mu_r = \frac{(2)(21) - 1}{3} = 13.67$$

$$\sigma_r = \sqrt{\frac{(16)(21) - 29}{90}} = 1.85$$

$$z = \frac{12 - 13.67}{1.85} = -.90$$

Since the absolute value $z = .90$ is lower than the tabled critical two-tailed value $z_{.05} = 1.96$ and the tabled critical one-tailed value $z_{.05} = 1.65$, the null hypothesis cannot be rejected, regardless of which alternative hypothesis is employed. The observed value $r = 12$ is well within chance expectation for the number of runs expected in a random distribution. The negative sign for z just indicates that the observed number of runs was less than the expected value.

The value $r = 12$ also does not achieve significance if one employs Zar's (1999) table of critical values. In the latter table, for $n = 21$, the critical two-tailed .05 values are respectively 9 and 18, and the critical one tailed .05 values are 10 and 18. In order to be significant, the obtained value of r must be equal to or less than the first number in each pair or equal to or greater than the second number in each pair. Since $r = 12$ is in between the limits which define the critical values, the null hypothesis is retained. Thus, regardless of whether we employ Equation 10.7 or Zar's (1999) table of critical values, we cannot conclude that the series for dispensing milk is not random.

It should be noted that the same set of data employed for Example 10.2 is evaluated with the **single-sample runs test** (see Example 10.5), also yielding a nonsignificant result. However, as is the case with the **single-sample runs test**, the **runs test for serial randomness** has the limitation that it can yield a nonsignificant result with data that is clearly nonrandom. Schmidt and Taylor (1970) provide an example in which the values of all the observations in the first half of a series fall below the median, while the values of all the observations in the second half of the series fall above the median value (a pattern that would not be expected in a random series). Yet in spite of the latter, the number of runs in the series falls within chance expectation if the data are analyzed with the **runs test for serial randomness**. As a general rule, **runs tests** are not the most stringent tests with respect to evaluating a hypothesis regarding randomness.

A discussion of the power of the **runs test for serial randomness** can be found in Levene (1952). Additional tests of randomness involving the analysis of runs can be found in Banks and Carson (1984), Phillips *et al.* (1976), and Schmidt and Taylor (1970). The latter sources describe tests which evaluate the observed versus expected frequency distribution of the length of runs for series evaluated with the **runs test for serial randomness**, as well as the **single-sample runs test**.

VII. Additional Discussion of the Single-Sample Runs Test

1. Additional discussion of the concept of randomness It is important to note that a distinction is made between a **random** and **pseudorandom** series of numbers. It is assumed that if a series is **random** there is no algorithm that will allow one to predict at above chance which of the possible outcomes will occur on a given trial. Truly random processes are event sequences that occur within a natural context (i.e., real world phenomena such as the radioactive decay of atomic nuclei and Browning molecular motion). A **pseudorandom** series, however, is generated through use of a computer program that employs a deterministic algorithm. As a result of this, if one is privy to the rule stated by the algorithm, one will be able to correctly predict all of the numbers in the pseudorandom series in the order in which they are generated. Pseudorandom series are often employed to simulate naturally occurring random events, since their use in the latter context provides researchers with a mechanism for studying phenomena that otherwise would be impossible or problematical to evaluate. Research employing pseudorandom number series is commonly referred to as **Monte Carlo research**. Peterson (1998) notes that the use of the term Monte Carlo grew out of the work of Stanislaw Ulam and John von Neumann, two brilliant mathematicians, who in the late 1940s using the earliest computers, conducted seminal research on the simulation of random processes.[6] In addition to using pseudorandom numbers for the latter, they are commonly employed to generate outcomes in slot machines as well as in presenting random stimuli in computer software (such as in video games). More recently, the popularity of data-driven statistical methods (discussed in Section IX (the **Addendum**) of the **Mann–Whitney *U* test (Test 12)**) has increased the demand for reliable random number generators.

To employ pseudorandom numbers effectively within the framework of simulation, it is essential to demonstrate that any mathematically generated series is, in fact, random. Yet the latter is easier said than done. In the second volume of his classic book *Seminumerical Algorithms* (1969, 1981, 1997), Donald Knuth notes that a number of mathematicians have suggested as a definition of randomness that a series of numbers should be able to pass each of the statistical tests which have been developed for evaluating randomness. Yet Knuth (1969, 1981, 1997) notes that it is virtually impossible to identify or generate a random series that will pass each and every statistical test which has been developed. Even if one could find such a series, it is all but certain that within the series there will be one or more sequences of numbers (often quite long in duration) which by themselves will fail one or more of the statistical tests for randomness. Peterson (1998) provides an interesting discussion on the limitations of employing pseudorandom numbers to simulate naturally occurring random processes. One of the examples he cites involves what is considered to be an excellent random number generator developed by Marsaglia and Zaman (1994), which is able to pass the most demanding tests of randomness. However, the Marsaglia–Zaman random number generator (i.e., an algorithm that generates a random sequence) yielded incorrect results in a computer-simulated study of magnetism. Thus, even an excellent random number generator may be characterized by peculiarities which may compromise its usefulness in simulating certain natural processes. It should be emphasized, however, that in spite of the latter, excellent random number generators are available for effectively simulating virtually all naturally occurring random processes.

The **single-sample runs test** is only one of many tests which have been developed for assessing randomness. In point of fact, many sources do not consider **runs tests** to be particularly effective mechanisms for assessing randomness. (For example, Conover (1999) states that **runs tests** leave a lot to be desired as tests of randomness, since they are very low in statistical power.). Section IX (the **Addendum**) describes alternative tests for randomness, as well as presenting some algorithms for generating pseudorandom numbers.

VIII. Additional Examples Illustrating the Use of the Single-Sample Runs Test

As is the case with Example 10.1, Examples 10.3–10.6 all involve series in which $N = 20$, $n_1 = n_2 = 10$, and $r = 11$. By virtue of employing identical data, the latter examples all yield the same result as Example 10.1. Example 10.6 illustrates the application of the **single-sample runs test** to a design involving two independent samples.[7] In Examples 10.1 and 10.3–10.5 it is implied that if the series involved are, in fact, random, it is probably reasonable to assume in the underlying population $\pi_1 = \pi_2 = .5$ (in other words, that each alternative has an equal likelihood of occurring in the underlying population, even if the latter is not reflected in the sample data). Example 10.7 illustrates the application of the **single-sample runs test** to a design in which it is known that in the underlying population $\pi_1 \neq \pi_2 \neq .5$.

Example 10.3 *A meteorologist conducts a study to determine whether or not humidity levels recorded at 12 noon for 20 consecutive days in July 1995 are distributed randomly with respect to whether they are above or below the average humidity recorded during the month of July during the years 1990 through 1994. Recorded below is a listing of whether the humidity for 20 consecutive days is above (+) or below (−) the July average.*

$$+ + + - - - + + - - + - + - + - - - + +$$

Do the data indicate that the series of temperature readings is random?

Example 10.4 *The gender of 20 consecutive patients who register at the emergency room of a local hospital is recorded below (where:* **M** = Male; **F** = Female).

F F F M M M F F M M F M F M F M M M F F

Do the data suggest that the gender distribution of entering patients is random?

Example 10.5 *A quality control study is conducted on a machine that pours milk into containers. The amount of milk (in liters) dispensed by the machine into* 21 *consecutive containers follows:* 1.90, 1.99, 2.00, 1.78, 1.77, 1.76, 1.98, 1.90, 1.65, 1.76, 2.01, 1.78, 1.99, 1,76, 1.94, 1.78, 1.67, 1.87, 1.91, 1.91, 1.89. *If the median number of liters the machine is programmed to dispense is* 1.89, *is the distribution random with respect to the amount of milk poured above versus below the median value?*

In Example 10.5 it can be assumed that if the process is random the scores should be distributed evenly throughout the series, and that there should be no obvious pattern with respect to scores above versus below the median. Thus, initially we list the 21 scores in sequential order with respect to whether they are above (+) or below (−) the median. Since one of the scores (that of the last container) is at the median, it is eliminated from the analysis. The latter protocol is employed for all scores equal to the median when the **single-sample runs test** is used within this context. The relationship of the first 20 scores to the median is recorded below.

$$+ \; + \; + \; - \; - \; - \; + \; + \; - \; - \; + \; - \; + \; - \; + \; - \; - \; - \; + \; +$$

Since the above sequence of runs is identical to the sequence observed for Examples 10.1, 10.3, and 10.4, it yields the same result. Thus, there is no evidence to indicate that the distribution is not random. Presence of a nonrandom pattern due to a defect in the machine can be reflected in a small number of large cycles (i.e., each cycle consists of many trials). Thus, one might observe 10 consecutive containers that are overfilled followed by 10 consecutive containers that are underfilled. A nonrandom pattern can also be revealed by an excess of runs attributed to multiple small cycles (i.e., each cycle consists of few trials).

The reader should take note of the fact that although the **Wilcoxon signed-ranks test (Test 6)** can also be employed to evaluate the data for Example 10.5, it is not appropriate to employ the latter test for evaluating a hypothesis regarding randomness. The **Wilcoxon signed-ranks test** can be used to evaluate whether or not the data indicate that the true median value for the machine is some value other than 1.89. It does not provide information concerning the ordering of the data. It should also be noted that the data for Example 10.5 are identical to that employed for Example 10.2. Note that in the latter example the **runs test for serial randomness** was employed to evaluate the hypothesis of randomness in reference to increments and decrements of liters on successive trials, and in the case of Example 10.2 it was concluded that the evidence did not suggest a lack of randomness.

Example 10.6   *In a study on the efficacy of an antidepressant drug, each of 20 clinically depressed patients is randomly assigned to one of two treatment groups. For 6 months one group is given the antidepressant drug and the other group is give a placebo. After 6 months have elapsed, subjects in both groups are rated for depression by a panel of psychiatrists who are blind with respect to group membership. Each subject is rated on a 100 point scale (the higher the rating, the greater the level of depression). The depression ratings for the two groups follow.*

> **Drug group:**   20, 25, 30, 48, 50, 60, 70, 80, 95, 98
> **Placebo group:**   35, 40, 42, 52, 55, 62, 72, 85, 87, 90

Do the data indicate there is a difference between the groups?

Since it is less powerful than alternative procedures for evaluating the same design (which typically contrast groups with respect to a measure of central tendency), the **single-sample runs test** is not commonly employed in evaluating a design involving two independent samples. Example 10.6 will, nevertheless, be used to illustrate its application to such a situation. In order to implement the runs test, the scores of the 20 subjects are arranged ordinally with respect to group membership as shown below (Where **D** represents the drug group and **P** represents the placebo group).

20	25	30	35	40	42	48	50	52	55	60	62	70	72	80	85	87	90	95	98
D	D	D	P	P	P	D	D	P	P	D	P	D	P	D	P	P	P	D	D

Runs are evaluated as in previous examples. In this instance, the two categories employed in the series represent the two groups from which the scores are obtained. When the scores are arranged ordinally, if there is a difference between the groups it is expected that most of the scores in one group will fall to the left of the series, and that most of the scores in the other group will fall to the right of the series. More specifically, if the drug is effective one will predict that the majority of the scores in the drug group will fall to the left of the series. Such an outcome will result in a small number of runs. Thus, if the number of runs is equal to or less than the tabled critical lower limit at the prespecified level of significance, one can conclude that the pattern of the data is nonrandom. Such an outcome will allow the researcher to conclude there is a significant difference between the groups. Since $n_1 = n_2 = 10$ and $r = 11$, the data for Example 10.6 are identical to those obtained for Examples 10.1, 10.3, and 10.4. Analysis of the data does not indicate that the series is nonrandom, and thus one cannot conclude that the groups differ from one another (i.e., represent two different populations).

Let us now consider two other possible patterns for Example 10.6. The first pattern depicted below contains $r = 2$ runs. It yields a significant result since in **Table A8**, for $n_1 = n_2 = 10$, any number of runs equal to or less than 6 is significant at the .05 level. Thus, the pattern depicted below will lead the researcher to conclude that the groups represent two different populations.

D D D D D D D D D D P P P P P P P P P P

Consider next the following pattern:

D D D D D P P P P P P P P P P D D D D D

Since the above pattern contains $r = 3$ runs, it is also significant at the .05 level. Yet inspection of the pattern suggests that a test which compares the mean or median values of the two groups will probably not result in a significant difference. This is based on the observation that the group receiving the drug contains the five highest and five lowest scores. Thus, if the performance of the drug group is summarized with a measure of central tendency, such a value will probably be close to the analogous value obtained for the placebo group (whose scores cluster in the middle of the series). Nevertheless, the pattern of the data certainly suggests that a difference with respect to the variability of scores exists between the groups. In other words, half of the people receiving the drug respond to it favorably, while the other half respond to it poorly. Most of the people in the placebo group, on the other hand, obtain scores in the middle of the distribution. The above example illustrates the fact that in certain situations the **single-sample runs test** may provide more useful information regarding two independent samples than other tests which are more commonly used for such a design — specifically, the *t* **test for two independent samples (Test 11)** and the **Mann–Whitney *U* test**, both of which evaluate measures of central tendency. The pattern of data depicted for the series under discussion is more likely to be identified by a test that contrasts the variability of two independent samples. In addition to the **single-sample runs test**, other procedures (discussed later in the book) which are better suited to identify differences with respect to group variability are the **Siegel–Tukey test of equal variability (Test 14)**, the **Moses test for equal variability (Test 15)**, and a number of tests described in Section VI of both the *t* **test for two independent samples** and the **single-factor between-subjects analysis of variance (Test 21)**.

Example 10.7 *A quality control engineer is asked by the manager of a factory to evaluate a machine which packages glassware. The manager informs the engineer that 90% of the glassware processed by the machine remains intact, while the remaining 10% of the glassware is cracked during the packaging process. It is suspected that some cyclical environmental* *condition may be causing the machine to produce breakages at certain points in time. In order to assess the situation, the quality control engineer records a series comprised of 1000 pieces of glassware packaged by the machine over a two-week period. It is determined that within the series 890 pieces of glassware remain intact and that 110 are cracked. It is also determined that within the series there are only 4 runs. Do the data indicate the series is nonrandom?*

In Example 10.7, $N = 1000$, $n_1 = 890$, $n_2 = 110$, and $r = 4$. Employing these values in Equation 10.1, the value $z = -3.12$ is computed. In employing the latter equation, the computed value for the expected number of runs is $\mu_r = 196.8$, which is well in excess of the observed value $r = 4$.

$$z = \frac{4 - \left[\dfrac{(2)(890)(110)}{890 + 110} + 1\right]}{\sqrt{\dfrac{(2)(890)(110)[(2)(890)(110) - 890 - 110]}{(890 + 110)^2(890 + 110 - 1)}}} = -3.12$$

Employing **Table A8**, we determine that the absolute value $z = 3.12$ is greater than the tabled critical two-tailed values $z_{.05} = 1.96$ and $z_{.01} = 2.58$, and the tabled critical one-tailed values $z_{.05} = 1.65$ and $z_{.01} = 2.33$. Thus, the nondirectional alternative hypothesis is supported at both .05 and .01 levels. Since the obtained value of z is negative, the directional alternative hypothesis predicting too few runs is supported at both the .05 and .01 levels. Obviously, the directional alternative hypothesis predicting too many runs is not supported.

Note that in Example 10.7 the plant manager informs the quality control engineer that the likelihood of a piece of glassware remaining intact is $\pi_1 = .9$, whereas the likelihood of it cracking is $\pi_2 = .1$. Thus, each of the two alternatives (**Intact** versus **Cracked**) is not equally likely to occur on a given trial. The observed proportion of cases in each of the two categories $p_1 = 890/1000 = .89$ and $p_2 = 110/1000 = .11$ are quite close to the values $\pi_1 = .9$ and $\pi_2 = .1$. As noted earlier in the discussion of the **single-sample runs test**, if the observed proportions are substantially different from the values assumed for the population proportions, the test will not identify such a difference and the analysis of runs will be based on the observed values of the proportions in the series, regardless of whether or not they are consistent with the underlying population proportions. The question of whether the proportions computed for the sample data are consistent with the population proportions is certainly relevant to the issue of whether or not the series is random. However, the **binomial sign test for a single sample (Test 9)** and the **chi-square goodness-of-fit test (Test 8)** are the appropriate tests to employ to evaluate the latter question.

IX. Addendum

1. The generation of pseudorandom numbers   In Section VII the distinction between random and pseudorandom numbers was discussed. In this section a number of pseudorandom number generators will be described. Various sources (e.g., Banks and Carson (1984), Phillips et al. (1976), and Schmidt and Taylor (1970)) note that an effective random number generator is characterized by the following properties: a) The numbers generated should conform as closely as possible to a **uniform distribution**. The latter means that each of the possible values in the distribution must have an equal likelihood of occurring. Thus, if each of the random numbers is an integer value between 0 and 9, each of the ten digits 0, 1, 2, 3, 4, 5, 6, 7, 8, 9 must have an equal likelihood of occurring; b) The random number generator should have a long **period**. The

period is the number of random numbers which are generated before the sequence begins to repeat. The term **cycling** is used when a sequence of numbers begins to repeat itself; c) A good generator should not **degenerate**. Degeneracy is when a random number generator at some point continually produces the same number; d) Since a researcher may wish to repeat the same experiment with the same set of numbers, a good random number generator should allow one to reproduce the same sequence of numbers, as well as having the ability to produce a unique sequence of numbers each time it is run; and e) The structure of a random number generator should be such that it can be executed quickly by a computer, and not utilize an excessive amount of computer memory.

Most random number generators begin with initial values called a **seed**, **constants**, and/or a **modulus**. Certain characteristics of these values (such as their magnitude or whether or not they are a prime number) may be critical in determining the quality of the random series that will be produced by a generator.[8] At this point a number of different types of random number generators will be described. Keep in mind that since quality random number generators involve the use of very large numbers and employ an excessive number of **iterations**, they require the use of a computer. (An **iteration** is the repetitive use of the same equation or set of mathematical operations.)

The midsquare method The first method for generating pseudorandom numbers was the **midsquare method** developed by John von Neumann in 1946 for simulation purposes. The midsquare method begins with a **seed** number, which is an initial value that the user arbitrarily selects. The first step of the midsquare method requires that the seed is squared, and the first random number will be the middle r digits of the resulting number. Each subsequent number is obtained by squaring the previous number, and once again employing the middle r digits of the resulting number as the next random number. The process is continued until the sequence of numbers cycles or degenerates.

To illustrate the midsquare method, assume that the seed number is 4931. The square of 4931 is 24,314,761. The middle four digits, 3147, will represent the first random number. When we square 3147, we obtain 9,903,609. The middle four digits of the latter number, 0360, will represent the next random number. When we square 0360, we obtain 129,600. The middle four digits of the latter number, 2960, will represent the next random number. As noted above, the process is continued until the sequence of numbers cycles or degenerates. Since the midsquare method tends to degenerate rapidly (resulting in a short period), it is seldom used today.

The midproduct method Although the **midproduct method** is superior to the midsquare method, it has a relatively short period compared to the best of the random number generators that are employed today. In the midproduct method one starts with two seed numbers (to be designated m_1 and m_2), each number containing the same number of digits. The values m_1 and m_2 are multiplied, and the middle r digits of the resulting value, designated m_3, are used to represent the first random number. m_3 is now multiplied by m_2, and the middle r digits of the resulting number are designated as the second random number. The process is continued until the sequence of numbers cycles or degenerates.

To illustrate the midproduct method, assume we begin with the two seed numbers 4931 and 7737, which when multiplied yield 38,151,147. The middle four digits of the latter number are 1511, which will represent the first random number. We next multiply the second seed (7737) by 1511, which yields 11,690,607. The middle four digits of the latter number are 6906, which will represent the next random number. We next multiply 1511 by 6906, yielding 10,434,966. The middle four digits of the latter number are 4349, which represent the next random number. As noted above, the process is continued until the sequence of numbers cycles or degenerates.

A variant of the midproduct method employs just one seed and another value called a **constant multiplier**. The seed is multiplied by the constant multiplier, and the first random number is the middle r digits of the resulting product. That value is then multiplied by the constant multiplier, and the second random number is the middle r digits of the resulting product, and so on. The process is continued until the sequence of numbers cycles or degenerates.

To illustrate this variant of the midproduct method, assume we begin with the seed 4931 and the constant multiplier 7737. When we multiply these two values we obtain 38,151,147. The middle four digits of the latter number are 1511, which will represent the first random number. We next multiply 1511 by the constant multiplier 7737, obtaining 11,690,607. The middle four digits of the latter number are 6906, which will represent the next random number. The value 6906 is multiplied by the constant multiplier 7737, yielding 53,431,722. The middle four digits of the latter number are 4317, which will represent the next random number. As noted above, the process is continued until the sequence of numbers cycles or degenerates.

The linear congruential method   **Congruential methods** are the most commonly used mechanisms employed today for generating random numbers. Congruential random number generators employ **modular arithmetic**, which means that they employ as a random number the remainder which results after dividing one number by another. To describe the use of modular arithmetic within the context of the congruential method, we will let the notation *mod* represent **modulus**. The notation y *mod* m means that some number designated by the symbol y is divided by the modulus which is represented by the value m. Whatever remainder results from this division will be employed as a random number.

The **linear congruential method**, which was developed by the mathematician Derrick Lehmer, is probably the most commonly used of the random number generators that are based on the congruential method. The linear congruential method produces random numbers which fall in the range 0 to $(m - 1)$. It is based on the following **recursive relationship** (a **recursive relationship** is one in which each result is computed by employing the information from the previous result): $x_{i+1} = (ax_i + c)$ mod m. In the aforementioned equation, a is a **constant multiplier**, c is referred to as the **additive constant** or **increment**, and m is the **modulus**. All of the aforementioned values remain unchanged each time the equation is employed. The initial value of x_i (which we will refer to as x_0) will be the **seed**. When the equation is employed the first time, the seed (x_0) is multiplied by the constant multiplier (a) and the additive constant (c) is added to the product. The latter value is divided by the modulus (m), and the remainder after the division represents the value on the left side of the equation (x_{i+1}). This latter value will represent the first random number (x_1). The equation is then employed again, using the value x_1 in the right side of the equation to represent x_i. The resulting value will represent the second random number. This process is continued until the sequence of numbers cycles or degenerates.

To illustrate the linear congruential method, assume we begin with the following values: **Seed** = x_0 = 47; **Constant multiplier** = a = 17; **Additive constant** = c = 79; **Modulus** = 100. Our initial equation is thus, x_1 = [(17)(47) + 79] mod 100 . The computed value of x_1 = 78, since (17)(47) + 79 = 878, which when divided by 100 yields 8 with a remainder of 78. Thus 78 will represent the first random number. The equation is employed again using the value x_1 = 78 to represent x_i on the right side of the equation. Thus, x_2 = [(17)(78) + 79] mod 100 , which yields 1405 divided by the modulus of 100. When 1405 is divided by 100, we obtain 14 with a remainder of 5. Thus, 5 is our second random number. The equation is then employed again using the value x_2 = 5 to represent x_i on the right side of the equation. Thus, x_3 = [(17)(5) + 79] mod 100 , which yields 164 divided by the modulus of 100. When 164 is divided by 100, we obtain 1 with a remainder of 64. Thus, 64 is our third random number. As noted above, this process is continued until the sequence of numbers cycles or degenerates. It should be noted that in order to generate a high quality sequence of random numbers with a congruential generator, the value of the modulus must be quite large. Bennett (1998) notes that Lehmer suggested the value 2,147,483,647 (which is equivalent to $(2^{31} - 1)$) for the modulus of congruential generators, and the latter value is, in fact, employed in many linear congruential generators. The values of the constant multiplier and additive constant vary from generator to generator. Bang *et al.* (1998) note that when the additive constant $c = 0$, the term **multiplicative congruential generator** is employed in reference to a linear congruential generator. In point of fact, most of the commonly used congruential random number generators are multiplicative congruential (i.e., since $c = 0$, the congruential equation becomes $x_{i+1} = ax_i \bmod m$).

One strategy which can further increase the likelihood that a series of numbers will be random is to **shuffle** the output of a random number generator. Specifically, a series of random digits is generated, and then a second series of random digits is generated. The latter values are then employed to select the random numbers to be used from the first series which was generated. As an example, the following series (comprised of 10 integer values which fall within the range 0 to 9) is generated: 2, 3, 3, 2, 1, 7, 1, 8, 9, 1. A second series of 10 values is then generated: 4, 2, 5, 6, 6, 9, 9, 1, 8, 3. The final values to be employed to represent the random sequence are selected as follows. Since the first digit in the second series is a 4, the fourth digit in the first series (2) is employed as the first digit. Since the second digit in the second series is 2, the second digit in the first series (3) is employed as the second digit. Since the third digit in the second series is 5, the fifth digit in the first series (1) is employed as the third digit, and so on.

Note that in all of the random number generators which have been described in this section, an integer number has been employed to represent each random number. In the case of the midsquare and midproduct methods, each value generated was a four-digit number that fell in the range 0000 to 9999. If we had wanted to, we could have selected the middle two or three digits or just one digit (instead of the middle four digits) from the product which was derived from multiplication. Also, by increasing the size of the seed(s) and/or the constant multiplier, a larger product can be obtained allowing one to select the middle six, seven, eight, etc. digits as a random number. Often when random generators are employed, the range of random numbers one desires may be very limited. Let us assume that we generate four-digit numbers as illustrated above, but want our sequence of random numbers to be comprised of two-digit numbers. The latter can be easily achieved by breaking each four-digit number into two numbers comprised of two digits, each number falling in the range 00 to 99. If one-digit numbers are desired, four one-digit numbers can be extracted from the four-digit number. If one is interested in a binary series of numbers, in which the only values employed are 0 and 1, each odd digit can be employed to represent one alternative and each even digit the other alternative. If one only wants three values, the digits 0, 1, 2 can be employed to represent one value, 3, 4, 5, a second value, and 6, 7, 8 a third value (the digit 9 would just be ignored when it occurs). By using the aforementioned logic, the number(s) generated can be formatted to represent as many alternatives as one wants to employ within a sequence of random numbers.

In actuality, most random number generators return a value that falls within the range 0 to 1. If one wished to convert any of the random numbers derived in this section into a value which falls within that range, each value can be divided by 10 raised to the appropriate power. In other words, if we take the random number 1511 generated earlier and divide it by 10,000, we obtain .1511. If we break 1511 into the two numbers 15 and 11, the latter values can be converted into .15 and .11 by dividing them by 100. Congruential generators typically return decimal values to represent random numbers. Thus, instead of expressing the remainder as an integer (as was done above), a decimal format is employed. For example, consider the value 78 used to represent the first random number generated by a linear congruential generator. The latter value was the remainder when 878 was divided by 100. The usual way of expressing the result of the division

878/100 is 8.78. The decimal part of the result, .78, can be employed to represent a random number in the range 0 to 1.

A modification of an algorithm noted by Dowdy (1986, p. 327) can be employed with random numbers which fall within the range $0 < RND < 1$ (where RND represents the value of the random number) to convert them into numbers which conform to a uniform discrete distribution that has a range of values between A and B (where A and B are integer values). Specifically, the algorithm to employ is: $X = A + (C - A) RND$, where $C = B + 1$. In the aforementioned algorithm the notation RND represents a random number which is some value greater than 0 but less than 1, and X is a value which will fall within the range A to C. To illustrate, assume that in order to simulate a series of rolls of a single die we wish to generate a series of random numbers which fall between the values $A = 1$ and $B = 6$. If the first value generated by a random number generator is $RND = .9888$, since $C = 6 + 1 = 7$, then $X = 1 + (7 - 1)(.9888) = 6.9328$. Since in simulating the roll of a die we are attempting to create a uniform discrete distribution which consists of the integer values 1, 2, 3, 4, 5, and 6, the value $X = 6.9328$ must be converted into an integer value between 1 and 6. The latter is easily accomplished by employing the integer portion of X (i.e., the value to the left of the decimal). Thus, $X = 6$. If the next random number generated is .0321, $X = 1 + (7 - 1)(.0321) = 1.1926$. X is set equal to 1, since the latter represents the integer portion of the computed value.

If one wished to generate a series of random numbers which represents a normally distributed variable, Dowdy (1986, p. 330) notes the following algorithm developed by Box and Muller (1958) can be employed: $z = \cos[(6.2832)(RND)] \sqrt{(-2)[\ln(RND)]}$. The following should be noted with respect to the algorithm: a) The cosine function in the algorithm employs radians to express the value of an angle. (**Radians** are discussed under the discussion of data transformation (within the context of an **arcsine transformation**) in Section VII of the *t* **test for two independent samples**.) Thus, if one employs a calculator to compute the cosine component of the equation, the calculator must be set in the radian mode; b) The logarithmic function employs natural logarithms (discussed in Endnote 14 in the **Introduction**). Consequently, if a calculator is employed the *ln* key should be used to compute the logarithm (and not the *log* key). The above algorithm converts a random number which falls between 0 and 1 into a *z* value (i.e., a standard deviation score), and a series of *z* values based on the algorithm will be normally distributed. To illustrate the above algorithm, if a random number is .5164453, the value derived for z will equal -1.143464, since $z = \cos[(6.2832)(.5164453)] \sqrt{(-2)[\ln(.5164453)]}$ = $(-.9946655)(1.149596) = -1.143464$.

Bang *et al.* (1998) note that at the current time there are three types of random number generators which are commonly employed. The first kind are congruential random number generators which were discussed earlier. The other two types of random number generators that are frequently used are: a) The **shift-register generator** (also known as the **Tausworthe generator**), which employs the binary structure of computers; and b) The **Fibonacci generator** (also known as the **additive generator**). Representative of the latter type of generator is the **Marsaglia–Zaman method**, which employs the **Fibonacci sequence**. The latter is a sequence of numbers in which, except for the first value, every number is the sum of the previous two numbers. Thus: 1, 1, 2, 3, 5, 8, 13, 21, etc. Peterson (1998) describes how the Fibonacci sequence was utilized by Marsaglia and Zaman to generate a random numbers series with an exceptionally long period. Bang *et al.* (1998), Banks and Carson (1984), Bennett (1998), Gentle (1998), Gruenberger and Jaffray (1965), James (1990), Knuth (1969, 1981, 1997), Peterson (1998), Phillips *et al.* (1976), and Schmidt and Taylor (1970) are sources which can provide the reader with a more detailed description and/or critique of random number generators.

A distinction can be made between a **software** versus a **hardware random number generator**. The generators described in this section are **software generators**, since they generate pseudorandom numbers through use of a computer algorithm/rule. **Hardware generators**, on the other hand, are generally viewed as producing genuine random numbers. The latter type of generators are physical devices which plug into a computer. Such devices generate random numbers by utilizing some physical process (such as emission of particles from a radioactive source or electrical noise from a resistor or semiconductor) which is converted into random units of information (such as digits) by the computer.

2. Alternative tests of randomness[9]   The **single-sample runs test** is one of many tests that have been developed for assessing randomness. Most of the alternative procedures for assessing randomness allow one to evaluate series in which, on each trial, there are two or more possible outcomes. A general problem with tests of randomness is that they do not employ the same criteria for assessing randomness. As a result of this some tests are more stringent than others, and thus it is not uncommon that a series of numbers may meet the requirements of one or more of the available tests of randomness, yet not meet the requirements of one or more of the other tests. This section will discuss some of the more commonly employed tests for evaluating random number sequences.

Test 10b: The frequency test   The **frequency test** (also known as the **equidistribution test**), which is probably the least demanding of the tests of randomness, assesses randomness on the basis of whether *k* or more equally probable alternatives occur an equal number of times within a series. The data for the **frequency test** are evaluated with the **chi-square goodness-of-fit test**, and when *k* = 2 the **binomial sign test for a single sample** (as well as the large sample normal approximation) can be employed. The **Kolmogorov–Smirnov goodness-of-fit test for a single sample (Test 7)** can also be employed to assess the uniformity of the scores in a distribution. The interested reader should consult Banks and Carson (1984) and Schmidt and Taylor (1970) for a description of how the latter test is employed within this context.

Since the **frequency test** only assesses a series with respect to the frequency of occurrence of each of the outcomes, it is insensitive to systematic patterns that may exist within a series. To illustrate this limitation of the **frequency test**, consider the following two binary series consisting of Heads (H) and Tails (T), where $\pi_1 = \pi_2 = .5$.

Series A: H H H T H T T T H T H H H T H T T T H T H
Series B: H T H T H T H T H T H T H T H T H T H T

Inspection of the data indicates that both **Series A** and **B** are comprised of 10 Heads and 10 Tails. Since the number of times each of the alternatives occurs is at the chance level (i.e., each occurs in 50% of the trials), if one elects to analyze either series employing either the **chi-square goodness-of-fit test** or the **binomial sign test for a single sample**, both series will meet the criterion for being random. However, visual inspection of the two series clearly suggests that, as opposed to **Series A**, **Series B** is characterized by a systematic pattern involving the alternation of Heads and Tails. This latter observation clearly suggests that **Series B** is not random (although it is theoretically possible for the latter sequence to occur in a random series).

At this point, the **frequency test** and the **single-sample runs test** will be applied to the same set of data. Specifically, both tests will be employed to evaluate whether or not the series below (which consists of *N* = 30 trials) is random. In the series, each of the runs has been underlined.

<u>H H H H H</u> <u>T</u> <u>H</u> <u>T</u> <u>H H H</u> <u>T</u> <u>H</u> <u>T</u> <u>H</u> <u>T</u> <u>H H H</u> <u>T T T T</u> <u>H H H H H H</u>

Since $k = 2$, the **binomial sign test for a single sample** will be employed to represent the **frequency test**. When the **binomial sign test** is employed, the null hypothesis that is evaluated is: H_0: $\pi_1 = .5$ (since it is assumed that $\pi_1 = \pi_2 = .5$). For both the **binomial sign test for a single sample** and the **single-sample runs test,** it will be assumed that a nondirectional alternative hypothesis is evaluated.

In the series that is being evaluated, the number of Heads is $n_1 = 21$ and the number of Tails is $n_2 = 9$. Employing Equation 9.9 (which is the continuity-corrected normal approximation for the **binomial sign test for a single sample**), the value $z = 2.01$ is computed.

$$z = \frac{\mid 21 - (30)(.5) \mid - .5}{\sqrt{(30)(.5)(.5)}} = 2.01$$

Since the obtained value $z = 2.01$ is greater than the tabled critical two-tailed value $z_{.05} = 1.96$, the nondirectional alternative hypothesis H_1: $\pi_1 \neq .5$ is supported. Thus, based on the above analysis with the **binomial sign test for a single sample**, one can conclude that the series is not random.

In evaluating the same series with the **single-sample runs test**, we determine that there are $r = 13$ runs in the data. The expected number of runs is $[[(2)(21)(9)]/(21 + 9)] + 1 = 13.6$, which is barely above the observed value $r = 13$. Employing Equation 10.1, the value $z = -.27$ is computed.[10]

$$z = \frac{13 - \left[\dfrac{(2)(21)(9)}{21 + 9} + 1 \right]}{\sqrt{\dfrac{(2)(21)(9)[(2)(21)(9) - 21 - 9]}{(21 + 9)^2(21 + 9 - 1)}}} = -.27$$

Since the obtained absolute value $z = .27$ is less than the tabled critical two-tailed value $z_{.05} = 1.96$, the nondirectional alternative hypothesis for the runs test is not supported. Thus, based on the above analysis with the **single-sample runs test**, one can conclude that the series is random.

That a significant result is obtained when the **binomial sign test for a single sample** is employed to evaluate the series reflects the fact that the latter test only takes into account the number of observations in each of the two categories, but does not take into consideration the ordering of the data. The **single-sample runs test**, on the other hand, is sensitive to the ordering of the data, yet will not always identify a nonrandom series if nonrandomness is a function of the number of outcomes for each of the alternatives deviating from the expected value predicted from the population parameter (in this instance the population parameter is $\pi_1 = .5$).

There will also be instances where one may conclude a series is random based on an analysis of the data with the **single-sample runs test**, yet not conclude the series is random if the **binomial sign test for a single sample** is employed for the analysis. Such a series, consisting of 15 Heads and 15 Tails, is depicted below.

H H H H H H H H H H H H H H H   T T T T T T T T T T T T T T T

When Equation 9.7 (the normal approximation of the **binomial sign test for a single sample**) is employed, it yields the following result: $z = [15 - (30)(.5)]/\sqrt{(30)(.5)(.5)} = 0$. Since the latter values is less than the tabled critical two-tailed value $z_{.05} = 1.96$, the result is not significant, and thus one can conclude that the series is random. (Equation 9.9, the continuity corrected equation (which is intended to provide a more conservative test), is not employed since it results in an absolute value larger than $z = 0$.)

When the same series, which consists of only two runs, is evaluated with Equation 10.1 (the equation for the normal approximation of the **single-sample runs test**), it yields the following result: $z = (r - \mu_r)/\sigma_r = (2 - 16)/2.69 = -5.20$. Since the absolute value $z = 5.20$ is greater than the tabled critical two-tailed values $z_{.05} = 1.96$ and $z_{.01} = 2.58$, the result is significant at both the .05 and .01 levels. Thus, if one employs the **single-sample runs test**, one can conclude that the series is not random.

One final set of data will be evaluated with the **frequency test**. The data to be presented (which were generated by a computer program) will also be evaluated with three other tests of randomness which will be presented in this section — specifically, the **gap test**, the **poker test**, and the **maximum test**. Keep in mind that in computer simulation research, the number of digits which comprise a series of random numbers will typically be more than 120 digits used in the data set to be presented. When random number generators are evaluated, millions or even billions of digits are generated and analyzed. One should keep in mind that within a random series of millions of numbers, there will undoubtedly be sequences of shorter duration that in and of themselves would not pass a test for randomness. In any event, the series presented below consists of 120 integer digits which fall within the range 0–9.

> 8, 9, 3, 7, 2, 3, 0, 2, 3, 1, 4, 7, 8, 5. 6, 2, 0, 9, 6, 8, 7, 5, 3, 0, 7, 8, 9, 6, 3, 5,
> 9, 9, 8, 4, 6, 3, 7, 9, 1, 0, 8, 3, 7, 6, 1, 0, 0, 3, 8, 9, 5, 6, 6, 7, 4, 1, 2, 0, 3, 6,
> 7, 8, 8, 8, 9, 9, 4, 5, 3, 3, 1, 1, 1, 6, 0, 0, 8, 7, 7, 3, 9, 7, 5, 2, 0, 3, 8, 6, 0, 4,
> 6, 3, 0, 2, 8, 6, 7, 0, 0, 1, 2, 5, 0, 5, 7, 9, 0, 8, 6, 4, 3, 2, 5, 8, 9, 6, 1, 0, 7, 8

The data are evaluated with the **chi-square goodness-of-fit test**. There are 10 categories, one corresponding to each of the 10 digits. Since it is assumed that each digit is equally likely to occur on a given trial, $\pi_i = .1$. Each digit followed by its observed frequency in the 120 digit series is presented: **0** (17); **1** (9); **2** (8); **3** (15); **4** (6); **5** (9); **6** (14); **7** (14); **8** (16); **9** (12). If (as it is employed in Equation 8.1) n represents the total number of observations, the expected frequency for each digit is $E_i = n\pi_i = (120)(.1) = 12$. Since the chi-square analysis will be based on $k = 10$ categories/cells, the degrees of freedom will be $df = k - 1 = 10 - 1 = 9$. When the data are evaluated with Equation 8.2, the value $\chi^2 = 10.65$ is computed. Table 10.1 provides a summary of the analysis.

Employing **Table A4 (Table of the Chi-Square Distribution)** in the **Appendix**, for $df = 9$, the tabled critical values are $\chi^2_{.05} = 16.92$ and $\chi^2_{.01} = 21.67$. Since the obtained value $\chi^2 = 10.65$ is less than $\chi^2_{.05} = 16.92$, the null hypothesis is retained. Thus, the data are consistent with the series being random.

Test 10c: The gap test Described in Banks and Carson (1984), Gruenberger and Jaffray (1965), Knuth (1969, 1981, 1997), Phillips *et al.* (1976), and Schmidt and Taylor (1970), the **gap test** evaluates the number of **gaps** between the appearance of a digit in a series and the reappearance of the same digit. Thus, if we have $k = 10$ digits and each of the digits is equally likely to occur, it would be expected that if the distribution of digits in a series consisting of n digits is random, the average gap/interval for the reoccurrence of each digit will equal $k = 10$. A gap for any digit can be determined by selecting that digit and counting until the next appearance of the same digit. To illustrate the concept of a gap, consider the following series of digits: 0121046720. For the digit 0 we can count two gaps of lengths 3 and 4 respectively. This

Table 10.1 Summary of Chi-Square Analysis for Frequency Test

Cell/Digit	Observed Frequency (O)	Expected Frequency (E)	$\dfrac{(O - E)^2}{E}$
0	17	12	2.08
1	9	12	.75
2	8	12	1.33
3	15	12	.75
4	6	12	3.00
5	9	12	.75
6	14	12	.33
7	14	12	.33
8	16	12	1.33
9	12	12	.00
Sums	120	120	$\chi^2 = 10.65$

is the case, since the number of digits between the first 0 and the second 0 is 3, and the number of digits between the second 0 and the third 0 is 4. In conducting the **gap test**, all of the gaps for each of the digits in the series are counted, after which the computed gap values are evaluated. The analysis of the data for a series within the framework of the **gap test** can employ one or more statistical tests which have been discussed in this book: a) The **single-sample z test (Test 1)** can be employed to contrast the computed mean gap value versus the expected mean gap value for each digit; b) The **single-sample chi-square test for a population variance (Test 3)** can be employed to contrast the observed versus expected variance of the gap values for each digit; and c) The **chi-square goodness-of-fit test** can be employed to compare the observed versus expected gap lengths of a specific value for each or all of the digits separately or together.

To illustrate some of the analyses which can be conducted on gap values, we will employ the same 120 digit series evaluated earlier with the **frequency test**. The number of gaps for any digit in a series will be one less than the frequency of occurrence of that digit in the series. Consequently the total number of gaps in the series will be $n - k$ (i.e., the total number of digits generated less the number of digit categories employed). Thus, in the case of a 120 digit series employing 10 digit values, the total number of gaps in the series will equal $120 - 10 = 110$. It was noted in the **frequency test** analysis of the 120-digit series, that the digit 0 appears 17 times. Thus, there are 16 gaps for the value **0**. Inspection of the series will reveal that the first gap length is 9, the second 6, and so on. The 16 gap lengths for the digit **0** are as follows: 4 gaps of length 3, 3 gaps of length 0, 2 gaps of length 10, and 1 gap of lengths 4, 5, 6, 8, 9, 15, and 16. The computed mean (through use of Equation I.1) and estimated population variance (through use of Equation I.8) of the 16 gap lengths are $\bar{X} = 5.94$ and $\tilde{s}^2 = 25.00$. The expected value for the mean (μ) is equal to k, the number of digit categories. Thus, $\mu = k = 10$. The expected value for the variance of the gaps for a digit is $\sigma^2 = k(k - 1)$. Thus, for each of the ten digits the expected variance is $\sigma^2 = 10(10 - 1) = 90$.

With respect to the data for the digit **0**, the **single-sample z test** (employing Equation 1.3) will be used to evaluate the null hypothesis that, for the digit **0**, the mean gap value for the sample is consistent with a population that has a mean gap value of 10. The **single-sample chi-square test for a population variance** (employing Equation 3.2) will be used to evaluate the null hypothesis that, for the digit **0**, the variance of the gap values in the sample is consistent with a population that has a gap value variance of 90. In a random distribution both of the aforementioned null hypotheses would not be rejected, not only in the case of the digit 0, but also in the case of any of the other nine digits.

When the **single-sample z test** is employed to evaluate the hypothesis about the mean value of the gaps for the digit 0, it yields the value $z = -1.69$. Note that the value 9.49 employed in Equation 1.3 represents the square root of the expected population variance (i.e., $\sqrt{\sigma^2} = \sqrt{90} = 9.49$. The value $n = 16$ employed in computing $\sigma_{\bar{X}}$ with Equation 1.2 represents the number of gaps.

$$ z = \frac{\bar{X} - \mu}{\sigma_{\bar{X}}} = \frac{5.94 - 10}{\frac{9.49}{\sqrt{16}}} = -1.69 $$

Since the absolute value $z = 1.69$ is greater than the tabled critical one-tailed value $z_{.05} = 1.65$, the directional alternative hypothesis predicting that the sample came from a population with a mean gap value less than 10 (since the value of z is negative) is supported at the .05 level, but not at the .01 level (since $z = 1.69$ is less than the tabled critical one-tailed value $z_{.01} = 2.33$). Since the absolute value $z = 1.69$ is less than the tabled critical two-tailed value $z_{.05} = 1.96$, the nondirectional alternative hypothesis predicting that the sample came from a population with mean gap value other than 10 is not supported. However, the fact that the one-tailed analysis yields a significant result suggests that the average gap value of $\bar{X} = 5.94$ for the digit 0 is inconsistent with what one would expect in a random distribution of digits.

When the **single-sample chi-square test for a population variance** is employed to evaluate the hypothesis about the variance of the gaps for the digit 0, (after computing an estimated population variance of $\tilde{s}^2 = 25.00$ through use of the sample gap values) Equation 3.2 is employed to compute the value $\chi^2 = 4.17$.

$$ \chi^2 = \frac{(n - 1)\tilde{s}^2}{\sigma^2} = \frac{(16 - 1)(25.00)}{90} = 4.17 $$

Employing **Table A4**, for $df = n - 1 = 15$, the tabled critical one-tailed values in the lower tail of the chi-square distribution (used to evaluate H_1: $\sigma^2 < 90$) are $\chi^2_{.05} = 7.26$ and $\chi^2_{.01} = 5.23$, and the tabled critical two tailed values (used to evaluate H_1: $\sigma^2 \neq 90$) are $\chi^2_{.05} = 6.26$ and $\chi^2_{.01} = 4.60$ (the lower tail is employed since the value computed for the estimated population variance is less than the hypothesized population variance). Since the obtained value $\chi^2 = 4.17$ is less than all of the aforementioned critical values, the null hypothesis is rejected, regardless of which of the above noted alternative hypotheses is employed. The data seem to clearly suggest that the sample came from a population with a variance gap value less than 90. Thus, the result of the evaluation of the mean and variance of the digit 0 is inconsistent with what would be expected in a random distribution of digits.

The above described analyses can also be employed to evaluate the means and estimated variances of the other nine digits. However, as noted in Section VI of the **chi-square goodness-of-fit test**, when a large number of tests are conducted on a set of data, the likelihood of committing at least one Type I error increases dramatically. (The general issue of conducting multiple tests on the same set of data is discussed in detail in Section VI of the **single-factor between-subjects analysis of variance**.) Nevertheless, it is likely that tests for multiple digits will yield significant results in the case of analyzing the means and/or variances of gap values for specific digits in a distribution that is not random.

A common test for randomness which evaluates gaps contrasts the observed frequencies of different gap values with their expected frequencies. In a distribution in which there are $k =$

10 digit categories, the probability that a gap of length r will occur (for any digit) is $P(r) = (.1)(.9)^r$. Thus, the likelihood of finding for any of the ten digits a gap of length 0 will be $(.1)(.9)^0 = .1$. The likelihood of finding a gap of length 1 will be $(.1)(.9)^1 = .09$. The likelihood of finding a gap of length 2 will be $(.1)(.9)^2 = .081$, and so on. The **chi-square goodness-of-fit test**, which will be demonstrated below, can be employed to compare the values of the observed and expected gap lengths for the whole series. If a significant difference is obtained, it would warrant the conclusion that the series is not random. The **Kolmogorov–Smirnov goodness-of-fit test for a single sample** can also be employed to assess the distribution of gap lengths within a series. The interested reader should consult Banks and Carson (1984) and Schmidt and Taylor (1970) for a description of how the latter test is employed within this context.

Table 10.2   Summary of Chi-Square Analysis for Gap Test

Cell/Gap Length	Observed number of gaps (O)	Expected number of gaps (E) & (P(r))	$\frac{(O - E)^2}{E}$
0–2	18	29.81 (.271)	4.68
3–5	21	21.67 (.197)	.02
6–8	25	15.84 (.144)	5.30
9–11	15	11.55 (.105)	1.03
12–14	13	8.36 (.076)	2.58
15–17	6	6.27 (.057)	.01
18–20	5	4.51 (.041)	.05
21–23	2	3.30 (.030)	.51
24–26	3	2.31 (.021)	.21
27–29	1	1.76 (.016)	.33
30–32	0	1.21 (.011)	1.21
33–35	0	.99 (.009)	.99
36–38	0	.66 (.006)	.66
39–41	1	.44 (.004)	.71
42–44	0	.33 (.003)	.33
45–47	0	.33 (.003)	.33
48–50	0	.33 (.003)	.33
51 or greater	0	.33 (.003)	.33
Sums	110	110 (1)	$\chi^2 = 19.61$

Table 10.2 summarizes the analysis of data of the 120 digit series with the **chi-square goodness-of-fit test**. The first column of the table lists 17 gap-length categories, which correspond to the 17 cells/categories in the chi-square table. It should be noted that the grouping of gaps into 17 categories is arbitrary. It is unlikely (although not impossible) that an alternative grouping format (i.e., a fewer or greater number of categories) will yield a substantially different result for the chi-square analysis. Each row in Table 10.2 lists in Column 2 the observed number of gaps in the data with lengths that correspond to the values listed in Column 1 of that row. The expected number of gaps for the lengths listed are noted in Column 3, followed in parentheses by the probability of obtaining gap values of the designated lengths. The latter probabilities were obtained by adding up the probability values for each of the gap lengths specified in a given row. Thus, the value .271 in the first row is the result of adding the values .1, .09, and .081, which are the values computed above (using the equation $P(r) = (.1)(.9)^r$) for the gap lengths 0, 1, and 2. The expected frequency in each row is computed by multiplying the expected row probability $(P(r))$ by 110 (which is the total number of gaps). Thus, in the case of the first row, $(110)(.271) = 29.81$.

The degrees of freedom for the chi-square table equals 17, since it is one less than the number of cells (which equals 18). In order to reject the null hypothesis, and thus conclude that the distribution is not random, it is required that the computed value of chi-square be equal to or greater than the tabled critical value at the prespecified level of significance. Employing **Table A4** for $df = 17$, $\chi^2_{.05} = 27.59$ and $\chi^2_{.01} = 33.41$. Since the obtained value $\chi^2 = 19.61$ is less than $\chi^2_{.05} = 27.59$, the null hypothesis is retained. Thus, the data are consistent with the series being random. Note, however, that the results of the above **gap test** are not consistent with the **gap test** analysis conducted previously (just on the digit **0**) with the **single-sample z test** and the **single-sample chi-square test for a population variance** (on the gap values of the mean and variance for the digit **0**). It is, however, consistent with the **frequency test** on the same series of numbers, which also did not find evidence of nonrandomness.

Test 10d: The poker test Described in Banks and Carson (1984), Gruenberger and Jaffray (1965), Knuth (1969, 1981, 1997), Phillips *et al.* (1976), and Schmidt and Taylor (1970), the **poker test** conceptualizes a series of digits as a set of hands in the game of poker. Starting with the first five digits in a series of n digits, the five digits are considered as the initial poker hand. In the analysis to be conducted, flushes are not possible, straights will not be employed, but five of a kind can occur. The analysis is repeated, employing as the second hand digits 6 through 10, and then as the third hand digits 11 through 15, and so on.[11] The **chi-square goodness-of-fit test** can be employed to evaluate the results of the test by comparing the observed frequencies for each of the possible hands with their theoretical/expected frequencies. Within the framework of tests of randomness that have been developed, the **poker test** is among the most stringent. Series of digits which are able to meet the criteria of other tests of randomness will often fail the **poker test**.

The same 120 digit series which was evaluated earlier with the **frequency test** and the **gap test** will be employed to illustrate the **poker test**. Each of the seven possible poker hands is listed, followed in parentheses by a five digit hand illustrating it, as well as the probability of obtaining that hand: a) **All five digits different** (12345; $p = .3024$); b) **One pair** (11234; $p = .5040$); c) **Two pair** (11223; $p = .1080$); d) **Three of a kind** (11123; $p = .0720$); d) **Four of a kind** (11112; $p = .0045$); e) **Five of a kind** (11111; $p = .0001$); f) **Full house** (11122; $p = .0090$).[12] (Note that the sum of the probabilities for the seven possible hands equals one.) Table 10.3 summarizes the chi-square analysis of the data for the 120 digit series. The latter series yields only 24 poker hands, since $120/5 = 24$. Although, in reality, a much larger number of hands should be employed in using the **poker test** to assess randomness, for purposes of demonstration we will employ the 24 available hands. In Table 10.3, the expected number of observations for each type of hand was computed by multiplying the total number of hands (24) by the probability of that hand occurring. Thus, in the case of the hand **All different**, the expected probability of 7.2576 was obtained as follows: $(24)(.3024) = 7.2576$.

The degrees of freedom for the chi-square table equals 6, since it is one less than the number of cells/categories (which equals 7). In order to reject the null hypothesis, and thus conclude that the distribution is not random, it is required that the computed value of chi-square is equal to or greater than the tabled critical value at the prespecified level of significance. Employing **Table A4** for $df = 6$, $\chi^2_{.05} = 12.59$ and $\chi^2_{.01} = 16.81$. Since the obtained value $\chi^2 = 8.29$ is less than $\chi^2_{.05} = 12.59$, the null hypothesis is retained. Thus, the data are consistent with the series being random.

Test 10e: The maximum test Described in Gruenberger and Jaffray (1965), the **maximum test** evaluates strings of three consecutive digits and records the number of cases in which the middle digit is higher than either of the outside two digits (e.g., in the string 152, the 5 is larger

Table 10.3 Summary of Chi-Square Analysis for Poker Test

Cell/Poker Hand	Observed number of hands (O)	Expected number of hands (E)	$\dfrac{(O - E)^2}{E}$
All different	13	7.2576	4.54
One pair	9	12.096	.79
Two pair	0	2.592	2.59
Three of a kind	2	1.728	.04
Four of a kind	0	.108	.11
Five of a kind	0	.0024	.00
Full house	0	.216	.22
Sums	24	24	$\chi^2 = 8.29$

than the 1 and 2). Gruenberger and Jaffray (1965) note that the likelihood of the latter occurring is .285. The **binomial sign test for a single sample** can be employed to evaluate the data, with n representing the total number of three digit strings analyzed in a sequence, and π_1 = .285 and π_2 = .715 respectively representing the likelihood of a **hit** versus a **miss** (where a **hit** is the middle digit being greater than the two outside digits).

The same series of 120 digits which has been evaluated with the other tests described in this section will now be evaluated with the **maximum test**. Beginning with the first digit and moving sequentially, 40 three-digit strings can be demarcated. Thus, the first string in the series which consists of the digits 893 is a hit, since the middle digit 9 is greater than the two outside digits 8 and 3. The second string 723 is a miss, since the middle digit 2 is not higher than both of the outside digits. Altogether, in 10 of the 40 strings ($p = 10/40 = .225$) the middle digit is larger than the two outside digits. The expected number of strings where the middle digit will be higher than the outside digit is $\mu = (\pi_1)(n) = (.285)(40) = 11.4$. Employing Equation 9.7, (the normal approximation of the **binomial sign test for a single sample**), the value $z = -.49$ is computed. The negative sign for the z value indicates that the observed number of strings is less than the expected number of strings. (Since Equation 9.9, the continuity-corrected equation, yields a value that is closer to zero (specifically, $z = -.32$), it will not be employed.)

$$z = \frac{x - m\pi_1}{\sqrt{n\pi_1\pi_2}} = \frac{10 - 11.4}{\sqrt{(40)(.285)(.715)}} = -.49$$

Since the absolute value $z = .49$ is lower than the tabled critical two-tailed value $z_{.05} = 1.96$ and the tabled critical one-tailed value $z_{.05} = 1.65$, the null hypothesis cannot be rejected, regardless of which alternative hypothesis is employed. The observed value $x = 10$ is well within chance expectation. Thus, we cannot conclude that the series is not random.

With the exception of the version of the **gap test**, which was only conducted on the mean and variance for the digit 0, the 120 digit series passed all of the other tests of randomness (i.e., the **frequency test**, the **gap test** (for the version of the test evaluating gap lengths for the entire distribution), the **poker test**, and the **maximum test**).

Test 10f: The coupon collector's test Imagine that a company inserts one of k different kinds of coupons in every box of a product it sells, and that the likelihood of any of the coupons being in a given box is $1/k$. The **coupon collector's test** addresses the question of how many boxes of the product a person would have to purchase in order to have a set which includes each of the k coupons. Within the context of evaluating numbers, the **coupon collector's test** evaluates the number of digits which will have to be generated in order to make a complete set of k digits in which there is at least one case of all the integer values 1 through k.

Mosteller (1965, p. 15) notes that in a random series, the average number of digits required (i.e., the expected value) to obtain a set which includes all k digits can be computed through use of the harmonic series $k[(1/k) + 1/(k-1) + 1/(k-2) + \cdots + 1/2 + 1]$. Thus, if the integers 1 through 5 are employed in a series of random numbers, the expected number of trials which will be required to have a set that includes all five digits is $5(1/5 + 1/4 + 1/3 + 1/2 + 1) = 11.42$. The average number of digits can also be approximated quite accurately by employing the following equation: $k \ln k + .577k + 1/2$. When the latter equation is solved for $k = 5$ digits, it results in the value 11.43 (i.e., $5\ln 5 + (.577)(5) + 1/2 = 11.43$).[13]

The **single-sample z test** or **single sample t test (Test 2)** (depending upon whether the expected or estimated value for the population variance is employed) can be used to contrast the predicted/expected average number of digits (as computed with the equation noted in the previous paragraph) with the observed average number of digits required for all of the complete sets comprised of k digits in a series comprised of m digits.

Blom *et al.* (1994, pp. 85–86), Feller (1968, pp. 60–61) and Knuth (1969, 1981, 1997) describe methods for computing probabilities within the context of the **coupon collector's test**. Blom *et al.* (1994) note that Equation 10.8 can be employed to compute the probability that all k digits will be included in a string of consecutive digits the length of which is equal to or less than n digits.

$$P(N \leq n) = \sum_{j=0}^{k}(-1)^j \binom{k}{j}\left(1 - \frac{j}{k}\right)^n \qquad \textbf{(Equation 10.8)}$$

To demonstrate the use of Equation 10.8, we will assume that we have a series of random numbers which is comprised of the digits 1, 2, 3, 4, and 5. Thus, $k = 5$. The notation in Equation 10.8 indicates that the operations recorded after the summation sign should initially be carried out with the value of j set equal to zero, and subsequently be employed for all integer values of j up to and including the value of k (i.e., if $k = 5$, for values of j equal to 0, 1, 2, 3, 4, 5). For each value of j the following product is obtained: -1 raised to the jth power, multiplied by the combination of k things taken j at a time, multiplied by the value $[1 - (j/k)]$ raised to the nth power. The sum of the six products obtained when j is set equal to 0, 1, 2, 3, 4, and 5 represents the probability that all k digits will be included in a string which is comprised of n or less consecutive digits. Initially we will compute the likelihood that all five digits will be included in a string which is comprised of $n = 5$ consecutive digits. Note that a minimum of a five digit string is required in order to include all five digits.

$$j = 0 \qquad (-1)^0 \binom{5}{0}\left(1 - \frac{0}{5}\right)^5 = +1$$

$$j = 1 \qquad (-1)^1 \binom{5}{1}\left(1 - \frac{1}{5}\right)^5 = -1.6384$$

$$j = 2 \qquad (-1)^2 \binom{5}{2}\left(1 - \frac{2}{5}\right)^5 = +.7776$$

$$j = 3 \qquad (-1)^3 \binom{5}{3}\left(1 - \frac{3}{5}\right)^5 = -.1024$$

$$j = 4 \qquad (-1)^4 \binom{5}{4}\left(1 - \frac{4}{5}\right)^5 = +.0016$$

$$j = 5 \quad (-1)^5 \binom{5}{5}\left(1 - \frac{5}{5}\right)^5 = 0$$

$$\sum = .0384$$

The value .0384 represents the probability that all five digits will be included in a string which is comprised of $n = 5$ consecutive digits.

We will now employ Equation 10.8 to compute the likelihood that all five digits will be included in a string which is comprised of $n = 6$ or less consecutive digits.

$$j = 0 \quad (-1)^0 \binom{5}{0}\left(1 - \frac{0}{5}\right)^6 = +1$$

$$j = 1 \quad (-1)^1 \binom{5}{1}\left(1 - \frac{1}{5}\right)^6 = -1.31072$$

$$j = 2 \quad (-1)^2 \binom{5}{2}\left(1 - \frac{2}{5}\right)^6 = +.46656$$

$$j = 3 \quad (-1)^3 \binom{5}{3}\left(1 - \frac{3}{5}\right)^6 = -.04096$$

$$j = 4 \quad (-1)^4 \binom{5}{4}\left(1 - \frac{4}{5}\right)^6 = +.00032$$

$$j = 5 \quad (-1)^5 \binom{5}{5}\left(1 - \frac{5}{5}\right)^6 = 0$$

$$\sum = .1152$$

Thus, the probability of including all five digits in a string which is comprised of $n = 6$ or less consecutive digits is .1152. Since it was previously determined that .0384 is the probability of obtaining all five digits in a five digit string, if we subtract .0384 from .1152 we can determine the probability that all five digits will be included in a string which is comprised of exactly $n = 6$ consecutive digits. Since $.1152 - .0384 = .0768$, the likelihood that all five digits will be included in a string which is comprised of exactly six consecutive digits is .0768.

The probability values computed above can be employed within the framework of the **chi-square goodness-of-fit test**. As an example, if we divide a large series comprised of m digits (e.g., a series comprised of 60,000 digits) into strings comprised of $n = 5$ consecutive digits (e.g., 60,000 digits would yield 12,000 five digit strings, since 60,000/5 = 12,000), we would expect 460.8 of the five digit strings to include all five digits, since (12,000)(.0384) = 460.8. Through use of the **chi-square test**, the latter expected value would be evaluated in relation to the observed number of five digit strings which contain all five digits. The **chi-square test** would involve two cells, one cell representing strings which included all five digits, and a second cell representing strings which did not include all five digits. The expected probability for the second cell would be $1 - .0384 = .9616$, and thus (12,000)(.9616) = 11,539.2 would be the number of strings not expected to contain all five digits. Another analysis could be conducted for six digit strings. Specifically, since we can obtain 10,000 six digit strings from a 60,000 digit series (since 60,000/6 = 10,000), we would expect 1152 of the six digit strings to include all five digits (i.e., (10,000)(.1152) = 1152). (The five digits could occur as a consecutive string of the digits or within the context of a nonconsecutive string.) The same procedure can be carried out for additional values of n. The reader should keep in mind, however, that (as noted in Section VI of the **chi-square goodness-of-fit test**) unless an adjustment is employed for the Type I error rate, the greater the number of analyses a researcher conducts, the higher the likelihood that a significant result may represent a Type I error.

An alternative application of the **chi-square goodness-of-fit test** requires that initially the procedure described for computing probabilities for a given value of n be carried out for all values of n (beginning with the lowest possible value for n) until the sum obtained for a specific value of n equals (or is extremely close to) 1. Upon doing the latter, one should select the first digit in a long series comprised of m digits, and count the number of consecutive digits required before a full set of five digits is obtained. At that point the procedure is repeated until another full set of the five digits is obtained. The procedure should be repeated as many times as necessary until there are no digits remaining in the series. The number of 5, 6, 7, ... , etc. digit strings which are obtained containing all five digits are employed to represent the observed frequencies, which are contrasted (through use of the **chi-square test**) with the expected frequencies computed by the method for computing probabilities described earlier.

The tests thus far described in this section are employed, for the most part, to evaluate series which are comprised of data which represent a discrete variable. The next test that will be described is designed to evaluate continuous data (i.e., data that are not constrained to assume a limited number of values).

Test 10g: The mean square successive difference test (for serial randomness) (The reader should note that this test evaluates interval/ratio data.) Attributed to Bellinson *et al.* (1941) and von Neumann (1941) and described in Bennett and Franklin (1954), Chou (1989), and Zar (1999), the **mean square successive difference test** contrasts the mean of the squares of the differences of $(n-1)$ successive differences in a series of n numbers with the variance of the n numbers. Within the framework of the test, the mean of the squares of the successive differences is conceptualized as an alternative measure of variance that is contrasted with the estimated population variance (which is computed with Equation I.8). The **mean square successive difference test** can be employed to evaluate whether or not a sequence of continuous interval/ratio scores is random. Zar (1999, p. 586) notes that an assumption of the latter test is that the data are derived from a normally distributed population.

Using a methodology described by Zar (1999, pp. 586–587), the **mean square successive difference test** will be employed to evaluate Example 10.2 (which employs the same data as Example 10.5), which was previously evaluated with the **runs test for serial randomness (Test 10a)**. The serial distribution of the 21 scores for Example 10.2 is presented below.

1.90, 1.99, 2.00, 1.78, 1.77, 1.76, 1.98, 1.90, 1.65, 1.76, 2.01, 1.78, 1.99, 1,76, 1.94, 1.78, 1.67, 1.87, 1.91, 1.91, 1.89

The initial value that is computed in conducting the **mean square successive difference test** is the estimated population variance. Employing Equation I.8, the latter value is computed to be $\hat{s}^2 = .0122$ for the $n = 21$ scores in the series (the mean score in the series is $\bar{X} = 1.857$). Equation 10.9 is now employed to compute the **mean of the squares of the successive differences** (more specifically, the unbiased estimate of that value in the population). The latter value will be represented by the symbol $\tilde{s}_{ms}^2$.

$$\tilde{s}_{ms}^2 = \frac{\sum_{i=1}^{n-1}(X_{i+1} - X_i)^2}{2(n-1)} \qquad \text{(Equation 10.9)}$$

The numerator of Equation 10.9 indicates that each score in the series is subtracted from the score that comes after it, each of the difference scores is squared, and the squared difference scores are summed. The sum of the squared difference scores is divided by $2(n-1)$, yielding the value of $\tilde{s}_{ms}^2$. The computation of the value $\tilde{s}_{ms}^2 = .0128$ is demonstrated below.

$$\tilde{s}_{ms}^2 = \frac{(1.99-1.90)^2 + (2.00-1.99)^2 + \cdots + (1.91-1.91)^2 + (1.89-1.91)^2}{2(21-1)} = \frac{.5107}{40} = .0128$$

Equation 10.10 is employed to compute the test statistic, which Young (1941) designated as C. Employing the latter equation, the value $C = -.049$ is computed. Note that the value of C will be positive when $\tilde{s}_{ms}^2 < \tilde{s}^2$, negative when $\tilde{s}_{ms}^2 > \tilde{s}^2$, and equal to 0 when the two values are equal.

$$C = 1 - \frac{\tilde{s}_{ms}^2}{\tilde{s}^2} = 1 - \frac{.0128}{.0122} = -.049 \qquad \text{(Equation 10.10)}$$

In order to reject the null hypothesis and conclude that the distribution is nonrandom, the absolute value of C must be equal to or greater than the tabled critical value of the C statistic at the prespecified level of significance. A large absolute C value indicates there is a large discrepancy between the values $\tilde{s}^2$ and $\tilde{s}_{ms}^2$. A table of critical values can be found in Zar (1999), who developed his table based on the work of Young (1941). Another table of the sampling distribution for this test (although not C values) was derived by Hart (1942), and can be found in Bennett and Franklin (1954). The computed absolute value $C = .049$ is less than Zar's (1999) tabled critical values $C_{.05} = .343$ and $C_{.01} = .470$. We thus retain the null hypothesis. In other words, the evidence does not indicate the distribution is not random.

In lieu of the exact table for the sampling distribution of C, we will employ a large sample normal approximation for the C statistic which is computed with Equation 10.11. Employing the latter equation, the value $z = -.24$ is computed.

$$z = \frac{C}{\sqrt{\dfrac{n-2}{n^2-1}}} = \frac{-.049}{\sqrt{\dfrac{21-2}{(21)^2-1}}} = -.24 \qquad \text{(Equation 10.11)}$$

Zar (1999) notes that one-tailed probabilities should be employed for the above analysis. To reject the null hypothesis and conclude that the distribution is nonrandom, the absolute value of z must be equal to or greater than the tabled critical value at the prespecified level of significance. Since the absolute value $z = .24$ is less than the tabled critical one-tailed value $z_{.05} = 1.65$, the null hypothesis cannot be rejected. Thus, the evidence does not indicate the distribution is nonrandom. This is consistent with the conclusion which was reached when the same set of data was evaluated with the **runs test for serial randomness**.

Additional tests of randomness

Gentle (1998, p. 155), who provides a detailed discussion of random number generators and tests of randomness, notes that the number of tests of randomness which can be developed is only limited by researchers' imaginations. Many tests of randomness (such as the **poker test**) are based on games of chance, while others are based on some other type of random process. Some of the more commonly employed alternative tests of randomness are summarized below.

1) **Autocorrelation** (also known as **serial correlation**) This procedure can be employed with series in which in each trial there are two or more possible categorical outcomes, or with continuous serial data (such as the data that were evaluated with the **runs test for serial randomness** and the **mean square successive difference test**). Within the framework of autocorrelation, one can conclude that a series is random if the correlation coefficient between successive numbers in the series is equal to zero. The **Durbin–Watson test (Test 28h)** (1950, 1951, 1971) is the most commonly employed procedure for detecting autocorrelation. The latter test is described in Section VII of the **Pearson product-moment correlation coefficient (Test 28)** within the context of a more extensive discussion of the general topic of autocorrelation. Within the context of the latter discussion, the use of the **single-sample runs test** is also discussed as an alternative, albeit less sensitive, procedure for detecting autocorrelation.

2) **The serial test** The **serial test** evaluates the occurrence of each two-digit combination ranging from 00 to kk, where k is the largest digit that can occur in the series. The **serial test**, which can be generalized to groups consisting of combinations of three or more digits, is described in more detail in Emshoff and Sisson (1970).

3) **The d^2 test of random numbers** Described in Gruenberger and Jaffray (1965), the d^2 **test of random numbers** conceptualizes random numbers as coordinates on a graph, and addresses the following question: If two points are chosen at random within a one unit square grid, what is the likelihood that the distance between the two points is greater than a specific value (e.g., that the value of d is greater than .5 and thus d^2 is greater than .25)? The d^2 **test of random numbers** is a stringent test which can be employed to evaluate random numbers that are in the range 0 to 1, since each point is represented by two numbers which fall in the latter range of values.

The d^2 **test of random numbers** can be classified as a **two-dimensional test of randomness**, since it involves evaluating the placement of numbers which are represented as points in two dimensional space. A simple variant of the latter test would be to construct a graph on which the values 0 to 1 are represented along both the abscissa and the ordinate. On each axis the values 0, .1, .2, .3, .4, .5, .6, .7, .8, .9, and 1 are recorded. The latter graph/two-dimensional plane can be viewed as a 10×10 grid comprised of 100 squares. Pairs of random numbers which fall between the values 0 and 1 are generated, and for each pair of values a point is placed in one of the 100 squares. For example, the pair of values .4536 and .6789 would yield a point that would be placed in the square which has the lower and upper boundaries respectively on the abscissa of .4 and .5, and the lower and upper boundaries respectively on the ordinate of .6 and .7. If a series of numbers is random, it would be expected that an equal number of points would fall within each of the 100 squares which comprise the 10×10 grid. The latter hypothesis could be evaluated through use of a **chi-square goodness-of-fit test**.

It is also possible to conduct a three or higher dimensional test of randomness. In the case, of a **three-dimensional test**, number triplets instead of number pairs are generated. Each triplet of numbers is employed to represent a point in a three-dimensional space which is conceptualized as a $10 \times 10 \times 10$ cube comprised of 1000 smaller cubes. As an example, the triplet of values .4536, .6789, .0232 would yield a point which would be placed in the small cube which has the lower and upper boundaries respectively on the horizontal plane of .4 and .5, the lower and upper boundaries respectively on the vertical plane of .6 and .7, and the lower and upper boundaries respectively on the plane representing depth of 0 and .1. Once again a **chi-square goodness-of-fit test** could be employed to evaluate the hypothesis of whether or not there are an equal number of points in all of the small cubes which comprise the $10 \times 10 \times 10$ cube.

4) **Tests of trend analysis/time series analysis**   Economists often refer to a set of observations that are measured over a period of time as a **time series**. The pattern of the data in a time series may be random or may instead be characterized by patterns or trends. **Trend analysis** and **time series analysis** are terms which are used to describe a variety of statistical procedures (such as those that have been described in this section) that are employed for analyzing such data. Among the other tests which are used for trend analysis that can be employed to identify a nonrandom series is the **Cox–Stuart test for trend** (developed by Cox and Stuart (1955) and described in Conover (1999) and Daniel (1990)), which is a modification of the **binomial sign test for a single sample**. Other tests of time series and trend analysis are commonly described in books on business and economic statistics (e.g., Chou (1989), Hoel and Jessen (1982), Montgomery and Peck (1992), and Netter *et al.* (1983)).

5) George Marsaglia (1985, 1995) developed a battery of tests of randomness called the **DIEHARD** (which is available on a CD-ROM). Gentle (1998, pp. 155–157) provides a succinct summary of the tests which are included in the **DIEHARD** battery.

References

Bang, J. W., Schumacker, R. E., & Schlieve, P. L. (1998).· Random-number generator validity in simulation studies: An investigation of normality. **Educational and Psychological Measurement**, 58, 430–450.

Banks, J. & Carson, J.S. (1984). **Discrete-event system simulation**. Englewood Cliffs, NJ: Prentice Hall.

Bellinson, H. R., Von Neumann, J., Kent, R. H., & Hart, B. I. (1941). The mean square successive difference. **Annals of Mathematical Statistics**, 12, 153–162.

Bennett, C. A. & Franklin, N. L. (1954). **Statistical analysis in chemistry and the chemical industry**. New York: John Wiley & Sons, Inc.

Bennett, D. J. (1998). **Randomness**. Cambridge, MA: Harvard University Press.

Beyer, W. H. (1968). **Handbook of tables for probability and statistics** (2nd ed.). Cleveland, OH: The Chemical Rubber Company.

Blom, G., Holst, L., Sandell, D. (1994). **Problems and snapshots from the world of probability**. New York: Springer–Verlag.

Box, G. E. & Muller, M. E. (1958). A note on the generation of random normal deviates. **Annals of Mathematical Statistics**, 29, 610–611.

Chou, Y. (1989). **Statistical analysis for business and economics**. New York: Elsevier.

Conover, W. J. (1999). **Practical nonparametric statistics** (3rd ed.). New York: John Wiley & Sons.

Cox, D. R. & Stuart, A. (1955). Some quick tests for trend in location and dispersion. **Biometrika**, 42, 80–95.

Daniel, W. (1990). **Applied nonparametric statistics** (2nd ed.). Boston: PWS–Kent Publishing Company.

Dowdy, S. (1986). **Statistical experiments using BASIC**. Boston: Duxbury Press.

Durbin, J. & Watson, G. S. (1950). Testing for serial correlation in least squares regression I. **Biometrika**, 37, 409–438.

Durbin, J. & Watson, G. S. (1951). Testing for serial correlation in least squares regression II. **Biometrika**, 38, 159–178.

Durbin, J. & Watson, G. S. (1971). Testing for serial correlation in least squares regression III. **Biometrika**, 58, 1–19.

Emshoff, J. R. & Sisson, R. L. (1970). **Design and use of computer simulation models**. New York: Macmillan Publishing Company.

Feller, W. (1968). **An introduction to probability theory and its applications (Volume I)** (3rd ed.). New York: John Wiley & Sons.

Gentle, J. E. (1998). **Random number generation and Monte Carlo methods**. New York: Springer–Verlag.

Gruenberger, F. & Jaffray, G. (1965). **Problems for computer solution**. New York: John Wiley & Sons.

Hart, B. I. (1942). Significance levels for the ratio of the mean square successive difference to the variance. **Annals of Mathematical Statistics**, 13, 445–447.

Hoel, P. G. & Jessen, R. J. (1982). **Basic statistics for business and economics** (3rd ed.). New York: John Wiley & Sons.

Hogg, R. V. & Tanis, E. A. (1988). **Probability and statistical inference** (3rd ed.). New York: Macmillan Publishing Company.

James, F. (1990). A review of pseudorandom number generators. **Computer Physics Communications**, 60, 329-344.

Knuth, D. (1969). **Semi-numerical algorithms: The art of computer programming (Vol. 2)**. Reading, MA: Addison–Wesley.

Knuth, D. (1981). **Semi-numerical algorithms: The art of computer programming (Vol. 2)** (2nd ed.). Reading, MA: Addison–Wesley.

Knuth, D. (1997). **Semi-numerical algorithms: The art of computer programming (Vol. 2)** (3rd ed.). Reading, MA: Addison–Wesley.

Larsen, R. J. & Marx, M. L. (1985). **An introduction to probability and its applications**. Englewood Cliffs, NJ: Prentice Hall.

Levene, H. (1952). On the power function of tests of randomness based on runs up and down. **Annals of Mathematical Statistics**, 23, 34–56.

Marsaglia, G. (1985). A current view of random number generators. In L. Billard (Ed.), **Computer science and statistics: 16th symposium on the interface** (pp. 3–10). Amsterdam: North–Holland.

Marsaglia, G. (1995). **The Marsaglia random number generator CDROM, including the DIEHARD battery of tests of randomness**. Department of Statistics, Florida State University, Tallahassee, FL.

Marsaglia, G. & Zaman, A. (1994). Some portable very-long period random number generators. **Computers in Physics**, 8, 117–121.

Montgomery, D. C. & Peck, E. A. (1992). **Introduction to linear regression analysis** (2nd ed.). New York: John Wiley & Sons, Inc.

Mosteller, F. (1965). **Fifty challenging problems in probability with solutions**. New York: Dover Publications, Inc.

Netter, J., Wasserman, W., & Kutner, M. H. (1983). **Applied linear regression models** (3rd ed.). Homewood, IL: Richard D. Irwin, Inc.

O'Brien, P. C. (1976). A test for randomness. **Biometrics**, 32, 391–401.

O'Brien, P. C. & Dyck, P. J. (1985). A runs test based on runs lengths. **Biometrics**, 41, 237–244.

Peterson, I. (1998). **The jungles of randomness**. New York: John Wiley & Sons.

Phillips, D. T., Ravindran, A., & Solberg, J. T. (1976). **Operations research: Principles and practice**. New York: John Wiley & Sons.

Schmidt, J. W. & Taylor, R. E. (1970). **Simulation and analysis of industrial systems.** Homewood, IL: Richard D. Irwin.

Siegel, S. (1956). **Nonparametric statistics for the behavioral sciences** (1st ed.). New York: McGraw–Hill Book Company.

Siegel, S. & Castellan, N. J., Jr. (1988). **Nonparametric statistics for the behavioral sciences** (2nd ed.). New York: McGraw–Hill Book Company.

Stirzaker, D. (1999). **Probability and random variables: A beginner's guide**. New York: Cambridge University Press.

Swed, F. S. & Eisenhart, C. (1943). Tables for testing randomness of grouping in a sequence of alternatives. **Annals of Mathematical Statistics**, 14, 66–87.

von Neumann, J. (1941). Distribution of the ratio of the mean square successive difference to the variance. **Annals of Mathematical Statistics**, 12, 307–395.

Wald, A. & Wolfowitz, J. (1940). On a test whether two samples are from the same population. **Annals of Mathematical Statistics**, 11, 147–162.

Wallis, W. A. & Moore, G. H. (1941). A significance test for time series analysis. **Journal of the American Statistical Association**, 36, 401–409.

Wallis, W. A. & Roberts, H. V. (1956). **Statistics: A new approach**. Glencoe, IL: Free Press.

Young, L. C. (1941). Randomness in ordered sequences. **Annals of Mathematical Statistics**, 12, 293–300.

Zar, J. H. (1999). **Biostatistical analysis** (4th ed.). Upper Saddle River, NJ: Prentice Hall.

Endnotes

1. An alternate definition of randomness employed by some sources is that in a random series, each of k possible alternatives is equally likely to occur on any trial, and that the outcome on each trial is independent of the outcome on any other trial. The problem with the latter definition is that it cannot be applied to a series in which on each trial there are two or more alternatives which do not have an equal likelihood of occurring (the stipulation regarding independence does, however, also apply to a series involving alternatives that do not have an equal likelihood of occurring on each trial). In point of fact, it is possible to apply the concept of randomness to a series in which $\pi_1 \neq \pi_2$. To illustrate the latter, consider the following example. Assume we have a series consisting of N trials involving a binomially distributed variable for which there are two possible outcomes **A** and **B**. The theoretical probabilities in the underlying population for each of the outcomes are $\pi_A = .75$ and $\pi_B = .25$. If a series involving the two alternatives is in fact random, on each trial the respective likelihoods of alternative **A** versus alternative **B** occurring will not be $\pi_A = \pi_B = .5$, but instead will be $\pi_A = .75$ and $\pi_B = .25$. If such a series is random it is expected that alternative **A** will occur approximately 75% of the time and alternative **B** will occur approximately 25% of the time. However, it is important to note that one cannot conclude that the above series is random purely on the basis of the relative frequencies of the two alternatives. To illustrate this, consider the following series consisting of 28 trials, which is characterized by the presence of an invariant pattern: **AAABAAABAAABAAABAAABAAABAAAB**. If one is attempting to predict the outcome on the 29th trial, and if, in fact, the periodicity of the pattern that is depicted is invariant, the likelihood that alternative **A** will occur on the next trial is not .75, but is, in fact, 1. This is the case, since the occurrence of events in the series can be summarized by the simple algorithm that the series is comprised of 4 trial cycles, and within each cycle alternative **A** occurs on the first 3 trials and alternative **B** on the fourth trial. The point to be made here is that it is entirely possible to have a random series, even if each of the

alternatives is not equally likely to occur on every trial. However, if the occurrence of the alternatives is consistent with their theoretical frequencies, the latter in and of itself does not insure that the series is random.

2. It should be pointed out that, in actuality, each of the three series depicted in Figure 10.1 has an equal likelihood of occurring. However, in most instances where a consistent pattern is present which persists over a large number of trials, such a pattern is more likely to be attributed to a nonrandom factor than it is to chance.

3. The computation of the values in **Table A8** is based on the following logic. If a series consists of N trials and alternative 1 occurs n_1 times and alternative 2 occurs n_2 times, the number of possible combinations involving alternative 1 occurring n_1 times and alternative 2 occurring n_2 times will be $\binom{N}{n_1} = N!/(n_1!n_2!)$. Thus, if a coin is tossed $N = 4$ times, since $\binom{4}{2} = 4!/(2!\ 2!) = 6$, there will be 6 possible ways of obtaining $n_1 = 2$ **Heads** and $n_2 = 2$ **Tails**. Specifically, the 6 ways of obtaining 2 **Heads** and 2 **Tails** are: HHTT, TTHH, THHT, HTTH, THTH, HTHT. Each of the 6 aforementioned sequences constitutes a series, and the likelihood of each of the series occurring is equal. The two series HHTT and TTHH are comprised of 2 runs, the two series THHT and HTTH are comprised of 3 runs, and the two series THTH and HTHT are comprised of 4 runs. Thus, the likelihood of observing 2 runs will equal $2/6 = .33$, the likelihood of observing 3 runs will equal $2/6 = .33$, and the likelihood of observing 4 runs will equal $2/6 = .33$. The likelihood of observing 3 or more runs will equal .67, and the likelihood of observing 2 or more runs will equal 1. A thorough discussion of the derivation of the sampling distribution for the **single-sample runs test**, which is attributed in some sources to Wald and Wolfowitz (1940), is described in Hogg and Tanis (1988).

4. Some of the cells in **Table A8** only list a lower limit. For the sample sizes in question, there is no maximum number of runs (upper limit) which will allow the null hypothesis to be rejected.

5. A general discussion of the correction for continuity can be found in Section VI of the **Wilcoxon signed-ranks test (Test 6)**. The reader should take note of the fact that the correction for continuity described in this section is intended to provide a more conservative test of the null hypothesis (i.e., make it more difficult to reject). However, when the absolute value of the numerator of Equation 10.1 is equal to or very close to zero, the z value computed with Equation 10.2 will be further removed from zero than the z value computed with Equation 10.1. Since the continuity-corrected z value will be extremely close to zero, this result is of no practical consequence (i.e., the null hypothesis will still be retained). Zar (1999, p. 493), however, notes that in actuality the correction for continuity should not be applied if it increases rather than decreases the absolute value of the test statistic. This observation regarding the correction for continuity can be generalized to the continuity-correction described in the book for other nonparametric tests.

6. The term Monte Carlo derives from the fact that Ulam had an uncle with a predilection for gambling who often frequented the casinos at **Monte Carlo**.

7. The application of the **single-sample runs test** to a design involving two independent samples is described in Siegel (1956) under the **Wald–Wolfowitz (1940) runs test**.

8. A number is prime if it has no divisors except for itself and the value 1. In other words, if a prime number is divided by any number except itself or 1, it will yield a remainder. Examples of prime numbers are 3, 7, 11, 13, 17, etc.

9. The author is indebted to Ted Sheskin for providing some of the reference material employed in this section.

10. Although Equation 10.2 (the continuity-corrected equation for the **single-sample runs test**) yields a slightly smaller absolute z value for this example, it leads to identical conclusions with respect to the null hypothesis.

11. Gruenberger and Jaffray (1965) note than an even more stringent variant of the poker test employs digits 2 through 6 as the second hand, digits 3 through 7 as the third hand, digits 4 through 8 as the fourth hand, and so on. The analysis is carried on until the end of the series (which will be the point at which a five-digit hand is no longer possible). The total of $(n - 4)$ possible hands can be evaluated with the **chi-square goodness-of-fit test**. The use of the latter test, however, is problematical, since as described above, the hands are not actually independent of one another (because they contain overlapping data), and consequently the assumption of independence for the **chi-square goodness-of-fit test** is violated.

12. a) Phillips *et al.* (1976) and Schmidt and Taylor (1970) describe the computation of the probabilities that are listed for the **poker test** for a **five digit hand**; b) Although it is generally employed with groups of five digits, the **poker test** can be applied to groups which consist of more or less than five digits. The **poker test** probabilities (for $k = 10$ digits) for a **four digit hand** (Schmidt and Taylor (1970)) and a **three digit hand** (Banks and Carson (1984)), along with a sample hand, are as follows: *Four digit hand*: **All four digits different** (1234; $p = .504$); **One pair** (1123; $p = .432$); **Two pair** (1122; $p = .027$); **Three of a kind** (1112; $p = .036$); **Four of a kind** (1111; $p = .001$.). *Three digit hand*: **All three digits different** (123; $p = .72$); **One pair** (112; $p = .27$); **Three of a kind** (111; $p = .01$).

13. Larsen and Marx (1985, p. 246) also discuss the methodology for computing the expected number of digits required for a complete set. Stirzaker (1999, p. 268) notes that the equation $\sum_{r=1}^{k} k/(k - r + 1)$ is an alternative way of expressing the harmonic series $k[(1/k) + 1/(k\text{-}1) + 1/(k\text{-}2) + \cdots + 1/2 + 1]$.

Inferential Statistical Tests Employed with Two Independent Samples (and Related Measures of Association/Correlation)

Test 11

The *t* Test for Two Independent Samples
(Parametric Test Employed with Interval/Ratio Data)

I. Hypothesis Evaluated with Test and Relevant Background Information

Hypothesis evaluated with test Do two independent samples represent two populations with different mean values?[1]

Relevant background information on test The *t* test for two independent samples, which is employed in a hypothesis testing situation involving two independent groups of subjects, is one of a number of inferential statistical tests that are based on the *t* distribution (which is discussed in detail under the **single-sample *t* test (Test 2)**). Two or more samples are independent of one another if each of the samples is comprised of different subjects.[2] In addition to being referred to as an **independent samples design**, a design involving two or more independent samples is also referred to as an **independent groups design**, a **between-subjects design**, a **between-groups design**, and a **randomized-groups design**. In order to eliminate the possibility of **confounding** in an **independent samples design**, each subject should be randomly assigned to one of the k (where $k \geq 2$) experimental conditions.

In conducting the *t* **test for two independent samples**, the two sample means (represented by the notations $\bar{X}_1$ and $\bar{X}_2$) are employed to estimate the values of the means of the populations (μ_1 and μ_2) from which the samples are derived. If the result of the *t* **test for two independent samples** is significant, it indicates the researcher can conclude there is a high likelihood the samples represent populations with different mean values. It should be noted that the *t* **test for two independent samples** is the appropriate test to employ for contrasting the means of two independent samples when the values of the underlying population variances are unknown. In instances where the latter two values are known, the appropriate test to employ is the *z* **test for two independent samples (Test 11d)**, which is described in Section VI.

The *t* **test for two independent samples** is employed with interval/ratio data, and is based on the following assumptions: a) Each sample has been randomly selected from the population it represents; b) The distribution of data in the underlying population from which each of the samples is derived is normal; and c) The third assumption, which is referred to as the **homogeneity of variance** assumption, states that the variance of the underlying population represented by Sample 1 is equal to the variance of the underlying population represented by Sample 2 (i.e., $\sigma_1^2 = \sigma_2^2$). The homogeneity of variance assumption is discussed in detail in Section VI. If any of the aforementioned assumptions are saliently violated, the reliability of the *t* test statistic may be compromised.

II. Example

Example 11.1 *In order to assess the efficacy of a new antidepressant drug, ten clinically depressed patients are randomly assigned to one of two groups. Five patients are assigned to Group 1, which is administered the antidepressant drug for a period of six months. The other five* *patients are assigned to* Group 2, *which is administered a placebo during the same six-month period. Assume that prior to introducing the experimental treatments, the experimenter confirmed that the level of depression in the two groups was equal. After six months elapse all ten subjects are rated by a psychiatrist (who is blind with respect to a subject's experimental condition) on their level of depression. The psychiatrist's depression ratings for the five subjects in each group follow (the higher the rating the more depressed a subject):* **Group 1:** 11, 1, 0, 2, 0; **Group 2:** 11, 11, 5, 8, 4. *Do the data indicate that the antidepressant drug is effective?*

III. Null versus Alternative Hypotheses

Null hypothesis H_0: $\mu_1 = \mu_2$

(The mean of the population Group 1 represents equals the mean of the population Group 2 represents.)

Alternative hypothesis H_1: $\mu_1 \neq \mu_2$

(The mean of the population Group 1 represents does not equal the mean of the population Group 2 represents. This is a **nondirectional alternative hypothesis** and it is evaluated with a **two-tailed test**. In order to be supported, the absolute value of t must be equal to or greater than the tabled critical two-tailed t value at the prespecified level of significance. Thus, either a significant positive t value or a significant negative t value will provide support for this alternative hypothesis.)

                                         or

                                H_1: $\mu_1 > \mu_2$

(The mean of the population Group 1 represents is greater than the mean of the population Group 2 represents. This is a **directional alternative hypothesis** and it is evaluated with a **one-tailed test**. It will only be supported if the sign of t is positive, and the absolute value of t is equal to or greater than the tabled critical one-tailed t value at the prespecified level of significance.)

                                         or

                                H_1: $\mu_1 < \mu_2$

(The mean of the population Group 1 represents is less than the mean of the population Group 2 represents. This is a **directional alternative hypothesis** and it is evaluated with a **one-tailed test**. It will only be supported if the sign of t is negative, and the absolute value of t is equal to or greater than the tabled critical one-tailed t value at the prespecified level of significance.)

    Note: Only one of the above noted alternative hypotheses is employed. If the alternative hypothesis the researcher selects is supported, the null hypothesis is rejected.[3]

IV. Test Computations

The data for Example 11.1 are summarized in Table 11.1. In the example there are $n_1 = 5$ subjects in Group 1 and $n_2 = 5$ subjects in Group 2. In Table 11.1 each subject is identified by a two digit number. The first digit before the comma indicates the subject's number within the group, and the second digit indicates the group identification number. Thus, Subject i, j is the ith subject in Group j. The scores of the 10 subjects are listed in the columns of Table 11.1 labeled X_1 and X_2. The adjacent columns labeled X_1^2 and X_2^2 contain the square of each subject's score.

Table 11.1 Data for Example 11.1

	Group 1			Group 2	
	X_1	X_1^2		X_2	X_2^2
Subject 1,1	11	121	Subject 1,2	11	121
Subject 2,1	1	1	Subject 2,2	11	121
Subject 3,1	0	0	Subject 3,2	5	25
Subject 4,1	2	4	Subject 4,2	8	64
Subject 5,1	0	0	Subject 5,2	4	16
	$\Sigma X_1 = 14$	$\Sigma X_1^2 = 126$		$\Sigma X_2 = 39$	$\Sigma X_2^2 = 347$

Employing Equations I.1 and I.8, the mean and estimated population variance for each sample are computed below.

$$\bar{X}_1 = \frac{\Sigma X_1}{n_1} = \frac{14}{5} = 2.8 \qquad \tilde{s}_1^2 = \frac{\Sigma X_1^2 - \frac{(\Sigma X_1)^2}{n_1}}{n_1 - 1} = \frac{126 - \frac{(14)^2}{5}}{5 - 1} = 21.7$$

$$\bar{X}_2 = \frac{\Sigma X_2}{n_2} = \frac{39}{5} = 7.8 \qquad \tilde{s}_2^2 = \frac{\Sigma X_2^2 - \frac{(\Sigma X_2)^2}{n_2}}{n_2 - 1} = \frac{347 - \frac{(39)^2}{5}}{5 - 1} = 10.7$$

When there are an **equal number of subjects** in each sample, Equation 11.1 can be employed to compute the test statistic for the *t* **test for two independent samples**.[4]

$$t = \frac{\bar{X}_1 - \bar{X}_2}{\sqrt{\frac{\tilde{s}_1^2}{n_1} + \frac{\tilde{s}_2^2}{n_2}}} \qquad \textbf{(Equation 11.1)}$$

Employing Equation 11.1, the value $t = -1.96$ is computed.

$$t = \frac{2.8 - 7.8}{\sqrt{\frac{21.7}{5} + \frac{10.7}{5}}} = \frac{-5}{2.55} = -1.96$$

Equation 11.2 is an alternative way of expressing Equation 11.1.

$$t = \frac{\bar{X}_1 - \bar{X}_2}{\sqrt{s_{\bar{X}_1}^2 + s_{\bar{X}_2}^2}} \qquad \textbf{(Equation 11.2)}$$

Note that in Equation 11.2 the values $s_{\bar{X}_1}^2$ and $s_{\bar{X}_2}^2$ represent the squares of the **standard error of the means** of the two groups. Employing the square of the value computed with Equation 2.2 (presented in Section IV of the **single-sample** *t* **test**), the squared standard error of the means of the two samples are computed: $s_{\bar{X}_1}^2 = \tilde{s}_1^2/n_1 = 21.7/5 = 4.34$ and $s_{\bar{X}_2}^2 = \tilde{s}_2^2/n_2 = 10.7/5 = 2.14$. When the values $s_{\bar{X}_1}^2 = 4.34$ and $s_{\bar{X}_2}^2 = 2.14$ are substituted in Equation 11.2, they yield the value $t = -1.96$: $t = (2.8 - 7.8)/\sqrt{4.34 + 2.14} = -1.96$.

The reader should take note of the fact that the values $\tilde{s}_1^2$, $\tilde{s}_2^2$, $s_{\bar{X}_1}^2$, and $s_{\bar{X}_2}^2$ (all of which are estimates of either the variance of a population or the variance of a sampling distribution) can never be negative numbers. If a negative value is obtained for any of the aforementioned values, it indicates a computational error has been made.

Equation 11.3 is a general equation for the *t* **test for two independent samples** which can be employed for both **equal and unequal sample sizes** (when $n_1 = n_2$, Equation 11.3 becomes equivalent to Equations 11.1/11.2). With respect to an **independent samples design** (which can have two or more experimental conditions), the term **balanced design** is sometimes employed to reflect the fact that there are an equal number of subjects in each group/experimental condition, while the term **unbalanced design** is employed to reflect that the sample sizes are not equivalent for all of the groups.

$$t = \frac{\bar{X}_1 - \bar{X}_2}{\sqrt{\left[\frac{(n_1 - 1)\tilde{s}_1^2 + (n_2 - 1)\tilde{s}_2^2}{n_1 + n_2 - 2}\right]\left[\frac{1}{n_1} + \frac{1}{n_2}\right]}} \qquad \textbf{(Equation 11.3)}$$

In the case of Example 11.1, Equation 11.3 yields the identical value $t = -1.96$ obtained with Equations 11.1/11.2.

$$t = \frac{2.8 - 7.8}{\sqrt{\left[\frac{(5 - 1)(21.7) + (5 - 1)(10.7)}{5 + 5 - 2}\right]\left[\frac{1}{5} + \frac{1}{5}\right]}} = -1.96$$

The left element inside the radical of the denominator of Equation 11.3 represents a weighted average (based on the values of n_1 and n_2) of the estimated population variances of the two groups. This weighted average is referred to as a **pooled variance estimate**, represented by the notation $\tilde{s}_{pooled}^2$.[5] Thus: $\tilde{s}_{pooled}^2 = [(n_1 - 1)\tilde{s}_1^2 + (n_2 - 1)\tilde{s}_2^2]/(n_1 + n_2 - 2)$. It should be noted that if Equations 11.1/11.2 are applied to data where $n_1 \neq n_2$, the absolute value of *t* will be slightly higher than the value computed with Equation 11.3. Thus, use of Equations 11.1/11.2 when $n_1 \neq n_2$ makes it easier to reject the null hypothesis, and consequently inflates the likelihood of committing a Type I error. The application of Equations 11.1–11.3 to a set of data when $n_1 \neq n_2$ is illustrated in Section VII.

Regardless of which equation is employed, the denominator of the *t* **test for two independent samples** is referred to as the **standard error of the difference**. This latter value, which can be summarized with the notation $s_{\bar{X}_1 - \bar{X}_2}$, represents an estimated standard deviation of difference scores for two populations. Thus in Example 11.1, $s_{\bar{X}_1 - \bar{X}_2} = 2.55$. If $s_{\bar{X}_1 - \bar{X}_2}$ is employed as the denominator of the equation for the *t* **test for two independent samples**, the equation can be written as follows: $t = (\bar{X}_1 - \bar{X}_2)/s_{\bar{X}_1 - \bar{X}_2}$.[6]

It should be noted that in some sources the numerator of Equations 11.1–11.3 is written as follows: $[(\bar{X}_1 - \bar{X}_2) - (\mu_1 - \mu_2)]$. The latter notation is only necessary if in stating the null hypothesis, a researcher stipulates that the difference between μ_1 and μ_2 is some value other than zero. When the null hypothesis is H_0: $\mu_1 = \mu_2$, the value $(\mu_1 - \mu_2)$ reduces to zero, leaving the term $(\bar{X}_1 - \bar{X}_2)$ as the numerator of the *t* **test** equation. The application of the *t* **test for two independent samples** to a hypothesis testing situation in which a value other than zero is stipulated in the null hypothesis is illustrated with Example 11.2 in Section VI.

V. Interpretation of the Test Results

The obtained value $t = -1.96$ is evaluated with **Table A2 (Table of Student's t Distribution)** in the **Appendix**. The degrees of freedom for the t **test for two independent samples** are computed with Equation 11.4.[7]

$$df = n_1 + n_2 - 2 \qquad \textbf{(Equation 11.4)}$$

Employing Equation 11.4, the value $df = 5 + 5 - 2 = 8$ is computed. Thus, the tabled critical t values which are employed in evaluating the results of Example 11.1 are the values recorded in the cells of **Table A2** that fall in the row for $df = 8$, and the columns with probabilities that correspond to the two-tailed and one-tailed .05 and .01 values. (The protocol for employing **Table A2** is described in Section V of the **single-sample t test**.) The critical t values for $df = 8$ are summarized in Table 11.2.

Table 11.2   Tabled Critical .05 and .01 t Values $df = 8$

	$t_{.05}$	$t_{.01}$
Two-tailed values	2.31	3.36
One-tailed values	1.86	2.90

The following guidelines are employed in evaluating the null hypothesis for the t **test for two independent samples**.

a) If the nondirectional alternative hypothesis H_1: $\mu_1 \neq \mu_2$ is employed, the null hypothesis can be rejected if the obtained absolute value of t is equal to or greater than the tabled critical two-tailed value at the prespecified level of significance.

b) If the directional alternative hypothesis H_1: $\mu_1 > \mu_2$ is employed, the null hypothesis can be rejected if the sign of t is positive, and the value of t is equal to or greater than the tabled critical one-tailed value at the prespecified level of significance.

c) If the directional alternative hypothesis H_1: $\mu_1 < \mu_2$ is employed, the null hypothesis can be rejected if the sign of t is negative, and the absolute value of t is equal to or greater than the tabled critical one-tailed value at the prespecified level of significance.

Employing the above guidelines, the null hypothesis can only be rejected (and only at the .05 level) if the directional alternative hypothesis H_1: $\mu_1 < \mu_2$ is employed. This is the case, since the obtained value $t = -1.96$ is a negative number, and the absolute value $t = 1.96$ is greater than the tabled critical one-tailed .05 value $t_{.05} = 1.86$. This outcome is consistent with the prediction that the group which receives the antidepressant will exhibit a lower level of depression than the placebo group. Note that the alternative hypothesis H_1: $\mu_1 < \mu_2$ is not supported at the .01 level, since the obtained absolute value $t = 1.96$ is less than the tabled critical one-tailed .01 value $t_{.01} = 2.90$.

The nondirectional alternative hypothesis H_1: $\mu_1 \neq \mu_2$ is not supported, since the obtained absolute value $t = 1.96$ is less than the tabled critical two-tailed .05 value $t_{.05} = 2.31$.

The directional alternative hypothesis H_1: $\mu_1 > \mu_2$ is not supported, since the obtained value $t = -1.96$ is a negative number. In order for the alternative hypothesis H_1: $\mu_1 > \mu_2$ to be supported, the computed value of t must be a positive number (as well as the fact that the absolute value of t must be equal to or greater than the tabled critical one-tailed value at the prespecified level of significance). It should be noted that it is not likely the researcher would employ the latter alternative hypothesis, since it predicts the placebo group will exhibit a lower level of depression than the group which receives the antidepressant.

A summary of the analysis of Example 11.1 with the *t* test for two independent samples follows: It can be concluded that the average depression rating for the group which receives the antidepressant medication is significantly less than the average depression rating for the placebo group. This conclusion can only be reached if the directional alternative hypothesis H_1: $\mu_1 < \mu_2$ is employed, and the prespecified level of significance is $\alpha = .05$. This result can be summarized as follows: $t(8) = 1.96$, $p < .05$.[8]

VI. Additional Analytical Procedures for the *t* Test for Two Independent Samples and/or Related Tests

1. The equation for the *t* test for two independent samples when a value for a difference other than zero is stated in the null hypothesis In some sources Equation 11.5 is presented as the equation for the *t* test for two independent samples.

$$t = \frac{(\bar{X}_1 - \bar{X}_2) - (\mu_1 - \mu_2)}{s_{\bar{X}_1 - \bar{X}_2}}$$
(Equation 11.5)

It is only necessary to employ Equation 11.5 if, in stating the null hypothesis, a researcher stipulates that the difference between μ_1 and μ_2 is some value other than zero. When the null hypothesis is H_0: $\mu_1 = \mu_2$ (which as noted previously can also be written as H_0: $\mu_1 - \mu_2 = 0$), the value of $(\mu_1 - \mu_2)$ reduces to zero, and thus what remains of the numerator in Equation 11.5 is $(\bar{X}_1 - \bar{X}_2)$, which constitutes the numerator of Equations 11.1–11.3. Example 11.2 will be employed to illustrate the use of Equation 11.5 in a hypothesis testing situation in which some value other than zero is stipulated in the null hypothesis.

Example 11.2   *The Accusharp Battery Company claims that the hearing aid battery it manufactures has an average life span which is two hours longer than the average life span of a battery manufactured by the Keenair Battery Company. In order to evaluate the claim, an independent researcher measures the life span of five randomly selected batteries from the stock of each of the two companies, and obtains the following values:* **Accusharp**: *10, 8, 10, 9, 11;* **Keenair**: *8, 9, 8, 7, 9. Do the data support the claim of the Accusharp Company?*

Since the Accusharp Company (which will be designated as Group 1) specifically predicts that the life span of its battery is 2 hours longer, the null hypothesis can be stated as follows: H_0: $\mu_1 - \mu_2 = 2$. The alternative hypothesis if stated nondirectionally is H_1: $\mu_1 - \mu_2 \neq 2$. If the computed absolute value of *t* is equal to or greater than the tabled critical two-tailed *t* value at the prespecified level of significance, the nondirectional alternative hypothesis is supported. If stated directionally, H_1: $\mu_1 - \mu_2 < 2$ is the appropriate alternative hypothesis to employ. The latter directional alternative hypothesis (which predicts a negative *t* value) is employed, since in order for the data to contradict the claim of the Accusharp Company (and thus reject the null hypothesis), the life span of the latter's battery can be any value that is less than 2 hours longer than that of the Keenair battery. The alternative hypothesis H_1: $\mu_1 - \mu_2 > 2$ (which is only supported with a positive *t* value) predicts that the superiority of the Accusharp battery is greater than 2 hours. If the latter alternative hypothesis is employed, the null hypothesis can only be rejected if the life span of the Accusharp battery is greater than 2 hours longer than that of the Keenair battery.

The analysis for Example 11.2 is summarized below.

$$\Sigma X_1 = 48 \qquad \bar{X}_1 = \frac{48}{5} = 9.6 \qquad \Sigma X_1^2 = 466$$

$$\Sigma X_2 = 41 \qquad \bar{X}_2 = \frac{41}{5} = 8.2 \qquad \Sigma X_2^2 = 339$$

$$\tilde{s}_1^2 = \frac{466 - \dfrac{(48)^2}{5}}{5 - 1} = 1.3 \qquad \tilde{s}_2^2 = \frac{339 - \dfrac{(41)^2}{5}}{5 - 1} = .7$$

$$t = \frac{(9.6 - 8.2) - 2}{\sqrt{\dfrac{1.3}{5} + \dfrac{.7}{5}}} = \frac{-.6}{.63} = -.95$$

Since $df = 5 + 5 - 2 = 8$, the tabled critical values in Table 11.2 can be employed to evaluate the results of the analysis. Since the obtained absolute value $t = .95$ is less than the tabled critical two-tailed value $t_{.05} = 2.31$, the null hypothesis is retained. Thus, the non-directional alternative hypothesis H_1: $\mu_1 - \mu_2 \neq 2$ is not supported. It is also true that the directional alternative hypothesis H_1: $\mu_1 - \mu_2 < 2$ is not supported. This is the case since although, as predicted, the sign of the computed t value is negative, the absolute value $t = .95$ is less than the tabled critical one-tailed value $t_{.05} = 1.86$. The directional alternative hypothesis H_1: $\mu_1 - \mu_2 > 2$ is not supported, since in order for the latter directional alternative hypothesis to be supported the sign of t must be positive, and the absolute value of t must be equal to or greater than tabled critical one-tailed value at the prespecified level of significance.

Thus, irrespective of whether a nondirectional or directional alternative hypothesis is employed, the data are consistent with the claim of the Accusharp company that it manufactures a battery which has a life span that is at least two hours longer than that of the Keenair battery. In other words, the obtained difference $(\bar{X}_1 - \bar{X}_2) = 1.4$ in the numerator of Equation 11.5 is not small enough to support the directional alternative hypothesis H_1: $\mu_1 - \mu_2 < 2$.

If H_0: $\mu_1 = \mu_2$ and H_1: $\mu_1 \neq \mu_2$ are employed as the null hypothesis and nondirectional alternative hypothesis for Example 11.2, analysis of the data yields the following result: $t = (9.6 - 8.2)/.63 = 2.22$. Since the obtained value $t = 2.22$ is less than the tabled critical two-tailed value $t_{.05} = 2.31$, the nondirectional alternative hypothesis H_1: $\mu_1 \neq \mu_2$ is not supported at the .05 level. Thus, one cannot conclude that the life span of the Accusharp battery is significantly different from the life span of the Keenair battery. The directional alternative hypothesis H_1: $\mu_1 > \mu_2$ (which can also be written as H_1: $\mu_1 - \mu_2 > 0$) is supported at the .05 level, since $t = 2.22$ is a positive number which is larger than the tabled critical one-tailed value $t_{.05} = 1.86$. Thus, if the null hypothesis H_0: $\mu_1 = \mu_2$ and the directional alternative hypothesis H_1: $\mu_1 > \mu_2$ are employed, the researcher is able to conclude that the life span of the Accusharp battery is significantly longer than the life span of the Keenair battery.

The evaluation of Example 11.2 in this section reflects that fact that the conclusions one reaches can be affected by how a researcher states the null and alternative hypotheses. In the case of Example 11.2, the fact that the nondirectional alternative hypothesis H_1: $\mu_1 - \mu_2 \neq 2$ is not supported suggests that there is a two-hour difference in favor of Accusharp (since the null hypothesis H_0: $\mu_1 - \mu_2 = 2$ is retained). Yet, if the nondirectional alternative hypothesis H_1: $\mu_1 \neq \mu_2$ is employed, the fact that it is not supported suggests there is no difference between the two brands of batteries.

Further discussion of the general subject of evaluating a hypothesis involving a difference for some value other than zero can be found in Section VII under the discussion of **clinical trials and tests of equivalence**. The latter discussion focuses on the fact that within the framework of the **classical hypothesis testing model**, a null hypothesis can never be accepted. Consequently, when a *t* **test for two independent samples** is conducted, retention of the null hypothesis only indicates that any difference detected between two treatments is not large enough to allow a researcher to conclude that the treatments are different from one another. Although the latter statement might suggest that retention of the null hypothesis is commensurate with concluding the treatments are equivalent to one another, in point of fact, it is not. In contrast to the **classical hypothesis testing model** (where the alternative hypothesis states there is a difference between treatments), in a **test of equivalence** the alternative hypothesis states that the treatments are, in fact, equivalent. Conversely, in a **test of equivalence** the null hypothesis states that a difference exists between the treatments. Since it is not mathematically feasible to establish an alternative hypothesis which states exact equality (i.e., a difference of zero) between experimental conditions, when one conducts a **test of equivalence**, prior to conducting a study a researcher stipulates a value which reflects a maximum difference which will be tolerated between two treatments in order for the researcher to conclude that the treatments are equivalent to one another. Any difference equal to or less than the stipulated value would be viewed so small as to be inconsequential. Thus, in a **test of equivalence** if a difference which is equal to or less than the value stipulated by the researcher is detected, the null hypothesis (stating that a difference does exist) can be rejected, and the alternative hypothesis (which stipulates equivalence) can be accepted. For a full discussion of this topic the reader is referred to the relevant material in Section VII.

2. Test 11a: Hartley's F_{max} test for homogeneity of variance/F test for two population variances: Evaluation of the homogeneity of variance assumption of the t test for two independent samples It is noted in Section I that one assumption of the *t* test for two independent samples is **homogeneity of variance**. Specifically, the homogeneity of variance assumption evaluates whether or not there is evidence to indicate that an inequality exists between the variances of the populations represented by the two experimental samples. When the latter condition exists, it is referred to as **heterogeneity of variance**. (Some sources (e.g., Grissom and Kim (2005) refer to **homogeneity of variance** as **homoscedasticity** and **heterogeneity of variance** as **heteroscedasticity**.) The null and alternative hypotheses employed in evaluating the homogeneity of variance assumption are as follows.

Null hypothesis $\qquad\qquad H_0: \sigma_1^2 = \sigma_2^2$

(The variance of the population Group 1 represents equals the variance of the population Group 2 represents.)

Alternative hypothesis $\qquad\qquad H_1: \sigma_1^2 \neq \sigma_2^2$

(The variance of the population Group 1 represents does not equal the variance of the population Group 2 represents. This is a **nondirectional alternative hypothesis** and it is evaluated with a **two-tailed test**. In evaluating the homogeneity of variance assumption for the *t* **test for two independent samples**, a nondirectional alternative hypothesis is always employed.)

One of a number of procedures that can be used to evaluate the homogeneity of variance hypothesis is **Hartley's F_{max} test** (Hartley (1940,1950)), which can be employed with a design involving two or more independent samples.[9] Although the F_{max} test assumes an equal number of subjects per group, among others, Kirk (1982, 1995) and Winer *et al.* (1991) note that if $n_1 \neq n_2$, but are approximately the same size, one can let the value of the larger sample size represent n when interpreting the F_{max} test statistic. The latter sources, however, note that using the larger n will result in a slight increase in the Type I error rate for the F_{max} test.

The test statistic for Hartley's F_{max} test is computed with Equation 11.6.

$$F_{max} = \frac{\tilde{s}_L^2}{\tilde{s}_S^2}$$ **(Equation 11.6)**

Where: $\tilde{s}_L^2$ = The larger of the two estimated population variances

$\tilde{s}_S^2$ = The smaller of the two estimated population variances

Employing Equation 11.6 with the estimated population variances computed for Example 11.1, the value F_{max} = 2.03 is computed. The reader should take note of the fact that the computed value for F_{max} will always be a positive number which is greater than 1 (unless $\tilde{s}_L^2 = \tilde{s}_S^2$, in which case F_{max} = 1).

$$F_{max} = \frac{21.7}{10.7} = 2.03$$

The computed value F_{max} = 2.03 is evaluated with **Table A9 (Table of the F_{max} Distribution)** in the **Appendix**. The tabled critical values for the F_{max} distribution are listed in reference to the values $(n - 1)$ and k, where n represents the number of subjects per group, and k represents the number of groups. In the case of Example 11.1, the value of $n = n_1 = n_2 = 5$. Thus, $n - 1 = 5 - 1 = 4$. Since there are two groups, $k = 2$.

In order to reject the null hypothesis and conclude that the homogeneity of variance assumption has been violated, the obtained F_{max} value must be equal to or greater than the tabled critical value at the prespecified level of significance. All values listed in **Table A9** are two-tailed values. Inspection of **Table A9** indicates that for $n - 1 = 4$ and $k = 2$, $F_{max_{.05}}$ = 9.6 and $F_{max_{.01}}$ = 23.2. Since the obtained value F_{max} = 2.03 is less than $F_{max_{.05}}$ = 9.6, the homogeneity of variance assumption is not violated — in other words the data do not suggest that the variances of the populations represented by the two groups are unequal. Thus, the null hypothesis is retained.

There are a number of additional points which should be made with regard to the above analysis that are discussed below.

Some sources employ Equation 11.7 or Equation 11.8 in lieu of Equation 11.6 to evaluate the homogeneity of variance assumption. When Equation 11.8 is employed to contrast two variances, it is often referred to as an ***F* test for two population variances**.

$$F = \frac{\tilde{s}_L^2}{\tilde{s}_S^2}$$ **(Equation 11.7)**

$$F = \frac{\tilde{s}_1^2}{\tilde{s}_2^2}$$ **(Equation 11.8)**

Both Equations 11.7 and 11.8 compute an F ratio, which is based on the F distribution. Critical values for the latter distribution are presented in **Table A10 (Table of the *F* Distribution)** in the **Appendix**. The F distribution (which is discussed in greater detail under the

single-factor between-subjects analysis of variance (Test 21)) is, in fact, the sampling distribution upon which the F_{max} distribution is based. In **Table A10**, critical values are listed in reference to the number of degrees of freedom associated with the numerator and the denominator of the F ratio. In employing the F distribution in reference to Equation 11.7, the degrees of freedom for the numerator of the F ratio is $df_{num} = n_L - 1$ (where n_L represents the number of subjects in the group with the larger estimated population variance), and the degrees of freedom for the denominator is $df_{den} = n_S - 1$ (where n_S represents the number of subjects in the group with the smaller estimated population variance). The tabled $F_{.975}$ value is employed to evaluate a two-tailed alternative hypothesis at the .05 level, and the tabled $F_{.995}$ value is employed to evaluate it at the .01 level.[10] The reason for employing the tabled $F_{.975}$ and $F_{.995}$ values instead of $F_{.95}$ (which in this analysis represents the two-tailed .10 value and the one-tailed .05 value) and $F_{.99}$ (which in this analysis represents the two-tailed .02 value and the one-tailed .01 value), is that both tails of the distribution are used in employing the F distribution to evaluate a hypothesis about two population variances. Thus, if one is conducting a two-tailed analysis with $\alpha = .05$, .025 (i.e., .05/2 = .025) represents the proportion of cases in the extreme left of the left tail of the F distribution, as well as the proportion of cases in the extreme right of the right tail of the distribution. With respect to a two-tailed analysis with $\alpha = .01$, .005 (i.e., .01/2 = .005) represents the proportion of cases in the extreme left of the left tail of the distribution, as well as the proportion of cases in the extreme right of the right tail of the distribution.

In point of fact, if $df_L = df_{num} = 4$ and $df_S = df_{den} = 4$ (which are the values employed in Example 11.1), the tabled critical two-tailed F values employed for $\alpha = .05$ and $\alpha = .01$ are $F_{.975} = 9.6$ and $F_{.995} = 23.15$. These are the same critical .05 and .01 values that are employed for the F_{max} test.[11] Thus, Equation 11.7 employs the same critical values and yields an identical result to that obtained with Equation 11.6. It should be noted, however, that if $n_1 \neq n_2$, Equation 11.6 and Equation 11.7 will employ different critical values, since Equation 11.7 (which uses the value of n for each group in determining degrees of freedom) can accommodate unequal sample sizes.

When Group 1 has a larger estimated population variance than Group 2, everything that has been said with respect to Equation 11.7 applies to Equation 11.8. However, when $\tilde{s}_2^2 > \tilde{s}_1^2$, the value of F computed with Equation 11.8 will be less than 1. In such an instance, one can do either of the following: a) Designate the group with the larger variance as Group 1 and the group with the smaller variance as Group 2. Upon doing this, divide the larger variance by the smaller variance (as is done in Equation 11.7), and thus obtain the same F value derived with Equation 11.7; or b) Use Equation 11.8, and employ the tabled critical $F_{.025}$ value to evaluate a two-tailed alternative hypothesis at the .05 level of significance, and the tabled critical $F_{.005}$ value to evaluate a two-tailed alternative hypothesis at the .01 level of significance.[12] In such a case, in order to be significant the computed F value must be equal to or less than the tabled critical $F_{.025}$ value (if $\alpha = .05$) or the tabled critical $F_{.005}$ value (if $\alpha = .01$). Both of the aforementioned methods will yield the same conclusions with respect to retaining or rejecting the null hypothesis at a given level of significance.

Equation 11.8 can also be used to test a directional alternative hypothesis concerning the relationship between two population variances. If a researcher specifically predicts that the variance of the population represented by Group 1 is larger than the variance of the population represented by Group 2 (i.e., H_1: $\sigma_1^2 > \sigma_2^2$), or that the variance of the population represented by Group 1 is smaller than the variance of the population represented by Group 2 (i.e., H_1: $\sigma_1^2 < \sigma_2^2$), a directional alternative hypothesis is evaluated. In such a case, the tabled critical one-tailed F value for $(n_1 - 1), (n_2 - 1)$ degrees of freedom at the prespecified level of significance is employed.

To illustrate the analysis of a one-tailed alternative hypothesis, let us assume that the alternative hypothesis H_1: $\sigma_1^2 > \sigma_2^2$ is evaluated. If $\alpha = .05$, in order for the result to be significant the computed value of F must be greater than 1. In addition, the tabled critical F value which is employed in evaluating the above alternative hypothesis is $F_{.95}$. To be significant, the obtained F value must be equal to or greater than the tabled critical $F_{.95}$ value. If $\alpha = .01$, the tabled critical F value that is employed in evaluating the alternative hypothesis is $F_{.99}$. To be significant, the obtained F value must be equal to or greater than the tabled critical $F_{.99}$ value.

Now let us assume that the alternative hypothesis being evaluated is H_1: $\sigma_1^2 < \sigma_2^2$. If $\alpha = .05$, in order for the result to be significant the computed value of F must be less than 1. The tabled critical F value which is employed in evaluating the above alternative hypothesis is $F_{.05}$. To be significant, the obtained F value must be equal to or less than the tabled critical $F_{.05}$ value. If $\alpha = .01$, the tabled critical F value that is employed in evaluating the alternative hypothesis is $F_{.01}$. To be significant, the obtained F value must be equal to or less than the tabled critical $F_{.01}$ value.

If Equation 11.8 is employed with a one-tailed alternative hypothesis with reference to two groups consisting of five subjects per group (as is the case in Example 11.1), the following tabled critical values listed for $(n_1 - 1) = 4$, $(n_2 - 1) = 4$ are employed: a) If $\alpha = .05$, $F_{.95} = 6.39$, and $F_{.05} = .157$; and b) If $\alpha = .01$, $F_{.99} = 15.98$, and $F_{.01} = .063$.[13] To illustrate this in a situation where an F value less than 1 is computed, assume for the moment that we employ the alternative hypothesis H_1: $\sigma_1^2 < \sigma_2^2$ for the two groups described in Example 11.1, and that the values of the two group variances are reversed — i.e., $\tilde{s}_1^2 = 10.7$ and $\tilde{s}_2^2 = 21.7$. Employing Equation 11.8 with this data, $F = 10.7/21.7 = .49$. The obtained value $F = .49$ is not significant, since to be significant the computed value of F must be equal to or less than $F_{.05} = .157$.

It should be noted that when the general procedure discussed in this section is employed to evaluate the homogeneity of variance assumption with reference to Example 11.1, in order to reject the null hypothesis for $\alpha = .05$, the larger of the estimated population variances must be more than 9 times the magnitude of the smaller variance, and for $\alpha = .01$ the larger variance must be more than 23 times the magnitude of the smaller variance. Within the framework of the F_{max} test, such a large discrepancy between the estimated population variances is tolerated when the number of subjects per group is small (since it is assumed that sampling error may be substantial). Inspection of **Table A9** reveals that as sample size increases (which is associated with a decrease in sampling error), the magnitudes of the tabled critical F_{max} values decrease. Thus, the larger the sample size, the smaller the difference between the variances which will be acceptable.

Two assumptions common to Equations 11.6–11.8 (i.e., all of the equations that can be employed to evaluate the homogeneity of variance assumption) are: a) Each sample has been randomly selected from the population it represents; and b) The distribution of data in the underlying population from which each of the samples is derived is normal. Violation of these assumptions can compromise the reliability of the F_{max} test statistic, which many sources note is extremely sensitive to violation of the normality assumption. Various sources (e.g., Keppel (1991)) point out that when the F_{max} test is employed to evaluate the homogeneity of variance hypothesis, it is not as powerful (i.e., likely to detect heterogeneity of variance when it is present) as some alternative but computationally more involved procedures. The consequence of not detecting heterogeneity of variance is that it increases the likelihood of committing a Type I error in conducting the *t* **test for two independent samples**. Additional discussion of the homogeneity of variance assumption and alternative procedures which can be used to evaluate it can be found in Section VI of the **single-factor between-subjects analysis of variance**.

In the event the homogeneity of variance assumption is violated, a number of different strategies (which yield similar but not identical results) are recommended with reference to conducting the **t test for two independent samples**. Since heterogeneity of variance increases the likelihood of committing a Type I error, all of the strategies that are recommended result in a more conservative t test (i.e., making it more difficult for the test to reject the null hypothesis). Such strategies compute either: a) An adjusted critical t value which is larger than the unadjusted critical t value; or b) An adjusted degrees of freedom value which is smaller than the value computed with Equation 11.4. By decreasing the degrees of freedom, a larger tabled critical t value is employed in evaluating the computed t value.

Before describing one of the procedures that can be employed when the homogeneity of variance assumption is violated, it should be pointed out that the existence of heterogeneity of variance in a set of data may in itself be noteworthy. It is conceivable that although the analysis of the data for an experiment may indicate there is no difference between the group means, one cannot rule out the possibility that there may be a significant difference between the variances of the two groups. This latter finding may be of practical importance in clarifying the relationship between the variables under study. This general issue is addressed in Section VIII of the **single-sample runs test (Test 10)** within the framework of the discussion of Example 10.6. In the latter discussion, an experiment is described in which subjects in an experimental group who receive an antidepressant either improve dramatically or regress while on the drug. In contrast, the scores of a placebo/control group exhibit little variability, and fall in between the two extreme sets of scores in the group which receives the antidepressant. Analysis of such a study with the **t test for two independent samples** will in all likelihood not yield a significant result, since the two groups will probably have approximately the same mean depression score. The fact that the group receiving the drug exhibits greater variability than the group receiving the placebo indicates that the effect of the drug is not consistent for all people who are depressed. Such an effect can be identified through use of a test such as the F_{max} test, which contrasts the variability of two groups.

Two statisticians (Behrens and Fisher) developed a sampling distribution for the t statistic when the homogeneity of variance assumption is violated. The latter sampling distribution is referred to as the t' distribution. Since tables of critical values developed by Behrens and Fisher (available in Fisher and Yates (1953)) can only be employed for a limited number of sample sizes, Cochran and Cox (1957) developed a methodology that allows one to compute critical values of t' for all values of n_1 and n_2. Equation 11.9 summarizes the computation of t'.

$$t' = \frac{t_1 \left[\frac{\tilde{s}_1^2}{n_1} \right] + t_2 \left[\frac{\tilde{s}_2^2}{n_2} \right]}{\frac{\tilde{s}_1^2}{n_1} + \frac{\tilde{s}_2^2}{n_2}}$$ (Equation 11.9)

Where: t_1 = The tabled critical t value at the prespecified level of significance for $df = n_1 - 1$.
t_2 = The tabled critical t value at the prespecified level of significance for $df = n_2 - 1$.

Equation 11.9 will be employed with the data for Example 11.1. For purposes of illustration, it will be assumed that the homogeneity of variance assumption has been violated, and that the nondirectional alternative hypothesis H_1: $\mu_1 \neq \mu_2$ is evaluated, with $\alpha = .05$. Since

$n_1 = n_2 = n = 5$, $df_1 = df_2 = n - 1 = 4$. Employing **Table A2**, we determine that for $\alpha = .05$ and $df = 4$, the tabled critical two-tailed .05 value is $t_{.05} = 2.78$. Thus, the values $t_1 = 2.78$ and $t_2 = 2.78$ are substituted in Equation 11.9, along with the values of the estimated population variances and the sample sizes.[14]

$$t' = \frac{2.78\left[\dfrac{21.7}{5}\right] + 2.78\left[\dfrac{10.7}{5}\right]}{\dfrac{21.7}{5} + \dfrac{10.7}{5}} = 2.78$$

Note that the computed value $t' = 2.78$ is larger than the tabled critical two-tailed .05 value $t_{.05} = 2.31$, which is employed if the homogeneity of variance adjustment is not violated. Since the value of t' will always be larger than the tabled critical t value at the prespecified level of significance for $df = n_1 + n_2 - 2$ (except for the instance noted in Endnote 14), use of the t' statistic will result in a more conservative test. In our hypothetical example, use of the t' statistic is designed to insure that the Type I error rate will conform to the prespecified value $\alpha = .05$. If there is heterogeneity of variance and the homogeneity of variance adjustment is not employed, the actual alpha level will be greater than $\alpha = .05$. Since in our example the computed value $t' = 2.78$ is larger than the computed absolute value $t = 1.96$ obtained for the t test, the null hypothesis cannot be rejected if the two-tailed alternative hypothesis $H_1: \mu_1 \neq \mu_2$ is employed. Recollect that when in Section V the homogeneity of variance assumption was assumed not to be violated, the two-tailed alternative hypothesis $H_1: \mu_1 \neq \mu_2$ was also not supported. If, on the other hand, the researcher evaluates the one-tailed alternative hypothesis $H_1: \mu_1 < \mu_2$, since $df_1 = df_2 = n - 1 = 4$, the tabled critical one-tailed .05 value $t_{.05} = 2.13$ is employed in Equation 11.9 which yields the value $t' = 2.13$. Since $t' = 2.13$ is larger than the computed absolute value $t = 1.96$ obtained for the t test, the null hypothesis cannot be rejected. Recollect that in Section V when the homogeneity of variance assumption was assumed not to be violated, the one-tailed alternative hypothesis $H_1: \mu_1 < \mu_2$ was supported.

The methodology described in this section for dealing with heterogeneity of variance provides for a slightly more conservative *t* **test for two independent samples** than do alternative strategies developed by Satterthwaite (1946) and Welch (1947) (which are described in Howell (2002, pp. 213–215)) and Winer *et al.* (1991)). Another strategy for dealing with heterogeneity of variance is to employ, in lieu of the *t* **test**, a nonparametric test which does not assume homogeneity of variance.

3. Computation of the power of the *t* test for two independent samples and the application of Test 11b: Cohen's *d* index In this section two methods for computing power, which are extensions of the methods presented for computing the power of the **single-sample *t* test**, will be described. Prior to reading this section the reader should review the discussion of power in Section VI of the latter test.

Method 1 for computing the power of the *t* test for two independent samples The first procedure to be described is a graphical method which reveals the logic underlying the power computations for the *t* **test for two independent samples**. In the discussion to follow, it will be assumed that the null hypothesis is identical to that employed for Example 11.1 (i.e., $H_0: \mu_1 - \mu_2 = 0$, which, as previously noted, is another way of writing $H_0: \mu_1 = \mu_2$). It will also be assumed that the researcher wants to evaluate the power of the *t* **test for two independent samples** in reference to the following alternative hypothesis: $H_1: |\mu_1 - \mu_2| \geq 5$ (which is the difference obtained between the sample means in Example 11.1). In other words, it is predicted that the absolute value of the difference between the two means is equal to or greater than 5. The latter alternative hypothesis is employed in lieu of H_0: $\mu_1 - \mu_2 \neq 0$ (which can also be written as H_1: $\mu_1 \neq \mu_2$), since in order to compute the power of the test, a specific value must be stated for the difference between the population means. Note that, as stated, the alternative hypothesis stipulates a nondirectional analysis, since it does not specify which of the two means will be the larger value. It will be assumed that $\alpha = .05$ is employed in the analysis.

Figure 11.1, which provides a visual summary of the power analysis, is comprised of two overlapping sampling distributions of difference scores. The distribution on the left, which will be designated as Distribution A, is a sampling distribution of difference scores that has a mean value of zero (i.e., $\mu_D = \mu_{\bar{X}_1 - \bar{X}_2} = 0$). This latter value will be represented by $\mu_{D_0} = 0$ in Figure 11.1. Distribution A represents the sampling distribution that describes the distribution of difference scores if the null hypothesis is true. The distribution on the right, which will be designated as Distribution B, is a sampling distribution of difference scores that has a mean value of 5 (i.e., $\mu_D = \mu_{\bar{X}_1 - \bar{X}_2} = 5$). This latter value will be represented by $\mu_{D_1} = 5$ in Figure 11.1. Distribution B represents the sampling distribution which describes the distribution of difference scores if the alternative hypothesis is true. It will be assumed that each of the sampling distributions has a standard deviation that is equal to the value computed for the standard error of the difference in Example 11.1 (i.e., $s_{\bar{X}_1 - \bar{X}_2} = 2.55$), since the latter value provides the best estimate of the standard deviation of the difference scores for the underlying populations.

In Figure 11.1, area (///) delineates the proportion of Distribution A that corresponds to the value $\alpha/2$, which equals .025. This is the case, since $\alpha = .05$ and a two-tailed analysis is conducted. Area ($\equiv$) delineates the proportion of Distribution B which corresponds to the probability of committing a Type II error (β). Area (\\\) delineates the proportion of Distribution B that represents the power of the test (i.e., $1 - \beta$).

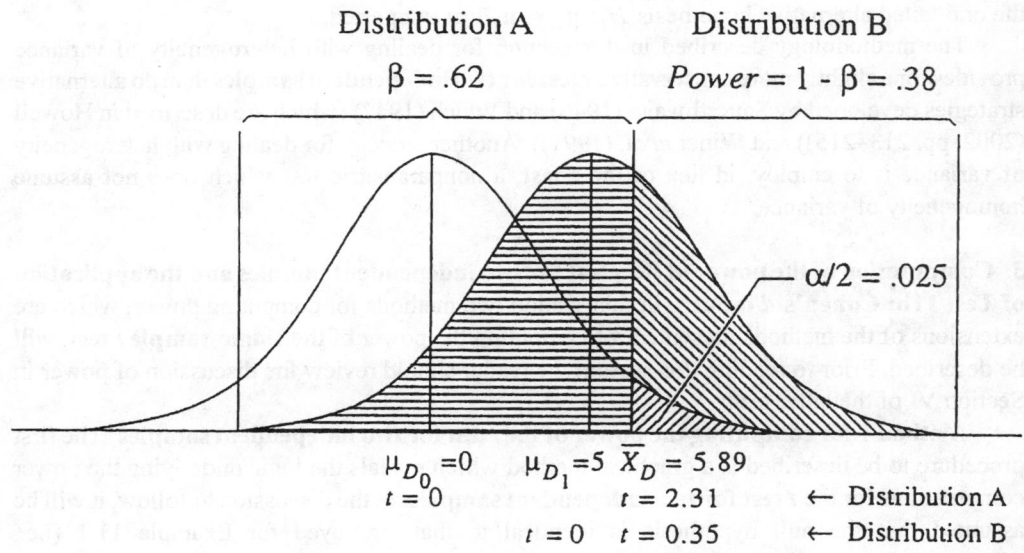

Distribution A

$\beta = .62$

Distribution B

$Power = 1 - \beta = .38$

$\alpha/2 = .025$

$\mu_{D_0} = 0$ $\mu_{D_1} = 5$ $\bar{X}_D = 5.89$

$t = 0$ $t = 2.31$ $\leftarrow$ Distribution A

$t = 0$ $t = 0.35$ $\leftarrow$ Distribution B

Figure 11.1 Visual Representation for Power for Example 11.1

The procedure for computing the proportions documented in Figure 11.1 will now be described. The first step in computing the power of the test requires one to determine how large a difference there must be between the sample means in order to reject the null hypothesis. In order to do this, we algebraically transpose the terms in Equation 11.1, using $s_{\bar{X}_1 - \bar{X}_2}$ to summarize the denominator of the equation, and $t_{.05}$ (the tabled critical two-tailed .05 t value) to represent t. Thus: $\bar{X}_1 - \bar{X}_2 = (t_{.05})(s_{\bar{X}_1 - \bar{X}_2})$. By substituting the values $t_{.05} = 2.31$ and $s_{\bar{X}_1 - \bar{X}_2} = 2.55$ in the latter equation, we determine that the minimum required difference is $\bar{X}_1 - \bar{X}_2 = (2.31)(2.55) = 5.89$. Thus, any difference between the two population means which is equal to or greater than 5.89 will allow the researcher to reject the null hypothesis at the .05 level.

The next step in the analysis requires one to compute the area in Distribution B that falls between the mean difference $\mu_{D_1} = 5$ (i.e., the mean of Distribution B) and a mean difference equal to 5.89 (represented by the notation $\bar{X}_D = 5.89$ in Figure 11.). This is accomplished by employing Equation 11.1. In using the latter equation, the value of $\bar{X}_1$ is represented by 5.89 and the value of $\bar{X}_2$ by $\mu_{D_1} = 5$.

$$t = \frac{\bar{X}_1 - \bar{X}_2}{s_{\bar{X}_1 - \bar{X}_2}} = \frac{5.89 - 5}{2.55} = .35$$

By interpolating the values listed in **Table A2** for $df = 8$, we determine that the proportion of Distribution B which lies to the right of a t score of .35 (which corresponds to a mean difference of 5.89) is approximately .38. The latter area corresponds to area (\\\\) in Distribution B. Note that the left boundary of area (\\\\) is also the boundary delineating the extreme 2.5% of Distribution A (i.e., $\alpha/2 = .025$, which is the rejection zone for the null hypothesis). Since area (\\\\) in Distribution B overlaps the rejection zone in Distribution A, area (\\\\) represents the power of the test — i.e., it represents the likelihood of rejecting the null hypothesis if the alternative hypothesis is true. The likelihood of committing a Type II error (β) is represented by area ($\equiv$), which comprises the remainder of Distribution B. The proportion of Distribution B that constitutes this latter area is determined by subtracting the value .38 from 1. Thus: $\beta = 1 - .38 = .62$.

Based on the results of the power analysis, we can state that if the alternative hypothesis H_1: $|\mu_1 - \mu_2| \geq 5$ is true, the likelihood the null hypothesis will be rejected is .38, and at the same time there is a .62 likelihood it will be retained. If the researcher considers the computed value for power too low (which in actuality should be determined prior to conducting a study), she can increase the power of the test by employing a larger sample size.

Method 2 for computing the power of the t test for two independent samples employing Test 11b: Cohen's d index (and the sample analogue g index)   Method 2, the quick computational method described for computing the power of the **single sample t test**, can be extended to the t **test for two independent samples**. In using the latter method, the researcher must stipulate an **effect size** (d), which in the case of the t **test for two independent samples** is computed with Equation 11.10. The effect size index computed with Equation 11.10 was developed by Cohen (1977, 1988), and is known as **Cohen's d index**. Further discussion of **Cohen's d index** can be found in the next section dealing with magnitude of treatment effect, as well as in Section IX (the **Appendix**) of the **Pearson product-moment correlation coefficient (Test 28)** under the discussion of **meta-analysis and related topics**.

$$d = \frac{|\mu_1 - \mu_2|}{\sigma} \qquad \textbf{(Equation 11.10)}$$

The numerator of Equation 11.10 represents the hypothesized difference between the two population means. As is the case with the graphical method described previously, when a power analysis is conducted after the mean of each sample has been obtained, the difference between the two sample means (i.e., $\overline{X}_1 - \overline{X}_2$) is employed as an estimate of the value of $|\mu_1 - \mu_2|$ (although the method allows a researcher to test for any value as the hypothesized difference between the means). It is assumed that the value of the standard deviation for the variable being measured is the same in each of the populations, and the latter value is employed to represent σ in the denominator of Equation 11.10 (i.e., $\sigma = \sigma_1 = \sigma_2$). In instances where the standard deviations of the two populations are not known or cannot be estimated, the latter value can be estimated from the sample data. Because of the fact that $\tilde{s}_1$ will usually not equal $\tilde{s}_2$, a pooled estimated population standard deviation ($\tilde{s}_{pooled}$) can be computed with Equation 11.11 (which is the square root of $\tilde{s}^2_{pooled}$ discussed in Section IV with reference to Equation 11.3).

$$\tilde{s}_{pooled} = \sqrt{\frac{(n_1 - 1)\tilde{s}_1^2 + (n_2 - 1)\tilde{s}_2^2}{n_1 + n_2 - 2}} \qquad \textbf{(Equation 11.11)}$$

Since the effect size computed with Equation 11.10 is based on population parameters, it is necessary to convert the value of d into a measure that takes into account the size of the samples (which is a relevant variable in determining the power of the test). This measure, as noted in the discussion of the **single-sample t test**, is referred to as the **noncentrality parameter** (See Endnote 10 of the **single-sample t test** for further clarification of the **noncentrality parameter**). Equation 11.12 is employed to compute the **noncentrality parameter** φ (which is a symbol for the lower case Greek letter **phi**) for the **t test for two independent samples**. When the sample sizes are equal, the value of n in Equation 11.12 will be $n = n_1 = n_2$. When the sample sizes are unequal, the value of n will be represented by the **harmonic mean** of the sample sizes, which is described later in this section (as well as in the **Introduction**).

$$\varphi = d\sqrt{\frac{n}{2}} \qquad \textbf{(Equation 11.12)}$$

The power of the **t test for two independent samples** will now be computed using the data for Example 11.1. For purposes of illustration, it will be assumed that the minimum difference between the population means the researcher is trying to detect is the observed 5 point difference between the two sample means — i.e., $|\overline{X}_1 - \overline{X}_2| = |2.8 - 7.8| = 5 = |\mu_1 - \mu_2|$. The value of σ employed in Equation 11.10 is estimated by computing a pooled value for the standard deviation using Equation 11.11. Substituting the relevant values from Example 11.1 in Equation 11.11, the value $\tilde{s}_{pooled} = 4.02$ is computed.

$$\tilde{s}_{pooled} = \sqrt{\frac{(5 - 1)(21.7) + (5 - 1)(10.7)}{5 + 5 - 2}} = 4.02$$

Substituting $|\mu_1 - \mu_2| = 5$ and $\sigma = 4.02$ in Equation 11.10, the value $d = 1.24$ is computed.

$$d = \frac{5}{4.02} = 1.24$$

Cohen (1977; 1988, pp. 24–27) has proposed the following (admittedly arbitrary) d values as criteria for identifying the magnitude of an effect size: a) A **small effect size** is one that is greater than .2 but not more than .5 standard deviation units; b) A **medium effect size** is one that is greater than .5 but not more than .8 standard deviation units; and c) A **large effect size** is greater than .8 standard deviation units. Employing Cohen's (1977, 1988) guidelines, the value $d = 1.24$ (which represents 1.24 standard deviation units) is categorized as a large effect size.[15]

Along with the value $n = 5$ (since $n_1 = n_2 = n = 5$), the value $d = 1.24$ is substituted in Equation 11.12, resulting in the value $\varphi = 1.96$.

$$\varphi = 1.24 \sqrt{\frac{5}{2}} = 1.96$$

The value $\varphi = 1.96$ is evaluated with **Table A3 (Power Curves for Student's t Distribution)** in the **Appendix**. We will assume that for the example under discussion a two-tailed test is conducted with $\alpha = .05$, and thus **Table A3-C** is the appropriate set of power curves to employ for the analysis. Since there is no curve for $df = 8$, the power of the test will be based on a curve that falls in between the $df = 6$ and $df = 12$ power curves. Through interpolation, the power of the **t test for two independent samples** is determined to be approximately .38 (which is the same value that is obtained with the graphical method). Thus, by employing 5 subjects in each group the researcher has a probability of .38 of rejecting the null hypothesis if the true difference between the population means is equal to or greater than 1.24 standard deviation units (which in Example 11.1 is equivalent to a 5 point difference between the means). It should be noted that in employing Equation 11.12, the smaller the value of n the smaller the computed value of φ, and consequently the lower the power of the test.

It was noted earlier in the discussion that when the sample sizes are unequal, the value of n in Equation 11.12 is represented by the **harmonic mean** of the sample sizes (which will be represented by the notation $\bar{n}_h$). The harmonic mean (the computation of which is also described in the **Introduction** in reference to a distribution of scores — see Equation I.5) is computed with Equation 11.13.[16]

$$\bar{n}_h = \frac{k}{\displaystyle\sum_{j=1}^{k} \left[\frac{1}{n_j}\right]}$$

(Equation 11.13)

Where: $\quad k$ = The number of groups
$\qquad\quad n_j$ = The number of subjects in the jth group

The use of Equation 11.13 will be illustrated for a case in which there is a total of 10 subjects, but there is an unequal number of subjects in each group. Thus, let us assume that $n_1 = 7$ and $n_2 = 3$. The harmonic mean can be computed as follows: $\bar{n}_h = 2/[(1/7) + (1/3)]$ $= 4.20$. The reader should take note of the fact that the value $\bar{n}_h = 4.20$ computed for the harmonic mean is lower than the average number of subjects per group ($\bar{n}$), which is computed to be $\bar{n} = (7 + 3)/2 = 5$. In point of fact, $\bar{n}_h$ will always be lower than $\bar{n}$ unless $n_1 = n_2$, in which case $\bar{n}_h = \bar{n}$. Since, when $n_1 \neq n_2$, $\bar{n}_h$ will always be less than $\bar{n}$, it follows that when $n_1 \neq n_2$, the value computed for the power of the test when the harmonic mean is employed in Equation 11.12 will be less than the value that is computed for the power of the test if $\bar{n}$ is employed to represent the value of n. This translates into the fact that for a specific total sample size (i.e., $n_1 + n_2$), the power of the **t test for two independent samples** will be maximized when $n_1 = n_2$.

As is the case with power computations for the **single-sample *t* test**, as long as a researcher knows or is able to estimate (from the sample data) the population standard deviation, by employing trial and error she can substitute various values of *n* in Equation 11.12 until the computed value of φ corresponds to the desired power value for the ***t* test for two independent samples** for a given effect size. This process can be facilitated by employing tables developed by Cohen (1977, 1988), which allow one to determine the minimum sample size necessary in order to achieve a specific level of power in reference to a given effect size.

Among others, Rosner (2000, pp. 306–309) notes that Equation 11.14 provides an alternative way of estimating the sample size required in order to conduct the ***t* test for two independent samples** which has a power of a specified value.

(Equation 11.14)

$$n_1 = \frac{\left(\sigma_1^2 + \dfrac{\sigma_2^2}{k} \right) (z_{\alpha/2} + z_\beta)^2}{\delta^2}$$

Where:　n_1 represents the number of subjects in Group 1.

k is the ratio of the number of subjects in **Group 2** relative to the number of subjects in **Group 1** (i.e., $k = n_1/n_2$). When $n_1 = n_2$, $k = 1$. If, on the other hand, it is expected that there will be twice as many subjects in **Group 2** than **Group 1**, $k = 2$, and so on.

δ represents $|\mu_1 - \mu_2|$, the numerator of Equation 11.10. Although it will not be employed in illustrating the use of Equation 11.14, many sources often employ the value $| \bar{X}_1 - \bar{X}_2 |$, which is assumed to be the best estimate of $|\mu_1 - \mu_2|$ (the true difference between the population means) to represent the value of δ. $| \bar{X}_1 - \bar{X}_2 |$ is commonly employed to represent the value of δ, when after obtaining a nonsignificant result a researcher wants to determine if the power of the test conducted had been set at a specific value, how large a sample size would have been required in order for the ***t* test for two independent samples** to identify the observed difference between the two group means to be significant.

σ_1^2 and σ_2^2 represent the population variances, which are often represented in Equation 11.14 by $\tilde{s}_1^2$ and $\tilde{s}_2^2$. Further discussion of the use of $\tilde{s}_1^2$ and $\tilde{s}_2^2$ in Equation 11.14 can be found below.

$z_{\alpha/2}$ represents the relevant tabled critical **two-tailed** *z* value in **Table A1**. It is important to note that $z_{\alpha/2}$ is employed in Equation 11.14 when the test of the alternative hypothesis is **nondirectional**. If, on the other hand, a **directional** alternative hypothesis is evaluated, z_α is employed in place of $z_{\alpha/2}$ — i.e., the relevant tabled critical **one-tailed** *z* value in **Table A1** should be employed.

z_β is the *z* value listed in **Table A1** for which a proportion (probability) is recorded that identifies the Type II error rate (i.e., β) employed in the analysis. The protocol to employ to determine the value of z_β is as follows: a) If the **power** stipulated for the test (i.e., the value $(1 - \beta)$), is **greater than .5**, z_β is the *z* value for which the proportion recorded in **Column 3** of **Table A1** is equal to the value **stipulated for β**. (It is also the *z* value for which the proportion recorded in **Column 2** of **Table A1** is equal to the value (**Power** − .5).) When the **power** stipulated for the test is **greater than .5**, the sign of z_β will always be **positive**; b) If the **power** stipulated for the test is **less than .5**, z_β is the *z* value for which the proportion recorded in **Column 3** of **Table A1** is equal to the **power** of the test (i.e., the value $(1 - \beta)$). (It is also the *z* value for which the proportion recorded in **Column 2** of **Table A1** is equal to the value (β – .5).) When the **power** stipulated for the test is **less than .5**, the sign of z_β will always be **negative**; c) If the **power** stipulated for the test is .5, $z_\beta = 0$. (In **Table A1** the value $\beta = .5000$ is listed in **Column 3**, and the value .0000 in **Column 2** represents the difference between β and .5 (i.e., $\beta - .5 = 0$, with **Power** $= 1 - \beta = .5$).)

The reader should take note of the following with regard to Equation 11.14: If the values of σ_1^2 and σ_2^2 are not known (as will often be the case), $\tilde{s}_1^2$ and $\tilde{s}_2^2$ should be employed in Equation 11.14 to represent σ_1^2 and σ_2^2. (When, in fact, the values of σ_1^2 and σ_2^2 are known the *t* **test for two independent samples** becomes *z* **test for two independent samples** (**Test 11e**), which is discussed later in this section.) Theoretically, when $\tilde{s}_1^2$ and $\tilde{s}_2^2$ are employed in Equation 11.14 instead of σ_1^2 and σ_2^2, the values $t_{\alpha/2}$ and t_β should be employed in the equation in place of $z_{\alpha/2}$ and z_β, and both of the values $t_{\alpha/2}$ and t_β will be larger than their respective analogs $z_{\alpha/2}$ and z_β. However, if one does not have access to exact probabilities for the *t* distribution, it will be difficult to estimate the appropriate value for t_β. Because of the latter, many sources employ $\tilde{s}_1^2$ and $\tilde{s}_2^2$ in Equation 11.14 along with the relevant values for $z_{\alpha/2}$ and z_β. As a general rule, the value of n_1 computed under the latter circumstances will underestimate the required sample size (unless the appropriate value for n_1 is greater than 200). The underestimation of the sample size when $\tilde{s}_1^2$ and $\tilde{s}_2^2$ are employed with $z_{\alpha/2}$ and z_β in Equation 11.14 results from the fact that the normal distribution allows for a more powerful test of an alternative hypothesis than the *t* distribution. Consequently, if the *t* distribution is the appropriate distribution to employ, it will require a slightly larger sample size to detect a difference than an analysis based on the normal distribution will indicate. It should also be noted that for a specified level of power, the total sample size (i.e., $n_1 + n_2$) will be larger when $n_1 \neq n_2$ than when $n_1 = n_2$.

Equation 11.14 will be employed to demonstrate the computation of an appropriate sample size for Example 11.1. We will assume that the researcher plans to evaluate the data with a *t* **test for independent samples**, and that a two-tailed analysis will be conducted with alpha set at .05. In addition, the researcher plans to have an equal number of subjects in each group, and wants the power of the test to be .80. In conducting the latter analysis, the following values are employed in Equation 11.14 to compute the number of subjects that will be required in **Group 1** (which will also be the number of subjects the researcher will need for **Group 2**, since when there are an equal number of subjects per group, $k = 1$): $\tilde{s}_1^2 = 21.7$, $\tilde{s}_2^2 = 10.7$, $\delta = |\bar{X}_1 - \bar{X}_2| = |2.8 - 7.8| = 5$, $z_{\alpha/2} = 1.96$ (If a one-tailed analysis is conducted $z_{\alpha/2} = 1.65$ should be employed.), $z_\beta = .84$. The rationale for employing the value $z_\beta = .84$ is as follows: Since **Power** $= 1 - \beta = .80$, $\beta = .20$. Employing the guidelines for using Equation 11.14 documented above to compute a sample size when **Power** $> .5$, in **Table A1**, $z = .84$ is the z value for which the proportion (probability) recorded in **Column 3** (.2005) is closest to the value $\beta = .20$. (It is also the case that for $z = .84$, if .5 is added to the value recorded in **Column 2** (i.e., .2995 + .5 = .7995), the resulting value is closest to the value .80 stipulated for the power of the test (and thus, **Power** $- .5 = .7995 - .5 = .2995$).) Since the power stipulated for the test is greater than .5, the sign of z_β will be positive. As noted above, the observed absolute difference between the two proportions $|\bar{X}_1 - \bar{X}_2| = 5$ will be employed to represent the value of δ.

When the appropriate values are substituted in Equation 11.14, the value $n_1 = 10.16$ is computed, indicating that in order for the power of the *t* **test for two independent samples** to be .80, the researcher must have at least 11 subjects per group (which is the next greatest integer value than integer element of the value computed for n_1), which is about twice the number of subjects per group that were actually employed in Example 11.1.[17]

$$n_1 = \frac{\left(21.7 + \frac{10.7}{1}\right)(1.96 + .84)^2}{(7.8 - 2.8)^2} = 10.16$$

If we stipulate that the power of the test should be .38 (which was the power value estimated earlier in this section for Example 11.1) the value $z_\beta = -.30$ is employed in Equation 11.14. The rationale for employing the value $z_\beta = -.30$ is as follows: We wish to compute the sample size if the power of the test is .38. Thus, if **Power** $= 1 - \beta = .38$, $\beta = .62$. Employing the guidelines for using Equation 11.14 documented earlier to compute a sample size when **Power** $< .5$, in **Table A1**, $z = .30$ is the z value for which the proportion (probability) recorded in **Column 3** (.3821) is closest to the value **Power** $= .38$. (It is also the case that for $z = .30$, if .5 is added to the value recorded in **Column 2** (i.e., .1179 + .5 = .6179), the resulting value is closest to the value .62 stipulated for the value of β (and thus, $\beta - .5 = .6179 - .5 = .1179$).) Since the power stipulated for the test is less than .5, the sign of z_β will be negative. When $z_\beta = -.30$ is employed in Equation 11.14, $n_1 = 3.57$, indicating that in order for the power of the test to equal .38 there must be at least 4 subjects per group. This result is generally consistent with the results obtained earlier when the alternative methods were employed to evaluate the power of the test. The slight discrepancy is attributable to the fact that instead of the exact values $t_{\alpha/2}$ and t_β, the values $z_{\alpha/2}$ and z_β were employed in Equation 11.14 (which results in an underestimation of the sample size, since with small sample sizes the z **test** is more powerful than the t **test**). In addition, it is possible that since interpolation was employed for the earlier methods, the latter could have resulted in some inaccuracy. The computations below (employing z scores) suggests that the power of the test employed in Example 11.1 was, in fact, somewhere within the range .38 to .42 (since if $z_\beta = -.20$, which is the appropriate value to use when the power equals .42, is substituted in Equation 11.14 the value $n_1 = 4.01$ is computed (which, since it is greater than 4, would be incremented to $n_1 = 5$).

$$n_1 = \frac{\left(21.7 + \frac{10.7}{1}\right)[1.96 + (-.30)]^2}{(7.8 - 2.8)^2} = 3.57$$

4. Measures of magnitude of treatment effect for the t test for two independent samples: Omega squared (Test 11c) and Eta squared (Test 11d) At the conclusion of an experiment a researcher may want to determine the proportion of the variability on the dependent variable which is associated with the experimental treatments (i.e., the independent variable). This latter value is commonly referred to as the **treatment effect**.[18] Unfortunately, the t value computed for the t **test for two independent samples** does not in itself provide information regarding the magnitude of a treatment effect. The reason for this is that the absolute value of t is not only a function of the treatment effect, but is also a function of the size of the sample employed in an experiment. Since the power of a statistical test is directly related to sample size, the larger the sample size, the more likely a significant t value will be obtained if there is any difference between the means of the underlying populations. Regardless of how small a treatment effect is present, the magnitude of the absolute value of t will increase as the size of the sample employed to detect that effect increases. Thus, a t value that is significant at any level (be it .05, .01, .001, etc.) can result from the presence of a large, medium, or small treatment effect.

Before describing measures of treatment effect for the t **test for two independent samples**, the distinction between **statistical significance** and **practical significance** (which is discussed briefly in the **Introduction**) will be clarified. Whereas **statistical significance** only indicates that a difference between the two group means has been detected and that the difference is unlikely to be the result of chance, **practical significance** refers to the practical implications of the obtained difference. As just noted, by employing a large sample size, a researcher will be able to detect a differences between two means which is extremely small. Although in some instances a small treatment effect can be of practical significance, more often than not a minimal difference will be of little or no practical value (other than perhaps allowing a researcher to get a study published, since significant results are more likely to be published than nonsignificant results). On the other hand, the larger the magnitude of treatment effect computed for an experiment, the more likely the results have practical implications. To go even further, when a researcher employs a small sample size, it is possible to have a moderate or large treatment effect present, yet not obtain a statistically significant result. Obviously, in such an instance (which represents an example of a Type II error) the computed *t* value is misleading with respect to the truth regarding the relationship between the variables under study.

A number of indices for measuring magnitude of treatment effect have been developed. Unlike the computed *t* value, measures of magnitude of treatment effect provide an index of the degree of relationship between the independent and dependent variables which is independent of sample size. (It should be noted, however, that the degree of error associated with a measure of treatment effect will decrease as the size of the sample employed to compute the measure increases.) A major problem with measures of treatment effect is that for most experimental designs two or more such measures are available which are not equivalent to one another, and researchers are often not in agreement with respect to which measure is appropriate to employ. In the case of the *t* **test for two independent samples**, the most commonly employed measure of treatment effect is **omega squared**. The statistic that is computed from the sample data to estimate the value of **omega squared** is represented by the notation $\tilde{\omega}^2$ (ω is the lower case Greek letter **omega**). This latter value provides an estimate of the underlying population parameter ω^2, which represents the proportion of variability on the dependent variable that is associated with the independent variable in the underlying population. The value of $\tilde{\omega}^2$ is computed with Equation 11.15.

$$\tilde{\omega}^2 = \frac{t^2 - 1}{t^2 + n_1 + n_2 - 1} \qquad \textbf{(Equation 11.15)}$$

Although the value of $\tilde{\omega}^2$ will generally fall in the range between 0 and 1, when $|t| < 1$, $\tilde{\omega}^2$ will be a negative number. The closer $\tilde{\omega}^2$ is to 1, the stronger the association between the independent and dependent variables, whereas the closer $\tilde{\omega}^2$ is to 0, the weaker the association between the two variables. A $\tilde{\omega}^2$ value equal to or less than 0 indicates there is no association between the variables.[19]

Employing Equation 11.15 with the data for Example 11.1, the value $\tilde{\omega}^2 = .22$ is computed.

$$\tilde{\omega}^2 = \frac{(-1.96)^2 - 1}{(-1.96)^2 + 5 + 5 - 1} = .22$$

The value $\tilde{\omega}^2 = .22$ indicates that 22% (or a proportion equal to .22) of the variability on the dependent variable (the depression ratings of the subjects) is associated with variability on the levels of the independent variable (the drug versus placebo conditions). To say it another way, 22% of the variability on the depression scores can be accounted for on the basis of which group a subject is a member.

Cohen (1977; 1988, pp. 285–288) has suggested the following (admittedly arbitrary) values, which are employed in psychology and a number of other disciplines, as guidelines for interpreting $\tilde{\omega}^2$: a) A **small effect size** is one that is greater than .0099 but not more than .0588;

b) A **medium effect size** is one that is greater than .0588 but not more than .1379; and c) A **large effect size** is greater than .1379.[20] In the case of Example 11.1, if one employs Cohen's (1977, 1988) guidelines, the obtained value $\tilde{\omega}^2$ = .22 indicates the presence of a large treatment effect.

Keppel (1991) and Keppel *et al.* (1992) note that in the experimental literature in the discipline of psychology it is unusual for a $\tilde{\omega}^2$ value to exceed .25 — indeed, one review of the psychological literature yielded an average $\tilde{\omega}^2$ value of .06. The inability of researchers to control experimental error with great precision is the most commonly cited reason for the low value obtained for $\tilde{\omega}^2$ in most psychological studies. The latter undoubtedly applies to studies in other disciplines (e.g., education, the social sciences, etc.).

Test 11d: Eta squared ($\tilde{\eta}^2$) is an alternative but less commonly used measure of association (since some sources note that it is a more biased measure than **omega squared**) which can also be employed to evaluate the magnitude of a treatment effect for a *t* **test for two independent samples**. (η is the lower case Greek letter **eta**.) The value of $\tilde{\eta}^2$ (which is an estimate of the value of **eta squared** (η^2) in the underlying population) is computed with Equation 11.16. For Example 11.1 the value $\tilde{n}^2$ = .32 is computed. Note that the latter value is not the same as $\tilde{\omega}^2$ = .22 computed earlier. The latter illustrates the fact that for the same set of data $\tilde{\omega}^2$ and $\tilde{\eta}^2$ will not yield identical values.

$$\tilde{\eta}^2 = \frac{t^2}{t^2 + df} = \left(\frac{-1.96)^2}{(-1.96)^2 + 8}\right) = .32 \qquad \textbf{(Equation 11.16)}$$

The **eta squared** statistic is discussed in greater detail in Sections VI and VII of the **single-factor between-subjects analysis of variance**, and in Section IX (the **Addendum**) of the **Pearson product-moment correlation coefficient** under the discussion of the **point-biserial correlation coefficient (Test 28i)**. In the latter discussion, it is demonstrated that **eta squared** and the square of the **point-biserial correlation coefficient** are equivalent measures when they are used to evaluate magnitude of treatment effect for a design involving two independent samples.

In closing the discussion of magnitude of treatment effect, it should be noted that at the present time many sources recommend that in reporting the results of an experimental analysis with an inferential statistical test (such as the *t* **test for two independent samples**), in addition to reporting the computed test statistic (e.g., a *t* value), the researcher should also present a measure of the magnitude of treatment effect (e.g., $\tilde{\omega}^2$), since, by including the latter value one is providing additional information which can further clarify the nature of the relationship between the variables under study. A general discussion of the latter issue, as well as an additional discussion of measures of treatment effect can be found in Section IX (the **Addendum**) of the **Pearson product-moment correlation coefficient** under the discussion of **meta-analysis and related topics**.

5. Computation of a confidence interval for the *t* test for two independent samples Prior to reading this section the reader should review the discussion on the computation of confidence intervals in Section VI of the **single-sample *t* test**. When interval/ratio data are available for two independent samples, a confidence interval can be computed that identifies a range of values with respect to which one can be confident to a specified degree contains the true difference between the two population means. Equation 11.17 is the general equation for computing the confidence interval for the difference between the means of two independent populations.

$$CI_{(1 - \alpha)} = (\bar{X}_1 - \bar{X}_2) \pm (t_{\alpha/2})(s_{\bar{X}_1 - \bar{X}_2}) \qquad \textbf{(Equation 11.17)}$$

Where: $t_{\alpha/2}$ represents the tabled critical two-tailed value in the t distribution, for $df = n_1 + n_2 - 2$, below which a proportion (percentage) equal to $[1 - (\alpha/2)]$ of the cases falls. If the proportion (percentage) of the distribution that falls within the confidence interval is subtracted from 1 (100%), it will equal the value of α.

Employing Equation 11.17, the 95% interval for Example 11.1 is computed below. In employing Equation 11.17, $(\bar{X}_1 - \bar{X}_2)$ represents the obtained difference between the group means (which is the numerator of the equation used to compute the value of t), $t_{.05}$ represents the tabled critical two-tailed .05 value for $df = n_1 + n_2 - 2$, and $s_{\bar{X}_1 - \bar{X}_2}$ represents the standard error of the difference (which is the denominator of the equation used to compute the value of t).

$$CI_{.95} = (\bar{X}_1 - \bar{X}_2) \pm (t_{.05})(s_{\bar{X}_1 - \bar{X}_2}) = -5 \pm (2.31)(2.55) = -5 \pm 5.89$$

$$-10.89 \le (\mu_1 - \mu_2) \le .89$$

This result indicates the researcher can be 95% confident (or the probability is .95) that the interval -10.89 to .89 contains the true difference between the population means. If the aforementioned interval contains the true difference between the population means, the mean of the population Group 2 represents is no more than 10.89 points higher than the mean of the population Group 1 represents, and the mean of the population Group 1 represents is no more than .89 points higher than the mean of the population Group 2 represents.[21]

Note that in using the above notation, when a confidence interval range involves both a negative and positive limit (as is the case in the above example), it indicates it is possible for either of the two population means to be the larger value. If, on the other hand, both limits identified by the confidence interval are positive values, the mean of Population 1 will always be greater than the mean of Population 2. If both limits identified by the confidence interval are negative values, the mean of Population 2 will always be greater than the mean of Population 1.

The 99% confidence interval for Example 11.1 will also be computed to illustrate that the range of values which defines a 99% confidence interval is always larger than the range that defines a 95% confidence interval.

$$CI_{.99} = (\bar{X}_1 - \bar{X}_2) \pm (t_{.01})(s_{\bar{X}_1 - \bar{X}_2}) = -5 \pm (3.36)(2.55) = -5 \pm 8.57$$

$$-13.57 \le (\mu_1 - \mu_2) \le 3.57$$

Thus, the researcher can be 99% confident (or the probability is .99) that the interval -13.57 to 3.57 contains the true difference between the population means. If the aforementioned interval contains the true difference between the population means, the mean of the population Group 2 represents is no more than 13.57 points higher than the mean of the population Group 1 represents, and the mean of the population Group 1 represents is no more than 3.57 points higher than the mean of the population Group 2 represents.

The reader should take note of the fact that the reliability of Equation 11.17 will be compromised if one or more of the assumptions of the **t test for two independent samples** are saliently violated. In instances where the homogeneity of variance assumption is violated, the t value employed in Equation 11.17 should be adjusted to reflect the degrees of freedom adjustment described earlier under the discussion of homogeneity of variance in this section. Wilcox (1996, 1997, 2003) and Grissom and Kim (2005, pp. 32–40, 64) provide detailed discussions of modified procedures for computing a confidence interval when there is reason to believe that one or more of the assumptions underlying the *t* test have been saliently violated.

In closing the discussion of confidence intervals, it is worth noting that the broad range of values which define the confidence intervals computed in this section will not allow a researcher to estimate with great precision the actual difference between the means of the underlying populations. The locus of the difference between means, however, can be further clarified through use of Smithson's (2003) methodology of employing confidence intervals in lieu of tests of significance (which is described in Section VI of the **single sample *t* test**).

6. Test 11e: The *z* test for two independent samples There are occasions (albeit infrequent) when a researcher wants to compare the means of two independent samples, and happens to know the variances of the two underlying populations. In such a case, the *z* test for two independent samples** should be employed to evaluate the data instead of the *t* **test for two independent samples**. As is the case with the latter test, the *z* **test for two independent samples** assumes that the two samples are randomly selected from populations which have normal distributions. The effect of violation of the normality assumption on the test statistic decreases as the size of the samples employed in an experiment increases. The homogeneity of variance assumption noted for the *t* **test for two independent samples** is not an assumption of the *z* **test for two independent samples**.

The null and alternative hypotheses employed for the *z* **test for two independent samples** are identical to those employed for the *t* **test for two independent samples**. Equation 11.18 is employed to compute the test statistic for the *z* **test for two independent samples**.

$$z = \frac{\bar{X}_1 - \bar{X}_2}{\sqrt{\dfrac{\sigma_1^2}{n_1} + \dfrac{\sigma_2^2}{n_2}}}$$

(Equation 11.18)

The only differences between Equation 11.18 and Equation 11.1 (the equation for the *t* test for two independent samples)** are: a) In the denominator of Equation 11.18 the population variances σ_1^2 and σ_2^2 are employed instead of the estimated population variances $\tilde{s}_1^2$ and $\tilde{s}_2^2$ (which are employed in Equation 11.1); and b) Equation 11.18 computes a *z* score which is evaluated with the normal distribution, while Equation 11.1 derives a *t* score which is evaluated with the *t* distribution. Unlike Equation 11.1, Equation 11.18 can be used with both equal and unequal sample sizes.[22]

If it is assumed that the two population variances are known in Example 11.1, and that $\sigma_1^2 = 21.7$ and $\sigma_2^2 = 10.7$, Equation 11.18 can be used to evaluate the data. Note that the obtained value $z = -1.96$ is identical to the value computed for *t* when Equation 11.1 is employed.

$$z = \frac{2.8 - 7.8}{\sqrt{\dfrac{21.7}{5} + \dfrac{10.7}{5}}} = -1.96$$

Except for the fact that the normal distribution is used in lieu of the *t* distribution, the same guidelines are employed in evaluating the computed *z* value as those stipulated for evaluating a *t* value in Section V. The obtained value $z = -1.96$ is evaluated with **Table A1 (Table of the Normal Distribution)** in the **Appendix**. In **Table A1** the tabled critical two-tailed .05 and .01 values are $z_{.05} = 1.96$ and $z_{.01} = 2.58$, and the tabled critical one-tailed .05 and .01 values are $z_{.05} = 1.65$ and $z_{.01} = 2.33$. Since the computed absolute value $z = 1.96$ is equal to the tabled critical two-tailed value $z_{.05}$ = 1.96, the nondirectional alternative hypothesis H_1: $\mu_1 \neq \mu_2$ is supported at the .05 level. Since the computed value $z = -1.96$ is a negative number and the absolute value of z is greater than the tabled critical one-tailed .05 value $z_{.05}$ = 1.65, the directional alternative hypothesis H_1: $\mu_1 < \mu_2$ is also supported at the .05 level.

When the same set of data is evaluated with the **t test for two independent samples**, although the directional alternative hypothesis H_1: $\mu_1 < \mu_2$ is supported at the .05 level, the nondirectional alternative hypothesis H_1: $\mu_1 \neq \mu_2$ is not supported. This latter fact illustrates that if the **z test for two independent samples** and the **t test for two independent samples** are employed to evaluate the same set of data (except when the value of $n_1 + n_2 - 2$ is extremely large), the latter test will provide a more conservative test of the null hypothesis (i.e., make it more difficult to reject H_0). This is the case, since the tabled critical values listed for the **z test for two independent samples** will always correspond to the tabled critical values listed in **Table A2** for $df = \infty$ (which are the lowest tabled critical values listed for the t distribution).

The final part of the discussion of the **z test for two independent samples** will describe a special case of the test in which it is employed to evaluate the difference between the average performance of two samples for whom scores have been obtained on a binomially distributed variable. Example 11.3, which is used to illustrate this application of the test, is an extension of Example 9.6 (discussed under the **z test for a population proportion (Test 9a)**) to a design involving two independent samples.

Example 11.3 *An experiment is conducted in which the performance of two groups is contrasted on a test of extrasensory perception. The two groups are comprised of five subjects who believe in extrasensory perception (Group 1) and five subjects who do not believe in it (Group 2). The researcher employs as test stimuli a list of 200 binary digits (specifically, the values 0 and 1) which have been randomly generated by a computer. During the experiment an associate of the researcher concentrates on each of the digits in the order it appears on the list. While the associate does this, each of the ten subjects, all of whom are in separate rooms, attempts to guess the value of the number for each of 200 trials. The number of correct guesses for the two groups of subjects follow:* **Group 1**: *105, 120, 130, 115, 110;* **Group 2**: *104, 99, 90, 100, 107. Is there a difference in the performance of the two groups?*

The null and alternative hypotheses evaluated in Example 11.3 are identical to those evaluated in Example 11.1. Example 11.3 is evaluated with Equation 11.19, which is the form Equation 11.17 assumes when $\sigma_1^2 = \sigma_2^2$. Note that in Equation 11.19, m_j is employed to represent the number of subjects in the jth group, since the notation n is employed with a binomially distributed variable to designate the number of trials each subject is tested. Thus, $m_1 = m_2 = 5$.

$$z = \frac{\bar{X}_1 - \bar{X}_2}{\sigma \sqrt{\dfrac{1}{m_1} + \dfrac{1}{m_2}}} \qquad \textbf{(Equation 11.19)}$$

In employing Equation 11.19, we first compute the average score of each of the groups: $\bar{X}_1 = 580/5 = 116$ and $\bar{X}_2 = 500/5 = 100$. Since scores on the binary guessing task described in Example 11.3 are assumed to be binomially distributed, as is the case in Example 9.6, the following is true: $n = 200$, $\pi_1 = .5$, and $\pi_2 = .5$. The computed value for the population standard deviation for the binomially distributed variable is $\sigma = \sqrt{n\pi_1\pi_2} = \sqrt{(200)(.5)(.5)} = 7.07$. (The computation of the latter values is discussed in Section I of the **binomial sign test for a single sample (Test 9)**.) When the appropriate values are substituted in Equation 11.19, the value $z = 3.58$ is computed.

$$z = \frac{116 - 100}{7.07\sqrt{\frac{1}{5} + \frac{1}{5}}} = 3.58$$

Since the computed absolute value $z = 3.58$ is greater than the tabled critical two-tailed values $z_{.05} = 1.96$ and $z_{.01} = 2.58$, the nondirectional alternative hypothesis H_1: $\mu_1 \neq \mu_2$ is supported at both the .05 and .01 levels. Since the computed value $z = 3.58$ is a positive number which is greater than the tabled critical one-tailed values $z_{.05} = 1.65$ and $z_{.01} = 2.33$, the directional alternative hypothesis H_1: $\mu_1 > \mu_2$ is supported at both the .05 and .01 levels. Thus, it can be concluded that the average score of Group 1 is significantly larger than the average score of Group 2.

After employing Equation 11.19 to evaluate an experiment such as the one described in Example 11.3, a researcher may want to determine whether or not either of the group averages is above or below the expected mean value (which, in Example 11.3, is $\mu = n\pi_1 = (200)(.5) = 100$). Equation 9.12 is employed to evaluate the performance of a single group. The null hypothesis that is employed for evaluating a single group is H_0: $\pi_1 = .5$ (which for Example 11.3 is commensurate with H_0: $\mu_j = 100$), and the nondirectional alternative hypothesis is H_1: $\pi_1 \neq .5$ (which for Example 11.3 is commensurate with H_1: $\mu_j \neq 100$). Since it is obvious from inspection of the data that the performance of Group 2 is at chance expectancy, the performance of Group 1, which is above chance, will be evaluated with Equation 9.12.

$$z = \frac{\bar{X} - \mu}{\frac{\sigma}{\sqrt{m}}} = \frac{116 - 100}{\frac{7.07}{\sqrt{5}}} = 5.06$$

Since the computed value $z = 5.06$ is greater than the tabled critical two-tailed values $z_{.05} = 1.96$ and $z_{.01} = 2.58$, the nondirectional alternative hypothesis H_1: $\pi_1 \neq .5$ is supported at both the .05 and .01 levels. Since the computed value $z = 5.06$ is a positive number which is greater than the tabled critical one-tailed values $z_{.05} = 1.65$ and $z_{.01} = 2.33$, the directional alternative hypothesis H_1: $\pi_1 > .5$ is also supported at both the .05 and .01 levels. Thus, it can be concluded that the average performance of Group 1 is significantly above chance.

VII. Additional Discussion of the *t* Test for Two Independent Samples

1. Unequal sample sizes   In Section IV it is noted that if Equation 11.1/11.2 (which can only be used when $n_1 = n_2$) is applied to data where $n_1 \neq n_2$, the absolute value of t will be larger than the value computed with Equation 11.3. To illustrate this point, Equations 11.1 and 11.3 will be applied to a modified form of Example 11.1. Specifically, one of the scores in Group 2 will be eliminated from the data. If the score of Subject 1 (i.e., 11) is eliminated, the four scores which remain are: 11, 5, 8, 4. Employing the latter values, $n_2 = 4$, $\Sigma X_2 = 28$, $\Sigma X_2^2 = 226$, $\bar{X}_2 = 7$, $\tilde{s}_2^2 = [226 - ((28)^2/4)]/(4 - 1) = 10$. For Group 1 the values $n_1 = 5$, $\bar{X}_1 = 2.8$, and $\tilde{s}_1^2 = 21.7$ remain unchanged. The relevant values are substituted in Equation 11.1.

$$t = \frac{2.8 - 7}{\sqrt{\frac{21.7}{5} + \frac{10}{4}}} = -1.61$$

The same information is now substituted in Equation 11.3.

$$t = \frac{2.8 - 7}{\sqrt{\left[\frac{(5 - 1)(21.7) + (4 - 1)(10)}{5 + 4 - 2}\right]\left[\frac{1}{5} + \frac{1}{4}\right]}} = -1.53$$

Note that the absolute value $t = 1.61$ computed with Equation 11.1 is larger than the absolute value $t = 1.53$ computed with Equation 11.3, and thus Equation 11.3 provides a more conservative test of the null hypothesis. It so happens that in this instance neither of the computed t values allows the null hypothesis to be rejected since, for $df = 5 + 4 - 2 = 7$, both absolute values are below the tabled critical two-tailed values $t_{.05} = 2.37$ and $t_{.01} = 3.50$, and the tabled critical one-tailed values $t_{.05} = 1.90$ and $t_{.01} = 3.00$.

2. Robustness of the *t* test for two independent samples[23] Some statisticians believe that if one or more of the assumptions of a parametric test (such as the *t* test for two independent samples) are saliently violated, the test results will be unreliable (see Wilcox (1887, 1996, 1997, 2003)), and because of this under such conditions it is more prudent to employ the analogous nonparametric test, which will generally have fewer or less rigorous assumptions than its parametric analog. In the case of the *t* test for two independent samples the most commonly employed analogous nonparametric tests are the **Mann–Whitney *U* test (Test 12)** and the **chi-square test for *r* × *c* tables (Test 16)**. Use of the **Mann–Whitney *U* test** (which is most likely to be recommended when the normality assumption of the *t* test for two independent samples** is saliently violated) requires that the original interval/ ratio scores be transformed into a rank-order format. By virtue of rank-ordering the data, information is sacrificed (since rank-orderings do not provide information regarding the magnitude of the differences between adjacent ranks). Given the fact that the **Mann–Whitney *U* test** employs less information, many researchers if given the choice will still elect to employ the *t* test for two independent samples, even if there is reason to believe that the normality assumption of the latter test is violated. Under such conditions, however, most researchers would probably conduct a more conservative *t* test in order to avoid inflating the likelihood of committing a Type I error (i.e., one might employ the tabled critical $t_{.01}$ value to represent the $t_{.05}$ value instead of the actual value listed for $t_{.05}$). In the unlikely event that a researcher elects to employ the **chi-square test for *r* × *c* tables** in place of the *t* test for two independent samples, he must convert the original interval/ratio data into a categorical format. This latter type of a transformation will result in an even greater loss of information than is the case with the **Mann–Whitney *U* test**.

The justification for using a parametric test in lieu of its nonparametric analog, even when one or more of the assumptions of the former test are violated, is that the results of numerous empirical sampling studies have demonstrated that under most conditions a parametric test like the *t* test for two independent samples is reasonably **robust**. A **robust** test is one which still provides reliable information about the underlying sampling distribution, in spite of the fact that one or more of the test's assumptions have been violated. In addition, researchers who are reluctant to employ nonparametric tests argue that parametric tests, such as the *t* test for two independent samples**, are more powerful than their nonparametric analogs. Proponents of nonparametric tests counter with the argument that the latter group of tests are almost equivalent in power to their parametric analogs, and because of this, they state that it is preferable to use the appropriate nonparametric test if any of the assumptions of a parametric test have been saliently violated. Throughout this book it is demonstrated that in most instances when the same set of data is evaluated with both a parametric and a nonparametric test (especially a nonparametric test employing rank-order data), the two tests yield comparable results. As a general rule, in instances where only one of the two tests is significant, the parametric test is the one that is more likely to be significant. However, in most cases where a parametric test achieves significance and the nonparametric test does not, the latter test will fall just short of being significant. In instances where both tests are significant, the alpha level at which the result is significant will generally be lower for the parametric test.

3. Outliers and data transformation (Test 11f: Median absolute deviation test for identifying outliers and Test 11g: Extreme Studentized deviate test for identifying outliers) An **outlier** is an observation (or subset of observations) in a set of data which does not appear to be consistent with the rest of the data. In most instances inconsistency is reflected in the magnitude of an observation (i.e., it is either much higher or much lower than any of the other observations). A researcher may be tempted to discard such an observation from a set of data, inferring that its presence is the result of some error on the part of the subject or experimenter. Yet what appears to be an inconsistent/extreme score to one observer may not appear to be so to another. Barnett and Lewis (1994) emphasize that a defining characteristic of an outlier is that it elicits genuine surprise in an observer. To illustrate the fact that what may surprise one observer may not surprise another, we will consider an example cited by Barnett and Lewis (1994, p. 15). The latter authors present data described by Fisher, Corbet, and Williams (1943), which represents the number of moths of a specific species which were caught in light-traps mounted in a geographical locale in England. The following 15 observations were obtained.

$$3, 3, 4, 5, 7, 11, 12, 15, 18, 24, 51, 54, 84, 120, 560$$

Barnett and Lewis (1994) point out that although the value 560 might appear to be an observation which would surprise most observers, in point of fact, it is not an anomaly. The reason why 560 would not be classified as an outlier is because an experienced entomologist would be privy to the fact that the distribution under study is characterized by a marked skewness, and consequently an occasional extreme score in the upper tail such as the value 560 is a matter-of-fact occurrence. Thus, a researcher familiar with the phenomenon under study would not classify 560 an outlier.[24] Another example of an outlier which should not be ignored is the person who exhibit an atypical/catastrophic response to an immunization shot. Certainly a physiological response measure for such an individual would be atypical when compared with the responses of the overwhelming majority of the population.

If at all possible, one should determine the source of any observation in a set of data that is viewed as anomalous, since the presence of one or more outliers can dramatically influence the values of both the mean and variance of a distribution. As a result of the latter, any test statistic computed for the data can be unreliable. As an example, assume that a researcher is comparing two groups with respect to their scores on a dependent variable, and that all the subjects in both groups except for one subject in Group 2 obtain a score between 0 and 20. The subject in Group 2 thought to represent an outlier obtains a score of 200. It should be obvious that the presence of this one score (even if the size of the sample for Group 2 is relatively large) will inflate the mean and variance of Group 2 relative to that of Group 1, and because of this either one or both of the following consequences may result: a) A significant difference between the two group means may be obtained which would not have been obtained if the outlier score was not included in the data; and/or b) The homogeneity of variance assumption will be violated due to the higher estimated population variance computed for Group 2. By virtue of adjusting the *t* **test** statistic for violation of the homogeneity of variance assumption, a more conservative test will be conducted, thus making it more difficult to reject the null hypothesis.[25]

Stevens (1996) notes that there are basically two strategies which can be used in dealing with outliers. One strategy is to develop and employ procedures for identifying outliers. Within the framework of the latter strategy, criteria should be established for determining under what conditions one or more scores which are identified as outliers should be deleted from a set of data. A second approach in dealing with outliers is to develop statistical procedures that are not influenced (or only minimally affected) by the presence of outliers. Such procedures are commonly called **robust statistical procedures** — the term **robust**, as noted earlier, referring to procedures which are not overly dependent on critical assumptions underlying a population distribution. The discussion of outliers within the framework of robustness is predicated on the fact that their presence may lead to violation of one or more assumptions underlying a statistical test. In the case of the *t* **test for two independent samples**, the assumptions of concern are those of normality and homogeneity of variance. Although a number of inferential statistical tests have been developed for identifying outliers, Sprent (1993, 1998) notes that ironically many of these tests lack robustness, and because of the latter may lack power with respect to their ability to identify the presence of one or more outliers. Sprent (1998), Barnett and Lewis (1994), Rosner (2000), and Wilcox (2003) discuss, among other things relating to outliers, the **masking effect**, which refers to the fact that a test's power in identifying a specific outlier may be compromised if there are one or more additional outliers in the data. Anderson (2001) is another source that provides additional insights on issues relating to outliers.

Barnett and Lewis (1994), who represent the most comprehensive source on the subject, describe a large number of tests for identifying outliers — they describe 48 tests alone for detection of outliers in data which are assumed to be drawn from a normal distribution. Some of the tests for identifying outliers are designed to identify a single outlier, some are designed for identification of multiple outliers, and some tests are specific with respect to identifying outliers in one or both tails of a distribution. Additionally, tests are described for detecting outliers in data that are assumed to be drawn from any number of a variety of nonnormal distributions (e.g., binomial, Poisson, gamma, exponential, etc.). Given the large number of tests which are available for detecting outliers, it is not unusual that two or more tests applied to the same set of data may not agree with one another with regard to whether or not a specific observation should be classified as an outlier.

Procedures for identifying outliers (Box-and-whisker plot criteria; Standard deviation score criterion; Test 11f: Median absolute deviation test for identifying outliers;Test 11g: Extreme Studentized deviate test for identifying outliers)

Box-and-whisker plot criteria   A visual approach for detecting outliers is described in the **Introduction** in the section on **visual methods for displaying data** under the discussion of **box-and-whisker plots**.

Standard deviation score criterion   The Russian mathematician Chebyshev demonstrated that regardless of the distribution of data, at least $[1 - (1/k^2)](100)$ percent of the observations in a set of data must fall within k standard deviations of the mean (where k is any value greater than 1). This stated relationship between the probability of a score and the deviation of the score from the mean in standard deviation units is referred to as **Chebyshev's inequality** (some sources employ the spelling Tchebycheff). Stevens (1996, p. 17) notes that by applying the latter, we can determine that when $k = 2$, at least 75% of the observations will fall within +2 and –2 standard deviations from the mean. The percentages for when k is equal to 3, 4, and 5 standard deviation units are respectively, 88.89%, 93.75%, and 96%. These values suggest that scores which are beyond two standard deviations from the mean are rare; that scores beyond three standard deviations are even rarer; and so on. Thus, scores which yield relatively high standard deviation values should be considered as possible outliers. Since the latter statement is rather general, two procedures will be described in this section which provide more definitive guidelines for classifying one or more scores in a distribution as outliers.

It should be noted that employing the magnitude of a standard deviation score (i.e., a z score) as a criterion for classifying outliers can often lead to misleading conclusions. Based on **Chebyshev's inequality**, Shiffler (1988) notes that the largest possible absolute z value which can occur in a set of data is defined by the limit $(n - 1)/\sqrt{n}$. For example, if $n = 15$, the absolute value of z cannot exceed $(15 - 1)/(\sqrt{15}) = 3.62$. Inspection of the equation $(n - 1)/\sqrt{n}$ reveals that the maximum possible absolute value z may attain is a direct function of the sample size — in other words, the larger the sample size, the larger the limiting value of z. In order to appreciate how the use of a z value with small sample sizes can lead to misleading conclusions if it is employed as a criterion for classifying outliers, consider the following examples presented by Shiffler (1988): a) A set of data consists of the following five scores: 0, 0, 0, 0, 1 million. On inspection, some researchers might immediately view the score of 1 million as an outlier. Yet when the sample mean and estimated population standard deviation are computed for the five scores, the z score associated with the score of 1 million is only $z = 1.79$; and b) A set of data is comprised of 18 scores, with 17 of the scores equal to 0 and the remaining score equal to 1. When the sample mean and estimated population standard deviation are computed for the 18 scores, the z score associated with the score of 1 is $z = 4.01$. The magnitude of the latter z value might suggest to some researchers that the score of 1 is an outlier. Thus, as noted above, when the sample size is relatively small, the theoretical limits imposed on the value of a z score may make it impractical to employ it as a criterion for classifying outliers.

Test 11f: Median absolute deviation test for identifying outliers Sprent (1993, 1998) describes a relatively robust procedure for identifying outliers, to be referred to as the **median absolute deviation test for identifying outliers**. Equation 11.20, which allows one to determine whether or not a score in a sample of n observations should be classified as an outlier, is employed to compute the test statistic for the **median absolute deviation test**.

$$\frac{|X_i - M|}{MAD} > Max \qquad \text{(Equation 11.20)}$$

Where: X_i represents any of the n scores being evaluated with respect to whether it is an outlier.

M is the median of the n scores in the sample

MAD is the **median absolute deviation**

Max is the critical value the result to the left of the inequality must exceed in order to conclude the value X_i is an outlier

In the left side of Equation 11.20, the numerator value $|X_i - M|$ represents a difference score. The denominator value MAD represents a measure of dispersion.[26] The equation is designed to yield a standard deviation score such as that obtained by Equation I.38 (which is the equation for computing a standard deviation score for a normally distributed variable). The value

Max on the right side of Equation 11.20 represents an extreme standard deviation score which is employed as a criterion for classifying an outlier.

The value of *MAD* in Equation 11.20 is determined as follows: a) Upon computing the sample median, obtain an **absolute deviation score** from the sample median for each of the n scores. The latter is done by computing the absolute value of the difference between the median and each score; b) When all n absolute deviation scores have been computed, arrange them ordinally (i.e., from lowest to highest); and c) Find the median of the n absolute deviation scores. The latter value represents the value of *MAD* to employ in Equation 11.20.

To determine whether or not any of the n scores in the sample is an outlier, the following protocol is employed: a) Select the score that deviates by the greatest amount from the median to initially represent the value of X_i; b) Subtract the median from the value of X_i, and divide the difference (which will always be a positive value since it is an absolute value) by *MAD*; c) If the value obtained in b) is greater than *Max*, one can conclude that X_i is an outlier. If the value obtained in b) is equal to or less than *Max*, one cannot conclude that X_i is an outlier; d) If it is concluded that the score selected to represent X_i is not an outlier, terminate the analysis. If it is concluded that the score selected to represent X_i is an outlier, select the score which has the second greatest deviation from the median and repeat steps b) and c); and e) Continue substituting X_i values in Equation 11.20 until an X_i value is identified which is not an outlier. The reader should take note of the fact that in most cases it is assumed that few if any observations within a given sample will be identified as outliers.

Sprent (1993, p. 278) notes that the selection of the value to represent *Max* in Equation 11.20 is somewhat arbitrary. He recommends that 5 is a reasonable value to employ, since if one assumes the data are derived from an approximately normally distributed population, the value *Max* = 5 will be extremely likely to identify scores which deviate from the mean by more than three standard deviations. (From **Column 3** of **Table A1** we can determine that the likelihood of a score being greater or less than three standard deviations from the mean is .0026.) It should be noted that if there is reason to believe the sample data are derived from a population which is not normally distributed, the choice of what value to employ to represent *Max* becomes more problematical. The reader should consult sources which discuss outliers in greater detail for a more in-depth discussion of the latter issue (e.g., Barnett and Lewis (1994) and Sprent (1993, 1998)).

To illustrate the application of Equation 11.20, assume we have a sample consisting of the following five scores: 2, 3, 4, 7, 18. Following the protocol described above: a) We determine that the median (i.e., the middle score) of the sample is 4; b) We compute the absolute deviation of each score from the median: $|2 - 4| = 2$, $|3 - 4| = 1$, $|4 - 4| = 0$, $|7 - 4| = 3$, $|18 - 4| = 14$. Arranging the five deviation scores ordinally (0, 1, 2, 3, 14), we determine the median of the five deviation scores is 2. The latter value will represent *MAD* in Equation 11.20; c) Since the only value we would suspect to be an outlier is the score of 18, we employ that value to represent X_i in Equation 11.20; d) Since we will assume the data are derived from a normal distribution, we employ the value *Max* = 5; e) Substituting the appropriate values in the left side of Equation 11.20, we compute $|18 - 4|/2 = 7$. Since the obtained value of 7 is greater than *Max* = 5, we conclude that the score of 18 is an outlier. Additional discussion of the *MAD* can be found in Grissom and Kim (2005, p. 17–18).

Test 11g: Extreme Studentized deviate test for identifying outliers   The **extreme Studentized deviate test**, developed by Rosner (1983) (and described in Rosner (1995, pp. 277–282; 2000, pp. 300–306)) represents a procedure which is recommended for data derived from a normal distribution that can detect one or more outliers in a sample of data. The test is most reliable when $n \geq 25$. The **extreme Studentized deviate test**, which is based on the t

distribution, controls the Type I error rate (i.e., erroneously detecting an outlier when, in fact, none are present in a set of data), as well as having good power in detecting the presence of one or more outliers in a sample. In order to demonstrate the **extreme Studentized deviate test**, consider the distribution of $n = 25$ scores noted below.

4, 4, 5, 6, 6, 7, 10, 11, 12, 14, 15, 17, 18, 22, 24, 32, 40, 42, 50, 54, 60, 62, 84, 120, 560

The test statistic for the **extreme Studentized deviate test** is computed with Equation 11.21. Note that the test statistic represents a measure of the distance between the score suspected of being an outlier (X_i) and the mean of the sample in standard deviation units.

$$ESD = \frac{|X_i - \bar{X}|}{\tilde{s}} \qquad \textbf{(Equation 11.21)}$$

Let us assume we want to determine whether or not the score of 560 — the most extreme score in the distribution — is an outlier. We initially compute the values $\bar{X} = 51.16$ (with Equation I.1) and $\tilde{s} = 109.83$ (with Equation I.11/2.1) for the 25 scores which comprise the distribution. Employing the value $X_i = 560$ in Equation 11.21, the value $ESD = 4.63$ is computed below.

$$ESD = \frac{|560 - 51.16|}{109.83} = 4.63$$

The computed value $ESD = 4.63$ is evaluated with **Table A26 (Table of Extreme Studentized Deviate Outlier Statistic)** in the **Appendix**. The tabled critical .05 and .01 values for the ESD statistic are recorded in reference to the value of n. The reader should take note of the following with regard to **Table A26**. As the size of a sample (which is assumed to be derived from a large population) increases, the greater the likelihood that one or more of the extreme scores which fall within the population will be included in the sample. Because a large population will inevitably have some extreme scores, such scores may not, in fact, represent outliers. Because of the latter, the larger the value of n the less likely it is that an extreme score should be classified as an outlier. This accounts for the fact that in **Table A26**, the larger the value of n, the larger the tabled ESD critical value at a given level of significance.

In order to reject the null hypothesis (which is that the score X_i is not an outlier) and conclude that the alternative hypothesis is supported (which is that the score X_i is an outlier), the obtained ESD value must be equal to or greater than the tabled critical value at the prespecified level of significance. Inspection of **Table A26** indicates that for $n = 25$, $ESD_{.05} = 2.82$ and $ESD_{.01} = 3.14$. Since the obtained value $ESD = 4.63$ is larger than both of the aforementioned critical values, the null hypothesis can be rejected at both the .05 or .01 levels. Thus, the test allows the researcher to conclude that 560 represents an outlier.

Earlier in the discussion of outliers it was noted that the term **masking effect** refers to the fact that a test's power in identifying a specific outlier may be compromised if there are one or more additional outliers in a sample. **Masking** can make it problematical to use Equation 11.21 as described above when it is suspected there may be multiple outliers in the sample. The problem arises from the fact that the presence of multiple outliers may inflate the value of the estimated population standard deviation (i.e., $\tilde{s}$), thus making it difficult to detect the presence of a single outlier, let alone additional outliers in the sample. Because of the latter, when the **extreme Studentized deviate test** is employed with a distribution in which a researcher suspects the presence of multiple outliers (especially if the multiple outliers are approximately the same distance from the mean), the values employed in Equation 11.21 for the mean and estimated population standard deviation should be recomputed, so as to not take into account any scores that are more extreme than the specific score which is being evaluated with regard to whether or not it is an outlier.

When the presence of more than one outlier is suspected, Rosner (1983, 1995, 2000) notes that it is advisable for a researcher to stipulate an upper boundary for the maximum number of outliers (to be designated n_{max}) which can be detected in a sample. He recommends that n_{max} be the largest integer value which is equal to or less than the value computed with the equation $n/10$. Since in the case of the distribution under discussion, $n/10 = 25/10 = 2.5$, n_{max} would be set equal to two. Consequently, only the two most extreme scores in the distribution can be considered as possible outliers. Rosner (1995, p. 280; 2000, p. 304) notes that when the value of n_{max} is determined to be more than five and, in fact, an analysis of individual scores with the **extreme Studentized deviate test** indicates there are more than five outliers in a sample, unless the sample is extremely large, it is likely that one is dealing with a nonnormal distribution (which will compromise the reliability of the **extreme Studentized deviate test**).

In the event we wish to employ the **extreme Studentized deviate test** to detect more than one outlier (such as in the case of the distribution under discussion, where $n_{max} = 2$ outliers), we must compute a test statistic for all of the suspected outliers (which in this instance are the two most extreme scores of 560 and 120). We initially evaluate the most extreme score of 560 as was done previously — i.e., we compute the values $\overline{X} = 51.16$ and $\tilde{s} = 109.83$ for the $n = 25$ scores, and use Equation 11.21 to compute the value $ESD = 4.63$. At this point we would not evaluate whether or not the latter value is statistically significant.

The second most extreme score of 120 is evaluated next. In evaluating the latter score the value 560 is omitted from the analysis, and thus the mean and estimated standard deviation of the distribution are computed for the remaining $n = 24$ scores. The latter values are computed to be $\overline{X} = 29.96$ and $\tilde{s} = 29.34$. Note how much smaller these values are (especially the value of $\tilde{s}$) than the analogous values computed when the score of 560 was included in the distribution. When the latter values, along with the value $X_i = 120$, are substituted in Equation 11.21, the value $ESD = 3.07$ is computed.

$$ESD = \frac{|\,120 - 29.96\,|}{29.34} = 3.07$$

Employing **Table A26** for $n = 24$, we determine that $ESD_{.05} = 2.80$ and $ESD_{.01} = 3.11$. Since the obtained value $ESD = 3.07$ is larger than $ESD_{.05} = 2.80$, the null hypothesis can be rejected at the .05 but not at the .01 level (since $ESD = 3.07$ is less than $ESD_{.01} = 3.11$). Thus, the test allows the researcher to conclude that the score of 120 represents an outlier if .05 had been the prestipulated value for alpha. We would also declare any values more extreme than 120 an outlier (in this case the score of 560), regardless of whether or not the ESD value computed for that score achieved significance (i.e., if the score is evaluated in reference to the value of n for which an ESD value was computed). In the event the researcher prestipulated .01 as the value for alpha, the score of 120 could not be declared an outlier. In such a case the ESD value computed for the more extreme score of 560 would be evaluated in reference to the appropriate $ESD_{.01}$ value for $n = 25$. As noted earlier, since the value $ESD = 4.63$ computed for $X_i = 560$ is larger than $ESD_{.01} = 3.14$, the score of 560 would be declared an outlier.

Since the value of n_{max} was set equal to two, the protocol described above would not permit the researcher to evaluate the next most extreme score in the distribution — specifically, the score of 84. However, for illustrative purposes it will be evaluated, and thus, for the moment, we will assume that the researcher had elected to set the value of n_{max} equal to three instead of two. In the event of the latter, the mean and estimated standard deviation of the sample would be computed for the remaining $n = 23$ scores (i.e., the scores 560 and 120 would be omitted from the analysis). When the latter values, which are computed to be $\bar{X} = 26.04$ and $\tilde{s} = 22.70$, along with the value $X_i = 84$ are substituted in Equation 11.21, the value $ESD = 2.55$ is computed.

$$ESD = \frac{|\,84 - 26.04\,|}{22.70} = 2.55$$

Employing **Table A26** for $n = 23$, we determine that $ESD_{.05} = 2.78$ and $ESD_{.01} = 3.08$. Since the obtained value $ESD = 2.55$ is less than $ESD_{.05} = 2.78$, the null hypothesis cannot be rejected at the .05 level. Thus, the test would not allow the researcher to conclude that the score of 84 represents an outlier if .05 had been the prestipulated value for alpha. We would then examine the next most extreme score (i.e., 120) and determine whether or not the ESD value computed for that score allowed the researcher to declare it an outlier, which, in fact, it does. Consequently, the score of 120 and the even more extreme score of 560 would both be declared as outliers. The reader should take note of the fact that if the value of n_{max} had been set equal to three, and it had turned out, in fact, that the ESD value obtained for the score of 84 was significant at the .05 level, the researcher would then be able to declare the score of 84 an outlier, and, in addition, any score more extreme than 84 would also automatically be declared an outlier (i.e., the scores of 120 and 560).

The rule to be employed when the **extreme Studentized deviate test** is used to detect the presence of more than one outlier in a sample can be summarized as follows: The test statistic should be computed for each of the n_{max} most extreme scores. If a significant ESD value is obtained for the least extreme of the n_{max} scores, all of the n_{max} scores are declared to be outliers. If a nonsignificant result is obtained for the least extreme score, the ESD value for the next most extreme score is evaluated. If the result for the next most extreme of the n_{max} scores is significant, that score and any scores more extreme than it are declared to be outliers. If a nonsignificant result is obtained for the next most extreme of the n_{max} scores, the result for next most extreme score is evaluated, and so on.

It should be emphasized that because of the **masking effect**, it will sometimes be the case that when there are multiple outliers in a distribution, use of the **extreme Studentized deviate test** on the most extreme score or even the second most extreme score will not yield a significant result due to the inflated value of $\tilde{s}$. However, when a less extreme score, which is also suspected of being an outlier, is evaluated, it may yield a significant result. In such a case, the latter score and the two more extreme scores are all declared to be outliers. The reader should also take note of the fact that in the example under discussion, the most extreme scores all happen to fall above the mean of the distribution. However, it is also possible for all of the n_{max} most extreme scores to fall below the mean, or that simultaneously, one or more of the n_{max} most extreme scores can fall above the mean and one or more of the n_{max} most extreme scores can fall below the mean.

Although the assumption of normality with respect to the data distribution under discussion may be questionable, the same set of data will now be evaluated through use of the **median absolute deviation test**. We initially determine that the **median** of the sample (i.e., the middle score, which is the score in the 13th ordinal position if the 25 scores are arranged ordinally) is 18. Starting with the lowest score of 4 and ascending to the highest score of 560, the absolute deviations of the 25 scores from the median are noted below.

Absolute deviation scores

14, 14, 13, 12, 12, 11, 8, 7, 6, 4, 3, 1, 0, 4, 6, 14, 22, 24, 32, 36, 42, 44, 66, 102, 542

When the above noted 25 absolute deviation scores are arranged ordinally, the **median of the absolute deviation scores** is determined to be 13 (since when the 25 absolute deviation scores are arranged ordinally, 13 is the absolute deviation score in the middle/13th ordinal position).

Absolute deviation scores arranged ordinally

0, 1, 3, 4, 4, 6, 6, 7, 8, 11, 12, 12, <u>13</u>, 14, 14, 14, 22, 24, 32, 36, 42, 44, 66, 102, 542

Thus, the value 13 will be employed to represent *MAD* in Equation 11.20. It turns out that the three most extreme scores — specifically, the scores of 560, 120, and 84 — can all be declared as outliers, since when the latter three scores are substituted in Equation 11.20 they yield a test statistic greater than $Max = 5$: a) $|560 – 18|/13 = 41.69$; b) $|120 – 18|/13 = 7.85$; c) $|84 – 18|/13 = 5.08$. Note that this result is not totally consistent with that obtained when the **extreme Studentized deviate test** was applied to the same set of data. Use of the latter test only allowed for the two most extreme values 560 and 120 to be declared as outliers. As noted earlier in the discussion, it is not uncommon that when two or more of the available tests for evaluating outliers are applied to the same set of data, the results may not be totally consistent with one another. In view of the latter, the reader may want to consult more detailed sources (such as Barnett and Lewis (1994)) to determine the most appropriate test for detecting outliers in a given set of data. Ultimately, however, it will be up to the researcher to make the final decision with regard to which if any scores in a sample should be declared as outliers.

Tabachnick and Fidell (1989, 1996) note that outliers present a researcher with a statistical problem if one elects to include them in the data analysis, and a moral problem if one is trying to decide whether or not to eliminate them from the analysis. In essence, how a researcher deals with outliers should depend on how one accounts for their presence. In instances where a researcher is reasonably sure that an outlier is due to any one of the following, Tabachnick and Fidell (1989, 1996) state that one has a strong argument for dropping such a score from the data: a) There is reason to believe an error was made in recording the score in question (either human recording error or instrumentation error); b) There is reason to believe the score is the result of failure on the part of a subject to follow instructions, or other behavior on the part of the subject indicating a lack of cooperation and/or attention to the experiment; and c) There is reason to believe the score is the result of failure on the part of the experimenter to employ the correct protocol in obtaining data from a subject.

One disadvantage of removing outliers from data is that it reduces the sample size, and the latter can result in a decrease in statistical power. Of course, if one is conducting a parametric test such as a *t* test, one may counteract the aforementioned loss in power by virtue of the fact that removal of an outlier will decrease the variability in the data. Tietjen (1986) notes that rejecting an outlier on a purely statistical basis may only be an indication that the data are derived from a population other than the one the researcher assumes to be the parent population. An example cited by Kruskal (1960) is employed by Tietjen (1986) to demonstrate that whether or not an observation should be considered an outlier will be a function of the hypothesis under study. The example used involves a study in which the concentration of a specific chemical in a mixture is being evaluated. Measures of the chemical are derived from five different samples, but one of the measures is way out of line in relation to the others. Let us assume that the latter measure is the result of instrumentation error. If the purpose of the study is to estimate the concentration of the chemical in the mixture, the researcher may elect to label the atypical measure an outlier, and only use the remaining four measures to compute an average value. If, instead, the purpose of the study is to assess the reliability of the instrument which is employed to measure the concentration of the chemical, the outlier observation is relevant and should be retained. Kruskal (1960) emphasizes the fact that the presence of one or more outliers in a set of data may indicate something that is of practical or theoretical relevance, and because of the latter, he recommends it should be standard protocol for researchers to report the presence of any outliers, regardless of whether or not they elect to employ them in the analysis of the data.

In the final analysis, excluding data which are deemed to be outliers is an extreme measure which entails the risk of mistakenly eliminating valid information about an underlying population.[27] At the other extreme, a researcher can include in the analysis all of the data, including observations which are viewed as outliers. Obviously, the latter strategy increases the likelihood the sample will be contaminated, and therefore what may emerge from the analysis may be a gross distortion of what is true regarding the underlying population under study.

When a researcher has reservations about employing either extreme — removal of outliers versus their inclusion — an alternative strategy known as **accommodation** may be employed. The latter involves the use of a procedure which utilizes all the data, but at the same time minimizes the influence of outliers. Two obvious options within the framework of accommodation which reduce the impact of outliers are: a) Use of the median in lieu of the mean as a measure of central tendency; b) Employing an inferential statistical test that uses rank-orders instead of interval/ratio data.

Accommodation is often described within the context of employing a robust statistical procedure (e.g., a procedure that assigns weights to the different observations when calculating the sample mean). Two commonly used methods for dealing with outliers (which are sometimes discussed within the context of accommodation) are **trimming** and **Winsorization**.

Trimming data   **Trimming** involves removing a fixed percentage of extreme scores from each of the tails of any of the distributions that are involved in the data analysis. As an example, in an experiment involving two groups, one might decide to omit the two highest and two lowest scores in each group (since by doing this, any outliers in either group would be eliminated). Sprent (1993) notes that common trimming levels are the top and bottom deciles (i.e., the extreme 10% from each tail of the distribution) and the first and fourth quartiles (i.e., the extreme 25% from each tail). The latter trimmed mean (i.e., the mean computed by only taking into account scores that fall in the middle 50% of the distribution) is often referred to as the **interquartile mean**. In addition to using trimming for reducing the impact of outliers, it is also employed when a distribution has **heavy or long tails** (i.e., a relatively flat distribution with a disproportionate number of observations falling in the tails). Other sources which provide guidelines to employ in trimming are Anderson (2001), Lee and Fung (1983) (who suggest a general trimming proportion of 15%), Rocke *et al.* (1982), and Wilcox (2003). Grissom and Kim (2005, p. 36–40) describe a method developed by Yuen (1974) for computing a confidence interval for the difference between two trimmed means.

Winsorization   Sprent (1993, p. 69) notes that the rationale underlying **Winsorization** (which Grissom and Kim (2005, p. 16) note is named for the statistician Charles Winsor) is that the outliers may provide some useful information concerning the magnitude of scores in the distribution, but at the same time may unduly influence the results of the analysis unless some adjustment is made. Winsorization involves replacing a fixed number of extreme scores with the score that is closest to them in the tail of the distribution in which they occur. As an example, in the distribution 0, 1, 18, 19, 23, 26, 26, 28, 33, 35, 98, 654 (which has a mean value of 80.08), one can substitute a score of 18 for both the 0 and 1 (which are the two lowest scores), and a score of 35 for the 98 and 654 (which are the two highest scores). Thus, the Winsorized distribution will be: 18, 18, 18, 19, 23, 26, 26, 28, 33, 35, 35, 35. The mean of the Winsorized distribution will be 26.17.

Barnett and Lewis (1994) note that the critical problem associated with trimming and Winsorization is selecting the number of scores that are trimmed or Winsorized. Grissom and Kim (2005, p. 17) note that, in the case of Winsorization, the most typical proportions employed for determining the number of scores to be deleted from each end of the distribution are .1, .2, or .25. If r is assumed to represent the actual number of scores to be trimmed or Winsorized in the right tail, and that l represents the number of scores to be trimmed or Winsorized in the left tail, if $r = l$ the trimming or Winsorization process is considered **symmetric**. If $r \neq l$, the trimming or Winsorization process is considered **asymmetric**. The issue of whether or not a researcher believes one or both tails of a sample distribution may be contaminated is one of a number of considerations which have to be taken into account in determining the most appropriate procedure to use. In Chapter 5 of their book, Barnett and Lewis (1994) address such issues, as well as modified trimming and Winsorization procedures. Wilcox (2003) also provides a good overview of the latter subjects.

Data transformation   Data transformation involves performing a mathematical operation on each of the scores in a set of data, and thereby converting the data into a new set of scores which are then employed to analyze the results of an experiment. In addition to being able to reduce the impact of outliers, a data transformation can be employed to equate heterogeneous group variances, as well as to normalize a nonnormal distribution. In point of fact, a transformation which results in homogeneity of variance, at the same time often normalizes data. Kirk (1995) and Winer *et al.* (1991) note that another reason for employing a data transformation is to insure that certain factorial designs are based on an **additive** model. A discussion of **factorial designs** and the concept of **additivity** can be found in Section V of the **between-subjects factorial analysis of variance (Test 27)**.

It is not uncommon that, as a result of a data transformation, data which will not yield a significant effect may be modified so as to be significant (or vice versa). Because of the latter, one might view a data transformation as little more than a convenient mechanism for "cooking" data until it allows a researcher to achieve a specific goal. While the latter may be true, when used judiciously, data transformation can be a valuable tool. One should consider the fact that the selection of the unit of measurement for a dependent variable will always be somewhat arbitrary. Howell (2002) cites numerous examples of transformations which are employed within various experimental settings because of the practical or theoretical advantage they provide. Illustrative of such transformations are the use of decibels in measuring sound intensity and the Richter scale in measuring the magnitude of energy for an earthquake (both of which are based on logarithmic data transformations). Another example of a data transformation is a set of test scores which are converted into percentiles rather than reporting the original raw scores (i.e., number of items correct) for subjects.

Although, as noted earlier, a data transformation can be misused to distort data to support a hypothesis favored by an experimenter, when employed judiciously, it can be of value. Specifically, under certain circumstances it can allow a researcher to provide a more accurate picture of the populations under study than will the analysis of untransformed data. Among those sources which describe and/or discuss data transformations in varying degrees of detail are Anderson (2001), Howell (2002), Kirk (1995), Myers and Well (1995), Tabachnick and Fidell (1989, 1996), Thöni (1967), and Winer *et al.* (1991). Articles by Games (1983, 1984) and Levine and Dunlap (1982, 1983) address some of the controversial issues surrounding data transformation.

Among the most commonly employed data transformations, all of which can reduce the impact of outliers (as well as normalize skewed data and/or produce homogeneity of variance), are converting scores into their **square root**, **logarithm** (a general discussion of logarithms can be found in Endnote 14 in the **Introduction**), **reciprocal**, and **arcsine**. In employing data transformations, a researcher may find it necessary to compare one or more different transformation procedures until he finds the one which best accomplishes the goal he is trying to achieve. Common sense would suggest that selection of a data transformation procedure can be based upon what it previously has been demonstrated to be successful at. In addition to the latter, Kirk (1995) recommends the following protocol for selecting a data transformation procedure: a) Apply each of the available transformation procedures to the largest and smallest scores in each of the experimental treatments/groups; b) Determine the range of values within each treatment, and compute within each treatment the ratio of the largest to the smallest value; and c) Employ the transformation procedure which yields the smallest ratio.

The most commonly employed data transformation procedures will now be discussed and demonstrated.

Square-root transformation   A **square-root transformation** may be useful when the mean is proportional to the variance (i.e., the proportion between the mean of a treatment and the variance of a treatment is approximately the same for all of the treatments). Under such circumstances the square-root transformation can be effective in normalizing distributions that have a moderate positive skew, as well as making the treatment variances more homogeneous. This is the case since a square-root scale will reduce the magnitude of difference between the two tails of a positively skewed distribution by pulling the right side of the distribution in toward the middle. Reaction time is a good example of a measure in psychology which characteristically exhibits a strong positive skew (i.e., there are many fast or relatively fast reactors but there are also a few slow reactors). Consequently a square-root transformation may be able to normalize a set of reaction time data. (If the square-root transformation is not successful, a logarithmic or reciprocal transformation discussed below may be more suitable for achieving this goal.) In such a case the square-root transformation can reduce skewness and stabilize distributional variance. Data taken from a Poisson distribution (discussed in Section IX (the **Addendum**) of the **binomial sign test for a single sample**) are sometimes effectively normalized with a square-root transformation. Such data typically consist of frequencies of randomly occurring objects or events that have a small probability of occurring over many trials or over a long period of time. In Poisson distributed data, the mean and variance are proportional (in fact, they are equal).

The square-root transformation is obtained through use of the equation $Y = \sqrt{X}$, where X is the original score and Y represents the transformed score. However, when any value of X is less than 10, any or all of the following equations are recommended by various sources in place of $Y = \sqrt{X}$, since they are more likely to result in homogeneous variances: a) $Y = \sqrt{X + .5}$; b) $Y = \sqrt{X} + \sqrt{X + 1}$; and c) $Y = \sqrt{X + .375}$.

Logarithmic transformation   A **logarithmic transformation** may be useful when the mean is proportional to the standard deviation (i.e., the proportion between the mean of a treatment and the standard deviation of a treatment is approximately the same for all of the treatments). Under such circumstances the logarithmic transformation can be effective in normalizing distributions that have a moderate positive skew, as well as making the treatment variances more homogeneous. Since a logarithmic transformation makes a more extreme adjustment than a square-root transformation, it can be employed to normalize distributions which have a more severe positive skew. On a logarithmic scale, the distance between adjacent points on the scale will be less than the distance between the corresponding points on the original scale of measurement. As is the case with the square-root transformation, the logarithmic transformation is often useful for normalizing a dependent variable that is a measure of response time.

The logarithmic transformation is obtained through use of the equation $Y = \log X$. Since a logarithm cannot be computed for the value zero, when one or more zeros or positive numbers close to zero are present in a set of data, the following equation is employed: $Y = \log(X + 1)$. Since a logarithm cannot be computed for a negative number, a constant (the value of which is a positive number that is minimally greater than one unit above the absolute value of the lowest negative number) can be added to all of the values in a set of data, to insure that each value will be a positive number. In employing a logarithmic transformation, it does not matter what base value is employed for the logarithm — some sources employ the base 10 while others use the base $e = 2.71828$ (see Endnote 14 in the **Introduction** for a clarification of what a base value of a logarithm represents).

Reciprocal transformation A **reciprocal transformation** (also referred to as an **inverse transformation**) may be useful when the square of the mean is proportional to the standard deviation (i.e., the proportion between the square of the mean of a treatment and the standard deviation of a treatment is approximately the same for all of the treatments). Under such circumstances the reciprocal transformation can be effective in normalizing distributions that have a moderate positive skew, as well as making the treatment variances more homogeneous. Since it exerts the most extreme adjustment with regard to normality, the reciprocal transformation is often able to normalize data that the square-root and logarithmic transformations are unable to normalize. Tabachnick and Fidell (1996) recommend the reciprocal transformation for normalizing a J-shaped distribution (i.e., a distribution which looks like the letter J or its mirror image — specifically, an extremely skewed unimodal distribution that is peaked without a tail at one end, and with a tail falling off toward the other end).

The reciprocal transformation is obtained through use of the equation $Y = 1/X$. If any of the scores are equal to zero, the equation $Y = 1/(X + 1)$ should be employed. Additional comments on the reciprocal transformation can be found at the end of the discussion on data transformations.

At this point some hypothetical data will be employed to demonstrate the **square-root**, **logarithmic**, and **reciprocal transformations**. To illustrate the application of the **square-root transformation**, assume we have two groups, with five subjects per group. The interval/ratio scores of the subjects in the two groups follow: **Group 1**: 2, 3, 4, 6, 10; **Group 2**: 10, 20, 20, 25, 25. Employing Equations I.1, I.11, and I.8, the sample means, estimated population standard deviations, and estimated population variances are computed to be $\bar{X}_1 = 5$, $\tilde{s}_1 = 3.16$, $\tilde{s}_1^2 = 10$; $\bar{X}_2 = 20$, $\tilde{s}_2 = 6.12$, $\tilde{s}_2^2 = 37.5$. Note that in each group the estimated population variance is approximately two times as large as the group mean. Let us assume we wish to make the variances in the two groups more homogeneous, since the variance of Group 2 is almost four times as large as the variance of Group 1.[28] Since the estimated population variances and means are proportional, we elect to employ a square root transformation. Since some of the scores in Group 1 are less than 10, we will employ the equation $Y = \sqrt{X + .5}$ to convert each score. The resulting corresponding transformed scores for the subjects in the two groups are: **Group 1**: 1.581, 1.871, 2.121, 2.550, 3.240; **Group 2**: 3.240, 4.528, 4.528, 5.050, 5.050. Employing Equations I.1 and I.8 with the transformed scores, the sample means (employing the notation $\bar{Y}_j$) and estimated population variances are computed to be $\bar{Y}_1 = 2.27$, $\tilde{s}_1^2 = .42$; $\bar{Y}_2 = 4.48$, $\tilde{s}_2^2 = .55$. Note that the estimated population variances are now almost equal — specifically, the variance of Group 2 is only 1.31 times larger than the variance of Group 1. When the **logarithmic transformation** is applied to the same set of data (using the equation $Y = \log X$), it also results in more homogeneous variances, with the variance of Group 1 being 2.70 times larger than the variance of Group 2. On the other hand, when the **reciprocal transformation** is applied to the data, it increases heterogeneity of variance (with the reciprocal transformed variance of Group 1 being 88.89 times larger than the reciprocal transformed variance of Group 2). It would appear that the square-root transformation is the most effective in equating the variances.

To illustrate the application of the **logarithmic transformation**, assume we have two groups, with five subjects per group. The interval/ratio scores of the subjects in the two groups follow: **Group 1**: 12, 14, 16, 18, 20; **Group 2**: 28, 32, 40, 50, 50. Employing Equations I.1, I.11 and I.8, the sample means, estimated population standard deviations, and estimated population variances are computed to be $\overline{X}_1 = 16$, $\tilde{s}_1 = 3.16$, $\tilde{s}_1^2 = 10$; $\overline{X}_2 = 40$, $\tilde{s}_2 = 10.10$, $\tilde{s}_2^2 = 102$. Note that in each group the estimated population standard deviation is between one-fourth to one-fifth the size of the group mean. Let us assume we wish to make the variances in the two groups more homogeneous, since the variance of Group 2 is about ten times as large as the variance of Group 1. Since the estimated population standard deviations and means are proportional, we elect to employ a logarithmic transformation (using the equation $Y = \log X$, which employs the base 10 for the logarithm). The resulting corresponding transformed scores for the subjects in the two groups are: **Group 1**: 1.079, 1.146, 1.204, 1.255, 1.301; **Group 2**: 1.447, 1.505, 1.602, 1.699, 1.699. Employing Equations I.1 and I.8 with the transformed scores, the sample means (employing the notation $\overline{Y}_j$) and estimated population variances are computed to be $\overline{Y}_1 = 1.197$, $\tilde{s}_1^2 = .008$; $\overline{Y}_2 = 1.590$, $\tilde{s}_2^2 = .013$. Note that the ratio of the larger variance ($\tilde{s}_2^2 = .013$) to the smaller variance ($\tilde{s}_1^2 = .008$) is now only 1.68. When the **square-root transformation** is applied to the same set of data (using the equation $Y = \sqrt{X}$, since none of the scores is less than 10), it also results in more homogeneous variances, with the variance of Group 2 being 4.13 times larger than the variance of Group 1. The **reciprocal transformation** applied to the same data also results in more homogeneous variances, with the variance of Group 1 being 3.47 times larger than the variance of Group 2. It would appear that although all three transformations make the variances more homogeneous, the logarithmic transformation is the most effective.

To illustrate the application of the **reciprocal transformation**, assume we have two groups, with five subjects per group. The interval/ratio scores of the subjects in the two groups follow: **Group 1**: 2, 3, 4, 6, 10; **Group 2**: 1, 1, 3, 5, 90. Employing Equations I.1, I.11 and I.8, the sample means, estimated population standard deviations, and estimated population variances are computed to be $\overline{X}_1 = 5$, $\tilde{s}_1 = 3.16$, $\tilde{s}_1^2 = 10$; $\overline{X}_2 = 20$, $\tilde{s}_2 = 39.17$, $\tilde{s}_2^2 = 1534$. Note that in each group the square of the mean is between 8 to 10 times the size of the estimated population standard deviation. Let us assume we wish to make the variances in the two groups more homogeneous, since the variance of Group 2 is approximately 153 times as large as the variance of Group 1. Since the square of the means and the estimated standard deviations are proportional, we elect to employ a reciprocal transformation (using the equation $Y = 1/X$). The resulting corresponding transformed scores for the subjects in the two groups are: **Group 1**: .5, .333, .25, .167, .1; **Group 2**: 1, 1, .333, .2, .011. Employing Equations I.1 and I.8 with the transformed scores, the sample means (employing the notation $\overline{Y}_j$) and estimated population variances are computed to be $\overline{Y}_1 = .27$, $\tilde{s}_1^2 = .024$; $\overline{Y}_2 = .51$, $\tilde{s}_2^2 = .21$. Note that although the estimated population variances are still not equal, they are considerably closer than the values computed for the untransformed data — specifically, as a result of the transformation the variance of Group 2 is 8.92 times larger than the variance of Group 1. When the **square-root transformation** (using the equation $Y = \sqrt{X + .5}$, since some of the scores are less than 10) is applied to the same set of data, the variance of Group 2 is 29.92 times larger than the variance of Group 1. When the **logarithmic transformation** is applied to the same set of data (using the equation $Y = \log X$), the variance of Group 2 is 8.8 times larger than the variance of Group 1. It would appear that although none of the three transformations is able to result in homogeneous group variances, the logarithmic and reciprocal transformations come closest to achieving that goal.

In the case of this latter example, the score of 90 in Group 2 would appear to be a possible outlier, and, as such, might not be included in the data or have its impact altered in some way through use of some method of accommodation designed to reduce the impact of an outlier. It should also be noted that use of the degree of proportionality among the means, squared means, standard deviations, and/or variances will not be the only factor that can be employed to determine what transformation will be most effective. Often, through use of trial and error, a researcher can investigate which, if any, transformation will best achieve the desired goal.

Arcsine (arcsin) transformation An **arcsine transformation** (also referred to as an **angular** or **inverse sine transformation**) involves the use of a trigonometric function which is able to transform a proportion between 0 and 1 (or percentage between 0% and 100%) into an angle expressed in radians (1 radian = 57.3 degrees, which is equal to $180°/\pi$, and one degree equals .01745 radians). The arcsine of a number is the angle whose sine is that number. Although some books contain tables of arcsine values, an arcsine can be computed on many calculators through use of the sin^{-1} key. An arcsine transformation may be useful for normalizing distributions when the means and variances are proportional, and the distributions are binomially distributed. Howell (2002) notes that although both the square-root and arcsine transformations are suitable when the means and variances are proportional, whereas the square-root transformation compresses the upper tail of the distribution, the arcsine transformation flattens the distribution by stretching out both tails.

The arcsine transformation is obtained through use of the equation $Y = 2 \arcsin\sqrt{X}$, where X will be a proportion between 0 and 1. Based on a paper by Bartlett (1947), sources recommend that the equation $Y = 2 \arcsin\sqrt{X + (1/2n)}$ (or $Y = 2 \arcsin\sqrt{X + (1/4n)}$) be employed when the value of X is equal to 0, and that the equation $Y = 2 \arcsin\sqrt{X - (1/2n)}$ (or $Y = 2 \arcsin\sqrt{X - (1/4n)}$) be employed when the value of X is equal to 1 (where n is the number of observations in the treatment for which a proportion is computed). All of the aforementioned equations yield a value of Y that is expressed in radians. To illustrate the arcsine transformation, assume we have the following five values which represent the proportion of bulbs that bloom in each of five flower beds: .25, .39, .5, .68, .75. When the latter values are employed in the equation $Y = 2 \arcsin\sqrt{X}$, the following values (in radians) are computed for Y: 1.0472, 1.3490, 1.5708, 1.9391, 2.0944. The possible range of values which Y can equal through use of any of the equations noted above is 0 radians (for a proportion of zero) to 3.1416 radians (for a proportion of 1). (Note that the value 3.1416 is equal to pi.)

Some sources (e.g., Myers and Well (1995), Rao (1998) and Zar (1999)) employ the following alternative arcsine transformation equation which expresses the value of Y in degrees: $Y = \arcsin\sqrt{X}$. At the conclusion of the data analysis, the transformed values can be converted back into the original proportions through use of the equation $X = (\sin Y)^2$. To illustrate the arcsine transformation using the alternative equation, assume we have the same five values representing the proportion of bulbs that bloom in each of five flower beds (i.e., .25, .39, .5, .68, .75). When the latter values are employed in the equation $Y = \arcsin\sqrt{X}$, the following values (in degrees) are computed for Y: 30, 38.65, 45, 55.55, 60.[29] Zar (1999) notes additional alternative equations for computing the value of Y in degrees. The possible range of values that Y can equal through use of any of the equations which compute an arcsine in degrees is $0°$ (for a proportion of zero) to $90°$ (for a proportion of 1). Additional information on the arcsine transformation, as well as examples illustrating its application, can be found in Rao (1998) and Zar (1999).

$Y = X^2$ **transformation** Zar (1999) notes that if there is an inverse relationship between the treatment standard deviations and the treatment means (i.e., as the standard deviations increase, the means decrease) and/or a distribution is negatively skewed, the following transformation may be useful for normalizing data and creating homogeneity of variance: $Y = X^2$.

Final comments on data transformation Some data transformations, such as a reciprocal transformation, may result in a reversal in the direction of the scores. In other words, if we have two scores a and b with $a > b$, if we obtain the reciprocal of both, the reciprocal of b will be greater than the reciprocal of a. The process of reversing the direction to restore the original ordinal relationship between the scores is called **reflection**. Reflection can also be used to convert negatively skewed data into positively skewed data. Tabachnick and Fidell (1996) note the latter can be accomplished by doing the following: a) Create a constant that is larger than all of the scores in a distribution by adding the value 1 to the largest score in the distribution; b) Subtract each score in the distribution from the constant value. At this point the converted data will have a positive skew and the appropriate transformation for normalizing positively skewed data (e.g., square-root, logarithmic, and reciprocal transformations) can be employed. After the appropriate statistical test has been employed with the normalized data, the researcher must remember to take into account the reversal in the direction of scoring in interpreting the results.

Although it is not generally discussed within the context of data transformation, the conversion of interval/ratio data into ranks (as is commonly done within the framework of many nonparametric tests) can be viewed as representing another example of transforming data. The latter point is noted by Anderson (2001, pp. 358–359) and Conover and Inman (1981), who point out that parametric tests conducted on data which have been converted to ranks generally yield results which are comparable to the results obtained when the same parametric test is employed with the original interval/ratio data.

In the final analysis, as with anything else, a data transformation should be judged on the basis of its practical consequences. Specifically, if through use of a data transformation a significant result is obtained and that result can be consistently replicated employing the same data transformation, a researcher can conclude that one is dealing with a reliable phenomenon. If the data obtained through use of a data transformation proves to be useful in a practical or theoretical sense, it is as valuable as data which when analyzed do not require any sort of transformation.

4. Missing data A problem which researchers are commonly confronted with is the absence of data for one or more subjects. The appropriate way to deal with missing data should ultimately be a function of whatever reasons a researcher believes account for the lack of data. Among the more common reasons for missing data are: a) Clerical error in recording data; b) Inability to collect all of the data for an experiment, due to the fact that some subjects are unavailable at the time data are collected; and c) Some subjects refuse to provide a researcher with a full set of data. The latter may be due to the fact that a subject's response on a variable may be viewed by the subject as an invasion of one's privacy. A subject might also elect not to cooperate with an experimenter for a variety of other reasons — e.g., resentment at having to participate in an experiment against one's will, a desire to escape an experimental situation as quickly as possible and consequently not responding conscientiously, thereby generating an atypical score (which may often represent an outlier within a set of data). In certain situations, such as in survey research, a subject may not provide an experimenter with a response, since she may not have an opinion or lacks the necessary information needed to provide a response.

Prior to doing anything about missing data a researcher should determine whether or not data that are missing are random in nature or whether they are characterized by a systematic pattern. If a researcher determines that subjects for whom data are missing are different in some respect from subjects who generate a full set of data, the ensuing statistical analysis of whatever data are available may be biased. As an example, if all subjects who have missing data are members of a specific demographic group (e.g., senior citizens), by omitting them from the analysis a researcher may arrive at different conclusions with respect to the relationship between the independent and dependent variables than would be the case if data for the latter group of subjects had been available.

A number of strategies will now be described for dealing with missing data. The simplest and most obvious strategy is to omit from an analysis any subject for whom all data are not available. In instances where each subject is required to contribute two or more pieces of data, the researcher may elect to eliminate from the analysis any subject who does not provide a full set of data, or, instead, evaluate for each variable whatever data are available (regardless of whether or not each subject has provided a full set of data). When the overall amount of data is large (especially when the pattern of missing data is random), omission of a few subjects/pieces of data from an analysis will generally not cause serious problems. However, as the amount of data omitted from an analysis increases, in addition to increasing the likelihood of generating unreliable results, there is a corresponding increase in the reduction of the power of the analysis due to a reduction in sample size. Harlow (2005, p. 32) notes that deleting data for all participants who have any missing data is the most extreme and least desirable approach to dealing with the problem. She advises against the latter strategy unless the sample size is very large (e.g., $n > 500$) and the percent of missing data is less than 5%.

Another approach for dealing with missing data is to employ one of a number of available **imputation methods**. The latter represent procedures which allow a researcher to estimate values for missing data. The method employed for estimating missing data can be based on a statistic computed for a specific variable through use of all available data on that variable for the sample employed in an experiment. Alternatively, utilization of preexisting relevant information on the relevant variable and/or the subject for whom data are missing can be employed as a basis for estimating the missing scores. Among the **imputation methods** available are the following (which are discussed in greater detail in sources such as Hair *et al.* (1995, 1998) and Tabachnick and Fidell (1996)).

Case substitution This method involves identifying a subject not employed in the original sample who is similar to the subject for whom data are missing. The necessary data are then obtained from the substitute subject, and the latter values are employed to represent any data not provided by the original subject

Mean substitution Tabachnick and Fidell (1996) note that use of the mean value (which Harlow (2005, p. 33) says is probably the most common method of dealing with missing data) represents a conservative strategy for estimating missing data. This method involves employing the mean value obtained for a variable from all available data in a sample to represent any missing data in the sample. An even more conservative alternative is to employ the population mean for a variable, if the latter value is available. A problem with the mean substitution method is that the use of the mean to represent all missing data in a sample will result in an underestimation of the population variance, as well as compromising the accuracy of any visual distribution of the data a researcher constructs. Because it underestimates the variance of a variable, the mean substitution method also reduces the magnitude of the correlation between that variable and other variables. (A full discussion of the concept of correlation can be found in the chapter on the **Pearson product-moment correlation coefficient.**)

Substitution of some value other than the mean A researcher may elect to employ some value other than the mean to represent missing data for a given variable. For instance, in the case of a skewed distribution or one in which outliers are present, a researcher might elect to employ the median instead of the mean. Alternatively, use of some value other than the mean might be predicated on the fact that preexisting information available to the researcher (based on research or some other source) indicates that a such a value might be optimal for estimating a subject's score on the relevant variable. Tabachnick and Fidell (1996) note that when prior knowledge is used to predict missing data, a researcher may elect to downgrade the level of measurement on a given variable, if the latter allows one to predict with a greater degree of confidence. For example, a researcher may elect to convert all of the interval/ratio scores on a variable into a categorical format if she will have more confidence in categorical estimates of missing data for that variable. Of course, by downgrading the level of measurement, the researcher sacrifices information on a variable and, among other things, compromises the power of a subsequent statistical analysis.

Use of regression analysis to predict missing data This method allows a researcher to estimate the score of a subject on a missing variable through use of an equation (referred to as a **regression equation**) which employs a previously determined statistical relationship between that variable and one or more additional variables (which are referred to as **predictor variables**). When the latter method is employed, it assumes that data are available on the predictor variable(s) for a subject for whom the researcher wishes to estimate data. The stronger the correlation (i.e., statistical relationship) between the missing data variable and the predictor variable(s), the more accurate the estimates will be for missing data. When there is little or no correlation between the missing data variable and the predictor variable(s), use of regression analysis will, for all practical purposes, yield the same results obtained when the mean is employed to represent missing data. A general discussion of regression analysis can be found in Section VI of the **Pearson product-moment correlation coefficient**.

Resampling procedures Another option for replacing missing data is through use of the **bootstrap**, which is a **resampling procedure** described in Section IX (the **Addendum**) of the **Mann–Whitney *U* test**. **Resampling** is a process in which subsets of scores are selected from an original set of data. Within this context, the computer would employ the available data from a given sample to represent the population from which scores are randomly selected to replace the missing data in that sample.

An alternative option is to have the computer generate values for missing data which are based on preexisting information available on one or more population parameters, or are based on one or more statistics computed from the sample data. To illustrate, the following two-step procedure can be employed: a) If the researcher does not know the population mean and/or standard deviation, employing all available data within a sample, the researcher computes the mean and/or estimated population standard deviation for the sample. If the population mean and/or standard deviation are known, the latter values are employed in lieu of the computed sample statistics; b) A random number generator (discussed in Section IX (the **Addendum**) of the **single-sample runs test**) is employed to compute a random value to represent each missing datum. The population distribution from which the random numbers are generated will have a mean and standard deviation equivalent to the mean and estimated population standard deviation computed from the sample data (or the actual population mean and standard deviation, if the latter are known). Unless the researcher has reason to believe otherwise, the distribution of the population distribution for which data are generated is assumed to be normal. If the researcher has reason to believe that the distribution is other than normal, the appropriate distribution is employed to generate the data.

It would certainly be useful to conduct computer based simulation studies (discussed in Section VII of the **single-sample runs test**) to contrast the performance of the above described methods for dealing with missing data. As an example, the following methodology can be employed in conducting a simulation study. Initially, the researcher stipulates the mean and standard deviation of one or more hypothetical populations, and the value(s) of the sample size(s) to employ in a simulation study. Initially, a full set of data is generated for the simulated experiment, and the latter data are evaluated through use of the appropriate statistical test(s). One or more pieces of datum are then randomly deleted from the original data. Employing each of the above described methods for dealing with missing data, the revised sets of data are once again evaluated with the appropriate statistical test(s). The results of the analyses adjusted for missing data are then compared with the results of the initial analysis involving the full set of data.

In closing this section, the reader should take note of the following regarding missing data: a) Bohj (1978) has developed an interesting methodology (which is described in Wilcox (1987)) for dealing with missing data, when one or more scores are not available in a design involving two dependent samples (i.e., a design in which two scores on a dependent variable are obtained from each subject). Bohj's (1978) methodology involves the use of both a **t test for two independent samples** and a **t test for two dependent samples (Test 17)**, and pooling the results of the two tests; b) **Censored data**, which can also be viewed as a form of missing data, is discussed in Section IX (the **Addendum**) of the **Mann–Whitney U test**; c) Hair *et al.* (1995, 1998) and Tabachnick and Fidell (1996) provide excellent discussions of issues and strategies involved in dealing with missing data within the framework of **multivariate analysis** (which refers to procedures that evaluate experimental designs in which there are multiple independent variables and/or multiple dependent variables). The latter authors note that the issue of missing data becomes an even more serious problem in multivariate analysis, since as the number of variables employed in a study increases, the likelihood increases that a full set of data will not be available for all subjects; d) Little and Rubin (1987, 2002) provide the most comprehensive coverage on the subject of missing data.

5. Clinical trials A **clinical trial** refers to an experiment involving human subjects which is designed to evaluate a health related issue (such as the relative efficacy of two types of therapy). Two common experimental designs which are utilized in conducting **clinical trials** are the **parallel group design** and the **crossover design**. The term **parallel group design** is often employed within the context of **clinical trials** to describe an **independent samples design** — in other words, it is a design in which each of n subjects is randomly assigned to one of k treatments. A **crossover design** is a type of **dependent samples design** (such designs are discussed in the chapter on the *t* **test for two dependent samples**) in which each of n subjects receives each of k experimental treatments. In the **two-treatment crossover design** (which is the simplest type of **crossover design**), every subject receives each of two treatments, with half the subjects receiving Treatment 1 first followed by Treatment 2, and the other half receiving the treatments in reverse order. Typically, when such studies are concerned with evaluating drug efficacy, there is a **washout period** for each subject intervening between a subject's two treatments. During the washout period (the duration of which can be anywhere from a few days to a few months) the subject receives no treatment, and it is assumed that during the washout period any residual biological effects derived from the previous treatment dissipate. A more detailed discussion of the **crossover design** (as well as the issues involved in analyzing such designs) can be found in Section VII of the *t* **test for two dependent samples** and in Section IX (the **Addendum**) of the **between-subjects factorial analysis of variance (Test 27)** under the discussion of the **factorial analysis of variance for a mixed design (Test 27i)**.

Participants in clinical trials are typically volunteers who are provided with **informed consent** prior to participating in a study. **Informed consent** provides a subject with the essential facts of a study (e.g., its purpose, duration, procedures involved, risks, and benefits) in order that the subject may decide whether he or she wishes to be a participant. If a subject agrees to participate in a clinical trial, the subject signs an informed consent document. The latter, however, is not a contract, since a subject is allowed to withdraw from a clinical trial at any time.

The purpose of a typical clinical trial is to evaluate an experimental treatment with regard to its efficacy in treating a specific illness. In a simple clinical trial an experimental treatment is compared with a control treatment, which may either be a **placebo** or the most current standard treatment for the illness in question. A **placebo** is an inactive substance (in the form of a pill or liquid) that has no physiological therapeutic impact on the illness in question. Among the types of clinical trials which are conducted are: a) **Treatment trials**, which assess new treatments for medical conditions; b) **Prevention trials**, which search for better ways of preventing people who have never had a disease from contracting it. In such trials, healthy subjects are exposed to one or more potential preventative treatments which are compared with control subjects (who generally receive a no treatment (or a placebo) or make no adjustments in their lifestyle); c) **Diagnostic trials**, which are employed in order to discover superior tests or procedures for diagnosing a medical condition. In such trials subjects who are not believed to have the condition are evaluated with an experimental diagnostic procedure and are compared with control subjects who are evaluated with the current standard diagnostic procedure.

Jennison and Turnbull (2000) note that a clinical trial evaluating a new treatment may last as long as ten years, since, among other things, it may take researchers a prolonged period of time to establish the safety and efficacy of a treatment. Other problems which can prolong a study are turnover in personnel, loss of subjects or difficulty finding subjects who meet eligibility criteria, and flaws identified in the design and/or execution of a study. Economic considerations can also play a major role in influencing the duration of a study. Obviously, if early results indicate a treatment is ineffective, a trial can be terminated in order to limit financial losses. Yet under other conditions, economic factors may motivate a company to extend a study, since if a treatment can be demonstrated to be effective the potential economic gains will be substantial. Within the framework of clinical trials (most frequently treatment trials) the following terminology is employed to describe different stages of a trial.

Phase I trials are exploratory insofar as they are concerned with controlling for the potential toxicity of a treatment. Phase I trials typically employ healthy volunteers who are given an experimental treatment in order to identify side effects and the safe dosage range which avoids serious side effects. Jennison and Turnbull (2000) note that typically between 20 and 80 subjects (who are often employees of a pharmaceutical company or medical students) are employed in this phase.

Phase II trials are a pilot/preliminary study which involves 100 to 300 people afflicted with the illness for which a treatment is being developed. During this phase the new treatment is given to subjects in order to evaluate its efficacy and to further evaluate it safety. In Phase II, as well as Phases III and IV, the term **endpoint** is often employed to designate the dependent variable, which typically involves some measure of therapeutic efficacy. This response can simply be binary in nature (such as cured versus not cured) or may involve a physiological measure which is indicative of the status of a person's health (e.g., the concentration of a specific chemical in a patient's blood). Jennison and Turnbull (2000) note that the term **survival endpoint** is often employed to refer to the time it takes for some event of interest to occur, such as the death of a patient or a relapse of an illness. It is not uncommon, however, that during a clinical trial one or more patients will fail to reach such an endpoint. As noted in Endnote 27, the term **censoring** is employed in reference to data which are not available for some subjects, since it is either not possible or desirable to follow each subject until the conclusion of a study. A detailed discussion of censoring and the analysis of censored data can be found in Section IX (the **Addendum**) of the **Mann–Whitney *U* test** under the discussion of **survival statistics**.

Phase III trials are conducted in order to provide a definitive evaluation of a new treatment. Of the four phases, Phase III (which typically involves 1000 to 3000 subjects) is the one where attention to experimental design and statistical analysis is most critical. In Phase III an experimental group is administered the new treatment and is compared with one or more control groups (which are administered the current standard treatment(s) and/or a placebo).

Phase III trials can last three to five years or longer, and within this phase, along with efficacy, long term side effects are monitored. It should be noted that not all treatment trials are designed to demonstrate the superiority of a new treatment when compared with the current standard treatment for a disease. In many instances a drug company may only want to demonstrate that a new treatment is **equivalent** to the standard treatment with regard to efficacy — the latter is most common when two formulations of the same drug are compared (e.g., comparing a generic version of a drug with the brand name). It is important to make the distinction between concluding that two drugs are equivalent to one another with regard to efficacy versus concluding there are no differences in efficacy between the two drugs. The latter conclusion (which derives from not obtaining a significant difference on the dependent variable for two drugs) could result from low statistical power, and in itself is not commensurate with establishing equivalency (which is discussed in detail later in this section).

One question which must be addressed during Phase III trials is the feasibility and/or desirability of randomly assigning subjects to treatment conditions. Fleiss *et al.* (2003, pp. 175–180) note that researchers are not in agreement with regard to whether or not a **double-blind study** involving random assignment of subjects to treatments should always be the preferred method for conducting a clinical trial. (In a **double-blind study** neither the patient or physician knows which treatment a patient receives. In a **single-blind study**, only the patient does not know which treatment he or she receives.) One situation in which a researcher might be reluctant or unable to randomize the assignment of treatments (as well as not inform subjects of the treatment they will receive) is when patients who are untreated will not survive. In such a case, potential participants in a treatment trial may refuse to participate unless they receives the experimental treatment. Such patients will view assignment to a placebo group or, for that matter, treatment with the current standard method (which may not substantially extend one's chances of survival) unacceptable. Another situation that may militate against randomization is when very few patients are available to participate in a study. Under such circumstances a researcher may elect to employ the experimental treatment with all of the available patients and use archival information on previous patients (who are deceased or cured) as a control. (The latter type of control is often referred to as a **historical control**.)

Another issue which must often be addressed during Phase III is whether to randomly assign subjects to treatments or to **match** subjects on prognostic factors such as age, sex, duration of illness, etc. (A detailed discussion of matching can be found in Section VII of the *t* **test for two dependent samples**.) Matching subjects requires that a researcher initially identify one or more variables (besides the independent variable) which she believes are positively correlated with the dependent variable employed in a study. Such a variable can be used as a matching variable. Each subject who is assigned to one of the k experimental conditions is matched with one subject in each of the other $(k - 1)$ experimental conditions. In matching subjects it is essential that any cohort of subjects who are matched with one another are equivalent (or reasonably comparable) with respect to any matching variables employed in the study. By matching subjects a researcher is able to conduct a more powerful statistical analysis than will be the case if subjects in the k conditions are not matched with one another. Although matching may be desirable it can be time consuming and expensive. Additionally, in studies where matching is employed, there may be differential loss of subjects assigned to different treatments thereby neutralizing any gains obtained as a result of matching.

Phase IV trials are conducted after a drug has been accepted for marketing, and during this phase additional information is accumulated on a new treatment with respect to its safety, efficacy, and usage profile. Phase IV trials are often referred to as **post-marketing surveillance**.

For a more in depth discussion on the general subject of clinical trials the reader should consult sources such as Armitage *et al.* (2002), Chow and Liu (2000) and Chow and Liu (2004).

6. Tests of equivalence: Test 11h: The Westlake–Schuirmann test of equivalence of two independent treatments and **Test 11i: Tryon's test of equivalence of two independent treatments** Within the framework of the **classical hypothesis testing model**, a null hypothesis can never be accepted. Consequently, when a *t* **test for two independent samples** is conducted, retention of the null hypothesis only indicates that any difference detected between two treatments is not large enough to allow a researcher to conclude that the treatments are different from one another. Although the latter statement might suggest retention of the null hypothesis is commensurate with concluding the treatments are equivalent to one another, in point of fact, it is not. As Wellek (2003, p. 2) notes, a nonsignificant difference must not be confused with significant homogeneity (i.e., equivalence of two or more experimental conditions). In order to demonstrate the equivalence of two or more experimental conditions, it is required that a researcher conduct what is referred to as a **test of equivalence**. The latter type of test is most commonly conducted within the framework of a clinical trial.

Wellek (2003) notes that the impetus for interest in tests of equivalence was the approval of generic drugs by the Food and Drug Administration in the United States (as well as by analogous authorities in other countries). With the introduction of generic drugs, researchers were expected to demonstrate their **bioequivalence** with a brand name drug. (**Bioequivalence** refers to equivalence of different formulations of the same drug (Westlake, 1981).) Three common experimental situations in which a researcher might wish to employ a **test of equivalence** are the following: a) The comparison of two mean values, one mean representing the bioavailability of a brand name drug, and the other the bioavailability of a generic drug. If the two drugs are equal with respect to bioavailability they are bioequivalent; b) The comparison of two proportions (or some other measure of efficacy expressed in a form other than a proportion), one proportion representing the success rate associated with a brand name drug, and the other the success rate associated with a generic drug; and c) A **test of equivalence** for goodness-of-fit, which is not the same as the goodness-of-fit tests described earlier in the book (e.g., the **Kolmogorov-Smirnov goodness-of-fit test (Test 7)** and the **chi-square goodness-of-fit test (Test 8)**). Wellek (2003, p. 3) notes the latter types of test are, in fact, designed to address the reverse problem of establishing lack of fit rather than equivalence of fit. Thus, if, in fact, the intent of a goodness-of-fit test is to demonstrate equality of distributions, rejection of the null hypothesis should result in support for the existence of equivalence rather than support for a lack of equivalence (the latter being the case when the **Kolmogorov-Smirnov** and **chi-square goodness-of-fit tests** are evaluated within the context of the **classical hypothesis testing model**).

The discussion of **tests of equivalence** in this book will be limited in scope insofar it will focus on a limited number of procedures that can be employed to evaluate experiments in which two independent means or proportions are contrasted with one another. Readers requiring comprehensive coverage of **tests of equivalence** are referred to sources such as Chow and Liu (2004) and Wellek (2003) (who, to date, provides the most comprehensive, albeit mathematically sophisticated, discussion of this subject). The **Journal of Articles in Support of the Null Hypothesis** (first published in 2002) presents research which fails to find significant differences between experimental conditions. Because of the latter, it is expected that this journal should provide numerous applications involving **tests of equivalence**.

The most obvious difference between the **classical hypothesis testing model** and the hypothesis testing model employed for **tests of equivalence** is reversal of the null and alternative hypotheses. In contrast to the **classical hypothesis testing model** (where the alternative hypothesis states there is a difference between treatments), in a **test of equivalence** the alternative hypothesis states that the treatments are, in fact, equivalent. Conversely, in a **test of**

equivalence the null hypothesis states that a difference exists between the treatments (in contrast to the null hypothesis employed within the framework of the **classical hypothesis testing model**, which states there is no difference between the treatments). Since Wellek (2003, p. 2) and others note it is not mathematically feasible to establish an alternative hypothesis which states exact equality (i.e., a difference of zero) between experimental conditions, when one conducts a **test of equivalence**, a parameter (to be designated ζ, which represents the lower case Greek letter **zeta**) is employed to reflect a maximum difference which will be tolerated between two treatments in order that a researcher might conclude the treatments are equivalent to one another. Any difference less than the value of ζ would be viewed so small as to be inconsequential (i.e., of no practical significance). Thus, in a **test of equivalence** if a difference less than the value of ζ is detected, the null hypothesis (stating a difference does exist) can be rejected, and the alternative hypothesis (which stipulates equivalence) can be accepted. In the discussion to follow, it will be emphasized that since in most instances rejection of the null hypothesis for a **test of equivalence** will involve detection of a substantially smaller effect size than the effect size a researcher is typically concerned with detecting within the framework of the **classical hypothesis testing model**, a **test of equivalence** will usually require greater statistical power, and consequently the use of a larger sample size, when contrasted with tests conducted within the framework of the **classical hypothesis testing model**. The smaller the value a researcher designates for ζ, the larger the sample size which will be required in order to reject the null hypothesis and thus conclude the treatments are equivalent.

Feinstein (2002, p. 519) notes that the **two-zone decision space** employed in the **classical hypothesis testing model** is not consistent with the realities of medical research. Specifically, the two-zone model stipulates a value which will be designated as δ (which, as noted earlier, is the lower case Greek letter **delta**) as the boundary, beyond which a difference between two statistics allows one to declare statistical significance. Thus, any absolute difference equal to or greater than δ will be statistically significant, while any absolute difference less than δ will not be statistically significant. Feinstein (2002) notes that, realistically, clinical decisions should probably be predicated on a **three-zone decision space** such as that illustrated in Figure 11.2. (A similar discussion of such a three-zone decision space, suggested by Keppel (1991), can be found in Section VI of the **single-factor between-subjects analysis of variance (Test 21)**.) If a researcher employs the three-zone decision space depicted in Figure 11.2, the intent of a **test of equivalence** will be to demonstrate that any difference detected between treatments falls in the leftmost zone, which will represent a zone of equivalence. (The idea of a three zone model utilizing a value such as ζ to indicate a lack of **practical significance** (as opposed to **statistical significance**) is also employed within the context of the **minimum-effect hypothesis testing model** (which is discussed in Section IX (the **Addendum**) of the **Pearson product-moment correlation coefficient**.)

Feinstein (2002, p. 520) notes that the issue of whether to conduct a directional or nondirectional analysis within the framework of a **test of equivalence** will be a function of the specific question a researcher wishes to answer. If the researcher is asking whether or not two treatments are equivalent to one another, then a **two-tailed/nondirectional** analysis is in order. In such a case the researcher would not be concerned with the direction of any trivial differences one detects between the two treatments. In other words, it will be irrelevant whether the score on the dependent variable for Treatment A is slightly larger than the score on the dependent variable for Treatment B, or vice versa. On the other hand, it is often the case the researcher expects to demonstrate that one treatment is perhaps slightly better than another, but not better enough to be of practical consequence. As an example, one might wish to establish that a generic drug is almost as good as a brand name drug, and that if the generic drug is not exactly equivalent to the brand name drug, the difference between them is of no practical consequence. In such a case the

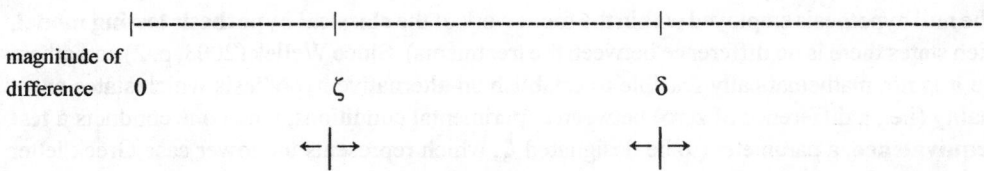

| Zone of trivial difference commensurate with no differences between conditions (i.e., equivalence), since the observed difference is deemed of no practical consequence. This is the zone of equivalence. Because a difference in this zone is so small, when the classical hypothesis testing model is employed the power of a statistical test will have to be extremely high (thus requiring a large sample size) in order for such a difference to be identified as statistically significant. Any difference equal to or less than the value of ζ specified by the researcher will fall within this zone. A test of equivalence employs the value ζ in evaluating the hypothesis of equivalence. | Zone of moderate difference yielding inconclusive results. Note that although within the framework of the classical hypothesis testing model a difference in this zone may achieve statistical significance, it is debatable whether or not such a difference is of any practical consequence. Any difference which is greater than the value of ζ but less than the value of δ specified by the researcher falls within this zone. Some sources refer to this zone as a zone of statistical indeterminancy. | Zone of substantial difference which is of practical consequence. Within the framework of the classical hypothesis testing model, a difference in this zone will often be statistically significant when a relatively small sample size is employed in a study, and because of this the power of a statistical test may not have to be large to detect such a difference. Any difference equal to or greater than the value of δ specified by the researcher will fall within this zone. |

Figure 11.2   Three-Zone Decision Model

researcher would only examine the degree to which the brand name drug might be superior to the generic drug, and would not concern oneself with whether the generic drug might prove slightly superior to the brand name drug.

In the discussion to follow, it will be assumed that the null and alternative hypotheses employed for a **test of equivalence** will be in reference to two treatments which are being compared with one another. One of the treatments, to be designated **Treatment 1**, will represent a **brand name drug**, while the other treatment, to be designated **Treatment 2**, will represent a **generic drug**. The goal of a study will be to establish the equivalence of the two treatments. In other words, the researcher wants to demonstrate that the generic drug is equivalent to the brand name drug with respect to bioavailability, efficacy or some other criterion. It will be assumed that a *t* **test for two independent samples** (employing Equation 11.5), a *z* **test for two independent samples** (employing Equation 11.5, but using the normal distribution in lieu of the *t* distribution), or the *z* **test for two independent proportions** (employing Equation 16.11) will be employed to contrast the two conditions. (The discussion of **tests of equivalence for two independent proportions** can be found in Section VII of the **chi-square test for *r* × *c* tables**.) In the numerator of all three of the aforementioned equations, the mean (in Equation 11.5) or proportion (in Equation 16.11) of **Treatment 2** will be subtracted from the mean or proportion of **Treatment 1**. In the **test of equivalence** the symbol θ (which represents the lower case Greek letter **theta**) will be employed in the null and alternative hypotheses to represent the true difference in the underlying populations between the two means ($\mu_1 - \mu_2$) or two proportions ($\pi_1 - \pi_2$). Thus, $\theta = \mu_1 - \mu_2$ or $\theta = \pi_1 - \pi_2$.

The specific numerical value for defining equivalence stipulated by the researcher in the null and alternative hypotheses will be designated as ζ. If a **two-tailed/nondirectional** analysis is conducted, the value of ζ stipulated by the researcher will represent a limiting value which the difference between the two means or proportions must not exceed, regardless of the direction of the difference. Thus, as long as the test allows one to conclude that the difference between the means or proportions in the underlying populations does not exceed the value stipulated for ζ, a researcher will be able to declare the two treatments equivalent. Consequently, if $\theta = \mu_1 - \mu_2$ (or in the case of proportions, $\theta = \pi_1 - \pi_2$) is a **positive number that is less than the value of** ζ, or $\theta = \mu_1 - \mu_2$ (or in the case of proportions, $\theta = \pi_1 - \pi_2$) is a **negative number that is greater than the value** $-\zeta$, the researcher will be able to conclude that the two treatments are equivalent. Put simply, **if the absolute value of θ is less than ζ, equivalency has been established**.

Often the value of ζ is defined as 10% (or some other relatively low percentage) of the sample mean of the brand name treatment. Thus, if a generic drug is being evaluated in reference to a brand name drug, the value stipulated for ζ may be 10% of the value obtained on the dependent variable for the sample mean of the brand name drug. In the case of proportions, the value of ζ may be a proportion equal to 10% (or some other relatively low percentage) of the success rate of the brand name drug.

The hypotheses employed for a **two-tailed/nondirectional test of equivalence** can be either **symmetrical** or **nonsymmetrical**. A **symmetrical analysis** is illustrated in Figure 11.3. The horizontal line in Figure 11.3 represents the range of possible values for the true difference in the underlying populations between the means/proportions of the two treatments (where $\theta = \mu_1 - \mu_2$ or $\theta = \pi_1 - \pi_2$). In a **symmetrical analysis** the absolute value hypothesized for ζ will be the same regardless of whether the difference between the two treatment means/proportions yields a positive or negative value. To illustrate, assume that in an analysis comparing the means of the two treatments, the value stipulated for ζ is 2. In such a case, the researcher will be able to declare equivalence if the data indicate that in the underlying populations the mean of **Treatment 1** is not greater than the mean of **Treatment 2** by more than 2 units, or that the mean of **Treatment 2** is not greater than the mean of **Treatment 1** by 2 units. Thus, the numerical values -2 and 2 would be substituted for the symbols $-\zeta$ and ζ in Figure 11.3. Note that the latter values represent boundaries beyond which equivalence will not exist. Thus, the zone of equivalence in Figure 11.3 is the area which falls between the values $-\zeta$ and ζ.

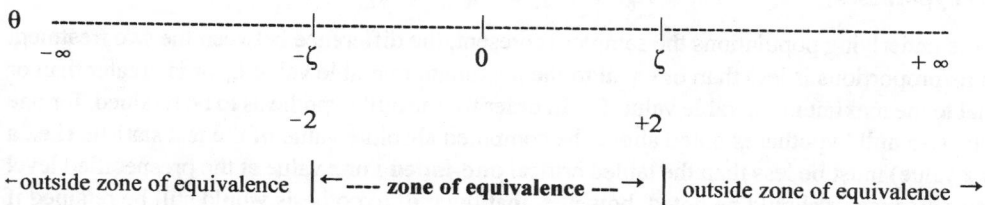

Figure 11.3 Symmetrical Analysis for Equivalence (Two-Tailed)

In the case of a **nonsymmetrical analysis**, which is illustrated in Figure 11.4, the absolute value hypothesized for ζ will depend upon whether or not the difference between the two treatment means/proportions yields a positive or negative value. To illustrate, assume that in an analysis comparing the means of the two treatments, the researcher is willing to tolerate the population mean of **Treatment 1** being 3 units or less greater than the population mean of **Treatment 2** in order to declare equivalence. However, he is only willing to tolerate the population mean of **Treatment 2** being 2 units or less greater than the population mean of **Treatment 1** in order to declare equivalence. In such a case, the researcher will be able to declare equivalence if the data indicate that in the underlying populations the mean of **Treatment 1** is not greater than the mean of **Treatment 2** by more than 3 units, or that the mean of **Treatment 2** is not greater than the mean of **Treatment 1** by 2 units. Thus, in Figure 11.4 the numerical values -2 and 3 are employed to represent $-\zeta$ and ζ (which can also be designated respectively as ζ_1 and ζ_2). The reader should take note of the fact that in the discussion to follow, in lieu of symbols $-\zeta$ and ζ, the values ζ_1 and ζ_2 will respectively be used to identify the boundaries employed within the framework of a **test of equivalence**, regardless of whether a **symmetrical** or **nonsymmetrical analysis** is conducted. Thus, in Figure 11.4 the zone of equivalence is the area which falls between the values ζ_1 and ζ_2.

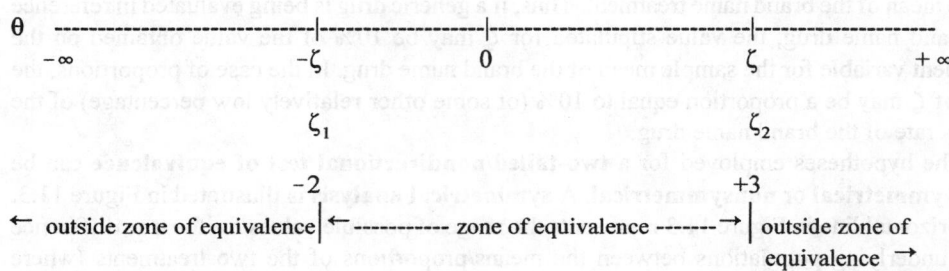

Figure 11.4 Nonsymmetrical Analysis for Equivalence (Two-Tailed)

If in a **test of equivalence** either a **symmetrical** or **nonsymmetrical two-tailed /nondirectional** analysis is conducted, the null and alternative hypotheses can be stated using the format noted below. As previously indicated, θ will be employed to represent the true difference in the underlying populations between the means/proportions of the two treatments, and the values ζ_1 and ζ_2 are used to demarcate the boundaries for equivalence. The reader should take note of the fact that since the statistical hypotheses involve both tails of the distribution, the null and alternative hypotheses will both be comprised of two directional elements.

Null hypotheses H_0: $\theta \le \zeta_1$ or H_0: $\theta \ge \zeta_2$

(In the underlying populations the samples represent, the difference between the two treatment means/proportions is less than or equal to the maximum tolerable value ζ_1 or is greater than or equal to the maximum tolerable value ζ_2. In order for the null hypothesis to be retained, for one of the two null hypotheses noted above the computed absolute value of the test statistic (i.e., a t or z value) must be less than the tabled critical **one-tailed** t or z value at the prespecified level of significance. It should be noted, however, that the null hypothesis would still be retained if **both** of the following conditions are met: a) The **sign** of the computed value of the test statistic is identical for the test of **both** of the null hypotheses noted above (i.e., **both** t (or z) tests yield a result with a positive sign or **both** tests yield a result with a negative sign); and b) In the case of **both** null hypotheses which are evaluated, the absolute value of the test statistic is equal to or greater than the tabled critical **one-tailed** t or z value at the prespecified level of significance. The null hypothesis is the **hypothesis of nonequivalence**.)

Alternative hypotheses $\qquad H_1: \theta > \zeta_1$ and $H_1: \theta < \zeta_2$

(In the underlying populations the samples represent, the difference between the two treatment means/proportions is greater than the maximum tolerable value ζ_1 and is less than the maximum tolerable value ζ_2. In order for the alternative hypothesis to be supported (and thus reject the null hypothesis) for **both** of the alternative hypotheses noted above the computed absolute value of the test statistic (i.e., a t or z value) must be equal to or greater than the tabled critical **one-tailed** t or z value at the prespecified level of significance. It should be noted, however, that in order for the alternative hypothesis to be supported, the **signs** of the computed value of the test statistics for the tests of **both** of the null hypotheses noted above must be **different** (in other words, one test yields a result with a positive sign and the other test yields a result with a negative sign). The alternative hypothesis is the **hypothesis of equivalence**.)

Rogers *et al.* (1993, pp. 554–555) note that in the case of a **two-tailed analysis**, a researcher is in actuality conducting two **one-tailed analyses**, and although in order to test for equivalence it is only necessary to conduct only one statistical analysis, each of the **null hypotheses** noted above must be rejected in order to demonstrate equivalence. In order to do the latter, it is only necessary to conduct an analysis involving one of the two values ζ_1 versus ζ_2 which is the shortest distance between the difference between the sample means (i.e., $\overline{X}_1 - \overline{X}_2$). This is the case, since the latter analysis will yield the smallest absolute value for the test statistic, and consequently of the two tests that are possible to conduct (one involving the value stipulated for ζ_1 and one involving the value stipulated for ζ_2), it will yield the larger probability value. Because of the latter, of the two possible tests evaluating the null hypotheses, the test involving the value ζ_1 or ζ_2 which is closest to the difference between the two sample means will have the lowest likelihood of establishing equivalence. If the analysis which yields the larger probability value allows the researcher to reject the specific null hypothesis it is evaluating, it follows that an analysis of value ζ_1 or ζ_2 stipulated in the other null hypothesis will yield a smaller probability value. Consequently if the larger of the two probability values for both of the elements which comprise the null hypotheses is equal to or less than the prespecified value of **alpha**, both of the probability values must be equal to or less than alpha, thus allowing the **null hypothesis of nonequivalence** to be rejected. In the evaluation of the two null hypotheses noted above, the likelihood of committing a Type I will be equal to the prestipulated value of alpha for the analysis yielding the larger probability value. Thus, the alpha level for the one-sided analysis that the researcher conducts will represent the likelihood of a Type I error for the **test of equivalency**. Figure 11.5 (which is based on Rogers *et al.* (1993, pp. 558) provides a visual depiction of the null and alternative hypotheses evaluated in a **test of equivalence**.[30]

It is also possible to conduct a single **one-tailed/directional test of equivalence**. In the latter case only one boundary is stipulated for equivalence. The most common scenario in which a researcher might elect to conduct a **one-tailed analysis** might be a study comparing a brand name drug with a generic drug. In such a study a researcher might only be willing to declare a lack of equivalence if it is demonstrated that the generic drug is inferior to the brand name drug, yet would not concern oneself with the unlikely possibility that the generic drug will prove superior to the brand name drug. If, in fact, the latter did occur, the researcher would still consider equivalence to have been established by virtue of the fact that the generic drug was not inferior to the brand name drug. To illustrate, assume that in an analysis comparing the means

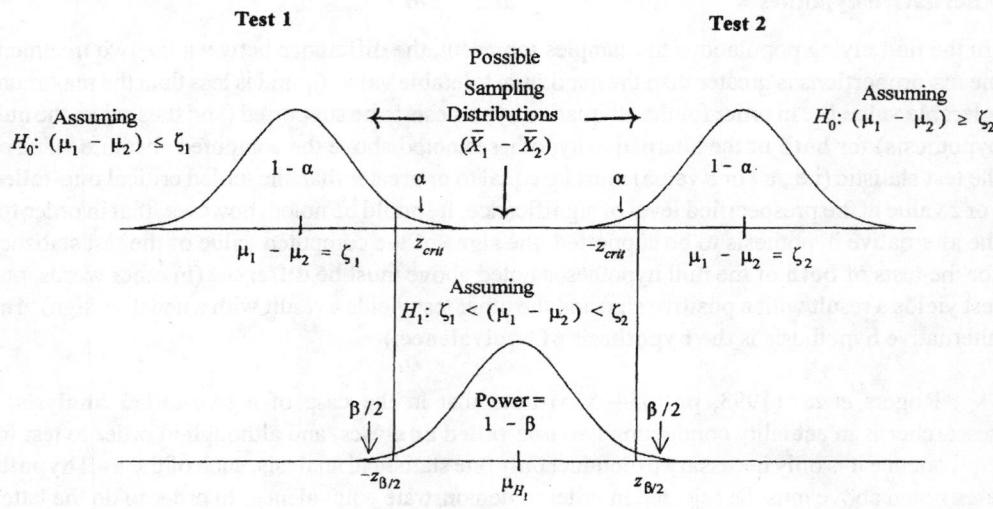

Figure 11.5   Visual Depiction of Null and Alternative Hypotheses in Test of Equivalence

of the two treatments (**Treatment 1** which represents a **brand name drug** versus **Treatment 2** which represents a **generic drug**) the value stipulated for ζ_2 is 2 and no value is stipulated for ζ_1. In such a case, the researcher will be able to declare equivalence if the data indicate that in the underlying populations the mean of **Treatment 1** is not greater than the mean of **Treatment 2** by more than 2 units. If, in fact, it turns out that the mean of **Treatment 2** is greater than the mean of **Treatment 1**, regardless of the magnitude of the difference, equivalence would still be declared, since the researcher is only focusing on the issue of whether or not the generic drug is inferior to the brand name drug. Figure 11.6 illustrates this latter situation.

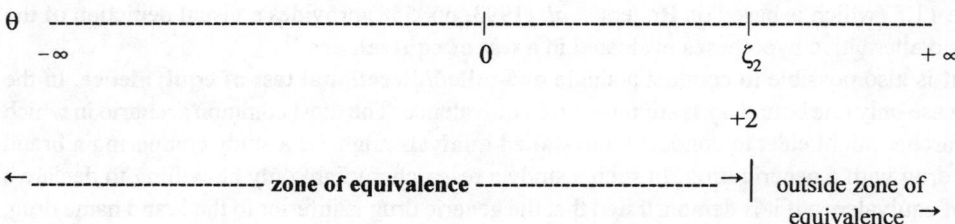

Figure 11.6   Directional Test of Equivalence (One-Tailed)

The null and alternative hypotheses for a **one-tailed analysis** are stated below.

Null hypothesis $H_0: \theta \geq \zeta_2$

(In the underlying populations the samples represent, the difference between the two treatment means/proportions is greater than or equal to the maximum tolerable value (i.e., ζ_2). This is a **directional null hypothesis** and it is evaluated with a **one-tailed test**. In order to be retained, the computed absolute value of t or z must be less than or equal to the tabled critical **one-tailed** t or z value at the prespecified level of significance. The null hypothesis is the **hypothesis of nonequivalence**.)

Alternative hypothesis $H_1: \theta < \zeta_2$

(In the underlying populations the samples represent, the difference between the two treatment means is less than the maximum tolerable value (i.e., ζ_2). This is a **directional alternative hypothesis** and it is evaluated with a **one-tailed test**. In order to be supported (and thus reject the null hypothesis), the computed absolute value of t or z must be greater than the tabled critical **one-tailed** t value at the prespecified level of significance, and the sign of t or z must be consistent with the directional prediction stipulated in the alternative hypothesis. The alternative hypothesis is the **hypothesis of equivalence**.)

This section will describe two procedures for evaluating the equivalence of two independent treatments/samples, which will be referred to as the **Westlake–Schuirmann test of equivalence of two independent treatments (Test 11h)** and **Tryon's test of equivalence of two independent treatments (Test 11i)**. Prior to illustrating the procedures for evaluating equivalence, the reader is asked to consider Examples 11.4 and 11.5.

Example 11.4   *The manufacturer of a generic drug for treating a disease conducts a study comparing it with the brand name drug for treating the disease. The major criterion for diagnosing the disease is that patients diagnosed with the disease have an abnormally low concentration of a specific factor in their blood. The criterion for comparing the generic drug with the brand name drug is the relative concentration of the criterion factor in the blood of patients. The study employs ten people who are afflicted with the disease (each patient exhibiting an abnormally low concentration of the criterion factor). Each of the subjects is randomly assigned to receive either the brand name drug or the generic drug for a period of six months (with five subjects per treatment). After six months have elapsed a measure of the criterion factor in the patients' blood is obtained. The criterion factor scores for the ten patients follow:* **Group 1 (Brand name drug)**: *10, 8, 10, 9, 11;* **Group 2 (Generic drug)**: *8, 9, 8, 7, 9. Do the data indicate there is a significant difference between the scores(i.e., concentrations of the criterion factor) of patients receiving the brand name drug versus the generic drug?*

The reader should take note of the fact that the data for Example 11.4 are identical to that employed for Example 11.2. Consequently, **Group 1**, the group which receives the **brand name drug** is analogous to the **Accusharp** battery condition, while **Group 2** which receives the **generic drug** is analogous to the **Keenair** battery condition. Thus, $\bar{X}_1 = 9.6$, $\bar{X}_2 = 8.2$, $\tilde{s}_1^2 = 1.3$, and $\tilde{s}_2^2 = .7$, and $n_1 = n_2 = 5$. If in Example 11.4 $H_0: \mu_1 = \mu_2$ and $H_1: \mu_1 \neq \mu_2$ are employed as the null hypothesis and nondirectional alternative hypothesis, analysis of the data yields the following result (as was the case for Example 11.2): $t = (9.6 - 8.2)/.63 = 2.22$. Since the obtained value $t = 2.22$ is less than the tabled critical two-tailed value $t_{.05} = 2.31$, the nondirectional alternative hypothesis H_1: $\mu_1 \neq \mu_2$ is not supported at the .05 level. Thus, one cannot conclude that the concentration of the criterion factor in **Group 1** (the **brand name drug**) is significantly different from the concentration of the criterion factor in **Group 2** (the **generic drug**). However, the fact that the null hypothesis H_0: $\mu_1 = \mu_2$ cannot be rejected since the alternative hypothesis H_1: $\mu_1 \neq \mu_2$ is not supported, does not provide sufficient grounds for the researcher declaring that the two treatments are equivalent.[31] (The directional alternative hypothesis H_1: $\mu_1 > \mu_2$ (which can also be written as H_1: $\mu_1 - \mu_2 > 0$) is supported at the .05 level, since $t = 2.22$ is a positive number which is larger than the tabled critical one-tailed value $t_{.05} = 1.86$. Thus, if the null hypothesis H_0: $\mu_1 = \mu_2$ and the directional alternative hypothesis H_1: $\mu_1 > \mu_2$ are employed, the researcher is able to conclude that the concentration of the criterion factor in **Group 1** (**brand name drug**) is significantly greater than the concentration of the criterion factor in **Group 2** (**generic drug**).)

Employing Equation 11.17, the computation of a 95% confidence interval for Example 11.4 is computed below, and visually represented in Figure 11.7.

$$CI_{.95} = (\bar{X}_1 - \bar{X}_2) \pm (t_{.05})(s_{\bar{X}_1 - \bar{X}_2}) = 1.4 \pm (2.306)(.63) = 1.4 \pm 1.45$$

$$-.05 \leq (\mu_1 - \mu_2) \leq 2.85$$

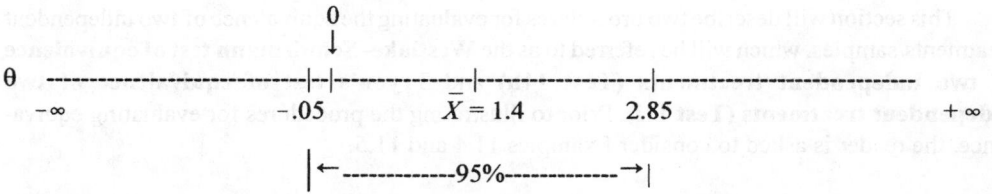

Figure 11.7   95% Confidence Interval For Example 11.4

This result indicates the researcher can be 95% confident (or the probability is .95) that the interval $-.05$ to 2.85 contains the true difference between the population means. If the afore-mentioned interval contains the true difference between the population means, the mean of the population **Group 2** (**generic drug**) represents is no more than .05 units higher than the mean of the population **Group 1** (**brand name drug**) represents, and the mean of the population **Group 1** represents is no more than 2.85 units higher than the mean of the population **Group 2** represents. Note the fact that the value zero falls within the above computed confidence interval is consistent with the result of the two-tailed analysis when alpha equals .05. In other words, when H_0: $\mu_1 = \mu_2$, H_1: $\mu_1 \neq \mu_2$, and alpha $= .05$, the 95% confidence interval will include the value zero if the two-tailed .05 analysis is not significant. In the same respect, when H_0: $\mu_1 = \mu_2$, H_1: $\mu_1 \neq \mu_2$, and alpha $= .01$, the 99% confidence interval will include the value zero if the two-tailed .01 analysis is not significant. However, when H_0: $\mu_1 = \mu_2$, H_1: $\mu_1 > \mu_2$, and alpha $= .05$, the 90% confidence interval will not include the value zero, since the one-tailed .05 analysis would be significant. The latter is illustrated below in Figure 11.8. It should be noted, however, that later in this section (under **Tryon's test of equivalence of two independent treatments (Test 11i)**) a more reliable method employing confidence intervals will be described for determining whether or not there is a significant difference between two treatments.

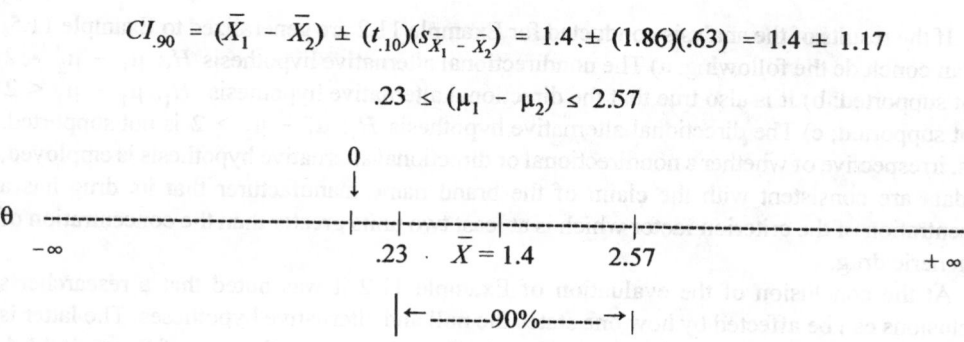

$$CI_{.90} = (\bar{X}_1 - \bar{X}_2) \pm (t_{.10})(s_{\bar{X}_1 - \bar{X}_2}) = 1.4 \pm (1.86)(.63) = 1.4 \pm 1.17$$

$$.23 \le (\mu_1 - \mu_2) \le 2.57$$

Figure 11.8   90% Confidence Interval For Example 11.4

Example 11.5   *The manufacturer of a brand name drug for treating a disease wants to empirically demonstrate it is superior to a generic drug which has been approved to treat the same disease. The major criterion for diagnosing the disease in question is that patients diagnosed with the disease have an abnormally low concentration of a specific factor in their blood. The manufacturer of the brand name drug claims that treatment with the latter drug results in a mean concentration of the criterion factor that is two units above the mean concentration of the criterion factor in patients treated with the generic drug. In order to demonstrate the latter, a study is conducted in which ten people afflicted with the disease (each patient exhibiting an abnormally low concentration of the criterion factor) are randomly assigned to receive either the brand name drug or the generic drug for a period of six months (with five subjects per treatment). After six months have elapsed a measure of the criterion factor is obtained for each patient. The criterion factor scores for the ten patients follow:* **Group 1 (Brand name drug):** 10, 8, 10, 9, 11; **Group 2 (Generic drug):** 8, 9, 8, 7, 9. *Do the data support the claim of the manufacturer of the brand name drug that it yields a mean concentration which is two units greater than the generic drug?*

The reader should take note of the fact that the data employed for Example 11.5 are identical to that employed for Example 11.2. Consequently, as was the case in Example 11.4, **Group 1**, the group which takes the **brand name drug** is analogous to the **Accusharp** battery condition, while **Group 2** which takes the **generic drug** is analogous to the **Keenair** battery condition. Since the manufacturer of the brand name drug predicts it will yield a mean score 2 units greater than the mean score of the generic drug, the null hypothesis can be stated as follows: H_0: $\mu_1 - \mu_2 = 2$ (which can also be written as H_0: $\theta = 2$). The alternative hypothesis if stated nondirectionally is H_1: $\mu_1 - \mu_2 \ne 2$ (which can also be written as H_1: $\theta \ne 2$). If stated directionally, the appropriate alternative hypothesis to employ is H_1: $\mu_1 - \mu_2 < 2$ (which can also be written as H_1: $\theta < 2$). The latter directional alternative hypothesis (which predicts a negative t value) is employed, since in order for the data to contradict the claim of the manufacturer of the brand name drug (and thus reject the null hypothesis), the score on the criterion factor for the latter drug can be any concentration which is 2 units or less than the concentration of the generic drug. The alternative hypothesis H_1: $\mu_1 - \mu_2 > 2$ (which can also be written as H_1: $\theta > 2$ and will only be supported with a positive t value) predicts that the superiority of the brand name drug is greater than 2 units. If the latter alternative hypothesis is employed, the null hypothesis can only be rejected if the concentration of the brand name drug is more than 2 units greater than the concentration of the generic drug. If, in fact, the latter alternative hypothesis is supported, it would further reinforce the claim of the brand name manufacturer that its drug is superior to the generic drug.

If the results of the analysis conducted for Example 11.2 are generalized to Example 11.5, we can conclude the following: a) The nondirectional alternative hypothesis H_1: $\mu_1 - \mu_2 \neq 2$ is not supported; b) It is also true that the directional alternative hypothesis H_1: $\mu_1 - \mu_2 < 2$ is not supported; c) The directional alternative hypothesis H_1: $\mu_1 - \mu_2 > 2$ is not supported. Thus, irrespective of whether a nondirectional or directional alternative hypothesis is employed, the data are consistent with the claim of the brand name manufacturer that its drug has a concentration of the criterion factor which is at least two units greater than the concentration of the generic drug.

At the conclusion of the evaluation of Example 11.2 it was noted that a researcher's conclusions can be affected by how one states the null and alternative hypotheses. The latter is once again applicable to Examples 11.4 and 11.5. Recollect that in the case of Example 11.4, when H_0: $\mu_1 = \mu_2$ and H_1: $\mu_1 \neq \mu_2$ are employed as the null hypothesis and nondirectional alternative hypothesis (for the same set of data evaluated for Example 11.5), the nondirectional alternative hypothesis H_1: $\mu_1 \neq \mu_2$ is not supported, and that the latter result indicates there is no significant difference between the two treatments. Yet in the case of Example 11.5, the fact that the nondirectional alternative hypothesis H_1: $\mu_1 - \mu_2 \neq 2$ is not supported suggests that, in fact, there is a difference between the treatments — specifically, a two unit difference in favor of the brand name drug (by virtue of the fact that the null hypothesis H_0: $\mu_1 - \mu_2 = 2$ is retained). In the final analysis, neither the analysis of Example 11.4 or 11.5 constitutes an appropriate methodology for answering the question of whether or not the two drugs are equivalent.

Test 11h: The Westlake–Schuirmann test of equivalence of two independent treatments
In order to directly address the question of equivalence, consider Example 11.6, which employs the same data as Examples 11.4 and 11.5. The reader should take note of the fact that although for illustrative purposes a small sample size is employed in Example 11.6 (as well as in Examples 11.4 and 11.5), substantially larger sample sizes would be expected in actual studies which evaluate drug efficacy.

Example 11.6 *The manufacturer of a generic drug for treating a disease wants to empirically demonstrate it is equivalent to a brand name drug, which up to the present time has been the only approved treatment for the disease. The major criterion for diagnosing the disease is that patients diagnosed with the disease have an abnormally low concentration of a specific factor in their blood. Treatment with the brand name drug results in a substantial increase in the concentration of the latter factor. The Federal Drug Administration declares that equivalency of a generic drug with the brand name drug can be established if the mean concentration of the disease factor in patients treated with the generic drug is within 2 units of the mean concentration of the factor obtained for patients treated with the brand name drug. To demonstrate equivalency of the two drugs, a study is conducted in which ten people afflicted with the disease (each patient exhibiting an abnormally low concentration of the disease factor) are randomly assigned to receive either the brand name drug or the generic drug for a period of six months. After six months have elapsed a measure of the disease factor is obtained for each patient. The results obtained from each of the groups follow:* **Group 1 (Brand name drug)**: 10, 8, 10, 9, 11; **Group 2 (Generic drug)**: 8, 9, 8, 7, 9. *Do the data indicate that the generic drug is equivalent to the brand name drug?*

The information provided in Example 11.6 allows for either a two-tailed or one-tailed analysis. If a researcher employs as a criterion for equivalence that the concentration of the brand name drug is no more than two units greater than the concentration of the generic drug, as well as the requirement that the concentration of the generic drug is no more than two units greater than the concentration of the brand name drug, a two-tailed analysis would be conducted. On the other hand, if the researcher only employs as a criterion for equivalence that the concentration of the brand name drug is no more than two units greater than the concentration of the generic drug, a one-tailed analysis would be conducted. Both the two-tailed and one-tailed analyses will be demonstrated. The two-tailed analysis is represented by Figure 11.3, and thus the values employed for ζ_1 and ζ_2 are respectively -2 and 2. A one-tailed analysis is represented by Figure 11.6, and thus the value 2 is employed for ζ_2, while no value is designated for ζ_1. Equation 11.22 will be employed for both analyses. Note that the latter equation is identical to Equation 11.5, except that in Equation 11.22 the value ζ is employed to represent the term $(\mu_1 - \mu_2)$ in the numerator of Equation 11.5. The value employed for ζ will be the appropriate value employed for ζ_1 or ζ_2 in the analysis being conducted.

$$ t = \frac{(\bar{X}_1 - \bar{X}_2) - \zeta}{s_{\bar{X}_1 - \bar{X}_2}} \qquad \textbf{(Equation 11.22)} $$

We will first consider a two-tailed analysis for Example 11.6. The null and alternative hypotheses for a **two-tailed analysis** are noted below.

Null hypotheses $\qquad H_0\colon \theta \leq -2$ or $H_0\colon \theta \geq 2$

(In the underlying populations the samples represent, the difference between the two treatment means is less than or equal to -2 or is greater than or equal to 2. This is the **hypothesis of nonequivalence.**)

Alternative hypotheses $\qquad H_1\colon \theta > -2$ and $H_1\colon \theta < 2$

(In the underlying populations the samples represent, the difference between the two treatment means is greater than -2 and is less than 2. This is the **hypothesis of equivalence.**)

The reader should take note of the fact that in a **test of equivalence** the alternative hypothesis stipulates a bounded range of values which are designated by the researcher, whereas in evaluating an alternative hypothesis within the framework of the **classical hypothesis testing model** (which was employed in evaluating Example 11.4), the alternative hypothesis has no boundaries assigned to it by the researcher, but instead is comprised of all values other than zero. The two-sided **test of equivalence** to be described in this section is referred to as **the Westlake–Schuirmann test of equivalence of two independent treatments** since in different sources it is attributed to Westlake (1976, 1981, 1988) and/or Schuirmann (1987). An excellent description of the rationale underlying the test can be found in Rogers *et al.* (1993).

In point of fact, the two-tailed analysis to be described requires that **two one-tailed tests** be conducted. The two one-tailed tests will now be conducted employing the value alpha = .05. Initially, the following null and alternative hypotheses are evaluated: $H_0\colon \theta \leq -2$ versus $H_1\colon \theta > -2$. Employing Equation 11.22, the value $t = 5.40$ is computed. Note that in the analysis the value $\zeta_1 = -2$ is employed in the numerator of the equation, since we are evaluating the left boundary for equivalence depicted in Figure 11.3.

$$ \tilde{s}_1^2 = \frac{466 - \dfrac{(48)^2}{5}}{5 - 1} = 1.3 \qquad \tilde{s}_2^2 = \frac{339 - \dfrac{(41)^2}{5}}{5 - 1} = .7 $$

$$t = \frac{(9.6 - 8.2) - (-2)}{\sqrt{\dfrac{1.3}{5} + \dfrac{.7}{5}}} = \frac{3.4}{.63} = 5.40$$

With $df = 5 + 5 - 2 = 8$, the tabled critical values Table **A2** in the **Appendix** (as well as Table 11.2) can be employed to evaluate the results of the analysis. Since the obtained absolute value $t = 5.40$ is greater than the tabled critical one-tailed value $t_{.05} = 1.86$, the null hypothesis H_0: $\theta \leq -2$ designating nonequivalence can be rejected. The probability value associated with this result is well below the .05 value which was prestipulated for alpha. Thus, the first of the null hypotheses designating nonequivalence can be rejected. However, it is important to note that in order to declare equivalence we must reject **both** of the null hypotheses designating nonequivalence, and the two test statistics computed must have opposite signs. In reference to the *t* test conducted above, the data are consistent with the alternative hypotheses H_1: $\theta > -2$, which stipulates that when the mean of **Group 2 (generic drug)** is subtracted from the mean of **Group 1 (brand name drug)** the difference is greater than -2 (which translates into the fact that if the mean of the **generic drug** is greater than the mean of the **brand name drug**, it is not greater than the latter value by more than two units).

We now evaluate the second null hypothesis designating nonequivalence. Specifically, a one-tailed test which evaluates the following null and alternative hypotheses is conducted: H_0: $\theta \geq 2$ versus H_1: $\theta < 2$. Note that this analysis (which yields the result obtained for Example 11.2) employs the same values as in the previous analysis, except for the fact that the value $\zeta_2 = 2$ is employed in the numerator of Equation 11.22.

$$t = \frac{(9.6 - 8.2) - (2)}{\sqrt{\dfrac{1.3}{5} + \dfrac{.7}{5}}} = \frac{-.6}{.63} = -.95$$

Since the obtained absolute value $t = -.95$ is less than the tabled critical one-tailed value $t_{.05} = 1.86$, the null hypothesis cannot be rejected. Thus, the second null hypothesis H_0: $\theta \geq 2$ designating nonequivalence cannot be rejected. The probability value associated with this result is well above .05. Since, as noted earlier, in order to declare equivalence we must reject **both** of the null hypotheses designating nonequivalence, equivalence cannot be established since the larger of the probability values computed for the two analyses exceeds .05. Specifically, the data are consistent with the null hypothesis H_0: $\theta \geq 2$, which stipulates that the mean of **Group 1** (the **brand name drug**) is two units or greater than the mean of **Group 2** (the **generic drug**).[32]

Rogers *et al.* (1993, pp. 554–555) note that when the value of alpha is set equal to .05 for each of the two analyses conducted above, the equivalency confidence interval is set at the $(1 - 2\alpha)$ (i.e., 90%) level of certainty/confidence rather than at the traditional $(1 - \alpha)$ (i.e., 95%) level of certainty/confidence. When both null hypotheses which are evaluated are rejected, the values employed for both ζ_1 and ζ_2 will fall outside of the $(1 - 2\alpha)$ % confidence interval. If either of the null hypotheses is retained, the value employed for ζ_1 or ζ_2 will fall within the limits of the confidence interval. **To be more specific, if equivalence is to be established, the values for both ζ_1 and ζ_2 must fall outside the limits of the confidence interval, with one of the values ζ_1 positioned to the left of the lower limit of the confidence interval and the other value ζ_2 positioned to the right of the upper limit of the confidence interval. Note, however, that if both ζ_1 and ζ_2 fall outside the limits of the confidence interval, and both ζ_1 and ζ_2 are below the lower limit of the confidence interval or both ζ_1 and ζ_2 are above the**

upper limit of the confidence interval, it would indicate a lack of equivalence. In the case of Example 11.6, in order to compute the 90% confidence interval, the tabled critical two-tailed .10 t value (which, as noted earlier, corresponds to the tabled critical one-tailed .05 t value) is employed in Equation 11.17. Employing Equation 11.17, the 90% confidence interval for Example 11.6 is computed below, and visually represented in Figure 11.9.[33]

$$CI_{.90} = (\bar{X}_1 - \bar{X}_2) \pm (t_{.10})(s_{\bar{X}_1 - \bar{X}_2}) = 1.4 \pm (1.86)(.63) = 1.4 \pm 1.17$$

$$.23 \leq (\mu_1 - \mu_2) \leq 2.57$$

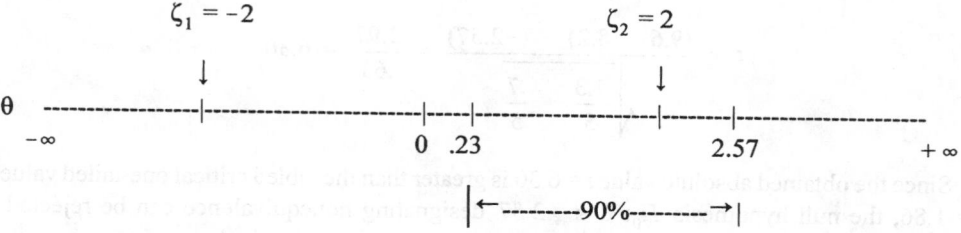

Figure 11.9 90% Confidence Interval For Example 11.6

This result indicates the researcher can be 90% confident (or the probability is .90) that the interval .23 to 2.57 contains the true difference between the population means. If the aforementioned interval contains the true difference between the population means, the mean of the population **Group 1 (Brand name drug)** represents is between .23 and 2.57 units greater than the mean of the population **Group 2 (Generic drug)** represents. Note that although the value $\zeta_1 = -2$ falls outside of the computed confidence interval, the value $\zeta_2 = 2$ falls inside it. Equivalency cannot be established, since the upper range of the confidence interval falls above the maximum tolerable difference between the means of the two conditions. In order for equivalency to have been established, the result for the second analysis would had to have been statistically significant at the .05 level, and if the latter had occurred the upper limit for the confidence interval would have been some value less than 2.

The reader should take note of the fact that simply by employing Equation 11.22, as was done earlier in this section to compute two t values, one can obtain the answer to the question of whether or not two treatments are equivalent. Alternatively, computation of the above confidence interval, in and of itself, will provide the answer to the same question. (Because of the latter Rosner (2000, pp. 638–640) refers to the **Westlake–Schuirmann test** as **inference based on confidence interval estimation**.) Use of both methods merely provides the same answer to the same question in different formats.

In point of fact, given the sample sizes employed in Example 11.6, we can easily determine the maximum absolute value which would have to be designated for ζ_1 and ζ_2 in order to establish equivalence. Specifically, we solve Equation 11.22 for ζ when the tabled critical one-tailed .05 value $t_{.05} = -1.86$ is obtained for the result of the analysis. The value $t_{.05} = -1.86$ is employed, since of the two t values computed above with Equation 11.22, the one which yielded the absolute t value closest to the critical value $t_{.05} = 1.86$ yielded a negative t value. When we solve for ζ, we obtain the value $\zeta = 2.57$ (which is the upper limit of the confidence interval computed earlier). Thus, the absolute value 2.57 is the maximum value that would have to be designated for ζ_1 and ζ_2 in order to establish equivalence.

$$t = \frac{(\bar{X}_1 - \bar{X}_2) - \zeta}{s_{\bar{X}_1 - \bar{X}_2}} = \frac{(9.6 - 8.2) - \zeta}{.63} = \frac{1.4 - \zeta}{.63} = -1.86$$

$$\zeta = 2.57$$

Thus any absolute value designated for ζ which is equal to or greater than 2.57 will yield a significant result for both of the *t* tests which are employed in testing for equivalence. To illustrate, if the values -2.57 and $+2.57$ are employed for ζ_1 and ζ_2, the two *t* tests for the **test of equivalence** yield the results below.

$$t = \frac{(9.6 - 8.2) - (-2.57)}{\sqrt{\dfrac{1.3}{5} + \dfrac{.7}{5}}} = \frac{3.97}{.63} = 6.30$$

Since the obtained absolute value $t = 6.30$ is greater than the tabled critical one-tailed value $t_{.05} = 1.86$, the null hypothesis H_0: $\theta \leq -2.57$ designating nonequivalence can be rejected. We now evaluate the second null hypothesis designating nonequivalence.

$$t = \frac{(9.6 - 8.2) - (2.57)}{\sqrt{\dfrac{1.3}{5} + \dfrac{.7}{5}}} = \frac{-1.17}{.63} = -1.86$$

Since the obtained absolute value $t = 1.86$ is equal to the tabled critical one-tailed value $t_{.05} = 1.86$, the null hypothesis can be rejected. Thus, the second null hypothesis H_0: $\theta \geq 2.57$ designating nonequivalence can also be rejected. Consequently, the data are consistent with the alternative hypotheses H_1: $\theta > -2.57$ and H_1: $\theta < 2.57$ which stipulate that the mean of **Group 1 (Brand name drug)** is within 2.57 units of the mean of **Group 2 (Generic drug)**. Note that in Figure 11.9, both of the values -2.57 and $+2.57$ would lie outside of the 90% confidence interval if they were employed in lieu of -2 and 2 for ζ_1 and ζ_2.[34]

Rogers *et al.* (1993, pp. 561–562) note that a **test of equivalence**, such as that conducted above, and a conventional test of significance conducted within the framework of the **classical hypothesis testing model** are not mutually exclusive of one another. In point of fact, if both of the latter types of test are conducted on the same set of data, they may or may not yield results which are consistent with one another. To be more specific consider the four possible outcomes noted below. We will assume that any analysis which is conducted will have sufficient power (and, consequently, a sufficiently large sample size) to evaluate any effect size which is present in the underlying populations that is equal to or greater than the absolute values a researcher specifies for ζ_1 and/or ζ_2.

1) A conventional test of significance conducted within the framework of the **classical hypothesis testing model** (as conducted with Example 11.4) yields a nonsignificant result suggesting no differences between the two treatments. A **test of equivalence** (as conducted with Example 11.6) yields a significant result suggesting equivalence between the two treatments. The two aforementioned outcomes would clearly suggest there are no meaningful differences between the two treatments.

2) A conventional test of significance conducted within the framework of the **classical hypothesis testing model** yields a significant result suggesting there is difference between the

two treatments. A **test of equivalence** does not yield a significant result suggesting nonequivalence between the two treatments. The two aforementioned outcomes would clearly suggest there are meaningful differences between the two treatments.

3) A conventional test of significance conducted within the framework of the **classical hypothesis testing model** yields a significant result suggesting there is difference between the two treatments. A **test of equivalence** yields a significant result suggesting equivalence between the two treatments. The latter outcomes suggest that although a difference between the two treatments was detected with the conventional significance test, the magnitude of the difference (which the conventional test only indicates is some value other than zero) falls within the boundaries for defining equivalence. Thus, the difference which was detected with the conventional significance test would be viewed as of no practical significance, thus allowing a researcher to conclude that the two treatments are equivalent. Obviously, if an extremely large sample size were employed in a study, the power of the conventional statistical test would be such that it would detect a difference which a researcher would deem as trivial in that it was less than the absolute values of both ζ_1 and ζ_2.

4) A conventional test of significance conducted within the framework of the **classical hypothesis testing model** yields a nonsignificant result suggesting there is no difference between the two treatments. A **test of equivalence** does not yield a significant result suggesting nonequivalence between the two treatments. The two aforementioned outcomes would suggest that the data are inconclusive and that further research on the hypothesis under investigation is in order. Obviously, the lack of a significant result for the conventional test of significance could result from the fact that the sample size employed in the study was too small, and thus the analysis lacked sufficient power to detect a difference of practical significance. In the same respect the sample size could have been too small such that a **test of equivalence** did not have sufficient power to detect equivalence.

If the above noted four categories are considered with respect to the analyses of Examples 11.4–11.6, the final judgement regarding the comparison of the **brand name drug** with the **generic drug** would seem to fall in Outcome 2, which suggests that the two drugs are not equivalent. Specifically, it would be conceptualized under Outcome 2 since: a) The conventional test of significance conducted within the framework of the **classical hypothesis testing model** yields a significant result for Example 11.4 (if the prespecified value of alpha is .05 and the directional alternative hypothesis H_1: $\mu_1 > \mu_2$ is employed), suggesting there is a difference between the two treatments; and b) The **test of equivalence** does not yield a significant result (when the values $|\zeta_1| = \zeta_2 = 2$ are employed) suggesting nonequivalence between the two treatments; c) Analysis of Example 11.5 resulted in all of the following alternative hypotheses not being supported: H_1: $\mu_1 - \mu_2 \neq 2$, H_1: $\mu_1 - \mu_2 < 2$, and H_1: $\mu_1 - \mu_2 > 2$. Thus, the analysis of Example 11.5 is consistent with the result of the **test of equivalence** in that it suggests that the **brand name drug** has a mean concentration of the criterion factor which is at least two units greater than the mean concentration of the **generic drug** — suggesting nonequivalence.

One could also argue that the analyses of Examples 11.4–11.6 could be conceptualized under Outcome 4 since: a) The conventional test of significance conducted within the framework of the **classical hypothesis testing model** yields a nonsignificant result for Example 11.4 if the prespecified value of alpha is .05 and the nondirectional alternative hypothesis H_1: $\mu_1 \neq \mu_2$ is employed. Although the latter result does indicate there is no difference between the two treatments, the lack of a significant result could easily be attributed to the test having low power because of the small sample size employed in the study; b) A **test of equivalence** does not yield a significant result (when the values $|\zeta_1| = \zeta_2 = 2$ are employed) suggesting nonequivalence between the two treatments; c) As noted earlier, analysis of Example 11.5 suggests the two treatments are not equivalent. The aforementioned outcomes might suggest to some researchers that the data are inconclusive, and thus further research on the hypothesis under investigation is in order.

It should be noted that many researchers would conceptualize Example 11.6 within the framework of the model depicted in Figure 11.6. In such a case a researcher would conduct only a single **one-tailed test** which employed the null and alternative hypotheses noted below.

Null hypothesis H_0: $\theta \geq 2$

(In the underlying populations the samples represent, the difference between the two treatment means is greater than or equal to 2. This is the **hypothesis of nonequivalence**.)

Alternative hypothesis H_1: $\theta < 2$

(In the underlying populations the samples represent, the difference between the two treatment means is less than 2. This is the **hypothesis of equivalence**.)

If the above analysis is conducted, only the second of the two *t* tests conducted previously in the analysis of Example 11.6 would be employed. Specifically:

$$t = \frac{(9.6 - 8.2) - (2)}{\sqrt{\dfrac{1.3}{5} + \dfrac{.7}{5}}} = \frac{-.6}{.63} = -.95$$

Since, as noted previously, the obtained absolute value $t = -.95$ is less than the tabled critical one-tailed value $t_{.05} = 1.86$, the null hypotheses H_0: $\theta \geq 2$ designating nonequivalence cannot be rejected. In this instance, as noted in Figure 11.10, only an upper boundary would be computed for a confidence interval. Specifically:

$$CI_{.95} = (\bar{X}_1 - \bar{X}_2) + (t_{.05})(s_{\bar{X}_1 - \bar{X}_2}) = 1.4 + (1.86)(.63) = 1.4 + 1.17$$

$$-\infty \leq (\mu_1 - \mu_2) \leq 2.57$$

$$\zeta_2 = 2$$
$$\downarrow$$

θ ------------------------------------|------------------------------|--------|-----------------------

$-\infty$ 0   2.57

|←--------------------------------------95%--------------------------------------→|

Figure 11.10   95% Confidence Interval For One-Tailed Analysis

The above result indicates the researcher can be 95% confident (or the probability is .95) that the interval $-\infty$ to 2.57 contains the true difference between the population means. If the aforementioned interval contains the true difference between the population means, the mean of the population **Group 1 (brand name drug)** represents is no more than 2.57 units higher than the mean of the population **Group 2 (generic drug)** represents. Since the value $\zeta_2 = 2$ falls inside the confidence interval, equivalency cannot be established.

It was noted earlier that since in most instances rejection of the null hypothesis for a **test of equivalence** will involve detection of a substantially smaller effect size than the effect size a researcher is typically concerned with detecting within the framework of the **classical hypothesis testing model**, a **test of equivalence** will usually require greater statistical power, and consequently the use of a larger sample size, when contrasted with tests conducted within the framework of the **classical hypothesis testing model**. The smaller the value a researcher designates for ζ, the more power will be required for the statistical analysis.[35]

Procedure for computing sample size in reference to the power of the Westlake–Schuirmann test of equivalence of two independent treatments Atherton Skaff *et al.* (2004) note that Equation 11.23 can be employed to estimate the sample size required in order to conduct a **Westlake–Schuirmann test of equivalence of two independent treatments** which has a power of a specified value. Note that the numerator of Equation 11.23 is identical to the numerator of Equation 11.14. The values of n_1, k, σ_1^2, σ_2^2, $z_{\alpha/2}$, z_β in Equation 11.23 are obtained in the identical manner they are obtained in Equation 11.14. The term ζ in the denominator of Equation 11.23 represents the limiting absolute value which the difference between the two means must not exceed in order for the researcher to declare the two treatments equivalent. The values μ_1 and μ_2 in the denominator of Equation 11.23 are the values of the population means (which are typically estimated with the values of the sample means $\bar{X}_1$ and $\bar{X}_2$).

(Equation 11.23)

$$n_1 = \frac{\left(\sigma_1^2 + \dfrac{\sigma_2^2}{k} \right) (z_{\alpha/2} + z_\beta)^2}{\left[|\mu_1 - \mu_2| - \zeta \right]^2}$$

Equation 11.23 will be employed within the framework of Example 11.6 to determine the sample size required in order for the power of the **Westlake–Schuirmann test** to equal .80 when the value $\zeta = 2$ is stipulated by the researcher. In conducting the latter analysis, the following values are employed in Equation 11.23 to compute the number of subjects that will be required in **Group 1**(which will also be the number of subjects the researcher will need for **Group 2**, since when there are an equal number of subjects per group, $k = 1$): $\tilde{s}_1^2 = 1.3$, $\tilde{s}_2^2 = .7$, $\zeta = 2$, $z_{\alpha/2} = 1.96$ (If a one-tailed analysis is conducted $z_{\alpha/2} = 1.65$ should be employed.), and $z_\beta = .84$. (The rationale for employing the values $z_{\alpha/2} = 1.96$ and $z_\beta = .84$ is provided in the discussion of the determination of the latter values for Equation 11.14.)

When the appropriate values are substituted in Equation 11.23, the value $n_1 = 43.56$ is computed, indicating that in order for the power of the **Westlake–Schuirmann test of two independent treatments** to be .80, the researcher must have at least 44 subjects per group (which is the next greatest integer value than integer element of the value computed for n_1).[36]

$$n_1 = \frac{\left(1.3 + \dfrac{.7}{1} \right) (1.96 + .84)^2}{\left[|9.6 - 8.2| - 2 \right]^2} = 43.56$$

To illustrate that a test of equivalence will usually require a larger sample size than a conventional test of significance, Equation 11.14 will be employed in reference to Example 11.4 to compute the required sample size for the power of the *t* test for two independent samples to equal .80.

$$n_1 = \frac{\left(1.3 + \frac{.7}{1}\right)(1.96 + .84)^2}{(9.6 - 8.2)^2} = 8$$

Thus, for a test with Power = .80 to identify the difference 9.6 − 8.2 = 2 as statistically significant at the .05 level, a sample size of 8 subjects per group (which is substantially less than the sample size value 43.56 per group computed for the test of equivalence) is required.

It should be noted that when a test of equivalence is used within the framework of drug assessment research, it is common practice for a one-tailed analysis to be employed — specifically, an evaluation of the directional alternative hypothesis that the efficacy of the generic drug will not be less than the efficacy of the brand name drug by more than the value stipulated for ζ. In such a case, the value $z_{\alpha/2} = 1.65$ would be employed in Equation 11.23. When the latter value is employed below in Equation 11.23, the required sample size (for a test with Power = .80) is reduced from 44 to 35 subjects per group.

$$n_1 = \frac{\left(1.3 + \frac{.7}{1}\right)(1.65 + .84)^2}{[|9.6 - 8.2| - 2]^2} = 34.45$$

Test 11i: Tryon's test of equivalence of two independent treatments   Tryon (2001) derived an alternative albeit more conservative methodology for evaluating equivalence, which involves both an alternative but algebraically equivalent approach for evaluating a null hypothesis within the framework of the **classical hypothesis testing model** and a **test of equivalence**. Tryon (2001) notes it is common practice when a study yields a significant result for a researcher to state that the study is significant at the actual probability value obtained for the outcome of a study (which is the probability value commonly printed out when data are evaluated with statistical software) rather than at some prestipulated probability value (such as .05 or .01). For example, a researcher might state a result is significant at the .004 level (which we will assume is the value computed with statistical software) as opposed to the .05 level, even though prior to conducting the study the researcher's intent was to reject the null hypothesis if the probability associated with the result was equal to or less than .05. Tryon (2001), among others, notes that if a researcher strictly adheres to the **classical hypothesis testing model**, it is expected that one will report the prestipulated probability value rather than the exact probability value computed for the data analysis. To insure that researchers employ a prestipulated probability value, Tryon (2001) proposes the use of confidence intervals for testing significance, since the latter will leave one with no choice but to use a prestipulated probability value.

Theoretically one would expect that when a statistically significant difference exists between two treatments the confidence intervals computed for the means of each of the treatments will not overlap. If the intent of a researcher is to conduct a two-tailed analysis with alpha set at .05, the two-tailed .05 critical value would be employed in computing the 95% confidence interval for each of the means. Thus in the case of Example 11.4, in order to evaluate the null hypothesis H_0: $\mu_1 = \mu_2$ at the .05 level, in lieu of conducting a **t test for two independent samples**, employing Equation 2.8/2.9 a 95% confidence interval can be computed for each of the two groups. The computation of the confidence intervals is presented below. Note that the values $s_{\bar{X}_1} = \sqrt{1.3/5} = .51$, $s_{\bar{X}_2} = \sqrt{.7/5} = .37$, and $t_{.05} = 2.776$ (the tabled critical two-tailed .05 value for $df = n - 1 = 5 - 1 = 4$) are employed in computing the two 95% confidence intervals. Since the two confidence intervals overlap (i.e., the upper limit of the confidence interval of Group 2 (9.23) is larger than the lower limit of the confidence interval of Group 1 (8.18)), the researcher cannot conclude there is a significant difference between the two treatments. This result is consistent with the result obtained when the *t* **test for two independent samples** was conducted. Recollect that when the latter test was employed to evaluate Example 11.4, the nondirectional alternative hypothesis H_1: $\mu_1 \neq \mu_2$ was not supported when alpha = .05 was employed.

$$CI_{.95} = \bar{X}_1 \pm (t_{.05})(s_{\bar{X}_1}) = 9.6 \pm (2.776)(.51) = 9.6 \pm 1.42$$

$$8.18 \leq \mu_1 \leq 11.02$$

$$CI_{.95} = \bar{X}_2 \pm (t_{.05})(s_{\bar{X}_2}) = 8.2 \pm (2.776)(.37) = 8.2 \pm 1.03$$

$$7.17 \leq \mu_2 \leq 9.23$$

Unfortunately, use of the above computed confidence intervals will not always yield a reliable result. To be more specific, Tryon (2001) notes that Goldstein and Healy (1995) demonstrated it is possible for two 95% confidence intervals to overlap when, in fact, a *t* **test for two independent samples** on the same set of data yields a significant difference at the .05 level. van Belle (2002, pp. 39–40), in fact, states that the overlap can be as much as 29%, and a good rule of thumb is to assume that overlaps of 25% or less suggest statistical significance. Schenker and Gentleman (2001, p. 182) summarize the latter issue in noting that evaluation of data through examination of confidence interval overlap provides for a more conservative test when the null hypothesis is true (i.e., rejects the null hypothesis less often) when contrasted with employing the results obtained from conducting the *t* test, as well as the fact that evaluation through use of confidence interval overlap is less powerful since it mistakenly fails to reject the null hypothesis as frequently as does the use of the *t* test when the null hypothesis is false. To insure that the latter inaccuracies will not occur, instead of computing the confidence intervals noted above (which will be referred to as **descriptive confidence intervals**), Tryon (2001) (based on Goldstein and Healy (1995)) describes the computation of modified confidence intervals which will be referred to as **inferential confidence intervals**. The latter confidence intervals not only provide for a more rigorous test of whether or not a statistically significant difference exists between the two treatments, but also allow a researcher to test for **statistical equivalence** (as well as **statistical indeterminacy**, which will exist when the analyses for statistical significance and statistical equivalence are inconsistent with one another).

Computation of an **inferential confidence interval** results in a smaller confidence interval for each of the treatment means. The latter is due to the fact a modified critical *t* value is employed in computing an **inferential confidence interval**. The modified critical *t* value employed in computing an **inferential confidence interval** will be smaller than the critical *t* value employed in computing the **descriptive confidence intervals** obtained above. **Inferential confidence intervals can be employed to determine whether or not the difference between two treatments is statistically significant, since when a statistically significant difference exists the inferential confidence intervals computed for the means of each of the treatments will not overlap.**

The modified critical *t* value employed in computing the **inferential confidence intervals** can be obtained by multiplying the tabled critical *t* value one would employ for computing the **descriptive confidence interval** by a ratio labeled *E* which is computed with Equation 11.24.[37]

$$E = \frac{\sqrt{s_{\bar{X}_1}^2 + s_{\bar{X}_2}^2}}{s_{\bar{X}_1} + s_{\bar{X}_2}}$$

(Equation 11.24)

The value $E = .7160$ is computed for Example 11.4 below. When the latter value is multiplied by the tabled critical two-tailed $.05$ value $t_{.05} = 2.776$ (for $df = n - 1 = 4$), the modified critical value $t_{mod} = 1.99$ is obtained. Employing the latter value for $t_{.05}$ in Equation 2.8/2.9, the two **inferential confidence intervals** are computed below.

$$E = \frac{\sqrt{(.51)^2 + (.37)^2}}{.51 + .37} = \frac{.63}{.88} = .7160$$

$$t_{mod} = (t_{.05})(E) = (2.776)(.7160) = 1.99$$

$$CI_{.95} = \bar{X}_1 \pm (t_{mod})(s_{\bar{X}_1}) = 9.6 \pm (1.99)(.51) = 9.6 \pm 1.01$$

$$8.59 \leq \mu_1 \leq 10.61$$

$$CI_{.95} = \bar{X}_2 \pm (t_{mod})(s_{\bar{X}_2}) = 8.2 \pm (1.99)(.37) = 8.2 \pm .74$$

$$7.46 \leq \mu_2 \leq 8.94$$

Note that, as was the case when the **descriptive confidence intervals** were computed earlier, the two confidence intervals overlap (since the upper limit of the confidence interval of Group 2 (8.94) is larger than the lower limit of the confidence interval of Group 1 (8.59)). Thus, the researcher cannot conclude there is a significant difference between the two treatments (when alpha = .05 is employed). The reader, however, should note the following: a) The width of each of the **inferential confidence intervals** is smaller than the width of the corresponding **descriptive confidence interval** which was computed earlier; b) The **inferential confidence intervals** overlap to a lesser degree than do the **descriptive confidence intervals**. Consequently, a researcher is more likely to declare a significant difference (by virtue of nonoverlapping confidence intervals) through use of **inferential confidence intervals** than through use of **descriptive confidence intervals**.

The inferential confidence intervals can also be employed to determine whether or not the two treatments are equivalent. Specifically, **equivalence will exist if the difference between the lower limit of the confidence interval computed for the smaller mean and the upper limit of the confidence interval computed for the larger mean is less than the absolute value of ζ stipulated by the researcher.** In our example the lower limit of the confidence interval computed for the smaller mean is 7.46 and the upper limit of the confidence interval computed for the larger mean is 10.61. Thus, we compute the difference $10.61 - 7.46 = 3.15$. Since the value 3.15 is greater than the absolute value $\zeta = 2$, we cannot conclude that the treatments are equivalent (when alpha = .05 is employed).

Tryon (2001) notes that when using his methodology, a researcher should conduct both analyses described above employing the **inferential confidence intervals**. In other words the researcher should use the **inferential confidence intervals** to determine if there is a statistically significant difference between the means of the two treatments, as well as using the **inferential confidence intervals** to test for equivalence. To be more specific, the following guidelines should be employed in determining what conclusions to reach regarding the hypothesis being evaluated.

1) A researcher is only able to declare a statistically significant difference between the treatments at the prestipulated alpha level if both of the following conditions are met: a) The **inferential confidence intervals** do not overlap; and b) The **test of equivalence** indicates that the difference between the lower limit of the confidence interval computed for the smaller mean and the upper limit of the confidence interval computed for the larger mean is greater than or equal to the absolute value of ζ.

2) A researcher is only able to declare equivalence at the prestipulated alpha level if both of the following conditions are met: a) The **inferential confidence intervals** overlap; and b) The **test of equivalence** indicates that the difference between the lower limit of the confidence interval computed for the smaller mean and the upper limit of the confidence interval computed for the larger mean is less than the absolute value of ζ.

3) A condition of **statistical indeterminancy** would exist if the results of the two analyses are inconsistent with one another. **Statistical indeterminancy** would exist if either of the following conditions is present: a) The difference between the means of the two treatments is not statistically significant (since the **inferential confidence intervals** overlap), yet at the same time the **test of equivalence** is not significant (since the difference between the lower limit of the confidence interval computed for the smaller mean and the upper limit of the confidence interval computed for the larger mean is greater than or equal to the absolute value of ζ); or b) The difference between the means of the two treatments is statistically significant (since the **inferential confidence intervals** do not overlap), yet at the same time the **test of equivalence** is significant (since the difference between the lower limit of the confidence interval computed for the smaller mean and the upper limit of the confidence interval computed for the larger mean is less than the absolute value of ζ). When **statistical indeterminancy** exists further research should be conducted in order to clarify the status of the hypothesis under study.

When Example 11.4 is evaluated with Tryon's methodology (employing a two-tailed analysis with alpha = .05), the final results would be categorized under **statistical indeterminancy**. The latter is the case, since a statistically significant difference is not obtained between the treatments (since the **inferential confidence intervals** overlap), yet the **test of equivalence** does not indicate the two treatments are equivalent (since the difference between the lower limit of the confidence interval computed for the smaller mean and the upper limit of the confidence interval computed for the larger mean is greater than $\zeta = 2$).

When Tryon's (2001) methodology is employed to evaluate Example 11.4 it yields a conclusion which is consistent with that reached when the **Westlake–Schuirmann test of equivalence of two independent treatments** was employed to evaluate the same set of data (assuming a two-tailed analysis with alpha = .05 is conducted). However, it should be noted that the two methodologies for evaluating equivalence will not always yield results which are consistent with one another. In other words, there will be instances in which one of the methodologies will allow the researcher to declare equivalence while the other will not. The author of the book has determined that Tryon's (2001) methodology provides a more conservative test of equivalence. Consequently, when the results of the two tests are inconsistent, the **Westlake–Schuirmann test of equivalence of two independent treatments** is more likely to allow the researcher to declare equivalence than **Tryon's test of equivalence of two independent treatments**. (The latter was confirmed in a simulation study conducted by Tryon (2005).) When such a circumstance exists, a researcher should consider whether a larger value could have been employed for ζ, so that if Tryon's (2001) methodology is employed one can conclude equivalence exists. If not, further research should be conducted in order to clarify the status of the hypotheses under study.

In concluding this section Example 11.7 will be evaluated in order to demonstrate a situation in which both the **Westlake–Schuirmann test of equivalence of two independent**

treatments and **Tryon's test of equivalence of two independent treatments** will allow a researcher to conclude that two treatments are equivalent.

Example 11.7 *The manufacturer of a generic drug for treating a disease wants to empirically demonstrate it is equivalent to a brand name drug, which up to the present time has been the only approved treatment for the disease. The major criterion for diagnosing the disease is that patients diagnosed with the disease have an abnormally low concentration of a specific factor in their blood. Treatment with the brand name drug results in a substantial increase in the concentration of the latter factor. The Federal Drug Administration declares that equivalency of a generic drug with the brand name drug can be established if the concentration of the disease factor in patients treated with the generic drug is within 3 units of the concentration of the factor obtained for patients treated with the brand name drug. To demonstrate equivalency of the two drugs, a study is conducted in which 20 people afflicted with the disease (each patient exhibiting an abnormally low concentration of the disease factor) are randomly assigned to receive either the brand name drug or the generic drug for a period of six months. After six months have elapsed a measure of the disease factor is obtained for each patient. The results obtained from each of the groups follow:* **Group 1 (Brand name drug):** *5, 5, 10, 11, 7, 8, 5, 5, 8, 9;* **Group 2 (Generic drug):** *7, 6, 3, 8, 8, 5, 7, 4, 10, 9. Do the data indicate that the generic drug is equivalent to the brand name drug?*

The null and alternative hypotheses for a test of equivalence (employing a two-tailed analysis) are stated below. Note that in Example 11.7 the absolute value of $\zeta = 3$ is established by the researcher.

Null hypotheses $\qquad\qquad H_0: \theta \leq -3$ or $H_0: \theta \geq 3$

(In the underlying populations the samples represent, the difference between the two treatment means is less than or equal to -3 or is greater than or equal to 3. This is the **hypothesis of nonequivalence**.)

Alternative hypotheses $\qquad H_1: \theta > -3$ and $H_1: \theta < 3$

(In the underlying populations the samples represent, the difference between the two treatment means is greater than -3 and is less than 3. This is the **hypothesis of equivalence**.)

The following summary data are computed for Example 11.7: **Group 1:** $\bar{X}_1 = 7.3$; $\tilde{s}_1 = 2.26$; $\tilde{s}_1^2 = 5.12$; $s_{\bar{X}_1} = .715$; $s_{\bar{X}_1}^2 = .511$; $n_1 = 10$; **Group 2:** $\bar{X}_2 = 6.7$; $\tilde{s}_2 = 2.21$; $\tilde{s}_2^2 = 4.88$; $s_{\bar{X}_2} = .699$; $s_{\bar{X}_2}^2 = .488$; $n_2 = 10$; $s_{\bar{X}_1 - \bar{X}_2} = 1.00$. We will initially assume that $H_0: \mu_1 = \mu_2$ and $H_1: \mu_1 \neq \mu_2$ are evaluated with alpha $= .05$. A t **test for two independent samples** is conducted below. With $df = 10 + 10 - 2 = 18$, we determine that the tabled critical .05 two-tailed t value in Table **A2** is $t_{.05} = 2.101$. Since the obtained value $t = .6$ is less than $t_{.05} = 2.101$, the null hypothesis $H_0: \mu_1 = \mu_2$ cannot be rejected.

$$t = \frac{7.3 - 6.7}{\sqrt{\dfrac{5.12}{10} + \dfrac{4.88}{10}}} = \frac{.6}{1.00} = .6$$

Within the framework of the **Westlake–Schuirmann test of equivalence of two independent treatments**, the two t tests for two independent samples will now be conducted.

Initially, the following null and alternative hypotheses are evaluated with a one-tailed test: $H_0: \theta \le -3$ versus $H_1: \theta > -3$.

$$t = \frac{(7.3 - 6.7) - (-3)}{\sqrt{\dfrac{5.12}{10} + \dfrac{4.88}{10}}} = \frac{3.6}{1.00} = 3.6$$

With $df = 10 + 10 - 2 = 18$, we determine that the tabled critical one-tailed .05 value in Table **A2** is $t_{.05} = 1.734$. Since the obtained value $t = 3.6$ is greater than $t_{.05} = 1.734$, the null hypothesis $H_0: \theta \le -3$ designating nonequivalence can be rejected.

We now evaluate the second null hypothesis designating nonequivalence. Specifically, a one-tailed test which evaluates the following null and alternative hypotheses is conducted: $H_0: \theta \ge 3$ versus $H_1: \theta < 3$.

$$t = \frac{(7.3 - 6.7) - (3)}{\sqrt{\dfrac{5.12}{10} + \dfrac{4.88}{10}}} = \frac{-2.4}{1.00} = -2.4$$

Since the obtained absolute value $t = 2.4$ is greater than the tabled critical one-tailed value $t_{.05} = 1.734$, the second of the null hypotheses designating nonequivalence ($H_0: \theta \ge 3$) can be rejected. Since in order to declare equivalence **both** of the null hypotheses designating non-equivalence must be rejected (with opposite signs required for the computed t values), and, in fact, both have been rejected with opposite signs, equivalence has been established.

Employing Equation 11.17, the computation of a 90% confidence interval for Example 11.7 is computed below.

$$CI_{.90} = (\bar{X}_1 - \bar{X}_2) \pm (t_{.10})(s_{\bar{X}_1 - \bar{X}_2}) = .6 \pm (1.734)(1.00) = .6 \pm 1.734$$

$$1.134 \le (\mu_1 - \mu_2) \le 2.234$$

This result indicates the researcher can be 90% confident (or the probability is .90) that the interval -2.334 to 1.134 contains the true difference between the population means. If the aforementioned interval contains the true difference between the population means, the mean of the population **Group 1 (Brand name drug)** represents is between 1.134 and 2.234 units greater than the mean of the population **Group 2 (Generic drug)** represents. Note that both of the values $\zeta_1 = -3$ and $\zeta_2 = 3$ fall outside the confidence interval. Thus, analysis of the data with the **Westlake–Schuirmann test of equivalence of two independent treatments** allows the researcher to conclude that the treatments are equivalent.

Example 11.7 will now be evaluated with **Tryon's test of equivalence of two independent treatments**. Employing Equation 11.24, the value $E = .7069$ is computed. When the latter value is multiplied by the tabled critical two-tailed .05 value $t_{.05} = 2.262$ (for $df = n - 1 = 9$), the modified critical value $t_{mod} = 1.60$ is obtained. Employing the latter value for $t_{.05}$ in Equation 2.8/2.9, the two **inferential confidence intervals** are computed below.

$$E = \frac{\sqrt{(.715)^2 + (.699)^2}}{.715 + .699} = .7069$$

$$t_{mod} = (t_{.05})(E) = (2.262)(.7069) = 1.60$$

$$CI_{.95} = \bar{X}_1 \pm (t_{mod})(s_{\bar{X}_1}) = 7.3 \pm (1.60)(.715) = 7.3 \pm 1.14$$

$$6.16 \le \mu_1 \le 8.44$$

$$CI_{.95} = \bar{X}_2 \pm (t_{mod})(s_{\bar{X}_2}) = 6.7 \pm (1.60)(.699) = 6.7 \pm 1.12$$

$$5.58 \le \mu_2 \le 7.82$$

Since the two **inferential confidence intervals** overlap, we can conclude there is not a statistically significant difference between the two treatments. Employing the inferential confidence intervals to determine whether or not the two treatments are equivalent, we determine that the difference between the lower limit of the confidence interval computed for the smaller mean and the upper limit of the confidence interval computed for the larger mean is $8.44 - 5.58 = 2.86$. Since the value 2.86 is less than the absolute value $\zeta = 3$, we can conclude that the treatments are equivalent. Thus, if $\zeta = 3$ is stipulated as the value of ζ, through use of either **Tryon's test of equivalence of two independent treatments** or the **Westlake–Schuirmann test of equivalence of two independent treatments**, the researcher can conclude that the difference between the treatment means $\bar{X}_1 = 7.3$ and $\bar{X}_2 = 6.7$ is of no practical consequence, and that the two treatments are equivalent to one another.

In closing this discussion, it should be noted that when two or more methods yield inconsistent results with regard to the presence or absence of equivalence, such inconsistencies can generally be resolved by conducting replication studies (the results of which can be evaluated through the use of **meta-analysis**, which is discussed in Section IX (the **Addendum**) of the **Pearson product-moment correlation coefficient**.), and basing one's final conclusions on the overall body of evidence. The reader should keep in mind that in the final analysis the criterion employed for equivalence will be some standard imposed by an outside agency (e.g., the FDA), or some alternative subjective criterion stipulated by a specific researcher. Ultimately, the final decision regarding the issue of equivalence should be based on multiple studies which yield a consistent pattern of results involving minimal differences between treatments that fall within a range of values which most practitioners in a discipline would consider to be reasonable.

7. Hotelling's T^2   The **multivariate** analog of the *t* test for two independent samples is **Hotelling's T^2** (Hotelling (1931)), which is one of a number multivariate statistical procedures discussed in the book. As noted previously, the term **multivariate analysis** is employed in reference to procedures that evaluate experimental designs in which there are multiple independent variables and/or multiple dependent variables (although some sources limit its use to analyses in which there are two or more dependent variables). **Hotelling's T^2 (Test 34)**, which is a special case of the **multivariate analysis of variance (Test 35)**, can be employed to analyze the data for an experiment that involves a single independent variable comprised of two levels and multiple dependent variables. With regard to the latter, instead of a single score, each subject produces scores on two or more dependent variables. To illustrate, let us assume that in Example 11.1 two scores are obtained for each subject. One score represents a subject's level of depression and a second score represents the subject's level of anxiety, and when taken together the two scores are viewed as representing a generalized measure of emotional stability. Within the framework of **Hotelling's T^2**, a composite mean based on both the depression and anxiety scores of subjects is computed for each group. The latter composite means are referred to as **mean vectors** or **centroids**. As is the case with the *t* **test for two independent samples**, the means (in this case composite) for the two groups are then compared with one another. **Hotelling's T^2** is described in detail in **Chapter 34**.

VIII. Additional Examples Illustrating the Use of the *t* Test for Two Independent Samples

Two additional examples which can be evaluated with the *t* **test for two independent samples** are presented in this section. Since Examples 11.8 and 11.9 employ the same data set as that employed in Example 11.1, they yield the identical result.

Example 11.8 *A researcher wants to assess the relative effect of two different kinds of punishment (loud noise versus a blast of cold air) on the emotionality of mice. Each of ten mice is randomly assigned to one of two groups. During the course of the experiment each mouse is sequestered in an experimental chamber. While in the chamber, each of the five mice in Group 1 is periodically presented with a loud noise, and each of the five mice in Group 2 is periodically presented with a blast of cold air. The presentation of the punitive stimulus for each of the animals is generated by a machine which randomly presents the stimulus throughout the duration of the time the mouse is in the chamber. The dependent variable of emotionality employed in the study is the number of times each mouse defecates while in the experimental chamber. The number of episodes of defecation for the 10 mice follow:* **Group 1**: *11, 1, 0, 2, 0;* **Group 2**: *11, 11, 5, 8, 4. Do mice exhibit differences in emotionality under the different experimental conditions?*

In Example 11.8, if the one-tailed alternative hypothesis H_1: $\mu_1 < \mu_2$ is employed it can be concluded that the group presented the blast of cold air (Group 2) obtains a significantly higher emotionality score than the group presented with loud noise (Group 1). This is the case, since the computed value $t = -1.96$ indicates the average defecation score of Group 2 is significantly higher than the average defecation score of Group 1. As is the case in Example 11.1, the nondirectional alternative hypothesis H_1: $\mu_1 \neq \mu_2$ is not supported, and thus, if the latter alternative hypothesis is employed the researcher cannot conclude that the blast of cold air results in higher emotionality.

Example 11.9 *Each of two companies that manufacture the same size precision ball bearing claims it has better quality control than its competitor. A quality control engineer conducts a study in which he compares the precision of ball bearings manufactured by the two companies. The engineer randomly selects five ball bearings from the stock of Company A and five ball bearings from the stock of Company B. He measures how much the diameter of each of the ten ball bearings deviates from the manufacturer's specifications. The deviation scores (in micrometers) for the ten ball bearings manufactured by the two companies follow:* **Company A**: *11, 1, 0 , 2, 0;* **Company B**: *11, 11, 5, 8, 4. What can the engineer conclude about the relative quality control of the two companies?*

In Example 11.9, if the one-tailed alternative hypothesis H_1: $\mu_1 < \mu_2$ is employed it can be concluded that Company B obtains a significantly higher deviation score than Company A. This will allow the researcher to conclude that Company A has superior quality control. As is the case in Example 11.1, the nondirectional alternative hypothesis H_1: $\mu_1 \neq \mu_2$ is not supported, and thus, if the latter alternative hypothesis is employed the researcher cannot conclude that Company A has superior quality control.

References

Anderson, N. N. (2001). **Empirical direction in design and analysis**. Mahwah, NJ: Lawrence Erlbaum Associates.

Anderson, S. & Hauck. W. W. (1983). A new procedure for testing equivalence in comparative bioavailability and other clinical trials. **Communication in Statistics — Theory and Methods**, 12, 2263–2692.

Armitage, P., Berry, G., & Matthews, J. (2002). **Statistical methods in medical research**. Malden, MA: Blackwell Science.

Atherton Skaff, P. J. & Sloan, J. A. (2004). Design and analysis of equivalence clinical trials via the SAS system. Website: **http://www2.sas.com/ proceedings/sugi23/Stats/p218. pdf.**

Barnett, V. & Lewis, T. (1994). **Outliers in statistical data** (3rd ed.). Chichester: John Wiley & Sons.

Bartlett, M. S. (1947). The use of transformations. **Biometrics**, 3, 39–52.

Bohj, D. S. (1978). Testing equality of means of correlated variates with missing data on both responses. **Biometrika**, 65, 225–228.

Chou, Y. (1989). **Statistical analysis for business and economics**. New York: Elsevier.

Chow, S. C. & Liu, J. P. (2000). **Design and analysis of clinical trials** (2nd ed.). New York: Wiley–Interscience.

Chow, S. C. & Liu, J. P. (2004). **Design and analysis of bioavailability and bioequivalence studies** (2nd ed., revised and expanded). New York: Dekker.

Cochran, W. G. & Cox, G. M. (1957). **Experimental designs** (2nd ed.). New York: John Wiley & Sons.

Cohen, B. H. (2001). **Explaining psychological statistics** (2nd ed.). New York: John Wiley & Sons.

Cohen, J. (1977). **Statistical power analysis for the behavioral sciences**. New York: Academic Press.

Cohen, J. (1988). **Statistical power analysis for the behavioral sciences** (2nd ed.). Hillsdale, NJ: Lawrence Erlbaum Associates, Publishers.

Conover, W. J. & Inman, R. L. (1981). Rank transformation as a bridge between parametric and nonparametric statistics. **The American Statistician**, 35, 124-129.

Cribbie, R. A., Gruman, J. A., & Arpin-Cribbie, C. A. (2004). Recommendations for applying tests of equivalence. **Journal of Clinical Psychology**, 60, 1–10.

Dunnett, C. W. & Gent, M. (1977). Significance testing to establish equivalence between treatments with special reference to data in the form of 2×2 tables. Biometrics, 33, 593–602.

Everitt, B. S. (2002). **Cambridge dictionary of statistics**. London: Cambridge University Press.

Fleiss, J. L., Levin, B. & Paik, M. C. (2003). **Statistical methods for rates and proportions** (3rd ed.). New York: John Wiley & Sons.

Feinstein, A.R. (2002). **Principles of medical statistics**. Boca Raton, FL: Chapman & Hall/CRC.

Fisher, R. A., Corbet, A. S., & Williams, C. B. (1943). The relation between the number of species and the number of individuals in a random sample of an animal population. **Journal of Animal Ecology**, 12, 42–57.

Fisher, R. A. & Yates, F. (1953). **Statistical tables for biological, agricultural, and medical research** (4th ed.). Edinburgh: Oliver & Boyd.

Games, P. (1983). Curvilinear transformation of the dependent variable. **Psychological Bulletin**, 93, 382–387.

Games, P. (1984). Data transformations, power, and skew: A rebuttal to Levine and Dunlap. **Psychological Bulletin**, 95, 345–347.

Glass, G. (1976). Primary, secondary and meta-analysis of research. **Educational Research**, 1976, 5, 3–8.

Goldstein, H. & Healy, M. J. R. (1995). The graphical presentation of a collection of means. **Journal of the Royal Statistical Society**, 158A, Part 1, 175–177.

Good, P. (1994). **Permutation tests: A practical guide to resampling methods for testing hypotheses**. New York: Springer.

Grissom, R. J. & Kim, J. J. (2005). **Effect sizes for research: A broad practical approach**. Mahwah, NJ: Lawrence Erlbaum Associates Publishers.

Guenther, W. C. (1965). **Concepts of statistical inference**. New York: McGraw–Hill Book Company.

Hair, J. F., Anderson, R. E., Tatham, R. L., & Black, W. C. (1995). **Multivariate data analysis with readings** (4th ed.). Upper Saddle River, NJ: Prentice Hall, Inc.

Hair, J. F., Anderson, R. E., Tatham, R. L., & Black, W. C. (1998). **Multivariate data analysis** (5th ed.). Upper Saddle River, NJ: Prentice Hall, Inc.

Harlow, L. L. (2005). **The essence of multivariate thinking**. Mahwah, NJ: Lawrence Erlbaum Associates.

Hartley, H. O. (1940). Testing the homogeneity of a set of variances. **Biometrika**, 31, 249–255.

Hartley, H. O. (1950). The maximum F-ratio as a shortcut test for heterogeneity of variance. **Biometrika**, 37, 308–312.

Hedges, L. V. (1981). Distribution theory for Glass's estimator of effect size and related estimators. **Journal of Educational Statistics**, 6, 107–128.

Hedges, L. V. (1982). Estimation of effect size from a series of independent experiments. **Psychological Bulletin**, 92, 490–499.

Hedges, L. V. & Olkin, I. (1985). **Statistical methods for meta-analysis**. Orlando, FL: Academic Press.

Hollander, M. & Wolfe, D.A. (1999). **Nonparametric statistical methods**. New York: John Wiley & Sons.

Hotelling, H. (1931). The generalization of Student's ratio. **Annals of Mathematical Statistics**, 2, 361–378.

Howell, D. C. (2002). **Statistical methods for psychology** (5th ed.). Pacific Grove, CA.: Duxbury.

Huber, P. J. (1981). **Robust Statistics**. New York: John Wiley & Sons.

Hunt, M. (1997). **How science takes stock**. New York: Russell Sage Foundation.

Jennison, C. & Turnbull, B. W. (2000) **Group sequential methods with applications to clinical trials**. Boca Raton: CRC Press.

Kaplan, E. L. & Meier, P. (1958). Nonparametric estimation from incomplete observations. **Journal of the American Statistical Association**, 53, 457–481.

Keppel, G. (1991). **Design and analysis: A researcher's handbook** (3rd ed.). Englewood Cliffs, NJ: Prentice Hall.

Keppel, G., Saufley, W. H. Jr., & Tokunaga, H. (1992). **Introduction to design and analysis: A student's handbook**. New York: W. H. Freeman & Company.

Kirk, R. E. (1982). **Experimental design: Procedures for the behavioral sciences** (2nd ed.). Belmont, CA: Brooks/Cole Publishing Company.

Kirk, R. (1995). **Experimental design: Procedures for the behavioral sciences**. Pacific Grove, CA: Brooks/Cole Publishing Company.

Kruskal, W. H. (1960). Some remarks on wild observations. **Technometrics**, 2, 1–3.

Lee, H. & Fung, K. Y. (1983). Robust procedures for multi-sample location problems with unequal group variances. **Journal of Statistical Computation and Simulation**, 18, 125–143.

Levine, D. W. & Dunlap, W. P. (1982). Power of the *F* test with skewed data: Should one transform or not? **Psychological Bulletin**, 92, 272–280.

Levine, D. W. & Dunlap, W. P. (1983). Data transformation, power, and skew: A rejoinder to Games. **Psychological Bulletin**, 93, 599–596.

Little, R. J. A. & Rubin, D. R. (1987). **Statistical analysis with missing data**. New York: John Wiley & Sons.

Little, R. J. A. & Rubin, D. R. (2002). **Statistical analysis with missing data**. (2nd ed.) Hoboken, N. J. :John Wiley & Sons.

Maxwell, S. E. & Delaney, H. D. (2004). **Designing experiments and analyzing data: A model comparison perspective** (2nd ed.). Mahwah, NJ: Lawrence Erlbaum Associates.

Miller, I. & Miller, M. (1999). **John E. Freund's mathematical statistics** (6th ed.). Upper Saddle River, NJ: Prentice Hall.

Myers, J. L. & Well, A. D. (1995). **Research design and statistical analysis**. Hillsdale, N.J.: Lawrence Erlbaum Associates, Publishers.

Pagano, M. & Gauvreau, K. (1993). **Principles of biostatistics**. Belmont, CA: Duxbury Press.

Rao, P. V. (1998). **Statistical research methods in the life sciences**. Pacific Grove, CA: Duxbury Press.

Rocke, D. M., Downs, G. W., & Rocke, A. J. (1982). Are robust estimators really necessary? **Technometrics**, 24, 95–110.

Rogers, J. L., Howard, K. I., & Vessey, J. T. (1993). Using significance tests to evaluate equivalence between two experimental groups. **Psychological Bulletin**, 113, 553–565.

Rosenthal, R., Rosnow, R. L., & Rubin, D. B. (2000). **Contrasts and effect sizes in behavioral research: A correlational approach**. Cambridge, UK: Cambridge University Press.

Rosner, B. (1983). Percentage points for a generalized ESD many-outlier procedure. **Technometrics,** 25(2), 165–172.

Rosner, B. (1995). **Fundamentals of biostatistics** (4th ed.). Belmont, CA: Duxbury Press.

Rosner, B. (2000). **Fundamentals of biostatistics** (5th ed.). Pacific Grove, CA: Duxbury Press.

Rosnow, R. L., Rosenthal, R & Rubin, D. B. (2000). Contrasts and correlations in effect-size estimation. **Psychological Science**, 11, 446–453.

Satterthwaite, F. E. (1946). An approximate distribution of estimates of variance components. **Biometrics Bulletin**, 2, 110–114.

Schenker, N. & Gentleman, J. F. (2001). On judging the significance of difference by examining the overlap between confidence intervals means. **The American Statistician**, 55, 182–186.

Schuirmann, D. J. (1987). A comparison of the two one-sided tests procedure and the power approach for assessing equivalence of average bioavailability. **Journal of Pharmacokinetics and Biopharmaceutics**, 15, 657–680.

Seaman, M. A. & Serlin, R. C. (1998). Equivalence confidence intervals for two-group comparisons of means. **Psychological Methods**, 3, 403–411.

Shiffler, R. (1988). Maximum *z* scores and outliers. **American Statistician**, 42, 79–80.

Smithson, M. (2003). **Confidence intervals**. Thousand Oaks, CA: Sage Publications.

Sprent, P. (1993). **Applied nonparametric statistical methods** (2nd ed.). London: Chapman & Hall.

Sprent, P. (1998). **Data driven statistical methods**. London: Chapman & Hall.

Staudte R. G. & Sheather S. J. (1990). **Robust estimation and testing**. New York: John Wiley & Sons.

Stevens, J. (1986). **Applied multivariate statistics for the social sciences**. Hillsdale, NJ: Lawrence Erlbaum Associates, Publishers.

Stevens, J. (1996). **Applied multivariate statistics for the social sciences** (3rd ed.). Mahwah, NJ: Lawrence Erlbaum Associates, Publishers.

Tabachnick, B.G. & Fidell, L. S. (1989). **Using multivariate statistics** (2nd ed.). New York: HarperCollins Publishers.

Tabachnick, B. G. & Fidell, L. S. (1996). **Using multivariate statistics** (3rd ed.). New York: HarperCollins College Publishers.

Thöni, H. (1967). Transformation of variables used in the analysis of experimental and observational data. A review. Technical Report Number 7, Statistical Laboratory, Iowa State University. Ames, Iowa, 61 pp.

Tietjen, G. L. (1986). The analysis and detection of outliers. In R. B. D'Agostino & M. A. Stephens (Eds.), **Goodness-of-fit techniques**. New York: Marcel Dekker, Inc.

Tryon, W. W. (2001). Evaluating statistical difference, equivalence, and indeterminacy using inferential confidence intervals: An integrated alternative method of conducting null hypothesis statistical tests. **Psychological Methods**, 6, 371–386.

Tryon, W. W. (2005). Evaluating the inferential confidence interval method of establishing statistical equivalence. Unpublished manuscript.

van Belle, G. (2002). **Statistical rules of thumb**. New York: John Wiley & Sons.

Welch, B. L. (1947). The generalization of Student's problem when several different population variances are involved. **Biometrika**, 34, 28–35.

Wellek, S. (2003). **Testing Statistical hypotheses of equivalence**. Boca Raton, FL: Chapman & Hall/CRC.

Westlake, W. J. (1976). Symmetrical confidence intervals for bioequivalence trials. **Biometrics**, 32, 741–744.

Westlake, W. J. (1981). Response to T.B. L. Kirkwood: Bioequivalence testing — a need to rethink. **Biometrics**, 37, 589–594.

Westlake, W. J. (1988). Bioavailability and bioequivalence of pharmaceutical formulations. In K. E. Peace (Ed.), **Biopharamaceutical statistics for drug development** (pp. 329–352). New York: Marcel Decker.

Wilcox, R. R. (1987). **New statistical procedures for the social sciences**. Hillsdale, NJ: Erlbaum.

Wilcox, R. R. (1996). **Statistics for the social sciences**. San Diego, CA: Academic Press.

Wilcox, R. R. (1997). **Introduction to robust estimation and hypothesis testing**. San Diego, CA: Academic Press.

Wilcox, R. R., (2003). **Applying contemporary statistical techniques**. San Diego, CA: Academic Press.

Winer, B. J., Brown, D., & Michels, K. (1991). **Statistical principles in experimental design** (3rd ed.). New York: McGraw–Hill Publishing Company.

Yuen, K. K. (1974). The two sample trimmed *t* for unequal population variances. **Biometrika**, 61, 165–170.

Zar, J. (1999). **Biostatistical analysis** (4th ed.). Upper Saddle River, NJ: Prentice Hall.

Zimmerman, D. W. & Zumbo, B. D. (1993). The relative power of parametric and non-parametric statistical methods. In Keren, G. & Lewis, C. (Eds.), **A handbook for data analysis in the behavioral sciences: Methodological issues**. Hillsdale, NJ: Lawrence Erlbaum Associates, Publishers.

Endnotes

1. Alternative terms which are commonly used to describe the different samples employed in an experiment are **groups, experimental conditions**, and **experimental treatments**.

2. It should be noted that there is a design in which different subjects serve in each of the k experimental conditions that is categorized as a **dependent samples design**. In a **dependent samples design** each subject either serves in all of the k experimental conditions, or else is matched with a subject in each of the other $(k-1)$ experimental conditions. When subjects are matched with one another they are equated on one or more variables which are believed to be correlated with scores on the dependent variable. The concept of matching and a general discussion of the **dependent sample design** can be found under the *t* **test for two dependent samples**.

3. An alternative but equivalent way of writing the null hypothesis is H_0: $\mu_1 - \mu_2 = 0$. The analogous alternative but equivalent ways of writing the alternative hypotheses in the order they are presented are: H_1: $\mu_1 - \mu_2 \neq 0$, H_1: $\mu_1 - \mu_2 > 0$, H_1: $\mu_1 - \mu_2 < 0$.

4. In order to be solvable, an equation for computing the *t* statistic requires that there is variability in the scores of at least one of the two groups. If all subjects in Group 1 have the same score and all subjects in Group 2 have the same score, the values computed for the estimated population variances will equal zero (i.e., $\tilde{s}_1^2 = \tilde{s}_2^2 = 0$). If the latter is true the denominator of any of the equations to be presented for computing the value of *t* will equal zero, thus rendering a solution impossible.

5. When $n_1 = n_2$, $\tilde{s}_{pooled} = \sqrt{(\tilde{s}_1^2 + \tilde{s}_2^2)/2}$.

6. The actual value that is estimated by $s_{\bar{X}_1 - \bar{X}_2}$ is $\sigma_{\bar{X}_1 - \bar{X}_2}$, which is the standard deviation of the sampling distribution of the difference scores for the two populations. The meaning of the **standard error of the difference** can be best understood by considering the following procedure for generating an empirical sampling distribution of difference scores: a) Obtain a random sample of n_1 scores from Population 1 and a random sample of n_2 scores from Population 2; b) Compute the mean of each sample; c) Obtain a difference score by subtracting the mean of Sample 2 from the mean of Sample 1 — i.e., $\bar{X}_1 - \bar{X}_2 = D$; and d) Repeat steps a) through c) *m* times. At the conclusion of this procedure one will have obtained *m* difference scores. The **standard error of the difference** represents the standard deviation of the *m* difference scores, and can be computed by using Equation I.11/2.1. Thus: $s_{\bar{X}_1 - \bar{X}_2} = \sqrt{[(\Sigma D^2 - ((\Sigma D)^2/m)]/[m - 1]}$. The standard deviation that is computed with the aforementioned equation is an estimate of $\sigma_{\bar{X}_1 - \bar{X}_2}$.

7. Equation 11.4 can also be written in the form $df = (n_1 - 1) + (n_2 - 1)$, which reflects the number of degrees of freedom for each of the groups.

8. The absolute value of *t* is employed to represent *t* in the summary statement.

9. a) The F_{max} test is one of a number of statistical procedures that are named after the English statistician Ronald Fisher. Among Fisher's contributions to the field of statistics was the development of a sampling distribution referred to as the F distribution (which bears the

first letter of his surname). The values in the F_{max} distribution are derived from the F distribution; b) Alternative tests (**Levene's test for homogeneity of variance (Test 21g)** and **The Brown–Forsythe test for homogeneity of variance (Test 21h)**) for evaluating homogeneity of variance (which like the F_{max} test can be employed for contrasting $k \geq 2$ experimental conditions), as well as a more in depth discussion of the pros and cons of the available tests of homogeneity of variance, can be found in Section VI of the **single-factor between-subjects analysis of variance**. Additionally, Grissom and Kim (2005, pp. 12–21) provide a detailed discussion of the limitations of the more commonly employed methodologies for assessing homogeneity of variance.

10. A tabled $F_{.975}$ value is the value below which 97.5% of the F distribution falls and above which 2.5% of the distribution falls. A tabled $F_{.995}$ value is the value below which 99.5% of the F distribution falls and above which .5% of the distribution falls.

11. In **Table A9** the value $F_{max_{.01}} = 23.2$ is the result of rounding off $F_{.995} = 23.15$.

12. A tabled $F_{.025}$ value is the value below which 2.5% of the F distribution falls and above which 97.5% of the distribution falls. A tabled $F_{.005}$ value is the value below which .5% of the F distribution falls and above which 99.5% of the distribution falls.

13. a) Most sources only list values in the upper tail of the F distribution. The values $F_{.05} = .157$ and $F_{.01} = .063$ are obtained from Guenther (1965). It so happens that when $df_{num} = df_{den}$, the value of $F_{.05}$ can be obtained by dividing 1 by the value of $F_{.95}$. Thus: $1/6.39 = .157$. In the same respect the value of $F_{.01}$ can be obtained by dividing 1 by the value of $F_{.99}$. Thus: $1/15.98 = .063$; b) Endnote 52 of the **single-factor between-subjects analysis of variance** describes the procedure for computing a confidence interval for the ratio of two variances, which allows for an alternative way of conducting the **F test for two population variances**.

14. When $n = n_1 = n_2$ and $t' = t_1 = t_2$, the t' value computed with Equation 11.9 will equal the tabled critical t value for $df = n - 1$. When $n_1 \neq n_2$, the computed value of t' will fall in between the values of t_1 and t_2. It should be noted that the effect of violation of the homogeneity of variance assumption on the t test statistic decreases as the number of subjects employed in each of the samples increases. This can be demonstrated in relation to Equation 11.9, in that if there are a large number of subjects in each group the value which is employed for both t_1 and t_2 in Equation 11.9 is $t_{.05} = 1.96$. The latter tabled critical two-tailed .05 value, which is also the tabled critical two-tailed .05 value for the normal distribution, is the value that is computed for t'. Thus, in the case of large sample sizes the tabled critical value for $df = n_1 + n_2 - 2$ will be equivalent to the value computed for $df = n_1 - 1$ and $df = n_2 - 1$.

15. a) The values .2, .5. and .8, which are respectively employed as the lower limits of a small, medium, and large effect size for **Cohen's d index**, can also be viewed with respect to the degree of overlap between the sampling distributions of the means for the two population distributions involved in an analysis. When $d = .20$ the degree of overlap between the two distributions is 85%, when $d = .50$ the degree of overlap is 67%, and when $d = .80$ the degree of overlap is only 53%. Obviously, the greater the similarity between the two distributions, the greater the degree of overlap between them and the smaller the effect size;

b) Since **Cohen's *d* index** is based on population parameters, sample analogues of the *d* **index** have been developed by and/or are described in Glass (1976), Hedges (1981, 1982), and Hedges and Olkin (1985). Equation 11.25 is most commonly employed for computation of a sample analogue index which is commonly referred to as the *g* **index**. The latter value is often employed as a measure of the effect size for a study. Note that the value $\tilde{s}_p$ computed with Equation 11.11 is employed in the denominator of Equation 11.25. Hedges (1981), Cohen (2001), and Maxwell and Delaney (2004, p. C-5) note that the value computed with Equation 11.25 (as well as the value computed with Equation 11.10 when sample statistics are employed to estimate the parameters $\bar{X}_1$, $\bar{X}_2$, and σ) is a slightly positively biased estimate (i.e., overestimates) of the true value of *d* in the underlying population. However, the degree of bias becomes minimal with the use of large sample sizes.

When the values $\bar{X}_1 = 2.8$, $\bar{X}_2 = 5.8$, and $\tilde{s}_{pooled} = 4.02$, (computed for Example 11.1) are substituted in Equation 11.25, the value $g = 1.24$ is obtained: $g = |2.8 - 5.8| / 4.02 = 1.24$. Note that this is the same value computed previously for *d*. *g* values are interpreted through use of the same guidelines employed to interpret *d* values. Thus, since $g = 1.24$ is greater than .8 it represents a large effect size.

$$g = \frac{|\bar{X}_1 - \bar{X}_2|}{\tilde{s}_p}$$ **(Equation 11.25)**

Various sources (e.g., Cohen (2001), Rosenthal *et al.* (2000), and Rosnow *et al.* (2000)) note that Equations 11.26 and 11.27 can respectively be employed to transform a *g* value into a *t* value and a *t* value into a *g* value. (In the aforementioned equations, the notation *N* represents the total sample size employed in a study while *n* represents the number of subjects per group.)

$$t = g\sqrt{\frac{n}{2}}$$ **(Equation 11.26)**

$$g = t\sqrt{\frac{2}{n}} = \frac{2t}{\sqrt{N}}$$ **(Equation 11.27)**

Note that when the value $g = 1.24$ computed for Example 11.1 is substituted in Equation 11.26, the value $t = 1.24\sqrt{5/2} = 1.96$ is obtained (which is the absolute value computed for *t* with Equation 11.1). Conversely, when the absolute value $t = 1.96$ is substituted in Equation 11.27, it yields the value $g = 1.24$ ($g = 1.96\sqrt{2/5} = 1.24$ or $g = (2)(1.96)/\sqrt{10} = 1.24$); c) Cohen (2001, p. 237) notes that approximate values for a confidence interval can be computed for a *g* value by dividing the upper and lower limits computed for the **confidence interval for the difference between the means** by the value of $\tilde{s}_p$. (A full description of the protocol for computing a confidence interval for the difference between the means can be found in the latter part of Section VI.) To illustrate, in the latter part of Section VI it is determined that the limiting values for the 95% confidence interval are -10.89 and .89. This result indicates the researcher can be 95% confident (or the probability is .95) the interval -10.89 to .89 contains the true difference between the population means. If the aforementioned interval contains the true difference between the population means, the mean of the population Group 2 represents is no more than 10.89 points higher than the mean of the population Group 1 represents, and the mean of the population Group 1 represents is no more than .89 points higher than the mean of the population Group 2 represents. If the absolute values of the two limits are divided by $\tilde{s}_p = 4.02$, the resulting values will define the limits for the 95% confidence interval for the **g index**. Thus, the upper limit for $g = 10.89/4.02 = 2.71$ and the lower limit for $g = .89/4.02 = .22$. (Note, however, that the two effect size values are in different directions.) Thus, the researcher can be 95% confident that the interval .22 to 2.71 (which is expressed in standard deviation units) contains the true effect size value in the underlying populations. Cohen (2001, p. 237) (also see Grissom and Kim (2005, p. 64)) notes that the latter values are biased estimates of the actual value of the confidence interval in the underlying populations, and that computation of the exact confidence interval requires access to the **noncentral *t* distribution** (which is a theoretical sampling distribution for the *t* statistic in which the mean is some value other than zero); d) Further discussion of the **g index** can be found in Section IX (the **Addendum**) of the **Pearson product-moment correlation coefficient** under the discussion of **meta-analysis and related topics**.

16. A comprehensive discussion on the appropriate use of the harmonic mean versus the arithmetic mean as a measure of central tendency can be found in Chou (1989, pp. 110–113). In the discussion Chou (1989, p. 112) notes that the use of the harmonic mean is best suited for situations in which the scores in a distribution are expressed inversely to what is desired to be reflected by the average score computed for the distribution.

17. van Belle (2002, pp. 31–33) notes that Equation 11.28 can be employed to estimate the sample size required in order to conduct a ***t* test for two independent samples**, if a nondirectional analysis is conducted with $\alpha = .05$. The value of k in the numerator of the latter equation will depend on the desired power of the test, and for the power values .50, .80, .90, .95, and .975 the values to employ for k are respectively 8, 16, 21, 26, and 31. Equation 11.28 will be demonstrated in reference to computing the necessary sample size for the power of the test to equal .80, if as previously the researcher wants to employ the latter power for evaluating the alternative hypothesis: H_1: $|\mu_1 - \mu_2| \geq 5$. If the power of the test is to equal .80, the value, $k = 16$ is employed in the numerator of Equation 11.28. Equation 11.10 is employed to compute the value $d = 1.24$ (which was computed earlier by dividing the 5 point difference stipulated in the alternative hypotheses by the value $\tilde{s}_{pooled} = 4.02$, which represents the best estimate of σ). When the appropriate values are substituted in Equation 11.28, the value $n_j = 10.41$ is computed (where n_j represents the number of subjects in the jth group). The latter result indicates that in order for the power of the test to equal .80, each group should have approximately 11 subjects.

$$n_j = \frac{k}{d^2} = \frac{16}{(1.24)^2} = 10.41 \qquad \textbf{(Equation 11.28)}$$

18. The **treatment effect** described in this section is not the same thing as **Cohen's *d* index** (the **effect size** computed with Equation 11.10). However, if a hypothesized effect size is present in a set of data, the computed value of *d* can be used as a measure of treatment effect (in point of fact, the sample analogue **g index** is usually employed to represent effect size). In such an instance, the value of *d* (or the **g index**) will be positively correlated with the value of the treatment effect described in this section. Cohen (1988, pp. 24–27) describes how the *d* **index** can be converted into one of the correlational treatment effect measures which are discussed in this section. Endnote 20 discusses the relationship between the *d* **index** and the **omega squared** statistic presented in this section in more detail.

19. The reader familiar with the concept of correlation can think of a measure of treatment effect as a correlational measure which provides information analogous to that provided by the **coefficient of determination** (designated by the notation r^2), which is the square of the **Pearson product-moment correlation coefficient**. The **coefficient of determination** (which is discussed in more detail in Section V of the **Pearson product-moment correlation coefficient**) measures the degree of variability on one variable which can be accounted for by variability on a second variable. This latter definition is consistent with the definition that is provided in this section for a treatment effect.

20. a) In actuality, Cohen (1977, 1988) employs the notation for **eta squared** (which is discussed briefly in the next paragraph and in greater detail in Section VI of the **single-factor between-subjects analysis of variance**) in reference to the aforementioned effect size values. Endnote 58 in the **single-factor between-subjects analysis of variance** clarifies Cohen's (1977, 1988) use of **eta squared** and **omega squared** to represent the same measure; b) Cohen (1977, 1988, pp. 23–27) states that the small, medium, and large effect size values of .0099, .0588, and .1379 are equivalent to the values .2, .5, and .8 for his **d index** (which was discussed previously in the section on statistical power). In point of fact, the values .2, .5, and .8 represent the minimum values for a small, medium, and large effect size for **Cohen's d index**. The conversion of an **omega squared/eta squared** value into the corresponding **Cohen's d index** value is described in Section IX (the **Addendum**) of the **Pearson product-moment correlation coefficient** under the discussion of **meta-analysis and related topics**; c) Grissom and Kim (2005, p. 5) note it is not uncommon for measures of effect size to overestimate the actual size of an effect in an underlying population. The latter is referred to as **positive** or **upward bias**. The latter authors provide information regarding recommended adjustments for some of the measures of effect size discussed in this book.

21. This result can also be written as: $-.89 \leq (\mu_2 - \mu_1) \leq 10.89$.

22. In instances where, in stating the null hypothesis, a researcher stipulates that the difference between the two population means is some value other than zero, the numerator of Equation 11.18 is the same as the numerator of Equation 11.5. The protocol for computing the value of the numerator is identical to that employed for Equation 11.5.

23. The general issues discussed in this section are relevant to any case in which a parametric and nonparametric test can be employed to evaluate the same set of data.

24. Barnett and Lewis (1994) note that the presence of an outlier may not always be obvious as a result of visual inspection of data. Typically, the more complex the structure of data the more difficult it becomes to visually detect outliers. **Multivariate analysis** (which is described in the last section of the book) often involves data for which visual detection of outliers is difficult.

25. In this latter instance, if, as a result of the presence of one or more outliers, the difference between the group means is also inflated, the use of a more conservative test will, in part, compensate for the heterogeneity of variance. The impact of outliers on the **t test for two independent samples** is discussed by Zimmerman and Zumbo (1993). The latter authors note that the presence of outliers in a sample may decrease the power of the *t* test to such a degree that the **Mann-Whitney U test** (which is the rank-order nonparametric analog of

the *t* **test for two independent samples**) will be a more powerful test for comparing two independent samples.

26. Barnett and Lewis (1994, p. 84) note that the use of the **median absolute deviation** as a measure of dispersion/variability can be traced back to the 19th century to the great German mathematician Johann Karl Friedrich Gauss. Barnett and Lewis (1994, p. 156) state that although the **median absolute deviation** is a less efficient measure of dispersion than the standard deviation, it is a more robust estimator (especially for nonnormally distributed data).

27. Samples in which data have been deleted or modified are sometimes referred to as **censored samples** (Barnett and Lewis (1994, p. 78). The term **censoring**, however, is most commonly employed in reference to data that are not available for some subjects, since it is either not desirable or possible to follow each subject until the conclusion of a study. This latter type of censored data is most commonly encountered in medical research when subjects no longer make themselves available for study, or a researcher is unable to locate subjects beyond a certain period of time. Good (1994, p. 117) notes that another example of censoring occurs when, within the framework of evaluating a variable, the measurement breaks down at some point on the measurement continuum (usually at an extreme point). Consequently, one must employ approximate scores instead of exact scores to represent the observations which cannot be measured with precision. Two obvious options that can be employed to negate the potential impact of censored data are: a) Use of the median in lieu of the mean as a measure of central tendency; and b) Employing an inferential statistical test which uses rank-orders instead of interval/ratio scores. A detailed discussion on the general subject of censored data can be found in Section IX (the **Addendum**) of the **Mann–Whitney *U* test**. Within the context of the latter discussion a number of methods for evaluating censored data are described including **the Kaplan–Meir estimate (Test 12d), Gehan's test for censored data (Test 12e)**, and the **log-rank test (Test 12f)**. Censored data is also briefly discussed later this section within the context of the discussion of clinical trials.

28. In this example, as well as other examples in this section, use of the F_{max} test may not yield a significant result (i.e., it may not result in the conclusion that the population variances are heterogeneous). The intent of the examples, however, is only to illustrate the variance stabilizing properties of the transformation methods.

29. If the relationship 1 radian = 57.3 degrees is applied for a specific proportion, the number of degrees computed with the equation $Y = \arcsin\sqrt{X}$ will not correspond to the number of radians computed with the equation $Y = 2\arcsin\sqrt{X}$. Nevertheless, if the transformed data derived from the two equations are evaluated with the same inferential statistical test, the same result is obtained. In point of fact, if the equation $Y = \arcsin\sqrt{X}$ is employed to derive the value of Y in radians, and the resulting value is multiplied by 57.3, it will yield the same number of degrees obtained when that equation is used to derive the value of Y in degrees. Since the multiplication of $\arcsin\sqrt{X}$ by 2 in the equation $Y = 2\arcsin\sqrt{X}$ does not alter the value of the ratio for the difference between means versus pooled variability (or other relevant parameters being estimated within the framework of a statistical test), it yields the same test statistic regardless of which equation is employed. The author is indebted to Jerrold Zar for clarifying the relationship between the arcsine equations.

30. Rogers *et al.* (1993, pp. 554–555) note that within the framework of the **classical hypothesis testing model**, a conventional two-tailed analysis (i.e., where the null hypothesis states equivalence of conditions) can be conceptualized as involving two one-tailed tests. In such an analysis, the null hypothesis is rejected if either one (but not both) of the two one-tailed analyses is significant. Because in such an analysis a Type I error occurs if by chance either of the two one-tailed tests is significant, the alpha level for each of the one-tailed tests must be added to compute the overall likelihood of committing a Type I error. As an example, if a two-tailed/nondirectional analysis is conducted with alpha = .05, it is commensurate with conducting two one-tailed tests employing alpha =.025 for each test. On the other hand, in the case of a **test of equivalence**, both one-tailed analyses stipulated in the null hypothesis must be significant in order to reject the null hypothesis. In the case of a **test of equivalence**, the likelihood that an analysis evaluating the larger of the two absolute values ζ_1 versus ζ_2 will yield a significant result, given the fact that the smaller of the two absolute values ζ_1 versus ζ_2 did yield a significant result, is in fact equal to 1. Because of the latter, the overall probability of committing a Type I error in such an analysis is the prestipulated alpha level employed for the **smaller of the two absolute values** ζ_1 and ζ_2 multiplied by 1, which, in fact, is equal to the prestipulated alpha level employed for analysis of the **smaller of the two absolute values** ζ_1 versus ζ_2.

31. Among others, Seaman and Serlin (1998) note that if an analysis evaluating the null hypothesis H_0: $\mu_1 = \mu_2$ is not significant, the latter result should perhaps motivate a researcher to go a step further and conduct a **test of equivalence**, since as noted earlier retention of the null hypothesis H_0: $\mu_1 = \mu_2$ is not commensurate with demonstrating equivalence.

32. Note that in a **symmetrical** analysis, the absolute value of ζ employed in Equation 11.22 for each of the two *t* **tests** conducted within the framework of the **test of equivalence** will always be the same. If, on the other hand, a **nonsymmetrical** analysis (which will not be demonstrated here) is conducted, the absolute value of ζ employed in each of the two *t* **tests** will not be the same, since by definition $|\zeta_1| \neq |\zeta_2|$.

33. Tryon (2001) notes that Dunnett and Gent (1977) employ the tabled critical two-tailed .05 critical value in computing a confidence interval within the framework of a test of equivalence. If the latter value $t_{.05} = 2.306$ (for $df = 8$) is employed for Example 11.6, the width of the confidence interval increases, thus resulting in a less conservative test. Specifically, the confidence interval becomes $1.4 \pm (2.306)(.63) = 1.4 \pm 1.45$, which results in the confidence interval $-.05 \leq (\mu_1 - \mu_2) \geq 2.85$. Thus any absolute value designated for ζ which is greater than 2.85 will yield a significant result for both of the two *t* **tests** that are employed in testing for equivalence. The test is less conservative since the researcher can specify a larger value for ζ in order to establish equivalence.

34. a) Strictly speaking, since the values of ζ_1 and ζ_2 should lie outside the confidence interval, the absolute value of θ employed in Equation 11.22 should be slightly larger than 2.57 (e.g., 2.5700001); b) Technically, if the statement is made that the mean of **Group 1** (**Brand name drug**) is within 2.57 units of the mean of **Group 2** (**Generic drug**), the alternative hypotheses would be written as follows: H_1: $\theta \geq -2.57$ and H_1: $\theta \leq 2.57$. Since the equals sign should not be included in the null hypotheses, the latter should be written as H_0: $\theta < -2.57$ or H_0: $\theta > 2.57$.

35. a) Cribbie *et al.* (2004) conducted a simulation study contrasting the ability of the *t* **test for two independent samples** when used within the framework of the **classical hypothesis testing model** (i.e., a conventional test of significance as employed with Example 11.4) versus the **Westlake–Schuirmann test of equivalence of two independent treatments** (employed to evaluate Example 11.6) in evaluating whether or not two treatments were equivalent to one another. In other words, when the *t* **test for two independent samples** was used within the framework of the **classical hypothesis testing model**, a nonsignificant result was interpreted as evidence of equivalence. Their analysis suggested that the **Westlake–Schuirmann test of equivalence of two independent treatments** is more effective than the conventional *t* **test** at detecting equivalence of population means when large sample sizes are employed in a study. However, the conventional *t* **test** was more effective in detecting equivalence of population means when small sample sizes were employed and/or when group variances were large; b) The reader should take note of the fact that in conducting a test of equivalence a researcher may elect to employ a smaller (or, for that matter, larger) alpha value than those used in the examples in this section.

36. If, in fact, there is no difference between the means of the two populations (i.e., $\mu_1 - \mu_2 = 0$), Equation 11.23 becomes Equation 11.29. Under the latter circumstances in order for the power of the test to be .80 only $n = 4$ subjects are required per group.

(Equation 11.29)

$$n_1 = \frac{\left(\sigma_1^2 + \dfrac{\sigma_2^2}{k}\right)(z_{\alpha/2} + z_\beta)^2}{\zeta^2} = \frac{\left(1.3 + \dfrac{.7}{1}\right)(1.96 + .84)^2}{(2)^2} = 3.92$$

37. a) The interested reader can find the rationale and derivation of E (which is the standard error of the difference employed for the *t* **test for two independent samples** divided by the sum of the two standard error of the means) in Tryon (2001, pp. 374–375); b) Tryon (2001, p. 375) notes that the value of E increases toward 1 as the standard errors of the two treatments become increasingly unequal. In spite of the latter, even when the standard error of one treatment is five times that of the other treatment, the value of $E = .8498$; c) If there are an unequal number of subjects in each group, a separate t_{mod} value must be computed for each group by multiplying the value of E by the appropriate tabled critical two-tailed .05 value (for $df = n - 1$, where n represents the number of subjects in a group) for a given group.

Test 12

The Mann–Whitney U Test
(Nonparametric Test Employed with Ordinal Data)

I. Hypothesis Evaluated with Test and Relevant Background Information

Hypothesis evaluated with test Do two independent samples represent two populations with different median values (or different distributions with respect to the rank-orderings of the scores in the two underlying population distributions)?

Relevant background information on test The **Mann–Whitney U test** is employed with ordinal (rank-order) data in a hypothesis testing situation involving a design with two independent samples. If the result of the **Mann–Whitney U test** is significant, it indicates there is a significant difference between the two sample medians, and as a result of the latter the researcher can conclude there is a high likelihood the samples represent populations with different median values.

Two versions of the test to be described under the label of the **Mann–Whitney U test** were independently developed by Mann and Whitney (1947) and Wilcoxon (1949). The version to be described here is commonly identified as the **Mann–Whitney U test**, while the version developed by Wilcoxon (1949) is usually referred to as the **Wilcoxon–Mann–Whitney test**.[1] Although they employ different equations and different tables, the two versions of the test yield comparable results. In employing the **Mann–Whitney U test**, one of the following is true with regard to the rank-order data that are evaluated: a) The data are in a rank-order format, since it is the only format in which scores are available; or b) The data have been transformed into a rank-order format from an interval/ratio format, since the researcher has reason to believe that the normality assumption (as well as, perhaps, the homogeneity of variance assumption) of the **t test for two independent samples (Test 11)** (which is the parametric analog of the **Mann–Whitney U test**) is saliently violated. It should be noted that when a researcher elects to transform a set of interval/ratio data into ranks, information is sacrificed. This latter fact accounts for the reluctance among some researchers to employ nonparametric tests such as the **Mann–Whitney U test**, even if there is reason to believe that one or more of the assumptions of the **t test for two independent samples** have been violated.

Various sources (e.g., Conover (1980, 1999), Daniel (1990), and Marascuilo and McSweeney (1977)) note that the **Mann–Whitney U test** is based on the following assumptions: a) Each sample has been randomly selected from the population it represents; b) The two samples are independent of one another; c) The original variable observed (which is subsequently ranked) is a continuous random variable. In truth, this assumption, which is common to many nonparametric tests, is often not adhered to, in that such tests are often employed with a dependent variable which represents a discrete random variable; and d) The underlying distributions from which the samples are derived are identical in shape. (Grissom and Kim (2005, p. 111) state that the term **nonhomomerity** indicates inequality in the shape of two or more distributions.) The shapes of the underlying population distributions, however, do not have to be normal. Maxwell and Delaney (1990) point out the assumption of identically shaped distributions implies equal dispersion of data within each distribution. Because of this, they (as well as Zimmerman and Zumbo (1993a; 1993b) and Grissom and Kim (2005, p. 100–101)) note that like the *t* **test for two independent samples**, the **Mann–Whitney** *U* **test** assumes homogeneity of variance with respect to the underlying population distributions. Because the latter assumption is not generally acknowledged for the **Mann–Whitney** *U* **test**, it is not uncommon for sources to state that violation of the homogeneity of variance assumption justifies use of the **Mann–Whitney** *U* **test** in lieu of the *t* **test for two independent samples**. It should be pointed out, however, that there is some empirical evidence which suggests that the sampling distribution for the **Mann–Whitney** *U* **test** is not as affected by violation of the homogeneity of variance assumption as is the sampling distribution for *t* **test for two independent samples**. One reason cited by various sources for employing the **Mann–Whitney** *U* **test** is that by virtue of ranking interval/ratio data, a researcher will be able to reduce or eliminate the impact of **outliers**. As noted in Section VII of the *t* **test for two independent samples**, since **outliers** can dramatically influence variability, they can be responsible for heterogeneity of variance between two or more samples. In addition, **outliers** can have a dramatic impact on the value of a sample mean.

II. Example

Example 12.1 is identical to Example 11.1 (which is evaluated with the *t* **test for two independent samples**). In evaluating Example 12.1 it will be assumed that the interval/ratio data are rank-ordered, since one or more of the assumptions of the *t* **test for two independent samples** have been saliently violated.

Example 12.1　*In order to assess the efficacy of a new antidepressant drug, ten clinically depressed patients are randomly assigned to one of two groups. Five patients are assigned to Group 1, which is administered the antidepressant drug for a period of six months. The other five patients are assigned to Group 2, which is administered a placebo during the same six-month period. Assume that prior to introducing the experimental treatments, the experimenter confirmed that the level of depression in the two groups was equal. After six months elapse all ten subjects are rated by a psychiatrist (who is blind with respect to a subject's experimental condition) on their level of depression. The psychiatrist's depression ratings for the five subjects in each group follow (the higher the rating, the more depressed a subject):* **Group 1**: 11, 1, 0, 2, 0; **Group 2**: 11, 11, 5, 8, 4. *Do the data indicate that the antidepressant drug is effective?*

III. Null versus Alternative Hypotheses

Null hypothesis　　　　　　　　　$H_0: \theta_1 = \theta_2$

(The median of the population Group 1 represents equals the median of the population Group 2 represents. With respect to the sample data, when both groups have an equal sample size, this translates into the sum of the ranks of Group 1 being equal to the sum of the ranks of Group 2 (i.e., $\Sigma R_1 = \Sigma R_2$). A more general way of stating this, which also encompasses designs involving unequal sample sizes, is that the means of the ranks of the two groups are equal (i.e., $\bar{R}_1 = \bar{R}_2$).

Alternative hypothesis　　　　　　$H_1: \theta_1 \neq \theta_2$

(The median of the population Group 1 represents does not equal the median of the population Group 2 represents. With respect to the sample data, when both groups have an equal sample size, this translates into the sum of the ranks of Group 1 not being equal to the sum of the ranks of Group 2 (i.e., $\Sigma R_1 \neq \Sigma R_2$). A more general way of stating this, which also encompasses designs involving unequal sample sizes, is that the means of the ranks of the two groups are not equal (i.e., $\bar{R}_1 \neq \bar{R}_2$). This is a **nondirectional alternative hypothesis** and it is evaluated with a **two-tailed test**.)

or

$$H_1: \theta_1 > \theta_2$$

(The median of the population Group 1 represents is greater than the median of the population Group 2 represents. With respect to the sample data, when both groups have an equal sample size (so long as a rank of 1 is given to the lowest score), this translates into the sum of the ranks of Group 1 being greater than the sum of the ranks of Group 2 (i.e., $\Sigma R_1 > \Sigma R_2$). A more general way of stating this (which also encompasses designs involving unequal sample sizes) is that the mean of the ranks of Group 1 is greater than the mean of the ranks of Group 2 (i.e., $\bar{R}_1 > \bar{R}_2$). This is a **directional alternative hypothesis** and it is evaluated with a **one-tailed test**.)

or

$$H_1: \theta_1 < \theta_2$$

(The median of the population Group 1 represents is less than the median of the population Group 2 represents. With respect to the sample data, when both groups have an equal sample size (so long as a rank of 1 is given to the lowest score), this translates into the sum of the ranks of Group 1 being less than the sum of the ranks of Group 2 (i.e., $\Sigma R_1 < \Sigma R_2$). A more general way of stating this (which also encompasses designs involving unequal sample sizes) is that the mean of the ranks of Group 1 is less than the mean of the ranks of Group 2 (i.e., $\bar{R}_1 < \bar{R}_2$). This is a **directional alternative hypothesis** and it is evaluated with a **one-tailed test**.)

Note: Only one of the above noted alternative hypotheses is employed. If the alternative hypothesis the researcher selects is supported, the null hypothesis is rejected.

IV. Test Computations

The data for Example 12.1 are summarized in Table 12.1. The total number of subjects employed in the experiment is $N = 10$. There are $n_1 = 5$ subjects in Group 1 and $n_2 = 5$ subjects in Group 2. The original interval/ratio scores of the subjects are recorded in the columns labeled X_1 and X_2. The adjacent columns R_1 and R_2 contain the rank-order assigned to each of the scores. The rankings for Example 12.1 are summarized in Table 12.2. The ranking protocol for the **Mann–Whitney U test** is described in this section. Note that in Table 12.1 and Table 12.2 each subject's identification number indicates the order in Table 12.1 in which a subject's score appears in a given group, followed by his/her group. Thus, Subject i, j is the ith subject in Group j.

The following protocol, summarized in Table 12.2, is used in assigning ranks.

a) All $N = 10$ scores are arranged in order of magnitude (irrespective of group membership), beginning on the left with the lowest score and moving to the right as scores increase. This is done in the second row of Table 12.2.

b) In the third row of Table 12.2, all $N = 10$ scores are assigned a rank. Moving from left to right, a rank of 1 is assigned to the score that is furthest to the left (which is the lowest score), a rank of 2 is assigned to the score that is second from the left (which, if there are no ties, will be the second lowest score), and so on until the score at the extreme right (which will be the highest score) is assigned a rank equal to N (if there are no ties for the highest score).

Table 12.1 Data for Example 12.1

	Group 1			Group 2	
	X_1	R_1		X_2	R_2
Subject 1,1	11	9	Subject 1,2	11	9
Subject 2,1	1	3	Subject 2,2	11	9
Subject 3,1	0	1.5	Subject 3,2	5	6
Subject 4,1	2	4	Subject 4,2	8	7
Subject 5,1	0	1.5	Subject 5,2	4	5
		$\Sigma R_1 = 19$			$\Sigma R_2 = 36$

$$\bar{R}_1 = \frac{\Sigma R_1}{n_1} = \frac{19}{5} = 3.8 \qquad \bar{R}_2 = \frac{\Sigma R_2}{n_2} = \frac{36}{5} = 7.2$$

Table 12.2 Rankings for the Mann–Whitney U Test for Example 12.1

Subject identification number	3,1	5,1	2,1	4,1	5,2	3,2	4,2	1,1	1,2	2,2
Depression score	0	0	1	2	4	5	8	11	11	11
Rank prior to tie adjustment	1	2	3	4	5	6	7	8	9	10
Tie-adjusted rank	1.5	1.5	3	4	5	6	7	9	9	9

c) The ranks in the third row of Table 12.2 must be adjusted when there are tied scores present in the data. Specifically, in instances where two or more subjects have the same score, the average of the ranks involved is assigned to all scores tied for a given rank. This adjustment is made in the fourth row of Table 12.2. To illustrate: Both Subjects 3,1 and 5,1 have a score of 0. Since the two scores of 0 are the lowest scores out of the total of ten scores, in assigning ranks to these scores we can arbitrarily assign one of the 0 scores a rank of 1 and the other a rank of 2. However, since both of these scores are identical it is more equitable to give each of them the same rank. To do this, we compute the average of the ranks involved for the two scores. Thus, the two ranks involved prior to adjusting for ties (i.e., the ranks 1 and 2) are added up and divided by two. The resulting value $(1 + 2)/2 = 1.5$ is the rank assigned to each of the subjects who is tied for 0. There is one other set of ties present in the data which involves three subjects. Subjects 1,1, 1,2, and 2,2 all obtain a score of 11. Since the ranks assigned to these three scores prior to adjusting for ties are 8, 9, and 10, the average of the three ranks $(8 + 9 + 10)/3 = 9$ is assigned to the scores of each of the three subjects who obtain a score of 11.

Although it is not the case in Example 12.1, it should be noted that any time each set of ties involves subjects in the same group, the tie adjustment will result in the identical sum and average for the ranks of the two groups which will be obtained if the tie adjustment is not employed. Because of this, under these conditions the computed test statistic will be identical regardless of whether or not one uses the tie adjustment. On the other hand, when one or more sets of ties involve subjects from both groups, the tie-adjusted ranks will yield a value for the test statistic that will be different from that which will be obtained if the tie adjustment is not employed. In Example 12.1, although the two subjects who obtain a score of zero happen to be in the same group, in the case of the three subjects who have a score of 11, one subject is in Group 1 and the other two subjects are in Group 2.

If the ranking protocol described in this section is used with Example 12.1, and the researcher elects to employ a one-tailed alternative hypothesis, the directional alternative hypothesis H_1: $\theta_1 < \theta_2$ is employed. The latter directional alternative hypothesis is employed, since it predicts that Group 1, the group which receives the antidepressant, will have a lower median score, and thus a lower sum of ranks/average rank (both of which are indicative of a lower level of depression) than Group 2.

It should be noted that it is permissible to reverse the ranking protocol described in this section. Specifically, one can assign a rank of 1 to the highest score, a rank of 2 to the second highest score, and so on, until reaching the lowest score which is assigned a rank equal to the value of N. Although this reverse ranking protocol will yield the identical **Mann–Whitney test** statistic as the ranking protocol described in this section, it will result in ranks which are the opposite of those obtained in Table 12.2. If the protocol employed in ranking is taken into account in interpreting the results of the **Mann–Whitney U test**, both ranking protocols will lead to identical conclusions. Since it is less likely to cause confusion in interpreting the test statistic, it is recommended that the original ranking protocol described in this section be employed — i.e., assigning a rank of 1 to the lowest score and a rank equivalent to the value of N to the highest score. In view of this, in all future discussion of the **Mann–Whitney U test**, as well as other tests that involve rank-ordering data, it will be assumed (unless otherwise stipulated) that the ranking protocol employed assigns a rank of 1 to the lowest score and a rank of N to the highest score.

Once all of the subjects have been assigned a rank, the sum of the ranks for each of the groups is computed. These values, $\Sigma R_1 = 19$ and $\Sigma R_2 = 36$, are computed in Table 12.1. Upon determining the sum of the ranks for both groups, the values U_1 and U_2 are computed employing Equations 12.1 and 12.2.

$$U_1 = n_1 n_2 + \frac{n_1(n_1 + 1)}{2} - \Sigma R_1 \qquad \text{(Equation 12.1)}$$

$$U_2 = n_1 n_2 + \frac{n_2(n_2 + 1)}{2} - \Sigma R_2 \qquad \text{(Equation 12.2)}$$

Employing Equations 12.1 and 12.2, the values $U_1 = 21$ and $U_2 = 4$ are computed.

$$U_1 = (5)(5) + \frac{5(5 + 1)}{2} - 19 = 21$$

$$U_2 = (5)(5) + \frac{5(5 + 1)}{2} - 36 = 4$$

Note that U_1 and U_2 can never be negative values. If a negative value is obtained for either, it indicates an error has been made in the rankings and/or calculations.

Equation 12.3 can be employed to confirm that the correct values have been computed for U_1 and U_2.

$$n_1 n_2 = U_1 + U_2 \qquad \text{(Equation 12.3)}$$

If the relationship in Equation 12.3 is not confirmed, it indicates an error has been made in ranking the scores or in the computation of the U values. The relationship described by Equation 12.3 is confirmed below for Example 12.1.

$$(5)(5) = 21 + 4 = 25$$

V. Interpretation of the Test Results

The smaller of the two values U_1 versus U_2 is designated as the obtained U statistic. Since $U_2 = 4$ is smaller than $U_1 = 21$, the value of $U = 4$. The value of U is evaluated with **Table A11 (Table of Critical Values for the Mann–Whitney U Statistic)** in the **Appendix**. In **Table A11**, the critical U values are listed in reference to the number of subjects in each group. For $n_1 = 5$ and $n_2 = 5$, the tabled critical two-tailed .05 and .01 values are $U_{.05} = 2$ and $U_{.01} = 0$, and the tabled critical one-tailed .05 and .01 values are $U_{.05} = 4$ and $U_{.01} = 1$. In order to be significant, the obtained value of U must be **equal to or less than** the tabled critical value at the prespecified level of significance.[2]

Since the obtained value $U = 4$ must be equal to or less than the aforementioned tabled critical values, the null hypothesis can only be rejected if the directional alternative hypothesis $H_1: \theta_1 < \theta_2$ is employed. The directional alternative hypothesis $H_1: \theta_1 < \theta_2$ is supported at the .05 level, since $U = 4$ is equal to the tabled critical one-tailed value $U_{.05} = 4$. The data are consistent with the directional alternative hypothesis $H_1: \theta_1 < \theta_2$, since the average of the ranks in Group 1 is less than the average of the ranks in Group 2 (i.e., $\bar{R}_1 < \bar{R}_2$).[3] The directional alternative hypothesis $H_1: \theta_1 < \theta_2$ is not supported at the .01 level, since the obtained value $U = 4$ is greater than the tabled critical one-tailed value $U_{.01} = 1$.

The nondirectional alternative hypothesis $H_1: \theta_1 \neq \theta_2$ is not supported, since the obtained value $U = 4$ is greater than the tabled critical two-tailed value $U_{.05} = 2$.

Since the data are not consistent with the directional alternative hypothesis $H_1: \theta_1 > \theta_2$, the latter alternative hypothesis is not supported. In order for the directional alternative hypothesis $H_1: \theta_1 > \theta_2$ to be supported, the average of the ranks in Group 1 must be greater than the average of the ranks in Group 2 (i.e., $\bar{R}_1 > \bar{R}_2$) (as well as the fact that the computed value of U must be equal to or less than the tabled critical one-tailed value at the prespecified level of significance).

The results of the **Mann–Whitney U test** are consistent with those obtained when the **t test for two independent samples** is employed to evaluate Example 11.1 (which employs the same set of data as Example 12.1). In both instances, the null hypothesis can only be rejected if the researcher employs a directional alternative hypothesis which predicts a lower degree of depression in the group which receives the antidepressant medication (Group 1).

VI. Additional Analytical Procedures for the Mann–Whitney U Test and/or Related Tests

1. The normal approximation of the Mann–Whitney U statistic for large sample sizes If the sample size employed in a study is relatively large, the normal distribution can be employed to approximate the **Mann–Whitney U statistic**. Although sources do not agree on the value of the sample size which justifies employing the normal approximation of the Mann–Whitney distribution (see Fahoome (2002) and Grissom and Kim (2005, p. 102)), they generally state that it should be employed for sample sizes larger than those documented in the exact table of the U distribution contained within the source. Equation 12.4 provides the normal approximation of the **Mann–Whitney U test** statistic.

$$z = \frac{U - \dfrac{n_1 n_2}{2}}{\sqrt{\dfrac{n_1 n_2 (n_1 + n_2 + 1)}{12}}}$$

(Equation 12.4)

In the numerator of Equation 12.4 the term $(n_1 n_2)/2$ is often summarized with the notation U_E, which represents the **expected (mean) value** of U if the null hypothesis is true. In other words, if in fact the two groups are equivalent, it is expected that $\bar{R}_1 = \bar{R}_2$. If the latter is true then $U_1 = U_2$, and both of the values U_1 and U_2 will equal $(n_1 n_2)/2$. The denominator of Equation 12.4 represents the expected **standard deviation** of the sampling distribution for the normal approximation of the U statistic (which can be summarized with the notation $\sqrt{Var(U)}$), and, consequently, the square of the denominator represents the **variance** $(Var(U))$ of the sampling distribution for the normal approximation of the U statistic.

Although Example 12.1 involves only $N = 10$ scores (a value most sources would view as too small to use with the normal approximation), it will be employed to illustrate Equation 12.4. The reader will see that in spite of employing Equation 12.4 with a small sample size, it yields a result which is consistent with the result obtained when the exact table for the **Mann–Whitney** U distribution is employed. It should be noted that since the smaller of the two values U_1 versus U_2 is selected to represent U, the value of z will always be negative (unless $U_1 = U_2$, in which case $z = 0$). This is the case, since by selecting the smaller value U will always be less than the expected value $U_E = (n_1 n_2)/2$.

Employing Equation 12.4, the value $z = -1.78$ is computed. Note that in the computations below $U_E = 12.5$ and $\sqrt{Var(U)} = 4.79$.[4]

$$z = \frac{4 - \dfrac{(5)(5)}{2}}{\sqrt{\dfrac{(5)(5)(5 + 5 + 1)}{12}}} = = \frac{-8.5}{4.79} = -1.78$$

The obtained value $z = -1.78$ is evaluated with **Table A1 (Table of the Normal Distribution)** in the **Appendix**. In order to be significant, the obtained absolute value of z must be equal to or greater than the tabled critical value at the prespecified level of significance. The tabled critical two-tailed .05 and .01 values are $z_{.05} = 1.96$ and $z_{.01} = 2.58$, and the tabled critical one-tailed .05 and .01 values are $z_{.05} = 1.65$ and $z_{.01} = 2.33$. The following guidelines are employed in evaluating the null hypothesis.

a) If the nondirectional alternative hypothesis $H_1: \theta_1 \neq \theta_2$ is employed, the null hypothesis can be rejected if the obtained absolute value of z is equal to or greater than the tabled critical two-tailed value at the prespecified level of significance.

b) If a directional alternative hypothesis is employed, one of the two possible directional alternative hypotheses is supported if the obtained absolute value of z is equal to or greater than the tabled critical one-tailed value at the prespecified level of significance. The directional alternative hypothesis which is supported is the one that is consistent with the data.

Employing the above guidelines with Example 12.1, the following conclusions are reached.

Since the obtained absolute value $z = 1.78$ must be equal to or greater than the tabled critical value at the prespecified level of significance, the null hypothesis can only be rejected if the directional alternative hypothesis $H_1: \theta_1 < \theta_2$ is employed. The directional alternative hypothesis $H_1: \theta_1 < \theta_2$ is supported at the .05 level, since the absolute value $z = 1.78$ is greater than the tabled critical one-tailed value $z_{.05} = 1.65$. As noted in Section V, the data are consistent with the directional alternative hypothesis $H_1: \theta_1 < \theta_2$. The directional alternative hypothesis $H_1: \theta_1 < \theta_2$ is not supported at the .01 level, since the obtained absolute value $z = 1.78$ is less than the tabled critical one-tailed value $z_{.01} = 2.33$.

The nondirectional alternative hypothesis $H_1: \theta_1 \neq \theta_2$ is not supported, since the obtained absolute value $z = 1.78$ is less than the tabled critical two-tailed .05 value $z_{.05} = 1.96$.

Since the data are not consistent with the directional alternative hypothesis H_1: $\theta_1 > \theta_2$, the latter alternative hypothesis is not supported. As noted in Section V, in order for the latter directional alternative hypothesis to be supported, the following condition must be met: $\bar{R}_1 > \bar{R}_2$.

It turns out that the above conclusions based on the normal approximation are identical to those reached when the exact table of the **Mann–Whitney U distribution** is employed.

It should be noted that, in actuality, either U_1 or U_2 can be employed in Equation 12.4 to represent the value of U. This is the case, since either value yields the same absolute value for z. Thus, if for Example 12.1 $U_1 = 21$ is employed in Equation 12.4, the value $z = 1.78$ is computed. Since the decision with respect to the status of the null hypothesis is a function of the absolute value of z, the value $z = 1.78$ leads to the same conclusions which are reached when $z = -1.78$ is employed. The decision with regard to a directional alternative hypothesis is not affected, since the data are still consistent with the directional alternative hypothesis H_1: $\theta_1 < \theta_2$.

2. The correction for continuity for the normal approximation of the Mann–Whitney U test[5] Although not used in most sources, Siegel and Castellan (1988) employ a correction for continuity for the normal approximation of the **Mann–Whitney test** statistic. Marascuilo and McSweeney (1977, p. 275) note that the correction for continuity is generally not employed, unless the computed absolute value of z is close to the prespecified tabled critical value. The correction for continuity, which reduces the absolute value of z, requires that .5 be subtracted from the absolute value of the numerator of Equation 12.4 (as well as the absolute value of the numerator of the alternative equation described in Endnote 4). The continuity-corrected version of Equation 12.4 is provided by Equation 12.5.

$$z = \frac{\left| U - \dfrac{n_1 n_2}{2} \right| - .5}{\sqrt{\dfrac{n_1 n_2 (n_1 + n_2 + 1)}{12}}} \qquad \textbf{(Equation 12.5)}$$

If the correction for continuity is employed with Example 12.1, the value computed for the numerator of Equation 12.5 is 8 (in contrast to the value 8.5 computed with Equation 12.4). Employing Equation 12.5 with Example 12.1, the value $z = 1.67$ is computed.

$$z = \frac{\left| 4 - \dfrac{(5)(5)}{2} \right| - .5}{\sqrt{\dfrac{(5)(5)(5 + 5 + 1)}{12}}} = \frac{8}{4.79} = 1.67$$

Since the absolute value $z = 1.67$ is greater than the tabled critical one-tailed value $z_{.05} = 1.65$, the directional alternative hypothesis H_1: $\theta_1 < \theta_2$ is still supported at the .05 level. (Note that in actuality the sign of z is still negative, since $U = 4$ is less than $U_E = 8$.) The reader should take note of the fact that the absolute value of the continuity-corrected z value will always be less than the absolute value computed when the correction for continuity is not used, except when the uncorrected value of $z = 0$ (in which case it will have no impact on the result of the analysis).

3. Tie correction for the normal approximation of the Mann–Whitney U statistic Some sources recommend that when an excessive number of ties is present in the data, a tie correction should be introduced into Equation 12.4. Equation 12.6 is the tie-corrected equation for the normal approximation of the **Mann–Whitney U distribution**. The latter equation results in a slight increase in the absolute value of z.[6]

$$z = \frac{U - \dfrac{n_1 n_2}{2}}{\sqrt{\dfrac{n_1 n_2 (n_1 + n_2 + 1)}{12} - \dfrac{n_1 n_2 \left[\displaystyle\sum_{i=1}^{s} (t_i^3 - t_i)\right]}{12(n_1 + n_2)(n_1 + n_2 - 1)}}}$$

(Equation 12.6)

The only difference between Equations 12.4 and 12.6 is the term to the right of the element $[n_1 n_2 (n_1 + n_2 + 1)]/12$ in the denominator. The result of this subtraction reduces the value of the denominator, thereby resulting in the slight increase in the absolute value of z. The term $\sum_{i=1}^{s}(t_i^3 - t_i)$ in the denominator of Equation 12.6 computes a value based on the number of ties in the data. In Example 12.1 there are $s = 2$ sets of ties. Specifically, there is a set of ties involving two subjects with the score 0, and a set of ties involving three subjects with the score 11. The notation $\sum_{i=1}^{s}(t_i^3 - t_i)$ indicates the following: a) For each set of ties, the number of ties in the set is subtracted from the cube of the number of ties in that set; and b) The sum of all the values computed in part a) is obtained. Thus, for Example 12.1:

$$\sum_{i=1}^{s} (t_i^3 - t_i) = [(2)^3 - 2] + [(3)^3 - 3] = 30$$

The tie-corrected value $z = -1.80$ is now computed employing Equation 12.6. Note that in the computations below $\sqrt{Var(U)} = 4.71$, which is slightly less than the value $\sqrt{Var(U)} = 4.79$ computed with Equation 12.4.

$$z = \frac{4 - \dfrac{(5)(5)}{2}}{\sqrt{\dfrac{(5)(5)(5 + 5 + 1)}{12} - \dfrac{(5)(5)(30)}{12(5 + 5)(5 + 5 - 1)}}} = \frac{-8.5}{4.71} = -1.80$$

The difference between $z = -1.80$ and the uncorrected value $z = -1.78$ is trivial, and consequently the decision the researcher makes with respect to the null hypothesis is not affected, regardless of which alternative hypothesis is employed.[7]

4. Computation of a confidence interval for a difference between the medians of two independent populations Prior to reading this section the reader should review the discussion of the computation of a confidence interval in Section VI of both the **single-sample t test** and the **t test for two independent samples**. The methods to be described in this section for computing a confidence interval for the **Mann–Whitney U test** (i.e., a confidence interval for a difference between the medians of two independent populations) are described in Higgins (2004)). Identical or similar methodologies for computing such a confidence interval can be found in Conover (1990), Daniel (1990), Hollander and Wolfe (1999) and Marascuilo and McSweeney (1977).

The first method to be described should be employed for samples sizes documented in **Table A11**, whereas the second method to be described is a large sample approximation based on the normal distribution. Both methods require constructing a table such as Table 12.3, which documents all pairwise differences between the scores in different groups. Note that in the latter table the scores in Group 1 are arranged ordinally in the top row, and the scores in Group 2 are arranged ordinally in the left column. Each of the scores in the cells of Table 12.3 are difference scores which are the result of subtracting the score in the extreme left of the row a cell appears (i.e., the corresponding Group 2 score) from the score at the top of the column in which the cell appears (i.e., the corresponding Group 1 score). In Table 12.4 the $n_1 n_2 = 25$ difference scores (to be referred to as **pairwise differences**) have been arranged with respect to ordinal position with from smallest to largest. In the latter table **OP** represents the ordinal position of a **pairwise difference (PWD)**. Note that the lowest pairwise difference of -11 is assigned an ordinal position 1 and the highest pairwise difference 7 is assigned the highest ordinal position 25.

Table 12.3 Pairwise Differences for Example 12.1

		Group 1 Scores			
		0	0	1	2	11
	4	-4	-4	-3	-2	7
	5	-5	-5	-4	-3	6
Group 2	8	-8	-8	-7	-6	3
Scores	11	-11	-11	-10	-9	0
	11	-11	-11	-10	-9	0

Table 12.4 Pairwise Differences Arranged with Respect to Ordinal Position for Example 12.1

OP	1	2	3	4	5	6	7	8	9	10	11	12	13
PWD	-11	-11	-11	-11	-10	-10	-9	-9	-8	-8	-7	-6	-5

OP	14	15	16	17	18	19	20	21	22	23	24	25
PWD	-5	-4	-4	-4	-3	-3	-2	0	0	3	6	7

In order to determine the 95% confidence interval we must identify the tabled critical two-tailed .05 value in **Table A11**, which as noted in Section V is $U_{.05} = 2$. (The tabled critical two-tailed .01 value is employed to compute the 99% confidence interval, while the tabled critical one-tailed .05 and .01 values are respectively employed to compute the 90% and 98% confidence intervals.) In order to compute a confidence interval we must stipulate a value to be designated as k which will always be **one unit above the relevant tabled critical value**. For the example under discussion, the value of k will equal 3 since the relevant tabled critical value equals 2 — i.e., $k = (U_{.05} = 2) + 1 = 3$. The value of k indicates the **ordinal position of the pairwise difference** (to be designated OP_L) **which will define the lower limit of the confidence interval**. The **ordinal position of the pairwise difference** (to be designated OP_U) **which will define upper limit of the confidence interval** will equal $OP_U = n_1 n_2 - OP_L + 1$. Since $k = 3$, the **ordinal position** of the pairwise difference which will define the lower limit of the 95% confidence interval is **3**. The pairwise difference of -11 in the **third ordinal position** is thus employed as the lower limit of the confidence interval. Since $OP_U = n_1 n_2 - OP_L + 1 = 25 - 3 + 1 = 23$, the pairwise difference of 3 in the **23rd ordinal position** is employed to represent the upper limit of the confidence interval. (The latter upper limit could also be determined through use of the value $k = 3$ by selecting the pairwise difference in the **third ordinal position**, if ordinal position is determined starting with the largest pairwise difference and moving downward to the smallest pairwise difference. If the latter protocol is employed, the pairwise difference in the **third ordinal position** in Table 12.4 will equal **3**.)

The notation $-11 \le (\theta_1 - \theta_2) \le 3$ summarizes the values which define the 95% confidence interval. This result indicates the researcher can be 95% confident (or the probability is .95) that the interval -11 to 3 contains the true difference between the population medians. If the aforementioned interval contains the true difference between the population medians, the median of the population Group 2 represents is no more than 11 points higher than the median of the population Group 1 represents, and the median of the population Group 1 represents is no more than 3 points higher than the median of the population Group 2 represents.

For sample sizes not documented in **Table A11**, Equations 12.7 and 12.8 can be employed to compute the values of OP_L and OP_U. Upon computing OP_L with Equation 12.7, the rounding off protocol for determining the final value of OP_U with Equation 12.8 should insure that the following relationship exists: $OP_L + OP_U = n_1 n_2 + 1$.

$$OP_{L_{.95}} = U_E - z_{.05} \sqrt{Var(U)} \qquad \textbf{(Equation 12.7)}$$

$$OP_{L_{.95}} = \frac{(5)(5)}{2} - (1.96)(4.79) = 3.11$$

$$OP_{U_{.95}} = 1 + U_E + z_{.05} \sqrt{Var(U)} \qquad \textbf{(Equation 12.8)}$$

$$OP_{U_{.95}} = 1 + \frac{(5)(5)}{2} + (1.96)(4.79) = 22.89$$

Rounding off to the nearest integer we set the values for computing the confidence interval at $OP_{L_{.95}} = 3$ and $OP_{U_{.95}} = 23$. Note that $[(OP_{L_{.95}} = 3) + (OP_{U_{.95}} = 23)] = [(n_1 n_2 = 25) + 1] = 26$. (If the tie corrected value $\sqrt{Var(U)} = 4.71$ is employed, $OP_{L_{.95}} = 3.27$ and $OP_{U_{.95}} = 22.73$, which when rounded off also yields $OP_{L_{.95}} = 3$ and $OP_{U_{.95}} = 23$.) Thus, the pairwise difference in the **third ordinal position** (the score of -11) will define the **lower limit of the confidence interval** and the pairwise difference in the **23rd ordinal position** (the score of **3**) will define the **upper limit of the confidence interval**. This result is identical to that obtained earlier when the critical value $U_{.05} = 2$ in **Table A11** was employed. Note that with respect to Examples 12.1 and 11.1, when the range of values which define the 95% confidence interval for a difference between the medians of two independent populations is contrasted with the range of values which define the confidence interval for a difference between the means of two independent populations (for which $-10.89 \le (\mu_1 - \mu_2) \le .89$ is computed in Section VI of the *t* test for two independent samples), the lower limit for both confidence intervals is almost identical, but there is a discrepancy with respect to the upper limit. Assuming the absence of outliers and that the assumptions underlying the *t* test for two independent samples are not saliently violated, in most instances in studies employing large sample sizes (which is not the case for Examples 11.1/12.1) the boundaries of a confidence interval computed for the difference between two means will be close to the boundaries of a confidence interval computed for the difference between two medians.

In closing this section, it should be noted that the **median of all pairwise differences** is referred to as the **Hodges–Lehman estimate**. From Table 12.4 it can be determined that the latter value equals -5.5 (which is the average of the two middle pairwise differences — i.e., the average of the pairwise differences in the 12th and 13th ordinal positions: $[(-6) + (-5)] / 2 = -5.5$). Higgins (2004) notes that the **Hodges–Lehman estimate** can be employed as a nonparametric alternative to the differences between the means (i.e., the value computed for the numerator of the *t* **test for two independent samples**, which in the case of Example 11.1 equals -5).

Additional Discussion of the Mann–Whitney *U* Test

1. Power-efficiency of the Mann–Whitney *U* test When the underlying population distributions are normal, the **asymptotic relative efficiency** (which is discussed in Section VII of the **Wilcoxon signed-ranks test (Test 6)**) of the **Mann–Whitney *U* test** is .955 (when contrasted with the *t* **test for two independent samples**). For population distributions that are not normal, the asymptotic relative efficiency of the **Mann–Whitney *U* test** is generally equal to or greater than 1. As a general rule, proponents of nonparametric tests take the position that when a researcher has reason to believe the normality assumption of the *t* **test for two independent samples** has been saliently violated, the **Mann–Whitney *U* test** provides a powerful test of the comparable alternative hypothesis.

2. Equivalency of the normal approximation of the Mann–Whitney *U* test and the *t* test for two independent samples with rank-orders Conover (1980, 1999), Conover and Iman (1981), and Zimmerman and Zumbo (1993a) note that the large sample normal approximation of the **Mann–Whitney *U* test** (i.e., Equation 12.4) yields a result which is equivalent (with respect to the exact alpha level computed for the data) to that which will be obtained if the *t* **test for two independent samples** is conducted on the same set of rank-orders. Even if the normal approximation is not employed, the results obtained from the **Mann–Whitney *U* test** through use of Equations 12.1 and 12.2 will be extremely close (in terms of the alpha value) to those that will be obtained if the *t* **test for two independent samples** is conducted on the same set of rank-orders.

In point of fact, if the *t* **test for two independent samples** is employed to contrast the ranks of the two groups recorded in Table 12.1, it leads to the same conclusion reached when the **Mann–Whitney *U* test** is employed to evaluate the ranks. Specifically, when the *t* **test for two independent samples** is employed with the ranks in Table 12.1, it yields the value $t = -2.13$. As is the case when the original unranked scores are evaluated with the *t* test, the result of the analysis is only significant at the .05 level for a one-tailed analysis (supporting the directional alternative hypothesis H_1: $\theta_1 < \theta_2$). This is the case, since the absolute value $t = 2.13$ is greater than the tabled critical one-tailed .05 value $t_{.05} = 1.86$ (for $df = 8$). The alternative hypothesis H_1: $\theta_1 < \theta_2$ is not supported at the .01 level, since the absolute value $t = 2.13$ is less than the tabled critical one-tailed .01 value $t_{.01} = 2.90$. The nondirectional alternative hypothesis H_1: $\theta_1 \neq \theta_2$ is not supported, since the absolute value $t = 2.13$ is less than the tabled critical two-tailed .05 value $t_{.05} = 2.31$. (In actuality, if the *t* test is employed to evaluate the ranks, the symbols employed within the framework of the null and alternative hypotheses should represent the mean values of the ranks in the underlying populations.)

3. Alternative nonparametric rank-order procedures for evaluating a design involving two independent samples In addition to the **Mann–Whitney *U* test** a number of other nonparametric procedures for two independent samples have been developed which can be employed with ordinal data. Among the more commonly cited alternative procedures are the following: a) Fligner and Policello (1981) describe a methodology which can be employed for computing an adjusted U statistic if the homogeneity of variance assumption underlying the **Mann–Whitney U test** is violated. The **Fligner–Policello test** is described in Hollander and Wolfe (1999, pp. 135–139). Delaney and Vargha (2002) and Grissom and Kim (2005, p. 101) also discuss robust alternatives to the **Mann–Whitney U test**; b) **The Kolmogorov–Smirnov test for two independent samples (Test 13)** (Kolmogorov (1933) and Smirnov (1939)), which is described in the next chapter; c) **The van der Waerden normal-scores test for k independent samples (Test 23)** (Van der Waerden (1952/1953)), which is described later in the book, as well as alternative **normal-scores tests** developed by Bell and Doksum (1965) and Terry and Hoeffding (Terry (1952)); d) **Tukey's quick test** (Tukey (1959)); e) **The median test for independent samples (Test 16e)**, which involves dichotomizing two samples with respect to their median values, and evaluating the data with the **chi-square test for r × c tables (Test 16)**; f) **The Wald–Wolfowitz runs test** (Wald and Wolfowitz (1940)) (briefly discussed in Endnote 7 of the single-sample runs test (Test 10)); and g) **Wilks' empty-cell test for identical populations** (Wilks (1961)). In addition to various books which specialize in nonparametric statistics, Sheskin (1984) describes these tests in greater detail.

VIII. Additional Examples Illustrating the Use of the Mann–Whitney U Test

The **Mann–Whitney U test** can be employed with any of the additional examples noted for the *t* **test for two independent samples**. Since Examples 11.4 and 11.5 use the same data as that employed in Example 12.1, they will yield the identical result. Examples 11.2 and 11.3 can also be evaluated with the **Mann–Whitney U test**, but employ different data than Example 12.1. The interval/ratio scores in all of the aforementioned examples have to be rank-ordered in order to employ the **Mann–Whitney U test**.

Example 12.2 provides one additional example which can be evaluated with the **Mann–Whitney U test**. It differs from Example 12.1 in the following respects: a) Since the original scores are in a rank-order format, there is no need to transform the scores into ranks from an interval/ratio format (as is the case in Example 12.1). It should be noted though, it is implied in Example 12.2 that the ranks are based on an underlying interval/ratio scale; and b) The sample sizes are unequal, with $n_1 = 6$ and $n_2 = 7$.

Example 12.2 *Doctor Radical, a math instructor at Logarithm University, has two classes in advanced calculus. There are six students in* Class 1 *and seven students in* Class 2. *The instructor uses a programmed textbook in* Class 1 *and a conventional textbook in* Class 2. *At the end of the semester, in order to determine if the type of text employed influences student performance, Dr. Radical has another math instructor, Dr. Root, rank the 13 students in the two classes with respect to math ability. The rankings of the students in the two classes follow:* **Class 1**: 1, 3, 5, 7, 11, 13; **Class 2**: 2, 4, 6, 8, 9, 10, 12 *(assume the lower the rank the better the student).*

Employing the **Mann–Whitney U test** with Example 12.2 the following values are computed.

$$\Sigma R_1 = 40 \qquad \Sigma R_2 = 51$$

$$U_1 = (6)(7) + \frac{6(6 + 1)}{2} - 40 = 23$$

$$U_2 = (6)(7) + \frac{7(7 + 1)}{2} - 51 = 19$$

$$(U_1 = 23) + (U_2 = 19) = (n_1 = 6)(n_2 = 7) = 42$$

$$z = \frac{19 - \dfrac{(6)(7)}{2}}{\sqrt{\dfrac{(6)(7)(6 + 7 + 1)}{12}}} = -.29$$

Since the value $U_2 = 19$ is less than $U_1 = 23$, $U = 19$. Employing **Table A11**, for $n_1 = 6$ and $n_2 = 7$, the tabled critical two-tailed values are $U_{.05} = 6$ and $U_{.01} = 3$, and the tabled critical one-tailed values are $U_{.05} = 8$ and $U_{.01} = 4$. Since in order to be significant the obtained value $U = 19$ must be equal to or less than the tabled critical value, the null hypothesis H_0: $\theta_1 = \theta_2$ cannot be rejected regardless of which alternative hypothesis is employed. The use of Equation 12.4 for the normal approximation confirms this result, since the absolute value $z = .29$ is less than the tabled critical two-tailed values $z_{.05} = 1.96$ and $z_{.01} = 2.58$, and the tabled critical one-tailed values $z_{.05} = 1.65$ and $z_{.01} = 2.33$.

IX. Addendum

1. Computer-intensive tests[8]   During the past 20 years the availability of computers has allowed for the use of hypothesis testing procedures which involve such an excessive amount of computation that in most instances it would be impractical to conduct an analysis by hand or even with a conventional calculator. The first statisticians who discussed such tests were Fisher (1935) and Pitman (1937a, 1937b, 1938), who described procedures known as **randomization** or **permutation tests**. Aside from their computer friendliness, another reason for the increased popularity of the **computer-intensive procedures** (also referred to as **data-driven procedures** (Sprent, 1998)) to be discussed in this section is that they have associated with them few if any of the distributional assumptions which underlie parametric tests, as well as certain nonparametric tests. As it is employed in this book, the term **computer-intensive test/procedure** (which is discussed briefly in Section VII of the *t* **test for two independent samples** under the discussion of **missing data**) is used to describe any of a variety of procedures which are computer-dependent.[9]

The computer dependency of such tests reflects the fact that in addition to carrying out numerous and/or complex computations, many of these tests employ the computer as a mechanism for repeated **resampling** of data. Julian Simon (1969) was among those who first discussed the advantages of employing computer-based resampling in the analysis of data. Resampling is a process in which subsets of scores are selected from an original set of data. In the final analysis, the distinguishing feature between the various computer-intensive procedures which employ resampling is the specific protocol employed in selecting subsamples from the original set of data. Two resampling procedures that will be described in this section are the **bootstrap** (developed by Efron (Efron (1979) and Efron and Tibshirani (1993)) and the **jackknife** (developed by Quenouille (1949) and named by Tukey (1958)).[10] The intent of the discussion to follow is to present the reader with the basic principles underlying computer-intensive procedures. Those who are interested in a more in-depth discussion of this complex topic should consult sources such as Manly (1997), Sprent (1998), and Sprent and Smeeton (2000), or other relevant references which are cited in the discussion to follow.

Randomization and permutation tests The terms **permutation test, randomization test, rerandomization test**, and **exact test** are employed interchangeably by many sources (Good, (1994)). Within the context of such tests, the word **permutation** is used to represent a specific configuration/arrangement of scores. Such tests yield results which are exclusively a function of the observed data, and are not based on any assumptions regarding an underlying population distribution. The use of the term **randomization test** reflects the fact that in most instances such tests contrast two or more groups to which it is assumed subjects have been randomly assigned. Although the data such tests are employed to evaluate need not be a random sample, if random selection cannot be assumed, it will limit the generalizability of the test's results.

Randomization tests evaluate the data obtained in an experiment within the framework of the distribution of all possible random arrangements which can be obtained for that set of data. In a randomization test, instead of evaluating the outcome of an experiment in reference to some underlying theoretical population distribution (e.g., the normal, t, F distributions, etc.), the data are employed to construct the relevant sampling distribution. By constructing a sampling distribution based on the data, the researcher is not restricted by any assumptions which might be associated with an underlying theoretical distribution. The fact that it is a requirement of a randomization test to compute a separate sampling distribution is the reason why, in almost all instances, a computer is needed to conduct such a test. In this section the principles underlying randomization tests will be demonstrated in reference to evaluating an experiment involving two independent samples. In the example presented, a randomization test will be employed to do the following: a) Evaluate the interval/ratio scores of two independent groups; and b) Evaluate the rank-orderings of the same set of interval/ratio scores. In the case of the latter analysis, it will be demonstrated that it is equivalent to the analysis conducted for the **Mann–Whitney U test** with the same set of ranks. This is the case since the **Mann–Whitney U test** (as well as many other rank-order tests) represents an example of a randomization/permutation test which is based upon permutations of ranks (i.e., configurations/arrangements of ranks).

Test 12a: The randomization test for two independent samples Among the other names that are employed for the test to be described in this section are **Fisher's randomization test for two independent samples** and the **Fisher–Pitman test** (since the procedure was first described by Fisher (1935) and Pitman (1937a, 1937b, 1938)). If a researcher has serious doubts regarding the normality and/or homogeneity of variance assumptions underlying the *t* test for two independent samples, the **randomization test for two independent samples** provides one with a viable alternative for evaluating the data. The data will be represented by the interval/ratio scores of two independent samples with n_1 scores in Group 1 and n_2 scores in Group 2, with $n_1 + n_2 = N$. Example 12.3 will be employed to illustrate the use of the **randomization test for two independent samples.**[11]

Example 12.3 *Each of six subjects is randomly assigned to one of two groups, with n_1 = 3 subjects in Group 1 and n_2 = 3 subjects in Group 2. While attempting to solve a mechanical problem, the subjects in Group 1 are exposed to a high concentration of atmospheric ozone. The subjects in Group 2 are required to solve the same mechanical problem under normal atmospheric conditions. The number of minutes it takes each of the subjects in the two groups to solve the problem follows:* **Group 1**: 15, 18, 21; **Group 2**: 7, 10, 11. *Do the data indicate that there are differences between the two groups?*

The null and alternative hypotheses evaluated with the **randomization test for two independent samples** are stated below. Note that, as stated, the null and alternative hypotheses do not include any reference to a population parameter. If a researcher included a population parameter in H_0 and H_1, there would be certain assumptions about the underlying population which would have to be met.

Null hypothesis H_0: There is no difference in the performance of the two groups. If the null hypothesis is supported, one can conclude that the two groups represent the same population.

Alternative hypothesis H_1: The performance of the two groups is not equivalent. If the alternative hypothesis is supported, one can conclude that the two groups do not represent the same population.

The alternative hypothesis stated above is **nondirectional**, and is evaluated with a **two-tailed test**. Either of the **directional alternative hypotheses** noted below can also be employed. If a **directional alternative hypothesis** is employed, a **one-tailed test** is conducted.

Alternative hypothesis H_1: The scores of subjects in Group 1 are higher than the scores of subjects in Group 2.

Alternative hypothesis H_1: The scores of subjects in Group 1 are lower than the scores of subjects in Group 2.

The general question addressed by the **randomization test for two independent samples** is as follows: If the scores of all N subjects are collapsed into a single group, how likely is it that the specific configuration of scores obtained in the experiment will be obtained if we randomly select n_1 scores and assign them to Group 1 and assign the remaining $N - n_1 = n_2$ scores to Group 2? Common sense suggests that if chance is operating, we will expect an equivalent distribution of scores in the two groups. On the other hand, if there are differences between the groups, it is expected that the distributions will not be equivalent. Thus, if one group has a preponderance of high scores and the other group has a preponderance of low scores, we will want to determine the exact likelihood of obtaining such an outcome purely as a result of chance. By constructing a sampling distribution from the data, the **randomization test for two independent samples** allows us to do this.

In order to evaluate the data for Example 12.3, we must first answer the following question: How many ways can six subjects be assigned to two groups with three subjects per group? This is equivalent to asking, what are the number of combinations of six things taken three at a time? (The reader may find it useful to review the discussion of **combinations** in the **Introduction** and/or in Section IV of the **binomial sign test for a single sample (Test 9)**.) Employing Equation I.48/Equation 9.4, we determine that the number of combinations of six things taken three at a time is equal to 20: $\binom{6}{3} = \frac{6!}{3!\,3!} = 20$. This result tells us that there are 20 possible ways six subjects can be assigned to two groups with three subjects per group.[12]

Table 12.5 summarizes the 20 possible ways in which the six scores 7, 10, 11, 15, 18, and 21 can be distributed between two groups with $n_1 = 3$ and $n_2 = 3$. The left side of Table 12.5 (Columns 2 and 3) lists the 20 possible ways three scores can be randomly assigned to Group 1 (or Group 2), as well as the sum of the three scores for each arrangement. The last two columns

Table 12.5 Possible Arrangements for Example 12.3 Data

Arrangement	Scores in Group 1	Sum of Scores in Group 1	Ranks in Group 1	Sum of Ranks in Group 1
1	7, 10, 11	28	1, 2, 3	6
2	7, 10, 15	32	1, 2, 4	7
3	7, 11, 15	33	1, 3, 4	8
4	7, 10, 18	35	1, 2, 5	8
5	7, 11, 18	36	1, 3, 5	9
6	10, 11, 15	36	2, 3, 4	9
7	7, 10, 21	38	1, 2, 6	9
8	7, 11, 21	39	1, 3, 6	10
9	10, 11, 18	39	2, 3, 5	10
10	7, 15, 18	40	1, 4, 5	10
11	10, 11, 21	42	2, 3, 6	11
12	7, 15, 21	43	1, 4, 6	11
13	10, 15, 18	43	2, 4, 5	11
14	11, 15, 18	44	3, 4, 5	12
15	10, 15, 21	46	2, 4, 6	12
16	7, 18, 21	46	1, 5, 6	12
17	11, 15, 21	47	3, 4, 6	13
18	10, 18, 21	49	2, 5, 6	13
19	11, 18, 21	50	3, 5, 6	14
20	15, 18, 21	54	4, 5, 6	15

Table 12.6 Sampling Distribution for Sums of Scores in Example 12.3

Sum of Scores	Frequency	Probability	Cumulative Probability
28	1	.05	.05
32	1	.05	.10
33	1	.05	.15
35	1	.05	.20
36	2	.10	.30
38	1	.05	.35
39	2	.10	.45
40	1	.05	.50
42	1	.05	.55
43	2	.10	.65
44	1	.05	.70
46	2	.10	.80
47	1	.05	.85
49	1	.05	.90
50	1	.05	.95
54	1	.05	1.00
Sums	**20**	**1.00**	

of the table are based on the assumption that the six scores have been converted into ranks, and lists the rank-orders for the three interval/ratio scores in a given row (Column 4), and the sum of the rank-orders for that row (Column 5). Within the framework of the 20 arrangements which are listed, Column 4 contains each of the possible ways three rank-orderings within a set of six rank-orderings can be randomly assigned to Group 1 (or Group 2).[13]

The values that are computed for the sums of the three interval/ratio scores for each of the 20 arrangements would represent the points on the abscissa (X-axis) if a graph of the sampling distribution for the sums was constructed. Since some of the 20 arrangements yield the same sum, there are a total of 16 possible values the sum of three scores can assume. The frequency or likelihood of occurrence for each of the 16 sums would be represented on the ordinate (*Y*-axis) of a graph of the sampling distribution of the sums. Table 12.6 summarizes the probabilities for the 16 sums of scores. Note that in Table 12.6 the sums of the scores are arranged in increasing order of magnitude, and that the sum of the column labeled **Frequency** is equal to 20, which is the total number of arrangements for the interval/ratio scores for our data. The probability distribution in Columns 3 and 4 of Table 12.6 represents the sampling distribution which is employed within the framework of the **randomization test for two independent samples** to evaluate the null hypothesis. Column 3 lists the probability for the sum of scores in each row, whereas Column 4 lists the cumulative probability for the sum of scores in that row (cumulative probabilities are discussed in the **Introduction** and in the discussion of the **Kolmogorov–Smirnov goodness-of-fit test for a single sample (Test 7)**).

Table 12.7   Sampling Distribution for Sums of Ranks in Example 12.3

Sum of Ranks	Frequency	Probability	Cumulative Probability
6	1	.05	.05
7	1	.05	.10
8	2	.10	.20
9	3	.15	.35
10	3	.15	.50
11	3	.15	.65
12	3	.15	.80
13	2	.10	.90
14	1	.05	.95
15	1	.05	1.00
Sums	**20**	**1.00**	

Table 12.7 summarizes the sampling distribution for the sums of ranks. It turns out that for our data there are only 10 possible values the sum of three ranks can equal. The frequency or likelihood of occurrence for each of the 10 sums of ranks is summarized in Table 12.7. Note that in Table 12.7 the sums of the ranks are arranged in increasing order of magnitude, and that the sum of Column 2 (labeled **Frequency**) is equal to 20, which is the total number of arrangements of ranks for our data. The probability distribution in Table 12.7 (Column 3 lists the probability for the sum of ranks in each row, whereas Column 4 lists the cumulative probability for the sum of ranks in that row) represents the sampling distribution which can be employed within the framework of the **randomization test for two independent samples** to evaluate the null hypothesis. It should be emphasized, however, that the original test developed by Fisher (1935) and Pitman (1937a, 1937b, 1938) evaluated the sums of interval/ratio scores and not summed values which were based on ranks. Thus, when summed values based on the ranks of two independent samples are evaluated using the **Fisher–Pitman randomization procedure**, the test is not generally referred to as the **randomization test for two independent samples**. Instead it is referred to as the **Mann–Whitney *U* test**.[14]

In order to reject the null hypothesis, the sum of scores/ranks (and consequently the arrangement which yields a specific sum) will have to be one that is highly unlikely to occur as a result of chance. In the case of a **two-tailed alternative hypothesis** this will translate into an arrangement with a sum of scores/ranks that is either very high or very low. In the case of a **one-tailed alternative hypothesis**, the following will apply: a) In order for the **one-tailed alternative hypothesis** predicting that the scores of the subjects in Group 1 are higher than the scores of the subjects in Group 2 to be supported, it will require an arrangement where the sum of scores/ranks for Group 1 is very high; b) In order for the **one-tailed alternative hypothesis** predicting that the scores of the subjects in Group 1 are lower than the scores of the subjects in Group 2 to be supported, it will require an arrangement where the sum of scores/ranks for Group 1 is very low.

In point of fact, the sum of scores for the observed data in Group 1 is 54 (based on the three scores 15, 18, and 21), which is the largest possible sum in Tables 12.5 and 12.6. In the same respect, the sum of the ranks for the observed ranks in Group 1 is 15 (based on the ranks 4, 5, and 6 for the three scores 15, 18, and 21), which is the largest possible sum of ranks in Tables 12.5 and 12.7. In both instances the likelihood of obtaining a sum of that magnitude is .05 (or 5%).[15] The latter value delineates the upper 5% of the sampling distribution we have derived for the data. This result allows us to reject the null hypothesis at the .05 level, but only for the **one-tailed alternative hypothesis** predicting that the scores of the subjects in Group 1 are higher than the scores of the subjects in Group 2. The latter alternative hypothesis is supported since there is only a 5% likelihood that a sum of scores or ranks equal to or greater than the one observed could have occurred as a result of chance.

The null hypothesis cannot be rejected if the **one-tailed alternative hypothesis** predicting that the scores of the subjects in Group 1 are lower than the scores of the subjects in Group 2 is employed. This is the case, since the data are inconsistent with the latter alternative hypothesis.

The null hypothesis cannot be rejected at the .05 level if the **two-tailed alternative hypothesis** is employed. The reason for this is that even though the observed data result in the highest possible sum of scores/ranks, the total number of possible arrangements is 20. As a result of the latter, the lowest probability possible for the most extreme arrangement in either direction (the highest or lowest sum) is $1/20 = .05$. In order for a **two-tailed alternative hypothesis** to be supported, it will require that the observed data fall within the extreme 5% of the cases involving both tails of the distribution. One-half of that 5% will have to come from the upper tail of the sampling distribution, and the other half from the lower tail. Thus, in order for the two-tailed alternative hypothesis to be supported, it will require that the probability associated with the highest sum (as well as the lowest sum) be equal to or less than $.05/2 = .025$. In our example, the small sample size (which results in 20 arrangements) does not allow us to evaluate a two-tailed alternative hypothesis at the .05 level, since it is not possible for the highest sum (or lowest sum) to have a probability as low as .025. In order to evaluate the two-tailed alternative hypothesis at that level, there would have to be a minimum of 40 arrangements (since $1/40 = .025$). It should be evident that because of our small sample size, it is not possible to evaluate the one-tailed or two-tailed alternative hypotheses at the .01 level.

It was noted earlier in the discussion of randomization tests that the **Mann–Whitney U test** represents an example of a randomization test which is based upon permutations of ranks (i.e., configurations/arrangements of ranks). What this translates into is that, if the data for Example 12.3 are evaluated with the **Mann–Whitney U test**, they should yield a result identical to that obtained with the **randomization test for two independent samples** when the latter procedure is applied to ranks. This will now be demonstrated by evaluating the data for Example 12.3 with the **Mann–Whitney U test**.

From Table 12.5 we know that the sum of the ranks for Group 1 is $\Sigma R_1 = 15$ (which is the sum of the ranks 4, 5, and 6 for the three scores 15, 18, and 21). Since the three scores for Group 2 are the three lowest scores, the sum of the ranks for Group 2 will equal $\Sigma R_2 = 6$ (which is the sum of the ranks 1, 2, and 3 for the three scores 7, 10, and 11). Employing Equations 12.1 and 12.2, we obtain $U_1 = 0$ and $U_2 = 9$.

$$U_1 = (3)(3) + \frac{3(3+1)}{2} - 15 = 0$$

$$U_2 = (3)(3) + \frac{3(3+1)}{2} - 6 = 9$$

Since the smaller of the two values U_1 versus U_2 is designated as the U statistic, $U = 0$. Note that in **Table A11** no tabled critical two-tailed .05 U value is listed for $n_1 = 3$ and $n_2 = 3$. The tabled critical one-tailed .05 value listed for $n_1 = 3$ and $n_2 = 3$ is $U_{.05} = 0$. No tabled critical .01 U values are listed due to the small sample sizes. Since the obtained value $U = 0$ is equal to the tabled critical one-tailed .05 value $U_{.05} = 0$, the one-tailed alternative hypothesis predicting that the scores of the subjects in Group 1 are higher than the scores of the subjects in Group 2 is supported at the .05 level. The one-tailed alternative hypothesis predicting that the scores of the subjects in Group 1 are lower than the scores of the subjects in Group 2 is not supported, since it is inconsistent with the observed data. Because no critical two-tailed values are listed in **Table A11** for $n_1 = 3$ and $n_2 = 3$, the two-tailed alternative hypothesis cannot be evaluated at the .05 level. The results of the **Mann–Whitney U test** are thus identical to those obtained when the data for Example 12.3 are evaluated with the **randomization test for two independent samples** (in which case the one-tailed alternative hypothesis that is consistent with the data is supported at the .05 level).

When the size of the samples for which a randomization test is employed becomes large, the number of possible combinations which can be computed for the data may become excessive to the point that it even becomes impractical for a computer to construct an exact sampling distribution based on every possible arrangement. In such a situation an **approximate randomization test** may be used. In the latter test, the computer constructs a sampling distribution based on randomly selecting a large number (but not all) of the possible arrangements for the data. The resulting sampling distribution is employed to evaluate the sample data.

Good (1994, p. 114) notes that under certain conditions a randomization test may provide a more powerful test of an alternative hypothesis than a parametric procedure. Sprent (1998, p. 52), however, points out a limitation of the **randomization test for two independent samples** is that if the test is employed with interval/ratio data containing one or more **outliers**, it behaves very much like the *t* **test for two independent samples** (and thus may be unreliable). Conover (1999, p. 408) notes that when the **randomization test for two independent samples** is applied to interval/ratio data, under certain conditions (e.g., outliers present in the data, skewed distributions) it provides a less powerful test of an alternative hypothesis than an analogous parametric procedure or a nonparametric rank-order procedure (which as noted in this section may be a randomization test on rank-orders). Lundbrook and Dudley (1998) provide a good history of randomization tests, and discuss the merits of employing such tests in biomedical research. Manly (1997) provides a comprehensive discussion of randomization tests.

Test 12b: The bootstrap Sprent (1998, p. 28) notes that although the philosophy underlying the bootstrap is different from that upon which permutation tests are based, it employs a similar methodology and often yields results that are concordant with those which will be obtained with a permutation test. The bootstrap is based on the general assumption that a random sample can be used to determine the characteristics of the underlying population from which the sample is derived. However, instead of using a sample statistic (e.g., the sample standard deviation) to estimate a population parameter (e.g., the population standard deviation), as is done within the framework of conventional parametric statistical tests, the bootstrap uses multiple samples derived from the original data to provide what in some instances may be a more accurate measure of the population parameter.[16] The most common application of the bootstrap involves estimating a population standard error and/or confidence interval.

The more ambiguous the information available to a researcher regarding an underlying population distribution, the more likely it is that the bootstrap may prove useful. Sprent (1998) notes that unlike permutation tests, which are able to provide a researcher with exact probability values, the use of the bootstrap only leads to approximate results. Nevertheless, in this case approximate results may, in the final analysis, be considerably more accurate than the results derived from an analysis which is based on an invalid theoretical model. For this reason, proponents of the bootstrap justify its use in circumstances where there is reasonable doubt regarding the characteristics of the underlying population distribution from which a sample is drawn. The most frequent justification for using the bootstrap is when there is reason to believe that data may not be derived from a normally distributed population. Another condition which might merit the use of the bootstrap is one involving a sample that contains one or more **outliers** (outliers are discussed in Section VII of the *t* **test for two independent samples**). If, as a result of outliers being present in the data, a researcher elects to trim scores from the tails of a sample distribution, many sources claim there is no equation available for providing the researcher with an unbiased estimate of certain population parameters (e.g., the standard error).[17]

Efron and Tibshirani (1993, p. 393) note that the bootstrap implements familiar statistical calculations (e.g., computing standard errors, confidence intervals, etc.) in an unfamiliar way — specifically, through use of computer-driven methods, as opposed to the use of mathematical equations. They state that, even though it employs a different methodology, the bootstrap is based on mathematical theory which insures its compatibility with traditional theories of statistical inference. Sprent (1993, p. 291) notes that bootstrapping can be a valuable technique when there is no clear analytic theory to obtain a measure of accuracy of an estimator. According to Efron and Tibshirani (1993, p. 394), at present the accuracy of the bootstrap is optimal in the estimation of values such as standard errors and confidence intervals, and it is weakest in evaluating those hypotheses which are typically evaluated with the more conventional inferential statistical tests.

At this point the methodology for the bootstrap will be demonstrated with a simple example. It is assumed that within the framework of actual research, the procedure to be described below would be carried out with a computer. Additionally, the sample size employed in a study will generally be larger than the value $n = 5$ employed in the example. Obviously, if a sample is randomly drawn for a population, the larger the value of n (i.e., the sample size) the more likely it is that the sample will be representative of the population. It is important to note that as a result of chance factors, even a random sample can be unrepresentative of an underlying population. To the degree that a sample is not representative of a population, the bootstrap will not provide an accurate estimate of the parameter under investigation. The general issue of what size a sample should be in order to employ the bootstrap is subject to debate. Mooney and Duval (1993) note in most instances a random sample comprised of 30 to 50 observations will be sufficient to provide a good bootstrap estimate. Although Manly (1997) states that the theory underlying the bootstrap insures it will work well in certain situations involving large samples, he notes that in actuality a substantial amount of published research does not employ large samples. Under such circumstances Manly (1997) states the use of the bootstrap becomes more problematical. To illustrate the latter, he provides an example involving data derived from an exponential distribution which requires a sample size of more than 100 in order for the bootstrap to provide an accurate estimate of a confidence interval.[18]

An example will now be presented to illustrate the application of the bootstrap methodology in computing a confidence interval.

Example 12.4 *Assume that five diamonds are randomly selected from a population of diamonds manufactured by one of the world's largest distributors of precious stones. For each of the n = 5 diamonds, the number of imperfections observed on its surface is recorded. Let us assume there is reason to believe the distribution of imperfections in the underlying population of diamonds is not normal. The number of imperfections observed in each of the five diamonds which comprise our sample follows: 12, 7, 8, 2, 4. A quality control supervisor in the company requests a 99% confidence interval for the population standard deviation for the number of imperfections per diamond. Employ the bootstrap to compute a 99% confidence interval for the population standard deviation.*

The methodology of the bootstrap requires that **sampling with replacement** be employed to select a large number of subsamples from the sample of five scores. (Endnote 1 of the **binomial sign test for a single sample (Test 9)** explains the distinction between **sampling with replacement** versus **sampling without replacement**.) We will let m represent the number of subsamples. Each of the m subsamples will be referred to as a **bootstrap sample**. As is the case with our original sample, the sample size for each of the bootstrap samples will be $n = 5$. Since we will employ sampling with replacement, within any bootstrap sample any of the original five scores can occur any number of times ranging from zero to five (since within any bootstrap sample, each score selected will always be randomly drawn from the pool of five scores that constitute the original sample). Let us assume we obtain $m = 1000$ bootstrap samples (all of which are randomly generated by a computer algorithm that selects random samples through use of sampling with replacement). The scores for the first two and last of the 1000 bootstrap samples are noted below.

Bootstrap sample 1: 2, 2, 7, 8, 12

Bootstrap sample 2: 8, 12, 7, 7, 12

. . . . .

Bootstrap sample 1000: 2, 4, 12, 8, 12

Devore and Farnum (1999, p. 317) note that, as a general rule, resampling methods such as the bootstrap work since random subsamples derived from a random sample are also random samples from a population. Consequently, each of the 1000 bootstrap samples obtained for Example 12.4 can be viewed as representing genuine random samples of size $n = 5$ obtained through use of sampling with replacement from the underlying population. Because of the latter, any statistic derived from such a set of bootstrap samples can be employed to construct a sampling distribution.

Employing Equation I.11, the computer calculates the unbiased estimate of the standard deviation for each of the $m = 1000$ bootstrap samples. The computation of $\tilde{s}$ is demonstrated for **Bootstrap sample 1** (where $\Sigma X_1 = 31$ and $\Sigma X_1^2 = 265$).

$$\tilde{s}_1 = \sqrt{\left[265 - \left[(31)^2/5\right]\right]/(5-1)} = 4.27$$

The computed values of $\tilde{s}$ for **Bootstrap sample 2** and **Bootstrap sample 1000** are $\tilde{s}_2 = 2.59$ and $\tilde{s}_{1000} = 4.56$. Once we have computed the standard deviation values for all 1000 bootstrap samples, the values are arranged ordinally — i.e., from lowest to highest. At this point we can employ the 1000 $\tilde{s}$ values to determine the 99% confidence interval. In the case of the 99% confidence interval, the researcher can be 99% confident that the computed interval contains the true value of the population standard deviation (or stated probabilistically, there is a .99 probability the computed confidence interval contains the true value of the population standard deviation). One-half of one per cent (.5% or .005 expressed as a proportion) of the scores fall to the left of the lower bound of the 99% confidence interval, and one-half of one per cent (.5%) of the scores fall to the right of the upper bound of the 99% confidence interval. The scores in between the two bounds fall within the 99% confidence interval. Thus, in the case of our $m = 1000$ bootstrap sample standard deviations, the middle 99% will fall within the 99% confidence interval. The extreme .5% in the left tail (i.e., the .5% lowest $\tilde{s}$ values computed) and the extreme .5% in the right tail (i.e., the .5% highest $\tilde{s}$ values computed) will fall outside the 99% confidence interval. Thus, the lower bound of the 99% confidence interval will be the score at the sixth ordinal position. We obtain the value of the latter ordinal position by multiplying $m = 1000$ (the total number of scores in our sampling distribution of standard deviation scores) by .005 (i.e., .5%) and adding 1 (i.e., $m(.005) + 1 = (1000)(.005) + 1 = 6$). Five of the 1000 scores will fall below that point. The upper bound of the 99% confidence interval will be the score in the 995th ordinal position. We obtain the latter ordinal position value by multiplying $m = 1000$ by .005, and subtracting the resultant value of 5 from 1000 (i.e., $m - m(.005) = 1000 - (1000)(.005) = 995$). Five of the 1000 scores will fall above that point. Assume that of the total 1000 bootstrap standard deviations computed, the six lowest and six highest values are listed below.

$$0 \quad 0 \quad .45 \quad .45 \quad .89 \quad .89 \quad \ldots\ldots\ldots\ldots\ldots \quad 4.27 \quad 4.47 \quad 4.47 \quad 4.47 \quad 4.67 \quad 4.67$$

Since the sixth score from the bottom is .89 and the sixth score from the top (which corresponds ordinally to the 995th score) is 4.27, the latter values represent the bounds that define the 99% confidence interval. Thus, we can be 99% confident (or the probability is .99) that the interval .89 to 4.27 contains the true value of the population standard deviation. This result can be summarized as follows: $.89 \leq \sigma \leq 4.27$.

It should be emphasized that the use of the bootstrap in Example 12.4 for computing a confidence interval will be predicated on the fact that the researcher is unaware of any acceptable mathematical method for accurately determining the confidence interval. Efron and Tibshirani (1993, p. 52) note that while estimating a standard error rarely necessitates more than 200 bootstrap samples, the computation of a confidence interval generally requires a minimum of 1000 bootstraps, and often considerably more depending upon the nature of the problem being evaluated (see Efron and Tibshirani (1993, Ch. 12–14) for a more detailed discussion).

The bootstrap can be applied to numerous hypothesis testing situations (although Kline (2004, pp. 279) expresses reservations about employing it for inferential statistical testing). As an example, Efron and Tibshirani (1993) and Sprent (1998) discuss the use of the bootstrap in evaluating the null hypothesis that two samples are derived from the same population. One approach to the two-sample hypothesis testing situation is to use the bootstrap in a manner analogous to that employed for a permutation test. However, instead of employing sampling without replacement, as is the case with a permutation test (since for each arrangement, n_1 scores are placed in one group, and the remaining $N - n = n_2$ scores by default constitute the other group), sampling with replacement is employed for the bootstrap. To illustrate, assume that we have two samples comprised of n_1 subjects in Group 1 and n_2 subjects in Group 2, with $n_1 + n_2 = N$. Using the total of N scores, employing sampling with replacement, we use the computer to randomly select a large number of bootstrap samples, each sample being comprised of N scores. Within each bootstrap sample, the first n_1 scores are employed to represent the scores for Group 1 and the remaining $N - n_1 = n_2$ scores are employed to represent the scores for Group 2. For each bootstrap sample, a difference score between the two group means is computed. The empirical sampling distribution of the difference scores is employed to evaluate the null hypothesis. Specifically, it is determined how likely it is to have obtained the observed difference in the original data when it is considered within the framework of a large number of bootstrap sample difference scores. This is essentially what is done in a permutation test, and Efron and Tibshirani (1993, p. 221) note that when a large number of bootstrap samples (e.g., 1000) are employed, the bootstrap and permutation test applied to the same set of data yield similar results. However, Efron and Tibshirani (1993, p. 220) make the general statement that although bootstrap tests are more widely applicable than permutation tests within the framework of hypothesis testing situations, they are not as accurate. As noted earlier in this section, the bootstrap is more accurate in estimating values such as standard errors and confidence intervals than it is in the area of hypothesis testing.

In concluding this discussion, it should be emphasized that the bootstrap is a relatively new methodology. Consequently, at the present time the method itself is the subject of considerable research which focuses on both its theoretical underpinnings and practical applications. Manly (1997) notes that although the bootstrap is based on a very simple concept, over the years the theory behind it has become increasingly complex. (Many of the applications of the bootstrap, as well as the theory behind them, are considerably more involved than what has been discussed in this section.) One could conjecture that if a researcher does not understand the theory or operations underlying a statistical procedure, he may become increasingly reluctant to use it, or, because of his ignorance, use it inappropriately. Manly (1997) recommends that in the future, appropriate user-friendly applications of the bootstrap be integrated into standard statistical software. Other sources which discuss the bootstrap in greater detail are Chernick (1999) and Davison and Hinckley (1997).

Test 12c: The jackknife   Another computer-intensive procedure which was developed (by Quenouille (1949)) prior to the bootstrap is the **jackknife**. Like the bootstrap, the jackknife is an alternative methodology available to the researcher that can be employed for point/interval estimation. Specifically, under certain conditions the jackknife can be employed to reduce the degree of bias associated with point estimation (i.e., increase accuracy in estimating a population parameter). Like the bootstrap, the jackknife might be considered as a viable alternative in situations where there is no clear analytic theory to obtain a measure of accuracy of an estimator. At this point a simple example will be presented to demonstrate how the jackknife can be employed to estimate the degree of bias associated with an estimator of a parameter. The data for Example 12.4 will be employed to demonstrate the use of the jackknife in estimating bias in the following situations: a) Employing $\bar{X}$, the sample mean (which is known to be an unbiased estimator of the population mean (μ)), to estimate the value of the population mean; and b) Employing the sample variance (s^2) as computed with Equation I.7 (which is known to be a biased estimator of the population variance (σ^2)) to estimate the value of the population variance.

Employing Equation I.1, we compute the sample mean for the data of Example 12.4 (i.e., the $n = 5$ scores 12, 7, 8, 2, 4) to be $\bar{X} = 6.6$. Employing Equation I.7, the sample variance (s^2) (which is known to be a biased estimate of σ^2) is computed to be $s^2 = 11.84$ ($s^2 = [277 - [(33)^2/5]]/5 = 11.84$). Employing Equation I.8, the estimated population variance ($\hat{s}^2$) (which is known to be an unbiased estimate of σ^2) is computed to be $\hat{s}^2 = 14.8$ ($\hat{s}^2 = [277 - [(33)^2/5]]/4 = 14.8$).

The methodology of the jackknife requires that n subsamples, to be designated **jackknife samples**, be derived from the original sample of n scores. Each of the jackknife samples will be comprised of $(n - 1)$ scores. In each of the n jackknife samples one of the original n scores is omitted. From the original set of 5 scores 12, 7, 8, 2, 4, we can derive the following five jackknife samples, comprised of four scores per sample: **Jackknife sample 1**: 7, 8, 2, 4;

Jackknife sample 2: 12, 8, 2, 4; **Jackknife sample 3**: 12, 7, 2, 4; **Jackknife sample 4**: 12, 7, 8, 4; **Jackknife sample 5**: 12, 7, 8, 2. Note that in **Jackknife sample 1** the first of the five scores listed is omitted, and in **Jackknife sample 2** the second of the five scores listed is omitted, and so on. Thus, each of the samples is comprised of four of the original five scores.

Employing Equations I.1 and I.7 we compute the mean ($\bar{X}$) and sample variance (s^2) for each of the five jackknife samples. These values are summarized below.

$$\bar{X}_1 = 5.25 \quad \bar{X}_2 = 6.5 \quad \bar{X}_3 = 6.25 \quad \bar{X}_4 = 7.75 \quad \bar{X}_5 = 7.25$$

$$s_1^2 = 5.6875 \quad s_2^2 = 14.75 \quad s_3^2 = 14.1875 \quad s_4^2 = 8.1875 \quad s_5^2 = 12.6875$$

Sprent (1993, p. 285) notes that Equation 12.9 is the general equation for computing a **jackknife estimate of a specific parameter** θ .(θ in this case is not the population median, which it was employed to represent earlier in this chapter.) Equation 12.10 is the general equation for the **jackknife estimate of the degree of bias** associated with the estimate, which will be represented with value B.

$$\theta_j = n\theta' - (n - 1)\theta_* \qquad \text{(Equation 12.9)}$$

Where: θ_j represents the jackknife estimate of the parameter
θ' represents the value of the statistic computed for the parameter from the original sample data comprised of n scores
θ_* represents the average value of the sample statistic computed for the n jackknife samples. (The same statistic employed to compute θ' should be employed to compute the estimate of the parameter for the n jackknife samples.)

$$B = (n - 1)(\theta_* - \theta') \qquad \text{(Equation 12.10)}$$

In employing the jackknife estimator of the population mean (μ), the symbol θ will be used to represent the population mean. The following values will be employed in Equations 12.9 and 12.10: a) $n = 5$; b) $\theta' = \bar{X} = 6.6$, since that is the mean value of the five scores 12, 7, 8, 2, 4; and c) $\theta_* = 6.6$, since that is the mean of five jackknife samples (i.e., (5.25 + 6.5 + 6.25 + 7.75 + 7.25)/5 = 6.6). The appropriate values are substituted in Equations 12.9 and 12.10.

$$\theta_j = 5(6.6) - (5 - 1)(6.6) = 6.6$$

$$B = (5 - 1)(6.6 - 6.6) = 0$$

The computed value $\theta_j = 6.6$ is the jackknife estimate of the population mean. The computed value $B = 0$ indicates that the degree of bias associated with the sample statistic $\bar{X}$ in estimating the population mean is zero. Since in the case of the population mean it is known that $\bar{X}$ provides an unbiased estimate of μ, it is expected that the value zero will be computed for B.

In employing the jackknife estimator of the population variance (σ^2), the symbol θ will be used to represent the population variance. The following values will be employed in Equations 12.9 and 12.10: a) $n = 5$; b) $\theta' = s^2 = 11.84$, since that is the value computed with Equation I.7 for the sample variance for the five scores 12, 7, 8, 2, 4; and c) $\theta_* = 11.1$, since that is the mean of the variances of the five jackknife samples (i.e., (5.6875 + 14.75 + 14.1875 + 8.1875 + 12.6875)/5 = 11.1). The appropriate values are substituted in Equations 12.9 and 12.10.

$$\theta_j = 5(11.84) - (5 - 1)(11.1) = 14.8$$

$$B = (5 - 1)(11.1 - 11.84) = -2.96$$

The computed value $\theta_j = 14.8$ is the jackknife estimate of the population variance. The computed value $B = -2.96$ indicates that the degree of bias associated with the sample statistic s^2 in estimating the population variance is -2.96 (the negative sign indicating underestimation of the population variance by 2.96 units through use of the sample statistic s^2). Note that the computed value $\theta_j = 14.8$ is 2.96 units above the value $s^2 = 11.84$, and that $\theta_j = 14.8$ is in fact equivalent to the unbiased estimate of the population variance $\tilde{s}^2 = 14.8$ computed with Equation I.8. In point of fact, if the value $\tilde{s}^2 = 14.8$ is employed in Equations 12.9 and 12.10 to represent θ', Equation I.8 is employed to compute the five jackknife variances (and the average of the five jackknife $\tilde{s}^2$ values is $\theta_* = 14.8$). In such a case, the jackknife estimate of the variance computed with Equation 12.9 remains unchanged, yielding the value $\theta_j = 14.8$. However, the resulting value computed with Equation 12.10 will equal zero, since, as is the case when $\bar{X}$ is employed to estimate μ, $\tilde{s}^2 = 14.8$ is an unbiased estimate of the population variance.

Sprent (1993, 1998) and Efron and Tibshirani (1993, p. 148) note that the jackknife will be ineffective in reducing bias for some parameters. As an example, both of the aforementioned authors demonstrate the ineffectiveness of the jackknife as an estimator of the population median. These authors, as well as Hollander and Wolfe (1999) discuss a modified jackknife procedure referred to as the **delete-d jackknife**. In the latter modification of the standard jackknife, d scores are omitted from each of the jackknife samples, where d is some integer value greater than 1. Under certain conditions the latter procedure may provide a more accurate estimate of a population parameter than use of the conventional jackknife where $d = 1$. The reader should take note of the fact that for a set of n scores, the number of jackknife samples will be limited in number and be a function of the value of n, while the number of bootstrap samples which can be employed for a set of n scores is unlimited. Within this context, Efron and Tibshirani (1993) note that since the jackknife employs less data than the bootstrap, it is less efficient, and in the final analysis, what the jackknife provides is an approximation of the results obtained with the bootstrap. Manly (1997) and Mooney and Duval (1993) (both of whom present detailed discussions of the jackknife and the bootstrap) describe analytical situations in which they state or demonstrate that the jackknife provides more accurate information than the bootstrap.

Final comments on computer-intensive procedures Procedures such as the bootstrap and jackknife are commonly discussed in books which address the general subject of **robust statistical procedures** (e.g., Huber (1981), Sprent (1993, 1998), Staudte and Sheather (1990)) — the latter term referring to statistical procedures which are not overly dependent on critical assumptions regarding an underlying population distribution (the concept of robustness is discussed in Section VII of the *t* **test for two independent samples**).[19] Conover (1999, p. 116) notes that the term **robustness** is most commonly applied to methods that are employed when the normality assumption underlying an inferential statistical test is violated. He points out in spite of the fact that when sample sizes are reasonably large certain tests such as the **single-sample** *t* **test** and the *t* **test for two independent samples** are known to be robust with respect to violation of the normality assumption (i.e., the accuracy of the tabled critical alpha values for the test statistics are not compromised), if the underlying distribution is not normal the power of such tests may still be appreciably reduced. Robust statistical procedures are directly related to the subject of **resistant indicators** (which is discussed in the **Introduction**). **Resistant indicators** (examples of which are the **median**, the **median absolute deviation** (discussed in the chapter on the *t* **test for two independent samples**, etc.) are measures which are relatively insensitive to data contamination (e.g., the presence of outliers), and which do not have associated with them strong assumptions regarding the shape of the underlying population distribution. Detailed discussion of robust statistical procedures and resistant indicators can be found in Wilcox (1987, 1996, 1997, 2001, 2003, 2005).

Another characteristic of data which is often discussed within the framework of robust statistical procedures is the subject of **outliers** alluded to earlier in this section. (For a more comprehensive discussion of outliers, the reader should consult Section VII of the *t* **test for two independent samples**.) Research has shown that a single outlier can substantially compromise the power of a parametric statistical test. (Staudte and Sheather (1990) provide an excellent example of this involving the **single-sample *t* test**.) Various sources suggest that when one or more outliers are present in a set of data, a computer-intensive procedure (such as the bootstrap or jackknife) may provide a researcher with more accurate information regarding the underlying population(s) than a parametric procedure.

Most recently, among others, Kline (2004) and Wilcox (1996, 1997, 2001, 2003, 2005) have been critical of conventional researchers ignorance of and/or reluctance to employ computer-dependent robust statistical procedures which have been developed during the past thirty years. The latter author takes software manufacturers to task for their failure to integrate computer-dependent robust statistical procedures into standard statistical packages. Wilcox (1997) employs the term **contaminated/mixed normal distribution** to represent a normal-shaped distribution which exhibits a slight departure from normality as a result of being comprised of elements which represent more than one subpopulation whose parameters have different values. He notes it is not uncommon for such distributions to be **heavy-tailed** (which more often than not can be attributed to the presence of outliers) — the tails of a heavy-tailed distribution are situated at a higher point on the ordinate than those of a standard normal distribution. In spite of the fact that in such distributions there is only a slight departure from normality, the latter can dramatically inflate a measure of variability, and by virtue of the latter compromise the reliability of conventional parametric statistical procedures. Wilcox (1997) notes that the use of a logarithmic transformation (described in Section VII of the *t* **test for two independent samples**) is not even able to effectively counteract the impact of such contamination on variability attributable to the presence of outliers. Under such circumstances a computer-intensive procedure might be the most prudent one to employ.

Staudte and Sheather (1990, p. 14) paint a bleak picture regarding the power of commonly employed **goodness-of-fit tests for normality**. Specifically, these authors suggest that unless a sample size is relatively large, goodness-of-fit tests for normality (such as the **Kolmogorov–Smirnov goodness-of-fit test for a single sample** or the **chi-square goodness-of-fit test (Test 8)**) will generally not result in rejection of the null hypothesis of normality, unless the fit with respect to normality is dramatically violated. Consequently, they conclude that most goodness-of-fit tests are ineffective mechanisms for providing confirmation for the normal distribution assumption which more often than not researchers assume characterizes an underlying population. Staudte and Sheather (1990) argue that as a result of the failure of goodness-of-fit tests to reject the normal distribution model, procedures based on the assumption of normality all too often are employed with data which are derived from nonnormal populations. In instances where the normality assumption is violated, Staudte and Sheather (1990) encourage a researcher to consider employing a **robust statistical procedure** (such as the bootstrap) to analyze the data. In accordance with this view, Sprent (1998) notes that the bootstrap will often yield a more accurate result for a nonnormal population than will analysis of the data with a statistical test which assumes normality.

2. Survival analysis The term **survival analysis** refers to a set of statistical procedures which are designed to evaluate the amount of time that elapses between an initial observation/diagnosis of a subject (or object) and the occurrence of a specific event. Within the framework of survival analysis a **survival function** (which can be in the form of a table or graph) is constructed. A survival function provides an estimate of the likelihood a subject or object will survive beyond a specified time period. Conover (1999, p. 89) notes that the variable of interest in a survival function is the lifetime (until death) of an organism, a product, a machine, a structure, or else it can be the time before the occurrence of a specific event (such as a cure for a disease, end of a marriage, attainment of a goal, loss of a job, arrival or departure of someone or something, etc.).[20] Survival analysis is most commonly employed in medical research where its primary focus is estimating the probability a subject will survive for a specific length of time. Among other disciplines where survival analysis may be used is the field of engineering where a researcher may want to determine the amount of time before a machine or structure will fail.

The term **survival endpoint** (Jennison and Turnbull (2000)) is often employed to refer to the time it takes for some event of interest to occur (such as the death of a patient, relapse of an illness, the failure of a machine, etc.). However, in some research (e.g., medical research during **clinical trials** (which is discussed in Section VII of the *t* **test for two independent samples**)) it is not uncommon that one or more subjects or objects will fail to reach such an endpoint. The term **censoring** (which is briefly discussed in Endnote 27 of the *t* **test for two independent samples**) is employed in reference to data which are not available for some subjects, since it is either not possible or desirable to follow each subject until the conclusion of a study. In other words, there are instances when a researcher evaluates data before the event of interest/endpoint has occurred in all of the subjects. In the case of subjects who the researcher has access to but for whom the event of interest has not occurred, the researcher only knows that such subjects did survive the duration of the study, but exactly how long they will ultimately survive is unknown. Alternatively, some subjects may have dropped out of a study or have proven impossible to contact at the time data are collected. A major goal of survival analysis is to provide an unbiased mechanism for evaluating censored data.

With regard to censored data, a distinction is made between data that are **right censored** versus **left censored**. Data which are **right censored** identify subjects (or objects) that do not manifest the event of interest prior to the termination of a study (and includes subjects who no longer make themselves available for the study, as well as subjects a researcher is unable to locate beyond a certain period in time). Another example of right censored data are the scores of subjects who die during the course of a study, yet their death is attributable to some disease other than the one that is being treated in the study. Although right censored data are typically encountered in survival analysis, data that are **left censored** are the exception rather than the rule. In left censored data, the start of the time interval during which a subject or object is observed begins prior to the initial recorded observation of the subject or object. An example of **left censored** data would be if a researcher wanted to record the exact duration of time a group of patients exhibited the symptoms of a disease, yet in the case of a specific patient the researcher was not sure when the symptoms of the disease first appeared. Consequently, the precise time duration for that patient would not be known, since the researcher could only document the presence of symptoms for that patient from the point in time the researcher first observed the patient. Wright (2000, p. 365) notes that most techniques described for survival analysis are not adequate for the analysis of left censored data.

Within the context of **right-censored data**, a distinction is often made with respect to a number of different types of censoring. In **random censoring** it is assumed subjects drop out of a study at random points of time and that any subject's reason for leaving the study is unrelated to the experimental treatment which he or she has been assigned. **Random censoring** is commonly assumed to be operating in medical studies, and, consequently, in such studies it is assumed that loss of subjects can be due to factors such as subjects dying as a result of circumstances unrelated to the treatments or their unwillingness or inability to continue in a study — but again, not resulting from the effects of the treatments.

It should be noted that if **random censoring** is assumed, it is not unreasonable to expect there may be a disparity between the number of censored scores in each of the experimental conditions. As an example, let us assume a clinical trail is conducted to evaluate two treatments with respect to their ability to prolong the survival time of patients diagnosed with pancreatic cancer. One hundred patients are assigned to each of the treatments. Prior to initiating the study the researcher decides data will be collected for a period of five years, at which point the study will be terminated. Common sense suggests that, regardless of the treatment one receives, at some point during a clinical trial one or more patients may become unable or unwilling to fulfill the long term time commitment required by the study. Let us assume that Treatment 1 is significantly more effective than Treatment 2 — to be more specific, only 10 of the 100 patients receiving Treatment 2 survive beyond two years from the point at which the study commences, whereas 65 of the 100 patients receiving Treatment 1 survive three years or longer. In such a case, by virtue of its prolonged impact on survival time, a researcher could expect a greater number of censored observations for Treatment 1, even though any side effects or other difficulties associated with the latter treatment are no more noxious than those associated with Treatment 2.

Another distinction made with respect to censoring is that between **Type I**, **Type II**, and **Type III** censoring. In **Type I censoring** (also referred to as **fixed-time censoring**) all of the subjects/objects are present at the beginning of a study, and the point at which the study will be terminated is predetermined by the experimenter. Any subjects/objects for whom the event in question (e.g., death) has not occurred at the conclusion of the study are assigned a survival time which corresponds to the length of the study. In **Type II censoring** all of the subjects/objects are present at the beginning of a study, but the point at which the study will be terminated is not initially fixed by the experimenter. The study is carried out until the event in question (e.g., death) has occurred for a prespecified proportion of the subjects/objects. Note that in **Type I censoring** the duration of the study is fixed beforehand while the number of subjects/objects surviving represents a random variable. In **Type II censoring** the number of subjects/objects surviving is fixed beforehand while the duration of the study represents a random variable. The term **Type III censoring** is employed in reference to studies where subjects/objects enter the study at different points in time.

Higgins (2004, p.226) notes that censored data may be present in research dealing with questions other than issues relating to survival. For example, a study is conducted in which the variable of interest is the concentration of a chemical in the air, yet the machine responsible for measuring the chemical is not able to detect concentrations which fall below the value a or above the value b. Because of the machines limitations, any value under a would be unmeasurable and consequently referred to as **lower-censored data** (which is analogous to left-censored data) and any value above b would be referred to as **upper-censored data** (which is analogous to right-censored data). As noted in Endnote 27 of the *t* **test for two independent samples**, another type of censored data is described by Good (1994, p. 117), who notes that on occasion the measurement of a variable may break down at some point on the measurement continuum (usually at an extreme point). In such a case, a researcher must employ approximate scores instead of exact scores to represent those observations which cannot be measured with precision.

Test 12d: The Kaplan–Meier estimate The **Kaplan–Meier estimate** (also known as the **product-limit method** or **product-limit estimator**) (Kaplan and Meier (1958)) is a commonly employed nonparametric procedure for estimating a **survival function** (i.e., a set of survival probabilities for specific time periods). Examples 12.5 and 12.6 will be employed to illustrate the application of the **Kaplan–Meier estimate**. Although both examples employ the same 16 scores, in Example 12.5 all of the scores are uncensored, whereas in Example 12.6 five of the scores are censored. As you will see, the latter will result in a different survival function for Example 12.5 versus Example 12.6.

Example 12.5 *The following survival times (in days) for 16 patients who are treated with an experimental drug designed to arrest the spread of a malignant brain tumor are obtained by a researcher: 28, 49, 54, 80, 80, 102, 120, 120, 120, 167, 200, 200, 200, 340, 500, 500. Construct a survival function for the data.*

Example 12.6 *The following survival times (in days) for 16 patients who are treated with an experimental drug designed to arrest the spread of a malignant brain tumor are obtained by a researcher: 28, 49, 54, 80, 80, 102⁺, 120, 120⁺, 120⁺, 167, 200, 200, 200⁺, 340, 500, 500⁺. The precise survival times for five of the patients are censored (as indicated with the ⁺ superscript. Specifically: a) The patient with a survival time of 102 days refused to provide any further data for the study beyond that point in time; b) The researcher was unable to contact two of the three patients with a survival time of 120 days beyond that point in time; c) One of the three patients with a survival time of 200 days committed suicide at that point in time; d) One of the two patients with a survival time of 500 days was alive at the end of the study. Construct a survival function for the data.*

The information derived from Example 12.5 for computing a survival function is summarized in Table 12.8, and the information for Example 12.6 is summarized in Table 12.9. In the latter tables the score for each patient indicates how many days the patient survived. Thus, in both tables the first patient survived for 28 days and died on the 29th day, while the second patient survived for 49 days and died on the 50th day, etc. Note that since there are no censored scores in Example 12.5, all of the information for scores of the same value are recorded in a single row in Table 12.8. In the case of Example 12.6, where censored scores are present, a separate row is employed in Table 12.9 for each of the 16 scores. The latter is done in order to facilitate understanding of the computation of the values in a survival function table when dealing with censored data. The actual estimated survival function for Examples 12.5 and 12.6 are recorded in **Column 6** of Tables 12.8 and 12.9, as well as in Figures 12.1 and 12.2, which depict the survival functions graphically. The data summarized in the aforementioned tables and graphs are referred to as **estimated survival functions**, since based on the analysis of the available data they provide the best estimate of survival probabilities in the underlying population.

The following information should be employed in interpreting the information in Tables 12.8 and 12.9.

1) **Column 1** contains the ordinal position (represented with the notation i) for each of the $n = 16$ scores (**uncensored** and **censored**) which comprise the set of data. In both Tables 12.8 and 12.9 the ordinal position of the lowest score is $i = 1$ and the ordinal position of the highest score is $i = 16$. As noted earlier, when the same score occurs more than once, a single row lists all of the ordinal positions represented by that score in Table 12.8, whereas in Table 12.9 a separate row is employed for each of the ordinal positions which represent that score.

2) Each of the rows in **Column 2** contains the **Score** (i.e., the number of days a patient survived) for each of the patients who participated in the study. In Table 12.9 any score which is **censored** has the superscript + recorded at the upper right of that score. Note that in Table 12.9, when two or more patients are assigned the same score, those patients with that score who scores are **censored** are listed below those patients with that score whose scores are **uncensored**. To illustrate, note that in Table 12.9 there are three scores of 120, and that the two **censored** **scores** of 120 are recorded below the single uncensored score of 120.

Table 12.8 Estimated Kaplan–Meier Survival Function for Example 12.5

i (Ordinal Position)	Score (Day)	Number at Risk	Number Died	Conditional Probability	Cumulative Probability (Survival Probability)	Standard Error	95% Confidence Interval (LL–UL)
1	28	16	1	15/16 = .9375	(1)(.9375) = .9375	.0605	.8189 – 1.000
2	49	15	1	14/15 = .9333	(.9375)(.9333) = .8750	.0827	.7129 – 1.000
3	54	14	1	13/14 = .9286	(.8750)(.9286) = .8125	.0976	.6212 – 1.000
4,5	80	13	2	11/13 = .8462	(.8125)(.8462) = .6875	.1159	.4603 – .9147
6	102	11	1	10/11 = .9091	(.6875)(.9091) = .6250	.1210	.3878 – .8622
7,8,9	120	10	3	7/10 = .7000	(.6250)(.7000) = .4375	.1240	.1945 – .6805
10	167	7	1	6/7 = .8571	(.4375)(.8571) = .3750	.1210	.1378 – .6122
11,12,13	200	6	3	3/6 = .5000	(.3750)(.5000) = .1875	.0976	0.000 – .3788
14	340	3	1	2/3 = .6667	(.1875)(.6667) = .1250	.0827	0.000 – .2871
15,16	500	2	2	0/2 = 0	(.1250)(0) = 0	.0000	

Table 12.9 Estimated Kaplan–Meier Survival Function for Example 12.6

i (Ordinal Position)	Score (Day)	Number at Risk	Number Died	Conditional Probability	Cumulative Probability (Survival Probability)	Standard Error	95% Confidence Interval (LL–UL)
1	28	16	1	15/16 = .9375	(1)(.9375) = .9375	.0605	.8189 – 1.000
2	49	15	1	14/15 = .9333	(.9375)(.9333) = .8750	.0827	.7129 – 1.000
3	54	14	1	13/14 = .9286	(.8750)(.9286) = .8125	.0976	.6212 – 1.000
4	80	13	2	11/13 = .8462	(.8125)(.8462) = .6875	.1159	.4603 – .9147
5	80+						
6	102+						
7	120	10	1	9/10 = .9000	(.6875)(.9000) = .6188	.1230	.3777 – .8591
8	120+						
9	120+						
10	167	7	1	6/7 = .8571	(.6188)(.8571) = .5304	.1335	.2687 – .7921
11	200	6	2	4/6 = .6667	(.5304)(.6667) = .3536	.1354	.0882 – .6190
12	200						
13	200+						
14	340	3	1	2/3 = .6667	(.3536)(.6667) = .2357	.1319	0.000 – .4942
15	500	2	1	1/2 = .5000	(.2357)(.5000) = .1179	.1063	0.000 – .3262
16	500+						

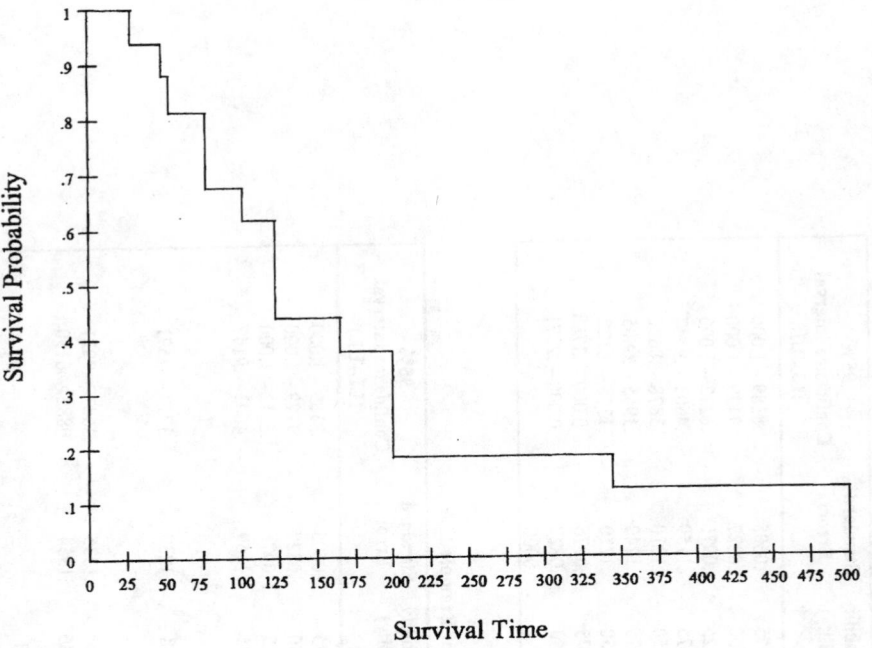

Figure 12.1   Kaplan–Meier Estimate of Survival Function for Example 12.5

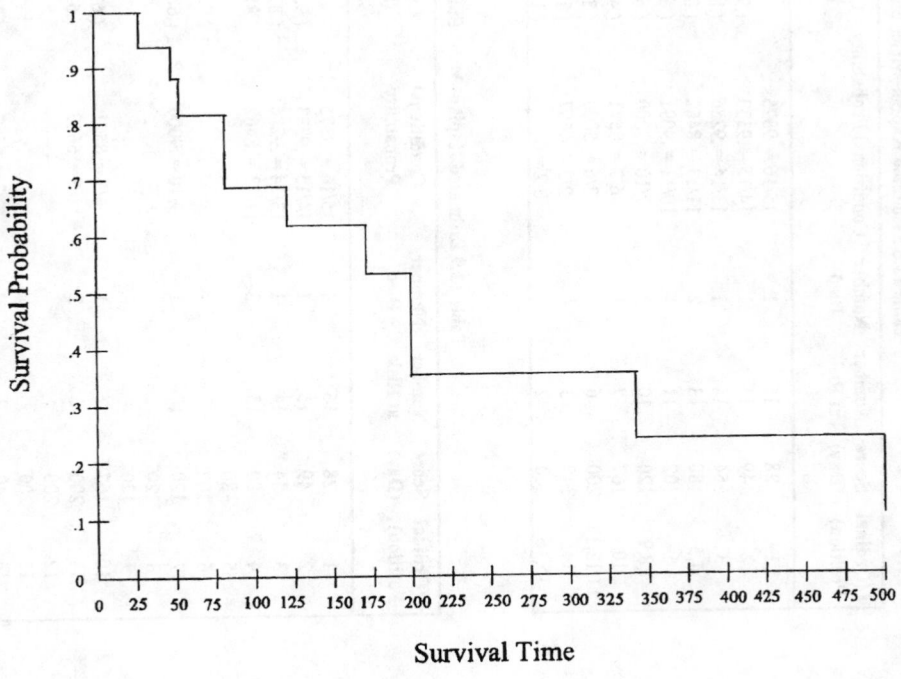

Figure 12.2   Kaplan–Meier Estimate of Survival Function for Example 12.6

3) The value entered in each of the rows of **Column 3**, which is labeled **Number at Risk**, is equal to the total number of scores (**uncensored** and **censored**) that are equal to or greater than the score in that row at that point in time. The value in **Column 3** represents the number of patients at the point in time designated by a given row who are still at risk of succumbing to the brain tumor. Note that in Table 12.9 **number at risk** values are not recorded for **censored scores**, and that when there are two or more instances of the identical **uncensored score**, the **number at risk** value is recorded in the row documenting the first instance of that **uncensored score** value. (If one chooses, the same **number at risk** value could also be recorded in any rows below it documenting other instances of that same **uncensored score** value.)

4) The value entered in each of the rows of **Column 4**, which is labeled **Number Died**, is the total number of **uncensored scores** that are equal to the score in that row. For a given row, the value in **Column 4** represents the number of patients who succumb to the brain tumor at that point in time (note that we do not include any patient who has a **censored score** equal to the **uncensored score** in that row, and consequently we do not record a **number died** value in any row that represents a **censored scores**). Note that when there are two or more instances of the identical **uncensored score**, the **number died** value is recorded in the row documenting the first instance of that **uncensored score** value. (If one chooses, the same **number died** value could also be recorded in any rows below it documenting other instances of that same **uncensored score** value.)

5) The value entered in each of the rows of **Column 5**, which is labeled **Conditional Probability**, represents the **conditional probability** for patients who obtained the score in that row. (The concept of **conditional probability** is discussed in the **Introduction** as well as in Section IX (the **Addendum**) of the **binomial sign test for a single sample (Test 9)**.) The **conditional probability** value recorded in a row indicates the probability a patient survived up to the time period designated for that row, given the fact the patient survived all of the earlier time periods documented in the table. A **conditional probability** is only computed for **uncensored scores**. Consequently, in Table 12.8 (in which all of the scores are **uncensored**), a **conditional probability** has been computed for each score. In Table 12.9 a **conditional probability** is only computed for 11 of the 16 scores, since five of the scores are **censored**. Note that if all instances of a score are **censored**, a **conditional probability** is not computed for that score (e.g., the score of 102 in Table 12.9). Also take note of the fact that in Table 12.9, when there are two or more instances of the identical **uncensored score**, the **conditional probability** value is recorded in the row documenting the first instance of that **uncensored score** value. (If one chooses, the same **conditional probability** value could also be recorded in any rows below it documenting other instances of that same **uncensored score** value.) Although **conditional probabilities** are not recorded for any of the **censored scores** in Table 12.9, some sources (e.g., Conover (1999, p. 91)) may elect to record a **conditional probability** for a **censored score** which is identical to the **conditional probability** recorded for the **uncensored score** of the same value.

The following guidelines should be used to construct the proportion employed to compute a **conditional probabilities** for an **uncensored score** in a given row: a) The **denominator of the proportion** will always be the **number at risk** value recorded for the score in that row; b) In the case of a score that has no **censored score(s)** of the same value, the **numerator of the proportion** will be the total number of scores (**uncensored** and **censored**) which are greater than the score in that row. In the case of a score that has at least one **censored score** of the same value, the **numerator of the proportion** will be the total number of scores (**uncensored** and **censored**) which are greater than the score in that row plus the number of **censored scores** that are equal to the score in that row. Put simply, the numerator for the proportion can be obtained by subtracting the **Number died** value for a row from the **Number at risk** value for that row — i.e., *Number at risk - Number died*; c) If there are any **censored scores** which have a value that is lower than the first **uncensored score** in a set of data, the **conditional probability** (if a source elects to record one) of those **censored scores** will always equal 1. This is the case since both the **numerator** and **denominator of the proportion** for computing the latter **conditional probability** will be equal to the total number of scores (**uncensored** and **censored**) which are greater than the score in that row; d) The probability in the last row of the **survival function table** will always equal **zero** (which is the case in Table 12.8) unless there are one or more **censored scores** equal to or greater than the last **uncensored score** (which is the case in Table 12.9). In the latter case, **censored scores** that are above the value of the last **uncensored score**, if assigned a **conditional probability** (most sources do not record one), would be assigned the same **conditional probability** assigned to the **uncensored score** that is less than but closest to the **censored score(s)** in question. In such a case the **conditional probability** in the last row of the table will not equal **zero**.

6) The value entered in each of the rows of **Column 6**, which is labeled **Cumulative Probability**, is employed to represent the **Survival Probability** for each of the time periods (i.e., day of survival values) listed in **Column 2**. The **cumulative probability** or **survival probability** is the likelihood that an individual survives longer than the time designated in a given row. A table or graph of the **cumulative probabilities** is employed to represent the **survival function** for a set of data. In both Figures 12.1 and 12.2, the scores (survival times) of patients are recorded on the abscissa (*X*-axis), while the **cumulative probability/survival probability** values are recorded on the ordinate (*Y*-axis).

A **cumulative probability** is only computed for **uncensored scores**. Consequently, in Table 12.8, in which all of the scores are **uncensored**, a **cumulative probability** has been computed for each score. In Table 12.9 a **cumulative probability** is only computed for 11 of the 16 scores, since five of the scores are **censored**. Note that in Table 12.9, when there are two or more instances of the identical **uncensored score** (e.g., there are two scores of 200 which are **uncensored**), the **cumulative probability** value is recorded in the row documenting the first instance of that **uncensored score** value. (Some sources (e.g., Conover (1999, p. 91)) record the same **cumulative probability** in any rows documenting other instances of that same **uncensored score** value.) If the same score occurs more than once, and at least one instance of that score is **uncensored**, any **censored scores** of that value are assigned the same **cumulative probability** as the value computed for the **uncensored score** of that value. If, on the other hand, all instances of a score are **censored**, a **cumulative probability** is not computed for that score (e.g., the score of 102 in Table 12.9). However, in the graph of the survival function, the latter scores are assigned the same **cumulative probability** as that of the highest **uncensored score** which falls below it (e.g., the score of 102 is assigned the **cumulative probability** of .6875 assigned to the score 80). Note that in Figures 12.1 and 12.2 the **survival probability/cumulative probability** recorded for an **uncensored score** is also the **survival probability** recorded for all of the scores/time periods greater than the **uncensored score** in question until the time period during which a new **uncensored score** is recorded, at which point the survival function decreases (i.e. moves to a lower point on the ordinate of the graph).

In order to compute a **cumulative probability**, the **multiplication rule** (which is described in the **Introduction** (in the section on **Basic principles of probability**)) must be employed. The **multiplication rule** allows one to compute *P(AB)* (which can also be written as *P(A∩B)*). Specifically, *P(AB)* = *P(A)P(B/A)*, which states the probability that two events *A* and *B* will occur simultaneously is equal to the probability of *A* times the conditional probability *P(B/A)*.[21]

In order to illustrate the multiplication rule, it will be applied to rows 2 and 3 (which respectively document ordinal positions 2 and 3) in Tables 12.8 and 12.9. In the discussion to follow I will refer to ordinal positions 2 and 3 respectively as Times 2 and 3, since the latter ordinal positions represent the second and third lowest survival times of the 16 subjects. If A represents the event that a patient is alive at 49 days (i.e., survives until Time 2) and B represents the event that a patient is alive at 54 days (i.e., survives until Time 3), employing the multiplication rule, the probability that a patient survives longer than 54 days is $P(AB) = P(A)P(B/A)$, which can also be written as noted below. (Note that P (Surviving to Time 3) represents $P(AB)$, since it represents the simultaneous occurrence of a person having been alive at both Times 2 and 3 (i.e., 49 days and 54 days).)

$$P(Surviving\ to\ Time\ 3) = [P\ (Surviving\ to\ Time\ 2)]\ [P(Surviving\ Time\ 3/Survived\ Time\ 2)]$$

The above equation indicates that to determine the **cumulative probability** for a given row, we multiply the **cumulative probability computed for the previous row in the table** by the **conditional probability computed for that row**. Applying the multiplication rule in both Tables 12.8 and 12.9, the **cumulative probability** for 54 days (Time 3) is (.8750)(.9286) = .8125. In other words, we multiply .8750 (which is the **cumulative probability** for Time 2 = 49 days) by .9286 (which is the **conditional probability** for 54 days = Time 3). Thus, the likelihood a patient survives longer than 54 days is .8125. It should be noted that tables or graphs of a survival function may include a time period prior to that designated by the first row of Tables 12.8 and 12.9 (i.e., a time period when all of the patients are alive). If such a row were included in the latter tables (e.g., a row for 27 days) it would be assigned both a **conditional** and **cumulative probability** of 1, since the likelihood of surviving to that time would equal 1 by virtue of the fact that all of the patients in the study were alive at that time.

The reader should take note of the fact that unlike Figures 12.1 and 12.2, survival functions which are based on a large number subjects typically assume the shape of a curve, since virtually every time period between the lowest and highest survival times will include at least one uncensored score (thus resulting in a continuous decrease of the line describing the function). Survival functions which are based on survival data for a large number of people are commonly referred to as **life tables**. An example of the latter would be a life expectancy table employed by an insurance company, which documents the likelihood of a person surviving a specific number of years. The total number of rows in such a table could be the maximum number of years a person in the population might be expected to survive — employing one row for each year (the latter values being recorded in **Column 1** of the table). Some life tables may elect to group survival times in intervals of fixed length. Thus, each of the rows of **Column 1** might represent a five year interval. Consequently the first row could contain information for subjects who survived between 1 and 5 years, the second row would contain information for subjects who survived between 6 and ten years, and so on.

7) The value entered in each of the rows of **Column 7**, which is labeled **Standard Error**, represents the **standard error** for the **cumulative/survival probability** estimated in each row. Note that a **standard error** is only computed for **uncensored scores**, and thus in Table 12.8, in which all of the scores are **uncensored**, a **standard error** has been computed for each score. In Table 12.9 a **standard error** is only computed for 11 of the 16 scores, since five of the scores are censored. Note that if all instances of a score are **censored**, a **standard error** is not computed for that score (e.g., the score of 120 in Table 12.9). Also take note of the fact that in Table 12.9, when there are two or more instances of the identical **uncensored score**, the **standard error** value is recorded in the row documenting the first instance of that **uncensored score** value. (If one chooses, the same **standard error** value could also be recorded in any rows below it documenting other instances of that same **uncensored score** value.) Although **standard errors** are not recorded for any of the **censored scores** in Table 12.9, some sources may elect

to record a **standard error** for a **censored score** which is identical to the **standard error** recorded for the **uncensored score** of the same value.

The **standard error** value listed in a given row is the square root of the **variance** of the **cumulative/survival probability** for that row. Although computer software is generally employed to generate the values of the **standard errors** (because of the complexity of the computations), the protocol for their computation is described below.

a) Multiply the **conditional probability** in a given row by the **conditional probabilities** of all the rows that are listed above it in the table (i.e., all rows which identify a lower survival time than the row in question). Upon doing the latter, square the resulting value. The obtained squared value will be designated as a.

b) For the row in question divide the number of **uncensored scores** in the sample which obtain the score in that row (i.e., the **Number died** value recorded for that row) by the product of the **numerator** (i.e., the **Number at risk** – **Number died**) and **denominator** (i.e., **Number at risk**) of the proportion that was employed to compute the **conditional probability** for the score in that row. Repeat the latter for all rows which are listed above that row in the table (i.e., all rows which identify a lower survival time than the row in question). Upon doing the latter, add up the values computed for each row. We will designate the obtained sum as b.

c) The **variance** for the score in that row is the product of a and b — i.e., $Variance = (a)(b)$. The square root of the **variance** is the **standard error** (s_e). Thus, $s_e = \sqrt{Variance} = \sqrt{(a)(b)}$. The latter computations are illustrated below for the scores 28 through 120.

$$s_e(28) = \sqrt{[.9375]^2 \left[\frac{1}{(15)(16)}\right]} = .0605$$

$$s_e(49) = \sqrt{[(.9333)(.9375)]^2 \left[\frac{1}{(14)(15)} + \frac{1}{(15)(16)}\right]} = .0827$$

$$s_e(54) = \sqrt{[(.9286)(.9333)(.9375)]^2 \left[\frac{1}{(13)(14)} + \frac{1}{(14)(15)} + \frac{1}{(15)(16)}\right]} = .0976$$

$$s_e(80) = \sqrt{[(.8462)(.9286)(.9333)(.9375)]^2 \left[\frac{2}{(11)(13)} + \frac{1}{(13)(14)} + \frac{1}{(14)(15)} + \frac{1}{(15)(16)}\right]} = .1159$$

$$s_e(120) = \sqrt{[(.9091)(.8462)(.9286)(.9333)(.9375)]^2 \left[\frac{1}{(9)(10)} + \frac{2}{(11)(13)} + \frac{1}{(13)(14)} + \frac{1}{(14)(15)} + \frac{1}{(15)(16)}\right]} = .1230$$

8) The values entered in each of the rows of **Column 8** represent the upper and lower limits of the **95% confidence interval** for the **cumulative probability** computed for a given row. A **confidence interval** is only computed for **uncensored scores**. Consequently, in Table 12.8, in which all of the scores are uncensored, a **confidence interval** has been computed for each score. In Table 12.9 a **confidence interval** is only computed for 11 of the 16 scores, since five of the scores are censored. Note that if all instances of a score are **censored**, a **confidence interval** is not computed for that score (e.g., the score of 102 in Table 12.9). Also take note of the fact that

in Table 12.9 when there are two or more instances of the identical **uncensored score**, the **confidence interval** value is recorded in the row documenting the first instance of that **uncensored score** value. (If one chooses, the same **confidence interval** value could also be recorded in any rows below it documenting other instances of that same **uncensored score** value.) Although **confidence intervals** are not recorded for any of the **censored scores** in Table 12.9, some sources may elect to record a **confidence interval** for a **censored score** which is identical to the **confidence interval** recorded for the **uncensored score** of the same value.

The 95% **confidence interval** is computed with Equation 8.6 (described in Section VI of the **chi-square goodness-of-fit test (Test 8)**), which is the equation for computing a confidence interval for a population proportion. To illustrate, the confidence interval for the **cumulative probability** of .8750 computed for a survival time of 49 days is computed below. Note that in Equation 8.6, the standard error is represented by the value $\sqrt{(p_1 p_2)/n}$.

$$\left[p_1 - z_{\alpha/2} \sqrt{\frac{p_1 \, p_2}{n}} \right] \leq \pi_1 \leq \left[p_1 + z_{\alpha/2} \sqrt{\frac{p_1 \, p_2}{n}} \right]$$

$$.8750 - (1.96)(.0827) \leq \pi_1 \leq .8750 + (1.96)(.0827)$$

$$\pi_1 = .8750 \pm .1621$$

$$.7129 \leq \pi_1 \leq 1.0371$$

which becomes

$$.7129 \leq \pi_1 \leq 1$$

Thus, the researcher can be 95% confident the interval .7129 to 1 contains the true proportion of cases in the underlying population who survive longer than 49 days. (The value 1 is employed in lieu of the computed value 1.0371, since a proportion cannot be greater than 1. In the same respect, if a lower limit for a confidence interval is computed to be a negative value (i.e., below zero), the lower limit of the interval is set at zero, since that is the lowest value a proportion can achieve.) Stated in probabilistic terms, there is a probability/likelihood of .95 that the interval .7129 to 1 contains the true value of the population proportion for a person surviving longer than 49 days.

In closing the discussion of the survival functions depicted in Table 12.8/Figure 12.1 and Table 12.9/Figure 12.2 the following should be noted: a) The **steepness** of a survival function at any point on the graph/curve can indicate time periods when a patient is at greatest risk for dying. Specifically, time periods during which the survival function remains relatively flat (i.e., remains for the most part on a horizontal plane) indicate that a patient is not at great risk of dying (since at such points in time a patient's probability of survival only decreases minimally). On the other hand, time periods associated with precipitous drops in the curve are indicative of times a patient is at greatest risk (since at such points in time a patient's probability of survival will decrease dramatically); b) Wright (2000, p. 367) notes that the measure of central tendency most commonly employed in survival analysis is the **median**. The median for a survival function will correspond to the time period which corresponds to a .5 survival probability. The latter value can be obtained by erecting a perpendicular line from the ordinate to the survival curve. At the point the latter line intersects the curve, a second line is drawn from the curve perpendicular to the abscissa. The point the latter line intersects the abscissa will correspond to the median time of survival; c) Wright (2000, p. 368) notes that the **interquartile range** (which is discussed in the **Introduction**) can be employed to stipulate a range of time a "typical" patient (i.e., a patient who is not at the lower or upper extreme times for survival) would be expected to survive. Since the boundaries which define the interquartile range are the 25th and 75th percentiles, the time periods on the survival curve which correspond to the probabilities .25 and .75 will represent the upper and lower limits of the interquartile range. The latter time periods can be obtained from a survival curve in the same manner as the median.

3. Procedures for evaluating censored data in a design involving two independent samples
An inferential statistical test for evaluating censored data is designed to provide a mechanism for minimizing bias which might result from the inclusion of censored scores in an evaluation of a null hypothesis. One could argue that by virtue of including the latter type of data in an analysis, such an inferential procedure will only be able to provide a general approximation with regard to whatever it is a researcher is evaluating with respect to one or more underlying populations, and that the approximation provided by such a test will be less accurate than that provided by an inferential test employed with uncensored data.

Under certain conditions a parametric test such as the *t* **test for two independent samples** can be employed to evaluate censored data. Specifically, if the censored observations fall at the upper and/or lower ends of the distributions, a procedure for dealing with outliers (such as trimming and Winsorization, which are discussed in Section VII of the latter test) can be employed to eliminate the censored scores from the analysis (which would employ trimmed or Winsorized values for the means and variances). Since, however, censored observations often fall at other than the endpoints of a distribution, the latter will usually not be feasible.

As a general rule, censored data are more amenable to be evaluated with nonparametric procedures. The latter generally involve evaluating the ordinal relationships between observations. This section will describe the following three nonparametric tests for evaluating censored data in a design involving two independent samples: a) A **permutation test based on the median**; b) **Gehan's test for censored data (Test 12e)**; and c) The **log-rank test (Test 12f)**.

Permutation test based on the median   In the **Introduction** (as well as earlier in this **Addendum**) it was noted that a **permutation** is an ordered arrangement of objects. As noted in Section IX, within the framework of inferential statistical testing, a **permutation test** involves the evaluation of ordered arrangements of scores. The discussion to follow will describe how a permutation test can be employed to evaluate a design involving two independent samples in which censored data are present.

Among others, Higgins (2004, p. 229–230) emphasizes the fact that permutation tests are only useful if the censoring mechanism in a study is random. Recollect that in **random censoring** it is assumed that subjects drop out of a study at random points in time, and that a subject's reason for leaving a study is unrelated to (i.e., independent of) the experimental treatment to which he or she has been assigned. To clarify the latter, assume that two drugs are equally effective in treating a disease, yet one of the drugs has a nasty side effect profile which increases the likelihood a patient will drop out of a study before the end of a clinical trial. Under such circumstances the censoring process will not be random, and by virtue of the latter a primary assumption underlying a permutation test is violated — specifically, that all possible permutations for the data are equally likely to occur. Because of the latter it would not be appropriate to employ a permutation test to evaluate the aforementioned study.

Example 12.7 illustrates a scenario where a permutation test could be employed to evaluate a set of data containing censored observations.

Example 12.7 Table 12.10 *contains the survival times (in days) of 20 cancer patients, each of whom is administered one of two treatments. Do the data indicate differences in the survival distributions for the two treatments?*

Table 12.10 Survival Times of Cancer Patients Exposed to Two Treatments

Treatment 1	134	167	201	233	344	356	378	399	567⁺	663⁺
Treatment 2	125	198	244	256	289	333	399	402	403	444

Note that in Table 12.10, within each of the treatments, the survival times of the patients who served in that treatment are listed in ascending order with respect to the number of days survived. Since two of the 20 scores (both of which are in Treatment 1) in Table 12.10 are censored, it is not possible to compute the mean length of survival for the two treatments (unless, of course, one elects to compute trimmed or Winsorized means). The medians for the treatments, however, can be computed since the censored scores are the two highest survival times among the 20 patients. Thus we can determine that the medians for Treatments 1 and 2 are respectively M_1 = 350 (i.e., $(344 + 356)/2 = 350$) and $M_2 = 311$ (i.e., $(289 + 333)/2 = 311$). The difference between the two median values is $350 - 311 = 39$. If we assume **random censoring**, a permutation test can be applied to evaluate the likelihood of obtaining a difference between the two median values equal to or larger than 39 days. Such a test would be impractical to conduct without the use of a computer because of the laborious computations involved. The methodology described earlier in this **Addendum** for conducting a **randomization test** (which, as noted earlier, is an alternative term for a **permutation test**) is employed to conduct such a test. Specifically, we can determine that the number of combinations of 20 things taken 10 at a time is equal to 184,756: $\binom{20}{10} = \frac{20!}{10!\,10!} = 184{,}756$. The latter indicates there are 184,766 possible ways that 20 subjects can be assigned to two treatments with 10 subjects per treatment. The magnitude of the value 184,756 might make it impractical for some computers to construct an exact sampling distribution based on every possible arrangement. As noted earlier, in such a situation an **approximate randomization test** can be conducted, wherein the computer constructs a sampling distribution based on randomly selecting a large number (but not all) of the possible arrangements for the data. The resulting sampling distribution is then employed to evaluate the sample data. As will be noted below, the use of an **approximate randomization test** in this context involves repeated **resampling** of the original data.

Let us assume we elect to employ an **approximate randomization test** to evaluate the data in Table 12.10. In conducting the latter test the computer is instructed to conduct a large number of simulations involving the survival scores of the 20 patients. For purposes of illustration let us assume that 10,000 simulations are conducted. In each simulation 10 of the 20 scores in Table 12.10 are randomly selected and assigned to Treatment 1 and the remaining 10 scores are assigned to Treatment 2. The median for each sample is computed, as well as the difference between the two sample medians. At the conclusion of the analysis there will be 10,000 difference scores. The frequency distribution of these difference scores represents a sampling distribution which is employed to evaluate the data in Table 12.10. Specifically, through use of the sampling distribution of difference scores a determination can be made with respect to the likelihood of obtaining a difference score equal to or more extreme than the median difference of 39 days obtained for the two treatments in Table 12.10.[22]

Test 12e: Gehan's test for censored data (also referred to as the **Gehan–Wilcoxon test for censored data**) The test to be described was developed by Gehan (1965) to compare two survival functions (representing two independent samples) in which there are one or more censored observations. In point of fact, **Gehan's test for censored data** is a permutation test, and if no censored observations are present the test is equivalent to the **Mann–Whitney U test**. The null hypothesis evaluated with **Gehan's test for censored data** is that the survival time distributions in the underlying populations the two treatments represent are identical. The nondirectional alternative hypothesis is that the survival time distributions for the two underlying populations are different, while a directional alternative hypothesis specifies that the distribution for the population representing Treatment 1 has longer or shorter survival times than the distribution for the population representing Treatment 2.

     Gehan's test for censored data will be illustrated with Example 12.8.

Example 12.8 Table 12.11 *contains the survival times (in days) of eight cancer patients, each of whom is administered one of two treatments. Do the data indicate differences in the survival distributions for the two treatments?*

     In Table 12.11 there are $N = 8$ patients, with $n_1 = n_2 = 4$. Note that in the latter table the configuration of the censored scores do not make the data amenable to being evaluated with the **permutation test** which was employed to evaluate the data in Table 12.10 (since use of the resampling procedure described for the latter test will not always result in samples for which an exact median value can be computed). Two methods will be described in this section for computing the **Gehan test statistic** for the data in Table 12.11.

Table 12.11   Survival Times of Cancer Patients Exposed to Two Treatments

Treatment 1	122^+	299	302	363^+
Treatment 2	199	367	403	403^+

Table 12.12   Computation of Gehan Scores

Observation	X_1	Y_1	X_2	X_3	X_4	Y_2	Y_3	Y_4
Score	122^+	199	299	302	363^+	367	403	403^+
L, G *	0, 0	0, 6	1, 5	2, 4	3, 0	3, 2	4, 1	5, 0
Gehan Score	0	−6	−4	−2	3	1	3	5

* L = Definitely less than; G = Definitely greater than

Method 1 for Gehan's test for censored data The first method to be described requires that initially the survival times of the $N = 8$ patients be rearranged in the format employed in Table 12.12. Note that in the first row of the latter table, X_i represents the ith observation in Treatment 1 and Y_i represents the ith observation in Treatment 2. The identification number of any observation within a treatment is a function of the ordinal position of its survival time in that treatment, with the lower the identification number the lower survival time for that observation within the treatment. In the second row of Table 12.12, the $N = 8$ survival times/scores are ordered from left to right with respect to magnitude. In conducting **Gehan's test for censored data**, a value to be referred to as a **Gehan score** is computed for each of the $N = n_1 + n_2 = 8$

observations in Tables 12.11/12.12. The sum of the **Gehan scores** for either treatment is designated U_G (since both sums will have the same absolute value), and is employed to represent the test statistic.

The following protocol is employed to compute a **Gehan score**. Beginning with the leftmost score in Table 12.12, each score is designated as a **target score**. The **target score** (to be designated A) is evaluated in reference to each of the other scores in the table with respect to whether the other score (to be designated B) is **definitely less than** the target score, **definitely greater than** the target score, or whether the ordering of the **target score** and the score with which it is being contrasted is **indeterminate**. The latter determinations are made as follows.

If two scores A and B are contrasted with one another, it can be concluded that score B is **definitely less than** score A if either of the following two conditions is met: 1) Both B and A are uncensored observations and $B < A$; or 2) If B is an uncensored observation and A is a censored observation, but $B \le A^+$. The value -1 is employed for each instance in which a score B is **definitely less than** score A. (Note that if A is censored, and the values of A^+ and B are identical, B is considered less than A, since although the exact survival time of A is not known, it is known that the survival time of A must be at least one day longer than the survival time of B.)

It can be concluded that score B is **definitely greater than** score A if either of the following two conditions is met: 1) Both B and A are uncensored observations and $B > A$; or 2) If B is a censored observation and A is an uncensored observation, but $B^+ \ge A$. The value $+1$ is employed for each instance in which a score B is **definitely greater than** score A. (Note that if B is censored, and the values of B^+ and A are identical, B is considered greater than A, since although the exact survival time of B is not known, it is known that the survival time of B must be at least one day longer than the survival time of A.)

The ordering of the scores is considered **indeterminate** if any of the following conditions is met: 1) If B is censored and A is uncensored and $B^+ < A$; 2) B is uncensored and A is censored and $B > A^+$; 3) Both B and A are censored but have different values (i.e., $B^+ \ne A^+$); 4) If A and B are tied (i.e., $A = B$ or $A^+ = B^+$). The value **0** is recorded for any pair of scores for which the ordering is considered **indeterminate**.

In the third row of Table 12.12 in the column for each of the $N = 8$ scores, the following information is recorded for when the score in that column is designated as the target score: The number of scores which are determined to be **definitely less than** the target score, followed by the number of scores which are determined to be **definitely greater than** the target score.

Employing the above criteria with the scores in Table 12.12, we can conclude the following. When the censored score $X_1 = 122^+$ (the first observation in Treatment 1) is designated as the **target score**, it can be determined that 0 scores are **definitely less than** $X_1 = 122^+$, and that 0 scores are **definitely greater than** $X_1 = 122^+$. When the score $Y_1 = 199$ (the first observation in Treatment 2) is designated as the **target score**, it can be determined that 0 scores are **definitely less than** $Y_1 = 199$, and that 6 scores are **definitely greater than** $Y_1 = 199$ (specifically, $X_2 = 299$, $X_3 = 302$, $X_4 = 363^+$, $Y_2 = 367$, $Y_3 = 403$, $Y_4 = 403^+$). When the score $X_2 = 299$ (the second observation in Treatment 1) is designated as the **target score**, it can be determined that 1 score is **definitely less than** $X_2 = 299$ (specifically, $Y_1 = 199$) and that 5 scores are **definitely greater than** $X_2 = 199$ (specifically, $X_3 = 302$, $X_4 = 363^+$, $Y_2 = 367$, $Y_3 = 403$, $Y_4 = 403^+$). When the score $X_3 = 302$ (the third observation in Treatment 1) is designated as the **target score**, it can be determined that 2 scores are **definitely less than** $X_3 = 302$ (specifically, $Y_1 = 199$, $X_2 = 299$) and that 4 scores are **definitely greater than** $X_3 = 302$ (specifically, $X_4 = 363^+$, $Y_2 = 367$, $Y_3 = 403$, $Y_4 = 403^+$). When the score $X_4 = 363^+$ (the fourth observation in Treatment 1) is designated as the **target score**, it can be determined that 3 scores are **definitely less than** $X_4 = 363^+$ (specifically, $Y_1 = 199$, $X_2 = 299$, $X_3 = 302$) and that 0 scores are **definitely greater than** $X_4 = 363^+$. When the score $Y_2 = 367$ (the second observation in Treatment 2) is designated as the **target score**, it can be determined that 3 scores are **definitely less than** $Y_2 = 367$ (specifically, $Y_1 = 199$, $X_2 = 299$, $X_3 = 302$) and that 2 scores are **definitely greater than** $Y_2 = 367$ (specifically, $Y_3 = 403$, $Y_4 = 403^+$). When the score $Y_3 = 403$ (the third observation in Treatment 2) is designated as the **target score**, it can be determined that 4 scores are **definitely less than** $Y_3 = 403$ (specifically, $Y_1 = 199$, $X_2 = 299$, $X_3 = 302$, $Y_2 = 367$) and that 1 score is **definitely greater than** $Y_3 = 403$ (specifically, $Y_4 = 403^+$). When the score $Y_4 = 403^+$ (the fourth observation in Treatment 2) is designated as the **target score**, it can be determined that 5 scores are **definitely less than** $Y_4 = 403^+$ (specifically, $Y_1 = 199$, $X_2 = 299$, $X_3 = 302$, $Y_2 = 367$, $Y_3 = 403$) and that 0 scores are **definitely greater than** $Y_4 = 403^+$.

The values in each column of the last row of Table 12.12, represent the **Gehan score** for the observation in that column. A **Gehan score** for an observation (which for each observation will be represented by the notation $\Sigma U_{i(G)}$) is computed as follows: The **number of observations which are definitely less than that observation minus the number of observations which are definitely greater than that observation**. Note that the sum of the N Gehan scores will always equal zero, and the absolute value of the sum of the n_1 Gehan scores for Treatment 1 will always equal the absolute value of the sum of the n_2 Gehan scores for Treatment 2. Thus, in Table 12.12, the sum of the $n_1 = 4$ **Gehan scores** in **Treatment 1** (which will be designated with the notation U_{G_1}) is $0 + (-4) + (-2) + 3 = -3$, and the sum of the $n_2 = 4$ scores **Gehan scores** in **Treatment 2** (which will be designated with the notation U_{G_2}) is $-6 + 1 + 3 + 5 = 3$. Either of the values $U_{G_1} = -3$ or $U_{G_2} = 3$ can be employed in Equation 12.12 (which is the equation for computing the **Gehan test statistic**) to represent the value U_G. The sign of U_G indicates the direction of the difference between the two treatment survival distributions.

Equation 12.12 is a normal approximation which is employed to compute the **Gehan test statistic** (which increases in accuracy as the sample sizes increase). If the null hypothesis is true, the expected mean value of U_G (represented with the notation $E(U_G)$) for the permutation distribution will equal zero. Note that since $E(U_G)$ will always equal zero, sources usually only employ the value U_G in the numerator of Equation 12.12. The estimated variance of the permutation distribution is computed with Equation 12.11. The notation $\Sigma U_{i(G)}^2$ in the latter equation indicates that each of the N Gehan scores in Table 12.12 is squared, and that the N squared Gehan scores are summed. Thus, $\Sigma U_{i(G)}^2$ can be referred to as the **sum of the squared Gehan scores**. Employing the values in last row of Table 12.12, $\Sigma U_{i(G)}^2 = (0)^2 + (-6)^2 + (-4)^2 + (-2)^2 + (3)^2 + (1)^2 + (3)^2 + (5)^2 = 100$. When the value $\Sigma U_{i(G)}^2 = 100$ along with the appropriate samples size information ($N = 8$, $n_1 = n_2 = 4$) is substituted in Equation 12.11, the value $Var(U_G) = 28.57$ is computed. Employing the latter value along with $U_{G_1} = -3$ in Equation 12.12 yields the value $z = -.56$. (Use of $U_{G_2} = 3$ in Equation 12.12 will yield the same absolute value, specifically, $z = .56$).

(Equation 12.11)

$$Var(U_G) = \frac{n_1 n_2}{N(N-1)} \Sigma U_{i(G)}^2 = \frac{(4)(4)}{8(8-1)}(100) = 28.57$$

(Equation 12.12)

$$z = \frac{U_G - E(U_G)}{\sqrt{Var(U_G)}} = \frac{U_G}{\sqrt{Var(U_G)}} = \frac{-3}{\sqrt{28.57}} = -.56$$

In order to reject the null hypothesis, the obtained absolute value of z must be equal to or greater than the tabled critical value at the prespecified level of significance. Since the obtained absolute value $z = .56$ is less than the tabled critical one- and two-tailed values $z_{.05} = 1.65$ and $z_{.05} = 1.96$, neither a directional or a nondirectional alternative hypothesis is supported. Thus, the data do not allow one to conclude that the underlying survival distributions for the two treatments are different. Although inspection of the data in Table 12.12 suggests longer survival times for patients in Treatment 2, the small sample size employed in the study resulted in a test with relatively low power, and, because of the latter, unlikely to detect a difference in the underlying populations if, in fact, one exists. One criticism of **Gehan's test of censored data** is it may not provide for as powerful a test as some alternative procedures for evaluating the same design involving censored data.

Although sources generally do not employ it, Gehan (1965, p. 208) stated that a more accurate approximation can be obtained if a correction for continuity is used with Equation 12.12, especially when the sample sizes are small. He recommended when there are relatively few tied and censored observations the continuity correction in the denominator of Equation 12.12 should be ± 1, and in other instances it should be ±½.

Method 2 for Gehan's test for censored data   At the beginning of the discussion of **Gehan's test for censored data** it was noted that if no censored observations are present the test is equivalent to the **Mann–Whitney U test**. If, however, the latter test is employed to evaluate a set of data containing one or more censored observations, it will not yield a result equivalent to that obtained with **Gehan's test**. Specifically, when the normal approximation of the **Mann–Whitney U test** (Equation 12.4) is employed to evaluate a design involving two independent samples containing censored data, it allows for a less powerful test of an alternative hypothesis when contrasted with **Gehan's test** (which will yield a higher absolute z value than that obtained with **Mann–Whitney** normal approximation). The methodology for the **Mann–Whitney test statistic** described Endnote 14 (and summarized in Table 12.21) will be employed to illustrate the latter, as well as to demonstrate an alternative methodology for computing the **Gehan test** statistic U_G computed above. Specifically, through use of Table 12.13 the number of **inversions** in the data will be computed, and the latter will be employed to determine the values U_1 and U_2 within the framework of the **Mann–Whitney U test**. The criterion for determining an inversion, however, will be modified to accommodate the presence of censored data.

In Table 12.13 each score in Treatment 2 is contrasted with each score in Treatment 1 with respect to ordinal position. Employing the criteria noted earlier, the values recorded in the cells of Table 12.13 indicate whether a score in Treatment 2 is **definitely greater than** or **definitely less than** a score in Treatment 1, or if the ordering for a pair of scores is **indeterminate**. Specifically, Table 12.13 contrasts each score in Treatment 2 with each score in Treatment 1 as follows: a) If it can be determined that a score in Treatment 2 is **definitely greater than** a score in Treatments 1, the entry + 1 appears in the cell that is the intersection of the two scores; b) If it can be determined that a score in Treatment 2 is **definitely less than** a score in Treatments 1, the entry – 1 appears in the cell that is the intersection of the two scores; c) If the ordering of the two scores being compared is **indeterminate**, the entry 0 appears in the cell that is the intersection of the two scores.

A number of comments are in order regarding similarities and differences between Tables 12.13 and 12.21 (which is identical to Table 12.3): a) In Table 12.21 the **sign in each cell** reflects the **direction of the difference** if the score in each row (i.e., Group 2 score) is subtracted from the score in the corresponding column (i.e., Group 1 score) (unless two scores are tied, in which case a 0 is entered in the relevant cell). Although, in addition to the sign, the magnitude of the difference between pairs of observations is also recorded in Table 12.21, the latter information is not necessary for determining the number of inversions in the table; b) In Table 12.13 the **sign in each cell** reflects the **direction of the difference** if the score in each row (i.e., Treatment 1 score) is subtracted from the score in the corresponding column (i.e., Treatment 2 score) (unless censoring or tied scores renders it impossible to specify the direction of the difference between a pair of observations, in which case a 0 is entered in the relevant cell); c) Note that in Table 12.21, 21 inversions were computed (since there were 20 negative signs and two zeros, with each zero counting as ½ point), and the latter value was employed to represent U_1. It was also noted that the value $U_2 = 4$ inversions can be obtained for Table 12.21 if the latter table is reconstructed so that the score in each column (i.e., Group 1 score) is subtracted from the score in the corresponding row (i.e., Group 2 score). Note that for Table 12.21, $(U_1 = 21) + (U_2 = 4) = n_1 n_2 = (5)(5) = 25$.

Table 12.13   Computation of U value for Gehan Test Statistic

		Treatment 2 Scores			
		199	367	403	403⁺
	122⁺	0	0	0	0
Treatment 1	299	−1	+1	+1	+1
Scores	302	−1	+1	+1	+1
	363⁺	−1	0	0	0

Employing the protocol described in Endnote 14 for determining the number of inversions, we compute there are 6½ inversions in Table 12.13. Specifically, there are 3 negative difference scores plus the seven zeros (which count at ½ a point each) which yield an additional 3½ inversions. Thus $3 + 3½ = 6½$. The latter value will be designated as U_1. Note that if Table 12.13 were reconstructed so that the score in each column (i.e., Treatment 2 score) is subtracted from the score in the corresponding row (i.e., Treatment 1 score) there would be a reversal of all of the signs in the table, yielding a table with 6 negative signs, 3 positive signs, and 7 zeros. The total number of inversions would equal 9½, which is the sum the 6 negative signs plus ½ point for each of the 7 zeros. Thus, $U_2 = 6 + 3½ = 9½$. Note that for Table 12.13 $(U_1 = 6½) + (U_2 = 9½) = n_1 n_2 = (4)(4) = 16$.

If we employ the same guidelines described in Section V for interpreting a U value obtained for the **Mann–Whitney U test**, the smaller of the two values U_1 versus U_2 is designated as the obtained U statistic. Since $U_1 = 6½$ is smaller than $U_2 = 9½$, the value of $U = 6½$. In **Table A11**, for $n_1 = 4$ and $n_2 = 4$, the tabled critical two-tailed .05 value is 0, but because of the small sample size no value is recorded for the tabled critical two-tailed .01 value. The tabled critical one-tailed .05 value is 1, but because of the small sample size no value is recorded for the tabled critical one-tailed .01 value. As noted in the guidelines for interpreting the result of the **Mann–Whitney U test**, in order to be significant the obtained value of U must be **equal to or less than** the tabled critical value at the prespecified level of significance. Since the obtained value $U = 6½$ is greater than all of the aforementioned tabled critical values, the null hypothesis (i.e., that the two underlying survival distributions are identical, or if stated within the context of the **Mann–Whitney U test**, H_0: $\theta_1 = \theta_2$) cannot be rejected. In other words, there is no evidence indicating a difference between the two treatments. (i.e., neither a nondirectional or directional alternative hypothesis is supported). Although the larger number of positive signs in Table 12.13 reflects longer survival times for patients in Treatment 2 (which is consistent with the alternative hypothesis H_1: $\theta_1 < \theta_2$), the small sample size employed in the study resulted in a test with relatively low power, and thus unlikely to detect a difference in the underlying populations if, in fact, one exists.[23]

In point of fact, the critical values recorded in **Table A11** will not be accurate if they are employed in evaluating a U value obtained for **Gehan's test for censored data**. Recollect that the inversion method employed in evaluating Table 12.21 uses a score of zero in a cell for tied observations, yet within the framework of the **Gehan test** a zero is also employed in cells where the ordering for a pair of observations is indeterminate. This latter modification in defining an inversion is what accounts for the breakdown in the accuracy of the tables. Deshpande *et al.* (1995, p. 194) note that the exact distribution for the **Gehan statistic** cannot be determined unless distributions of the censoring variables are known, which is generally not going to be the case.

Sprent and Smeeton (2000, p. 165) note that when substantial censoring is present in a set of data, alternative procedures (such as the **log-rank test**) may allow for a more powerful test of the alternative hypothesis. Pyke and Thompson (1986) note that one criticism of **Gehan's test for censored data** is that when the survival times for censored observations are less than the survival times of uncensored observations, the test tends to give greater weight to earlier deaths. Other sources which discuss **Gehan's test for censored data** are Higgins (2004, pp. 233–235), Deshpande *et al.* (1995, pp. 191–195), and Desu and Raghavarao (2003, pp. 114–116).

Test 12f: The log-rank test   The test to be described (developed by Mantel (1966) and also known as the **Mantel two-sample test for censored data** (Hollander and Wolfe (1999, 550–555)) compares two survival functions representing two independent samples in which there are one or more censored observations.[24] The null hypothesis evaluated with the **log-rank test** is that the survival time distributions in the underlying populations the two treatments represent are identical. The nondirectional alternative hypothesis is that the survival time distributions for the two underlying populations are different, while a directional alternative hypothesis specifies that the distribution for the population representing Treatment 1 has longer or shorter survival times than the distribution for the population representing Treatment 2.

The **log-rank test** will be illustrated with Example 12.9.

Example 12.9   Table 12.14 *contains the survival times of 42 cancer patients, each of whom is administered one of two treatments. Each treatment is comprised of 21 patients. Do the data indicate differences in the survival distributions for the two treatments?*

Table 12.14   Survival Times of 42 Cancer Patients

Treatment 1	60, 60, 60, 60$^+$, 70, 90$^+$, 100, 100$^+$, 110$^+$, 130, 160, 170$^+$, 190$^+$, 200$^+$, 220, 230, 250$^+$, 320$^+$, 320$^+$, 340$^+$, 350$^+$
Treatment 2	10, 10, 20, 20, 30, 40, 40, 50, 50, 80, 80, 80, 80, 110, 110, 120, 120, 150, 170, 220$^+$, 230$^+$

Table 12.15, which represents a 2×2 contingency table, summarizes the notation that will be used in conducting the **log-rank test**. A 2×2 contingency table is employed to summarize the frequency of observations in each of 4 cells which are formed as a result of the intersection of different levels/categories of two variables. One of the variables is represented by the two rows of the table, while the other variable is represented by the two columns. In the case of Table 12.15, the row variable represents the two treatments and the column variable represents the survival status of a patient. Note that although it categorizes a patient with respect to the

individual **dying** versus **surviving**, the column variable employed for the **log-rank test** can more generally be defined in terms of specifying an **event of note occurring** (e.g., death) versus the **event of note not occurring** (e.g., survival). Other examples which could be employed to represent the column variable in a 2×2 contingency table such as Table 12.15 are: a) The event of note could be a reformed alcoholic resuming drinking versus a reformed alcoholic remaining abstinent (which would represent the nonoccurrence of the event); b) The event of note could be the performance of a machine falling below a specified level of precision versus the performance of the machine remaining above the specified level of precision (which would represent the nonoccurrence of the event). It should also be noted that the basis for categorization on the row variable in Table 12.15 can be something other than differential treatments. Thus, one might employ **males** versus **females** as the row variable in order to determine whether there are gender differences in survival, or employ **Factory 1** versus **Factory 2** in order to determine whether there are differences in the breakdown rate of machines manufactured in two factories.

Table 12.15 Survival Table for Analysis with Log-Rank Test

	Die/Event of note occurred	Survive/Event of note did not occur	At Risk
Treatment 1	d_{1j}	$n_{1j} - d_{1j}$	n_{1j}
Treatment 2	d_{2j}	$n_{2j} - d_{2j}$	n_{2j}
Column sums	d_j	$n_j - d_j$	n_j

If one assumes the null hypothesis of no difference between the two underlying survival distributions is true, a patient's survival status will be independent of the treatment the patient receives. In Table 12.15, n_{ij} represents the number of patients in Treatment i at risk at the beginning of Time j, and $n_{ij} - d_{ij}$ represents the number of patients in Treatment i who are still alive at the end of Time j. Thus, n_{1j} represents the number of patients in Treatment 1 at risk at the beginning of Time j, and $n_{1j} - d_{1j}$ represents the number of patients in Treatment 1 who are still alive at the end of Time j. n_{2j} represents the number of patients in Treatment 2 at risk at the beginning of Time j, and $n_{2j} - d_{2j}$ represents the number of patients in Treatment 2 who are still alive at the end of Time j. n_j represents the total number of patients at risk in both Treatment 1 and Treatment 2 at the beginning of Time j, and $n_j - d_j$ represents the number of patients in both Treatment 1 and Treatment 2 who are still alive at the end of Time j. Note that $n_j = n_{1j} + n_{2j}$. d_{ij} represents the number of patients in Treatment i who die during Time j. Thus, d_{1j} represents the number of patients in Treatment 1 who die during Time j, and d_{2j} represents the number of patients in Treatment 2 who die during Time j. The notation d_j represents the total number of deaths which are known to have occurred in both Treatment 1 and Treatment 2 during Time j. Note that $d_j = d_{1j} + d_{2j}$.

The **log-rank test** compares the observed value of d_{1j} (or d_{2j}) with its expected value. Under the assumption of fixed marginal frequencies, the distribution of d_{1j} (or d_{2j}) in Table 12.15 will be hypergeometric with an expected value E_{1j} and variance V_{1j} (or in the case of Treatment 2 an expected value E_{2j} and variance V_{2j} (it will be noted later that $V_{1j} = V_{2j}$)). The normal (as well as the chi-square) distribution can be employed to provide an excellent approximation of the exact test statistic.

In order to conduct the **log-rank test**, Table 12.16 is constructed through use of the data in Table 12.14. The reader should take note of the following with respect to Table 12.16.

Table 12.16 Computation of Log-Rank Test Statistic for Example 12.9

j	d_{1j}	d_{2j}	d_j	n_{1j}	n_{2j}	n_j	$e_{1j} = (d_j/n_j)\,n_{1j}$	$v_{1j} = \dfrac{n_{1j}\,n_{2j}\,d_j\,(n_j - d_j)}{n_j^2(n_j - 1)}$
10	0	2	2	21	21	42	1.0000	.4878
20	0	2	2	21	19	40	1.0500	.4860
30	0	1	1	21	17	38	.5526	.2472
40	0	2	2	21	16	37	1.1351	.4772
50	0	2	2	21	14	35	1.2000	.4659
60	3	0	3	21	12	33	1.9091	.6508
70	1	0	1	17	12	29	.5862	.2426
80	0	4	4	16	12	28	2.2857	.8707
100	1	0	1	15	8	23	.6522	.2268
110	0	2	2	13	8	21	1.2381	.4481
120	0	2	2	12	6	18	1.3333	.4183
130	1	0	1	12	4	16	.7500	.1875
150	0	1	1	11	4	15	.7333	.1956
160	1	0	1	11	3	14	.7857	.1684
170	0	1	1	10	3	13	.7692	.1775
220	1	0	1	7	2	9	.7778	.1728
230	1	0	1	6	1	7	.8571	.1224
$\Sigma d_{1j} = 9$							$\Sigma e_{1j} = 17.6154$	$\Sigma v_{1j} = 6.0456$

The notation j at the top of Column 1 is employed to represent each of the **distinct failure times** in Table 12.14. A **distinct failure time** is a survival time for which there is **at least one uncensored value**. The values in Column 1 of Table 12.16 were obtained by examining the 42 survival times in Table 12.14 and identifying any times for which it is known that at least one patient died. Note that the **distinct failure times** are arranged ordinally from smallest to largest in Column 1.

As noted previously, the notation d_{1j} in Column 2 represents the number of deaths that are known to have occurred in Treatment 1 during Time j. The notation d_{2j} in Column 3 represents the number of deaths that are known to have occurred in Treatment 2 during Time j. The notation d_j in Column 4 represents the total number of deaths that are known to have occurred in both Treatment 1 and Treatment 2 during Time j. Note that for a given row/distinct failure time,

$d_j = d_{1j} + d_{2j}$. The notation n_{1j} in Column 5 represents the number of patients at risk in Treatment 1 at the beginning of Time j. More specifically, n_{1j} represents the number of patients in Treatment 1 who have not died or been censored prior to Time j. The notation n_{2j} in Column 6 represents the number of patients at risk in Treatment 2 at the beginning of Time j. More specifically, n_{2j} represents the number of patients in Treatment 2 who have not died or been censored prior to Time j. The notation n_j in Column 7 represents the total number of patients at risk in both Treatment 1 and Treatment 2 at the beginning of Time j. More specifically, n_j represents the total number of patients in Treatment 1 and Treatment 2 who have not died or been censored prior to Time j. Note that for a given row/survival time, $n_j = n_{1j} + n_{2j}$.

The reader should take note of the fact that for a given value of j, the values computed for n_{1j} and n_{2j} include all patients who have not died or been censored **prior to Time j**. Consequently in computing the values n_{1j} and n_{2j}, any patient who dies or has his score censored during Time j is included. For example, consider the value $j = 60$ where $n_{1j} = 21$. Note that even though three patients in Treatment 1 die on day 60 and the score of one of the patients is censored on that day, all of the patients who receive Treatment 1 are included in computing the value $n_{1j} = 21$, since none of the patients in Treatment 1 died prior to day 60. Note that when $j = 70$, $n_{1j} = 17$, since the three patients in Treatment 1 who are known to have died prior to day 70 along with the one patient whose score was censored prior to that day are no longer considered to be at risk of dying. The one patient in Treatment 1, however, who dies on day 70 is included in computing the value $n_{1j} = 17$, since he did not die prior to day 70.

The values e_{1j} and v_{1j} in Columns 8 and 9 of Table 12.16 are the expected value and variance of d_{1j} for the distinct failure Time j. Equations 12.13 and 12.14 are respectively employed to compute the value of e_{1j} and v_{1j} in each row. To illustrate the computation of an e_{1j} value, for $j = 10$, $e_{1,10} = (2/42)(21) = 1$; for $j = 20$, $e_{1,20} = (2/40)(21) = 1.0500$; for $j = 30$, $e_{1,30} = (1/38)(21) = .5526$, etc. To illustrate the computation of a v_{1j} value, for $j = 10$, $v_{1,10} = [(21)(21)(2)(42 - 2)] / [(42)^2(42 - 1)] = .4878$; $v_{1,20} = [(21)(19)(2)(40 - 2)] / [(40)^2(40 - 1)] = .4860$; $v_{1,30} = [(21)(17)(1)(38 - 1)] / [(38)^2(38 - 1)] = .2472$, etc.

$$e_{1j} = \left(\frac{d_j}{n_j}\right) n_{1j} \qquad \textbf{(Equation 12.13)}$$

$$v_{1j} = \frac{n_{1j} n_{2j} d_j (n_j - d_j)}{n_j^2 (n_j - 1)} \qquad \textbf{(Equation 12.14)}$$

The test statistic for the **log-rank test** (to be designated z_M), which as noted earlier is based on a normal approximation, is computed with Equation 12.15. Note that the values Σd_{1j}, E_{1j} and V_{1j} in the latter equation are computed in last row of Table 12.16. Specifically, $\Sigma d_{1j} = 9$ is the sum of the d_{1j} values in Column 2; the expected value $E_{1j} = 17.6154$ is the sum of the e_{1j} values in Column 8 (i.e., $E_{1j} = \Sigma e_{1j}$); and the variance $V_{1j} = 6.0456$ is the sum of the v_{1j} values in Column 9 (i.e., $V_{1j} = \Sigma v_{1j}$). When the latter values are substituted in Equation 12.15, $z_M = -3.50$ is obtained.

In order to reject the null hypothesis, the obtained absolute value of z_M must be equal to or greater than the tabled critical value at the prespecified level of significance. Since the absolute value $z_M = 3.50$ is greater than the tabled critical one- and two-tailed values $z_{.01} = 2.33$ and $z_{.01}$

= 2.58, the directional alternative hypothesis stating longer survival times for Treatment 1 is supported at the .01 level, and the nondirectional alternative hypothesis stating that the survival time distributions for the two treatments are different is also supported at the .01 level. Note that the sign of z_M only provides information regarding the direction of any difference between the two survival distributions — in this case the negative sign indicates longer survival times for Treatment 1. Thus, the researcher can conclude that the survival times for Treatment 1 are longer than the survival times for Treatment 2.

$$z_M = \frac{\Sigma d_{1j} - E_{1j}}{\sqrt{V_{1j}}} = \frac{9 - 17.6154}{\sqrt{6.0456}} = -3.50 \qquad \textbf{(Equation 12.15)}$$

In point of fact, the same absolute value will be obtained for z_M if Treatment 2 is employed as the reference treatment instead of Treatment 1. If Treatment 2 is employed as the reference treatment, the values e_{2j} (where $e_{2j} = (d_j/n_j)\, n_{2j}$) and v_{2j} are computed in Columns 8 and 9 for each row/distinct failure time in Table 12.16. Since the value v_{2j} is also computed with Equation 12.14, for a given row/distinct failure time, $v_{2j} = v_{1j}$, and consequently ($V_{2j} = \Sigma v_{2j}$) = ($V_{1j} = \Sigma v_{1j}$). The values $\Sigma d_{2j} = 19$ (which is the sum of Column 3 of Table 12.16), E_{2j} (which equals $\Sigma e_{2j} = 10.3846$) and V_{2j} (which equals $\Sigma v_{2j} = 6.0456$) are employed in Equation 12.15 in place of E_{1j} and V_{1j}. When the latter values are substituted in Equation 12.15, $z_M = 3.50$ is obtained for the test statistic. Note that the absolute value of the test statistic remains unchanged, and that the positive sign is indicative of the lower survival times for Treatment 2.

$$z_M = \frac{\Sigma d_{2j} - E_{2j}}{\sqrt{V_{2j}}} = \frac{19 - 10.3846}{\sqrt{6.0456}} = 3.50$$

Some sources employ Equation 12.16 to compute the test statistic for the **log-rank test**. In point of fact, Equation 12.16, which employs the chi-square distribution, provides an equivalent result to that obtained with Equation 12.15. In Equation 12.16, O_1 represents the observed number of deaths in Treatment 1 (i.e., $O_1 = \Sigma d_{1j}$); E_1 represents the expected number of deaths in Treatment 1 (i.e., $E_1 = E_{1j} = \Sigma e_{1j}$); and $V = V_{1j} = \Sigma v_{1j}$ represents the variance. When the appropriate values are substituted in Equation 12.16, the value $\chi^2_M = 12.28$ is computed. One degree of freedom is employed in interpreting the latter chi-square value. Employing **Table A4 (Table of the Chi-Square Distribution)** in the **Appendix**, it can be determined that the tabled critical one and two-tailed chi-square values for $df = 1$ are respectively $\chi^2_{01} = 5.43$ and $\chi^2_{01} = 6.63$.[25] Since $\chi^2_M = 12.28$ is greater than both of the aforementioned critical values, as was the case when Equation 12.15 was employed, the directional alternative hypothesis stating longer survival times for Treatment 1 is supported at the .01 level, and the nondirectional alternative hypothesis stating that the survival time distributions for the two treatments are different is also supported at the .01 level.

The reader should take note of the fact that the square root of the chi-square value computed with Equation 12.16 will always be equal to the absolute z_M value computed with Equation 12.15, and the probability value associated with the result of both equations will be identical. Note that $[\chi^2_M = 12.28] = [(z_M = 3.50)^2]$ (the minimal difference is due to rounding off error).[26]

$$\chi^2_M = \frac{(O_1 - E_1)^2}{V} = \frac{(9 - 17.6154)^2}{6.0456} = 12.28 \qquad \text{(Equation 12.16)}$$

If Example 12.9 is evaluated with **Gehan's test for censored data**, the absolute value U_{G_1} = U_{G_2} = 258 is computed. When the value $\Sigma U^2_{i(G)}$ = 21,958 (which is the sum of the squared Gehan scores for the data) along with the appropriate samples size information ($N = 42$, n_1 = n_2 = 21) is substituted in Equation 12.11, the value $Var(U_G) = 5623.39$ is obtained. Substituting the values $U_G = 258$ and $Var(U_G) = 5623.39$ in Equation 12.12 yields the value $z = 3.44$.

$$Var(U_G) = \frac{n_1 n_2}{N(N-1)} \Sigma U^2_{i(G)} = \frac{(21)(21)}{42(42-1)}(21,958) = 5623.39$$

$$z = \frac{U_G - E(U_G)}{\sqrt{Var(U_G)}} = \frac{U_G}{\sqrt{Var(U_G)}} = \frac{258}{\sqrt{5623.39}} = 3.44$$

Since the obtained value $z = 3.44$ is greater than the tabled critical one- and two-tailed values $z_{.01} = 2.33$ and $z_{.01} = 2.58$, the directional alternative hypothesis stating longer survival times for Treatment 1 is supported at the .01 level, and the nondirectional alternative hypothesis is also supported at the .01 level. Thus, as is the case with the **log-rank test**, the **Gehan test** allows one to conclude that the survival times in the underlying distribution for Treatment 1 are longer than the survival times in the underlying distribution for Treatment 2. Although analysis of the same set of data with the **log-rank** and **Gehan tests** will not always yield an equivalent result, in this instance the absolute value $z = 3.44$ obtained for the **Gehan test** is almost identical to the value $z_M = 3.50$ obtained for the **log-rank test**.

Equivalency of the log-rank test and the Mantel–Haenszel test of association (Test 16I-c) (The reader should read the relevant material in the chapter on the **chi-square test for** $r \times c$ **tables** prior to reading this section) Various sources note that the label **log-rank test** is misleading with respect to the underlying nature of the test described in this section, since the **log-rank test** really does not involve the use of either logarithms or ranks. In spite of the fact the normal approximation represented by Equation 12.15 (or the equivalent chi-square analysis using Equation 12.16) is commonly employed to compute the **log-rank test** statistic, in actuality the **log-rank test** can be conceptualized as a set of chi-square analyses comparing the observed versus expected number of deaths at each of the distinct failure times documented in a survival table (such as Table 12.16). In point of fact, the **log-rank test** is equivalent to the **Mantel–Haenszel test of association (Test 16I-c)** (discussed in Section VI of the **chi-square test for** r $\times c$ **tables**), a procedure which is commonly employed to pool the results of multiple 2×2 contingency tables. Table 12.17 summarizes the general model for a 2×2 contingency table employed in most sources (as well as in Chapter 16 — e.g., see Table 16.6). Note the only difference between Table 12.17 and the model for a 2×2 table depicted previously with Table 12.15 is the notation employed. Table 12.18 summarizes the equivalencies with respect to notation with reference to the two aforementioned tables.

In the case of Example 12.9, the data for each of the distinct failure times listed in Column 1 of Table 12.16 can be reconfigured as 17 2×2 contingency tables. The **Mantel–Haenszel test of association** (specifically, Equation 16.46) can be employed to pool the data contained in the latter set of 17 contingency tables. The test statistic computed for the **Mantel–Haenszel test of association** is a chi-square value which will be equivalent to the chi-square value obtained for

Table 12.17 Model for 2 × 2 Contingency Table

	Column 1	Column 2	Row sums
Row 1	a	b	$a + b$
Row 2	c	d	$c + d$
Column sums	$a + c$	$b + d$	n

Table 12.18 Survival Table for Analysis with Log-Rank Test

	Die/Event of note occurred	Survive/Event of note did not occur	At Risk
Treatment 1	$a = d_{1j}$	$b = n_{1j} - d_{1j}$	$a + b = n_{1j}$
Treatment 2	$c = d_{2j}$	$d = n_{2j} - d_{2j}$	$c + d = n_{2j}$
Column sums	$a + c = d_j$	$b + d = n_j - d_j$	$n = n_j$

the **log-rank test** with Equation 12.16 (as well as being equal to the square of the z_M value computed with Equation 12.15).

Example 12.10 will be employed to illustrate the use of the **Mantel–Haenszel test of association** in computing the test statistic for the **log-rank test**. Note that in Example 12.10 there are only four distinct failure times (specifically, **Time 1**: January 1, 2002 to December 31, 2002; **Time 2**: January 1, 2003 to December 31, 2003; **Time 3**: January 1, 2004 to December 31, 2004; **Time 4**: January 1, 2005 to December 31, 2005). Consistent with the definition of a distinct failure time, the designation of the years 2002, 2003, 2004, and 2005 as distinct failure times requires that a minimum of at least one patient must have died during each of the four years. Table 12.20 summarizes the analysis of the data with the **log-rank test** through use of Equations 12.15 and 12.16.

Example 12.10 *A researcher is interested in comparing the survival rates of 100 male versus 100 female cancer patients who are treated with a specific drug. Table 12.19 summarizes the survival data over a four year period for the 200 patients. Each of the four contingency tables which comprise Table 12.19 respectively summarizes the survival information at the end of the first, second, third, and fourth year of the study. Do the data indicate differences in the survival distributions of males versus females?*

It should be noted that it is not unusual for sources to simply employ the label **die** for the column labeled **die or censored** in the tables which comprise Table 12.19. In such a case one would not have information regarding whether the data for any of the patients was, in fact, censored. However, as is the case for Tables 12.15/12.16, it is assumed that in each of the tables which comprise Table 12.19, for a given value of j, the values computed for n_{1j} (which corresponds to the row sum $a + b$ in Table 12.17) and n_{2j} (which corresponds to the row sum $c + d$ in Table 12.17) include all patients who have not died or been censored **prior to the beginning of Time j** (i.e., the first day of the year in question). Consequently, in computing the values n_{1j} and n_{2j}, any patient who dies or has his score censored during Time j is included. For example, consider the value $j = 1$ (i.e., Time 1: January 1, 2002 to December 31, 2002) where $n_{1j} = 100$. Note that even though 12 males die or may have had their score censored during the year 2002, all of the male patients are included in computing the value $n_{1j} = 100$, since it is assumed that none of them died or were censored prior to January 1, 2002. Although

Table 12.19 Survival Data for Example 12.10

Table 12.19a   Survival Data for Year 1
(Time 1: January 1, 2002 to December 31, 2002)

	Die or censored	Survive	At Risk
Males	12	88	100
Females	18	82	100
Column sums	30	170	200

Table 12.19b   Survival Data for Year 2
(Time 2: January 1, 2003 to December 31, 2003)

	Die or censored	Survive	At Risk
Males	14	74	88
Females	20	62	82
Column sums	34	136	170

Table 12.19c   Survival Data for Year 3
(Time 3: January 1, 2004 to December 31, 2004)

	Die or censored	Survive	At Risk
Males	16	58	74
Females	28	34	62
Column sums	44	92	136

Table 12.19d   Survival Data for Year 4
(Time 4: January 1, 2005 to December 31, 2005)

	Die or censored	Survive	At Risk
Males	21	37	58
Females	28	6	34
Column sums	49	43	92

it is impossible to determine from looking at Table 12.19a the exact number of patients who died as opposed to being censored in 2002, it is not necessary to have the latter information so long as the protocol described above for determining frequencies is adhered to.[27] Note that when $j = 2$, $n_{1j} = 88$, since the 12 males who are known to have died or had their scores censored prior to the beginning of Time 2 (i.e., January 1, 2003) are no longer considered to be at risk of dying. The 14 males who die during 2003 are included in computing the value $n_{1j} = 88$, since they did not die prior to the first day of Time 2 (January 1, 2003).

Table 12.20 summarizes the data in Table 12.19 when it is converted into the format employed in Table 12.16 for computing the values necessary to conduct the **log-rank test**. Equations 12.15 and 12.16 are employed to compute the test statistic for the **log-rank test**. Note that $(|z| = 4.785)^2 = (\chi^2 = 22.897)$.

$$z_M = \frac{\Sigma d_{1j} - E_{1j}}{\sqrt{V_{1j}}} = \frac{63 - 87.4325}{\sqrt{26.0711}} = -4.785$$

$$\chi^2_M = \frac{(O_1 - E_1)^2}{V} = \frac{(63 - 87.4325)^2}{26.0711} = 22.897$$

Table 12.20   Computation of Log-Rank Test Statistic for Example 12.10

j^*	d_{1j}	d_{2j}	d_j	n_{1j}	n_{2j}	n_j	$e_{1j} = (d_j/n_j)\,n_{1j}$	$v_{1j} = \dfrac{n_{1j}n_{2j}d_j(n_j - d_j)}{n_j^2(n_j - 1)}$
1	12	18	30	100	100	200	15.0000	6.4070
2	14	20	34	88	82	170	17.6000	6.8317
3	16	28	44	74	62	136	23.9412	7.4379
4	21	28	49	58	34	92	30.8913	5.3945

$\Sigma d_{1j} = 63$ $\qquad\qquad\qquad\qquad\qquad\qquad\qquad$ $\Sigma e_{1j} = 87.4325$ $\qquad$ $\Sigma v_{1j} = 26.0711$

* The values of j respectively represent distinct failure times/years 1 (2002), 2 (2002), 3 (2004), and 4 (2005).

Since the absolute value $z_M = 4.785$ is greater than the tabled critical one and two-tailed values $z_{.01} = 2.33$ and $z_{.01} = 2.58$, the directional alternative hypothesis stating longer survival times for males is supported at the .01 level, and the nondirectional alternative hypothesis stating that the survival time distributions for the two genders are different is also supported at the .01 level. Equivalently, since $\chi^2 = 12.28$ is greater than $\chi^2_{.01} = 5.43$ and $\chi^2_{.01} = 6.63$ (for $df = 1$), the aforementioned directional and nondirectional alternative hypotheses are supported at the .01 level. Note that $[\chi^2_M = 22.897] = [(z_M = -4.785)^2]$.

It should be noted that if the notation in Table 12.17 for a 2×2 contingency table is substituted in the equations for computing the **log-rank test** statistic, the term Σd_{1j} in Tables 12.15 and 12.16 can be written as Σa_j (the subscript j is employed to designate the distinct failure time). Additionally, Equations 12.17 and 12.18 (both of which employ the notation in Table 12.17) provide an alternative way of writing Equations 12.13 and 12.14.

$$e_{1j} = \left(\frac{d_j}{n_j}\right)n_{1j} = \frac{(a_j + c_j)(a_j + b_j)}{n_j} \qquad \textbf{(Equation 12.17)}$$

(Equation 12.18)

$$v_{1j} = \frac{n_{1j}n_{2j}d_j(n_j - d_j)}{n_j^2(n_j - 1)} = \frac{(a_j + b_j)(c_j + d_j)(a_j + c_j)(b_j + d_j)}{n_j^2(n_j - 1)}$$

Using the above notation reveals that Equation 12.19 is equivalent to Equation 12.16. In point of fact, Equation 12.19 is equivalent to Equation 16.46, which is the test statistic for the **Mantel–Haenszel test of association**, when a correction for continuity is not employed.[28]

$$\chi_M^2 = \frac{(O - E)^2}{V} = \frac{(\Sigma d_{1j} - \Sigma e_{ij})^2}{\Sigma v_{1j}} =$$

(Equation 12.19)

$$\frac{\left(\Sigma a_j - \Sigma \left[\frac{(a_j + b_j)(a_j + c_j)}{n_j} \right] \right)^2}{\Sigma \left[\frac{(a_j + b_j)(c_j + d_j)(a_j + c_j)(b_j + d_j)}{n_j^2(n_j - 1)} \right]}$$

Further discussion of the **log-rank test** and/or other procedures for evaluating censored data can be found in Deshpande *et al.* (1995), Desu and Raghavarao (2003), Good (1994), Grimm and Yarnold (2000), Higgins (2004), Hollander and Wolfe (1999), and Rosner (2000).

References

Barnett, V. & Lewis, T. (1994). **Outliers in statistical data** (3rd ed.). Chichester: John Wiley & Sons.

Bell, C. B. & Doksum, K. A. (1965). Some new distribution-free statistics. **Annals of Mathematical Statistics**, 36, 203–214.

Bergmann, R., Lundbrook, J. & Spooren, W. P. (2000). Different outcomes of the Wilcoxon–Mann–Whitney test from different statistical packages. **The American Statistician**, 54, 72–77.

Chernick, M. R. (1999). **Bootstrap methods: A practitioner's guide**. New York: John Wiley & Sons.

Conover, W. J. (1980). **Practical nonparametric statistics** (2nd ed.). New York: John Wiley & Sons.

Conover, W. J. (1999). **Practical nonparametric statistics** (3rd ed.). New York: John Wiley & Sons.

Conover, W. J. & Iman, R. L. (1981). Rank transformations as a bridge between parametric and nonparametric statistics. **The American Statistician**, 35, 124–129.

Cox, D. B. (1972). Regression models and life tables (with discussion). **Journal of the Royal Statistical Society**, 26b, 103–110.

Cox D. B. & Oakes, D. (1984). **Analysis of survival data**. New York: Chapman & Hall.

Daniel, W. W. (1990). **Applied Nonparametric statistics** (2nd ed.). Boston: PWS–Kent Publishing Company.

Davison, A. C. & Hinckley, D. V. (1997). **Bootstrap methods and their applications**. Cambridge, England: Cambridge University Press.

Delaney, H. D. & Vargha, A. (2002). Comparing several robust tests of stochastic equality with ordinally scaled variables and small to moderate sized samples. **Psychological Methods**, 7, 485–503.

Deshpande, J. V., Gore, A. P. & Shanubhogue, A. (1995). **Statistical analysis of nonnormal data**. New Delhi: New Age International Publishers Limited/Wiley Eastern Limited.

Desu, M. M. & Raghavarao, D. (2003). **Nonparametric statistical methods for complete and censored data**. Boca Raton, FL: Chapman & Hall/CRC.

Devore, J. & Farnum, N. (1999). **Applied statistics for engineers and scientists**. Pacific Grove, CA: Duxbury Press.

Efron, B. (1979). Bootstrap methods: another look at the jackknife. **Annals of Mathematical Statistics**, 7, 1–26.

Efron, B. & Tibshirani R. J. (1993). **An introduction to the bootstrap**. New York: Chapman & Hall.

Fahoome, G. (2002). Twenty nonparametric statistics and their large-sample approximations. **Journal of Modern Applied Statistical Methods**, 2, 248–268.

Fisher, R. A. (1935). **The design of experiments.** Edinburgh: Oliver & Boyd.

Fligner, M. A. & Policello, II, G. E. (1981). Robust rank procedures for the Behrens-Fisher problem. **Journal of the American Statistical Association**, 76, 162–174.

Gehan, E. A. (1965a). A generalized Wilcoxon test for comparing arbitrarily singly censored samples. **Biometrika**, 52, 203–223.

Gehan, E. A. (1965b). A generalized two-sample Wilcoxon test for doubly censored data. **Biometrika**, 52, 650–653.

Good, P. (1994). **Permutation tests: A practical guide to resampling methods for testing hypotheses**. New York: Springer.

Grimm, L.G. & Yarnold, P.R. (Eds.) (2000). **Reading and understanding more multivariate statistics**. Washington, D.C.: American Psychological Association.

Grissom, R. J. (1994). Probability of the superior outcome of one treatment over another. **Journal of Applied Psychology**, 79, 314–316.

Grissom , R. J. & Kim, J. J. (2005). **Effect sizes for research: A broad practical approach**. Mahwah, NJ: Lawrence Erlbaum Associates, Publishers.

Hartigan, J. A. (1969). Using subsample values as typical values. **Journal of the American Statistical Association**, 64, 1303–1317.

Higgins, J. J. (2004). **An introduction to modern nonparametric statistics**. Belmont, CA: Duxbury Press.

Hollander, M. & Wolfe, D.A. (1999). **Nonparametric statistical methods**. New York: John Wiley & Sons.

Huber, P. J. (1981). **Robust Statistics**. New York: John Wiley & Sons.

Jennison, C. & Turnbull, B. (2000). **Group sequential methods with applications to clinical trials**. Boca Raton: CRC Press.

Kaplan, E. L. & Meier, P. (1958). Nonparametric estimation for incomplete observations. **Journal of the American Statistical Association**, 53, 457–481.

Keller-McNulty, S. & Higgins, J. (1987). Effect of tail weight and outliers on power and Type-I error of robust permutation test for location. **Communication in Statistics: Simulations and Computations**, 16(1), 17–36.

Kline, R. B. (2004). **Beyond significance testing: Reforming data analysis methods in behavioral research**. Washington, D. C.: American Psychological Association.

Kruskal, W. H, (1957). Historical notes on the Wilcoxon unpaired two-sample test. **Journal of the American Statistical Association**, 52, 356–360.

Kolmogorov, A. N. (1933). Sulla determinazione empiraca di una legge di distribuzione. **Giorn dell'Inst. Ital. degli. Att.**, 4, 89–91.

Lundbrook, J. & Dudley, H. (1998). Why permutation tests are superior to the t and F tests in biomedical research. **The American Statistician**, 52, 127–132.

Manly, B. F. J. (1997). **Randomization, bootstrap and Monte Carlo methods in biology** (2nd ed.). London: Chapman & Hall.

Mann, H. & Whitney, D. (1947). On a test of whether one of two random variables is stochastically larger than the other. **Annals of Mathematical Statistics**, 18, 50–60.

Mantel, N. (1966). Evaluation of survival data and two new rank order statistics arising in its consideration. **Cancer Chemotherapy Report**, 50, 113–170.

Marascuilo, L. A. & McSweeney, M. (1977). **Nonparametric and distribution-free methods for the social sciences**. Monterey, CA: Brooks/Cole Publishing Company.

Maxwell, S. E. & Delaney, H. (1990). **Designing experiments and analyzing data**. Monterey, CA: Wadsworth Publishing Company.

Mooney, C. Z. & Duval, D. (1993). **Bootstrapping: A nonparametric approach to statistical inference**. Newbery Park, CA: Sage Publications.

Norušis, N. J. (2004). **SPSS 13.0 advanced statistical procedures companion**. Upper Saddle River, NJ: Prentice Hall.

Pagano, M. & Gauvreau, K. (1993). **Principles of biostatistics**. Belmont, CA: Duxbury Press

Pitman, E. J. G. (1937a). Significance tests that may be applied to samples from any population. **Journal of the Royal Statistical Society: Supplement**, 4, 119–130.

Pitman, E. J. G. (1937b). Significance tests that may be applied to samples from any population, fII. The correlation coefficient test. **Journal of the Royal Statistical Society: Supplement**, 4, 225–232.

Pitman, E. J. G. (1938). Significance tests that may be applied to samples from any population, III. The analysis of variance test. **Biometrika**, 29, 322–335.

Pyke, D. A. & Thompson, H. (1986). Statistical analysis of survival and removal rate experiments. **Ecology**, 67, 240–245.

Quenouille, M. H. (1949). Approximate tests of correlation in time series. **Journal of the Royal Statistical Society, B**, 11, 18–84.

Rosner, B. (2000). **Fundamentals of biostatistics** (5th ed.). Pacific Grove, CA: Duxbury Press.

Selvin, S. (1995). **Practical biostatistical methods**. Belmont, CA: Duxbury.

Sheskin, D. J. (1984). **Statistical tests and experimental design: A guidebook**. New York: Gardner Press.

Siegel, S. & Castellan, N. J., Jr. (1988). **Nonparametric statistics for the behavioral sciences** (2nd ed.). New York: McGraw–Hill Book Company.

Simon, J. L. (1969). **Basic research methods in social science**. New York: Random House.

Smirnov, N. V. (1939). On the estimation of the discrepancy between empirical curves of distributions for two independent samples. **Bulletin University of Moscow**, 2, 3–14.

Sprent, P. (1993). **Applied nonparametric statistical methods** (2nd ed.). London: Chapman & Hall.

Sprent, P. (1998). **Data driven statistical methods**. London: Chapman & Hall.

Sprent, P. & Smeeton, N. C. (2000). **Applied nonparametric statistical methods** (3rd ed.). London: Chapman & Hall.

Staudte R. G. & Sheather S. J. (1990). **Robust estimation and testing**. New York: John Wiley & Sons.

Terry, M. E. (1952). Some rank-order tests, which are most powerful against specific parametric alternatives. **Annals of Mathematical Statistics**, 23, 346–366.

Tukey, J. W. (1958). Bias and confidence in not quite large samples (Abstract). **Annals of Mathematical Statistics**, 29, 614.

Tukey, J. W. (1959). A quick, compact, two-sample test to Duckworth's specifications. **Technometrics**, 1, 31–48.

Van der Waerden, B. L. (1952/1953). Order tests for the two-sample problem and their power. **Proceedings Koninklijke Nederlandse Akademie van Wetenshappen** (A), 55 (**Indagationes Mathematicae** 14), 453–458, and 56 (**Indagationes Mathematicae**, 15), 303–316 (corrections appear in Vol. 56, p. 80).

Wald, A. & Wolfowitz, J. (1940). On a test whether two samples are from the same population. **Annals of Mathematical Statistics**, 11, 147–162.

Wilcox, R. R. (1987). **New statistical procedures for the social sciences**. Hillsdale, NJ: Erlbaum.

Wilcox, R. R. (1996). **Statistics for the social sciences**. San Diego, CA: Academic Press.

Wilcox, R. R. (1997). **Introduction to robust estimation and hypothesis testing**. San Diego, CA: Academic Press.

Wilcox, R. R., (2001). **Fundamentals of modern statistical methods: Substantially increasing power and accuracy**. New York: Springer.

Wilcox, R. R., (2003). **Applying contemporary statistical techniques**. San Diego, CA: Academic Press.

Wilcox, R. R. (2005). **Robust estimation and hypothesis testing** (2nd ed.) Amsterdam: Elsevier Academic Press.

Wilcoxon, F. (1949). **Some rapid approximate statistical procedures**. Stamford, CT: Stamford Research Laboratories, American Cyanamid Corporation.

Wilks, S. S. (1961). A combinatorial test for the problems of two samples from continuous distributions. In J. Neyman (Ed.), **Proceedings of the Fourth Berkeley Symposium on Mathematical Statistics and Probability**. Berkeley & Los Angeles: University of California Press, Vol. I, 707–717.

Wright, R. E. (2000). Survival analysis. In Grimm, L. G. & Yarnold, P. R. (Eds.), **Reading and understanding more multivariate statistics** (pp. 363–407). Washington, D.C.: American Psychological Association.

Yuen, K. K. (1974). The two-sample trimmed *t* for unequal population variances. **Biometrika**, 64, 165–179.

Zimmerman, D. W. & Zumbo, B. D. (1993a). The relative power of parametric and nonparametric statistical methods. In Keren, G. & Lewis, C. (Eds.), **A handbook for data analysis in the behavioral sciences: Methodological issues**. Hillsdale, NJ: Lawrence Erlbaum Associates, Publishers.

Zimmerman, D. W. & Zumbo, B. D. (1993b). Rank transformations and the power of the Student *t* test and Welch *t'* test. **Canadian Journal of Experimental Psychology**, 47, 523–529.

Endnotes

1. a) The test to be described in this chapter is also referred to as the **Wilcoxon rank-sum test** and the **Mann–Whitney–Wilcoxon test**; b) Grissom and Kim (2005, p. 105) note that Kruskal (1957) provides an overview of the development of the ideas underlying the test prior to its formal introduction, and that alternative versions of the test are discussed in Bergmann, Lundbrook and Spooren (2000); c) The population median will be represented by θ, which is the lower case Greek letter **theta**.

2. The reader should take note of the following with respect to the table of critical values for the **Mann–Whitney U distribution**: a) No critical values are recorded in the **Mann–Whitney table** for very small sample sizes, since a level of significance of .05 or less cannot be achieved for sample sizes below a specific minimum value; b) The critical values published in **Mann–Whitney tables** by various sources may not be identical. Such differences are trivial (usually one unit), and are the result of rounding off protocol; and c) The distribution of the **Mann–Whitney U statistic** is two-tailed, and **Table A11** only lists critical values from the lower tail of the distribution. The result of the **Mann–Whitney U test** can also be evaluated through use of critical values in the upper tail of the distribution. In such a case one designates the larger of the two values U_1 versus U_2 as the obtained U

statistic. In the latter instance, the tabled critical value to employ in evaluating the latter U statistic can be computed from **Table A11** by subtracting the critical value recorded in the relevant cell for the two sample sizes in the table from the product $n_1 n_2$. Thus if $U_1 = 21$ is designated as the value of U, since $n_1 n_2 = 25$, the tabled critical two-tailed .05 and .01 values are $U_{.05} = 25 - 2 = 23$ and $U_{.01} = 25 - 0 = 25$, and the tabled critical one-tailed .05 and .01 values are $U_{.05} = 25 - 4 = 21$ and $U_{.01} = 25 - 1 = 24$. In order to be significant, the obtained value of U (i.e., when the larger of the two values U_1 versus U_2 represents U) must be **equal to or greater than** the tabled critical value at the prespecified level of significance. Note that since the obtained value $U = 21$ is equal to the tabled critical one-tailed .05 value $U_{.05} = 21$ (but less than the other three above noted critical values), as was the case when $U_2 = 4$ was employed as the U statistic, the null hypothesis can only be rejected if the directional alternative hypothesis $H_1: \theta_1 < \theta_2$ is employed; d) The table for the alternative version of the **Mann–Whitney U test** (which was developed by Wilcoxon (1949)) contains critical values that are based on the sampling distribution of the sums of ranks, which differ from the tabled critical values contained in **Table A11** (which represents the sampling distribution of U values).

3. Although for Example 12.1 we can also say that since $\Sigma R_1 < \Sigma R_2$ the data are consistent with the directional alternative hypothesis $H_1: \theta_1 < \theta_2$, the latter will not necessarily always be the case when $n_1 \neq n_2$. Since the relationship between the average of the ranks will always be applicable to both equal and unequal sample sizes, it will be employed in describing the hypothesized relationship between the ranks of the two groups.

4. Some sources employ an alternative normal approximation equation which yields the same result as Equation 12.4. The alternative equation is noted below.

$$z = \frac{\Sigma R_1 - n_1 \left[\dfrac{n_1 + n_2 + 1}{2} \right]}{\sqrt{\dfrac{n_1 n_2 (n_1 + n_2 + 1)}{12}}}$$

Note that the only difference between the above equation and Equation 12.4 is with respect to the numerator. If the value $\Sigma R_1 = 19$ from Example 12.1 is substituted in the above equation, it yields the value -8.5 for the numerator (which is the value for the numerator computed with Equation 12.4), and consequently the value $z = -1.78$. If ΣR_2 is employed in the numerator of the above equation in lieu of ΣR_1, the numerator of the equation assumes the form $\Sigma R_2 - [[n_2(n_1 + n_2 + 1)]/2]$. If $\Sigma R_2 = 36$ is substituted in the revised numerator, the value 8.5 is computed for the numerator, which results in the value $z = 1.78$. (Later on in this discussion it will be noted that the same conclusions regarding the null hypothesis are reached with the values $z = 1.78$ and $z = -1.78$.) The above equation is generally employed in sources which describe the version of the **Mann–Whitney U test** developed by Wilcoxon (1949). The latter version of the test only requires that the sum of the ranks be computed for each group, and does not require the computation of U values. As noted in Endnote 2, the table of critical values for the alternative version of the test is based on the sampling distribution of the sums of the ranks.

5. A general discussion of the correction for continuity can be found in Section VI of the **Wilcoxon signed-ranks test**.

6. Some sources employ the term below for the denominator of Equation 12.6. It yields the identical result.

$$\sigma_U = \sqrt{\frac{n_1 n_2}{N(N-1)}\left[\frac{N^3 - N}{12} - \sum_{i=1}^{s}\frac{(t_i^3 - t_i)}{12}\right]}$$

7. A correction for continuity can be used by subtracting .5 from the absolute value computed for the numerator of Equation 12.6. The continuity correction yields the absolute value z = 8/4.71 = 1.70 (which in actuality is a negative number since $U = 4$ is less than $U_E = 8$).

8. The rationale for discussing computer-intensive procedures in the **Addendum** of the **Mann–Whitney U test** is that the **Mann–Whitney test** (as well as many other rank-order procedures) can be conceptualized as an example of a **randomization** or **permutation test**, which is the first of the computer based procedures to be described in the **Addendum**.

9. Another application of a computer-intensive procedure is **Monte Carlo research** which is discussed in Section VII of the **single-sample runs test**.

10. The term **bootstrap** is derived from the saying that one lifts oneself up by one's bootstraps. Within the framework of the statistical procedure, bootstrapping indicates a single sample is used as a basis for generating multiple additional samples — in other words, one makes the most out of what little resources one has. Manly (1997) notes that the use of the term **jackknife** is based on the idea that a jackknife is a multipurpose tool which can be used for many tasks, in spite of the fact that for any single task it is seldom the best tool.

11. The reader should take note of the fact that although Example 12.3 involves two independent samples, by applying the basic methodology to be described in this section, randomization tests can be employed to evaluate virtually any type of experimental design.

12. a) Suppose in the above example we have an unequal number of subjects in each group. Specifically, let us assume two subjects are randomly assigned to Group 1 and four subjects to Group 2. The total number of possible arrangements will be the combination of six things taken two at a time, which is equivalent to the combination of six things taken four at a time. This results in 15 different arrangements: $\binom{6}{2} = \binom{6}{4} = \frac{6!}{2!\,4!} = 15$; b) To illustrate how large the total number of arrangements can become, suppose we have a total of 40 subjects and randomly assign 15 subjects to Group 1 and 25 subjects to Group 2. The total number of possible arrangements is the combination of 40 things taken 15 at a time, which is equivalent to the combination of 40 things taken 25 at a time. The total number of possible arrangements will be $\binom{40}{15} = \binom{40}{25} = \frac{40!}{15!\,25!} = 40{,}225{,}345{,}060$. Obviously, without the aid of a computer it will be impossible to evaluate such a large number of arrangements; b) In actuality, Equations I.48 and 9.4 are the equations for computing the **number of combinations of n things taken x at a time**. Although the test discussed in this section is referred to as a **permutation test**, it is actually based on the computation of **combinations** rather than **permutations**. As noted in the **Introduction**, in computing **combinations** one is not interested in the ordering of the elements within an arrangement, whereas in computing **permutations** one is interested in the ordering.

13. The reader should take note of the following: a) If the 20 corresponding arrangements for Group 2 are listed in Table 12.5, the same 20 arrangements which are listed for Group 1 will appear in the table, but in different rows. To illustrate, the first arrangement in Table 12.5 for Group 1 is comprised of the scores 7, 10, 11. The corresponding Group 2 arrangement is 15, 18, 21 (which are the three remaining scores in the sample). In the last row of Table 12.5 the scores 15, 18, 21 are listed. The corresponding arrangement for Group 2 will be the remaining scores, which are 7, 10, 11. If we continue this process for the remaining 18 Group 1 arrangements, the final distribution of arrangements for Group 2 will be comprised of the same 20 arrangements obtained for Group 1; b) When $n_1 \neq n_2$, the distribution of the arrangements of the scores in the two groups will not be identical, since all the arrangements in Group 1 will always have n_1 scores and all the arrangements in Group 2 will always have n_2 scores, and $n_1 \neq n_2$. Nevertheless, computation of the appropriate sampling distribution for the data only requires that a distribution be computed which is based on the arrangements for one of the two groups. Employing the distribution for the other group will yield the identical result for the analysis.

14. a) Although the result obtained with the **Mann–Whitney U test** is equivalent to the result which will be obtained with a randomization/permutation test conducted on the rank-orders, only the version of the test which was developed by Wilcoxon (1949) directly evaluates the permutations of the ranks. Marascuilo and McSweeney (1977, pp. 270–272) and Sprent (1998, pp. 85–86) note that the version of the test described by Mann and Whitney (1947) actually employs a statistical model which evaluates the number of **inversions** in the data. An **inversion** is defined as follows: Assume that we begin with the assumption that all the scores in one group (designated Group 1) are higher than all the scores in the other group (designated Group 2). If we compare all the scores in Group 1 with all the scores in Group 2, an inversion is any instance in which a score in Group 1 is not higher than a score in Group 2. It turns out an inversion-based model yields a result which is equivalent to that obtained when the permutations of the ranks are evaluated (as is done in Wilcoxon's (1949) version of the **Mann–Whitney U test**). Employing the data from Example 12.1, consider Table 12.21 (which is identical to Table 12.3) where the scores in Group 1 are arranged ordinally in the top row, and the scores in Group 2 are arranged ordinally in the left column.

Table 12.21 Inversion Model of Mann–Whitney U test

		Group 1 Scores			
		0	0	1	2	11
	4	−4	−4	−3	−2	7
	5	−5	−5	−4	−3	6
Group 2	8	−8	−8	−7	−6	3
Scores	11	−11	−11	−10	−9	0
	11	−11	−11	−10	−9	0

Each of the scores in the cells of Table 12.21 are difference scores that are the result of subtracting the score in the extreme left of the row a cell appears (i.e., the corresponding Group 2 score) from the score at the top of the column in which the cell appears (i.e., the corresponding Group 1 score). Note that any negative difference score recorded in the table meets the criterion established for an **inversion**. Ties result in a zero value for a cell. The total number of inversions in the data is based on the number of negative difference scores and the number of ties. One point is allocated for each negative difference score, and one-half a point for each zero/tie (since the latter is viewed as a less extreme inversion than one associated with a negative difference score). Employing the aforementioned protocol, the number of inversions in the Table 12.21 is the 20 negative difference scores plus the two zeros, which sums to 21 inversions. Note that the latter value corresponds to the value computed for U_1 with Equation 12.1. The value $U_2 = 4$ inversions will be obtained if Table 12.21 is reconstructed so that the score in each column (i.e., Group 1 score) is subtracted from the score in the corresponding row (i.e., Group 2 score). In point of fact, if the table is reconstructed all of the signs in the table will be reversed, yielding a table with 20 positive signs, three negative signs, and two zeros. The three negative signs plus ½ a point for each zero yield a total of $U_2 = 4$ inversions; b) Grissom and Kim (2005, pp. 98–103) discuss a statistic which Grissom (1994) states can be employed as a measure of **effect size** for ranked data in the case of two independent samples. The latter measure, which Grissom (1994) labels the **probability of superiority** (*PS*), is computed with Equation 12.20. The **probability of superiority** is a direct function of the number of inversions in a set of data. Specifically, it is the proportion of times (or probability) that one group has a higher score (or lower score if the latter is indicative of superior performance) than the other group. In the case of Example 12.1, in computing the **probability of superiority** the value $U_1 = 21$ is employed to indicate the number of times subjects in Group 1 obtain a superior score to subjects in Group 2 (In Example 12.1, the lower a subject's score on the dependent variable, the less depressed the subject.) The latter value is computed below to equal *PS* = .84. Additionally, the value $U_2 = 4$ can be employed to compute an alternative **probability of superiority** to indicate the number of times subjects in Group 2 obtain a superior score to subjects in Group 1. The latter value is computed below to equal *PS* = .16. Note that Equation 12.20 simply involves dividing the number of inversions by the total number of possible pairwise comparisons, which is represented by the term $n_1 n_2$ in the denominator (which as noted in Equation 12.3 will always equal the sum of U_1 and U_2) and is visually displayed in Tables 12.3/12.21. A **probability of superiority** value will always fall between a maximum *PS* = 1 and minimum of *PS* = 0, with the value *PS* = .5 indicating that members of the two groups outscore each other equally often. Employing the number of inversions in the denominator of Equation 12.20 insures that when ties are present, one half of the ties will be allocated to each group. Additionally, the sum of the two possible *PS* values which can be computed for two independent samples will equal 1 (as is the case below since .84 + .16 = 1).

$$PS = \frac{U}{n_1 n_2} \qquad \textbf{(Equation 12.20)}$$

$$PS_{Gp\ 1 > Gp\ 2} = \frac{U_1}{n_1 n_2} = \frac{21}{(5)(5)} = .84$$

$$PS_{Gp\ 2 > Gp\ 1} = \frac{U_2}{n_1 n_2} = \frac{4}{(5)(5)} = .16$$

15. Although in this example the identical probability is obtained for the highest sum of scores and highest sum of ranks, this will not always be the case.

16. Efron and Tibshirani (1993, p. 394) note that the bootstrap differs from more conventional simulation procedures (discussed briefly in Section VII of the **single-sample runs test**), in that in conventional simulation, data are generated through use of a theoretical model (such as sampling from a theoretical population such as a normal distribution for which the

mean and standard deviation have been specified). In the bootstrap the simulation is data-based. Specifically, multiple samples are drawn from a sample of data which is derived in an experiment. One problem with the bootstrap is that since it involves drawing random subsamples from a set of data, two or more researchers conducting a bootstrap may not reach the same conclusions due to differences in the random subsamples generated.

17. a) Sprent (1998) notes that when the bootstrap is employed correctly in situations where the normality assumption is not violated, it generally yields conclusions which will be consistent with those derived from the use of conventional parametric and nonparametric tests, as well as permutation tests; b) Wilcox (2003) describes a procedure developed by Yuen (1974) which can be employed for comparing trimmed means.

18. The **exponential distribution** (which is discussed in Section IX (the **Addendum**) of the **binomial sign test for a single sample**) is a continuous probability distribution that is often useful in investigating reliability theory and stochastic processes. Manly (1997) recommends that prior to employing the bootstrap for inferential purposes, it is essential to evaluate its performance with small samples derived from various theoretical probability distributions (such as the exponential distribution). Monte Carlo studies (i.e., computer simulations involving the derivation of samples from theoretical distributions for which the values of the relevant parameters have been specified) can be employed to evaluate the reliability of the bootstrap.

19. a) Huber (1981) notes that a statistic is defined as robust if it is **efficient** — specifically, if the variance of the statistic is not dramatically increased in situations where the assumptions for the underlying population are violated. As an example, whereas the variance of the sample mean is sensitive to deviations from the normality assumption, the variance of the sample median is relatively unaffected by departures from normality.

20. a) In contrast to a survival function, Rosner (1995, p. 609) notes that a **hazard function** (which documents **hazard rates**) indicates the "instantaneous probability of having an event at time t given that one has survived up to time t" — i.e., a hazard function is a mathematical relationship describing changes in the risk that an event will occur over time. Selvin (1995, p. 437) states that a hazard function "measures the rate of change in a survival function at time t relative to the probability of surviving beyond time t." A large hazard indicates events occur at a fast rate, while a small hazard indicates a slower rate of occurrence. Wright (2000, p. 380) notes that unlike a survival function, a hazard function can increase, decrease, or fluctuate over time. As an example, the latter author notes the hazard function for human mortality is U-shaped, since the probability of a person dying is highest among newborns and the elderly and lower during the intervening years. On the other hand, the survival function describing human mortality ranges from 1 down to 0, with the probability of survival decreasing with the progression of time. **Cox regression** (Cox (1972) and Cox and Oates (1984)) (which is described in many books on biostatistics) is a procedure commonly employed to evaluate hazard functions. Among others, Wright (2000, p. 398) notes that the latter procedure is commonly classified as a **semiparametric (or partially) parametric procedure**. **Semiparametric procedures** or **models** are characterized by having multiple components, some of which are parametric and others nonparametric, and consequently such procedures or models do not clearly fall within the realm of being parametric or nonparametric (Sprent and Smeeton (2002)). Norušis (2004)

is an excellent reference on **Cox regression**, as well as on the use of *SPSS* for conducting a survival analysis.

21.   As noted in the **Introduction**, it is also true that $P(AB) = P(B)P(A/B)$.

22.   a) Research by Keller-McNulty and Higgins (1987) (summarized in Higgins (2004, pp. 63–64)) indicates that although the default number of simulations implemented by *StatXact* (which is a statistical software package commonly utilized in conducting **permutation tests**) is 10,000, in actuality, in order to obtain a reliable result it is not necessary to conduct more than 1600 simulations; b) Most contemporary high speed computers could compute the complete sampling distribution (i.e., all 184,766 possible combinations) in a relatively short period of time.

23.   If the value $U_1 = 6\frac{1}{2}$ is substituted in Equation 12.4 (the equation for the normal approximation of the **Mann–Whitney U statistic**), it yields the absolute value $z = .43$. The continuity-corrected Equation 12.5 yields the absolute value $z = .29$. Neither of the latter values achieves statistical significance at the .05 level.

24.   The **log-rank test** (also spelled in many sources as **logrank test**) can also be employed to contrast two survival functions in which no data are censored.

25.   Clarification with respect to determining a one-tailed chi-square critical value can be found in Section V of the **single-sample chi-square test for a population variance**.

26.   a) Although most sources do not employ a correction for continuity for the **log-rank test**, the latter can be applied to Equations 12.15 and 12.16 resulting in Equations 12.21 and 12.22. As noted below, use of the latter equations will result in a slightly lower absolute value for z_M and χ^2_M.

(Equation 12.21)

$$z_M = \frac{|\Sigma d_{1j} - E_{1j}| - .5}{\sqrt{V_{1j}}} = \frac{|9 - 17.6154| - .5}{\sqrt{6.0456}} = 3.30$$

(Equation 12.22)

$$\chi^2_M = \frac{(|O_1 - E_1| - .5)^2}{V} = \frac{(|9 - 17.6154| - .5)^2}{6.0456} = 10.89$$

27.   It should be noted that **Gehan's test for censored data** cannot be employed to evaluate Example 12.10/Table 12.19, since the available information does not allow one to determine the number of patients who **died** versus had their scores **censored** during the four time periods (which one must know in order to conduct the latter test).

28.   Equation 16.45, which employs a correction for continuity, will yield a result that is equivalent to that obtained with Equations 12.21 and 12.22.

Test 13

The Kolmogorov–Smirnov Test for Two Independent Samples
(Nonparametric Test Employed with Ordinal Data)

I. Hypothesis Evaluated with Test and Relevant Background Information

Hypothesis evaluated with test Do two independent samples represent two different populations?

Relevant background information on test The **Kolmogorov–Smirnov test for two independent samples** was developed by Smirnov (1939). Daniel (1990) notes that because of the similarity between Smirnov's test and a goodness-of-fit test developed by Kolmogorov (1933) (the **Kolmogorov–Smirnov goodness-of-fit test for a single sample (Test 7)**), the test to be discussed in this chapter is often referred to as the **Kolmogorov–Smirnov test for two independent samples** (although other sources (Conover (1980, 1999)) simply refer to it as the **Smirnov test**).

Daniel (1990), Marascuilo and McSweeney (1977), and Siegel and Castellan (1988) note that when a nondirectional/two-tailed alternative hypothesis is evaluated, the **Kolmogorov–Smirnov test for two independent samples** is sensitive to any kind of distributional difference (i.e., a difference with respect to location/central tendency, dispersion/variability, skewness, and kurtosis). When a directional/one-tailed alternative hypothesis is evaluated, the test evaluates the relative magnitude of the scores in the two distributions.

As is the case with the **Kolmogorov–Smirnov goodness-of-fit test for a single sample** discussed earlier in the book, computation of the test statistic for the **Kolmogorov–Smirnov test for two independent samples** involves the comparison of two **cumulative frequency distributions**. Whereas the **Kolmogorov–Smirnov goodness-of-fit test for a single sample** compares the cumulative frequency distribution of a single sample with a hypothesized theoretical or empirical cumulative frequency distribution, the **Kolmogorov–Smirnov test for two independent samples** compares the cumulative frequency distributions of two independent samples. If, in fact, the two samples are derived from the same population, the two cumulative frequency distributions would be expected to be identical or reasonably similar to one another. The test protocol for the **Kolmogorov–Smirnov test for two independent samples** is based on the principle that if there is a significant difference at any point along the two cumulative frequency distributions, the researcher can conclude there is a high likelihood the samples are derived from different populations.

The **Kolmogorov–Smirnov test for two independent samples** is categorized as a test of ordinal data because it requires that cumulative frequency distributions be constructed (which requires that within each distribution scores be arranged in order of magnitude). Further clarification of the defining characteristics of a cumulative frequency distribution can be found in the **Introduction**, and in Section I of the **Kolmogorov–Smirnov goodness-of-fit test for a**

single sample. Since the **Kolmogorov–Smirnov test for two independent samples** represents a nonparametric alternative to the *t* **test for two independent samples (Test 11)**, the most common situation in which a researcher might elect to employ the **Kolmogorov–Smirnov test** to evaluate a hypothesis about two independent samples (where the dependent variable represents interval/ratio measurement) is when there is reason to believe that the normality and/or homogeneity of variance assumption of the *t* test have been saliently violated. The **Kolmogorov–Smirnov test for two independent samples** is based on the following assumptions: a) All of the observations in the two samples are randomly selected and independent of one another; and b) The scale of measurement is at least ordinal.

II. Example

Example 13.1 is identical to Examples 11.1/12.1 (which are evaluated with the *t* **test for two independent samples** and the **Mann–Whitney** *U* **test (Test 12)**).

Example 13.1 *In order to assess the efficacy of a new antidepressant drug, ten clinically depressed patients are randomly assigned to one of two groups. Five patients are assigned to Group 1, which is administered the antidepressant drug for a period of six months. The other five patients are assigned to Group 2, which is administered a placebo during the same six-month period. Assume that prior to introducing the experimental treatments, the experimenter confirmed that the level of depression in the two groups was equal. After six months elapse all ten subjects are rated by a psychiatrist (who is blind with respect to a subject's experimental condition) on their level of depression. The psychiatrist's depression ratings for the five subjects in each group follow (the higher the rating, the more depressed a subject):* **Group 1**: 11, 1, 0, 2, 0; **Group 2**: 11, 11, 5, 8, 4. *Do the data indicate that the antidepressant drug is effective?*

III. Null versus Alternative Hypotheses

Prior to reading the null and alternative hypotheses to be presented in this section, the reader should take note of the following: a) The protocol for the **Kolmogorov–Smirnov test for two independent samples** requires that a cumulative probability distribution be constructed for each of the samples. The test statistic is defined by the point which represents the greatest vertical distance at any point between the two cumulative probability distributions; and b) Within the framework of the null and alternative hypotheses, the notation $F_j(X)$ represents the population distribution from which the jth sample/group is derived. $F_j(X)$ can also be conceptualized as representing the cumulative probability distribution for the population from which the jth sample/group is derived.

Null hypothesis $\qquad H_0: F_1(X) = F_2(X)$ for all values of X

(The distribution of data in the population that Sample 1 is derived from is consistent with the distribution of data in the population that Sample 2 is derived from. Another way of stating the null hypothesis is as follows: At no point is the greatest vertical distance between the cumulative probability distribution for Sample 1 (which is assumed to be the best estimate of the cumulative probability distribution of the population from which Sample 1 is derived) and the cumulative probability distribution for Sample 2 (which is assumed to be the best estimate of the cumulative probability distribution of the population from which Sample 2 is derived) larger than what would be expected by chance, if the two samples are derived from the same population.)

Alternative hypothesis H_1: $F_1(X) \neq F_2(X)$ for at least one value of X

(The distribution of data in the population that Sample 1 is derived from is not consistent with the distribution of data in the population that Sample 2 is derived from. Another way of stating this alternative hypothesis is as follows: There is at least one point where the greatest vertical distance between the cumulative probability distribution for Sample 1 (which is assumed to be the best estimate of the cumulative probability distribution of the population from which Sample 1 is derived) and the cumulative probability distribution for Sample 2 (which is assumed to be the best estimate of the cumulative probability distribution of the population from which Sample 2 is derived) is larger than what would be expected by chance, if the two samples are derived from the same population. At the point of maximum deviation separating the two cumulative probability distributions, the cumulative probability for Sample 1 is either significantly greater or less than the cumulative probability for Sample 2. This is a **nondirectional alternative hypothesis** and it is evaluated with a **two-tailed test**.)

or

$$H_1: F_1(X) > F_2(X) \text{ for at least one value of } X$$

(The distribution of data in the population that Sample 1 is derived from is not consistent with the distribution of data in the population that Sample 2 is derived from. Another way of stating this alternative hypothesis is as follows: There is at least one point where the greatest vertical distance between the cumulative probability distribution for Sample 1 (which is assumed to be the best estimate of the cumulative probability distribution of the population from which Sample 1 is derived) and the cumulative probability distribution for Sample 2 (which is assumed to be the best estimate of the cumulative probability distribution of the population from which Sample 2 is derived) is larger than what would be expected by chance, if the two samples are derived from the same population. At the point of maximum deviation separating the two cumulative probability distributions, the cumulative probability for Sample 1 is significantly greater than the cumulative probability for Sample 2. This is a **directional alternative hypothesis** and it is evaluated with a **one-tailed test**.)

or

$$H_1: F_1(X) < F_2(X) \text{ for at least one value of } X$$

(The distribution of data in the population that Sample 1 is derived from is not consistent with the distribution of data in the population that Sample 2 is derived from. Another way of stating this alternative hypothesis is as follows: There is at least one point where the greatest vertical distance between the cumulative probability distribution for Sample 1 (which is assumed to be the best estimate of the cumulative probability distribution of the population from which Sample 1 is derived) and the cumulative probability distribution for Sample 2 (which is assumed to be the best estimate of the cumulative probability distribution of the population from which Sample 2 is derived) is larger than what would be expected by chance, if the two samples are derived from the same population. At the point of maximum deviation separating the two cumulative probability distributions, the cumulative probability for Sample 1 is significantly less than the cumulative probability for Sample 2. This is a **directional alternative hypothesis** and it is evaluated with a **one-tailed test**.)

Note: Only one of the above noted alternative hypotheses is employed. If the alternative hypothesis the researcher selects is supported, the null hypothesis is rejected.

IV. Test Computations

As noted in Sections I and III, the test protocol for the **Kolmogorov–Smirnov test for two independent samples** contrasts the two sample cumulative probability distributions with one another. Table 13.1 summarizes the steps which are involved in the analysis. There are a total of $n = 10$ scores, with $n_1 = 5$ scores in Group 1 and $n_2 = 5$ scores in Group 2.

Table 13.1 Calculation of Test Statistic for Kolmogorov–Smirnov Test for Two Independent Samples for Example 13.1

A (X_1)	B $S_1(X)$	C (X_2)	D $S_2(X)$	E $S_1(X) - S_2(X)$
0,0	2/5 = .40	–	0	.40 – 0 = .40
1	3/5 = .60	–	0	.60 – 0 = .60
2	4/5 = .80	–	0	.80 – 0 = .80 = M
–	4/5 = .80	4	1/5 = .20	.80 – .20 = .60
–	4/5 = . 80	5	2/5 = .40	.80 – .40 = .40
–	4/5 = .80	8	3/5 = .60	.80 – .60 = .20
11	5/5 = 1.00	11, 11	5/5 = 1.00	1.00 – 1.00 = .00

The values represented in the columns of Table 13.1 are summarized below.

The values of the psychiatrist's depression ratings for the subjects in Group 1 are recorded in **Column A**. Note that there are five scores recorded in **Column A**, and that if the same score is assigned to more than one subject in Group 1, each of the scores of that value is recorded in the same row in **Column A**.

Each value in **Column B** represents the cumulative proportion associated with the value of the X score recorded in **Column A**. The notation $S_1(X)$ is commonly employed to represent the cumulative proportions for Group/Sample 1 recorded in **Column B**. The value in **Column B** for any row is obtained as follows: a) The Group 1 cumulative frequency for the score in that row (i.e., the frequency of occurrence of all scores in Group 1 equal to or less than the score in that row) is divided by the total number of scores in Group 1 ($n_1 = 5$). To illustrate, in the case of **Row 1**, the score 0 is recorded twice in **Column A**. Thus, the cumulative frequency is equal to 2, since there are 2 scores in Group 1 that are equal to 0 (a depression rating score cannot be less than 0). Thus, the cumulative frequency 2 is divided by $n_1 = 5$, yielding 2/5 = .40. The value .40 in **Column B** represents the cumulative proportion in Group 1 associated with a score of 0. It means that the proportion of scores in Group 1 which is equal to 0 is .40. The proportion of scores in Group 1 that is larger than 0 is .60 (since $1 - .40 = .60$). In the case of **Row 2**, the score 1 is recorded in **Column A**. The cumulative frequency is equal to 3, since there are 3 scores in Group 1 that are equal to or less than 1 (2 scores of 0 and a score of 1). Thus, the cumulative frequency 3 is divided by $n_1 = 5$, yielding 3/5 = .60. The value .60 in **Column B** represents the cumulative proportion in Group 1 associated with a score of 1. It means that the proportion of scores in Group 1 which is equal to or less than 1 is .60. The proportion of scores in Group 1 that is larger than 1 is .40 (since $1 - .60 = .40$). In the case of **Row 3**, the score 2 is recorded in **Column A**. The cumulative frequency is equal to 4, since there are 4 scores in Group 1 that are equal to or less than 2 (two scores of 0, a score of 1, and a score of 2). Thus, the cumulative frequency 4 is divided by $n_1 = 5$, yielding 4/5 = .80. The value .80 in **Column B** represents the cumulative proportion in Group 1 associated with a score of 2. It means that the proportion of scores in Group 1 which is equal to or less than 2 is .80. The proportion of scores in Group 1 that is larger than 2 is .20 (since $1 - .80 = .20$). Note that the value of the cumulative proportion in **Column B** remains .8 in **Rows 4, 5**, and **6**, since until a new score is recorded in **Column A**, the cumulative proportion recorded in **Column B** will remain the same. In the case of **Row 7**, the score 11 is recorded in **Column A**. The cumulative frequency is equal to 5, since there are 5 scores in Group 1 that are equal to or less than 11 (i.e., all of the scores in Group 1 are equal to or less than 11). Thus, the cumulative frequency 5 is divided by $n_1 = 5$, yielding $5/5 = 1$. The value 1 in **Column B** represents the cumulative proportion in Group 1 associated with a score of 11. It means that the proportion of scores in Group 1 which is equal to or less than 11 is 1. The proportion of scores in Group 1 that is larger than 11 is 0 (since $1 - 1 = 0$).

The values of the psychiatrist's depression ratings for the subjects in Group 2 are recorded in **Column C**. Note that there are five scores recorded in **Column C**, and if the same score is assigned to more than one subject in Group 2, each of the scores of that value is recorded in the same row in **Column C**.

Each value in **Column D** represents the cumulative proportion associated with the value of the X score recorded in **Column C**. The notation $S_2(X)$ is commonly employed to represent the cumulative proportions for Group/Sample 2 recorded in **Column D**. The value in **Column D** for any row is obtained as follows: a) The Group 2 cumulative frequency for the score in that row (i.e., the frequency of occurrence of all scores in Group 2 equal to or less than the score in that row) is divided by the total number of scores in Group 2 ($n_2 = 5$). To illustrate, in the case of **Rows 1, 2**, and **3**, no score is recorded in **Column C**. Thus, the cumulative frequencies for each of those rows are equal to 0, since up to that point in the analysis there are no scores recorded for Group 2. Consequently, for each of the first three rows, the cumulative frequency 0 is divided by $n_2 = 5$, yielding $0/5 = 0$. In each of the first three rows, the value 0 in **Column D** represents the cumulative proportion for Group 2 up to that point in the analysis. For each of those rows, the proportion of scores in Group 2 that remain to be analyzed is 1 (since $1 - 0 = 1$). In the case of **Row 4**, the score 4 is recorded in **Column C**. The cumulative frequency is equal to 1, since there is 1 score in Group 2 that is equal to or less than 4 (i.e., the score 4 in that row). Thus, the cumulative frequency 1 is divided by $n_2 = 5$, yielding $1/5 = .20$. The value .20 in **Column D** represents the cumulative proportion in Group 2 associated with a score of 4. It means that the proportion of scores in Group 2 which is equal to or less than 4 is .20. The proportion of scores in Group 2 that is larger than 4 is .80 (since $1 - .20 = .80$). In the case of **Row 5**, the score 5 is recorded in **Column C**. The cumulative frequency is equal to 2, since there are 2 scores in Group 2 that are equal to or less than 5 (the scores of 4 and 5). Thus, the cumulative frequency 2 is divided by $n_2 = 5$, yielding $2/5 = .40$. The value .40 in **Column D** represents the cumulative proportion in Group 2 associated with a score of 5. It means that the proportion of scores in Group 2 which is equal to or less than 5 is .40. The proportion of scores in Group 2 that is larger than 5 is .60 (since $1 - .40 = .60$). In the case of **Row 6**, the score 8 is recorded in **Column C**. The cumulative frequency is equal to 3, since there are 3 scores in Group 2 that are equal to or less than 8 (the scores of 4, 5, and 8). Thus, the cumulative frequency 3 is divided by $n_2 = 5$, yielding $3/5 = .60$. The value .60 in **Column D** represents the cumulative proportion in Group 2 associated with a score of 8. It means that the proportion of scores in Group 2 which is equal to or less than 8 is .60. The proportion of scores in Group 2 that is larger than 8 is .40 (since $1 - .60 = .40$). In the case of **Row 7**, the score 11 is recorded twice in **Column C**. The cumulative frequency is equal to 5, since there are 5 scores in Group 2 that are equal to or less than 11 (i.e., all of the scores in Group 2 are equal to or less than 11). Thus, the cumulative frequency 5 is divided by $n_2 = 5$, yielding $5/5 = 1$. The value 1 in **Column D** represents the cumulative proportion in Group 2 associated with a score of 11. It means that the proportion of scores in Group 2 which is equal to or less than 11 is 1. The proportion of scores in Group 2 that is larger than 11 is 0 (since $1 - 1 = 0$).

The values in **Column E** are difference scores between the cumulative proportions recorded in **Row B** for Group 1 and **Row D** for Group 2. Thus, for **Row 1** the entry in **Column E** is .40, which represents the **Column B** cumulative proportion of .40 for Group 1, minus 0, which represents the **Column D** cumulative proportion for Group 2. For **Row 2** the entry in **Column E** is .60, which represents the **Column B** cumulative proportion of .60 for Group 1, minus 0, which represents the **Column D** cumulative proportion for Group 2. The same procedure is employed with the remaining five rows in the table.

As noted in Section III, the test statistic for the **Kolmogorov–Smirnov test for two independent samples** is defined by the greatest vertical distance at any point between the two cumulative probability distributions. The largest absolute value obtained in **Column E** will represent the latter value. The notation M will be employed for the test statistic. In Table 13.1 the largest absolute value is .80 (which is recorded in **Row 3**). Therefore, $M = .80$.[1]

V. Interpretation of the Test Results

The test statistic for the **Kolmogorov–Smirnov test for two independent samples** is evaluated with **Table A23 (Table of Critical Values for the Kolmogorov–Smirnov test for two independent samples)** in the **Appendix**. If, at any point along the two cumulative probability distributions, the greatest distance (i.e., the value of M) is equal to or greater than the tabled critical value recorded in **Table A23**, the null hypothesis is rejected. The critical values in Table **A23** are listed in reference to the values of n_1 and n_2. For n_1 = 5 and n_2 = 5, the tabled critical two-tailed .05 and .01 values are $M_{.05}$ = .800 and $M_{.01}$ = .800, and the tabled critical one-tailed .05 and .01 values are $M_{.05}$ = .600 and $M_{.01}$ = .800.[2]

The following guidelines are employed in evaluating the null hypothesis for the **Kolmogorov–Smirnov test for two independent samples**.

a) If the nondirectional alternative hypothesis H_1: $F_1(X) \neq F_2(X)$ is employed, the null hypothesis can be rejected if the computed absolute value of the test statistic is equal to or greater than the tabled critical two-tailed M value at the prespecified level of significance.

b) If the directional alternative hypothesis H_1: $F_1(X) > F_2(X)$ is employed, the null hypothesis can be rejected if the computed absolute value of the test statistic is equal to or greater than the tabled critical one-tailed M value at the prespecified level of significance. Additionally, the difference between the two cumulative probability distributions must be such that in reference to the point which represents the test statistic, the cumulative probability for Sample 1 must be larger than the cumulative probability for Sample 2 (which will result in a positive sign for the value of M).

c) If the directional alternative hypothesis H_1: $F_1(X) < F_2(X)$ is employed, the null hypothesis can be rejected if the computed absolute value of the test statistic is equal to or greater than the tabled critical one-tailed M value at the prespecified level of significance. Additionally, the difference between the two cumulative probability distributions must be such that in reference to the point which represents the test statistic, the cumulative probability for Sample 1 must be less than the cumulative probability for Sample 2 (which will result in a negative sign for the value of M).

The above guidelines will now be employed in reference to the computed test statistic $M = .80$.

a) If the nondirectional alternative hypothesis H_1: $F_1(X) \neq F_2(X)$ is employed, the null hypothesis can be rejected at both the .05 and .01 levels, since the absolute value $M = .80$ is equal to the tabled critical two-tailed values $M_{.05}$ = .800 and $M_{.01}$ = .800.

b) If the directional alternative hypothesis H_1: $F_1(X) > F_2(X)$ is employed, the null hypothesis can be rejected at both the .05 and .01 levels, since the absolute value $M = .80$ is greater than or equal to the tabled critical one-tailed values $M_{.05}$ = .600 and $M_{.01}$ = .800. Additionally, since in **Row 3** of Table 13.1 $[S_1(X) = .80] > [S_2(X) = 0]$, the data are consistent with the alternative hypothesis H_1: $F_1(X) > F_2(X)$. In other words, in computing the value of M, the cumulative proportion for Sample 1 is larger than the cumulative proportion for Sample 2 (which results in a positive sign for the value of M).

c) If the directional alternative hypothesis H_1: $F_1(X) < F_2(X)$ is employed, the null hypothesis cannot be rejected, since in order for the latter alternative hypothesis to be supported, in computing the value of M, the cumulative proportion for Sample 2 must be larger than the cumulative proportion for Sample 1 (which would result in a negative sign for the value of M— which is not the case in **Row 3** of Table 13.1).

A summary of the analysis of Example 13.1 with the **Kolmogorov–Smirnov test for two independent samples** follows: It can be concluded there is a high likelihood the two groups are derived from different populations. More specifically, the data indicate that the depression ratings for Group 1 (i.e., the group that receives the antidepressant medication) are significantly less than the depression ratings for Group 2 (the placebo group).

When the same set of data is evaluated with the *t* **test for two independent samples** and the **Mann–Whitney** *U* **test** (i.e., Examples 11.1/12.1), in the case of both of the latter tests, the null hypothesis can only be rejected (and only at the .05 level) if the researcher employs a directional alternative hypothesis which predicts a lower level of depression for Group 1. The latter result is consistent with the result obtained with the **Kolmogorov–Smirnov test**, in that the directional alternative hypothesis H_1: $F_1(X) > F_2(X)$ is supported. Note, however, that the latter directional alternative hypothesis is supported at both the .05 and .01 levels when the **Kolmogorov–Smirnov test** is employed. In addition, the nondirectional alternative hypothesis is supported at both the .05 and .01 levels with the **Kolmogorov–Smirnov test**, but is not supported when the *t* **test** and **Mann–Whitney** *U* **test** are used. Although the results obtained with the **Kolmogorov–Smirnov test for two independent samples** are not identical with the results obtained with the *t* **test for two independent samples** and the **Mann–Whitney** *U* **test**, they are reasonably consistent.

It should be noted that in most instances the **Kolmogorov–Smirnov test for two independent samples** and the *t* **test for two independent samples** are employed to evaluate the same set of data, the **Kolmogorov–Smirnov test** will provide a less powerful test of an alternative hypothesis. Thus, although it did not turn out to be the case for Examples 11.1/13.1, if a significant difference is present, the *t* **test** will be the more likely of the two tests to detect it. Siegel and Castellan (1988) note that when compared with the *t* **test for two independent samples**, the **Kolmogorov–Smirnov test** has a power efficiency (which is defined in Section VII of the **Wilcoxon signed-ranks test (Test 6)**) of .95 for small sample sizes, and a slightly lower power efficiency for larger sample sizes.

VI. Additional Analytical Procedures for the Kolmogorov–Smirnov Test for Two Independent Samples and/or Related Tests

1. Graphical method for computing the Kolmogorov–Smirnov test statistic Conover (1980, 1999) employs a graphical method for computing the **Kolmogorov–Smirnov test** statistic which is based on the same logic as the graphical method that is briefly discussed for computing the test statistic for the **Kolmogorov–Smirnov goodness-of-fit test for a single sample**. The method involves constructing a graph of the cumulative probability distribution for each sample and measuring the point of maximum distance between the two cumulative probability distributions. The latter graph is similar to the one depicted in Figure 7.1. Daniel (1990) describes a graphical method which employs a graph referred to as a **pair chart** as an alternative way of computing the **Kolmogorov–Smirnov test** statistic. The latter method is attributed to Hodges (1958) and Quade (1973) (who cites Drion (1952) as having developed the pair chart).

2. Computing sample confidence intervals for the Kolmogorov–Smirnov test for two independent samples The same procedure that is described for computing a confidence interval for cumulative probabilities for the sample distribution which is evaluated with the **Kolmogorov– Smirnov goodness-of-fit test for a single sample** can be employed to compute a confidence interval for cumulative probabilities for either one of the samples that are evaluated with the **Kolmogorov–Smirnov test for two independent samples**. Specifically, Equation 7.1 is employed to compute the upper and lower limits for each of the points in a confidence interval. Thus, for each sample, $M_{\alpha/2}$ is added to and subtracted from each of the $S_j(X)$ values. Note that the value of $M_{\alpha/2}$ employed in constructing a confidence interval for each of the samples is derived from **Table A21 (Table of Critical Values for the Kolmogorov–Smirnov Goodness-of-Fit Test for a Single Sample)** in the **Appendix**. Thus, if one is computing a 95% confidence interval for each of the samples, the tabled critical two-tailed value $M_{.05} = .563$ for $n_j = n_1 = n_2 = 5$ is employed to represent $M_{\alpha/2}$ in Equation 7.1.

Note the notation $S_j(X)$ is used to represent the points on a cumulative probability distribution for the **Kolmogorov–Smirnov test for two independent samples**, while the notation $S(X_i)$ is used to represent the points on the cumulative probability distribution for the sample evaluated with the **Kolmogorov–Smirnov goodness-of-fit test for a single sample**. In the case of the latter test, there is only one sample for which a confidence interval can be computed, while in the case of the **Kolmogorov–Smirnov test for two independent samples**, a confidence interval can be constructed for each of the independent samples.

3. Large sample chi-square approximation for a one-tailed analysis of the Kolmogorov–Smirnov test for two independent samples Siegel and Castellan (1988) note that Goodman (1954) has shown that Equation 13.1 (which employs the chi-square distribution) can provide a good approximation for large sample sizes when a one-tailed/directional alternative hypothesis is evaluated.[3]

$$\chi^2 = 4M^2\left(\frac{n_1\,n_2}{n_1 + n_2}\right) \qquad \text{(Equation 13.1)}$$

The computed value of chi-square is evaluated with **Table A4 (Table of the Chi-Square Distribution)** in the **Appendix**. The degrees of freedom employed in the analysis will always be $df = 2$. The tabled critical one-tailed .05 and .01 chi-squared values in **Table A4** for $df = 2$ are $\chi^2_{.05} = 5.99$ and $\chi^2_{.01} = 9.21$. If the computed value of chi-square is equal to or greater than either of the aforementioned values, the null hypothesis can be rejected at the appropriate level of significance (i.e., the directional alternative hypothesis that is consistent with the data will be supported). Although our sample size is too small for the large sample approximation, for purposes of illustration we will use it. When the appropriate values for Example 13.1 are substituted in Equation 13.1, the value $\chi^2 = 6.4$ is computed. Since $\chi^2 = 6.4$ is larger than $\chi^2_{.05} = 5.99$ but less than $\chi^2_{.01} = 9.21$, the null hypothesis can be rejected, but only at the .05 level. Thus, the directional alternative hypothesis H_1: $F_1(X) > F_2(X)$ is supported at the .05 level. Note than when the tabled critical values in **Table A23** are employed, the latter alternative hypothesis is also supported at the .01 level. The latter is consistent with the fact that Siegel and Castellan (1988) note that when Equation 13.1 is employed with small sample sizes, it tends to yield a conservative result (i.e., it is less likely to reject a false null hypothesis).

$$\chi^2 = 4(.80)^2\left(\frac{(5)(5)}{5 + 5}\right) = 6.4$$

VII. Additional Discussion of the Kolmogorov–Smirnov Test for Two Independent Samples

1. Additional comments on the Kolmogorov–Smirnov test for two independent samples
a) Daniel (1990) states that if for both populations a continuous dependent variable is evaluated, the **Kolmogorov–Smirnov test for two independent samples** yields exact probabilities. He notes, however, that Noether (1963, 1967) has demonstrated that if a discrete dependent variable is evaluated, the test tends to be conservative (i.e., is less likely to reject a false null hypothesis);
b) Sprent (1993) notes the **Kolmogorov–Smirnov test for two independent samples** may not be as powerful as tests which focus on whether or not there is a difference on a specific distributional characteristic such as a measure of central tendency and/or variability. Siegel and Castellan (1988) state that the **Kolmogorov–Smirnov test for two independent samples** is more powerful than the **chi-square test for** $r \times c$ **tables (Test 16)** and the **median test for independent samples (Test 16e)**. They also note that for small sample sizes, the **Kolmogorov–Smirnov test** has a higher power efficiency than the **Mann–Whitney** U **test**, but as the sample size increases the opposite becomes true with regard to the power efficiency of the two tests; and
c) Conover (1980, 1999) and Hollander and Wolfe (1999) provide a more detailed discussion of the theory underlying the **Kolmogorov–Smirnov test for two independent samples**.

VIII. Additional Examples Illustrating the Use of the Kolmogorov–Smirnov Test for Two Independent Samples

Since Examples 11.4 and 11.5 in Section VIII of the *t* **test for two independent samples** employ the same data as Example 13.1, the **Kolmogorov–Smirnov test for two independent samples** will yield the same result if employed to evaluate the latter two examples. In addition, the **Kolmogorov–Smirnov test** can be employed to evaluate Examples 11.2 and 11.3. Since different data are employed in the latter examples, the result obtained with the **Kolmogorov–Smirnov test** will not be the same as that obtained for Example 13.1. Example 11.2 is evaluated below with the **Kolmogorov–Smirnov test for two independent samples**. Table 13.2 summarizes the analysis.

Table 13.2 Calculation of Test Statistic for Kolmogorov–Smirnov Test for Two Independent Samples for Example 11.2

A (X_1)	B $S_1(X)$	C (X_2)	D $S_2(X)$	E $S_1(X) - S_2(X)$
–	0	7	$1/5 = .20$	$0 - .20 = -.20$
8	$1/5 = .20$	8, 8	$3/5 = .60$	$.20 - .60 = -.40$
9	$2/5 = .40$	9, 9	$5/5 = 1.00$	$.40 - 1.00 = -.60 = M$
10, 10	$4/5 = .80$	–	$5/5 = 1.00$	$.80 - 1.00 = -.20$
11	$5/5 = 1.00$	–	$5/5 = 1.00$	$1.00 - 1.00 = .00$

The obtained value of test statistic is $M = .60$, since .60 is the largest absolute value for a difference score recorded in **Column E** of Table 13.2. Since $n_1 = 5$ and $n_2 = 5$, we employ the same critical values used in evaluating Example 13.1. If the nondirectional alternative hypothesis H_1: $F_1(X) \neq F_2(X)$ is employed, the null hypothesis cannot be rejected at the .05 level, since $M = .60$ is less than the tabled critical two-tailed value $M_{.05} = .800$. The data are consistent with the directional alternative hypothesis H_1: $F_1(X) < F_2(X)$, since in **Row 3** of Table 13.2 $[S_1(X) = .40] < [S_2(X) = 1]$. In other words, in computing the value of M, the cumulative proportion for Sample 2 is larger than the cumulative proportion for

Sample 1 (resulting in a negative sign for the computed value of M). The directional alternative hypothesis H_1: $F_1(X) < F_2(X)$ is supported at the .05 level, since $M = .60$ is equal to the tabled critical one-tailed value $M_{.05} = .600$. It is not, however, supported at the .01 level, since $M = .60$ is less than the tabled critical one-tailed value $M_{.01} = .800$. The directional alternative hypothesis H_1: $F_1(X) > F_2(X)$ is not supported, since it is not consistent with the data (i.e., the sign of the value computed for M is not positive).

When the null hypothesis H_0: $\mu_1 = \mu_2$ is evaluated with the t test for two independent samples, the only alternative hypothesis which is supported (but only at the .05 level) is the directional alternative hypothesis H_1: $\mu_1 > \mu_2$. The latter result (indicating higher scores in Group 1) is consistent with the result obtained when the Kolmogorov–Smirnov test for two independent samples is employed to evaluate the same set of data.

References

Conover, W. J. (1980). **Practical nonparametric statistics** (2nd ed.). New York: John Wiley & Sons.

Conover, W. J. (1999). **Practical nonparametric statistics** (3rd ed.). New York: John Wiley & Sons.

Daniel, W. W. (1990). **Applied nonparametric statistics** (2nd ed.). Boston: PWS–Kent Publishing Company.

Drion, E. F. (1952). Some distribution-free tests for the difference between two empirical cumulative distribution functions. **Annals of Mathematical Statistics**, 23, 563–574.

Goodman, L. A. (1954). Kolmogorov–Smirnov tests for psychological research. **Psychological Bulletin**, 51, 160–168.

Hodges, J. L., Jr. (1958). The significance probability of the Smirnov two-sample test. **Ark. Mat.**, 3, 469–486.

Hollander, M. & Wolfe, D. A. (1999). **Nonparametric statistical methods**. New York: John Wiley & Sons.

Kolmogorov, A. N. (1933). Sulla determinazione empirica di una legge di distribuzione. **Giorn dell'Inst. Ital. degli. Att.**, 4, 89–91.

Marascuilo, L. A. & McSweeney, M. (1977). **Nonparametric and distribution-free methods for the social sciences.** Monterey, CA: Brooks/Cole Publishing Company.

Massey, F. J., Jr. (1952). Distribution tables for the deviation between two sample cumulatives. **Annals of Mathematical Statistics**, 23, pp. 435–441.

Noether, G. E. (1963). Note on the Kolmogorov statistic in the discrete case. **Metrika**, 7, 115–116.

Noether, G. E. (1967). **Elements of nonparmetric statistics**. New York: John Wiley & Sons.

Quade, D. (1973). The pair chart. **Statistica Neerlandica**, 27, 29–45.

Siegel, S. & Castellan, N. J., Jr. (1988). **Nonparametric statistics for the behavioral sciences** (2nd ed.). New York: McGraw–Hill Book Company.

Smirnov, N. V. (1936). Sur la distribution de W^2 (criterium de M. R. v. Mises). **Comptes Rendus** (Paris), 202, 449–452.

Smirnov, N. V. (1939). Estimate of deviation between empirical distribution functions in two independent samples (Russian), **Bull Moscow Univ.**, 2, 3–16.

Sprent, P. (1993). **Applied nonparametric statistics** (2nd ed). London: Chapman & Hall.

Zar, J. H. (1999). **Biostatistical analysis** (4th ed.). Upper Saddle River, NJ: Prentice Hall.

Endnotes

1. Marasucilo and McSweeney (1977) employ a modified protocol which can result in a larger absolute value for M in **Column E** than the one obtained in Table 13.1. The latter protocol employs a separate row for the score of each subject when the same score occurs more than once within a group. If the latter protocol is employed in Table 13.1, the first two rows of the table will have the score of 0 in **Column A** for the two subjects in Group 1 who obtain that score. The first 0 will be in the first row, and have a cumulative proportion in **Column B** of $1/5 = .20$. The second 0 will be in the second row, and have a cumulative proportion in **Column B** of $2/5 = .40$. In the same respect the first of the two scores of 11 (obtained by two subjects in Group 2) will be in a separate row in **Column C**, and have a cumulative proportion in **Column D** of $4/5 = .80$. The second score of 11 will be in the last row of the table, and have a cumulative proportion in **Column D** of $5/5 = 1$. In the case of Example 13.1, the outcome of the analysis will not be affected if the aforementioned protocol is employed. In some instances, however, it can result in a larger M value. The protocol employed by Marasucilo and McSweeney (1977) is used by sources who argue that when there are ties present in the data (i.e., the same score occurs more than once within a group), the protocol described in this chapter (which is used in most sources) results in an overly conservative test (i.e., makes it more difficult to reject a false null hypothesis).

2. When the values of n_1 and n_2 are small, some of the .05 and .01 critical values listed in **Table A23** are identical to one another.

3. The last row in **Table A23** can also be employed to compute a critical M value for large sample sizes.

Test 14

The Siegel–Tukey Test for Equal Variability
(Nonparametric Test Employed with Ordinal Data)

I. Hypothesis Evaluated with Test and Relevant Background Information

Hypothesis evaluated with test   Do two independent samples represent two populations with different variances?

Relevant background information on test   Developed by Siegel and Tukey (1960), the **Siegel–Tukey test for equal variability** is employed with ordinal (rank-order) data in a hypothesis testing situation involving two independent samples. If the result of the **Siegel–Tukey test for equal variability** is significant, it indicates there is a significant difference between the sample variances, and as a result of the latter the researcher can conclude there is a high likelihood that the samples represent populations with different variances.

The **Siegel–Tukey test for equal variability** is one of a number of tests of **dispersion** (also referred to as tests of **scale** or **spread**) which have been developed for contrasting the variances of two independent samples. A discussion of alternative nonparametric tests of dispersion can be found in Section VII. Some sources recommend the use of nonparametric tests of dispersion for evaluating the homogeneity of variance hypothesis when there is reason to believe that the normality assumption of the appropriate parametric test for evaluating the same hypothesis is violated. Sources that are not favorably disposed toward nonparametric tests recommend the use of Hartley's F_{max} **test for homogeneity of variance**/F **test for two population variances (Test 11a)** or one of the alternative parametric tests described in Section VI of the **single-factor between subjects analysis of variance (Test 21)** which are available for evaluating homogeneity of variance, regardless of whether or not the normality assumption of the parametric test is violated. Such sources do, however, recommend that in employing a parametric test a researcher employ a lower significance level to compensate for the fact that violation of the normality assumption can inflate the Type I error rate associated with the test. When there is no evidence to indicate that the normality assumption of the parametric test has been violated, sources are in general agreement that such a test is preferable to the **Siegel–Tukey test for equal variability** (or an alternative nonparametric test of dispersion), since a parametric test (which uses more information than a nonparametric test) provides a more powerful test of an alternative hypothesis.

Since nonparametric tests are not assumption free, the choice of which of the available tests of dispersion to employ will primarily depend on what assumptions a researcher is willing to make with regard to the underlying distributions represented by the sample data. The **Siegel–Tukey test for equal variability** is based on the following assumptions: a) Each sample has been randomly selected from the population it represents; b) The two samples are independent of one another; c) The level of measurement the data represent is at least ordinal; and d) The two populations from which the samples are derived have equal medians. If the latter assumption is violated, but the researcher does know the values of the population medians, the scores in the groups can be adjusted so as to allow the use of the **Siegel–Tukey test for equal variability**. When, however, the population medians are unknown and one is unwilling to assume they are equal, the **Siegel–Tukey test for equal variability** is not the appropriate nonparametric test of dispersion to employ. The assumption of equality of population medians presents a practical problem, in that when evaluating two independent samples a researcher will often have no prior knowledge regarding the population medians. In point of fact, most hypothesis testing addresses the issue of whether or not the medians (or means) of two or more populations are equal. In view of this, sources such as Siegel and Castellan (1988) note that if the latter values are not known, it is not appropriate to estimate them with the sample medians.

In employing the **Siegel–Tukey test for equal variability**, one of the following is true with regard to the rank-order data that are evaluated: a) The data are in a rank-order format, since it is the only format in which scores are available; or b) The data have been transformed to a rank-order format from an interval/ratio format, since the researcher has reason to believe that the normality assumption of the analogous parametric test is saliently violated.

II. Example

Example 14.1 *In order to assess the effect of two antidepressant drugs, 12 clinically depressed patients are randomly assigned to one of two groups. Six patients are assigned to Group 1, which is administered the antidepressant drug Elatrix for a period of six months. The other six patients are assigned to Group 2, which is administered the antidepressant drug Euphyria during the same six-month period. Assume that prior to introducing the experimental treatments, the experimenter confirmed that the level of depression in the two groups was equal. After six months elapse, all 12 subjects are rated by a psychiatrist (who is blind with respect to a subject's experimental condition) on their level of depression. The psychiatrist's depression ratings for the six subjects in each group follow (the higher the rating, the more depressed a subject):*
Group 1: *10, 10, 9, 1, 0, 0;* **Group 2**: *6, 6, 5, 5, 4, 4.*

The fact that the mean and median of each group are equivalent (specifically, both values equal 5) is consistent with prior research which suggests that there is no difference in efficacy for the two drugs (when the latter is based on a comparison of group means and/or medians). Inspection of the data does suggest, however, that there is much greater variability in the depression scores of subjects in Group 1. To be more specific, the data suggest that the drug Elatrix may, in fact, decrease depression in some subjects, yet increase it in others. The researcher decides to contrast the variability within the two groups through use of the Siegel–Tukey test for equal variability. The use of the latter nonparametric test is predicated on the fact that there is reason to believe that the distributions of the posttreatment depression scores in the underlying populations are not normal (which is why the researcher is reluctant to evaluate the data with a parametric test of dispersion). Do the data indicate there is a significant difference between the variances of the two groups?

III. Null versus Alternative Hypotheses

Null hypothesis $H_0: \sigma_1^2 = \sigma_2^2$

(The variance of the population Group 1 represents equals the variance of the population Group 2 represents. With respect to the sample data, when both groups have an equal sample size, this translates into the sum of the ranks of Group 1 being equal to the sum of the ranks of Group 2 (i.e., $\Sigma R_1 = \Sigma R_2$). A more general way of stating this, which also encompasses designs involving unequal sample sizes, is that the means of the ranks of the two groups are equal (i.e., $\bar{R}_1 = \bar{R}_2$)).

Alternative hypothesis $\qquad\qquad H_1: \sigma_1^2 \neq \sigma_2^2$

(The variance of the population Group 1 represents does not equal the variance of the population Group 2 represents. With respect to the sample data, when both groups have an equal sample size, this translates into the sum of the ranks of Group 1 not being equal to the sum of the ranks of Group 2 (i.e., $\Sigma R_1 \neq \Sigma R_2$). A more general way of stating this, which also encompasses designs involving unequal sample sizes, is that the means of the ranks of the two groups are not equal (i.e., $\bar{R}_1 \neq \bar{R}_2$). This is a **nondirectional alternative hypothesis** and it is evaluated with a **two-tailed test**.)

or

$$H_1: \sigma_1^2 > \sigma_2^2$$

(The variance of the population Group 1 represents is greater than the variance of the population Group 2 represents. With respect to the sample data, when both groups have an equal sample size (so long as a rank of 1 is given to the lowest score), this translates into the sum of the ranks of Group 1 being less than the sum of the ranks of Group 2 (i.e., $\Sigma R_1 < \Sigma R_2$). A more general way of stating this, which also encompasses designs involving unequal sample sizes, is that the mean of the ranks of Group 1 is less than the mean of the ranks of Group 2 (i.e., $\bar{R}_1 < \bar{R}_2$). This is a **directional alternative hypothesis** and it is evaluated with a **one-tailed test**.)

or

$$H_1: \sigma_1^2 < \sigma_2^2$$

(The variance of the population Group 1 represents is less than the variance of the population Group 2 represents. With respect to the sample data, when both groups have an equal sample size (so long as a rank of 1 is given to the lowest score), this translates into the sum of the ranks of Group 1 being greater than the sum of the ranks of Group 2 (i.e., $\Sigma R_1 > \Sigma R_2$). A more general way of stating this, which also encompasses designs involving unequal sample sizes, is that the mean of the ranks of Group 1 is greater than the mean of the ranks of Group 2 (i.e., $\bar{R}_1 > \bar{R}_2$). This is a **directional alternative hypothesis** and it is evaluated with a **one-tailed test**.)

Note: Only one of the above noted alternative hypotheses is employed. If the alternative hypothesis the researcher selects is supported, the null hypothesis is rejected.

IV. Test Computations

The total number of subjects employed in the experiment is $N = 12$. There are $n_1 = 6$ subjects in Group 1 and $n_2 = 6$ subjects in Group 2. The data for the analysis are summarized in Table 14.1. The original interval/ratio scores of the subjects are recorded in the columns labeled X_1 and X_2. The adjacent columns R_1 and R_2 contain the rank-order assigned to each of the scores. The rankings for Example 14.1 are summarized in Table 14.2. The ranking protocol for the **Siegel–Tukey test for equal variability** is described in this section. Note that in Table 14.1 and Table 14.2 each subject's identification number indicates the order in Table 14.1 in which a subject's score appears in a given group, followed by his/her group. Thus, Subject i, j is the ith subject in Group j.

The computational procedure for the **Siegel–Tukey test for equal variability** is identical to that employed for the **Mann–Whitney U test (Test 12)**, except for the fact that the two

Table 14.1   Data for Example 14.1

	Group 1			Group 2	
	X_1	R_1		X_2	R_2
Subject 1,1	10	2.5	Subject 1,2	6	8.5
Subject 2,1	10	2.5	Subject 2,2	6	8.5
Subject 3,1	9	6	Subject 3,2	5	11.5
Subject 4,1	1	5	Subject 4,2	5	11.5
Subject 5,1	0	2.5	Subject 5,2	4	8.5
Subject 6,1	0	2.5	Subject 6,2	4	8.5
		$\Sigma R_1 = 21$			$\Sigma R_2 = 57$

$$\bar{R}_1 = \frac{\Sigma R_1}{n_1} = \frac{21}{6} = 3.5 \qquad \bar{R}_2 = \frac{\Sigma R_2}{n_2} = \frac{57}{6} = 9.5$$

Table 14.2   Rankings for the Siegel–Tukey Test for Equal Variability for Example 14.1

Subject identification number	5,1	6,1	4,1	5,2	6,2	3,2	4,2	1,2	2,2	3,1	1,1	2,1
Depression score	0	0	1	4	4	5	5	6	6	9	10	10
Rank prior to tie adjustment	1	4	5	8	9	12	11	10	7	6	3	2
Tie-adjusted rank	2.5	2.5	5	8.5	8.5	11.5	11.5	8.5	8.5	6	2.5	2.5

tests employ a different ranking protocol. Recollect that in the description of the alternative hypotheses for the **Siegel–Tukey test for equal variability**, it is noted that when a directional alternative hypothesis is supported, the average of the ranks of the group with the larger variance will be **less** than the average of the ranks of the group with the smaller variance. On the other hand, when a directional hypothesis for the **Mann–Whitney *U* test** is supported, the average of the ranks of the group with the larger median will be **greater** than the average of the ranks of the group with the smaller median. The difference between the two tests with respect to the ordinal position of the average ranks reflects the fact that the tests employ different ranking protocols. Whereas the ranking protocol for the **Mann–Whitney *U* test** is designed to identify differences with respect to central tendency (specifically, the median values), the ranking protocol for the **Siegel–Tukey test for equal variability** is designed to identify differences with respect to variability. The ranking protocol for the **Siegel–Tukey test for equal variability** is based on the premise that within the overall distribution of *N* scores, the distribution of scores in the group with the higher variance will contain more extreme values (i.e., scores that are very high and scores that are very low) than the distribution of scores in the group with the lower variance.

The following protocol, which is summarized in Table 14.2, is used in assigning ranks.

a) All $N = 12$ scores are arranged in order of magnitude (irrespective of group membership), beginning on the left with the lowest score and moving to the right as scores increase. This is done in the second row of Table 14.2.

b) Ranks are now assigned in the following manner: A rank of 1 is assigned to the lowest score (0). A rank of 2 is assigned to the highest score (10), and a rank of 3 is assigned to the second highest score (10). A rank of 4 is assigned to the second lowest score (0), and a rank of 5 is assigned to the third lowest score (1). A rank of 6 is assigned to the third highest score (9), and a rank of 7 is assigned to the fourth highest score (6). A rank of 8 is assigned to the fourth lowest score (4), and a rank of 9 is assigned to the fifth lowest score (4). A rank of 10 is assigned to the fifth highest score (6), and a rank of 11 is assigned to the sixth highest score (5). A rank of 12 is assigned to the sixth lowest score (5). Note that the ranking protocol assigns ranks to the distribution of $N = 12$ scores by alternating from one extreme of the distribution to the other. The ranks assigned employing this protocol are listed in the third row of Table 14.2

c) The ranks in the third row of Table 14.2 must be adjusted when there are tied scores present in the data. The same procedure for handling ties which is described for the **Mann–Whitney U test** is also employed for the **Siegel–Tukey test for equal variability**. Specifically, in instances where two or more subjects have the same score, the average of the ranks involved is assigned to all scores tied for a given rank. This adjustment is made in the fourth row of Table 14.2. To illustrate: Both Subjects 5,1 and 6,1 have a score of 0. Since the ranks assigned to the scores of these two subjects are, respectively, 1 and 4, the average of the two ranks $(1 + 4)/2 = 2.5$ is assigned to the score of both subjects. Both Subjects 1,1 and 2,1 have a score of 10. Since the ranks assigned to the scores of these two subjects are, respectively, 3 and 2, the average of the two ranks $(3 + 2)/2 = 2.5$ is assigned to the score of both subjects. For the remaining three sets of ties (which all happen to fall in Group 2) the same averaging procedure is employed.

It should be noted that in Example 14.1 each set of tied scores involves subjects who are in the same group. Any time each set of ties involves subjects in the same group, the tie adjustment will result in the identical sum and average for the ranks of the two groups which will be obtained if the tie adjustment is not employed. Because of this, under these conditions the computed test statistic will be identical regardless of whether or not one uses the tie adjustment. On the other hand, when one or more sets of ties involve subjects from both groups, the tie-adjusted ranks will yield a value for the test statistic which is different from that which will be obtained if the tie adjustment is not employed.

It should be noted that it is permissible to reverse the ranking protocol described in this section. Specifically, one can assign a rank of 1 to the highest score, a rank of 2 to the lowest score, a rank of 3 to the second lowest score, a rank of 4 to the second highest score, a rank of 5 to the third highest score, and so on. This reverse-ranking protocol will result in the same test statistic and, consequently, the same conclusion with respect to the null hypothesis as the ranking protocol described in this section.[1]

Once all of the subjects have been assigned a rank, the sum of the ranks for each of the groups is computed. These values, $\Sigma R_1 = 21$ and $\Sigma R_2 = 57$, are computed in Table 14.1. Upon determining the sum of the ranks for both groups, the values U_1 and U_2 are computed employing Equations 12.1 and 12.2, which are used for the **Mann–Whitney U test**. The basis for employing the same equations and the identical distribution as that used for the **Mann–Whitney U test** is predicated on the fact that both tests employ the same sampling distribution.

$$U_1 = n_1 n_2 + \frac{n_1(n_1 + 1)}{2} - \Sigma R_1 = (6)(6) + \frac{6(6 + 1)}{2} - 21 = 36$$

$$U_2 = n_1 n_2 + \frac{n_2(n_2 + 1)}{2} - \Sigma R_2 = (6)(6) + \frac{6(6 + 1)}{2} - 57 = 0$$

Note that U_1 and U_2 can never be negative values. If a negative value is obtained for either, it indicates an error has been made in the rankings and/or calculations.

As is the case for the **Mann–Whitney U test**, Equation 12.3 can be employed to verify the calculations. If the relationship in Equation 12.3 is not confirmed, it indicates an error has been made in ranking the scores or in the computation of the U values. The relationship described by Equation 12.3 is confirmed below for Example 14.1.

$$n_1 n_2 = U_1 + U_2$$

$$(6)(6) = 36 + 0 = 36$$

V. Interpretation of the Test Results

The smaller of the two values U_1 versus U_2 is designated as the obtained U statistic. Since $U_2 = 0$ is smaller than $U_1 = 36$, the value of $U = 0$. The value of U is evaluated with **Table A11 (Table of Critical Values for the Mann–Whitney U Statistic)** in the **Appendix**. In order to be significant, the obtained value of U must be **equal to or less than** the tabled critical value at the prespecified level of significance. For $n_1 = 6$ and $n_2 = 6$, the tabled critical two-tailed values are $U_{.05} = 5$ and $U_{.01} = 2$, and the tabled critical one-tailed values are $U_{.05} = 7$ and $U_{.01} = 3$.[2]

Since the obtained value $U = 0$ is less than the tabled critical two-tailed values $U_{.05} = 5$ and $U_{.01} = 2$, the nondirectional alternative hypothesis $H_1: \sigma_1^2 \neq \sigma_2^2$ is supported at both the .05 and .01 levels. Since the obtained value of U is less than the tabled critical one-tailed values $U_{.05} = 7$ and $U_{.01} = 3$, the directional alternative hypothesis $H_1: \sigma_1^2 > \sigma_2^2$ is also supported at both the .05 and .01 levels. The latter directional alternative hypothesis is supported since $\bar{R}_1 < \bar{R}_2$, which indicates that the variability of scores in Group 1 is greater than the variability of scores in Group 2. The directional alternative hypothesis $H_1: \sigma_1^2 < \sigma_2^2$ is not supported, since in order for the latter alternative hypothesis to be supported, $\bar{R}_1$ must be greater than $\bar{R}_2$ (which indicates that the variability of scores in Group 2 is greater than the variability of scores in Group 1).

Based on the results of the **Siegel–Tukey test for equal variability**, the researcher can conclude there is greater variability in the depression scores of the group which receives the drug Elatrix (Group 1) than the group which receives the drug Euphyria (Group 2).

VI. Additional Analytical Procedures for the Siegel–Tukey Test for Equal Variability and/or Related Tests

1. The normal approximation of the Siegel–Tukey test statistic for large sample sizes As is the case with the **Mann–Whitney U test**, the normal distribution can be employed with large sample sizes to approximate the **Siegel–Tukey test** statistic. Equation 12.4, which is employed for the large sample approximation of the **Mann–Whitney distribution**, can also be employed for the large sample approximation of **Siegel–Tukey test** statistic. As is noted in Section VI of the **Mann–Whitney U test**, the large sample approximation is generally used for sample sizes larger than those documented in the exact table contained within the source one is employing.

In the discussion of the **Mann–Whitney U test**, it is noted that the term $(n_1 n_2)/2$ in the numerator of Equation 12.4 represents the expected (mean) value of U if the null hypothesis is true. This is also the case when the normal distribution is employed to approximate the **Siegel–Tukey test statistic**. Thus, if the two population variances are in fact equal, it is expected that $\bar{R}_1 = \bar{R}_2$ and, consequently, $U_1 = U_2 = (n_1 n_2)/2$.

Although Example 14.1 involves only $N = 12$ scores (a value most sources would view as too small to use with the normal approximation), it will be employed to illustrate Equation 12.4. The reader will see that in spite of employing Equation 12.4 with a small sample size, it yields a result which is consistent with the result obtained when the exact table for the **Mann–Whitney U distribution** is employed. As is noted in Section VI of the **Mann–Whitney U test**, since the smaller of the two values U_1 versus U_2 is selected to represent U, the value of z will always be negative (unless $U_1 = U_2$, in which case $z = 0$).

Employing Equation 12.4, the value $z = -2.88$ is computed.

$$z = \frac{U - \dfrac{n_1 n_2}{2}}{\sqrt{\dfrac{n_1 n_2 (n_1 + n_2 + 1)}{12}}} = \frac{0 - \dfrac{(6)(6)}{2}}{\sqrt{\dfrac{(6)(6)(6 + 6 + 1)}{12}}} = -2.88$$

The obtained value $z = -2.88$ is evaluated with **Table A1 (Table of the Normal Distribution)** in the **Appendix**. To be significant, the obtained absolute value of z must be equal to or greater than the tabled critical value at the prespecified level of significance. The tabled critical two-tailed .05 and .01 values are $z_{.05} = 1.96$ and $z_{.01} = 2.58$, and the tabled critical one-tailed .05 and .01 values are $z_{.05} = 1.65$ and $z_{.01} = 2.33$. The following guidelines are employed in evaluating the null hypothesis.

a) If the nondirectional alternative hypothesis $H_1: \sigma_1^2 \neq \sigma_2^2$ is employed, the null hypothesis can be rejected if the obtained absolute value of z is equal to or greater than the tabled critical two-tailed value at the prespecified level of significance.

b) If a directional alternative hypothesis is employed, one of the two possible directional alternative hypotheses is supported if the obtained absolute value of z is equal to or greater than the tabled critical one-tailed value at the prespecified level of significance. The directional alternative hypothesis which is supported is the one that is consistent with the data.

Employing the above guidelines with Example 14.1, the following conclusions are reached.

Since the obtained absolute value $z = 2.88$ is greater than the tabled critical two-tailed values $z_{.05} = 1.96$ and $z_{.01} = 2.58$, the nondirectional alternative hypothesis $H_1: \sigma_1^2 \neq \sigma_2^2$ is supported at both the .05 and .01 levels. Since the obtained absolute value $z = 2.88$ is greater than the tabled critical one-tailed values $z_{.05} = 1.65$ and $z_{.01} = 2.33$, the directional alternative hypothesis $H_1: \sigma_1^2 > \sigma_2^2$ is supported at both the .05 and .01 levels. The latter directional alternative hypothesis is supported since it is consistent with the data. The directional alternative hypothesis $H_1: \sigma_1^2 < \sigma_2^2$ is not supported, since it is not consistent with the data. Note that the conclusions reached with reference to each of the possible alternative hypotheses are consistent with those reached when the exact table of the U distribution is employed.

As is the case when normal approximation is used with the **Mann–Whitney U test**, either U_1 or U_2 can be employed in Equation 12.4 to represent the value of U, since both values will yield the same absolute value for z.

2. The correction for continuity for the normal approximation of the Siegel–Tukey test for equal variability

Although not described in most sources, the correction for continuity employed for the normal approximation of the **Mann–Whitney U test** can also be applied to the **Siegel–Tukey test for equal variability**. Employing Equation 12.5 (the **Mann–Whitney** continuity-corrected equation) with the data for Example 14.1, the value $z = -2.80$ is computed.

$$z = \frac{\left| U - \dfrac{n_1 n_2}{2} \right| - .5}{\sqrt{\dfrac{n_1 n_2 (n_1 + n_2 + 1)}{12}}} = \frac{\left| 0 - \dfrac{(6)(6)}{2} \right| - .5}{\sqrt{\dfrac{(6)(6)(6 + 6 + 1)}{12}}} = -2.80$$

The obtained absolute value $z = 2.80$ is greater than the tabled critical two-tailed .05 and .01 values $z_{.05} = 1.96$ and $z_{.01} = 2.58$, and the tabled critical one-tailed .05 and .01 values $z_{.05} = 1.65$ and $z_{.01} = 2.33$. Thus, as is the case when the correction for continuity is not employed, both the nondirectional alternative hypothesis $H_1: \sigma_1^2 \neq \sigma_2^2$ and the directional alternative hypothesis $H_1: \sigma_1^2 > \sigma_2^2$ are supported at both the .05 and .01 levels. The reader should take note of the fact that the absolute value of the continuity-corrected z value will always be less than the absolute value computed when the correction for continuity is not used, except when the uncorrected value of $z = 0$ (in which case it will have no impact on the result of the analysis).

3. Tie correction for the normal approximation of the Siegel–Tukey test statistic It is noted in the discussion of the normal approximation of the **Mann–Whitney U test** that some sources recommend that Equation 12.4 be modified when an excessive number of ties is present in the data. Since the identical sampling distribution is involved, the same tie correction (which results in a slight increase in the absolute value of z) can be employed for the normal approximation of the **Siegel–Tukey test for equal variability**. Employing Equation 12.6 (the **Mann–Whitney** tie correction equation), the tie-corrected value $z = -2.91$ is computed for Example 14.1. Note that in Example 14.1 there are $s = 5$ sets of ties, each set involving two ties. Thus, in Equation 12.6 the term $\sum_{i=1}^{s}(t_i^3 - t_i) = 5[(2)^3 - 2] = 30$.

$$z = \cfrac{U - \cfrac{n_1 n_2}{2}}{\sqrt{\cfrac{n_1 n_2 (n_1 + n_2 + 1)}{12} - \cfrac{n_1 n_2 \sum_{i=1}^{s}(t_i^3 - t_i)}{12(n_1 + n_2)(n_1 + n_2 - 1)}}}$$

$$= \cfrac{0 - \cfrac{(6)(6)}{2}}{\sqrt{\cfrac{(6)(6)(6 + 6 + 1)}{12} - \cfrac{(6)(6)(30)}{12(6 + 6)(6 + 6 - 1)}}} = -2.91$$

The difference between $z = -2.91$ and the uncorrected value $z = -2.88$ is trivial, and consequently the decision the researcher makes with respect to the null hypothesis is not affected, regardless of which alternative hypothesis is employed.

4. Adjustment of scores for the Siegel–Tukey test for equal variability when $\theta_1 \neq \theta_2$ It is noted in Section I that if the values of the population medians are known but are not equal, in order to employ the **Siegel–Tukey test for equal variability** it is necessary to adjust the scores. In such a case, prior to ranking the scores the difference between the two population medians is subtracted from each of the scores in the group which represents the population with the higher median (or added to each of the scores in the group which represents the population with the lower median). This adjustment procedure will be demonstrated with Example 14.2.

Example 14.2 *In order to evaluate whether or not two teaching methods result in different degrees of variability with respect to performance, a mathematics instructor employs two methods of instruction with different groups of students. Prior to initiating the study it is determined that the two groups are comprised of students of equal math ability. Group 1, which* *is comprised of five subjects, is taught through the use of lectures and a conventional textbook* (Method A). *Group 2, which is comprised of six subjects, is taught through the use of a computer software package* (Method B). *At the conclusion of the course the final exam scores of the two groups are compared. The final exam scores follow (the maximum possible score on the final exam is 15 points and the minimum 0):* **Group 1**: 7, 5, 4, 4, 3; **Group 2**: 13, 12, 7, 7, 4, 3. *The researcher elects to rank-order the scores of the subjects, since she does not believe the data are normally distributed in the underlying populations. If the* **Siegel–Tukey test for equal variability** *is employed to analyze the data, is there a significant difference in within-groups variability?*

From the sample data we can determine that the median score of Group 1 is 4, and the median score of Group 2 is 7. Although the computations will not be shown here, in spite of the three-point difference between the medians of the groups, if the **Mann–Whitney** U **test** is employed to evaluate the data, the null hypothesis H_0: $\theta_1 = \theta_2$ (i.e., that the medians of the underlying populations are equal) cannot be rejected at the .05 level, regardless of whether a nondirectional or directional alternative hypothesis is employed. The fact that the null hypothesis cannot be rejected is largely the result of the small sample size, which limits the power of the **Mann–Whitney** U **test** to detect a difference between underlying populations, if, in fact, one exists.

Table 14.3 Data for Example 14.2 Employing Adjusted X_2 Scores

	Group 1			Group 2	
	X_1	R_1		X_2	R_2
Subject 1,1	7	6	Subject 1,2	10	2
Subject 2,1	5	7	Subject 2,2	9	3
Subject 3,1	4	9.5	Subject 3,2	4	9.5
Subject 4,1	4	9.5	Subject 4,2	4	9.5
Subject 5,1	3	5	Subject 5,2	1	4
			Subject 6,2	0	1
		$\Sigma R_1 = 37$			$\Sigma R_2 = 29$

$$\bar{R}_1 = \frac{\Sigma R_1}{n_1} = \frac{37}{5} = 7.4 \qquad \bar{R}_2 = \frac{\Sigma R_2}{n_2} = \frac{29}{6} = 4.83$$

Table 14.4 Rankings for the Siegel–Tukey Test for Equal Variability for Example 14.2

Subject identification number	6,2	5,2	5,1	3,1	4,1	3,2	4,2	2,1	1,1	2,2	1,2
Exam score	0	1	3	4	4	4	4	5	7	9	10
Rank prior to tie adjustment	1	4	5	8	9	11	10	7	6	3	2
Tie-adjusted rank	1	4	5	9.5	9.5	9.5	9.5	7	6	3	2

Let us assume, however, that based on prior research there is reason to believe the median of the population represented by Group 2 is, in fact, three points higher than the median of the population represented by Group 1. In order to employ the **Siegel–Tukey test for equal varia-bility** to evaluate the null hypothesis H_0: $\sigma_1^2 = \sigma_2^2$, the groups must be equated with respect to their median values. This can be accomplished by subtracting the difference between the population medians from each score in the group with the higher median. Thus, in Table 14.3 three points have been subtracted from the score of each of the subjects in Group 2.[3] The scores in Table 14.3 are ranked in accordance with the **Siegel–Tukey test** protocol. The ranks are summarized in Table 14.4.

Equations 12.1 and 12.2 are employed to compute the values $U_1 = 8$ and $U_2 = 22$.

$$U_1 = n_1 n_2 + \frac{n_1(n_1 + 1)}{2} - \Sigma R_1 = (5)(6) + \frac{5(5 + 1)}{2} - 37 = 8$$

$$U_2 = n_1 n_2 + \frac{n_2(n_2 + 1)}{2} - \Sigma R_2 = (5)(6) + \frac{6(6 + 1)}{2} - 29 = 22$$

Employing Equation 12.3, we confirm the relationship between the sample sizes and the computed values of U_1 and U_2.

$$n_1 n_2 = U_1 + U_2$$

$$(5)(6) = 8 + 22 = 30$$

Since $U_1 = 8$ is smaller than $U_2 = 22$, the value of $U = 8$. Employing **Table A11** for $n_1 = 5$ and $n_2 = 6$, we determine that the tabled critical two-tailed .05 and .01 values are $U_{.05} = 3$ and $U_{.01} = 1$, and the tabled critical one-tailed .05 and .01 values are $U_{.05} = 5$ and $U_{.01} = 2$. Since the obtained value $U = 8$ is greater than all of the aforementioned tabled critical values, the null hypothesis cannot be rejected at either the .05 or .01 level, regardless of whether a nondirectional or directional alternative hypothesis is employed.

If the normal approximation for the **Siegel–Tukey test for equal variability** is employed with Example 14.2, it is also the case that the null hypothesis cannot be rejected, regardless of which alternative hypothesis is employed. The latter is the case, since the computed absolute value $z = 1.28$ is less than the tabled critical .05 and .01 two-tailed values $z_{.05} = 1.96$ and $z_{.01} = 2.58$, and the tabled critical .05 and .01 one-tailed values $z_{.05} = 1.65$ and $z_{.01} = 2.33$.

$$z = \frac{8 - \dfrac{(5)(6)}{2}}{\sqrt{\dfrac{(5)(6)(5 + 6 + 1)}{12}}} = -1.28$$

Thus, the data do not indicate the two teaching methods represent populations with different variances. Of course, as is the case when the **Mann–Whitney U test** is employed with the same set of data, it is entirely possible that a difference does exist in the underlying populations but is not detected because of the small sample size employed in the study.

VII. Additional Discussion of the Siegel–Tukey Test for Equal Variability

1. Analysis of the homogeneity of variance hypothesis for the same set of data with both a parametric and nonparametric test, and the power-efficiency of the Siegel–Tukey test for equal variability As noted in Section I, the use of the **Siegel–Tukey test for equal variability** would most likely be based on the fact that a researcher has reason to believe that the data in the underlying populations are not normally distributed. If, however, in the case of Example 14.1 the normality assumption is not an issue, or if it is but in spite of it a researcher prefers to use a parametric procedure such as **Hartley's F_{max} test for homogeneity of variance/F test for two**

population variances, she can still reject the null hypothesis H_0: $\sigma_1^2 = \sigma_2^2$ at both the .05 or .01 levels, regardless of whether a nondirectional or directional alternative hypothesis is employed. This is demonstrated by employing Equation 11.6 (the equation for **Hartley's F_{max} test for homogeneity of variance**) with the data for Example 14.1.

$$\Sigma X_1 = 30 \quad \Sigma X_1^2 = 282 \quad \Sigma X_2 = 30 \quad \Sigma X_2^2 = 154$$

$$\hat{s}_1^2 = \frac{282 - \frac{(30)^2}{6}}{6 - 1} = 26.4 \quad \hat{s}_2^2 = \frac{154 - \frac{(30)^2}{6}}{6 - 1} = .8$$

$$F_{max} = \frac{\hat{s}_L^2}{\hat{s}_S^2} = \frac{26.4}{.8} = 33$$

Table A9 (Table of the F_{max} Distribution) in the **Appendix** is employed to evaluate the computed value $F_{max} = 33$. For $k = 2$ groups and $(n - 1) = (6 - 1) = 5$ (since $n_1 = n_2 = n = 6$), the appropriate tabled critical values for a nondirectional analysis are $F_{max_{.05}} = 7.15$ and $F_{max_{.01}} = 14.9$. Since the obtained value $F_{max} = 33$ is greater than both of the aforementioned tabled critical values, the nondirectional alternative hypothesis H_1: $\sigma_1^2 \neq \sigma_2^2$ is supported at both the .05 and .01 levels.[4]

In the case of a directional analysis, the appropriate tabled critical one-tailed .05 and .01 values must be obtained from **Table A10 (Table of the F Distribution)** in the **Appendix**. In **Table A10**, the values for $F_{.95}$ and $F_{.99}$ for $df_{num} = n_1 - 1 = 6 - 1 = 5$ and $df_{den} = n_2 - 1 = 6 - 1 = 5$ are employed. The appropriate values derived from **Table A10** are $F_{.95} = 5.05$ and $F_{.99} = 10.97$. Since the obtained value $F_{max} = 33$ is greater than both of the aforementioned tabled critical values, the directional alternative hypothesis H_1: $\sigma_1^2 > \sigma_2^2$ is supported at both the .05 and .01 levels.

Note that the difference between the computed F_{max} (or F) value and the appropriate tabled critical value is more pronounced in the case of **Hartley's F_{max} test for homogeneity of variance/F test for two population variances** than the difference between the computed test statistic and the appropriate tabled critical value for the **Siegel–Tukey test for equal variability** (when either the exact U distribution or the normal approximation is employed). The actual probability associated with the outcome of the analysis is, in fact, less than .01 for both the F_{max} and **Siegel–Tukey tests,** but is even further removed from .01 in the case of the F_{max} test. This latter observation is consistent with the fact that when both a parametric and nonparametric test are applied to the same set of data, the former test will generally provide a more powerful test of an alternative hypothesis.[5]

The above outcome is consistent with the fact that various sources (e.g., Marascuilo and McSweeney (1977) and Siegel and Castellan (1988)) note that the **asymptotic relative efficiency** (discussed in Section VII of the **Wilcoxon signed-ranks test (Test 6)**) of the **Siegel–Tukey** **test** relative to the F_{max} test is only .61. However, the asymptotic relative efficiency of the **Siegel–Tukey test** may be considerably higher when the underlying population distributions are not normal.

2. Alternative nonparametric tests of dispersion In Section I it is noted that the **Siegel–Tukey test for equal variability** is one of a number of nonparametric tests for ordinal data which have been developed for evaluating the hypothesis that two populations have equal variances. The determination with respect to which of these tests to employ is generally based on the specific assumptions a researcher is willing to make about the underlying population distributions. Other factors that can determine which test a researcher elects to employ are the relative power efficiencies of the tests under consideration, and the complexity of the computations required for a specific test. This section will briefly summarize a few of the alternative procedures which evaluate the same hypothesis as the **Siegel–Tukey test for equal variability**. One or more of these procedures are described in detail in various books which specialize in nonparametric statistics (e.g., Conover (1980, 1999), Daniel (1990), Hollander and Wolfe (1999), Marascuilo and McSweeney (1977), Siegel and Castellan (1988), and Sprent (1993)). In addition, Sheskin (1984) provides a general overview and bibliography of nonparametric tests of dispersion.

The **Ansari–Bradley test** (Ansari and Bradley (1960) and Freund and Ansari (1957)) evaluates the same hypothesis as the **Siegel–Tukey test for equal variability**, as well as sharing its assumptions. The **Moses test for equal variability (Test 15)** (Moses, 1963), which is described in the next chapter, can also be employed to evaluate the same hypothesis. However, the **Moses test** is more computationally involved than the two aforementioned tests. Unlike the **Siegel–Tukey test** and **Ansari–Bradley test**, the **Moses test** assumes that the data evaluated represent at least interval level measurement. In addition, the **Moses test** does not assume that the two populations have equal medians. Among other nonparametric tests of dispersion are procedures developed by Conover (Conover and Iman (1978), Conover (1980, 1999)), Klotz (1962), and Mood (1954). Of the tests just noted, the **Siegel–Tukey test for equal variability**, the **Klotz test**, and the **Mood test** can be extended to designs involving more than two independent samples. In addition to all of the aforementioned procedures, **tests of extreme reactions** developed by Moses (1952) (the **Moses test of extreme reactions** is described in Siegel (1956)) and Hollander (1963) can be employed to contrast the variability of two independent groups. Since there is extensive literature on nonparametric tests of dispersion, the interested reader should consult sources which specialize in nonparametric statistics for a more comprehensive discussion of the subject.

VIII. Additional Examples Illustrating the Use of the Siegel–Tukey Test for Equal Variability

The **Siegel–Tukey test for equal variability** can be employed to evaluate the null hypothesis H_0: $\sigma_1^2 = \sigma_2^2$ with any of the examples noted for the *t* **test for two independent samples (Test 11)** and the **Mann–Whitney *U* test**. In order to employ the **Siegel–Tukey test for equal variability** with any of the aforementioned examples, the data must be rank-ordered employing the protocol described in Section IV. Example 14.3 is an additional example which can be evaluated with the **Siegel–Tukey test for equal variability**. It is characterized by the fact that (unlike Examples 14.1 and 14.2) in Example 14.3 subjects are rank-ordered without initially obtaining scores which represent interval/ratio level measurement. Although it is implied that the ranks in Example 14.3 are based on an underlying interval/ratio scale, the data are never expressed in such a format.

Example 14.3 *A company determines there is no difference with respect to enthusiasm for a specific product after people are exposed to a monochromatic versus a polychromatic advertisement for the product. The company, however, wants to determine whether or not different degrees of variability are associated with the two types of advertisement. To answer the question, a study is conducted employing twelve subjects who as a result of having no knowledge of the product are neutral towards it. Six of the subjects are exposed to a monochromatic* advertisement for the product (Group 1), *and the other six are exposed to a polychromatic version of the same advertisement* (Group 2). *One week later each subject is interviewed by a market researcher who is blind with respect to which advertisement a subject was exposed. Upon interviewing all 12 subjects, the market researcher rank-orders them with respect to their level of enthusiasm for the product. The rank-orders of the subjects in the two groups follow (assume that the lower the rank-order, the lower the level of enthusiasm for the product):*

Group 1:   **Subject 1,1**: 12; **Subject 2,1**: 2; **Subject 3,1**: 4; **Subject 4,1**: 6;
           Subject 5,1: 3; **Subject 6,1**: 10
Group 2:   **Subject 1,2**: 7; **Subject 2,2**: 5; **Subject 3,2**: 9; **Subject 4,2**: 8;
           Subject 5,2: 11; **Subject 6,2**: 1

Is there a significant difference in the degree of variability within each of the groups?

Employing the ranking protocol for the **Siegel–Tukey test for equal variability** with the above data, the ranks of the two groups are converted into the following new set of ranks (i.e., assigning a rank of 1 to the lowest rank, a rank of 2 to the highest rank, a rank of 3 to the second highest rank, etc.).

Group 1:   **Subject 1,1**: 2; **Subject 2,1**: 4; **Subject 3,1**: 8; **Subject 4,1**: 12;
           Subject 5,1: 5; **Subject 6,1**: 6
Group 2:   **Subject 1,2**: 11; **Subject 2,2**: 9; **Subject 3,2**: 7; **Subject 4,2**: 10;
           Subject 5,2: 3; **Subject 6,2**: 1

Employing the above set of ranks, $\Sigma R_1 = 37$ and $\Sigma R_2 = 41$. Through use of Equations 12.1 and 12.2, the values of U_1 and U_2 are computed to be $U_1 = (6)(6) + [[6(6 + 1)]/2] - 37 = 20$ and $U_2 = (6)(6) + [[6(6 + 1)]/2 - 41] = 16$. Since $U_2 = 16$ is less than $U_1 = 20$, $U = 16$. In **Table A11**, for $n_1 = 6$ and $n_2 = 6$, the tabled critical two-tailed .05 and 01 values are $U_{.05} = 5$ and $U_{.01} = 2$, and the tabled critical one-tailed .05 and .01 values are $U_{.05} = 7$ and $U_{.01} = 3$. Since $U = 16$ is greater than all of the aforementioned critical values, the null hypothesis H_0: $\sigma_1^2 = \sigma_2^2$ cannot be rejected, regardless of whether a nondirectional or directional alternative hypothesis is employed. Thus, there is no evidence to indicate that the two types of advertisements result in different degrees of variability.

References

Ansari, A. R. & Bradley, R. A. (1960). Rank-sum tests for dispersions. **Annals of Mathematical Statistics**, 31, 1174–1189.

Conover, W. J. (1980). **Practical nonparametric statistics** (2nd. ed.). New York: John Wiley & Sons.

Conover, W. J. (1999). **Practical nonparametric statistics** (3rd. ed.). New York: John Wiley & Sons.

Conover, W. J. & Iman, R. L. (1978). Some exact tables for the squared ranks test. **Communication in Statistics: Simulation and Computation**, B7 (5), 491–513.

Daniel, W. W. (1990). **Applied nonparametric statistics** (2nd ed.). Boston: PWS–Kent Publishing Company.

Freund, J. E. & Ansari, A. R. (1957). Two-way rank-sum test for variances. Technical Report Number 34, Virginia Polytechnic Institute and State University, Blacksburg, VA.

Hollander, M. (1963). A nonparametric test for the two-sample problem. **Psychometrika**, 28, 395–403.

Hollander, M. & Wolfe, D. A. (1999). **Nonparametric statistical methods**. New York: John Wiley & Sons.

Klotz, J. (1962). Nonparametric tests for scale. **Annals of Mathematical Statistics**, 33, 498–512.

Marascuilo, L. A. & McSweeney, M. (1977). **Nonparametric and distribution-free methods for the social sciences**. Monterey, CA: Brooks/Cole Publishing Company.

Mood, A. M. (1954). On the asymptotic efficiency of certain nonparametric two-sample tests. **Annals of Mathematical Statistics**, 25, 514–522.

Moses, L. E. (1952). A two-sample test. **Psychometrika**, 17, 234–247.

Moses, L. E. (1963). Rank tests of dispersion. **Annals of Mathematical Statistics**, 34, 973–983.

Sheskin, D. J. (1984). **Statistical tests and experimental design**. New York: Gardner Press.

Siegel, S. (1956). **Nonparametric statistics**. New York: McGraw Hill Book Company.

Siegel, S. & Castellan, N. J., Jr. (1988). **Nonparametric statistics for the behavioral sciences** (2nd ed.). New York: McGraw–Hill Book Company.

Siegel, S. & Tukey, J. W. (1960). A nonparametric sum of ranks procedure for relative spread in unpaired samples. **Journal of the American Statistical Association**, 55, 429–445.

Sprent, P. (1993). **Applied nonparametric statistical methods** (2nd ed.). London: Chapman & Hall.

Endnotes

1. a) As is the case with the **Mann–Whitney U test**, if the reverse ranking protocol is employed, the values of U_1 and U_2 are reversed (i.e. U_1 becomes U_2 and U_2 becomes U_1). Since, by virtue of the latter, the value of U, which represents the test statistic, is the lower of the two values U_1 versus U_2, the value designated U will be the same U value obtained with the original ranking protocol; b) If, on the other hand, the reverse ranking protocol is employed and the values of U_1 and U_2 are not reversed, the researcher must designate the larger of the two values U_1 versus U_2 as the value of U. The protocol for interpreting the U value under the latter circumstances is described in Endnote 2 of the **Mann–Whitney U test**.

2. As is the case with the **Mann–Whitney U test**, in describing the **Siegel–Tukey test for equal variability** some sources do not compute a U value, but rather provide tables which are based on the smaller and/or larger of the two sums of ranks. The equation for the normal approximation (to be discussed in Section VI) in these sources is also based on the sums of the ranks.

3. As previously noted, we can instead add three points to each score in Group 1.

4. If one employs Equation 11.7, and thus uses **Table A10**, the same tabled critical values are listed for $F_{.975}$ and $F_{.995}$ for $df_{num} = 6 - 1 = 5$ and $df_{den} = 6 - 1 = 5$. Thus, $F_{.975} = 7.15$ and $F_{.995} = 14.94$. (The latter value is only rounded off to one decimal place in **Table A9**.) The use of **Table A10** in evaluating homogeneity of variance is discussed in Section VI of the *t* test for two independent samples.

5. The **Levene test for homogeneity of variance (Test 21g)** and the **Brown–Forsythe test for homogeneity of variance (Test 21h)** (both of which are described in Section VI of the

single-factor between subjects analysis of variance) represent two parametric tests which provide for an even more powerful test of the alternative hypothesis specifying heterogeneity of variance than **Hartley's** F_{max} **test for homogeneity of variance**/F **test for two population variances**.

Test 15

The Moses Test for Equal Variability
(Nonparametric Test Employed with Ordinal Data)

I. Hypothesis Evaluated with Test and Relevant Background Information

Hypothesis evaluated with test Do two independent samples represent two populations with different variances?

Relevant background information on test Developed by Moses (1963), the **Moses test for equal variability** is a nonparametric procedure which can be employed in a hypothesis testing situation involving two independent samples. If the result of the **Moses test for equal variability** is significant, it indicates there is a significant difference between the sample variances, and as a result of the latter the researcher can conclude there is a high likelihood that the samples represent populations with different variances.

The **Moses test for equal variability** is one of a number of tests of **dispersion** (also referred to as tests of **scale** or **spread**) which have been developed for contrasting the variances of two independent samples. A discussion of alternative nonparametric tests of dispersion can be found in Section VII. Some sources recommend the use of nonparametric tests of dispersion for evaluating the homogeneity of variance hypothesis when there is reason to believe that the normality assumption of the appropriate parametric test for evaluating the same hypothesis is violated. Sources that are not favorably disposed toward nonparametric tests recommend the use of Hartley's F_{max} test for homogeneity of variance/F test for two population variances (Test 11a) or one of the alternative parametric tests described in Section VI of the single-factor between subjects analysis of variance (Test 21) which are available for evaluating homogeneity of variance, regardless of whether or not the normality assumption of the parametric test is violated. Such sources do, however, recommend that in employing a parametric test a researcher employ a lower significance level to compensate for the fact that violation of the normality assumption can inflate the Type I error rate associated with the test. When there is no evidence to indicate that the normality assumption of the parametric test has been violated, sources are in general agreement that such a test is preferable to the **Moses test for equal variability** (or an alternative nonparametric test of dispersion), since a parametric test (which uses more information than a nonparametric test) provides a more powerful test of an alternative hypothesis.

Since nonparametric tests are not assumption free, the choice of which of the available tests of dispersion to employ will primarily depend on what assumptions a researcher is willing to make with regard to the underlying distributions represented by the sample data. The **Moses test for equal variability** is based on the following assumptions: a) Each sample has been randomly selected from the population it represents; b) The two samples are independent of one another; c) The original scores obtained for each of the subjects are in the format of interval/ratio data, and the dependent variable is a **continuous** variable. (A **continuous variable** is characterized by the fact that a given score can assume any value within the range of values which define the limits of that variable.); and d) The underlying populations from which the samples are derived are similar in shape.

It is important to note that a major difference between the **Moses test for equal variability** and the previously discussed **Siegel–Tukey test for equal variability (Test 14)** is that the **Moses test** does not assume the two populations from which the samples are derived have equal medians (which is an assumption underlying the **Siegel-Tukey test**).

It should be noted that all of the other tests in this text which rank data (with the exception of the **Wilcoxon signed-ranks test (Test 6)** and the **Wilcoxon matched-pairs signed-ranks test (Test 18)**) rank the original interval/ratio scores of subjects. The **Moses test for equal variability**, however, does not rank the original interval/ratio scores, but instead ranks the sums of squared difference/deviation scores. For this reason, some sources (e.g., Siegel and Castellan (1988)) categorize the **Moses test for equal variability** as a test of interval/ratio data. In this book, however, the **Moses test for equal variability** is categorized as a test of ordinal data, by virtue of the fact that a ranking procedure constitutes a critical part of the test protocol.

II. Example

Example 15.1 is identical to Example 14.1 (which is evaluated with the **Siegel–Tukey test for equal variability**). Although Example 15.1 suggests that the underlying population medians are equal, as noted above, the latter is not an assumption of the **Moses test for equal variability**.

Example 15.1 *In order to assess the effect of two antidepressant drugs, 12 clinically depressed patients are randomly assigned to one of two groups. Six patients are assigned to Group 1, which is administered the antidepressant drug Elatrix for a period of six months. The other six patients are assigned to Group 2, which is administered the antidepressant drug Euphyria during the same six-month period. Assume that prior to introducing the experimental treatments, the experimenter confirmed that the level of depression in the two groups was equal. After six months elapse, all 12 subjects are rated by a psychiatrist (who is blind with respect to a subject's experimental condition) on their level of depression. The psychiatrist's depression ratings for the six subjects in each group follow (the higher the rating, the more depressed a subject):*
Group 1: 10, 10, 9, 1, 0, 0; **Group 2**: 6, 6, 5, 5, 4, 4.

The fact that the mean and median of each group are equivalent (specifically, both values equal 5) is consistent with prior research which suggests that there is no difference in efficacy for the two drugs (when the latter is based on a comparison of group means and/or medians). Inspection of the data does suggest, however, that there is much greater variability in the depression scores of subjects in Group 1. To be more specific, the data suggest that the drug Elatrix may, in fact, decrease depression in some subjects, yet increase it in others. The researcher decides to contrast the variability within the two groups through use of the Siegel–Tukey test for equal variability. The use of the latter nonparametric test is predicated on the fact that there is reason to believe that the distributions of the posttreatment depression scores in the underlying populations are not normal (which is why the researcher is reluctant to evaluate the data with a parametric test of dispersion. Do the data indicate there is a significant difference between the variances of the two groups?

III. Null versus Alternative Hypotheses

The test statistic for the **Moses test for equal variability** is computed with the **Mann–Whitney U test (Test 12)**. In order to understand the full text of the null and alternative hypotheses presented in this section, the reader will have to read the protocol involved in conducting the **Moses test for equal variability**, which is described in Section IV.

Null hypothesis $$H_0: \sigma_1^2 = \sigma_2^2$$

(The variance of the population Group 1 represents equals the variance of the population Group 2 represents. When, within the framework of the **Mann–Whitney** U **test** analysis to be described, both groups have an equal sample size, this translates into the sum of the ranks of the sums of the squared difference scores of Group 1 being equal to the sum of the ranks of the sums of the squared difference scores of Group 2. A more general way of stating this, which also encompasses designs involving unequal sample sizes, is that the means of the ranks of the sums of the squared difference scores of the two groups are equal.)

Alternative hypothesis $$H_1: \sigma_1^2 \neq \sigma_2^2$$

(The variance of the population Group 1 represents does not equal the variance of the population Group 2 represents. When, within the framework of the **Mann–Whitney** U **test** analysis to be described, both groups have an equal sample size, this translates into the sum of the ranks of the sums of the squared difference scores of Group 1 not being equal to the sum of the ranks of the sums of the squared difference scores of Group 2. A more general way of stating this, which also encompasses designs involving unequal sample sizes, is that the means of the ranks of the sums of the squared difference scores of the two groups are not equal. This is a **nondirectional alternative hypothesis** and it is evaluated with a **two-tailed test**.)

or

$$H_1: \sigma_1^2 > \sigma_2^2$$

(The variance of the population Group 1 represents is greater than the variance of the population Group 2 represents. When, within the framework of the **Mann–Whitney** U **test** analysis to be described, both groups have an equal sample size (so long as a rank of 1 is given to the lowest score), this translates into the sum of the ranks of the sums of the squared difference scores of Group 1 being greater than the sum of the ranks of the sums of the squared difference scores of Group 2. A more general way of stating this, which also encompasses designs involving unequal sample sizes, is that the mean of the ranks of the sums of the squared difference scores of Group 1 is greater than the mean of the ranks of the sums of the squared difference scores of Group 2. This is a **directional alternative hypothesis** and it is evaluated with a **one-tailed test**.)

or

$$H_1: \sigma_1^2 < \sigma_2^2$$

(The variance of the population Group 1 represents is less than the variance of the population Group 2 represents. When, within the framework of the **Mann–Whitney** U **test** analysis to be described, both groups have an equal sample size (so long as a rank of 1 is given to the lowest score), this translates into the sum of the ranks of the sums of the squared difference scores of Group 1 being less than the sum of the ranks of the sums of the squared difference scores of Group 2. A more general way of stating this, which also encompasses designs involving unequal sample sizes, is that the mean of the ranks of the sums of the squared difference scores of Group 1 is less than the mean of the ranks of the sums of the squared difference scores of Group 2. This is a **directional alternative hypothesis** and it is evaluated with a **one-tailed test**.)

Note: Only one of the above noted alternative hypotheses is employed. If the alternative hypothesis the researcher selects is supported, the null hypothesis is rejected.

IV. Test Computations

The protocol described below is employed for computing the test statistic for the **Moses test for equal variability**. In employing the protocol the following values are applicable: a) The total number of subjects employed in the experiment is $N = 12$; and b) There are $n_1 = 6$ subjects in Group 1 and $n_2 = 6$ subjects in Group 2.

　　a) The protocol for the **Moses test for equal variability** requires that the original interval/ratio scores be broken down into subsamples. A subsample is a set of scores derived from a sample, with the number of scores in a subsample being less than the total number of scores in the sample.

　　b) Divide the n_1 scores in Group 1 into m_1 subsamples (where $m_1 > 1$), with each subsample being comprised of k scores. Selection of the k scores for each of the m_1 subsamples should be random. **Sampling without replacement** (which is defined in Endnote 1 of the **binomial sign test for a single sample (Test 9)**) is employed in forming the subsamples. In other words, each of the n_1 scores in Group 1 is employed in only one of the m_1 subsamples.

　　c) Divide the n_2 scores in Group 2 into m_2 subsamples (where $m_2 > 1$), with each subsample being comprised of k scores. Selection of the k scores for each of the m_2 subsamples should be random. **Sampling without replacement** is employed in forming the subsamples. In other words, each of the n_2 scores in Group 2 is employed in only one of the m_2 subsamples.[1]

　　d) Note that regardless of which group a subsample is derived from, all subsamples will be comprised of the same number of scores (i.e., k scores). The number of subsamples derived from each group, however, need not be equal. In other words, the values of m_1 and m_2 do not have to be equivalent. The number of scores in each subsample should be such that the products $(m_1)(k)$ and $(m_2)(k)$ include as many of the scores as possible. Although the optimal situation would be if $(m_1)(k) = n_1$ and $(m_2)(k) = n_2$, it will often not be possible to achieve the latter (i.e., include all N scores in the $m_1 + m_2$ subsamples).

　　To illustrate the formation of subsamples, let us assume that $n_1 = 20$ and $n_2 = 20$. Employing the data from Group 1, we can form $m_1 = 4$ subsamples comprised of $k = 5$ scores per subsample. Thus, each of the $n_1 = 20$ scores in Group 1 will be included in one of the subsamples. Employing the data from Group 2, we can form $m_2 = 4$ subsamples comprised of $k = 5$ scores per subsample. Thus, each of the $n_2 = 20$ scores in Group 2 will be included in one of the subsamples.

　　Now let us assume there are 18 subjects in Group 1 and 20 subjects in Group 2 (i.e., $n_1 = 18$ and $n_2 = 20$). If we still employ $k = 5$ scores per subsample, we can only form $m_1 = 3$ subsamples (which include 15 of the 18 scores in Group 1). In such a case, three scores in Group 1 will have to be omitted from the analysis (which will employ $m_1 = 3$ subsamples comprised of $k = 5$ scores per subsample, and $m_2 = 4$ subsamples comprised of $k = 5$ scores per subsample). In order to include more subjects in the total analysis with $n_1 = 18$ and $n_2 = 20$, we can employ $k = 4$ scores per subsample, in which case we will have $m_1 = 4$ subsamples comprised of $k = 4$ scores per subsample, and $m_2 = 5$ subsamples comprised of $k = 4$ scores per subsample. In the latter case, only two scores in Group 1 will have to be omitted from the analysis. Obviously, if $n_1 = 18$ and $n_2 = 20$, we can only include all $N = 20$ subjects in the analysis, if we have $k = 2$ scores per subsample. In such a case we will have $m_1 = 9$ subsamples comprised of $k = 2$ scores per subsample, and $m_2 = 10$ subsamples comprised of $k = 2$ scores per subsample.

　　Daniel (1990) notes that Shorack (1969) recommends the following criteria in determining the values of k, m_1, and m_2: 1) k should be as large as possible, but not more than 10; and 2) The values of m_1 and m_2 should be large enough to derive meaningful results. The latter translates into employing values for m_1 and m_2 which meet the minimum sample size requirements for the **Mann–Whitney U test**, which is employed to compute the test statistic for the **Moses test for equal variability**. This latter point will be clarified in Section V.

d) Compute the mean of each of the m_1 subsamples derived from Group 1. Within each subsample do the following: 1) Subtract the mean of the subsample from each of the k scores in the subsample; 2) Square each of the difference scores; and 3) Obtain the sum of the k squared difference scores. The notation ΣD_{1i}^2 will represent the sum of the squared difference scores for the ith subsample in Group 1. There will be a total of m_1 ΣD_{1i}^2 scores for Group 1.

e) Compute the mean of each of the m_2 subsamples derived from Group 2. Within each subsample do the following: 1) Subtract the mean of the subsample from each of the k scores in the subsample; 2) Square each of the difference scores; and 3) Obtain the sum of the k squared difference scores. The notation ΣD_{2i}^2 will represent the sum of the squared difference scores for the ith subsample in Group 2. There will be a total of m_2 ΣD_{2i}^2 scores for Group 2.

f) The reader may want to review the protocol for the **Mann–Whitney U test** (in Section IV of the latter test) prior to continuing this section, since at this point in the analysis the **Mann–Whitney U test** is employed to compute the test statistic for the **Moses test for equal variability**. Specifically, within the framework of the **Mann–Whitney U test** model, each of the m_1 sums of the squared difference scores in Group 1 (i.e., the m_1 ΣD_{1i}^2 scores) is conceptualized as one of the n_1 scores in Group 1, and each of the m_2 sums of the squared difference scores in Group 2 (i.e., the m_2 ΣD_{2i}^2 scores) is conceptualized as one of the n_2 scores in Group 2.

If the null hypothesis is true, it is expected that the rank-orders for the sums of the squared difference scores in Groups 1 and 2 will be evenly dispersed and, consequently, the sum of the ranks for the Group 1 ΣD_{1i}^2 scores will be equal or close to the sum of the ranks for the Group 2 ΣD_{2i}^2 scores. If, on the other hand, there is greater variability in one of the groups, the rank-orderings of the sums of the squared difference scores for that group will be higher than the rank-orderings of the sums of the squared difference scores for the other group.

The protocol described in this section for the **Moses test for equal variability** will now be employed to evaluate Example 15.1. Table 15.1 summarizes the initial part of the analysis, which requires that subsamples be selected from each of the groups. As a result of the small sample size employed in the study, each subsample will be comprised of $k = 2$ scores, and thus there will be $m_1 = 3$ subsamples for Group 1 and $m_2 = 3$ subsamples for Group 2. Since $(m_1)(k) = (3)(2) = 6 = n_1$ and $(m_2)(k) = (3)(2) = 6 = n_2$, the scores of all $N = 12$ subjects are employed in the analysis. As noted earlier, the assignment of scores to subsamples is random. In this instance the author used a table of random numbers to select the scores for each of the subsamples.[2]

The information in Table 15.1 can be summarized as follows: a) Column 1 lists the two scores in each subsample. Each row contains a separate subsample; b) Column 2 lists the mean ($\bar{X}$) of each subsample; c) Column 3 lists the difference scores ($(X - \bar{X})$) obtained when the mean of a subsample is subtracted from each score in the subsample; d) Column 4 lists the squared difference scores ($(X - \bar{X})^2$) for each subsample (which are the squares of the scores in Column 3); and e) Column 5 lists the sum of the squared difference scores ($\Sigma(X - \bar{X})^2 = \Sigma D_{ji}^2$) for each of the subsamples (i.e., within each row/subsample, the values in Column 5 are the sum of the values in Column 4). Note that the notation ΣD_{ji}^2 represents the sum of the squared difference scores of the ith subsample in Group j. The values in Column 5 are evaluated in Table 15.2 with the **Mann–Whitney U test**.

Table 15.1 Summary of Analysis of Example 15.1

	Group 1			
Subsample	$\bar{X}_i$	$(X - \bar{X}_i)$	$(X - \bar{X}_i)^2$	$\Sigma(X - \bar{X}_i)^2 = \Sigma D_{1i}^2$
1) 1, 10	5.5	−4.5, 4.5	20.25, 20.25	40.5
2) 10, 0	5	5, −5	25, 25	50
3) 9, 0	4.5	4.5, −4.5	20.25, 20.25	40.5
	Group 2			
Subsample	$\bar{X}_i$	$(X - \bar{X}_i)$	$(X - \bar{X}_i)^2$	$\Sigma(X - \bar{X}_i)^2 = \Sigma D_{2i}^2$
1) 4, 4	4	0, 0	0, 0	0
2) 5, 6	5.5	−.5, .5	.25, .25	.5
3) 5, 6	5.5	−.5, .5	.25, .25	.5

15.2 Analysis of Example 15.1 with Mann–Whitney U Test

Group 1		**Group 2**	
ΣD_{1i}^2	Rank	ΣD_{2i}^2	Rank
40.5	4.5	0	1
50	6	.5	2.5
40.5	4.5	.5	2.5
	$\Sigma R_1 = 15$		$\Sigma R_2 = 6$

$$U_1 = n_1 n_2 + \frac{n_1(n_1 + 1)}{2} - \Sigma R_1 = (3)(3) + \frac{3(3 + 1)}{2} - 15 = 0$$

$$U_2 = n_1 n_2 + \frac{n_2(n_2 + 1)}{2} - \Sigma R_2 = (3)(3) + \frac{3(3 + 1)}{2} - 6 = 9$$

V. Interpretation of the Test Results

The smaller of the two values U_1 versus U_2 is designated as the obtained U statistic. Since $U_1 = 0$ is smaller than $U_2 = 9$, the value of $U = 0$. The value of U is evaluated with **Table A11 (Table of Critical Values for the Mann–Whitney U Statistic)** in the **Appendix**. In the case of Example 15.1, there are three scores in each group (which are the sums of the squared difference scores for the three subsamples which comprise that group). Thus, $n_1 = 3$ and $n_2 = 3$. Because of the small sample size, **Table A11** does not list critical two-tailed .05 and .01 values, nor does it list a critical one-tailed .01 value. It does, however, list the critical one-tailed .05 value $U_{.05} = 0$. In order to be significant, the obtained value of U must be **equal to or less than** the tabled critical value at the prespecified level of significance. Since $U = 0$ is equal to $U_{.05} = 0$, the directional alternative hypothesis that is consistent with the data is supported at the .05 level. The latter alternative hypothesis is H_1: $\sigma_1^2 > \sigma_2^2$.

In Section III it is noted that when both groups have an equal sample size, if the directional alternative H_1: $\sigma_1^2 > \sigma_2^2$ is supported, the sum (as well as average) of the ranks of the sums of the squared difference scores of Group 1 will be greater than the sum (as well as average) of the ranks of the sums of the squared difference scores of Group 2 (i.e., $\Sigma R_1 > \Sigma R_2$ and $\bar{R}_1 > \bar{R}_2$). Since the latter is the case in Example 15.1 ($\Sigma R_1 = 15 > \Sigma R_2 = 6$ and $\bar{R}_1 = 5 > \bar{R}_2 = 2$), the directional alternative hypothesis H_1: $\sigma_1^2 > \sigma_2^2$ is supported at the .05 level.

For the directional alternative hypothesis H_1: $\sigma_1^2 < \sigma_2^2$ to be supported, the sum (as well as average) of the ranks of the sums of the squared difference scores of Group 2 must be greater than the sum (as well as average) of the ranks of the sums of the squared difference scores of Group 1 (i.e., $\Sigma R_1 < \Sigma R_2$ and $\bar{R}_1 < \bar{R}_2$). Since, as noted above, the opposite is true, the directional alternative hypothesis H_1: $\sigma_1^2 < \sigma_2^2$ is not supported.

For the nondirectional alternative hypothesis H_1: $\sigma_1^2 \neq \sigma_2^2$ to be supported, the sum (as well as average) of the ranks of the sums of the squared difference scores of Group 1 must not be equal to the sum (as well as average) of the ranks of the sums of the squared difference scores of Group 2 (i.e., $\Sigma R_1 \neq \Sigma R_2$ and $\bar{R}_1 \neq \bar{R}_2$). In point of fact, the latter is true, and the computed value $U = 0$ is the smallest possible U value which can be obtained. However, as noted earlier, because of the small sample size, no two-tailed .05 and .01 critical values are listed in **Table A11** for $n_1 = 3$ and $n_2 = 3$.

Based on the results of the **Moses test for equal variability**, the researcher can conclude there is greater variability in the depression scores of the group which receives the drug Elatrix (Group 1) than the group that receives the drug Euphyria (Group 2).

When the same data are evaluated with the **Siegel–Tukey test for equal variability**, as well as with **Hartley's F_{max} test for homogeneity of variance/F test for two population variances**, both the nondirectional alternative hypothesis H_1: $\sigma_1^2 \neq \sigma_2^2$ and the directional alternative hypothesis H_1: $\sigma_1^2 > \sigma_2^2$ are supported at both the .05 and .01 levels.[3] The fact that the latter two alternative procedures for evaluating the null hypothesis H_0: $\sigma_1^2 = \sigma_2^2$ yield a more significant result than the **Moses test** is consistent with the fact that of the three procedures, the one with the lowest statistical power is the **Moses test** (the issue of the power of the **Moses test for equal variability** is discussed in greater detail in Section VII).

VI. Additional Analytical Procedures for the Moses Test for Equal Variability and/or Related Tests

1. The normal approximation of the Moses test statistic for large sample sizes Although the sample size is too small to employ the large sample normal approximation of the **Mann–Whitney U test** statistic, for demonstration purposes the latter value is computed below with Equation 12.4. As is noted in Section VI of the **Mann–Whitney U test**, the large sample normal approximation is generally used for sample sizes larger than those documented in the exact table for the **Mann–Whitney U test** contained within the source one is employing. Employing Equation 12.4, the absolute value $z = 1.96$ is computed.

The obtained absolute value $z = 1.96$ is evaluated with **Table A1 (Table of the Normal Distribution)** in the **Appendix**. In order to be significant, the obtained absolute value of z must be equal to or greater than the tabled critical value at the prespecified level of significance. The tabled critical two-tailed .05 and .01 values are $z_{.05} = 1.96$ and $z_{.01} = 2.58$, and the tabled critical one-tailed .05 and .01 values are $z_{.05} = 1.65$ and $z_{.01} = 2.33$.

$$z = \frac{U - \dfrac{n_1 n_2}{2}}{\sqrt{\dfrac{n_1 n_2 (n_1 + n_2 + 1)}{12}}} = \frac{0 - \dfrac{(3)(3)}{2}}{\sqrt{\dfrac{(3)(3)(3 + 3 + 1)}{12}}} = -1.96$$

The following guidelines are employed in evaluating the null hypothesis.

a) If the nondirectional alternative hypothesis H_1: $\sigma_1^2 \neq \sigma_2^2$ is employed, the null hypothesis can be rejected if the obtained absolute value of z is equal to or greater than the tabled critical two-tailed value at the prespecified level of significance.

   b) If a directional alternative hypothesis is employed, one of the two possible directional alternative hypotheses is supported if the obtained absolute value of z is equal to or greater than the tabled critical one-tailed value at the prespecified level of significance. The directional alternative hypothesis which is supported is the one that is consistent with the data.

   Since the computed absolute value $z = 1.96$ is equal to $z_{.05} = 1.96$ but less than $z_{.01} = 2.58$, the nondirectional alternative hypothesis H_1: $\sigma_1^2 \neq \sigma_2^2$ is supported at the .05 level, but not at the .01 level. Since the computed absolute value $z = 1.96$ is greater than $z_{.05} = 1.65$ but less than $z_{.01} = 2.33$, the directional alternative hypothesis H_1: $\sigma_1^2 > \sigma_2^2$ is supported at the .05 level, but not at the .01 level.

   The continuity-corrected Equation 12.5 yields the slightly lower absolute value $z = 1.75$. Since the computed absolute value $z = 1.75$ is greater than $z_{.05} = 1.65$ but less than $z_{.01} = 2.33$, the directional alternative hypothesis H_1: $\sigma_1^2 > \sigma_2^2$ is supported, but only at the .05 level. Since the computed absolute value $z = 1.75$ is less than $z_{.05} = 1.96$ and $z_{.01} = 2.58$, the nondirectional alternative hypothesis H_1: $\sigma_1^2 \neq \sigma_2^2$ is not supported. Thus, the result obtained with the continuity-corrected equation is exactly the same as the result obtained when the values in **Table A11** are employed.[4]

$$z = \frac{\left| U - \dfrac{n_1 n_2}{2} \right| - .5}{\sqrt{\dfrac{n_1 n_2 (n_1 + n_2 + 1)}{12}}} = \frac{\left| 0 - \dfrac{(3)(3)}{2} \right| - .5}{\sqrt{\dfrac{(3)(3)(3 + 3 + 1)}{12}}} = -1.75$$

VII. Additional Discussion of the Moses Test for Equal Variability

1. Power-efficiency of the Moses test for equal variability   Daniel (1990) and Siegel and Castellan (1988) note the power-efficiency of the **Moses test for equal variability** relative to a parametric procedure such as **Hartley's F_{max} test for homogeneity of variance/F test for two population variances** is a function of the size of the subsamples. With small subsamples the **asymptotic relative efficiency** (which is discussed in Section VII of the **Wilcoxon signed-ranks test**) of the test is relatively low if the underlying population distributions are normal (e.g., the power efficiency is .50 when $k = 3$). Although as the value of k increases, the power efficiency approaches an upper limit of .95, the downside of employing a large number of scores in each subsample is that as k increases, the number of subsamples which are available for analysis will decrease (and the latter will compromise the power of the **Mann–Whitney** analysis employed to compute the **Moses test** statistic). Thus, in deciding whether or not to employ the **Moses test for equal variability**, the researcher must weigh the test's relatively low power efficiency against the following factors: a) The extreme sensitivity of **Hartley's F_{max} test for homogeneity of variance/F test for two population variances** (as well as other parametric tests for evaluating homogeneity of variance discussed in Section VI of the **single-factor between-subjects analysis of variance**) to violations of the assumption of normality in the underlying populations (which is not an assumption of the **Moses test**); and b) The assumption of equal population medians associated with the **Siegel–Tukey test for equal variability**. Since the power of a statistical test is directly related to sample size, for the same set of data, the

Siegel–Tukey test for equal variability will have higher power than the **Moses test for equal variability**. The latter is true, since the sample size for the test statistic with the **Siegel–Tukey test** will always be the values n_1 and n_2, while with the **Moses test** the sample size will be m_1 and m_2 (the values of which will always be less than n_1 and n_2).

2. Issue of repetitive resampling An obvious problem associated with the **Moses test for equal variability** is that its result is dependent on the configuration of the data in each of the random subsamples employed in the analysis. It is entirely possible that an analysis based on one set of subsamples may yield a different result than an analysis based on a different set of subsamples. Because of the latter, if a researcher has a bias in favor of obtaining a significant (or perhaps nonsignificant) result, she can continue to select subsamples until she obtains a set of subsamples which yields the desired result. Obviously, the latter protocol is inappropriate and would compromise the integrity of one's results. Put simply, the **Moses test for equal variability** should be run one time, with the researcher accepting the resulting outcome. If the researcher has reason to believe the outcome does not reflect the truth regarding the populations in question, a replication study should be conducted.

3. Alternative nonparametric tests of dispersion In Section I it is noted that the **Moses test for equal variability** is one of a number of nonparametric tests which have been developed for evaluating the hypothesis that two populations have equal variances. The determination with respect to which of these tests to employ is generally based on the specific assumptions a researcher is willing to make about the underlying population distributions. Other factors that can determine which test a researcher elects to employ are the relative power efficiencies of the tests under consideration, and the complexity of the computations required for a specific test. As noted in Section VI of the **Siegel–Tukey test for equal variability**, among the other procedures that are available for evaluating a hypothesis about equality of population variances are the **Ansari–Bradley test** (Ansari and Bradley (1960) and Freund and Ansari (1957)), and nonparametric tests of dispersion developed by Conover (Conover and Iman (1978), Conover (1980, 1999)), Hollander (1963), Klotz (1962), Mood (1954), and Moses (1952). Of the aforementioned tests, the **Siegel–Tukey test for equal variability**, the **Klotz test**, and the **Mood test** can be extended to designs involving more than two independent samples.

Since there is extensive literature on nonparametric tests of dispersion, the interested reader should consult sources which specialize in nonparametric statistics for a more comprehensive discussion of the subject. One or more of these procedures are described in detail in various books that specialize in nonparametric statistics (e.g., Conover (1980, 1999), Daniel (1990), Hollander and Wolfe (1999), Marascuilo and McSweeney (1977), Siegel and Castellan (1988), and Sprent (1993)). In addition, Sheskin (1984) provides a general overview and bibliography of nonparametric tests of dispersion.

VIII. Additional Examples Illustrating the Use of the Moses Test for Equal Variability

In the discussion of the **Siegel–Tukey test for equal variability**, Example 14.2 was employed to illustrate an example in which a researcher is not able to assume that the two population medians are equal. Although the latter is a situation where the **Moses test for equal variability** would be more appropriate to employ than the **Siegel–Tukey test** (since the **Moses test** does not assume equal population medians), because of the small sample size ($n_1 = 5$ and $n_2 = 6$) the **Moses test** cannot be employed. In the case of Example 14.2, the only way to obtain more than one subsample per group is to set $k = 2$, and have $m_1 = 2$ subsamples in Group 1 (for which the five scores are 7, 5, 4, 4, 3), and m_2 = 3 subsamples in Group 2 (for which the six scores are 13, 12, 7, 7, 4, 3). However, since there are no critical values listed in the **Mann–Whitney** table (i.e., **Table A11**) for n_1 = 2 and n_2 = 3 (which are the number of subsamples/sums of squared difference scores in each group), the probability level associated with the result of the **Mann–Whitney** U test will always be above .05, regardless of whether a one-tailed or two-tailed analysis is employed.

Example 15.2 is an additional problem that will be evaluated with the **Moses test for equal variability**.

Example 15.2 *A researcher wants to determine whether or not a group of subjects who are given a low dose of a stimulant drug exhibit more variability with respect to the number of errors they make on a test of eye-hand coordination than a group of subjects who are given a placebo. There are* n_1 = 12 *subjects in the group administered the drug and* n_2 = 17 *subjects in the placebo group. The scores of the N = 29 subjects are listed below.*

> **Group 1:** 8, 5, 4, 3, 2, 9, 6, 1, 14, 18, 8, 8
> **Group 2:** 7, 7, 7, 8, 9, 7, 8, 9, 8, 8, 7, 10, 11, 12, 7, 9, 5

Is there a significant difference between the degree of variability within each of the groups?

In evaluating the data with the **Moses test for equal variability** we will employ $k = 3$ subjects per subsample, which means that we will derive m_1 = 4 subsamples for Group 1 (since n_1 = 12 divided by 3 equals 4), and m_2 = 5 subsamples for Group 1 (since n_2 = 17 divided by 3 equals 5, with a remainder of 2). Since n_2 = 17 is not evenly divisible by 3, two of the scores in Group 2 will not be included in the Group 2 subsamples (specifically, the score of one of the six subjects who obtained a 7 and the score of the subject who obtained a 5 were not selected for inclusion in the Group 2 subsamples during the random selection process). Tables 15.3 and 15.4 summarize the analysis. In Table 15.4, within the framework of the **Mann–Whitney** U test, the values n_1 =4 and n_2 = 5 are employed to represent the m_1 = 4 sums of squared difference scores for Group 1 and the m_2 = 5 sums of squared difference scores for Group 2.

Table 15.3   Summary of Analysis of Example 15.2

Group 1				
Subsample	$\bar{X}$	$(X - \bar{X})$	$(X - \bar{X})^2$	$\Sigma(X - \bar{X})^2 = \Sigma D_{1i}^2$
1) 5, 6, 4	5	0, 1, –1	0, 1, 1	2
2) 8, 18, 1	9	–1, 9, –8	1, 81, 64	146
3) 8, 14, 9	10.33	–2.33, 3.67, –1.33	5.43, 13.47, 1.77	20.67
4) 3, 8, 2	4.33	–1.33, 3.67, –2.33	1.77, 13.47, 5.43	20.67

Group 2				
Subsample	$\bar{X}$	$(X - \bar{X})$	$(X - \bar{X})^2$	$\Sigma(X - \bar{X})^2 = \Sigma D_{2i}^2$
1) 12, 7, 7	8.67	3.33, –1.67, –1.67	11.09, 2.79, 2.79	16.67
2) 9, 9, 8	8.67	.33, .33, –.67	.11, .11, .45	.67
3) 10, 11, 7	9.33	.67, 1.67, –2.33	.45, 2.79, 5.43	8.67
4) 8, 8, 7	7.67	.33, .33, –.67	.11, .11, .45	.67
5) 8, 7, 9	8	0, –1, 1	0, 1, 1	2

Table 15.4   Analysis of Example 15.2 with Mann–Whitney U Test

Group 1		Group 2	
ΣD_{1i}^2	Rank	ΣD_{2i}^2	Rank
2	3.5	16.67	6
146	9	.67	1.5
20.67	7.5	8.67	5
20.67	7.5	.67	1.5
		2	3.5
$\Sigma R_1 = 27.5$		$\Sigma R_2 = 17.5$	

$$U_1 = n_1 n_2 + \frac{n_1(n_1 + 1)}{2} - \Sigma R_1 = (4)(5) + \frac{4(4 + 1)}{2} - 27.5 = 2.5$$

$$U_2 = n_1 n_2 + \frac{n_2(n_2 + 1)}{2} - \Sigma R_2 = (4)(5) + \frac{5(5 + 1)}{2} - 17.5 = 17.5$$

Table 15.5   Analysis of Example 15.2 with Siegel–Tukey Test for Equal Variability

Group 1		Group 2	
ΣD_{1i}^2	Rank	ΣD_{2i}^2	Rank
8	24.86	7	20.5
5	10.5	7	20.5
4	8	7	20.5
3	5	8	24.86
2	4	9	14.5
9	14.5	7	20.5
6	13	8	24.86
1	1	9	14.5
14	3	8	24.86
18	2	8	24.86
8	24.86	7	20.5
8	24.86	10	10
		11	7
		12	6
		7	20.5
		9	14.5
		5	10.5
$\Sigma R_1 = 135.58$		$\Sigma R_2 = 299.44$	

$$U_1 = n_1 n_2 + \frac{n_1(n_1 + 1)}{2} - \Sigma R_1 = (12)(17) + \frac{12(12 + 1)}{2} - 135.58 = 146.42$$

$$U_2 = n_1 n_2 + \frac{n_2(n_2 + 1)}{2} - \Sigma R_2 = (12)(17) + \frac{17(17 + 1)}{2} - 299.44 = 57.56$$

Since $U_1 = 2.5$ is smaller than $U_2 = 17.5$, the value of $U = 2.5$. Employing **Table A11**, for the values $n_1 = 4$ and $n_2 = 5$, the tabled critical two-tailed .05 value is $U_{.05} = 1$. Because of the small sample size, no tabled critical two-tailed .01 value is listed. The tabled critical one-tailed values are $U_{.05} = 2$ and $U_{.01} = 0$. Since the computed value $U_1 = 2.5$ is larger than all of the tabled critical values, the null hypothesis cannot be rejected, regardless of which alternative hypothesis is employed. Thus, the researcher cannot conclude that there are differences in the variances of the two groups.

The data for Example 15.2 are consistent with the directional alternative hypothesis H_1: $\sigma_1^2 > \sigma_2^2$, since the average of the ranks for Group 1 ($\bar{R}_1 = 6.875$) is larger than the average of the ranks for Group 2 ($\bar{R}_2 = 3.5$). Support for the latter directional alternative hypothesis falls just short of being significant, since $U = 2.5$ is only .5 units above the tabled critical one-tailed value $U_{.05} = 2$.

It turns out that when the data for Example 15.2 are evaluated with the **Siegel–Tukey test for equal variability**, the computed value of the test statistic (through use of the **Mann–Whitney U test**) is $U = 57.56$.[5] Table 15.5 summarizes the analysis with the **Siegel–Tukey test**. In the case of the **Siegel–Tukey test**, the original sample size values $n_1 = 12$ and $n_2 = 17$ are employed in obtaining critical values from **Table A11**. For $n_1 = 12$ and $n_2 = 17$, the tabled critical two-tailed .05 and 01 values are $U_{.05} = 57$ and $U_{.01} = 44$, and the tabled critical one-tailed .05 and .01 values are $U_{.05} = 64$ and $U_{.01} = 49$. Since $U = 57.56$ is greater (albeit barely) than the tabled critical two-tailed value $U_{.05} = 57$, the nondirectional alternative hypothesis H_1: $\sigma_1^2 \neq \sigma_2^2$ is not supported. The directional alternative hypothesis H_1: $\sigma_1^2 > \sigma_2^2$ is supported, but only at the .05 level, since $U = 57.56$ is less than the tabled critical one-tailed value $U_{.05} = 64$. Thus, when the **Siegel–Tukey test for equal variability** is employed, the researcher can conclude there is a greater degree of variability in the scores of Group 1.

When the data for Example 15.2 are evaluated with **Hartley's F_{max} test for homogeneity of variance/F test for two population variances**, the nondirectional alternative hypothesis H_1: $\sigma_1^2 \neq \sigma_2^2$ and the directional alternative hypothesis H_1: $\sigma_1^2 > \sigma_2^2$ are supported at both the .05 and .01 levels. The computations for the latter test are shown below. Equation 11.6 is employed to compute the value $F_{max} = 8.39$.

$$\tilde{s}_1^2 = \frac{\Sigma X_1^2 - \dfrac{(\Sigma X_1)^2}{n_1}}{n_1 - 1} = \frac{884 - \dfrac{(86)^2}{12}}{12 - 1} = 24.33$$

$$\tilde{s}_2^2 = \frac{\Sigma X_2^2 - \dfrac{(\Sigma X_2)^2}{n_2}}{n_2 - 1} = \frac{1183 - \dfrac{(139)^2}{17}}{17 - 1} = 2.90$$

$$F_{max} = \frac{\tilde{s}_L^2}{\tilde{s}_S^2} = \frac{24.33}{2.90} = 8.39$$

The computed value $F_{max} = 8.39$ is evaluated with **Table A9 (Table of the F_{max} Distribution)** in the **Appendix**. The tabled critical values for the F_{max} distribution are listed in reference to the values $(n - 1)$ and k, where n represents the number of subjects per group, and k represents the number of groups. In the case of Example 15.2, the computed value $F_{max} = 8.39$ is larger than the tabled critical values in **Table A9** for $k = 2$ and $n = 12$. (With unequal sample sizes, for the most conservative test of the null hypothesis, we employ the smaller of the two sample size values $n_1 = 12$ and $n_2 = 17$.)[6] The tabled critical values for $k = 2$ and $n = 12$ are $F_{max_{.05}} = 3.28$ and $F_{max_{.01}} = 4.91$. Since the obtained value $F_{max} = 8.39$ is larger than both of the aforementioned critical values, we can reject the null hypothesis at both the .05 and .01 levels. Thus, the nondirectional alternative hypothesis H_1: $\sigma_1^2 \neq \sigma_2^2$ is supported. If the latter nondirectional alternative hypothesis is supported at both the .05 and.01 levels, the directional alternative hypothesis H_1: $\sigma_1^2 > \sigma_2^2$ will also be supported at both .05 and .01 levels, since the critical values for the latter alternative hypothesis will be lower than the critical two-tailed values noted above. Thus, when **Hartley's F_{max} test for homogeneity of variance/F test for two population variances** is employed, the researcher can conclude there is a higher degree of variability in the scores of Group 1. As noted earlier, the latter test will generally provide a more powerful test of an alternative hypothesis than either the **Moses test** (which does not yield a significant result for Example 15.2) or the **Siegel–Tukey test** (which does yield a significant result for Example 15.2, but only for a one-tailed alternative hypothesis).

References

Ansari, A. R. & Bradley, R. A. (1960). Rank-sum tests for dispersions. **Annals of Mathematical Statistics**, 31, 1174–1189.

Conover, W. J. (1980). **Practical nonparametric statistics** (2nd ed.). New York: John Wiley & Sons.

Conover, W. J. (1999). **Practical nonparametric statistics** (3rd ed.). New York: John Wiley & Sons.

Conover, W. J. & Iman, R. L. (1978). Some exact tables for the squared ranks test. **Communication in Statistics: Simulation and Computation**, B7 (5), 491–513.

Daniel, W. W. (1990). **Applied nonparametric statistics** (2nd ed.). Boston: PWS–Kent Publishing Company.

Freund, J. E. & Ansari, A. R. (1957). Two-way rank-sum test for variances. Technical Report Number 34, Virginia Polytechnic Institute and State University, Blacksburg, VA.

Hollander, M. (1963). A nonparametric test for the two-sample problem. **Psychometrika**, 28, 395–403.

Hollander, M. & Wolfe, D. A. (1999). **Nonparametric statistical methods**. New York: John Wiley & Sons.

Klotz, J. (1962). Nonparametric tests for scale. **Annals of Mathematical Statistics**, 33, 498–512.

Marascuilo, L. A. & McSweeney, M. (1977). **Nonparametric and distribution-free methods for the social sciences**. Monterey, CA: Brooks/Cole Publishing Company.

Mood, A. M. (1954). On the asymptotic efficiency of certain nonparametric two-sample tests. **Annals of Mathematical Statistics**, 25, 514–522.

Moses, L. E. (1952). A two-sample test. **Psychometrika**, 17, 234–247.

Moses, L. E. (1963). Rank tests of dispersion. **Annals of Mathematical Statistics**, 34, 973–983.

Sheskin, D. J. (1984). **Statistical tests and experimental design**. New York: Gardner Press.

Shorack, G. R. (1969). Testing and estimating ratios of scale parameters. **Journal of the American Statistical Association**, 64, 999–1013.

Siegel, S. & Castellan, N. J., Jr. (1988). **Nonparametric statistics for the behavioral sciences** (2nd ed.). New York: McGraw–Hill Book Company.

Siegel, S. & Tukey, J. W. (1960). A nonparametric sum of ranks procedure for relative spread in unpaired samples. **Journal of the American Statistical Association**, 55, 429–445.

Sprent, P. (1993). **Applied nonparametric statistical methods** (2nd ed.). London: Chapman & Hall.

Endnotes

1. One could argue that the use of random subsamples for the **Moses test for equal variability** allows one to conceptualize the test within the framework of the general

category of **resampling procedures**, which are discussed in Section IX (the **Addendum**) of the **Mann–Whitney *U* test**.

2. A typical table of random numbers is a computer-generated series of random digits which fall within the range 0 through 9. If there are six scores in a group, the sequential appearance of the digits 1 through 6 in the table can be used to form subsamples. For example, let us assume that the following string of digits appears in a random number table: 235223945567590091293737394940404. For Group 1, we will form three subsamples with two scores per subsample. Since the first digit which appears in the random number table is 2, the second score listed for the group will be the first score assigned to a Subsample 1. Since 3 is the next digit, the third score listed becomes the second score in Subsample 1. Since 5 is the next digit, the fifth score listed becomes the first score in the Subsample 2. We ignore the next four digits (2, 2, 3 and 9) since: a) We have already selected the second and third scores from the group; and b) The digit 9 indicates that we should select the ninth score. The latter score, however, does not exist, since there are only six scores in the group. Since the next digit is 4, the fourth score in the group becomes the second score in the Subsample 2. By default, the two scores which remain in the group (the first and sixth scores) will comprise Subsample 3. Continuing with the string of random numbers, the procedure is then repeated for the n_2 scores in Group 2, forming three more subsamples with each subsample being comprised of two scores.

3. The **Levene test for homogeneity of variance (Test 21g)** and the **Brown–Forsythe test for homogeneity of variance (Test 21h)** (both of which are described in Section VI of the **single-factor between subjects analysis of variance**) represent two parametric tests that provide for an even more powerful test of the alternative hypothesis specifying heterogeneity of variance than **Hartley's F_{max} test for homogeneity of variance/*F* test for two population variances**.

4. Equation 12.6 (the tie-corrected **Mann–Whitney** normal approximation equation) can be employed if ties are present in the data (i.e., there are one or more identical values for squared difference score sums).

5. For purposes of illustration we will assume that the medians of the populations the two groups represent are equal (which is an assumption of the **Siegel–Tukey test for equal variability**). In actuality, the sample medians computed for Groups 1 and 2 are respectively 7 and 8.

6. In other words, when $n_1 \neq n_2$, the smaller sample size is employed when using **Table A9** in order to minimize the likelihood of committing a Type I error.

Test 16

The Chi-Square Test for $r \times c$ Tables
(Nonparametric Test Employed with Categorical/Nominal Data)

I. Hypothesis Evaluated with Test and Relevant Background Information

Hypothesis evaluated with test In the underlying population(s) represented by the sample(s) in a contingency table, are the observed cell frequencies different from the expected frequencies?

Relevant background information on test The **chi-square test for $r \times c$ tables** is one of a number of tests described in this book for which the chi-square distribution is the appropriate sampling distribution.[1] The **chi-square test for $r \times c$ tables** is an extension of the **chi-square goodness-of-fit test (Test 8)** to two-dimensional tables. Whereas the latter test can only be employed with a single sample categorized on a single dimension (the single dimension is represented by the k cells/categories that comprise the frequency distribution table), the **chi-square test for $r \times c$ tables** can be employed to evaluate designs which summarize categorical data in the form of an $r \times c$ table (which is often referred to as a **contingency table**). An $r \times c$ table consists of r rows and c columns. Both the values of r and c are integer numbers that are equal to or greater than 2. The total number of cells in an $r \times c$ table is obtained by multiplying the value of r by the value of c. The data contained in each of the cells of a contingency table represent the number of observations (i.e., subjects or objects) which are categorized in the cell.

Table 16.1 presents the general model for an $r \times c$ contingency table. There are a total of n observations in the table. Note that each cell is identified with a subscript that consists of two elements. The first element identifies the row in which the cell falls and the second element identifies the column in which the cell falls. Thus, the notation O_{ij} represents the number of observations in the cell that is in the ith row and the jth column. $O_{i.}$ represents the number of observations in the ith row and $O_{.j}$ represents the number of observations in the jth column.

Table 16.1 General Model for an $r \times c$ Contingency Table

			Column variable					Row sums
		C_1	C_2	$\cdots$	C_j	$\cdots$	C_c	
	R_1	O_{11}	O_{12}	$\cdots$	O_{1j}	$\cdots$	O_{1c}	$O_{1.}$
	R_2	O_{21}	O_{22}	$\cdots$	O_{2j}	$\cdots$	O_{2c}	$O_{2.}$
	$\vdots$	$\vdots$	$\vdots$		$\vdots$		$\vdots$	$\vdots$
Row variable	R_i	O_{i1}	O_{i2}	$\cdots$	O_{ij}	$\cdots$	O_{ic}	$O_{i.}$
	$\vdots$	$\vdots$	$\vdots$		$\vdots$		$\vdots$	$\vdots$
	R_r	O_{r1}	O_{r2}	$\cdots$	O_{rj}	$\cdots$	O_{rc}	$O_{r.}$
Column sums		$O_{.1}$	$O_{.2}$	$\cdots$	$O_{.j}$	$\cdots$	$O_{.c}$	n

In actuality, there are two chi-square tests that can be conducted with an $r \times c$ table. The two tests which will be described are the **chi-square test for homogeneity (Test 16a)** and the **chi-square test of independence (Test 16b)**. The general label **chi-square test for $r \times c$ tables** will be employed to refer to both of the aforementioned tests, since the two tests are computationally identical. Although, in actuality, the **chi-square test for homogeneity** and the **chi-square test of independence** evaluate different hypotheses, a generic hypothesis can be stated that is applicable to both tests. A brief description of the two tests follows.

The chi-square test for homogeneity (Test 16a)   The **chi-square test for homogeneity** is employed when r independent samples (where $r \geq 2$) are categorized on a single dimension which consists of c categories (where $c \geq 2$). The data for the r independent samples (which are generally represented by the r rows of the contingency table) are recorded with reference to the number of observations in each of the samples that fall within each of c categories (which are generally represented by the c columns of the contingency table). It is assumed that each of the samples is randomly drawn from the underlying population it represents. The **chi-square test for homogeneity** evaluates whether or not the r samples are homogeneous with respect to the proportion of observations in each of the c categories. To be more specific, if the data are homogeneous, the proportion of observations in the jth category will be equal in all of the r populations. The **chi-square test for homogeneity** assumes that the sums of the r rows (which represent the number of observations in each of the r samples) are determined by the researcher prior to the data collection phase of a study. Example 16.1 in Section II is employed to illustrate the **chi-square test for homogeneity**.

The chi-square test of independence (Test 16b)   The **chi-square test of independence** is employed when a single sample is categorized on two dimensions/variables. It is assumed that the sample is randomly selected from the population it represents. One of the dimensions/variables is comprised of r categories (where $r \geq 2$) which are represented by the r rows of the contingency table, while the second dimension/variable is comprised of c categories (where $c \geq 2$) which are represented by the c columns of the contingency table. The **chi-square test of independence** evaluates the general hypothesis that the two variables are independent of one another. Another way of stating that two variables are independent of one another is to say that there is a zero correlation between them. A zero correlation indicates there is no way to predict at above chance in which category an observation will fall on one of the variables, if it is known which category the observation falls on the second variable. (For an overview of the concept of correlation, the reader should consult the **Introduction** and/or Section I of the **Pearson product-moment correlation coefficient (Test 28)**.) The **chi-square test of independence** assumes that neither the sums of the r rows (which represent the number of observations in each of the r categories for Variable 1) or the sums of the c columns (which represent the number of observations in each of the c categories for Variable 2) are predetermined by the researcher prior to the data collection phase of a study. Example 16.2 in Section II is employed to illustrate the **chi-square test of independence**.

The **chi-square test for $r \times c$ tables** (i.e., both the **chi-square test for homogeneity** and the **chi-square test of independence**) is based on the following assumptions: a) Categorical/nominal data (i.e., frequencies) for $r \times c$ mutually exclusive categories are employed in the analysis; b) The data that are evaluated represent a random sample comprised of n independent observations. This assumption reflects the fact that each subject or object can only be represented once in the data; and c) The expected frequency of each cell in the contingency table is 5 or greater. When the expected frequency of one or more cells is less than 5, the probabilities in the chi-square distribution may not provide an accurate estimate of the underlying sampling distribution. As is the case for the **chi-square goodness-of-fit test**, sources are not in agreement with respect to the minimum acceptable value for an expected frequency. Many sources employ criteria suggested by Cochran (1952), who stated that none of the expected frequencies should be less than 1, and that no more than 20% of the expected frequencies should be less than 5. However, many sources suggest the latter criteria may be overly conservative. In instances where a researcher believes that one or more expected cell frequencies are too small, two or more cells can be combined with one another to increase the values of the expected frequencies.

In actuality, the chi-square distribution only provides an approximation of the exact sampling distribution for a contingency table.[2] The accuracy of the chi-square approximation increases as the size of the sample increases and, except for instances involving small sample sizes, the chi-square distribution provides an excellent approximation of the exact sampling distribution. One case for which an exact probability is often computed is a 2 × 2 contingency table involving a small sample size. In the latter instance, an exact probability can be computed through use of the hypergeometric distribution. The computation of an exact probability for a 2 × 2 table using the hypergeometric distribution is described under the **Fisher exact test (Test 16c)** in Section VI.

II. Examples

Example 16.1 *A researcher conducts a study in order to evaluate the effect of noise on altruistic behavior. Each of the 200 subjects who participate in the experiment is randomly assigned to one of two experimental conditions. Subjects in both conditions are given a one-hour test which is ostensibly a measure of intelligence. During the test the 100 subjects in Group 1 are exposed to continual loud noise, which they are told is due to a malfunctioning generator. The 100 subjects in Group 2 are not exposed to any noise during the test. Upon completion of this stage of the experiment, each subject on leaving the room is confronted by a middle-aged man whose arm is in a sling. The man asks the subject if she would be willing to help him carry a heavy package to his car. In actuality, the man requesting help is an experimental confederate (i.e., working for the experimenter). The number of subjects in each group who help the man is recorded. Thirty of the 100 subjects who were exposed to noise elect to help the man, while 60 of the 100 subjects who were not exposed to noise elect to help the man. Do the data indicate that altruistic behavior is influenced by noise?*

The data for Example 16.1, which can be summarized in the form of a 2 × 2 contingency table, are presented in Table 16.2.

Table 16.2   Summary of Data for Example 16.1

	Helped the confederate	Did not help the confederate	Row sums
Noise	30	70	100
No noise	60	40	100
Column sums	90	110	Total observations 200

The appropriate test to employ for evaluating Example 16.1 is the **chi-square test for homogeneity**. This is the case, since the design of the study involves the use of categorical data (i.e., frequencies are represented in each of the $r \times c$ cells in the contingency table) with multiple independent samples (specifically two) that are categorized on a single dimension (altruism). To be more specific, the differential treatments to which the two groups are exposed (i.e., **noise** versus **no-noise**) constitute the independent variable. The latter variable is the row variable, since it is represented by the two rows in Table 16.2. Note that the researcher assigns 100 subjects to each of the two levels of the independent variable prior to the data collection phase of the study. This is consistent with the fact that when the **chi-square test for homogeneity** is employed, the sums for the row variable are predetermined prior to collecting the data. The dependent variable is whether or not a subject exhibits altruistic behavior. The latter variable is represented by the two categories **helped the confederate** versus **did not help the confederate**. The dependent variable is the column variable, since it is represented by the two columns in Table 16.2. The hypothesis which is evaluated with the **chi-square test for homogeneity** is whether there is a difference between the two groups with respect to the proportion of subjects who help the confederate.

Example 16.2 *A researcher wants to determine if there is a relationship between the personality dimension of introversion-extroversion and political affiliation. Two hundred people are recruited to participate in the study. All of the subjects are given a personality test on the basis of which each subject is classified as an introvert or an extrovert. Each subject is then asked to indicate whether he or she is a Democrat or a Republican. The data for Example 16.2, which can be summarized in the form of a 2 × 2 contingency table, are presented in Table 16.3. Do the data indicate there is a significant relationship between one's political affiliation and whether or not one is an introvert versus an extrovert?*

Table 16.3 Summary of Data for Example 16.2

	Democrat	Republican	Row sums
Introvert	30	70	100
Extrovert	60	40	100
Column sums	90	110	Total observations 200

The appropriate test to employ for evaluating Example 16.2 is the **chi-square test of independence**. This is the case since: a) The study involves a single sample which is categorized on two dimensions; and b) The data are comprised of frequencies for each of the $r \times c$ cells in the contingency table. To be more specific, a sample of 200 subjects is categorized on the following two dimensions, with each dimension being comprised of two mutually exclusive categories: a) **introvert** versus **extrovert**; and b) **Democrat** versus **Republican**. In Example 16.2 the **introvert–extrovert** dimension is the row variable and the **Democrat–Republican** dimension is the column variable.[3] Note that in selecting the sample of 200 subjects, the researcher does not determine beforehand the number of **introverts, extroverts, Democrats,** and **Republicans** to include in the study.[4] Thus, in Example 16.2 (consistent with the use of the **chi-square test of independence**) the sums of the rows and columns (which are referred to as the **marginal sums**) are not predetermined. The hypothesis which is evaluated with the **chi-square test of independence** is whether the two dimensions are independent of one another.

III. Null versus Alternative Hypotheses

Even though the hypotheses evaluated with the **chi-square test for homogeneity** and the **chi-square test of independence** are not identical, generic null and alternative hypotheses employing common symbolic notation can be used for both tests. The generic null and alternative hypotheses employ the observed and expected cell frequencies in the underlying population(s) represented by the sample(s). The observed and expected cell frequencies for the population(s) are represented respectively by the lower case Greek letters **omicron** (o) and **epsilon** (ε). Thus, o_{ij} and ε_{ij} respectively represent the observed and expected frequency of Cell$_{ij}$ in the underlying population.

Null hypothesis
$$H_0: o_{ij} = \varepsilon_{ij} \text{ for all cells}$$

(This notation indicates that in the underlying population(s) the sample(s) represent(s), for each of the $r \times c$ cells the observed frequency of a cell is equal to the expected frequency of the cell. With respect to the sample data, this translates into the observed frequency of each of the $r \times c$ cells being equal to the expected frequency of the cell.)

Alternative hypothesis
$$H_1: o_{ij} \neq \varepsilon_{ij} \text{ for at least one cell}$$

(This notation indicates that in the underlying population(s) the sample(s) represent(s), for at least one of the $r \times c$ cells the observed frequency of a cell is not equal to the expected frequency of the cell. With respect to the sample data, this translates into the observed frequency of at least one of the $r \times c$ cells not being equal to the expected frequency of the cell. This notation should not be interpreted as meaning that in order to reject the null hypothesis there must be a discrepancy between the observed and expected frequencies for all $r \times c$ cells. Rejection of the null hypothesis can be the result of a discrepancy between the observed and expected frequencies for one cell, two cells, ..., or all $r \times c$ cells.)

Although it is possible to employ a directional alternative hypothesis for the **chi-square test for $r \times c$ tables**, in the examples used to describe the test it will be assumed that the alternative hypothesis will always be stated nondirectionally. A discussion of the use of a directional alternative hypothesis can be found in Section VI.

The null and alternative hypotheses for each of the two tests which are described under the **chi-square test for $r \times c$ tables** can also be expressed within the framework of a different format. The alternative format for stating the null and alternative hypotheses employs the proportion of observations in the cells of the $r \times c$ contingency table. Before presenting the hypotheses in the latter format, the reader should take note of the following with respect to Tables 16.2 and 16.3. In both Tables 16.2 and 16.3 four cells can be identified: a) Cell$_{11}$ is the upper left cell in each table (i.e., in Row 1 and Column 1 the cell with the observed frequency $O_{11} = 30$). In the case of Example 16.1, $O_{11} = 30$ represents the number of subjects exposed to **noise** who **helped the confederate**. In the case of Example 16.2, $O_{11} = 30$ represents the number of **introverts** who are **Democrats**; b) Cell$_{12}$ is the upper right cell in each table (i.e., in Row 1 and Column 2 the cell with the observed frequency $O_{12} = 70$). In the case of Example 16.1, $O_{12} = 70$ represents the number of subjects exposed to **noise** who **did not help the confederate**. In the case of Example 16.2, $O_{12} = 70$ represents the number of **introverts** who are **Republicans**; c) Cell$_{21}$ is the lower left cell in each table (i.e., in Row 2 and Column 1 the cell with the observed frequency $O_{21} = 60$). In the case of Example 16.1, $O_{21} = 60$ represents the number of subjects exposed to **no noise** who **helped the confederate**. In the case of Example 16.2, $O_{21} = 60$ represents the number of **extroverts** who are **Democrats**; d) Cell$_{22}$ is the lower right cell in each table (i.e., in Row 2 and Column 2 the cell with the observed frequency $O_{22} = 40$). In the case of Example 16.1, $O_{22} = 40$ represents the number of subjects exposed to **no noise** who **did not help the confederate**. In the case of Example 16.2, $O_{22} = 40$ represents the number of **extroverts** who are **Republicans**.

Alternative way of stating the null and alternative hypotheses for the chi-square test for homogeneity If the independent variable (which represents the different groups) is employed as the row variable, the null and alternative hypotheses can be stated as follows:

H_0: In the underlying populations the samples represent, all of the proportions in the same column of the $r \times c$ table are equal.

H_1: In the underlying populations the samples represent, all of the proportions in the same column of the $r \times c$ table are not equal for at least one of the columns.

Viewing the above hypotheses in relation to the sample data in Table 16.2, the null hypothesis states that there are an equal proportion of observations in $Cell_{11}$ and $Cell_{21}$. With respect to the sample data, the proportion of observations in $Cell_{11}$ is the proportion of subjects who are exposed to **noise** who **helped the confederate** (which equals $O_{11}/O_{1.} = 30/100 = .3$). The proportion of observations in $Cell_{21}$ is the proportion of subjects who are exposed to **no noise** who **helped the confederate** (which equals $O_{21}/O_{2.} = 60/100 = .6$). The null hypothesis also requires an equal proportion of observations in $Cell_{12}$ and $Cell_{22}$. The proportion of observations in $Cell_{12}$ is the proportion of subjects who are exposed to **noise** who **did not help the confederate** (which equals $O_{12}/O_{1.} = 70/100 = .7$). The proportion of observations in $Cell_{22}$ is the proportion of subjects who are exposed to **no noise** who **did not help the confederate** (which equals $O_{22}/O_{2.} = 40/100 = .4$).

Alternative way of stating the null and alternative hypotheses for the chi-square test of independence The null and alternative hypotheses for the **chi-square test of independence** can be stated as follows:

$$H_0: \pi_{ij} = (\pi_{i.})(\pi_{.j}) \text{ for all } r \times c \text{ cells.}$$

$$H_1: \pi_{ij} \neq (\pi_{i.})(\pi_{.j}) \text{ for at least one cell.}$$

Where: π represents the value of a proportion in the population

The above notation indicates that if the null hypothesis is true, in the underlying population represented by the sample, for each of the $r \times c$ cells the proportion of observations in a cell will equal the proportion of observations in the row in which the cell appears multiplied by the proportion of observations in the column in which the cell appears. This will now be illustrated with respect to Example 16.2. In illustrating the relationship described in the null hypothesis, the notation p is employed to represent the relevant proportions obtained for the sample data. If the null hypothesis is true, in the case of $Cell_{11}$ it is required that the proportion of observations in $Cell_{11}$ is equivalent to the product of the proportion of observations in Row 1 (which equals $p_{1.} = O_{1.}/n = 100/200 = .5$) and the proportion of observations in Column 1 (which equals $p_{.1} = O_{.1}/n = 90/200 = .45$). The result of multiplying the row and column proportions is $(p_{1.})(p_{.1}) = (.5)(.45) = .225$. Thus, if the null hypothesis is true, the proportion of observations in $Cell_{11}$ must equal $p_{11} = .225$.[5] Consequently, if the value .225 is multiplied by 200, which is the total number of observations in Table 16.3, the resulting value $(p_{11})(n) = (.225)(200) = 45$ is the number of observations that is expected in $Cell_{11}$ if the null hypothesis is true. The same procedure can be used for the remaining three cells to determine the number of observations that are required in each cell in order for the null hypothesis to be supported. In Section IV these values, which in actuality correspond to the expected frequencies of the cells, are computed for each of the four cells in Table 16.3 (as well as Table 16.2).

IV. Test Computations

The computations for the **chi-square test for r × c tables** will be described for Example 16.1. The procedure to be described in this section when applied to Example 16.2 yields the identical result since: a) The computational procedure for the **chi-square test of independence** is identical to that employed for the **chi-square test for homogeneity**; and b) The identical data are employed for Examples 16.1 and 16.2. Table 16.4 summarizes the data and computations for Example 16.1.

Table 16.4 Chi-Square Summary Table for Example 16.1

Cell	O_{ij}	E_{ij}	$(O_{ij} - E_{ij})$	$(O_{ij} - E_{ij})^2$	$\dfrac{(O_{ij} - E_{ij})^2}{E_{ij}}$
Cell $_{11}$ — Noise/Helped the confederate	30	45	−15	225	5.00
Cell $_{12}$ — Noise/Did not help the confederate	70	55	15	225	4.09
Cell $_{21}$ — No noise/Helped the confederate	60	45	15	225	5.00
Cell $_{22}$ — No noise/Did not help the confederate	40	55	−15	225	4.09

$$\Sigma O_{ij} = 200 \quad \Sigma E_{ij} = 200 \quad \Sigma(O_{ij} - E_{ij}) = 0 \qquad \chi^2 = 18.18$$

The observed frequency of each cell (O_{ij}) is listed in Column 2 of Table 16.4. Column 3 contains the expected cell frequencies (E_{ij}). In order to conduct the **chi-square test for r × c tables**, the observed frequency for each cell must be compared with its expected frequency. In order to determine the expected frequency of a cell, the data should be arranged in a contingency table which employs the format of Table 16.2. The following protocol is then employed to determine the expected frequency of a cell: a) Multiply the sum of the observations in the row in which the cell appears by the sum of the observations in the column in which the cell appears; b) Divide n, the total number of observations, into the product that results from multiplying the row and column sums for the cell.

The computation of an expected cell frequency can be summarized by Equation 16.1.

$$E_{ij} = \frac{(O_{i.})(O_{.j})}{n} \qquad\qquad \textbf{(Equation 16.1)}$$

Applying Equation 16.1 to Cell$_{11}$ in Table 16.2 (i.e., **noise/helped the confederate**), the expected cell frequency can be computed as follows. The row sum is the total number of subjects who were exposed to **noise**. Thus, $O_{1.} = 100$. The column sum is the total number of subjects who **helped the confederate**. Thus, $O_{.1} = 90$. Employing Equation 16.1, the expected frequency for Cell$_{11}$ can now be computed: $E_{11} = [(O_{1.})(O_{.1})]/n = [(100)(90)]/200 = 45$. The expected frequencies for the remaining three cells in the 2 × 2 contingency table that summarizes the data for Example 16.1 are computed below.

$$\text{Cell}_{12} = [(O_{1.})(O_{.2})/n = [(100)(110)]/200 = 55$$

$$\text{Cell}_{21} = [(O_{2.})(O_{.1})/n = [(100)(90)]/200 = 45$$

$$\text{Cell}_{22} = [(O_{2.})(O_{.2})/n = [(100)(110)]/200 = 55$$

Upon determining the expected cell frequencies, the test statistic for the **chi-square test for** $r \times c$ **tables** is computed with Equation 16.2.[6]

$$\chi^2 = \sum_{i=1}^{r} \sum_{j=1}^{c} \left[\frac{(O_{ij} - E_{ij})^2}{E_{ij}} \right] \qquad \text{(Equation 16.2)}$$

The operations described by Equation 16.2 (which are the same as those described for computing the chi-square statistic for the **chi-square goodness-of-fit test**) are as follows: a) The expected frequency of each cell is subtracted from its observed frequency (summarized in Column 4 of Table 16.4); b) For each cell, the difference between the observed and expected frequency is squared (summarized in Column 5 of Table 16.4); c) For each cell, the squared difference between the observed and expected frequency is divided by the expected frequency of the cell (summarized in Column 6 of Table 16.4); and d) The value of chi-square is computed by summing all of the values in Column 6. For Example 16.1, Equation 16.2 yields the value $\chi^2 = 18.18$.[7]

Note that in Table 16.4 the sums of the observed and expected frequencies are identical. This must always be the case, and any time these sums differ from one another, it indicates a computational error has been made. It is also required that the sum of the differences between the observed and expected frequencies equals zero (i.e., $\Sigma(O_{ij} - E_{ij}) = 0$). Any time the latter value does not equal zero, it indicates an error has been made. Since all of the $(O_{ij} - E_{ij})$ values are squared in Column 5, the sum of Column 6, which represents the value of χ^2, must always be a positive number. If a negative value for chi-square is obtained, it indicates an error has been made. The only time χ^2 will equal zero is when $O_{ij} = E_{ij}$ for all $r \times c$ cells.

V. Interpretation of the Test Results

The obtained value $\chi^2 = 18.18$ is evaluated with **Table A4 (Table of the Chi-Square Distribution)** in the **Appendix**. A general discussion of the values in **Table A4** can be found in Section V of the **single-sample chi-square test for a population variance (Test 3)**. When the chi-square distribution is employed to evaluate the **chi-square test for** $r \times c$ **tables**, the degrees of freedom employed for the analysis are computed with Equation 16.3.

$$df = (r - 1)(c - 1) \qquad \text{(Equation 16.3)}$$

The tabled critical values in **Table A4** for the **chi-square test for** $r \times c$ **tables** are always derived from the right tail of the distribution. The critical chi-square value for a specific value of alpha is the tabled value at the percentile which corresponds to the value $(1 - \alpha)$. Thus, the tabled critical .05 chi-square value (to be designated $\chi^2_{.05}$) is the tabled value at the 95th percentile. In the same respect, the tabled critical .01 chi-square value (to be designated $\chi^2_{.01}$) is the tabled value at the 99th percentile. In order to reject the null hypothesis, the obtained value of chi-square must be equal to or greater than the tabled critical value at the prespecified level of significance.

The aforementioned guidelines for determining tabled critical chi-square values are employed when the alternative hypothesis is stated nondirectionally (which, as noted earlier, is generally the case for the **chi-square test for $r \times c$ tables**). The determination of tabled critical chi-square values in reference to a directional alternative hypothesis is discussed in Section VI.

The guidelines for a nondirectional analysis will now be applied to Example 16.1. Since $r = 2$ and $c = 2$, the degrees of freedom are computed to be $df = (2 - 1)(2 - 1) = 1$. The tabled critical .05 chi-square value for $df = 1$ is $\chi^2_{.05} = 3.84$, which as noted above is the tabled chi-square value at the 95th percentile. The tabled critical .01 chi-square value for $df = 1$ is $\chi^2_{.01} = 6.63$, which as noted above is the tabled chi-square value at the 99th percentile. Since the computed value $\chi^2 = 18.18$ is greater than both of the aforementioned critical values, the null hypothesis can be rejected at both the .05 and .01 levels. Rejection of the null hypothesis at the .01 level can be summarized as follows: $\chi^2(1) = 18.18$, $p < .01$.

The significant chi-square value obtained for Example 16.1 indicates that subjects who served in the **noise** condition **helped the confederate** significantly less than subjects who served in the **no noise** condition. This can be confirmed by visual inspection of Table 16.2, which reveals that twice as many subjects who served in the **no noise** condition **helped the confederate** than subjects who served in the **noise** condition.

As noted previously, the chi-square analysis described in this section also applies to Example 16.2, since the latter example employs the same data as Example 16.1. Thus, with respect to Example 16.2, the significant $\chi^2 = 18.18$ value allows the researcher to conclude that a subject's categorization on the **introvert–extrovert** dimension is associated with (i.e., not independent of) one's political affiliation. This can be confirmed by visual inspection of Table 16.3, which reveals that **introverts** are more likely to be **Republicans** whereas **extroverts** are more likely to be **Democrats**.

It is important to note that Example 16.2 represents a correlational study, and as such does not allow a researcher to draw any conclusions with regard to cause and effect.[8] To be more specific, the study does not allow one to conclude that a subject's categorization on the personality dimension **introvert–extrovert** is the cause of one's political affiliation (**Democrat** versus **Republican**), or vice versa (i.e., that political affiliation causes one to be an **introvert** versus an **extrovert**). Although it is possible that the two variables employed in a correlational study are causally related to one another, such studies do not allow one to draw conclusions regarding cause and effect, since they fail to control for the potential influence of confounding variables. Because of this, when studies which are evaluated with the **chi-square test of independence** (such as Example 16.2) yield a significant result, one can only conclude that in the underlying population the two variables have a correlation with one another which is some value other than zero (which is not commensurate with saying that one variable causes the other).

Studies such as that represented by Example 16.2 can also be conceptualized within the framework of a **natural experiment** (also referred to as an **ex post facto study**) which is discussed in the **Introduction** of the book. In the latter type of study, one of the two variables is designated as the independent variable, and the second variable as the dependent variable. The independent variable is (in contrast to the independent variable in a **true experiment**) a non-manipulated variable. A subject's score (or category in the case of Example 16.2) on a non-manipulated independent variable is based on some preexisting subject characteristic, rather than being a direct result of some manipulation on the part of the experimenter. Thus, if in Example 16.2 the **introvert–extrovert** dimension is designated as the independent variable, it represents a nonmanipulated variable, since the experimenter does not determine whether or not a subject becomes an **introvert** or an **extrovert**. Which of the two aforementioned categories a subject falls into is determined beforehand by "nature" (thus the term **natural experiment**). The same logic also applies if political affiliation is employed as the independent variable, since, like **introvert–extrovert**, the **Democrat–Republican** dichotomization is a preexisting subject characteristic.

In Example 16.1, however, the independent variable, which is whether or not a subject is exposed to **noise**, is a manipulated variable. This is the case, since the experimenter randomly determines those subjects who are assigned to the **noise** condition and those who are assigned to the **no noise** condition. As noted in the **Introduction**, an experiment in which the researcher manipulates the level of the independent variable to which a subject is assigned is referred to as a **true experiment**. In the latter type of experiment, by virtue of randomly assigning subjects to the different experimental conditions, the researcher is able to control for the effects of potentially confounding variables. Because of this, if a significant result is obtained in a **true experiment**, a researcher is justified in drawing conclusions with regard to cause and effect.

VI. Additional Analytical Procedures for the Chi-Square Test for $r \times c$ Tables and/or Related Tests

1. Yates' correction for continuity In Section I it is noted that, in actuality, the **chi-square test for $r \times c$ tables** employs a continuous distribution to approximate a discrete probability distribution. Under such conditions, some sources recommend that a correction for continuity be employed. As noted previously in the book, the correction for continuity is based on the premise that if a continuous distribution is employed to estimate a discrete distribution, such an approximation will inflate the Type I error rate. By employing the correction for continuity, the Type I error rate is ostensibly adjusted to be more compatible with the prespecified alpha level designated by the researcher. Sources which recommend the correction for continuity for the **chi-square test for $r \times c$ tables** only recommend that it be employed in the case of 2×2 contingency tables. Equation 16.4 (which was developed by Yates (1934)) is the continuity-corrected chi-square equation for 2×2 tables.

$$\chi^2 = \sum_{i=1}^{r} \sum_{j=1}^{c} \left[\frac{(\mid O_{ij} - E_{ij} \mid - .5)^2}{E_{ij}} \right] \qquad \textbf{(Equation 16.4)}$$

Note that by subtracting .5 from the absolute value of the difference between each set of observed and expected frequencies, the chi-square value obtained with Equation 16.4 will be lower than the value computed with Equation 16.2.

Statisticians are not in agreement with respect to whether or not it is prudent to employ the correction for continuity described by Equation 16.4 with a 2×2 contingency table. To be more specific, various sources take the following positions with respect to what the most effective strategy is for evaluating 2×2 tables: a) Most sources agree that when the sample size for a 2×2 table is small (generally less than 20), the **Fisher exact test** (which is described later in this section) should be employed instead of the **chi-square test for $r \times c$ tables**. Cochran (1952, 1954) stated that in the case of 2×2 tables, the **chi-square test for $r \times c$ tables** should not be employed when $n < 20$, and that when $20 < n < 40$ the test should only be employed if all of the expected frequencies are at least equal to 5. Additionally, when $n > 40$ all expected frequencies should be equal to or greater than 1; b) Some sources recommend that for small sample sizes **Yates' correction for continuity** be employed. This recommendation assumes that the size of the sample is at least equal to 20 (since, when $n < 20$, the **Fisher exact test** should be employed), but less than some value which defines the maximum size of a small sample with respect to the use of Yates' correction. Sources do not agree on what value of n defines the upper limit beyond which Yates' correction is not required; c) Some sources recommend that **Yates' correction for continuity** should always be employed with 2 × 2 tables, regardless of the sample size; d) To further confuse the issue, many sources take the position that **Yates' correction for continuity** should never be used, since the chi-square value computed with Equation 16.4 results in an overcorrection — i.e., it results in an overly conservative test; and e) Haber (1980, 1982) argues that alternative continuity correction procedures (including one developed by Haber) are superior to **Yates' correction for continuity**. Haber's (1980, 1982) continuity correction procedure is described in Zar (1999, pp. 494–495).

Table 16.5 illustrates the application of **Yates' correction for continuity** with Example 16.1. By employing Equation 16.4, the obtained value of chi-square is reduced to 16.98 (in contrast to the value $\chi^2 = 18.18$ obtained with Equation 16.2). Since the obtained value $\chi^2 = 16.98$ is greater than both $\chi^2_{.05} = 3.84$ and $\chi^2_{.01} = 6.83$, the null hypothesis can still be rejected at both the .05 and .01 levels. Thus, in this instance, **Yates' correction for continuity** leads to the same conclusions as those reached when Equation 16.2 is employed.

Table 16.5   Chi-Square Summary Table for Example 16.1 Employing Yates' Correction for Continuity

Cell	O_{ij}	E_{ij}	$(\|O_{ij} - E_{ij}\| - .5)$	$(\|O_{ij} - E_{ij}\| - .5)^2$	$\dfrac{(\|O_{ij} - E_{ij}\| - .5)^2}{E_{ij}}$
Cell$_{11}$ — Noise/ Helped the confederate	30	45	14.5	210.25	4.67
Cell$_{12}$ — Noise/ Did not help the confederate	70	55	14.5	210.25	3.82
Cell$_{21}$ — No noise/ Helped the confederate	60	45	14.5	210.25	4.67
Cell$_{22}$ — No noise/ Did not help the confederate	40	55	14.5	210.25	3.82
$\Sigma O_{ij} = 200$ $\Sigma E_{ij} = 200$					$\chi^2 = 16.98$

2. Quick computational equation for a 2 × 2 table   Equation 16.5 is a quick computational equation which can be employed for the **chi-square test for r × c tables** in the case of a 2 × 2 table. Unlike Equation 16.2, it does not require that the expected cell frequencies be computed. The notation employed in Equation 16.5 is based on the model for a 2 × 2 contingency table summarized in Table 16.6.

Table 16.6   Model for 2 × 2 Contingency Table

	Column 1	Column 2	Row sums
Row 1	a	b	$a + b = n_1$
Row 2	c	d	$c + d = n_2$
Column sums	$a + c$	$b + d$	n

$$\chi^2 = \frac{n(ad - bc)^2}{(a + b)(c + d)(a + c)(b + d)} \qquad \textbf{(Equation 16.5)}$$

Where: $a, b, c,$ and d represent the number of observations in the relevant cell

Using the model depicted in Table 16.6, by employing the appropriate observed cell frequencies for Examples 16.1 and 16.2, we know that $a = 30$, $b = 70$, $c = 60$, and $d = 40$. Substituting these values in Equation 16.5, the value $\chi^2 = 18.18$ is computed (which is the same chi-square value that is computed with Equation 16.2).

$$\chi^2 = \frac{200[(30)(40) - (70)(60)]^2}{(30 + 70)(60 + 40)(30 + 60)(70 + 40)} = 18.18$$

If **Yates' correction for continuity** is applied to a 2 × 2 table, Equation 16.6 is the continuity-corrected version of Equation 16.5.

$$\chi^2 = \frac{n(|ad - bc| - .5n)^2}{(a + b)(c + d)(a + c)(b + d)} \qquad \textbf{(Equation 16.6)}$$

Substituting the data for Examples 16.1 and 16.2 in Equation 16.6, the value $\chi^2 = 16.98$ is computed (which is the same continuity-corrected chi-square value computed with Equation 16.4).

$$\chi^2 = \frac{200[|(30)(40) - (70)(60)| - (.5)(200)]^2}{(30 + 70)(60 + 40)(30 + 60)(70 + 40)} = 16.98$$

3. Evaluation of a directional alternative hypothesis in the case of a 2 × 2 contingency table
In the case of 2 × 2 contingency tables it is possible to employ a directional/one-tailed alternative hypothesis. Prior to reading this section, the reader may find it useful to review the relevant material on this subject in Section VII of the **chi-square goodness-of-fit test**.

In the case of a 2 × 2 contingency table, it is possible to make two directional predictions. In stating the null and alternative hypotheses, the following notation (in reference to the sample data) based on the model for a 2 × 2 contingency table described in Table 16.6 will be employed.

$$p_1 = \frac{a}{a + b} = \frac{a}{n_1} \qquad p_2 = \frac{c}{c + d} = \frac{c}{n_2}$$

The value p_1 represents the proportion of observations in Row 1 that falls in Cell a, while the value p_2 represents the proportion of observations in Row 2 that falls in Cell c. The analogous proportions in the underlying populations which correspond to p_1 and p_2 will be represented by the notation π_1 and π_2. Thus, π_1 represents the proportion of observations in Row 1 in the underlying population that falls in Cell a, while the proportion π_2 represents the proportion of observations in Row 2 in the underlying population that falls in Cell c. Employing the aforementioned notation, it is possible to make either of the two following directional predictions for a 2 × 2 contingency table.

a) In the underlying population(s) the sample(s) represent, the proportion of observations in Row 1 that falls in Cell a is greater than the proportion of observations in Row 2 that falls in Cell c. The null hypothesis and directional alternative hypothesis for this prediction are stated as follows: $H_0: \pi_1 = \pi_2$ versus $H_1: \pi_1 > \pi_2$. With respect to Example 16.1, the latter alternative hypothesis predicts that a larger proportion of subjects in the **noise** condition will **help the confederate** than subjects in the **no noise** condition. In Example 16.2, the alternative hypothesis predicts that a larger proportion of **introverts** will be **Democrats** than **extroverts**.

b) In the underlying population(s) the sample(s) represent, the proportion of observations in Row 1 that falls in Cell a is less than the proportion of observations in Row 2 that falls in Cell c. The null hypothesis and directional alternative hypothesis for this prediction are stated as follows: H_0: $\pi_1 = \pi_2$ versus H_1: $\pi_1 < \pi_2$. With respect to Example 16.1, the latter alternative hypothesis predicts that a larger proportion of subjects in the **no noise** condition will **help the confederate** than subjects in the **noise** condition. In Example 16.2, the alternative hypothesis predicts that a larger proportion of **extroverts** will be **Democrats** than **introverts**.

As is the case for the **chi-square goodness-of-fit test**, if a researcher wants to evaluate a one-tailed alternative hypothesis at the .05 level, the appropriate critical value to employ is $\chi^2_{.90}$, which is the tabled chi-square value at the .10 level of significance. The latter value is represented by the tabled chi-square value at the 90th percentile (which demarcates the extreme 10% in the right tail of the chi-square distribution). If a researcher wants to evaluate a one-tailed/directional alternative hypothesis at the .01 level, the appropriate critical value to employ is $\chi^2_{.98}$, which is the tabled chi-square value at the .02 level of significance. The latter value is represented by the tabled chi-square value at the 98th percentile (which demarcates the extreme 2% in the right tail of the chi-square distribution).

If a one-tailed alternative hypothesis is evaluated for Examples 16.1 and 16.2, from **Table A4** it can be determined that for $df = 1$ the relevant tabled critical one-tailed .05 and .01 values are $\chi^2_{.90} = 2.71$ and $\chi^2_{.98} = 5.43$.[9] Note that when one employs a one-tailed alternative hypothesis it is easier to reject the null hypothesis, since the one-tailed .05 and .01 critical values are less than the two-tailed .05 and .01 values (which for $df = 1$ are $\chi^2_{.05} = 3.84$ and $\chi^2_{.01} = 6.63$).[10] In conducting a one-tailed analysis, it is important to note, however, if the obtained value of chi-square is equal to or greater than the tabled critical value at the prespecified level of significance, only one of the two possible alternative hypotheses can be supported. The alternative hypothesis that is supported is the one which is consistent with the data.

Since for Examples 16.1 and 16.2, the computed value $\chi^2 = 18.18$ is greater than both of the one-tailed critical values $\chi^2_{.90} = 2.71$ and $\chi^2_{.98} = 5.43$, the null hypothesis can be rejected at both the .05 and .01 levels, but only if the directional alternative hypothesis H_1: $\pi_1 < \pi_2$ is employed. If the directional hypothesis H_1: $\pi_1 > \pi_2$ is employed, the null hypothesis cannot be rejected, since the data are not consistent with the latter alternative hypothesis.

When evaluating contingency tables in which the number of rows and/or columns is greater than two, it is possible to run a **multi-tailed test** (which as noted in the discussion of the **chi-square goodness-of-fit test** is the term that is sometimes used when there are more than two possible directional alternative hypotheses). It would be quite unusual to encounter the use of a multi-tailed analysis for an $r \times c$ table, which requires that a researcher determine all possible directional patterns/ordinal configurations for a set of data, and then predict one or more of the specific patterns which will occur. In the event one elects to conduct a multi-tailed analysis, the determination of the appropriate critical values is based on the same guidelines which are discussed for multi-tailed tests under the **chi-square goodness-of-fit test**.

4. Test 16c: The Fisher exact test In Section I it is noted that the chi-square distribution provides an approximation of the exact sampling distribution for a contingency table. In the case of 2×2 tables, the chi-square distribution is employed to approximate the hypergeometric distribution which will be discussed in this section. (The hypergeometric distribution is discussed in detail in Section IX (the **Addendum**) of the **binomial sign test for a single sample (Test 9)**.) As noted earlier, when $n < 20$ most sources recommend that the **Fisher exact test** (which employs exact hypergeometric probabilities) be employed to evaluate a 2×2 contingency table. Table 16.6, which is used earlier in this section to summarize a 2×2 table, will be employed to describe the model for the hypergeometric distribution upon which the **Fisher exact test** is based.

According to Daniel (1990) the **Fisher exact test,** which is also referred to as the **Fisher–Irwin test,** was simultaneously described by Fisher (1934, 1935), Irwin (1935), and Yates (1934). The test shares the same assumptions as those noted for the **chi-square test for $r \times c$ tables,** with the exception of the assumption regarding small expected frequencies (which reflects the limitations of the latter test with small sample sizes). An additional assumption of the **Fisher exact test** is that of **fixed marginal frequencies,** which means that both the row and column sums of a 2×2 contingency table are predetermined by the researcher. Daniel (1990, p. 121) notes that the assumption of fixed marginal frequencies is only rarely encountered in practice, and the truth of the matter is that the **Fisher exact test** is used with 2×2 contingency tables involving small sample sizes when one or neither of the marginal sums is predetermined by the researcher. Among others, Daniel (1990, p. 124) cites sources which discuss the impact of employing the **Fisher exact test** with data when the assumption of fixed marginal frequencies is violated. The **Fisher exact test** is more commonly employed with the model described for the **chi-square test of homogeneity** than it is with the model described for the **chi-square test of independence.**

Equation 16.7, which is the equation for a hypergeometrically distributed variable, allows for the computation of the exact probability (P) of obtaining a specific set of observed frequencies in a 2×2 contingency table. Equation 16.7, which uses the notation for the **Fisher exact test** model, is equivalent to Equation 9.17, which is more commonly employed to represent the general equation for a hypergeometrically distributed variable. Since Equation 16.7 involves the computation of combinations, the reader may find it useful to review the discussion of combinations in the **Introduction** and/or in Section IV of the **binomial sign test for a single sample.**

$$P = \frac{\binom{a + c}{a}\binom{b + d}{b}}{\binom{n}{a + b}} \qquad \textbf{(Equation 16.7)}$$

Equation 16.8 is a computationally more efficient form of Equation 16.7 which yields the same probability value.

$$P = \frac{(a + c)! \ (b + d)! \ (a + b)! \ (c + d)!}{n! \ a! \ b! \ c! \ d!} \qquad \textbf{(Equation 16.8)}$$

Example 16.3 (which is a small sample size version of Example 16.1) will be employed to illustrate the **Fisher exact test.**

Example 16.3 *A researcher conducts a study in order to evaluate the effect of noise on altruistic behavior. Each of the 12 subjects who participate in the experiment is randomly assigned to one of two experimental conditions. Subjects in both conditions are given a one-hour test which is ostensibly a measure of intelligence. During the test the six subjects in Group 1 are exposed to continual loud noise, which they are told is due to a malfunctioning generator. The six subjects in Group 2 are not exposed to any noise during the test. Upon completion of this stage of the experiment, each subject on leaving the room is confronted by a middle-aged man whose arm is in a sling. The man asks the subject if she would be willing to help him carry a heavy package to his car. In actuality, the man requesting help is an experimental confederate (i.e., working for the experimenter). The number of subjects in each group who help the man is recorded. One of the six subjects who were exposed to noise elects to help the man, while five of the six subjects who were not exposed to noise elect to help the man. Do the data indicate that altruistic behavior is influenced by noise?*

The data for Example 16.3, which can be summarized in the form of a 2 × 2 contingency table, are presented in Table 16.7.

Table 16.7 Summary of Data for Example 16.3

	Helped the confederate	Did not help the confederate	Row sums
Noise	$a = 1$	$b = 5$	$a + b = n_1 = 6$
No noise	$c = 5$	$d = 1$	$c + d = n_2 = 6$
Column sums	$a + c = 6$	$b + d = 6$	$n = 12$

The null and alternative hypotheses for the **Fisher exact test** are most commonly stated using the format described in the discussion of the evaluation of a directional alternative hypothesis for a 2 × 2 contingency table. Thus, the null hypothesis and nondirectional alternative hypothesis are as follows:

$$H_0: \pi_1 = \pi_2$$

(In the underlying populations the samples represent, the proportion of observations in Row 1 (the **noise** condition) that falls in Cell a is equal to the proportion of observations in Row 2 (the **no noise** condition) that falls in Cell c.)

$$H_1: \pi_1 \neq \pi_2$$

(In the underlying populations the samples represent, the proportion of observations in Row 1 (the **noise** condition) that falls in Cell a is not equal to the proportion of observations in Row 2 (the **no noise** condition) that falls in Cell c.)

The alternative hypothesis can also be stated directionally, as described in the discussion of the evaluation of a directional alternative hypothesis for a 2 × 2 contingency table — i.e., $H_1: \pi_1 > \pi_2$ or $H_1: \pi_1 < \pi_2$.[11]

Employing Equations 16.7 and 16.8, the probability of obtaining the specific set of observed frequencies in Table 16.7 is computed to be $P = .039$.

Equation 16.7:

$$P = \frac{\binom{6}{1}\binom{6}{5}}{\binom{12}{6}} = \frac{\left[\dfrac{6!}{1!\,5!}\right]\left[\dfrac{6!}{5!\,1!}\right]}{\dfrac{12!}{6!\,6!}} = .039$$

Equation 16.8:

$$P = \frac{6!\ 6!\ 6!\ 6!}{12!\ 1!\ 5!\ 5!\ 1!} = .039$$

In order to evaluate the null hypothesis, in addition to the probability $P = .039$ (which is the probability of obtaining the set of observed frequencies in Table 16.7) it is also necessary to compute the probabilities for any sets of observed frequencies which are even more extreme than the observed frequencies in Table 16.7. The only result that is more extreme than the result summarized in Table 16.7 is if all six subjects in the **no noise** condition **helped the confederate**, while all six subjects in the **noise** condition **did not help the confederate**. Table 16.8 summarizes the observed frequencies for the latter result.

Table 16.8 Most Extreme Possible Set of Observed Frequencies for Example 16.3

	Helped the confederate	Did not help the confederate	Row sums
Noise	$a = 0$	$b = 6$	$a + b = n_1 = 6$
No noise	$c = 6$	$d = 0$	$c + d = n_2 = 6$
Column sums	$a + c = 6$	$b + d = 6$	$n = 12$

Employing Equations 16.7 and 16.8, the probability of obtaining the set of observed frequencies in Table 16.8 is computed to be $P = .001$.

Equation 16.7:

$$P = \frac{\binom{6}{0}\binom{6}{6}}{\binom{12}{6}} = \frac{\left[\dfrac{6!}{0!\ 6!}\right]\left[\dfrac{6!}{6!\ 0!}\right]}{\dfrac{12!}{6!\ 6!}} = .001$$

Equation 16.8:

$$P = \frac{6!\ 6!\ 6!\ 6!}{12!\ 0!\ 6!\ 6!\ 0!} = .001$$

When $P = .001$ (the probability of obtaining the set of observed frequencies in Table 16.8) is added to $P = .039$, the resulting probability represents the likelihood of obtaining a set of observed frequencies that is equal to or more extreme than the set of observed frequencies in Table 16.7. The notation P_T will be used to represent the latter value. Thus, in our example, $P_T = .039 + .001 = .04$.[12]

The following guidelines are employed for Example 16.3 in evaluating the null hypothesis for the **Fisher exact test**.

a) If the nondirectional alternative hypothesis H_1: $\pi_1 \neq \pi_2$ is employed, the value of P_T (i.e., the probability of obtaining a set of observed frequencies equal to or more extreme than the set obtained in the study) must be equal to or less than $\alpha/2$. Thus, if the prespecified value of alpha is $\alpha = .05$, the obtained value of P_T must be equal to or less than $.05/2 = .025$. If the prespecified value of alpha is $\alpha = .01$, the obtained value of P_T must be equal to or less than $.01/2 = .005$.

b) If a directional alternative hypothesis is employed, the observed set of frequencies for the study must be consistent with the directional alternative hypothesis, and the value of P_T must be equal to or less than the prespecified value of alpha. Thus, if the prespecified value of alpha is $\alpha = .05$, the obtained value of P_T must be equal to or less than $.05$. If the prespecified value of alpha is $\alpha = .01$, the obtained value of P_T must be equal to or less than $.01$.

Employing the above guidelines, the following conclusions can be reached.

If $\alpha = .05$, the nondirectional alternative hypothesis H_1: $\pi_1 \neq \pi_2$ is not supported, since the obtained value $P_T = .04$ is greater than $.05/2 = .025$.

The directional alternative hypothesis H_1: $\pi_1 < \pi_2$ is supported, but only at the .05 level. This is the case, since the data are consistent with the latter alternative hypothesis, and the obtained value $P_T = .04$ is less than $\alpha = .05$. The latter alternative hypothesis is not supported at the .01 level, since $P_T = .04$ is greater than $\alpha = .01$.

The directional alternative hypothesis H_1: $\pi_1 > \pi_2$ is not supported, since it is not consistent with the data. In order for the data to be consistent with the alternative hypothesis H_1: $\pi_1 > \pi_2$, it is required that a larger proportion of subjects in the **noise** condition **helped the confederate** than subjects in the **no noise** condition.

To further clarify how to interpret a directional versus a nondirectional alternative hypothesis, consider Table 16.9 which presents all seven possible outcomes of observed cell frequencies for $n = 12$ in which the marginal sums (i.e., the row and column sums) equal six (which are the values for the marginal sums in Example 16.3).

The sum of the probabilities for the seven outcomes presented in Table 16.9 equals 1. This is the case, since the seven outcomes represent all the possible outcomes for the cell frequencies if the marginal sum of each row and column equals six. As noted earlier, if a researcher evaluates the directional alternative hypothesis H_1: $\pi_1 < \pi_2$ for Example 16.3, he will only be interested in **Outcomes 1 and 2**. The combined probability for the latter two outcomes is $P_T = .04$, which is less than the one-tailed value $\alpha = .05$ (which represents the extreme 5% of the sampling distribution in one of the two tails of the distribution). Since the data are consistent with the directional alternative hypothesis H_1: $\pi_1 < \pi_2$ and $P_T = .04$ is less than $\alpha = .05$, the latter alternative hypothesis is supported.

If, however, the nondirectional alternative hypothesis H_1: $\pi_1 \neq \pi_2$ is employed, in addition to considering **Outcomes 1 and 2**, the researcher must also consider **Outcomes 6 and 7**, which are the analogous extreme outcomes in the opposite tail of the distribution. The latter set of outcomes also has a combined probability of .04. If the probability associated with **Outcomes 1 and 2** and **Outcomes 6 and 7** are summed, the resulting value $P_T = .04 + .04 = .08$ represents the likelihood in both tails of the distribution of obtaining an outcome equal to or more extreme than the outcome observed in Table 16.7. Since the value $P_T = .08$ is greater than the two-tailed value $\alpha = .05$, the nondirectional alternative hypothesis H_1: $\pi_1 \neq \pi_2$ is not supported. This is commensurate with saying that in order for the latter alternative hypothesis to be supported at the .05 level, in each of the tails of the distribution the maximum permissible probability value for outcomes which are equivalent to or more extreme than the observed outcome cannot be greater than the value $.05/2 = .025$. As noted earlier, since the computed probability in the relevant tail of the distribution equals .04 (which is greater than $.05/2 = .025$), the nondirectional alternative hypothesis is not supported.

If the directional alternative hypothesis H_1: $\pi_1 > \pi_2$ is employed the researcher is interested in **Outcomes 6 and 7**. As is the case for **Outcomes 1 and 2**, the combined probability for these two outcomes is $P_T = .04$, which is less than the one-tailed value $\alpha = .05$ (which represents the extreme 5% of the sampling distribution in the other tail of the distribution). However, since the data are not consistent with the directional alternative hypothesis H_1: $\pi_1 > \pi_2$, it is not supported.

To compare the results of the **Fisher exact test** with those that will be obtained if Example 16.3 is evaluated with the **chi-square test for r × c tables**, Equation 16.2 will be employed to evaluate the data in Table 16.7. Table 16.10 summarizes the chi-square analysis. Note that the expected frequency of each cell is 3, since when Equation 16.1 is employed, the value $E_{ij} = [(6)(6)]/12 = 3$ is computed for all $r \times c = 4$ cells.

Since for $df = 1$, the obtained value $\chi^2 = 5.32$ is greater than the tabled critical two-tailed .05 value $\chi^2_{.05} = 3.84$, the nondirectional alternative hypothesis H_1: $\pi_1 \neq \pi_2$ is supported at the .05 level. It is not, however, supported at the .01 level, since $\chi^2 = 5.32$ is less than the tabled critical two-tailed .01 value $\chi^2_{.01} = 6.63$.

The directional alternative hypothesis H_1: $\pi_1 < \pi_2$ is supported at the .05 level, since the obtained value $\chi^2 = 5.32$ is greater than the tabled critical one-tailed .05 value $\chi^2_{.90} = 2.71$. The

**Table 16.9 Possible Outcomes for Observed Cell Frequencies
If All Marginal Sums Equal 6, When $n = 12$**

Outcome 1: $P = .001$

	Col. 1	Col. 2	Row sums
Row 1	0	6	6
Row 2	6	0	6
Column sums	6	6	12

Outcome 2: $P = .039$

	Col. 1	Col. 2	Row sums
Row 1	1	5	6
Row 2	5	1	6
Column sums	6	6	12

Outcome 3: $P = .243$

	Col. 1	Col. 2	Row sums
Row 1	2	4	6
Row 2	4	2	6
Column sums	6	6	12

Outcome 4: $P = .433$

	Col. 1	Col. 2	Row sums
Row 1	3	3	6
Row 2	3	3	6
Column sums	6	6	12

Outcome 5: $P = .243$

	Col. 1	Col. 2	Row sums
Row 1	4	2	6
Row 2	2	4	6
Column sums	6	6	12

Outcome 6: $P = .039$

	Col. 1	Col. 2	Row sums
Row 1	5	1	6
Row 2	1	5	6
Column sums	6	6	12

Outcome 7: $P = .001$

	Col. 1	Col. 2	Row sums
Row 1	6	0	6
Row 2	0	6	6
Column sums	6	6	12

Table 16.10 Chi-Square Summary Table for Example 16.3

Cell	O_{ij}	E_{ij}	$(O_{ij} - E_{ij})$	$(O_{ij} - E_{ij})^2$	$\dfrac{(O_{ij} - E_{ij})^2}{E_{ij}}$
Cell$_{11}$ — Noise/Helped the confederate	1	3	−2	4	1.33
Cell$_{12}$ — Noise/Did not help the confederate	5	3	2	4	1.33
Cell$_{21}$ — No noise/Helped the confederate	5	3	2	4	1.33
Cell$_{22}$ — No noise/Did not help the confederate	1	3	−2	4	1.33
	$\Sigma O_{ij} = 12$	$\Sigma E_{ij} = 12$	$\Sigma (O_{ij} - E_{ij}) = 0$		$\chi^2 = 5.32$

latter directional alternative hypothesis falls just short of being supported at the .01 level, since $\chi^2 = 5.32$ is less than the tabled critical one-tailed .01 value $\chi^2_{.98} = 5.43$.

Note that when the **Fisher exact test** is employed, the nondirectional alternative hypothesis H_1: $\pi_1 \neq \pi_2$ is not supported at the .05 level, yet it is supported at the .05 level when the **chi-square test for $r \times c$ tables** is used. Both the **Fisher exact test** and **chi-square test for $r \times c$ tables** allow the researcher to reject the null hypothesis at the .05 level if the directional alternative hypothesis H_1: $\pi_1 < \pi_2$ is employed. However, whereas the latter alternative hypothesis falls just short of significance at the .01 level when the **chi-square test for $r \times c$ tables** is used, it is further removed from being significant when the **Fisher exact test** is employed. The discrepancy between the two tests when they are applied to the same set of data involving a small sample size suggests that the chi-square approximation underestimates the actual probability associated with the observed frequencies and, consequently, increases the likelihood of committing a Type I error.

On the other hand, the result for the **chi-square test for $r \times c$ tables** for Example 16.3 will be totally consistent with the result obtained with the **Fisher exact test** if **Yates' correction for continuity** is employed in evaluating the data. If **Yates' correction** is employed, all of the $(O_{ij} - E_{ij})$ values in Table 16.10 become 1.5, which when squared equals 2.25. When the latter value is divided by the expected cell frequency of 3, it yields .75 which will be the entry for each cell in the last column of Table 16.10. The sum of the values in the latter column is $\chi^2 = 3$, which is the continuity-corrected chi-square value. Since $\chi^2 = 3$ is less than the tabled critical two-tailed values $\chi^2_{.05} = 3.84$ and $\chi^2_{.01} = 6.63$, the nondirectional alternative hypothesis is not supported. Since $\chi^2 = 3$ is greater than the tabled critical one-tailed .05 value $\chi^2_{.90} = 2.71$ but less than the tabled critical .01 one-tailed value $\chi^2_{.98} = 5.43$, the directional alternative hypothesis that is consistent with the data (H_1: $\pi_1 < \pi_2$) is supported, but only at the .05 level. This is the same conclusion which is reached when the **Fisher exact test** is employed.

5. Test 16d: The z test for two independent proportions

The **z test for two independent proportions** is an alternative large sample procedure for evaluating a 2×2 contingency table. In point of fact, the **z test for two independent proportions** yields a result which is equivalent to that obtained with the **chi-square test for $r \times c$ tables**. Later in this discussion it will be demonstrated that if both the **z test for two independent proportions** (which is based on the normal distribution) and the **chi-square test for $r \times c$ tables** are applied to the same set of data, the square of the z value obtained for the former test will equal the chi-square value obtained for the latter test.

The **z test for two independent proportions** is most commonly employed to evaluate the null and alternative hypotheses which are described for the **Fisher exact test** (which for a 2×2 contingency table are equivalent to the null and alternative hypotheses presented in Section III for the **chi-square test for homogeneity**). Thus, in reference to Example 16.1, employing the model for a 2×2 contingency table summarized by Table 16.6, the null hypothesis and non-directional alternative hypothesis for the **z test for two independent proportions** are as follows.

$$H_0: \pi_1 = \pi_2$$

(In the underlying populations the samples represent, the proportion of observations in Row 1 (the **noise** condition) that falls in Cell a is equal to the proportion of observations in Row 2 (the **no noise** condition) that falls in Cell c.)

$$H_1: \pi_1 \neq \pi_2$$

(In the underlying populations the samples represent, the proportion of observations in Row 1 (the **noise** condition) that falls in Cell a is not equal to the proportion of observations in Row 2 (the **no noise** condition) that falls in Cell c.)

An alternate but equivalent way of stating the above noted null hypothesis and nondirectional alternative hypothesis is as follows: H_0: $\pi_1 - \pi_2 = 0$ versus H_1: $\pi_1 - \pi_2 \neq 0$. As is the case with the **Fisher exact test** (as well as the **chi-square test for $r \times c$ tables**), the alternative hypothesis can also be stated directionally. Thus, the following two directional alternative hypotheses can be employed: H_1: $\pi_1 > \pi_2$ (which can also be stated as H_1: $\pi_1 - \pi_2 > 0$) or H_1: $\pi_1 < \pi_2$ (which can also be stated as H_1: $\pi_1 - \pi_2 < 0$).

Equation 16.9 is employed to compute the test statistic for the z **test for two independent proportions**.

$$z = \frac{p_1 - p_2}{\sqrt{p(1-p)\left[\dfrac{1}{n_1} + \dfrac{1}{n_2}\right]}}$$ (**Equation 16.9**)

Where: n_1 represents the number of observations in Row 1
n_2 represents the number of observations in Row 2
$p_1 = a/(a+b) = a/n_1$ represents the proportion of observations in Row 1 that falls in Cell a. It is employed to estimate the population proportion π_1.
$p_2 = c/(c+d) = c/n_2$ represents the proportion of observations in Row 2 that falls in Cell c. It is employed to estimate the population proportion π_2.
$p = (a+c)/(n_1 + n_2) = (a+c)/n$. p is a pooled estimate of the proportion of observations in Column 1 in the underlying population.

The denominator of Equation 16.9, which represents a standard deviation of a sampling distribution of differences between proportions, is referred to as the **standard error of the difference between two proportions** (which is often summarized with the notation $s_{p_1 - p_2}$). This latter value is analogous to the **standard error of the difference** ($s_{\bar{X}_1 - \bar{X}_2}$), which is the denominator of Equations 11.1, 11.2, 11.3, and 11.5 (which are all described in reference to the t **test for two independent samples (Test 11)**). Whereas the latter value is a standard deviation of a sampling distribution of difference scores between the means of two populations, the **standard error of the difference between two proportions** is a standard deviation of a sampling distribution of difference scores between proportions for two populations.

For Example 16.1 we either know or can compute the following values.[13]

$$a = 30 \qquad c = 60 \qquad n_1 = 100 \qquad n_2 = 100$$

$$p_1 = \frac{a}{n_1} = \frac{30}{100} = .3 \qquad p_2 = \frac{c}{n_2} = \frac{60}{100} = .6$$

$$p = \frac{a+c}{n_1 + n_2} = \frac{30 + 60}{100 + 100} = .45 \qquad 1 - p = 1 - .45 = .55$$

Employing the above values in Equation 16.9, the value $z = -4.26$ is computed.

$$z = \frac{.3 - .6}{\sqrt{(.45)(.55)\left[\dfrac{1}{100} + \dfrac{1}{100}\right]}} = -4.26$$

The obtained value $z = -4.26$ is evaluated with **Table A1 (Table of the Normal Distribution)** in the **Appendix**. In **Table A1** the tabled critical two-tailed .05 and .01 values are $z_{.05} = 1.96$ and $z_{.01} = 2.58$, and the tabled critical one-tailed .05 and .01 values are $z_{.05} = 1.65$ and $z_{.01} = 2.33$. The following guidelines are employed in evaluating the null hypothesis:

a) If the nondirectional alternative hypothesis $H_1: \pi \neq \pi_2$ is employed, the null hypothesis can be rejected if the obtained absolute value of z is equal to or greater than the tabled critical two-tailed value at the prespecified level of significance.

b) If the directional alternative hypothesis H_1: $\pi_1 > \pi_2$ is employed, the null hypothesis can be rejected if $p_1 > p_2$, and the obtained value of z is a positive number which is equal to or greater than the tabled critical one-tailed value at the prespecified level of significance.

c) If the directional alternative hypothesis H_1: $\pi_1 < \pi_2$ is employed, the null hypothesis can be rejected if $p_1 < p_2$, and the obtained value of z is a negative number with an absolute value which is equal to or greater than the tabled critical one-tailed value at the prespecified level of significance.

Employing the above guidelines, the following conclusions can be reached.

Since the obtained absolute value $z = 4.26$ is greater than the tabled critical two-tailed values $z_{.05} = 1.96$ and $z_{.01} = 2.58$, the nondirectional alternative hypothesis H_1: $\pi \neq \pi_2$ is supported at both the .05 and .01 levels.

Since the obtained value of z is a negative number and absolute value $z = 4.26$ is greater than the tabled critical one-tailed values $z_{.05} = 1.65$ and $z_{.01} = 2.33$, the directional alternative hypothesis H_1: $\pi_1 < \pi_2$ is supported at both the .05 and .01 levels.

The directional alternative hypothesis H_1: $\pi_1 > \pi_2$ is not supported, since as previously noted, in order for it to be supported the following must be true: $p_1 > p_2$.

Note that the above conclusions for the analysis of Example 16.1 (as well as Example 16.2) with the **z test for two independent proportions** are consistent with the conclusions that are reached when the **chi-square test for r × c tables** is employed to evaluate the same set of data. In point of fact, the square of the z value obtained with Equation 16.9 will always be equal to the value of chi-square computed with Equation 16.2. This relationship can be confirmed by the fact that in the example under discussion, $(z = -4.26)^2 = (\chi^2 = 18.18)$. It is also the case that the square of a tabled critical z value at a given level of significance will equal the tabled critical chi-square value at the corresponding level of significance. This is confirmed for the tabled critical two-tailed z and χ^2 values at the .05 and .01 levels of significance: $(z_{.05} = 1.96)^2 = (\chi^2_{.05} = 3.84)$ and $(z_{.01} = 2.58)^2 = (\chi^2_{.01} = 6.63)$.[14]

Yates' correction for continuity can also be applied to the **z test for two independent proportions**. Equation 16.10 is the continuity-corrected equation for the **z test for two independent proportions**.

$$z = \frac{[p_1 - p_2] \pm \left[\frac{1}{2}\right]\left[\frac{1}{n_1} + \frac{1}{n_2}\right]}{\sqrt{p(1 - p)\left[\frac{1}{n_1} + \frac{1}{n_2}\right]}}$$

(Equation 16.10)

The following protocol is employed with respect to the numerator of Equation 16.10: a) If $(p_1 - p_2)$ is a positive number, the term $[1/2][(1/n_1) + (1/n_2)]$ is subtracted from $(p_1 - p_2)$; and b) If $(p_1 - p_2)$ is a negative number, the term $[1/2][(1/n_1) + (1/n_2)]$ is added to $(p_1 - p_2)$. An alternative way of computing the value of the numerator of Equation 16.10 is to subtract $[1/2][(1/n_1) + (1/n_2)]$ from the absolute value of $(p_1 - p_2)$, and then restore the original sign of the latter value.

Employing Equation 16.10, the continuity-corrected value $z = -4.12$ is computed.

$$z = \frac{[.3 - .6] + \left[\frac{1}{2}\right]\left[\frac{1}{100} + \frac{1}{100}\right]}{\sqrt{(.45)(.55)\left[\frac{1}{100} + \frac{1}{100}\right]}} = -4.12$$

Note that the absolute value of the continuity-corrected z value computed with Equation 16.10 will always be smaller than the absolute z value computed with Equation 16.9. As is the case when Equation 16.9 is employed, the square of the continuity-corrected z value will equal the continuity-corrected chi-square value computed with Equation 16.4. Thus, $(z = -4.12)^2 = (\chi^2 = 16.98)$.

By employing Equation 16.10, the obtained absolute value of z is reduced to 4.12 (when contrasted with the absolute value $z = 4.26$ computed with Equation 16.9). Since the absolute value $z = 4.12$ is greater than both $z_{.05} = 1.96$ and $z_{.01} = 2.58$, the nondirectional alternative hypothesis H_1: $\pi_1 \neq \pi_2$ is still supported at both the .05 and .01 levels. The directional alternative hypothesis H_1: $\pi_1 < \pi_2$ is also still supported at both the .05 and .01 levels, since the absolute value $z = 4.12$ is greater than both $z_{.05} = 1.65$ and $z_{.01} = 2.33$.

The protocol which has been described for the z **test for two independent proportions** assumes that the researcher employs the null hypothesis H_0: $\pi_1 = \pi_2$. If, in fact, the null hypothesis stipulates a difference other than zero between the two values π_1 and π_2, Equation 16.11 is employed to compute the test statistic for the z **test for two independent proportions.**[15]

$$z = \frac{(p_1 - p_2) - (\pi_1 - \pi_2)}{\sqrt{\dfrac{p_1(1 - p_1)}{n_1} + \dfrac{p_2(1 - p_2)}{n_2}}} \qquad \textbf{(Equation 16.11)}$$

The use of Equation 16.11 is based on the assumption that $\pi_1 \neq \pi_2$. Whenever the latter is the case, instead of computing a pooled estimate for a common population proportion, it is appropriate to estimate a separate value for the proportion in each of the underlying populations (i.e., π_1 and π_2) by using the values p_1 and p_2. This is in contrast to Equation 16.9, which computes a pooled p value (which represents the pooled estimate of the proportion of observations in Column 1 in the underlying populations). In the case of Equation 16.9, the computation of a pooled p value is based on the assumption that $\pi_1 = \pi_2$.

To illustrate the application of Equation 16.11, let us assume that the following null hypothesis and nondirectional alternative hypothesis are employed for Example 16.1: H_0: $\pi_1 - \pi_2 = -.1$ (which can also be written H_0: $\pi_2 - \pi_1 = .1$) versus H_1: $\pi_1 - \pi_2 \neq -.1$ (which can also be written H_1: $\pi_2 - \pi_1 \neq .1$). The null hypothesis states that in the underlying populations represented by the samples, the difference between the proportion of observations in Row 1 (the **noise** condition) that falls in Cell a and the proportion of observations in Row 2 (the **no noise** condition) that falls in Cell c is $-.1$. The alternative hypothesis states that in the underlying populations represented by the samples, the difference between the proportion of observations in Row 1 (the **noise** condition) that falls in Cell a and the proportion of observations in Row 2 (the **no noise** condition) that falls in Cell c is some value other than $-.1$.

In order for the nondirectional alternative hypothesis H_1: $\pi_1 - \pi_2 \neq -.1$ to be supported, the obtained absolute value of z must be equal to or greater than the tabled critical two-tailed value at the prespecified level of significance. The directional alternative hypothesis which is consistent with the data is H_1: $\pi_1 - \pi_2 < -.1$. In order for the latter alternative hypothesis to be supported, the sign of the obtained value of z must be negative, and the obtained absolute value of z must be equal to or greater than the tabled one-tailed critical value at the prespecified level of significance. The directional alternative hypothesis H_1: $\pi_1 - \pi_2 > -.1$ is not consistent with the data. Either a significant positive z value (in which case $p_1 - p_2 > 0$) or a significant negative z value (in which case $0 > (p_1 - p_2) > -.1$) is required to support the latter alternative hypothesis.

Employing Equation 16.11 for the above analysis, the value $z = -2.99$ is computed.

$$z = \frac{(.3 - .6) - (-.1)}{\sqrt{\dfrac{(.3)(.7)}{100} + \dfrac{(.6)(.4)}{100}}} = -2.99$$

Since the obtained absolute value $z = 2.99$ is greater than the tabled critical two-tailed .05 and .01 values $z_{.05} = 1.96$ and $z_{.01} = 2.58$, the nondirectional alternative hypothesis $H_1: \pi_1 - \pi_2 \neq -.1$ is supported at both the .05 and .01 levels. Thus, one can conclude that in the underlying populations the difference $(\pi_1 - \pi_2)$ is some value other than $-.1$. The directional alternative hypothesis $H_1: \pi_1 - \pi_2 < -.1$ is supported at both the .05 and .01 levels, since the obtained value $z = -2.99$ is a negative number and the absolute value $z = -2.99$ is greater than the tabled critical one-tailed values $z_{.05} = 1.65$ and $z_{.01} = 2.33$. Thus, if the latter alternative hypothesis is employed, one can conclude that in the underlying populations the difference $(\pi_1 - \pi_2)$ is some value that is less than $-.1$ (i.e., is a negative number with an absolute value larger than .1). As noted earlier, the directional alternative hypothesis $H_1: \pi_1 - \pi_2 > -.1$ is not supported, since it is not consistent with the data.

Yates' correction for continuity can also be applied to Equation 16.11 by employing the same correction factor in the numerator which is employed in Equation 16.10. Using the correction for continuity, the numerator of Equation 16.11 becomes:

$$(p_1 - p_2) - (\pi_1 - \pi_2) \pm \left[\frac{1}{2}\right]\left[\frac{1}{n_1} + \frac{1}{n_2}\right]$$

Without the correction for continuity, the value of the numerator of Equation 16.11 is $-.2$. Since $[1/2][(1/100) + (1/100)] = .01$, using the guidelines outlined previously, the value of the numerator becomes $-2 + .01 = -.19$. When the latter value is divided by the denominator (which for the example under discussion equals .067), it yields the continuity-corrected value $z = -2.84$. Thus, by employing the correction for continuity, the absolute value of z is reduced from $z = 2.99$ to $z = 2.84$. Since the absolute value $z = 2.84$ is greater than $z_{.05} = 1.96$ and $z_{.01} = 2.58$, the nondirectional alternative hypothesis $H_1: \pi_1 - \pi_2 \neq -.1$ is still supported at both the .05 and .01 levels. The directional alternative hypothesis $H_1: \pi_1 - \pi_2 < -.1$ is also still supported at both the .05 and .01 levels, since the absolute value $z = 2.84$ is greater than $z_{.05} = 1.65$ and $z_{.01} = 2.33$.

Procedure for computing sample size in reference to the power of the z test for two independent proportions Equation 16.12 can be employed to estimate the sample size required in order to conduct a z test for two independent proportions which has a power of a specified value.

(Equation 16.12)

$$n_1 = \frac{\left[z_{\alpha/2}\sqrt{p(1 - p)\left(1 + \dfrac{1}{k}\right)} + z_\beta\sqrt{p_1(1 - p_1) + \left(\dfrac{p_2(1 - p_2)}{k}\right)}\right]^2}{\delta^2}$$

Where: n_1 represents the number of subjects in Group 1.

k is the ratio of the number of subjects in **Group 2** relative to the number of subjects in **Group 1**. When $n_1 = n_2$, $k = 1$. If, on the other hand, it is expected that there will be twice as many subjects in **Group 2** than **Group 1**, $k = 2$, and so on.

The terms p, p_1 and p_2 are employed to represent the best estimates of the population proportions π, π_1 and π_2. Although some sources may employ the notation π, π_1 and π_2 in Equation 16.12 in place of p, p_1 and p_2, it is common practice to employ the latter values in computing the sample size.

δ represents the minimum effect size (in proportion units) the researcher wants to be able to identify. Many sources employ the value $|p_1 - p_2|$, which is assumed to be the best estimate of $|\pi_1 - \pi_2|$ (the true difference between the population proportions), to represent the value of δ. $|p_1 - p_2|$ is commonly employed to represent the value of δ, when after obtaining a nonsignificant result a researcher wants to determine if the power of the test conducted had been set at a specific value, how large a sample size would have been required in order for the z **test for two independent proportions** to identify the observed difference between the two proportions to be significant. $z_{\alpha/2}$ represents the relevant tabled critical two-tailed z value in **Table A1** employed in the analysis.

z_β is the z value listed in **Table A1** for which a proportion (probability) is recorded which identifies the Type II error rate (i.e., β) employed in the analysis. The protocol to employ to determine the value of z_β is as follows: a) If the **power** stipulated for the test (i.e., the value $(1 - \beta)$), is **greater than .5**, z_β is the z value for which the proportion recorded in **Column 3** of **Table A1** is equal to the value **stipulated for β**. (It is also the z value for which the proportion recorded in **Column 2** of **Table A1** is equal to the value (**Power** – .5).) When the **power** stipulated for the test is **greater than .5**, the sign of z_β will always be **positive**; b) If the **power** stipulated for the test is **less than .5**, z_β is the z value for which the proportion recorded in **Column 3** of **Table A1** is equal to the **power** of the test (i.e., the value $(1 - \beta)$). (It is also the z value for which the proportion recorded in **Column 2** of **Table A1** is equal to the value (β – .5).) When the **power** stipulated for the test is **less than .5**, the sign of z_β will always be **negative**; c) If the **power** stipulated for the test is .5, $z_\beta = 0$. (In **Table A1** the value $\beta = .5000$ is listed in **Column 3**, and the value .0000 in **Column 2** represents the difference between β and .5 (i.e., β – .5 = 0, with **Power** = 1 – β = .5).)

Equation 16.12 will be employed to demonstrate the computation of an appropriate sample size for the study summarized in Table 16.2 in which subjects in the **noise** condition (**Group 1**) are compared with subjects in the **no noise** condition (**Group 2**). We will assume that the researcher plans to evaluate the data with a z **test for two independent proportions**, and that a two-tailed analysis will be conducted with alpha set at .05. In addition, the researcher plans to have an equal number of subjects in each group, and wants the power of the test to be .80. The following values will be employed in Equation 16.12 to compute the number of subjects that will be required in **Group 1**(which will also be the number of subjects the researcher will need for **Group 2**, since when there are an equal number of subjects per group, $k = 1$): $p_1 = .3$, $(1 - p_1)$ = .7, $p_2 = .6$, $(1 - p_2) = .4$, $p = .45$, $(1 - p) = .55$, $z_{\alpha/2} = 1.96$ (If a one-tailed analysis is conducted $z_{\alpha/2} = 1.65$ should be employed.), $z_\beta = .84$. The rationale for employing the value $z_\beta = .84$ is as follows: Since **Power** = 1 – β = .80, $\beta = .20$. Employing the guidelines for using Equation 16.12 documented above to compute a sample size when **Power** > .5, in **Table A1**, $z = .84$ is the z value for which the proportion (probability) recorded in **Column 3** (.2005) is closest to the value β = .20. (It is also the case that for $z = .84$, if .5 is added to the value recorded in **Column 2** (i.e., .2995 + .5 = .7995), the resulting value is closest to the value .80 stipulated for the power of the test (and thus, **Power** – .5 = .7995 – .5 = .2995).) Since the power stipulated for the test is greater than .5, the sign of z_β will be positive. The observed absolute difference between the two proportions $|p_1 - p_2| = |.3 - .6| = .3$ will be employed to represent the value of δ.

When the appropriate values are substituted in Equation 16.12, the value $n_1 = 41.92$ is computed, indicating that in order for the power of the z **test for two independent proportions** to be .80, the researcher must have at least 42 subjects per group (which is the next integer value that is greater than the value computed for n_1). Since, in fact, 100 subjects were employed in the study summarized in Table 16.2, the power of the test was well above .80.

$$n_1 = \frac{\left[1.96\sqrt{(.45)(.55)(2)} + .84\sqrt{(.3)(.7) + (.6)(.4)}\right]^2}{(.3 - .6)^2} = 41.92$$

A number of sources state that because it does not employ a correction for continuity, Equation 16.12 may underestimate the required sample size. Fleiss *et al.* (2003, p. 72) and Zar (1999, p. 560) note that Equation 16.13 may provide a more accurate estimate of the required sample size (where n' represents the value computed for n_1 with Equation 16.12). For the problem under discussion, use of Equation 16.13 would indicate that the sample size per group should be increased from approximately 42 to 49 in order to insure that the power of the test is .80. Further discussion of evaluating the power of the z **test for two independent proportions**, as well as alternative methodologies, can be found in Feinstein (2002, p. 503), Fleiss *et al.* (2003, Chapter 4), Rosner (2000, pp. 384–387) and Zar (1999, pp. 558–563).

(Equation 16.13)

$$n_1 = \frac{n'}{4}\left(1 + \sqrt{1 + \frac{4}{n'\delta}}\right)^2 = \frac{41.92}{4}\left(1 + \sqrt{1 + \frac{4}{(41.92)(|.3 - .6|)}}\right)^2 = 48.36$$

In addition to the sources cited above, Cohen (1977, 1988) has developed a statistic called the h **index** which can be employed to compute the power of the z **test for two independent proportions**. The value h is an **effect size** index reflecting the difference between two population proportions. The concept of effect size is discussed in Section VI of the **single-sample t test (Test 2)**. It is discussed in greater detail in Section VI of the t **test for two independent samples**, and in Section IX (the **Addendum**) of the **Pearson product-moment correlation coefficient** under the discussion of **meta-analysis and related topics**.

The equation for the h **index** is $h = \varphi_1 - \varphi_2$ (where φ_1 and φ_2 are the arcsine transformed values for the proportions). Cohen (1977; 1988, Ch. 6) has derived tables which allow a researcher through use of the h **index** to determine the appropriate sample size to employ if one wants to test a hypothesis about the difference between two population proportions at a specified level of power. Cohen (1977; 1988, pp. 184–185) has proposed the following (admittedly arbitrary) h values as criteria for identifying the magnitude of an effect size: a) A **small effect size** is one that is greater than .2 but not more than .5; b) A **medium effect size** is one that is greater than .5 but not more than .8; and c) A **large effect size** is greater than .8.

6. Computation of a confidence interval for a difference between two proportions With **large sample sizes**, a confidence interval can be computed which identifies a range of values within which one can be confident to a specified degree that the true difference lies between the two population proportions π_1 and π_2. Equation 16.14, which employs the normal distribution, is the large sample approximation for computing the confidence interval for the difference between two population proportions. The notation employed in Equation 16.14 is identical to that used in the discussion of the z **test for two independent proportions**.

$$CI_{(1 - \alpha)} = (p_1 - p_2) \pm (z_{\alpha/2})(s_{p_1 - p_2}) \qquad \textbf{(Equation 16.14)}$$

Where: $s_{p_1 - p_2} = \sqrt{[[p_1(1 - p_1)]/n_1] + [[p_2(1 - p_2)]/n_2]}$

$z_{\alpha/2}$ represents the tabled critical two-tailed value in the normal distribution, below which a proportion (percentage) equal to $[1 - (\alpha/2)]$ of the cases falls. If the proportion (percentage) of the distribution that falls within the confidence interval is subtracted from 1 (100%), it will equal the value of α.

Employing Equation 16.14, the 95% confidence interval for Examples 16.1/16.2 is computed below. In employing Equation 16.14, $(p_1 - p_2)$ (which is the numerator of Equation 16.9) represents the obtained difference between the sample proportions, $z_{.05}$ represents the tabled critical two-tailed .05 value $z_{.05} = 1.96$, and $s_{p_1 - p_2}$ (which is the denominator of Equation 16.11) represents the **standard error of the difference between two proportions.**[16]

$$p_1 - p_2 = .3 - .6 = -.3 \qquad s_{p_1 - p_2} = \sqrt{\frac{(.3)(.7)}{100} + \frac{(.6)(.4)}{100}} = .067$$

$$CI_{.95} = (p_1 - p_2) \pm (z_{.05})(s_{p_1 - p_2}) = -.3 \pm (1.96)(.067) = -.3 \pm .131$$

$$-.169 \geq (\pi_1 - \pi_2) \geq -.431$$

This result indicates the researcher can be 95% confident (or the probability is .95) that the interval $-.431$ to $-.169$ contains the true difference between the population proportions. If the aforementioned interval contains the true difference between the population proportions, the proportion for Population 1 (π_1) is less than the proportion for Population 2 (π_2) by a value that is less than or equal to .169 but not less than .431. The result can also be written as $.169 \leq (\pi_2 - \pi_1) \leq .431$, which indicates that if the aforementioned interval contains the true difference between the population proportions, the proportion for Population 2 (π_2) is larger than the proportion for Population 1 (π_1) by a value that is greater than or equal to .169 but not greater than .431.

The 99% confidence interval, which results in a broader range of values, is computed below employing the tabled critical two-tailed .01 value $z_{.01} = 2.58$ in Equation 16.14 in lieu of $z_{.05} = 1.96$.

$$CI_{.99} = (p_1 - p_2) \pm (z_{.01})(s_{p_1 - p_2}) = -.3 \pm (2.58)(.067) = -.3 \pm .173$$

$$-.127 \geq (\pi_1 - \pi_2) \geq -.473$$

Thus, the researcher can be 99% confident (or the probability is .99) that the interval $-.127$ to $-.473$ contains the true difference between the population proportions. If the aforementioned interval contains the true difference between the population proportions, the proportion for Population 1 (π_1) is less than the proportion for Population 2 (π_2) by a value that is less than or equal to .127 but not less than .473.

Some sources (Fleiss *et al.* (2003, pp. 60–61) and Zar (1999, pp. 555–557)) employ a **correction for continuity** in computing the confidence interval. In the latter case, the width of the computed confidence interval is wider than the width of the confidence interval computed with Equation 16.14. The continuity corrected confidence interval is computed with Equation 16.15.

(Equation 16.15)

$$(\pi_1 - \pi_2) \geq (p_1 - p_2) - (z_{\alpha/2})\,(s_{p_1 - p_2}) - \frac{1}{2}\left(\frac{1}{n_1} + \frac{1}{n_2}\right)$$

and

$$(\pi_1 - \pi_2) \leq (p_1 - p_2) + (z_{\alpha/2})\,(s_{p_1 - p_2}) + \frac{1}{2}\left(\frac{1}{n_1} + \frac{1}{n_2}\right)$$

Since $\frac{1}{2}(1/n_1 + 1/n_2) = \frac{1}{2}[(1/100) + (1/100)] = .01$, the latter value is subtracted from the lower limit of the confidence interval computed with Equation 16.14 and added to the upper limit of the confidence interval computed with the latter equation. Thus, in the case of the 95% confidence interval, the continuity corrected interval is $-.159 \geq (\pi_1 - \pi_2) \geq -.441$, which as noted earlier is slightly larger than the confidence interval computed with Equation 16.14.

Grissom and Kim (2005, p. 183) discuss alternative methods for computing a confidence interval for a difference between two proportions.

7. Test 16e: The median test for independent samples The model for the **median test for independent samples** assumes there are k independent groups, and that within each group each observation is categorized with respect to whether it is **above** or **below** a composite median value. In actuality, the **median test for independent samples** is a label which is commonly used when the **chi-square test for r × c tables**, the **z test for two independent proportions**, or the **Fisher exact test** is employed to evaluate the hypothesis that in each of k groups there is an equal proportion of observations which are **above** versus **below** a composite median. With large sample sizes, the **median test for independent samples** is computationally identical to the **chi-square test for r × c tables** (when $k \geq 2$) and the **z test for two independent proportions** (when $k = 2$). In the case of small sample sizes, the test is computationally identical to the **Fisher exact test** (when $k = 2$). Table 16.6, which is used to summarize the model for the three aforementioned tests, can also be applied to the **median test for independent samples**. The two rows are employed to represent the two groups, and the two columns are used to represent the two categories on the dependent variable — specifically, whether a score falls **above** versus **below the median**. Example 16.4 will be employed to illustrate the **median test for two independent samples**.

Example 16.4 *A study is conducted to determine whether or not five-year-old females are more likely than five-year-old males to score above the population median on a standardized test of eye-hand coordination. One hundred randomly selected females and 100 randomly selected males are administered the test of eye-hand coordination, and categorized with respect to whether they score above or below the overall population median (i.e., the 50th percentile for both males and females). Table 16.11 summarizes the results of the study. Do the data indicate there are gender differences in performance?*

Table 16.11 Summary of Data for Example 16.4

	Above the median	Below the median		Row sums
Males	30	70		100
Females	60	40		100
Column sums	90	110	Total observations	200

Since Example 16.4 involves a large sample size, it can be evaluated with Equation 16.2, the equation for the **chi-square test for $r \times c$ tables** (as well as with Equation 16.9, the equation for the z **test for two independent proportions**). The reader should take note of the fact that the study described in Example 16.4 conforms to the model for the **chi-square test for homogeneity**. This is the case, since the row sums (i.e., the number of **males** and **females**) are predetermined by the researcher. Since it is consistent with the model for the latter test, the null and alternative hypotheses which are evaluated with the **median test for independent samples** are identical to those evaluated with the **chi-square test for homogeneity**, the **Fisher exact test**, and the z **test for two independent proportions**.[17]

Since the data for Example 16.4 are identical to the data for Examples 16.1/16.2, analysis of Example 16.4 with Equation 16.2 yields the value $\chi^2 = 18.18$ (which is the value obtained for Examples 16.1/16.2). Since (as noted earlier) $\chi^2 = 18.18$ is significant at both the .05 and .01 levels, one can conclude that **females** are more likely than **males** to score **above the median**.

In the case of the **median test for independent samples**, in the event that one or more subjects obtain a score which is equal to the population median, the following options are available for handling such scores: a) If the number of subjects who obtain a score which equals the median is reasonably large, a strong argument can be made for adding a third column to Table 16.11 for subjects who scored **at the median**. In such a case the contingency table is transformed into a 2×3 table; b) If a minimal number of subjects obtain a score at the median, such subjects can be dropped from the data; and c) Within each group, half the scores which fall at the median value are assigned to the **above the median** category and the other half to the **below the median** category. In the final analysis, the critical thing the researcher must be concerned with in employing any of the aforementioned strategies is to make sure that the procedure he employs does not lead to misleading conclusions regarding the distribution of scores in the underlying populations.

The **median test for independent samples** can be extended to designs involving more than two groups. As an example, in Example 16.4, instead of evaluating the number of **males** and **females** who score **above** versus **below the median**, four groups of children representing different ethnic groups (e.g., **Caucasian, Asian-American, African-American, Native-American**) could be evaluated with respect to whether they score **above** versus **below the median**. In such a case, the data are summarized in the form of a 4×2 contingency table, with the four rows representing the four ethnic groups and the two columns representing **above** versus **below the median**.

It should be noted that some sources categorize the **median test for independent samples** as a test of ordinal data, since categorizing scores with respect to whether they are **above** versus **below the median** involves placing scores in one of two ordered categories. The reader should also be aware of the fact that it is possible to categorize scores on more than two ordered categories. As an example, **male** and **female** children can be categorized with respect to whether they score in the **first quartile** (lower 25%), **second quartile** (25% to 50%), **third quartile** (50% to 75%), or **fourth quartile** (upper 25%) on the test of eye–hand coordination. In such a case, the data are summarized in the form of a 2×4 contingency table, with the two rows representing **males** and **females** and the four columns representing the four quartiles. It is also possible to have a design involving more than two groups of subjects (e.g., the four ethnic groups discussed above), and a dependent variable involving more than two ordered categories (e.g., four quartiles). The data for such a design are summarized in the form of a 4×4 contingency table. Finally, it is possible to have contingency tables in which both the row and the column variables are ordered. Although the **chi-square test for $r \times c$ tables** can be employed to evaluate a design in which both variables are ordered, alternative procedures may provide the researcher with more information about the relationship between the two variables. An example of such an alternative procedure is **Goodman and Kruskal's gamma (Test 32)** discussed later in the book.

8. Extension of the chi-square test for r × c tables to contingency tables involving more than two rows and/or columns, and associated comparison procedures It is noted in the previous section that the **chi-square test for r × c tables** can be employed with tables involving more than two rows and/or columns. In this section larger contingency tables will be discussed, and within the framework of the discussion additional analytical procedures which can be employed with such tables will be described. Example 16.5 will be employed to illustrate the use of the **chi-square test for r × c tables** with a larger contingency table — specifically, a 4 × 3 table.

Example 16.5 *A researcher conducts a study in order to determine if there are differences in the frequency of biting among different species of laboratory animals. He selects random samples of four laboratory species from the stock of various animal supply companies. Sixty mice, 50 gerbils, 90 hamsters, and 80 guinea pigs are employed in the study. Each of the animals is handled over a two-week period, and categorized into one of the following three categories with respect to biting behavior: not a biter, mild biter, flagrant biter.* Table 16.12 *summarizes the data for the study. Do the data indicate there are interspecies differences in biting behavior?*

Table 16.12   Summary of Data for Example 16.5

	Not a biter	Mild biter	Flagrant biter		Row sums
Mice	20	16	24		60
Gerbils	30	10	10		50
Hamsters	50	30	10		90
Guinea pigs	19	11	50		80
Column sums	119	67	94	Total observations	280

The study described in Example 16.5 conforms to the model for the **chi-square test for homogeneity**. This is the case, since the row sums (i.e., the number of animals representing each of the four species) are predetermined by the researcher. The row variable represents the independent variable in the study. The independent variable, which is comprised of four levels, is nonmanipulated, since it is based on a preexisting subject characteristic (i.e., species). The reader should take note of the fact that it is not necessary to have an equal number of subjects in each of the groups/categories which constitute the row variable. The column variable, which is comprised of three categories, is the biting behavior of the animals. The latter variable represents the dependent variable in the study. Note that the marginal sums for the column variable are not predetermined by the researcher.

As is the case with a 2 × 2 contingency table, Equation 16.1 is employed to compute the expected frequency for each cell. As an example, the expected frequency of Cell$_{11}$ (**mice/not a biter**) is computed as follows: $E_{11} = [(O_{1.})(O_{.1})]/n = [(60)(119)]/280 = 25.5$. In employing Equation 16.1, the value 60 represents the sum for Row 1 (which represents the total number of **mice**), the value 119 represents the sum for Column 1 (which represents the total number of animals categorized as **not a biter**), and 280 represents the total number of subjects/observations in the study.

After employing Equation 16.1 to compute the expected frequency for each of the 4 × 3 = 12 cells in the contingency table, Equation 16.2 is employed to compute the value $\chi^2 = 59.16$. The analysis is summarized in Table 16.13.

Table 16.13 Chi-Square Summary Table for Example 16.5

Cell	O_{ij}	E_{ij}	$(O_{ij} - E_{ij})$	$(O_{ij} - E_{ij})^2$	$\dfrac{(O_{ij} - E_{ij})^2}{E_{ij}}$
Mice/Not a biter	20	25.50	−5.50	30.25	1.19
Mice/Mild biter	16	14.36	1.64	2.69	.19
Mice/Flagrant biter	24	20.14	3.86	14.90	.74
Gerbils/Not a biter	30	21.25	8.75	76.56	3.60
Gerbils/Mild biter	10	11.96	−1.96	3.84	.32
Gerbils/Flagrant biter	10	16.79	−6.79	46.10	2.75
Hamsters/Not a biter	50	38.25	11.75	138.06	3.61
Hamsters/Mild biter	30	21.54	8.46	71.57	3.32
Hamsters/Flagrant biter	10	30.21	−20.21	408.44	13.52
Guinea pigs/Not a biter	19	34.00	−15.00	225.00	6.62
Guinea pigs/Mild biter	11	19.14	−8.14	66.26	3.36
Guinea pigs/Flagrant biter	50	26.86	23.14	535.46	19.94
Column sums	280	280.00	0		$\chi^2 = 59.16$

Substituting the values $r = 4$ and $c = 3$ in Equation 16.3, the number of degrees of freedom for the analysis are $df = (4 - 1)(3 - 1) = 6$. Employing **Table A4**, the tabled critical .05 and .01 chi-square values for $df = 6$ are $\chi^2_{.05} = 12.59$ and $\chi^2_{.01} = 16.81$. Since the computed value $\chi^2 = 59.16$ is greater than both of the aforementioned critical values, the null hypothesis can be rejected at both the .05 and .01 levels. By virtue of rejecting the null hypothesis, the researcher can conclude that the four species are not homogeneous with respect to biting behavior, or to be more precise, that at least two of the species are not homogeneous.

In the case of a 2 × 2 contingency table, a significant result indicates that the two groups employed in the study are not homogeneous with respect to the dependent variable. However, in the case of a larger contingency table, although a significant result indicates that at least two of the r groups are not homogeneous, the chi-square analysis does not indicate which of the groups differ from one another or which of the cells are responsible for the significant effect. Visual inspection of Tables 16.12 and 16.13 suggests that the significant effect in Example 16.5 is most likely attributable to the disproportionately large number of **flagrant biters** among **guinea pigs**, and the disproportionately small number of **flagrant biters** among **hamsters**. In lieu of visual inspection (which is not a precise method for identifying the cells which are primarily responsible for a significant effect), the following two types of comparisons are among those that can be conducted.

Simple comparisons A **simple comparison** is a comparison between two of the r rows of an $r \times c$ contingency table (or two of the c columns). Table 16.14 summarizes a simple comparison which contrasts the biting behavior of **mice** with the biting behavior of **guinea pigs**. Note that in the simple comparison, the data for the other two species employed in the study (i.e., **gerbils** and **hamsters**) are not included in the analysis.

It should be noted that a simple comparison does not have to involve all of the columns of the contingency table. Thus, one can compare the two species **mice** and **guinea pigs**, but limit the comparison within those species to only those animals who are classified **not a biter** and **flagrant biter**. The resulting 2 × 2 contingency table for such a comparison will differ from Table 16.14 in that the second column (**mild biter**) is not included. As a result of omitting the latter column, the row sum for **mice** is reduced to 44 and the row sum for **guinea pigs** is reduced to 69. The total number of observations for the comparison is 113.

Table 16.14 Simple Comparison for Example 16.5

	Not a biter	Mild biter	Flagrant biter	Row sums
Mice	20	16	24	60
Guinea pigs	19	11	50	80
Column sums	39	27	74	Total observations 140

Complex comparisons A **complex comparison** is a comparison between two or more of the *r* rows of an *r* × *c* contingency table with one of the other rows or two or more of the other rows of the table.[18] Table 16.5 summarizes an example of a complex comparison which contrasts the biting behavior of **guinea pigs** with the combined biting behavior of the other three species employed in the study (i.e., **mice, gerbils,** and **hamsters**).

As is the case for a simple comparison, a complex comparison does not have to include all of the columns on which the groups are categorized. Thus, one can compare the **mice, gerbils, and hamsters** with **guinea pigs,** but limit the comparison to only those animals who are classified **not a biter** and **flagrant biter**. The resulting 2 × 2 contingency table for such a comparison will differ from Table 16.15, in that the second column (**mild biter**) will not be included. As a result of omitting the latter column, the row sum for **mice, gerbils, and hamsters** is reduced to 144 and the row sum for **guinea pigs** is reduced to 69. The total number of observations for the comparison is 213.

The null and alternative hypotheses for simple and complex comparisons are identical to those which are employed for evaluating the original *r* × *c* table, except for the fact that they are stated in reference to the specific cells involved in the comparison.

Table 16.15 Complex Comparison for Example 16.5

	Not a biter	Mild biter	Flagrant biter	Row sums
Mice, Gerbils, and Hamsters	100	56	44	200
Guinea pigs	19	11	50	80
Column sums	119	67	94	Total observations 280

Sources are not in agreement with respect to what protocol is most appropriate to employ in conducting simple and/or complex comparisons following the computation of a chi-square value for an *r* × *c* contingency table. Among those sources which describe comparison procedures for *r* × *c* contingency tables are Keppel and Saufley (1980), Keppel *et al.* (1992), Marascuilo and McSweeney (1977), and Siegel and Castellan (1988). Keppel *et al.* (1992) note that Wickens (1989) states that, as a general rule, the different protocols which have been developed for conducting comparisons yield comparable results. In discussing the general issue of partitioning contingency tables (i.e., breaking down the table for the purposes of conducting comparisons), most sources emphasize that whenever feasible, a researcher should plan a limited number of simple and/or complex comparisons prior to the data collection phase of a study, and that any comparisons one conducts should be meaningful at a theoretical and/or practical level.[19] In the discussion of comparisons in Section VI of the **chi-square goodness-of-fit test,** it is noted that when a limited number of comparisons are planned beforehand, most sources take the position that a researcher is not obliged to control the overall Type I error rate. However, when comparisons are not planned, there is general agreement that in order to avoid inflating the Type I error rate, the latter value should be adjusted. One way of adjusting the Type I error rate is to divide the maximum overall Type I error rate one is willing to tolerate by the total number of comparisons to be conducted. The resulting probability value can then be employed as the alpha level for each comparison that is conducted. To illustrate, if one intends to conduct three comparisons and does not want the overall Type I error rate for all of the comparisons to be greater than $\alpha = .05$, the alpha level to employ for each comparison is α/number of comparisons $= .05/3 = .0167$. There are those who would argue the latter adjustment is too severe, since it substantially reduces the power associated with each comparison. In the final analysis, the researcher must be the one who decides what per comparison alpha level strikes an equitable balance in terms of the likelihood of committing a Type I error and the power associated with a comparison. Obviously, if a researcher employs a severely reduced alpha value, it may become all but impossible to detect actual differences which exist between the underlying populations. Before continuing this section, the reader may find it useful to review the discussion of comparisons for the **chi-square goodness-of-fit test**. A thorough discussion of the issues involved in conducting comparisons can be found in Section VI of the **single-factor between-subjects analysis of variance (Test 21)**. In the remainder of this section, two comparison procedures for an $r \times c$ contingency table (based on Keppel *et al.* (1992) and Keppel and Saufley (1980)) will be described. The procedures to be described can be employed for both planned and unplanned comparisons.

Method 1 The first procedure to be described (which is derived from Keppel *et al.* (1992)) employs for both simple and complex comparisons the same protocol to evaluate a comparison contingency table as the protocol which is employed when the complete $r \times c$ table is evaluated. In the case of Example 16.5, Equation 16.2 is employed to evaluate the simple comparison summarized by the 2×3 contingency table in Table 16.14. The analysis assumes there is a total of 140 observations. Equation 16.1 is employed to compute the expected frequency of each cell — i.e., the sum of the observations in the row in which the cell appears is multiplied by the sum of the observations in the column in which the cell appears, with the resulting product divided by the total number of observations in the 2×3 contingency table. As an example, the expected frequency of Cell$_{11}$ (**mice/not a biter**) is computed as follows: $E_{11} = [(O_{1.})(O_{.1})]/n = [(60)(39)]/140 = 16.71$. In employing Equation 16.1, the value 60 represents the sum for Row 1 (which represents the total number of **mice**), the value 39 represents the sum for Column 1 (which represents the total number of **mice** and **guinea pigs** categorized as **not a biter**), and 140 represents the total number of observations in the table (i.e., **mice** and **guinea pigs**).

The sums for the two rows/species involved in the simple comparison under discussion are identical to the row sums for those species in the original 4×3 contingency table. The column sums, however, only represent the sums of the columns for the two species involved in the comparison, and are thus different from the column sums for all four species which are computed when the original 4×3 contingency table is evaluated. Table 16.16 summarizes the computation of the value $\chi^2 = 7.39$ (with Equation 16.2) for the simple comparison summarized in Table 16.14.

Substituting the values $r = 2$ and $c = 3$ in Equation 16.3, the number of degrees of freedom for the comparison are $df = (2-1)(3-1) = 2$. Employing **Table A4**, the tabled critical .05 and .01 chi-square values for $df = 2$ are $\chi_{.05}^2 = 5.99$ and $\chi_{.01}^2 = 9.21$. Since the computed value $\chi^2 = 7.39$ is greater than $\chi_{.05}^2 = 5.99$, the null hypothesis can be rejected at the .05 level. It cannot, however, be rejected at the .01 level, since $\chi^2 = 7.39$ is less than $\chi_{.01}^2 = 9.21$. By virtue of rejecting the null hypothesis the researcher can conclude that **mice** and **guinea pigs**, the two species involved in the comparison, are not homogeneous with respect to biting behavior. Inspection of Table 16.14 reveals that the significant difference can primarily be attributed to the fact there are a disproportionately large number of **flagrant biters** among **guinea pigs**.

Table 16.16 Chi-Square Summary Table for Simple Comparison in Table 16.14 Employing Method 1

Cell	O_{ij}	E_{ij}	$(O_{ij} - E_{ij})$	$(O_{ij} - E_{ij})^2$	$\dfrac{(O_{ij} - E_{ij})^2}{E_{ij}}$
Mice/Not a biter	20	16.71	3.29	10.82	.65
Mice/Mild biter	16	11.57	4.43	19.62	1.70
Mice/Flagrant biter	24	31.71	−7.71	59.44	1.87
Guinea pigs/Not a biter	19	22.29	−3.29	10.82	.49
Guinea pigs/Mild biter	11	15.43	−4.43	19.62	1.27
Guinea pigs/Flagrant biter	50	42.29	7.71	59.44	1.41
Column sums	140	140.00	0		$\chi^2 = 7.39$

Method 2   An alternative and somewhat more cumbersome method for conducting both simple and complex comparisons described by Bresnahan and Shapiro (1966) and Castellan (1965) is presented in Keppel and Saufley (1980). Although most statisticians would consider the method to be described in this section as preferable to **Method 1**, in most cases the latter method will result in similar conclusions with regard to the hypothesis under study. The method to be described in this section is identical to **Method 1**, except for the fact that in computing the chi-square value, for each cell in the comparison contingency table the value $(O_{ij} - E_{ij})^2$ is divided by the expected frequency computed for the cell when the original $r \times c$ contingency table is evaluated (instead of the expected frequency for the cell based on the row and column sums in the comparison contingency table). Specifically, for each of the cells in the comparison contingency table, the following values are computed: a) The expected frequency for each cell is computed as it is for **Method 1** (i.e., using the row and column sums for the comparison contingency table and employing the number of observations in the comparison contingency table to represent the value of *n*); b) The values $(O_{ij} - E_{ij})$ and $(O_{ij} - E_{ij})^2$ are computed for each cell just as they are in **Method 1**; and c) In computing the values in the last column of the chi-square summary table, instead of dividing $(O_{ij} - E_{ij})^2$ by E_{ij}, it is divided by the expected frequency of the cell when all $r \times c$ cells in the original table are employed to compute the expected cell frequency. In the analysis to be described, the latter expected for each cell is represented by the notation E_{ij}'. The steps noted above are illustrated in Table 16.17, which summarizes the application of **Method 2** for computing the chi-square statistic for the simple comparison presented in Table 16.14. Note that the value of E_{ij}' for each cell is the same as the value of the expected frequency for that cell in Table 16.13.

As is the case when **Method 1** is employed, the null hypothesis can be rejected at the .05 level but not at the .01 level, since for $df = 2$, the computed value $\chi^2 = 8.12$ is greater than $\chi^2_{.05} = 5.99$, but less than $\chi^2_{.01} = 9.21$. Although the computed value $\chi^2 = 8.12$ is slightly larger than the value $\chi^2 = 7.39$ computed with **Method 1**, the difference between the two chi-square values is minimal. As noted previously, the two methods will generally yield approximately the same value.

Method 1 and **Method 2** are also employed to evaluate the complex comparison summarized in Table 16.15. The results of these analyses are summarized in Tables 16.18 and 16.19. Although in the case of the complex comparison, **Method 1** and **Method 2** both yield the value $\chi^2 = 42.03$, the two methods will not always yield the same value. Substituting the values $r = 2$ and $c = 3$ in Equation 16.3, the number of degrees of freedom for the comparison are $df = (2 - 1)(3 - 1) = 2$. Employing **Table A4**, the tabled critical .05 and .01 chi-square values

**Table 16.17 Chi-Square Summary Table for Simple Comparison
in Table 16.14 Employing Method 2**

Cell	O_{ij}	E_{ij}	$(O_{ij} - E_{ij})$	$(O_{ij} - E_{ij})^2$	E'_{ij}	$\dfrac{(O_{ij} - E_{ij})^2}{E'_{ij}}$
Mice/Not a biter	20	16.71	3.29	10.82	25.50	.42
Mice/Mild biter	16	11.57	4.43	19.62	14.36	1.37
Mice/Flagrant biter	24	31.71	−7.71	59.44	20.14	2.95
Guinea pigs/Not a biter	19	22.29	−3.29	10.82	34.00	.32
Guinea pigs/Mild biter	11	15.43	−4.43	19.62	19.14	1.03
Guinea pigs/Flagrant biter	50	42.29	7.71	59.44	26.86	2.03
Column sums	140	140.00	0			$\chi^2 = 8.12$

**Table 16.18 Chi-Square Summary Table for Complex Comparison
in Table 16.15 Employing Method 1**

Cell	O_{ij}	E_{ij}	$(O_{ij} - E_{ij})$	$(O_{ij} - E_{ij})^2$	$\dfrac{(O_{ij} - E_{ij})^2}{E_{ij}}$
Mice, Gerbils, Hamsters/Not a biter	100	85.00	15.00	225.00	2.65
Mice, Gerbils, Hamsters/Mild biter	56	47.86	8.14	66.26	1.38
Mice, Gerbils, Hamsters/Flagrant biter	44	67.14	−23.14	535.46	7.98
Guinea pigs/Not a biter	19	34.00	−15.00	225.00	6.62
Guinea pigs/Mild biter	11	19.14	−8.14	66.26	3.46
Guinea pigs/Flagrant biter	50	26.86	23.14	535.46	19.34
Column sums	280	280.00	0		$\chi^2 = 42.03$

**Table 16.19 Chi-Square Summary Table for Complex Comparison
in Table 16.15 Employing Method 2**

Cell	O_{ij}	E_{ij}	$(O_{ij} - E_{ij})$	$(O_{ij} - E_{ij})^2$	E'_{ij}	$\dfrac{(O_{ij} - E_{ij})^2}{E'_{ij}}$
Mice, Gerbils, Hamsters/Not a biter	100	85.00	15.00	225.00	85.00	2.65
Mice, Gerbils, Hamsters/Mild biter	56	47.86	8.14	66.26	47.86	1.38
Mice, Gerbils, Hamsters/Flagrant biter	44	67.14	−23.14	535.46	67.14	7.98
Guinea pigs/Not a biter	19	34.00	−15.00	225.00	34.00	6.62
Guinea pigs/Mild biter	11	19.14	−8.14	66.26	19.14	3.46
Guinea pigs/Flagrant biter	50	26.86	23.14	535.46	26.86	19.94
Column sums	280	280.00	0			$\chi^2 = 42.03$

for $df = 2$ are $\chi^2_{.05} = 5.99$ and $\chi^2_{.01} = 9.21$. Since the computed value $\chi^2 = 42.03$ is greater than both of the aforementioned critical values, the null hypothesis can be rejected at both the .05 and .01 levels. By virtue of rejecting the null hypothesis, the researcher can conclude that a combined population of **mice, gerbils, and hamsters** is not homogeneous with a population of **guinea pigs** with respect to biting behavior. Inspection of Table 16.15 reveals the latter can primarily be attributed to the discrepancy between the number of **guinea pigs** in the **flagrant biters** category and the number of **mice, gerbils, and hamsters** in the **not a biter** category.

As previously noted, if the two comparisons summarized in Tables 16.14 and 16.15 are not planned prior to collecting the data, most sources would argue that each comparison should be evaluated at a lower alpha level. If one does not want the likelihood of committing at least one

Type I error in the set of two comparisons to be greater than .05, one can adjust the alpha level as follows: Adjusted α level = α/number of comparisons = .05/2 = .025. Thus, in evaluating each of the comparisons, the tabled critical chi-square value at the .025 level is employed instead of the tabled critical .05 value $\chi^2_{.05} = 5.99$ (although as noted earlier some sources might consider the latter adjustment to be too severe). In **Table A4**, for the appropriate degrees of freedom, the tabled critical .025 value corresponds to the value listed under $\chi^2_{.975}$. In the case of $df = 2$, $\chi^2_{.975} = 7.38$. Note that for the simple comparison discussed in this section, the obtained chi-square value $\chi^2 = 7.39$ obtained with **Method 1** barely achieves significance if the latter critical value is used, whereas without the adjustment the latter result is significant at the .05 level by a comfortable margin. This example should serve to illustrate that by employing a lower alpha level, in addition to decreasing the likelihood of committing a Type I error, one also (by virtue of reducing the power of the test) decreases the likelihood of rejecting a false null hypothesis.

   The rationale for presenting two comparison methods in this section is to demonstrate that although there is not a consensus among different sources with respect to what procedure should be employed for conducting comparisons, as a general rule, if a significant effect is present it will be identified regardless of which method one employs. Although **Method 1** is the simpler of the two methods described in this section, as noted earlier, most sources would probably take the position that **Method 1** is more subject to challenge on statistical grounds. Although in some instances there will be differences with respect to the precise probability values associated with the two methods (as well as the probabilities associated with other available methods), in most cases such differences will be trivial, and will thus be of little importance in terms of their practical and/or theoretical implications. In the final analysis, the per comparison alpha level one elects to employ (rather than the use of a different comparison procedure) is the most likely reason why two or more researchers may reach dramatically different conclusions with respect to a specific comparison. In such an instance, a replication study is the best available option for clarifying the status of the null hypothesis. Alternative procedures for conducting comparisons with contingency tables are described in Marascuilo and McSweeney (1977), Marascuilo and Serlin (1988), and Siegel and Castellan (1988).

9. The analysis of standardized residuals   As noted in Section VI of the **chi-square goodness-of-fit test**, an alternative procedure for conducting comparisons (developed by Haberman (1973) and cited in Siegel and Castellan (1988)) involves the computation of **standardized residuals**. Within the context of a chi-square analysis, the term **residual** is employed to represent the absolute difference between the expected and observed cell frequencies. A **standardized residual** is a residual expressed within a format of a **standard deviation score** (i.e., a z score). By virtue of standardizing difference scores, standardized residuals allow one to compare two or more contingency tables with one another (independent of the sample sizes employed in the different contingency tables) with respect to the overall degree of difference between expected and observed cell frequencies.

   By computing standardized residuals, one is able to determine which cells are the major contributors to a significant chi-square value. The computation of residuals can be useful in reinforcing or clarifying information derived from the comparison procedures described in the previous section, as well as providing additional information on the data contained in an r × c table. Through use of Equation 16.16, a standardized residual ($z_{res_{ij}}$) can be computed for each cell in an r × c contingency table.

$$z_{res_{ij}} = \frac{(O_{ij} - E_{ij})}{\sqrt{E_{ij}}}$$ **(Equation 16.16)**

The value computed for a residual with Equation 16.16, which is interpreted as a normally distributed variable, is evaluated with **Table A1**. Any residual with an absolute value which is equal to or greater than the tabled critical two-tailed .05 value $z_{.05}$ = 1.96 is significant at the .05 level. Any residual which is equal to or greater than the tabled critical two-tailed .01 value $z_{.01}$ = 2.58 is significant at the .01 level. Any cell in a contingency table that has a significant residual makes a significant contribution to the obtained chi-square value. For any cell which has a significant residual, one can conclude that the observed frequency of the cell differs significantly from its expected frequency. The sign of the standardized residual indicates whether the observed frequency of the cell is above (+) or below (−) the expected frequency. The sum of the squared residuals for all $r \times c$ cells will equal the obtained value of chi-square. The analysis of the residuals for Example 16.5 is summarized in Table 16.20.

Table 16.20 Analysis of Residuals for Example 16.5

Cell	O_{ij}	E_{ij}	$(O_{ij} - E_{ij})$	$z_{res_{ij}} = \dfrac{(O_{ij} - E_{ij})}{\sqrt{E_{ij}}}$	$z^2_{res_{ij}} = \left[\dfrac{(O_{ij} - E_{ij})}{\sqrt{E_{ij}}}\right]^2$
Mice/Not a biter	20	25.50	−5.50	−1.09	1.19
Mice/Mild biter	16	14.36	1.64	.43	.19
Mice/Flagrant biter	24	20.14	3.86	.86	.74
Gerbils/Not a biter	30	21.25	8.75	1.90	3.61
Gerbils/Mild biter	10	11.96	−1.96	−.57	.33
Gerbils/Flagrant biter	10	16.79	−6.79	−1.66	2.76
Hamsters/Not a biter	50	38.25	11.75	1.90	3.61
Hamsters/Mild biter	30	21.54	8.46	1.82	3.32
Hamsters/Flagrant biter	10	30.21	−20.21	−3.68**	13.52
Guinea pigs/Not a biter	19	34.00	−15.00	−2.57*	6.62
Guinea pigs/Mild biter	11	19.14	−8.14	−1.86	3.46
Guinea pigs/Flagrant biter	50	26.86	23.14	4.46**	19.93
Column sums	280	280.00	0		$\Sigma z^2_{res_{ij}} = \chi^2 = 59.28$

*Significant at the .05 level.
**Significant at the .01 level.

Inspection of Table 16.20 indicates that the residual computed for the following cells is significant: a) **guinea pigs/flagrant biters** — Since the absolute value $z_{res_{ij}}$ = 4.46 computed for the residual is greater than $z_{.05}$ = 1.96 and $z_{.01}$ = 2.58, the residual is significant at both the .05 and .01 levels. The positive value of the residual indicates that the observed frequency of the cell is significantly above its expected frequency. The value of the residual for this cell is consistent with the fact that in the comparisons discussed in the previous section, the cell **guinea pigs/flagrant biters** appears to play a critical role in the significant effect that is detected; b) **hamsters/flagrant biter** — Since the absolute value $z_{res_{ij}}$ = 3.68 computed for the residual is greater than $z_{.05}$ = 1.96 and $z_{.01}$ = 2.58, the residual is significant at both the .05 and .01 levels. The negative value of the residual indicates that the observed frequency of the cell is significantly below its expected frequency. The value of the residual for this cell is consistent with the fact that in the comparisons discussed in the previous section the cell **hamsters/flagrant biter** appears to play a critical role in the significant effect that is detected; and c) **guinea pigs/not a biter** — since the absolute value $z_{res_{ij}}$ = 2.57 computed for the residual is greater than $z_{.05}$ = 1.96, but is less (albeit barely) than $z_{.01}$ = 2.58, the residual is significant at the .05 level. The negative value of the residual indicates that the observed frequency of the cell is significantly below its expected frequency.

Four other cells in Table 16.20 approach being significant at the .05 level (**gerbils/not a biter, hamsters/not a biter, guinea pigs/mild biter, hamsters/mild biter**). The absolute value of the residual for all of the aforementioned cells is close to the tabled critical value $z_{.05} = 1.96$. Note that in Table 16.20, the sum of the squared residuals is essentially equal to the chi-square value computed in Table 16.13 for Example 16.5 (the minimal discrepancy is the result of rounding off error).

Everitt (2001, p. 276) states that the analysis for computing standardized residuals described in this section may be too conservative in estimating the contribution of cells in the computation of the chi-square value. He provides an alternative adjusted residual equation also developed by Haberman (1974).

10. Sources for computing the power of the chi-square test for r × c tables Cohen (1977, 1988) has developed a statistic called the **w index** which can be employed to compute the power of the **chi-square test for r × c tables** (as well as the **chi-square goodness-of-fit test**). The value *w* is an **effect size** index reflecting the difference between expected and observed frequencies. The concept of effect size is discussed in Section VI of the **single-sample *t* test**. It is discussed in greater detail in Section VI of the *t* **test for two independent samples**, and in Section IX (the **Addendum**) of the **Pearson product-moment correlation coefficient** under the discussion of **meta-analysis and related topics**.

The equation for the **w index** is $w = \sqrt{\Sigma[(P_{alt} - P_{null})^2/P_{null}]}$. The latter equation indicates the following: a) For each of the cells in the chi-square table, the proportion of cases hypothesized in the null hypothesis is subtracted from the proportion of cases hypothesized in the alternative hypothesis; b) The obtained difference in each cell is squared, and then divided by the proportion hypothesized in the null hypothesis for that cell; c) All of the values obtained for the cells in part b) are summed; and d) *w* represents the square root of the sum obtained in part c).

Cohen (1977, 1988, Ch. 7) has derived tables which allow a researcher to determine, through use of the **w index**, the appropriate sample size to employ if one wants to test a hypothesis about the difference between observed and expected frequencies in a chi-square table at a specified level of power. Cohen (1977, 1988, pp. 224–226) has proposed the following (admittedly arbitrary) *w* values as criteria for identifying the magnitude of an effect size: a) A **small effect size** is one that is greater than .1 but not more than .3; b) A **medium effect size** is one that is greater than .3 but not more than .5; and c) A **large effect size** is greater than .5.

11. Measures of association for r × c contingency tables Prior to reading this section the reader may find it useful to review the discussion of **magnitude of treatment effect** (which is a measure of the degree of association between two or more variables) in Section VI of the *t* **test for two independent samples**. The computed chi-square value for an *r × c* contingency table does not provide a researcher with precise information regarding the size of the **treatment effect** present in the data. The reason why a chi-square value computed for a contingency table is not an accurate index of the degree of association between the two variables is because the chi-square value is a function of both the total sample size and the proportion of observations in each of the *r × c* cells of the table. The degree of association, on the other hand, is independent of the total sample size, and is only a function of the cell proportions. In point of fact, the magnitude of the computed chi-square statistic is directly proportional to the total sample size employed in a study.

To illustrate the latter point, consider Table 16.21 which presents a different set of data for Example 16.1. In actuality, the effect size for the data presented in Table 16.21 is the same as the effect size in Table 16.2, and the only difference between the two tables is that in Table 16.21

the total sample size and the number of observations in each of the cells are one-half of the corresponding values listed in Table 16.2. If Equation 16.2 is applied to the data presented in Table 16.21, the value $\chi^2 = 9.1$ is computed. Note that the latter value is one-half of $\chi^2 = 18.18$, which is the value computed for Table 16.2. Thus, even though the same proportion of observations appears in the corresponding cells of the two tables, the chi-square value computed for each table is directly proportional to the total sample size.

Table 16.21 Summary of Data for Example 16.1 with Reduced Sample Size

	Helped the confederate	Did not help the confederate	Row sums
Noise	15	35	50
No noise	30	20	50
Column sums	45	55	Total observations 100

A number of different measures of association/correlation which are independent of sample size can be employed as indices of the magnitude of a treatment effect for an $r \times c$ contingency table. In this section the following measures of association will be described: a) **Test 16f: The contingency coefficient**; b) **Test 16g: The phi coefficient**; c) **Test 16h: Cramér's phi coefficient**; d) **Test 16i: Yule's Q**; and e) **Test 16j: The odds ratio**. This section also contains a description of **Cohen's kappa (Test 16k)**, which is a chance-corrected measure of association that can be employed with certain contingency tables to indicate the degree of agreement between two judges with respect to assigning each of n subjects/objects to one of k mutually exclusive categories.

As a general rule (although there are some exceptions), the value computed for a measure of association/correlation will usually fall within a range of values between 0 and +1 or between −1 and +1. Whereas a value of 0 indicates no relationship between the two variables, an absolute value of 1 indicates a maximum relationship between the variables. Consequently, the closer the absolute value of a measure of association is to 1, the stronger the relationship between the variables. As noted above, some measures of association can assume values in the range between −1 and +1. In such cases, the absolute value of the measure indicates the strength of the relationship, and the sign of the measure indicates the direction of the relationship — i.e., the pattern of observations among the cells of a contingency table. Before continuing this section the reader may find it useful to read the material in the **Introduction** on correlation and/or Section I of the **Pearson product-moment correlation coefficient**, both of which provide a general discussion of correlation/association.

Measures of association for $r \times c$ contingency tables can be evaluated with respect to statistical significance. In the case of a number of the measures of association to be discussed in this section, if the computed chi-square value for the contingency table is statistically significant at a given level of significance, the measure of association computed for the contingency table will be significant at the same level of significance. In most instances, the null hypothesis and nondirectional alternative hypothesis evaluated with reference to a measure of association are as follows:[20]

Null hypothesis The correlation/degree of association between the two variables in the underlying population is zero.
Alternative hypothesis The correlation/degree of association between the two variables in the underlying population is some value other than zero.

The same guidelines discussed earlier in reference to employing a directional alternative hypothesis for the **chi-square test for r × c tables** can also be applied if one wants to state the alternative hypothesis for a measure of association directionally.

Since the different measures of association which can be computed for a contingency table do not employ the same criteria in measuring the strength of the relationship between the two variables, if two or more measures are applied to the same set of data they may not yield comparable coefficients of association. Although in the material to follow, there will be some discussion of factors that can be taken into account in considering which of the various measures of association to employ, in most cases one measure is not necessarily superior to another in terms of providing information about a contingency table. Indeed, Conover (1980, 1999) notes that the choice of which measure to employ is based more on prevailing tradition than it is on statistical considerations.

Test 16f: The contingency coefficient (*C*) The **contingency coefficient** (also known as **Pearson's contingency coefficient**) is a measure of association that can be computed for an *r × c* contingency table of any size. The value of the **contingency coefficient**, which will be represented with the notation *C*, is computed with Equation 16.17.

$$C = \sqrt{\frac{\chi^2}{\chi^2 + n}}$$ **(Equation 16.17)**

Where: χ^2 is the computed chi-square value for the contingency table
n is the total number of observations in the contingency table

Since *n* can never equal zero, the value of *C* can never equal 1. Consequently, the range of values *C* may assume is $0 \leq C < +1$. One limitation of the contingency coefficient is that its upper limit (i.e., the highest value it can attain) is a function of the number of rows and columns in the *r × c* contingency table. The upper limit of *C* (represented by C_{max}) can be determined with Equation 16.18.

$$C_{max} = \sqrt{\frac{k - 1}{k}}$$ **(Equation 16.18)**

Where: *k* represents the smaller of the two values of *r* and *c* in the contingency table

Employing Equation 16.18 for a 2 × 2 contingency table, we can determine that the maximum value for such a table is $C_{max} = \sqrt{(2 - 1)/2} = .71$.

Employing Equation 16.19, the value *C* = .29 is computed as the value of the **contingency coefficient** for Examples 16.1/16.2.

$$C = \sqrt{\frac{18.18}{18.18 + 200}} = .29$$

Note that the value *C* = .29 is also obtained for Table 16.21 (which as noted earlier has the same effect size but half the sample size as Tables 16.2/16.3): $C = \sqrt{(9.1)/(9.1 + 100)} = .29$.

As noted in the introductory remarks on measures of association, the computed value *C* = .29 will be statistically significant at both the .05 and .01 levels, since the computed value $\chi^2 = 18.18$ for Tables 16.2/16.3 (as well as the computed value $\chi^2 = 9.1$ for Table 16.21) is significant at the aforementioned levels of significance. Thus, one can conclude that in the underlying population the **contingency coefficient** between the two variables is some value other than zero.

The reader should take note of the fact that if the value of n is reduced, but the effect size present in Tables 16.1/16.2/16.21 is maintained, at some point the chi-square value will not achieve significance. Although in such a case Equation 16.17 will yield the value $C = .29$, the latter value will not be statistically significant. This is the case, since in order for the value of C to be significant, the computed value of chi-square must be significant.

Ott *et al.* (1992) note that among the disadvantages associated with the **contingency coefficient** is that it will always be less than 1, even when the two variables are totally dependent on one another. In addition, **contingency coefficients** which have been computed for two or more tables can only be compared with one another if all of the tables have the same number of rows and columns. One suggestion that is made to counteract the latter problems is to employ Equation 16.19 to compute an adjusted value for the **contingency coefficient**.

$$C_{adj} = \frac{C}{C_{max}}$$ **(Equation 16.19)**

By employing Equation 16.19, if a perfect association between the variables exists, the value of C_{adj} will equal 1. If Equation 16.19 is employed for Examples 16.1/16.2 and Table 16.21, the value $C_{adj} = .29/.71 = .41$ is computed. It should be pointed out that although the use of C_{adj} allows for better comparison between tables of unequal size, it still does not allow one to compare such tables with complete accuracy.

In the section which discusses the computation of the power of the **chi-square test for** *r* × *c* **tables**, it notes that Cohen (1977, 1988) has developed a measure of effect size for the **chi-square test for** *r* × *c* **tables** called the *w* **index**. It also notes that Cohen (1977, 1988; pp. 224–226) has proposed the following (admittedly arbitrary) *w* values as criteria for identifying the magnitude of an effect size: a) A **small effect size** is one that is greater than .1 but not more than .3; b) A **medium effect size** is one that is greater than .3 but not more than .5; and c) A **large effect size** is greater than .5.

Cohen (1977; 1988, p. 222) states that the value computed for a **contingency coefficient** can be converted into the *w* **index** through use of the following equation: $w = \sqrt{C^2/(1 - C^2)}$. When the computed value $C = .29$ is substituted in the latter equation, the resulting value is $w = \sqrt{(.29)^2/[1 - (.29)^2]} = .30$. Employing Cohen's (1977, 1988) criteria for the *w* **index**, the value .30 is at the lower limit for a medium effect size. Since the upper limiting value the **contingency coefficient** approaches is 1, and in the case of a 2 × 2 contingency table the maximum value C can attain is .71, one can argue that in lieu of C, C_{adj} should be employed in computing w. If $C_{adj} = .41$ is employed to compute w, the value $w = \sqrt{(.41)^2/[1 - (.41)^2]} = .45$ is obtained, which based on Cohen's (1977, 1988) criteria still falls within the limits of a medium effect size.

Smithson (2003, pp. 72–73) describes a procedure for computing a confidence interval for the **contingency coefficient** through use of the **noncentral chi-square distribution** (which in order to employ requires the use of the appropriate computer software). The chi-square distribution (for which **Table A4** displays probability values and corresponding chi-square values) employed throughout this book is sometimes referred to as the **central chi-square distribution**. The latter distribution is based on the assumption that a null hypothesis is correct. The **noncentral chi-square distribution**, on the other hand, is not based on the latter assumption. Among others, Kline (2004, p. 35) and Smithson (2003) note that the **noncentral chi-square distribution** is actually a family of distributions (of which the **central chi-square**

distribution represents a special case), and that the shape of a specific noncentral distribution will be a function of the value of an additional parameter called the **noncentrality parameter** (which is also discussed in earlier chapters in reference to the **noncentral** *t* **distribution**). The **noncentrality parameter** (which in the case of the **central chi-square distribution** equals zero) essentially reflects the degree of deviation from the null hypothesis.

Test 16g: The phi coefficient The **phi coefficient** (represented by the notation ϕ, which is the lower case Greek letter **phi**) is a measure of association that can only be employed with a 2 × 2 contingency table. The **phi coefficient**, which is discussed further in Section VII of the **Pearson product-moment correlation coefficient**, is, in actuality, a special case of the latter correlation coefficient. Specifically, it is a **Pearson product-moment correlation coefficient** which is computed if the values 0 and 1 are employed to represent the levels of two **dichotomous variables**. (A **dichotomous variable** is comprised of two mutually exclusive categories.) Although the value of **phi** can fall within the range −1 to +1, the latter statement must be qualified, since the lower and upper limits of **phi** are dependent on certain conditions. Carroll (1961) and Guilford (1965) note that in order for **phi** to equal −1 or +1, the following two conditions must be met with respect to the 2 × 2 contingency table described by the model in Table 16.6: $(a + b) = (c + d)$ and $(a + c) = (b + d)$.

Employing the notation in Table 16.6 for a 2 × 2 contingency table, the value of **phi** is computed with Equation 16.20.

$$\phi = \frac{ad - bc}{\sqrt{(a + b)(c + d)(a + c)(b + d)}}$$ **(Equation 16.20)**

Since $\phi^2 = \chi^2/n$, many sources compute the value of **phi** through use of Equation 16.21 (which can be derived from the equation $\phi^2 = \chi^2/n$). Note that the result of Equation 16.21 will always be the absolute value of **phi** derived with Equation 16.20.[21]

$$\phi = \sqrt{\frac{\chi^2}{n}}$$ **(Equation 16.21)**

Employing Equation 16.20, the value $\phi = -.30$ is computed below for Examples 16.1/16.2 (it will yield the same value for the data in Table 16.21).

$$\phi = \frac{(30)(40) - (70)(60)}{\sqrt{(30 + 70)(60 + 40)(30 + 60)(70 + 40)}} = -.30$$

If Equation 16.21 is employed, the absolute value of **phi** is computed to be $\phi = \sqrt{18.18/200} = .30$. As noted in the introductory remarks on measures of association, the computed absolute value $\phi = .30$ will be statistically significant at both the .05 and .01 levels, since the computed value $\chi^2 = 18.18$ is significant at the aforementioned levels of significance. Thus, one can conclude that in the underlying population the **phi** coefficient between the two variables is some value other than zero.[22] In addition to the nondirectional alternative hypothesis being supported, the directional alternative hypothesis which is consistent with the data is also supported. This is the case since the computed value of chi-square is significant at the .10 and .02 levels (i.e., the obtained value $\chi^2 = 18.18$ is greater than $\chi^2_{.90} = 2.71$ and $\chi^2_{.98} = 5.41$).

It turns out the absolute value $\phi = .30$ computed for Examples 16.1/16.2 is almost identical to $C = .29$, the value of the **contingency coefficient** computed for the same set of data. As a general rule, although the two values will not be identical, the absolute value of **phi** and the **contingency coefficient** will be close to one another.

Cohen (1977; 1988, p. 223) notes that **phi** is identical in value to the **w index** discussed in the previous section. Thus, Cohen (1977, 1988) has proposed the following (admittedly arbitrary) **phi** values as criteria for identifying the magnitude of an effect size (which are identical to the criteria stated earlier for the **w index**): **small effect size**: $.10 \leq \phi < .30$; **medium effect size**: $.30 \leq \phi < .50$; **large effect size**: $\phi \geq .50$. Employing Cohen's (1977, 1988) guidelines, the observed effect size for Examples 16.1/16.2 ($\phi = .30$) is at the lower limit of a medium effect size — i.e., there is a moderate relationship between the two variables employed in the study.

The use of the **phi coefficient** is most commonly endorsed in the case of 2×2 contingency tables involving two variables which are dichotomous in nature. Because of this, among others, Guilford (1965) and Fleiss (1981) note that one of the most useful applications of **phi** is for determining the intercorrelation between the responses of subjects on two dichotomous test items.[23] Siegel and Castellan (1988) note that when the two variables being correlated are ordered variables, computation of the **phi** coefficient sacrifices information, and because of this under such conditions it is preferable to employ alternative measures of association which are designed for ordered tables (such as **Goodman and Kruskal's gamma** which is discussed later in the book). Grissom and Kim (2005, pp. 175–176) discuss the limitations of using **phi** as a measure of effect size. For a more thorough overview of the **phi coefficient**, the reader should consult Guilford (1965) and Fleiss (1981), who, among other things, discuss various sources which argue in favor of employing measures other than **phi** with 2×2 tables.

Test 16h: Cramér's phi coefficient Developed by Cramér (1946), **Cramér's phi coefficient** (which will be represented by the notation ϕ_C) is an extension of the **phi coefficient** to contingency tables that are larger than 2×2 tables. **Cramér's phi coefficient**, which can assume a value between 0 and +1, is computed with Equation 16.22.

$$\phi_C = \sqrt{\frac{\chi^2}{n(k-1)}} \qquad \text{(Equation 16.22)}$$

Where: k represents the smaller of the two values of r and c in the contingency table

The derivation of **Cramér's phi coefficient** is based on the fact that the maximum value chi-square can attain for a set of data is $\chi^2_{max} = n(k-1)$. Thus, the value ϕ_C is the square root of a proportion which represents the computed value of chi-square divided by the maximum possible chi-square value for a set of data. When the computed chi-square value for a set of data equals χ^2_{max}, the value $\phi_C = 1$ will be obtained, which indicates maximum dependency between the two variables.

Since for 2×2 tables **Cramér's phi** and the **phi coefficient** are equivalent (i.e., when $k = 2$, $\phi_C = \phi = \sqrt{\chi^2/n}$), they will both yield the absolute value of .30 for Examples 16.1/16.2. **Cramér's phi** is computed below for the 4×3 contingency table presented in Example 16.5. Since $c = 3$ is less than $r = 4$, $k = c = 3$.

$$\phi_C = \sqrt{\frac{59.16}{(280)(3-1)}} = .325$$

The computed value $\phi_C = .325$ is significant at both the .05 and .01 levels, since the computed value $\chi^2 = 59.16$ is significant at the aforementioned levels of significance. Thus, one can conclude that in the underlying population, the value of **Cramér's phi coefficient** is some value other than zero.

Cohen (1977; 1988, p. 223) states that the value computed for **Cramér's phi coefficient** can be converted into the previously discussed **w index** through use of the following equation: $w = \phi_C \sqrt{k - 1}$. (Note that when $k = 2$, $\phi_C = \phi = w$.) When the computed value $\phi_C = .325$ is substituted in the latter equation, the resulting value is $w = (.325)\sqrt{3 - 1} = .46$. Employing Cohen's (1977, 1988) criteria for the **w index**, the value $w = .46$ falls within the upper region of the range of values listed for a medium effect size.

Daniel (1990) notes that when a contingency table is square (i.e., $r = c$) and $\phi_C = 1$, there is a perfect correlation between the two variables (which will be reflected by the fact that all of the observations will be in the cells of one of the diagonals of the table). When $r \neq c$ and $\phi_C = 1$, however, the two variables will not be perfectly correlated in the same manner as is the case with a square contingency table. Conover (1980, 1999) notes that although under all conditions the possible range of values for ϕ_C will be between 0 and +1, its interpretation will depend on the values of r and c. He states there is a tendency for the value of χ^2 (and consequently the value of ϕ_C) to increase as the values of r and c become larger. For this reason Conover (1980, 1999) suggests that ϕ_C may not be completely accurate for comparing the degree of association in different size tables. Daniel (1990) notes that when $r = c = 2$, the value of ϕ_C is equal to the square of the tie-adjusted value of **Kendall's tau (Test 30)** (discussed later in the book) computed for the same set of data. As is the case with the **phi coefficient**, when ordered categories are employed for both variables, sources do not recommend employing **Cramér's phi** since it sacrifices information. In designs involving variables with ordered categories, it is preferable to employ an alternative measure of association such as **Goodman and Kruskal's gamma**.

Smithson (2003, pp. 39–41) describes a procedure for computing a confidence interval for **Cramér's phi** through use of the **noncentral chi-square distribution** (which, as noted earlier, requires access to the appropriate computer software).

Test 16i: Yule's Q   **Yule's Q** (Yule (1900)) is a measure of association for a 2 × 2 contingency table. It is presented in this section to illustrate that if two or more measures of association are computed for the same set of data, they may not yield comparable values. Since it employs less information than the **phi coefficient**, **Yule's Q** is less frequently recommended than **phi** as a measure of association for 2 × 2 tables. **Yule's Q** is actually a special case of **Goodman and Kruskal's gamma** (although unlike **gamma**, which is only used with ordered contingency tables, **Yule's Q** can be used for both ordered and unordered tables).

Employing the notation in Table 16.6 for a 2 × 2 contingency table, Equation 16.23 is employed to compute the value of **Yule's Q**.

$$Q = \frac{ad - bc}{ad + bc}$$ **(Equation 16.23)**

Sources which discuss **Yule's Q** generally note that it tends to inflate the degree of association in the underlying population. Ott *et al.* (1992) note that an additional limitation is that if the absolute value of Q equals 1, it does not necessarily mean there is a perfect association between the two variables. In point of fact, if the observed frequency of any of the four cells in a 2 × 2 contingency table equals 0, the value of **Yule's Q** will equal either −1 or +1 (which are the lower and upper limits that **Yule's Q** may fall within). For this reason, the meaning of Q can be quite misleading in cases where the frequency of one of the cells is equal to 0. Because of the latter, it is not recommended that **Yule's Q** be employed when there is a small number of frequencies in any of the four cells of a 2 × 2 contingency table.

Employing Equation 16.22, the value $Q = -.56$ is computed for Examples 16.1/16.2 (as well as for the data in Table 16.21).

$$Q = \frac{(30)(40) - (70)(60)}{(30)(40) + (70)(60)} = -.56$$

Note that although both the values of Q and **phi** suggest the presence of a negative association between the two variables, the absolute value $Q = .56$ is almost twice the value $\phi = .30$ computed previously for the same set of data. When Cohen's (1977, 1988) criteria are applied to the computed value of **phi** it indicates the presence of a medium effect size, yet if the same criteria are applied to $Q = .56$, it suggests the presence of a large effect size. (It should be noted, however, that Cohen does not endorse the use of the criterion values he lists for **phi** with **Yule's Q**.)

Ott *et al.* (1992) note that the significance of **Yule's Q** can be evaluated with Equation 16.24.[24]

$$z = \frac{Q}{\sqrt{\frac{1}{4}(1 - Q^2)^2 \left[\frac{1}{a} + \frac{1}{b} + \frac{1}{c} + \frac{1}{d} \right]}} \qquad \textbf{(Equation 16.24)}$$

Employing Equation 16.24 with the data for Examples 16.1/16.2, the value $z = -5.46$ is computed.

$$z = \frac{-.56}{\sqrt{\left[\frac{1}{4} \right]\left[1 - (.56)^2\right]^2 \left[\frac{1}{30} + \frac{1}{70} + \frac{1}{60} + \frac{1}{40} \right]}} = -5.46$$

The obtained value $z = -5.46$ is evaluated with **Table A1**. Since the obtained absolute value $z = 5.46$ is greater than the tabled critical two-tailed .05 and .01 values $z_{.05} = 1.96$ and $z_{.01} = 2.58$, the nondirectional alternative hypothesis is supported at both the .05 and .01 levels. Since $z = 5.46$ is greater than the tabled critical one-tailed .05 and .01 values $z_{.05} = 1.65$ and $z_{.01} = 2.33$, the directional alternative hypothesis which is consistent with the data is also supported at both the .05 and .01 levels.

Test 16j: The odds ratio (and the concept of relative risk) As is the case with **Yule's Q** (but unlike C, ϕ, and ϕ_C), the **odds ratio** (which is attributed to Cornfield (1951)) is a measure of association employed with contingency tables that is not a function of chi-square. Although the **odds ratio**, which will be represented by the notation o, can be applied to any size contingency table, it is easiest to interpret in the case of 2×2 tables. The **odds ratio** expresses the degree of association between the two variables in a different numerical format than all of the previously discussed measures of association. In some respects it provides a more straightforward way of interpreting the results of a contingency table than the correlational measures of association discussed previously.

The **odds ratio** is one of two measures which are commonly employed in epidemiological research to indicate the risk of a person contracting a disease. Although **relative risk**, which is the second of the two measures, is intuitively easier to understand, the **odds ratio** is more useful when employed within the context of statistical analysis. In this section both of the aforementioned measures will be examined. In order to do this, consider the data in Table 16.22. We will assume that the latter table represents the results of a hypothetical study which evaluates if someone who washes her hands immediately after handling a diseased animal is less likely to contract the disease than someone who does not wash her hands.

Note that Table 16.22 is identical to Table 16.2, except for the following: a) In the case of the row variable, the **noise** condition has been replaced by a **washes hands** condition, and the **no noise** condition has been replaced by a **does not wash** condition; b) In the case of the column variable, the **helped the confederate** response category has been replaced by the **contracts the disease** category, and the **did not help the confederate** response category has been replaced by the **does not contract the disease** category. Thus, the **independent variable** in the study summarized in Table 16.22 is **whether or not a person washes her hands**, and the **dependent variable** is **whether or not the person contracts the disease**.

Table 16.22 Summary of Hand Washing Study

	Contracts the disease	Does not contract the disease	Row sums
Washes hands	30	70	**100**
Does not wash	60	40	**100**
Column sums	**90**	**110**	**Total observations** **200**

Relative risk allows a researcher to compare the relative probabilities of contracting a disease. Specifically, **relative risk** is the probability of contracting a disease if you are a member of one group (typically the group which is considered to have the higher risk) divided by the probability of contracting the disease if you are a member of the other group (typically the group which is considered to have the lower risk). In the case of our example, the probability of contracting the disease in the **does not wash** group (which will be considered the high risk group) is $60/100 = .6$. The latter probability is simply the number of people in the **does not wash** group who **contract the disease** (60) divided by the total number of people who constitute the **does not wash** group (100). In the same respect, the probability of contracting the disease in the **washes hands** group (which will be considered the low risk group) is $30/100 = .3$. The latter probability is simply the number of people in the **washes hands** group who **contract the disease** (30) divided by the total number of people who constitute the **washes hands** group (100). Through use of the notation in Table 16.6, the above can be expressed as follows.

$$p(Contracts\ disease/Does\ not\ wash) = c/(c + d)$$
$$p(Contracts\ disease/Washes\ hands) = a/(a + b)$$

The relative risk (RR) is computed with Equation 16.25.

$$RR = \frac{c/(c + d)}{a/(a + b)} = \frac{60/(60 + 40)}{30/(30 + 70)} = 2 \qquad \textbf{(Equation 16.25)}$$

or

$$RR = \frac{ac + bc}{ac + ad} = \frac{(30)(60) + (70)(60)}{(30)(60) + (30)(40)} = 2$$

The value $RR = 2$ computed for **relative risk** means that a person who does not wash her hands is 2 times more likely to contract the disease than a person who washes her hands. If the numerator and denominator of Equation 16.25 are reversed (which computes the relative risk for someone who washes relative to someone who does not wash), the value $RR = .5$ is computed.

The value $RR = .5$ indicates that a person who washes her hands has half (.5) the likelihood of contracting the disease than a person who does not wash her hands.

If we apply the computations for **relative risk** to the variables employed in Example 16.1, we can state that someone in the **no noise** condition is 2 times more likely to **help the confederate** than someone in the **noise** condition.

At this point we will turn our attention to the **odds ratio**. Before considering the latter measure it will be useful to clarify the concept of **odds**. The odds that an event (to be designated X) will occur are computed by dividing the probability that the event will occur by the probability that the event will not occur. The latter is described by Equation 16.26.

$$Odds(X) = \frac{p(X\ will\ occur)}{p(X\ will\ not\ occur)} \qquad \textbf{(Equation 16.26)}$$

The following information regarding **odds** was noted previously in Section IX (the **Addendum**) of the **binomial sign test for a single sample** under the discussion of **Bayesian hypothesis testing**: a) The computed value for odds can fall anywhere in the range 0 to infinity; b) When the value computed for odds is greater than 1, it indicates that the probability of an event occurring is better than one-half ($\frac{1}{2}$). Thus, the larger the odds of an event occurring, the higher the probability the event will occur; c) When the value computed for odds is less than 1, it indicates that the probability of an event occurring is less than one-half ($\frac{1}{2}$). Thus, the smaller the odds of an event occurring, the lower the probability the event will occur. The lowest value that can be computed for odds is 0, which indicates the probability the event will occur is 0; d) When the value computed for odds equals 1, it indicates that the probability of the event occurring is one half ($\frac{1}{2}$) — i.e., there is a 50-50 chance of the event occurring.[25]

Odds will now be computed for the data in Tables 16.2/16.22.

$$Odds(Help/No\ noise) = \frac{p(Help/No\ noise)}{p(Not\ help/No\ noise)} = \frac{60/100}{40/100} = 1.5$$

$$Odds(Help/Noise) = \frac{p(Help/Noise)}{p(Not\ help/Noise)} = \frac{30/100}{70/100} = .429$$

$$Odds(Contracts\ disease/Does\ not\ wash) = \frac{p(Contracts\ disease/Does\ not\ wash)}{p(Does\ not\ contract\ disease/Does\ not\ wash)}$$

$$= \frac{60/100}{40/100} = 1.5$$

$$Odds(Contracts\ disease/Washes\ hands) = \frac{p(Contracts\ disease/Washes\ hands)}{p(Does\ not\ contract\ disease/Washes\ hands)}$$

$$= \frac{30/100}{70/100} = .429$$

The above results indicate that the odds of **helping** in the **no noise** condition are 1.5 to 1. Since the value 1.5 is greater than 1, it indicates the probability of a subject **helping** in the **no noise** condition is greater than $\frac{1}{2}$. The odds of **helping** in the **noise** condition are .429 to 1. Since the value .429 is less than 1, it indicates the probability of a subject **helping** in the **noise** condition is less than $\frac{1}{2}$. In the same respect the odds of **contracting the disease** in the **does not wash** condition are 1.5 to 1, while the odds of **contracting the disease** in the **washes hands** condition are .429 to 1.

The **odds ratio** is simply the ratio between the two odds computed for a contingency table. Typically the larger number is divided by the smaller number. Thus $o = 1.5/.429 = 3.5$. In the case of Example 16.1, the latter value indicates that the odds of **helping** in the **no noise** condition are 3.5 times larger than the odds of **helping** in the **noise** condition. In the case of the data in Table 16.22, the value $o = 3.5$ indicates that the odds of **contracting the disease** in the **does not wash** condition are 3.5 times larger than the odds of **contracting the disease** in the **washes hands** condition. One can also divide .439 by 1.5 and obtain the value $o' = .29$. The latter value, which is often referred to as the **inverse odds ratio** (and will be represented with the notation o'), indicates that the odds of **helping** in the **noise** condition are .29 times as large as the odds of **helping** in the **no noise** condition. In the case of the data in Table 16.22, the value $o' = .29$ indicates that the odds of **contracting the disease** in the **washes hands** condition are .29 times as large as the odds of **contracting the disease** in the **does not wash** condition.[26]

It can be algebraically demonstrated that Equations 16.27 and 16.28 (which employ the notation in Table 16.6) provide a simple method for computing the two **odds ratio** values $o = 3.5$ and $o' = .29$ noted in the previous paragraph. The latter equation is employed below to obtain the values $o = 3.5$ and $o' = .29$ for our data.[27]

$$o = \frac{bc}{ad} = \frac{(70)(60)}{(30)(40)} = 3.5 \qquad \textbf{(Equation 16.27)}$$

$$o' = \frac{ad}{bc} = \frac{(30)(40)}{(70)(60)} = .29 \qquad \textbf{(Equation 16.28)}$$

The reader should take note of the fact that many sources employ Equation 16.28 as the equation for the **odds ratio**. In the final analysis, the value computed for an **odds ratio** for a 2 × 2 contingency table will be a function of how the rows and columns of the table are configured. Nevertheless, regardless of which of the two possible odds ratio values one elects to compute for a 2 × 2 table (i.e., o versus o'), a researcher will reach the same conclusions regarding the relationship between the variables in the table. For the example under discussion, if the rows in Tables 16.2 and 16.22 are reversed (i.e., the category **no noise/does not wash hands** is employed as the first row in the contingency table and the category **noise/washes hands** is employed as the second row), then $a = 60$, $b = 40$, $c = 30$, and $d = 70$. When the latter values are employed in Equations 16.27 and 16.28, the values $o = .29$ and $o' = 3.5$ are obtained for the **odds ratio** and **inverse odds ratio**. Note that the latter values are the identical two values obtained previously for odds ratios, but the original **odds ratio** is now designated as the **inverse odds ratio**, while the original **inverse odds ratio** is now designated as the **odds ratio**. Nevertheless, the odds associated with the variables as well as the conclusions with respect to the relationship between the variables remains the same.

Various sources (e.g., Pagano and Gauvreau (1993)) note that when the probability of an event occurring is very low, the values in cells a and c of Tables 16.6/16.22 will be very small. When the latter is true, the values computed for the **relative risk** and **odds ratio** will be very close together, since if the values of a and c are close to 0, the equation $RR = (ac + bc)/(ac + ad)$ for computing relative risk essentially becomes Equation 16.27. Although the latter is not true for the example under discussion, a case where the **relative risk** and **odds ratio** are, in fact, almost identical is described in Section IX (the **Appendix**) of the **Pearson product-moment correlation coefficient** under the discussion of **meta-analysis and related topics**.

Although, as noted earlier, the **odds ratio** can be extended beyond 2 × 2 tables, it becomes more difficult to interpret with larger contingency tables. However, in instances where there are more than two rows but only two columns, its interpretation is still relatively straightforward.

To illustrate, assume that in Example 16.1 we have three noise conditions instead of two — specifically, **loud noise, moderate noise,** and **no noise.** Table 16.23 depicts a hypothetical set of data which summarizes the results of such a study.

Table 16.23 Summary of Data for a 3 × 2 Contingency Table

	Helped the confederate	Did not help the confederate	Row sums
Loud noise	30	70	100
Moderate noise	50	50	100
No noise	80	20	100
Column sums	160	140	Total observations 300

Within any of the three noise conditions, we can determine the odds that someone in that condition will **help the confederate.** This is accomplished by dividing the proportion of subjects in a given condition who **helped the confederate** by the proportion of subjects in the condition who **did not help the confederate.** Thus, for the **loud noise** condition the odds that someone will **help the confederate** are $(30/100)/(70/100) = .43$. For the **moderate noise** condition the odds that someone will **help the confederate** are $(50/100)/(50/100) = 1$. For the **no noise** condition the odds that someone will **help the confederate** are $(80/100)/(20/100) = 4$. From these values we can compute the following three odds ratios: a) The **odds ratio** of someone in the **no noise** condition **helping the confederate** versus someone in the **loud noise** condition: $o = 4/.43 = 9.3$; b) The **odds ratio** of someone in the **no noise** condition **helping the confederate** versus someone in the **moderate noise** condition: $o = 4/1 = 4$; and c) The **odds ratio** of someone in the **moderate noise** condition **helping the confederate** versus someone in the **loud noise** condition: $o = 1/.43 = 2.33$.

Thus, the odds of someone in the **no noise** condition **helping the confederate** are 9.3 times larger than the odds of someone in the **loud noise** condition **helping the confederate,** and 4 times larger than the odds of someone in the **moderate noise** condition **helping the confederate.** The odds of someone in the **moderate noise** condition **helping the confederate** are 2.33 times larger than the odds of someone in the **loud noise** condition **helping the confederate.**

A discussion of the use of the concepts of **relative risk** and the **odds ratio** as measures of effect size can be found in Grissom and Kim (2005, pp. 183–198).

Test 16j-a: Test of significance for an odds ratio and computation of a confidence interval for an odds ratio   Christensen (1990, 1999) and Pagano and Gauvreau (1993) note that Equation 16.29 can be employed to evaluate the null hypothesis that the true value of the **odds ratio** in the underlying population is equal to 1 (i.e., that the probability an event will occur is equal to the probability the event will not occur). Since the sampling distribution of the **odds ratio** is positively skewed, a logarithmic scale transformation is employed in computing the test statistic, which is a standard normal deviate (i.e., a z score). (Logarithmic scale transformations are discussed in Section VII of the *t* **test for two independent samples.**) Without the logarithmic transformation, the numerator of Equation 16.29 would be $(o - 1)$, where o is the computed value of the **odds ratio,** and 1 is the expected value of the **odds ratio** if the probabilities for the event occurring and not occurring are equal. By virtue of the logarithmic transformation, the numerator of Equation 16.29 becomes the natural logarithm (which is defined in Endnote 14 in the **Introduction**) of the **odds ratio** minus 0, which is the natural logarithm of 1.

$$z = \frac{\ln(o) - \ln(1)}{s_o} = \frac{\ln(o) - 0}{s_o} \qquad \textbf{(Equation 16.29)}$$

Where: $\ln(o)$ represents the natural logarithm of the computed value of the **odds ratio**
s_o represents the standard error, which is the estimated standard deviation of the sampling distribution. The standard error is computed as follows:[28]

$$s_o = \sqrt{\frac{1}{a} + \frac{1}{b} + \frac{1}{c} + \frac{1}{d}}$$

If we evaluate the null hypothesis regarding the **odds ratio** in reference to the data for Example 16.1, the following values are computed: a) Through use of the appropriate tables or a calculator (using the *ln* key), the natural logarithm of the **odds ratio** ($o = 3.5$) is determined to be 1.2528; and b) The value of the standard error is computed to be $SE = .2988$.

$$s_o = \sqrt{\frac{1}{30} + \frac{1}{70} + \frac{1}{60} + \frac{1}{40}} = .2988$$

Substituting the values $\ln(o) = \ln(3.5) = 1.2528$ and $SE = .2988$ in Equation 16.29, the value $z = 4.19$ is computed.

$$z = \frac{1.2528 - 0}{.2988} = 4.19$$

The value $z = 4.19$ is evaluated with **Table A1** in the **Appendix**. In **Table A1** the tabled critical two-tailed .05 and .01 values are $z_{.05} = 1.96$ and $z_{.01} = 2.58$, and the tabled critical one-tailed .05 and .01 values are $z_{.05} = 1.65$ and $z_{.01} = 2.33$. If a nondirectional alternative hypothesis is employed, in order to reject the null hypothesis the absolute value of z must be equal to or greater than the tabled critical two-tailed value at the prespecified level of significance. If a directional alternative hypothesis is employed, in order to reject the null hypothesis the following must be true: a) The absolute value of z must be equal to or greater than the tabled critical one-tailed value at the prespecified level of significance; and b) If the alternative hypothesis stipulates that the population **odds ratio** is less than 1, the sign of the computed z value must be negative (since a positive number less than 1 will yield a negative value for a logarithm). If the alternative hypothesis stipulates that the underlying population **odds ratio** is greater than 1, the sign of the computed z value must be positive (since a number greater than 1 will yield a positive value for a logarithm). Since the computed value $z = 4.19$ is greater than the two-tailed .05 and .01 values $z_{.05} = 1.96$ and $z_{.01} = 2.58$, the nondirectional alternative hypothesis is supported at both the .05 and .01 levels. Since $z = 4.19$ is a positive number that is greater than the tabled critical one-tailed .05 and .01 values $z_{.05} = 1.65$ and $z_{.01} = 2.33$, the directional alternative hypothesis stipulating that the underlying population **odds ratio** is greater than 1 is supported at both the .05 and .01 levels. Thus, we can conclude that the population **odds ratio** is not equal to 1.

The reader should also take note of the fact that if the natural logarithm of the **inverse odds ratio** ($o' = .29$) is substituted in Equation 16.29, except for the sign of z, the identical result will be obtained. Specifically, if the natural logarithm of the **inverse odds ratio** is computed to 9 decimal places (i.e., $o' = .285714286$), $\ln(o') = -1.2528$. The absolute value of the latter number is identical to the value of the natural logarithm obtained for $o = 3.5$. When the value -1.2528 is substituted in Equation 16.29, it yields the identical absolute value $z = 4.19$ obtained earlier when $o = 3.5$ was evaluated. Thus, the identical absolute z value will be obtained regardless of whether the natural logarithm of the **odds ratio** or the **inverse odds ratio** is employed in Equation 16.29. Consequently, the same conclusions will be reached with respect to the whether or not the underlying population **odds ratio** or **inverse odds ratio** is equal to 1.

$$z = \frac{-1.2528 - 0}{.2988} = -4.19$$

Christensen (1990, 1999) and Pagano and Gauvreau (1993) describe the computation of a confidence interval for the **odds ratio**. The latter computation initially requires that a product based on multiplying the standard error by the appropriate z value for the confidence interval be obtained. Thus, if one is computing the 95% confidence interval, the standard error is multiplied by the tabled critical two-tailed .05 value $z_{.05} = 1.96$. If one is computing the 99% confidence interval, the standard error is multiplied by the tabled critical two-tailed .01 value $z_{.01} = 2.58$. The latter product for the desired confidence interval is then added to and subtracted from the natural logarithm of the odds ratio. The antilogarithms (which are the original numbers for which the logarithms were computed) of the resulting two values are determined. The latter values define the limits of the confidence interval.

To illustrate, the 95% confidence interval will be computed. We first multiply the standard error $s_o = .2988$ by the tabled critical two-tailed .05 value $z_{.05} = 1.96$, and obtain $(.2988)(1.96) = .5856$. The latter value is added to and subtracted from the natural logarithm of the odds ratio, which was previously computed to be 1.2528. Thus $1.2528 \pm .5856$ yields the two values .6672 and 1.8384. Through use of the appropriate tables or a calculator (using the e^x key), the antilogarithms of the latter two values are determined to be 1.9488 and 6.2865. These values represent the limits which define the 95% confidence interval for the **odds ratio**. In other words, we can be 95% confident (or the probability is .95) the interval 1.9488 to 6.2865 contains the true value of the odds ratio in the population (represented by the notation o_p). The computations are summarized below.

$$CI_{.95} = \ln(o) \pm (z_{.05})(s_o) = 1.2528 \pm (1.96)(.2988) = 1.2528 \pm .5856$$

Compute antilogarithms of $1.2528 - .5856 = $ **.6672** and $1.2528 + .5856 = $ **1.8384** which are **1.9488** and **6.2865**. Thus, $1.9488 \leq o_{pop} \leq 6.2865$

We can also compute the 95% confidence interval for the **inverse odds ratio** $o' = .29$ by adding and subtracting .5856 from -1.2528 (which, as noted earlier, is the natural logarithm of o' carried out to nine decimal places), and then computing the antilogarithms of the latter values. Thus, $-1.2528 \pm .5856$ yields the two values $-.6672$ and -1.8384. The antilogarithms of the latter two values are respectively .5131 and .1591. These values represent the limits which define the 95% confidence interval for the **inverse odds ratio**. In other words, we can be 95% confident (or the probability is .95) the interval .1591 to .5131 contains the true value of the **inverse odds ratio** in the population. Thus, $.1591 \leq o'_{pop} \leq .5131$.

Pagano and Gauvreau (1993, pp. 325) note that like Equation 16.29, a confidence interval can be employed to evaluate the null hypothesis that the population **odds ratio** is equal to 1 against the alternative hypothesis that the population **odds ratio** is equal to some value other than 1. Specifically, if the range of values which define a confidence interval for an **odds ratio** includes the value 1, the above null hypothesis is retained. On the other hand, if the confidence interval does not include the value 1, the null hypothesis can be rejected. Rejection of the null hypothesis is viewed as evidence that a significant association exists between the variables in the contingency table.

In the case of a 95% confidence interval, the above null hypothesis is evaluated with a two-tailed test with alpha set equal to .05. Since in our example, the range 1.9488 to 6.2865 obtained for the **odds ratio** confidence interval does not include the value 1, we can reject the null hypothesis, and thus the alternative hypothesis stating that the population **odds ratio** is some value other than 1 is supported. The null hypothesis can also be rejected in the case of the **inverse odds ratio**, since the range .1591 to .5131 obtained for the **inverse odds ratio** confidence interval also does not include the value 1.

Test 16k: Cohen's kappa Developed by Cohen (1960), **kappa** is a chance-corrected measure of association which can be employed with certain contingency tables to indicate the degree of agreement between two judges with respect to assigning each of n subjects/objects to one of k mutually exclusive categories. It should be noted, however, that although **kappa** is computed for data which can be summarized in the form of a contingency table, the **kappa** statistic is not based on the chi-square distribution. **Cohen's kappa** is one of a number of measures of association described in this book which can be employed to measure **interobserver reliability**. The term **interobserver reliability** (which also referred to as **interjudge reliability**) refers to the degree of agreement between two or more observers/judges with respect to their categorization of n subjects/objects. In addition to **Cohen's kappa** the following measures of association, which are described later in the book, are also employed as measures of **interobserver reliability**: a) The **intraclass correlation coefficient (Test 24i)** (employed with interval/ratio data); b) The **Pearson product-moment correlation coefficient** (employed with interval/ratio data); c) **Spearman's rank-order correlation coefficient (Test 29)** (employed with ordinal data); d) **Kendall's tau** (employed with ordinal data); e) **Kendall's coefficient of concordance (Test 31)** (employed with ordinal data). Two features which distinguish **Cohen's kappa** from the aforementioned measures of association are: a) **kappa** is employed with categorical data; b) **kappa** corrects for chance agreement which may occur within the framework of the categorization process by two observers.

Example 16.6 will be employed to illustrate the computation of **Cohen's kappa**.

Example 16.6 *In order to determine the degree of agreement between doctors with respect to diagnosing psychiatric patients, two staff psychiatrists are asked to assign each of 100 patients admitted to a hospital to one of the following three categories: schizophrenia, bipolar disorder, other. Table 16.24 summarizes the diagnostic decisions of the two psychiatrists. The value recorded in each of the nine cells of the 3 × 3 summary table represents the number of patients who were assigned by Doctor 1 to the category recorded at the top of the column in which the cell appears and assigned by Doctor 2 to the category recorded at the left of the row in which the cell appears.*

Table 16.24 Summary Table for Example 16.6

		Doctor 1			Row sums
		Schizophrenia	Bipolar	Other	
	Schizophrenia	31	4	2	37
Doctor 2	Bipolar	6	29	8	43
	Other	10	7	3	20
	Column sums	47	40	13	100

If one does not employ a correction for chance agreement between the two psychiatrists, a measure of agreement can be obtained by merely summing the three values in the diagonal of

Table 16.24, and dividing the sum by the total number of patients evaluated in the table (which is $n = 100$). In other words, the sum $31 + 29 + 3 = 63$ (which is the number of patients for whom the two psychiatrists are in agreement with respect to their diagnosis) is divided by the total of $n = 100$ patients evaluated by the psychiatrists. Thus, $63/100 = .63$, which expressed as a percentage indicates a 63% agreement rate.

Equation 16.30 is employed to compute **Cohen's kappa**, which as noted above, provides a chance-corrected agreement rate. **Cohen's kappa** is represented by the notation κ (which is the lower case Greek letter **kappa**).

$$\kappa = \frac{\Sigma O_{ij} - \Sigma E_{ij}}{n - \Sigma E_{ij}} \qquad \textbf{(Equation 16.30)}$$

Where: $i = j$
  O_{ij} represents the observed frequency of Cell ij
  E_{ij} represents the expected frequency of Cell ij
  n represents the total number of observations/judgements recorded in the summary table

The notation O_{ij} and E_{ij} in Equation 16.30 respectively represent the observed and expected frequencies of the cell which is in the ith row and j^{th} column. It is important to note that the only cells employed in Equation 16.30 are those cells in which $i = j$ — i.e., the cells which comprise the diagonal of the summary table. Thus, in the case of Example 16.6, the three cells employed in Equation 16.30 will represent: a) The number of cases in which both doctors are in agreement that a patient is schizophrenic (31); b) The number of cases in which both doctors are in agreement that a patient has bipolar disorder (29); and c) The number of cases in which both doctors are in agreement that a patient should be assigned to some other category (3).

In the numerator of Equation 16.30, the number of agreements which would be expected by chance (i.e., ΣE_{ij}) is subtracted from the total number of agreements in the summary table (i.e., ΣO_{ij}). In the denominator of the equation, the total of $n = 100$ patients who are diagnosed is also reduced by the number of agreements which would be expected by chance (i.e., ΣE_{ij}). Consequently, **kappa** is a ratio of the two chance-corrected values computed in the numerator and denominator of Equation 16.30.

Although the **kappa** statistic is not based on the chi-square distribution, an expected frequency for a cell is computed in an identical manner to that employed for computing an expected frequency for a cell in a chi-square summary table. Thus, Equation 16.1 is employed to compute the expected cell frequencies for Table 16.24. Through use of Equation 16.1 the expected frequencies for the three diagonal cells are computed below.

$$E_{11} = \frac{(37)(47)}{100} = 17.39$$

$$E_{22} = \frac{(43)(40)}{100} = 17.20$$

$$E_{33} = \frac{(20)(13)}{100} = 2.60$$

The notation ΣE_{ij} in Equation 16.30 represents the sum of the expected frequencies of the cells in the diagonal of the summary table. Thus:

$$\Sigma E_{ij} = 17.39 + 17.20 + 2.60 = 37.19$$

The notation ΣO_{ij} in Equation 16.30 represents the sum of the observed frequencies of the cells in the diagonal of the summary table. Thus:

$$\Sigma O_{ij} = 31 + 29 + 3 = 63$$

Substituting the appropriate values in Equation 16.30, the value $\kappa = .41$ is computed.

$$\kappa = \frac{63 - 37.19}{100 - 37.19} = \frac{25.81}{62.81} = .41$$

Expressed as a percentage (we move the decimal place two places to the right), the computed value $\kappa = .41$ indicates an agreement rate of 41% between the two psychiatrists. Note that the chance-corrected 41% agreement rate computed for **Cohen's kappa** is substantially less than the 63% agreement rate computed earlier which did not employ the chance-correction.

Cohen (1960) notes that Equation 16.31 provides a good approximation of the **standard deviation** (which can also be referred to as the **standard error**) of **kappa**, and that σ_κ can be employed to compute a confidence interval for **kappa**.

$$\sigma_\kappa = \sqrt{\frac{\Sigma O_{ij}(n - \Sigma O_{ij})}{n(n - \Sigma E_{ij})^2}} \qquad \textbf{(Equation 16.31)}$$

Employing Equation 16.31, the value $\sigma_\kappa = .077$ is computed.

$$\sigma_\kappa = \sqrt{\frac{63(100 - 63)}{100(100 - 37.19)^2}} = \sqrt{\frac{2331}{394509.61}} = .077$$

In order to compute a confidence interval for **kappa**, a product is obtained based on multiplying the standard error by the z value for the appropriate confidence interval. Thus, if one is computing the 95% confidence interval, the standard error is multiplied by the tabled critical two-tailed .05 value $z_{.05} = 1.96$. If one is computing the 99% confidence interval, the standard error is multiplied by the tabled critical two-tailed .01 value $z_{.01} = 2.58$. The product for the desired confidence interval is then added to and subtracted from the value computed for **kappa** with Equation 16.31. The latter values define the limits of the confidence interval.

To illustrate, the 95% confidence interval will be computed. We first multiply the standard error $\sigma_\kappa = .077$ by the tabled critical two-tailed .05 value $z_{.05} = 1.96$, and obtain $(.077)(1.96) = .151$. The latter value is added to and subtracted from $\kappa = .41$, yielding the values .561 and .259. These values represent the limits which define the 95% confidence interval for **kappa**. In other words, we can be 95% confident (or the probability is .95) the interval .259 to .561 contains the true value of **kappa** in the population.

Test 16k-a: Test of significance for Cohen's kappa Cohen (1960) notes that Equation 16.32 can be employed to evaluate the null hypothesis H_0: $\kappa_p = 0$ (where κ_p represents the true value of **kappa** in the underlying population). In other words, Equation 16.32 allows a researcher to evaluate whether in the underlying population the true value of **kappa** is some value other than zero. The alternative hypothesis evaluated will generally be the nondirectional alternative hypothesis H_1: $\kappa_p \neq 0$ or the directional alternative hypothesis H_1: $\kappa_p > 0$. Cohen (1960) states that although **kappa** can assume a negative value (specifically, when the degree of agreement between judges is less than chance expectancy) the latter is generally interpreted as a zero correlation, since it indicates there is a lack of agreement between the judges — consequently, the directional alternative hypothesis H_1: $\kappa_p < 0$ is generally not considered.

$$z = \frac{\kappa}{\sigma_{\kappa_p}}$$ **(Equation 16.32)**

Where: $\sigma_{\kappa_p} = \sqrt{\dfrac{\Sigma E_{ij}}{n(n - \Sigma E_{ij})}}$, which represents the **standard error of the sampling distribution**

The null hypothesis H_0: $\kappa_p = 0$ is evaluated below in reference to the value $\kappa = .41$ computed earlier. It should be noted that although the value $\sigma_{\kappa_p} = .077$ computed below is identical to $\sigma_k = .077$ computed earlier, the two values will generally not be equivalent.

$$\sigma_{\kappa_p} = \sqrt{\frac{37.19}{100(100 - 37.19)}} = .077$$

$$z = \frac{.41}{.077} = 5.32$$

The obtained value $z = 5.32$ is evaluated with **Table A1**. Since $z = 5.32$ is greater than the tabled critical two-tailed .05 and .01 values $z_{.05} = 1.96$ and $z_{.01} = 2.58$, the nondirectional alternative hypothesis H_1: $\kappa_p \neq 0$ is supported at both the .05 and .01 levels. Since $z = 5.32$ is greater than the tabled critical one-tailed .05 and .01 values $z_{.05} = 1.65$ and $z_{.01} = 2.33$, the directional alternative hypothesis H_1: $\kappa_p > 0$ is also supported at both the .05 and .01 levels. Cohen (1960) notes the above test is generally of little practical value, since a relatively low value of **kappa** can yield a significant result. In other words, a value such as $\kappa = .41$ (in spite of the fact it is statistically significant) may be deemed by a researcher to be too low a level of reliability (i.e., degree of agreement between judges) to be of practical value in a clinical setting.

Test 16k-b: Test of significance for two independent values of Cohen's kappa   Cohen (1960) notes that Equation 16.33 can be employed to evaluate whether or not there is a significant difference between two independent values of **kappa** (i.e., evaluate the null hypothesis H_0: $\kappa_1 = \kappa_2$). Let us assume that a second set of judges evaluates the same (or a different) set of patients with respect to the three categories summarized in Table 16.24. We will also assume that the computed value for **kappa** is $\kappa = .56$, with a standard error of $\sigma_\kappa = .062$. Employing Equation 16.33, the value $\kappa_2 = .56$ is contrasted with the value $\kappa_1 = .41$ originally computed for Example 16.6.

$$z = \frac{\kappa_1 - \kappa_2}{\sqrt{\sigma_{\kappa_1}^2 + \sigma_{\kappa_2}^2}}$$ **(Equation 16.33)**

$$z = \frac{.41 - .56}{\sqrt{(.077)^2 + (.062)^2}} = \frac{-.15}{.099} = -1.52$$

The obtained value $z = -1.52$ is evaluated with **Table A1**. Since the absolute value $z = 1.52$ is less than the tabled critical two-tailed .05 and .01 values $z_{.05} = 1.96$ and $z_{.01} = 2.58$, the nondirectional alternative hypothesis H_1: $\kappa_1 \neq \kappa_2$ is not supported. Note that the sign of z is negative, since $\kappa_1 < \kappa_2$. However, since the absolute value $z = 1.52$ is less than the tabled critical one-tailed .05 and .01 values $z_{.05} = 1.65$ and $z_{.01} = 2.33$, the directional alternative hypothesis H_1: $\kappa_1 < \kappa_2$ is not supported. The directional alternative hypothesis H_1: $\kappa_1 > \kappa_2$ is not supported, since in order for the latter alternative hypothesis to be supported κ_1 must be greater than κ_2, as well as the fact that the positive value obtained for z must be equal to or greater than the prespecified tabled critical one-tailed value.

Additional discussion of **Cohen's kappa** can be found in Everitt (2001) who, among other things, provides references for extending **kappa** to more than two judges.

12. Combining the results of multiple 2 × 2 contingency tables: Heterogeneity chi-square analysis for a 2 × 2 contingency table and Test 16l: The Mantel–Haenszel test (and Test 16l-a: Test of homogeneity of odds ratios for Mantel–Haenszel analysis, Test 16l-b: Summary odds ratio for Mantel–Haenszel analysis, and Test 16l-c: Mantel–Haenszel test of association) This section will describe two procedures for pooling the results from multiple 2 × 2 contingency tables which summarize the results of m independent experiments which evaluate the same hypothesis. The first procedure (**heterogeneity chi-square analysis**), which was included in the first three editions of this book, can lead to erroneous conclusions if not employed judiciously. The second procedure (the **Mantel–Haenszel analysis**) is the method of choice in most sources, since it is less likely to lead to erroneous conclusions, as well as the fact that it is used in conjunction with computing a **summary odds ratio** for m 2 × 2 contingency tables.

Heterogeneity chi-square analysis for a 2 × 2 contingency table Let us suppose that a researcher conducts m independent studies (where $m \geq 2$), each of which evaluates the same hypothesis. In order to illustrate the analysis to be described in this section, we will assume that the null hypothesis evaluated in each study is that a particular type of kidney disease is equally likely to occur in males and females. The data for each study are summarized in a 2 × 2 contingency table, and each of the contingency tables is evaluated with the **chi-square test for homogeneity**. Although none of the m analyses yield a statistically significant result, visual inspection of the data suggests to the researcher that the disease occurs more frequently in females than males. The researcher suspects that because of the relatively small sample size employed in each study, the absence of significant results may be due to a lack of statistical power. In order to increase the power of the analysis, the researcher wants to combine the data for the m studies into one 2 × 2 contingency table, and evaluate the latter table with the **chi-square test for homogeneity**. This section will present a procedure (described in Zar (1999, pp. 500–504)) for determining whether or not a researcher is justified in pooling data under such conditions. The procedure to be presented, which is referred to as **heterogeneity chi-square analysis**, was previously employed in reference to a one-dimensional chi-square table in Section VI of the **chi-square goodness-of-fit test**.

The null and alternative hypotheses which are evaluated with a **heterogeneity chi-square analysis** are as follows.

Null hypothesis H_0: The m samples are derived from the same population (i.e., population homogeneity).

Alternative hypothesis H_1: At least two of the m samples are not derived from the same population (i.e., population heterogeneity).

Table 16.25 summarizes the **heterogeneity chi-square analysis** conducted on $m = 3$ 2 × 2 contingency tables. Each table summarizes the results of a study evaluating the frequency

Table 16.25 Heterogeneity Chi-Square Analysis for Three 2 × 2 Contingency Tables

A. Chi-square analysis of three individual studies

Cell	O_{ij}	E_{ij}	$(O_{ij} - E_{ij})$	$(O_{ij} - E_{ij})^2$	$\dfrac{(O_{ij} - E_{ij})^2}{E_{ij}}$
			Study 1		
(a) Male/Disease	12	14.5	−2.5	6.25	.43
(b) Male/No disease	17	14.5	2.5	6.25	.43
(c) Female/Disease	18	15.5	2.5	6.25	.40
(d) Female/No disease	13	15.5	−2.5	6.25	.40
	$\Sigma O_{ij} = 60$ $\Sigma E_{ij} = 60$ $\Sigma(O_{ij} - E_{ij}) = 0$				$\chi_1^2 = 1.69$
			Study 2		
(a) Male/Disease	10	13	−3	9	.69
(b) Male/No disease	16	13	3	9	.69
(c) Female/Disease	15	12	3	9	.75
(d) Female/No disease	9	12	−3	9	.75
	$\Sigma O_{ij} = 50$ $\Sigma E_{ij} = 50$ $\Sigma(O_{ij} - E_{ij}) = 0$				$\chi_2^2 = 2.88$
			Study 3		
(a) Male/Disease	6	8.55	−2.55	6.50	.76
(b) Male/No disease	12	9.45	2.55	6.50	.69
(c) Female/Disease	13	10.45	2.55	6.50	.62
(d) Female/No disease	9	11.55	−2.55	6.50	.56
	$\Sigma O_{ij} = 40$ $\Sigma E_{ij} = 40$ $\Sigma(O_{ij} - E_{ij}) = 0$				$\chi_3^2 = 2.63$

Sum of chi-square values for three studies = $\chi_{sum}^2 = 1.69 + 2.88 + 2.63 = 7.20$

B. Chi-square analysis of pooled data

Pooled data for $m = 3$ studies

Cell	O_{ij}	E_{ij}	$(O_{ij} - E_{ij})$	$(O_{ij} - E_{ij})^2$	$\dfrac{(O_{ij} - E_{ij})^2}{E_{ij}}$
(a) Male/Disease	28	36.01	−8.01	64.16	1.78
(b) Male/No disease	45	36.99	8.01	64.16	1.73
(c) Female/Disease	46	37.99	8.01	64.16	1.69
(d) Female/No disease	31	39.01	−8.01	64.16	1.64
	$\Sigma O_{ij} = 150$ $\Sigma E_{ij} = 60$ $\Sigma(O_{ij} - E_{ij}) = 0$				$\chi_{pooled}^2 = 6.84$

C. Heterogeneity of chi-square analysis

Heterogeneity chi-square = Sum of chi-square values for four studies − Pooled chi-square value

$$\chi_{het}^2 = (\chi_{sum}^2 = 7.20) - (\chi_{pooled}^2 = 6.84) = .36$$

D. Continuity-corrected chi-square analysis of pooled data

Pooled data for $m = 3$ studies (continuity-corrected)

Cell	O_{ij}	E_{ij}	$|O_{ij} - E_{ij}| -5$	$(|O_{ij} - E_{ij}| -5)^2$	$\dfrac{(|O_{ij} - E_{ij}| -5)^2}{E_{ij}}$
(a) Male/Disease	28	36.01	7.51	56.40	1.57
(b) Male/No disease	45	36.99	7.51	56.40	1.52
(c) Female/Disease	46	37.99	7.51	56.40	1.48
(d) Female/No disease	31	39.01	7.51	56.40	1.45
	$\Sigma O_{ij} = 150$ $\Sigma E_{ij} = 60$ $\Sigma(O_{ij} - E_{ij}) = 0$				$\chi_{pooled}^2 = 6.02$

of occurrence of the kidney disease in males and females. Part A of Table 16.25 presents the analysis of each study with the **chi-square test for homogeneity**. Column 2 for each of the studies contains the observed frequency for the relevant cell of the 2 × 2 contingency table. Thus, using the notation in Table 16.6, the following cells are represented in the contingency tables: **Cell a:** Males who develop the disease; **Cell b:** Males who do not develop the disease; **Cell c:** Females who develop the disease; **Cell d:** Females who do not develop the disease.

The following protocol is employed in the **heterogeneity chi-square analysis**: a) Employing the **chi-square test for homogeneity**, a chi-square value is computed for each of the individual studies. Zar (1999) notes that even though each table has one degree of freedom, the correction for continuity should not be employed in analyzing the individual contingency tables at this point in the analysis; b) The sum of the m chi-square values obtained in a) for the individual studies is computed. The latter value itself represents a chi-square value, and will be designated χ^2_{sum}. In addition, the sum of the degrees of freedom for the m studies is computed. The latter degrees of freedom value, which will be designated df_{sum}, is obtained by summing the df values for each of the individual studies. Since in the case of a 2 × 2 contingency table $df = (r-1)(c-1) = 1$, the value of df_{sum} will equal the number of studies that are evaluated; c) The data for the m studies are combined into one table, and through use of the **chi-square test for homogeneity** a chi-square value, which will be designated χ^2_{pooled}, is computed for the pooled data. The degrees of freedom for the table with the pooled data, which will be designated df_{pooled}, is equal to $df = 1$, since df for a 2 × 2 contingency table is $df = (r-1)(c-1) = 1$. Even though the table has one degree of freedom, the correction for continuity is not used in analyzing the table with the pooled data at this point in the analysis; d) The **heterogeneity chi-square analysis** is based on the premise that if the m samples are in fact homogeneous, the sum of the m individual chi-square values (χ^2_{sum}) should be approximately the same value as the chi-square value computed for the pooled data (χ^2_{pooled}). In order to determine the latter, the absolute value of the difference between the sum of the m chi-square values (obtained in b)) and the pooled chi-square value (obtained in c)) is computed. The obtained difference, which is itself a chi-square value, is the **heterogeneity chi-square value**, which will be designated χ^2_{het}. Thus, $\chi^2_{het} = |\chi^2_{sum} - \chi^2_{pooled}|$. The null hypothesis will be rejected when there is a large difference between the values of χ^2_{sum} and χ^2_{pooled}. The value χ^2_{het}, which represents the test statistic, is evaluated with a degrees of freedom value that is the sum of the degrees of freedom for the m individual studies (df_{sum}) less the degrees of freedom obtained for the table with the pooled data (df_{pooled}). Thus, $df_{het} = df_{sum} - df_{pooled}$. In order to reject the null hypothesis, the value χ^2_{het} must be equal to or greater than the tabled critical value at the prespecified level of significance for df_{het}; and e) If the null hypothesis is rejected the data cannot be pooled. If, however, the null hypothesis is retained, the data can be pooled, and the computed value for χ^2_{pooled} is employed to evaluate the goodness-of-fit hypothesis. Zar (1999) notes, however, that the table for the pooled data should be reevaluated employing the correction for continuity, and the continuity-corrected χ^2_{pooled} value (which will be a little lower than the original χ^2_{pooled} value) should be employed to evaluate the relevant hypothesis for the contingency table.

The computed chi-square values for the three studies (summarized in **Part A** of Table 16.25) are $\chi^2_1 = 1.69$, $\chi^2_2 = 2.88$, and $\chi^2_3 = 2.63$. The total number of degrees of freedom employed for the three studies is $df_{sum} = (3)(1) = 3$ (i.e., the number of studies (3) multiplied by the number of degrees of freedom per study (1)). Since a single 2 × 2 contingency table is evaluated for the pooled data, the degrees of freedom for the latter table is $df_{pooled} = 1$. By summing the chi-square values for the three studies, we compute the value $\chi^2_{sum} = 1.69 + 2.88 + 2.63 = 7.20$. Since in **Part B** of Table 16.25 we compute $\chi^2_{pooled} = 6.84$, the value for heterogeneity chi-square (computed in **Part C** of Table 16.25) is $\chi^2_{het} = |(\chi^2_{sum} = 7.20) - (\chi^2_{pooled} = 6.84)| = .36$. The degrees of freedom employed to evaluate the latter chi-square value are $df_{het} = (df_{sum} = 3) - (df_{pooled} = 1) = 2$. The tabled critical .05 and .01 chi-square values in **Table A4** for $df = 2$ are $\chi^2_{.05} = 5.99$ and $\chi^2_{.01} = 9.21$. Since the computed value $\chi^2_{het} = .36$ is less than $\chi^2_{.05} = 5.99$, the null hypothesis is retained. In other words we can conclude the three samples are homogeneous (i.e., come from the same population), and thus we can justify pooling the data into a single table.

As noted earlier, the correction for continuity is not employed in computing the value $\chi^2_{pooled} = 6.84$ for the pooled data in **Part B** of Table 16.25. The data for the pooled contingency table are now reevaluated with the **chi-square test for homogeneity**, employing the correction for continuity. The continuity-corrected chi-square value (obtained in **Part D** of Table 16.25) is $\chi^2_{pooled} = 6.02$. The degrees of freedom for the pooled contingency table are $df = (r-1)(c-1) = 1$. The tabled critical .05 and .01 values in **Table A4** for $df = 1$ are $\chi^2_{.05} = 3.84$ and $\chi^2_{.01} = 6.63$. Since the value $\chi^2_{pooled} = 6.02$ is larger than $\chi^2_{.05} = 3.84$, the null hypothesis for the data in the contingency table can be rejected at the .05 level (but not at the .01 level). In other words, with respect to the pooled data, we can conclude that in the case of at least one of the cells, there is a difference between its observed and expected frequency. Inspection of the data suggests that the disease occurs more frequently in females than it does in males. A more detailed discussion of the **heterogeneity chi-square analysis** can be found in Zar (1999).

It should be emphasized that a researcher should employ common sense in applying the heterogeneity chi-square analysis described in this section. To be more specific, there may be occasions when even though the computed value of χ^2_{het} is not significant, in spite of the latter it would not be recommended the researcher pool the data from two or more smaller tables. Put simply, one should not pool data from two or more tables employing small sample sizes (which, when evaluated individually, fail to yield a significant chi-square value) in order to obtain a significant pooled chi-square value, when there is an obvious inconsistency in the cell proportions for two or more of the tables. In other words, when the data from m tables are pooled, the proportion of cases in the cells of each of the m tables should be approximately the same.

Test 16l: The Mantel–Haenszel analysis (Test 16l-a: Test of homogeneity of odds ratios for Mantel–Haenszel analysis, Test 16l-b: Summary odds ratio for Mantel–Haenszel analysis, and Test 16l-c: Mantel-Haenszel test of association)   Because of the limitations of the heterogeneity chi-square analysis described in this section, among others, Everitt (1977, 1992), Fleiss (1981), Fleiss *et al.* (2003, Chapter 10), and Hollander and Wolfe (1999) recommend alternative procedures for pooling the data from multiple 2 × 2 contingency tables. The most commonly recommended procedure (as well as the most frequently employed, according to Hollander (1999)) is the **Mantel–Haenszel analysis** (Mantel and Haenszel (1959)). The latter methodology (as well as other procedures for evaluating 2 × 2 contingency tables) is often employed to pool 2 × 2 contingency tables based on different subgroups that are part of a larger population. Such subgroups, which are commonly referred to as **strata** (which is discussed in the section on **sampling methodologies** in the **Introduction**), are differentiated from one another based on a potential confounding variable which a researcher wants to take into consideration prior to pooling the data from the different studies. To illustrate, let us assume that in the three studies summarized in Table 16.25 (in which the researcher evaluates a hypothesis regarding the relationship between a kidney disease and gender) the researcher suspects that exposure to cigarette smoke during one's adult lifetime might represent a potential confounding variable — i.e., the amount of smoke a person has been exposed to may be associated with the likelihood of one developing the disease. Consequently, each of the three studies conducted might have employed three distinct (i.e., homogeneous) subgroups (i.e., strata) of subjects who had been differentially exposed to cigarette smoke during their adult lifetimes. As an example, the 60 subjects in Study 1 could be individuals who had smoked all of their adult lives; the 50 subjects in Study 2 could be individuals who did not smoke themselves, yet lived with a smoker for the duration of their adult lives; and the 40 subjects in Study 3 could be subjects who neither smoked or lived with a smoker during their adult lives. If, in fact, the researcher determined that the data for the three subgroups/strata were homogeneous (i.e., that the statistical relationship between a person developing the kidney disease and one's gender was consistent regardless of the degree to which one was exposed to cigarette smoke during one's adult lifetime), the data for the three 2 × 2 contingency tables could be pooled.

The **Mantel–Haenszel analysis** can be broken down into two stages which should be conducted in following sequential order: a) Initially the researcher should compute an **odds ratio** for each of the *m* 2 × 2 contingency tables, and then conduct a **test of homogeneity** on the *m* **odds ratios**. If it is determined that the *m* **odds ratios** are homogeneous, the researcher is then justified in pooling the data from the *m* contingency tables. If, however, the analysis indicates that the *m* **odds ratios** are not homogeneous, the data cannot be pooled and the analysis should be terminated at that point; b) If during Stage 1 it is determined that the *m* **odds ratios** are homogeneous, the researcher can now compute a **summary odds ratio** for the *m* contingency tables. Upon doing the latter, a confidence interval can be computed for the **summary odds ratio**. If the range of values which define the 95% confidence interval (assuming the .05 level of significance is employed) does not include the value 1, the researcher can conclude that a significant association exists between the variables in the *m* contingency tables. The latter can also be demonstrated through use of the **Mantel–Haenszel test of association**, which is a statistical test (that employs either the chi-square or normal distributions).

Test 16l-a: Test of homogeneity of odds ratios for Mantel–Haenszel analysis The *m* = 3 independent 2 × 2 contingency tables summarized in Table 16.25 will be employed in the discussion of the **test of homogeneity of odds ratios** (as well as in the discussion of the other procedures described in this section). In assessing homogeneity of one or more **odds ratios**, one initially computes an **odds ratio** for each of the 2 × 2 contingency tables. As noted earlier in this chapter (under the discussion of the **odds ratio**), it is possible to compute two **odds ratios** for a 2 × 2 contingency table (with the second **odds ratio** identified as the **inverse odds ratio**). In the case of the 2 × 2 tables summarized in Table 16.25, Equation 16.27 can be employed to compute the **odds ratio** which indicates how much greater the odds are of being a female having the disease than being a male having the disease. Equation 16.28, on the other hand, can be employed to compute the **odds ratio** which indicates how much greater the odds are of being a male having the disease than being a female having the disease.

Employing Equation 16.27, an **odds ratio** is computed for each of the three contingency tables in Table 16.25. The three values computed for the **odds ratios** indicate the following: a) In Study 1, the odds of a female having the disease are 1.96 times greater than the odds of a male having the disease; b) In Study 2, the odds of a female having the disease are 2.67 times greater than the odds of a male having the disease; c) In Study 3, the odds of a female having the disease are 2.89 times greater than odds of a male having the disease.

$$o_1 = \frac{b_1 c_1}{a_1 d_1} = \frac{(17)(18)}{(12)(13)} = 1.96$$

$$o_2 = \frac{b_2 c_2}{a_2 d_2} = \frac{(16)(15)}{(10)(9)} = 2.67$$

$$o_3 = \frac{b_3 c_3}{a_3 d_3} = \frac{(12)(13)}{(6)(9)} = 2.89$$

The **test of homogeneity of odds ratios for Mantel–Haenszel analysis** (which is discussed in Pagano and Gauvreau (1993, pp. 343–346)) will now be conducted to determine whether or not the three **odds ratio** (which represent estimates of the **odds ratios** in the underlying populations) are homogenous. Specifically, the null and alternative hypotheses evaluated with the test are as follows.

H_0: In the underlying populations represented by the three samples (i.e., the data in each of the three contingency tables) the **odds ratios** are identical.

H_1: In the underlying populations represented by the three samples (i.e., the data in each of the three contingency tables) the **odds ratios** are not identical in at least two of the three populations.

The test statistic for the **test of homogeneity of odds ratios for Mantel–Haenszel analysis** is computed with Equation 16.34.

$$\chi^2 = \sum_{i=1}^{m} w_i (y_i - \bar{Y})^2 \qquad \text{(Equation 16.34)}$$

Where: y_i represents the natural logarithm of the estimated **odds ratio** in the ith contingency table.

w_i represents the weight assigned to the ith contingency table, based on the number of observations in the contingency table.

$\bar{Y} = \sum_{i=1}^{m} [(w_i y_i) / \sum_{i=1}^{m} w_i]$ (which will be designated as Equation 16.36) represents the weighted average of the m **odds ratios** expressed as a natural logarithm. If we conclude that the m individual **odds ratios** are homogeneous, the antilogarithm of $\bar{Y}$ is employed to represent the **summary odds ratio** for the m contingency tables.

We first compute the natural logarithm of the three **odds ratios** computed earlier with Equation 16.27: $\ln(o_1) = \ln(1.96) = .6729$; $\ln(o_2) = \ln(2.67) = .9821$; $\ln(o_3) = \ln(2.89) = 1.0613$. Thus, $y_1 = .6729$, $y_2 = .9821$, and $y_3 = .1.0613$. Equation 16.35 is employed to compute the weight to assign to each of the **odds ratios**, based on the total number of observations in the contingency table for which the **odds ratio** was computed. Employing the latter equation, following weights are computed: $w_1 = 3.6417$, $w_2 = 2.9386$, $w_3 = 2.2831$.

$$w_i = \frac{1}{\dfrac{1}{a_i} + \dfrac{1}{b_i} + \dfrac{1}{c_i} + \dfrac{1}{d_i}} \qquad \text{(Equation 16.35)}$$

$$w_1 = \frac{1}{\dfrac{1}{12} + \dfrac{1}{17} + \dfrac{1}{18} + \dfrac{1}{13}} = 3.6417$$

$$w_2 = \frac{1}{\dfrac{1}{10} + \dfrac{1}{16} + \dfrac{1}{15} + \dfrac{1}{9}} = 2.9386$$

$$w_3 = \frac{1}{\dfrac{1}{6} + \dfrac{1}{12} + \dfrac{1}{13} + \dfrac{1}{9}} = 2.2831$$

Employing Equation 16.36, the value $\bar{Y} = .8755$ is computed below.

(Equation 16.36)

$$\bar{Y} = \frac{\sum\limits_{i=1}^{m} w_i y_i}{\sum\limits_{i=1}^{m} w_i} = \frac{(3.6417)(.6729) + (2.9386)(.9821) + (2.2831)(1.0613)}{3.6417 + 2.9386 + 2.2831} = .8755$$

The antilogarithm of $\bar{Y} = .8755$ is 2.40, which as noted earlier will be employed to represent the **summary odds ratio** (which will be represented by the notation o_{MH}) if the **odds ratios** for the three contingency tables are homogeneous.

Equation 16.34 is now employed to evaluate the null hypothesis of homogeneity of the three odds ratios.

$$\chi^2 = (3.6417)(.6729 - .8755)^2 + (2.9386)(.9821 - .8755)^2 + (2.2831)(1.0613 - .8755)^2 = .2617$$

The degrees of freedom employed to evaluate the chi-square value computed with Equation 16.34 are $df = m - 1$. Thus, $df = 3 - 1 = 2$. Employing **Table A4**, the tabled critical .05 and .01 chi-square values for $df = 2$ are $\chi^2_{.05} = 5.99$ and $\chi^2_{.01} = 9.21$. Since the computed value $\chi^2 = .2617$ is less than $\chi^2_{.05} = 5.99$, the null hypothesis is retained, and we can conclude that the **odds ratios** in the three contingency tables are, in fact, homogeneous. Because of the latter we are justified in pooling the data for the three contingency tables.

It should be noted that if it is determined the **odds ratios** computed for m independent contingency tables are homogeneous, the same result will be obtained if the **inverse odds ratios** are evaluated for homogeneity. Thus, if Equation 16.28 were employed to compute the inverse **odds ratio** for each of the three contingency tables, the same chi-square value will be obtained with Equation 16.34. The **inverse odds ratios** (which can also be obtained by computing the reciprocal of an **odds ratio** — i.e., dividing the latter value into 1) for the example under discussion are $o_1' = 1/1.96 = .51$; $o_2' = 1/2.67 = .37$; $o_3' = 1/2.89 = .35$).

Test 16l-b: Summary odds ratio for Mantel–Haenszel analysis It was noted earlier that if the m **odds ratios** are homogeneous, the antilogarithm of $\bar{Y}$ will be the **summary odds ratio**, and thus o_{MH} = antilog $(.8755) = 2.40$. Equation 16.37 can also be employed to compute the **summary odds ratio**. Equation 16.38 can be employed to compute the **summary inverse odds ratio** $o'_{MH} = .42$ (represented with the notation o'_{MH}) for the set of $m = 3$ contingency tables. The obtained value $o_{MH} = 2.40$ indicates that if the pooled data from the three contingency tables are employed, the odds of a female having the disease are 2.40 times greater than a male having the disease. The obtained value $o'_{MH} = .42$ indicates that if the pooled data from the three contingency tables are employed, the odds of a male having the disease are .42 times as large as a female having the disease. Note also that since $o'_{MH} = 1/o_{MH}$, $o'_{MH} = 1/2.40 = .42$.

$$o_{MH} = \frac{\sum\limits_{i=1}^{m} (b_i c_i)/n_i}{\sum\limits_{i=1}^{m} (a_i d_i)/n_i}$$

(Equation 16.37)

$$o_{MH} = \frac{[(17)(18)]/60 + [(16)(15)]/50 + [(12)(13)]/40}{[(12)(13)]/60 + [(10)(9)]/50 + [(6)(9)]/40} = 2.40$$

$$o'_{MH} = \frac{\sum\limits_{i=1}^{m} (a_i d_i)/n_i}{\sum\limits_{i=1}^{m} (b_i c_i)/n_i}$$

(Equation 16.38)

$$o'_{MH} = \frac{[(12)(13)]/60 + [(10)(9)]/50 + [(6)(9)]/40}{[(17)(18)]/60 + [(16)(15)]/50 + [(12)(13)]/40} = .42$$

Computation of a confidence interval for the summary odds ratio Pagano and Gauvreau (1993, p. 347) recommend that prior to computing a confidence interval for a **summary odds ratio**, Equations 16.39–16.42 be employed to determine whether or not the overall sample size for the m contingency tables is large enough to insure that a reliable value will be computed for the confidence interval. In the case of all four equations (each of which computes a pooled expected frequency for one of the four cells that comprise a pooled 2×2 contingency table), the computed expected frequency should be equal to or greater than 5. The notation employed in all four equations stipulates that for the indicated cell, the expected frequency is computed for each of the $m = 3$ contingency tables, and upon doing the latter, the $m = 3$ expected frequencies are summed, yielding an overall expected frequency for that cell. Employing Equations 16.39–16.42, the values $E(a_{pooled}) = 36.05$, $E(b_{pooled}) = 36.95$, $E(c_{pooled}) = 37.95$, and $E(d_{pooled}) = 39.05$ are computed.

(Equation 16.39)

$$E(a_{pooled}) = \sum_{i=1}^{m} \left[\frac{(a_i + b_i)(a_i + c_i)}{n_i} \right] = \frac{(29)(30)}{60} + \frac{(26)(25)}{50} + \frac{(18)(19)}{40} = 36.05$$

(Equation 16.40)

$$E(b_{pooled}) = \sum_{i=1}^{m} \left[\frac{(a_i + b_i)(b_i + d_i)}{n_i} \right] = \frac{(29)(30)}{60} + \frac{(26)(25)}{50} + \frac{(18)(21)}{40} = 36.95$$

(Equation 16.41)

$$E(c_{pooled}) = \sum_{i=1}^{m} \left[\frac{(c_i + d_i)(a_i + c_i)}{n_i} \right] = \frac{(31)(30)}{60} + \frac{(24)(25)}{50} + \frac{(22)(19)}{40} = 37.95$$

(Equation 16.42)

$$E(d_{pooled}) = \sum_{i=1}^{m} \left[\frac{(c_i + d_i)(b_i + d_i)}{n_i} \right] = \frac{(31)(30)}{60} + \frac{(24)(25)}{50} + \frac{(22)(21)}{40} = 39.05$$

Since all of the computed values for the expected frequencies for the pooled data are larger than 5, we can proceed to compute a confidence interval for the **summary odds ratio**. As is the case with the computation of a confidence interval for an **odds ratio**, in the case of a **summary odds ratio** the sampling distribution of the relative odds will also be positively skewed. Since the sampling distribution of the natural logarithm of the **odds ratio** provides a better approximation of a normally distributed variable, a confidence interval is initially computed for the natural logarithm of the **summary odds ratio**. In order to do the latter, Equation 16.43 must be employed to compute an **estimated standard error** for $\bar{Y}$, which as noted earlier is the weighted average of the m **odds ratios** expressed as a natural logarithm. The notation in Equation 16.43 indicates that to compute the **standard error** (which will be represented by the notation $s_{\bar{Y}_{MH}}$), the number 1 is divided by the square root of the sum of the w_i values (which, as noted earlier, represent the weights assigned to each of the contingency tables). The value $s_{\bar{Y}_{MH}} = .3359$ is computed below.

(Equation 16.43)

$$s_{\bar{Y}_{MH}} = \frac{1}{\sqrt{\sum_{i=1}^{m} w_i}} = \frac{1}{\sqrt{3.6417 + 2.9386 + 2.2831}} = .3359$$

Equation 16.44 is employed to compute the confidence interval for the **summary odds ratio**. The 95% confidence interval is computed below with the latter equation.

$$CI_{(1-\alpha)} = \bar{Y} \pm (z_{\alpha/2})(s_{\bar{y}_{MH}})$$ **(Equation 16.44)**

$$CI_{.95} = \bar{Y} \pm (z_{.05})(s_{\bar{Y}_{MH}}) = .8755 \pm (1.96)(.3359) = .8755 \pm .6584$$

When .6584 is subtracted from and added to .8755, the values .2171 and 1.5339 are obtained. The antilogarithms of the latter values will represent the lower and upper boundaries of the 95% confidence interval of the **summary odds ratio**. Thus, antilog (.2171) = 1.2425 and antilog (1.5339) = 4.6362. Thus, we can be 95% confident (or the probability is .95) the interval 1.2425 to 4.6362 contains the true value of the **summary odds ratio** in the population. We can also compute the 95% confidence interval for $o'_{MH} = .42$, the **summary inverse odds ratio**, by first computing the natural logarithm of the latter value: ln (.42) = −.8675. The value .6584 is then subtracted from and added to −.8675, yielding −1.5259 and −.2091. The antilogarithms of the latter values will represent the lower and upper boundaries of the 95% confidence interval of the **summary inverse odds ratio**. Thus, antilog (−1.5259) = .2174 and antilog (−.2091) = .8113. Thus, we can be 95% confident (or the probability is .95) the interval .2174 to .8113 contains the true value of the **summary inverse odds ratio** in the population.

Among others, Pagano and Gauvreau (1993, pp. 350) note that a confidence interval can be employed to evaluate the null hypothesis that the population **summary odds ratio** is equal to 1 against the alternative hypothesis that the population **summary odds ratio** is equal to some value other than 1. Specifically, if the range of values which define a confidence interval for an **odds ratio** includes the value 1, the above null hypothesis is retained. On the other hand, if the confidence interval does not include the value 1, the null hypothesis can be rejected. Rejection of the null hypothesis is viewed as evidence that a significant association exists between the variables in the m contingency tables. Because of the latter, the above null and alternative hypotheses can also be stated as follows: H_0: When the data from the $m = 3$ 2×2 contingency tables are pooled, there is no statistical association between the two variables which comprise the contingency table (i.e., in the example under discussion, kidney disease and gender); H_1: There is a significant statistical association between the two variables.

In the case of a 95% confidence interval, the above null hypothesis is evaluated with a two-tailed test with alpha set equal to .05. Since in our example, the range 1.2425 to 4.6362 obtained for the **summary odds ratio** confidence interval does not include the value 1, we can reject the null hypothesis, and thus the alternative hypothesis stating that the population **summary odds ratio** is some value other than 1 is supported. The null hypothesis can also be rejected in the case of the **summary inverse odds ratio**, since the range .2174 to .8113 obtained for the **summary inverse odds ratio** confidence interval also does not include the value 1. As noted in the previous paragraph, rejection of the null hypothesis is commensurate with concluding that a significant association exists between the variables in the m contingency tables — in the case of the pooled data for the three studies summarized in Table 16.25, there is a significant relationship (i.e., statistical association) between gender and whether or not a person has the kidney disease, with females being more likely to have the disease than males.

Test 16l-c: The Mantel–Haenszel test of association   It was noted at the beginning of the discussion of the **Mantel–Haenszel analysis**, that the **Mantel–Haenszel test of association** evaluates the same hypothesis which can be evaluated by determining whether or not the range of values for a confidence interval includes the value 1. Thus, the null hypothesis evaluated with the **Mantel–Haenszel test of association** is that when the data from the $m = 3$ 2×2 contingency tables are pooled, there is no statistical association between the two variables which comprise the table, versus the alternative hypothesis that there is a significant statistical association between the two variables. Alternatively, the null and alternative hypotheses can be stated as follows: H_0: The population **summary odds ratio** is equal to 1; H_1: The population **summary odds ratio** is equal to some value other than 1.

Employing the notation in Table 16.6, the null hypothesis can also be stated as follows:

H_0: If the m contingency tables are viewed as a whole, the likelihood of an observation in row 1 being in column 1 equals the likelihood of an observation in row 2 being in column 1. Employed in reference to the notation in Table 16.6, it states that $a/(a + b) = c/(c + d)$. In reference to Table 16.25, the latter states the likelihood of a male having the disease is equal to the likelihood of a female having the disease.

H_1: If the m contingency tables are viewed as a whole, the likelihood of an observation in row 1 being in column 1 is greater than the likelihood of an observation in row 2 being in column 1, or the likelihood of an observation in row 1 being in column 1 is less than the likelihood of an observation in row 2 being in column 1. Employing the notation in Table 16.6, it states that $a/(a + b) > c/(c + d)$ or $a/(a + b) < c/(c + d)$. In reference to Table 16.25, the latter states the likelihood of a male having the disease is greater than the likelihood of a female having the disease or the likelihood of a male having the disease is less than the likelihood of a female having the disease. As stated, the alternative hypothesis is **nondirectional**, which is evaluated with the appropriate tabled critical two-tailed value. The alternative hypothesis can also be stated **directionally** by stipulating a specific directional difference for the pooled data. A directional alternative hypothesis is evaluated with the appropriate tabled critical one-tailed value.

Equation 16.45 is employed to compute the test statistic (represented by $\chi^2_{MH_c}$) for the **Mantel–Haenszel test of association** when a correction for continuity is employed. Equation

16.46 is employed to compute the test statistic (represented by χ^2_{MH}) for the **Mantel–Haenszel test of association** when a correction for continuity is not employed. Equations 16.45 and 16.46 are both based on the assumption that the marginal sums (i.e., the sums of the rows and columns) of the *m* contingency tables are fixed (or nonrandom) (which as noted earlier in this chapter is an underlying assumption of the **Fisher exact test**). Conover (1999, p. 193) notes that if one is unwilling to accept the latter assumption (and consequently assumes that the values of the row and column sums are randomly determined), Equation 16.49 (to be discussed later in this section) should be employed, since the use of the latter equation will result in a more powerful test of the alternative hypothesis. If the chi-square value computed with Equations 16.45 or 16.46 is statistically significant, it indicates that when the data for the *m* independent 2 × 2 contingency tables are pooled there is a significant statistical association between the two variables.

$$\chi^2_{MH_c} = \sum_{i=1}^{m} \left[\frac{(|O - E| - .5)^2}{V} \right]$$

(Equation 16.45)

Where: *m* represents the number of contingency tables (i.e., independent studies conducted) $O = \sum_{i=1}^{m} O_i$, where $O_i = a_i$. Specifically, employing the notation in Table 16.6, a_i represents the number of observations in cell *a* of the ith contingency table in the set of *m* contingency tables. Thus, *O* is the sum of the observed frequencies in cell *a* of the *m* contingency tables.

$E = \sum_{i=1}^{m} E_i$, where $E_i = [(a_i + b_i)(a_i + c_i)] / n_i$. Specifically, employing the notation in Table 16.6, in the set of *m* contingency tables, a_i represents the number of observations in cell *a* of the ith contingency table, b_i represents number of observations in cell *b* of the ith contingency table, c_i represents the number of observations in cell *c* of the ith contingency table, d_i represents the number of observations in cell *d* of the ith contingency table and n_i represents total number of observations in the ith contingency table. Inspection of the equation for E_i should reveal that E_i is the expected frequency for cell *a* in the ith contingency table. Consequently, *E* is the sum of the expected frequencies for cell *a* in the *m* contingency tables.

V represents the variance of the observed frequencies. The value of *V* is computed as follows: $V = \sum_{i=1}^{m} V_i = \sum_{i=1}^{m} [[(a_i + b_i)(c_i + d_i)(a_i + c_i)(b_i + d_i)] / [n_i^2(n_i - 1)]]$. The latter equation indicates that for each of the *m* contingency tables, the product of the marginal sums (i.e., the sums of the two rows and two columns) is obtained and divided by $[n_i^2(n_i - 1)]$, yielding the value V_i. *V* is the sum of the *m* V_i values. Rosner (1995, p. 407) notes that the **Mantel–Haenszel test of association** should only be conducted if $V \geq 5$.

Equation 16.45 will now be employed to evaluate the data in the three 2 × 2 contingency tables documented in Table 16.25. The values $O = 28$, $E = 36.05$, and $V = 9.5251$ are computed as follows: a) $O_1 = a_1 = 12$; $O_2 = a_2 = 10$; $O_3 = a_3 = 6$. Thus, $O = \sum_{i=1}^{m} O_i = 12 + 10 + 6 = 28$; b) $E_1 = E(a_1) = [(29)(30)]/60 = 14.5$; $E_2 = E(a_2) = [(26)(25)]/50 = 13$; $E_3 = E(a_3) = [(18)(19)]/40 = 8.55$. Thus, $E = \sum_{i=1}^{m}(E_i) = 14.5 + 13 + 8.55 = 36.05$.

$$V = \frac{(29)(31)(30)(30)}{(60)^2(59)} + \frac{(26)(24)(25)(25)}{(50)^2(49)} + \frac{(18)(22)(19)(21)}{(40)^2(39)} =$$

$$3.8093 + 3.1837 + 2.5321 = 9.5251$$

$$\chi^2_{MH_c} = \sum_{i=1}^{m} \left[\frac{(|O - E| - .5)^2}{V} \right] = \left[\frac{(|28 - 36.05| - .5)^2}{9.5251} \right] = \frac{57.0025}{9.5251} = 5.984$$

The obtained $\chi^2_{MH_c}$ value is always evaluated employing one degree of freedom (i.e., $df =$ 1). If we conduct a two-tailed/nondirectional analysis, the tabled critical two-tailed .05 and .01 chi-square values for $df = 1$ in **Table A4** are $\chi^2_{.05} = 3.84$ and $\chi^2_{.01} = 6.63$. Since the computed value $\chi^2 = 5.984$ is greater than $\chi^2_{.05} = 3.84$, the null hypothesis can be rejected at the .05 level, but not at the .01 level (since $\chi^2_{MH_c} = 5.984$ is less than $\chi^2_{.01} = 6.63$).

If Equation 16.46 is employed, which does not employ the correction for continuity, the value $\chi^2_{MH} = 6.803$ is obtained. Since the computed value $\chi^2_{MH} = 6.803$ is greater than $\chi^2_{.05} = 3.84$ and $\chi^2_{.01} = 6.63$, the null hypothesis can be rejected at both the .05 and .01 levels.

(Equation 16.46)

$$\chi^2_{MH} = \sum_{i=1}^{m} \left[\frac{(O - E)^2}{V} \right] = \left[\frac{(28 - 36.05)^2}{9.5251} \right] = \frac{64.8025}{9.5251} = 6.803$$

Regardless of which equation is employed, the researcher can conclude there is a significant statistical relationship between gender and whether or not one develops the kidney disease. As was the case with the **heterogeneity chi-square analysis** conducted earlier, inspection of the contingency tables indicates that the disease occurs more frequently in females than it does in males.

Equations 16.47–16.49 (all of which employ the normal distribution) provide an alternative but equivalent method for computing the **Mantel–Haenszel test** statistic. Equation 16.47 yields a result equivalent to that obtained with Equation 16.45, and Equation 16.48 yields a result equivalent to that obtained with Equation 16.46. Note that Equation 16.47, which employs a correction for continuity, yields a slightly lower z value than Equation 16.48, which does not employ the latter correction. Conover (1999, pp. 192–195) states that Equations 16.47 and 16.48 are based on the assumption that the marginal sums (i.e., the sums of the rows and columns) of the m contingency tables are fixed (which is as noted earlier an underlying assumption of the **Fisher exact test**). The square of the continuity corrected z value computed with Equation 16.47 will equal $\chi^2_{MH_c}$ value computed with Equation 16.45, while square of the z value computed with Equation 16.48 will equal χ^2_{MH} value computed with Equation 16.46. Equation 16.49 does not assume that the marginal sums are fixed, and should be employed if it is assumed that the values of the row and column sums are determined randomly. Equation 16.49 will yield a slightly larger z value than both Equations 16.47 and 16.48, and thus provide for a more powerful test of the alternative hypothesis. In the final analysis, the values obtained for a test statistic through use of Equations 16.47–16.49 will be quite close to one another, and in instances where one equation yields a significant result and another does not, further research may be necessary in order to clarify the status of the hypothesis under study.

(Equation 16.47)

$$z_{MH_c} = \frac{\left| \sum_{i=1}^{m} a_i - \sum_{i=1}^{m} \left(\frac{(a_i + b_i)(a_i + c_i)}{n_i} \right) \right| - .5}{\sqrt{\sum_{i=1}^{m} \left(\frac{(a_i + b_i)(c_i + d_i)(a_i + c_i)(b_i + d_i)}{n_i^2(n_i - 1)} \right)}}$$

$$z_{MH} = \frac{\left| \sum_{i=1}^{m} a_i - \sum_{i=1}^{m} \left(\frac{(a_i + b_i)(a_i + c_i)}{n_i} \right) \right|}{\sqrt{\sum_{i=1}^{m} \left(\frac{(a_i + b_i)(c_i + d_i)(a_i + c_i)(b_i + d_i)}{n_i^2(n_i - 1)} \right)}}$$ **(Equation 16.48)**

$$z_{MH} = \frac{\left| \sum_{i=1}^{m} a_i - \sum_{i=1}^{m} \left(\frac{(a_i + b_i)(a_i + c_i)}{n_i} \right) \right|}{\sqrt{\sum_{i=1}^{m} \left(\frac{(a_i + b_i)(c_i + d_i)(a_i + c_i)(b_i + d_i)}{n_i^3} \right)}}$$ **(Equation 16.49)**

In Equations 16.47–16.49, the notation $\sum_{i=1}^{m}$ indicates that the operations noted should be carried out with regard to the relevant cells noted for each of the m 2 × 2 contingency tables, after which the obtained values for each of the m tables is summed. The computations for 16.47–16.49 are presented below, respectively yielding the values $z_{MH_c} = 2.45$, $z_{MH} = 2.61$, and $z_{MH} = 2.635$.

$$z_{MH_c} = \frac{\left| (12 + 10 + 6) - \left(\frac{(29)(30)}{60} + \frac{(26)(25)}{50} + \frac{(18)(19)}{40} \right) \right| - .5}{\sqrt{\frac{(29)(31)(30)(30)}{(60)^2(59)} + \frac{(26)(24)(25)(25)}{(50)^2(49)} + \frac{(18)(22)(19)(21)}{(40)^2(39)}}} = \frac{7.55}{3.08} = 2.45$$

$$z_{MH} = \frac{\left| (12 + 10 + 6) - \left(\frac{(29)(30)}{60} + \frac{(26)(25)}{50} + \frac{(18)(19)}{40} \right) \right|}{\sqrt{\frac{(29)(31)(30)(30)}{(60)^2(59)} + \frac{(26)(24)(25)(25)}{(50)^2(49)} + \frac{(18)(22)(19)(21)}{(40)^2(39)}}} = \frac{8.05}{3.08} = 2.61$$

$$z_{MH} = \frac{\left| (12 + 10 + 6) - \left(\frac{(29)(30)}{60} + \frac{(26)(25)}{50} + \frac{(18)(19)}{40} \right) \right|}{\sqrt{\frac{(29)(31)(30)(30)}{(60)^3} + \frac{(26)(24)(25)(25)}{(50)^3} + \frac{(18)(22)(19)(21)}{(40)^3}}} = \frac{8.05}{3.055} = 2.635$$

If a two-tailed/nondirectional analysis is conducted, the tabled critical two-tailed .05 and .01 values $z_{.05} = 1.96$ and $z_{.01} = 2.58$ in **Table A1** are employed. Since the value $z_{MH_c} = 2.45$ obtained with Equation 16.47 is greater than $z_{.05} = 1.96$, the null hypothesis can be rejected at the .05 level, but not at the .01 level (since $z_{MH_c} = 2.45$ is less than $z_{.01} = 2.58$). Since the values $z_{MH} = 2.61$ and $z_{MH} = 2.635$ obtained with Equations 16.48 and 16.49 are larger than $z_{.05} = 1.96$ and $z_{.01} = 2.58$, in the latter instances the null hypothesis can be rejected at both the .05 and .01 levels. Note that the square of the value $z_{MH_c} = 2.45$ obtained with Equation 16.47 is equal to the value $\chi^2_{MH_c} = 5.984$ obtained with Equation 16.45 (i.e. ($z_{MH_c} = 2.45)^2 = (\chi^2_{MH_c} = 5.984)$ — the minimal difference is due to rounding off error), and the square of the value $z_{MH} = 2.61$ obtained with Equation 16.48 is equal to the value $\chi^2_{MH} = 6.803$ obtained with Equation 16.46 (i.e. (z_{MH} = 2.61)^2 = (χ^2_{MH} = 6.803) — the minimal difference is due to rounding off error). Thus, the **Mantel–Haenszel test** statistics obtained through use of the chi-square versus normal distributions are identical, and in both cases allow the researcher to conclude that females are more likely to develop the kidney disease than males.

Before concluding the discussion of the **Mantel–Haenszel test of association** a few final comments are in order.

1) If a researcher conducts a one-tailed/directional analysis, in order for a result to be significant a computed **Mantel–Haenszel test** statistic must be equal to or greater than the relevant tabled critical one-tailed value, and the data must be consistent with the directional prediction stipulated in alternative hypothesis. In the case of Equations 16.47–16.49, the absolute value notation can be omitted in the numerator of the latter equations. If the obtained z value is negative it would indicate that in the pooled contingency table, the probability of an observation in cell a is less than the probability of an observation in cell c. On the other hand, if the obtained z value is positive it would indicate that in the pooled contingency table, the probability of an observation in cell c is less than the probability of an observation in cell a. In the case of a one tailed analysis, in order to be significant the sign of the obtained z value must be consistent with the prediction stipulated in the alternative hypothesis, and the absolute value of z must be equal to or greater than the prestipulated tabled critical z value.

2) Note that in Equations 16.45–16.49 cell a is always stipulated as the **target cell**. Specifically, the **target cell** refers to the cell in the 2×2 contingency table for which a **summary odds ratio** will be stated in reference to. As you will see, based on how the three contingency tables summarized in Table 16.25 are configured, either cell a (which contains the number of males who have the disease) or cell c (which contains the number of females who have the disease) could be employed as the target cell for the **Mantel–Haenszel test of association**. Note that when cell a is the target cell, the relevant **summary odds ratio** indicates how many times greater the odds are that a male will have the disease than that a female will have the disease (which we previously determined is $o'_{MH} = .42$, the value computed earlier for the **summary inverse odds ratio**). On the other hand, if cell c is stipulated as the target cell, the relevant **summary odds ratio** will indicate how many times greater the odds are that a female will have the disease than that a male will have the disease (which we previously determined is $o_{MH} = 2.40$, the value computed earlier for the **summary odds ratio**). If, in fact, we wish to stipulate cell c as the target cell, the numerators of Equations 16.45–16.49 would have to be modified. Upon making the appropriate modifications, the chi-square or z values computed with Equations 16.45–16.49 would be identical to those obtained when cell a was stipulated as the target cell. Specifically, if cell c is stipulated as the target cell, the values O and E in the numerator of Equations 16.45 and 16.46 should be changed as follows.

a) $O = \sum_{i=1}^m O_i$, where $O_i = c_i$. Thus, O becomes the sum of the observed frequencies in cell c of the m contingency tables — i.e., the sum of the observed frequencies in the target cell (in place of the sum of the observed frequencies in cell a of the m contingency tables).

b) $E = \sum_{i=1}^m E_i$, where $E_i = [(c_i + d_i)(a_i + c_i)] / n_i$. Thus, E becomes the sum of the expected frequencies for cell c in the m contingency tables — i.e., the sum of the expected frequencies in the target cell (in place of the sum of the observed frequencies in cell a of the m contingency tables).

If cell c is stipulated as the target cell, the numerator of Equations 16.47–16.49 should be changed as follows.

a) The term $\sum_{i=1}^m c_i$ is employed in place of $\sum_{i=1}^m a_i$. Note that in Equations 16.47–16.49, the term $\Sigma_{i=1}^{m} a_i$ is equivalent to the term $O = \Sigma_{i=1}^{m} O_i$ in Equations 16.45 and 16.46 — i.e., it is the sum of the observed frequencies in the target cell a. Thus, if we employ $\Sigma_{i=1}^{m} c_i$ in Equations 16.47–16.49 in place of $\Sigma_{i=1}^{m} a_i$, we are stipulating the sum of the observed frequencies in target cell c of the m contingency tables.

   b) The term $\Sigma_{i=1}^{m}[[(a_i + b_i)(a_i + c_i)] / n_i]$ (which is the sum of the expected frequencies in the target cell a) is replaced by $\Sigma_{i=1}^{m}[[(c_i + d_i)(a_i + c_i)] / n_i]$ (which is the sum of the expected frequencies in the target cell c). Note that in Equations 16.47–16.49, the term $\Sigma_{i=1}^{m}[[(a_i + b_i)(a_i + c_i)] / n_i]$ is equivalent to the term $E = \Sigma_{i=1}^{m} E_i$ in Equations 16.45 and 16.46 — i.e., it is the sum of the expected frequencies in the target cell a. Thus, if we employ $\Sigma_{i=1}^{m}[[(c_i + d_i)(a_i + c_i)] / n_i]$ in Equations 16.47–16.49 in place of $\Sigma_{i=1}^{m}[[(a_i + b_i)(a_i + c_i)] / n_i]$, we are stipulating the sum of the expected frequencies in target cell c of the m contingency tables.

   To illustrate that the same result is obtained when cell c is employed as the target cell in place of cell a, the modified form of Equation 16.46 is employed below. Note that obtained value $\chi^2 = 6.80$ is identical to that computed previously with the latter equation when cell a was the target cell.

$$O = \sum_{i=1}^{m} c_i = 18 + 15 + 13 = 46.$$

$$E = \sum_{i=1}^{m} E_i = \frac{(31)(30)}{60} + \frac{(24)(25)}{50} + \frac{(22)(19)}{40} = 37.95$$

$$\chi^2 = \frac{(46 - 37.95)^2}{9.525} = 6.80$$

   3) If .8755 (the absolute value of the natural logarithm of both the **summary odds ratio** and **summary inverse odds ratio** if the latter values are carried out to nine decimal places) is substituted in Equation 16.29 (with the **standard error** of the **summary odds ratio** employed in lieu of s_o), the value $z = 2.61$ is computed. Note that the latter value is identical to that obtained with Equation 16.48 (which yields a result equivalent to that obtained with Equation 16.46). Thus, Equation 16.29 can also be employed to evaluate the hypothesis of whether or not there is a significant statistical association between the two variables in the pooled contingency table.

$$z = \frac{\ln(o) - \ln(1)}{s_{\bar{Y}_{MH}}} = \frac{\ln(o) - 0}{s_{\bar{Y}_{MH}}} = \frac{.8755}{.3359} = 2.61$$

   4) The use of the **Mantel–Haenszel test of association** for evaluating survival tables (which include or do not include censored data) in the form of contingency tables is described in Section IX (the **Addendum**) of the **Mann–Whitney U test**. The latter discussion describes the use of the **Mantel–Haenszel test of association** within the context of the **log-rank test (Test 12f)**, and demonstrates the equivalency of Equation 16.46 with Equations 12.16 and 12.19.

   5) Fleiss *et al.* (2003, Chapter 10) provide a detailed discussion of the **Mantel–Haenszel analysis** as well as alternative procedures for evaluating 2 × 2 contingency tables.

   Meta-analysis (which is discussed in Section IX (the **Addendum**) of the **Pearson product-moment correlation coefficient**) can be also employed to obtain a combined result for multiple 2 × 2 contingency tables which evaluate the same general hypothesis (although **meta-analysis** does not pool all of the data into a single 2 × 2 table). In point of fact, one of the alternative procedures documented in Everitt (1977, pp. 25–27; 1992) is described in the discussion of **meta-analysis**. The latter alternative procedure is referred to as the **Stouffer**

procedure for obtaining a combined significance level (*p* value) for *k* studies (Test 28m). In the discussion of the **Stouffer procedure**, the data for the three contingency tables in **Part A** of Table 16.25 are pooled through use of the latter method. If the reader consults the latter discussion she will discover that the analysis with the **Stouffer method** yields essentially the same conclusions which are reached when the **heterogeneity chi-square analysis** methodology and the **Mantel–Haenszel analysis** described in this section are employed to evaluate the three 2×2 contingency tables. In addition to computing a combined probability value through use of the **Stouffer procedure**, the discussion of **meta-analysis** provides a test of homogeneity of effect size for the three studies (which is described in the discussion of the **procedure for comparing *k* studies with respect to effect size (Test 28n))**, as well as providing an average effect size value for the studies (which is described in the discussion of the **procedure for obtaining a combined effect size for *k* studies (Test 28o)**).

VII. Additional Discussion of the Chi-Square Test for $r \times c$ Tables

1. Equivalency of the chi-square test for $r \times c$ tables when $c = 2$ with the *t* test for two independent samples (when $r = 2$) and the single-factor between-subjects analysis of variance (when $r \geq 2$)   In evaluating the data in a 2×2 contingency table, the *t* test for two independent samples (which is discussed earlier in the book) will yield a result equivalent to that obtained with the **chi-square test for $r \times c$ tables**. To generalize even further, it can be stated that in the case of any $r \times 2$ contingency table, a result equivalent to that obtained with the **chi-square test for $r \times c$ tables** can be obtained through use of the **single-factor between-subjects analysis of variance** (which is discussed later in the book). The *t* **test for two independent samples** and the **single-factor between-subjects analysis of variance** are categorized as parametric tests involving independent samples employed with interval/ratio data. Whereas the *t* **test for two independent samples** can only be employed with two independent samples, the **single-factor between-subjects analysis of variance** can be employed with two or more independent samples. A detailed discussion of the conditions of equivalency between the **chi-square test for $r \times c$ tables** and the *t* **test for two independent samples/single-factor between-subjects analysis of variance** can be found in Section VII of the **single-factor between-subjects analysis of variance**.

2. Test of equivalence for two independent proportions: Test 16m: The Westlake–Schuirmann test of equivalence of two independent proportions   Prior to reading this section the reader should review the discussion of **tests of equivalence** in Section VII of the *t* **test for two independent samples**. In the latter discussion it was noted that within the framework of the classical hypothesis testing model, a null hypothesis can never be accepted. In contrast to the **classical hypothesis testing model** (where the alternative hypothesis states that there is a difference between treatments), in a **test of equivalence** the alternative hypothesis states that the treatments are, in fact, equivalent. Conversely, in a **test of equivalence** the null hypothesis states that a difference exists between the treatments (in contrast to a test of significance within the framework of the **classical hypothesis testing model**, where the null hypothesis states that there is no difference between the treatments). Since it is not mathematically feasible to establish an alternative hypothesis which states exact equality (i.e., a difference of zero) between experimental conditions, when one conducts a **test of equivalence**, a parameter (to be designated ζ, which represents the lower case Greek letter **zeta**) is employed to reflect a maximum difference which will be tolerated between two treatments in order that a researcher might conclude that the treatments are equivalent to one another. Any difference less than the value of ζ would be viewed so small as to be inconsequential (i.e., of no practical significance). Thus, in a **test of equivalence** if a difference less than the value of ζ is detected, the null hypothesis (stating that a difference does exist) can be rejected, and the alternative hypothesis (which stipulates equivalence) can be accepted. In the discussion to follow, it will be emphasized that since in most instances rejection of the null hypothesis for a **test of equivalence** will involve detection of a substantially smaller effect size than the effect size a researcher is typically concerned with detecting within the framework of the **classical hypothesis testing model**, a **test of equivalence** will usually require greater statistical power, and consequently the use of a larger sample size, when contrasted with tests conducted within the framework of the **classical hypothesis testing model**. The smaller the value a researcher designates for ζ, the larger the sample size which will be required in order to reject the null hypothesis and thus conclude the treatments are equivalent.

With respect to evaluating a difference between two independent proportions, assume that the *z* **test for two independent proportions** described earlier in this chapter is employed to evaluate whether or not the proportion of patients afflicted with a specific disease who respond to a brand name drug is equal to the proportion of patients who respond to a generic drug. Assume that the researcher employs the null hypothesis H_0: $\pi_1 = \pi_2$ and the nondirectional alternative hypothesis H_1: $\pi_1 \neq \pi_2$, and that analysis of the data indicates there is not a significant difference between the two treatments. In such a case, retention of the null hypothesis only indicates that any difference detected between two treatments is not large enough to allow the researcher to conclude that the treatments are different from one another. Although the latter statement might suggest that retention of the null hypothesis is commensurate with concluding that the treatments are equivalent to one another, in point of fact, it is not. In order to demonstrate the equivalence of two or more experimental conditions, it is required that a researcher conducts what is referred to as a **test of equivalence**.

In this section a procedure for evaluating the **equivalence of two independent proportions** will be described which is an analog of the procedures described earlier in the book for evaluating the equivalence of the means of two independent samples (described in Section VII of the *t* **test for two independent samples**). The procedure to be described in this section will be referred to as the **Westlake–Schuirmann test of equivalence of two independent proportions (Test 16m)** (Westlake (1976, 1981, 1988) and Schuirmann (1987); also discussed in Chen *et al.* (2000)). Prior to demonstrating the latter methodology for evaluating equivalence, the reader is asked to consider Example 16.7.

Example 16.7 *The manufacturer of a generic drug for treating a disease wants to determine whether or not there is a significant difference in the response of patients to the generic drug when compared with the response of patients to a brand name drug, which up to the present time has been the only approved treatment for the disease. A study is conducted employing 200 people afflicted with the disease who receive either the brand name drug or generic drug for a period of six months. Of the 200 patients, 98 receive the brand name drug and 102 receive the generic drug. A panel of physicians, who are blind with respect to the treatment each patient receives, categorize a patient with respect to whether he or she responded successfully to the treatment. Eighty-two of the 98 patients who received the brand name drug (Treatment 1) were categorized as a success while 80 of the 102 patients who received the generic drug (Treatment 2) were categorized as a success. The data for the study are summarized in Table 16.26. Do the data indicate there is a significant difference between the two treatments?*

Assume that the null hypothesis H_0: $\pi_1 = \pi_2$ and the nondirectional alternative hypothesis H_1: $\pi_1 \neq \pi_2$ are employed with alpha = .05. For Example 16.7 we either know or can compute the following values.

Table 16.26 Summary of Data for Example 16.7

	Success	Failure	Row sums
Brand name drug	82	16	98
Generic drug	80	22	102
Column sums	162	38	Total observations 200

$$a = 82 \qquad c = 80 \qquad n_1 = 98 \qquad n_2 = 102$$

$$p_1 = \frac{a}{n_1} = \frac{82}{98} = .837 \qquad p_2 = \frac{c}{n_2} = \frac{80}{102} = .784$$

$$p = \frac{a + c}{n_1 + n_2} = \frac{82 + 80}{98 + 102} = .81 \qquad 1 - p = 1 - .81 = .19$$

Employing the above values in Equation 16.9, the value $z = .96$ is computed. (If the correction for continuity is used $z = .78$.)[29]

$$z = \frac{.837 - .784}{\sqrt{(.81)(.19)\left[\dfrac{1}{98} + \dfrac{1}{102}\right]}} = \frac{.053}{.055} = .96$$

Since the computed value $z = .96$ is less than $z_{.05} = 1.96$ (the tabled critical two-tailed .05 value in **Table A1**), the null hypothesis H_0: $\pi_1 = \pi_2$ cannot be rejected. Thus, the researcher can conclude that the differences between the proportion of successes for the brand name drug versus the generic drug is not statistically significant. However, as noted earlier, the latter is not commensurate with the conclusion that the two treatments are equivalent to one another. In order to demonstrate the latter the researcher must conduct a **test of equivalence**. Consequently, at this point Example 16.7 will be reconceptualized as Example 16.8 in order that a **test of equivalence** can be conducted. Example 16.8 will be evaluated with the **Westlake–Schuirmann test of equivalence of two independent proportions**.

Example 16.8 *The manufacturer of a generic drug for treating a disease wants to empirically demonstrate that it is equivalent to a brand name drug, which up to the present time has been the only approved treatment for the disease. To demonstrate equivalency of the two drugs, a study is conducted employing 200 people afflicted with the disease who receive either the brand name drug or the generic drug for a period of six months. Of the 200 patients, 98 receive the brand name drug and 102 receive the generic drug. A panel of physicians, who are blind with respect to the treatment each patient receives, categorize a patient with respect to whether he or she responded successfully to the treatment. Eighty-two of the 98 patients who received the brand name drug (Treatment 1) were categorized as a success while 80 of the 102 patients who received the generic drug (Treatment 2) were categorized as a success. The data for the study are summarized in Table 16.26. Do the data indicate that the generic drug is equivalent to the brand name drug if the researcher is willing to declare equivalency if the difference between the treatment proportions is less than 10%?*

The information provided in Example 16.8 allows for either a two-tailed or one-tailed **test of equivalence**. If a researcher employs as a criterion for equivalence that the success rate of the brand name drug is no more than 10% greater than the success rate of the generic drug, as well as the requirement that success rate of the generic drug is no more than 10% greater than the success rate of the brand name drug, a two-tailed analysis would be conducted. (The value 10% is equivalent to a proportional difference of .1.) On the other hand, if the researcher only employs as a criterion for equivalence that the success rate of the brand name drug is no more than 10% greater than the success rate of the generic drug, a one-tailed analysis would be conducted.

The two-tailed analysis can be represented by Figure 16.1 where the values employed for ζ_1 and ζ_2 are respectively $-.10$ and $.10$ (which represent a difference between the two treatment proportions). The two-tailed analysis depicted in Figure 16.1 is symmetrical since the absolute values of ζ_1 (which is equal to $-\zeta$) and ζ_2 (which is equal to ζ) are equivalent to one another. As noted in the discussion of **tests of equivalence** in Section VII of the *t* test for two independent samples, it is also possible to conduct a nonsymmetrical two-tailed analysis, in which case the absolute values of ζ_1 and ζ_2 are different.

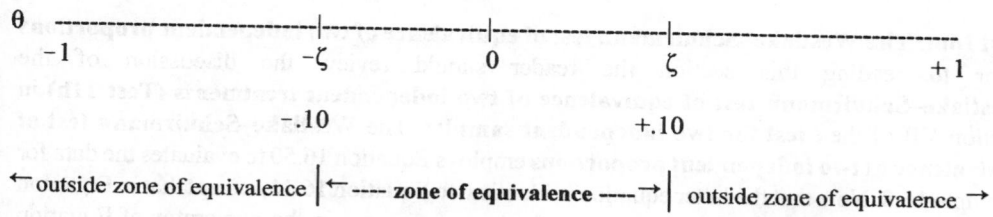

Figure 16.1 Symmetrical Analysis for Equivalence (Two-Tailed)

A one-tailed analysis would be represented by Figure 16.2, and thus the value .10 is employed for ζ_2, while no value is designated for ζ_1. Note that in Figures 16.1 and 16.2, the values along the horizontal line represent differences between proportions, whereas the values recorded along the horizontal line in the analogous figures (Figures 11.3 and 11.6) in the discussion of **tests of equivalence** in Section VII of the *t* test for two independent samples represent differences between means.

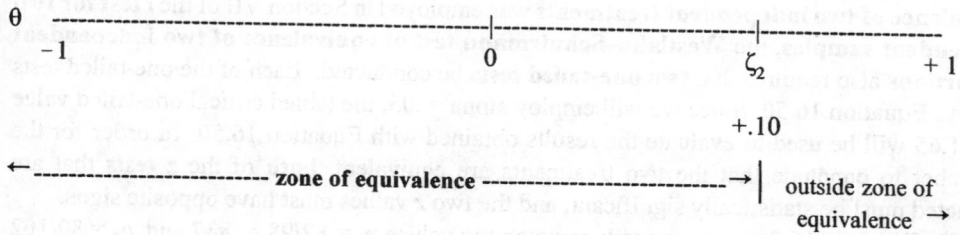

Figure 16.2 Directional Test of Equivalence (One-Tailed)

In the case of Example 16.8 we will assume that the researcher elects to conduct a two-tailed analysis with alpha set equal to .05. A full discussion of the protocol for a one-tailed analysis can be found in the discussion of **tests of equivalence** in Section VII of the *t* test for two independent samples. A one-tailed analysis will always provide a less conservative **test of equivalence** than will a two-tailed analysis — i.e., for a given level of alpha, it will always be easier to declare equivalence when a one-tailed analysis is employed as opposed to a two-tailed analysis.

The null and alternative hypotheses for a **two-tailed analysis** are noted below. The notation θ (which is the lower case Greek letter **theta**) in the null and alternative hypotheses represents the difference between the two proportions in the underlying populations. It is assumed that the estimate of that difference is based on subtracting the proportion of successes for the generic drug from the proportion of successes for the brand name drug.

Null hypotheses H_0: $\theta \leq -.10$ or H_0: $\theta \geq .10$

(In the underlying populations the samples represent, the difference between the two treatment proportions is less than or equal to $-.10$ or is greater than or equal to .10. This is the **hypothesis of nonequivalence**.)

Alternative hypotheses H_1: $\theta > -.10$ and H_1: $\theta < .10$

(In the underlying populations the samples represent, the difference between the two treatment proportions is greater than $-.10$ and is less than .10. This is the **hypothesis of equivalence**.)

Test 16m: The Westlake–Schuirmann test of equivalence of two independent proportions
Prior to reading this section the reader should review the discussion of the **Westlake–Schuirmann test of equivalence of two independent treatments (Test 11h)** in Section VII of the *t* **test for two independent samples**. The **Westlake–Schuirmann test of equivalence of two independent proportions** employs Equation 16.50 to evaluates the data for Example 16.8. Note that the latter equation is identical to Equation 16.11, except that in Equation 16.50 the value ζ is employed to represent the term $(\pi_1 - \pi_2)$ in the numerator of Equation 16.11. The value employed for ζ will be the appropriate value employed for ζ_1 or ζ_2 in the analysis being conducted.

$$z = \frac{(p_1 - p_2) - \zeta}{\sqrt{\dfrac{p_1(1 - p_1)}{n_1} + \dfrac{p_2(1 - p_2)}{n_2}}} \qquad \textbf{(Equation 16.50)}$$

As was the case for a two-tailed analysis when the **Westlake–Schuirmann test of equivalence of two independent treatments** was employed in Section VII of the *t* **test for two independent samples**, the **Westlake–Schuirmann test of equivalence of two independent proportions** also requires that **two one-tailed tests** be conducted. Each of the one-tailed tests employs Equation 16.50. Since we will employ alpha = .05, the tabled critical one-tailed value $z_{.05} = 1.65$ will be used to evaluate the results obtained with Equation 16.50. In order for the researcher to conclude that the two treatments are equivalent, **both** of the *z* tests that are conducted must be statistically significant, and the two *z* values must have opposite signs.

For Example 16.8, we once again compute the values $p_1 = 82/98 = .837$ and $p_2 = 80/102 = .784$ which will be employed in Equation 16.50. Employing the latter equation, the following null and alternative hypotheses will be initially evaluated: H_0: $\theta \leq -.10$ versus H_1: $\theta > -.10$. Note that in this analysis the value $\zeta_1 = -.10$ is employed in the numerator of Equation 16.50, since we are evaluating the left boundary for equivalence depicted in Figure 16.1.

$$z = \frac{(.837 - .784) - (-.10)}{\sqrt{\dfrac{(.837)(.163)}{98} + \dfrac{(.784)(.216)}{102}}} = \frac{.153}{\sqrt{.00139 + .00166}} = \frac{.153}{.0552} = 2.77$$

Since the obtained value $z = 2.77$ is greater than the tabled critical one-tailed value $z_{.05} = 1.65$, the null hypothesis H_0: $\theta \leq -.10$ designating nonequivalence can be rejected. (If the correction for continuity (i.e., $\pm [1/2][(1/n_1) + (1/n_2)]$) is employed in the numerator of the above equation, .01 is subtracted from the numerator of the equation yielding the value $z = 2.59$). The probability value associated with this result (with or without the correction for continuity) is well below the .05 value which was prestipulated for alpha. Thus, the first of the null hypotheses designating nonequivalence can be rejected. However, it once again should be emphasized that in order to declare equivalence we must reject **both** of the null hypotheses designating nonequivalence, and the two z computed values must have opposite signs. In reference to the z **test** conducted above, the data are consistent with the alternative hypotheses H_1: $\theta > -.10$, which stipulates that when the proportion of successes for **Group 2 (Generic drug)** is subtracted from the proportion of successes for **Group 1 (Brand name drug)** the difference is greater than $-.10$ (which translates into the fact that if the proportion of successes for the **generic drug** is greater than the proportion of successes for the **brand name drug**, it is not greater than the latter value by more than .10).

We now evaluate the second null hypothesis designating nonequivalence. Specifically, a one-tailed test which evaluates the following null and alternative hypotheses is conducted: H_0: $\theta \geq .10$ versus H_1: $\theta < .10$. Note that this analysis employs the same values as in the previous analysis, except for the fact that the value $\zeta_2 = .10$ is employed in the numerator of Equation 16.50.

$$z = \frac{(.837 - .784) - (.10)}{\sqrt{\dfrac{(.837)(.163)}{98} + \dfrac{(.784)(.216)}{102}}} = \frac{-.047}{\sqrt{.00139 + .00166}} = \frac{-.047}{.0552} = -.85$$

Since the obtained absolute value $z = -.85$ is less than the tabled critical one-tailed value $z_{.05} = 1.65$, the null hypothesis cannot be rejected. (If the correction for continuity (i.e.,$[1/2][(1/n_1) + (1/n_2)]$) is employed in the numerator of the above equation, .01 is added to the numerator of the equation yielding the value $z = -.67$). Thus, the second null hypothesis H_0: $\theta \geq .10$ designating nonequivalence cannot be rejected. The probability value associated with this result is well above .05 (with or without the correction for continuity).Since, as noted earlier, in order to declare equivalence we must reject **both** of the null hypotheses designating nonequivalence (as well as the fact that the z values must have different signs), equivalence cannot be established since the larger of the probability values computed for the two analyses exceeds .05. Specifically, the data are consistent with the null hypothesis H_0: $\theta \geq 10$, which stipulates that the proportion of success for **Group 1 (Brand name drug)** is .1 or greater than the proportion of successes for **Group 2 (Generic drug)**.

Rogers *et al.* (1993, pp. 554–555) note that when the value of alpha is set equal to .05 for each of the two analyses conducted above, the equivalency confidence interval is set at the (1 -2α) (i.e., 90%) level of certainty/confidence rather than at the traditional $(1 - \alpha)$ (i.e., 95%) level of certainty/confidence. When both null hypotheses which are evaluated are rejected, the values employed for both ζ_1 and ζ_2 will fall outside of the $(1 - 2\alpha)$ % confidence interval. If either of the null hypotheses is retained, the value employed for ζ_1 or ζ_2 will fall within the limits of the confidence interval. **To be more specific, if equivalence is to be established, the values for both ζ_1 and ζ_2 must fall outside the limits of the confidence interval, with one of the values ζ_1 positioned to the left of the lower limit of the confidence interval and the other value ζ_2 positioned to the right of the upper limit of the confidence interval. Note, however, that if both ζ_1 and ζ_2 fall outside the limits of the confidence interval, and both ζ_1 and ζ_2 are below the lower limit of the confidence interval or both ζ_1 and ζ_2 are above the**

upper limit of the confidence interval, it would indicate a lack of equivalence. In the case of Example 16.8, in order to compute the 90% confidence interval, the tabled critical two-tailed .10 z value (which corresponds to the tabled critical one-tailed .05 z value) is employed in Equation 16.14. Employing Equation 16.14, the 90% confidence interval for Example 16.8 is computed below, and visually represented in Figure 16.3.

$$p_1 - p_2 = .837 - .784 = .053 \qquad s_{p_1 - p_2} = \sqrt{\frac{(.837)(.163)}{98} + \frac{(.784)(.216)}{102}} = .0552$$

$$CI_{90} = (p_1 - p_2) \pm (z_{.10})(s_{p_1 - p_2}) = .053 \pm (1.65)(.0552) = .053 \pm .091$$

$$-.038 \geq (\pi_1 - \pi_2) \geq .144$$

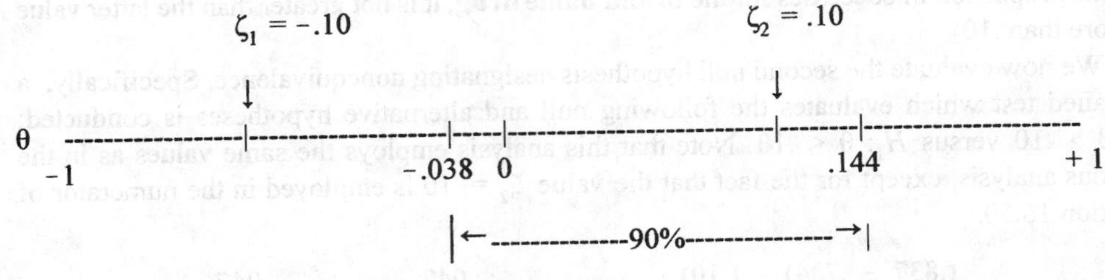

$$\zeta_1 = -.10 \qquad\qquad\qquad\qquad\qquad \zeta_2 = .10$$

$$\downarrow \qquad\qquad\qquad\qquad\qquad\qquad\qquad \downarrow$$

$$\theta \quad \text{-------------------|--------------------|----|--------------------|----------------|-------------}$$
$$-1 \qquad\qquad\qquad\qquad -.038 \;\; 0 \qquad\qquad\qquad\qquad .144 \qquad\qquad +1$$

$$|\leftarrow\text{----------------}90\%\text{----------------}\rightarrow|$$

Figure 16.3   Confidence Interval For Example 16.8 when $\zeta = .10$

    This result indicates the researcher can be 90% confident (or the probability is .90) that the interval −.038 to .144 contains the true difference between the population proportions. (If Equation 16.15, which employs the correction for continuity (i.e., $\pm .5 [(1/n_1) + (1/n_2)]$), is employed, .01 is subtracted from the lower limit and added to the upper limit, yielding the confidence interval −.048 to .154.) If the interval derived without using the correction for continuity contains the true difference between the population proportions, the proportion of successes for the population **Group 1 (Brand name drug)** represents is not greater than the proportion of successes for the population **Group 2 (Generic drug)** represents than the proportion value .144. Additionally, the proportion of successes for the population **Group 2 (Generic drug)** represents is not greater than the proportion of successes for the population **Group 1 (Brand name drug)** represents than the proportion value .038. Note that although the value $\zeta_1 = -.10$ falls outside of the computed confidence interval, the value $\zeta_2 = .10$ falls inside it. Equivalency cannot be established, since the upper range of the confidence interval falls above the maximum tolerable difference between the proportions for the two treatments. In order for equivalency to have been established, the result for the second analysis would had to have been statistically significant at the .05 level, and if the latter had occurred the upper limit for the confidence interval would have been some value less than .10.

    The reader should take note of the fact that just by employing Equation 16.50, as was done earlier in this section to compute two z values, one can obtain the answer to the question of whether or not two treatments are equivalent. Alternatively, computation of the above confidence interval, in and of itself, will provide the answer to the same question. Use of both methods merely provides the answer to the same question in different formats.

    It is interesting to note that the analyses of Examples 16.7 and 16.8 are inconsistent with one another. Whereas Example 16.7 implies no differences between the treatments, Example 16.8 leads to the conclusion that the treatments are not equivalent. Rogers *et al.* (1993, pp. 561–562) note that when a conventional test of significance conducted within the framework of the **classical hypothesis testing model** (as was the case in evaluating Example 16.7) yields a nonsignificant result, suggesting there is no difference between the two treatments, yet a **test of equivalence** does not yield a significant result, suggesting nonequivalence between the two treatments, further research is warranted on the hypothesis under investigation. Obviously, the lack of a significant result for the conventional test of significance could be attributed to the fact that the sample size employed in the study was extremely small, and thus the analysis lacked sufficient power to detect a small difference between the population proportions. In the same respect, the small sample size could also be employed in rationalizing why the **test of equivalence** did not detect equivalence between the two treatments.

In point of fact, given the sample sizes employed in Example 16.8, we can easily determine the maximum absolute value that would have to be designated for ζ_1 and ζ_2 in order to establish equivalence. Specifically, we solve Equation 16.50 for ζ when the tabled critical one-tailed .05 value $z_{.05} = -1.65$ is obtained for the result of the analysis. The value $z_{.05} = -1.65$ is employed, since of the two z values computed above with Equation 16.50, the one which yielded the absolute z value closest to the critical value $z_{.05} = 1.65$ yielded a negative z value. When we solve for ζ, we obtain the value $\zeta = .144$ (which is the upper limit of the confidence interval computed earlier). Thus, the absolute value .144 is the maximum value that would have to be designated for ζ_1 and ζ_2 in order to establish equivalence.

$$z = \frac{(p_1 - p_2) - \zeta}{\sqrt{\dfrac{p_1(1 - p_1)}{n_1} + \dfrac{p_2(1 - p_2)}{n_2}}} = \frac{(.837 - .784) - \zeta}{.0552} = \frac{.053 - \zeta}{.0552} = -1.65$$

$$\zeta = .144$$

Thus any absolute value designated for ζ which is equal to or greater than .144 will yield a significant result for both of the z tests which are employed in testing for equivalence. To illustrate, if the values $-.144$ and $+.144$ (which are just above the absolute value .144) are employed for ζ_1 and ζ_2, the two t tests for the **test of equivalence** yield the results below.

$$z = \frac{(.837 - .784) - (-.144)}{\sqrt{\dfrac{(.837)(.163)}{98} + \dfrac{(.784)(.216)}{102}}} = \frac{.197}{.0552} = 3.57$$

Since the obtained value $z = 3.57$ is greater than the tabled critical one-tailed value $z_{.05} = 1.65$, the null hypothesis H_0: $\theta \leq -.145$ designating nonequivalence can be rejected. (If the correction for continuity is used $z = 3.39$.)

We now evaluate the second null hypothesis designating nonequivalence.

$$z = \frac{(.837 - .784) - (.144)}{\sqrt{\dfrac{(.837)(.163)}{98} + \dfrac{(.784)(.216)}{102}}} = \frac{-.091}{.0552} = -1.65$$

Since the obtained absolute value $z = 1.65$ is equal to the tabled critical one-tailed value $z_{.05} = 1.65$, the null hypothesis can be rejected. Thus, the second null hypothesis H_0: $\theta \geq .144$ designating nonequivalence can also be rejected. Consequently, the data are consistent with the alternative hypotheses H_1: $\theta > -.144$ and H_1: $\theta < .144$ which stipulate that the difference between the proportion of successes between the two groups is within a proportion value of .144.[30]

It should be noted that many researchers would elect to employ a one-tailed analysis for the data in Example 16.8 — specifically, stipulating as the alternative hypothesis that the proportion of successes for **Treatment 1 (the brand name drug)** is not more than a proportion equal to .10 greater than the proportion of successes for **Treatment 2 (the generic drug)**. In such a case a researcher would conduct only a single **one-tailed test** which employed the null and alternative hypotheses noted below.

Null hypothesis H_0: $\theta \geq .10$

(In the underlying populations the samples represent, the difference between the two treatment proportions is greater than or equal to .10. This is the **hypothesis of nonequivalence**.)

Alternative hypothesis H_1: $\theta < .10$

(In the underlying populations the samples represent, the difference between the two treatment proportions is less than .10. This is the **hypothesis of equivalence**.)

If the above analysis is conducted, only the second of the two z-tests conducted previously in the analysis of Example 16.8 would be employed. Specifically:

$$z = \frac{(.837 - .784) - (.10)}{\sqrt{\dfrac{(.837)(.163)}{98} + \dfrac{(.784)(.216)}{102}}} = \frac{-.047}{\sqrt{.00139 + .00166}} = \frac{-.047}{.0552} = -.85$$

Since, as noted previously, the obtained absolute value $z = -.85$ is less than the tabled critical one-tailed value $z_{.05} = 1.65$, the null hypotheses H_0: $\theta \geq .10$ designating nonequivalence cannot be rejected. In this instance, as noted in Figure 16.2, only the upper boundary would be determined for the confidence interval computed previously, with the requirement for equivalence being that the value of ζ would have to fall above the upper limit of the confidence interval (which $\zeta = .10$ does not).

Since in most instances rejection of the null hypothesis for a **test of equivalence** will involve detection of a substantially smaller effect size than the effect size a researcher is typically concerned with detecting within the framework of the **classical hypothesis testing model**, a **test of equivalence** will usually require greater statistical power, and consequently the use of a larger sample size, when contrasted with tests conducted within the framework of the **classical model**. The smaller the value a researcher designates for ζ, the more power will be required for the statistical analysis. In order to demonstrate the latter, the sample sizes required in order to conduct a test of significance which has a power of .80 will be computed for Examples 16.7 and 16.8.

Procedure for computing sample size in reference to the power of the Westlake–Schuirmann test of equivalence of two independent proportions   Atherton Skaff *et al.* (2004) and Rosner (2000, pp. 639–640) note that Equation 16.51 can be employed to estimate the sample size required in order to conduct a **Westlake–Schuirmann test of equivalence of two independent proportions** which has a power of a specified value. The notation in Equation 16.51 represents the same values stipulated for Equation 16.12, except for the fact that ζ (which is employed in lieu of δ) represents the limiting absolute value which the difference between the two proportions must not exceed in order for the researcher to declare the two treatments equivalent.

(Equation 16.51)

$$n_1 = \frac{\left(p_1(1 - p_1) + \frac{p_2(1 - p_2)}{k} \right) (z_{\alpha/2} + z_\beta)^2}{\left[|p_1 - p_2| - \zeta \right]^2}$$

Equation 16.51 will be employed within the framework of Example 16.8 to determine the sample size required in order for the power of the **Westlake–Schuirmann test** to equal .80 when the value $\zeta = .10$ is stipulated by the researcher. In conducting the latter analysis, the following values are employed in Equation 16.51 to compute the number of subjects that will be required in **Group 1**(which will also be the number of subjects the researcher will need for **Group 2**, since when there are an equal number of subjects per group, $k = 1$): $p_1 = .837$, $(1 - p_1) = .163$, $p_2 = .784$, $(1 - p_2) = .216$, $\zeta = .10$, $z_{\alpha/2} = 1.96$ (If a one-tailed analysis is conducted $z_{\alpha/2} = 1.65$ should be employed.), $z_\beta = .84$. The rationale for employing the value $z_\beta = .84$ is identical to that employed for its use in demonstrating Equation 16.12.

When the appropriate values are substituted in Equation 16.51, the value $n_1 = 1085.25$ is computed, indicating that in order for the power of the test **Westlake–Schuirmann test of equivalence of two independent proportions** to be .80, the researcher must have at least 1086 subjects per group (which is the next integer value that is greater than the value computed for n_1).[31]

$$n_1 = \frac{\left((.837)(.163) + \frac{(.784)(.216)}{1} \right)(1.96 + .84)^2}{[|.837 - .784| - .1]^2} = 1085.23$$

Equation 16.13 (employed in Section VI within the framework of computing the sample size for the *z* **test for two independent proportions**) can be employed to compute a continuity corrected value for n_1 (since some sources would argue that Equation 16.51 underestimates the required sample size). In the latter equation the value $\zeta = .1$ is employed to represent the value of δ. For the problem under discussion, use of Equation 16.13 would indicate that the sample size per group should be increased from approximately 1086 to 1106 in order to insure that the power of the test is .80.

$$n_1 = \frac{n'}{4} \left(1 + \sqrt{1 + \frac{4}{n'\delta}} \right)^2 = \frac{1085.23}{4} \left(1 + \sqrt{1 + \frac{4}{(1085.23)(.1)}} \right)^2 = 1105.14$$

To illustrate that a test of equivalence will usually require a larger sample size than a conventional test of significance, Equation 16.12 will be employed in reference to Example 16.7 to compute the required sample size for the power of the *z* **test for two independent proportions** to equal .80. Note that in the computations below $p = [(.837)(.784)]/2 = .811$ and $(1 - p) = [(.163)(.216)]/2 = .190$. Thus, for a test with Power = .80 to identify the difference .837 − .784 = .053 as statistically significant at the .05 level will require a sample size of 859 subjects per group (which is less than the sample size value 1086 per group computed for the test of equivalence).

$$n_1 = \frac{\left[1.96\sqrt{(.811)(.190)(2)} + .84\sqrt{(.837)(.163) + (.784)(.216)} \right]^2}{(.837 - .784)^2} = 858.12$$

It should be noted that when a test of equivalence is used within the framework of drug assessment research, it is common practice for a one-tailed analysis to be employed — specifically, an evaluation of the directional alternative hypothesis that the efficacy of the generic drug will not be less than the efficacy of the brand name drug by more than the value stipulated for ζ. In such a case, the value $z_{\alpha/2} = 1.65$ would be employed in Equation 16.51. When the latter value is employed below in Equation 16.51, the required sample size (for a test with Power = .80) is reduced from 1086 to 859.

$$n_1 = \frac{\left((.837)(.163) + \frac{(.784)(.216)}{1} \right)(1.65 + .84)^2}{[|.837 - .784| - .1]^2} = 858.23$$

3. Test 16n: The log-likelihood ratio Prior to reading this section the reader should review Endnote 14 in the **Introduction** which discusses the general subject of logarithms. Two alternative approaches for evaluating one- and two-dimensional tables are the **log-likelihood ratio** and **log-linear analysis**. If either of the latter methods is employed to evaluate a one- or two-dimensional table it generally yields comparable results to those obtained with a conventional chi-square analysis, and because of the latter most sources state there is no advantage to using the above noted alternative methods (both which are computationally more involved than a chi-square analysis). The discussion in this section will focus on the computation of the **log-likelihood ratio** and the statistic based upon it which is referred to as the **likelihood ratio statistic**.

For a one-dimensional table the **log-likelihood ratio** (which Zar (1999, p. 473) notes was first proposed by Wilks (1935)) is expressed by Equation 16.52, which is equivalent to Equation 16.53. For a two-dimensional table the **log-likelihood ratio** (to be summarized with the notation LLR) is expressed by Equation 16.54, which is equivalent to Equation 16.55. Note that **natural logarithms** (represented with the notation **ln**) are employed in all of the aforementioned equations.

$$LLR = \sum_{i=1}^{r} \left[O_i \ln\left(\frac{O_i}{E_i} \right) \right] \qquad \textbf{(Equation 16.52)}$$

$$LLR = \sum_{i=1}^{r} (O_i)(\ln O_i) - \sum_{i=1}^{r} (O_i)(\ln E_i) \qquad \textbf{(Equation 16.53)}$$

$$LLR = \sum_{i=1}^{r}\sum_{j=1}^{c} \left[O_{ij} \ln\left(\frac{O_{ij}}{E_{ij}} \right) \right] \qquad \textbf{(Equation 16.54)}$$

$$LLR = \sum_{i=1}^{r}\sum_{j=1}^{c} (O_{ij})(\ln O_{ij}) - \sum_{i=1}^{r}\sum_{j=1}^{c} (O_{ij})(\ln E_{ij}) \qquad \textbf{(Equation 16.55)}$$

The notation G, which is commonly referred to as the **likelihood ratio statistic**, is employed to represent twice the value of the **log-likelihood ratio**. (Some sources employ the notation G^2 in lieu of G.) Equation 16.56 summarizes the computation of G for a one-dimensional table by multiplying the values computed for *LLR* with Equations 16.52 and 16.53 by 2. Equation 16.57 summarizes the computation of G for a two-dimensional table by multiplying the values computed for *LLR* with Equations 16.54 and 16.55 by 2.

(Equation 16.56)

$$G = 2\sum_{i=1}^{r}\left[O_i \ln\left(\frac{O_i}{E_i}\right)\right] = 2\left[\sum_{i=1}^{r}(O_i)(\ln O_i) - \sum_{i=1}^{r}(O_i)(\ln E_i)\right]$$

(Equation 16.57)

$$G = 2\left(\sum_{i=1}^{r}\sum_{j=1}^{c}\left[O_{ij} \ln\left(\frac{O_{ij}}{E_{ij}}\right)\right]\right) = 2\left[\sum_{i=1}^{r}\sum_{j=1}^{c}(O_{ij})(\ln O_{ij}) - \sum_{i=1}^{r}\sum_{j=1}^{c}(O_{ij})(\ln E_{ij})\right]$$

Zar (1999, p. 474) notes that **common logarithms** (which employ the notation **log**) can be employed in Equations 16.56 and 16.57 if the multiplier 2 at the extreme left of the latter equations is replaced by 4.60517. The latter is illustrated below for a two dimensional table in Equation 16.58.

(Equation 16.58)

$$G = 4.60517\left(\sum_{i=1}^{r}\sum_{j=1}^{c}\left[O_{ij} \log\left(\frac{O_{ij}}{E_{ij}}\right)\right]\right) = 4.60517\left[\sum_{i=1}^{r}\sum_{j=1}^{c}(O_{ij})(\log O_{ij}) - \sum_{i=1}^{r}\sum_{j=1}^{c}(O_{ij})(\log E_{ij})\right]$$

The **likelihood ratio statistic** (i.e., G) is evaluated with the chi-square distribution with the same degrees of freedom employed for a conventional chi-square analysis (i.e., $df = k - 1$ for a one-dimensional table and $df = (r - 1)(c - 1)$ for a two-dimensional table). To illustrate the computation of a **likelihood ratio statistic**, the latter value is computed below through use of Equation 16.57 for Examples 16.1/16.2 (i.e., the data in Tables 16.2/16.3). Note that the computed value $G = 18.48$ is extremely close to the value $\chi^2 = 18.18$ computed in Section IV for the same set of data. As noted above, G is interpreted as a chi-square value with $df = (r - 1)(c - 1)$. Since $r = 2$ and $c = 2$, $df = (2 - 1)(2 - 1) = 1$, and thus $\chi^2_{.05} = 3.84$ and $\chi^2_{.01} = 6.63$. Since the computed value $G = 18.48$ is greater than both of the aforementioned critical values, the null hypothesis can be rejected at both the .05 and .01 levels. Thus, the conclusions regarding the data are identical to those reached when the conventional chi-square analysis was employed.

$$G = 2\left(\sum_{i=1}^{r}\sum_{j=1}^{c}\left[O_{ij} \ln\left(\frac{O_{ij}}{E_{ij}}\right)\right]\right) =$$

$$2\left[30 \ln\left(\frac{30}{45}\right) + 70 \ln\left(\frac{70}{55}\right) + 60 \ln\left(\frac{60}{45}\right) + 40 \ln\left(\frac{40}{55}\right)\right] =$$

$$2[(30)(-.4055) + (70)(.2412) + (60)(.2877) + (40)(-.3185)] = (2)(9.24) = 18.48$$

or

$$2\left[\sum_{i=1}^{r}\sum_{j=1}^{c}(O_{ij})(\ln O_{ij}) - \sum_{i=1}^{r}\sum_{j=1}^{c}(O_{ij})(\ln E_{ij})\right] =$$

$$2([30\ln(30) + 70\ln(70) + 60\ln(60) + 40\ln(40)] - [30\ln(45) + 70\ln(55) + 60\ln(45) + 40\ln(55)]) =$$

$$2[792.646 - 783.406] = 2(9.24) = 18.48$$

In the case of a two-dimensional table, when $df = 1$ (as is the case for Examples 16.1/16.2) a correction for continuity (which will yield a slightly smaller value for G) can be employed where each observed frequency is made closer to its expected frequency by .5. In the case of Examples 16.1/16.2 the values of the observed frequencies 30, 70, 60, 40 are changed to 30.5, 69.5, 59.5, 40.5, while the values for the expected frequencies 45, 55, 45, 55 remained unchanged. When the adjusted corrected for continuity observed frequencies are employed in Equation 16.57, the value $G = 17.25$ is computed (which is also significant at both the .05 and .01 levels). As noted above, most sources state that the use of a conventional chi-square analysis versus a **log-likelihood analysis** of a contingency table will generally yield comparable results. Zar (1999, p. 475), however, notes that Williams (1976) recommends the **likelihood ratio statistic** whenever for any cell $|O_i - E_i| \geq E_i$ in the case of a one-dimensional table or $|O_{ij} - E_{ij}| \geq E_{ij}$ in the case of a two-dimensional table.

4. Simpson's paradox **Simpson's paradox** is where either the direction or magnitude of the relationship between two variables (to be designated X and Y) is influenced by a third variable (to be designated Z). When the relationship between X and Y is summarized in a two-dimensional contingency table, the direction and/or the magnitude of the relationship between the two variables is different from what it appears to be when the data are summarized in a three-dimensional contingency table which takes into account all three variables.

To illustrate, let us assume that a study is conducted which evaluates the efficacy of two surgical treatments (to be designated Treatment A and Treatment B) for a seizure disorder. The study is conducted at two hospitals to be designated as Hospital 1 and Hospital 2. The neurosurgeons who perform the surgery in Hospital 1 are extremely experienced in using both of the surgical techniques. The neurosurgeons at Hospital 2, on the other hand, are relatively inexperienced in using the surgical procedures. However, the researcher who designs the study is unaware of differences in experience between the surgeons at the two hospitals. Consequently, the researcher conceptualizes the study as having one independent variable, which is the type of surgical treatment a patient receives. The dependent variable (which constitutes the second variable in the study) is the measure of the efficacy of the treatments — specifically, the categorization of a patient as a success or a failure. For purposes of illustration we will assume that 550 patients participate in the study, and that half of the patients receive Treatment A and the other half Treatment B. Of the 550 patients, 370 are treated at Hospital 1, and 180 are treated at Hospital 2. Table 16.27 summarizes the results of the study in the form of a 2×2 contingency table.

Table 16.27 Data for Neurosurgery Study in 2 × 2 Contingency Table

| | | Response to treatment | | |
		Success	Failure	Totals
Treatment	Treatment A	160	115	275
	Treatment B	160	115	275
	Totals	320	230	550

It should be apparent from inspection of Table 16.27 that Treatment A and Treatment B yield the identical number of successes and failures, and thus there is no difference in the efficacy of the two treatments. The latter can be confirmed by the fact that if a **chi-square test for homogeneity** is employed to evaluate the data (without using the correction for continuity), it yields the value $\chi^2 = 0$ (since the expected and observed frequency for each cell will be equal). We can also make the following additional statements about the data: a) A patient has a $160/275 = .58$ probability of responding favorably to Treatment A, and a $160/275 = .58$ probability of responding favorably to Treatment B; b) If the **odds ratio** is computed (through use of Equation 16.27), the value $o = [(115)(160)]/[(160)(115)] = 1$ is computed. An **odds ratio** of 1 in reference to Table 16.27 indicates that the odds of Treatment A being successful are equal to the odds of Treatment B being successful.

Table 16.28 summarizes the results of the study in the format of a three-dimensional contingency table. Note that in addition to the two variables taken into account in Table 16.27, Table 16.28 includes as a third variable the hospital at which a patient received the treatment. Inspection of Table 16.28 reveals that if the hospital which administered the treatment is taken into account, one reaches a different conclusion regarding the efficacy of the two treatments.

Table 16.28 Data for Neurosurgery Study in Three-Dimensional Contingency Table

| | | Hospital 1 | | Hospital 2 | | |
		Success	Failure	Success	Failure	Totals
Treatment	Treatment A	150	75	10	40	275
	Treatment B	120	25	40	90	275
	Totals	270	100	50	130	550

Inspection of Table 16.28 reveals the following: a) In Hospital 1 a patient has a $120/145 = .83$ probability of having a successful response to Treatment B, but only a $150/225 = .67$ probability of having a successful response to Treatment A. If the **odds ratio** (through use of Equation 16.27) is computed for the Hospital 1 data, the value $o = [(75)(120)]/[(150)(25)] = 2.4$ is computed. The latter value indicates that the odds of Treatment B being successful are 2.4 times larger than the odds of Treatment A being successful. If a **chi-square test for homogeneity** is employed to evaluate the 2 × 2 table for Hospital 1, it yields the value $\chi^2 = 11.58$, which is significant at both the .05 and .01 levels (since for $df = 1$, $\chi^2_{.05} = 3.84$ and $\chi^2_{.01} = 6.63$). The values computed in Hospital 1 clearly indicate that Treatment B is more successful than Treatment A; b) In Hospital 2 a patient has a $40/130 = .31$ probability of having a successful response to Treatment B, but only a $10/50 = .20$ probability of having a successful response to Treatment A. If the **odds ratio** (through use of Equation 16.27) is computed for the Hospital 2 data, the value $o = [(40)(40)]/[(10)(90)] = 1.78$ is computed. The latter value indicates that the odds of Treatment B being successful are 1.78 times larger than the odds of Treatment A being successful. If a **chi-square test for homogeneity** is employed to evaluate the 2 × 2 table for Hospital 2, it yields the value $\chi^2 = 2.09$ (which is not significant, since $\chi^2 = 2.09$ is less than $\chi^2_{.05} = 3.84$).[32] The fact remains, however, that the success rate in Hospital 2 for Treatment B is higher than it is for Treatment A. The lack of a significant result for the Hospital 2 data may be a function of the relatively small sample size, which limits the power of the analysis. Thus, there is a suggestion that Treatment B may also be more successful than Treatment A in Hospital 2.

Analysis of the treatments within each of the two hospitals strongly suggests that Treatment B is more successful than Treatment A. Yet when the data are pooled and expressed within the format of a 2 × 2 contingency table, there is no apparent difference between the two treatments.

This is a classic example of **Simpson's paradox**, in that both the magnitude and direction of the relationship between two variables (the type of treatment administered and a patient's response to the treatment) are influenced by a third variable (i.e., the hospital at which a treatment was administered). The fact that the success rate for both of the treatments is higher in Hospital 1 than it is in Hospital 2 reflects the fact that the doctors at Hospital 1 are more skilled in using the surgical treatments than the doctors at Hospital 2 (although there is the possibility that the lower success rate at Hospital 2 could be due to the fact that a disproportionate number of the more difficult cases were treated at that hospital). The superiority of Treatment B over Treatment A is masked in Table 16.27, because the latter table ignores the differential outcome rates at the two hospitals. The two factors which are responsible for the different conclusions reached through use of the data in Table 16.27 versus Table 16.28 are: a) In Hospital 1 fewer patients received Treatment B than Treatment A, while in Hospital 2 a greater number of patients received Treatment B than Treatment A; and b) The doctors at Hospital 1 are more experienced (and thus probably more successful in conducting the surgery for both treatments, but especially Treatment B) than the doctors at Hospital 2. The joint influence of these two factors is what is responsible for the absence of any apparent differences between the two treatments when the data are summarized in a 2 × 2 contingency table (i.e., Table 16.27). Thus, **Simpson's paradox** illustrates that when data based on three variables (typically two independent variables and one dependent variable) are collapsed into a 2 × 2 table, it can dramatically distort what really occurred in a study. In point of fact, **Simpson's paradox** is the result of a phenomenon referred to as an **interaction**, which is not revealed in the data contained in Table 16.27 but is revealed in Table 16.28. The concept of **interaction** is discussed in the next section.

5. Analysis of multidimensional contingency tables through use of a chi-square analysis
A multidimensional contingency table (such as Table 16.28) contains information on three or more variables. This section will focus on the generalization of the **chi-square test for $r \times c$ tables** in evaluating a three-dimensional table. The next section will describe the use of an alternative methodology called **log-linear analysis (Test 16o)** in evaluating multidimensional tables. The reader should keep in mind that the three-dimensional contingency table to be evaluated in this section and the next will represent the simplest multidimensional table which can be constructed — specifically, a table with three variables, with each variable comprised of two levels. As the number of variables (as well as the number of levels/categories per variable) increases, the analysis of a multidimensional table becomes increasingly complex. In addition, the more variables that are included in a study, the more difficult it becomes to obtain a clear interpretation of the results of an analysis.

The analysis of **multidimensional contingency tables** shares a number of things in common with the analysis of **factorial designs** with an **analysis of variance** (factorial designs are discussed in the chapter on the **between-subjects factorial analysis of variance (Test 27)**). Both of the aforementioned analyses can be employed to evaluate data when there are two or more independent variables and one dependent variable. In designs in which there are multiple independent variables, within the framework of conducting an omnibus test on the complete body of data, it is necessary to determine whether or not any **interactions** are present. An **interaction** is present in a set of data when the performance of subjects on one independent variable is not consistent across all the levels of another independent variable.

If we employ the definition of an **interaction** in reference to the example employed to illustrate **Simpson's paradox**, we can conclude there is an interaction between the type of treatment a person receives and the hospital at which he or she is treated — or to put it another way, the success rate for Treatment A versus Treatment B (the two treatments representing one of the independent variables) is not consistent with respect to Hospital 1 versus Hospital 2 (the two hospitals representing a second independent variable). In the case of multidimensional contingency tables, the concept of interaction can also be generalized to designs where no clear distinction is made with respect to whether variables are independent or dependent variables. Although the concept of interaction will be discussed further later in this section, it is discussed in greater detail in Section V of the **between-subjects factorial analysis of variance**.

As noted earlier, it is possible to have a three-dimensional table which summarizes the data for two independent variables and a dependent variable.[33] A summary table of the data for Example 16.9 (which is similar to Example 16.1, except for the fact that it includes a second independent variable) will yield such a multidimensional table. It was also noted that it is possible to have a three-dimensional table where no clear-cut distinction is made between the independent and dependent variables. Example 16.10 (which is similar to Example 16.2, except for the fact that it describes a study involving three variables instead of two variables) will result in a multidimensional table of this type. Examples 16.9 and 16.10 will be employed to illustrate the generalization of the **chi-square test for r × c tables** to a three-dimensional contingency table.

Example 16.9 *A researcher conducts a study on altruistic behavior. All subjects who participate in the experiment are males. Each of the 160 male subjects is given a one-hour test which is ostensibly a measure of intelligence. During the test 65 subjects are exposed to continual loud noise, which they are told is due to a malfunctioning generator. Another 95 subjects are not exposed to any noise during the test. Upon completion of this stage of the experiment, each subject on leaving the room is confronted by a middle-aged man or woman whose arm is in a sling. The latter individual asks the subject if she would be willing to help him/her carry a heavy package to his/her car. 80 of the subjects are confronted by a male who asks for help, while the other 80 are confronted by a female. In actuality, the person requesting help is an experimental confederate (i.e., working for the experimenter). The dependent variable in the experiment is whether or not a subject helps the person carry the package. Table 16.29 summarizes the data for the experiment. Do the data indicate that altruistic behavior is influenced by noise and/or the gender of the person requesting help?*

Example 16.10 *A researcher wants to determine if there is a relationship between the following three variables/dimensions: a) A person's political affiliation — specifically, whether a person is a Democrat or a Republican; b) A person's categorization on the personality dimension of introversion–extroversion; and c) A person's gender (male versus female). One hundred and sixty people are recruited to participate in the study. All of the subjects are given a personality test, on the basis of which each subject is classified as an introvert or an extrovert. Each subject is then asked to indicate whether he or she is a Democrat or a Republican. The data for Example 16.10, which can be summarized in the form of a three-dimensional contingency table, are presented in Table 16.29. Do the data indicate that the three variables are independent of one another?*

Table 16.29 is a three-dimensional contingency table which simultaneously summarizes the data for Examples 16.9 and 16.10. In a three-dimensional table, one of the variables is designated as the **row** variable, a second variable is designated as the **column** variable, and the third variable is designated as the **layer** variable. (Zar (1999) employs the term **tier** variable to designate the third variable.) In a three-dimensional table there will be a total of $r \times c \times l$ cells, where r represents the number of row categories, c the number of column categories, and l the number of layer categories.

As noted above, the experiment described in Example 16.9 has two independent variables. One of the independent variables is the **noise manipulation**, which is comprised of the two levels **noise** versus **no noise**. In Table 16.29 the **noise manipulation** independent variable is designated as the **row variable**. The second independent variable is whether a subject is confronted by a **male** or a **female** confederate. In Table 16.29 the independent variable represented by the **gender of the confederate** is designated as the **column variable**. The dependent variable is whether or not a subject **helped the confederate**. In Table 16.29 **whether or not a subject helped,** which has the two levels **helped** and **did not help,** is designated as the **layer variable**.

In the case of Example 16.10, where no clear-cut distinction is made with respect to a variable being an independent or dependent variable, the **row variable** will be the categorization of a person on the **introversion-extroversion dimension**, the **column variable** will be the **gender of the subject (male-female)**, and the **layer variable** will be a person's **political affiliation (Democrat-Republican)**.

Table 16.29　Summary Data for Examples 16.9/16.10

	Helped/Democrat		Did not help/Republican		
	Male	Female	Male	Female	Totals
Noise/Introvert	10	15	25	15	65
No Noise/Extrovert	25	45	20	5	95
Totals	35	60	45	20	160

Sums:	Row 1 (Noise/Introvert)	$= O_{1..} = R_1 = 65$
	Row 2 (No Noise/Extrovert)	$= O_{2..} = R_2 = 95$
	Column 1 (Male)	$= O_{.1.} = C_1 = 80$
	Column 2 (Female)	$= O_{.2.} = C_2 = 80$
	Layer 1 (Helped/Democrat)	$= O_{..1} = L_1 = 95$
	Layer 2 (Did not help/Republican	$= O_{..2} = L_2 = 65$

Christensen (1990, pp. 63–64; 1999) notes that among the possible ways of conceptualizing the relationship between the variables in a three-way contingency table are the following: a) Rows, columns, and layers are all independent of one another; b) Rows are independent of columns and layers (with columns and layers not necessarily being independent of one another); c) Columns are independent of rows and layers (with rows and layers not necessarily being independent of one another); d) Layers are independent of rows and columns (with rows and columns not necessarily being independent of one another); e) Given any specific level of a row, columns and layers are independent of one another; f) Given any specific level of a column, rows and layers are independent of one another; g) Given any specific level of a layer, rows and columns are independent of one another.

Each of the aforementioned ways of conceptualizing a contingency table is often referred to as a **model**. Of the seven models noted above, **Model a** is referred to as the **model of complete (or mutual) independence**, **Models b, c,** and **d** as **models of partial independence**, and **Models e, f,** and **g** as **models of conditional independence**. The discussion to follow will be limited to a description of the analysis of the models of **complete** and **partial independence**.

Test of model of complete independence　The initial analysis which will be conducted on Table 16.29 will evaluate whether or not in each of the studies all three variables are independent of one another. This analysis (which, as noted earlier, assesses **complete independence**) will be evaluated with what will be referred to as the **omnibus chi-square analysis**. Within the framework of the latter analysis, the following null and alternative hypotheses will be evaluated.

Null hypothesis H_0: In Example 16.9, in the underlying population the three variables (exposure to noise, gender of confederate, and helping behavior) are all independent of one another. In Example 16.10, in the underlying population the three variables (introversion–extroversion, gender, and political affiliation) are all independent of one another. The notation H_0: $O_{ijk} = \varepsilon_{ijk}$ *for all cells* can also be employed, which means that in the underlying population(s) the sample(s) represent(s), for each of the $r \times c \times l$ cells, the observed frequency of a cell is equal to the expected frequency of the cell. With respect to the sample data, this translates into the observed frequency of each of the $r \times c \times l$ cells being equal to the expected frequency of the cell.

Alternative hypothesis H_1: In Example 16.9, in the underlying population the three variables (exposure to noise, gender of confederate, and helping behavior) are not all independent of one another. In Example 16.10, in the underlying population the three variables (introversion-extroversion, gender, and political affiliation) are not all independent of one another. The notation H_1: $O_{ijk} \neq \varepsilon_{ijk}$ *for at least one cell* can also be employed, which means that in the underlying population(s) the sample(s) represent(s), for at least one of the $r \times c \times l$ cells, the observed frequency of a cell is not equal to the expected frequency of the cell. With respect to the sample data, this translates into the observed frequency of at least one of the $r \times c \times l$ cells not being equal to the expected frequency of the cell. It will be assumed that the alternative hypothesis is nondirectional (although it is possible to have a directional alternative hypothesis).

Equation 16.59 (which is a generalization of Equation 16.2 to a three-dimensional table) is employed to compute the test statistic for a three-dimensional table.

$$\chi^2 = \sum_{i=1}^{r} \sum_{j=1}^{c} \sum_{k=1}^{l} \left[\frac{(O_{ijk} - E_{ijk})^2}{E_{ijk}} \right] \qquad \textbf{(Equation 16.59)}$$

Note that the notation k in the ijk subscript for the observed and expected frequencies in Equation 16.59 (as well as in the null and alternative hypotheses) represents the kth layer of the layer variable. Thus, in Equation 16.59 the notation E_{ijk} means the expected frequency for the cell in Row i, Column j, and Layer k. Note that just as there are r levels on the row variable and c levels on the column variable, there are l levels on the layer variable.

The operations described by Equation 16.59 (which are the same as those described for computing the chi-square statistic for the **chi-square goodness-of-fit test** and the **chi-square test for r × c tables**) are as follows: a) The expected frequency of each cell is subtracted from its observed frequency; b) For each cell, the difference between the observed and expected frequency is squared; c) For each cell, the squared difference between the observed and expected frequency is divided by the expected frequency of the cell; and d) The value of chi-square is computed by summing all of the values computed in part c).

Note that in contrast to Equation 16.59, there is only one summation sign in the analogous equation for the **chi-square goodness-of-fit test** (Equation 8.2), since the latter test has only a single variable (designated as the row variable). In the same respect there are two summation signs for the analogous equation for the **chi-square test for r × c tables** (Equation 16.2), since the latter test has two variables — a row variable and a column variable. Since there are three variables, the three-dimensional equation requires summing over all three dimensions. (If there are four dimensions the chi-square equation will have four summation signs, since summing will have to be done over all four dimensions. Five dimensions will require five summation signs, and so on.)

The protocol for conducting the chi-square analysis is identical to that employed for a two-dimensional table. The only aspect of the analysis which is different is the computation of the expected frequency of a cell, which in the case of a three-dimensional table is computed with Equation 16.60. Note that Equation 16.60 is written two ways. The first representation of the equation is consistent with the format used in Equation 16.1, the equation for computing the expected frequency of a two-dimensional table. Thus, $O_{i..}$ represents the number of observations in the ith row, which can be represented more simply as R_i. $O_{.j.}$ represents the number of observations in the jth column, which can be represented more simply as C_j. $O_{..k}$ represents the number of observations in the kth layer, which can be represented more simply as L_k. The totals for the rows, columns, and layers are recorded at the bottom of Table 16.29.

$$E_{ijk} = \frac{O_{i..} O_{.j.} O_{..k}}{n^2} = \frac{R_i C_j L_k}{n^2} \qquad \text{(Equation 16.60)}$$

The notation in Equation 16.60 indicates that to compute the expected frequency of a cell, the observed frequency of the row the cell is in is multiplied by the observed frequency of the column the cell is in, and the resulting value is multiplied by the observed frequency of the layer the cell is in. The resulting product is divided by the square of the total number of observations in the contingency table.

To illustrate, we will compute the expected frequency for the cell in the upper left of Table 16.29. The latter cell, which is Cell$_{111}$ (i.e., the cell in Row 1, Column 1, and Layer 1), represents subjects in the **noise/introvert** category, the **male** category, and the **helped/Democrat** category. We thus multiply the total number of observations in Row 1 (which is 65) by the total number of observations in Column 1 (which is 80) by the total number of observations in Layer 1 (which is 95). The product is divided by the square of the total number of observations in the contingency table (the total number of observations being 160). Thus: $E_{111} = [(65)(80)(95)]/(160)^2 = 19.30$. The expected frequencies of the remaining seven cells are computed below.

$$E_{121} = [(65)(80)(95)]/(160)^2 = 19.30$$
$$E_{112} = [(65)(80)(65)]/(160)^2 = 13.20$$
$$E_{122} = [(65)(80)(65)]/(160)^2 = 13.20$$
$$E_{211} = [(95)(80)(95)]/(160)^2 = 28.20$$
$$E_{221} = [(95)(80)(95)]/(160)^2 = 28.20$$
$$E_{212} = [(95)(80)(65)]/(160)^2 = 19.30$$
$$E_{222} = [(95)(80)(65)]/(160)^2 = 19.30$$

The chi-square analysis from this point on is identical to that employed for a two-dimensional table. The analysis is summarized in Table 16.30. The cell identification codes in the table are **N/I** = Noise/Introvert; **NN/E** = No noise/Extrovert; **M** = Male; **F** = Female; **H/D** = Helped/Democrat; **DNH/R** = Did not help/Republican.

The degrees of freedom employed in the omnibus analysis of a three-dimensional contingency table are $df = rcl - r - c - l + 2$.[34] Thus, for our example, since $r = c = l = 2$, $df = (2)(2)(2) - 2 - 2 - 2 + 2 = 4$. Employing **Table A4**, the tabled critical .05 and .01 chi-square values for $df = 4$ are $\chi^2_{.05} = 9.49$ and $\chi^2_{.01} = 13.28$. Since the computed value $\chi^2 = 37.24$ is greater than both of the aforementioned critical values, the null hypothesis can be rejected at both the .05 and .01 levels. By virtue of rejecting the null hypothesis, the researcher can conclude that in Examples 16.9/16.10 the three variables are not independent of one another.

Table 16.30 Chi-Square Summary Table for Omnibus Analysis of Examples 16.9/16.10

Cell	O_{ijk}	E_{ijk}	$(O_{ijk} - E_{ijk})$	$(O_{ijk} - E_{ijk})^2$	$\dfrac{(O_{ijk} - E_{ijk})^2}{E_{ijk}}$
111 (N/I,M,H/D)	10	19.30	−9.3	86.49	4.48
121 (N/I,F,H/D)	15	19.30	−4.3	18.49	.96
112 (N/I,M,DNH/R)	25	13.20	11.8	139.24	10.55
122 (N/I,F,DNH/R)	15	13.20	1.8	3.24	.25
211 (NN/E,M,H/D)	25	28.20	−3.2	10.24	.36
221 (NN/E,F,H/D)	45	28.20	16.8	282.24	10.01
212 (NN/E,M,DNH/R)	20	19.30	.7	.49	.03
222 (NN/E,F,DNH/R)	5	19.30	−14.3	204.49	10.60
Sums	160	160	0		$\chi^2 = 37.24$

Test of models of partial independence If the evaluation of the model of **complete independence** through use of the omnibus chi-square analysis yields a significant result, a researcher should conduct additional analyses in order to further clarify the nature of the relationship among the three variables. As noted earlier, among the analyses that can be conducted are those for **partial independence**, which determine whether one variable is independent of the other two variables. Thus, we can determine the following: a) Whether rows are independent of columns and layers; b) Whether columns are independent of rows and layers; and c) Whether layers are independent of rows and columns. The latter three analyses will now be conducted.

Test of independence of rows versus columns and layers Equation 16.59 is employed to determine whether or not the row variable is independent of the column and layer variables. The only difference in the analysis to be described in this section from the omnibus analysis conducted in the previous section is the computation of the expected frequency for each cell. Equation 16.61 is employed to compute the expected frequency of a cell.

$$E_{ijk} = \frac{(O_{i..})(O_{.j.}O_{..k})}{n} = \frac{(R_i)(C_j L_k)}{n} \qquad \textbf{(Equation 16.61)}$$

The notation in Equation 16.61 indicates that to compute the expected frequency of a cell, the number of observations for the row in which that cell appears is multiplied by the number of observations that are in both the column and layer designated for that cell. The resulting product is divided by the total number of observations in the contingency table.

To illustrate, we will compute the expected frequency for the cell in the upper left of Table 16.29. The latter cell, which is Cell$_{111}$ (i.e., the cell in Row 1, Column 1, and Layer 1), represents subjects in the **noise/introvert** category, the **male** category, and the **helped/Democrat** category. We thus multiply the total number of observations in Row 1 (**noise/introvert**) (which is 65) by the total number of observations in both Column 1 and Layer 1 (i.e., observations in the **male** column that are in the **helped/Democrat** layer) (which is 35). The product is divided by the total number of observations in the contingency table (which is 160). Thus: $E_{111} = [(65)(35)]/(160)$ = 14.22. The expected frequencies of the remaining seven cells are computed below.

$$E_{121} = [(65)(60)]/(160) = 24.38$$
$$E_{112} = [(65)(45)]/(160) = 18.28$$
$$E_{122} = [(65)(20)]/(160) = 8.13$$

$$E_{211} = [(95)(35)]/(160) = 20.78$$
$$E_{221} = [(95)(60)]/(160) = 35.63$$
$$E_{212} = [(95)(45)]/(160) = 26.72$$
$$E_{222} = [(95)(20)]/(160) = 11.88$$

The chi-square analysis to determine whether or not the row variable is independent of the column and layer variables is summarized in Table 16.31.

Table 16.31   Chi-Square Summary Table for Rows versus Columns/Layers Analysis of Examples 16.9/16.10

Cell	O_{ijk}	E_{ijk}	$(O_{ijk} - E_{ijk})$	$(O_{ijk} - E_{ijk})^2$	$\dfrac{(O_{ijk} - E_{ijk})^2}{E_{ijk}}$
111 (N/I,M,H/D)	10	14.22	−4.22	17.18	1.25
121 (N/I,F,H/D)	15	24.38	−9.38	87.98	3.61
112(N/I,M,DNH/R)	25	18.28	6.72	45.16	2.47
122(N/I,F,DNH/R)	15	8.13	6.87	47.20	5.81
211(NN/E,M,H/D)	25	20.78	4.22	17.81	.86
221(NN/E,F,H/D)	45	35.63	9.37	87.80	2.46
212(NN/E,M,DNH/R)	20	26.72	−6.72	45.16	1.69
222(NN/E,F,DNH/R)	5	11.88	−6.88	47.33	3.98
Sums	160	160	0		$\chi^2 = 22.13$

The degrees of freedom employed in the analysis are $df = rcl - cl - r + 1 = (2)(2)(2) - (2)(2) - 2 + 1 = 3$.[35] Employing **Table A4**, the tabled critical .05 and .01 chi-square values for $df = 3$ are $\chi^2_{.05} = 7.81$ and $\chi^2_{.01} = 11.34$. Since the computed value $\chi^2 = 22.13$ is greater than both of the aforementioned critical values, the null hypothesis can be rejected at both the .05 and .01 levels. By virtue of rejecting the null hypothesis, the researcher can conclude that the row variable (**noise manipulation/introversion-extroversion**) is not independent of the column (**gender**) and layer (**helping/political affiliation**) variables.

Test of independence of columns versus rows and layers   Equation 16.59 is employed to determine whether or not the column variable is independent of the row and layer variables. The only difference in the analysis to be described in this section and the previous analyses described is the computation of the expected frequency for each cell. The data have been reorganized in Table 16.32 to facilitate the computation of the expected frequencies for the analysis of independence of columns versus rows and layers.

Table 16.32   Summary of Data for Examples 16.9/16.10 for Columns versus Rows/Layers Analysis

	Helped/Democrat		Did not help/Republican		
	Noise/Introvert	No Noise/Extrovert	Noise/Introvert	No Noise/Extrovert	Totals
Male	10	25	25	20	80
Female	15	45	15	5	80
Totals	25	70	40	25	160

Equation 16.62 is employed to compute the expected frequency of a cell.

$$E_{ijk} = \frac{(O_{.j.})(O_{i..}O_{..k})}{n} = \frac{(C_j)(R_i L_k)}{n}$$ **(Equation 16.62)**

The notation in Equation 16.62 indicates that to compute the expected frequency of a cell, the number of observations for the column in which that cell appears is multiplied by the number of observations which are in both the row and layer designated for that cell. The resulting product is divided by the total number of observations in the contingency table.

To illustrate, we will compute the expected frequency for the cell in the upper left of Table 16.32. The latter cell, which is Cell$_{111}$ (i.e., the cell in Row 1, Column 1, and Layer 1) represents subjects in the **noise/introvert** category, the **male** category, and the **helped/Democrat** category. We thus multiply the total number of observations in Column 1 (**males**) (which is 80) by the total number of observations in both Row 1 and Layer 1 (i.e., observations in the **noise/introvert** column that are in the **helped/Democrat** layer) (which is 25). The product is divided by the total number of observations in the contingency table (which is 160). Thus: $E_{111} = [(80)(25)]/(160)$ = 12.5. The expected frequencies of the remaining seven cells are computed below.

$$E_{121} = [(80)(25)]/(160) = 12.5$$
$$E_{112} = [(80)(40)]/(160) = 20$$
$$E_{122} = [(80)(40)]/(160) = 20$$
$$E_{211} = [(80)(70)]/(160) = 35$$
$$E_{221} = [(80)(70)]/(160) = 35$$
$$E_{212} = [(80)(25)]/(160) = 12.5$$
$$E_{222} = [(80)(25)]/(160) = 12.5$$

The chi-square analysis to determine whether or not the column variable is independent of the row and layer variables is summarized in Table 16.33.

Table 16.33 Chi-Square Summary Table for Columns versus Rows/Layers Analysis of Examples 16.9/16.10

Cell	O_{ijk}	E_{ijk}	$(O_{ijk} - E_{ijk})$	$(O_{ijk} - E_{ijk})^2$	$\dfrac{(O_{ijk} - E_{ijk})^2}{E_{ijk}}$
111 (N/I,M,H/D)	10	12.50	−2.5	6.25	.50
121 (N/I,F,H/D)	15	12.50	2.5	6.25	.50
112(N/I,M,DNH/R)	25	20.00	5.0	25.00	1.25
122(N/I,F,DNH/R)	15	20.00	−5.0	25.00	1.25
211(NN/E,M,H/D)	25	35.00	−10.0	100.00	2.86
221(NN/E,F,H/D)	45	35.00	10.0	100.00	2.86
212(NN/E,M,DNH/R)	20	12.50	7.5	56.25	4.50
222(NN/E,F,DNH/R)	5	12.50	−7.5	56.25	4.50
Sums	160	160	0		$\chi^2 = 18.22$

The degrees of freedom employed in the analysis are $df = rcl - rl - c + 1 = (2)(2)(2) - (2)(2) - 2 + 1 = 3$.[36] Employing **Table A4**, the tabled critical .05 and .01 chi-square values for $df = 3$ are $\chi^2_{.05} = 7.81$ and $\chi^2_{.01} = 11.34$. Since the computed value $\chi^2 = 18.22$ is greater than both of the aforementioned critical values, the null hypothesis can be rejected at both the .05 and .01 levels. By virtue of rejecting the null hypothesis, the researcher can conclude that the column variable (**gender**) is not independent of the row (**noise manipulation/introversion-extroversion**) and layer (**helping/ political affiliation**) variables.

Test of independence of layers versus rows and columns   Equation 16.59 is employed to determine whether or not the layer variable is independent of the row and column variables. The only difference in the analysis to be described in this section and the previous analyses described is the computation of the expected frequency for each cell. The data have been reorganized in Table 16.34 to facilitate the computation of the expected frequencies for the analysis of independence of layers versus rows and columns.

Table 16.34   Summary of Data for Examples 16.9/16.10 for Layers versus Rows/Columns Analysis

| | Noise/Introvert | | No Noise/Extrovert | | |
	Male	Female	Male	Female	Totals
Helped/Democrat	10	15	25	45	95
Did not Help/Republican	25	15	20	5	65
Totals	35	30	45	50	160

Equation 16.63 is employed to compute the expected frequency of a cell.

$$E_{ijk} = \frac{(O_{..k})(O_{i..}O_{.j.})}{n} = \frac{(L_k)(R_i C_j)}{n} \qquad \textbf{(Equation 16.63)}$$

The notation in Equation 16.63 indicates that to compute the expected frequency of a cell, the number of observations for the layer in which that cell appears is multiplied by the number of observations that are in both the row and column designated for that cell. The resulting product is divided by the total number of observations in the contingency table.

To illustrate, we will compute the expected frequency for the cell in the upper left of Table 16.34. The latter cell, which is $Cell_{111}$ (i.e., the cell in Row 1, Column 1, and Layer 1), represents subjects in the **noise/introvert** category, the **male** category, and the **helped/Democrat** category. We thus multiply the total number of observations in Layer 1 (**helped/Democrat**) (which is 95) by the total number of observations in both Row 1 and Column 1 (i.e., observations in the **noise/ introvert** row that are in the **male** column) (which is 35). The product is divided by the total number of observations in the contingency table (which is 160). Thus: $E_{111} = [(95)(35)]/(160)$ = 20.78. The expected frequencies of the remaining seven cells are computed below.

$$E_{121} = [(95)(30)]/(160) = 17.81$$
$$E_{112} = [(65)(35)]/(160) = 14.22$$
$$E_{122} = [(65)(30)]/(160) = 12.19$$
$$E_{211} = [(95)(45)]/(160) = 26.72$$
$$E_{221} = [(95)(50)]/(160) = 29.69$$
$$E_{212} = [(65)(45)]/(160) = 18.28$$
$$E_{222} = [(65)(50)]/(160) = 20.31$$

The chi-square analysis to determine whether or not the layer variable is independent of the row and column variables is summarized in Table 16.35. The degrees of freedom employed in the analysis are $df = rcl - rc - l + 1 = (2)(2)(2) - (2)(2) - 2 + 1 = 3$.[37] Employing **Table A4**, the tabled critical .05 and .01 chi-square values for $df = 3$ are $\chi^2_{.05} = 7.81$ and $\chi^2_{.01} = 11.34$. Since the computed value $\chi^2 = 34.55$ is greater than both of the aforementioned critical values, the null hypothesis can be rejected at both the .05 and .01 levels. By virtue of rejecting the null hypothesis, the researcher can conclude that the layer variable (**helping/political affiliation**) is not independent of the row (**noise manipulation/ introversion-extroversion**) and column (**gender**) variables.

Table 16.35 Chi-Square Summary Table for Layers versus Rows/Columns Analysis of Examples 16.9/16.10

Cell	O_{ijk}	E_{ijk}	$(O_{ijk} - E_{ijk})$	$(O_{ijk} - E_{ijk})^2$	$\dfrac{(O_{ijk} - E_{ijk})^2}{E_{ijk}}$
111 (N/I,M,H/D)	10	20.78	−10.78	116.21	5.59
121 (N/I,F,H/D)	15	17.81	−2.81	7.90	.44
112 (N/I,M,DNH/R)	25	14.22	10.78	116.21	8.17
122 (N/I,F,DNH/R)	15	12.19	2.81	7.90	.65
211 (NN/E,M,H/D)	25	26.72	−1.72	2.96	.11
221 (NN/E,F,H/D)	45	29.69	15.31	234.40	7.89
212 (NN/E,M,DNH/R)	20	18.28	1.72	2.96	.16
222 (NN/E,F,DNH/R)	5	20.31	−15.31	234.40	11.54
Sums	160	160	0		$\chi^2 = 34.55$

To clarify the nature of the relationship between the variables with greater precision, it will be necessary to conduct additional analyses on the data. To illustrate this, we will just examine the data in reference to Example 16.9 in greater detail. Recollect that in the latter study there are two independent variables, the **noise manipulation** and **gender**, and a dependent variable, which is the **helping behavior of subjects**. Let us assume that prior to the study the experimenter predicted the following: a) Subjects exposed to the **no noise** condition will be more likely to **help the confederate** than subjects exposed to the **noise condition**; and b) Subjects exposed to a **female** confederate will be more likely to **help** than subjects exposed to a **male** confederate. The two aforementioned hypotheses are predicting what is referred to as a **main effect** on both of the independent variables. The term **main effect** (which is discussed in greater detail in Sections I and V of the **between-subjects factorial analysis of variance**) describes the effect of one independent variable (also referred to as a factor) on the dependent variable, ignoring any effect any of the other independent variables/factors might have on the dependent variable. If the researcher considers each independent variable separately, two 2 × 2 contingency tables can be constructed to summarize the data. The latter is done with Tables 16.36 and 16.37, which respectively summarize the data when the **noise manipulation** independent variable is considered by itself and when the **gender** independent variable is considered by itself.

With regard to the predicted main effects, consider the following information that can be derived from Tables 16.36 and 16.37.

a) Without employing a test of significance on the data (which in this case would be the **chi-square test for homogeneity/z test for two independent proportions**), it appears that a subject is more likely to **help** in the **no noise** condition than the **noise** condition. This is the case, since the proportion of subjects who **helped** the confederate in the **no noise** condition is 70/95 = .74, while the proportion of subjects who helped the confederate in the **noise** condition is only 25/65 = .38. This clearly suggests the presence of a **main effect** on the **noise manipulation** independent variable. In other words, if in analyzing the data the researcher does not bother to consider **gender** as a second independent variable, but considers the **noise manipulation** as the only independent variable, the researcher will conclude that subjects are more likely to **help the confederate** in the **no noise** condition rather than in the **noise** condition.

Table 16.36 Summary of Data for Example 16.9 Employing Only Noise Manipulation Independent Variable

	Helped the Confederate	Did not Help the Confederate	Totals
Noise	25	40	65
No Noise	70	25	95
Totals	95	65	160

Table 16.37 Summary of Data for Example 16.9 Employing Only Gender Independent Variable

	Helped the Confederate	Did not Help the Confederate	Totals
Male	35	45	80
Female	60	20	80
Totals	95	65	160

b) Without employing a test of significance on the data (which in this case would be the **chi-square test for homogeneity/z test for two independent proportions**), it appears that a subject is more likely to **help** a **female** confederate than a **male** confederate. This is the case, since the proportion of subjects who **helped** the **female** confederate is 60/80 = .75, while the proportion of subjects who helped the **male** confederate is only 35/80 = .44. This clearly suggests the presence of a **main effect** on the **gender** independent variable. In other words, if in analyzing the data the researcher does not bother to consider the **noise manipulation** as a second independent variable, but considers **gender** as the only independent variable, the researcher will conclude that subjects are more likely to **help the female confederate** rather than the **male confederate**.

It was noted earlier in the discussion of **Simpson's paradox** that when data based on three variables are collapsed into two 2 × 2 contingency tables, a distorted picture of what actually occurred in a study may result. In the case of Example 16.9, there appears to be a definite interaction between the **independent variables** of **noise** and **gender**. Although for Example 16.9 the conclusions which will be reached if one employs two 2 × 2 contingency tables (i.e., Tables 16.36 and 16.37) are not as skewed as in the example employed to demonstrate **Simpson's paradox**, the use of two 2 × 2 tables still does not present an entirely accurate picture of what occurred in the study. To be more specific, the 2 × 2 tables are unable to reveal the interaction between the two independent variables. Specifically, consider the following, all of which are summarized in Table 16.38: a) The proportion of subjects who **helped** who were exposed to **noise** and a **male** confederate is 10/35 = .29; b) The proportion of subjects who **helped** who were exposed to **noise** and a **female** confederate is 15/30 = .50; c) The proportion of subjects who **helped** who were exposed to **no noise** and a **male** confederate is 25/45 = .56; and d) The proportion of subjects who **helped** who were exposed to **no noise** and a **female** confederate is 45/50 = .90. Note that in Table 16.38, the proportion .44 for **males** in Column 1 is the proportion of all subjects exposed to a **male** confederate who helped (35/80 = .44). The proportion .75 for **females** in Column 2 is the proportion of all subjects exposed to a **female** confederate who helped (60/80 = .75). The proportion .38 for **noise** in Row 1 is the proportion of all subjects exposed to **noise** who helped (25/65 = .38). The proportion .74 for **no noise** in Row 2 is the proportion of all subjects exposed to **no noise** who helped (70/95 = .74).

As noted earlier, an interaction is present in a set of data when the performance of subjects on one independent variable is not consistent across all the levels of another independent variable. Examination of Table 16.38 clearly suggests the presence of an interaction. Specifically, the following appears to be the case: Subjects are more likely to help a **female** confederate than a **male** confederate, but the proportion of **females** helped relative to the proportion of **males** helped is larger in the **no noise** condition than in the **noise** condition. We can also say that subjects are more likely to help the confederate in the **no noise** condition than the **noise** condition, but the proportion of subjects who help in the **no noise** condition relative to the **noise** condition is larger when the confederate is a **female** as opposed to a **male**.

Table 16.38 Summary of Interaction for Example 16.9:
Proportions of Helping across Both Independent Variables

	Male	Female	Row proportions
Noise	.29	.50	.38
No Noise	.56	.90	.74
Column proportions	.44	.74	

Tables such as Table 16.38, as well as graphs (such as Figures 27.1 and 27.2 employed to illustrate an interaction for a **between-subjects factorial analysis of variance**), can be extremely useful in providing a researcher with visual information regarding whether or not an interaction is present in a set of data.[38] It should be emphasized that in order to definitively establish the presence of an interaction, it is required that the appropriate inferential statistical statistic be conducted, and that the latter test yields a significant result for the interaction in question.

The analysis of multidimensional contingency tables is a complex topic which is discussed further in the next section. Alternative sources which discuss the analysis of multidimensional tables are Fienberg (1980), Marascuilo and McSweeney (1977), Marascuilo and Serlin (1988), Wickens (1989), and Zar (1999).

6. Test 16o: Analysis of multidimensional contingency tables with log-linear analysis

General overview of log-linear analysis As noted earlier, an alternative methodology for evaluating multidimensional tables is **log-linear analysis**, which, in fact, is the most common method employed to evaluate such tables.[39] Since it involves the analysis of more than two variables, **log-linear analysis** is sometimes discussed within the framework of **multivariate statistics** (which is discussed in detail in the latter part of the book). Because of the complexity of the computations involved in **log-linear analysis**, it is impractical to implement (as well as other multivariate procedures) without the aid of computer software.

The name **log-linear** derives from the fact that through use of logarithms a multiplicative relationship is transformed into a linear relationship. Specifically, in the case of a two-dimensional table, the null hypothesis for an $r \times c$ contingency table states that the probability for any cell in a table is the product of the probability of the row a cell falls within and the probability of the column the cell falls within. If the aforementioned probabilities are expressed as logarithms, the null hypothesis now states that the logarithm of the probability for any cell in the table is the sum of logarithms of the probabilities for the row and column the cell falls within. By using a logarithmic transformation, **log-linear analysis** can evaluate the multiplicative model represented by a contingency table as an additive model — in other words, in a manner analogous to that employed in evaluating the **general linear model** (defined in Section VII of the **single-factor between-subjects analysis of variance** and Endnote 2 in the chapter on **canonical correlation (Test 38)**). By doing the latter, the analysis becomes conceptually similar to what is done in a **factorial analysis of variance** — specifically, each of the categorical variables can be evaluated with respect to main effects and any interactions which may be present. The output generated by computer software for **log-linear analysis** documents the goodness-of-fit for each of a number of alternative models, and in the process identifies the model which best fits a set of data — i.e., the model which best accounts for the observed frequencies in a multidimensional table. Each model that is evaluated specifies the presence (or absence) of one or more main effects as well as the presence (or absence) of one or more interactions among the variables. The method to be described in this section for evaluating alternative models is referred to as **hierarchical log-linear analysis**. The latter type of analysis initially evaluates the most complex model (i.e., the model containing the highest-order interaction, all lower order interactions, and all main effects), and then continues to evaluate other models in descending order of complexity. This methodology is referred to as **backward elimination** (which is also discussed in Section I of the chapter on **multiple regression (Test 33))**.

Although the remainder of this discussion will focus on some basic guidelines for interpreting a **log-linear analysis** through use of *SPSS*, the reader should keep in mind that **log-linear analysis** is a complex topic which goes well beyond the scope of the material to be presented in this section. Those requiring more detailed coverage of **log-linear analysis** are referred to Agresti (1984, 1990), Christensen (1990, 1999), Everitt (1977, 1992), Field (2005), Fienberg (1980), Howell (2002), Kennedy (1983), Norušis (2004), Reynolds (1977a, 1977b), Selvin (1995), Tabachnick and Fidell (1996, 2001), and Wickens (1989).

At this point in the discussion the essentials of **log-linear analysis** will be demonstrated with respect to Examples 16.9 and 16.10. A distinction regarding multidimensional contingency tables which is relevant to the latter examples is that which is made between **symmetric** versus **asymmetric relationships**. When all of the variables are treated as dependent variables (or some sources simply state that no distinction is made between the independent and dependent variables), the relationship between the variables are viewed as **symmetric**, and in such a case the researcher will be interested in examining the relationships among all of the variables. This type of **symmetric** relationship is illustrated by Example 16.10 in which the row (introversion/extroversion), column (male/female), and layer (Democrat/Republican) variables all represent dependent variables, and consequently the researcher will be interested in examining all possible interactions involving any of the three variables.

On the other hand, when among three or more variables one is designated as the dependent variable and the others as independent variables, the relationships of interest will be those between the dependent variable and the independent variables. Such a relationship is said to be **asymmetric**, and in the case of such an analysis the researcher is not interested in examining the relationships among independent variables. An **asymmetric** relationship is illustrated by Example 16.9 in which the row (noise/no noise) and column (male/female) variables are independent variables, and the layer variable (help/did not help) represents a dependent variable. In the case of Example 16.9 the researcher will only be interested in examining the possible interactions the independent variables might have with the dependent variable.

Within the framework of **log-linear analysis** Reynolds (1977b, p. 57) defines a **model** as a hypothesis about distributions and interrelationships among the variables in a contingency table. Put simply, in **log-linear analysis** a model will be represented by a linear equation that describes the configuration of relationships between a set of variables. The goal in such an analysis will be to identify the simplest yet optimally predictive model with respect to the variables in question. In the case of a three-dimensional table, the methodology of **hierarchical log-linear analysis** will be employed to evaluate the models noted below in the order in which they are listed.

1) The model that will always serve as a baseline against which an initial alternative model will be compared is referred to as the **saturated** (or **full**) **model** — the latter being the model that fits the data perfectly, and consequently has no prediction error associated with it. More specifically, in a **saturated model** the expected frequencies in a contingency table are equal to the observed frequencies. The goal of **log-linear analysis** is to identify a simpler model (i.e., an **unsaturated model**) which when contrasted with the **saturated model** does not result in a significant decrease in predictive efficiency. Within the context of a three-dimensional contingency table the **saturated model** can be summarized with Equation 16.64. In the latter equation, the term $\ln(F_{rcl})$ is the sum of all the terms (commonly referred to as **effect parameters**) on the right side of the equation. The right side of the equation contains a **constant** represented by the notation θ (which is the lower case Greek letter **theta**). Some sources refer to θ as the **intercept** or as the **mean** (μ) since, in fact, it is the mean of the natural logarithms of all of the observed frequencies in a table. The remaining terms on the right side of the equation represent all possible effects that can be present in a three-dimensional table. Each effect is represented by the notation λ (which represents the lower case Greek letter **lambda**). Thus, λ_r, λ_c, and λ_l respectively represent a **main effect** (often referred to as a **first-order effect**) on the row, column, and layer variables; λ_{rc}, λ_{rl}, and λ_{cl} respectively represent the three possible **two-way interactions** (often referred to as **second-order effects**) which can be written as **rows × columns**, **rows × layers**, and **columns × layers**; and λ_{rcl} represents the **three-way interaction** (often referred to as a **third-order effect**) between the rows, columns, and layers which can be written as **rows × columns × layers**.

(Equation 16.64)

$$\ln(F_{rcl}) = \theta + \lambda_r + \lambda_c + \lambda_l + \lambda_{rc} + \lambda_{rl} + \lambda_{cl} + \lambda_{rcl}$$

2) In a three-variable **log-linear analysis**, the initial model contrasted with the **saturated model** is the model that omits the three-way interaction — specifically, the model described by Equation 16.65 which only contains the two-way interactions, the main effects, and the constant. This model can be further refined by removing one or two of the two-way interactions (and in the latter case one of the main effects).

(Equation 16.65)

$$\ln(F_{rcl}) = \theta + \lambda_r + \lambda_c + \lambda_l + \lambda_{rc} + \lambda_{rl} + \lambda_{cl}$$

3) The next model that can be evaluated is described by Equation 16.66 which involves only the main effects (λ_r, λ_c, and λ_l) and the constant. This model can be further refined by removing one or two of the main effects. In reality, in most instances, a model which only involves one or more main effects and the constant will not be of any practical value to a researcher.

(Equation 16.66)

$$\ln(F_{rcl}) = \theta + \lambda_r + \lambda_c + \lambda_l$$

4) The last model that can be evaluated is the **no-effects model** — in other words, the model described by Equation 16.67 which only includes the constant θ. In the case of this model, all three variables are independent of one another, and the row, column, and layers variables are equally probable (Reynolds (1977b, p. 59)).

(Equation 16.67)

$$\ln(F_{rcl}) = \theta$$

SPSS log-linear analysis of Examples 16.9/16.10 Output generated by *SPSS* for a log-linear analysis of Examples 16.9 and 16.10 is displayed in Table 16.39.[40] Readers should take

Table 16.39 *SPSS* Output for Linear-Analysis of Examples 16.9/16.10

```
* * * * * * * * H I E R A R C H I C A L   L O G   L I N E A R * * * * * * * *

DESIGN 1 has generating class

    ROWVAR*COLVAR*LAYERVAR

Note: For saturated models  .500 has been added to all observed cells.
This value may be changed by using the CRITERIA = DELTA subcommand.


The Iterative Proportional Fit algorithm converged at iteration 1.
The maximum difference between observed and fitted marginal totals is    .000
and the convergence criterion is    .250

- - - - - - - - - - - - - - - - - - - - - - - - - - - - - - - - - - - - - - -


Observed, Expected Frequencies and Residuals.

     Factor        Code       OBS count   EXP count   Residual   Std Resid

  ROWVAR          1
   COLVAR          1
    LAYERVAR        1          10.5        10.5        .00         .00
    LAYERVAR        2          25.5        25.5        .00         .00
   COLVAR          2
    LAYERVAR        1          15.5        15.5        .00         .00
    LAYERVAR        2          15.5        15.5        .00         .00

  ROWVAR          2
   COLVAR          1
    LAYERVAR        1          25.5        25.5        .00         .00
    LAYERVAR        2          20.5        20.5        .00         .00
   COLVAR          2
    LAYERVAR        1          45.5        45.5        .00         .00
    LAYERVAR        2           5.5         5.5        .00         .00

- - - - - - - - - - - - - - - - - - - - - - - - - - - - - - - - - - - - - - -


Goodness-of-fit test statistics

      Likelihood ratio chi square =      .00000   DF = 0   P = -INF
              Pearson chi square =      .00000   DF = 0   P = -INF

- - - - - - - - - - - - - - - - - - - - - - - - - - - - - - - - - - - - - - -


* * * * * * * * H I E R A R C H I C A L   L O G   L I N E A R * * * * * * * *

Tests that K-way and higher order effects are zero.

   K    DF    L.R. Chisq   Prob   Pearson Chisq    Prob    Iteration

   3     1       1.957    .1618       1.936       .1641        4
   2     4      38.994    .0000      37.206       .0000        2
   1     7      50.311    .0000      52.500       .0000        0

- - - - - - - - - - - - - - - - - - - - - - - - - - - - - - - - - - - - - - -


Tests that K-way effects are zero.

   K    DF    L.R. Chisq   Prob   Pearson Chisq    Prob    Iteration

   1     3      11.317    .0101      15.294       .0016        0
   2     3      37.037    .0000      35.270       .0000        0
   3     1       1.957    .1618       1.936       .1641        0
```

Table 16.39   *SPSS* Output for Linear-Analysis of Examples 16.9/16.10 (continued)

```
* * * * * * * *  H I E R A R C H I C A L   L O G   L I N E A R  * * * * * * * *

Backward Elimination (p = .050) for DESIGN 1 with generating class

     ROWVAR*COLVAR*LAYERVAR

Likelihood ratio chi square =        .00000   DF = 0  P =  -INF
- - - - - - - - - - - - - - - - - - - - - - - - - - - - - - - - - - - - - - - -


If Deleted Simple Effect is               DF   L.R. Chisq Change   Prob  Iter

  ROWVAR*COLVAR*LAYERVAR                    1               1.957   .1618   4

Step 1

   The best model has generating class

        ROWVAR*COLVAR
        ROWVAR*LAYERVAR
        COLVAR*LAYERVAR

   Likelihood ratio chi square =     1.95716   DF = 1  P =  .162
- - - - - - - - - - - - - - - - - - - - - - - - - - - - - - - - - - - - - - - -


If Deleted Simple Effect is               DF   L.R. Chisq Change   Prob  Iter

  ROWVAR*COLVAR                             1                .483   .4868   2
  ROWVAR*LAYERVAR                           1              19.864   .0000   2
  COLVAR*LAYERVAR                           1              16.360   .0001   2

Step 2

   The best model has generating class

        ROWVAR*LAYERVAR
        COLVAR*LAYERVAR

   Likelihood ratio chi square =     2.44065   DF = 2  P =  .295
- - - - - - - - - - - - - - - - - - - - - - - - - - - - - - - - - - - - - - - -


If Deleted Simple Effect is               DF   L.R. Chisq Change   Prob  Iter

  ROWVAR*LAYERVAR                           1              20.029   .0000   2
  COLVAR*LAYERVAR                           1              16.525   .0000   2


Step 3

   The best model has generating class

        ROWVAR*LAYERVAR
        COLVAR*LAYERVAR

   Likelihood ratio chi square =     2.44065   DF = 2  P =  .295
- - - - - - - - - - - - - - - - - - - - - - - - - - - - - - - - - - - - - - - -

   The final model has generating class

        ROWVAR*LAYERVAR
        COLVAR*LAYERVAR

   The Iterative Proportional Fit algorithm converged at iteration 0.
   The maximum difference between observed and fitted marginal totals is      .000
   and the convergence criterion is     .250

- - - - - - - - - - - - - - - - - - - - - - - - - - - - - - - - - - - - - - - -
```

Table 16.39   *SPSS* Output for Linear-Analysis of Examples 16.9/16.10 (continued)

* * * * * * * * H I E R A R C H I C A L   L O G   L I N E A R * * * * * * * *

Tests of PARTIAL associations.

Effect Name	DF	Partial Chisq	Prob	Iter
ROWVAR*COLVAR	1	.483	.4868	2
ROWVAR*LAYERVAR	1	19.864	.0000	2
COLVAR*LAYERVAR	1	16.360	.0001	2
ROWVAR	1	5.658	.0174	2
COLVAR	1	.000	1.0000	2
LAYERVAR	1	5.658	.0174	2

- -

Observed, Expected Frequencies and Residuals.

Factor	Code	OBS count	EXP count	Residual	Std Resid
ROWVAR	1				
COLVAR	1				
LAYERVAR	1	10.0	9.2	.79	.26
LAYERVAR	2	25.0	27.7	-2.69	-.51
COLVAR	2				
LAYERVAR	1	15.0	15.8	-.79	-.20
LAYERVAR	2	15.0	12.3	2.69	.77
ROWVAR	2				
COLVAR	1				
LAYERVAR	1	25.0	25.8	-.79	-.16
LAYERVAR	2	20.0	17.3	2.69	.65
COLVAR	2				
LAYERVAR	1	45.0	44.2	.79	.12
LAYERVAR	2	5.0	7.7	-2.69	-.97

- -

Goodness-of-fit test statistics

```
    Likelihood ratio chi square =    2.44065    DF = 2  P =  .295
              Pearson chi square =    2.35721    DF = 2  P =  .308
```

- -

note of the fact that only selected output for the analysis is presented in the latter table, and more detailed displays of output for a **log-linear analysis** can be found in some of the sources noted earlier.

a) The section of the Table 16.39 labeled **Observed, Expected Frequencies and Residuals** displays the observed and expected frequencies (which as noted earlier will be identical) for the **saturated model**. The observed frequencies in this table are identical to those in Table 16.29, except for the fact that .5 has been added to each frequency.[41]

b) The overall goodness-of-fit for the most appropriate model is determined on the basis of the similarity of the observed frequencies with the expected frequencies for a specific model. The goodness-of-fit of a model is assessed through use of the **likelihood ratio statistic** (also

Table 16.39 *SPSS* **Output for Linear-Analysis of Examples 16.9/16.10 (continued)**

Estimates for Parameters.

ROWVAR*COLVAR*LAYERVAR

Parameter	Coeff.	Std. Err.	Z-Value	Lower 95 CI	Upper 95 CI
1	.1259259341	.09315	1.35192	-.05664	.30849

ROWVAR*COLVAR

Parameter	Coeff.	Std. Err.	Z-Value	Lower 95 CI	Upper 95 CI
1	-.0785336573	.09315	-.84312	-.26110	.10403

ROWVAR*LAYERVAR

Parameter	Coeff.	Std. Err.	Z-Value	Lower 95 CI	Upper 95 CI
1	-.4023151243	.09315	-4.31918	-.58488	-.21975

COLVAR*LAYERVAR

Parameter	Coeff.	Std. Err.	Z-Value	Lower 95 CI	Upper 95 CI
1	-.3477517328	.09315	-3.73339	-.53032	-.16519

ROWVAR

Parameter	Coeff.	Std. Err.	Z-Value	Lower 95 CI	Upper 95 CI
1	-.0887287499	.09315	-.95257	-.27130	.09384

COLVAR

Parameter	Coeff.	Std. Err.	Z-Value	Lower 95 CI	Upper 95 CI
1	.1056270727	.09315	1.13399	-.07694	.28819

LAYERVAR

Parameter	Coeff.	Std. Err.	Z-Value	Lower 95 CI	Upper 95 CI
1	.1804893256	.09315	1.93770	-.00208	.36306

- -

referred to as a **likelihood ratio chi-square statistic**, which as noted earlier in Section VII is often referred to as G or G^2) that is evaluated with the chi-square distribution. The value of the **likelihood ratio statistic** (as well as the value of an alternative statistic, the **Pearson chi-square statistic**, which will essentially yield an equivalent result) for the **saturated model** will always equal 0 (since the ratio of the observed to expected frequencies for the latter model will always equal 1, and $\ln(1) = 0$). The result of the latter analysis is displayed in the middle of the first page of Table 16.39 in the section labeled **Goodness-of-fit test statistic**. Note the values **Likelihood ratio chi square** = 0 and **Pearson chi square** = 0. The value $P = -\mathbf{INF}$, which in some software may be written as $\mathbf{P = 1}$, indicates that the **saturated model** provides a **perfect fit** for the data.

c) As noted above, the goal of **log-linear analysis** is to identify an **unsaturated model** which provides the best fit for the data — in other words, a model that provides a fit which does not deviate significantly from the fit provided by the **saturated model**. In order to do the latter a **backward hierarchical analysis** is conducted beginning with the next most complex model, which in the case of three variables is a model that contains all of the two-way interactions and all of the main effects.[42] The result of the latter analysis is displayed at the bottom of the first page of Table 16.39 in the sections labeled **Tests that K-way and higher order effects are zero** and **Tests that K-way effects are zero**. Whereas the former table displays the value of the **likelihood ratio statistic** computed for each of the order effects (as well as any higher-order effects), the latter table displays the magnitude of change when a model containing specific order-effects is contrasted with a previously considered alternative model.

Each row of the table labeled **Tests that K-way and higher-order effects are zero** can be viewed as **evaluating whether the order effect identified by the number in a given row and all potential higher-order effects are equal to zero**. A nonsignificant probability for a given row indicates that the order effect identified by the number in that row, as well as any potential higher-order effects, should be omitted from the final model. Thus, the value 1.957 in the row labeled 3 in the column labeled **L. R. Chisq** represents the **likelihood ratio** obtained for the **third-order effect** (i.e., the three-way interaction). Since the probability .1618 in the column labeled **Prob** is less than .05, it indicates that the third-order interaction is not significant, and thus should not be included in the final model. The value **L. R. Chisq** = 38.994 in the row labeled 2 represents the **likelihood ratio** obtained for the **second-order effect** (i.e., the two-way interactions). The significant probability **Prob** = .0000 indicates there is a significant second-order effect, and thus at least one of the two-way interactions (and any main effects which are components of a second-order interaction) should be included in the final model. If the probability for the row labeled 2 was nonsignificant (i.e., **Prob** was greater than .05), it would indicate that both the second and third-order effects should be omitted from the final model. The value **L. R. Chisq** = 50.311 in the row labeled 1 represents the **likelihood ratio** obtained for the **first-order effect** (i.e., the main effects). The significant probability **Prob** = .0000 indicates there is a significant first-order effect, and thus at least of the one main effects should be included in the final model.[43]

Among others, Howell (2002, pp. 675–677) presents the following alternative way to interpret the information in the table labeled **Tests that K-way and higher-order effects are zero** (which yields the same conclusions as above). The question addressed in the row labeled 1 is whether a model which only includes the constant provides an adequate fit for the data. More specifically, the null hypothesis evaluated in that row is that all the main effects (λ_r, λ_c, λ_l), all the two-way interactions (λ_{rc}, λ_{rl}, λ_{cl}) and the three-way interaction (λ_{rcl}) equal zero. The fact that the value 50.311 computed for the **likelihood ratio** in the column labeled **L. R. Chisq** is significant (since .0000 in the column labeled **Prob** is less than .05) indicates our model should include at least one of the aforementioned effects. The question addressed in the row labeled 2 is whether a model which only includes one or more of the main effects and the constant provides an adequate fit for the data. More specifically, the null hypothesis evaluated in that row is that all the two-way interactions (λ_{rc}, λ_{rl}, λ_{cl}) and the three-way interaction (λ_{rcl}) equal zero. The fact that the value **L. R. Chisq** = 38.994 is significant (since **Prob** = .0000 is less than .05) indicates our model should include at least one of the aforementioned effects. The question addressed in the row labeled 3 is whether a model which only includes the constant, one or more of the main effects, and one or more of the two-way interactions provides an adequate fit for the data. More specifically, the null hypothesis evaluated in the row is that the three-way interaction (λ_{rcl}) equals zero. The fact that the value **L. R. Chisq** = 1.957 is not significant (since

Prob = .1618 is greater than .05) indicates the null hypothesis cannot be rejected, and thus it is not necessary for our model to include the three-way interaction.[44]

Each row of the table labeled **Tests that K-way effects are zero** can be viewed as **evaluating whether or not the order effect identified by the number listed in a given row is equal to zero**. For a given row, the **likelihood ratio** value in the column labeled **L. R. Chisq** indicates the change in the **likelihood ratio** if a higher-order effect is added to the model containing all lower-order effects. Thus, the value **L. R. Chisq** =11.317 in the row labeled **1** for main effects indicates that if the main effects are removed from the model leaving only the constant, the change in the likelihood ratio is 11.317. A significant probability value (i.e., .05 or less) in the column labeled **Prob** indicates that the order effect for the row in question is not equal to zero, and because of the latter it will be of value to include that order effect in the final model. Thus, the probability .0101 in the row labeled **1** indicates a significant first-order effect (i.e., one or more of the main effects is significant). The value **L. R. Chisq** =37.037 in the row labeled **2** for second-order effects indicates that if second-order effects are removed from the model leaving only main effects and the constant, the change in the likelihood ratio is 37.037. The significant probability .0000 in the column labeled **Prob** indicates that second-order effects are not equal to zero (i.e., one or more of the two-way interactions are significant), and because of the latter it will be of value to include second-order effects in the final model. The value **L. R. Chisq** =1.957 in the row labeled **3** for a third-order effect indicates that if the third-order effect is removed from the model leaving only second-order effects, main effects and the constant, the change in the likelihood ratio is 1.957. The nonsignificant probability .1618 in the column labeled **Prob** indicates that the third-order effect is equal to zero (or to put it another way, we cannot reject the null hypothesis that the third-order effect is equal to zero). Because of the latter, the third-order interaction will not be included in the final model. Note that these results are consistent with those obtained for the table labeled **Tests that K-way and higher-order effects are zero**.

The reader should take note of the following with respect to values of **L. R. Chisq** in the table labeled **Tests that K-way and higher-order effects are zero** versus the value of **L. R. Chisq** in the table labeled **Tests that K-way effects are zero**. Whereas the table labeled **Tests that K-way and higher-order effects are zero** displays the value of the **likelihood ratio** computed for each order effect as well as potential higher-order effects, the table labeled **Tests that K-way effects are zero** displays the **change** in the **likelihood ratio** for a given model relative to the previous model tested. With respect to the latter, one can easily determine whether a new model has changed the **likelihood ratio** by subtracting the **likelihood ratio** computed for the previous model from the **likelihood ratio** computed for a new model (the same use of subtraction can also be employed with respect to computing the degrees of freedom associated with each model). Thus, if in the table labeled **Tests that K-way and higher-order effects are zero** we subtract the **likelihood ratio** of 1.957 obtained for the model with a third-order effect (in the row labeled **3**) from 38.994 (the **likelihood ratio** obtained for the model with second-order effects in the row labeled **2**), we obtain $38.994 - 1.957 = 37.037$, which is the value in the row labeled **2** in the column labeled **L.R. Chisq** in the table labeled **Tests that K-way effects are zero** (note also that $[(df = 4) - (df = 1)] = (df = 3)$). The difference 37.037 is a measure of how much the **likelihood ratio** increases when second-order effects are included, and the value .0000 (which is less than .05) in the column labeled **Prob** (in the table labeled **Tests that K-way effects are zero**) indicates that the researcher can reject the null hypothesis of second-order effects being equal to zero. Or to put it another way, if second-order effects are removed it will have a detrimental effect on the final model.

In the same respect, if in the table labeled **Tests that K-way and higher-order effects are zero** we subtract the **likelihood ratio** of 38.994 obtained for the model with second-order effects (in the row labeled 2) from 50.311 (the **likelihood ratio** obtained for the model with first-order effects in the row labeled 1), we obtain 50.311 − 38.994 = 11.317, which is the value in the row labeled 1 in the column labeled **L.R. Chisq** in the table labeled **Tests that K-way effects are zero** (note also that $[(df = 7) - (df = 4)] = (df = 3)$). The difference 11.317 is a measure of how much the **likelihood ratio** increases when first-order effects are included, and the value .0101 (which is less than .05) in the column labeled **Prob** (in the table labeled **Tests that K-way effects are zero**) indicates that the researcher can reject the null hypothesis that first-order effects equal zero. Or to put it another way, if first-order effects are removed it will have a detrimental effect on the final model.

When there are three variables, the **L. R. Chisq** value in the row labeled **3** in the table labeled **Tests that K-way effects are zero** will always equal the **L. R. Chisq** value in the row labeled **3** in the table labeled **Tests that K-way and higher-order effects are zero**, since zero (the value of the **likelihood ratio** for the **saturated model**) will have been subtracted from the **L. R. Chisq** value in the row labeled 3 in the latter table (i.e., 1.957 − 0 = 1.957). The difference 1.957 is a measure of how much the **likelihood ratio** increases when the third-order effect is included, and the value .1618 (which is greater than .05) in the column labeled **Prob** (in the table labeled **Tests that K-way effects are zero**) indicates the researcher cannot reject the null hypothesis that the **third-order effect** equals zero. Or to put it another way, if the third-order effect is not included it will not have a detrimental effect on the final model.

In summary, the results from the tables labeled **Tests that K-way and higher-order effects are zero** and **Tests that K-way effects are zero** indicate that at least one main effect and one second-order effect should be included in the final model.[45] As will be noted later, when any higher-order effect (in this case a second-order effect) is retained, any lower-order effect which is a component of a retained higher-order effect must also be employed in the final model. Consequently, our final model (which will not be significantly inferior to the **saturated model**) will include one or more second-order effects and any main effects which are a component of any retained second-order effects.

The second page of Table 16.39 documents the final results of the **backward elimination procedure**. First take note of **ROWVAR*LAYERVAR** and **COLVAR*LAYERVAR** displayed at the bottom of the page under the heading **The final model has generating class**. The latter indicates that the **rows × layers** and **columns × layers** interactions are significant. The probabilities associated with the latter interactions are respectively .0000 and .0000, which are printed in the section just above **Step 3**. The fact that the value 20.029 recorded for **L. R. Chisq Change** for **ROWVAR*LAYERVAR** is slightly larger than the value 16.525 for **COLVAR* LAYERVAR** suggests that the **rows × layers** interaction is slightly more important than the **columns × layers** interaction. The latter is also suggested on the third page of Table 16.39 in the section labeled **Tests of PARTIAL associations** which contains what are referred to as **partial likelihood ratios** (which are measures of the unique contribution of each effect). The latter table breaks the model into all of its components — specifically, the three two-way interactions and the three main effects. By examining the **likelihood ratio** (in the column labeled **Partial Chisq**) computed for each component we can once again determine which of the two-way interactions is significant. Note that for **ROWVAR*LAYERVAR** and **COLVAR* LAYERVAR** the probabilities .0000 (for **Partial Chisq** = 19.864) and .0001 (for the **Partial Chisq** = 16.360) are significant, indicating that the latter two-way interactions are significant. On the other hand, the probability .4868 (for **Partial Chisq** = .483) for **ROWVAR*COLVAR** is not significant, and thus indicates the **row × column** interaction is not significant.[46]

With respect to Example 16.9 the latter interactions are the same as noted in the previous section (when the three-way table was evaluated with the **chi-square test for r × c tables**) with regard to the relationship between the row variable noise/no noise with the layer variable of helping behavior, and the relationship between the column variable of gender of the confederate with the layer variable of helping behavior. In the case of Example 16.10 the above noted interactions are between the row variable introversion/extroversion with the layer variable Democrat/ Republican and the column variable male/female with the layer variable Democrat/ Republican. The latter interactions can be scrutinized more closely by breaking the original three-dimensional table into smaller two-dimensional tables comprised of the two variables involved in a specific interaction.

Further examination of the **Tests of PARTIAL associations** section of Table 16.39 reveals that the **row** and **layer main effects** are significant, since **Prob** = .0174 (for **Partial Chisq** = 5.658) for **ROWVAR** and **Prob** = .0174 (for **Partial Chisq** = 5.658) for **LAYERVAR**. The latter main effects are probably not of interest to the researcher since all they indicate is that: a) With respect to the row variable in the case of Example 16.9, significantly more subjects are in the no noise category (95) than the noise category (65), and in the case of Example 16.10 there are significantly more extroverts (95) than introverts (65); b) With respect to the layer variable in the case of Example 16.9, significantly more subjects helped the confederate (95) than did not help (65), and in the case of Example 16.10 there are significantly more Democrats (95) than Republicans (65). Note that the main effect for the column variable is not significant since **Prob** = 1 (for **Partial Chisq** = 0). The latter is consistent with the fact that in both Examples 16.9 and 16.10 there are an equal number of subjects in the male (80) versus female (80) categories. It is important to note, however, that even if a main effect is nonsignificant it must be retained in the final model if it is a component of a significant higher-order interaction that will be retained in the model. Thus, the main effect for the column variable will be included in the final model, since it is part of the **rows × layers** and **columns × layers** interactions. As noted in Endnote 45, we can state that latter two interactions represent the **generating class** for the final model, since the elements in the equation for the final model will be determined by those interactions.

The final model derived from the **log-linear analysis** can be summarized with Equation 16.68. Note that along with the constant, the model includes the significant **rows × layers** and **columns × layers** interactions as well as all three main effects, since, as noted above, the row, column, and layer variables are included in one or more of the aforementioned interactions.[47]

(Equation 16.68)

$$\ln{(F_{rcl})} = \theta + \lambda_r + \lambda_c + \lambda_l + \lambda_{rl} + \lambda_{cl}$$

The table at the bottom of the third page of Table 16.39 displays the observed frequencies for the data along with the expected frequencies computed for the final model (along with additional information regarding goodness-of-fit for the model). Note that the probability **P** = .295 associated with **Likelihood ratio chi-square** = 2.44065 at the bottom of the page is nonsignificant. The latter indicates that the final model provides a **good fit** for the data — in other words, it indicates there is not a significant discrepancy between the observed frequencies and the expected frequencies computed for the model. A significant result would indicate significant deviations between expected frequencies computed for the model versus observed frequencies — an indicator of a less than adequate model.

The same conclusions regarding the final model are also obtained by examining the values computed for the **effect parameters** (which represent values of λ in Equation 16.68). The values computed for the parameters (which represent estimates of the underlying population parameters) for Example 16.9/16.10 (along with their standard errors and confidence intervals) are displayed in the column labeled **Coeff.** on the last page of Table 16.39. Since parameters are standardized they can be evaluated as z scores, allowing for the null hypothesis $H_0: \lambda = 0$ to be evaluated for each parameter. Rejection of a null hypothesis that the value of a parameter is equal to zero is associated with a significant effect. The z value computed for each parameter is obtained by dividing the value of a parameter by its standard error. Inspection of the z-**Values** in Column 4 reveals that the parameters for the **ROWVAR*LAYERVAR** and **COLVAR*LAYERVAR** interactions are significant. The latter is the case since the obtained absolute values $z = 4.32$ and $z = 3.73$ are greater than the tabled critical two-tailed .05 and .01 values $z_{.05} = 1.96$ and $z_{.01} = 2.58$ (although it should be noted that as a result of employing multiple tests of significance, in order to avoid an inflated Type I error rate use of a more conservative alpha value is often recommended in evaluating parameters). Note that use of the parameters once again leads to the conclusion that the **rows × layers** and **columns × layers** interactions represent the **generating class** for the final model.

Norušis (2004, pp. 23–24) notes that a researcher should only view the final model as a "suggested" model, in that there may be alternative models which provide an equivalent fit for the observed data and may be easier to interpret. Any decision made with respect to a satisfactory model should ultimately be based on multiple studies evaluating data from independent samples derived from the same population. It should also be emphasized that a **log-linear analysis** becomes increasingly difficult to interpret as the number of dimensions in a multidimensional table increase.

As noted earlier, the information which can be displayed by computer software for a **log-linear analysis** can go well beyond the output which has been presented in this section. For example, among the measures that are available which can provide additional useful information regarding the relationships among the variables are: a) **Residuals** computed by *SPSS* and other software can be employed to evaluate goodness-of-fit for each of the cells in the final model. Through use of residuals a researcher can identify specific cells which do not provide a good fit for the derived model, and possibly indicate that the model is unacceptable; c) **Measures of marginal association** are measures of effects based on marginal totals, collapsing across the other variables; and d) **Odds ratios**, which can be employed to measure effect size.

Tables 16.40–16.45 present some additional hypothetical data for three-way contingency tables — specifically, tables involving **no-effects** versus a **first-order effect** versus a **third-order effect**. Tables 16.41, 16.43 and 16.45 display the *SPSS* output for the tables labeled **Tests that K-way and higher order effects are zero** and **Tests that K-way effects are zero**.

Table 16.40 displays a hypothetical set of data which conforms to a model containing **no-effects**, and Table 16.41 displays the *SPSS* output for the tables labeled **Tests that K-way and higher order effects are zero** and **Tests that K-way effects are zero**. Note that in the former table all of the probabilities (in the rows labeled **3**, **2**, and **1**) are above .05 (specifically, they all equal 1, indicating that the model containing **no-effects** is not significantly less predictive than the **saturated model**).

Table 16.42 displays a hypothetical set of data which conforms to a model containing **first-order effects** (specifically, a main effect on the column variable), and Table 16.43 displays the *SPSS* output for the tables labeled **Tests that K-way and higher order effects are zero** and **Tests that K-way effects are zero**. Note that in the former table the probabilities for the **second** and **third-order effects** (in the rows labeled **2** and **3**) are above .05 (specifically, they both equal 1, indicating that the model containing **first-order effects** is not significantly less predictive than the **saturated model**). Note, however, that since in the case of Example 16.10, the **first-order effect** in Table 16.42 only indicates that all of the subjects are female, it would in all likelihood be of no interest to the researcher (except, of course, if he or she was unaware of this characteristic with respect to the sample of 160 subjects). In the case of Example 16.9, a researcher would have control over the number of male and female confederates employed in the study, and would thus be aware of the fact that all the confederates were females.

Table 16.44 displays a hypothetical set of data which conforms to a model containing a **third-order effect/three-way interaction**. A **three-way interaction** means that the relationship between any two variables changes with changes in the level of the third variable. Table 16.45 displays the *SPSS* output for the tables labeled **Tests that K-way and higher order effects are zero** and **Tests that K-way effects are zero**. Note that in the former table the probabilities for the **third**, **second** and **first-order effects** (in the rows labeled **3, 2** and **1**) are all less than .05, indicating that the model containing a **third-order effect** should be retained, since removing any of the lower-order effects will have a detrimental effect on the final model. One should also take note of the fact that when a **three-way interaction** is present, it takes precedence over any lower-order effects, since the latter must viewed within the context of the **three-way interaction**.

Table 16.40 Summary Data for No Effects

| | Helped/Democrat | | Did not help/Republican | | |
	Male	Female	Male	Female	Totals
Noise/Introvert	20	20	20	20	80
No Noise/Extrovert	20	20	20	20	80
Totals	40	40	40	40	160

Table 16.41 Partial *SPSS* Output for a Model Containing No-Effects

```
* * * * * * * *  H I E R A R C H I C A L   L O G   L I N E A R  * * * * * * * *

Tests that K-way and higher order effects are zero.

    K    DF   L.R. Chisq    Prob   Pearson Chisq    Prob    Iteration

    3    1        .000    1.0000           .000   1.0000        1
    2    4        .000    1.0000           .000   1.0000        1
    1    7        .000    1.0000           .000   1.0000        0

- - - - - - - - - - - - - - - - - - - - - - - - - - - - - - - - - - - - - - - -

Tests that K-way effects are zero.

    K    DF   L.R. Chisq    Prob   Pearson Chisq    Prob    Iteration

    1    3        .000    1.0000           .000   1.0000        0
    2    3        .000    1.0000           .000   1.0000        0
    3    1        .000    1.0000           .000   1.0000        0
```

Table 16.42 Summary Data for Main Effect on Column Variable

| | Helped/Democrat | | Did not help/Republican | | |
	Male	Female	Male	Female	Totals
Noise/Introvert	0	40	0	40	80
No Noise/Extrovert	0	40	0	40	80
Totals	0	80	0	80	160

Table 16.43 Partial *SPSS* Output for a Model Containing a First-Order Effect

```
* * * * * * * *  H I E R A R C H I C A L   L O G   L I N E A R  * * * * * * * *

Tests that K-way and higher order effects are zero.

     K   DF   L.R. Chisq    Prob   Pearson Chisq   Prob   Iteration

     3    1      .000     1.0000        .000     1.0000       2
     2    4      .000     1.0000        .000     1.0000       2
     1    7    221.807     .0000     160.000      .0000       0

- - - - - - - - - - - - - - - - - - - - - - - - - - - - - - - - - - - - - - - -

Tests that K-way effects are zero.

     K   DF   L.R. Chisq    Prob   Pearson Chisq   Prob   Iteration

     1    3    221.807     .0000     160.000      .0000       0
     2    3      .000     1.0000        .000     1.0000       0
     3    1      .000     1.0000        .000     1.0000       0
```

Table 16.44 Summary Data for Three-Way Interaction

	Helped/Democrat		Did not help/Republican		
	Male	**Female**	**Male**	**Female**	**Totals**
Noise/Introvert	10	5	10	15	40
No Noise/Extrovert	5	65	20	30	120
Totals	15	70	30	45	160

Table 16.45 Partial *SPSS* Output for a Model Containing a Third-Order Effect

```
* * * * * * * *  H I E R A R C H I C A L   L O G   L I N E A R  * * * * * * * *

Tests that K-way and higher order effects are zero.

     K   DF   L.R. Chisq    Prob   Pearson Chisq   Prob   Iteration

     3    1     15.212     .0001      15.311      .0001       4
     2    4     39.300     .0000      35.175      .0000       2
     1    7    113.471     .0000     140.000      .0000       0

- - - - - - - - - - - - - - - - - - - - - - - - - - - - - - - - - - - - - - - -

Tests that K-way effects are zero.

     K   DF   L.R. Chisq    Prob   Pearson Chisq   Prob   Iteration

     1    3     74.171     .0000     104.825      .0000       0
     2    3     24.088     .0000      19.864      .0002       0
     3    1     15.212     .0001      15.311      .0001       0
```

VIII. Additional Examples Illustrating the Use of the Chi-Square Test for $r \times c$ Tables

Examples 16.11–16.14 are additional examples which can be evaluated with the **chi-square test for $r \times c$ tables**.

Example 16.11 *A researcher conducts a study to evaluate the relative problem-solving ability of male versus female adolescents. One hundred males and 80 females are randomly selected from a population of adolescents. Each subject is given a mechanical puzzle to solve. The dependent variable is whether or not a person is able to solve the puzzle. Sixty out of the 100 male subjects are able to solve the puzzle, while only 30 out of the 80 female subjects are able to solve the puzzle. Is there a significant difference between males and females with respect to their ability to solve the puzzle?*

Table 16.46 summarizes the data for Example 16.11. Example 16.11 conforms to the requirements of the **chi-square test for homogeneity**. This is the case, since there are two independent samples/groups (**males** versus **females**) which are dichotomized with respect to the following two categories on the dimension of problem-solving ability: **solved puzzle** versus **did not solve puzzle**. The grouping of subjects on the basis of gender represents a nonmanipulated independent variable, while the problem-solving performance of subjects represents the dependent variable. Note that the number of people for each gender is predetermined by the experimenter. Also note that it is not necessary to have an equal number of observations in the categories of the row variable, which represents the independent variable. Since the independent variable is nonmanipulated, if the chi-square analysis is significant it will only allow the researcher to conclude that a significant association exists between gender and one's ability to solve the puzzle. The researcher cannot conclude that gender is the direct cause of any observed differences in problem-solving ability between males and females. Employing Equation 16.2, the obtained chi-square value for Table 16.46 is $\chi^2 = 9$, which for $df = 1$ is greater than $\chi^2_{.05} = 3.84$ and $\chi^2_{.01} = 6.63$. Thus, the null hypothesis can be rejected at both the .05 and .01 levels. Inspection of Table 16.46 reveals that a larger proportion of **males** is able to solve the puzzle than **females**.

Table 16.46 Summary of Data for Example 16.11

	Solved puzzle	Did not solve puzzle		Row sums
Males	60	40		100
Females	30	50		80
Column sums	90	90	Total observations	180

Example 16.12 *A pollster conducts a survey to evaluate whether or not Caucasians and African-Americans differ in their attitude toward gun control. Five hundred people are randomly selected from a telephone directory and called at 8 P.M. in the evening. An interview is conducted with each individual, at which time a person is categorized with respect to both race and whether one supports or opposes gun control. Table 16.47 summarizes the results of the survey. Is there evidence of racial differences with respect to attitude toward gun control?*

This example conforms to the requirements of the **chi-square test of independence**, since a single sample is categorized on two dimensions. The two dimensions with respect to which subjects are categorized are race, for which there are the two categories **Caucasian** versus **African-American**, and attitude toward gun control, for which there are the two categories **supports gun control** versus **opposes gun control**. Note that neither the number of **Caucasians** or **African-Americans** (i.e., the sums of the rows), nor the people who **support gun control** or **oppose gun control** (i.e., the sums of the columns) are predetermined prior to the pollster conducting the survey. The pollster selects a single sample of 500 subjects and categorizes them on both dimensions after the data are collected. The obtained chi-square value for Table 16.47 is $\chi^2 = 59.91$, which for $df = 1$ is greater than $\chi^2_{.05} = 3.84$ and $\chi^2_{.01} = 6.63$. Thus, the null hypothesis can be rejected at both the .05 and .01 levels. Inspection of Table 16.47 reveals that a larger proportion of **Caucasians opposes gun control**, whereas a larger proportion of **African-Americans supports gun control**.

Table 16.47 Summary of Data for Example 16.12

	Supports gun control	Opposes gun control	Row sums
Caucasians	120	170	290
Afro-Americans	160	50	210
Column sums	280	220	Total observations 500

Example 16.13 *A researcher conducts a study on a college campus to examine the relationship between a student's class standing and the number of times a student visits a physician during the school year. Table 16.48 summarizes the responses of a random sample of 280 students employed in the study. Do the data indicate that the number of visits a student makes to a physician is independent of his or her class standing?*

Table 16.48 Summary of Data for Example 16.13

	0 visits	1–5 visits	More than 5 visits	Row sums
Freshman	20	16	24	60
Sophomore	30	10	10	50
Junior	50	30	10	90
Senior	19	11	50	80
Column sums	119	67	94	Total observations 280

This example conforms to the requirements of the **chi-square test of independence**, since a single sample is categorized on two dimensions. The two dimensions with respect to which subjects are categorized are class standing, for which there are the four categories: **Freshman, Sophomore, Junior, Senior**, and the number of visits to a physician, for which there are the three categories: **0 visits, 1–5 visits, more than 5 visits**. Note that the sums of neither the rows or columns are predetermined by the researcher. The researcher randomly selects 280 subjects and categorizes each subject on both dimensions after the data are collected. Since the data for Example 16.13 are identical to those employed in Example 16.5, they yield the same result. The null hypothesis can be rejected, since the obtained value $\chi^2 = 59.16$ is significant at both the .05 and .01 levels. Thus, the researcher can conclude that a student's class standing and the number of visits one makes to a physician are not independent of one another (i.e., the two dimensions seem to be associated/correlated with one another). As is the case with Example 16.5, a more detailed analysis of the data can be conducted through use of the comparison procedures described in Section VI.

Example 16.14 *A researcher conducts a study to evaluate whether or not a nationally acclaimed astrologer is able to match subjects with their correct sun sign. The astrologer and researcher agree to employ a format in which the astrologer views a five minute videotape of a subject who verbally responds to five open ended questions of a personal nature. Upon viewing the videotape, the astrologer indicates which of the 12 sun signs he believes the person's birth data fall within. Over a three-month period the astrologer views videotapes for 718 subjects. Table 16.49 summarizes the results of the study. Since each column of the table corresponds to a subject's actual sun sign, and each row the sun sign selected by the astrologer, all correct responses by the astrologer appear in the 12 cells that constitute the diagonal of the table (i.e., any cell in which the row and column labels are identical). Does the performance of the astrologer indicate that he can reliably identify a person's sun sign?*

Table 16.49 Astrology Study Data

Selected sun sign	Actual Sun Sign												Column sums
	Aquarius	Pisces	Aries	Taurus	Gemini	Cancer	Leo	Virgo	Libra	Scorpio	Sagittarius	Capricorn	
Aquarius	6	2	4	8	5	3	9	7	2	8	6	4	64
Pisces	3	2	0	5	7	1	9	9	4	2	8	7	57
Aries	0	9	3	1	7	6	2	1	9	8	9	4	59
Taurus	9	3	4	3	4	8	1	8	5	6	2	6	59
Gemini	8	8	2	6	3	9	4	6	4	1	3	9	63
Cancer	6	6	4	7	7	7	2	4	6	3	8	1	61
Leo	7	7	1	4	9	9	3	9	2	1	4	6	62
Virgo	2	4	6	8	8	7	1	0	3	5	1	9	54
Libra	8	8	8	8	3	6	2	9	0	6	2	4	64
Scorpio	5	1	9	0	0	5	7	9	9	0	6	6	57
Sagittarius	0	4	7	7	8	3	6	1	9	2	9	0	56
Capricorn	7	0	5	9	5	2	0	7	9	8	6	4	62
Row sums	61	54	53	66	66	66	46	70	62	50	64	60	718

Table 16.49 summarizes the results of the study in the format of a 12 × 12 contingency table. Since the table is based on a single sample of subjects who are categorized two times (i.e., each subject is categorized by the astrologer with regard to sun sign and categorized on the basis of one's actual birth date), the **chi-square test of independence** can be employed to evaluate the body of data contained within the whole table. However, the latter analysis will really not address the question of primary interest with any degree of precision. The simplest and most straightforward analysis will be to determine whether or not the number of correct responses by the astrologer is significantly above chance expectation. Thus, instead of evaluating the data with the **chi-square test of independence**, the **binomial sign test for a single sample** will be employed to evaluate the number of correct responses in the diagonal of the table.

Equation 9.7 will be employed to evaluate the data.[48] If the astrologer is just guessing a person's sun sign, he has a one in 12 chance of being correct, and an 11 in 12 chance of being incorrect.[49] Thus, the expected number of correct responses will be the total number of responses (which corresponds to the total number of subjects/observations) multiplied by 1/12. Employing Equation 9.1, the expected number of correct responses is computed to be $\mu = n\pi_1 = (718)(1/12) = 59.83$. Equation 9.2 is employed to compute the standard deviation of the binomially distributed variable: $\sigma = \sqrt{n\pi_1\pi_2} = \sqrt{(718)(1/12)(11/12)} = 7.41$. Since the sum of the 12 values in the diagonal of the Table 16.49 equals 40, the value $x = 40$ is employed to represent the number of correct responses in Equation 9.7. Substituting the appropriate values in the latter equation, the value $z = -2.68$ is computed. The negative z value is consistent with the fact that the number of correct responses is below chance expectation. Certainly the latter, in and of itself, invalidates the astrologer's claim that he can reliably identify a person's sun sign.

$$z = \frac{x - n\pi_1}{\sqrt{n\pi_1\pi_2}} = \frac{40 - (718)(1/12)}{\sqrt{(718)(1/12)(11/12)}} = \frac{40 - 59.83}{7.41} = -2.68$$

The obtained value $z = -2.68$ is evaluated with **Table A1** in the **Appendix**. In **Table A1** the tabled critical two-tailed .05 and .01 values are $z_{.05} = 1.96$ and $z_{.01} = 2.58$, and the tabled critical one-tailed .05 and .01 values are $z_{.05} = 1.65$ and $z_{.01} = 2.33$. Since the absolute value $z = 2.68$ is larger than all of the aforementioned critical values, both the nondirectional alternative hypothesis and the directional alternative hypothesis which is consistent with the data are supported. Put simply, the researcher will conclude that the astrologer's performance is significantly below chance.

Given the fact that Table 16.49 is a 12 × 12 contingency table, there are numerous other analyses which can be conducted on the data. For instance, the researcher can examine the accuracy of the astrologer's responses within each of the sun signs. As an example, with respect to the sun sign Sagittarius, the astrologer correctly identified 9 of the 64 subjects who are, in fact, a Sagittarius. Since the chance probability of a correct response within a given sun sign is also 1/12, it turns out that a score of 9 is significantly above chance (if a one-tailed analysis is conducted). On the other hand, the astrologer's scores of 0 for Virgo, Libra, and Scorpio are all significantly below chance. The point to be made here is that if one sifts through a large body of data, just by chance some results will be significant, and some of the significant results will be in the direction toward which a researcher is biased. Thus, if there is reason to believe that a specific element of the data will be significant, it should be specified beforehand. If the latter then turns out to be significant, that is quite different from finding the same significant difference after the fact.

References

Agresti, A (1984). **Analysis of ordinal categorical data**. New York: John Wiley & Sons

Agresti, A (1990). **Categorical data analysis**. New York: John Wiley & Sons

Atherton Skaff, P. J. & Sloan, J. A. (2004). Design and analysis of equivalence clinical trials via the SAS system. Website: **http://www2.sas.com/ proceedings/sugi23/Stats/p218. pdf.**

Beyer, W. H. (Ed.) (1968). **CRC handbook of tables for probability and statistics** (2nd ed.). Boca Raton, FL: CRC Press.

Bresnahan, J. L. & Shapiro, M. M. (1966). A general equation and technique for the exact partitioning of chi-square contingency tables. **Psychological Bulletin**, 66, 252–262.

Carroll, J. B. (1961). The nature of data, or how to choose a correlation coefficient. **Psychometrika**, 26, 347–372.

Castellan, N. J., Jr. (1965). On the partitioning of contingency tables. **Psychological Bulletin**, 64, 330–338.

Chen, J. J., Tsong, Y., & Kang, S. H. (2000). Tests for equivalence or noninferiority between two proportions. **Drug Information Journal**, 34, 569–578.

Christensen, R. (1990). **Log-linear models**. New York: Springer–Verlag.

Christensen, R. (1999). **Log-linear models** (2nd ed.). New York: Springer–Verlag.

Cochran, W. G. (1952). The chi-square goodness-of-fit test. **Annals of Mathematical Statistics**, 23, 315–345.

Cochran, W. G. (1954). Some methods for strengthening the common chi-square tests. **Biometrics**, 10, 417–451.

Cohen, J. (1960). A coefficient of agreement for nominal scales. **Educational and Psychological Measurement**, 10, 37–46.

Cohen, J. (1977). **Statistical power analysis for the behavioral sciences**. New York: Academic Press.

Cohen, J. (1988). **Statistical power analysis for the behavioral sciences** (2nd ed.). Hillsdale, NJ: Lawrence Erlbaum Associates, Publishers.

Conover, W. J. (1980). **Practical nonparametric statistics** (2nd ed.). New York: John Wiley & Sons.

Conover, W. J. (1999). **Practical nonparametric statistics** (3rd ed.). New York: John Wiley & Sons.

Cornfield, J. (1951). A method of estimating comparative rates from clinical data. Applications to cancer of the lung, breast and cervix. **Journal of the National Cancer Institute**, 11, 1229– 1275.

Cramér, H. (1946). **Mathematical models of statistics**. Princeton, NJ: Princeton University Press.

Daniel, W. W. (1990). **Applied nonparametric statistics** (2nd ed.). Boston: PWS–Kent Publishing Company.

Everitt, B. S. (1977). **The analysis of contingency tables**. New York: Chapman & Hall.

Everitt, B. S. (1992). **The analysis of contingency tables** (2nd ed.). New York: Chapman & Hall.

Everitt, B. S. (2001). **Statistics for psychologists: An intermediate course**. Mahwah, NJ: Lawrence Erlbaum Associates.

Feinstein, A.R. (2002). **Principles of medical statistics**. Boca Raton, FL: Chapman & Hall/CRC.

Field, A. (2005). **Discovering statistics using SPSS** (2nd ed.). Sage Publications: London.

Fienberg, S. E. (1980). **The analysis of cross-classified categorical data** (2nd ed.). Cambridge, MA: MIT Press.

Fisher, R. A. (1934). **Statistical methods for research workers** (5th ed.). Edinburgh: Oliver & Boyd.

Fisher, R. A. (1935). The logic of inductive inference. **Journal of the Royal Statistical Society, Series A**, 98, 39–54.

Fleiss, J. L. (1981). **Statistical methods for rates and proportions** (2nd ed.). New York: John Wiley & Sons.

Fleiss, J. L., Levin, B. & Paik, M. C. (2003). **Statistical methods for rates and proportions** (3rd ed.). New York: John Wiley & Sons.

Garson, D. G. (2006). Statistics: Topics in multivariate analysis. Website: **http://www2.chas.ncsu.edu/garson/pa765/statnote.htm.**

Goldstein, H. & Healy, M. J. R. (1995). The graphical presentation of a collection of means. **Journal of the Royal Statistical Society**, 158A, Part 1, 175–177.

Goodman, L. A. (1970). The multivariate analysis of qualitative data. **Journal of the American Statistical Association**, 65, 226–256.

Grissom, R. J. & Kim, J. J. (2005). **Effect sizes for research: A broad practical approach**. Mahwah, NJ: Lawrence Erlbaum Associates Publishers.

Guilford, J. P. (1965). **Fundamental statistics in psychology and education** (4th ed.). New York: McGraw–Hill Book Company.

Haber, M. (1980). A comparison of some continuity corrections for the chi-squared test on 2×2 tables. **Journal of the American Statistical Association**, 75, 510–515.

Haber, M. (1982). The continuity correction and statistical testing. **International Statistical Review**, 50, 135–144.

Haberman, S. J. (1973). The analysis of residuals in cross-classified tables. **Biometrics**, 29, 205–220.

Haberman, S. J. (1974). **The analysis of frequency data.** Chicago: University of Chicago Press.

Hollander, M. & Wolfe, D. A. (1999). **Nonparametric statistical methods.** New York: John Wiley & Sons.

Howell, D. C. (1992). **Statistical methods for psychology** (3rd ed.). Boston: PWS–Kent Publishing Company.

Howell, D. C. (2002). **Statistical methods for psychology** (5th ed.). Pacific Grove, CA.: Duxbury.

Irwin, J. O. (1935). Tests of significance for differences between percentages based on small numbers. **Metron**, 12, 83–94.

Kennedy, J. J. (1983). **Analyzing quantitative data: Introductory loglinear analysis for behavioral research.** New York: Prager.

Keppel, G. & Saufley, W. H., Jr. (1980). **Introduction to design and analysis: A student's handbook.** San Francisco: W. H. Freeman & Company.

Keppel, G., Saufley, W. H. Jr., & Tokunaga, H. (1992). **Introduction to design and analysis: A student's handbook** (2nd ed.). New York: W. H. Freeman & Company.

Kline, R. B. (2004). **Beyond significance testing: Reforming data analysis methods in behavioral research.** Washington, D. C.: American Psychological Association.

Mantel, N. & Haenszel, W. (1959). Statistical aspects of the analysis of data from retrospective studies of disease. **Journal of the National Cancer Institute**, 22, 719–748.

Marascuilo, L. A. & McSweeney, M. (1977). **Nonparametric and distribution-free methods for the social sciences.** Belmont, CA: Brooks/Cole Publishing Company.

Marascuilo, L. A. & Serlin, R. C. (1988). **Statistical methods for the social and behavioral sciences.** New York: W.H. Freeman & Company.

Norušis, N. J. (2004). **SPSS 13.0 advanced statistical procedures companion.** Upper Saddle River, NJ: Prentice Hall.

Ott, R. L., Larson, R., Rexroat, C., and Mendenhall, W. (1992). **Statistics: A tool for the social sciences** (5th ed.). Boston: PWS–Kent Publishing Company.

Owen, D. B. (1962). **Handbook of statistical tables.** Reading, MA: Addison–Wesley.

Pagano, M. & Gauvreau, K. (1993). **Principles of biostatistics.** Belmont, CA: Duxbury Press.

Reynolds, H. T. (1977a). **The analysis of cross-classifications.** New York: The Free Press.

Reynolds, H. T. (1977b). **Analysis of nominal data.** Beverly Hills, CA: Sage Publications.

Rogers, J. L., Howard, K. I., & Vessey, J. T. (1993). Using significance tests to evaluate equivalence between two experimental groups. **Psychological Bulletin**, 113, 553–565.

Rosner, B. (1995). **Fundamental of biostatistics** (4th ed.) Belmont, CA: Duxbury Press.

Rosner, B. (2000). **Fundamental of biostatistics** (5th ed.) Pacific Grove, CA: Duxbury Press.

Schenker, N. & Gentleman, J. F. (2001). On judging the significance of difference by examining the overlap between confidence intervals means. **The American Statistician**, 55, 182–186.

Schuirmann, D. J. (1987). A comparison of the two one-sided tests procedure and the power approach for assessing equivalence of average bioavailability. **Journal of Pharmacokinetics and Biopharmaceutics**, 15, 657–680.

Selvin, S. (1995). **Practical biostatistical methods**. Belmont, CA: Duxbury.

Siegel, S. & Castellan, N. J., Jr. (1988). **Nonparametric statistics for the behavioral sciences** (2nd ed.) New York: McGraw–Hill Book Company.

Smithson, M. (2003). **Confidence intervals**. Thousand Oaks, CA: Sage Publications.

Sprent, P. (1993). **Applied nonparametric statistical methods** (2nd ed.). London: Chapman & Hall.

Tabachnick, B. G. & Fidell, L. S. (1996). **Using multivariate statistics** (3rd ed.). New York: Harper Collins Publishers.

Tabachnick, B. G. & Fidell, L. S. (2001). **Using multivariate statistics** (4th ed.). Boston: Allyn & Bacon.

Tryon, W. W. (2001). Evaluating statistical difference, equivalence, and indeterminacy using inferential confidence intervals: An integrated alternative method of conducting null hypothesis statistical tests. **Psychological Methods**, 6, 371–386.

Tryon, W. W. (2005). Evaluating proportions for statistical difference, equivalence, and indeterminacy using inferential confidence intervals, Unpublished manuscript.

Westlake, W. J. (1976). Symmetrical confidence intervals for bioequivalence trials. **Biometrics**, 32, 741–744.

Westlake, W. J. (1981). Response to T.B. L. Kirkwood: Bioequivalence testing — a need to rethink. **Biometrics**, 37, 589–594.

Westlake, W. J. (1988). Bioavailability and bioequivalence of pharmaceutical formulations. In K. E. Peace (Ed.), **Biopharamaceutical statistics for drug development** (pp. 329–352). New York: Marcel Decker.

Wickens, T. (1989). **Multiway contingency table analysis for the social sciences**. Hillsdale, NJ: Lawrence Erlbaum Associates, Publishers.

Wilks, S. S. (1935). The likelihood test of independence in a contingency table. **Annals of Mathematical Statistics**, 6, 190–196.

Williams, K. (1976). The failure of Pearson's goodness of fit statistic. **Statistician**, 25, 49.

Yates, F. (1934). Contingency tables involving small numbers and the chi-square test. **Journal of the Royal Statistical Society**, 1, 217–235.

Yule, G. (1900). On the association of the attributes in statistics: With illustrations from the material of the childhood society, &c. **Philosophical Transactions of the Royal Society**, Series A, 194, 257–319.

Zar, J. H. (1999). **Biostatistical analysis** (4th ed.). Upper Saddle River, NJ: Prentice Hall.

Endnotes

1. A general discussion of the chi-square distribution can be found in Sections I and V of the **single-sample chi-square test for a population variance**.

2. The use of the chi-square approximation (which employs a continuous probability distribution to approximate a discrete probability distribution) is based on the fact that the computation of exact probabilities requires an excessive amount of calculations.

3. In the case of both the **chi-square test of independence** and the **chi-square test of homogeneity**, the same result will be obtained regardless of which of the variables is designated as the row variable versus the column variable.

4. It is just coincidental that the number of **introverts** equals the number of **extroverts**.

5. In the context of the discussion of the **chi-square test of independence**, the proportion of observations in $Cell_{11}$ refers to the number of observations in $Cell_{11}$ divided by the total number of observations in the 2×2 table. In the discussion of the hypothesis for the **chi-square test for homogeneity**, the proportion of observations in $Cell_{11}$ refers to the number of observations in $Cell_{11}$ divided by the total number of observations in Row 1 (i.e., the row in which $Cell_{11}$ appears).

6. Equation 16.2 is an extension of Equation 8.2 (which is employed to compute the value of chi-square for the **chi-square goodness-of-fit test**) to a two-dimensional table. In Equation 16.2, the use of the two summation expressions $\sum_{i=1}^{r} \sum_{j=1}^{c}$ indicates that the operations summarized in Table 16.4 are applied to all of the cells in the $r \times c$ table. In contrast, the single summation expression $\sum_{i=1}^{k}$ in Equation 8.2 indicates that the operations summarized in Table 8.2 are applied to all k cells in a one-dimensional table.

7. The same chi-square value will be obtained if the row and column variables are reversed — i.e., the helping variable represents the row variable and the noise variable represents the column variable.

8. Correlational studies are discussed in the **Introduction** and in the chapter on the **Pearson product-moment correlation coefficient**.

9. The value $\chi^2_{.98} = 5.43$ is determined by interpolation. It can also be derived by squaring the tabled critical one-tailed .01 value $z_{.01} = 2.33$, since the square of the latter value is equivalent to the chi-square value at the 98th percentile. The use of z values in reference to a 2×2 contingency table is discussed later in this section under the **z test for two independent proportions**.

10. Within the framework of this discussion, the value $\chi^2_{.05} = 3.84$ represents the tabled chi-square value at the 95th percentile (which demarcates the extreme 5% in the right tail of the chi-square distribution). Thus, using the format employed for the one-tailed .05 and .01 values, the notation identifying the two-tailed .05 value can also be written as $\chi^2_{.95} = 3.84$. In the same respect, the value $\chi^2_{.01} = 6.63$ represents the tabled chi-square value at the 99th percentile (which demarcates the extreme 1% in the right tail of the chi-square distribution). Thus, using the format employed for the one-tailed .05 and .01 values, the notation identifying the two-tailed .01 value can also be written as $\chi^2_{.99} = 6.63$.

11. The null and alternative hypotheses presented for the **Fisher exact test** in this section are equivalent to the alternative form for stating the null and alternative hypotheses for the **chi-square test of homogeneity** presented in Section III (if the hypotheses in Section III are applied to a 2×2 contingency table).

12. Sourcebooks documenting statistical tables (e.g., Owen (1962) and Beyer (1968)), as well as many books which specialize in nonparametric statistics (e.g., Daniel (1990), Marascuilo and McSweeney (1977), and Siegel and Castellan (1988)) contain tables of the hypergeometric distribution which can be employed with 2×2 contingency tables. Such tables eliminate the requirement of employing Equations 16.7/16.8 to compute the value of P_T.

13. The value $(1 - p)$, which is often represented by the notation q, can also be computed as follows: $q = (1 - p) = (b + d)/(n_1 + n_2) = (b + d)/n$. The value q is a pooled estimate of the proportion of observations in Column 2 in the underlying population.

14. Due to rounding off error there may be a minimal discrepancy between the square of a z value and the corresponding chi-square value.

15. The logic for employing Equation 16.11 in lieu of Equation 16.9 is the same as that discussed in reference to the *t* **test for two independent samples**, when in the case of the latter test the null hypothesis stipulates a value other than zero for the difference between the population means (and Equation 11.5 is employed to compute the test statistic in lieu of Equations 11.1/11.2/11.3).

16. The denominator of Equation 16.11 is employed to compute $s_{p_1 - p_2}$ instead of the denominator of Equation 16.9, since in computing a confidence interval it cannot be assumed that $\pi_1 = \pi_2$ (which is assumed in Equation 16.9, and serves as the basis for computing a pooled p value in the latter equation).

17. The **median test for independent samples** can also be employed within the framework of the model for the **chi-square test of independence**. To illustrate this, assume that Example 16.4 is modified so that the researcher randomly selects a sample of 200 subjects, and does not specify beforehand that the sample is comprised of 100 females and 100 males. If it just happens by chance that the sample is comprised of 100 females and 100 males, one can state that neither the sum of the rows nor the sum of the columns is predetermined by the researcher. As noted in Section I, when neither of the marginal sums is predetermined, the design conforms to the model for the **chi-square test of independence**.

18. The word **column** can be interchanged with the word **row** in the definition of a complex comparison.

19. Another consideration which should be mentioned with respect to conducting comparisons is that two or more comparisons for a set of data can be **orthogonal** (which means they are independent of one another), or comparisons can overlap with respect to the information they provide. As a general rule, when a limited number of comparisons is planned, it is most efficient to conduct orthogonal comparisons. The general subject of orthogonal comparisons is discussed in greater detail in Section VI of the **single-factor between-subjects analysis of variance**.

20. The null and alternative hypotheses stated below do not apply to the **odds ratio**.

21. a) In some sources (and in the previous edition of this book) the **phi coefficient** is represented by the symbol φ; b) Some sources note that the **phi coefficient** can only assume a range of values between 0 and +1. In these sources, the term $|ad - bc|$ is employed in the numerator of Equation 16.20. By employing the absolute value of the term in the numerator of Equation 16.20, the value of **phi** will always be a positive number. Under the latter condition the following will be true: $\phi = \sqrt{\chi^2/n}$.

22. In the case of small sample sizes, the results of the **Fisher exact test** are employed as the criterion for determining whether or not the computed value of **phi** is significant.

23. In such a case, the data are summarized in the form of a 2 × 2 contingency table documenting the proportion of subjects who answer in each of the categories of two dichotomous variables (e.g., **True** versus **False** for both variables/test items).

24. The reason why the result of the **chi-square test for $r \times c$ tables** is not employed to assess the significance of Q is because Q is not a function of chi-square. It should be noted that since Q is a special case of **Goodman and Kruskal's gamma**, it can be argued that Equation 32.2 (the significance test for **gamma**) can be employed to assess whether or not Q is significant. However, Ott *et al*. (1992) state that a different procedure is employed for evaluating the significance of Q versus **gamma**. Equation 32.2 will not yield the same result as that obtained with Equation 16.24 when it is applied to a 2×2 table. If the **gamma** statistic is computed for Examples 16.1/16.2 it yields the absolute value $\gamma = .56$ (γ is the lower case Greek letter **gamma**), which is identical to the value of Q computed for the same set of data. (The absolute value is employed since the contingency table is not ordered, and thus, depending upon how the cells are arranged, a value of either $+.56$ or $-.56$ can be derived for **gamma**.) However, when Equation 32.2 is employed to assess the significance of $\gamma = .56$, it yields the absolute value $z = 3.51$, which although significant at both the .05 and .01 levels is lower than the absolute value $z = 5.46$ obtained with Equation 16.24.

25. A more detailed discussion of odds can be found in Section IX (the **Addendum**) of the **binomial sign test for a single sample** under the discussion of **Bayesian hypothesis testing**.

26. a) As noted earlier, when odds are employed what is often stated are the odds that an event will not occur. Using this definition, the odds that a person in the **noise** condition **did not help the confederate** (or that someone who **washes her hands** does **not contract the disease**) are 2.33:1 (since $(70/100)/(30/100) = 2.33$). The odds that a person in the **no noise** condition **did not help the confederate** (or that someone who **does not wash her hands** does **not contract the disease**) are .667:1 or 2:3 (since $(40/100)/(60/100) = .667$). These values yield the same **odds ratio**, since $2.33/.667 = 3.49$; b) It is also the case that $o = 1/o'$ $= 1/.29 = 3.45$ and $o' = 1/o = 1/3.5 = .286$ (the minimal differences are due to rounding off error); c) van Belle (2002, pp. 78–82) notes the following with respect to the **odds ratio** and the **relative risk** (both of which are discussed in Section VI of the **chi-square test for $r \times c$ tables**): 1) The value of the **odds ratio** is always further removed from the value 1 than the value of the **relative risk**; 2) When the baserate is .05 or less, the difference between the **odds ratio** and the **relative risk** will be minimal; 3) The **relative risk** can never be greater than $1/baserate$; 4) When both the **odds ratio** and the **relative risk** are less than 1, the difference between the two values will be less pronounced; d) Endnotes 41 and 42 in the chapter on the **Pearson product-moment correlation coefficient** provide additional discussion on the relationship between the **odds ratio** and **relative risk**.

27. Equations 16.69 and 16.70 are alternate equations for computing the **odds ratios** $o = 3.5$ and $o' = .29$.

$$o = \frac{p_b\,p_c}{p_a\,p_d} \qquad \textbf{(Equation 16.69)}$$

$$o' = \frac{p_a\,p_d}{p_b\,p_c} \qquad \textbf{(Equation 16.70)}$$

From Tables 16.2/16.22, we can determine that $p_a = a/n = 30/200 = .15$, $p_b = b/n$ $= 70/200 = .35$, $p_c = c/n = 60/200 = .3$, and $p_d = d/n = 40/200 = .2$. Employing Equations 16.69 and 16.70 with the data for Example 16.1, the values $o = 3.5$ and $o' = .29$ are computed.

$$o = \frac{(.35)(.3)}{(.15)(.2)} = 3.5$$

$$o' = \frac{(.15)(.2)}{(.35)(.3)} = .29$$

28. Pagano and Gauvreau (1993) note that if the expected frequencies for any of the cells in the contingency table are less than 5, the equation below should be employed to compute the standard error.

$$SE = \sqrt{\frac{1}{a + .5} + \frac{1}{b + .5} + \frac{1}{c + .5} + \frac{1}{d + .5}}$$

29. If the **continuity corrected** equation (Equation 16.10) is employed for the examples in this section which evaluate whether or not there is a difference between two independent proportions, a lower absolute z value will be computed for any analysis that is conducted. The lower absolute z value will not, however, affect the conclusions reached for any analysis. Later in this section Equation 16.14 is employed to compute a confidence interval for a difference between two proportions. If in lieu of Equation 16.14 the **continuity corrected equation** (Equation 16.15) is employed, a slightly wider confidence is computed. (In the relevant example, .01 is subtracted from the lower limit of a confidence interval and .01 is added to the upper limit of a confidence interval.) The latter, however, does not affect the conclusions reached for the analysis.

30. a) If the correction for continuity is used to compute the second z value, $z = -1.44$ is obtained. The latter illustrates that if the correction of continuity is employed, the absolute value stipulated for ζ must be greater than .154, which is the upper limit of the continuity corrected confidence interval. Employing $\zeta = .154$, the continuity corrected equation becomes $z = [(.837 - .784) - .154 + .01] / .0552 = -1.65$; b) Technically, if the statement is made that the difference between the proportion of successes between the two groups is within a proportion value of .144, the alternative hypotheses should be written as follows: $H_1: \theta \geq -.144$ and $H_1: \theta \leq .144$. Since the equals sign should not be included in the null hypotheses, the latter should be written as $H_0: \theta < -.144$ or $H_0: \theta > .144$.

31. Rosner (2000, p. 640) notes that if, in fact, there is no difference between the sample proportions (i.e., $p_1 = p_2 = p$ and we let $q = p - 1$) and $k = 1$, Equation 16.51 becomes Equation 16.71. If, for purposes of illustration we assume that $p_1 = p_2 = .837$, under the latter circumstances in order for the power of the test to be .80 the number of subjects required per group is reduced to $n = 214$.

(Equation 16.71)

$$n_1 = \frac{2pq\,(z_{\alpha/2} + z_\beta)^2}{\zeta^2} = \frac{[(2)(.837)(.163)][1.96 + .84]^2}{(.1)^2} = 213.92$$

32. **Yates' correction for continuity** was not used to compute the values $\chi^2 = 11.58$ and $\chi^2 = 2.09$ for the two hospitals. If **Yates' correction** is used, the computed chi-square values will be a little lower.

33. Some sources (e.g., Christensen (1990)) employ the term **factors** (which is the term that is commonly employed within the framework of a **factorial analysis of variance**) to identify the different independent variables.

34. Zar (1999) notes that the degrees of freedom are the sum of the degrees of freedom for all of the interactions. Specifically $df = rcl - r - c - l + 2 = (r-1)(c-1)(l-1) + (r-1)(c-1) + (r-1)(l-1) + (c-1)(l-1)$.

35. Zar (1999) notes that the degrees of freedom are the sum of the following: $df = rcl - cl - r + 1 = (r-1)(c-1)(l-1) + (r-1)(c-1) + (r-1)(l-1)$.

36. Based on Endnote 32, it logically follows that the degrees of freedom are the sum of the following: $df = rcl - rl - c + 1 = (r-1)(c-1)(l-1) + (c-1)(r-1) + (c-1)(l-1)$.

37. Based on Endnote 32, it logically follows that the degrees of freedom are the sum of the following: $df = rcl - rc - l + 1 = (r-1)(c-1)(l-1) + (l-1)(r-1) + (l-1)(c-1)$.

38. Zar (1999) discusses and cites reference on **mosaic displays** for contingency tables, which represent an alternative to conventional graphs for visually summarizing the data in a contingency table.

39. a) Readers who are not familiar with factorial designs and the concepts of main effects and interactions are advised to read Sections I–V of the **between-subjects factorial analysis of variance (Test 27)** prior to reading the discussion of **log-linear analysis**; b) Tabachnick and Fidell (1996, 2001) refer to **log-linear analysis** as **multiway frequency analysis**.

40. The following *SPSS* command sequence was employed in this chapter in conducting a **log-linear analysis**: a) Click **Analyze**; b) Click **Loglinear**; c) Click **Model Selection**; d) Highlight the categorization variables (i.e., the **row**, **column**, and **layer variables** in Examples 16,9/16.10) one at a time, and move each one separately into the **Factor(s)** window. After moving a categorization variable into the **Factor(s)** window, click **Define Range** and enter 1 for **Minimum** and enter the number of categories for that variable for **Maximum** (which will be 2 for all three categorization variables employed in Examples 16.9/16.10), and then click **Continue** (After entering the categorization variables and defining the range for each one, take note of the fact that under **Model Building** the default setting is **Use backward elimination**, which was the option employed for Examples 16.9/16.10. Additionally, if you Click on **Model**, under **Specify Model**, note that **Saturated** (which is the default setting) is checked, and then click **Continue** to return to the main window); g) Click **Options** and check off desired information (e.g., **Parameter estimates, Association Table**), and then click **Continue**; h) Click **OK** to obtain the output for the analysis.

41. a) Reynolds (1977b, p. 66) notes that Goodman (1970), who pioneered **log-linear analysis**, recommended .5 be added to each of the observed frequencies in order to avoid having to compute a logarithm for zero (which is undefined); b) As is the case for the **chi-square test for $r \times c$ tables**, one assumption underlying **log-linear analysis** is that the expected frequencies of all cells are greater than 1, and that no more than 20% of the cells have an expected frequency less than 5. Reynolds (1977b, p. 78) notes that, as a general rule, the sample size for a **log-linear analysis** will be adequate if the result of dividing the number

of subjects in a contingency table by the total number of cells which comprise the table is greater than five; c) As an alternative to employing **backward hierarchical analysis** described in this section, it is possible for the researcher to stipulate one or more specific models she wants to evaluate in relation to the **saturated model**; d) The total degrees of freedom for the **saturated model** under discussion where $r = 2$, $c = 2$, and $l = 2$ are $df = 8$. The latter value is computed as follows: $df_{Total} = [(df_\theta = 1) + (df_r = r - 1 = 1) + (df_c = c - 1 = 1) + (df_l = l - 1 = 1) + (df_{rc} = (r - 1)(c - 1) = 1) + (df_{rl} = (r - 1)(l - 1) = 1) + (c - 1)(l - 1) = 1) + (df_{rcl} = (r - 1)(c - 1)(l - 1) = 1)] = 8$ (Christensen (1990, p. 87).

42. In the case of a three-way table, a model involving two-way interactions is said to be **nested** within the **saturated model**. In other words a less complex model will always be nested within a more complex model. Consequently, a model involving main effects will be nested within both the **saturated model** and a model involving two-way interactions.

43. Norušis (2004, p. 13) notes that the **Pearson Chisq** values and associated probabilities in Table 16.39 are based on evaluating the data with the **chi-square test for $r × c$ tables**. For large sample sizes the latter test statistic is equivalent to the **likelihood ratio chi-square statistic**. The reason the analysis focuses on the **likelihood ratio statistic** is because it can be subdivided into components, each of which can be interpreted, and the sum of the components will add up to the total value of the statistic.

44. Unlike in most analyses where a researcher typically wishes to reject the null hypothesis, and is thus looking to obtain a p value equal to or less than .05, within the context of the method being described for interpreting the value of a **likelihood ratio statistic**, the researcher wants to retain the null hypothesis that an alternative model is not significantly inferior to the **saturated model**, and in order to do the latter a p value above .05 is required.

45. a) With respect to the tables labeled **Tests that K-way and higher-order effects are zero** and **Tests that K-way effects are zero**, Tabachnick and Fidell (1996, p. 280) suggest that in the case of a small sample size in order to increase the power of an analysis, a researcher might employ a p value larger than .05 as a criterion for significance; b) The reader need not be concerned with the values in the columns labeled **Iteration** in the tables labeled **Tests that K-way and higher-order effects are zero** and **Tests that K-way effects are zero**. An **iteration** is a set of operations which is sequentially repeated until the best possible approximation is computed for a value in question. In the case of **log-linear analysis** multiple calculations are required in order to compute the best possible estimate for the expected cell frequencies. The values in the **Iteration** column just indicate the number of calculations required for each model.

46. a) Norušis (2004, p. 14) notes that the **generating class** of a model refers to the highest-order interaction involved in a model. Thus, if the **generating class** for a model is the second-order interactions $r × l$ and $c × l$, the model will contain the latter interactions plus all of the lower-order "relatives" of those latter interactions (i.e., the main effects r, c, and l, which are components of the aforementioned second-order interactions); b) Norušis (2006, personal communication) notes that it is possible for the **final model** (displayed under the heading **The final model has generating class** at the bottom of page 2 of Table 16.39) to contain a higher-order interaction than what is indicated by the results for the tables labeled **Tests that K-way and higher order effects are zero** and **Tests that K-way**

effects are zero. She notes that the tests in the latter tables "are testing sequential hypotheses about models with **all** interactions terms of a particular order absent or present. When you remove the restriction that **all** interaction terms of a particular order must be present or absent and require only that resulting models be hierarchical, the significance of individual effects may change." Thus, although it was not the case for the example under discussion, it is theoretically possible that the final model could have contained a three-way interaction; c) In point of fact, the magnitude of difference between the probabilities associated with the computed **L.R. Chisq Change** values for the **ROWVAR* LAYERVAR** and **COLVAR*LAYERVAR** interactions is trivial and could easily be attributable to sampling error. Consequently, before drawing at any conclusions regarding the relative importance of the two interactions the researcher would be required to conduct further studies on other samples derived from the same population; d) The author is indebted to Marija Norušis for clarifying some of the information displayed by *SPSS* for log-linear analysis.

47. Within the context of discussing **log-linear analysis**, sources (e.g., Christensen (1990), Garson (2006), and Selvin (1995)) employ the following terminology with respect to identifying models in reference to a three-dimensional contingency table: a) A model in which all of the two-way interactions are significant is referred to as the **homogeneous association model**; b) A model in which two of the three possible two-way interactions are significant is referred to as a **conditional independence model**; c) A model in which one of the three possible two-way interactions is significant is referred to as a **partial independence model** or **one-factor independence model**; and d) A model in which all of the factors are unrelated to one another is referred to as a **complete independence model** or **equiprobability model**. Other sources, however, may employ different terminology in reference to one or more of the latter models (e.g., Howell (2002, p. 665)).

48. Since Equations 9.6 and 9.7 are equivalent, either one can be employed for the analysis. In addition, since our analysis involves a binary situation (i.e., the two response categories of **being correct** or **incorrect**), the data can also be evaluated with the **chi-square goodness-of-fit test**. The chi-square value obtained with the latter test will be equal to the square of the z value obtained with Equations 9.6/9.7.

49. In the interest of precision, the number of days in each sun sign is not one-twelfth of the total number of days in a year (since 365 cannot be divided evenly by 12, it logically follows that the number of days in each sun sign will not be equal). In spite of the latter, for all practical purposes the value 1/12 can be accurately employed to represent the probability for each sun sign.

Inferential Statistical Tests Employed with Two Dependent Samples (and Related Measures of Association/Correlation)

Test 17

The *t* Test for Two Dependent Samples
(Parametric Test Employed with Interval/Ratio Data)

I. Hypothesis Evaluated with Test and Relevant Background Information

Hypothesis evaluated with test Do two dependent samples represent two populations with different mean values?

Relevant background information on test The *t* test for two dependent samples, which is employed in a hypothesis testing situation involving two dependent samples, is one of a number of inferential statistical tests that are based on the *t* distribution (which is described in detail under the **single-sample *t* test (Test 2))**. Throughout the discussion of the *t* test for two dependent samples, the term **experimental conditions** will also be employed to represent the dependent samples employed in a study. In a **dependent samples design**, each subject either serves in all of the *k* (where $k \geq 2$) experimental conditions, or else is matched with a subject in each of the other $(k-1)$ experimental conditions (matching is discussed in Section VII).[1] In designs which are evaluated with the *t* test for two dependent samples, the value of *k* will always equal 2.

In conducting the *t* test for two dependent samples, the means of the two experimental conditions (represented by the notations $\bar{X}_1$ and $\bar{X}_2$) are employed to estimate the values of the means of the populations (μ_1 and μ_2) the conditions represent. If the result of the *t* test for two dependent samples is significant, it indicates the researcher can conclude there is a high likelihood the two experimental conditions represent populations with different mean values. It should be noted that the *t* test for two dependent samples is the appropriate test to employ for contrasting the means of two dependent samples when the values of the underlying population variances are unknown. In instances where the latter two values are known, the appropriate test to employ is the *z* test for two dependent samples (Test 17e), which is described in Section VI.

The *t* test for two dependent samples is employed with interval/ratio data, and is based on the following assumptions: a) The sample of *n* subjects has been randomly selected from the population it represents; b) The distribution of data in the underlying populations each of the experimental conditions represents is normal; and c) The third assumption, which is referred to as the **homogeneity of variance** assumption, states that the variance of the underlying population represented by Condition 1 is equal to the variance of the underlying population represented by Condition 2 (i.e., $\sigma_1^2 = \sigma_2^2$). It should be noted that the *t* test for two dependent samples is more sensitive to violation of the homogeneity of variance assumption (which is discussed in Section VI) than is the *t* test for two independent samples (Test 11). If any of the aforementioned assumptions of the *t* test for two dependent samples are saliently violated, the reliability of the test statistic may be compromised.

When a study employs a dependent samples design, the following two issues related to experimental control must be taken into account: a) In a dependent samples design in which each subject serves in both experimental conditions, it is essential that the experimenter control for **order effects**. An **order effect** is where an obtained difference on the dependent variable is a direct result of the order of presentation of the experimental conditions, rather than being due to the independent variable manipulated by the experimenter. Order effects can be controlled through the use of a technique called **counterbalancing**, which is discussed in Section VII; and b) When a dependent samples design employs matched subjects, within each pair of matched subjects each of the two subjects must be randomly assigned to one of the two experimental conditions. Nonrandom assignment of subjects to the experimental conditions can compromise the **internal validity** of a study.[2] A more thorough discussion of **matching** can be found in Section VII.

II. Example

Example 17.1 *A psychologist conducts a study to determine whether or not people exhibit more emotionality when they are exposed to sexually explicit words than when they are exposed to neutral words. Each of ten subjects is shown a list of 16 randomly arranged words, which are projected onto a screen one at a time for a period of five seconds. Eight of the words on the list are sexually explicit and eight of the words are neutral. As each word is projected on the screen, a subject is instructed to say the word softly to himself or herself. As a subject does this, sensors attached to the palms of the subject's hands record galvanic skin response (GSR), which is used by the psychologist as a measure of emotionality. The psychologist computes two scores for each subject, one score for each of the experimental conditions:* **Condition 1**: *GSR/Explicit — The average GSR score for the eight sexually explicit words;* **Condition 2**: *GSR/Neutral — The average GSR score for the eight neutral words. The GSR/Explicit and the GSR/Neutral scores of the ten subjects follow. (The higher the score, the higher the level of emotionality.)* **Subject 1** *(9, 8);* **Subject 2** *(2, 2);* **Subject 3** *(1, 3);* **Subject 4** *(4, 2);* **Subject 5** *(6, 3);* **Subject 6** *(4, 0);* **Subject 7** *(7, 4);* **Subject 8** *(8, 5);* **Subject 9** *(5, 4);* **Subject 10** *(1, 0).*[3] *Do subjects exhibit differences in emotionality with respect to the two categories of words?*

III. Null versus Alternative Hypotheses

Null hypothesis $H_0: \mu_1 = \mu_2$

(The mean of the population Condition 1 represents equals the mean of the population Condition 2 represents.)

Alternative hypothesis $H_1: \mu_1 \neq \mu_2$

(The mean of the population Condition 1 represents does not equal the mean of the population Condition 2 represents. This is a **nondirectional alternative hypothesis** and it is evaluated with a **two-tailed test**. In order to be supported, the absolute value of t must be equal to or greater than the tabled critical two-tailed t value at the prespecified level of significance. Thus, either a significant positive t value or a significant negative t value will provide support for this alternative hypothesis.)

or

$$H_1: \mu_1 > \mu_2$$

(The mean of the population Condition 1 represents is greater than the mean of the population Condition 2 represents. This is a **directional alternative hypothesis** and it is evaluated with a

one-tailed test. It will only be supported if the sign of t is positive, and the absolute value of t is equal to or greater than the tabled critical one-tailed t value at the prespecified level of significance.)

or

$$H_1: \mu_1 < \mu_2$$

(The mean of the population Condition 1 represents is less than the mean of the population Condition 2 represents. This is a **directional alternative hypothesis** and it is evaluated with a **one-tailed test**. It will only be supported if the sign of t is negative, and the absolute value of t is equal to or greater than the tabled critical one-tailed t value at the prespecified level of significance.)

Note: Only one of the above noted alternative hypotheses is employed. If the alternative hypothesis the researcher selects is supported, the null hypothesis (H_0) is rejected.[4]

IV. Test Computations

Two methods can be employed to compute the test statistic for the *t* **test for two dependent samples**. The method to be described in this section, which is referred to as the **direct-difference method**, allows for the quickest computation of the t statistic. In Section VI, a computationally equivalent but more tedious method for computing t is described.

Table 17.1 Data for Example 17.1

Subject	Condition 1 X_1	Condition 2 X_2	D	D^2
1	9	8	1	1
2	2	2	0	0
3	1	3	−2	4
4	4	2	2	4
5	6	3	3	9
6	4	0	4	16
7	7	4	3	9
8	8	5	3	9
9	5	4	1	1
10	1	0	1	1
	$\Sigma X_1 = 47$	$\Sigma X_2 = 31$	$\Sigma D- = -2$ $\Sigma D+ = 18$	$\Sigma D^2 = 54$
			$\Sigma D = 16$	
	$\bar{X}_1 = \dfrac{47}{10} = 4.7$	$\bar{X}_2 = \dfrac{31}{10} = 3.1$		

The data for Example 17.1 and the preliminary computations for the **direct-difference method** are summarized in Table 17.1. Note that there are $n = 10$ subjects, and that there is a total of $2n = (2)(10) = 20$ scores, since each subject has two scores. The two scores of the 10 subjects are listed in the columns of Table 17.1 labeled X_1 and X_2. The score of a subject in the column labeled X_1 is the average GSR score of the subject for the eight sexually explicit words (Condition 1), while the score of a subject in the column labeled X_2 is the average GSR score of the subject for the eight neutral words (Condition 2). Column 4 of Table 17.1 lists a difference score for each subject (designated by the notation D), which is computed by subtracting a subject's X_2 score from his X_1 score (i.e., $D = X_1 - X_2$). Column 5 of the table lists a D^2 score for each subject, which is obtained by squaring a subject's D score.

In Column 4 of Table 17.1, the summary value $\Sigma D = 16$ is obtained by adding $\Sigma D+ = 18$, the sum of the positive difference scores (i.e., all those difference scores with a + sign), and $\Sigma D- = -2$, the sum of the negative difference scores (i.e., all those difference scores with a – sign). The reader should take note of the fact that whenever $\Sigma X_1 > \Sigma X_2$ (and consequently $\bar{X}_1 > \bar{X}_2$), the value ΣD will be a positive number, whereas whenever $\Sigma X_1 < \Sigma X_2$ (and consequently $\bar{X}_1 < \bar{X}_2$), the value ΣD will be a negative number.

Equation 17.1 is the direct-difference equation for computing the test statistic for the **t test for two dependent samples**.

$$t = \frac{\bar{D}}{s_{\bar{D}}}$$ **(Equation 17.1)**

Where:　$\bar{D}$ represents the mean of the difference scores
　　　　　$s_{\bar{D}}$ represents the **standard error of the mean difference**

The **mean of the difference scores** is computed with Equation 17.2.

$$\bar{D} = \frac{\Sigma D}{n}$$ **(Equation 17.2)**

Employing Equation 17.2, the value $\bar{D} = 1.6$ is computed.

$$\bar{D} = \frac{16}{10} = 1.6$$

Equation 17.3 is employed to compute $\tilde{s}_D$, which represents the **estimated population standard deviation of the difference scores**.[5]

$$\tilde{s}_D = \sqrt{\frac{\Sigma D^2 - \frac{(\Sigma D)^2}{n}}{n-1}}$$ **(Equation 17.3)**

Employing Equation 17.3, the value $\tilde{s}_D = 1.78$ is computed. (Note that $\tilde{s}_D^2 = (1.78)^2 = 3.17$. The latter value, which is the **estimated population variance of the difference scores**, will be employed at a later point in the discussion of the **t test for two dependent samples**.)

$$\tilde{s}_D = \sqrt{\frac{54 - \frac{(16)^2}{10}}{10-1}} = 1.78$$

Equation 17.4 is employed to compute the value $s_{\bar{D}}$. The value $s_{\bar{D}}$ represents the **standard error of the mean difference**, which is an estimated population standard deviation of mean difference scores.[6]

$$s_{\bar{D}} = \frac{\tilde{s}_D}{\sqrt{n}}$$ **(Equation 17.4)**

Employing Equation 17.4, the value $s_{\bar{D}} = .56$ is computed.

$$s_{\bar{D}} = \frac{1.78}{\sqrt{10}} = .56$$

Substituting $\bar{D} = 1.6$ and $s_{\bar{D}} = .56$ in Equation 17.1, the value $t = 2.86$ is computed.[7]

$$t = \frac{1.6}{.56} = 2.86$$

The reader should take note of the fact that the values $\tilde{s}_D$ and $s_{\bar{D}}$, which are respectively estimates of a population standard deviation and a standard deviation of a sampling distribution, can never be a negative number. If a negative value is obtained for either of the aforementioned values, it indicates a computational error has been made.

V. Interpretation of the Test Results

The obtained value $t = 2.86$ is evaluated with **Table A2 (Table of Student's t Distribution)** in the **Appendix**. The degrees of freedom for the **t test for two dependent samples** are computed with Equation 17.5.

$$df = n - 1$$ **(Equation 17.5)**

Employing Equation 17.5, the value $df = 10 - 1 = 9$ is computed. The tabled critical two-tailed and one-tailed .05 and .01 t values for $df = 9$ are summarized in Table 17.2. (For a review of the protocol for employing **Table A2**, the reader should review Section V of the **single-sample t test**.)

Table 17.2   Tabled Critical .05 and .01 t Values for df = 9

	$t_{.05}$	$t_{.01}$
Two-tailed values	2.26	3.25
One-tailed values	1.83	2.82

The following guidelines are employed in evaluating the null hypothesis for the **t test for two dependent samples**.

a) If the nondirectional alternative hypothesis H_1: $\mu_1 \neq \mu_2$ is employed, the null hypothesis can be rejected if the obtained absolute value of t is equal to or greater than the tabled critical two-tailed value at the prespecified level of significance.

b) If the directional alternative hypothesis H_1: $\mu_1 > \mu_2$ is employed, the null hypothesis can be rejected if the sign of t is positive, and the value of t is equal to or greater than the tabled critical one-tailed value at the prespecified level of significance.

c) If the directional alternative hypothesis H_1: $\mu_1 < \mu_2$ is employed, the null hypothesis can be rejected if the sign of t is negative, and the absolute value of t is equal to or greater than the tabled critical one-tailed value at the prespecified level of significance.

Employing the above guidelines, the following conclusions can be reached.

The nondirectional alternative hypothesis H_1: $\mu_1 \neq \mu_2$ is supported at the .05 level, since the computed value $t = 2.86$ is greater than the tabled critical two-tailed value $t_{.05} = 2.26$. The latter alternative hypothesis, however, is not supported at the .01 level, since $t = 2.86$ is less than the tabled critical two-tailed value $t_{.01} = 3.25$.

The directional alternative hypothesis H_1: $\mu_1 > \mu_2$ is supported at both the .05 and .01 levels, since the obtained value $t = 2.86$ is a positive number which is greater than the tabled critical one-tailed values $t_{.05} = 1.83$ and $t_{.01} = 2.82$. Note that when the directional alternative hypothesis H_1: $\mu_1 > \mu_2$ is supported, it is required that $\bar{X}_1 > \bar{X}_2$.

The directional alternative hypothesis H_1: $\mu_1 < \mu_2$ is not supported, since the obtained value $t = 2.86$ is a positive number. In order for the directional alternative hypothesis H_1: $\mu_1 < \mu_2$ to be supported, the computed value of t must be a negative number (as well as the fact that the absolute value of t must be equal to or greater than the tabled critical one-tailed value at the prespecified level of significance). In order for the data to be consistent with the directional alternative hypothesis H_1: $\mu_1 < \mu_2$, it is required that $\bar{X}_1 < \bar{X}_2$.

A summary of the analysis of Example 17.1 with the *t* **test for two dependent samples** follows: It can be concluded that the average GSR (emotionality) score for the sexually explicit words is significantly higher than the average GSR score for the neutral words. This result can be summarized as follows (if $\alpha = .05$ is employed): $t(9) = 2.86$, $p < .05$.[8]

VI. Additional Analytical Procedures for the *t* Test for Two Dependent Samples and/or Related Tests

1. Alternative equation for the *t* test for two dependent samples   Equation 17.6 is an alternative equation which can be employed to compute the test statistic for the *t* **test for two dependent samples**.[9]

$$t = \frac{\bar{X}_1 - \bar{X}_2}{\sqrt{s_{\bar{X}_1}^2 + s_{\bar{X}_2}^2 - 2(r_{X_1 X_2})(s_{\bar{X}_1})(s_{\bar{X}_2})}} \qquad \text{(Equation 17.6)}$$

The computation of t with Equation 17.6 requires more computations than does Equation 17.1 (the direct difference method equation). Equation 17.6, unlike Equation 17.1, requires that the estimated population variance be computed for each of the samples (in Section VI it is noted that the latter values are required in order to evaluate the homogeneity of variance assumption of the *t* **test for two dependent samples**). Since a total understanding of Equation 17.6 requires an understanding of the concept of correlation, the reader may find it useful to review the relevant material on correlation in the **Introduction** as well as Section I of the **Pearson product-moment correlation coefficient (Test 28)** prior to continuing this section.

Except for the last term in the denominator of Equation 17.6 (i.e., $2(r_{X_1 X_2})(s_{\bar{X}_1})(s_{\bar{X}_2})$), the latter equation is identical to Equation 11.2 (the equation for the *t* **test for two independent samples** when $n_1 = n_2$). The value $r_{X_1 X_2}$ represents the coefficient of correlation between the two scores of subjects (or matched pairs of subjects) on the dependent variable. It is expected that as a result of using the same subjects in both conditions (or by employing matched subjects), a positive correlation will exist between pairs of scores (i.e., scores that are in the same row of Table 17.1). The closer the value of $r_{X_1 X_2}$ is to $+1$, the stronger the association between the scores of subjects on the dependent variable. (When $r = +1$, subjects who have a high score in Condition 1 will have a comparably high score in Condition 2, and subjects who have a low score in Condition 1 will have a comparably low score in Condition 2.) As the value of $r_{X_1 X_2}$ approaches $+1$, the value of the denominator of Equation 17.6 decreases, which will result in an increase in the absolute value computed for t. Note that if $r_{X_1 X_2} = 0$, the denominator of Equation 17.6

becomes identical to the denominator of Equation 11.2. Thus, if the scores of the n subjects under the two experimental conditions are not correlated with one another, the equation for the **t test for two dependent samples** (as represented by Equation 17.6) reduces to Equation 11.2.

The intent of the above discussion is to illustrate that one advantage of employing a design which can be evaluated with the **t test for two dependent samples**, as opposed to a design that is evaluated with the **t test for two independent samples**, is that if there is a substantial positive correlation between pairs of scores, the former test will provide a more powerful test of an alternative hypothesis than will the latter test. The greater power associated with the **t test for two dependent samples** is a direct result of the lower value which will be computed for the denominator of Equation 17.6 when contrasted with the denominator that will be computed for Equation 11.2 for the same set of data. In the case of both equations, the denominator is an estimated measure of variability in a sampling distribution. By employing pairs of scores which are positively correlated with one another, the estimated variability in the sampling distribution will be less than will be the case if the scores are not correlated with one another.[10]

The computation of t with Equation 17.6 will now be illustrated. Note that in order to compute t with the latter equation, the following values are required: $\bar{X}_1$, $\bar{X}_2$, $s_{\bar{X}_1}$, $s_{\bar{X}_2}$, $s_{\bar{X}_1}^2$, $s_{\bar{X}_2}^2$, $r_{X_1 X_2}$. In order to compute the estimated population variances and standard deviations, the values ΣX_1^2 and ΣX_2^2 are required. The latter values are computed in Table 17.3. Employing the summary information provided in Table 17.1 and the values $\Sigma X_1^2 = 293$ and $\Sigma X_2^2 = 147$, all of the above noted values, with the exception of $r_{X_1 X_2}$, are computed.

$$\bar{X}_1 = \frac{\Sigma X_1}{n} = \frac{47}{10} = 4.7 \quad \bar{X}_2 = \frac{\Sigma X_2}{n} = \frac{31}{10} = 3.1$$

$$\tilde{s}_1 = \sqrt{\frac{\Sigma X_1^2 - \frac{(\Sigma X_1)^2}{n}}{n-1}} = \sqrt{\frac{293 - \frac{(47)^2}{10}}{10-1}} = 2.83 \quad \tilde{s}_2 = \sqrt{\frac{\Sigma X_2^2 - \frac{(\Sigma X_2)^2}{n}}{n-1}} = \sqrt{\frac{147 - \frac{(31)^2}{10}}{10-1}} = 2.38$$

$$s_{\bar{X}_1} = \frac{\tilde{s}_1}{\sqrt{n}} = \frac{2.83}{\sqrt{10}} = .89 \quad s_{\bar{X}_2} = \frac{\tilde{s}_2}{\sqrt{n}} = \frac{2.38}{\sqrt{10}} = .75$$

$$s_{\bar{X}_1}^2 = (.89)^2 = .79 \quad s_{\bar{X}_2}^2 = (.75)^2 = .56$$

Equation 17.7 is employed to compute the value $r_{X_1 X_2}$.

$$r_{X_1 X_2} = \frac{\Sigma X_1 X_2 - \frac{(\Sigma X_1)(\Sigma X_2)}{n}}{\sqrt{\left[\Sigma X_1^2 - \frac{(\Sigma X_1)^2}{n}\right]\left[\Sigma X_2^2 - \frac{(\Sigma X_2)^2}{n}\right]}} \qquad \textbf{(Equation 17.7)}$$

The only value in Equation 17.7 that is required to compute $r_{X_1 X_2}$ which has not been computed for Example 17.1 is the term $\Sigma X_1 X_2$ in the numerator. The latter value, which is computed in Table 17.3, is obtained as follows: Each subject's X_1 score is multiplied by the subject's X_2 score. The resulting score represents an $X_1 X_2$ score for the subject. The n $X_1 X_2$ scores are summed, and the resulting value represents the term $\Sigma X_1 X_2$ in Equation 17.7.

Employing Equation 17.7, the value $r_{X_1 X_2} = .78$ is computed.

$$r_{X_1 X_2} = \frac{193 - \frac{(47)(31)}{10}}{\sqrt{\left[293 - \frac{(47)^2}{10}\right]\left[147 - \frac{(31)^2}{10}\right]}} = .78$$

Table 17.3 Computation of ΣX_1^2, ΣX_2^2, and $\Sigma X_1 X_2$ for Example 17.1

	Condition 1		Condition 2		
Subject	X_1	X_1^2	X_2	X_2^2	$X_1 X_2$
1	9	81	8	64	72
2	2	4	2	4	4
3	1	1	3	9	3
4	4	16	2	4	8
5	6	36	3	9	18
6	4	16	0	0	0
7	7	49	4	16	28
8	8	64	5	25	40
9	5	25	4	16	20
10	1	1	0	0	0
	$\Sigma X_1 = 47$	$\Sigma X_1^2 = 293$	$X_2 = 31$	$X_2^2 = 147$	$\Sigma X_1 X_2 = 193$

When the relevant values are substituted in Equation 17.6, the value $t = 2.86$ is computed (which is the same value computed for t with Equation 17.1). Note that the value of the numerator of Equation 17.6 is $\bar{X}_1 - \bar{X}_2 = \bar{D} = 1.6$.[11]

$$t = \frac{4.7 - 3.1}{\sqrt{.79 + .56 - 2(.78)(.89)(.75)}} = 2.86$$

In order to illustrate that the *t* **test for two dependent samples** provides a more powerful test of an alternative hypothesis than the *t* **test for two independent samples**, the data for Example 17.1 will be evaluated with Equation 11.2 (which is the equation for the latter test). In employing Equation 11.2, the positive correlation which exists between the scores of subjects under the two experimental conditions will not be taken into account. Use of Equation 11.2 with the data for Example 17.1 assumes that in lieu of having $n = 10$ subjects, there are instead two independent groups, each group being comprised of 10 subjects. Thus, $n_1 = 10$, and the 10 X_1 scores in Table 17.1 represent the scores of the 10 subjects in Group 1, and $n_2 = 10$, and the 10 X_2 scores in Table 17.1 represent the scores of the 10 subjects in Group 2. Since the values $\bar{X}_1$, $\bar{X}_2$, $s_{\bar{X}_1}^2$, and $s_{\bar{X}_2}^2$ have already been computed, they can be substituted in Equation 11.2. When the relevant values are substituted in Equation 11.2, the value $t = 1.37$ is computed.

$$t = \frac{\bar{X}_1 - \bar{X}_2}{\sqrt{s_{\bar{X}_1}^2 + s_{\bar{X}_2}^2}} = \frac{4.7 - 3.1}{\sqrt{.79 + .56}} = 1.37$$

Employing Equation 11.4, the degrees of freedom for Equation 11.2 are $df = 10 + 10 - 2 = 18$. In **Table A2**, the tabled critical two-tailed .05 and .01 values for $df = 18$ are $t_{.05} = 2.10$ and $t_{.01} = 2.88$, and the tabled critical one-tailed .05 and .01 values are $t_{.05} = 1.73$ and $t_{.01} = 2.55$.

Since the obtained value $t = 1.37$ is less than all of the aforementioned critical values, the null hypothesis cannot be rejected, regardless of whether a nondirectional or directional alternative hypothesis is employed. Recollect that when the same set of data is evaluated with Equations 17.1/17.6, the nondirectional alternative hypothesis H_1: $\mu_1 \neq \mu_2$ is supported at the .05 level and the directional alternative hypothesis H_1: $\mu_1 > \mu_2$ is supported at both the .05 and .01 levels.

Note that by employing twice the degrees of freedom, the tabled critical values employed for Equation 11.2 will always be smaller than those employed for Equations 17.1/17.6. However, if there is a reasonably high positive correlation between the pairs of scores, the lower critical values associated with the *t* **test for two independent samples** will be offset by the fact that the *t* value computed with Equation 11.2 will be substantially smaller than the value computed with Equations 17.1/17.6.

When the *t* **test for two dependent samples** is employed to evaluate a dependent samples design, it is assumed that a positive correlation exists between the scores of subjects in the two experimental conditions. It is, however, theoretically possible (although unlikely) that the two scores of subjects will be negatively correlated. If, in fact, the correlation between the X_1 and X_2 scores of subjects is negative, the value of the denominator of Equation 17.6 will actually be larger than will be the case if Equation 11.2 is employed to evaluate the same set of data (since in the denominator of Equation 17.6, if $r_{X_1 X_2}$ is a negative number, the product $2(r_{X_1 X_2})(s_{\bar{X}_1})(s_{\bar{X}_2})$ will be added to instead of subtracted from $s_{\bar{X}_1}^2 + s_{\bar{X}_2}^2$). In addition, if the correlation between the two scores is a very low positive value which is close to 0, the slight increment in the value of *t* computed with Equation 17.6 may be offset by the loss of degrees of freedom (and the consequent increase in the tabled critical *t* value), so as to allow Equation 11.2 (the *t* **test for two independent samples**) to provide a more powerful test of an alternative hypothesis than Equation 17.6 (the *t* **test for two dependent samples**).

In the unlikely event that in a dependent samples design there is a substantial negative correlation between subjects' scores in the two experimental conditions, it is very unlikely that evaluation of the data with Equation 17.6 will yield a significant result. However, the presence of a significant negative correlation, in and of itself, can certainly be of statistical importance. When a negative correlation is present, subjects who obtain a high score in one experimental condition will obtain a low score in the other experimental condition, and vice versa. The closer the negative correlation is to -1, the more pronounced the tendency for a subject's scores in the two conditions to be in the opposite direction. To illustrate the presence of a negative correlation, consider the following example. Assume that employing a dependent samples design, each of five subjects who serve in two experimental conditions obtains the following scores: **Subject 1** (1, 5); **Subject 2** (2, 4); **Subject 3** (3, 3); **Subject 4** (4, 2); **Subject 5** (5, 1). In this hypothetical example, it turns out that the correlation between the five pairs of scores is $r_{X_1 X_2} = -1$, which is the strongest possible negative correlation. Since, however, the mean and median value for both of the experimental conditions is equal to 3, evaluation of the data with the *t* **test for two dependent samples**, as well as the more commonly employed nonparametric procedures employed for a design involving two dependent samples (such as the **Wilcoxon matched-pairs signed-ranks test (Test 18)** and the **binomial sign test for two dependent samples (Test 19)**), will lead one to conclude there is no difference between the scores of subjects under the two conditions. This is the case, since such tests base the comparison of conditions on an actual or implied measure of central tendency. If it is assumed that in the above example the sample data accurately reflect what is true with respect to the underlying populations, it appears that the higher a subject's score in Condition 1, the lower the subject's score in Condition 2, and vice versa. In such a case, it can be argued that if the coefficient of correlation is statistically significant, that in itself can indicate the presence of a significant treatment effect (albeit an unusual one), in that there is a significant association between the two sets of scores. The determination of whether or not a correlation coefficient is significant is discussed under the **Pearson product-moment correlation coefficient**, as well as in the discussion of a number of the other correlational procedures in the book.[12]

2. The equation for the *t* test for two dependent samples when a value for a difference other than zero is stated in the null hypothesis If in stating the null hypothesis for the *t* test for two dependent samples a researcher stipulates that the difference between μ_1 and μ_2 is some value other than zero, Equation 17.8 is employed to evaluate the null hypothesis in lieu of Equation 17.6.[13]

$$t = \frac{(\bar{X}_1 - \bar{X}_2) - (\mu_1 - \mu_2)}{\sqrt{s_{\bar{X}_1}^2 + s_{\bar{X}_2}^2 - 2(r_{X_1 X_2})(s_{\bar{X}_1})(s_{\bar{X}_2})}} \qquad \textbf{(Equation 17.8)}$$

When the null hypothesis is H_0: $\mu_1 = \mu_2$ (which as noted previously can also be written as H_0: $\mu_1 - \mu_2 = 0$), the value of $(\mu_1 - \mu_2)$ in Equation 17.8 reduces to zero, and thus what remains of the numerator in Equation 17.8 is $(\bar{X}_1 - \bar{X}_2)$ (which represents the numerator of Equation 17.6). In evaluating the value of *t* computed with Equation 17.8, the same protocol is employed that is described for evaluating a *t* value for the **t test for two independent samples** when the difference stated in the null hypothesis for the latter test is some value other than zero (in which case Equation 11.5 is employed to compute *t*).

3. Test 17a: The *t* test for homogeneity of variance for two dependent samples: Evaluation of the homogeneity of variance assumption of the *t* test for two dependent samples Prior to reading this section, the reader should review the discussion of homogeneity of variance in Section VI of the *t* test for two independent samples. As is the case with an independent samples design, in a dependent samples design the homogeneity of variance assumption evaluates whether or not there is evidence to indicate that an inequality exists between the variances of the populations represented by the two experimental conditions. The null and alternative hypotheses employed in evaluating the homogeneity of variance assumption are as follows.

Null hypothesis H_0: $\sigma_1^2 = \sigma_2^2$

(The variance of the population Condition 1 represents equals the variance of the population Condition 2 represents.)

Alternative hypothesis H_1: $\sigma_1^2 \neq \sigma_2^2$

(The variance of the population Condition 1 represents does not equal the variance of the population Condition 2 represents. This is a **nondirectional alternative hypothesis** and it is evaluated with a **two-tailed test**. In evaluating the homogeneity of variance assumption, a nondirectional alternative hypothesis is always employed.)

The test which will be described in this section for evaluating the homogeneity of variance assumption is referred to as the *t* **test for homogeneity of variance for two dependent samples**. The reader should take note of the fact that the F_{max} **test/F test for two population variances (Test 11a)** (employed in evaluating the homogeneity of variance assumption for the *t* **test for two independent samples**) is not appropriate to use with a dependent samples design, since it does not take into account the correlation between subjects' scores in the two experimental conditions (i.e., $r_{X_1 X_2}$). Equation 17.9 is the equation for the *t* test for **homogeneity of variance for two dependent samples**.

$$t = \frac{(\tilde{s}_L^2 - \tilde{s}_S^2)\sqrt{(n - 2)}}{\sqrt{4\tilde{s}_L^2 \tilde{s}_S^2(1 - r_{X_1 X_2}^2)}}$$ **(Equation 17.9)**

Where: $\tilde{s}_L^2$ is the larger of the two estimated population variances

$\tilde{s}_S^2$ is the smaller of the two estimated population variances

Since for Example 17.1 it has already been determined that $\tilde{s}_1 = 2.83$ and $\tilde{s}_2 = 2.38$, by squaring the latter values we can determine the values of the estimated population variances. Thus: $\tilde{s}_1^2 = (2.83)^2 = 8.01 = \tilde{s}_L^2$ and $\tilde{s}_2^2 = (2.38)^2 = 5.66 = \tilde{s}_S^2$. Substituting the appropriate values in Equation 17.9, the value $t = .79$ is computed.

$$t = \frac{(8.01 - 5.66)\sqrt{(10 - 2)}}{\sqrt{4(8.01)(5.66)(1 - (.78)^2)}} = .79$$

The degrees of freedom to employ for evaluating the *t* value computed with Equation 17.9 are computed with Equation 17.10.

$$df = n - 2$$ **(Equation 17.10)**

Employing Equation 17.10, the degrees of freedom for the analysis are $df = 10 - 2 = 8$. For $df = 8$, the tabled critical two-tailed .05 and .01 values in **Table A2** are $t_{.05} = 2.31$ and $t_{.01} = 3.35$. In order to reject the null hypothesis the obtained value of *t* must be equal to or greater than the tabled critical value at the prespecified level of significance. Since the value $t = .79$ is less than both of the aforementioned critical values, the null hypothesis cannot be rejected. Thus, the homogeneity of variance assumption is not violated.

There are a number of additional points which should be noted with respect to the *t* **test for homogeneity of variance for two dependent samples**.

a) Unless $\tilde{s}_L^2 = \tilde{s}_S^2$ (in which case $t = 0$), Equation 17.9 will always yield a positive *t* value. This is the case, since in the numerator of the equation the smaller variance is subtracted from the larger variance.

b) In some sources Equation 17.9 is written in the form of Equation 17.11, which employs the notation $\tilde{s}_1^2$ and $\tilde{s}_2^2$ in place of $\tilde{s}_L^2$ and $\tilde{s}_S^2$.

$$t = \frac{(\tilde{s}_1^2 - \tilde{s}_2^2)\sqrt{(n - 2)}}{\sqrt{4\tilde{s}_1^2 \tilde{s}_2^2(1 - r_{X_1 X_2}^2)}}$$ **(Equation 17.11)**

If Equation 17.11 is employed, the computed value of *t* can be a negative number. Specifically, *t* will be negative when $\tilde{s}_2^2 > \tilde{s}_1^2$. In point of fact, the sign of *t* is irrelevant, unless one is evaluating a directional alternative hypothesis. Since the homogeneity of variance assumption involves evaluation of a nondirectional alternative hypothesis, when Equation 17.11 is employed, the researcher is only interested in the absolute value of *t*.

c) As is the case for a test of homogeneity of variance for two independent samples, it is possible to use the *t* **test for homogeneity of variance for two dependent samples** to evaluate a directional alternative hypothesis regarding the relationship between the variances of two populations. Thus, if a researcher specifically predicts that the variance of the population represented by Condition 1 is larger than the variance of the population represented by Condition 2 (i.e., H_1: $\sigma_1^2 > \sigma_2^2$), or that the variance of the population represented by Condition 1 is smaller than the variance of the population represented by Condition 2 (i.e., H_1: $\sigma_1^2 < \sigma_2^2$), the latter pair of directional alternative hypotheses can be evaluated with Equation 17.11. In such a case, the sign of the computed t value is relevant. If one employs Equation 17.11 to evaluate a directional alternative hypothesis, the following guidelines are employed in evaluating the null hypothesis.

1) If the directional alternative hypothesis H_1: $\sigma_1^2 > \sigma_2^2$ is employed, the null hypothesis can be rejected if the obtained absolute value of t is positive, and the value of t is equal to or greater than the tabled critical one-tailed value at the prespecified level of significance.

2) If the directional alternative hypothesis H_1: $\sigma_1^2 < \sigma_2^2$ is employed, the null hypothesis can be rejected if the sign of t is negative, and the absolute value of t is equal to or greater than the tabled critical one-tailed value at the prespecified level of significance.

If the directional alternative hypothesis H_1: $\sigma_1^2 > \sigma_2^2$ is evaluated for Example 17.1, the tabled critical one-tailed .05 and .01 values employed for the analysis are $t_{.05} = 1.86$ and $t_{.01} = 2.90$ (which respectively correspond to the tabled values at the 95th and 99th percentiles). Although the data are consistent with the directional alternative hypothesis H_1: $\sigma_1^2 > \sigma_2^2$, the null hypothesis cannot be rejected, since the obtained value $t = .79$ is less than the aforementioned one-tailed critical values. ($t = .79$ is obtained with Equation 17.11, since as previously noted, $\tilde{s}_1^2 = \tilde{s}_L^2 = 8.01$ and $\tilde{s}_2^2 = \tilde{s}_S^2 = 5.66$.)

d) Equation 17.12 is an alternative but equivalent form of Equation 17.11.

$$t = \frac{(F - 1)\sqrt{(n - 2)}}{2\sqrt{F(1 - r_{X_1 X_2}^2)}} \qquad \textbf{(Equation 17.12)}$$

In Equation 17.12, the value of F is computed with Equation 11.8 ($F = \tilde{s}_1^2 / \tilde{s}_2^2$, which is described under the *t* **test for two independent samples**). Substituting the appropriate values in Equation 17.12, the value $t = .79$ is computed.

$$F = \frac{8.01}{5.66} = 1.42 \qquad t = \frac{(1.42 - 1)\sqrt{(10-2)}}{2\sqrt{(1.42(1 - .78^2)}} = .79$$

Note that Equation 17.12 can only yield a positive t value.

e) Equation 17.13 is an alternative but equivalent form of Equation 17.9 which can only be employed to evaluate a nondirectional alternative hypothesis. Since $\tilde{s}_1^2 = \tilde{s}_L^2 = 8.01$ and $\tilde{s}_2^2 = \tilde{s}_S^2 = 5.66$, Equation 17.13 yields the value $t = .79$ (since $F_{max} = 8.01/5.66 = 1.42$).

$$\textbf{(Equation 17.13)}$$

$$t = \frac{(F_{max} - 1)\sqrt{(n - 2)}}{2\sqrt{F_{max}(1 - r_{X_1 X_2}^2)}} = \frac{(1.42 - 1)\sqrt{10 - 2}}{2\sqrt{(1.42)[1 - (.78)^2]}} = .79$$

Where: $F_{max} = \tilde{s}_L^2 / \tilde{s}_S^2$ (which is Equation 11.6)

All of the equations noted in this section for the *t* **test for homogeneity of variance for two dependent samples** are based on the following two assumptions: a) The samples have been randomly drawn from the populations they represent; and b) The distribution of data in the underlying population each of the samples represents is normal. It is noted in the discussion of homogeneity of variance in Section VI of the *t* **test for two independent samples** that violation of the normality assumption can severely compromise the reliability of certain tests of homogeneity of variance. The *t* **test for homogeneity of variance for two dependent samples** is among those tests whose reliability can be compromised if the normality assumption is violated.

The problems associated with the use of the *t* **test for two independent samples** when the homogeneity of variance assumption is violated are also applicable to the *t* **test for two dependent samples**. Thus, if the homogeneity of variance assumption is violated, it will generally inflate the Type I error rate associated with the *t* **test for two dependent samples**. The reader should take note of the fact that when the homogeneity of variance assumption is violated with a dependent samples design, its effect on the Type I error rate will be greater than for an independent samples design. In the event the homogeneity of variance assumption is violated for a dependent samples design, either of the following strategies can be employed: a) In conducting the *t* **test for two dependent samples**, the researcher can run a more conservative test. Thus, if the researcher does not want the Type I error rate to be greater than .05, instead of employing $t_{.05}$ as the tabled critical value, she can employ $t_{.01}$ to represent the latter value; or b) In lieu of the *t* **test for two dependent samples**, a nonparametric test which does not assume homogeneity of variance can be employed to evaluate the data (such as the **Wilcoxon matched-pairs signed-ranks test**). A more detailed discussion of violation of the homogeneity of variance assumption for a dependent samples design can be found in Section VI (under violation of the sphericity assumption) of the **single-factor within-subjects analysis of variance (Test 24)**, which is an alternative parametric procedure that can be employed to evaluate the data for Example 17.1.

4. Computation of the power of the *t* test for two dependent samples and the application of Test 17b: Cohen's *d* index In this section the two methods for computing power which are described for computing the power of the *t* **test for two independent samples** will be extended to the *t* **test for two dependent samples**. Prior to reading this section, the reader may find it useful to review the discussion of power for both the **single-sample *t* test** and the *t* **test for two independent samples**.

The first procedure to be described is the graphical method, which reveals the logic underlying the power computations for the *t* **test for two dependent samples**. In the discussion to follow, it will be assumed that the null hypothesis is identical to that employed for Example 17.1 (i.e., H_0: $\mu_1 - \mu_2 = 0$, which, as previously noted, is another way of writing H_0: $\mu_1 = \mu_2$). It will also be assumed that the researcher wants to evaluate the power of the *t* **test for two dependent samples** in reference to the following alternative hypothesis: H_1: $|\mu_1 - \mu_2| \geq 1.6$ (which is the difference obtained between the means of the two experimental conditions in Example 17.1). In other words, it is predicted that the absolute value of the difference between the two means is equal to or greater than 1.6. The latter alternative hypothesis is employed in lieu of H_1: $\mu_1 - \mu_2 \neq 0$ (which can also be written as H_1: $\mu_1 \neq \mu_2$), since in order to compute the power of the test, a specific value must be stated for the difference between the population means. Note that, as stated, the alternative hypothesis stipulates a nondirectional analysis, since it does not specify which of the two means will be the larger value. It will be assumed that $\alpha = .05$ is employed in the analysis.

Figure 17.1, which provides a visual summary of the power analysis, is comprised of two overlapping sampling distributions of difference scores. The distribution on the left, which will be designated as Distribution A, is a sampling distribution of difference scores that has a mean value of zero (i.e., $\mu_{\bar{D}} = \mu_{\bar{X}_1 - \bar{X}_2} = 0$). This latter value will be represented by $\mu_{D_0} = 0$ in

Figure 17.1. Distribution A represents the sampling distribution which describes the distribution of difference scores if the null hypothesis is true. The distribution on the right, which will be designated as Distribution B, is a sampling distribution of difference scores which has a mean value of 1.6 (i.e., $\mu_D = \mu_{\bar{X}_1 - \bar{X}_2} = 1.6$). This latter value will be represented by $\mu_{D_1} = 1.6$ in Figure 17.1. Distribution B represents the sampling distribution that describes the distribution of difference scores if the alternative hypothesis is true. Each of the sampling distributions has a standard deviation which is equal to the value computed for $s_{\bar{D}} = .56$, the estimated standard error of the mean difference, since the latter value provides the best estimate of the standard deviation of the mean difference in the underlying populations.

In Figure 17.1, area (///) delineates the proportion of Distribution A that corresponds to the value $\alpha/2$, which equals .025. This is the case, since $\alpha = .05$ and a two-tailed analysis is conducted. Area ($\equiv$) delineates the proportion of Distribution B which corresponds to the probability of committing a Type II error (β). Area (\\\) delineates the proportion of Distribution B that represents the power of the test (i.e., $1 - \beta$).

The procedure for computing the proportions documented in Figure 17.1 will now be described. The first step in computing the power of the test requires one to determine how large a difference there must be between the sample means in order to reject the null hypothesis. In order to do this, we algebraically transpose the terms in Equations 17.1/17.6, using $s_{\bar{D}}$ to summarize the denominator of the equation, and $t_{.05}$ (the tabled critical two-tailed .05 t value) to represent t. Thus: $\bar{X}_1 - \bar{X}_2 = (t_{.05})(s_{\bar{D}})$. By substituting the values $t_{.05} = 2.26$ and $s_{\bar{D}} = .56$ in the latter equation, we determine that the minimum required difference is $\bar{X}_1 - \bar{X}_2 = (2.26)(.56) = 1.27$ (which is represented by the notation $\bar{X}_D = 1.27$ in Figure 17.1). Thus, any difference between the two population means which is equal to or greater than 1.27 will allow the researcher to reject the null hypothesis at the .05 level.

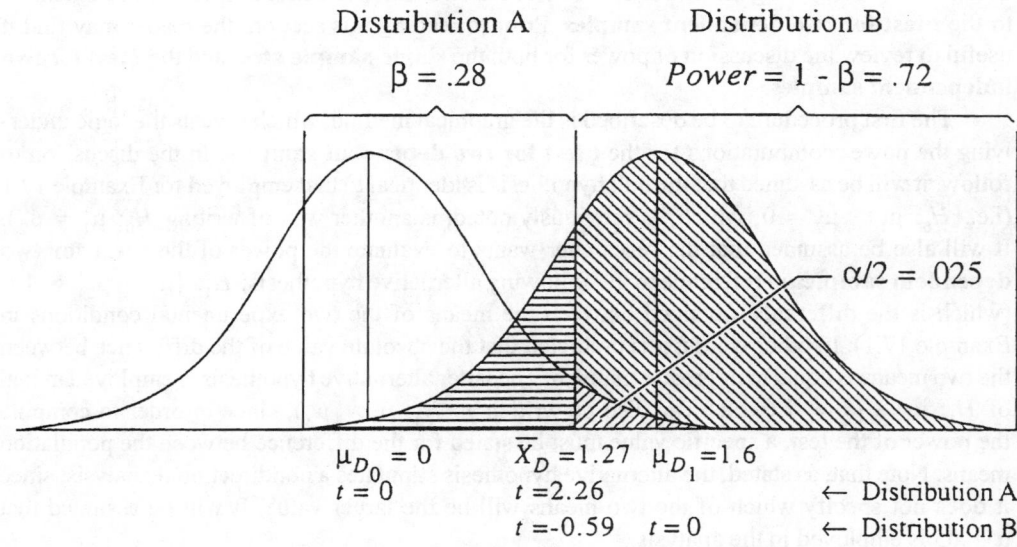

Distribution A   Distribution B

$\beta = .28$ $Power = 1 - \beta = .72$

$\alpha/2 = .025$

$\mu_{D_0} = 0$ $\bar{X}_D = 1.27$ $\mu_{D_1} = 1.6$
$t = 0$ $t = 2.26$ ← Distribution A
                  $t = -0.59$ $t = 0$ ← Distribution B

Figure 17.1   Visual Representation of Power for Example 17.1

The next step in the analysis requires one to compute the area in Distribution B that falls between the mean difference μ_{D_1} = 1.6 (i.e., the mean of Distribution B) and a mean difference equal to 1.27 (represented by the notation $\bar{X}_D$ = 1.27 in Figure 17.1). This is accomplished by employing Equations 17.1/17.6. In using the latter equation, the value of $\bar{X}_1$ is represented by 1.27 and the value of $\bar{X}_2$ by μ_{D_1} = 1.6.

$$ t = \frac{\bar{X}_1 - \bar{X}_2}{s_{\bar{D}}} = \frac{1.27 - 1.6}{.56} = -.59 $$

By interpolating the values listed in **Table A2** for $df = 9$ (since $n = 10$), we determine that the proportion of Distribution B which falls between the mean and a t score of $-.59$ (which corresponds to a mean difference of 1.27) is approximately .22. The latter area plus the 50% of Distribution B to the right of the mean corresponds to area (\\\\) in Distribution B. Note that the left boundary of area (\\\\) is also the boundary delineating the extreme 2.5% of Distribution A (i.e., $\alpha/2$ = .025, which is the rejection zone for the null hypothesis). Since area (\\\\) in Distribution B overlaps the rejection zone in Distribution A, area (\\\\) represents the power of the test — i.e., it represents the likelihood of rejecting the null hypothesis if the alternative hypothesis is true. The power of the test is obtained by adding .22 and .5. Thus, the power of the test equals .72. The likelihood of committing a Type II error (β) is represented by area ($\equiv$), which comprises the remainder of Distribution B. The proportion of Distribution B which constitutes this latter area is determined by subtracting the value .72 from 1. Thus: $\beta = 1 - .72 = .28$.

Based on the results of the power analysis, we can state that if the alternative hypothesis H_1: $|\mu_1 - \mu_2| \geq 1.6$ is true, the likelihood that the null hypothesis will be rejected is .72, and at the same time, there is a .28 likelihood that it will be retained. If the researcher considers the computed value for power too low (which in actuality should be determined prior to conducting a study), she can increase the power of the test by employing a larger sample size.

Method 2 for computing the power of the t test for two dependent samples employing Test 17b: Cohen's d index Method 2 described for computing the power of the t **test for two independent samples** can also be extended to the t **test for two dependent samples**. In using this latter method, the researcher must stipulate an **effect size** (d), which in the case of the t **test for two dependent samples** is computed with Equation 17.14. The effect size index computed with Equation 17.14 was developed by Cohen (1977, 1988), and is known as **Cohen's d index**. Further discussion of **Cohen's d index** (as well as the sample analogue referred to as the **g index**) can be found in Endnote 15 of the t **test for two independent samples**, and in Section IX (the **Appendix**) of the **Pearson product-moment correlation coefficient** under the discussion of **meta-analysis and related topics**.

$$ d = \frac{|\mu_1 - \mu_2|}{\sigma_D} \qquad \textbf{(Equation 17.14)} $$

The numerator of Equation 17.14 represents the hypothesized difference between the two population means. As is the case with the graphical method described previously, when a power analysis is conducted after the mean of each sample has been obtained, the difference between the two sample means (i.e., $\bar{X}_1 - \bar{X}_2$) is employed as an estimate of the value of $|\mu_1 - \mu_2|$ (although the method allows a researcher to test for any value as the hypothesized difference between the means). In Equation 17.14, the value of σ_D represents the standard deviation of the difference scores in the population. In order to compute the power of the t **test for two dependent samples**, the latter value must either be known or be estimated by the researcher. If power is computed after the sample data have been collected, one can employ the value computed for $\tilde{s}_D$ to estimate the value of σ_D. Thus, in the case of Example 17.1 we can employ $\tilde{s}_D = 1.78$ as an estimate of σ_D.

It should be noted that if one computes the power of a test prior to collecting the data (which is what a researcher should ideally do) most researchers will have great difficulty coming up with a reasonable estimate for the value of σ_D. Since a researcher is more likely to be able to estimate the values of σ_1 and σ_2 (i.e., the population standard deviation for each of the experimental conditions), if it can be assumed that $\sigma_1 = \sigma_2$ (which is true if the population variances are homogeneous) the value of σ_D can be estimated with Equation 17.15.[14]

$$\sigma_D = \sigma\sqrt{2(1 - \rho_{X_1 X_2})} \qquad \textbf{(Equation 17.15)}$$

Where: $\sigma = \sigma_1 = \sigma_2$ is the population variance for each of the experimental conditions
$\rho_{X_1 X_2}$ is the correlation between the two variables in the underlying populations

Since the effect size computed with Equation 17.14 is only based on population parameters, it is necessary to convert the value of d into a measure that takes into account the size of the sample (which is a relevant variable in determining the power of the test). This measure, as noted in the discussions of the **single-sample t test** and the **t test for two independent samples**, is referred to as the **noncentrality parameter**. Equation 17.16 is employed to compute the **noncentrality parameter** φ (which is a symbol for the lower case Greek letter **phi**) for the **t test for two dependent samples**.

$$\varphi = d\sqrt{n} \qquad \textbf{(Equation 17.16)}$$

The power of the **t test for two dependent samples** will now be computed using the data for Example 17.1. For purposes of illustration, it will be assumed that the minimum difference between the population means the researcher is trying to detect is the observed 1.6 point difference between the two sample means — i.e., $|\bar{X}_1 - \bar{X}_2| = |4.7 - 3.1| = 1.6 = |\mu_1 - \mu_2|$. The value of σ_D that will be employed in Equation 17.14 is $\tilde{s}_D = 1.78$ (which is the estimated value of the population parameter computed for the sample data). Substituting $|\mu_1 - \mu_2| = 1.6$ and $\sigma_D = 1.78$ in Equation 17.14, the value $d = .90$ is computed.

$$d = \frac{1.6}{1.78} = .90$$

Cohen (1977; 1988, pp. 24–27) has proposed the following (admittedly arbitrary) d values as criteria for identifying the magnitude of an effect size: a) A **small effect size** is one that is greater than .2 but not more than .5 standard deviation units; b) A **medium effect size** is one that is greater than .5 but not more than .8 standard deviation units; and c) A **large effect size** is greater than .8 standard deviation units. Employing Cohen's (1977, 1988) guidelines, the value $d = .90$ (which represents .90 standard deviation units) is categorized as a large effect size.

Along with the value $n = 10$, the value $d = .90$ is substituted in Equation 17.16, resulting in the value $\delta = 2.85$.

$$\varphi = .90\sqrt{10} = 2.85$$

The value $\varphi = 2.85$ is evaluated with **Table A3 (Power Curves for Student's t Distribution)** in the **Appendix**. We will assume that for the example under discussion a two-tailed test is conducted with $\alpha = .05$, and thus **Table A3-C** is the appropriate set of power curves to employ for the analysis. Since there is no curve for $df = 9$, the power of the test will be based on a curve that falls between the $df = 6$ and $df = 12$ power curves. Through interpolation, the power of the **t test for two dependent samples** is determined to be approximately .72 (which is the same value which is obtained with the graphical method). Thus, by employing 10 subjects the researcher has a probability of .72 of rejecting the null hypothesis if the true difference between the population means is equal to or greater than .90 σ_D units (which in Example 17.1 is equivalent to a 1.6 point difference between the means).

As long as a researcher knows or is able to estimate the value of σ_D, by employing trial and error she can substitute various values of n in Equation 17.16, until the computed value of φ corresponds to the desired power value for the **t test for two dependent samples** for a given effect size. This process can be facilitated by employing tables developed by Cohen (1977, 1988) which allow one to determine the minimum sample size necessary in order to achieve a specific level of power in reference to a given effect size.

Among others, Rosner (2000, pp. 309–312) and Zar (1999, p. 164) note that Equation 17.17 provides an alternative way of estimating the sample size (n) required in order to conduct the **t test for two dependent samples** which has a power of a specified value.

(Equation 17.17)

$$n = \frac{\sigma_D^2 \, (z_{\alpha/2} + z_{\beta})^2}{\delta^2}$$

Where: δ represents $|\mu_1 - \mu_2|$, the numerator of Equation 17.14. Although it will not be employed in illustrating the use of Equation 17.17, many sources often employ the value $| \bar{X}_1 - \bar{X}_2 |$, which is assumed to be the best estimate of $|\mu_1 - \mu_2|$ (the true difference between the population means) to represent the value of δ. $| \bar{X}_1 - \bar{X}_2 |$ is commonly employed to represent the value of δ, when after obtaining a nonsignificant result a researcher wants to determine if the power of the test conducted had been set at a specific value, how large a sample size would have been required in order for the **t test for two dependent samples** to identify the observed difference between the two treatment/condition means to be significant.

σ_D^2 represents the **variance of the difference scores for the population**, which is often represented in Equation 17.17 by s_D^2. Further discussion of the use of s_D^2 in Equation 17.17 can be found below.

$z_{\alpha/2}$ represents the relevant tabled critical **two-tailed** z value in **Table A1**. It is important to note that $z_{\alpha/2}$ is employed in Equation 17.17 when the test of the alternative hypothesis is **nondirectional**. If, on the other hand, a **directional** alternative hypothesis is evaluated, z_{α} is employed in place of $z_{\alpha/2}$ — i.e., the relevant tabled critical **one-tailed** z value in **Table A1** should be employed.

z_{β} is the z value listed in **Table A1** for which a proportion (probability) is recorded that identifies the Type II error rate (i.e., β) employed in the analysis. The protocol to employ to determine the value of z_{β} is as follows: a) If the **power** stipulated for the test (i.e., the value $(1 - \beta)$), is **greater than .5**, z_{β} is the z value for which the proportion recorded in **Column 3** of **Table A1** is equal to the value **stipulated for β**. (It is also the z value for which the proportion recorded in **Column 2** of **Table A1** is equal to the value (**Power** − .5).) When the **power** stipulated for the test is **greater than .5**, the sign of z_{β} will always be **positive**; b) If the **power** stipulated for the test is **less than .5**, z_{β} is the z value for which the proportion recorded in **Column 3** of **Table A1** is equal to the **power** of the test (i.e., the value $(1 - \beta)$). (It is also the z value for which the proportion recorded in **Column 2** of **Table A1** is equal to the value (β − .5).) When the **power** stipulated for the test is **less than .5**, the sign of z_{β}

will always be **negative**; c) If the **power** stipulated for the test is .5, $z_\beta = 0$. (In **Table A1** the value $\beta = .5000$ is listed in **Column 3**, and the value .0000 in **Column 2** represents the difference between β and .5 (i.e., $\beta - .5 = 0$, with **Power** $= 1 - \beta = .5$).)

The reader should take note of the following with regard to Equation 17.17: a) The value σ_D^2 represents the actual value of the **variance of the difference scores** for the population. σ_D^2 can also be written as follows: $\sigma_D^2 = \sigma_1^2 + \sigma_2^2 - 2(r_{X_1 X_2})(\sigma_1)(\sigma_2)$; b) If the value of σ_D^2 is not known (as will often be the case), s_D^2 (which can also be written as $s_D^2 = \tilde{s}_1^2 + \tilde{s}_2^2 - 2(r_{X_1 X_2})(\tilde{s}_1)(\tilde{s}_2)$) must be computed and employed in Equation 17.17 in place of it. (When, in fact, the value of σ_D^2 is known the *t* **test for two dependent samples** becomes *z* **test for two dependent samples** (**Test 11e**), which is discussed later in this section.) Theoretically, when s_D^2 is employed in Equation 17.17 instead of σ_D^2, the values $t_{\alpha/2}$ and t_β should be employed in the equation in place of $z_{\alpha/2}$ and z_β, and both of the values $t_{\alpha/2}$ and t_β will be larger than their respective analogs $z_{\alpha/2}$ and z_β. However, if one does not have access to exact probabilities for the *t* distribution, it will be difficult to estimate the appropriate value for t_β. Because of the latter, many sources employ s_D^2 in Equation 17.17 along with the relevant values for $z_{\alpha/2}$ and z_β. As a general rule, the value of *n* computed under the latter circumstances will underestimate the required sample size (unless the appropriate value for *n* is greater than 200). The underestimation of the sample size when s_D^2 is employed with $z_{\alpha/2}$ and z_β in Equation 17.17 results from the fact that the normal distribution allows for a more powerful test of an alternative hypothesis than the *t* distribution. Consequently, if the *t* distribution is the appropriate distribution to employ, it will require a slightly larger sample size to detect a difference than an analysis based on the normal distribution will indicate; c) Note that Equation 17.17 is identical to Equation 2.7 (which is employed to compute the sample size for the **single-sample *t* test**), except for the fact that σ_D^2 is employed in Equation 17.17 in place of σ^2. The commonality of the equations results from the fact that Equation 17.17 can be conceptualized as computing a sample size for a single sample of difference scores.

Equation 17.17 will be employed to demonstrate the computation of an appropriate sample size for Example 17.1. We will assume that the researcher plans to evaluate the data with a *t* **test for dependent samples**, and that a two-tailed analysis will be conducted with alpha set at .05. In addition, the researcher wants the power of the test to be .72 (which was the value computed for the power of the test earlier in this section for Example 17.1 using the alternative methods). In conducting the latter analysis, the following values are employed in Equation 17.17 to compute the number of subjects that will be required: $s_D^2 = 3.17$, $\delta = |\bar{X}_1 - \bar{X}_2| = |4.7 - 3.1| = 1.6$, $z_{\alpha/2} = 1.96$ (If a one-tailed analysis is conducted $z_{\alpha/2} = 1.65$ should be employed.), $z_\beta = .58$. The rationale for employing the value $z_\beta = .58$ is as follows: Since **Power** $= 1 - \beta = .72$, $\beta = .28$. Employing the guidelines for using Equation 17.17 documented above to compute a sample size when **Power** $> .5$, in **Table A1**, $z = .58$ is the *z* value for which the proportion (probability) recorded in **Column 3** (.2810) is closest to the value $\beta = .28$. (It is also the case that for $z = .58$, if .5 is added to the value recorded in **Column 2** (i.e., $.2190 + .5 = .7190$), the resulting value is closest to the value .72 stipulated for the power of the test (and thus, **Power** $- .5 = .7190 - .5 = .2190$).) Since the power stipulated for the test is greater than .5, the sign of z_β will be positive. As noted above, the observed absolute difference between the two means $|\bar{X}_1 - \bar{X}_2| = 1.6$ will be employed to represent the value of δ.

When the appropriate values are substituted in Equation 17.17, the value $n = 7.99$ is computed, indicating that in order for the power of the *t* **test for two dependent samples** to be .72, the researcher must employ at least 8 subjects (which is the next greatest integer value than the integer element computed for *n*). (As noted above, the value $s_D^2 = 3.17$ employed in Equation

17.17 is equal to $s_D^2 = \tilde{s}_1^2 + \tilde{s}_2^2 - 2\,(r_{X_1 X_2})(\tilde{s}_1)(\tilde{s}_2))$. Specifically, since $\tilde{s}_1^2 = (2.83)^2 = 8.01$ and $\tilde{s}_2^2$ $= (2.38)^2 = 5.66$, $s_D^2 = 8.01 + 5.66 - 2(.78)(2.83)(2.38) = 3.17$.)

$$n = \frac{(3.17)\,(1.96 + .58)^2}{(4.7 - 3.1)^2} = 7.99$$

This result (which is a little lower than the actual samples size employed, $n = 10$) is generally consistent with the results obtained earlier when a power value of .72 was computed for the analysis of Example 17.1 with the **t test for two dependent samples**. The slight discrepancy is attributable to the fact that instead of the exact values $t_{\alpha/2}$ and t_β, the values $z_{\alpha/2}$ and z_β were employed in Equation 17.17 (which results in an underestimation of the sample size, since with small sample sizes the **z test** is more powerful than the **t test**). In addition, it is possible that since interpolation was employed for the earlier methods, the latter could have resulted in some inaccuracy.

In order to demonstrate that at a specified level of power a **t test for two dependent samples** will require substantially fewer subjects than a **t test for two independent samples**, let us assume that the two experimental conditions in Example 17.1 represent two independent samples. In such a case, Equation 11.14 would be employed to compute the sample size required to conduct a **t test for two independent samples**. Once again we will set the value of the power at .72 and employ $\delta = 1.6$. We will assume there will be an equal number of subjects per group. The values $\tilde{s}_1^2 = (2.83)^2 = 8.01$, $\tilde{s}_2^2 = (2.38)^2 = 5.66$ (the estimated population variances for each of the experimental conditions) are employed in Equation 11.14 in place of s_D^2. The same values as before will be used for $z_{\alpha/2}$ and z_β (although it should be noted that if the actual values of $t_{\alpha/2}$ and t_β were used in Equation 11.14 they would be different, due to the fact that the degrees of freedom for the independent samples analysis is two times the degrees of freedom for the dependent samples analysis). The result of the analysis of the data for Example 17.1 when viewed within the framework of an independent samples analysis indicates that in order to have a test with power = .72, approximately 42 subjects are required per group. Note how much larger the latter value is than the value $n = 8$ computed for the dependent samples analysis.

$$n_1 = \frac{\left(8.01 + \dfrac{5.66}{1}\right)(1.96 + .84)^2}{(4.7 - 3.1)^2} = 41.86$$

5. Measure of magnitude of treatment effect for the t test for two dependent samples: Omega squared (Test 17c)   Prior to reading this section, the reader should review the discussion of **magnitude of treatment effect** and the **omega squared statistic** in Section VI of the **t test for two independent samples**. In the latter discussion, it is noted that the computation of a **t** value only provides a researcher with information concerning the likelihood of the null hypothesis being false, but does not provide information on the magnitude of any treatment effect which is present. As noted in the discussion of the **t test for two independent samples**, a treatment effect is defined as the proportion of the variability on the dependent variable that is associated with the experimental treatments/conditions. As is the case with the **t test for two independent samples**, the magnitude of a treatment effect for the **t test for two dependent samples** can be estimated with the **omega squared** statistic ($\hat{\omega}^2$).

Equation 11.15, which was employed to compute the value of omega squared for the **t test for two independent samples**, can be modified to compute the same statistic for the **t test for**

two dependent samples. Specifically, Equation 17.18 is employed to compute **omega squared** for the *t* **test for two dependent samples**. Employing Equation 17.18 with the data for Example 17.1, the value $\tilde{\omega}^2 = .264$ is computed. The value $\tilde{\omega}_p^2 = .264$ indicates 26.4% of the variability on the dependent variable (galvanic skin response) is associated with variability on the different levels of the independent variable (sexually explicit versus neutral words).

$$\tilde{\omega}^2 = \frac{t^2 - 1}{t^2 + 2n - 1} = \frac{(2.86)^2 - 1}{(2.86)^2 + (2)(10) - 1} = .264 \quad \text{(Equation 17.18)}$$

In the previous editions of this book a different but equivalent equation was employed to compute the value of **omega squared** for the *t* **test for two dependent samples**. In order to use the alternative equation it was necessary to obtain additional values which have not been computed for Example 17.1. The latter values are obtained within the framework of conducting a **single-factor within-subjects analysis of variance**, which, as noted earlier, is an alternative parametric procedure that can be employed to evaluate the data for Example 17.1 (yielding equivalent results). The derivation of the relevant values for computing **omega squared** (which are summarized in Table 17.4) is described in the discussion of the **single-factor within-subjects analysis of variance**.[15]

Table 17.4　Summary Table of Single-Factor Within-Subjects Analysis
of Variance for Example 17.1

Source of variation	SS	df	MS	F
Between-subjects	108.8	9	12.09	
Between-conditions	12.8	1	12.80	8.11
Residual	14.2	9	1.58	
Total	135.8	19		

Keppel (1991) and Kirk (1995) note that there is disagreement with respect to which of the components derived in the analysis of variance should be employed in computing **omega squared** for a within-subjects design. One method of computing **omega squared** (which computes a value referred to as **standard omega squared**) was employed in the first edition of this book. The latter method expresses treatment (i.e., between-conditions) variability as a proportion of the sum of all the elements which account for variability in a within-subjects design. Another method of computing **omega squared** (discussed in the second edition of the book) is referred to as **partial omega squared**. The latter measure, which Keppel (1991) and Kirk (1995) view as more meaningful than **standard omega squared**, ignores between-subjects variability, and expresses treatment (i.e., between-conditions) variability as a proportion of the sum of between-conditions, and residual variability. For a set of data, the value computed for **partial omega squared** will always be larger than the value computed for **standard omega squared**.

Equation 17.19 is employed to compute **partial omega squared** ($\tilde{\omega}_p^2$). Since in Equation 17.19 *k* equals the number of experimental conditions, in the case of the *t* **test for two dependent samples** *k* will always equal 2. (Since, $t^2 = F$ and $k = 2$, if the aforementioned values are substituted in Equation 17.19, Equation 17.18 will result.)

$$\tilde{\omega}_p^2 = \frac{(k - 1)(F_{BC} - 1)}{(k - 1)(F_{BC} - 1) + nk}$$

(Equation 17.19)

$$\tilde{\omega}_p^2 = \frac{(2 - 1)(8.11 - 1)}{(2 - 1)(8.11 - 1) + (10)(2)} = .262$$

Note that the value $\tilde{\omega}_p^2 = .262$ computed for **partial omega squared** with Equation 17.19 is equivalent to the value computed for **omega squared** with Equation 17.18 (the small difference is due to rounding off error). In the first edition of this book the value computed for **omega squared** was .08, which, as noted earlier, represents **standard omega squared**. The fact that .08 is less that .262 is consistent with the fact that the value of **standard omega squared** will always be smaller than the value of **partial omega squared**.

It is noted in an earlier discussion of **omega squared** (in Section VI of the *t* **test for two independent samples**) that Cohen (1977; 1988, pp. 285–288) has suggested the following (admittedly arbitrary) values, which are employed in psychology and a number of other disciplines, as guidelines for interpreting $\tilde{\omega}^2$: a) A **small effect size** is one that is greater than .0099 but not more than .0588; b) A **medium effect size** is one that is greater than .0588 but not more than .1379; and c) A **large effect size** is greater than .1379. If Cohen's (1977, 1988) guidelines are employed, the value $\tilde{\omega}_p^2 = .262$ is categorized as a large effect size. If the value .08 computed for **standard omega squared** is employed, it is categorized as a medium effect size.

A full discussion of the computation of an **omega squared** value for a within-subjects design can be found in Section VI of the **single-factor within-subjects analysis of variance**, as well as in Keppel (1991) and Kirk (1995).

6. Computation of a confidence interval for the *t* test for two dependent samples Prior to reading this section the reader should review the discussion of the computation of confidence intervals in Section VI of the **single-sample *t* test** and the *t* **test for two independent samples**. When interval/ratio data are available for two dependent samples, a confidence interval can be computed that identifies a range of values with respect to which one can be confident to a specified degree contains the true difference between the two population means. Equation 17.20 is the general equation for computing the confidence interval for the difference between two dependent population means.[16]

$$CI_{(1 - \alpha)} = (\overline{X}_1 - \overline{X}_2) \pm (t_\alpha)(s_{\overline{D}})$$

(Equation 17.20)

Where: $t_{\alpha/2}$ represents the tabled critical two-tailed value in the *t* distribution, for $df = n - 1$, below which a proportion (percentage) equal to $[1 - (\alpha/2)]$ of the cases falls. If the proportion (percentage) of the distribution that falls within the confidence interval is subtracted from 1 (100%), it will equal the value of α.

Employing Equation 17.20, the 95% interval for Example 17.1 is computed below. In employing Equation 17.20, $(\overline{X}_1 - \overline{X}_2)$ represents the obtained difference between the means of the two conditions (which is the numerator of the equation used to compute the value of *t*), $t_{.05}$ represents the tabled critical two-tailed .05 value for $df = n - 1$, and $s_{\overline{D}}$ represents the standard error of the mean difference (which is the denominator of the equation used to compute the value of *t*).

$$CI_{.95} = (\bar{X}_1 - \bar{X}_2) \pm (t_{.05})(s_{\bar{D}}) = 1.6 \pm (2.26)(.56) = 1.6 \pm 1.27$$

$$.33 \leq (\mu_1 - \mu_2) \leq 2.87$$

This result indicates the researcher can be 95% confident (or the probability is .95) that the interval .33 to 2.87 contains the true difference between the population means. If the aforementioned interval contains the true difference between the population means, the mean of the population Condition 1 represents is greater than the mean of the population Condition 2 represents by at least .33 GSR units but not by more than 2.87 GSR units.

The 99% confidence interval for Example 17.1 will also be computed to illustrate that the range of values that define a 99% confidence interval is always larger than the range which defines a 95% confidence interval.

$$CI_{.99} = (\bar{X}_1 - \bar{X}_2) \pm (t_{.01})(s_{\bar{D}}) = 1.6 \pm (3.25)(.56) = 1.6 \pm 1.82$$

$$-.22 \leq (\mu_1 - \mu_2) \leq 3.42$$

Thus the researcher can be 99% confident (or the probability is .99) that the interval -.22 to 3.42 contains the true difference between the population means. If the aforementioned interval contains the true difference between the population means, the mean of the population Condition 2 represents is no more than .22 GSR units higher than the mean of the population Condition 1 represents, and the mean of the population Condition 1 represents is no more than 3.42 GSR units higher than the mean of the population Condition 2 represents. The reader should take note of the fact that the reliability of Equation 17.20 will be compromised if one or more of the assumptions of the *t* **test for two dependent samples** are saliently violated.

7. Test 17d: Sandler's *A* test Sandler (1955) derived a computationally simpler procedure, referred to as **Sandler's *A* test**, which is mathematically equivalent to the *t* **test for two dependent samples**. The test statistic for **Sandler's *A* test** is computed with Equation 17.21.

$$A = \frac{\Sigma D^2}{(\Sigma D)^2} \qquad \text{(Equation 17.21)}$$

Note that in Equation 17.21, ΣD and ΣD^2 are the same elements computed in Table 17.1 which are employed for the direct difference method for the *t* **test for two dependent samples**. When Equation 17.21 is employed for Example 17.1, the value $A = .211$ is computed.

$$A = \frac{54}{(16)^2} = .211$$

The reader should take note of the fact that except for when $\Sigma D = 0$, the value of A must be a positive number. If a negative value is obtained for A, it indicates that a computational error has been made. If $\Sigma D = 0$ (which indicates that the means of the two conditions are equal), Equation 17.21 becomes unsolvable.

The obtained value $A = .211$ is evaluated with **Table A12 (Table of Sandler's *A* Statistic)** in the **Appendix**. As is the case for the *t* **test for two dependent samples**, the degrees of freedom employed for **Sandler's *A* test** are computed with Equation 17.5. Thus, $df = 10 - 1 = 9$. The tabled critical two-tailed and one-tailed .05 and .01 values for $df = 9$ are summarized in Table 17.5.

Table 17.5 Tabled Critical .05 and .01 A Values for df = 9

	$t_{.05}$	$t_{.01}$
Two-tailed values	.276	.185
One-tailed values	.368	.213

The following guidelines are employed in evaluating the null hypothesis for **Sandler's A test**.

a) If the nondirectional alternative hypothesis H_1: $\mu_1 \neq \mu_2$ is employed, the null hypothesis can be rejected if the obtained value of A is equal to or less than the tabled critical two-tailed value at the prespecified level of significance.

b) If the directional alternative hypothesis H_1: $\mu_1 > \mu_2$ is employed, the null hypothesis can be rejected if the sign of ΣD is positive (i.e., $\bar{X}_1 > \bar{X}_2$), and the value of A is equal to or less than the tabled critical one-tailed value at the prespecified level of significance.

c) If the directional alternative hypothesis H_1: $\mu_1 < \mu_2$ is employed, the null hypothesis can be rejected if the sign of ΣD is negative (i.e., $\bar{X}_1 < \bar{X}_2$), and the value of A is equal to or less than the tabled critical one-tailed value at the prespecified level of significance.

Employing the above guidelines, the following conclusions can be reached.

The nondirectional alternative hypothesis H_1: $\mu_1 \neq \mu_2$ is supported at the .05 level, since the computed value $A = .211$ is less than the tabled critical two-tailed value $A_{.05} = .276$. The latter alternative hypothesis, however, is not supported at the .01 level, since $A = .211$ is greater than the tabled critical two-tailed value $A_{.01} = .185$.

The directional alternative hypothesis H_1: $\mu_1 > \mu_2$ is supported at both the .05 and .01 levels, since $\Sigma D = 16$ is a positive number, and the obtained value $A = .211$ is less than the tabled critical one-tailed values $A_{.05} = .368$ and $A_{.01} = .213$.

The directional alternative hypothesis H_1: $\mu_1 < \mu_2$ is not supported, since $\Sigma D = 16$ is a positive number. In order for the directional alternative hypothesis H_1: $\mu_1 < \mu_2$ to be supported, the value of ΣD must be a negative number (as well as the fact that the computed value of A must be equal to or less than the tabled critical one-tailed value at the prespecified level of significance).

Note that the results obtained for **Sandler's A test** are identical to those obtained when the *t* **test for two dependent samples** is employed to evaluate Example 17.1. Equation 17.22 describes the relationship between **Sandler's A** statistic and the *t* value computed for the *t* **test for two dependent samples**.[17]

$$A = \frac{n-1}{nt^2} + \frac{1}{n} \qquad \textbf{(Equation 17.22)}$$

It is demonstrated below that when $t = 2.86$ (the value computed with Equations 17.1/17.6) is substituted in Equation 17.22, it yields the value $A = .211$ computed with Equation 17.21.

$$A = \frac{10-1}{(10)(2.86)^2} + \frac{1}{10} = .21$$

8. Test 17e: The z test for two dependent samples There are occasions (albeit infrequent) when a researcher wants to compare the means of two dependent samples, and happens to know the variances of the two underlying populations. In such a case, the z **test for two dependent samples** should be employed to evaluate the data instead of the *t* **test for two dependent samples**. As is the case with the latter test, the z **test for two dependent samples** assumes that the two samples are randomly selected from populations which have normal distributions. The effect of violation of the normality assumption on the test statistic decreases as the size of the sample employed in an experiment increases. The homogeneity of variance assumption noted for the *t* test for two dependent samples is not an assumption of the *z* test for two dependent samples.

The null and alternative hypotheses employed for the *z* test for two dependent samples are identical to those employed for the *t* test for two dependent samples. Equation 17.23 is employed to compute the test statistic for the *z* test for two dependent samples.[18]

$$z = \frac{\bar{X}_1 - \bar{X}_2}{\sqrt{\sigma_{\bar{X}_1}^2 + \sigma_{\bar{X}_2}^2 - 2(r_{X_1 X_2})(\sigma_{\bar{X}_1})(\sigma_{\bar{X}_2})}} \qquad \textbf{(Equation 17.23)}$$

Where:	$\sigma_{\bar{X}_i}^2 = \sigma_i^2/n$ and $\sigma_{\bar{X}_i} = \sigma_i/\sqrt{n}$

The only differences between Equation 17.23 and Equation 17.6 (the equation for the *t* test for two dependent samples) are: a) In the denominator of Equation 17.23, in computing the standard error of the mean for each condition, the population standard deviations σ_1 and σ_2 are employed instead of the estimated population standard deviations $\tilde{s}_1$ and $\tilde{s}_2$ (which are employed in Equation 17.6); and b) Equation 17.23 computes a *z* score which is evaluated with the normal distribution, while Equation 17.6 derives a *t* score which is evaluated with the *t* distribution.

If it is assumed that the two population variances are known for Example 17.1, and that $\sigma_1^2 = 8.01$ and $\sigma_2^2 = 5.66$, Equation 17.23 can be employed to evaluate the data. Note that the obtained value $z = 2.86$ is identical to the value which was computed for *t* when Equation 17.6 was employed.

$$z = \frac{4.7 - 3.1}{\sqrt{.79 + .56 - 2(.78)(.89)(.75)}} = 2.86$$

Except for the fact that the normal distribution is employed in lieu of the *t* distribution, the same guidelines are employed in evaluating the computed *z* value as those stipulated for evaluating a *t* value in Section V. The obtained value $z = 2.86$ is evaluated with **Table A1 (Table of the Normal Distribution)** in the **Appendix**. In **Table A1** the tabled critical two-tailed .05 and .01 values are $z_{.05} = 1.96$ and $z_{.01} = 2.58$, and the tabled critical one-tailed .05 and .01 values are $z_{.05} = 1.65$ and $z_{.01} = 2.33$. Since the computed value $z = 2.86$ is greater than the tabled critical two-tailed values $z_{.05} = 1.96$ and $z_{.01} = 2.58$, the nondirectional alternative hypothesis $H_1: \mu_1 \neq \mu_2$ is supported at both the .05 and .01 levels. Since the computed value $z = 2.86$ is a positive number which is greater than the tabled critical one-tailed values $z_{.05} = 1.65$ and $z_{.01} = 2.33$, the directional alternative hypothesis $H_1: \mu_1 > \mu_2$ is also supported at both the .05 and .01 levels.

When the same set of data is evaluated with the *t* test for two dependent samples, although the directional alternative hypothesis $H_1: \mu_1 > \mu_2$ is supported at both the .05 and .01 levels, the nondirectional alternative hypothesis $H_1: \mu_1 \neq \mu_2$ is only supported at the .05 level. This latter fact illustrates that if the *z* test for two dependent samples and the *t* test for two dependent samples are employed to evaluate the same set of data (unless the value of *n* is extremely large), the latter test will provide a more conservative test of the null hypothesis (i.e., make it more difficult to reject H_0). This is the case, since the tabled critical values listed for the *z* **test for two dependent samples** will always correspond to the tabled critical values listed in **Table A2** for *df* = ∞ (which are the lowest tabled critical values listed for the *t* distribution).

The final part of the discussion of the *z* **test for two dependent samples** will describe a special case of the test in which it is employed to evaluate the difference between the average performance of two conditions, when the scores of subjects are based on a binomially distributed variable. Example 17.2, which is used to illustrate this application of the test, is the dependent samples analog of Example 11.3 (which illustrates the analysis of a binomially distributed variable with the *z* **test for two independent samples**). The null and alternative hypotheses evaluated in Example 17.2 are identical to those evaluated in Example 17.1.

Example 17.2 *An experiment is conducted in which each of five subjects is tested for extrasensory perception under two experimental conditions. In* Condition 1 *a subject listens to a relaxation training tape, after which the subject is tested while in a relaxed state of mind. In* Condition 2 *each subject is tested while in a normal state of mind. Assume that the order of presentation of the two experimental conditions is counterbalanced, although not completely, since to do the latter would require that an even number of subjects be employed in the study. Thus, three of the five subjects initially serve in* Condition 1 *followed by* Condition 2, *while the remaining two subjects initially serve in* Condition 2 *followed by* Condition 1. *(The concept of* **counterbalancing** *is discussed in* Section VII.)*

In each experimental condition a subject is tested for 200 trials. In each condition the researcher employs as stimuli a different list of 200 binary digits (specifically, the values 0 and 1) which have been randomly generated by a computer. On each trial, an associate of the researcher concentrates on a digit in the order it appears on the list for that condition. While the associate does this, a subject is required to guess the value of the number which is employed as the stimulus for that trial. The number of correct guesses for subjects under the two experimental conditions follow. (The first score for each subject is the number of correct responses in Condition 1, *and the second score is the number of correct responses in* Condition 2.): **Subject 1** (105, 90); **Subject 2** (120, 104); **Subject 3** (130, 107); **Subject 4** (115, 100); **Subject 5** (110, 99). Table 17.6 *summarizes the data for the experiment.*

Table 17.6 Data for Example 17.2

Subject	Condition 1		Condition 2		
	X_1	X_1^2	X_2	X_2^2	$X_1 X_2$
1	105	11025	90	8100	9450
2	120	14400	104	10816	12480
3	130	16900	107	11449	13919
4	115	13225	100	10000	11500
5	110	12100	99	9801	10890
	$\Sigma X_1 = 580$	$\Sigma X_1^2 = 67650$	$\Sigma X_2 = 500$	$\Sigma X_2^2 = 50166$	$\Sigma X_1 X_2 = 58230$

$$\bar{X}_1 = \frac{580}{5} = 116 \qquad \bar{X}_2 = \frac{500}{5} = 100$$

Note that in Example 17.2, the five scores in each of the two experimental conditions are identical to the five scores employed in the two experimental conditions in Example 11.3. The only difference between the two examples is the order in which the scores are listed. Specifically, in Example 17.2 the scores have been arranged so that the two scores in each row (i.e., the two scores of each subject) have a high positive correlation with one another. Through use of Equation 17.7, it is demonstrated that the correlation between subjects' scores in the two experimental conditions is $r_{X_1 X_2} = .93$.

$$r_{X_1 X_2} = \frac{58230 - \frac{(580)(500)}{5}}{\sqrt{\left[67650 - \frac{(580)^2}{5}\right]\left[50166 - \frac{(500)^2}{5}\right]}} = .93$$

Example 17.2 will be evaluated with Equation 17.24, which is the form Equation 17.23 assumes when $\sigma_1^2 = \sigma_2^2$.

$$z = \frac{\overline{X}_1 - \overline{X}_2}{\sigma\sqrt{\frac{1}{m} + \frac{1}{m} - \frac{2(r_{X_1 X_2})}{m}}}$$ **(Equation 17.24)**

Note that in Equation 17.24, m is employed to represent the number of subjects, since the notation n is employed with the binomial variable to designate the number of trials each subject is tested. Since scores on the binary guessing task described in Example 17.2 are assumed to be binomially distributed, as is the case in Example 11.3, the following is true: $n = 200$, $\pi_1 = .5$, and $\pi_2 = .5$. The computed value for the population standard deviation for the binomially distributed variable is $\sigma = \sqrt{n\pi_1\pi_2} = \sqrt{(200)(.5)(.5)} = 7.07$. (The computation of the latter values is discussed in Section I of the **binomial sign test for a single sample**.) When the appropriate values are substituted in Equation 17.24, the value $z = 13.52$ is computed.

$$z = \frac{116 - 100}{7.07\sqrt{\frac{1}{5} + \frac{1}{5} - \frac{(2)(.93)}{5}}} = 13.52$$

Since the computed value $z = 13.52$ is greater than the tabled critical two-tailed values $z_{.05} = 1.96$ and $z_{.01} = 2.58$, the nondirectional alternative hypothesis $H_1: \mu_1 \neq \mu_2$ is supported at both the .05 and .01 levels. Since the computed value $z = 13.52$ is a positive number which is greater than the tabled critical one-tailed values $z_{.05} = 1.65$ and $z_{.01} = 2.33$, the directional alternative hypothesis $H_1: \mu_1 > \mu_2$ is supported at both the .05 and .01 levels. Thus, it can be concluded that the average score in Condition 1 is significantly larger than the average score in Condition 2. Note that when Equation 11.19 is employed with the same set of data, it yields the value $z = 3.58$. The fact that the value $z = 13.52$ obtained with Equation 17.24 is larger than the value $z = 3.58$ obtained with Equation 11.19 illustrates if there is a positive correlation between the scores of subjects employed in a dependent samples design, a z **test for two dependent samples** will provide a more powerful test of an alternative hypothesis than will a z **test for two independent samples** (due to the lower value of the denominator for the former test).

VII. Additional Discussion of the t Test for Two Dependent Samples

1. The use of matched subjects in a dependent samples design   It is noted in Section I that the t **test for two dependent samples** can be applied to a design involving matched subjects. Matching subjects requires that a researcher initially identify one or more variables (besides the independent variable) which she believes are positively correlated with the dependent variable employed in a study. Such a variable can be employed as a matching variable. Each subject who is assigned to one of the k experimental conditions is matched with one subject in each of the other $(k-1)$ experimental conditions. In matching subjects it is essential that any cohort of subjects who are matched with one another are equivalent (or reasonably comparable) with respect to any matching variables employed in the study. In a design employing matched subjects there will be n cohorts (also referred to as **blocks**) of matched subjects, and within each cohort there will be k subjects. Each of the k subjects should be randomly assigned to one of the k experimental conditions/levels of the independent variable. Thus, when $k = 2$, each of the two subjects within the n pairs/cohorts will be randomly assigned to one of the two experimental conditions.

By matching subjects a researcher is able to conduct a more powerful statistical analysis than will be the case if subjects in the two conditions are not matched with one another (i.e., if an independent samples design is employed).[19] The more similar the cohorts of subjects are on the matched variable(s), the greater the power of the statistical test. In actuality, the most extreme case of matching is when each subject is matched with himself or herself, and thus serves in each of the k experimental conditions. Within this framework, the design employed for Example 17.1 can be viewed as a matched-subjects design. However, as the term **matching** is most commonly employed, within each row of the data summary table (i.e., Table 17.1) different subjects serve in each of the k experimental conditions. When $k = 2$, the most extreme case of matching involving different subjects in each condition is when n pairs of identical twins are employed as subjects. By virtue of their common genetic makeup, identical twins allow an experimenter to match a subject with his or her "clone." In Example 17.3 identical twins are employed as subjects in the same experiment described by Example 17.1. Analysis of Example 17.3 with the *t* **test for two dependent samples** yields the same result as that obtained for Example 17.1, since both examples employ the same set of data.

Example 17.3 *A psychologist conducts a study to determine whether or not people exhibit more emotionality when they are exposed to sexually explicit words than when they are exposed to neutral words. Ten sets of identical twins are employed as subjects. Within each twin pair, one of the twins is randomly assigned to* Condition 1, *in which the subject is shown a list of eight sexually explicit words, while the other twin is assigned to* Condition 2, *in which the subject is shown a list of eight neutral words. As each word is projected on the screen, a subject is instructed to say the word softly to himself or herself. As a subject does this, sensors attached to the palms of the subject's hands record galvanic skin response (GSR), which is used by the psychologist as a measure of emotionality. The psychologist computes two scores for each pair of twins to represent the emotionality score for each of the experimental conditions:* **Condition 1**: *GSR/Explicit — The average GSR score for the twin presented with the eight sexually explicit words;* **Condition 2**: *GSR/Neutral — The average GSR score for the twin presented with the eight neutral words. The GSR/Explicit and the GSR/Neutral scores of the ten pairs of twins follow. (The first score for each twin pair represents the score of the twin presented with the sexually explicit words, and the second score represents the score of the twin presented with the neutral words. The higher the score, the higher the level of emotionality.)* **Twin pair 1** (9, 8); **Twin pair 2** (2, 2); **Twin pair 3** (1, 3); **Twin pair 4** (4, 2); **Twin pair 5** (6, 3); **Twin pair 6** (4, 0); **Twin pair 7** (7, 4); **Twin pair 8** (8, 5); **Twin pair 9** (5, 4); **Twin pair 10** (1, 0). *Do subjects exhibit differences in emotionality with respect to the two categories of words?*

In the event $k = 3$ and a researcher wants to use identical siblings, identical triplets can be employed in a study. If $k = 4$, identical quadruplets can be used, and so on. If the critical variable(s) with respect to which the researcher wants to match subjects are believed to be influenced by environmental factors, the suitability of employing identical siblings as matched subjects will be compromised to the degree that within each set of siblings the members of the set do not share common environmental experiences. Realistically, the number of available identical siblings in a human population will be quite limited. Thus, with the exception of identical twins, it would be quite unusual to encounter a study which employs identical siblings. Because of the low frequency of identical siblings in the general population, in matching subjects a researcher may elect to employ biological relatives who share less in common with one another or employ people who are not blood relatives. Example 17.4 illustrates the latter type of matching. Analysis of Example 17.4 with the *t* test for two dependent samples yields the same result as that obtained for Examples 17.1 and 17.3, since all three experiments employ the same set of data.

Example 17.4 *A psychologist conducts a study to determine whether or not people exhibit more emotionality when they are exposed to sexually explicit words than when they are exposed to neutral words. Based on previous research, the psychologist has reason to believe that the following three variables are highly correlated with the dependent variable of emotionality: a) gender; b) autonomic hyperactivity (which is measured by a series of physiological measures); and c) repression–sensitization (which is measured by a pencil and paper personality test). Ten pairs of matched subjects who are identical (or very similar) on the three aforementioned variables are employed in the study. Within each pair, one person is randomly assigned to a condition in which the subject is shown a list of eight sexually explicit words, while the other person is assigned to a condition in which the subject is shown a list of eight neutral words. As each word is projected on the screen, a subject is instructed to say the word softly to himself or herself. As a subject does this, sensors attached to the palms of the subject's hands record galvanic skin response (GSR), which is used by the psychologist as a measure of emotionality. The psychologist computes two scores for each pair of matched subjects to represent the emotionality score for each of the experimental conditions:* **Condition 1**: *GSR/Explicit — The average GSR score for the subject presented with the eight sexually explicit words;* **Condition 2**: *GSR/Neutral — The average GSR score for the subject presented with the eight neutral words. The GSR/Explicit and the GSR/Neutral scores of the ten pairs of subjects follow. (The first score for each pair represents the score of the person presented with the sexually explicit words, and the second score represents the score of the person presented with the neutral words. The higher the score, the higher the level of emotionality.)* **Pair 1** (9, 8); **Pair 2** (2, 2); **Pair 3** (1, 3); **Pair 4** (4, 2); **Pair 5** (6, 3); **Pair 6** (4, 0); **Pair 7** (7, 4); **Pair 8** (8, 5); **Pair 9** (5, 4); **Pair 10** (1, 0). *Do subjects exhibit differences in emotionality with respect to the two categories of words?*

One reason a researcher may elect to employ matched subjects (as opposed to employing each subject in all *k* experimental conditions) is because in many experiments it is not feasible to have a subject serve in more than one condition. Specifically, a subject's performance in one or more of the conditions might be influenced by his or her experience in one or more of the conditions which precede it. In some instances counterbalancing can be employed to control for such effects, but in other cases even counterbalancing does not provide the necessary control.

In spite of the fact that it can increase the power of a statistical analysis, matching is not commonly employed in experiments involving human subjects. The reason for this is that matching requires a great deal of time and effort on the part of a researcher. Not only is it necessary to identify one or more matching variables which are correlated with the dependent variable, but it is also necessary to identify and obtain the cooperation of a sufficient number of matched subjects to participate in an experiment. The latter does not present as much of a problem in animal research, where litter mates can be employed as matched cohorts. Example 17.5, which is evaluated in the next section, illustrates a design that employs animal litter mates as subjects.

Example 17.5 *A researcher wants to assess the relative effect of two different kinds of punishment (loud noise versus a blast of cold air) on the emotionality of mice. Five pairs of mice derived from five separate litters are employed as subjects. Within each pair, one of the litter mates is randomly assigned to one of two experimental conditions. During the course of the experiment each mouse is sequestered in an experimental chamber. While in the chamber, each of the five mice in* Condition 1 *is periodically presented with a loud noise, and each of the five mice in* Condition 2 *is periodically presented with a blast of cold air. The presentation of the punitive stimulus for each of the animals is generated by a machine which randomly presents the stimulus throughout the duration of the time an animal is in the chamber. The dependent variable of emotionality employed in the study is the number of times each mouse defecates while in the experimental chamber. The number of episodes of defecation for the five pairs of mice follows. (The first score represents the litter mate exposed to noise and the second score represents the litter mate exposed to cold air.)* **Litter 1** (11, 11); **Litter 2** (1, 11); **Litter 3** (0, 5); **Litter 4** (2, 8); **Litter 5** (0, 4). *Do subjects exhibit differences in emotionality under the different experimental conditions?*

2. Relative power of the *t* test for two dependent samples and the *t* test for two independent samples Example 17.5 will be employed to illustrate that the *t* **test for two dependent samples** provides a more powerful test of an alternative hypothesis than does the *t* **test for two independent samples**. Except for the fact that it employs a dependent samples design involving matched subjects, Example 17.5 is identical to Example 11.4 (which employs an independent samples design). Both examples employ the same set of data and evaluate the same null and alternative hypotheses. The summary values for evaluating Example 17.5 with the *t* **test for two dependent samples** (using either Equation 17.1 or 17.6) are noted below. Some of the values listed can also be found in Table 11.1 (which summarizes the same set of data for analysis with the *t* **test for two independent samples**).

$$\Sigma X_1 = 14 \qquad \Sigma X_1^2 = 126 \qquad \Sigma X_2 = 39 \qquad \Sigma X_2^2 = 347 \qquad \bar{X}_1 = 2.8 \qquad \bar{X}_2 = 7.8$$

$$\Sigma X_1 X_2 = 148 \qquad \tilde{s}_2^2 = 21.7 \qquad \tilde{s}_1^2 = 10.7 \qquad r_{X_1 X_2} = .64$$

$$\Sigma D = -25 \qquad \Sigma D^2 = 177 \qquad \bar{D} = -5 \qquad \tilde{s}_D = 3.61 \qquad s_{\bar{D}} = 1.61$$

$$t = \frac{\bar{D}}{s_{\bar{D}}} = \frac{-5}{1.61} = -3.10$$

Since $n = 5$ in Example 17.5, $df = 5 - 1 = 4$. In **Table A2**, for $df = 4$, the tabled critical two-tailed .05 and .01 values are $t_{.05} = 2.78$ and $t_{.01} = 4.60$, and the tabled critical one-tailed .05 and .01 values are $t_{.05} = 2.13$ and $t_{.01} = 3.75$.

The nondirectional alternative hypothesis H_1: $\mu_1 \neq \mu_2$ is supported at the .05 level, since the computed absolute value $t = 3.10$ is greater than the tabled critical two-tailed value $t_{.05} = 2.78$. It is not, however, supported at the .01 level, since the absolute value $t = 3.10$ is less than the tabled critical two-tailed value $t_{.01} = 4.60$.

The directional alternative hypothesis H_1: $\mu_1 < \mu_2$ is supported at the .05 level, since the computed value $t = -3.10$ is a negative number, and the absolute value $t = 3.10$ is greater than the tabled critical one-tailed value $t_{.05} = 2.13$. It is not, however, supported at the .01 level since the absolute value $t = 3.10$ is less than the tabled critical one-tailed value $t_{.01} = 3.75$.

The directional alternative hypothesis H_1: $\mu_1 > \mu_2$ is not supported, since the computed value $t = -3.10$ is a negative number.

Note that the absolute value $t = 3.10$ computed for Example 17.5 is substantially higher than the absolute value $t = 1.96$ computed for Example 11.4 (which has the same data as Example 11.1) with the **t test for two independent samples**. In the case of Example 11.4 (as well as Example 17.5), the directional alternative hypothesis H_1: $\mu_1 < \mu_2$ is supported at the .05 level. However, the nondirectional alternative hypothesis H_1: $\mu_1 \neq \mu_2$ (which is supported at the .05 level in the case of Example 17.5) is not supported in Example 11.4 when the data are evaluated with the **t test for two independent samples**. The difference in the conclusions reached with the two tests reflects the fact that the **t test for two dependent samples** provides a more powerful test of an alternative hypothesis (assuming there is a positive correlation (which in the example under discussion is $r_{X_1 X_2} = .64$) between the scores of subjects in the two experimental conditions). In closing this discussion, it is worth noting that designs involving independent samples are more commonly employed in research than designs involving dependent samples. The reason for this is that over and above the fact that a dependent samples design allows for a more powerful test of an alternative hypothesis, it presents more practical problems in its implementation (e.g., controlling for problems which might result from subjects serving in multiple conditions; the difficulty of identifying matching variables; identifying and obtaining the cooperation of an adequate number of matched subjects).

3. Counterbalancing and order effects   When each of the n subjects in an experiment serves in all k experimental conditions, it is often necessary to control for the order of presentation of the conditions. The latter is done in order to control for what is commonly referred to as **order effects**.[20] Anderson (2001, p. 415) notes that there are two types of **order effects, position effects** and **carryover effects**.

Position effects are assumed to only be a function of the **serial position** of a treatment (which is the order in which the treatment is presented), and are not affected by what particular treatment(s) occurred in earlier serial positions. The most common **position effects** are **practice effects** in which the performance of subjects improves with each subsequent treatment, and fatigue and boredom in which each subsequent treatment results in a decrease in the performance of subjects. In other words, if all n subjects are administered Condition 1 first followed by Condition 2 (or vice versa), factors such as practice or fatigue, which are a direct function of the order of presentation of the conditions, can differentially affect subjects' scores on the dependent variable. Specifically, subjects may perform better in Condition 2 due to practice effects or subjects may perform worse in Condition 2 as a result of fatigue or boredom.

A **carryover effect**, which represent the second type of order effect, is where the effect of a specific treatment is not independent of a treatment which precedes it. As an example, assume within the framework of a within-subjects design the following three treatments are administered on separate days to each of n subjects in order to assess each treatment's effect on eye-hand coordination: Drug A, Drug B, a placebo. Assume that a side effect of Drug A is that it makes a person drowsy, and that the latter adversely influences performance. To compound matters, it requires 36 hours before Drug A completely washes out of a person's body. Because of the latter, if either of the other two treatments is administered to a subject the day after he or she receives Drug A, the subject's performance under that treatment will also be adversely affected due to the carryover effect of residual drowsiness deriving from Drug A.

As noted earlier, **counterbalancing** is a procedure which allows a researcher to control for order effects. In **complete counterbalancing** all possible orders for presenting the experimental conditions are represented an equal number of times with respect to the total number of subjects employed in a study. Thus, if a study with $n = 10$ subjects and $k = 2$ conditions is completely counterbalanced, five subjects will initially serve in Condition 1 followed by Condition 2, while the other five subjects will initially serve in Condition 2 followed by Condition 1. If the number of experimental conditions is $k = 3$, there will be $k! = 3! = 6$ possible presentation orders (i.e., 1,2,3; 1,3,2; 2,1,3; 2,3,1; 3,1,2; 3,2,1). Under such conditions a minimum of six subjects will be required in order to employ complete counterbalancing. If a researcher wants to assign two subjects to each of the presentation orders, $6 \times 2 = 12$ subjects must be employed. It should be obvious that to completely counterbalance the order of presentation of the experimental conditions, the number of subjects must equal the value of $k!$ or be some value that is evenly divisible by it.

As the number of experimental conditions increases, **complete counterbalancing** becomes more difficult to implement, since the number of subjects required increases substantially. Specifically, if there are $k = 5$ experimental conditions, there are $5! = 120$ presentation orders — thus requiring a minimum of 120 subjects (which can be a prohibitively large number for a researcher to use) in order that one subject serves in each of the possible presentation orders. When it is not possible to completely counterbalance the order of presentation of the conditions, alternative less complete counterbalancing procedures are available. One experimental design which allows for **incomplete counterbalancing** (i.e., where not all possible presentation orders are employed in a study) is the **Latin square design**, which is employed more frequently in agricultural and industrial research than in research in education and the behavioral sciences. If an independent variable is comprised of only a few levels, it is feasible to use a Latin square design to provide **complete counterbalancing**. However, as the number of levels of the independent variable increase, it is more likely that the Latin square design will be employed to control for order effects through use of **incomplete counterbalancing**. The Latin square design is discussed briefly in Section VII of the **single-factor within-subjects analysis of variance** and in detail in Section IX (the **Addendum**) of the **between-subjects factorial analysis of variance (Test 27)** under the **analysis of variance for a Latin square design (Test 27j)**.

The general subject of **order effects** is often discussed within the framework of a **crossover design** (which as noted in Endnote 1 is alternative term some sources employ for a **dependent samples design**). The term **crossover design** is employed most frequently within the context of **clinical trials** (which is discussed in Section VII of the *t* **test for two independent samples**). A **clinical trial** refers to an experiment involving human subjects that is designed to evaluate a health related issue (such as the relative efficacy of two or more types of therapy — most commonly two drugs or a drug versus a placebo). In a **two-treatment crossover design**, every subject receives each of two treatments, with half of the subjects receiving Treatment 1 first followed by Treatment 2, and the other half receiving the treatments in reverse order. Typically, in studies evaluating the efficacy of two drugs, there is a **washout period** for each subject intervening between a subject's two treatments. During the washout period (the duration of which can be anywhere from a few days to a few months) the subject receives no treatment, and it is assumed that any residual biological effects from previous treatment dissipate. An order effect is present when, in fact, there is a residual biological effect from a previous treatment. As noted previously, the latter type of order effect is commonly referred to as a carryover effect. The analysis of both treatment and order effects in a crossover design is described in Section IX (the **Addendum**) of the **between-subjects factorial analysis of variance** under the discussion of the **factorial analysis of variance for a mixed design (Test 27i)**. In the latter discussion it is noted that the standard *t* **test for two dependent samples** is not the most powerful method for evaluating a treatment effect in a **two-treatment crossover design**.

Rosner (2000, p. 648–649) notes it is advantageous to employ a crossover design rather than a **parallel group design** if no carryover effects are present in a set of data. (The term **parallel group design** is employed within the context of clinical trials for an **independent samples design** in which subjects are randomly assigned to one of k groups, with each group receiving a different treatment.) Specifically, if there is a consistent increase or decrease in the responses of *n* subjects associated with the order of presentation of *k* treatments, it would only be appropriate for the researcher to employ the results of the first treatment period in assessing the effect of the treatment. However, for a sample of *n* subjects, by limiting the analysis to comparing the main treatment with the other treatments during the first treatment period, the researcher will employ a less powerful statistical test than would be the case if a **parallel group design** were used instead (since the total number of subjects employed for the latter design would be substantially larger in that it would be comprised of *k* treatments with *n* subjects per treatment). Rosner (2000) also notes that a crossover design is most appropriate for evaluating treatments which are expected to culminate in a clearly defined **endpoint** that occurs within a relatively short period of time after the treatment is initiated (an **endpoint** is some measure of therapeutic efficacy), as well as treatments which do not have a long lasting residual effect once the treatment is terminated. Consequently, long term studies, such as **phase III clinical trials** (discussed in Section VII of the *t* **test for independent samples**) are better suited to be evaluated with a **parallel group design**.

4. Analysis of a one-group pretest-posttest design with the *t* test for two dependent samples In a **one-group pretest-posttest design** (discussed in the **Introduction**) *n* subjects are administered a pretest on a dependent variable.[21] After the pretest, all *n* subjects are exposed to the experimental treatment. Subjects are then administered a posttest on the same dependent variable. The *t* **test for two dependent samples** can be employed to determine if there is a significant difference between the pretest versus posttest scores of subjects. Although there are published studies which employ the *t* **test for two dependent samples** to evaluate the aforementioned design, it is important to note that since it lacks a control group, a **one-group pretest-posttest design** does not allow a researcher to conclude that the experimental treatment is responsible for a significant difference. If a significant result is obtained in such a study, it only allows the researcher to conclude that there is a significant statistical association/correlation between the experimental treatment and the dependent variable. Since correlational information does not allow one to draw conclusions with regard to cause and effect, a researcher cannot conclude that the treatment is directly responsible for the observed difference. Although it is possible that the treatment is responsible for the difference, it is also possible that the difference is due to one or more other variables (discussed in the **Introduction**) which intervened between the pretest and the posttest.

To modify a **one-group pretest-posttest design** to insure adequate experimental control, it is required that two groups of subjects be employed. In such a modification, pretest and posttest scores are obtained for both groups, but only one of the groups (the experimental group) is exposed to the experimental treatment in the time period that intervenes between the pretest and the posttest. By virtue of employing a control group which is not exposed to the experimental treatment, the researcher is able to rule out the potential influence of confounding variables. Thus, order effects, as well as other factors in the environment that subjects may have been exposed to between the pretest and the posttest, can be ruled out through use of a control group. Example 17.6 illustrates a study which employs a **one-group pretest-posttest design** without the addition of the necessary control group.

Example 17.6 *In order to assess the efficacy of electroconvulsive therapy (ECT), a psychiatrist evaluates ten clinically depressed patients before and after a series of ECT treatments. A standardized interview is used to operationalize a patient's level of depression, and on the basis of the interview each patient is assigned a score ranging from 0 to 10 with respect to his or her level of depression prior to (pretest score) and after (posttest score) the administration of ECT.*

The higher a patient's score, the more depressed the patient. The pretest and posttest scores of the ten patients follow: **Patient 1** (9, 8); **Patient 2** (2, 2); **Patient 3** (1, 3); **Patient 4** (4, 2); **Patient 5** (6, 3); **Patient 6** (4, 0); **Patient 7** (7, 4); **Patient 8** (8, 5); **Patient 9** (5, 4); **Patient 10** (1, 0). *Do the data indicate that ECT is effective?*

Since the data for Example 17.6 are identical to that employed in Example 17.1, the same result is obtained. Thus, analysis of the data with the *t* **test for two dependent samples** indicates there is a significant decrease in depression following the ECT. However, as previously noted, because there is no control group, the psychiatrist cannot conclude that ECT is responsible for the decrease in depression. Inclusion of a "sham" ECT group (which is analogous to a placebo group) can provide the necessary control to evaluate the impact of ECT. Such a group would be comprised of ten additional patients for whom pretest and posttest depression scores are obtained. Between the pretest and posttest, the patients in the control group undergo all of the preparations involved in ECT, but are only administered a simulated ECT treatment (i.e., they are not actually administered the shock treatment). Only by including such a control group can one rule out the potential role of extraneous variables that might also be responsible for the lower level of depression during the posttest. An example of such an extraneous variable would be if all of the subjects who receive ECT are in psychotherapy throughout the duration of the experiment. Without a control group, a researcher cannot determine whether or not a lower posttest depression score is the result of the ECT, the psychotherapy, the ECT and psychotherapy interacting with one another, or some other variable of which the researcher is unaware. By including a control group, it is assumed (although not insured) that if any extraneous variables are present, by virtue of randomly assigning subjects to two groups, the groups will be equated on such variables.[22]

When the **one-group pretest-posttest design** is modified by the addition of the appropriate control group, the resulting design is referred to as a **pretest-posttest control group design** (discussed in the **Introduction**). Unfortunately, researchers are not in agreement with respect to what statistical analysis is most appropriate for the latter design. Among the analytical procedures which have been recommended are the following: a) The difference scores of the two groups can be contrasted with a *t* **test for two independent samples** (or a **single-factor between-subjects analysis of variance (Test 21)**, which will yield an equivalent result when there are two experimental conditions); b) The results can be evaluated by employing a **factorial analysis of variance for a mixed design**. A factorial design has two or more independent variables (which are also referred to as factors). Thus, if the appropriate control group is employed in Example 17.6, the resulting **pretest-posttest control group design** can be conceptualized as being comprised of two independent variables. One of the independent variables is represented by the ECT versus sham ECT manipulation. The second independent variable is the pretest-posttest dichotomy. In a mixed factorial design involving two factors, one of the independent variables is a between-subjects variable (i.e., different subjects serve under different levels of that independent variable). Thus, in the example under discussion, the ECT versus sham ECT manipulation represents a between-subjects independent variable. The other independent variable in a mixed factorial design is a within-subjects variable (i.e., each subject serves under all levels of that independent variable). In the example under discussion, the pretest-posttest dichotomy represents a within-subjects independent variable; and c) The **single-factor between-subjects analysis of covariance (Test 21I)** (a procedure which is discussed in Section IX (the **Addendum**) of the **single-factor between-subjects analysis of variance**) can also be employed to evaluate a **pretest-posttest control group design**. In conducting an analysis of covariance, the pretest scores of subjects are employed as the **covariate** (which is defined in the discussion of the analysis of covariance).

Weinfurt (2000) (who provides a comprehensive discussion of the three procedures noted above for evaluating a pretest-post design involving two groups) recommends the use of the analysis of covariance when the design to be evaluated involves random assignment of subjects to the two experimental conditions. On the other hand, when a **nonequivalent control group design** (i.e., a quasi-experimental design (discussed in the **Introduction**) which employs intact groups to represent the two levels of the independent variable) is employed, he notes that the use of the analysis of covariance can lead to an erroneous conclusion regarding the effect of the independent variable. Because of the latter, when a **nonequivalent control group design** is employed, Weinfurt (2000) recommends that the difference scores be evaluated with the *t* **test for two independent samples/single-factor between-subjects analysis of variance**.

Cohen (2001, p. 491) also recommends the use of the analysis of covariance in evaluating a pretest-post design, since it allows for a more powerful test when contrasted with the alternative procedures. Cohen (2001, p. 491) notes that when the difference scores for two groups are evaluated with the *t* **test for two independent samples**, the computed value of *t* will be equal to the *F* value obtained for the interaction when the same data are evaluated with a **factorial analysis of variance for a mixed design**. It should be noted that the **pretest-posttest control group design** can be modified to include more than two groups, and that such designs can also be evaluated through use of the analytical procedures described above. Cohen (2001) notes that if pretest-posttest difference scores for more than two groups are evaluated with a **single-factor between-subjects analysis of variance**, the obtained *F* value will be equal to the *F* value obtained for the interaction when the same data are evaluated with a **factorial analysis of variance for a mixed design**. Within the framework of the discussion of the merits of contrasting difference scores versus use of the **factorial analysis of variance for a mixed design**, citing a paper by Huck and McLean (1975), Cohen (2001) notes that when the data are evaluated with the latter analysis of variance, the *F* value obtained for the interaction should be employed to assess whether or not a significant effect can be attributed to the independent variable, rather than the *F* value obtained for the main effect on the independent variable. For clarification of the latter *F* values, the reader should read the discussion of the **factorial analysis of variance for a mixed design** in Section IX (the **Addendum**) of the **between-subjects factorial analysis of variance**.

5. Tests of equivalence: Test 17f: The Westlake–Schuirmann test of equivalence of two dependent treatments and **Test 17g: Tryon's test of equivalence of two dependent treatments** Prior to reading this section the reader should review the material on **tests of equivalence** in Section VII of the *t* **test for two independent samples**. The latter section describes procedures for assessing equivalence which can be generalized to assess the equivalence of two dependent samples. Specifically, the **Westlake–Schuirmann test of equivalence of two independent treatments (Test 11h)** (Westlake (1976, 1981, 1988) and/or Schuirmann (1987)) and **Tryon's test of equivalence of two independent treatments (Test 11i)** (Tryon (2001)) can be modified to accommodate designs involving two dependent samples. In the latter cases the tests are referred to as the **Westlake–Schuirmann test of equivalence of two dependent treatments** and **Tryon's test of equivalence of two dependent treatments**.

The discussion of tests of equivalence in Section VII of the *t* **test for two independent samples** notes that in contrast to the **classical hypothesis testing model** (where the alternative hypothesis states that there is a difference between treatments), in a **test of equivalence** the alternative hypothesis states that the treatments are, in fact, equivalent. Conversely, in a **test of equivalence** the null hypothesis states that a difference exists between the treatments (in contrast to a test of significance within the framework of the **classical hypothesis testing model**, where the null hypothesis states that there is no difference between the treatments). Since it is not mathematically feasible to establish an alternative hypothesis which states exact equality (i.e., a difference of zero) between experimental conditions, when one conducts a **test of equivalence**, a parameter (to be designated ζ, which represents the lower case Greek letter **zeta**) is employed to reflect a maximum difference which will be tolerated between two treatments in order that a researcher might conclude that the treatments are equivalent to one another. Any difference less than the value of ζ would be viewed so small as to be inconsequential (i.e., of no practical significance). Thus, in a **test of equivalence** if a difference less than the value of ζ is detected, the null hypothesis (stating that a difference does exist) can be rejected, and the alternative hypothesis (which stipulates equivalence) can be accepted. In the discussion to follow, it will be emphasized that since in most instances rejection of the null hypothesis for a **test of equivalence** will involve detection of a substantially smaller effect size than the effect size a researcher is typically concerned with detecting within the framework of the **classical hypothesis testing model**, a **test of equivalence** will usually require greater statistical power, and consequently the use of a larger sample size, when contrasted with tests conducted within the framework of the **classical hypothesis testing model**. The smaller the value a researcher designates for ζ, the larger the sample size which will be required in order to reject the null hypothesis and thus conclude the treatments are equivalent.

Test 11f The Westlake–Schuirmann test of equivalence of two dependent treatments In order to directly address the question of equivalence consider Example 17.7. The reader should take note of the fact that although for illustrative purposes a small sample size is employed in Example 17.7, a substantially larger sample size would be expected in an actual study designed to evaluate drug efficacy.

Example 17.7 *The manufacturer of a generic drug for treating a disease wants to empirically demonstrate it is equivalent to a brand name drug, which up to the present time has been the only approved treatment for the disease. The major criterion for diagnosing the disease is that patients diagnosed with the disease have an abnormally low concentration of a specific factor in their blood. Treatment with the brand name drug results in a substantial increase in the concentration of the latter factor. The Federal Drug Administration declares that equivalency of a generic drug with the brand name drug can be established if the mean concentration of the disease factor in patients treated with the generic drug is within 2 units of the mean concentration of the factor obtained for patients treated with the brand name drug. To demonstrate equivalency of the two drugs, a study is conducted in which ten people afflicted with the disease (each patient exhibiting an abnormally low concentration of the disease factor) are treated with both the brand name and generic drugs. Five of the subjects initially receive the brand name drug which they take for six months, after which a measure of the disease factor is obtained for each patient. After a two month washout period, the five subjects receive the generic drug for six months, after which a measure of the disease factor is once again obtained for each patient. The other five subjects in the study receive the treatments in the reverse order. The blood concentrations for the disease factor for the ten subjects under the two treatments follow (The first score represents a subject's blood concentration under the brand name drug, and the second score represents a subject's blood concentration under the generic drug):* **Subject 1** (9, 8); **Subject 2** (2, 2); **Subject 3** (1, 3); **Subject 4** (4, 2); **Subject 5** (6, 3); **Subject 6** (4, 0); **Subject 7** (7, 4); **Subject 8** (8, 5); **Subject 9** (5, 4); **Subject 10** (1, 0). *Do the data indicate that the generic drug is equivalent to the brand name drug?*

Note that Example 17.7 (which employs the same data as Example 17.1) stipulates that in order for the two treatments to be declared equivalent, the difference between the treatment means can not be greater than two units. Thus, we will assume that the researcher elects to conduct a two-tailed analysis with alpha set equal to .05. (A full discussion of the protocol for a one-tailed analysis can be found in the discussion of **tests of equivalence** in Section VII of the *t* **test for two independent samples**.) Equation 17.25 is employed to conduct two *t*-**tests for two dependent samples** which together comprise the **Westlake–Schuirmann test of equivalence of two dependent treatments**. Note that the latter equation is identical to Equation 17.8, except that in Equation 17.25 the value ζ is employed to represent the term $(\mu_1 - \mu_2)$ in the numerator of Equation 17.8. Because two units was stipulated as the maximum acceptable difference, the absolute value of ζ will be 2, and the values $\zeta_1 = -2$ and $\zeta_2 = 2$ will be employed in the two *t* tests that will be conducted.

$$t = \frac{(\bar{X}_1 - \bar{X}_2) - \zeta}{s_{\bar{D}}} \qquad \textbf{(Equation 17.25)}$$

The null and alternative hypotheses for a **two-tailed analysis** are noted below. The notation θ (which is the lower case Greek letter **theta**) in the null and alternative hypotheses represents the difference between the two treatment means in the underlying populations. It is assumed that difference is based on subtracting the population mean for the generic drug from the population mean for the brand name drug.

Null hypotheses $H_0: \theta \leq -2$ or $H_0: \theta \geq 2$

(In the underlying populations the treatments represent, the difference between the two treatment means is less than or equal to -2 or is greater than or equal to 2. This is the **hypothesis of nonequivalence**.)

Alternative hypotheses $H_1: \theta > -2$ and $H_1: \theta < 2$

(In the underlying populations the samples represent, the difference between the two treatment means is greater than -2 and is less than 2. This is the **hypothesis of equivalence**.)

As was the case for a two-tailed analysis when the **Westlake–Schuirmann test of equivalence of two independent treatments** was employed in Section VII of the *t* **test for two independent samples**, the **Westlake–Schuirmann test of equivalence of two dependent treatments** requires that **two one-tailed tests** be conducted. In order for the researcher to conclude that the two treatments are equivalent, **both** of the *t* tests that are conducted must be statistically significant, and the two *t* values must have opposite signs.

The two one-tailed tests will now be conducted employing the value alpha = .05. Since the data for Example 17.7 are identical to that employed for Example 17.1, in both analyses the values $\bar{X}_1 - \bar{X}_2 = 1.6$ and $s_{\bar{D}} = .56$ will be employed in Equation 17.25. Initially, the following null and alternative hypotheses are evaluated: $H_0: \theta \leq -2$ versus $H_1: \theta > -2$. Note that in the analysis the value $\zeta_1 = -2$ is employed in the numerator of the equation, since we are evaluating the null hypothesis that the mean for the generic drug is 2 or more units greater than the mean for the brand name drug (i.e., $\bar{X}_{brand\ name} - \bar{X}_{generic} \leq -2$). Employing Equation 17.25, the value $t = 6.43$ is computed.

$$t = \frac{(\bar{X}_1 - \bar{X}_2) - \zeta}{s_{\bar{D}}} = \frac{(4.7 - 3.1) - (-2)}{.56} = \frac{3.6}{.56} = 6.43$$

For $df = n - 1 = 9$, the tabled critical one-tailed value $t_{.05} = 1.83$ in Table **A2** in the **Appendix** is employed to evaluate the result. Since the obtained absolute value $t = 6.43$ is greater than the tabled critical one-tailed value $t_{.05} = 1.83$, the null hypothesis H_0: $\theta \leq -2$ designating nonequivalence can be rejected. Thus, the first of the null hypotheses designating nonequivalence can be rejected. However, it is important to note that in order to declare equivalence we must reject **both** of the null hypotheses designating nonequivalence, and the two test statistics computed must have opposite signs. In reference to the **t test** conducted above, the data are consistent with the alternative hypotheses H_1: $\theta > -2$, which stipulates that when the mean of **Treatment 2 (generic drug)** is subtracted from the mean of **Treatment 1 (brand name drug)** the difference is greater than -2 (which translates into the fact that if the mean of the **generic drug** is greater than the mean of the **brand name drug**, it is not greater than the latter value by more than two units).

We now evaluate the second null hypothesis designating nonequivalence. Specifically, a one-tailed test which evaluates the following null and alternative hypotheses is conducted: H_0: $\theta \geq 2$ versus H_1: $\theta < 2$. This analysis employs the same values as in the previous analysis, except for the fact that the value $\zeta_2 = 2$ is employed in the numerator of the equation, since we are evaluating the null hypothesis that the mean for the brand name drug is 2 or more units greater than the mean for the generic drug (i.e., $\bar{X}_{brand\ name} - \bar{X}_{generic} \geq -2$).

$$t = \frac{(\bar{X}_1 - \bar{X}_2) - \zeta}{s_{\bar{D}}} = \frac{(4.7 - 3.1) - (2)}{.56} = \frac{-.4}{.56} = -.71$$

Since the obtained absolute value $t = -.71$ is less than the tabled critical one-tailed value $t_{.05} = 1.83$, the null hypothesis cannot be rejected. Thus, the second null hypothesis H_0: $\theta \geq 2$ designating nonequivalence cannot be rejected. The probability value associated with this result is well above .05. Since, as noted earlier, in order to declare equivalence we must reject **both** of the null hypotheses designating nonequivalence, equivalence cannot be established since the larger of the probability values computed for the two analyses exceeds .05. Specifically, the data are consistent with the null hypothesis H_0: $\theta \geq 2$, which stipulates that the mean of **Treatment 1 (the brand name drug)** is two units or greater than the mean of **Treatment 2 (the generic drug)**.

Rogers *et al.* (1993, pp. 554–555) note that when the value of alpha is set equal to .05 for each of the two analyses conducted above, the equivalency confidence interval is set at the $(1 - 2\alpha)$ (i.e., 90%) level of certainty/confidence rather than at the traditional $(1 - \alpha)$ (i.e., 95%) level of certainty/confidence. When both null hypotheses which are evaluated are rejected, the values employed for both ζ_1 and ζ_2 will fall outside of the $(1 - 2\alpha)$ % confidence interval. If either of the null hypotheses is retained, the value employed for ζ_1 or ζ_2 will fall within the limits of the confidence interval. **To be more specific, if equivalence is to be established, the values for both ζ_1 and ζ_2 must fall outside the limits of the confidence interval, with one of the values ζ_1 positioned to the left of the lower limit of the confidence interval and the other value ζ_2 positioned to the right of the upper limit of the confidence interval. Note, however, that if both ζ_1 and ζ_2 fall outside the limits of the confidence interval, and both ζ_1 and ζ_2 are below the lower limit of the confidence interval or both ζ_1 and ζ_2 are above the upper limit of the confidence interval, it would indicate a lack of equivalence.** In the case of Example 17.7, in order to compute the 90% confidence interval, the tabled critical two-tailed .10 t value (which corresponds to the tabled critical one-tailed .05 t value) is employed in Equation 17.20. Employing Equation 17.20, the 90% confidence interval for Example 17.7 is computed below, and visually represented in Figure 17.2.

$$CI_{(1 - \alpha)} = (\bar{X}_1 - \bar{X}_2) \pm (t_\alpha)(s_{\bar{D}}) = (4.7 - 3.1) \pm (1.83)(.56) = 1.6 \pm 1.02$$

$$.58 \leq (\mu_1 - \mu_2) \leq 2.62$$

This result indicates the researcher can be 90% confident (or the probability is .90) that the interval .58 to 2.62 contains the true difference between the population means. If the aforementioned interval contains the true difference between the population means, the mean of the population **Treatment 1 (Brand name drug)** represents is between .58 and 2.62 units greater than the mean of the population **Treatment 2 (Generic drug)** represents. Note that although the value $\zeta_1 = -2$ falls outside of the computed confidence interval, the value $\zeta_2 = 2$ falls inside it. Equivalency cannot be established, since the upper range of the confidence interval falls above the maximum tolerable difference between the means of the two treatments. In order for equivalency to have been established, the result for the second analysis would had to have been statistically significant at the .05 level, and if the latter had occurred the upper limit for the confidence interval would had to have been some value less than 2.

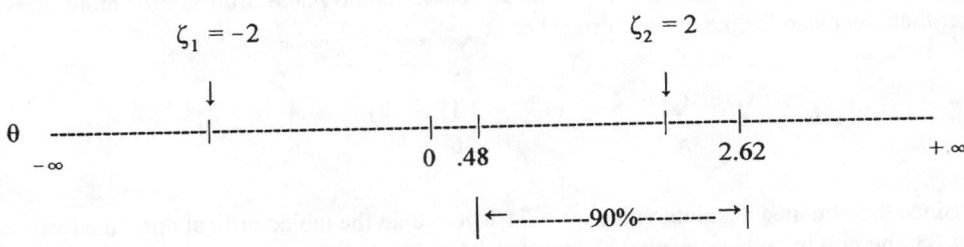

Figure 17.2   90% Confidence Interval For Example 17.7

The reader should take note of the fact that just by employing Equation 17.25, as was done earlier in this section in conducting the two *t* tests, one can obtain the answer to the question of whether or not the two treatments are equivalent. Alternatively, computation of the above confidence interval, in and of itself, will provide the answer to the same question. Use of both methods merely provides the same answer to the same question in different formats.

It should be noted that many researchers would elect to employ a one-tailed analysis for the data in Example 17.7 — specifically, stipulating as the alternative hypothesis that the mean of **Treatment 1** (the **brand name drug**) is not more than 2 units greater than the mean of **Treatment 2** (the **generic drug**). In such a case a researcher would conduct only a single **one-tailed test** which employed the null and alternative hypotheses noted below.

Null hypothesis $H_0: \theta \geq 2$

(In the underlying populations the samples represent, the difference between the two treatment means is greater than or equal to 2. This is the **hypothesis of nonequivalence**.)

Alternative hypothesis $H_1: \theta < 2$

(In the underlying populations the samples represent, the difference between the two treatment means is less than 2. This is the **hypothesis of equivalence**.)

If the above analysis is conducted, only the second of the two *t*-tests conducted previously would be employed. Specifically:

$$t = \frac{(\bar{X}_1 - \bar{X}_2) - \zeta}{s_{\bar{D}}} = \frac{(4.7 - 3.1) - (2)}{.56} = \frac{-.4}{.56} = -.71$$

Since, as noted previously, the obtained absolute value $t = -.71$ is less than the tabled critical one-tailed value $t_{.05} = 1.83$, the null hypotheses H_0: $\theta \geq 2$ designating nonequivalence cannot be rejected. In this instance, as noted in Figure 17.3, only the upper boundary would be determined for the confidence interval computed previously, with the requirement for equivalence being that the value of ζ_2 (which represents ζ) would have to fall above the upper limit of the confidence interval (which $\zeta = 2$ does not).

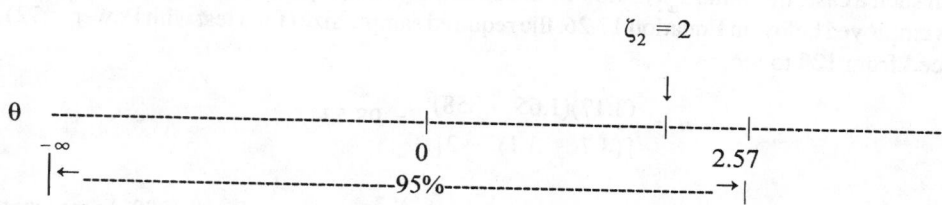

Figure 17.3 95% Confidence Interval For One-Tailed Analysis

Procedure for computing sample size in reference to the power of the Westlake–Schuirmann test of equivalence of two dependent treatments Equation 17.26 can be employed to estimate the sample size (n) required in order to conduct a **Westlake–Schuirmann test of equivalence of two dependent treatments** which has a power of a specified value. The values in the numerator of Equation 17.26 and $|\mu_1 - \mu_2|$ in the denominator of the latter equation are obtained in the identical manner they are obtained in Equation 17.17. The term ζ in the denominator of Equation 17.26 represents the limiting absolute value which the difference between the two means must not exceed in order for the researcher to declare the two treatments equivalent.

$$n = \frac{\sigma_D^2 \left(z_{\alpha/2} + z_\beta\right)^2}{\left[|\mu_1 - \mu_2| - \zeta\right]^2} \qquad \textbf{(Equation 17.26)}$$

Equation 17.26 will be employed within the framework of Example 17.7 to determine the sample size required in order for the power of the **Westlake–Schuirmann test** to equal .72 when the value $\zeta = 2$ is stipulated by the researcher. In conducting the latter analysis, the following values are employed in Equation 17.26 to compute the number of subjects that will be required: $s_D^2 = 3.17$, $|\mu_1 - \mu_2| = 1.6$, $\delta = 2$, $z_{\alpha/2} = 1.96$ (If a one-tailed analysis is conducted $z_{\alpha/2} = 1.65$ should be employed.), $z_\beta = .58$. (The rationale for employing the values $z_{\alpha/2} = 1.96$ and $z_\beta = .58$ is provided in the discussion of the determination of the latter values for Equation 17.17.)

When the appropriate values are substituted in Equation 17.26, the value $n = 127.82$ is computed, indicating that in order for the power of the test **Westlake–Schuirmann test of two dependent treatments** to be .72, the researcher must have at least 128 subjects (which is the next greatest integer value than integer element of the value computed for n).

$$n = \frac{(3.17)(1.96 + .58)^2}{[(4.7 - 3.1) - 2]^2} = 127.82$$

To illustrate that a test of equivalence will usually require a larger sample size than a conventional test of significance, recollect that when Equation 17.17 was employed with the same set of data in Section VI to compute the required sample size for the power of the *t* test for **two dependent samples** to equal .72, the much smaller value $n =7.99$ was computed for the sample size.

It should be noted that when a test of equivalence is used within the framework of drug assessment research, it is common practice for a one-tailed analysis to be employed — specifically, an evaluation of the directional alternative hypothesis that the efficacy of the generic drug will not be less than the efficacy of the brand name drug by more than the value stipulated for ζ. In such a case, the value $z_{\alpha/2} = 1.65$ would be employed in Equation 17.26. When the latter value is employed below in Equation 17.26, the required sample size (for a test with Power $= .72$) is reduced from 128 to 99.

$$ n = \frac{(3.17)(1.65 + .58)^2}{[(4.7 - 3.1) - 2]^2} = 98.53 $$

Test 17g: Tryon's test of equivalence of two dependent treatments Tryon (2001) notes that it is possible for the 95% confidence intervals computed for the means of two treatments to overlap when, in fact, a *t* **test for two dependent samples** yields a significant difference at the .05 level. Because of the latter, Schenker and Gentleman (2001, p. 182) summarize the latter issue in noting that evaluation of data through examination of confidence interval overlap provides for a more conservative test when the null hypothesis is true (i.e., rejects the null hypothesis less often) when contrasted with employing the results obtained from conducting the *t*-test, as well as the fact that evaluation through use of confidence interval overlap is less powerful since it mistakenly fails to reject the null hypothesis as frequently as does the use of the *t*-test when the null hypothesis is false. To insure that the latter inaccuracies will not occur, instead of computing the confidence intervals of the treatment means with Equation 2.8/2.9(which Tryon (2001) refers to as **descriptive confidence intervals**), modified confidence intervals referred to as **inferential confidence intervals** are computed. The latter confidence intervals not only provide for a more rigorous test of whether or not a statistically significant difference exists between the two treatments, but also allow a researcher to test for **statistical equivalence** (as well as **statistical indeterminancy**, which will exist when the analyses for statistical significance and statistical equivalence are inconsistent with one another).

As is the case with **Tryon's test of equivalence of two independent treatments**, **Tryon's test of equivalence of two dependent treatments** requires that a modified critical *t* value be computed in order to determine an **inferential confidence interval** for each of the treatment means. Within the framework of **Tryon's test of equivalence of two dependent treatments**, the modified critical *t* value employed in computing the **inferential confidence intervals** for the two treatment means can be obtained by multiplying the appropriate tabled critical *t* value for computing the **descriptive confidence interval** for each of the treatment means (using Equation 2.8/2.9) by the ratio E, which is computed with Equation 17.27. Note that in contrast to Equation 11.24 (the analogous equation used within the framework of **Tryon's test of equivalence of two independent treatments**) in which the **standard error of the difference** comprises the numerator, in Equation 17.27 the **standard error of the mean difference** comprises the numerator.

$$ E = \frac{\sqrt{s_{\bar{X}_1}^2 + s_{\bar{X}_2}^2 - 2(r_{X_1 X_2})(s_{\bar{X}_1})(s_{\bar{X}_2})}}{s_{\bar{X}_1} + s_{\bar{X}_2}} = \frac{s_{\bar{D}}}{s_{\bar{X}_1} + s_{\bar{X}_2}} \qquad \textbf{(Equation 17.27)} $$

The value $E = .3415$ is computed for Example 17.7 below. The reader should take note of the fact that the value of E computed for a dependent samples analysis will be substantially lower than the E value computed for an independent samples analysis. Because of the latter the range of values which define an **inferential confidence interval** in a dependent samples analysis will be narrower than the range of values obtained for an independent samples analysis. When the value $E = .3415$ is multiplied by the tabled critical two-tailed .05 value $t_{.05} = 2.262$ (for $df = n - 1 = 9$, since there are $n = 10$ scores per treatment), the modified critical value $t_{mod.} = .7725$ is obtained. Employing the latter value for $t_{.05}$ in Equation 2.8, an **inferential confidence interval** is computed below for each of the treatment means.

$$E = \frac{.56}{.89 + .75} = .3415$$

$$t_{mod} = (t_{.05})(E) = (2.262)(.3415) = .7725$$

$$CI_{.95} = \bar{X}_1 \pm (t_{mod})(s_{\bar{X}_1}) = 4.7 \pm (.7725)(.89) = 4.7 \pm .69$$

$$4.01 \leq \mu_1 \leq 5.39$$

$$CI_{.95} = \bar{X}_2 \pm (t_{mod})(s_{\bar{X}_2}) = 3.1 \pm (.7725)(.75) = 3.1 \pm .58$$

$$2.52 \leq \mu_2 \leq 3.68$$

Note that the two inferential confidence intervals do not overlap. Thus, the researcher can conclude there is a significant difference between the two treatments (when alpha = .05 is employed). The inferential confidence intervals can also be employed to determine whether or not the two treatments are equivalent. As was done previously within the framework of conducting the **Westlake–Schuirmann test of equivalence of two dependent treatments**, the value $\zeta = 2$ will be employed to stipulate the maximum tolerable difference between the two means. In conducting **Tryon's test of equivalence of two dependent treatments, equivalence will exist if the difference between the lower limit of the confidence interval computed for the smaller mean and the upper limit of the confidence interval computed for the larger mean is less than the absolute value of ζ stipulated by the researcher**. In our example the lower limit of the confidence interval computed for the smaller mean is 2.52 and the upper limit of the confidence interval computed for the larger mean is 5.39. Thus, we compute the difference $5.39 - 2.52 = 2.87$. Since the value 2.87 is greater than the absolute value $\zeta = 2$, we cannot conclude the treatments are equivalent (when alpha = .05 is employed).

Tryon (2001) notes that when using his methodology, a researcher should conduct both analyses described above employing the **inferential confidence intervals**. In other words the researcher should use the **inferential confidence intervals** to determine if there is a statistically significant difference between the means of the two treatments, as well as using the **inferential confidence intervals** to test for equivalence. To be more specific, the following guidelines should be employed in determining what conclusions to reach regarding the hypotheses being evaluated.

1) A researcher is only able to declare a statistically significant difference between the treatments at the prestipulated alpha level if both of the following conditions are met: a) The **inferential confidence intervals** do not overlap; b) The **test of equivalence** indicates that the difference between the lower limit of confidence interval computed for the smaller mean and the upper limit of the confidence interval computed for the larger mean is greater than or equal to the absolute value of ζ.

2) A researcher is only able to declare equivalence at the prestipulated alpha level if both of the following conditions are met: a) The **inferential confidence intervals** overlap; b) The **test of equivalence** indicates that the difference between the lower limit of the confidence interval computed for the smaller mean and the upper limit of the confidence interval computed for the larger mean is less than the absolute value of ζ.

3) A condition of **statistical indeterminancy** would exist if the results of the two analyses are inconsistent with one another. **Statistical indeterminancy** would exist if either of the following conditions is present: a) The difference between the means of the two treatments is not statistically significant (since the **inferential confidence intervals** overlap), yet at the same time the **test of equivalence** is not significant (since the difference between the lower limit of the confidence interval computed for the smaller mean and the upper limit of the confidence interval computed for the larger mean is greater than or equal to the absolute value of ζ); or b) The difference between the means of the two treatments is statistically significant (since the **inferential confidence intervals** do not overlap), yet at the same time the **test of equivalence** is significant (since the difference between the lower limit of the confidence interval computed for the smaller mean and the upper limit of the confidence interval computed for the larger mean is less than the absolute value of ζ). When **statistical indeterminancy** exists further research should be conducted in order to clarify the status of the hypotheses under study.

When Example 17.7 is evaluated with Tryon's methodology (employing a two-tailed analysis with alpha = .05), the final results fall in the first category noted above — in other words, the researcher can declare that there is a statistically significant difference between the two treatment means, since: a) The **inferential confidence intervals** do not overlap, and b) The **test of equivalence** indicates the difference of 2.87 units between the lower limit of confidence interval computed for the smaller mean and the upper limit of the confidence interval computed for the larger mean is greater $\zeta = 2$.

When Tryon's (2001) methodology is employed to evaluate Example 17.7 it yields a conclusion which is consistent with that reached when the **Westlake–Schuirmann test of equivalence of two dependent treatments** was employed to evaluate the same set of data (assuming a two-tailed analysis with alpha = .05 is conducted). However, it should be noted that the two methodologies for evaluating equivalence will not always yield results which are consistent with one another. In other words, there will be instances in which one of the methodologies will allow the researcher to declare equivalence while the other will not. The author of the book has determined that Tryon's (2001) methodology provides a more conservative test of equivalence. Consequently, when the results of the two tests are inconsistent, the **Westlake–Schuirmann test of equivalence of two dependent treatments** may allow the researcher to declare equivalence, while the **Tryon test of equivalence of two dependent treatments** will not. When such a circumstance exists, a researcher should consider whether a larger value might have been employed for ζ, so that if Tryon's (2001) methodology is employed one can conclude equivalence exists.

In closing this discussion, it should be noted that when two or more methods yield inconsistent results with regard to the presence or absence of equivalence, such inconsistencies should be resolved by conducting replication studies (the results of which can be evaluated through the use of **meta-analysis**), and basing one's final conclusions on the overall body of evidence. The reader should keep in mind that in the final analysis the criterion employed for equivalence will be some standard imposed by an outside agency (e.g., the FDA), or some alternative subjective criterion stipulated by a specific researcher. Ultimately, the final decision regarding the issue of equivalence should be based on multiple studies which yield a consistent pattern of results involving minimal differences between treatments that fall within a range of values which most practitioners in a discipline would consider to be reasonable.

VIII. Additional Example Illustrating the Use of the *t* Test for Two Dependent Samples

Example 17.8 is an additional example which can be evaluated with the *t* test for two dependent samples. Since Example 17.8 employs the same data as Example 17.1, it yields the same result. Note that in Example 17.8 complete counterbalancing is employed in order to control for order effects.

Example 17.8 *A study is conducted to evaluate the relative efficacy of two drugs (Clearoxin and Lesionoxin) on chronic psoriasis. Ten subjects afflicted with chronic psoriasis participate in the study. Each subject is exposed to both drugs for a six-month period, with a three-month hiatus between treatments. Five subjects are treated with Clearoxin initially, after which they are treated with Lesionoxin. The other five subjects are treated with Lesionoxin first and then with Clearoxin. The dependent variable employed in the study is a rating of the severity of a subject's lesions under the two drug conditions. The higher the rating the more severe a subject's psoriasis. The scores of the ten subjects under the two treatment conditions follow. (The first score represents the Clearoxin condition (which represents* Condition 1*), and the second score the Lesionoxin condition (which represents* Condition 2*).)* **Subject 1** *(9, 8);* **Subject 2** *(2, 2);* **Subject 3** *(1, 3);* **Subject 4** *(4, 2);* **Subject 5** *(6, 3);* **Subject 6** *(4, 0);* **Subject 7** *(7, 4);* **Subject 8** *(8, 5);* **Subject 9** *(5, 4);* **Subject 10** *(1, 0). Do the data indicate that subjects respond differently to the two types of medication?*

The significant result $t(9) = 2.86$, $p < .05$ indicates that the mean severity rating for patients treated with Clearoxin/Condition 1 ($\overline{X}_1 = 4.7$) is significantly higher than the mean rating for patients treated with Lesionoxin/Condition 2 ($\overline{X}_2 = 3.1$).

References

Anderson, N. N. (2001). **Empirical direction in design and analysis**. Mahwah, NJ: Lawrence Erlbaum Associates.

Cohen, B. H. (2001). **Explaining psychological statistics** (2nd ed.). New York, NY: John Wiley & Sons.

Cohen, J. (1977). **Statistical power analysis for the behavioral sciences**. New York: Academic Press.

Cohen, J. (1988). **Statistical power analysis for the behavioral sciences** (2nd ed.). Hillsdale, NJ: Lawrence Erlbaum Associates, Publishers.

Fisher, R. A. (1935). **The design of experiments** (7th ed.). Edinburgh–London: Oliver & Boyd.

Huck, S. W. & McLean, R. A. (1975). Using a repeated measures ANOVA to analyze the data from a pretest-posttest design: A potentially confusing task. **Psychological Bulletin**, 82, 511–518.

Keppel, G. (1991) **Design and analysis: A researcher's handbook** (3rd ed.). Englewood Cliffs, NJ: Prentice Hall.

Kirk, R. E. (1982). **Experimental design: Procedures for the behavioral sciences** (2nd. ed.). Belmont, CA: Brooks/Cole Publishing Company.

Kirk, R. E. (1995). **Experimental design: Procedures for the behavioral sciences** (3rd ed.). Pacific Grove, CA: Brooks/Cole Publishing Company.

Rogers, J. L., Howard, K. I., & Vessey, J. T. (1993). Using significance tests to evaluate equivalence between two experimental groups. **Psychological Bulletin**, 113, 553–565.

Rosner, B. (2000). **Fundamental of biostatistics** (5th ed.) Pacific Grove, CA: Duxbury Press.

Sandler, J. (1955) A test of the significance of difference between the means of correlated measures based on a simplification of Student's *t*. **British Journal of Psychology**, 46, 225–226.

Schenker, N. & Gentleman, J. F. (2001). On judging the significance of difference by examining the overlap between confidence intervals means. **The American Statistician**, 55, 182–186.

Schuirmann, D. J. (1987). A comparison of the two one-sided tests procedure and the power approach for assessing equivalence of average bioavailability. **Journal of Pharmacokinetics and Biopharmaceutics**, 15, 657–680.

Seaman, M. A. & Serlin, R. C. (1998). Equivalence confidence intervals for two-group comparisons of means. **Psychological Methods**, 3, 403–411.

Tryon, W. W. (2001). Evaluating statistical difference, equivalence, and indeterminacy using inferential confidence intervals: An integrated alternative method of conducting null hypothesis statistical tests. **Psychological Methods**, 6, 371–386.

van Belle, G. (2002). **Statistical rules of thumb**. New York: John Wiley & Sons.

Weinfurt, K.P. (2000). Repeated measures analysis: ANOVA, MANOVA and HLM. Grimm, L. G. & Yarnold, P. R. (Eds), **Reading and understanding more multivariate statistics** (pp. 317–361). Washington, D.C.: American Psychological Association.

Wellek, S. (2003). **Testing Statistical hypotheses of equivalence**. Boca Raton, FL: Chapman & Hall/CRC.

Westlake, W. J. (1976). Symmetrical confidence intervals for bioequivalence trials. **Biometrics**, 32, 741–744.

Westlake, W. J. (1981). Response to T.B. L. Kirkwood: Bioequivalence testing — a need to rethink. **Biometrics**, 37, 589–594.

Westlake, W. J. (1988). Bioavailability and bioequivalence of pharmaceutical formulations. In K. E. Peace (Ed.), **Biopharamaceutical statistics for drug development** (pp. 329–352). New York: Marcel Decker.

Endnotes

1. a) Alternative terms which are employed in describing a **dependent samples design** and the *t*-test described in this chapter for such a design are **repeated measures design/*t*-test, within-subjects design/*t*-test, paired-samples design/*t*-test, treatment-by-subjects design/*t*-test, correlated samples design/*t*-test, matched-subjects design/*t*-test, matched-pairs design/*t*-test, crossover design, randomized-blocks design**, and **split-plot design**. The use of the terms **blocks** within the framework of a dependent samples design is discussed in Endnote 1 of the **single-factor within-subjects analysis of variance**; b) A **dependent samples design** (in which there are two or more experimental conditions) is considered a **balanced design** if there is no missing data — i.e., a score is obtained from each subject for each of the experimental conditions. In the event at least one subject is missing data for one or more the experimental conditions a **dependent samples design** would be categorized as an **unbalanced design**.

2. As noted in the **Introduction**, a study has **internal validity** to the extent that observed differences between the experimental conditions on the dependent variable can be unambiguously attributed to a manipulated independent variable. Random assignment of subjects to the different experimental conditions is the most effective way to optimize the

likelihood of achieving **internal validity** (by eliminating the possible influence of confounding/extraneous variables). In contrast to **internal validity, external validity** refers to the degree to which the results of an experiment càn be generalized. The results of an experiment can only be generalized to a population of subjects, as well as environmental conditions, which are comparable to those that are employed in the experiment.

3. In actuality, when galvanic skin response (which is a measure of skin resistance) is measured, the higher a subject's GSR the less emotional the subject. In Example 17.1, it is assumed that the GSR scores have been transformed so that the higher a subject's GSR score, the greater the level of emotionality.

4. An alternative but equivalent way of writing the null hypothesis is H_0: $\mu_1 - \mu_2 = 0$. The analogous alternative but equivalent ways of writing the alternative hypotheses in the order they are presented are: H_1: $\mu_1 - \mu_2 \neq 0$; H_1: $\mu_1 - \mu_2 > 0$; and H_1: $\mu_1 - \mu_2 < 0$.

5. Note that the basic structure of Equation 17.3 is the same as Equations I.11/2.1 (the equation for the estimated population standard deviation which is employed within the framework of the **single-sample t test**). In Equation 17.3 a standard deviation is computed for n D scores, whereas in Equations I.11/2.1 a standard deviation is computed for n X scores.

6. The actual value that is estimated by $s_{\bar{D}}$ is $\sigma_{\bar{D}}$, which is the standard deviation of the sampling distribution of mean difference scores for the two populations. The meaning of the **standard error of the mean difference** can be best understood by considering the following procedure for generating an empirical sampling distribution of difference scores: a) Obtain n difference scores for a random sample of n subjects; b) Compute the mean difference score $(\bar{D})$ for the sample; and c) Repeat steps a) and b) m times. At the conclusion of this procedure one will have obtained m mean difference scores. The **standard error of the mean difference** represents the standard deviation of the m mean difference scores, and can be computed by substituting the term $\bar{D}$ for D in Equation 17.3. Thus: $s_{\bar{D}} = \sqrt{[\Sigma\bar{D}^2 - ((\Sigma\bar{D})^2/m)]/[m - 1]}$. The standard deviation which is computed with Equation 17.4 is an estimate of $\sigma_{\bar{D}}$.

7. In order for Equation 17.1 to be soluble, there must be variability in the n difference scores. If each subject produces the same difference score, the value of $\tilde{s}_D$ computed with Equation 17.3 will equal 0. As a result of the latter, Equation 17.4 will yield the value $s_{\bar{D}} = 0$. Since $s_{\bar{D}}$ is the denominator of Equation 17.1, when the latter value equals zero the t **test** equation will be insoluble.

8. The same result for the t **test** will be obtained if in obtaining a difference score for each subject a subject's X_1 score is subtracted from his X_2 score (i.e., $D = X_2 - X_1$). Employing the latter protocol will only result in a change in the sign of the value computed for the t statistic. In the case of Example 17.1 the aforementioned protocol will yield the value $t = -2.86$. The obtained value of t is interpreted in the same manner as noted at the beginning of this section, except for the fact that in order for the directional alternative hypothesis H_1: $\mu_1 > \mu_2$ to be supported the sign of t must be negative, and for the directional alternative hypothesis H_1: $\mu_1 < \mu_2$ to be supported the sign of t must be positive.

9. The numerator of Equation 17.6 will always equal $\bar{D}$ (i.e., the numerator of Equation 17.1). In the same respect the denominator of Equation 17.6 will always equal $s_{\bar{D}}$ (the denominator of Equation 17.1). The denominator of Equation 17.6 can also be written as follows:

$$\sqrt{\frac{\tilde{s}_1^2}{n} + \frac{\tilde{s}_2^2}{n} - 2(r_{X_1 X_2})\left[\frac{\tilde{s}_1}{\sqrt{n}}\right]\left[\frac{\tilde{s}_2}{\sqrt{n}}\right]}$$

10. a) The reader should take note of the fact that if the correlation between the scores of the n subjects under the two experimental conditions is a low positive value, the **t test for two dependent samples** may actually provide a less powerful test of an alternative hypothesis than would be the case if the same set of data were evaluated with a **t test for two independent samples**. The reason for the latter is that the degrees of freedom employed for the **t test for two independent samples** are larger than the degrees of freedom employed for the **t test for two dependent samples**. Use of a larger degrees of freedom value allows a researcher to employ a smaller critical t value for the **t test for two independent samples**, which might offset any power advantage resulting from the lower error term in the denominator of the **t test for two dependent samples**; b) Note that in the case of Example 11.1 (which is employed to illustrate the **t test for two independent samples**), it is reasonable to assume that scores in the same row of Table 11.1 (which summarizes the data for the study) will not be correlated with one another (by virtue of the fact that two independent samples are employed in the study). When independent samples are employed, it is assumed that random factors determine the values of any pair of scores in the same row of a table summarizing the data, and consequently it is assumed that the correlation between pairs of scores in the same row will be equal to (or close to) 0.

11. a) Due to rounding off error, there may be a slight discrepancy between the value of t computed with Equations 17.1 and 17.6; b) If instead of $(\bar{X}_1 - \bar{X}_2)$, $(\bar{X}_2 - \bar{X}_1)$ is employed in the numerator of Equation 17.6, the obtained absolute value of t will be identical but the sign of t will be reversed. Thus, in the case of Example 17.1 it will yield the value $t = -2.86$.

12. A noncorrelational procedure which allows a researcher to evaluate whether or not a treatment effect is present in the above described example is **Fisher's randomization procedure** (Fisher (1935)), which is generally categorized as a **permutation test**. The **randomization test for two independent samples (Test 12a)**, which is an example of a test that is based on **Fisher's randomization procedure**, is described in Section IX (the **Addendum**) of the **Mann–Whitney U test (Test 12)**. **Fisher's randomization procedure** requires that all possible score configurations which can be obtained for the value of the computed sum of the difference scores be determined. Upon computing the latter information, one can determine the likelihood of obtaining a configuration of scores which is equal to or more extreme than the one obtained for a set of data.

13. Equation 17.1 can be modified as follows to be equivalent to Equation 17.8:
$t = [\bar{D} - (\mu_1 - \mu_2)]/s_{\bar{D}}$.

14. Although Equation 17.15 is intended for use prior to collecting the data, it should yield the same value for σ_D if the values computed for the sample data are substituted in it. Thus, if

we employ the values $\sigma = 2.60$ (which is the average of the values $\tilde{s}_1 = 2.83$ and $\tilde{s}_2 = 2.38$) and $\rho_{X_1 X_2} = .78$ (which is the population correlation coefficient estimated by the value $r_{X_1 X_2} = .78$), and substitute them in Equation 17.15, the value $\sigma_D = 1.72$ is computed, which is quite close to the computed value $\tilde{s}_D = 1.78$. The slight discrepancy between the two values can be attributed to the fact that the estimated population standard deviations are not identical.

15. In contrast to the *t* **test for two dependent samples** (which can only be employed with two dependent samples), the **single-factor within-subjects analysis of variance** can be used with a dependent samples design involving interval/ratio data in which there are k samples, where $k \geq 2$.

16. Note that the basic structure of Equation 17.20 is the same as Equation 11.17 (which is employed for computing a confidence interval for the *t* **test for two independent samples**), except that the latter equation employs $s_{\bar{X}_1 - \bar{X}_2}$ in place of $s_{\bar{D}}$.

17. It was noted earlier that if all n subjects obtain the identical difference score, Equations 17.1/17.6 become unsolvable. In the case of Equation 17.21, for a given value of n, if all n subjects obtain the same difference score the same A value will always be computed, regardless of the magnitude of the identical difference score obtained by each of the n subjects. If the value of A computed under such conditions is substituted in the equation $t = \sqrt{(n-1)/(An-1)}$ (which is algebraically derived from Equation 17.22), the latter equation becomes unsolvable (since the value $(An-1)$ will always equal zero). The conclusion that results from this observation is that Equation 17.21 is insensitive to the magnitude of the difference between experimental conditions when all subjects obtain the same difference score.

18. a) Equation 17.23 can also be written as follows:

$$ z = \frac{\bar{X}_1 - \bar{X}_2}{\sqrt{\dfrac{\sigma_1^2}{n} + \dfrac{\sigma_2^2}{n} - 2(r_{X_1 X_2})\left[\dfrac{\sigma_1}{\sqrt{n}}\right]\left[\dfrac{\sigma_2}{\sqrt{n}}\right]}} $$

In instances when a researcher stipulates in the null hypothesis that the difference between the two population means is some value other than zero, the numerator of Equation 17.22 is the same as the numerator of Equation 17.8. The protocol for computing the value of the numerator is identical to that employed for Equation 17.8; b) Another way of writing Equation 17.23 is $z = \bar{D}/\sigma_{\bar{D}}$, where $\sigma_{\bar{D}} = \sigma_D/\sqrt{n}$.

19. van Belle (2002, p. 61–63) notes that in order for matching to be of value, the correlation for the matched pairs on the dependent variable (i.e., $r_{X_1 X_2}$) should be .5 or greater.

20. a) In Example 17.1, the order of presentation of the conditions is controlled by randomly distributing the sexually explicit and neutral words throughout the 16 word list presented to each subject; b) In some sources the term **order effects** is employed synonymously with the terms **carryover effects**, or **sequencing**.

21. In the previous editions of this book the **one-group pretest-posttest design** was referred to as a **before-after design**.

22. A doctor conducting such a study might justify the absence of a control group on ethical grounds, based on one's belief that patients in such a group would be deprived of a potentially beneficial treatment.

Test 18

The Wilcoxon Matched-Pairs Signed-Ranks Test
(Nonparametric Test Employed with Ordinal Data)

I. Hypothesis Evaluated with Test and Relevant Background Information

Hypothesis evaluated with test Do two dependent samples represent two different populations?

Relevant background information on test The **Wilcoxon matched-pairs signed-ranks test** (Wilcoxon (1945, 1949)) is a nonparametric procedure employed in a hypothesis testing situation involving a design with two dependent samples. Whenever one or more of the assumptions of the *t* **test for two dependent samples (Test 17)** are saliently violated, the **Wilcoxon matched-pairs signed-ranks test** (which has less stringent assumptions) may be preferred as an alternative procedure. Prior to reading the material on the **Wilcoxon matched-pairs signed-ranks test**, the reader may find it useful to review the general information regarding a dependent samples design contained in Sections I and VII of the *t* **test for two dependent samples**.

The **Wilcoxon matched-pairs signed-ranks test** is essentially an extension of the **Wilcoxon signed-ranks test (Test 6)** (which is employed for a single sample design) to a design involving two dependent samples. In order to employ the **Wilcoxon matched-pairs signed-ranks test**, it is required that each of *n* subjects (or *n* pairs of matched subjects) has two interval/ratio scores (each score having been obtained under one of the two experimental conditions). A difference score is computed for each subject (or pair of matched subjects) by subtracting a subject's score in Condition 2 from his score in Condition 1. The hypothesis evaluated with the **Wilcoxon matched-pairs signed-ranks test** is whether or not in the underlying populations represented by the samples/experimental conditions, the median of the difference scores (which will be represented by the notation θ_D) equals zero. If a significant difference is obtained, it indicates there is a high likelihood the two samples/conditions represent two different populations.

The **Wilcoxon matched-pairs signed-ranks test** is based on the following assumptions:[1] a) The sample of *n* subjects has been randomly selected from the population it represents; b) The original scores obtained for each of the subjects are in the format of interval/ratio data; and c) The distribution of the difference scores in the populations represented by the two samples is symmetric about the median of the population of difference scores.

As is the case for the *t* **test for two dependent samples**, in order for the **Wilcoxon matched-pairs signed-ranks test** to generate valid results, the following guidelines should be adhered to: a) To control for order effects, the presentation of the two experimental conditions should be random or, if appropriate, be counterbalanced; and b) If matched samples are employed, within each pair of matched subjects each of the subjects should be randomly assigned to one of the two experimental conditions.

As is the case with the *t* **test for two dependent samples**, the **Wilcoxon matched-pairs signed-ranks test** can also be employed to evaluate a **one-group pretest-posttest design**. The limitations of the **one-group pretest-posttest design** (which are discussed in Section VII of the *t* **test for two dependent samples** and the **Introduction**) are also applicable when it is evaluated with the **Wilcoxon matched-pairs signed-ranks test**.

It should be noted that all of the other tests in this text which rank data (with the exception of the **Wilcoxon signed-ranks test** and the **Moses test for equal variability (Test 15)**), rank the original interval/ratio scores of subjects. The **Wilcoxon matched-pairs signed-ranks test**, however, does not rank the original interval/ratio scores, but instead ranks the interval/ratio difference scores of subjects (or matched pairs of subjects). For this reason, some sources categorize the **Wilcoxon matched-pairs signed-ranks test** as a test of interval/ratio data. Most sources, however (including this book), categorize the **Wilcoxon matched-pairs signed-ranks test** as a test of ordinal data, by virtue of the fact that a ranking procedure is part of the test protocol.

II. Example

Example 18.1 is identical to Example 17.1 (which is evaluated with the *t* **test for two dependent samples**). In evaluating Example 18.1 it will be assumed that the ratio data are rank-ordered, since one or more of the assumptions of the *t* **test for two dependent samples** have been saliently violated.

Example 18.1 *A psychologist conducts a study to determine whether or not people exhibit more emotionality when they are exposed to sexually explicit words than when they are exposed to neutral words. Each of ten subjects is shown a list of 16 randomly arranged words, which are projected onto a screen one at a time for a period of five seconds. Eight of the words on the list are sexually explicit and eight of the words are neutral. As each word is projected on the screen, a subject is instructed to say the word softly to himself or herself. As a subject does this, sensors attached to the palms of the subject's hands record galvanic skin response (GSR), which is used by the psychologist as a measure of emotionality. The psychologist computes two scores for each subject, one score for each of the experimental conditions:* **Condition 1**: *GSR/Explicit — The average GSR score for the eight sexually explicit words;* **Condition 2**: *GSR/Neutral — The average GSR score for the eight neutral words. The GSR/Explicit and the GSR/Neutral scores of the ten subjects follow. (The higher the score, the higher the level of emotionality.)* **Subject 1** *(9, 8);* **Subject 2** *(2, 2);* **Subject 3** *(1, 3);* **Subject 4** *(4, 2);* **Subject 5** *(6, 3);* **Subject 6** *(4, 0);* **Subject 7** *(7, 4);* **Subject 8** *(8, 5);* **Subject 9** *(5, 4);* **Subject 10** *(1, 0). Do subjects exhibit differences in emotionality with respect to the two categories of words?*

III. Null versus Alternative Hypotheses

Null hypothesis　　　　　　　　　　　　H_0: $\theta_D = 0$

(In the underlying populations represented by Condition 1 and Condition 2, the median of the difference scores equals zero. With respect to the sample data, this translates into the sum of the ranks of the positive difference scores being equal to the sum of the ranks of the negative difference scores (i.e., $\Sigma R+ = \Sigma R-$).

Alternative hypothesis　　　　　　　　　H_1: $\theta_D \neq 0$

(In the underlying populations represented by Condition 1 and Condition 2, the median of the difference scores is some value other than zero. With respect to the sample data, this translates into the sum of the ranks of the positive difference scores not being equal to the sum of the ranks of the negative difference scores (i.e., $\Sigma R+ \neq \Sigma R-$). This is a **nondirectional alternative hypothesis** and it is evaluated with a **two-tailed test**.)

or

$$H_1:\ \theta_D > 0$$

(In the underlying populations represented by Condition 1 and Condition 2, the median of the difference scores is some value that is greater than zero. With respect to the sample data, this translates into the sum of the ranks of the positive difference scores being greater than the sum of the ranks of the negative difference scores (i.e., $\Sigma R+ > \Sigma R-$). The latter result indicates that the scores in Condition 1 are higher than the scores in Condition 2. This is a **directional alternative hypothesis** and it is evaluated with a **one-tailed test**.)

or

$$H_1:\ \theta_D < 0$$

(In the underlying populations represented by Condition 1 and Condition 2, the median of the difference scores is some value that is less than zero (i.e., a negative number). With respect to the sample data, this translates into the sum of the ranks of the positive difference scores being less than the sum of the ranks of the negative difference scores (i.e., $\Sigma R+ < \Sigma R-$). The latter result indicates that the scores in Condition 2 are higher than the scores in Condition 1. This is a **directional alternative hypothesis** and it is evaluated with a **one-tailed test**.)

Note: Only one of the above noted alternative hypotheses is employed. If the alternative hypothesis the researcher selects is supported, the null hypothesis is rejected.

IV. Test Computations

The data for Example 18.1 are summarized in Table 18.1. Note that there are 10 subjects and that each subject has two scores.

Table 18.1 Data for Example 18.1

Subject	X_1	X_2	$D = X_1 - X_2$	Rank of $\lvert D \rvert$	Signed rank of $\lvert D \rvert$
1	9	8	1	2	2
2	2	2	0	–	–
3	1	3	–2	4.5	–4.5
4	4	2	2	4.5	4.5
5	6	3	3	7	7
6	4	0	4	9	9
7	7	4	3	7	7
8	8	5	3	7	7
9	5	4	1	2	2
10	1	0	1	2	2
				$\Sigma R+$ = 40.5	
				$\Sigma R-$ = 4.5	

In Table 18.1, X_1 represents each subject's score in Condition 1 (sexually explicit words) and X_2 represents each subject's score in Condition 2 (neutral words). In Column 4 of Table 18.1 a D score is computed for each subject by subtracting a subject's score in Condition 2 from the subject's score in Condition 1 (i.e., $D = X_1 - X_2$). In Column 5 the D scores have been ranked with respect to their absolute values. Since the ranking protocol employed for the **Wilcoxon matched-pairs signed-ranks test** is identical to that employed for the **Wilcoxon signed-ranks test**, the reader may find it useful to review the ranking protocol described in Section IV of the latter test. To reiterate, the following guidelines should be adhered to when ranking the difference scores for the **Wilcoxon matched-pairs signed-ranks test**.

a) The **absolute values** of the difference scores ($|D|$) are ranked (i.e., the sign of a difference score is not taken into account). Because absolute values are employed to represent the difference scores, $D = X_2 - X_1$ can also be employed to compute the value of D.

b) Any difference score that equals zero is not ranked. This translates into eliminating from the analysis any subject who yields a difference score of zero.

c) When there are tied scores present in the data, the average of the ranks involved is assigned to all scores tied for a given rank.

d) As is the case with the **Wilcoxon signed-ranks test**, when ranking difference scores for the **Wilcoxon matched-pairs signed-ranks test** it is essential that a rank of 1 be assigned to the difference score with the lowest absolute value, and that a rank of n be assigned to the difference score with the highest absolute value (where n represents the number of signed ranks — i.e., difference scores which have been ranked).[2]

Upon ranking the absolute values of the difference scores, the sign of each difference score is placed in front of its rank. The signed ranks of the difference scores are listed in Column 6 of Table 18.1. Note that although 10 subjects participated in the experiment there are only $n = 9$ signed ranks, since Subject 2 had a difference score of zero which was not ranked. Table 18.2 summarizes the rankings of the difference scores for Example 18.1.

Table 18.2 Ranking Procedure for Wilcoxon Matched-Pairs Signed-Ranks Test

Subject number	2	1	9	10	3	4	5	7	8	6
Subject's difference score	0	1	1	1	−2	2	3	3	3	4
Absolute value of difference score	−	1	1	1	2	2	3	3	3	4
Rank of $|D|$	−	2	2	2	4.5	4.5	7	7	7	9

The sum of the ranks that have a positive sign (i.e., $\Sigma R+ = 40.5$) and the sum of the ranks that have a negative sign (i.e., $\Sigma R- = 4.5$) are recorded at the bottom of Column 6 in Table 18.1. Equation 18.1 (which is identical to Equation 6.1) allows one to check the accuracy of these values. If the relationship indicated by Equation 18.1 is not obtained, it indicates an error has been made in the calculations.

$$\Sigma R+ \; + \; \Sigma R- \; = \; \frac{n(n + 1)}{2} \qquad \text{(Equation 18.1)}$$

Employing the values $\Sigma R+ = 40.5$ and $\Sigma R- = 4.5$ in Equation 18.1, we confirm that the relationship described by the equation is true.

$$40.5 \; + \; 4.5 \; = \; \frac{(9)(10)}{2} \; = \; 45$$

V. Interpretation of the Test Results

As noted in Section III, if the sample is derived from a population in which the median of the difference scores equals zero, the values of $\Sigma R+$ and $\Sigma R-$ will be equal to one another. When

$\Sigma R+$ and $\Sigma R-$ are equivalent, both of these values will equal $[n(n + 1)]/4$, which in the case of Example 18.1 will be $[(9)(10)]/4 = 22.5$. This latter value is commonly referred to as the **expected value** of the **Wilcoxon T statistic**.

If the value of $\Sigma R+$ is significantly greater than the value of $\Sigma R-$, it indicates there is a high likelihood that Condition 1 represents a population with higher scores than the population represented by Condition 2. On the other hand, if $\Sigma R-$ is significantly greater than $\Sigma R+$, it indicates there is a high likelihood that Condition 2 represents a population with higher scores than the population represented by Condition 1. Table 18.1 reveals that $\Sigma R+ = 40.5$ is greater than $\Sigma R- = 4.5$, and thus the data are consistent with the directional alternative hypothesis H_1: $\theta_D > 0$ (i.e., it indicates that subjects obtained higher scores in Condition 1 than Condition 2). The question is, however, whether the difference is significant — i.e., whether it is large enough to conclude that it is unlikely to be the result of chance.

The absolute value of the **smaller** of the two values $\Sigma R+$ versus $\Sigma R-$ is designated as the **Wilcoxon T test statistic**. Since $\Sigma R- = 4.5$ is smaller than $\Sigma R+ = 40.5$, $T = 4.5$. The T value is interpreted by employing **Table A5 (Table of Critical T Values for Wilcoxon's Signed-Ranks and Matched-Pairs Signed-Ranks Tests)** in the **Appendix**. **Table A5** lists the critical one- and two-tailed .05 and .01 T values in relation to the number of signed ranks in a set of data. In order to be significant, the obtained value of T must be **equal to or less than** the tabled critical T value at the prespecified level of significance.[3] Table 18.3 summarizes the tabled critical one- and two-tailed .05 and .01 Wilcoxon T values for $n = 9$ signed ranks.

Table 18.3   Tabled Critical Wilcoxon T Values for $n = 9$ Signed Ranks

	$T_{.05}$	$T_{.01}$
Two-tailed values	5	1
One-tailed values	8	3

Since the null hypothesis can only be rejected if the computed value $T = 4.5$ is equal to or less than the tabled critical value at the prespecified level of significance, we can conclude the following.

In order for the nondirectional alternative hypothesis H_1: $\theta_D \neq 0$ to be supported, it is irrelevant whether $\Sigma R+ > \Sigma R-$ or $\Sigma R- > \Sigma R+$. In order for the result to be significant, the computed value of T must be equal to or less than the tabled critical two-tailed value at the prespecified level of significance. Since the computed value $T = 4.5$ is less than the tabled critical two-tailed .05 value $T_{.05} = 5$, the nondirectional alternative hypothesis H_1: $\theta_D \neq 0$ is supported at the .05 level. It is not, however, supported at the .01 level, since $T = 4.5$ is greater than the tabled critical two-tailed .01 value $T_{.01} = 1$.

In order for the directional alternative hypothesis H_1: $\theta_D > 0$ to be supported, $\Sigma R+$ must be greater than $\Sigma R-$. Since $\Sigma R+ > \Sigma R-$, the data are consistent with the directional alternative hypothesis H_1: $\theta_D > 0$. In order for the result to be significant, the computed value of T must be equal to or less than the tabled critical one-tailed value at the prespecified level of significance. Since the computed value $T = 4.5$ is less than the tabled critical one-tailed .05 value $T_{.05} = 8$, the directional alternative hypothesis H_1: $\theta_D > 0$ is supported at the .05 level. It is not, however, supported at the .01 level, since $T = 4.5$ is greater than the tabled critical one-tailed .01 value $T_{.01} = 3$.

In order for the directional alternative hypothesis H_1: $\theta_D < 0$ to be supported, the following two conditions must be met: a) $\Sigma R-$ must be greater than $\Sigma R+$; and b) The computed value of T must be equal to or less than the tabled critical one-tailed value at the prespecified level of significance. Since the first of these conditions is not met, the directional alternative hypothesis H_1: $\theta_D < 0$ is not supported.

A summary of the analysis of Example 18.1 with the **Wilcoxon matched-pairs signed-ranks test** follows: It can be concluded that subjects exhibited higher GSR (emotionality) scores with respect to the sexually explicit words than the neutral words.

The results obtained with the **Wilcoxon matched-pairs signed-ranks test** are reasonably consistent with those obtained when the *t* **test for two dependent samples** is employed to evaluate the same set of data. In the case of both tests, the analogous nondirectional alternative hypotheses H_1: $\theta_D \neq 0$ and H_1: $\mu_1 \neq \mu_2$ are supported, but only at the .05 level. In the case of the **Wilcoxon matched-pairs signed-ranks test**, the directional alternative hypothesis H_1: $\theta_D > 0$ is only supported at the .05 level, whereas the analogous directional alternative hypothesis H_1: $\mu_1 > \mu_2$ is supported at both the .05 and .01 levels when the data are evaluated with the *t* **test for two dependent samples**. The latter discrepancy between the two tests reflects the fact that when a parametric and nonparametric test are applied to the same set of data, the parametric test will generally provide a more powerful test of an alternative hypothesis. In most instances, however, similar conclusions will be reached if the same data are evaluated with the *t* **test for two dependent samples** and the **Wilcoxon matched-pairs signed-ranks test**.

VI. Additional Analytical Procedures for the Wilcoxon Matched-Pairs Signed-Ranks Test and/or Related Tests

1. The normal approximation of the Wilcoxon *T* statistic for large sample sizes As is the case with the **Wilcoxon signed-ranks test**, if the sample size employed in a study is relatively large, the normal distribution can be employed to approximate the Wilcoxon *T* statistic. Although sources do not agree on the value of the sample size which justifies employing the normal approximation of the Wilcoxon distribution, they generally state that it should be employed for sample sizes larger than those documented in the Wilcoxon table contained within the source. Equation 18.2 (which is identical to Equation 6.2) provides the normal approximation for Wilcoxon *T*. In the equation *T* represents the computed value of Wilcoxon *T*, which for Example 18.1 is $T = 4.5$. *n*, as noted previously, represents the number of signed ranks. Thus, in our example, $n = 9$. Note that in the numerator of Equation 18.2, the term $[n(n + 1)]/4$ represents the expected value of *T* (often summarized with the symbol T_E), which is defined in Section V. The denominator of Equation 18.2 represents the expected standard deviation of the sampling distribution of the *T* statistic.

$$z = \frac{T - \dfrac{n(n + 1)}{4}}{\sqrt{\dfrac{n(n + 1)(2n + 1)}{24}}} \qquad \textbf{(Equation 18.2)}$$

Although Example 18.1 involves only nine signed ranks (a value most sources would view as too small to use with the normal approximation), it will be employed to illustrate Equation 18.2. The reader will see that in spite of employing Equation 18.2 with a small sample size, it will yield essentially the same result as that obtained when the exact table of the Wilcoxon distribution is employed. When the values $T = 4.5$ and $n = 9$ are substituted in Equation 18.2, the value $z = -2.13$ is computed.

$$z = \frac{4.5 - \dfrac{(9)(10)}{4}}{\sqrt{\dfrac{(9)(10)(19)}{24}}} = -2.13$$

The obtained value $z = -2.13$ is evaluated with **Table A1 (Table of the Normal Distribution)** in the **Appendix**. In **Table A1** the tabled critical two-tailed .05 and .01 values are $z_{.05} = 1.96$ and $z_{.01} = 2.58$, and the tabled critical one-tailed .05 and .01 values are $z_{.05} = 1.65$ and $z_{.01} = 2.33$.

Since the smaller of the two values $\Sigma R+$ versus $\Sigma R-$ is selected to represent T, the value of z computed with Equation 18.2 will always be a negative number (unless $\Sigma R+ = \Sigma R-$, in which case z will equal zero). This is the case, since by selecting the smaller value T will always be less than the expected value T_E. As a result of this, the following guidelines are employed in evaluating the null hypothesis.

a) If a nondirectional alternative hypothesis is employed, the null hypothesis can be rejected if the obtained absolute value of z is equal to or greater than the tabled critical two-tailed value at the prespecified level of significance.

b) When a directional alternative hypothesis is employed, one of the two possible directional alternative hypotheses will be supported if the obtained absolute value of z is equal to or greater than the tabled critical one-tailed value at the prespecified level of significance. Which alternative hypothesis is supported depends on the prediction regarding which of the two values $\Sigma R+$ versus $\Sigma R-$ is larger. The null hypothesis can only be rejected if the directional alternative hypothesis that is consistent with the data is supported.

Employing the above guidelines, when the normal approximation is employed with Example 18.1 the following conclusions can be reached.

The nondirectional alternative hypothesis $H_1: \theta_D \neq 0$ is supported at the .05 level. This is the case, since the computed absolute value $z = 2.13$ is greater than the tabled critical two-tailed .05 value $z_{.05} = 1.96$. The nondirectional alternative hypothesis $H_1: \theta_D \neq 0$ is not supported at the .01 level, since the absolute value $z = 2.13$ is less than the tabled critical two-tailed .01 value $z_{.01} = 2.58$. This decision is consistent with the decision that is reached when the exact table of the Wilcoxon distribution is employed to evaluate the nondirectional alternative hypothesis $H_1: \theta_D \neq 0$.

The directional alternative hypothesis $H_1: \theta_D > 0$ is supported at the .05 level. This is the case, since the data are consistent with the latter alternative hypothesis (i.e., $\Sigma R+ > \Sigma R-$), and the computed absolute value $z = 2.13$ is greater than the tabled critical one-tailed .05 value $z_{.05} = 1.65$. The directional alternative hypothesis $H_1: \theta_D > 0$ is not supported at the .01 level, since the obtained absolute value $z = 2.13$ is less than the tabled critical one-tailed .01 value $z_{.01} = 2.33$. This decision is consistent with the decision that is reached when the exact table of the Wilcoxon distribution is employed to evaluate the directional alternative hypothesis $H_1: \theta_D > 0$.

The directional alternative hypothesis $H_1: \theta_D < 0$ is not supported, since the data are not consistent with the latter alternative hypothesis (which requires that $\Sigma R- > \Sigma R+$).

It should be noted that, in actuality, either $\Sigma R+$ or $\Sigma R-$ can be employed to represent the value of T in Equation 18.2. Either value will yield the same absolute value for z. The smaller of the two values will always yield a negative z value, and the larger of the two values will always yield a positive z value (which in this instance will be $z = 2.13$ if $\Sigma R+ = 40.5$ is employed to represent T). In evaluating a nondirectional alternative hypothesis the sign of z is irrelevant. In the case of a directional alternative hypothesis, one must determine whether the data are consistent with the alternative hypothesis that is stipulated. If the data are consistent, one then determines whether the absolute value of z is equal to or greater than the tabled critical one-tailed value at the prespecified level of significance.

2. The correction for continuity for the normal approximation of the Wilcoxon matched-pairs signed-ranks test
As noted in the discussion of the **Wilcoxon signed-ranks test**, a

correction for continuity can be employed for the normal approximation of the Wilcoxon test statistic. The same correction for continuity can be applied to the **Wilcoxon matched-pairs signed-ranks test**. The correction for continuity (which results in a slight reduction in the absolute value computed for z) requires that .5 be subtracted from the absolute value of the numerator of Equation 18.2. Thus, Equation 18.3 (which is identical to Equation 6.3) represents the continuity-corrected normal approximation of the Wilcoxon test statistic.

$$z = \frac{\left| T - \frac{n(n+1)}{4} \right| - .5}{\sqrt{\frac{n(n+1)(2n+1)}{24}}} \qquad \textbf{(Equation 18.3)}$$

Employing Equation 18.3, the continuity-corrected value $z = 2.07$ is computed. Note that as a result of the absolute value conversion, the numerator of Equation 18.3 will always be a positive number, thus yielding a positive z value.

$$z = \frac{\left| 4.5 - \frac{(9)(10)}{4} \right| - .5}{\sqrt{\frac{(9)(10)(19)}{24}}} = 2.07$$

The result of the analysis with Equation 18.3 leads to the same conclusions that are reached with Equation 18.2 (i.e., when the correction for continuity is not employed). Specifically, since the absolute value $z = 2.07$ is greater than the tabled critical two-tailed .05 value $z_{.05} = 1.96$, the nondirectional alternative hypothesis $H_1: \theta_D \neq 0$ is supported at the .05 level (but not at the .01 level). Since the absolute value $z = 2.07$ is greater than the tabled critical one-tailed .05 value $z_{.05} = 1.65$, the directional alternative hypothesis $H_1: \theta_D > 0$ is supported at the .05 level (but not at the .01 level).

3. Tie correction for the normal approximation of the Wilcoxon test statistic Equation 18.4 (which is identical to Equation 6.4) is an adjusted version of Equation 18.2 that is recommended in some sources (e.g., Daniel (1990) and Marascuilo and McSweeney (1977)) when tied difference scores are present in the data. The tie correction (which is identical to the one described for the **Wilcoxon signed-ranks test**) results in a slight increase in the absolute value of z. Unless there are a substantial number of ties, the difference between the values of z computed with Equations 18.2 and 18.4 will be minimal.

$$z = \frac{T - \frac{n(n+1)}{4}}{\sqrt{\frac{n(n+1)(2n+1)}{24} - \frac{\Sigma t^3 - \Sigma t}{48}}} \qquad \textbf{(Equation 18.4)}$$

Table 18.4 illustrates the application of the tie correction with Example 18.1. In the data for Example 18.1 there are three sets of tied ranks: Set 1 involves three subjects (Subjects 1, 9, and 10); Set 2 involves two subjects (Subjects 3 and 4); Set 3 involves three subjects (Subjects 5, 7, and 8). The number of subjects involved in each set of tied ranks represents the values of t in the third column of Table 18.4. The three t values are cubed in the last column of the table, after which the values Σt and Σt^3 are computed. The appropriate values are now substituted in Equation 18.4.[4]

Table 18.4 Correction for Ties with Normal Approximation

Subject	Rank	t	t^3
1	2		
9	2	3	27
10	2		
3	4.5	2	8
4	4.5		
5	7		
7	7	3	27
8	7		
6	9		
		$\Sigma t = 8$	$\Sigma t^3 = 62$

$$z = \frac{4.5 - \dfrac{(9)(10)}{4}}{\sqrt{\dfrac{(9)(10)(19)}{24} - \dfrac{62 - 8}{48}}} = -2.15$$

The absolute value $z = 2.15$ is slightly larger than the absolute value $z = 2.13$ obtained without the tie correction. The difference between the two methods is trivial, and in this instance, regardless of which alternative hypothesis is employed, the decision the researcher makes with respect to the null hypothesis is not affected.[5]

Conover (1980, 1999) and Daniel (1990) discuss and/or cite sources on the subject of alternative ways of handling tied difference scores. Conover (1980, 1999) also notes that in some instances retaining and ranking zero difference scores may actually provide a more powerful test of an alternative hypothesis than the more conventional method employed in this book (which eliminates zero difference scores from the data).

4. Computation of a confidence interval for a median difference between two dependent populations Prior to reading this section the reader should review the section on the computation of a confidence interval for a population median in Section VII of the **Wilcoxon signed-ranks test**, as well as the discussion of the computation of a confidence interval for a population mean in Section VI of the **single-sample *t* test**. The procedure which was described for computing a confidence interval for a population median in Section VII of the **Wilcoxon signed-ranks test** can also be used to compute a confidence interval for a median difference between two dependent populations. Equations 18.5 and 16.6, which are identical to Equations 6.5 and 6.6, can be employed to compute a large sample approximation for a confidence interval for a median difference between to dependent populations.[6]

$$U = \frac{n + 1}{2} + \frac{z_{\alpha/2} \sqrt{n}}{2} \qquad \textbf{(Equation 18.5)}$$

$$L = n - U + 1 \qquad \textbf{(Equation 18.6)}$$

Where: $z_{\alpha/2}$ represents the tabled critical two-tailed value in the normal distribution below which a proportion (percentage) equal to $[1 - (\alpha/2)]$ of the cases falls. If the proportion (percentage) of the distribution that falls within the confidence interval is subtracted from 1 (100%), it will equal the value of α.

U will represent the **ordinal position** of the score which represents the upper limit of the $[1 - (\alpha/2)]^{th}$ confidence interval. Any fractional value is rounded off to the next highest integer number.

L will represent the **ordinal position** of the score which represents the lower limit of the $[1 - (\alpha/2)]^{th}$ confidence interval.

The data for the sample employed in Example 18.1 will be used to demonstrate the use of Equations 18.5 and 18.6 to compute the 95% confidence interval. The difference scores of the ten subjects who comprise the sample are arranged below from lowest to highest moving from left to right. Note that any difference score equal to zero is employed in computing a confidence interval.

$$-2, 0, 1, 1, 1, 2, 3, 3, 3, 4$$

The values $n = 10$ and $z_{.05} = 1.96$ (which represents the tabled critical two-tailed .05 z value in **Table A1**) are employed in Equation 18.5 to compute the upper boundary of the 95% confidence interval. Substituting $n = 10$ and $z_{.05} = 1.96$ in Equation 18.5 yields the value $U = 8.60$, which rounded off to the next highest integer value is set equal to 9. When $U = 9$ is substituted in Equation 18.6 it yields the value $L = 2$. The result indicates that the score in the **ninth ordinal position** (which is a score of 3 in the above distribution) represents the upper limit of the confidence interval, while the score in the **second ordinal position** (which is the score of **0** in the above distribution) represents the lower limit of the confidence interval.

$$U \geq \frac{10 + 1}{2} + \frac{1.96\sqrt{10}}{2} = 5.5 + 3.10 = 8.60$$

$$L = 10 - 9 + 1 = 2$$

This result indicates the researcher can be 95% confident (or the probability is .95) the interval 0 to 3 contains the true difference between the population medians. If the aforementioned interval contains the true difference between the population medians, the median of the population Condition 1 represents is at least equal to the median of the population Condition 2 represents but not by more than 3 GSR units. It should be noted that the fact zero is computed for the lower limit of the confidence interval can be viewed as inconsistent with the result obtained earlier for the **Wilcoxon matched-pairs signed-ranks test**, insofar that if, in fact, there is a significant difference between the two population medians, one would not expect 0 to be included in the confidence interval for the difference between the medians.

Note that the range of values $0 \leq \theta_D \leq 3$, which define the limits of the above computed confidence interval, are close (but not identical) to the range of values $.33 \leq (\mu_1 - \mu_2) \leq 2.87$ computed in Section VI of the *t* **test for two dependent samples** for the confidence interval for the difference between the two population means for the same set of data. As noted in Section VII of the **Wilcoxon signed-ranks test**, the confidence interval based on the normal distribution computed in this section is in actuality an approximation of an exact confidence interval which can be computed through use of the **binomial distribution**. The methodology for computing a confidence interval for a median difference between two dependent populations employing the binomial distribution (which employs the probabilities in **Table A6 (Table of the Binomial Distribution, Individual Probabilities)** in the **Appendix**) is described in Higgins (2004), Marascuilo and McSweeney (1977), and Zar (1999). Conover (1980, 1999) and Daniel (1990) describe alternative procedures for computing the large sample normal approximation of the confidence interval.

VII. Additional Discussion of the Wilcoxon Matched-Pairs Signed-Ranks Test

1. **Power-efficiency of the Wilcoxon matched-pairs signed-ranks test**[7] When the underlying population distributions are normal, the **asymptotic relative efficiency** (which is discussed in Section VII of the **Wilcoxon signed-ranks test**) of the **Wilcoxon matched-pairs signed-ranks test** is .955 (when contrasted with the *t* **test for two dependent samples**). For population distributions that are not normal, the asymptotic relative efficiency of the **Wilcoxon matched-pairs signed-ranks test** is generally equal to or greater than 1. As a general rule, proponents of nonparametric tests take the position that when a researcher has reason to believe that the normality assumption of the *t* **test for two dependent samples** has been saliently violated, the **Wilcoxon matched-pairs signed-ranks test** provides a powerful test of the comparable alternative hypothesis.

2. **Probability of superiority as a measure of effect size** Endnote 14 of the **Mann–Whitney** *U* **test (Test 12)** discusses a measure of effect size referred to as **probability of superiority** (*PS*) for ranked data in the case of two independent samples. Grissom and Kim (2005, pp. 114–115) note that the latter measure can be extended to a dependent samples design involving ranked data, where *PS* is simply the proportion of cases in which a subject's score is higher in one experimental condition as opposed to the other (ties are dropped from the analysis). Thus, in the case of Example 18.1, if Subject 2 (whose scores are tied) is omitted, 8 of the 9 subjects had a higher score in Condition 1 than Condition 2, yielding the value $PS = 8/9 = .89$. Alternatively, we can compute a second probability of superiority measure for Condition 2 — specifically, $PS = 1/9 = .11$, which is indicative of the fact that 1 out of the 9 subjects had a higher score in Condition 2 than in Condition 1. Note that a **probability of superiority** value will always fall between a maximum $PS = 1$ and minimum of $PS = 0$, with the value $PS = .5$ indicating an equal proportion of superiority for each of the experimental conditions. The sum of the two *PS* values which can be computed for two dependent samples will always equal 1 (as is the case for Example 18.1, since $.89 + .11 = 1$). In point of fact, the proportion represented by a **probability of superiority** measure is evaluated more rigorously within the framework of the **binomial sign test for two dependent samples (Test 19)** (which is described in the next chapter).

3. **Alternative nonparametric procedures for evaluating a design involving two dependent samples** In addition to the **Wilcoxon matched-pairs signed-ranks test**, the **binomial sign test for two dependent samples** can be employed to evaluate a design involving two dependent samples. Marascuilo and McSweeney (1977) describe the extension of the **van der Waerden normal-scores test for** *k* **independent samples (Test 23)** (Van der Waerden (1952/1953)) (which is discussed later in the book) to a design involving *k* dependent samples. (**Normal-scores tests** are procedures which involve trans-formation of ordinal data through use of the normal distribution.) Conover (1980, 1999) notes that a normal-scores test developed by Bell and Doksum (1965) can be extended to a dependent samples design. Another procedure which can be employed with a dependent samples design is **Fisher's randomization procedure** (Fisher (1935)) (which is described in Conover (1980, 1999), Marascuilo and McSweeney (1977) and Siegel and Castellan (1988)). **The randomization test for two independent samples (Test 12a)** (which is described in Section IX (the **Addendum**) of the **Mann–Whitney** *U* **test**) illustrates the use of **Fisher's randomization procedure** with two independent samples. Additional nonparametric procedures that can be employed with a *k* dependent samples design are either discussed or referenced in Conover (1980, 1999), Daniel (1990), Hollander and Wolfe (1999), Marascuilo and McSweeney (1977), and Sheskin (1984).

VIII. Additional Examples Illustrating the Use of the Wilcoxon Matched-Pairs Signed-Ranks Test

The **Wilcoxon matched-pairs signed-ranks test** can be employed to evaluate any of the additional examples noted for the *t* **test for two dependent samples** (i.e., Examples 17.2–17.7). In all instances in which the **Wilcoxon matched-pairs signed-ranks test** is employed, difference scores are obtained for subjects (or pairs of matched subjects). All difference scores are then ranked and evaluated in accordance with the ranking protocol described in Section IV.

References

Bell, C. B. & Doksum, K. A. (1965). Some new distribution-free statistics. **Annals of Mathematical Statistics**, 36, 203–214.

Conover, W. (1980). **Practical nonparametric statistics** (2nd ed.). New York: John Wiley & Sons.

Conover, W. (1999). **Practical nonparametric statistics** (3rd ed.). New York: John Wiley & Sons.

Daniel, W. (1990). **Applied nonparametric statistics** (2nd ed.). Boston: PWS–Kent Publishing Company.

Fisher, R. A. (1935). **The design of experiments** (7th ed.). Edinburgh-London: Oliver & Boyd.

Glass, G. V. & Hopkins, K. D. (1996). **Statistical methods in education and psychology** (3rd ed.) Boston: Allyn & Bacon.

Grissom , R. J. & Kim, J. J. (2005). **Effect sizes for research: A broad practical approach**. Mahwah, NJ: Lawrence Erlbaum Associates, Publishers.

Higgins, J. J. (2004). **An introduction to modern nonparametric statistics**. Belmont, CA: Duxbury Press.

Hollander, M. & Wolfe, D. A. (1999). **Nonparametric statistical methods**. New York: John Wiley & Sons.

Marascuilo, L. & McSweeney, M. (1977). **Nonparametric and distribution-free methods for the social sciences**. Monterey, CA: Brooks/Cole Publishing Company.

Sheskin, D. J. (1984). **Statistical tests and experimental design: A guidebook**. New York: Gardner Press.

Siegel, S. & Castellan, N., Jr. (1988). **Nonparametric statistics for the behavioral sciences** (2nd ed.). New York: McGraw–Hill Book Company.

Snedecor, G. W. & Cochran, W. G. (1980). **Statistical methods** (8th ed.). Ames, IA: Iowa State University Press.

Van der Waerden, B. L. (1952/1953). Order tests for the two-sample problem and their power. **Proceedings Koninklijke Nederlandse Akademie van Wetenshappen** (A), 55 (**Indagationes Mathematicae** 14), 453–458, and 56 (**Indagationes Mathematicae**, 15), 303–316 (corrections appear in Vol. 56, p. 80).

Wilcoxon, F. (1945). Individual comparisons by ranking methods. **Biometrics**, 1, 80–83.

Wilcoxon, F. (1949). **Some rapid approximate statistical procedures**. Stamford, CT: Stamford Research Laboratories, American Cyanamid Corporation.

Zar, J. H. (1999). **Biostatistical analysis** (4th ed.). Upper Saddle River, NJ: Prentice Hall.

Endnotes

1. Some sources note that one assumption of the **Wilcoxon matched-pairs signed-ranks test** is that the variable being measured is based on a continuous distribution. In practice, however, this assumption is often not adhered to.

2. When there are tied scores for either the lowest or highest difference scores, as a result of averaging the ordinal positions of the tied scores, the rank assigned to the lowest difference score will be some value greater than 1, and the rank assigned to the highest difference score will be some value less than n.

3. A more thorough discussion of **Table A5** can be found in Section V of the **Wilcoxon signed-ranks test**.

4. The term $(\Sigma t^3 - \Sigma t)$ in Equation 18.4 can also be written as $\Sigma_{i=1}^{s}(t_i^3 - t_i)$. The latter notation indicates the following: a) For each set of ties, the number of ties in the set is subtracted from the cube of the number of ties in that set; and b) the sum of all the values computed in part a) is obtained. Thus, in the example under discussion (in which there are $s = 3$ sets of ties):

$$\sum_{i=1}^{s} (t_i^3 - t_i) = [(3)^3 - 3] + [(2)^3 - 2] + [(3)^3 - 3] = 54$$

   The computed value of 54 is the same as the corresponding value $(\Sigma t^3 - \Sigma t) = 62 - 8 = 54$ computed in Equation 18.4 through use of Table 18.4.

5. A correction for continuity can be used in conjunction with the tie correction by subtracting .5 from the absolute value computed for the numerator of Equation 18.4. Use of the correction for continuity will reduce the tie-corrected absolute value of z.

6. Sources are not in agreement with respect to the minimum sample size for which the latter equations should be employed.

7. The concept of **power-efficiency** is discussed in Section VII of the **Wilcoxon signed-raks test**.

Test 19

The Binomial Sign Test for Two Dependent Samples
(Nonparametric Test Employed with Ordinal Data)

I. Hypothesis Evaluated with Test and Relevant Background Information

Hypothesis evaluated with test Do two dependent samples represent two different populations?

Relevant background information on test The **binomial sign test for two dependent samples** is essentially an extension of the **binomial sign test for a single sample (Test 9)** to a design involving two dependent samples. Since a complete discussion of the binomial distribution (which is the distribution upon which the test is based) is contained in the discussion of the **binomial sign test for a single sample**, the reader is advised to read the material on the latter test prior to continuing this section. Whenever one or more of the assumptions of the *t* test **for two dependent samples (Test 17)** or the **Wilcoxon matched-pairs signed-ranks test (Test 18)** are saliently violated, the **binomial sign test for two dependent samples** can be employed as an alternative procedure. The reader should review the assumptions of the aforementioned tests, as well as the information on a dependent samples design discussed in Sections I and VII of the *t* **test for two dependent samples**.

To employ the **binomial sign test for two dependent samples**, it is required that each of n subjects (or n pairs of matched subjects) has two scores (each score having been obtained under one of the two experimental conditions). The two scores are represented by the notations X_1 and X_2. For each subject (or pair of matched subjects), a determination is made with respect to whether a subject obtains a higher score in Condition 1 or Condition 2. Based on the latter, a signed difference ($D+$ or $D-$) is assigned to each pair of scores. The sign of the difference assigned to a pair of scores will be positive if a higher score is obtained in Condition 1 (i.e., $D+$ if $X_1 > X_2$), whereas the sign of the difference will be negative if a higher score is obtained in Condition 2 (i.e., $D-$ if $X_2 > X_1$). The hypothesis the **binomial sign test for two dependent samples** evaluates is whether or not in the underlying population represented by the sample, the proportion of subjects who obtain a positive signed difference (i.e., obtain a higher score in Condition 1) is some value other than .5. If the proportion of subjects who obtain a positive signed difference (which, for the underlying population, is represented by the notation $\pi+$) is some value that is either significantly above or below .5, it indicates there is a high likelihood the two dependent samples represent two different populations.

The **binomial sign test for two dependent samples** is based on the following assumptions:[1] a) The sample of n subjects has been randomly selected from the population it represents; and b) The format of the data is such that within each pair of scores the two scores can be rank-ordered.

As is the case for the *t* **test for two dependent samples** and the **Wilcoxon matched-pairs signed-ranks test**, in order for the **binomial sign test for two dependent samples** to generate valid results, the following guidelines should be adhered to: a) To control for order effects, the presentation of the two experimental conditions should be random or, if appropriate, be counterbalanced; and b) If matched samples are employed, within each pair of matched subjects each of the subjects should be randomly assigned to one of the two experimental conditions.

As is the case with the *t* **test for two dependent samples** and the **Wilcoxon matched-pairs signed-ranks test**, the **binomial sign test for two dependent samples** can also be employed to evaluate a **one-group pretest-posttest design**. The limitations of the **one-group pretest-posttest design** (which are discussed in Section VII of the *t* **test for two dependent samples** and the **Introduction**) are also applicable when it is evaluated with the **binomial sign test for two dependent samples**.

II. Example

Example 19.1 is identical to Examples 17.1 and 18.1 (which are respectively evaluated with the *t* **test for two dependent samples** and the **Wilcoxon matched-pairs signed-ranks test**). In evaluating Example 19.1 it will be assumed that the **binomial sign test for two dependent samples** is employed, since one or more of the assumptions of the *t* **test for two dependent samples** and the **Wilcoxon matched-pairs signed-ranks test** have been saliently violated.

Example 19.1 *A psychologist conducts a study to determine whether or not people exhibit more emotionality when they are exposed to sexually explicit words than when they are exposed to neutral words. Each of ten subjects is shown a list of 16 randomly arranged words which are projected onto a screen one at a time for a period of five seconds. Eight of the words on the list are sexually explicit in nature and eight of the words are neutral. As each word is projected on the screen, a subject is instructed to say the word softly to himself or herself. As a subject does this, sensors attached to the palms of the subject's hands record galvanic skin response (GSR), which is used by the psychologist as a measure of emotionality. The psychologist computes two scores for each subject, one score for each of the experimental conditions:* **Condition 1**: *GSR/Explicit — The average GSR score for the eight sexually explicit words;* **Condition 2**: *GSR/Neutral – The average GSR score for the eight neutral words. The GSR/Explicit and the GSR/Neutral scores of the ten subjects follow. (The higher the score, the higher the level of emotionality.)* **Subject 1** *(9, 8);* **Subject 2** *(2, 2);* **Subject 3** *(1, 3);* **Subject 4** *(4, 2);* **Subject 5** *(6, 3);* **Subject 6** *(4, 0);* **Subject 7** *(7, 4);* **Subject 8** *(8, 5);* **Subject 9** *(5, 4);* **Subject 10** *(1, 0). Do subjects exhibit differences in emotionality with respect to the two categories of words?*

III. Null versus Alternative Hypotheses

Null hypothesis H_0: $\pi+ = .5$

(In the underlying population the sample represents, the proportion of subjects who obtain a positive signed difference (i.e., a higher score in Condition 1 than Condition 2) equals .5.)

Alternative hypothesis H_1: $\pi+ \neq .5$

(In the underlying population the sample represents, the proportion of subjects who obtain a positive signed difference (i.e., a higher score in Condition 1 than Condition 2) does not equal .5. This is a **nondirectional alternative hypothesis**, and it is evaluated with a **two-tailed test**. In order to be supported, the observed proportion of positive signed differences in the sample data (which will be represented with the notation $p+$) can be either significantly larger than the hypothesized population proportion $\pi+ = .5$ or significantly smaller than $\pi+ = .5$.)

or

$$H_1: \pi+ > .5$$

(In the underlying population the sample represents, the proportion of subjects who obtain a positive signed difference (i.e., a higher score in Condition 1 than Condition 2) is greater than .5. This is a **directional alternative hypothesis**, and it is evaluated with a **one-tailed test**. In order to be supported, the observed proportion of positive signed differences in the sample data must be significantly larger than the hypothesized population proportion $\pi+ = .5$.)

or

$$H_1: \pi+ < .5$$

(In the underlying population the sample represents, the proportion of subjects who obtain a positive signed difference (i.e., a higher score in Condition 1 than Condition 2) is less than .5. This is a **directional alternative hypothesis**, and it is evaluated with a **one-tailed test**. In order to be supported, the observed proportion of positive signed differences in the sample data must be significantly smaller than the hypothesized population proportion $\pi+ = .5$.)

Note: Only one of the above noted alternative hypotheses is employed. If the alternative hypothesis the researcher selects is supported, the null hypothesis is rejected.[2]

IV. Test Computations

The data for Example 19.1 are summarized in Table 19.1. Note that there are 10 subjects and that each subject has two scores.

Table 19.1 Data for Example 19.1

Subject	X_1	X_2	$D = X_1 - X_2$	Signed Difference
1	9	8	1	+
2	2	2	0	0
3	1	3	-2	-
4	4	2	2	+
5	6	3	3	+
6	4	0	4	+
7	7	4	3	+
8	8	5	3	+
9	5	4	1	+
10	1	0	1	+

$$\Sigma D+ = 8$$
$$\Sigma D- = 1$$

The following information can be derived from Table 19.1: a) Eight subjects (Subjects 1, 4, 5, 6, 7, 8, 9, 10) yield a difference score with a positive sign — i.e., a positive signed difference; b) One subject (Subject 3) yields a difference score with a negative sign — i.e., a negative signed difference; and c) One subject (Subject 2) obtains the identical score in both conditions, and as a result of this yields a difference score of zero.

As is the case with the **Wilcoxon matched-pairs signed-ranks test**, in employing the **binomial sign test for two dependent samples**, any subject who obtains a zero difference score is eliminated from the data analysis. Since Subject 2 falls in this category, the size of the sample is reduced to $n = 9$, which is the same number of signed ranks employed when the **Wilcoxon matched-pairs signed-ranks test** is employed to evaluate the same set of data.

The sampling distribution of the signed differences represents a binomially distributed variable with an expected probability of .5 for each of the two mutually exclusive categories (i.e., positive signed difference versus negative signed difference). The logic underlying the **binomial sign test for two dependent samples** is that if the two experimental conditions represent equivalent populations, the signed differences should be randomly distributed. Thus, assuming that subjects who obtain a difference score of zero are eliminated from the analysis, if the remaining signed differences are, in fact, randomly distributed, one-half of the subjects should obtain a positive signed difference and one-half of the subjects should obtain a negative signed difference. In Example 19.1 the observed proportion of positive signed differences is $p+$ = 8/9 = .89 and the observed proportion of negative signed differences is $p-$ = 1/9 = .11.

Equation 19.1 (which is identical to Equation 9.5, except for the fact that $\pi+$ and $\pi-$ are used in place of π_1 and π_2) is employed to determine the probability of obtaining $x = 8$ or more positive signed differences in a set of $n = 9$ scores.

$$P(\geq x) = \sum_{r=x}^{n} \binom{n}{x} (\pi+)^x (\pi-)^{(n-x)} \qquad \textbf{(Equation 19.1)}$$

Where: $\pi+$ and $\pi-$, respectively, represent the hypothesized values for the proportion of positive and negative signed differences
n represents the number of signed differences
x represents the number of positive signed differences

In employing Equation 19.1 with Example 19.1, the following values are employed: a) $\pi+$ = .5 and $\pi-$ = .5, since if the null hypothesis is true, the proportion of positive and negative signed differences should be equal. Note that the sum of $\pi+$ and $\pi-$ must always equal 1; b) $n = 9$, since there are 9 signed differences; and c) $x = 8$, since 8 subjects obtain a positive signed difference.

The notation $\sum_{r=x}^{n}$ in Equation 19.1 indicates that the probability of obtaining a value of x equal to the observed number of positive signed differences must be computed, as well as the probability for all values of x greater than the observed number of positive signed differences up through and including the value of n. Thus, in the case of Example 19.1, the binomial probability must be computed for the values $x = 8$ and $x = 9$. Equation 19.1 is employed below to compute the latter probability. The obtained value .0195 represents the likelihood of obtaining 8 or more positive signed differences in a set of $n = 9$ signed differences.

$$P(x \geq 8) = \binom{9}{8} (.5)^8 (.5)^1 + \binom{9}{9} (.5)^9 (.5)^0 = .0195$$

An even more efficient way of obtaining the probability $P(8$ or $9/9) = .0195$ is through use of **Table A7 (Table of the Binomial Distribution, Cumulative Probabilities)** in the **Appendix**. In employing **Table A7** we find the section for $n = 9$, and locate the cell that is the intersection of the row $x = 8$ and the column $\pi = .5$. The entry .0195 in that cell represents the probability of obtaining 8 or more (i.e., 8 and 9) positive signed differences, if there are a total of 9 signed differences.[3]

Equation 19.2 (which is identical to Equation 9.3 employed for the **binomial sign test for a single sample**, except for the fact that $\pi+$ and $\pi-$ are employed in place of π_1 and π_2) can be employed to compute each of the individual probabilities that are summed in Equation 19.1.

$$P(x) = \binom{n}{x} (\pi+)^x \, (\pi-)^{(n-x)} \qquad \textbf{(Equation 19.2)}$$

Since the computation of binomial probabilities can be quite tedious, in lieu of employing Equation 19.2, **Table A6 (Table of the Binomial Distribution, Individual Probabilities)** in the **Appendix** can be used to determine the appropriate probabilities. In employing **Table A6** we find the section for $n = 9$, and locate the cell that is the intersection of the row $x = 8$ and the column $\pi = .5$. The entry .0176 in that cell represents the probability of obtaining exactly 8 positive signed differences, if there are a total of 9 signed differences. Additionally, we locate the cell that is the intersection of the row $x = 9$ and the column $\pi = .5$. The entry .0020 in that cell represents the probability of obtaining exactly 9 positive signed differences, if there are a total of 9 signed differences. Summing the latter two values yields the value $P(8 \text{ or } 9/9) = .0196$, which is the likelihood of observing 8 or 9 positive signed differences in a set of $n = 9$ signed differences.[4] For a comprehensive discussion on the computation of binomial probabilities and the use of **Tables A6** and **A7**, the reader should review Section IV of the **binomial sign test for a single sample**.

V. Interpretation of the Test Results

The following guidelines are employed in evaluating the null hypothesis.

a) If a nondirectional alternative hypothesis is employed, the null hypothesis can be rejected if the probability of obtaining a value equal to or more extreme than x is equal to or less than $\alpha/2$ (where α represents the prespecified value of α). The reader should take note of the fact that if the proportion of positive signed differences in the data (i.e., $p+$) is greater than $\pi+ = .5$, a value which is more extreme than x will be any value that falls above the observed value of x, whereas if the proportion of positive signed differences in the data is less than $\pi+ = .5$, a value which is more extreme than x will be any value that falls below the observed value of x.

b) If a directional alternative hypothesis is employed which predicts that the underlying population proportion is above the hypothesized value $\pi+ = .5$, in order to reject the null hypothesis both of the following conditions must be met: 1) The proportion of positive signed differences must be greater than the value $\pi+ = .5$ stipulated in the null hypothesis; and 2) The probability of obtaining a value equal to or greater than x is equal to or less than the prespecified value of α.

c) If a directional alternative hypothesis is employed which predicts that the underlying population proportion is below the hypothesized value $\pi+ = .5$, in order to reject the null hypothesis both of the following conditions must be met: 1) The proportion of positive signed differences must be less than the value $\pi+ = .5$ stipulated in the null hypothesis; and 2) The probability of obtaining a value equal to or less than x is equal to or less than the prespecified value of α.

Applying the above guidelines to Example 19.1, we can conclude the following.

The nondirectional alternative hypothesis H_1: $\pi+ \neq .5$ is supported at the $\alpha = .05$ level, since the obtained probability .0195 is less than $\alpha/2 = .05/2 = .025$. The nondirectional alternative hypothesis H_1: $\pi+ \neq .5$ is not supported at the $\alpha = .01$ level, since the probability .0195 is greater than $\alpha/2 = .01/2 = .005$.[5]

The directional alternative hypothesis H_1: $\pi+ > .5$ is supported at the $\alpha = .05$ level. This is the case because: a) The data are consistent with the directional alternative hypothesis H_1: $\pi+ > .5$, since $p+ = .89$ is greater than the value $\pi+ = .5$ stated in the null hypothesis; and b) The obtained probability .0195 is less than $\alpha = .05$. The directional alternative hypothesis H_1: $\pi+ > .5$ is not supported at the $\alpha = .01$ level, since the probability .0195 is greater than $\alpha = .01$.

The directional alternative hypothesis H_1: $\pi+ < .5$ is not supported, since the data are not consistent with it. Specifically, $p+ = .89$ does not meet the requirement of being less than the value $\pi+ = .5$ stated in the null hypothesis.

A summary of the analysis of Example 19.1 with the **binomial sign test for two dependent samples** follows: It can be concluded that subjects exhibited higher GSR (emotionality) scores with respect to the sexually explicit words than the neutral words.

When the **binomial sign test for two dependent samples** and the **Wilcoxon matched-pairs signed-ranks test** are applied to the same set of data, the two tests yield identical conclusions. Specifically, both tests support the nondirectional alternative hypothesis and the directional alternative hypothesis that is consistent with the data at the .05 level. Although it is not immediately apparent from this example, as a general rule, when applied to the same data the **binomial sign test for two dependent samples** tends to be less powerful than the **Wilcoxon matched-pairs signed-ranks test**. This is the case, since by not considering the magnitude of the difference scores, the **binomial sign test for two dependent samples** employs less information than the **Wilcoxon matched-pairs signed-ranks test**. As is the case with the **Wilcoxon matched-pairs signed-ranks test**, the **binomial sign test for two dependent samples** utilizes less information than the ***t* test for two dependent samples**, and thus in most instances, it will provide a less powerful test of an alternative hypothesis than the latter test. In point of fact, in the case of Example 19.1, if the data are evaluated with the ***t* test for two dependent samples**, the directional alternative hypothesis which is consistent with the data is supported at both the .05 and .01 levels. It should be noted, however, that if the normality assumption of the ***t* test for two dependent samples** is saliently violated, in some instances the **binomial sign test for two dependent samples** may provide a more powerful test of an analogous alternative hypothesis.

VI. Additional Analytical Procedures for the Binomial Sign Test for Two Dependent Samples and/or Related Tests

1. The normal approximation of the binomial sign test for two dependent samples with and without a correction for continuity With large sample sizes the normal approximation for the binomial distribution (which is discussed in Section VI of the **binomial sign test for a single sample**) can provide a large sample approximation for the **binomial sign test for two dependent samples**. As a general rule, most sources recommend employing the normal approximation for sample sizes larger than those documented in the table of the binomial distribution contained in the source. Equation 19.3 (which is equivalent to Equation 9.7) is the normal approximation equation for the **binomial sign test for two dependent samples**. When a **correction for continuity** is used, Equation 19.4 (which is equivalent to Equation 9.9) is employed.[6]

$$z = \frac{x - (n)(\pi+)}{\sqrt{(n)(\pi+)(\pi-)}} \qquad \textbf{(Equation 19.3)}$$

$$z = \frac{|x - (n)(\pi+)| - .5}{\sqrt{(n)(\pi+)(\pi-)}} \qquad \textbf{(Equation 19.4)}$$

Although Example 19.1 involves only nine signed ranks (a value most sources would view as too small to use with the normal approximation), it will be employed to illustrate Equations 19.3 and 19.4. The reader will see that in spite of employing the normal approximation with a small sample size, it yields essentially the same results as those obtained with the exact binomial probabilities.

Employing Equation 19.3, the value $z = 2.33$ is computed.

$$z = \frac{8 - (9)(.5)}{\sqrt{(9)(.5)(.5)}} = 2.33$$

Employing Equation 19.4, the value $z = 2.00$ is computed.

$$z = \frac{|8 - (9)(.5)| - .5}{\sqrt{(9)(.5)(.5)}} = 2.00$$

The obtained z values are evaluated with **Table A1 (Table of the Normal Distribution)** in the **Appendix**. In **Table A1** the tabled critical two-tailed .05 and .01 values are $z_{.05} = 1.96$ and $z_{.01} = 2.58$, and the tabled critical one-tailed .05 and .01 values are $z_{.05} = 1.65$ and $z_{.01} = 2.33$. The following guidelines are employed in evaluating the null hypothesis.

a) If a nondirectional alternative hypothesis is employed, the null hypothesis can be rejected if the obtained absolute value of z is equal to or greater than the tabled critical two-tailed value at the prespecified level of significance.

b) If a directional alternative hypothesis is employed, only the directional alternative hypothesis which is consistent with the data can be supported. With respect to the latter alternative hypothesis, the null hypothesis can be rejected if the obtained absolute value of z is equal to or greater than the tabled critical one-tailed value at the prespecified level of significance.

Employing the above guidelines, we can conclude the following.

Since the value $z = 2.33$ computed with Equation 19.3 is greater than the tabled critical two-tailed value $z_{.05} = 1.96$ but less than the tabled critical two-tailed value $z_{.01} = 2.58$, the non-directional alternative hypothesis H_1: $\pi+ \neq .5$ is supported, but only at the .05 level. Since the value $z = 2.33$ is greater than the tabled critical one-tailed value $z_{.05} = 1.65$ and equal to the tabled critical one-tailed value $z_{.01} = 2.33$, the directional alternative hypothesis H_1: $\pi+ > .5$ is supported at both the .05 and .01 levels. Note that when the exact binomial probabilities are employed, both the nondirectional alternative hypothesis H_1: $\pi+ \neq .5$ and the directional alternative hypothesis H_1: $\pi+ > .5$ are supported, but in both instances, only at the .05 level.

When the correction for continuity is employed, the value $z = 2.00$ computed with Equation 19.4 is greater than the tabled critical two-tailed value $z_{.05} = 1.96$ but less than the tabled critical two-tailed value $z_{.01} = 2.58$. Thus, the nondirectional alternative hypothesis H_1: $\pi+ \neq .5$ is only supported at the .05 level. Since the value $z = 2.00$ is greater than the tabled critical one-tailed value $z_{.05} = 1.65$ but less than the tabled critical one-tailed value $z_{.01} = 2.33$, the directional alternative hypothesis H_1: $\pi+ > .5$ is also supported, but only at the .05 level. The results with the correction for continuity are identical to those obtained when the exact binomial probabilities are employed. Note that the continuity-corrected normal approximation provides a more conservative test of the null hypothesis than does the uncorrected normal approximation.

The **chi-square goodness-of-fit test (Test 8)** (either with or without the correction for continuity) can also be employed to provide a large sample approximation of the **binomial sign test for two dependent samples**. The **chi-square goodness-of-fit test**, which evaluates the relationship between the observed and expected frequencies in the two categories (i.e., positive signed difference versus negative signed difference), will yield a result which is equivalent to that obtained with the normal approximation. The computed chi-square value will equal the square of the z value derived with the normal approximation.

Equation 8.2 is employed for the chi-square analysis, without using a correction for continuity. Equation 8.5 is the continuity-corrected equation. Table 19.2 summarizes the analysis of Example 19.1 with Equation 8.2.

Table 19.2 Chi-Square Summary Table for Example 19.1

Cell	O_i	E_i	$(O_i - E_i)$	$(O_i - E_i)^2$	$\dfrac{(O_i - E_i)^2}{E_i}$
Positive signed differences	8	4.5	3.5	12.25	2.72
Negative signed differences	1	4.5	−3.5	12.25	2.72
	$\Sigma O_i = 9$	$\Sigma E_i = 9$	$\Sigma(O_i - E_i) = 0$		$\chi^2 = 5.44$

In Table 19.2, the expected frequency $E_i = 4.5$ for each cell is computed by multiplying the hypothesized population proportion for the cell (.5 for both cells) by $n = 9$. Since $k = 2$, the degrees of freedom employed for the chi-square analysis are $df = k - 1 = 2$. The obtained value $\chi^2 = 5.44$ is evaluated with **Table A4 (Table of the Chi-Square Distribution)** in the **Appendix**. For $df = 1$, the tabled critical .05 and .01 chi-square values are $\chi^2_{.05} = 3.84$ (which corresponds to the chi-square value at the 95th percentile) and $\chi^2_{.01} = 6.63$ (which corresponds to the chi-square value at the 99th percentile). Since the obtained value $\chi^2 = 5.44$ is greater than $\chi^2_{.05} = 3.84$ but less than $\chi^2_{.01} = 6.63$, the nondirectional alternative hypothesis H_1: $\pi+ \neq .5$ is supported, but only at the .05 level. Since $\chi^2 = 5.44$ is greater than the tabled critical one-tailed .05 value $\chi^2_{.05} = 2.71$ (which corresponds to the chi-square value at the 90th percentile) and the tabled critical one-tailed .01 value $\chi^2_{.01} = 5.43$ (which corresponds to the chi-square value at the 98th percentile), the directional alternative hypothesis H_1: $\pi+ > .5$ is supported at both the .05 and .01 levels.[7] The aforementioned conclusions are identical to those reached when Equation 19.3 is employed.

As noted previously, if the z value obtained with Equation 19.3 is squared, it will always equal the chi-square value computed for the same data. Thus, in the current example where $z = 2.33$ and $\chi^2 = 5.44$, $(z = 2.33)^2 = (\chi^2 = 5.43)$. (The minimal discrepancy is the result of rounding off error.)

Equation 8.5 (which, as noted previously, is the continuity-corrected equation for the **chi-square goodness-of-fit test**) is employed below, and yields an equivalent result to that obtained with Equation 19.4. In employing Equation 8.5, the value $(|O_i - E_i| - .5) = 3$ is employed for each cell. Thus:

$$\chi^2 = \sum_{i=1}^{k} \left[\frac{(|O_i - E_i| - .5)^2}{E_i} \right] = \frac{(3)^2}{4.5} + \frac{(3)^2}{4.5} = 4$$

Note that the obtained value $\chi^2 = 4$ is equal to the square of the value $z = 2.00$ obtained with Equation 19.4. Since $\chi^2 = 4$ is greater than $\chi^2_{.05} = 3.84$ but less than $\chi^2_{.01} = 6.63$, the nondirectional alternative hypothesis H_1: $\pi+ \neq .5$ is supported, but only at the .05 level. Since $\chi^2 = 4$ is greater than the tabled critical one-tailed .05 value $\chi^2_{.05} = 2.71$ but less than the tabled critical .01 value $\chi^2_{.01} = 5.43$, the directional alternative hypothesis H_1: $\pi+ > .5$ is also supported, but only at the .05 level. The aforementioned conclusions are identical to those reached when Equation 19.4 is employed.

2. Computation of a confidence interval for the binomial sign test for two dependent samples

Equation 19.5 (which is equivalent to Equation 8.6, except for the fact that $p+$, $p-$, and $\pi+$ are employed in place of p_1, p_2, and π_1) can be used to compute a confidence interval for the **binomial sign test for two dependent samples**. Since Equation 19.5 is based on the normal approximation, it should be employed with large sample sizes. It will, however, be used here with the data for Example 19.1. The 95% confidence interval computed with Equation 19.5 estimates an interval for which there is a 95% likelihood it contains the true proportion of positive signed differences in the underlying population.

$$\left[p+ - z_{(\alpha/2)}\sqrt{\frac{(p+)(p-)}{n}} \right] \leq \pi+ \leq \left[p+ + z_{(\alpha/2)}\sqrt{\frac{(p+)(p-)}{n}} \right] \quad \textbf{(Equation 19.5)}$$

Employing the values computed for Example 19.1, Equation 19.5 is employed to compute the 95% confidence interval.

$$\left[.89 - (1.96)\sqrt{\frac{(.89)(.11)}{9}} \right] \leq \pi+ \leq \left[.89 + (1.96)\sqrt{\frac{(.89)(.11)}{9}} \right]$$

$$\pi+ = .89 \pm .204$$

$$.686 \leq \pi+ \leq 1.094$$

Thus, the researcher can be 95% confident (or the probability is .95) the interval .686 to 1.094 contains the true proportion of positive signed differences in the underlying population. Obviously, since a proportion cannot be greater than 1, the range of values identified by the confidence interval will fall between .686 and 1.[8]

It should be noted that the above method for computing a confidence interval ignores the presence of any zero difference scores. Consequently, the range of values computed for the confidence interval assumes there are no zero difference scores in the underlying population. If, in fact, there are zero difference scores in the population, the above computed confidence interval only identifies proportions which are relevant to the total number of cases in the population that are not zero difference scores. In point of fact, when one or more zero difference scores are present in the sample data, a researcher may want to assume that zero difference scores are present in the underlying population. If the researcher makes such an assumption and employs the sample data to estimate the proportion of zero difference scores in the population, the value employed for $p+$ in Equation 19.5 will represent the number of positive signed differences in the sample divided by the total number of scores in the sample, including any zero difference scores. Thus, in the case of Example 19.1, the value $p+ = 8/10 = .8 = .8$ is computed by dividing 8 (the number of positive signed differences) by $n = 10$. The value $p-$ in Equation 19.5 will no longer represent just the negative signed differences, but will represent all signed differences that are not positive (i.e., both negative signed differences and zero difference scores). Thus, in the case of Example 19.1, $p- = 2/10 = .2$, since there is one negative signed difference and one zero difference score. If the values $n = 10$, $p+ = .8$, and $p- = .2$ are employed in Equation 19.5, the confidence interval $.552 \leq \pi+ \leq 1.048$ is computed.

$$\left[.8 - (1.96) \sqrt{\frac{(.8)(.2)}{10}} \right] \leq \pi+ \leq \left[.8 + (1.96) \sqrt{\frac{(.8)(.2)}{10}} \right]$$

$$\pi+ = .8 \pm .248$$

$$.552 \leq \pi+ \leq 1.048$$

Thus, the researcher can be 95% confident (or the probability is .95) the interval .552 to 1.048 contains the true proportion of positive signed differences in the underlying population. Since, as noted earlier, a proportion cannot be greater than 1, the range identified by the confidence interval will fall between .552 and 1. If the researcher wants to employ the same method for computing a confidence interval for the proportion of minus signed differences in the population (i.e., $\pi-$), the product $z_{(\alpha/2)} \sqrt{[(p+)(p-)]/n}$ is added to and subtracted from $p- = .1$. The values $p- = 1/10 = .1$ and $p+ = 9/10 = .9$ are employed in the confidence interval equation, since there is one negative signed difference and nine signed differences which are not negative (i.e., eight positive signed differences and one zero difference score). The reader should take note of the fact that although a computed proportion for a limiting value of a confidence interval may result in a negative value, the lower limit assigned to a negative proportion is zero.

3. Sources for computing the power of the binomial sign test for two dependent samples, and comments on asymptotic relative efficiency of the test　　Cohen (1977, 1988) has developed a statistic called the **g index** which can be employed to compute the power of the **binomial sign test for a single sample** when H_0: $\pi_i = .5$ is evaluated. The latter effect size index can be generalized to compute the power of the **binomial sign test for two dependent samples** (in reference to the values $\pi+$ or $\pi-$). The **g index** represents the distance in units of proportion from the value .50. The equation Cohen (1977, 1988) employs for the **g index** is $g = P - .5$ when H_1 is directional, and $g = |P - .5|$ when H_1 is nondirectional (where P represents the hypothesized value of the population proportion stated in the alternative hypothesis) — in employing the **g index** it is assumed that the researcher has stated a specific value in the alternative hypothesis as an alternative to the value that is stipulated in the null hypothesis.

Cohen (1977; 1988, Ch. 5) has derived tables which allow a researcher, through use of the **g index**, to determine the appropriate sample size to employ if one wants to test a hypothesis about the distance of a proportion from the value .5 at a specified level of power. Cohen (1977; 1988, pp. 147–150) has proposed the following (admittedly arbitrary) g values as criteria for identifying the magnitude of an effect size: a) A **small effect size** is one that is greater than .05 but not more than .15; b) A **medium effect size** is one that is greater than .15 but not more than .25; and c) A **large effect size** is greater than .25.

Marascuilo and McSweeney (1977) note that if the underlying population distribution is normal, the **asymptotic relative efficiency** (which is discussed in Section VII of the **Wilcoxon signed-ranks test (Test 6)**) of the **binomial sign test** is .637, in contrast to an asymptotic relative efficiency of .955 for the **Wilcoxon matched-pairs signed-ranks test** (with both asymptotic relative efficiencies being in reference to the *t* **test for two dependent samples**). When the underlying population distribution is not normal, in most cases, the asymptotic relative efficiency of the **Wilcoxon matched-pairs signed-ranks test** will be higher than the analogous value for the **binomial sign test for two dependent samples**.

VII. Additional Discussion of the Binomial Sign Test for Two Dependent Samples

1. The problem of an excessive number of zero difference scores   When there is an excessive number of subjects who have a zero difference score in a set of data, a substantial amount of information is sacrificed if the **binomial sign test for two dependent samples** is employed to evaluate the data. Under such conditions, it is advisable to evaluate the data with the *t* **test for two dependent samples** (assuming the interval/ratio scores of subjects are available). If one or more of the assumptions of the latter test are saliently violated, the alpha level employed for the *t* test should be adjusted.

2. Equivalency of the binomial sign test for two dependent samples and the Friedman two-way analysis variance by ranks when *k* = 2   In Section VII of the **Friedman two-way analysis of variance by ranks (Test 25)**, it is demonstrated that when there are two dependent samples and there are no zero difference scores, the latter test (which can be employed for two or more dependent samples) is equivalent to the chi-square approximation of the **binomial sign test for two dependent samples** (i.e., it will yield the same chi-square value computed with Equation 8.2). When employing the **Friedman two-way analysis of variance by ranks** with two dependent samples, the two scores of each subject (or pair of matched subjects) are rank-ordered. The data for Example 19.1 can be expressed in a rank-order format, if for each subject a rank of 1 is assigned to the lower of the two scores and a rank of 2 is assigned to the higher score (or vice versa). If a researcher only has such rank-order information, it is still possible to assign a signed difference to each subject, since the ordering of a subject's two ranks provides sufficient information to determine whether the difference between the two scores of a subject would yield a positive or negative value if the interval/ratio scores of the subject were available. Consequently, under such conditions one can still conduct the **binomial sign test for two dependent samples**. On the other hand, if a researcher only has the sort of rank-order information noted above, one will not be able to evaluate the data with either the *t* **test for two dependent samples** or the **Wilcoxon matched-pairs signed-ranks test**, since the latter two tests require the interval/ ratio scores of subjects.

VIII. Additional Examples Illustrating the Use of the Binomial Sign Test for Two Dependent Samples

The **binomial sign test for two dependent samples** can be employed with any of the additional examples noted for the *t* **test for two dependent samples** and the **Wilcoxon matched-pairs signed-ranks test**. In each of the examples, a signed difference must be computed for each subject (or pair of matched subjects). The signed differences are then evaluated employing the protocol for the **binomial sign test for two dependent samples**.

References

Cohen, J. (1977). **Statistical power analysis for the behavioral sciences**. New York: Academic Press.

Cohen, J. (1988). **Statistical power analysis for the behavioral sciences** (2nd ed.). Hillsdale, NJ: Lawrence Erlbaum Associates, Publishers.

Conover, W. J. (1980). **Practical nonparametric statistics** (2nd ed.). New York: John Wiley & Sons.

Conover, W. J. (1999). **Practical nonparametric statistics** (3rd ed.). New York: John Wiley & Sons.

Daniel, W. W. (1990). **Applied nonparametric statistics** (2nd ed.). Boston: PWS–Kent Publishing Company.

Marascuilo, L. A. & McSweeney, M. (1977). **Nonparametric and distribution-free methods for the social sciences**. Monterey, CA: Brooks/Cole Publishing Company.

Siegel, S. & Castellan, N. J., Jr. (1988). **Nonparametric statistics for the behavioral sciences** (2nd ed.). New York: McGraw–Hill Book Company.

Endnotes

1. Some sources note that one assumption of the **binomial sign test for two dependent samples** is that the variable being measured is based on a continuous distribution. In practice, however, this assumption is often not adhered to.

2. Another way of stating the null hypothesis is that in the underlying population the sample represents, the proportion of subjects who obtains a positive signed difference is equal to the proportion of subjects who obtains a negative signed difference. The null and alternative hypotheses can also be stated with respect to the proportion of people in the population who obtains a higher score in Condition 2 than Condition 1, thus yielding a negative difference score. The notation $\pi-$ represents the proportion of the population who yield a difference with a negative sign (referred to as a negative signed difference). Thus, H_0: $\pi- = .5$ can be employed as the null hypothesis, and the following nondirectional and directional alternative hypotheses can be employed: H_1: $\pi- \neq .5$; H_1: $\pi- > .5$; H_1: $\pi- < .5$.

3. It is also the likelihood of obtaining 8 or 9 negative signed differences in a set of 9 signed differences.

4. Due to rounding off protocol, the value computed with Equation 19.1 will be either .0195 or .0196, depending upon whether one employs **Table A6** or **Table A7**.

5. An equivalent way of determining whether or not the result is significant is by doubling the value of the cumulative probability obtained from **Table A7**. In order to reject the null hypothesis, the resulting value must not be greater than the value of α. Since $2 \times .0195 = .039$ is less than $\alpha = .05$, we confirm that the nondirectional alternative hypothesis is supported when $\alpha = .05$. Since .039 is greater than $\alpha = .01$, it is not supported at the .01 level.

6. Equations 9.6 and 9.8 are respectively alternate but equivalent forms of Equations 19.3 and 19.4. Note that in Equations 9.6–9.9, π_1 and π_2 are employed in place of $\pi+$ and $\pi-$ to represent the two population proportions.

7. A full discussion of the protocol for determining one-tailed chi-square values can be found in Section VII of the **chi-square goodness-of-fit test**.

8. Alternative equations for computing a confidence interval (e.g., Equations 8.7–8.11 and Equations 8.14 and 8.15) described in the chapter on the **chi-square goodness-of-fit test** will yield approximately the same limits for a confidence interval as those obtained with Equation 19.5.

Test 20

The McNemar Test
(Nonparametric Test Employed with Categorical/Nominal Data)

I. Hypothesis Evaluated with Test and Relevant Background Information

Hypothesis evaluated with test Do two dependent samples represent two different populations?

Relevant background information on test It is recommended that before reading the material on the **McNemar test**, the reader review the general information on a dependent samples design contained in Sections I and VII of the *t* **test for two dependent samples (Test 17)**. The **McNemar test** (McNemar, 1947) is a nonparametric procedure for categorical data employed in a hypothesis testing situation involving a design with two dependent samples. In actuality, the **McNemar test** is a special case of the **Cochran Q test (Test 26)**, which can be employed to evaluate a *k* dependent samples design involving categorical data, where $k \geq 2$. The **McNemar test** is employed to evaluate an experiment in which a sample of *n* subjects (or *n* pairs of matched subjects) is evaluated on a **dichotomous** dependent variable (i.e., scores on the dependent variable must fall within one of two mutually exclusive categories). The **McNemar test** assumes that each of the *n* subjects (or *n* pairs of matched subjects) contributes two scores on the dependent variable. The test is most commonly employed to analyze data derived from the two types of experimental designs described below.

a) The **McNemar test** can be employed to evaluate categorical data obtained in a **true experiment** (i.e., an experiment involving a manipulated independent variable).[1] In such an experiment, the two scores of each subject (or pair of matched subjects) represent a subject's responses under the two levels of the independent variable (i.e., the two experimental conditions). A significant result allows the researcher to conclude there is a high likelihood the two experimental conditions represent two different populations. As is the case with the *t* **test for two dependent samples**, the **Wilcoxon matched-pairs signed-ranks test (Test 18)**, and the **binomial sign test for two dependent samples (Test 19)**, when the **McNemar test** is employed to evaluate the data for a **true experiment**, in order for the test to generate valid results, the following guidelines should be adhered to: 1) In order to control for order effects, the presentation of the two experimental conditions should be random or, if appropriate, be counterbalanced; and 2) If matched samples are employed, within each pair of matched subjects each of the subjects should be randomly assigned to one of the two experimental conditions.

b) The **McNemar test** can be employed to evaluate a **one-group pretest-posttest design** (which is described in the **Introduction** as well as in Section VII of the *t* **test for two dependent samples**). In applying the **McNemar test** to a **one-group pretest-posttest design**, *n* subjects are administered a pretest on a dichotomous dependent variable. Following the pretest, all of the subjects are exposed to an experimental treatment, after which they are administered a posttest on the same dichotomous dependent variable. The hypothesis evaluated with a **one-group pretest-posttest design** is whether or not there is a significant difference between the pretest and posttest scores of subjects on the dependent variable. The reader is advised to review the discussion of the **one-group pretest-posttest design** in the **Introduction** and in Section VII of the *t* **test for two dependent samples**, since the limitations noted for the design also apply when it is evaluated with the **McNemar test**.

The 2 × 2 table depicted in Table 20.1 summarizes the **McNemar test** model. The entries for Cells *a*, *b*, *c*, and *d* in Table 20.1 represent the number of subjects/observations in each of four possible categories which can be employed to summarize the two responses of a subject (or matched pair of subjects) on a dichotomous dependent variable. Each of the four response category combinations represents the number of subjects/observations whose response in Condition 1/Pretest falls in the response category for the row in which the cell falls, and whose response in Condition 2/Posttest falls in the response category for the column in which the cell falls. Thus, the entry in Cell *a* represents the number of subjects who respond in Response category 1 in both Condition 1/Pretest and Condition 2/Posttest. The entry in Cell *b* represents the number of subjects who respond in Response category 1 in Condition 1/Pretest and in Response category 2 in Condition 2/Posttest. The entry in Cell *c* represents the number of subjects who respond in Response category 2 in Condition 1/Pretest and in Response category 1 in Condition 2/Posttest. The entry in Cell *d* represents the number of subjects who respond in Response category 2 in both Condition 1/Pretest and Condition 2/Posttest.

Table 20.1　Model for the McNemar Test

		Condition 2/Posttest		
		Response category 1	Response category 2	Row sums
Condition 1/Pretest	Response category 1	a	b	$a + b = n_1$
	Response category 2	c	d	$c + d = n_2$
	Column sums	$a + c$	$b + d$	n

The **McNemar test** is based on the following assumptions: a) The sample of *n* subjects has been randomly selected from the population it represents; b) Each of the *n* observations in the contingency table is independent of the other observations; c) The scores of subjects are in the form of a dichotomous categorical measure involving two mutually exclusive categories; and d) Most sources state that the **McNemar test** should not be employed with extremely small sample sizes. Although the chi-square distribution is generally employed to evaluate the **McNemar test** statistic, in actuality the latter distribution is used to provide an approximation of the exact sampling distribution which is, in fact, the binomial distribution. When the sample size is small, in the interest of accuracy, the exact binomial probability for the data should be computed. Sources do not agree on the minimum acceptable sample size for computing the **McNemar test** statistic (i.e., using the chi-square distribution). Some sources endorse the use of a correction for continuity with small sample sizes (discussed in Section VI), in order to insure that the computed chi-square value provides a more accurate estimate of the exact binomial probability.

II. Examples

Since, as noted in Section I, the **McNemar test** is employed to evaluate a **true experiment** and a **one-group pretest-posttest design**, two examples, each representing one of the aforementioned designs, will be presented in this section. Since the two examples employ identical data, they will result in the same conclusion with respect to the null hypothesis. Example 20.1

describes a **true experiment** and Example 20.2 describes a study which employs a **one-group pretest-posttest design**.

Example 20.1 *A psychologist wants to compare a drug for treating enuresis (bed-wetting) with a placebo. One hundred enuretic children are administered both the drug (Endurin) and a placebo in a double blind study conducted over a six month period. During the duration of the study, each child has six drug and six placebo treatments, with each treatment lasting one week. To insure that there are no carryover effects from one treatment to another, during the week following each treatment a child is not given either the drug or the placebo. The order of presentation of the 12 treatment periods for each child is randomly determined. The dependent variable in the study is a parent's judgement with respect to whether or not a child improves under each of the two experimental conditions. Table 20.2 summarizes the results of the study. Do the data indicate the drug was effective?*

Table 20.2 Summary of Data for Example 20.1

		Favorable response to drug		
		Yes	No	Row sums
Favorable response	Yes	10	13	23
to placebo	No	41	36	77
	Column sums	51	49	100

Note that the data in Table 20.2 indicate the following: a) 10 subjects respond favorably to both the drug and the placebo; b) 13 subjects do not respond favorably to the drug but do respond favorably to the placebo; c) 41 subjects respond favorably to the drug but do not respond favorably to the placebo; and d) 36 subjects do not respond favorably to either the drug or the placebo. Of the 100 subjects, 51 respond favorably to the drug, while 49 do not. 23 of the 100 subjects respond favorably to the placebo, while 77 do not.

Example 20.2 *A researcher conducts a study to investigate whether or not a weekly television series which is highly critical of the use of animals as subjects in medical research influences public opinion. One hundred randomly selected subjects are administered a pretest to determine their attitude concerning the use of animals in medical research. Based on their responses, subjects are categorized as pro-animal research or anti-animal research. Following the pretest, all of the subjects are instructed to watch the television series (which lasts two months). At the conclusion of the series each subject's attitude toward animal research is reassessed. The results of the study are summarized in Table 20.3. Do the data indicate that a shift in attitude toward animal research occurred after subjects viewed the television series?*

Table 20.3 Summary of Data for Example 20.2

		Posttest		
		Anti	Pro	Row sums
Pretest	Anti	10	13	23
	Pro	41	36	77
	Column sums	51	49	100

Note that the data in Table 20.3 indicate the following: a) 10 subjects express an anti-animal research attitude on both the pretest and the posttest; b) 13 subjects express an anti-animal research attitude on the pretest but a pro-animal research attitude on the posttest; c) 41 subjects express a pro-animal research attitude on the pretest but an anti-animal research attitude on the posttest; and d) 36 subjects express a pro-animal research attitude on both the pretest and the posttest. Of the 100 subjects, 23 are anti on the pretest and 77 are pro on the pretest. Of the 100 subjects, 51 are anti on the posttest, while 49 are pro on the posttest.

Table 20.4 summarizes that data for Examples 20.1 and 20.2.

Table 20.4　Summary of Data for Examples 20.1 and 20.2

		Favorable response to drug/Posttest		
		Yes/Anti	No/Pro	Row sums
Favorable response	Yes/Anti	$a = 10$	$b = 13$	23
to placebo/Pretest	No/Pro	$c = 41$	$d = 36$	77
	Column sums	51	49	100

III.　Null versus Alternative Hypotheses

In conducting the **McNemar test**, the cells of interest in Table 20.4 are Cells b and c, since the latter two cells represent those subjects who respond in different response categories under the two experimental conditions (in the case of a **true experiment**) or in the pretest versus posttest (in the case of a **one-group pretest-posttest design**). In Example 20.1, the frequencies recorded in Cells b and c, respectively, represent subjects who respond **favorably to the placebo/ unfavorably to the drug** and **favorably to the drug/unfavorably to the placebo**. If the drug is more effective than the placebo, one would expect the proportion of subjects in Cell c to be larger than the proportion of subjects in Cell b. In Example 20.2, the frequencies recorded in Cells b and c, respectively, represent subjects who are **anti-animal research in the pretest/pro-animal research in the posttest** and **pro-animal research in the pretest/anti-animal research in the posttest**. If there is a shift in attitude from the pretest to the posttest (specifically from **pro-animal research** to **anti-animal research**), one would expect the proportion of subjects in Cell c to be larger than the proportion of subjects in Cell b.

It will be assumed that in the underlying population, π_b and π_c represent the following proportions: $\pi_b = b/(b + c)$ and $\pi_c = c/(b + c)$. If there is no difference between the two experimental conditions (in the case of a **true experiment**) or between the pretest and the posttest (in the case of a **one-group pretest-posttest design**), the following will be true: $\pi_b = \pi_c = .5$. With respect to the sample data, the values π_b and π_c are estimated with the values p_b and p_c, which in the case of Examples 20.1 and 20.2 are $p_b = b/(b + c) = 13/(13 + 41) = .24$ and $p_c = c/(b + c) = 41/(13 + 41) = .76$.

Employing the above information the null and alternative hypotheses for the **McNemar test** can now be stated.[2]

Null hypothesis　　　　　　　　　　H_0: $\pi_b = \pi_c$

(In the underlying population the sample represents, the proportion of observations in Cell b equals the proportion of observations in Cell c.)

Alternative hypothesis H_1: $\pi_b \neq \pi_c$

(In the underlying population the sample represents, the proportion of observations in Cell b does not equal the proportion of observations in Cell c. This is a **nondirectional alternative hypothesis** and it is evaluated with a two-tailed test. In order to be supported, the proportion of observations in Cell b (p_b) can be either significantly larger or significantly smaller than the proportion of observations in Cell c (p_c). In the case of Example 20.1, this alternative hypothesis will be supported if the proportion of subjects who respond **favorably to the placebo/ unfavorably to the drug** is significantly greater than the proportion of subjects who respond **favorably to the drug/unfavorably to the placebo**, or the proportion of subjects who respond **favorably to the drug/unfavorably to the placebo** is significantly greater than the proportion of subjects who respond **favorably to the placebo/unfavorably to the drug**. In the case of Example 20.2, this alternative hypothesis will be supported if, in the pretest versus posttest, a significantly larger proportion of subjects shifts its response from **pro-animal research** to **anti-animal research** or a significantly larger proportion of subjects shifts its response from **anti-animal research** to **pro-animal research**.)

or

H_1: $\pi_b > \pi_c$

(In the underlying population the sample represents, the proportion of observations in Cell b is greater than the proportion of observations in Cell c. This is a **directional alternative hypothesis** and it is evaluated with a one-tailed test. In order to be supported, the proportion of observations in Cell b (p_b) must be significantly larger than the proportion of observations in Cell c (p_c). In the case of Example 20.1, this alternative hypothesis will be supported if the proportion of subjects who respond **favorably to the placebo/unfavorably to the drug** is significantly greater than the proportion of subjects who respond **favorably to the drug/ unfavorably to the placebo**. In the case of Example 20.2, this alternative hypothesis will be supported if, in the pretest versus posttest, a significantly larger proportion of subjects shifts its response from **anti-animal research** to **pro-animal research**.)

or

H_1: $\pi_b < \pi_c$

(In the underlying population the sample represents, the proportion of observations in Cell b is less than the proportion of observations in Cell c. This is a **directional alternative hypothesis** and it is evaluated with a one-tailed test. In order to be supported, the proportion of observations in Cell b (p_b) must be significantly smaller than the proportion of observations in Cell c (p_c). In the case of Example 20.1, this alternative hypothesis will be supported if the proportion of subjects who respond **favorably to the drug/unfavorably to the placebo** is significantly greater than the proportion of subjects who respond **favorably to the placebo/unfavorably to the drug**. In the case of Example 20.2, this alternative hypothesis will be supported if, in the pretest versus posttest, a significantly larger proportion of subjects shifts its responses from **pro-animal research** to **anti-animal research**.)

Note: Only one of the above noted alternative hypotheses is employed. If the alternative hypothesis the researcher selects is supported, the null hypothesis is rejected.

IV. Test Computations

The test statistic for the **McNemar test**, which is based on the chi-square distribution, is computed with Equation 20.1.[3]

$$\chi^2 = \frac{(b - c)^2}{b + c}$$
                                                                          (Equation 20.1)

Where: b and c represent the number of observations in Cells b and c of the **McNemar test** summary table

Substituting the appropriate values in Equation 20.1, the value $\chi^2 = 14.52$ is computed for Examples 20.1/20.2.

$$\chi^2 = \frac{(13 - 41)^2}{13 + 41} = 14.52$$

The computed chi-square value must always be a positive number. If a negative value is obtained, it indicates that an error has been made. The only time the value of chi-square will equal zero is when $b = c$.

V. Interpretation of the Test Results

The obtained value $\chi^2 = 14.52$ is evaluated with **Table A4 (Table of the Chi-Square Distribution)** in the **Appendix**.[4] The degrees of freedom employed in the analysis are $df = 1$.[5] Employing **Table A4**, for $df = 1$ the tabled critical two-tailed .05 and .01 chi-square values are $\chi^2_{.05} = 3.84$ (which corresponds to the chi-square value at the 95th percentile) and $\chi^2_{.01} = 6.63$ (which corresponds to the chi-square value at the 99th percentile). The tabled critical one-tailed .05 and .01 values are $\chi^2_{.05} = 2.71$ (which corresponds to the chi-square value at the 90th percentile) and $\chi^2_{.01} = 5.43$ (which corresponds to the chi-square value at the 98th percentile).[6]

The following guidelines are employed in evaluating the null hypothesis for the **McNemar test**.

a) If the nondirectional alternative hypothesis $H_1: \pi_b \neq \pi_c$ is employed, the null hypothesis can be rejected if the obtained chi-square value is equal to or greater than the tabled critical two-tailed value at the prespecified level of significance.

b) If a directional alternative hypothesis is employed, only the directional alternative hypothesis which is consistent with the data can be supported. With respect to the latter alternative hypothesis, the null hypothesis can be rejected if the obtained chi-square value is equal to or greater than the tabled critical one-tailed value at the prespecified level of significance.

Applying the above guidelines to Examples 20.1/20.2, we can conclude the following.

Since the obtained value $\chi^2 = 14.52$ is greater than the tabled critical two-tailed values $\chi^2_{.05} = 3.84$ and $\chi^2_{.01} = 6.63$, the nondirectional alternative hypothesis $H_1: \pi_b \neq \pi_c$ is supported at both the .05 and .01 levels. Since $\chi^2 = 14.52$ is greater than the tabled critical one-tailed values $\chi^2_{.05} = 2.71$ and $\chi^2_{.01} = 5.43$, the directional alternative hypothesis $H_1: \pi_b < \pi_c$ is supported at both the .05 and .01 levels (since $p_b = .24$ is less than $p_c = .76$).

A summary of the analysis of Examples 20.1 and 20.2 with the **McNemar test** follows:

Example 20.1: It can be concluded that the proportion of subjects who respond favorably to the drug is significantly greater than the proportion of subjects who respond favorably to the placebo.

Example 20.2: It can be concluded that following exposure to the television series, there is a significant change in attitude toward the use of animals as subjects in medical research. The direction of the change is from pro-animal research to anti-animal research. It is important to note, however, that since Example 20.2 is based on a **one-group pretest-posttest design**, the researcher is not justified in concluding that the change in attitude is a direct result of subjects watching the television series. This is the case because (as noted in the **Introduction** and in Section VII of the *t* **test for two dependent samples**) a **one-group pretest-posttest design** is an incomplete experimental design. Specifically, in order to be an adequately controlled experimental design, a **one-group pretest-posttest design** requires the addition of a control group which is administered the identical pretest and posttest at the same time periods as the group described in Example 20.2. The control group, however, would not be exposed to the television series between the pretest and the posttest. Without inclusion of such a control group, it is not possible to determine whether an observed change in attitude from the pretest to the posttest is due to the experimental treatment (i.e., the television series), or is the result of one or more extraneous variables that may also have been present during the intervening time period between the pretest and the posttest.

VI. Additional Analytical Procedures for the McNemar Test and/or Related Tests

1. Alternative equation for the McNemar test statistic based on the normal distribution Equation 20.2 is an alternative equation that can be employed to compute the **McNemar test** statistic. It yields a result which is equivalent to that obtained with Equation 20.1.

$$z = \frac{b - c}{\sqrt{b + c}}$$
(Equation 20.2)

The sign of the computed z value is only relevant insofar as it indicates the directional alternative hypothesis with which the data are consistent. Specifically, the z value computed with Equation 20.2 will be a positive number if the number of observations in Cell b is greater than the number of observations in Cell c, and it will be a negative number if the number of observations in Cell c is greater than the number of observations in Cell b. Since in Examples 20.1/20.2 $c > b$, the computed value of z will be a negative number. Substituting the appropriate values in Equation 20.2, the value $z = -3.81$ is computed.

$$z = \frac{13 - 41}{\sqrt{13 + 41}} = -3.81$$

The square of the z value obtained with Equation 20.2 will always equal the chi-square value computed with Equation 20.1. This relationship can be confirmed by the fact that $(z = -3.81)^2 = (\chi^2 = 14.52)$. It is also the case that the square of a tabled critical z value at a given level of significance will equal the tabled critical chi-square value at the corresponding level of significance.

The obtained z value is evaluated with **Table A1 (Table of the Normal Distribution)** in the **Appendix**. In **Table A1** the tabled critical two-tailed .05 and .01 values are $z_{.05} = 1.96$ and $z_{.01} = 2.58$, and the tabled critical one-tailed .05 and .01 values are $z_{.05} = 1.65$ and $z_{.01} = 2.33$. In interpreting the z value computed with Equation 20.2, the following guidelines are employed.

a) If the nondirectional alternative hypothesis H_1: $\pi_b \neq \pi_c$ is employed, the null hypothesis can be rejected if the obtained absolute value of z is equal to or greater than the tabled critical two-tailed value at the prespecified level of significance.

b) If a directional alternative hypothesis is employed, only the directional alternative hypothesis that is consistent with the data can be supported. With respect to the latter alternative hypothesis, the null hypothesis can be rejected if the obtained absolute value of z is equal to or greater than the tabled critical one-tailed value at the prespecified level of significance.

Employing the above guidelines with Examples 20.1/20.2, we can conclude the following. Since the obtained absolute value $z = 3.81$ is greater than the tabled critical two-tailed values $z_{.05} = 1.96$ and $z_{.01} = 2.58$, the nondirectional alternative hypothesis H_1: $\pi_b \neq \pi_c$ is supported at both the .05 and .01 levels. Since the obtained absolute value $z = 3.81$ is greater than the tabled critical one-tailed values $z_{.05} = 1.65$ and $z_{.01} = 2.33$, the directional alternative hypothesis H_1: $\pi_b < \pi_c$ is supported at both the .05 and .01 levels. These conclusions are identical to those reached when Equation 20.1 is employed to evaluate the same set of data.

2. The correction for continuity for the McNemar test　Since the **McNemar test** employs a continuous distribution to approximate a discrete probability distribution, some sources recommend that a correction for continuity be employed in computing the test statistic. Sources which recommend such a correction either recommend it be limited to small sample sizes or that it be used in all instances.[7] Equations 20.3 and 20.4 are the continuity-corrected versions of Equations 20.1 and 20.2.[8]

$$\chi^2 = \frac{(|b - c| - 1)^2}{b + c} \qquad \textbf{(Equation 20.3)}$$

$$z = \frac{|b - c| - 1}{\sqrt{b + c}} \qquad \textbf{(Equation 20.4)}$$

Substituting the appropriate values in Equations 20.3 and 20.4, the values $\chi^2 = 13.5$ and $z = 3.67$ are computed for Examples 20.1/20.2.

$$\chi^2 = \frac{(|13 - 41| - 1)^2}{13 + 41} = 13.5$$

$$z = \frac{|13 - 41| - 1}{\sqrt{13 + 41}} = 3.67$$

As is the case without the continuity correction, the square of the z value obtained with Equation 20.4 will always equal the chi-square value computed with Equation 20.3. This relationship can be confirmed by the fact that $(z = 3.67)^2 = (\chi^2 = 13.5)$. Note that the chi-square value computed with Equation 20.3 will always be less than the value computed with Equation 20.1. In the same respect, the absolute value of z computed with Equation 20.4 will always be less than the absolute value of z computed with Equation 20.2. The lower absolute values computed for the continuity-corrected statistics reflect the fact that the latter analysis provides a more conservative test of the null hypothesis than does the uncorrected analysis. In this instance, the decision the researcher makes with respect to the null hypothesis is not affected by the correction for continuity, since the values $\chi^2 = 13.5$ and $z = 3.67$ are both greater than the relevant tabled critical one- and two-tailed .05 and .01 values. Thus, the nondirectional alternative hypothesis H_1: $\pi_b \neq \pi_c$ and the directional alterative hypothesis H_1: $\pi_b < \pi_c$ are supported at both the .05 and .01 levels.[9]

3. Computation of the exact binomial probability for the McNemar test model with a small sample size In Section I it is noted that the exact probability distribution for the **McNemar test** model is the binomial distribution, and that the chi-square distribution is employed to approximate the latter distribution. Although for large sample sizes the chi-square distribution provides an excellent approximation of the binomial distribution, many sources recommend that for small sample sizes the exact binomial probabilities be computed. In order to demonstrate the computation of an exact binomial probability for the **McNemar test** model, assume that Table 20.5 is a revised summary table for Examples 20.1/20.2.

Table 20.5   Revised Summary Table for Examples 20.1 and 20.2 for Binomial Analysis

		Favorable response to drug/Posttest		
		Yes/Anti	No/Pro	Row sums
Favorable response to	Yes/Anti	$a = 10$	$b = 2$	12
placebo/Pretest	No/Pro	$c = 8$	$d = 36$	44
	Column sums	18	38	56

Note that although the frequencies for Cells a and d in Table 20.5 are identical to those employed in Table 20.4, different frequencies are employed for Cells b and c. Although in Table 20.5 the total sample size of $n = 56$ is reasonably large, the total number of subjects in Cells b and c is quite small, and, in the final analysis, it is when the sum of the frequencies of the latter two cells is small that computation of the exact binomial probability is recommended. The fact that the frequencies of Cells a and d are not taken into account represents an obvious limitation of the **McNemar test**. In point of fact, the frequencies of Cells a and d could be 0 and 0 instead of 10 and 36, and the same result will be obtained when the **McNemar test** statistic is computed. In the same respect, frequencies of 1000 and 3600 for Cells a and d will also yield the identical result. Common sense suggests, however, that the difference $|b - c| = |2 - 8| = 6$ will be considered more important if the total sample size is small (which will be the case if the frequencies of Cells a and d are 0 and 0) than if the total sample size is very large (which will be the case if the frequencies of Cells a and d are 1000 and 3600). What the latter amounts to is that a significant difference between Cells b and c may be of little or no practical significance if the total number of observations in all four cells is very large.

Employing Equations 20.1 and 20.2 with the data in Table 20.5, the values $\chi^2 = 3.6$ and $z = 1.90$ are computed for the **McNemar test** statistic.

$$\chi^2 = \frac{(2 - 8)^2}{2 + 8} = 3.6$$

$$z = \frac{2 - 8}{\sqrt{2 + 8}} = 1.90$$

Employing Equations 20.3 and 20.4, the values $\chi^2 = 2.5$ and $z = 1.58$ are the continuity-corrected values computed for the **McNemar test** statistic.

$$\chi^2 = \frac{(|2 - 8| - 1)^2}{2 + 8} = 2.5$$

$$z = \frac{|2 - 8| - 1}{\sqrt{2 + 8}} = 1.58$$

Employing **Table A1**, we determine that the exact one-tailed probability for the value $z = 1.90$ computed with Equation 20.2 (as well as for $\chi^2 = 3.6$ computed with Equation 20.1) is .0287. We also determine that the exact one-tailed probability for the value $z = 1.58$ computed with Equation 20.4 (as well as for $\chi^2 = 2.5$ computed with Equation 20.3) is .0571.[10] Note that since the continuity-correction results in a more conservative test, the probability associated with the continuity-corrected value will always be higher than the probability associated with the uncorrected value. Without the continuity correction, the directional alternative hypothesis H_1: $\pi_b < \pi_c$ is supported at the .05 level, since $\chi^2 = 3.6/z = 1.90$ are greater than the tabled critical one-tailed values $\chi^2_{.05} = 2.71/z_{.05} = 1.65$. The nondirectional alternative hypothesis H_1: $\pi_b \neq \pi_c$ is not supported, since $\chi^2 = 3.6/z = 1.90$ are less than the tabled critical two-tailed values $\chi^2_{.05} = 3.84/z_{.05} = 1.96$. When the continuity correction is employed, the directional alternative hypothesis H_1: $\pi_b < \pi_c$ fails to achieve significance at the .05 level, since $\chi^2 = 2.5/z = 1.58$ are less than $\chi^2_{.05} = 2.71/z_{.05} = 1.65$. The nondirectional alternative hypothesis H_1: $\pi_b \neq \pi_c$ is not supported, since $\chi^2 = 2.5/z = 1.58$ are less than the tabled critical two-tailed values $\chi^2_{.05} = 3.84/z_{.05} = 1.96$.

At this point, the exact binomial probability will be computed for the same set of data. As is the case with the equations for the **McNemar test** which are based on the chi-square and normal distributions, the binomial analysis only considers the frequencies of Cells b and c. Since only two cells are taken into account, the binomial analysis becomes identical to the analysis described for the **binomial sign test for a single sample (Test 9)**.[11]

Equation 20.5 is the binomial equation that is employed to determine the likelihood of obtaining a frequency of 8 or larger in one of the two cells (or 2 or less in one of the two cells) in the **McNemar model** summary table, if the total frequency in the two cells is 10 (where $m = b + c = 2 + 8 = 10$). Note that Equation 20.5 is identical to Equation 9.5, except for the fact that π_b and π_c are used in place of π_1 and π_2, and the value m, which represents $b + c$, is used in place of n.

$$P(\geq x) = \sum_{r=x}^{m} \binom{m}{x} (\pi_b)^x (\pi_c)^{(m-x)} \qquad \textbf{(Equation 20.5)}$$

In evaluating the data in Table 20.5, the following values are employed in Equation 20.5: $\pi_b = \pi_c = .5$ (which will be the case if the null hypothesis is true), $m = 10$, $x = 8$.

$$P(x \geq 8) = \binom{10}{8} (.5)^8 (.5)^2 + \binom{10}{9} (.5)^9 (.5)^1 + \binom{10}{10} (.5)^{10} (.5)^0 = .0547$$

The computed probability .0547 is the likelihood of obtaining a frequency of 8 or greater in one of the two cells (as well as the likelihood of obtaining a frequency of 2 or less in one of the two cells). The value .0547 can also be obtained from **Table A7 (Table of the Binomial Distribution, Cumulative Probabilities)** in the **Appendix**. In using **Table A7** we find the section for $m = 10$ (which is represented by $n = 10$ in the table), and locate the cell which is the intersection of the row $x = 8$ and the column $\pi = .5$. The entry for the latter cell is .0547. The value .0547 computed for the exact binomial probability is quite close to the continuity-corrected probability of .0571 obtained with Equations 20.3/20.4 (which suggests that even when the sample size is small, the continuity-corrected chi-square/normal approximation provides an excellent estimate of the exact probability). As is the case when the data are evaluated with

Equations 20.3/20.4, the directional alternative hypothesis H_1: $\pi_b < \pi_c$ is not supported if the binomial analysis is employed. This is the case, since the probability .0547 is greater than $\alpha = .05$. In order for the directional alternative hypothesis H_1: $\pi_b < \pi_c$ to be supported, the tabled probability must be equal to or less than $\alpha = .05$. The nondirectional alternative hypothesis H_1: $\pi_b \neq \pi_c$ is also not supported, since the probability .0547 is greater than $\alpha/2 = .05/2 = .025$.[12]

4. Computation of the power of the McNemar test   Connett *et al.* (1987) describe a procedure for computing the power of the **McNemar test** (which is also described in Zar (1999, pp. 171–172)) through use of Equation 20.6. The precision of the latter equation, which employs the normal distribution, increases as the value of n increases.

$$z_\beta = \frac{\sqrt{n}\sqrt{p}\,(\psi - 1) - z_\alpha\sqrt{\psi + 1}}{\sqrt{(\psi + 1) - p(\psi - 1)^2}}$$ **(Equation 20.6)**

Where: n is the sample size/number of matched pairs employed in the study
   p is the smaller of the two ratios b/n versus c/n (which is generally estimated in a pilot study)
   ψ (which is the lower case Greek letter **psi**) is the minimum acceptable effect size (specifically, the magnitude of the larger of the two ratios *(b/n) / (c/n)* versus *(c/n) / (b/n)*)
   z_α is the tabled critical z value at the prespecified level of significance to be employed in the study
   z_β will represent the z value in **Table A1** employed to compute the Type II error rate for the study (i.e., the value of β)

   To illustrate the application of Equation 20.6, we will assume that the data summarized in Table 20.4 represent a **pilot study**. (A **pilot study** is a preliminary study conducted prior to conducting a final study which will be employed to evaluate a specific hypothesis.) We will also assume that in the final study which will be conducted to evaluate the hypothesis in question, the researcher plans to use $n = 64$ subjects. Employing the data in Table 20.4, the value of p is estimated by computing the two ratios $b/n = 13/100 = .13$, and $c/n = 41/100 = .41$. Since $b/n = .13$ is the smaller of the two values, $p = .13$. The researcher elects to specify $\psi = 3$ as the minimum acceptable effect size. The latter indicates that in order for the result of the study to be of practical significance, the researcher requires that either of the two ratios *(b/n) / (c/n)* versus *(c/n) / (b/n)* is equal to or greater than 3. For the purposes of illustration we will assume that the prespecified tabled critical value $z = 1.96$ (which allows for a one-tailed analysis at the .025 level and a two-tailed analysis at the .05 level) will be employed in Equation 20.6 to represent z_α. In the case of both Examples 20.1/20.2 one could argue that the researcher is primarily interested in a one-tailed analysis with respect to power. Consequently, if the latter is the case and $\psi = 3$, the directional prediction being made is *(c/n) / (b/n)* ≥ 3 (which implies that at least three times as many people will respond favorably to the drug than the placebo (Example 20.1), and at least three times as many people will shift from pro to anti as opposed to anti to pro following exposure to the television series (Example 20.2)). Substituting the appropriate values in Equation 20.6, the value $z_\beta = .992$ is computed.

$$z_\beta = \frac{\sqrt{64}\,\sqrt{.13}\,(3-1)-(1.96)\sqrt{3+1}}{\sqrt{(3+1)-(.13)(3-1)^2}} = .992$$

The computed value $z_\beta = .992$ represents the z value associated with the Type II error rate (i.e., the value of β) in the study. The value of β will be the proportion recorded in **Column 3** of **Table A1** for the appropriate z value. For the value $z = .99$ (which is closest to .992) the proportion recorded in **Column 3** is .1611. Since **Power = 1 − β**, the power of the test will be equal to $1-.1611 = .8389$. In other words, in reference to the values employed in Equation 20.6, the likelihood of rejecting the null hypothesis, given the latter is incorrect, will be approximately .84.

Equation 20.7 (which is algebraically derived from Equation 20.6) allows a researcher to compute the sample size required in order to conduct the **McNemar test** at a specified level of power. The latter equation is employed below to compute the sample size value $n = 64$ (used in Equation 20.6) if prior to conducting the study the researcher specifies a power of .8389. Specifically, $n = 64$ is employed in Equation 20.7 along with the values $\psi = 3$, $p = .13$, $z_{\alpha/2} = 1.96$, and $z_\beta = .992$ (which are the same values employed or computed above with Equation 20.6). Note that the value $z_\beta = .992$ (which was computed with Equation 20.6) must be obtained from **Table A1** based on the value of β (which in our example is .1611). As noted above, β is the proportion in **Column 3** of **Table A1**, and it is the result of the difference $1 - \textbf{Power} = \beta$. Thus, if prior to conducting the study, the researcher specifies .8389 as the desired power for the **McNemar test** (which is the power value determined earlier through use of Equation 20.6), $\beta = 1 - .8389 = .1611$. Consequently the value $z = .99$ associated with $\beta = .1611$ is employed in Equation 20.7 to represent z_β.

$$n = \frac{\left[z_{\alpha/2}\sqrt{\psi+1}+z_\beta\sqrt{(\psi+1)-p(\psi-1)^2}\right]^2}{p(\psi-1)^2} \qquad \textbf{(Equation 20.7)}$$

$$n = \frac{\left[(1.96)\sqrt{3+1}+(.992)\sqrt{(3+1)-(.13)(3-1)^2}\right]^2}{(.13)(3-1)^2} = 64.2$$

It should be noted that Equations 20.6 and 20.7 do not employ a correction for continuity. If a researcher elects to employ the correction for continuity for the **McNemar test**, the above equations would have to be adjusted. If a correction for continuity is employed in the above equations, it will result in a lower power value than that computed with Equation 20.6 and a larger sample size than that computed with Equation 20.7. Alternative forms of Equations 20.6 and 20.7 can be found in Rosner (2000, pp. 384–389). Alternative methodologies for determining sample size for the **McNemar test** are discussed in Fleiss *et al.* (2003, 394–399).

The reader should take note of the fact that although the number of observations in cells *a* and *d* is not included in the computation of the test statistic (i.e., χ^2 or z value) for the **McNemar test** (through use of Equations 20.2–20.4), the latter cell values are relevant in the computation of power (since they are included in the total sample size (n) computed with equation 20.7). Daniel (1990) and Fleiss (1981) provide additional references which discuss the power of the **McNemar test** relative to alternative procedures (such as the **Gart test for order effects** which is discussed later in this section).

5. Computation of a confidence interval for the McNemar test A procedure for computing a confidence interval for the difference between the marginal probabilities (i.e., $[(a + b)/n] - [(a + c)/n]$) in a **McNemar test** summary table will be presented in this section. As noted in Endnote

2, we can allow π_1 to represent the proportion of observations in the underlying population that responds in Response category 1 in Condition 1/Pretest, and π_2 to represent the proportion of observations in the underlying population that responds in Response category 1 in Condition 2/Posttest. With respect to the sample data, the values p_1 and p_2 are employed to estimate π_1 and π_2, where $p_1 = (a + b)/n$ and $p_2 = (a + c)/n$. Employing the latter notation in reference to Table 20.4, we will compute a confidence interval for $\pi_1 - \pi_2$ through use of Equation 20.8. As noted in Endnote 9, the value $\sqrt{(b + c)/n^2}$ (which does not take into account the frequencies in cells a and d of Table 20.1/20.2/20.3) should not be used to represent the standard error in computing a confidence interval. Instead Equation 20.9 (which does take into account the frequencies in cells a and d) is employed to compute the value of the standard error of the difference between the proportions p_1 and p_2.[13]

$$CI_{(1-\alpha)} = (p_1 - p_2) \pm (z_{\alpha/2})(s_{p_1 - p_2}) \qquad \textbf{(Equation 20.8)}$$

$$s_{p_1 - p_2} = \sqrt{\frac{(a + d)(b + c) + 4bc}{n^3}} \qquad \textbf{(Equation 20.9)}$$

In the case of Examples 20.1 and 20.2, $p_1 = (10 + 13)/100 = .23$ and $p_2 = (10 + 41)/100 = .51$. The value $s_{p_1 - p_2} = .0679$ is computed below.

$$s_{p_1 - p_2} = \sqrt{\frac{(10 + 36)(13 + 41) + (4)(13)(41)}{(100)^3}} = .0679$$

Substituting the appropriate values in Equation 20.8 the 95% confidence interval (which employs the tabled critical two-tailed value $z_{.05} = 1.96$ in **Table A1**) is computed below.

$$CI_{(1-\alpha)} = (.23 - .51) \pm (1.96)(.0679) = -.28 \pm .13$$

$$-.15 \geq (\pi_1 - \pi_2) \geq -.41$$

This result indicates the researcher can be 95% confident (or the probability is .95) that the interval $-.41$ to $-.15$ contains the true difference between the population proportions. If the aforementioned interval contains the true difference between the population proportions, the proportion for Condition 1 (π_1) (i.e., **placebo/pretest**) is less than the proportion for Condition 2 (π_2) (i.e., **drug/posttest**) by a value that is less than or equal to .15 but not less than .41. The result can also be written as $.15 \leq (\pi_2 - \pi_1) \leq .41$, which indicates that if the aforementioned interval contains the true difference between the population proportions, the proportion for Condition 2 (π_2) is larger than the proportion for Condition 1 (π_1) by a value that is greater than or equal to .15 but not greater than .41. Further discussion of the computation of a confidence interval for the **McNemar test** can be found in Marascuilo and McSweeney (1977), Fleiss (1981), Fleiss *et al.* (2003), and Selvin (1995, 2004).

6. Computation of an odds ratio for the McNemar test Prior to reading this section the reader may want to review the material on the **odds ratio** in Section VI of the **chi-square test for $r \times c$ tables (Test 16)**. Equations 20.10 and 20.11 can respectively be employed to compute an **odds ratio** (o_m) and an **inverse odds ratio** (o'_m) for the **McNemar test**. Employing the latter equations with the data in Table 20.4, the values $o_m = 3.15$ and $o'_m = .32$ are computed below.

$$o_m = \frac{c}{b} = \frac{41}{13} = 3.15 \qquad \text{(Equation 20.10)}$$

$$o'_m = \frac{b}{c} = \frac{13}{41} = .32 \qquad \text{(Equation 20.11)}$$

The value $o_m = 3.15$ indicates that the odds of a person responding to the drug are 3.15 times greater than the odds of a person responding to the placebo, whereas value $o'_m = .32$ indicates that the odds of a person responding to the placebo are .32 times as large as the odds of a person responding to the drug.[14]

7. Additional analytical procedures for the McNemar test

a) Fleiss (1981, p. 118) and Fleiss *et al.* (2003, pp. 379–380) note that if the null hypothesis is rejected in a study such as that described by Example 20.1, Equation 20.12 can be employed to determine the **relative difference** (represented by the notation p_e) between the two treatments. The latter concept is based on the assumption that the drug can only benefit subjects who fail to respond to the placebo, and thus the relative value of the drug can be estimated by p_e. Equation 20.12 is employed with the data for Example 20.1 to compute the value $p_e = .36$.

$$p_e = \frac{c - b}{c + d} = \frac{41 - 13}{41 + 36} = .36 \qquad \text{(Equation 20.12)}$$

The computed value .36 indicates that in a sample of 100 patients who do not respond favorably to the placebo, $(.36)(100) = 36$ would be expected to respond favorably to the drug. Fleiss (1981) and Fleiss *et al.* (2003, pp. 379–380) describe the computation of the estimated standard error for the value of the relative difference computed with Equation 20.12, as well as the procedure for computing a confidence interval for the relative difference.

b) In Section IX (the **Addendum**) of the **Pearson product-moment correlation coefficient (Test 28)**, the use of the **phi coefficient (Test 16g** described in Section VI of the **chi-square test for $r \times c$ tables**) as a measure of association for the **McNemar test** model is discussed within the context of the **tetrachoric correlation coefficient (Test 28k)**.

8. Test 20a: The Gart test for order effects

In some experiments it is possible to expose a group of subjects to each of two experimental treatments simultaneously. For example, the relative effects of two topical creams for treating psoriasis can be evaluated by employing one of the creams on the right side of each subject's body and the other cream on the left side of each subject's body. (Note, however, that it could be argued that the simultaneous presence of the two creams on a person's body could result in a unique physiological state which is not commensurate with the physiological state associated with the presence of either one of the two creams by itself.) Other experiments, however, require that two treatments be sequentially applied to the same set of subjects, since it is not feasible to expose each subject to the two treatment conditions simultaneously. For example, the effect of two allergy medications may be evaluated on a group of subjects, with each subject taking each of the medicines during separate three month periods. In experiments involving sequential presentation of experimental treatments, the potential impact of order effects (which, as previously noted, is also referred to as carryover effects) is often controlled for through use of counterbalancing or randomizing the presentation of treatments. When, for whatever reason, a researcher is unable to implement the latter methodologies or views them as unsatisfactory, a test developed by Gart (1969) can be employed to evaluate a set of data for order effects.

The **Gart test for order effects** is employed to evaluate a design with two dependent samples involving categorical data to determine whether or not the order of presentation of two treatments influences subjects' responses to the treatments. For small sample sizes the **Gart test for order effects** involves the use of the **Fisher exact test (Test 16c),** and for large sample sizes the test involves the use of the **chi-square test for** $r \times c$ **tables**. In both instances the **Gart test for order effects** involves the analysis of two 2×2 contingency tables which summarize the responses of subjects in relation to the differential treatments, as well as with respect to the order of presentation of the treatments. The examples to be employed below to illustrate the **Gart test for order effects** will assume a reasonably large sample size, and consequently will be evaluated through use of the **chi-square test for** $r \times c$ **tables**. In employing the latter test a correction for continuity will not be employed. (If a correction for continuity is employed, the chi-square values in this section will be slightly lower than the values computed.)

The **Gart test for order effects** will initially be employed to evaluate Example 20.3, which illustrates the use of a **crossover design** (which is discussed in Section VII of the *t* **test for two dependent samples** as well as in Section IX (the **Addendum**) of the **between-subjects factorial analysis of variance (Test 27)** under the discussion of the **factorial analysis of variance for a mixed design (Test 27i)**. The reader should take note of the fact that the numerical values employed in Example 20.3 (which are identical to those used in Table 20.2) will only be employed in demonstrating the **Gart test** in reference to the data displayed in Tables 20.6 and 20.7. In Tables 20.8 and 20.9 at this end of this section, different numerical values (than those used in Example 20.3/Table 20.2) will be employed in illustrating the use of the **Gart test**. It should also be noted that, ideally, in a two-treatment **crossover design** there will be an equal number of subjects initially assigned to each of the two different presentation orders. Due to practical and ethical issues, however, it is often not possible to achieve the latter. Even in instances where the number of subjects assigned to the different presentation orders is equal at the beginning of a study, due to differential subject loss, more often than not it is no longer the case at the conclusion of the study.

Example 20.3 *A clinical trial is conducted in order to evaluate the efficacy of a drug for treating attention deficit disorder relative to the impact of a placebo. In a double blind study, 54 children are administered the drug for a duration of three months and a placebo for a duration of three months. Twenty-six of the 54 children receive the drug as their first treatment followed by the placebo, while the other 28 children receive the placebo first followed by the drug. To insure that there are no carryover effects from one treatment to another, there is a one month washout period at the conclusion of the first treatment a child receives. The dependent variable in the study is a parent's judgement with respect to whether or not a child improves under each of the two treatment conditions.*

Unlike the 2×2 data summary table employed for the **McNemar test**, the 2×2 data summary tables employed for the **Gart test for order effects** only document the data of those subjects who respond favorably to either the placebo or the drug, but not both treatments (i.e., subjects in cells *b* and *c* of Table 20.2). The following two analyses are conducted within the framework of the **Gart test**: a) An analysis is conducted on a 2×2 contingency table in which the row variable represents a favorable response to only the first of the two treatments administered versus a favorable response to only the second of the two treatments administered. The column variable represents the order *DP* with respect to the presentation of the two treatments versus the order *PD* with respect to the presentation of the two treatments (assuming *P* and *D* respectively represent the placebo versus drug treatments). The analysis of this 2×2

contingency table informs the researcher whether or not there is a significant treatment effect. A statistically significant result indicates that a significant treatment effect is present in the data; b) An analysis is conducted on a 2 × 2 contingency table in which the row variable represents a favorable response only to the drug versus a favorable response only to the placebo. The column variable represents the order *DP* with respect to the presentation of the two treatments versus the order *PD* with respect to the presentation of the two treatments. The analysis of this 2 × 2 contingency table informs the researcher whether or not the order of presentation of the two treatments influences the responses of subjects. A statistically significant result indicates that a significant order effect is present in the data.

At this point in the discussion it will be assumed that the results of Example 20.3 are summarized in Tables 20.6 and 20.7 (both of which employ the data in Table 20.2). The **Gart test for order effects** will be employed to evaluate the data in both of the aforementioned tables. Table 20.6 will be employed to illustrate a set of data which results in a significant treatment effect and a nonsignificant order effect. Table 20.7 will be employed to illustrate a set of data which results in both a significant treatment effect and a significant order effect. Later in this section Tables 20.8 and 20.9 will present a different set of data which results in a nonsignificant treatment effect but a significant order effect.

The null hypothesis and nondirectional alternative hypothesis evaluated with part (a) of all of the tables to be presented in describing the **Gart test for order effects** are as follows:

Null hypothesis H_0: There is no treatment effect in the underlying populations represented by the data. (The null hypothesis can also be stated as follows: There is no difference between the two treatments with respect to the number of subjects who respond favorably.)

Alternative hypothesis H_1: There is a treatment effect in the underlying population represented by the data. (The alternative hypothesis can also be stated as follows: There is a difference between the two treatments with respect to the number of subjects who respond favorably.)

The null and alternative hypotheses evaluated with part (b) of all the tables to be presented in describing the **Gart test for order effects** are as follows:

Null hypothesis H_0: There is no order effect with respect to the presentation of the two treatments in the underlying populations represented by the data. (The null hypothesis can also be stated as follows: The order of presentation of the two treatments does not affect the number of subjects who respond favorably.)

Alternative hypothesis H_1: There is an order effect with respect to the presentation of the two treatments in the underlying population represented by the data. (The order of presentation of the two treatments does affect the number of subjects who respond favorably.)

Tables 20.6 and 20.7 are based on the responses of the 54 subjects in cells *b* and *c* of Table 20.2. Thirteen of those 54 subjects (i.e., the subjects in cell *b*) responded favorably to the placebo but not to the drug, while 41 of those 54 subjects responded favorably to the drug but not to the placebo.

In the case of Table 20.6 (which is comprised of two parts: (a) and (b)), for purposes of illustration, it will be assumed that 27 of the 54 subjects responded to only the first of the two treatments, while the remaining 27 subjects responded to only the second of the two treatments. The cell entries in Table 20.6a indicate the following. Of the 27 subjects who responded to the first of the two treatments, 20 responded to the drug since the order of presentation of the two treatments for those 20 subjects was *DP*. The latter 20 subjects are represented in cell *a* of Table 20.6a. Of the 27 subjects who responded to the first of the two treatments, 7 responded to the placebo since the order of presentation of the two treatments for those 7 subjects was *PD*. The latter 7 subjects are represented in cell *b* of Table 20.6a. Of the 27 subjects who responded to the second of the two treatments, 6 responded to the placebo since the order of presentation of the two treatments for those 6 subjects was *DP*. The latter 6 subjects are represented in cell *c* of Table 20.6a. Of the 27 subjects who responded to the second of the two treatments, 21 responded to the drug since the order of presentation of the two treatments for those 21 subjects was *PD*. The latter 21 subjects are represented in cell *d* of Table 20.6a.

The cell entries in Table 20.6b indicate the following: Of the 41 subjects who responded to the drug, 20 responded to the drug when it was the first of the two treatment conditions, since the order of presentation of the two treatments for those 20 subjects was *DP*. The latter 20 subjects are represented in cell *a* of Table 20.6b. Of the 41 subjects who responded to the drug, 21 responded to the drug when it was the second of the two treatments, since the order of presentation of the two treatments for those 21 subjects was *PD*. The latter 21 subjects are represented in cell *b* of Table 20.6b. Of the 13 subjects who responded to the placebo, 6 responded to the placebo when it was the second of the two treatments, since the order of presentation of the two treatments for those 6 subjects was *DP*. The latter 6 subjects are represented in cell *c* of Table 20.6b. Of the 13 subjects who responded to the placebo, 7 responded to the placebo when it was the first of the two treatments, since the order of presentation of the two treatments for those 7 subjects was *PD*. The latter 7 subjects are represented in cell *d* of Table 20.6b.

The data in Tables 20.6 (a) and (b) can be evaluated with the **Fisher exact test** (which yields an exact probability value) or the **chi-square test for *r* × *c* tables**.(which provides an approximation of the exact **Fisher exact test** probability). Since for $n = 54$ subjects the **chi-square test** provides an excellent approximation of the exact probability, the latter test will be employed in all of the computations to follow. Note that the **Fisher exact test** or **chi-square test** is employed to evaluate the **Gart test for order effects** data tables, since the response of each of the 54 subjects summarized in each **Gart test** table is intended to only summarize one response per subject (rather than two responses per subject, as is the case with the data evaluated in Table 20.2 with the **McNemar test**).

Employing Equation 16.5 (which yields the same result as Equation 16.2) (both of which are the equations for the **chi-square test for *r* × *c* tables**), the value $\chi^2 = 14.54$ is computed for Table 20.6a.

$$\chi^2 = \frac{54\,[(20)(21) - (7)(6)]^2}{(20 + 7)(6 + 21)(20 + 6)(7 + 21)} = 14.54$$

Equation 16.3 is employed to compute the degrees of freedom, yielding the value $df = (2 - 1)(2 - 1) = 1$). The tabled critical .05 and .01 chi-square values in **Table A4** for $df = 1$ are $\chi^2_{.05} = 3.84$ and $\chi^2_{.01} = 6.63$. Since the computed value $\chi^2 = 14.54$ is larger than both of the aforementioned critical values, the result is significant at both the .05 and .01 levels. Thus, we can conclude there is a significant treatment effect present in the data — specifically, that a larger number of subjects exhibit a favorable response to the drug than to the placebo (which in Table 20.6a is a function of the difference in the cell frequencies for cells *a* and *d* versus cells *b* and *c*).

Table 20.6   Summary of Data for Gart Test for Order Effects

Table 20.6a   Table for Evaluating Treatment Effect

		Order of presentation of treatments		
		DP	PD	Row sums
Outcome	Favorable response to first treatment	20	7	27
	Favorable response to second treatment	6	21	27
	Column sums	26	28	54

Table 20.6b   Table for Evaluating Order Effect

		Order of presentation of treatments		
		DP	PD	Row sums
Outcome	Favorable response to drug	20	21	41
	Favorable response to placebo	6	7	13
	Column sums	26	28	54

Employing Equation 16.5 (or 16.2), the value $\chi^2 = .03$ is computed for Table 20.6b.

$$\chi^2 = \frac{54\,[\,(20)(7) - (21)(6)\,]^2}{(20 + 21)(6 + 7)(20 + 6)(21 + 7)} = .03$$

As is the case for Table 20.6a, $df = 1$, $\chi^2_{.05} = 3.84$, and $\chi^2_{.01} = 6.63$. Since the computed value $\chi^2 = .03$ is less than $\chi^2_{.05} = 3.84$, we can conclude there is not a significant order effect present in the data. In other words, the order of presentation of the two treatments has no effect on how subjects responded. The lack of an order effect is revealed in Table 20.6b in that approximately the same number of subjects responded favorably to the drug in the *DP* order of presentation as in the *PD* order of presentation (i.e., cells *a* and *b*). In the same respect, approximately the same number of subjects responded favorably to the placebo in the *DP* order of presentation as in the *PD* order of presentation (i.e., cells *c* and *d*).

As noted earlier, Table 20.7 will be employed to illustrate a set of data which results in both a significant treatment effect and a significant order effect. In the case of Table 20.7 (which is comprised of two parts: (a) and (b)), for purposes of illustration, it will be assumed that 12 of the 54 subjects responded to only the first of the two treatments, while the remaining 42 subjects responded to only the second of the two treatments. The cell entries in Table 20.7a indicate the following. Of the 12 subjects who responded to the first of the two treatments, 10 responded to the drug, since the order of presentation of the two treatments for those 10 subjects was *DP*. The latter 10 subjects are represented in cell *a* of Table 20.7a. Of the 12 subjects who responded to the first of the two treatments, 2 responded to the placebo, since the order of presentation of the two treatments for those 2 subjects was *PD*. The latter 2 subjects are represented in cell *b* of

Table 20.7a. Of the 42 subjects who responded to the second of the two treatments, 11 responded to the placebo, since the order of presentation of the two treatments for those 11 subjects was *DP*. The latter 11 subjects are represented in cell *c* of Table 20.7a. Of the 42 subjects who responded to the second of the two treatments, 31 responded to the drug, since the order of presentation of the two treatments for those 31 subjects was *PD*. The latter 31 subjects are represented in cell *d* of Table 20.7a.

The cell entries in Table 20.7b indicate the following: Of the 41 subjects who responded to the drug, 10 responded to the drug when it was the first of the two treatments, since the order of presentation of the two treatments for those 10 subjects was *DP*. The latter 10 subjects are represented in cell *a* of Table 20.7b. Of the 41 subjects who responded to the drug, 31 responded to the drug when it was the second of the two treatments, since the order of presentation of the two treatments for those 31 subjects was *PD*. The latter 31 subjects are represented in cell *b* of Table 20.7b. Of the 13 subjects who responded to the placebo, 11 responded to the placebo when it was the second of the two treatments, since the order of presentation of the two treatments for those 11 subjects was *DP*. The latter 11 subjects are represented in cell *c* of Table 20.7b. Of the 13 subjects who responded to the placebo, 2 responded to the placebo when it was the first of the two treatments, since the order of presentation of the two treatments for those 2 subjects was *PD*. The latter 2 subjects are represented in cell *d* of Table 20.7b.

Employing Equation 16.5 (or 16.2), the value $\chi^2 = 12.82$ is computed for Table 20.7a.

$$\chi^2 = \frac{54\,[\,(10)(31) - (2)(11)\,]^2}{(10 + 2)(11 + 31)(10 + 11)(2 + 31)} = 12.82$$

Table 20.7   Summary of Data for Gart Test for Order Effects

Table 20.7a   Table for Evaluating Treatment Effect

		Order of presentation of treatments		
		DP	PD	Row sums
Outcome	Favorable response to first treatment	10	2	12
	Favorable response to second treatment	11	31	42
	Column sums	21	33	54

Table 20.7b   Table for Evaluating Order Effect

		Order of presentation of treatments		
		DP	PD	Row sums
Outcome	Favorable response to drug	10	31	41
	Favorable response to placebo	11	2	13
	Column sums	21	33	54

As is the case for Table 20.6, the values $df = 1$, $\chi^2_{.05} = 3.84$, and $\chi^2_{.01} = 6.63$ are employed to evaluate the computed chi-square value. Since the computed value $\chi^2 = 12.82$ is larger than both of the aforementioned critical values, we can conclude there is a significant treatment effect present in the data — specifically, that a larger number of subjects exhibit a favorable response to the drug than to the placebo (which is a function of the difference in the cell frequencies for cells a and d versus cells b and c).

Employing Equation 16.5 (or 16.2), the value $\chi^2 = 15.06$ is computed for Table 20.6b.

$$\chi^2 = \frac{54\,[\,(10)(2) - (31)(11)\,]^2}{(10 + 31)(11 + 2)(10 + 11)(31 + 2)} = 15.06$$

As is the case for Table 20.7a, $df = 1$, and thus $\chi^2_{.05} = 3.84$, and $\chi^2_{.01} = 6.63$. Since the computed value $\chi^2 = 15.06$ is larger than both of the aforementioned critical values, the result is significant at both the .05 and .01 levels. Thus, we can conclude there is a significant order effect present in the data. In other words, the order of presentation of the two treatments appears to have an effect on how subjects responded. The presence of an order effect is revealed in Table 20.7b in that approximately three times as many subjects responded favorably to the drug in the *PD* order of presentation as opposed to the *DP* order of presentation (i.e., cells b and a). In the case of the placebo, approximately five times the number of subjects responded favorably to the placebo in the *DP* order of presentation as opposed to the *PD* order of presentation (i.e., cells c and d). Thus, it would appear that regardless of whether one is using the drug or the placebo, the second of the two treatments administered is more likely to elicit a favorable response than the first treatment administered. In spite of the latter, however, the drug does appear to be more efficacious than the placebo.

Earlier in this section it was noted that Tables 20.8 and 20.9 will be employed to illustrate a set of data which result in a nonsignificant treatment effect but a significant order effect. In order to demonstrate the latter, Table 20.8 (which is a modified **McNemar test** summary table for Example 20.3) is presented below. Note that the difference between Table 20.8 and Table 20.2 is the frequencies for cells b and c, which in Table 20.8 are respectively 22 and 32. The latter two entries indicate that 54 subjects responded to only one of the two treatments, and 22 of those subjects responded to the placebo but not the drug, while 32 of the subjects responded to the drug but not the placebo.

A **McNemar test** will now be employed to evaluate the data summarized in Table 20.8. Substituting the appropriate values in Equation 20.1, the value $\chi^2 = 1.85$ is computed for the test statistic.

$$\chi^2 = \frac{(22 - 32)^2}{22 + 32} = 1.85$$

Since the computed value $\chi^2 = 1.85$ is less than the tabled critical values (for $df = 1$) $\chi^2_{.05} = 3.84$, and $\chi^2_{.01} = 6.63$, the result is not significant. Thus, there is no evidence for a treatment

Table 20.8 Modified McNemar Test Summary Table for Example 20.3

		Favorable response to drug		
		Yes	No	Row sums
Favorable response	Yes	10	22	32
to placebo	No	32	36	68
	Column sums	42	58	100

Table 20.9   Summary of Data for Gart Test for Order Effects

Table 20.9a   Table for Evaluating Treatment Effect

| | | Order of presentation of treatments | | |
		DP	PD	Row sums
Outcome	**Favorable response to first treatment**	26	17	43
	Favorable response to second treatment	5	6	11
	Column sums	31	23	54

Table 20.9b   Table for Evaluating Order Effect

| | | Order of presentation of treatments | | |
		DP	PD	Row sums
Outcome	**Favorable response to drug**	26	6	32
	Favorable response to placebo	5	17	22
	Column sums	31	23	54

effect. Use of the **Gart test for order effects** will, however, reveal that there is a significant order effect present in the data. In order to demonstrate the latter, consider Table 20.9 which provides us with more information on the hypothetical study summarized in Table 20.8.

In Table 20.9 it will be assumed that 31 of the 54 subjects in cells *b* and *c* of Table 20.8 were administered the treatments in the order *DP*, while the other 23 subjects were administered the treatments in the order *PD*. Inspection of Table 20.9a reveals that 43 of the 54 subjects responded to only the first of the two treatments, while the remaining 11 subjects responded to only the second of the two treatments. The cell entries in Table 20.9a indicate the following. Of the 43 subjects who responded to the first of the two treatments, 26 responded to the drug, since the order of presentation of the two treatments for those 26 subjects was *DP*. The latter 26 subjects are represented in cell *a* of Table 20.9a. Of the 43 subjects who responded to the first of the two treatments, 17 responded to the placebo, since the order of presentation of the two treatments for those 17 subjects was *PD*. The latter 17 subjects are represented in cell *b* of Table 20.9a. Of the 11 subjects who responded to the second of the two treatments, 5 responded to the placebo, since the order of presentation of the two treatments for those 5 subjects was *DP*. The latter 5 subjects are represented in cell *c* of Table 20.9a. Of the 11 subjects who responded to the second of the two treatments, 6 responded to the drug, since the order of presentation of the two treatments for those 6 subjects was *PD*. The latter 6 subjects are represented in cell *d* of Table 20.9a.

The cell entries in Table 20.9b indicate the following: Of the 32 subjects who responded to the drug, 26 responded to the drug when it was the first of the two treatments, since the order of presentation of the two treatments for those 26 subjects was *DP*. The latter 26 subjects are represented in cell a of Table 20.9b. Of the 32 subjects who responded to the drug, 6 responded to the drug when it was the second of the two treatments, since the order of presentation of the two treatments for those 6 subjects was *PD*. The latter 6 subjects are represented in cell b of Table 20.9b. Of the 22 subjects who responded to the placebo, 5 responded to the placebo when it was the second of the two treatments, since the order of presentation of the two treatments for those 5 subjects was *DP*. The latter 5 subjects are represented in cell c of Table 20.9b. Of the 22 subjects who responded to the placebo, 17 responded to the placebo when it was the first of the two treatments, since the order of presentation of the two treatments for those 17 subjects was *PD*. The latter 17 subjects are represented in cell d of Table 20.9b.

Employing Equation 16.5 (or 16.2), the value $\chi^2 = .81$ is computed for Table 20.9a.

$$\chi^2 = \frac{54\,[\,(26)(6)\, -\, (17)(5)\,]^2}{(26\, +\, 17)(5\, +\, 6)(26\, +\, 5)(17\, +\, 6)} = .81$$

As is the case for the previous tables discussed in this section, the values $df = 1$, $\chi^2_{.05} = 3.84$, and $\chi^2_{.01} = 6.63$ are employed to evaluate the computed chi-square value. Since the computed value $\chi^2 = .81$ is less than $\chi^2_{.05} = 3.84$, we can conclude there is not a significant treatment effect present in the data. Although a larger number of subjects exhibit a favorable response to the drug than the placebo (which is a function of the difference in the cell frequencies for cells a and d versus cells b and c), the magnitude of the difference is not significant.

Employing Equation 16.5 (or 16.2), the value $\chi^2 = 18.26$ is computed for Table 20.9b.

$$\chi^2 = \frac{54\,[\,(26)(17)\, -\, (6)(5)\,]^2}{(26\, +\, 6)(5\, +\, 17)(26\, +\, 5)(6\, +\, 17)} = 18.26$$

Since the computed value $\chi^2 = 18.26$ is larger than the tabled critical values (for $df = 1$) $\chi^2_{.05} = 3.84$, and $\chi^2_{.01} = 6.63$, the result is significant at both the .05 and .01 levels. Thus, we can conclude there is a significant order effect present in the data. In other words, the order of presentation of the two treatments appears to have an effect on how subjects responded. The presence of an order effect is revealed in Table 20.9b in that more than four times as many subjects responded favorably to the drug in the *DP* order of presentation as opposed to the *PD* order of presentation (i.e., cells a and b). In the case of the placebo, more than three times as many subjects responded favorably to the placebo in the *PD* order of presentation as opposed to the *DP* order of presentation (i.e., cells c and d). Thus, it would appear that regardless of whether one is using the drug or a placebo, the first of the two treatments presented is more likely to elicit a favorable response than is the second treatment.

Other sources which discuss the **Gart test for order effects** are Everitt (1977, pp. 22–26; 1992), Fleiss (1981, pp. 104–105), Fleiss *et al.* (2003, p. 164), Leach (1979, pp. 136–140) and Zar (1999, pp. 173–175). Leach (1979, pp. 139–140) also describes the extension of the **Gart test for order effects** to designs involving ordinal data.

VII. Additional Discussion of the McNemar Test

1. Alternative format for the McNemar test summary table and modified test equation Although in this book Cells b and c are designated as the two cells in which subjects are inconsistent with respect to their response categories, the **McNemar test** summary table can be rearranged so that Cells a and d become the relevant cells. Table 20.10 represents such a rearrangement with respect to the response categories employed in Examples 20.1/20.2.

Table 20.10 Alternative Format for Summary of Data for Examples 20.1 and 20.2

		Favorable response to drug/Posttest		
		Yes/Anti	No/Pro	Row sums
Favorable response to placebo/Pretest	No/Pro	$a = 41$	$b = 36$	77
	Yes/Anti	$c = 10$	$d = 13$	23
	Column sums	51	49	100

Note that in Table 20.10 Cells a and d are the key cells, since subjects who are inconsistent with respect to response categories are represented in these two cells. If Table 20.10 is employed as the summary table for the **McNemar test**, Cells a and d will, respectively, replace Cells c and b in stating the null and alternative hypotheses. In addition, Equations 20.13 and 20.14 will, respectively, be employed in place of Equations 20.1 and 20.2. When the appropriate values are substituted in the aforementioned equations, Equation 20.13 yields the value $\chi^2 = 14.52$ computed with Equation 20.1, and Equation 20.14 yields the value $z = -3.81$ computed with Equation 20.2.

$$\chi^2 = \frac{(d - a)^2}{d + a} = \frac{(13 - 41)^2}{13 + 41} = 14.52 \qquad \textbf{(Equation 20.13)}$$

$$z = \frac{d - a}{\sqrt{d + a}} = \frac{13 - 41}{\sqrt{13 + 41}} = -3.81 \qquad \textbf{(Equation 20.14)}$$

2. The effect of disregarding matching Selvin (2004, pp. 309–311) notes that on occasion a researcher may elect to evaluate data accumulated within the framework of a matched-pairs design within the context of an independent samples design. In computing the test statistic, the latter type of analysis will take into consideration the frequencies of all four cells of the 2×2 contingency table (which, as noted earlier, is not the case with the **McNemar test** which only evaluates data for two of the four cells). Selvin (2004, pp. 310–311) notes that although analysis of data for a matched-pairs design within the context of an independent samples design allows for a more powerful test of the null hypothesis, the resulting **odds ratios** will be biased (since the matching variable is a confounding variable). He further notes, however, that in most instances any differences between the two methods with respect to the obtained chi-square values and **odds ratios** will be unremarkable.

The analysis of matched-pairs data within the context of an independent sample design will be demonstrated in reference to Example 20.1. Table 20.11 represents a reconfiguration of the data for the latter example if it is conceptualized within the framework of an independent samples design. Within the latter context, the marginal sums in Table 20.1 (i.e., the sums of the rows and columns) are employed as the cell frequencies in Table 20.11, which results in a 2×2 contingency table comprised of $n = 200$ scores. Thus, 23 scores indicate a favorable response to the placebo, 51 scores a favorable response to the drug, 77 scores an unfavorable response to the placebo, and 49 scores an unfavorable response to the drug.

Analysis of the data in Table 20.11 with Equation 16.2 yields the value $\chi^2 = 16.92$. (If a correction for continuity is employed, Equation 16.4 yields $\chi^2 = 15.64$.) Since $r = 2$ and $c = 2$, $df = (2 - 1)(2 - 1) = 1$, and for $df = 1$, $\chi^2_{.05} = 3.84$ and $\chi^2_{.01} = 6.63$. Since $\chi^2 = 16.92$ is greater than $\chi^2_{.01} = 6.63$, the null hypothesis can be rejected at both the .05 and .01 levels. Inspection of Table 20.11 reveals that a larger proportion of subjects responded favorably to the drug than the placebo. Note that the above computed chi-square values $\chi^2 = 16.92$ and $\chi^2 = 15.64$ are close to the analogous chi-square values $\chi^2 = 14.52$ and $\chi^2 = 13.5$ computed earlier with the **McNemar test** with Equations 20.1 and 20.3 for the matched-pairs analysis.

Table 20.11   Analysis of Example 20.1 as an Independent Samples Design

	Favorable response	Unfavorable response		Row sums
Drug	51	49		100
Placebo	23	77		100
			Total	
Column sums	74	126	observations	200

Equations 16.27 and 16.28 (which are used to compute the **odds ratio** and **inverse odds ratio** for a 2×2 contingency table involving an independent samples design) are employed below with the data in Table 20.11, yielding the values $o = .29$ and $o' = 3.48$.

$$o = \frac{bc}{ad} = \frac{(49)(23)}{(51)(77)} = .29$$

$$o' = \frac{ad}{bc} = \frac{(51)(77)}{(49)(23)} = 3.48$$

The value $o' = 3.48$ indicates that the odds of a person responding to the drug are 3.48 times greater than the odds of a person responding to the placebo, whereas value $o = .29$ indicates that the odds of a person responding to the placebo are .29 times as large as the odds of a person responding to the drug. Note that the latter **odds ratios** are quite close to the analogous **odds ratio** values 3.15 and .32 computed earlier with Equations 20.10 and 20.11 for the matched data within the context of the **McNemar test**. The latter is consistent with Selvin's (2004, pp. 310–311) observation that in most instances any difference between the matched-pairs versus independent samples analysis will be minimal.

3. Alternative nonparametric procedures for evaluating a design with two dependent samples involving categorical data The **McNemar test** model has been extended by Bowker (1948) (**The Bowker test of internal symmetry (Test 20c)**) and Maxwell (1970) and Stuart (1955, 1957) (the **Stuart–Maxwell test of marginal homogeneity (Test 20d)**) to dependent samples designs in which the dependent variable is a categorical measure which is comprised of more than two categories. The latter two tests are discussed in Section IX (the **Addendum**).

4. Tests of equivalence for two dependent proportions: Test 20b: The Westlake–Schuirmann test of equivalence of two dependent proportions Prior to reading this section the reader should review the discussion of **tests of equivalence** in Section VII of the *t* **test for two independent samples**. In the latter discussion it was noted that within the framework of the **classical hypothesis testing model**, a null hypothesis can never be accepted. In contrast to the **classical hypothesis testing model** (where the alternative hypothesis states that there is a difference between treatments), in a **test of equivalence** the alternative hypothesis states that the treatments are, in fact, equivalent. Conversely, in a **test of equivalence** the null hypothesis states that a difference exists between the treatments (in contrast to a test of significance within the framework of the **classical hypothesis testing model,** where the null hypothesis states that there is no difference between the treatments). Since it is not mathematically feasible to establish an alternative hypothesis which states exact equality (i.e., a difference of zero) between experimental conditions, when one conducts a **test of equivalence,** a parameter (to be designated ζ, which represents the lower case Greek letter **zeta**) is employed to reflect a maximum difference which will be tolerated between two treatments in order that a researcher might conclude that the treatments are equivalent to one another. Any difference less than the value of ζ would be viewed so small as to be inconsequential (i.e., of no practical significance). Thus, in a **test of equivalence** if a difference less than the value of ζ is detected, the null hypothesis (stating that a difference does exist) can be rejected, and the alternative hypothesis (which stipulates equivalence) can be accepted. Since in most instances rejection of the null hypothesis for a **test of equivalence** will involve detection of a substantially smaller effect size than the effect size a researcher is typically concerned with detecting within the framework of the **classical hypothesis testing model,** a **test of equivalence** will usually require greater statistical power, and consequently the use of a larger sample size, when contrasted with tests conducted within the framework of the **classical model.** The smaller the value a researcher designates for ζ, the larger the sample size which will be required in order to reject the null hypothesis and thus conclude the treatments are equivalent.

In this section a methodology attributed to Westlake (1976, 1981, 1988) and Schuirmann (1987) described earlier in the book for evaluating the equivalence of the means of two independents samples (described in Section VII of the *t* **test for two independent samples**), two independent proportions (described in Section VII of the **chi-square test for** *r* × *c* **tables**), and two dependent samples (described in Section VII of the *t* **test for two dependent samples**) will be extended to evaluate the **equivalence of two dependent proportions.** In point of fact, a number of different methods have been suggested for evaluating equivalence in a design involving two dependent proportions, and although the available methods will generally yield consistent results with regard to the issue of equivalence, there may be occasions where the latter is not true. Such inconsistencies can generally be resolved by conducting replication studies (the results of which can be evaluated through the use of **meta-analysis**), and basing one's final conclusions on the overall body of evidence. (**Meta-analysis** is described in Section IX (the **Addendum**) of the **Pearson product-moment correlation coefficient.**) The reader should keep in mind that in the final analysis the criterion employed for equivalence will be some standard imposed by an outside agency (e.g., the FDA), or some alternative subjective criterion stipulated by a specific researcher. Ultimately, the final decision regarding the issue of equivalence should be based on multiple studies which yield a consistent pattern of results involving minimal differences between treatments that fall within a range of values which most practitioners in a discipline would consider to be reasonable.

Descriptions of alternative methodologies for evaluating equivalence of two dependent proportions can be found in Chow and Liu (2004), Dunnet and Gent (1977), Liu *et al.* (2002), Liu and Cumberland (2001), Nam (1997), Selwyn *et al.* (1981), Selwyn and Hall (1984), Tango (1998, 1999), and Wellek (2003).[15] It should be noted that the methodology to be described in this section would not be recommended for analysis of equivalence in a crossover design, unless prior analysis of the data with the **Gart test for order effects** indicated the absence of order effects.

Prior to demonstrating the application of the **Westlake–Schuirmann test of equivalence of two dependent proportions,** the reader is asked to consider Example 20.4, which describes a study in which a researcher is required to determine whether or not there is a significant difference between two treatments. The latter, however, is not commensurate with being asked to establish whether or not two treatments are equivalent to one another. Since the purpose of the study is to merely assess whether or not a statistically significant difference is present, the data for Example 20.4 would be evaluated with the **McNemar test**. (The **Gart test** could be employed if more detailed information were available regarding performance under the different presentation orders.)

Example 20.4 *A study is conducted in order to determine whether or not there is a difference in efficacy with respect to a brand name versus a generic drug in treating a specific disease. In the study 100 people afflicted with the disease are treated with both the brand name and generic drugs. Fifty of the subjects initially receive the brand name drug which they take for six months, after which each subject is categorized with respect to whether he or she exhibited a favorable response to the brand name drug. After a two month washout period, the 50 subjects receive the generic drug for six months, after which each subject is categorized with respect to whether he or she exhibited a favorable response to the generic drug. The remaining 50 subjects are also administered the two treatments but in the reverse order (with the two month washout period intervening between the two treatments). The results of the study are summarized in Table 20.12.*

Table 20.12 Summary of Data for Example 20.4

		Favorable response to drug		
		Yes	No	Row sums
Favorable response	Yes	10	29	39
to placebo	No	42	19	61
	Column sums	52	48	100

When $b = 29$ and $c = 42$ are substituted in Equations 20.1 and 20.2, the values $\chi^2 = 2.38$ and $z = -1.54$ are respectively computed for the **McNemar test** statistic. (The continuity corrected Equations 20.3 and 20.4 yield the values $\chi^2 = 2.03$ and $z = 1.42$.) Since the obtained absolute values $\chi^2 = 2.38$ and $z = 1.54$ (as well as the continuity corrected values $\chi^2 = 2.03$ and $z = 1.42$ computed with Equations 20.3 and 20.4) are less than the tabled critical two-tailed values $\chi^2_{.05} = 3.84$ and $z_{.05} = 1.96$, the nondirectional alternative hypothesis $H_1: \pi_b \neq \pi_c$ is not supported at the .05 level. Additionally, since the obtained absolute values $\chi^2 = 2.38$ and $z = 1.54$ (as well as the continuity corrected values $\chi^2 = 2.03$ and $z = 1.42$) are less than the tabled critical one-tailed values $\chi^2_{.05} = 2.71$ and $z_{.05} = 1.65$, the directional alternative hypothesis $H_1: \pi_b < \pi_c$ is not supported at the .05 level. Thus, regardless of whether one conducts a two- or one-tailed analysis, the null hypothesis $H_0: \pi_b = \pi_c$ cannot be rejected. Consequently, the research cannot conclude there is a difference in efficacy between the two treatments.

$$\chi^2 = \frac{(29 - 42)^2}{29 + 42} = 2.38$$

$$z = \frac{29 - 42}{\sqrt{29 + 42}} = -1.54$$

$$\chi^2 = \frac{(|29 - 42| - 1)^2}{29 + 42} = 2.03$$

$$z = \frac{|29 - 42| - 1}{\sqrt{29 + 42}} = 1.42$$

In Endnote 2 it is noted that if the values p_1 and p_2 (where $p_1 = (a + b)/n$ and $p_2 = (a + c)/n$) are employed to estimate π_1 and π_2, an alternative way of stating the null and alternative hypotheses for the **McNemar test** is as follows: **Null hypothesis:** H_0: $\pi_1 = \pi_2$ (which can also be written as H_0: $\pi_1 - \pi_2 = 0$); **Nonndirectional alternative hypothesis:** H_1: $\pi_1 \neq \pi_2$ (which can also be written as H_1: $\pi_1 - \pi_2 \neq 0$); **Directional alternative hypotheses:** H_1: $\pi_1 > \pi_2$ or H_1: $\pi_1 < \pi_2$ (which can also be written respectively as H_1: $\pi_1 - \pi_2 > 0$ and H_1: $\pi_1 - \pi_2 < 0$). As noted in Endnote 9, Equations 20.18 and 20.19 are alternative equations which employ the latter notation that can be used to compute the **McNemar test** statistic. When the values $p_1 = (10 + 29)/100 = .39$ and $p_2 = (10 + 42)/100 = .52$ are employed in the Equations 20.18 and 20.19, the values $z = -1.54$ and $z = 1.42$ are obtained which are identical to the values previously computed with Equations 20.2 and 20.4. Thus, the null hypothesis H_0: $\pi_1 = \pi_2$ (which is equivalent to H_0: $\pi_b = \pi_c$) cannot be rejected.

$$z = \frac{p_1 - p_2}{\sqrt{\dfrac{b + c}{n^2}}} = \frac{.39 - .52}{\sqrt{\dfrac{29 + 42}{(100)^2}}} = -1.54$$

$$z = \frac{(|p_1 - p_2|) - \dfrac{1}{n}}{\sqrt{\dfrac{b + c}{n^2}}} = \frac{(|.39 - .52|) - \dfrac{1}{100}}{\sqrt{\dfrac{29 + 42}{(100)^2}}} = 1.42$$

As noted earlier, when the result obtained for the **McNemar test** does not achieve statistical significance, the latter does not justify concluding that the two treatments are equivalent to one another. In order to address the issue of equivalence, consider Example 20.5 which employs the same data as Example 20.4.

Example 20.5 *The manufacturer of a generic drug for treating a disease wants to empirically demonstrate it is equivalent to a brand name drug, which up to the present time has been the only approved treatment for the disease. To demonstrate equivalency of the two drugs, a study is conducted in which 100 people afflicted with the disease are treated with both the generic drug (Treatment 1) and the brand name drug (Treatment 2). Fifty of the subjects initially receive the generic drug which they take for six months, after which each subject is categorized with respect to whether he or she exhibited a favorable response to the generic drug. After a two month washout period, the 50 subjects receive the brand name drug for six months, after which each subject is categorized with respect to whether he or she exhibited a favorable response to the brand name drug. The remaining 50 subjects are also administered the two treatments but in the reverse order (with the two month washout period intervening between the two treatments). The results of the study are summarized in Table 20.12. Do the data indicate that the generic drug is equivalent to the brand name drug if the FDA is willing to declare equivalency if the difference between the treatment proportions is less than 10%?*

The information provided in Example 20.5 allows for either a two- or one-tailed **test of equivalence**. If a researcher employs as a criterion for equivalence that the success rate of the brand name drug is no more than 10% greater than the success rate of the generic drug, as well as the requirement that success rate of the generic drug is no more than 10% greater than the success rate of the brand name drug, a two-tailed analysis would be conducted. On the other hand, if the researcher only employs as a criterion for equivalence that the success rate of the brand name drug is not more than 10% greater than the success rate of the generic drug, a one-tailed analysis would be conducted.

Since 10% expressed as a proportion is equivalent to .10, the value $\zeta = .1$ will be employed in the test of equivalence. The two-tailed analysis can be represented by Figure 20.1 where the values employed for ζ_1 and ζ_2 are respectively $-.10$ and $.10$ (which represent a difference between the two treatment proportions). The two-tailed analysis depicted in Figure 20.1 is symmetrical since the absolute values of ζ_1 (which is equal to $-\zeta$) and ζ_2 (which is equal to ζ) are equivalent to one another. As noted in the discussion of **tests of equivalence** in Section VII of the *t* **test for two independent samples**, it is also possible to conduct a nonsymmetrical two-tailed analysis, in which case the absolute values of ζ_1 and ζ_2 are different.

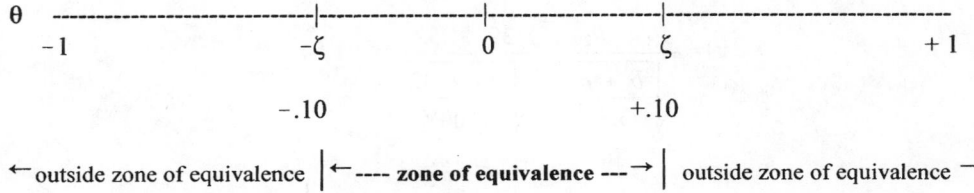

Figure 20.1 Symmetrical Analysis for Equivalence (Two-Tailed)

A one-tailed analysis would be represented by Figure 20.2, and thus the value $-.10$ is employed for ζ_1, while no value is designated for ζ_2. Note that the value $\zeta_1 = -.10$ is employed as the boundary for the zone of equivalence since if the brand name drug is superior to the generic drug $p_1 - p_2$ will be a negative value that is less than $-.10$.

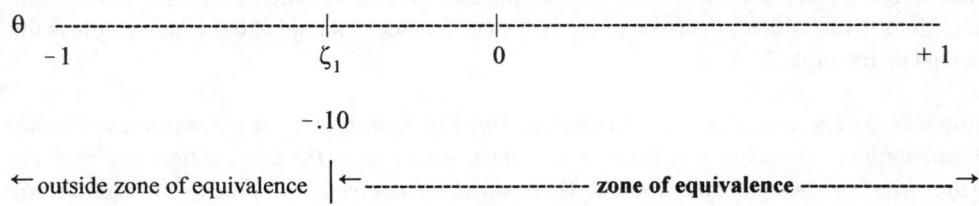

Figure 20.2 Directional Test of Equivalence (One-Tailed)

In the case of Example 20.5 we will assume that the researcher elects to conduct a two-tailed analysis with alpha set equal to .05. The null and alternative hypotheses for a **two-tailed analysis** are noted below. The notation θ (which is the lower case Greek letter **theta**) in the null and alternative hypotheses represents the difference between the two proportions in the underlying populations. It is assumed that the estimate of that difference is based on subtracting the proportion of successes for the brand name drug (p_2) from the proportion of successes for the generic drug (p_1).

Null hypotheses H_0: $\theta \leq -.10$ or H_0: $\theta \geq .10$

(In the underlying populations the samples represent, the difference between the two treatment proportions is less than or equal to $-.10$ or is greater than or equal to $.10$. This is the **hypothesis of nonequivalence**.)

Alternative hypotheses H_1: $\theta > -.10$ and H_1: $\theta < .10$

(In the underlying populations the samples represent, the difference between the two treatment proportions is greater than $-.10$ and is less than $.10$. This is the **hypothesis of equivalence**.)

The **Westlake–Schuirmann test of equivalence of two dependent proportions** employs Equation 20.15 to evaluate the data for Example 20.5. Except for the addition of the term ζ, the numerator of Equation 20.15 is identical to the numerator of Equation 20.18. The value employed for ζ will be the appropriate value employed for ζ_1 or ζ_2 in the analysis being conducted. The denominator of Equation 20.15 is the **standard error** $s_{p_1 - p_2}$ computed with Equation 20.9. In the discussion of the computation of a confidence interval for the **McNemar test** in Section VI it was noted that the value $\sqrt{(b + c)/n^2}$ employed as the standard error in Equation 20.18 does not take into account the frequencies in cells a and d, and consequently does not provide the most accurate estimate of the standard error for computing a confidence interval. The standard error computed with Equation 20.9 (which does take into account the frequencies in cells a and d) is the appropriate measure of variability to employ in computing the test statistic in assessing equivalence.[16]

$$z = \frac{(p_1 - p_2) - \zeta}{\sqrt{\dfrac{(a + d)(b + c) + 4bc}{n^3}}} \qquad \textbf{(Equation 20.15)}$$

As was the case for a two-tailed analysis when the **Westlake–Schuirmann test of equivalence of two independent treatments** was employed in Section VII of the *t* **test for two independent samples**, the **Westlake–Schuirmann test of equivalence of two dependent proportions** also requires that **two one-tailed tests** be conducted. Each of the one-tailed tests employs Equation 20.15. Since we will employ alpha = .05, the tabled critical one-tailed value $z_{.05} = 1.65$ will be used to evaluate the results obtained with Equation 20.15. In order for the researcher to conclude that the two treatments are equivalent, **both** of the z **tests** that are conducted must be statistically significant, and the two z values must have opposite signs.

For Example 20.5, we once again compute the values $p_1 = (10 + 29)/100 = .39$ and $p_2 = (10 + 42)/100 = .52$ which will be employed in Equation 20.15. Employing the latter equation, the following null and alternative hypotheses will be evaluated first: H_0: $\theta \geq .10$ versus H_1: $\theta < .10$. Note that in this analysis the value $\zeta_2 = .10$ is employed in the numerator of Equation 20.15, since we are evaluating the right boundary for equivalence depicted in Figure 20.1.

$$z = \frac{(.39 - .52) - (.10)}{\sqrt{\dfrac{(10 + 19)(29 + 42) + (4)(29)(42)}{(100)^3}}} = \frac{-.23}{.083} \cong -2.77$$

Since the obtained absolute value $z = 2.77$ is greater than the tabled critical one-tailed value $z_{.05} = 1.65$, the null hypothesis H_0: $\theta \geq .10$ designating nonequivalence can be rejected.

(If the correction for continuity (i.e., $- 1/n$) is employed in the numerator of the above equation, $1/100 = .01$ is subtracted from the numerator of the equation yielding the absolute value $z = 2.65$). The probability value associated with this result (with or without the correction for continuity) is well below the .05 value that was prestipulated for alpha. Thus, the first of the null hypotheses designating nonequivalence can be rejected. However, it once again should be emphasized that in order to declare equivalence we must reject **both** of the null hypotheses designating nonequivalence, and the two z values computed must have opposite signs. In reference to the z **test** conducted above, the data are consistent with the alternative hypotheses H_1: $\theta < .10$, which stipulates that when the proportion of successes for the **brand name drug** is subtracted from the proportion of successes for the **generic drug** the difference is less than .10 (which translates into the fact that if the proportion of successes for the **generic name drug** is greater than the proportion of successes for the **brand name drug**, it is not greater than the latter value by more than .10).

We now evaluate the second null hypothesis designating nonequivalence. Specifically, a one-tailed test which evaluates the following null and alternative hypotheses is conducted: H_0: $\theta \le -.10$ versus H_1: $\theta > -.10$. Note that this analysis employs the same values as in the previous analysis, except for the fact that the value $\zeta_1 = -.10$ is employed in the numerator of Equation 20.15, since we are evaluating the left boundary for equivalence depicted in Figure 20.1.

$$z = \frac{(.39 - .52) - (-.10)}{\sqrt{\dfrac{(10 + 19)(29 + 42) + (4)(29)(42)}{(100)^3}}} = \frac{-.03}{.083} \cong -.36$$

Since the obtained absolute value $z = .36$ is less than the tabled critical one-tailed value $z_{.05} = 1.65$, the null hypothesis H_0: $\theta \le -.10$ designating nonequivalence cannot be rejected. (If the correction for continuity (i.e., $- 1/n$) is employed in the numerator of the above equation, .01 is subtracted from the numerator of the equation yielding the absolute value $z = .24$). The probability value associated with this result (with or without the correction for continuity) is well above the .05 value which was prestipulated for alpha. Thus, the second of the null hypotheses designating nonequivalence cannot be rejected. It was noted above that in order to declare equivalence **both** of the null hypotheses designating nonequivalence must be rejected. Since the null hypothesis H_0: $\theta \le -.10$ evaluated above cannot be rejected, we cannot declare that the two drugs are equivalent to one another. In reference to the z **test** conducted above, the data are consistent with the null hypotheses H_0: $\theta \le -.10$, which stipulates that when the proportion of successes for the **brand name drug** is subtracted from the proportion of successes for the **generic drug** the difference is less than or equal to $-.10$ (which translates into the fact that if the proportion of successes for the **brand name drug** is greater than the proportion of successes for the **generic name drug**, it is equal to or greater than .10).

Rogers *et al.* (1993, pp. 554–555) note that when the value of alpha is set equal to .05 for each of the two analyses conducted above, the equivalency confidence interval is set at the $(1 -2\alpha)$ (i.e., 90%) level of certainty/confidence rather than at the traditional $(1 - \alpha)$ (i.e., 95%) level of certainty/confidence. When both null hypotheses which are evaluated are rejected, the values employed for both ζ_1 and ζ_2 will fall outside of the $(1 -2\alpha)$ % confidence interval. If either of the null hypotheses is retained, the value employed for ζ_1 or ζ_2 will fall within the limits of the confidence interval. **To be more specific, if equivalence is to be established, the values for both ζ_1 and ζ_2 must fall outside the limits of the confidence interval, with one of the values ζ_1 positioned to the left of the lower limit of the confidence interval and the other value ζ_2 positioned to the right of the upper limit of the confidence interval. Note,** however, that if both ζ_1 and ζ_2 fall outside the limits of the confidence interval, and both ζ_1 and ζ_2 are below the lower limit of the confidence interval or both ζ_1 and ζ_2 are above the upper limit of the confidence interval, it would indicate a lack of equivalence. In the case of Example 20.5, in order to compute the 90% confidence interval, the tabled critical two-tailed .10 z value (which corresponds to the tabled critical one-tailed .05 z value) is employed in Equation 20.8. Note that in computing the confidence interval Equation 20.9 is employed to compute $s_{p_1 - p_2}$. Employing Equation 20.8, the 90% confidence interval for Example 20.5 is computed below, and visually represented in Figure 20.3.

$$p_1 - p_2 = .39 - .52 = -.13 \qquad s_{p_1 - p_2} = \sqrt{\frac{(10 + 19)(29 + 42) + (4)(29)(42)}{(100)^3}} = .083$$

$$CI_{.90} = (p_1 - p_2) \pm (z_{.10})(s_{p_1 - p_2}) = .-.13 \pm (1.65)(.083) = -.13 \pm .137$$

$$-.267 \leq (\pi_1 - \pi_2) \leq .007$$

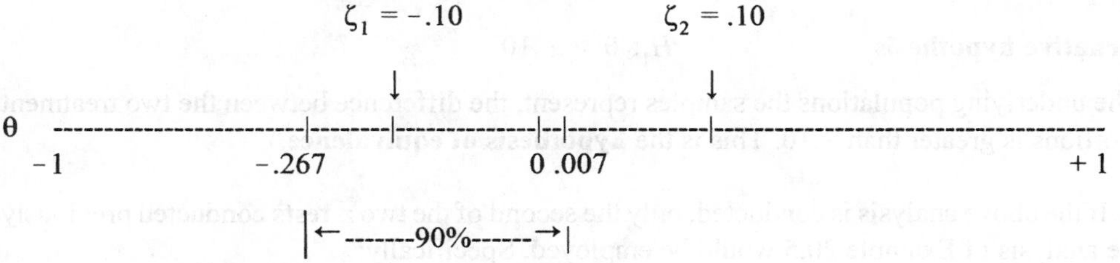

$$\zeta_1 = -.10 \qquad\qquad \zeta_2 = .10$$

$$\downarrow \qquad\qquad\qquad \downarrow$$

θ ----------------------------|--------|------------|--|------------|------------------------------------

-1 $-.267$ 0 $.007$ $+1$

$|\leftarrow$ -------90%------ $\rightarrow|$

Figure 20.3 90% Confidence Interval For Example 20.5 when $\zeta = .10$

This result indicates that the researcher can be 90% confident (or the probability is .90) that the interval $-.267$ to $.007$ contains the true difference between the population proportions. If the aforementioned interval contains the true difference between the population proportions, the proportion of successes for the population the **generic drug** represents is not less than the proportion of successes for the population the **brand name drug** represents than the proportion value .267. Additionally, the proportion of successes for the population the **generic drug** represents is not more than the proportion of successes for the population the **brand name drug** represents than the proportion value .007. The result can also be written as $-.007 \leq (\pi_2 - \pi_1)$ $\leq .267$, which indicates that if the aforementioned interval contains the true difference between the population proportions, the proportion of successes for the population the **brand name drug** represents is larger than the proportion of successes for the population the **generic drug** represents by a proportion value that is not greater than .267. Additionally, the proportion of successes for the population the **brand name drug** represents is not less than the proportion of successes for the population the **generic drug** represents than the proportion value .007. Note that although the value $\zeta_2 = .10$ falls outside of the computed confidence interval, the value ζ_1 $= -.10$ falls inside it. Equivalency cannot be established, since the lower range of the confidence interval falls below the minimum tolerable difference between the proportions for the two treatments. In order for equivalency to have been established, the result for the second analysis would had to have been statistically significant at the .05 level, and if the latter had occurred the lower limit for the confidence interval would have been some value greater than $-.10$.

The reader should take note of the fact that just by employing Equation 20.15, as was done earlier in this section to compute two z values, one can obtain the answer to the question of whether or not two treatments are equivalent. Alternatively, computation of the above confidence interval, in and of itself, will provide the answer to the same question. Use of both methods merely provides the same answer to the question in different formats.

It should be noted that many researchers would elect to employ a one-tailed analysis for the data in Example 20.5. A one-tailed analysis will always provide a less conservative **test of equivalence** than will a two-tailed analysis — i.e., for a given level of alpha, it will always be easier to declare equivalence when a one-tailed analysis is employed as opposed to a two-tailed analysis. If a one-tailed analysis were employed for Example 20.5 it would stipulate as the alternative hypothesis that the proportion of successes for the **brand name drug** is not more than a proportion equal to .10 greater than the proportion of successes for the **generic drug**. In such a case a researcher would conduct only a single **one-tailed test** which employed the null and alternative hypotheses noted below.

Null hypothesis H_0: $\theta \leq -.10$

(In the underlying populations the samples represent, the difference between the two treatment proportions is less than or equal to $-.10$. This is the **hypothesis of nonequivalence**.)

Alternative hypothesis H_1: $\theta > -.10$

(In the underlying populations the samples represent, the difference between the two treatment proportions is greater than $-.10$. This is the **hypothesis of equivalence**.)

If the above analysis is conducted, only the second of the two z-tests conducted previously in the analysis of Example 20.5 would be employed. Specifically:

$$z = \frac{(.39 - .52) - (-.10)}{\sqrt{\dfrac{(10 + 19)(29 + 42) + (4)(29)(42)}{(100)^3}}} = \frac{-.03}{.083} \cong -.36$$

Since, as noted previously, the obtained absolute value $z = .36$ is less than the tabled critical one-tailed value $z_{.05} = 1.65$, the null hypotheses H_0: $\theta \leq -.10$ designating nonequivalence cannot be rejected. In this instance, as noted in Figure 20.3, only the lower boundary would be determined for the confidence interval computed previously, with the requirement for equivalence being that the value of ζ would have to fall below the lower limit of the confidence interval (which $\zeta_1 = -.10$ does not).

In closing the discussion of tests of equivalence for two dependent proportions, in order to demonstrate a scenario where equivalence can be established, the data for Examples 20.4/20.5 will be reevaluated employing a larger value for ζ, specifically $\zeta = .30$. Since the original data for Examples 20.4/20.5 have not been altered, the use of the **McNemar test** (through use of Equation 20.2) still yields the absolute value $z = 1.54$. As noted earlier, although the value $z = 1.54$ does not allow the researcher to reject the null hypothesis that there is no difference between the two treatments, the latter is not commensurate with the conclusion that the two treatments are equivalent to one another. In order to determine the latter, the data will now be reevaluated with the **Westlake–Schuirmann test of equivalence of two dependent proportions** employing the more liberal criterion $\zeta = .30$ for equivalence in place of the original value employed $\zeta = .10$.

In other words, the researcher is now using as the criterion for equivalence that the success rate of the brand name drug is no more than 30% greater than the success rate of the generic drug, as well as the requirement that success rate of the generic drug is no more than 30% greater than the success rate of the brand name drug. The reader should take note of the fact that $\zeta = .30$ is used here for illustrative purposes, and, in reality, would most likely be considered too large a value to employ in an actual equivalency study.

Employing Equation 20.15 within the framework of the **Westlake–Schuirmann test of equivalence of two dependent proportions**, the two z tests are conducted below. Initially, the following null and alternative hypotheses are evaluated with a one-tailed test: H_0: $\theta \geq .30$ versus H_1: $\theta < .30$.

$$z = \frac{(.39 - .52) - (.30)}{\sqrt{\dfrac{(10 + 19)(29 + 42) + (4)(29)(42)}{(100)^3}}} = \frac{-.43}{.083} \equiv -5.18$$

Since the obtained absolute value $z = 5.18$ is greater than the tabled critical one-tailed value $z_{.05} = 1.65$, the null hypothesis H_0: $\theta \geq .30$ designating nonequivalence can be rejected. (The correction for continuity yields the absolute value $z = 5.06$.) The probability value associated with this result (with or without the correction for continuity) is well below the .05 value that was prestipulated for alpha. Thus, the first of the null hypotheses designating nonequivalence can be rejected. However, it once again should be emphasized that in order to declare equivalence we must reject **both** of the null hypotheses designating nonequivalence, and the two z values computed must have opposite signs.

We now evaluate the second null hypothesis designating nonequivalence. Specifically, the following null and alternative hypotheses are evaluated employing a one-tailed test: H_0: $\theta \leq -.30$ versus H_1: $\theta > -.30$.

$$z = \frac{(.39 - .52) - (-.30)}{\sqrt{\dfrac{(10 + 19)(29 + 42) + (4)(29)(42)}{(100)^3}}} = \frac{.17}{.083} \equiv 2.05$$

Since the obtained absolute value $z = 2.05$ is greater than the tabled critical one-tailed value $z_{.05} = 1.65$, the null hypothesis H_0: $\theta \leq .-30$ designating nonequivalence can be rejected. (The correction for continuity yields the absolute value $z = 1.93$.) The probability value associated with this result (with or without the correction for continuity) is well below the .05 value that was prestipulated for alpha. Since in order to declare equivalence **both** of the null hypotheses designating nonequivalence must be rejected (with opposite signs required for the computed z values), and, in fact, both have been rejected with opposite signs, equivalence has been established.

Since the 90% confidence interval previously computed for the **Westlake–Schuirmann test** is not affected by the value the researcher stipulates for ζ, the researcher can still be 90% confident (or the probability is .90) that the interval $-.267$ to $.007$ contains the true difference between the population proportions. Inspection of Figure 20.4 reveals that both $\zeta_1 = -.30$ and $\zeta_2 = .30$ fall outside of the computed confidence interval. The latter is consistent with the fact that in order for equivalency to be established the values of both ζ_1 and ζ_2 must fall outside the limits of the confidence interval, with one of the values ζ_1 positioned to the left of the lower limit of the confidence interval and the other value ζ_2 positioned to the right of the upper limit of the confidence interval. The latter is the case in Figure 20.4 where $\zeta = .30$ is employed (in contrast to Figure 20.3 when $\zeta = .10$ was employed, and equivalence was not established).

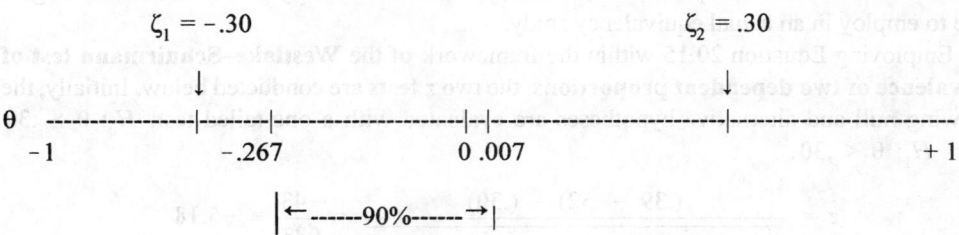

Figure 20.4   90% Confidence Interval For Example 20.5 when $\zeta = .30$

VIII. Additional Examples Illustrating the Use of the McNemar Test

Three additional examples which can be evaluated with the **McNemar test** are presented in this section. Since Examples 20.6–20.8 employ the same data employed in Example 20.1, they yield the identical result.

Example 20.6   *In order to determine if there is a relationship between schizophrenia and enlarged cerebral ventricles, a researcher evaluates* 100 *pairs of identical twins who are discordant with respect to schizophrenia (i.e., within each twin pair, only one member of the pair has schizophrenia). Each subject is evaluated with a CAT scan to determine whether or not there is enlargement of the ventricles. The results of the study are summarized in* Table 20.13. *Do the data indicate there is a statistical relationship between schizophrenia and enlarged ventricles?*

Table 20.13   Summary of Data for Example 20.6

		Schizophrenic twin		Row sums
		Enlarged ventricles	Normal ventricles	
Normal twin	Enlarged ventricles	10	13	23
	Normal ventricles	41	36	77
	Column sums	51	49	100

Since Table 20.13 summarizes categorical data derived from $n = 100$ pairs of matched subjects, the **McNemar test** is employed to evaluate the data. The reader should take note of the fact that since the independent variable employed in Example 20.6 is **nonmanipulated** (specifically, it is whether a subject is **schizophrenic** or **normal**), analysis of the data will only provide correlational information, and thus will not allow the researcher to draw conclusions with regard to cause and effect. In other words, although the study indicates that schizophrenic subjects are significantly more likely than normal subjects to have enlarged ventricles, one cannot conclude that enlarged ventricles cause schizophrenia or that schizophrenia causes enlarged ventricles. Although either of the latter is possible, the design of the study only allows one to conclude that the presence of enlarged ventricles is associated with schizophrenia.

Example 20.7   *A company which manufactures an insecticide receives complaints from its employees about premature hair loss. An air quality analysis reveals a large concentration of a vaporous compound emitted by the insecticide within the confines of the factory. In order to determine whether or not the vaporous compound (which is known as Acherton) is related to hair loss, the following study is conducted. Each of 100 mice is exposed to air containing high concentrations of Acherton over a two-month period. The same mice are also exposed to air which is uncontaminated with Acherton during another two-month period. Half of the mice are initially exposed to the Acherton-contaminated air followed by the uncontaminated air, while the other half are initially exposed to the uncontaminated air followed by the Acherton-contaminated air. The dependent variable in the study is whether or not a mouse exhibits hair loss during each of the experimental conditions.* Table 20.14 *summarizes the results of the study. Do the data indicate a relationship between Acherton and hair loss?*

Table 20.14   Summary of Data for Example 20.7

		Acherton contaminated air		Row sums
		Hair loss	**No hair loss**	
Uncontaminated air	**Hair loss**	10	13	23
	No hair loss	41	36	77
	Column sums	51	49	100

Analysis of the data in Table 20.14 reveals that the mice are significantly more likely to exhibit hair loss when exposed to Acherton as opposed to when they are exposed to uncontaminated air. Although the results of the study suggest that Acherton may be responsible for hair loss, one cannot assume that the results can be generalized to humans.

Example 20.8   *A market research firm is hired to determine whether or not a debate between the two candidates who are running for the office of Governor influences voter preference. The gubernatorial preference of 100 randomly selected voters is determined before and after a debate between the two candidates, Edgar Vega and Vera Myers.* Table 20.15 *summarizes the results of the voter preference survey. Do the data indicate that the debate influenced voter preference?*

Table 20.15   Summary of Data for Example 20.8

		Voter preference before debate		Row sums
		Edgar Vega	**Vera Meyers**	
Voter preference	**Edgar Vega**	10	13	23
after debate	**Vera Meyers**	41	36	77
	Column sums	51	49	100

When the data for Example 20.8 (which represents a **one-group pretest-posttest design**) are evaluated with the **McNemar test**, the result indicates that following the debate there is a significant shift in voter preference in favor of Vera Myers. As noted in Section V, since a **one-group pretest-posttest design** does not adequately control for the potential influence of extraneous variables, one cannot rule out the possibility that some factor other than the debate is responsible for the shift in voter preference.

IX. Addendum

Extension of the McNemar test model beyond 2 × 2 contingency tables The **McNemar test** model has been extended by Bowker (1948) and Stuart (1955, 1957) to a dependent samples design in which the dependent variable is a categorical measure that is comprised of more than two categories. In the test models for the **Bowker test of internal symmetry (Test 20c)** and the **Stuart–Maxwell test of marginal homogeneity (Test 20d)** (both of which will be described in this section), a $k \times k$ (i.e., square) contingency table (where k is the number of response categories, and $k \geq 3$) is employed to categorize n subjects (or n pairs of matched subjects) on a dependent variable under two conditions (or two time periods). The **Bowker test of internal symmetry** evaluates differences with respect to the joint probability distributions, or to put it more simply, whether the data are distributed symmetrically about the main diagonal of the table. The **Stuart–Maxwell test of marginal homogeneity**, on the other hand, evaluates differences between marginal probabilities. When either of the two tests yields a significant result, the **Bowker test of internal symmetry** is more specific in the information it provides about the differences detected in the $k \times k$ table than is the **Stuart–Maxwell test of marginal homogeneity**. This latter point will be clarified in greater detail within the framework of describing the two tests.

1. Test 20c: The Bowker test of internal symmetry Bowker (1948) developed a test to evaluate whether or not the data in a $k \times k$ contingency table are distributed symmetrically about the main diagonal of the table. Note that in a $k \times k$ contingency table, $k = r = c$ (i.e., the number of rows and columns are equal). A lack of symmetry in a $k \times k$ table is interpreted to mean that there is a difference in the distribution of the data under the two experimental conditions/time periods. In the case of a 2×2 table, the **Bowker test of internal symmetry** becomes equivalent to the **McNemar test**. Example 20.9 will be employed to illustrate the use of the **Bowker test of internal symmetry**.

Example 20.9 *Two drugs that are believed to have mood-altering effects are tested on 276 subjects. The order of presentation of the drugs is counterbalanced so that half the subjects receive Drug A followed by Drug B, while the reverse sequence of presentation is used for the other subjects. Based on his response to each drug a subject is assigned to one of the following three response categories: No change in mood (NC); Moderate mood alteration (MA); Dramatic mood alteration (DA). Table 20.16 represents a joint distribution which summarizes the responses of the 276 subjects to both of the drugs. (The value in any cell in Table 20.16 represents the number of subjects whose response to Drug A corresponds to the column category for that cell, and whose response to Drug B corresponds to the row category for that cell.) Do the data indicate that the two drugs differ with respect to their mood-altering properties?*

Table 20.16 Summary of Data for Example 20.9

		Response to Drug A		Row sums
	NC	MA	DA	
Response to Drug B NC	26	10	16	52
MA	20	50	14	84
DA	30	40	70	140
Column sums	76	100	100	276

The null and alternative hypotheses evaluated with the **Bowker test of internal symmetry** are as follows.

Null hypothesis $H_0: p_{ij} = p_{ji}$ (where $j > i$)

(In the $k \times k$ contingency table, the probabilities are equal for each of the off-diagonal/ symmetric pairs. The latter translates into the fact that the distribution of data above the main diagonal of the $k \times k$ contingency table is the same as the distribution of data below the main diagonal. With respect to Example 20.9, the null hypothesis is stating that the response distributions of subjects for the two drugs will be the same.)

Alternative hypothesis $H_1: p_{ij} \neq p_{ji}$ for at least one cell (where $j > i$)

(In the $k \times k$ contingency table, the probabilities are not equal for at least one pair of the total off-diagonal/symmetric pairs. The latter translates into the fact that the distribution of data above the main diagonal of the $k \times k$ contingency table is not the same as the distribution of data below the main diagonal. With respect to Example 20.9, the null hypothesis is stating that the response distributions of subjects for the two drugs will not be the same. The alternative hypothesis is **nondirectional.**)[17]

Equation 20.16 is employed to compute the test statistic for the **Bowker test of internal symmetry**.

$$\chi^2 = \sum_{i=1}^{r} \sum_{j>i} \left[\frac{(n_{ij} - n_{ji})^2}{n_{ij} + n_{ji}} \right]$$ **(Equation 20.16)**

The notation in Equation 20.16 indicates the following: a) Each frequency above the main diagonal which is in Row i and Column j (i.e., n_{ij}, where $j > i$) is paired with the frequency below the main diagonal which is in Row j and Column i (i.e., n_{ji}, where $j < i$). The latter pair is referred to as an **off-diagonal** or **symmetric pair**. Within each of the off-diagonal pairs the following is done: a) The difference between n_{ij} and n_{ji} is obtained; b) The difference is squared; c) The squared difference is divided by the sum of n_{ij} and n_{ji}; and d) All of the values computed in part c) are summed, and the resulting value represents the test statistic, which is a chi-square value.

The number of off-diagonal pairs in a table will be equal to $\binom{k}{2}$, which is the number of combinations of k things taken two at a time. (A discussion of **combinations** can be found in the **Introduction** in the section on the basic principles of probability.) The number of off-diagonal pairs is also equal to the value $(k \times k - k)/2$, which is equal to $[k(k-1)]/2$ (which is the number of degrees of freedom employed for the analysis). Thus, if $k = 3$ (i.e., $r = c = 3$), there will be three pairs, since $\binom{3}{2} = 3$, or $(3 \times 3 - 3)/2 = [3(3-1)]/2 = 3$. Specifically, the three pairs will involve the following combinations of cell subscripts: 1,2; 1,3; and 2, 3. Thus, the following pairs of cells will be contrasted through use of Equation 20.16 (where the first digit represents the row (i) in which the cell appears, and the second digit represents the column (j) in which the cell appears): Cell$_{12}$ versus Cell$_{21}$; Cell$_{13}$ versus Cell$_{31}$, Cell$_{23}$ versus Cell$_{32}$. Note that for the first cell listed in each pair, $j > i$, and for the second cell in each pair, $j < i$.

The data for Example 20.9 are evaluated below with Equation 20.16.

$$\chi^2 = \frac{(n_{12} - n_{21})^2}{(n_{12} + n_{21})} + \frac{(n_{13} - n_{31})^2}{(n_{13} + n_{31})} + \frac{(n_{23} - n_{32})^2}{(n_{23} + n_{32})}$$

$$\chi^2 = \frac{(10 - 20)^2}{(10 + 20)} + \frac{(16 - 30)^2}{(16 + 30)} + \frac{(14 - 40)^2}{(14 + 40)} = 20.11$$

As noted earlier, the degrees of freedom employed for the **Bowker test of internal symmetry** analysis are $[k(k - 1)]/2$. Since $k = 3$, $df = [3(3–1)]/2 = 3$. Employing **Table A4**, the tabled critical .05 and .01 chi-square values for $df = 3$ are $\chi^2_{.05} = 7.81$ and $\chi^2_{.01} = 11.34$. In order to reject the null hypothesis, the computed value of chi-square must be equal to or greater than the tabled critical value at the prespecified level of significance. Since the computed value $\chi^2 = 20.11$ is greater than both of the aforementioned critical values, the null hypothesis can be rejected at both the .05 and .01 levels.

Inspection of Table 20.16 clearly suggests that Drug B is more likely to be associated with a mood change than Drug A. To be more specific, the largest absolute deviation value in the denominator of Equation 20.16 (i.e., the largest $(n_{ij} - n_{ji})$ value) is the difference $(n_{23} - n_{32})$ = |14 - 40| = 26. By virtue of being the largest absolute deviation value, the latter indicates that the biggest difference between the two drugs is attributable to a larger number of subjects (i.e., 26 subjects) exhibiting a dramatic mood alteration (DA) to Drug B versus a moderate mood alteration (MA) to Drug A, when contrasted with the number of subjects who exhibit a dramatic mood alteration (DA) to Drug A versus a moderate mood alteration (MA) to Drug B. To a lesser extent, the difference between the two drugs can be attributed to the absolute difference $(n_{13} - n_{31})$ = |16 - 30| = 14. The latter indicates that a larger number of subjects (i.e., 14 subjects) exhibit a dramatic mood alteration (DA) to Drug B versus no change in mood (NC) to Drug A, when contrasted with the number of subjects who exhibit a dramatic mood alteration (DA) to Drug A versus no change in mood (NC) to Drug B. Finally, the absolute difference $(n_{12} - n_{21})$ = |10 - 20| = 10 indicates that a larger number of subjects (i.e., 10 subjects) exhibit a moderate mood alteration (MA) to Drug B versus no change in mood (NC) to Drug A, when contrasted with the number of subjects who exhibit a moderate mood alteration (MA) to Drug A versus no change in mood (NC) to Drug B.

When a significant result is obtained for either the **Bowker test of internal symmetry** or the **Stuart–Maxwell test of marginal homogeneity** (discussed in the next section), a 3×3 contingency table can be broken down into smaller 2×2 contingency tables. The **McNemar test** can then be employed with the 2×2 tables to compare specific combinations of cells with one another. To illustrate, the **McNemar test** is employed below to evaluate whether or not the absolute deviation score $(n_{23} - n_{32})$ = |14 - 40| = 26 is statistically significant. The latter test is based on an analysis of the data in Table 20.17, which isolate those cells in the original 3×3 table that summarizes the responses of moderate mood alteration (MA) and dramatic mood alteration (DA) to the two drugs.

Table 20.17 Summary of Data for McNemar Test to Evaluate $(n_{23} - n_{32})$

		Response to Drug A		
		MA	DA	Row sums
Response to Drug B	MA	50	14	64
	DA	40	70	110
	Column sums	90	84	174

Substituting the appropriate values in Equation 20.1, the value $\chi^2 = 12.52$ is computed for the data in Table 20.17.

$$\chi^2 = \frac{(14 - 40)^2}{14 + 40} = 12.52$$

Since the computed value $\chi^2 = 12.52$ is larger than the tabled critical values (for $df = 1$) $\chi^2_{.05} = 3.84$, and $\chi^2_{.01} = 6.63$, the result is significant at both the .05 and .01 levels. Thus, we can conclude that the absolute deviation score $(n_{23} - n_{32}) = |14 - 40| = 26$ represents a statistically significant difference.

Tables 20.18 and 20.19 summarize the data for the other two absolute deviation scores. The appropriate **McNemar test** is noted below each table.

Table 20.18 Summary of Data for McNemar Test to Evaluate $(n_{13} - n_{31})$

		Response to Drug A		
		NC	DA	Row sums
Response to Drug B	NC	26	16	42
	DA	30	70	100
	Column sums	56	86	142

$$\chi^2 = \frac{(16 - 30)^2}{16 + 30} = 4.26$$

Table 20.19 Summary of Data for McNemar Test to Evaluate $(n_{12} - n_{21})$

		Response to Drug A		
		NC	MA	Row sums
Response to Drug B	NC	26	10	36
	MA	20	50	70
	Column sums	46	60	106

$$\chi^2 = \frac{(10 - 20)^2}{10 + 20} = 3.33$$

In the case of Table 20.18, the **McNemar test** on $(n_{13} - n_{31}) = |16 - 30| = 14$ yields the value $\chi^2 = 4.26$, which is significant at only the .05 level (since $\chi^2 = 4.26$ is greater than $\chi^2_{.05} = 3.84$), but not at the .01 level. It should be noted, however, that when the correction for continuity (Equation 20.3) is employed, the computed chi-square value is $\chi^2 = 3.67$, which is not significant at the .05 level.

In the case of Table 20.19, the **McNemar test** on $(n_{12} - n_{21}) = |10 - 20| = 10$ yields the value $\chi^2 = 3.33$, which is not significant. Thus, the only clearly significant difference appears to be the absolute deviation $(n_{23} - n_{32}) = |14 - 40| = 26$.

When more than one **McNemar test** is employed to conduct comparisons within the framework of a 3 × 3 contingency table (such as those illustrated above), an adjustment of the Type I error rate may be in order. The protocol for the latter adjustment is discussed in Section VI of the **chi-square goodness-of-fit test (Test 8)** and in Section VI of the **single-factor between-subjects analysis of variance (Test 21)**.

The reader should take note of the fact that, as is the case with the **McNemar test,** Equation 20.16 employed for the **Bowker test of internal symmetry** only takes into account those subjects who fall in a different category in each of the experimental conditions. In other words, the chi-square value computed with Equation 20.16 is independent of the number of subjects who fall in the diagonal of Table 20.16. Further discussion of the **Bowker test of internal symmetry** can be found in Everitt (1977, pp. 114–115; 1992), Marascuilo and McSweeney (1977), Marascuilo and Serlin (1988), Sprent (1993) and Zar (1999).

2. Test 20d: The Stuart–Maxwell test of marginal homogeneity Everitt (1977, 1992) notes that if on the basis of the **Bowker test of internal symmetry** the hypothesis of symmetry (which will generally be the hypothesis of primary interest, since it provides the most meaningful information) is rejected, a researcher may employ the **Stuart–Maxwell test of marginal homogeneity** (also referred to in some sources as the **Stuart test** or the **Maxwell test**) to further clarify the distribution of the data. The null and alternative hypotheses evaluated with the **Stuart–Maxwell test of marginal homogeneity** are as follows.

Null hypothesis H_0: $p_{i.} = p_{.j}$ (for all integer values from 1 to k, where $i = j$, and $p_{i.}$ is the population probability for the ith row and $p_{.j}$ is the population probability for the jth column)

(In the $k \times k$ contingency table, for all integer values from 1 to k where $i = j$, the marginal probability of the ith row is equal to the marginal probability of the jth column (the latter condition represents **marginal homogeneity**). With respect to Example 20.9, the null hypothesis is stating that the row and column marginal response distributions of subjects to the two drugs will be the same.)

Alternative hypothesis H_1: $p_{i.} \neq p_{.j}$ for at least one set of marginal probabilities

(In the $k \times k$ contingency table, for all integer values from 1 to k where $i = j$, for at least one set of marginal probabilities the marginal probability of the ith row will not be equal to the marginal probability of the jth column (which violates the condition of **marginal homogeneity**). With respect to Example 20.9, the null hypothesis is stating that at least one of the row and column marginal response distributions of subjects to the two drugs will not be the same. The alternative hypothesis is **nondirectional**.)

Although complex mathematics involving matrix algebra is required to compute the test statistic for the **Stuart–Maxwell test of marginal homogeneity,** Fleiss and Everitt (1971) have derived Equation 20.17 to compute the test statistic when $k = 3$.

$$\chi^2 = \frac{\bar{n}_{23} d_1^2 + \bar{n}_{13} d_2^2 + \bar{n}_{12} d_3^2}{2 (\bar{n}_{12} \bar{n}_{13} + \bar{n}_{12} \bar{n}_{23} + \bar{n}_{13} \bar{n}_{23})} \qquad \textbf{(Equation 20.17)}$$

Where: $\bar{n}_{ij} = \dfrac{n_{ij} + n_{ji}}{2}$

$d_i = n_{i.} - n_{.i}$ (with $i = j$)

The data for Example 20.9 are evaluated below with Equation 20.17.

$$d_1 = n_1. - n._1 = 52 - 76 = -24$$

$$d_2 = n_2. - n._2 = 84 - 100 = -16$$

$$d_3 = n_3. - n._3 = 140 - 100 = 40$$

$$\bar{n}_{12} = \frac{n_{12} + n_{21}}{2} = \frac{10 + 20}{2} = 15$$

$$\bar{n}_{13} = \frac{n_{13} + n_{31}}{2} = \frac{16 + 30}{2} = 23$$

$$\bar{n}_{23} = \frac{n_{23} + n_{32}}{2} = \frac{14 + 40}{2} = 27$$

Substituting the above values in Equation 20.17, the value $\chi^2 = 16.57$ is computed.

$$\chi^2 = \frac{(27)(-24)^2 + (23)(-16)^2 + (15)(40)^2}{2[(15)(23) + (15)(27) + (23)(27)]} = 16.57$$

Everitt (1977, p. 116) notes that for the above analysis, $df = (k - 1)$. Employing **Table A4**, the tabled critical chi-square values for $df = (3-1) = 2$ are $\chi^2_{.05} = 5.99$ and $\chi^2_{.01} = 9.21$. In order to reject the null hypothesis, the computed value of chi-square must be equal to or greater than the tabled critical value at the prespecified level of significance. Since the computed value $\chi^2 = 16.57$ is greater than both of the aforementioned critical values, the null hypothesis can be rejected at both the .05 and .01 levels. Thus we can conclude that the marginal probabilities are not homogeneous.

Table 20.20   Summary of Marginal Probabilities for Example 20.9

		Response			Row sums
		NC	MA	DA	
Drug	Drug A	76/276=.275	100/276=.362	100/276=.362	1
	Drug B	52/276=.188	84/276 =.304	140/276=.507	1

Further clarification of the above result can be obtained from inspection of Table 20.20, which summarizes the marginal probabilities in Table 20.16. Table 20.20 indicates the following: a) Subjects are more likely to exhibit a dramatic mood alteration (DA) to Drug B ($p = .507$) when contrasted with a dramatic mood alternation (DA) to Drug A ($p = .362$); b) Subjects are more likely to exhibit a moderate mood alteration (MA) to Drug A ($p = .362$) when contrasted with a moderate mood alteration (MA) to Drug B ($p = .304$); and c) Subjects are more likely to exhibit no change in mood (NC) to Drug A ($p = .275$) when contrasted with no change in mood (NC) to Drug B ($p = .188$). In the final analysis, however, the most meaningful explanation of the differences between the two drugs is contained in the analysis of the internal structure of the 3 × 3 table with the **Bowker test of internal symmetry** — specifically, that the major difference with respect to the mood altering properties of the two drugs can be attributed to a larger proportion of subjects who exhibit a dramatic mood alteration (DA) to Drug B relative to the proportion of subjects who exhibit a dramatic mood alteration (DA) to Drug A.

Fleiss (1981, p. 121) and Fleiss and Everitt (1971) describe additional procedures which can be employed to conduct a more detailed analysis of a $k \times k$ contingency table within the framework of the **Stuart–Maxwell test of marginal homogeneity**. As noted earlier under the discussion of the **Bowker test of internal symmetry**, one or more **McNemar tests** can be conducted on 2×2 contingency tables derived from the original 3×3 table. Marascuilo and Serlin (1988, p. 392) note that the **Bowker test of internal symmetry** and the **Stuart–Maxwell test of marginal homogeneity** are not the optimal tests to employ for $k \times k$ contingency tables when the categories in the table are ordered.

Sources which discuss the **Stuart–Maxwell test of marginal homogeneity** are Everitt (1977, pp. 115–116; 1992), Fleiss (1981, pp. 119–123), Fleiss *et al.* (2003, pp. 381–384), Hinkle *et al.* (1998), Marascuilo and McSweeney (1977), and Marascuilo and Serlin (1988).

References

Bowker, A. H. (1948). A test for symmetry in contingency tables. **Journal of the American Statistical Association**, 43, 572–574.

Chow, S. C. & Liu, J. P. (2004). **Design and analysis of clinical trials: Concepts and methodologies** (2nd ed.). Hoboken, NJ: John Wiley & Sons.

Connett, J. E., Smith, J. A., & McHugh, R. B. (1987). Sample size and power for pair-matched case-control studies. **Statistics in Medicine**, 6, 53–59.

Daniel, W. W. (1990). **Applied nonparametric statistics** (2nd ed.). Boston: PWS–Kent Publishing Company.

Dunnet, C. W. & Gent, M. (1977). Significance testing to establish equivalence between treatments with special reference to data in the form of 2×2 tables. **Biometrics**, 33, 593–602.

Edwards, A. L. (1948). Note on the "correction for continuity" in testing the significance of the difference between correlated proportions. **Psychometrika**, 13, 185–187.

Everitt, B. S. (1977). **The analysis of contingency tables**. London: Chapman & Hall.

Everitt, B. S. (1992). **The analysis of contingency tables** (2nd ed.). New York: Chapman & Hall.

Fleiss, J. L. (1981). **Statistical methods for rates and proportions** (2nd ed.). New York: John Wiley & Sons.

Fleiss, J. L. & Everitt, B. S. (1971). Comparing the marginal totals of square contingency tables. **British Journal of Math. Statist. Psychol.**, 24, 117–123.

Fleiss, J. L., Levin, B. & Paik, M. C. (2003). **Statistical methods for rates and proportions** (3rd ed.). New York: John Wiley & Sons.

Gart, J. J. (1969). An exact test for comparing matched proportions in crossover designs. **Biometrika**, 56, 75–80.

Hinkle, D. E., Wiersma, W., & Jurs, S.G. (1998). **Applied statistics for the behavioral sciences**) (4th ed.). Boston: Houghton Mifflin Company.

Leach, C. (1979). **Introduction to statistics: A nonparametric approach for the social sciences**. Chichester, England: John Wiley & Sons.

Liu, J. P., Hsueh, H. M. Hiseh, E, & Chen, J. J. (2002). Tests for equivalence or non-inferiority for paired binary data. **Statistics in Medicine**, 21, 231–245.

Liu, K. J., & Cumberland, W. G. (2001). A test procedure of equivalence in ordinal data with matched-pairs. **Biometrical Journal**, 43, 977–983.

Lu, Y. & Bean, J. A. (1995). On the sample size for one-sided equivalence of senitivities based upon McNemar's test. **Statistics in Medicine**, 14, 1831–1839.

Marascuilo, L. A. & McSweeney, M. (1977). **Nonparametric and distribution-free methods for the social sciences.** Monterey, CA: Brooks/Cole Publishing Company.

Marascuilo, L. A. & Serlin, R. C. (1988). **Statistical methods for the social and behavioral sciences.** New York: W. H. Freeman & Company.

Maxwell, A. E. (1970). Comparing the classification of subjects by two independent judges. **British Journal of Psychiatry**, 116, 651–655.

McNemar, Q. (1947). Note on the sampling error of the difference between correlated proportions or percentages. **Psychometrika**, 12, 153–157.

Morikawa, T. &, Yanagawa, T. (1995). Equivalence testing for paired dichotomous data. **Proceedings of Annual Conference of Biometric Society of Japan**, 123–126.

Nam, J. (1997). Establishing equivalence of two treatments and sample size requirements in matched-pairs designs. **Biometrics**, 53, 1422–1430.

Rosner, B. (2000). **Fundamentals of biostatistics** (5th ed.). Pacific Grove, CA: Duxbury Press.

Schuirmann, D. J. (1987). A comparison of the two one-sided tests procedure and the power approach for assessing equivalence of average bioavailability. **Journal of Pharmacokinetics and Biopharmaceutics**, 15, 657–680.

Selvin, S. (1995). **Practical biostatistical methods**. Belmont, CA: Duxbury.

Selvin, S. (2004). **Statistical analysis of epidemiological data** (3rd ed.). Oxford: Oxford University Press.

Selwyn, M. R., Dempster, A. P. & Hall, N. R. (1981). A Bayesian approach to bioequivalence for the 2 × 2 changeover design. **Biometrics**, 37, 11–21.

Selwyn, M. R. & Hall, N. R. (1984). On Baysian methods for bioequivalence. **Biometrics**, 40, 1103–1108.

Siegel, S. & Castellan, N. J., Jr. (1988). **Nonparametric statistics for the behavioral sciences** (2nd ed.). New York: McGraw–Hill Book Company.

Sprent, P. (1993). **Applied nonparametric statistical methods** (2nd ed.). London: Chapman & Hall.

Stuart, A. A. (1955). A test for homogeneity of the marginal distributions in a two way classification. **Biometrika**, 42, 412–416.

Stuart, A. A. (1957). The comparison of frequencies in matched samples. **British Journal of Statistical Psychology**, 10, 29–32.

Tango, T. (1998). Equivalence test and confidence interval for the difference in proportions for the paired-sample design. **Statistics in Medicine**, 17, 891–908.

Tango, T. (1999). Improved confidence intervals for the difference between binomial proportions based on paired data. **Statistics in Medicine**, 18, 3511–3513.

Wellek, S. (2003). **Testing Statistical hypotheses of equivalence**. Boca Raton, FL: Chapman & Hall/CRC.

Westlake, W. J. (1976). Symmetrical confidence intervals for bioequivalence trials. **Biometrics**, 32, 741–744.

Westlake, W. J. (1981). Response to T.B. L. Kirkwood: Bioequivalence testing — a need to rethink. **Biometrics**, 37, 589–594.

Westlake, W. J. (1988). Bioavailability and bioequivalence of pharmaceutical formulations. In K. E. Peace (Ed.), **Biopharamaceutical statistics for drug development** (pp. 329–352). New York: Marcel Decker.

Zar, J. H. (1999). **Biostatistical analysis** (4th ed.). Upper Saddle River, NJ: Prentice Hall.

Endnotes

1. The distinction between a **true experiment** and a **natural experiment** is discussed in more detail in the **Introduction**.

2. a) The reader should take note of the following with respect to the null and alternative hypotheses stated in this section:

    a) If n represents the total number of observations in Cells a, b, c, and d, the proportion of observations in Cells b and c can also be expressed as follows: b/n and c/n. The latter two values, however, are not equivalent to the values p_b and p_c which are used to estimate the values π_b and π_c employed in the null and alternative hypotheses.

    b) Many sources employ an alternative but equivalent way of stating the null and alternative hypotheses for the **McNemar test**. Assume that π_1 represents the proportion of observations in the underlying population that responds in Response category 1 in Condition 1/Pretest, and π_2 represents the proportion of observations in the underlying population that responds in Response category 1 in Condition 2/Posttest. With respect to the sample data, the values p_1 and p_2 are employed to estimate π_1 and π_2, where $p_1 = (a + b)/n$ and $p_2 = (a + c)/n$. In the case of Examples 20.1 and 20.2, $p_1 = (10 + 13)/100 = .23$ and $p_2 = (10 + 41)/100 = .51$. If there is no difference in the proportion of observations in Response category 1 in Condition 1/Pretest versus the proportion of observations in Response category 1 in Condition 2/Posttest, p_1 and p_2 would be expected to be equal, and if the latter is true one can conclude that in the underlying population $\pi_1 = \pi_2$. If, however, $p_1 \neq p_2$ (and consequently in the underlying population $\pi_1 \neq \pi_2$), it indicates a difference between the two experimental conditions in the case of a **true experiment**, and a difference between the pretest and the posttest responses of subjects in the case of a **one-group pretest-posttest design**. Employing this information, the null hypothesis can be stated as follows: H_0: $\pi_1 = \pi_2$ (which can also be written as H_0: $\pi_1 - \pi_2 = 0$). The null hypothesis H_0: $\pi_1 = \pi_2$ is equivalent to the null hypothesis H_0: $\pi_b = \pi_c$. The **nondirectional alternative hypothesis** can be stated as H_1: $\pi_1 \neq \pi_2$ (which can also be written as H_1: $\pi_1 - \pi_2 \neq 0$). The nondirectional alternative hypothesis H_1: $\pi_1 \neq \pi_2$ is equivalent to the nondirectional alternative hypothesis H_1: $\pi_b \neq \pi_c$. The two **directional alternative hypotheses** which can be employed are H_1: $\pi_1 > \pi_2$ or H_1: $\pi_1 < \pi_2$ (which can also be written respectively as H_1: $\pi_1 - \pi_2 > 0$ and H_1: $\pi_1 - \pi_2 < 0$). The directional alternative hypothesis H_1: $\pi_1 > \pi_2$ is equivalent to the directional alternative hypothesis H_1: $\pi_b > \pi_c$. The directional alternative hypothesis H_1: $\pi_1 < \pi_2$ is equivalent to the directional alternative hypothesis H_1: $\pi_b < \pi_c$.

3. It can be demonstrated algebraically that Equation 20.1 is equivalent to Equation 8.2 (which is the equation for the **chi-square goodness-of-fit test**). Specifically, if Cells a and d are eliminated from the analysis, and the **chi-square goodness-of-fit test** is employed to evaluate the observations in Cells b and c, $n = b + c$. If the expected probability for each of the cells is .5, Equation 8.2 reduces to Equation 20.1. As will be noted in Section VI, a limitation of the **McNemar test** (which is apparent from inspection of Equation 20.1) is that it only employs the data for two of the four cells in the contingency table.

4. A general overview of the chi-square distribution and interpretation of the values listed in **Table A4** can be found in Sections I and V of the **single-sample chi-square test for a population variance (Test 3)**.

5. The degrees of freedom are based on Equation 8.3, which is employed to compute the degrees of freedom for the **chi-square goodness-of-fit test**. In the case of the **McNemar test**, $df = k - 1 = 2 - 1 = 1$, since only the observations in Cells b and c (i.e., $k = 2$ cells) are evaluated.

6. A full discussion of the protocol for determining one-tailed chi-square values can be found in Section VII of the **chi-square goodness-of-fit test**.

7. A general discussion of the correction for continuity can be found in Section VI of the **Wilcoxon signed-ranks test (Test 6)**. Fleiss (1981) notes that the correction for continuity for the **McNemar test** was recommended by Edwards (1948).

8. The numerator of Equation 20.4 is sometimes written as $(b - c) \pm 1$. In using the latter format, 1 is added to the numerator if the term $(b - c)$ results in a negative value, and 1 is subtracted from the numerator if the term $(b - c)$ results in a positive value. Since we are only interested in the absolute value of z, it is simpler to employ the numerator in Equation 20.4, which results in the same absolute value that is obtained when the alternative form of the numerator is employed. If the alternative form of the numerator is employed for Examples 20.1/20.2, it yields the value $z = -3.67$.

9. Equations 20.18 and 20.19 are alternative equations which can be employed to compute the **McNemar test** statistic. Whereas Equation 20.18 does not employ a correction for continuity, Equation 20.19 does. As noted in Endnote 2, the values p_1 versus p_2 in the latter equations are computed as follows: $p_1 = (a + b)/n = (10 + 13)/100 = .23$; $p_2 = (a + c)/n = (10 + 41)/100 = .51$. Equations 20.18 and 20.19 are employed below to compute the values $z = -3.81$ and $z = 3.67$ which are identical to the values previously computed with Equations 20.2 and 20.4.

(Equation 20.18)

$$z = \frac{p_1 - p_2}{\sqrt{\dfrac{b + c}{n^2}}} = \frac{.23 - .51}{\sqrt{\dfrac{13 + 41}{(100)^2}}} = \frac{-.28}{.0735} = -3.81$$

(Equation 20.19)

$$z = \frac{(|p_1 - p_2|) - \dfrac{1}{n}}{\sqrt{\dfrac{b + c}{n^2}}} = \frac{(|.23 - .51|) - \dfrac{1}{100}}{\sqrt{\dfrac{13 + 41}{(100)^2}}} = \frac{.28}{.0735} = 3.67$$

Fleiss *et al.* (2003, pp. 375–378) note that the denominator of Equations 20.18 and 20.19, $\sqrt{(b + c)/n^2}$ (which does not take into account the frequencies in cells a and d), represents a measure of the standard error of the difference between the two proportions p_1 and p_2 when the two underlying population proportions π_1 and π_2 are hypothesized to be equal. Various sources (e.g., Fleiss *et al.* (2003, p. 375) and Selvin (1995, p. 265)) note, however, that the value $\sqrt{(b + c)/n^2}$ should not be employed as the value of the standard error when the two underlying population proportions π_1 and π_2 are not hypothesized to be equal. Because of the latter, $\sqrt{(b + c)/n^2}$ should not be employed for the standard error in the computation of a confidence interval for the difference $p_1 - p_2$ (since computation of a confidence interval is not predicated on the assumption that the two proportions are equal). In computing a confidence interval a modified value which does, in fact, take into account the frequencies in cells a and d is employed for the standard error. The latter value is computed with Equation 20.9 in part 5 of Section VI.

10. The values .0287 and .0571 respectively represent the proportion of the normal distribution which falls above the values $z = 1.90$ and $z = 1.58$.

11. In point of fact, it can also be viewed as identical to the analysis conducted with the **binomial sign test for two dependent samples**. In Section VII of the **Cochran Q test**, it is demonstrated that when the **McNemar test** (as well as the **Cochran Q test** when $k = 2$) and the **binomial sign test for two dependent samples** are employed to evaluate the same set of data, they yield equivalent results.

12. For a comprehensive discussion of the computation of binomial probabilities and the use of **Table A7**, the reader should review Section IV of the **binomial sign test for a single sample**.

13. a) The reader should take note of the fact that if the value $\sqrt{(b + c)/n^2}$ is employed to compute the standard error, it results in the slightly larger value $\sqrt{(13 + 41)/(100)^2} = .0735$; b) There are a number of alternative but equivalent equations for computing the value of the standard error computed with Equation 20.9 (which is also employed in Selvin (1995, p. 265)). Equations 20.20 and 20.21 (Fleiss *et al.*(2003, p. 378)) and Equation 20.22 (Marascuilo and McSweeney (1977, pp. 170–171)) represent three alternative but equivalent equations for computing the standard error.

(Equation 20.20)

$$s_{p_1 - p_2} = \frac{\sqrt{n(b + c) - (b - c)^2}}{n\sqrt{n}} = \frac{\sqrt{100(13 + 41) - (13 - 41)^2}}{100\sqrt{100}} = .0679$$

(Equation 20.21)

$$s_{p_1 - p_2} = \frac{\sqrt{(a + d)(b + c) + 4bc}}{n\sqrt{n}} = \frac{\sqrt{(10 + 36)(13 + 41) + (4)(13)(41)}}{100\sqrt{100}} = .0679$$

(Equation 20.22)

$$s_{p_1 - p_2} = \sqrt{\frac{\left(\dfrac{a + c}{n}\right)\left(\dfrac{b + d}{n}\right) + \left(\dfrac{a + b}{n}\right)\left(\dfrac{c + d}{n}\right) - \dfrac{2a}{n} + 2\left(\dfrac{a + b}{n}\right)\left(\dfrac{a + c}{n}\right)}{n}} =$$

$$\sqrt{\frac{\left(\dfrac{10 + 41}{100}\right)\left(\dfrac{13 + 36}{100}\right) + \left(\dfrac{10 + 13}{100}\right)\left(\dfrac{41 + 36}{100}\right) - \dfrac{(2)(10)}{100} + 2\left(\dfrac{10 + 13}{100}\right)\left(\dfrac{10 + 41}{100}\right)}{100}} = .0679$$

14. a) Fleiss *et al.* (2003, p. 376) notes that the **standard error** of an **odds ratio** (s_{o_m}) can be computed with Equation 20.23 (which is analogous to the equation used to compute the standard error for an **odds ratio** employed in Equation 16.29). For the data in Table 20.4 the value $s_{o_m} = 1$ is computed.

$$s_{o_m} = o_m\sqrt{\frac{1}{b} + \frac{1}{c}} = 3.15\sqrt{\frac{1}{13} + \frac{1}{41}} = 1.00 \qquad \textbf{(Equation 20.23)}$$

b) Sources are not in agreement with respect to what method allows for optimal computation of a confidence interval for an **odds ratio** in the case of a 2×2 contingency table involving dependent samples. Although, one method described by Rosner (2000, pp. 605–606) will be presented below, readers should be aware of the fact that alternative methods for approximating a confidence interval (described in sources such as Fleiss *et al.* (2003, pp. 376–377) and Selvin (2004, pp. 305–307)) will not yield the same values as the method described below.

The following notation will be employed within the framework of computing a confidence interval for the **odds ratio** for the data in Table 20.4: $m = b + c = 13 + 41 = 54$, $p = b/m = 13/54 = .241$ and $1 - p = c/m = 41/54 = .759$. Equation 20.24 is employed below to compute the value $s_{o_m} = .3182$, which represents an estimate of the standard error to be employed in the equation for computing the **odds ratio**. (Note, however, that Equation 20.24 does not compute the same standard error computed with Equation 20.23.)

(Equation 20.24)

$$s_{o_m} = \sqrt{\frac{1}{mp(1 - p)}} = \sqrt{\frac{1}{(54)(.241)(.759)}} = .3182$$

Equation 20.25 is employed to derive a confidence interval for the **odds ratio**. The latter equation is employed below in reference to $o_m = 3.15$ to compute the 95% confidence interval (which employs the tabled critical two-tailed value $z_{.05} = 1.96$ in **Table A1**). (In the computations below the natural logarithm is based on the value $o_m = 3.15$ being computed to nine decimal places.)

$$CI_{1 - \alpha} = \ln(o_m) \pm (z_{\alpha/2})(s_{o_m})$$

(Equation 20.25)

$$CI_{.95} = \ln(o_m) \pm (z_{.05})(s_{o_m})$$

$$= \ln(3.15) \pm (1.96)(.3182) = 1.1486 \pm (1.96)(.3182) = 1.1486 \pm .6237$$

When .6237 is subtracted from and added to 1.1486, the values .5249 and 1.7723 are obtained. The antilogarithms of the latter values will represent the lower and upper boundaries of the 95% confidence interval of the **odds ratio**. Thus, antilog (.5249) = 1.6903 and antilog (1.7723) = 5.8844. The result indicates we can be 95% confident (or the probability is .95) the interval 1.6903 to 5.8844 contains the true value of the **odds ratio** in the population.

15. The methodology to be described in this section for evaluating equivalence is essentially the same as methodologies described by Lu and Bean (1995) and Morikawa and Yanagawa (1995) which are described in Tango (1998).

16. If the denominator of Equation 20.18 (i.e., the standard error which only takes into account the frequencies in cells b and c) is employed as the denominator of Equation 20.15, the absolute value computed for z will be slightly larger than the value computed with Equation 20.15. In other words, use of standard error which ignores the frequencies for cells a and d will result in a slightly more conservative test of equivalence.

17. Marascuilo and McSweeney (1977) note that it is only possible to state the alternative hypothesis directionally when the number of degrees of freedom employed for the test is 1, which will always be the case for a 2 × 2 table.

Inferential Statistical Tests Employed with Two or More Independent Samples (and Related Measures of Association/Correlation)

Test 21: The Single-Factor Between-Subjects Analysis of Variance

Test 22: The Kruskal–Wallis One-Way Analysis of Variance by Ranks

Test 23: The van der Waerden Normal-Scores Test for k Independent Samples

Test 21

The Single-Factor Between-Subjects Analysis of Variance
(Parametric Test Employed with Interval/Ratio Data)

I. Hypothesis Evaluated with Test and Relevant Background Information

Hypothesis evaluated with test In a set of k independent samples (where $k \geq 2$), do at least two of the samples represent populations with different mean values?

Relevant background information on test The term **analysis of variance** (for which the acronym **ANOVA** is often employed) describes a group of inferential statistical procedures developed by the British statistician Ronald Fisher. Analysis of variance procedures are employed to evaluate whether or not there is a difference between at least two means in a set of data for which two or more means can be computed. The test statistic computed for an analysis of variance is based on the F distribution (which is named after Fisher), which is a continuous theoretical probability distribution. A computed F value (commonly referred to as an **F ratio**) will always fall within the range $0 \leq F \leq \infty$. As is the case with the t and chi-square distributions discussed earlier in the book, there are an infinite number of F distributions — each distribution being a function of the number of degrees of freedom employed in the analysis (with degrees of freedom being a function of both the number of samples and the number of subjects per sample). A more thorough discussion of the F distribution can be found in Section V and Endnote 11.

The **single-factor between-subjects analysis of variance** is the most basic of the analysis of variance procedures.[1] It is employed in a hypothesis testing situation involving k independent samples. In contrast to the **t test for two independent samples (Test 11)**, which only allows for a comparison between the means of two independent samples, the **single-factor between-subjects analysis of variance** allows for a comparison of two or more independent samples. The **single-factor between-subjects analysis of variance** is also referred to as the **completely randomized single-factor analysis of variance**, the **simple analysis of variance**, the **one-way analysis of variance**, and the **single-factor analysis of variance**.

In conducting the **single-factor between-subjects analysis of variance**, each of the k sample means is employed to estimate the value of the mean of the population a sample represents. If the computed test statistic is significant, it indicates there is a significant difference between at least two of the sample means in the set of k means. As a result of the latter, the researcher can conclude there is a high likelihood at least two of the samples represent populations with different mean values.

In order to compute the test statistic for the **single-factor between-subjects analysis of variance**, the **total variability** in the data is divided into **between-groups variability** and **within-groups variability**. **Between-groups variability** (which is also referred to as **treatment variability**) is essentially a measure of the variance of the means of the k samples. **Within-groups variability** (which is essentially an average of the variance within each of the k samples) is variability which is attributable to chance factors that are beyond the control of a researcher. Since such chance factors are often referred to as experimental error, **within-groups variability** is also referred to as **error** or **residual variability**. The *F* **ratio**, which is the test statistic for the **single-factor between-subjects analysis of variance**, is obtained by dividing **between-groups variability** by **within-groups variability**. Since **within-groups variability** is employed as a baseline measure of the variability in a set of data which is beyond a researcher's control, it is assumed that if the *k* samples are derived from a population with the same mean value, the amount of variability between the sample means (i.e., **between-groups variability**) will be approximately the same value as the amount of variability within any single sample (i.e., **within-groups variability**). If, on the other hand, **between-groups variability** is significantly larger than **within-groups variability** (in which case the value of the *F* ratio will be larger than 1), it is likely that something in addition to chance factors is contributing to the amount of variability between the sample means. In such a case, it is assumed that whatever it is which differentiates the groups from one another (i.e., the independent variable/experimental treatments) accounts for the fact that **between-groups variability** is larger than **within-groups variability**.[2] A thorough discussion of the logic underlying the **single-factor between-subjects analysis of variance** can be found in Section VII.

The **single-factor between-subjects analysis of variance** is employed with interval/ratio data and is based on the following assumptions: a) Each sample has been randomly selected from the population it represents; b) The distribution of data in the underlying population from which each of the samples is derived is normal; and c) The third assumption, which is referred to as the **homogeneity of variance** assumption, states that the variances of the *k* underlying populations represented by the *k* samples are equal to one another. The homogeneity of variance assumption is discussed in detail in Section VI.[3] If any of the aforementioned assumptions of the **single-factor between-subjects analysis of variance** are saliently violated, the reliability of the computed test statistic may be compromised.

II. Example

Example 21.1   *A psychologist conducts a study to determine whether or not noise can inhibit learning. Each of 15 subjects is randomly assigned to one of three groups. Each subject is given 20 minutes to memorize a list of 10 nonsense syllables, which she is told she will be tested on the following day. The five subjects assigned to* **Group 1**, *the* **no noise** *condition, study the list of nonsense syllables while they are in a quiet room. The five subjects assigned to* **Group 2**, *the* **moderate noise** *condition, study the list of nonsense syllables while listening to classical music. The five subjects assigned to* **Group 3**, *the* **extreme noise** *condition, study the list of nonsense syllables while listening to rock music. The number of nonsense syllables correctly recalled by the 15 subjects follows:* **Group 1**: *8, 10, 9, 10, 9;* **Group 2**: *7, 8, 5, 8, 5;* **Group 3**: *4, 8, 7, 5, 7. Do the data indicate that noise influenced subjects' performance?*

III. Null versus Alternative Hypotheses

Null hypothesis $H_0: \mu_1 = \mu_2 = \mu_3$

(The mean of the population Group 1 represents equals the mean of the population Group 2 represents equals the mean of the population Group 3 represents.)

Alternative hypothesis $\qquad\qquad\qquad$ H_1: Not H_0

(This indicates there is a difference between at least two of the $k = 3$ population means. It is important to note that the alternative hypothesis should not be written as follows: H_1: $\mu_1 \neq \mu_2 \neq \mu_3$. The reason why the latter notation for the alternative hypothesis is incorrect is because it implies that all three population means must differ from one another in order to reject the null hypothesis. In this book it will be assumed (unless stated otherwise) that the alternative hypothesis for the analysis of variance is stated **nondirectionally**.[4] In order to reject the null hypothesis, the obtained F value must be equal to or greater than the tabled critical F value at the prespecified level of significance.)

IV. Test Computations

The test statistic for the **single-factor between-subjects analysis of variance** can be computed with either **computational** or **definitional equations**. Although definitional equations reveal the underlying logic behind the analysis of variance, they involve considerably more calculations than the computational equations. Because of the latter, computational equations will be employed in this section to demonstrate the computation of the test statistic. The definitional equations for the **single-factor between-subjects analysis of variance** are described in Section VII.

$\qquad$ The data for Example 21.1 are summarized in Table 21.1. The scores of the $n_1 = 5$ subjects in Group 1 are listed in the column labeled X_1, the scores of the $n_2 = 5$ subjects in Group 2 are listed in the column labelled X_2, and the scores of the $n_3 = 5$ subjects in Group 3 are listed in the column labeled X_3. Since there are an equal number of subjects in each group, the notation n is employed to represent the number of subjects per group. In other words, $n = n_1 = n_2 = n_3$. The columns labeled X_1^2, X_2^2, and X_3^2 list the squares of the scores of the subjects in each of the three groups.

Table 21.1 Data for Example 21.1

Group 1		Group 2		Group 3	
X_1	X_1^2	X_2	X_2^2	X_3	X_3^2
8	64	7	49	4	16
10	100	8	64	8	64
9	81	5	25	7	49
10	100	8	64	5	25
9	81	5	25	7	49
$\Sigma X_1 = 46$	$\Sigma X_1^2 = 426$	$\Sigma X_2 = 33$	$X_2^2 = 227$	$X_3 = 31$	$X_3^2 = 203$
$\bar{X}_1 = \dfrac{\Sigma X_1}{n_1} = \dfrac{46}{5} = 9.2$		$\bar{X}_2 = \dfrac{\Sigma X_2}{n_2} = \dfrac{33}{5} = 6.6$		$\bar{X}_3 = \dfrac{\Sigma X_3}{n_3} = \dfrac{31}{5} = 6.2$	

The notation N represents the total number of subjects employed in the experiment. Thus:

$$N = n_1 + n_2 + \cdots + n_k$$

Since there are $k = 3$ groups:

$$N = n_1 + n_2 + n_3 = 5 + 5 + 5 = 15$$

The value ΣX_T represents the total sum of the scores of the N subjects who participate in the experiment. Thus:

$$\Sigma X_T = \Sigma X_1 + \Sigma X_2 + \cdots + \Sigma X_k$$

Since there are $k = 3$ groups, $\Sigma X_T = 110$.

$$\Sigma X_T = \Sigma X_1 + \Sigma X_2 + \Sigma X_3 = 46 + 33 + 31 = 110$$

$\bar{X}_T$ represents the grand mean, where $\bar{X}_T = \Sigma X_T / N$. Thus, $\bar{X}_T = 110/15 = 7.33$. Although $\bar{X}_T$ is not employed in the computational equations to be described in this section, it is employed in some of the definitional equations described in Section VII.

The value ΣX_T^2 represents the total sum of the squared scores of the N subjects who participate in the experiment. Thus:

$$\Sigma X_T^2 = \Sigma X_1^2 + \Sigma X_2^2 + \cdots + \Sigma X_k^2$$

Since there are $k = 3$ groups, $\Sigma X_T^2 = 856$.

$$\Sigma X_T^2 = \Sigma X_1^2 + \Sigma X_2^2 + \Sigma X_3^2 = 426 + 227 + 203 = 856$$

Although the group means are not required for computing the analysis of variance test statistic, it is recommended they be computed since visual inspection of the group means can provide the researcher with a general idea of whether or not it is reasonable to expect a significant result. To be more specific, if two or more of the group means are far removed from one another, it is likely that the analysis of variance will be significant (especially if the number of subjects in each group is reasonably large). Another reason for computing the group means is that they are required for comparing individual groups with one another, something that is often done following the analysis of variance on the full set of data. The latter types of comparisons are described in Section VI.

As noted in Section I, in order to compute the test statistic for the **single-factor between-subjects analysis of variance**, the **total variability** in the data is divided into **between-groups variability** and **within-groups variability**. In order to do this, the following values are computed: a) The **total sum of squares** which is represented by the notation SS_T; b) The **between-groups sum of squares** which is represented by the notation SS_{BG}. The **between-groups sum of squares** is the numerator of the equation which represents **between-groups variability** (i.e., the equation that represents the amount of variability between the means of the k groups); and c) The **within-groups sum of squares** which is represented by the notation SS_{WG}. The **within-groups sum of squares** is the numerator of the equation which represents **within-groups variability** (i.e., the equation that represents the average amount of variability within each of the k groups, which, as noted earlier, represents error variability).

Equation 21.1 describes the relationship between SS_T, SS_{BG}, and SS_{WG}.

$$SS_T = SS_{BG} + SS_{WG} \qquad \text{(Equation 21.1)}$$

Equation 21.2 is employed to compute SS_T.

$$SS_T = \Sigma X_T^2 - \frac{(\Sigma X_T)^2}{N} \qquad \text{(Equation 21.2)}$$

Employing Equation 21.2, the value SS_T = 49.33 is computed.

$$SS_T = 856 - \frac{(110)^2}{15} = 49.33$$

Equation 21.3 is employed to compute SS_{BG}. In Equation 21.3, the notation n_j and ΣX_j, respectively, represent the values of n and ΣX for the jth group/sample.

$$SS_{BG} = \sum_{j=1}^{k} \left[\frac{(\Sigma X_j)^2}{n_j} \right] - \frac{(\Sigma X_T)^2}{N} \qquad \textbf{(Equation 21.3)}$$

The notation $\sum_{j=1}^{k} [(\Sigma X_j)^2/n_j]$ in Equation 21.3 indicates that for each group the value $(\Sigma X_j)^2/n_j$ is computed, and the latter values are summed for all k groups. When there are an equal number of subjects in each group (as is the case in Example 21.1), the notation n can be employed in Equation 21.3 in place of n_j.[5]

With reference to Example 21.1, Equation 21.3 can be rewritten as follows:

$$SS_{BG} = \left[\frac{(\Sigma X_1)^2}{n_1} + \frac{(\Sigma X_2)^2}{n_2} + \frac{(\Sigma X_3)^2}{n_3} \right] - \frac{(\Sigma X_T)^2}{N}$$

Substituting the appropriate values from Example 21.1 in Equation 21.3, the value SS_{BG} = 26.53 is computed.[6]

$$SS_{BG} = \left[\frac{(46)^2}{5} + \frac{(33)^2}{5} + \frac{(31)^2}{5} \right] - \frac{(110)^2}{15} = 833.2 - 806.67 = 26.53$$

By algebraically transposing the terms in Equation 21.1, the value of SS_{WG} can be computed with Equation 21.4.

$$SS_{WG} = SS_T - SS_{BG} \qquad \textbf{(Equation 21.4)}$$

Employing Equation 21.4, the value SS_{WG} = 22.80 is computed.

$$SS_{WG} = 49.33 - 26.53 = 22.80$$

Since the value obtained with Equation 21.4 is a function of the values obtained with Equations 21.2 and 21.3, if the computations for either of the latter two equations are incorrect Equation 21.4 will not yield the correct value for SS_{WG}. For this reason, one may prefer to compute the value of SS_{WG} with Equation 21.5.

$$SS_{WG} = \sum_{j=1}^{k} \left[\Sigma X_j^2 - \frac{(\Sigma X_j)^2}{n_j} \right] \qquad \textbf{(Equation 21.5)}$$

The summation sign $\sum_{j=1}^{k}$ in Equation 21.5 indicates that for each group the value $\Sigma X_j^2 - [(\Sigma X_j)^2/n_j]$ is computed, and the latter values are summed for all k groups. With reference to Example 21.1, Equation 21.5 can be written as follows:

$$SS_{WG} = \left[\sum X_1^2 - \frac{(\sum X_1)^2}{n_1}\right] + \left[\sum X_2^2 - \frac{(\sum X_2)^2}{n_2}\right] + \left[\sum X_3^2 - \frac{(\sum X_3)^2}{n_3}\right]$$

Employing Equation 21.5, the value $SS_{WG} = 22.80$ is computed, which is the same value computed with Equation 21.4.[7]

$$SS_{WG} = \left[426 - \frac{(46)^2}{5}\right] + \left[227 - \frac{(33)^2}{5}\right] + \left[203 - \frac{(31)^2}{5}\right] = 22.80$$

The reader should take note of the fact that the values SS_T, SS_{BG}, and SS_{WG} must always be positive numbers. If a negative value is obtained for any of the aforementioned values, it indicates a computational error has been made.

At this point the values of the **between-groups variance** and the **within-groups variance** can be computed. In the **single-factor between-subjects analysis of variance**, the **between-groups variance** is referred to as the **mean square between-groups**, which is represented by the notation MS_{BG}. MS_{BG} is computed with Equation 21.6. (The term **mean square** reflects the fact that the definition of the **variance** is that it is the **mean/average of the squared difference scores**.)

$$MS_{BG} = \frac{SS_{BG}}{df_{BG}} \qquad \text{(Equation 21.6)}$$

The **within-groups variance** is referred to as the **mean square within-groups**, which is represented by the notation MS_{WG}. MS_{WG} is computed with Equation 21.7.[8]

$$MS_{WG} = \frac{SS_{WG}}{df_{WG}} \qquad \text{(Equation 21.7)}$$

Note that a total mean square is not computed.

In order to compute MS_{BG} and MS_{WG}, it is required that the values df_{BG} and df_{WG} (the denominators of Equations 21.6 and 21.7) be computed. df_{BG}, which represents the **between-groups degrees of freedom**, are computed with Equation 21.8.

$$df_{BG} = k - 1 \qquad \text{(Equation 21.8)}$$

df_{WG}, which represents the **within-groups degrees of freedom**, are computed with Equation 21.9.[9]

$$df_{WG} = N - k \qquad \text{(Equation 21.9)}$$

Although it is not required in order to determine the F ratio, the **total degrees of freedom** are generally computed, since it can be used to confirm the df values computed with Equations 21.8 and 21.9, as well as the fact that it is employed in the analysis of variance summary table. The total degrees of freedom (represented by the notation df_T) are computed with Equation 21.10.[10]

$$df_T = N - 1 \qquad \text{(Equation 21.10)}$$

The relationship between df_{BG}, df_{WG}, and df_T is described by Equation 21.11.

$$df_T = df_{BG} + df_{WG} \qquad \textbf{(Equation 21.11)}$$

Employing Equations 21.8–21.10, the values $df_{BG} = 2$, $df_{WG} = 12$, and $df_T = 14$ are computed. Note that $df_T = df_{BG} + df_{WG} = 2 + 12 = 14$.

$$df_{BG} = 3 - 1 = 2 \qquad df_{WG} = 15 - 3 = 12 \qquad df_T = 15 - 1 = 14$$

Employing Equations 21.6 and 21.7, the values $MS_{BG} = 13.27$ and $MS_{WG} = 1.9$ are computed.

$$MS_{BG} = \frac{26.53}{2} = 13.27 \qquad MS_{WG} = \frac{22.8}{12} = 1.9$$

The F ratio, which is the test statistic for the **single-factor between-subjects analysis of variance**, is computed with Equation 21.12.

$$F = \frac{MS_{BG}}{MS_{WG}} \qquad \textbf{(Equation 21.12)}$$

Employing Equation 21.12, the value $F = 6.98$ is computed.

$$F = \frac{13.27}{1.9} = 6.98$$

The reader should take note of the fact that the values MS_{BG}, MS_{WG}, and F must always be positive numbers. If a negative value is obtained for any of the aforementioned values, it indicates a computational error has been made. If $MS_{WG} = 0$, Equation 21.12 will be insoluble. The only time $MS_{WG} = 0$ is when within each group all subjects obtain the same score (i.e., there is no within-groups variability). If all of the groups have the identical mean value, $MS_{BG} = 0$, and if the latter is true, $F = 0$.

V. Interpretation of the Test Results

It is common practice to summarize the results of a **single-factor between-subjects analysis of variance** with the summary table represented by Table 21.2.

Table 21.2 Summary Table of Analysis of Variance for Example 21.1

Source of variation	SS	df	MS	F
Between-groups	26.53	2	13.27	6.98
Within-groups	22.80	12	1.90	
Total	**49.33**	**14**		

The obtained value $F = 6.98$ is evaluated with **Table A10 (Table of the F Distribution)** in the **Appendix**. In **Table A10** critical values are listed in reference to the number of degrees of freedom associated with the numerator and the denominator of the F ratio (i.e., df_{num} and df_{den}).[11] In employing the F distribution in reference to Example 21.1, the degrees of freedom for the numerator are $df_{BG} = 2$ and the degrees of freedom for the denominator are $df_{WG} = 12$. In **Table A10** the tabled $F_{.95}$ and $F_{.99}$ values are, respectively, employed to evaluate the nondirectional alternative hypothesis H_1: Not H_0 at the .05 and .01 levels. Throughout the discussion of the analysis of variance the notation $F_{.05}$ is employed to represent the tabled critical F value at the .05 level. The latter value corresponds to the relevant tabled $F_{.95}$ value in **Table A10**. In the same respect, the notation $F_{.01}$ is employed to represent the tabled critical F value at the .01 level, and corresponds to the relevant tabled $F_{.99}$ value in **Table A10**.

For $df_{num} = 2$ and $df_{den} = 12$, the tabled $F_{.95}$ and $F_{.99}$ values are $F_{.95} = 3.89$ and $F_{.99} = 6.93$. Thus, $F_{.05} = 3.89$ and $F_{.01} = 6.93$. In order to reject the null hypothesis, the obtained F value must be equal to or greater than the tabled critical value at the prespecified level of significance. Since $F = 6.98$ is greater than $F_{.05} = 3.89$ and $F_{.01} = 6.93$, the alternative hypothesis is supported at both the .05 and .01 levels.

A summary of the analysis of Example 21.1 with the **single-factor between-subjects analysis of variance** follows: It can be concluded there is a significant difference between at least two of the three groups exposed to different levels of noise. This result can be summarized as follows: $F(2, 12) = 6.98$, $p < .01$.

VI. Additional Analytical Procedures for the Single-Factor Between-Subjects Analysis of Variance and/or Related Tests

1. Comparisons following computation of the omnibus F value for the single-factor between-subjects analysis of variance The F value computed with the analysis of variance is commonly referred to as the **omnibus F value**. The latter term implies that the obtained F value is based on an evaluation of all k group means. Recollect that in order to reject the null hypothesis, it is only required that at least two of the k group means differ significantly from one another. As a result of this, the omnibus F value does not indicate whether just two or, in fact, more than two groups have mean values which differ significantly from one another. In order to answer this question it is necessary to conduct additional tests, which are referred to as **comparisons** (since they involve comparing the means of two or more groups with one another).

Researchers are not in total agreement with respect to the appropriate protocol for conducting comparisons.[12] The basis for the disagreement revolves around the fact that each comparison one conducts increases the likelihood of committing at least one Type I error within a set of comparisons. For this reason, it can be argued that a researcher should employ a lower Type I error rate per comparison to insure that the overall likelihood of committing at least one Type I error in the set of comparisons does not exceed a prespecified alpha value which is reasonably low (e.g., $\alpha = .05$). At this point in the discussion the following two terms are defined: a) The **familywise Type I error rate** (represented by the notation α_{FW}) is the likelihood that there will be at least one Type I error in a **set of c comparisons**;[13] and b) The **per comparison Type I error rate** (represented by the notation α_{PC}) is the likelihood that any single comparison will result in a Type I error.

Equation 21.13 defines the relationship between the **familywise Type I error rate** and the **per comparison Type I error rate**, where c = the number of comparisons.[14]

$$\alpha_{FW} = 1 - (1 - \alpha_{PC})^c \qquad \text{(Equation 21.13)}$$

Let us assume that upon computing the value $F = 6.98$ for Example 21.1, the researcher decides to compare each of the three group means with one another — i.e., $\bar{X}_1$ versus $\bar{X}_2$; $\bar{X}_1$ versus $\bar{X}_3$; and $\bar{X}_2$ versus $\bar{X}_3$. The three aforementioned comparisons can be conceptualized as a **family/set** of comparisons, with $c = 3$. If for each of the comparisons the researcher establishes the value $\alpha_{PC} = .05$, employing Equation 21.13 it can be determined that the **familywise Type I error rate** will equal $\alpha_{FW} = .14$. This result tells the researcher that the likelihood of committing at least one Type I error in the set of three comparisons is .14.

$$\alpha_{FW} = 1 - (1 - .05)^3 = .14$$

Equation 21.14, which is computationally more efficient than Equation 21.13, can be employed to provide an approximation of α_{FW}.[15]

$$\alpha_{FW} = (c)(\alpha_{PC}) \qquad \textbf{(Equation 21.14)}$$

Employing Equation 21.14, the value $\alpha_{FW} = .15$ is computed.

$$\alpha_{FW} = (3)(.05) = .15$$

Note that the **familywise Type I error rate** $\alpha_{FW} = .14$ is almost three times the value of the **per comparison Type I error rate** $\alpha_{PC} = .05$. Of greater importance is the fact that the value $\alpha_{FW} = .14$ is considerably higher than .05, the usual maximum value permitted for a Type I error rate in hypothesis testing. Thus, some researchers would consider the value $\alpha_{FW} = .14$ to be excessive, and if, in fact, a maximum **familywise Type I error rate** of $\alpha_{FW} = .05$ is stipulated by a researcher, it is required that the Type I error rate for each comparison (i.e., α_{PC}) be reduced. Through use of Equation 21.15 or Equation 21.16 (which are, respectively, the algebraic transpositions of Equation 21.13 and Equation 21.14), it can be determined that in order to have $\alpha_{FW} = .05$, the value of α_{PC} must equal .017.[16]

$$\alpha_{PC} = 1 - \sqrt[c]{1 - \alpha_{FW}} = 1 - \sqrt[3]{1 - .05} = .017 \qquad \textbf{(Equation 21.15)}$$

$$\alpha_{PC} = \frac{\alpha_{FW}}{c} = \frac{.05}{3} = .0167 \qquad \textbf{(Equation 21.16)}$$

The reader should take note of the fact that although a reduction in the value of α_{FW} reduces the likelihood of committing a Type I error, it increases the likelihood of committing a Type II error (i.e., not rejecting a false null hypothesis). Thus, as one reduces the value of α_{FW}, the power associated with each of the comparisons which is conducted is reduced. In view of this, it should be apparent that if a researcher elects to adjust the value of α_{FW}, he must consider the impact it will have on the Type I versus Type II error rates for all of the comparisons which are conducted within the set of comparisons.

A number of different strategies have been developed with regard to what a researcher should do about adjusting the **familywise Type I error rate**. These strategies are employed within the framework of the following two types of comparisons which can be conducted following an omnibus F test: **planned comparisons** versus **unplanned comparisons**. The distinction between **planned** and **unplanned comparisons** follows.

Planned comparisons (also known as *a priori* **comparisons**) **Planned comparisons** are comparisons a researcher plans prior to collecting the data for a study. In a well designed experiment one would expect that a researcher will probably predict differences between specific groups prior to conducting the study. As a result of this, there is general agreement that following the computation of an omnibus F value, a researcher is justified in conducting any comparisons which have been planned beforehand, regardless of whether or not the omnibus F value is significant. In point of fact, in experiments involving an independent variable which is comprised of three of more levels, some researchers may not even bother to conduct the omnibus F test, but only conduct one or more of the comparisons which were planned prior to the study. Although most sources state that when planned comparisons are conducted it is not necessary to adjust the **familywise Type I error rate**, under certain conditions (such as when there are a large number of planned comparisons) an argument can be made for adjusting the value of α_{FW}.

In actuality, there are two types of planned comparisons (as well as unplanned comparisons) that can be conducted, which are referred to as **simple comparisons** versus **complex comparisons**. A **simple comparison** is any comparison in which two groups are compared with one another. For instance: Group 1 versus Group 2 (i.e., $\bar{X}_1$ versus $\bar{X}_2$, which allows one to evaluate the null hypothesis H_0: $\mu_1 = \mu_2$). Simple comparisons are often referred to as **pairwise comparisons**. Note that the total number of simple/pairwise comparisons which can be conducted is the **combination of k things taken two at a time** (where k represents the number of groups). The latter can be expressed as $\binom{n}{k}$, which is Equation I.48 in the **Introduction**.

Thus, in the case Example 21.1, where $k = 3$, $\binom{3}{2} = \dfrac{3!}{2!\,(3-2)!} = 3$. The three possible pairwise comparisons that can be conducted are $\bar{X}_1$ versus $\bar{X}_2$, $\bar{X}_1$ versus $\bar{X}_3$, and $\bar{X}_2$ versus $\bar{X}_3$,

A **complex comparison** is any comparison in which the combined performance of two or more groups is compared with the performance of one of the other groups or the combined performance of two or more of the other groups. For instance: Group 1 versus the average of Groups 2 and 3 (i.e., $\bar{X}_1$ versus $(\bar{X}_2 + \bar{X}_3)/2$, which evaluates the null hypothesis H_0: $\mu_1 = (\mu_2 + \mu_3)/2$). If there are four groups, one can conduct a complex comparison involving the average of Groups 1 and 2 versus the average of Groups 3 and 4 (i.e., $(\bar{X}_1 + \bar{X}_2)/2$ versus $(\bar{X}_3 + \bar{X}_4)/2$, which evaluates the null hypothesis H_0: $(\mu_1 + \mu_2)/2 = (\mu_3 + \mu_4)/2$).

It should be noted that if the omnibus F value is significant, it indicates there is at least one significant difference among all of the possible comparisons which can be conducted. Grissom and Kim (2005, p. 119), Kirk (1982, 1995) and Maxwell and Delaney (1990, 2000), among others, note that in such a situation it is theoretically possible that none of the simple comparisons are significant (especially if a conservative comparison procedure such as the **Bonferroni–Dunn** or **Scheffé** tests described later in the section is employed), and that the one (or perhaps more than one) significant comparison is a complex comparison. (Grissom and Kim (2005, p. 130) note it is also possible to obtain a nonsignificant omnibus F value, yet obtain a significant result for one or more comparisons.) It is important to note that regardless of what types of comparisons a researcher conducts, all comparisons should be meaningful within the context of the problem under study, and, as a general rule, comparisons should not be redundant with respect to one another.

Unplanned comparisons (also known as **post hoc**, **multiple**, or *a posteriori* **comparisons**) An **unplanned comparison** (which can be either a simple or complex comparison) is a comparison a researcher decides to conduct after collecting the data for a study. In conducting **unplanned comparisons**, following the data collection phase of a study a researcher examines the values of the k group means, and at that point decides which groups to compare with one another. Although for many years most researchers argued that unplanned comparisons should not be conducted unless the omnibus F value is significant, more recently many researchers (including this author) have adopted the viewpoint that it is acceptable to conduct unplanned comparisons regardless of whether or not a significant F value is obtained. Although there is general agreement among researchers that the **familywise Type I error rate** should be adjusted when unplanned comparisons are conducted, there is a lack of consensus with regard to the degree of adjustment that is required. For example, Anderson (2001, p. 536) notes that some sources consider an α_{FW} value of .05 or even .10 to be too stringent, and endorses employing a value as large as $\alpha_{FW} = .25$. Certainly the magnitude of α_{FW} a researcher employs should be a function of the number of experimental treatments employed in a study. Consequently, one would have a hard time justifying the use of $\alpha_{FW} = .25$ when $k = 3$, yet might have a convincing argument if $k = 10$.

The lack of consensus with respect to what value should be employed for α_{FW} is reflected in the fact that a variety of unplanned comparison procedures have been developed, each of which employs a different method for adjusting the value of α_{FW}. More often than not, when unplanned comparisons are conducted, a researcher will compare each of the k groups with all of the other $(k - 1)$ groups (i.e., all possible comparisons between pairs of groups are made). This "shotgun" approach, which maximizes the number of comparisons conducted, represents the classic situation for which most sources argue it is imperative to control the value of α_{FW}.

The rationale behind the argument that it is more important to adjust the value of α_{FW} in the case of unplanned comparisons as opposed to planned comparisons will be illustrated with a simple example.[17] Let us assume that a set of data is evaluated with an analysis of variance, and a significant omnibus F value is obtained. Let us also assume that within the whole set of data it is possible to conduct 20 comparisons between pairs of means and/or combinations of means. If the truth were known, however, no differences exist between the means of any of the populations being compared. However, in spite of the fact that none of the population means differ, within the set of 20 possible comparisons, one comparison (specifically, the one involving $\bar{X}_1$ versus $\bar{X}_2$) results in a significant difference at the .05 level. In this example we will assume that the difference $\bar{X}_1 - \bar{X}_2$ is the largest difference between any pair of means or combination of means in the set of 20 possible comparisons. If, in fact, $\mu_1 = \mu_2$, a significant result obtained for the comparison $\bar{X}_1$ versus $\bar{X}_2$ will represent a Type I error.

Let us assume that the comparison $\bar{X}_1$ versus $\bar{X}_2$ is planned beforehand, and it is the only comparison the researcher intends to make. Since there are 20 possible comparisons, the researcher has only a 1 in 20 chance (i.e., .05) of conducting the comparison Group 1 versus Group 2, and in the process commit a Type I error. If, on the other hand, the researcher does not plan any comparisons beforehand, but after computing the omnibus F value decides to make all 20 possible comparisons, he has a 100% chance of making a Type I error, since it is certain he will compare Groups 1 and 2. Even if the researcher decides to make only one unplanned comparison — specifically, the one involving the largest difference between any pair of means or combination of means — he will also have a 100% chance of committing a Type I error, since the comparison he will make will be $\bar{X}_1$ versus $\bar{X}_2$. This example illustrates that from a probabilistic viewpoint, the **familywise Type I error rate** associated with a set of unplanned comparisons will be higher than the rate for a set of planned comparisons.

The remainder of this section will describe the most commonly recommended comparison procedures. Specifically, the following procedures will be described: a) **Linear contrasts**; b) **Multiple t tests/Fisher's LSD test**; c) **The Bonferroni–Dunn test**; d) **Tukey's HSD test**; e) **The Newman–Keuls test**; f) **The Scheffé test**; and g) **The Dunnett test**. Although any of the aforementioned comparison procedures can be used for both planned and unplanned comparisons, **linear contrasts** and **multiple t tests/Fisher's LSD test** (which do not control the value of α_{FW}) are generally described within the context of planned comparisons. Some sources, however, do employ the latter procedures for unplanned comparisons. The **Bonferroni–Dunn test**, **Tukey's HSD test**, the **Newman–Keuls test**, the **Scheffé test**, and the **Dunnett test** are generally described as unplanned comparison procedures which are employed when a researcher wants to control the value of α_{FW}. The point that will be emphasized throughout the discussion to follow is that the overriding issue in selecting a comparison procedure is whether or not the researcher wants to control the value of α_{FW}, and if so, to what degree. This latter issue is essentially what determines the difference between the various comparison procedures to be described in this section.

Linear contrasts Many sources employ the terms **contrast** and **comparison** synonymously. The term **linear contrast** refers a **planned comparison** which evaluates whether or not the difference between two or more means or linear combinations involving two or more means is some value other than zero. Another way to define a **contrast** is that it represents a **weighted sum of population means** (employing sample means to estimate the population means). As will be demonstrated later in this section, the value of a **contrast** (which is represented by the lower case Greek letter **psi** (ψ)) can be computed by summing the products of contrast weights and estimated population means.

In the case of a simple comparison, a **linear contrast** consists of comparing two of the group means with one another (e.g., $\bar{X}_1$ versus $\bar{X}_2$). In the case of a complex comparison, the combined performance of two or more groups is compared with the performance of one of the other groups or the combined performance of two or more of the other groups (e.g., $\bar{X}_1$ versus $(\bar{X}_2 + \bar{X}_3)/2$). In conducting both simple and complex comparisons, the researcher must assign weights, which are referred to as **coefficients**, to all of the group means. These coefficients reflect the relative contribution of each of the group means to the two mean values which are being contrasted with one another in the comparison. In the case of a complex comparison, at least one of the two mean values that are contrasted in the comparison will be based on a weighted combination of the means of two or more of the groups.

The use of **linear contrasts** will be described for both simple and complex comparisons. In the examples which will be employed to illustrate **linear contrasts**, it will be assumed that all comparisons are planned beforehand, and that the researcher is making no attempt to control the value of α_{FW}. Thus, for each comparison to be conducted it will be assumed that $\alpha_{PC} = .05$. All of the comparisons (both simple and complex) to be described in this section are referred to as **single degree of freedom (df) comparisons**. This is the case, since one degree of freedom is always employed in the numerator of the F ratio (which represents the test statistic for a comparison).[18]

Linear contrast of a planned simple comparison   Let us assume that prior to obtaining the data for Example 21.1, the experimenter hypothesizes there will be a significant difference between Group 1 (**no noise**) and Group 2 (**moderate noise**). After conducting the omnibus F test, the simple planned comparison $\bar{X}_1$ versus $\bar{X}_2$ is conducted to compare the performance of the two groups. The null and alternative hypotheses for the comparison follow: $H_0: \mu_1 = \mu_2$ versus $H_1: \mu_1 \neq \mu_2$.[19] Table 21.3 summarizes the information required to conduct the planned comparison $\bar{X}_1$ versus $\bar{X}_2$.

Table 21.3   Planned Simple Comparison: Group 1 versus Group 2

Group	$\bar{X}_1$	Coefficient (c_j)	Product $(c_j)(\bar{X}_j)$	Squared Coefficient (c_j^2)
1	9.2	+1	$(+1)(9.2) = +9.2$	1
2	6.6	−1	$(−1)(6.6) = −6.6$	1
3	6.2	0	$(0)(6.2) =$ 0	0
		$\Sigma c_j = 0$	$\Sigma(c_j)(\bar{X}_j) =$ 2.6	$\Sigma c_j^2 = 2$

The following should be noted with respect to Table 21.3: a) The rows of Table 21.3 represent data for each of the three groups employed in the experiment. Even though the comparison involves two of the three groups, the data for all three groups are included to illustrate how the group which is not involved in the comparison is eliminated from the calculations; b) Column 2 contains the mean score of each of the groups; c) In Column 3 each of the groups is assigned a **coefficient**, represented by the notation c_j. The value of c_j assigned to each group is a weight that reflects the proportional contribution of the group to the comparison. Any group not involved in the comparison (in this instance Group 3) is assigned a coefficient of zero. Thus, $c_3 = 0$. When only two groups are involved in a comparison, one of the groups (it does not matter which one) is assigned a coefficient of +1 (in this instance Group 1 is assigned the coefficient $c_1 = +1$) and the other group a coefficient of -1 (i.e., $c_2 = -1$). Note that Σc_j, the sum of the coefficients (which is the sum of Column 3), must always equal zero (i.e., $c_1 + c_2 + c_3 = (+1) + (-1) + 0 = 0$); d) In Column 4 a product is obtained for each group. The product for a group is obtained by multiplying the mean of the group $(\bar{X}_j)$ by the coefficient which has been assigned to that group (c_j). Although it may not be immediately apparent from looking at the table, the sum of Column 4, $\Sigma(c_j)(\bar{X}_j)$ (which is often referred to as the **sum of the weighted treatment means**) is, in fact, the difference between the two means being compared (i.e., $\Sigma(c_j)(\bar{X}_j)$ is equal to $\bar{X}_1 - \bar{X}_2 = 9.2 - 6.6 = 2.6$), and the latter value defines the contrast being evaluated — i.e., $\psi = \Sigma(c_j)(\bar{X}_j)$. As was noted, $\Sigma(c_j)(\bar{X}_j)$, equals the difference between the two mean values stipulated in the null hypothesis for a **single degree of freedom** comparison. The information in Column 4 can be summarized as $\psi = \Sigma(c_j)(\bar{X}_j) = (+1)(9.2) + (-1)(6.6) + (0)(6.2) = 2.6$; and e) Σc_j^2, the sum of Column 5, is the sum of the squared coefficients.[20]

The test statistic for the comparison is an F ratio, represented by the notation F_{comp}. In order to compute the value F_{comp}, a **sum of squares** (SS_{comp}), a **degrees of freedom value** (df_{comp}), and a **mean square** (MS_{comp}) for the comparison must be computed. The **comparison sum of squares** (SS_{comp}) is computed with Equation 21.17. Note that Equation 21.17 assumes there are an equal number of subjects (n) in each group.[21]

$$SS_{comp} = \frac{n\left[\Sigma(c_j)(\bar{X}_j)\right]^2}{\Sigma c_j^2} \qquad \text{(Equation 21.17)}$$

Substituting the appropriate values from Example 21.1 in Equation 21.17, the value $SS_{comp} = 16.9$ is computed.

$$SS_{comp} = \frac{5(2.6)^2}{2} = 16.9$$

The **comparison mean square** (MS_{comp}) is computed with Equation 21.18. MS_{comp} represents a measure of between-groups variability which takes into account just the two group means involved in the comparison.

$$MS_{comp} = \frac{SS_{comp}}{df_{comp}} \qquad \text{(Equation 21.18)}$$

In a **single degree of freedom comparison**, df_{comp} will always equal 1 since the number of mean values being compared in such a comparison will always be $k_{comp} = 2$, and $df_{comp} = k_{comp} - 1 = 2 - 1 = 1$. Substituting the values $SS_{comp} = 16.9$ and $df_{comp} = 1$ in Equation 21.18, the value $MS_{comp} = 16.9$ is computed. Note that since in a **single degree of freedom comparison** the value of df_{comp} will always equal 1, the values SS_{comp} and MS_{comp} will always be equivalent.

$$MS_{comp} = \frac{16.9}{1} = 16.9$$

The test statistic F_{comp} is computed with Equation 21.19. F_{comp} is a ratio which is comprised of the variability of the two means involved in the comparison divided by the within-groups variability employed for the omnibus F test.[22]

$$F_{comp} = \frac{MS_{comp}}{MS_{WG}} \qquad \text{(Equation 21.19)}$$

Substituting the values $MS_{comp} = 16.9$ and $MS_{WG} = 1.9$ in Equation 21.19, the value $F_{comp} = 8.89$ is computed.

$$F_{comp} = \frac{16.9}{1.9} = 8.89$$

The value $F_{comp} = 8.89$ is evaluated with **Table A10**. Employing **Table A10**, the appropriate degrees of freedom value for the numerator is $df_{num} = 1$. This is the case, since the numerator of the F_{comp} ratio is MS_{comp}, and the degrees of freedom associated with the latter value is $df_{comp} = 1$. The denominator degrees of freedom will be the value of df_{WG} employed for the omnibus F test, which for Example 21.1 is $df_{WG} = 12$.

For $df_{num} = 1$ and $df_{den} = 12$, the tabled critical .05 and .01 F values are $F_{.05} = 4.75$ and $F_{.01} = 9.33$. Since the obtained value $F_{comp} = 8.89$ is greater than $F_{.05} = 4.75$, the nondirectional alternative hypothesis H_1: $\mu_1 \neq \mu_2$ is supported at the .05 level. Since $F_{comp} = 8.89$ is less than $F_{.01} = 9.33$, the latter alternative hypothesis is not supported at the .01 level. Thus, if the value $\alpha = .05$ is employed, the researcher can conclude that Group 1 recalled a significantly greater number of nonsense syllables than Group 2.

With respect to Example 21.1, it is possible to conduct the following additional simple comparisons: $\overline{X}_1$ versus $\overline{X}_3$; $\overline{X}_2$ versus $\overline{X}_3$ (both of which would represent **linear contrasts** if they were planned comparisons). The latter two simple comparisons will be conducted later employing **multiple t tests/Fisher's LSD test** (which, as will be noted, are computationally equivalent to the **linear contrast** procedure described in this section).

Linear contrast of a planned complex comparison   Let us assume that prior to obtaining the data for Example 21.1, the experimenter hypothesizes there will be a significant difference between the performance of Group 3 (**extreme noise**) and the combined performance of Group 1 (**no noise**) and Group 2 (**moderate noise**). Such a comparison is a complex comparison, since it involves a single group being contrasted with two other groups. As is the case with the simple comparison $\overline{X}_1$ versus $\overline{X}_2$, two means are also contrasted within the framework of the complex comparison. However, one of the two means is a composite mean which is based upon the combined performance of two groups. The complex comparison represents a single degree of freedom comparison. This is the case, since two means are being contrasted with one another — specifically, the mean of Group 3 with the composite mean of Groups 1 and 2. The fact that it is a single degree of freedom comparison is also reflected in the fact that there is one equals sign (=) in both the null and alternative hypotheses for the comparison. The null and alternative hypotheses for the complex comparison are H_0: $\mu_3 = (\mu_1 + \mu_2)/2$ versus H_1: $\mu_3 \neq (\mu_1 + \mu_2)/2$.[23] Table 21.4 summarizes the information required to conduct the planned complex comparison $\overline{X}_3$ versus $(\overline{X}_1 + \overline{X}_2)/2$.

Table 21.4 Planned Complex Comparison: Group 3 versus Groups 1 and 2

Group	$(\bar{X}_j)$	Coefficient (c_j)	Product $(c_j)(\bar{X}_j)$	Squared Coefficient (c_j^2)
1	9.2	$-\frac{1}{2}$	$\left(-\frac{1}{2}\right)(9.2) = -4.6$	$\frac{1}{4}$
2	6.6	$-\frac{1}{2}$	$\left(-\frac{1}{2}\right)(6.6) = -3.3$	$\frac{1}{4}$
3	6.2	$+1$	$(+1)(6.2) = +6.2$	1
		$\Sigma c_j = 0$	$\Sigma(c_j)(\bar{X}_j) = -1.7$	$\Sigma c_j^2 = 1.5$

Note that the first two columns of Table 21.4 are identical to the first two columns of Table 21.3. The different values in the remaining columns of Table 21.4 result from the fact that different coefficients are employed for the complex comparison. The absolute value $\Sigma(c_j)(\bar{X}_j)$ $= -1.7$ represents the difference between the two sets of means contrasted in the null hypothesis — specifically, the difference between $\bar{X}_3 = 6.2$ and the composite mean of Group 1 ($\bar{X}_1 = 9.2$) and Group 2 ($\bar{X}_2 = 6.6$). The latter composite mean will be represented with the notation $\bar{X}_{1/2}$. Since $\bar{X}_{1/2} = (9.2 + 6.6)/2 = 7.9$, the difference between the two means evaluated with the comparison is $6.2 - 7.9 = -1.7$ (which is the same as the value $\Sigma(c_j)(\bar{X}_j) = -1.7$, the sum of Column 4 in Table 21.4).

Before conducting the computations for the complex comparison, a general protocol will be described for assigning coefficients to the groups involved in either a simple or complex comparison. Within the framework of describing the protocol, it will be employed to determine the coefficients for the complex comparison under discussion.

1) Write out the null hypothesis (i.e., H_0: $\mu_3 = (\mu_1 + \mu_2)/2$). Any group not involved in the comparison (i.e., not noted in the null hypothesis) is assigned a coefficient of zero. Since all three groups are included in the present comparison, none of the groups receives a coefficient of zero.

2) On each side of the equals sign of the null hypothesis write the number of group means designated in the null hypothesis. Thus:

$$H_0: \mu_3 = \frac{\mu_1 + \mu_2}{2}$$

1 mean   2 means

3) To obtain the coefficient for each of the groups included in the null hypothesis, employ Equation 21.20. The latter equation, which is applied to each side of the null hypothesis, represents the reciprocal of the number of means on a specified side of the null hypothesis.[24]

$$\text{Coefficient} = \frac{1}{\text{Number of group means}} \qquad \textbf{(Equation 21.20)}$$

Since there is only one mean to the left of the equals sign (μ_3), employing Equation 21.20 we determine that the coefficient for that group (Group 3) equals $\frac{1}{1} = 1$. Since there are two means to the right of the equals sign (μ_1 and μ_2), using Equation 21.20 we determine the coefficient for both of the groups to the right of the equals sign (Groups 1 and 2) equals $\frac{1}{2}$. Notice that all groups on the same side of the equals sign receive the same coefficient.[25]

4) The coefficient(s) on one side of the equals sign are assigned a positive sign, and the coefficient(s) on the other side of the equals sign are assigned a negative sign. Equivalent results will be obtained irrespective of which side of the equals sign is assigned positive versus negative coefficients. In the complex comparison under discussion, a positive sign is assigned to the coefficient to the left of the equals sign, and negative signs are assigned to coefficients to the right of the equals sign. Thus, the values of the coefficients are: $c_1 = -1/2$; $c_2 = -1/2$; $c_3 = +1$. Note that the sum of the coefficients must always equal zero (i.e., $c_1 + c_2 + c_3 = (-\frac{1}{2}) + (-\frac{1}{2}) + (+1)$).[26]

Equations 21.17–21.19, which are employed for the simple comparison, are also used to evaluate a complex comparison. Substituting the appropriate information from Table 21.4 in Equation 21.17, the value $SS_{comp} = 9.63$ is computed.

$$SS_{comp} = \frac{5(-1.7)^2}{1.5} = 9.63$$

Employing Equation 21.18, the value $MS_{comp} = 9.63$ is computed. Note that since the complex comparison is a single degree of freedom comparison, $df_{comp} = 1$.

$$MS_{comp} = \frac{9.63}{1} = 9.63$$

Employing Equation 21.19, the value $F_{comp} = 5.07$ is computed.

$$F_{comp} = \frac{9.63}{1.9} = 5.07$$

The protocol for evaluating the value $F_{comp} = 5.07$ computed for the complex comparison is identical to that employed for the simple comparison. In determining the tabled critical F value in **Table A10**, the same degrees of freedom values are employed. This is the case since the numerator degrees of freedom for any single degree of freedom comparison is $df_{comp} = 1$. The denominator degrees of freedom is df_{WG}, which, as in the case of the simple comparison, is the value of df_{WG} employed for the omnibus F test. Thus, the appropriate degrees of freedom for the complex comparison are $df_{num} = 1$ and $df_{den} = 12$. The tabled critical .05 and .01 F values in **Table A10** for the latter degrees of freedom are $F_{.05} = 4.75$ and $F_{.01} = 9.33$. Since the obtained value $F = 5.07$ is greater than $F_{.05} = 4.75$, the nondirectional alternative hypothesis H_1: $\mu_3 \neq (\mu_1 + \mu_2)/2$ is supported at the .05 level. Since $F = 5.07$ is less than $F_{.01} = 9.33$, the latter alternative hypothesis is not supported at the .01 level. Thus, if the value $\alpha = .05$ is employed, the researcher can conclude that Group 3 recalled a significantly fewer number of nonsense syllables than the average number recalled when the performance of Groups 1 and 2 are combined.

Orthogonal comparisons[27]   Most sources agree that intelligently planned studies involve a limited number of meaningful comparisons which the researcher plans prior to collecting the data. As a general rule, any comparisons that are conducted should address critical questions underlying the general hypothesis under study. Some researchers believe that it is not even necessary to obtain an omnibus F value if, in fact, the critical information one is concerned with is contained within the framework of the planned comparisons. It is generally recommended that the maximum number of planned comparisons one conducts should not exceed the value of df_{BG} employed for the omnibus F test. If the number of planned comparisons is equal to or less than df_{BG}, sources generally agree that a researcher is not obliged to adjust the value of α_{FW}. When, however, the number of planned comparisons exceeds df_{BG}, many sources recommend that the value of α_{FW} be adjusted. Applying this protocol to Example 21.1, since $df_{BG} = 2$, one can conduct two planned comparisons without being obliged to adjust the value of α_{FW}.

The subject of **orthogonal comparisons** is relevant to the general question of how many (and specifically which) comparisons a researcher should conduct. Orthogonal comparisons are defined as comparisons which are independent of one another. In other words, such comparisons are not redundant in that they do not overlap with respect to the information they provide. In point of fact, the two comparisons which have been conducted (i.e., the simple comparison of Group 1 versus Group 2, and the complex comparison of Group 3 versus the combined performance of Groups 1 and 2) are orthogonal comparisons. This can be demonstrated by employing Equation 21.21, which defines the relationship that will exist between two comparisons if they are orthogonal to one another. In Equation 21.21, c_{j1} is the coefficient assigned to Group j in Comparison 1 and c_{j2} is the coefficient assigned to Group j in Comparison 2. If, in fact, two comparisons are orthogonal, the sum of the products of the coefficients of all of k groups will equal zero.

$$\sum_{j=1}^{k} (c_{j_1})(c_{j_2}) = 0 \qquad \textbf{(Equation 21.21)}$$

Equation 21.21 is employed below with the two comparisons which have been conducted in this section. Notice that for each group, the first value in the parentheses is the coefficient for that group for the simple comparison (Group 1 versus Group 2), while the second value in parentheses is the coefficient for that group for the complex comparison (Group 3 versus Groups 1 and 2).

$$\text{Group 1} \qquad \text{Group 2} \qquad \text{Group 3}$$
$$(+1)\left(-\frac{1}{2}\right) \; + \; (-1)\left(-\frac{1}{2}\right) \; + \; (0)(+1) \; = \left(-\frac{1}{2}\right) + \left(+\frac{1}{2}\right) + 0 = 0$$

If there are k treatments, there will be $(k-1)$ (which corresponds to df_{BG} employed for the omnibus F test) orthogonal comparisons (also known as **orthogonal contrasts**) within each complete orthogonal set. This is illustrated by the fact that two comparisons comprise the orthogonal set demonstrated above — specifically, Group 1 versus Group 2, and Group 3 versus Groups 1 and 2. Actually, when (as is the case in Example 21.1) there are $k = 3$ treatments, there are three possible sets of orthogonal comparisons — each set being comprised of one simple comparison and one complex comparison. In addition to the set noted above, the following two additional orthogonal sets can be formed: a) Group 1 versus Group 3; Group 2 versus Groups 1 and 3; b) Group 2 versus Group 3; Group 1 versus Groups 2 and 3.

When $k > 3$ there will be more than 2 contrasts in a set of orthogonal contrasts. Within that full set of orthogonal contrasts, if the coefficients from any two of the contrasts are substituted in Equation 21.21, they will yield a value of zero. It should also be noted that the number of possible sets of contrasts will increase as the value of k increases. It is important to note, however, that when all possible sets of contrasts are considered, most of them will not be orthogonal. With respect to determining those contrasts which are orthogonal, Howell (2002) describes a simple protocol that can be employed to derive most (although not all) orthogonal contrasts in a body of data. The procedure described by Howell (2002) is summarized in Figure 21.1.

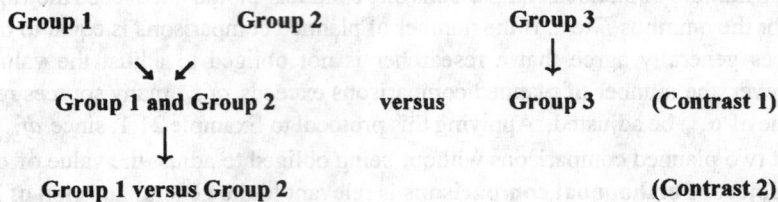

Figure 21.1   Tree Diagram for Determining Orthogonal Contrasts

In employing Figure 21.1, initially two blocks of groups are formed employing all k groups. A block can be comprised of one or more of the groups. In Figure 21.1, the first block is comprised of Groups 1 and 2, and the second block of Group 3. This will represent the first contrast, which corresponds to the complex comparison which is described in this section. Any blocks that remain which are comprised of two or more groups are broken down into smaller blocks. Thus, the block comprised of Group 1 and Group 2 is broken down into two blocks, each consisting of one group. The contrast of these two groups (Group 1 versus Group 2) represents the second contrast in the orthogonal set.

Figure 21.1 can also be employed to derive the other two possible orthogonal sets for Example 21.1. To illustrate, the initial two blocks derived can be a block consisting of Groups 1 and 3 and a second block consisting of Group 2. This represents the complex comparison of Group 2 versus Groups 1 and 3. The remaining block consisting of Groups 1 and 3 can be broken down into two blocks consisting of Group 1 and Group 3. The comparison of these two groups, which represents a simple comparison, constitutes the second comparison in that orthogonal set.

Note that once a group has been assigned to a block, and that block is compared to an adjacent block, from that point on any other comparisons involving that group will be with other groups that fall within its own block. Thus, in our example, if the first comparison is Groups 1 and 2 versus Group 3, the researcher cannot use the comparison Group 1 versus Group 3 as the second comparison for that set, since the two groups are in different blocks. If the latter comparison is conducted, the sum of the products of the coefficients of all k groups for the two comparisons will not equal zero, and thus not constitute an orthogonal set. To illustrate this latter fact, the coefficients of the three groups for the simple comparison depicted in Table 21.3 are rearranged as follows: $c_1 = +1$, $c_2 = 0$, and $c_3 = -1$. The latter coefficients are employed if the simple comparison Group 1 versus Group 3 is conducted. Equation 21.21 is now employed to demonstrate that the Group 1 versus Group 3 comparison is not orthogonal to the complex comparison Group 3 versus Groups 1 and 2 summarized in Table 21.4.

Group 1   Group 2   Group 3

$$(+1)\left(-\tfrac{1}{2}\right) \;+\; (0)\left(-\tfrac{1}{2}\right) \;+\; (-1)(+1) \;=\; \left(-\tfrac{1}{2}\right) + 0 + (-1) \;=\; -1\tfrac{1}{2}$$

It should be pointed out that when more than one set of orthogonal comparisons is conducted, since the different sets are not orthogonal to one another, many of the comparisons one conducts will not be independent of one another. For this reason, a researcher who does not want to conduct any nonindependent comparisons should only conduct those comparisons involving the orthogonal set which provide the most meaningful information with regard to the hypothesis under study. It should be noted, however, that there is no immutable rule which states that a researcher can only conduct orthogonal comparisons. Many sources point out there are times when the questions addressed by nonorthogonal comparisons can often contribute to a researcher's understanding of the general hypothesis under study.

Another characteristic of orthogonal comparisons is that the sum of squares for all comparisons which comprise a set of orthogonal comparisons equals the value of SS_{BG} for the omnibus F test. This reflects the fact that the variability in a set of orthogonal contrasts will account for all the between-groups variability in the full set of data. Employing the data for the simple comparison summarized in Table 21.3 and the complex comparison summarized in Table 21.4, it is confirmed below that the sum of squares for the latter two comparisons (which comprise an orthogonal set) equals the value $SS_{BG} = 26.53$ obtained with Equation 21.3.

$$SS_{BG} = 26.53 = SS_{\text{simple comparison}} + SS_{\text{complex comparison}} = 16.9 + 9.63 = 26.53$$

Test 21a: Multiple *t* tests/Fisher's LSD test   One option a researcher has available after computing an omnibus F value is to run **multiple *t* tests** (specifically, the *t* **test for two independent samples**), in order to determine whether there is a significant difference between any of the pairs of means which can be contrasted within the framework of either simple or complex comparisons. In point of fact, it can be algebraically demonstrated that in the case of both simple and complex comparisons, the use of **multiple *t* tests** will yield a result which is equivalent to that obtained with the protocol described for conducting **linear contrasts**. When **multiple *t* tests** are discussed as a procedure for conducting comparisons, most sources state that: a) **Multiple *t* tests** should only be employed for planned comparisons; b) Since **multiple *t* tests** are only employed for planned comparisons, they can be conducted regardless of whether or not the omnibus F value is significant; and c) In conducting **multiple *t* tests** for planned comparisons, the researcher is not required to adjust the value of α_{FW}, as long as a limited number of comparisons are conducted. (As noted earlier, most sources state the number of planned comparisons should not exceed df_{BG}.) All of the aforementioned stipulations noted for **multiple *t* tests** also apply to **linear contrasts** (since as noted above, **multiple *t* tests** and **linear contrasts** are computationally equivalent).

When, on the other hand, comparisons are unplanned and **multiple *t* tests** are employed to compare pairs of means, the use of **multiple *t* tests** within the latter context is referred to as **Fisher's LSD test** (the term **LSD** is an abbreviation for **least significant difference**). Since the use of **Fisher's LSD** test is only justified if the omnibus F value is significant, it attempts to control the value of α_{FW} by substantially reducing the number of unplanned comparisons that would be conducted if, in fact, the null hypothesis is true. When **Fisher's LSD test** is compared with other unplanned comparison procedures, it provides the most powerful test with respect to identifying differences between pairs of means, since it only minimally attempts to control the value of α_{FW} (and thus, among the unplanned comparison methods, it requires the smallest difference between two means in order to conclude that a difference is significant). However, **Fisher's LSD test** has the highest likelihood of committing one or more Type I errors in a set/family of comparisons. In the discussion to follow, since **multiple *t* tests** and **Fisher's LSD method** are computationally equivalent (as well as equivalent to **linear contrasts**), the term **multiple *t* tests/Fisher's LSD test** will refer to a computational procedure, which can be employed for both planned and unplanned comparisons, that does not adjust the value of α_{FW}.

Equation 21.22 can be employed to compute the test statistic (which employs the *t* distribution) for **multiple *t* tests/Fisher's LSD test**. Whereas Equation 21.22 can only be employed for simple comparisons, Equation 21.23 is a generic equation which can be employed for both simple and complex comparisons. It will be assumed that the null and alternative hypotheses for any comparisons conducted are as follows: H_0: $\mu_a = \mu_b$ versus H_1: $\mu_a \neq \mu_b$. In the case of a **simple comparison**, the value Σc_j^2 in Equation 21.23 will always equal 2, thus resulting in Equation 21.22.

$$t = \frac{\bar{X}_a - \bar{X}_b}{\sqrt{\dfrac{2MS_{WG}}{n}}} \qquad \textbf{(Equation 21.22)}$$

Where: $\bar{X}_a$ and $\bar{X}_b$ represent the two means contrasted in the comparison

$$t = \frac{\bar{X}_a - \bar{X}_b}{\sqrt{\dfrac{(\Sigma c_j^2)(MS_{WG})}{n}}} \qquad \textbf{(Equation 21.23)}$$

The degrees of freedom employed in evaluating the t value computed with Equations 21.22 and 21.23 is the value of df_{WG} computed for the omnibus F test. Thus, in the case of Example 21.1, the value $df_{WG} = 12$ is employed. Note that in Equations 21.22/21.23, the value MS_{WG} computed for the omnibus F test is employed in computing the **standard error of the difference** in the denominator of the t test equation, as opposed to the value $\sqrt{(\tilde{s}_a^2/n_a) + (\tilde{s}_b^2/n_b)}$, which is employed in Equation 11.1 (the equation for the t **test for two independent samples** when $n_a = n_b$). This is the case, since MS_{WG} is a pooled estimate of the population variance based on the full data set (i.e., the k groups for which the omnibus F value is computed).[28]

Equation 21.22 is employed below to conduct the simple comparison of Group 1 versus Group 2.

$$t = \frac{9.2 - 6.6}{\sqrt{\dfrac{(2)(1.9)}{5}}} = 2.99$$

The obtained value $t = 2.99$ is evaluated with **Table A2 (Table of Student's t Distribution)** in the **Appendix**. For $df = 12$, the tabled critical two-tailed .05 and .01 values are $t_{.05} = 2.18$ and $t_{.01} = 3.06$. Since $t = 2.99$ is greater than $t_{.05} = 2.18$, the nondirectional alternative hypothesis H_1: $\mu_a \neq \mu_b$ is supported at the .05 level. Since $t = 2.99$ is less than $t_{.01} = 3.06$, the latter alternative hypothesis is not supported at the .01 level. Thus, if $\alpha = .05$, the researcher can conclude that Group 1 recalled a significantly greater number of nonsense syllables than Group 2. This result is consistent with that obtained in the previous section using the protocol for **linear contrasts**.

Equation 21.24 employs **multiple t tests/Fisher's LSD test** to compute the minimum required difference in order for two means to differ significantly from one another at a prespecified level of significance. The latter value is represented by the notation CD_{LSD}, with CD being the abbreviation for **critical difference**. Whereas Equation 21.24 only applies to simple comparisons, Equation 21.25 is a generic equation which can be employed for both simple and complex comparisons.[29]

$$CD_{LSD} = \sqrt{F_{(1, WG)}} \sqrt{\frac{2MS_{WG}}{n}} \qquad \textbf{(Equation 21.24)}$$

$$CD_{LSD} = \sqrt{F_{(1,WG)}} \sqrt{\frac{(\Sigma c_j^2)(MS_{WG})}{n}} \qquad \textbf{(Equation 21.25)}$$

Where: $F_{(1,WG)}$ is the tabled critical F value for $df_{num} = 1$ and $df_{den} = df_{WG}$ at the prespecified level of significance[30]

Employing the appropriate values from Example 21.1 in Equation 21.24, the value $CD_{LSD} = 1.90$ is computed.[31]

$$CD_{LSD} = \sqrt{4.75}\sqrt{\frac{(2)(1.9)}{5}} = 1.90$$

Thus, in order to differ significantly at the .05 level, the means of any two groups must differ from one another by at least 1.90 units. Employing Table 21.5, which summarizes the differences between pairs of means involving all three experimental groups, it can be seen that the following simple comparisons are significant at the .05 level if **multiple *t* tests/Fisher's LSD test** are employed: $\bar{X}_1 - \bar{X}_2 = 2.6$; $\bar{X}_1 - \bar{X}_3 = 3$. The difference $\bar{X}_2 - \bar{X}_3 = .4$ is not significant, since it is less than $CD_{LSD} = 1.90$.

Within the framework of the discussion to follow, it will be demonstrated that the CD value computed with **multiple *t* tests/Fisher's LSD test** is the smallest CD value which can be computed with any of the comparison methods that can be employed for the analysis of variance.

Table 21.5 Differences between Pairs of Means in Example 21.1

$$\bar{X}_1 - \bar{X}_2 = 9.2 - 6.6 = 2.6$$
$$\bar{X}_1 - \bar{X}_3 = 9.2 - 6.2 = 3.0$$
$$\bar{X}_2 - \bar{X}_3 = 6.6 - 6.2 = 0.4$$

Multiple *t* tests/Fisher's LSD test will now be demonstrated for a complex comparison. Equations 21.23 and 21.25 are employed to evaluate the complex comparison involving the mean of Group 3 versus the composite mean of Groups 1 and 2. Employing Equation 21.23, the absolute value $t = 2.25$ is computed.[32] The value $CD_{LSD} = 1.65$ is computed with Equation 21.25. Note that in computing the values of t and CD_{LSD}, the value $\Sigma c_j^2 = 1.5$ is employed in Equations 21.23 and 21.25, as opposed to $\Sigma c_j^2 = 2$, which is employed in Equations 21.22 and 21.24. This latter fact accounts for why the value $CD_{LSD} = 1.65$ computed for the complex comparison is smaller than the value $CD_{LSD} = 1.90$ computed for the simple comparison.

$$t = \frac{6.2 - 7.9}{\sqrt{\frac{(1.5)(1.9)}{5}}} = -2.25$$

$$CD_{LSD} = \sqrt{4.75}\sqrt{\frac{(1.5)(1.9)}{5}} = 1.65$$

Since the obtained absolute value $t = 2.25$ is greater than $t_{.05} = 2.18$ (which is the tabled critical two-tailed .05 t value for $df_{WG} = 12$), the nondirectional alternative hypothesis H_1:

$\mu_3 \neq (\mu_1 + \mu_2)/2$ is supported at the .05 level. Since $t = 2.25$ is less than the tabled critical two-tailed value $t_{.01} = 3.06$, the latter alternative hypothesis is not supported at the .01 level. Thus, if $\alpha = .05$, the researcher can conclude that Group 3 recalled a significantly fewer number of nonsense syllables than the average number recalled when the performances of Groups 1 and 2 are combined. This result is consistent with that obtained in the previous section using the protocol for **linear contrasts**.

The fact that the obtained absolute value of the difference $|\overline{X}_3 - \overline{X}_{1/2}| = |6.2 - 7.9| = 1.7$ is larger than the computed value, $CD_{LSD} = 1.65$, is consistent with the fact that the difference for the **complex comparison** is significant at the .05 level. The computed value $CD_{LSD} = 1.65$ indicates that for any complex comparison involving the set of coefficients employed in Table 21.4, the minimum required difference in order for the two mean values stipulated in the null hypothesis to differ significantly from one another at the .05 level is $CD_{LSD} = 1.65$.

Test 21b: The Bonferroni–Dunn test   First formally described by Dunn (1961), the **Bonferroni–Dunn test** (which Anderson (2001, p. 527) credits to Fisher, who employed the term **splitting** for the procedure) is based on the **Bonferroni inequality**, which states that the probability of the occurrence of a set of events can never be greater than the sum of the individual probabilities for each event. Although the **Bonferroni–Dunn test** (often just referred to as the **Bonferroni test/procedure**) is identified in most sources as a planned comparison procedure, it can also be employed for unplanned comparisons. In actuality, the **Bonferroni–Dunn test** is computationally identical to **multiple t tests/Fisher's LSD test/linear contrasts**, except for the fact that the equation for the test statistic employs an adjustment in order to reduce the value of α_{FW}. By virtue of reducing α_{FW}, the power of the **Bonferroni–Dunn test** will always be less than the power associated with **multiple t tests/Fisher's LSD test/linear contrasts** (since the latter procedure does not adjust the value of α_{FW}). As a general rule, whenever the **Bonferroni–Dunn test** is employed to conduct all possible pairwise comparisons in a set of data, it provides the least powerful test of an alternative hypothesis of all the available comparison procedures.

The **Bonferroni–Dunn test** requires a researcher to initially stipulate the highest **family-wise Type I error rate** he is willing to tolerate. For purposes of illustration, let us assume that in conducting comparisons in reference to Example 21.1 the researcher does not want the value of α_{FW} to exceed .05 (i.e., he does not want more than a 5% chance of committing at least one Type I error in a set of comparisons). Let us also assume he either plans beforehand or decides after computing the omnibus F value that he will compare each of the group means with one another. This will result in the following three simple comparisons: Group 1 versus Group 2; Group 1 versus Group 3; Group 2 versus Group 3. To insure that the **familywise Type I error rate** does not exceed .05, $\alpha_{FW} = .05$ is divided by $c = 3$, which represents the number of comparisons which comprise the set. The resulting value $\alpha_{PC} = \alpha_{FW}/c = .05/3 = .0167$ represents, for each of the comparisons that are conducted, the likelihood of committing a Type I error. Thus, even if a Type I error is made for all three of the comparisons, the overall **familywise Type I error rate** will not exceed .05 (since $(3)(.0167) = .05$).

In the case of a simple comparison, Equation 21.26 is employed to compute $CD_{B/D}$, which will represent the **Bonferroni–Dunn test** statistic. $CD_{B/D}$ is the minimum required difference in order for two means to differ significantly from one another, if the familywise Type I error rate is set at a prespecified level.[33]

$$CD_{B/D} = t_{B/D} \sqrt{\frac{2MS_{WG}}{n}} \qquad \textbf{(Equation 21.26)}$$

The value $t_{B/D}$ in Equation 21.26 represents the tabled critical t value at the level of significance that corresponds to the value of α_{PC} (which in this case equals $t_{.0167}$) for df_{WG} (which for Example 21.1 is $df_{WG} = 12$). The value of $t_{B/D}$ can be obtained from detailed tables of the t distribution prepared by Dunn (1961), or can be computed with Equation 21.27 (which is described in Keppel (1991)).

$$t_{B/D} = z + \frac{z^3 + z}{4(df_{WG} - 2)} \qquad \textbf{(Equation 21.27)}$$

The value of z in Equation 21.27 is derived from **Table A1 (Table of the Normal Distribution)** in the **Appendix**. It is the z value above which $\alpha_{PC/2}$ of the cases in the normal distribution falls. Since in our example $\alpha_{PC} = .0167$, we look up the z value above which $.0167/2 = .00835$ of the distribution falls.[34] From **Table A1** we can determine that the z value which corresponds to the proportion .00835 is $z = 2.39$ (i.e., the value in Column 3 which is closest to .00835). Substituting $z = 2.39$ in Equation 21.27, the value $t_{B/D} = 2.79$ is computed.

$$t_{B/D} = 2.39 + \frac{(2.39)^3 + 2.39}{4(12 - 2)} = 2.79$$

Substituting the value $t_{B/D} = 2.79$ in Equation 21.26, the value $CD_{B/D} = 2.43$ is computed.

$$CD_{B/D} = 2.79 \sqrt{\frac{(2)(1.9)}{5}} = 2.43$$

Thus, in order to be significant, the difference between any pair of means contrasted in a simple comparison must be at least 2.43 units. Referring to Table 21.5, we can determine that (as is the case with **multiple t tests/Fisher's LSD test**) the following comparisons are significant: Group 1 versus Group 2; Group 1 versus Group 3 (since the difference between the means of the aforementioned groups is larger than $CD_{B/D} = 2.43$). Note that the value $CD_{B/D} = 2.43$ is larger than the value $CD_{LSD} = 1.90$, computed with Equation 21.24. The difference between the two CD values reflects the fact that in the case of the **Bonferroni–Dunn test**, for the set of $c = 3$ simple comparisons the value of α_{FW} is .05, whereas in the case of **multiple t tests/Fisher's LSD test**, the value of α_{FW} (which is computed with Equation 21.13) is .14. By virtue of adjusting the value of α_{FW}, the **Bonferroni–Dunn test** will always result in a larger CD value than the value computed with **multiple t tests/Fisher's LSD test**.[35]

The **Bonferroni–Dunn test** can be used for both simple and complex comparisons. Earlier in this section, it was noted that both simple and complex comparisons involving two sets of means (where each set of means in a complex comparison consists of a single mean or a combination of means) represent single degree of freedom comparisons. Keppel (1991, p. 167) notes that the number of possible single degree of freedom comparisons (to be designated $c_{(1\ df)}$) in a set of data can be computed with Equation 21.28.

$$c_{(1\ df)} = 1 + \frac{(3^k - 1)}{2} - 2^k \qquad \textbf{(Equation 21.28)}$$

Employing Equation 21.28 with Example 21.1 (where $k = 3$), the value $c_{(1\ df)} = 6$ is computed.

$$c_{(1\ df)} = 1 + \frac{(3^3 - 1)}{2} - 2^3 = 6$$

The $c_{(1\ df)} = 6$ possible single degree of freedom comparisons which are possible when $k = 3$ follow: Group 1 versus Group 2; Group 1 versus Group 3; Group 2 versus Group 3; Group 1 versus Groups 2 and 3; Group 2 versus Groups 1 and 3; Group 3 versus Groups 1 and 2. Thus, if the **Bonferroni–Dunn test** is employed to conduct all six possible single degree of freedom comparisons with $\alpha_{FW} = .05$, the per comparison error rate will be $\alpha_{PC} = .05/6 = .0083$. Since $\alpha_{PC}/2 = .0083/2 = .00415$, employing **Table A1** we can determine the z value above which .00415 proportion of cases falls is approximately 2.635. Substituting $z = 2.635$ in Equation 21.27, the value $t_{B/D} = 3.16$ is computed.

$$t_{B/D} = 2.635 + \frac{(2.635)^3 + 2.635}{4(12 - 2)} = 3.16$$

Substituting $t_{B/D} = 3.16$ in Equation 21.26, the value $C_{B/D} = 2.75$ is computed.

$$C_{B/D} = 3.16\sqrt{\frac{(2)(1.9)}{5}} = 2.75$$

Since Equation 21.26 is only valid for a simple comparison, the value $C_{B/D} = 2.75$ only applies to the three simple comparisons which are possible within the full set of six single degree of freedom comparisons. Thus, in order to be significant, the difference between any two means contrasted in a simple comparison must be at least 2.75 units. Note that the latter value is larger than $CD_{B/D} = 2.43$ computed for a set of $c = 3$ comparisons.

If one or more complex comparisons are conducted, the $t_{B/D}$ value a researcher employs will depend on the total number of comparisons (both simple and complex) being conducted. As is the case with **multiple t tests/Fisher's LSD test**, the computed value of $CD_{B/D}$ will also be a function of the coefficients employed in a comparison. Equation 21.29 is a generic equation which can be employed for both simple and complex comparisons to compute the value of $CD_{B/D}$. In the case of a simple comparison the term $\Sigma c_j^2 = 2$, thus resulting in Equation 21.26.

$$CD_{B/D} = t_{B/D}\sqrt{\frac{(\Sigma c_j^2)(MS_{WG})}{n}} \qquad \textbf{(Equation 21.29)}$$

Equation 21.29 will be employed for the complex comparison of Group 3 versus the combined performance of Groups 1 and 2. On the assumption that the six possible single degree of freedom comparisons are conducted, the value $t_{B/D} = 3.16$ will be employed in the equation.

$$t_{B/D} = 3.16\sqrt{\frac{(1.5)(1.9)}{5}} = 2.39$$

Thus, in order to be significant, the absolute value of the difference $\bar{X}_3 - [(\bar{X}_1 + \bar{X}_2)/2]$ must be at least 2.39 units. Since the obtained absolute difference of 1.7 is less than the latter value, the nondirectional alternative hypothesis H_1: $\mu_3 \neq (\mu_1 + \mu_2)/2$ is not supported. Recollect that when **multiple t tests/Fisher's LSD test** are employed for the same comparison, the latter alternative hypothesis is supported. The difference between the results of the two comparison procedures illustrates that the **Bonferroni–Dunn test** is a more conservative/less powerful procedure.

It should be noted that in conducting comparisons (especially if they are planned) a researcher may not elect to conduct all possible comparisons between pairs of means. Obviously, the fewer comparisons that are conducted, the higher the value for α_{PC} which can be employed for the **Bonferroni–Dunn test**. Nevertheless, some researchers consider the **Bonferroni–Dunn** adjustment to be too severe, regardless of how many comparisons are conducted. In view of this, some sources recommend a procedure recommended by Šidák (1967), who proposed a modification which provides for a slight increase in the power of the test. Šidák (1967) recommended that instead of setting the per comparison error rate at $\alpha_{PC} = \alpha_{FW}/c$ (computed with Equation 21.16) it be set at the exact value $\alpha_{PC} = 1 - (1 - \alpha_{FW})^{1/c}$ (computed with Equation 21.15). Use of the latter equation provides for a more powerful test per comparison, since $1 - (1 - \alpha_{FW})^{1/c} > \alpha_{FW}/c$. Thus, in the example under discussion, α_{PC} would be set at $1 - (1 - \alpha_{FW})^{1/c} = 1 - (1 - .05)^{1/3} = .01695$, which is slightly larger than the value $\alpha_{FW}/c = .05/3 = .167$ employed for the **Bonferroni–Dunn test**. More detailed discussion of the **Šidák–Bonferroni correction** (including tables for evaluating the results of the test) can be found in Keppel and Wickens (2004) and Kirk (1995). An alternative modification of the **Bonferroni–Dunn test** has been proposed by Holm (1979). The latter modification, which is described in Howell (2002, pp. 386–388), also allows for a more powerful test of an alternative hypothesis.

It should also be pointed out that in conducting the **Bonferroni–Dunn test**, as well as any other comparison procedures which result in a reduction of the value of α_{PC}, it is not necessary that each comparison be assigned the same α_{PC} value. As long as the sum of the α_{PC} values adds up to the value stipulated for α_{FW}, the α_{PC} values can be distributed in any way the researcher deems prudent. Thus, if certain comparisons are considered more important than others, the researcher may be willing to tolerate a higher α_{PC} rate for such comparisons. As an example, assume a researcher conducts three comparisons and sets $\alpha_{FW} = .05$. If the first of the comparisons is considered to be the one of most interest and the researcher wants to maximize the power of that comparison, the α_{PC} rate for that comparison can be set equal to .04, and the α_{PC} rate for each of the other two comparisons can be set at .005. Note that since the sum of the three values is .05, the value $\alpha_{FW} = .05$ is maintained.

Test 21c: Tukey's HSD test (The term **HSD** is an abbreviation for **honestly significant difference**)[36] **Tukey's HSD test** (Tukey, 1953) is generally recommended for unplanned comparisons when a researcher wants to make **all possible pairwise comparisons** (i.e., simple comparisons) in a set of data. The total number of pairwise comparisons (c) which can be conducted for a set of data can be computed with the following equation: $c = [k(k-1)]/2$. Thus, if $k = 3$, the total number of possible pairwise comparisons is $c = [3(3-1)]/2 = 3 = 3$ (which in the case of Example 21.1 are Group 1 versus Group 2, Group 1 versus Group 3, and Group 2 versus Group 3).

Tukey's HSD test controls the **familywise Type I error rate** so that it will not exceed the prespecified alpha value employed in the analysis. Many sources view it as a good compromise among the available unplanned comparison procedures, in that it maintains an acceptable level for α_{FW}, without resulting in an excessive decrease in power.[37] **Tukey's HSD test** is one of a number of comparison procedures which are based on the **Studentized range statistic** (which is represented by the notation q). Like the t distribution, the distribution for the **Studentized range statistic** is also employed to compare pairs of means. When the total number of groups/ treatments involved in an experiment is greater than two, a tabled critical q value will be higher than the corresponding tabled critical t value for **multiple t tests/Fisher's LSD test** for the same comparison.[38] For a given degrees of freedom value, the magnitude of a tabled critical q value increases as the number of groups employed in an experiment increases. **Table A13** in the **Appendix** (which is discussed in more detail later in this section) is the **Table of the Studentized Range Statistic**.

Equation 21.30 is employed to compute the q statistic, which represents the test statistic for **Tukey's HSD test**.

$$q = \frac{\bar{X}_a - \bar{X}_b}{\sqrt{\dfrac{MS_{WG}}{n}}}$$

(Equation 21.30)

Equation 21.30 will be employed with Example 21.1 to conduct the simple comparison of Group 1 versus Group 2. When the latter comparison is evaluated with Equation 21.30, the value $q = 4.22$ is computed.

$$q = \frac{9.2 - 6.6}{\sqrt{\dfrac{1.9}{5}}} = 4.22$$

The obtained value $q = 4.22$ is evaluated with **Table A13**. The latter table contains the two-tailed .05 and .01 critical q values which are employed to evaluate a nondirectional alternative hypothesis at the .05 and 01 levels, or a directional alternative hypothesis at the .10 and .02 levels. (It should be noted that Kirk (1995, p. 144) states that the **Tukey procedure** and all other unplanned comparison procedures should only be employed to evaluate a nondirectional alternative hypothesis.) As is the case with previous comparisons, the analysis will be in reference to the nondirectional alternative hypothesis H_1: $\mu_1 \neq \mu_2$, with $\alpha_{FW} = .05$. Employing the section of **Table A13** for the .05 critical values, we locate the q value that is in the cell which is the intersection of the column for $k = 3$ means (which represents the total number of groups upon which the omnibus F value is based) and the row for $df_{error} = 12$ (which represents the value $df_{WG} = df_{error} = 12$ computed for the omnibus F test). The tabled critical $q_{.05}$ value for $k = 3$ means and $df_{error} = 12$ is $q_{.05} = 3.77$. In order to reject the null hypothesis, the obtained absolute value of q must be equal to or greater than the tabled critical value at the prespecified level of significance.[39] Since the obtained value $q = 4.22$ is greater than the tabled critical two-tailed value $q_{.05} = 3.77$, the nondirectional alternative hypothesis H_1: $\mu_1 \neq \mu_2$ is supported at the .05 level (where .05 represents the value of α_{FW}). It is not supported at the .01 level, since $q = 4.22$ is less than the tabled critical value $q_{.01} = 5.05$.

Equation 21.31, which is algebraically derived from Equation 21.30, can be employed to compute the minimum required difference (CD_{HSD}) in order for two means to differ significantly from one another at a prespecified level of significance.[40]

$$CD_{HSD} = q_{(k, df_{WG})} \sqrt{\frac{MS_{WG}}{n}}$$

(Equation 21.31)

Where: $q_{(k, df_{WG})}$ is the tabled critical q value for k groups/means and df_{WG} is the value employed for the omnibus F test

Employing the appropriate values in Equation 21.31, the value CD_{HSD} = 2.32 is computed.

$$CD_{HSD} = (3.77)\sqrt{\frac{1.9}{5}} = 2.32$$

Thus, in order to be significant, the difference $\bar{X}_1 - \bar{X}_2$ must be at least 2.32 units. Note that the value CD_{HSD} = 2.32 is greater than CD_{LSD} = 1.90, but less than $CD_{B/D}$ = 2.43. This reflects the fact that in conducting all pairwise comparisons, **Tukey's HSD test** will provide a more powerful test of an alternative hypothesis than the **Bonferroni–Dunn test**. It also indicates that **Tukey's HSD test** is less powerful than **multiple *t* tests/Fisher's LSD test** (which does not control the value of α_{FW}).

With respect to conducting all three pairwise comparisons, **Tukey's HSD test** indicates the following comparisons are significant at the .05 level (since the difference between the means is greater than CD_{HSD} = 2.32): $\bar{X}_1 - \bar{X}_2$ = 2.6 and $\bar{X}_1 - \bar{X}_3$ = 3.0. The difference $\bar{X}_2 - \bar{X}_3$ = .4 is not significant, since it is less than CD_{HSD} = 2.32.

The use of Equations 21.30 and 21.31 for **Tukey's HSD test** is based on the following assumptions: a) The distribution of data in the underlying population from which each of the samples is derived is normal; b) The variances of the k underlying populations represented by the k samples are equal to one another (i.e., homogeneous); and c) The sample sizes of each of the groups being compared are equal. Kirk (1982, 1995), among others, discusses a number of modifications of **Tukey's HSD test** which are recommended when there is reason to believe that all or some of the aforementioned assumptions are violated.[41]

Hayter (1986) suggested a modification of the use of the **Studentized range statistic** which allows for a more powerful test than **Tukey's HSD test** in conducting all possible pairwise comparisons. The **Fisher–Hayter test** (which Keppel and Wickens (2004) and Kirk (1995) note can be viewed as a modification of **Fisher's LSD test**) is identical to **Tukey's HSD test**, except for the fact that it employs the value $q_{(k-1,df_{WG})}$ in Equation 21.31 in place of $q_{(k,df_{WG})}$. Since $q_{(k-1,df_{WG})}$ will always be smaller than $q_{(k,df_{WG})}$, the **critical difference** computed for the **Fisher–Hayter test statistic** (to be designated $CD_{F/H}$) will always be smaller than CD_{HSD}. With respect to the example under discussion, to compute $CD_{F/H}$, the value $q_{(k-1,df_{WG})} = q_{2,12} = 3.08$ in **Table A13** is employed in Equation 21.31, yielding the value $CD_{F/H} = (3.08)\sqrt{1.9/5} = 1.90$, which is less than the previously computed value CD_{HSD} = 2.32. (Although $CD_{F/H}$ = 1.90 is equal to the value computed for CD_{LSD}, it will usually be greater than the latter value.) Although the result for the **Fisher–Hayter test** (which sources state should only be employed if the omnibus F test is significant) does not alter our conclusions with regard to the significance or lack of for the difference between the means for the three possible pairwise comparisons, it does provide for a more powerful test than **Tukey's HSD test**. More detailed discussion of the **Fisher–Hayter test** can be found in Keppel and Wickens (2004) and Kirk (1995). Keppel and Wickens (2004, p. 125) note that a number of simulation studies have demonstrated that the **Fisher–Hayter test** provides excellent control over the familywise Type I error rate.

Test 21d: The Newman–Keuls test The **Newman–Keuls test** (Keuls (1952), Newman (1939)) is another procedure for pairwise unplanned comparisons that employs the **Studentized range statistic**. (Anderson (2001, p. 546) notes that Newman gives Student credit for the idea behind the test.) Although the **Newman–Keuls test** is more powerful than **Tukey's HSD test**, unlike the latter test it does not insure that in a set of pairwise comparisons the **familywise Type I error rate** will not exceed a prespecified alpha value. Equations 21.30 and 21.31 (which are employed for **Tukey's HSD test**) are also employed for the **Newman–Keuls test**. When, however, the latter equations are used for the **Newman–Keuls test**, the appropriate tabled critical q value will be a function of how far apart the two means being compared are from one another.

To be more specific, the k means are arranged ordinally (i.e., from lowest to highest). Thus, in the case of Example 21.1, the $k = 3$ means are arranged in the following order:

$$\bar{X}_3 = 6.2 \quad \bar{X}_2 = 6.6 \quad \bar{X}_1 = 9.2$$

For each pairwise comparison which can be conducted, the number of **steps** between the two means involved is determined. Because of the fact that the tabled critical q value is a function of the number of steps or **layers** that separate two mean values, the **Newman–Keuls test** is often referred to as a **stepwise** or **layered test**. The number of steps (which will be represented by the notation s) between any two means is determined as follows: Starting with the lower of the two mean values (which will be the mean to the left), count until the higher of the two mean values is reached. Each mean value employed in counting from the lower to the higher value in the pair represents one step. Thus, $s = 2$ steps are involved in the simple comparison of Group 1 versus Group 2, since we start at the left with $\bar{X}_2 = 6.6$ (the lower of the two means involved in the comparison) which is step 1, and move right to the adjacent value $\bar{X}_1 = 9.2$ (the higher of the two means involved in the comparison), which represents step 2. If Group 1 and Group 3 are compared, $s = 3$ steps separate the two means, since we start at the left with $\bar{X}_3 = 6.2$ (the lower of the two means involved in the comparison), move to the right counting $\bar{X}_2 = 6.6$ as step 2, and then move to $\bar{X}_1 = 9.2$ (the higher of the two means involved in the comparison), which is step 3.

The **Newman–Keuls test** protocol requires that the pairwise comparisons be conducted in a specific order. The first comparison conducted is between the two means which are separated by the largest number of steps (which will represent the largest absolute difference between any two means). If the latter comparison is significant, any comparisons involving the second largest number of steps are conducted. If all the comparisons in the latter subset of comparisons are significant, the subset of comparisons for the next largest number of steps is conducted, and so on. The basic rule upon which the protocol is based is that if at any point in the analysis a comparison fails to yield a significant result, no further comparisons are conducted on pairs of means which are separated by fewer steps than the number of steps involved in the nonsignificant comparison. Employing this protocol with Example 21.1, the first comparison that is conducted is $\bar{X}_1$ versus $\bar{X}_3$, which involves $s = 3$ steps. If that comparison is significant, the comparisons $\bar{X}_1$ versus $\bar{X}_2$ and $\bar{X}_2$ versus $\bar{X}_3$, both of which involve $s = 2$ steps, are conducted.

In employing **Table A13** to determine the tabled critical q value to employ with Equations 21.30 and 21.31, instead of employing the column for k (the total number of groups/means in the set of data), the column which is used is the one that corresponds to the number of steps between the two means involved in a comparison. Thus, if Group 1 and Group 3 are compared, the column for $k = 3$ groups in **Table A13** is employed in determining the value of q. Since the value of df_{error} remains $df_{WG} = 12$, the value $q_{.05} = 3.77$ is employed in Equation 21.31. Since the latter value is identical to the q value employed for **Tukey's HSD test**, the **Newman–Keuls** value computed for the minimum required difference for two means ($CD_{N/K}$) will be the same as the value CD_{HSD} computed for **Tukey's HSD test**.[42] Thus, $CD_{N/K} = (3.77)\sqrt{1.9/5} = 2.32$. The latter result indicates that if $\alpha = .05$, the minimum required difference between two means for the comparison Group 1 versus Group 3 is $CD_{N/K} = 2.32$.[43] Since the absolute difference $|\bar{X}_1 - \bar{X}_3| = 3$ is greater than $CD_{N/K} = 2.32$, the comparison is significant.

Since the result of the $s = 3$ step comparison is significant, the **Newman–Keuls test** is employed to compare Group 1 versus Group 2 and Group 2 versus Group 3, which represent $s = 2$ step comparisons. The value $q_{.05}$ = 3.08 is employed in Equation 21.31, since the latter value is the tabled critical $q_{.05}$ value for $k = 2$ means and df_{error} = 12. Substituting $q_{.05}$ = 3.08 in Equation 21.31 yields the value $CD_{N/K}$ = $(3.08)\sqrt{1.9/5}$ = 1.90. Since the absolute difference $|\bar{X}_1 - \bar{X}_2|$ = 2.6 is greater than $CD_{N/K}$ = 1.90, there is a significant difference between the means of Groups 1 and 2. Since the absolute difference $|\bar{X}_2 - \bar{X}_3|$ = .4 is less than $CD_{N/K}$ = 1.90, the researcher cannot conclude there is a significant difference between the means of Groups 2 and 3. Note that the value $CD_{N/K}$ = 1.90 computed for a two-step analysis is smaller than the value $CD_{N/K}$ = 2.32 computed for the three-step analysis. The general rule is that the fewer the number of steps, the smaller the computed $CD_{N/K}$ value.

The astute reader will observe that $CD_{N/K}$ = 1.90 is identical to CD_{LSD} = 1.90 obtained with Equation 21.24. The latter result illustrates that when two steps are involved in a **Newman–Keuls** comparison, it will always yield a value which is identical to CD_{LSD} (i.e., the value computed with **multiple t tests/Fisher's LSD test**). In point of fact, when $s = 2$, for a given degrees of freedom value, the relationship between the values of q and t is as follows: q = $t\sqrt{2}$ and t = $q/\sqrt{2}$. If the value $q_{.05}$ = 3.08 is employed in the equation t = $q/\sqrt{2}$, t = $(3.08)/\sqrt{2}$ = 2.18. Note that $(t_{.05}$ = $2.18)^2$ = $(F_{.05}$ = 4.75), and that $F_{.05}$ = 4.75 is the tabled critical value employed in Equation 21.24 which yields the value CD_{LSD} = 1.90.

The value of α_{FW} associated with the **Newman–Keuls test** will be higher than the value of α_{FW} for **Tukey's HSD test**. This is a direct result of the fact that the tabled critical **Studentized range** values employed for the **Newman–Keuls test** are smaller than those employed for **Tukey's HSD test** (with the exception of the comparison contrasting the lowest and highest means in the set of k means). Within any subset of comparisons which are an equal number of steps apart from one another, the overall Type I error rate for the **Newman–Keuls test** within that subset will not exceed the prespecified value of alpha. The latter, however, does not insure that α_{FW} for all the possible pairwise comparisons will not exceed the prespecified value of alpha. Because of its higher α_{FW} rate, the **Newman–Keuls test** is generally not held in high esteem as an unplanned comparison procedure. Excellent discussions of the **Newman–Keuls test** can be found in Maxwell and Delaney (1990, 2000) and Howell (2002).

Test 21e: The Scheffé test The **Scheffé test** (Scheffé, 1953), which is employed for unplanned comparisons, is commonly described as the most conservative of the unplanned comparison procedures. Since each hypothesis evaluated with the **Scheffé test** is stated in terms of a linear combination involving a set of coefficients and population means, the latter combination is commonly referred to as a **contrast**. Because of the latter some sources refer to the **Scheffé test** as the **Scheffé multiple contrast procedure**. The **Scheffé test** maintains a fixed value for α_{FW}, regardless of how many simple and complex comparisons are conducted. By virtue of controlling for a large number of potential comparisons, the error rate for any single comparison (i.e., α_{PC}) will be lower than the error rate associated with any of the other comparison procedures (assuming an alternative procedure employs the same value for α_{FW}).[44] Since in conducting unplanned pairwise comparisons **Tukey's HSD test** provides a more powerful test of an alternative hypothesis than the **Scheffé test**, most sources note that it is not prudent to employ the **Scheffé test** if only simple comparisons are being conducted. The **Scheffé test** is, however, recommended whenever a researcher wants to maintain a specific α_{FW} level, regardless of how many simple and complex comparisons are conducted. Sources note that the **Scheffé test** can accommodate unequal sample sizes and is quite robust with respect to violations of the assumptions underlying the analysis of variance (i.e., homogeneity of variance and normality of the underlying population distributions). Because of the low value the **Scheffé test** imposes on the value of α_{PC} and the consequent loss of power associated with each comparison, some sources recommend that in using the test a researcher employ a larger value for α_{FW} than would ordinarily be the case. Thus, one might employ $\alpha_{FW} = .10$ for the **Scheffé test**, instead of $\alpha_{FW} = .05$ which might be employed for a less conservative comparison procedure.

Equation 21.32 is employed to compute the minimum required difference for the **Scheffé test** (CD_S) in order for two means to differ significantly from one another at a prespecified level of significance. Whereas Equation 21.32 only applies to simple comparisons, Equation 21.33 is a generic equation which can be employed for both simple and complex comparisons.

$$CD_S = \sqrt{(k - 1)(F_{(df_{BG}, df_{WG})})}\sqrt{\frac{2MS_{WG}}{n}} \qquad \textbf{(Equation 21.32)}$$

$$CD_S = \sqrt{(k - 1)(F_{(df_{BG}, df_{WG})})}\sqrt{\frac{(\Sigma c_j^2)(MS_{WG})}{n}} \qquad \textbf{(Equation 21.33)}$$

In Equations 21.32 and 21.33, the value $F_{(df_{BG}, df_{WG})}$ is the tabled critical value that is employed for the omnibus F test for a value of alpha which corresponds to the value of α_{FW}. Thus, in Example 21.1, if $\alpha_{FW} = .05$, $F_{.05} = 3.89$ for $df = 2,12$ is used in Equations 21.32/21.33. Employing Equation 21.32, the value $CD_S = 2.43$ is computed for the simple comparison of Group 1 versus Group 2.

$$CD_S = \sqrt{(3 - 1)(3.89)}\sqrt{\frac{(2)(1.9)}{5}} = 2.43$$

Thus, in order to be significant, the difference $\bar{X}_1 - \bar{X}_2$ (as well as the difference between the means of any other two groups) must be at least 2.43 units. Since the absolute difference $|\bar{X}_1 - \bar{X}_2| = 2.6$ is greater than $CD_S = 2.43$, the nondirectional alternative hypothesis $H_1: \mu_1 \neq \mu_2$ is supported. Note that for the same comparison, the CD value computed for the **Scheffé test** is larger than the previously computed values $CD_{LSD} = CD_{N/K} = 1.90$ and $CD_{HSD} = 2.32$, but is equivalent to $CD_{B/D} = 2.43$. The fact that $CD_S = CD_{B/D}$ for the comparison under discussion illustrates that although for simple comparisons the **Scheffé test** is commonly described as the most conservative of the unplanned comparison procedures, when the **Bonferroni–Dunn test** is employed for simple comparisons it may be as or more conservative than the **Scheffé test**. Maxwell and Delaney (1990, 2000) note that in instances where a researcher conducts a small number of comparisons, the **Bonferroni–Dunn test** will provide a more powerful test of an alternative hypothesis than the **Scheffé test**. However, as the number of comparisons increases, at some point the **Scheffé test** will become more powerful than the **Bonferroni–Dunn test**. In general (although there are some exceptions), the **Bonferroni–Dunn test** will be more powerful than the **Scheffé** test when the number of comparisons conducted is less than $[k(k - 1)]/2$.

The **Scheffé test** is most commonly recommended when at least one complex comparison is conducted in a set of unplanned comparisons. Although the other comparison procedures discussed in this section can be employed for unplanned complex comparisons, the **Scheffé test** is viewed as a more desirable alternative by most sources. It will now be demonstrated how the **Scheffé test** can be employed for the complex comparison $\bar{X}_3$ versus $(\bar{X}_1 + \bar{X}_2)/2$.

Substituting the value $\Sigma c_j^2 = 1.5$ (which is computed in Table 21.4) and the other relevant values in Equation 21.33, the value $CD_S = 2.11$ is computed.

$$CD_S = \sqrt{(3 - 1)(3.89)}\sqrt{\frac{(1.5)(1.9)}{5}} = 2.11$$

Thus, in order to be significant, the difference $\bar{X}_3 - [(\bar{X}_1 + \bar{X}_2)/2]$ (as well as the difference between any set of means in a complex comparison for which $\Sigma c_j^2 = 1.5$) must be equal to or greater than 2.11 units. Since the absolute difference $|\bar{X}_3 - [(\bar{X}_1 + \bar{X}_2)/2]| = 1.7$ is less than $CD_S = 2.11$, the nondirectional alternative hypothesis $H_1: \mu_3 \neq (\mu_1 + \mu_2)/2$ is not supported. Note that the CD value computed for the **Scheffé test** is larger than the previously computed value $CD_{LSD} = 1.65$, but less than $CD_{B/D} = 2.39$ computed for the same complex comparison. This reflects the fact that for the complex comparison $\bar{X}_3$ versus $(\bar{X}_1 + \bar{X}_2)/2$, the **Scheffé test** is not as powerful as **multiple t tests/Fisher's LSD test**, but is more powerful than the **Bonferroni–Dunn test**.[45]

Equation 21.34 can also be employed for the **Scheffé test** for both simple and complex comparisons.

$$F_S = (k - 1)F_{(BG, WG)} \qquad \textbf{(Equation 21.34)}$$

Where: $F_{(BG, WG)}$ is the tabled critical value at the prespecified level of significance employed in the omnibus F test

In order to use Equation 21.34 it is necessary to first employ Equations 21.17–21.19 to compute the value of F_{comp} for the comparison being conducted. The value computed for F_{comp} will serve as the test statistic for the **Scheffé test**. Earlier in this section (under the discussion of **linear contrasts**) the value $F_{comp} = 5.07$ was computed for the complex comparison $\bar{X}_3$ versus $(\bar{X}_1 + \bar{X}_2)/2$. When F_{comp} is used as the test statistic for the **Scheffé test**, the critical value employed to evaluate it is different from the critical value employed in evaluating a **linear contrast**. Equation 21.34 is used to determine the **Scheffé test** critical value. In order for a comparison to be significant, the computed value of F_{comp} must be equal to or greater than the critical F value computed with Equation 21.34 (which is represented by the notation F_S). In employing Equation 21.34 to compute F_S, the tabled critical value employed for the omnibus F test (which in the case of Example 21.1 is $F_{.05} = 3.89$) is multiplied by $(k - 1)$. Obviously, the resulting value will be higher than the tabled critical value employed for the **linear contrast** for the same comparison. When the appropriate values for Example 21.1 are substituted in Equation 21.34, the value $F_S = 7.78$ is computed.

$$F_S = (3 - 1)(3.89) = 7.78$$

Since the computed value $F_{comp} = 5.07$ is less than the critical value $F_S = 7.78$ computed with Equation 21.34, it indicates that the nondirectional alternative hypothesis $H_1: \mu_3 \neq (\mu_1 + \mu_2)/2$ is not supported. Note that the value $F_S = 7.78$ is larger than the value $F_{.05} = 4.75$ (which is the tabled critical value for $df_{num} = 1$ and $df_{den} = df_{WG} = 12$) which is employed for the linear contrast for the same comparison. Recollect that the alternative hypothesis $H_1: \mu_3 \neq (\mu_1 + \mu_2)/2$ is supported when a **linear contrast** is conducted.

In closing the discussion of the **Scheffé test** some general comments will be made regarding the value of α_{FW} for the **Scheffé test**, the **Bonferroni–Dunn test**, and **Tukey's HSD test**. In the discussion to follow it will be assumed that upon computation of an omnibus F value, a researcher wishes to conduct a series of unplanned comparisons for which the **familywise error rate** does not exceed $\alpha_{FW} = .05$.[46]

a) If all possible comparisons (simple and complex) are conducted with the **Scheffé test**, the value of α_{FW} will equal exactly .05. When $k \geq 3$ there are actually an infinite number of comparisons which can be made (Maxwell and Delaney (1990, p. 190; 2000)). Marascuilo and Serlin (1988, p. 366) note that Scheffé determined that if the omnibus null hypothesis could be rejected, it indicated that at least one of the possible infinite number of comparisons was statistically significant. To illustrate the fact that an infinite number of comparisons are possible, assume $k = 3$. Besides the three pairwise/simple comparisons (Group 1 versus Group 2; Group 1 versus Group 3; Group 2 versus Group 3) and the three apparent complex comparisons (The average of Groups 1 and 2 versus Group 3; the average of Groups 1 and 3 versus Group 2; the average of Groups 2 and 3 versus Group 1), in the case of a complex comparison it is possible to combine two groups so that one group contributes more to the composite mean representing the two groups than does the other group. As an example, in comparing Groups 1 and 2 with Group 3, a coefficient of 7/8 can be assigned to Group 1 and a coefficient of 1/8 to Group 2. Employing these coefficients, a composite mean value can be computed to represent the mean of the two groups which is contrasted with Group 3. It should be obvious that if one can stipulate any combination of two coefficients/weights which add up to 1, there are potentially an infinite number of coefficient combinations that can be assigned to any two groups, and therefore an infinite number of possible comparisons can result from coefficient combinations involving the comparison of two groups with a third group. If fewer than all possible comparisons are conducted with the **Scheffé test**, the value of α_{FW} will be less than .05, thus making it an overly conservative test (since the value of α_{PC} will be lower than is necessary for α_{FW} to equal .05).

b) If all possible comparisons are conducted employing the **Bonferroni–Dunn test**, the value of α_{FW} will be less than .05. As noted earlier, as the number of comparisons conducted increases, at some point the value of α_{FW} for the **Bonferroni–Dunn test** will be less than α_{FW} for the **Scheffé test**, and thus at that point the **Bonferroni–Dunn test** will be even more conservative (and thus less powerful) than the **Scheffé test**. The decrease in the value of α_{FW} for the **Bonferroni–Dunn test** results from the fact that within the set of comparisons conducted, not all comparisons will be orthogonal with one another. Winer *et al.* (1991) note that when comparisons conducted with the **Bonferroni–Dunn test** are orthogonal, the following is true: $\alpha_{FW} = c(\alpha_{PC})$. However, when some of the comparisons conducted are not orthogonal, $\alpha_{FW} < c(\alpha_{PC})$. Thus, by virtue of some of the comparisons being nonorthogonal, the **Bonferroni–Dunn test** becomes a more conservative test (i.e., $\alpha_{FW} < .05$).

c) If **Tukey's HSD test** is employed for conducting all possible pairwise comparisons, the value of α_{FW} will be exactly .05, even though the full set of pairwise comparisons will not constitute an orthogonal set. As noted above, if the **Bonferroni–Dunn test** is employed for the full set of pairwise comparisons, due to the presence of nonorthogonal comparisons, the value of α_{FW} will be less than .05, and thus the value of $CD_{B/D}$ will be larger than the value of CD_{HSD}. If in addition to conducting all pairwise comparisons with **Tukey's HSD test**, complex comparisons are also conducted, the value of α_{FW} will exceed .05. As the number of complex comparisons conducted increases, the value of α_{FW} increases. When complex comparisons are conducted, **Tukey's HSD test** is not as powerful as the **Scheffé test**.

Test 21f: The Dunnett test The **Dunnett test** (1955, 1964) is a comparison procedure, only employed for simple comparisons, which is designed to compare a control group with the other $(k-1)$ groups in a set of data. Under such conditions the **Dunnett test** provides a more powerful test of an alternative hypothesis than do the **Bonferroni–Dunn test**, **Tukey's HSD test**, and the **Scheffé test**. This is the case, since for the same value of α_{FW}, the α_{PC} value associated with the **Dunnett test** will be higher than the α_{PC} values associated with the aforementioned procedures (and, by virtue of this, provides a more powerful test of an alternative hypothesis). The larger α_{PC} value for the **Dunnett test** is predicated on the fact that by virtue of limiting the comparisons to contrasting a control group with the other groups, the **Dunnett test** statistic is based on the assumption that fewer comparisons are conducted than will be the case if all pairwise comparisons are conducted. Consequently, if a researcher specifies that $\alpha_{FW} = .05$, and the mean of the control group is contrasted with the means of each of the other $(k-1)$ groups, the **Dunnett test** insures that the **familywise Type I error rate** will not exceed .05. It should be noted that since the control group is involved in each of the comparisons which are conducted, the comparisons will not be orthogonal to one another. In illustrating the computation of the **Dunnett test** statistic, we will assume that in Example 21.1 Group 1 (the group that is not exposed to noise) is a control group, and that Groups 2 and 3 (both of which are exposed to noise) are experimental groups. Thus, employing the **Dunnett test**, the following two comparisons will be conducted with $\alpha_{FW} = .05$: Group 1 versus Group 2; Group 1 versus Group 3.

The test statistic for the **Dunnett test** (t_D) is computed with Equation 21.35, which, except for the fact that a t_D value is computed, is identical to Equation 21.22 (which is employed to compute the test statistic for **multiple t tests/Fisher's LSD test**).[47]

$$t_D = \frac{\bar{X}_a - \bar{X}_b}{\sqrt{\dfrac{2MS_{WG}}{n}}} \qquad \text{(Equation 21.35)}$$

Equation 21.35 is employed below to compute the value $t_D = 2.99$ for the simple comparison of Group 1 versus Group 2.

$$t_D = \frac{9.2 - 6.6}{\sqrt{\dfrac{(2)(1.9)}{5}}} = 2.99$$

The computed value $t_D = 2.99$ is evaluated with **Table A14 (Table of Dunnett's Modified t Statistic for a Control Group Comparison)** in the **Appendix**. The latter table, which contains both two- and one-tailed .05 and .01 critical values, is based on a modified t distribution derived by Dunnett (1955, 1964). Dunnett (1955) computed one-tailed critical values, since in comparing one or more treatments with a control group a researcher is often interested in the direction of the difference. The tabled critical t_D values are listed in reference to k, the total number of groups/treatments employed in the experiment, and the value of $df_{error} = df_{WG}$ computed for the omnibus F test.

For $k = 3$ and $df_{error} = df_{WG} = 12$, the tabled critical two-tailed .05 and .01 values are $t_{D_{.05}} = 2.50$ and $t_{D_{.01}} = 3.39$, and the tabled critical one-tailed .05 and .01 values are $t_{D_{.05}} = 2.11$ and $t_{D_{.01}} = 3.01$. The computed value $t_D = 2.99$ is greater than the tabled critical two-one-tailed .05 t_D values but less than the tabled critical two- and one-tailed .01 t_D values. Thus, for α_{FW} = .05 (but not α_{FW} = .01), the nondirectional alternative hypothesis H_1: $\mu_1 \neq \mu_2$ and the directional alternative hypothesis H_1: $\mu_1 > \mu_2$ are supported. The second comparison, involving the control group (Group 1) versus Group 3, yields the value t_D = (9.2 – 6.2)/ $\sqrt{[(2)(1.9)]/5}$ = 3.45. Since the value t_D = 3.45 is greater than the tabled critical two- and one-tailed .05 and .01 t_D values, the nondirectional alternative hypothesis H_1: $\mu_1 \neq \mu_3$ and the directional alternative hypothesis H_1: $\mu_1 > \mu_3$ are supported for both α_{FW} = .05 and α_{FW} = .01.

Equation 21.36 is employed to compute the minimum required difference for the **Dunnett test** (designated CD_D) in order for two means to differ significantly from one another at a pre-specified level of significance.

$$CD_D = t_{D_{(k, df_{WG})}} \sqrt{\frac{2MS_{WG}}{n}}$$　　　　　　　　**(Equation 21.36)**

Where:　$t_{D_{(k, df_{WG})}}$ is the tabled critical value for **Dunnett's modified t statistic** for k groups and df_{WG} at the prespecified value of α_{FW}

In employing Equation 21.36, α_{FW} = .05 will be employed, and it will be assumed that a nondirectional alternative hypothesis is evaluated for both comparisons. Substituting the two-tailed .05 value $t_{D_{.05}}$ = 2.50 in Equation 21.36, the value CD_D = 2.18 is computed.

$$CD_D = 2.50 \sqrt{\frac{2(1.9)}{5}} = 2.18$$

Thus, in order to be significant, the differences $\bar{X}_1 - \bar{X}_2$ and $\bar{X}_L - \bar{X}_3$ must be at least 2.18 units. Since the absolute differences $|\bar{X}_1 - \bar{X}_2|$ = 2.6 and $|\bar{X}_1 - \bar{X}_3|$ = 3 are greater than CD_D = 2.18, the nondirectional alternative hypothesis is supported for both comparisons. Note that CD_D = 2.18 computed for the **Dunnett test** is larger than CD_{LSD} = 1.90 computed for **multiple t tests/Fisher's LSD test** (for a simple comparison), but is less than the CD values computed for a simple comparison for the **Bonferroni–Dunn test** ($CD_{B/D}$ = 2.43), **Tukey's HSD test** (CD_{HSD} = 2.32), and the **Scheffé test** (CD_S = 2.43).

Additional discussion of comparison procedures and final recommendations The accuracy of the comparison procedures described in this section may be compromised if the homogeneity of variance assumption underlying the analysis of variance (the evaluation of which is described later in Section VI) is violated. This is the case, since in such an instance MS_{WG} may not provide the best measure of error variability for a given comparison. Violation of the homogeneity of variance assumption can either increase or decrease the Type I error rate associated with a comparison, depending upon whether MS_{WG} over or underestimates the pooled variability of the groups involved in a specific comparison. It is also the case that when the homogeneity of variance assumption is violated, the accuracy of a comparison may be even further compromised when there is not an equal number of subjects in each group. Sources that discuss these general issues (e.g., Howell (2002), Keppel and Wickens (2004), Kirk (1982, 1995), Maxwell and Delaney (1990, 2000), Winer et al. (1991)) provide alternative equations which are recommended when the homogeneity of variance assumption is violated and/or sample sizes are unequal.

As a general rule, the measure of within-groups variability that is employed in equations which are recommended when there is heterogeneity of variance is based on the pooled within-groups variability of just those groups which are involved in a specific comparison. Since the latter measure has a smaller degrees of freedom associated with it than MS_{WG}, the tabled critical value for the analysis will be based on fewer degrees of freedom. Although the loss of degrees of freedom can reduce the power of the test, it may be offset if the revised measure of within-groups variability is less than MS_{WG}. A full discussion of the subject of violation of the homogeneity of variance assumption with comparisons is beyond the scope of this book. The reader who refers to sources which discuss the subject in greater detail will discover that there is a lack of agreement with respect to what procedure is most appropriate to employ when the assumption is violated.

Numerous other multiple comparison procedures have been developed in addition to those described in this section. Dayton (2003), Grissom and Kim (2005, pp. 129–134), Howell (2002), Hsu (1996), Keppel and Wickens (2004), Kirk (1995), Maxwell and Delaney (2004), Toothaker (1991), and Wilcox (2003), among others, describe and/or discuss a number of alternative procedures. Howell (2002) describes a procedure developed by Ryan (1960), which is a compromise between **Tukey's HSD test** and the **Newman–Keuls test**. To be more specific, Ryan's (1960) procedure maintains the value of α_{FW} at the desired level, but at the same time allows the critical difference required between pairs of means to vary as a function of step size. Kirk (1995, Ch. 4) describes the merits and limitations of 22 multiple comparison procedures. Within the framework of his discussion, Kirk (1995) notes that in conducting comparisons, a researcher's priority should be to guard against inflation of the Type I error rate, yet at the same time to employ a procedure which maximizes power (i.e., has a high likelihood of identifying a significant effect). Rosenthal *et al.* (2000) discuss alternative methodologies for contrasting group means.

At this point the author will present some general recommendations regarding the use of comparison procedures for the analysis of variance.[48] From what has been said, it should be apparent that in conducting comparisons the minimum value required for two means to differ significantly from one another can vary dramatically, depending upon which comparison procedure a researcher employs. Although some recommendations have been made with respect to when it is viewed most prudent to employ each of the comparison procedures, in the final analysis the use of any of the procedures does not insure that a researcher will determine the truth regarding the relationship between the variables under study. Aside from the fact that researchers do not agree among themselves on which comparison procedure to employ (due largely to the fact that they do not concur with respect to the maximum acceptable value for α_{FW}), there is also the problem that one is not always able to assume with a high degree of confidence that all of the assumptions underlying a specific comparison procedure have, in fact, been met. In view of this, any probability value associated with a comparison may always be subject to challenge. Although most of the time a probability value may not be compromised to that great a degree, when one considers the fact that researchers may quibble over whether one should employ $\alpha_{FW} = .05$ versus $\alpha_{FW} = .10$, a minimal difference with respect to a probability value can mean a great deal, since it ultimately may determine whether a researcher elects to retain or reject a null hypothesis. If, in fact, the status of the null hypothesis is based on the result of a single study, it would seem that a researcher is obliged to arrive at a probability value in which he and others can have a high degree of confidence.[49]

In view of everything which has been discussed, this writer believes that a general strategy for conducting comparisons suggested by Keppel (1991) is the most prudent to employ. Keppel (1991) suggests that in hypothesis testing involving **unplanned comparisons**, instead of just employing the two decision categories of **retaining the null hypothesis** versus **rejecting the null hypothesis**, a third category, **suspend judgement** be added. Specifically, Keppel (1991) recommends the following:

a) If the obtained difference between two means is less than the value of CD_{LSD}, **retain the null hypothesis**. Since the value of CD_{LSD} will be the smallest CD value computed with any of the available comparison procedures, it allows for the most powerful test of an alternative hypothesis.

b) If the obtained difference between two means is equal to or greater than the CD value associated with the comparison procedure which results in the **largest α_{FW} value** one is willing to tolerate, **reject the null hypothesis**. Based on the procedures described in this book (depending upon the number of comparisons that are conducted), the largest CD value will be generated through use of either the **Bonferroni–Dunn test** or the **Scheffé test**. However, the largest α_{FW} value a researcher may be willing to tolerate may be larger than the α_{FW} value associated with either of the aforementioned procedures. In such a case, the minimum CD value required to reject the null hypothesis will be smaller than $CD_{B/D}$ or CD_S.

c) If the obtained difference between two means is greater than or equal to CD_{LSD}, but less than the CD value associated with the largest α_{FW} value one is willing to tolerate, **suspend judgement**. It is recommended that one or more replication studies be conducted employing the relevant groups/treatments for any comparisons which fall in the **suspend judgement** category.

The above guidelines will now be applied to Example 21.1. Let us assume that three simple comparisons are conducted, and that the maximum **familywise Type I error rate** the researcher is willing to tolerate is $\alpha_{FW} = .05$. Employing the above guidelines, the null hypothesis can be rejected for the comparisons of Group 1 versus Group 2 and Group 1 versus Group 3. This is the case, since $\bar{X}_1 - \bar{X}_2 = 2.6$ and $\bar{X}_1 - \bar{X}_3 = 3$, and in both instances the obtained difference between the means of the two groups is greater than $CD_{HSD} = 2.32$ (which within the framework of **Tukey's HSD test** insures that α_{FW} will not exceed .05). On the other hand, for the comparison of Group 2 versus Group 3, since the difference $\bar{X}_2 - \bar{X}_3 = .4$ is less than $CD_{LSD} = 1.90$ (which is the lowest of the computed CD values), the null hypothesis cannot be rejected. Thus, in the case of Example 21.1, none of the comparisons falls in the **suspend judgement** category. In point of fact, even if the α_{FW} value associated with the **Bonferroni–Dunn** and/or **Scheffé** tests is employed as the maximum acceptable α_{FW} rate, none of the comparisons falls in the **suspend judgement** category, since the absolute difference between the means for each comparison is either greater than or less than the relevant CD values.

As noted earlier, in the case of **planned comparisons** the issue of whether or not a researcher should control the value of α_{FW} is subject to debate. It can be argued that the strategy described for unplanned comparisons should also be employed with planned comparisons — with the stipulation that the largest value for α_{FW} which one is willing to tolerate for planned comparisons be higher than the value employed for unplanned comparisons. Nevertheless, one can omit the latter stipulation and argue that the same criterion be employed for both unplanned and planned comparisons. The rationale for the latter position is as follows. Assume that two researchers independently conduct the identical study. Researcher 1 has the foresight to plan c comparisons beforehand. Researcher 2, on the other hand, conducts the same set of c comparisons, but does not plan them beforehand. The truth regarding the populations involved in the comparisons is totally independent of who conducts the study. As a result of this, one can argue that the same criterion be applied, regardless of who conducts the investigation. If Researcher 1 is allowed to conduct a less conservative analysis (i.e., tolerate a higher α_{FW} rate) than Researcher 2, it is commensurate with giving Researcher 1 a bonus for having a bit more acumen than Researcher 2 (if we consider allowing one greater latitude with respect to rejecting the null hypothesis to constitute a bonus). It would seem that if, in the final analysis, the issue at hand is the truth concerning the populations under study, each of the two researchers should be expected to adhere to the same criterion, regardless of their expectations prior to conducting the study. If one accepts this line of reasoning, it would seem that the guidelines described in this section for unplanned comparisons should also be employed for planned comparisons.

In the final analysis, regardless of whether one has conducted planned or unplanned comparisons, when there is reasonable doubt in the mind of the researcher or there is reason to believe that there will be reasonable doubt among those who scrutinize the results of a study, it is always prudent to replicate a study. In other words, anytime the result of an analysis falls within the **suspend judgement** category (or perhaps even if the result is close to falling within it), a strong argument can be made for replicating a study. Thus, regardless of which comparison procedure one employs, if the result of a comparison is not significant, yet would have been had another comparison procedure been employed, it would seem logical to conduct one or more replication studies in order to clarify the status of the null hypothesis. There is also the case where the result of a comparison turns out to be significant, yet would not have been with a more conservative comparison procedure. If in such an instance the researcher (or others who are familiar with the relevant literature) has reason to believe that a Type I error may have been committed, it would seem prudent to reevaluate the null hypothesis. In the final analysis, multiple replications of a result provide the most powerful evidence regarding the status of a null hypothesis. An effective tool that can be employed to pool the results of multiple studies which evaluate the same hypothesis is a methodology called **meta-analysis**, which is discussed in Section IX (the **Addendum**) of the **Pearson product-moment correlation coefficient (Test 28)**.

At this point in the discussion it is worth reiterating the difference between **statistical** and **practical significance** (which is discussed in both the **Introduction** and in Section VI of the *t* **test for two independent samples**). By virtue of employing a large sample size, virtually any test evaluating a difference between the means of two populations will turn out to be significant. This results from the fact that two population means are rarely identical. However, in most instances a minimal difference between two means is commensurate with there being no difference at all, since the magnitude of such a difference is of no practical or theoretical value. In conducting comparisons (or, for that matter, conducting any statistical test) one must decide what magnitude of difference is of practical and/or theoretical significance. To state it another way, one must determine the magnitude of the **effect size** one is attempting to identify.[50] If a researcher is able to stipulate a meaningful effect size prior to collecting the data for a study, he can design the study so that the test which is employed to evaluate the null hypothesis is sufficiently powerful to identify the desired effect size. As a general rule, a researcher is best able to control the power of a statistical test by employing a sample size which exceeds some minimal value. In the latter part of Section VI the computation of power for the **single-factor between-subjects analysis of variance** is discussed in reference to both the omnibus *F* test as well as for comparison procedures.

In the final analysis, when the magnitude of the obtained difference between the means involved in any comparison is deemed too small to be of practical or theoretical significance, it really becomes irrelevant whether a result is statistically significant. In such an instance if two or more comparison procedures yield conflicting results, replication of a study is not in order.

The computation of a confidence interval for a comparison   The computation of a **confidence interval** allows a researcher to identify a range of values with respect to which one can be confident to a specified degree contains the true difference between the means of two populations. Computation of a confidence interval for a comparison is a straightforward procedure which can be easily implemented following the computation of a *CD* value.[51] Specifically, to compute the range of values which define the 95% confidence interval for any comparison, one should do the following: Add to and subtract the computed value of CD from the obtained difference between the two means involved in the comparison. As an example, let us assume **Tukey's HSD test** is employed to compute the value $CD_{HSD} = 2.32$ for the comparison involving Group 1 versus Group 2. To compute the 95% confidence interval, the value 2.32 is added to and subtracted from 2.6, which is the difference between the two means. Thus, $CI_{HSD_{.95}} = 2.6 \pm 2.32$, which can also be written as $.28 \leq (\mu_1 - \mu_2) \leq 4.92$. In other words, the researcher can be 95% confident (or the probability is .95) the interval .28 to 4.92 contains the true difference between the means of the populations represented by Groups 1 and 2. If the aforementioned interval contains the true difference between the population means, the mean of the population Group 1 represents is between .28 to 4.92 units larger than the mean of the population Group 2 represents. If the researcher wants to compute the 99% confidence interval for a comparison, the same procedure can be used, except for the fact that in computing the CD value for the comparison, the relevant .01 tabled critical q value is employed (or the tabled critical .01 t, F, or t_D value if the comparison procedure happens to employ either of the aforementioned distributions).

It should be emphasized that the range of values which define a confidence interval will be a function of the tabled critical value for the relevant test statistic that a researcher elects to employ in the analysis. As noted earlier, the tabled critical value one employs will be a function of the value established for α_{FW}. In the case of the **Bonferroni–Dunn test**, the magnitude of the critical $t_{B/D}$ value one employs (and, consequently, the range of values which defines a confidence interval) increases as the number of comparisons one conducts increases.

If one has reason to believe that the homogeneity of variance assumption underlying the analysis of variance is violated, one can argue that the measure of error variability employed in computing a confidence interval should be a pooled measure of variability based only on the groups involved in a specific comparison. Under such circumstances, one can also argue that it is acceptable to employ Equation 11.17 (employed to compute a confidence interval for the *t* **test for two independent samples**) to compute the confidence interval for a comparison. In using the latter equation for a simple comparison, one can argue that $df = n_a + n_b - 2$ (the degrees of freedom used for the *t* **test for two independent samples**) should be employed for the analysis as opposed to df_{WG}. The use of Equation 11.17 can also be justified in circumstances where: a) The researcher views the groups involved in a comparison as distinct (with respect to both the value of μ and σ^2) from the other groups involved in the study; and b) The researcher is not attempting to control the value of α_{FW}.

In view of everything which has been said, it should be apparent that the value of a confidence interval can vary dramatically depending upon which of the comparison procedures a researcher employs, and what assumptions one is willing to make with reference to the underlying populations involved in a study. For this reason, two researchers may compute substantially different confidence intervals as a result of employing different comparison procedures. In the final analysis, however, each of the researchers may be able to offer a persuasive argument in favor of the methodology he employs.

Smithson (2003, pp. 11–12) notes that (as is the case with regard to multiple comparisons) the probability value associated with a confidence interval for a comparison will be affected by the number of comparisons conducted. As an example, if a researcher computes more than one 95% confidence interval, the actual confidence level for each comparison will be some value less than 95% (depending upon the number of comparisons conducted). Through use of a **Bonferroni correction**, if a researcher desires a familywise confidence interval of $(1 - \alpha)$, then the value for each of the confidence intervals computed should be equal to $(1 - \alpha/c)$.

2. Comparing the means of three or more groups when $k \geq 4$ Within the framework of a **single-factor between-subjects analysis of variance** involving $k = 4$ or more groups, a researcher may wish to evaluate a general hypothesis with respect to the means of a subset of groups, where the number of groups in the subset is some value less than k. Although the latter type of situation is not commonly encountered in research, this section will describe the protocol for conducting such an analysis.

To illustrate, assume that a fourth group is added to Example 21.1. Assume that the scores of the five subjects who serve in Group 4 are as follows: 3, 2, 1, 4, 5. Thus, $\Sigma X_4 = 15$, $\bar{X}_4 = 3$, and $\Sigma X_4^2 = 55$. If the data for Group 4 are integrated into the data for the other three groups whose performance is summarized in Table 21.1, the following summary values are computed: $N = nk = (5)(4) = 20$, $\Sigma X_T = 125$, $\Sigma X_T^2 = 911$. Substituting the revised values for $k = 4$ groups in Equations 21.2, 21.3, and 21.4/21.5, the following sum of squares values are computed: $SS_T = 129.75$, $SS_{BG} = 96.95$, $SS_{WG} = 32.8$. Employing the values $k=4$ and $N=20$ in Equations 21.8 and 21.9, the values $df_{BG} = 4 - 1 = 3$ and $df_{WG} = 20 - 4 = 16$ are computed. Substituting the appropriate values for the sum of squares and degrees of freedom in Equations 21.6 and 21.7, the values $MS_{BG} = 96.95/3 = 32.32$ and $MS_{WG} = 32.8/16 = 2.05$ are computed. Equation 21.12 is employed to compute the value $F = 32.32/2.05 = 15.77$. Table 21.6 is the summary table of the analysis of variance.

Table 21.6 Summary Table of Analysis of Variance for Example 21.1 When $k = 4$

Source of variation	SS	df	MS	F
Between-groups	96.95	3	32.32	15.77
Within-groups	32.80	16	2.05	
Total	129.75	19		

Employing $df_{num} = 3$ and $df_{den} = 16$, the tabled critical .05 and .01 values are $F_{.05} = 3.24$ and $F_{.01} = 5.29$. Since the obtained value $F = 15.77$ is greater than both of the aforementioned critical values, the null hypothesis (which for $k = 4$ is H_0: $\mu_1 = \mu_2 = \mu_3 = \mu_4$) can be rejected at both the .05 and .01 levels.

Let us assume that prior to the above analysis the researcher has reason to believe that Groups 1, 2, and 3 may be distinct from Group 4. However, before he contrasts the composite mean of Groups 1, 2, and 3 with the mean of Group 4 (i.e., conducts the complex comparison which evaluates the null hypothesis H_0: $(\mu_1 + \mu_2 + \mu_3)/3 = \mu_4$), he decides to evaluate the null hypothesis H_0: $\mu_1 = \mu_2 = \mu_3$. If the latter null hypothesis is retained, he will assume that the three groups share a common mean value and, on the basis of this, he will compare their composite mean with the mean of Group 4. In order to evaluate the null hypothesis H_0: $\mu_1 = \mu_2 = \mu_3$, it is necessary for the researcher to conduct a separate analysis of variance which just involves the data for the three groups identified in the null hypothesis. The latter analysis of variance has already been conducted, since it is the original analysis of variance that is employed for Example 21.1 — the results of which are summarized in Table 21.2.

Upon conducting an analysis of variance on the data for all $k = 4$ groups as well as an analysis of variance on the data for the subset comprised of $k_{subset} = 3$ groups, the researcher has the necessary information to compute the appropriate F ratio (which will be represented with the notation $F_{(1/2/3)}$) for evaluating the null hypothesis H_0: $\mu_1 = \mu_2 = \mu_3$. In computing the F ratio to evaluate the latter null hypothesis, the following values are employed: a) $MS_{BG} = 26.53$ (which is the value of MS_{BG} computed for the analysis of variance in Table 21.2 that involves only the three groups identified in null hypothesis H_0: $\mu_1 = \mu_2 = \mu_3$) is employed as the numerator of the F ratio; and b) $MS_{WG} = 2.05$ (which is the value of MS_{WG} computed in Table 21.6 for the omnibus F test when the data for all $k = 4$ groups are evaluated) is employed as the denominator of the F ratio. The reason for employing the latter value instead of $MS_{WG} = 1.9$ (which is the value of MS_{WG} computed for the analysis of variance in Table 21.2 that only employs the data for the three groups identified in the null hypothesis H_0: $\mu_1 = \mu_2 = \mu_3$) is because $MS_{WG} = 2.05$ is a pooled estimate of all $k = 4$ population variances. If, in fact, the populations represented by all four groups have equal variances, this latter value will provide the most accurate estimate of MS_{WG}. Thus:

$$F_{(1/2/3)} = \frac{MS_{BG_{(1/2/3)}}}{MS_{WG_{(1/2/3/4)}}} = \frac{13.27}{2.05} = 6.47$$

The degrees of freedom employed for the analysis are based on the mean square values employed in computing the $F_{(1/2/3)}$ ratio. Thus: $df_{num} = k_{subset} - 1 = 3 - 1 = 2$ (where $k_{subset} = 3$ groups) and $df_{den} = df_{WG_{(1/2/3/4)}} = 16$ (which is df_{WG} for the omnibus F test involving all $k = 4$ groups). For $df_{num} = 2$ and $df_{den} = 16$, $F_{.05} = 3.63$ and $F_{.01} = 6.23$. Since the obtained value $F = 6.47$ is greater than both of the aforementioned critical values, the null hypothesis can be rejected at both the .05 and .01 levels. Thus, the data do not support the researcher's hypothesis that Groups 1, 2, and 3 represent a homogenous subset. In view of this, the researcher would not conduct the contrast $(\overline{X}_1 + \overline{X}_2 + \overline{X}_3)/3$ versus $\overline{X}_4$.

It should be noted that if the researcher has reason to believe that Groups 1, 2, and 3 have homogeneous variances, and the variance of Group 4 is not homogenous with the variance of the latter three groups, it would be more appropriate to employ $MS_{WG_{(1/2/3)}} = 1.9$ as the denominator of the F ratio as opposed to $MS_{WG_{(1/2/3/4)}} = 2.05$. With respect to the problem under discussion, since the two MS_{WG} values are almost equivalent, using either value produces essentially the same result (i.e., if $MS_{WG_{(1/2/3)}} = 1.9$ is employed to compute the $F_{(1/2/3)}$ ratio, $F_{(1/2/3)} = 13.27/1.9 = 6.98$, which is greater than both the tabled critical .05 and .01 F values).

3. Evaluation of the homogeneity of variance assumption of the single-factor between-subjects analysis of variance: Test 11a: Hartley's F_{max} test, Test 21g: The Levene test for homogeneity of variance and Test 21h: The Brown–Forsythe test for homogeneity of variance In Section I it is noted that one assumption of the **single-factor between-subjects analysis of variance** is homogeneity of variance. As noted in the discussion of the *t* **test for two independent samples**, when there are $k = 2$ groups the homogeneity of variance assumption evaluates whether or not there is evidence to indicate an inequality exists between the variances of the populations represented by the two samples/groups. When there are two or more groups, the homogeneity of variance assumption evaluates whether there is evidence to indicate that an inequality exists between at least two of the population variances represented by the k samples/groups. When the latter condition exists it is referred to as **heterogeneity of variance**. In reference to Example 21.1, the null and alternative hypotheses employed in evaluating the homogeneity of variance assumption are as follows:

Null hypothesis H_0: $\sigma_1^2 = \sigma_2^2 = \sigma_3^2$

(The variance of the population Group 1 represents equals the variance of the population Group 2 represents equals the variance of the population Group 3 represents.)

Alternative hypothesis H_1: Not H_0

(This indicates that there is a difference between at least two of the three population variances.)

One of a number of procedures that can be employed to evaluate the homogeneity of variance hypothesis is **Hartley's F_{max} test (Test 11a)**, which is also employed to evaluate homogeneity of variance for the *t* **test for two independent samples**. The reader is advised to review the discussion of the F_{max} test in Section VI of the *t* **test for two independent samples** prior to continuing this section.

Equation 21.37 (which is identical to Equation 11.6) is employed to compute the F_{max} test statistic.

$$F_{max} = \frac{\tilde{s}_L^2}{\tilde{s}_S^2} \qquad \text{(Equation 21.37)}$$

Where: $\tilde{s}_L^2$ = The largest of the estimated population variances of the k groups
$\tilde{s}_S^2$ = The smallest of the estimated population variances of the k groups

Employing Equation I.8, the estimated population variances are computed for the three groups.

$$\tilde{s}_1^2 = \frac{426 - \frac{(46)^2}{5}}{5 - 1} = .7 \qquad \tilde{s}_2^2 = \frac{227 - \frac{(33)^2}{5}}{5 - 1} = 2.3 \qquad \tilde{s}_3^2 = \frac{203 - \frac{(31)^2}{5}}{5 - 1} = 2.7$$

The largest and smallest estimated population variances which are employed in Equation 21.37 are $\tilde{s}_L^2 = \tilde{s}_3^2 = 2.7$ and $\tilde{s}_S^2 = \tilde{s}_1^2 = .7$. Substituting the latter values in Equation 21.37, the value $F_{max} = 3.86$ is computed.

$$F_{max} = \frac{2.7}{.7} = 3.86$$

The value of F_{max} will always be a positive number that is greater than 1 (unless $\tilde{s}_L^2 = \tilde{s}_S^2$, in which case $F_{max} = 1$). The F_{max} value obtained with Equation 21.37 is evaluated with **Table A9 (Table of the F_{max} Distribution)** in the **Appendix**. Since in Example 21.1, there are $k = 3$ groups and $n = 5$ subjects per group, the tabled critical values in **Table A9** which are employed are the values in the cell that is the intersection of the row $n - 1 = 4$ and the column $k = 3$. In order to reject the null hypothesis, and thus conclude that the homogeneity of variance assumption is violated, the obtained F_{max} value must be equal to or greater than the tabled critical value at the prespecified level of significance. Inspection of **Table A9** indicates that for $n - 1 = 4$ and $k = 3$, $F_{max_{.05}} = 15.5$ and $F_{max_{.01}} = 37$. Since the obtained value $F_{max} = 3.86$ is less than $F_{max_{.05}} = 15.5$, the null hypothesis cannot be rejected. In other words, the alternative hypothesis indicating the presence of heterogeneity of variance is not supported.

Two assumptions of the F_{max} test are: a) Each of the samples has been randomly selected from the population it represents; and b) The distribution of data in the underlying populations from which each of the samples is derived is normal. Various sources (e.g., Keppel (1991), Keppel and Wickens (2004), Maxwell and Delaney (1990, 2000), and Winer *et al.* (1991)) note that when the normality assumption is violated, the accuracy of the F_{max} test may be severely compromised. This problem becomes exacerbated when violation of the normality assumption occurs within the framework of an analysis involving small and/or unequal sample sizes. As noted in Section VI of the *t* **test for two independent samples**, the F_{max} **test** assumes there are an equal number of subjects per group. However, if the sample sizes of the groups being compared are unequal, but are approximately the same value, the value of the larger sample size can be employed to represent *n* in evaluating the test statistic. Kirk (1982, 1995) and Winer *et al.* (1991) note that using the larger *n* will result in a slight increase in the Type I error rate for the test.

An alternative strategy involving the use of the F_{max} **test** suggested by Keppel *et al.* (1992, pp. 119–120) (also discussed in Keppel (1991, p. 220)) is that regardless of the values of *n* or *k*, if $F_{max} > 3$ a lower alpha level should be employed in evaluating the results of the analysis of variance in order to avoid inflating the Type I error rate associated with the latter analysis. The aforementioned strategy is based on research which indicates that when $F_{max} > 3$, there is an increased likelihood that the accuracy of the tabled critical values in the *F* distribution will be compromised when an omnibus *F* value is evaluated.

One criticism of the F_{max} **test** is that it is less powerful than some alternative but computationally more involved procedures for evaluating the homogeneity of variance assumption. Among the more commonly cited alternatives to the F_{max} **test** are tests developed by Bartlett (1937), Cochran (1941), Levene (1960), and Brown and Forsythe (1974a). The tests attributed to Levene (1960) and Brown and Forsythe, (1974a), both of which are described below, are recommended by a number of sources since they are not as sensitive to violations of the normality assumption as is the F_{max} **test**, as well as the fact that they provide for a more powerful test of the alternative hypothesis than the latter test.[52]

Test 21g: The Levene test for homogeneity of variance The test developed by Levene (1960) (which is employed by the statistical software package *SPSS* within the framework of conducting an analysis of variance) involves determining within each group an absolute deviation score from the **group mean** for each of the *n* scores in a group. An **absolute deviation score** is the absolute value of the difference between a score and the group mean. A **single-factor between-subjects analysis of variance** is then employed to contrast the means of the deviation scores computed for the *k* groups. The null hypothesis of homogeneity of variance is rejected if a significant result is obtained.

Table 21.7 contains the values computed for the **absolute deviation scores** for each of the 15 subjects employed in Example 21.1. For each of the three groups, the absolute deviation scores of the five subjects within a group are recorded in the column labeled X_j. Within each group the latter values are obtained by subtracting the group mean from each of the 5 scores in the group, and employing the absolute values of the five difference/deviation scores. In Table 21.1, the values $\bar{X}_1 = 9.2$, $\bar{X}_2 = 6.6$, and $\bar{X}_3 = 6.2$ are recorded for the means of the three groups. Thus, in the case of Group 1, the value $\bar{X}_1 = 9.2$ is subtracted from each of the fives scores 8, 10, 9, 10, 9, yielding the following five absolute deviation scores $|8 - 9.2| = 1.2$, $|10 - 9.2| = .8$, $|9 - 9.2| = .2$, $|10 - 9.2| = .8$, $|9 - 9.2| = .2$. In the case of Group 2, the value $\bar{X}_2 = 6.6$ is subtracted from each of the fives scores 7, 8, 5, 8, 5, yielding the following five absolute deviation scores $|7 - 6.6| = .4$, $|8 - 6.6| = 1.4$, $|5 - 6.6| = 1.6$, $|8 - 6.6| = 1.4$, $|5 - 6.6| = 1.6$. In the case of Group 3, the value $\bar{X}_3 = 6.2$ is subtracted from each of the fives scores 4, 8, 7, 5, 7, yielding the following five absolute deviation scores $|4 - 6.2| = 2.2$, $|8 - 6.2| = 1.8$, $|7 - 6.2| = .8$, $|5 - 6.2| = 1.2$, $|7 - 6.2| = .8$.

In order to conduct the **Levene test for homogeneity of variance**, a **single-factor between-subjects analysis of variance** is employed to evaluate the null hypothesis of equality of the three group means $\bar{X}_1 = .64$, $\bar{X}_2 = 1.28$, and $\bar{X}_3 = 1.36$ computed for the absolute deviation

Table 21.7 Absolute Deviation Scores for Levene Test for Example 21.1

Group 1		Group 2		Group 3	
X_1	X_1^2	X_2	X_2^2	X_3	X_3^2
1.2	1.44	.4	.16	2.2	4.84
.8	.64	1.4	1.96	1.8	3.24
.2	.04	1.6	2.56	.8	.64
.8	.64	1.4	1.96	1.2	1.44
.2	.04	1.6	2.56	.8	.64
$\Sigma X_1^2 = 3.2$	$\Sigma X_1^2 = 2.80$	$\Sigma X_2^2 = 6.4$	$X_2^2 = 9.20$	$\Sigma X_3^2 = 6.8$	$X_3^2 = 10.80$

$$\bar{X}_1 = \frac{\Sigma X_1}{n_1} = \frac{3.2}{5} = .64 \qquad \bar{X}_2 = \frac{\Sigma X_2}{n_2} = \frac{6.4}{5} = 1.28 \qquad \bar{X}_3 = \frac{\Sigma X_3}{n_3} = \frac{6.8}{5} = 1.36$$

scores in Table 21.7. The **Levene test** is based on the premise that if the homogeneity of variance assumption is violated a significant F value will be obtained, thus allowing the researcher to reject the null hypothesis — or to put it another way, if a significant F value is obtained between the mean absolute deviation scores of the k groups, it indicates a significant difference is present with respect to within-group variability between at least two of the three groups.

Initially, we compute the following summary values for the data in Table 21.7.

$$\Sigma X_T = \Sigma X_1 + \Sigma X_2 + \Sigma X_3 = 3.2 + 6.4 + 6.8 = 16.4$$
$$\Sigma X_T^2 = \Sigma X_1^2 + \Sigma X_2^2 + \Sigma X_3^2 = 2.8 + 9.2 + 10.8 = 22.8$$

Employing Equation 21.2, the value $SS_T = 4.87$ is computed.

$$SS_T = 22.8 - \frac{(16.4)^2}{15} = 4.87$$

Employing Equation 21.3, the value $SS_{BG} = 1.56$ is computed.

$$SS_{BG} = \left[\frac{(3.2)^2}{5} + \frac{(6.4)^2}{5} + \frac{(6.8)^2}{5} \right] - \frac{(16.4)^2}{15} = 19.49 - 17.93 = 1.56$$

Employing Equation 21.4, the value $SS_{WG} = 3.31$ is computed.

$$SS_{WG} = 4.87 - 1.56 = 3.31$$

The values $df_{BG} = 2$, $df_{WG} = 12$, and $df_T = 14$ computed originally with Equations 21.8–21.10 in the analysis of Example 21.1 are also employed for the analysis of the data in Table 21.7. Employing Equations 21.6 and 21.7, the values $MS_{BG} = .78$ and $MS_{WG} = .28$ are computed.

$$MS_{BG} = \frac{1.56}{2} = .78 \qquad MS_{WG} = \frac{3.31}{12} = .28$$

Employing Equation 21.12, the value $F = 2.79$ is computed.

$$F = \frac{.78}{.28} = 2.79$$

Table 21.8 is a summary table of the analysis of variance.

**Table 21.8   Levene Test Analysis of Variance Summary Table
for Example 21.1**

Source of variation	SS	df	MS	F
Between-groups	1.56	2	.78	2.79
Within-groups	3.31	12	.28	
Total	**4.87**	**14**		

As was the case for the original analysis of Example 21.1, in **Table A10** the tabled critical values for df_{num} = 2 and df_{den} = 12 are $F_{.05}$ = 3.89 and $F_{.01}$ = 6.93. Since $F = 2.79$ is less than $F_{.05}$ = 3.89, the null hypothesis can not be rejected. Thus, the result of the **Levene test** indicates that the homogeneity of variance assumption is not violated, which is consistent with the result obtained for the F_{max} **test**. Note, however, that the value $F = 2.79$ obtained for the analysis of variance in Table 21.8 is closer to the critical value $F_{.05}$ = 3.89, than the value F_{max} = 3.86 computed for the F_{max} **test** is to the critical value $F_{max_{.05}}$ = 15.5. To be more specific, the result of the **Levene test** comes closer to allowing the researcher to reject the null hypothesis than does the result of the F_{max} **test**. The latter is consistent with the fact that the **Levene test** is more sensitive (i.e., has greater power) than the F_{max} **test** with respect to detecting differences in within-groups variability.

Test 21h: The Brown–Forsythe test for homogeneity of variance   The test developed by Brown and Forsythe (1974a) involves determining within each group an absolute deviation score from the **group median** for each of the n scores in a group. A **single-factor between-subjects analysis of variance** is then employed to contrast the means of the deviation scores computed for the k groups. The null hypothesis of homogeneity of variance is rejected if a significant result is obtained.

Table 21.9   Absolute Deviation Scores for Brown–Forsythe Test for Example 21.1

Group 1		Group 2		Group 3	
X_1	X_1^2	X_2	X_2^2	X_3	X_3^2
1	1	0	0	3	9
1	1	1	1	1	1
0	0	2	4	0	0
1	1	1	1	2	4
0	0	2	4	0	0
$\Sigma X_1^2 = 3$	$\Sigma X_1^2 = 3$	$\Sigma X_2^2 = 6$	$X_2^2 = 10$	$\Sigma X_3^2 = 6$	$X_3^2 = 14$

$$\bar{X}_1 = \frac{\Sigma X_1}{n_1} = \frac{3}{5} = .6 \qquad \bar{X}_2 = \frac{\Sigma X_2}{n_2} = \frac{6}{5} = 1.2 \qquad \bar{X}_3 = \frac{\Sigma X_3}{n_3} = \frac{6}{5} = 1.2$$

Table 21.9 contains the values computed for the **absolute deviation scores** for each of the 15 subjects employed in Example 21.1. For each of the three groups, the absolute deviation scores of the five subjects within a group are recorded in the column labeled X_j. Within each group the latter values are obtained by subtracting the **group median** from each of the 5 scores in the group, and employing the absolute values of the five difference/deviation scores. Employing the guidelines for computing a median described in the **Introduction**, the medians computed for the three groups are respectively $M_1 = 9$, $M_2 = 7$, and $M_3 = 7$. (The latter values are obtained by arranging the scores in each group in order of magnitude and selecting within each group the middle score, which for all of the groups is the score that falls in the third ordinal position.) In the case of Group 1, the value $M_1 = 9$ is subtracted from each of the fives scores 8, 10, 9, 10, 9, yielding the following five absolute deviation scores $|8 - 9| = 1$, $|10 - 9| = 1$, $|9 - 9| = 0$, $|10 - 9| = 1$, $|9 - 9| = 0$. In the case of Group 2, the value $M_2 = 7$ is subtracted from each of the fives scores 7, 8, 5, 8, 5, yielding the following five absolute deviation scores $|7 - 7| = 0$, $|8 - 7| = 1$, $|5 - 7| = 2$, $|8 - 7| = 1$, $|5 - 7| = 2$. In the case of Group 3, the value $M_3 = 7$ is subtracted from each of the fives scores 4, 8, 7, 5, 7, yielding the following five absolute deviation scores $|4 - 7| = 3$, $|8 - 7| = 1$, $|7 - 7| = 0$, $|5 - 7| = 2$, $|7 - 7| = 0$.

In order to conduct the **Brown–Forsythe test**, a **single-factor between-subjects analysis of variance** is employed to evaluate the null hypothesis of equality of the three group means $\bar{X}_1 = .6$, $\bar{X}_2 = 1.2$, and $\bar{X}_3 = 1.2$ computed for the absolute deviation scores in Table 21.9. The **Brown–Forsythe test** is based on the premise that if the homogeneity of variance assumption is violated a significant F value will be obtained, thus allowing the researcher to reject the null hypothesis — or to put it another way, if a significant F value is obtained between the mean absolute deviation scores of the k groups, it indicates that a significant difference is present with respect to within-group variability between at least two of the three groups.

Initially, we compute the following summary values for the data in Table 21.9.

$$\Sigma X_T = \Sigma X_1 + \Sigma X_2 + \Sigma X_3 = 3 + 6 + 6 = 15$$
$$\Sigma X_T^2 = \Sigma X_1^2 + \Sigma X_2^2 + \Sigma X_3^2 = 3 + 10 + 14 = 27$$

Employing Equation 21.2, the value $SS_T = 12$ is computed.

$$SS_T = 27 - \frac{(15)^2}{15} = 12$$

Employing Equation 21.3, the value $SS_{BG} = 1.2$ is computed.

$$SS_{BG} = \left[\frac{(3)^2}{5} + \frac{(6)^2}{5} + \frac{(6)^2}{5} \right] - \frac{(15)^2}{15} = 16.2 - 15 = 1.2$$

Employing Equation 21.4, the value $SS_{WG} = 10.8$ is computed.

$$SS_{WG} = 12 - 1.2 = 10.8$$

The values $df_{BG} = 2$, $df_{WG} = 12$, and $df_T = 14$ computed originally with Equations 21.8–21.10 in the analysis of Example 21.1 are also employed for the analysis of the data in Table 21.9. Employing Equations 21.6 and 21.7, the values $MS_{BG} = .6$ and $MS_{WG} = .9$ are computed.

$$MS_{BG} = \frac{1.2}{2} = .6 \qquad MS_{WG} = \frac{10.8}{12} = .9$$

Employing Equation 21.12, the value $F = .67$ is computed.

$$F = \frac{.6}{.9} = .67$$

Table 21.10 is a summary table of the analysis of variance. As was the case for the original analysis of Example 21.1, in **Table A10** the tabled critical values for $df_{num} = 2$ and $df_{den} = 12$ are $F_{.05} = 3.89$ and $F_{.01} = 6.93$. Since $F = .67$ is less than $F_{.05} = 3.89$, the null hypothesis can

Table 21.10 Brown–Forsythe Test for Homogeneity of Variance
Analysis of Variance Summary Table for Example 21.1

Source of variation	SS	df	MS	F
Between-groups	1.2	2	.6	.67
Within-groups	10.8	12	.9	
Total	**12.0**	**14**		

not be rejected. Thus, the result of the **Brown– Forsythe test for homogeneity of variance** indicates that the homogeneity of variance assumption is not violated, which is consistent with the result obtained for both the F_{max} **test** and the **Levene test**. As is the case with the latter test, the **Brown–Forsythe test** is more sensitive (i.e., has greater power) than the F_{max} **test** with respect to detecting differences in within-groups variability. Keppel and Wickens (2004, p. 152) note that in order to insure that the **Brown–Forsythe test** has sufficient power, some researchers recommend that an alpha level of .10 or greater be employed in lieu of the standard .05 value.

Although alternative tests for evaluating homogeneity of variance do use more information than the F_{max} **test**, they may also be subject to distortion when the underlying populations are not normally distributed. Winer *et al.* (1991) discuss tests developed by Box (1953) and Scheffé (1959) which, like the **Levene** and **Brown–Forsythe tests**, are not as likely to be affected by violation of the normality assumption as the F_{max} **test**. Alternative parametric procedures developed by Brown and Forsythe (1974b), James (1951), and Welch (1951) are discussed in various sources. Keppel (1991) notes that the Brown and Forsythe (1974b) and Welch procedures are not acceptable when $k > 4$, and that James' procedure is too computationally involved for conventional use. In addition, some sources believe that the aforementioned procedures have not been sufficiently researched to justify their use as an alternative to the analysis of variance, even if the homogeneity of variance assumption of the latter test is violated. Keppel (1991), Keppel and Wickens (2004, p. 151), and Winer *et al.* (1991) note that in a review of 56 tests of homogeneity of variance, Conover *et al.* (1981) recommend the **Brown–Forsythe test** described in this section. Howell (2002) and Maxwell and Delaney (1990, 2000), on the other hand, endorse the use of a test developed by O'Brien (1981).

When, in fact, the homogeneity of variance assumption is violated, there is an increased likelihood of committing a Type I error in conducting the analysis of variance evaluating the k group means. Factors which influence the degree to which the Type I error rate for the analysis of variance will be larger than the prespecified value of alpha are the size of the samples and the shapes of the underlying population distributions. Sources generally agree that the effect of violation of the homogeneity of variance assumption on the accuracy of the tabled critical F values is exacerbated when there are not an equal number of subjects in each group.

A variety of strategies have been suggested regarding how a researcher should deal with heterogeneity of variance. Among the procedures which have been suggested are the following: a) Keppel (1991) states that one option available to the researcher is to employ an adjusted tabled critical F value in evaluating the analysis of variance. Specifically, one can employ a tabled critical value associated with a lower alpha level than the prespecified alpha level, so as to provide a more accurate estimate of the latter value (i.e., employ $F_{.025}$ or $F_{.01}$ to estimate $F_{.05}$). The problem with this strategy is that if too low an alpha value is employed, the power of the omnibus F test may be compromised to an excessive degree. Loss of power, however, can be offset by employing a large sample size; b) The data can be evaluated with a procedure other than an analysis of variance. Thus, one can employ a rank-order nonparametric procedure such as the

Kruskal–Wallis one-way analysis of variance by ranks (Test 22). However, by virtue of rank-ordering the data, the latter test will usually provide a less powerful test of an alternative hypothesis than the analysis of variance. Another option is to employ the **van der Waerden normal scores test for k independent samples (Test 23)**, which is a nonparametric procedure which under certain conditions can be as or more powerful than the analysis of variance; and c) Another option available to the researcher is to equate the estimated population variances by employing a data transformation procedure (discussed in Section VII of the *t* **test for two independent samples**), and to conduct an analysis of variance on the transformed data.

4. Computation of the power of the single-factor between-subjects analysis of variance
Prior to reading this section the reader may find it useful to review the discussion on power in Section VI of both the **single-sample *t* test (Test 2)** and the *t* **test for two independent samples**. Before conducting an analysis of variance a researcher may want to determine the minimum sample size required in order to detect a specific effect size. To conduct such a power analysis, the researcher will have to estimate the means of all of the populations that are represented by the experimental treatments/groups. Additionally, he will have to estimate a standard deviation value which it will be assumed represents the standard deviation of all k populations. Understandably, the accuracy of one's power calculations will be a function of the researcher's ability to come up with good approximations for the means and standard deviations of the populations that are involved in the study. The basis for making such estimates will generally be prior research concerning the hypothesis under study.

Equation 21.38 is employed for computing the power of the **single-factor between-subjects analysis of variance**. The test statistic φ represents what is more formally known as the **noncentrality parameter**, and is based on the **noncentral F distribution**.[53]

$$\varphi = \sqrt{n \left[\frac{\Sigma(\mu_j - \mu_T)^2}{k\sigma_{WG}^2} \right]}$$ **(Equation 21.38)**

Where: μ_j = The estimated mean of the population represented by Group j
μ_T = The grand mean, which is the average of the k estimated population means
σ_{WG}^2 = The estimated population variance for each of the k groups
n = The number of subjects per group
k = The number of groups

The computation of the minimum acceptable sample size required to achieve a specified level of power is generally determined prior to collecting the data for an experiment. Using trial and error, a researcher can determine what value of n (based on the assumption that there are an equal number of subjects per group) when substituted in Equation 21.38 will yield the desired level of power. To illustrate the use of Equation 21.38 with Example 21.1, let us assume that prior to conducting the study the researcher estimates that the means of the populations represented by the three groups are as follows: $\mu_1 = 10$, $\mu_2 = 8$, $\mu_3 = 6$. Additionally, it will be assumed that he estimates the variance for each of the three populations the groups represent is $\sigma_{WG}^2 = 2.5$.[54] Based on this information, the value $\mu_T = 8$ can be computed: $\mu_T = (\mu_1 + \mu_2 + \mu_3)/k = (10 + 8 + 6)/3 = 8$. The appropriate values are now substituted in Equation 21.38.

$$\varphi = \sqrt{n \left[\frac{(10 - 8)^2 + (8 - 8)^2 + (6 - 8)^2}{(3)(2.5)} \right]} = \sqrt{1.07n} = 1.03\sqrt{n}$$

At this point, **Table A15 (Graphs of the Power Function for the Analysis of Variance)** in the **Appendix** can be employed to determine the necessary sample size required in order to have the power stipulated by the experimenter. **Table A15** is comprised of sets of power curves which were derived by Pearson and Hartley (1951). Each set of curves is based on a different value for df_{num}, which in the case of a **single-factor between-subjects analysis of variance** is df_{BG} employed for the omnibus F test. Within each set of curves, for a given value of df_{num} there are power functions for both $\alpha = .05$ and $\alpha = .01$. For our analysis (for which it will be assumed that $\alpha = .05$) the appropriate set of curves to employ is the set for $df_{num} = df_{BG} = 2$. Let us assume we want the omnibus F test to have a power of at least .80. We now substitute what we consider to be a reasonable value for n in the equation $\varphi = 1.03\sqrt{n}$ (which is the result obtained with Equation 21.38). To illustrate, the value $n = 5$ (the sample size employed for Example 21.1) is substituted in the equation. The resulting value is $\varphi = 1.03\sqrt{5} = 2.30$.

The value $\varphi = 2.30$ is located on the abscissa (X-axis) of the relevant set of curves in **Table A15** — specifically, the set for $df_{num} = 2$. At the point corresponding to $\varphi = 2.30$, a perpendicular line is erected from the abscissa which intersects with the power curve that corresponds to the value of $df_{den} = df_{WG}$ employed for the omnibus F test. Since $df_{WG} = 12$, the curve for the latter value is employed.[55] At the point the perpendicular intersects the curve $df_{WG} = 12$, a second perpendicular line is drawn in relation to the ordinate (Y-axis). The point at which this perpendicular intersects the ordinate indicates the power of the test. Since $\varphi = 2.30$, we determine that the power is approximately .89. Thus, if we employ five subjects per group, there is a probability of .89 of detecting an effect size equal to or larger than the one stipulated by the researcher (which is a function of the estimated values for the population means relative to the value estimated for the variance of a population). Since the probability of committing a Type II error is $\beta = 1 - $ Power, $\beta = 1 - .89 = .11$. The latter value represents the likelihood of not detecting an effect size equal to or greater than the one stipulated.

Cohen (1977, 1988), who provides a detailed discussion of power computations for the analysis of variance, describes a measure of effect size for the analysis of variance based on standard deviation units which is comparable to the **d index** computed for the different types of **t tests**. In the case of the analysis of variance, d is the difference between the smallest and largest of the estimated population means divided by the standard deviation of the populations. In other words, $d = (\mu_L - \mu_S)/\sigma$. In our example the largest estimated population mean is $\mu_1 = 10$ and the smallest is $\mu_3 = 6$. The value of σ is $\sigma_{WG} = \sqrt{\sigma_{WG}^2} = \sqrt{2.5} = 1.58$. Thus, $d = (10 - 6)/1.58 = 2.53$. This result tells us that if $n = 5$, a researcher has a .89 probability of detecting a difference of about two and one-half standard deviation units.

It is also possible to conduct a power analysis for comparisons which are conducted following the computation of the omnibus F value. In fact, Keppel (1991) recommends that the sample size employed in an experiment be based on the minimum acceptable power necessary to detect the smallest effect size among all of the comparisons the researcher plans before collecting the data. The value φ_{comp} (described by McFatter and Gollob (1986)), which is computed with Equation 21.39, is employed to determine the power of a comparison.

$$\varphi_{comp} = \sqrt{n\left[\frac{(\mu_a - \mu_b)^2}{2(\sigma_{WG}^2)(\Sigma c_j^2)}\right]}$$ **(Equation 21.39)**

Equation 21.39 can be used for both simple and complex single degree of freedom comparisons. As a general rule, the equation is used for planned comparisons. Although it can be extended to unplanned comparisons, published power tables for the analysis of variance generally only apply to per comparison error rates of $\alpha = .05$ and $\alpha = .01$. In the case of planned and especially unplanned comparisons which involve α_{PC} rates other than .05 or .01, more detailed tables are required.[56]

For single degree of freedom comparisons, the power curves in **Table A15** for $df_{num} = 1$ are always employed. The use of Equation 21.39 will be illustrated for the simple comparison Group 1 versus Group 2 (summarized in Table 21.3). Since $\Sigma c_j^2 = 2$, and we have estimated $\mu_a = \mu_1 = 10$, $\mu_b = \mu_2 = 8$, and $\sigma_{WG}^2 = 2.5$, the following result is obtained.

$$\varphi_{comp} = \sqrt{n\left[\frac{(10 - 8)^2}{(2)(2.5)(2)}\right]} = \sqrt{.4n} = .63\sqrt{n}$$

Substituting $n = 5$ in the equation $\varphi_{comp} = .63\sqrt{n}$, we obtain $\varphi_{comp} = .63\sqrt{5} = 1.41$. Employing the power curves for $df_{num} = 1$ with $\alpha = .05$, we use the curve for $df_{WG} = 12$ (the df_{WG} employed for the omnibus F test), and determine that when $\varphi_{comp} = 1.41$, the power of the test is approximately .44.

It should be noted that if the methodology described for computing the power of the *t* test **for two independent samples** is employed for the above comparison or any other simple comparison, it will produce the identical result. To demonstrate that Equation 21.39 produces a result which is equivalent to that obtained with the protocol for computing the power of the *t* test **for two independent samples**, Equations 11.10 and 11.12 will be employed to compute the power of the comparison Group 1 versus Group 2. Employing Equation 11.10:

$$d = \frac{\mu_1 - \mu_2}{\sigma} = \frac{10 - 8}{1.58} = 1.27$$

Substituting the value of d in Equation 11.12, the value $\delta = 2.01$ is computed.

$$\delta = d\sqrt{\frac{n}{2}} = 1.27\sqrt{\frac{5}{2}} = 2.01$$

The value of $\delta = 2.01$ is evaluated with **Table A3 (Power Curves for Student's *t* Distribution)** in the **Appendix**. A full description of how to employ the power curves in **Table A3** can be found in Section VI of the **single-sample *t* test**. Employing **Table A3-C** (which is the set of curves for a two-tailed analysis with $\alpha = .05$), we use the curve for $df_{WG} = 12$. It is determined that when $\delta = 2.01$, the power of the comparison is approximately .44.

A final note regarding Equation 21.39: The value of Σc_j^2 computed for a complex comparison will always be lower than the value Σc_j^2 computed for a simple comparison (in the case of a simple comparison, Σc_j^2 will always equal 2). Because of this, for a fixed value of n the computed value of φ_{comp} will always be larger for a complex comparison and, consequently, the power of a complex comparison will always be greater than the power of a simple comparison.

5. Measures of magnitude of treatment effect for the single-factor between-subjects analysis of variance: Omega squared (Test 21i), eta squared (Test 21j), and Cohen's f index (Test 21k) Prior to reading this section the reader should review the discussion of the measures of magnitude of treatment effect in Section VI of the *t* test for two independent samples. As is the case with the *t* value computed for the latter test, the omnibus *F* value computed for the **single-factor between-subjects analysis of variance** only provides a researcher with information regarding whether or not the null hypothesis can be rejected — i.e., whether or not a significant difference is present between at least two of the experimental treatments. The *F* value (as well as the level of significance with which it is associated), however, does not provide the researcher with any information regarding the size of any treatment effect which is present. As is the case with a *t* value, an *F* value is a function of both the difference between the means of the experimental treatments and the sample size. The measures described in this section are variously referred to as **measures of effect size, measures of magnitude of treatment effect, measures of association**, and **correlation coefficients**.

Omega squared (Test 21i)   A number of measures of the **magnitude of treatment effect** have been developed which can be employed for the **single-factor between-subjects analysis of variance**. Such measures, which are independent of sample size, provide an estimate of the proportion of variability on the dependent variable that is associated with the independent variable/experimental treatments. Although sources are not in agreement with respect to which measure of treatment effect is most appropriate for the **single-factor between-subjects analysis of variance**, one commonly employed measure is the **omega squared** statistic ($\tilde{\omega}^2$), which provides an estimate of the underlying population parameter ω^2. The value of $\tilde{\omega}^2$ is computed with Equation 21.41. The $\tilde{\omega}^2$ value computed with Equation 21.41 is the best estimate of the proportion of variability in the data that is attributed to the experimental treatments. It is obtained by dividing treatment variability (σ_{BG}^2) by total variability (which equals $\sigma_{BG}^2 + \sigma_{W}^2$). Thus, Equation 21.40 represents the population parameter estimated by Equation 21.41.

$$\omega^2 = \frac{\sigma_{BG}^2}{\sigma_{BG}^2 + \sigma_{WG}^2} \qquad \textbf{(Equation 21.40)}$$

$$\tilde{\omega}^2 = \frac{SS_{BG} - (k - 1)MS_{WG}}{SS_T + MS_{WG}} \qquad \textbf{(Equation 21.41)}$$

Although the value of $\tilde{\omega}^2$ will generally fall in the range between 0 and 1, when $F < 1$, $\tilde{\omega}^2$ will be a negative number. The closer $\tilde{\omega}^2$ is to 1, the stronger the association between the independent and dependent variables, whereas the closer $\tilde{\omega}^2$ is to 0, the weaker the association between the two variables. A $\tilde{\omega}^2$ value equal to or less than 0 indicates that there is no association between the variables. Keppel (1991) notes that in behavioral science research (which is commonly characterized by a large amount of error variability) the value of $\tilde{\omega}^2$ will rarely be close to 1.

Employing Equation 21.41 with the data for Example 21.1, the value $\tilde{\omega}^2 = .44$ is computed.

$$\tilde{\omega}^2 = \frac{26.53 - (3 - 1)(1.9)}{49.33 + 1.9} = .44$$

Equation 21.42 is an alternative equation for computing the value of $\tilde{\omega}^2$ that yields the same value as Equation 21.41.[57]

$$\tilde{\omega}^2 = \frac{(k - 1)(F - 1)}{(k - 1)(F - 1) + nk} \qquad \textbf{(Equation 21.42)}$$

$$\tilde{\omega}^2 = \frac{(2)(6.98 - 1)}{(2)(6.98 - 1) + (5)(3)} = .44$$

The value $\tilde{\omega}^2 = .44$ indicates that 44% (or a proportion equal to .44) of the variability on the dependent variable (the number of nonsense syllables correctly recalled) is associated with variability on the levels of the independent variable (noise). To say it another way, 44% of the variability on the recall scores of subjects can be accounted for on the basis of which group a subject is a member. As noted in the discussion of **omega squared** in Section VI of the *t* **test for two independent samples,** Cohen (1977; 1988, pp. 284–287) has suggested the following (admittedly arbitrary) values, which are employed in psychology and a number of other disciplines, as guidelines for interpreting $\tilde{\omega}^2$: a) A **small effect size** is one that is greater than .0099 but not more than .0588; b) A **medium effect size** is one that is greater than .0588 but not more than .1379; c) A **large effect size** is greater than .1379. If one employs Cohen's (1977, 1988) guidelines for magnitude of treatment effect, $\tilde{\omega}^2 = .44$ represents a large treatment effect.

Eta squared (Test 21j) Another measure of treatment effect employed for the **single-factor between-subjects analysis of variance** is the **eta squared** statistic ($\tilde{\eta}^2$) (which estimates the underlying population parameter η^2). $\tilde{\eta}^2$ is computed with Equation 21.43.

$$\tilde{\eta}^2 = \frac{SS_{BG}}{SS_T} \qquad \textbf{(Equation 21.43)}$$

Employing Equation 21.43 with the data for Example 21.1, the value $\tilde{\eta}^2 = .54$ is computed.

$$\tilde{\eta}^2 = \frac{26.53}{49.33} = .54$$

Note that the value $\tilde{\eta}^2 = .54$ is larger than $\tilde{\omega}^2 = .44$ computed with Equations 21.41/21.42. Sources (e.g., Grissom and Kim (2005, p. 121)) note that $\tilde{\eta}^2$ tends to be positively biased (i.e., it tends to overestimate true value of η^2), and that $\tilde{\eta}^2$ is a more biased estimate of the magnitude of treatment effect in the underlying population than is $\tilde{\omega}^2$, since $\tilde{\eta}^2$ employs the values SS_{BG} and SS_T, which by themselves are biased estimates of population variability. Darlington and Carlson (1987) note that Equation 21.44 can be employed to compute a less biased estimate of the population parameter that is estimated by $\tilde{\eta}^2$.

$$\textit{Adjusted } \tilde{\eta}^2 = 1 - \frac{MS_{WG}}{MS_T} \qquad \textbf{(Equation 21.44)}$$

$$\textbf{Where: } MS_T = \frac{SS_T}{N - 1}$$

Since $MS_T = 49.33/14 = 3.52$, the adjusted $\tilde{\eta}^2 = .46$ falls in between $\tilde{\omega}^2 = .44$ (computed with Equations 21.41/21.42) and $\tilde{\eta}^2 = .54$ (computed with Equation 21.43).

$$\text{Adjusted } \tilde{\eta}^2 = 1 - \frac{1.9}{3.52} = .46$$

When $\tilde{\omega}^2$ and $\tilde{\eta}^2$ are based on a small sample size, their standard error (which is a measure of error variability) will be large, and, consequently, the reliability of the measures of treatment effect will be relatively low. The latter will be reflected in the fact that under such conditions a confidence interval computed for a measure of treatment effect (the computation of which will not be described in this book) will have a wide range. It should be emphasized that a measure of magnitude of treatment effect is a measure of association/correlation and, in and of itself, it is not a test of significance. The significance of $\tilde{\omega}^2$ and $\tilde{\eta}^2$ is based on whether or not the omnibus F value is significant.[58]

Cohen's f index (Test 21k)   Cohen (1977, 1988) describes an index of effect size that can be employed with the **single-factor between-subjects analysis of variance** which he designates as f. **Cohen's f index** is a generalization of his d **effect size index** in the case of three or more means. The d **index (Test 2a)** was employed to compute the power of the **single-sample t test**, the t **test for two independent samples**, and the t **test for two dependent samples (Test 17)**. Kirk (1995, p. 181) notes that the value of f can be computed with either Equation 21.45 or Equation 21.46. When the latter equations are employed for Example 21.1, the value $f = .89$ is obtained.[59]

$$f = \sqrt{\frac{\tilde{\omega}^2}{1 - \tilde{\omega}^2}} = \sqrt{\frac{.44}{1 - .44}} = .89 \qquad \text{(Equation 21.45)}$$

$$\text{(Equation 21.46)}$$

$$f = \sqrt{\frac{\frac{k-1}{nk}[MS_{BG} - MS_{WG}]}{MS_{WG}}} = \sqrt{\frac{\left(\frac{3-1}{(5)(3)}\right)[13.27 - 1.90]}{1.90}} = .89$$

It should be noted that although Cohen (1977; 1988, p. 284) employs the notation for **eta squared** in Equation 21.45 in place of **omega squared**, the definition of the statistic he uses is consistent with the definition of **omega squared**. If, for Example 21.1, one elects to employ the values $\tilde{\eta}^2 = .54$ or *Adjusted* $\tilde{\eta}^2 = .46$ in Equation 21.45, the values $f = 1.08$ and $f = .92$ are computed.

Cohen (1977; 1988, p. 275–276) defines an f value as a "standard deviation of standardized means," and notes that f will equal 0 when the k treatment means are equal, and continue to increase as the ratio of between-groups variability to within-groups variability gets larger. Cohen (1977; 1988, pp. 284–288) has proposed the following (admittedly arbitrary) f values as criteria for identifying the magnitude of an effect size: a) A **small effect size** is one that is greater than .1 but not more than .25; b) A **medium effect size** is one that is greater than .25 but not more than .4; and c) A **large effect size** is greater than .4. Employing Cohen's criteria, the value $f = .89$ represents a large effect size. The f effect size index is commonly employed in computing the power of the **single-factor between-subjects analysis of variance** (as well as the power of the analysis of variance when it is employed with other designs). Cohen (1977; 1988, Chapter 8) contains power tables for the analysis of variance which employ f as a measure of effect size.

Equation 21.47 allows one to convert an f value into an $\tilde{\omega}^2$ or **eta squared** value. (As was the case with Equation 21.45, although Cohen (1977; 1988, p. 281) employs the notation for **eta squared** in Equation 21.47 in place of **omega squared**, the definition of the statistic he uses is consistent with the definition of **omega squared**.)

$$\tilde{\omega}^2 = \frac{f^2}{1 + f^2} = \frac{(.89)^2}{1 - (.89)^2} = .44 \qquad \textbf{(Equation 21.47)}$$

If the values $f = 1.08$ and $f = .92$ are substituted in Equation 21.47, the values $\tilde{\eta}^2 = .54$ and *Adjusted* $\tilde{\eta}^2 = .46$ are computed. The choice of employing **omega squared** or **eta squared** in Equation 21.45 is up to the researcher's discretion.

Many sources recommend that in summarizing the results of an experiment, in addition to reporting the omnibus test statistic (e.g., an F or t value), a measure of magnitude of treatment effect also be included, since the latter can be useful in further clarifying the nature of the relationship between the independent and dependent variables. It is important to note that if the value of a measure of magnitude of treatment effect is small, it does not logically follow that the relationship between the independent and the dependent variables is trivial. There are instances when a small treatment effect may be of practical and/or theoretical value (an illustration of this is provided in Section IX (the **Addendum**) of the **Pearson product-moment correlation coefficient**, under the discussion of **meta-analysis and related topics**). It should be noted that when the independent variable is a nonmanipulated variable, it is possible that any treatment effect which is detected may be due to some variable other than the independent variable. Such studies (referred to as **ex post facto** studies) do not allow a researcher to adequately control for the potential effects of extraneous variables on the dependent variable. As a result of this, even if a large treatment effect is present, a researcher is not justified in drawing conclusions with regard to cause and effect — specifically, the researcher cannot conclude that the independent variable is responsible for group differences on the dependent variable. Other sources which discuss measures of magnitude of treatment effect for the **single-factor between-subjects analysis of variance** are Cortina and Nouri (2001), Howell (2002), Keppel (1991), Kirk (1982, 1995), Maxwell and Delaney (1990, 2000, 2004), and Winer *et al.* (1991). Rosenthal *et al.* (2000) and Rosnow *et al.* (2000) provide alternative methodologies for assessing effect size for a **single-factor between-subjects analysis of variance**. Further discussion of the indices of treatment effect discussed in this section and the relationship between effect size and statistical power can be found in Section IX (the **Addendum**) of the **Pearson product-moment correlation coefficient** under the discussion of **meta-analysis and related topics**.

Final comments on measures of effect size During the past decade researchers have become increasingly critical of tests of statistical significance and the use of the probability value which results from such a test. In lieu of tests of significance, greater emphasis has been put on the importance in published research of reporting measures of effect size and confidence intervals (which are discussed in the next section). This writer does not share the disillusionment which some researchers have toward tests of significance, since he is of the opinion that if used intelligently such tests can be extremely powerful tools within the armamentarium of the researcher. It is the opinion of this writer that the problems which result from the use of tests of significance can be attributed to confusion among researchers who use them, rather than the tests themselves. Put simply, it is a fact of academic training that individuals who employ statistical analysis are often ill informed, and because of the latter are prone to misuse and/or misinterpret tests of significance.

In the case of measures of effect size, a problem associated with such measures is that there is a general lack of agreement with respect to which of the numerous available measures is the best to employ. To go even further, in the case of the **omega squared** statistic, for some analysis of variance procedures (e.g., the **single-factor within-subjects analysis of variance (Test 24)** and the **between-subjects factorial analysis of variance (Test 27)**) there are alternative equations yielding different values which are recommended by different sources for computing the value of **omega squared**. In view of the latter, the use of **omega squared** may result in more confusion than clarification with respect to how a set of data should be interpreted (also see Anderson (2001, p. 557)). Alternative measures of effect size discussed elsewhere in the book (e.g., the **odds ratio (Test 16j)** and the **binomial effect size display (Test 28p)**) are probably more informative and easier to interpret than the measures of effect size discussed in this section for the analysis of variance. Grissom and Kim (2005, p. 124–127) provide a good discussion of criticisms which have been directed towards the measures of effect size discussed in this chapter. Among the criticisms noted by the latter authors is that the measures of effect size which reflect the percentage of explained variance (i.e., $\tilde{\omega}^2$, $\tilde{\eta}^2$, $\tilde{\varepsilon}^2$) are global in nature (i.e., reflect overall effect size among all k groups), and because of the latter, when $k > 2$, it might be more meaningful to compute two or more measures of effect size for specific pairs of groups.

In the case of confidence intervals, although the recent emphasis on the importance of reporting of such intervals certainly has merit, at the same time the broad range of values that are generally computed for a confidence interval limit their practical value. As noted in the **Introduction**, more recently critics of significance testing (e.g., Hunter and Schmidt (1990, 2004) and Kline (2004), Smithson (2003), and Thompson (2002)) have argued that researchers should utilize confidence intervals in decision making, as well as the fact that all published research should include confidence intervals for any relevant statistics. In some instances, those who vigorously advocate the use of confidence intervals argue that the test of statistical significance should no longer be employed for hypothesis testing, and that instead decisions should be made on the basis of values computed for confidence intervals. The position of this writer is that, like tests of significance, measures of effect size and confidence intervals are limited with respect to the information they provide, and, as such, can only be useful if utilized intelligently by researchers.

6. Computation of a confidence interval for the mean of a treatment population Prior to reading this section the reader may find it useful to review the discussion of confidence intervals in Section VI of the **single-sample t test**. Equation 21.48 can be employed to compute a confidence interval for the mean of a population represented by one of the k treatments for which an omnibus F value has been computed.[60]

$$CI = \bar{X}_j \pm t_{df_{WG}} \sqrt{\frac{MS_{WG}}{n}} \qquad \textbf{(Equation 21.48)}$$

Where: $t_{df_{wg}}$ represents the tabled critical two-tailed value in the t distribution, for df_{WG}, below which a proportion (percentage) equal to $[1 - (\alpha/2)]$ of the cases falls. If the proportion (percentage) of the distribution that falls within the confidence interval is subtracted from 1 (100%), it will equal the value of α.

The use of the tabled critical t value for df_{WG} in Equation 21.48 instead of a degrees of freedom value based on the number of subjects who served in each group (i.e., $df = n - 1$) is predicated on the fact that the estimated population variance is based on the pooled variance of the k groups. If, however, there is reason to believe that the homogeneity of variance assumption is violated, it is probably more prudent to employ Equation 2.8 to compute the confidence interval. The latter equation employs the estimated population standard deviation for a specific group for which a confidence interval is computed, rather than the pooled variability (which is represented by MS_{WG}). If the standard deviation of the group is employed rather than the pooled variability, $df = n - 1$ is used for the analysis. It can also be argued that it is preferable to employ Equation 2.8 in computing the confidence interval for a population mean in circumstances where a researcher has reason to believe that the group in question represents a population which is distinct from the populations represented by the other $(k - 1)$ groups. Specifically, if the mean value of a group is significantly above or below the means of the other groups, it can be argued that one can no longer assume that the group shares a common variance with the other groups. In view of this, one can take the position that the variance of the group is the best estimate of the variance of the population represented by that group (as opposed to the pooled variance of all of the groups involved in the study). The point to be made here is that, as is the case in computing a confidence interval for a comparison, depending upon how one conceptualizes the data, more than one methodology can be employed for computing a confidence interval for a population mean.

The computation of the 95% confidence interval for the mean of the population represented by Group 1 is illustrated below employing Equation 21.48. The value $t_{.05} = 2.18$ is the tabled critical two-tailed $t_{.05}$ value for $df_{WG} = 12$.

$$CI_{95} = 9.2 \pm 2.18 \sqrt{\frac{1.9}{5}} = 9.2 \pm 1.34$$

Thus, the researcher can be 95% confident (or the probability is .95) the interval 7.86 to 10.54 includes the actual value of the mean of the population represented by Group 1. Stated symbolically: $7.86 \leq u_1 \leq 10.54$.

If, on the other hand, the researcher elects to employ Equation 2.8 (and more specifically Equation 2.9) to compute the 95% confidence interval for the mean of the population represented by Group 1, the standard deviation of Group 1 is employed in lieu of the pooled variability of all the groups. In addition, the tabled critical two-tailed $t_{.05}$ value for $df = n - 1 = 4$ is employed in Equation 2.8. The computations are shown below.

$$\tilde{s}_1 = \sqrt{\frac{\sum X_1^2 - \frac{(\sum X_1)^2}{n}}{n - 1}} = \sqrt{\frac{426 - \frac{(46)^2}{5}}{4}} = .837$$

$$CI_{95} = \bar{X}_1 \pm t_{.05} \left(\frac{\tilde{s}_1}{\sqrt{n}} \right) = 9.2 \pm 2.78 \left(\frac{.837}{\sqrt{5}} \right) = 9.2 \pm 1.04$$

Thus, employing Equation 2.8, the range for CI_{95} is $8.16 \leq \mu_1 \leq 10.24$.[61] Note that although the ranges of values computed with Equation 21.48 and Equation 2.8 are reasonably close to one another, the range of the confidence interval computed with Equation 21.48 is wider. Depending on the values of $t_{df_{WG}}$ versus $t_{df_{(n-1)}}$ and $\sqrt{MS_{WG}/n}$ versus $s_{\bar{x}_j}$, there will usually be a discrepancy between the confidence interval computed with the two equations. When the estimated population variances of all k groups are equal (or reasonably close to one another),

Equation 2.8 will yield a wider confidence interval than Equation 21.48, since $\sqrt{MS_{WG}/n}$ will equal $s_{\bar{X}_g}$ and $t_{df_{WG}} < t_{df_{(n-1)}}$. In the example illustrated in this section, the reason why use of Equation 2.8 yields a smaller confidence interval than Equation 21.48 (in spite of the fact that a larger t value is employed in Equation 2.8) is because $\tilde{s}^2 = .7$ (the estimated variance of the population represented by Group 1) is substantially less than $MS_{WG} = 1.9$ (the pooled estimate of within-groups variability).

7. Trend analysis The comparison methods described in the first part of Section VI are most useful when a **qualitative independent variable** is employed in an experiment. When, however, the independent variable is **quantitative**, a methodology referred to as **trend analysis** can provide more useful information regarding the relationship between the independent and dependent variables. More specifically, trend analysis allows a researcher to derive a mathematical function which describes the nature of the relationship between a quantitative independent variable and a dependent variable.

A **qualitative independent variable** is one in which the k levels represent k different **kinds** of something (i.e., whatever it is that the independent variable represents). For example, the independent variable employed in Example 21.1 represents a **qualitative independent variable** insofar as the three levels (**no noise, moderate noise/classical music**, and **extreme noise/rock music**) are not defined as specific quantitative values, but instead are defined by different kinds/qualities of music. On the other hand, a **quantitative independent variable** is one in which the k levels represent k different **amounts** of something. Thus, if in Example 21.1, the three levels **no noise, moderate noise** and **extreme noise**, were replaced with exposure to a sound with an amplitudes of **0, 40 or 80 decibels** (a decibel is a measure of sound intensity), the independent variable would be quantitative. To be more specific, assume that the **no noise** condition is replaced by a subject being exposed to **zero decibels**, **moderate noise** by a subject being exposed to a continuous **40 decibel tone**, and **extreme noise** by a subject being exposed to a continuous **80 decibel tone**. In such an experiment, the three levels of the independent variable are all expressed within the context of the same numerical scale, and thus the independent variable is quantitative in nature. Other examples of a quantitative independent variable (which typically differentiates the k levels of the independent variable by employing numerical values based upon some interval/ratio scale of measurement) would be the dosage of a drug administered to subjects in k experimental groups, or the amount of time k groups of subjects are allowed to study a passage of verbal material prior to being tested on its contents. By definition, no trend is present in a set of data when the means of the k experimental conditions are equal to one another.

It is not required that an omnibus analysis of variance be conducted on the k group means prior to conducting a trend analysis, since the latter type of analysis can be viewed as a special case of planned comparisons. Nevertheless, a researcher is likely to conduct an analysis of variance prior to a trend analysis in order to determine if any significant differences are present in the data. If a quantitative independent variable is employed in an experiment and the result of the analysis of variance does not allow the researcher to reject the null hypothesis, a trend analysis on the data will also yield nonsignificant results. On the other hand, if the result of the omnibus F test is significant, in lieu of (or in addition to) employing the comparison procedures described earlier, the researcher can conduct a trend analysis to determine the mathematical function which best describes the relationship between the different quantitative levels of the independent variable and the dependent variable.

The results of a trend analysis can be summarized both numerically and visually. Numerical summary of a trend analysis involves computing one or more F values through use of the same methodology described earlier for the evaluation of **linear contrasts**. Additionally, the mathematical function which provides the best description of the relationship between the two variables is computed. Visual summary of a trend analysis is accomplished by a constructing a graph which depicts the pattern of the relationship between the different quantitative levels of the independent variable (the values of which are represented on the abscissa/X-axis) and the k group means for the dependent variable (the values of which are represented on the ordinate/Y-axis).

It should be emphasized that in contrast to the comparison procedures described earlier, which address differences between pairs of means, trend analysis focuses on the pattern of the relationship among the complete set of k means. To be more specific, trend analysis attempts to determine the mathematical function that best describes the relationship between the quantitative values which define the levels of the independent variable and the magnitude of scores on the dependent variable. Generally speaking, a mathematical function can be either linear or nonlinear (i.e., curvilinear). The mathematical functions most commonly employed in trend analysis are referred to as **polynomials**. By definition, a **polynomial** is an equation that is comprised of the sum of two or more terms which involve a sum of powers multiplied by coefficients in one or more of the terms. Two examples of polynomials are $Y = 3X + 3$ (which is a linear polynomial — or more simply, the equation of a straight line) and $Y = 2X^2 + X - 1$ (which is a quadratic polynomial, which is one type of curvilinear function).

When an ordered quantitative independent variable is comprised of $k = 2$ levels, only a **linear trend** can describe the relationship between the two variables. However, when such an independent variable is comprised of $k = 3$ or more levels, a researcher can test for the presence of a **linear** as well as one or more different types of **curvilinear trends**. The number of curvilinear trends that can be evaluated will be a direct function of the number of levels of the independent variable. At this point in the discussion the concept of a linear trend as well as three different kinds of curvilinear trends will be illustrated through use of Figure 21.2.

The methodology in trend analysis is to initially determine whether or not a straight line can be employed to describe the relationship between the independent and dependent variables. If, in fact, the latter turns out to be the case, one can conclude that a **linear trend** (which represents the simplest relationship between the variables) is present. As an example, let us assume that in Example 21.1 the independent variable had five levels, and that each level represented one of the five following decibel values which subjects in each of the five groups were exposed to while they were studying the list of 10 nonsense syllables: **Group 1: 0 decibels**; **Group 2: 20 decibels**; **Group 3: 40 decibels**; **Group 4: 60 decibels**; **Group 5: 80 decibels**. Let us also assume that the average number of nonsense syllables recalled the following day by subjects in each of the groups were as follows: $\bar{X}_1 = 8$, $\bar{X}_2 = 6$, $\bar{X}_3 = 4$, $\bar{X}_4 = 2$, $\bar{X}_5 = 0$. (It was previously noted that when the result of a trend analysis is summarized graphically, subjects' scores on the dependent variable are represented on the ordinate/Y-axis and the values of the levels of the quantitative independent variable are represented on the abscissa/X-axis. In order to be consistent with the use of the notation X and Y representing the two variables on the axes of a graph, in the discussion to follow the notation $\bar{Y}_j$ will be employed in place of $\bar{X}_j$ to represent the mean of the jth group. Thus, for the example under discussion (as well as in other examples discussed in this section) the notation $\bar{Y}_1 = 8$, $\bar{Y}_2 = 6$, $\bar{Y}_3 = 4$, $\bar{Y}_4 = 2$, $\bar{Y}_5 = 0$ will be employed to represent the five group means.)

By utilizing the values 0, 20, 40, 60 and 80 for the levels of the independent variable in combination with the values noted above for the group means, a graph can be constructed employing the following five data points (with each data point corresponding to the horizontal position of a decibel level measured along the abscissa and the vertical position of a group mean

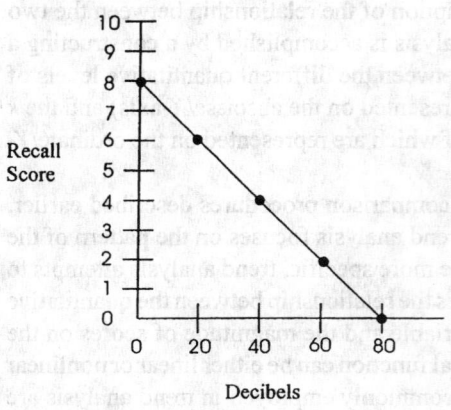

Figure 21.2a   Linear Trend

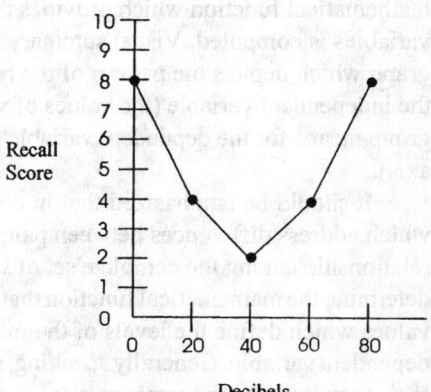

Figure 21.2b   Quadratic Trend

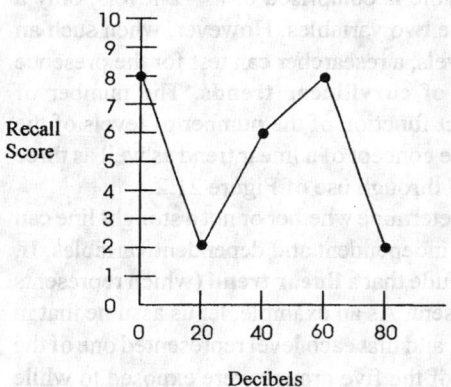

Figure 21.2c   Cubic Trend

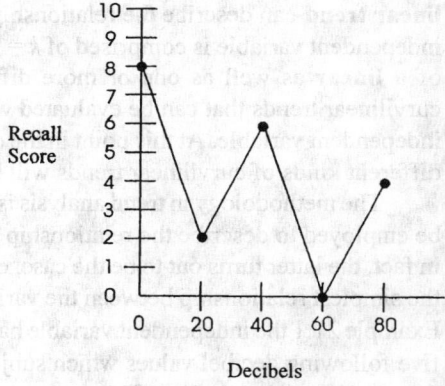

Figure 21.2d   Quartic Trend

Figure 21.2   Examples of Linear and Curvilinear Trends

measured along the ordinate): (0, 8), (20, 6), (40, 4), (60, 2), (80, 0). When the latter data points are connected they yield the negatively sloped line depicted in Figure 21.2a, which indicates that an increase in decibel level is associated with a proportional decrease in recall. (A **negatively sloped line** is characterized by the upper part of the line being positioned at the left of the graph with the line slanting downward as one moves to the right. Further clarification of the **slope** of a line can be found in Section VI of the **Pearson product-moment correlation coefficient**.) Because of the latter, it would appear that a **linear trend** provides an accurate description of the relationship between the independent and dependent variables. Note that a linear trend describes a function which increases or decreases at the same rate as one moves from left to right along the horizontal plane that defines the quantitative levels of the independent variable. To put it another way, when a linear trend is present the graph of the function which defines the relationship between the independent and dependent variable **does not at any point reverse its direction**. An equation referred to as a **linear polynomial** can be computed for the straight line depicted in Figure 21.2a. The general form of the equation for a straight line is $Y = a + bX$ (where a represents a constant (also referred to as the Y intercept, since it is the point at which the line intersects the ordinate), and b is a numerical value representing the **slope** of the line). The actual equation for the straight line in Figure 21.2a is $Y = 8 - .1X$. Note that if a value for X representing one of the five decibel levels employed to define the quantitative independent variable is substituted in the latter equation, it yields a value for Y which corresponds to the mean recall score obtained by the group exposed to that decibel level. (The protocol for determining the equation of a straight line is discussed later in this section, as well as in Section VI of the **Pearson product-moment correlation coefficient**.)

It is often the case that a set of data may be characterized by a trend which is **nonlinear**. The three most common **nonlinear trends** a researcher is likely to consider are a **quadratic trend** (also referred to as a **second order polynomial** — a linear function can also be referred to as a **first order polynomial**), **cubic trend** (also referred to as a **third order polynomial**), and a **quartic trend** (also referred to as a **fourth order polynomial**), all of which can be described by polynomial equations for curvilinear functions. When an ordered quantitative independent variable has at least $k = 3$ levels, a researcher can test for the presence of either a **linear** or **quadratic trend**. Within the framework of the graph which describes the relationship between the variables, a **quadratic trend** is present if there is **one reversal** in the direction of the line that connects the k data points (the curve which describes such a trend is often referred to as being **concave** — being characterized by a single bend in either an upward or downward direction). A **quadratic trend** is illustrated in Figure 21.2b. Note that in the latter figure, when the five data points $(0, 8)$, $(20, 4)$, $(40, 2)$, $(60, 4)$, and $(80, 8)$ are connected, they do not yield a straight line, but instead a line which reverses its direction at the 40 decibel level. The **quadratic trend** in Figure 21.2b indicates that the number of nonsense syllables recalled decreases as the number of decibels increase from 0 to 40, but beyond that the number of nonsense syllables recalled increases as the number of decibels increase up to the value 80. The equation that can be computed for the curvilinear function depicted in Figure 21.2b is referred to as a **quadratic polynomial**. The general form of such an equation is $Y = a + b_1 X + b_2 X^2$ (where a represents a constant, and b_1 and b_2 respectively represent numerical values referred to as coefficients for the linear and quadratic components of the equation).

When a quantitative independent variable is comprised of $k = 4$ or more levels, it is possible to evaluate whether a **linear**, **quadratic** or a **cubic trend** is present. Within the framework of the graph which describes the relationship between the variables, a **cubic trend** is present if there are **two reversals** in the direction of the line that connects the k data points. A **cubic trend** is illustrated in Figure 21.2c. Note that in the latter figure, when the five data points $(0, 8)$, $(20, 2)$, $(40, 6)$, $(60, 8)$, and $(80, 2)$ are connected, they do not yield a straight line, but instead a line which reverses its direction twice, initially at the 20 decibel level and again at the 60 decibel level. The **cubic trend** in Figure 21.2c indicates that the number of nonsense syllables recalled decreases as the number of decibels increase from 0 to 20, then increases up to the 60 decibel level, after which recall once again decreases. The equation that can be computed for the curvilinear function depicted in Figure 21.2c is referred to as a **cubic polynomial**. The general form of such an equation is $Y = a + b_1 X + b_2 X^2 + b_3 X^3$ (where a represents a constant, and b_1, b_2, and b_3 respectively represent the coefficients for the linear, quadratic, and cubic components of the equation).

When a quantitative independent variable is comprised of $k = 5$ or more levels, it is possible to evaluate whether a **linear**, **quadratic**, **cubic** or a **quartic trend** is present. Within the framework of the graph which describes the relationship between the variables, a **quartic trend** is present if there are **three reversals** in the direction of the line that connects the k data points. A **quartic trend** is illustrated in Figure 21.2d. Note that in the latter figure, when the five data points (0, 8), (20, 2), (40, 6), (60, 0), and (80, 4) are connected, they do not yield a straight line, but instead a line which reverses its direction three times, initially at the 20 decibel level as well as subsequently at the 40 and 60 decibel levels. The quartic trend in Figure 21.2d indicates that the number of nonsense syllables recalled decreases as the number of decibels increases from 0 to 20, then increases up to the 40 decibel level, then decreases up to the 60 decibel level, after which recall once again increases. The equation that can be computed for the curvilinear function depicted in Figure 21.2d is referred to as a **quartic polynomial**. The general form of such an equation is $Y = a + b_1 X + b_2 X^2 + b_3 X^3 + b_4 X^4$ (where a represents a constant, and b_1, b_2, b_3, and b_4 respectively represent the coefficients for the linear, quadratic, cubic, and quartic components of the equation).

Although the discussion of nonlinear trends in this section (as well as the discussion of trend analysis in most texts) will focus on polynomial functions, one should be aware of the fact that other curvilinear trends can be identified. Examples of nonpolynomial curvilinear trends are **exponential** and **logarithmic functions**. Another type of trend analysis which will not be discussed in this section is **monotonic trend analysis** (described in Keppel and Wickens (2004, pp. 108–109)), which can be employed when the spacing of the levels of an independent variable is viewed as being arranged ordinally as opposed to being defined by specific interval/ratio values.

As noted earlier, although strictly speaking, prior to conducting a trend analysis it is not required that one conducts an omnibus analysis of variance on the k group means, most researchers will probably conduct an omnibus analysis of variance for the following reasons: a) The presence of a linear or curvilinear trend will be a function of one or more significant differences among the k group means; and b) In order to conduct a trend analysis it is necessary to compute MS_{WG}, which is computed within the context of the analysis of variance. It should be emphasized, that in conducting the analysis of variance, rejection of the null hypothesis, in and of itself, does not provide specific information regarding any trend(s) which may characterize the relationship between the means values on the dependent variable and the quantitative levels of an independent variable.

In Endnote 18 of the discussion of **linear contrasts** it was noted that a **contrast** can be viewed as a **trend** which is hypothesized with respect to the means of the treatment conditions. The discussion of linear contrasts focused on contrasts (which were also referred to as **single degree of freedom comparisons**) involving the means of two groups or the means of two sets of groups. Trend analysis is merely an extension of the methodology employed for conducting **single degree of freedom comparisons** in order to examine the configuration of the means of all k groups. Through use of Equations 21.17–21.19, in each stage of a trend analysis a researcher can determine whether or not the configuration of k means represents a linear trend or one or more of the possible curvilinear trends which can characterize a set of k means. In order to conduct a trend analysis, one must have access to a set of coefficients which reflect the presence of the specific trend one is evaluating. Tables of coefficients for trend analysis (which can be found in books of statistical tables such as Beyer (1968), as well as in sources on the analysis of variance such as Keppel and Wickens (2004) and Kirk (1995)) list the values of the coefficients for evaluating the presence of a specific type of trend in relation to a specific number of levels for an independent variable.

To further clarify the role of coefficients in trend analysis, recollect it was previously noted that when an independent variable is comprised of $k = 3$ levels, it is only possible to evaluate for a linear or quadratic trend. One set of coefficients (comprised of $k = 3$ numerical values) is employed to test whether or not a linear trend is present, whereas a different set of coefficients (also comprised of $k = 3$ numerical values) is employed to test for the presence of a quadratic trend. More specifically, the coefficients -1, 0, 1 are employed in evaluating a linear trend, while the coefficients 1, -2, 1 are employed in evaluating a quadratic trend. Note that the sum of each set of coefficients equals zero. The aforementioned coefficients provide idealized versions of the mathematical function represented by the trend in question.

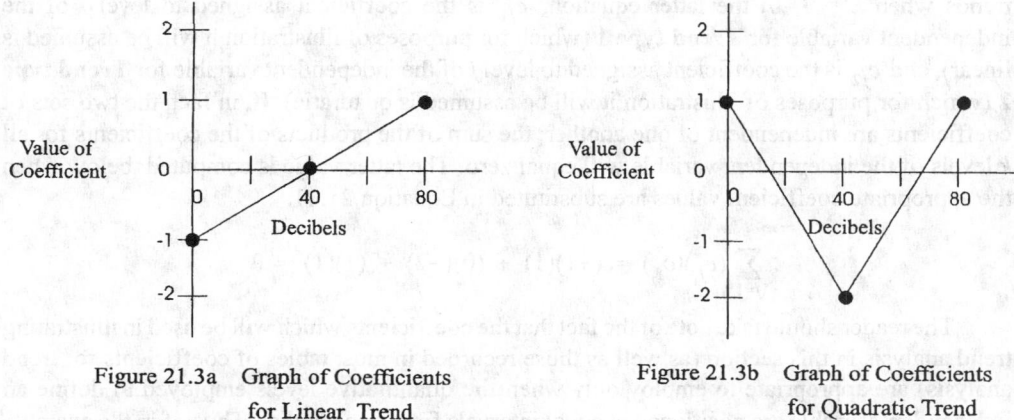

Figure 21.3a Graph of Coefficients
for Linear Trend

Figure 21.3b Graph of Coefficients
for Quadratic Trend

Figure 21.3 Graphs of Coefficients for Linear and Quadratic Trends

To further clarify the concept of a coefficient, k data points can be derived if coefficients are viewed within the context of a graph and are represented along the ordinate, with the quantitative levels of the independent variable represented along the abscissa. The k data points are derived by beginning with the first coefficient in a set comprised of k coefficients, and pairing each coefficient with the quantitative level of the independent variable which is in the same ordinal position as the coefficient. The latter will be illustrated through use of Figure 21.3. Let us assume an independent variable is comprised of three levels of noise corresponding to 0, 40, and 80 decibels. Note that in Figure 21.3a, when each of the coefficients -1, 0, 1 used for evaluating a linear trend is matched with the corresponding level of the independent variable in the same ordinal position, the following three data points are derived: (0, -1), (40, 0), (80, 1). When the latter data points are connected they yield a straight line. On the other hand, in Figure 21.3b, when the coefficients 1, -2, 1 employed to evaluate a quadratic trend are matched with the corresponding level of the independent variable, we derive the three data points (0, 1), (40, -2), (80, 1), which when connected yield a pattern consistent with a quadratic function.

The coefficients most commonly employed for trend analysis are commonly referred to as **orthogonal polynomials**. Use of the latter coefficients insures that evaluation of two or more types of trend in a set of data will be independent of one another. Consequently, in the case of an independent variable comprised of $k = 3$ levels, by virtue of employing **orthogonal polynomial coefficients** the analysis for linear trend will not be contaminated by any quadratic trend which might also be present in the data or vice versa. In the event a quantitative independent variable is comprised of four levels (which allows for assessment of linear, quadratic and cubic trends), the **coefficients of orthogonal polynomials** employed would provide for a measure of each trend independent of either of the other two trends. To be more specific, when a quantitative independent variable has $k = 4$ levels, the coefficients employed for evaluating the three possible types of trend are: Linear (-3, -1, 1, 3); Quadratic (1, -1, -1, 1); Cubic (-1, 3,

-3, 1). Note that the number of values which comprise any set of coefficients will always be equal to the number of levels of the independent variable.

Equation 21.21, which was previously employed to describe an orthogonal relationship between two comparisons (through use of the coefficients), can also be utilized to define an orthogonal relationship between two sets of coefficients in this context. The latter equation will be employed to demonstrate the independence of the sets of coefficients for the linear (specifically, the coefficients $-1, 0, 1$) versus quadratic (specifically, the coefficients $1, -2, 1$) trends when $k = 3$. In the latter equation, c_{j_1} is the coefficient assigned to level j of the independent variable for **Trend type 1** (which for purposes of illustration it will be assumed is linear), and c_{j_2} is the coefficient assigned to level j of the independent variable for **Trend type 2** (which for purposes of illustration it will be assumed is quadratic). If, in fact, the two sets of coefficients are independent of one another, the sum of the products of the coefficients for all k levels of the independent variable will equal zero. The latter value is computed below when the appropriate coefficient values are substituted in Equation 21.21.

$$\sum_{j=1}^{k} (c_{j_1})(c_{j_2}) = (-1)(1) + (0)(-2) + (1)(1) = 0$$

The reader should take note of the fact that the coefficients which will be used in illustrating trend analysis in this section (as well as those recorded in most tables of coefficients for trend analysis) are appropriate to employ only when the quantitative levels employed to define an independent variable are positioned at equal intervals from one another. Thus, if in the example under discussion instead of the decibel levels 0, 40, and 80 (where each successive value is 40 decibels greater than the one which precedes it), the values 0, 20, and 80 had been employed, coefficients other than $-1, 0, 1$ would have to be employed. For a description of the protocol for determining coefficients under the latter circumstances (which can be implemented by most standardized statistical software programs), the reader should consult sources which discuss analysis of variance procedures in greater depth.

Illustration of trend analysis At this point a trend analysis will be conducted on the data for Example 21.1. However, it will be assumed that the three decibel values 0, 40, and 80 will represent the three levels of the independent variable in lieu of original levels **no noise**, **moderate noise**, and **extreme noise**. The respective mean recall scores obtained by the groups serving under the 0, 40, and 80 decibel levels are $\bar{Y}_1 = 9.2$, $\bar{Y}_2 = 6.6$, and $\bar{Y}_3 = 6.2$. Initially, we will determine whether or not the configuration of the k mean values on the dependent variable indicates the presence of a linear trend.

The protocol for evaluating a linear trend is identical to that employed for the **single degree of freedom simple comparison** summarized in Table 21.3. We initially compute the **sum of the weighted treatment means**, which is the value computed for the sum of Column 4 of the latter table. Before proceeding further, the reader should take note of the following with regard to the information to be discussed in reference to Column 4 of Table 21.3: a) The notation ψ (along with the appropriate subscript) will be employed to represent the sum of the weighted treatments for a specific trend; b) Each of the coefficients employed for a specific trend is multiplied by the mean of the group which is in the corresponding ordinal position; c) As noted earlier, the notation $\bar{Y}_j$ is employed to represent the group means in lieu of the notation $\bar{X}_j$ used in Table 21.3 and Equation 21.17.

The **sum of the weighted treatment means** for a linear trend (ψ_{linear}) is computed below for Example 21.1.

$$\psi_{\text{linear}} = \Sigma(c_j)(\bar{Y}_j) = (-1)(9.2) + (0)(6.6) + (+1)(6.2) = -3$$

Equation 21.17 is now employed to compute a sum of squares for the linear trend, to be designated SS_{linear} (which is employed in the latter equation in place of SS_{comp}). Note that Σc_j^2, which represents the sum of the squared coefficients, is $(-1)^2 + (0)^2 + (+1)^2 = 2$. The reader should take note of the fact that the same value will be computed for SS_{linear} if the signs of the coefficients are reversed (i.e., the values $+1, 0, -1$ are employed in place of $-1, 0, +1$). Reversal of the signs does not affect the value of SS_{linear}, since the absolute value of ψ_{linear} (which is squared in Equation 21.17) remains the same.

$$SS_{linear} = \frac{n\left[\Sigma(c_j)(\overline{Y}_j)\right]^2}{\Sigma c_j^2} = \frac{n(\psi_{linear})^2}{\Sigma c_j^2} = \frac{(5)(-3)^2}{2} = 22.5$$

The **mean square** for the linear trend (MS_{linear}), which is a measure of the degree of variability among the treatment means that can be summarized through use of a straight line, is computed with Equation 21.18 (employing the subscript **linear** in place of **comp**). As is the case for any single degree of freedom comparison, the degrees of freedom employed in evaluating a linear trend (df_{linear}) (or for that matter the degrees of freedom for evaluating any higher order trend) will always be equal to 1.

$$MS_{linear} = \frac{SS_{linear}}{df_{linear}} = \frac{22.5}{1} = 22.5$$

Equation 21.19 is now employed to conduct an F test for linear trend (once again employing the subscript **linear** in place of **comp**). F_{linear} is a ratio which is comprised of the variability attributable to a linear trend divided by the within-groups variability computed for the full set of data. If the value of F_{linear} computed with Equation 21.19 is significant, it indicates the presence of a significant linear trend.

$$F_{linear} = \frac{MS_{linear}}{MS_{WG}} = \frac{22.5}{1.9} = 11.84$$

In **Table A10**, for $df_{num} = df_{linear} = 1$ and $df_{den} = df_{WG} = 12$ the tabled critical .05 and .01 values are $F_{.05} = 4.75$ and $F_{.01} = 9.33$. Since the obtained value $F_{linear} = 11.84$ is greater than $F_{.01} = 9.33$, we can reject the null hypothesis (of no linear trend being present) at the .01 level of significance. Thus, we can conclude there is evidence of a linear trend.

Through use of Equation 21.49 in which the sum of squares for the linear component is subtracted from the between-groups sum of squares, one can determine the degree of variability between the group means which can be explained by a higher order trend (which, since $k = 3$, can only be a quadratic trend). The result obtained with Equation 21.49 is referred to as the **residual sum of squares** (represented by the notation SS_{res}).

$$SS_{res} = SS_{BG} - SS_{linear} = 26.53 - 22.5 = 4.03 \qquad \textbf{(Equation 21.49)}$$

Equation 21.52 can be employed to determine whether or not a significant amount of residual variability (i.e., variability attributed to a higher order trend) remains after taking into account variability attributable to the linear trend. If a significant result is obtained for Equation 21.52 it would indicate that additional variability is present in the data which can be explained through use of a higher order trend — specifically, a quadratic trend. Note that Equations 21.50 and 21.51 are respectively employed to compute the values $df_{res} = 1$ and $MS_{res} = 4.03$, which are required for computing the value $F_{residual} = 2.12$ obtained with Equation 21.52.

$$df_{res} = df_{BG} - df_{linear} = 2 - 1 = 1 \qquad \textbf{(Equation 21.50)}$$

$$MS_{res} = \frac{SS_{res}}{df_{res}} = \frac{4.03}{1} = 4.03 \qquad \textbf{(Equation 21.51)}$$

$$F_{res} = \frac{MS_{res}}{MS_{WG}} = \frac{4.03}{1.90} = 2.12 \qquad \textbf{(Equation 21.52)}$$

In **Table A10**, for $df_{num} = df_{res} = 1$ and $df_{den} = df_{WG} = 12$ the tabled critical .05 and .01 values are $F_{.05} = 4.75$ and $F_{.01} = 9.33$. Since the obtained value $F_{res} = 2.12$ is less than $F_{.05} = 4.75$, we cannot reject the null hypothesis (of no higher order trend being present). It thus appears that a linear trend is sufficient to describe the relationship between levels of the independent variable and the group means.

Although there is no indication of the presence of a quadratic trend, if the result obtained through use of Equation 21.52 had been statistically significant, Equations 21.17–21.19 would once again be employed to evaluate for the presence of the latter type of trend. For purposes of illustration the analysis of a quadratic trend is demonstrated below employing the coefficients 1, –2, 1 (which as noted earlier are the **coefficients of orthogonal polynomials** employed to evaluate a quadratic trend when $k = 3$).

$$\varphi_{quadratic} = \Sigma(c_j)(\bar{Y}_j) = (+1)(9.2) + (-2)(6.6) + (+1)(6.2) = 2.2$$

$$SS_{quadratic} = \frac{n\left[\Sigma(c_j)(\bar{Y}_j)\right]^2}{\Sigma c_j^2} = \frac{n(\psi_{quadratic})^2}{\Sigma c_j^2} = \frac{(5)(2.2)^2}{6} = 4.03$$

$$MS_{quadratic} = \frac{SS_{quadratic}}{df_{quadratic}} = \frac{4.03}{1} = 4.03$$

$$F_{quadratic} = \frac{MS_{quadratic}}{MS_{WG}} = \frac{4.03}{1.9} = 2.12$$

In **Table A10**, for $df_{num} = df_{quadratic} = 1$ and $df_{den} = df_{WG} = 12$ the tabled critical .05 and .01 values are $F_{.05} = 4.75$ and $F_{.01} = 9.33$. Since the obtained value $F_{quadratic} = 2.12$ is less than $F_{.05} = 4.75$, we cannot reject the null hypothesis (of no quadratic trend being present). Note that the sum of the variability in the data attributable to the linear and quadratic contributions is equal to the between-groups variability: $SS_{BG} = SS_{linear} + SS_{quadratic} = 22.5 + 4.03 = 26.53$.

It should be noted that if the values F_{res} and $F_{quadratic}$ computed with Equations 21.52 and 21.19 had been significant, one would have concluded that the form of the relationship between decibel level and recall is quadratic, with a significant linear component. Although linear and quadratic trends are the most likely ones to be encountered in research, as noted earlier, higher order trends may also be detected when a quantitative independent variable is comprised of four or more levels. In instances where both significant linear and quadratic trends are present, the procedure described above following the detection of a specific trend should be repeated. Specifically, if we assume an independent variable is comprised of $k = 4$ levels and that significant linear and quadratic trends have been detected, Equation 21.53 would be employed to determine the residual sum of squares. Equation 21.54, followed by use of Equations 21.51 and 21.52 allows us to determine whether or not the magnitude of the residual variability is significant. If the value F_{res} computed with Equation 21.52 is significant, Equations 21.17–21.19 are employed to determine whether or not there is evidence for a cubic trend.

$$SS_{res} = SS_{BG} - SS_{linear} - SS_{quadratic} \qquad \textbf{(Equation 21.53)}$$

$$df_{res} = df_{BG} - df_{linear} - df_{quadratic} \qquad \textbf{(Equation 21.54)}$$

With regard to Equations 21.53 and 21.54, the reader should also take note of the following: a) The value $df_{quadratic} = 1$ is used in Equation 21.54; b) In the event higher order trends are evaluated for an independent variable comprised of five or more levels, Equations 21.53 and 21.54 are extended to accommodate the specific type of trend being evaluated. Specifically, in Equation 21.53 the sum of squares value computed for each of the trends is subtracted from SS_{BG}. In Equation 21.54 the degrees of freedom value for each of the trends is subtracted from df_{BG}. Regardless of the specific higher order trend under consideration, the df value for a trend will always equal 1.

Keppel and Zedeck (1989, pp. 496–499) note that Equations 21.55 and 21.56 (both of which represent a square of a correlation coefficient/measure of association) can respectively be employed to determine the proportion of variability on the group means which can be explained on the basis of each of the possible trend components. The result for Equation 21.55 indicates that the proportion of variability on the group means which can be explained on the basis of a linear trend is .456, while the result for Equation 21.56 indicates the proportion of variability which can be explained on the basis of a quadratic trend is .082.

$$r^2_{linear} = \frac{SS_{linear}}{SS_T} = \frac{22.5}{49.33} = .456 \qquad \textbf{(Equation 21.55)}$$

$$r^2_{quadratic} = \frac{SS_{quadratic}}{SS_T} = \frac{4.03}{49.33} = .082 \qquad \textbf{(Equation 21.56)}$$

Winer *et al.* (1991, p. 209) state that the squared correlation $\eta^2 = .538$ computed with Equation 21.57 indicates the proportion of variability on the group means which can be explained on the basis of a polynomial of any degree. Note that for the example under discussion: $[(r^2_{linear} = .456) + (r^2_{quadratic} = .082)] = [\eta^2 = .538]$. Kirk (1995, p. 198) notes that if the variables are related linearly, r_{linear} (which in our example equals $\sqrt{r^2_{linear}} = \sqrt{.456} = .675$) will equal η (which in our example equals $\sqrt{\eta^2} = \sqrt{.538} = .733$), but as the data depart from a perfect linear relationship the value of η will increase relative to the absolute value of r. (In actuality, the value of r_{linear} is $-.675$, since there is an inverse linear relationship between the values of the levels of the independent variable and the mean scores on the dependent variable.)

$$\eta^2 = \frac{SS_{BG}}{SS_T} = \frac{26.53}{49.33} = .538 \qquad \textbf{(Equation 21.57)}$$

Derivation of the line of best fit for a linear trend   A major goal of trend analysis is to identify the mathematical function which best describes the relationship between scores on the dependent variable and the quantitative levels of the independent variable. If (as is the case in the example under discussion) it is determined that a linear trend describes the relationship between the variables, a polynomial equation, referred to as the **linear regression equation**, should be derived. The latter equation can be represented graphically as the straight line (referred to as a **regression line** or **line of best fit**) which comes closest to passing through all k data points. At this point in the discussion, the derivation of the **linear regression equation** for the example under discussion will be described. The reader interested in a more detailed discussion of the methodology to be described in this section should consult Section VI of the **Pearson product-moment correlation coefficient**.

It was noted earlier that the general form of the equation for a straight line is $Y = a + bX$ (where a represents a constant (also referred to as the Y **intercept**, since it is the point at which the line intersects the ordinate), and b is a numerical value representing the **slope** of the line). Although there is an alternative way of deriving the line of best fit for a linear trend, the methodology which will be described in this section is identical to that employed for linear regression analysis in Section VI of the **Pearson product-moment correlation coefficient**. The reader should take note of the fact that the equation $Y = a + bX$ corresponds to Equation 28.4 in the latter chapter, and Equations 21.58 and 21.59 (which are respectively employed to compute the values of b and a) correspond to Equations 28.9 and 28.11 the aforementioned chapter. The derivation of the regression line is based on the values of the $n = 3$ data points $(0, 9.2)$, $(40, 6.6)$, $(80, 6.2)$ (each of which represents a level of the independent variable paired with the mean recall score of the group which served under that level). Table 21.11 summarizes the computation of the values which are required to compute the values of b and a. It should also be noted that thorough use of Equation 28.1 (which is the same as Equation 17.7), it can be determined that the **Pearson product moment correlation** (which is discussed briefly in the **Introduction** and in detail in Chapter 28) between the three decibel levels (X variable) and the mean group recall scores (Y variable) is $r = -.921$. Although the latter value (which should not be confused with the value $r_{linear} = -.675$ computed earlier) indicates a strong inverse linear relationship between the variables, in and of itself, it is not sufficient to establish a linear trend between the two variables.

Substituting the values $a = 8.83$ and $b = -.0375$ in the equation $Y = a + bX$, we determine that the equation for the straight line for the data in Example 21.1 is $Y = 8.83 - .0375X$. Since two points can be employed to construct a straight line, we can select two values for X and substitute each value in the equation $Y = 8.83 - .0375X$, and solve for the value which would be obtained for Y. Each pair of values which is comprised of an X score and the resulting Y value will represent one point on the line of best fit. To demonstrate this, if the value $X = 0$ is substituted in the regression equation, it yields the value $Y = 8.83$ (which equals the value of a): $Y = 8.83 - (-.0375)(0) = 8.83$. Thus, the first point that will be employed in constructing the regression line is $(0, 8.83)$. If we next substitute the value $X = 80$ in the regression equation, it yields the value $Y = 5.83$: $Y = 8.83 + (-.0375)(80) = 5.83$. Thus, the second point to be used in constructing the regression line is $(80, 5.83)$. The regression line that results from connecting the points $(0, 8.83)$ and $(80, 5.83)$ is depicted in Figure 21.4. If through use of the regression line the researcher wanted to predict a score on the Y variable by employing a value for a level of the independent variable, the appropriate X value would be substituted in the equation $Y = 8.83 - .0375X$. (In Chapter 28 the notation Y' is employed in the linear regression equation to indicate a predicted Y value.) Alternatively, through use of Figure 21.4, one could make the same prediction as follows: Erect a perpendicular line from the point on the abscissa which corresponds to the relevant decibel level, and at the point where the perpendicular intersects the regression line, a second perpendicular is drawn in relation to the ordinate. The point at which the latter line intersects the ordinate delineates the predicted score on the dependent variable.

Table 21.11   Summary of Data for Example 21.1

Data point	X	X^2	Y	Y^2	XY
1	0	0	9.2	84.64	0
2	40	1600	6.6	43.56	264
3	80	6400	6.2	38.44	496
	$\Sigma X = 120$	$\Sigma X^2 = 8000$	$\Sigma Y = 22$	$\Sigma Y^2 = 166.64$	$\Sigma XY = 760$

$$\bar{X} = \Sigma X/n = 120/3 = 40 \qquad \bar{X} = \Sigma Y/n = 22/3 = 7.33$$

$$r = \frac{\Sigma XY - \dfrac{(\Sigma X)(\Sigma Y)}{n}}{\sqrt{\left[\Sigma X^2 - \dfrac{(\Sigma X)^2}{n}\right]\left[\Sigma Y^2 - \dfrac{(\Sigma Y)^2}{n}\right]}} = \frac{760 - \dfrac{(120)(22)}{3}}{\sqrt{\left[8000 - \dfrac{(120)^2}{3}\right]\left[166.64 - \dfrac{(22)^2}{3}\right]}} = -.921$$

$$b = \frac{\Sigma XY - \dfrac{(\Sigma X)(\Sigma Y)}{n}}{\Sigma X^2 - \dfrac{(\Sigma X)^2}{n}} = \frac{760 - \dfrac{(120)(22)}{3}}{8000 - \dfrac{(120)^2}{3}} = .0375 \qquad \textbf{(Equation 21.58)}$$

$$a_Y = \bar{Y} - b\bar{X} = 7.33 - (-.0375)(40) = 8.83 \qquad \textbf{(Equation 21.59)}$$

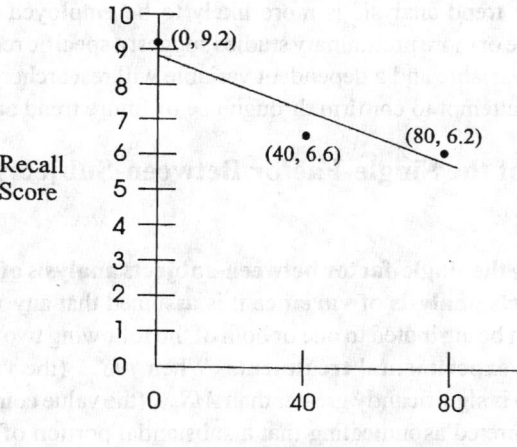

Figure 21.4   Regression Line for Example 21.1

Final comments on trend analysis In the event the trend analysis described above indicated the presence of a quadratic trend, the researcher would determine the quadratic function which provided the best fit for the data. Because of the complexity of the computations involved in the derivation of equations for curvilinear trends (which are generally derived with statistical software), such procedures will not be described in this book.

In closing the discussion of trend analysis it should be noted that if prior to a study a researcher plans to conduct a trend analysis, the specific values to be employed for the levels of the quantitative independent variable should be carefully chosen. Cohen (2001, p. 606) and Keppel and Wickens (2004) note that although equally spaced quantitative levels are the easiest to analyze, they may not always be optimal in allowing one to identify the precise nature of a trend which defines a set of data. If prior to a study the researcher has a theoretical model which predicts a specific relationship between the quantitative levels of the independent variable and scores on the dependent variable, the nature of such a model will dictate the specific quantitative values that will be employed to define the levels of the independent variable. In instances where a researcher predicts a curvilinear trend, the number of intervals employed as well as the numerical distance of the intervals from one another will be dictated by the precise shape of the curvilinear function one is predicting. As an example, if in Figure 21.2d the researcher had only employed the 0, 40, and 80 decibel values, he would have erroneously concluded that the data conformed to a linear trend (since a straight line connects the three data points (0, 80), (40, 2), and (80, 4). However, by also including the data points (20, 2) and (60, 0), the researcher was able to identify the presence of curvilinear trend — specifically a quartic trend. Put simply, the fewer the number of levels employed to define a quantitative independent variable, the lower the order of the trends one will be able to detect.

In order to allow for the possibility of identifying one or more higher order curvilinear trends, among others, Keppel (1991, p. 154) notes that researchers should probably employ between 5 to 7 values for defining the levels of a quantitative independent variable. If prior to a study a researcher hypothesizes a specific theoretical model, the specific values selected for the levels of the independent variable should allow him to determine whether or not the pattern of the data is consistent with that model. Within the context of employing the analysis of variance, **model driven trend analysis** may be the exception rather than the rule. (By **model driven trend analysis**, I am referring to the use of trend analysis in evaluating the tenability of a specific linear or curvilinear model which is hypothesized by a researcher prior to conducting a study.) In the initial stages of a research program, trend analysis is more likely to be employed within an exploratory context, and only after one or more preliminary studies suggest a specific relationship between a quantitative independent variable and a dependent variable will researchers arrive at a model that they may subsequently attempt to confirm through use of future trend analyses.

VII. Additional Discussion of the Single-Factor Between-Subjects Analysis of Variance

1. Theoretical rationale underlying the single-factor between-subjects analysis of variance In the **single-factor between-subjects analysis of variance** it is assumed that any variability between the means of the k groups can be attributed to one or both of the following two elements: a) **Experimental error**; and b) **The experimental treatments**. When MS_{BG} (the value computed for between-groups variability) is significantly greater than MS_{WG} (the value computed for within-groups variability), it is interpreted as indicating that a substantial portion of between-groups variability is due to a treatment effect. The rationale for this is as follows.

Experimental error is random variability in the data which is beyond the control of the researcher. In an independent groups design the average amount of variability within each of the

k groups is employed to represent experimental error. Thus, the value computed for MS_{WG} is the normal amount of variability that is expected between the scores of different subjects who serve in the same group. Within this framework, within-groups variability is employed as a baseline to represent variability which results from factors that are beyond an experimenter's control. The experimenter assumes that since such uncontrollable factors are responsible for within-groups differences, it is logical to assume that they can produce differences of a comparable magnitude between the means of the k groups. As long as the variability between the group means (MS_{BG}) is approximately the same as within-groups variability (MS_{WG}), the experimenter can attribute any between-groups variability to experimental error. When, however, between-groups variability is substantially greater than within-groups variability, it indicates that something over and above error variability is contributing to the variability between the k group means. In such a case, it is assumed that a treatment effect is responsible for the larger value of MS_{BG} relative to the value of MS_{WG}. In essence, if within-groups variability is subtracted from between-groups variability, any remaining variability can be attributed to a treatment effect. If there is no treatment effect, the result of the subtraction will be zero. Of course, one can never completely rule out the possibility that if MS_{BG} is larger than MS_{WG}, the larger value for MS_{BG} is entirely due to error variability. However, since the latter is unlikely, when MS_{BG} is significantly larger than MS_{WG}, it is interpreted as indicating the presence of a treatment effect.

Tables 21.12 and 21.13 will be used to illustrate the relationship between between-groups variability and within-groups variability in the analysis of variance. Assume that both tables contain data for two hypothetical studies employing an independent groups design involving $k = 3$ groups and $n = 3$ subjects per group. In the hypothetical examples below, even though it is not employed as the measure of variability in the analysis of variance, the range (the difference between the lowest and highest score) will be used as a measure of variability. The range is employed since: a) It is simpler to employ for purposes of illustration than the variance; and b) What is derived with respect to the range from this example can be generalized to the variance/mean squares as they are employed within the framework of the analysis of variance.

Table 21.12 presents a set of data where there is no treatment effect, since between-groups variability equals within-groups variability. In Table 21.12, in order to assess within-groups variability we do the following: a) Compute the range of each group. This is done by subtracting the lowest score in each group from the highest score in that group; and b) Compute the average range of the three groups. Since the range for all three groups equals 2, the average range equals 2 (i.e., $(2 + 2 + 2)/3 = 2$). This value will be used to represent within-groups variability (which is a function of individual differences in performance among members of the same group).

Table 21.12   Data Illustrating No Treatment Effect

Group 1	Group 2	Group 3
3	4	5
4	5	6
5	6	7
$\bar{X}_1 = 4$	$\bar{X}_2 = 5$	$\bar{X}_3 = 6$

Between-groups variability is obtained by subtracting the lowest group mean ($\bar{X}_1 = 4$) from highest group mean ($\bar{X}_3 = 6$). Thus, since $\bar{X}_3 - \bar{X}_1 = 6 - 4 = 2$, between-groups variability equals 2. As noted previously, if between-groups variability is significantly larger than within-groups variability, it is interpreted as indicating the presence of a treatment effect. Since, in the example under discussion, both between-groups and within-groups variability equal 2, there is no evidence of a treatment effect. To put it another way, if within-groups variability is subtracted from between-groups variability, the difference is zero. In order for there to be a treatment effect, a positive difference should be present.

If, for purposes of illustration, we employ the range in the F ratio to represent variability, the value of the F will equal 1 since:

$$F = \frac{\text{Between-groups variability}}{\text{Within-groups variability}} = \frac{2}{2} = 1$$

As noted previously, when $F = 1$, there is no evidence of a treatment effect, and thus the null hypothesis is retained.[62]

Table 21.13 presents a set of data where a treatment effect is present as a result of between-groups variability being greater than within-groups variability. In Table 21.13, if we once again use the range to assess variability, within-groups variability equals 2. This is the case since the range of each group equals 2, yielding an average range of 2. The between-groups variability, on the other hand, equals 27. The latter value is obtained by subtracting the smallest of the three group means $\bar{X}_1 = 4$ from the largest of the group means $\bar{X}_3 = 31$. Thus, $\bar{X}_3 - \bar{X}_1 = 31 - 4 = 27$. Because the value 27 is substantially larger than 2 (which is a baseline measure of error variability that will be tolerated between the group means), there is strong evidence that a treatment effect is present. Specifically, since we will assume that only 2 of the 27 units which comprise between-groups variability can be attributed to experimental error, the remaining $27 - 2 = 25$ units can be assumed to represent the contribution of the treatment effect.

Table 21.13   Data Illustrating Treatment Effect

Group 1	Group 2	Group 3
3	10	30
4	11	31
5	12	32
$\bar{X}_1 = 4$	$\bar{X}_2 = 11$	$\bar{X}_3 = 31$

If once again, for purposes of illustration, the values of the range are employed to compute the F ratio, the resulting value will be $F = 27/2 = 13.5$. As noted previously, when the value of F is substantially greater than 1, it is interpreted as indicating the presence of a treatment effect, and consequently the null hypothesis is rejected.

2. Definitional equations for the single-factor between-subjects analysis of variance In the description of the computational protocol for the **single-factor between-subjects analysis of variance** in Section IV, Equations 21.2, 21.3, and 21.5 are employed to compute the values SS_T, SS_{BG}, and SS_{WG}. The latter set of computational equations was employed, since it allows for the most efficient computation of the sum of squares values. As noted in Section IV, computational equations are derived from definitional equations which reveal the underlying logic involved in the derivation of the sums of squares. This section will describe the definitional equations for the **single-factor between-subjects analysis of variance**, and apply them to Example 21.1 in order to demonstrate that they yield the same values as the computational equations.

As noted previously, the total sum of squares (SS_T) is made up of two elements, the between-groups sum of squares (SS_{BG}) and the within-groups sum of squares (SS_{WG}). The contribution of any single subject's score to the total variability in the data can be expressed in terms of a between-groups component and a within-groups component. When the between-groups component and the within-groups component are added, the sum reflects that subject's total contribution to the overall variability in the data. The contribution of all N subjects to the total variability (SS_T) and the elements which comprise it (SS_{BG} and SS_{WG}) are summarized in Table 21.14. The definitional equations described in this section employ the following notation: X_{ij} represents the score of the ith subject in the jth group; $\bar{X}_T$ represents the grand mean (which is $\bar{X}_T = (\sum_{j=1}^{k}\sum_{i=1}^{n} X_{ij})/N = 110/15 = 7.33$); and $\bar{X}_j$ represents the mean of the jth group.

Table 21.14 Computation of Sums of Squares for Example 21.1 with Definitional Equations

	X_{ij}	$SS_{WG} = \sum_{j=1}^{k}\sum_{i=1}^{n}(X_{ij} - \bar{X}_j)^2$	$SS_{BG} = n\sum_{j=1}^{k}(\bar{X}_j - \bar{X}_T)^2$	$SS_T = \sum_{j=1}^{k}\sum_{i=1}^{n}(X_{ij} - \bar{X}_T)^2$
	8	$(8-9.2)^2 = 1.44$	$(9.2-7.33)^2 = 3.497$	$(8-7.33)^2 = .449$
	10	$(10-9.2)^2 = .64$	$(9.2-7.33)^2 = 3.497$	$(10-7.33)^2 = 7.129$
Group 1	9	$(9-9.2)^2 = .04$	$(9.2-7.33)^2 = 3.497$	$(9-7.33)^2 = 2.789$
	10	$(10-9.2)^2 = .64$	$(9.2-7.33)^2 = 3.497$	$(10-7.33)^2 = 7.129$
	9	$(9-9.2)^2 = .04$	$(9.2-7.33)^2 = 3.497$	$(9-7.33)^2 = 2.789$
	7	$(7-6.6)^2 = .16$	$(6.6-7.33)^2 = .533$	$(7-7.33)^2 = .109$
	8	$(8-6.6)^2 = 1.96$	$(6.6-7.33)^2 = .533$	$(8-7.33)^2 = .449$
Group 2	5	$(5-6.6)^2 = 2.56$	$(6.6-7.33)^2 = .533$	$(5-7.33)^2 = 5.429$
	8	$(8-6.6)^2 = 1.96$	$(6.6-7.33)^2 = .533$	$(8-7.33)^2 = .449$
	5	$(5-6.6)^2 = 2.56$	$(6.6-7.33)^2 = .533$	$(5-7.33)^2 = 5.429$
	4	$(4-6.2)^2 = 4.84$	$(6.2-7.33)^2 = 1.277$	$(4-7.33)^2 = 11.089$
	8	$(8-6.2)^2 = 3.24$	$(6.2-7.33)^2 = 1.277$	$(8-7.33)^2 = .449$
Group 3	7	$(7-6.2)^2 = .64$	$(6.2-7.33)^2 = 1.277$	$(7-7.33)^2 = .109$
	5	$(5-6.2)^2 = 1.44$	$(6.2-7.33)^2 = 1.277$	$(5-7.33)^2 = 5.429$
	7	$(7-6.2)^2 = .64$	$(6.2-7.33)^2 = 1.277$	$(7-7.33)^2 = .109$
		$SS_{WG} = 22.80$	$SS_{BG} = 26.535$	$SS_T = 49.335$

Equation 21.60 is the definitional equation for the **total sum of squares**.[63]

$$SS_T = \sum_{j=1}^{k} \sum_{i=1}^{n} (X_{ij} - \bar{X}_T)^2 \qquad \textbf{(Equation 21.60)}$$

In employing Equation 21.60 to compute SS_T, the grand mean ($\bar{X}_T$) is subtracted from each of the N scores and each of the N difference scores is squared. The total sum of squares (SS_T) is the sum of the N squared difference scores. Equation 21.60 is computationally equivalent to Equation 21.2.

Equation 21.61 is the definitional equation for the **between-groups sum of squares**.

$$SS_{BG} = n\sum_{j=1}^{k} (\bar{X}_j - \bar{X}_T)^2 \qquad \textbf{(Equation 21.61)}$$

In employing Equation 21.61 to compute SS_{BG}, the following operations are carried out for each group. The grand mean ($\bar{X}_T$) is subtracted from the group mean ($\bar{X}_j$). The difference score is squared, and the squared difference score is multiplied by the number of subjects in the group (n). After this is done for all k groups, the values which have been obtained for each group as a result of multiplying the squared difference score by the number of subjects in a group are summed. The resulting value represents the between-groups sum of squares (SS_{BG}). Equation 21.61 is computationally equivalent to Equation 21.3. An alternative but equivalent method of obtaining SS_{BG} (which is employed in deriving SS_{BG} in Table 21.14) is as follows: Within each group, for each of the n subjects the grand mean is subtracted from the group mean, each difference score is squared, and upon doing this for all k groups, the N squared difference scores are summed.

Equation 21.62 is the definitional equation for the **within-groups sum of squares**.

$$SS_{WG} = \sum_{j=1}^{k} \sum_{i=1}^{n} (X_{ij} - \bar{X}_j)^2 \qquad \textbf{(Equation 21.62)}$$

In employing Equation 21.62 to compute SS_{WG}, the following operations are carried out for each group. The group mean ($\bar{X}_j$) is subtracted from each score in the group. The difference scores are squared, after which the sum of the squared difference scores is obtained. The sum of the sum of the squared difference scores for all k groups represents the within-groups sum of squares. Equation 21.62 is computationally equivalent to Equations 21.4/21.5.

Table 21.14 illustrates the use of Equations 21.60, 21.61, and 21.62 with the data for Example 21.1.[64] The resulting values of SS_T, SS_{BG}, and SS_{WG} are identical to those obtained with the computational equations (Equations 21.2, 21.3, and 21.5). Any minimal discrepancies are the result of rounding off error.

3. Equivalency of the single-factor between-subjects analysis of variance and the t test for two independent samples when $k = 2$ Interval/ratio data for an experiment involving $k = 2$ independent groups can be evaluated with either a **single-factor between-subjects analysis of variance** or a t **test for two independent samples**. When both of the aforementioned tests are employed to evaluate the same set of data, they will yield the same result. Specifically, the following will always be true with respect to the relationship between the computed F and t values for the same set of data: $F = t^2$ and $t = \sqrt{F}$. It will also be the case that the square of the tabled critical t value at a prespecified level of significance for $df = n_1 + n_2 - 2$ will be equal to the tabled critical F value at the same level of significance for $df_{BG} = 1$ and df_{WG} (which will be $df_{WG} = N - k = N - 2$, which is equivalent to the value $df = n_1 + n_2 - 2$ employed for the t **test for two independent samples**).

To illustrate the equivalency of the results obtained with the **single-factor between-subjects analysis of variance** and the t **test for two independent samples** when $k = 2$, an F value will be computed for Example 11.1. The value $t = -1.96$ ($t = -1.964$ if carried out to 3 decimal places) is obtained for the latter example when the t **test for two independent samples** is employed. When the same set of data is evaluated with the **single-factor between-subjects analysis of variance**, the value $F = 3.86$ is computed. Note that $(t = -1.964)^2 = (F = 3.86)$. Equations 21.2, 21.3, and 21.4 are employed below to compute the values SS_T, SS_{BG}, and SS_{WG} for Example 11.1. Since $k = 2$, $n = 5$, and $nk = N = 10 = 10$, $df_{BG} = 2 - 1 = 1$, $df_{WG} = N - k = 10 - 2 = 8$, and $df_T = N - 1 = 10 - 1 = 9$. The full analysis of variance is summarized in Table 21.15.

$$SS_T = 473 - \frac{(53)^2}{10} = 192.1 \qquad SS_{BG} = \left[\frac{(14)^2}{5} + \frac{(39)^2}{5} \right] - \frac{(53)^2}{10} = 62.5$$

$$SS_{WG} = 192.1 - 62.5 = 129.6$$

Table 21.15 Summary Table of Analysis of Variance for Example 11.1

Source of variation	SS	df	MS	F
Between-groups	62.5	1	62.5	3.86
Within-groups	129.6	8	16.2	
Total	192.1	9		

For df_{BG} = 1 and df_{WG} = 8, the tabled critical .05 and .01 values are $F_{.05}$ = 5.32 and $F_{.01}$ = 11.26 (which are appropriate for a nondirectional analysis). Note that (if one takes into account rounding off error) the square roots of the aforementioned tabled critical values are (for df = 8) the tabled critical two-tailed values $t_{.05}$ = 2.31 and $t_{.01}$ = 3.36 which are employed in Example 11.1 to evaluate the value t = –1.96. Since the obtained value F = 3.86 is less than the tabled critical .05 value $F_{.05}$ = 5.32, the nondirectional alternative hypothesis H_1: $\mu_1 \neq \mu_2$ is not supported. The directional alternative hypothesis H_1: $\mu_1 < \mu_2$ is supported at the .05 level, since F = 3.86 is greater than the tabled critical one-tailed .05 value $F_{.90}$ = 3.46 (the square root of which is the tabled critical one-tailed .05 value $t_{.05}$ = 1.86 employed in Example 11.1). The directional alternative hypothesis H_1: $\mu_1 < \mu_2$ is not supported at the .01 level, since F = 3.86 is less than the tabled critical one-tailed .01 value $F_{.98}$ = 8.41 (the square root of which is the tabled critical one-tailed .01 value $t_{.01}$ = 2.90 employed in Example 11.1).[65] The conclusions derived from the **single-factor between-subjects analysis of variance** are identical to those reached when the data are evaluated with the **t test for two independent samples**.

4. Robustness of the single-factor between-subjects analysis of variance   The general comments made with respect to the robustness of the **t test for two independent samples** (in Section VII of the latter test) are applicable to the **single-factor between-subjects analysis of variance**. Most sources state that the **single-factor between-subjects analysis of variance** is robust with respect to violation of its assumptions. Nevertheless, when either the normality and/ or homogeneity of variance assumption is violated, it is recommended that a more conservative analysis be conducted (i.e., employ the tabled $F_{.01}$ value or even a tabled value for a lower alpha level to represent the $F_{.05}$ value). When the violation of one or both assumptions is extreme, some sources recommend that a researcher consider employing an alternative procedure. Alternative procedures are discussed earlier in this section under the homogeneity of variance assumption. Keppel (1991) and Maxwell and Delaney (1990, 2000) provide comprehensive discussions on the general subject of the robustness of the **single-factor between-subjects analysis of variance**.

5. Equivalency of the single-factor between-subjects analysis of variance and the t test for two independent samples with the chi-square test for r × c tables when c = 2   It was brought to my attention by a colleague, David Boynton, that in evaluating the data in a 2 × 2 contingency table, the **t test for two independent samples** will yield a result equivalent to that obtained with the **chi-square test for r × c tables (Test 16)**. On an even more general level, it was noted in Section VII of the **chi-square test for r × c tables** that in the case of any r × 2 contingency table, a result equivalent to that obtained with the **chi-square test for r × c tables** can be obtained through use of the **single-factor between-subjects analysis of variance**. The aforementioned relationship between the **chi-square test for r × c tables** and the **t test for two independent samples/single-factor between-subjects analysis of variance** assumes that the contingency table in question summarizes data which involve r groups (i.e., r can be any integer value), and that the $c = 2$ columns represent scores on a **dichotomous** dependent variable. A dichotomous variable involves two categories, and, as a general rule, the values 0 and 1 are used when a numerical value is employed to represent the scores or categorization of subjects on such a variable.

The following will be now demonstrated in reference to the above: a) The equivalency between the **chi-square test for $r \times c$ tables** (when $r = 2$) and the t **test for two independent samples**; b) The equivalency between the **chi-square test for $r \times c$ tables** (when $r \geq 2$) and the **single-factor between-subjects analysis of variance**.

The **chi-square test for $r \times c$ tables** was demonstrated through use of Example 16.1. At this point the latter example will be evaluated with the t **test for two independent samples**. (The analysis to be presented is also applicable to Example 16.2 if the **introvert** versus **extrovert** categorization is employed to represent which of two groups a subject is a member, and the **Democrat** versus **Republican** dichotomy is employed to represent a subject's score/ categorization on the dependent variable.) Table 21.16 (as well as Table 16.2) summarizes the data for Example 16.1. In the latter tables, the row variable represents the independent variable, which is comprised of two levels represented by the two rows. Thus, subjects exposed to the **noise** versus **no noise** conditions represent two independent groups. The column variable, which is represented by the two columns, represents the dependent variable. In order to evaluate Example 16.1 with the t **test for two independent samples**, any subject who **helped** the confederate will be assigned a score of 1 and any subject who **did not help** the confederate will be assigned a score of 0. (The same result for the analysis to be described will be obtained if a score of 0 is assigned to all subjects who **helped** the confederate and a score of 1 is assigned to all subjects who **did not help** the confederate.).

From Table 21.16 we can determine the summary values required for using Equation 11.1 (the equation which will be employed for the t **test for two independent samples**). Note that each of the two groups is comprised of 100 subjects. Thus, $n_1 = 100$ and $n_2 = 100$. Since 30 of the 100 subjects in Group 1 obtain a score of 1 (i.e., **helped**) and 70 of the subjects obtain a score of 0 (i.e., **did not help**), the sum of the scores for Group 1 is $\Sigma X_1 = 30$. When each of the 30 X_1 scores of 1 is squared (yielding 1 for each squared score), and the 30 squared X_1 values of 1 are summed, the value $\Sigma X_1^2 = 30$ is obtained. (The sum of the 70 squared X_1 values of 0 equals 0.) Since 60 of the 100 subjects in Group 2 obtain a score of 1 (i.e., **helped**) and 40 of the subjects obtain a score of 0 (i.e., **did not help**), the sum of the scores for Group 2 is $\Sigma X_2 = 60$. When each of the 60 X_2 scores of 1 is squared (yielding 1 for each squared score), and the 60 squared X_2 values are summed, the value $\Sigma X_2^2 = 60$ is obtained. (The sum of the 40 squared X_2 values of 0 equals 0.)

Table 21.16   Summary of Data for Example 16.1

| | | Column variable (Dependent variable) | | Row sums |
		1 = Helped	0 = Did not help	
Row variable (Independent variable)	Group 1: Noise	30	70	100
	Group 2: No noise	60	40	100
	Column sums	90	110	Total = 200

With respect to the data in Table 21.16, the following computations are noted: a) Equations I.1 and I.8 are employed to compute the mean and estimated population variance for each of the groups; b) Employing Equation 11.1, the value $t = -4.45$ is computed; c) Employing Equation 11.4, $df = 198$ is the value computed for the degrees of freedom for the analysis of the data in Table 21.16 with the *t* test.

$$\bar{X}_1 = \frac{\Sigma X_1}{n_1} = \frac{30}{100} = .3 \qquad \tilde{s}_1^2 = \frac{\Sigma X_1^2 - \frac{(\Sigma X_1)^2}{n_1}}{n_1 - 1} = \frac{30 - \frac{(30)^2}{100}}{100 - 1} = .212$$

$$\bar{X}_2 = \frac{\Sigma X_2}{n_2} = \frac{60}{100} = .6 \qquad \tilde{s}_2^2 = \frac{\Sigma X_2^2 - \frac{(\Sigma X_2)^2}{n_2}}{n_2 - 1} = \frac{60 - \frac{(60)^2}{100}}{100 - 1} = .242$$

$$t = \frac{\bar{X}_1 - \bar{X}_2}{\sqrt{\frac{\tilde{s}_1^2}{n_1} + \frac{\tilde{s}_2^2}{n_2}}} = \frac{.3 - .6}{\sqrt{\frac{.212}{100} + \frac{.242}{100}}} = -.4.45$$

$$df = n_1 + n_2 - 2 = 100 + 100 - 2 = 198.$$

The obtained absolute value $t = 4.45$ is evaluated with **Table A2**. (The absolute value of *t* is employed since, as is the case with the **chi-square test**, a nondirectional alternative hypothesis is evaluated.) The tabled critical two-tailed .05 and .01 values for $df = \infty$ can be employed for $df = 198$. Thus, $t_{.05} = 1.96$ and $t_{.01} = 2.58$. Since the absolute value $t = 4.45$ is greater than $t_{.01} = 2.58$, the nondirectional alternative hypothesis $H_1: \mu_1 \neq \mu_2$ is supported at the .01 level. To be more precise, the likelihood of obtaining an absolute *t* value equal to or greater than 4.45 is less than .0001, which is also the likelihood of obtaining a chi-square value equal to or greater than $\chi^2 = 18.18$ for the same set of data — the latter being the value obtained when the data were evaluated with the **chi-square test for $r \times c$ tables**.

Table 21.17 Summary Table of Analysis of Variance for Example 16.1

Source of variation	SS	df	MS	F
Between-groups	4.50	1	4.500	19.82
Within-groups	45.00	198	.227	
Total	49.50	199		

It was noted earlier in this section that when there are $k = 2$ independent groups, the *t* test for two independent samples and the single-factor between-subjects analysis of variance will yield identical results for the same set of data. This is confirmed below in Table 21.17, the summary table of an analysis of variance conducted on the data in Table 21.16. The summary values $n_1 = 100$, $n_2 = 100$, $N = 200$, $\Sigma X_1 = 30$, $\Sigma X_1^2 = 30$, $\Sigma X_2 = 60$, $\Sigma X_2^2 = 60$, $\Sigma X_T = 90$, $\Sigma X_T^2 = 90$ are employed to compute the between-groups, within-groups, and total sum of squares values in Table 21.17. Note that the obtained value $F = 19.82$ is the square of the previously computed absolute value $t = 4.45$ (which reflects the relationship that $F = t^2$). As is the case with the value $t = 4.45$, the value $F = 19.82$ is significant (with $df = 1, 198$), and the likelihood of obtaining an F value equal to or greater than 19.82 is less than .0001.

Further evidence of the equivalency of the **chi-square test for $r \times c$ tables** with the t **test for two independent samples** and the **single-factor between-subjects analysis of variance** is provided by the fact that an identical magnitude of treatment effect is computed for all three of the aforementioned tests. The latter will always be the case for a 2×2 contingency table when the square of the **phi coefficient** is employed to compute the magnitude of treatment effect for the **chi-square test for $r \times c$ tables**, and **eta squared** is employed to compute the magnitude of treatment effect for the t **test for two independent samples** and the **single-factor between-subjects analysis of variance**. Employing Equation 16.21 and squaring the result, the value .0909 is computed below for **phi squared** (ϕ^2). The identical value is also computed below through use of Equations 11.16 and 21.43 for **eta squared**.

$$\phi = \sqrt{\frac{\chi^2}{n}} = \sqrt{\frac{18.18}{200}} = .3015 \qquad \phi^2 = (.3015)^2 = .0909$$

$$\tilde{\eta}^2 = \frac{t^2}{t^2 + df} = \frac{(4.45)^2}{(4.45)^2 + 198} = .0909 \qquad \tilde{\eta} = \sqrt{.0909} = .3015$$

$$\tilde{\eta}^2 = \frac{SS_{BG}}{SS_T} = \frac{4.5}{49.5} = .0909 \qquad \tilde{\eta} = \sqrt{.0909} = .3015$$

The reader should note the following with respect to the values computed above: a) **phi squared** and **eta squared** (i.e., ϕ^2 and η^2) are commonly employed measures of magnitude of treatment effect employed to determine the proportion of variability on the dependent variable which is associated with variability on the independent variable; b) The square roots of **phi squared** and **eta squared** (i.e., ϕ and η) are referred to as **correlation coefficients** or **measures of association**. The latter measures indicate the degree of relationship between two variables (in this case an independent and a dependent variable). The closer the absolute value of a correlation coefficient is to 1 the stronger the relationship between the variables, and the closer the absolute value of a correlation coefficient is to 0 the weaker the relationship between the variables. A detailed discussion of correlation coefficients can be found in the chapter on the **Pearson product-moment correlation coefficient**; c) In Section IX (the **Addendum**) of the **Pearson product-moment correlation coefficient**, under the discussion of the **point-biserial correlation coefficient (Test 28h)**, it is noted that **eta** is equivalent to the **point-biserial correlation coefficient** (which is represented by the notation r_{pb}). Consequently, if Example 16.1 is conceptualized as it has been in this section and it is evaluated with the t **test for two independent samples**, within the latter context a **point-biserial correlation coefficient** can be computed for the data. If the latter is done, the resulting value of the **point-biserial correlation coefficient** equals .3015 — i.e., the same value computed above for **eta** (as well as the fact that $r_{pb} = \eta = \phi$ and $r_{pb}^2 = \eta^2 = \phi^2$).

The data in Table 16.23, a 3×2 contingency table, will be employed to illustrate the equivalency between the **chi-square test for $r \times c$ tables** and the **single-factor between-subjects analysis of variance** when $r > 2$. The analysis of the latter table with the **chi-square test for $r \times c$ tables** yields a chi-square value of $\chi^2 = 50.89$. For $df = (r - 1)(c - 1) = 2$, the result is significant at a probability of less than .0001.The computed values of **Cramér's phi coefficient (Test 16h)** (which is the analog of **phi** for larger contingency tables and which employs Equation 16.22) and **Cramér's phi squared** are respectively $\phi_C = \sqrt{50.89/300} =$ .4118 and $\phi_C^2 = .1696$.

Table 21.18 is the summary table for a **single-factor between-subjects analysis of variance** employed with the same set of data. The summary values $n_1 = 100$, $n_2 = 100$, $n_3 = 100$, $N = 300$, $\Sigma X_1 = 30$, $\Sigma X_1^2 = 30$, $\Sigma X_2 = 50$, $\Sigma X_2^2 = 50$, $\Sigma X_3 = 80$, $\Sigma X_3^2 = 80$, $\Sigma X_T = 160$, $\Sigma X_T^2 = 160$ are employed to compute the between-groups, within-groups, and total sum of squares values in Table 21.18.

Table 21.18 Summary Table of Analysis of Variance for Table 16.24

Source of variation	SS	df	MS	F
Between-groups	12.67	2	6.33	30.31
Within-groups	62.00	297	.21	
Total	74.67	299		

For $df = 2, 297$, the result is significant at a probability of less than .0001. The computed values of **eta squared** (through use of Equation 21.43) and **eta** are respectively $\eta^2 = SS_{BG}/SS_T$ $= 12.67/74.67 = .1696$ and $\eta = .4118$. Thus, it can be seen that the **chi-square test for $r \times c$ tables** and the **single-factor between-subjects analysis of variance** yield equivalent results with respect to the data in Table 16.23. As noted previously, the two tests will always yield equivalent results for an $r \times 2$ contingency table.

6. The general linear model A large number of statistical methods including the **analysis of variance** and *t* **test** can be viewed within an overall conceptual framework referred to as the **general linear model (GLM)**. The latter model is a generalization of the model employed for **simple linear regression** (which is the basis for bivariate correlation). The **simple linear regression model,** which evaluates one independent variable and one dependent variable, is discussed in the chapter on the **Pearson product-moment correlation coefficient**. The extension of the **simple linear regression model** to designs involving multiple independent variables and one dependent variable is discussed in the chapter on **multiple regression (Test 33)**, while its extension to designs involving multiple independent variables and multiple dependent variables is discussed in the chapter on **canonical correlation (Test 38)**. All of the aforementioned analyses fall within the framework of the **general linear model**, which can be employed to evaluate any hypothesis regarding the relationship between one or more independent (i.e., predictor) variables and one or more dependent (i.e., criterion) variables, assuming that the latter are measured on an interval/ratio scale. Two questions which are addressed within the framework of the **general linear model** are: a) How well does a linear model fit a set of data? and b) Which of one or more independent variables is/are significantly associated with one or more dependent variables?

Field (2005, p. 304) notes that the **general linear model** can be succinctly summarized by the equation *Dependent variable = Model + error variability*. The simplest form of the latter equation is the equation for a straight line (discussed earlier in the section on trend analysis), which is derived in **simple linear regression**. In the latter equation (see Equation 28.4), the term *Model* is represented by the terms to the right of the equal sign — i.e., $a + bX$. Without getting into a complex mathematical explanation, it is sufficient to say that the **general linear model** states that parametric statistical procedures (such as the **single-factor between-subjects analysis of variance**) can be summarized in the form of one or more linear equations, and in the final analysis a correlational format can be employed to summarize all such procedures (or, to put it another way, all parametric procedures can be conceptualized as special cases of **linear regression analysis**). Equation 21.86 in Endnote 2 represents the linear function which describes the model (as well as error variability) for the **single-factor between-subjects analysis of variance**.

Cohen and Cohen (1983), Licht (1995, p. 20), and Tatsuoka (1975) note that in developing the **analysis of variance** Fisher originally employed **linear regression analysis**. Ultimately, however, he used the more familiar computational procedures described for the **analysis of variance** rather than **regression analysis**, because of the complexity of the computations required for the latter. The point to be made is that the **analysis of variance** represents a special case of **regression analysis**. Since computer software (such as *SPSS*) often employs **regression analysis** within the framework of conducting the **analysis of variance**, it is common to see reference to the **general linear model** and relevant terminology employed in **regression analysis** (e.g., intercept) in the menus and output associated with such software.

An example of the equivalency of the *t* **test for two independent samples/single-factor between-subjects analysis of variance** and bivariate correlation is demonstrated later in the book in the discussion of the **point-biserial correlation coefficient (Test 28i)** (discussed in Section IX (the **Addendum**) of the **Pearson product-moment correlation coefficient**). In the latter discussion the independent variable is represented as a **dummy variable** — the latter being a dichotomous variable which employs two values (typically 0 and 1) to indicate membership in one of two categories. The use of **dummy variables** in **regression analysis** (which is pivotal in expressing an inferential procedure as a correlational analysis) is discussed in Endnote 4 in the chapter on **logistic regression (Test 39)**. Further discussion of the **general linear model** can be found in Endnote 1 in the chapter on **multiple regression** and Endnote 2 in the chapter on **canonical correlation**.

7. Fixed-effects versus random-effects models for the single-factor between-subjects analysis of variance The terms **fixed-** versus **random-effects models** refer to the way in which a researcher selects the levels of the independent variables which are employed in an experiment. Whereas a **fixed-effects model** assumes that the levels of the independent variable are the same levels that will be employed in any attempted replication of the experiment, a **random-effects model** assumes that the levels have been randomly selected from the overall population of all possible levels which can be employed for the independent variable. The discussion of the **single-factor between-subjects analysis of variance** in this book assumes a **fixed-effects model**. With the exception of the computation of measures of magnitude of treatment effect, the equations employed for the fixed-effects model are identical to those that are employed for a random-effects model. However, in the case of more complex designs (within-subjects and factorial designs), the computational procedures for fixed- versus random-effects models may differ.[66] The degree to which a researcher may generalize the results of an experiment will be a direct function of which model is employed. Specifically, if a fixed-effects model is employed, one can only generalize the results of an experiment to the specific levels of the independent variable which are used in the experiment. On the other hand, if a random-effects model is employed, one can generalize the results to all possible levels of the independent variable.

8. Multivariate analysis of variance (commonly identified by the acronym **MANOVA**) The **multivariate** analog of the **single-factor between-subjects analysis of variance** is the **multivariate analysis of variance (Test 35)**, which is one of a number of multivariate statistical procedures discussed in the book. As noted under the discussion of **Hotelling's T²** (**Test 34**) (discussed in Section VII of the *t* **test for two independent samples**), the term multivariate is employed in reference to procedures that evaluate experimental designs in which there are multiple independent variables and/or multiple dependent variables. The **multivariate analysis of variance** is a generalization of **Hotelling's T²** to experimental designs involving more than two groups. In point of fact, **Hotelling's T²** represents a special case of the **multivariate analysis of variance**. The **multivariate analysis of variance** can be employed to analyze the data for an experiment which involves a single independent variable comprised of two or more levels and multiple dependent variables. With regard to the latter, instead of a single score, each subject produces scores on two or more dependent variables. To illustrate, let us assume that in Example 21.1 two scores are obtained for each subject. One score represents the number of correct responses, and a second score represents response latency (i.e., speed of response). Within the framework of the **multivariate analysis of variance**, a composite mean based on both the number of correct responses and response latency scores is computed for each group. The latter composite means are referred to as **mean vectors** or **centroids**. As is the case with the **single-factor between-subjects analysis of variance**, the means (in this case, composite) for the three groups are then compared with one another.

Stevens (1996, 2002) notes the following reasons why a researcher should consider employing a **multivariate analysis of variance** instead of an analysis of variance procedure which evaluates just a single dependent variable: a) Most treatments which have an effect on subjects will impact them in a variety of ways. Consequently, an experimental design that evaluates multiple effects accruing from a treatment will provide a more comprehensive picture of the overall impact of the treatment; b) As noted in Section VI, conducting multiple statistical tests can inflate the overall Type I error rate associated with a body of data. By employing multiple dependent variables within the framework of a single study, a researcher can avoid the inflated Type I error rate which can result from conducting separate studies, with each one requiring an analysis of a single dependent variable; c) Univariate analysis of variance (i.e., the analysis of variance described in this chapter) does not take into consideration that two or more potential dependent variables may be correlated with one another, whereas a **multivariate analysis of variance** takes such intercorrelations into account; and d) The effect of one or more dependent variables by themselves may not be strong enough to yield a significant result, yet when employed together within the framework of a design that is evaluated with a **multivariate analysis of variance**, their combined effect may be significant.

Like most multivariate procedures, the mathematics involved in conducting a **multivariate analysis of variance** is quite complex, and for this reason it becomes laborious if not impractical to implement without the aid of a computer. A detailed description of the **multivariate analysis of variance** can be found in Chapter 35.

VIII. Additional Examples Illustrating the Use of the Single-Factor Between-Subjects Analysis of Variance

Since the **single-factor between-subjects analysis of variance** can be employed to evaluate interval/ratio data for any independent groups design involving two or more groups, it can be used to evaluate examples which are evaluated with the *t* **test for two independent samples** (e.g., Examples 11.1, 11.8, and 11.9). Examples 21.2, 21.3, and 21.4 in this section are extensions of Examples 11.1, 11.8, and 11.9 to a design involving $k = 3$ groups. Since the data for all of the examples are identical to the data employed in Example 21.1, they yield the same result.

Example 21.2 *In order to assess the efficacy of a new antidepressant drug, 15 clinically depressed patients are randomly assigned to one of three groups. Five patients are assigned to Group 1, which is administered the antidepressant drug for a period of six months. Five patients are assigned to Group 2, which is administered a placebo during the same six-month period. Five patients are assigned to Group 3, which does not receive any treatment during the six-month period. Assume that prior to introducing the experimental treatments, the experimenter confirmed that the level of depression in the three groups was equal. After six months elapse, all 15 subjects are rated by a psychiatrist (who is blind with respect to a subject's experimental condition) with respect to their mood. The psychiatrist's mood ratings for the five subjects in each group follow. (The higher the rating, the less depressed a subject.)* **Group 1**: *8, 10, 9, 10, 9;* **Group 2**: *7, 8, 5, 8, 5;* **Group 3**: *4, 8, 7, 5, 7. Do the data indicate that the antidepressant drug is effective?*

Example 21.3 *A researcher wants to assess the relative effect of three different kinds of punishment on the emotionality of mice. Each of 15 mice is randomly assigned to one of three groups. During the course of the experiment each mouse is sequestered in an experimental chamber. While in the chamber, each of the five mice in Group 1 is periodically presented with a loud noise, each of the five mice in Group 2 is periodically presented with a blast of cold air, and each of the mice in Group 3 is periodically presented with an electric shock. The presentation of the punitive stimulus for each of the animals is generated by a machine which randomly presents the stimulus throughout the duration of the time it is in the chamber. The dependent variable of emotionality employed in the study is the number of times each mouse defecates while in the experimental chamber. The number of episodes of defecation for the 15 mice follow:* **Group 1**: *8, 10, 9, 10, 9;* **Group 2**: *7, 8, 5, 8, 5;* **Group 3**: *4, 8, 7, 5, 7. Do mice exhibit differences in emotionality under the different experimental conditions?*

Example 21.4 *Each of three companies that manufacture the same size precision ball bearing claims it has better quality control than its competitor. A quality control engineer conducts a study in which he compares the precision of ball bearings manufactured by the three companies. The engineer randomly selects five ball bearings from the stock of Company A, five ball bearings from the stock of Company B, and five ball bearings from the stock of Company C. He measures how much the diameter of each of the 15 ball bearings deviates from the manufacturer's specifications. The deviation scores (in micrometers) for the 15 ball bearings manufactured by the three companies follow:* **Company A**: *8, 10, 9, 10, 9;* **Company B**: *7, 8, 5, 8, 5;* **Company C**: *4, 8, 7, 5, 7. What can the engineer conclude about the relative quality control of the three companies?*

IX. Addendum

Test 21l: The single-factor between-subjects analysis of covariance Analysis of covariance (for which the acronym **ANCOVA** is commonly employed) is an analysis of variance procedure that employs a statistical adjustment (involving regression analysis, which is discussed under the **Pearson product-moment correlation coefficient**) to control for the effect of one or more extraneous variables on a dependent variable. Although it is possible to employ multiple extraneous variables within the framework of an analysis of covariance (which Winer *et al.* (1991) note was originally described by Fisher (1932)), in this section the **single-factor between-subjects analysis of covariance** involving one extraneous variable will be discussed. Analysis of covariance is an analysis of variance procedure which utilizes data on an extraneous variable that has a linear correlation with the dependent variable. Such an extraneous variable is referred to as a **covariate variable** (also referred to as a **concomitant, moderating, or mediating variable**). By utilizing the correlation between the covariate and the dependent variable, the researcher is able to remove variability on the dependent variable which is attributable to the covariate. The effect of the latter is a reduction in the error variability employed in computing the *F* ratio, thereby resulting in a more powerful test of the alternative hypothesis. A second potential effect of an analysis of covariance is that by utilizing the correlation between scores on a covariate and the dependent variable, the mean scores of the different groups can be adjusted for any pre-existing differences on the dependent variable which are present prior to the administration of the experimental treatments. Thus, one component of the analysis of covariance computations involves computing adjusted mean values for each of the *k* treatment means.

When analysis of covariance is employed, the most commonly used covariates are pretest scores on the dependent variable or subject variables such as intelligence, anxiety, weight, etc. Thus, in Example 21.1, if a researcher believes that the number of nonsense syllables a subject learns is a function of verbal intelligence, and if there is a linear correlation between verbal intelligence and one's ability to learn nonsense syllables, verbal intelligence can be employed as a covariate. In order to employ it as a covariate, it is necessary to have a verbal intelligence score for each subject who participates in the experiment. By employing the latter scores the researcher can use an analysis of covariance to determine whether or not there are performance differences on the dependent variable between the three groups which are independent of verbal intelligence. Later in this section an analysis of covariance will be employed to evaluate the data for Example 21.1 using verbal intelligence as a covariate.

The optimal design for which an analysis of covariance can be employed is an experiment involving a manipulated independent variable in which subjects are randomly assigned to groups (ideally, there should be an equal number of subjects in each group), and in which scores on the covariate are measured prior to the introduction of the experimental treatments. In such a design the analysis of covariance is able to remove variability on the dependent variable that is attributed to differences between the groups on the covariate. However, when, prior to introducing the experimental treatments it is known that a strong linear correlation exists between the covariate and the dependent variable, and that the groups are not equal with respect to the covariate, the following two options are available to a researcher: a) Subjects can be randomly reassigned to groups, after which the researcher can check that the resulting groups are equivalent with respect to the covariate; or b) The covariate can be integrated into the study as a second independent variable. Some sources endorse the use of either of the aforementioned strategies as preferable to employing an analysis of covariance on groups that are not equal with respect to the covariate.

A less ideal situation for employing an analysis of covariance is an experiment involving a manipulated independent variable in which subjects are randomly assigned to groups, but in which scores on the covariate are not measured until after the experimental manipulation. The latter situation is more problematical with respect to using the analysis of covariance, since subjects' scores on the covariate could have been influenced by the experimental treatments. Stevens (2002, p. 347) notes that if the covariate is influenced by the treatments, the change in the covariate may be correlated with the dependent variable, and under the latter circumstances part of the treatment effect will be removed. Simply put, if a covariate is not independent of the treatments it will be more difficult to interpret the results of an analysis.

An even more problematical use of the analysis of covariance is for a design in which subjects are not randomly assigned to groups. (In the **Introduction** such designs are discussed under **quasi-experimental designs**.) The latter can involve the use of intact groups (such as two different classes at a school) who are exposed to a manipulated independent variable, or two groups which are formed on the basis of some pre-existing subject characteristic (i.e., an **ex post facto study** involving a nonmanipulated independent variable such as gender, race, etc.). It should be noted that if a substantial portion of between-groups variability can be explained on the basis of an extraneous variable (i.e., a potential covariate), it implies there was probably some sort of systematic bias involved in forming the experimental groups. Because of the latter, it is reasonable to expect that if a researcher identifies one extraneous variable which has a substantial correlation with the dependent variable, there are probably other extraneous variables whose effects will not be controlled for, even if one evaluates the data with an analysis of covariance. In view of this, some sources (Keppel and Zedeck (1989) and Lord (1967, 1969)) argue that if subjects are not randomly assigned to groups, a researcher will not be able to adequately control for the potential effects of other pre-existing extraneous variables on the dependent variable. More specifically, they argue that the analysis of covariance will not be able to produce the necessary statistical control to allow one to unambiguously interpret the effect of the independent variable on the dependent variable. Thus, in designs in which subjects are not randomly assigned to groups, sources either state that the analysis of covariance should never be employed, or that if it is employed, the results should be interpreted with extreme caution.

Sources (e.g., Hinkle *et al.* (1998) and Keppel (1991)) note that in order for the analysis of covariance to provide effective statistical control the following two requirements must be met: a) A linear relationship must exist between the dependent variable and the covariate (since if they are not linearly related, the adjusted mean values computed for the k experimental conditions will be biased). Anderson (2001, p. 384) and Harlow (2005, p. 64) suggest that in order for an analysis of covariance to be worthwhile there should be at least a moderate correlation between the covariate and the dependent variable (with Harlow stipulating a minimal correlation of .30 and Anderson .40). Harlow (2005, p. 64) also notes that the reliability of a covariate (i.e., correlation between two measures of the covariate for each subject) should be substantial — specifically, at least $r = .70$; and b) The covariate should be independent of (i.e., not be influenced by) the experimental treatments/independent variable. The simplest way to evaluate this latter requirement is to conduct an analysis of variance using the covariate as the dependent variable. If the null hypothesis of equality between the covariate treatment means is rejected, it will indicate a lack of independence between the covariate and the independent variable.

Maxwell and Delaney (1990; pp. 380–385; 2000) provide an excellent description of the conditions which can result in a lack of independence between the covariate and independent variable, and the impact the latter will have on interpreting the results of an analysis of covariance. When a lack of independence is present, the results will be most difficult to interpret when assignment of subjects to groups is nonrandom, and/or when the treatments affect the covariate in a situation where the covariate is measured after the administration of the treatments. Even in the ideal situation where the researcher randomly assigns subjects to the k experimental conditions and measures the covariate before introducing the experimental treatment, a significant difference among the groups with respect to their means on the covariate may be detected. The latter situation could merely be representative of a Type I error — in other words, the mean differences on the covariate represent a fluke of chance variation. Under such circumstances, to insure minimal ambiguity in interpreting the results of an analysis of covariance, the most prudent strategy would probably be to randomly reassign subjects until the k covariate treatment means are approximately the same. Maxwell and Delaney (1990; 2000) note that within the framework of an analysis of covariance, the accuracy of the adjusted means which are computed for the dependent variable decreases as the group means on the covariate deviate from the grand mean on the covariate. What the latter translates into is (assuming the covariate is, in fact, independent of the independent variable) that when the treatments differ with respect to covariate means, the adjusted values of the dependent variable will be less likely to result in a significant difference — in other words, the power of the analysis of covariance to detect a significant effect between the treatment means on the dependent variable will be reduced. It is worth noting that some sources (e.g., Maxwell and Delaney (1990, 2000)) take the position that although problems in interpretation will result if the covariate is not independent of the experimental treatments, it does not necessarily mean that under such conditions an analysis of covariance cannot yield useful information.

Weinfurt (2000, pp. 339–340), however, states that in the case of a **quasi-experimental design** in which subjects are not randomly assigned to the different levels of the independent variable, an analysis of covariance will for all practical purposes be useless if the groups do not have the same mean scores on the covariate. Weinfurt (2000, p. 340) notes that in the case of a **nonequivalent control group design** (which is described in the **Introduction**), comparing the posttest scores of two groups will be of no value if the groups are not equivalent with respect to their pretest scores. (In an analysis of covariance the pretest scores would be employed to represent the covariate.) Under the latter circumstances, instead of employing an analysis of covariance a researcher should employ a *t* **test for two independent samples** (or a **single-factor between-subjects analysis of variance**) to contrast the mean difference scores of the two groups. (In such a case a difference score is obtained for each subject, which is the difference between a subject's pretest versus posttest score.)

Kachigan (1986), among others, notes that one should never use an analysis of covariance to adjust for between-groups differences with respect to a covariate which are attributable to normal sampling error. Aside from the fact that such variability is a part of expected error variability within the framework of conducting an analysis of variance, a major reason for not employing an analysis of covariance (for which the test statistic is also an omnibus F value) is that the number of within-groups degrees of freedom required for the analysis will be one less than df_{WG} required for an analysis of variance on the same set of data. In such a case, any reduction in error variance associated with the analysis of covariance may be offset by the fact that it will require a larger critical F value to reject the null hypothesis than will an analysis of variance on the original data.

A final point which should be made is that it is not uncommon for researchers to employ multiple covariates in a study. It is important that covariates are not highly correlated with one another, since, as Steven's (2002, pp. 345–346) notes, highly correlated covariates will be removing much of the same error variance from the dependent variable (and consequently will not substantially reduce error variability). If, on the other hand, the correlation between two covariates is minimal, they will each remove relatively distinct portions of error variability, and the latter will result in a substantially lower error variance for the analysis of covariance. (The term **collinearity** is employed when two covariates are intercorrelated with one another. **Collinearity** is discussed in greater detail in the chapter on **multiple regression (Test 33)**.) In addition, Harlow (2005, p 67) notes that when multiple covariates are employed, since each covariate uses up one degree of freedom, the more degrees of freedom used up by the covariates, the less powerful will be the test of the null hypothesis. In such a case the analysis of covariance may be no more powerful (or perhaps even less powerful) than a simple analysis of variance involving just the independent variable.[67]

Example 21.5 will be employed to illustrate the use of the **single-factor between-subjects analysis of covariance**. Example 21.5 is identical to Example 21.1, except for the fact that the covariate of verbal intelligence is included in the analysis. It is because of the latter that the data are evaluated with the **single-factor between-subjects analysis of covariance**.

Example 21.5 *A psychologist conducts a study to determine whether or not noise can inhibit learning. Each of 15 subjects is randomly assigned to one of three groups. Each subject is given 20 minutes to memorize a list of 10 nonsense syllables, which she is told she will be tested on the following day. The five subjects assigned to* **Group 1**, *the* **no noise** *condition, study the list of nonsense syllables while they are in a quiet room. The five subjects assigned to* **Group 2**, *the*

moderate noise *condition, study the list of nonsense syllables while listening to classical music. The five subjects assigned to* **Group 3***, the* **extreme noise** *condition, study the list of nonsense syllables while listening to rock music. The number of nonsense syllables correctly recalled by the 15 subjects follow:* **Group 1**: 8, 10, 9, 10, 9; **Group 2**: 7, 8, 5, 8, 5; **Group 3**: 4, 8, 7, 5, 7. *From previous research it is known that a subject's ability to learn nonsense syllables is highly correlated with one's verbal intelligence. As a result of the latter, a test of verbal intelligence (for which the maximum possible score is 20) is administered to each subject prior to introducing the experimental treatments. The verbal intelligence scores of the subjects follow:* **Group 1**: 14, 16, 15, 16, 15; **Group 2**: 15, 17, 15, 17, 15; **Group 3**: 14, 17, 16, 16, 16. *Do the data indicate that noise influenced subjects' performance?*

The **null** and **alternative hypotheses** employed for the **single-factor between-subjects analysis of covariance** are identical to those employed for the **single-factor between-subjects analysis of variance**. The treatment means, however, which are contrasted in the analysis of covariance are adjusted values. As noted earlier, the means are adjusted for the effect of the covariate. In view of the fact that the analysis of covariance evaluates adjusted mean values, the notation μ_j' will be employed to represent the adjusted value of the mean of the population the jth group represents. The null and alternative hypotheses for the analysis of covariance in reference to Example 21.5 are as follows.

Null hypothesis H_0: $\mu_1' = \mu_2' = \mu_3'$

(The adjusted mean of the population Group 1 represents equals the adjusted mean of the population Group 2 represents equals the adjusted mean of the population Group 3 represents.)

Alternative hypothesis H_1: Not H_0

(This indicates there is a difference between at least two of the $k = 3$ adjusted population means.)

Computational procedures for the single-factor between-subjects analysis of covariance This section will describe a number of computational procedures which are used within the framework of an analysis of covariance. The reader should keep in mind that the analysis of covariance can be employed with experimental designs other than a between-subjects design, and that in conducting an analysis of covariance there can be more than one covariate. All of the procedures to be described in this section, however, are in reference to the **single-factor between-subjects analysis of covariance** involving a single covariate.

The data analysis in this section will evaluate Example 21.5. In the latter example the number of nonsense syllables correctly recalled represents the dependent variable (which will be the Y variable), while the verbal intelligence test scores of subjects represent the covariate (which will be the X variable). Initially, an analysis of variance will be conducted on the covariate. The latter is generally done in order to demonstrate independence between the covariate and the experimental treatments. In the case of Example 21.5, one could argue that it is not necessary to conduct an analysis of variance on the covariate, since subjects are randomly assigned to groups, and the covariate is measured prior to the introduction of the experimental treatments. However, random assignment in and of itself does not insure that the group means on the covariate will be equal. As noted earlier, Maxwell and Delaney (1990, 2000) state that the accuracy of the adjusted means which are computed for the dependent variable will decrease as the group means on the covariate deviate from the grand mean on the covariate. Most sources agree that if the assignment of subjects to groups is nonrandom, and/or the covariate is measured before or during the administration of treatments, an analysis of variance on the covariate should be conducted. Although a nonsignificant result for the latter analysis does not guarantee that the analysis of covariance (as well as the computed values for the adjusted treatments means) will be unbiased, it makes it unlikely.

After establishing there are no significant differences between the covariate means, an analysis of covariance will be conducted. After doing the latter, adjusted treatment means on the dependent variable will be computed for the three groups. The remainder of the section will describe or discuss the following: a) Conducting multiple comparisons on the adjusted treatment means; b) Evaluating the homogeneity of regression assumption underlying the analysis of covariance; c) Computing measures of magnitude of treatment effect for the analysis of covariance; and d) Computing power for the analysis of covariance.

Table 21.19 summarizes the data for Example 21.5. As noted above, X represents the scores on the covariate, and Y represents the scores on the dependent variable. In the last column of Table 21.19 an XY score is computed for each subject. A subject's XY score is obtained by multiplying the subject's X score by the subject's Y score. The latter values are employed in computing the sum of products, which is represented with notation SP. (The sum of products is discussed in greater detail in Section VII of the **Pearson product-moment correlation coefficient** under the discussion of **covariance**.) The general equation for the sum of products is $SP_{XY} = \Sigma XY - [(\Sigma X)(\Sigma Y)]/n$. All terms in Table 21.19 with the subscript j (e.g., ΣX_j, ΣY_j, etc., where j equals a specific group number) are based on the scores of the $n_j = 5$ subjects in the jth group, while all terms with the subscript T (e.g., ΣX_T, ΣY_T, etc.) are based on the scores of all $N = 15$ subjects.

Analysis of variance on the covariate The procedure for the analysis of variance on the covariate is identical to the procedure employed for the analysis of variance for Example 21.1, except for the fact that the covariate scores (X) are evaluated instead of the scores on the dependent variable (Y). The null and alternative hypotheses employed are identical to those employed for Example 21.1, except for the fact they are stated in reference to the population means for the covariate. Thus, H_0: $\mu_{X_1} = \mu_{X_2} = \mu_{X_3}$ and H_1: Not H_0. Computation of the sums of squares for the analysis of variance on the covariate are summarized below.

$$SS_T = \Sigma X_T^2 - \frac{(\Sigma X_T)^2}{N} = 3664 - \frac{(234)^2}{15} = 13.6$$

$$SS_{BG} = \sum_{j=1}^{k} \left[\frac{(\Sigma X_j)^2}{n_j} \right] - \frac{(\Sigma X_T)^2}{N} = \left[\frac{(76)^2}{5} + \frac{(79)^2}{5} + \frac{(79)^2}{5} \right] - \frac{(234)^2}{15} = 1.2$$

$$SS_{WG} = SS_T - SS_{BG} = 13.6 - 1.2 = 12.4$$

Table 21.20 summarizes the analysis of variance on the covariate. The number of degrees of freedom employed are the same as those employed for Example 21.1, since the identical number of groups and subjects are employed in the analysis. Thus, $df_{BG} = 2$ and $df_{WG} = 12$. The tabled critical values employed in **Table A10** are $F_{.95} = 3.89$ and $F_{.99} = 6.93$. Since the computed value $F = .58$ is less than $F_{.05} = 3.89$, the null hypothesis cannot be rejected at the .05 level. Thus, we can conclude there is no difference between the groups with respect to their mean scores on the covariate (which are $\bar{X}_1 = 15.2$, $\bar{X}_2 = 15.8$, $\bar{X}_3 = 15.8$). As noted earlier, this is generally interpreted as indicating that the covariate is independent of the experimental treatments.

Table 21.19 Data for Example 21.5

Group 1

X_1	X_1^2	Y_1	Y_1^2	$X_1 Y_1$
14	196	8	64	112
16	256	10	100	160
15	225	9	81	135
16	256	10	100	160
15	225	9	81	135

$$\Sigma X_1 = 76 \qquad \Sigma X_1^2 = 1158 \qquad \Sigma Y_1 = 46 \qquad \Sigma Y_1^2 = 426 \qquad \Sigma X_1 Y_1 = 702$$
$$\overline{X}_1 = 15.2 \qquad\qquad\qquad \overline{Y}_1 = 9.2$$

$$SS_{X_1} = \Sigma X_1^2 - \frac{(\Sigma X_1)^2}{n_1} = 1158 - \frac{(76)^2}{5} = 2.8 \qquad SS_{Y_1} = \Sigma Y_1^2 - \frac{(\Sigma Y_1)^2}{n_1} = 426 - \frac{(46)^2}{5} = 2.8$$

$$SP_{1(XY)} = SP_{X_1 Y_1} = \Sigma X_1 Y_1 - \frac{(\Sigma X_1)(\Sigma Y_1)}{n_1} = 702 - \frac{(76)(46)}{5} = 2.8$$

Group 2

X_2	X_2^2	Y_2	Y_2^2	$X_2 Y_2$
15	225	7	49	105
17	289	8	64	136
15	225	5	25	75
17	289	8	64	136
15	225	5	25	75

$$\Sigma X_2 = 79 \qquad \Sigma X_2^2 = 1253 \qquad \Sigma Y_2 = 33 \qquad \Sigma Y_2^2 = 227 \qquad \Sigma X_2 Y_2 = 527$$
$$\overline{X}_2 = 15.8 \qquad\qquad\qquad \overline{Y}_2 = 6.6$$

$$SS_{X_2} = \Sigma X_2^2 - \frac{(\Sigma X_2)^2}{n_2} = 1253 - \frac{(79)^2}{5} = 4.8 \qquad SS_{Y_2} = \Sigma Y_2^2 - \frac{(\Sigma Y_2)^2}{n_2} = 227 - \frac{(33)^2}{5} = 9.2$$

$$SP_{2(XY)} = SP_{X_2 Y_2} = \Sigma X_2 Y_2 - \frac{(\Sigma X_2)(\Sigma Y_2)}{n_2} = 527 - \frac{(79)(33)}{5} = 5.6$$

Group 3

X_3	X_3^2	Y_3	Y_3^2	$X_3 Y_3$
14	196	4	16	56
17	289	8	64	136
16	256	7	49	112
16	256	5	25	80
16	256	7	49	112

$$\Sigma X_3 = 79 \qquad \Sigma X_3^2 = 1253 \qquad \Sigma Y_3 = 31 \qquad \Sigma Y_3^2 = 203 \qquad \Sigma X_3 Y_3 = 496$$
$$\overline{X}_3 = 15.8 \qquad\qquad\qquad \overline{Y}_3 = 6.2$$

$$SS_{X_3} = \Sigma X_3^2 - \frac{(\Sigma X_3)^2}{n_3} = 1253 - \frac{(79)^2}{5} = 4.8 \qquad SS_{Y_3} = \Sigma Y_3^2 - \frac{(\Sigma Y_3)^2}{n_3} = 203 - \frac{(31)^2}{5} = 10.8$$

$$SP_{3(XY)} = SP_{X_3 Y_3} = \Sigma X_3 Y_3 - \frac{(\Sigma X_3)(\Sigma Y_3)}{n_3} = 496 - \frac{(79)(31)}{5} = 6.2$$

$$\Sigma X_T = 234 \qquad \Sigma X_T^2 = 3664 \qquad \Sigma Y_T = 110 \qquad \Sigma Y_T^2 = 856 \qquad \Sigma(XY)_T = 1725$$

$$SS_{T(X)} = \Sigma X_T^2 - \frac{(\Sigma X_T)^2}{N} = 3664 - \frac{(234)^2}{5} = 13.6 \qquad SS_{T(Y)} = \Sigma Y_T^2 - \frac{(\Sigma Y_T)^2}{N} = 856 - \frac{(110)^2}{5} = 49.33$$

$$SP_{T(XY)} = \Sigma(XY)_T - \frac{(\Sigma X_T)(\Sigma Y_T)}{N} = 1725 - \frac{(234)(110)}{15} = 9$$

**Table 21.20 Summary Table of Analysis of Variance
on Covariate for Example 21.5**

Source of variation	SS	df	MS	F
Between-groups	1.20	2	.60	.58
Within-groups	12.40	12	1.03	
Total	13.60	14		

The analysis of covariance In conducting the analysis of covariance we will employ the sum of squares values which were previously computed in the analysis of variance on the dependent variable and in the analysis of variance on the covariate. The values that will be employed are listed below. Note that the notation for each of the sum of squares values listed indicates whether a sum of squares value is for the X variable (the covariate) or the Y variable (the dependent variable): a) $SS_{BG(X)} = 1.2$, which represents the between-groups sum of squares for the covariate; b) $SS_{WG(X)} = 12.4$, which represents the within-groups sum of squares for the covariate; c) $SS_{T(X)} = 13.6$, which represents the total sum of squares for the covariate; d) $SS_{BG(Y)} = 26.53$, which represents the between-groups sum of squares for the dependent variable; e) $SS_{WG(Y)} = 22.8$, which represents the within-groups sum of squares for the dependent variable; and f) $SS_{T(Y)} = 49.33$, which represents the total sum of squares for the dependent variable. In addition to employing the aforementioned values, we must compute the values noted below.

a) A **between-groups sum of products**, represented by the notation $SP_{BG(XY)}$, is computed with Equation 21.63. The notation $\sum_{j=1}^{k}[[(\Sigma X_j)(\Sigma Y_j)]/n_j]$ on the right side of Equation 21.63 indicates that for each group the sum of the scores on the X variable is multiplied by the sum of the scores on the Y variable, and the product is divided by the number of subjects in the group. Upon doing the latter for all k groups, the k values that have been obtained are summed. The value $[(\Sigma X_T)(\Sigma Y_T)]/N$ is subtracted from the resulting sum. The value $[(\Sigma X_T)(\Sigma Y_T)]/N$ is obtained by multiplying the total sum of the scores on the X variable (ΣX_T) by the total sum of the scores on the Y variable (ΣY_T), and dividing the product by the total number of subjects (N). Note that unlike a sum of squares value, a sum of products value can be a negative number.

(Equation 21.63)

$$SP_{BG(XY)} = \sum_{j=1}^{k}\left[\frac{(\Sigma X_j)(\Sigma Y_j)}{n_j}\right] - \frac{(\Sigma X_T)(\Sigma Y_T)}{N}$$

$$SP_{BG(XY)} = \left[\frac{(76)(46)}{5} + \frac{(79)(33)}{5} + \frac{(79)(31)}{5}\right] - \frac{(234)(110)}{15} = 1710.4 - 1716 = -5.6$$

b) A **within-groups sum of products**, represented by the notation $SP_{WG(XY)}$, is computed with Equation 21.64. As noted earlier, the notation $\sum_{j=1}^{k}[[(\Sigma X_j)(\Sigma Y_j)]/n_j]$ on the right side of Equation 21.64 indicates that for each group the sum of the scores on the X variable is multiplied by the sum of the scores on the Y variable, and the product is divided by the number of subjects in the group. Upon doing the latter for all k groups, the k values that have been obtained are summed. The resulting value is subtracted from the term $\Sigma(XY)_T$, which is the sum of the XY scores of all $N = 15$ subjects.

(Equation 21.64)

$$SP_{WG(XY)} = \Sigma(XY)_T - \sum_{j=1}^{k}\left[\frac{(\Sigma X_j)(\Sigma Y_j)}{n_j}\right]$$

$$SP_{WG(XY)} = 1725 - \left[\frac{(76)(46)}{5} + \frac{(79)(33)}{5} + \frac{(79)(31)}{5}\right] = 1725 - 1710.4 = 14.6$$

c) A **total sum of products**, represented by the notation $SP_{T(XY)}$, is computed with Equation 21.65. In computing the total sum of products, the value $[(\Sigma X_T)(\Sigma Y_T)]/N$ (which, as noted earlier, is obtained by multiplying the total sum of the scores on the X variable (ΣX_T) by the total sum of the scores on the Y variable (ΣY_T), and dividing the product by the total number of subjects (N)) is subtracted from the term $\Sigma(XY)_T$, which, as noted earlier, is the sum of the XY scores of all $N = 15$ subjects.

(Equation 21.65)

$$SP_{T(XY)} = \Sigma(XY)_T - \frac{(\Sigma X_T)(\Sigma Y_T)}{N}$$

$$SP_{T(XY)} = 1725 - \frac{(234)(110)}{15} = 1725 - 1716 = 9$$

As is the case with the analysis of variance, sum of squares values are computed for the analysis of covariance. However, the sums of squares computed for the analysis of covariance are adjusted for the effects of the covariate. The analysis of covariance requires that the following three adjusted sum of squares values be computed: a) The **adjusted total sum of squares**, which will be represented by the notation $SS_{T(adj)}$; b) The **adjusted within-groups sum of squares**, which will be represented by the notation $SS_{WG(adj)}$; and c) The **adjusted between-groups sum of squares**, which will be represented by the notation $SS_{BG(adj)}$.

The **adjusted total sum of squares** is computed with Equation 21.66.

(Equation 21.66)

$$SS_{T(adj)} = SS_{T(Y)} - \frac{(SP_{T(XY)})^2}{SS_{T(X)}}$$

$$SS_{T(adj)} = 49.33 - \frac{(9)^2}{13.6} = 43.37$$

The **adjusted within-groups sum of squares** is computed with Equation 21.67.[68]

(Equation 21.67)

$$SS_{WG(adj)} = SS_{WG(Y)} - \frac{(SP_{WG(XY)})^2}{SS_{WG(X)}}$$

$$SS_{WG(adj)} = 22.8 - \frac{(14.6)^2}{12.4} = 5.61$$

The **adjusted between-groups sum of squares** is computed with Equation 21.68.

(Equation 21.68)

$$SS_{BG(adj)} = SS_{T(adj)} - SS_{WG(adj)}$$

$$SS_{BG(adj)} = 43.37 - 5.61 = 37.76$$

Table 21.21 summarizes the analysis of covariance.

Table 21.21 Summary Table of Analysis of Covariance for Example 21.5

Source of variation	SS_{adj}	df_{adj}	MS_{adj}	F
Between-groups	37.76	2	18.88	37.02
Within-groups	5.61	11	.51	
Total	43.37	13		

Note that variability for the analysis of covariance is partitioned into between-groups variability and within-groups variability, the same two components that variability is partitioned into in the analysis of variance. As is the case for an analysis of variance, a mean square is computed for each source of variability by dividing the sum of squares by its respective degrees of freedom. Equations 21.69–21.71 summarize the computation of the degrees of freedom values. Note that in the analysis of covariance, the within-groups (error) degrees of freedom are $df_{WG(adj)} = N - k - 1$ as opposed to the value $df_{WG} = N - k$ computed for the analysis of variance. The loss of one degree of freedom in the analysis of covariance reflects the use of the covariate as a statistical control. Equations 21.72 and 21.73 summarize the computation of the mean square values, and Equation 21.74 computes the F ratio for the analysis of covariance.

$$df_{BG(adj)} = k - 1 = 3 - 1 = 2 \qquad \textbf{(Equation 21.69)}$$

$$df_{WG(adj)} = N - k - 1 = 15 - 3 - 1 = 11 \qquad \textbf{(Equation 21.70)}$$

$$df_{T(adj)} = N - 2 \qquad \textbf{(Equation 21.71)}$$

$$MS_{BG(adj)} = \frac{SS_{BG(adj)}}{df_{BG(adj)}} = \frac{37.76}{2} = 18.88 \qquad \textbf{(Equation 21.72)}$$

$$MS_{WG(adj)} = \frac{SS_{WG(adj)}}{df_{WG(adj)}} = \frac{5.61}{11} = .51 \qquad \textbf{(Equation 21.73)}$$

$$F = \frac{MS_{BG(adj)}}{MS_{WG(adj)}} = \frac{18.88}{.51} = 37.02 \qquad \textbf{(Equation 21.74)}$$

The obtained value $F = 37.02$ is evaluated with **Table A10**, employing as the numerator and denominator degrees of freedom $df_{num} = df_{BG(adj)} = 2$ and $df_{den} = df_{WG(adj)} = 11$. For $df_{num} = 2$ and $df_{den} = 11$, the tabled $F_{.95}$ and $F_{.99}$ values are $F_{.05} = 3.98$ and $F_{.01} = 7.21$. In order to reject the null hypothesis, the obtained F value must be equal to or greater than the tabled critical value at the prespecified level of significance. Since $F = 37.02$ is greater than $F_{.05} = 3.98$ and $F_{.01} = 7.21$, the alternative hypothesis is supported at both the .05 and .01 levels.

Note that the value $F = 37.02$ computed for the analysis of covariance is substantially larger than the value $F = 6.98$ computed for the analysis of variance in Example 21.1. The larger F value for the analysis of covariance can be attributed to the following two factors: a) The reduction in error variability in the analysis of covariance. The latter is reflected by the fact that the **within-groups mean square** value for the analysis of covariance $MS_{WG(adj)} = .51$ is substantially less than the analogous value $MS_{WG} = 1.90$ computed for the analysis of variance; and b) The larger **between-groups mean square** computed for the analysis of covariance. With respect to the latter, in the case of the analysis of covariance $MS_{BG(adj)} = 18.88$, whereas for the analysis of variance $MS_{BG} = 13.27$.

Thus, when we contrast the result of the analysis of covariance for Example 21.5 with the result of the analysis of variance for Example 21.1, we can state that by virtue of controlling for variation attributable to the covariate we are able to conduct a more sensitive analysis with respect to the impact of the treatments on the dependent variable. As is the case for the analysis of variance, it can be concluded that there is a significant difference between at least two of the three groups exposed to different levels of noise. This result can be summarized as follows: $F(2,11) = 37.02, p < .01$.

Some sources employ the format in Table 21.22 to summarize the results of an analysis of covariance.

Table 21.22   Alternative Summary Table of Analysis of Covariance for Example 21.5

Source of variation	SS_{adj}	df_{adj}	MS_{adj}	F
Covariate	5.96	1	5.96	11.65
Between-groups	37.76	2	18.88	37.02
Within-groups	5.61	11	.51	
Total	49.33	14		

Note that there are two differences between Tables 21.21 and 21.22: a) Table 21.22 contains a row for variability attributed to the covariate; and b) The total sum of squares and total degrees of freedom in the last row of Table 21.22 are different from the values listed in Table 21.21. In point of fact, the value for total variability in Table 21.22 ($SS_T = 49.33$) is equal to the total variability computed for Table 21.2 (the table for the original analysis of variance, which did not include the covariate in the study). In the same respect, the value for total degrees of freedom in Table 21.22 ($df_T = 14$) is equal to the total degrees of freedom computed for Table 21.2. In Table 21.22 one degree of freedom is employed for the covariate, and consequently the total degrees of freedom are equal to $df = N - 1$ (which is the same value employed for the total degrees of freedom for the analysis of variance in Table 21.2). The value $SS_{cov} = 5.96$ is computed with Equation 21.75, and Equations 21.76 and 21.77 are employed to compute the values $MS_{cov} = 5.96$ and $F = 11.65$ for the covariate.

$$SS_{cov} = SS_{T(Y)} - SS_{T(adj)} = 49.33 - 43.37 = 5.96 \qquad \textbf{(Equation 21.75)}$$

$$MS_{cov} = \frac{SS_{cov}}{df_{cov}} = \frac{5.96}{1} = 5.96 \qquad \textbf{(Equation 21.76)}$$

$$F = \frac{MS_{cov}}{MS_{WG(adj)}} = \frac{5.96}{.51} = 11.65 \qquad \textbf{(Equation 21.77)}$$

Note that the F ratio computed for the covariate with Equation 21.77 is not the same F ratio computed for the covariate in Table 21.20. Whereas the latter F ratio does not take the dependent variable into account, the value $F = 11.65$ computed with Equation 21.77 assumes the presence of both a dependent variable and a covariate in the data. Hinkle *et al.* (1998) note that the value $F = 11.65$ computed for the covariate can be employed to evaluate the null hypothesis

H_0: ρ_T = 0, which states that in the underlying population the sample represents, the correlation between the scores of subjects on the covariate and the dependent variable equals 0 (the notation ρ is the lower case Greek letter **rho**, which is employed to represent the population correlation). Or to put it another way, the null hypothesis is stating that there is no linear relationship between the covariate and the dependent variable. The alternative hypothesis (which is nondirectional) that is evaluated is H_1: ρ_T ≠ 0. The latter alternative hypothesis states that in the underlying population the sample represents, the correlation between the scores of subjects on the covariate and the dependent variable does not equal 0. Or to put it another way, there is a linear relationship between the covariate and the dependent variable.

Since an assumption of the analysis of covariance is that the dependent variable and covariate are linearly related, we want to reject the null hypothesis. If the computed F ratio computed with Equation 21.77 is statistically significant, the null hypothesis H_0: ρ_T = 0 can be rejected.[69] The numerator and denominator degrees of freedom employed in the analysis of the covariate F ratio are df_{num} = df_{cov} = 1 and df_{den} = $df_{WG(adj)}$ = 11. In **Table A10**, for df_{num} = 1 and df_{den} = 11, the tabled $F_{.95}$ and $F_{.99}$ values are $F_{.05}$ = 4.84 and $F_{.01}$ = 9.65. Since F = 11.65 is greater than $F_{.05}$ = 4.84 and $F_{.01}$ = 9.65, the alternative hypothesis is supported at both the .05 and .01 levels. In other words, we can conclude there is a significant linear relationship between the dependent variable and the covariate.

It is possible to compute the following three correlation coefficients from the data in Table 21.19: a) The **overall correlation coefficient** based on the total number of scores in the three groups. The latter value, represented by the notation r_T, is computed in Endnote 69 to be r_T = .347; b) The **within-groups correlation coefficient** (represented by the notation r_{WG}), which is a weighted average correlation coefficient between the covariate and the dependent variable within each of the k = 3 groups. The within-groups correlation coefficient is computed with Equation 21.78 to be r_{WG} = .868. The greater the absolute value of r_{WG}, the greater the precision of the analysis of covariance;[70] and c) the **between-groups correlation coefficient** (represented by the notation r_{BG}), which is a correlation coefficient between the k = 3 treatment means on the covariate and the k = 3 treatment means on the dependent variable. The between-groups correlation coefficient is computed (employing Equation 28.1) to be r_{BG} = −.992. In computing the latter value, the three pairs of X and Y scores, which are substituted in Equation 28.1, with n = 3, are as follows: ($\overline{X}_1$ = 15.2, $\overline{Y}_1$ = 9.2), ($\overline{X}_2$ = 15.8, $\overline{Y}_2$ = 6.6), and ($\overline{X}_3$ = 15.8, $\overline{Y}_3$ = 6.2).

(Equation 21.78)

$$r_{WG} = \frac{SP_{WG(XY)}}{\sqrt{SS_{WG(X)})(SS_{WG(Y)})}} = \frac{14.6}{\sqrt{(12.4)(22.8)}} = .868$$

Kirk (1995, p. 719) notes that if the value of r_{BG} is larger than the value of r_{WG}, any reduction in error variability has a high likelihood of being offset by a reduction in between-groups variability. As a result of the latter, the F value computed for the analysis of covariance may actually be lower than the F value computed for the analysis of variance on the dependent variable. However, if r_{BG} is negative and r_{WG} is positive (as is the case in our example), the F value computed for the analysis of covariance will be greater than the F value computed for the analysis of variance on the dependent variable. Winer *et al.* (1991) note that the higher the value of r_{WG}, the lower the value of the error term which will be computed for the analysis of covariance relative to value of the error term that will be computed if an analysis of variance is conducted on the dependent variable. Thus, the greater the absolute value of r_{WG}, the greater the precision of the analysis of covariance.

Computing the adjusted group/treatment means　The analysis of covariance is an analysis of any variability on the dependent variable which is not accounted for by the covariate. If the result of the analysis of covariance is significant, it indicates that two or more treatment means differ significantly from one another. However, prior to comparing treatment means it is necessary that the latter values be adjusted for the effects of the covariate. In computing the adjusted treatment means, we are determining what the scores on the dependent variable would be if the groups did not differ on the covariate. Maxwell and Delaney (1990, p. 378; 2000) note that the adjusted treatment means on the dependent variable can be viewed as estimates of the values that would be obtained if the mean of the covariate for each of the groups was equal to mean of all N subjects on the covariate.

Equation 21.79 is the general equation for computing an adjusted treatment/group mean.

$$\overline{Y}_j' = \overline{Y}_j - b_{WG}(\overline{X}_j - \overline{X}_T) \qquad \textbf{(Equation 21.79)}$$

In Equation 21.79, the adjusted mean on the dependent variable for the jth group is represented by the notation $\overline{Y}_j'$. The notation $\overline{Y}_j$ is the unadjusted mean for the jth group on the dependent variable (i.e., in the case of Example 21.5, the unadjusted means are the $\overline{Y}_j$ values in Table 21.19 which are the same as the group means computed for Example 21.1). Thus, $\overline{Y}_1 = 9.2$, $\overline{Y}_2 = 6.6$, and $\overline{Y}_3 = 6.2$. The value $\overline{X}_j$ represents the mean of the jth group on the covariate. Thus, as noted in Table 21.19, $\overline{X}_1 = 15.2$, $\overline{X}_2 = 15.8$, and $\overline{X}_3 = 15.8$. The value $\overline{X}_T$ is the mean of the $N = 15$ subjects on the covariate. Thus, $\overline{X}_T = (\Sigma X_1 + \Sigma X_2 + \Sigma X_3)/15$ $= (76 + 79 + 79)/15 = 15.6$. The value b_{WG}, which is referred to as the **within-groups regression coefficient**, is computed with either Equation 21.80 or Equation 21.81. For Example 21.5, the value $b_{WG} = 1.18$ is computed.[71]

$$b_{WG} = \frac{SP_{WG(XY)}}{SS_{WG(X)}} = \frac{14.6}{12.4} = 1.18 \qquad \textbf{(Equation 21.80)}$$

$$b_{WG} = r_{WG}\sqrt{\frac{MS_{WG(Y)}}{MS_{WG(X)}}} = .868\sqrt{\frac{1.90}{1.03}} = 1.18 \qquad \textbf{(Equation 21.81)}$$

The adjusted group means are computed for the $k = 3$ groups below.

$$\overline{Y}_1' = 9.2 - 1.18(15.2 - 15.6) = 9.672$$

$$\overline{Y}_2' = 6.6 - 1.18(15.8 - 15.6) = 6.364$$

$$\overline{Y}_3' = 6.2 - 1.18(15.8 - 15.6) = 5.964$$

Conducting comparisons among the adjusted group/treatment means　The same types of comparisons for contrasting group means on the dependent variable which are described in Section VI can be employed in contrasting the adjusted group means. Visual inspection of the adjusted and unadjusted group means reveals that in the case of Groups 1 and 2 the adjusted means for the latter groups are further removed from one another than the unadjusted means are from one another. Employing the same protocol used in Section VI, the **planned simple comparison** on the adjusted means of Groups 1 and 2 is summarized in Table 21.23. The test statistic is then computed employing the appropriate equations presented in Section VI. Note that

Table 21.23 Planned Simple Comparison: Group 1 versus Group 2

Group	$\bar{Y}_j'$	Coefficient (c_j)	Product $(c_j)(\bar{Y}_j')$	Squared Coefficient $(c_j)^2$
1	9.672	+1	$(+1)(9.672) = +9.672$	1
2	6.364	−1	$(-1)(6.364) = -6.364$	1
3	5.964	0	$(0)(5.964) = 0$	0
		$\Sigma(c_j) = 0$	$\Sigma(c_j)(\bar{Y}_j') = 3.308$	$\Sigma c_j^2 = 2$

in the latter equations, the notation appropriate for comparing the adjusted mean values is substituted for the notation employed in Section VI.

Equation 21.17 is employed to compute the value $SS_{comp(adj)} = 27.36$.

$$SS_{comp(adj)} = \frac{n\left[\Sigma(c_j)(\bar{Y}_j')\right]^2}{\Sigma c_j^2} = \frac{5(3.308)^2}{2} = 27.36$$

Equation 21.18 is employed to compute the value $MS_{comp(adj)} = 27.36$.

$$MS_{comp(adj)} = \frac{SS_{comp(adj)}}{df_{comp}} = \frac{27.36}{1} = 27.36$$

The test statistic F_{comp} is computed with Equation 21.83. Keppel (1990) notes that instead of employing $MS_{WG(adj)}$ as the denominator in Equation 21.83, the value computed with Equation 21.82 for MS_{WG}' should be employed.[72]

(Equation 21.82)

$$MS_{WG}' = MS_{WG(adj)}\left[1 + \frac{MS_{BG(X)}}{SS_{WG(X)}}\right] = .51\left[1 + \frac{.6}{12.4}\right] = .535$$

$$F_{comp(adj)} = \frac{MS_{comp(adj)}}{MS_{WG}'} = \frac{27.36}{.535} = 51.14 \qquad \textbf{(Equation 21.83)}$$

Substituting $MS_{WG}' = .535$ in Equation 21.83, the value $F_{comp(adj)} = 51.14$ is computed. The numerator and denominator degrees of freedom employed for evaluating the computed F value for the comparison are $df_{num} = 1$ (since it is a single degree of freedom comparison), and $df_{den} = df_{WG(adj)} = 11$. In **Table A10**, for $df_{num} = 1$ and $df_{den} = 11$, the tabled $F_{.95}$ and $F_{.99}$ values are $F_{.05} = 4.84$ and $F_{.01} = 9.65$. Since $F_{comp(adj)} = 51.14$ is greater than $F_{.05} = 4.84$ and $F_{.01} = 9.65$, the alternative hypothesis is supported at both the .05 and .01 levels. In other words, we can conclude there is a significant difference between the adjusted means of Groups 1 and 2. Note how much larger the value $F_{comp(adj)} = 51.14$ is than the value $F_{comp} = 8.89$ computed for the unadjusted Group 1 versus Group 2 comparison in Section VI.

Equations 21.30 and 21.31 (which are employed for **Tukey's HSD test**) will be used to illustrate an **unplanned comparison** on the adjusted means of Groups 1 and 2. When Equation 21.30 is employed, the value $q = 10.12$ is computed.

$$q = \frac{\overline{Y}'_1 - \overline{Y}'_2}{\sqrt{\dfrac{MS'_{WG}}{n}}} = \frac{9.672 - 6.364}{\sqrt{\dfrac{.535}{5}}} = 10.12$$

The obtained value $q = 10.12$ is evaluated with **Table A13**. In the latter table, for the prespecified probability level, we locate the q value that is in the cell which is the intersection of the column for $k = 3$ means (which represents the total number of groups upon which the omnibus F value for the analysis of covariance is based), and the row for $df_{error} = 11$ (which represents the value $df_{WG(adj)} = 11$ computed for the analysis of covariance). The tabled critical $q_{.05}$ and $q_{.01}$ values for $k = 3$ means and $df_{error} = 11$ are $q_{.05} = 3.82$ and $q_{.01} = 5.15$. Since the obtained value $q = 10.12$ is greater than the aforementioned tabled critical two-tailed values, the nondirectional alternative hypothesis H_1: $\mu'_1 \neq \mu'_2$ is supported at both the .05 and .01 levels. Note that when the unadjusted means of Groups 1 and 2 are evaluated with **Tukey's HSD test** in Section VI, the alternative hypothesis is only supported at the .05 level.

When Equation 21.31 is employed to compute the minimum required difference (CD_{HSD}), in order for two means to differ significantly from one another at a prespecified level of significance, the CD_{HSD} value computed for the adjusted means is less than the value computed in Section VI for the unadjusted means. It is demonstrated below that the value $CD_{HSD} = 1.25$ computed at the .05 level for the adjusted means is less than the value $CD_{HSD} = 2.32$ computed in Section VI for the unadjusted means. It should be emphasized, however, that employing a covariate in and of itself does not guarantee a more sensitive analysis. The increased sensitivity of the analysis in the case of Example 21.5 is the result of the covariate having a substantial linear relationship with the dependent variable.

$$CD_{HSD} = q_{.05} \sqrt{\frac{MS'_{WG}}{n}} = (3.82)\sqrt{\frac{.535}{5}} = 1.25$$

Additional procedures which can be employed for conducting comparisons for an analysis of covariance can be found in Kirk (1995) and Winer *et al.* (1991).

Evaluation of the homogeneity of regression assumption   The **homogeneity of regression assumption** of the analysis of covariance is that within each of the k groups there is a linear correlation between the dependent variable and the covariate, and that the k group regression lines have the same slope (i.e., are parallel to one another). A full discussion of the concept of linear regression and the procedure for determining the slope of a regression line can be found in Section VI of the **Pearson product-moment correlation coefficient**. The reader who is unfamiliar with the concept of regression is advised to review the latter discussion before continuing this section.

In order to evaluate the homogeneity of regression assumption, the following null and alternative hypotheses (which are evaluated with Equation 21.85) are employed. (The notation β_j in the null hypothesis represents the slope of the regression line of Y on X for the jth group.)

Null hypothesis H_0: $\beta_1 = \beta_2 = \beta_3$

(In the underlying populations represented by the three groups, the slope of the regression line of Y on X for Group 1 equals the slope of the regression line of Y on X for Group 2 equals the slope of the regression line of Y on X for Group 3.)

Alternative hypothesis H_1: Not H_0

(This indicates there is a difference between the slopes of at least two of the $k = 3$ regression lines. In order to reject the null hypothesis, the F value computed with Equation 21.85 must be equal to or greater than the tabled critical F value at the prespecified level of significance.)

If the test of the homogeneity of regression assumption results in rejection of the null hypothesis, it means that the assumption is violated. Hinkle *et al.* (1998) and Tabachnick and Fidell (1996) note that if the null hypothesis for the homogeneity of regression assumption is rejected, it suggests that an interaction is present with regard to the relationship between the covariate and the dependent variable. The presence of an interaction means that the relationship between the covariate and the dependent variable is not consistent across the different treatments/groups. Consequently, a different adjustment will be required for each of the treatment means — in other words, Equation 21.79 cannot be employed to compute the adjusted means for each of the groups. Although Keppel (1991, p. 317) cites studies which suggest that the analysis of covariance is reasonably robust with respect to violation of the homogeneity of regression assumption, other sources argue that violation of the assumption can seriously compromise the reliability of an analysis of covariance.

Equations 21.84 and 21.85 are employed to evaluate the homogeneity of regression assumption. Through use of Equation 21.84, a **within-groups regression sum of squares** (which is explained in Endnote 73) is computed. The latter value is then employed in Equation 21.85 to compute the test statistic.

$$SS_{wgreg} = \sum_{j=1}^{k}\left[SS_{Y_j}(1 - r_{j(XY)}^2) \right] \qquad \textbf{(Equation 21.84)}$$

Where: $SS_{Y_j} = \Sigma Y_j^2 - [(\Sigma Y_j)^2/n_j]$. The latter value indicates that the sum of squares for the Y variable (the dependent variable) is computed for each group.

$r_{j(XY)}^2$ is the square of the correlation between the X variable (the covariate) and the Y variable (the dependent variable) within the jth group.

The notation in Equation 21.84 indicates that for each group, the square of the correlation between the covariate and dependent variable is subtracted from 1, and the difference is multiplied by the sum of squares for the Y variable within that group. The value SS_{wgreg} is the sum of the values computed for the k groups.

The values $r_{1(XY)} = 1$, $r_{2(XY)} = .843$, and $r_{3(XY)} = .861$ are computed for each group by employing in Equation 28.1 the appropriate summary values for the five X and Y scores for that group, or through use of the equation $r = SP_{j(XY)}/\sqrt{SS_{X_j}SS_{Y_j}}$ (which is another way of expressing Equation 28.1). Employing the relevant summary values in Table 21.19, the three $r_{j(XY)}$ values are computed:

$$r_{1(XY)} = SP_{1(XY)}/\sqrt{SS_{X_1}SS_{Y_1}} = 2.8/\sqrt{(2.8)(2.8)} = 1$$

$$r_{2(XY)} = SP_{2(XY)}/\sqrt{SS_{X_2}SS_{Y_2}} = 5.6/\sqrt{(4.8)(9.2)} = .843$$

$$r_{3(XY)} = SS_{3(XY)}/\sqrt{SS_{X_3}SS_{Y_3}} = 6.2/\sqrt{(4.8)(10.8)} = .861$$

Employing Equation 21.84, the value $SS_{wgreg} = 5.46$ is computed.

$$SS_{wgreg} = \left[\left(426 - \frac{(46)^2}{5}\right)(1 - (1)^2)\right] + \left[\left(227 - \frac{(33)^2}{5}\right)(1 - (.843)^2)\right]$$

$$+ \left[\left(203 - \frac{(31)^2}{5}\right)(1 - (.861)^2)\right] = 5.46$$

Equation 21.85 is employed to compute the test statistic for the homogeneity of regression assumption. Two forms of the latter equation are presented below.

(Equation 21.85)

$$F = \frac{(SS_{WG(adj)} - SS_{wgreg})/(k - 1)}{SS_{wgreg}/k(n_j - 2)} = \frac{k(n_j - 2)(SS_{WG(adj)} - SS_{wgreg})}{(k - 1)SS_{wgreg}}$$

Employing Equation 21.85, the value $F = .124$ is computed.

$$F = \frac{(5.61 - 5.46)/(3 - 1)}{5.46/[(3)(5 - 2)]} = \frac{3(5 - 2)(5.61 - 5.46)}{(3 - 1)5.46} = .124$$

The obtained value $F = .124$ is evaluated with **Table A10**, employing as the numerator and denominator degrees of freedom $df_{num} = k - 1 = 2$ and $df_{den} = k(n_j - 2) = 9$. For $df_{num} = 2$ and $df_{den} = 9$, the tabled $F_{.95}$ and $F_{.99}$ values are $F_{.05} = 4.26$ and $F_{.01} = 8.02$. In order to reject the null hypothesis, the obtained F value must be equal to or greater than the tabled critical value at the prespecified level of significance. Since $F = .124$ is less than $F_{.05} = 4.26$, the null hypothesis cannot be rejected. Thus, we can conclude that the three regression lines are homogeneous.[73]

Estimating the magnitude of treatment effect for the single-factor between-subjects analysis of covariance Values for the **omega squared** and **eta squared statistics** and **Cohen's f index** computed in Section VI can also be computed for the **single-factor between-subjects analysis of covariance**. The appropriate adjusted values from the analysis of covariance can be substituted in any of the Equations 21.41–21.47. Through use of Equations 21.41, 21.43, and 21.45, the values of **omega squared**, **eta squared**, and **Cohen's f index** are computed for the analysis of covariance. Notice that the computed values $\tilde{\omega}^2 = .84$, $\tilde{\eta}^2 = .87$, and $f = 2.29$ are substantially larger than the analogous values computed for the magnitude of the treatment effect for the analysis of variance. The latter reflects the fact that when the data are adjusted for the effects of the covariate, a substantially larger proportion of the variability on the dependent variable is associated with the independent variable.

$$\tilde{\omega}^2_{adj} = \frac{SS_{BG(adj)} - (k - 1)MS_{WG(adj)}}{SS_{T(adj)} + MS_{WG(adj)}} = \frac{37.76 - (3 - 1)(.51)}{43.37 + .51} = .84$$

$$\tilde{\eta}^2 = \frac{SS_{BG(adj)}}{SS_{T(adj)}} = \frac{37.76}{43.37} = .87$$

$$f = \sqrt{\frac{\tilde{\omega}^2}{1 - \tilde{\omega}^2}} = \sqrt{\frac{.84}{1 - .84}} = 2.29$$

Computation of the power of the single-factor between-subjects analysis of covariance
Equation 21.38, which is employed to compute the power of the **single-factor between-subjects analysis of variance**, can also be used to compute the power of the **single-factor between-subjects analysis of covariance**. The value $\sigma^2_{WG(\text{adj})}$ is employed in Equation 21.38 in place of σ^2_{WG}. If prior data are available from previous studies, the value of $MS_{WG(\text{adj})}$ in such studies can be used as an estimate of $\sigma^2_{WG(\text{adj})}$. Cohen (1977, 1988) and Keppel (1991) discuss power computations for the analysis of covariance in more detail.

Final remarks on the analysis of covariance   In closing the discussion of analysis of covariance, it should be emphasized that the procedure should not be used indiscriminately, since a considerable investment of time and effort is required to obtain subjects' scores on a covariate, as well as the fact that the analysis requires laborious computations (although at this point in time the latter will not be an issue, since one will probably have access to the appropriate computer software). Kirk (1995, p. 710) suggests the following with regard to determining the appropriateness of employing an analysis of covariance: a) If a researcher is aware of one or more extraneous variables which cannot be controlled experimentally that may influence the dependent variable, they can be employed as covariates; and b) The measures obtained for the covariate(s) should not be influenced by the experimental treatments under study. The latter condition will be met if any of the following is true: 1) The measures of the covariate are obtained prior to introducing the experimental treatment; 2) The measures of the covariate are obtained after the introduction of the experimental treatment, but before the treatment can have an impact on the covariate; and 3) Based on prior information, the researcher can assume the covariate is unaffected by the experimental treatment. In such a case, the covariate can be measured after the administration of the experimental treatment.

Winer *et al.* (1991, pp. 787–788) discuss the advantages and disadvantages of employing an analysis of covariance versus the use of a factorial design in which the covariate is employed as a second independent variable. Factorial designs are discussed in detail in the discussion of the **between-subjects factorial analysis of variance**. Winer *et al.* (1991) note that one advantage of employing a factorial design (which is evaluated with the appropriate factorial analysis of variance) is that its assumptions are less restrictive than the assumptions for an analysis of covariance.[74] Anderson (2001), Keppel (1991, p. 326) and Kirk (1995, p. 739) discuss the pros and cons of employing stratification/blocking, and then conducting the appropriate analysis of variance (commonly referred to as a **randomized-blocks design**) as an alternative to the analysis of covariance. Stratification involves employing the covariate to form homogeneous blocks of subjects, in order to reduce the measure of error variability employed in computing the F ratio. The subject of blocking is discussed in Section I of the **single-factor within-subjects analysis of variance**. The use of blocking within the context of a **randomized-blocks design** is discussed in Section VII of the **between-subjects factorial analysis of variance**.

For a more detailed discussion of analysis of covariance the reader should consult such sources as Anderson (2001), Hinkel *et al.* (1998), Howell (2002), Keppel (1991), Keppel and Zedeck (1989), Kirk (1995), Marascuilo and Levin (1983), Marascuilo and Serlin (1988), Maxwell and Delaney (1990, 2000), Myers and Well (1995), Tabachnick and Fidell (1996), and Winer *et al.* (1991).

References

Anderson, N. N. (2001). **Empirical direction in design and analysis**. Mahwah, NJ: Lawrence Erlbaum Associates.

Bartlett, M. S. (1937). Some examples of statistical methods of research in agriculture and applied biology. **Journal of the Royal Statistical Society Supplement**, 4, 137–170.

Beyer, W. H. (1968). **Handbook of tables for probability and statistics** (2nd ed.). Cleveland, OH: CRC Press.

Box, G. E. (1953). Non-normality and tests on variance. **Biometrika**, 40, 318–335.

Brown, M. B. & Forsythe, A. B. (1974a). Robust tests for equality of variances. **Journal of the American Statistical Association**, 69, 364–367.

Brown, M. B. & Forsythe, A. B. (1974b). The small sample behavior of some statistics which test the equality of several means. **Technometrics**, 16, 129–132.

Brown, M. B. & Forsythe, A. B. (1974c). The ANOVA and multiple comparisons for data with heterogenous variances. **Biometrics**, 30, 719–724.

Cochran, W. G. (1941). The distribution of the largest of a set of estimated variances as a fraction of their total. **Annals of Eugenics**, 11, 47–52.

Cochran, W. G. & Cox, G. M. (1957). **Experimental designs** (2nd ed.). New York: John Wiley & Sons.

Cohen, B. H. (2001). **Explaining psychological statistics** (2nd ed.). New York: John Wiley & Sons.

Cohen, J. (1977). **Statistical power analysis for the behavioral sciences**. New York: Academic Press.

Cohen, J. (1988). **Statistical power analysis for the behavioral sciences** (2nd ed.). Hillsdale, NJ: Lawrence Erlbaum Associates, Publishers.

Cohen, J. & Cohen, P. (1983). **Applied multiple regression/correlation analysis for the behavioral sciences** (2nd ed.). Hillsdale, NJ: Lawrence Erlbaum Associates, Publishers.

Conover, W. J., Johnson, M. E., & Johnson, M. M. (1981). A comparative study of tests of homogeneity of variance with applications to the outer continental shelf bidding data. **Technometrics**, 23, 351–361.

Cortina, J. M. & Nouri, H. (2000). **Effect sizes for ANOVA designs**. Thousand Oaks, CA: Sage Publications.

Darlington, R. B. & Carlson, P. M. (1987). **Behavioral statistics: Logic and methods**. New York: The Free Press.

Dayton, C. M. (2003). Information criteria for pairwise comparisons. **Psychological methods**, 8, 61–71.

Dunn, O. J. (1961). Multiple comparisons among means. **Journal of the American Statistical Association**, 56, 52–64.

Dunnett, C. W. (1955). A multiple comparison procedure for comparing several treatments with a control. **Journal of the American Statistical Association**, 50, 1096–1121.

Dunnett, C. W. (1964). New tables for multiple comparisons with a control. **Biometrics**, 20, 482–491.

Fisher, R. A. (1932). **Statistical methods for research workers**. Edinburgh: Oliver & Boyd.

Fisher, R. A. (1935). **The design of experiments**. Edinburgh & London: Oliver & Boyd.

Grissom, R. J. & Kim, J. J. (2005). **Effect sizes for research: A broad practical approach**. Mahwah, NJ: Lawrence Erlbaum Associates Publishers.

Harlow, L. L. (2005). **The essence of multivariate thinking**. Mahwah, NJ: Lawrence Erlbaum Associates.

Hartley, H. O. (1940). Testing the homogeneity of a set of variances. **Biometrika**, 31, 249–255.

Hartley, H. O. (1950). The maximum *F* ratio as a shortcut test for heterogeneity of variance. **Biometrika**, 37, 308–312.

Hayter, A. J. (1986). The maximum familywise error rate of Fisher's least significant difference test. **Journal of the American Statistical Association**, 81, 1000–1004.

Hinkle, D. E., Wiersma, W., & Jurs, S. G. (1998). **Applied statistics for the behavioral sciences**. (4th ed.). Boston: Houghton Mifflin Company.

Holm, S. (1979). A simple sequentially rejective multiple test procedure. **Scandanavian Journal of Statistics**, 6, 65–70.

Howell, D. C. (2002). **Statistical methods for psychology** (5th ed.). Pacific Grove, CA: Duxbury Press.

Hsu, J. C. (1996). **Multiple comparisons: Theory and methods**. New York: Chapman & Hall.

Huitema, B. (1980). **The analysis of covariance and alternatives**. New York: John Wiley & Sons.

Hunter, J. E. & Schmidt, F. L. (1990). **Methods of meta-analysis: Correcting error and bias in research findings** (1st ed.). Newbury Park, CA: Sage Publications.

Hunter, J. E. & Schmidt, F. L. (2004). **Methods of meta-analysis: Correcting error and bias in research findings** (2nd ed.). Thousand Oaks, CA: Sage Publications.

James, G. S. (1951). The comparison of several groups of observations when the ratios of the population variances are unknown. **Biometrika**, 38, 324–329.

Kachigan, S. K. (1986). **Statistical analysis: An interdisciplinary introduction to univariate and multivariate methods**. New York: Radius Press.

Keppel, G. (1991) **Design and analysis: A researcher's handbook** (3rd ed.). Englewood Cliffs, NJ: Prentice Hall.

Keppel, G., Saufley, W. H. & Tokunaga, H. (1992). **Introduction to design and analysis: A student's handbook** (2nd ed.). New York: W. H. Freeman & Company.

Keppel, G. & Wickens, T. D. (2004). **Design and analysis: A researcher's handbook** (4th ed.). Upper Saddle River, NJ: Pearson/Prentice Hall.

Keppel, G. & Zedeck, S. (1989) **Data analysis for research designs**. New York: W. H. Freeman & Company.

Keuls, M. (1952). The use of studentized range in connection with an analysis of variance. **Euphytica**, 1, 112–122.

Kirk, R. E. (1982). **Experimental design: Procedures for the behavioral sciences** (2nd ed.). Belmont, CA: Brooks/Cole Publishing Company.

Kirk, R. E. (1995). **Experimental design: Procedures for the behavioral sciences** (3rd ed.). Pacific Grove, CA: Brooks/Cole Publishing Company.

Kline, R. B. (2004). **Beyond significance testing: Reforming data analysis methods in behavioral research**. Washington, D. C.: American Psychological Association.

Levene, H. (1960). Robust tests for the equality of variance. In I. Olkin (Ed.) **Contributions to probability and statistics** (pp. 278–292). Palo Alto, CA: Stanford University Press.

Licht, M. H. (1995). Multiple regression and correlation. In Grimm, L. G. & Yarnold, P. R. (Eds.) (pp. 100–136). **Reading and understanding multivariate statistics**. Washington, D.C.: American Psychological Association.

Lord, F. M. (1967). A paradox in the interpretation of group comparisons. **Psychological Bulletin**, 68, 304–305.

Lord, F. M. (1969). Statistical adjustments when comparing pre-existing groups. **Psychological Bulletin**, 72, 336–337.

Marascuilo, L. A. & Levin, J. R. (1983). **Multivariate statistics in the social sciences**. Monterey, CA: Brooks/Cole Publishing Company.

Marascuilo, L. A. & Serlin, R. C. (1988). **Statistical methods for the social and behavioral**

sciences. New York: W. H. Freeman & Company.

Maxwell, S. E. & Delaney, H. D. (1990) **Designing experiments and analyzing data**. Belmont, CA: Wadsworth Publishing Company.

Maxwell, S. E. & Delaney, H. D. (2000) **Designing experiments and analyzing data**. Mahwah, NJ: Lawrence Erlbaum Associates.

Maxwell, S. E. & Delaney, H. D. (2004) **Designing experiments and analyzing data: A model comparison perspective** (2nd ed.). Mahwah, NJ: Lawrence Erlbaum Associates.

McFatter, R. M. & Gollob, H. F. (1986). The power of hypothesis tests for comparisons. **Educational and Psychological Measurement**, 46, 883–886.

Myers, J. L. & Well, A. D. (1995). **Research design and statistical analysis**. Hillsdale, NJ: Lawrence Erlbaum Associates, Publishers.

Myers, J. L. & Well, A. D. (2003). **Research design and statistical analysis** (2nd ed.). Mahwah, NJ: Lawrence Erlbaum Associates, Publishers.

Newman, D. (1939). The distribution of the range in samples from a normal population, expressed in terms of an independent estimate of standard deviation. **Biometrika**, 31, 20–30.

O'Brien, R. G. (1981). A simple test for variance effects in experimental designs. **Psychological Bulletin**, 89, 570–574.

Pearson, E. S. & Hartley, H. O. (1951). Charts of the power function for analysis of variance, derived from the non-central F distribution. **Biometrika**, 38, 112–130.

Rosenthal, R., Rosnow, R. L., & Rubin, D. B. (2000). **Contrasts and effect sizes in behavioral research: A correlational approach**. Cambridge, UK: Cambridge University Press.

Rosnow, R. L., Rosenthal, R., & Rubin, D. B. (2000). Contrasts and correlations in effect-size estimation. **Psychological Science**, 11, 446–453.

Ryan, T. A. (1960). Significance tests for multiple comparisons of proportions, variances, and other statistics. **Psychological Bulletin**, 57, 318–328.

Scheffé, H. A. (1953). A method for judging all possible contrasts in the analysis of variance. **Biometrika**, 40, 87–104.

Scheffé, H. A. (1959). **The analysis of variance**. New York: John Wiley & Sons.

Šidák Z. (1967). Rectangular confidence regions for the means of multivariate normal distributions. **Journal of the American Statistical Association**, 62, 626–633.

Smithson, M. (2003). **Confidence intervals**. Thousand Oaks, CA: Sage Publications.

Stevens, J. (1996). **Applied multivariate statistics for the social sciences**. Hillsdale, NJ: Lawrence Erlbaum Associates, Publishers.

Stevens, J. (2002). **Applied multivariate statistics for the social sciences** (4th ed.). Hillsdale, NJ: Lawrence Erlbaum Associates, Publishers.

Tabachnick, B. G. & Fidell, L. S. (1989). **Using multivariate statistics** (2nd ed.). New York: Harper Collins Publishers.

Tabachnick, B. G. & Fidell, L. S. (1996). **Using multivariate statistics** (3rd ed.). New York: Harper Collins.

Tatsuoka, M. M. (1975). **The general linear model: A new trend in analysis of variance**. Champaign, IL: Institute for Personality & Ability Testing.

Thompson, B. (2002). What future quantitative social science research could look like: Confidence intervals for effect sizes. **Educational Researcher**, 31, 25–32.

Tiku, M. L. (1967). Tables of the power of the F test. **Journal of the American Statistical Association**, 62, 525–539.

Toothaker, L. E. (1991). **Multiple comparisons for researchers**. Newbury Park, CA: Sage.

Tukey, J. W. (1953). The problem of multiple comparisons. Unpublished paper, Princeton University, Princeton, NJ.

Weinfurt, K.P. (2000). Repeated measures analysis: ANOVA, MANOVA and HLM. Grimm, L. G. & Yarnold, P. R. (Eds.). **Reading and understanding more multivariate statistics** (pp. 317–361). Washington, D.C.: American Psychological Association.

Welch, B. L. (1951). On the comparison of several mean values: An alternative approach. **Biometrika**, 38, 330–336.

Wilcox, R. R., (2003). **Applying contemporary statistical techniques**. San Diego, CA: Academic Press.

Winer, B. J., Brown, D. R., & Michels, K. M. (1991). **Statistical principles in experimental design** (3rd ed.). New York: McGraw–Hill Publishing Company.

Endnotes

1. The term **single-factor** refers to the fact that the design for which the analysis of variance is employed involves a single independent variable. Since factor and independent variable mean the same thing, **multifactor designs** (more commonly called **factorial designs**) which are evaluated with the analysis of variance involve more than one independent variable. Multifactor analysis of variance procedures are discussed in the chapter on the **between-subjects factorial analysis of variance**.

2. a) It should be noted that if an experiment is confounded, one cannot conclude that a significant portion of **between-groups variability** is attributed to the independent variable. This is the case, since if one or more confounding variables systematically vary with the levels of the independent variable, a significant difference can be due to a confounding variable rather than the independent variable; b) The model for the **single-factor between-subjects analysis of variance** can be summarized by the Equation 21.86, which is a **linear** (or **additive**) **function** of the relevant parameters involved in the analysis of variance. The latter equation describes the elements which contribute to the score of any subject on the dependent variable.

    $$X_{ij} = \mu_T + \tau_i + \varepsilon_{ij} \qquad \text{(Equation 21.86)}$$

    Where: X_{ij} represents the score of the ith subject in Level j of the independent variable
    μ_T represents the grand mean of all the treatment populations
    τ_i represents the part of X_{ij} attributable to the experimental treatment (i.e., the independent variable)
    ε_{ij} represents the part of X_{ij} attributable chance/error variability

3. The homogeneity of variance assumption is also discussed in Section VI of the **t test for two independent samples** in reference to a design involving two independent samples.

4. Although it is possible to conduct a directional analysis, such an analysis will not be described with respect to the analysis of variance. A discussion of a directional analysis when $k = 2$ can be found under the **t test for two independent samples**. In addition, a discussion of one-tailed F values can be found in Section VI of the latter test under the discussion of **Hartley's F_{max} test for homogeneity of variance/F test for two population variances**. A discussion of the evaluation of a directional alternative hypothesis when $k \geq$ 3 can be found in Section VII of the **chi-square goodness-of-fit test (Test 8)**. Although the latter discussion is in reference to analysis of a k independent samples design involving categorical data, the general principles regarding the analysis of a directional alternative hypothesis when $k \geq 3$ are applicable to the analysis of variance.

5. Some sources present an alternative method for computing SS_{BG} when the number of subjects in each group is not equal. Whereas Equation 21.3 weighs each group's contribution based on the number of subjects in the group, the alternative method (which is not generally recommended) weighs each group's contribution equally, irrespective of sample size. Keppel (1991), who describes the latter method, notes that as a general rule the value it computes for SS_{BG} is close to the value obtained with Equation 21.3, except when the sample sizes of the groups differ substantially from one another.

6. Since there are an equal number of subjects in each group, the Equation for SS_{BG} can also be written as follows:

$$SS_{BG} = \frac{[(46)^2 + (33)^2 + (31)^2]}{5} - \frac{(110)^2}{15} = 26.53$$

7. SS_{WG} can also be computed with the following equation:

$$SS_{WG} = \Sigma X_T^2 - \sum_{j=1}^{k}\left[\frac{(\Sigma X_j)^2}{n_j}\right] = 856 - \left[\frac{(46)^2}{5} + \frac{(33)^2}{5} + \frac{(31)^2}{5}\right] = 22.80$$

   Since there are an equal number of subjects in each group, n can be used in place of n_j in the above equation, as well as in Equation 21.5. The numerators of the term in the brackets can be combined and written over a single denominator which equals the value $n = 5$.

8. When $n_1 = n_2 = n_3$, $MS_{WG} = (\tilde{s}_1^2 + \tilde{s}_2^2 + \tilde{s}_3^2)/k$. Thus, since $\tilde{s}_1^2 = .7$, $\tilde{s}_2^2 = 2.3$, and $\tilde{s}_3^2 = 2.7$, $MS_{WG} = (.7 + 2.3 + 2.7)/3 = 1.9$.

9. Equation 21.9 can be employed if there are an equal or unequal number of subjects in each group. The following equation can also be employed when the number of subjects in each group is equal or unequal: $df_{WG} = (n_1 - 1) + (n_2 - 1) + \cdots + (n_k - 1)$. The equation $df_{WG} = k(n - 1) = nk - k$ can be employed to compute df_{WG}, but only when the number of subjects in each group is equal.

10. When there are an equal number of subjects in each group, since $N = nk$, $df_T = nk - 1$.

11. There is a separate F distribution for each combination of df_{num} and df_{den} values. Figure 21.5 depicts the F distribution for three different sets of degrees of freedom values. Note that in each of the distributions, 5% of the distribution falls to the right of the tabled critical $F_{.05}$ value. Most tables of the F distribution do not include tabled critical values for all possible values of df_{num} and df_{den}. The protocol which is generally employed for determining a critical F value for a df value that is not listed is to either employ interpolation or to employ the df value closest to the desired df value. Some sources qualify the latter by stating that in order to insure that the Type I error rate does not exceed the prespecified value of alpha, one should employ the df value which is closest to but not above the desired df value.

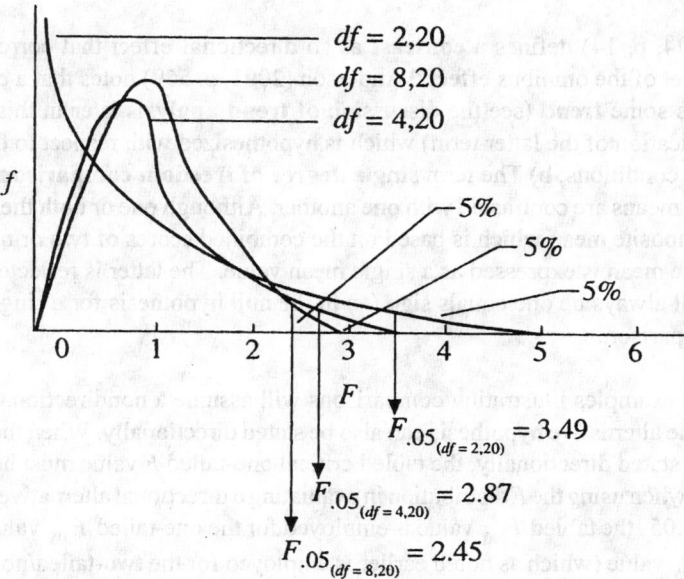

Figure 21.5 Representative *F* Distributions for Different *df* Values

12. Although the discussion of comparison procedures in this section will be limited to the analysis of variance, the underlying general philosophy can be generalized to any inferential statistical test for which comparisons are conducted.

13. 1) The terms **family** and **set** are employed synonymously throughout this discussion; 2) When all of the comparisons in a family/set of comparisons comprise the sum total of the comparisons conducted for an experiment, the error rate for the latter is commonly referred to as the **experimentwise Type I error rate**.

14. The accuracy of Equation 21.13 will be compromised if all of the comparisons are not independent of one another. Independent comparisons, which are commonly referred to as **orthogonal comparisons**, are discussed later in the section.

15. Equation 21.14 tends to overestimate the value of α_{FW}. The degree to which it overestimates α_{FW} increases as either the value of c or α_{PC} increases. For larger values of c and α_{PC}, Equation 21.14 is not very accurate. Howell (2002, p. 371) notes that the limits on α_{FW} are $\alpha_{PC} \leq \alpha_{FW} \leq (c)(\alpha_{PC})$.

16. Equation 21.16 provides a computationally quick approximation of the value computed with Equation 21.15. The value computed with Equation 21.16 tends to underestimate the value of α_{PC}. The larger the value of α_{FW} or c, the greater the degree α_{PC} will be underestimated with the latter equation. When $\alpha = .05$, however, the two equations yield values which are almost identical.

17. This example is based on a discussion of this issue in Howell (2002) and Maxwell and Delaney (1990, 2000).

18. a) Kline (2004, p. 14) defines a **contrast** as "a directional effect that corresponds to a particular facet of the omnibus effect." Anderson (2001, p. 559) notes that a **contrast** can be defined as some **trend** (see the discussion of **trend analysis** later in this section for further clarification of the latter term) which is hypothesized with respect to the means of the treatment conditions; b) The term **single degree of freedom comparison** reflects the fact that $k = 2$ means are contrasted with one another. Although one or both the $k = 2$ means may be a composite mean which is based on the combined scores of two or more groups, any composite mean is expressed as a single mean value. The latter is reflected in the fact that there will always be one equals sign (=) in the null hypothesis for a single degree of freedom comparison.

19. Although the examples illustrating comparisons will assume a nondirectional alternative hypothesis, the alternative hypothesis can also be stated directionally. When the alternative hypothesis is stated directionally, the tabled critical one-tailed F value must be employed. Specifically, when using the F distribution in evaluating a directional alternative hypothesis, when $\alpha_{PC} = .05$, the tabled $F_{.90}$ value is employed for the one-tailed $F_{.05}$ value instead of the tabled $F_{.95}$ value (which as noted earlier is employed for the two-tailed/nondirectional $F_{.05}$ value). When $\alpha_{PC} = .01$, the tabled $F_{.98}$ value is employed for the one-tailed $F_{.01}$ value instead of the tabled $F_{.99}$ value (which is employed for the two-tailed/nondirectional $F_{.01}$ value).

20. If the coefficients of Groups 1 and 2 are reversed (i.e., $c_1 = -1$ and $c_2 = +1$), the value of $\psi = \Sigma(c_j)(\bar{X}_j)$ will equal -2.6. The fact that the sign of the latter value is negative will not affect the test statistic for the comparison. This is the case, since, in computing the F value for the comparison, the value $\psi = \Sigma(c_j)(\bar{X}_j)$ is squared and, consequently, it becomes irrelevant whether $\psi = \Sigma(c_j)(\bar{X}_j)$ is a positive or negative number.

21. When the sample sizes of all k groups are not equal, the value of the **harmonic mean** (which is discussed in the **Introduction** and in Section VI of the *t* **test for two independent samples**) is employed to represent n in Equation 21.17. However, when the harmonic mean is employed, if there are large discrepancies between the sizes of the samples, the accuracy of the analysis may be compromised.

22. MS_{WG} is employed as the estimate of error variability for the comparison, since, if the homogeneity of variance assumption is not violated, the pooled within-groups variability employed in computing the omnibus F value will provide the most accurate estimate of error (i.e., within-groups) variability.

23. As is the case with simple comparisons, the alternative hypothesis for a complex planned comparison can also be evaluated nondirectionally.

24. A reciprocal of a number is the value 1 divided by that number.

25. When there are an equal number of subjects in each group it is possible, though rarely done, to assign different coefficients to two or more groups on the same side of the equals sign. In such an instance the composite mean reflects an unequal weighting of the groups. On the other hand, when there are an unequal number of subjects in any of the groups on the same

side of the equals sign, any groups which do not have the same sample size will be assigned a different coefficient. The coefficient a group is assigned will reflect the proportion it contributes to the total number of subjects on that side of the equals sign. Thus, if Groups 1 and 2 are compared with Group 3, and there are 4 subjects in Group 1 and 6 subjects in Group 2, there are a total of 10 subjects involved on that side of the equals sign. The absolute value of the coefficient assigned to Group 1 will be $\frac{4}{10} = \frac{2}{5}$, whereas the absolute value assigned to Group 2 will be $\frac{6}{10} = \frac{3}{5}$.

26. When any of the coefficients are fractions, in order to simplify calculations some sources prefer to convert the coefficients into integers. In order to do this, each coefficient must be multiplied by a **least common denominator**. A **least common denominator** is the smallest number (excluding 1) which is divisible by all of the denominators of the coefficients. With respect to the complex comparison under discussion, the least common denominator is 2, since 2 is the smallest number that can be divided by 1 and 2 (which are the denominators of the coefficients $1/1 = 1$ and ½). If all of the coefficients are multiplied by 2, the coefficients are converted into the following values: $c_1 = -1$, $c_2 = -1$, $c_3 = +2$. If the latter coefficients are employed in the calculations which follow, they will produce the same end result as that obtained through use of the coefficients employed in Table 21.4. It should be noted, however, that if the converted coefficients are employed, the value $\Sigma(c_j)(\bar{X}_j) = -1.7$ in Table 21.4 will become twice the value that is presently listed (i.e., it becomes 3.4). As a result of this, $\Sigma(c_j)(\bar{X}_j)$ will no longer represent the difference between the two sets of means contrasted in the null hypothesis. Instead, it will be a multiple of that value — specifically, the multiple of the value by which the coefficients are multiplied.

27. A special case of orthogonal contrasts/comparisons can be summarized through use of the term **trend analysis** which is discussed at the end of this section.

28. a) If $n_a \neq n_b$, $\sqrt{(MS_{WG}/n_a) + (MS_{WG}/n_b)}$ is employed as the denominator of Equation 21.22 and $\sqrt{([(\Sigma c_a^2)(MS_{WG})]/n_a) + ([(\Sigma c_b^2)(MS_{WG})]/n_b)}$ is employed as the denominator of Equation 21.23; b) If the value $\sqrt{(\tilde{s}_a^2/n_a) + (\tilde{s}_b^2/n_b)}$ is employed as the denominator of Equations 21.22/21.23, or if the degrees of freedom for Equations 21.22/21.23 are computed with Equation 11.4 ($df = n_a + n_b - 2$), a different result will be obtained since: 1) Unless the variance for all of the groups is equal, it is unlikely that the computed t value will be identical to the one computed with Equations 21.22/21.23; and 2) If $df = n_a + n_b - 2$ is used, the tabled critical t value employed will be larger than the tabled critical value employed for Equations 21.22/21.23. This is the case, since the df value associated with MS_{WG} is larger than the df value associated with $df = n_a + n_b - 2$. The larger the value of df, the lower the corresponding tabled critical t value. Consequently, **Fisher's LSD method** will provide an even more powerful test of an alternative hypothesis than the conventional t test.

29. Equations 21.24 and 21.25 are, respectively, derived from Equations 21.22 and 21.23. When two groups are compared with one another, the data may be evaluated with either the t distribution or the F distribution. The relationship between the two distributions is $F = t^2$. In view of this, the term $\sqrt{F_{(1, WG)}}$ in Equations 21.24 and 21.25 can also be written as $t_{df_{WG}}$. In other words, Equation 21.24 can be written as $CD_{LSD} = t_{df_{WG}}\sqrt{(2MS_{WG})/n}$ (and,

in the case of Equation 21.25, the same equation is employed, except for the fact that, inside the radical, (Σc_j^2) is employed in place of 2). Thus, in the computations to follow, if a nondirectional alternative hypothesis is employed with $\alpha = .05$, one can employ either $F_{.05} = 4.75$ (for $df_{num} = 1$, $df_{den} = 12$) or $t_{.05} = 2.18$ (for $df = 12$), since $(t_{.05} = 2.18)^2 = (F_{.05} = 4.75)$.

30. As is noted in reference to **linear contrasts**, the $df = 1$ value for the numerator of the F ratio for a single degree of freedom comparison is based on the fact that the comparison involves $k = 2$ groups with $df = k - 1$. In the same respect, in the case of **complex comparisons** there are two sets of means, and thus $df = k_{comp} - 1$.

31. As indicated in Endnote 29, if a t value is substituted in Equation 21.24 it will yield the same result. Thus, if $t_{.05} = 2.18$ is employed in the equation $CD_{LSD} = t_{df_{WG}}\sqrt{(2MS_{WG})/n}$, $CD_{LSD} = 2.18\sqrt{[(2)(1.9)]/5} = 1.90$.

32. The absolute value of t is employed, since a nondirectional alternative hypothesis is evaluated.

33. When $n_1 \neq n_2$, $\sqrt{(MS_{WG}/n_a) + (MS_{WG}/n_b)}$ is employed in Equation 21.26 in place of $\sqrt{(2MS_{WG})/n}$.

34. Since, in a two-tailed analysis we are interested in a proportion which corresponds to the most extreme .0167 cases in the distribution, one-half of the cases (i.e., 00835) falls in each tail of the distribution.

35. The only exception to this will be when one comparison is being made (i.e., $c = 1$), in which case both methods yield the identical CD value.

36. Some sources employ the abbreviation **WSD** for **wholly significant difference** instead of **HSD**.

37. The value of α_{FW} for **Tukey's HSD test** is compared with the value of α_{FW} for other comparison procedures within the framework of the discussion of the **Scheffé test** later in this section.

38. When the tabled critical q value for $k = 2$ treatments is employed, the **Studentized range statistic** will produce equivalent results to those obtained with **multiple t tests/Fisher's LSD test**. This will be demonstrated later in reference to the **Newman–Keuls test**, which is another comparison procedure that employs the **Studentized range statistic**.

39. As is the case with a t value, the sign of q will only be relevant if a directional alternative hypothesis is evaluated. When a directional alternative hypothesis is evaluated, in order to reject the null hypothesis, the sign of the computed q value must be in the predicted direction, and the absolute value of q must be equal to or greater than the tabled critical q value at the prespecified level of significance.

40. Although Equations 21.30 and 21.31 can be employed for both simple and complex comparisons, sources generally agree that **Tukey's HSD test** should only be employed for

simple comparisons. Its use with only simple comparisons is based on the fact that in the case of complex comparisons it provides an even less powerful test of an alternative hypothesis than does the **Scheffé test** (which is an extremely conservative procedure) discussed later in this section.

41. a) When $n_a \neq n_b$ and/or the homogeneity of variance assumption is violated, some sources recommend using the following modified form of Equation 21.31 (referred to as the **Tukey–Kramer procedure**) for computing CD_{HSD}.

$$CD_{HSD} = q_{(k, df_{WG})}\sqrt{MS_{WG}[[(1/n_a) + (1/n_b)]/2]}$$

   Other sources, however, do not endorse the use of the latter equation and provide alternative approaches for dealing with unequal sample sizes; b) In the case of unequal sample sizes, some sources (e.g., Winer *et al.*, 1991) recommend employing the **harmonic mean** of the sample sizes to compute the value of n for Equations 21.30 and 21.31 (especially if the sample sizes are approximately the same value). The harmonic mean is described in the **Introduction** as well as in Section VI of the *t* **test for two independent samples**.

42. The only comparison for which the minimum required difference computed for the **Newman–Keuls test** will equal CD_{HSD} will be the comparison contrasting the smallest and largest means in the set of k means. One exception to this is a case in which two or more treatments have the identical mean value, and that value is either the lowest or highest of the treatment means. Although the author has not seen such a configuration discussed in the literature, one can argue that in such an instance the number of steps between the lowest and highest mean should be some value less than k. One can conceptualize all cases of a tie as constituting one step, or perhaps, for those means which are tied, one might employ the average of the steps that would be involved if all those means are counted as separate steps. The larger the step value that is employed, the more conservative the test.

43. If Equation 21.30 is employed with the framework of the **Newman–Keuls test**, it uses the same values which are employed for **Tukey's HSD test**, and thus yields the identical q value. Thus, $q = (9.2 - 6.6)/\sqrt{1.9/5} = 4.22$.

44. One exception to this involving the **Bonferroni–Dunn test** will be discussed later in this section.

45. Recollect that the **Bonferroni–Dunn test** assumed that a total of 6 comparisons are conducted (3 simple comparisons and 3 complex comparisons). Thus, as noted earlier, since the number of comparisons exceeds $[k(k - 1)]/2 = [3(3 - 1)]/2 = 3$, the **Scheffé test** will provide a more powerful test of an alternative hypothesis than the **Bonferroni– Dunn test**.

46. The author is indebted to Scott Maxwell for his input on the content of the discussion to follow.

47. Dunnett (1964) developed a modified test procedure (described in Winer *et al.*, (1991)) to be employed in the event there is a lack of homogeneity of variance between the variance of the control group and the variance of the experimental groups with which it is contrasted.

48.   The philosophy to be presented here can be generalized to any inferential statistical analysis.

49.   Although the difference between α_{FW} = .05 and α_{FW} = .10 may seem trivial, it is a fact of scientific life that a result which is declared statistically significant is more likely to be submitted and/or accepted for publication than a nonsignificant result.

50.   The reader may find it useful to review the discussion of **effect size** in Section VI of the **single sample *t* test** and the ***t* test for two independent samples**. Effect size indices are also discussed in Section IX (the **Addendum**) of the **Pearson product-moment correlation coefficient** under the discussion of **meta-analysis and related topics**.

51.   The reader may want to review the discussion of confidence intervals in Section VI of both the **single sample *t* test** and the ***t* test for two independent samples**.

52.   a) The ***F* test for two population variances** (discussed in conjunction with the F_{max} test as **Test 11a** in Section VI of the ***t* test for two independent samples**) can only be employed to evaluate the homogeneity of variance hypothesis when $k = 2$; b) Smithson (2003, pp. 25–27) provides Equation 21.87 for computing a confidence interval for the ratio of two variances. He notes that the latter equation allows for computation of a one-tailed confidence interval which will have an upper limit of ∞. The lower limit of the latter confidence interval will be the result obtained with Equation 21.87. To illustrate, to compute a one-tailed 95% confidence interval, the lower limit will be the result obtained when the ratio of the two variances is divided by the appropriate tabled critical $F_{.05}$ value for $(n_L - 1)$, $(n_S - 1)$ degrees of freedom in **Table A10** (specifically, the tabled critical value for $(n_L - 1)$, $(n_S - 1)$ degrees of freedom in **Table A10** for $F_{.95}$). An upper limit for a one-tailed 95% confidence interval with a lower limit of zero could also be computed if one had access to tabled critical values for the lower tail of the F distribution. The latter .05 value would then be divided into the ratio of the two variances in order to obtain the upper limit of the confidence interval. (The reader should consult Endnote 13 of the ***t* test for two independent samples** for clarification of the latter.)

      To compute a two-tailed 95% confidence interval for the ratio of two variances, one would employ the tabled critical .025 values for $(n_L - 1)$, $(n_S - 1)$ degrees of freedom . In computing the lower limit, the relevant tabled critical $F_{.975}$ value in **Table A10** would be divided into the ratio of the two variances. In computing the upper limit, one would need access to tables of the F distribution which listed the .025 critical values in the lower tail of the distribution. Division of the ratio of the two variances by the latter critical value will yield the upper limit of the two-tailed 95% confidence interval.

      Smithson (2003, p. 27) notes that a one-sided confidence interval will always be consistent with the result of the F test contrasting the two variances — specifically, if the F test is significant at the .05 level, the lower limit of the confidence interval will always exclude the value 1 (i.e., the lower limit of the confidence interval will be greater than 1, which is consistent with the presence of heterogeneity of variance). However, a two-tailed test will not always be consistent with the result of the F test — more specifically a two-tailed test will require a confidence interval of $100(1 - 2\alpha)\%$ in order for it to always be consistent with the result of the ***F* test for two population variances**.

$$CI_{(1-\alpha)} = \frac{\tilde{s}_L^2}{\tilde{s}_S^2} \quad \text{(Equation 21.87)}$$
$$\frac{}{F_{(n_L - 1),(n_S - 1)}}$$

53. a) See Endnote 10 of the **single-sample *t* test** for further clarification of the **noncentrality parameter**; b)In Endnote 21 of the **chi-square test for *r* × *c* tables**, it was noted that φ is an alternative form of the lower case Greek letter **phi**. This alternative form of **phi** is employed here (as well as in other chapters) to represent the noncentrality parameter for the F distribution. The notation φ should not be confused with the notation ϕ, which is employed for the **phi coefficient** in various places in the book.

54. If a power analysis is conducted after the analysis of variance, the value computed for MS_{WG} can be employed to represent σ_{WG}^2 if the researcher has reason to believe it is a reliable estimate of error/with-groups variability in the underlying population. If prior to conducting a study reliable data are available from other studies, the MS_{WG} values derived in the latter studies can be used as a basis for estimating σ_{WG}^2.

55. When there is no curve which corresponds exactly to df_{WG}, one can either interpolate or employ the df_{WG} value closest to it.

56. Tiku (1967) has derived more detailed power tables that include the alpha values .025 and .005.

57. a) When $k = 2$, Equations 21.41/21.42 and Equation 11.15 will yield the same $\tilde{\omega}^2$ value; b) Grissom and Kim (2005, p. 122–123) note the following with respect to $\tilde{\omega}^2$: 1) The equation for computing $\tilde{\omega}^2$ assumes equal sample sizes and homogeneity of variance; 2) A statistically significant F value indicates the value of $\tilde{\omega}^2$ is significantly different from zero; 3) When a value computed for a treatment effect is less than zero it should be reported for purposes of computing a confidence interval (which Grissom and Kim (2005, p. 122) provide a reference for) and for use in **meta-analysis**; 4) Equation 21.88 can be employed for computing an effect size (through use of $\tilde{\omega}^2$) for a simple comparison.

$$\tilde{\omega}_{comp}^2 = \frac{SS_{comp} - MS_{WG}}{SS_T + MS_{WG}} \quad \text{(Equation 21.88)}$$

5) Grissom and Kim (2005, pp. 127–128) also note that a standardized difference effect size for a comparison (i.e., a d score) can be computed by dividing the difference between the two means involved in a comparison by $\sqrt{MS_{WG}}$ (which is the square root of the pooled variance) (if the homogeneity of variance assumption has not been violated with respect to the k groups) or the standard error of the difference for the two comparisons (in the event that the homogeneity of variance assumption has been violated with respect to the k groups). It should be noted that although a number of different measures have been developed to determine the magnitude of treatment effect for comparison procedures, researchers are not in agreement with respect to which of the available measures is most appropriate to employ. Keppel (1991) provides a comprehensive discussion of this subject.

58. The following should be noted with respect to the **eta squared** statistic: a) Earlier in this section it was noted that Cohen (1977; 1988, pp. 284–287) employs the values .0099, .0588, and .1379 as the lower limits for defining a small versus medium versus large effect size for the **omega squared** statistic. In actuality, Cohen (1977; 1988, pp. 284–287) employs the notation for **eta squared** (i.e., η^2) in reference to the aforementioned effect size values. However, the definition Cohen (1977; 1988. p. 281) provides for **eta squared** is identical to Equation 21.40 (which is the equation for **omega squared**). For the latter reason, various sources (e.g., Keppel (1991) and Kirk (1995)) employ the values .0099, .0588, and .1379 in reference to the **omega squared** statistic; b) Equation 21.44 (the equation for *Adjusted* $\tilde{\eta}^2$) is essentially equivalent to Equation 21.40 (the definitional equation for the population parameter **omega squared**); c) When $k = 2$, the **eta squared** statistic is equivalent to r_{pb}, which represents the **point-biserial correlation coefficient** (**Test 28i**). Under the discussion of r_{pb} the equivalency of $\tilde{\eta}^2$ and r_{pb} is demonstrated; d) Grissom and Kim (2005, p. 121) note that when the independent variable is quantitative, η represents the correlation between the independent and dependent variables, but unlike a **Pearson product-moment correlation** between the two variables, η reflects a curvilinear as well as a linear relationship; e) Some sources employ the notation R^2 for the **eta squared** statistic. The statistic represented by R^2 or $\tilde{\eta}^2$ is commonly referred to as the **correlation ratio**, which is the squared **multiple correlation coefficient** which is computed when **multiple regression (Test 33)** is employed to predict subjects' scores on the dependent variable based on group membership. (Grissom and Kim (2005, p. 121) note that originally **eta** (η) was used to represent the **correlation ratio** but the latter term has since been employed to refer to $\tilde{\eta}^2$.) The use of R^2 in this context reflects the fact that the analysis of variance can be conceptualized within the framework of a **multiple regression** model. The value of $R^2 = \tilde{\eta}^2$ (which as noted can be computed with Equation 21.43) can also be computed with Equation 21.89 (which is similar to but not identical with Equation 21.42 employed to compute $\tilde{\omega}^2$).

(Equation 21.89)

$$R^2 = \eta^2 = \frac{(k - 1)F}{(k - 1)F + k(n - 1)} = \frac{(3 - 1)(6.98)}{(3 - 1)(6.98) + 3(5 - 1)} = .54$$

f) Grissom and Kim (2005, p. 121) note that **epsilon squared** ($\tilde{\varepsilon}^2$) (ε is the lower case Greek letter **epsilon**) is an alternative measure of effect size which is less biased than $\tilde{\eta}^2$ but more biased than $\tilde{\omega}^2$. The value $\tilde{\varepsilon}^2 = .46$ (which is larger than $\tilde{\omega}^2 = .44$ but smaller than $\tilde{\eta}^2 = .56$) is computed below for Example 21.1 with Equation 21.90.

(Equation 21.90)

$$\varepsilon^2 = \frac{SS_{BG} - (k - 1)MS_{WG}}{SS_T} = \frac{26.53 - (3 - 1)1.90}{49.33} = .46$$

59. Further discussion of f can be found in Grissom and Kim (2005, pp. 119–120) who note that Equation 21.91 can also be employed to compute the value of f.

(Equation 21.91)

$$f = \sqrt{\frac{(k - 1)}{N}(F - 1)} = \sqrt{\frac{(3 - 1)}{15}(6.98 - 1)} = .89$$

60. Some sources employ $\sqrt{F_{(1, df_{WG})}}$ in Equation 21.48 instead of $t_{df_{WG}}$. Since $t = \sqrt{F}$, the two values produce equivalent results. The tabled critical two-tailed t value at a prespecified level of significance will be equivalent to the square root of the tabled critical F value at the same level of significance for $df_{num} = 1$ and $df_{den} = df_{WG}$.

61. In using Equation 2.7, $s_1/\sqrt{n}$ is equivalent to $s_{\bar{X}_1}$.

62. In the interest of accuracy, Keppel and Zedeck (1989, p. 98) note that although when the null hypothesis is true, the median of the sampling distribution for the value of F equals 1, the mean of the sampling distribution of F is slightly above one. Specifically, the expected value of $F = df_{WG}/df_{(WG - 2)}$. It should also be noted that although it rarely occurs, it is possible for $MS_{WG} > MS_{BG}$. In such a case, the value of F will be less than 1 and, obviously, if $F < 1$, the result cannot be significant.

63. In employing double (or even more than two) summation signs such as $\sum_{j=1}^{k} \sum_{i=1}^{n}$, the mathematical operations specified are carried out beginning with the summation sign that is farthest to the right and continued sequentially with those operations specified by summation signs to the left. To illustrate, if $k = 3$ and $n = 5$, the notation $\sum_{j=1}^{k} \sum_{i=1}^{n} X_{ij}$ indicates that the sum of the n scores in Group 1 is computed, after which the sum of the n scores in Group 2 is computed, after which the sum of the n scores in Group 3 is computed. The leftmost notation sign indicates that final result will be the sum of the three sums which have been computed for each of the $k = 3$ groups.

64. For each of the $N = 15$ subjects in Table 21.14, the following is true with respect to the contribution of a subject's score to the total variability in the data:

$$(X_{ij} - \bar{X}_T) = (\bar{X}_j - \bar{X}_T) + (X_{ij} - \bar{X}_j)$$

Total deviation score = BG deviation score + WG deviation score

65. In evaluating a directional alternative hypothesis, when $k = 2$ the tabled $F_{.90}$ and $F_{.98}$ values (for the appropriate degrees of freedom) are respectively employed as the one-tailed .05 and .01 values. Since the values for $F_{.90}$ and $F_{.98}$ are not listed in **Table A10**, the values $F_{.90} = 3.46$ and $F_{.98} = 8.41$ can be obtained by squaring the tabled critical one-tailed values $t_{.05} = 1.86$ and $t_{.01} = 2.90$, or by employing more extensive tables of the F distribution available in other sources, or through interpolation.

66. In the case of a factorial design it is also possible to have a **mixed-effects model**. In a **mixed-effects model** it is assumed that at least one of the independent variables is based on a fixed-effects model and that at least one is based on a random-effects model.

67. Stevens (2002, p. 346) notes that when a researcher elects to employ more than one covariate, Huitema (1980, p. 161) recommends limiting the number of covariates such that the ratio [*Number of covariates* + (*Number of groups* – 1)] / *Total sample size* is less than .10.

68. The equations below are alternative equations which can be employed to compute the values $SS_{T(adj)}$ and $SS_{WG(adj)}$. The slight discrepancy in values is the result of rounding off

error. The computation of the values r_T and r_{WG} is described at a later point in the discussion of the analysis of covariance.

$$SS_{T(\text{adj})} = SS_{T(Y)}(1 - r_T^2) = 49.33[1 - (.347)^2] = 43.39$$

$$SS_{WG(\text{adj})} = SS_{WG(Y)}(1 - r_{WG}^2) = 22.8[1 - (.868)^2] = 5.62$$

69.   a) Through use of Equation 28.1, the sample correlation between the covariate and the dependent variable is computed to be $r_T = .347$. In the computations below, N is employed to represent the total number of subjects (in lieu of n, which is employed in Equation 28.1). The values $\Sigma X_T^2 = 3664$ and $\Sigma Y_T^2 = 856$ are the sums of the ΣX_j^2 and ΣY_j^2 scores for the $k = 3$ groups.

$$r = \frac{\Sigma(XY)_T - \dfrac{(\Sigma X_T)(\Sigma Y_T)}{N}}{\sqrt{\left[\Sigma X_T^2 - \dfrac{(\Sigma X_T)^2}{N}\right]\left[\Sigma Y_T^2 - \dfrac{(\Sigma Y_T)^2}{N}\right]}} = \frac{1725 - \dfrac{(234)(110)}{15}}{\sqrt{\left[3664 - \dfrac{(234)^2}{15}\right]\left[856 - \dfrac{(110)^2}{15}\right]}} = .347$$

The value $r = .347$ can be computed through use of the equation $r = SP_{XY}/\sqrt{SS_X SS_Y}$ (which is discussed at the end of Section IV of the **Pearson product-moment correlation coefficient**). When the values $SP_{T(XY)} = 9$, $SS_{T(X)} = 13.6$, and $SS_{T(Y)} = 49.33$ (computed in the bottom of Table 21.19) are substituted in the latter equation, the resulting value is $r = 9/\sqrt{(13.6)(49.33)} = .347$; b) It should be noted that the data for Example 21.5 are fictitious, and, as such, were configured by the author to yield an F value for the analysis of covariance which is associated with a lower probability than the F value computed for the analysis of variance conducted on Example 21.1. Inspection of Table 21.19 reveals that the range of values of subjects' scores on the covariate is quite small (the lowest score being 14 and the highest score 17). Since the possible range of values for a score a subject can obtain on the covariate (the test of verbal intelligence) is 0–20, the use of covariate scores in the restricted range 14–17 may yield a value for r_T which, in fact, is really not indicative of the true value of the population correlations between the covariate and the dependent variable (since the latter would be based on the full range of values for the covariate). It should be emphasized that if the latter is true, the results obtained for the analysis of covariance would be viewed as questionable.

70.   The equation noted below is an alternative way of computing r_{WG} through use of the elements employed in Equation 28.1. The relevant values within each group are pooled in the numerator and denominator of the equation.

$$r = \frac{\displaystyle\sum_{j=1}^{k}\left[\Sigma X_j Y_j - \dfrac{(\Sigma X_j)(\Sigma Y_j)}{n_j}\right]}{\sqrt{\displaystyle\sum_{j=1}^{k}\left[\Sigma X_j^2 - \dfrac{(\Sigma X_j)^2}{n_j}\right]\sum_{j=1}^{k}\left[\Sigma Y_j^2 - \dfrac{(\Sigma Y_j)^2}{n_j}\right]}}$$

$$r = \frac{\left[702 - \frac{(76)(46)}{5}\right] + \left[527 - \frac{(79)(33)}{5}\right] + \left[496 - \frac{(79)(31)}{5}\right]}{\sqrt{\left[\left[1158 - \frac{(76)^2}{5}\right] + \left[1253 - \frac{(79)^2}{5}\right] + \left[1253 - \frac{(79)^2}{5}\right]\right]\left[\left[426 - \frac{(46)^2}{5}\right] + \left[227 - \frac{(33)^2}{5}\right] + \left[203 - \frac{(31)^2}{5}\right]\right]}}$$

$$= .868$$

71. If the homogeneity of regression assumption for the analysis of covariance is violated, the value computed for b_{WG} will result in biased adjusted mean values (i.e., the adjusted mean values for the groups will not be accurate estimates of their true values in the underlying populations). Evaluation of the homogeneity of regression assumption is discussed later in this section.

72. Equation 21.82 can also be written as follows:

$$MS'_{WG} = MS_{WG(\text{adj})} + MS_{WG(\text{adj})}\left[\frac{MS_{BG(X)}}{SS_{WG(X)}}\right] = .51 + .51\left[\frac{.6}{12.4}\right] = .535$$

73. An alternative way to evaluate the homogeneity of regression assumption is presented by Keppel (1991, pp. 317–320), who notes that the **adjusted within-groups sum of squares** ($SS_{WG(\text{adj})}$) can be broken down into the following two components: a) The **between-groups regression sum of squares** (SS_{bgreg}), which is a source of variability that represents the degree to which the group regression coefficients deviate from the average regression coefficient for all of the data; and b) The **within-groups regression sum of squares** (SS_{wgreg}), which is a source of variability that represents the degree to which the scores of individual subjects deviate from the regression line of the group of which a subject is a member. Since $SS_{WG(\text{adj})}$ is the sum of SS_{bgreg} and SS_{wgreg}, the value of SS_{bgreg} can be expressed as follows: $SS_{bgreg} = SS_{WG(\text{adj})} - SS_{wgreg}$.

Although we have already computed the value of SS_{wgreg}, an alternative way to compute SS_{wgreg} is with the equation 21.92 noted below.

$$SS_{wgreg} = \sum_{j=1}^{k}\left[SS_{Y_j} - \frac{(SP_{j(XY)})^2}{SS_{X_j}}\right] \qquad \textbf{(Equation 21.92)}$$

The notation in Equation 21.92 indicates that within each group, the result of dividing the square of the sum of products (i.e., $SP_{j(XY)}$ is squared) by the sum of squares for the X variable/the covariate (SS_{X_j}) is subtracted from the sum of squares for the Y variable/the dependent variable (SS_{Y_j}). The resulting values for the $k = 3$ groups are then summed. The values for SS_{Y_j}, SS_{X_j}, and $SP_{j(XY)}$ are listed in Table 21.19 for each group at the bottom of the summary information for that group. The computation of SS_{wgreg} is demonstrated below.

$$SS_{wgreg} = \left[2.8 - \frac{(2.8)^2}{2.8}\right] + \left[9.2 - \frac{(5.6)^2}{4.8}\right] + \left[10.8 - \frac{(6.2)^2}{4.8}\right] = 5.46$$

Employing the values $SS_{WG(\text{adj})} = 5.61$ and $SS_{wgreg} = 5.46$, we can compute that the value of $SS_{bgreg} = .15$.

$$SS_{bgreg} = SS_{WG(adj)} - SS_{wgreg} = 5.61 - 5.46 = .15$$

A mean square is computed for the between-groups regression element and the within-groups regression element as noted below. The degrees of freedom for the between-groups regression element is $df_{bgreg} = k - 1 = 3 - 1 = 2$, and the degrees of freedom for the within-groups regression element is $df_{wgreg} = k(n - 2) = 3(5 - 2) = 9$.

$$MS_{bgreg} = \frac{SS_{bgreg}}{df_{bgreg}} = \frac{.15}{2} = .075 \quad MS_{wgreg} = \frac{SS_{wgreg}}{df_{wgreg}} = \frac{5.46}{9} = .606$$

The F ratio for evaluating the homogeneity of regression assumption is computed with the equation below. The degrees of freedom employed in evaluating the F ratio are $df_{num} = df_{bgreg} = 2$ and $df_{den} = df_{wgreg} = 9$. For $df_{num} = 2$ and $df_{den} = 9$, the tabled $F_{.95}$ and $F_{.99}$ values are $F_{.05} = 4.26$ and $F_{.01} = 8.02$. Since $F = .124$ is less than $F_{.05} = 4.26$, the null hypothesis cannot be rejected. The result is identical to that obtained with the other method.

$$F = \frac{MS_{bgreg}}{MS_{wgreg}} = \frac{.075}{.606} = .124$$

74. In addition to the assumptions already noted for the analysis of covariance, it also has the usual assumptions for the analysis of variance (i.e., normality of the underlying population distributions and homogeneity of variance).

Test 22

The Kruskal–Wallis One-Way Analysis of Variance by Ranks

(Nonparametric Test Employed with Ordinal Data)

I. Hypothesis Evaluated with Test and Relevant Background Information

Hypothesis evaluated with test In a set of k independent samples (where $k \geq 2$), do at least two of the samples represent populations with different median values?

Relevant background information on test The **Kruskal–Wallis one-way analysis of variance by ranks** (Kruskal (1952) and Kruskal and Wallis (1952)) is employed with ordinal (rank-order) data in a hypothesis testing situation involving a design with two or more independent samples. The test is an extension of the **Mann–Whitney U test (Test 12)** to a design involving more than two independent samples and, when $k = 2$, the **Kruskal–Wallis one-way analysis of variance by ranks** will yield a result that is equivalent to that obtained with the **Mann–Whitney U test**. If the result of the **Kruskal–Wallis one-way analysis of variance by ranks** is significant, it indicates there is a significant difference between at least two of the sample medians in the set of k medians. As a result of the latter, the researcher can conclude there is a high likelihood at least two of the samples represent populations with different median values.

In employing the **Kruskal–Wallis one-way analysis of variance by ranks** one of the following is true with regard to the rank-order data that are evaluated: a) The data are in a rank-order format, since it is the only format in which scores are available; or b) The data have been transformed into a rank-order format from an interval/ratio format, since the researcher has reason to believe that one or more of the assumptions of the **single-factor between-subjects analysis of variance (Test 21)** (which is the parametric analog of the **Kruskal–Wallis test**) are saliently violated. It should be noted that when a researcher elects to transform a set of interval/ratio data into ranks, information is sacrificed. This latter fact accounts for why there is reluctance among some researchers to employ nonparametric tests such as the **Kruskal– Wallis one-way analysis of variance by ranks**, even if there is reason to believe that one or more of the assumptions of the **single-factor between-subjects analysis of variance** have been violated.

Various sources (e.g., Conover (1980, 1999), Daniel (1990), and Marascuilo and McSweeney (1977)) note that the **Kruskal–Wallis one-way analysis of variance by ranks** is based on the following assumptions: a) Each sample has been randomly selected from the population it represents; b) The k samples are independent of one another; c) The dependent variable (which is subsequently ranked) is a continuous random variable. In truth, this assumption, which is common to many nonparametric tests, is often not adhered to, in that such tests are often employed with a dependent variable which represents a discrete random variable; and d) The underlying distributions from which the samples are derived are identical in shape. The shapes of the underlying population distributions, however, do not have to be normal. Maxwell and

Delaney (1990; 2004, pp. 137–143) point out that the assumption of identically shaped distributions implies equal dispersion of data within each distribution. Because of this, they note that, like the **single-factor between-subjects analysis of variance**, the **Kruskal–Wallis one-way analysis of variance by ranks** assumes homogeneity of variance with respect to the underlying population distributions. Because the latter assumption is not generally acknowledged for the **Kruskal–Wallis one-way analysis of variance by ranks**, it is not uncommon for sources to state that violation of the homogeneity of variance assumption justifies use of the **Kruskal–Wallis one-way analysis of variance by ranks** in lieu of the **single-factor between-subjects analysis of variance**. It should be pointed out, however, that there is some empirical research which suggests that the sampling distribution for the **Kruskal–Wallis test** statistic is not as affected by violation of the homogeneity of variance assumption as is the F distribution (which is the sampling distribution for the **single-factor between-subjects analysis of variance**). One reason cited by various sources for employing the **Kruskal–Wallis one-way analysis of variance by ranks** is that by virtue of ranking interval/ratio data a researcher can reduce or eliminate the impact of **outliers**. As noted in Section VII of the *t* **test for two independent samples**, since **outliers** can dramatically influence variability, they can be responsible for heterogeneity of variance between two or more samples. In addition, **outliers** can have a dramatic impact on the value of a sample mean. Maxwell and Delaney (2004, pp. 141–142), however, note that under certain conditions (e.g., unequal sample sizes) the **Kruskal–Wallis test** is not robust (i.e., its reliability is compromised) with respect to violation of the homogeneity of variance assumption, and that Delaney and Vargha (2002) have proposed alternative versions of the **Kruskal–Wallis test** which should be considered when heterogeneity of variance is suspected in studies employing unequal sample sizes.

Zimmerman and Zumbo (1993) note that the result obtained with the **Kruskal–Wallis one-way analysis of variance by ranks** is equivalent (in terms of the derived probability value) to that which will be obtained if the rank-orders employed for the **Kruskal–Wallis test** are evaluated with a **single-factor between-subjects analysis of variance**.

II. Example

Example 22.1 is identical to Example 21.1 (which is evaluated with the **single-factor between-subjects analysis of variance**). In evaluating Example 22.1 it will be assumed that the ratio data (i.e., the number of nonsense syllables correctly recalled) are rank-ordered, since one or more of the assumptions of the **single-factor between-subjects analysis of variance** have been saliently violated.

Example 22.1   *A psychologist conducts a study to determine whether or not noise can inhibit learning. Each of 15 subjects is randomly assigned to one of three groups. Each subject is given 20 minutes to memorize a list of 10 nonsense syllables which she is told she will be tested on the following day. The five subjects assigned to* **Group 1,** *the* **no noise** *condition, study the list of nonsense syllables while they are in a quiet room. The five subjects assigned to* **Group 2,** *the* **moderate noise** *condition, study the list of nonsense syllables while listening to classical music. The five subjects assigned to* **Group 3,** *the* **extreme noise** *condition, study the list of nonsense syllables while listening to rock music. The number of nonsense syllables correctly recalled by the 15 subjects follows:* **Group 1**: 8, 10, 9, 10, 9; **Group 2**: 7, 8, 5, 8, 5; **Group 3**: 4, 8, 7, 5, 7. *Do the data indicate that noise influenced subjects' performance?*

III. Null versus Alternative Hypotheses

Null hypothesis H_0: $\theta_1 = \theta_2 = \theta_3$

(The median of the population Group 1 represents equals the median of the population Group 2 represents equals the median of the population Group 3 represents. With respect to the sample data, when there are an equal number of subjects in each group, the sums of the ranks will be equal for all k groups — i.e., $\Sigma R_1 = \Sigma R_2 = \Sigma R_3$. A more general way of stating this (which also encompasses designs involving unequal sample sizes) is that the means of the ranks of the k groups will be equal (i.e., $\bar{R}_1 = \bar{R}_2 = \bar{R}_3$).)

Alternative hypothesis H_1: Not H_0

(This indicates that there is a difference between at least two of the $k = 3$ population medians. It is important to note that the alternative hypothesis should not be written as follows: H_1: $\theta_1 \neq \theta_2 \neq \theta_3$. The reason why the latter notation for the alternative hypothesis is incorrect is because it implies that all three population medians must differ from one another in order to reject the null hypothesis. With respect to the sample data, if there are an equal number of subjects in each group, when the alternative hypothesis is true the sums of the ranks of at least two of the k groups will not be equal. A more general way of stating this (which also encompasses designs involving unequal sample sizes) is that the means of the ranks of at least two of the k groups will not be equal. In this book it will be assumed (unless stated otherwise) that the alternative hypothesis for the **Kruskal–Wallis one-way analysis of variance by ranks** is stated **nondirectionally**.)[1]

IV. Test Computations

The data for Example 22.1 are summarized in Table 22.1. The total number of subjects employed in the experiment is $N = 15$. There are $n_j = n_1 = n_2 = n_3 = 5$ subjects in each group. The original interval/ratio scores of the subjects are recorded in the columns labeled X_1, X_2, and X_3. The adjacent columns R_1, R_2, and R_3 contain the rank-order assigned to each of the scores. The rankings for Example 22.1 are summarized in Table 22.2.

The ranking protocol employed for the **Kruskal–Wallis one-way analysis of variance by ranks** is the same as that employed for the **Mann–Whitney U test**. In Table 22.2 the two-digit subject identification number indicates the order in which a subject's score appears in Table 22.1 followed by his/her group. Thus, Subject i, j is the ith subject in Group j.

Table 22.1   Data for Example 22.1

Group 1		Group 2		Group 3	
X_1	R_1	X_2	R_2	X_3	R_3
8	9.5	7	6	4	1
10	14.5	8	9.5	8	9.5
9	12.5	5	3	7	6
10	14.5	8	9.5	5	3
9	12.5	5	3	7	6
$\Sigma R_1 = 63.5$		$\Sigma R_2 = 31$		$\Sigma R_3 = 25.5$	
$\bar{R}_1 = \dfrac{\Sigma R_1}{n_1} = \dfrac{63.5}{5} = 12.7$		$\bar{R}_2 = \dfrac{\Sigma R_2}{n_2} = \dfrac{31}{5} = 6.2$		$\bar{R}_3 = \dfrac{\Sigma R_3}{n_3} = \dfrac{25.5}{5} = 5.1$	

Table 22.2　Rankings for the Kruskal–Wallis Test for Example 22.1

Subject identification number	1,3	3,2	5,2	4,3	1,2	3,3	5,3	1,1	2,2	4,2	2,3	3,1	5,1	2,1	4,1
Number correct	4	5	5	5	7	7	7	8	8	8	8	9	9	10	10
Rank prior to tie adjustment	1	2	3	4	5	6	7	8	9	10	11	12	13	14	15
Tie-adjusted rank	1	3	3	3	6	6	6	9.5	9.5	9.5	9.5	12.5	12.5	14.5	14.5

A brief summary of the ranking protocol employed in Table 22.2 follows:

a) All $N = 15$ scores are arranged in order of magnitude (irrespective of group membership), beginning on the left with the lowest score and moving to the right as scores increase. This is done in the second row of Table 22.2.

b) In the third row of Table 22.2, all $N = 15$ scores are assigned a rank. Moving from left to right, a rank of 1 is assigned to the score that is furthest to the left (which is the lowest score), a rank of 2 is assigned to the score that is second from the left (which, if there are no ties, will be the second lowest score), and so on until the score at the extreme right (which will be the highest score) is assigned a rank equal to N (if there are no ties for the highest score).

c) In instances where two or more subjects have the same score, the average of the ranks involved is assigned to all scores tied for a given rank. The tie-adjusted ranks are listed in the fourth row of Table 22.2. For a comprehensive discussion of how to handle tied ranks, the reader should review the description of the ranking protocol in Section IV of the **Mann–Whitney U test**.

It should be noted that, as is the case with the **Mann–Whitney U test**, it is permissible to reverse the ranking protocol described above. Specifically, one can assign a rank of 1 to the highest score, a rank of 2 to the second highest score, and so on, until reaching the lowest score which is assigned a rank equal to the value of N. This reverse ranking protocol will yield the same value for the **Kruskal–Wallis test** statistic as the protocol employed in Table 22.2.

Upon rank-ordering the scores of the $N = 15$ subjects, the sum of the ranks is computed for each group. In Table 22.1 the sum of the ranks of the jth group is represented by the notation ΣR_j. Thus, $\Sigma R_1 = 63.5$, $\Sigma R_2 = 31$, $\Sigma R_3 = 25.5$.

The chi-square distribution is used to approximate the **Kruskal–Wallis test** statistic. Equation 22.1 is employed to compute the chi-square approximation of the **Kruskal–Wallis test** statistic (which is represented in most sources by the notation H).

$$H = \frac{12}{N(N + 1)} \sum_{j=1}^{k} \left[\frac{(\Sigma R_j)^2}{n_j} \right] - 3(N + 1) \qquad \textbf{(Equation 22.1)}$$

Note that in Equation 22.1, the term $\sum_{j=1}^{k}[(\Sigma R_j)^2/n_j]$ indicates that for each of the k groups, the sum of the ranks is squared and then divided by the number of subjects in the group. Upon doing this for all k groups, the resulting values are summed. Substituting the appropriate values from Example 22.1 in Equation 22.1, the value $H = 8.44$ is computed.

$$H = \frac{12}{(15)(15 + 1)} \left[\frac{(63.5)^2}{5} + \frac{(31)^2}{5} + \frac{(25.5)^2}{5} \right] - (3)(15 + 1) = 8.44$$

V. Interpretation of the Test Results

In order to reject the null hypothesis, the computed value $H = \chi^2$ must be equal to or greater than the tabled critical chi-square value at the prespecified level of significance. The computed chi-square value is evaluated with **Table A4 (Table of the Chi-Square Distribution)** in the **Appendix**. For the appropriate degrees of freedom, the tabled $\chi^2_{.95}$ value (which is the chi-square value at the 95th percentile) and the tabled $\chi^2_{.99}$ value (which is the chi-square value at the 99th percentile) are employed as the .05 and .01 critical values for evaluating a nondirectional alternative hypothesis. The number of degrees of freedom employed in the analysis are computed with Equation 22.2. Thus, $df = 3 - 1 = 2$.

$$df = k - 1 \qquad \text{(Equation 22.2)}$$

For $df = 2$, the tabled critical .05 and .01 chi-square values are $\chi^2_{.05} = 5.99$ and $\chi^2_{.01} = 9.21$. Since the computed value $H = 8.44$ is greater than $\chi^2_{.05} = 5.99$, the alternative hypothesis is supported at the .05 level. Since, however, $H = 8.44$ is less than $\chi^2_{.01} = 9.21$, the alternative hypothesis is not supported at the .01 level.[2] A summary of the analysis of Example 22.1 with the **Kruskal–Wallis one-way analysis of variance by ranks** follows: It can be concluded that there is a significant difference between at least two of the three groups exposed to different levels of noise. This result can be summarized as follows: $H = \chi^2(2) = 8.44$, $p < .05$.

It should be noted that when the data for Example 22.1 are evaluated with a **single-factor between-subjects analysis of variance**, the null hypothesis can be rejected at both the .05 and .01 levels (although it barely achieves significance at the latter level and, in the case of the **Kruskal–Wallis test**, the result just falls short of significance at the .01 level). The slight discrepancy between the results of the two tests reflects the fact that, as a general rule (assuming that none of the assumptions of the analysis of variance are saliently violated), the **Kruskal–Wallis one-way analysis of variance by ranks** provides a less powerful test of an alternative hypothesis than the **single-factor between-subjects analysis of variance**.

VI. Additional Analytical Procedures for the Kruskal–Wallis One-Way Analysis of Variance by Ranks and/or Related Tests

1. Tie correction for the Kruskal–Wallis one-way analysis of variance by ranks Some sources recommend that if there is an excessive number of ties in the overall distribution of N scores, the value of the **Kruskal–Wallis test** statistic be adjusted. The tie correction results in a slight increase in the value of H (thus providing a slightly more powerful test of the alternative hypothesis). Equation 22.3 is employed to compute the value C, which represents the tie correction factor for the **Kruskal–Wallis one-way analysis of variance by ranks**.

$$C = 1 - \frac{\sum\limits_{i=1}^{s} (t_i^3 - t_i)}{N^3 - N} \qquad \text{(Equation 22.3)}$$

Where: s = The number of sets of ties

t_i = The number of tied scores in the ith set of ties

The notation $\sum_{i=1}^{s} (t_i^3 - t_i)$ indicates the following: a) For each set of ties, the number of ties in the set is subtracted from the cube of the number of ties in that set; and b) The sum of all the values computed in part a) is obtained. The correction for ties will now be computed for Example 22.1. In the latter example there are $s = 5$ sets of ties (i.e., three scores of 5, three scores of 7, four scores of 8, two scores of 9, and two scores of 10). Thus:

$$\sum_{i=1}^{s} (t_i^3 - t_i) = [(3)^3 - 3] + [(3)^3 - 3] + [(4)^3 - 4] + [(2)^3 - 2] + [(2)^3 - 2] = 120$$

Employing Equation 22.3, the value $C = .964$ is computed.

$$C = 1 - \frac{120}{(15)^3 - 15} = .964$$

H_C, which represents the tie-corrected value of the **Kruskal–Wallis test** statistic, is computed with Equation 22.4.

$$H_C = \frac{H}{C} \qquad\qquad \textbf{(Equation 22.4)}$$

Employing Equation 22.4, the tie-corrected value $H_C = 8.76$ is computed.

$$H_C = \frac{8.44}{.964} = 8.76$$

As is the case with $H = 8.44$ computed with Equation 22.1, the value $H_C = 8.76$ computed with Equation 22.4 is significant at the .05 level (since it is greater than $\chi^2_{.05} = 5.99$), but is not significant at the .01 level (since it is less than $\chi^2_{.01} = 9.21$). Although Equation 22.4 results in a slightly less conservative test than Equation 22.1, in this instance the two equations lead to identical conclusions with respect to the null hypothesis.

2. Pairwise comparisons following computation of the test statistic for the Kruskal–Wallis one-way analysis of variance by ranks Prior to reading this section the reader should review the discussion of comparisons in Section VI of the **single-factor between-subjects analysis of variance**. As is the case with the omnibus F value computed for the **single-factor between-subjects analysis of variance**, the H value computed with Equation 22.1 is based on an evaluation of all k groups. When the value of H is significant, it does not indicate whether just two or, in fact, more than two groups differ significantly from one another. In order to answer the latter question, it is necessary to conduct comparisons contrasting specific groups with one another. This section will describe methodologies which can be employed for conducting **unplanned simple/pairwise comparisons** following the computation of a significant H value.[3]

In conducting a simple comparison, the null hypothesis and nondirectional alternative hypothesis are as follows: H_0: $\theta_a = \theta_b$ versus H_1: $\theta_a \neq \theta_b$. In the aforementioned hypotheses, θ_a and θ_b represent the medians of the populations represented by the two groups involved in the comparison. The alternative hypothesis can also be stated directionally as follows: H_1: $\theta_a > \theta_b$ or H_1: $\theta_a < \theta_b$.

Various sources (e.g., Daniel (1990) and Siegel and Castellan (1988)) describe a comparison procedure recommended for use with large sample sizes for the **Kruskal–Wallis one-way analysis of variance by ranks** (described by Dunn (1964)), which is essentially the application of the **Bonferroni–Dunn method** described in Section VI of the **single-factor between-subjects analysis of variance** to the **Kruskal–Wallis test** model. Through use of

Equation 22.5, the procedure allows a researcher to identify the minimum required difference between the means of the ranks of any two groups (designated as CD_{KW}) in order for them to differ from one another at the prespecified level of significance.[4]

$$CD_{KW} = z_{adj} \sqrt{\frac{N(N+1)}{12} \left(\frac{1}{n_a} + \frac{1}{n_b} \right)}$$ **(Equation 22.5)**

Where: n_a and n_b represent the number of subjects in each of the groups involved in the simple comparison

The value of z_{adj} is obtained from **Table A1 (Table of the Normal Distribution)** in the **Appendix**. In the case of a nondirectional alternative hypothesis, z_{adj} is the z value above which a proportion of cases corresponding to the value $\alpha_{FW}/2c$ falls (where c is the total number of comparisons that are conducted). In the case of a directional alternative hypothesis, z_{adj} is the z value above which a proportion of cases corresponding to the value α_{FW}/c falls. When all possible pairwise comparisons are made $c = [k(k-1)]/2$, and thus, $2c = k(k-1)$. In Example 22.1 the number of pairwise/simple comparisons that can be conducted is $c = [3(3-1)]/2 = 3$ — specifically, Group 1 versus Group 2, Group 1 versus Group 3, and Group 2 versus Group 3.

The value of z_{adj} will be a function of both the maximum **familywise Type I error rate** (α_{FW}) the researcher is willing to tolerate and the total number of comparisons that are conducted. When a limited number of comparisons are planned prior to collecting the data, most sources take the position that a researcher is not obliged to control the value of α_{FW}. In such a case, the **per comparison Type I error rate** (α_{PC}) will be equal to the prespecified value of alpha. When α_{FW} is not adjusted, the value of z_{adj} employed in Equation 22.5 will be the tabled critical z value which corresponds to the prespecified level of significance. Thus, if a nondirectional alternative hypothesis is employed and $\alpha = \alpha_{PC} = .05$, the tabled critical two-tailed .05 value $z_{.05} = 1.96$ is used to represent z_{adj} in Equation 22.5. If $\alpha = \alpha_{PC} = .01$, the tabled critical two-tailed .01 value $z_{.01} = 2.58$ is used in Equation 22.5. In the same respect, if a directional alternative hypothesis is employed, the tabled critical .05 and .01 one-tailed values $z_{.05} = 1.65$ and $z_{.01} = 2.33$ are used for z_{adj} in Equation 22.5.

When comparisons are not planned beforehand, it is generally acknowledged that the value of α_{FW} must be controlled so as not to become excessive. The general approach for controlling the latter value is to establish a **per comparison Type I error rate** which insures that α_{FW} will not exceed some maximum value stipulated by the researcher. One method for doing this (described under the **single-factor between-subjects analysis of variance** as the **Bonferroni–Dunn method**) establishes the **per comparison Type I error rate** by dividing the maximum value one will tolerate for the **familywise Type I error rate** by the total number of comparisons conducted. Thus, in Example 22.1, if one intends to conduct all three pairwise comparisons and wants to insure that α_{FW} does not exceed .05, $\alpha_{PC} = \alpha_{FW}/c = .05/3 = .0167$. The latter proportion is used to determine the value of z_{adj}. As noted earlier, if a directional alternative hypothesis is employed for a comparison, the value of z_{adj} employed in Equation 22.5 is the z value above which a proportion equal to $\alpha_{PC} = \alpha_{FW}/c$ of the cases falls. In **Table A1**, the z value that corresponds to the proportion .0167 is $z = 2.13$. By employing $z_{adj} = 2.13$ in Equation 22.5, one can be assured that within the "family" of three pairwise comparisons, α_{FW} will not exceed .05 (assuming all of the comparisons are directional). If a nondirectional alternative hypothesis is employed for all of the comparisons, the value of z_{adj} will be the z value above which a proportion equal to $\alpha_{FW}/2c = \alpha_{PC}/2$ of the cases falls. Since $\alpha_{PC}/2 = .0167/2 = .0083, z = 2.39$. By employing $z_{adj} = 2.39$ in Equation 22.5, one can be assured that α_{FW} will not exceed .05.[5]

In order to employ the CD_{KW} value computed with Equation 22.5, it is necessary to determine the mean rank for each of the k groups, and then compute the absolute value of the difference between the mean ranks of each pair of groups that are compared.[6] In Table 22.1 the following values for the mean ranks of the groups are computed: $\bar{R}_1 = 12.7$, $\bar{R}_2 = 6.2$, $\bar{R}_3 = 5.1$. Employing the latter values, Table 22.3 summarizes the difference scores between pairs of mean ranks.

Table 22.3   Difference Scores between Pairs of Mean Ranks for Example 22.1

$$|\bar{R}_1 - \bar{R}_2| = |12.7 - 6.2| = 6.5$$
$$|\bar{R}_1 - \bar{R}_3| = |12.7 - 5.1| = 7.6$$
$$|\bar{R}_2 - \bar{R}_3| = |6.2 - 5.1| = 1.1$$

If any of the differences between mean ranks is equal to or greater than the CD_{KW} value computed with Equation 22.5, a comparison is declared significant. Equation 22.5 will now be employed to evaluate the nondirectional alternative hypothesis $H_1: \theta_a \neq \theta_b$ for all three pairwise comparisons. Since it will be assumed that the comparisons are unplanned and the researcher does not want the value of α_{FW} to exceed .05, the value $z_{adj} = 2.39$ will be used in computing CD_{KW}.

$$CD_{KW} = (2.39)\sqrt{\frac{(15)(15 + 1)}{12}\left(\frac{1}{5} + \frac{1}{5}\right)} = (2.39)(2.83) = 6.76$$

The obtained value $CD_{KW} = 6.76$ indicates that any difference between the mean ranks of two groups which is equal to or greater than 6.76 is significant. With respect to the three pairwise comparisons, the only difference between mean ranks that is greater than $CD_{KW} = 6.76$ is $|\bar{R}_1 - \bar{R}_3| = 7.6$. Thus, we can conclude there is a significant difference between the performance of Group 1 and Group 3. Note that although $|\bar{R}_1 - \bar{R}_2| = 6.5$ is close to $CD_{KW} = 6.76$, it is not statistically significant unless the researcher is willing to tolerate a **familywise error rate** slightly above .05.[7]

An alternative strategy which can be employed for conducting pairwise comparisons for the **Kruskal–Wallis test** model is to use the **Mann–Whitney U test** for each comparison. Use of the latter test requires that the data for each pair of groups to be compared be rank-ordered, and that a separate U value be computed for that comparison. The exact distribution of the **Mann–Whitney test** statistic can only be used when the value of α_{PC} is equal to one of the probabilities documented in **Table A11 (Table of Critical Values for the Mann–Whitney U Statistic)** in the **Appendix**. When α_{PC} is a value other than those listed in **Table A11**, the normal approximation of the **Mann–Whitney U test** statistic must be employed.

When the **Mann–Whitney U test** is employed for the three pairwise comparisons, the following U values are computed: a) Group 1 versus Group 2: $U = 1$; b) Group 1 versus Group 3: $U = .5$; and c) Group 2 versus Group 3: $U = 10$. When the aforementioned U values are substituted in Equations 12.4 and 12.5 (the uncorrected and continuity-corrected normal approximations for the **Mann–Whitney U test**), the following absolute z values are computed: a) Group 1 versus Group 2: $z = 2.40$ and $z = 2.30$; b) Group 1 versus Group 3: $z = 2.51$ and $z = 2.40$;

and c) Group 2 versus Group 3: $z = .52$ and $z = .42$. If we want to evaluate a nondirectional alternative hypothesis and insure that α_{FW} does not exceed .05, the value of α_{PC} is set equal to .0167. **Table A11** cannot be employed, since it does not list two-tailed critical U values for $\alpha_{.0167}$. In order to evaluate the result of the normal approximation, we identify the tabled critical two-tailed .0167 z value in **Table A1**. In employing Equation 22.5 earlier in this section, we determined that the latter value is $z_{.0167} = 2.39$. Since the uncorrected values $z = 2.40$ (for the comparison Group 1 versus Group 2) and $z = 2.51$ (for the comparison Group 1 versus Group 3) computed with Equation 12.4 are greater than $z_{.0167} = 2.39$, the latter two comparisons are significant if we wish to insure that α_{FW} does not exceed .05. If the correction for continuity is employed, only the value $z = 2.40$ (for the comparison Group 1 versus Group 3), computed with Equation 12.5 is significant, since it exceeds $z_{.0167} = 2.39$. The Group 1 versus Group 2 comparison falls short of significance, since $z = 2.30$ is less than $z_{.0167} = 2.39$. Recollect that when Equation 22.5 is employed to conduct the same set of comparisons, the Group 1 versus Group 3 comparison is significant, whereas the Group 1 versus Group 2 comparison falls just short of significance. Thus, the result obtained with Equation 22.5 is identical to that obtained when the continuity-corrected normal approximation of the **Mann–Whitney U test** is employed.

In the event the researcher elects not to control the value of α_{FW} and employs $\alpha_{PC} = .05$ in evaluating the three pairwise comparisons (once again assuming a nondirectional analysis), both the Group 1 versus Group 2 and Group 1 versus Group 3 comparisons are significant at the .05 level, regardless of which comparison procedure is employed. Specifically, both the uncorrected and corrected normal approximations are significant, since $z = 2.40$ and $z = 2.30$ (computed for the comparison Group 1 versus Group 2) and $z = 2.51$ and $z = 2.40$ (computed for the comparison Group 1 versus Group 3) are greater than the tabled critical two-tailed value $z_{.05} = 1.96$. Employing **Table A11**, we also determine that both the Group 1 versus Group 2 and Group 1 versus Group 3 comparisons are significant at the .05 level, since the computed values $U = 1$ and $U = .5$ are less than the tabled critical two-tailed .05 value $U_{.05} = 2$ (based on $n_1 = 5$ and $n_2 = 5$). If Equation 22.5 is employed for the same set of comparisons, $CD_{KW} = (1.96)(2.83) = 5.55$.[8] Thus, if the latter equation is employed, the Group 1 versus Group 2 and Group 1 versus Group 3 comparisons are significant, since in both instances the difference between the mean ranks is greater than $CD_{KW} = 5.55$.

The above discussion of comparisons illustrates that, generally speaking, the results obtained with Equation 22.5 and the **Mann–Whitney U test** will be reasonably consistent with one another. As noted in Endnote 7 (as well as in the discussion of comparisons in Section VI of the **single-factor between-subjects analysis of variance**), in instances where two or more comparison procedures yield inconsistent results, the most effective way to clarify the status of the null hypothesis is to replicate a study one or more times. In the final analysis, the decision regarding which of the available comparison procedures to employ is usually not the most important issue facing the researcher conducting comparisons. The main issue is what maximum value one is willing to tolerate for α_{FW}. Additional sources on comparison procedures for the **Kruskal–Wallis test** model are Marascuilo and McSweeney (1977) (who describe a methodology for conducting complex comparisons), Wike (1978) (who provides a comparative analysis of a number of different procedures), and Hollander and Wolfe (1999).

The same logic employed for computing a confidence interval for a comparison described in Section VI of the **single-factor between-subjects analysis of variance** can be employed to compute a confidence interval for the **Kruskal–Wallis test** model. Specifically: Add and subtract the computed value of CD_{KW} to and from the obtained difference between the two mean ranks involved in a comparison. Thus, $CI_{.95}$ (based on $\alpha_{FW} = .05$) for the comparison Group 1 versus Group 3 is computed as follows: $CI_{.95}$ = 7.6 ± 6.76. In other words, the researcher can be 95% confident (or the probability is .95) the interval .84 to 14.36 contains the true difference between the means of the ranks of the populations represented by Groups 1 and 3. If the aforementioned interval contains the true difference between the means of the ranks of the two populations, the mean of the ranks of the population Group 1 represents is between .84 to 14.36 units larger than the mean of the ranks of the population Group 3 represents. Marascuilo and McSweeney (1977) provide a more detailed discussion of the computation of a confidence interval for the **Kruskal–Wallis test** model.

VII. Additional Discussion of the Kruskal–Wallis One-Way Analysis of Variance by Ranks

1. Exact tables of the Kruskal–Wallis distribution Although an exact probability value can be computed for obtaining a configuration of ranks which is equivalent to or more extreme than the configuration observed in the data evaluated with the **Kruskal–Wallis one-way analysis of variance by ranks**, the chi-square distribution is generally employed to estimate the latter probability. As the values of k and N increase, the chi-square distribution provides a more accurate estimate of the exact Kruskal–Wallis distribution. Although most sources employ the chi-square approximation regardless of the values of k and N, some sources recommend that exact tables be employed under certain conditions. Beyer (1968), Daniel (1990), and Siegel and Castellan (1988) provide exact Kruskal–Wallis probabilities for whenever $k = 3$ and the number of subjects in any of the samples is five or less. Use of the chi-square distribution for small sample sizes will generally result in a slight decrease in the power of the test (i.e., there is a higher likelihood of retaining a false null hypothesis). Thus, for small sample sizes the tabled critical chi-square value should, in actuality, be a little lower than the value listed in **Table A4**.

In point of fact, the exact tabled critical H values for $k = 3$ and $n_j = 5$ are $H_{.05} = 5.78$ and $H_{.01} = 7.98$ (versus the values $\chi^2_{.05} = 5.99$ and $\chi^2_{.01} = 9.21$ used earlier). If the critical values from the exact tables are employed, the value $H = 8.44$ computed for Example 22.1 is significant at both the .05 and .01 levels (which was also the case when the data were evaluated with the **single-factor between-subjects analysis of variance**), since $H = 8.44$ is greater than both $H_{.05} = 5.78$ and $H_{.01} = 7.98$. Thus, in this instance the exact tables and the chi-square approximation do not yield identical results.

2. Equivalency of the Kruskal–Wallis one-way analysis of variance by ranks and the Mann–Whitney U test when $k = 2$ In Section I it is noted that when $k = 2$ the **Kruskal–Wallis one-way analysis of variance by ranks** will yield a result which is equivalent to that obtained with the **Mann–Whitney U test**. To be more specific, the **Kruskal–Wallis test** will yield a result which is equivalent to the normal approximation for the **Mann–Whitney U test** when the correction for continuity is not employed (i.e., the result obtained with Equation 12.4). In order to demonstrate the equivalency of the two tests, Equation 22.1 is employed below to analyze the data for Example 12.1, which were previously evaluated with the **Mann–Whitney U test**.

$$H = \frac{12}{(10)(10 + 1)}\left[\frac{(19)^2}{5} + \frac{(36)^2}{5}\right] - (3)(10 + 1) = 3.15$$

Employing Equation 22.2, $df = 2 - 1 = 1$. For $df = 1$, the tabled critical .05 and .01 chi-square values are $\chi^2_{.05} = 3.84$ and $\chi^2_{.01} = 6.63$. Since the obtained value $H = 3.15$ is less than $\chi^2_{.05} = 3.84$, the null hypothesis cannot be rejected.

Equation 12.4 yields the value $z = -1.78$ for the same set of data. When the latter value is squared, it yields a value which is equal to the H (chi-square) value computed with Equation 22.1 (i.e., $(z = -1.78)^2 = (\chi^2 = 3.15)$).[9] It is also the case that the square of the tabled critical z value employed for the normal approximation of the **Mann–Whitney U test** will always be equal to the tabled critical chi-square value employed for the **Kruskal–Wallis test** at the same level of significance. Thus, the square of the tabled critical two-tailed value $z_{.05} = 1.96$ employed for the normal approximation of the **Mann–Whitney U test** equals $\chi^2_{.05} = 3.84$ employed for the **Kruskal–Wallis test** (i.e., $(z = 1.96)^2 = (\chi^2 = 3.84)$).

3. Power-efficiency of the Kruskal–Wallis one-way analysis of variance by ranks When the underlying population distributions are normal, the **asymptotic relative efficiency** (which is discussed in Section VII of the **Wilcoxon signed-ranks test (Test 6)**) of the **Kruskal–Wallis one-way analysis of variance by ranks** is .955 (when contrasted with the **single-factor between-subjects analysis of variance**). For population distributions which are not normal, the asymptotic relative efficiency of the **Kruskal–Wallis test** is generally equal to or greater than 1. As a general rule, proponents of nonparametric tests take the position that when a researcher has reason to believe that the normality assumption of the **single-factor between-subjects analysis of variance** has been saliently violated, the **Kruskal–Wallis one-way analysis of variance by ranks** provides a powerful test of the comparable alternative hypothesis. In point of fact, Maxwell and Delaney (2004, p. 142) note that for certain nonnormal distributions — more specifically, symmetric but heavy tailed distributions (i.e., distributions characterized by the presence of more extreme scores/outliers), as well as when distributions are identical but skewed — the **Kruskal–Wallis test** is considerably more powerful than the **single-factor between-subjects analysis of variance**.

4. Alternative nonparametric rank-order procedures for evaluating a design involving k independent samples In addition to the **Kruskal–Wallis one-way analysis of variance by ranks**, a number of other nonparametric procedures for two or more independent samples have been developed that can be employed with ordinal data. Among the more commonly cited alternative procedures are the following: a) The **van der Waerden normal-scores test for k independent samples (Test 23)** (Van der Waerden (1953/1953)), which is described in the next chapter, as well as alternative **normal-scores tests** developed by Terry and Hoeffding (Terry (1952)) and Bell and Doksum (1965); b) The **Jonckheere–Terpstra test for ordered alternatives (Test 22a)** (Jonckheere (1954); Terpstra (1952)) (which is described in Section IX (the **Addendum**)) can be employed when the alternative hypothesis for a k independent samples design specifies the rank-order of the k population medians; c) **The median test for independent samples (Test 16e)** (discussed in Section VI of the **chi-square test for $r \times c$ tables (Test 16)**) can be extended to three or more independent samples by dichotomizing k samples with respect to their median values; and d) **The Kolmogorov–Smirnov test for two independent samples (Test 13)** (Kolmogorov (1933) and Smirnov (1939)) can be extended to three or more independent samples. The use of the latter test with more than two samples is discussed in Bradley (1968) and Conover (1980, 1999). Sheskin (1984) describes some of the aforementioned tests in greater detail, as well as citing additional procedures which can be employed for k independent samples designs involving rank-order data.

VIII. Additional Examples Illustrating the Use of the Kruskal–Wallis One-Way Analysis of Variance by Ranks

The **Kruskal–Wallis one-way analysis of variance by ranks** can be employed to evaluate any of the additional examples noted for the **single-factor between-subjects analysis of variance**, if the data for the latter examples are rank-ordered. In addition, the **Kruskal–Wallis test** can be used to evaluate the data for any of the additional examples noted for the *t* **test for two independent samples (Test 11)** and the **Mann–Whitney *U* test**. Examples 22.2 and 22.3 are two additional examples which can be evaluated with the **Kruskal–Wallis one-way analysis of variance by ranks**. Example 22.2 is an extension of Example 12.2 (evaluated with the **Mann–Whitney *U* test**) to a design involving $k = 3$ groups. In Example 22.2 (as well as Example 22.3) the original data are presented as ranks, rather than in an interval/ratio format.[10] Since the rank-orderings for Example 22.2 are identical to those employed in Example 22.1, it yields the same result. Example 22.3 (which is also an extension of Example 12.2) illustrates the use of the **Kruskal–Wallis one-way analysis of variance by ranks** with a design involving $k = 4$ groups and unequal sample sizes.

Example 22.2 *Doctor Radical, a math instructor at Logarithm University, has three classes in advanced calculus. There are five students in each class. The instructor uses a programmed textbook in* Class 1, *a conventional textbook in* Class 2, *and his own printed notes in* Class 3. *At the end of the semester, in order to determine if the type of instruction employed influences student performance, Dr. Radical has another math instructor, Dr. Root, rank the 15 students in the three classes with respect to math ability. The rankings of the students in the three classes follow:* **Class 1**: 9.5, 14.5, 12.5, 14.5, 12.5; **Class 2**: 6, 9.5, 3, 9.5, 3; *and* **Class 3**: 1, 9.5, 6, 3, 6 *(assume the lower the rank, the better the student).*

Note that whereas in Example 22.1, Group 1 (the group with the highest sum of ranks) has the best performance, in Example 22.2, Class 3 (the class with the lowest sum of ranks) is evaluated as the best class. This is the case, since in Example 22.2 the lower a student's rank the better the student, whereas in Example 22.1 the lower a subject's rank, the poorer the subject performed.

Example 22.3 *Doctor Radical, a math instructor at Logarithm University, has four classes in advanced calculus. There are six students in* Class 1, *seven students in* Class 2, *eight students in* Class 3, *and six students in* Class 4. *The instructor uses a programmed textbook in* Class 1, *a conventional textbook in* Class 2, *his own printed notes in* Class 3, *and no written instructional material in* Class 4. *At the end of the semester, in order to determine if the type of instruction employed influences student performance, Dr. Radical has another math instructor, Dr. Root, rank the 27 students in the four classes with respect to math ability. The rankings of the students in the four classes follow:* **Class 1**: 1, 2, 4, 6, 8, 9; **Class 2**: 10, 14, 18, 20, 21, 25, 26; **Class 3**: 3, 5, 7, 11, 12 16, 17, 22; **Class 4**: 13, 15, 19, 23, 24, 27 *(assume the lower the rank, the better the student).*

Example 22.3 provides us with the following information: $n_1 = 6$, $n_2 = 7$, $n_3 = 8$, $n_4 = 6$, and $N = 27$. The sums of the ranks for the four groups are: $\Sigma R_1 = 30$, $\Sigma R_2 = 134$, $\Sigma R_3 = 93$, $\Sigma R_4 = 121$. We can also determine that $\bar{R}_1 = 5$, $\bar{R}_2 = 19.14$, $\bar{R}_3 = 11.63$, $\bar{R}_4 = 20.17$. Substituting the appropriate values in Equation 22.1, the value $H = 14.99$ is computed.

$$H = \frac{12}{(27)(27+1)}\left[\frac{(30)^2}{6} + \frac{(134)^2}{7} + \frac{(93)^2}{8} + \frac{(121)^2}{6}\right] - (3)(27+1) = 14.99$$

Employing Equation 22.2, $df = 4 - 1 = 3$. For $df = 3$, the tabled critical .05 and .01 chi-square values are $\chi^2_{.05} = 7.81$ and $\chi^2_{.01} = 11.34$. Since the obtained value $H = 14.99$ is greater than both of the aforementioned critical values, the null hypothesis can be rejected at both the .05 and .01 levels. Thus, one can conclude that the rankings for at least two of the four classes differed significantly from one another. Although multiple comparisons will not be conducted for this example, visual inspection of the data suggests that the rankings for Class 1 are dramatically superior to those for the other three classes. It also appears that the rankings for Group 3 are superior to those for Groups 2 and 4.

IX. Addendum

Test 22a: The Jonckheere–Terpstra test for ordered alternatives In experiments involving two or more groups/treatments a researcher may predict *a priori* (i.e., prior to conducting the study) the ordering of the groups with respect to their scores on the dependent variable. A common example of the latter type of experiment is one in which it is hypothesized that the greater the therapeutic dosage of a drug the greater the degree of its efficacy. In the case where there are two treatment levels, such a hypothesis of **ordered alternatives** (which is the term employed for a prediction involving the exact ordinal relationship between two or more treatments) can be evaluated through use of a one-tailed analysis (as opposed to a two-tailed analysis). The latter option, however, is not available when there are three or more treatments. Jonckheere (1954) and Terpstra (1952) independently developed the test to be described in this section, which can be employed when the alternative hypothesis for a k independent samples design (where $k \geq 2$) specifies the rank-order of the k population medians.

The **Jonckheere–Terpstra test for ordered alternatives** can be employed under the same conditions a researcher might consider using the **Kruskal–Wallis one-way analysis of variance by ranks** as an alternative to the **single-factor between-subjects analysis of variance**. Daniel (1990) and Jonckheere (1954) note that the **Jonckheere–Terpstra test for ordered alternatives** is based on the following assumptions: a) Each sample has been randomly selected from the population it represents; b) The k samples are independent of one another; c) The dependent variable is a continuous random variable; d) The level of measurement for the dependent variable is at least ordinal; and e) The k underlying populations are identical except for possible differences in location parameters (i.e., measures of central tendency).

The **Jonckheere–Terpstra test for ordered alternatives** will be illustrated through use of Example 22.1.[11] In the latter example, it will be assumed that prior to conducting the study the experimenter predicted the order of the three groups with respect to the number of nonsense syllables which would be recalled. Specifically, the experimenter hypothesized that **Group 1 (No noise)** would recall the largest number of nonsense syllables, followed by **Group 2 (Moderate noise),** followed by **Group 3 (Extreme noise)** (which would recall the smallest number of nonsense syllables).

The null and alternative hypotheses employed for the above prediction when it is made within the framework of the **Jonckheere–Terpstra test for ordered alternatives** are as follows.

Null hypothesis $H_0: \theta_1 = \theta_2 = \theta_3$

(The median of the population Group 1 represents equals the median of the population Group 2 represents equals the median of the population Group 3 represents.)

Alternative hypothesis $H_1: \theta_1 \geq \theta_2 \geq \theta_3$, with at least one strict inequality

(With respect to at least one of the inequalities, the median of the population Group 1 represents is greater than or equal to the median of the population Group 2 represents which is greater than or equal to the median of the population Group 3 represents. With respect to Example 22.1, in order for the alternative hypothesis to be supported, at least one of the ordinal relationships stated in H_1 must be supported. It should be noted, however, that although the notation $\geq$ is employed by all sources in stating the alternative hypothesis, the use of the term **strict** within the context of stating the alternative hypothesis requires that in order for it to be supported, in at least one case the median of one of the populations must be greater than the median of one of the other populations in the order stipulated.)

Notes: 1) The alternative hypothesis stipulated above is directional. Leach (1979, p. 174) notes that the **Jonckheere–Terpstra test for ordered alternatives** can also be employed to evaluate a two-tailed alternative hypothesis. In the case of Example 22.1, a two-tailed/nondirectional alternative hypothesis will be supported if the data are consistent with either of the following two predictions: a) $\theta_1 \geq \theta_2 \geq \theta_3$, with at least one strict inequality; or b) $\theta_1 \leq \theta_2 \leq \theta_3$, with at least one strict inequality. Note that reversal of the inequalities stipulated for the alternative hypothesis allows for evaluation of the analogous alternative hypothesis (i.e., $H_1: \theta_1 \leq \theta_2 \leq \theta_3$, with at least one strict inequality) in the opposite tail of the sampling distribution; 2) Some sources in stating the null and alternative hypotheses employ notation designating equalities or inequalities in reference to treatment effects, as opposed to the population medians which have been used above.

Table 22.4   Arrangement of Data for Example 22.1 for Jonckheere–Terpstra Test for Ordered Alternatives

		Levels of noise			
	Extreme noise (Group 3)			Moderate noise (Group 2)	No noise (Group 1)
a	3	3		2	
			Score		Score	Score
b	1	2		1	
	5	5	4	5	5	8
	5	4	5	5	5	9
	5	2.5	7	5	7	9
	5	2.5	7	4.5	8	10
	4.5	1	8	4.5	8	10
U_{ab}	24.5	15		24		

The data for Example 22.1 are summarized in Table 22.4. Note that the ordering of the groups in Table 22.4 is such that the group hypothesized to have the lowest median value on the dependent variable (Group 3) is positioned at the extreme left of the table, followed by the group hypothesized to have the second lowest median value (Group 2), followed on the extreme right by the group hypothesized to have the highest median value (Group 1). For each of the groups, the original ratio scores of subjects on the dependent variable are recorded in the columns labeled **Score**. Note that the $n_j = 5$ scores within each group are arranged ordinally, such that the lowest score in a group is recorded at the top of the **Score** column, followed by the next lowest score, and so on until the highest score in the group is recorded at the bottom of the **Score** column.

The **Jonckheere–Terpstra test for ordered alternatives** requires one to count the number of times each score in the ath group is lower than a score in the bth group (where a and b represent the identification numbers of the two groups being compared with one another). The latter values are summarized in the columns labeled $\frac{a}{b}$, and indicate count values between the following pairs of groups: Group 3 and Group 1; Group 3 and Group 2; Group 2 and Group 1. Each of the values recorded in the columns labeled $\frac{a}{b}$ indicates for the relevant score in Group a the number of scores in Group b it falls below. For instance, the column labeled $\frac{3}{1}$ specifies for each of the scores in Group 3 the number of scores in Group 1 it falls below. Ties between a relevant score and a score in another group are assigned a count value of .5. To illustrate, the first score recorded for Group 3 in the **Score** column of Table 22.4 is 4. The value 5 in the column labeled $\frac{3}{1}$ indicates that the score of 4 in Group 3 is less than all 5 scores in Group 1 (i.e., it falls below the scores 8, 9, 9, 10, 10, resulting in a count of 5). Thus, the total count recorded in the $\frac{3}{1}$ column for the score of 4 in Group 3 is 5. In the case of the second score recorded for Group 3 (the score of 5), the value 5 in the column labeled $\frac{3}{1}$ indicates that the score of 5 is less than all 5 scores in Group 1 (i.e., it falls below the scores 8, 9, 9, 10, 10, resulting in a count of 5). Thus, the total count recorded in the $\frac{3}{1}$ column for the score of 5 in Group 3 is 5. In the case of both the third and fourth scores recorded for Group 3 (the two scores of 7), the value 5 in the column labeled $\frac{3}{1}$ indicates that the score of 7 is less than all 5 scores in Group 1 (i.e., it falls below the scores 8, 9, 9, 10, 10, resulting in a count of 5). Thus, the total count recorded in the $\frac{3}{1}$ column for each score of 7 in Group 3 is 5. In the case of last score recorded for Group 3 (the score of 8), the value 4.5 in the column labeled $\frac{3}{1}$ indicates that the score of 8 is less than 4 of the scores in Group 1 (i.e., it falls below the scores 9, 9, 10, 10) (which results in a count of 4), while 1 score in Group 1 is equal to 8 (which results in a count of .5, which, as noted earlier, is the count value assigned for a tie). Thus, the total count recorded in the $\frac{3}{1}$ column for the score of 8 in Group 3 is $4 + .5 = 4.5$. The sum of the values recorded in the $\frac{3}{1}$ column, $U_{31} = 24.5$, indicates the total number of times a score in Group 3 is less than a score in Group 1.

The above described counting procedure is next employed for the 5 scores in Group 3 in reference to the 5 scores in Group 2. The latter count values are recorded in the column labeled $\frac{3}{2}$. Thus, the value 5 at the top of the column labeled $\frac{3}{2}$ indicates that the score of 4 in Group 3 is less than all 5 scores in Group 2 (i.e., it falls below the scores 5, 5, 7, 8, 8, resulting in a count of 5). Thus, the total count recorded in the $\frac{3}{2}$ column for the score of 5 in Group 3 is 5. In the case of the second score recorded for Group 3 (the score of 5), the value 4 in the column labeled $\frac{3}{2}$ indicates that the score of 5 is less than 3 of the scores in Group 2 (i.e., it falls below the scores 7, 8, 8) (which results in a count of 3), while 2 scores in Group 2 are equal to 5 (which results in a count of $.5 + .5 = 1$). Thus, the total count recorded in the $\frac{3}{2}$ column for the score of 5 in Group 3 is $3 + 1 = 4$. In the case of the third and fourth scores recorded for Group 3 (the two scores of 7), the value 2.5 in the column labeled $\frac{3}{2}$ indicates that the score of 7 is less than 2 of the scores in Group 2 (i.e., it falls below the scores 8, 8) (which results in a count of 2), while 1 score in Group 2 is equal to 7 (which results in a count of .5). Thus, the total count recorded in the $\frac{3}{2}$ column for each score of 7 in Group 3 is 2 + .5 = 2.5. In the case of the last score recorded for Group 3 (the score of 8), the value 1 in the column labeled $\frac{3}{2}$ indicates that the score of 8 is not less than any of the scores in Group 2 (which results in a count of 0), while 2 scores in Group 2 are equal to 8 (which results in a count of .5 + .5 = 1). Thus, the total count recorded in the $\frac{3}{2}$ column for the score of 8 in Group 3 is 0 + 1 = 1. The sum of the values recorded in the $\frac{3}{2}$ column, $U_{32} = 15$, indicates the total number of times a score in Group 3 is less than a score in Group 2.

Upon completion of the $\frac{3}{2}$ set of counts, the counting procedure is employed a final time for a set of counts contrasting Group 2 with Group 1. Thus, the column labeled $\frac{2}{1}$ contains the count values for the number of times a score in Group 2 is less than a score in Group 1. Note that all of the scores in Group 2 are less than the scores in Group 1, except for the 2 scores of 8 in Group 2 which are tied with 1 of the scores in Group 1. Consequently, the scores 5, 5, 7 in Group 2 are all assigned a count of 5 in the $\frac{2}{1}$ column (since they all fall below the scores 8, 9, 9, 10, 10 in Group 1). The 2 scores of 8 in Group 2 are both assigned a count of 4.5 in the $\frac{2}{1}$ column, since the latter count value indicates that the score of 8 is less than 4 of the scores in Group 1 (9, 9, 10, 10) and equal to 1 score in Group 1 (thus, 4 + .5 = 4.5). The sum of the values recorded in the $\frac{2}{1}$ column, $U_{21} = 24$, indicates the total number of times a score in Group 2 is less than a score in Group 1.

Note that the number of sets of counts (i.e., pairs of groups) for which the counting procedure must be employed is $[k(k-1)]/2$. In the case of Example 22.1, where $k = 3$, there are $[3(3-1)]/2 = 3$ sets of counts (specifically $\frac{3}{1}, \frac{3}{2}, \frac{2}{1}$).

The test statistic for the **Jonckheere–Terpstra test for ordered alternatives** is computed with Equation 22.6. The latter equation indicates that the test statistic, designated by the notation J, is the sum of the U_{ab} values recorded at the bottom of each of the $\frac{a}{b}$ columns in Table 22.4. The value J represents the total number of instances a score in Group a is less than a score in Group b, when all possible pairs of groups are compared with one another.[12]

$$J = \Sigma U_{ab} \qquad\qquad \text{(Equation 22.6)}$$

The value of the test statistic computed for Example 22.1 is $J = 63.5$.

$$J = 24.5 + 15 + 24 = 63.5$$

In order to reject the null hypothesis, the computed value of J must be equal to or greater than the tabled critical value at the prespecified level of significance in **Table A24 (Table of critical values for the Jonckheere–Terpstra Test Statistic)** in the **Appendix**. Inspection of the latter table reveals that the tabled critical **one-tailed .05 and .01 values** in **Table A24** for the sample sizes (5, 5, 5) are $J_{.05} = 54$ and $J_{.01} = 60$.[13] Since the computed value $J = 63.5$ is larger than both of the aforementioned critical values, the null hypothesis can be rejected at both the .05 and .01 levels. Thus, the alternative hypothesis is supported — i.e., the ordering of the population medians is $\theta_1 \geq \theta_2 \geq \theta_3$.

The reader should take note of the fact that the order in which the sample sizes are listed in **Table A24** does not assume that the first value listed in a row for k sample sizes corresponds to the number of subjects in Group 1, the second value listed corresponds to the number of subjects in Group 2, etc. The sample size values in each row merely represent the composite set of k values for the sample sizes employed in a study. To clarify the latter, assume in an experiment involving three groups there are $n_1 = 5$ subjects in Group 1, $n_2 = 4$ subjects in Group 2, and $n_3 = 5$ subjects in Group 3. The relevant sample size value to employ in **Table A24** will be the row which designates one group comprised of 4 subjects and two groups comprised of 5 subjects. Thus, the row with the sample size values 4, 5, 5 should be employed.

The normal approximation of the Jonckheere–Terpstra test for ordered alternatives If the composite sample size employed in a study is relatively large, the normal distribution can be employed to approximate the test statistic for the **Jonckheere–Terpstra test for ordered alternatives**. Sources generally recommend that the normal approximation should be used when the sample sizes employed in a study are larger than those documented in **Table A24**. Equation 22.7 provides the normal approximation of the **Jonckheere–Terpstra test** statistic.

$$z = \frac{J - \dfrac{\left[N^2 - \sum\limits_{j=1}^{k} n_j^2\right]}{4}}{\sqrt{\dfrac{N^2(2N + 3) - \sum\limits_{j=1}^{k} n_j^2(2n_j + 3)}{72}}}$$
(Equation 22.7)

In the numerator of Equation 22.7, the term subtracted from J (i.e., $[N^2 - \sum\limits_{j=1}^{k} n_j^2]/4$) represents the expected (mean) value of the sampling distribution of J (which can be represented with the notation J_E), while the denominator represents the expected standard deviation of the sampling distribution of J (which can be represented with the notation SD_J). Thus, Equation 22.7 can be summarized as follows: $z = (J - J_E) / SD_J$.

With respect to the notation in Equation 22.7 the reader should note the following:

1) The element $[N^2 - \sum\limits_{j=1}^{k} n_j^2]/4$ in the numerator of Equation 22.7 indicates the following a) For each group the value of n_j is squared, and the sum of all the squared values is computed; b) The result obtained in part a) is subtracted from the square of the value of N (i.e., the total number of subjects in all k groups); c) The result of part b) is divided by 4.

2) The element $\sum\limits_{j=1}^{k} n_j^2(2n_j + 3)$ in the denominator of Equation 22.7 indicates the following a) For each group the value of n_j is squared, and multiplied by the value $(2n_j + 3)$; b) The sum of all the k values obtained in part a) is computed

Although Example 22.1 employs a sample size which most sources would view as too small to use within the framework of the normal approximation, the latter approximation will be employed to illustrate Equation 22.7. The reader will see that although Equation 22.7 is used with a relatively small sample size, it still yields a result which is consistent with that obtained when the exact distribution for the **Jonckheere–Terpstra** test statistic is employed. Employing Equation 22.7, the value $z = 2.75$ is computed.[14]

The obtained value $z = 2.75$ is evaluated with **Table A1**. In order to reject the null hypothesis, the computed value of z must be equal to or greater than the tabled critical one-tailed value at the prespecified level of significance. The tabled critical one-tailed .05 and .01 values are $z_{.05} = 1.65$ and $z_{.01} = 2.33$. Since $z = 2.75$ is larger than both of the aforementioned critical values, the null hypothesis can be rejected at both the .05 and .01 levels. Thus, as is the case when the exact sampling distribution is employed, the alternative hypothesis is supported.[15]

$$z = \frac{63.5 - \left[\dfrac{15^2 - (5^2 + 5^2 + 5^2)}{4}\right]}{\sqrt{\dfrac{15^2[(2)(15) + 3] - [(5^2)[(2)(5) + 3] + (5^2)[(2)(5) + 3] + (5^2)[(2)(5) + 3]]}{72}}}$$

$$= \frac{63.5 - 37.5}{9.46} = 2.75$$

Note that regardless of whether one employs the exact critical values for the sampling distribution of the *J* statistic in **Table A24** or computes the normal approximation, the result of the **Jonckheere–Terpstra test for ordered alternatives** is generally consistent with the result obtained when the same set of data was evaluated with both the **Kruskal–Wallis one-way analysis of variance by ranks** and the **single-factor between-subjects analysis of variance** (neither of which, however, specifies the ordering of the groups in the alternative hypothesis). When the **Kruskal–Wallis one-way analysis of variance by ranks** was employed with the same set of data without specifying the ordering of groups, the null hypothesis of equal population medians was rejected at the .05 level. When the **single-factor between-subjects analysis of variance** was employed with the same set of data without specifying the ordering of groups, the null hypothesis of equal population means was rejected at both the .05 and .01 levels.

The fact that the null hypothesis can be rejected at the .01 level on the basis of the result of the **Jonckheere–Terpstra test for ordered alternatives**, yet only at the .05 level when the data are evaluated with the **Kruskal–Wallis one-way analysis of variance by ranks**, is consistent with a statement made by Hollander and Wolfe (1999, p. 210) that the **Jonckheere– Terpstra test for ordered alternatives** is more powerful than the **Kruskal–Wallis one-way analysis of variance by ranks** when the alternative hypothesis is ordered. Siegel and Castellan (1988, p. 222) note that when contrasted with the **single-factor between-subjects analysis of variance**, the **asymptotic relative efficiency** of the **Jonckheere–Terpstra test for ordered alternatives** for normally distributed data is .955 (which is also the case for the **Kruskal–Wallis one-way analysis of variance by ranks**).

Equivalency of the Jonckheere–Terpstra test for ordered alternatives and a one-tailed analysis with the Mann–Whitney *U* test when *k* = 2   When there are *k* = 2 groups, the result obtained for the **Jonckheere–Terpstra test for ordered alternatives** will be equivalent to that obtained for the **Mann–Whitney *U* test**, when a one-tailed analysis is employed for the latter test (Hollander, personal communication). When *k* = 2, the *J* statistic computed for the **Jonckheere–Terpstra test for ordered alternatives** will be equal to the number of inversions computed for the **Mann–Whitney *U* test** (the latter value being computed through use of the methodology described in Endnote 14 of the **Mann–Whitney *U* test**), as well as the value of either U_1 or U_2.

Table 22.5 illustrates the computation of the **Jonckheere–Terpstra** test statistic for Example 12.1 (which was employed to illustrate the **Mann–Whitney *U* test**). The column labeled $\frac{1}{2}$ in Table 22.5 specifies for each of the scores in Group 1 the number of scores in Group 2 it falls below. Note that the computed value $J = U_{ab} = U_{21} = 21$ is equivalent to the value U_1 = 21 computed for Example 12.1 (where $U_1 = 21$ and $U_2 = 4$). Additionally, $J = 21$ is equal to

Table 22.5 Arrangement of Data for Example 12.1 for the Jonckheere–Terpstra Test for Ordered Alternatives

		Treatment	
		Drug (Group 1)	Placebo (Group 2)
a	1		
b	2	Score	Score
	5	0	4
	5	0	5
	5	1	8
	5	2	11
	1	11	11
$U_{ab} = 21$			

$$J = \Sigma U_{ab} = 21$$

the number of inversions computed through use of the methodology described in Endnote 14 of the **Mann–Whitney U test**. In order to employ the critical values recorded in **Table A11 (Table of Critical Values for the Mann–Whitney U Statistic)**, the smaller of the two values U_1 versus U_2 is designated as U, and consequently $U = 4$. In point of fact, either U_1 or U_2 can be employed to represent the U statistic, depending upon how tables of critical values for the U statistic are configured (the values U_1 and U_2 represent U values in each of the two tails of the sampling distribution of the U statistic). Thus, in the final analysis, the test statistic computed for Example 12.1 with the **Jonckheere–Terpstra test for ordered alternatives** is identical to that computed for the **Mann–Whitney U test**. Consequently, the result of a one-tailed analysis for the latter test will be identical to that obtained for a one-tailed analysis with the **Jonckheere–Terpstra test for ordered alternatives**. Note that when $k = 2$, **Table A24** cannot be employed for the **Jonckheere–Terpstra test for ordered alternatives**, since the latter table only contains critical values for J when $k \geq 3$.

Equation 22.7 is employed below to compute the value $z = 1.78$, the normal approximation of the J statistic.

$$z = \frac{21 - \left[\dfrac{10^2 - (5^2 + 5^2)}{4} \right]}{\sqrt{\dfrac{10^2[(2)(10) + 3] - [(5^2)[(2)(5) + 3] + (5^2)[(2)(5) + 3]]}{72}}} = \frac{8.5}{4.787} = 1.78$$

The value $z = 1.78$ is identical to absolute z value computed for the same set of data with Equation 12.4 (which is employed to compute the normal approximation for the **Mann–Whitney U test**)). Note that since the sign of z indicates the direction of the difference, both of the U values computed for Example 12.1 (i.e., $U_1 = 21$ and $U_2 = 4$) will yield the absolute value $z = 1.78$ when employed in Equation 12.4. Obviously, since both Equations 22.7 and 12.4 yield the same absolute value, they will result in the same conclusions when a one-tailed analysis is conducted, (i.e., the direction of the difference will be the same regardless of whether one employs the **Jonckheere–Terpstra test for ordered alternatives** or the **Mann–Whitney U test**).[16]

Additional comments on the Jonckheere–Terpstra test for ordered alternatives a) A number of sources (e.g., Leach (1979) and Hollander and Wolfe (1999, p. 204)) describe a rather computationally involved tie-correction which is recommended when there are an extensive number of ties in a set of data. The tie correction reduces the value of the standard deviation of Equation 22.7, and results in a slight increase in the computed absolute value of z. Unless the number of ties in a set of data is excessive, the difference between the tie-corrected z value versus the uncorrected z value will be minimal; b) The pairwise comparison procedures described earlier in this chapter for the **Kruskal–Wallis one-way analysis of variance by ranks** can also be employed for evaluating the **Jonckheere–Terpstra test for ordered alternatives**. One-tailed critical values should be used in Equation 22.5 in conducting pairwise comparisons for the **Jonckheere–Terpstra test for ordered alternatives** if a directional alternative hypothesis is employed for a comparison; c) Terpstra (1952) and Conover (1999, p. 325) discuss the equivalency of the result obtained with the **Jonckheere–Terpstra test for ordered alternatives** with the result obtained when the same set of data is evaluated with **Kendall's tau (Test 30)** (a measure of association discussed later in the book) under certain conditions; d) Additional discussion of the **Jonckheere–Terpstra test for ordered alternatives** can be found in the following sources: Conover (1999), Daniel (1990), Gibbons (1997), Hollander and Wolfe (1999), Leach (1979), Lehman (1975), Siegel and Castellan (1988), and Sprent (1993).

References

Bell, C. B. and Doksum, K. A. (1965). Some new distribution-free statistics. **Annals of Mathematical Statistics**, 36, 203–214.

Beyer, W. H. (1968). **Handbook of tables for probability and statistics** (2nd ed.). Cleveland, OH: The Chemical Rubber Company.

Bradley, J. V. (1968). **Distribution-free statistical tests**. Englewood Cliffs, NJ: Prentice Hall.

Conover, W. J. (1980). **Practical nonparametric statistics** (2nd ed.). New York: John Wiley & Sons.

Conover, W. J. (1999). **Practical nonparametric statistics** (3rd ed.). New York: John Wiley & Sons.

Daniel, W. W. (1990). **Applied nonparametric statistics** (2nd ed.). Boston: PWS–Kent Publishing Company.

Delaney, H. D. & Vargha, A. (2002). Comparing several robust tests of stochastic equality with ordinally scaled variables and small to moderate sized samples. **Psychological Methods**, 7, 485–503.

Dunn, O. J. (1964). Multiple comparisons using rank sums. **Technometrics**, 6, 241–252.

Gibbons, J. D. (1997). **Nonparametric methods for quantitative analysis** (3rd ed.). Columbus, OH: American Sciences Press.

Hollander, M. & Wolfe, D. A. (1999). **Nonparametric statistical methods**. New York: John Wiley & Sons.

Jonckheere, A. R. (1954). A distribution-free k sample test against ordered alternatives. **Biometrika**, 41, 133–145.

Kolmogorov, A. N. (1933). Sulla determinazione empiraca di una legge di distribuzione. **Giorn dell'Inst. Ital. degli. Att.**, 4, 89–91.

Kruskal, W. H. (1952). A nonparametric test for the several sample problem. **Annals of Mathematical Statistics**, 23, 525–540.

Kruskal, W. H. & Wallis, W. A. (1952). Use of ranks in one-criterion variance analysis. **Journal of the American Statistical Association**, 47, 583–621.

Leach, C. (1979). **Introduction to statistics: A nonparametric approach for the social sciences**. Chichester: John Wiley & Sons.

Lehman, E. L. (1975). **Nonparametric statistical methods based on ranks**. San Francisco: Holden–Day.

Marascuilo, L. A. & McSweeney, M. (1977). **Nonparametric and distribution-free methods for the social sciences**. Monterey, CA: Brooks/Cole Publishing Company.

Maxwell, S. & Delaney, H. (1990). **Designing experiments and analyzing data**. Belmont, CA: Wadsworth Publishing Company.

Maxwell, S. E. & Delaney, H. D. (2004). **Designing experiments and analyzing data: A model comparison perspective** (2nd ed.). Mahwah, NJ: Lawrence Erlbaum Associates.

Odeh, R. E. (1971). Jonckheere's k-sample test against ordered alternatives. **Technometrics**, 13, 912–918.

Sheskin, D. J. (1984). **Statistical tests and experimental design: A guidebook.** New York: Gardner Press.

Siegel, S. & Castellan, N. J., Jr. (1988). **Nonparametric statistics for the behavioral sciences** (2nd ed.). New York: McGraw–Hill Book Company.

Smirnov, N. V. (1939). On the estimation of the discrepancy between empirical curves of distributions for two independent samples. **Bulletin University of Moscow**, 2, 3–14.

Sprent, P. (1993). **Applied nonparametric statistical methods** (2nd ed.). London: Chapman & Hall.

Terpstra, T. J. (1952). The asymptotic normality and consistency of Kendall's test against trend, when ties are present in one ranking. **Indagationes Mathematicae**, 14, 327–333.

Terry, M. E. (1952). Some rank-order tests, which are most powerful against specific parametric alternatives. **Annals of Mathematical Statistics**, 23, 346–366.

Van der Waerden, B. L. (1952/1953). Order tests for the two-sample problem and their power. **Proceedings Koninklijke Nederlandse Akademie van Wetenshappen** (A), 55 (**Indagationes Mathematicae** 14), 453–458, & 56 (**Indagationes Mathematicae**, 15), 303–316 (corrections appear in Vol. 56, p. 80).

Wike, E. L. (1978). A Monte Carlo investigation of four nonparametric multiple-comparison tests for k independent samples. **Bulletin of the Psychonomic Society**, 11, 25–28.

Wike, E. L. (1985). **Numbers: A primer of data analysis**. Columbus, OH: Charles E. Merrill Publishing Company.

Zar, J. H. (1999). **Biostatistical analysis** (4th ed.). Upper Saddle River, NJ: Prentice Hall.

Zimmerman, D. W. & Zumbo, B. D. (1993). The relative power of parametric & nonparametric statistical methods. In Keren, G. & Lewis, C. (Eds.), **A handbook for data analysis in the behavioral sciences: Methodological issues** (pp. 481–517). Hillsdale, NJ: Lawrence Erlbaum Associates, Publishers.

Endnotes

1. Although it is possible to conduct a directional analysis, such an analysis will not be described with respect to the **Kruskal–Wallis one-way analysis of variance by ranks**. A discussion of a directional analysis when $k = 2$ can be found under the **Mann–Whitney U test**. A discussion of the evaluation of a directional alternative hypothesis when $k \geq 3$ can be found in Section VII of the **chi-square goodness-of-fit test (Test 8)**. Although the latter

discussion is in reference to analysis of a k independent samples design involving categorical data, the general principles regarding analysis of a directional alternative hypothesis when $k \geq 3$ are applicable to the **Kruskal–Wallis one-way analysis of variance by ranks**.

2. As noted in Section IV, the chi-square distribution provides an approximation of the **Kruskal–Wallis test** statistic. Although the chi-square distribution provides an excellent approximation of the Kruskal–Wallis sampling distribution, some sources recommend the use of exact probabilities for small sample sizes. Exact tables of the Kruskal–Wallis distribution are discussed in Section VII.

3. In the discussion of comparisons under the **single-factor between-subjects analysis of variance**, it is noted that a **simple** (also known as a **pairwise**) **comparison** is a comparison between any two groups in a set of k groups.

4. Note that in Equation 22.5, as the value of N increases the value computed for CD_{KW} will also increase because of the greater number (range of values) of rank-orderings required for the data.

5. The rationale for the use of the proportions .0167 and .0083 in determining the appropriate value for z_{adj} is as follows. In the case of a one-tailed/directional analysis, the relevant probability/proportion employed is based on only one of the two tails of the normal distribution. Consequently, the proportion of the normal curve which is used to determine the value of z_{adj} will be a proportion that is equal to the value of α_{PC} in the appropriate tail of the distribution (which is designated in the alternative hypothesis). The value $z = 2.13$ is employed, since the proportion of cases that falls above $z = 2.13$ in the right tail of the distribution is .0167, and the proportion of cases that falls below $z = -2.13$ in the left tail of the distribution is .0167 (i.e., the value $z = 2.13$ has an entry in column 3 of **Table A1** which is closest to .0167). In the case of a two-tailed/nondirectional analysis, the relevant probability/proportion employed is based on both tails of the distribution. Consequently, the proportion of the normal curve which is used to determine the value of z_{adj} will be a proportion that is equal to the value of $\alpha_{PC}/2$ in each tail of the distribution. The proportion $\alpha_{PC}/2 = .0167/2 = .0083$ is employed for a two-tailed/nondirectional analysis, since one-half of the proportion which comprises $\alpha_{PC} = .0167$ comes from the left tail of the distribution and the other half from the right tail. Consequently, the value $z = 2.39$ is employed, since the proportion of cases that falls above $z = 2.39$ in the right tail of the distribution is .0083, and the proportion of cases that falls below $z = -2.39$ in the left tail of the distribution is .0083 (i.e., the value $z = 2.39$ has an entry in column 3 of **Table A1** which is closest to .0083).

6. It should be noted that when a directional alternative hypothesis is employed, the sign of the difference between the two mean ranks is relevant in that it must be consistent with the prediction stated in the directional alternative hypothesis. When a nondirectional alternative hypothesis is employed, the direction of the difference between two mean ranks is irrelevant.

7. a) Many researchers would probably be willing to tolerate a somewhat higher familywise Type I error rate than .05. In such a case the difference $|\bar{R}_1 - \bar{R}_2| = 6.5$ will be significant, since the value of z_{adj} employed in Equation 22.5 will be less than $z = 2.39$, thus resulting in a lower value for CD_{KW}; b) When there are a large number of ties in the

data, a modified version of Equation 22.5 is recommended by some sources (e.g., Daniel (1990)), which reduces the value of CD_{KW} by a minimal amount. Marascuilo and McSweeney (1977, p. 318) recommend that when ties are present in the data, the tie correction factor $C = .964$ computed with Equation 22.3 be multiplied by the term in the radical of Equation 22.5. When the latter is done with the data for Example 22.1 (as noted below), the value $CD_{KW} = 6.64$ is obtained. As is the case with Equation 22.5, only the Group 1 versus Group 3 pairwise difference is significant.

$$CD_{KW} = (2.39)\sqrt{(.964)\left(\frac{(15)(15+1)}{12}\right)\left(\frac{1}{5}+\frac{1}{5}\right)} = (2.39)(2.77) = 6.64$$

c) In contrast to Equation 22.5, Conover (1980, 1999) employs Equation 22.8 for conducting pairwise comparisons. Conover (1980, 1999) states that Equation 22.8 is recommended when there are no ties. Although he provides a tie correction equation, he notes that the result obtained with the tie correction equation will be very close to that obtained with Equation 22.8 when there are only a few ties in the data. Because of the latter, Equation 22.8 will be employed in the discussion to follow. The latter equation will yield a substantially lower CD_{KW} value than Equation 22.5. Equation 22.8 is analogous to Equation 23.4, which Conover (1980, 1999) recommends for conducting comparisons for the **van der Waerden normal-scores test for k independent samples** (which is discussed in the next chapter). Note that if the element $[(N - 1 - H)/(N - k)]$ is omitted from the radical in Equation 22.8, it becomes Equation 22.5. It is demonstrated below that when Equation 22.8 is employed with the data for Example 22.1, the value $CD_{KW} = 4.60$ is obtained. If the latter CD_{KW} value is used, both the Group 1 versus Group 3 and Group 1 versus Group 2 comparisons are significant (the latter being the case, since the obtained difference of 6.5 for the Group 1 versus Group 2 comparison is greater than $CD_{KW} = 4.60$). ($CD_{KW} = 4.47$ is obtained if the tie corrected value $H_C = 8.76$ is employed in Equation 22.8.) Since Equation 22.8 will always yield a smaller value for CD_{KW}, it allows for a more powerful test of an alternative hypothesis than Equation 22.5.

(Equation 22.8)

$$CD_{KW} = z_{adj}\sqrt{\left(\frac{N(N+1)}{12}\right)\left(\frac{N-1-H}{N-k}\right)\left(\frac{1}{n_a}+\frac{1}{n_b}\right)}$$

$$= (2.39)\sqrt{\left(\frac{15(15+1)}{12}\right)\left(\frac{15-1-8.44}{15-3}\right)\left(\frac{1}{5}+\frac{1}{5}\right)} = 4.60$$

In a personal communication Conover (2005) informed the author that Equation 22.8 is, in fact, equivalent to Equation 21.24 (employed for **Multiple t tests/Fisher's LSD test (Test 21a)**), if the value of MS_{WG} in the latter equation is determined on the basis of a **single-factor between-subjects analysis of variance** conducted on the ranks of the subjects (i.e., not on the original unranked scores as was done for Example 21.1). (In other words, MS_{WG} will represent $MS_{WG_{ranks}}$.) Within the latter context, Equation 22.8 can be expressed in the form $CD_{KW} = z_{adj}\sqrt{(2\,MS_{WG_{ranks}})/n}$. (Note that z_{adj} in the latter equation is analogous to $\sqrt{F_{(1,WG)}}$ in Equation 21.24, since $\sqrt{F_{(1,WG)}} = \sqrt{t^2}$, which in Equation 22.8 can be viewed as $\sqrt{z_{adj}^2}$.) The author demonstrated the validity of the latter by conducting a **single-factor between-subjects analysis of variance** on the ranks, the result of which is summarized in Table 22.6. (Again, the reader should take note of the fact that the value $MS_{WG} = 8.44$ in Table 22.6 represents $MS_{WG_{ranks}}$.)

Table 22.6 Summary Table of Analysis of Variance on Ranks for Example 22.1

Source of variation	SS	df	MS	F
Between-groups	168.70	2	84.35	9.92*
Within-groups	101.30	12	8.44	
Total	270.00	14		

*Significant at .01 level

Before continuing with the discussion of the comparisons, it should be noted that Conover indicated to the author, that the information derived from the analysis of variance on the ranks can be employed to compute the omnibus **Kruskal–Wallis test** statistic as follows: $H = [(SS_{BG})(N - 1)] / SS_T$. When the appropriate values recorded in Table 22.6 are substituted in the latter equation, it yields $H = [(168.7)(15 - 1)] / 270 = 8.75$, which is, in fact, equivalent (if rounding off error is ignored) to the tie corrected value $H = 8.76$ obtained with Equation 22.4.

When $F_{(1,WG)}$ is employed in Equation 21.24 to compute the minimum required difference between the mean ranks of any pair of groups, it yields the value $CD_{LSD} = 4.00$. If, however, the square of $(z_{adj} = 2.39)^2 = 5.71$ is employed as the F value in the latter equation, the value $CD_{LSD} = 4.39$ is obtained. When rounding off error is taken into account, Equations 21.24 and 22.8 will essentially yield the same value for CD_{KW}, and the latter value will always be lower than the value computed with Equation 22.5.

$$CD_{LSD} = \sqrt{F_{(1,WG)}} \sqrt{\frac{2\,MS_{WG}}{n}} = \sqrt{4.75} \sqrt{\frac{(2)(8.44)}{5}} = 4.00$$

$$CD_{LSD} = \sqrt{z_{adj}^2} \sqrt{\frac{2\,MS_{WG}}{n}} = \sqrt{5.71} \sqrt{\frac{(2)(8.44)}{5}} = 4.39$$

The author also computed the CD values noted below for the **Bonferroni–Dunn test** (**Test 21b**) (using Equation 21.26), **Tukey's HSD test** (**Test 21c**) (using Equation 21.31), and the **Scheffé test** (**Test 21e**) (using Equation 21.32). (Note that the identical values are employed in all of the equations used to compute CD values as those used for the comparisons with Example 21.1, except for the value of MS_{WG}, which in the case of the analysis of variance of the ranks equals $MS_{WG_{ranks}} = 8.44$.) It is interesting to note that all of the values $CD_{LSD} = 4.00$ or 4.39, $CD_{KW} = 4.47$ (using Equation 22.8), $CD_{HSD} = 4.90$, $CD_{B/D} = 5.13$, and $CD_S = 5.12$ are less than $CD_{KW} = 6.75$ computed with Equation 22.5.

$$CD_{B/D} = t_{B/D} \sqrt{\frac{2MS_{WG}}{n}} \equiv 2.79 \sqrt{\frac{(2)(8.44)}{5}} = 5.13$$

$$CD_{HSD} = q_{(k, df_{WG})} \sqrt{\frac{MS_{WG}}{n}} = = (3.77)\sqrt{\frac{8.44}{5}} = 4.90$$

$$CD_S = \sqrt{(k - 1)(F_{(df_{BG}, df_{WG})})} \sqrt{\frac{2MS_{WG}}{n}} = \sqrt{(3 - 1)(3.89)}\sqrt{\frac{(2)(8.44)}{5}} = 5.12$$

In point of fact, Dunn (1964, p. 250) acknowledges that Equation 22.5 (as well as the use of multiple **Mann–Whitney tests** which employ an adjusted alpha level) provides for a conservative test of the alternative hypothesis. One could argue that the analysis should indeed be more conservative, since when an analysis of variance is conducted on the ranks, the normality assumption underlying the latter test is violated (as well as being violated with respect to all of the multiple comparison procedures employed above). In the final analysis, a justification can be made for employing Equation 22.5 or, for that matter, Equation 22.8. In view of the fact that sources do not agree on the methodology for conducting pairwise comparisons, if two or more methods yield dramatically different results for a specific comparison, one or more replication studies employing reasonably large sample sizes should clarify whether or not an obtained difference is reliable, as well as the magnitude of the difference (if, in fact, one exists).

8. In Equation 22.5 the value $z_{.05} = 1.96$ is employed for z_{adj}, and the latter value is multiplied by 2.83, which is the value computed for the term in the radical of the equation for Example 22.1.

9. The slight discrepancy is due to rounding off error, since the actual absolute value of z computed with Equation 12.4 is 1.7756.

10. In accordance with one of the assumptions noted in Section I for the **Kruskal–Wallis one-way analysis of variance by ranks**, in both Examples 22.2 and 22.3 it is assumed that Dr. Radical implicitly or explicitly evaluates the N students on a continuous interval/ratio scale prior to converting the data into a rank-order format.

11. As noted, one of the assumptions stipulated for both the **Kruskal–Wallis one-way analysis of variance by ranks** and the **Jonckheere–Terpstra test for ordered alternatives** is that the dependent variable is a continuous random variable, yet, in practice, this assumption (which is also common to a number of other nonparametric tests employed with ordinal data) is often not adhered to. In the case of Example 22.1, the dependent variable (number of nonsense syllables recalled) would be conceptualized by most people as a discrete rather than a continuous variable. On the basis of the latter one could argue that the accuracy of both tests described in this chapter may be compromised if they are employed to evaluate the data for Example 22.1. On the other hand, if in the description of Example 22.1 it was stated that the experimenter had the option of awarding a subject partial credit with respect to whether or not he or she was successful in recalling a nonsense syllable, the dependent variable could then be conceptualized as continuous rather than discrete.

12. The U_{ab} values are commonly referred to as U counts, since the methodology employed in the counting procedure described for the **Jonckheere–Terpstra test for ordered alternatives** is the same as the counting procedure described for the **Mann–Whitney U test** in Endnote 14 of the latter test.

13. It should be noted that the critical values listed in **Table A24** are based on probabilities which have been computed to five decimal places, and thus a given tabled critical J value will be associated with a probability which when computed to five decimal places is equal to or less than the designated alpha value. There may be a one-unit discrepancy between the critical values listed in **Table A24** and those found in other sources when the values listed in such sources are based on rounding off probabilities to fewer than five decimal places. More detailed tables for the **Jonckheere–Terpstra test** statistic can be found in Daniel (1990) and Hollander and Wolfe (1999).

14. Jonckheere (1954, p. 142) notes that a correction for continuity can be employed for Equation 22.7 by subtracting the value 1 from the absolute difference obtained for the numerator of the equation. Sources that describe the **Jonckheere–Terpstra test for ordered alternatives**, however, do not recommend using the correction for continuity.

15. It should be noted that it is possible to obtain a negative z value for Equation 22.7 if the data are consistent with the analogous alternative hypothesis in the other tail of the sampling distribution. Put simply, if the results are the exact opposite of what is predicted, the value of z will be negative. Specifically, if the data were consistent with the alternative hypothesis $H_1: \theta_1 \leq \theta_2 \leq \theta_3$, the value computed for J would be less than the expected value of J computed in the numerator of Equation 22.7 (which in our example was $J_E = 37.5$). When the latter is true, the value computed for z will be negative. In instances where a researcher is employing a two-tailed alternative hypothesis, the absolute value of z is employed, since the sign of z has no bearing on whether or not a result is significant.

   To illustrate the computation of a negative z value, if the analogous alternative hypothesis in the other tail of the sampling distribution is employed (i.e., $H_1: \theta_1 \leq \theta_2 \leq \theta_3$), Equation 22.6 will yield the value $J = 11.5$ for Example 22.1. This is the case since the U_{ab} values for the three $\frac{a}{b}$ sets are 1 for the set $\frac{1}{2}$ (i.e., the count value for the number of times a score in Group 1 is less than a score in Group 2 is 1), .5 for the set $\frac{1}{3}$ (i.e., the count value for the number of times a score in Group 1 is less than a score in Group 3 is .5), and 10 for the set $\frac{2}{3}$ (i.e., the count value for the number of times a score in Group 2 is less than a score in Group 3 is 10). Thus, the sum of the 3 U_{ab} values is $1 + .5 + 10 = 11.5$. When the value $J = 11.5$ is substituted in Equation 22.7, the value $z = -2.75$ is computed, which is identical to the absolute value computed when the alternative hypothesis $H_1: \theta_1 \geq \theta_2 \geq \theta_3$ is employed. Thus, the sign of z merely indicates the direction of a one-tailed analysis. (**Table A24**, however, cannot be employed to evaluate the value $J = 11.5$ within the framework of $H_1: \theta_1 \leq \theta_2 \leq \theta_3$, since the latter alternative hypothesis is inconsistent with Equation 22.6 in reference to Table 22.4.)

16. When $k = 2$, the result obtained for the **Jonckheere–Terpstra test for ordered alternatives** will also be equivalent to the one-tailed result obtained for the **Kruskal–Wallis one-way analysis of variance by ranks**, since when $k = 2$ the latter test is equivalent to the **Mann–Whitney U test**.

Test 23

The van der Waerden Normal-Scores Test for *k* Independent Samples
(Nonparametric Test Employed with Ordinal Data)

I. Hypothesis Evaluated with Test and Relevant Background Information

Hypothesis evaluated with test Are *k* independent samples derived from identical population distributions?

Relevant background information on test The **van der Waerden normal-scores test for *k* independent samples** (van der Waerden (1952/1953)) is employed with ordinal (rank-order) data in a hypothesis testing situation involving a design with two or more independent samples. The **van der Waerden test** is one of a number of normal-scores tests which have been developed for evaluating data. **Normal-scores tests** transform a set of rank-orders into a set of standard deviation scores (i.e., *z* scores) based on the standard normal distribution. Marascuilo and McSweeney (1977, p. 280) note that normal-scores tests are often described as **distribution-free tests**, insofar as the shape of the underlying population distribution(s) for the original data has little effect on the results of such tests. Because of their minimal assumptions, normal-scores tests are categorized as nonparametric tests.

Just as converting a set of interval/ratio data into rank-orders transforms the data into what is essentially a uniform distribution, transforming a set of ranks into normal-scores transforms the ranks into essentially a normal distribution. Although the normal-scores transformation results in some loss of the information contained in the original data, much of the information is retained, albeit in a different format. Conover (1980, 1999) notes that when the underlying population distributions for the original data are normal, by virtue of employing a normal-scores transformation the statistical power of the resulting normal-scores test will generally be equal to that of the analogous parametric test, and that when the underlying population distributions for the original data are not normal, the power of the normal-scores test may actually be higher than that of the analogous parametric test.[1]

If the result of the **van der Waerden normal-scores test for *k* independent samples** is significant, it indicates there is a significant difference between at least two of the samples in the set of *k* samples. Consequently, the researcher can conclude there is a high likelihood that at least two of the *k* samples are not derived from the same population. Because of the latter, there is a high likelihood that the magnitude of the scores in one distribution is greater than the magnitude of the scores in the other distribution.

As is the case with the **Kruskal–Wallis one-way analysis of variance by ranks (Test 22)**, in employing the **van der Waerden normal-scores test for *k* independent samples**, one of the following is true with regard to the rank-order data that are evaluated: a) The data are in a rank-order format, since it is the only format in which scores are available; or b) The data have been transformed into a rank-order format from an interval/ratio format, since the researcher has reason to believe that one or more of the assumptions of the **single-factor between-subjects analysis of variance (Test 21)** (which is the parametric analog of the **Kruskal–Wallis test** and the **van der Waerden test**) are saliently violated. It should be noted that when a researcher elects to transform a set of interval/ratio data into ranks, information is sacrificed. This latter fact accounts for why there is reluctance among some researchers to employ nonparametric tests such as the **Kruskal–Wallis one-way analysis of variance by ranks** and the **van der Waerden normal-scores test for k independent samples**, even if there is reason to believe that one or more of the assumptions of the **single-factor between-subjects analysis of variance** have been violated.

Conover (1980, 1999) notes that the **van der Waerden normal-scores test for k independent samples** is based on the same assumptions as the **Kruskal–Wallis one-way analysis of variance by ranks**, which are as follows: a) Each sample has been randomly selected from the population it represents; b) The k samples are independent of one another; c) The dependent variable (which is subsequently ranked) is a continuous random variable. In truth, this assumption, which is common to many nonparametric tests, is often not adhered to, in that such tests are often employed with a dependent variable that represents a discrete random variable; and d) The underlying distributions from which the samples are derived are identical in shape. The shapes of the underlying population distributions, however, do not have to be normal.

II. Example

Example 23.1 is identical to Examples 21.1/22.1 (which are respectively evaluated with the **single-factor between-subjects analysis of variance** and the **Kruskal–Wallis one-way analysis of variance by ranks**). In evaluating Example 23.1 it will be assumed that the ratio data (i.e., the number of nonsense syllables correctly recalled) are rank-ordered, since one or more of the assumptions of the **single-factor between-subjects analysis of variance** have been saliently violated.

Example 23.1 *A psychologist conducts a study to determine whether or not noise can inhibit learning. Each of 15 subjects is randomly assigned to one of three groups. Each subject is given 20 minutes to memorize a list of 10 nonsense syllables which she is told she will be tested on the following day. The five subjects assigned to* **Group 1,** *the* **no noise** *condition, study the list of nonsense syllables while they are in a quiet room. The five subjects assigned to* **Group 2,** *the* **moderate noise** *condition, study the list of nonsense syllables while listening to classical music. The five subjects assigned to* **Group 3,** *the* **extreme noise** *condition, study the list of nonsense syllables while listening to rock music. The number of nonsense syllables correctly recalled by the 15 subjects follows:* **Group 1:** *8, 10, 9, 10, 9;* **Group 2:** *7, 8, 5, 8, 5;* **Group 3:** *4, 8, 7, 5, 7. Do the data indicate that noise influenced subjects' performance?*

III. Null versus Alternative Hypotheses

Null hypothesis H_0: The $k = 3$ groups are derived from the same population.

(If the null hypothesis is supported, the averages of the normal-scores for each of the $k = 3$ groups will be equal (i.e., $\bar{z}_1 = \bar{z}_2 = \bar{z}_3$). If the latter is true, it indicates that the three underlying populations are equivalent with respect to the magnitude of the scores in each of the distributions.)

Alternative hypothesis H_1: At least two of the $k = 3$ groups are not derived from the same population.

(If the alternative hypothesis is supported, the averages of the normal-scores computed for at least two of the $k = 3$ groups will not be equal to one another. It is important to note that the alternative hypothesis does not require that $\bar{z}_1 \neq \bar{z}_2 \neq \bar{z}_3$, since the latter implies that all three normal-scores means must differ from one another. If the alternative hypothesis is supported, it indicates that at least two of the three populations are not equivalent with respect to the magnitude of the scores in each of the distributions. In this book it will be assumed (unless stated otherwise) that the alternative hypothesis for the **van der Waerden normal-scores test for k independent samples** is stated **nondirectionally**.)[2]

IV. Test Computations

The total number of subjects employed in the experiment is $N = 15$, and there are $n_j = n_1 = n_2 = n_3 = 5$ subjects in each group. The use of the **van der Waerden normal-scores test for k independent samples** assumes that the ratio scores of the $N = 15$ subjects have been rank-ordered in accordance with the rank-ordering procedure described for the **Kruskal–Wallis one-way analysis of variance by ranks** (the aforementioned rank-ordering procedure is described in Section IV of the latter test). Table 23.1 summarizes the data for Example 23.1. The table lists the following values for each of the $n_j = 5$ subjects in the $k = 3$ groups (where j indicates the jth group): a) The original ratio scores of the subjects (i.e., X_j, the number of nonsense syllables correctly recalled); b) The rank-order of each score (R_j) within the framework of the rank-ordering procedure described for the **Kruskal–Wallis one-way analysis of variance by ranks**; and c) The normal score value (z_j) computed for each rank-order.

Table 23.1 Summary of Data for Example 23.1

Group 1			Group 2			Group 3		
X_1	R_1	z_1	X_2	R_2	z_2	X_3	R_3	z_3
8	9.5	.24	7	6	−.32	4	1	−1.53
10	14.5	1.32	8	9.5	.24	8	9.5	.24
9	12.5	.78	5	3	−.89	7	6	−.32
10	14.5	1.32	8	9.5	.24	5	3	−.89
9	12.5	.78	5	3	−.89	7	6	−.32
	$\Sigma z_1 = 4.44$			$\Sigma z_2 = -1.62$			$\Sigma z_3 = -2.82$	
	$\bar{z}_1 = .888$			$\bar{z}_2 = -.324$			$\bar{z}_3 = -.564$	

As noted above, in order to conduct the **van der Waerden normal-scores test for k independent samples**, each of the rank-orders must be converted into a normal score. Since the latter conversion is based on cumulative proportions (or percentiles) for the normal distribution, at this point in the discussion the reader may want to review the following material: a) The material in the **Introduction** on percentiles and cumulative proportions (within the context of a cumulative frequency distribution), and the material on the normal distribution; and b) The material on cumulative proportions in Section I of the **Kolmogorov–Smirnov goodness-of-fit test for a single sample (Test 7)**.

The following protocol is employed to convert a rank-order into a normal score: Divide the rank-order by the value $(N + 1)$. The resulting value will be a proportion (which will be designated as P) that is greater than 0 but less than 1. The proportion which is computed is conceptualized as the percentile for that score (when the decimal point is moved two places to the right). The standard normal score (i.e., z value) which corresponds to that percentile will represent the normal score for that rank-order. The latter z value is obtained through use of **Table A1 (Table of the Normal Distribution)** in the **Appendix**.

To illustrate, the computation of the normal score of $z = .24$ for Subject 1 in Group 1 will now be explained. The original score for the subject in question is 8, which is assigned a rank-order of 9.5 within the framework of the **Kruskal–Wallis** ranking procedure. Employing the procedure described above for converting a rank-order into a proportion, the rank-order 9.5 is divided by $N + 1 = 15 + 1 = 16$. Thus, $P = 9.5/16 = .5938$. The latter proportion is conceptualized as a cumulative proportion or percentile in a normal distribution (by moving the decimal point two places to the right, the proportion .5938 is converted into 59.38%). This result tells us that the normal score for the rank-order 9.5 will be the z value which corresponds to the cumulative proportion .5938 in the normal distribution (or expressed within the framework of a percentile, the z value that falls at the 59.38[th] percentile). The point in question corresponds to the point on the normal distribution below which 59.38% of the cases fall (or to state it proportionally, .5938 is the proportion of cases which fall below that point). Since the latter proportion/percentile is above .50/50% (i.e., above the mean of the normal distribution), **the sign of the z value will be positive**. In order to identify the appropriate z value, we look in **Column 2** of **Table A1** for the value that is closest to $.5938 - .5000 = .0938$. The latter value is $z = .24$, since the entry .0948 in **Column 2** for $z = .24$ is closest to .0938.[3]

The general rule for determining the normal score for any proportion (P) that is greater than .5000 is to find in **Column 2** of **Table A1** the proportion (which we will designate as Q) that is closest to the difference between the value of P and .5000 (i.e., $P - .5000 = Q$). The z value in the row which corresponds to the value of Q will be the normal score for the proportion P.

The computation of a normal score for a proportion that is less than .5000 will now be explained. To demonstrate this we will use the normal score of $z = -1.53$ for Subject 1 in Group 3. The original score for the subject in question is 4, which is assigned a rank-order of 1 within the framework of the **Kruskal–Wallis** ranking procedure. Employing the protocol for converting a rank-order into a proportion, the rank-order 1 is divided by $N + 1 = 15 + 1 = 16$. Thus, $P = 1/16 = .0625$. The latter proportion is conceptualized as a cumulative proportion or percentile in a normal distribution (by moving the decimal point two places to the right, the proportion .0625 is converted into 6.25%). This result tells us that the normal score for the rank-order 1 will be the z value which corresponds to the cumulative proportion .0625 in the normal distribution (or expressed within the framework of a percentile, the z value that falls at the 6.25[th] percentile). The point in question corresponds to the point on the normal distribution below which 6.25% of the cases fall (or to state it proportionally, .0625 is the proportion of cases that fall below that point). Since the latter proportion/percentile is below .50/50% (i.e., below the mean of the normal distribution), **the sign of the z value will be negative**. In order to identify the appropriate z value, we look in **Column 3** of **Table A1** for the value that is closest to .0625. The latter value is $z = 1.53$, since the entry .0630 in **Column 3** for $z = 1.53$ is closest to .0625.

The general rule for determining the normal score for any proportion (P) that is less than .5000 is to find in **Column 3** of **Table A1** the proportion which is closest to the value of P. The z value in the row that corresponds to the value of P will be the normal score for the proportion P. A negative sign is assigned to the obtained z value.

After computing a normal score for each of the 15 rank-orders in Table 23.1, the sum of the normal-scores is computed for each group. The latter values in Table 23.1 are $\Sigma z_1 = 4.44$, $\Sigma z_2 = -1.62$, $\Sigma z_3 = -2.82$. An average normal score for each group is computed by dividing the sum of the normal-scores for the group by the number of subjects in the group. Thus: $\bar{z}_1 = \Sigma z_1/n_1 = 4.44/5 = .888$, $\bar{z}_2 = \Sigma z_2/n_2 = -1.62/5 = -.324$, and $\bar{z}_3 = \Sigma z_3/n_3 = -2.82/5 = -.564$. In order to compute the test statistic for the **van der Waerden normal-scores test for**

k independent samples, it is necessary to compute the estimated population variance ($\bar{s}^2$) of the normal-scores. The latter is computed with Equation 23.1.

$$\bar{s}^2 = \frac{\sum\limits_{j=1}^{k} \sum\limits_{i=1}^{n} z_{ij}^2}{N - 1}$$ **(Equation 23.1)**

The notation in Equation 23.1 indicates the following: a) If we let z_{ij} represent the normal score for the ith subject in Group j, then $\sum_{j=1}^{k} \sum_{i=1}^{n} z_{ij}^2$ indicates that each of the N (i.e., $N = (k)(n)$) normal-scores is squared, and the N squared normal-scores are summed; and b) The sum of the squared normal-scores is divided by $(N - 1)$, which yields the value of the variance. Employing Equation 23.1, the value $\bar{s}^2 = .7482$ is computed. Conover (1980, 1999) notes that the computed value for the variance will generally be close to 1.[4]

$$\bar{s}^2 = \frac{(.24)^2 + (1.32)^2 + ... + (-.89)^2 + (-.32)^2}{15 - 1} = .7482$$

The chi-square distribution provides an excellent estimate of the exact probability distribution for the **van der Waerden** test statistic. The chi-square estimate of the test statistic (to be designated χ^2_{vdw}) is computed with Equation 23.2.[5]

$$\chi^2_{vdw} = \frac{\sum\limits_{j=1}^{k} n_j (\bar{z}_j)^2}{\bar{s}^2}$$ **(Equation 23.2)**

The notation in Equation 23.2 indicates the following: a) In the numerator of the equation, the number of subjects in each group (n_j) is multiplied by the square of the mean of the normal-scores for that group. Upon doing the latter for all k groups, the k resulting values are summed; and b) The final sum obtained in part a) is divided by the estimated population variance computed with Equation 23.1.

When Equation 23.2 is employed to compute the test statistic for the **van der Waerden normal-scores test for** k **independent samples**, the value $\chi^2_{vdw} = 8.10$ is obtained.

$$\chi^2_{vdw} = \frac{(5)(.888)^2 + (5)(-.324)^2 + (5)(-.564)^2}{.7482} = 8.10$$

V. Interpretation of the Test Results

In order to reject the null hypothesis, the computed value χ^2_{vdw} must be equal to or greater than the tabled critical chi-square value at the prespecified level of significance. The computed chi-square value is evaluated with **Table A4 (Table of the Chi-Square Distribution)** in the **Appendix**. For the appropriate degrees of freedom, the tabled $\chi^2_{.95}$ value (which is the chi-square value at the 95th percentile) and the tabled $\chi^2_{.99}$ value (which is the chi-square value at the 99th percentile) are employed as the .05 and .01 critical values for evaluating a nondirectional alternative hypothesis. The number of degrees of freedom employed in the analysis are computed with Equation 23.3. Thus, $df = 3 - 1 = 2$.

$$df = k - 1$$ **(Equation 23.3)**

For $df = 2$, the tabled critical .05 and .01 chi-square values are $\chi^2_{.05} = 5.99$ and $\chi^2_{.01} = 9.21$. Since the computed value $\chi^2_{vdw} = 8.10$ is greater than $\chi^2_{.05} = 5.99$, the alternative hypothesis is supported at the .05 level. Since, however, $\chi^2_{vdw} = 8.10$ is less than $\chi^2_{.01} = 9.21$, the alternative hypothesis is not supported at the .01 level. A summary of the analysis of Example 23.1 with the **van der Waerden normal-scores test for k independent samples** follows: It can be concluded there is a significant difference between at least two of the three groups exposed to different levels of noise. This result can be summarized as follows: $\chi_{vdw}(2) = 8.10, p < .05$.

It should be noted that when the data for Example 23.1 are evaluated with the **Kruskal–Wallis one-way analysis of variance by ranks**, the identical conclusions are reached (i.e., the null hypothesis can be rejected at the .05 level, but not at the .01 level — although the **Kruskal–Wallis test** results in a slightly larger chi-square value). When the data are evaluated with **single-factor between-subjects analysis of variance**, the null hypothesis can be rejected at both the .05 and .01 levels (although it barely achieves significance at the latter level). The slight discrepancy between the results of the **van der Waerden test** (as well as the **Kruskal–Wallis test**) and the **analysis of variance** suggests that in the case of Examples 21.1/22.1/23.1, it would appear that the **analysis of variance** provides a slightly more powerful test of the alternative hypothesis than either the **van der Waerden** or **Kruskal–Wallis tests**.

VI. Additional Analytical Procedures for the van der Waerden Normal-Scores Test for k Independent Samples and/or Related Tests

1. Pairwise comparisons following computation of the test statistic for the van der Waerden normal-scores test for k independent samples   Prior to reading this section the reader should review the discussion of comparisons in Section VI of the **single-factor between-subjects analysis of variance**. As is the case with the omnibus F value computed for the **single-factor between-subjects analysis of variance**, the χ^2_{vdw} value computed with Equation 23.2 is based on an evaluation of all k groups. When the value of χ^2_{vdw} is significant, it does not indicate whether just two or, in fact, more than two groups differ significantly from one another. In order to answer the latter question, it is necessary to conduct comparisons contrasting specific groups with one another. This section will describe a methodology which can be employed for conducting **simple/pairwise comparisons** following the computation of an χ^2_{vdw} value.[6]

In conducting a simple comparison between any two groups to be designated a and b, the null hypothesis and nondirectional alternative hypothesis are as follows: H_0: Groups a and b are derived from identical population distributions; H_1: Groups a and b are not derived from identical population distributions. The alternative hypothesis can also be stated directionally insofar as the researcher can predict that the magnitude of the scores in one distribution is greater than the magnitude of the scores in the other distribution. As is the case with the omnibus null hypothesis, the decision made with regard to the null hypothesis for a comparison will be a function of the magnitude of the difference between the averages of the normal-scores for the groups that are involved in the comparison.

Conover (1980, 1999) describes the use of Equation 23.4 to conduct comparisons for the **van der Waerden normal-scores test for k independent samples**. The latter equation allows a researcher to identify the minimum required difference between the means of the normal-scores of any two groups (designated as CD_{vdw}) in order for them to differ from one another at the prespecified level of significance.

$$CD_{vdw} = t_{adj} \sqrt{\tilde{s}^2 \left(\frac{N - 1 - \chi^2_{vdw}}{N - k} \right) \left(\frac{1}{n_a} + \frac{1}{n_b} \right)} \qquad \textbf{(Equation 23.4)}$$

Where: n_a and n_b represent the number of subjects in each of the groups involved in the simple comparison

The value of t_{adj} is obtained from the **Table A2 (Table of Student's t Distribution)** in the **Appendix**. In the case of a nondirectional alternative hypothesis, t_{adj} is the t value for $df = N - k$, above which a proportion of cases corresponding to the value $\alpha_{FW}/2c$ falls (where c is the total number of comparisons that are conducted). In Example 23.1, $df = 15 - 3 = 12$. In the case of a directional alternative hypothesis, t_{adj} is the t value above which a proportion of cases corresponding to the value α_{FW}/c falls. When all possible pairwise comparisons are made $c = [k(k - 1)]/2$, and thus, $2c = k(k - 1)$. In Example 23.1 the number of pairwise/simple comparisons that can be conducted is $c = [3(3 - 1)]/2 = 3$ — specifically, Group 1 versus Group 2, Group 1 versus Group 3, and Group 2 versus Group 3.

The value of t_{adj} will be a function of both the maximum **familywise Type I error rate** (α_{FW}) the researcher is willing to tolerate and the total number of comparisons that are conducted. When a limited number of comparisons are planned prior to collecting the data, most sources take the position that a researcher is not obliged to control the value of α_{FW}. In such a case, the **per comparison Type I error rate** (α_{PC}) will be equal to the prespecified value of alpha. When α_{FW} is not adjusted, the value of t_{adj} employed in Equation 23.4 will be the tabled critical t value which corresponds to the prespecified level of significance. Thus, if a nondirectional alternative hypothesis is employed and $\alpha = \alpha_{PC} = .05$, for $df = 12$, the tabled critical two-tailed .05 value $t_{.05} = 2.179$ is used to represent t_{adj} in Equation 23.4. If $\alpha = \alpha_{PC} = .01$, the tabled critical two-tailed .01 value $z_{.01} = 3.055$ is used in Equation 23.4. In the same respect, if a directional alternative hypothesis is employed, the tabled critical .05 and .01 one-tailed values $t_{.05} = 1.782$ and $t_{.01} = 2.681$ are used for t_{adj} in Equation 23.4.

When comparisons are not planned beforehand, it is generally acknowledged that the value of α_{FW} must be controlled so as not to become excessive. The general approach for controlling the latter value is to establish a **per comparison Type I error rate** which insures that α_{FW} will not exceed some maximum value stipulated by the researcher. One method for doing this (described under the **single-factor between-subjects analysis of variance** as the **Bonferroni–Dunn method**) establishes the **per comparison Type I error rate** by dividing the maximum value one will tolerate for the **familywise Type I error rate** by the total number of comparisons conducted. Thus, in Example 23.1, if one intends to conduct all three pairwise comparisons and wants to insure that α_{FW} does not exceed .05, $\alpha_{PC} = \alpha_{FW}/c = .05/3 = .0167$. The latter proportion is used to determine the value of t_{adj}. As noted earlier, if a directional alternative hypothesis is employed for a comparison, the value of t_{adj} employed in Equation 23.4 is the t value above which a proportion equal to $\alpha_{PC} = \alpha_{FW}/c$ of the cases falls. In **Table A2**, by interpolation (since the exact value is not listed) the t value which corresponds approximately to the proportion .0167 (for $df = 12$) is $t = 2.35$. By employing $t_{adj} = 2.35$ in Equation 23.4, one can be assured that within the "family" of three pairwise comparisons, α_{FW} will not exceed .05 (assuming all of the comparisons are directional). If a nondirectional alternative hypothesis is employed for all of the comparisons, the value of t_{adj} will be the t value above which a proportion equal to

$\alpha_{FW}/2c = \alpha_{PC}/2$ of the cases falls. Since $\alpha_{PC}/2 = .0167/2 = .0083$, $t = 2.75$. By employing $t_{adj} = 2.75$ in Equation 23.4, one can be assured that α_{FW} will not exceed .05.[7]

In order to employ the CD_{vdw} value computed with Equation 23.4, it is necessary to determine the mean normal score for each of the k groups (which we have already done), and then compute the absolute value of the difference between the mean normal-scores of each pair of groups that are compared.[8] In Table 23.1 the following values for the mean normal-scores of the groups are computed: $\bar{z}_1 = .888$, $\bar{z}_2 = -.324$, $\bar{z}_3 = -.564$. Employing the latter values, Table 23.2 summarizes the difference scores between pairs of mean normal-scores.

Table 23.2 Difference Scores between Pairs of Mean Normal-Scores for Example 23.1

$$|\bar{z}_1 - \bar{z}_2| = |.888 - (-.324)| = 1.212$$
$$|\bar{z}_1 - \bar{z}_3| = |.888 - (-.564)| = 1.452$$
$$|\bar{z}_2 - \bar{z}_3| = |(-.324) - (-.564)| = .24$$

If any of the differences between mean normal-scores is equal to or greater than the CD_{vdw} value computed with Equation 23.4, a comparison is declared significant. Equation 23.4 will now be employed to evaluate the nondirectional alternative hypothesis for all three pairwise comparisons. Since it will be assumed that the comparisons are unplanned and that the researcher does not want the value of α_{FW} to exceed .05, the value $t_{adj} = 2.75$ will be used in computing CD_{vdw}.

$$CD_{vdw} = (2.75)\sqrt{(.7482)\left(\frac{15 - 1 - 8.10}{15 - 3}\right)\left(\frac{1}{5} + \frac{1}{5}\right)} = (2.75)(.3836) = 1.055$$

The obtained value $CD_{vdw} = 1.055$ indicates that any difference between the mean normal-scores of two groups which is equal to or greater than 1.055 is significant. With respect to the three pairwise comparisons, the differences $|\bar{z}_1 - \bar{z}_2| = 1.212$ and $|\bar{z}_1 - \bar{z}_3| = 1.452$ are greater than $CD_{vdw} = 1.055$. Thus, we can conclude there is a significant difference between the performance of Group 1 and Group 2 and between the performance of Group 1 and Group 3.[9]

The same logic employed for computing a confidence interval for a comparison described in Section VI of the **single-factor between-subjects analysis of variance** can be employed to compute a confidence interval for the **van der Waerden** test model. Specifically: Add and subtract the computed value of CD_{vdw} to and from the obtained difference between the two mean normal-scores involved in a comparison. Thus, $CI_{.95}$ (based on $\alpha_{FW} = .05$) for the comparison Group 1 versus Group 2 is computed as follows: $CI_{.95} = 1.212 \pm 1.055$. In other words, the researcher can be 95% confident (or the probability is .95) the interval .157 to 2.267 contains the true difference between the normal-score means of the populations represented by Groups 1 and 2. If the aforementioned interval contains the true difference between the normal score means, the mean of the normal-scores in the population represented by Group 1 is between .157 to 2.267 standard deviation units larger than the mean of the normal-scores in the population represented by Group 2.

VII. Additional Discussion of the van der Waerden Normal-Scores Test for k Independent Samples

1. Alternative normal-scores tests Alternative normal-scores procedures have been developed by Terry and Hoeffding (Terry (1952)) and Bell and Doksum (1965). The latter procedures are described in Marascuilo and McSweeney (1977), who also discuss the extension of normal-scores tests to other experimental designs (e.g., within-subjects designs). The alternative normal-scores tests employ a different procedure from the one described for the **van der Waerden test** for determining a normal score for each of the rank-orders in a sample. For example, the **Bell–Doksum normal-scores test** (1965) obtains N random normal deviates (which are typically generated with a pseudorandom number generator, which is discussed in Section IX (the **Addendum**) of the **single-sample runs test (Test 10)**), ordinally arranges the N random normal deviates (i.e., the N z scores), and then pairs each of the random deviates (z scores) with the rank-order in the same ordinal position.

VIII. Additional Examples Illustrating the Use of the van der Waerden Normal-Scores Test for k Independent Samples

The **van der Waerden normal-scores test for k independent samples** can be employed to evaluate any of the additional examples noted for the **single-factor between-subjects analysis of variance**, if the data for the latter examples are rank-ordered, and the rank-orders are then converted into normal-scores. The **van der Waerden test** can also be employed to evaluate Examples 22.2 and 22.3, which are presented in Section VIII of the **Kruskal–Wallis one-way analysis of variance by ranks**.

As noted earlier, the **van der Waerden normal-scores test for k independent samples** can be employed to evaluate a design involving $k = 2$ independent samples/groups. Thus, the test can be used to evaluate the data for any of the examples noted for the t **test for two independent samples (Test 11)** and the **Mann–Whitney U test (Test 12)** (assuming the interval/ratio scores are rank-ordered, and the ranks are then converted into normal-scores). Example 23.2 is identical to Example 11.1/12.1, which is employed to illustrate the use of the t **test for two independent samples** and the **Mann–Whitney U test**. The example will be employed to illustrate the use of the **van der Waerden normal-scores test for k independent samples** when there are $k = 2$ independent samples/groups.

Example 23.2 *In order to assess the efficacy of a new antidepressant drug, ten clinically depressed patients are randomly assigned to one of two groups. Five patients are assigned to* Group 1, *which is administered the antidepressant drug for a period of six months. The other five patients are assigned to* Group 2, *which is administered a placebo during the same six-month period. Assume that prior to introducing the experimental treatments, the experimenter confirmed that the level of depression in the two groups was equal. After six months elapse, all ten subjects are rated by a psychiatrist (who is blind with respect to a subject's experimental condition) on their level of depression. The psychiatrist's depression ratings for the five subjects in each group follow (the higher the rating, the more depressed a subject):* **Group 1**: 11, 1, 0, 2, 0; **Group 2**: 11, 11, 5, 8, 4. *Do the data indicate that the antidepressant drug is effective?*

Table 23.3 summarizes the analysis of the data with the **van der Waerden normal-scores test for k independent samples**. The null hypothesis evaluated with the test is: H_0: The $k = 2$ groups are derived from the same population. The nondirectional alternative hypothesis evaluated with the test is: H_1: The $k = 2$ groups are not derived from the same population. If a directional alternative hypothesis is employed, it will predict that the two groups are derived from different populations, but more specifically, that the magnitude of the scores in Group 2 will be larger than the magnitude of the scores in Group 1.

The computed value for the **van der Waerden test statistic** is $\chi^2_{vdw} = 3.16$. Since $k = 2$, $df = 2 - 1 = 1$. Employing **Table A4**, we determine that for $df = 1$, the tabled critical two-tailed values that are employed to evaluate a nondirectional alternative hypothesis are $\chi^2_{.05} = 3.84$ and $\chi^2_{.01} = 6.63$. Since the computed value $\chi^2_{vdw} = 3.16$ is less than $\chi^2_{.05} = 3.84$, the nondirectional alternative hypothesis is not supported at the .05 level. The tabled critical one-tailed values which are employed for evaluating a directional alternative hypothesis are $\chi^2_{.05} = 2.71$ and $\chi^2_{.01} = 5.10$.[10] Since the computed value $\chi^2_{vdw} = 3.16$ is greater than $\chi^2_{.05} = 2.71$, but less than $\chi^2_{.01} = 5.10$, the directional alternative hypothesis predicting that Group 1 (the group which received the drug) will have a lower level of depression is supported at the .05 level, but not at the .01 level. This result is consistent with that obtained when the same data are evaluated with the *t* **test for two independent samples** and the **Mann–Whitney** *U* **test** — in other words, in the case of the latter tests, the same directional alternative hypothesis is supported, but only at the .05 level.

Table 23.3   Summary of Data for Example 23.2

Group 1			Group 2		
X_1	R_1	z_1	X_2	R_2	z_2
11	9	.91	11	9	.91
1	3	−.60	11	9	.91
0	1.5	−1.10	5	6	.11
2	4	−.35	8	7	.35
0	1.5	−1.10	4	5	−.11

$$\Sigma z_1 = -2.24 \qquad\qquad \Sigma z_2 = 2.17$$
$$\bar{z}_1 = -.448 \qquad\qquad \bar{z}_2 = .434$$

$$\tilde{s}^2 = \frac{(.91)^2 + (-.60)^2 + \ldots + (.35)^2 + (-.11)^2}{10 - 1} = .6148$$

$$\chi^2_{vdw} = \frac{(5)(-.448)^2 + (5)(.434)^2}{.6148} = 3.16$$

References

Bell, C. B. & Doksum, K. A. (1965). Some new distribution-free statistics. **Annals of Mathematical Statistics**, 36, 203–214.

Conover, W. J. (1980). **Practical nonparametric statistics** (2nd ed.). New York: John Wiley & Sons.

Conover, W. J. (1999). **Practical nonparametric statistics** (3rd ed.). New York: John Wiley & Sons.

Daniel, W. W. (1990). **Applied nonparametric statistics** (2nd ed.). Boston: PWS–Kent Publishing Company.

Marascuilo, L. A. & McSweeney, M. (1977). **Nonparametric and distribution-free methods for the social sciences**. Monterey, CA: Brooks/Cole Publishing Company.

Sheskin, D. J. (1984). **Statistical tests and experimental design: A guidebook**. New York: Gardner Press.

Siegel, S. & Castellan, N. J., Jr. (1988). **Nonparametric statistics for the behavioral sciences** (2nd ed.). New York: McGraw–Hill Book Company.

Sprent, P. (1993). **Applied nonparametric statistical methods** (2nd ed.). London: Chapman & Hall

Terry, M. E. (1952). Some rank-order tests, which are most powerful against specific parametric alternatives. **Annals of Mathematical Statistics, 23**, 346–366.

Van der Waerden, B. L. (1952/1953). Order tests for the two-sample problem and their power. **Proceedings Koninklijke Nederlandse Akademie van Wetenshappen** (A), 55 (**Indagationes Mathematicae** 14), 453–458, & 56 (**Indagationes Mathematicae,** 15), 303–316 (corrections appear in Vol. 56, p. 80).

Endnotes

1. Conover (1980, 1999) and Marascuilo and McSweeney (1977) note that normal-scores tests have power equal to or greater than their parametric analogs. The latter sources state that the **asymptotic relative efficiency** (which is discussed in Section VII of the **Wilcoxon signed-ranks test (Test 6)**) of a normal-scores test is equal to 1 when the underlying population distribution(s) are normal, and often greater than 1 when the underlying population distribution(s) are something other than normal. What the latter translates into is that for a given level of power, a normal-scores test will require an equal number of or even fewer subjects than the analogous parametric test in evaluating an alternative hypothesis.

2. Although it is possible to conduct a directional analysis when $k \geq 3$, such an analysis will not be described with respect to the **van der Waerden normal-scores test for k independent samples**. A discussion of a directional analysis when $k = 2$ can be found in Section VIII where the **van der Waerden test** is employed to evaluate the data for Examples 11.1/12.1 (which are employed to illustrate the *t* **test for two independent samples** and the **Mann–Whitney U test**). A discussion of the evaluation of a directional alternative hypothesis when $k \geq 3$ can be found in Section VII of the **chi-square goodness-of-fit test (Test 8)**. Although the latter discussion is in reference to analysis of a *k* independent samples design involving categorical data, the general principles regarding analysis of a directional alternative hypothesis when $k \geq 3$ are applicable to the **van der Waerden normal-scores test for k independent samples**.

3. The proportion of cases in the normal distribution that falls below the mean is .5000. The value .0948 in **Column 2** represents the proportion of cases which falls between the mean and the value $z = .24$. Thus, $.5000 + .0948 = .5948$ represents the proportion of cases that falls below the value $z = .24$.

4. Conover (1980, 1999) notes that if there are no ties, the mean of the N z_{ij} scores (i.e., the mean of all N normal-scores) will equal zero, and be extremely close to zero when there are ties. Thus, if the mean equals zero, the equation $\tilde{s}^2 = \Sigma(X - \bar{X})^2/(n - 1)$ (which is Equation I.8, the definitional equation for computing the unbiased estimate of a population variance) reduces to $\tilde{s}^2 = \Sigma X^2/(n - 1)$. If z is employed in place of X and N in place of n (since in Equation 23.1 the variance of N z scores is computed), we obtain Equation 23.1, $\tilde{s}^2 = \Sigma z_{ij}^2/(N - 1)$.

5. When two or more groups do not have the same sample size, the value of n_j for a given group is used to represent the group sample size in any of the equations which require the group sample size.

6. In the discussion of comparisons under the **single-factor between-subjects analysis of variance**, it is noted that a **simple** (also known as a **pairwise**) **comparison** is a comparison between any two groups in a set of k groups.

7. The rationale for the use of the proportions .0167 and .0083 is explained more thoroughly in Endnote 5 of the **Kruskal–Wallis one-way analysis of variance**. In the case of the latter test, since the normal distribution is employed in the comparison procedure, the explanation of the proportions is in reference to a standard normal deviate (i.e., a z value). The same rationale applies when the t distribution is employed, with the only difference being that for a corresponding probability level, the t values which are used are different from the z values employed for the **Kruskal–Wallis** comparison procedure.

    The values $t_{.0167} = 2.35$ and $t_{.0083} = 2.75$ are based on interpolating the t values in **Table A2** (since exact values are not listed for $t_{.0167}$ and $t_{.0083}$). The value $t_{.0167} = 2.35$ is the best estimate of the t value at the 98.33^{th} percentile (since $1 - .9833 = .0167$). The value $t_{.0083} = 2.75$ is the best estimate of the t value at the 99.17^{th} percentile (since $1 - .9917 = .0083$).

8. It should be noted that when a directional alternative hypothesis is employed, the sign of the difference between the two mean normal-scores is relevant in that it must be consistent with the prediction stated in the directional alternative hypothesis. When a nondirectional alternative hypothesis is employed, the direction of the difference between two mean normal-scores is irrelevant.

9. Equation 23.4 is analogous to Equation 22.8, which Conover (1980, 1999) employs in conducting a comparison for the **Kruskal–Wallis one-way analysis of variance**. If the element $[(N - 1 - \chi^2_{vdw})/(N - k)]$ is omitted from the radical in Equation 23.4, it becomes Equation 23.5, which is analogous to Equation 22.5. Equation 23.5 will yield a larger CD_{vdw} value than Equation 23.4. It is demonstrated below that when Equation 23.5 is employed with the data for Example 23.1, the value $CD_{vdw} = 1.504$ is obtained. If the latter CD_{vdw} value is employed, none of the pairwise comparisons are significant, since no difference is equal to or greater than 1.504. As noted in the discussion of comparisons in the **Kruskal–Wallis one-way analysis of variance**, an equation in the form of Equation 23.5 conducts a less powerful/more conservative comparison.

    (Equation 23.5)

$$CD_{vdw} = t_{adj.}\sqrt{\tilde{s}^2\left(\frac{1}{n_a} + \frac{1}{n_b}\right)} = (2.75)\sqrt{(.7482)\left(\frac{1}{5} + \frac{1}{5}\right)} = 1.504$$

    In Section VI of the **Kruskal–Wallis one-way analysis of variance** it is noted that sources do not agree on the methodology for conducting pairwise comparisons for the latter test. The comments regarding comparisons in Endnote 7 of the **Kruskal–Wallis test** can be generalized to the **van der Waerden normal-scores test for k independent samples** (since sources do not agree on the comparison protocol for the **van der Waerden test**). Thus, if two or more methods for conducting comparisons yield dramatically different results for a specific comparison, one or more replication studies employing reasonably large sample sizes should clarify whether or not an obtained difference is reliable, as well as the magnitude of the difference (if, in fact, one exists).

10. The tabled critical one-tailed .05 and .01 values are, respectively, the tabled chi-square values at the 90th and 98th percentiles/quantiles of the chi-square distribution. For clarification regarding the latter values, the reader should review the material on the evaluation of a directional hypothesis involving the chi-square distribution in Section VII of the **chi-square goodness-of-fit test**, and the discussion of **Table A4** in Section IV of the **single-sample chi-square test for a population variance (Test 3)**. Since the chi-square value at the 98th percentile is not in **Table A4**, the value $\chi^2_{.01} = 5.10$ is an approximation of the latter value.

Inferential Statistical Tests Employed with Two or More Dependent Samples (and Related Measures of Association/Correlation)

Test 24

The Single-Factor Within-Subjects Analysis of Variance
(Parametric Test Employed with Interval/Ratio Data)

I. Hypothesis Evaluated with Test and Relevant Background Information

Hypothesis evaluated with test In a set of k dependent samples (where $k \geq 2$), do at least two of the samples represent populations with different mean values?

Relevant background information on test Prior to reading this section the reader should review the general comments on the analysis of variance in Section I of the **single-factor between-subjects analysis of variance (Test 21)**. In addition, the general information regarding a dependent samples design contained in Sections I and VII of the **t test for two dependent samples (Test 17)** should be reviewed. The **single-factor within-subjects analysis of variance** (which is also referred to as the **single-factor repeated-measures analysis of variance** and the **randomized-blocks one-way analysis of variance**)[1] is employed in a hypothesis testing situation involving k dependent samples. In contrast to the **t test for two dependent samples**, which only allows for a comparison of the means of two dependent samples, the **single-factor within-subjects analysis of variance** allows for comparison of two or more dependent samples. When the number of dependent samples is $k = 2$, the **single-factor within-subjects analysis of variance** and the **t test for two dependent samples** yield equivalent results.

In conducting the **single-factor within-subjects analysis of variance**, each of the k sample means is employed to estimate the value of the mean of the population the sample represents. If the computed test statistic is significant, it indicates there is a significant difference between at least two of the sample means in the set of k means. As a result of the latter, the researcher can conclude there is a high likelihood at least two of the samples represent populations with different mean values.

In order to compute the test statistic for the **single-factor within-subjects analysis of variance**, the following two variability components (which are part of the **total variability**) are contrasted with one another: **between-conditions variability** and **residual variability**. **Between-conditions variability** (which is also referred to as **treatment variability**) is essentially a measure of the variance of the means of the k experimental conditions. **Residual variability** is the amount of variability within the k scores of each of the n subjects which cannot be accounted for on the basis of a treatment effect. **Residual variability** is viewed as variability which results from chance factors that are beyond the control of a researcher, and since chance factors are often referred to as experimental error, **residual variability** is also referred to as **error variability**. The F **ratio**, which is the test statistic for the **single-factor within-subjects analysis of variance**, is obtained by dividing **between-conditions variability** by **residual variability**. Since **residual variability** is employed as a baseline measure of the variability in a set of data which is beyond a researcher's control, it is assumed that if the k experimental conditions represent populations with the same mean value, the amount of variability between the means of the k experimental conditions (i.e., **between-conditions variability**) will be approximately the same value as the **residual variability**. If, on the other hand, **between-conditions variability** is significantly larger than **residual variability** (in which case the value of the F ratio will be larger than 1), it is likely that something in addition to chance factors is contributing to the amount of variability between the means of the experimental conditions. In such a case, it is assumed that whatever it is which differentiates the experimental conditions from one another (i.e., the independent variable/experimental treatments) accounts for the fact that **between-conditions variability** is larger than **residual variability**.[2] A thorough discussion of the logic underlying the **single-factor within-subjects analysis of variance** can be found in Section VII.

The **single-factor within-subjects analysis of variance** is employed with interval/ratio data and is based on the following assumptions: a) The sample of n subjects has been randomly selected from the population it represents; b) The distribution of data in the underlying populations each of the experimental conditions represents is normal; and c) The third assumption, which is referred to as the **sphericity assumption**, is the analog of the homogeneity of variance assumption of the **single-factor between-subjects analysis of variance**. The assumption of sphericity, which is mathematically more complex than the homogeneity of variance assumption, essentially revolves around the issue of whether or not the underlying population variances and covariances are equal. A full discussion of the sphericity assumption (as well as the concept of **covariance**) can be found in Section VI. It should also be noted that the **single-factor within-subjects analysis of variance** is more sensitive to violations of its assumptions than is the **single-factor between-subjects analysis of variance**.

As is the case for the *t* **test for two dependent samples**, in order for the **single-factor within-subjects analysis of variance** to generate valid results the following guidelines should be adhered to: a) In order to control for **order effects** (which are discussed in Section VII under the Latin square design) the presentation of the k experimental conditions should be random or, if appropriate, be counterbalanced; and b) If matched samples are employed, within each set of matched subjects each of the subjects should be randomly assigned to one of the k experimental conditions.

As is the case with the *t* **test for dependent samples**, when $k = 2$ the **single-factor within-subjects analysis of variance** can be employed to evaluate a **one-group pretest-posttest design**, as well as extensions of the latter design which involve more than two measurement periods. The limitations of the **one-group pretest-posttest design** (which are discussed in the **Introduction** and in Section VII of the *t* **test for dependent samples**) are also applicable when it is evaluated with the **single-factor within-subjects analysis of variance**.

The reader should take note of the fact that there are certain advantages associated with employing a within-subjects design as opposed to a between-subjects design. If within-subjects and between-subjects designs which evaluate the same hypothesis and involve the same number of scores in each of the experimental conditions are compared with one another, the number of subjects required for the within-subjects analysis is a fraction (specifically, $1/k^{th}$) of the number required for the between-subjects analysis. Another advantage of a within-subjects analysis is that it provides for a more powerful test of an alternative hypothesis. The latter can be attributed to the fact that the error variability associated with a within-subjects analysis is less than that associated with a between-subjects analysis.[3] In spite of the aforementioned advantages of employing a within-subjects design, the between-subjects design is more commonly employed in research, since in many experiments it is impractical for a subject to serve in more than one experimental condition.

II. Example

Example 24.1 *A psychologist conducts a study to determine whether or not noise can inhibit learning. Each of six subjects is tested under three experimental conditions. In each of the experimental conditions a subject is given 20 minutes to memorize a list of 10 nonsense syllables, which the subject is told she will be tested on the following day. The three experimental conditions each subject serves under are as follows:* **Condition 1,** *the* **no noise** *condition, requires subjects to study the list of nonsense syllables in a quiet room.* **Condition 2,** *the* **moderate noise** *condition, requires subjects to study the list of nonsense syllables while listening to classical music.* **Condition 3,** *the* **extreme noise** *condition, requires subjects to study the list of nonsense syllables while listening to rock music. Although in each of the experimental conditions subjects are presented with a different list of nonsense syllables, the three lists are comparable with respect to those variables that are known to influence a person's ability to learn nonsense syllables. To control for order effects, the order of presentation of the three experimental conditions is completely counterbalanced.*[4] *The number of nonsense syllables correctly recalled by the six subjects under the three experimental conditions follows. (Subjects' scores are listed in the order* **Condition 1, Condition 2, Condition 3.**) **Subject 1:** 9, 7, 4; **Subject 2:** 10, 8, 7; **Subject 3:** 7, 5, 3; **Subject 4:** 10, 8, 7; **Subject 5:** 7, 5, 2; **Subject 6:** 8, 6, 6. *Do the data indicate that noise influenced subjects' performance?*

III. Null versus Alternative Hypotheses

Null hypothesis H_0: $\mu_1 = \mu_2 = \mu_3$

(The mean of the population Condition 1 represents equals the mean of the population Condition 2 represents equals the mean of the population Condition 3 represents.)

Alternative hypothesis H_1: Not H_0

(This indicates there is a difference between at least two of the $k = 3$ population means. It is important to note that the alternative hypothesis should not be written as follows: H_1: $\mu_1 \neq \mu_2 \neq \mu_3$. The reason why the latter notation for the alternative hypothesis is incorrect is because it implies that all three population means must differ from one another in order to reject the null hypothesis. In this book it will be assumed (unless stated otherwise) that the alternative hypothesis for the analysis of variance is stated **nondirectionally**.[5] In order to reject the null hypothesis, the obtained F value must be equal to or greater than the tabled critical F value at the prespecified level of significance.)

IV. Test Computations

The test statistic for the **single-factor within-subjects analysis of variance** can be computed with either **computational** or **definitional equations**. Although definitional equations reveal the underlying logic behind the analysis of variance, they involve considerably more calculations than do the computational equations. Because of the latter, computational equations will be employed in this section to demonstrate the computation of the test statistic. The definitional equations for the **single-factor within-subjects analysis of variance** are described in Section VII.

The data for Example 24.1 are summarized in Table 24.1. In Table 24.1 the $k = 3$ scores of the $n = 6$ subjects in Conditions 1, 2, and 3 are, respectively, listed in the columns labeled X_1,

X_2, and X_3. The notation n is employed to represent the number of scores in each of the experimental conditions. Since there are $n = 6$ scores in each condition, $n = n_1 = n_2 = n_3 = 6$. The columns labeled X_1^2, X_2^2, and X_3^2 list the squares of the scores of the six subjects in each of the three experimental conditions. The last column labeled ΣS_i lists for each of the six subjects the sum of a subject's $k = 3$ scores. Thus, the value ΣS_i is the sum of the scores for Subject i under Conditions 1, 2, and 3.

Table 24.1   Data for Example 24.1

	Condition 1		Condition 2		Condition 3		
	X_1	X_1^2	X_2	X_2^2	X_3	X_3^2	ΣS_i
Subject 1	9	81	7	49	4	16	20
Subject 2	10	100	8	64	7	49	25
Subject 3	7	49	5	25	3	9	15
Subject 4	10	100	8	64	7	49	25
Subject 5	7	49	5	25	2	4	14
Subject 6	8	64	6	36	6	36	20
	$\Sigma X_1 = 51$	$\Sigma X_1^2 = 443$	$\Sigma X_2 = 39$	$\Sigma X_2^2 = 263$	$\Sigma X_3 = 29$	$\Sigma X_3^2 = 163$	$\Sigma X_T = 119$

$$\bar{X}_1 = \frac{\Sigma X_1}{n_1} = \frac{51}{6} = 8.5 \qquad \bar{X}_2 = \frac{\Sigma X_2}{n_2} = \frac{39}{6} = 6.5 \qquad \bar{X}_3 = \frac{\Sigma X_3}{n_3} = \frac{29}{6} = 4.83$$

The notation N represents the total number of scores in the experiment. Since there are $n = 6$ subjects and each subject has $k = 3$ scores, there are a total of $nk = N = (6)(3) = 18$ scores. The value ΣX_T represents the sum of the $N = 18$ scores (i.e., the total sum of scores). Thus:

$$\Sigma X_T = \Sigma X_1 + \Sigma X_2 + \cdots + \Sigma X_k$$

Since there are $k = 3$ experimental conditions, $\Sigma X_T = 119$.

$$\Sigma X_T = \Sigma X_1 + \Sigma X_2 + \Sigma X_3 = 51 + 39 + 29 = 119$$

The value ΣX_T can also be computed by adding up the ΣS_i scores computed for the n subjects. Thus:

$$\Sigma X_T = \Sigma S_1 + \Sigma S_2 + \cdots + \Sigma S_n$$

$$\Sigma X_T = \Sigma S_1 + \Sigma S_2 + \Sigma S_3 + \Sigma S_4 + \Sigma S_5 + \Sigma S_6$$

$$= 20 + 25 + 15 + 25 + 14 + 20 = 119$$

$\bar{X}_T$ represents the grand mean, where $\bar{X}_T = \Sigma X_T / N$. Thus, $\bar{X}_T = 119/18 = 6.61$. Although $\bar{X}_T$ is not employed in the computational equations to be described in this section, it is employed in some of the definitional equations described in Section VII.

The value ΣX_T^2 represents the total sum of the N squared scores. Thus:

$$\Sigma X_T^2 = \Sigma X_1^2 + \Sigma X_2^2 + \cdots + \Sigma X_k^2$$

Since there are $k = 3$ experimental conditions, $\Sigma X_T^2 = 869$.

$$\Sigma X_T^2 = \Sigma X_1^2 + \Sigma X_2^2 + \Sigma X_3^2 = 443 + 263 + 163 = 869$$

Although the means for each of the experimental conditions are not required for computing the analysis of variance test statistic, it is recommended that they be computed since visual inspection of the condition means can provide the researcher with a general idea of whether or not it is reasonable to expect a significant result. To be more specific, if two or more of the condition means are far removed from one another, it is likely that the analysis of variance will be significant (especially if there are a relatively large number of subjects employed in the experiment). Another reason for computing the condition means is that they are required for comparing individual conditions with one another, which is something that is often done following the analysis of variance on the full set of data. The latter types of comparisons are described in Section VI.

In order to compute the test statistic for the **single-factor within-subjects analysis of variance**, the **total variability** in the data is divided into a number of different components. Specifically, the following variability components are computed: a) The **total sum of squares** which is represented by the notation SS_T; b) The **between-conditions sum of squares** which is represented by the notation SS_{BC}. The **between-conditions sum of squares** is the numerator of the equation which represents **between-conditions variability** (i.e., the equation that represents the amount of variability between the means of the k conditions); c) The **between-subjects sum of squares** is represented by the notation SS_{BS}. The **between-subjects sum of squares** is the numerator of the equation that represents **between-subjects variability**, which is the amount of variability between the mean scores of the n subjects (the mean of each subject being the average of a subject's k scores); and d) The **residual sum of squares** is represented by the notation SS_{res}. The **residual sum of squares** is the numerator of the equation which represents **residual variability** (i.e., error variability that is beyond the researcher's control).

Equation 24.1 describes the relationship between SS_T, SS_{BC}, SS_{BS}, and SS_{res}.[6]

$$SS_T = SS_{BC} + SS_{BS} + SS_{res} \qquad \textbf{(Equation 24.1)}$$

Equation 24.2 is employed to compute SS_T.

$$SS_T = \Sigma X_T^2 - \frac{(\Sigma X_T)^2}{N} \qquad \textbf{(Equation 24.2)}$$

Employing Equation 24.2, the value $SS_T = 82.28$ is computed.

$$SS_T = 869 - \frac{(119)^2}{18} = 869 - 786.72 = 82.28$$

Equation 24.3 is employed to compute SS_{BC}. In Equation 24.3 the notation ΣX_j represents the sum of the n scores in the jth condition. Note that in Equation 24.3 the notation n_j can be employed in place of n, since there are $n = n_j$ scores in the jth condition.

$$SS_{BC} = \sum_{j=1}^{k} \left[\frac{(\Sigma X_j)^2}{n} \right] - \frac{(\Sigma X_T)^2}{N} \qquad \textbf{(Equation 24.3)}$$

The notation $\sum_{j=1}^{k}[(\Sigma X_j)^2/n]$ in Equation 24.3 indicates that for each condition the value $(\Sigma X_j)^2/n$ is computed, and the latter values are summed for all k conditions.

With reference to Example 24.1, Equation 24.3 can be rewritten as follows:

$$SS_{BC} = \left[\frac{(\Sigma X_1)^2}{n} + \frac{(\Sigma X_2)^2}{n} + \frac{(\Sigma X_3)^2}{n} \right] - \frac{(\Sigma X_T)^2}{N}$$

Substituting the appropriate values from Example 24.1 in Equation 24.3, the value $SS_{BC} = 40.45$ is computed.[7]

$$SS_{BC} = \left[\frac{(51)^2}{6} + \frac{(39)^2}{6} + \frac{(29)^2}{6} \right] - \frac{(119)^2}{18} = 827.17 - 786.72 = 40.45$$

Equation 24.4 is employed to compute SS_{BS}.

$$SS_{BS} = \sum_{i=1}^{n} \left[\frac{(\Sigma S_i)^2}{k} \right] - \frac{(\Sigma X_T)^2}{N} \qquad \textbf{(Equation 24.4)}$$

The notation $\sum_{i=1}^{n}\left[(\Sigma S_i)^2/k\right]$ in Equation 24.4 indicates that for each subject the value $(\Sigma S_i)^2/k$ is computed, and the latter values are summed for all n subjects. With reference to Example 24.1, Equation 24.4 can be rewritten as follows.

$$SS_{BS} = \left[\frac{(\Sigma S_1)^2}{k} + \frac{(\Sigma S_2)^2}{k} + \frac{(\Sigma S_3)^2}{k} + \frac{(\Sigma S_4)^2}{k} + \frac{(\Sigma S_5)^2}{k} + \frac{(\Sigma S_6)^2}{k} \right] - \frac{(\Sigma X_T)^2}{N}$$

Substituting the appropriate values from Example 24.1 in Equation 24.4, the value $SS_{BS} = 36.93$ is computed.[8]

$$SS_{BS} = \left[\frac{(20)^2}{3} + \frac{(25)^2}{3} + \frac{(15)^2}{3} + \frac{(25)^2}{3} + \frac{(14)^2}{3} + \frac{(20)^2}{3} \right] - \frac{(119)^2}{18} = 823.65 - 786.72 = 36.93$$

By algebraically transposing the terms in Equation 24.1, the value of SS_{res} can be computed with Equation 24.5.

$$SS_{res} = SS_T - SS_{BC} - SS_{BS} \qquad \textbf{(Equation 24.5)}$$

Employing Equation 24.5, the value $SS_{res} = 4.9$ is computed.

$$SS_{res} = 82.28 - 40.45 - 36.93 = 4.9$$

Equation 24.6 is a computationally more complex equation for computing the value of SS_{res}.

$$SS_{res} = \Sigma X_T^2 - \sum_{j=1}^{k} \left[\frac{(\Sigma X_j)^2}{n} \right] - \sum_{i=1}^{n} \left[\frac{(\Sigma S_i)^2}{k} \right] + \frac{(\Sigma X_T)^2}{N} \qquad \textbf{(Equation 24.6)}$$

Since $\Sigma X_T^2 = 869$, $\sum_{j=1}^{k}[(\Sigma X_j)^2/n] = 827.17$, $\sum_{i=1}^{n}[(\Sigma S_i)^2/k] = 823.65$, and $(\Sigma X_T)^2/N = 786.72$, employing Equation 24.6, the value $SS_{res} = 4.9$ is computed.

$$SS_{res} = 869 - 827.17 - 823.65 + 786.72 = 4.9$$

The reader should take note of the fact that the values SS_T, SS_{BC}, SS_{BS}, and SS_{res} must always be positive numbers. If a negative value is obtained for any of the aforementioned values, it indicates a computational error has been made.

At this point the values of the **between-conditions variance**, **between-subjects variance**, and the **residual variance** can be computed. In the **single-factor within-subjects analysis of variance**, the **between-conditions variance** is referred to as the **mean square between-conditions**, which is represented by the notation MS_{BC}. MS_{BC} is computed with Equation 24.7.

$$MS_{BC} = \frac{SS_{BC}}{df_{BC}} \qquad \textbf{(Equation 24.7)}$$

The **between-subjects variance** is referred to as the **mean square between-subjects**, which is represented by the notation MS_{BS}. MS_{BS} is computed with Equation 24.8.

$$MS_{BS} = \frac{SS_{BS}}{df_{BS}} \qquad \textbf{(Equation 24.8)}$$

The **residual variance** is referred to as the **mean square residual**, which is represented by the notation MS_{res}. MS_{res} is computed with Equation 24.9.

$$MS_{res} = \frac{SS_{res}}{df_{res}} \qquad \textbf{(Equation 24.9)}$$

Note that a total mean square is not computed.

In order to compute MS_{BC}, MS_{BS}, and MS_{res}, it is required that the values df_{BC}, df_{BS}, and df_{res} (the denominators of Equations 24.7–24.9) be computed. df_{BC}, which represents the **between-conditions degrees of freedom**, are computed with Equation 24.10.

$$df_{BC} = k - 1 \qquad \textbf{(Equation 24.10)}$$

df_{BS}, which represents the **between-subjects degrees of freedom**, are computed with Equation 24.11.

$$df_{BS} = n - 1 \qquad \textbf{(Equation 24.11)}$$

df_{res}, which represents the **residual degrees of freedom**, are computed with Equation 24.12.

$$df_{res} = (n - 1)(k - 1) \qquad \textbf{(Equation 24.12)}$$

Although it is not required in order to determine the F ratio, the **total degrees of freedom** are generally computed, since it can be used to confirm the df values computed with Equations 24.10–24.12, as well as the fact it is employed in the analysis of variance summary table. The total degrees of freedom (represented by the notation df_T) are computed with Equation 24.13.

$$df_T = nk - 1 = N - 1 \qquad \textbf{(Equation 24.13)}$$

The relationship between df_{BC}, df_{BS}, df_{res}, and df_T is described by Equation 24.14.

$$df_T = df_{BC} + df_{BS} + df_{res} \qquad \textbf{(Equation 24.14)}$$

Employing Equations 24.10–24.13, the values $df_{BC} = 2$, $df_{BS} = 5$, $df_{res} = 10$, and $df_T = 17$ are computed. Note that $df_T = df_{BC} + df_{BS} + df_{res} = 2 + 5 + 10 = 17$.

$$df_{BC} = 3 - 1 = 2 \qquad df_{BS} = 6 - 1 = 5$$

$$df_{res} = (6 - 1)(3 - 1) = 10 \qquad df_T = 18 - 1 = 17$$

Employing Equations 24.7–24.9, the values $MS_{BC} = 20.23$, $MS_{BS} = 7.39$, and $MS_{res} = .49$ are computed.

$$MS_{BC} = \frac{40.45}{2} = 20.23 \qquad MS_{BS} = \frac{36.93}{5} = 7.39 \qquad MS_{res} = \frac{4.9}{10} = .49$$

The F ratio, which is the test statistic for the **single-factor within-subjects analysis of variance**, is computed with Equation 24.15.

$$F = \frac{MS_{BC}}{MS_{res}} \qquad \textbf{(Equation 24.15)}$$

Employing Equation 24.15, the value $F = 41.29$ is computed.

$$F = \frac{20.23}{.49} = 41.29$$

The reader should take note of the fact that the values MS_{BC}, MS_{BS}, and MS_{res} must always be positive numbers. If a negative value is obtained for any of the aforementioned values, it indicates a computational error has been made. If $MS_{res} = 0$, Equation 24.15 will be insoluble. If all of the conditions have the identical mean value, $MS_{BC} = 0$, and if the latter is true, $F = 0$.

V. Interpretation of the Test Results

It is common practice to summarize the results of a **single-factor within-subjects analysis of variance** with the summary table represented by Table 24.2.

Table 24.2 Summary Table of Analysis of Variance for Example 24.1

Source of variation	SS	df	MS	F
Between-subjects	36.93	5	7.39	
Between-conditions	40.45	2	20.23	41.29
Residual	4.90	10	.49	
Total	**82.28**	**17**		

The obtained value $F = 41.29$ is evaluated with **Table A10 (Table of the F Distribution)** in the **Appendix**. In **Table A10** critical values are listed in reference to the number of degrees of freedom associated with the numerator and the denominator of the F ratio (i.e., df_{num} and df_{den}). In employing the F distribution in reference to Example 24.1, the degrees of freedom for the numerator are $df_{BC} = 2$ and the degrees of freedom for the denominator are $df_{res} = 10$. In **Table A10** the tabled $F_{.95}$ and $F_{.99}$ values are, respectively, employed to evaluate the nondirectional alternative hypothesis H_1: Not H_0 at the .05 and .01 levels. As is the case for the **single-factor between-subjects analysis of variance**, the notation $F_{.05}$ is employed to represent the tabled critical F value at the .05 level. The latter value corresponds to the relevant tabled $F_{.95}$ value in **Table A10**. In the same respect, the notation $F_{.01}$ is employed to represent the tabled critical F value at the .01 level, and corresponds to the relevant tabled $F_{.99}$ value in **Table A10**.

For df_{num} = 2 and df_{den} = 10, the tabled $F_{.95}$ and $F_{.99}$ values are $F_{.95}$ = 4.10 and $F_{.99}$ = 7.56. Thus, $F_{.05}$ = 4.10 and $F_{.01}$ = 7.56. In order to reject the null hypothesis, the obtained F value must be equal to or greater than the tabled critical value at the prespecified level of significance. Since F = 41.29 is greater than both $F_{.05}$ = 4.10 and $F_{.01}$ = 7.56, the alternative hypothesis is supported at both the .05 and .01 levels.

A summary of the analysis of Example 24.1 with the **single-factor within-subjects analysis of variance** follows: It can be concluded there is a significant difference between at least two of the three experimental conditions (i.e., different levels of noise). This result can be summarized as follows: $F(2,10) = 41.29, p < .01$.

VI. Additional Analytical Procedures for the Single-Factor Within-Subjects Analysis of Variance and/or Related Tests

1. Comparisons following computation of the omnibus F value for the single-factor within-subjects analysis of variance Prior to reading this section the reader should review the discussion of comparison procedures in Section VI of the **single-factor between-subjects analysis of variance**. As is the case with the latter test, the omnibus F value computed for a **single-factor within-subjects analysis of variance** is based on a comparison of the means of all k experimental conditions. Thus, in order to reject the null hypothesis, it is only required that the means of at least two of the k conditions differ significantly from one another.[9]

The same procedures which are employed for conducting comparisons for the **single-factor between-subjects analysis of variance** can be used for the **single-factor within-subjects analysis of variance**. Thus, the following comparison procedures discussed under the latter test can be employed for conducting comparisons within the framework of the **single-factor within-subjects analysis of variance**: **Test 24a: Multiple t tests/Fisher's LSD test** (which is equivalent to **linear contrasts**); **Test 24b: The Bonferroni–Dunn test**; **Test 24c: Tukey's HSD test**; **Test 24d: The Newman–Keuls test**; **Test 24e: The Scheffé test**; **Test 24f: The Dunnett test**. The only difference in applying the aforementioned comparison procedures to the analysis of variance under discussion is that a different measure of error variability is employed. Recollect that in the case of the **single-factor between-subjects analysis variance**, a pooled measure of within-groups variability (MS_{WG}) is employed as the measure of experimental error. In the case of the **single-factor within-subjects analysis of variance**, MS_{res} is employed as the measure of error variability. Consequently, in conducting comparisons for a **single-factor within-subjects analysis of variance**, MS_{res} is employed in place of MS_{WG} as the error term in the comparison equations described in Section VI of the **single-factor between-subjects analysis variance**. It should be noted, however, that if the sphericity assumption (which, as noted in Section I, is based on homogeneity of the underlying population variances and covariances) of the **single-factor within-subjects analysis of variance** is violated, MS_{res} may not provide the most accurate measure of error variability to employ in conducting comparisons. Because of this, an alternative measure of error variability (which is not influenced by violation of the sphericity assumption) will be presented later in this section.

At this point, employing MS_{res} as the error term, the following two **single degree of freedom comparisons** will be conducted: a) The **simple comparison** Condition 1 versus Condition 2, which is summarized in Table 24.3; and b) The **complex comparison** Condition 3 versus the combined performance of Conditions 1 and 2, which is summarized in Table 24.4.

Table 24.3 Planned Simple Comparison: Condition 1 versus Condition 2

Condition	$\bar{X}_j$	Coefficient (c_j)	Product $(c_j)(\bar{X}_j)$	Squared Coefficient (c_j^2)
1	8.5	+1	(+1)(8.5) = +8.5	1
2	6.5	−1	(−1)(6.5) = −6.5	1
3	4.83	0	(0)(4.83) = 0	0
		$\Sigma c_j = 0$	$\Sigma(c_j)(\bar{X}_j) = 2$	$\Sigma c_j^2 = 2$

Since it will be assumed that the above comparisons are planned prior to collecting the data, **linear contrasts** will be conducted in which no attempt is made to control the value of the **familywise Type I error rate** (α_{FW}). The null hypothesis and nondirectional alternative hypothesis for the above comparisons are identical to those employed when the analogous comparisons are conducted in Section VI for the **single-factor between-subjects analysis variance** (i.e., H_0: $\mu_1 = \mu_2$ versus H_1: $\mu_1 \neq \mu_2$ for the simple comparison, and H_0: $\mu_3 = (\mu_1 + \mu_2)/2$ versus H_1: $\mu_3 \neq (\mu_1 + \mu_2)/2$ for the complex comparison).

Table 24.4 Planned Complex Comparison: Condition 3 versus Conditions 1 and 2

Condition	$\bar{X}_j$	Coefficient (c_j)	Product $(c_j)(\bar{X}_j)$	Squared Coefficient (c_j^2)
1	8.5	$-\frac{1}{2}$	$\left(-\frac{1}{2}\right)(8.5) = -4.25$	$\frac{1}{4}$
2	6.5	$-\frac{1}{2}$	$\left(-\frac{1}{2}\right)(6.5) = -3.25$	$\frac{1}{4}$
3	4.83	+1	(+1)(4.83) = +4.83	1
		$\Sigma c_j = 0$	$\Sigma(c_j)(\bar{X}_j) = -2.67$	$\Sigma c_j^2 = 1.5$

Equations 21.17 and 21.18, which are employed to compute SS_{comp} and MS_{comp} for linear contrasts for the **single-factor between-subjects analysis of variance**, are also employed for linear contrasts for the **single-factor within-subjects analysis of variance**. The latter equations are employed below for the simple comparison Condition 1 versus Condition 2.

$$SS_{comp} = \frac{n\left[\Sigma(c_j)(\bar{X}_j)\right]^2}{\Sigma c_j^2} = \frac{6(2)^2}{2} = 12$$

$$MS_{comp} = \frac{SS_{comp}}{df_{comp}} = \frac{12}{1} = 12$$

Equation 24.16 (which is identical to Equation 21.19, except for the fact that it employs MS_{res} as the error term) is used to compute the value of F_{comp}.

$$F_{comp} = \frac{MS_{comp}}{MS_{res}} = \frac{12}{.49} = 24.49 \qquad \textbf{(Equation 24.16)}$$

The degrees of freedom employed in evaluating the obtained value $F_{comp} = 24.49$ are $df_{num} = df_{comp} = 1$ (since in a **single degree of freedom comparison** df_{comp} will always equal 1) and $df_{den} = df_{res} = 10$. For $df_{num} = 1$ and $df_{den} = 10$, the tabled critical .05 and .01 F values in **Table A10** are $F_{.05} = 4.96$ and $F_{.01} = 10.04$. Since the obtained value $F_{comp} = 24.49$ is greater than both of the aforementioned critical values, the nondirectional alternative hypothesis $H_1: \mu_1 \neq \mu_2$ is supported at both the .05 and .01 levels.

Applying Equations 21.17, 21.18, and 24.16 to the complex comparison Condition 3 versus Conditions 1 and 2, the following result is obtained.

$$SS_{comp} = \frac{n\left[\Sigma(c_j)(\bar{X}_j)\right]^2}{\Sigma c_j^2} = \frac{6(-2.67)^2}{1.5} = 28.52$$

$$MS_{comp} = \frac{SS_{comp}}{df_{comp}} = \frac{28.52}{1} = 28.52$$

$$F_{comp} = \frac{MS_{comp}}{MS_{res}} = \frac{28.52}{.49} = 58.20$$

As is the case for the simple comparison, the degrees of freedom employed for the complex comparison in evaluating the value $F_{comp} = 58.20$ are $df_{num} = df_{comp} = 1$ and $df_{den} = df_{res} = 10$. Since the obtained value $F_{comp} = 58.20$ is greater than the tabled critical values $F_{.05} = 4.96$ and $F_{.01} = 10.04$, the nondirectional alternative hypothesis $H_1: \mu_3 \neq (\mu_1 + \mu_2)/2$ is supported at both the .05 and .01 levels.

For both of the comparisons which have been conducted, a CD value can be computed that identifies a minimum required difference in order for two means (or sets of means) to differ from one another at a prespecified level of significance. For both the simple comparison Condition 1 versus Condition 2 and the complex comparison Condition 3 versus Conditions 1 and 2, a CD value will be computed employing **multiple *t* tests/Fisher's LSD test** (which is equivalent to a **linear contrast** in which the value of α_{FW} is not controlled) and the **Scheffé test**. Whereas **multiple *t* tests/Fisher's LSD test** allow a researcher to determine the minimum CD value (designated CD_{LSD}) which can be computed through use of any of the comparison procedures that are described for the **single-factor between-subjects analysis of variance**, the **Scheffé test** generally results in the maximum CD value (designated CD_S) which can be computed by the available procedures. Thus, if at a prespecified level of significance the obtained difference for a comparison is equal to or greater than CD_S, it will be significant regardless of which comparison procedure is employed. On the other hand, if the obtained difference for a comparison is less than CD_{LSD}, it will not achieve significance, regardless of which comparison procedure is employed. In illustrating the computation of CD values for both simple and complex comparisons, it will be assumed that the total number of comparisons conducted is $c = 3$.[10] It will also be assumed that when the researcher wants to control the value of the **familywise Type I error rate**, the value $\alpha_{FW} = .05$ is employed irrespective of which comparison procedure is used.

The equations employed for computing the values CD_{LSD} and CD_S are essentially the same as those used for the **single-factor between-subjects analysis of variance**. Equations 24.17 and 24.18 are employed below to compute the values of CD_{LSD} and CD_S for the simple comparison Condition 1 versus Condition 2. Note that Equations 24.17 and 24.18 are identical to Equations 21.24 and 21.32, except for the fact that MS_{res} is employed as the error term in both equations, and in Equation 24.18 df_{BC} is used in place of df_{BG} in determining the numerator degrees of freedom. In employing Equation 24.18, the tabled critical value $F_{.05} = 4.10$ is employed since $df_{num} = df_{BC} = 2$ and $df_{den} = df_{res} = 10$.

$$CD_{LSD} = \sqrt{F_{(1.res)}} \sqrt{\frac{2MS_{res}}{n}} = \sqrt{4.96}\sqrt{\frac{(2)(.49)}{6}} = .90 \quad \textbf{(Equation 24.17)}$$

$$CD_S = \sqrt{(k-1)(F_{(df_{BC}, df_{res})})} \sqrt{\frac{2MS_{res}}{n}}$$

$$\textbf{(Equation 24.18)}$$

$$= \sqrt{(3-1)(4.10)}\sqrt{\frac{(2)(.49)}{6}} = 1.16$$

Thus, if one employs **multiple t tests/Fisher's LSD test** (i.e., conducts linear contrasts with α_{FW} not adjusted), in order to differ significantly at the .05 level, the means of any two conditions must differ from one another by at least .90 units. If, on the other hand, the **Scheffé test** is employed, in order to differ significantly the means of any two conditions must differ from one another by at least 1.16 units. Since the difference score for the comparison Condition 1 versus Condition 2 equals $\bar{X}_1 - \bar{X}_2 = 2$, the comparison is significant at the .05 level, regardless of which comparison procedure is employed.

Table 24.5 summarizes the differences between pairs of means involving all of the experimental conditions. Since the difference scores for all three simple/pairwise comparisons are greater than $CD_S = 1.16$, all of the comparisons are significant at the .05 level, regardless of which comparison procedure is employed.

Table 24.5 Differences between Pairs of Means in Example 24.1

$$\bar{X}_1 - \bar{X}_2 = 8.5 - 6.5 = 2$$
$$\bar{X}_1 - \bar{X}_3 = 8.5 - 4.83 = 3.67$$
$$\bar{X}_2 - \bar{X}_3 = 6.5 - 4.83 = 1.67$$

The complex comparison Condition 3 versus Conditions 1 and 2 is illustrated below employing Equations 24.19 and 24.20. The latter two equations (which are the generic forms of Equations 24.17 and 24.18 that can be used for both simple and complex comparisons) are identical to Equations 21.25 and 21.33 (except for the use of MS_{res} as the error term, and the use of df_{BC} in place of df_{BG} in Equation 24.20).

$$CD_{LSD} = \sqrt{F_{(1.res)}} \sqrt{\frac{(\Sigma c_j^2)(MS_{res})}{n}}$$

$$\textbf{(Equation 24.19)}$$

$$= \sqrt{4.96}\sqrt{\frac{(1.5)(.49)}{6}} = .78$$

$$CD_S = \sqrt{(k - 1)(F_{(df_{BC}, df_{res})})} \sqrt{\frac{(\Sigma c_j^2)(MS_{res})}{n}}$$

(Equation 24.20)

$$= \sqrt{(3 - 1)(4.10)} \sqrt{\frac{(1.5)(.49)}{6}} = 1.00$$

Thus, if one employs **multiple *t* tests/Fisher's LSD test** (i.e., conducts a linear contrast with α_{FW} not adjusted), in order to differ significantly the difference between $\bar{X}_3$ and $(\bar{X}_1 + \bar{X}_2)/2$ must be at least .78 units. If, on the other hand, the **Scheffé test** is employed, in order to differ significantly the two sets of means must differ from one another by at least 1.00 unit. Since the obtained difference of 2.67 is greater than $CD_S = 1.00$, the nondirectional alternative hypothesis $H_1: \mu_3 \neq (\mu_1 + \mu_2)/2$ is supported, regardless of which comparison procedure is employed.

The computation of a confidence interval for a comparison The procedure which is described for computing a confidence interval for a comparison for the **single-factor between-subjects analysis of variance** can also be used with the **single-factor within-subjects analysis of variance**. Thus, in the case of the latter analysis of variance, a confidence interval for a comparison is computed by adding to and subtracting the relevant CD value for the comparison from the obtained difference between the means involved in the comparison. As an example, let us assume the **Scheffé test** is employed to compute the value $CD_S = 1.16$ for the comparison Condition 1 versus Condition 2. To compute the 95% confidence interval, the value 1.16 is added to and subtracted from 2, which is the absolute value of the difference between the two means. Thus, $CI_{.95} = 2 \pm 1.16$, which can also be written as $.84 \leq (\mu_1 - \mu_2) \leq 3.16$. In other words, the researcher can be 95% confident (or the probability is .95) that the interval .84 to 3.16 contains the true difference between the means of the populations represented by Conditions 1 and 2. If the aforementioned interval contains the true difference between the population means, the mean of the population Condition 1 represents is between .84 to 3.16 units larger than the mean of the population Condition 2 represents.

Alternative methodology for computing MS_{res} for a comparison Earlier in this section it is noted that if the sphericity assumption underlying the **single-factor within-subjects analysis of variance** is violated, MS_{res} may not provide an accurate measure of error variability for a specific comparison. Because of this, many sources recommend that when the sphericity assumption is violated a separate measure of error variability be computed for each comparison which is conducted. The procedure that will be discussed in this section (which is described in Keppel (1991)) can be employed any time a researcher has reason to believe that MS_{res} employed in computing the omnibus F value is not representative of the actual error variability for the experimental conditions involved in a specific comparison. The procedure (which will be demonstrated for both a simple and complex comparison) requires that a **single-factor within-subjects analysis of variance** be conducted employing only the data for those experimental conditions involved in a comparison. In the case of a simple comparison, the scores of subjects in the two comparison conditions are evaluated with the analysis of variance. In the case of a complex comparison, a weighted score must be computed for each subject for any composite mean that is a combination of two or more experimental conditions.

With respect to a simple planned comparison, the procedure will be employed to evaluate the difference between Condition 1 and Condition 3. The reason it will not be used for the Condition 1 versus Condition 2 comparison is because the error variability associated with the latter comparison is $MS_{res} = 0$. The reason why $MS_{res} = 0$ for the latter comparison is revealed by inspection of the scores of the six subjects in Table 24.1. Observe that the score of each of the six subjects in Condition 1 is two units higher than it is in Condition 2. Anytime all of the subjects in a within-subjects design involving two treatments obtain identical difference scores, the measure of error variability will equal zero. Consequently, whenever the value of $MS_{res} = 0$, the value of F_{comp} will be indeterminate, since the denominator of the F ratio will equal zero. In such an instance, one can either use the MS_{res} value employed in computing the omnibus F value, or elect to use the smallest MS_{res} value which can be computed for any of the other simple comparisons in the set of data.[11]

Table 24.6 summarizes the data for the comparison Condition 1 versus Condition 3. Note that the values $\Sigma X_T = 80$ and $\Sigma X_T^2 = 606$ differ from those computed in Table 24.1, since in this instance they only include the data for two of the three experimental conditions.

Table 24.6 Data for Comparison of Condition 1 versus Condition 3

	Condition 1		Condition 3		
	X_1	X_1^2	X_3	X_3^2	ΣS_i
Subject 1	9	81	4	16	13
Subject 2	10	100	7	49	17
Subject 3	7	49	3	9	10
Subject 4	10	100	7	49	17
Subject 5	7	49	2	4	9
Subject 6	8	64	6	36	14
	$\Sigma X_1 = 51$	$\Sigma X_1^2 = 443$	$\Sigma X_2 = 29$	$\Sigma X_2^2 = 163$	$\Sigma X_T = 80$
	$\Sigma X_T = 51 + 29 = 80$		$\Sigma X_T^2 = 443 + 163 = 606$		

The sum of squares values which are required for the analysis of variance are computed with Equations 24.2–24.5. Note that since we are only dealing with two conditions, $k = 2$, and thus, $N = nk = (6)(2) = 12$.

$$SS_T = \Sigma X_T^2 - \frac{(\Sigma X_T)^2}{N} = 606 - \frac{(80)^2}{12} = 72.67$$

$$SS_{BC} = \left[\frac{(\Sigma X_1)^2}{n} + \frac{(\Sigma X_3)^2}{n}\right] - \frac{(\Sigma X_T)^2}{N} = \left[\frac{(51)^2}{6} + \frac{(29)^2}{6}\right] - \frac{(80)^2}{12} = 40.34$$

$$SS_{BS} = \left[\frac{(\Sigma S_1)^2}{k} + \cdots + \frac{(\Sigma S_6)^2}{k}\right] - \frac{(\Sigma X_T)^2}{N} = \left[\frac{(13)^2}{2} + \frac{(17)^2}{2} + \cdots + \frac{(14)^2}{2}\right] - \frac{(80)^2}{12} = 28.67$$

$$SS_{res} = SS_T - SS_{BC} - SS_{BS} = 72.67 - 40.34 - 28.67 = 3.66$$

Upon computing the sum of squares values, the appropriate degrees of freedom (employing Equations 24.10–24.13) and mean square values (employing Equations 24.7–24.9) are computed. Employing Equation 24.15, the value $F = F_{comp} = 55.26$ is computed for the comparison. The analysis of variance is summarized in Table 24.7. Note that in computing the degrees of freedom for the analysis, the values $n = 6$ and $k = 2$ are employed.

Table 24.7 Summary Table for Analysis of Variance for Comparison of Condition 1 versus Condition 3

Source of variation	SS	df	MS	F
Between-subjects	28.67	5	5.73	
Between-conditions	40.34	1	40.34	55.26
Residual	3.66	5	.73	
Total	72.67	11		

Employing $df_{num} = df_{BC} = df_{comp} = 1$ and $df_{den} = df_{res} = 5$, the tabled critical values employed in **Table A10** are $F_{.05} = 6.61$ and $F_{.01} = 16.26$. Since the obtained value $F = 55.26$ is greater than both of the aforementioned critical values, the nondirectional alternative hypothesis H_1: $\mu_1 \neq \mu_3$ is supported at both the .05 and .01 levels. The reader should note that the value $df_{res} = 5$ employed for the comparison of Condition 1 versus Condition 3 is less than the value $df_{res} = 10$ employed when the same comparison was conducted earlier in this section employing the value $MS_{res} = .49$ obtained for the omnibus F test. The lower df_{res} value associated with the method under discussion results in a less powerful test of the alternative hypothesis. In some instances the loss of power associated with this method may be offset if the value of MS_{res} for the comparison conditions is smaller than the value of MS_{res} obtained for the omnibus F test.

Employing Equations 24.17 and 24.18, a CD_{LSD} and CD_S value can be computed for the Condition 1 versus Condition 3 comparison.

$$CD_{LSD} = \sqrt{F_{(1.res)}} \sqrt{\frac{2MS_{res}}{n}} = \sqrt{6.61}\sqrt{\frac{(2)(.73)}{6}} = 1.27$$

$$CD_S = \sqrt{(k-1)(F_{(df_{BC}, df_{res})})} \sqrt{\frac{2MS_{res}}{n}} = \sqrt{(3-1)(4.10)}\sqrt{\frac{(2)(.73)}{6}} = 1.41$$

In computing $CD_{LSD} = 1.27$, the tabled critical F value employed in Equation 24.17 is based on $df_{num} = df_{BC} = df_{comp} = 1$, and $df_{den} = df_{res} = 5$. Note that $df_{den} = df_{res} = 5$ is only based on the data for the two experimental conditions employed in the comparison. In computing $CD_S = 1.41$, however, the tabled critical F value employed in Equation 24.18 is based on all three conditions employed in the experiment, and thus $df_{BC} = 2$ and $df_{res} = 10$. Note that the values $CD_{LSD} = 1.27$ and $CD_S = 1.41$ are larger than the corresponding values $CD_{LSD} = .90$ and $CD_S = 1.16$, which are computed for simple comparisons when $MS_{res} = .49$ is employed. It should be obvious through inspection of Equations 24.17 and 24.18 that the larger the value of MS_{res}, the greater the magnitude of the computed CD value.

At the beginning of this section it is noted that when a complex comparison is conducted employing a residual variability which is based only on the specific conditions involved in the comparison, a weighted score must be computed for each subject for any composite mean which is a combination of two or more experimental conditions. This will now be illustrated for the comparison of Condition 3 versus the combined performance of Conditions 1 and 2.

Table 24.8 contains the scores of the six subjects in Condition 3 and a weighted score for each subject for Conditions 1 and 2. The weighted score of a subject for a combination of two or more conditions is obtained as follows: a) A subject's score in each condition is multiplied by the absolute value of the coefficient for that condition (based on the coefficients for the comparison in Table 24.4); and b) The subject's weighted score is the sum of the products obtained in part a).

To clarify how the aforementioned procedure is employed to compute the weighted scores in Table 24.8, the computation of weighted scores for Subjects 1 and 2 will be described. Since the scores of Subject 1 in Conditions 1 and 2 are 9 and 7, each score is multiplied by the absolute value of the coefficient for the corresponding condition. Employing the absolute values of the coefficients noted in Table 24.4 for Conditions 1 and 2, each score is multiplied by ½, yielding $(9)(½) = 4.5$ and $(7)(½) = 3.5$. The weighted score for Subject 1 is obtained by summing the latter two values. Thus, the weighted score for Subject 1 is $4.5 + 3.5 = 8$. In the case of Subject 2, $(10)(½) = 5$ and $(8)(½) = 4$. Thus, the weighted score for Subject 2 is $5 + 4 = 9$. The same procedure is used for the remaining four subjects.[12]

Table 24.8   Data for Comparison of Condition 3 versus Conditions 1 and 2

	Condition 3		Conditions 1 & 2		
	X_3	X_3^2	$X_{1/2}$	$X_{1/2}^2$	ΣS_i
Subject 1	4	16	8	64	12
Subject 2	7	49	9	81	16
Subject 3	3	9	6	36	9
Subject 4	7	49	9	81	16
Subject 5	2	4	6	36	8
Subject 6	6	36	7	49	13
	$\Sigma X_3 = 29$	$\Sigma X_3^2 = 163$	$\Sigma X_{1/2} = 45$	$\Sigma X_{1/2}^2 = 347$	$\Sigma X_T = 74$
	$\Sigma X_T = 29 + 45 = 74$		$\Sigma X_T^2 = 163 + 347 = 510$		

The sum of squares values which are required for the analysis of variance are computed with Equations 24.2–24.5. Note that since we are only dealing with two sets of means, $k = 2$, and thus, $N = nk = (6)(2) = 12$.

$$SS_T = \Sigma X_T^2 - \frac{(\Sigma X_T)^2}{N} = 510 - \frac{(74)^2}{12} = 53.67$$

$$SS_{BC} = \left[\frac{(\Sigma X_3)^2}{n} + \frac{(\Sigma X_{1/2})^2}{n}\right] - \frac{(\Sigma X_T)^2}{N} = \left[\frac{(29)^2}{6} + \frac{(45)^2}{6}\right] - \frac{(74)^2}{12} = 21.34$$

$$SS_{BS} = \left[\frac{(\Sigma S_1)^2}{k} + \cdots + \frac{(\Sigma S_6)^2}{k}\right] - \frac{(\Sigma X_T)^2}{N} = \left[\frac{(12)^2}{2} + \frac{(16)^2}{2} + \cdots + \frac{(13)^2}{2}\right] - \frac{(74)^2}{12} = 28.67$$

$$SS_{res} = SS_T - SS_{BC} - SS_{BS} = 53.67 - 21.34 - 28.67 = 3.66$$

Upon computing the sum of squares values, the appropriate degrees of freedom (employing Equations 24.10–24.13) and mean square values (employing Equations 24.7–24.9) are computed. Employing Equation 24.15, the value $F = F_{comp} = 29.23$ is computed for the comparison. The analysis of variance is summarized in Table 24.9. Note that in computing the degrees of freedom for the analysis, the values $n = 6$ and $k = 2$ are employed.

Employing $df_{num} = df_{BC} = df_{comp} = 1$ and $df_{den} = df_{res} = 5$, the tabled critical values employed in **Table A10** are $F_{.05} = 6.61$ and $F_{.01} = 16.26$. Since the obtained value $F = 29.23$ is greater than both of the aforementioned tabled critical values, the nondirectional alternative hypothesis $H_1: \mu_3 \neq (\mu_1 + \mu_2)/2$ is supported at both the .05 and .01 levels.

**Table 24.9 Summary Table for Analysis of Variance
for Comparison of Condition 3 versus Conditions 1 and 2**

Source of variation	SS	df	MS	F
Between-subjects	28.67	5	5.73	
Between-conditions	21.34	1	21.34	29.23
Residual	3.66	5	.73	
Total	53.67	11		

Employing Equations 24.19 and 24.20, a CD_{LSD} and CD_S value can be computed for the complex comparison.

$$CD_{LSD} = \sqrt{F_{(1,\text{res})}} \sqrt{\frac{(\Sigma c_j^2)(MS_{\text{res}})}{n}} = \sqrt{6.61}\sqrt{\frac{(1.5)(.73)}{6}} = 1.10$$

$$CD_S = \sqrt{(k-1)(F_{(df_{BC}, df_{\text{res}})})} \sqrt{\frac{(\Sigma c_j^2)(MS_{\text{res}})}{n}}$$

$$= \sqrt{(3-1)(4.10)}\sqrt{\frac{(1.5)(.73)}{6}} = 1.22$$

Note that the values $CD_{LSD} = 1.10$ and $CD_S = 1.22$ are larger than the corresponding values $CD_{LSD} = .78$ and $CD_S = 1.00$, which are computed for the same complex comparison when $MS_{\text{res}} = .49$ is used.

Hsu (1996), Maxwell and Delaney (2004) and Rosenthal *et al.* (2000) discuss alternative methodologies for contrasting means within the framework of the **single-factor within-subjects analysis of variance**.

2. Comparing the means of three or more conditions when $k \geq 4$ Within the framework of a **single-factor within-subjects analysis of variance** involving $k = 4$ or more conditions, a researcher may wish to evaluate a general hypothesis with respect to the means of a subset of conditions, where the number of conditions in the subset is some value less than k. Although the latter type of situation is not commonly encountered in research, this section will describe the protocol for conducting such an analysis. Specifically, the protocol described for the analogous analysis for a **single-factor between-subjects analysis of variance** will be extended to the **single-factor within-subjects analysis of variance**.

To illustrate, assume that a fourth experimental condition is added to Example 24.1. Assume that the scores of Subjects 1–6 in Condition 4 are respectively: 3, 8, 2, 6, 4, 6. Thus, $\Sigma X_4 = 29$, $\bar{X}_4 = 4.83$, and $\Sigma X_4^2 = 165$. If the data for Condition 4 are integrated into the data for the other three conditions (which are summarized in Table 24.1), the following summary values are computed: $N = nk = (6)(4) = 24$, $\Sigma X_T = 148$, $\Sigma X_T^2 = 1034$. Substituting the revised values for $k = 4$ conditions in Equations 24.2–24.5, the following sum of squares values are computed: $SS_T = 121.33$, $SS_{BC} = 54.66$, $SS_{BS} = 54.33$, $SS_{\text{res}} = 12.34$. Employing the values $k = 4$ and $n = 6$ in Equations 24.10–24.12, the values $df_{BC} = 4 - 1 = 3$, $df_{BS} = 6 - 1 = 5$, and $df_{\text{res}} = (6-1)(4-1) = 15$ are computed. Substituting the appropriate values for the sums of squares and degrees of freedom in Equations 24.7–24.9, the values $MS_{BC} = 54.66/3 = 18.22$, $MS_{BS} = 54.33/5 = 10.87$, and $MS_{\text{res}} = 12.34/15 = .82$ are computed. Equation 24.15 is employed to compute the value $F = 18.22/.82 = 22.22$. Table 24.10 is the summary table of the analysis of variance.

Table 24.10 Summary Table of Analysis of Variance for Example 24.1 When $k = 4$

Source of variation	SS	df	MS	F
Between-subjects	54.33	5	10.87	
Between-conditions	54.66	3	18.22	22.22
Residual	12.34	15	.82	
	121.33	23		

Employing $df_{num} = 3$ and $df_{den} = 15$, the tabled critical .05 and .01 F values are $F_{.05} = 3.29$ and $F_{.01} = 5.42$. Since the obtained value $F = 22.22$ is greater than both of the aforementioned critical values, the null hypothesis (which for $k = 4$ is H_0: $\mu_1 = \mu_2 = \mu_3 = \mu_4$) can be rejected at both the .05 and .01 levels.

Let us assume that prior to the above analysis the researcher has reason to believe that Conditions 1, 2, and 3 may be distinct from Condition 4. However, before he contrasts the composite mean of Conditions 1, 2, and 3 with the mean of Condition 4 (i.e., conducts the complex comparison which evaluates the null hypothesis H_0: $(\mu_1 + \mu_2 + \mu_3)/3 = \mu_4$), he decides to evaluate the null hypothesis H_0: $\mu_1 = \mu_2 = \mu_3$. If the latter null hypothesis is retained, he will assume that the three conditions share a common mean value, and on the basis of this he will compare their composite mean with the mean of Condition 4. In order to evaluate the null hypothesis H_0: $\mu_1 = \mu_2 = \mu_3$, it is necessary for the researcher to conduct a separate analysis of variance which just involves the data for the three conditions identified in the null hypothesis. The latter analysis of variance has already been conducted, since it is the original analysis of variance that is employed for Example 24.1 — the results of which are summarized in Table 24.2.

Upon conducting an analysis of variance on the data for all $k = 4$ conditions as well as an analysis of variance on the data for the subset comprised of $k_{subset} = 3$ conditions, the researcher has the necessary information to compute the appropriate F ratio (which will be represented with the notation $F_{(1/2/3)}$) for evaluating the null hypothesis H_0: $\mu_1 = \mu_2 = \mu_3$. If we apply the same logic employed when the analogous analysis is conducted in reference to the **single-factor between-subjects analysis of variance**, the following values are employed to compute the F ratio to evaluate the latter null hypothesis: a) $MS_{BC} = 20.23$ (which is the value of MS_{BC} computed for the analysis of variance in Table 24.2 which involves only the three conditions identified in the null hypothesis H_0: $\mu_1 = \mu_2 = \mu_3$) is employed as the numerator of the F ratio; and b) $MS_{res} = .82$ (which is the value of MS_{res} computed in Table 24.10 for the omnibus F test when the data for all $k = 4$ conditions are evaluated) is employed as the denominator of the F ratio. The use of $MS_{res} = .82$ instead of $MS_{res} = .49$ (which is the value of MS_{res} computed for the analysis of variance in Table 24.2) as the denominator of the F ratio is predicated on the assumption that $MS_{res} = .82$ provides a more accurate estimate of error variability than $MS_{res} = .49$. If the latter assumption is made, the value $F_{(1/2/3)} = 24.67$ is computed.

$$F_{(1/2/3)} = \frac{MS_{BC_{(1/2/3)}}}{MS_{res_{(1/2/3/4)}}} = \frac{20.23}{.82} = 24.67$$

The degrees of freedom employed for the analysis are based on the mean square values employed in computing the $F_{(1/2/3)}$ ratio. Thus: $df_{num} = k_{subset} - 1 = 3 - 1 = 2$ (where $k_{subset} = 3$ conditions) and $df_{den} = df_{res_{(1/2/3/4)}} = 15$ (which is df_{res} for the omnibus F test involving all $k = 4$ conditions). For $df_{num} = 2$ and $df_{den} = 15$, $F_{.05} = 3.68$ and $F_{.01} = 6.36$. Since the obtained value $F = 24.67$ is greater than both of the aforementioned critical values, the null hypothesis can be rejected at both the .05 and .01 levels. Thus, the data do not support the researcher's hypothesis that Conditions 1, 2, and 3 represent a homogeneous subset. In view of this, the researcher would not conduct the contrast $(\bar{X}_1 + \bar{X}_2 + \bar{X}_3)/3$ versus $\bar{X}_4$.

If a researcher is not willing to assume that $MS_{res} = .82$ provides a more accurate estimate of error variability than $MS_{res} = .49$, the latter value can be employed as the denominator term in computing the $F_{(1/2/3)}$ ratio. Thus, if for some reason a researcher believes that by virtue of adding a fourth experimental condition experimental error is either increased or decreased, one can justify employing the value $MS_{res} = .49$ (computed for the $k = 3$ conditions) as the denominator term in computing the value $F_{(1/2/3)}$. If $MS_{res} = .49$ is employed to compute the F ratio, the value $F_{(1/2/3)} = 20.23/.49 = 41.29$ is computed. Since the latter value is greater than $F_{.05} = 3.68$ and $F_{.01} = 6.36$, the researcher can still reject the null hypothesis at both the .05 and .01 levels. However, the fact that $F_{(1/2/3)} = 41.29$ is substantially larger than $F_{(1/2/3)} = 24.67$ illustrates that depending upon which of the two error terms is employed, it is possible that they may lead to different conclusions regarding the status of the null hypothesis. The determination of which error term to use will be based on the assumptions a researcher is willing to make concerning the data. In the final analysis, in instances where the two error terms yield inconsistent results, it may be necessary for a researcher to conduct one or more replication studies in order to clarify the status of a null hypothesis.

3. Evaluation of the sphericity assumption underlying the single-factor within-subjects analysis of variance In Section I it is noted that one of the assumptions underlying the **single-factor within-subjects analysis of variance** is the existence of a condition referred to as **sphericity**. Sphericity exists when there is homogeneity of variance among the populations of difference scores. The latter can be explained as follows: Assume that for each of the n subjects who serve under all k experimental conditions, a difference score is calculated for all pairs of conditions. The number of difference scores which can be computed for each subject will equal $[k(k-1)]/2$. When $k = 3$, three sets of difference scores can be computed. Specifically: a) A set of difference scores that is the result of subtracting each subject's score in Condition 2 from the subject's score in Condition 1; b) A set of difference scores that is the result of subtracting each subject's score in Condition 3 from the subject's score in Condition 1; and c) A set of difference scores that is the result of subtracting each subject's score in Condition 3 from the subject's score in Condition 2.[13]

The sphericity assumption states that if the estimated population variances for the three sets of difference scores are computed, the values of the variances should be equal. The derivation of the three sets of difference scores for Example 24.1, and the computation of their estimated population variances are summarized in Table 24.11. Note that for each set of difference scores, a D value is computed for each subject. The estimated population variance of the D values (which is computed with Equation I.8) represents the estimated population variance of a set of difference scores.

Visual inspection of the estimated population variances of the difference scores reveals that the three variances are quite close to one another. This latter fact suggests that the sphericity assumption is unlikely to have been violated. Unfortunately, the tests which are discussed in this

Table 24.11　Computation of Estimated Population Variances of Difference Scores

		Condition 1 versus Condition 2		
	X_1	X_2	D	D^2
Subject 1	9	7	2	4
Subject 2	10	8	2	4
Subject 3	7	5	2	4
Subject 4	10	8	2	4
Subject 5	7	5	2	4
Subject 6	8	6	2	4
			$\Sigma D = 12$	$\Sigma D^2 = 24$

$$\tilde{s}^2_{(X_1 - X_2)} = \frac{\Sigma D^2 - \frac{(\Sigma D)^2}{n}}{n - 1} = \frac{24 - \frac{(12)^2}{6}}{6 - 1} = 0$$

		Condition 1 versus Condition 3		
	X_1	X_3	D	D^2
Subject 1	9	4	5	25
Subject 2	10	7	3	9
Subject 3	7	3	4	16
Subject 4	10	7	3	9
Subject 5	7	2	5	25
Subject 6	8	6	2	4
			$\Sigma D = 22$	$\Sigma D^2 = 88$

$$\tilde{s}^2_{(X_1 - X_3)} = \frac{\Sigma D^2 - \frac{(\Sigma D)^2}{n}}{n - 1} = \frac{88 - \frac{(22)^2}{6}}{6 - 1} = 1.47$$

		Condition 2 versus Condition 3		
	X_2	X_3	D	D^2
Subject 1	7	4	3	9
Subject 2	8	7	1	1
Subject 3	5	3	2	4
Subject 4	8	7	1	1
Subject 5	5	2	3	9
Subject 6	6	6	0	0
			$\Sigma D = 10$	$\Sigma D^2 = 24$

$$\tilde{s}^2_{(X_2 - X_3)} = \frac{\Sigma D^2 - \frac{(\Sigma D)^2}{n}}{n - 1} = \frac{24 - \frac{(10)^2}{6}}{6 - 1} = 1.47$$

book for evaluating homogeneity of variance are not appropriate for comparing the variances of the difference scores within the framework of evaluating the sphericity assumption. The procedures that have been developed for evaluating sphericity require the use of matrix algebra and are generally conducted with the aid of a computer. Further reference to such procedures will be made later in this discussion.

Sources on analysis of variance (e.g., Myers and Well (1995, 2003)) note there is another condition known as **compound symmetry** which is sufficient, although not necessary, in order for sphericity to exist. Compound symmetry, which represents a special case of sphericity, exists when both of the following conditions have been met: a) **Homogeneity of variance** — All of the populations that are represented by the k experimental conditions have equal variances; and b) **Homogeneity of covariance** — All of the population **covariances** are equal to one another. When the latter is true, the correlations between the scores of subjects on each pair of levels of the independent variable will be equal. Specifically, in the case of Example 24.1, if homogeneity of covariance exists, the correlation between the scores of subjects in Condition 1 versus Condition 2 will be equal to the correlation between the scores of subjects in Condition 1 versus Condition 3, which will be equal to the correlation between the scores of subjects in Condition 2 versus Condition 3. (For a detailed clarification of the concept of correlation, the reader should consult the **Introduction** as well as the chapter on the **Pearson product-moment correlation coefficient (Test 28)**.)

At this point we will examine the variances of the three experimental conditions, as well as the covariances of each pair of conditions. Employing Equation I.8, the estimated population variance for each of the three experimental conditions is computed.

$$\tilde{s}_1^2 = \frac{\sum X_1^2 - \frac{(\sum X_1)^2}{n}}{n-1} = \frac{443 - \frac{(51)^2}{6}}{5} = 1.9$$

$$\tilde{s}_2^2 = \frac{\sum X_2^2 - \frac{(\sum X_2)^2}{n}}{n-1} = \frac{263 - \frac{(39)^2}{6}}{5} = 1.9$$

$$\tilde{s}_3^2 = \frac{\sum X_3^2 - \frac{(\sum X_3)^2}{n}}{n-1} = \frac{163 - \frac{(29)^2}{6}}{5} = 4.57$$

Although Equation 17.9 is not actually employed to evaluate homogeneity of variance within the framework of the sphericity assumption, for illustrative purposes it will be used. The latter equation is employed to evaluate the homogeneity of variance assumption for the *t* **test for two dependent samples** by contrasting the **highest estimated population variance** (which in Example 24.1 is $\tilde{s}_3^2 = 4.57$) and the **lowest estimated population variance** (which in Example 24.1 is $\tilde{s}_1^2 = \tilde{s}_2^2 = 1.9$). In order to employ Equation 17.9 it is necessary to compute the correlation between subjects' scores in Condition 1 (which will be used to represent the lowest variance) and Condition 3. The value of the correlation coefficient is computed with Equation 17.7. Employing Equation 17.7, the value $r_{X_1 X_3} = .85$ is computed.[14]

$$r_{X_1 X_3} = \frac{\sum X_1 X_3 - \frac{(\sum X_1)(\sum X_3)}{n}}{\sqrt{\left[\sum X_1^2 - \frac{(\sum X_1)^2}{n}\right]\left[\sum X_3^2 - \frac{(\sum X_3)^2}{n}\right]}} = \frac{259 - \frac{(51)(29)}{6}}{\sqrt{\left[443 - \frac{(51)^2}{6}\right]\left[163 - \frac{(29)^2}{6}\right]}} = .85$$

Substituting the appropriate values in Equation 17.9, the value $t = 1.72$ is computed.

$$t = \frac{(\tilde{s}_L^2 - \tilde{s}_S^2)\sqrt{n-2}}{\sqrt{4\tilde{s}_L^2 \tilde{s}_S^2(1 - r_{X_1 X_3}^2)}} = \frac{(4.57 - 1.9)\sqrt{(6-2)}}{\sqrt{(4)(4.57)(1.9)(1 - (.85)^2)}} = 1.72$$

The degrees of freedom associated with the t value computed with Equation 17.9 are $df = n - 2 = 4$. Since the computed value $t = 1.72$ is less than the tabled critical two-tailed value $t_{.05} = 2.78$ (for $df = 4$), the null hypothesis H_0: $\sigma_L^2 = \sigma_S^2$ (which states there is homogeneity of variance) is retained. Thus, there is no evidence to suggest that the homogeneity of variance assumption is violated.

Earlier in the discussion it was noted that the sphericity assumption assumes equal population covariances. Whereas variance is a measure of variability of the scores of n subjects on a single variable, covariance (which is discussed in more detail in Section VII of the **Pearson product-moment correlation coefficient**) is a measure that represents the degree to which two variables vary together. A positive covariance is associated with variables that are positively correlated with one another, and a negative covariance is associated with variables that are negatively correlated with one another.

Equation 24.21 is the general equation for computing **covariance**. The value computed with Equation 24.21 represents the estimated covariance between **Population a** and **Population b**.

$$\text{cov}_{X_a X_b} = \frac{\sum X_a X_b - \frac{(\sum X_a)(\sum X_b)}{n}}{n - 1} \qquad \textbf{(Equation 24.21)}$$

Since a covariance can be computed for any pair of experimental conditions, in Example 24.1 three covariances can be computed — specifically, the covariance between Conditions 1 and 2, the covariance between Conditions 1 and 3, and the covariance between Conditions 2 and 3. To illustrate the computation of the covariance, the covariance between Conditions 1 and 2 ($\text{cov}_{X_1 X_2}$) will be computed employing Equation 24.21. Table 24.12, which reproduces the data for Conditions 1 and 2, summarizes the values employed in the calculation of the covariance.

Table 24.12 Data Required for Computing Covariance of Condition 1 and Condition 2

	X_1	X_2	$X_1 X_2$
Subject 1	9	7	63
Subject 2	10	8	80
Subject 3	7	5	35
Subject 4	10	8	80
Subject 5	7	5	35
Subject 6	8	6	48
	$\sum X_1 = 51$	$\sum X_2 = 39$	$\sum X_1 X_2 = 341$

Employing Equation 24.21 the value $\text{cov}_{X_1 X_2} = 1.9$ is computed.

$$\text{cov}_{X_1 X_2} = \frac{341 - \frac{(51)(39)}{6}}{6 - 1} = 1.9$$

If the relevant data for the other two sets of scores are employed, the values $\text{cov}_{X_1 X_3} = 2.5$ and $\text{cov}_{X_2 X_3} = 2.5$ are computed.

$$\text{cov}_{X_1 X_3} = \frac{259 - \frac{(51)(29)}{6}}{6 - 1} = 2.5$$

$$\text{cov}_{X_2 X_3} = \frac{201 - \frac{(39)(29)}{6}}{6 - 1} = 2.5$$

Since the three values for covariance are extremely close to one another, on the basis of visual inspection it would appear that the data are characterized by homogeneity of covariance. Coupled with the fact that homogeneity of variance also appears to exist, it would seem reasonable to conclude that the assumptions underlying compound symmetry (and thus of sphericity) are unlikely to have been violated.

The conditions necessary for compound symmetry (which, as previously noted, is not required in order for sphericity to exist) are, in fact, more stringent than the general requirement of sphericity (i.e., that there be homogeneity of variance among the populations of difference scores). Whenever data are characterized by compound symmetry, homogeneity of variance will exist among the populations of difference scores. However, it is possible to have homogeneity of variance among the populations of difference scores, yet not have compound symmetry.

At this point some general comments are in order regarding the consequences of violating the sphericity assumption. In the discussion of the *t* **test for two dependent samples** it is noted that the latter test is much more sensitive to violation of the homogeneity of variance assumption than is the *t* **test for two independent samples (Test 11)**. Since this observation can be generalized to designs involving more than two treatments, the **single-factor within-subjects analysis of variance** is more sensitive to violation of the sphericity assumption than is the **single-factor between-subjects analysis of variance** to violation of its assumption of homogeneity of variance. In point of fact, there is general agreement that the **single-factor within-subjects analysis of variance** is extremely sensitive to violations of the sphericity assumption, and that when the latter assumption is violated the tabled critical values in **Table A10** will not be accurate. Specifically, when the sphericity assumption is violated, the tabled critical F value associated with the appropriate degrees of freedom for the analysis of variance will be too low (i.e., the Type I error rate for the analysis will actually be higher than the prespecified value). One option proposed by Geisser and Greenhouse (1958) is to employ the tabled critical F value associated with $df_{num} = 1$ and $df_{den} = n - 1$ instead of the tabled critical F value associated with the usual degrees of freedom values (i.e., $df_{num} = df_{BC} = k - 1$ and $df_{den} = df_{res} = (n - 1)(k - 1)$). However, since the Geisser–Greenhouse method tends to overcorrect the value of F (i.e., it results in too high a critical value), some sources recommend an alternative but computationally more involved method developed by Box (1954) which does not result in as severe an adjustment of the critical F value as the Geisser–Greenhouse method.

An alternative methodology employed to evaluate the sphericity assumption is based on a statistic referred to as ε (which is the lower case Greek letter **epsilon**), which was derived independently by Greenhouse and Geisser (1959) and Huynh and Feldt (1976). Unfortunately the latter two methodologies described for computing ε usually do not result in the same value. Weinfurt (2000, p. 330) notes that the value of ε computed through use of the Greenhouse –Geiser method is thought to underestimate the true value of the underlying population parameter, while the Huynh–Feldt method is thought to overestimate it. When $\varepsilon = 1$, the sphericity assumption has been met, whereas when $\varepsilon = 1/(k - 1)$ the worst possible violation of the sphericity assumption is present. Weinfurt (2000) notes that although there is not a consensus among researchers with respect to the minimum acceptable value for ε, there is general agreement that when $\varepsilon > .90$ it is reasonably safe to assume that the sphericity assumption has not been violated. When the sphericity assumption is violated, an adjustment of the tabled critical F value is made by multiplying the numerator and denominator degrees of freedom employed in the computation of the F ratio by ε. The latter allows for a more conservative test, since it provides for a larger critical value than would be the case otherwise. Further discussion of ε can be found in Everitt (2001, pp. 142–143) and Howell (2002, pp. 486 –488).

Additional discussion of the tests which are employed to evaluate the sphericity assumption underlying the **single-factor within-subjects analysis of variance** can be found in selected texts that specialize in analysis of variance (e.g., Kirk (1982, 1995)). Keppel (1991), among others, notes, however, that tests which have been developed to evaluate the sphericity assumption have their own assumptions, and when the assumptions of the latter tests are violated (which may be more often than not), their reliability will be compromised. In view of this, Keppel (1991) questions the wisdom of employing such tests for evaluating the sphericity assumption.

Keppel (1991) notes that it is quite common for the sphericity assumption to be violated in experiments which utilize a within-subjects design. In view of this he recommends that in employing the **single-factor within-subjects analysis of variance** to evaluate the latter design, it is probably always prudent to run a more conservative test in order to insure that the Type I error rate is adequately controlled. Some sources suggest that the value most affected by violation of the sphericity assumption will not necessarily be the omnibus F value, but instead will be the results computed for comparisons/contrasts (which employ the error term computed for the analysis of variance) conducted after the F test (Weinfurt (2000, p. 331) and Howell (2002, p. 519)). With regard to the latter, Anderson (2001, p. 593) recommends always computing a separate error term for each comparison/contrast conducted, regardless of whether the sphericity assumption is violated.

In recent years many sources have argued that when there is reason to believe the sphericity assumption underlying the **single-factor within-subjects analysis of variance** is violated, the data should be evaluated with a **multivariate analysis of variance (Test 35)**. The latter analysis is described in Section VI of the chapter on **Hotelling's T^2 (Test 34)** (which is a special case of the **multivariate analysis of variance**) in the discussion of **the use of the single-sample Hotelling's T^2 to evaluate a dependent samples design (Test 34b)**. Sources (e.g., Weinfurt (2000), Howell (2002), and Maxwell and Delaney (1990; 2000; 2004, Ch. 13)) which discuss the use of the **multivariate analysis of variance** in evaluating a within-subjects design note the main advantage of the latter procedure is that it does not assume sphericity. Stevens (2002, p. 509) provides a thoughtful discussion on the merits and liabilities of the use of the **single-factor within-subjects analysis of variance** versus the **multivariate analysis of variance** for a within-subjects design. He notes that when the sphericity assumption is not violated, the **single-factor within-subjects analysis of variance** provides a more powerful test of an alternative hypothesis than does a **multivariate analysis of variance**, and if the sphericity assumption is violated and the epsilon correction is employed, since the Type I error rate will be controlled adequately, there is no reason to employ the multivariate approach. Yet other sources are not in agreement with respect to which analysis to employ when the sphericity assumption is violated. Maxwell and Delaney (1990, pp. 602–604; 2004) note that, as a general rule, for small sample sizes the **multivariate analysis of variance** provides a less powerful test of an alternative hypothesis than does the **single-factor within-subjects analysis of variance**, and within the latter context they suggest that the multivariate analysis should not be employed if the value of n is less than (k + 10). McCulloch (2005) argues that since the **single-factor within-subjects analysis of variance** is prone to yielding inaccurate results under certain circumstances (e.g., scores for one or more subjects are not available for all *k* conditions), it is his recommendation that researchers employ alternative procedures, such as maximum likelihood methodology (which will not be described in this book for such an analysis), in evaluating a within-subjects design. Further discussion of the use of the **multivariate analysis of variance** in evaluating a within-subjects design can be found in **Chapter 34**.

It should be apparent from the discussion in this section that there is lack of agreement with respect to the most appropriate methodology for dealing with violation of the sphericity assumption.[15] A cynic might conclude that regardless of which method one employs there will always be reason to doubt the accuracy of the probability value associated with the outcome of a study. As noted throughout this book, in situations where there are doubts concerning the reliability of an analysis, the most powerful tool the researcher has at her disposal is replication. In the final analysis, the truth regarding a hypothesis will ultimately emerge if one or more researchers conduct multiple studies which evaluate the same hypothesis. In instances where replication studies have been conducted, **meta-analysis** (discussed in Section IX (the **Addendum**) of the **Pearson product-moment correlation coefficient**) can be employed to derive a pooled probability value for all of the published studies. It should be noted, however, that if the accuracy of the probabilities associated with the outcome of one or more studies is subject to challenge, the accuracy of a pooled probability will be compromised. One can argue, however, that if enough replication studies are conducted, probability inaccuracies in one direction will most likely be balanced by probability inaccuracies in the opposite direction.

4. Computation of the power of the single-factor within-subjects analysis of variance Prior to reading this section the reader should review the discussion of power in Section VI of the **single-factor between-subjects analysis of variance**, since basically the same procedure is employed to determine the power of the **single-factor within-subjects analysis of variance**. The power of the **single-factor within-subjects analysis of variance** is computed with Equation 24.22, which is identical to Equation 21.38 (which is the equation used for computing the power of the **single-factor between-subjects analysis of variance**), except for the fact that the estimated value of σ^2_{res} is employed as the measure of error variability in place of σ^2_{WG} (McFatter and Gollob (1986)).

$$\varphi = \sqrt{n \left[\frac{\Sigma(\mu_j - \mu_T)^2}{k\sigma^2_{res}} \right]}$$ **(Equation 24.22)**

Where: μ_j = The estimated mean of the population represented by Condition *j*
  μ_T = The grand mean, which is the average of the *k* estimated population means
  σ^2_{res} = The estimated measure of error variability
  n = The number of subjects
  k = The number of experimental conditions

To illustrate the use of Equation 24.22 with Example 24.1, let us assume that prior to conducting the study the researcher estimates that the means of the populations represented by the three conditions are as follows: $\mu_1 = 8$, $\mu_2 = 6$, $\mu_3 = 4$. Additionally, it will be assumed that he estimates the population error variance associated with the analysis will equal $\sigma^2_{res} = 1$. Based on this information, the value μ_T can be computed: $\mu_T = (\mu_1 + \mu_2 + \mu_3)/k = (8 + 6 + 4)/3 = 6$. The appropriate values are now substituted in Equation 24.22.

$$\varphi = \sqrt{n\left[\frac{(8 - 6)^2 + (6 - 6)^2 + (4 - 6)^2}{(3)(1)}\right]} = \sqrt{2.67n} = 1.63\sqrt{n}$$

At this point **Table A15 (Graphs of the Power Function for the Analysis of Variance)** in the **Appendix** can be employed to determine the necessary sample size required in order to have the power stipulated by the experimenter. For our analysis (for which it will be assumed $\alpha = .05$) the appropriate set of curves to employ is the set for $df_{num} = df_{BC} = 2$. Let us assume we want the omnibus F test to have a power of at least .80. We now substitute what we consider to be a reasonable value for n in the equation $\varphi = 1.63\sqrt{n}$ (which is the result obtained with Equation 24.22). To illustrate, the value $n = 6$ (the sample size employed in Example 24.1) is substituted in the equation. The resulting value is $\varphi = 1.63\sqrt{6} = 3.99$.

The value $\varphi = 3.99$ is located on the abscissa (X-axis) of the relevant set of curves in **Table A15** — specifically, the set for $df_{num} = 2$. At the point corresponding to $\varphi = 3.99$, a perpendicular line is erected from the abscissa which intersects with the power curve that corresponds to $df_{den} = df_{res}$ employed for the omnibus F test. Since $df_{res} = 10$, the curve for the latter value is employed (or closest to it if a curve for the exact value is not available). At the point the perpendicular intersects the curve $df_{res} = 10$, a second perpendicular line is drawn in relation to the ordinate (Y-axis). The point at which this perpendicular intersects the ordinate indicates the power of the test. Since $\varphi = 3.99$, we determine the power equals 1.[16] Thus, if we employ six subjects in a within-subjects design, there is a 100% likelihood (which corresponds to a probability of 1) of detecting an effect size equal to or larger than the one stipulated by the researcher (which is a function of the estimated values for the population means relative to the value estimated for error variability). Since the probability of committing a Type II error is $\beta = 1 - power$, $\beta = 1 - 1 = 0$. This value represents the likelihood of not detecting an effect size equal to or greater than the one stipulated.

Equation 24.23 (described in McFatter and Gollob (1986)) can be employed to conduct a power analysis for a comparison associated with a **single-factor within-subjects analysis of variance**. Equation 24.23 is identical to Equation 21.39 (which is the equation for evaluating the power of a comparison for the **single-factor between-subjects analysis of variance**), except for the fact that σ_{res}^2 is employed as the measure of error variability in place of σ_{WG}^2.

$$\varphi_{comp} = \sqrt{n\left[\frac{(\mu_a - \mu_b)^2}{2(\sigma_{res}^2)(\Sigma c_j^2)}\right]} \qquad \textbf{(Equation 24.23)}$$

As is the case for a **single-factor between-subjects analysis of variance**, Equation 24.23 can be used for both simple and complex single degree of freedom comparisons. As a general rule, the equation is used for planned comparisons. As noted in the discussion of the **single-factor between-subjects analysis of variance**, although the equation can be extended to unplanned comparisons, published power tables for the analysis of variance generally only apply to per comparison error rates of $\alpha = .05$ and $\alpha = .01$. In the case of planned and especially unplanned comparisons which involve α_{PC} rates other than .05 or .01, more detailed tables are required.

For single degree of freedom comparisons, the power curves in **Table A15** for $df_{num} = 1$ are always employed. The use of Equation 24.23 will be illustrated for the simple comparison Condition 1 versus Condition 2 (summarized in Table 24.3). Since $\Sigma c_j = 2$, and we have estimated $\mu_a = \mu_1 = 8$, $\mu_b = \mu_2 = 6$ and $\sigma_{res}^2 = 1$, the following result is obtained:

$$\varphi_{comp} = \sqrt{n \left[\frac{(8 - 6)^2}{(2)(1)(2)} \right]} = \sqrt{n}$$

Substituting $n = 6$ in the equation $\varphi_{comp} = \sqrt{n}$, we obtain $\varphi_{comp} = \sqrt{6} = 2.45$. Employing the power curves for $df_{num} = 1$ with $\alpha = .05$, we use the curve for $df_{res} = 10$ (df_{res} employed for the omnibus F test) and determine that when $\varphi_{comp} = 2.45$, the power of the test is approximately .88.

5. Measures of magnitude of treatment effect for the single-factor within-subjects analysis of variance: Omega squared (Test 24g) and Cohen's f index (Test 24h)

Prior to reading this section the reader should review the discussion of measures of magnitude of treatment effect in Section VI of both the *t* test for two independent samples and the single-factor between-subjects analysis of variance. The discussion for the latter test notes that the computation of an omnibus F value only provides a researcher with information regarding whether or not the null hypothesis can be rejected — i.e., whether a significant difference exists between at least two of the experimental conditions. The F value (as well as the level of significance with which it is associated), however, does not provide the researcher with any information regarding the size of any treatment effect that is present. As is noted in earlier discussions of treatment effect, the latter is defined as the proportion of the variability on the dependent variable which is associated with the independent variable/experimental conditions. The measures described in this section are variously referred to as **measures of effect size**, **measures of magnitude of treatment effect**, **measures of association**, and **correlation coefficients**.

Omega squared (Test 24g) The **omega squared** statistic is a commonly computed measure of treatment effect for the **single-factor within-subjects analysis of variance**. Keppel (1991) and Kirk (1995) note there is disagreement with respect to which variance components should be employed in computing **omega squared** for a within-subjects design. One method of computing **omega squared** (which computes a value referred to as **standard omega squared**) was employed in the first edition of this book. The latter method expresses treatment (i.e., between-conditions) variability as a proportion of the sum of all the elements which account for variability in a within-subjects design. Equation 24.25, which is presented in Myers and Well (1995), can be employed to compute **standard omega squared** ($\tilde{\omega}_s^2$).[17] The **omega squared** value computed with Equation 24.25 is an estimate of the proportion of variability in the data that is attributed to the experimental treatments (σ_{BC}^2) divided by the sum total of variability in the data (i.e., treatment variability (σ_{BC}^2) plus between-subjects variability (σ_{BS}^2) plus residual variability (σ_{res}^2)). Thus, Equation 24.24 represents the population parameter (ω_s^2) estimated by Equation 24.25. Employing Equation 24.25 with the data for Example 24.1, the value $\tilde{\omega}_s^2 = .44$ is computed.

$$\omega_s^2 = \frac{\sigma_{BC}^2}{\sigma_{BC}^2 + \sigma_{BS}^2 + \sigma_{res}^2}$$

(Equation 24.24)

(Equation 24.25)

$$\tilde{\omega}_s^2 = \frac{(k - 1)(MS_{BC} - MS_{res})}{(k - 1)(n - 1)MS_{res} + (k - 1)MS_{BC} + nMS_{BS}}$$

$$\tilde{\omega}_s^2 = \frac{(3 - 1)(20.23 - .49)}{(3 - 1)(6 - 1)(.49) + (3 - 1)(20.23) + (6)(7.39)} = .44$$

The value $\tilde{\omega}_s^2 = .44$ indicates that 44% (or a proportion of .44) of the variability on the dependent variable (the number of nonsense syllables correctly recalled) is associated with variability on the levels of the independent variable (noise).

A second method for computing **omega squared** (discussed in the second and third editions of this book) computes what is referred to as **partial omega squared**. The latter measure, which Keppel (1991) and Kirk (1995) view as more meaningful than **standard omega squared**, ignores between-subjects variability, and expresses treatment (i.e., between-conditions) variability as a proportion of the sum of between-conditions and residual variability. Grissom and Kim (2005, p. 136) (also see Maxwell and Delaney (2004, Ch. 7) for a contrary viewpoint) note that **partial omega square** when computed for a **within-subjects design** is comparable to an **omega squared** value for a **between-subjects design** by virtue of eliminating subject variability from the total variability. Equation 24.27 is employed to compute **partial omega squared** ($\tilde{\omega}_p^2$). Equation 24.26 represents the population parameter (ω_p^2) estimated by Equation 24.27. Employing Equation 24.27, the value $\tilde{\omega}_p^2 = .82$ is computed.

$$\omega_p^2 = \frac{\sigma_{BC}^2}{\sigma_{BC}^2 + \sigma_{res}^2}$$ **(Equation 24.26)**

$$\tilde{\omega}_p^2 = \frac{\tilde{\sigma}_{BC}^2}{\tilde{\sigma}_{BC}^2 + \tilde{\sigma}_{res}^2}$$ **(Equation 24.27)**

Where:

$$\tilde{\sigma}_{BC}^2 = \frac{df_{BC}(MS_{BC} - MS_{res})}{nk} = \frac{(2)(20.23 - .49)}{(6)(3)} = 2.19$$

$$\tilde{\sigma}_{res}^2 = MS_{res} = .49$$

Thus: $\tilde{\omega}_p^2 = \dfrac{\tilde{\sigma}_{BC}^2}{\tilde{\sigma}_{BC}^2 + \tilde{\sigma}_{res}^2} = \dfrac{2.19}{2.19 + .49} = .82$

Equation 24.28 can also be employed to compute the value of **partial omega squared**.

$$\tilde{\omega}_p^2 = \frac{(k - 1)(F - 1)}{(k - 1)(F - 1) + nk}$$

(Equation 24.28)

$$\tilde{\omega}_p^2 = \frac{(3 - 1)(41.29 - 1)}{(3 - 1)(41.29 - 1) + (6)(3)} = .82$$

The value $\tilde{\omega}_p^2 = .82$ computed for **partial omega squared** indicates that 82% (or a proportion of .82) of the variability on the dependent variable (the number of nonsense syllables correctly recalled) is associated with variability on the levels of the independent variable (noise). Note that because it does not take into account between-subjects variability, **partial omega squared** yields a much higher value than **standard omega squared**.

It was noted in an earlier discussion of **omega squared** (in Section VI of the *t* **test for two independent samples**) that Cohen (1977; 1988, pp. 284–287) has suggested the following (admittedly arbitrary) values, which are employed in psychology and a number of other disciplines, as guidelines for interpreting $\tilde{\omega}^2$: a) A **small effect size** is one that is greater than .0099 but not more than .0588; b) A **medium effect size** is one that is greater than .0588 but not more than .1379; and c) A **large effect size** is greater than .1379. If one employs Cohen's (1977,1988) guidelines for magnitude of treatment effect, both $\tilde{\omega}_s^2 = .44$ and $\tilde{\omega}_p^2 = .82$ represent a large treatment effect.[18]

Cohen's *f* index (Test 24h)   If the value of **partial omega squared** is substituted in Equation 21.45, **Cohen's *f* index** can be computed. In Section VI of the **single-factor between-subjects analysis of variance**, it was noted that **Cohen's *f* index** is an alternate measure of effect size which can be employed for an analysis of variance. The computation of **Cohen's *f* index** with Equation 21.45 yields the value $f = 2.13$.

$$f = \sqrt{\frac{\tilde{\omega}^2}{1 - \tilde{\omega}^2}} = \sqrt{\frac{.82}{1 - .82}} = 2.13$$

In the discussion of **Cohen's *f* index** in Section VI of the **single-factor between-subjects analysis of variance**, it was noted that Cohen (1977; 1988, pp. 284–288) employed the following (admittedly arbitrary) *f* values as criteria for identifying the magnitude of an effect size: a) A **small effect size** is one that is greater than .1 but not more than .25; b) A **medium effect size** is one that is greater than .25 but not more than .4; and c) A **large effect size** is greater than .4. Employing Cohen's criteria, the value $f = 2.13$ represents a large effect size.

Sources which discuss computation of a measure of magnitude of treatment effect for a **single-factor within-subjects analysis of variance** are Cortina and Nouri (2000), Grissom and Kim (2005), Keppel (1991), Kirk (1995), Kline (2004), Maxwell and Delaney (2004), Rosenthal *et al.* (2000) and Rosnow *et al.* (2000). Further discussion of the indices of treatment effect discussed in this section and the relationship between effect size and statistical power can be found in Section IX (the **Addendum**) of the **Pearson product-moment correlation coefficient** under the discussion of **meta-analysis and related topics**.

6. Computation of a confidence interval for the mean of a treatment population   Prior to reading this section the reader should review the discussion of confidence intervals in Section VI of both the **single-sample *t* test (Test 2)** and the **single-factor between-subjects analysis of variance**. The same procedure employed to compute a confidence interval for a treatment population for the **single-factor between-subjects analysis of variance** is employed for computing the confidence interval for the mean of a treatment population for a **single-factor within-subjects analysis of variance**. In other words, in order to compute a confidence interval for any experimental treatment/condition, one must conceptualize a within-subjects design as if it was a between-subjects design. The reason for this is that a confidence interval for any single condition will be a function of the variability of the scores of subjects who serve within that condition. Since MS_{res}, the measure of error variability for the repeated-measures analysis of variance, is a measure of within-subjects variability which is independent of any treatment effect, it cannot be employed to estimate the error variability for a specific treatment if one wants to compute a confidence interval for the mean of a treatment population. As is the case with the **single-factor between-subjects analysis of variance**, one can employ either of the following two strategies in computing the confidence interval for a treatment.

　　a) If one assumes that all of the treatments represent a population with the same variance, Equation 21.48 can be employed to compute the confidence interval (in our example we will assume the 95% confidence interval is being computed). In order to employ Equation 21.48 it is necessary to compute the value of MS_{WG}, which in the case of the **single-factor within-subjects analysis of variance** can be conceptualized as a **within-conditions mean square** (MS_{WC}). In order to compute MS_{WC}, it is necessary to first compute the **within-conditions sum of squares** (SS_{WC}). The latter value is computed employing Equation 24.29 (which is identical to Equation 21.5, except it employs the subscript WC in place of WG).

$$SS_{WC} = \sum_{j=1}^{k} \left[\sum X_j^2 - \frac{(\sum X_j)^2}{n} \right] \qquad \textbf{(Equation 24.29)}$$

$$= \left[443 - \frac{(51)^2}{6} \right] + \left[263 - \frac{(39)^2}{6} \right] + \left[163 - \frac{(29)^2}{6} \right] = 41.83$$

　　The within-conditions degrees of freedom is computed in an identical manner as is the within-groups degrees of freedom for the **single-factor between-subjects analysis of variance**. Thus, using Equation 21.9 (using the subscript WC in place of WG), $df_{WC} = N - k = 18 - 3 = 15$. The **within-conditions mean square** can now be computed: $MS_{WC} = SS_{WC}/df_{WC} = 41.83/15 = 2.79$. Employing Equation 21.48, the 95% confidence interval for the mean of the population represented by Condition 1 is computed. The value $t_{.05} = 2.13$ is the tabled critical two-tailed $t_{.05}$ value for $df_{WC} = 15$.

$$CI_{.95} = \bar{X}_j \pm t_{df_{WC}} \sqrt{\frac{MS_{WC}}{n}} = 8.5 \pm 2.13 \sqrt{\frac{2.79}{6}} = 8.5 \pm 1.45$$

　　Thus, the researcher can be 95% confident (or the probability is .95) the interval 7.05 to 9.95 includes the actual value of the mean of the population represented by Condition 1. Stated symbolically: $7.05 \le \mu_1 \le 9.95$.

　　b) If one has reason to believe that the treatment in question is distinct from the other treatments, Equation 2.8 (and more specifically Equation 2.9) can be employed to compute the 95% confidence interval. Specifically, if the mean value of a treatment is substantially above or below the means of the other treatments, it can be argued that one can no longer assume that the treatment shares a common variance with the other treatments. In view of this, one can take the position that the variance of the treatment is the best estimate of the variance of the population represented by that treatment (as opposed to the pooled variability of all of the treatments involved in the study). This position can also be taken even if the means of the k treatments are equal, but the treatments have substantially different estimated population variances.

　　If Equation 2.8 is employed to compute the 95% confidence interval for the population mean of Condition 1, the estimated variance of the population Condition 1 represents is employed in lieu of the pooled within-conditions variability. In addition, the tabled critical two-tailed $t_{.05}$ value for $df = n - 1 = 5$ (which is $t_{.05} = 2.57$) is employed in the equation. The computation of the confidence interval is illustrated below. Initially, the estimated population standard deviation is computed, which is then substituted in Equation 2.8. The result obtained with Equation 2.8 indicates that the range for $CI_{.95}$ is $7.05 \le \mu_1 \le 9.95$.[19] Note that in the case of Condition 1, the confidence intervals computed with Equations 21.48 and 2.8 are identical. This will not always be true, especially when the within-condition variability of the treatment for which the confidence interval is computed is substantially different from the within-condition variability of the other treatments.

$$\tilde{s}_1 = \sqrt{\frac{\Sigma X_1^2 - \frac{(\Sigma X_1)^2}{n}}{n-1}} = \sqrt{\frac{443 - \frac{(51)^2}{6}}{5}} = 1.38$$

$$CI_{.95} \quad \overline{X}_1 \pm t_{.05}\left(\frac{\tilde{s}_1}{\sqrt{n}}\right) = 8.5 \pm 2.57\left(\frac{1.38}{\sqrt{6}}\right) = 8.5 \pm 1.45$$

7. Test 24i: The intraclass correlation coefficient Prior to reading this section the reader may want to review the material in the **Introduction** on correlation. A measure of association referred to as the **intraclass correlation coefficient** can be computed for the model upon which the **single-factor within-subjects analysis of variance** is based. The latter correlation coefficient is generally employed when a researcher desires a measure of **interjudge reliability** with respect to k judges rating n subjects/objects on some variable. (**Interjudge reliability** is the degree to which two or more judges are in agreement with one another.) A number of measures of association which can be used to assess interjudge reliability are described in this book. Specifically: a) **Cohen's kappa (Test 16k)** is described in reference to categorical data when two judges are involved; b) **Spearman's rank-order correlation coefficient (Test 29)** and **Kendall's tau (Test 30)** are described in reference to data that are rank-ordered by two judges; c) The **Pearson product-moment correlation coefficient** is described in reference to interval/ratio data when two judges are involved; and d) **Kendall's coefficient of concordance (Test 31)** is described in reference to data that are rank-ordered by more than two judges. The **intraclass correlation coefficient** (represented by the notation r_{IC}) can be employed to assess interjudge reliability in reference to interval/ratio data when more than two judges are involved.

When $k = 3$ or more judges evaluate n subjects/objects, a **Pearson product-moment correlation coefficient** between the ratings of each pair of judges can be computed, and the average correlation for the set of judges can be determined (through use of Equation 28.25). However, a limitation of the latter approach is that it does not take into account the differences between the ratings of the judges. To be more specific, if one judge consistently uses higher ratings than another judge, the product-moment correlation will not take the latter into account. As an example, assume that $k = 2$ judges are instructed to rate subjects with respect to a variable (e.g., attractiveness) on a ten point scale. If one judge rates all of the subjects between 7 and 10 and a second judge rates all of the subjects between 1 and 4, as long as the ratings of the judges are proportional with respect to the subjects/objects being rated, the value for the correlation coefficient will equal 1. In other words, as long as there is **relative agreement** between the judges, the value of the correlation coefficient will be high (which is also the case for the rank-order correlation coefficients referred to in the previous paragraph). If, on the other hand, one is concerned about the **absolute differences** between the judges' ratings, the use of the **intraclass correlation coefficient** is recommended, since if there is a substantial difference between the mean ratings of k judges the **intraclass correlation coefficient** will reflect such differences (in that the value of the correlation will be close to zero). In point of fact, the **intraclass correlation coefficient** can only achieve the maximum value of $+1$ if all of the judges assign the exact same ratings to the subjects/objects. Put simply, if most of the variability in the data is due to between-subjects variability (which would be reflected in substantial differences between the mean ratings assigned to each of the n subjects) with very low between-judges variability (which would be reflected in minimal differences between the mean ratings assigned by each of the *k* judges), the value for the **intraclass correlation** will be close to one. On the other hand, if most of the variability is due to a large degree of interjudge variability (which would be reflected in substantial differences between the mean ratings of the *k* judges), the **intraclass correlation** will be low. Among others, Bartko (1976) notes that an **intraclass correlation** may fall within the following range of values: $-1/(k-1) < r_{IC} < +1$. Although the latter indicates it is theoretically possible to obtain a negative value for r_{IC}, a negative correlation is generally treated as $r_{IC} = 0$. It should be noted that the analysis to be described below for computing an **intraclass correlation** assumes a **random-effects model** (as opposed to a **fixed-effects model**, which requires a different computational equation).

To illustrate the computation of an r_{IC} value, let us assume that $k = 3$ judges rate $n = 6$ subjects with respect to physical attractiveness, and that an **intraclass correlation coefficient** will be employed to assess interjudge reliability. The data are summarized in Table 24.13.

Table 24.13 Data for Intraclass Correlation

	Judge 1	Judge 2	Judge 3
Subject 1	9	7	4
Subject 2	10	8	7
Subject 3	7	5	3
Subject 4	10	8	7
Subject 5	7	5	2
Subject 6	8	6	6

In point of fact, Table 24.13 is identical to Table 24.1, except that the column variable in Table 24.13 is represented by the judges (as opposed to it representing the levels of noise (i.e., the independent variable) as is the case in Table 24.1). Equation 24.30 is employed to compute the value of the **intraclass correlation coefficient**. The values required for computing r_{IC} are determined by conducting a **single-factor within-subjects analysis of variance** on the data. Since the data in Table 24.13 are identical to those contained in Table 24.1, the **within-subjects analysis of variance** yields the results summarized in Table 24.2. In Equation 24.30 the between-judges variability (MS_{BJ}) is equivalent to the between-conditions variability (MS_{BC}) computed in the original analysis. Since the value $r_{IC} = .38$ is substantially less than 1, it indicates poor interjudge reliability (i.e., poor agreement among the judges). The low value computed for the **intraclass correlation** can be attributed to the fact that there are considerable differences between the mean ratings of the three judges (specifically, the mean ratings of Judges 1, 2, and 3 are respectively 8.5, 6.5, 4.83).

Shrout and Fleiss (1979, p. 424) note that in order to evaluate the null hypothesis H_0: $\rho_{IC} = 0$, the following *F* ratio (which was not computed in Table 24.2) is employed: $F_{BS} = MS_{BS}/MS_{res}$. Employing the mean square values for Example 24.1, $F_{BS} = 7.39/.49 = 15.08$. The latter value is evaluated with $df_{BS} = n - 1 = 6 - 1 = 5$, and $df_{res} = (n-1)(k-1) = (6-1)(3-1) = 10$. For $df_{num} = 5$ and $df_{den} = 10$, the tabled $F_{.95}$ and $F_{.99}$ values are $F_{.95} = 3.33$ and $F_{.99} = 5.64$. Since $F = 15.08$ is greater than both of the aforementioned critical values, the null hypothesis can be rejected at both the .05 and .01 levels. Thus one can conclude that the underlying population **intraclass correlation** (ρ_{IC}) is some value other than zero (which, it should be emphasized, in itself does not indicate high interjudge reliability — the latter usually requiring a value of .7 or greater). The value $r_{IC} = .38$ is in sharp contrast to the average of the three **Pearson product-moment correlation coefficients** for the ratings of each possible pair of judges. The latter average is computed to be $r = .951$ (through use of Equation 28.25) or $r = .9$ (if the three *r* values of 1, .85, and .85 obtained earlier in the discussion of the sphericity assumption are averaged). Further discussion of the **intraclass correlation coefficient** can be found in Section VII of **Kendall's coefficient of concordance**, as well as in Cohen (2001), Everitt (2001), Howell (2002), Maxwell and Delaney (2004, pp. 563–567), Shrout and Fleiss (1979), and Wilcox (2003).

(Equation 24.30)

$$r_{IC} = \frac{MS_{BS} - MS_{res}}{MS_{BS} + (k - 1)\,MS_{res} + \frac{k}{n}\left(MS_{BJ} - MS_{res}\right)} = \frac{7.39 - .49 + (3 - 1)(.49) + \left(\frac{3}{6}\right)(20.23 - .49)}{7.39} = .38$$

VII. Additional Discussion of the Single-Factor Within-Subjects Analysis of Variance

1. Theoretical rationale underlying the single-factor within-subjects analysis of variance
In the **single-factor within-subjects analysis of variance** the total variability can be partitioned into the following two elements: a) **Between-subjects variability** (which is represented by MS_{BS}) represents the variability between the mean scores of the n subjects. In other words, a mean value for each subject who has served in each of the k experimental conditions is computed, and MS_{BS} represents the variance of the n subject means; and b) **Within-subjects variability** (which will be represented by the notation MS_{WS}) represents variability within the k scores of each of the n subjects. In other words, for each subject the variance for that subject's k scores is computed, and the average of the n variances represents within-subjects variability. **Within-subjects variability** can itself be partitioned into two elements: **between-conditions variability** (which is represented by MS_{BC}) and **residual variability** (which is represented by MS_{res}). Between-conditions variability is essentially a measure of variance of the means of the k experimental conditions. In the **single-factor within-subjects analysis of variance**, it is assumed that any variability between the means of the conditions can be attributed to one or both of the following two elements: a) **The experimental treatments**; and b) **Experimental error**. When MS_{BC} (the value computed for between-conditions variability) is significantly greater than MS_{res} (the value computed for error variability), it is interpreted as indicating that a substantial portion of between-conditions variability is due to a treatment effect. The rationale for this is as follows.

Experimental error is random variability in the data which is beyond the control of the researcher. In a within-subjects design the average amount of variability within the k scores of each of the n subjects that cannot be accounted for on the basis of a treatment effect is employed to represent experimental error. Thus, the value computed for MS_{res} is the normal amount of variability which is expected for any subject who serves in each of k experimental conditions, if the conditions are equivalent to one another. Within this framework, residual variability is employed as a baseline to represent variability that results from factors which are beyond an experimenter's control. The experimenter assumes that even if no treatment effect is present, since such uncontrollable factors are responsible for within-subjects variability, it is logical to assume that they can produce differences of a comparable magnitude between the means of the k experimental conditions. As long as the variability between the condition means (MS_{BC}) is approximately the same as residual variability (MS_{res}), the experimenter can attribute any between-conditions variability present to experimental error. When, however, between-conditions variability is substantially greater than residual variability, it indicates that something over and above error variability is contributing to the variability between the k condition means. In such a case, it is assumed that a treatment effect is responsible for the larger value of MS_{BC} relative to the value of MS_{res}. In essence, if residual variability is subtracted from within-subjects variability, any remaining variability within the scores of subjects can be attributed to a treatment effect. If there is no treatment effect, the result of the subtraction will be zero. Of course, one can never completely rule out the possibility that if MS_{BC} is larger than MS_{res}, the larger value for MS_{BC} is entirely due to error variability. However, since the latter is unlikely, when MS_{BC} is significantly larger than MS_{res}, it is interpreted as indicating the presence of a treatment effect.

Table 24.14 Alternative Summary Table for Analysis of Variance for Example 24.1

Source of variation	SS	df	MS	F
Between-subjects	36.93	5	7.39	
Within-subjects	45.35	12	3.78	
Between-conditions	40.45	2	20.23	41.29
Residual	4.9	10	.49	
Total	**82.28**	17		

In some sources a table employing the format depicted in Table 24.14 is used to summarize the results of a **single-factor within-subjects analysis of variance**. In contrast to Table 24.2, which does not include a row documenting within-subjects variability, Table 24.14 includes the latter variability, which is partitioned into between-conditions variability and residual variability. In point of fact, it is not necessary to compute the information documented in the row for within-subjects variability in order to compute the F ratio. (The reader should also take note of the fact that in some sources the terms **interaction** or **conditions × subjects interaction** are employed to represent the **residual variability** for the **single-factor within-subjects analysis of variance.**)

Note that in Table 24.14 the following relationships will always be true: a) $SS_T = SS_{BS} + SS_{WS}$; b) $df_T = df_{BS} + df_{WS}$; c) $SS_{WS} = SS_{BC} + SS_{res}$; and d) $df_{WS} = df_{BC} + df_{res}$. The values SS_{WS}, df_{WS}, and MS_{WS} in Table 24.14 are, respectively, computed with Equations 24.31, 24.32, and 24.33.

$$SS_{WS} = \Sigma X_T^2 - \sum_{i=1}^{n} \left[\frac{(\Sigma S_i)^2}{k} \right] \qquad \textbf{(Equation 24.31)}$$

$$df_{WS} = n(k - 1) \qquad \textbf{(Equation 24.32)}$$

$$MS_{WS} = \frac{SS_{WS}}{df_{WS}} \qquad \textbf{(Equation 24.33)}$$

The values $SS_{WS} = 45.35$, $df_{WS} = 12$, and $MS_{WS} = 3.78$ are computed below for Example 24.1. Note that $SS_{WS} = SS_{BC} + SS_{res} = 40.45 + 4.9 = 45.35$ and $df_{WS} = df_{BC} + df_{res} = 2 + 10 = 12$.

$$SS_{WS} = 869 - \left[\frac{(20)^2}{3} + \frac{(25)^2}{3} + \frac{(15)^2}{3} + \frac{(25)^2}{3} + \frac{(14)^2}{3} + \frac{(20)^2}{3} \right] = 45.35$$

$$df_{WS} = 6(3 - 1) = 12 \qquad MS_{WS} = \frac{45.35}{12} = 3.78$$

Other sources employ the format in Table 24.15 to summarize the results of a **single-factor within-subjects analysis of variance**. As is the case for the **single-factor between-subjects analysis of variance**, in the latter table variability is partitioned into **between-conditions variability** (which is the same as **between-groups variability**) and **within-conditions variability** (which is the same as **within-groups variability**). Note, however, that in the **single-factor within-subjects analysis of variance** the **within-conditions variability** is partitioned into **between-subjects variability** and **residual variability**, since variability within any of the k experimental conditions will be a function of individual differences among subjects and error/residual variability. Whereas in the **single-factor between-subjects analysis of variance** the variability attributable to individual differences is considered part of error variability (since the two cannot be separated from one another), in the **single-factor within-subjects analysis of variance** variability attributable to individual differences and error variability can be separated from one another. The latter is accomplished by subtracting **between-subjects variability** from **within-conditions variability**, with the difference representing **error/residual variability**.

Table 24.15 Alternative Summary Table for Analysis of Variance for Example 24.1

Source of variation	SS	df	MS	F
Between-conditions	40.45	2	20.23	41.29
Within-conditions	41.83	15	2.79	
Between-subjects	36.93	5	7.39	
Residual	4.9	10	.49	
Total	**82.28**	**17**		

Except for the **within-conditions variability**, the values for all of the sources of variability listed in Table 24.15 have already been computed. The relevant values for within-conditions variability are computed below through use of Equations 21.5, 21.9, and 21.7 (which are, respectively, the equations for SS_{WC}, df_{WC}, and MS_{WC}, which are the same as the equations for SS_{WG}, df_{WG}, and MS_{WG}). Employing the latter equations, the values $SS_{WC} = 41.83$, $df_{WC} = 15$, and $MS_{WC} = 2.79$ are computed. Note that $SS_{res} = SS_{WC} - SS_{BS} = 41.83 - 36.93 = 4.9$, and $df_{res} = df_{WC} - df_{BS} = 15 - 5 = 10$. As is the case in Tables 24.2 and 24.14, the F ratio is computed with Equation 24.15.

$$SS_{WC} = SS_{WG} = \sum_{j=1}^{k} \left[\sum X_j^2 - \frac{(\sum X_j)^2}{n_j} \right]$$

Which for Example 24.1 indicates:

$$SS_{WC} = SS_{WG} = \left[\sum X_1^2 - \frac{(\sum X_1)^2}{n_1} \right] + \left[\sum X_2^2 - \frac{(\sum X_2)^2}{n_2} \right] + \left[\sum X_3^2 - \frac{(\sum X_3)^2}{n_3} \right]$$

Thus:

$$SS_{WC} = SS_{WG} = \left[443 - \frac{(51)^2}{6} \right] + \left[263 - \frac{(39)^2}{6} \right] + \left[163 - \frac{(29)^2}{6} \right] = 41.83$$

$$df_{WC} = df_{WG} = N - k = 18 - 3 = 15$$

$$MS_{WC} = MS_{WG} = \frac{41.83}{15} = 2.79$$

2. Definitional equations for the single-factor within-subjects analysis of variance In the description of the computational protocol for the **single-factor within-subjects analysis of variance**, Equations 24.2–24.5 are employed to compute the values SS_T, SS_{BC}, SS_{BS}, and SS_{res}. The latter set of computational equations were employed, since they allow for the most efficient computation of the sum of squares values. As noted in Section IV, computational equations are derived from definitional equations which reveal the underlying logic involved in the derivation of the sums of squares. This section will describe the definitional equations for the **single-factor within-subjects analysis of variance**, and apply them to Example 24.1 in order to demonstrate that they yield the same values as the computational equations.

As noted previously, the total sum of squares (SS_T) is made up of two elements, the between-subjects sum of squares (SS_{BS}) and the within-subjects sum of squares (SS_{WS}), and that the latter sum of squares can be partitioned into the between-conditions sum of squares (SS_{BC}) and the residual sum of squares (SS_{res}). The contribution of any single subject's score to the total variability in the data can be expressed in terms of a between-subjects component and a within-subjects component. When the between-subjects component and the within-subjects component are added, the sum reflects that subject's total contribution to the overall variability in the data. The contribution of all N scores to the total variability (SS_T) and the elements that comprise it (SS_{BS} and SS_{WS}, and SS_{BC} and SS_{res} which comprise the latter) are summarized in Table 24.16. The definitional equations described in this section employ the following notation: X_{ij} represents the score of the ith subject in the jth condition, $\bar{X}_T$ represents the grand mean (which is $\bar{X}_T = (\sum_{j=1}^{k}\sum_{i=1}^{n}X_{ij})/N = 119/18 = 6.61$), $\bar{X}_j$ represents the mean of the jth condition, and $\bar{S}_i$ represents the mean of the k scores of the ith subject.

Equation 24.34 is the definitional equation for the **total sum of squares**.[20]

$$SS_T = \sum_{j=1}^{k} \sum_{i=1}^{n} (X_{ij} - \bar{X}_T)^2 \qquad \text{(Equation 24.34)}$$

In employing Equation 24.34 to compute SS_T, the grand mean $(\bar{X}_T)$ is subtracted from each of the N scores and each of the N difference scores is squared. The total sum of squares (SS_T) is the sum of the N squared difference scores. Equation 24.34 is computationally equivalent to Equation 24.2.

Equation 24.35 is the definitional equation for the **between-subjects sum of squares**.

$$SS_{BS} = k \sum_{i=1}^{n} (\bar{S}_i - \bar{X}_T)^2 \qquad \text{(Equation 24.35)}$$

In employing Equation 24.35 to compute SS_{BS}, the following operations are carried out for each of the n subjects. The grand mean $(\bar{X}_T)$ is subtracted from the mean of the subject's k scores. The difference score is squared and the squared difference score is multiplied by the number of experimental conditions (k). After this is done for all n subjects, the values which have been obtained for each subject as a result of multiplying the squared difference score by k are summed. The resulting value represents the between-subjects sum of squares (SS_{BS}). Equation 24.35 is computationally equivalent to Equation 24.4.

Equation 24.36 is the definitional equation for the **within-subjects sum of squares**.

$$SS_{WS} = \sum_{i=1}^{n} \sum_{j=1}^{k} (X_{ij} - \bar{S}_i)^2 \qquad \textbf{(Equation 24.36)}$$

In employing Equation 24.36 to compute SS_{WS}, the following operations are carried out for the k scores of each of the n subjects. The mean of a subject's k scores ($\bar{S}_i$) is subtracted from each of the subject's scores, and the k difference scores for that subject are squared. The sum of the k squared difference scores for all n subjects (i.e., the sum total of N squared difference scores) represents the within-subjects sum of squares (SS_{WS}). Equation 24.36 is computationally equivalent to Equation 24.31.

Equation 24.37 is the definitional equation for the **between-conditions sum of squares**.

$$SS_{BC} = n \sum_{j=1}^{k} (\bar{X}_j - \bar{X}_T)^2 \qquad \textbf{(Equation 24.37)}$$

In employing Equation 24.37 to compute SS_{BC}, the following operations are carried out for each experimental condition. The grand mean ($\bar{X}_T$) is subtracted from the condition mean ($\bar{X}_j$). The difference score is squared, and the squared difference score is multiplied by the number of scores in that condition (n). After this is done for all k conditions, the values which have been obtained for each condition as a result of multiplying the squared difference score by the number of subjects in the condition are summed. The resulting value represents the between-conditions sum of squares (SS_{BC}). Equation 24.37 is computationally equivalent to Equation 24.3. An alternative but equivalent method of obtaining SS_{BC} (which is employed in deriving SS_{BC} in Table 24.16) is as follows: Within each condition, for each of the n subjects the grand mean is subtracted from the condition mean, each difference score is squared, and upon doing this for all k conditions, the N squared difference scores are summed.

Equation 24.38 is the definitional equation for the **residual sum of squares**.

$$SS_{res} = \sum_{j=1}^{k} \sum_{i=1}^{n} \left[(X_{ij} - \bar{X}_T) - (\bar{S}_i - \bar{X}_T) - (\bar{X}_j - \bar{X}_T) \right]^2 \qquad \textbf{(Equation 24.38)}$$

In employing Equation 24.38 to compute SS_{res}, the following operations are carried out for each of the N scores: a) The grand mean ($\bar{X}_T$) is subtracted from the score (X_{ij}); b) The grand mean ($\bar{X}_T$) is subtracted from the mean of the k scores for that subject ($\bar{S}_i$); and c) The grand mean ($\bar{X}_T$) is subtracted from the mean of the condition from which the score is derived ($\bar{X}_j$). The value of the difference score obtained in b) is subtracted from the value of the difference score obtained in a), and the difference score obtained in c) is subtracted from the resulting difference. The resulting value is squared, and the sum of the squared values for all N scores represents the residual sum of squares (SS_{res}). Note that in Equation 24.38, for each subject a between-subjects and between-conditions component of variability is subtracted from the subject's contribution to the total variability, resulting in the subject's contribution to the residual variability. Equation 24.38 is computationally equivalent to Equations 24.5/24.6.

Table 24.16 illustrates the use of Equations 24.34–24.38 with the data for Example 24.1.[21] In the computations summarized in Table 24.16, the following S_i values are employed: $\bar{S}_1$ = 20/3 = 6.67, $\bar{S}_2$ = 25/3 = 8.33, $\bar{S}_3$ = 15/3 = 5, $\bar{S}_4$ = 25/3 = 8.33, $\bar{S}_5$ = 14/3 = 4.67, $\bar{S}_6$ = 20/3

Table 24.16 Computation of Sums of Squares for Example 24.1 with Definitional Equations

(Subject, Condition)	X_{ij}	$SS_T = \sum\limits_{j=1}^{k}\sum\limits_{i=1}^{n}(X_{ij} - \bar{X}_T)^2$	$SS_{BS} = k\sum\limits_{j=1}^{n}(\bar{S}_i - \bar{X}_T)^2$	$SS_{WS} = \sum\limits_{j=1}^{n}\sum\limits_{i=1}^{k}(X_{ij} - \bar{S}_i)^2$
Condition 1				
(1,1)	9	$(9.00-6.61)^2 = 5.71$	$(6.67-6.61)^2 = .00$	$(9.00-6.67)^2 = 5.43$
(2,1)	10	$(10.00-6.61)^2 = 11.49$	$(8.33-6.61)^2 = 2.96$	$(10.00-8.33)^2 = 2.79$
(3,1)	7	$(7.00-6.61)^2 = .15$	$(5.00-6.61)^2 = 2.59$	$(7.00-5.00)^2 = 4.00$
(4,1)	10	$(10.00-6.61)^2 = 11.49$	$(8.33-6.61)^2 = 2.96$	$(10.00-8.33)^2 = 2.79$
(5,1)	7	$(7.00-6.61)^2 = .15$	$(4.67-6.61)^2 = 3.76$	$(7.00-4.67)^2 = 5.43$
(6,1)	8	$(8.00-6.61)^2 = 1.93$	$(6.67-6.61)^2 = .00$	$(8.00-6.67)^2 = 1.77$
Condition 2				
(1,2)	7	$(7.00-6.61)^2 = .15$	$(6.67-6.61)^2 = .00$	$(7.00-6.67)^2 = .11$
(2,2)	8	$(8.00-6.61)^2 = 1.93$	$(8.33-6.61)^2 = 2.96$	$(8.00-8.33)^2 = .11$
(3,2)	5	$(5.00-6.61)^2 = 2.59$	$(5.00-6.61)^2 = 2.59$	$(5.00-5.00)^2 = .00$
(4,2)	8	$(8.00-6.61)^2 = 1.93$	$(8.33-6.61)^2 = 2.96$	$(8.00-8.33)^2 = .11$
(5,2)	5	$(5.00-6.61)^2 = 2.59$	$(4.67-6.61)^2 = 3.76$	$(5.00-4.67)^2 = .11$
(6,2)	6	$(6.00-6.61)^2 = .37$	$(6.67-6.61)^2 = .00$	$(6.00-6.67)^2 = .45$
Condition 3				
(1,3)	4	$(4.00-6.61)^2 = 6.81$	$(6.67-6.61)^2 = .00$	$(4.00-6.67)^2 = 7.13$
(2,3)	7	$(7.00-6.61)^2 = .15$	$(8.33-6.61)^2 = 2.96$	$(7.00-8.33)^2 = 1.77$
(3,3)	3	$(3.00-6.61)^2 = 13.03$	$(5.00-6.61)^2 = 2.59$	$(3.00-5.00)^2 = 4.00$
(4,3)	7	$(7.00-6.61)^2 = .15$	$(8.33-6.61)^2 = 2.96$	$(7.00-8.33)^2 = 1.77$
(5,3)	2	$(2.00-6.61)^2 = 21.25$	$(4.67-6.61)^2 = 3.76$	$(2.00-4.67)^2 = 7.13$
(6,3)	6	$(6.00-6.61)^2 = .37$	$(6.67-6.61)^2 = .00$	$(6.00-6.67)^2 = .45$
		$SS_T = 82.24$	$SS_{BS} = 36.81$	$SS_{WS} = 45.35$

(Subject, Condition)	X_{ij}	$SS_{BC} = n\sum\limits_{j=1}^{k}(\bar{X}_j - \bar{X}_T)^2$	$SS_{res} = \sum\limits_{j=1}^{k}\sum\limits_{i=1}^{n}[(X_{ij} - \bar{X}_T) - (\bar{S}_i - \bar{X}_T) - (\bar{X}_j - \bar{X}_T)]^2$
Condition 1			
(1,1)	9	$(8.50-6.61)^2 = 3.57$	$[(9.00-6.61)-(6.67-6.61)-(8.50-6.61)]^2 = .19$
(2,1)	10	$(8.50-6.61)^2 = 3.57$	$[(10.00-6.61)-(8.33-6.61)-(8.50-6.61)]^2 = .05$
(3,1)	7	$(8.50-6.61)^2 = 3.57$	$[(7.00-6.61)-(5.00-6.61)-(8.50-6.61)]^2 = .01$
(4,1)	10	$(8.50-6.61)^2 = 3.57$	$[(10.00-6.61)-(8.33-6.61)-(8.50-6.61)]^2 = .05$
(5,1)	7	$(8.50-6.61)^2 = 3.57$	$[(7.00-6.61)-(4.67-6.61)-(8.50-6.61)]^2 = .19$
(6,1)	8	$(8.50-6.61)^2 = 3.57$	$[(8.00-6.61)-(6.67-6.61)-(8.50-6.61)]^2 = .31$
Condition 2			
(1,2)	7	$(6.50-6.61)^2 = .01$	$[(7.00-6.61)-(6.67-6.61)-(6.50-6.61)]^2 = .19$
(2,2)	8	$(6.50-6.61)^2 = .01$	$[(8.00-6.61)-(8.33-6.61)-(6.50-6.61)]^2 = .05$
(3,2)	5	$(6.50-6.61)^2 = .01$	$[(5.00-6.61)-(5.00-6.61)-(6.50-6.61)]^2 = .01$
(4,2)	8	$(6.50-6.61)^2 = .01$	$[(8.00-6.61)-(8.33-6.61)-(6.50-6.61)]^2 = .05$
(5,2)	5	$(6.50-6.61)^2 = .01$	$[(5.00-6.61)-(4.67-6.61)-(6.50-6.61)]^2 = .19$
(6,2)	6	$(6.50-6.61)^2 = .01$	$[(6.00-6.61)-(6.67-6.61)-(6.50-6.61)]^2 = .31$
Condition 3			
(1,3)	4	$(4.83-6.61)^2 = 3.17$	$[(4.00-6.61)-(6.67-6.61)-(4.83-6.61)]^2 = .79$
(2,3)	7	$(4.83-6.61)^2 = 3.17$	$[(7.00-6.61)-(8.33-6.61)-(4.83-6.61)]^2 = .20$
(3,3)	3	$(4.83-6.61)^2 = 3.17$	$[(3.00-6.61)-(5.00-6.61)-(4.83-6.61)]^2 = .05$
(4,3)	7	$(4.83-6.61)^2 = 3.17$	$[(7.00-6.61)-(8.33-6.61)-(4.83-6.61)]^2 = .20$
(5,3)	2	$(4.83-6.61)^2 = 3.17$	$[(2.00-6.61)-(4.67-6.61)-(4.83-6.61)]^2 = .79$
(6,3)	6	$(4.83-6.61)^2 = 3.17$	$[(6.00-6.61)-(6.67-6.61)-(4.83-6.61)]^2 = 1.23$
		$SS_{BC} = 40.50$	$SS_{res} = 4.86$

$= 6.7$. The resulting values of SS_T, SS_{BS}, SS_{WS}, SS_{BC}, and SS_{res} are identical to those obtained with the computational equations (Equations 24.2, 24.4, 24.31, 24.3, and 24.5/24.6). Any minimal discrepancies are the result of rounding off error.

3. Relative power of the single-factor within-subjects analysis of variance and the single-factor between-subjects analysis of variance The use of MS_{res} as the measure of error variability (as opposed to MS_{WC}) for the **single-factor within-subjects analysis of variance** provides for an optimally powerful test of an alternative hypothesis.[22] The reason why MS_{res} allows for a more powerful test of an alternative hypothesis than MS_{WC} is because when no treatment effect is present in the data, it is expected that the average variability of the k scores of n subjects will be less than the average variability of the scores of n different subjects who serve in any single experimental condition (in an experiment involving k experimental conditions).

To illustrate this point, let us assume that the data for Example 24.1 are obtained in an experiment employing an independent groups/between-subjects design, and as a result of the latter $MS_{WC} = MS_{WG}$ is employed as the measure of error variability. Thus, we will assume that each of $N = 18$ subjects is randomly assigned to one of $k = 3$ experimental conditions, resulting in $n = 6$ scores per condition. The data for such an experiment will be evaluated with a **single-factor between-subjects analysis of variance**. In conducting the computations for the latter analysis, the value of SS_T is computed with Equation 21.2 (which, in fact, is identical to Equation 24.2, which is employed to compute SS_T when Example 24.1 is evaluated with a **single-factor within-subjects analysis of variance**). Thus, $SS_T = 82.28$. Equation 21.3, which is employed to compute the between-groups sum of squares (SS_{BG}) is, in fact, identical to Equation 24.3 (which is employed in Section IV to compute the between-conditions sum of squares (SS_{BC})). Thus, $SS_{BG} = SS_{BC} = 40.45$. The within-groups sum of squares (SS_{WG}) can be computed with Equation 21.4. Thus, $SS_{WG} = SS_T - SS_{BG} = 82.28 - 40.45 = 41.83$. Note that the latter value is identical to the value computed with Equation 24.29 (which as noted earlier is computationally equivalent to Equation 21.5, which yields the same value as Equation 21.4). Employing the values $k = 3$ and $N = 18$ in Equations 21.8–21.10, the values $df_{BG} = 2$, $df_{WG} = 15$, and $df_T = 17$ are computed. Substituting the appropriate degrees of freedom in Equations 21.6 and 21.7, the values $MS_{BG} = 40.45/2 = 20.23$ and $MS_{WG} = 41.83/15 = 2.79$ are computed. Using Equation 21.12, $F = 20.23/2.79 = 7.25$. Table 24.17 is the summary table of the analysis of variance.

Since $df_{num} = df_{BG} = 2$ and $df_{den} = df_{WG} = 15$, $F_{.05} = 3.68$ and $F_{.01} = 6.36$ are the critical values in **Table A10** which are employed to evaluate the nondirectional alternative hypothesis. Since the obtained value $F = 7.25$ is greater than both of the aforementioned critical values, the alternative hypothesis is supported at both the .05 and .01 levels. Note, however, that the value $F = 7.25$ is substantially less than the value $F = 41.29$, which is obtained when the same set of data is evaluated with the **single-factor within-subjects analysis of variance**. Although the value $F = 7.25$ obtained for a between-subjects analysis is significant at both the .05 and .01 levels, $F = 7.25$ is not very far removed from the tabled critical value $F_{.01} = 6.36$. The value $F = 41.29$, on the other hand, is well above the tabled critical value $F_{.01} = 7.56$ (which is the tabled critical .01 value employed for the **single-factor within-subjects analysis of variance** for $df_{BC} = 2$ and $df_{res} = 10$). The fact that the difference between the computed F value and the tabled critical $F_{.01}$ value is much larger when the **single-factor within-subjects analysis of variance** is employed illustrates that a within-subjects analysis provides a more powerful test of an alternative hypothesis than a between-subjects analysis.[23]

**Table 24.17   Summary Table of Single-Factor
Between-Subjects Analysis of Variance for Example 24.1**

Source of variation	SS	df	MS	F
Between-groups	40.45	2	20.23	7.25
Within-groups	41.83	15	2.79	
Total	82.28	17		

It should be noted that for the same set of data, the tabled critical F value at a given level of significance for a **single-factor between-subjects analysis of variance** will always be lower than the tabled critical F value for a **single-factor within-subjects analysis of variance** (unless there is an extremely large number of scores in each condition, in which case the tabled critical F values for both analyses will be equivalent). This is the case since (as long as n is not extremely large) the number of degrees of freedom associated with the denominator of the F ratio will always be larger for a **single-factor between-subjects analysis of variance** (assuming the values of $n = n_j$ and k for both analyses are equal) than for a **single-factor within-subjects analysis of variance** — i.e., $df_{WG} > df_{res}$. It is important to note, however, that any loss of degrees of freedom associated with a within-subjects analysis will more than likely be offset as a result of employing MS_{res} as the error term in the computation of the F ratio. A final point which should be made is that, if in a within-subjects design, subjects' scores in the k experimental conditions are not correlated with one another (which is highly unlikely), a **single-factor within-subjects analysis of variance** and a **single-factor between-subjects analysis of variance** (as well as a *t* **test for two dependent samples** and a *t* **test for two independent samples** when $k = 2$) will yield comparable results.

4. Equivalency of the single-factor within-subjects analysis of variance and the *t* test for two dependent samples when $k = 2$ Interval/ratio data for an experiment involving $k = 2$ dependent samples can be evaluated with either a **single-factor within-subjects analysis of variance** or a *t* **test for two dependent samples**. When both of the aforementioned tests are employed to evaluate the same set of data they will yield the same result. Specifically, the following will always be true with respect to the relationship between the computed F and t values for the same set of data: $F = t^2$ and $t = \sqrt{F}$. It will also be the case that the square of the tabled critical t value at a prespecified level of significance for $df = n - 1$ will be equal to the tabled critical F value at the same level of significance for $df_{BC} = 1$ and df_{res} (which will be $df_{res} = (n - 1)(k - 1) = (n - 1)(2 - 1) = n - 1$, which is equivalent to the value $df = n - 1$ employed for the *t* **test for two dependent samples**).

To illustrate the equivalency of the results obtained with the **single-factor within-subjects analysis of variance** and the *t* **test for two dependent samples** when $k = 2$, an F value will be computed for Example 17.1. The value $t = 2.86$ is obtained (a more precise value $t = 2.848$ is obtained if all computations are carried out to 3 decimal places) for the latter example when the *t* **test for two dependent samples** is employed. When the same set of data is evaluated with the **single-factor within-subjects analysis of variance**, the value $F = 8.11$ is computed. Note that $(t = 2.848)^2 = (F = 8.11)$. Equations 24.2–24.5 are employed below to compute the values SS_T, SS_{BC}, SS_{BS}, and SS_{res} for Example 17.1. Since $k = 2$, $n = 10$, and $nk = N = 20$, $df_{BC} = 2 - 1 = 1$, $df_{BS} = 10 - 1 = 9$, $df_{res} = (10 - 1)(2 - 1) = 9$, and $df_T = 20 - 1 = 19$. The full analysis of variance is summarized in Table 24.18.

For $df_{BC} = 1$ and $df_{res} = 9$, the tabled critical .05 and .01 values are $F_{.05} = 5.12$ and $F_{.01} = 10.56$ (which are appropriate for a nondirectional analysis). Note that (if one takes into account rounding off error) the square roots of the aforementioned tabled critical values are (for $df = 9$) the tabled critical two-tailed values $t_{.05} = 2.26$ and $t_{.01} = 3.25$ which are employed in Example 17.1 to evaluate the value $t = 2.86$. Since the obtained value $F = 8.11$ is greater than $F_{.05} = 5.12$ but less than $F_{.01} = 10.56$, the nondirectional alternative hypothesis H_1: $\mu_1 \neq \mu_2$ is supported, but only at the .05 level. The directional alternative hypothesis H_1: $\mu_1 > \mu_2$ is supported at both the .05 and .01 levels, since $F = 8.11$ is greater than the tabled critical one-tailed .05 and .01 values $F_{.05} = 3.36$ and $F_{.01} = 7.95$ (the square roots of which are the tabled critical one-tailed .05 and .01 values $t_{.05} = 1.83$ and $t_{.01} = 2.82$ employed for Example 17.1).[24] The conclusions derived from the **single-factor within-subjects analysis of variance** are identical to those reached when the data are evaluated with the *t* **test for two dependent samples**.

$$SS_T = 440 - \frac{(78)^2}{20} = 135.8 \qquad SS_{BC} = \left[\frac{(47)^2 + (31)^2}{10}\right] - \frac{(78)^2}{20} = 12.8$$

$$SS_{BS} = \left[\frac{(17)^2 + (4)^2 + (4)^2 + (6)^2 + (9)^2 + (4)^2 + (11)^2 + (13)^2 + (9)^2 + (1)^2}{2}\right] - \frac{(78)^2}{20} = 108.8$$

$$SS_{res} = 135.8 - 12.8 - 108.8 = 14.2$$

**Table 24.18 Summary Table of Analysis of Variance
for Example 17.1**

Source of variation	SS	df	MS	F
Between-subjects	108.8	9		
Between-conditions	12.8	1	12.80	8.11
Residual	14.2	9	1.58	
Total	135.8	19		

5. The Latin square design In the discussion of **counterbalancing and order effects** in Section VII of the *t* **test for two dependent samples**, it is noted that there are two types of **order effects**, **position effects** and **carryover effects**. **Position effects** are assumed to only be a function of the **serial position** of a treatment (which is the order in which the treatment is presented), and are not affected by what particular treatment(s) occurred in earlier serial positions. The most common **position effects** are **practice effects** in which the performance of subjects improves with each subsequent treatment, and fatigue and boredom in which each subsequent treatment results in a decrease in the performance of subjects. A **carryover effect** is where the effect of a specific treatment is not independent of a treatment which precedes it. As an example, assume within the framework of a within-subjects design the following three treatments are administered on separate days to each of *n* subjects in order to assess each treatment's effect on eye-hand coordination: Drug A, Drug B, a placebo. Assume that a side effect of Drug A is that it makes a person drowsy, and that the latter adversely influences performance. To compound matters, it requires 36 hours before Drug A completely washes out of a person's body. Because of the latter, if either of the other two treatments is administered to a subject the day after he or she receives Drug A, the subject's performance under that treatment will also be adversely affected due to the carryover effect of residual drowsiness deriving from Drug A.

Counterbalancing is a procedure which allows a researcher to control for **order effects**. In **complete counterbalancing** all possible orders for presenting the experimental conditions are represented an equal number of times with respect to the total number of subjects employed in a study. Thus, if a study with $n = 10$ subjects and $k = 3$ conditions is completely counterbalanced, there will be $k! = 3! = 6$ possible presentation orders for the three treatments (i.e., 1,2,3; 1,3,2; 2,1,3; 2,3,1; 3,1,2; 3,2,1). Under such conditions a minimum of six subjects will be required in order to employ complete counterbalancing. If a researcher wants to assign two subjects to each of the presentation orders, $6 \times 2 = 12$ subjects must be employed. It should be obvious that to completely counterbalance the order of presentation of the experimental conditions, the number of subjects must equal the value of $k!$ or be some value that is evenly divisible by it.

As the number of experimental conditions increases, **complete counterbalancing** becomes more difficult to implement, since the number of subjects required increases substantially. Specifically, if there are $k = 5$ experimental conditions, there are $5! = 120$ presentation orders — thus requiring a minimum of 120 subjects (which can be a prohibitively large number for a researcher to use) in order that one subject serves in each of the possible presentation orders. When it is not possible to completely counterbalance the order of presentation of the conditions, alternative less complete counterbalancing procedures are available. One experimental design which allows for **incomplete counterbalancing** (i.e., where not all possible presentation orders are employed in a study) is the **Latin square design**, which is employed more frequently in agricultural and industrial research than in research in education and the behavioral sciences. If an independent variable is comprised of only a few levels, it is feasible to use a Latin square design to provide **complete counterbalancing**. However, as the number of levels of the independent variable increase, it is more likely that the Latin square design will be employed to control for order effects through use of **incomplete counterbalancing**.

If we conceptualize a table summarizing a within-subjects design as being comprised of n rows (corresponding to each of the n subjects) and p columns (corresponding to each of p presentation orders employed for the k treatments, where $n = p = k$), we can define a Latin square design as one in which each treatment appears only one time in each row and only one time in each column. Figure 24.1 represents a 4×4 **Latin square** since there are $n = 4$ subjects/rows, $p = 4$ presentation orders/columns, and $k = 4$ treatments (which are identified by the letters **A**, **B**, **C**, and **D**). In the square depicted in Figure 24.1, Subject 1 receives the treatments in the order **A, B, C, D**, Subject 2 receives the treatments in the order **C, A, D, B**, and so on. As noted above, a Latin square does not typically employ **complete counterbalancing**. When there are $k = 4$ treatments, there are, in fact, $k! = 4! = 24$ possible presentation orders. Consequently, a minimum of 24 subjects would be required in order to have complete counterbalancing.

		Presentation order			
		1	2	3	4
	1	A	B	C	D
	2	B	D	A	C
Subject	**3**	C	A	D	B
	4	D	C	B	A

Figure 24.1 4 × 4 Latin Square Design

The Latin square design is discussed in detail in Section IX (the **Addendum**) of the **between-subjects analysis of variance (Test 27)** under the **analysis of variance for a Latin square design (Test 27j)**.

VIII. Additional Examples Illustrating the Use of the Single-Factor Within-Subjects Analysis of Variance

Since the **single-factor within-subjects analysis of variance** can be employed to evaluate interval/ratio data for any dependent samples design involving two or more experimental conditions, it can be used to evaluate any of the examples that are evaluated with the *t* **test for two dependent samples** (with the exception of Example 17.2). Examples 24.2–24.6 are, respectively, extensions of Examples 17.1, 17.3, 17.5, 17.6, and 17.8. As is the case with Examples 17.3 and 17.5, Examples 24.3 and 24.4 employ matched subjects, and are thus evaluated as a within-subjects design. Examples 24.6 and 24.7 represent extensions of the **one-group pretest-posttest design** to a design involving $k = 3$ experimental conditions. Since the data for all of the examples are identical to the data employed in Example 24.1, they yield the same result.

Example 24.2 *A psychologist conducts a study to determine whether or not people exhibit more emotionality when they are exposed to sexually explicit words, aggressively toned words, or neutral words. Each of six subjects is shown a list of 15 randomly arranged words, which are projected on a screen one at a time for a period of five seconds. Five of the words on the list are sexually explicit, five of the words are aggressively toned, and five of the words are neutral. As each word is projected on the screen, a subject is instructed to say the word softly to himself or herself. As a subject does this, sensors attached to the palms of the subject's hands record galvanic skin response (GSR), which is used by the psychologist as a measure of emotionality. The psychologist computes the following three scores for each subject, one score for each of the three experimental conditions:* **Condition 1**: *GSR/Sexually explicit — The average GSR score for the five sexually explicit words;* **Condition 2**: *GSR/Aggressively toned — The average GSR score for the five aggressively toned words;* **Condition 3**: *GSR/Neutral — The average GSR score for the five neutral words. The GSR/Sexually explicit, GSR/Aggressively toned, and GSR/Neutral scores of the six subjects follow. (The higher the score, the higher the level of emotionality.)* **Subject 1** (9, 7, 4); **Subject 2** (10, 8, 7); **Subject 3** (7, 5, 3); **Subject 4** (10, 8, 7); **Subject 5** (7, 5, 2); **Subject 6** (8, 6, 6). *Do subjects exhibit differences in emotionality with respect to the three categories of words?*

Example 24.3 *A psychologist conducts a study in order to determine whether people exhibit more emotionality when they are exposed to sexually explicit words, aggressively toned words, or neutral words. Six sets of identical triplets are employed as subjects and within each set of triplets one member of the set is treated as follows: a) One of the triplets is randomly assigned to* Condition 1, *in which the subject is shown a list of five sexually explicit words; b) One of the triplets is randomly assigned to* Condition 2, *in which the subject is shown a list of five aggressively toned words; and c) One of the triplets is randomly assigned to* Condition 3, *in which the subject is shown a list of five neutral words. As each word is projected on the screen, a subject is instructed to say the word softly to himself or herself. As a subject does this, sensors attached to the palms of the subject's hands record galvanic skin response (GSR), which is used by the psychologist as a measure of emotionality. The psychologist computes the following three scores for each set of triplets to represent the emotionality score for each of the experimental conditions:* **Condition 1**: *GSR/Sexually explicit — The average GSR score for the subject presented with the five sexually explicit words;* **Condition 2**: *GSR/Aggressively toned — The average GSR score for the subject presented with the five aggressively toned words;* **Condition 3**: *GSR/Neutral — The average GSR score for the subject presented with the five neutral words. The GSR/Sexually explicit, GSR/Aggressively toned, and GSR/Neutral scores of the six sets of triplets follow. (The first score for each triplet set represents the score of the subject presented* with the sexually explicit words, the second score represents the score of the subject presented with the aggressively toned words, and the third score represents the score of the subject presented with the neutral words. The higher the score, the higher the level of emotionality.) **Triplet set 1** (9, 7, 4); **Triplet set 2** (10, 8, 7); **Triplet set 3** (7, 5, 3); **Triplet set 4** (10, 8, 7); **Triplet set 5** (7, 5, 2); **Triplet set 6** (8, 6, 6). *Do subjects exhibit differences in emotionality with respect to the three categories of words?*

Example 24.4 *A researcher wants to assess the impact of different types of punishment on the emotionality of mice. Six sets of mice derived from six separate litters are employed as subjects. Within each set, one of the litter mates is randomly assigned to one of the three experimental conditions. During the course of the experiment each mouse is sequestered in an experimental chamber. While in the chamber, each of the six mice in* Condition 1 *is periodically presented with a loud noise, and each of the six mice in* Condition 2 *is periodically presented with a blast of cold air. The six mice in* Condition 3 *(which is a no-treatment control condition) are not exposed to any punishment. The presentation of the punitive stimulus for the animals in* Conditions 1 *and* 2 *is generated by a machine that randomly presents the stimulus throughout the duration of the time an animal is in the chamber. The dependent variable of emotionality employed in the study is the number of times each mouse defecates while in the experimental chamber. The number of episodes of defecation for the six sets of mice follows. (The higher the score, the higher the level of emotionality.)* **Litter 1** (9, 7, 4); **Litter 2** (10, 8, 7); **Litter 3** (7, 5, 3); **Litter 4** (10, 8, 7); **Litter 5** (7, 5, 2); **Litter 6** (8, 6, 6). *Do subjects exhibit differences in emotionality under the different experimental conditions?*

Example 24.5 *A study is conducted to evaluate the relative efficacy of two drugs (Clearoxin and Lesionoxin) and a placebo on chronic psoriasis. Six subjects afflicted with chronic psoriasis participate in the study. Each subject is exposed to both drugs and the placebo for a six-month period, with a three-month hiatus between treatments. Within the six subjects, the order of presentation of the experimental treatments is completely counterbalanced. The dependent variable employed in the study is a rating of the severity of a subject's lesions under the three experimental conditions. The lower the rating the more severe a subject's psoriasis. The scores of the six subjects under the three treatment conditions follow. (The first score represents the Clearoxin condition (which represents* Condition 1), *the second score the Lesionoxin condition (which represents* Condition 2), *and the third score the placebo condition (which represents* Condition 3).) **Subject 1** (9, 7, 4); **Subject 2** (10, 8, 7); **Subject 3** (7, 5, 3); **Subject 4** (10, 8, 7); **Subject 5** (7, 5, 2); **Subject 6** (8, 6, 6). *Do the data indicate differences in subjects' responses under the three experimental conditions?*

Example 24.6 *In order to assess the efficacy of electroconvulsive therapy (ECT), a psychiatrist evaluates six clinically depressed patients who receive a series of ECT treatments. Each patient is evaluated at the following three points in time: a) One day prior to the first treatment in the ECT series; b) The day following the final treatment in the ECT series; and c) Six months after the final treatment in the ECT series. During each evaluation period a standardized interview is used to operationalize a patient's level of depression, and on the basis of the interview a patient is assigned a score ranging from 0 to 10. The higher a patient's score, the more depressed the patient. The depression scores of the six patients during each of the three time periods follow:* **Patient 1** (9, 7, 4); **Patient 2** (10, 8, 7); **Patient 3** (7, 5, 3); **Patient 4** (10, 8, 7); **Patient 5** (7, 5, 2); **Patient 6** (8, 6, 6). *Do the data indicate that the ECT is effective, and, if so, is the effect maintained six months after the treatment?*

Although, as described, Example 24.6 can be evaluated with a **single-factor within-subjects analysis of variance**, the design of the study does not allow one to rule out the potential impact of confounding variables. To be more specific, Example 24.6 (which represents an extension of a **one-group pretest-posttest design** to more than two measurement periods) does not allow a researcher to draw definitive conclusions with respect to whether any observed changes in mood are, in fact, due to the ECT treatments.[25] Thus, even if there is a significant decrease in subjects' depression scores following the final ECT treatment, and the effect is still present six months later, factors other than ECT can account for such a result. As an example, all of the patients may have been depressed about a problem related to the economy, and if, in fact, during the course of the study the economy improves dramatically, the observed changes in mood can be attributed to the improved economy rather than the ETC. In order for the design of the above study to be suitable, it is necessary to include a control group — specifically, a comparable group of depressed patients who are not given ECT (or are given "sham" ECT treatments). By contrasting the depression scores of the control group with those of the treatment group, one can determine whether or not any observed differences across the three time periods are, in fact, attributable to the ETC. Inclusion of such a control group would require that the design of the above study be modified into a **mixed factorial design**. The latter design and the analysis of variance employed to evaluate it are discussed in Section IX of the **between-subjects factorial analysis of variance**.

Example 24.7 *In order to assess the efficacy of a drug which a pharmaceutical company claims is effective in treating hyperactivity, six hyperactive children are evaluated during the following three time periods: a) One week prior to taking the drug; b) After a child has taken the drug for six consecutive months; and c) Six months after the drug is discontinued. The children are observed by judges who employ a standardized procedure for evaluating hyperactivity. During each time period a child is assigned a score between 0 and 10, in which the higher the score, the higher the level of hyperactivity. During the evaluation process, the judges are blind with respect to whether or not a child is taking medication at the time he or she is evaluated. The hyperactivity scores of the six children during the three time periods follow:* **Child 1**: *(9, 7, 4);* **Child 2**: *(10, 8, 7);* **Child 3**: *(7, 5, 3);* **Child 4**: *(10, 8, 7);* **Child 5**: *(7, 5, 2);* **Child 6**: *(8, 6, 6).* *Do the data indicate that the drug is effective?*

Since it lacks a control group, Example 24.7 is subject to the same criticism which is noted for Example 24.6. Because of the lack of a control group (i.e., a group of hyperactive children who do not receive medication), any observed differences in hyperactivity between two or more of the measurement periods can be the result of extraneous factors in the external environment or physiological/maturational changes in the children that are independent of whether a child is taking the drug. In spite of its limitations, it is not unusual to encounter the use of the design employed in Example 24.7 (which is commonly referred to as an **ABA design**) in behavior modification research. Such designs are most commonly employed with individual subjects in order to assess the efficacy of a treatment protocol. (The latter types of designs, which are referred to as **single-subject designs**, are discussed in the **Introduction**.) The letters **A** and **B** in an **ABA design** refer to whether or not a treatment is in effect during a specific time period. In Example 24.7, Time period 1 is designated **A** since no treatment is in effect. This initial measure of the subject's behavior provides the researcher with a baseline measure of hyperactivity. During Time period 2, which is designated by the letter **B**, the treatment is in effect. If the treatment is effective, a decrease in hyperactivity in Time period 2 relative to Time period 1 is expected. Time period 3 is once again designated **A**, since the treatment is no longer employed. If, in fact, the treatment is effective it is expected that a subject's level of hyper-activity during Time period 3 will be higher than in Time period 2 (which, in actuality, did not occur), and, in fact, return to the baseline level obtained during Time period 1 (unless, of course, the drug has a permanent residual effect). When an **ABA design** is employed with an individual subject, the format of the data resulting from such a study is not suitable for evaluation with an analysis of variance.

References

Anderson, N. N. (2001). **Empirical direction in design and analysis**. Mahwah, NJ: Lawrence Erlbaum Associates.

Bartko, J. J. (1976). On various intraclass correlation reliability coefficients. **Psychological Bulletin**, 83, 762–765.

Box, G. E. P. (1954). Some theorems on quadratic forms applied in the study of analysis of variance problems, II. Effect of inequality of variances and correlation between error in two-way classification. **Annals of Mathematical Statistics**, 25, 484–498.

Cohen, B. H. (2001). **Explaining psychological Statistics** (2nd ed.). New York: John Wiley & Sons.

Cohen, J. (1977). **Statistical power analysis for the behavioral sciences**. New York: Academic Press.

Cohen, J. (1988). **Statistical power analysis for the behavioral sciences** (2nd ed.). Hillsdale, NJ: Erlbaum.

Cortina, J. M. & Nouri, H. (2000). **Effect sizes for ANOVA designs**. Thousand Oaks, CA: Sage Publications.

Everitt, B. S. (2001). **Statistics for psychologists: An intermediate course**. Mahwah, NJ: Lawrence Erlbaum Associates.

Geisser, S. & Greenhouse, S. W. (1958). An extension of Box's results to the use of the F distribution in multivariate analysis. **Annals of Mathematical Statistics**, 29, 885–891.

Greenhouse, S. W. & Geisser (1959). On the methods in the analysis of profile data. **Psychometrika**, 24, 95–112.

Grissom, R. J. & Kim, J. J. (2005). **Effect sizes for research: A broad practical approach**. Mahwah, NJ: Lawrence Erlbaum Associates Publishers.

Howell, D. C. (2002). **Statistical methods for psychology** (5th ed.). Pacific Grove, CA: Duxbury Press.

Hsu, J. C. (1996). **Multiple comparisons: Theory and methods**. New York: Chapman & Hall.

Huynh, H. & Feldt, L. S. (1976). Estimates of the correction for degrees of freedom for sample data in randomized block and split-plot designs. **Journal of Educational Statistics**, 1, 69–82.

Keppel, G. (1991) **Design and analysis: A researcher's handbook** (3rd ed.). Englewood Cliffs, NJ: Prentice Hall.

Keppel, G. & Wickens, T. D. (2004). **Design and analysis: A researcher's handbook** (4th ed.). Upper Saddle River, NJ: Pearson/Prentice Hall.

Keppel, G. & Zedeck, S. (1989) **Data analysis for research designs**. New York: W. H. Freeman & Company.

Kirk, R. E. (1982) **Experimental design: Procedures for the behavioral sciences** (2nd ed.). Belmont, CA: Brooks/Cole Publishing Company.

Kirk, R. (1995). **Experimental design: Procedures for the behavioral sciences**. Pacific Grove, CA: Brooks/Cole Publishing Company.

Kline, R. B. (2004). **Beyond significance testing: Reforming data analysis methods in behavioral research**. Washington, D. C.: American Psychological Association.

McFatter, R. M. & Gollob, H. F. (1986). The power of hypothesis tests for comparisons. **Educational and Psychological Measurement**, 46, 883–886.

Mauchly, J. W. (1940). Significance test for sphericity of a normal *n*-variate distribution. **Annals of Mathematical Statistics**, 11, 204–209.

Maxwell, S. E. & Delaney, H. D. (2000) **Designing experiments and analyzing data**. Mahwah, NJ: Lawrence Erlbaum Associates.

Maxwell, S. E. & Delaney, H. D. (2004) **Designing experiments and analyzing data: A model comparison perspective** (2nd ed.). Mahwah, NJ: Lawrence Erlbaum Associates.

McCulloch, C. E. (2005). Repeated measures ANOVA, RIP? **Chance**, 18, 29–33.

McFatter, R. M. & Gollob, H. F. (1986). The power of hypothesis tests for comparisons. **Educational and Psychological Measurement**, 46, 883–886.

Myers, J. L. & Well, A. D. (1995). **Research design and statistical analysis**. Hillsdale, NJ: Lawrence Erlbaum Associates, Publishers.

Myers, J. L. & Well, A. D. (2003). **Research design and statistical analysis** (2nd ed.) Mahwah, NJ: Lawrence Erlbaum Associates, Publishers.

Rosenthal, R., Rosnow, R. L., & Rubin, D. B. (2000). **Contrasts and effect sizes in behavioral research: A correlational approach**. Cambridge, UK: Cambridge University Press.

Rosnow, R. L., Rosenthal, R. & Rubin, D. B. (2000). Contrasts and correlations in effect-size estimation. **Psychological Science**, 11, 446–453.

Shrout, P. E. & Fleiss, J. L. (1979). Intraclass correlations: Use in assessing rater reliability. **Psychological Bulletin**, 86, 420–428.

Stevens, J.P. (2002). **Applied multivariate statistics for the social scieences** (4th ed.). Mahwah, NJ: Lawrence Erlbaum Associates.

Tabachnick, B. C. & Fidell, L. S. (1989). **Using multivariate statistics** (2nd ed.). New York: HarperCollins Publishers.

Weinfurt, K.P. (2000). Repeated measures analysis: ANOVA, MANOVA and HLM. Grimm, L. G. & Yarnold, P. R. (Eds.), **Reading and understanding more multivariate statistics** (pp. 317–361). Washington, D.C.: American Psychological Association.

Wilcox, R. R., (2003). **Applying contemporary statistical techniques**. San Diego, CA: Academic Press.

Winer, B. J., Brown, D. R., & Michels, K. M. (1991). **Statistical principles in experimental design** (3rd ed.). New York: McGraw–Hill Publishing Company.

Endnotes

1. A **within-subjects/repeated-measures design** in which each subject serves under each of the *k* levels of the independent variable is often described as a special case of a **randomized-blocks design**. The term **randomized-blocks design** is commonly employed to describe a dependent samples design involving matched subjects. As an example, assume that 10 sets of identical triplets are employed in a study to determine the efficacy of two drugs when compared with a placebo. Within each set of triplets one of the members is randomly assigned to each of the three experimental conditions. Such a design is described in various sources as a **matched-subjects/samples design**, a **dependent samples design**, a **correlated-subjects design**, or a **randomized-blocks design**. Within the usage of the term **randomized-blocks design**, each set of triplets constitutes a **block**, and consequently, 10 blocks are employed in the study with three subjects in each block. Further discussion of the **randomized-blocks design** can be found in Section VII of the **between-subjects factorial analysis of variance**.

2. It should be noted that if an experiment is confounded, one cannot conclude that a significant portion of **between-conditions** variability is attributed to the independent variable.

This is the case, since if one or more confounding variables systematically vary with the levels of the independent variable, a significant difference can be due to a confounding variable rather than the independent variable.

3. Keppel (1991, p. 350) notes that if substantial **practice effects** (which represent one type of **order effect**) are present in a within-subjects design and are not taken into account in conducting the analysis of variance, by virtue of the latter the error variability computed for the latter analysis of variance may become inflated to the point that any advantages which result from employing a within-subjects design over a between-subjects design are negated. Consequently, **counterbalancing** is required in order to control for practice effects in a within-subjects design. Counterbalancing is discussed in Section VII of the *t* **test for two dependent samples** as well as in Section VII of this chapter.

4. In other words, each subject is tested under one of the six possible presentation orders for the three experimental conditions, and within the sample of six subjects each of the presentation orders is presented once. Specifically, the following six presentation orders are employed: 1,2,3; 1,3,2; 2,1,3; 2,3,1; 3,1,2; 3,2,1.

5. Although it is possible to conduct a directional analysis, such an analysis will not be described with respect to the **single-factor within-subjects analysis of variance**. A discussion of a directional analysis when $k = 2$ can be found under the *t* **test for two dependent samples**. Endnote 4 under the **single-factor between-subjects analysis of variance** indicates where the reader may find relevant information related to a directional analysis.

6. In Section VII it is noted that the sum of **between-conditions variability** and **residual variability** represents what is referred to as **within-subjects variability**. The sum of squares of **within-subjects variability** (SS_{WS}) is the sum of **between-conditions variability** and **residual variability** — i.e., $SS_{WS} = SS_{BC} + SS_{res}$.

7. Since there is an equal number of scores in each condition, the Equation for SS_{BC} can also be written as follows:

$$SS_{BC} = \left[\frac{(\Sigma X_1)^2 + (\Sigma X_2)^2 + \cdots + (\Sigma X_k)^2}{n} \right] - \frac{(\Sigma X_T)^2}{N}$$

Thus:
$$SS_{BC} = \frac{[(51)^2 + (39)^2 + (29)^2]}{6} - \frac{(119)^2}{18} = 40.45$$

8. The equation for SS_{BS} can also be written as follows:

$$SS_{BS} = \frac{[(\Sigma S_1)^2 + (\Sigma S_2)^2 + \cdots + (\Sigma S_n)^2]}{k} - \frac{(\Sigma X_T)^2}{n}$$

$$= \frac{[(20)^2 + (25)^2 + (15)^2 + (14)^2 + (20)^2]}{3} - \frac{(119)^2}{18} = 36.93$$

9. In the interest of accuracy, as is the case with the **single-factor between-subjects analysis of variance**, a significant omnibus F value indicates that there is at least one significant difference among all possible comparisons that can be conducted. Thus, it is theoretically possible that none of the simple/pairwise comparisons are significant, and that the significant difference (or differences) involves one or more complex comparisons.

10. As noted in Section VI of the **single-factor between-subjects analysis of variance**, in some instances the $CD_{B/D}$ value associated with the **Bonferroni–Dunn test** will be larger than the CD_S value associated with the **Scheffé test**. However, when there are $c = 3$ comparisons, CD_S will be greater than $CD_{B/D}$.

11. One can, of course, conduct a replication study and base the estimate of MS_{res} on the value of MS_{res} obtained for the comparison in the latter study. In point of fact, one or more replication studies can serve as a basis for obtaining the best possible estimate of error variability to employ for any comparison conducted following an analysis of variance.

12. If the means of each of the conditions for which a composite mean is computed are weighted equally, an even simpler method for computing the composite score of a subject is to add the subject's scores and divide the sum by the number of conditions which are involved. Thus, the composite score of Subject 1 can be obtained by adding 9 and 7 and dividing by 2. The averaging procedure will only work if all of the means are weighted equally. The protocol described in Section VI must be employed in instances where a comparison involves unequal weighting of means.

13. The same result is obtained if (for the three difference scores) the score in the first condition noted is subtracted from the score in the second condition noted (i.e., Condition 2 – Condition 1; Condition 3 – Condition 1; Condition 3 – Condition 2).

14. If the variance of Condition 2 is employed to represent the lowest variance, $r_{X_2 X_3}$ also equals .85.

15. a) The statistical software package *SPSS* evaluates the sphericity assumption with **Mauchly's test of sphericity** (Mauchly (1940)) (which is displayed in a separate table labeled **Mauchly's Test of Sphericity**). If the probability value for the latter test (which is displayed in a column labeled **Sig.**) is less than .05, the sphericity assumption is violated. Adjusted F values (based on Geisser and Greenhouse (1958) and Huynh and Feldt (1976)) (along with their associated probabilities) are printed out which should be employed when the sphericity assumption is violated; b) Kline (2004) notes the following two additional options to those noted in this section for dealing with the violation of the sphericity assumption: a) Conduct specific comparisons between pairs of conditions instead of conducting an omnibus F test; b) Use statistical resampling (specifically the bootstrap method (discussed in Section IV (the Addendum) of the **Mann–Whitney U test (Test 12)**) to conduct the **single factor within-subjects analysis of variance**.

16. Inspection of the $df_{res} = 10$ curve reveals that for $df_{res} = 10$, a value of approximately $\varphi = 3.1$ or greater will be associated with a power of 1.

17. A number of different alternative equations have been proposed for computing **standard omega squared**. Although a slightly different equation was employed in the first edition of this book, it yields approximately the same result that is obtained with Equation 24.25. Grissom and Kim (2005, p. 135) note that Equation 24.39 can be employed to compute **standard omega squared**. The latter Equation is equivalent to Equation 25.25.

(**Equation 24.39**)

$$\tilde{\omega}_s^2 = \frac{(k - 1)(MS_{BC} - MS_{res})}{SS_T + MS_{BS}} = \frac{(3 - 1)(20.23 - .49)}{82.28 + 7.39} = .44$$

18. The **eta squared** statistic ($\tilde{\eta}^2$) (also represented by R^2, and commonly referred to as the **correlation ratio**) computed for the **single-factor between-subjects analysis of variance** can also be computed for the **single-factor within-subjects analysis of variance**. The **partial** version of the latter statistic can be computed with Equation 24.40. Employing the latter equation, the value $\tilde{\eta}^2 = .89$ is computed below. Keppel and Wickens (2004, p. 362) note that the value computed with Equation 24.40 tends to overestimate the degree of relationship between the independent and dependent variables in the underlying population.

$$\tilde{\eta}^2 = R^2 = \frac{SS_{BC}}{SS_{BC} + SS_{res}} = \frac{40.45}{40.45 + 4.9} = .89 \qquad \text{(\textbf{Equation 24.40})}$$

19. In using Equation 2.8, $\tilde{s}_1/\sqrt{n}$ is equivalent to $s_{\bar{X}_1}$.

20. In employing double (or even more than two) summation signs such as $\sum_{j=1}^{k} \sum_{i=1}^{n}$, the mathematical operations specified are carried out beginning with the summation sign that is farthest to the right and continued sequentially with those operations specified by summation signs to the left. Specifically, if $k = 3$ and $n = 6$, the notation $\sum_{j=1}^{k} \sum_{i=1}^{n} X_{ij}$ indicates that the sum of the n scores in Condition 1 is computed, after which the sum of the n scores in Condition 2 is computed, after which the sum of the n scores in Condition 3 is computed. The final result will be the sum of all the aforementioned values which have been computed. On the other hand, the notation $\sum_{i=1}^{n} \sum_{j=1}^{k} X_{ij}$ indicates that the sum of the $k = 3$ scores of Subject 1 is computed, after which the sum of the $k = 3$ scores of Subject 2 is computed, and so on until the sum of the $k = 3$ scores of Subject 6 is computed. The final result will be the sum of all the aforementioned values which have been computed. In this example the final value computed for $\sum_{j=1}^{k} \sum_{i=1}^{n} X_{ij}$ will be equal to the final value computed for $\sum_{i=1}^{n} \sum_{j=1}^{k} X_{ij}$. In obtaining the final value, however, the order in which the operations are conducted is reversed. Specifically, in computing $\sum_{j=1}^{k} \sum_{i=1}^{n} X_{ij}$, the sums of the k columns are computed and summed in order to arrive at the grand sum, while in computing $\sum_{i=1}^{n} \sum_{j=1}^{k} X_{ij}$, the sums of the n rows are computed and summed in order to arrive at the grand sum.

21. For each of the $N = 18$ scores in Table 24.16, the following is true with respect to the contribution of any score to the total variability in the data.

$$(X_{ij} - \bar{X}_T) = (\bar{S}_i - \bar{X}_T) + (X_{ij} - \bar{S}_i)$$

Total deviation score = *BS* deviation score + *WS* deviation score

and

$$(X_{ij} - \bar{S}_i) = (\bar{X}_j - \bar{X}_T) + [(X_{ij} - \bar{X}_T) - (\bar{S}_i - \bar{X}_T) - (\bar{X}_j - \bar{X}_T)]$$

WS deviation score = *BC* deviation score + Residual deviation score

22. As noted in Section VI under the discussion of computation of a confidence interval, MS_{WC} is equivalent to MS_{WG} (which is the analogous measure of variability for the **single-factor between-subjects analysis of variance**).

23. An issue discussed by Keppel (1991) which is relevant to the power of the **single-factor within-subjects analysis of variance** is that even though counterbalancing is an effective procedure for distributing practice effects evenly over the k experimental conditions in a within-subjects design, if practice effects are, in fact, present in the data, the value of MS_{res} will be inflated, and because of the latter the power of the **single-factor within-subjects analysis of variance** will be reduced. Keppel (1991) describes a methodology for computing an adjusted measure of MS_{res}, which is independent of practice effects, that allows for a more powerful test of an alternative hypothesis.

24. In evaluating a directional alternative hypothesis, when $k = 2$ the tabled $F_{.90}$ and $F_{.98}$ values (for the appropriate degrees of freedom) are respectively employed as the one-tailed .05 and .01 values. Since the values for $F_{.90}$ and $F_{.98}$ are not listed in **Table A10**, the values $F_{.90} = 3.36$ and $F_{.98} = 7.95$ can be obtained by squaring the tabled critical one-tailed values $t_{.05} = 1.83$ and $t_{.01} = 2.82$, or by employing more extensive tables of the F distribution available in other sources, or through interpolation.

25. Example 24.6 (as well as Example 24.7) can also be viewed as an example of what is commonly referred to as a **time-series design** (although time-series designs typically involve more measurement periods than are employed in the latter example). The latter design is essentially a **one-group pretest-posttest design** involving one or more measurement periods prior to an experimental treatment, and one or more measurement periods following the experimental treatment.

Test 25

The Friedman Two-Way Analysis of Variance by Ranks
(Nonparametric Test Employed with Ordinal Data)

I. Hypothesis Evaluated with Test and Relevant Background Information

Hypothesis evaluated with test In a set of k dependent samples (where $k \geq 2$), do at least two of the samples represent populations with different median values?

Relevant background information on test Prior to reading the material on the **Friedman two-way analysis of variance by ranks**, the reader may find it useful to review the general information regarding a dependent samples design contained in Sections I and VII of the *t* test **for two dependent samples (Test 17)**. The **Friedman two-way analysis of variance by ranks** (Friedman (1937)) is employed with ordinal (rank-order) data in a hypothesis testing situation involving a design with two or more dependent samples. The test is an extension of the **binomial sign test for two dependent samples (Test 19)** to a design involving more than two dependent samples, and when $k = 2$ the **Friedman two-way analysis of variance by ranks** will yield a result which is equivalent to that obtained with the **binomial sign test for two dependent samples**.[1] If the result of the **Friedman two-way analysis of variance by ranks** is significant, it indicates there is a significant difference between at least two of the sample medians in the set of k medians. As a result of the latter, the researcher can conclude there is a high likelihood at least two of the samples represent populations with different median values.

In employing the **Friedman two-way analysis of variance by ranks**, one of the following is true with regard to the rank-order data that are evaluated: a) The data are in a rank-order format, since it is the only format in which scores are available; or b) The data have been transformed into a rank-order format from an interval/ratio format, since the researcher has reason to believe that one or more of the assumptions of the **single-factor within-subjects analysis of variance (Test 24)** (which is the parametric analog of the **Friedman test**) are saliently violated. It should be pointed out that when a researcher elects to transform a set of interval/ratio data into ranks, information is sacrificed. This latter fact accounts for why there is reluctance among some researchers to employ nonparametric tests such as the **Friedman two-way analysis of variance by ranks**, even if there is reason to believe that one or more of the assumptions of the **single-factor within-subjects analysis of variance** have been violated.

Various sources (e.g., Conover (1980, 1999), Daniel (1990)) note that the **Friedman two-way analysis of variance by ranks** is based on the following assumptions: a) The sample of n subjects has been randomly selected from the population it represents; and b) The dependent variable (which is subsequently ranked) is a continuous random variable. In truth, this assumption, which is common to many nonparametric tests, is often not adhered to, in that such tests are often employed with a dependent variable which represents a discrete random variable.

As is the case for other tests which are employed to evaluate data involving two or more dependent samples, in order for the **Friedman two-way analysis of variance by ranks** to generate valid results the following guidelines should be adhered to:[2] a) To control for order effects, the presentation of the *k* experimental conditions should be random or, if appropriate, be counterbalanced; and b) If matched samples are employed, within each set of matched subjects each of the subjects should be randomly assigned to one of the *k* experimental conditions.

As is noted with respect to other tests which are employed to evaluate a design involving two or more dependent samples, the **Friedman two-way analysis of variance by ranks** can also be used to evaluate a **one-group pretest-posttest design**, as well as extensions of the latter design that involve more than two measurement periods. The limitations of the **one-group pretest-posttest design** (which are discussed in the **Introduction** and in Section VII of the *t* **test for two dependent samples**) are also applicable when it is evaluated with the **Friedman two-way analysis of variance by ranks**.

II. Example

Example 25.1 is identical to Example 24.1 (which is evaluated with the **single-factor within-subjects analysis of variance**). In evaluating Example 25.1 it will be assumed that the ratio data (i.e., the number of nonsense syllables correctly recalled) are rank-ordered, since one or more of the assumptions of the **single-factor within-subjects analysis of variance** have been saliently violated.

Example 25.1 *A psychologist conducts a study to determine whether or not noise can inhibit learning. Each of six subjects is tested under three experimental conditions. In each of the experimental conditions a subject is given 20 minutes to memorize a list of 10 nonsense syllables, which the subject is told she will be tested on the following day. The three experimental conditions each subject serves under are as follows:* **Condition 1**, *the* **no noise** *condition, requires subjects to study the list of nonsense syllables in a quiet room.* **Condition 2**, *the* **moderate noise** *condition, requires subjects to study the list of nonsense syllables while listening to classical music.* **Condition 3**, *the* **extreme noise** *condition, requires subjects to study the list of nonsense syllables while listening to rock music. Although in each of the experimental conditions subjects are presented with a different list of nonsense syllables, the three lists are comparable with respect to those variables that are known to influence a person's ability to learn nonsense syllables. To control for order effects, the order of presentation of the three experimental conditions is completely counterbalanced. The number of nonsense syllables correctly recalled by the six subjects under the three experimental conditions follow. (Subjects' scores are listed in the order* **Condition 1**, **Condition 2**, **Condition 3**.*)* **Subject 1**: 9, 7, 4; **Subject 2**: 10, 8, 7; **Subject 3**: 7, 5, 3; **Subject 4**: 10, 8, 7; **Subject 5**: 7, 5, 2; **Subject 6**: 8, 6, 6. *Do the data indicate that noise influenced subjects' performance?*

III. Null versus Alternative Hypotheses

Null hypothesis H_0: $\theta_1 = \theta_2 = \theta_3$

(The median of the population Condition 1 represents equals the median of the population Condition 2 represents equals the median of the population Condition 3 represents. With respect to the sample data, when the null hypothesis is true the sums of the ranks (as well as the mean ranks) of all *k* conditions will be equal.)

Alternative hypothesis H_1: Not H_0

(This indicates there is a difference between at least two of the $k = 3$ population medians. It is important to note that the alternative hypothesis should not be written as follows: H_1: $\theta_1 \neq \theta_2 \neq \theta_3$. The reason why the latter notation for the alternative hypothesis is incorrect is because it implies that all three population medians must differ from one another in order to reject the null hypothesis. With respect to the sample data, if the alternative hypothesis is true, the sum of the ranks (as well as the mean ranks) of at least two of the k conditions will not be equal. In this book it will be assumed (unless stated otherwise) that the alternative hypothesis for the **Friedman two-way analysis of variance by ranks** is stated **nondirectionally**.)[3]

IV. Test Computations

The data for Example 25.1 are summarized in Table 25.1. The number of subjects employed in the experiment is $n = 6$, and thus within each condition there are $n = n_1 = n_2 = n_3 = 6$ scores. The original interval/ratio scores of the six subjects are recorded in the columns labeled X_1, X_2, and X_3. The adjacent columns R_1, R_2, and R_3 note the rank-order assigned to each of the scores.

Table 25.1   Data for Example 25.1

	Condition 1		Condition 2		Condition 3	
	X_1	R_1	X_2	R_2	X_3	R_3
Subject 1	9	3	7	2	4	1
Subject 2	10	3	8	2	7	1
Subject 3	7	3	5	2	3	1
Subject 4	10	3	8	2	7	1
Subject 5	7	3	5	2	2	1
Subject 6	8	3	6	1.5	6	1.5
	$\Sigma R_1 = 18$		$\Sigma R_2 = 11.5$		$\Sigma R_3 = 6.5$	

$$\bar{R}_1 = \frac{\Sigma R_1}{n_1} = \frac{18}{6} = 3 \qquad \bar{R}_2 = \frac{\Sigma R_2}{n_2} = \frac{11.5}{6} = 1.92 \qquad \bar{R}_3 = \frac{\Sigma R_3}{n_3} = \frac{6.5}{6} = 1.08$$

The ranking procedure employed for the **Friedman two-way analysis of variance by ranks** requires that each of the k scores of a subject be ranked within that subject.[4] Thus, in Table 25.1, for each subject a rank of 1 is assigned to the subject's lowest score, a rank of 2 to the subject's middle score, and a rank of 3 to the subject's highest score. In the event of tied scores, the same protocol described for handling ties for other rank-order tests (discussed in detail in Section IV of the **Mann–Whitney U test (Test 12))** is employed. Specifically, the average of the ranks involved is assigned to all scores tied for a given rank. The only example of tied scores in Example 25.1 is in the case of Subject 6 who has a score of 6 in both Conditions 2 and 3. In Table 25.1 both of these scores are assigned a rank of 1.5, since if the scores of Subject 6 in Conditions 2 and 3 were not identical but were still less than the subject's third score (which is 8 in Condition 1), one of the two scores that are, in fact, tied would receive a rank of 1 and the other a rank of 2. The average of these two ranks (i.e., $(1 + 2)/2 = 1.5$) is thus assigned to each of the two tied scores.

It should be noted that it is permissible to reverse the ranking protocol described above. Specifically, one can assign a rank of 1 to a subject's highest score, a rank of 2 to the subject's middle score, and a rank of 3 to the subject's lowest score. This reverse ranking protocol will yield the same value for the **Friedman test** statistic as the protocol employed in Table 25.1.

Upon rank-ordering the scores of the $n = 6$ subjects, the sum of the ranks is computed for each of the experimental conditions. In Table 25.1 the sum of the ranks of the jth condition is represented by the notation ΣR_j. Thus, $\Sigma R_1 = 18$, $\Sigma R_2 = 11.5$, $\Sigma R_3 = 6.5$. Although they are not required for the **Friedman test** computations, the mean rank ($\bar{R}_j$) for each of the conditions is also noted in Table 25.1.

The chi-square distribution is used to approximate the **Friedman test** statistic. Equation 25.1 is employed to compute the chi-square approximation of the **Friedman test** statistic (which is represented in most sources by the notation χ_r^2).

$$\chi_r^2 = \frac{12}{nk(k + 1)}\left[\sum_{j=1}^{k} (\Sigma R_j)^2\right] - 3n(k + 1) \qquad \text{(Equation 25.1)}$$

Note that in Equation 25.1 the term $[\sum_{j=1}^{k}(\Sigma R_j)^2]$ indicates that the sum of the ranks for each of the k experimental conditions is squared, and that the squared sums of ranks are summed. Substituting the appropriate values from Example 25.1 in Equation 25.1, the value $\chi_r^2 = 11.08$ is computed.

$$\chi_r^2 = \frac{12}{(6)(3)(3 + 1)}[(18)^2 + (11.5)^2 + (6.5)^2] - (3)(6)(3 + 1) = 11.08$$

V. Interpretation of the Test Results

In order to reject the null hypothesis, the computed value χ_r^2 must be equal to or greater than the tabled critical chi-square value at the prespecified level of significance. The computed chi-square value is evaluated with **Table A4 (Table of the Chi-Square Distribution)** in the **Appendix**. For the appropriate degrees of freedom, the tabled $\chi_{.95}^2$ value (which is the chi-square value at the 95th percentile) and the tabled $\chi_{.99}^2$ value (which is the chi-square value at the 99th percentile) are employed as the .05 and .01 critical values for evaluating a nondirectional alternative hypothesis. The number of degrees of freedom employed in the analysis are computed with Equation 25.2. Thus, $df = 3 - 1 = 2$.

$$df = k - 1 \qquad \text{(Equation 25.2)}$$

For $df = 2$, the tabled critical .05 and .01 chi-square values are $\chi_{.05}^2 = 5.99$ and $\chi_{.01}^2 = 9.21$. Since the computed value $\chi_r^2 = 11.08$ is greater than $\chi_{.05}^2 = 5.99$ and $\chi_{.01}^2 = 9.21$, the alternative hypothesis is supported at both the .05 and .01 levels.[5] A summary of the analysis of Example 25.1 with the **Friedman two-way analysis of variance by ranks** follows: It can be concluded there is a significant difference between at least two of the three experimental conditions exposed to different levels of noise. This result can be summarized as follows: $\chi_r^2(2) = 11.08$, $p < .01$.

It should be noted that when the data for Example 25.1 are evaluated with a **single-factor within-subjects analysis of variance**, the null hypothesis can also be rejected at both the .05 and .01 levels. The reader should note, however, that the difference between the value $\chi_r^2 = 11.08$ (obtained for the **Friedman test**) and $\chi_{.01}^2 = 9.21$ (the .01 tabled critical value for the **Friedman test**) is much smaller than the difference between $F = 41.29$ (obtained for the analysis of variance) and $F_{.01} = 7.56$ (the .01 tabled critical value for the analysis of variance). The smaller difference between the computed test statistic and the tabled critical value in the case of the **Friedman test** reflects the fact that, as a general rule (assuming that none of the assumptions of the analysis of variance are saliently violated), it provides a less powerful test of an alternative hypothesis than the analysis of variance.

VI. Additional Analytical Procedures for the Friedman Two-Way Analysis of Variance by Ranks and/or Related Tests

1. Tie correction for the Friedman two-way analysis of variance by ranks Some sources recommend that if there is an excessive number of ties in the overall distribution of scores, the value of the **Friedman test** statistic be adjusted. The tie correction results in a slight increase in the value of χ_r^2 (thus providing a slightly more powerful test of the alternative hypothesis). Equation 25.3 (based on a methodology described in Daniel (1990) and Marascuilo and McSweeney (1977)) is employed in computing the value C, which represents the tie correction factor for the **Friedman two-way analysis of variance by ranks**.

$$C = 1 - \frac{\sum\limits_{i=1}^{s} (t_i^3 - t_i)}{n(k^3 - k)} \qquad \text{(Equation 25.3)}$$

Where: s = the number of sets of ties
t_i = the number of tied scores in the ith set of ties

The notation $\sum_{i=1}^{s}(t_i^3 - t_i)$ indicates the following: a) For each set of ties, the number of ties in the set is subtracted from the cube of the number of ties in that set; and b) The sum of all the values computed in part a) is obtained. The tie correction will now be computed for Example 25.1. In the latter example there is $s = 1$ set of ties in which there are $t_i = 2$ ties (i.e., the two scores of 6 for Subject 6 under Conditions 2 and 3). Thus:

$$\sum\limits_{i=1}^{s} (t_i^3 - t_i) = [(2)^3 - 2] = 6$$

Employing Equation 25.3, the value $C = .958$ is computed.

$$C = 1 - \frac{6}{6[(3)^3 - 3]} = .958$$

$\chi_{r_C}^2$, which represents the tie-corrected value of the **Friedman test** statistic, is computed with Equation 25.4.

$$\chi_{r_C}^2 = \frac{\chi_r^2}{C} \qquad \text{(Equation 25.4)}$$

Employing Equation 25.4, the tie-corrected value $\chi_{r_C}^2 = 11.57$ is computed.

$$\chi_{r_C}^2 = \frac{11.08}{.958} = 11.57$$

As is the case with $\chi_r^2 = 11.08$ computed with Equation 25.1, the value $\chi_{r_C}^2 = 11.57$ computed with Equation 25.4 is significant at both the .05 and .01 levels (since it is greater than $\chi_{.05}^2 = 5.99$ and $\chi_{.01}^2 = 9.21$). Although Equation 25.4 results in a slightly less conservative test than Equation 25.1, in this instance the two equations lead to identical conclusions with respect to the null hypothesis.

2. Pairwise comparisons following computation of the test statistic for the Friedman two-way analysis of variance by ranks Prior to reading this section the reader should review the discussion of comparisons in Section VI of the **single-factor between-subjects analysis of variance (Test 21)**. As is the case with the omnibus F value computed for an analysis of variance, the χ_r^2 value computed with Equation 25.1 is based on an evaluation of all k experimental conditions. When the value of χ_r^2 is significant, it does not indicate whether just two or, in fact, more than two conditions differ significantly from one another. In order to answer the latter question, it is necessary to conduct comparisons contrasting specific conditions with one another. This section will describe methodologies which can be employed for conducting **unplanned simple/pairwise comparisons** following the computation of a significant χ_r^2 value.[6]

In conducting a simple comparison, the null hypothesis and nondirectional alternative hypothesis are as follows: H_0: $\theta_a = \theta_b$ versus H_1: $\theta_a \neq \theta_b$. In the aforementioned hypotheses, θ_a and θ_b represent the medians of the populations represented by the two conditions involved in the comparison. The alternative hypothesis can also be stated directionally as follows: H_1: $\theta_a > \theta_b$ or H_1: $\theta_a < \theta_b$.

Various sources (e.g., Daniel (1990) and Siegel and Castellan (1988)) describe a comparison procedure recommended for use with large sample sizes for the **Friedman two-way analysis of variance by ranks** (which is essentially the application of the **Bonferroni-Dunn method** described in Section VI of the **single-factor between-subjects analysis of variance** to the **Friedman test** model). Through use of Equation 25.5, the procedure allows a researcher to identify the minimum required difference between the sums of the ranks of any two conditions (designated as CD_F) in order for them to differ from one another at the prespecified level of significance.[7]

$$CD_F = z_{adj} \sqrt{\frac{nk(k+1)}{6}} \qquad \text{(Equation 25.5)}$$

The value of z_{adj} is obtained from **Table A1 (Table of the Normal Distribution)** in the **Appendix**. In the case of a nondirectional alternative hypothesis, z_{adj} is the z value above which a proportion of cases corresponding to the value $\alpha_{FW}/2c$ falls (where c is the total number of comparisons that are conducted). In the case of a directional alternative hypothesis, z_{adj} is the z value above which a proportion of cases corresponding to the value α_{FW}/c falls. When all possible pairwise comparisons are made, $c = [k(k-1)]/2$, and thus, $2c = k(k-1)$. In Example 25.1 the number of pairwise/simple comparisons that can be conducted is c [3(3 − 1)]/2 = 3 — specifically, Condition 1 versus Condition 2, Condition 1 versus Condition 3, and Condition 2 versus Condition 3.

The value of z_{adj} will be a function of both the maximum **familywise Type I error rate** (α_{FW}) the researcher is willing to tolerate and the total number of comparisons that are conducted. When a limited number of comparisons are planned prior to collecting the data, most sources take the position that a researcher is not obliged to control the value of α_{FW}. In such a case, the **per comparison Type I error rate** (α_{PC}) will be equal to the prespecified value of alpha. When α_{FW} is not adjusted, the value of z_{adj} employed in Equation 25.5 will be the tabled critical z value which corresponds to the prespecified level of significance. Thus, if a nondirectional alternative hypothesis is employed and $\alpha = \alpha_{PC} = .05$, the tabled critical two-tailed .05 value $z_{.05} = 1.96$ is used to represent z_{adj} in Equation 25.5. If $\alpha = \alpha_{PC} = .01$, the tabled critical two-tailed .01 value $z_{.01} = 2.58$ is used in Equation 25.5. In the same respect, if a directional alternative hypothesis is employed, the tabled critical .05 and .01 one-tailed values $z_{.05} = 1.65$ and $z_{.01} = 2.33$ are used for z_{adj} in Equation 25.5.

When comparisons are not planned beforehand, it is generally acknowledged that the value of α_{FW} must be controlled so as not to become excessive. The general approach for controlling the latter value is to establish a **per comparison Type I error rate** which insures that α_{FW} will not exceed some maximum value stipulated by the researcher. One method for doing this (described under the **single-factor between-subjects analysis of variance** as the **Bonferroni–Dunn method**) establishes the **per comparison Type I error rate** by dividing the maximum value one will tolerate for the **familywise Type I error rate** by the total number of comparisons conducted. Thus, in Example 25.1, if one intends to conduct all three pairwise comparisons and wants to insure that α_{FW} does not exceed .05, $\alpha_{PC} = \alpha_{FW}/c = .05/3 = .0167$. The latter proportion is used to determine the value of z_{adj}. As noted earlier, if a directional alternative hypothesis is employed for a comparison, the value of z_{adj} employed in Equation 25.5 is the z value above which a proportion equal to $\alpha_{PC} = \alpha_{FW}/c$ of the cases falls. In **Table A1**, the z value that corresponds to the proportion .0167 is $z = 2.13$. By employing $z_{adj} = 2.13$ in Equation 25.5, one can be assured that within the "family" of three pairwise comparisons, α_{FW} will not exceed .05 (assuming all of the comparisons are directional). If a nondirectional alternative hypothesis is employed for all of the comparisons, the value of z_{adj} will be the z value above which a proportion equal to $\alpha_{FW}/2c = \alpha_{PC}/2$ of the cases falls. Since $\alpha_{PC}/2 = .0167/2 = .0083$, $z = 2.39$. By employing $z_{adj} = 2.39$ in Equation 25.5, one can be assured that α_{FW} will not exceed .05.[8]

In order to employ the CD_F value computed with Equation 25.5, it is necessary to determine the absolute value of the difference between the sums of the ranks of each pair of experimental conditions that are compared.[9] Table 25.2 summarizes the difference scores between pairs of sums of ranks.

Table 25.2 Difference Scores between Pairs of Sums of Ranks for Example 25.1

$$|\Sigma R_1 - \Sigma R_2| = |18 - 11.5| = 6.5$$
$$|\Sigma R_1 - \Sigma R_3| = |18 - 6.5| = 11.5$$
$$|\Sigma R_2 - \Sigma R_3| = |11.5 - 6.5| = 5$$

If any of the differences between the sums of ranks is equal to or greater than the CD_F value computed with Equation 25.5, a comparison is declared significant. Equation 25.5 will now be employed to evaluate the nondirectional alternative hypothesis $H_1: \theta_a \neq \theta_b$ for all three pairwise comparisons. Since it will be assumed that the comparisons are unplanned and the researcher does not want the value of α_{FW} to exceed .05, the value $z_{adj} = 2.39$ will be used in computing CD_F.

$$CD_F = (2.39)\sqrt{\frac{(6)(3)(3 + 1)}{6}} = (2.39)(3.46) = 8.28$$

The obtained value $CD_F = 8.28$ indicates that any difference between the sums of ranks of two conditions which is equal to or greater than 8.28 is significant. With respect to the three pairwise comparisons, the only difference between the sum of ranks of two conditions which is greater than $CD_F = 8.28$ is $|\Sigma R_1 - \Sigma R_3| = 11.5$. Thus, we can conclude there is a significant difference between Condition 1 and Condition 3. We cannot conclude that the difference between any other pair of conditions is significant.[10]

An alternative strategy which can be employed for conducting pairwise comparisons for the **Friedman test** model is to use one of the tests that are described for evaluating a dependent samples design involving $k = 2$ samples. Specifically, one can employ either the **Wilcoxon matched-pairs signed-ranks test (Test 18)** or the **binomial sign test for two dependent samples**. Whereas the **binomial sign test** only takes into consideration the direction of the difference of subjects' scores in the two experimental conditions, the **Wilcoxon test** rank-orders the interval/ratio difference scores of subjects. Because of the latter, the **Wilcoxon test** employs more information than the **binomial sign test**, and, consequently, will provide a more powerful test of an alternative hypothesis. Both the **Wilcoxon test** and **binomial sign test** will be used to conduct the three pairwise comparisons for Example 25.1.[11]

Use of the **Wilcoxon matched-pairs signed-ranks test** requires that for each comparison which is conducted the difference scores of subjects in the two experimental conditions be rank-ordered, and that the Wilcoxon T statistic be computed for that comparison. The exact distribution of the **Wilcoxon test** statistic can only be used when the value of α_{PC} is equal to one of the probabilities documented in **Table A5 (Table of Critical T Values for Wilcoxon's Signed-Ranks and Matched-Pairs Signed-Ranks Tests)** in the **Appendix**. When α_{PC} is a value other than those listed in **Table A5**, the normal approximation of the **Wilcoxon test** statistic must be employed.

When the **Wilcoxon matched-pairs signed-ranks test** is employed for the three pairwise comparisons, the following T values are computed: a) Condition 1 versus Condition 2: $T = 0$; b) Condition 1 versus Condition 3: $T = 0$; and c) Condition 2 versus Condition 3: $T = 0$.[12] When the aforementioned T values are substituted in Equations 18.2 and 18.3 (the uncorrected and continuity-corrected normal approximations for the **Wilcoxon test**), the following absolute z values are computed: a) Condition 1 versus Condition 2: $z = 2.20$ and $z = 2.10$; b) Condition 1 versus Condition 3: $z = 2.20$ and $z = 2.10$; and c) Condition 2 versus Condition 3: $z = 2.02$ and $z = 1.89$. If we want to evaluate a nondirectional alternative hypothesis and insure that α_{FW} does not exceed .05, the value of α_{PC} is set equal to .0167. **Table A5** cannot be employed, since it does not list two-tailed critical T values for $\alpha_{.0167}$. In order to evaluate the result of the normal approximation, we identify the tabled critical two-tailed .0167 z value in **Table A1**. In employing Equation 25.5 earlier in this section, we determined that the latter value is $z_{.0167} = 2.39$. Since none of the z values computed for the normal approximation is equal to or greater than $z_{.0167} = 2.39$, none of the pairwise comparisons is significant. This result is not identical to that obtained with Equation 25.5, in which case a significant difference is computed for the comparison Condition 1 versus Condition 3. Although the latter comparison (as well as the Condition 1 versus Condition 2 comparison) comes close when it is evaluated with the **Wilcoxon test**, it falls just short of achieving significance.

In the event the researcher elects not to control the value of α_{FW} and employs $\alpha_{PC} = .05$ in evaluating the three pairwise comparisons (once again assuming a nondirectional analysis), both the Condition 1 versus Condition 2 and Condition 1 versus Condition 3 comparisons are significant at the .05 level if the **Wilcoxon test** is employed. Specifically, both the uncorrected and corrected normal approximations are significant, since $z = 2.20$ and $z = 2.10$ (computed for both the Condition 1 versus Condition 2 and Condition 1 versus Condition 3 comparisons) are greater than the tabled critical two-tailed value $z_{.05} = 1.96$. The Condition 2 versus Condition 3 comparison is also significant, but only if the uncorrected value $z = 2.02$ is employed. Employing **Table A5**, we also determine that both the Condition 1 versus Condition 2 and Condition 1 versus Condition 3 comparisons are significant at the .05 level, since the computed value $T = 0$ for both comparisons is equal to the tabled critical two-tailed .05 value $T_{.05} = 0$ (for $n = 6$). The Condition 2 versus Condition 3 comparison is not significant, since no two-tailed .05 critical value is listed in **Table A5** for $n = 5$. If Equation 25.5 is employed for the same set of comparisons, however, only the Condition 1 versus Condition 3 comparison is significant. This is the case, since $CD_F = (1.96)(3.46) = 6.78$, and only the difference $|\Sigma R_1 - \Sigma R_3| = 11.5$ is greater than $CD_F = 6.78$.[13] The difference $|\Sigma R_1 - \Sigma R_2| = 6.5$ (which is significant with the **Wilcoxon test**) just falls short of achieving significance. Although the result obtained with Equation 25.5 is not identical to that obtained with the **Wilcoxon test**, the two analyses are reasonably consistent with one another.

In the event the **binomial signed test for two dependent samples** is employed to conduct comparisons, a researcher must determine for each comparison the number of subjects who yield positive versus negative difference scores. With the exception of Subject 6 in the Condition 2 versus Condition 3 comparison, all of the difference scores for the three pairwise comparisons are positive (since all subjects obtain a higher score in Condition 1 than Condition 2, in Condition 1 than Condition 3, and in Condition 2 than Condition 3). For the Condition 1 versus Condition 2 and Condition 1 versus Condition 3 comparisons, we must compute $P(x = 6)$ for $n = 6$. For the Condition 2 versus Condition 3 comparison (which does not include Subject 6 in the analysis, since the latter subject has a zero difference score), we must compute $P(x = 5)$ for $n = 5$. For all three pairwise comparisons $\pi+ = \pi- = .5$.

Employing **Table A6 (Table of the Binomial Distribution, Individual Probabilities)** (or **Table A7**) in the **Appendix** we can determine that when $n = 6$, $P(x = 6) = .0156$. Thus, the computed two-tailed probability for the Condition 1 versus Condition 2 and Condition 1 versus Condition 3 comparisons is $(2)(.0156) = .0312$. When $n = 5$, $P(x = 5) = .0312$. The computed two-tailed probability for the Condition 2 versus Condition 3 comparison is $(2)(.0312) = .0624$.

As before, if we want to evaluate a nondirectional alternative hypothesis and insure that α_{FW} does not exceed .05, the value of α_{PC} is set equal to .0167. Thus, in order to reject the null hypothesis the computed two-tailed binomial probability for a comparison must be equal to or less than .0167. Since the computed two-tailed probabilities .0312 (for the Condition 1 versus Condition 2 and the Condition 1 versus Condition 3 comparisons) and .0624 (for the Condition 2 versus Condition 3 comparison) are greater than .0167, none of the pairwise comparisons is significant.

In the event the researcher elects not to control the value of α_{FW} and employs $\alpha_{PC} = .05$ for evaluating the three pairwise comparisons (once again assuming a nondirectional analysis), both the Condition 1 versus Condition 2 and Condition 1 versus Condition 3 comparisons are significant at the .05 level, since the computed two-tailed probability .0312 (for the Condition 1 versus Condition 2 and the Condition 1 versus Condition 3 comparisons) is less than .05. The Condition 2 versus Condition 3 comparison is not significant, since the computed two-tailed probability .0624 for the latter comparison is greater than .05.

When the results obtained with the **binomial sign test** are compared with those obtained with Equation 25.5 and the **Wilcoxon test**, it would appear that of the three procedures the

binomial sign test results in the most conservative test (and thus, as noted previously, the least powerful test). However, if one takes into account the obtained binomial probabilities, they are, in actuality, not far removed from the probabilities obtained when Equation 25.5 and the **Wilcoxon test** are used.

In the case of Example 25.1, regardless of which comparison procedure one employs, it would appear that unless one uses a very low value for α_{PC}, the Condition 1 versus Condition 3 comparison is significant. There is some suggestion that the Condition 1 versus Condition 2 comparison may also be significant, but some researchers would recommend conducting additional studies in order to clarify whether or not the two conditions represent different populations. Although, based on the analyses which have been conducted, the Condition 2 versus Condition 3 comparison does not appear to be significant, it is worth noting that if the researcher uses the **Wilcoxon test** (specifically, the normal approximation not corrected for continuity) to evaluate the directional alternative hypothesis H_1: $\theta_2 > \theta_3$ with $\alpha_{PC} = .05$, the latter comparison also yields a significant result. Thus, further studies might be in order to clarify the relationship between the populations represented by Conditions 2 and 3.

Although the intent of presenting three comparison procedures in this section is to illustrate that the results obtained with the different comparison procedures are reasonably consistent with one another, as noted in Endnote 10, other sources recommend alternative procedures which may yield different results. As is noted in the discussion of comparisons in Section VI of both the **single-factor between-subjects analysis of variance** and the **Kruskal–Wallis one-way analysis of variance by ranks**, in instances where two or more comparison procedures yield inconsistent results, the most effective way to resolve such a problem is by conducting one or more replication studies. By doing the latter, a researcher should be able to clarify whether or not an obtained difference is reliable, as well as the magnitude of the difference (if, in fact, one exists).

It is also noted throughout the book that, in the final analysis, the decision regarding which of the available comparison procedures to employ is usually not the most important issue facing the researcher conducting comparisons. The main issue is what maximum value one is willing to tolerate for α_{FW}. Additional discussion of comparison procedures for the **Friedman test** model can be found in Church and Wike (1979) (who provide a comparative analysis of a number of different comparison procedures), Daniel (1990) (who describes a methodology for comparing $(k - 1)$ conditions with a control group, as well as a methodology for estimating the size of a difference between the medians of any pair of experimental conditions), Marascuilo and McSweeney (1977) (who within the framework of a comprehensive discussion of the **Friedman test** model describe a methodology for conducting complex comparisons), Marascuilo and Serlin (1988), Siegel and Castellan (1988) (who also describe the methodology for comparing $(k - 1)$ conditions with a control group), and Zar (1999).

Marascuilo and McSweeney (1977) also discuss the computation of a confidence interval for a comparison for the **Friedman test** model. One approach for computing a confidence interval is to add and subtract the computed value of CD_F to and from the obtained difference between the sums of ranks (or mean ranks, if the equation in Endnote 7 is employed) involved in the comparison. The latter approach is based on the same logic employed for computing a confidence interval for a comparison in Section VI of the **single-factor between-subjects analysis of variance**.

VII. Additional Discussion of the Friedman Two-Way Analysis of Variance by Ranks

1. Exact tables of the Friedman distribution Although an exact probability value can be computed for obtaining a configuration of ranks which is equivalent to or more extreme than the configuration observed in the data evaluated with the **Friedman two-way analysis of variance by ranks**, the chi-square distribution is generally employed to estimate the latter probability. Although most sources employ the chi-square approximation regardless of the values of k and n, some sources recommend that exact tables be employed when the values of n and/or k are small. The exact sampling distribution for the **Friedman two-way analysis of variance by ranks** is based on the use of **Fisher's method of randomization** (which is discussed in Section IX (the **Addendum**) of the **Mann-Whitney U test**).

Tables of exact critical values, which can be viewed as adjusted chi-square values, can be found in Marascuilo and McSweeney (1977) and Siegel and Castellan (1988) (who list critical values for various values of n between 5 and 13 when the value of k is between 3 and 5). Depending upon the values of k and n, exact critical values may be either slightly larger or smaller than the critical chi-square values in **Table A4**. In point of fact, for $k = 3$ and $n = 6$ the exact tabled critical .05 and .01 values for the **Friedman test** statistic are respectively $\chi^2_{r.05} = 7.00$ and $\chi^2_{r.01} = 9.00$. Since the value $\chi^2_r = 11.08$ computed for Example 25.1 is greater than both of the aforementioned critical values, the null hypothesis can still be rejected at both the .05 and .01 levels. Although the conclusions with respect to Example 25.1 are the same regardless of whether one employs the exact critical values or the chi-square values in **Table A4**, inspection of the two sets of values indicates that the exact .05 critical value is larger than the corresponding critical value $\chi^2_{.05} = 5.99$ derived from **Table A4**, while the reverse is true with respect to the exact .01 critical value, which is less than the corresponding value $\chi^2_{.01} = 9.21$ in **Table A4**. It should be noted that for a given value of k, as the value of n increases, the exact critical value approaches the tabled chi-square value in **Table A4**. An additional point of interest relevant to evaluating the **Friedman test** statistic is that Daniel (1990), Conover (1980; 1999, pp. 370–373) and Zar (1999, pp. 264–265) cite a study by Iman and Davenport (1980) which suggests that the F distribution can be used to approximate the sampling distribution for the **Friedman test**, and that the latter approximation may be more accurate than the more commonly employed chi-square approximation.[14]

2. Equivalency of the Friedman two-way analysis of variance by ranks and the binomial sign test for two dependent samples when $k = 2$ In Section I it is noted that when $k = 2$ the **Friedman two-way analysis of variance by ranks** will yield a result which is equivalent to that obtained with the **binomial sign test for two dependent samples**. To be more specific, the **Friedman test** will yield a result which is equivalent to the normal approximation of the **binomial sign test for two dependent samples** when the correction for continuity is not employed (i.e., the result obtained with Equation 19.3).[15] It should be noted, however, that the two tests will only yield an equivalent result when none of the subjects has the same score in the two experimental conditions. In the case of the **binomial sign test**, any subject who has the same score in both conditions is eliminated from the data analysis. In the case of the **Friedman test**, however, such subjects are included in the analysis. In order to demonstrate the equivalency of the two tests, Equation 25.1 will be employed to analyze the data for Example 19.1 (which was previously evaluated with the **binomial sign test for two dependent samples**). In using Equation 25.1 the data for Subject 2 are not included, since the latter subject has identical scores in both conditions. Thus, in our analysis $n = 9$ and $k = 2$. Table 25.3 summarizes the rank-ordering of data for Example 19.1 within the framework of the **Friedman test** model.

Table 25.3 Summary of Data for Example 19.1 for Friedman Test Model

	Condition 1		Condition 2	
	X_1	R_1	X_2	R_2
Subject 1	9	2	8	1
Subject 2	2	1.5	2	1.5
Subject 3	1	1	3	2
Subject 4	4	2	2	1
Subject 5	6	2	3	1
Subject 6	4	2	0	1
Subject 7	7	2	4	1
Subject 8	8	2	5	1
Subject 9	5	2	4	1
Subject 10	1	2	0	1
		$\Sigma R_1 = 18.5$		$\Sigma R_2 = 11.5$

Since the data for Subject 2 are not included in the analysis, the rank-orders for Subject 2 under the two conditions are subtracted from the values $\Sigma R_1 = 18.5$ and $\Sigma R_2 = 11.5$, yielding the revised values $\Sigma R_1 = 17$ and $\Sigma R_2 = 10$ which are employed in Equation 25.1. Employing the latter equation, the value $\chi_r^2 = 5.44$ is computed.

$$\chi_r^2 = \frac{12}{(9)(2)(2+1)} [(17)^2 + (10)^2] - (3)(9)(3) = 5.44$$

Employing Equation 25.2, $df = 2 - 1 = 1$. For $df = 1$, the tabled critical .05 and .01 chi-square values are $\chi_{.05}^2 = 3.84$ and $\chi_{.01}^2 = 6.63$. Since the obtained value $\chi_r^2 = 5.44$ is greater than $\chi_{.05}^2 = 3.84$, the alternative hypothesis is supported at the .05 level. It is not, however, supported at the .01 level, since $X_r^2 = 5.44$ is less than $\chi_{.01}^2 = 6.63$.[16]

Equation 19.3 yields the value $z = 2.33$ for the same set of data. Since the square of a z value will equal the corresponding chi-square value computed for the same set of data, z^2 should equal χ_r^2. In point of fact, $(z = 2.33)^2 = (\chi_r^2 = 5.44)$ (the minimal discrepancy is due to rounding off error). It is also the case that the square of the tabled critical z value employed for the normal approximation of the **binomial sign test for two dependent samples** will always equal the tabled critical chi-square value employed for the **Friedman test** at the same level of significance. Thus, the square of the tabled critical two-tailed value $z_{.05} = 1.96$ employed for the normal approximation of the **binomial sign test** equals $\chi_{.05}^2 = 3.84$ employed for the **Friedman test** (i.e., $(z = 1.96)^2 = (\chi^2 = 3.84)$).

3. Power-efficiency of the Friedman two-way analysis of variance by ranks   Daniel (1990) notes that Noether (1967) states when the underlying population distributions are normal, the **asymptotic relative efficiency** (which is discussed in Section VII of the **Wilcoxon signed-ranks test (Test 6)**) of the **Friedman two-way analysis of variance by ranks** (relative to the **single-factor within-subjects analysis of variance**) is $.955k/(k + 1)$. Thus, when $k = 2$ the asymptotic relative efficiency of the **Friedman test** is .64, but when $k = 10$ it equals .87. For a uniform distribution, the asymptotic relative efficiency of the **Friedman test** is $k/(k + 1)$.

4. Alternative nonparametric rank-order procedures for evaluating a design involving k dependent samples   In addition to the **Friedman two-way analysis of variance by ranks**, a number of other nonparametric procedures for two or more dependent samples have been developed which can be employed with ordinal data. Among the more commonly cited alternative procedures are the following: a) Marascuilo and McSweeney (1977) describe the extension of the **van der Waerden normal-scores test for** k **independent samples (Test 23)** (Van der Waerden (1952/1953)) to a design involving k dependent samples. Conover (1980, 1999) notes that the normal-scores test developed by Bell and Doksum (1965) can also be extended to the latter design; b) The **Page test for ordered alternatives** (Page (1963)) (which is described in Section IX (The **Addendum**)) can be employed with k dependent samples to evaluate an ordered alternative hypothesis. Specifically, in stating the alternative hypothesis, the ordinal position of the treatment effects is stipulated (as opposed to just stating that a difference exists between at least two of the k experimental conditions).

5. Relationship between the Friedman two-way analysis of variance by ranks and Kendall's coefficient of concordance The **Friedman two-way analysis of variance by ranks** and **Kendall's coefficient of concordance (Test 31)** (which is one of a number of measures of association that are described in this book) are based on the same statistical model. The latter measure of association is employed with three or more sets of ranks when rankings are based on the **Friedman test** protocol. A full discussion of the relationship between the **Friedman two-way analysis of variance by ranks** and **Kendall's coefficient of concordance** (which can be used as a measure of effect size for the **Friedman test**) can be found in Section VII of **Kendall's coefficient of concordance**. When there are $n = 2$ subjects/sets of matched subjects, **Spearman's rank-order correlation coefficient (Test 29)**, which is linearly related to **Kendall's coefficient of concordance**, can be conceptualized within the framework of the **Friedman test** model. The latter relationship is discussed in Section VII of **Spearman's rank-order correlation coefficient**.

VIII. Additional Examples Illustrating the Use of the Friedman Two-Way Analysis of Variance by Ranks

The **Friedman two-way analysis of variance by ranks** can be employed to evaluate any of the additional examples noted for the **single-factor within-subjects analysis of variance**, if the data for the latter examples are rank-ordered. In addition, the **Friedman test** can be used to evaluate the data for any of the additional examples noted for the *t* **test for two dependent samples/ binomial sign test for two dependent samples/Wilcoxon matched-pairs signed-ranks test**. Example 25.2 is an additional example that can be evaluated with the **Friedman two-way analysis of variance by ranks**. In Example 25.2 there is no need to rank-order interval/ratio data, since the results of the study are summarized in a rank-order format.[17]

Example 25.2 *Six horses are rank-ordered by a trainer with respect to their racing form on three different surfaces. Specifically,* Track A *has a cement surface,* Track B *a clay surface, and* Track C *a grass surface. Except for the surface, the three tracks are comparable to one another in all other respects.* Table 25.4 *summarizes the rankings of the horses on the three tracks. (In the case of* Horse 6, *the rank of* 1.5 *for both the clay and grass tracks reflects the fact that the horse was perceived to have equal form on both surfaces.) Do the data indicate that the form of a horse is related to the surface on which it is racing?*

Table 25.4   Data for Example 25.2

	Track A (Cement)	Track B (Clay)	Track C (Grass)
Horse 1	3	2	1
Horse 2	3	2	1
Horse 3	3	2	1
Horse 4	3	2	1
Horse 5	3	2	1
Horse 6	3	1.5	1.5

Since the ranks employed in Example 25.2 are identical to those employed for Example 25.1, the **Friedman test** will yield the identical result. Since most people would probably be inclined to employ a rank of 1 to represent a horse's best surface and a rank of 3 to represent a horse's worst surface, using such a ranking protocol, the track with the lowest sum of ranks (Track C) is associated with the best racing form and the track with the highest sum of ranks (Track A) is associated with the worst racing form.

IX. Addendum

Test 25a: The Page test for ordered alternatives   In experiments involving two or more experimental conditions a researcher may predict *a priori* (i.e., prior to conducting the study) the ordering of the experimental conditions with respect to their scores on the dependent variable. A common example of the latter type of experiment involving a dependent samples design is one in which it is hypothesized that the greater the therapeutic dosage of a drug the greater the degree of its efficacy. In the case where subjects serve in each of two experimental conditions (e.g., a drug condition versus a placebo condition), such a hypothesis of **ordered alternatives** (which is the term employed for a prediction involving the exact ordinal relationship between two or more experimental conditions) can be evaluated through use of a one-tailed analysis (as opposed to a two-tailed analysis). The latter option, however, is not available when there are three or more experimental conditions. Page (1963) developed the test to be described in this section, which can be employed when the alternative hypothesis for a k dependent samples design (where $k \geq 2$) specifies the rank-order of the k population medians.

The **Page test for ordered alternatives** can be employed under the same conditions a researcher might consider using the **Friedman two-way analysis of variance by ranks** as an alternative to the **single-factor within-subjects analysis of variance**. Daniel (1990) notes that the **Page test for ordered alternatives** is based on the same assumptions underlying the **Friedman two-way analysis of variance by ranks**.

The **Page test for ordered alternatives** will be illustrated through use of Example 25.1. In the latter example, it will be assumed that prior to conducting the study the experimenter predicted the order of the three experimental conditions with respect to the number of nonsense syllables which would be recalled. Specifically, the experimenter hypothesized that subjects would recall the largest number of nonsense syllables under **Condition 1 (No noise)**, followed by **Condition 2 (Moderate noise)**, followed by **Condition 3 (Extreme noise)** (in which the smallest number of nonsense syllables would be recalled).

The null and alternative hypotheses employed for the above prediction when it is made within the framework of the **Page test for ordered alternatives** are as follows.

Null hypothesis H_0: $\theta_1 = \theta_2 = \theta_3$

(The median of the population Condition 1 represents equals the median of the population Condition 2 represents equals the median of the population Condition 3 represents.)

Alternative hypothesis H_1: $\theta_1 \geq \theta_2 \geq \theta_3$, with at least one strict inequality

(With respect to at least one of the inequalities, the median of the population Condition 1 represents is greater than or equal to the median of the population Condition 2 represents which is greater than or equal to the median of the population Condition 3 represents. With respect to Example 25.1, in order for the alternative hypothesis to be supported, at least one of the ordinal relationships stated in H_1 must be supported. It should be noted, however, that although the notation $\geq$ is employed by all sources in stating the alternative hypothesis, the use of the term **strict** within the context of stating the alternative hypothesis requires that in order for it to be supported, in at least one case the median of one of the populations must be greater than the median of one of the other populations in the order stipulated.)

Notes: 1) The alternative hypothesis stipulated above is directional. The **Page test for ordered alternatives** can also be employed to evaluate a two-tailed alternative hypothesis. In the case of Example 25.1, a two-tailed/nondirectional alternative hypothesis will be supported if the data are consistent with either of the following two predictions: a) $\theta_1 \geq \theta_2 \geq \theta_3$, with at least one strict inequality; or b) $\theta_1 \leq \theta_2 \leq \theta_3$, with at least one strict inequality. Note that reversal of the inequalities stipulated for the alternative hypothesis allows for evaluation of the analogous alternative hypothesis (i.e., H_1: $\theta_1 \leq \theta_2 \leq \theta_3$, with at least one strict inequality) in the opposite tail of the sampling distribution; 2) Some sources in stating the null and alternative hypotheses employ notation designating equalities or inequalities in reference to treatment effects, as opposed to the population medians which have been used above.

The test statistic (designated with the letter L) for the **Page test for ordered alternatives** is computed with Equation 25.6.

$$L = \sum_{j=1}^{k} jR_j = R_1 + 2R_2 + \dots kR_k \qquad \textbf{(Equation 25.6)}$$

Where R_1 represents the sum of the ranks of the experimental condition predicted to have the **smallest sum of ranks**, R_2 represents the sum of ranks of the experimental condition predicted to have the **second smallest sum of ranks**, and so on, and with R_k representing the sum of ranks of the experimental condition predicted to have the **largest sum of ranks**.

The notation in Equation 25.6 indicates that the sum of the ranks for each of the experimental conditions (the values of which are summarized in Table 25.1) is multiplied by its predicted rank order if all k of the sum of ranks values are arranged ordinally. In arranging the sum of ranks values ordinally, the value 1 is employed to designate the rank order for the sum of ranks value that is predicted to be the smallest of the k sum of ranks values, the value 2 is employed to designate the rank order for the sum of ranks value that is predicted to be the second smallest of the k sum of ranks values, and so on, until the value of k is employed to designate the rank order for the sum of ranks value which is predicted to be the largest of the k sum of ranks values.

If, in fact, the data are consistent with the ordering stipulated for the sum of ranks values in the alternative hypothesis, then the sum of ranks value R_a will be larger than the sum of ranks value R_b, when $b < a$. As a result of the fact that the sum of ranks values for each of the k experimental conditions is assigned a weight which indicates its ordinal position as specified in the alternative hypothesis, the value computed for L with Equation 25.6 will be large if the alternative hypothesis is supported.

Note that in reference to Example 25.1, the alternative hypothesis stated for the **Page test for ordered alternatives** earlier in this section predicts the following: Condition 3 will have the smallest median value (and consequently it will have the smallest sum of ranks), Condition 2 will have the second smallest median value (and consequently it will have the second smallest sum of ranks), and Condition 1 will have the largest median value (and consequently it will have the largest sum of ranks). Consequently, the sum of the ranks for the condition predicted to have the smallest sum of ranks (i.e., Condition 3) is multiplied by 1 (where the value 1 designates the rank for the smallest predicted sum of ranks value); the sum of the ranks for the condition predicted to have the second smallest sum of ranks (i.e., Condition 2) is multiplied by 2 (where the value 2 designates the rank for the second smallest predicted sum of ranks value); and the sum of ranks for the condition predicted to have the third smallest (i.e., the largest) sum of ranks (i.e., Condition 1) is multiplied by 3 (where the value 3 (which is equal to the total number of experimental conditions $k = 3$) designates the rank for the third smallest predicted sum of ranks value). Thus in the case of Example 25.1, Equation 25.6 can be expressed in the form of the equation noted below. It is important to note that in the equation below, the subscripts employed correspond to the original subscripts designated for the three experimental conditions (i.e., **Condition 1 = No noise; Condition 2 = Moderate noise; Condition 3 = Loud noise**).

$$L = R_3 + 2R_2 + 3R_1$$

Employing the above equation with the data for Example 25.1, the value $L = 83.5$ is computed for the **Page test statistic**.

$$J = (1)(6.5) + (2)(11.5) + (3)(18) = 6.5 + 23 + 54 = 83.5$$

In order to reject the null hypothesis, the computed value of L must be equal to or greater than the tabled critical value at the prespecified level of significance in **Table A25 (Table of critical values for the Page Test Statistic)** in the **Appendix**. Inspection of the latter table reveals that the tabled critical .05 and .01 values in **Table A25** for $k = 3$ experimental conditions with $n = 6$ subjects are $L_{.05} = 79$ and $L_{.01} = 81$. Since the computed value $L = 83.5$ is larger than both of the aforementioned critical values, the null hypothesis can be rejected at both the .05 and .01 levels. Thus, the alternative hypothesis is supported — i.e., the ordering of the population medians is $\theta_1 \geq \theta_2 \geq \theta_3$.

The normal approximation of the Page test for ordered alternatives If the composite sample size (i.e., the values of n and/or k) employed in a study is relatively large, the normal distribution can be employed to approximate the test statistic for the **Page test for ordered alternatives**. Sources generally recommend that the normal approximation should be used when values of n and/or k employed in a study are larger than those documented in **Table A25**. Equation 25.7 provides the normal approximation of the **Page test** statistic.

$$z = \frac{L - \left| \dfrac{nk(k + 1)^2}{4} \right|}{\sqrt{\dfrac{n(k^3 - k)^2}{144(k - 1)}}}$$ **(Equation 25.7)**

In the numerator of Equation 25.7, the term subtracted from L (i.e., $[nk(k + 1)^2]/4$) represents the expected (mean) value of the sampling distribution of L (which can be represented with the notation L_E), while the denominator represents the expected standard deviation of the sampling distribution of L (which can be represented with the notation SD_L). Thus, Equation 25.7 can be summarized as follows: $z = (L - L_E)/SD_L$.

Although Example 25.1 employs a sample size which most sources would view as too small to use within the framework of the normal approximation, the latter approximation will be employed to illustrate Equation 25.7. The reader will see that although Equation 25.7 is used with a relatively small sample size, it still yields a result which is consistent with that obtained when the exact distribution for the **Page test** statistic is employed. Employing Equation 25.7, the value $z = 3.32$ is computed.

$$z = \frac{83.5 - \left| \dfrac{(6)(3)(3 + 1)^2}{4} \right|}{\sqrt{\dfrac{(6)[(3)^3 - 3]^2}{144(3 - 1)}}} = \frac{83.5 - 72}{3.46} = 3.32$$

The obtained value $z = 3.32$ is evaluated with **Table A1**. In order to reject the null hypothesis, the computed value of z must be equal to or greater than the tabled critical one-tailed value at the prespecified level of significance. The tabled critical one-tailed .05 and .01 values are $z_{.05} = 1.65$ and $z_{.01} = 2.33$. Since $z = 3.32$ is larger than both of the aforementioned critical values, the null hypothesis can be rejected at both the .05 and .01 levels. Thus, as is the case when the exact sampling distribution is employed, the alternative hypothesis is supported.[18]

Note that, regardless of whether one employs the exact critical values for the sampling distribution of the L statistic in **Table A25** or computes the normal approximation, the result of the **Page test for ordered alternatives** is consistent with the result obtained when the same set of data was evaluated with both the **Friedman two-way analysis of variance by ranks** and the **single-factor within-subjects analysis of variance** (neither of which, however, specified the ordering of the conditions in the alternative hypothesis). When the **Friedman two-way analysis of variance by ranks** was employed with the same set of data without specifying the ordering of the conditions, the null hypothesis of equal population medians was rejected at both the .05 and .01 levels. When the **single-factor within-subjects analysis of variance** was employed with the same set of data without specifying the ordering of conditions, the null hypothesis of equal population means was rejected at both the .05 and .01 levels. Siegel and Castellan (1988, p. 188) note that when the ordering of the conditions is not stipulated in the alternative hypothesis, the **Page test for ordered alternatives** is equal in power to the **Friedman two-way analysis of variance by ranks** when the two tests are evaluated in relation to the **single-factor within-subjects analysis of variance** and the *t* **test for two dependent samples** (i.e., the **Friedman two-way analysis of variance by ranks** and the **Page test for ordered alternatives** are equal with respect to their **asymptotic relative efficiency**). However, when the alternative hypothesis stipulates the order of the population medians, the **Page test for ordered alternatives** is more powerful than the **Friedman two-way analysis of variance by ranks** when evaluated in relation to the **single-factor within-subjects analysis of variance** and the *t* **test for two dependent samples**.

Equivalency of the Page test for ordered alternatives and a one-tailed analysis with both the Friedman two-way analysis of variance by ranks and the binomial sign test for two dependent samples when $k = 2$ When there are $k = 2$ conditions and a one-tailed alternative hypothesis is employed, the result obtained for the **Page test for ordered alternatives** will be equivalent to the result obtained with both the **binomial sign test for two dependent samples** and the **Friedman two-way analysis of variance by ranks**. (Recollect that earlier in this chapter it was noted that the latter two tests are equivalent to one another when $k = 2$.) It should be noted, however, that the equivalency of the **Page test** (and **Friedman test**) with the **binomial sign test** will only apply when there are no tied ranks in the data.

The data in Table 25.3, which summarizes Example 19.1 (which was employed to illustrate the **binomial sign test for two dependent samples**, as well as the equivalency of the latter test with the **Friedman two-way analysis of variance by ranks** when $k = 2$), can be employed to illustrate the equivalency with the **Page test for ordered alternatives**. Let us assume that the directional alternative hypothesis H_1: $\theta_1 \geq \theta_2$ is employed for Example 19.1. As was the case when the latter example was employed to illustrate the equivalency between the **binomial sign test for two dependent samples** and the **Friedman two-way analysis of variance by ranks**, the data for Subject 2 will not be included in the analysis (since the latter subject has the same score in both experimental conditions). Since the data for Subject 2 are not included in the analysis, the rank-orders for Subject 2 under the two conditions are subtracted from the values $\Sigma R_1 = 18.5$ and $\Sigma R_2 = 11.5$, yielding the revised values $\Sigma R_1 = 17$ and $\Sigma R_2 = 10$. In addition, our revised sample size value will be $n = 9$. Employing Equation 25.6, the value $L = 44$ is computed, since $L = (1)(10) + (2)(17) = 44$. Note that in Equation 25.6, ΣR_1 represents the sum of the ranks for Condition 2, since it is the condition predicted to have the smallest sum of ranks, and thus is the condition whose sum of ranks value is multiplied by 1. ΣR_2 represents the sum of the ranks for Condition 1, since it is the condition predicted to have the second smallest (and in this case largest) sum of ranks, and thus is the condition whose sum of ranks value is multiplied by 2. Since **Table A25** does not list critical values for $k = 2$, the normal approximation must be employed to evaluate the computed value $L = 44$. Substituting $L = 44$ in Equation 25.7 yields the value $z = 2.33$.

$$z = \frac{44 - \left[\dfrac{(9)(2)(2 + 1)^2}{4}\right]}{\sqrt{\dfrac{(9)[(2)^3 - 2]^2}{144(2 - 1)}}} = \frac{44 - 40.5}{1.5} = 2.33$$

The obtained value $z = 2.33$ is evaluated with **Table A1**. Since $z = 2.33$ is larger than the tabled critical one-tailed value $z_{.05} = 1.65$, and equal to the tabled critical one-tailed value $z_{.01} = 2.33$, the null hypothesis H_0: $\theta_1 = \theta_2$ can be rejected at both the .05 and .01 levels. Thus, when the **Page test for ordered alternatives** is employed with Example 19.1, the one-tailed/directional alternative hypothesis H_1: $\theta_1 \geq \theta_2$ is supported at both the .05 and .01 levels. The value $z = 2.33$ is identical to that obtained when the same set of data is evaluated with Equation 19.3 (the equation for the normal approximation of the **binomial sign test for two dependent samples**) and the **Friedman two-way analysis of variance by ranks** (in which case Equation 25.1 yielded the value $\chi^2 = 5.44$ (which as noted earlier is equal to the square of the z value computed with Equation 19.3 — i.e., $z^2 = (2.33)^2 = \chi^2 = 5.44$)).

Additional comments on the Page test for ordered alternatives a) The presence of ties in the data will result in an overly conservative test statistic for the **Page test for ordered alternatives**. In other words, both the computed value of L (through use of Equation 25.6), as well as the z value computed with Equation 25.7, will be slightly lower when ties are present in the data (thus making it more difficult to reject H_0). Sources, however, do not recommend the use of a tie correction for the **Page test for ordered alternatives**; b) The pairwise comparison procedures described earlier in this chapter for the **Friedman two-way analysis of variance by ranks** can also be employed for evaluating the **Page test for ordered alternatives**. One-tailed critical values should be used in Equation 25.5 in conducting pairwise comparisons for the **Page test for ordered alternatives** if a directional alternative hypothesis is employed for a comparison; c) Various sources (e.g., Conover (1999), Hollander and Wolfe (1999), Marascuilo and McSweeney (1977), and Siegel and Castellan (1988)) state that the L statistic computed for the **Page test for ordered alternatives** is related to **Spearman's rank-order correlation coefficient**. Conover (1999, p. 380) notes that the L statistic is a monotonic function of the latter correlation coefficient if there are no ties within the blocks/rows of data.[19] Hollander and Wolfe (1999, p. 292) provide an equation which specifies the exact nature of the relationship between the L statistic and **Spearman's rank-order correlation coefficient**; d) Additional discussion of the **Page test for ordered alternatives** can be found in Daniel (1990), Hollander and Wolfe (1999), Marascuilo and McSweeney (1977), Sheskin (1984), and Siegel and Castellan (1988). An excellent discussion of additional distribution-free tests which can be employed to evaluate a hypothesis about ordered alternatives when there are k dependent samples can be found in Daniel (1990).

References

Bell, C. B. & Doksum, K. A. (1965). Some new distribution-free statistics. **Annals of Mathematical Statistics**, 36, 203–214.

Church, J. D. & Wike, E. L. (1979). A Monte Carlo study of nonparametric multiple-comparison tests for a two-way layout. **Bulletin of the Psychonomic Society**, 14, 95–98.

Conover, W. J. (1980). **Practical nonparametric statistics** (2nd ed.). New York: John Wiley & Sons.

Conover, W. J. (1999). **Practical nonparametric statistics** (3rd ed.). New York: John Wiley & Sons.

Daniel, W. W. (1990). **Applied nonparametric statistics** (2nd ed.). Boston: PWS–Kent Publishing Company.

Friedman, M. (1937). The use of ranks to avoid the assumption of normality implicit in the analysis of variance. **Journal of the American Statistical Association**, 32, 675–701.

Hollander, M. & Wolfe, D. A. (1999). **Nonparametric statistical methods**. New York: John Wiley & Sons.

Iman, R. L. & Davenport, J. M. (1980). Approximations of the critical region of the Friedman statistic. **Communication in Statistics — Theory and Methods**, 9, 571–595.

Marascuilo, L. A. & McSweeney, M. (1977). **Nonparametric and distribution-free methods for the social sciences**. Monterey, CA: Brooks/Cole Publishing Company.

Marascuilo, L. A. & Serlin, R. C. (1988). **Statistical methods for the social and behavioral sciences**. New York: W. H. Freedman & Company.

Noether, G. E. (1967). **Elements of nonparametric statistics**. New York: John Wiley & Sons.

Page, E. B. (1963). Ordered hypotheses for multiple treatments: A significance test for linear ranks. **Journal of the American Statistical Association**, 58, 216–230.

Sheskin, D. J. (1984). **Statistical tests and experimental design: A guidebook**. New York: Gardner Press.

Siegel, S. & Castellan, N. J., Jr. (1988). **Nonparametric statistics for the behavioral sciences** (2nd ed.). New York: McGraw–Hill Book Company.

Sprent, P. & Smeeton, N. C. (2000). **Applied nonparametric statistical methods** (3rd ed.). London: Chapman & Hall.

Van der Waerden, B. L. (1952/1953). Order tests for the two-sample problem and their power. **Proceedings Koninklijke Nederlandse Akademie van Wetenshappen** (A), 55 (**Indagationes Mathematicae** 14), 453–458, & 56 (**Indagationes Mathematicae**, 15), 303–316 (corrections appear in Vol. 56, p. 80).

Wike, E. L. (1985). **Numbers: A primer of data**. Columbus, OH: Charles E. Merrill Publishing Company.

Zar, J. H. (1999). **Biostatistical analysis** (4th ed.). Upper Saddle River, NJ: Prentice Hall.

Endnotes

1. The reader should take note of the fact that when there are $k = 2$ dependent samples, the **Wilcoxon matched-pairs signed-ranks test** (which is also described in this book as a nonparametric test for evaluating ordinal data) will not yield a result equivalent to that obtained with the **Friedman two-way analysis of variance by ranks**. Since the **Wilcoxon test** (which rank-orders interval/ratio difference scores) employs more information than the **Friedman test/binomial sign test**, it provides a more powerful test of an alternative hypothesis than the latter tests.

2. A more detailed discussion of the guidelines noted below can be found in Sections I and VII of the *t* **test for two dependent samples**.

3. Although it is possible to conduct a directional analysis, such an analysis will not be described with respect to the **Friedman two-way analysis of variance by ranks**. A discussion of a directional analysis when $k = 2$ can be found under the **binomial sign test for two dependent samples**. A discussion of the evaluation of a directional alternative hypothesis when $k \geq 3$ can be found in Section VII of the **chi-square goodness-of-fit test (Test 8)**. Although the latter discussion is in reference to analysis of a k independent samples design involving categorical data, the general principles regarding analysis of a directional alternative hypothesis when $k \geq 3$ are applicable to the **Friedman two-way analysis of variance by ranks**.

4. Note that this ranking protocol differs from that employed for other rank-order procedures discussed in the book. In other rank-order tests, the rank assigned to each score is based on the rank-order of the score within the overall distribution of $nk = N$ scores.

5. As noted in Section IV, the chi-square distribution provides an approximation of the **Friedman test** statistic. Although the chi-square distribution provides an excellent approximation of the Friedman sampling distribution, some sources recommend the use of exact probabilities for small sample sizes. Exact tables of the Friedman distribution are discussed in Section VII.

6. In the discussion of comparisons in reference to the analysis of variance, it is noted that a **simple** (also known as a **pairwise**) **comparison** is a comparison between any two groups/ conditions in a set of k groups/conditions.

7. Equation 25.8 is an alternative form of the comparison equation, which identifies the minimum required difference between the **means of the ranks** of any two conditions in order for them to differ from one another at the prespecified level of significance. If the CD_F value computed with Equation 25.5 is divided by n, it yields the value $CD_{F_{(\bar{R}_a - \bar{R}_b)}}$ computed with the Equation 25.8.

    (Equation 25.8)

    $$CD_{F_{(\bar{R}_a - \bar{R}_b)}} = z_{adj}\sqrt{\frac{k(k + 1)}{6n}}$$

8. The method for deriving the value of z_{adj} for the **Friedman two-way analysis of variance by ranks** is based on the same logic that is employed in Equation 22.5 (which is used for conducting comparisons for the **Kruskal–Wallis one-way analysis of variance by ranks**). A rationale for the use of the proportions .0167 and .0083 in determining the appropriate value for z_{adj} in Example 25.1 can be found in Endnote 5 of the **Kruskal–Wallis one-way analysis of variance by ranks.**

9. It should be noted that when a directional alternative hypothesis is employed, the sign of the difference between the two sums of ranks must be consistent with the prediction stated in the directional alternative hypothesis. When a nondirectional alternative hypothesis is employed, the direction of the difference between two sums of ranks is irrelevant.

10. Unfortunately sources are not in agreement with respect to what equation is most appropriate to employ for a comparison. Among the alternative procedures recommended are the following: a) Hollander and Wolfe (1999, p. 296) and Zar (1999, p. 267) recommend Equation 25.9 (which employs the **Studentized range statistic** discussed in Section VI of the **single-factor between-subjects analysis of variance**) for unplanned comparisons with large samples. (Zar (1999, p. 267) states that Equation 25.5 should only be employed if a number of groups are compared one at a time with a control group — i.e., that Equation 25.5 is employed within the framework of the **Dunnet** methodology described in Section VI of the **single-factor between-subjects analysis of variance**.) The q value in Equation 25.9 is obtained from **Table A13 (Table of the Studentized Range Statistic)** in the **Appendix** (where k represents the number of experimental conditions and the value ∞ is employed for df_{error}). Use of Equation 25.9 with Example 25.1 (employing the $q_{.05}$ value (for 3, ∞)) yields the value $CD_F = 8.11$, which indicates that in order to declare a difference between two conditions significant (with the familywise error rate adjusted to .05) the difference between the sums of ranks of any two conditions must be equal to or greater than 8.11 (which, as is the case when Equation 25.5 is employed, is only true for $|\Sigma R_1 - \Sigma R_3| = 8.11$).

    (Equation 25.9)

    $$CD_F = q_{(k, \infty)}\sqrt{\frac{nk(k + 1)}{12}} = (3.31)\sqrt{\frac{(6)(3)(3 + 1)}{12}} = 8.11$$

    b) Conover (1999, pp. 370–373) and Sprent and Smeeton (1988, p. 222) recommend Equation 25.10 (which is based on Fisher's **least significant difference** methodology discussed in Section VI of the **single-factor between-subjects analysis of variance**) for unplanned comparisons if the value computed for χ_r^2 is statistically significant. In the latter equation: 1) $\Sigma R_{ij} = 83.5$ represents the sum of the squared ranks in Table 25.1 (i.e., each of the N ranks is squared and the N squared values are summed). Conover (1999, p. 371) notes that if there are no ties $\Sigma R_{ij} = [nk(k + 1)(2k + 1)]/6$; 2) ΣR_j^2 (which equals 83.5) indicates that the sum of the ranks for each of the k experimental conditions is squared, and that the squared sums of ranks are summed; and c) $t_{(n - 1)(k - 1)}$ indicates use of the tabled critical t value (for $df = (n - 1)(k - 1)$) in **Table A2 (Table of Student's t Distribution)** in the **Appendix**. Although Conover (1999, pp. 370–373) does not adjust the familywise error rate, use of Equation 25.10 with Example 25.1 will employ the value $t_{\alpha = .05 \text{ familywise}} = 3.5$, which is the approximate critical t value, for $df = 10$, that allows for a two-tailed test .0167 per comparison alpha value and a familywise alpha level of .05. The value $CD_F = 1.75$ is obtained, which indicates that in order to declare a difference between two conditions significant at the .05 level the difference between the sums of ranks of any two conditions must be equal to or greater than 1.75. Note that the value $CD_F = 1.75$, which is dramatically lower than all previously computed CD_F values, allows the researcher to declare all three pairwise comparisons significant.

(Equation 25.10)

$$CD_F = t_{(n - 1)(k - 1)} \sqrt{\frac{2(n\Sigma R_{ij}^2 - \Sigma R_j^2)}{(n - 1)(k - 1)}} = (3.5)\sqrt{\frac{2[(6)(83.5) - 498.5]}{(6 - 1)(3 - 1)}} = 1.75$$

c) Marascuilo and McSweeney (1977, pp. 362–366) describe an unplanned comparison procedure based on the **Scheffé** methodology described in Chapter 21, as well as a procedure for planned comparisons. Marascuilo and Serlin (1988, 581–583) also describe a procedure for planned comparisons along with the procedure for unplanned comparisons described by Equation 25.9.

11.   a) In the case of both the **Wilcoxon matched-pairs signed-ranks test** and the **binomial sign test for two dependent samples**, it is assumed that for each pairwise comparison a subject's score in the second condition which is listed for a comparison is subtracted from the subject's score in the first condition that is listed for the comparison. In the case of both tests, reversing the order of subtraction will yield the same result; b) Marascuilo and McSweeney (1977, p. 369) state that the use of the **Wilcoxon matched-pairs signed-ranks test** is acceptable for comparisons as long as it is restricted to planned pairwise comparisons.

12.   The value $n = 6$ is employed for the Condition 1 versus Condition 2 and Condition 1 versus Condition 3 comparisons, since no subject has the same score in both experimental conditions. On the other hand, the value $n = 5$ is employed in the Condition 2 versus Condition 3 comparison, since Subject 6 has the same score in Conditions 2 and 3. The use of $n = 5$ is predicated on the fact that in conducting the **Wilcoxon matched-pairs signed-ranks test**, subjects who have a difference score of zero are not included in the computation of the test statistic.

13.   In Equation 25.5 the value $z_{.05} = 1.96$ is employed for z_{adj}, and the latter value is multiplied by 3.46, which is the value computed for the term in the radical of the equation for Example 25.1.

14. Iman and Davenport (1980) argued that Equation 25.1 allows for too conservative a test of the alternative hypothesis (i.e., inflates the likelihood of committing a Type II error), and that Equation 25.11, which employs the F distribution, allows for a more powerful test of the **Friedman two-way analysis of variance by ranks** test statistic. The F value obtained with Equation 25.11 is evaluated with **Table A10 (Table of the F Distribution)** in the **Appendix**. The degrees of freedom employed for the analysis are $df_{num} = k - 1$ and $df_{den} = (k-1)(n-1)$. In the case of Example 25.1, $df_{num} = 3 - 1 = 2$ and $df_{den} = (3-1)(6-1) = 10$, and the tabled $F_{.95}$ and $F_{.99}$ values for $df_{num} = 2$ and $df_{den} = 10$ are $F_{.95} = 4.10$ and $F_{.99} = 7.56$. In order to reject the null hypothesis, the obtained F value must be equal to or greater than the tabled critical value at the prespecified level of significance.

    When the appropriate values for Example 25.1 are substituted in Equation 25.11 the value $F = 6.02$ is obtained. Since the latter value is greater than $F_{.05} = 4.10$ but less than $F_{.01} = 7.56$, the alternative hypothesis is supported, but at only the .05 level. The latter analysis is, in fact, more conservative than the original analysis (when the value $X_r^2 = 11.08$ was evaluated) when the alternative hypothesis was supported at both the .05 and .01 levels.

$$F = \frac{(n - 1)\chi_r^2}{n(k - 1) - \chi_r^2} = \frac{(6 - 1)(11.08)}{(6)(3 - 1) - 11.08} = 6.02 \quad \textbf{(Equation 25.11)}$$

15. It is also the case that the exact binomial probability for the **binomial sign test for two dependent samples** will correspond to the exact probability for the **Friedman test statistic**.

16. If Subject 2 is included in the analysis, Equation 25.1 yields the value $\chi_r^2 = 4.9$, which is also significant at the .05 level.

17. In Section I it is noted that in employing the **Friedman test** it is assumed that the variable which is ranked is a continuous random variable. Thus, it would be assumed that the racing form of a horse was at some point either explicitly or implicitly expressed as a continuous interval/ratio variable.

18. It should be noted that it is possible to obtain a negative z value for Equation 25.7 if the data are consistent with the analogous hypothesis in the other tail of the sampling distribution. Put simply, if the results are the exact opposite of what is predicted, the value of z will be negative. Specifically, if the data were consistent with the alternative hypothesis H_1: $\theta_1 \leq \theta_2 \leq \theta_3$, the value computed for L would be less than the expected value of L computed in the numerator of Equation 25.7 (which in our example was $L_E = 72$). When the latter is true, the value computed for z will be negative. In instances where a researcher is employing a two-tailed alternative hypothesis, the absolute value of z is employed, since the sign of z has no bearing on whether or not a result is significant. In the case of a two-tailed alternative hypothesis, tabled critical two-tailed values are employed.

    To illustrate the computation of a negative z value, if the analogous alternative hypothesis in the other tail of the sampling distribution is employed (i.e., H_1: $\theta_1 \leq \theta_2 \leq \theta_3$), Equation 25.6 will yield the value $L = 60.5$ for Example 25.1, since $L = (1)(18) + (2)(11.5) + (3)(6.5) = 60.5$. When the value $L = 60.5$ is substituted in Equation 25.7, the value $z = -3.32$ is computed, which is identical to the absolute value computed when the alternative hypothesis $H_1: \theta_1 \geq \theta_2 \geq \theta_3$ is employed. Thus, the sign of z merely indicates the direction of a one-tailed analysis. (**Table A25**, however, cannot be employed to evaluate the value $L = 60.5$ within the framework of $H_1: \theta_1 \leq \theta_2 \leq \theta_3$, since the latter alternative hypothesis is not consistent with the configuration of Equation 25.6.)

19. A **monotonic** relationship is defined in Section I of **Spearman's rank-order correlation coefficient**.

Test 26

The Cochran Q Test
(Nonparametric Test Employed with Categorical/Nominal Data)

I. Hypothesis Evaluated with Test and Relevant Background Information

Hypothesis evaluated with test In a set of k dependent samples (where $k \geq 2$), do at least two of the samples represent different populations?

Relevant background information on test It is recommended that before reading the material on the **Cochran Q test**, the reader review the general information on a dependent samples design contained in Sections I and VII of the *t* **test for two dependent samples (Test 17)**. The **Cochran Q test** (Cochran (1950)) is a nonparametric procedure for categorical data employed in a hypothesis testing situation involving a design with $k = 2$ or more dependent samples. The test is employed to evaluate an experiment in which a sample of n subjects (or n sets of matched subjects) is evaluated on a dichotomous dependent variable (i.e., scores on the dependent variable must fall within one of two mutually exclusive categories). The test assumes that each of the n subjects (or n sets of matched subjects) contributes k scores on the dependent variable. The **Cochran Q test** is an extension of the **McNemar test (Test 20)** to a design involving more than two dependent samples, and when $k = 2$ the **Cochran Q test** will yield a result which is equivalent to that obtained with the **McNemar test**. If the result of the **Cochran Q test** is significant, it indicates there is a high likelihood at least two of the k experimental conditions represent different populations.

The **Cochran Q test** is based on the following assumptions: a) The sample of n subjects has been randomly selected from the population it represents; and b) The scores of subjects are in the form of a dichotomous categorical measure involving two mutually exclusive categories.

Although the chi-square distribution is generally employed to evaluate the **Cochran test** statistic, in actuality the latter distribution is used to provide an approximation of the exact sampling distribution. Sources on nonparametric analysis (e.g., Daniel (1990), Marascuilo and McSweeney (1977), and Siegel and Castellan (1988)) recommend that for small sample sizes exact tables of the Q distribution derived by Patil (1975) be employed. Use of exact tables is generally recommended when $n < 4$ and/or $nk < 24$.

As is the case for other tests which are employed to evaluate data involving two or more dependent samples, in order for the **Cochran Q test** to generate valid results the following guidelines should be adhered to:[1] a) To control for order effects, the presentation of the k experimental conditions should be random or, if appropriate, be counterbalanced; and b) If matched samples are employed, within each set of matched subjects each of the subjects should be randomly assigned to one of the k experimental conditions.

As is noted with respect to other tests which are employed to evaluate a design involving two or more dependent samples, the **Cochran Q test** can also be used to evaluate a **one-group pretest-posttest design**, as well as extensions of the latter design that involve more than two measurement periods. The limitations of the **one-group pretest-posttest design** (which are discussed in the **Introduction** and in Section VII of the *t* **test for two dependent samples**) are also applicable when it is evaluated with the **Cochran Q test**.

II. Example

Example 26.1 *A market researcher asks 12 female subjects whether or not they would purchase an automobile manufactured by three different companies. Specifically, subjects are asked whether they would purchase a car manufactured by the following automobile manufacturers: Chenesco, Howasaki, and Gemini (which respectively correspond to* **Condition 1**, **Condition 2**, *and* **Condition 3***). The responses of the 12 subjects follow:* **Subject 1** *said she would purchase a Chenesco and a Howasaki but not a Gemini;* **Subject 2** *said she would only purchase a Howasaki;* **Subject 3** *said she would purchase all three makes of cars;* **Subject 4** *said she would only purchase a Howasaki;* **Subject 5** *said she would only purchase a Howasaki;* **Subject 6** *said she would purchase a Howasaki and a Gemini but not a Chenesco;* **Subject 7** *said she would not purchase any of the automobiles;* **Subject 8** *said she would only purchase a Howasaki;* **Subject 9** *said she would purchase a Chenesco and a Howasaki but not a Gemini;* **Subject 10** *said she would only purchase a Howasaki;* **Subject 11** *said she would not purchase any of the automobiles; and* **Subject 12** *said she would only purchase a Gemini. Can the market researcher conclude that there are differences with respect to car preference based on the responses of subjects?*

III. Null versus Alternative Hypotheses

In stating the null and alternative hypotheses the notation π_j will be employed to represent the proportion of **Yes** responses in the population represented by the jth experimental condition. Stated more generally, π_j represents the proportion of responses in one of the two response categories in the population represented by the jth experimental condition.

Null hypothesis $H_0: \pi_1 = \pi_2 = \pi_3$

(The proportion of **Yes** responses in the population represented by Condition 1 equals the proportion of **Yes** responses in the population represented by Condition 2 equals the proportion of **Yes** responses in the population represented by Condition 3.)

Alternative hypothesis H_1: Not H_0

(This indicates that in at least two of the underlying populations represented by the $k = 3$ conditions, the proportion of **Yes** responses is not equal. It is important to note that the alternative hypothesis should not be written as follows: $H_1: \pi_1 \neq \pi_2 \neq \pi_3$. The reason why the latter notation for the alternative hypothesis is incorrect is because it implies that all three population proportions must differ from one another in order to reject the null hypothesis. In this book it will be assumed (unless stated otherwise) that the alternative hypothesis for the **Cochran Q test** is stated **nondirectionally**.)[2]

IV. Test Computations

The data for Example 26.1 are summarized in Table 26.1. The number of subjects employed in the experiment is $n = 12$, and thus within each condition there are $n = n_1 = n_2 = n_3 = 12$ scores. The values **1** and **0** are employed to represent the two response categories in which a subject's response/categorization may fall. Specifically, a score of **1** indicates a **Yes** response and a score of **0** indicates a **No** response.

The following summary values are computed in Table 26.1 which will be employed in the analysis of the data:

a) The value ΣC_j represents the number of **Yes** responses in the jth condition. Thus, the number of **Yes** responses in Conditions 1, 2, and 3 are, respectively, $\Sigma C_1 = 3$, $\Sigma C_2 = 9$, and $\Sigma C_3 = 3$.

b) The value $(\Sigma C_j)^2$ represents the square of the ΣC_j value computed for the jth condition. The sum of the $k = 3$ $(\Sigma C_j)^2$ scores can be represented by the notation $\Sigma(\Sigma C_j)^2$. Thus, for Example 26.1, $\Sigma(\Sigma C_j)^2 = (\Sigma C_1)^2 + (\Sigma C_2)^2 + (\Sigma C_3)^2 = 9 + 81 + 9 = 99$.

c) The value R_i represents the sum of the $k = 3$ scores of the ith subject (i.e., the number of **Yes** responses for the ith subject). Note that an R_i value is computed for each of the $n = 12$ subjects. The sum of the n R_i scores is ΣR_i. Thus, for Example 26.1, $\Sigma R_i = 2 + 1 + \cdots + 0 + 1 = 15$. Note that the value of ΣR_i will always equal the value $\Sigma(\Sigma C_j)$. In the case of Example 26.1, both of the aforementioned values equal 15 (since $\Sigma R_i = \Sigma(\Sigma C_j) = (\Sigma C_1 = 3) + (\Sigma C_2 = 9) + (\Sigma C_3 = 3) = 15$).

d) The value R_i^2 represents the square of the R_i score of the ith subject. The sum of the n R_i^2 scores is ΣR_i^2. Thus, for Example 26.1, $\Sigma R_i^2 = 4 + 1 + \cdots + 0 + 1 = 27$.

e) The value p_j represents the proportion of **Yes** responses in the jth condition. The value of p_j is computed as follows: $p_j = \Sigma C_j / n_j$. Thus, in Table 26.1 the values of p_j for Conditions 1, 2, and 3 are, respectively, $p_1 = .25$, $p_2 = .75$, and $p_3 = .25$.

Table 26.1 Data for Example 26.1

	Chenesco C_1	Howasaki C_2	Gemini C_3	R_i	R_i^2
Subject 1	1	1	0	2	4
Subject 2	0	1	0	1	1
Subject 3	1	1	1	3	9
Subject 4	0	1	0	1	1
Subject 5	0	1	0	1	1
Subject 6	0	1	1	2	4
Subject 7	0	0	0	0	0
Subject 8	0	1	0	1	1
Subject 9	1	1	0	2	4
Subject 10	0	1	0	1	1
Subject 11	0	0	0	0	0
Subject 12	0	0	1	1	1
	$\Sigma C_1 = 3$	$\Sigma C_2 = 9$	$\Sigma C_3 = 3$	$\Sigma R_i = 15$	$\Sigma R_i^2 = 27$
	$(\Sigma C_1)^2 = (3)^2 = 9$	$(\Sigma C_2)^2 = (9)^2 = 81$	$(\Sigma C_3)^2 = (3)^2 = 9$		
	$p_1 = \dfrac{\Sigma C_1}{n_1} = \dfrac{3}{12} = .25$	$p_2 = \dfrac{\Sigma C_2}{n_2} = \dfrac{9}{12} = .75$	$p_3 = \dfrac{\Sigma C_3}{n_3} = \dfrac{3}{12} = .25$		

Equation 26.1 is employed to calculate the test statistic for the **Cochran Q test**. The Q value computed with Equation 26.1 is interpreted as a chi-square value. In Equation 26.1 the following notation is employed with respect to the summary values noted in this section: a) C is employed to represent the value computed for $\Sigma(\Sigma C_j)^2$; b) T is employed to represent the value computed for ΣR_i (which as noted above is equal to $\Sigma(\Sigma C_j)$); and c) R is employed to represent the value computed for ΣR_i^2. Thus, for Example 26.1, $C = 99$, $T = 15$, and $R = 27$.[3]

$$Q = \frac{(k-1)[(k)(C) - (T)^2]}{(k)(T) - R}$$ (Equation 26.1)

Substituting the appropriate values from Example 26.1 in Equation 26.1, the value $Q = 8$ is computed.[4]

$$Q = \frac{(3-1)[(3)(99) - (15)^2]}{(3)(15) - 27} = 8$$

V. Interpretation of the Test Results

In order to reject the null hypothesis, the computed value $Q = \chi^2$ must be equal to or greater than the tabled critical chi-square value at the prespecified level of significance. The computed chi-square value is evaluated with **Table A4 (Table of the Chi-Square Distribution)** in the **Appendix**. For the appropriate degrees of freedom, the tabled $\chi^2_{.95}$ value (which is the chi-square value at the 95th percentile) and the tabled $\chi^2_{.99}$ value (which is the chi-square value at the 99th percentile) are employed as the .05 and .01 critical values for evaluating a nondirectional alternative hypothesis. The number of degrees of freedom employed in the analysis are computed with Equation 26.2. Thus, $df = 3 - 1 = 2$.

$$df = k - 1$$ (Equation 26.2)

For $df = 2$, the tabled critical .05 and .01 chi-square values are $\chi^2_{.05} = 5.99$ and $\chi^2_{.01} = 9.21$. Since the computed value $Q = 8$ is greater than $\chi^2_{.05} = 5.99$, the alternative hypothesis is supported at the .05 level. Since, however, $Q = 8$ is less than $\chi^2_{.01} = 9.21$, the alternative hypothesis is not supported at the .01 level. A summary of the analysis of Example 26.1 with the **Cochran Q test** follows: It can be concluded there is a significant difference in subjects' preferences for at least two of the three automobiles. This result can be summarized as follows: $Q(2) = 8, p < .05$.

VI. Additional Analytical Procedures for the Cochran Q Test and/or Related Tests

1. Pairwise comparisons following computation of the test statistic for the Cochran Q test Prior to reading this section the reader should review the discussion of comparisons in Section VI of the **single-factor between-subjects analysis of variance (Test 21)**. As is the case with the omnibus F value computed for an analysis of variance, the Q value computed with Equation 26.1 is based on an evaluation of all k experimental conditions. When the value of Q is significant, it does not indicate whether just two or, in fact, more than two conditions differ significantly from one another. In order to answer the latter question, it is necessary to conduct comparisons contrasting specific conditions with one another. This section will describe methodologies which can be employed for conducting **unplanned simple/pairwise comparisons** following the computation of a significant Q value.[5]

In conducting a simple comparison, the null hypothesis and nondirectional alternative hypothesis are as follows: $H_0: \pi_a = \pi_b$ versus $H_1: \pi_a \neq \pi_b$. In the aforementioned hypotheses, π_a and π_b represent the proportion of **Yes** responses in the populations represented by the two conditions involved in the comparison. The alternative hypothesis can also be stated directionally as follows: $H_1: \pi_a > \pi_b$ or $H_1: \pi_a < \pi_b$.

A number of sources (e.g., Fleiss (1981) and Marascuilo and McSweeney (1977)) describe comparison procedures for the **Cochran Q test**. The procedure to be described in this section, which is recommended for large sample sizes, is one of two procedures described in Marascuilo and McSweeney (1977). It is essentially the application of the **Bonferroni–Dunn method** described in Section VI of the **single-factor between-subjects analysis of variance** to the **Cochran Q test** model. Through use of Equation 26.3, the procedure allows a researcher to identify the minimum required difference between the observed proportion of **Yes** responses for any two experimental conditions (designated as CD_C) in order for them to differ from one another at the prespecified level of significance.

$$CD_C = z_{adj} \sqrt{2 \left[\frac{(k)(T) - R}{(n^2)(k)(k - 1)} \right]}$$ **(Equation 26.3)**

The value of z_{adj} is obtained from **Table A1 (Table of the Normal Distribution)** in the **Appendix**. In the case of a nondirectional alternative hypothesis, z_{adj} is the z value above which a proportion of cases corresponding to the value $\alpha_{FW}/2c$ falls (where c is the total number of comparisons that are conducted). In the case of a directional alternative hypothesis, z_{adj} is the z value above which a proportion of cases corresponding to the value α_{FW}/c falls. When all possible pairwise comparisons are made $c = [k(k - 1)]/2$, and thus, $2c = k(k - 1)$. In Example 26.1 the number of pairwise/simple comparisons that can be conducted is $c = [3(3 - 1)]/2 = 3$ — specifically, Condition 1 versus Condition 2, Condition 1 versus Condition 3, and Condition 2 versus Condition 3.

The value of z_{adj} will be a function of both the maximum **familywise Type I error rate** (α_{FW}) the researcher is willing to tolerate and the total number of comparisons that are conducted. When a limited number of comparisons are planned prior to collecting the data, most sources take the position that a researcher is not obliged to control the value of α_{FW}. In such a case, the **per comparison Type I error rate** (α_{PC}) will be equal to the prespecified value of alpha. When α_{FW} is not adjusted, the value of z_{adj} employed in Equation 26.3 will be the tabled critical z value which corresponds to the prespecified level of significance. Thus, if a nondirectional alternative hypothesis is employed and $\alpha = \alpha_{PC} = .05$, the tabled critical two-tailed .05 value $z_{.05} = 1.96$ is used to represent z_{adj} in Equation 26.3. If $\alpha = \alpha_{PC} = .01$, the tabled critical two-tailed .01 value $z_{.01} = 2.58$ is used in Equation 26.3. In the same respect, if a directional alternative hypothesis is employed, the tabled critical .05 and .01 one-tailed values $z_{.05} = 1.65$ and $z_{.01} = 2.33$ are used for z_{adj} in Equation 26.3.

When comparisons are not planned beforehand, it is generally acknowledged that the value of α_{FW} must be controlled so as not to become excessive. The general approach for controlling the latter value is to establish a **per comparison Type I error rate** which insures that α_{FW} will not exceed some maximum value stipulated by the researcher. One method for doing this (described under the **single-factor between-subjects analysis of variance** as the **Bonferroni–Dunn method**) establishes the **per comparison Type I error rate** by dividing the maximum value one will tolerate for the **familywise Type I error rate** by the total number of comparisons conducted. Thus, in Example 26.1, if one intends to conduct all three pairwise comparisons and wants to insure that α_{FW} does not exceed .05, $\alpha_{PC} = \alpha_{FW}/c = .05/3 = .0167$. The latter proportion is used in determining the value of z_{adj}. As noted earlier, if a directional alternative hypothesis is employed for a comparison, the value of z_{adj} employed in Equation 26.3 is the

z value above which a proportion equal to $\alpha_{PC} = \alpha_{FW}/c$ of the cases falls. In **Table A1**, the z value that corresponds to the proportion .0167 is $z = 2.13$. By employing z_{adj} in Equation 26.3, one can be assured that within the "family" of three pairwise comparisons, α_{FW} will not exceed .05 (assuming all of the comparisons are directional). If a nondirectional alternative hypothesis is employed for all of the comparisons, the value of z_{adj} will be the z value above which a proportion equal to $\alpha_{FW}/2c = \alpha_{PC}/2$ of the cases falls. Since $\alpha_{PC}/2 = .0167/2 = .0083$, $z = 2.39$. By employing z_{adj} in Equation 26.3, one can be assured that α_{FW} will not exceed .05.[6]

In order to employ the CD_C value computed with Equation 26.3, it is necessary to determine the absolute value of the difference between the proportion of **Yes** responses for each pair of experimental conditions which are compared.[7] Table 26.2 summarizes the difference scores between pairs of proportions.

Table 26.2   Difference Scores between Pairs of Proportions for Example 26.1

$\lvert p_1 - p_2 \rvert = \lvert .25 - .75 \rvert = .50$ $\lvert p_1 - p_3 \rvert = \lvert .25 - .25 \rvert = 0$ $\lvert p_2 - p_3 \rvert = \lvert .75 - .25 \rvert = .50$	

If any of the differences between two proportions is equal to or greater than the CD_C value computed with Equation 26.3, a comparison is declared significant. Equation 26.3 will now be employed to evaluate the nondirectional alternative hypothesis H_1: $\pi_a \neq \pi_b$ for all three pairwise comparisons. Since it will be assumed that the comparisons are unplanned and that the researcher does not want the value of α_{FW} to exceed .05, the value $z_{adj} = 2.39$ will be used in computing CD_C.

$$CD_C = (2.39)\sqrt{2\left[\frac{(3)(15) - 27}{(12)^2(3)(3 - 1)}\right]} = (2.39)(.204) = .49$$

The obtained value $CD_C = .49$ indicates that any difference between a pair of proportions which is equal to or greater than .49 is significant. With respect to the three pairwise comparisons, the difference between Condition 1 and Condition 2 (which equals .50) and the difference between Condition 2 and Condition 3 (which also equals .50) are significant, since they are both greater than $CD_C = .49$. We cannot conclude that the difference between Condition 1 and Condition 3 is significant, since $\lvert p_1 - p_3 \rvert = 0$ is less than $CD_C = .49$.

An alternative strategy which can be employed for conducting pairwise comparisons for the **Cochran Q test** is to use the **McNemar test** for each comparison. In employing the **McNemar test** one can employ either the chi-square or normal approximation of the test statistic for each comparison (the continuity-corrected value generally providing a more accurate estimate for a small sample size), or compute the exact binomial probability for the comparison. (It will be demonstrated in Section VII that the computed chi-square value computed with Equation 20.1 for the **McNemar test** yields the same $Q = \chi^2$ value which is obtained if a **Cochran Q test** (Equation 26.1) is employed to compare the same set of experimental conditions.) In this section the exact binomial probabilities for the three pairwise comparisons for the **McNemar test** model will be computed. In order to compute the exact binomial probabilities (or, for that matter, the chi-square or normal approximations of the test statistic), the data for each comparison must be placed within a 2×2 table like Table 20.1 (which is the table for the **McNemar test** model). To illustrate this, the data for the Condition 1 versus Condition 2 comparison are recorded in Table 26.3. Note that of the $n = 12$ subjects involved in the comparison, only 6 of the subjects' scores are actually taken into account in computing the test statistic, since the other 6 subjects have the same score in both conditions.

Table 26.3 McNemar Test Model for Binomial Analysis of Condition 1 versus Condition 2 Comparison

		Condition 1		Row sums
		Yes (1)	No (0)	
Condition 2	Yes (1)	$a = 3$	$b = 6$	9
	No (0)	$c = 0$	$d = 3$	3
	Column sums	3	9	12

Employing **Table A6 (Table of the Binomial Distribution, Individual Probabilities)** in the **Appendix** to compute the binomial probability, we determine that when $n = 6$, $P(x = 6)$ = .0156. Thus, the two-tailed binomial probability for the Condition 1 versus Condition 2 comparison is $(2)(.0156) = .0312$. In the case of the Condition 1 versus Condition 3 comparison, the frequencies for Cells a, b, c, and d are, respectively, 1, 2, 2, and 7. Since the frequency of both Cells b and c is 2 (and thus, $n = 4$), the Condition 1 versus Condition 3 comparison results in no difference. In the case of the Condition 2 versus Condition 3 comparison, the frequencies for Cells a, b, c, and d are, respectively, 2, 1, 7, and 2. For the latter comparison, since $n = 8$, the frequencies for Cells b and c are 1 and 7. Using **Table A6** (or **Table A7** which is the **Table of the Binomial Distribution, Cumulative Probabilities**), we determine that when $n = 8$, $P(x \geq 7) = .0352$. Thus, the two-tailed binomial probability for the Condition 2 versus Condition 3 comparison is $(2)(.0352) = .0704$. Note that the binomial probabilities computed for the Condition 1 versus Condition 2 and Condition 2 versus Condition 3 comparisons are not identical, since by virtue of eliminating subjects who respond in the same category in both conditions from the analysis, the two comparisons employ different values for n (which is the sum of the frequencies for Cells b and c).

As before, if we wish to evaluate a nondirectional alternative hypothesis and insure that α_{FW} does not exceed .05, the value of α_{PC} is set equal to .0167. Thus, in order to reject the null hypothesis the computed two-tailed binomial probability for a comparison must be equal to or less than .0167. Since the computed two-tailed probabilities .0312 (for the Condition 1 versus Condition 2 comparison) and .0704 (for the Condition 2 versus Condition 3 comparison) are greater than .0167, none of the pairwise comparisons is significant.

In the event the researcher elects not to control the value of α_{FW} and employs $\alpha_{PC} = .05$ in evaluating the three pairwise comparisons (once again assuming a nondirectional analysis), only the Condition 1 versus Condition 2 comparison is significant, since the computed two-tailed binomial probability .0312 is less than .05. The Condition 2 versus Condition 3 comparison falls short of significance, since the computed two-tailed binomial probability .0704 is greater than .05. It should be noted that both of the aforementioned comparisons are significant if the directional alternative hypothesis which is consistent with the data is employed (since the one-tailed probabilities .0156 and .0352 are less than .05). If Equation 26.3 is employed for the same set of comparisons, $CD_C = (1.96)(.204) = .40$.[8] Thus, employing the latter equation, the Condition 1 versus Condition 2 and Condition 2 versus Condition 3 comparisons are significant, since in both instances the difference between the two proportions is greater than $CD_C = .40$.

Although the binomial probabilities for the **McNemar test** for the Condition 1 versus Condition 2 and Condition 2 versus Condition 3 comparisons are larger than the probabilities associated with the use of Equation 26.3, both comparison procedures yield relatively low probability values for the two aforementioned comparisons. Thus, in the case of Example 26.1, depending upon which comparison procedure one employs (as well as the value of α_{PC} and whether one evaluates a nondirectional or directional alternative hypothesis) it would appear that there is a high likelihood the Condition 1 versus Condition 2 and Condition 2 versus Condition 3 comparisons are significant. The intent of presenting two different comparison procedures in this section is to illustrate that, generally speaking, the results obtained with different procedures will be reasonably consistent with one another.[9] As is noted in the discussion of comparisons in Section VI of the **single-factor between-subjects analysis of variance**, in instances where two or more comparison procedures yield inconsistent results, the most effective way to clarify the status of the null hypothesis is to replicate a study one or more times. It is also noted throughout the book that, in the final analysis, the decision regarding which of the available comparison procedures to employ is usually not the most important issue facing the researcher conducting comparisons. The main issue is what maximum value one is willing to tolerate for α_{FW}.

Marascuilo and McSweeney (1977) discuss the computation of a confidence interval for a comparison for the **Cochran Q test** model. One approach for computing a confidence interval is to add and subtract the computed value of CD_C to and from the obtained difference between the proportions involved in the comparison. The latter approach is based on the same logic employed for computing a confidence interval for a comparison in Section VI of the **single-factor between-subjects analysis of variance**.

VII. Additional Discussion of the Cochran Q Test

1. Issues relating to subjects who obtain the same score under all of the experimental conditions

a) Cochran (1950) noted that since the value computed for Q is not affected by the scores of any subject (or any row, if matched subjects are employed) who obtains either all 0s or all 1s in each of the experimental conditions, the scores of such subjects can be deleted from the data analysis. If the latter is done with respect to Example 26.1, the data for Subjects 3, 7, and 11 can be eliminated from the analysis (since Subjects 7 and 11 obtain all 0s, and Subject 3 obtains all 1s). It is demonstrated below that if the scores of Subjects 3, 7, and 11 are eliminated from the analysis, the value $Q = 8$ is still obtained when the revised summary values are substituted in Equation 26.1.

$$\Sigma C_1 = 2 \quad \Sigma C_2 = 8 \quad \Sigma C_3 = 2 \quad (\Sigma C_1)^2 = 4 \quad (\Sigma C_2)^2 = 64 \quad (\Sigma C_3)^2 = 4$$

$$\Sigma(\Sigma C_j)^2 = 4 + 64 + 4 = 72 \quad \Sigma R_i = 12 \quad \Sigma R_i^2 = 18$$

$$\text{Thus:} \quad C = 72 \quad T = 12 \quad R = 18$$

$$Q = \frac{(3 - 1)[(3)(72) - (12)^2]}{(3)(12) - 18} = 8$$

It is noted in Section VI of the **McNemar test** that the latter test essentially eliminates from the analysis any subject who obtains the same score under both experimental conditions, and that this represents a limitation of the test. What was said with regard to the **McNemar test** in this respect also applies to the **Cochran Q test**. Thus, it is entirely possible to obtain a significant Q value even if the overwhelming majority of the subjects in a sample obtain the same score in each of the experimental conditions. To illustrate this, the value $Q = 8$ (obtained for Example 26.1) can be obtained for a sample of 1009 subjects, if 1000 of the subjects obtained a score of 1 in all three experimental conditions, and the remaining nine subjects had the same scores as Subjects 1, 2, 4, 5, 6, 8, 9, 10, and 12 in Example 26.1. Since the computation of the Q value in such an instance will be based on a sample size of 9 rather than on the actual sample size of 1009, it is reasonable to assume that such a result, although statistically significant, will not be of any practical significance from the perspective of the three automobile manufacturers. The latter statement is based on the fact that since all but 9 of the 1009 subjects said they would buy all three automobiles, there really does not appear to be any difference in preference which will be of any economic consequence to the manufacturers.

b) In Section I it is noted that when $n < 4$ and/or $nk < 24$, it is recommended that tables for the exact **Cochran test** statistic (derived by Patil (1975)) be employed instead of the chi-square approximation. In making such a determination, the value of n which should be used should not include any subjects who obtain all 0s or all 1s in each of the experimental conditions. Thus, in Example 26.1, the value $n = 9$ is employed, and not the value $n = 12$. Consequently $nk = (9)(3)$ $= 27$. Since $n > 4$ and $nk > 24$, it is acceptable to employ the chi-square approximation.

c) Note that Equation 26.3 (the equation employed for conducting comparisons) employs the value of n for the total number of subjects, irrespective of whether a subject obtains the same score in all k experimental conditions. The use of the latter n value maximizes the power of the comparison procedure. Certainly one could make an argument for employing as n in Equation 26.3 the number of subjects who have at least two different scores in the k experimental conditions. In most instances, the latter n value will be less than the total number of subjects employed in an experiment, and, subsequently, if the smaller n value is employed in Equation 26.3, the comparison procedure will be more conservative (since it will result in a higher value for CD_C).

2. Equivalency of the Cochran Q test and the McNemar test when $k = 2$

In Section I it is noted that when $k = 2$ the **Cochran Q test** yields a result which is equivalent to that obtained with the **McNemar test**. To be more specific, the **Cochran Q test** will yield a result which is equivalent to the **McNemar test** statistic when the correction for continuity is not employed for the latter test (i.e., the result obtained with Equation 20.1). In order to demonstrate the equivalency of the two tests, Example 26.2 will be evaluated with both tests.

Example 26.2 *A market researcher asks 10 female subjects whether or not they would purchase an automobile manufactured by two different companies. Specifically, subjects are asked whether they would purchase an automobile manufactured by Chenesco and Howasaki. Except for Subjects 2 and 3, all of the subjects said they would purchase a Chenesco but would not purchase a Howasaki. Subject 2 said she would not purchase either car, while Subject 3 said she would purchase a Howasaki but not a Chenesco. Based on the responses of subjects, can the market researcher conclude that there are differences with respect to car preference?*

Tables 26.4 and 26.5 respectively summarize the data for the study within the framework of the **Cochran Q test** model and the **McNemar test** model. Example 26.2 is evaluated below employing both the **Cochran Q test** and the **McNemar test**. Note the computed value $Q = \chi^2 = 5.44$ is equivalent to $\chi^2 = 5.44$ computed for the **McNemar test**.

Table 26.4　Summary of Data for Analysis of Example 26.2 with Cochran Q Test

	Chenesco C_1	Howasaki C_2	R_i	R_i^2
Subject 1	1	0	1	1
Subject 2	0	0	0	0
Subject 3	0	1	1	1
Subject 4	1	0	1	1
Subject 5	1	0	1	1
Subject 6	1	0	1	1
Subject 7	1	0	1	1
Subject 8	1	0	1	1
Subject 9	1	0	1	1
Subject 10	1	0	1	1
	$\Sigma C_1 = 8$	$\Sigma C_2 = 1$	$\Sigma C_3 = 9$	$\Sigma R_i = 9$

Table 26.5　Summary of Data for Analysis of Example 26.2 with McNemar Test

		Condition 1 (Chenesco)		Row sums
		No (0)	Yes (1)	
Condition 2	No (0)	$a = 1$	$b = 8$	9
(Howasaki)	Yes (1)	$c = 1$	$d = 0$	1
	Column sums	2	8	10

Cochran Q test:

$$\Sigma C_1 = 8 \quad \Sigma C_2 = 1 \quad (\Sigma C_1)^2 = 64 \quad (\Sigma C_2)^2 = 1$$

$$\Sigma(\Sigma C_j)^2 = 64 + 1 = 65 \quad \Sigma R_i = 9 \quad \Sigma R_i^2 = 9$$

Thus:　$C = 65 \quad T = 9 \quad R = 9$

$$Q = \frac{(2 - 1)[(2)(65) - (9)^2]}{(2)(9) - 9} = 5.44$$

McNemar test:

$$\chi^2 = \frac{(b - c)^2}{(b + c)} = \frac{(8 - 1)^2}{(8 + 1)} = 5.44$$

In the case of both tests, $df = 1$ (since in the case of the **Cochran Q test** $df = k - 1 = 2 - 1 = 1$, and in the case of the **McNemar test** the number of degrees of freedom is always $df = 1$). The tabled critical .05 and .01 chi-square values for $df = 1$ are $\chi^2_{.05} = 3.84$ and $\chi^2_{.01} = 6.63$. Since the obtained value $\chi^2 = 5.44$ is greater than $\chi^2_{.05} = 3.84$, the nondirectional alternative hypothesis is supported at the .05 level. Since $\chi^2 = 5.44$ is less than $\chi^2_{.01} = 6.63$, it is not supported at the .01 level.

In point of fact, the data in Table 26.4 are based on Example 19.1, which is employed to illustrate the **binomial sign test for two dependent samples (Test 19)**. If we assume that in the case of Example 19.1 a subject is assigned a score of 1 in the condition in which she has a higher score and a score of 0 in the condition in which she has a lower score, plus the fact that a subject is assigned a score of 0 (or 1) in both conditions if she has the same score, the data in Table 19.1

will be identical to those presented in Table 26.4.[10] When Equation 19.3 (the uncorrected (for continuity) normal approximation for the **binomial sign test for two dependent samples**) is employed to evaluate Example 19.1, it yields the value $z = 2.33$ which, if squared, equals the obtained chi-square value for Example 26.2 — i.e., $(z = 2.33)^2 = (\chi^2 = 5.44)$.[11] Thus, when $k = 2$ the **McNemar test/Cochran Q test** are equivalent to the **binomial sign test for two dependent samples**. It should also be noted that the exact binomial probability computed for the **binomial sign test for two dependent samples** will be equivalent to the exact binomial probability computed when the **McNemar test/Cochran Q test** is employed to evaluate the same data. For Examples 19.1/26.2, the two-tailed binomial probability is $P(x \geq 8) = (2)(.0196) = .0392$ (for $n = 9$).

3. Alternative nonparametric procedures with categorical data for evaluating a design involving k dependent samples Daniel (1990) and Fleiss (1981) note that alternative procedures for comparing k or more matched samples have been developed by Bennett (1967, 1968) and Shah and Claypool (1985). Chou (1989) describes a median test that can be employed to evaluate more than two dependent samples. The latter test, which employs the chi-square distribution to approximate the exact sampling distribution, employs subject/block medians as reference points in determining whether or not two or more of the treatment conditions represent different populations. The test described by Chou (1989) assumes that subjects' original scores are in an interval/ratio format and are converted into categorical data.

VIII. Additional Examples Illustrating the Use of the Cochran Q Test

Since the **Cochran Q test** can be employed to evaluate any dependent samples design involving two or more experimental conditions, it can also be used to evaluate any of the examples discussed under the **McNemar test**. Examples 26.3–26.7 are additional examples which can be evaluated with the **Cochran Q test**. Examples 26.6 and 26.7 represent extensions of a **one-group pretest-posttest design** to a design involving $k = 3$ experimental conditions. Since the data for all of the examples in this section (with the exception of Example 26.7) are identical to the data employed in Example 26.1, they yield the same result.

Table 26.6 Data for Example 26.3

	Brand of paint		
	Brightglow	**Colorfast**	**Prismalong**
Block 1	1	1	0
Block 2	0	1	0
Block 3	1	1	1
Block 4	0	1	0
Block 5	0	1	0
Block 6	0	1	1
Block 7	0	0	0
Block 8	0	1	0
Block 9	1	1	0
Block 10	0	1	0
Block 11	0	0	0
Block 12	0	0	1

Example 26.3 *A researcher wants to assess the relative likelihood of three brands of house paint fading within two years of application. In order to make this assessment he applies the*

following three brands of house paint which are identical in hue to a sample of houses that have cedar shingles: Brightglow, Colorfast, and Prismalong. In selecting the houses the researcher identifies 12 neighborhoods which vary with respect to geographical conditions, and within each neighborhood he randomly selects 3 houses. Within each block of three houses, one of the houses is painted with Brightglow, a second house with Colorfast, and a third house with Prismalong. Thus, a total of 36 houses are painted in the study. Two years after the houses are painted, an independent judge categorizes each house with respect to whether or not the paint on its shingles has faded. A house is assigned the number 1 if there is evidence of fading and the number 0 if there is no evidence of fading. Table 26.6 *summarizes the results of the study. Do the data indicate differences among the three brands of house paint with respect to fading?*

Note that in Example 26.3 the 12 blocks, comprised of 3 houses per block, are analogous to the use of 12 sets of matched subjects with 3 subjects per set/block. The brands of house paint represent the three levels of the independent variable, and the judge's categorization for each house with respect to fading (i.e., 1 versus 0) represents the dependent variable. Based on the analysis conducted for Example 26.1, there is a strong suggestion that Colorfast paint is perceived as more likely to fade than the other two brands.

Example 26.4 *Twelve male marines are administered a test of physical fitness which requires that an individual achieve the minimum criterion noted for the following three tasks: a) Climb a 100 ft. rope; b) Do 25 chin-ups; and c) Run a mile in under six minutes. Within the sample of 12 subjects, the order of presentation of the three tasks is completely counterbalanced (i.e., each of the six possible presentation orders for the tasks is presented to two subjects). For each of the tasks a subject is assigned a score of 1 if he achieves the minimum criterion and a score of 0 if he does not.* Table 26.7 *summarizes the results of the testing. Do the data indicate there is a difference among the three tasks with respect to subjects achieving the criterion?*

Table 26.7 Data for Example 26.4

	Task		
	Rope climb	Chin-ups	Mile run
Subject 1	1	1	0
Subject 2	0	1	0
Subject 3	1	1	1
Subject 4	0	1	0
Subject 5	0	1	0
Subject 6	0	1	1
Subject 7	0	0	0
Subject 8	0	1	0
Subject 9	1	1	0
Subject 10	0	1	0
Subject 11	0	0	0
Subject 12	0	0	1

Based on the analysis conducted for Example 26.1, the data suggest that subjects are more likely to achieve the criterion for chin-ups than the criteria for the other two tasks.

Example 26.5 *A horticulturist working at a university is hired to evaluate the effectiveness of three different kinds of weed killer (Zapon, Snuffout, and Shalom). Twelve athletic fields of equal* size are selected as test sites. The researcher divides each athletic field into three equally sized areas, and within each field (based on random determination) he applies one kind of weed killer to one third of the field, a second kind of weed killer to another third of the field, and the third kind of weed killer to the remaining third of the field. This procedure is employed for all 12 athletic fields, resulting in 36 separate areas to which weed killer is applied. Six months after application of the weed killer, an independent judge evaluates the 36 areas with respect to weed growth. The judge employs the number **1** to indicate that an area has evidence of weed growth and the number **0** to indicate that an area does not have evidence of weed growth. Table 26.8 summarizes the judge's categorizations. Do the data indicate there is a difference in the effectiveness among the three kinds of weed killer?

Note that in Example 26.5 the 12 athletic fields are analogous to 12 subjects who are evaluated under three experimental conditions. The three brands of weed killer represent the levels of the independent variable, and the judge's categorization of each area with respect to weed growth (i.e., **1** versus **0**) represents the dependent variable. Based on the analysis conducted for Example 26.1, the data suggest that Snuffout is less effective than the other two brands of weed killer.

Table 26.8   Data for Example 26.5

	Weed killer		
	Zapon	**Snuffout**	**Shalom**
Field 1	1	1	0
Field 2	0	1	0
Field 3	1	1	1
Field 4	0	1	0
Field 5	0	1	0
Field 6	0	1	1
Field 7	0	0	0
Field 8	0	1	0
Field 9	1	1	0
Field 10	0	1	0
Field 11	0	0	0
Field 12	0	0	1

Example 26.6 *A social scientist conducts a study assessing the impact of a federal gun control law on rioting in large cities. Assume that as a result of legislative changes the law in question, which severely limits the public's access to firearms, was not in effect between the years 1985–1989, but was in effect during the five years directly preceding and following that time period (i.e., the gun control law was in effect during the periods 1980–1984 and 1990–1994). In conducting the study, the social scientist categorizes 12 large cities with respect to whether or not there was a major riot within each of the three designated time periods. Thus, each city is categorized with respect to whether or not a riot occurred during: a) 1980–1984, during which time the gun control law was in effect (Time 1); b) 1985–1989, during which time the gun control law was not in effect (Time 2); and c) 1990–1994, during which time the gun control law was in effect (Time 3). A code of **1** is employed to indicate the occurrence of at least one major riot during a specified five-year time period, and a code of **0** is employed to indicate the absence of a major riot during a specified time period. Table 26.9 summarizes the results of the study. Do the data indicate the gun control law had an effect on rioting?*

Note that in Example 26.6 the 12 cities are analogous to 12 subjects who are evaluated during three time periods. Example 26.6 can be conceptualized as representing what is referred

Table 26.9   Data for Example 26.6

	Time period		
	Time 1 (1980–1984)	Time 2 (1985–1989)	Time 3 (1990–1994)
New York	1	1	0
Chicago	0	1	0
Detroit	1	1	1
Philadelphia	0	1	0
Los Angeles	0	1	0
Dallas	0	1	1
Houston	0	0	0
Miami	0	1	0
Washington	1	1	0
Boston	0	1	0
Baltimore	0	0	0
Atlanta	0	0	1

to as a **time-series design**. As noted in the **Introduction** (under the discussion of experimental design), a **time-series design** is essentially a **one-group pretest-posttest design** in which one or more blocks are evaluated one or more times both prior to and following an experimental treatment. In Example 26.6 each of the cities represents a block. **Time-series** designs are most commonly employed in the social sciences when a researcher wants to evaluate social change through analysis of archival data (i.e., public records). The internal validity of a **time-series design** is limited insofar as the treatment is not manipulated by the researcher, and thus any observed differences across time periods with respect to the dependent variable may be due to extraneous variables over which the researcher has no control. Thus, although when Example 26.6 is evaluated with the **Cochran Q test** the obtained Q value is significant (suggesting that more riots occurred when the gun control law was not in effect), other circumstances (such as the economy, race relations, etc.) may have varied across time periods, and such factors (as opposed to the gun control law) may have been responsible for the observed effect.[12] Another limitation of the time-series design described by Example 26.6 is that since the membership of the blocks is not the result of random assignment, the various blocks will not be directly comparable to one another. In closing the discussion of Example 26.6, it should be noted that in practice the **Cochran Q test** is not commonly employed to evaluate a **time-series design**. [13]

Example 26.7 *In order to assess the efficacy of a drug which a pharmaceutical company claims is effective in treating hyperactivity, 12 hyperactive children are evaluated during the following three time periods: a) One week prior to taking the drug; b) After a child has taken the drug for six consecutive months; and c) Six months after the drug is discontinued. The children are observed by judges who employ a standardized procedure for evaluating hyperactivity. The procedure requires that during each time period a child be assigned a score of 1 if he is hyperactive and a score of 0 if he is not hyperactive. During the evaluation process, the judges are blind with respect to whether a child is taking medication at the time he or she is evaluated. Table 26.10 summarizes the results of the study. Do the data indicate the drug is effective?*

Example 26.7 employs the same experimental design to evaluate the hypothesis which is evaluated in Example 24.7 (in Section VIII of the **single-factor within-subjects analysis of**

variance (Test 24)). In Example 26.7, however, categorical data are employed to represent the dependent variable. Evaluation of the data with the **Cochran Q test** yields the value $Q = 14.89$.

$$\Sigma C_1 = 12 \quad \Sigma C_2 = 3 \quad \Sigma C_3 = 10 \quad (\Sigma C_1)^2 = 144 \quad (\Sigma C_2)^2 = 9 \quad (\Sigma C_3)^2 = 100$$

$$\Sigma(\Sigma C_j)^2 = 144 + 9 + 100 = 253 \quad \Sigma R_i = 25 \quad \Sigma R_i^2 = 57$$

$$\text{Thus:} \quad C = 253 \quad T = 25 \quad R = 57$$

$$Q = \frac{(3 - 1)[(3)(253) - (25)^2]}{(3)(25) - 57} = 14.89$$

Table 26.10 Data for Example 26.7

	Time Period		
	Time 1	Time 2	Time 3
Child 1	1	0	1
Child 2	1	0	1
Child 3	1	0	1
Child 4	1	0	0
Child 5	1	0	0
Child 6	1	0	1
Child 7	1	1	1
Child 8	1	0	1
Child 9	1	0	1
Child 10	1	0	1
Child 11	1	1	1
Child 12	1	1	1

Since $k = 3$, $df = 2$. The tabled critical .05 and .01 values for $df = 2$ are $\chi^2_{.05} = 5.99$ and $\chi^2_{.01} = 9.21$. Since the obtained value $Q = \chi^2 = 14.89$ is greater than both of the aforementioned critical values, the alternative hypothesis is supported at both the .05 and .01 levels. Inspection of Table 26.10 strongly suggests that the significant effect is due to the lower frequency of hyperactivity during the time subjects are taking the drug (Time period 2). The latter, of course, can be confirmed by conducting comparisons between pairs of time periods.

As noted in the discussion of Example 24.7, the design of the latter study does not adequately control for the effects of extraneous/confounding variables. The same comments noted in the aforementioned discussion also apply to Example 26.7. Example 26.7 can also be conceptualized within the context of a time-series design.

References

Bennett, B. M. (1967). Tests of hypotheses concerning matched samples. **Journal of the Royal Statistical Society**, Ser. B., 29, 468–474.

Bennett, B. M. (1968). Notes on χ^2 tests for matched samples. **Journal of the Royal Statistical Society**, Ser. B., 30, 368–370.

Chou, Y. (1989). **Statistical analysis for business and economics**. New York: Elsevier.

Cochran, W. G. (1950). The comparison of percentages in matched samples. **Biometrika**, 37, 256–266.

Conover, W. J. (1980). **Practical nonparametric statistics** (2nd ed.). New York: John Wiley & Sons.

Conover, W. J. (1999). **Practical nonparametric statistics** (3rd ed.). New York: John Wiley & Sons.

Daniel, W. J. (1990). **Applied nonparametric statistics** (2nd ed.). Boston: PWS-Kent Publishing Company.

Fleiss, J. L. (1981). **Statistical methods for rates and proportions** (2nd ed.). New York: John Wiley & Sons.

Marascuilo, L. A. & McSweeney, M. (1977). **Nonparametric and distribution-free methods for the social sciences**. Monterey, CA: Brooks/Cole Publishing Company.

Patil, K. D. (1975). Cochran's Q test: Exact distribution. **Journal of the American Statistical Association**, 70, 186–189.

Shah, A. K. & Claypool, P. L. (1985). Analysis of binary data in the randomized complete block design. **Communications in Statistics — Theory and Methods**, 14, 1175–1179.

Sheskin, D. J. (1984). **Statistical tests and experimental design: A guidebook**. New York: Gardner Press.

Siegel, S. & Castellan, N. J., Jr. (1988). **Nonparametric statistics for the behavioral sciences** (2nd ed.). New York: McGraw–Hill Book Company.

Winer, B. J., Brown, D. R., & Michels, K. M. (1991). **Statistical principles in experimental design** (3rd ed.). New York: McGraw–Hill, Inc.

Endnotes

1. A more detailed discussion of the guidelines noted below can be found in Sections I and VII of the *t* **test for two dependent samples**.

2. Although it is possible to conduct a directional analysis, such an analysis will not be described with respect to the **Cochran Q test**. A discussion of a directional analysis when $k = 2$ can be found under the **McNemar test**. A discussion of the evaluation of a directional alternative hypothesis when $k \geq 3$ can be found in Section VII of the **chi-square goodness-of-fit test (Test 8)**. Although the latter discussion is in reference to analysis of a *k* independent samples design involving categorical data, the general principles regarding analysis of a directional alternative hypothesis when $k \geq 3$ are applicable to the **Cochran Q test**.

3. The use of Equation 26.1 to compute the **Cochran Q test** statistic assumes that the columns in the summary table (i.e., Table 26.1) are employed to represent the k levels of the independent variable, and that the rows are employed to represent the n subjects/matched sets of subjects. If the columns and rows are reversed (i.e., the columns are employed to represent the subjects/matched sets of subjects, and the rows the levels of the independent variable), Equation 26.1 cannot be employed to compute the value of Q.

4. The same Q value is obtained if the frequencies of **No** responses (0) are employed in computing the summary values used in Equation 26.1 instead of the frequencies of **Yes** (1) responses. To illustrate this, the data for Example 26.1 are evaluated employing the frequencies of **No** (0) responses.

$$\Sigma C_1 = 9 \quad \Sigma C_2 = 3 \quad \Sigma C_3 = 9 \quad (\Sigma C_1)^2 = 81 \quad (\Sigma C_2)^2 = 9 \quad (\Sigma C_3)^2 = 81$$

$$\Sigma(\Sigma C_j)^2 = 81 + 9 + 81 = 171 \quad \Sigma R_i = 21 \quad \Sigma R_i^2 = 45$$

Thus: $C = 171 \quad T = 21 \quad R = 45$

$$Q = \frac{(3 - 1)[(3)(171) - (21)^2]}{(3)(21) - 45} = 8$$

5. In the discussion of comparisons in reference to the analysis of variance, it is noted that a **simple** (also known as a **pairwise**) **comparison** is a comparison between any two groups/conditions in a set of k groups/conditions.

6. The method for deriving the value of z_{adj} for the **Cochran Q test** is based on the same logic that is employed in Equation 22.5 (which is used for conducting comparisons for the **Kruskal–Wallis one-way analysis of variance by ranks (Test 22)**. A rationale for the use of the proportions .0167 and .0083 in determining the appropriate value for z_{adj} in Example 26.1 can be found in Endnote 5 of the **Kruskal–Wallis one-way analysis of variance by ranks**.

7. It should be noted that when a directional alternative hypothesis is employed, the sign of the difference between the two proportions must be consistent with the prediction stated in the directional alternative hypothesis. When a nondirectional alternative hypothesis is employed, the direction of the difference between the two proportions is irrelevant.

8. In Equation 26.3 the value $z_{.05} = 1.96$ is employed for z_{adj}, and the latter value is multiplied by .204, which is the value computed for the term in the radical of the equation for Example 26.1.

9. In point of fact, Equation 26.3 employs more information than the **McNemar test**, and thus provides a more powerful test of an alternative hypothesis than the latter test (assuming both tests employ the same value for α_{PC}). The lower power of the **McNemar test** is directly attributed to the fact that for a given comparison, it only employs the scores of those subjects who obtain different scores under the two experimental conditions.

10. In conducting the **binomial sign test for two dependent samples**, what is relevant is in which of the two conditions a subject has a higher score, which is commensurate with assigning a subject to one of two response categories. As is the case with the **McNemar test** and the **Cochran Q test**, the analysis for the **binomial sign test for two dependent samples** does not include subjects who obtain the same score in both conditions.

11. The value $\chi^2 = 5.44$ is also obtained for Example 19.1 through use of Equation 8.2, which is the equation for the **chi-square goodness-of-fit test**. In the case of Example 19.1, the latter equation produces an equivalent result to that obtained with Equation 19.3 (the normal approximation). The result of the binomial analysis of Example 19.1 with the **chi-square goodness-of-fit test** is summarized in Table 19.2.

12. Within the framework of a time-series design, one or more blocks can be included to serve as controls. Specifically, in Example 26.6 additional cities might have been selected in which the gun control law was always in effect (i.e., in effect during Time 2 as well as during Times 1 and 3). Differences on the dependent variable during Time 2 between the control cities and the cites in which the law was nullified between 1985–1989 could be contrasted to further evaluate the impact of the gun control law. Unfortunately, if the law in question is national, such control cities would not be available in the nation in which the study is conducted. The reader should note, however, that even if such control cities were available, the internal validity of such a study would still be subject to challenge, since it would still not ensure adequate control over all potential extraneous variables.

13. a) Cochran (1950) and Winer *et al.* (1991) note that if a **single-factor within-subjects analysis of variance** is employed to evaluate the data in a **Cochran Q test** summary table (e.g., Tables 26.1–26.9), it generally leads to similar conclusions as those reached when the data are evaluated with Equation 26.1. However, the question of whether it is appropriate to employ an analysis of variance to evaluate the categorical data in the **Cochran Q test** summary table is an issue on which researchers do not agree; b) It might be more worthwhile to conceptualize a study such as that represented by Example 26.6 within the framework of a **mixed factorial design** (which will always involve at least two independent variables). In a **mixed factorial design** involving two independent variables, one independent variable is a between-subjects variable (i.e., each subject/block is evaluated under only one level of that independent variable), while the other independent variable is a within-subjects variable (i.e., each subject/block is evaluated under all levels of that independent variable). In the latter type of analysis, the dependent variable is generally represented by interval/ratio level data. If a study such as Example 26.6 were conceptualized as a **mixed factorial design**, different *types* of cities might be employed to represent the between-subjects independent variable, and the three time periods would represent a within-subjects independent variable. If the study was conceptualized within the framework of a mixed factorial design, each block (i.e., row in Table 26.9) would be comprised of two or more cities. For example, each row might be comprised of two or more cities with populations that fell within a specified range. Thus, Row 1/Block 1 might be comprised of five cities with a population greater than five million; Row 2/Block 2 by five cities with a population between one and five million; Row 3/Block 3 by five cities with a population between 500,000 and one million, and so on. Such a design is typically evaluated with the **factorial analysis of variance for a mixed design (Test 27i)** which is discussed in Section IX (the **Addendum**) of the **between-subjects factorial analysis of variance (Test 27)**.

Inferential Statistical Test Employed with a Factorial Design (and Related Measures of Association/Correlation)

Test 27: The Between-Subjects Factorial Analysis of Variance

Test 27

The Between-Subjects Factorial Analysis of Variance
(Parametric Test Employed with Interval/Ratio Data)

I. Hypothesis Evaluated with Test and Relevant Background Information

The **between-subjects factorial analysis of variance** is one of a number of analysis of variance procedures which are employed to evaluate a **factorial design**. A factorial design is employed to simultaneously evaluate the effect of two or more independent variables on a dependent variable. Each of the independent variables is referred to as a **factor**. Each of the factors has two or more levels, which refer to the number of groups/experimental conditions which comprise that independent variable. If a factorial design is not employed to assess the effect of multiple independent variables on a dependent variable, separate experiments must be conducted to evaluate the effect of each of the independent variables. One major advantage of a factorial design is that it allows the same set of hypotheses to be evaluated at a comparable level of power by using only a fraction of the subjects which would be required if separate experiments were conducted to evaluate the relevant hypotheses for each of the independent variables. Another advantage of a factorial design is that it permits a researcher to evaluate whether or not there is an **interaction** between two or more independent variables — something which cannot be determined if only one independent variable is employed in a study. An **interaction** is present in a set of data when the performance of subjects on one independent variable is not consistent across all the levels of another independent variable. The concept of interaction is discussed in detail in Section V.

The **between-subjects factorial analysis of variance** (also known as a **completely randomized factorial analysis of variance**) is an extension of the **single-factor between-subjects analysis of variance (Test 21)** to experiments involving two or more independent variables. Although the **between-subjects factorial analysis of variance** can be used for more than two factors, the computational procedures described in this book will be limited to designs involving two factors. One of the factors will be designated by the letter **A**, and will have p levels, and the second factor will be designated by the letter **B**, and will have q levels. As a result of this, there will be a total of $p \times q$ groups. A $p \times q$ **between-subjects/completely randomized factorial design** requires that each of the $p \times q$ groups is comprised of different subjects who have been randomly assigned to that group. Each group serves under one of the p levels of Factor A and one of the q levels of Factor B, with no two groups serving under the same combination of levels of the two factors. All possible combinations of the levels of Factor A and Factor B are represented by the total $p \times q$ groups.

The **between-subjects factorial analysis of variance** evaluates the following hypotheses:

a) With respect to Factor A: In the set of p independent samples (where $p \geq 2$), do at least two of the samples represent populations with different mean values? The latter hypothesis can also be stated as follows: Do at least two of the levels of Factor A represent populations with different mean values?

b) With respect to Factor B: In the set of q independent samples (where $q \geq 2$), do at least two of the samples represent populations with different mean values? The latter hypothesis can also be stated as follows: Do at least two of the levels of Factor B represent populations with different mean values?

c) In addition to evaluating the above hypotheses (which assess the presence or absence of what are referred to as **main effects**), the **between-subjects factorial analysis of variance** evaluates the hypothesis of whether there is a significant **interaction** between the two factors/independent variables.[1]

A discussion of the theoretical rationale underlying the evaluation of the three sets of hypotheses for the **between-subjects factorial analysis of variance** can be found in Section VII.

The **between-subjects factorial analysis of variance** is employed with interval/ratio data and is based on the following assumptions: a) Each sample has been randomly selected from the population it represents; b) The distribution of data in the underlying population from which each of the samples is derived is normal; and c) The third assumption, which is referred to as the **homogeneity of variance** assumption, states that the variances of the $p \times q$ underlying populations represented by the $p \times q$ groups are equal to one another. The homogeneity of variance assumption (which is discussed earlier in the book in reference to the *t* **test for two independent samples (Test 11)**, the *t* **test for two dependent samples (Test 17)**, the **single-factor between-subjects analysis of variance** and the **single-factor within-subjects analysis of variance (Test 24)**) is discussed in greater detail in Section VI. If any of the aforementioned assumptions of the **between-subjects factorial analysis of variance** are saliently violated, the reliability of the computed test statistic may be compromised.

II. Example

Example 27.1 *A study is conducted to evaluate the effect of humidity (to be designated as Factor A) and temperature (to be designated as Factor B) on mechanical problem-solving ability. The experimenter employs a 2×3 between-subjects factorial design. The two levels which comprise Factor A are A_1: Low humidity; A_2: High humidity. The three levels which comprise Factor B are B_1: Low temperature; B_2: Moderate temperature; B_3: High temperature. The study employs 18 subjects, each of whom is randomly assigned to one of the six experimental groups (i.e., $p \times q = 2 \times 3 = 6$), resulting in three subjects per group. Each of the six experimental groups represents a different combination of the levels that comprise the two factors. The number of mechanical problems solved by the three subjects in each of the six experimental conditions/groups follow. (The notation Group AB_{jk} indicates the group that served under Level j of Factor A and Level k of Factor B.) Group AB_{11}: Low humidity/Low temperature (11, 9, 10); Group AB_{12}: Low humidity/Moderate temperature (7, 8, 6); Group AB_{13}: Low humidity/High temperature (5, 4, 3); Group AB_{21}: High humidity/Low temperature (2, 4, 3); Group AB_{22}: High humidity/Moderate temperature (4, 5, 3); Group AB_{23}: High humidity/ High temperature (0, 1, 2). Do the data indicate that either humidity or temperature influences mechanical problem-solving ability?*

III. Null versus Alternative Hypotheses

A **between-subjects factorial analysis of variance** involving two factors evaluates three sets of hypotheses. The first set of hypotheses evaluates the effect of Factor A on the dependent variable, the second set evaluates the effect of Factor B on the dependent variable, and the third set evaluates whether or not there is an interaction between the two factors.

Set 1: Hypotheses for Factor A

Null hypothesis $$H_0: \mu_{A_1} = \mu_{A_2}$$

(The mean of the population Level 1 of Factor A represents equals the mean of the population Level 2 of Factor A represents.)

Alternative hypothesis $$H_1: \mu_{A_1} \neq \mu_{A_2}$$

(The mean of the population Level 1 of Factor A represents does not equal the mean of the population Level 2 of Factor A represents. This is a nondirectional alternative hypothesis. In the discussion of the **between-subjects factorial analysis of variance** it will be assumed (unless stated otherwise) that an alternative hypothesis is stated **nondirectionally**.[2] In order for the alternative hypothesis for Factor A to be supported, the obtained F value for Factor A (designated by the notation F_A) must be equal to or greater than the tabled critical F value at the prespecified level of significance.)

Set 2: Hypotheses for Factor B

Null hypothesis $$H_0: \mu_{B_1} = \mu_{B_2} = \mu_{B_3}$$

(The mean of the population Level 1 of Factor B represents equals the mean of the population Level 2 of Factor B represents equals the mean of the population Level 3 of Factor B represents.)

Alternative hypothesis $$H_1: \text{Not } H_0$$

(This indicates there is a difference between at least two of the $q = 3$ population means. It is important to note that the alternative hypothesis should not be written as follows: $H_1: \mu_{B_1} \neq \mu_{B_2} \neq \mu_{B_3}$. The reason why the latter notation for the alternative hypothesis is incorrect is because it implies that all three population means must differ from one another in order to reject the null hypothesis. In order for the alternative hypothesis for Factor B to be supported, the obtained F value for Factor B (designated by the notation F_B) must be equal to or greater than the tabled critical F value at the prespecified level of significance.)

Set 3: Hypotheses for interaction

H_0: There is no interaction between Factor A and Factor B.

H_1: There is an interaction between Factor A and Factor B.

Although it is possible to state the null and alternative hypotheses for the interaction symbolically, such a format will not be employed since it requires a considerable amount of notation. It should be noted that in predicting an interaction, a researcher may be very specific with respect to the pattern of the interaction that is predicted. As a general rule, however, such predictions are not reflected in the statement of the null and alternative hypotheses. In order for the alternative hypothesis for the interaction to be supported, the obtained F value for the interaction (designated by the notation F_{AB}) must be equal to or greater than the tabled critical F value at the prespecified level of significance.

IV. Test Computations

The test statistics for the **between-subjects factorial analysis of variance** can be computed with either **computational** or **definitional equations**. Although definitional equations reveal the underlying logic behind the analysis of variance, they involve considerably more calculations than do the computational equations. Because of the latter, computational equations will be employed in this section to demonstrate the computation of the test statistic. The definitional equations for the **between-subjects factorial analysis of variance** are described in Section VII.

The data for Example 27.1 are summarized in Table 27.1. In the latter table the following notation is employed.

N represents the total number of subjects who serve in the experiment. In Example 27.1, $N = 18$.

ΣX_T represents the total sum of the scores of the $N = 18$ subjects who serve in the experiment.

$\bar{X}_T$ represents the mean of the scores of the $N = 18$ subjects who serve in the experiment. $\bar{X}_T$ will be referred to as the **grand mean**.

ΣX_T^2 represents the total sum of the squared scores of the $N = 18$ subjects who serve in the experiment.

$(\Sigma X_T)^2$ represents the square of the total sum of scores of the $N = 18$ subjects who serve in the experiment.

$n_{AB_{jk}}$ represents the number of subjects who serve in Group AB_{jk}. In Example 27.1, $n_{AB_{jk}} = 3$. In some of the equations that follow, the notation n is employed to represent the value $n_{AB_{jk}}$.

$\Sigma X_{AB_{jk}}$ represents the sum of the scores of the $n_{AB_{jk}} = 3$ subjects who serve in Group AB_{jk}.

$\bar{X}_{AB_{jk}}$ represents the mean of the scores of the $n_{AB_{jk}} = 3$ subjects who serve in Group AB_{jk}.

$\Sigma X_{AB_{jk}}^2$ represents the sum of the squared scores of the $n_{AB_{jk}} = 3$ subjects who serve in Group AB_{jk}.

$(\Sigma X_{AB_{jk}})^2$ represents the square of the sum of scores of the $n_{AB_{jk}} = 3$ subjects who serve in Group AB_{jk}.

n_{A_j} represents the number of subjects who serve in level j of Factor A. In Example 27.1, $n_{A_j} = (n_{AB_{jk}})(q) = (3)(3) = 9$.

ΣX_{A_j} represents the sum of the scores of the $n_{A_j} = 9$ subjects who serve in Level j of Factor A.

$\bar{X}_{A_j}$ represents the mean of the scores of the $n_{A_j} = 9$ subjects who serve in level j of Factor A.

$\Sigma X_{A_j}^2$ represents the sum of the squared scores of the $n_{A_j} = 9$ subjects who serve in Level j of Factor A.

$(\Sigma X_{A_j})^2$ represents the square of the sum of scores of the $n_{A_j} = 9$ subjects who serve in Level j of Factor A.

n_{B_k} represents the number of subjects who serve in level k of Factor B. In Example 27.1, $n_{B_k} = (n_{AB_{jk}})(p) = (3)(2) = 6$.

ΣX_{B_k} represents the sum of the scores of the $n_{B_k} = 6$ subjects who serve in Level k of Factor B.

$\bar{X}_{B_k}$ represents the mean of the scores of the $n_{B_k} = 6$ subjects who serve in Level k of Factor B.

$\Sigma X_{B_k}^2$ represents the sum of the squared scores of the $n_{B_k} = 6$ subjects who serve in Level k of Factor B.

$(\Sigma X_{B_k})^2$ represents the square of the sum of scores of the $n_{B_k} = 6$ subjects who serve in Level k of Factor B.

Table 27.1 Data for Example 27.1

Factor A (Humidity)	Factor B (Temperature)			Row sums
	B_1 (Low)	B_2 (Moderate)	B_3 (High)	
	Group AB_{11}	**Group AB_{12}**	**Group AB_{13}**	**Level A_1**
	$X_{AB_{11}}$ $X^2_{AB_{11}}$	$X_{AB_{12}}$ $X^2_{AB_{12}}$	$X_{AB_{13}}$ $X^2_{AB_{13}}$	
	11 121	7 49	5 25	
	9 81	8 64	4 16	
	10 100	6 36	3 9	
A_1 (Low)	$n_{AB_{11}} = 3$	$n_{AB_{12}} = 3$	$n_{AB_{13}} = 3$	$n_{A_1} = 9$
	$\Sigma X_{AB_{11}} = 30$	$\Sigma X_{AB_{12}} = 21$	$\Sigma X_{AB_{13}} = 12$	$\Sigma X_{A_1} = 63$
	$\bar{X}_{AB_{11}} = \dfrac{\Sigma X_{AB_{11}}}{n_{AB_{11}}} = \dfrac{30}{3} = 10$	$\bar{X}_{AB_{12}} = \dfrac{\Sigma X_{AB_{12}}}{n_{AB_{12}}} = \dfrac{21}{3} = 7$	$\bar{X}_{AB_{13}} = \dfrac{\Sigma X_{AB_{13}}}{n_{AB_{13}}} = \dfrac{12}{3} = 4$	$\bar{X}_{A_1} = \dfrac{\Sigma X_{A_1}}{n_{A_1}} = \dfrac{63}{9} = 7$
	$\Sigma X^2_{AB_{11}} = 302$	$\Sigma X^2_{AB_{12}} = 149$	$\Sigma X^2_{AB_{13}} = 50$	$\Sigma X^2_{A_1} = 501$
	$(\Sigma X_{AB_{11}})^2 = (30)^2 = 900$	$(\Sigma X_{AB_{12}})^2 = (21)^2 = 441$	$(\Sigma X_{AB_{13}})^2 = (12)^2 = 144$	$(\Sigma X_{A_1})^2 = (63)^2 = 3969$
	Group AB_{21}	**Group AB_{22}**	**Group AB_{23}**	**Level A_2**
	$X_{AB_{21}}$ $X^2_{AB_{21}}$	$X_{AB_{22}}$ $X^2_{AB_{22}}$	$X_{AB_{23}}$ $X^2_{AB_{23}}$	
	2 4	4 16	0 0	
	4 16	5 25	1 1	
	3 9	3 9	2 4	
A_2 (High)	$n_{AB_{21}} = 3$	$n_{AB_{22}} = 3$	$n_{AB_{23}} = 3$	$n_{A_2} = 9$
	$\Sigma X_{AB_{21}} = 9$	$\Sigma X_{AB_{22}} = 12$	$\Sigma X_{AB_{23}} = 3$	$\Sigma X_{A_2} = 24$
	$\bar{X}_{AB_{21}} = \dfrac{\Sigma X_{AB_{21}}}{n_{AB_{21}}} = \dfrac{9}{3} = 3$	$\bar{X}_{AB_{22}} = \dfrac{\Sigma X_{AB_{22}}}{n_{AB_{22}}} = \dfrac{12}{3} = 4$	$\bar{X}_{AB_{23}} = \dfrac{\Sigma X_{AB_{23}}}{n_{AB_{23}}} = \dfrac{3}{3} = 1$	$\bar{X}_{A_2} = \dfrac{\Sigma X_{A_2}}{n_{A_2}} = \dfrac{24}{9} = 2.67$
	$\Sigma X^2_{AB_{21}} = 29$	$\Sigma X^2_{AB_{22}} = 50$	$\Sigma X^2_{AB_{23}} = 5$	$\Sigma X^2_{A_2} = 84$
	$(\Sigma X_{AB_{21}})^2 = (9)^2 = 81$	$(\Sigma X_{AB_{22}})^2 = (12)^2 = 144$	$(\Sigma X_{AB_{23}})^2 = (3)^2 = 9$	$(\Sigma X_{A_2})^2 = (24)^2 = 576$
	Level B_1	**Level B_2**	**Level B_3**	**Grand Total**
	$n_{B_1} = 6$	$n_{B_2} = 6$	$n_{B_3} = 6$	$N = 18$
	$\Sigma X_{B_1} = 39$	$\Sigma X_{B_2} = 33$	$\Sigma X_{B_3} = 15$	$\Sigma X_T = 87$
Column sums	$\bar{X}_{B_1} = \dfrac{\Sigma X_{B_1}}{n_{B_1}} = \dfrac{39}{6} = 6.5$	$\bar{X}_{B_2} = \dfrac{\Sigma X_{B_2}}{n_{B_2}} = \dfrac{33}{6} = 5.5$	$\bar{X}_{B_3} = \dfrac{\Sigma X_{B_3}}{n_{B_3}} = \dfrac{15}{6} = 2.5$	$\bar{X}_T = \dfrac{\Sigma X_T}{N} = \dfrac{87}{18} = 4.83$
	$\Sigma X^2_{B_1} = 331$	$\Sigma X^2_{B_2} = 199$	$\Sigma X^2_{B_3} = 55$	$\Sigma X^2_T = 585$
	$(\Sigma X_{B_1})^2 = (39)^2 = 1521$	$(\Sigma X_{B_2})^2 = (33)^2 = 1089$	$(\Sigma X_{B_3})^2 = (15)^2 = 225$	$(\Sigma X_T)^2 = (87)^2 = 7569$

As is the case for the **single-factor between-subjects analysis of variance**, the total variability for the **between-subjects factorial analysis of variance** can be divided into **between-groups variability** and **within-groups variability**. The **between-groups variability** can be divided into the following: a) Variability attributable to Factor A; b) Variability attributable to Factor B; and c) Variability attributable to any interaction that is present between Factors A and B (which will be designated as AB variability). For each of the variability components involved in the **between-subjects factorial analysis of variance**, a sum of squares is computed. Thus, the following sum of squares values are computed: a) SS_T, the **total sum of squares**; b) SS_{BG}, the **between-groups sum of squares**; c) SS_A, the **sum of squares for Factor A** (which can also be referred to as the **row sum of squares**, since in Table 27.1 Factor A is the row variable); d) SS_B, the **sum of squares for Factor B** (which can also be referred to as the **column sum of squares**, since in Table 27.1 Factor B is the column variable); e) SS_{AB}, the **interaction sum of squares**; and f) SS_{WG}, the **within-groups sum of squares**, which is also referred to as the **error sum of squares** or **residual sum of squares**, since it represents variability that is due to chance factors which are beyond the control of the researcher. Each of the aforementioned sum of squares values represents the numerator in the equation which is employed to compute the variance for that variability component (which is referred to as the **mean square** for that component).

Equations 27.1–27.3 summarize the relationship between the sum of squares components for the **between-subjects factorial analysis of variance**. Equation 27.1 summarizes the relationship between the between-groups, the within-groups, and the total sums of squares.

$$SS_T = SS_{BG} + SS_{WG} \qquad \text{(Equation 27.1)}$$

Because of the relationship noted in Equation 27.2, Equation 27.1 can also be written in the form of Equation 27.3.

$$SS_{BG} = SS_A + SS_B + SS_{AB} \qquad \text{(Equation 27.2)}$$

$$SS_T = SS_A + SS_B + SS_{AB} + SS_{WG} \qquad \text{(Equation 27.3)}$$

In order to compute the sums of squares for the **between-subjects factorial analysis of variance**, the following summary values are computed with Equations 27.4–27.8 which will be employed as elements in the computational equations: $[XS]$, $[T]$, $[A]$, $[B]$, $[AB]$.[3] The reader should take note of the fact that in the equations that follow, the following is true: a) $n_{AB_{jk}} = n = 3$; b) $N = npq$, and thus, $N = (3)(2)(3) = 18$; c) $n_{A_j} = nq$, and thus, $n_{A_j} = (3)(3) = 9$; d) $n_{B_k} = np$, and thus, $n_{B_k} = (3)(2) = 6$.

The summary value $[XS] = 585$ is computed with Equation 27.4.

$$\text{(Equation 27.4)}$$
$$[XS] = \Sigma X_T^2 = (11)^2 + (9)^2 + (10)^2 + \cdots + (0)^2 + (1)^2 + (2)^2 = 585$$

The summary value $[T] = 420.5$ is computed with Equation 27.5.

$$[T] = \frac{(\Sigma X_T)^2}{N} = \frac{(87)^2}{18} = 420.5 \qquad \text{(Equation 27.5)}$$

The summary value $[A] = 505$ is computed with Equation 27.6.

$$[A] = \sum_{j=1}^{p} \left[\frac{(\Sigma X_{A_j})^2}{n_{A_j}} \right] = \frac{(63)^2}{9} + \frac{(24)^2}{9} = 505 \qquad \textbf{(Equation 27.6)}$$

The notation $\sum_{j=1}^{p}[(\Sigma X_{A_j})^2/n_{A_j}]$ in Equation 27.6 indicates that for each level of Factor A, the scores of the $n_{A_j} = 9$ subjects who serve under that level of the factor are summed, the resulting value is squared, and the obtained value is divided by $n_{A_j} = 9$. The values obtained for each of the $p = 2$ levels of Factor A are then summed.

The summary value $[B] = 472.5$ is computed with Equation 27.7.

$$[B] = \sum_{k=1}^{q} \left[\frac{(\Sigma X_{B_k})^2}{n_{B_k}} \right] = \frac{(39)^2}{6} + \frac{(33)^2}{6} + \frac{(15)^2}{6} = 472.5 \qquad \textbf{(Equation 27.7)}$$

The notation $\sum_{k=1}^{q}[(\Sigma X_{B_k})^2/n_{B_k}]$ in Equation 27.7 indicates that for each level of Factor B, the scores of the $n_{B_k} = 6$ subjects who serve under that level of the factor are summed, the resulting value is squared, and the obtained value is divided by $n_{B_k} = 6$. The values obtained for each of the $q = 3$ levels of Factor B are then summed.

The summary value $[AB] = 573$ is computed with Equation 27.8.

$$[AB] = \sum_{k=1}^{q} \sum_{j=1}^{p} \left[\frac{(\Sigma X_{AB_{jk}})^2}{n_{AB_{jk}}} \right] = \frac{(30)^2}{3} + \frac{(9)^2}{3}$$

$$\textbf{(Equation 27.8)}$$

$$+ \frac{(21)^2}{3} + \frac{(12)^2}{2} + \frac{(12)^2}{3} + \frac{(3)^2}{3} = 573$$

The notation $\sum_{k=1}^{q} \sum_{j=1}^{p}[(\Sigma X_{AB_{jk}})^2/n_{AB_{jk}}]$ in Equation 27.8 indicates that for each of the $pq = 6$ groups, the scores of the $n_{AB_{jk}} = 3$ subjects who serve in that group are summed, the resulting value is squared, and the obtained value is divided by $n_{AB_{jk}} = 3$. The values obtained for each of the $pq = 6$ groups are then summed.

Employing the summary values computed with Equations 27.4–27.8, Equations 27.9–27.14 can be employed to compute the values SS_T, SS_{BG}, SS_A, SS_B, SS_{AB}, and SS_{WG}.

Equation 27.9 is employed to compute the value $SS_T = 164.5$.

$$SS_T = [XS] - [T] = 585 - 420.5 = 164.5 \qquad \textbf{(Equation 27.9)}$$

Equation 27.10 is employed to compute the value $SS_{BG} = 152.5$.

$$SS_{BG} = [AB] - [T] = 573 - 420.5 = 152.5 \qquad \textbf{(Equation 27.10)}$$

Equation 27.11 is employed to compute the value $SS_A = 84.5$.

$$SS_A = [A] - [T] = 505 - 420.5 = 84.5 \qquad \textbf{(Equation 27.11)}$$

Equation 27.12 is employed to compute the value $SS_B = 52$.

$$SS_B = [B] - [T] = 472.5 - 420.5 = 52 \qquad \textbf{(Equation 27.12)}$$

Equation 27.13 is employed to compute the value $SS_{AB} = 16$.

(Equation 27.13)

$$SS_{AB} = [AB] - [A] - [B] + [T] = 573 - 505 - 472.5 + 420.5 = 16$$

Equation 27.14 is employed to compute the value $SS_{WG} = 12.$[4]

$$SS_{WG} = [XS] - [AB] = 585 - 573 = 12 \qquad \text{(Equation 27.14)}$$

Note that $SS_{BG} = SS_A + SS_B + SS_{AB} = 84.5 + 52 + 16 = 152.5$ and $SS_T = SS_{BG} + SS_{WG} = 152.5 + 12 = 164.5$. The reader should take note of the fact that the values SS_T, SS_{BG}, SS_A, SS_B, SS_{AB}, and SS_{WG} must always be positive numbers. If a negative value is obtained for any of the aforementioned values, it indicates a computational error has been made.

At this point the **mean square** values (which as previously noted represent variances) for the above components can be computed. In order to compute the test statistics for the **between-subjects factorial analysis of variance**, it is only required that the following mean square values be computed: MS_A, MS_B, MS_{AB}, and MS_{WG}.

MS_A is computed with Equation 27.15.

$$MS_A = \frac{SS_A}{df_A} \qquad \text{(Equation 27.15)}$$

MS_B is computed with Equation 27.16.

$$MS_B = \frac{SS_B}{df_B} \qquad \text{(Equation 27.16)}$$

MS_{AB} is computed with Equation 27.17.

$$MS_{AB} = \frac{SS_{AB}}{df_{AB}} \qquad \text{(Equation 27.17)}$$

MS_{WG} is computed with Equation 27.18.

$$MS_{WG} = \frac{SS_{WG}}{df_{WG}} \qquad \text{(Equation 27.18)}$$

In order to compute MS_A, MS_B, MS_{AB}, and MS_{WG}, it is required that the following **degrees of freedom** values be computed: df_A, df_B, df_{AB}, and df_{WG} (which are the denominators of Equations 27.15–27.18).

df_A are computed with Equation 27.19.

$$df_A = p - 1 \qquad \text{(Equation 27.19)}$$

df_B are computed with Equation 27.20.

$$df_B = q - 1 \qquad \text{(Equation 27.20)}$$

df_{AB} are computed with Equation 27.21.

$$df_{AB} = (p - 1)(q - 1)$$ **(Equation 27.21)**

df_{WG} (i.e., the **within-groups degrees of freedom**) are computed with Equation 27.22. As noted earlier, the value n is equivalent to the value $n_{AB_{jk}}$. The use of n in any of the equations for the **between-subjects factorial analysis of variance** assumes that there are an equal number of subjects in each of the pq groups.

$$df_{WG} = pq(n - 1)$$ **(Equation 27.22)**

Although they are not required in order to compute the F ratios for the **between-subjects factorial analysis of variance**, the **between-groups degrees of freedom** (df_{BG}), and the **total degrees of freedom** (df_T) are generally computed, since they can be used to confirm the df values computed with Equations 27.19–27.22, as well as the fact that they are employed in the analysis of variance summary table.

df_{BG} are computed with Equation 27.23.

$$df_{BG} = pq - 1$$ **(Equation 27.23)**

df_T are computed with Equation 27.24.

$$df_T = N - 1$$ **(Equation 27.24)**

The relationships between the various degrees of freedom values are described below.

$$df_{BG} = df_A + df_B + df_{AB} \qquad df_T = df_{BG} + df_{WG}$$

Employing Equations 27.19–27.24, the values $df_A = 1$, $df_B = 2$, $df_{AB} = 2$, $df_{WG} = 12$, $df_{BG} = 5$, and $df_T = 17$ are computed.

$$df_A = 2 - 1 = 1 \qquad df_B = 3 - 1 = 2 \qquad df_{AB} = (2 - 1)(3 - 1) = 2$$

$$df_{WG} = (2)(3)(3 - 1) = 12 \qquad df_{BG} = [(2)(3)] - 1 = 5 \qquad df_T = 18 - 1 = 17$$

Note that $df_{BG} = df_A + df_B + df_{AB} = 1 + 2 + 2 = 5$ and $df_T = df_{BG} + df_{WG} = 5 + 12 = 17$.

Employing Equations 27.15–27.18, the following values are computed: $MS_A = 84.5$, $MS_B = 26$, $MS_{AB} = 8$, $MS_{WG} = 1$.

$$MS_A = \frac{84.5}{1} = 84.5 \qquad MS_B = \frac{52}{2} = 26 \qquad MS_{AB} = \frac{16}{2} = 8 \qquad MS_{WG} = \frac{12}{12} = 1$$

The F ratio is the test statistic for the **between-subjects factorial analysis of variance**. Since, however, there are three sets of hypotheses to be evaluated, it is required that three F ratios be computed — one for each of the components which comprise the between-groups variability. Specifically, an F ratio is computed for Factor A, for Factor B, and for the AB interaction. Equations 27.25–27.27 are, respectively, employed to compute the three F ratios.

$$F_A = \frac{MS_A}{MS_{WG}}$$ **(Equation 27.25)**

$$F_B = \frac{MS_B}{MS_{WG}}$$ **(Equation 27.26)**

$$F_{AB} = \frac{MS_{AB}}{MS_{WG}}$$ **(Equation 27.27)**

Employing Equations 27.25–27.27, the values $F_A = 84.5$, $F_B = 26$, and $F_{AB} = 8$ are computed.

$$F_A = \frac{84.5}{1} = 84.5 \qquad F_B = \frac{26}{1} = 26 \qquad F_{AB} = \frac{8}{1} = 8$$

The reader should take note of the fact that any value computed for a mean square or an F ratio must always be a positive number. If a negative value is obtained for any mean square or F ratio, it indicates a computational error has been made. If $MS_{WG} = 0$, Equations 27.25–27.27 will be insoluble. The only time $MS_{WG} = 0$ is when, within each of the pq groups, all subjects obtain the same score (i.e., there is no within-groups variability). If the mean values for all of the levels of any factor are identical, the mean square value for that factor will equal zero, and, if the latter is true, the F value for that factor will also equal zero.

V. Interpretation of the Test Results

It is common practice to summarize the results of a **between-subjects factorial analysis of variance** with the summary table represented by Table 27.2.

Table 27.2 Summary Table of Analysis of Variance for Example 27.1

Source of variation	SS	df	MS	F
Between-groups	152.5	5		
A	84.5	1	84.5	84.5
B	52	2	26	26
AB	16	2	8	8
Within-groups	12	12	1	
Total	164.5	17		

The obtained F values are evaluated with **Table A10 (Table of the F Distribution)** in the **Appendix**. In **Table A10** critical values are listed in reference to the number of degrees of freedom associated with the numerator and the denominator of an F ratio. Thus, in the case of Example 27.1 the values of df_A, df_B, and df_{AB} are employed for the numerator degrees of freedom for each of the three F ratios, while df_{WG} is employed as the denominator degrees of freedom for all three F ratios. As is the case in the discussion of other analysis of variance procedures discussed in the book, the notation $F_{.05}$ is employed to represent the tabled critical F value at the .05 level. The latter value corresponds to the tabled $F_{.95}$ value in **Table A10**. In the same respect, the notation $F_{.01}$ will be employed to represent the tabled critical F value at the .01 level, and the latter value will correspond to the relevant tabled $F_{.99}$ value in **Table A10**.

The following tabled critical values are employed in evaluating the three F ratios computed for Example 27.1: a) **Factor A**: For $df_{num} = df_A = 1$ and $df_{den} = df_{WG} = 12$, $F_{.05} = 4.75$ and $F_{.01} = 9.33$; b) **Factor B**: For $df_{num} = df_B = 2$ and $df_{den} = df_{WG} = 12$, $F_{.05} = 3.89$ and $F_{.01} = 6.93$; c) **AB interaction**: For $df_{num} = df_{AB} = 2$ and $df_{den} = df_{WG} = 12$, $F_{.05} = 3.89$ and $F_{.01} = 6.93$.

In order to reject the null hypothesis in reference to a computed F ratio, the obtained F value must be equal to or greater than the tabled critical value at the prespecified level of significance. Since the computed value $F_A = 84.5$ is greater than $F_{.05} = 4.75$ and $F_{.01} = 9.33$, the alternative hypothesis for Factor A is supported at both the .05 and .01 levels. Since the computed value $F_B = 26$ is greater than $F_{.05} = 3.89$ and $F_{.01} = 6.93$, the alternative hypothesis for Factor B is supported at both the .05 and .01 levels. Since the computed value $F_{AB} = 8$ is greater than $F_{.05} = 3.89$ and $F_{.01} = 6.93$, the alternative hypothesis for an interaction between Factors A and B is supported at both the .05 and .01 levels. The aforementioned results can be summarized as follows: $F_A(1,12) = 84.5$, $p < .01$; $F_B(2,12) = 26$, $p < .01$; $F_{AB}(2,12) = 8$, $p < .01$.

The analysis of the data for Example 27.1 allows the researcher to conclude that both humidity (Factor A) and temperature (Factor B) have a significant impact on problem-solving scores. Thus, both main effects are significant. As previously noted, a main effect describes the effect of one factor/independent variable on the dependent variable, ignoring any effect any of the other factors/independent variables might have on the dependent variable. There is also, however, a significant interaction present in the data. As noted in Section I, the latter indicates that the effect of one factor is not consistent across all the levels of the other factor. It is important to note that the presence of an interaction can, in some instances, render one or more significant main effects meaningless, since it may require that the relationship described by a main effect be qualified. This is the case in Example 27.1, where the presence of an interaction reflects the fact that the nature of the relationship between the levels of a factor on which a significant main effect is detected will depend upon which level of the second factor is considered. Table 27.3, which summarizes the data for Example 27.1, will be used to illustrate this point. The six cells in the Table 27.3 contain the means of the $pq = 6$ groups. The values in the margins of the rows and columns of the table, respectively, represent the means of the levels of Factor A and Factor B. In Table 27.3 the average of any row or column can be obtained by adding all of the values in that row or column, and dividing the sum by the number of cells in that row or column.[5]

Table 27.3 Group and Marginal Means for Example 27.1

		Factor B (Temperature)			
		B_1 (Low)	B_2 (Moderate)	B_3 (High)	Row averages
Factor A (Humidity)	A_1 (Low)	10	7	4	7
	A_2 (High)	3	4	1	2.67
Column averages		6.5	5.5	2.5	Grand mean = 4.83

In Table 27.3 the main effect for Factor A (Humidity) indicates that as humidity increases the number of problems solved decreases (since $(\bar{X}_{A_1} = 7) > (\bar{X}_{A_2} = 2.67)$). Similarly, the main effect for Factor B (Temperature) indicates that as temperature increases, the number of problems solved decreases (since $(\bar{X}_{B_1} = 6.5) > (\bar{X}_{B_2} = 5.5) > (\bar{X}_{B_3} = 2.5)$). However, closer inspection of the data reveals that the effects of the factors on the dependent variable are not as straightforward as the main effects suggest. Specifically, the ordinal relationship depicted for the main effect on Factor B is only applicable to Level 1 of Factor A. Although under the low humidity condition (A_1) the number of problems solved decreases as temperature increases, the latter is not true for the high humidity condition (A_2). Under the latter condition the number of problems solved increases from 3 to 4 as temperature increases from low to moderate but then decreases to 1 under the high temperature condition. Thus, the main effect for Factor B is misleading, since it is based on the result of averaging the data from two rows which do not contain consistent patterns of information. In the same respect, if one examines the main effect on Factor A, it suggests that as humidity increases, performance decreases. Table 27.3, however, reveals that although this ordinal relationship is observed for all three levels of Factor B, the effect is much more pronounced for Level 1 (low temperature) than it is for either Level 2 (moderate temperature) or Level 3 (high temperature). Thus, even though the ordinal relationship described by the main effect is consistent across the three levels of Factor B, the magnitude of the relationship varies depending upon which level of Factor B is considered.

Figure 27.1 summarizes the information presented in Table 27.3 in a graphical format. Each of the points depicted in the graphs described by Figures 27.1a and 27.1b represents the average score of the group which corresponds to the level of the factor represented by the line on which that point falls and the level of the factor on the abscissa (X-axis) above which the point falls. An interaction is revealed on either graph when two or more of the lines are not equidistant from one another throughout the full length of the graph, as one moves from left to right. When two or more lines on a graph intersect with one another, as is the case in Figure 27.1a, or two or more lines diverge substantially from one another, as is the case in Figure 27.1b, it indicates the presence of an interaction. The ultimate determination, however, with respect to whether or not a significant interaction is present should always be based on the computed value of the F_{AB} ratio.

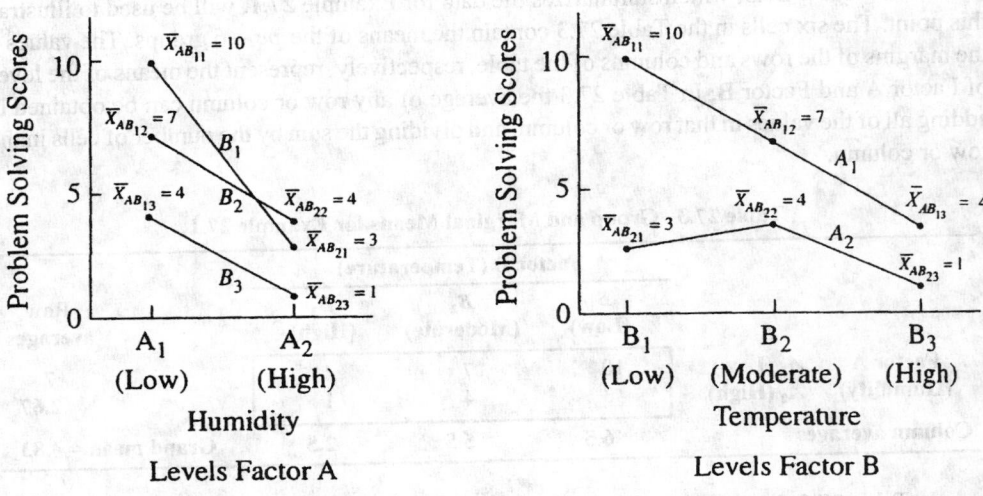

Figure 27.1   Graphical Summary of Results of Example 27.1

In an experiment in which there are two factors, either of two graphs can be employed to summarize the results of the study. In Figure 27.1a the levels of Factor A are represented on the abscissa, and three lines are employed to represent subjects' performance on each of the levels of Factor B (with reference to the specific levels of Factor A). In Figure 27.1b the levels of Factor B are represented on the abscissa, and two lines are employed to represent subjects' performance on each of the levels of Factor A (with reference to the specific levels of Factor B). As noted earlier, the fact that an interaction is present is reflected in Figures 27.1a and 27.1b, since the lines are not equidistant from one another throughout the length of both graphs.[6]

Table 27.4 and Figure 27.2 summarize a hypothetical set of data (for the same experiment described by Example 27.1) in which no interaction is present. For purposes of illustration it will be assumed that in this example the computed values F_A and F_B are significant, while F_{AB} is not.

Table 27.4 Hypothetical Values for Group and Marginal Means When There Is No Interaction

		Factor B (Temperature)			
		B_1 (Low)	B_2 (Moderate)	B_3 (High)	Row averages
Factor A (Humidity)	A_1 (Low)	10	8	6	8
	A_2 (High)	6	4	2	4
Column averages		8	6	4	Grand mean = 6

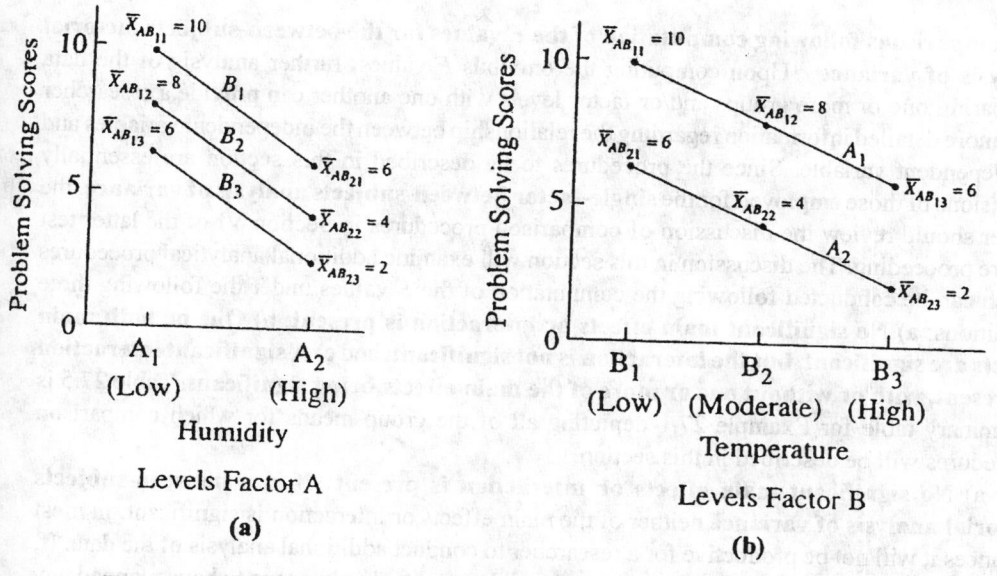

Figure 27.2 Graphical Summary of Results Described in Table 27.4

Inspection of Table 27.4 and Figure 27.2 indicates the presence of a main effect on both Factors A and B and the absence of an interaction. The presence of a main effect on Factor A is reflected in the fact that there is a reasonably large difference between $\overline{X}_{A_1} = 8$ and $\overline{X}_{A_2} = 4$. In the same respect, the significant main effect on Factor B is reflected in the discrepancy between the mean values $\overline{X}_{B_1} = 8$, $\overline{X}_{B_2} = 6$, and $\overline{X}_{B_3} = 4$. The conclusion that there is no interaction is based on the fact that the relationship described by each of the main effects is consistent across all of the levels of the second factor. To illustrate this, consider the main effect described for Factor A. In Table 27.4, the main effect for Factor A indicates that subjects solve 4 more problems under the low humidity condition than under the high humidity condition, and since this is the case regardless of which level of Factor B one considers, it indicates there is no interaction between the two factors. The absence of an interaction is reflected in Figure 27.2a, since the three lines are equidistant from one another.[7] In Table 27.4 the main effect for Factor B indicates that the number of problems solved decreases in steps of 2 as one progresses from low to moderate to high temperature. This pattern is consistent across both of the levels of Factor A. The absence of an interaction is also reflected in Figure 27.2b, since the two lines representing each of the levels of Factor A are equidistant from one another (as well as being parallel) throughout the length of the graph.

The term **additive model** is commonly employed to describe an analysis of variance in which there is no interaction (whereas the term **nonadditive** is employed when there is an interaction). The use of the term **additive** within this context reflects the fact that the mean of any of the $p \times q$ cells can be obtained by adding the row and column effects for that cell to the grand mean (Myers and Well, 1995). The concept of **additivity** is discussed further in Section IX (the Addendum) under the discussion of the **analysis of variance for a Latin square design (Test 27j)**.

VI. Additional Analytical Procedures for the Between-Subjects Factorial Analysis of Variance and/or Related Tests

1. Comparisons following computation of the F values for the between-subjects factorial analysis of variance Upon computing the omnibus F values, further analysis of the data comparing one or more groups and/or factor levels with one another can provide a researcher with more detailed information regarding the relationship between the independent variables and the dependent variable. Since the procedures to be described in this section are essentially extensions of those employed for the **single-factor between-subjects analysis of variance**, the reader should review the discussion of comparison procedures in Section VI of the latter test before proceeding. The discussion in this section will examine additional analytical procedures which can be conducted following the computation of the F values under the following three conditions: a) **No significant main effects or interaction is present**; b) **One or both main effects are significant, but the interaction is not significant**; and c) **A significant interaction is present, with or without one or more of the main effects being significant**. Table 27.5 is a summary table for Example 27.1 depicting all of the group means for which comparison procedures will be described in this section.

a) **No significant main effects or interaction is present** If in a **between-subjects factorial analysis of variance** neither of the main effects or interaction is significant, in most instances it will not be productive for a researcher to conduct additional analysis of the data. If, however, prior to the data collection phase of a study a researcher happens to have planned any of the specific types of analyses to be discussed later in this section, he can still conduct them

Table 27.5 Summary Table of Means for Example 27.1

		Factor B (Temperature)			Row averages
		B_1	B_2	B_3	
Factor A (Humidity)	A_1	$\bar{X}_{AB_{11}}$	$\bar{X}_{AB_{12}}$	$\bar{X}_{AB_{13}}$	$\bar{X}_{A_1}$
	A_2	$\bar{X}_{AB_{21}}$	$\bar{X}_{AB_{22}}$	$\bar{X}_{AB_{23}}$	$\bar{X}_{A_2}$
Column averages		$\bar{X}_{B_1}$	$\bar{X}_{B_2}$	$\bar{X}_{B_3}$	

regardless of whether or not any of the F values is significant (and not be obliged to control the value of α_{FW}). Although one can also justify conducting additional analytical procedures which are unplanned, in such a case most statisticians believe that a researcher should control the familywise Type I error rate (α_{FW}), in order that it not exceed what would be considered to be a reasonable level.

b) **One or both main effects are significant, but the interaction is not significant** When at least one of the F values is significant, the first question the researcher must ask prior to conducting any additional analytical procedures is whether or not the interaction is significant. When the interaction is not significant, a factorial design can essentially be conceptualized as being comprised of two separate single factor experiments. As such, both **simple** and **complex comparisons** can be conducted contrasting different means or sets of means which represent the levels of each of the factors. Such comparisons involve contrasting within a specific factor the **marginal means** (i.e., the means of the p rows and the means of the q columns). In the case of Example 27.1, a simple comparison can be conducted in which two of the three levels of Factor B are compared with one another (i.e., $\bar{X}_{B_1}$ versus $\bar{X}_{B_2}$), or a complex comparison in which a composite mean involving two levels of Factor B is compared with the mean of the third level of Factor B (i.e., $(\bar{X}_{B_2} + \bar{X}_{B_3})/2$ versus $\bar{X}_{B_1}$). If Factor B has four levels, a complex comparison contrasting two sets of composite means (each set representing a composite mean of two of the four levels) can be conducted (i.e., $(\bar{X}_{B_1} + \bar{X}_{B_2})/2$ versus $(\bar{X}_{B_3} + \bar{X}_{B_4})/2$). Since there are only two levels of Factor A, no additional comparisons are possible involving the means of the levels of that factor (i.e., the omnibus F value for Factor A represents the comparison $\bar{X}_{A_1}$ versus $\bar{X}_{A_2}$). As is the case for the **single-factor between-subjects analysis of variance**, in designs in which one or both of the factors are comprised of more than three levels, it is possible to conduct an omnibus F test comparing the means of three or more of the levels of a specific factor. In addition to all of the aforementioned comparisons, within a given level of a specific factor, simple and complex comparisons can be conducted which contrast the means of specific groups that are a combination of both factors (i.e., a simple comparison such as $\bar{X}_{AB_{11}}$ versus $\bar{X}_{AB_{12}}$, or a complex comparison such as $\bar{X}_{AB_{11}}$ versus $(\bar{X}_{AB_{12}} + \bar{X}_{AB_{13}})/2$).[8] It is worth reiterating that, whenever possible, comparisons should be planned prior to the data collection phase of a study, and that any comparisons which are conducted should address important theoretical and/or practical questions that underlie the hypotheses under study. In addition, the total number of comparisons that are conducted should be limited in number, and should not be redundant with respect to the information they provide.

c) **A significant interaction is present with or without one or more of the main effects being significant** As noted previously, when the interaction is significant the main effects may essentially be rendered meaningless, since any main effects will have to be qualified in reference to the levels of a second factor. Thus, any comparison which involves the levels of a specific factor (e.g., $\bar{X}_{B_1}$ versus $\bar{X}_{B_2}$) will reflect both the contribution of that factor, as well as the interaction between that factor and the second factor. For this reason, the most logical strategy to employ if a significant interaction is obtained is to test for what are referred to as **simple effects**. A test of a simple effect is essentially an analysis of variance evaluating all of the levels of one factor across only one level of the other factor. In the case of Example 27.1, two simple effects can be evaluated for Factor B. Specifically, an F test can be conducted which evaluates the scores of subjects on Factor B, but only for those subjects who serve under Level 1 of Factor A (i.e., an F ratio is computed for Groups AB_{11}, AB_{12}, and AB_{13}). A second simple effect for Factor B can be evaluated by contrasting the scores of subjects on Factor B, but only for those subjects who serve under Level 2 of Factor A (i.e., Groups AB_{21}, AB_{22}, and AB_{23}). In the case of Factor A, there are three possible simple effects that can be evaluated. Specifically, separate F tests can be conducted which evaluate the scores of subjects on Factor A for only those subjects who serve under: a) Level 1 of Factor B (i.e., Groups AB_{11} and AB_{21}); b) Level 2 of Factor B (i.e., Groups AB_{12} and AB_{22}); and c) Level 3 of Factor B (i.e., Groups AB_{13} and AB_{23}). In the event that one or more of the simple effects are significant, additional simple and complex comparisons contrasting specific groups within a given level of a factor can be conducted (e.g., a simple comparison such as $\bar{X}_{AB_{11}}$ versus $\bar{X}_{AB_{12}}$ or a complex comparison such as $\bar{X}_{AB_{11}}$ versus $(\bar{X}_{AB_{12}} + \bar{X}_{AB_{13}})/2$).

Description of analytical procedures (Including the following comparison procedures that are described for the **single-factor between-subjects analysis of variance** which, in this section, are described in reference to the **between-subjects factorial analysis of variance**: Test 27a: **Multiple t tests/Fisher's LSD test** (which is equivalent to **linear contrasts**); Test 27b: **The Bonferroni–Dunn test**; Test 27c: **Tukey's HSD test**; Test 27d: **The Newman–Keuls test**; Test 27e: **The Scheffé test**; Test 27f: **The Dunnett test**)

Comparisons between the marginal means The equations that are employed in conducting simple and complex comparisons involving the marginal means are basically the same equations which are employed for conducting comparisons for the **single-factor between-subjects analysis of variance**. Thus, in comparing two marginal means or two sets of marginal means (in the case of complex comparisons), **linear contrasts** can be conducted when no attempt is made to control the value of α_{FW} (which will generally be the case for planned comparisons).[9] In the case of either planned or unplanned comparisons where the value of α_{FW} is controlled, any of the multiple comparison procedures discussed under the **single-factor between-subjects analysis of variance** can be employed (i.e., the **Bonferroni–Dunn test**, **Tukey's HSD test**, the **Newman–Keuls test**, the **Scheffé test**, and the **Dunnett test**). The only difference in employing any of the latter comparison procedures with a factorial design is that the sample size employed in a comparison equation will reflect the number of subjects in each of the levels of the relevant factor. Thus, any comparison involving the marginal means of Factor A will involve nq subjects per group (in Example 27.1, $nq = (3)(3) = 9$), and any comparison involving the marginal means of Factor B will involve np subjects per group (in Example 27.1, $np = (3)(2) = 6$).

    As an example, assume we want to compare the scores of subjects on two of the levels of Factor B — specifically Level 1 versus Level 3 (i.e., $\bar{X}_{B_1}$ versus $\bar{X}_{B_3}$). If no attempt is made to control the value of α_{FW}, Equations 27.28–27.30 (which are the analogs of Equations 21.17–21.19 employed for conducting **linear contrasts** for the **single-factor between-subjects analysis of variance**) are employed to conduct a **linear contrast** (which is equivalent to **Fisher's LSD test**) comparing the two levels of Factor B (which within the framework of the comparison are conceptualized as two groups). Note that in Equation 27.28 the value np

represents the number of subjects who served under each level of Factor B, and $[\Sigma(c_{B_k})(\bar{X}_{B_k})]^2$ will equal the squared difference between the means of the two levels of Factor B that are being compared (i.e., in the case of the comparison under discussion, $[\Sigma(c_{B_k})(\bar{X}_{B_k})]^2 = (\bar{X}_{B_1} - \bar{X}_{B_3})^2$).

$$SS_{B\ comp} = \frac{np\left[\Sigma(c_{B_k})(\bar{X}_{B_k})\right]^2}{\Sigma c_{B_k}^2} \qquad \textbf{(Equation 27.28)}$$

$$MS_{B\ comp} = \frac{SS_{B\ comp}}{df_{B\ comp}} \qquad \textbf{(Equation 27.29)}$$

$$F_{B\ comp} = \frac{MS_{B\ comp}}{MS_{WG}} \qquad \textbf{(Equation 27.30)}$$

The data from Example 27.1 are now employed in Equations 27.28–27.30 to conduct the comparison $\bar{X}_{B_1}$ versus $\bar{X}_{B_3}$. Note that since Levels 1 and 3 of Factor B constitute the groups that are involved in the comparison, the coefficients for the comparison are $c_{B_1} = +1$, $c_{B_2} = 0$, $c_{B_3} = -1$. Thus, $\Sigma c_{B_k}^2 = 2$ and $[\Sigma(c_{B_k})(\bar{X}_{B_k})]^2 = (\bar{X}_{B_1} - \bar{X}_{B_3})^2 = (6.5 - 2.5)^2 = (4)^2 = 16$. Substituting the appropriate values in Equation 27.28, the value $SS_{B\ comp} = 48$ is computed.

$$SS_{B\ comp} = \frac{(3)(2)(4)^2}{2} = 48$$

Since all **linear contrasts** represent a **single degree of freedom comparison**, $df_{B\ comp} = 1$. Employing Equations 27.29 and 27.30, the values $MS_{B\ comp} = 48$ and $F_{B_{comp}} = 48$ are computed. Note that the value $MS_{WG} = 1$ computed for the omnibus F test is employed in the denominator of Equation 27.30.

$$MS_{B\ comp} = \frac{48}{1} = 48$$

$$F_{B\ comp} = \frac{48}{1} = 48$$

The value $F_{B\ comp} = 48$ is evaluated with **Table A10**. Employing the latter table, the appropriate degrees of freedom for the numerator and denominator are $df_{num} = df_{B\ comp} = 1$ and $df_{den} = df_{WG} = 12$. For $df_{num} = 1$ and $df_{den} = 12$, the tabled critical .05 and .01 values are $F_{.05} = 4.75$ and $F_{.01} = 9.33$. Since the obtained value $F_{comp} = 48$ is greater than the aforementioned critical values, the nondirectional alternative hypothesis $H_1: u_{B_1} \neq \mu_{B_3}$ is supported at both the .05 and .01 levels.

Equations 27.31–27.33 are employed to evaluate comparisons involving the levels of Factor A.

$$SS_{A\ comp} = \frac{nq[\Sigma(c_{A_j})(\bar{X}_{A_j})]^2}{\Sigma c_{A_j}^2} \qquad \textbf{(Equation 27.31)}$$

$$MS_{A\ comp} = \frac{SS_{A\ comp}}{df_{A\ comp}} \qquad \textbf{(Equation 27.32)}$$

$$F_{A\ comp} = \frac{MS_{A\ comp}}{MS_{WG}} \qquad \textbf{(Equation 27.33)}$$

Note that in Equation 27.31 nq represents the sample size, which in this case is the number of subjects who serve in each level of Factor A. The value $[\Sigma(c_{A_j})(\bar{X}_{A_j})]^2$ is equal to $(\bar{X}_{A_x} - \bar{X}_{A_y})^2$, which is the squared difference between the two means involved in the comparison (where x and y represent the levels of Factor A that are employed in the comparison).

As is the case with comparisons conducted for a **single-factor between-subjects analysis of variance**, a CD value can be computed for any comparison. Recollect that a CD value represents the minimum required difference in order for two means to differ significantly from one another. To demonstrate this, two CD values will be computed for the comparison $\bar{X}_{B_1}$ versus $\bar{X}_{B_3}$. Specifically, CD_{LSD} and CD_S will be computed. CD_{LSD} (which is the CD value associated with the **linear contrast** which is conducted with Equations 27.28–27.30) is the lowest possible difference that can be computed with any of the available comparison procedures. CD_S (the value for the **Scheffé test**), on the other hand, computes the largest CD value from the methods that are available. If the obtained difference for a comparison is less than CD_{LSD}, the null hypothesis will be retained, whereas if it is larger than CD_S it will be rejected. For the purpose of this discussion, it will be assumed that an obtained difference which is larger than CD_{LSD} but less than CD_S will be relegated to the **suspend judgement** category.[10]

Equation 27.34 is employed to compute the value $CD_{LSD} = 1.25$ for the **simple comparison** $\bar{X}_{B_1}$ versus $\bar{X}_{B_3}$, with $\alpha = .05$. In point of fact, the CD_{LSD} value computed with Equation 27.34 applies to all three simple comparisons which can be conducted with respect to Factor B (i.e., $\bar{X}_{B_1} - \bar{X}_{B_2} = 6.5 - 5.5 = 1$; $\bar{X}_{B_1} - \bar{X}_{B_3} = 6.5 - 2.5 = 4$; $\bar{X}_{B_2} - \bar{X}_{B_3} = 5.5 - 2.5 = 3$).

$$CD_{LSD} = \sqrt{F_{(1,WG)}}\sqrt{\frac{2MS_{WG}}{np}} = \sqrt{4.75}\sqrt{\frac{(2)(1)}{(3)(2)}} = 1.25 \quad \textbf{(Equation 27.34)}$$

In order to differ significantly at the .05 level, the means of any two levels of Factor B must differ from one another by at least 1.25 units. Thus, the differences $\bar{X}_{B_1} - \bar{X}_{B_3} = 4$ and $\bar{X}_{B_2} - \bar{X}_{B_3} = 3$ are significant, while the difference $\bar{X}_{B_1} - \bar{X}_{B_2} = 1$ is not.

Note that Equation 27.34 is identical to Equation 21.24 employed for computing the CD_{LSD} value for the **single-factor between-subjects analysis of variance**, except for the fact that in Equation 27.34, np subjects are employed per group/level of Factor B. In Equation 27.34, the value $F_{(1,WG)} = 4.75$ is the tabled critical .05 F value for df_{num} and df_{den}, which represent the degrees of freedom associated with the $F_{B\ comp}$ value computed with Equation 27.30.

Equations 27.35 and 27.36, which are analogous to Equation 21.25 (which is the generic equation for both simple and complex comparisons for CD_{LSD} for the **single-factor between-subjects analysis of variance**), are, respectively, the generic equations for Factors A and B for computing CD_{LSD}.

$$CD_{LSD} = \sqrt{F_{(1,WG)}}\sqrt{\frac{(\Sigma c_{A_j}^2)(MS_{WG})}{nq}} \qquad \textbf{(Equation 27.35)}$$

$$CD_{LSD} = \sqrt{F_{(1, WG)}} \sqrt{\frac{(\Sigma c_{B_k}^2)(MS_{WG})}{np}}$$ **(Equation 27.36)**

At this point the **Scheffé test** will be employed to conduct the simple comparison $\bar{X}_{B_1}$ versus $\bar{X}_{B_3}$. Equation 27.37, which is analogous to Equation 21.32 (which is the equation for simple comparisons for CD_S for the **single-factor between-subjects analysis of variance**), is employed to compute the value $CD_S = 1.61$, with $\alpha_{FW} = .05$. The value $F_{(B, WG)} = 3.89$ used in Equation 27.37 is the tabled critical .05 F value employed in evaluating the main effect for Factor B in the omnibus F test.

(Equation 27.37)

$$CD_S = \sqrt{(q - 1)F_{df_{(B, WG)}}} \sqrt{\frac{2MS_{WG}}{np}} = \sqrt{(3 - 1)(3.89)} \sqrt{\frac{(2)(1)}{(3)(2)}} = 1.61$$

Thus, in order to differ significantly at the .05 level, the means of any two levels of Factor B must differ from one another by at least 1.61 units. As is the case when $CD_{LSD} = 1.25$ is computed, the differences $\bar{X}_{B_1} - \bar{X}_{B_3} = 4$ and $\bar{X}_{B_2} - \bar{X}_{B_3} = 3$ are significant, while the difference $\bar{X}_{B_1} - \bar{X}_{B_2} = 1$ is not.

Equations 27.38 and 27.39, which are analogous to Equation 21.33 (which is the generic equation for both simple and complex comparisons for CD_S for the **single-factor between-subjects analysis of variance**), are respectively the generic equations for Factors A and B for computing CD_S. Note that in conducting comparisons involving the levels of Factor A, the value $F_{(A, WG)}$ employed in Equation 27.38 is the tabled critical F value at the prespecified level of significance used in evaluating the main effect for Factor A in the omnibus F test.

$$CD_S = \sqrt{(p - 1)F_{df_{(A, WG)}}} \sqrt{\frac{(\Sigma c_{A_j}^2)(MS_{WG})}{nq}}$$ **(Equation 27.38)**

$$CD_S = \sqrt{(q - 1)F_{df_{(B, WG)}}} \sqrt{\frac{(\Sigma c_{B_k}^2)(MS_{WG})}{np}}$$ **(Equation 27.39)**

In closing the discussion of the **Scheffé test**, it should be noted that since Equation 27.37 only takes into account those comparisons which are possible involving the levels of Factor B, it may not be viewed as imposing adequate control over α_{FW} if one intends to conduct additional comparisons involving the levels of Factor A and/or specific groups that are a combination of both factors. Because of this, some sources make a distinction between the **familywise error rate** (α_{FW}) and the **experimentwise error rate**. Although in the case of a single factor experiment the two values will be identical, in a multifactor experiment, a **familywise error rate** can be computed for comparisons within each factor as well as for comparisons between groups that are based on combinations of the factors. The **experimentwise error rate** will be a composite error rate which will be the result of combining all of the **familywise error rates**. Thus, in the above example if one intends to conduct additional comparisons involving the levels of Factor A and/or groups which are combinations of both factors, one can argue that the **Scheffé test** as employed does not impose sufficient control over the value of the **experimentwise error rate**. Probably the simplest way to deal with such a situation is to conduct a more conservative test in evaluating any null hypotheses involving the levels of Factor A, Factor B, or groups that are combinations of both factors (i.e., evaluate a null hypothesis at the .01 level instead of at the .05 level).

Evaluation of an omnibus hypothesis involving more than two marginal means   If the interaction is not significant, it is conceivable that a researcher may wish to conduct an F test on three or more marginal means in a design where the factor involved has four or more levels. In other words, if in Example 27.1 there were four levels on Factor B instead of three, one might want to evaluate the null hypothesis H_0: $\mu_{B_1} = \mu_{B_2} = \mu_{B_3}$. The logic that is employed in conducting such an analysis for the **single-factor between-subjects analysis of variance** can be extended to a factorial design. Specifically, in the case of a 2×4 design, a **between-subjects factorial analysis of variance** employing all of the data is conducted initially. Upon determining that the interaction is not significant, a **single-factor between-subjects analysis of variance** can then be conducted employing only the data for the three levels of Factor B in which the researcher is interested (i.e., B_1, B_2, and B_3). The following F ratio is computed: $F_{(B_1/B_2/B_3)} = MS_{BG_{(B_1/B_2/B_3)}} / MS_{WG}$. Note that the mean square value in the numerator is based on the between-groups variability in the **single-factor between-subjects analysis of variance** which involves only the data for levels B_1, B_2, and B_3 of Factor B. The degrees of freedom associated with the numerator of the F ratio is 2, since it is based on the number of levels of Factor B evaluated with the **single-factor between-subjects analysis of variance** (i.e., $df_{(B_1/B_2/B_3)} = 3 - 1 = 2$). The mean square and degrees of freedom for the denominator of the F ratio are the within-groups mean square and degrees of freedom computed for the **between-subjects factorial analysis of variance** when the full set of data is employed (i.e., the data for all four levels of Factor B). For further clarification of the aforementioned procedure the reader should review Section VI of the **single-factor between-subjects analysis of variance**.

Comparisons between specific groups that are a combination of both factors   The procedures employed for comparing the marginal means can also be employed to evaluate differences between specific groups which are a combination of both factors (e.g., a comparison such as $\bar{X}_{AB_{11}}$ versus $\bar{X}_{AB_{12}}$). Such differences are most likely to be of interest when an interaction is present. It should be noted that these are not the only types of comparisons that can provide more specific information regarding the nature of an interaction. A more comprehensive discussion of further analysis of an interaction can be found in books which specialize in the analysis of variance. Keppel (1991), among others, provides an excellent discussion of this general subject.

    In comparing specific groups with one another, the same equations are essentially employed which are used for the comparison of marginal means, except for the fact that the equations must be modified in order to accommodate the sample size of the groups. Both simple and complex comparisons can be conducted. As an example, let us assume we want to conduct a **linear contrast** for the simple comparison $\bar{X}_{AB_{11}}$ versus $\bar{X}_{AB_{12}}$. Equation 27.40 is employed for conducting such a comparison. Note that the latter equation has the same basic structure as Equations 27.28 and 27.31, but is based on the sample size of n, which is the sample size of each of the $p \times q$ groups.

$$SS_{comp} = \frac{n[\Sigma(c_{AB_{jk}})(\bar{X}_{AB_{jk}})]^2}{\Sigma c_{AB_{jk}}^2}$$ **(Equation 27.40)**

In Equation 27.40 the value $[\Sigma(c_{AB_{jk}})(\bar{X}_{AB_{jk}})]^2$ is equal to the squared difference between the means of the two groups which are being compared (i.e., for the comparison under discussion it yields the same value as $(\bar{X}_{AB_{11}} - \bar{X}_{AB_{12}})^2$). Note that since only two of the $p \times q = 6$ groups are involved in the comparison, the coefficients for the comparison are $c_{AB_{11}} = +1$, $c_{AB_{12}} = -1$, and $c_{AB_{jk}} = 0$ for the remaining four groups. Thus, $[\Sigma(c_{AB_{jk}})(\bar{X}_{AB_{jk}})]^2 = (\bar{X}_{AB_{11}} - \bar{X}_{AB_{12}})^2 = (10 - 7)^2 = (3)^3 = 9$ and $\Sigma c_{AB_{jk}}^2 = 2$. Substituting the appropriate values in Equation 27.40, the value $SS_{comp} = 13.5$ is computed.

$$SS_{comp} = \frac{(3)(3)^2}{2} = 13.5$$

Employing Equations 21.18 and 21.19, the values $MS_{comp} = 13.5$ and $F_{comp} = 13.5$ are computed.

$$MS_{comp} = \frac{SS_{comp}}{df_{comp}} = \frac{13.5}{1} = 13.5$$

$$F_{comp} = \frac{MS_{comp}}{MS_{WG}} = \frac{13.5}{1} = 13.5$$

Employing **Table A10**, the appropriate degrees of freedom for the numerator and denominator are $df_{num} = df_{comp} = 1$ (since the comparison is a **single degree of freedom comparison**) and $df_{den} = df_{WG} = 12$. For $df_{num} = 1$ and $df_{den} = 12$, the tabled critical .05 and .01 values are $F_{.05} = 4.75$ and $F_{.01} = 9.33$. Since the obtained value $F_{comp} = 13.5$ is greater than the aforementioned critical values, the nondirectional alternative hypothesis H_1: $\mu_{AB_{11}} \neq \mu_{AB_{12}}$ is supported at both the .05 and .01 levels.

CD_{LSD} and CD_S values will now be computed for the above comparison, with $\alpha = .05$. CD_{LSD} is computed with Equation 27.41 (which is identical to Equation 21.24 employed to compute CD_{LSD} for the **single-factor between-subjects analysis of variance**). Note that the sample size employed in Equation 27.41 is $n = n_{AB_{jk}} = 3$. Substituting the appropriate values in Equation 27.41, the value $CD_{LSD} = 1.78$ is computed.

$$CD_{LSD} = \sqrt{F_{(1, WG)}}\sqrt{\frac{2MS_{WG}}{n}} = \sqrt{4.75}\sqrt{\frac{(2)(1)}{3}} = 1.78 \quad \textbf{(Equation 27.41)}$$

Since in order to differ significantly at the .05 level the means of any two groups must differ from one another by at least 1.78 units, the difference $\bar{X}_{AB_{11}} - \bar{X}_{AB_{12}} = 3$ is significant. If we conduct comparisons for all 15 possible differences between pairs of groups (i.e., all simple comparisons), any difference which is equal to or greater than 1.78 units is significant at the .05 level.[11] Recollect, though, that since the computation of a CD_{LSD} value does not control the value of α_{FW}, the **per comparison Type I error rate** will equal .05.

Equation 27.42 (which is analogous to Equation 21.25 employed for the **single-factor between-subjects analysis of variance**) is the generic form of Equation 27.41 which can be employed for both simple and complex comparisons.

$$CD_{LSD} = \sqrt{F_{(1, WG)}} \sqrt{\frac{(\Sigma c_{AB_{jk}}^2)(MS_{WG})}{n}} \qquad \textbf{(Equation 27.42)}$$

CD_S is computed with Equation 27.43 (which is analogous to Equation 21.32 employed to compute CD_S for the **single-factor between-subjects analysis of variance**). Substituting the appropriate values in Equation 27.43, the value $CD_S = 3.22$ is computed. The value $F_{df_{(BG, WG)}} = 3.11$ used in Equation 27.43 is the tabled critical .05 F value for $df_{num} = df_{BG} = pq - 1 = (2)(3) - 1 = 5$ and $df_{den} = df_{WG} = 12$ employed in the omnibus F test.

$$CD_S = \sqrt{(pq - 1)F_{df_{(BG, WG)}}} \sqrt{\frac{2MS_{WG}}{n}} \qquad \textbf{(Equation 27.43)}$$

$$= \sqrt{[(2)(3) - 1](3.11)} \sqrt{\frac{(2)(1)}{3}} = 3.22$$

Thus, in order for any pair of means to differ significantly, the difference between the two means must be equal to or greater than 3.22 units. Since the difference $\bar{X}_{AB_{11}} - \bar{X}_{AB_{12}} = 3$ is less than $CD_S = 3.22$, the null hypothesis cannot be rejected if the **Scheffé test** is employed. Thus, the nondirectional alternative hypothesis $H_1: \mu_{AB_{11}} \neq \mu_{AB_{12}}$ is not supported.

Equation 27.44 (which is analogous to Equation 21.33 employed for the **single-factor between-subjects analysis of variance**) is the generic form of Equation 27.43 that can be employed for both simple and complex comparisons.

$$CD_S = \sqrt{(pq - 1)F_{df_{(BG, WG)}}} \sqrt{\frac{(\Sigma c_{AB_{jk}}^2)(MS_{WG})}{n}} \qquad \textbf{(Equation 27.44)}$$

Since the **linear contrast** procedure yields a significant difference and the **Scheffé test** does not, one might want to **suspend judgement** with respect to the comparison $\bar{X}_{AB_{11}}$ versus $\bar{X}_{AB_{12}}$ until a replication study is conducted. However, it is certainly conceivable that many researchers might consider the **Scheffé test** to be too conservative a procedure. Thus, one might elect to use a less conservative procedure such as **Tukey's HSD test**. Equation 27.45 (which is analogous to Equation 21.31 employed to compute CD_{HSD} for the **single-factor between-subjects analysis of variance**) is employed to compute CD_{HSD}. The value $q_{(pq, df_{WG})} = 4.75$ in Equation 27.45 is the value of the **Studentized range statistic** in **Table A13 (Table of the Studentized Range Statistic)** in the **Appendix** for $k = pq = 6$ and $df_{WG} = 12$.

$$CD_{HSD} = q_{(pq, df_{WG})} \sqrt{\frac{MS_{WG}}{n}} = 4.75 \sqrt{\frac{1}{3}} = 2.74 \qquad \textbf{(Equation 27.45)}$$

Since $CD_{HSD} = 2.74$ is less than $\bar{X}_{AB_{11}} - \bar{X}_{AB_{12}} = 3$, we can conclude that the difference between the groups is significant.[12]

Hsu (1996), Maxwell and Delaney (2004) and Rosenthal *et al.* (2000) discuss alternative methodologies for contrasting means within the framework of the **between-subjects factorial analysis of variance**.

The computation of a confidence interval for a comparison The same procedure described for computing a confidence interval for a comparison for the **single-factor between-subjects analysis of variance** can also be employed for the **between-subjects factorial analysis of variance**. Specifically, the following procedure is employed for computing a confidence interval for any of the methods described in this section: The obtained CD value is added to and subtracted from the obtained difference between the two means (or sets of means in the case of a complex comparison). The resulting range of values defines the confidence interval. The 95% confidence interval will be associated with a computed $CD_{.05}$ value, and the 99% confidence interval will be associated with a computed $CD_{.01}$ value. To illustrate the computation of a confidence interval, the 95% confidence interval for the value $CD_{HSD} = 2.74$ computed for the comparison $\bar{X}_{AB_{11}}$ versus $\bar{X}_{AB_{12}}$ is demonstrated below.

$$CI_{.95} = (\bar{X}_{AB_{11}} - \bar{X}_{AB_{12}}) \pm CD_{HSD} = 3 \pm 2.74$$

Thus, the researcher can be 95% confident (or the probability is .95) the interval .26 to 5.74 contains the true difference between the means of the populations represented by Group AB_{11} and Group AB_{12}. If the aforementioned interval contains the true difference between the population means, the mean of the population Group AB_{11} represents is between .26 to 5.74 units larger than the mean of the population Group AB_{12} represents. This result can be stated symbolically as follows: $.26 \leq (\mu_{AB_{11}} - \mu_{AB_{12}}) \leq 5.74$.

Analysis of simple effects Earlier in this section it was noted that the most logical strategy to employ when a significant interaction is detected is to initially test for what is referred to as **simple effects**. A test of a **simple effect** is essentially an analysis of variance evaluating all of the levels of one factor across only one level of the other factor. The analysis of simple effects will be illustrated with Example 27.1 by evaluating the simple effects of Factor B. Specifically, an F test will be conducted to evaluate the scores of subjects on the three levels of Factor B, but only the nq subjects who served under Level 1 of Factor A (i.e., an F ratio will be computed evaluating the Groups AB_{11}, AB_{12}, and AB_{13}). This represents the analysis of the simple effect of Factor B at level A_1. An analysis of a second simple effect (which represents the analysis of the simple effect of Factor B at level A_2) will evaluate the scores of subjects on the three levels of Factor B, but only the $(n_{AB_{jk}})(q)$ subjects who served under Level 2 of Factor A (i.e., Groups AB_{21}, AB_{22}, and AB_{23}).

Although it will not be done in reference to Example 27.1, since an interaction is present a comprehensive analysis of the data would also involve evaluating the simple effects of Factor A. There are three simple effects of Factor A which can be evaluated, each one involving comparing the scores of subjects on Factor A, but employing only the $(n_{AB_{jk}})(p)$ subjects who served under one of the three levels of Factor B. The three simple effects of Factor A involve the following contrasts: 1) The simple effect of Factor A at Level B_1: $\bar{X}_{AB_{11}}$ versus $\bar{X}_{AB_{21}}$; 2) The simple effect of Factor A at Level B_2: $\bar{X}_{AB_{12}}$ versus $\bar{X}_{AB_{22}}$; and 3) The simple effect of Factor A at Level B_3: $\bar{X}_{AB_{13}}$ versus $\bar{X}_{AB_{23}}$.[13]

In order to evaluate a simple effect, it is necessary to initially compute a sum of squares for the specific effect. Thus, in evaluating the simple effects of Factor B it is necessary to compute a sum of squares for Factor B at Level 1 of Factor A ($SS_{B \text{ at } A_1}$) and a sum of squares for Factor B at Level 2 of Factor A ($SS_{B \text{ at } A_2}$). Upon computing all of the sums of squares for the simple effects for a specific factor, F ratios are computed for each of the simple effects by dividing the mean square for a simple effect (which is obtained by dividing the simple effect sum of squares by its degrees of freedom) by the within-groups mean square derived for the factorial analysis of variance. This procedure will now be demonstrated for the simple effects of Factor B.

Equation 27.46 (which is equivalent to Equation 21.3 employed for computing the **between-groups sum of squares** for the **single-factor between-subjects analysis of variance**) is employed to compute the sum of squares for each of the simple effects. If ΣX_{AB_j} represents the sum of the scores on Level j of Factor A of subjects who serve under a specific level of Factor B, the notation $\Sigma[(\Sigma X_{AB_j})^2 / n_{AB_j}]$ in Equation 27.46 indicates that the sum of the scores for each level of Factor B at a given level of Factor A is squared, divided by n_{AB_j}, and the q resulting values are summed. The notation $(\Sigma\Sigma X_{AB_j})^2$ represents the square of the sum of scores of the $(n_{AB_j})(q)$ subjects who serve under the specified level of Factor A.[14]

(Equation 27.46)

$$SS_{B \text{ at } A_j} = \Sigma \left[\frac{(\Sigma X_{AB_j})^2}{n_{AB_j}} \right] - \frac{(\Sigma\Sigma X_{AB_j})^2}{(n_{AB_j})(q)}$$

$$SS_{B \text{ at } A_1} = \Sigma \left[\frac{(\Sigma X_{AB_1})^2}{n_{AB_1}} \right] - \frac{(\Sigma\Sigma X_{AB_1})^2}{(n_{AB_1})(q)} = \left[\frac{(30)^2}{3} + \frac{(21)^2}{3} + \frac{(12)^2}{3} \right] - \frac{(63)^2}{(3)(3)} = 54$$

$$SS_{B \text{ at } A_2} = \Sigma \left[\frac{(\Sigma X_{AB_2})^2}{n_{AB_2}} \right] - \frac{(\Sigma\Sigma X_{AB_2})^2}{(n_{AB_2})(q)} = \left[\frac{(9)^2}{3} + \frac{(12)^2}{3} + \frac{(3)^2}{3} \right] - \frac{(24)^2}{(3)(3)} = 14$$

Table 27.6 summarizes the analysis of variance for the simple effects of Factor B. Note that for each of the simple effects, the degrees of freedom for the effect is $df_{B \text{ at } A_j} = q - 1 = 3 - 1 = 2$ (which equals df_B employed for the **between-subjects factorial analysis of variance**). The mean square for each simple effect is obtained by dividing the sum of squares for the simple effect by its degrees of freedom. The F value for each simple effect is obtained by dividing the mean square for the simple effect by $MS_{WG} = 1$ computed for the factorial analysis of variance. Thus, $F_{B \text{ at } A_1} = 27/1 = 27$ and $F_{B \text{ at } A_2} = 7/1 = 7$.

Table 27.6   Analysis of Simple Effects of Factor B

Source of variation	SS	df	MS	F
B at A_1	54	2	27	27
B at A_2	14	2	7	7
Within-groups	12	12	1	

Employing **Table A10**, the degrees of freedom used in evaluating each of the simple effects are $df_{num} = df_{B \text{ at } A_j} = 2$ and $df_{den} = df_{WG} = 12$. Since both of the obtained values $F_{B \text{ at } A_1} = 27$ and $F_{B \text{ at } A_2} = 7$ are greater than $F_{.05} = 3.89$ and $F_{.01} = 6.93$ (which are the tabled critical values for $df_{num} = 2$ and $df_{den} = 12$), each of the simple effects is significant at both the .05 and .01 levels. On the basis of the result of the analysis of the simple effects of Factor B, we can conclude that within each level of Factor A there is at least one simple or complex comparison involving the levels of Factor B that is significant.

As noted earlier, when one or more of the simple effects is significant, additional simple and complex comparisons contrasting specific groups can be conducted. Thus, for Level 1 of Factor A, simple comparisons between $\bar{X}_{AB_{11}}$, $\bar{X}_{AB_{12}}$, and $\bar{X}_{AB_{13}}$, as well as complex comparisons (such as $\bar{X}_{AB_{11}}$ versus $(\bar{X}_{AB_{12}} + \bar{X}_{AB_{13}})/2$) can clarify the locus of the significant simple effect.

If the homogeneity of variance assumption of the **between-subjects factorial analysis of variance** (which is discussed in the next section) is violated, in computing the F ratios for the simple effects a researcher can justify employing a MS_{WG} value that is just based on the groups involved in analyzing a specific simple effect, instead of the value of MS_{WG} computed for the factorial analysis of variance. If the latter is done, the within-groups degrees of freedom employed in the analysis of the simple effects of Factor B becomes $df_{WG} = q(n - 1)$ instead of $df_{WG} = pq(n - 1)$. Since the within-groups degrees of freedom is smaller if $df_{WG} = q(n - 1)$ is employed, the test will be less powerful than a test employing $df_{WG} = pq(n - 1)$. The loss of power can be offset, however, if the new value for MS_{WG} is lower than the value derived for the omnibus F test.[15]

The reader should take note of the fact that the variability within each of the simple effects is the result of contributions from both the main effect on the factor for which the simple effect is being evaluated (Factor B in our example), as well as any interaction between the two factors. For this reason, the total of the sum of squares for each of the simple effects for a given factor will be equal to the interaction sum of squares (SS_{AB}) plus the sum of squares for that factor (SS_B). This can be confirmed by the fact that in our example the following is true.

$$[(SS_{B \text{ at } A_1} = 54) + (SS_{B \text{ at } A_2} = 14)] = [(SS_{AB} = 16) + (SS_B = 52)] = 68$$

It should be noted that analysis of simple effects in and of itself cannot provide definitive evidence with regard to the presence or absence of an interaction. In point of fact, it is possible for only one of two simple effects to be significant, and yet the value of F_{AB} computed for the factorial analysis of variance may not be significant. For a full clarification of this issue the reader should consult Keppel (1991) or Keppel and Wickens (2004).

2. Evaluation of the homogeneity of variance assumption of the between-subjects factorial analysis of variance The homogeneity of variance assumption discussed in reference to the **single-factor between-subjects analysis of variance** is also an assumption of the **between-subjects factorial analysis of variance**. Since both tests employ the same protocol in evaluating this assumption, prior to reading this section the reader should review the relevant material for evaluating the homogeneity of variance assumption (through use of **Hartley's F_{max} test (Test 11a)**) in Section VI of the **single-factor between-subjects analysis of variance** (as well as the material on **Hartley's F_{max} test** in Section VI of the *t* **test for two independent samples**).

In the case of the **between-subjects factorial analysis of variance**, evaluation of the homogeneity of variance assumption requires the researcher to compute the estimated population variances for each of the pq groups. The latter values are computed with Equation I.8. As it turns out, the value of the estimated population variance for all six groups equals $\hat{s}^2_{AB_{jk}} = 1$. This is demonstrated below for Group AB_{11}.

$$\tilde{s}_{AB_{11}}^2 = \frac{\Sigma X_{AB_{11}}^2 - \dfrac{(\Sigma X_{AB_{11}})^2}{n_{AB_{11}}}}{n_{AB_{11}} - 1} = \frac{302 - \dfrac{(30)^2}{3}}{3 - 1} = 1$$

Upon determining that the value of both the largest and smallest of the estimated population variances equals 1, Equation 21.37 is employed to compute the value of the F_{max} statistic. Employing Equation 21.37, the value $F_{max} = 1$ is computed.

$$F_{max} = \frac{\tilde{s}_L^2}{\tilde{s}_S^2} = \frac{1}{1} = 1$$

In order to reject the null hypothesis (H_0: $\sigma_L^2 = \sigma_S^2$) and thus conclude that the homogeneity of variance assumption is violated, the obtained F_{max} value must be equal to or greater than the tabled critical value at the prespecified level of significance. Employing **Table A9 (Table of the F_{max} Distribution)** in the **Appendix**, we determine that the tabled critical F_{max} values for $n = n_{AB_{jk}} = 3$ and $k = pq = 6$ groups are $F_{max_{05}} = 266$ and $F_{max_{01}} = 1362$. Since the obtained value $F_{max} = 1$ is less than $F_{max_{05}} = 266$, the null hypothesis cannot be rejected. In other words, the alternative hypothesis indicating the presence of heterogeneity of variance is not supported. The latter should be obvious without the use of the F_{max} test, since the same value is computed for the variance of each of the groups.

In instances where the homogeneity of variance assumption is violated, the researcher should employ one of the strategies recommended for heterogeneity of variance which are discussed in Section VI of the **single-factor between-subjects analysis of variance**. The simplest strategy is to use a more conservative test (i.e., employ a lower α level) in evaluating the three sets of hypotheses for the factorial analysis of variance.[16]

3. Computation of the power of the between-subjects factorial analysis of variance Prior to reading this section the reader should review the procedure described for computing the power of the **single-factor between-subjects analysis of variance**, since the latter procedure can be generalized to the **between-subjects factorial analysis of variance**. In determining the appropriate sample size for a factorial design, a researcher must consider the predicted effect size for each of the factors, as well as the magnitude of any predicted interactions. Thus, in the case of Example 27.1, prior to the experiment a separate power analysis can be conducted with respect to the main effect for Factor A, the main effect for Factor B, and the interaction between the two factors. The sample size the researcher should employ will be the largest of the sample sizes derived from analyzing the predicted effects associated with the two factors and the interaction. As is the case for the **single-factor between-subjects analysis of variance**, such an analysis will require the researcher to estimate the means of all of the experimental groups, as well as the value of error/within-groups variability (i.e., σ_{WG}^2).

Equation 27.47, which contains the same basic elements that comprise Equation 21.38, is the general equation which is employed for determining the minimum sample size necessary in order to achieve a specified power with regard to either of the main effects or the interaction.

$$\varphi = \sqrt{(\text{number of observations}) \left[\frac{\Sigma d^2}{(df_{effect} + 1)(\sigma_{WG}^2)} \right]} \qquad \textbf{(Equation 27.47)}$$

The following should be noted with respect to Equation 27.47:

a) The value employed for the number of observations will equal nq for Factor A, np for Factor B, and n for the interaction.

b) Σd^2 represents the sum of the squared deviation scores. This value is obtained as follows: 1) For **Factor A,** p deviation scores are computed by subtracting the estimated grand mean (μ_G) from each of the estimated means of the levels of Factor A (i.e., $d_{A_j} = \mu_{A_j} - \mu_G$). Σd^2, the sum of the squared deviation scores, is obtained by squaring the p deviation scores and summing the resulting values; 2) For **Factor B,** q deviation scores are computed by subtracting the estimated grand mean from each of the estimated means of the levels of Factor B (i.e., $d_{B_k} = \mu_{B_k} - \mu_G$). The sum of the squared deviation scores is obtained by squaring the q deviation scores and summing the resulting values; and 3) For the interaction, pq deviation scores are computed — one for each of the groups. A deviation score is computed for each group by employing the following equation: $d_{AB_{jk}} = \mu_{AB_{jk}} - \mu_{A_j} - \mu_{B_k} + \mu_G$. The latter equation indicates the following: The mean of the group is estimated ($\mu_{AB_{jk}}$), after which both the estimated mean of the level of Factor A the group serves under (μ_{A_j}) and the estimated mean of the level of Factor B the group serves under (μ_{B_k}) are subtracted from the estimated mean of the group. The estimated grand mean (μ_G) is then added to this result. The resulting value represents the deviation score for that group. Upon computing a deviation score for each of the pq groups, the pq deviation scores are squared, after which the resulting squared deviation scores are summed. The resulting value equals Σd^2.

c) ($df_{\text{effect}} + 1$) for Factor A equals $df_A + 1 = p$. ($df_{\text{effect}} + 1$) for Factor B equals $df_B + 1 = q$. ($df_{\text{effect}} + 1$) for the interaction equals $df_{AB} + 1 = (p - 1)(q - 1) + 1$.

d) σ^2_{WG} is the estimate of the population variance for any one of the pq groups (which are assumed to have equal variances if the homogeneity of variance assumption is true). If a power analysis is conducted after the data collection phase of a study, it is logical to employ MS_{WG} as the estimate of σ^2_{WG}.

To illustrate the use of Equation 27.47, the power of detecting the main effect on Factor B will be computed. Let us assume that based on previous research, prior to evaluating the data, we estimate the following values: $\mu_{B_1} = 7$, $\mu_{B_2} = 5$, $\mu_{B_3} = 3$. Since we know that $\mu_G = (\mu_{B_1} + \mu_{B_2} + \mu_{B_3})/q$, $\mu_G = (7 + 5 + 3)/3 = 5$.[17] It will also be assumed that the estimated value for error variability is $\sigma^2_{WG} = 1.5$. The relevant values are now substituted in Equation 27.47.

$$\varphi = \sqrt{np\left[\frac{\Sigma(\mu_{B_k} - \mu_G)^2}{(q)(\sigma^2_{WG})}\right]} = \sqrt{n(2)\left[\frac{(7 - 5)^2 + (5 - 5)^2 + (3 - 5)^2}{(3)(1.5)}\right]} = 1.89\sqrt{n}$$

If we employ $n = 3$ subjects per group (as is the case in Example 27.1), the value $\varphi = 3.27$ is computed: $\varphi = 1.89\sqrt{3} = 3.27$. Employing **Table A15 (Graphs of the Power Function for the Analysis of Variance)** in the **Appendix**, we use the set of power curves for $df_{\text{num}} = df_{\text{effect}} = df_B = 2$, and within that set employ the curve for $df_{\text{den}} = df_{WG} = 12$, for $\alpha = .05$. Since a perpendicular line erected from the value $\varphi = 3.27$ on the abscissa to the curve for $df_{WG} = 12$ is beyond the highest point on the curve, the power of the test for the estimated effect on Factor B will be 1 if $n = 3$ subjects are employed per group. Thus, there is a 100% likelihood that an effect equal to or larger than the one stipulated by the values employed in Equation 27.47 will be detected.

Although it will not be demonstrated here, to conduct a thorough power analysis it is necessary to also determine the minimum required sample sizes required to achieve what a researcher would consider to be the minimum acceptable power for identifying the estimated effects for Factor A and the interaction. The largest of the values computed for n for each of the three power analyses is the sample size which should be employed for each of the pq groups in the study. For a more comprehensive discussion on computing the power of the **between-subjects factorial analysis of variance** the reader should consult Cohen (1977, 1988).

4. Measures of magnitude of treatment effect for the between-subjects factorial analysis of variance: Omega squared (Test 27g) and Cohen's f index (Test 27h) Prior to reading this section the reader should review the discussion of magnitude of treatment effect in Section VI of both the *t* **test for two independent samples** and the **single-factor between-subjects analysis of variance**. The discussion for the latter test notes that the computation of an omnibus F value only provides a researcher with information regarding whether or not the null hypothesis can be rejected — i.e., whether a significant difference exists between at least two of the experimental treatments within a given factor. An F value (as well as the level of significance with which it is associated), however, does not provide the researcher with any information regarding the size of any treatment effect that is present. As is noted in earlier discussions of treatment effect, the latter is defined as the proportion of the variability on the dependent variable which is associated with the independent variable/experimental treatments. The measures described in this section are variously referred to as **measures of effect size**, **measures of magnitude of treatment effect**, **measures of association**, and **correlation coefficients**.

Omega squared (Test 27g) The **omega squared** statistic is a commonly computed measure of treatment effect for the **between-subjects factorial analysis of variance**. Keppel (1991) and Kirk (1995) note that there is disagreement with respect to which variance components should be employed in computing **omega squared** for a factorial design. One method of computing **omega squared** (which computes a value referred to as **standard omega squared**) was employed in the first edition of this book. The latter method expresses treatment variability for each of the factors as a proportion of the sum of all the elements which account for variability in a between-subjects factorial design (i.e., the variability for a given factor is divided by the sum of variability for all of the factors, interactions, and within-groups variability). A second method for computing **omega squared** (discussed in the second edition of the book) computes what is referred to as **partial omega squared** (which was also computed in reference to the **single-factor within-subjects analysis of variance**). In computing the latter measure, which Keppel (1991) and Kirk (1995) view as more meaningful than **standard omega squared**, the proportion of variability for a given factor is divided by the sum of the proportion of variability for that factor and within-groups variability (i.e., variability attributable to other factors and interactions is ignored).

Equations 27.51–27.53, respectively, summarize the elements which are employed to compute **standard omega squared** ($\tilde{\omega}_s^2$) for Factors A and B and the AB interaction (i.e., $\tilde{\omega}_{sA}^2$, $\tilde{\omega}_{sB}^2$, and $\tilde{\omega}_{sAB}^2$ represent **standard omega squared** for Factors A and B and the AB interaction). Equations 27.48–27.50 represent the population parameters (ω_{sA}^2, ω_{sB}^2, and ω_{sAB}^2) estimated by Equations 27.51–27.53.

$$\omega_{sA}^2 = \frac{\sigma_A^2}{\sigma_A^2 + \sigma_B^2 + \sigma_{AB}^2 + \sigma_{WG}^2} \qquad \textbf{(Equation 27.48)}$$

$$\omega_{sB}^2 = \frac{\sigma_B^2}{\sigma_A^2 + \sigma_B^2 + \sigma_{AB}^2 + \sigma_{WG}^2} \qquad \textbf{(Equation 27.49)}$$

$$\omega_{sAB}^2 = \frac{\sigma_{AB}^2}{\sigma_A^2 + \sigma_B^2 + \sigma_{AB}^2 + \sigma_{WG}^2} \qquad \textbf{(Equation 27.50)}$$

$$\tilde{\omega}_{sA}^2 = \frac{\tilde{\sigma}_A^2}{\tilde{\sigma}_A^2 + \tilde{\sigma}_B^2 + \tilde{\sigma}_{AB}^2 + \tilde{\sigma}_{WG}^2} \qquad \textbf{(Equation 27.51)}$$

$$\tilde{\omega}_{sB}^2 = \frac{\tilde{\sigma}_B^2}{\tilde{\sigma}_A^2 + \tilde{\sigma}_B^2 + \tilde{\sigma}_{AB}^2 + \tilde{\sigma}_{WG}^2} \qquad \textbf{(Equation 27.52)}$$

$$\tilde{\omega}_{sAB}^2 = \frac{\tilde{\sigma}_{AB}^2}{\tilde{\sigma}_A^2 + \tilde{\sigma}_B^2 + \tilde{\sigma}_{AB}^2 + \tilde{\sigma}_{WG}^2} \qquad \textbf{(Equation 27.53)}$$

Equations 27.57–27.59, respectively, summarize the elements which are employed to compute **partial omega squared** ($\tilde{\omega}_p^2$) for Factors A and B and the AB interaction. Equations 27.54–27.56 represent the population parameters estimated by Equations 27.57–27.59.

$$\omega_{pA}^2 = \frac{\sigma_A^2}{\sigma_A^2 + \sigma_{WG}^2} \qquad \textbf{(Equation 27.54)}$$

$$\omega_{pB}^2 = \frac{\sigma_B^2}{\sigma_B^2 + \sigma_{WG}^2} \qquad \textbf{(Equation 27.55)}$$

$$\omega_{pAB}^2 = \frac{\sigma_{AB}^2}{\sigma_{AB}^2 + \sigma_{WG}^2} \qquad \textbf{(Equation 27.56)}$$

$$\tilde{\omega}_{pA}^2 = \frac{\tilde{\sigma}_A^2}{\tilde{\sigma}_A^2 + \tilde{\sigma}_{WG}^2} \qquad \textbf{(Equation 27.57)}$$

$$\tilde{\omega}_{pB}^2 = \frac{\tilde{\sigma}_B^2}{\tilde{\sigma}_B^2 + \tilde{\sigma}_{WG}^2} \qquad \textbf{(Equation 27.58)}$$

$$\tilde{\omega}_{pAB}^2 = \frac{\tilde{\sigma}_{AB}^2}{\tilde{\sigma}_{AB}^2 + \tilde{\sigma}_{WG}^2} \qquad \textbf{(Equation 27.59)}$$

Where:

$$\tilde{\sigma}_A^2 = \frac{df_A(MS_A - MS_{WG})}{npq} = \frac{(1)(84.5 - 1)}{(3)(2)(3)} = 4.64$$

$$\tilde{\sigma}_B^2 = \frac{df_B(MS_B - MS_{WG})}{npq} = \frac{(2)(26 - 1)}{(3)(2)(3)} = 2.78$$

$$\tilde{\sigma}^2_{AB} = \frac{df_{AB}(MS_{AB} - MS_{WG})}{npq} = \frac{(2)(8 - 1)}{(3)(2)(3)} = .78$$

$$\tilde{\sigma}^2_{WG} = MS_{WG} = 1$$

Thus:

$$\tilde{\omega}^2_{sA} = \frac{4.64}{4.64 + 2.78 + .78 + 1} = .50$$

$$\tilde{\omega}^2_{sB} = \frac{2.78}{4.64 + 2.78 + .78 + 1} = .30$$

$$\tilde{\omega}^2_{sAB} = \frac{.78}{4.64 + 2.78 + .78 + 1} = .08$$

$$\tilde{\omega}^2_{pA} = \frac{4.64}{4.64 + 1} = .82$$

$$\tilde{\omega}^2_{pB} = \frac{2.78}{2.78 + 1} = .74$$

$$\tilde{\omega}^2_{pAB} = \frac{.78}{.78 + 1} = .44$$

Equations 27.60–27.62 can also be employed to compute the values of **partial omega squared**.

$$\tilde{\omega}^2_{pA} = \frac{(p - 1)(F_A - 1)}{(p - 1)(F_A - 1) + npq}$$

(Equation 27.60)

$$= \frac{(2 - 1)(84.5 - 1)}{(2 - 1)(84.5 - 1) + (3)(2)(3)} = .82$$

$$\tilde{\omega}^2_{pB} = \frac{(q - 1)(F_B - 1)}{(q - 1)(F_B - 1) + npq}$$

(Equation 27.61)

$$= \frac{(3 - 1)(26 - 1)}{(3 - 1)(26 - 1) + (3)(2)(3)} = .74$$

$$\tilde{\omega}^2_{pAB} = \frac{(p - 1)(q - 1)(F_{AB} - 1)}{(p - 1)(q - 1)(F_{AB} - 1) + npq}$$

(Equation 27.62)

$$= \frac{(2 - 1)(3 - 1)(8 - 1)}{(2 - 1)(3 - 1)(8 - 1) + (3)(2)(3)} = .44$$

The results of the above analysis for **standard omega squared** indicate that 50% of the variability on the dependent variable is associated with Factor A (Humidity), 30% with Factor B (Temperature), and 8% with the AB interaction. Thus, 50% + 30% + 8% = 88% of the variability on the dependent variable (problem-solving scores) is associated with variability on the two factors/independent variables and the interaction between them. It should be noted that although in some instances a small value for **omega squared** may indicate that the contribution of a factor or the interaction is trivial, this will not always be the case. Thus, in the example under discussion, although the value $\tilde{\omega}^2_{sAB} = .08$ is small relative to the **omega squared** values computed for the main effects, inspection of the data clearly indicates that in order to understand the influence of temperature on problem-solving scores, it is imperative that one take into account the level of humidity, and vice versa.

The values computed for **partial omega squared** indicate that 82% of the variability on the dependent variable is associated with Factor A (Humidity), 74% with Factor B (Temperature), and 44% with the AB interaction. Note that since the value computed for **partial omega squared** for a given factor does not take into account variability on the other factors or the interaction, it yields a much higher value for that factor and the interaction than **standard omega squared** computed for the same factor and the interaction. You should also note that when **partial omega squared** is computed, the sum of the proportions/percentage values can exceed 1/100%.

It was noted in an earlier discussion of **omega squared** (in Section VI of the *t* test **for two independent samples**) that Cohen (1977; 1988, pp. 284–287) has suggested the following (admittedly arbitrary) values, which are employed in psychology and a number of other disciplines, as guidelines for interpreting $\tilde{\omega}^2$: a) A **small effect size** is one that is greater than .0099 but not more than .0588; b) A **medium effect size** is one that is greater than .0588 but not more than .1379; and c) A **large effect size** is greater than .1379. If one employs Cohen's (1977, 1988) guidelines for magnitude of treatment effect, all of the **omega squared** values computed in this section represent a large treatment effect, with the exception of $\tilde{\omega}^2_{sAB} = .08$, which represents a medium effect.[18]

Grissom and Kim (2005, p.142–143) note **omega squared** values computed for a factorial design should be interpreted with caution because of the following: a) It is not necessarily the case that an **omega squared** value for a specific effect whose corresponding *F* value is statistically significant will be larger than the **omega squared** value for another effect with a corresponding *F* value that is not statistically significant. The latter can result from the fact that the power of the two *F* tests was not equivalent; b) A larger significant *F* value for a specific effect does not necessarily indicate a larger **omega squared** value for that effect when compared with the **omega squared** value for another effect which is associated with a smaller *F* value; c) Because of the high degree of sampling variability associated with the **omega squared** statistic, it is prudent to compute confidence intervals for such values; d) Olejnik and Algina (2000) note that researchers should not compare **partial omega squared** values for different effects in the same study, since the denominators involved in computing the latter values may not be equivalent; e) Extreme caution should be employed in using ratios of **omega squared** values computed for different factors in the same study in order to assess the relative importance of the factors, since analogous ratios based on alternative effect size values (e.g., *d* values) can be quite different.

Cohen's *f* index (Test 27h) If, for a given factor, the value of **partial omega squared** is substituted in Equation 21.45, **Cohen's *f* index** can be computed. In Section VI of the **single-factor between-subjects analysis of variance**, it was noted that **Cohen's *f* index** is an alternate measure of effect size that can be employed for an analysis of variance. The computation of **Cohen's *f* index** with Equation 21.45 yields the following values: $f_A = 2.13$, $f_B = 1.69$, $f_{AB} = .89$.

$$f_A = \sqrt{\frac{\tilde{\omega}^2_{pA}}{1 - \tilde{\omega}^2_{pA}}} = \sqrt{\frac{.82}{1 - .82}} = 2.13$$

$$f_B = \sqrt{\frac{\tilde{\omega}_{pB}^2}{1 - \tilde{\omega}_{pB}^2}} = \sqrt{\frac{.74}{1 - .74}} = 1.69$$

$$f_{AB} = \sqrt{\frac{\tilde{\omega}_{pAB}^2}{1 - \tilde{\omega}_{pAB}^2}} = \sqrt{\frac{.44}{1 - .44}} = .89$$

In the discussion of **Cohen's f index** in Section VI of the **single-factor between-subjects analysis of variance**, it was noted that Cohen (1977; 1988, pp. 284–288) employed the following (admittedly arbitrary) f values as criteria for identifying the magnitude of an effect size: a) A **small effect size** is one that is greater than .1 but not more than .25; b) A **medium effect size** is one that is greater than .25 but not more than .4; and c) A **large effect size** is greater than .4. Employing Cohen's criteria, all of the values computed for f represent large effect sizes.

Sources which discuss computation of a measure of magnitude of treatment effect for a **between-subjects factorial analysis of variance** (as well as other factorial designs) are Cortina and Nouri (2000), Grissom and Kim (2005, Ch. 7), Keppel (1991), Kirk (1995), Kline (2004), Maxwell and Delaney (2004), Rosenthal *et al.* (2000) and Rosnow *et al.* (2000). Further discussion of the indices of treatment effect discussed in this section and the relationship between effect size and statistical power can be found in Section IX (the **Addendum**) of the **Pearson product-moment correlation coefficient** under the discussion of **meta-analysis and related topics**.

5. Computation of a confidence interval for the mean of a population represented by a group   The same procedure employed to compute a confidence interval for a treatment population for the **single-factor between-subjects analysis of variance** can be employed with the **between-subjects factorial analysis of variance** to compute a confidence interval for the mean of any population represented by the pq groups. Although it will not be demonstrated here, the computational procedure requires that the appropriate values be substituted in Equation 21.48. In the event a researcher wants to compute a confidence interval for the mean of one of the levels of any of the factors, the number of subjects in the denominator of the radical of Equation 21.48 is based on the number of subjects who served within each level of the relevant factor (i.e., nq in the case of Factor A and np in the case of Factor B).

VII. Additional Discussion of the Between-Subjects Factorial Analysis of Variance

1. Theoretical rationale underlying the between-subjects factorial analysis of variance As noted in Section IV, as is the case for the **single-factor between-subjects analysis of variance**, the total variability for the **between-subjects factorial analysis of variance** can be divided into **between-groups variability** and **within-groups variability**. Although it is not required in order to compute the F ratios, the value MS_{BG} (which represents between-groups variability) can be used to represent the variance of the means of the pq groups. MS_{BG} can be computed with the equation $MS_{BG} = SS_{BG}/df_{BG}$. As noted earlier, between-groups variability is comprised of the following elements: a) **Variability attributable to Factor A** (represented by the notation MS_A), which represents the variance of the means of the p levels of Factor A; b) **Variability attributable to Factor B** (represented by the notation MS_B), which represents the variance of the means of the q levels of Factor B; and c) **Variability attributable to any interaction that is present between Factors A and B** (represented by the notation MS_{AB}), which is a measure of variance that represents whatever remains of between-groups variability after the contributions of the main effects of Factors A and B have been subtracted from between-groups variability.

In computing the three F ratios for the **between-subjects factorial analysis of variance**, the values MS_A, MS_B, and MS_{AB} are contrasted with MS_{WG}, which serves as a baseline measure of error variability. In other words, MS_{WG} represents experimental error which results from circumstances that are beyond an experimenter's control. As is the case for the **single-factor between-subjects analysis of variance**, in the **between-subjects factorial analysis of variance**, MS_{WG} is the normal amount of variability which is expected between the scores of different subjects who serve under the same experimental condition. Thus, MS_{WG} represents the average of the variances computed for each of the pq groups. As long as any of the elements that comprise between-groups variability (MS_A, MS_B, or MS_{AB}) are approximately the same value as within-groups variability (MS_{WG}), the experimenter can attribute variability on a between-groups component to experimental error. When, however, any of the components which comprise between-groups variability are substantially greater than MS_{WG}, it indicates that something over and above error variability is contributing to that element of variability. In such a case it is assumed that the inflated level of variability for the between-groups component is the result of a treatment effect.

2. Definitional equations for the between-subjects factorial analysis of variance In the description of the computational protocol for the **between-subjects factorial analysis of variance** in Section IV, Equations 27.9–27.14 are employed to compute the values SS_T, SS_{BG}, SS_A, SS_B, SS_{AB}, and SS_{WG}. The latter set of computational equations were employed, since they allow for the most efficient computation of the sum of squares values. As noted in Section IV, computational equations are derived from definitional equations which reveal the underlying logic involved in the derivation of the sums of squares.

As noted previously, the total sum of squares (SS_T) can be broken down into two elements, the between-groups sum of squares (SS_{BG}) and the within-groups sum of squares (SS_{WG}). The contribution of any single subject's score to the total variability in the data can be expressed in terms of a between-groups component and a within-groups component. When the between-groups component and the within-groups component are added, the sum reflects that subject's total contribution to the overall variability in the data. Furthermore, the between-groups sum of squares can be broken down into the following three elements: a sum of squares for Factor A (SS_A), a sum of squares for Factor B (SS_B), and an interaction sum of squares (SS_{AB}). The contribution of any single subject's score to between-groups variability in the data can be expressed in terms of an A, a B, and an AB component. When the A, B, and AB components for a given subject are added, the sum reflects that subject's total contribution to between-groups variability in the data. The aforementioned information is reflected in the definitional equations which will now be described for computing the sums of squares.

Equation 27.63 is the definitional equation for the **total sum of squares**.[19] In Equation 27.63 the notation X_{ijk} is employed to represent the score of the ith subject in the group that serves under Level j of Factor A and Level k of Factor B.[20] When the notation $\sum_{k=1}^{q} \sum_{j=1}^{p} \sum_{i=1}^{n}$ precedes a term in parentheses, it indicates that the designated operation should be carried out for all $N = npq$ subjects.[21]

$$SS_T = \sum_{k=1}^{q} \sum_{j=1}^{p} \sum_{i=1}^{n} (X_{ijk} - \bar{X}_T)^2 \qquad \textbf{(Equation 27.63)}$$

In employing Equation 27.63 to compute SS_T, the grand mean ($\bar{X}_T$) is subtracted from each of the $N = npq$ scores and each of the N difference scores is squared. The total sum of squares (SS_T) is the sum of the N squared difference scores. Equation 27.63 is computationally equivalent to Equation 27.9.

Equation 27.64 is the definitional equation for the **between-groups sum of squares**.

$$SS_{BG} = n \sum_{k=1}^{q} \sum_{j=1}^{p} (\bar{X}_{AB_{jk}} - \bar{X}_T)^2 \qquad \textbf{(Equation 27.64)}$$

In employing Equation 27.64 to compute SS_{BG}, the following operations are carried out for each of the pq groups. The grand mean ($\bar{X}_T$) is subtracted from the group mean ($\bar{X}_{AB_{jk}}$). The difference score is squared, and the squared difference score is multiplied by the number of subjects in the group (n). After this is done for all pq groups, the values that have been obtained for each group as a result of multiplying the squared difference score by the number of subjects in a group are summed. The resulting value represents the between-groups sum of squares (SS_{BG}). Equation 27.64 is computationally equivalent to Equation 27.10.

Equation 27.65 is the definitional equation for the **sum of squares for Factor A**.

$$SS_A = nq \sum_{j=1}^{p} (\bar{X}_{A_j} - \bar{X}_T)^2 \qquad \textbf{(Equation 27.65)}$$

In employing Equation 27.65 to compute SS_A, the following operations are carried out for each of the p levels of Factor A. The grand mean ($\bar{X}_T$) is subtracted from the mean of that level of Factor A ($\bar{X}_{A_j}$). The difference score is squared, and the squared difference score is multiplied by the number of subjects in the level ($n_{A_j} = nq$). After this is done for all p levels of Factor A, the values that have been obtained for each level as a result of multiplying the squared difference score by the number of subjects in a level are summed. The resulting value represents the sum of squares for Factor A (SS_A). Equation 27.65 is computationally equivalent to Equation 27.11.

Equation 27.66 is the definitional equation for the **sum of squares for Factor B**.

$$SS_B = np \sum_{k=1}^{q} (\bar{X}_{B_k} - \bar{X}_T)^2 \qquad \textbf{(Equation 27.66)}$$

In employing Equation 27.66 to compute SS_B, the following operations are carried out for each of the q levels of Factor B. The grand mean ($\bar{X}_T$) is subtracted from the mean of that level of Factor B ($\bar{X}_{B_k}$). The difference score is squared, and the squared difference score is multiplied by the number of subjects in the level ($n_{B_k} = np$). After this is done for all q levels of Factor B, the values that have been obtained for each level as a result of multiplying the squared difference score by the number of subjects in a level are summed. The resulting value represents the sum of squares for Factor B (SS_B). Equation 27.66 is computationally equivalent to Equation 27.12.

Equation 27.67 is the definitional equation for the **interaction sum of squares**.

$$SS_{AB} = n\sum_{k=1}^{q} \sum_{j=1}^{p} (\bar{X}_{AB_{jk}} - \bar{X}_{A_j} - \bar{X}_{B_k} + \bar{X}_T)^2 \qquad \textbf{(Equation 27.67)}$$

In employing Equation 27.67 to compute SS_{AB}, the following operations are carried out for each of the pq groups. The mean of the level of Factor A the group represents ($\bar{X}_{A_j}$) and the mean of the level of Factor B the group represents ($\bar{X}_{B_k}$) are subtracted from the mean of the group ($\bar{X}_{AB_{jk}}$), and the grand mean ($\bar{X}_T$) is added to the resulting value. The result of the aforementioned operations is squared, and the squared difference score is multiplied by the number of subjects in that group ($n = n_{AB_{jk}}$). After this is done for all pq groups, the values that have been obtained for each group are summed, and the resulting value represents the sum of squares for the interaction (SS_{AB}).[22] Equation 27.67 is computationally equivalent to Equation 27.13.

Equation 27.68 is the definitional equation for the **within-groups sum of squares**.

$$SS_{WG} = \sum_{k=1}^{q} \sum_{j=1}^{p} \sum_{i=1}^{n} (X_{ijk} - \bar{X}_{AB_{jk}})^2 \qquad \textbf{(Equation 27.68)}$$

In employing Equation 27.68 to compute SS_{WG}, the following operations are carried out for each of the pq groups. The group mean ($\bar{X}_{AB_{jk}}$) is subtracted from each score in the group. The difference scores are squared, after which the sum of the squared difference scores is obtained. The sum of the sum of the squared difference scores for all pq groups represents the within-groups sum of squares. Equation 27.68 is computationally equivalent to Equation 27.14.

3. Unequal sample sizes The equations presented in this book for the **between-subjects factorial analysis of variance** assume there is an equal number of subjects in each of the pq groups (i.e., the value of $n_{AB_{jk}}$ is equal for each group). When the number of subjects per group is not equal, most sources recommend that adjusted sum of squares and sample size values be employed in conducting the analysis of variance. One approach to dealing with unequal sample sizes, which is generally referred to as the **unweighted means procedure**, employs the harmonic mean of the sample sizes of the pq groups to represent the value of $n = n_{AB_{jk}}$.[23] Based on the computed value of the harmonic mean (which will be designated $\bar{n}_h$), the sample size of each row and column, as well as the total sample size, are adjusted as follows: $n_{A_j} = (\bar{n}_h)(q)$, $n_{B_k} = (\bar{n}_h)(p)$, $N = (\bar{n}_h)(p)(q)$. In addition, the $\sum X_{AB_{jk}}$ score of each group is adjusted by multiplying the mean of the group derived from the original data by the value computed for the harmonic mean (i.e., $(\bar{X}_{AB_{jk}})(\bar{n}_h)$ = Adjusted value of $\sum X_{AB_{jk}}$). Employing the adjusted $\sum X_{AB_{jk}}$ values, the value of $\sum X_T$ and the values of the sums of the rows ($\sum X_{A_j}$) and columns ($\sum X_{B_k}$) are adjusted accordingly. The adjusted values of $\sum X_T$, $\sum X_{AB_{jk}}$, $\sum X_{A_j}$, $\sum X_{B_k}$, $n_{AB_{jk}}$, n_{A_j}, n_{B_k}, and N are substituted in Equations 27.9–27.13 to compute the values SS_T, SS_{BG}, SS_A, SS_B, and SS_{AB}. The value of SS_{WG}, on the other hand, is a pooled within-groups sum of squares which is based on the original unadjusted data. Thus, employing the original unadjusted values of $\sum X_{AB_{jk}}$ and $n_{AB_{jk}}$, the sum of squares is computed for each of the pq groups employing the following equation: $\sum X_{AB_{jk}}^2 - [(\sum X_{AB_{jk}})^2/n_{AB_{jk}}]$. The sum of the pq sum of squares values represents the pooled within-groups sum of squares. This value is, in fact, computed in Endnote 4. The values

MS_A, MS_B, and MS_{AB} are computed with Equations 27.15–27.17 by dividing the relevant sum of squares value by the appropriate degrees of freedom. The degrees of freedom for the aforementioned mean square values are computed with Equations 27.19–27.21. Although the value MS_{WG} is computed with Equation 27.18, the value df_{WG} in the denominator of Equation 27.18 is a pooled within-groups degrees of freedom. The latter degrees of freedom value is determined by computing the value $(n_{AB_{jk}} - 1)$ for each group, and summing the $(n_{AB_{jk}} - 1)$ values for each of the pq groups. When the resulting degrees of freedom value is divided into the pooled within-groups sum of squares, it yields the value MS_{WG} which is employed in computing the F ratios. Equations 27.25–27.27 are employed to compute the values F_A, F_B, and F_{AB}. Keppel (1991) notes that since the F values derived by the method described in this section may underestimate the likelihood of committing a Type I error, it is probably prudent to employ a lower tabled probability to represent the prespecified level of significance — i.e., employ the tabled critical $F_{.01}$ value to represent the tabled critical $F_{.05}$ value. Alternative methods for dealing with unequal sample sizes are described in Keppel (1991), Kirk (1982, 1995), and Winer et al. (1991).

4. The randomized-blocks design Let us assume a researcher wishes to evaluate the impact of the dosage level of a specific drug on the ability of subjects to learn verbal material. The researcher conducts an experiment employing an **independent samples design**. To be more specific, each of 27 subjects is randomly assigned one of three groups. The subjects in Group 1 are administered 10 milligrams of the drug, the subjects in Group 2 are administered 20 milligrams of the drug, and the subjects in Group 3 are administered 30 milligrams of the drug. Thus, there are $k = 3$ groups in the study with $n = 9$ subjects per group. A subject's score will represent the amount of verbal material learned, and the higher the score the greater the amount of learning. The data for the study are summarized in Table 27.7.

Table 27.7 Data for Independent Samples Design

Group 1 (10 mg.)	Group 2 (20 mg.)	Group 3 (30 mg.)
X_1	X_2	X_3
2	4	3
1	2	4
3	2	4
4	4	5
5	5	6
4	6	7
6	7	9
7	8	8
6	7	7
$\Sigma X_1 = 38$	$\Sigma X_2 = 45$	$\Sigma X_3 = 53$
$\overline{X}_1 = 4.22$	$\overline{X}_2 = 5.00$	$\overline{X}_3 = 5.89$

The data for the above described experiment are evaluated with a **single-factor between-subjects analysis of variance**. Table 27.8 is the summary table for the analysis of variance. Since $df_{num} = 2$ and $df_{den} = 24$, from **Table A10** we determine that the tabled critical value $F_{.05} = 3.40$ is employed to evaluate the null hypothesis. Since $F = 1.47$ is less than $F_{.05} = 3.40$

Table 27.8 Summary Table of Single-Factor Between-Subjects Analysis of Variance

Source of variation	SS	df	MS	F
Between-groups	12.52	2	6.26	1.47
Within-groups	102.44	24	4.27	
Total	114.96	26		

the null hypothesis cannot be rejected. Thus, the researcher cannot conclude there are any differences in learning among subjects who were administered different dosages of the drug.

It will now be illustrated how the design of the above described experiment can be modified in order that it can be conceptualized as a **randomized-blocks design**. The latter design (which will be evaluated with a **between-subjects factorial analysis of variance**) will allow us to conduct a much more powerful test of the alternative hypothesis (which is that in the underlying populations at least one of the $k = 3$ mean values is different from one of the other mean values). Although it is noted in Endnote 1 of the **single-factor within-subjects analysis of variance** that the term **randomized-blocks design** is commonly employed to describe a dependent samples design involving matched subjects, the term **randomized-blocks design** is also used to refer to a factorial design involving two factors in which one of the factors (which will be designated Factor A) represents k experimental treatments (which in the case of the experiment under discussion are the three dosage levels of the drug), and the other factor (which will be designated Factor B) is a subject/nonmanipulated independent variable which a researcher has reason to believe is correlated with the dependent variable. The latter variable will represent what is referred to as a **blocked variable**. In our example we will employ the subject variable of intelligence (which it is logical to believe is correlated with the dependent variable of verbal learning ability) as the second independent variable in the study. In order to employ intelligence as Factor B, we must obtain a score on intelligence for each subject prior to randomly assigning subjects to the experimental conditions. Blocking subjects requires that we initially rank the subjects on the blocked variable — i.e., we must rank the $N = 27$ subjects employed in the study with respect to intelligence. The researcher must decide on the number of levels he wishes to employ on the blocked variable. The number of levels on the blocked variable will be represented by the notation j. Once subjects have been ranked on the blocked variable, $1/j^{th}$ of the subjects are assigned to each of the levels of Factor B. Level 1 of Factor B is comprised of the $1/j^{th}$ of the subjects who rank the lowest on the blocked variable/intelligence. Level 2 of Factor B is comprised of the $1/j^{th}$ of the subjects who are ranked directly above the subjects assigned to Level 1 on the blocked variable/intelligence. Level 3 of Factor B is comprised of the $1/j^{th}$ of the subjects who are ranked directly above the subjects assigned to Level 2 on the blocked variable/intelligence, and so on. In the study to be described we will employ $j = 3$ levels for the blocked variable. Thus, $N/j = 27/3 = 9$ subjects will be assigned to each level of Factor B.[24]

Table 27.9 summarizes the data for our example when the variable of intelligence is employed as a second factor. Note that values of the scores in each of the levels of Factor A in Table 27.9 are identical to the values of the scores in each of the levels of the independent variable (which is equivalent to Factor A in Table 27.9) in Table 27.7. The only difference between the two tables is that in Table 27.9 subjects are also categorized with respect to Factor B — specifically, whether they fall in the lowest, middle, or highest category with respect to intelligence. By virtue of employing Factor B, which is comprised of $j = 3$ levels, the original data presented in Table 27.7 have been converted into a 3×3 between-subjects factorial design. Note that all together there are $pq = (3)(3) = 9$ independent groups of subjects employed in the

Table 27.9 Data for Randomized-Blocks Design

Factor B	Factor A			Row summary information
	A_1 (10 mg)	A_2 (20 mg)	A_3 (30 mg)	
B_1 (Lowest intelligence)	2 1 3	4 2 2	3 4 4	Level B_1 $\sum X_{B_1} = 25$ $\overline{X}_{B_1} = 2.78$
B_2 (Middle intelligence)	4 5 4	4 5 6	5 6 7	Level B_2 $\sum X_{B_2} = 46$ $\overline{X}_{B_2} = 5.11$
B_3 (Highest intelligence)	6 7 6	7 8 7	9 8 7	Level B_3 $\sum X_{B_3} = 65$ $\overline{X}_{B_3} = 7.22$
Column summary information	Level A_1 $\sum X_{A_1} = 38$ $\overline{X}_{A_1} = 4.22$	Level A_2 $\sum X_{A_2} = 45$ $\overline{X}_{A_2} = 5.00$	Level A_3 $\sum X_{A_3} = 53$ $\overline{X}_{A_3} = 5.89$	Grand Total $\sum X_T = 136$ $\overline{X}_T = 5.04$

Table 27.10 Summary Table of Analysis of Variance

Source of variation	SS	df	MS	F
Between-groups	101.63	8		
A	12.52	2	6.26	8.45
B	88.96	2	44.48	60.05
AB	.15	4	.04	.05
Within-groups	13.33	18	.74	
Total	114.96	26		

study, each group being categorized with respect to a different combination of levels of the two factors. The data for this design are evaluated with a **between-subjects factorial analysis of variance** in the same manner as was Example 27.1. Table 27.10 summarizes the analysis of variance on the data in Table 27.9.

Since the intent of the study is to determine whether or not the drug had an effect on subjects' learning we are interested in whether there is a significant main effect on Factor A. Employing **Table A10**, for $df_{num} = 2$ and $df_{den} = 18$, we determine that the tabled critical $F_{.05}$ and $F_{.01}$ values are $F_{.95} = 3.55$ and $F_{.01} = 6.01$. Since $F_A = 8.45$ is greater than both of the aforementioned values, the null hypothesis can be rejected at both the .05 and .01 levels. Thus, the researcher can conclude that there is at least one significant difference between the means of two of the three drug dosage conditions.

The fact that the same body of data which did not yield a significant result when it was evaluated with a **single-factor between-subjects analysis of variance** yields a significant result when it is evaluated with a **between-subjects factorial analysis of variance** can be attributed to the use of the blocking variable. Specifically, by virtue of employing a blocking variable, the **randomized-blocks design** matches subjects across treatments. Note that the values computed for MS_{BG} in Table 27.8 and MS_A in Table 27.10 are identical (i.e., $MS_{BG} = MS_A = 6.26$). On the other hand, the value $MS_{WG} = .74$ computed in Table 27.10 is substantially lower than the value $MS_{WG} = 4.27$ computed in Table 27.8. This latter fact accounts for the substantially higher F value computed for Factor A in Table 27.10 when compared with the F value computed in Table 27.8. To clarify this further, if the blocking variable had not been employed in the study, the within-groups sum of squares for the resulting **single-factor between-subjects analysis of variance** would be the value of the error term (i.e., SS_{WG}) computed in the factorial analysis of variance plus the values computed for SS_B and SS_{AB}. Note that the sum of the latter three sum of squares values in Table 27.10 is, in fact, equal to $SS_{WG} = 102.44$ computed in Table 27.8 (i.e., in Table 27.10 [($SS_B = 88.96$) + ($SS_{AB} = .15$) + ($SS_{WG} = 13.33$)] = [$SS_{WG} = 102.44$]). Put simply, because subjects' scores on Factor B (the blocking variable) are correlated with their scores on the dependent variable, the end result is that when the blocking variable is integrated into the analysis of variance, the error term (MS_{WG}) computed will be substantially lower than the error term (MS_{WG}) obtained when the blocking variable is not employed. The greater the correlation between the blocking variable and the dependent variable, the greater will be the reduction in the error term employed for the factorial analysis of variance. Anderson (2001, p. 412) notes that for blocking to result in a substantial loss in error variability, the correlation between the blocked variable and the dependent variable must be at least .35. He further notes that in practice, it is difficult to find a blocking variable which has that high a correlation with a dependent variable.

It should also be pointed out that although the value $df_{WG} = 18$ employed for the **between-subjects factorial analysis of variance** is smaller than the value $df_{WG} = 24$ employed for the **single-factor between-subjects analysis of variance** (resulting in a larger critical value being employed in evaluating the result of the factorial analysis of variance), the difference in degrees of freedom is not nearly enough to offset the impact of the lower error term derived for the factorial analysis of variance.

The reader should take note of the fact that in Table 27.10 a significant main effect is obtained for Factor B (which is evaluated with the same critical values as Factor A). The latter can be attributed to the fact that by virtue of blocking subjects on Factor B, the researcher will most likely create a significant effect on the blocking variable. Anderson (2001, p. 408) notes that in order to reduce the value of the error term for the factorial analysis of variance, the value of F_B need not be statistically significant, but that the value of MS_B should be substantially larger than the value of MS_{AB}. In the case of our example, the value of $MS_{AB} = .04$ is quite low and thus the interaction (which is evaluated with $F_{.05} = 2.93$, the tabled critical value for $df_{num} = 4$ and $df_{den} = 18$) is not significant.

It was noted in Section IX (the **Addendum**) of the **single-factor analysis of variance** that an **analysis of covariance (Test 21j)** is an analysis of variance procedure which utilizes data on an extraneous variable that has a linear correlation with the dependent variable. Such an extraneous variable is referred to as a **covariate**. By utilizing the correlation between the covariate and the dependent variable, the researcher is able to remove variability on the dependent variable which is attributable to the covariate. The effect of the latter is a reduction in the error variability employed in computing the F ratio, thereby resulting in a more powerful test of the alternative hypothesis. The use of blocking as described above also capitalizes on a linear correlation between an extraneous variable (in this case intelligence) and the dependent variable. Consequently, blocking can serve as an alternative to employing an analysis of covariance as a means of reducing error variability in a set of data. If an analysis of covariance had been used in our study, the intelligence scores of subjects would be employed to represent the covariate. If the actual intelligence scores of subjects were employed they would provide more information than blocking, insofar as blocking only resulted in grouping subjects in three categories based on their rank ordering on intelligence. Yet blocking is more reasonable to use when a researcher only has enough information on the blocked variable to assign subjects to mutually exclusive categories.

A discussion of the relative merits of employing an analysis of covariance versus blocking can be found in Anderson (2001, pp. 387–388) and Maxwell and Delaney (1990, pp. 398–400). In the latter discussion Anderson (2001, p. 388) notes that one advantage of analysis of covariance over blocking is that whereas in the case of the analysis of covariance subjects do not have to be measured on the covariate/blocked variable before randomly assigning them to groups, the latter is required if blocking is to be employed. He also notes that although analysis of covariance is slightly more effective than blocking when the underlying assumptions of the analysis of covariance have not been violated, in most instances blocking and analysis of covariance will yield comparable results.

5. Additional comments on the between-subjects factorial analysis of variance

   a) **Fixed-effects versus random-effects versus mixed-effects models** In Section VII of the **single-factor between-subjects analysis of variance** it is noted that one assumption underlying the analysis of variance is whether or not the levels of an independent variable are **fixed** or **random**. Whereas a **fixed-effects model** assumes that the levels of an independent variable are the same levels that will be employed in any attempted replication of the experiment, a **random-effects model** assumes that the levels have been randomly selected from the overall population of all possible levels which can be employed for the independent variable. The computational procedures for all of the analysis of variance procedures described in this book assume a fixed-effects model for all factors.

   In the case of factorial designs it is also possible to have a **mixed-effects model**, which is a combination of a fixed-effects and a random-effects model. Specifically, in the case of a two-factor design, a mixed-effects model assumes that one of the factors is based on a fixed-effects model while the second factor is based on a random-effects model. When there are three or more factors, a mixed-effects model assumes that one or more of the factors is based on a fixed-effects model and one or more of the factors is based on a random-effects model. Texts that specialize in the analysis of variance provide in-depth discussions of this general subject, as well as describing the modified equations which are appropriate for evaluating factorial designs that are based on random- and mixed-effects models.

   b) **Nested factors/hierarchical designs and designs involving more than two factors** In designing experiments that involve two or more factors, it is possible to employ what are referred to as **nested factors**. Nesting is present in an experimental design when different levels of one factor do not occur at all levels of another factor. To illustrate nesting, let us assume that a researcher wants to evaluate two teaching methods (which will comprise Factor A) in 10 different classes, each of which is unique with respect to the ethnic makeup of its students. The 10 different classes will comprise Factor B. Five of the classes (B_1 ... B_5) are taught by teaching method A_1 and the other five classes (B_6 ... B_{10}) by teaching method A_2. In such a case Factor B is nested under Factor A, since each level of Factor B serves under only one level of Factor A. Figure 27.3 outlines the design.[25]

$$A_1 \qquad\qquad\qquad\qquad A_2$$

$$B_1 \quad B_2 \quad B_3 \quad B_4 \quad B_5 \qquad\qquad B_6 \quad B_7 \quad B_8 \quad B_9 \quad B_{10}$$

Figure 27.3 Example of a Nested Design

Winer *et al.* (1991) note that in the above described design it is not possible to evaluate whether or not there is an interaction between the two factors. The reason for this is that in order to test for an interaction, it is necessary that all levels of Factor B must occur under all of the levels of Factor A. When the latter is true (as is the case in Example 27.1), the two factors are said to be **crossed**. The term **hierarchical design** is often employed to describe designs in which there are two or more nested factors. It is also possible to have a **partially hierarchical design**, in which at least one factor is nested and at least one factor is crossed. Since the statistical model upon which nested designs are based differs from the model which has been employed for the **between-subjects factorial analysis of variance**, the analysis of such designs requires the use of different equations.

As noted earlier, a **between-subjects factorial analysis of variance** (as well as a **factorial analysis of variance for a mixed design** and a **within-subjects factorial analysis of variance** discussed in Section IX) can involve more than two factors. To further complicate matters, designs with three or more factors can involve some factors that are nested and others which are crossed, plus a fixed-effects model may be assumed for some factors and a random-effects model for others. Honeck *et al.* (1983) is an excellent source for deriving the appropriate equations for the use of the analysis of variance with experimental designs involving nesting and/or the use of more than two factors. Other sources on this subject are Keppel (1991), Kirk (1982, 1995), Montgomery (2000) and Winer *et al.* (1991).

c) **Screening designs** Screening designs are typically employed in the initial stages of researching a multifactorial problem when a experimenter has reason to believe that many of the factors (i.e., independent variables) under initial consideration have little or no impact on a dependent variable. The goal of such a design is to "screen out" from the large set of factors those which have a major impact on the dependent variable, after which the latter factors are scrutinized more closely in subsequent studies (Montgomery (2001)). In the final analysis, a **screening design** is a multifactor experiment which is deliberately confounded in order that a researcher can efficiently evaluate a large number of factors employing a minimal number of subjects/observations.[26] Since screening designs are compromised with respect to their ability to test interactions, they are often referred to as **incomplete designs** (although not all incomplete designs would be considered screening designs, since the term incomplete design is often used generically to describe an inadequately controlled design). Except for the **Latin-square design** (which is discussed later in Section IX (the **Addendum**)), screening designs are rarely used in the behavioral or natural sciences, except perhaps within the context of pilot (i.e., preliminary) studies. On the other hand, in industrial settings such designs are frequently employed in the early stages of sequential experimentation, since a major goal of industrial research is to minimize cost by extracting the maximum amount of unbiased information regarding the factors affecting a production process from as few observations as possible.

In the behavioral and natural sciences, within the framework of conducting multifactor experiments involving more than two factors researchers are often concerned with **higher-order interactions** (i.e., interactions involving three or more independent variables), since the latter types of interactions may prove to be of theoretical or practical significance. In these disciplines it would be highly unusual to encounter a **full-factorial design** (i.e., a factorial design in which every possible combination of all the factors is included) which involves more than five factors, since a study with that many variables becomes problematic with respect to evaluating the data and/or interpreting the results. In contrast, research in industrial settings often requires that a large number of factors be evaluated simultaneously, and within the latter context higher-order interactions are often viewed as little more than a nuisance which prevent a researcher from detecting the presence of one or more main effects. Because of the latter, the industrial researcher may have occasion to employ a screening design to filter out important main effects, while at the same time having little or no concern with potential interactions.

A commonly employed screening design is the **fractional-factorial design**. Prior to describing the latter design, it should be noted that a major problem with full-factorial designs is that the number of combinations of different experimental conditions increases exponentially with the total number of factors involved in an experiment. Even if the number of factors is relatively small, the number of combinations involved in a full-factorial design can become very large. For example, if a full-factorial design has five factors with two levels per factor, $2^5 = 32$ combinations (often referred to as **runs** within the content of industrial research) of experimental conditions would be required. Thus, in the case of a completely randomized design, a minimum of 32 subjects would be required in order to have at least one subject in each of the treatment combinations. If the researcher wanted two subjects in each treatment combination, $32 \times 2 = 64$ subjects would be required, and so on. Imagine, however, if a researcher was interested in screening 10 independent variables. In such a case $2^{10} = 1024$ is the minimum number of subjects/observations that would be required in order to have one observation in each of the treatment combinations. Obviously, in the latter instance it would be impractical to employ a full-factorial design. Consequently, the most reasonable alternative is the use of a **fractional-factorial design** in which only a fraction of the treatment combinations required for a full-factorial experiment are selected to be run. Depending upon how many subjects are employed, fractional-factorial designs usually allow for testing of all or some of the two-way interactions. For example, if a study employing 256 subjects within the framework of a fractional-factorial design is conducted involving ten independent variables (each with two levels), only some of the two-way interactions (not to mention none of the higher-order interactions) could be assessed, since only a fraction of the $2^{10} = 1024$ possible treatment combinations would be included in the study.[27] Because fractional-factorial designs are difficult to hand generate (i.e., the determination of which specific treatment combinations to employ), they are usually constructed with the aid of computer software (see Tabachnick and Fidell (2001, pp. 599–606)).

The more interactions that can be evaluated within the framework of a fractional-factorial design, the higher the **resolution** of the design. **Resolution III designs** (which are sometimes called **low resolution designs**) can only test for main effects (use of such a design implies a researcher considers any interactions to be negligible). **Resolution IV** designs can test for main effects and some two-way interactions, while **Resolution V** designs can test for main effects and all two-way interactions (use of the latter two type of designs implies a researcher considers any higher-order interactions to be negligible). (See Montgomery (2001) and Tabachnick and Fidell (2001, pp. 570–571) for more details.)

Two of the more commonly employed types of screening designs are **Plackett–Burman designs** and **Taguchi designs**. **Plackett–Burman designs** (Placket and Burman(1946)) are fractional-factorial designs which are useful for screening a large number of independent variables, each with two levels, using the smallest possible number of combinations of experimental conditions. This type of design allows for testing all main effects but none of the interactions. Like **Plackett–Burman designs**, **Taguchi designs** (Taguchi (1986, 1993)) also provide tests for a large number of independent variables using a minimum number of treatment combinations (Tabachnick and Fidell (2001, p. 578)). The latter type of design was developed by Dr. Genichi Taguchi, a Japanese engineer who viewed experimental design as off-line quality control, since he believed it represents a methodology for ensuring good performance in the design stage of products or processes. **Taguchi designs** focus on identifying the combination of levels of the independent variables which produce optimal performance, and reflect Taguchi's goal of developing products that operate consistently and optimally over a broad spectrum of conditions (more specifically, products that are characterized by minimal variability in performance and minimal sensitivity to noise (i.e., extraneous nuisance variables)).

From the above discussion, it should be apparent that two shortcomings of screening designs are their limited (or total lack of) ability to evaluate interactions (which in some instances may incorrectly be assumed to be negligible), and the issue of questionable reliability for scores on the dependent variable due to the use of small sample sizes. Analytical procedures for generating and evaluating screening designs are complicated and usually require the aid of computer software. Since the general subject of screening designs is complex and goes well beyond the material presented in this section, readers requiring more detailed information should consult sources such as Box *et al.* (1978), Montgomery (2000), Tabachnick and Fidell (2001), as well as references on statistical applications in industry and engineering.

VIII. Additional Examples Illustrating the Use of the Between-Subjects Factorial Analysis of Variance

Examples 27.2 and 27.3 are two additional examples which can be evaluated with the **between-subjects factorial analysis of variance**. Since the data for both examples are identical to those employed in Example 27.1, they yield the same result. Note that whereas in Example 27.1 both Factor A and Factor B are manipulated independent variables, in Example 27.2 Factor B is manipulated while Factor A is nonmanipulated (i.e., is a subject/attribute independent variable). In Example 27.3 both factors are nonmanipulated independent variables.

Example 27.2 *A study is conducted in order to evaluate the impact of gender (to be designated as Factor A) and anxiety level (to be designated as Factor B) on affiliation. The experimenter employs a 2 × 3 between-subjects (completely-randomized) factorial design. The two levels that comprise Factor A are A_1: Male; A_2: Female. The three levels that comprise Factor B are B_1: Low Anxiety; B_2: Moderate Anxiety; B_3: High Anxiety. Each of nine males and nine females is randomly assigned to one of three experimental conditions. All of the subjects are told they are participants in a learning experiment which will require them to learn lists of words. Subjects in the low anxiety condition are told there will be no consequences for poor performance in the experiment. Subjects in the moderate anxiety condition are told if they perform below a certain level they will have to drink a distasteful beverage. Subjects in the high anxiety condition are told if they perform below a certain level they will be given a painful electric shock. All subjects are told that while waiting to be tested they can either wait by themselves or with other people. Each subject is asked to designate the number of people he or she would like in the room with him or her while waiting to be tested. This latter measure is employed to represent the dependent variable of affiliation. The experimenter assumes that the higher a subject is in affiliation, the more people the subject will want to be with while waiting. The affiliation scores of the three subjects in each of the six experimental groups/conditions (which result from the combinations of the levels which comprise the two factors) follow: Group AB_{11}: Male/Low anxiety (11, 9, 10); Group AB_{12}: Male/Moderate anxiety (7, 8, 6); Group AB_{13}: Male/High anxiety (5, 4, 3); Group AB_{21}: Female/Low anxiety (2, 4, 3); Group AB_{22}: Female/Moderate anxiety (4, 5, 3); Group AB_{23}: Female/High anxiety (0, 1, 2). Do the data indicate that either gender or anxiety level are related to affiliation?*

Example 27.3 *A study is conducted in order to evaluate if there is a relationship between ethnicity (to be designated as Factor A) and socioeconomic class (to be designated as Factor B), and the number of times a year a person visits a doctor. The experimenter employs a 2×3 between-subjects (completely randomized) factorial design. The two levels that comprise Factor A are A_1: Caucasian; A_2: African-American. The three levels that comprise Factor B are B_1: Lower socioeconomic class; B_2: Middle socioeconomic class; B_3: Upper socioeconomic class. Based on their occupation and income, each of nine Caucasians and nine African-Americans is categorized with respect to whether he or she is a member of the lower, middle, or upper socioeconomic class. Upon doing this the experimenter determines the number of times during the past year each of the subjects has visited a doctor. This latter measure represents the dependent variable in the study. The number of visits for the three subjects in each of the six experimental groups/conditions (which result from the combinations of the levels that comprise the two factors) follow:* **Group AB_{11}:** *Caucasian/Lower socioeconomic class* (11, 9, 10); **Group AB_{12}:** *Caucasian/Middle socioeconomic class* (7, 8, 6); **Group AB_{13}:** *Caucasian/Upper socioeconomic class* (5, 4, 3); **Group AB_{21}:** *African-American/Lower socioeconomic class* (2, 4, 3); **Group AB_{22}:** *African-American/Middle socioeconomic class* (4, 5, 3); **Group AB_{23}:** *African-American/Upper socioeconomic class* (0, 1, 2). *Do the data indicate that either ethnicity or socioeconomic class is related to how often a person visits a doctor?*

IX. Addendum

Discussion of and computational procedures for additional analysis of variance procedures for factorial designs This addendum describes the following analysis of variance procedures: The **factorial analysis of variance for a mixed design (Test 27i)**, the **analysis of variance for a Latin square design (Test 27j)**, and the **within-subjects factorial analysis of variance (Test 27k)**.

1. Test 27i: The factorial analysis of variance for a mixed design A mixed factorial design involves two or more independent variables/factors in which at least one of the independent variables is measured between-subjects (i.e., different subjects serve under each of the levels of that independent variable) and at least one of the independent variables is measured within-subjects (i.e., the same subjects or matched sets of subjects serve under all of the levels of that independent variable).[28] Although the **factorial analysis of variance for a mixed design** can be used with designs involving more than two factors, the computational protocol to be described in this section will be limited to the two-factor experiment. For purposes of illustration it will be assumed that Factor A is measured between-subjects (i.e., different subjects serve in each of the p levels of Factor A), and that Factor B is measured within-subjects (i.e., all subjects are measured on each of the q levels of Factor B). Since one of the factors is measured within-subjects, a **mixed factorial design** requires a fraction of the subjects that are needed to evaluate the same set of hypotheses with a **between-subjects factorial design** (assuming both designs employ the same number of scores in each of the pq experimental conditions). To be more specific, the fraction of subjects required is 1 divided by the number of levels of the within-subjects factor (i.e., $1/q$ if Factor B is the within-subjects factor). The advantages as well as the disadvantages of a within-subjects analysis (which are discussed under the *t* **test for two dependent samples** and the **single-factor within-subjects analysis of variance**) also apply to the within-subjects factor which is evaluated with the **factorial analysis of variance for a mixed design**. Probably the most notable advantage associated with the within-subjects factor is that it allows for a more powerful test of an alternative hypothesis when contrasted with the between-subjects factor. Example 27.4 is employed to illustrate the use of the **factorial analysis of variance for a mixed design**.

Example 27.4 *A study is conducted to evaluate the effect of humidity (to be designated as Factor A) and temperature (to be designated as Factor B) on mechanical problem-solving ability. The experimenter employs a 2×3 mixed factorial design. The two levels which comprise Factor A are A_1: Low humidity; A_2: High humidity. The three levels which comprise Factor B are B_1: Low temperature; B_2: Moderate temperature; B_3: High temperature. The study employs six subjects, three of whom are randomly assigned to Level 1 of Factor A and three of whom are randomly assigned to Level 2 of Factor A. Each subject is exposed to all three levels of Factor B. The order of presentation of the levels of Factor B is completely counterbalanced within the six subjects. The number of mechanical problems solved by the subjects in the six experimental conditions (which result from combinations of the levels of the two factors) follow:* **Condition AB_{11}**: *Low humidity/Low temperature (11, 9, 10);* **Condition AB_{12}**: *Low humidity/ Moderate temperature (7, 8, 6);* **Condition AB_{13}**: *Low humidity/High temperature (5, 4, 3);* **Condition AB_{21}**: *High humidity/Low temperature (2, 4, 3);* **Condition AB_{22}**: *High humidity/ Moderate temperature (4, 5, 3);* **Condition AB_{23}**: *High humidity/High temperature (0, 1, 2). Do the data indicate that either humidity or temperature influences mechanical problem-solving ability?*

Table 27.11   Data for Example 27.4 for Evaluation with the Factorial Analysis of Variance for a Mixed Design

	A_1			Subject sums (ΣS_i)
	B_1	B_2	B_3	
Subject 1	11	7	5	$\Sigma S_1 = 23$
Subject 2	9	8	4	$\Sigma S_2 = 21$
Subject 3	10	6	3	$\Sigma S_3 = 19$
Condition sums	$\Sigma X_{AB_{11}} = 30$	$\Sigma X_{AB_{12}} = 21$	$\Sigma X_{AB_{13}} = 12$	$\Sigma X_{A_1} = 63$
	$\Sigma X^2_{AB_{11}} = 302$	$\Sigma X^2_{AB_{12}} = 149$	$\Sigma X^2_{AB_{13}} = 50$	

	A_2			Subject sums (ΣS_i)
	B_1	B_2	B_3	
Subject 4	2	4	0	$\Sigma S_4 = 6$
Subject 5	4	5	1	$\Sigma S_5 = 10$
Subject 6	3	3	2	$\Sigma S_6 = 8$
Condition sums	$\Sigma X_{AB_{21}} = 9$	$\Sigma X_{AB_{22}} = 12$	$\Sigma X_{AB_{23}} = 3$	$\Sigma X_{A_2} = 24$
	$\Sigma X^2_{AB_{21}} = 29$	$\Sigma X^2_{AB_{22}} = 50$	$\Sigma X^2_{AB_{23}} = 5$	
	$\Sigma X_{B_1} = 39$	$\Sigma X_{B_2} = 33$	$\Sigma X_{B_3} = 15$	$\Sigma X_T = 87$
				$\Sigma X^2_T = 585$

The data for Example 27.4 are summarized in Table 27.11. Examination of Table 27.11 reveals that since the data employed for Example 27.4 are identical to those employed for Example 27.1, the summary values for the rows, columns, and *pq* experimental conditions are identical to those in Table 27.1. Thus, the following values in Table 27.11 are identical to those obtained in Table 27.1: $n_{A_j} = 9$ and the values computed for ΣX_{A_j} and $\Sigma X^2_{A_j}$ for each of the levels of Factor A; $n_{B_k} = 6$ and the values computed for ΣX_{B_k} and $\Sigma X_{B_k}^2$ for each of the levels of Factor B; $n_{AB_{jk}} = n = 3$ and the values computed for $\Sigma X_{AB_{jk}}$ and $\Sigma X_{AB_{jk}}^2$ for each of the pq experimental conditions that result from combinations of the levels of the two factors: $N = npq = 18$; $\Sigma X_T = 87$; $\Sigma X_T^2 = 585$.

Note that in both the **between-subjects factorial analysis of variance** and the **factorial analysis of variance for a mixed design**, the value $n_{AB_{jk}} = n = 3$ represents the number of scores in each of the pq experimental conditions. In the case of the **factorial analysis of variance for a mixed design**, the value $N = npq = 18$ represents the total number of scores in the set of data. Note, however, that the latter value does not represent the total number of subjects employed in the study, as it does in the case of the **between-subjects factorial analysis of variance**. The number of subjects employed for a **factorial analysis of variance for a mixed design** will always be the value of n multiplied by the number of levels of the between-subjects factor. Thus, in Example 27.4 the number of subjects is $np = (3)(2) = 6$.[29]

As is the case for the **between-subjects factorial analysis of variance**, the following three F ratios are computed for the **factorial analysis of variance for a mixed design**: F_A, F_B, F_{AB}. The equations required for computing the F ratios are summarized in Table 27.12. Table 27.13 summarizes the computations for the **factorial analysis of variance for a mixed design** when it is employed to evaluate Example 27.4. In order to compute the F ratios for the **factorial analysis of variance for a mixed design**, it is required that the following summary values (which are also computed for the **between-subjects factorial analysis of variance**) be computed: $[XS]$, $[T]$, $[A]$, $[B]$, and $[AB]$. Since the summary values computed in Table 27.11 are identical to those computed in Table 27.1 (for Example 27.1), the same summary values are employed in Tables 27.12 and 27.13 to compute the values $[XS]$, $[T]$, $[A]$, $[B]$, and $[AB]$ (which are, respectively, computed with Equations 27.4–27.8). Thus: $[XS] = 585$, $[T] = 420.5$, $[A] = 505$, $[B] = 472.5$, and $[AB] = 573$. Since the same set of data and the same equations are employed for the **factorial analysis of variance for a mixed design** and the **between-subjects factorial analysis of variance**, both analysis of variance procedures yield identical values for $[XS]$, $[T]$, $[A]$, $[B]$, and $[AB]$. Inspection of Table 27.12 also reveals that the **factorial analysis of variance for a mixed design** and the **between-subjects factorial analysis of variance** employ the same equations to compute the values SS_A, SS_B, SS_{AB}, SS_T, MS_A, MS_B, and MS_{AB}.

In order to compute a number of additional sum of squares values for the **factorial analysis of variance for a mixed design**, it is necessary to compute the element $[AS]$ (which is not computed for the **between-subjects factorial analysis of variance**). $[AS]$, which is computed with Equation 27.69, is employed in Tables 27.12 and 27.13 to compute the following values: $SS_{\text{Between-subjects}}$, $SS_{\text{Subjects } WG}$, $SS_{\text{Within-subjects}}$, $SS_{B \times \text{subjects } WG}$.

$$[AS] = \sum_{i=1}^{np} \left[\frac{(\Sigma S_i)^2}{q} \right] \qquad \textbf{(Equation 27.69)}$$

The notation $\sum_{i=1}^{np}[(\Sigma S_i)^2/q]$ in Equation 27.69 indicates that for each of the $np = 6$ subjects, the score of the subject is squared and divided by q. The resulting values obtained for the np subjects are summed, yielding the value $[AS]$. Employing Equation 27.69, the value $[AS] = 510.33$ is computed.

$$[AS] = \frac{(23)^2}{3} + \frac{(21)^2}{3} + \frac{(19)^2}{3} + \frac{(6)^2}{3} + \frac{(10)^2}{3} + \frac{(8)^2}{3} = 510.33$$

Table 27.12   Summary Table of Equations for the Factorial Analysis of Variance for a Mixed Design

Source of variation	SS	df	MS	F
Between-subjects	$[AS]-[T]$	$np-1$		
A	$[A]-[T]$	$p-1$	$\dfrac{SS_A}{df_A}$	$F_A = \dfrac{MS_A}{MS_{\text{Subjects }WG}}$
Subjects WG	$[AS]-[A]$	$p(n-1)$	$\dfrac{SS_{\text{Subjects }WG}}{df_{\text{Subjects }WG}}$	
Within-subjects	$[XS]-[AS]$	$np(q-1)$		
B	$[B]-[T]$	$q-1$	$\dfrac{SS_B}{df_B}$	$F_B = \dfrac{MS_B}{MS_{B\times\text{subjects }WG}}$
AB	$[AB]-[A]-[B]+[T]$	$(p-1)(q-1)$	$\dfrac{SS_{AB}}{df_{AB}}$	$F_{AB} = \dfrac{MS_{AB}}{MS_{B\times\text{subjects }WG}}$
B × subjects WG	$[XS]-[AB]-[AS]+[A]$	$p(q-1)(n-1)$	$\dfrac{SS_{B\times\text{subjects }WG}}{df_{B\times\text{subjects }WG}}$	
Total	$[XS]-[T]$	$N-1 = npq-1$		

Table 27.13   Summary Table of Computations for Example 27.4

Source of variation	SS	df	MS	F
Between-subjects	$510.33-420.5 = 89.83$	$(3)(2)-1 = 5$		
A	$505-420.5 = 84.5$	$2-1 = 1$	$MS_A = \dfrac{84.5}{1} = 84.5$	$F_A = \dfrac{84.5}{1.33} = 63.53$
Subjects WG	$510.33-505 = 5.33$	$2(3-1) = 4$	$MS_{\text{Subjects }WG} = \dfrac{5.33}{4} = 1.33$	
Within-subjects	$585-510.33 = 74.67$	$(3)(2)(3-1) = 12$		
B	$472.5-420.5 = 52$	$3-1 = 2$	$MS_B = \dfrac{52}{2} = 26$	$F_B = \dfrac{26}{.83} = 31.33$
AB	$573-505-472.5+420.5 = 16$	$(2-1)(3-1) = 2$	$MS_{AB} = \dfrac{16}{2} = 8$	$F_{AB} = \dfrac{8}{.83} = 9.64$
B×subjects WG	$585-573-510.33+505 = 6.67$	$2(3-1)(3-1) = 8$	$MS_{B\times\text{subjects }WG} = \dfrac{6.67}{8} = .83$	
Total	$585-420.5 = 164.5$	$18-1 = (3)(2)(3)-1 = 17$		

The reader should take note of the following relationships in Tables 27.12 and 27.13:

$$SS_{\text{Between-subjects}} = SS_A + SS_{\text{Subjects } WG}$$

$$SS_{\text{Within-subjects}} = SS_B + SS_{AB} + SS_{B \times \text{subjects } WG}$$

$$SS_T = SS_{\text{Between-subjects}} + SS_{\text{Within-subjects}}$$

$$df_{\text{Between-subjects}} = df_A + df_{\text{Subjects } WG}$$

$$df_{\text{Within-subjects}} = df_B + df_{AB} + df_{B \times \text{subjects } WG}$$

$$df_T = df_{\text{Between-subjects}} + df_{\text{Within-subjects}}$$

Inspection of Table 27.2 and Tables 27.12/27.13 reveals that if a **between-subjects factorial analysis of variance** and a **factorial analysis of variance for a mixed design** are employed with the same set of data, identical values are computed for the following: SS_A, SS_B, SS_{AB}, SS_T, df_A, df_B, df_{AB}, df_T, MS_A, MS_B, and MS_{AB}.

In Table 27.13 the error term $MS_{\text{Subjects } WG} = 1.33$ (employed in computing the value $F_A = 63.53$) is identical to the value MS_{WG} which would be obtained if Factor B was not taken into account in Example 27.4, and the data on Factor A were evaluated with a **single-factor between-subjects analysis of variance**. Although the error term $MS_{B \times \text{subjects } WG} = .83$, employed in computing the values $F_B = 31.33$ and $F_{AB} = 9.64$ is analogous to the error term employed for the **single-factor within-subjects analysis of variance**, it will not yield the same value that would be obtained for MS_{res} if one ignored Factor A and just conducted a **single-factor analysis of variance** on the q levels of Factor B. Keppel and Wickens (2004, p. 437) note that the error term $MS_{B \times \text{subjects } WG}$ is a combination (actually an average) of the separate $B \times subjects$ *interaction* at each level of Factor A, each of which is the appropriate error term for its portion of the data. For a thorough discussion of the derivation of the error terms for the **factorial analysis of variance for a mixed design**, the reader should consult books that discuss analysis of variance procedures in greater detail (e.g., Keppel (1991), Winer *et al.* (1991), and Keppel and Wickens (2004)).

The following tabled critical values derived from **Table A10** are employed in evaluating the three F ratios computed for Example 27.4: a) **Factor A**: For $df_{\text{num}} = df_A = 1$ and $df_{\text{den}} = df_{\text{Subjects } WG} = 4$, $F_{.05} = 7.71$ and $F_{.01} = 21.20$; b) **Factor B**: For $df_{\text{num}} = df_B = 2$ and $df_{\text{den}} = df_{B \times \text{subjects } WG} = 8$, $F_{.05} = 4.46$ and $F_{.01} = 8.65$; and c) **AB interaction**: For $df_{\text{num}} = df_{AB} = 2$ and $df_{\text{den}} = df_{B \times \text{subjects } WG} = 8$, $F_{.05} = 4.46$ and $F_{.01} = 8.65$.

The identical null and alternative hypotheses that are evaluated in Section III of the **between-subjects factorial analysis of variance** are evaluated in the **factorial analysis of variance for a mixed design**. In order to reject the null hypothesis in reference to a computed F ratio, the obtained F value must be equal to or greater than the tabled critical value at the prespecified level of significance. Since the computed value $F_A = 63.53$ is greater than $F_{.05} = 7.71$ and $F_{.01} = 21.20$, the alternative hypothesis for Factor A is supported at both the .05 and .01 levels. Since the computed value $F_B = 31.33$ is greater than $F_{.05} = 4.46$ and $F_{.01} = 8.65$, the alternative hypothesis for Factor B is supported at both the .05 and .01 levels. Since the computed value $F_{AB} = 9.64$ is greater than $F_{.05} = 4.46$ and $F_{.01} = 8.65$, the alternative hypothesis for an interaction between Factors A and B is supported at both the .05 and .01 levels.

The analysis of the data for Example 27.4 allows the researcher to conclude that both humidity (Factor A) and temperature (Factor B) have a significant impact on problem-solving scores. However, as is the case when the same set of data is evaluated with a **between-subjects factorial analysis of variance**, the relationships depicted by the main effects must be qualified because of the presence of a significant interaction. Comparison procedures following the computation of the omnibus *F* ratios (as well as the other analytical procedures for determining power, effect size, etc.) described in Section VI of the **between-subjects factorial analysis of variance** can be extended to the **factorial analysis of variance for a mixed design**. With the exception of one comparison procedure which will be employed within the framework of evaluating a **crossover design** (which will be described next), such procedures will not be described in this book. For a full description of comparison procedures for the **factorial analysis of variance for a mixed design**, the reader should consult texts that discuss analysis of variance procedures in greater detail (e.g., Keppel (1991), Keppel and Wickens (2004), and Winer *et al.* (1991)).

Analysis of a crossover design with a factorial analysis of variance for a mixed design   The **crossover design** is described in Section VII (in the discussion of **counterbalancing and order effects**) of the *t* **test for two dependent samples**. As noted in the latter discussion, the **crossover design** is employed most frequently within the context of **clinical trials** (which is discussed in Section VII of the *t* **test for two independent samples**). A **clinical trial** refers to an experiment involving human subjects which is designed to evaluate a health related issue (such as the relative efficacy of two or more types of therapy — most commonly two drugs or a drug versus a placebo). In a **two-treatment crossover design**, every subject receives each of two treatments, with half of the subjects receiving Treatment 1 first followed by Treatment 2, and the other half receiving the treatments in reverse order. Typically, in studies evaluating the efficacy of two drugs, there is a **washout period** for each subject intervening between a subject's two treatments. During the washout period (the duration of which can be anywhere from a few days to a few months) the subject receives no treatment, and it is assumed that during that period any residual biological effects derived from the previous treatment will dissipate. An **order effect** is present when, in fact, there is a residual biological effect from a previous treatment. The latter type of **order effect** is commonly referred to as a **carryover effect**. The **factorial analysis of variance for a mixed design** can be employed to simultaneously evaluate the efficacy of the two treatments, as well as whether or not **carryover effects** are present in the data.   Since in a **crossover design** each of the two possible presentation orders are presented to an equal number of subjects, the order of presentation of the treatments is **completely counterbalanced**. As a result of the latter, when such a design is evaluated with a **factorial analysis of variance for a mixed design**, any variability attributable to presentation order can be extracted as a separate sum of squares which can be eliminated from error variability (thus reducing the magnitude of the latter value). For this reason, the **factorial analysis of variance for a mixed design** provides for the most powerful test for assessing whether or not a treatment effect is present.

Example 27.5 will be employed to demonstrate the analysis of a **crossover design** with the **factorial analysis of variance for a mixed design**.

Example 27.5 *A clinical trial is conducted in order to evaluate the efficacy of a drug for treating migraines relative to the impact of a placebo. In a double blind study, 12 subjects are administered the drug for a duration of six months and a placebo for a duration of six months. Six of the 12 subjects receive the drug as their first treatment followed by the placebo (the* DP *condition), while the other 6 subjects receive the placebo first followed by the drug (the* PD *condition). To insure that there are no carryover effects from one treatment to another, there*

Table 27.14 Data for Example 27.5 for Evaluation with the Factorial Analysis of Variance for a Mixed Design

	A_1		Subject sums (ΣS_i)
	B_1	B_2	
Subject 1	6	5	$\Sigma S_1 = 11$
Subject 2	3	2	$\Sigma S_2 = 5$
Subject 3	5	4	$\Sigma S_3 = 9$
Subject 4	2	3	$\Sigma S_4 = 5$
Subject 5	3	3	$\Sigma S_5 = 6$
Subject 6	2	1	$\Sigma S_6 = 3$
Condition sums	$\Sigma X_{AB_{11}} = 21$	$\Sigma X_{AB_{12}} = 18$	$\Sigma X_{A_1} = 39$
	$\Sigma X^2_{AB_{11}} = 87$	$\Sigma X^2_{AB_{12}} = 64$	

	A_2		Subject sums (ΣS_i)
	B_1	B_2	
Subject 7	8	4	$\Sigma S_7 = 12$
Subject 8	6	3	$\Sigma S_8 = 9$
Subject 9	3	2	$\Sigma S_9 = 5$
Subject 10	4	1	$\Sigma S_{10} = 5$
Subject 11	5	3	$\Sigma S_{11} = 8$
Subject 12	4	2	$\Sigma S_{12} = 6$
Condition sums	$\Sigma X_{AB_{21}} = 30$	$\Sigma X_{AB_{22}} = 15$	$\Sigma X_{A_2} = 45$
	$\Sigma X^2_{AB_{21}} = 166$	$\Sigma X^2_{AB_{22}} = 43$	
	$\Sigma X_{B_1} = 51$	$\Sigma X_{B_2} = 33$	$\Sigma X_T = 84$
			$\Sigma X^2_T = 360$

Table 27.15 Group and Marginal Means for Example 27.5

		Factor B (Treatment)		Row averages
		B_1 (Drug)	B_2 (Placebo)	
Factor A (Order of Treatment)	A_1 (DP)	$\overline{AB}_{11} = 3.5$	$\overline{AB}_{12} = 3$	$\overline{A}_1 = 3.25$
	A_2 (PD)	$\overline{AB}_{21} = 5$	$\overline{AB}_{22} = 2.5$	$\overline{A}_2 = 3.75$
Column averages		$\overline{B}_1 = 4.25$	$\overline{B}_2 = 2.75$	Grand mean = 3.5

is a two month washout period at the conclusion of the first treatment a subject receives. The dependent variable in the study is a subject's rating on a ten point scale with respect to the degree to which a treatment is effective (with 0 indicating no improvement). Tables 27.14 and 27.15 summarize the results of the study.

The reason that a **factorial analysis of variance for a mixed design** is employed to evaluate Example 27.5 is predicated on the fact that a **crossover design** can be conceptualized as being comprised of two factors, one being a between-subjects factor and the other a within-subjects factor. In Example 27.5, **Factor A**, the order of presentation of the treatments, represents a **between-subjects factor**, since a subject does not serve under all levels of the latter factor. Factor A is comprised of the following $p = 2$ levels: A_1: **Drug followed by Placebo (DP)** and A_2: **Placebo followed by Drug (PD)**. Note that $n_{A_1} = 6$ of the total of 12 subjects serve under **Level 1 (DP)** of Factor A and the other $n_{A_2} = 6$ subjects serve under **Level 2 (PD)** of Factor A. **Factor B**, the specific treatment a subject is administered, represents a **within-subjects factor**. The latter factor is comprised of the following $q = 2$ levels: B_1: **Drug** and B_2: **Placebo**. Note that all 12 subjects serve under both levels of Factor B, since every subject is administered both the drug and the placebo (and thus, $n_{B_k} = n_{B_1} = n_{B_2} = 12$). Note that there are $n_{AB_{jk}} = 6$ scores in each of the $pq = (2)(2) = 4$ cells which comprise Table 27.15 (with each cell representing one of the $q = 2$ treatments and one of the $p = 2$ presentation orders). $N = (p)(q)(n_{AB_{jk}}) = (2)(2)(6) = 24$ scores comprise the total set of data.

In evaluating the data with a **factorial analysis of variance for a mixed design**, the main effect for Factor A (as well as the **AB interaction**) is employed in evaluating whether or not there is any evidence of **carryover effects**. The main effect for Factor B is employed to determine whether or not there is a significant difference in efficacy with respect to the drug versus the placebo.

The following logic will be employed in evaluating the results of Example 27.5. Let us assume that the drug is more effective than the placebo. If the latter is true, it would be expected that, regardless of the order of presentation of the two treatments, the mean efficacy score of subjects under the drug condition (i.e., Condition B_1) will be greater than the mean efficacy score of subjects under the placebo condition (i.e., Condition B_2). Additionally, let us assume that, given the drug is more effective than the placebo, there is a residual carryover effect deriving from taking the drug prior to the placebo. If the latter were true, it would be expected that the overall mean efficacy score of subjects who receive the placebo after receiving the drug (i.e., subjects in Group A_1) will be greater than the overall mean efficacy score of subjects who receive the placebo prior to taking the drug (i.e., subjects in Group A_2). In the latter instance, the mean efficacy for the placebo when it is presented following the drug (i.e., Condition AB_{12}) will provide a biased estimate for the effect of the placebo. Under such circumstances, only first period data should be employed to evaluate the relative efficacy of the drug versus the placebo (i.e., conduct a comparison of Condition AB_{11} versus Condition AB_{22}). Table 27.16 is a summary table of a hypothetical set of data which would reflect the carryover effect noted above — i.e., a main effect on both the treatment variable and the order of presentation of the treatments, with a higher average for the placebo when it is administered after the drug as opposed to it being the initial treatment received by a subject.

Table 27.16 Group and Marginal Means Illustrating Carryover Effect

		Factor B (Treatment)		
		B_1 (Drug)	B_2 (Placebo)	Row averages
Factor A (Order of Treatment)	A_1 (DP)	$\overline{AB}_{11} = 4.2$	$\overline{AB}_{12} = 3.6$	$\overline{A}_1 = 3.9$
	A_2 (PD)	$\overline{AB}_{21} = 4.4$	$\overline{AB}_{22} = 2$	$\overline{A}_2 = 3.2$
Column averages		$\overline{B}_1 = 4.3$	$\overline{B}_2 = 2.8$	Grand mean = 3.55

At this point the data for Example 27.5 will be evaluated with a **factorial analysis of variance for a mixed design**. Initially, the summary value $[XS] = 360$ is computed with Equation 27.4.

$$[XS] = \sum X_T^2 = (6)^2 + (3)^2 + (5)^2 + \cdots + (1)^2 + (3)^2 + (2)^2 = 360$$

The summary value $[T] = 294$ is computed with Equation 27.5.

$$[T] = \frac{(\sum X_T)^2}{N} = \frac{(84)^2}{24} = 294$$

The summary value $[A] = 295.5$ is computed with Equation 27.6.

$$[A] = \sum_{j=1}^{p} \left[\frac{(\sum X_{A_j})^2}{n_{A_j}} \right] = \frac{(39)^2}{12} + \frac{(45)^2}{12} = 295.5$$

The summary value $[B] = 307.5$ is computed with Equation 27.7.

$$[B] = \sum_{k=1}^{q} \left[\frac{(\sum X_{B_k})^2}{n_{B_k}} \right] = \frac{(51)^2}{12} + \frac{(33)^2}{12} = 307.5$$

The summary value $[AB] = 315$ is computed with Equation 27.8.

$$[AB] = \sum_{k=1}^{q} \sum_{j=1}^{p} \left[\frac{(\sum X_{AB_{jk}})^2}{n_{AB_{jk}}} \right] = \frac{(21)^2}{6} + \frac{(18)^2}{6} + \frac{(30)^2}{6} + \frac{(15)^2}{6} = 315$$

The summary value $[AS] = 336$ is computed with Equation 27.69.

$$[AS] = \sum_{i=1}^{np} \left[\frac{(\sum S_i)^2}{q} \right] = \frac{(11)^2}{2} + \frac{(5)^2}{2} + \frac{(9)^2}{2} + \frac{(5)^2}{2} + \frac{(6)^2}{2} + \frac{(3)^2}{2}$$

$$+ \frac{(12)^2}{2} + \frac{(9)^2}{2} + \frac{(5)^2}{2} + \frac{(5)^2}{2} + \frac{(8)^2}{2} + \frac{(6)^2}{2} = 336$$

Equation 27.9 is employed to compute the value $SS_T = 66$.

$$SS_T = [XS] - [T] = 360 - 294 = 66$$

Equation 27.70 is employed to compute the value $SS_{BS} = 42$.

$$SS_{BS} = [AS] - [T] = 336 - 294 = 42 \qquad \textbf{(Equation 27.70)}$$

Equation 27.11 is employed to compute the value $SS_A = 1.5$.

$$SS_A = [A] - [T] = 295.5 - 294 = 1.5$$

Equation 27.71 is employed to compute the value $SS_{Subjects\ WG} = 40.5$.

$$SS_{Subjects\ WG} = [AS] - [A] = 336 - 295.5 = 40.5 \quad \textbf{(Equation 27.71)}$$

Equation 27.72 is employed to compute the value $SS_{Within-subjects} = 24$.

$$SS_{Within-subjects} = [XS] - [AS] = 360 - 336 = 24 \quad \textbf{(Equation 27.72)}$$

Equation 27.12 is employed to compute the value $SS_B = 13.5$.

$$SS_B = [B] - [T] = 307.5 - 294 = 13.5$$

Equation 27.13 is employed to compute the value $SS_{AB} = 6$.

$$SS_{AB} = [AB] - [A] - [B] + [T] = 315 - 295.5 - 307.5 + 294 = 6$$

Equation 27.73 is employed to compute the value $SS_{B \times subjects\ WG} = 4.5$.

$$\textbf{(Equation 27.73)}$$
$$SS_{B \times subjects\ WG} = [XS] - [AB] - [AS] + [A] = 360 - 315 - 336 + 295.5 = 4.5$$

The relevant **mean square** values are computed below. Note that the **degrees of freedom** values employed for the denominator of the mean square calculations are based on the use of the values $n = 6$, $p = 2$, $q = 2$, and $N = npq = 24$.

$$MS_A = \frac{1.5}{1} = 1.5 \qquad MS_{Subjects\ WG} = \frac{40.5}{10} = 4.05$$

$$MS_B = \frac{13.5}{1} = 13.5 \qquad MS_{AB} = \frac{6}{1} = 6 \qquad MS_{B \times subjects\ WG} = \frac{4.5}{10} = .45$$

The F ratios for the **factorial analysis of variance for a mixed design** are computed below, and Table 27.17 is a summary table for the analysis of variance.

$$F_A = \frac{MS_A}{MS_{Subjects\ WG}} = \frac{1.5}{4.05} = .37$$

$$F_B = \frac{MS_B}{MS_{B \times Subjects\ WG}} = \frac{13.5}{.45} = 30$$

$$F_{AB} = \frac{MS_{AB}}{MS_{B \times Subjects\ WG}} = \frac{6}{.45} = 13.33$$

The following tabled critical values derived from **Table A10** are employed in evaluating the three F ratios computed for Example 27.5: a) **Factor A**: For $df_{num} = df_A = 1$ and $df_{den} = df_{Subjects\ WG} = 10$, $F_{.05} = 4.96$ and $F_{.01} = 10.04$; b) **Factor B**: For $df_{num} = df_B = 1$ and $df_{den} = df_{B \times subjects\ WG} = 10$, $F_{.05} = 4.96$ and $F_{.01} = 10.04$; and c) **AB interaction**: For $df_{num} = df_{AB} = 1$ and $df_{den} = df_{B \times subjects\ WG} = 10$, $F_{.05} = 4.96$ and $F_{.01} = 10.04$.

Table 27.17 Summary Table of Analysis of Variance for Example 27.5

Source of variation	SS	df	MS	F
Between-subjects	42	11		
A	1.5	1	1.5	.37
Subjects WG	40.5	10	4.05	
Within-subjects	24	12		
B	13.5	1	13.5	30
AB	6	1	6	13.33
B × Subjects WG	4.5	10	.45	
Total	66	23		

Since the computed value $F_A = .37$ is less than $F_{.05} = 4.96$, the alternative hypothesis for Factor A is not supported at the .05 level. The latter result indicates that in Table 27.15 there is no difference between the marginal row means of $\bar{A}_1 = 3.25$ and $\bar{A}_2 = 3.75$ (which represent the overall means for the two presentation orders). Based on the description of the logic discussed earlier for evaluating the results of the analysis, the fact that the value $F_A = .37$ is not significant appears to indicate that no **carryover effects** are present. However, as will be noted later, the presence of a significant interaction will require that the latter conclusion be qualified.

Since the computed value $F_B = 30$ is greater than $F_{.05} = 4.96$ and $F_{.01} = 10.04$, the alternative hypothesis for Factor B is supported at both the .05 and .01 levels. The latter result indicates that in Table 27.15 there is a significant difference between the marginal column means of $\bar{B}_1 = 4.25$ and $\bar{B}_2 = 2.75$ (which represent the overall means for the drug and placebo treatments). The latter result clearly indicates that the drug is rated more effective than the placebo. Once again, however, the presence of a significant interaction will require that the latter statement be qualified.

Since the computed value $F_{AB} = 13.33$ is greater than $F_{.05} = 4.96$ and $F_{.01} = 10.04$, the alternative hypothesis for an interaction between Factors A and B is supported at both the .05 and .01 levels. Visual examination of the mean values computed for the four cells which comprise Table 27.15 reveals that the superiority of the drug over the placebo appears to be greater in the **PD** presentation order (where the drug mean is $\bar{AB}_{21} = 5$ and the placebo mean is $\bar{AB}_{22} = 2.5$) than the **DP** presentation order (where the drug mean is $\bar{AB}_{11} = 3.5$ and the placebo mean is $\bar{AB}_{12} = 3$). Thus, the presence of a significant interaction suggests an **order effect** — specifically, that the drug is rated more effective than the placebo when the drug is the second treatment administered to a subject. Further confirmation of the latter can be obtained by an evaluation of the **simple effects** (which, as noted in Section VI, is an analysis of variance evaluating all of the levels of one factor across only one level of the other factor). However, prior to discussing the **simple effects** for Example 27.5, alternative methods (often documented in other sources) for evaluating the main effects summarized in Table 27.17 will be described.

Alternative methods for evaluating the main effects in a two-treatment crossover design
An alternative but equivalent methodology for evaluating the main effect on Factor A (i.e., the order of presentation of the treatments) is to first compute an average score for each subject (i.e., add up the score of each subject for the drug and placebo conditions and divide the sum by two). The latter values can easily be obtained for the $n = 12$ subjects by dividing the value recorded for a subject in the **Subject's sums** column of Table 27.14 by two. A **t test for two independent samples** is then employed to compare the average scores of the six subjects who served under **Level 1** of Factor A (i.e., the presentation order **DP**) with the average scores of the six subjects

who served under **Level 2** of Factor A (i.e., the presentation order **PD**). The average scores of the six subjects who served under the two levels of Factor A follow: A_1: 5.5, 2.5, 4.5, 2.5, 3, 1.5; A_2: 6, 4.5, 2.5, 2.5, 4, 3. In conducting the *t* **test for two independent samples** the following values are employed $\bar{A}_1 = 3.25$, $\bar{A}_2 = 3.75$, $\tilde{s}_{A_1} = 1.475$, $\tilde{s}_{A_2} = 1.369$, and $t = [\,|\,3.25 - 3.75\,|\,]\,/\,\sqrt{[(1.475)^2/6)] + [(1.369)^2/6)]} = .609$ (with $p > .05$ for $df = n_1 + n_2 - 2 = 10$). As is the case with the analysis of variance, the result is not significant, since $t = .609$ is less than $t_{.05} = 1.81$ and $t_{.05} = 2.23$ (which are the tabled critical one- and two-tailed $t_{.05}$ values for $df = 10$ in **Table A2 (Table of Student's *t* Distribution)** in the **Appendix**). Note that $t^2 = (.609)^2$ equals the value $F_A = .37$ computed in Table 27.17 for the main effect on Factor A. The latter confirms the fact that the above analysis is equivalent to the analysis of the main effect on Factor A obtained with the **factorial analysis of variance for a mixed design**.

In the case of Factor B (i.e., the two treatments) one might expect that by conducting a *t* **test for two dependent samples** which compares the average scores of the $n = 12$ subjects under the drug condition with their average scores under the placebo condition, a result will be obtained which is equivalent to that obtained for the main effect for Factor B in Table 27.17. If the latter analysis is equivalent, the obtained value of t should be $\sqrt{F_B} = \sqrt{30} = 5.48$.

Table 27.18   Analysis of Treatment Effect Only for Example 27.5

	Subject	Drug X_{B_1}	Placebo X_{B_2}	D	D^2
A_1	1	6	5	1	1
	2	3	2	1	1
	3	5	4	1	1
	4	2	3	−1	1
	5	3	3	0	0
	6	2	1	1	1
A_2	7	8	4	4	16
	8	6	3	3	9
	9	3	2	1	1
	10	4	1	3	9
	11	5	3	2	4
	12	4	2	2	4

$$\Sigma X_{B_1} = 51 \qquad \Sigma X_{B_2} = 33 \qquad \begin{array}{l}\Sigma D- = -1 \\ \Sigma D+ = 19 \end{array} \qquad \Sigma D^2 = 48$$

$$\Sigma D = 18$$

$$\bar{X}_{B_1} = \frac{51}{12} = 4.25 \qquad \bar{X}_{B_2} = \frac{33}{12} = 2.75 \qquad \bar{D} = \frac{18}{12} = 1.5$$

$$\tilde{s}_D = \sqrt{\frac{48 - \dfrac{(18)^2}{12}}{12 - 1}} = 1.382 \qquad s_{\bar{D}} = \frac{1.382}{\sqrt{12}} = .399 \qquad t = \frac{1.5}{.399} = 3.76$$

Table 27.18, which summarizes the data for Example 27.5, is formatted to only take into account the two treatments but not their order of presentation. When the *t* **test for two dependent samples** is employed to compare the two treatments, the following values are computed from Table 27.18 (through use of Equations 17.2, 17.3, 17.4 and 17.1): $\bar{D} = 18/12 = 1.5$, $\tilde{s}_D = 1.382$, $s_{\bar{D}} = .399$, $t = 1.5/.399 = 3.76$ (with $p < .01$ for $df = n - 1 = 11$). As is the case with the analysis of variance, the result is significant at the .01 level, since $t = 3.76$ is greater than $t_{.01} = 2.72$ and $t_{.01} = 3.11$ (which are the tabled critical one and two-tailed $t_{.01}$ values for $df = 11$ in **Table A2**). Although the result indicates that subjects' mean efficacy ratings for the drug are significantly greater than for the placebo, note that $t^2 = (3.76)^2 = 14.14$, which is substantially less than the value $F_B = 30$ computed for the main effect for Factor B in Table 27.17. The latter indicates that analysis of the treatment effects through use of the **t test for two dependent samples** as just illustrated does not yield a result which is equivalent to that obtained with the analysis of variance. In point of fact, it provide a less powerful test of the main effect for Factor B than the result obtained from the analysis of variance. As noted earlier in the discussion of the **factorial analysis of variance for a mixed design**, the within-subjects error term $MS_{B \times \text{subjects } WG} = .45$ employed in computing the values $F_B = 30$ and $F_{AB} = 13.33$ will not yield the same value that would be obtained for MS_{res} if one ignored Factor A and just conducted a **single-factor analysis of variance** on the q levels of Factor B (or, when $q = 2$, a **t test for two dependent samples**).

Fleiss (1986, pp. 26 7–268) and Rosner (2000, pp. 641–645) note that the appropriate **t test** to conduct in order to evaluate the main effect on Factor B is a test (which is conceptualized as a **t test for two independent samples**) which divides the **mean of the mean difference scores for Levels 1 and 2 of Factor A** by the **standard error of the mean difference for Levels 1 and 2 of Factor A**. Specifically, the value $t = 5.48$ is computed below through use of Equations 27.74–27.80. Note that $\Sigma D_{A_1} = 3$ and $\Sigma D_{A_1}^2 = 5$ are the sum of the difference scores and sum of the squared difference scores in Table 27.18 for the $n_{A_1} = 6$ subjects who served under Level 1 of Factor A. $\Sigma D_{A_2} = 15$ and $\Sigma D_{A_2}^2 = 43$ are the sum of the difference scores and sum of the squared difference scores in Table 27.18 for the $n_{A_2} = 6$ subjects who served under Level 2 of Factor A. The value $\bar{D} = 1.5$, which is computed with Equation 27.76, is identical to the value for $\bar{D}$ computed for the $n = n_{A_1} + n_{A_2} = 12$ difference scores in Table 27.18. The values $\tilde{s}_{D_{A_1}} = .837$ and $\tilde{s}_{D_{A_2}} = 1.049$, computed with Equation 27.77, are respectively the **estimated population standard deviations of the differences scores** of the subjects who served under Levels 1 and 2 of Factor A. The value $\tilde{s}_{D_{pooled}}^2 = .900$, computed with Equation 27.78, is the **pooled variance of the within-subjects differences**. The value $s_{\bar{D}} = .274$, computed with Equation 27.79, is the **standard error of the mean difference**. The reader should take note of the fact that the value computed with Equation 27.79 for $s_{\bar{D}}$ is different (specifically, it is smaller) than the value computed in Table 27.18 for $s_{\bar{D}}$. The smaller value computed for $s_{\bar{D}}$ with Equation 27.79 is the reason why a larger t value is computed with Equation 27.80 (since in both Table 27.18 and the computation of the value $t = 5.48$ with Equation 27.80, the same value (specifically, $\bar{D} = 1.5$) is employed for the numerator in the **t test** equation).

$$\bar{D}_{A_1} = \frac{\Sigma D_{A_1}}{n_{A_1}} = \frac{3}{6} = .5 \qquad \textbf{(Equation 27.74)}$$

$$\bar{D}_{A_2} = \frac{\Sigma D_{A_2}}{n_{A_2}} = \frac{15}{6} = 2.5 \qquad \textbf{(Equation 27.75)}$$

$$\bar{D} = \frac{\bar{D}_{A_1} + \bar{D}_{A_2}}{2} = \frac{.5 + 2.5}{2} = 1.5 \qquad \textbf{(Equation 27.76)}$$

$$\tilde{s}_{D_{A_j}} = \sqrt{\frac{\sum D_{A_j}^2 - \frac{(\sum D_{A_j})^2}{n_{A_j}}}{n_{A_j} - 1}} \qquad \textbf{(Equation 27.77)}$$

$$\tilde{s}_{D_{A_1}} = \sqrt{\frac{5 - \frac{(3)^2}{6}}{6 - 1}} = .837 \qquad \tilde{s}_{D_{A_2}} = \sqrt{\frac{43 - \frac{(15)^2}{6}}{6 - 1}} = 1.049$$

$$\tilde{s}_{D_{pooled}}^2 = \left[\frac{(n_{A_1} - 1)\tilde{s}_{D_{A_1}}^2 + (n_{A_2} - 1)\tilde{s}_{D_{A_2}}^2}{n_{A_1} + n_{A_2} - 2} \right] \qquad \textbf{(Equation 27.78)}$$

$$\tilde{s}_{D_{pooled}}^2 = \frac{(6 - 1)(.837)^2 + (6 - 1)(1.049)^2}{6 + 6 - 2} = .900$$

$$s_{\bar{D}} = \sqrt{\frac{\tilde{s}_{D_{pooled}}^2}{4} \left[\frac{1}{n_{A_1}} + \frac{1}{n_{A_2}} \right]} = \sqrt{\frac{.900}{4} \left[\frac{1}{6} + \frac{1}{6} \right]} = .274 \quad \textbf{(Equation 27.79)}$$

$$t = \frac{\bar{D}}{s_{\bar{D}}} = \frac{1.5}{.274} = 5.48 \qquad \textbf{(Equation 27.80)}$$

Since the analysis employs two groups of subjects representing the two levels of Factor A, the degrees of freedom employed in evaluating the value $t = 5.48$ are computed with $df = n_{A_1} + n_{A_2} - 2 = 10$. Since the absolute value $t = 5.48$ is greater than $t_{.01} = 2.76$ and $t_{.01} = 3.17$ (which are the tabled critical one- and two-tailed $t_{.01}$ values for $df = 10$ in **Table A2**), as is the case with the analysis of variance, the result is significant at the .01 level (for either a directional analysis predicting the drug as more effective than the placebo or a nondirectional analysis). Note that $t^2 = (5.48)^2$ equals the value $F_B = 30$ computed in Table 27.17 for the main effect on Factor B (the minimal difference is due to rounding off error). The latter confirms the fact that the above analysis is equivalent to the analysis of the main effect on Factor B obtained with the **factorial analysis of variance for a mixed design**.

A confidence interval for the mean difference ($\bar{D}$) can be obtained through use Equation 27.81. To illustrate, the latter equation is employed to compute the 95% confidence interval.

$$CI_{(1-\alpha)} = \bar{D} \pm (t_{\alpha/2})(s_{\bar{D}}) \qquad \textbf{(Equation 27.81)}$$

$$CI_{95} = 1.5 \pm (2.228)(.274) = 1.5 \pm .610$$

$$.89 \leq \mu_{\bar{D}} \leq 2.11$$

This result indicates the researcher can be 95% confident (or the probability is .95) that the interval .89 to 2.11 contains the true difference between the population treatment means $\mu_{\bar{B}_1}$ and $\mu_{\bar{B}_2}$ (where $u_{\bar{D}} = \mu_{\bar{B}_1} - \mu_{\bar{B}_2}$). If the aforementioned interval contains the true difference between the population means, the mean of the population the drug condition represents is greater than the mean of the population the placebo condition represents by at least .89 units but not by more than 2.11 units.

Analysis of simple effects Earlier in the discussion of the results obtained for Example 27.5 it was noted that the presence of a significant interaction suggests an order effect — specifically, that the drug is rated more effective than the placebo when the drug is the second treatment administered to a subject. Further confirmation of the latter can be obtained by an evaluation of the **simple effects** (which, as noted in Section VI, is an analysis of variance evaluating all of the levels of one factor across only one level of the other factor). Specifically, the following two comparisons will be conducted to evaluate the **simple effects of Factor A across each of the levels of Factor B**: 1) $\bar{AB}_{11} = 3.5$ versus $\bar{AB}_{21} = 5$; 2) $\bar{AB}_{12} = 3$ versus $\bar{AB}_{22} = 2.5$. In addition, the following two comparisons will evaluate the **simple effects of Factor B across each of the levels of Factor A**: 1) $\bar{AB}_{11} = 3.5$ versus $\bar{AB}_{12} = 3$; 2) $\bar{AB}_{21} = 5$ versus $\bar{AB}_{22} = 2.5$.

Initially, we will evaluate the **simple effects** of **Factor A across each of the levels of Factor B**. The latter analysis involves conducting a separate **single-factor between-subjects analysis of variance** across each of the two levels of Factor B. As noted in Endnote 14, Equation 27.105 (which is equivalent to Equation 21.3 employed for computing the **between-groups sum of squares** for the **single-factor between-subjects analysis of variance**) is employed to compute the **sum of squares for each of the simple effects** of Factor A. The appropriate computations are carried out below.

$$SS_{A \text{ at } B_k} = \sum \left[\frac{(\Sigma X_{AB_{.k}})^2}{n_{AB_{.k}}} \right] - \frac{(\Sigma\Sigma X_{AB_{.k}})^2}{(n_{AB_{.k}})(p)}$$

$$SS_{A \text{ at } B_1} = \sum \left[\frac{(\Sigma X_{AB_{.1}})^2}{n_{AB_{.1}}} \right] - \frac{(\Sigma\Sigma X_{AB_{.1}})^2}{(n_{AB_{.1}})(p)} = \left[\frac{(21)^2}{6} + \frac{(30)^2}{6} \right] - \frac{(51)^2}{(6)(2)} = 6.75$$

$$SS_{A \text{ at } B_2} = \sum \left[\frac{(\Sigma X_{AB_{.2}})^2}{n_{AB_{.2}}} \right] - \frac{(\Sigma\Sigma X_{AB_{.2}})^2}{(n_{AB_{.2}})(p)} = \left[\frac{(18)^2}{6} + \frac{(15)^2}{6} \right] - \frac{(33)^2}{(6)(2)} = .75$$

In addition to the above computations, Equation 27.82 will be needed to compute the appropriate error terms for evaluating the **simple effects** of Factor A.

(Equation 27.82)

$$SS \text{ Subjects WG at } B_k = \Sigma\Sigma X_{AB_{.k}}^2 - \sum \left[\frac{(\Sigma X_{AB_{.k}})^2}{n_{AB_{.k}}} \right]$$

$$SS \text{ Subjects WG at } B_1 = \Sigma\Sigma X^2_{AB_{.1}} - \Sigma\left[\frac{(\Sigma X_{AB_{.1}})^2}{n_{AB_{.1}}}\right]$$

$$SS \text{ Subjects WG at } B_1 = (87 + 166) - \left[\frac{(21)^2}{6} + \frac{(30)^2}{6}\right] = 253 - 223.5 = 29.5$$

$$SS \text{ Subjects WG at } B_2 = \Sigma\Sigma X^2_{AB_{.2}} - \Sigma\left[\frac{(\Sigma X_{AB_{.2}})^2}{n_{AB_{.2}}}\right]$$

$$SS \text{ Subjects WG at } B_2 = (64 + 43) - \left[\frac{(18)^2}{6} + \frac{(15)^2}{6}\right] = 107 - 91.5 = 15.5$$

Tables 27.19 and 27.20 summarize the analysis of variance for the simple effects of Factor A. The mean square for each simple effect is obtained by dividing the sum of squares for the simple effect by its degrees of freedom (where $df_{A \text{ at } B_k} = p - 1$ and $df_{\text{Subjects WG at } B_k} = p(n_{AB_{jk}} - 1)$). The F value for each simple effect is obtained by dividing the mean square for the simple effect by the error term **Subjects WG at B_k**.

Table 27.19 Analysis of Simple Effect of Factor A at B_1

Source of variation	SS	df	MS	F
A at B_1	6.75	1	6.75	2.29
Subjects WG at B_1	29.5	10	2.95	
Total	36.25	11		

Table 27.20 Analysis of Simple Effect of Factor A at B_2

Source of variation	SS	df	MS	F
A at B_2	.75	1	.75	.48
Subjects WG at B_2	15.5	10	1.55	
Total	16.25	11		

Employing **Table A10**, the degrees of freedom used in evaluating each of the simple effects are $df_{\text{num}} = df_{A \text{ at } B_k} = 1$ and $df_{\text{den}} = df_{\text{Subjects WG at } b_k} = 10$. Since both of the obtained values $F_{A \text{ at } B_1} = 2.29$ and $F_{A \text{ at } B_2} = .48$ are less than $F_{.05} = 4.96$ (which is the tabled critical .05 value for $df_{\text{num}} = 1$ and $df_{\text{den}} = 10$), neither of the simple effects is significant at the .05 level. Thus, neither of the comparisons $\overline{AB}_{11} = 3.5$ versus $\overline{AB}_{21} = 5$ or $\overline{AB}_{12} = 3$ versus $\overline{AB}_{22} = 2.5$ represents a significant difference. The nonsignificant result for the **simple effect of Factor A at B_1** indicates that regardless of whether it is the first or second treatment a subject receives, there is no difference with respect to the efficacy of the drug. The nonsignificant result for the **simple effect of Factor A at B_2** indicates that regardless of whether it is the first or second treatment a subject receives, there is no difference with respect to the efficacy of the placebo.

It should be noted that an equivalent result for the above analysis can be obtained if a *t* test **for two independent samples** is employed to evaluate the two above noted simple effects/comparisons. Specifically, if the *t* test for two independent samples is employed to compare the $n_{AB_{11}} = 6$ scores in Condition AB_{11} with the $n_{AB_{21}} = 6$ scores in Condition AB_{21}, the following values are computed $\bar{X}_{AB_{11}} = 3.5$, $\bar{X}_{AB_{21}} = 5$, $\tilde{s}_{AB_{11}} = 1.643$, $\tilde{s}_{AB_{21}} = 1.789$, and $t = [\,|\,3.5 - 5\,|\,] / \sqrt{[(1.643)^2/6)] + [(1.789)^2/6)]} = 1.51$ (with $p > .05$, since the latter *t* value is less than the tabled critical one- and two-tailed .05 values (for $df = n_1 + n_2 - 2 = 10$) $t_{.05} = 1.81$ and $t_{.05} = 2.23$). Note that $t^2 = (1.51)^2$ equals the value $F = 2.29$ computed for the same comparison in Table 27.19 (the minimal difference is due to rounding off error). It should be noted that the fact the difference of 1.5 units between $\bar{X}_{AB_{11}} = 3.5$ and $\bar{X}_{AB_{21}} = 5$ is nonsignificant might be attributed to low power (because of the small sample size).

Additionally, if the *t* test for two independent samples is employed to compare the $n_{AB_{12}} = 6$ scores in Condition AB_{12} with the $n_{AB_{22}} = 6$ scores in Condition AB_{22}, the following values are computed $\bar{X}_{AB_{12}} = 3$, $\bar{X}_{AB_{22}} = 2.5$, $\tilde{s}_{AB_{12}} = 1.414$, $\tilde{s}_{AB_{22}} = 1.049$, and $t = [\,|\,3 - 2.5\,|\,] / \sqrt{[(1.414)^2/6)] + [(1.049)^2/6)]} = .70$ (with $p > .05$, since the latter *t* value is less than the tabled critical one- and two-tailed .05 values (for $df = n_1 + n_2 - 2 = 10$) $t_{.05} = 1.81$ and $t_{.05} = 2.23$). Note that $t^2 = (.70)^2$ equals the value $F = .48$ computed for the same comparison in Table 27.20 (the minimal difference is due to rounding off error).

As previously noted in Section VI, the variability within each of the simple effects is the result of contributions from both the main effect on the factor for which the simple effect is being evaluated (Factor A in this instance), as well as any interaction between the two factors. For this reason, the total of the sum of squares for each of the simple effects for a given factor will be equal to the interaction sum of squares (SS_{AB}) plus the sum of squares for that factor (SS_A). This can be confirmed by the fact that for the calculations under discussion the following is true.

$$[(SS_{A \text{ at } B_1} = 6.75) + (SS_{A \text{ at } B_2} = .75)] = [(SS_{AB} = 6) + (SS_A = 1.5)] = 7.5$$

As previously noted, the analysis of simple effects in and of itself cannot provide definitive evidence with regard to the presence or absence of an interaction.

We can also evaluate the **simple effects of Factor B across each of the levels of Factor A**. The latter analysis involves conducting a separate **single-factor within-subjects analysis of variance** across each of the two levels of Factor A. Equation 27.46 (which is equivalent to Equation 24.3 employed for computing the **between-conditions sum of squares** for the **single-factor within-subjects analysis of variance**) is employed to compute the **sum of squares for each of the simple effects**.

$$SS_{B \text{ at } A_j} = \sum \left[\frac{(\Sigma X_{AB_j})^2}{n_{AB_j}} \right] - \frac{(\Sigma\Sigma X_{AB_j})^2}{(n_{AB_j})(q)}$$

$$SS_{B \text{ at } A_1} = \sum \left[\frac{(\Sigma X_{AB_1})^2}{n_{AB_1}} \right] - \frac{(\Sigma\Sigma X_{AB_1})^2}{(n_{AB_1})(q)} = \left[\frac{(21)^2}{6} + \frac{(18)^2}{6} \right] - \frac{(39)^2}{(6)(2)} = .75$$

$$SS_{B \text{ at } A_2} = \sum \left[\frac{(\Sigma X_{AB_2})^2}{n_{AB_2}} \right] - \frac{(\Sigma\Sigma X_{AB_2})^2}{(n_{AB_2})(q)} = \left[\frac{(30)^2}{6} + \frac{(15)^2}{6} \right] - \frac{(45)^2}{(6)(2)} = 18.75$$

In addition, for each **single-factor within-subjects analysis of variance** which is to be conducted, Equations 24.4 and 24.6 will respectively be employed to compute the **between-subjects** and **residual sums of squares**. Initially, Equations 24.4 and 24.6 will be employed to compute the **between-subjects** and **residual sums of squares** for the simple effect of **Factor B at A_1**. Note that in using Equations 24.4 and 24.6, $n = n_{A_1} = 6$, ΣS_i is computed for the $n_{A_1} = 6$ subjects in Level 1 of Factor A, and ΣX_T is computed for the $N = (n_{AB_1})(q) = 12$ scores in Level 1 of Factor A.

$$SS_{BS} = \sum_{i=1}^{n}\left[\frac{(\Sigma S_i)^2}{k}\right] - \frac{(\Sigma X_T)^2}{N}$$

$$SS_{BS} = \left[\frac{(11)^2}{2} + \frac{(5)^2}{2} + \frac{(9)^2}{2} + \frac{(5)^2}{2} + \frac{(6)^2}{2} + \frac{(3)^2}{2}\right] - \frac{(39)^2}{(6)(2)} =$$

$$148.5 - 126.75 = 21.75$$

$$SS_{res} = \Sigma X_T^2 - \sum_{j=1}^{k}\left[\frac{(\Sigma X_j)^2}{n}\right] - \sum_{i=1}^{n}\left[\frac{(\Sigma S_i)^2}{q}\right] + \frac{(\Sigma X_T)^2}{nq}$$

$$[87 + 64] - \left[\frac{(21)^2}{6} + \frac{(18)^2}{6}\right] - \left[\frac{(11)^2}{2} + \frac{(5)^2}{2} + \frac{(9)^2}{2} + \frac{(5)^2}{2} + \frac{(6)^2}{2} + \frac{(3)^2}{2}\right] + \left[\frac{(39)^2}{(6)(2)}\right] =$$

$$151 - 127.5 - 148.5 + 126.75 = 1.75$$

Table 27.21 summarizes the analysis of variance for the simple effect of **Factor B at A_1**. The mean square for each simple effect is obtained by dividing the sum of squares for the simple effect by its degrees of freedom (where $df_{B \ at \ A_j} = q - 1 = 2 - 1 = 1$, $df_{BS} = n_{A_j} - 1 = 6 - 1 = 5$, and $df_{res} = (n_{A_j} - 1)(q - 1) = (5)(1) = 5$). The F value for each simple effect is obtained by dividing the mean square for the simple effect by the value computed for MS_{res}.

Table 27.21   Analysis of Simple Effect of Factor B at A_1

Source of variation	SS	df	MS	F
Between-subjects	21.75	5	4.35	
B at A_1	.75	1	.75	2.14
Residual	1.75	5	.35	
Total	24.25	11		

Employing **Table A10**, the degrees of freedom used in evaluating the simple effect are $df_{num} = df_{B \ at \ A_1} = 1$ and $df_{den} = df_{res} = 5$. Since the obtained value $F_{B \ at \ A_1} = 2.14$ is less than $F_{.05} = 6.61$ (which is the tabled critical .05 value for $df_{num} = 1$ and $df_{den} = 5$), the simple effect of **Factor B at A_1** is not significant. Thus, the comparison $\overline{AB}_{11} = 3.5$ versus $\overline{AB}_{12} = 3$ does not represent a significant difference. The nonsignificant result for the **simple effect of Factor B at A_1** indicates that there is no difference in efficacy between the drug and placebo when the order of presentation of the treatments is the drug followed by the placebo.

Equations 24.4 and 24.6 will now be employed to compute the **between-subjects** and **residual sums of squares** for the simple effect of **Factor B at A_2**. Note that in using Equations 24.4 and 24.6, $n = n_{A_2} = 6$, ΣS_i is computed for the $n_{A_2} = 6$ subjects in Level 2 of Factor A, and ΣX_T is computed for the $N = (n_{AB_2})(q) = 12$ scores in Level 2 of Factor A.

$$SS_{BS} = \sum_{i=1}^{n} \left[\frac{(\Sigma S_i)^2}{k} \right] - \frac{(\Sigma X_T)^2}{N}$$

$$SS_{BS} = \left[\frac{(12)^2}{2} + \frac{(9)^2}{2} + \frac{(5)^2}{2} + \frac{(5)^2}{2} + \frac{(8)^2}{2} + \frac{(6)^2}{2} \right] - \frac{(45)^2}{(6)(2)} =$$

$$187.5 - 168.75 = 18.75$$

$$SS_{res} = \Sigma X_T^2 - \sum_{j=1}^{k} \left[\frac{(\Sigma X_j)^2}{n} \right] - \sum_{i=1}^{n} \left[\frac{(\Sigma S_i)^2}{q} \right] + \frac{(\Sigma X_T)^2}{nq}$$

$$[166 + 43] - \left[\frac{(30)^2}{6} + \frac{(15)^2}{6} \right] - \left[\frac{(12)^2}{2} + \frac{(9)^2}{2} + \frac{(5)^2}{2} + \frac{(5)^2}{2} + \frac{(8)^2}{2} + \frac{(6)^2}{2} \right] + \left[\frac{(45)^2}{(6)(2)} \right] =$$

$$209 - 187.5 - 187.5 + 168.75 = 2.75$$

Table 27.22 summarizes the analysis of variance for the simple effect of **Factor B at A_2**.

Table 27.22　Analysis of Simple Effect of Factor B at A_2

Source of variation	SS	df	MS	F
Between-subjects	18.75	5	3.75	
B at A_2	18.75	1	18.75	34.09
Residual	2.75	5	.55	
Total	40.25	11		

Employing **Table A10**, the degrees of freedom used in evaluating the simple effect are $df_{num} = df_{B \ at \ A_2} = 1$ and $df_{den} = df_{res} = 5$. Since the obtained value $F_{B \ at \ A_2} = 34.09$ is greater than $F_{.01} = 16.26$ (which is the tabled critical .01 value for $df_{num} = 1$ and $df_{den} = 5$), the simple effect of **Factor B at A_2** is significant at the .01 level. Thus, the comparison $\bar{AB}_{21} = 5$ versus $\bar{AB}_{22} = 2.5$ represents a significant difference. Specifically, subjects have a significantly higher efficacy score for the drug than the placebo when the drug is administered as the second treatment.

It should be noted that equivalent results for the **simple effects for Factor B** can be obtained if a *t* **test for two dependent samples** is employed to evaluate the two above described simple effects/comparisons. Specifically, if the *t* **test for two dependent samples** is employed to compare the scores of the $n_{A_1} = 6$ subjects for Condition AB_{11} versus Condition AB_{12}, the following values can be employed with either Equation 17.1 or 17.6 to compute the value $t = 1.46$ (with $p > .05$, since the latter *t* value is less than the tabled critical one- and two-tailed .05 values (for $df = n - 1 = 5$) $t_{.05} = 2.02$ and $t_{.05} = 2.57$): $\bar{D} = .5$, $\tilde{s}_D = .837$, $s_{\bar{D}} = .342$ (thus, $t = .5/ .342 = 1.46$), $\bar{X}_{AB_{11}} = 3.5$, $\bar{X}_{AB_{12}} = 3$, $\tilde{s}_{AB_{11}} = 1.643$, $\tilde{s}_{AB_{12}} = 1.414$, $r_{11 \ vs. \ 12} = .861$. Note that $t^2 = (1.46)^2$ equals the value $F = 2.14$ computed for the same comparison in Table 27.21 (the minimal difference is due to rounding off error).

If the *t* **test for two dependent samples** is employed to compare the scores of the $n_{A_2} = 6$ subjects for Condition AB_{21} versus Condition AB_{22}, the following values can be employed in either Equation 17.1 or 17.6 to compute the value $t = 5.84$ (with $p < .01$, since the latter *t* value is greater than the tabled critical one- and two-tailed .01 values (for $df = n - 1 = 5$) $t_{.01} = 3.37$

and $t_{.01} = 4.03$): $\bar{D} = 2.5$, $\tilde{s}_D = 1.049$, $s_{\bar{D}} = .428$ (thus, $t = 2.5/.428 = 5.84$), $\bar{X}_{AB_{21}} = 5$, $\bar{X}_{AB_{22}} = 2.5$, $\tilde{s}_{AB_{21}} = 1.789$, $\tilde{s}_{AB_{22}} = 1.049$, $r_{21 \ vs. \ 22} = .853$. Note that $t^2 = (5.84)^2$ equals the value $F = 34.09$ computed for the same comparison in Table 27.22 (the minimal difference is due to rounding off error).

As previously noted in Section VI, the variability within each of the simple effects is the result of contributions from both the main effect on the factor for which the simple effect is being evaluated (Factor B in this instance), as well as any interaction between the two factors. For this reason, the total of the sum of squares for each of the simple effects for a given factor will be equal to the interaction sum of squares (SS_{AB}) plus the sum of squares for that factor (SS_B). This can be confirmed by the fact that for the calculations under discussion the following is true.

$$[(SS_{B \ at \ A_1} = .75) + (SS_{B \ at \ A_2} = 18.75)] = [(SS_{AB} = 6) + (SS_B = 13.5)] = 19.5$$

When viewed together, the presence of the significant interaction in the omnibus analysis of variance (summarized in Table 27.17) along with the results of the analysis of the simple effects for Factor B, suggests that there is a significant relationship between the order of presentation of the treatments and the efficacy scores of subjects. Such a relationship would not be detected if a researcher merely elected to evaluate the main effect for Factor A through use of the **t test for two independent samples** described earlier. Keeping in mind that the sample size in Example 27.5 is far below that which would be required for an actual clinical trial, the available data indicate the drug is rated more effective than the placebo when the drug is the second treatment administered to a subject (as revealed by the significant simple effect for **Factor B at A_2**). Yet curiously (as revealed by the nonsignificant simple effect for **Factor B at A_1**), when the drug is the first treatment, it is not significantly superior to the placebo. The latter observation is inconsistent with the significant main effect obtained with the omnibus analysis of variance for Factor B — the latter indicating that the drug is superior to the placebo. Although the aforementioned inconsistency in the results may just be an anomaly and the drug is, in fact, superior to the placebo regardless of the order in which it is presented, it is apparent that further research is required in order to clarify whether the latter is, in fact, true.

In the event that a **residual carryover effect** resulting from taking the drug prior to the placebo is detected (thus, incrementing subjects' efficacy scores for the placebo when it is taken after the drug), and because of the latter the researcher is forced to only employ data from the first test period, a **t test for two independent samples** can be employed to compare the means for the drug and placebo when both treatments were presented during the first test period — in other words, the mean of Condition AB_{11} (during which the drug was the first treatment presented) would be contrasted with the mean of Condition AB_{22} (during which the placebo was the first treatment presented). In point of fact, when the latter analysis is conducted the difference between the drug and placebo is not significant. Specifically, if the **t test for two independent samples** is employed to compare the $n_{AB_{11}} = 6$ scores in Condition AB_{11} with the $n_{AB_{22}} = 6$ scores in Condition AB_{22}, the following values are computed $\bar{X}_{AB_{11}} = 3.5$, $\bar{X}_{AB_{22}} = 2.5$, $\tilde{s}_{AB_{11}} = 1.643$, $\tilde{s}_{AB_{22}} = 1.049$, and $t = [|3.5 - 2.5|]/\sqrt{[(1.643)^2/6]} + [(1.049)^2/6]} = 1.26$ (with $p > .05$, since the latter t value is less than the tabled critical one and two-tailed .05 values (for $df = n_1 + n_2 - 2 = 10$) $t_{.05} = 1.81$ and $t_{.05} = 2.23$).

The reader should take note of the fact that the discussion of the evaluation of a **crossover design** has involved a number of comparisons involving pairs of groups or conditions. As noted in Section VI of the **single-factor between-subjects analysis of variance**, unless comparisons are planned beforehand, a researcher should adjust the per comparison alpha level accordingly.

2. Test 27j: Analysis of variance for a Latin square design   In the discussion of **counter-balancing and order effects** in Section VII of both the *t* **test for two dependent samples** and the **single-factor within-subjects analysis of variance**, it is noted that there are the following two types of **order effects**: **position effects** and **carryover effects**. **Position effects** are assumed to only be a function of the **serial position** of a treatment (which is the order in which the treatment is presented), and are not affected by what particular treatment(s) occurred in earlier serial positions. The most common **position effects** are **practice effects** in which the performance of subjects improves with each subsequent treatment, and fatigue and boredom in which each subsequent treatment results in a decrease in the performance of subjects. As noted in the discussion of the crossover design, a **carryover effect** is where the effect of a specific treatment is not independent of a treatment which precedes it. As an example, assume within the framework of a within-subjects design the following three treatments are administered on separate days to each of *n* subjects in order to assess each treatments effect on eye-hand coordination: Drug A, Drug B, placebo. Assume that a side effect of Drug A is it makes a person drowsy, and that the latter adversely influences performance. To compound matters, it requires 36 hours before Drug A completely washes out of a person's body. Because of the latter, if either of the other two treatments is administered to a subject the day after he or she receives Drug A, the subject's performance under that treatment will also be adversely affected due to the carryover effect of residual drowsiness deriving from Drug A.

Counterbalancing is a procedure which allows a researcher to control for **order effects**. In **complete counterbalancing** all possible orders for presenting the experimental conditions are represented an equal number of times with respect to the total number of subjects employed in a study. Thus, if a study with *n* = 10 subjects and *k* = 3 conditions is completely counterbalanced, there will be *k*! = 3! = 6 possible presentation orders for the three treatments (i.e., 1,2,3; 1,3,2; 2,1,3; 2,3,1; 3,1,2; 3,2,1). Under such conditions a minimum of six subjects will be required in order to employ complete counterbalancing. If a researcher wants to assign two subjects to each of the presentation orders, 6 × 2 = 12 subjects must be employed. It should be obvious that to completely counterbalance the order of presentation of the experimental conditions, the number of subjects must equal the value of *k*! or be some value that is evenly divisible by it.

As the number of experimental conditions increases, **complete counterbalancing** becomes more difficult to implement, since the number of subjects required increases substantially. Specifically, if there are *k* = 5 experimental conditions, there are 5! = 120 presentation orders — thus requiring a minimum of 120 subjects (which can be a prohibitively large number for a researcher to use) in order that one subject serves in each of the possible presentation orders. When it is not possible to completely counterbalance the order of presentation of the conditions, alternative less complete counterbalancing procedures are available. One experimental design which allows for **incomplete counterbalancing** (i.e., where not all possible presentation orders are employed in a study) is the **Latin square design**[30], which is employed more frequently in agricultural and industrial research than in research in education and the behavioral sciences. If an independent variable is comprised of only a few levels, it is feasible to use a Latin square design to provide **complete counterbalancing**. However, as the number of levels of the independent variable increase, it is more likely that the Latin square design will be employed to control for order effects through use of **incomplete counterbalancing**.

If we conceptualize a within-subjects design as being comprised of *n* rows (corresponding to each of the *n* subjects) and *p* columns (corresponding to each of *p* presentation orders employed for the *k* treatments, where *n* = *p* = *k*), we can define a Latin square design as one in which each treatment appears only one time in each row and only one time in each column. It is also possible in a Latin square for each of the *n* rows to represent a block of *m* subjects (resulting in a total of (*n*)(*m*) subjects), and all of the subjects within a given block are presented the treatments in the order indicated for its row/block. Figure 27.4 represents a **4 × 4 Latin square** since there are $n = 4$ subjects/rows, $p = 4$ presentation orders/columns, and $k = 4$ treatments (which are identified by the Latin letters **A, B, C,** and **D**).[31] In the square depicted in Figure 27.4, Subject 1 receives the treatments in the order **A, B, C, D,** Subject 2 receives the treatments in the order **C, A, D, B,** and so on. Which of the four presentation orders a specific subject is assigned should be determined randomly.[32] As noted above, a Latin square does not typically employ **complete counterbalancing.** When there are $k = 4$ treatments, there are, in fact, $k! = 4! = 24$ possible presentation orders. Consequently, a minimum of 24 subjects would be required in order to have complete counterbalancing.

The Latin square depicted in Figure 27.4 is referred to as a **standard square,** which by definition is characterized by the fact that both the first row and first column are ordered alphabetically. A standard square can also be employed to construct one or more **nonstandard squares** by interchanging the rows and columns of the standard square. The total number of nonstandard squares which are possible will depend upon how many ways there are of interchanging the rows and columns of the total number of standard squares that can be constructed for a given value of $k = n = p$. Winer *et al.* (1991, p. 677) and Kirk (1995, 321) note that from any standard square it is possible to generate $[k! (k - 1)! - 1]$ additional nonstandard squares.

Presentation order

		1	2	3	4
	1	A	B	C	D
	2	B	D	A	C
Subject	3	C	A	D	B
	4	D	C	B	A

Figure 27.4   **4 × 4 Latin Square Design**

Presentation order

		1	2
Subject/	1	A	B
Block	2	B	A

Figure 27.5   **2 × 2 Latin Square Design**

Presentation order

		1	2	3
	1	A	B	C
Subject	2	B	C	A
	3	C	A	B

Figure 27.6   **3 × 3 Standard Latin Square**

When there are four or more treatments, it is possible to construct more than one standard Latin square. When the number of treatments equals $k = 2$ or $k = 3$, only one standard Latin square can be constructed. In actuality, when there are $k = 2$ treatments, the **2 × 2 crossover design** discussed earlier can be conceptualized as a Latin square design. Specifically, the **2 × 2 crossover design** can be viewed as the **2 × 2 Latin square design** depicted in Figure 27.5. When the number of treatments is $k = 3$, the only standard Latin square which can be constructed is depicted in Figure 27.6.

When $k = 4$, in addition to the standard Latin square depicted in Figure 27.4, it is also possible to construct the three additional standard Latin squares depicted in Figure 27.7.

Square 1

Presentation order

		1	2	3	4
	1	A	B	C	D
	2	B	A	D	C
Subject	3	C	D	A	B
	4	D	C	B	A

Square 2

Presentation order

		1	2	3	4
	1	A	B	C	D
	2	B	A	D	C
Subject	3	C	D	B	A
	4	D	C	A	B

Square 3

Presentation order

		1	2	3	4
	1	A	B	C	D
	2	B	C	D	A
Subject	3	C	D	A	B
	4	D	A	B	C

Figure 27.7   Additional Standard 4 × 4 Latin Squares

Keppel and Wickens (2004, p. 384) note that when k is greater than 4, the number of standard squares which can be constructed increases dramatically (e.g., when $k = 5$, 56 squares; when $k = 6$, 9,408 squares; and from then on increasing at an exponential rate). Among others, the latter authors recommend that when a researcher employs four or more treatments and elects to employ a Latin square design employing just one standard square, the selection of the single square should be made randomly from the total available standard squares for that value of k. In order to minimize the possibility of **carryover effects**, the use of what is commonly referred to as a **cyclic square** should be avoided, and instead one should select a square that is **diagram-balanced** (also referred to as **row-balanced**). In a **diagram-balanced** square (which is represented by the square depicted in Figure 27.4) each of the treatments follows each of the other treatments only once. The latter will not be the case in a **cyclic square** (which is represented by Square 3 in Figure 27.7), whose configuration is such that each new row is created by merely moving the first letter of the previous row to the position of the last letter of the row that will appear beneath it. Note that the cyclic square depicted in Figure 27.7 is susceptible to carryover effects, since Condition A follows Condition D three times, Condition B follows Condition A three times, Condition C follows Condition B three times, and Condition D follows Condition C three times. Consequently, any carryover effect from Condition D is completely confounded with the response to Condition A, any carryover effect from Condition A is completely confounded with the response to Condition B, any carryover effect from Condition B is completely confounded with the response to Condition C, and any carryover effect from Condition C is completely confounded with the response to Condition D. The latter problems resulting from potential carryover effects would be avoided if, instead, one employed the diagram-balanced square depicted in Figure 27.4. A full discussion of guidelines for selecting squares can be found in Kirk (1995), who also lists sources containing all Latin squares of size 7 × 7 and smaller.

If there are four presentation orders and a researcher wishes to have counterbalancing, in order to increase the total number of subjects the researcher could employ multiple subjects in each row of the same square as depicted in Figure 27.8. Note that in the latter square, each row is identified as a block. An equal number of subjects would be assigned to each block. Thus, if a researcher wanted to employ a total of 16 subjects, within the framework of the Latin square depicted in Figure 27.8 (which is the same Latin square depicted in Figure 27.4), the first block would be comprised of four subjects, all of whom would be presented the treatments in the order **ABCD**. The second, third, and fourth blocks would also be comprised of four subjects per block, and the order of presentation of the treatments to subjects in the second, third, and fourth blocks would respectively be **BDAC**, **CADB**, and **DCBA**. The latter use of the Latin square design in which multiple subjects serve in each of the presentation orders can be conceptualized as a **mixed factorial design**, and can be evaluated with a **factorial analysis of variance for a mixed design**.

Presentation order

		1	2	3	4
	Block 1	A	B	C	D
Block	Block 2	B	D	A	C
	Block 3	C	A	D	B
	Block 4	D	C	B	A

Figure 27.8   4 × 4 Latin Square Design

Anderson (2001, p. 418) and Keppel and Wickens (2004, p. 384) note that a limitation of a design such as that depicted in Figure 27.8 is that the Latin square (or for that matter, any of the other three possible standard Latin squares which could be employed when there are $k = 4$ treatments) might have idiosyncratic effects associated with the orders of presentation it contains. To avoid the latter, it would be preferable to employ all four of the possible standard Latin squares which can be constructed for $k = 4$ treatments. Since the different squares have different treatment sequences, by doing the latter, any carryover effects which are present are more likely to be balanced. Thus, if four different Latin squares were employed, the 16 subjects would be divided into four sets (also referred to as **replicates**) comprised of four subjects per set/replicate. Each set of four subjects would then be randomly assigned to one of the four standard Latin squares summarized in Figures 27.4 and 27.7.

The purpose of the Latin square design is to equally distribute any order effects which are present over the k experimental treatments. A Latin square design, however, will only provide effective control for order effects if there is **no interaction** between the experimental treatments and their order of presentation. In point of fact, the absence of an interaction between order of presentation and treatments is critical in any within-subjects design, since if an interaction is present it will not be possible to obtain a pure measure for any treatment effects which may be present.

As noted in Section V, an **additive model** for an analysis of variance is one which assumes there is no interaction between the independent variables, whereas a **nonadditive model** assumes the presence of an interaction. The use of a Latin square design thus assumes an **additive model**. The consequence of employing a Latin square design with data that is not additive (i.e., an interaction is present, which is sometimes referred to as **nonadditivity**), is that the power of the analysis of variance employed for the Latin square will compromised — i.e., the F ratio computed for the treatment effect will be spuriously reduced. The advantage of employing a Latin square design when an additive model is, in fact, applicable is that the analysis of variance for such a design provides for a more efficient analysis of whether or not a treatment effect is present when contrasted with the use a **single-factor within subjects analysis of variance** on the same set of data. Myers and Well (1991, 1995, p. 345) note that a Latin square design is often referred to as an **incomplete block design** since each subject or block of subjects is not tested under all possible combinations of the treatment variable and the column variable (which in the case of Figures 27.4–27.8 is presentation order), and it is because of the latter that the design does not allow for the evaluation of interactions.

Analysis of a single Latin square employing one subject per row   Many sources which discuss the Latin square design describe the analysis for a single Latin square employing a number of subjects equal to the number of rows/columns/treatments which comprise the Latin square. Although some sources (e.g., Miller and Freud (1965, p. 284)) note that, in reality, the latter design is rarely employed, since its residual degrees of freedom will be a minimal value — specifically, $df_{res} = (k - 1)(k - 2)$. The consequence of the latter is that the residual mean square will probably not be sufficiently low so as to allow a researcher to take advantage of the Latin square design's ability to substantially reduce error variance, thus providing for a more powerful test regarding the presence of a treatment effect. Because of the latter, Myers and Well (2003, p .461) recommend that the minimally sized square which should be used for one set of n subjects is a 5×5 square.

For purposes of illustration, Example 27.6 will be employed to demonstrate the use of a Latin square design comprised of one set of $n = 4$ subjects. The Latin square used for Example 27.6 is the 4×4 square depicted in Figure 27.4 which, as previously noted, is a **standard square** as well as being **diagram-balanced**. Note that when $k = 4$, the number of possible presentation orders is $4! = 24$. The Latin square employed in the study, however, will only use the following four presentation orders: **ABCD, BDAC, CADB, DCBA** (which correspond to the four rows of the square). Note that the treatment identification numbers **1, 2, 3**, and **4** employed in Example 27.6 respectively correspond to the letters **A, B, C**, and **D** in the Latin square. Inspection of the 4×4 square depicted in Figure 27.4 indicates that the four subjects are administered the four treatments in the following presentation orders: **Subject 1**: 1234, **Subject 2**: 2413; **Subject 3**: 3142; **Subject 4**: 4321.

Example 27.6 *A psychologist conducts a study to determine whether or not noise can inhibit learning. Each of four subjects is tested under four experimental conditions. In each of the experimental conditions a subject is given 10 minutes to memorize a list of 20 nonsense syllables, which the subject is told she will be tested on the following day. The four experimental conditions each subject serves under are as follows:* **Condition 1,** *the* **extreme noise** *condition, requires subjects to study the list of nonsense syllables while music is played in the background at an extremely high amplitude.* **Condition 2,** *the* **moderate noise** *condition, requires subjects to study the list of nonsense syllables while music is played in the background at a moderate amplitude.* **Condition 3,** *the* **low noise** *condition, requires subjects to study the list of nonsense syllables while music is played in the background at a low amplitude.* **Condition 4,** *the* **no noise** *condition, requires subjects to study the list of nonsense syllables in a quiet room. Although in each of the experimental conditions subjects are presented with a different list of nonsense syllables, the four lists are comparable with respect to those variables that are known to influence a person's ability to learn nonsense syllables. To control for order effects, the experimenter employs a 4 × 4 Latin square design employing the square depicted in Figure 27.4. The number of nonsense syllables correctly recalled by the four subjects under the four experimental conditions follows. (Subjects' scores are listed in the order* **Condition 1, Condition 2, Condition 3, Condition 4**) **Subject 1:** *5, 5, 7, 7;* **Subject 2:** *6, 13, 8, 12;* **Subject 3:** *7, 3, 9, 7;* **Subject 4:** *1, 5, 10, 18. The data for the study are summarized in Table 27.23. Do the data indicate that noise influenced subjects' performance?*

Table 27.23 Data for Example 27.6

Subject	Subject presentation order	Table A				Table B				
		Ordinal position of treatment				Conditions/ Treatments				
		p_1	p_2	p_3	p_4	1	2	3	4	Sum
1	1234	5	5	7	7	5	5	7	7	24
2	2413	13	12	6	8	6	13	8	12	39
3	3142	9	7	7	3	7	3	9	7	26
4	4321	18	10	5	1	1	5	10	18	34
Sum		45	34	25	19	19	26	34	44	123
Mean		11.25	8.5	6.25	4.75	4.75	6.5	8.5	11	

A separate **single-factor within-subjects analysis of variance** can be employed to evaluate each of the two tables which comprise Table 27.23. Note that in **Table B** the experimental conditions/treatments represent the column/independent variable, while in **Table A** the order of presentation of the treatments represents the column/independent variable.

Initially, a **single-factor within-subjects analysis of variance** will be employed to evaluate the data in **Table B**. Note that the latter analysis only examines whether or not a treatment effect is present, without taking into consideration the order of presentation of the treatments. The appropriate sums of squares for the analysis of variance on **Table B** are computed below.

With $\Sigma X_T^2 = 1199$, employing Equation 24.2, the value $SS_T = 253.44$ is computed.

$$SS_T = 1199 - \frac{(123)^2}{16} = 1199 - 945.56 = 253.44$$

Equation 24.3 is employed to compute $SS_{BC} = 86.69$.

$$SS_{BC} = \left[\frac{(19)^2}{4} + \frac{(26)^2}{4} + \frac{(34)^2}{4} + \frac{(44)^2}{4}\right] - \frac{(123)^2}{16} = 1032.25 - 945.56 = 86.69$$

Equation 24.4 is employed to compute $SS_{BS} = 36.69$.

$$SS_{BS} = \left[\frac{(24)^2}{4} + \frac{(39)^2}{4} + \frac{(26)^2}{4} + \frac{(34)^2}{4}\right] - \frac{(123)^2}{16} = 982.25 - 945.56 = 36.69$$

$SS_{res} = 130.06$ is computed with Equation 24.5.

$$SS_{res} = 253.44 - 86.69 - 36.69 = 130.06$$

Table 27.24 summarizes the **single-factor within-subjects analysis of variance** employed with the data in **Table B**, which yields the result $F = 2.00$. In **Table A10** for $df_{num} = 3$ and $df_{den} = 9$ (which are the appropriate degrees of freedom values to employ for Table 27.24), the tabled $F_{.95}$ and $F_{.99}$ values are $F_{.95} = 3.86$ and $F_{.99} = 6.99$. Since the value $F = 2.00$ is less than $F_{.05} = 3.86$, the null hypothesis for a treatment effect cannot be rejected. Thus, the data do not indicate there is a significant difference between the means of any of the four experimental conditions in Example 27.6.

Table 27.24 Summary Table of Analysis of Variance of Table B in Table 27.23 for Treatment Effect

Source of variation	SS	df	MS	F
Between-subjects	36.69	3	12.23	
Between-conditions	86.69	3	28.90	2.00
Residual	130.06	9	14.45	
Total	253.44	15		

A second **single-factor within-subjects analysis of variance** can be employed with the data in **Table A** in Table 27.23 in order to evaluate whether or not subjects' scores are affected by the order of presentation of the treatments. The latter analysis, however, will not address the question of whether or not a treatment effect is present in the data. The second analysis of variance to be conducted will evaluate the null hypothesis that in the underlying populations represented by sample, the means of the four presentation orders are equal to one another. The alternative hypothesis will be that the means of at least two of four presentation orders are different from one another.

The **total variability** for the **single-factor within-subjects analysis of variance** which will be conducted on the data in **Table A** to evaluate the effect of order of presentation is comprised of the following three sources: **between-subjects variability**, **residual variability**, and **between-presentation orders variability**. The following values (involving total and between-subjects variability) obtained previously in the analysis of **Table B** (which are recorded in Table 27.24) remain unchanged, and are also used in the analysis of **Table A**: $SS_T = 253.44$, $df_T = 15$, $SS_{BS} = 36.69$, $df_{BS} = 3$, and $MS_{BS} = 12.23$. The following values which were not employed in the analysis of **Table B** must be either computed or revised: a) For presentation order (which in the analysis of **Table A** represents the independent variable) we must compute a value for the sum of squares, degrees of freedom, and mean square — specifically, SS_{BP}, df_{BP} MS_{BP}; b) New values must be computed for SS_{res}, df_{res}, and MS_{res}.

Equation 27.83 is employed to compute the sum of squares associated with between-presentation orders.

$$SS_{BP} = \sum_{j=1}^{P} \left[\frac{(\Sigma P_j)^2}{n} \right] - \frac{(\Sigma X_T)^2}{N} \qquad \textbf{(Equation 27.83)}$$

The notation $\sum_{j=1}^{P} \left[(\Sigma P_j)^2 / n \right]$ in Equation 27.83 indicates that for each of the presentation orders the value $(\Sigma P_j)^2/n$ is computed, and the latter values are summed for all p presentation orders. With reference to Example 27.6, Equation 27.83 can be rewritten as follows.

$$SS_{BP} = \left[\frac{(\Sigma P_1)^2}{n} + \frac{(\Sigma P_2)^2}{n} + \frac{(\Sigma P_3)^2}{n} + \frac{(\Sigma P_4)^2}{n} \right] - \frac{(\Sigma X_T)^2}{N}$$

Substituting the appropriate values from **Table A** in Table 27.23 in Equation 27.83, the value $SS_{BP} = 96.19$ is computed.

$$SS_{BP} = \left[\frac{(45)^2}{4} + \frac{(34)^2}{4} + \frac{(25)^2}{4} + \frac{(19)^2}{4} \right] - \frac{(123)^2}{16} = 1041.75 - 945.56 = 96.19$$

The between-presentation orders degrees of freedom and mean square are respectively computed below with Equations 27.84 and 27.85.

$$df_{BP} = p - 1 = 4 - 1 = 3 \qquad \textbf{(Equation 27.84)}$$

$$MS_{BP} = \frac{SS_{BP}}{df_{BP}} = \frac{96.19}{3} = 32.06 \qquad \textbf{(Equation 27.85)}$$

SS_{res} for the analysis can be computed with either Equation 27.86 or Equation 27.87. Employing the latter equations, the value $SS_{res} = 49.34$ is computed.

$$SS_{res} = SS_T - SS_{BP} - SS_{BS} \qquad \textbf{(Equation 27.86)}$$

$$SS_{res} = 253.44 - 96.19 - 36.69 = 120.56$$

(Equation 27.87)

$$SS_{res} = \Sigma X_T^2 - \sum_{j=1}^{p}\left[\frac{(\Sigma P_j)^2}{n}\right] - \sum_{i=1}^{n}\left[\frac{(\Sigma S_i)^2}{k}\right] + \frac{(\Sigma X_T)^2}{N}$$

$$SS_{res} = 1199 - 1041.75 - 982.25 + 945.56 = 120.56$$

The residual degrees of freedom and mean square are respectively computed below with Equations 27.88 and 27.89.

$$df_{res} = (n - 1)(p - 1)$$

(Equation 27.88)

$$df_{res} = (4 - 1)(4 - 1) = 9$$

$$MS_{res} = \frac{SS_{res}}{df_{res}} = \frac{120.56}{9} = 13.40 \qquad \textbf{(Equation 27.89)}$$

The F ratio is computed with Equation 27.90.

$$F = \frac{MS_{BP}}{MS_{res}} = \frac{32.06}{13.40} = 2.39 \qquad \textbf{(Equation 27.90)}$$

Table 27.25 summarizes the **single-factor within-subjects analysis of variance** employed with the data in **Table A**. In **Table A10** for $df_{num} = 3$ and $df_{den} = 9$ (which are the appropriate degrees of freedom values to employ for Table 27.25), the tabled $F_{.95}$ and $F_{.99}$ values are $F_{.95} = 3.86$ and $F_{.99} = 6.99$ (which are the same critical values employed in the analysis of variance for the treatment effects summarized in Table 27.24). Since the value $F = 2.39$ is less than $F_{.05} = 3.86$, the null hypothesis for order effects cannot be rejected. Thus, the data do not indicate there is a significant difference between the means of any the four presentation orders employed in Example 27.6.

Table 27.25 Summary Table of Analysis of Variance of Table A in Table 27.23 for Order Effects

Source of variation	SS	df	MS	F
Between-subjects	36.69	3	12.23	
Between-presentation orders	96.19	3	32.06	2.39
Residual	120.56	9	13.40	
Total	**253.44**	**15**		

Inspection of the mean values recorded in **Table B** in Table 27.23 appears to indicate that there are substantial differences between the four treatment/condition means ($\bar{X}_1 = 4.75$, $\bar{X}_2 = 6.5$, $\bar{X}_3 = 8.5$, $\bar{X}_4 = 11$), suggesting that subjects' performance was affected by the experimental conditions. In addition, inspection of the mean values recorded in **Table A** appears to indicate that there are substantial differences between the three presentation order means ($\bar{X}_{P_1} = 11.25$, $\bar{X}_{P_2} = 8.5$, $\bar{X}_{P_3} = 6.25$, $\bar{X}_{P_4} = 4.75$), suggesting that the higher the ordinal position of the presentation order the poorer the performance of a subject.

In point of fact, the analyses summarized in Tables 27.24 and 27.25 do not represent the optimal analysis (insofar as maximizing power) for assessing treatment effects and/or presentation order. With regard to evaluating the presence of a treatment effect, Keppel (1991, p. 361) notes that although counterbalancing results in practice effects being equally distributed over the different treatments, it also deposits variability due to practice effects in residual variability (thus inflating the latter value).

In order to maximize power in the assessment of both presentation order and treatments, variability attributable to both between-conditions (i.e., variability attributed to treatments) and between-presentation orders should be integrated into a single analysis of variance which is summarized in Table 27.26. The latter can be done since, as Keppel and Wickens (2004, p. 388) note, in a Latin square design the three main effects (i.e., subjects (which can be conceptualized as a separate factor), treatments, and presentation order) are mutually orthogonal (i.e., independent of one another). Note that the values computed previously (in Tables 27.24 and 27.25) for $SS_T = 253.44$, $df_T = 15$, $SS_{BS} = 36.69$, $df_{BS} = 3$, $SS_{BC} = 86.69$, $df_{BC} = 3$, $SS_{BP} = 96.19$, and $df_{BP} = 3$ are also employed in Table 27.26. However, the values for SS_{res} and df_{res} in the latter table are not the same as the values computed for SS_{res} and df_{res} in either Table 27.24 or Table 27.25. In the case of Table 27.26, Equations 27.91 and 27.92 are respectively employed to compute the values $SS_{res} = 33.87$ and $df_{res} = 6$.

$$SS_{res} = SS_T - SS_{BP} - SS_{BS} - SS_{BC}$$

(Equation 27.91)

$$SS_{res} = 253.44 - 96.19 - 36.69 - 86.69 = 33.87$$

$$df_{res} = df_T - df_{BP} - df_{BS} - df_{BC}$$

(Equation 27.92)

$$df_{res} = 15 - 3 - 3 - 3 = 6$$

With $MS_{res} = SS_{res}/df_{res} = 33.87/6 = 5.65$, Equations 27.93 and 27.94 are employed to compute the F ratios for the treatment and presentation order sources of variation. Note that in computing the F ratios, the value $MS_{res} = 5.65$ is much lower than the value for MS_{res} employed in either Table 24.24 ($MS_{res} = 14.45$) or Table 27.25 ($MS_{res} = 13.40$). Consequently, the value $F_{BC} = 5.11$ is larger than the value $F_{BC} = 2.00$ computed in Table 24.24, and the value $F_{BP} = 5.67$ is larger than the value $F_{BP} = 2.39$ computed in Table 27.25.

$$F_{BC} = \frac{MS_{BC}}{MS_{res}} = \frac{28.90}{5.65} = 5.11$$

(Equation 27.93)

$$F_{BP} = \frac{MS_{BP}}{MS_{res}} = \frac{32.06}{5.65} = 5.67$$

(Equation 27.94)

Table 27.26 Summary Table of Analysis of Variance for Latin Square Analysis of Example 27.6

Source of variation	SS	df	MS	F
Between-subjects	36.69	3	12.23	
Between-presentation orders	96.19	3	32.06	5.67
Between-conditions	86.69	3	28.90	5.11
Residual	33.87	6	5.65	
Total	253.44	15		

In the analysis of variance summarized in Table 27.26 involving one set of $n = 4$ subjects, where n is equal to the number of rows which equals the number of columns as well as the number of treatments in the Latin square, the degrees of freedom for the analysis can also be computed with the Equations 27.95–27.99 noted below.

$$df_{BS} = df_{rows} = n - 1 = 4 - 1 = 4 \qquad \textbf{(Equation 27.95)}$$

$$df_{BP} = df_{columns} = n - 1 = 4 - 1 = 3 \qquad \textbf{(Equation 27.96)}$$

$$df_{BC} = n - 1 = 4 - 1 = 3 \qquad \textbf{(Equation 27.97)}$$

$$df_{res} = (n - 1)(n - 2) = (4 - 1)(4 - 2) = 6 \qquad \textbf{(Equation 27.98)}$$

$$df_T = n^2 - 1 = 4^2 - 1 = 15 \qquad \textbf{(Equation 27.99)}$$

The degrees of freedom employed in **Table A10** in evaluating the values $F_{BP} = 5.67$ and $F_{BC} = 5.11$ computed in Table 27.26 are $df_{num} = 3$ and $df_{den} = 6$. The tabled $F_{.95}$ and $F_{.99}$ values for the aforementioned degrees of freedom are $F_{.95} = 4.76$ and $F_{.99} = 9.78$. Since $F_{BC} = 5.11$ is greater than $F_{.95} = 4.76$, the null hypothesis for a treatment effect can be rejected at the .05 level (which was not the case in Table 27.24). The latter indicates there is a difference between the means of at least two of the four experimental conditions. Since $F_{BP} = 5.67$ is greater than $F_{.95} = 4.76$, the null hypothesis for presentation orders can also be rejected at the .05 level (which was not the case in Table 27.25). The latter indicates there is a difference between at least two of the four presentation orders means. More detailed analysis of the results can be obtained through conducting comparisons between the means of pairs of treatments and the means of pairs of presentation orders. The error term computed in Table 27.26 would be employed in conducting such comparisons.

Although the analysis summarized in Table 27.26 will usually provide a more powerful test of treatment and order effects than the analyses summarized in Tables 27.24 and 27.25, the reader should take note of the fact that, by virtue of employing smaller df values, the critical F values employed for Table 27.26 are larger than those employed for Tables 27.24 and 27.25. Although under certain conditions the latter could make it more difficult to reject a null hypothesis, it may be counteracted by the fact that the residual mean square computed for the analysis summarized in Table 27.26 will be substantially lower than the residual mean square values computed for the analyses summarized in Tables 27.24 and 27.25. With regard to both treatment and order effects, a researcher could conduct the three analyses summarized in Tables 27.24, 27.25, and 27.26. In the unlikely event the analysis summarized in Tables 27.24 and/or 27.25 provide a more powerful test than the analysis summarized in Table 27.26 (i.e., the probability values associated with a result in Table 27.24 and/or Table 27.25 is less than the probability value associated with the result in Table 27.26), the researcher can employ the F_{BC} and/or F_{BP} value associated with the lower probability.

Latin square designs can be employed to control for variables other than order effects, which up to this point in the discussion is the only extraneous variable that has been considered. Extraneous variables which a researcher might want to control for are often referred to as **nuisance variables**. In the final analysis, a Latin square design simultaneously controls for any effect two nuisance variables might have on the independent variable of interest by integrating the nuisance variables into the design of an experiment. As is the case with any Latin square, the two nuisance variables in Example 27.6 are represented by the rows and columns which comprise the Latin square — specifically, the between-subjects variable (which is the row variable in Figure 27.4) and presentation order (which is the column variable in Figure 27.4). By virtue of integrating the nuisance variables into its structure, a Latin square design can be used as an alternative to employing a factorial design involving three independent variables (which would be comprised of the independent variable of interest plus the two nuisance variables). The Latin square design is appropriate to employ when prior to conducting a study a researcher has reason to believe there are no interactions (or, at most, only a minimal interaction(s)) between the nuisance variables and the independent variable of interest.

As an example, suppose a researcher wanted to evaluate the effect of four dietary formulas (to be designated **A**, **B**, **C**, and **D**) on the performance of race horses, and had four race horses and four jockeys for use in a study. One available option would be to employ a three factor design — specifically, a $4 \times 4 \times 4$ factorial design. with the dietary formulas, horses, and jockeys representing the three factors. The latter design would require the researcher to obtain a total of $4 \times 4 \times 4 = 64$ scores involving all possible combinations of the three variables. Let us assume, however, that there is reason to believe there are no interactions between the dietary formulas and the jockeys, the dietary formulas and the race horses, or the jockeys and the race horses. In the final analysis, the researcher is only interested in estimating the effect of dietary formula on performance, or to state it another way, he is only interested in the **main effect** for dietary formula. Since a main effect refers to the effect of one independent variable on the dependent variable, while ignoring the effect any of the other independent variables may have on the dependent variable, the two nuisance variables represent the other independent variables. Although you are primarily interested in the impact of the dietary formulas, at the same time, you want to make sure the main effects for the two nuisance variables (the jockeys and race horses) do not bias your estimate of the main effect for the dietary formulas. The Latin square design depicted in Figure 27.9 would allow the researcher to control for any bias which might result from the effects of the jockey and the race horse variables.

Note that the square in Figure 27.9 is identical to the square depicted in Figure 27.4, except for the fact the four rows (i.e., $r = 4$, which is analogous to $n = 4$ in Figure 27.4) represent the four jockeys and the four columns (i.e., $c = 4$, which is analogous to $p = 4$ in Figure 27.4) represent the four horses. The row and column variables in a Latin square design are commonly referred to as **blocking variables**, since the data are categorized in reference to blocks which represent the levels of the latter nuisance variables. Notice that in our example, each jockey tests each dietary formula and that each dietary formula is tested with each horse. However, not all possible combinations of dietary formula/horse/jockey are present. The Latin Square depicted in Figure 27.9 would be evaluated in the same manner as Example 27.6.

		Horses			
		1	2	3	4
Jockeys	1	A	B	C	D
	2	B	D	A	C
	3	C	A	D	B
	4	D	C	B	A

Figure 27.9 4 × 4 Latin Square Design for Dietary Formula Example

Examples 27.7 and 27.8 (which respectively represent applications of a Latin square design to the fields of agriculture and industry) represent two studies which conform to the 4×4 Latin square design depicted in Figure 27.4 in which there are four rows, four columns, and four treatments. Figure 27.10 (which also contains the score for each cell) and Table 27.27 summarize the data for Examples 27.7 and 27.8.

		Column/ Time of day			
		1/8AM	2/11AM	3/2PM	4/4 PM
Row/Month	1/Jan	A (5)	B (5)	C (7)	D (7)
	2/May	B (13)	D (12)	A (6)	C (8)
	3/Aug	C (9)	A (7)	D (7)	B (3)
	4/Dec	D (18)	C (10)	B (5)	A (1)

Figure 27.10   4 × 4 Latin Square employed for Examples 27.7 and 27.8

Table 27.27   Data for Examples 27.7 and 27.8

Row/ Month	Treatment sequence	Table A Column/ Time of day				Table B Treatments				Sum
		1/ 8AM	2/ 11AM	3/ 2PM	4/ 4PM	A	B	C	D	
1/Jan	ABCD	5	5	7	7	5	5	7	7	24
2/May	BDAC	13	12	6	8	6	13	8	12	39
3/Aug	CADB	9	7	7	3	7	3	9	7	26
4/Dec	DCBA	18	10	5	1	1	5	10	18	34
Sum		45	34	25	19	19	26	34	44	123
Mean		11.25	8.5	6.25	4.75	4.75	6.5	8.5	11	

Example 27.7   *A study was conducted in order to examine the efficacy of four different formulas of plant nutrient on corn production. The four formulas of plant nutrient are designated as* **A**, **B**, **C**, *and* **D**. *To control for the possibility that the growth of corn could be differentially affected by the quality of the soil in different parts of the field, a four acre field was divided into four equally sized rows and columns, resulting in a grid comprised of 16 equally sized squares. In each of the squares one of the four formulas was applied to the soil in which corn has been planted. Within the 16 square grid, each formula was applied only once in each row and only once in each column. Figure 27.10 and Table 27.27 respectively summarize the 4 × 4 Latin square design and the data obtained for the study. The scores in each cell of Figure 27.10 and Table 27.27 represent a corn productivity index, with the higher the index the higher the level of productivity.*

Example 27.8   *A study was conducted in order to examine the impact of four industrial odors on the performance of factory workers. The four odors will be designated as* **A**, **B**, **C**, *and* **D**. *Based on research which suggests the impact of odor is a function of both the time of the year as well as the time of day it is inhaled, a 4 × 4 Latin square design was employed, with the row* factor corresponding to four different months during the year (specifically, the months of January, May, August and December), and the column factor corresponding to four different times during the work day (specifically, 8–9 AM, 11–12 noon, 2–3 PM and 4–5 PM). One factory was employed in the study, and the workers in the factory were exposed each of the four odors four times over the course of one year. Within the 16 cell Latin square, each odor was presented to the workers in the factory only once in a given month and only once at a given time period. Figure 27.10 and Table 27.27 respectively summarize the 4 × 4 Latin square design employed and data obtained for the study. The scores in each cell of Figure 27.10 and Table 27.27 represent a performance index for the workers in the factory, with the higher the score the higher the level of performance.

Since the data for Examples 27.7 and 27.8 are identical to that employed in Example 27.6, the results for Examples 27.7 and 27.8 can be summarized by the analysis of variance in Table 27.26. Note that in the Examples 27.27 and 27.28, Treatments **A**, **B**, **C**, and **D** are analogous to Conditions **1, 2, 3** and **4** employed in Example 27.6. The reader should also take note of the fact that in Examples 27.7 and 27.8 the researcher is interested in the between-rows/months F ratio (which was not the case for Example 27.6/Table 27.26, where the analogous F ratio was the between-subjects F ratio ($F_{BS} = MS_{BS}/MS_{res} = 12.23/5.65 = 2.16$)). The results for Examples 27.7 and 27.8 are summarized in Table 27.28.

Table 27.28 Summary Table of Analysis of Variance for Latin Square Analysis of Examples 27.7 and 27.8

Source of variation	SS	df	MS	F
Between-rows/months	36.69	3	12.23	2.16
Between-columns/times of day	96.19	3	32.06	5.67
Between-formulas/odors	86.69	3	28.90	5.11
Residual	33.87	6	5.65	
Total	253.44	15		

With respect to Example 27.7, the significant value $F_{BC} = 5.11$ for the treatment variable indicates there is a significant difference in productivity between at least two of the four plant nutrient formulas. The significant value $F_{BP} = F_{columns} = 5.67$ for the column variable (which corresponds to the presentation order variable in Example 27.6) indicates there is a significant difference in productivity between at least two of the four columns in which corn was planted. The nonsignificant value $F_{BS} = F_{rows} = 2.16$ for the row variable (which corresponds to the between-subjects variable in Example 27.6) indicates there is no significant difference in productivity between any of the four rows in which corn was planted.

With respect to Example 27.8, the significant value $F_{BC} = 5.11$ for the treatment variable indicates there is a significant difference in performance for at least two of the four odors. The significant value $F_{BP} = F_{columns} = 5.67$ for the column variable (which represents to the time of day the odor is presented, and corresponds to the presentation order variable in Example 27.6) indicates there is a significant difference in performance during at least two of the four times of day. The nonsignificant value $F_{BS} = F_{rows} = 2.16$ for the row variable (which represents the month of the year the odor was presented, and corresponds to the between-subjects variable in Example 27.6) indicates there is no significant difference in performance between any of the four months.

Analysis of multiple Latin squares Earlier in the discussion of the Latin square design it was noted that it is common practice to employ multiple subjects in each row of the same Latin square or employ different Latin squares with k subjects assigned to each square. In order to demonstrate the latter analysis, a methodology described by Cohen (2001, pp 500–502) and Keppel and Wickens (2004, pp. 386–393) will be employed. Example 24.1 (which was used to illustrate the **single-factor within-subjects analysis of variance**) will be reevaluated under the assumption that the order of presentation of the $k = 3$ experimental conditions/treatments is determined through the use of the two Latin squares depicted in Figure 27.11. When $k = 3$, the number of possible presentation orders is 3! = 6. Specifically, the six presentation orders are **ABC, BCA, CAB** (which comprise the three rows of **Square 1** in Figure 27.11), and **ACB, BAC, CBA** (which comprise the three rows of **Square 2** in Figure 27.11). Note that the treatment identification numbers **1, 2,** and **3** employed in reference to Example 24.1 respectively correspond to the letters **A, B,** and **C** in the two Latin squares. Inspection of the two squares depicted in Figure 27.11 indicates that the six subjects are administered the three treatments in the following presentation orders: **Subject 1**: 123; **Subject 2**: 231; **Subject 3**: 312; **Subject 4**: 132; **Subject 5**: 213; **Subject 6**: 321. Note that **Square 1** conforms to the definition of a **standard square**, whereas **Square 2** does not. **Square 1** is **cyclic** but not **diagram-balanced**, whereas **Square 2** does not conform to the definition of either. The two squares taken together, however, provide balance in the order of presentation since: a) Within the sample of six subjects employed in the study, each treatment appears exactly two times in each of the three presentation positions; b) Within the presentation orders for the six subjects: Condition A/1 is followed by Condition B/2 two times; Condition A/1 is followed by Condition C/3 two times; Condition B/2 is followed by Condition A/1 two times; Condition B/2 is followed by Condition C/3 two times; Condition C/3 is followed by Condition A/1 two times; Condition C/3 is followed by Condition B/2 two times. By virtue of the latter any position effects will be spread evenly among the treatments.

Square 1

Presentation order

		1	2	3
	1	A=1	B=2	C=3
Subject	2	B=2	C=3	A=1
	3	C=3	A=1	B=2

Square 2

Presentation order

		1	2	3
	4	A=1	C=3	B=2
Subject	5	B=2	A=1	C=3
	6	C=3	B=2	A=1

Figure 27.11 3 × 3 Latin Squares employed for Example 24.1

Table 27.29 summarizes the data for Example 24.1 when the two Latin squares depicted in Table 27.10 are employed. Note that, like Table 27.23, Table 27.29 is also comprised of two tables, **Tables A** and **B**, which respectively list the data in reference to each subject's score in the three orders of presentation and for each of the three treatments.

Table 27.29   Data for Example 24.1 Configured as Latin Square Design

Subject	Subject presentation order	Table A Ordinal position of treatment			Table B Treatments			Sum
		P_1	P_2	P_3	1	2	3	
1	123	9	7	4	9	7	4	20
2	231	8	7	10	10	8	7	25
3	312	3	7	5	7	5	3	15
4	132	10	7	8	10	8	7	25
5	213	5	7	2	7	5	2	14
6	321	6	6	8	8	6	6	20
Sum		41	41	37	51	39	29	119
Mean		6.83	6.83	6.17	8.50	6.50	4.83	

A separate **single-factor within-subjects analysis of variance** is employed to evaluate each of the two tables which comprise Table 27.29. **Table B** has already been evaluated with the **single-factor within-subjects analysis of variance** which was employed to evaluate Example 24.1 in Chapter 24. The latter analysis which is summarized again in Table 27.30 (which is identical to Table 24.2) yields the value $F = 41.29$. For $df_{num} = 2$ and $df_{den} = 10$ (which are the appropriate degrees of freedom to employ for the analysis), the tabled $F_{.95}$ and $F_{.99}$ values are $F_{.95} = 4.10$ and $F_{.99} = 7.56$. Since $F = 41.29$ is greater than both of the aforementioned critical values, the null hypothesis for a treatment effect can be rejected at both the .05 and .01 levels. Thus, the data indicate there is a significant difference between at least two of the three treatment means in Example 24.1.

Table 27.30   Summary Table of Analysis of Variance of Table B in Table 27.29 for Treatment Effect

Source of variation	SS	df	MS	F
Between-subjects	36.93	5	7.39	
Between-conditions	40.45	2	20.23	41.29
Residual	4.90	10	.49	
Total	82.28	17		

A second **single-factor within-subjects analysis of variance** will be employed with the data in **Table A** in Table 27.29 in order to evaluate whether or not subjects' scores are affected by the order of presentation of the treatments. As was the case with the analysis for Example 27.6, the latter analysis will not address the question of whether or not a treatment effect is present in the data. Visual inspection of the three presentation order means ($\bar{X}_{P_1} = 6.83$, $\bar{X}_{P_2} = 6.83$, $\bar{X}_{P_3} = 6.17$) would seem to suggest there are no differences between the three mean values. The analysis of variance, however, provides us with a more formal analysis.

The protocol for the **single-factor within-subjects analysis of variance** to be employed with the data in **Table A** of Table 27.29 to evaluate the effect of order of presentation is identical to that employed for Example 27.6. Thus, the following values (involving total and between-subjects variability) obtained previously in the analysis of **Table B** (which are recorded in Table 27.30/24.2) remain unchanged, and are also used in the analysis of **Table A**: $SS_T = 82.28$, $df_T = 17$, $SS_{BS} = 36.93$, $df_{BS} = 5$, and $MS_{BS} = 7.39$. The following values which were not employed in the analysis of **Table B** must be either computed or revised: a) For presentation order (which in the analysis of **Table A** represents the independent variable) we must compute values for the sum of squares, degrees of freedom, and mean square — specifically, SS_{BP}, df_{BP}, and MS_{BP}; b) New values must be computed for SS_{res}, df_{res}, and MS_{res}.

Employing Equation 27.83 the value $SS_{BP} = 1.78$ is computed.

$$SS_{BP} = \left[\frac{(41)^2}{6} + \frac{(41)^2}{6} + \frac{(37)^2}{6}\right] - \frac{(119)^2}{18} = 788.5 - 786.72 = 1.78$$

The between-presentation orders degrees of freedom and mean square are respectively computed below with Equations 27.84 and 27.85.

$$df_{BP} = p - 1 = 3 - 1 = 2$$

$$MS_{BP} = \frac{SS_{BP}}{df_{BP}} = \frac{1.78}{2} = .89$$

The value $SS_{res} = 43.57$ is computed below with both Equations 27.86 and Equation 27.87.

$$SS_{res} = SS_T - SS_{BP} - SS_{BS}$$

$$SS_{res} = 82.28 - 1.78 - 36.93 = 43.57$$

$$SS_{res} = \Sigma X_T^2 - \sum_{j=1}^{p}\left[\frac{(\Sigma P_j)^2}{n}\right] - \sum_{i=1}^{n}\left[\frac{(\Sigma S_i)^2}{k}\right] + \frac{(\Sigma X_T)^2}{N}$$

$$SS_{res} = 869 - 788.5 - 823.65 + 786.72 = 43.57$$

The residual degrees of freedom and mean square are respectively computed below with Equations 27.88 and 27.89.

$$df_{res} = (n - 1)(p - 1)$$

$$df_{res} = (6 - 1)(3 - 1) = 10$$

$$MS_{res} = \frac{SS_{res}}{df_{res}} = \frac{43.57}{10} = 4.357$$

The value $F = .20$ is computed with Equation 27.90.

$$F = \frac{MS_{BP}}{MS_{res}} = \frac{.89}{4.36} = .20$$

Table 27.31 is a summary table of the analysis of variance of order effects.

Table 27.31 Summary Table of Analysis of Variance of Table A in Table 27.29 for Order Effects

Source of variation	SS	df	MS	F
Between-subjects	36.93	5	7.39	
Between-presentation orders	1.78	2	.89	.20
Residual	43.57	10	4.36	
Total	82.28	17		

In **Table A10** for $df_{num} = 2$ and $df_{den} = 10$ (which are the appropriate degrees of freedom to employ for the analysis), the tabled $F_{.95}$ and $F_{.99}$ values are $F_{.95} = 4.10$ and $F_{.99} = 7.56$ (which are the same critical values employed in the analysis of variance for the treatment effects summarized in Table 27.30/24.2). Since $F = .20$ is less than $F_{.05} = 4.10$, the null hypothesis cannot be rejected. Thus, the data do not indicate any significant differences between the means for the three presentation orders.

As noted earlier, the analysis summarized in Table 27.31 does not represent the optimal analysis for assessing whether or not there is a significant effect due to presentation order, as well as the fact that analysis summarized in Table 27.30/24.2 may not represent the optimal analysis for assessing the presence of a treatment effect. In order to maximize power in the assessment of both presentation order and treatments, as was done in the analysis of Example 27.6, variability attributable to both between-conditions (i.e., variability attributed to treatments) and between-presentation orders will be integrated into a single analysis of variance which is summarized in Table 27.32. Note that the values computed previously (in Tables 27.30 and 27.31) for $SS_T = 82.28$, $df_T = 17$, $SS_{BS} = 36.93$, $df_{BS} = 5$, $SS_{BC} = 40.45$, $df_{BC} = 2$, $SS_{BP} = 1.78$, and $df_{BP} = 2$ are also employed in Table 27.32. However, the values for SS_{res} and df_{res} in the latter table are not the same as the values computed for SS_{res} and df_{res} in either Table 27.30/24.2 or Table 27.31. In the case of Table 27.32, Equations 27.91 and 27.92 are respectively employed to compute the values $SS_{res} = 3.12$ and $df_{res} = 8$.

$$SS_{res} = SS_T - SS_{BP} - SS_{BS} - SS_{BC}$$

$$SS_{res} = 82.28 - 1.78 - 36.93 - 40.45 = 3.12$$

$$df_{res} = df_T - df_{BP} - df_{BS} - df_{BC}$$

$$df_{res} = 17 - 2 - 5 - 2 = 8$$

With $MS_{res} = SS_{res}/df_{res} = 3.12/8 = .39$, Equations 27.93 and 27.94 are employed to compute the F ratios for the treatment and presentation orders sources of variation. Note that in computing the F ratios, the value $MS_{res} = .39$ is lower than the value for MS_{res} employed in either Table 27.30/24.2 ($MS_{res} = .49$) or Table 27.31 ($MS_{res} = 4.36$). Consequently, the value $F_{BC} = 51.87$ is larger than the value $F_{BC} = 41.29$ computed in Table 27.30/24.2, and the value $F_{BP} = 2.28$ is larger than the value $F_{BP} = .20$ computed in Table 27.31.

$$F_{BC} = \frac{MS_{BC}}{MS_{res}} = \frac{20.23}{.39} = 51.87$$

$$F_{BP} = \frac{MS_{BP}}{MS_{res}} = \frac{.89}{.39} = 2.28$$

**Table 27.32 Summary Table of Analysis of Variance
for Latin Square Analysis of Example 24.1**

Source of variation	SS	df	MS	F
Between-subjects	36.93	5	7.39	
Between-presentation orders	1.78	2	.89	2.28
Between-conditions	40.45	2	20.23	51.87
Residual	3.12	8	.39	
Total	82.28	17		

The degrees of freedom employed in **Table A10** in evaluating both $F_{BC} = 51.87$ and F_{BP} = 2.28 are df_{num} = 2 and df_{den} = 8. The tabled $F_{.95}$ and $F_{.99}$ values for the aforementioned degrees of freedom are $F_{.95}$ = 4.46 and $F_{.99}$ = 8.65. Since F_{BC} = 51.87 is greater than both of the aforementioned critical values, the null hypothesis for a treatment effect can be rejected at both the .05 and .01 levels (which was also the case in Table 27.30/24.2). Since F_{BP} = 2.28 is less than $F_{.05}$ = 4.46, the null hypothesis for presentations orders cannot be rejected (which was also the case in Table 27.31). Thus, even by reducing the residual variation in the analysis summarized in Table 27.32, we are still not able to conclude there is a significant effect attributable to the order of presentation of the three treatments.

If in the analysis of a Latin square design described above the researcher had wanted to employ more than six subjects, additional subjects would be added in multiples of six. One subject in each additional set of six subjects would be presented the three treatments in one of the six possible presentation orders. As an example, if six additional subjects were added to the study, there would be six more rows in Table 27.29. The six rows would represent the scores of Subjects 7–12, each of whom would respectively be presented with the three treatments in the following orders: 123, 231, 312, 132, 213, 321. The same equations as those employed above would be used for the analysis with n = 12 and the total number of scores $N = nk = (12)(3) = 36$.

It was noted earlier it is only recommended that a Latin square design be employed when there is reason to believe little or no interaction is present between the independent variable of interest and a nuisance variable such as presentation order. When there is reason to believe a significant interaction may be present between an independent variable of interest and the nuisance variable, a researcher can employ the nuisance variable as a second independent variable within the framework of a **mixed factorial design**.

To illustrate, assume that in Example 24.1 (as described within the context of the data in Table 27.29) presentation order is employed as a second independent variable. If the latter example were conceptualized as a **mixed factorial design**, presentation order would represent a between-subjects independent variable comprised of k levels, each of which corresponded to one of the k presentation orders. Within each presentation order there would be a block of m = 2 subjects. The within-subjects independent variable would be the $k = 3$ experimental conditions/ treatments. If both a treatment effect and orders effects are present, a **factorial analysis of**

variance for a mixed design will yield: a) A significant F value for the between-subjects variable (i.e., presentation order); and b) A significant F value for the within-subjects variable (i.e. the treatment effect). In addition, the F value obtained for the treatment effect with a **factorial analysis of variance for a mixed design** will in all likelihood be larger than the F value obtained if a **single-factor within-subjects analysis of variance** were employed to only evaluate the treatment effect (i.e., order effects were not included in the analysis). It should be noted, however, that the error degrees of freedom for the **factorial analysis of variance for a mixed design** will be smaller than the error degrees of freedom for the **single-factor within-subjects analysis of variance**. By virtue of the latter, the critical values for the former analysis of variance will be larger than those required for the latter analysis of variance, thus making it more difficult to reject the null hypothesis for the treatment effect. When an order effect is present, yet weak, the smaller degrees of freedom associated with the **factorial analysis of variance for a mixed design** may offset the larger F value it will be likely to yield for a treatment effect when compared with the F value obtained for the treatment effect through use of a **single-factor within-subjects analysis of variance**. Because of the latter, Cohen (2001, p. 500) recommends that if the data are evaluated with a **factorial analysis of variance for a mixed design**, and both the main effect attributable to order of presentation and the interaction between the treatments and order of presentation are weak, then the order of presentation should be omitted from the analysis. In such a case the data should be reevaluated with a **single-factor within-subjects analysis of variance**, and the resulting F value should be employed to assess the treatment effect. Myers and Well (1991, 1995, pp. 364–366) describe an alternative methodology for evaluating a Latin square design when it is conceptualized as a mixed factorial design.

In Section VII of the **Wilcoxon signed-ranks test** it was noted that the concept of **relative efficiency** (also referred to as **power efficiency**) is employed to indicate the power of two tests relative to one another. Various sources note that, when appropriate, because of its greater power (which is attributable to its lower residual variability), the **analysis of variance for a Latin-square design** is more efficient that the **single-factor within-subjects analysis of variance**. Myers and Well (1995, p. 351) note that Equation 27.100 can be employed to compute the **relative efficiency** (*RE*) of the **analysis of variance for a Latin-square design** when contrasted with the **single-factor within-subjects analysis of variance**.[33] In reference to the analysis of the treatment effect in Example 24.1, when the values MS_{res} = .49 and df_{res} = 10 recorded in Table 27.30/24.2 for **single-factor within-subjects analysis of variance** and MS_{res} = .39 and df_{res} = 8 recorded in Table 27.32 for **analysis of variance for a Latin-square design** are substituted in the latter equation, the value *RE* = 1.30 is computed. The value *RE* = 1.30 indicates that the efficiency of the **Latin-square design** relative to the **single-factor within-subjects design** is 1.3 (which expressed as a percentage is 130%). Alternatively, we can state that, if we employ a correction factor to compensate for the discrepancy in degrees of freedom employed in the two analyses, the residual variability for the **single-factor within-subjects analysis of variance** is 1.30 times larger than the residual variability for the **analysis of variance for a Latin-square design**. The latter, of course, accounts for the fact that **analysis of variance for a Latin-square design** provides for a more powerful test of the treatment effect.

(Equation 27.100)

$$RE = \left(\frac{MS_{res_{wsanova}}}{MS_{res_{lsanova}}} \right) \left(\frac{df_{res_{wsanova}} + 1}{df_{res_{wsanova}} + 3} \right) \left(\frac{df_{res_{lsanova}} + 3}{df_{res_{lsanova}} + 1} \right)$$

$$RE = \left(\frac{.49}{.39}\right)\left(\frac{10 + 1}{10 + 3}\right)\left(\frac{8 + 3}{8 + 1}\right) = 1.30$$

In concluding this section, the reader should be made aware of the fact that the discussion of the Latin square design has been limited in scope. In point of fact, alternative more complex analytical procedures are available when the design of a study involves more than one Latin square with one subject per row of each square, as well as for studies involving one or more Latin squares with two or more subjects employed in each row of each square. Detailed discussions of the latter models and alternative analytical procedures which can be employed to evaluate them can be found in books which focus on analysis of variance procedures such as Kirk (1995), Montgomery (2000), Myers and Well (1991, 1995), Tabachnick and Fidell (2001), and Winer *et al.* (1991).

Graeco–Latin square design The Latin square design described up to this point attempts to control for the effect of two nuisance variables. A **Graeco–Latin square design** allows for control of three nuisance variables by superimposing a second Latin square design (which uses Greek letters to identify k levels of a third nuisance variable) on a conventional Latin square design (which employs Latin letters to identify the k levels of the independent variable of interest). Figure 27.12 depicts a 4 × 4 Graeco–Latin square design which employs the Latin letters **A**, **B**, **C**, and **D** and the Greek letters **alpha**, **beta**, **gamma**, and **delta** (α, β, δ, and γ). Note that in Figure 27.12, each Latin letters appears one time in each row and one time in each column, and each Greek letter appears one time in each row and one time in each column. Additionally, each combination of Latin and Greek letters appears exactly once in each row and each column. The analysis of variance for the Graeco–Latin square design is similar to that employed earlier, except for the fact that a sum of squares (as well as degrees of freedom and a mean square) is computed for the nuisance variable represented by the Greek letters. Experimental error is further reduced by subtracting the variability attributable to the latter variable from total variability (in addition subtracting between-conditions variability and variability due to both the row and column nuisance variables from total variability).

	Presentation order			
	1	2	3	4
1	A α	B β	C γ	D δ
2	B δ	D γ	A β	C α
Subject 3	C β	A α	D δ	B γ
4	D γ	C δ	B α	A β

Figure 27.12   4 × 4 Graeco–Latin Square Design

Myers and Well (1991, 1995, pp. 371–372) note that by virtue of introducing an additional independent variable into an analysis, the residual degrees of freedom for a Graeco–Latin square design will be less than the residual degrees of freedom for a Latin square design involving only one independent variable of interest. The consequence of the latter will be that the analysis of variance for a Graeco–Latin square design will be **negatively biased**, which means that the power of the analysis of variance will be compromised — i.e., it will have a lower likelihood of detecting treatment effects on the independent variables of interest, since the lower the residual degrees of freedom, the larger the residual variability, and consequently the smaller the value that will be computed for an F ratio. Because of the latter, Myers and Well (1991, 1995) recommend that a Graeco–Latin square should be 6×6 or larger in order that it have an adequate number of residual degrees of freedom to minimize the likelihood of inadequate power. The aforementioned authors also note that the introduction of an additional variable in a Graeco–Latin square design increases the likelihood one or more interactions (which cannot be evaluated within the framework of such a design) may be confounded with the main effects on the independent variables of interest. Because of the latter problems, Myers and Well (1991, 1995, p. 372) recommend the use of some alternative design other than a Graeco–Latin square. More detailed discussions of the Graeco–Latin square design can be found in Kirk (1995), Montgomery (2000), and Winer *et al.* (1991).

3. Test 27k: The within-subjects factorial analysis of variance A **within-subjects factorial design** involves two or more factors, and all subjects are measured on each of the levels of all of the factors. The **within-subjects factorial analysis of variance** (also known as a **repeated-measures factorial analysis of variance**) is an extension of the **single-factor within-subjects analysis of variance** to experiments involving two or more independent variables/factors. Although the **within-subjects factorial analysis of variance** can be used with designs involving more than two factors, the computational protocol to be described in this section will be limited to the two-factor experiment. Within the framework of the **within-subjects factorial design**, each subject contributes pq scores (which result from the combinations of the levels that comprise the two factors). Since subjects serve under all pq experimental conditions, a **within-subjects factorial design** requires a fraction of the subjects that are needed to evaluate the same set of hypotheses with either the **between-subjects factorial design** or the **mixed factorial design** (assuming a given design employs the same number of scores in each of the pq experimental conditions). To be more specific, only $1/pq^{\text{th}}$ of the subjects are required for a **within-subjects factorial design** in contrast to a **between-subjects factorial design**. The requirement of fewer subjects, and the fact that a within-subjects analysis provides for a more powerful test of an alternative hypothesis than a between-subjects analysis, must to be weighed against the fact that it is often impractical or impossible to have subjects serve in multiple experimental conditions. In addition, a within-subjects analysis of variance is more sensitive to violations of its assumptions than a between-subjects analysis of variance. Example 27.9 is employed to illustrate the **within-subjects factorial analysis of variance**.

Example 27.9 *A study is conducted to evaluate the effect of humidity (to be designated as Factor A) and temperature (to be designated as Factor B) on mechanical problem-solving ability. The experimenter employs a 2×3 within-subjects factorial design. The two levels which comprise Factor A are A_1: Low humidity; A_2: High humidity. The three levels which comprise Factor B are B_1: Low temperature; B_2: Moderate temperature; B_3: High temperature. The study employs three subjects, all of whom serve under the two levels of Factor A and the three levels of Factor B. The order of presentation of the combinations of the two factors is incompletely counterbalanced.[34] The number of mechanical problems solved by the subjects in the six experimental conditions (which result from combinations of the levels of the two factors) follow:* **Condition AB_{11}:** *Low humidity/Low temperature* (11, 9, 10); **Condition AB_{12}:** *Low humidity/Moderate temperature* (7, 8, 6); **Condition AB_{13}:** *Low humidity/High temperature* (5, 4, 3); **Condition AB_{21}:** *High humidity/Low temperature* (2, 4, 3); **Condition AB_{22}:** *High humidity/Moderate temperature* (4, 5, 3); **Condition AB_{23}:** *High humidity/High temperature* (0, 1, 2). *Do the data indicate that either humidity or temperature influences mechanical problem-solving ability?*

Table 27.33 Data for Example 27.9 for Evaluation with the Within-Subjects Factorial Analysis of Variance

	A_1			A_2			Subject sums (ΣS_i)
	B_1	B_2	B_3	B_1	B_2	B_3	
Subject 1	11	7	5	2	4	0	$\Sigma S_1 = 29$
Subject 2	9	8	4	4	5	1	$\Sigma S_2 = 31$
Subject 3	10	6	3	3	3	2	$\Sigma S_3 = 27$
Condition Sums	$\Sigma X_{AB_{11}} = 30$	$\Sigma X_{AB_{12}} = 21$	$\Sigma X_{AB_{13}} = 12$	$\Sigma X_{AB_{21}} = 9$	$\Sigma X_{AB_{22}} = 12$	$\Sigma X_{AB_{23}} = 3$	$\Sigma X_T = 87$
	$\Sigma X_{AB_{11}}^2 = 302$	$\Sigma X_{AB_{12}}^2 = 149$	$\Sigma X_{AB_{13}}^2 = 50$	$\Sigma X_{AB_{21}}^2 = 29$	$\Sigma X_{AB_{22}}^2 = 50$	$\Sigma X_{AB_{23}}^2 = 5$	$\Sigma X_T^2 = 585$

Table 27.34 Scores of Subjects on Levels of Factor A for Example 27.9

	A_1	A_2	Subject sums (ΣS_i)
Subject 1	$S_{1_{A_1}} = 23$	$S_{1_{A_2}} = 6$	$\Sigma S_1 = 29$
Subject 2	$S_{2_{A_1}} = 21$	$S_{2_{A_2}} = 10$	$\Sigma S_2 = 31$
Subject 3	$S_{3_{A_1}} = 19$	$S_{3_{A_2}} = 8$	$\Sigma S_3 = 27$
Sums for levels of Factor A	$\Sigma X_{A_1} = 63$	$\Sigma X_{A_2} = 24$	$\Sigma X_T = 87$

Table 27.35 Scores of Subjects on Levels of Factor B for Example 27.9

	B_1	B_2	B_3	Subject sums (ΣS_i)
Subject 1	$S_{1_{B_1}} = 13$	$S_{1_{B_2}} = 11$	$S_{1_{B_3}} = 5$	$\Sigma S_1 = 29$
Subject 2	$S_{2_{B_1}} = 13$	$S_{2_{B_2}} = 13$	$S_{2_{B_3}} = 5$	$\Sigma S_2 = 31$
Subject 3	$S_{3_{B_1}} = 13$	$S_{3_{B_2}} = 9$	$S_{3_{B_3}} = 5$	$\Sigma S_3 = 27$
Sums for levels of Factor B	$\Sigma X_{B_1} = 39$	$\Sigma X_{B_2} = 33$	$\Sigma X_{B_3} = 15$	$\Sigma X_T = 87$

The data for Example 27.9 are summarized in Tables 27.33–27.35. In Table 27.34, $S_{i_{A_j}}$ represents the score of Subject i under Level j of Factor A. In Table 27.35, $S_{i_{B_k}}$ represents the score of Subject i under Level k of Factor B. Examination of Tables 27.33–27.35 reveals that since the data employed for Example 27.9 are identical to those employed for Examples 27.1 and 27.4, the summary values for the rows, columns, and pq experimental conditions are identical to those in Tables 27.1 and 27.11. Thus, the following values in Tables 27.33–27.35 are identical to those obtained in the tables for Examples 27.1 and 27.4: $n_{A_j} = 9$ and the values computed for ΣX_{A_j} and $\Sigma X_{A_j}^2$ for each of the levels of Factor A; $n_{B_k} = 6$ and the values computed for ΣX_{B_k} and $\Sigma X_{B_k}^2$ for each of the levels of Factor B; $n_{AB_{jk}} = n = 3$ and the values computed for $\Sigma X_{AB_{jk}}$ and $\Sigma X_{AB_{jk}}^2$ for each of the pq experimental conditions that result from combinations of the levels of the two factors: $N = npq = 18$; $\Sigma X_T = 87$; $\Sigma X_T^2 = 585$.

Note that in the **within-subjects factorial analysis of variance**, the **between-subjects factorial analysis of variance**, and the **factorial analysis of variance for a mixed design**, the

Table 27.36 Summary Table of Equations for the Within-Subjects Factorial Design

Source of variation	SS	df	MS	F
Between-subjects	$[S]-[T]$	$n-1$		
Within-subjects	$[XS]-[S]$	$n(pq-1)$		
A	$[A]-[T]$	$p-1$	$\dfrac{SS_A}{df_A}$	$F_A = \dfrac{MS_A}{MS_{A\times subjects}}$
B	$[B]-[T]$	$q-1$	$\dfrac{SS_B}{df_B}$	$F_B = \dfrac{MS_B}{MS_{B\times subjects}}$
AB	$[AB]-[A]-[B]+[T]$	$(p-1)(q-1)$	$\dfrac{SS_{AB}}{df_{AB}}$	$F_{AB} = \dfrac{MS_{AB}}{MS_{AB\times subjects}}$
A × subjects	$[AS]-[A]-[S]+[T]$	$(p-1)(n-1)$	$\dfrac{SS_{A\times subjects}}{df_{A\times subjects}}$	
B × subjects	$[BS]-[B]-[S]+[T]$	$(q-1)(n-1)$	$\dfrac{SS_{B\times subjects}}{df_{B\times subjects}}$	
AB × subjects	$[XS]-[AB]-[AS]-[BS]$ $+[A]+[B]+[S]-[T]$	$(p-1)(q-1)(n-1)$	$\dfrac{SS_{AB\times subjects}}{df_{AB\times subjects}}$	
Total	$[XS]-[T]$	$N-1 = npq-1$		

Table 27.37 Summary Table of Computations for Example 27.9

Source of variation	SS	df	MS	F
Between-subjects	$421.83-420.5 = 1.33$	$3-1 = 2$		
Within-subjects	$585-421.83 = 163.17$	$3[(2)(3)-1] = 15$		
A	$505-420.5 = 84.5$	$2-1 = 1$	$MS_A = \frac{84.5}{1} = 84.5$	$F_A = \frac{84.5}{2} = 42.25$
B	$472.5-420.5 = 52$	$3-1 = 2$	$MS_B = \frac{52}{2} = 26$	$F_B = \frac{26}{.67} = 38.81$
AB	$573-505-472.5+420.5 = 16$	$(2-1)(3-1) = 2$	$MS_{AB} = \frac{16}{2} = 8$	$F_{AB} = \frac{8}{1} = 8$
A × subjects	$510.33-505-421.83$ $+420.5 = 4$	$(2-1)(3-1) = 2$	$MS_{A\times subjects} = \frac{4}{2} = 2$	
B × subjects	$476.5-472.5-421.83$ $+420.5 = 2.67$	$(3-1)(3-1) = 4$	$MS_{B\times subjects} = \frac{2.67}{4} = .67$	
AB × subjects	$585-573-510.33-476.5+505$ $+472.5+421.83-420.5 = 4$	$(2-1)(3-1)(3-1) = 4$	$MS_{AB\times subjects} = \frac{4}{4} = 1$	
Total	$585-420.5 = 164.5$	$18-1=(3)(2)(3)-1 =17$		

value $n_{AB_{jk}} = n = 3$ represents the number of scores in each of the pq experimental conditions. In the case of the **within-subjects factorial analysis of variance**, the value $N = npq = 18$ represents the total number of scores in the set of data. Note, however, that the latter value does not represent the total number of subjects employed in the study as it does in the case of the **between-subjects factorial analysis of variance**. The number of subjects employed for a **within-subjects factorial analysis of variance** will always be the value of $n = n_{AB_{jk}}$. Thus, in Example 27.9 the number of subjects is $n = n_{AB_{jk}} = 3$.

As is the case for the **between-subjects factorial analysis of variance** and the **factorial analysis of variance for a mixed design**, the following three F ratios are computed for the **within-subjects factorial analysis of variance**: F_A, F_B, F_{AB}. The equations required for computing the F ratios are summarized in Table 27.36. Table 27.37 summarizes the computations for the **within-subjects factorial analysis of variance** when it is employed to evaluate Example 27.9. In order to compute the F ratios for the **within-subjects factorial analysis of variance**, it is required that the following summary values (which are also computed for the **between-subjects factorial analysis of variance** and the **factorial analysis of variance for a mixed design**) be computed: $[XS]$, $[T]$, $[A]$, $[B]$, and $[AB]$. Since the summary values computed in Tables 27.33–27.35 are identical to those computed in Tables 27.1 and 27.11 (for Example 27.1 and Example 27.4), the same summary values are employed in Tables 27.36 and 27.37 to compute the values $[XS]$, $[T]$, $[A]$, $[B]$, and $[AB]$ (which are, respectively, computed with Equations 27.4–27.8). Thus: $[XS] = 585$, $[T] = 420.5$, $[A] = 505$, $[B] = 472.5$, and $[AB] = 573$. Since the same set of data and the same equations are employed for the **within-subjects factorial analysis of variance**, the **between-subjects factorial analysis of variance**, and the **factorial analysis of variance for a mixed design**, all three analysis of variance procedures yield identical values for $[XS]$, $[T]$, $[A]$, $[B]$, and $[AB]$. Inspection of Table 27.36 also reveals that the **within-subjects factorial analysis of variance**, the **between-subjects factorial analysis of variance**, and the **factorial analysis of variance for a mixed design** employ the same equations to compute the values SS_A, SS_B, SS_{AB}, SS_T, MS_A, MS_B, and MS_{AB}.

In order to compute a number of additional sum of squares values for the **within-subjects factorial analysis of variance**, it is necessary to compute the following three elements which are not computed for the **between-subjects factorial analysis of variance**: $[S]$, $[AS]$, and $[BS]$.

$[S]$, which is computed with Equation 27.101, is employed in Tables 27.36 and 27.37 to compute the following values: $SS_{\text{Between-subjects}}$, $SS_{\text{Within-subjects}}$, $SS_{A \times \text{subjects}}$, $SS_{B \times \text{subjects}}$, $SS_{AB \times \text{subjects}}$.

$$[S] = \sum_{i=1}^{n} \left[\frac{(\Sigma S_i)^2}{pq} \right] \qquad \textbf{(Equation 27.101)}$$

The notation $\sum_{i=1}^{n} [(\Sigma S_i)^2 / pq]$ in Equation 27.101 indicates that for each of the $n = 3$ subjects, the sum of that subject's three scores (i.e., ΣS_i) is squared and divided by pq. The resulting values obtained for the $n = 3$ subjects are summed, yielding the value $[S]$. Employing Equation 27.101, the value $[S] = 421.83$ is computed.

$$[S] = \frac{(29)^2}{6} + \frac{(31)^2}{6} + \frac{(27)^2}{6} = 421.83$$

$[AS]$, which is computed with Equation 27.102, is employed in Tables 27.36 and 27.37 to compute the following values: $SS_{A \times \text{subjects}}$, $SS_{AB \times \text{subjects}}$.

$$[AS] = \sum_{i=1}^{n} \sum_{j=1}^{p} \left[\frac{(\Sigma S_{i_{A_j}})^2}{q} \right] \qquad \textbf{(Equation 27.102)}$$

The notation $\sum_{i=1}^{n} \sum_{j=1}^{p} [(\Sigma S_{i_{A_j}})^2/q]$ in Equation 27.102 indicates that each of the $p = 2$ $S_{i_{A_j}}$ scores of the $n = 3$ subjects is squared and divided by $q = 3$. The resulting $np = 6$ values are summed, yielding the value $[AS]$. Employing Equation 27.102, the value $[AS] = 510.33$ is computed (which is the same value computed for $[AS]$ when the same set of data is evaluated with the **factorial analysis of variance for a mixed design**).

$$[AS] = \frac{(23)^2}{3} + \frac{(6)^2}{3} + \frac{(21)^2}{3} + \frac{(10)^2}{3} + \frac{(19)^2}{3} + \frac{(8)^2}{3} = 510.33$$

$[BS]$, which is computed with Equation 27.103, is employed in Tables 27.36 and 27.37 to compute the following values: $SS_{B \times \text{subjects}}$, $SS_{AB \times \text{subjects}}$.

$$[BS] = \sum_{i=1}^{n} \sum_{k=1}^{q} \left[\frac{(\Sigma S_{i_{B_k}})^2}{p} \right] \qquad \textbf{(Equation 27.103)}$$

The notation $\sum_{i=1}^{n} \sum_{k=1}^{q} [(\Sigma S_{i_{B_k}})^2/p]$ in Equation 27.103 indicates that each of the $q = 3$ $S_{i_{B_k}}$ scores of the $n = 3$ subjects is squared and divided by $p = 2$. The resulting $nq = 9$ values are summed, yielding the value $[BS]$. Employing Equation 27.103, the value $[BS]$ 476.5 is computed.

$$[BS] = \frac{(13)^2}{2} + \frac{(11)^2}{2} + \frac{(5)^2}{2} + \frac{(13)^2}{2} + \frac{(13)^2}{2} + \frac{(5)^2}{2} + \frac{(13)^2}{2} + \frac{(9)^2}{2} + \frac{(5)^2}{2} = 476.5$$

The reader should take note of the following relationships in Tables 27.36 and 27.37:

$$SS_{\text{Within-subjects}} = SS_A + SS_B + SS_{AB} + SS_{A \times \text{subjects}} + SS_{B \times \text{subjects}} + SS_{AB \times \text{subjects}}$$

$$SS_T = SS_{\text{Between-subjects}} + SS_{\text{Within-subjects}}$$

$$df_{\text{Within-subjects}} = df_A + df_B + df_{AB} + df_{A \times \text{subjects}} + df_{B \times \text{subjects}} + df_{AB \times \text{subjects}}$$

$$df_T = df_{\text{Between-subjects}} + df_{\text{Within-subjects}}$$

Inspection of Table 27.2, Tables 27.36/27.37, and Tables 27.12/27.13 reveals that if a **between-subjects factorial analysis of variance**, a **within-subjects factorial analysis of variance**, and a **factorial analysis of variance for a mixed design** are employed with the same set of data, identical values are computed for the following: SS_A, SS_B, SS_{AB}, SS_T, df_A, df_B, df_{AB}, df_T, MS_A, MS_B, and MS_{AB}.

In Table 27.37, the error term $MS_{A \times \text{subjects}} = 2$, employed in computing the value $F_A = 42.25$, is analogous to the error term which would be obtained if, in evaluating the data for Example 27.9, Factor B was not taken into account, and the data on Factor A were evaluated with a **single-factor within-subjects analysis of variance**. The error term $MS_{B \times \text{subjects}} = .67$, employed in computing the value $F_B = 38.81$, is analogous to the error term that would be obtained if, in evaluating the data for Example 27.9, Factor A was not taken into account, and the data on Factor B were evaluated with a **single-factor within-subjects analysis of variance**. The value $MS_{AB \times subjects} = 1$, employed in computing the value $F_{AB} = 8$, is a measure of error variability specific to the AB interaction for the **within-subjects factorial analysis of variance**. For a thorough discussion of the derivation of the error terms for the **within-subjects factorial analysis of variance**, the reader should consult books which discuss analysis of variance procedures in greater detail (e.g., Keppel (1991) and Winer *et al.* (1991)).

The following tabled critical values derived from **Table A10** are employed in evaluating the three F ratios computed for Example 27.9: a) **Factor A**: For $df_{num} = df_A = 1$ and $df_{den} = df_{A \times subjects} = 2$, $F_{.05} = 18.51$ and $F_{.01} = 98.50$; b) **Factor B**: For $df_{num} = df_B = 2$ and $df_{den} = df_{B \times subjects} = 4$, $F_{.05} = 6.94$ and $F_{.01} = 18.00$; and c) **AB interaction**: For $df_{num} = df_{AB} = 2$ and $df_{den} = df_{AB \times subjects} = 4$, $F_{.05} = 6.94$ and $F_{.01} = 18.00$.

The identical null and alternative hypotheses that are evaluated in Section III of the **between-subjects factorial analysis of variance** are evaluated in the **within-subjects factorial analysis of variance**. In order to reject the null hypothesis in reference to a computed F ratio, the obtained F value must be equal to or greater than the tabled critical value at the prespecified level of significance. Since the computed value $F_A = 42.25$ is greater than $F_{.05} = 18.51$, the alternative hypothesis for Factor A is supported, but only at the .05 level. Since the computed value $F_B = 38.81$ is greater than $F_{.05} = 6.94$ and $F_{.01} = 18.00$, the alternative hypothesis for Factor B is supported at both the .05 and .01 levels. Since the computed value $F_{AB} = 8$ is greater than $F_{.05} = 6.94$, the alternative hypothesis for an interaction between Factors A and B is supported, but only at the .05 level.[35]

The analysis of the data for Example 27.9 allows the researcher to conclude that both humidity (Factor A) and temperature (Factor B) have a significant impact on problem-solving scores. However, as is the case when the same set of data is evaluated with a **between-subjects factorial analysis of variance**, the relationships depicted by the main effects must be qualified because of the presence of a significant interaction. Although the comparison procedures following the computation of the omnibus F ratios (as well as the other analytical procedures for determining power, effect size, etc.) described in Section VI of the **between-subjects factorial analysis of variance** can be extended to the **within-subjects factorial analysis of variance**, they will not be described in this book. For a full description of such procedures, the reader should consult texts which discuss analysis of variance procedures in greater detail (e.g., Keppel (1991) and Winer *et al.* (1991)).

4. Analysis of higher order factorial designs When a **full-factorial design** (i.e., one in which all possible combinations of all factors are represented) involves more than two factors it will have additional interactions that will have to be evaluated. In the case of a **three- factor design** involving Factors A, B, and C, the following interactions (referred to as **first-order interactions**) can be evaluated: AB, AC, BC. Additionally, the **second-order interaction** ABC can also be evaluated — the latter interaction evaluating the joint effect of all three variables. A **second-order interaction** is an example of a **higher-order interaction** — the latter being any interaction which involves three or more factors. Howell (2002, p. 457) notes that probably the simplest way of viewing a **second-order interaction** is to conceptualize it as the AB interaction interacting with factor C.

In a **four-factor design** there will be four main effects for each of the factors A, B, C, and D, six two-way interactions (AB, AC, AD, BC, BD, CD), four three-way interactions (ABC, ABD, ACD, BCD), and one four-way interaction (ABCD), with the latter representing a **third-order interaction**.

It should be noted that the complexity of the computations involved in evaluating factorial designs (be they between-subjects, mixed, or within-subjects) involving more than five factors made such analyses impractical up until the point in time computers became available for analysis. However, as noted in Section VII under the discussion of **screening designs**, even today, in most scientific disciplines one would not be likely to encounter a full-factorial design involving more than five factors, for the simple reason that the more factors involved in an analysis the more difficult it becomes to arrive at a clear interpretation of the results. Keppel (1991, Ch. 22), who provides a good discussion of the liabilities and benefits of higher-order factorial designs, emphasizes the fact that as the number of factors increase the number of treatment groups increase, and assuming the total number of subjects is held constant, the number of subjects in each treatment group will decrease, with the latter compromising the reliability of the treatment means as well as the power for detecting higher-order interactions.

Honeck *et al.* (1983) is an excellent source for deriving the appropriate equations for the use of the analysis of variance with higher-order factorial designs. Other sources that provide more detailed documentation of such designs are Keppel (1991), Kirk (1982, 1995), Montgomery (2000) and Winer *et al.* (1991).

References

Anderson, N. N. (2001). **Empirical direction in design and analysis**. Mahwah, NJ: Lawrence Erlbaum Associates.

Box, G. E. P., Hunter, W. G., & Hunter, J. S. (1978). **Statistics for experimenters**. New York: John Wiley & Sons.

Cohen, B. H. (2001). **Explaining psychological Statistics** (2nd ed.). New York: John Wiley & Sons.

Cohen, J. (1977). **Statistical power analysis for the behavioral sciences**. New York: Academic Press.

Cohen, J. (1988). **Statistical power analysis for the behavioral sciences** (2nd ed.). Hillsdale, NJ: Lawrence Erlbaum Associates, Publishers.

Cortina, J. M. & Nouri, H. (2000). **Effect sizes for ANOVA designs**. Thousand Oaks, CA: Sage Publications.

Edwards, A. L. (1985). **Experimental design in psychological research** (5th ed.). New York: Harper & Row.

Fleiss, J. L. (1986). **The design and analysis of clinical experiments**. New York: John Wiley & Sons.

Grissom, R. J. & Kim, J. J. (2005). **Effect sizes for research: A broad practical approach**. Mahwah, NJ: Lawrence Erlbaum Associates Publishers.

Honeck, R. P., Kibler, C. T., & Sugar, J. (1983). **Experimental design and analysis: A systematic approach.** Lanham, MD: University Press of America.

Howell, D. C. (1992). **Statistical methods for psychology** (3rd ed.). Boston: PWS–Kent Publishing Company.

Howell, D. C. (2002). **Statistical methods for psychology** (5th ed.). Pacific Grove, CA.: Duxbury.

Hsu, J. C. (1996). **Multiple comparisons: Theory and methods**. New York: Chapman & Hall.

Keppel, G. (1991). **Design and analysis: A researcher's handbook** (3rd ed.). Englewood Cliffs, NJ: Prentice Hall.

Keppel, G. & Wickens, T. D. (2004). **Design and analysis: A researcher's handbook** (4th ed.). Upper Saddle River, NJ: Pearson/Prentice Hall.

Keppel, G., Saufley, W. H., & Tokunaga, H. (1992). **Introduction to design and analysis: A student's handbook** (2nd ed.). New York: W. H. Freeman & Company.

Kirk, R. E. (1982). **Experimental design: Procedures for the behavioral sciences** (2nd ed.). Belmont, CA: Brooks/Cole Publishing Company.

Kirk, R. (1995). **Experimental design: Procedures for the behavioral sciences**. Pacific Grove, CA: Brooks/Cole Publishing Company.

Kline, R. B. (2004). **Beyond significance testing: Reforming data analysis methods in behavioral research**. Washington, D. C.: American Psychological Association.

Maxwell, S. E. & Delaney, H. D. (1990). **Designing experiments and analyzing data**. Belmont, CA: Wadsworth Publishing Company.

Maxwell, S. E. & Delaney, H. D. (2000) **Designing experiments and analyzing data**. Mahwah, NJ: Lawrence Erlbaum Associates.

Maxwell, S. E. & Delaney, H. D. (2004) **Designing experiments and analyzing data: A model comparison perspective** (2nd ed.). Mahwah, NJ: Lawrence Erlbaum Associates.

Miller, I. & Freund, J. E. (1965). **Probability and statistics for engineers**. Englewood-Cliffs, NJ: Prentic-Hall.

Montgomery, D. C. (2000). **Design and analysis of experiments** (5th ed.). New York: John Wiley & Sons.

Myers, J. L. & Well, A. D. (1991). **Research design and statistical analysis**. New York: Harper Collins.

Myers, J. L. & Well, A. D. (1995). **Research design and statistical analysis**. Hillsdale, NJ: Lawrence Erlbaum Associates, Publishers.

Myers, J. L. & Well, A. D. (2003). **Research design and statistical analysis** (2nd ed.) Mahwah, NJ: Lawrence Erlbaum Associates, Publishers.

Olejnik, S. & Algina, J. (2000). Measures of effect size for comparative studies: Applications, interpretations, and limitations. **Contemporary Educational Psychology**, 25, 241–286.

Plackett, R. L. & Burman, J. P. (1946). The design of optimal multifactorial experiments. **Biometrika**, 33, 305–325.

Rosenthal, R., Rosnow, R. L., & Rubin, D. B. (2000). **Contrasts and effect sizes in behavioral research: A correlational approach**. Cambridge, UK: Cambridge University Press.

Rosnow, R. L., Rosenthal, R. & Rubin, D. B. (2000). Contrasts and correlations in effect-size estimation. **Psychological Science**, 11, 446–453.

Tabachnick, B. G. & Fidell, L. S. (2001). **Computer-assisted research design and analysis**. Boston: Allyn & Bacon.

Taguchi, G. (1986). **Introduction to quality engineering**. White Plains: NY: Asian Productivity Organization, UNIPUB.

Taguchi, G. (1993). **Taguchi methods: Design of experiments**. Dearborn, MI: ASI Press.

Winer, B. J., Brown, D. R., & Michels, K. M. (1991). **Statistical principles in experimental design** (3rd ed.). New York: McGraw–Hill Publishing Company.

Endnotes

1. a) A **main effect** refers to the effect of one independent variable on the dependent variable, while ignoring the effect any of the other independent variables may have on the dependent variable; b) The model for the **between-subjects factorial analysis of variance** can be summarized by the Equation 27.104, which is a **linear** (or **additive**) **function** of the relevant parameters involved in the analysis of variance. The latter equation describes the elements which contribute to the score of any subject on the dependent variable.

$$X_{ijk} = \mu_T + \alpha_i + \beta_j + (\alpha\beta)_{ij} + \varepsilon_{ijk} \qquad \textbf{(Equation 27.104)}$$

Where: X_{ijk} represents the score of the ith subject in Level i of Factor A and Level j in Factor B

μ_T represents the grand mean of all the treatment populations

α_i represents the part of X_{ijk} attributable to Factor A

β_j represents the part of X_{ijk} attributable to Factor B

$(\alpha\beta)_{ij}$ represents the part of X_{ijk} attributable to an interaction between Factors A and B

ε_{ijk} represents the part of X_{ijk} attributable chance/error variability

2. Although it is possible to conduct a directional analysis, such an analysis will not be described with respect to a factorial analysis of variance. A discussion of a directional analysis when an independent variable is comprised of two levels can be found under the *t* **test for two independent samples**. In addition, a discussion of one-tailed *F* values can be found in Section VI of the latter test under the discussion of **Hartley's F_{max} test for homogeneity of variance/*F* test for two population variances**. A discussion of the evaluation of a directional alternative hypothesis when there are two or more groups can be found in Section VII of the **chi-square goodness-of-fit test (Test 8)**. Although the latter discussion is in reference to analysis of a *k* independent samples design involving categorical data, the general principles regarding the analysis of a directional alternative hypothesis are applicable to the analysis of variance.

3. The notational system employed for the factorial analysis of variance procedures described in this chapter is based on Keppel (1991).

4. The value $SS_{WG} = 12$ can also be computed employing the following equation:

$$SS_{WG} = \sum_{k=1}^{q} \sum_{j=1}^{p} \left[\sum X_{AB_{jk}}^2 - \frac{(\sum X_{AB_{jk}})^2}{n_{AB_{jk}}} \right]$$

$$= \left[302 - \frac{(30)^2}{3} \right] + \left[29 - \frac{(9)^2}{3} \right] + \left[149 - \frac{(21)^2}{3} \right] + \left[50 - \frac{(12)^2}{3} \right] + \left[50 - \frac{(12)^2}{3} \right] + \left[5 - \frac{(3)^2}{3} \right]$$

$$= 2 + 2 + 2 + 2 + 2 + 2 = 12$$

Note that in the above equation a within-groups sum of squares is computed for each of the $pq = 6$ groups, and $SS_{WG} = 12$ represents the sum of the six sum of squares values.

5. This averaging protocol only applies when there is an equal number of subjects in the groups represented in the specific row or column for which an average is computed.

6. a) If the factor represented on the abscissa is comprised of two levels (as is the case in Figure 27.1a), when no interaction is present the lines representing the different levels of the second factor will be parallel to one another by virtue of being equidistant from one another. When the abscissa factor is comprised of more than two factors, the lines can be equidistant but not parallel when no interaction is present; b) A distinction is often made between an **ordinal** versus a **disordinal interaction**. In an **ordinal interaction** the direction of the effect of one independent variable is consistent across the levels of a second

independent variable — in other words, as is the case for the two lines in Figure 27.1b, the means for each of the subgroups comprising one line are consistently higher than/above the means for the subgroups comprising the other line. In a **disordinal interaction** the direction of the effect of one independent variable is reversed across the levels of another independent variable — in other words, as is the case with the lines for levels B_1 and B_2 in Figure 27.1a, the lines in the graph at some point cross one another.

7. As noted earlier, the fact that the lines are parallel to one another is not a requirement if no interaction is present when the abscissa factor is comprised of three or more levels.

8. If no interaction is present, such comparisons should yield results which are consistent with those obtained when the means of the levels of that factor are contrasted.

9. As noted in Section VI of the **single-factor between-subjects analysis of variance**, a **linear contrast** is equivalent to **multiple t tests/Fisher's LSD test**.

10. Many researchers would elect to employ a comparison procedure which is less conservative than the **Scheffé test**, and thus would not require as large a value as CD_S in order to reject the null hypothesis.

11. The number of pairwise comparisons is $[k(k-1)]/2 = [6(6-1)]/2 = 15$, where $k = pq = (2)(3) = 6$ represents the number of groups.

12. If **Tukey's HSD test** is employed to contrast pairs or sets of marginal means for Factors A and B, the values $q_{(A, df_{WG})}$ and $q_{(B, df_{WG})}$ are, respectively, employed from **Table A13**. The sample sizes used in Equation 27.45 for Factors A and B are, respectively, nq and np.

13. When there are only two levels involved in analyzing the simple effects of a factor (as is the case with Factor A), the procedure to be described in this section will yield an F value for a simple effect which is equivalent to the F_{comp} value that can be computed by comparing the two groups employing the **linear contrast** procedure described earlier (i.e., the procedure for which Equation 27.40 is employed to compute SS_{comp}).

14. Equation 27.105 (which is equivalent to Equation 21.3 employed for computing the **between-groups sum of squares** for the **single-factor between-subjects analysis of variance**) is employed to compute the sum of squares for each of the simple effects of Factor A.

(Equation 27.105)

$$SS_{A \text{ at } B_k} = \sum \left[\frac{(\Sigma X_{AB_k})^2}{n_{AB_k}} \right] - \frac{(\Sigma\Sigma X_{AB_k})^2}{(n_{AB_k})(p)}$$

If ΣX_{AB_k} represents the sum of the scores on Level k of Factor B of subjects who serve under a specific level of Factor A, the notation $\Sigma[(\Sigma X_{AB_k})^2/n_{AB_k}]$ in the above equation indicates that the sum of the scores for each level of Factor A at a given level of Factor B is squared, divided by n_{AB_k}, and the p resulting values are summed. The notation $(\Sigma\Sigma X_{AB_k})^2$ represents the square of the sum of scores of the $(n_{AB_k})(p)$ subjects who serve under the specified level of Factor B.

15. In the case of the simple effects of Factor A, the modified degrees of freedom value is $df_{WG} = p(n - 1)$.

16. a) The fact that in the example under discussion the tabled critical values employed for evaluating F_{max} are extremely large is due to the small value of n. However, under the discussion of homogeneity of variance under the **single-factor between-subjects analysis of variance**, it is noted that Keppel (1991) suggests employing a more conservative test anytime the value of $F_{max} \geq 3$; b) As noted in Section VI of the **single-factor between-subjects analysis of variance**, alternative tests of homogeneity of variance, such as the **Levene test for homogeneity of variance (Test 21g)** and the **Brown–Forsythe test of homogeneity of variance (Test 21h)** provide for a more powerful test of the alternative hypothesis and are less sensitive to violations of the normality assumption underlying the **between-subjects factorial analysis of variance**.

17. The procedure described in this section assumes there is an equal number of subjects in each group. If the latter is true, it is also the case for Example 27.1 that $\mu_G = (\mu_{A_1} + \mu_{A_2})/2$ and $\mu_G = (\mu_{AB_{11}} + \mu_{AB_{12}} + \mu_{AB_{13}} + \mu_{AB_{21}} + \mu_{AB_{22}} + \mu_{AB_{23}})/6$.

18. a) Different but equivalent forms of Equations 27.51–27.53 were employed to compute **standard omega squared** in the first edition of this book; b) Equation 27.106 is a general equation which can also be employed to compute the values for **standard omega squared**. Note that in Equation 27.106 (as well as in Equations 27.107 and 27.108), the subscript **effect** refers to the main effect on Factor A, Factor B, or the interaction. Employing Equation 27.106, the values $\omega_{sA}^2 = .50$, $\omega_{sB}^2 = .30$, and $\omega_{sAB}^2 = .08$ obtained previously are computed below.

(Equation 27.106)

$$\tilde{\omega}_{s(effect)}^2 = \frac{df_{effect}(F_{effect} - 1)}{df_A(F_A - 1) + df_B(F_B - 1) + df_{AB}(F_{AB} - 1) + npq}$$

$$\tilde{\omega}_{sA}^2 = \frac{(1)(84.5 - 1)}{(1)(84.5 - 1) + (2)(26 - 1) + (2)(8 - 1) + (3)(2)(3)} = .50$$

$$\tilde{\omega}_{sB}^2 = \frac{(2)(26 - 1)}{(1)(84.5 - 1) + (2)(26 - 1) + (2)(8 - 1) + (3)(2)(3)} = .30$$

$$\tilde{\omega}_{sAB}^2 = \frac{(2)(8 - 1)}{(1)(84.5 - 1) + (2)(26 - 1) + (2)(8 - 1) + (3)(2)(3)} = .08$$

c) Equation 27.107 (Grissom and Kim (2005, p. 142) is an alternative equation for computing **partial omega squared** for a specific effect which will yield the same **partial omega squared** values obtained with Equations 27.60–27.62.

$$\tilde{\omega}_{partial}^2 = \frac{SS_{effect} - (df_{effect}\, MS_{WG})}{SS_{effect} + (N - df_{effect})\, MS_{WG}} \qquad \textbf{(Equation 27.107)}$$

d) The **eta squared** statistic ($\tilde{\eta}^2$) (also represented by R^2, and commonly referred to as the **correlation ratio**) computed for the **single-factor between-subjects analysis of variance** and the **single-factor within-subjects analysis of variance** can also be computed for the **between-subjects factorial analysis of variance**. Equations 27.108 and 27.109 are general equations which can be employed to respectively compute the **standard** ($\tilde{\eta}_s^2 = R_s^2$) and **partial** ($\tilde{\eta}_p^2 = R_p^2$) versions of the latter statistic.

$$\tilde{\eta}_s^2 = R_s^2 = \frac{df_{effect} F_{effect}}{df_A F_A + df_B F_B + df_{AB} F_{AB} + df_{WG}} \quad \text{(Equation 27.108)}$$

$$\tilde{\eta}_{sA}^2 = R_{sA}^2 = \frac{(1)(84.5)}{(1)(84.5) + (2)(26) + (2)(8) + 12} = .51$$

$$\tilde{\eta}_{sB}^2 = R_{sB}^2 = \frac{(2)(26)}{(1)(84.5) + (2)(26) + (2)(8) + 12} = .32$$

$$\tilde{\eta}_{sAB}^2 = R_{sAB}^2 = \frac{(2)(8)}{(1)(84.5) + (2)(26) + (2)(8) + 12} = .10$$

$$\tilde{\eta}_p^2 = R_p^2 = \frac{df_{effect} F_{effect}}{df_{effect} F_{effect} + df_{WG}} \quad \text{(Equation 27.109)}$$

$$\tilde{\eta}_{pA}^2 = R_{pA}^2 = \frac{(1)(84.5)}{(1)(84.5) + 12} = .88$$

$$\tilde{\eta}_{pB}^2 = R_{pB}^2 = \frac{(2)(26)}{(2)(26) + 12} = .81$$

$$\tilde{\eta}_{pAB}^2 = R_{pAB}^2 = \frac{(2)(8)}{(2)(8) + 12} = .57$$

19. For a clarification of the use of multiple summation signs, the reader should review Endnote 63 under the **single-factor between-subjects analysis of variance** and Endnote 19 under the **single-factor within-subjects analysis of variance**.

20. The notation X_{ijk} is a simpler form of the notation $X_{i_{AB_{jk}}}$, which is more consistent with the notational format used throughout the discussion of the **between-subjects factorial analysis of variance**.

21. The notation $\sum_{k=1}^{q} \sum_{j=1}^{p} \sum_{i=1}^{n} X_{ijk}$ is an alternative way of writing $\sum X_T$. $\sum_{k=1}^{q} \sum_{j=1}^{p} \sum_{i=1}^{n} X_{ijk}$ indicates that the scores of each of the $n = n_{AB_{jk}}$ subjects in each of the pq groups are summed.

22. Since the interaction sum of squares is comprised of whatever remains of between-groups variability after the contributions of the main effects for Factor A and Factor B have been removed, Equation 27.67 can be derived from the equation noted below which subtracts Equations 27.65 and 27.66 from Equation 27.64.

$$SS_{AB} = n \sum_{k=1}^{q} \sum_{j=1}^{p} (\bar{X}_{AB_{jk}} - \bar{X}_T)^2 - nq \sum_{j=1}^{p} (\bar{X}_{A_j} - \bar{X}_T)^2 - np \sum_{k=1}^{q} (\bar{X}_{B_k} - \bar{X}_T)^2$$

23. The computation of the harmonic mean is described in the **Introduction** and in Section VI of the *t* **test for two independent samples**.

24. Note that in the example to illustrate the **randomized-blocks design**, the letter *k* is employed to represent the number of levels on Factor A and *j* the number of levels on Factor B, whereas in Example 27.1, the letter *j* is employed as a subscript with respect to the levels of Factor A and the letter *k* as a subscript with respect to the levels of Factor B.

25. Some sources note that the subjects employed in such an experiment (or for that matter any experiment involving independent samples) are nested within the level of the factor to which they are assigned, since each subject serves under only one level of that factor.

26. Sources commonly employ the term **aliasing** to represent **confounding** within the context of screening designs.

27. a) A fractional-factorial design should be **balanced**, which means there should be an equal number of subjects in each of the treatment combinations employed in the design. Thus, in a study involving 256 subjects there could be 256 of the 1024 possible treatment combinations with one subject per combination (although realistically a researcher would employ more than one subject per combination), or 128 treatment combinations with two subjects per combination, and so on; b) Although in an experiment involving a large number of factors it is possible to have subjects serve in multiple conditions (see the **factorial analysis of variance for a mixed design** and the **within-subjects factorial analysis of variance** in Section IX (the **Addendum**)), as the number of factors increase it becomes increasingly difficult and/or impractical to employ the same subjects in multiple conditions.

28. The **mixed factorial design** is often referred to as a **split-plot design**.

29. The computational procedure for the **factorial analysis of variance for a mixed design** assumes there is an equal number of subjects in each of the levels of the between-subjects factor. When the latter is not true, adjusted equations should be employed which can be found in books which describe the **factorial analysis of variance for a mixed design** in greater detail.

30. a) The term Latin square was first used by the Swiss mathematician Lennard Euler (1707-1783; b) A Latin square design is categorized in some sources as a **screening design** (which is discussed in Section VII), since it does not allow interactions to be evaluated.

31. The letters employed in the English language alphabet are more formally referred to as Latin letters, since they are derived from the ancient language of Latin.

32. Many sources recommend that once selecting a Latin square to employ in a study, the researcher should do the following in the order noted: a) Randomly rearrange the ordinal position of the rows; b) Randomly rearrange the ordinal position of the columns; c) Randomly rearrange the notation for the experimental treatments.

    The resulting Latin square is the one which should be employed in the study. The aforementioned procedure is demonstrated below in reference to the Latin square depicted in Figure 27.4.

1. Rows of Figure 27.4 randomly rearranged in Figure 27.13a.

Presentation order

		1	2	3	4
	1	D	C	B	A
	2	C	A	D	B
Subject	3	A	B	C	D
	4	B	D	A	C

Figure 27.13a Random Rearrangement of Rows in Figure 27.4

2. Columns of Figure 27.13a randomly rearranged in Figure 27.13b.

Presentation order

		1	2	3	4
	1	B	D	A	C
	2	D	C	B	A
Subject	3	C	A	D	B
	4	A	B	C	D

Figure 27.13b Random Rearrangement of Columns in Figure 27.13a

Rearrange the treatments by randomly determining to replace A with D, B with A, C with B, and D with C within the cells of the table. The latter is done with respect to the square depicted in Figure 27.13b in the square depicted in Figure 27.14. The square depicted in Figure 27.14 is the one that would be used for the four subjects employed in the design.

Presentation order

		1	2	3	4
	1	A	C	D	B
	2	C	B	A	D
Subject	3	B	D	C	A
	4	D	A	B	C

Figure 27.14 Rearrangement of Square in Figure 27.13b by Randomly Interchanging Letters

33. Equation 27.100 expresses **relative efficiency/power efficiency** in a different format than it is expressed with Equation 6.7.

34. There are 12 possible presentation orders involving combinations of the two factors ($p!q! = 3!2! = 12$). The sequences for presentation of the levels of both factors are determined in the following manner: If A_1 is followed by A_2, presentation of the levels of Factor B can be in the six following sequences: 123, 132, 213, 231, 312, 321. If A_2 is followed by A_1, presentation of the levels of Factor B can be in the same six sequences noted previously. Thus, there are a total of 12 possible sequence combinations. Since there are only six subjects in Example 27.9, only six of the 12 possible sequence combinations can be employed.

35. If Factors A and B are both within-subjects factors and a significant effect is present for the main effects and the interaction, the **within-subjects factorial analysis of variance** would be the most likely of the three factorial analysis of variance procedures discussed to yield significant F ratios. The F_A, F_B, and F_{AB} values obtained in Examples 27.1 and 27.4 are significant at both the .05 and .01 levels when the data are respectively evaluated with a **between-subjects factorial analysis of variance** and a **factorial analysis of variance for a mixed design**. However, when Example 27.9 is evaluated with the **within-subjects factorial analysis of variance**, although F_B is significant at both the .05 and .01 levels, F_A and F_{AB} are only significant at the .05 level. This latter result can be attributed to the fact that the data set employed for the three examples is hypothetical, and is not based on the scores of actual subjects who were evaluated within the framework of a within-subjects factorial design. In point of fact, in the case of the **within-subjects factorial analysis of variance**, the lower value for df_{den} employed for a specific effect (in contrast to the values of df_{den} employed for the **between-subjects factorial analysis of variance** and the **factorial analysis of variance for a mixed design**) will be associated with a tabled critical F value that is larger than the values employed for the latter two tests. Thus, unless there is an actual correlation between subjects' scores under different conditions (which should be the case if a variable is measured within-subjects), the loss of degrees of freedom will nullify the increase in power associated with the **within-subjects factorial analysis of variance** (assuming the data are derived from the appropriate design). The superior power of the **within-subjects factorial analysis of variance** derives from the smaller *MS* error terms employed in evaluating the main effects and interaction.

Measures of
Association/Correlation

Test 28

The Pearson Product-Moment Correlation Coefficient
(Parametric Measure of Association/Correlation Employed with Interval/Ratio Data)

I. Hypothesis Evaluated with Test and Relevant Background Information

The **Pearson product-moment correlation coefficient** is one of a number of measures of correlation or association discussed in this book. Measures of correlation are not inferential statistical tests, but are, instead, descriptive statistical measures which represent the degree of relationship between two or more variables. Within the framework of computing a correlation coefficient additional analyses are conducted which are described by the general term **regression analysis** (which as Kachigan (1986, p. 239) notes could just as easily be called **prediction analysis**). The latter allows a researcher to derive an equation which can be employed to predict a subject's score on one variable (referred to as the **criterion variable**) from her score on the other variable (referred to as the **predictor variable**). Within the context of computing a correlation coefficient and conducting a regression analysis, it is common practice to employ a variety of inferential statistical tests in order to evaluate one or more hypotheses concerning a number of population parameters. The hypothesis stated below is always evaluated when a **Pearson product-moment correlation coefficient** is computed.

Hypothesis evaluated with test In the underlying population represented by a sample, is the correlation between subjects' scores on two variables some value other than zero? The latter hypothesis can also be stated in the following form: In the underlying population represented by the sample, is there a significant linear relationship between the two variables?

Relevant background information on test Developed by Pearson (1896, 1900), the **Pearson product-moment correlation coefficient** is employed with interval/ratio data to determine the degree to which two variables covary (i.e., vary in relation to one another). Any measure of correlation/association which assesses the degree of relationship between two variables is referred to as a **bivariate** measure of association. In evaluating the extent to which two variables covary, the **Pearson product-moment correlation coefficient** determines the degree to which a linear relationship exists between the variables. One variable (usually designated as the X variable) is referred to as the **predictor variable**, since if indeed a linear relationship does exist between the two variables, a subject's score on the predictor variable can be used to predict the subject's score on the second variable. The latter variable, which is referred to as the **criterion variable**, is usually designated as the Y variable.[1] The degree of accuracy with which a researcher will be able to predict a subject's score on the criterion variable from the subject's score on the predictor variable will depend upon the strength of the linear relationship between the two variables.

The statistic computed for the **Pearson product-moment correlation coefficient** is represented by the letter r. r is an estimate of ρ (the Greek letter **rho**), which is the correlation between the two variables in the underlying population. r can assume any value within the range of -1 to $+1$ (i.e., $-1 \leq r \leq +1$). Thus, the value of r can never be less than -1 (i.e., r cannot equal -1.2, -50, etc.) or be greater than $+1$ (i.e., r cannot equal 1.2, 50, etc.). The **absolute value** of r (i.e., $|r|$) indicates the **strength** of the relationship between the two variables. As the absolute value of r approaches 1, the degree of linear relationship between the variables becomes stronger, achieving the maximum when $|r| = 1$ (i.e., when r equals either $+1$ or -1). The closer the absolute value of r is to 1, the more accurately a researcher will be able to predict a subject's score on one variable from the subject's score on the other variable. The closer the absolute value of r is to 0, the weaker the linear relationship between the two variables. As the absolute value of r approaches 0, the degree of accuracy with which a researcher can predict a subject's score on one variable from the other variable decreases, until finally, when $r = 0$ there is no predictive relationship between the two variables. To state it another way, when $r = 0$ the use of the correlation coefficient to predict a subject's Y score from the subject's X score (or vice versa) will not be any more accurate than a prediction which is based on some random process (i.e., a prediction that is based purely on chance).

The **sign** of r indicates the nature or **direction** of the linear relationship which exists between the two variables. A positive sign indicates a **direct** linear relationship, whereas a negative sign indicates an **indirect** (or **inverse**) linear relationship. A direct linear relationship is one in which a change on one variable is associated with a change on the other variable in the same direction (i.e., an increase on one variable is associated with an increase on the other variable, and a decrease on one variable is associated with a decrease on the other variable). When there is a direct relationship, subjects who have a high score on one variable will have a high score on the other variable, and subjects who have a low score on one variable will have a low score on the other variable. The closer a positive value of r is to $+1$, the stronger the direct relationship between the two variables, whereas the closer a positive value of r is to 0, the weaker the direct relationship between the variables. Thus, when r is close to $+1$, most subjects who have a high score on one variable will have a comparably high score on the second variable, and most subjects who have a low score on one variable will have a comparably low score on the second variable. As the value of r approaches 0, the consistency of the general pattern described by a positive correlation deteriorates, until finally, when $r = 0$ there will be no consistent pattern which allows one to predict at above chance a subject's score on one variable if one knows the subject's score on the other variable.

An indirect/inverse relationship is one in which a change on one variable is associated with a change on the other variable in the opposite direction (i.e., an increase on one variable is associated with a decrease on the other variable, and a decrease on one variable is associated with an increase on the other variable). When there is an indirect linear relationship, subjects who have a high score on one variable will have a low score on the other variable, and vice versa. The closer a negative value of r is to -1, the stronger the indirect relationship between the two variables, whereas the closer a negative value of r is to 0, the weaker the indirect relationship between the variables. Thus, when r is close to -1, most subjects who have a high score on one variable will have a comparably low score on the second variable (i.e., as extreme a score in the opposite direction), and most subjects who have a low score on one variable will have a comparably high score on the second variable. As the value of r approaches 0, the consistency of the general pattern described by a negative correlation deteriorates, until finally, when $r = 0$ there will be no consistent pattern which allows one to predict at above chance a subject's score on one variable if one knows the subject's score on the other variable.

The use of the **Pearson product-moment correlation coefficient** assumes that a linear function best describes the relationship between the two variables. If, however, the relationship between the variables is better described by a curvilinear function, the value of *r* computed for a set of data may not indicate the actual extent of the relationship between the variables. In view of this, when a computed *r* value is equal to or close to 0, a researcher should always rule out the possibility that the two variables are related curvilinearly. One quick way of assessing the likelihood of the latter is to construct a **scatterplot** of the data. A scatterplot, which is described in Section VI, displays the data for a correlational analysis in a graphical format.

It is important to note that correlation does not imply causation. Consequently, if there is a strong correlation between two variables (i.e., the absolute value of *r* is close to 1), a researcher is not justified in concluding that one variable causes the other variable. Although it is possible that when a strong correlation exists one variable may, in fact, cause the other variable, the information employed in computing the **Pearson product-moment correlation coefficient** does not allow a researcher to draw such a conclusion. This is the case, since extraneous variables which have not been taken into account by the researcher can be responsible for the observed correlation between the two variables.

The **Pearson product-moment correlation coefficient** (as well as the subsequent regression analysis conducted on the data) is based on the following assumptions: a) The sample of *n* subjects for which the value *r* is computed is randomly selected from the population it represents; b) The level of measurement upon which each of the variables is based is interval or ratio data. Although this assumption is applicable to the conventional use of the **Pearson product-moment correlation coefficient**, there are special cases in which the equation for **Pearson *r*** can be employed with rank-order data (see Section VI of **Spearman's rank-order correlation coefficient (Test 29)**), and categorical data involving one or both variables (see the discussions of the **phi coefficient (Test 16g)** in Section VII, and the discussion of the **point-biserial correlation coefficient (Test 28i)** in Section IX (the **Addendum**)); c) The two variables have a **bivariate normal distribution**. The assumption of bivariate normality states that each of the variables and the linear combination of the two variables are normally distributed. With respect to the latter, if every possible pair of data points is plotted on a three-dimensional plane, the resulting surface (which will look like a mountain with a rounded peak) will be a three-dimensional normal distribution (i.e., a three-dimensional structure in which any cross-section is a standard normal distribution). Another characteristic of a bivariate normal distribution is that for any given value of the *X* variable, the scores on the *Y* variable will be normally distributed, and for any given value of the *Y* variable, the scores on the *X* variable will be normally distributed. In conjunction with the latter, the variances for the *Y* variable will be equal for each of the possible values of the *X* variable, and the variances for the *X* variable will be equal for each of the possible values of the *Y* variable;[2] d) Related to the bivariate normality assumption is the assumption of **homoscedasticity**. Homoscedasticity exists in a set of data if the relationship between the *X* and *Y* variables is of equal strength across the whole range of both variables. Tabachnick and Fidell (1989, 1996) note that when the assumption of bivariate normality is met, the two variables will be homoscedastic. The concept of homoscedasticity is discussed in Section VII; and e) Another assumption, which is discussed in detail in Section VII, is that the **residuals** are independent of one another. A **residual** is the degree of error with respect to prediction — more specifically, it is the difference between a subject's actual score versus his/her predicted score on a criterion variable (the latter being determined through use of the linear regression equation derived in the analysis).

Prior to computing a correlation coefficient and conducting a regression analysis a researcher should carefully scrutinize her data. Prior screening allows the researcher to do the following: a) Evaluate whether or not any of the above noted assumptions have been violated.

b) Correct any erroneous data entries; c) Determine whether or not **outliers** are present, and, if so, how they should be dealt with. (**Outliers,** which are atypically extreme scores, are discussed in the **Introduction** and in Section VII of the *t* **test for two independent samples (Test 11).**) Additionally, a number of diagnostic measures referred to as **distance, leverage,** and **influence** (discussed in Section VII and/or in Section IV/V of the chapter on **multiple regression (Test 33)),** which are available with most computer software, should be employed to identify any observations that will exert undue influence on statistics computed for a set of data.

Among those factors which can compromise the reliability of a correlation coefficient (i.e., result in an *r* value that does not accurately reflecting the actual correlation between two variables in the underlying population) are the following: a) If the range of scores on one or both variables is restricted (i.e., the range of scores for the sample is not representative of the range of scores in the underlying population), the absolute value of the computed correlation coefficient will be smaller than its actual value in the underlying population. Howell (2002, p. 283) notes that one exception to this in which the absolute value of *r* may actually be increased is if, by employing a restricted range of scores in a sample, an actual curvilinear relationship in the underlying population between the two variables is concealed; b) The presence of one or more outliers in a set of data can spuriously increase or decrease the value of *r* relative to what its actual value is in the underlying population; c) When the underlying population distributions for the *X* and *Y* variables are skewed in the opposite direction, the maximum possible absolute value for a correlation between the two variables will be less than 1. In such an instance a data transformation (discussed in Section VII of the *t* **test for two independent samples**) can be employed to normalize the shapes of the distributions; d) If in the underlying population the relationship between two variables is curvilinear, the degree of relationship reflected by the absolute value of *r* will underestimate the actual degree of the relationship between the two variables; e) Although the sample size employed to compute a correlation coefficient does not influence the magnitude of the absolute value of *r* (except when $n = 2$, in which case *r* can only equal 0 or ± 1), it does affect the accuracy of the *r* value (relative to what the actual value of the correlation between the two variables is in the underlying population). As is the case in computing any statistic, the larger the sample size the more confidence one can have that the computed value of a statistic provides an accurate estimate of the underlying population parameter.

As noted earlier, the use of correlational data for predictive purposes is summarized under the general subject of **regression analysis** (or more formally, **linear regression analysis,** since when prediction is discussed in reference to the **Pearson product-moment correlation coefficient** it is based on the degree of linear relationship between the two variables). The term **regression** (which was discussed briefly in the **Introduction** under the discussion of experimental design) was first employed in 1877 by the British scientist Francis Galton (1822-1911). Galton used the term regression within the context of a study he conducted in which he empirically demonstrated that tall fathers were more likely to have sons who were shorter than themselves, whereas short fathers were more likely to have sons who were taller than themselves. In point of fact, Galton discovered that the heights of the sons tended to move inward toward the mean height for all males in the general population, rather than outward in the direction of being even more extreme than their fathers. This phenomenon of the offsprings' height moving inward toward the mean is summarized by the expression **regression toward the mean**. Galton coined the term **regression** to designate the process of predicting one variable (in this case, the heights of the boys) from another variable (the heights of their fathers).

Subsequent research of the phenomenon identified by Galton revealed that when a score is measured on any variable, an organism which obtains an extreme score on that variable will be likely to yield a score which is closer to the mean value for that score in the general population if the score is remeasured. In other words, extreme scores regress toward the mean when remeasured. The degree to which an extreme score will regress toward the mean will be a function of the reliability of the measurement process. Put simply, the greater the degree of error in measurement, the more likely an organism that obtains an extreme score on a variable will obtain a score closer to the mean if the score is remeasured. In statistical terms, whenever the correlation between two variables (where scores on one variable are employed to predict scores on the other variable) is imperfect (i.e., the absolute value of r is some value other than 1), regression toward the mean occurs. One implication of this discussion is that whenever one or more organisms that obtain extreme scores on a variable are the focus of a study, if after exposure to some experimental treatment the scores of those organisms are closer to the population mean for that variable, the change in scores may not be the result of the experimental treatment but, in fact, may just be an example of regression toward the mean. A full discussion of **linear regression analysis** can be found in Section VI.

II. Example

Example 28.1 *A nutritionist conducts a study employing a sample of five children to determine whether or not there is a statistical relationship between the number of ounces of sugar a ten-year-old child eats per week (which will represent the X variable) and the number of cavities in a child's mouth (which will represent the Y variable). The two scores (ounces of sugar consumed per week and number of cavities) obtained for each of the five children follow:* **Child 1** (20, 7); **Child 2** (0, 0); **Child 3** (1, 2); **Child 4** (12, 5); **Child 5** (3, 3). *Is there a significant correlation between sugar consumption and the number of cavities?*

III. Null versus Alternative Hypotheses

Upon computing the **Pearson product-moment correlation coefficient**, it is common practice to determine whether the obtained absolute value of the correlation coefficient is large enough to allow a researcher to conclude that the underlying population correlation coefficient between the two variables is some value other than zero. Section V describes how the latter hypothesis, which is stated below, can be evaluated through use of tables of critical r values or through use of an inferential statistical test which is based on the t distribution.

Null hypothesis H_0: $\rho = 0$

(In the underlying population the sample represents, the correlation between the scores of subjects on Variable X and Variable Y equals 0.)

Alternative hypothesis H_1: $\rho \neq 0$

(In the underlying population the sample represents, the correlation between the scores of subjects on Variable X and Variable Y equals some value other than 0. This is a **nondirectional alternative hypothesis**, and it is evaluated with a **two-tailed test**. Either a significant positive r value or a significant negative r value will provide support for this alternative hypothesis. In order to be significant, the obtained absolute value of r must be equal to or greater than the tabled critical two-tailed r value at the prespecified level of significance.)

or

H_1: $\rho > 0$

(In the underlying population the sample represents, the correlation between the scores of subjects on Variable X and Variable Y equals some value greater than 0. This is a **directional alternative hypothesis**, and it is evaluated with a **one-tailed test**. Only a significant positive r value will provide support for this alternative hypothesis. In order to be significant (in addition to the requirement of a positive r value), the obtained absolute value of r must be equal to or greater than the tabled critical one-tailed r value at the prespecified level of significance.)

or

$$H_1: \rho < 0$$

(In the underlying population the sample represents, the correlation between the scores of subjects on Variable X and Variable Y equals some value less than 0. This is a **directional alternative hypothesis**, and it is evaluated with a **one-tailed test**. Only a significant negative r value will provide support for this alternative hypothesis. In order to be significant (in addition to the requirement of a negative r value), the obtained absolute value of r must be equal to or greater than the tabled critical one-tailed r value at the prespecified level of significance.)

Note: Only one of the above noted alternative hypotheses is employed. If the alternative hypothesis the researcher selects is supported, the null hypothesis is rejected.

IV. Test Computations

Table 28.1 summarizes the data for Example 28.1. The following should be noted with respect to Table 28.1: a) The number of subjects is $n = 5$. Each subject has an X score and a Y score, and thus there are five X scores and five Y scores; b) ΣX, ΣX^2, ΣY, and, ΣY^2, respectively, represent the sum of the five subjects' scores on the X variable, the sum of the five subjects' squared scores on the X variable, the sum of the five subjects' scores on the Y variable, and the sum of the five subjects' squared scores on the Y variable; and c) An XY score is obtained for each subject by multiplying a subject's X score by the subject's Y score. ΣXY represents the sum of the five subjects' XY scores. Each XY score represents a **cross-product** and ΣXY represents the **sum of the cross-products**.

Table 28.1   Summary of Data for Example 28.1

Subject	X	X^2	Y	Y^2	XY
1	20	400	7	49	140
2	0	0	0	0	0
3	1	1	2	4	2
4	12	144	5	25	60
5	3	9	3	9	9
	$\Sigma X = 36$	$\Sigma X^2 = 554$	$\Sigma Y = 17$	$\Sigma Y^2 = 87$	$\Sigma XY = 211$

Although they are not required for computing the value of r, the mean score ($\bar{X}$ and $\bar{Y}$) and the estimated population standard deviation ($\hat{s}_X$ and $\hat{s}_Y$) for each of the variables are computed (the latter values are computed with Equation I.11). These values are employed in Section VI to derive **regression equations**, which are used to predict a subject's score on one variable from the subject's score on the other variable.

$$\bar{X} = \frac{\Sigma X}{n} = \frac{36}{5} = 7.2 \qquad \bar{Y} = \frac{\Sigma Y}{n} = \frac{17}{5} = 3.4$$

$$\tilde{s}_X = \sqrt{\frac{\Sigma X^2 - \frac{(\Sigma X)^2}{n}}{n-1}} = \sqrt{\frac{554 - \frac{(36)^2}{5}}{5-1}} = 8.58$$

$$\tilde{s}_Y = \sqrt{\frac{\Sigma Y^2 - \frac{(\Sigma Y)^2}{n}}{n-1}} = \sqrt{\frac{87 - \frac{(17)^2}{5}}{5-1}} = 2.70$$

Equation 28.1 (which is identical to Equation 17.7, except for the fact that the notation X and Y is used in place of X_1 and X_2) is employed to compute the value of r.[3]

$$r = \frac{\Sigma XY - \frac{(\Sigma X)(\Sigma Y)}{n}}{\sqrt{\left[\Sigma X^2 - \frac{(\Sigma X)^2}{n}\right]\left[\Sigma Y^2 - \frac{(\Sigma Y)^2}{n}\right]}} \qquad \textbf{(Equation 28.1)}$$

Substituting the appropriate values in Equation 28.1, the value $r = .995$ is computed.

$$r = \frac{211 - \frac{(36)(17)}{5}}{\sqrt{\left[554 - \frac{(36)^2}{5}\right]\left[87 - \frac{(17)^2}{5}\right]}} = .955$$

The numerator of Equation 28.1, which is referred to as the **sum of products** (which is summarized with the notation SP_{XY}), will determine the sign of r. If the numerator is a negative value, r will be a negative number. If the numerator is a positive value, r will be a positive number. If the numerator equals zero, r will equal zero. In the case of Example 28.1, $SP_{XY} = \Sigma XY - [[(\Sigma X)(\Sigma Y)]/n] = 211 - [[(36)(17)]/5] = 88.6$. The denominator of Equation 28.1 is the square root of the product of the **sum of squares of the X scores** (which is summarized with the notation SS_X), and the **sum of squares of the Y scores** (which is summarized with the notation SS_Y). Thus, $SS_X = \Sigma X^2 - [(\Sigma X)^2/n] = 554 - [(36)^2/5] = 294.8$ and $SS_Y = \Sigma Y^2 - [(\Sigma Y)^2/n] = 87 - [(17)^2/5] = 29.2$. The aforementioned sum of squares values represent the numerator of the equation for computing the estimated population standard deviation of the X and Y scores (i.e., Equation I.11). Employing the notation for the sum of products and the sums of squares, the equation for the **Pearson product-moment correlation coefficient** can be expressed as follows: $r = SP_{XY}/\sqrt{SS_X SS_Y}$.

The reader should take note of the fact that each of the sum of squares values must be a positive number. If either of the sum of squares values is a negative number, it indicates that a computational error has been made. The only time a sum of squares value will equal zero will be if all of the subjects have the identical score on the variable for which the sum of squares is computed. Any time one or both of the sum of squares values equals zero, Equation 28.1 will be insoluble. It is noted in Section I that the computed value of r must fall within the range $-1 \le r \le +1$. Consequently, if the value of r is less than -1 or greater than $+1$, it indicates that a computational error has been made. One final point which should be made is that the sample size $n = 5$ used for Example 28.1 is extremely small, and that realistically such a small sample size would not be employed in a serious study. The use of the small sample size is to enable readers to follow the computational procedures to be described for correlational analysis more easily.

V. Interpretation of the Test Results

The obtained value $r = .995$ is evaluated with **Table A16 (Table of Critical Values for Pearson r)** in the **Appendix**. The degrees of freedom employed for evaluating the significance of r are computed with Equation 28.2.

$$df = n - 2 \qquad \textbf{(Equation 28.2)}$$

Employing Equation 28.2, the value $df = 5 - 3 = 2$ is computed. Using **Table A16**, it can be determined that the tabled critical two-tailed r values at the .05 and .01 levels of significance are $r_{.05} = .878$ and $r_{.01} = .959$, and the tabled critical one-tailed r values at the .05 and .01 levels of significance are $r_{.05} = .805$ and $r_{.01} = .934$.

The following guidelines are employed in evaluating the null hypothesis H_0: $\rho = 0$.

a) If the nondirectional alternative hypothesis H_1: $\rho \ne 0$ is employed, the null hypothesis can be rejected if the obtained absolute value of r is equal to or greater than the tabled critical two-tailed value at the prespecified level of significance.

b) If the directional alternative hypothesis H_1: $\rho > 0$ is employed, the null hypothesis can be rejected if the sign of r is positive, and the value of r is equal to or greater than the tabled critical one-tailed value at the prespecified level of significance.

c) If the directional alternative hypothesis H_1: $\rho < 0$ is employed, the null hypothesis can be rejected if the sign of r is negative, and the absolute value of r is equal to or greater than the tabled critical one-tailed value at the prespecified level of significance.

Employing the above guidelines, the nondirectional alternative hypothesis H_1: $\rho \ne 0$ is supported at the .05 level, since the computed value $r = .955$ is greater than the tabled critical two-tailed value $r_{.05} = .878$. It is not, however, supported at the .01 level, since $r = .955$ is less than the tabled critical two-tailed value $r_{.01} = .959$.

The directional alternative hypothesis H_1: $\rho > 0$ is supported at both the .05 and .01 levels, since the computed value $r = .955$ is a positive number which is greater than the tabled critical one-tailed values $r_{.05} = .805$ and $r_{.01} = .934$.

The directional alternative hypothesis H_1: $\rho < 0$ is not supported, since the computed value $r = .955$ is a positive number. In order for the alternative hypothesis H_1: $\rho < 0$ to be supported, the computed value of r must be a negative number (as well as the fact that the absolute value of r must be equal to or greater than the tabled critical one-tailed value at the prespecified level of significance).

It may seem surprising that such a large correlation (i.e., an r value which almost equals 1) is not significant at the .01 level. Inspection of **Table A16** reveals that when the sample size is small (as is the case in Example 28.1), the tabled critical r values are relatively large. The large critical values reflect the fact that the smaller the sample size, the higher likelihood of sampling error resulting in a spuriously inflated correlation. Further examination of **Table A16** reveals that as the value of n increases, the tabled critical values at a given level of significance decrease, until finally when n is quite large the tabled critical values are quite low. What this translates into is that when the sample size is extremely large, an absolute r value which is barely above zero will be statistically significant. Keep in mind, however, that the alternative hypothesis which is evaluated only stipulates that the underlying population correlation is some value other than zero. The distinction between **statistical** versus **practical significance** (which is discussed in the **Introduction** and in Section VI of the *t* **test for two independent samples**) is germane to this discussion, in that a small correlation may be statistically significant, yet not be of any practical and/or theoretical value. It should be noted, however, that in many instances a significant correlation which is close to zero may be of practical and/or theoretical significance.

As noted in Section I, the factors noted below are among those which can dramatically influence the value of r. In point of fact, these factors are more likely to be present and/or distort an r value when a small sample size is employed in a study: a) If the range of scores on either the X or Y variable is restricted, the absolute value of r will be reduced; b) A correlation based on a sample which is characterized by the presence of extreme scores on one or both of the variables (even though the scores are not extreme enough to be considered outliers, which are atypically extreme scores) may be spuriously high (i.e., the absolute value of r will be higher than the absolute value of ρ in the underlying population); and c) The presence of one or more outliers can grossly distort the absolute value of r, or even affect the sign of r.

Test 28a: Test of significance for a Pearson product-moment correlation coefficient In the event a researcher does not have access to **Table A16**, Equation 28.3, which employs the t distribution, provides an alternative way of evaluating the null hypothesis H_0: $\rho = 0$.

$$t = \frac{r\sqrt{n-2}}{\sqrt{1-r^2}}$$ **(Equation 28.3)**

Substituting the appropriate values in Equation 28.3, the value $t = 5.58$ is computed.

$$t = \frac{.955\sqrt{5-2}}{\sqrt{1-(.955)^2}} = 5.58$$

The computed value $t = 5.58$ is evaluated with **Table A2 (Table of Student's *t* Distribution)** in the **Appendix**. The degrees of freedom employed in evaluating Equation 28.3 are $df = n - 2$. Thus, $df = 5 - 2 = 3$. For $df = 3$, the tabled critical two-tailed .05 and .01 values are $t_{.05} = 3.18$ and $t_{.01} = 5.84$, and the tabled critical one-tailed .05 and .01 values are $t_{.05} = 2.35$ and $t_{.01} = 4.54$. Since the sign of the t value computed with Equation 28.3 will always be the same as the sign of r, the guidelines described earlier in reference to **Table A16** for evaluating an r value can also be applied in evaluating the t value computed with Equation 28.3 (i.e., substitute t in place of r in the text of the guidelines for evaluating r).

Employing the guidelines, the nondirectional alternative hypothesis H_1: $\rho \neq 0$ is supported at the .05 level, since the computed value $t = 5.58$ is greater than the tabled critical two-tailed value $t_{.05} = 3.18$. It is not, however, supported at the .01 level, since $t = 5.58$ is less than the tabled critical two-tailed value $t_{.01} = 5.84$.

The directional alternative hypothesis H_1: $\rho > 0$ is supported at both the .05 and .01 levels, since the computed value $t = 5.58$ is a positive number which is greater than the tabled critical one-tailed values $t_{.05} = 2.35$ and $t_{.01} = 4.54$.

The directional alternative hypothesis H_1: $\rho < 0$ is not supported, since the computed value $t = 5.58$ is a positive number. In order for the alternative hypothesis H_1: $\rho < 0$ to be supported, the computed value of t must be a negative number (as well as the fact that the absolute value of t must be equal to or greater than the tabled critical one-tailed value at the prespecified level of significance). (An alternative methodology for evaluating the null hypothesis H_0: $\rho = 0$, which yields a result equivalent to that obtained with Equation 28.3, can be found in the Section VII under the discussion of **residuals**.)

Note that the results obtained through use of Equation 28.3 are consistent with those that are obtained when **Table A16** is employed.[4] A summary of the analysis of Example 28.1 follows: It can be concluded there is a significant positive correlation between the number of ounces of sugar a ten-year-old child eats and the number of cavities in a child's mouth. This result can be summarized as follows (if it is assumed the nondirectional alternative hypothesis H_1: $\rho \neq 0$ is employed): $r = .955$, $p < .05$.

The coefficient of determination The square of a computed r value (i.e., r^2) is referred to as the **coefficient of determination**. r^2 represents the proportion of variance on one variable which can be accounted for by variance on the other variable.[5] The use of the term "accounted for" in the previous sentence should not be interpreted as indicating that a cause–effect relationship exists between the two variables. As noted in Section I, a substantial correlation between two variables does not allow one to conclude that one variable causes the other.

For Example 28.1 the coefficient of determination is computed to be $r^2 = (.955)^2 = .912$, which expressed as a percentage is 91.2%.[6] This indicates that 91.2% of the variation on the Y variable can be accounted for on the basis of variability on the X variable (or vice versa). Although it is possible that X causes Y (or that Y causes X), it is also possible that one or more extraneous variables which are related to X and/or Y that have not been taken into account in the analysis are the real reason for the strong relationship between the two variables. In order to demonstrate that the amount of sugar a child eats is the direct cause of the number of cavities he or she develops, a researcher would be required to conduct an experiment in which the amount of sugar consumed is a manipulated independent variable, and the number of cavities is the dependent variable. As noted in the **Introduction** of the book, an experiment in which the independent variable is directly manipulated by the researcher is often referred to as a **true experiment**. If a researcher conducts a true experiment to evaluate the relationship between the amount of sugar eaten and the number of cavities, such a study would require randomly assigning a representative sample of young children to two or more groups. By virtue of random assignment, it would be assumed that the resulting groups are comparable to one another. Each of the groups would be differentiated from one another on the basis of the amount of sugar the children within a group consume (which would be determined by the experimenter). Since the independent variable is manipulated, the amount of sugar consumed by each group is under the direct control of the experimenter. Any observed differences on the dependent variable between the groups at some later point in time could be attributed to the manipulated independent variable. Thus, if in fact significant group differences with respect to the number of cavities are observed, the researcher would have a reasonable basis for concluding that sugar consumption is responsible for such differences.

Whereas the correlational study represented by Example 28.1 is not able to control for potentially confounding variables, the true experiment described above is able control for such variables. Common sense suggests, however, that practical and ethical considerations would make it all but impossible to conduct the sort of experiment described above. Realistically, in a democratic society a researcher cannot force a parent to feed her child a specified amount of sugar if the parent is not naturally inclined to do so. Even if a researcher discovers that through the use of monetary incentives she can persuade some parents to feed their children different amounts of sugar than they deem prudent, the latter sort of inducement would most likely compromise a researcher's ability to randomly assign subjects to groups, not to mention the fact that it would be viewed as unethical by many people. Consequently, if a researcher is inclined to conduct a study evaluating the relationship between sugar consumption and the number of cavities, it is highly unlikely that sugar consumption would be employed as a manipulated independent variable. In order to assess what, if any, relationship there is between the two variables, it is much more likely that a researcher would solicit parents whose children ate large versus moderate versus small amounts of sugar, and use the latter as a basis for defining her groups. In such a study, the amount of sugar consumed would be a nonmanipulated independent variable (since it represents a preexisting subject characteristic). The information derived from this type of study (which is commonly referred to as an **ex post facto study** or a **natural experiment**) is correlational in nature. This is the case, since in any study in which the independent variable is not manipulated by the experimenter, one is not able to effectively control for the influence of potentially confounding variables. Thus, if in fact differences are observed between two or more groups in an **ex post facto study**, although such differences may be due to the independent variable, they can also be due to extraneous variables. Consequently, in the case of the example under discussion, any observed differences in the number of cavities between two or more groups can be due to extraneous factors such as maternal prenatal health care, different home environments, dietary elements other than sugar, socioeconomic and/or educational differences among the families that comprise the different groups, etc.

VI. Additional Analytical Procedures for the Pearson Product-Moment Correlation Coefficient and/or Related Tests

1. Derivation of a regression line The obtained value $r = .955$ suggests there is a strong direct relationship between sugar consumption (X) and the number of cavities (Y). The high positive value of the correlation coefficient suggests that as the number of ounces of sugar consumed increases, there is a corresponding increase in the number of cavities. This is confirmed in Figure 28.1 which is a **scatterplot** of the data for Example 28.1. A scatterplot depicts the data employed in a correlational analysis in a graphical format. Each subject's two scores are represented by a single point on the scatterplot. The point which depicts a subject's two scores is arrived at by moving horizontally on the abscissa (X-axis) the number of units that corresponds to the subject's X score, and moving vertically on the ordinate (Y-axis) the number of units that corresponds to the subject's Y score.

Employing the scatterplot, one can visually estimate the straight line which comes closest to passing through all of the data points. This line is referred to as the **regression line** (also known as the **line of best fit**). In actuality, there are two regression lines. The most commonly determined line is **the regression line of Y on X**. The latter line is employed to predict a subject's Y score (which represents the criterion variable) by employing the subject's X score (which represents the predictor variable). The second regression line, **the regression line of X on Y**, allows one to predict a subject's X score by employing the subject's Y score. As will be noted later in this discussion, the only time the two regression lines will be identical is when the absolute value of r equals 1. Because X is usually designated as the predictor variable and Y as the criterion variable, the **regression line of Y on X** is the more commonly determined of the two regression lines.

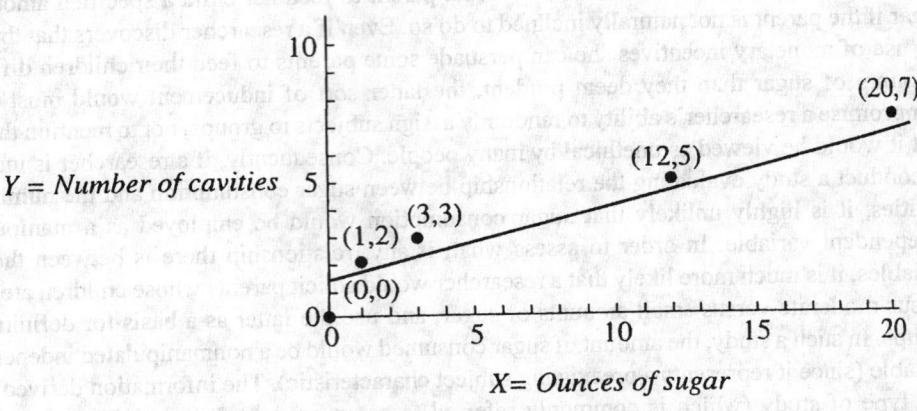

Figure 28.1 Scatterplot for Example 28.1

The **regression line of Y on X** (which, along with the **regression line of X on Y**, is determined mathematically later in this section) is displayed in Figure 28.1. Note that the line is positively sloped — i.e., the lowest part of the line is on the lower left of the graph with the line slanting upward to the right. A line which is positively sloped reflects the fact that a change on one variable in a specific direction is accompanied by a change in the other variable in the same direction. A positive correlation will always result in a positively sloped regression line. A negative correlation, on the other hand, will always result in a negatively sloped regression line. In a negatively sloped regression line, the upper part of the line is at the left of the graph and the line slants downward as one moves to the right. A line which is negatively sloped reflects the fact that a change on one variable in a specific direction is accompanied by a change in the other variable in the opposite direction.

Whereas the slope of the regression line indicates whether a computed r value is a positive or negative number, the magnitude of the absolute value of r reflects how close the n data points fall in relation to the regression line. When $r = +1$ or $r = -1$, all of the data points fall on the regression line. As the absolute value of r deviates from 1 and moves toward 0, the data points deviate further and further from the regression line. Figure 28.2 depicts a variety of hypothetical regression lines, which are presented to illustrate the relationship between the sign and absolute value of r and the regression line.

In Figure 28.2 the regression lines (a), (b), (c), and (d) are positively sloped, and are thus associated with a positive correlation. Lines (e), (f), (g), and (h), on the other hand, are negatively sloped, and are associated with a negative correlation. Note that in each graph, the closer the data points are to the regression line, the closer the absolute value of r is to one. Thus, in graphs (a) – (h), the strength of the correlation (i.e., maximum, strong, moderate, weak) is a function of how close the data points are to the regression line.

The use of the terms strong, moderate, and weak in relation to specific values of correlation coefficients is somewhat arbitrary. For the purpose of discussion, the following rough guidelines will be employed for designating the strength of a correlation coefficient: a) If $|r| \geq .7$, a correlation is considered to be strong; b) If $.3 \leq |r| < .7$, a correlation is considered to be moderate; and c) If $|r| < .3$, a correlation is considered to be weak. In point of fact, most statistically significant correlations in the scientific literature are in the weak to moderate range.

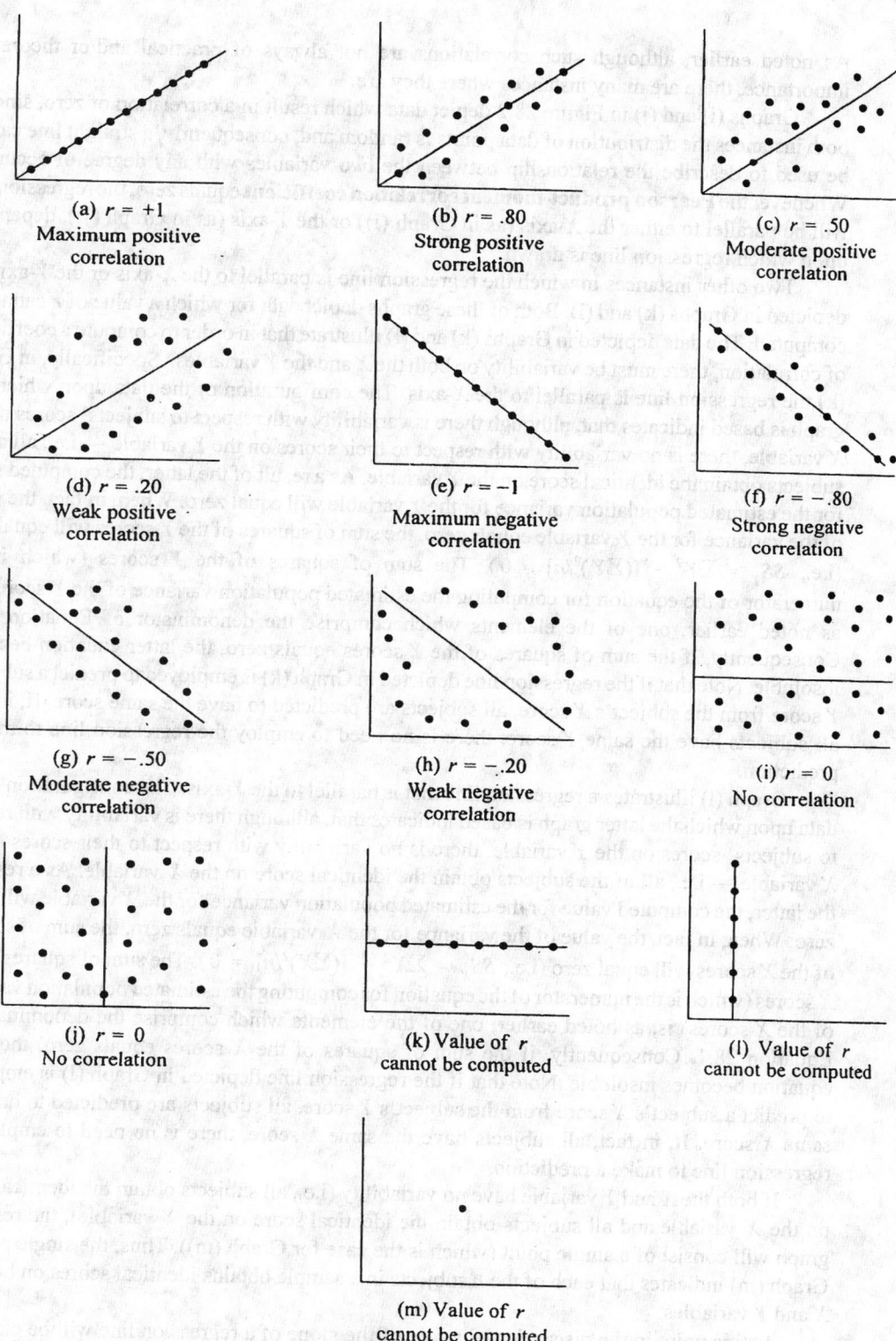

(a) $r = +1$
Maximum positive
correlation

(b) $r = .80$
Strong positive
correlation

(c) $r = .50$
Moderate positive
correlation

(d) $r = .20$
Weak positive
correlation

(e) $r = -1$
Maximum negative
correlation

(f) $r = -.80$
Strong negative
correlation

(g) $r = -.50$
Moderate negative
correlation

(h) $r = -.20$
Weak negative
correlation

(i) $r = 0$
No correlation

(j) $r = 0$
No correlation

(k) Value of r
cannot be computed

(l) Value of r
cannot be computed

(m) Value of r
cannot be computed

Figure 28.2 Hypothetical Regression Lines

As noted earlier, although such correlations are not always of practical and/or theoretical importance, there are many instances where they are.

Graphs (i) and (j) in Figure 28.2 depict data which result in a correlation of zero, since in both instances the distribution of data points is random and, consequently, a straight line cannot be used to describe the relationship between the two variables with any degree of accuracy. Whenever the **Pearson product-moment correlation coefficient** equals zero, the regression line will be parallel to either the X-axis (as in Graph (i)) or the Y-axis (as in Graph (j)), depending upon which regression line is drawn.

Two other instances in which the regression line is parallel to the X-axis or the Y-axis are depicted in Graphs (k) and (l). Both of these graphs depict data for which a value of r cannot be computed. The data depicted in Graphs (k) and (l) illustrate that in order to compute a coefficient of correlation, there must be variability on both the X and the Y variables. Specifically, in Graph (k) the regression line is parallel to the X-axis. The configuration of the data upon which this graph is based indicates that, although there is variability with respect to subjects' scores on the X variable, there is no variability with respect to their scores on the Y variable — i.e., all of the subjects obtain the identical score on the Y variable. As a result of the latter, the computed value for the estimated population variance for the Y variable will equal zero. When, in fact, the value of the variance for the Y variable equals zero, the sum of squares of the Y scores will equal zero (i.e., $SS_Y = \Sigma Y^2 - [(\Sigma Y)^2/n] = 0$). The sum of squares of the Y scores (which is the numerator of the equation for computing the estimated population variance of the Y scores) is, as noted earlier, one of the elements which comprise the denominator of Equation 28.1. Consequently, if the sum of squares of the Y scores equals zero, the latter equation becomes insoluble. Note that if the regression line depicted in Graph (k) is employed to predict a subject's Y score from the subject's X score, all subjects are predicted to have the same score. If, in fact, all subjects have the same Y score, there is no need to employ the regression line to make a prediction.

Graph (l) illustrates a regression line that is parallel to the Y-axis. The configuration of the data upon which the latter graph is based indicates that, although there is variability with respect to subjects' scores on the Y variable, there is no variability with respect to their scores on the X variable — i.e., all of the subjects obtain the identical score on the X variable. As a result of the latter, the computed value for the estimated population variance for the X variable will equal zero. When, in fact, the value of the variance for the X variable equals zero, the sum of squares of the X scores will equal zero (i.e., $SS_X = \Sigma X^2 - [(\Sigma X)^2/n] = 0$). The sum of squares of the X scores (which is the numerator of the equation for computing the estimated population variance of the X scores) is, as noted earlier, one of the elements which comprise the denominator of Equation 28.1. Consequently, if the sum of squares of the X scores equals zero, the latter equation becomes insoluble. Note that if the regression line depicted in Graph (l) is employed to predict a subject's X score from the subject's Y score, all subjects are predicted to have the same X score. If, in fact, all subjects have the same X score, there is no need to employ the regression line to make a prediction.

If both the X and Y variable have no variability (i.e., all subjects obtain the identical score on the X variable and all subjects obtain the identical score on the Y variable), the resulting graph will consist of a single point (which is the case for Graph (m)). Thus, the single point in Graph (m) indicates that each of the n subjects in a sample obtains identical scores on both the X and Y variables.

At this point in the discussion, the role of the slope of a regression line will be clarified. The slope of a line indicates the number of units the Y variable will change if the X variable is

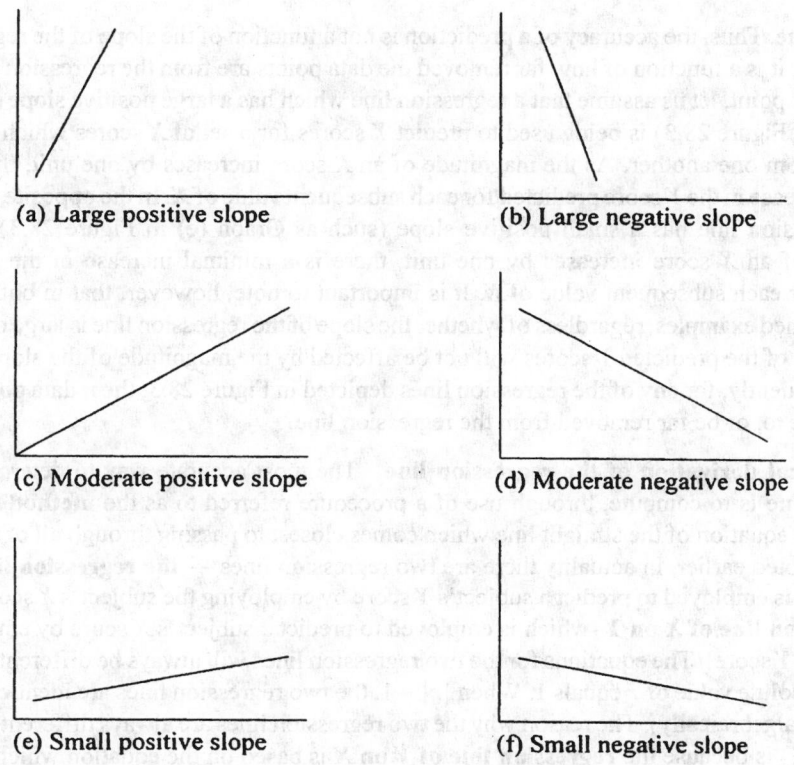

(a) Large positive slope   (b) Large negative slope

(c) Moderate positive slope   (d) Moderate negative slope

(e) Small positive slope   (f) Small negative slope

Figure 28.3 Hypothetical Regression Lines with Different Slopes

incremented by one unit. This definition for the slope is applicable to the **regression line of Y on X**. The slope of the **regression line of X on Y**, on the other hand, indicates the number of units the X variable will change if the Y variable is incremented by one unit. The discussion to follow will employ the definition of the slope in reference to the **regression line of Y on X**.

A line with a large positive slope or large negative slope is inclined upward (in one direction or the other) away from the X-axis — i.e., like a hill with a high grade. The more the magnitude of a positive slope or negative slope increases, the more the line approaches being parallel to the Y-axis. A line with a small positive slope or small negative slope has a minimal inclination in relation to the X-axis — i.e., like a hill with a low grade. The smaller the slope of a line, the more the line approaches being parallel to the X-axis. The graphs in Figure 28.3 reflect the following degrees of slope: Graphs (a) and (b) respectively depict lines with a large positive slope and a large negative slope; Graphs (c) and (d) respectively depict lines with a moderate positive slope and a moderate negative slope (i.e., the severity of the angle in relation to the X-axis is in between that of a line with a large slope and a small slope); Graphs (e) and (f) respectively depict lines with a small positive slope and a small negative slope.

It is important to keep in mind that although the slope of the **regression line of Y on X** plays a role in determining the specific value of Y which is predicted from the value of X, the magnitude of the slope is not related to the magnitude of the absolute value of the coefficient of correlation. A regression line with a large slope can be associated with a correlation coefficient that has a large, moderate, or small absolute value. In the same respect, a regression line with a small slope can be associated with a correlation coefficient which has a large, moderate, or small absolute value. Thus, the accuracy of a prediction is not a function of the slope of the regression line. Instead, it is a function of how far removed the data points are from the regression line. To illustrate this point, let us assume that a regression line which has a large positive slope (such as Graph (a) in Figure 28.3) is being used to predict Y scores for a set of X scores which are one unit apart from one another. As the magnitude of an X score increases by one unit, there is a sizeable increase in the Y score predicted for each subsequent value of X. In the opposite respect, if the regression line has a small positive slope (such as Graph (e) in Figure 28.3), as the magnitude of an X score increases by one unit, there is a minimal increase in the Y score predicted for each subsequent value of X. It is important to note, however, that in both of the aforementioned examples, regardless of whether the slope of the regression line is large or small, the accuracy of the predicted Y scores will not be affected by the magnitude of the slope of the line. Consequently, for any of the regression lines depicted in Figure 28.3, the n data points can fall on, close to, or be far removed from the regression line.

Mathematical derivation of the regression line   The most accurate way to determine the regression line is to compute, through use of a procedure referred to as the **method of least squares**, the equation of the straight line which comes closest to passing through all of the data points. As noted earlier, in actuality there are two regression lines — **the regression line of Y on X** (which is employed to predict a subject's Y score by employing the subject's X score), and **the regression line of X on Y** (which is employed to predict a subject's X score by employing the subject's Y score). The equations for the two regression lines will always be different, except when the absolute value of r equals 1. When $|r| = 1$, the two regression lines are identical (both visually and algebraically). The reason why the two regression lines are always different (except when $|r| = 1$) is because the **regression line of Y on X** is based on the equation which results in the **minimum squared distance** of all the data points from the line, when the distance of the points from the line is measured **vertically** (i.e., ↑ or ↓). On the other hand, the **regression line of X on Y** is based on the minimum squared distance of the data points from the line, when the distance of the points from the line is measured **horizontally** (i.e., → or ←).[7] Since when $|r| = 1$ all the points fall on the regression line, both the vertical and horizontal squared distance for each data point equals zero. Consequently, when $|r| = 1$ the two regression lines are identical.

The **regression line of Y on X** is determined with Equation 28.4.

$$Y' = a_Y + b_Y X \qquad \textbf{(Equation 28.4)}$$

Where: Y' represents the predicted Y score for a subject

X represents the subject's X score that is used to predict the value Y'

a_Y represents the Y intercept, which is the point at which the regression line crosses the Y-axis

b_Y represents the slope of the **regression line of Y on X**

In order to derive Equation 28.4, the values of b_Y (which is often referred to as the **regression coefficient**) and a_Y (which is often referred to as the **constant** in the regression equation) must be computed. Either Equation 28.5 or Equation 28.6 can be employed to compute the value b_Y. The latter equations are employed below to compute the value $b_Y = .30$.

$$b_Y = \frac{SP_{XY}}{SS_X} = \frac{\sum XY - \dfrac{(\sum X)(\sum Y)}{n}}{\sum X^2 - \dfrac{(\sum X)^2}{n}} = \frac{211 - \dfrac{(36)(17)}{5}}{554 - \dfrac{(36)^2}{5}} = .30 \quad \textbf{(Equation 28.5)}$$

$$b_Y = r\left(\frac{\tilde{s}_Y}{\tilde{s}_X}\right) = (.955)\left(\frac{2.70}{8.58}\right) = .30 \qquad \textbf{(Equation 28.6)}$$

Equation 28.7 is employed to compute the value a_Y. The latter equation is employed below to compute the value $a_Y = 1.24$.

$$a_Y = \overline{Y} - b_Y\overline{X} = 3.4 - (.30)(7.2) = 1.24 \qquad \textbf{(Equation 28.7)}$$

Substituting the values $a_Y = 1.24$ and $b_Y = .30$ in Equation 28.4, we determine that the equation for **regression line of Y on X** is $Y' = 1.24 + .3X$. Since two points can be used to construct a straight line, we can select two values for X and substitute each value in Equation 28.4, and solve for the values which would be predicted for Y'. Each set of values that is comprised of an X score and the resulting Y' value will represent one point on the regression line. Thus, if we plot any two points derived in this manner and connect them, the resulting line is the **regression line of Y on X**. To demonstrate this, if the value $X = 0$ is substituted in the regression equation, it yields the value $Y' = 1.24$ (which equals the value of a_Y): $Y' = 1.24 - (.30)(0) = 1.24$. Thus, the first point which will be employed in constructing the regression line is $(0, 1.24)$. If we next substitute the value $X = 5$ in the regression equation, it yields the value $Y' = 2.74$: $Y' = 1.24 + (.30)(5) = 2.74$. Thus, the second point to be used in constructing the regression line is $(5, 2.74)$. The regression line that results from connecting the points $(0, 1.24)$ and $(5, 2.74)$ is displayed in Figure 28.4.

If the researcher wants to predict a subject's score on the Y variable by employing the subject's score on the X variable, the predicted value Y' can be derived either from the regression equation or from Figure 28.4. If the regression equation is employed, the value Y' is derived by substituting a subject's X score in the equation (which is the same procedure that is employed to determine the two points which are used to construct the regression line). Thus, if a child consumes ten ounces of sugar per week, employing $X = 10$ in the regression equation, the predicted number of cavities for the child is $Y' = 1.24 + (.30)(10) = 4.24 = 4.24$.

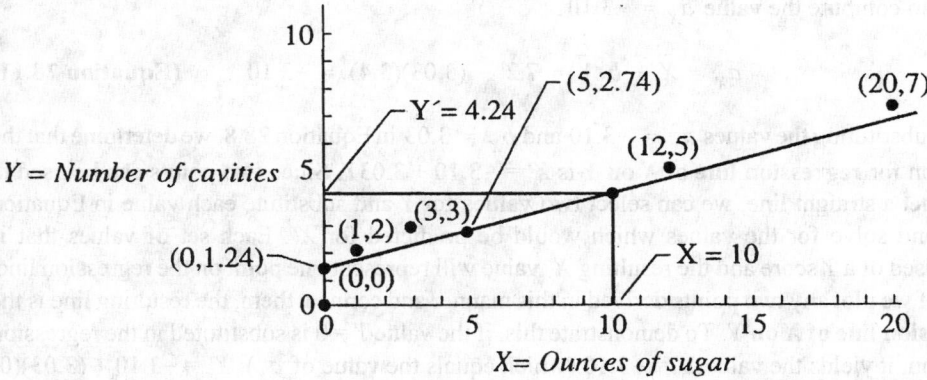

Figure 28.4   Regression Line of Y on X for Example 28.1

In using Figure 28.4 to predict the value of Y', we identify the point on the X-axis which corresponds to the subject's score on the X variable. A perpendicular line is erected from that point until it intersects the regression line. At the point the perpendicular line intersects the regression line, a second perpendicular line is dropped to the Y-axis. The point at which the latter perpendicular line intersects the Y-axis corresponds to the predicted value Y'. This procedure, which is illustrated in Figure 28.4, yields the same value, $Y' = 4.24$, that is obtained when the regression equation is employed.

The **regression line of X on Y** is determined with Equation 28.8.

$$X' = a_X + b_X Y \qquad \text{(Equation 28.8)}$$

Where: X' represents the predicted X score for a subject

Y represents the subject's Y score which is used to predict the value X'

a_X represents the X intercept, which is the point at which the regression line crosses the X-axis

b_X represents the slope of the **regression line of X on Y**

In order to derive Equation 28.8, the values of b_X (which is often referred to as the **regression coefficient**) and a_X (which is often referred to as the **constant** in the regression equation) must be computed. Either Equation 28.9 or Equation 28.10 can be employed to compute the value b_X. The latter equations are employed below to compute the value $b_X = 3.03$.

$$b_X = \frac{SP_{XY}}{SS_Y} = \frac{\sum XY - \dfrac{(\sum X)(\sum Y)}{n}}{\sum Y^2 - \dfrac{(\sum Y)^2}{n}} = \frac{211 - \dfrac{(36)(17)}{5}}{87 - \dfrac{(17)^2}{5}} = 3.03 \quad \text{(Equation 28.9)}$$

$$b_X = r\left(\frac{\tilde{s}_X}{\tilde{s}_Y}\right) = (.955)\left(\frac{8.58}{2.70}\right) = 3.03 \qquad \text{(Equation 28.10)}$$

Equation 28.11 is employed to compute the value a_X. The latter equation is employed below to compute the value $a_X = -3.10$.

$$a_X = \overline{X} - b_X \overline{Y} = 7.2 - (3.03)(3.4) = -3.10 \qquad \text{(Equation 28.11)}$$

Substituting the values $a_X = -3.10$ and $b_X = 3.03$ in Equation 28.8, we determine that the equation for **regression line of X on Y** is $X' = -3.10 + 3.03Y$. Since two points can be used to construct a straight line, we can select two values for Y and substitute each value in Equation 28.8, and solve for the values which would be predicted for X'. Each set of values that is comprised of a Y score and the resulting X' value will represent one point on the regression line. Thus, if we plot any two points derived in this manner and connect them, the resulting line is the **regression line of X on Y**. To demonstrate this, if the value $Y = 0$ is substituted in the regression equation, it yields the value $X' = -3.10$ (which equals the value of a_X): $X' = -3.10 + (3.03)(0)$ $= -3.10$. Thus, the first point which will be employed in constructing the regression line is $(-3.10, 0)$. If we next substitute the value $Y = 5$ in the regression equation, it yields the value X' $= 12.05$: $X' = -3.10 + (3.03)(5) = 12.05$. Thus, the second point to be used in constructing the regression line is $(12.05, 5)$. The regression line that results from connecting the points $(-3.10, 0)$ and $(12.05, 5)$ is displayed in Figure 28.5. Note that since the value $Y = 0$ results in a negative

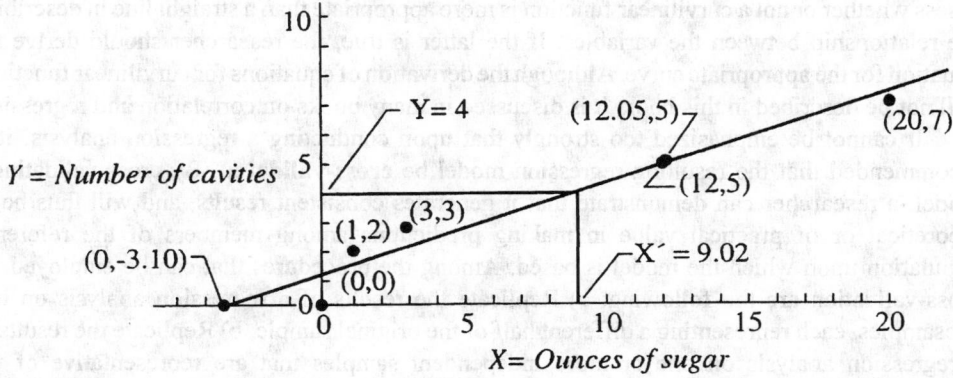

Figure 28.5 Regression Line of *X* on *Y* for Example 28.1

X value, the *X*-axis in Figure 28.5 must be extended to the left of the origin in order to accommodate the value $X = -3.10$.

If the researcher wants to predict a subject's score on the *X* variable by employing the subject's score on the *Y* variable, the predicted value X' can be derived either from the regression equation or from Figure 28.5. If the regression equation is employed, the value X' is derived by substituting a subject's *Y* score in the equation (which is the same procedure that is employed to determine the two points which are used to construct the regression line). Thus, if a child has four cavities, employing $Y = 4$ in the regression equation, the predicted number of ounces of sugar the child eats per week is $X' = -3.10 + (3.03)(4) = 9.02$.

In using Figure 28.5 to predict the value of X', we identify the point on the *Y*-axis which corresponds to the subject's score on the *Y* variable. A perpendicular line is erected from that point until it intersects the regression line. At the point the perpendicular line intersects the regression line, a second perpendicular line is dropped to the *X*-axis. The point at which the latter perpendicular line intersects the *X*-axis corresponds to the predicted value X'. This procedure, which is illustrated in Figure 28.5, yields the same value, $X' = 9.02$, that is obtained when the regression equation is employed.

The protocol described in this section for deriving a regression equation does not provide any information regarding the accuracy of prediction that will result from such an equation.[8] The **standard error of estimate**, which is discussed in the next section, is used as an index of accuracy in regression analysis. The standard error of estimate is a function of a set of *n* deviation scores that are referred to as **residuals**. A **residual** is the difference between the predicted value of the criterion variable for a subject (i.e., Y' or X'), and a subject's actual score on the criterion variable (i.e., *Y* or *X*). (A discussion of the role of residuals in regression analysis can be found in Section VII.)

In closing the discussion of the derivation of a regression line, it is important to emphasize that in some instances where the value of *r* is equal to or close to zero, there may actually be a curvilinear relationship between the two variables (which, if present, is most likely to be detected if scores on the predictor variable represent the full range of possible values). When the absolute value of *r* is such that there is a weak to moderate relationship between the variables, if in fact a curvilinear function best describes the relationship between the variables, it will provide a more accurate basis for prediction than will the straight line derived through use of the method of least squares. One advantage of constructing a scatterplot is that it allows a researcher to visually assess whether or not a curvilinear function is more appropriate than a straight line in describing the relationship between the variables. If the latter is true, the researcher should derive the equation for the appropriate curve. Although the derivation of equations for curvilinear functions will not be described in this book, it is discussed in many books on correlation and regression.

It cannot be emphasized too strongly that upon conducting a regression analysis, it is recommended that the resulting regression model be **cross-validated**. By cross-validating a model, a researcher can demonstrate that it generates consistent results, and will thus be of theoretical or of practical value in making predictions among members of the reference population upon which the model is based. Among the procedures that can be employed for cross-validation are the following: a) Replicate the results of a regression analysis on two subsamples, each representing a different half of the original sample; b) Replicate the results of a regression analysis on one or more independent samples that are representative of the population to which one wishes to apply the regression model; and c) A third method of cross-validation is to initially conduct a regression analysis on a sample of subjects. A second sample is then obtained from the same population, and the regression equation derived for the first sample is employed to predict the scores of subjects on the criterion variable in the second sample. The latter set of predicted scores are then correlated with the actual scores of the subjects on the criterion variable in the second sample. The obtained correlation coefficient will reflect the accuracy of prediction when the original regression equation is employed to predict scores on the criterion variable for subjects in the second sample. As Licht (1995, p. 23) notes, this method of cross-validation is not the same as the previously described methods. Whereas in the latter methods the results of a second regression analysis are compared with an initial regression analysis, in this type of cross-validation the results of the initial analysis are evaluated by employing them with data obtained from a second sample.

2. The standard error of estimate   The **standard error of estimate** is a standard deviation of the distribution of error scores employed in regression analysis. More specifically, it is an index of the difference between the predicted versus the actual value of the criterion variable. The standard error of estimate for the **regression line of Y on X** (which is represented by the notation $s_{Y.X}$) represents the standard deviation of the values of Y for a specific value of X. The standard error of estimate for the **regression line of X on Y** (which is represented by the notation $s_{X.Y}$) represents the standard deviation of the values of X for a specific value of Y. Thus, in Example 28.1, $s_{Y.X}$ represents the standard deviation for the number of cavities of any subject whose weekly sugar consumption is equal to a specific number of ounces. $s_{X.Y}$, on the other hand, represents the standard deviation for the number of ounces of sugar consumed by any subject who has a specific number of cavities.

The standard error of estimate can be employed to compute a confidence interval for the predicted value of Y (or X). The larger the value of a standard error of estimate, the larger will be the range of values which define the confidence interval and, consequently, the less likely it is that the predicted value Y' (or X') will equal or be close to the actual score of a given subject on that variable.

Equations 28.12 and 28.13 are, respectively, employed to compute the values $s_{Y.X}$ and $s_{X.Y}$ (which are estimates of the underlying population parameters $\sigma_{Y.X}$ and $\sigma_{X.Y}$).

$$s_{Y.X} = \tilde{s}_Y \sqrt{\left[\frac{n-1}{n-2}\right][1 - r^2]} \qquad\qquad \textbf{(Equation 28.12)}$$

$$s_{X.Y} = \tilde{s}_X \sqrt{\left[\frac{n-1}{n-2}\right][1 - r^2]} \qquad \textbf{(Equation 28.13)}$$

As the size of the sample increases, the value $(n - 1)/(n - 2)$ in the radical of Equations 28.12 and 28.13 approaches 1, and thus for large sample sizes the equations simplify to $s_{Y.X} = \tilde{s}_Y \sqrt{1 - r^2}$ and $s_{X.Y} = \tilde{s}_X \sqrt{1 - r^2}$. Note, however, that for small sample sizes the latter equations underestimate the values of $s_{Y.X}$ and $s_{X.Y}$.

Equations 28.12 and 28.13 are employed to compute the values $s_{Y.X} = .92$ and $s_{X.Y} = 2.94$.[9]

$$s_{Y.X} = 2.70 \sqrt{\left[\frac{5-1}{5-2}\right][1 - (.955)^2]} = .92$$

$$s_{X.Y} = 8.58 \sqrt{\left[\frac{5-1}{5-2}\right][1 - (.955)^2]} = 2.94$$

3. Computation of a confidence interval for the value of the criterion variable[10] It turns out that Equations 28.12 and 28.13 are not unbiased estimates of error throughout the full range of values the criterion variable may assume. What this translates into is that if a researcher wants to compute a confidence interval with respect to a specific subject's score on the criterion variable, in the interest of complete accuracy an adjusted standard error of estimate value should be employed. The adjusted standard error of estimate values will be designated $\tilde{s}_{Y.X}$ and $\tilde{s}_{X.Y}$. The values computed for $\tilde{s}_{Y.X}$ and $\tilde{s}_{X.Y}$ will always be larger than the values computed for $s_{Y.X}$ and $s_{X.Y}$. The larger the deviation between a subject's score on the predictor variable and the mean score for the predictor variable, the greater the difference between the values $s_{Y.X}$ versus $\tilde{s}_{Y.X}$ and $s_{X.Y}$ versus $\tilde{s}_{X.Y}$. Equations 28.14 and 28.15 are employed to compute the values $\tilde{s}_{Y.X}$ and $\tilde{s}_{X.Y}$.[11] In the latter equations, the values X and Y, respectively, represent the X and Y scores of the specific subject for whom the standard error of estimate is computed.

$$\tilde{s}_{Y.X} = s_{Y.X} \sqrt{1 + \frac{1}{n} + \frac{(X - \bar{X})^2}{SS_X}} \qquad \textbf{(Equation 28.14)}$$

$$\tilde{s}_{X.Y} = s_{X.Y} \sqrt{1 + \frac{1}{n} + \frac{(Y - \bar{Y})^2}{SS_Y}} \qquad \textbf{(Equation 28.15)}$$

At this point two confidence intervals will be computed employing the values $\tilde{s}_{Y.X}$ and $\tilde{s}_{X.Y}$. The two confidence intervals will be in reference to the two subjects for whom the values $Y' = 4.24$ and $X' = 9.02$ are predicted in the previous section (employing Equations 28.4 and 28.8). Initially, the use of Equation 28.14 will be demonstrated to compute a confidence interval for the subject who consumes 10 ounces of sugar (i.e., $X = 10$), and (through use of Equation 28.4) is predicted to have $Y' = 4.24$ cavities. Equation 28.16 is employed to compute a confidence interval for the predicted value of Y.

$$CI_{(1 - \alpha)} = Y' \pm (t_{\alpha/2})(\tilde{s}_{Y.X}) \qquad \textbf{(Equation 28.16)}$$

Where: $t_{\alpha/2}$ represents the tabled critical two-tailed value in the t distribution, for $df = n - 2$, below which a proportion (percentage) equal to $[1 - (\alpha/2)]$ of the cases falls. If the proportion (percentage) of the distribution that falls within the confidence interval is subtracted from 1 (100%), it will equal the value of α.

In the computation of a confidence interval, the predicted value Y' can be conceptualized as the mean value in a population of scores on the Y variable for a specific subject. When the sample size employed for the analysis is large (i.e., $n > 100$), one can assume that the shape of such a distribution for each subject will be normal, and in such a case the relevant tabled critical two-tailed z value (i.e., $z_{\alpha/2}$) can be employed in Equation 28.16 in place of the relevant tabled critical t value. For smaller sample sizes (as is the case for Example 28.1), however, the t distribution provides a more accurate approximation of the underlying population distribution. Use of the normal distribution with small sample sizes underestimates the range of values which define a confidence interval. Inspection of Equation 28.16 reveals that the range of values computed for a confidence interval is a function of the magnitude of the standard error of estimate and the tabled critical t value (the magnitude of the latter being inversely related to the sample size).

In order to use Equation 28.16, the value $\tilde{s}_{Y.X}$ must be computed with Equation 28.14. Employing Equation 28.14, the value $\tilde{s}_{Y.X} = 1.02$ is computed. Note that the latter value is slightly larger than the value $s_{Y.X} = .92$ computed with Equation 28.12.

$$\tilde{s}_{Y.X} = .92 \sqrt{1 + \frac{1}{5} + \frac{(10 - 7.2)^2}{294.8}} = 1.02$$

To demonstrate the use of Equation 28.16, the 95% confidence interval will be computed. The value $t_{.05} = 3.18$ is employed to represent $t_{\alpha/2}$, since in **Table A2** it is the tabled critical two-tailed .05 t value for $df = 3$. The appropriate values are now substituted in Equation 28.16 to compute the 95% confidence interval.

$$CI_{.95} = 4.24 \pm (3.18)(1.02) = 4.24 \pm 3.24$$

This result indicates the researcher can be 95% confident (or the probability is .95) that the interval 1.00 to 7.48 contains the actual number of cavities for the subject (i.e., $1.00 \leq Y \leq 7.48$).

A confidence interval will now be computed for the subject who has 4 cavities (i.e., $Y = 4$), and (through use of Equation 28.8) is predicted to eat 9.02 ounces of sugar. Equation 28.17 is employed to compute a confidence interval for the value of X.

$$CI_{(1 - \alpha)} = X' \pm (t_{\alpha/2})(\tilde{s}_{X.Y}) \qquad \textbf{(Equation 28.17)}$$

In order to use Equation 28.17 the value $\tilde{s}_{X.Y}$ must be computed with Equation 28.15. Employing Equation 28.15, the value $\tilde{s}_{X.Y} = 3.24$ is computed. Note that the latter value is slightly larger than the value $s_{X.Y} = 2.94$ computed with Equation 28.13.

$$\tilde{s}_{X.Y} = 2.94 \sqrt{1 + \frac{1}{5} + \frac{(4 - 3.4)^2}{29.2}} = 3.24$$

As is done in the previous example, the 95% confidence interval will be computed. Thus, the values $t_{.05} = 3.18$ and $\tilde{s}_{X.Y} = 3.24$ are substituted in Equation 28.17.

$$CI_{.95} = 9.02 \pm (3.18)(3.24) = 9.02 \pm 10.30$$

This result indicates the researcher can be 95% confident (or the probability is .95) that the interval -1.28 to 19.32 contains the number of ounces of sugar the subject actually eats (i.e., $-1.28 \le X \le 19.32$). Since it is impossible to have a negative number of ounces of sugar, the result translates into between 0 and 19.32 ounces of sugar (i.e., $0 \le X \le 19.32$).

4. Computation of a confidence interval for a Pearson product-moment correlation coefficient In order to compute a confidence interval for a computed value of the **Pearson product-moment correlation coefficient**, it is necessary to employ a procedure developed by Fisher (1921) referred to as **Fisher's z_r (or z) transformation**. The latter procedure transforms an r value to a scale which is based on the normal distribution. The rationale behind the use of **Fisher's z_r transformation** is that although the theoretical sampling distribution of the correlation coefficient can be approximated by the normal distribution when the value of a population correlation is equal to zero, as the value of the population correlation deviates from zero, the sampling distribution becomes more and more skewed. Thus, in computing confidence intervals (as well as in testing hypotheses involving one or more populations in which a hypothesized population correlation is some value other than zero), **Fisher's z_r transformation** is required to transform a skewed sampling distribution into a normalized format.

Equation 28.18 is employed to convert an r value into a **Fisher transformed value**, which is represented by the notation z_r.

$$z_r = \frac{1}{2} \ln \left[\frac{1 + r}{1 - r} \right] \qquad \textbf{(Equation 28.18)}$$

Where: ln represents the **natural logarithm** of a number (which is defined in Endnote 14 in the **Introduction**)

Although logarithmic values can be computed with a function key on most scientific calculators, if one does not have access to a calculator, **Table A17 (Table of Fisher's z_r Transformation)** in the **Appendix** provides an alternative way of deriving the Fisher transformed values. The latter table contains the z_r values which correspond to specific values of r. The reader should take note of the fact that in employing Equation 28.18 or **Table A17**, the sign assigned to a z_r value is always the same as the sign of the r value upon which it is based. Thus, a positive r value will always be associated with a positive z_r value, and a negative r value will always be associated with a negative z_r value. When $r = 0$, z_r will also equal zero.

Equation 28.19 is employed to compute the confidence interval for a computed r value.

$$CI_{z_{r(1-\alpha)}} = z_r \pm (z_{\alpha/2}) \sqrt{\frac{1}{n - 3}} \qquad \textbf{(Equation 28.19)}$$

Where: $z_{\alpha/2}$ represents the tabled critical two-tailed value in the normal distribution below which a proportion (percentage) equal to $[1 - (\alpha/2)]$ of the cases fall. If the proportion (percentage) of the distribution that falls within the confidence interval is subtracted from 1 (100%), it will equal the value of α.

The value $\sqrt{1/(n-3)}$ in Equation 28.19 represents the standard error of z_r. In employing Equation 28.19 to compute the 95% confidence interval, the product of the tabled critical two-tailed .05 z value and the standard error of z_r are added to and subtracted from the Fisher transformed value for the computed r value. The two resulting values, which represent z_r values, are then reconverted into correlation coefficients through use of **Table A17** or by reconfiguring Equation 28.18 to solve for r.[12] Use of **Table A17** for the latter is accomplished by identifying the r values which correspond to the computed z_r values. The resulting r values derived from the table identify the limits that define the 95% confidence interval.

Equation 28.19 will now be used to compute the 95% confidence interval for $r = .955$. From **Table A17** it is determined that the Fisher transformed value which corresponds to $r = .955$ is $z_r = 1.886$. The latter value can also be computed with Equation 28.18: $z_r = (1/2)\ln[(1 + .955)/(1 - .955)] = 1.886$. The appropriate values are now substituted in Equation 28.19.

$$CI_{z_{r(.95)}} = 1.886 \pm (1.96)\sqrt{\frac{1}{5-3}} = 1.886 \pm 1.386$$

Subtracting from and adding 1.386 to 1.886, yields the values .5 and 3.272. The latter values are now converted into r values through use of **Table A17**. By interpolating, we can determine that a z_r value of .5 corresponds to the value $r = .462$, which will define the lower limit of the confidence interval. Since the value $z_r = 3.272$ is substantially above the z value which corresponds to the largest tabled r value, it will be associated with the value $r = 1$. Thus, we can be 95% confident (or the probability is .95) the interval .462 to 1 contains the true value of the population correlation. Symbolically, this can be written as follows: $.462 \le \rho \le 1$. Note that because of the small sample size employed in the experiment, the range of values which define the confidence interval is quite large.

If the 99% confidence interval is computed, the tabled critical two-tailed .01 value $z_{.01} = 2.58$ is employed in Equation 28.19 in place of $z_{.05} = 1.96$. As is always the case in computing a confidence interval, the range of values that defines a 99% confidence interval will be larger than the range which defines a 95% confidence interval.

It should be noted that sources are not in agreement with respect to the best method to employ in computing a population confidence interval. Alternative methodologies are discussed in Grissom and Kim (2005, pp. 72), Hunter and Schmidt (2004), Smithson (2003), and Wilcox (2003).

5. Test 28b: Test for evaluating the hypothesis that the true population correlation is a specific value other than zero In certain instances, a researcher may want to evaluate whether an obtained correlation could have come from a population in which the true correlation between two variables is a specific value other than zero. The null and alternative hypotheses that are evaluated under such conditions are as follows.

$$H_0: \rho = \rho_0$$

(In the underlying population the sample represents, the correlation between the scores of subjects on Variable X and Variable Y equals ρ_0.)

$$H_1: \rho \ne \rho_0$$

(In the underlying population the sample represents, the correlation between the scores of subjects on Variable X and Variable Y equals some value other than ρ_0. The alternative hypothesis as stated is **nondirectional**, and is evaluated with a **two-tailed test**. It is also possible to state the alternative hypothesis directionally (H_1: $\rho > \rho_0$ or H_1: $\rho < \rho_0$), in which case it is evaluated with a one-tailed test.)

Equation 28.20 is employed to evaluate the null hypothesis H_0: $\rho = \rho_0$.

$$z = \frac{z_r - z_{\rho_0}}{\sqrt{\dfrac{1}{(n-3)}}} \qquad \textbf{(Equation 28.20)}$$

Where: z_r represents the Fisher transformed value of the computed value of r

z_{ρ_0} represents the Fisher transformed value of ρ_0, the hypothesized population correlation

Equation 28.20 will now be employed in reference to Example 28.1. Let us assume that we want to evaluate whether the true population correlation between the number of ounces of sugar consumed and the number of cavities is .80. Thus, the null hypothesis is H_0: $\rho = .80$, and the nondirectional alternative hypothesis is H_1: $\rho \neq .80$.

By employing **Table A17** (or Equation 28.18), we determine that the corresponding z_r values for the obtained correlation coefficient $r = .955$ and the hypothesized population correlation coefficient $\rho = .80$ are, respectively, $z_r = 1.886$ and $z_\rho = 1.099$ (the notation z_ρ is employed in place of z_r whenever the relevant element in an equation identifies a population correlation). Substituting the Fisher transformed values in Equation 28.20, the value $z = 1.11$ is computed.[13]

$$z = \frac{1.886 - 1.099}{\sqrt{\dfrac{1}{(5-3)}}} = 1.11$$

The computed value $z = 1.11$ is evaluated with **Table A1 (Table of the Normal Distribution)** in the **Appendix**. In order to reject the null hypothesis, the obtained absolute value of z must be equal to or greater than the tabled critical two-tailed value at the prespecified level of significance. Since $z = 1.11$ is less than the tabled critical two-tailed values $z_{.05} = 1.96$ and $z_{.01} = 2.58$, the null hypothesis cannot be rejected at either the .05 or .01 level. Thus, the null hypothesis that the true population correlation equals .80 is retained.

If the alternative hypothesis is stated directionally, in order to reject the null hypothesis the obtained absolute value of z must be equal to or greater than the tabled critical one-tailed value at the prespecified level of significance (i.e., $z_{.05} = 1.65$ or $z_{.01} = 2.33$). Since $z = 1.11$ is less than $z_{.05} = 1.65$, the directional alternative hypothesis H_0: $\rho > .80$ is not supported. Note that the sign of the value of z computed with Equation 28.20 will be positive when the computed value of r is greater than the hypothesized value ρ_0, and negative when the computed value of r is less than the hypothesized value ρ_0. Since $r = .955$ is a positive number, the directional alternative hypothesis H_0: $\rho < .80$ is inconsistent with the data, and is thus not supported.

6. Computation of power for the Pearson product-moment correlation coefficient Prior to collecting correlational data, a researcher can determine the likelihood of detecting a population correlation of a specific magnitude if a specific value of n is employed. As a result of such a power analysis, one can determine the minimum sample size required to detect a prespecified population correlation. To illustrate the computation of power, let us assume that prior to collecting the data for 25 subjects a researcher wants to determine the power associated with the analysis if the value of the population correlation he wants to detect is $\rho = .40$.[14] It will be assumed that a nondirectional analysis is conducted, with $\alpha = .05$.

Equation 28.21 (which is described in Guenther (1965, pp. 244–246)) is employed to compute the power of the analysis.

$$\varphi = |z_{\rho_0} - z_{\rho_1}| \sqrt{n - 3}$$ **(Equation 28.21)**

Where: z_{ρ_0} is the **Fisher transformed value** of the population correlation stipulated in the null hypothesis, and z_{ρ_1} is the **Fisher transformed value** of the population correlation the researcher wants to detect

Table A17 in the **Appendix** reveals that the Fisher transformed value associated with $\rho = 0$ is $z_\rho = z_{\rho_0} = 0$, and thus when the null hypothesis H_0: $\rho = 0$ is employed (which it will be assumed is the case), Equation 28.21 reduces to $\varphi = z_{\rho_1} \sqrt{n - 3}$. Employing **Table A17**, we determine that the Fisher transformed value for the population correlation of $\rho = .40$ is $z_\rho = .424$. Substituting the appropriate values in Equation 28.21, the value $\varphi = 1.99$ is computed.

$$\varphi = |0 - .424| \sqrt{25 - 3} = 1.99$$

The obtained value $\varphi = 1.99$ is evaluated with **Table A3 (Power Curves for Student's t Distribution)** in the **Appendix**. A full discussion on the use of **Table A3** (which is employed to evaluate the power of a number of different types of t **tests**) can be found in Section VI of the **single-sample t test** (**Test 2**). Employing the power curve for $df = \infty$ in **Table A3-C** (the appropriate table for a nondirectional/two-tailed analysis, with $\alpha = .05$), we determine the power of the correlational analysis to be approximately .52. (For the analysis described in this section the $df = \infty$ curve is employed for the relevant set of power curves since **Fisher's z_r transformation** is based on the normal distribution.) Thus, if the underlying population correlation is $\rho = .40$ and a sample size of $n = 25$ is employed, the likelihood of the researcher rejecting the null hypothesis is only .52. If this value is deemed too small, the researcher can substitute larger values of n in Equation 28.21 until a value is computed for φ which is associated with an acceptable level of power.

Equation 28.21 can also be employed if the value stated in the null hypothesis is some value other than $\rho = 0$. Assume that a number of studies suggest the population correlation between two variables is $\rho = .60$. A researcher, who has reason to believe that the latter value may over-estimate the true population correlation, wants to compute the power of a correlational analysis to determine if the true population correlation is, in fact, $\rho = .40$. In this example (for which the value $n = 25$ will be employed) H_0: $\rho = .60$. Since the researcher believes the true population correlation may be less than .60, the alternative hypothesis is stated directionally. Thus, H_1: $\rho < .60$.

Employing **Table A17**, we determine the Fisher transformed value for $\rho = 60$ is $z_\rho = .693$ and, from the previous analysis, we know that for $\rho = .40$, $z_\rho = .424$. Substituting the appropriate values in Equation 28.21, the value $\varphi = 1.26$ is computed.

$$\varphi = |.693 - .424| \sqrt{25 - 3} = 1.26$$

Employing the power curve for $df = \infty$ in **Table A3-D** (i.e., the curves for the one-tailed .05 value), we determine the power of the analysis to be approximately .37. Thus, if the underlying population correlation is $\rho = .40$ and a sample size of $n = 25$ is employed, the likelihood of the researcher rejecting the null hypothesis H_0: $\rho = .60$ is only .37.

Cohen (1977; 1988, Chapter 3) has derived tables that allow a researcher to determine the appropriate sample size to employ if one wants to evaluate an alternative hypothesis which designates a specific value for a population correlation (when the null hypothesis is H_0: $\rho = 0$). These tables can be employed as an alternative to the procedure described in this section in computing power for the **Pearson product-moment correlation coefficient**.[15]

7. Test 28c: Test for evaluating a hypothesis on whether there is a significant difference between two independent correlations There are occasions when a researcher will compute a correlation between the same two variables for two independent samples. In the event the correlation coefficients obtained for the two samples are not equal, the researcher may wish to determine whether the difference between the two correlations is statistically significant. The null and alternative hypotheses that are evaluated under such conditions are as follows.

$$H_0: \rho_1 = \rho_2$$

(In the underlying populations represented by the two samples, the correlation between the two variables is equal.)

$$H_1: \rho_1 \neq \rho_2$$

(In the underlying populations represented by the two samples, the correlation between the two variables is not equal. The alternative hypothesis as stated is **nondirectional**, and is evaluated with a **two-tailed test**. It is also possible to state the alternative hypothesis directionally (H_1: $\rho_1 > \rho_2$ or H_1: $\rho_1 < \rho_2$), in which case it is evaluated with a one-tailed test.)

To illustrate, let us assume that in Example 28.1 the correlation of $r = .955$ between the number of ounces of sugar eaten per week and the number of cavities is based on a sample of five ten-year-old boys (to be designated Sample 1). Let us also assume that the researcher evaluates a sample of five ten-year-old girls (to be designated Sample 2), and determines that in this second sample the correlation between the number of ounces of sugar eaten per week and the number of cavities is $r = .765$. Equation 28.22 can be employed to determine whether the difference between $r_1 = .955$ and $r_2 = .765$ is significant.

$$z = \frac{z_{r_1} - z_{r_2}}{\sqrt{\dfrac{1}{n_1 - 3} + \dfrac{1}{n_2 - 3}}}$$ **(Equation 28.22)**

Where: z_{r_1} represents the Fisher transformed value of the computed value of r_1 for Sample 1

z_{r_2} represents the Fisher transformed value of the computed value of r_2 for Sample 2

n_1 and n_2 are, respectively, the number of subjects in Sample 1 and Sample 2

Since there are five subjects in both samples, $n_1 = n_2 = 5$. From the analysis in the previous section we already know that the Fisher transformed value of $r_1 = .955$ is

z_{r_1} = 1.886. For the female sample, employing **Table A17** we determine that the Fisher transformed value of r_2 = .765 is z_{r_2} = 1.008. When the appropriate values are substituted in Equation 28.22, they yield the value z = .878.

$$z = \frac{1.886 - 1.008}{\sqrt{\dfrac{1}{5-3} + \dfrac{1}{5-3}}} = .878$$

The value z = .878 is evaluated with **Table A1**. In order to reject the null hypothesis, the obtained absolute value of z must be equal to or greater than the tabled critical two-tailed value at the prespecified level of significance. Since z = .878 is less than the tabled critical two-tailed values $z_{.05}$ = 1.96 and $z_{.01}$ = 2.58, the nondirectional alternative hypothesis H_1: $\rho_1 \neq \rho_2$ is not supported at either the .05 or .01 level. Thus, we retain the null hypothesis that there is an equal correlation between the two variables in each of the populations represented by the samples.

If the alternative hypothesis is stated directionally, in order to reject the null hypothesis the obtained absolute value of z must be equal to or greater than the tabled critical one-tailed value at the prespecified level of significance (i.e., $z_{.05}$ = 1.65 or $z_{.01}$ = 2.33). The sign of z will be positive when $r_1 > r_2$, and thus can only support the alternative hypothesis H_1: $\rho_1 > \rho_2$. The sign of z will be negative when $r_1 < r_2$, and thus can only support the alternative hypothesis H_1: $\rho_1 < \rho_2$. Since z = .878 is less than $z_{.05}$ = 1.65, the directional alternative hypothesis H_1: $\rho_1 > \rho_2$ is not supported.

Edwards (1984) notes that when the null hypothesis is retained, since the analysis suggests the two samples represent a single population, Equation 28.23 can be employed to provide a weighted estimate of the common population correlation.

$$\bar{z}_r = \frac{(n_1 - 3)z_{r_1} + (n_2 - 3)z_{r_2}}{(n_1 - 3) + (n_2 - 3)} \qquad \textbf{(Equation 28.23)}$$

Substituting the data in Equation 28.23, the Fisher transformed value $\bar{z}_r$ = 1.447 is computed.

$$\bar{z}_r = \frac{(5 - 3)(1.886) + (5 - 3)(1.008)}{(5 - 3) + (5 - 3)} = 1.447$$

Employing **Table A17**, we determine that the Fisher transformed value z_r = 1.447 corresponds to the value r = .895. Thus, r = .895 can be employed as the best estimate of the common population correlation. Note that the estimated common population correlation computed with Equation 28.23 is not the same value that is obtained if, instead, one calculates the weighted average of the two correlations (which, since the sample sizes are equal, is the average of the two correlations: $(.955 + .765)/2 = .86$). The fact that the weighted average of the two correlations yields a different value from the result obtained with Equation 28.23 can be attributed to the theoretical sampling distribution of the correlation coefficient becoming more skewed as the absolute value of r approaches 1.

Cohen (1977; 1988, Ch. 4) has developed a statistic referred to as the **q index** which can be employed for computing the power of the test comparing two independent correlation coefficients. A brief discussion of the *q* index can be found in Section IX (the **Addendum**) under the discussion of **meta-analysis and related topics**.

8. Test 28d: Test for evaluating a hypothesis on whether *k* independent correlations are homogeneous Test 28c can be extended to determine whether more than two independent correlation coefficients are homogeneous (in other words, can be viewed as representing the same population correlation, ρ). The null and alternative hypotheses that are evaluated under such conditions are as follows.

$$H_0: \rho_1 = \rho_2 = \cdots = \rho_k$$

(In the underlying populations represented by the *k* samples, the correlation between the two variables is equal.)

$$H_1: \text{Not } H_0$$

(In the underlying populations represented by the *k* samples, the correlation between the two variables is not equal in at least two of the populations. The alternative hypothesis as stated is **nondirectional**.)

To illustrate, let us assume that the correlation between the number of ounces of sugar eaten per week and the number of cavities is computed for three independent samples, each sample consisting of five children living in different parts of the country. The values of the correlations obtained for the three samples are as follows: $r_1 = .955$, $r_2 = .765$, $r_3 = .845$.

Equation 28.24 is employed to determine whether the $k = 3$ sample correlations are homogeneous. (Equation 28.91 in Section IX (the **Addendum**) is a different but equivalent version of Equation 28.24.) In Equation 28.24, wherever the summation sign $\sum_{j=1}^{k}$ appears it indicates that the operation following the summation sign is carried out for each of the $k = 3$ samples, and the resulting $k = 3$ values are summed.

$$\chi^2 = \sum_{j=1}^{k} [(n_j - 3) z_{r_j}^2] - \frac{\left[\sum_{j=1}^{k} (n_j - 3) z_{r_j} \right]^2}{\sum_{j=1}^{k} (n_j - 3)} \qquad \textbf{(Equation 28.24)}$$

Since there are five subjects in each sample, $n_1 = n_2 = n_3 = 5$. From the analysis in the previous section, we already know that the Fisher transformed values of $r_1 = .955$ and $r_2 = .765$ are, respectively, $z_{r_1} = 1.886$ and $z_{r_2} = 1.008$. Employing **Table A17**, we determine for Sample 3 the Fisher transformed value of $r_3 = .845$ is $z_{r_3} = 1.238$. When the appropriate values are substituted in Equation 28.24, they yield the value $\chi^2 = .83$.

$$\chi^2 = [(5 - 3)(1.886)^2 + (5 - 3)(1.008)^2 + (5 - 3)(1.238)^2]$$
$$- \frac{[(5 - 3)(1.886) + (5 - 3)(1.008) + (5 - 3)(1.238)]^2}{(5 - 3) + (5 - 3) + (5 - 3)} = .83$$

The value $\chi^2 = .83$ is evaluated with **Table A4 (Table of the Chi-Square Distribution)** in the **Appendix**. The degrees of freedom employed in evaluating the obtained chi-square value are $df = k - 1$. Thus, for the above example, $df = 3 - 1 = 2$. In order to reject the null hypothesis, the obtained value of χ^2 must be equal to or greater than the tabled critical value at the prespecified level of significance. Since $\chi^2 = .83$ is less than $\chi^2_{.05} = 5.99$ and $\chi^2_{.01} = 9.21$ (which are the tabled critical .05 and .01 values for $df = 2$ when a nondirectional alternative hypothesis is employed), the null hypothesis cannot be rejected at either the .05 or .01 level. Thus, we retain the null hypothesis that in the underlying populations represented by the $k = 3$ samples, the correlations between the two variables are equal.[16]

Edwards (1984) and Hedges and Olkin (1985) note that when the null hypothesis is retained, since the analysis suggests that the k samples represent a single population, Equation 28.25 can be employed to provide a weighted estimate of the common population correlation.[17]

$$\bar{z}_r = \frac{\sum_{j=1}^{k} (n_j - 3)z_{r_j}}{\sum_{j=1}^{k} (n_j - 3)} \qquad \textbf{(Equation 28.25)}$$

Substituting the data in Equation 28.25, the Fisher transformed value $\bar{z}_r = 1.377$ is computed.

$$\bar{z}_r = \frac{(5 - 3)(1.886) + (5 - 3)(1.008) + (5 - 3)(1.238)}{(5 - 3) + (5 - 3) + (5 - 3)} = 1.377$$

Employing **Table A17**, we determine that the Fisher transformed value $z_r = 1.377$ corresponds to the value $r = .88$. Thus, $r = .88$ can be employed as the best estimate of the common population correlation. Note that, as is the case when the same analysis is conducted for $k = 2$ samples, the value obtained for the common population correlation (using Equation 28.25) is not the same as the value that is obtained if the weighted average of the three correlation coefficients is computed (i.e., $\bar{r} = (.955 + .765 + .845)/3 = .855$).

9. Test 28e: Test for evaluating the null hypothesis H_0: $\rho_{XZ} = \rho_{YZ}$ There are instances when a researcher may want to evaluate if, within a specific population, one variable (X) has the same correlation with some criterion variable (Z) as does another variable (Y). The null and alternative hypotheses which are evaluated in such a situation are as follows.

$$H_0: \ \rho_{XZ} = \rho_{YZ}$$

(In the underlying population represented by the sample, the correlation between variables X and Z is equal to the correlation between variables Y and Z.)

$$H_1: \ \rho_{XZ} \neq \rho_{YZ}$$

(In the underlying population represented by the sample, the correlation between variables X and Z is not equal to the correlation between variables Y and Z. The alternative hypothesis as stated is **nondirectional**, and is evaluated with a **two-tailed test**. It is also possible to state the alternative hypothesis directionally (H_1: $\rho_{XZ} > \rho_{YZ}$ or H_1: $\rho_{XZ} < \rho_{YZ}$), in which case it is evaluated with a one-tailed test.)

To illustrate how one can evaluate the null hypothesis H_0: $\rho_{XZ} = \rho_{YZ}$, let us assume that the correlation between the number of ounces of sugar eaten per week and the number of cavities is computed for five subjects, and $r = .955$. Let us also assume that for the same five subjects we determine that the correlation between the number of ounces of salt eaten per week and the number of cavities is $r = .52$. We want to determine whether or not there is a significant difference in the correlation between the number of ounces of sugar eaten per week and the number of cavities versus the number of ounces of salt eaten per week and the number of cavities. Let us also assume that for the sample of five subjects, the correlation between the number of ounces of sugar eaten per week and the number of ounces of salt eaten per week is $r = .37$.

In the above example, within the framework of the hypothesis being evaluated we have two predictor variables — the number of ounces of sugar eaten per week and the number of ounces of salt eaten per week. These two predictor variables will respectively represent the X and Y variables in the analysis to be described. The number of cavities, which is the criterion variable, will be designated as the Z variable. Thus, $r_{XZ} = .955$, $r_{YZ} = .52$, $r_{XY} = .37$.

The test statistic for evaluating the null hypothesis, which is based on the t distribution, is computed with Equation 28.26. A more detailed description of the test statistic can be found in Steiger (1980), who notes that Equation 28.26 provides a superior test of the hypothesis being evaluated when compared with an alternative procedure developed by Hotelling (1940) (which is described in Lindeman *et al.* (1980)).

(Equation 28.26)

$$t = (r_{YZ} - r_{XZ}) \sqrt{\frac{(n-1)(1+r_{XY})}{2\left[\frac{(n-1)}{(n-3)}\right][1 - r_{YZ}^2 - r_{XZ}^2 - r_{XY}^2 + 2r_{YZ}r_{XZ}r_{XY}] + \left[\frac{r_{YZ}+r_{XZ}}{2}\right]^2 [1 - r_{XY}]^3}}$$

Substituting the values $n = 5$, $r_{XZ} = .955$, $r_{YZ} = .52$, and $r_{XY} = .37$ in Equation 28.26, the value $t = -1.78$ is computed.

$$t = (.52-.955)\sqrt{\frac{(5-1)(1+.37)}{2\frac{(5-1)}{(5-3)}[1-(.52)^2-(.955)^2-(.37)^2+2(.52)(.955)(.37)]+\left[\frac{(.52+.955)}{2}\right]^2[1-(.37)]^3}}$$

$$= -1.78$$

The value $t = -1.78$ is evaluated with **Table A2**. The degrees of freedom employed in evaluating the obtained t value are $df = n - 3$. Thus, for the above example, $df = 5 - 3 = 2 = 2$. In order to reject the null hypothesis, the obtained absolute value of t must be equal to or greater than the tabled critical value at the prespecified level of significance. Since the absolute value $t = 1.78$ is less than $t_{.05} = 4.30$ and $t_{.01} = 9.93$ (which are the tabled critical two-tailed values for $df = 2$), the null hypothesis cannot be rejected at either the .05 or .01 level. Thus, we retain the null hypothesis that the population correlation between variables X and Z is equal to the population correlation between variables Y and Z.

If the alternative hypothesis is stated directionally, in order to reject the null hypothesis, the obtained absolute value of t must be equal to or greater than the tabled critical one-tailed value at the prespecified level of significance (which for $df = 2$ is $t_{.05} = 2.92$ or $t_{.01} = 6.97$). The sign of t must be positive if $H_1: \rho_{XZ} < \rho_{YZ}$, and must be negative if $H_1: \rho_{XZ} > \rho_{YZ}$. Since the absolute value $t = 1.78$ is less than $t_{.05} = 2.92$, the directional alternative hypothesis

H_1: ρ_{XZ} > ρ_{YZ} (which is consistent with the data) is not supported. The directional alternative hypothesis H_1: ρ_{XZ} < ρ_{YZ} (which is not consistent with the data) is also not supported, since the sign of t is negative.

In the event the t value obtained with Equation 28.26 is significant, it indicates that the predictor variable which correlates highest with the criterion variable (i.e., the one with the highest absolute value) is the best predictor of subjects' scores on the latter variable. It should be noted that because the analysis discussed in this section represents a dependent samples analysis (since all three correlations are based on the same sample), Equation 28.22 (the equation for contrasting two independent correlations) is not appropriate to use to evaluate the null hypothesis H_0: ρ_{XZ} = ρ_{YZ}.

10. Tests for evaluating a hypothesis regarding one or more regression coefficients A number of tests have been developed which evaluate hypotheses concerning the slope of a regression line. (As noted earlier, the slope of a regression line is also referred to as a **regression coefficient**.) This section will present a brief description of such tests. In the statement of the null and alternative hypotheses of tests concerning a regression coefficient, the notation β is employed to represent the slope of the line in the underlying population represented by a sample. Thus, β_Y is the population regression coefficient of the **regression line of Y on X**, and β_X is the population regression coefficient of the **regression line of X on Y**.

Test 28f: Test for evaluating the null hypothesis H_0: β = 0 A test of significance can be conducted to evaluate the hypothesis of whether in the underlying population the value of the slope of a regression line is equal to zero. The null hypotheses which can be evaluated in reference to the two regression lines are H_0: β_Y = 0 and H_0: β_X = 0. In point of fact, the test of the generic null hypothesis H_0: β = 0 will always yield the same result as that obtained when the null hypothesis H_0: ρ = 0 is evaluated using **Test 28a** (which employs Equation 28.3). This is the case, since whenever, $\rho = 0$, the slope of a regression line in the underlying population will also equal zero. Equations 28.27 and 28.28 are respectively employed to evaluate the null hypotheses H_0: β_Y = 0 and H_0: β_X = 0. The equations are employed below with the data for Example 28.1 and, in both instances, yield the same t value as that obtained when the null hypothesis H_0: ρ = 0 is evaluated with Equation 28.3. The slight discrepancies between the t values computed with Equations 28.3, 28.27, and 28.28 are the result of rounding off error.

$$t = \frac{(b_Y)(\tilde{s}_X)\sqrt{n-1}}{s_{Y.X}} = \frac{(.30)(8.58)\sqrt{5-1}}{.92} = 5.60 \qquad \textbf{(Equation 28.27)}$$

$$t = \frac{(b_X)(\tilde{s}_Y)\sqrt{n-1}}{s_{X.Y}} = \frac{(3.03)(2.70)\sqrt{5-1}}{2.94} = 5.57 \qquad \textbf{(Equation 28.28)}$$

The t values computed with Equations 28.27 and 28.28 are evaluated with **Table A2**. The degrees of freedom employed in evaluating the obtained t values are $df = n - 2$. Since both t values are identical to the t value computed with Equation 28.3 and the same degrees of freedom are employed, interpretation of the t values leads to the same conclusions — except, in this case, the conclusions are in reference to the regression coefficients β_Y and β_X. Thus, the nondirectional alternative hypotheses H_1: β_Y ≠ 0 and H_1: β_X ≠ 0 are supported at the .05 level, since (for $df = 3$) the computed t values are greater than the tabled critical two-tailed value $t_{.05} = 3.18$.

The directional alternative hypotheses $H_1: \beta_Y > 0$ and $H_1: \beta_X > 0$ are supported at both the .05 and .01 levels, since the sign of t (as well as the sign of each of the regression coefficients) is positive, and the computed t values are greater than the tabled critical one-tailed values $t_{.05} = 2.35$ and $t_{.01} = 4.54$. (An alternative methodology for evaluating the null hypothesis $H_0: \beta = 0$, which yields a result equivalent to that obtained with Equation 28.27 or 28.28, can be found in the Section VII under the discussion of **residuals**.)

Equations 28.29 and 28.30 can respectively be employed to compute confidence intervals for the regression coefficients β_Y and β_X. The computation of the 95% confidence interval for the two regression coefficients is demonstrated below. The value $t_{.05} = 3.18$ (which is also employed in computing the confidence intervals derived with Equations 28.16 and 28.17) is employed in Equations 28.29 and 28.30 to represent $t_{\alpha/2}$, since in **Table A2** it is the tabled critical two-tailed .05 t value for $df = 3$ (which is computed with $df = n - 2$).

(Equation 28.29)

$$CI_{\beta_{Y_{(1 - \alpha)}}} = b_Y \pm (t_{\alpha/2}) \left[\frac{s_{Y.X}}{(\tilde{s}_X)\sqrt{n - 1}} \right] = .30 \pm (3.18) \left[\frac{.92}{8.58\sqrt{5 - 1}} \right] = .30 \pm .17$$

(Equation 28.30)

$$CI_{\beta_{X_{(1 - \alpha)}}} = b_X \pm (t_{\alpha/2}) \left[\frac{s_{X.Y}}{(\tilde{s}_Y)\sqrt{n - 1}} \right] = 3.03 \pm (3.18) \left[\frac{2.94}{2.70\sqrt{5 - 1}} \right] = 3.03 \pm 1.73$$

The above results indicate the following: a) There is a 95% likelihood the interval .13 to .47 contains the population regression coefficient β_Y (i.e., $.13 \leq \beta_Y \leq .47$); and b) There is a 95% likelihood the interval 1.30 to 4.76 contains the population regression coefficient β_X (i.e., $1.30 \leq \beta_X \leq 4.76$). Since the nondirectional alternative hypotheses $H_1: \beta_Y \neq 0$ and $H_1: \beta_X \neq 0$ are supported at the .05 level, it logically follows that the value zero will not fall within the range which defines either of the confidence intervals.

Test 28g: Test for evaluating the null hypothesis $H_0: \beta_1 = \beta_2$ A test of significance can be conducted to evaluate whether the slopes of two regression lines obtained from two independent samples are equal to one another. As is the case with **Test 28c**, it is assumed that the correlations for the independent samples are for the same two variables. The null hypotheses evaluated by the test for the **regression lines of Y on X** and the **regression lines of X on Y** are respectively $H_0: \beta_{Y_1} = \beta_{Y_2}$ and $H_0: \beta_{X_1} = \beta_{X_2}$ (where β_{Y_i} represents the slope of the **regression line of Y on X** in the underlying population represented by Sample i, and β_{X_i} represents the slope of the **regression line of X on Y** in the underlying population represented by Sample i). As a result of evaluating the null hypothesis in reference to two independent **regression lines of Y on X**, a researcher can determine if the degree of change on the Y variable when the X variable is incremented by one unit is equivalent in the two samples. In the case of two independent **regression lines of X on Y**, a researcher can determine if the degree of change on the X variable when the Y variable is incremented by one unit is equivalent in the two samples. It should be noted that the test employed to evaluate the generic null hypothesis $H_0: \beta_1 = \beta_2$ is not equivalent to **Test 28c**, which evaluates the null hypothesis $H_0: \rho_1 = \rho_2$. This is the case since (as is illustrated in Figure 28.2) it is entirely possible for the regression lines associated with two independent correlations to have identical slopes, yet be associated with dramatically different correlations.

Equations 28.31 and 28.32 are employed to evaluate the null hypotheses H_0: $\beta_{Y_1} = \beta_{Y_2}$ and H_0: $\beta_{X_1} = \beta_{X_2}$. The t values computed with Equations 28.31 and 28.32 are evaluated with **Table A2**. The degrees of freedom employed in evaluating the obtained t values are $df = (n_1 - 2) + (n_2 - 2) = n_1 + n_2 - 4$.[18]

$$t = \frac{b_{Y_1} - b_{Y_2}}{\sqrt{\dfrac{s_{Y.X_1}^2}{(\tilde{s}_{X_1}^2)(n_1 - 1)} + \dfrac{s_{Y.X_2}^2}{(\tilde{s}_{X_2}^2)(n_2 - 1)}}} \qquad \textbf{(Equation 28.31)}$$

$$t = \frac{b_{X_1} - b_{X_2}}{\sqrt{\dfrac{s_{X.Y_1}^2}{(\tilde{s}_{Y_1}^2)(n_1 - 1)} + \dfrac{s_{X.Y_2}^2}{(\tilde{s}_{Y_2}^2)(n_2 - 1)}}} \qquad \textbf{(Equation 28.32)}$$

Equations 28.31 and 28.32 can be employed to evaluate the regression coefficients associated with the two independent correlations described within the framework of the example employed to demonstrate **Test 28c**. If, for instance, the regression coefficient b_{Y_1} computed for boys (who will represent Sample 1) is larger than the regression coefficient b_{Y_2} computed for girls (who will represent Sample 2), Equation 28.31 can be employed to evaluate the null hypothesis H_0: $\beta_{Y_1} = \beta_{Y_2}$. Although the full analysis will not be done here, the following values would be used in Equation 28.31 for Sample 1/boys (whose data are the same as those employed in Example 28.1): $n_1 = 5$, $b_{Y_1} = .30$, $\tilde{s}_{X_1}^2 = (8.58)^2 = 73.62$, $s_{Y.X_1}^2 = (.92)^2 = .85$. If upon substituting the analogous values for a sample of five girls in Equation 28.31 the resulting t value is significant, the null hypothesis H_0: $\beta_{Y_1} = \beta_{Y_2}$ is rejected. The number of degrees of freedom employed for the analysis are $df = 5 + 5 - 4 = 6 = 6$. The tabled critical .05 and .01 two-tailed and one-tailed t values for $df = 6$ are respectively $t_{.05} = 2.45$ and $t_{.01} = 3.71$, and $t_{.05} = 1.94$ and $t_{.01} = 3.14$. If the nondirectional alternative hypothesis H_1: $\beta_{Y_1} \neq \beta_{Y_2}$ is employed, in order to be significant the obtained absolute value of t must be equal to or greater than the tabled critical two-tailed value at the prespecified level of significance. If the directional alternative hypothesis H_1: $\beta_{Y_1} > \beta_{Y_2}$ is employed, in order to be significant the computed t value must be a positive number that is equal to or greater than the tabled critical one-tailed value at the prespecified level of significance. If the directional alternative hypothesis H_1: $\beta_{Y_1} < \beta_{Y_2}$ is employed, in order to be significant, the computed t value must be a negative number that has an absolute value which is equal to or greater than the tabled critical one-tailed value at the prespecified level of significance.[19]

11. Additional correlational procedures The **Addendum** (Section IX) of this chapter describes the following additional correlational procedures which are directly or indirectly related to the **Pearson product-moment correlation coefficient**: a) **Test 28i: The point-biserial correlation coefficient**; b) **Test 28j: The biserial correlation coefficient**; and c) **Test 28k: The tetrachoric correlation coefficient**. The chapter on **multiple regression** describes the extension of the **Pearson product-moment correlation coefficient** to designs involving two or more predictor variables.

VII. Additional Discussion of the Pearson Product-Moment Correlation Coefficient

1. The definitional equation for the Pearson product-moment correlation coefficient

Although more computationally tedious than Equation 28.1, Equation 28.33 is a conceptually more meaningful equation for computing the **Pearson product-moment correlation coefficient**. Unlike Equation 28.1, which allows for the quick computation of r, Equation 28.33 reveals the fact that Pearson conceptualized the product-moment correlation coefficient as the average of the products of the paired z scores of subjects on the X and Y variables.

$$r = \frac{\sum_{i=1}^{n} z_{X_i} z_{Y_i}}{n-1}$$

(Equation 28.33)

Where: $z_{X_i} = (X_i - \bar{X})/\tilde{s}_X$ and $z_{Y_i} = (Y_i - \bar{Y})/\tilde{s}_Y$, with X_i and Y_i representing the scores of the ith subject on the X and Y variables

As noted above, the correlation coefficient is, in actuality, the mean of the product of each subject's X and Y scores, when the latter are expressed as z scores. Since the computed r value represents an average score, many books employ n as the denominator of Equation 28.33 instead of $(n-1)$. In point of fact, n can be employed as the denominator of Equation 28.33 if, in computing the z_{X_i} and z_{Y_i} scores, the sample standard deviations s_X and s_Y (computed with Equation I.10) are employed in place of the estimated population standard deviations $\tilde{s}_X$ and $\tilde{s}_Y$ (computed with Equation I.11). When the estimated population standard deviations are employed, however, $(n-1)$ is the appropriate value to employ in the denominator of Equation 28.33.

Table 28.2 Computation of r with Equation 28.33

X_i	$z_{X_i} = \dfrac{X_i - \bar{X}}{\tilde{s}_X}$	Y	$z_{Y_i} = \dfrac{Y_i - \bar{Y}}{\tilde{s}_Y}$	$z_{X_i} z_{Y_i}$
20	1.49	7	1.33	1.982
0	−.84	0	−1.26	1.058
1	−.72	2	−.52	.374
12	.56	5	.59	.330
3	−.49	3	−.15	.074

$$\sum z_{X_i} z_{Y_i} = 3.818$$

In employing Equation 28.33 to compute the value of r, initially the mean and estimated population standard deviation of the X and Y scores must be computed. Each X score is then converted into a z score by employing the equation for converting a raw score into a z score (i.e., $z_{X_i} = (X_i - \bar{X})/\tilde{s}_X$). Each Y score is also converted into a z score using the same equation with reference to the Y variable (i.e., $z_{Y_i} = (Y_i - \bar{Y})/\tilde{s}_Y$).[20] The product of each subject's z_{X_i} and z_{Y_i} score is obtained, and the sum of the products for the n subjects z scores is computed. The latter sum is divided by $(n-1)$, yielding the value of r which represents an average of the sum of the products.[21] The value of r computed with Equation 28.33 will be identical to that computed with Equation 28.1.

The computation of the value $r = .955$ with Equation 28.33 is demonstrated in Table 28.2. In deriving the z_{X_i} and z_{Y_i} scores, the following summary values are employed: $\bar{X} = 7.2$, $\tilde{s}_X = 8.58$, $\bar{Y} = 3.4$, $\tilde{s}_Y = 2.70$. Equation 28.33 yields the value $r = 3.818/4 = .955$ when the sum of the products in the last column of Table 28.2 is divided by $(n-1)$.

2. Covariance In Section IV it is noted that the numerator of Equation 28.1 is referred to as the **sum of products**. When the sum of products is divided by $(n-1)$, the resulting value represents a measure which is referred to as the **covariance**. Equation 28.34 is the computational equation for the covariance.

$$\text{cov}_{XY} = \frac{\Sigma XY - \dfrac{(\Sigma X)(\Sigma Y)}{n}}{n-1} \qquad \textbf{(Equation 28.34)}$$

Equation 28.35 is the definitional equation of the covariance, which reveals the fact that covariance is an index of the degree to which two variables covary (i.e., vary in relation to one another). Examination of Equation 28.35 reveals that covariance is the product of the average difference between one variable and its mean and a second variable and its mean. As Harlow (2005, p. 18) notes, correlation is covariance expressed in its standardized form reflecting the degree to which two variables are linearly related to one another.

$$\text{cov}_{XY} = \frac{\sum_{i=1}^{n}\left[(X_i - \bar{X})(Y_i - \bar{Y})\right]}{n-1} \qquad \textbf{(Equation 28.35)}$$

Each subject's contribution to the covariance is computed as follows: The difference between a subject's score on the X variable and the mean of the X variable, and the difference between a subject's score on the Y variable and the mean of the Y variable, are computed. The two resulting deviation scores are multiplied together. The resulting product represents that subject's contribution to the covariance. Upon obtaining a product for all n subjects, the sum of the n products (which is the numerator of Equation 28.1) is divided by $(n-1)$. The resulting value represents the covariance, which it can be seen is essentially the average of the products of the deviation scores. The reason why the sum of the products is divided by $(n-1)$ instead of n is because (as is also the case in computing the variance) division by the latter value provides a biased estimate of the population covariance. In the event one is computing a covariance for a sample and not using it as an estimate of the underlying population covariance, n is employed as the denominator of Equations 28.34 and 28.35.

Inspection of Equation 28.35 reveals that subjects who are above the mean on both variables or below the mean on both variables will contribute a positive product to the covariance. On the other hand, subjects who are above the mean on one of the variables but below the mean on the other variable will contribute a negative product to the covariance. If all or most of the subjects contribute positive products, the covariance will be a positive number. Since the value of r is a direct function of the sign of the covariance (which is a function of the sum of products), the resulting correlation coefficient will be a positive number. If all or most of the subjects contribute negative products, the covariance will be a negative number, and, consequently, the resulting correlation coefficient will also be negative. When among the n subjects the distribution of negative and positive products is such that they sum to zero, the sum of products will equal zero resulting in zero covariance, and r will equal zero. If, for one of the two variables, all subjects obtain the identical score, each subject will yield a product of zero, resulting in zero covariance (since the sum of products will equal zero). However, as noted in Section VI, since the value for the sum of squares will equal zero for a variable on which all subjects have the same score, Equation 28.1 becomes insoluble when all subjects have the same score on either of the variables. Based on what has been said with respect to the relationship between the sum of products and the covariance, the computation of r can be summarized by either of the following equations: $r = \text{cov}_{XY}/(\tilde{s}_X \tilde{s}_Y)$ and $r = SP_{XY}/\sqrt{SS_X SS_Y}$.

3. The homoscedasticity assumption of the Pearson product-moment correlation coefficient It is noted in Section I that one of the assumptions underlying the **Pearson product-moment correlation coefficient** is a condition referred to as **homoscedasticity** (**homo** means *same* and **scedastic** means *scatter*). Homoscedasticity exists in a set of data if the relationship between the X and Y variables is of equal strength across the whole range of both variables. Data that are not homoscedastic are **heteroscedastic**. When data are homoscedastic the accuracy of a prediction based on the regression line will be consistent across the full range of both variables. To illustrate, if data are homoscedastic and a strong positive correlation is computed between X and Y, the strong positive correlation will exist across all values of both variables. However, if for high values of X the correlation between X and Y is a strong positive one, but the strength of this relationship decreases as the value of X decreases, the data are heteroscedastic. As a general rule, if the distribution of one or both of the variables employed in a correlation is saliently skewed, the data are likely to be heteroscedastic. When, however, the data for both variables are distributed normally, the data will be homoscedastic.

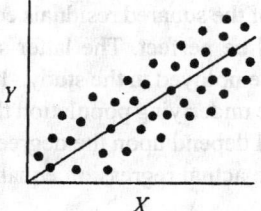 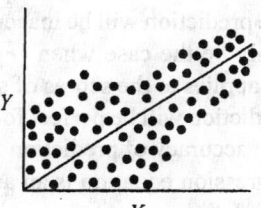

Figure 28.6 Homoscedastic Versus Heteroscedastic Data

Figure 28.6 presents two regression lines (which it will be assumed represent the **regression line of Y on X**) and the accompanying data points. Note that in Figure 28.6a, which represents homoscedastic data, the distance of the data points from the regression line is about the same along the entire length of the line. Figure 28.6b, on the other hand, represents heteroscedastic data, since the data are not dispersed evenly along the regression line. Specifically, in Figure 28.6b the data points are close to the line for high values of X, yet as the value of X decreases, the data points become further removed from the line. Thus, the strength of the positive correlation is much greater for high values of X than it is for low values. This translates into the fact that a subject's Y score can be predicted with a greater degree of accuracy if the subject has a high score on the X variable as opposed to a low score. Directly related to this is the fact that the value of the standard error of estimate computed with Equation 28.12 ($s_{Y \cdot X}$) will not be a representative measure of error variability for all values of X. Specifically, when $\tilde{s}_{Y.X}$ computed with Equation 28.14 (which is a function of the value $s_{Y.X}$) is employed to compute a confidence interval through use of Equation 28.16, the value of $\tilde{s}_{Y.X}$ will be larger for subjects who have a low score on the X variable, and thus the confidence interval associated with the predicted scores of such subjects will be larger than the confidence interval for subjects who have a high score on the X variable.

4. Residuals, analysis of variance for regression analysis and regression diagnostics In Section VI it is noted that a **residual** is the difference between the predicted value of a subject's score on the criterion variable and the subject's actual score on the criterion variable. Thus, a residual indicates the amount of error between a subject's actual and predicted scores. If e_i represents the amount of error for the ith subject, the residual for a subject can be defined as follows (assuming the **regression line of Y on X** is employed): $e_i = (Y_i - Y_i')$. In the least squares regression model, the sum of the residuals will always equal zero. Thus, $\sum_{i=1}^{n} e_i = \sum_{i=1}^{n}(Y_i - Y_i') = 0$. Since the sum of the residuals equals zero, the average of the residuals will also equal zero (i.e., $\bar{e}_i = (\sum_{i=1}^{n} e_i)/n = 0$). The latter reflects the fact that for some of the subjects the predicted value of Y_i' will be larger than Y_i, while for other subjects the predicted value of Y_i' will be smaller than Y_i (of course, for some subjects Y_i' may equal Y_i). It should be noted that if the sum of the squared distances of the data points from the regression line is not the minimum possible value, the sum of the residuals will be some value other than zero.

$\sum_{i=1}^{n} e_i^2 = \sum_{i=1}^{n}(Y_i - Y_i')^2$ (which is the sum of the squared residuals) provides an index of the accuracy of prediction that results from use of the regression equation (in reference to the sample data). When the sum of the squared residuals is small prediction will be accurate, but when it is large prediction will be inaccurate. When the sum of the squared residuals equals zero (which will only be the case when $|r| = 1$), prediction will be perfect. The latter statement, however, only applies to the scores of subjects in the sample employed in the study. It does not ensure that prediction will be perfect for other members of the underlying population the sample represents. The accuracy of prediction for the population will depend upon the degree to which the derived regression equation is an accurate estimate of the actual regression equation in the underlying population.

The partitioning of the variation on the criterion variable in the least squares regression model can be summarized by Equation 28.36.

$$\sum_{i=1}^{n} (Y_i - \bar{Y})^2 = \sum_{i=1}^{n} (Y_i' - \bar{Y})^2 + \sum_{i=1}^{n} (Y_i - Y_i')^2$$

(Equation 28.36)

Total variation = Explained variation + Error variation

Note that in Equation 28.36, the error (unexplained) variation is the sum of the squared residuals. When $|r| = 1$, $\sum_{i=1}^{n}(Y_i - Y_i')^2 = 0$, which as noted earlier results in perfect prediction. When, on the other hand, $r = 0$, $\sum_{i=1}^{n}(Y_i' - \bar{Y})^2 = 0$, and thus the value $\bar{Y}$ will be the predicted value of Y' for each subject (since, using Equations 28.6 and 28.7, if $r = 0$ then $b_Y = 0$ and $a_Y = \bar{Y}$. If $a_Y = \bar{Y}$, the value of Y' computed with Equation 28.4 is $Y' = \bar{Y}$).

Through use of the residuals, the **coefficient of determination** can be expressed with Equation 28.37.

$$r^2 = \frac{\text{Explained variation}}{\text{Total variation}} = \frac{\sum_{i=1}^{n} (Y_i' - \bar{Y})^2}{\sum_{i=1}^{n} (Y_i - \bar{Y})^2} \qquad \textbf{(Equation 28.37)}$$

Equation 28.38 (which is the square root of Equation 28.37) represents an alternative (albeit more tedious) way of computing the correlation coefficient.

$$r = \pm \sqrt{\frac{\sum_{i=1}^{n} (Y_i' - \bar{Y})^2}{\sum_{i=1}^{n} (Y_i - \bar{Y})^2}} \qquad \textbf{(Equation 28.38)}$$

The value $s_{Y.X}^2$ is the **residual variance** (i.e., the variance of the residuals). The residual variance (which is the square of the value $s_{Y.X}$ computed with Equation 28.12) can be defined by Equation 28.39. The denominator of Equation 28.39 represents the degrees of freedom employed in the analysis.

$$s_{Y.X}^2 = \frac{\sum_{i=1}^{n} (Y_i - Y_i')^2}{n - 2} \qquad \textbf{(Equation 28.39)}$$

Equation 28.40, which is the square root of Equation 28.39, is an alternative (albeit more tedious) way of computing the **standard error of estimate**. Inspection of Equation 28.40 reveals that the greater the sum of the squared residuals, the greater the value of $s_{Y.X}$.

$$s_{Y.X} = \sqrt{\frac{\sum_{i=1}^{n} (Y_i - Y_i')^2}{n - 2}} \qquad \textbf{(Equation 28.40)}$$

Everything that has been said about the residuals with reference to the **regression line of Y on X** can be generalized to the **regression line of X on Y** (in which case the residual for each subject is represented by $e_i = (X_i - X_i')$). Thus, all of the equations described in this section can be generalized to the second regression line by respectively employing the values X_i, X_i', and $\bar{X}$ in place of Y_i, Y_i', and $\bar{Y}$.

In Equation 28.36 the value $\sum_{i=1}^{n}(Y_i' - \bar{Y})^2$ computed for explained variation is commonly referred to as the **regression sum of squares** (which will be designated with the notation SS_{reg}), since it reflects the amount of variability on the Y variable which can be accounted for by the regression line. The value $\sum_{i=1}^{n}(Y_i - Y_i')^2$ is commonly referred to as the **residual sum of squares** (which will be designated with the notation SS_{res}), since it reflects unexplained variability. (The term **residual** implies *left over* variability — i.e., variability which cannot be explained through use of the regression line.) Since the latter two sum of squares values comprise the total variability on the Y variable, Equation 28.36 can be rewritten as noted below or in the symbolic form expressed by Equation 28.41.

Sum of squares Y variable = Total sum of squares = Regression sum of squares + Residual sum of squares

$$SS_Y = SS_{Total} = SS_{reg} + SS_{res} \qquad \text{(Equation 28.41)}$$

In point of fact, an analysis of variance (more specifically, a **single-factor between-subjects analysis of variance (Test 21)**) can be employed to determine how well the regression line fits the data. In reference to the **regression line of Y on X**, an F test can be conducted which simultaneously evaluates the null hypotheses H_0: $\rho = 0$ and H_0: $\beta_Y = 0$, which, as noted earlier, are respectively evaluated with Equations 28.3 and 28.27. It was also previously noted that the analysis of the two aforementioned null hypotheses will always yield an equivalent result. In point of fact, the square of the t value computed with Equations 28.3 and 28.27 will always be equal to the F value computed for the analysis of variance employed to evaluate the same hypotheses. (As will be demonstrated later in this section, the latter is also true with respect to the evaluation of the **regression line of X on Y** when the null hypothesis H_0: $\beta_X = 0$ is evaluated with Equation 28.28.) When the variability due to linear regression (i.e., the regression line) is substantially greater than the residual variability, the value of the F ratio will be significantly greater than 1, and thus allow a researcher to reject the null hypotheses H_0: $\rho = 0$ and H_0: $\beta_Y = 0$.

Equations 28.42–28.48 are employed below to compute the relevant sum of squares, degrees of freedom, and mean square values for the analysis of variance. (The values SS_X, SS_Y, and SP_{XY} are computed in Section IV.) Note that in the equations for computing the degrees of freedom, the notation k represents the number of predictor variables in the regression analysis. For the analysis under discussion, $k = 1$ since the only predictor variable in the analysis is the X variable. (The chapter on **multiple regression** discusses analyses which involve more than two predictor variables.)

$$SS_{reg} = \frac{(SP_{XY})^2}{SS_X} = \frac{(88.6)^2}{294.8} = 26.63 \qquad \text{(Equation 28.42)}$$

$$SS_{res} = SS_Y - SS_{reg} = 29.2 - 26.63 = 2.57 \qquad \text{(Equation 28.43)}$$

$$df_{Total} = df_Y = n - 1 = 5 - 1 = 4 \qquad \text{(Equation 28.44)}$$

$$df_{reg} = k = 1 \qquad \text{(Equation 28.45)}$$

$$df_{res} = n - k - 1 = 5 - 1 - 1 = 3 \qquad \text{(Equation 28.46)}$$

$$MS_{reg} = \frac{SS_{reg}}{df_{reg}} = \frac{26.63}{1} = 26.63 \qquad \text{(Equation 28.47)}$$

$$MS_{res} = \frac{SS_{res}}{df_{res}} = \frac{2.57}{3} = .86 \qquad \text{(Equation 28.48)}$$

The F ratio is computed with Equation 28.49. The result of the analysis is summarized in Table 28.3.

$$F = \frac{MS_{reg}}{MS_{res}} = \frac{26.63}{.86} = 31.06 \qquad \text{(Equation 28.49)}$$

Table 28.3 Summary Table of Analysis of Variance for Regression Analysis of Regression Line of Y on X

Source of variation	SS	df	MS	F
Regression	26.63	1	26.63	31.06
Residual	2.57	3	.86	
Total	29.20	4		

The obtained value $F = 31.06$ is evaluated with **Table A10 (Table of the F Distribution)** in the **Appendix**. The degrees of freedom employed for the analysis are the numerator and denominator degrees of freedom employed in computing the F ratio. Thus, $df_{num} = df_{reg} = 1$ and $df_{den} = df_{res} = 3$. For $df_{num} = 1$ and $df_{den} = 3$, the tabled $F_{.95}$ and $F_{.99}$ values are $F_{.95} = 10.13$ and $F_{.99} = 34.12$. In order to reject the null hypothesis, the obtained F value must be equal to or greater than the tabled critical value at the prespecified level of significance. Since $F = 31.06$ is greater than $F_{.95} = 10.13$ (which represents the tabled critical .05 F value), the null hypotheses $H_0: \rho = 0$ and $H_0: \beta_Y = 0$ can be rejected at the .05 level. They cannot, however, be rejected at the .01 level, since $F = 31.06$ is less than $F_{.99} = 34.12$ (which represents the tabled critical .01 F value). Note that this result is equivalent to that obtained when the same null hypotheses were evaluated through use of a *t* **test** (employing Equations 28.3 and 28.27). The square of the *t* value obtained with the aforementioned equations is equal to the value $F = 31.06$ obtained for the analysis of variance. (*t* was equal to approximately 5.6 — the slight discrepancy between the values of t^2 and F is due to rounding off error.)

It is also the case that r^2 will always equal the proportion of total variability (on the Y variable) attributable to the regression sum of squares. Specifically, the value of r^2 can be defined by the relationship described in Equation 28.50. Note that in Section V the value $r^2 = (.955)^2 = .912$ was previously determined to be the **coefficient of determination**.

$$r^2 = \frac{SS_{reg}}{SS_{Total}} = \frac{26.63}{29.2} = .912 \qquad \textbf{(Equation 28.50)}$$

In reference to the second regression line, the **regression line of X on Y**, an F test can be conducted (which will yield the same F value obtained above in Table 28.3) that simultaneously evaluates the null hypotheses $H_0: \rho = 0$ and $H_0: \beta_X = 0$. Thus, if X represents the criterion variable and Y represents the predictor variable, the total variability on the X variable can be expressed by Equation 28.51.

$$SS_X = SS_{reg} + SS_{res} \qquad \textbf{(Equation 28.51)}$$

Equations 28.52–28.55 are employed below to compute the relevant sum of squares and mean square values for the analysis of variance. As was the case for the previous analysis of variance, Equations 28.44–28.46 are employed to compute the degrees of freedom values (Equation 28.44, however, is now $df_{Total} = df_X = n - 1 = 4$).

$$SS_{reg} = \frac{(SP_{XY})^2}{SS_Y} = \frac{(88.6)^2}{29.2} = 268.83 \qquad \textbf{(Equation 28.52)}$$

$$SS_{res} = SS_X - SS_{reg} = 294.80 - 268.83 = 25.97 \quad \textbf{(Equation 28.53)}$$

$$MS_{reg} = \frac{SS_{reg}}{df_{reg}} = \frac{268.83}{1} = 268.83 \quad \textbf{(Equation 28.54)}$$

$$MS_{res} = \frac{SS_{res}}{df_{res}} = \frac{25.97}{3} = 8.66 \quad \textbf{(Equation 28.55)}$$

The F ratio is computed with Equation 28.49. The result of the analysis is summarized in Table 28.4.

$$F = \frac{MS_{reg}}{MS_{res}} = \frac{268.83}{8.65} = 31.06$$

Table 28.4 Summary Table of Analysis of Variance for Regression Analysis of Regression Line of X on Y

Source of variation	SS	DF	MS	F
Regression	268.83	1	268.83	31.06
Residual	25.97	3	8.66	
Total	**294.80**	**4**		

The obtained value $F = 31.06$ is evaluated with **Table A10**. For $df_{num} = df_{reg} = 1$ and $df_{den} = df_{res} = 3$, $F_{.95} = 10.13$ and $F_{.99} = 34.12$. Since $F = 31.06$ is greater than $F_{.95} = 10.13$, the null hypotheses, H_0: $\rho = 0$ and H_0: $\beta_X = 0$, can be rejected at the .05 level. They cannot, however, be rejected at the .01 level, since $F = 31.06$ is less than $F_{.99} = 34.12$. Note that, as was the case for the analysis of variance on the **regression line of Y on X**, this result is equivalent to that obtained when the same null hypotheses were evaluated through use of a t test (employing Equations 28.3 and 28.28).

Once again, it is also the case that r^2 equals the proportion of total variability (in this case on the X variable) attributable to the regression sum of squares.

$$r^2 = \frac{SS_{reg}}{SS_{Total}} = \frac{268.83}{294.80} = .912$$

Regression diagnostics through use of residual scatterplots Up to this point, the discussion of regression analysis has been predicated on the assumption that the relationship between the X and Y variables is best described by a linear model derived through use of the method of least squares. However, prior to employing the latter model for predictive purposes, a researcher should perform **regression diagnostics** — more specifically, conduct a more detailed analysis of the residuals in order to determine whether or not the derived linear relationship does, in fact, provide the most appropriate fit for the data.

A scatterplot of the residuals can provide a researcher with information that is generally not obvious from a simple scatterplot of subjects' scores (such as the scatterplot depicted in Figure 28.1). The distribution of the residuals can provide information regarding whether or not any of the assumptions (i.e., linearity, homoscedasticity, normality, and independence of error terms) underlying regression analysis has been violated. Although visual inspection of the residuals (which will be described in this section) represents an informal method for regression diagnostics, it can provide a researcher with important information with respect to whether or not the derived regression model is satisfactory for describing the nature of the relationship between the variables.

Table 28.5   Computation of Residuals for Example 28.1

Subject	X	Y	Y'	$e_i = (Y_i - Y_i')$	$e_i = (Y_i - Y_i')^2$	z_{res_i}
1	20	7	7.24	$-.24$	.0576	$-.2592$
2	0	0	1.24	-1.24	1.5376	-1.3392
3	1	2	1.54	.46	.2116	.4968
4	12	5	4.84	.16	.0256	.1728
5	3	3	2.14	.86	.7396	.9288
				$\Sigma e_i = 0$	$\Sigma e_i^2 = 2.572$	$\Sigma z_{res_i} = 0$

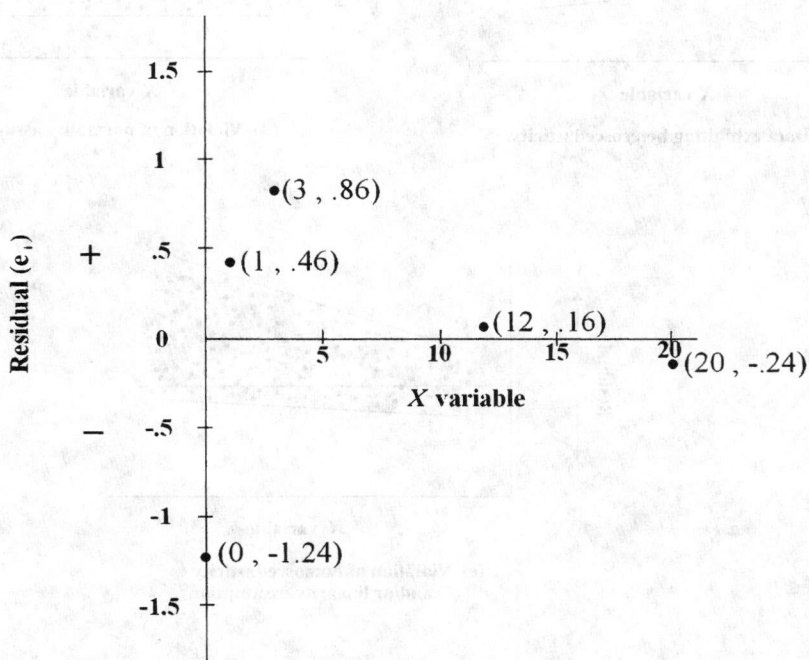

Figure 28.7   Scatterplot of Residuals for Example 28.1

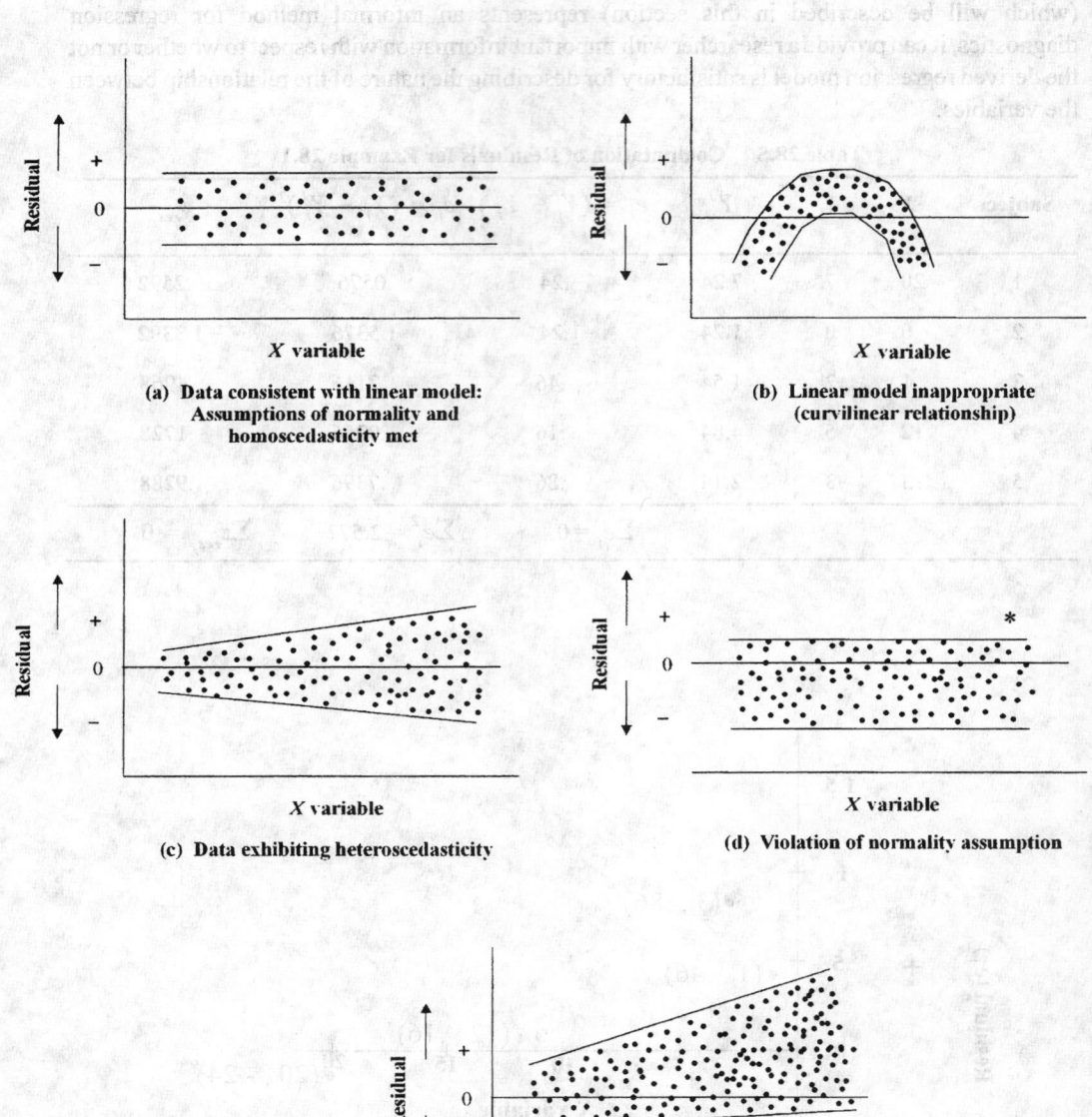

Figure 28.8　Scatterplot of Residuals

Table 28.5 summarizes the computation of the residuals (with reference to the **regression line of Y on X**), and Figure 28.7 is a scatterplot of the residuals for the data in Example 28.1. In plotting the residuals, each point on the original scatterplot (i.e., Figure 28.1) is transferred to a new plot in which the X axis remains intact, but the Y axis is modified to reflect the values of the residuals. Specifically, the Y axis is employed to indicate the extent to which a predicted value of Y (i.e., the value of Y' computed with the regression equation $Y' = 1.24 + .3X$ computed in Section VI) falls above or below the actual value of Y for a given value of X. Inspection of Table 28.5 reveals that the five data points in Figure 28.1 employed in the plot of the residuals for the subjects in Example 28.1 are: **Subject 1**: 20, – .24; **Subject 2**: 0, – 1.24; **Subject 3**: 1, .46; **Subject 4**: 12, .16; **Subject 5**: 3, .86.

Regression diagnostics can be facilitated by computing **standardized residuals**, which, among other things, can be useful in identifying outliers. In order to determine a standardized residual (which expresses a residual within the format of a standard deviation score — i.e., a z score) it is necessary to compute the **variance of the residuals** (s_{res}^2), which is computed below with Equation 28.56 (which is an alternative way of writing Equation 28.39).

$$s_{res}^2 = \frac{\sum e_i^2}{n - 2} = \frac{2.572}{5 - 2} = .8573 \qquad \textbf{(Equation 28.56)}$$

Equation 28.57 is employed to compute the value z_{res_i}, which represents the **standardized residual** for the ith observation.

$$z_{res_i} = \frac{e_i - \bar{e}_i}{\sqrt{s_{res}^2}} = \frac{e_i}{\sqrt{s_{res}^2}} \qquad \textbf{(Equation 28.57)}$$

To illustrate the computation of a standardized residual (the values of which are recorded in the last column of Table 28.5), the value $z_{res_1} = -.2592$ computed for Subject 1 in Table 28.5 is computed through use of Equation 28.57 as follows: $z_{res_1} = -.24 / \sqrt{.8573} = -.2592$. Note that the sum of the standardized residuals ($\sum z_{res_i}$) will always equal zero.

Because of the small sample size employed in Example 28.1, visual inspection of the residual scatterplot represented by Figure 28.7 does not provide enough information to allow one to draw any conclusions regarding the appropriateness of the regression model derived for the latter example. Consequently, the residual scatterplots depicted in Figure 28.8 (which contains five hypothetical sets of residuals — all of which contain a sufficient number of observations for detecting the presence of a pattern) will be employed to demonstrate the nature of the information that can be derived from a visual analysis of residuals. It should be noted that the use of a residual scatterplot does not insure a researcher will detect a violation of one or more of the assumptions underlying the linear regression. Shiffler and Adams (1995, p. 621) note that, as a general rule, it will be easier to assess whether or not any of the assumptions have been violated when such a scatterplot contains two or more observations for each value for X (which, in reality, is often not the case).

When the assumption of linearity, along with the other assumptions underlying linear regression have been met, the scatterplot of the residuals should approximate the form of Figure 28.8a — i.e., a rectangular shape with a concentration of scores around the horizontal centerline that is perpendicular to the vertical axis. Note that the horizontal centerline (i.e., the X axis) intercepts the Y axis at a residual value equal to zero (i.e., the point at which the predicted value of Y will equal the observed value of Y). Thus, the distribution of the points which comprise the scatterplot should be symmetrical in relation to the X axis — in other words, approximately the same number of points should fall both above and below the X axis. When the pattern of data is more appropriately described by a curvilinear function as opposed to a linear function, the distribution of the residuals will, in fact, be curvilinear as is the case in Figure 28.8b (which suggests the presence of a quadratic effect of a predictor variable). (A definition of a quadratic function (illustrated in Figure 21.2b) can be found in Section VI of the **single-factor between-subjects analysis of variance** under the discussion of **trend analysis**.) As a general rule, when a curvilinear relationship is appropriate for describing the relationship between the two variables, the latter will be more obvious from visual inspection of a plot of the residuals (such as the data depicted in Figure 28.8b) as opposed to inspection of a conventional scatterplot (such as Figure 28.1). Further discussion of curvilinear models can be found in the latter part of this section.

When one or more of the assumptions underlying the linear regression model are violated, transformation of the data (discussed in Section VII of the *t* **test for two independent samples**) on either the X and/or Y variables may result in data which conform to the assumptions of a simple linear model. Neter *et al.* (1990, pp. 142–145) note that if the assumption of linearity appears to be violated, yet both the normality and homoscedasticity assumptions (to be discussed below) have not been violated, one can attempt to achieve linearity by transforming the values of the X variable through use of a transformation involving square roots, squares, and/or reciprocals. They do not recommend transforming the values of the Y variable, since the latter may result in violation of the normality and/or homoscedasticity assumptions. Webster (1995, pp. 756–757) discusses the use of a logarithmic transformation on the X and/or Y variables in dealing with nonlinearity.

If the homoscedasticity assumption underlying linear regression has been violated (i.e., the data are heteroscedastic since the strength of the relationship between the X and Y variables is not consistent across the whole range of both variables), the residuals will be distributed unevenly above and below the centerline, and, more specifically, may assume a fan shaped pattern such as that depicted in Figure 28.8c (although other patterns depicting inconsistent variability can also indicate the presence of heteroscedasticity). The pattern depicted in Figure 28.8c indicates that as the value of X increases there is an increase in variability associated with the value of Y. Or to put it another way, as the value of X increases, there is an increase in residual variability. Tabachnick and Fidell (1996, p. 138) note that serious heteroscedasticity is present when the standard deviation of the residuals for the value of X associated with the greatest residual variability is at least three times the standard deviation of the residuals for the value of X associated with the lowest residual variability. An inferential statistical test developed by White (1980) can be employed to evaluate the homoscedasticity assumption. Neter *et al.* (1990, pp. 145–150) and Zar (1999, p. 356) note a logarithmic transformation of the data might resolve violation of the homoscedasticity assumption, which Neter *et al.* (1990, p. 145 note is often present when the normality assumption underlying linear regression is also violated).

Another assumption underlying linear regression is the normality assumption, which states that for every possible predicted Y score, the values of the residuals will be normally distributed about a value of zero. Tabachnick and Fidell (1996, p. 137) note that as a result of the latter, for each predicted value of Y, the residual scatterplot should reveal a concentration of residuals near the centerline and a normalized pattern of residuals trailing off symmetrically from the centerline (i.e., the further the distance from the centerline, the fewer the number of residual values in the scatterplot for a given value of X). When the normality assumption is violated, the distribution of the residuals will be skewed. The most serious violation of the normality assumption is likely to occur when one or more outliers are present in a set of observations. Sinich (1996, p. 746) notes that regression analysis is relatively robust with respect to violation of the normality assumption — i.e., minor to moderate departures from normality are not likely to seriously compromise any inferences derived from a regression analysis. Other sources, however, view mild to moderate departures from normality having a more serious impact.

Figure 28.8d represents a scatterplot of residuals when the normality assumption is violated (since the majority of the residuals fall below the centerline). A scatterplot of residuals can alert one to the presence of one or more outliers, insofar as the latter values will be far removed from the centerline. The asterisk in the upper right of Figure 28.8d represents a residual computed for an outlier. The inclusion of an outlier in a linear regression model (especially a model which is based on a small number of observations) may result in a drastic distortion of the regression line, since the inclination of the regression line will be shifted in the direction of the location of the outlier. Sinich (1996, p. 747) suggests that any residual with an absolute value which is three standard deviations or more from the predicted value of Y (i.e., $|Y_i - Y_i'| \geq 3s_{res}$) should be viewed as a potential outlier. Neter *et al.* (1990, pp. 121–122) do not recommend deleting an outlier from a regression analysis unless there is strong reason to believe the observation represents an error of some type on the part of the experimenter or the subject, since an outlier may convey significant information regarding the relationship between the variables. For example, an outlier can result from an interaction with an extraneous variable — in other words, another potential predictor variable which was not included in the regression model. In such a case, the model for predicting the criterion variable may be incomplete, insofar it may have omitted one or more additional critical predictor variables. In such a case it would be more appropriate to employ a **multiple regression model** (discussed in Chapter 33) to predict the Y variable.

It should be noted that even if there is no suggestion of the presence of one or more outliers in a set of data, the presence of an extraneous variable can compromise the validity of a regression model. The most common example of the latter is when data are collected over sequential periods of time — commonly referred to as a **time series**. Within a time series it is not uncommon that residuals computed for multiple adjacent observations conform to specific patterns with respect to sign and/or magnitude. When, in fact, the latter is true, a critical assumption underlying the linear regression model is likely to have been violated — specifically, the assumption of independence of the residuals. The most effective mechanism for determining whether or not a potentially critical variable such as time should be included in a regression model is to plot the residuals against the corresponding values of the variable in question. A full discussion of time as a potentially confounding variable in correlational research, as well as a discussion of the assumption of independence of the residuals underlying the linear regression model, can be found in the section on **autocorrelation** (which follows this section).

Sources on regression analysis describe a number of alternative graphical methods for assessing the normality assumption. Among the methods described are constructing a histogram, a stem-and-leaf display, and/or a box plot (all of which are discussed in the **Introduction**) of the residuals. The latter types of visual displays should yield patterns of residuals which approximate a normal distribution. It should be emphasized, however, that a reasonably large sample size is required in order for such visual displays to provide reliable information regarding the shape of the underlying distribution of residuals. Another visual method for assessing normality is a **normal probability plot of the residuals**. In the latter display, the expected values for the residuals when the distribution is normal are represented on the abscissa (X-axis), and the observed values of the residuals are recorded on the ordinate (Y-axis). If the residuals are distributed normally, the plot of the points should approximate a straight line. A good description of the protocol for constructing a normal probability plot can be found in Neter *et al.* (1990, pp. 125–128).

An alternative strategy for evaluating the normality assumption (also described in Neter *et al.* (1990, pp. 124–125)) is to evaluate the values computed for the standardized residuals with respect to goodness-of-fit for a normal distribution. The latter can be accomplished through use of one of goodness-of- fit tests described earlier in the book, such as the **Kolmogorov–Smirnov goodness-of-fit test for a single sample (Test 7)** (for which an analysis of goodness-of-fit for a normal distribution is demonstrated in Section IV), or the **chi-square goodness-of-fit test (Test 8)** (for which an analysis of goodness-of-fit for a normal distribution is demonstrated in Section VII). An alternative test developed by Lin and Mudholkar (1980) for assessing the normality assumption, which employs the **jacknife procedure** (described in Section IX (the **Addendum**) of the **Mann–Whitney U test (Test 12))**, is described in Shiffler and Adams (1995, pp. 624–626).

Neter *et al.* (1990, pp. 128; pp. 145–151) state that violation of the normality assumption may be related to violation of one or more of the other assumptions (specifically, the homoscedasticity and linearity assumptions) underlying linear regression. Kachigan (1986, p. 258) notes that a distribution of residuals such as that depicted in Figure 28.8e represents a situation in which there is violation of either the linearity or homoscedasticity assumptions, or quite possibly both. Because violation of the normality assumption may be associated with violations of other assumptions, Neter *et al.* (1990) suggest it is prudent to evaluate such assumptions prior to assessing normality. They also note that when the normality assumption is violated, one may be able to achieve normality (as well as homoscedasticity, if the latter assumption is also violated) through use of a data transformation (e.g., square root, logarithmic, or reciprocal) on the Y variable.

It should also be emphasized that a scatterplot of residuals should also be scrutinized for **influential observations**. An **influential observation** is one which exerts an undue influence on the orientation of the regression line. Although an outlier represents a typical example of what might be viewed as an influential observation, an influential observation need not necessarily be categorized as an outlier. Among others, Moore and McCabe (1993, p. 134) note that a simple way of evaluating the impact of a suspected **influential observation** is to visually inspect the scatterplot of the residuals, and to compute the regression line with and without the presence of the observation in question. The suspicion that an observation is influential will be confirmed if there is a substantial difference in the position of the regression line obtained when the observation in question is included in the analysis versus the regression line obtained when it is omitted. Berenson and Levine (1996, pp. 755–758) describe a number of computational procedures (attributed to Cook and Weisberg (1982) and Hoagalin and Welsch (1978)) which have been developed for identifying the presence of influential observations (these measures are described in Section IV/V in the chapter on **multiple regression**). Ultimately, the final decision with respect to whether or not to retain an observation deemed influential should be based on the judgement of the researcher rather than on visual or statistical analysis of the data. In other words, the researcher must decide whether the observation in question is the result of error or, in fact, is a valid observation from the underlying population.

When a linear model is not adequate for describing the data, the most commonly employed curvilinear model (although the tenability of alternative curvilinear models can also be assessed) is a **polynomial regression model**. The latter model involves fitting the data to a **polynomial function**. (A full discussion of polynomial functions can be found in Section VI of the **single-factor between-subjects analysis of variance** under the discussion of **trend analysis**.) Initially, one would determine whether a **quadratic function** (also referred to as a **second order polynomial** — a **linear function** being a **first order polynomial**) provides the best fit for the data. If not, the aptness of a **cubic function** (also referred to as a **third order polynomial**) would be evaluated, followed by a test of the aptness of a **quartic function** (also referred to as a **fourth order polynomial**), etc. Neter *et al.* (1990, p. 318) note, however, that in reality higher order polynomial regression models other than those for quadratic and cubic functions are rarely employed, since such models become problematic when extrapolating to the actual distribution of data in an underlying population. More specifically, a higher order curvilinear function is more likely to exhibit inconsistent behavior if generalized to values of the X variable which fall above and/or below the obtained values of X for the sample employed in computing the regression equation.

Among others, Webster (1995, p. 762) notes that a researcher may be able to derive two or more curvilinear models (in addition to perhaps a linear model), all of which provide reasonably good fits for a set of data. When, however, the different models are compared with one another, one may have a higher coefficient of determination (i.e., r^2 value) (which is desirable) than competing models, yet at the same time have a higher standard error of estimate (which is undesirable). Webster (1995, p. 762) suggests that the determination of which model to employ should be predicated, in part, on the purpose for which a model is intended. If a model is desired for strictly explanatory/theoretical purposes it is preferable to select the one with the higher coefficient of determination, whereas if one's primary goal is prediction, the model with the smaller standard error of estimate is to be preferred. However, it should be emphasized that trial and error exploration on the part of a researcher to find a suitable model can be problematical, insofar as one may ultimately arrive at a model which, in the final analysis, does not accurately extrapolate to the actual distribution of the data in an underlying population.

The discussion of regression diagnostics in this section has not addressed what many sources consider to be the most critical assumption underlying linear regression analysis — the assumption of independence of the error terms. The latter assumption is discussed in the next section within the context of the discussion of **autocorrelation**.

5. Autocorrelation (and **Test 28h: Durbin–Watson test**) As noted in the previous section, perhaps the most critical assumption underlying linear regression analysis is independence of the error terms (i.e., residuals). This latter assumption is most likely to be violated in applications in business and economics, where it is not uncommon for data to be collected over sequential periods of time with the ultimate goal of employing regression analysis for the prediction of future trends. As noted earlier in this section, data collected over sequential time periods are commonly referred to as a **time series**, and within a time series it is not unusual for residuals computed for multiple adjacent observations to conform to specific patterns with respect to sign and/or magnitude. For example, a positive residual obtained at Time i might have a high likelihood of being preceded by a positive residual at Time $(i - 1)$ and followed by a positive residual at Time $(i + 1)$, while a negative residual at Time i might be more likely to be preceded by a negative residual at Time $(i - 1)$ and followed by a negative residual at Time $(i + 1)$. The presence of the latter pattern involving similar adjacent residuals in a set of data represents an example of **autocorrelation** (also called to as **serial correlation** and **nonautoregression**) — more specifically, a positive autocorrelation.

The assumption of independence of the error terms in a linear regression model assumes the residuals are not autocorrelated, or, to put it another way, that the correlation between adjacent pairs of residuals in a set of data is $r_{res} = 0$ (which will be true if the correlation between adjacent scores in the data set equals zero). If, in fact, the latter is true with respect to the underlying population represented by the sample data, the null hypothesis H_0: $\rho_{res} = 0$ is supported (in other words, in the underlying population the correlation between the residuals is zero). If analysis of data leads a researcher to reject the null hypothesis H_0: $\rho_{res} = 0$, one of the following alternative hypotheses would be supported: H_1: $\rho_{res} \neq 0$ (which is nondirectional); H_1: $\rho_{res} > 0$ (which is directional, and indicates the presence of positive autocorrelation); H_1: $\rho_{res} < 0$ (which is directional, and indicates the presence of negative autocorrelation). If the null hypothesis H_0: ρ_{res} = 0 is rejected, it indicates the derived linear regression model is invalid due to a lack of independence of the residuals.

Figure 28.9 presents three sets of hypothetical data in which the residuals have been plotted against time (which is represented on the X axis). The numerical values for the time periods listed could represent successive days, month, or years, and the residual values could represent magnitude of error in predicting the values of variables such as unemployment rate, sales revenue, interest rate, gross national product, stock price, crop yield, crime rate, temperature, etc. Whereas in Figure 28.9a there is no apparent pattern when the residuals are plotted against time, in Figures 28.9b and 28.9c there are distinct patterns which suggest that the residuals are correlated with one another. The data depicted in Figure 28.9b, which is characterized by cyclical runs involving successive positive or negative residuals, represents an idealized distribution of residuals when there is a positive autocorrelation. The data depicted in Figure 28.9c, which is characterized by the alternation of positive and negative residuals for successive time periods, represents an idealized distribution of residuals when there is a negative autocorrelation.

Figure 28.10 (based on Webster (1995, p. 751)) is an alternative visual display which allows one to differentiate between positive and negative autocorrelation. In Figures 28.10a and 28.10b each residual (e_i) is plotted against the residual for the prior time period (e_{i-1}). In Figure 28.10a a positive autocorrelation is indicated by the fact that when e_i is a positive value e_{i-1} also tends to be a positive value, and when e_i is a negative value e_{i-1} also tends to be a negative value. In Figure 28.10b a negative autocorrelation is indicated by the fact that when e_i is a positive value e_{i-1} tends to be a negative value, and when e_i is a negative value e_{i-1} tends to be a positive value.

In reality, the presence of a positive autocorrelation is much more common than a negative autocorrelation. In point of fact, it is not unusual for the residuals for data collected in disciplines such as business and economics to exhibit a positive autocorrelation. The latter is commonly found on measures which exhibit cyclic trends, such as unemployment rate, sales, interest rates, or gross national product. To illustrate, if annual rates for unemployment are scrutinized over a long period of time (e.g., 20 years), it is likely there will be multiple consecutive months (if not years) characterized by high unemployment, followed by multiple consecutive months (if not years) during which unemployment is low. Shifts from high to low unemployment will generally not occur overnight, but instead will evolve gradually over a more extended period of time. Thus, if (relative to the long-term rate of unemployment for the full time series — e.g., twenty years) the unemployment rate for a specific month is unusually high, in all likelihood the unemployment rate will also be high during the adjacent months (i.e., the months preceding and following the month in question). In the same respect, if the unemployment rate for a specific month is unusually low, in all likelihood the unemployment rate will also be low during the adjacent months. Webster (1995, p. 590) notes that if unemployment rates are unusually high for a limited time period (e.g., 12 months), the prediction derived from the regression model will probably underestimate the rate of unemployment (i.e., the observed value of the Y variable), resulting in a positive residual (since if $Y' < Y, (Y - Y')$ will be a positive number). If, on the other hand, unemployment rates are unusually low for a limited time period, the regression model will probably overestimate the rate of unemployment, resulting in a negative residual (since if $Y' > Y, (Y - Y')$ will be a negative number). Such a pattern of multiple successive positive residuals followed by multiple successive negative residuals is indicative of a positive autocorrelation as that depicted in Figures 28.9b and 28.10a.

In the final analysis, if the residuals are plotted against time, they should be randomly distributed above and below a residual value of zero, with a nonrandom pattern suggesting lack of independence between the residuals (which indicates the data are autocorrelated). On the other hand (as noted in the previous section), if the residuals are plotted against the observed values

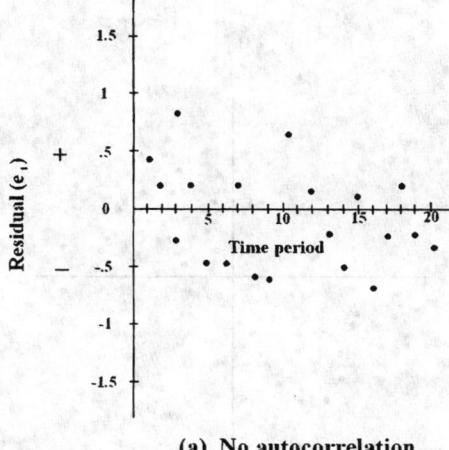

(a) No autocorrelation

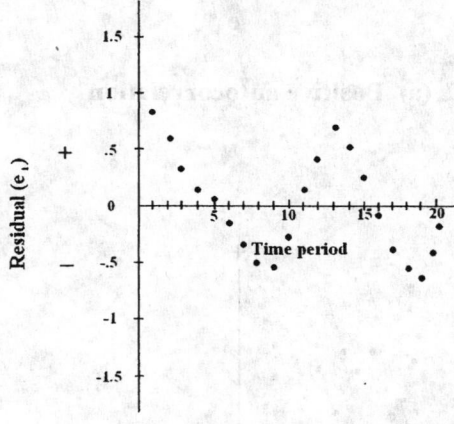

(b) Positive autocorrelation

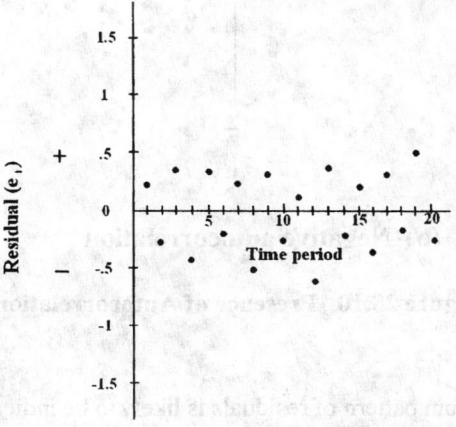

(c) Negative autocorrelation

Figure 28.9   Examples of Autocorrelation

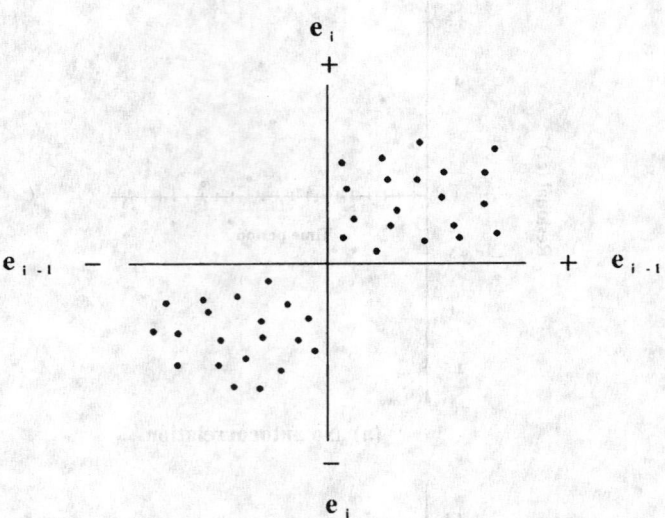

(a) Positive autocorrelation

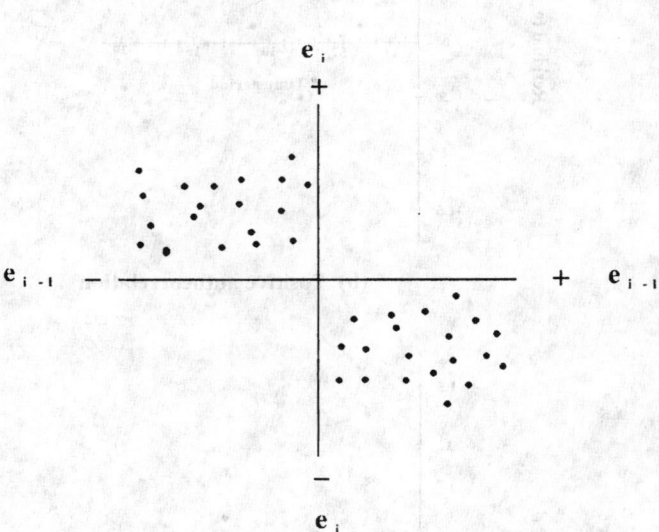

(b) Negative autocorrelation

Figure 28.10   Presence of Autocorrelation

of the X variable, a nonrandom pattern of residuals is likely to be indicative of a poor fit for the derived regression model (as is the case in Figure 28.8b), rather than a lack of independence between the residuals. Related to the latter is that if the n observations in a conventional scatterplot (e.g., Figure 28.1) are not randomly distributed about the derived regression line, such a pattern can also be viewed as evidence of a poor fit for the derived regression model.

Although the **single-sample runs test (Test 10)** represents a simple method for determining whether or not the distribution of the residuals is random, among others, Ryan (1997, p. 46) notes the latter test is limited in this context, since it is not able to detect certain patterns of nonrandomness. If, however, one elected to employ the **single-sample runs test** to evaluate the independence of residuals, the number of runs for residual values which fall above versus below zero are computed (when the residuals are plotted against time). Employing **Table A8 (Table of Critical Values for the Single-Sample Runs Test)** in the **Appendix**, a determination is made (with n representing the total number of time periods) whether the observed number of runs is fewer than the tabled critical lower limit (which would suggest the presence of a positive autocorrelation) or greater than the tabled critical upper limit (which would suggest the presence of a negative autocorrelation). If either of the latter conditions is met, the researcher can reject the null hypothesis of independence of residuals.

It is also the case that if the residuals are plotted against the scores on the X variable, the **single-sample runs test** can be employed to evaluate the aptness of fit of the regression model (with too few or too many runs suggesting lack of fit with the model). In the same respect, the **runs test** can be employed to evaluate the aptness of fit of a regression model through analysis of the distribution of the observations on a conventional scatterplot such as Figure 28.1. In the latter case, the number of runs involving observations which fall above and below the derived regression line are determined, with too few or too many runs suggesting lack of fit for the model.

Test 28h: Durbin–Watson test   When data are autocorrelated, the pattern of residuals is usually not as obvious as those depicted in Figures 28.9 and 28.10, and, as noted above, may not be detected through use of the **single-sample runs test**. The test of choice for determining whether or not residuals are, in fact, correlated with one another (i.e., for evaluating the null hypothesis H_0: $\rho_{res} = 0$) is the **Durbin–Watson test** (1950, 1951, 1971). The latter test evaluates what is referred to as a **first-order autocorrelation**, which is the computation of the correlation between adjacent residuals (i.e., neighboring residuals one time period apart). A more general approach for testing autocorrelation between residuals more than one time period apart (which will not be described in this book) is the **modified Box–Pierce statistic** (also known as the **Box–Ljung–Pierce statistic**) (Box and Pierce (1970); Ljung and Box (1978)). The term **lag value** (which will be discussed in more detail later in this section) is employed to identify the number of time periods between residuals. In **first-order autocorrelation** the lag value is 1 (often referred to as **lag (t – 1)**), since the computed correlation is between neighboring residuals one time period apart.

The use of the **Durbin–Watson test** for detecting autocorrelation will be demonstrated with Example 28.2. The reader should take note of the fact that in the interest of computational simplicity, the two variables employed in the latter example are quantified through the use of index numbers which are limited to low integer values. However, in reality, if such a study were conducted, the likelihood is that the variable of advertising would be quantified in dollars and the variable of sales volume would be quantified in dollars or number of units sold.

Example 28.2 *Data are available for a 16 year period documenting the relationship between the amount of money spent advertising a specific product and the volume of sales recorded for the product. A researcher is asked to derive a linear regression model describing the relationship between the two variables, and to employ the **Durbin–Watson test** to determine whether or not there is evidence of a positive autocorrelation for the sales residuals. In conducting the analysis, two 10 point indices (both employing the integer values 0 through 9) are used to indicate the amount of money spent on advertising (which represents the X variable)* and on volume of sales (which represented the Y variable). (The higher the annual advertising index number, the greater the amount expended on advertising, and the higher the annual sales index number, the greater the volume of sales.) Table 28.6 summarizes the data for the study along with the computations involved in conducting the **Durbin–Watson test**.

Table 28.6 Data for Example 28.2 and Computations for Durbin–Watson Test

Time	X_i	Y_i	Y_i'	$e_i = Y_i - Y_i'$	e_i^2	$e_i - e_{i-1}$	$(e_i - e_{i-1})^2$
1	1	2	1.364	.636	.404	—	—
2	2	3	2.273	.727	.529	.091	.0083
3	3	4	3.182	.818	.669	.091	.0083
4	4	5	4.091	.909	.826	.091	.0083
5	5	6	5.000	1.000	1.000	.091	.0083
6	6	7	5.909	1.091	1.190	.091	.0083
7	7	8	6.818	1.182	1.397	.091	.0083
8	8	9	7.727	1.273	1.621	.091	.0083
9	9	8	8.636	-.636	.404	-1.909	3.644
10	8	7	7.727	-.727	.529	-.091	.0083
11	7	6	6.818	-.818	.669	-.091	.0083
12	6	5	5.909	-.909	.826	-.091	.0083
13	5	4	5.000	-1.000	1.000	-.091	.0083
14	4	3	4.091	-1.091	1.190	-.091	.0083
15	3	2	3.182	-1.182	1.397	-.091	.0083
16	2	1	2.273	-1.273	1.621	-.091	.0083
Sums	80	80		$\Sigma = 0$	$\Sigma = 15.272$	$\Sigma = -1.909$	$\Sigma = 3.760$

Preliminary analysis of the data in Table 28.6 (which does not take into consideration the order in which the X and Y scores are listed in the table) yields a correlation coefficient of $r = .909$ between the amount of money spent on advertising and sales volume. The computations are demonstrated below through use of Equation 28.1 (employing the values $\Sigma X = \Sigma Y = 80$, $\Sigma X^2 = \Sigma Y^2 = 488$, and $\Sigma XY = 480$). Employing **Table A16** in the **Appendix**, for $df = 16 - 2 = 14$, the critical two-tailed .05 and .01 r values are $r_{.05} = .497$ and $r_{.01} = .623$. Since the computed value $r = .909$ is greater than $r_{.01} = .623$, it can be concluded at the .01 level of significance that the correlation between the variables in the underlying population is some value other than zero.

Analysis of the data yields the regression equation $Y' = .455 + .909X$ (with the values $b = .909$ and $a = .455$ respectively computed with Equations 28.5/28.6 and 28.7, with $\tilde{s}_X = \tilde{s}_Y = 2.42$).

$$r = \frac{480 - \dfrac{(80)(80)}{16}}{\sqrt{\left[488 - \dfrac{(80)^2}{16}\right]\left[488 - \dfrac{(80)^2}{16}\right]}} = \frac{80}{\sqrt{(88)(88)}} = .909$$

$$b_Y = \frac{SP_{XY}}{SS_X} = \frac{\Sigma XY - \dfrac{(\Sigma X)(\Sigma Y)}{n}}{\Sigma X^2 - \dfrac{(\Sigma X)^2}{n}} = \frac{80}{88} = .909$$

$$a_Y = \bar{Y} - b_Y \bar{X} = 5 - (.909)(5) = .455$$

At this point the **Durbin–Watson test** statistic will be computed for the data. The test statistic for the **Durbin–Watson test** (to be designated as d) is computed with Equation 28.58.

$$d = \frac{\sum\limits_{i=1}^{n} (e_i - e_{i-1})^2}{\sum\limits_{i=1}^{n} e_i^2} \qquad \text{(Equation 28.58)}$$

In order to compute the latter test statistic, the following values and/or operations are employed in Table 28.6: a) **Column 1** contains the 16 sequential time periods/years evaluated in the study; b) **Columns 2** and **3** respectively contain the index scores for the X (advertising expense) and Y (sales volume) variables during each time period. As noted above, the latter values for X and Y were employed to compute the value $r = .909$; c) The Y' values recorded in **Column 4** represent the predicted sales index during each of the 16 time periods. The value of Y' for a given time period is computed by substituting the value of X for that time period in the linear regression equation $Y' = .455 + .909X$ derived from the data for Example 28.2; d) The values $e_i = (Y_i - Y_i')$ recorded in **Column 5** represent the residuals computed for each of the 16 time periods. During each time period the predicted sales index (Y_i') is subtracted from the actual sales index (Y_i); e) In **Column 6** the value of each of the residuals in **Column 5** has been squared; f) The value $(e_i - e_{i-1})$ recorded in each row of **Column 7** is the difference between the residual value computed for that time period and the residual value for the time period which preceded it. Note that the latter is only done for $(n - 1)$ of the n time periods (since no value can be entered for time period 1 since no time period preceded it). Shiffler and Adams (1995, p. 630) note that when the computations in **Column 7** are correct, the sum of the values in Column 7 will equal the value obtained if the residual for time period 1 is subtracted from the residual for time period n (i.e., $\Sigma(e_i - e_{i-1}) = e_n - e_1$). The latter is the case in Table 28.6 where $\Sigma(e_i - e_{i-1}) = -1.909$, which is equal to $[(e_n = -1.273) - (e_1 = .636)] = -1.909$; g) The values $(e_i - e_{i-1})^2$ recorded in each row of **Column 8** (with the exception of the row for time period 1) are the squares of the difference scores recorded in **Column 7**.

When the values $\sum\limits_{i=1}^{n} (e_i - e_{i-1})^2 = 3.760$ and $e_i^2 = 15.272$ computed in Table 28.6 are substituted in Equation 28.58, the value $d = .246$ is obtained for the **Durbin–Watson test** statistic.

$$d = \frac{\sum\limits_{i=1}^{n}(e_i - e_{i-1})^2}{\sum\limits_{i=1}^{n} e_i^2} = \frac{3.760}{15.272} = .246$$

The obtained value of d is evaluated through use of **Table A27 (Table of Critical Values for the Durbin–Watson test statistic)** in the **Appendix**. Note that determination of whether the **Durbin–Watson test statistic** is significant requires the use of the following two values: a) n represents the number of subjects/time periods. (If n is less than 15 one must attempt to estimate the values which would define the lower and upper limits that define the critical values.); and b) k represents the number of predictor/X variables. Although the discussion in this section will only deal with an analysis involving $k = 1$ predictor variable, the **Durbin–Watson test** can be employed to evaluate a **multiple regression** model (in which there are two or more predictor variables). **Table A27** contains critical **one-tailed** .05 and .01 values. Note that at a given alpha level, for each combination of n and k two critical values d_L and d_U are listed, with d_L representing a lower limit and d_U representing an upper limit. For Example 28.2, where $n = 16$ and $k = 1$, the tabled critical .05 and .01 values are respectively $d_{L_{.05}} = 1.10$ and $d_{U_{.05}} = 1.37$, and $d_{L_{.01}} = .84$ and $d_{U_{.01}} = 1.09$.

It can be demonstrated that the value of the d statistic ranges from 0 to 4, and when the data are not autocorrelated the value of d will be close to 2. Analysis of a d value can result in any of the following three decisions: a) Rejection of the null hypothesis H_0: $\rho_{res} = 0$; b) Retention of the null hypothesis H_0: $\rho_{res} = 0$; or c) An inconclusive result. Table 28.7 (based on Wilson and Keating (1990, p. 134)) summarizes the decision protocol to employ in evaluating the d statistic. Interpreting the computed value $d = .246$ with respect to the guidelines in Table 28.7 indicates our decision should be based on Outcome 2. Specifically, since $d = .246$ is greater than zero but less than $d_{L_{.01}} = .84$, the null hypothesis H_0: $\rho_{res} = 0$ can be rejected at the .01 level, and the researcher can conclude the data are positively autocorrelated (i.e., the directional alternative hypothesis H_1: $\rho_{res} > 0$ is supported).

Table 28.7　Decision Guidelines for Interpreting the Durbin–Watson Test Statistic (d)

Outcome	Value computed for d	Decision
1	$d_L < d < d_U$	Result is inconclusive
2	$0 < d < d_L$	Positive autocorrelation present
3	$2 < d < (4 - d_U)$	No autocorrelation present
4	$d_U < d < 2$	No autocorrelation present
5	$(4 - d_L) < d < 4$	Negative autocorrelation present
6	$(4 - d_U) < d < (4 - d_L)$	Result is inconclusive

It should be noted that if a researcher elected to conduct a **two-tailed analysis** (i.e., evaluate the alternative hypothesis H_1: $\rho_{res} \neq 0$ (which is commensurate with simultaneously evaluating the two alternative hypotheses H_1: $\rho_{res} > 0$ and H_1: $\rho_{res} < 0$)), the Type I error rates associated with the .05 and .01 critical values listed in **Table A26** are respectively .10 and .02.

The positive autocorrelation obtained for Example 28.2 is consistent with the fact that the residuals for the first eight time periods in **Column 5** of Table 28.6 are all positive values while the residuals computed for the last eight time periods are all negative. One should take note of the fact that when adjacent residuals are positively correlated with one another, the difference scores in **Column 7** will be smaller than will be the case when adjacent residuals are not positively correlated, thus yielding a minimal sum for the squared differences of the adjacent residuals (i.e., a minimal sum for **Column 8** of Table 28.6, which represents the numerator of Equation 28.58). Note that the smaller the numerator of Equation 28.58, the closer d will be to zero, with a value of d close to zero indicating a positive autocorrelation. When data are negatively autocorrelated, the value of d will be in the upper range, since the difference scores for adjacent residuals recorded in **Column 7** of Table 28.6 will tend to be larger, thus resulting in a higher value for the sum of the squared differences of the adjacent residuals. As noted earlier, when the residuals are not autocorrelated, the value computed for d will be close to 2.

Table 28.8   Correlation of Adjacent Residuals for Example 28.2

Residual Pairs	$X = e_i$	$X^2 = e_i^2$	$Y = e_{i-1}$	$Y^2 = e_{i-1}^2$	$XY = (e_i)(e_{i-1})$
1	.727	.529	.636	.404	.462
2	.818	.669	.727	.529	.595
3	.909	.826	.818	.669	.744
4	1.000	1.000	.909	.826	.909
5	1.091	1.190	1.000	1.000	1.091
6	1.182	1.397	1.091	1.190	1.290
7	1.273	1.621	1.182	1.397	1.505
8	-.636	.404	1.273	1.621	-.810
9	-.727	.529	-.636	.404	.462
10	-.818	.669	-.727	.529	.595
11	-.909	.826	-.818	.669	.744
12	-1.000	1.000	-.909	.826	.909
13	-1.091	1.190	-1.000	1.000	1.091
14	-1.182	1.397	-1.091	1.190	1.290
15	-1.273	1.621	-1.182	1.397	1.505
Sums	$\Sigma X = -.636$	$\Sigma X^2 = 14.868$	$\Sigma Y = 1.273$	$\Sigma Y^2 = 13.651$	$\Sigma XY = 12.382$

$$r_{e_i/e_{i-1}} = \frac{12.382 - \dfrac{(-.636)(1.273)}{15}}{\sqrt{\left[14.868 - \dfrac{(-.636)^2}{15}\right]\left[13.651 - \dfrac{(1.273)^2}{15}\right]}} = .877$$

Various sources (e.g., Hildebrand (1986, p. 707) and Webster (1995, p. 753)) note that an approximation of the d statistic can be computed with Equation 28.59, where $r_{e_i/e_{i-1}}$ is the correlation between each residual value and the residual value for the preceding time period.

$$d = 2(1 - r_{e_i/e_{i-1}}) = 2(1 - .877) = .246 \qquad \textbf{(Equation 28.59)}$$

Table 28.8 summarizes the correlation of the value $r_{e_i/e_{i-1}} = .877$. Note that in the Table 28.8, beginning with time period 2, every e_i value recorded in each row of **Column 5** in Table 28.6 is employed to represent the X variable, and is paired with the e_i value recorded for the prior time period, which represents the Y variable. When the value $r_{e_i/e_{i-1}} = .877$ computed for the correlation between the adjacent residuals is substituted in Equation 28.59, it yields the value $d = .246$, which is the identical value computed with Equation 28.58 (although the value computed with Equation 28.59 will not usually be identical to that computed with Equation 28.58). It should also be noted that when the value $r_{e_i/e_{i-1}} = .877$ is evaluated in reference to the critical values in **Table A16** (for $df = 15 - 2 = 13$, the critical two-tailed .05 and .01 values are $r_{.05} = .514$ and $r_{.01} = .641$), the correlation between adjacent residuals is significant at the .01 level (since $r_{e_i/e_{i-1}} = .877$ is greater than $r_{.01} = .641$). The latter result is consistent with the presence of a positive autocorrelation.

A revised set of data for Example 28.2 in which there is no autocorrelation will now be considered. The latter data is summarized in Table 28.9. Inspection of the latter table reveals that the 16 pairs of scores employed in Table 28.9 are identical to those employed in Table 28.6. However, the order (i.e., the time period) in which the 16 pairs of scores are listed has been modified so as to eliminate the positive autocorrelation previously detected for the data in Table 28.6. Since the same values as those employed in Table 28.6 are employed in Table 28.9 for each pair of X (advertising) and Y (sales volume) scores, the correlation coefficient of $r = .909$ computed earlier with respect to Table 28.6 between the amount of money spent on advertising and sales volume is also computed for the data in Table 28.9. Similarly, the regression equation $Y' = .455 + .909X$ obtained for the data in Table 28.6 is also obtained for the data in Table 28.9. Note, however, that when the ordering of the paired scores is taken into account, and the data in Table 28.9 are evaluated for autocorrelation with Equation 28.58, the value $d = 1.614$ is computed for the **Durbin–Watson test** statistic.

$$d = \frac{\sum\limits_{i=1}^{n}(e_i - e_{i-1})^2}{\sum\limits_{i=1}^{n}e_i^2} = \frac{24.653}{15.272} = 1.614$$

Employing **Table A27**, the value $d = 1.614$ is evaluated with the same critical values for $n = 16$ and $k = 1$ employed previously — specifically, $d_{L_{.05}} = 1.10$ and $d_{U_{.05}} = 1.37$ and $d_{L_{.01}} = .84$ and $d_{U_{.01}} = 1.09$. Interpreting the computed value $d = 1.614$ with respect to the guidelines in Table 28.7 indicates our decision should be based on Outcome 4. Specifically, since $d = 1.614$ is greater than $d_{U_{.05}} = 1.37$ but less than 2, we can conclude the data are not autocorrelated. Consequently, the null hypothesis $H_0: \rho_{res} = 0$ cannot be rejected at the .05 level. The absence of autocorrelation for the data in Table 28.9 is consistent with the fact that the residuals in **Column 5** of the latter table do not appear to conform to a predictable pattern (which was not

Table 28.9 Revised Data for Example 28.2 and Computations for Durbin–Watson Test

Time	X_i	Y_i	Y_i'	$e_i = Y_i - Y_i'$	e_i^2	$e_i - e_{i-1}$	$(e_i - e_{i-1})^2$
1	8	9	7.727	1.273	1.621	—	—
2	3	2	3.182	−1.182	1.397	−2.455	6.027
3	3	4	3.182	.818	.669	2.000	4.000
4	2	3	2.273	.727	.529	−.091	.008
5	7	6	6.818	−.818	.669	−1.545	2.387
6	9	8	8.636	−.636	.404	.182	.033
7	4	3	4.091	−1.091	1.190	−.455	.207
8	2	1	2.273	−1.273	1.621	−.182	.033
9	8	7	7.727	−.727	.529	.546	.298
10	6	5	5.909	−.909	.826	−.182	.033
11	5	6	5.000	1.000	1.000	1.909	3.644
12	5	4	5.000	−1.000	1.000	−2.000	4.000
13	4	5	4.091	.909	.826	1.909	3.644
14	6	7	5.909	1.091	1.190	.182	.033
15	7	8	6.818	1.182	1.397	.091	.008
16	1	2	1.364	.636	.404	−.546	.298
Sums	80	80		$\Sigma = 0$	$\Sigma = 15.272$	$\Sigma = -1.909$	$\Sigma = 24.653$

the case for Table 28.6). Note also that in Table 28.9 the following relationship can be confirmed: $\Sigma(e_i - e_{i-1}) = -.637$ (i.e., the sum of the values in **Column 7**) is equal to $[(e_n = .636) - (e_1 = 1.273)] = -.637$.

Table 28.10 summarizes the correlation of the value $r_{e_i/e_{i-1}} = .133$ obtained for the data in Table 28.9. When the value $r_{e_i/e_{i-1}} = .133$ computed for the correlation between the adjacent residuals is substituted in Equation 28.59, it yields the value $d = 1.734$, which is close to the value $d = 1.614$ computed with Equation 28.58 (and like the latter d value indicates our decision should be based on Outcome 4 in Table 28.7). It should also be noted that when the value $r_{e_i/e_{i-1}} = .133$ is evaluated in reference to the critical values in **Table A16** (for $df = 15 - 2 = 13$, the critical two-tailed .05 and .01 values are $r_{.05} = .514$ and $r_{.01} = .641$), the correlation between adjacent residuals is not significant (since $r_{e_i/e_{i-1}} = .133$ is less than $r_{.05} = .514$). The latter result is consistent with the absence of autocorrelation.

$$d = 2(1 - r_{e_i/e_{i-1}}) = 2(1 - .133) = 1.734$$

Table 28.10 Correlation of Adjacent Residuals for Revised Data for Example 28.2

Residual Pairs	$X = e_i$	$X^2 = e_i^2$	$Y = e_{i-1}$	$Y^2 = e_{i-1}^2$	$XY = (e_i)(e_{i-1})$
1	−1.182	1.397	1.273	1.621	−1.505
2	.818	.669	−1.182	1.397	−.967
3	.727	.529	.818	.669	.595
4	−.818	.669	.727	.529	−.595
5	−.636	.404	−.818	.669	.520
6	−1.091	1.190	−.636	.404	.694
7	−1.273	1.621	−1.091	1.190	1.389
8	−.727	.529	−1.273	1.621	.925
9	−.909	.826	−.727	.529	.661
10	1.000	1.000	−.909	.826	−.909
11	−1.000	1.000	1.000	1.000	−1.000
12	.909	.826	−1.000	1.000	−.909
13	1.091	1.190	.909	.826	.992
14	1.182	1.397	1.091	1.190	1.290
15	.636	.404	1.182	1.397	.752
Sums	$\Sigma X = -1.273$	$\Sigma X^2 = 13.652$	$\Sigma Y = -.636$	$\Sigma Y^2 = 14.868$	$\Sigma XY = 1.933$

$$r_{e_i/e_{i-1}} = \frac{1.933 - \dfrac{(-1.273)(-.636)}{15}}{\sqrt{\left[13.652 - \dfrac{(-1.273)^2}{15}\right]\left[14.868 - \dfrac{(-.636)^2}{15}\right]}} = .133$$

When the result of the **Durbin–Watson test** is significant, thus indicating the presence of autocorrelation, a researcher should initially attempt to identify whether or not a key independent/predictor variable (such as the time variable) has been omitted from the regression model, and if so integrate the latter variable into the model — i.e., employ a **multiple regression** model. If the latter model is not successful at eliminating autocorrelation, there are a number of data transformations which can be attempted, which are described in Neter *et al.* (1990, pp. 495–504). The latter authors also note that when the **Durbin–Watson test** yields an inconclusive result, it indicates further analysis should be conducted employing a larger sample size, and if the latter is not feasible and/or successful in eliminating autocorrelation, the remedial measures noted above for eliminating the presence of autocorrelation should be employed. If the issue of an indeterminate result cannot be resolved, one should assume the presence of autocorrelation.

Among the problems that might occur when data are, in fact, autocorrelated are: a) Estimates of regression coefficients may be inaccurate, and, more specifically t values computed with Equations 28.27 and 28.28 for evaluating the null hypothesis H_0: $\beta = 0$ will be inflated, thus increasing the likelihood of committing a Type I error; b) The value of s_{res}^2 may underestimate the actual residual variability; c) Values computed for confidence intervals will be inaccurate; and d) Hildebrand (1986, p. 707) notes that even an autocorrelation with a relatively low value such as $r_{res} = .25$ can have a serious effect on any inferences drawn from a regression model.

When data are autocorrelated, the term **autoregression** is employed to describe the methodology for predicting values on the dependent variable (i.e., Y values) by employing values of that variable obtained during prior time periods — in other words, the dependent nature of time-sequenced errors is incorporated into the regression model. Thus, by employing the available information on the correlation between successive residuals, autoregression (which involves the use of **multiple regression**) allows one to optimize prediction. A full discussion of autoregression (which can be implemented with most standard statistical software) can be found in texts which specialize in regression analysis (e.g., Netter *et al.* (1990)) as well as in many books on business statistics (e.g., Black (2001, Chapter 15), and Canavos and Miller (1995, Chapter 11)). In addition to describing the use of regression methods in business forecasting, texts in business statistics typically describe alternative forecasting methods (i.e., methods which do not involve regression analysis) which are commonly referred to as time-series techniques.

Autocorrelation as a test of randomness   In Section IX (the **Addendum**) of the **single-sample runs test** a number of procedures are discussed which are employed in determining whether the ordering of a series of numbers is random. Among the procedures that are briefly discussed is **autocorrelation**, which is also referred to as **serial correlation**. In contrast to most of the tests of randomness discussed in Section IX of the **single-sample runs test** (which can only be employed with a discrete variable), autocorrelation can be employed to evaluate either a continuous or discrete variable for randomness.

The most basic methodology that can be employed for autocorrelation when it is used to evaluate whether or not a series of numbers is random is to pair each of the numbers in the series of n numbers with the number which follows (or precedes) it in the series. Upon doing this, the **Pearson product-moment correlation coefficient** between the resulting $(n - 1)$ pairs of numbers is computed. It is also possible to pair each number with the number whose ordinal position is some value other than one digit after (or before) it. In other words, each number can be paired with the number that is two digits after (or before) it, three digits after (or before) it, etc. in the series. In autocorrelation the number of digits which separates two values that are paired with one another is referred to as the **lag value**. In the example to be employed in this section the lag value +1 will be used, since each number will be paired with the number that is one ordinal position above it in the series. If, instead, each number is paired with the number which falls two ordinal positions above it in the series, the lag value is +2. If, on the other hand, each number is paired with the number that precedes it by one ordinal position in the series, the lag value is −1. The higher the absolute value of the lag value, the fewer the number of pairs which will be employed in computing the correlation. Thus, if in a series of ten digits each number is paired with the number that is above it by two ordinal positions, there will only be $n - 2 = 8$ pairs of X and Y scores. This is the case, since the first two numbers in the series can only be *X* scores, and the last two numbers in the series can only be *Y* scores. Regardless of the lag value employed in an autocorrelation, if the sequence of numbers in a series is random, the computed value of the correlation coefficient should equal zero.

One variant of the methodology which has been described above (which is referred to as **noncircular autocorrelation** (or **noncircular serial correlation**)) is a procedure referred to as **circular autocorrelation** (or **circular serial correlation**). In **circular autocorrelation** every number in a series of *n* numbers is paired with another number, including any number in the series which does not have a number following it. Numbers that are not followed by any number are sequentially paired with the numbers at the beginning of the series. Thus, if the lag value is +1, the last number in the series is paired with the first number in the series. If the lag value is +2, the $(n-1)^{th}$ number is paired with the first number in the series, and the nth number is paired with the second number in the series.

To illustrate the use of autocorrelation in evaluating randomness the following 17 digit series of numbers will be evaluated: 1, 2, 3, 4, 5, 6, 7, 8, 9, 8, 7, 6, 5, 4, 3, 2, 1. In Table 28.11 the 17 digits are arranged sequentially in Column A from top to bottom. The same 17 digits are arranged sequentially in Column B, except for the fact they are arranged so that each digit in Column B is adjacent to the digit in Column A which directly precedes it in the series. If each pair of adjacent values is treated as a set of scores, the value in Column A can be designated as an *X* score, and the value in Column B can be designated as a *Y* score. If the latter is done, each of the 17 digits in the series will at some point be designated as both an *X* score and a *Y* score, except for the first digit which will only be an *X* score and the last digit which will only be a *Y* score.

Table 28.11 Arrangement of Digits for Autcorrelation

Row	Column A (*X*)	Column B (*Y*)
1		1
2	1	2
3	2	3
4	3	4
5	4	5
6	5	6
7	6	7
8	7	8
9	8	9
10	9	8
11	8	7
12	7	6
13	6	5
14	5	4
15	4	3
16	3	2
17	2	1
18	1	

Note that if the data in Rows 2 – 17 of Table 28.11 (i.e., the 16 rows which contain a score both Column A and Column B) are viewed as listing a set of *X* scores in Column A and a set of *Y* scores in Column B, the data in the table become identical to that in Table 28.6, and consequently the same correlation coefficient $r = .909$ previously computed for Table 28.6 is obtained for the 16 pairs of X and Y scores in Table 28.11. Recollect that use of the **Durbin–Watson test** to evaluate whether or not the data in Table 28.6 are autocorrelated resulted in the value $d = .246$, indicating the presence of a significant positive autocorrelation. The same d value would be computed for Table 28.11, indicating a significant positive correlation between adjacent digits in the 17 digit series. Because of the latter the researcher would conclude that the distribution of the digits in the series is not random.

An alternative but not commonly employed methodology (described in earlier editions of this book) for determining whether or not data are autocorrelated will now be considered. As noted in the discussion of the analysis of Table 28.6, when the obtained correlation $r = .909$ is evaluated with the critical values in **Table A16**, the tabled critical two-tailed .05 and .01 r values employed (for $df = 16 - 2 = 14$) are $r_{.05} = .497$ and $r_{.01} = .623$. Since the computed value $r = .909$ (which is also obtained for Table 28.11) is greater than $r_{.01} = .623$, one might surmise it can be concluded at the .01 level of significance that the correlation between the variables in the underlying population is some value other than zero, and consequently that the 17 digit series is not random.

In point of fact, the tabled critical values in **Table A16** are not the most appropriate values to employ in evaluating the value $r = .909$ with respect to whether or not the 17 digit series in Table 28.11 indicates the presence of autocorrelation. This is the case, since the sampling distribution for a serial correlation coefficient is not identical to the sampling distribution upon which the critical values in **Table A16** are based. The sampling distribution upon which **Table A16** is based assumes that the n pairs of scores are independent of one another. Since in Table 28.11 all of the digits in the series (with the exception of the last digit) represent both an X and a Y variable, the latter assumption is violated. Because the pairs are not independent, the residuals derived from the data may also not be independent (the latter, as noted earlier, being a critical assumption of the least squares regression model.).[22]

Because of the fact the residuals may not be independent, a sampling distribution other than the one upon which the critical values in **Table A16** are based should be employed to evaluate the value $r = .909$ computed with Equation 28.1. Anderson (1942) demonstrated that in the sampling distribution for a serial correlation, the absolute value of a critical value at a prespecified level of significance is smaller than the corresponding critical value in **Table A16**. Furthermore, the limits which define a critical value at a prespecified level of significance are asymmetrical (i.e., the absolute value of a critical value will not be identical for a positive versus a negative r value). Anderson (1942) computed the critical two-tailed .05 and .01 values of r for values of n between 5 and 75 for lag +1. For large sample sizes he determined that Equation 28.60 (which employs the normal distribution) can be used to provide a good approximation of the critical values of r when the lag value is +1.[23]

$$r = \pm z \sqrt{\frac{n - 2}{(n - 1)^2} - \frac{1}{n - 1}}$$ **(Equation 28.60)**

Where: z represents the tabled critical value in the normal distribution which corresponds to the prespecified level of significance employed in evaluating r

n represents the total number of numbers in the series. Note that n is not the number of pairs of numbers employed in computing the coefficient of correlation.

Employing the values derived by Anderson (1942) for the exact sampling distribution of the serial correlation coefficient (which is not reproduced in this book), it can be determined (for

$\alpha = .05$ and $n = 17$) that in order to reject the null hypothesis H_0: $\rho = 0$, the computed value of r must be equal to or greater than $r_{.05} = .315$ or equal to or less than $r_{.05} = -.428$. (The value of n used in Anderson's (1942) table represents the total number of digits in the series and not the number of pairs of digits employed in computing the correlation.) Since $r = .909$ is greater than $r_{.05} = .315$, the null hypothesis can be rejected. Thus, regardless of whether one employs **Table A16** or Anderson's (1942) critical values, the null hypothesis can be rejected. Nevertheless, the difference between the critical values in the two tables is substantial.

Use of Anderson's (1942) tables or Equation 28.60 provide for a more powerful test of the alternative hypothesis H_1: $\rho \neq 0$ than do the critical values in **Table A16**. Although the degree of discrepancy between a critical value in **Table A16** and a critical value computed with Equation 28.60 decreases as the size of n increases, even for large sample sizes the absolute values in **Table A16** are noticeably higher. It should be noted that use of Equation 28.60 with small samples yields absolute critical values that are too high. In closing this discussion it should be emphasized that the **Durbin–Watson test** would represent the optimal method for assessing whether or not the data in Table 28.11 are autocorrelated, assuming the lag values $+1$ or -1 are employed. When the lag value has an absolute value greater than 1, one of the tests noted earlier at the beginning of the discussion of the **Durbin–Watson test** for evaluating higher lag values would be recommended.

In the discussion of tests of randomness in the chapter on the **single-sample runs test**, it is noted that it is not uncommon for two or more of the available tests for determining randomness to yield conflicting results. Although autocorrelation is not considered to be among the most rigorous tests for randomness, if one conducts multiple autocorrelations on a series (i.e., for the lag values $+1$, $+2$, $+3$, etc. and -1, -2, -3, etc.), and all or most lead to retention of the null hypothesis, such a protocol will provide a more authoritative analysis with respect to randomness than will the single analysis for lag $+1$ conducted in this section. It should be noted that if for a series of n numbers (where the value of n is large) an autocorrelation is conducted for every possible positive and negative lag value, just by chance it is expected that some of the computed serial correlations will be significant. Whatever prespecified alpha value the researcher employs will determine the proportion of significant correlations that can be obtained which will still allow one to retain the null hypothesis H_0: $\rho = 0$.

One limitation of autocorrelation as a test of randomness should be noted. Assume that a researcher is evaluating a series in which on any trial a number can assume any one of $k = 5$ possible values. For instance, consider a situation in which the integer values 1, 2, 3, 4, and 5 are the only possible values that can occur. In a truly random series of reasonable length, each of the five digits would be expected to occur approximately the same number of times. Yet, it is entirely possible to have a series of numbers in which one or more of the integer values does not even occur one time, yet the resulting autocorrelation is $r = 0$. For instance, a computer can be programmed to generate a series of 1000 pseudorandom numbers employing the integer values 1, 2, 3, 4, and 5. Yet it is theoretically possible for the computer algorithm to generate 1000 digits, all of which are either 1, 2, 3 or 4, yet no instances of the digit 5. If the autocorrelation between all of the digits generated is zero, it will suggest the sequence of numbers is random. Although it may be a random sequence for a population in which the only values the numbers may assume are the integer values 1, 2, 3, and 4, it is not a random series for a population in which the numbers may assume an integer value between 1 and 5. Whereas most of the other tests that are employed in evaluating randomness will identify this problem, autocorrelation will not.

6. The phi coefficient as a special case of the Pearson product-moment correlation coefficient A number of the correlational procedures discussed in this book represent special cases of the **Pearson product-moment correlation coefficient**. One of the procedures, the **phi coefficient**, is described in Section VI of the **chi-square test for $r \times c$ tables (Test 16)**. Another of the procedures, the **point-biserial correlation coefficient**, is described in Section IX (the **Addendum**). A third procedure, **Spearman's rank-order correlation coefficient**, is discussed in the next chapter.

In this section it will be demonstrated how the **phi coefficient** (represented by the notation ϕ) can be computed with Equation 28.1. In the discussion of the latter measure of association, it is noted that the value of **phi** is equivalent to the value of the **Pearson product-moment correlation coefficient** which will be obtained if the scores 0 and 1 are employed with reference to two dichotomous variables in a 2×2 contingency table. Using the data for Examples 16.1/16.2 (which employ a 2×2 contingency table), the scores 0 and 1 are employed for each of the categories on the two variables. Table 28.12 summarizes the data.

Table 28.12 Summary of Data for Examples 16.1/16.2

		X variable		Row sums
		0	1	
Y variable	0	$a = 30$	$b = 70$	100
	1	$c = 60$	$d = 40$	100
Column sums		90	110	Total = 200

Table 28.12 reveals the following: 30 subjects have both an X score and a Y score of 0; 70 subjects have an X score of 1 and a Y score of 0; 60 subjects have a X score of 0 and a Y score of 1; 40 subjects have both an X score and a Y score of 1. Employing this information we can determine $\Sigma X = 110$, $\Sigma X^2 = 110$, $\Sigma Y = 100$, $\Sigma Y^2 = 100$, $\Sigma XY = 40$. Substituting these values in Equation 28.1, the value $r = .30$ is computed. The latter value is identical to $\phi = .30$ computed for Examples 16.1/16.2 with Equations 16.20 and 16.21.

$$ r = \frac{40 - \frac{(110)(100)}{200}}{\sqrt{\left[110 - \frac{(110)^2}{200}\right]\left[100 - \frac{(100)^2}{200}\right]}} = .30 $$

7. Ecological correlation In investigating the relationship between two variables, a researcher may want to obtain correlational information regarding the behavior of individuals, yet only has access to summary data for two or more groups of which the individuals are members. A correlation coefficient which is based on averages, proportions, or percentages for **groups** is referred to as an **ecological correlation**. **Ecological inference** is a conclusion regarding statistical association with respect to individuals which is based upon variables that have been measured and evaluated on an aggregate level (i.e., based on grouped data). To illustrate the latter concepts, let us assume a researcher has access to the mean educational level (which will represent the X variable) and mean income (which will represent the Y variable) of the 50 states which comprise the United States. Employing the 50 pairs of educational level–income scores, the researcher determines that the correlation between the two variables is $r = .77$. The latter value represents an **ecological correlation**, since it was computed through use of summary values for group data. Let us assume the researcher subsequently obtains the actual scores for educational level and income for the 200 million people in United States for whom the state averages with respect to the latter variables were employed in computing the correlation $r = .77$. Will the correlation computed for 150 million pairs of scores be equal to or close to the previously computed ecological correlation of $r = .77$? The answer to the latter question is that there is a reasonable likelihood it would not. In point of fact, it is not at all unusual for an ecological correlation to have a different sign and/or absolute value from the sign and/or absolute value of the correlation coefficient computed from all of the individual scores. It would also not be unusual that within any of the 50 states the correlation between educational level and income for the residents of that state would not be equal or close to the ecological correlation $r = .77$.

The fact there may be a substantial discrepancy between a correlation coefficient based on summary values for group data versus a correlation coefficient based on the individual scores of all or some of those who comprised the groups can be explained by the **ecological fallacy** (often discussed within the context of **aggregation bias**). The latter refers to the erroneous assumption that an observed relationship at the aggregate level (i.e., a relationship based on groups) implies the same relationship exists at the individual level (i.e., the relationship of the scores of each of the individuals who comprise the relevant groups).

In a seminal paper on ecological correlation, Robinson (1950) illustrated the pitfalls of the methodology with an example involving the relationship between nativity and literacy (based on data for the year 1930). Robinson (1950) noted that since educational standards were lower for foreign born than for native born, it was expected that a correlational analysis would indicate foreign born tend to be less literate than native born. Employing data for approximately 100,000 subjects (10 years or older) who were categorized with respect to being native or foreign born (which represented the X variable) and whether a person was literate or illiterate (which represented the Y variable), a correlation (specifically a phi coefficient) of $r = .118$ was obtained between the two variables. The latter weak positive correlation indicated a slight trend toward greater literacy among native born. However, an ecological correlation coefficient (based on 48 pairs of scores obtained for the 48 states which comprised the United States at that point in time) between the percentage of foreign born people residing in a state (which represented the X variable) and the percentage of residents in a state who were literate (which represented the Y variable) was computed to be $-.526$, suggesting a moderate negative correlation between nativity and literacy (i.e., that foreign born are more likely to be literate than native born). It turns out that the substantial discrepancy between the values of the ecological correlation and the correlation based on individual scores can be explained on the basis of a confounding/intervening variable which the ecological correlation failed to take into account — specifically, the fact that foreign born residents were more likely to live in states where native born residents are more literate.

The simplest explanation for why ecological correlations can be misleading is that by virtue of computing a summary value, such as an average for each group, individual variation and covariation within a group (which are the key components in computing a correlation coefficient) is lost. The latter is illustrated in Figure 28.11 which summarizes the scores of 42 subjects in each of three groups A, B, and C (with 14 subjects per group) who have been measured on an X and a Y variable. In the latter figure, each occurrence of a letter represents the composite score of a subject in that group on the X and Y variables. Note that since the configuration of the 14 letters in each group is rectangular, the correlation between X and Y for a group will equal 0. Yet when the means of the groups are computed for each of the variables and plotted as the three points $\bar{A}$, $\bar{B}$, and $\bar{C}$, the arrangement of the latter values results in a positively sloped straight line which results in an ecological correlation of $r = 1$. Note that use of the group means to summarize the data totally obscures the variability and covariability within each of the groups, and consequently leads to totally different conclusions regarding the relationship between the *X* and *Y* variables. To further complicate matters, if the 42 subjects are collapsed into one group, depending upon how far removed the 14 letters in each group are from the positively sloped line depicted in Figure 28.11, the resulting correlation coefficient will be a moderate to high positive value (which will be less than the value $r = 1$ computed for the ecological correlation). Thus, aside from the fact that a substantial discrepancy exists between the values of the ecological and individual correlations, it appears that both correlations fail to provide an accurate summary of the nature of the relationship between the two variables within each of the groups. The above example illustrates the importance of obtaining a visual picture of data before coming to conclusions which are entirely based on the value of a single correlation coefficient.

In addition to plotting and carefully scrutinizing a complete set of data, the best remedy for avoiding the ecological inference problem (i.e., drawing erroneous conclusions regarding individual behavior based on analysis of grouped data) is that prior to conducting an analysis a researcher should attempt to identify potentially confounding variables which are relevant to the grouping process, and integrate the latter variables into an overall model. Only after doing the latter can the researcher determine the most suitable methodology for evaluating the data, and thus be in a position to correctly interpret the nature of the relationship between the variables of interest.

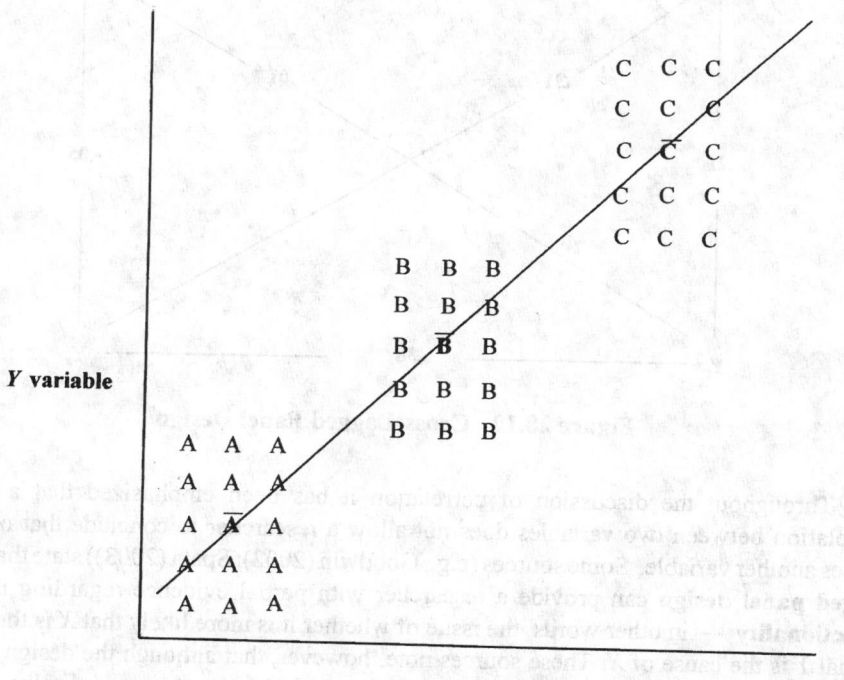

Figure 28.11 Ecological versus Individual Correlation

8. Cross-lagged panel and regression-discontinuity designs This section will describe two experimental designs (both of which are most commonly categorized as **quasi-experimental designs**) which some sources note when employed with correlational data are able to clarify issues relating to cause and effect. In point of fact, as will be noted below, only the latter of the two designs, the **regression-discontinuity design**, provides strong evidence with respect to clarifying issues of causality.

Cross-lagged panel design The **cross-lagged panel design** (which according to Cook and Campbell (1979, p. 309) was first introduced by Lazarsfeld (1947, 1948) and was subsequently popularized and/or further evaluated by, among others, Campbell (1963), Campbell and Stanley (1963), Pelz and Andrews (1964), Rozelle and Campbell (1969), and Kenny (1973, 1975)) is a **quasi-experimental design** — more specifically, a type of **longitudinal/panel design**. (A discussion of **quasi-experimental** and **longitudinal designs** can be found in the **Introduction** in the section on experimental design.) The **cross-lagged panel design** represents a **longitudinal design** since repeated measures on two variables are taken on a sample of subjects at two different points in time as depicted in Figure 28.12. The configuration of the four measures and the accompanying correlations depicted in the latter figure are commonly referred to as a **panel**.

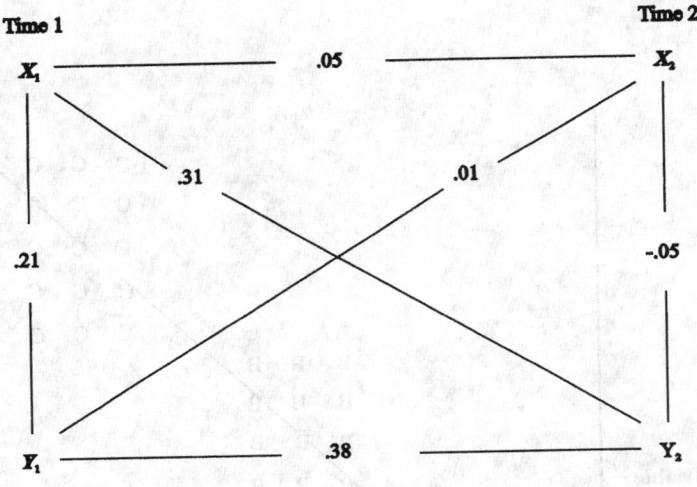

Figure 28.12 Cross-Lagged Panel Design

    Throughout the discussion of correlation it has been emphasized that a substantial correlation between two variables does not allow a researcher to conclude that one variable causes another variable. Some sources (e.g., Goodwin (2002); Spata (2003)) state that the **cross-lagged panel design** can provide a researcher with partial evidence regarding the issue of **directionality** — in other words, the issue of whether it is more likely that X is the cause of Y or that Y is the cause of X. These sources note, however, that although the design can help to clarify the issue of directionality, in the final analysis it cannot unequivocally establish it. Rogosa (1980, 1987), among others, argues that the **cross-lagged panel design** is not a reliable methodology for evaluating longitudinal/panel data, and consequently should never be used for that purpose. Instead, the latter author recommends the use of alternative more contemporary methods for evaluating panel data. (One such methodology, **path analysis** (which can be employed to clarify possible causal relationships between two or more variables), is discussed in Section VII of the chapter on **multiple regression**.)

In order to illustrate a **cross-lagged panel design** an oft cited study conducted by Eron *et al.* (1972) will be described. The latter study investigated the relationship between preference for violent television programs (which will be designated as the X variable) and aggression (which will be designated as the Y variable) in children at two points in time — specifically, when the children were in the third grade (**Time 1**) and ten years later (**Time 2**). The notation X_1 will represent a child's score on the X variable during **Time 1**; X_2 will represent a child's score on the X variable during **Time 2**; Y_1 will represent a child's score on the Y variable during **Time 1**; Y_2 will represent a child's score on the Y variable during **Time 2**. The results of the study are summarized in Figure 28.12. Note that when the children were in the third grade the correlation between preference for violent television programs and aggression was $r_{X_1 Y_1} = .21$, indicating a modest positive relationship. Ten years later the correlation between the two variables was only $r_{X_2 Y_2} = -.05$.

In addition to the latter two correlation coefficients, the use of a **crossed-lagged panel design** allowed the researchers to compute the following four additional correlations: $r_{X_1 X_2} = .05$, $r_{Y_1 Y_2} = .38$, $r_{X_1 Y_2} = .31$, and $r_{Y_1 X_2} = .01$. The two correlation coefficients $r_{X_1 Y_1} = .21$ and $r_{X_2 Y_2} = -.05$ on the vertical pane of the panel depicted in Figure 28.12 are not **time lagged**, since the measures on the X and Y variables are obtained during the same time period. On the other hand, the two correlation coefficients $r_{X_1 X_2} = .05$ and $r_{Y_1 Y_2} = .38$ on the horizontal plane of the panel are **time lagged**, since one of the variables was measured during **Time 1** while the other variable was measured during **Time 2**. Likewise, both of the correlation coefficients $r_{X_1 Y_2} = .31$ and $r_{Y_1 X_2} = .01$ on the diagonal of the panel are **time lagged**, since one of the variables was measured during **Time 1** while the other variable was measured during **Time 2**. The latter two diagonal correlations are said to be **cross-lagged**.

The **cross-lagged correlations** are the most critical ones evaluated within the framework of the **cross-lagged panel design**. The rationale for the latter is that since cause precedes effect, one would expect a variable which occurs at an earlier time period to have a moderate to high relationship with a variable which occurs at a later time period. The design assumes that if $r_{X_1 Y_2}$ is larger than $r_{Y_1 X_2}$ it would suggest that the X variable affects the Y variable, while, on the other hand if $r_{Y_1 X_2}$ is larger than $r_{X_1 Y_2}$ it would suggest that the Y variable affects the X variable. With respect to the study under discussion, the following two hypotheses can be evaluated: a) The amount of television violence children preferred in **Time 1** is causally related to their level of aggression at **Time 2**. More specifically, if the amount of television violence children preferred in the third grade was causally related to their level of aggression ten years later, the value of the correlation $r_{X_1 Y_2}$ should be moderate to high; b) The level of aggression in children in **Time 1** is causally related to the amount of television violence they preferred in **Time 2**. More specifically, if the aggression level of children during the third grade was causally related to the amount of television violence they preferred ten years later, the value of the correlation $r_{Y_1 X_2}$ should be moderate to high. The fact that $r_{X_1 Y_2} = .31 > r_{Y_1 X_2} = .01$ was interpreted by Eron *et al.* (1972) as supporting the hypothesis that the amount of television violence children preferred in the third grade was causally related to their level of aggression ten years later. The fact that the correlation $r_{Y_1 X_2} = .01$ (a minimal value) was interpreted as suggesting there is no causal relationship between the aggression level of children during the third grade and the amount of television violence they preferred ten years later.

Although most sources which discuss the **cross-lagged design** focus on the values of the cross-lagged correlations, the other four correlation coefficients in the panel are sometimes taken into consideration in evaluating the nature of the relationship between the variables. With respect to the latter, however, sources are not in agreement. Among the statements made regarding the other correlation coefficients in the panel are the following: a) Spata (2003, p. 213) notes that if there is a relationship between the two variables one would expect a substantial correlation in the same direction for both of the vertical panel correlations $r_{X_1 Y_1}$ and $r_{X_2 Y_2}$, with the value of $r_{X_2 Y_2}$ being larger than $r_{X_1 Y_1}$ — the latter indicating a stable relationship over time and the fact that the value obtained for $r_{X_1 Y_1}$ was not due to chance. Note, however, that the latter is not the case in Figure 28.12; b) Spata (2003, p. 213) also suggests if, in fact, the researcher hypothesizes that X is the cause of Y as opposed to Y being the cause of X, it is expected that the suspected cause should be more stable over time and therefore have a higher correlation — i.e., with respect to the horizontal panel correlations, $r_{X_1 X_2} > r_{Y_1 Y_2}$ (which is not the case in Figure 28.12). If, on the other hand, Y causes X, one would expect that $r_{Y_1 Y_2} > r_{X_1 X_2}$ (which is the case in Figure 28.12). Neale and Liebert (1980, p. 212), on the other hand, state that the reliability of the measures should not change between the two measurement periods (which is not the case in Figure 28.12), and that if one of the variables has a higher reliability coefficient than the other, the variable with the higher reliability coefficient is more likely to be the effect variable rather than the cause variable (which is the case in Figure 28.12).

An obvious limitation of **cross-lagged correlation** is that it does not eliminate the **third variable problem** — specifically, that a third variable (or possibly more than one additional variable) not included in the study can better explain the correlations obtained in the panel. (Such a variable is often referred to as a **moderator** or **mediating variable**.) Thus, even if the results suggest that X may be partially responsible for causing Y or vice versa, there is still the possibility that some uncontrolled for third variable Z, which is correlated with both X and Y, is causally related to both. One way of addressing the third variable problem is for a researcher to identify potentially confounding variables, and through the use of **partial correlation** (which is discussed in the chapter on **multiple regression**) evaluate the effects of such variables on the variables of interest. A **partial correlation coefficient** allows a researcher to measure the degree of association between two variables after any linear association one or more additional variables have with the other two variables has been removed. In the case of a third variable Z, **partial correlation** allows one to measure the relationship between X and Y with any effect that might be attributed to the variable Z controlled for. Consequently, if a **partial correlation** between X and Y (represented with the notation $r_{YX.Z}$) is essentially the same as the original correlation obtained between X and Y (i.e., r_{XY}), then Z can be ruled out as a possible influencing variable. In point of fact, Eron *et al.* (1972) used **partial correlation** to rule out the role of a number of potential extraneous variables to reinforce their conclusion that the amount of television violence children preferred in the third grade was likely to have had some effect on their level of aggression ten years later

In view of the criticisms and inconsistent predictions associated with the **crossed-lagged panel design**, anyone considering using it is advised to consult sources such as Neale and Liebert (1980), Rogosa (1980, 1987), and Shadish *et al.* (2002) (which provide more thorough documentation of its shortcomings) prior to employing it as a research design. In the final analysis, most sources would recommend that researchers interested in trying to clarify hypotheses regarding cause and effect within the framework of longitudinal research would be better advised to employ more contemporary methods such as **path analysis** and **structural equation modeling (SEM)** (both of which are discussed in the chapter on **multiple regression**).

Regression-discontinuity design In a **regression-discontinuity design** subjects are assigned to experimental conditions on the basis of a **cutoff score** on some **predetermined assignment variable**. The design (which was initially employed in research during the early 1960s) has been most commonly employed in educational research for the purpose of program evaluation. An example of a **regression-discontinuity design** might involve dividing third grade students into two groups based on whether they are above or below a state mandated target cutoff score on a reading achievement test. Upon forming the groups the children who score below the cutoff point are enrolled in a remedial training program (which represents the **independent variable** in the study), whereas no special treatment is given to the children who score above the cutoff point. Analysis of the **regression-discontinuity design** would allow a researcher to determine whether or not remedial training is effective in raising childrens' level of performance (on the same achievement test or on some other appropriate criterion measure of performance, such as a reading achievement test administered at the end of the fourth grade — the latter will be employed to represent the **dependent variable** in the study to be described in more detail below). It should be noted that a critical factor vital for insuring the integrity of the **regression-continuity design** is that the researcher has a sound theoretical basis for expecting a substantial correlation between the pretest and posttest measures employed in the study.

The appeal of the **regression-discontinuity design** when contrasted with the use of a **randomized-groups/independent samples design** for evaluating the same hypothesis is that the **regression-continuity design** does not require the use of the more typical type of control group. In the case of the experiment described above, the **randomized-groups design** would require a researcher to randomly assign a sample of subjects in need of remedial training to either an experimental or control group, with only the former group receiving remedial training. Thus, from an ethical standpoint the **regression-discontinuity design** might be deemed more appropriate, since it does not require a control group of subjects which score below the cutoff point who if employed within the context of a **randomized-groups design** would be denied remedial training.

An actual example that illustrates the use of a **regression-discontinuity design** is a study conducted by Berk and Rauma (1983) and Rauma and Berk (1987) (which is described in Shadish *et. al.* (2002, p. 207)) that assessed whether or not providing inmates with unemployment insurance upon their release from prison decreased the likelihood of recidivism. The study was based on a California law which provided unemployment insurance to inmates who had worked a minimum specified number of hours in prison during the year prior to their release. Inmates were divided into two groups based on whether or not they qualified for unemployment compensation based on the specified cutoff number of hours worked during the year prior to release. Evaluation of the data obtained within the context of a **regression-discontinuity design** indicated that the unemployment compensation program (the independent variable) was effective in reducing the rate of recidivism (the dependent variable).

Figure 28.13 summarizes a **regression-continuity design** — specifically, a **regression-continuity design** for the example described initially which involved remedial training for third grade children who score below the state mandated target in reading achievement. Inspection of the latter figure reveals that the design has similarities with the **pretest-posttest control group design** (Figure I.26) as well as the **nonequivalent control group design** (Figure I.19) — both of which are described in the **Introduction**. The feature which differentiates the **regression-discontinuity design** from the latter two designs is the method employed for group assignment. Note that in the **pretest-posttest control group design** subject assignment to groups is random, and in the **nonequivalent control group design** subject assignment to groups is based on membership in a preexisting intact group. In the **regression-discontinuity design** group assignment is based on a cutoff score on a specific measure obtained prior to initiating the program to be evaluated. The design depicted in Figure 28.13 is also applicable to the Berk and Rauma (1983)/Rauma and Berk (1987) prison inmate unemployment study, except for the fact that in the case of the latter study the group which scored above the cutoff point on the assignment variable (i.e., had worked above the required number of hours) would be the experimental group, while the control group would be the group that scored below the cutoff point.

	Time 1	Time 2	Time 3
Experimental group (Below cutoff on assignment variable)	Pretreatment score on assignment variable	Treatment	Posttreatment score on criterion variable
Control group (Above cutoff on assignment variable)	Pretreatment score on assignment variable	-------------	Posttreatment score on criterion variable

Figure 28.13   Regression-Discontinuity Design

Figure 28.14 illustrates hypothetical results for the study involving remedial training for third grade children who score below the state mandated target in reading achievement. Specifically, Figure 28.14a depicts an outcome if a treatment effect is not present (i.e., the remedial program was not effective), Figure 28.14b depicts an outcome consistent with the treatment being effective in raising student performance, and Figure 28.14c depicts an outcome (albeit unexpected) indicating a negative impact resulting from remedial training. Note that in the three graphs depicted in Figure 28.14, the X axis is employed for the pretreatment scores of subjects on the assignment variable (the achievement test) and the Y axis for the posttreatment scores on the criterion variable (which we will assume is a reading achievement test administered to all of the children at the conclusion of the fourth grade). It will be assumed that on both the pretreatment and posttreatment tests a child can obtain a score ranging from 0 to 100, and that the cutoff score on the assignment variable (i.e., the pretreatment test) is 60. Thus, any third grader who obtains a score below 60 is assigned to the experimental group, with the remaining students assigned to the control group. The vertical line perpendicular to the pretest score of 60 on the X axis demarcates the cutoff score on the third grade achievement test. In each of the scatterplots in Figure 28.14, a given dot to the left of the vertical line represents the two scores of a specific subject in the experimental group on the pretest and the posttreatment test, while a given dot to the right of the vertical line represents the two scores of a specific subject in the control group. For purposes of illustration, it will be assumed that in the case of a positive treatment effect (depicted in Figure 28.14a) the remedial program consistently raised the scores of subjects in the experimental group about ten points, and that in the case of a negative treatment effect (depicted in Figure 28.14c) the scores of subjects in the experimental group consistently dropped by about ten points.

The regression line depicted in Figure 28.14a indicates the absence of a treatment effect. Specifically, the graph indicates a strong positive relationship between the pretreatment and posttreatment scores of subjects — i.e., the higher a subject's pretreatment score the higher the subject's posttreatment score. Note that the regression line is **continuous** insofar as if one connects the most extreme point at the lower left of the portion of the regression line to the left of the vertical line demarcating the cutoff point with the most extreme point at the upper right of the portion of the regression line to the right of the vertical line, the end result is a continuous straight line. This latter pattern is consistent with the absence of a treatment effect.

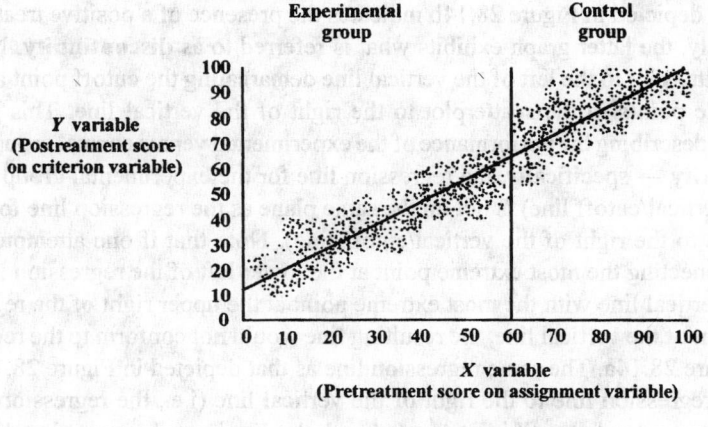

(a) **No treatment effect**

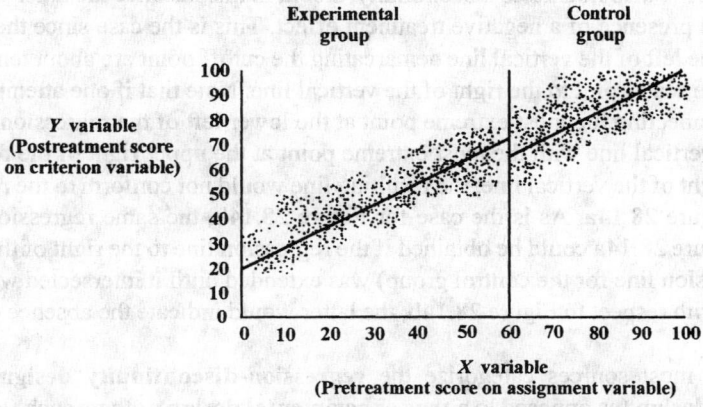

(b) **Positive treatment effect**

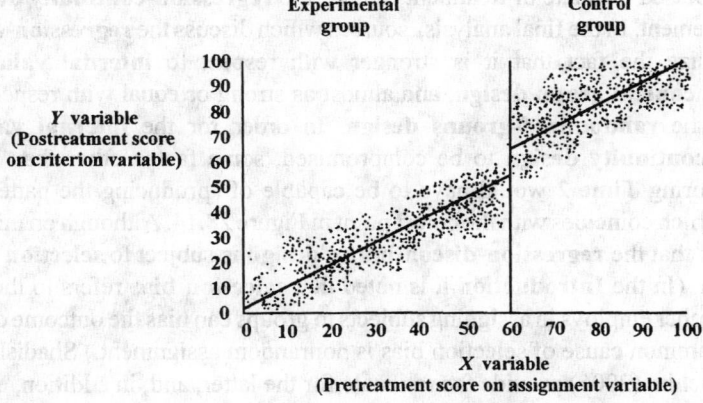

(c) **Negative treatment effect**

Figure 28.14   Hypothetical Outcomes for Regression-Discontinuity Design

The graph depicted in Figure 28.14b indicates the presence of a positive treatment effect. More specifically, the latter graph exhibits what is referred to as **discontinuity**. Note that the points in the scatterplot to the left of the vertical line demarcating the cutoff point are about ten points above the points in the scatterplot to the right of the vertical line. This break in the regression lines describing the performance of the experimental versus control groups is referred to as **discontinuity** — specifically, the regression line for the experimental group (which is to the left of the vertical/cutoff line) is not on the same plane as the regression line for the control group (which is to the right of the vertical/cutoff line). Note that if one attempted to draw a straight line connecting the most extreme point at the lower left of the regression line that is to the left of the vertical line with the most extreme point at the upper right of the regression line that is to the right of the vertical line, the resulting line would not conform to the regression line depicted in Figure 28.14a. The same regression line as that depicted in Figure 28.14a could be obtained if the regression line to the right of the vertical line (i.e., the regression line for the control group) was extended until it intersected with the Y axis, and, as noted earlier, the latter would indicate the absence of a treatment effect.

Figure 28.14c also illustrates discontinuity, albeit, in this instance the latter would indicate the unexpected presence of a negative treatment effect. This is the case since the points in the scatterplot to the left of the vertical line demarcating the cutoff point are about ten points below the points in the scatterplot to the right of the vertical line. Note that if one attempted to draw a straight line connecting the most extreme point at the lower left of the regression line that is to the left of the vertical line with the most extreme point at the upper right of the regression line that is to the right of the vertical line, the resulting line would not conform to the regression line depicted in Figure 28.14a. As is the case for Figure 28.14b, the same regression line as that depicted in Figure 28.14a could be obtained if the regression line to the right of the vertical line (i.e., the regression line for the control group) was extended until it intersected with the Y axis, and, as noted with respect to Figure 28.14b, the latter would indicate the absence of a treatment effect.

Although most sources categorize the **regression-discontinuity design** as a **quasi-experimental design** (as opposed to a **true experimental design**), others such (e.g., Mosteller (1990)) state it is more than just a **quasi-experimental design**. The rationale for the latter is that if a **true experimental design** is defined as any design which, if appropriately implemented, provides an unbiased estimate of treatment effects, the **regression-continuity design** satisfies the latter requirement. In the final analysis, sources which discuss the **regression-discontinuity design** emphasize the fact that it is stronger with respect to **internal validity** than the **nonequivalent control group design**, and almost as strong or equal with respect to **internal validity** than the **randomized-groups design**. In order for the **internal validity** of the **regression-discontinuity design** to be compromised, some factor other than the treatment administered during **Time 2** would have to be capable of producing the pattern of sudden discontinuity which coincides with the cutoff point in Figure 28.14. Although on initial reflection one might think that the **regression-discontinuity design** is subject to **selection bias**, in point of fact, it is not. (In the **Introduction** it is noted that **selection bias** refers to the fact that the method a researcher employs in assigning subjects to groups can bias the outcome of a study, and that the most common cause of selection bias is nonrandom assignment.) Shadish *et al.* (2002, p. 237) and Trochim (2005) provide the rationale for the latter, and, in addition, explain why it is highly implausible that any of the other threats to **internal validity** discussed in the **Introduction** (i.e., history, maturation, instrumentation, statistical regression, mortality) could be responsible for the specific pattern of discontinuity associated with the **regression-continuity design**. Shadish *et al.* (2002, p. 242), however, does note that the **regression-discontinuity design** has less power than a balanced randomized experiment (i.e., a **randomized-groups**

design in which $n_1 = n_2$), and consequently requires a larger sample size in order to evaluate the same hypothesis at a comparable level of power.

Statistical analysis of the **regression-discontinuity design** typically involves the use of **multiple regression** and/or **analysis of covariance (Test 21I)**. Readers interested in the latter should consult Trochim (2005) who, along with Cook and Campbell (1979) and Shadish *et al.* (2002), provides detailed discussion of the **regression-continuity design**.

VIII. Additional Examples Illustrating the Use of the Pearson Product-Moment Correlation Coefficient

Two additional examples which can be evaluated with the **Pearson product-moment correlation coefficient** are presented in this section. Since the data for Examples 28.3 and 28.4 are identical to the data employed in Example 28.1, they yield the same result.

Example 28.3 *The editor of an automotive magazine conducts a survey to see whether or not it is possible to predict the number of traffic citations one receives for speeding based on how often a person changes his or her motor oil. The responses of five subjects on the two variables follow: (For each subject, the first score represents the number of oil changes (which represents the X variable), and the second score the number of traffic citations (which represents the Y variable))* **Subject 1** (20, 7); **Subject 2** (0, 0); **Subject 3** (1, 2); **Subject 4** (12, 5); **Subject 5** (3, 3). *Do the data indicate there is a significant correlation between the two variables?*

Example 28.4 *A pediatrician speculates that the length of time an infant is breast fed may be related to how often a child becomes ill. In order to answer the question, the pediatrician obtains the following two scores for five three-year-old children: The number of months the child was breast fed (which represents the X variable) and the number of times the child was brought to the pediatrician's office during the current year (which represents the Y variable). The scores for the five children follow:* **Child 1** (20, 7); **Child 2** (0, 0); **Child 3** (1, 2); **Child 4** (12, 5); **Child 5** (3, 3). *Do the data indicate that the length of time a child is breast fed is related to the number of times a child is brought to the pediatrician?*

IX. Addendum

The **Addendum** will discuss two additional topics which are directly or indirectly related to the general subject of correlational analysis.

1) The first part of the **Addendum** describes three bivariate correlational measures that are related to the **Pearson product-moment correlation coefficient**. The three procedures which will be described are: a) The **point-biserial correlation coefficient (Test 28i)**; b) The **biserial correlation coefficient (Test 28j)**; and c) The **tetrachoric correlation coefficient (Test 28k)**.

2) The second part of the **Addendum** discusses **meta-analysis and related topics**. **Meta-analysis** is methodology for pooling the results of multiple studies which evaluate the same general hypothesis. A major component of meta-analysis involves evaluating measures of **effect size**, which are correlational measures. Within the framework of the discussion of meta-analysis, criticisms that have been directed at the conventional hypothesis testing model (which are discussed earlier in the book) will once again be considered. More specifically, the argument that measures of effect size are more meaningful indicators of the nature and strength of the relationship between experimental variables than statistical significance will be discussed in greater detail.

1. Bivariate measures of correlation that are related to the Pearson-product moment correlation coefficient This section of the **Addendum** will describe three bivariate correlational measures which are related to the **Pearson product-moment correlation coefficient**. Each of the correlation coefficients to be described assumes that the scores on at least one of the variables can be expressed within the format of interval/ratio data, and that the underlying distribution of these scores is continuous and normal. Two of the correlational procedures assume that the underlying interval/ratio scores on one or both of the variables have been converted into a dichotomous (two category) format. A brief description of the three procedures follows.

The point-biserial correlation coefficient (Test 28i) The **point-biserial correlation coefficient** (r_{pb}) (which is a special case of the **Pearson product-moment correlation coefficient**) is employed if one variable is expressed as interval/ratio data, and the other variable is represented by a dichotomous nominal/categorical scale (i.e., two categories).

The biserial correlation coefficient (Test 28j) The **biserial correlation coefficient** (r_b) is employed if both variables are based on an interval/ratio scale, but the scores on one of the variables have been transformed into a dichotomous nominal/categorical scale. It provides an estimate of the value that would be obtained for the **Pearson product-moment correlation coefficient** if, instead of the dichotomized variable, one employed the scores on the underlying interval/ratio scale which the latter variable represents.

The tetrachoric correlation coefficient (Test 28k) The **tetrachoric correlation coefficient** (r_{tet}) is employed if both variables are based on an interval/ratio scale, but the scores on both of the variables have been transformed into a dichotomous nominal/categorical scale. It provides an estimate of the value that would be obtained for the **Pearson product-moment correlation coefficient**, if, instead of the dichotomized variables, one employed the scores on the underlying interval/ratio scales which the latter variables represent.

Test 28i: The point-biserial correlation coefficient (r_{pb}) As noted earlier, the **point-biserial correlation coefficient** represents a special case of the **Pearson product-moment correlation coefficient**. The **point-biserial correlation coefficient** is employed if one variable is expressed as interval/ratio data and the other variable is represented by a dichotomous nominal/categorical scale. Examples of variables which constitute a dichotomous nominal/categorical scale are **male** versus **female** and **employed** versus **unemployed**. In using the **point-biserial correlation coefficient**, it is assumed that the dichotomous variable is not based on an underlying continuous interval/ratio distribution. If, in fact, the dichotomous variable is based on the latter type of distribution, the **biserial correlation coefficient (Test 28j)** is the appropriate measure to employ. Examples of variables that are expressed in a dichotomous format, but which are based on an underlying continuous interval/ratio distribution are **pass** versus **fail** and **above average intelligence** versus **below average intelligence**. Obviously, not everyone who passes (or fails) a test or a course performs at the same level. In the same respect, the distributions of intelligence of people who are **above average** or **below average** are not uniform. There are, of course, variables with respect to which it can be argued whether they are based on an underlying continuous distribution (such as, perhaps, handedness, which will be employed as a dichotomous variable in the example to be presented in this section). As is the case with the **Pearson product-moment correlation coefficient**, the range of values within which r_{pb} can fall are $-1 \leq r_{pb} \leq +1$. Example 28.5 will be employed to illustrate the use of the **point-biserial correlation coefficient**.

Example 28.5   *A study is conducted to determine whether there is a correlation between handedness and eye-hand coordination. Five right-handed and five left-handed subjects are administered a test of eye-hand coordination. The test scores of the subjects follow (the higher a subject's score, the better his or her eye-hand coordination):* **Right-handers**: 11, 1, 0, 2, 0; **Left-handers**: 11, 11, 5, 8, 4. *Is there a statistical relationship between handedness and eye-hand coordination?*

In the analysis handedness will represent the X variable, and the eye-hand coordination test scores will represent the Y variable. With respect to handedness (which is a dichotomous variable), all right-handed subjects will be assigned a score of 1 on the X variable, and all left-handed subjects will be assigned a score of 0. In this experiment, the X variable represents what is referred to as a **dummy variable**. A **dummy variable** is a dichotomous variable which employs two values (typically 0 and 1) to indicate membership in one of two categories. Table 28.13 summarizes the data for the ten subjects employed in the study.

Table 28.13 Data for Example 28.5

Subject	X	X^2	Y	Y^2	XY
1	1	1	11	121	11
2	1	1	1	1	1
3	1	1	0	0	0
4	1	1	2	4	2
5	1	1	0	0	0
6	0	0	11	121	0
7	0	0	11	121	0
8	0	0	5	25	0
9	0	0	8	64	0
10	0	0	4	16	0
	$\Sigma X = 5$	$\Sigma X^2 = 5$	$\Sigma Y = 53$	$\Sigma Y^2 = 473$	$\Sigma XY = 14$

Since the **point-biserial correlation coefficient** is a special case of the **Pearson product-moment correlation coefficient**, Equation 28.61 (which is identical to Equation 28.1) is employed to compute r_{pb}. Employing Equation 28.61, the value $r_{pb} = -.57$ is computed.

(Equation 28.61)

$$r_{pb} = \frac{\Sigma XY - \dfrac{(\Sigma X)(\Sigma Y)}{n}}{\sqrt{\left[\Sigma X^2 - \dfrac{(\Sigma X)^2}{n}\right]\left[\Sigma Y^2 - \dfrac{(\Sigma Y)^2}{n}\right]}} = \frac{14 - \dfrac{(5)(53)}{10}}{\sqrt{\left[5 - \dfrac{(5)^2}{10}\right]\left[473 - \dfrac{(53)^2}{10}\right]}} = -.57$$

Equation 28.62 is an alternative equation for computing the **point-biserial correlation coefficient**.

$$r_{pb} = \left[\frac{\overline{Y}_1 - \overline{Y}_0}{\tilde{s}_Y}\right]\sqrt{p_0 p_1}\sqrt{\frac{n}{n-1}}$$

(Equation 28.62)

$$= \left[\frac{2.8 - 7.8}{4.62}\right]\sqrt{(.5)(.5)}\sqrt{\frac{10}{10-1}} = -.57$$

Where: $\bar{Y}_0$ and $\bar{Y}_1$ are, respectively, the average scores on the Y variable for subjects who are categorized 0 versus 1 on the X variable

p_0 equals the proportion of subjects with an X score of 0

p_1 equals the proportion of subjects with an X score of 1

In employing Equation 28.62, $\bar{Y}_1 = 2.8$ and $\bar{Y}_0 = 7.8$. The value $\tilde{s}_Y = 4.62$, which represents the unbiased estimate of the population standard deviation for the Y variable (which is computed with Equation I.11), is computed below.

$$\tilde{s}_Y = \sqrt{\frac{\sum Y^2 - \frac{(\sum Y)^2}{n}}{n - 1}} = \sqrt{\frac{473 - \frac{(53)^2}{10}}{10 - 1}} = 4.62$$

It should be noted that some sources employ Equation 28.63 to compute the value of the **point-biserial correlation coefficient**.

$$r_{pb} = \left[\frac{\bar{Y}_1 - \bar{Y}_0}{s_Y} \right] \sqrt{p_0 p_1} = \left[\frac{2.8 - 7.8}{4.38} \right] \sqrt{(.5)(.5)} = -.57 \qquad \textbf{(Equation 28.63)}$$

Note that Equation 28.63 employs the sample standard deviation (computed with Equation I.9 — i.e., $s_Y = \sqrt{[\sum Y^2 - ((\sum Y)^2/n)]/n}$), which is a biased estimate of the population standard deviation. For Example 28.2, $s_Y = 4.38$. When $s_Y = 4.38$ is substituted in the Equation 28.63, it yields the value $r_{pb} = -.57$.

The reader should take note of the fact that the sign of r_{pb} is irrelevant unless the categories on the dichotomized variable are ordered (which is not the case for Example 28.5). The reason for employing the absolute value of r_{pb} is that the use of the scores 0 and 1 for the two categories is arbitrary, and does not indicate that one category is superior to the other. (If all right-handed subjects are assigned a score of 0 on the X variable and all left-handed subjects are assigned a score of 1, the value computed for $r_{pb} = +.57$, which is the same absolute value computed for the data in Table 28.13.) Since the categories are not ordered, from this point on in the discussion, the absolute value $r_{pb} = .57$ will be employed to represent the value of r_{pb}. In the event the categories are ordered, the score 1 should be employed for the category associated with higher performance/quality, and the score 0 should be employed for the category associated with lower performance/quality. In all likelihood, if the categories are ordered they are likely to be based on an underlying continuous distribution, and in such a case the appropriate correlational measure to employ is the **biserial correlation coefficient**.

The square of the **point-biserial correlation coefficient** represents the **coefficient of determination**, which as noted in Section VI indicates the amount of variability on the Y variable which can be accounted for by variability on the X variable. Since, $r_{pb}^2 = (.57)^2 = .325, 32.5\%$ of the variability on the test of eye-hand coordination can be accounted for on the basis of a person's handedness.

The data employed for Example 28.5 are identical to those employed for Example 11.1 (which is used to illustrate the *t* **test for two independent samples**). In point of fact, the **point-**

biserial correlation coefficient can be employed to measure the magnitude of treatment effect in an experiment which has been evaluated with a *t* **test for two independent samples,** if the grouping of the subjects is conceptualized as the dichotomous variable. Thus, in Example 11.1, if each subject in Group 1 (Drug Group) is assigned an *X* score of 1, and each subject in Group 2 (Placebo Group) is assigned an *X* score of 0, the data for the experiment can be summarized with Table 28.13. Since analysis of the data in Table 28.13 yields $r_{pb} = .57$ and $r_{pb}^2 = .325$, the researcher can conclude that 32.5% of the variability on the dependent variable (the depression ratings for subjects) can be accounted for on the basis of which group a subject is a member.

In Section VI of the *t* **test for two independent samples**, the measure of association that is employed to measure the magnitude of treatment effect is the **omega squared** ($\tilde{\omega}^2$) statistic. The value of **omega squared** computed for Example 11.1 is $\tilde{\omega}^2 = .22$. Since $\tilde{\omega}^2$ is interpreted in the same manner as r_{pb}^2, the value $\tilde{\omega}^2 = .22$ indicates that 22% of the variability on the dependent variable can be accounted for on the basis of which group a subject is a member. Obviously, the latter value is lower than the value $r_{pb}^2 = .325$ computed in this section. The discrepancy between the two values will be discussed further later in this section. In point of fact, $r_{pb}^2 = .325$ is equivalent to the **eta squared** ($\tilde{\eta}^2$) statistic, which is an alternative measure of association that some sources employ in assessing the magnitude of treatment effect for the *t* **test for two independent samples**. Marascuilo and Serlin (1988) note that both r_{pb}^2 and $\tilde{\eta}^2$ represent a **correlation ratio**. The **correlation ratio**, which can be defined within the framework of an analysis of variance, is the ratio of the explained sum of squares over the total sum of squares. To clarify the meaning of a correlation ratio, let us assume that in lieu of the *t* **test for two independent samples**, the data for Example 11.1 are evaluated with the **single-factor between-subjects analysis of variance**. Table 28.14 is the summary table of the analysis of variance for Example 11.1.

Table 28.14 Summary Table of Analysis of Variance for Example 11.1

Source of variation	SS	df	MS	F
Between-groups	62.5	1	62.5	3.86
Within-groups	129.6	8	16.2	
Total	**192.1**	9		

Within the framework of the **single-factor between-subjects analysis of variance**, the **correlation ratio** (which is computed with Equation 21.43) is $\tilde{\eta}^2 = SS_{BG}/SS_T$. Since both $\tilde{\eta}^2$ and r_{pb}^2 represent the correlation ratio, $\tilde{\eta}^2 = r_{pb}^2 = SS_{BG}/SS_T$. Thus, for Example 11.1, $\tilde{\eta}^2 = r_{pb}^2 = 62.5/192.1 = .325$.

Equation 28.64 (which is identical to Equation 11.16) can also be employed to compute the value $\tilde{\eta}^2 = r_{pb}^2$ (where $t = \sqrt{F} = \sqrt{3.86} = 1.964$).

$$\tilde{\eta}^2 = r_{pb}^2 = \frac{t^2}{t^2 + df} = \frac{(1.964)^2}{(1.964)^2 + 8} = .325 \qquad \textbf{(Equation 28.64)}$$

Where: $df = n_1 + n_2 - 2$ (which is the degrees of freedom for the *t* **test for two independent samples**)

The fact that different measures of magnitude of treatment effect may not yield the same value is discussed in Section VI of the **single-factor between-subjects analysis of variance**.

In the latter discussion it is noted that the computed value $\tilde{\eta}^2$ is a biased estimate of the underlying population parameter η^2, and an adjusted value (which is less biased) can be computed with Equation 21.44. The latter value is now computed for Example 11.1: Adjusted $\tilde{\eta}^2 = 1 - [MS_{WG}/MS_T] = 1 - [16.2/21.34] = .24$. (Where, $MS_T = SS_T/df_T = 192.1/9 = 21.34$.) Note that the value Adjusted $\tilde{\eta}^2 = .24$ is closer to the value $\tilde{\omega}^2 = .22$ than the previously computed value $\tilde{\eta}^2 = r_{pb}^2 = .325$.

Test 28i-a: Test of significance for a point-biserial correlation coefficient   The null hypothesis $H_0: \rho_{pb} = 0$ can be evaluated with Equation 28.65 (which is identical to Equation 28.3, which is employed to evaluate the same hypothesis with reference to the **Pearson product-moment correlation coefficient**). As is the case for Equation 28.3, the degrees of freedom for Equation 28.65 are $df = n - 2$. Employing Equation 28.65, the value $t = 1.96$ is computed.

$$t = \frac{r_{pb}\sqrt{n - 2}}{\sqrt{1 - r_{pb}^2}} = \frac{.57\sqrt{10 - 2}}{\sqrt{1 - (.57)^2}} = 1.96 \qquad \textbf{(Equation 28.65)}$$

It will be assumed that the nondirectional alternative hypothesis $H_1: \rho_{pb} \neq 0$ is evaluated. Employing **Table A2**, for $df = 10 - 2 = 8$, the tabled critical two-tailed .05 and .01 values are $t_{.05} = 2.31$ and $t_{.01} = 3.36$. Since the obtained value $t = 1.96$ is less than both of the aforementioned critical values, the null hypothesis $H_0: \rho_{pb} = 0$ cannot be rejected.

Since the value of the **point-biserial correlation coefficient** is a direct function of the difference between $\bar{Y}_0$ and $\bar{Y}_1$, a significant difference between the latter two mean values indicates the absolute value of the correlation between the two variables is significantly above zero. Thus, an alternative way of evaluating the null hypothesis $H_0: \rho_{pb} = 0$ is to conduct a *t* **test for two independent samples**, contrasting the two mean values $\bar{Y}_0$ and $\bar{Y}_1$. The fact that the latter analysis will yield a result which is equivalent to that obtained with Equation 28.65 can be confirmed by the fact that the value $t = 1.96$ computed above with Equation 28.65 is identical to the absolute *t* value computed for the same set of data with Equation 11.1 (for Example 11.1). Sources that provide additional discussion of the **point-biserial correlation coefficient** are Guilford (1965), Lindeman *et al.* (1980), and McNemar (1969).

Test 28j: The biserial correlation coefficient (r_b)   The **biserial correlation coefficient** is employed if both variables are based on an interval/ratio scale, but the scores on one of the variables have been transformed into a dichotomous nominal/categorical scale. An example of a situation where an interval/ratio variable would be expressed as a dichotomous variable is a test based on a normally distributed interval/ratio scale for which the only information available is whether a subject has passed or failed the test. The value computed for the **biserial correlation coefficient** represents an estimate of the value that would be obtained for the **Pearson product-moment correlation coefficient**, if, instead of employing a dichotomized variable, one had employed the scores on the underlying interval/ratio scale.

The **biserial correlation coefficient** is based on the assumption that the underlying distribution for both of the variables is continuous and normal. Since the accuracy of r_b is highly dependent upon the assumption of normality, it should not be employed unless there is empirical evidence to indicate that the distribution underlying the dichotomous variable is normal. If the underlying distribution of the dichotomous variable deviates substantially from normality, the computed value of r_b will not be an accurate approximation of the underlying population correlation r_b estimates. One consequence of the normality assumption being violated is that, under certain conditions, the absolute value computed for r_b may exceed 1. In point of fact, Lindeman *et al.* (1980) note that the theoretical limits of r_b are $-\infty \leq r_b \leq +\infty$.

In contrast to r_{pb}, the sign of the **biserial correlation coefficient** should be taken into account, since it clarifies the nature of the relationship between the two variables. This is the case, since the dichotomous variable will involve two ordered categories. In assigning scores to subjects on the ordered dichotomized variable, the score 1 should be employed for the category associated with higher performance/quality, and the score 0 should be employed for the category associated with lower performance/quality.

In order to illustrate the computation of the **biserial correlation coefficient**, let us assume a researcher wants to determine whether there is a statistical relationship between intelligence and eye-hand coordination. Ten subjects are categorized with respect to both variables. Although the evaluation of each subject's intelligence is based on an interval/ratio intelligence test score, we will assume that the only information available to the researcher is whether an individual is above or below average in intelligence. In view of this, intelligence, which will be designated the X variable, will have to be represented as a dichotomous variable. Subjects who are above average in intelligence will be assigned a score of 1, and subjects who are below average in intelligence will be assigned a score of 0. The scores on the eye-hand coordination test will represent the Y variable. Example 28.6, which employs the same set of data as Example 28.5, summarizes the above described experiment.

Example 28.6 *A study is conducted to determine whether there is a correlation between intelligence and eye-hand coordination. Five subjects who are above average in intelligence and five subjects who are below average in intelligence are administered a test of eye-hand coordination. The test scores of the subjects follow (the higher a subject's score, the better his or her eye-hand coordination):* **Above average intelligence**: 11, 1, 0, 2, 0; **Below average intelligence**: 11, 11, 5, 8, 4. *Is there a statistical relationship between intelligence and eye-hand coordination?*

The **biserial correlation coefficient** can be computed with either Equation 28.66 or 28.67. It can also be computed with Equation 28.68 if r_{pb} has been computed for same set of data. Note that except for h, all of the terms in the aforementioned equations are also employed in computing the **point-biserial correlation coefficient**. The value h represents the height (known more formally as the **ordinate**) of the standard normal distribution at the point which divides the proportions p_0 and p_1. Specifically, employing **Table A1**, the z value is identified which delineates the point on the normal curve that a proportion of the cases corresponding to the smaller of the two proportions p_0 versus p_1 falls above and the larger of the two proportions falls below. The tabled value of h (in **Column 4** of **Table A1**) associated with that z value is employed in whatever equation one employs for computing r_b. If, as is the case in our example, $p_0 = p_1 = .5$, the value of z will equal zero, and thus the corresponding value of $h = .3989$. When $h = .3989$ and the other appropriate values employed for Example 28.6 (which are summarized in Table 28.13) are substituted in Equations 28.66–28.68, the value $r_b = -.71$ is computed.

(Equation 28.66)

$$r_b = \left[\frac{\bar{Y}_1 - \bar{Y}_0}{\tilde{s}_Y}\right]\left[\frac{p_0 p_1}{h}\right]\sqrt{\frac{n}{n-1}} = \left[\frac{2.8 - 7.8}{4.62}\right]\left[\frac{(.5)(.5)}{.3989}\right]\sqrt{\frac{10}{10-1}} = -.71$$

$$r_b = \left[\frac{\bar{Y}_1 - \bar{Y}_0}{s_Y}\right]\left[\frac{p_0 p_1}{h}\right] = \left[\frac{2.8 - 7.8}{4.38}\right]\left[\frac{(.5)(.5)}{.3989}\right] = -.71 \quad \textbf{(Equation 28.67)}$$

$$r_b = \frac{r_{pb}\sqrt{p_0 p_1}}{h} = \frac{(-.57)\sqrt{(.5)(.5)}}{.3989} = -.71 \quad \textbf{(Equation 28.68)}$$

Note that for the same set of data, the absolute value r_b = .71 is larger than the absolute value r_{pb} = .57 computed for the **point-biserial correlation coefficient**. In point of fact, for the same set of data (except when $r_b = r_{pb} = 0$) the absolute value of r_b will always be larger than the absolute value of r_{pb}, since $\sqrt{p_0 p_1}/h$ will always be larger than 1. The closer together the values p_0 and p_1, the less the discrepancy between the values of r_b and r_{pb}. If there is reason to believe that the normality assumption for the dichotomous variable has been violated, most sources recommend computing r_{pb} instead of r_b, since r_b may be a spuriously inflated estimate of the underlying population correlation. When the latter is taken into consideration, along with the fact that by dichotomizing a continuous variable one sacrifices valuable information, it can be understood why the **biserial correlation coefficient** is infrequently employed within the framework of research. With regard to the latter, Hunter and Schmidt (2004) note that dichotomizing a continuous variable will almost always result in it having a lower correlation with another variable in contrast to the correlation that would be computed had it not been dichotomized.

Guilford (1965) notes that, given the normality assumption has not been violated, those conditions which optimize the likelihood of r_b providing a good estimate of the underlying population parameter ρ_b are as follows: a) The value of n is large; and b) The values of p_0 and p_1 are close together. It should be noted that (as is the case for Pearson r and r_{pb}) if the relationship between two variables is nonlinear, the computed value of r_b will only represent the degree of linear relationship between the variables.

Test 28j-a: Test of significance for a biserial correlation coefficient Lindeman *et al*. (1980) note that the null hypothesis H_0: ρ_b = 0 can be evaluated with Equation 28.69. Although the latter equation, which is based on the normal distribution, assumes a large sample size, it is employed to evaluate the value r_b = -.71 computed for Example 28.6. Note that the sign of z will always be the same as the sign of r_b.

$$z = \frac{h r_b}{\sqrt{\dfrac{p_0 p_1}{n}}} = \frac{(.3989)(-.71)}{\sqrt{\dfrac{(.5)(.5)}{10}}} = -1.79 \quad \textbf{(Equation 28.69)}$$

Employing **Table A1**, the tabled critical two-tailed values are $z_{.05}$ = 1.96 and $z_{.01}$ = 2.58, and the tabled critical one-tailed values are $z_{.05}$ = 1.65 and $z_{.01}$ = 2.33. The nondirectional alternative hypothesis H_1: $\rho_b \ne 0$ is not supported, since the absolute value z = 1.79 is less than the tabled critical two-tailed value $z_{.05}$ = 1.96. However, the directional alternative hypothesis H_1: $\rho_b < 0$ is supported at the .05 level, since z = -1.79 is a negative number with an absolute value that is greater than the tabled critical one-tailed value $z_{.05}$ = 1.65. The moderately strong negative correlation between the two variables indicates that subjects who score below average on the intelligence test perform better on the test of eye-hand coordination than do subjects who score above average on the intelligence test. The latter can be confirmed by the fact that $\overline{Y}_1 = 2.8$ is less than $\overline{Y}_0 = 7.8$.[24]

As is the case for the **point-biserial correlation coefficient**, since the value of the **biserial correlation coefficient** is a direct function of the difference between $\overline{Y}_0$ and $\overline{Y}_1$, a significant difference between the two mean values indicates that the absolute value of the correlation between the two variables is significantly above zero. Thus, an alternative way of evaluating the null hypothesis $H_0: \rho_b = 0$ is to contrast the means $\overline{Y}_0$ and $\overline{Y}_1$ with a *t* **test for two independent samples**. However, the result obtained with Equation 28.69 will not necessarily be consistent with the result obtained if the *t* **test for two independent samples** is employed to contrast $\overline{Y}_0$ versus $\overline{Y}_1$ (especially if the sample size is small). In point of fact, use of the *t* **test for two independent samples** to contrast $\overline{Y}_0$ versus $\overline{Y}_1$ assumes the use of r_{pb} as a measure of association. Within the context of employing the *t* **test**, the correlational example under discussion can be conceptualized as a study in which intelligence represents the independent variable and eye-hand coordination the dependent variable. The independent variable, which is nonmanipulated, is comprised of the two levels, **above average intelligence** versus **below average intelligence**. Sources that provide additional discussion of the **biserial correlation coefficient** are Guilford (1965), Lindeman *et al.* (1980), and McNemar (1969).

Test 28k: The tetrachoric correlation coefficient (r_{tet}) The **tetrachoric correlation coefficient** is employed if both variables are based on an interval/ratio scale, but the scores on both of the variables have been transformed into a dichotomous nominal/categorical scale. The value computed for the **tetrachoric correlation coefficient** represents an estimate of the value one would obtain for the **Pearson product-moment correlation coefficient** if, instead of employing dichotomized variables, one had used the scores on the underlying interval/ratio scales. The **tetrachoric correlation coefficient** (which was developed by Karl Pearson (1901)) is based on the assumption that the underlying distribution for both of the variables is continuous and normal. Among others, Cohen and Cohen (1983) note that caution should be employed in using both the **tetrachoric** and **biserial correlation coefficients**, since both measures are based on hypothetical underlying distributions which are not directly observed. Since the accuracy of r_{tet} is highly dependent upon the assumption of normality, it should not be employed unless there is empirical evidence to indicate that the distributions underlying the dichotomous variables are normal.

Since the magnitude of the standard error of estimate of r_{tet} is large relative to the standard error of estimate of Pearson *r*, in order to provide a reasonable estimate of *r* the sample size employed for computing r_{tet} should be quite large. Guilford (1965) and Lindeman *et al.* (1980) state that the value of *n* employed in computing r_{tet} should be at least two times that which would be employed to compute *r*. As is the case for the **Pearson product-moment correlation coefficient**, the following apply to the **tetrachoric correlation coefficient**: a) The range of values within which r_{tet} can fall is $-1 \leq r_{tet} \leq +1$; and b) If the relationship between two variables is nonlinear, the computed value of r_{tet} will only represent the degree of linear relationship between the variables.

Earlier in this section (as well as in the discussion of the **chi-square test for $r \times c$ tables**) it is noted that the **phi coefficient (ϕ)** is also employed as a measure of association for a 2×2 contingency table involving two dichotomous variables. The basic difference between r_{tet} and ϕ is that the latter measure is employed with two genuinely dichotomous variables (i.e., variables that are not based on an underlying distribution involving an interval/ratio scale). Cohen and

Cohen (1983) and McNemar (1969) note that the value of r_{tet} computed for a 2×2 contingency table will always be larger than the value of ϕ computed for the same data.[25]

A number of reasons account for the fact that the **tetrachoric correlation coefficient** is infrequently employed within the framework of research. One reason is that, in most instances, data on variables which represent an interval/ratio scale are available in the latter format, and thus there is no need to convert it into a dichotomous format. Another reason is the reluctance of researchers to accept the normality assumption with respect to variables for which only dichotomous information is available.

Without the aid of a computer or special tables (which can be found in Guilford (1965) and Lindeman *et al.* (1980)), the computation of the exact value of r_{tet} is both time consuming and tedious. There are, however, two equations that have been developed which provide reasonably good approximations of r_{tet} under most conditions. These equations will be employed to evaluate Example 28.7.

Example 28.7 *Two hundred subjects are asked whether they* **Agree** *(which will be assigned a score of 1) or* **Disagree** *(which will be assigned a score of 0) with the following two statements:* **Question 1**: *I believe that abortion should be legal.* **Question 2**: *I believe that murderers should be executed. The responses of the 200 subjects are summarized in* Table 28.15. *Is there a statistical relationship between subjects' responses to the two questions?*

Table 28.15   Summary of Data for Example 28.7

		X variable Question 1		
		0 Disagree	**1** Agree	Row Sums
Y variable Question 2	**0** Disagree	$a = 30$	$b = 70$	100
	1 Agree	$c = 60$	$d = 40$	100
	Column Sums	90	110	Total = 200

Subjects' responses to Question 1 will represent the X variable, and their responses to Question 2 will represent the Y variable. The use of the **tetrachoric correlation coefficient** in evaluating the data is based on the assumption that the permissible responses **Agree** versus **Disagree** represent two points that lie on a continuous scale. It will be assumed that if subjects are allowed to present their opinions to the questions with more precision, their responses can be quantified on an interval/ratio scale, and that the overall distribution of these responses in the underlying population will be normal. Thus, the responses of 0 and 1 on each variable are the result of dichotomizing information which is based on an underlying interval/ratio scale.

Equations 28.70 and 28.71 can be employed to compute reasonably good approximations of the value of r_{tet}. Lindeman *et al.* (1980) note that Equation 28.70 provides a good approximation of r_{tet} when $p_0 = p_1 = .5$ for both of the dichotomous variables. In other words, for the X variable both $p_{0_X} = (a + c)/n$ and $p_{1_X} = (b + d)/n$ will equal .5, and for the Y variable both $p_{0_Y} = (a + b)/n$ and $p_{1_Y} = (c + d)/n$ will equal .5 (where $n = a + b + c + d$). Equation 28.71, on the other hand, is recommended when the values of p_0 and p_1 are not equal. As the discrepancy between p_0 and p_1 increases, the less accurate the approximation provided by Equation 28.71 becomes.

In both Equations 28.70 and 28.71, *a* and *d* will always represent the frequencies of cells in which subjects provide the same response for both variables/questions, and *b* and *c* will always represent the frequencies of cells in which subjects provide opposite responses for the two variables/questions. Inspection of Table 28.15 (which is identical to Table 28.12, which is employed to illustrate that ϕ is a special case of *r*) indicates the following: 1) $a = 30$ subjects respond **Disagree** to both questions; 2) $d = 40$ subjects respond **Agree** to both questions; 3) $b = 70$ subjects respond **Agree** to Question 1 and **Disagree** to Question 2; and 4) $c = 60$ subjects respond **Disagree** to Question 1 and **Agree** to Question 2.

The configuration of the data is such that $p_0 = p_1$ for the *Y* variable, and the relationship is closely approximated for the *X* variable. Specifically, for the *Y* variable, $p_{0_Y} = (a + b)/n$ $= (30 + 70)/200 = .5$ and $p_{1_Y} = (c + d)/n = (60 + 40)/200 = .5$. In the case of the *X* variable, $p_{0_X} = (a + c)/n = (30 + 60)/200 = .45$ and $p_{1_X} = (b + d)/n = (70 + 40)/200 = .55$.

The appropriate values are substituted in Equations 28.70 and 28.71 below. The trigonometric functions in each of the equations can be easily calculated with one keystroke on most scientific calculators.

(Equation 28.70)

$$r_{tet} = \sin\left[90°\left(\frac{a + d - b - c}{n}\right)\right] = \sin\left[90°\left(\frac{30 + 40 - 70 - 60}{200}\right)\right] = \sin\ -27° = -.45$$

(Equation 28.71)

$$r_{tet} = \cos\left(\frac{180°}{1 + \sqrt{\dfrac{ad}{bc}}}\right) = \cos\left(\frac{180°}{1 + \sqrt{\dfrac{(30)(40)}{(70)(60)}}}\right) = \cos\ 117.30° = -.46$$

Since for both variables the condition $p_0 = p_1 = .5$ is present or approximated, the two equations result in almost identical values. The negative sign in front of the correlation coefficient reflects the fact that subjects who are in one response category on one variable are more likely to be in the other response category on the other variable. A positive correlation would indicate that subjects tend to be in the same response category on both variables. Note that the absolute value $r_{tet} = .45$ (or .46) is greater than the value $\phi = .30$ obtained for the same set of data (i.e., Table 28.12). This is consistent with what was noted earlier — that the value of r_{tet} computed for a 2×2 table will always be larger than the value of ϕ computed for the same data. The reader should take note of the fact, however, that unlike r_{tet}, ϕ is always expressed as a positive number. Thus, in comparing the two values for the same set of data, the absolute value of r_{tet} should be employed.

Test 28k-a: Test of significance for a tetrachoric correlation coefficient In order to evaluate the null hypothesis H_0: $\rho_{tet} = 0$, the standard error of estimate of r_{tet} must first be computed employing Equation 28.72. In the latter equation, the values h_X and h_Y are the height (ordinate) of the standard normal distribution at the point for each of the variables which divides the proportions p_0 and p_1. The protocol for determining the ordinate for each of the variables is identical to that employed for determining the ordinate for the **biserial correlation coefficient**. Employing **Table A1** for both the *X* and the *Y* variables, the *z* value is identified which delineates the point on the normal curve that a proportion of cases corresponding to the smaller of the two proportions p_0 versus p_1 falls above and the larger of the two proportions falls below. The corresponding ordinate (in **Column 4** of **Table A1**) is then determined. Thus, in the case of the X variable, $h_X = .3958$ (since 45% of cases fall above the corresponding value $z = .128$). (The values $z = .128$ and $h_X = .3958$ are interpolated from **Table A1**, since z falls between .12 and .13.) In the case of the Y variable, $h_Y = .3989$ (since 50% of cases fall above the corresponding value $z = 0$). When the appropriate values are substituted in Equation 28.72, the value $s_{r_{tet}} .111$ is computed.

$$s_{r_{tet}} = \frac{\sqrt{p_{0_X} p_{1_X} p_{0_Y} p_{1_Y}}}{h_X h_Y \sqrt{n}} = \frac{\sqrt{(.45)(.55)(.5)(.5)}}{(.3958)(.3989)\sqrt{200}} = .111 \quad \textbf{(Equation 28.72)}$$

The value $s_{r_{tet}} = .111$ is substituted in Equation 28.73, which is employed to evaluate the null hypothesis H_0: $r_{tet} = 0$. Use of the normal distribution in evaluating the null hypothesis assumes that the computation of r_{tet} is based on a large sample size (since, as noted earlier, r_{tet} will be extremely unreliable if it is based on a small sample). Employing Equation 28.73, the value $z = -4.14$ is computed. Note that the sign of z will always be the same as the sign of r_{tet}.

$$z = \frac{r_{tet}}{s_{r_{tet}}} = \frac{-.46}{.111} = -4.14 \quad \textbf{(Equation 28.73)}$$

It will be assumed that the nondirectional alternative hypothesis H_1: $\rho_{tet} \neq 0$ is evaluated. Employing **Table A1**, it is determined that the tabled critical two-tailed .05 and .01 values are $z_{.05} = 1.96$ and $z_{.01} = 2.58$. Since the obtained absolute value $z = 4.14$ is greater than both of the aforementioned critical values, the nondirectional alternative hypothesis H_1: $\rho_{tet} \neq 0$ is supported at both the .05 and .01 levels. Additional discussion of the **tetrachoric correlation coefficient** can be found in Guilford (1965), Lindeman *et al.* (1980), and McNemar (1969).

2. Meta-analysis and related topics **Meta-analysis** is methodology for pooling the results of multiple studies which evaluate the same general hypothesis. The purpose of meta-analysis is to allow a research community to come to some conclusion with respect to the validity of a hypothesis which is not based on one or two studies, but rather is based on a multitude of studies which have addressed the same research hypothesis. Various sources (e.g., Kline (2004, p. 251), Hunter and Schmidt (2004, p. 21) and Schulze (2004, p. 9)) note that a meta-analysis can evaluate summary statistics which are the result of **primary** or **secondary analyses** of research. A **primary analysis** is the analysis of data by the original researcher who conducts a study, whereas a **secondary analysis** involves reanalysis of data reported in the original study through use of a different methodology and/or additional analysis of the published data by a second party.

During the past 20 years the number of published studies employing meta-analyses has increased dramatically in disciplines such as education, the social and behavioral sciences, and medicine. To illustrate, within the field of psychology meta-analytic reviews are regularly published in the *Psychological Bulletin*, a prominent journal in the latter discipline, and Hunter and Schmidt (2004; p. 28) (citing the work of Hunt (1997) and Antman *et al.* (1992)) note the great impact of meta-analysis in the field of medicine citing the leadership of Thomas Chalmers (Chalmers *et al.* (1987)).

Hedges and Olkin (1985) and Lipsey and Wilson (2001) note that Karl Pearson in 1904 and Ronald Fisher (within the framework of analyzing agricultural research) in the 1930s are credited with developing the first meta-analytic procedures. However, the work of Gene Glass (an educational psychologist who initially employed meta-analysis to evaluate the efficacy of psychotherapy) and his associates (Glass (1976, 1977), Glass *et al.* (1981), Smith and Glass (1977)) are largely responsible for popularizing the use of meta-analysis within the scientific community. Two industrial/personnel psychologists, Frank Schmidt and John Hunter (Hunter and Schmidt (1977, 1990, 2004)), who primarily employ meta-analysis to evaluate predictive validity in personnel selection, also introduced meta-analytic methodologies about the same time as Glass and his associates. Robert Rosenthal (a psychologist) and Donald Rubin (a statistician) (Rosenthal and Rubin (1978)) made seminal as well as substantial later contributions toward developing meta-analytic methodologies. Larry Hedges and Ingram Olkin (two educational research methodologists trained in statistics) published a notable volume outlining statistical theory and methods involved in meta-analysis (Hedges and Olkin, 1985).

Pooling the results of multiple studies which evaluate the same hypothesis is not a simple and straightforward matter. More often than not there are differences between two or more studies which address the same general hypothesis. Rarely if ever are two studies identical with respect to the details of their methodology, the quality of their design, the soundness of execution, and the target populations which are evaluated. To further complicate matters, there is the issue of additional studies that may have evaluated the same hypothesis which were either never submitted for publication or were submitted but rejected. Rosenthal (1979, 1991, 1993) refers to this latter phenomenon (which will be discussed later in this section in greater detail) as the **file drawer problem**. In spite of the practical and theoretical difficulties involved in pooling the results of multiple studies, during the past 20 years numerous analytical procedures have been developed for this purpose.

Hedges and Olkin (1985) and Rosenthal (1991, 1993) note that two general approaches characterize meta-analytic research. One approach involves procedures which evaluate statistical significance for the combined results of multiple studies, while the second approach estimates treatment effects across studies. Hedges and Olkin (1985), Rosenthal (1991, 1993), and Wolf (1986) note that one method for evaluating the statistical significance of the combined results of multiple studies is the **vote-counting method**. The latter procedure involves identifying all of the studies which are believed to evaluate the same general hypothesis, and then determining the number of studies that yield a statistically significant result. The proportion of significant studies is then contrasted with the proportion of studies that are not significant (through use of a procedure such as the **binomial sign test for a single sample**). A variant of the vote-counting method statistically combines the probability values from two or more studies in order to compute a pooled probability value. Hedges and Olkin (1985) (also see Hunter and Schmidt (2004, p. 62 and Ch. 11)[26] state that in spite of the intuitive appeal of such a simple and straightforward approach as the **vote-counting method**, the latter procedure tends to be strongly biased toward the conclusion that there is no overall treatment effect for the variables under study. The latter bias is largely attributed to the relatively low power (due to the use of small sample sizes employed in studies in which a small to moderate effect size may be present) of research in certain disciplines such as the social and behavioral sciences. If one assumes low power, the vote-counting method will most likely only include as significant those studies in which the effect size is large, and fail to include studies where a weak or moderate effect size is present. Thus, the advantage of meta-analytic techniques which ignore the level of significance, but instead pool effect sizes, is that they circumvent the problem of low statistical power. Hedges and Olkin (1985) note that an optimal meta-analytic strategy should allow an investigator to compute an average treatment effect across all of the studies, as well as the consistency of the treatment effect across the studies. The general subject of treatment effects has been discussed throughout this book, often within the framework of the discussion of power. Specifically, various indices of treatment effect and measures of association (which are commonly employed as measures of treatment effect) are discussed in detail in Section VI of the following tests: The **single-sample *t* test**, the *t* **tests for two independent and dependent samples**, the **chi-square test for *r* × *c* tables**, and the tests which involve **analysis of variance** procedures. Prior to describing a number of meta-analytic procedures, the subject of effect size will be discussed in greater detail.

Measures of effect size   The discussion will begin by clarifying the relationship between statistical significance and effect size. Equation 28.74 is a general equation (presented by Rosenthal (1991, 1993) and discussed in Tatsuoka (1993)) which describes the relationship between effect size and a test statistic employed to measure statistical significance.

(Equation 28.74)

$$Effect\ size = Significance\ test\ statistic/Sample\ size$$

The *Effect size* (*ES*) value on the left side of Equation 28.74 can be any one of various measures of effect size discussed throughout this book. The value designated *Significance test statistic* will be the computed value for the inferential test statistic which is employed to determine statistical significance (e.g., a t, F, χ^2, etc. value). The number employed to represent the *Sample size* will be some index that reflects the overall size of a sample employed in a study (but will usually not correspond exactly to the total number of subjects employed in a study). The relationship in Equation 28.74 reflects the fact that if sample size varies, in order for the effect size to remain unchanged, there must be a direct relationship between the magnitude of the computed test statistic and the sample size (i.e., as the value of the sample size increases, the magnitude of the test statistic must increase).

This relationship was demonstrated earlier in reference to Example 16.1, which is employed to illustrate the **chi-square test for *r* × *c* tables**. In Section II of the latter test, Table 16.2 summarizes the data for Example 16.1. Analysis of the data in the latter table, which is comprised of 200 observations, yields a chi-square value of $\chi^2 = 18.18$. In Section VI of the **chi-square test for *r* × *c* tables** (under **measures of association for *r* × *c* contingency tables**), Table 16.21 summarizes the same experiment employing numbers (in the rows, columns, and cells of the summary table) that are half the value of those employed in Table 16.2. The number of observations in Table 16.21 is 100, and the computed test statistic is $\chi^2 = 9.1$. Since the identical degrees of freedom are employed in the analysis of both tables (i.e., $df = 1$, with $\chi^2_{.05} = 3.84$ and $\chi^2_{.01} = 6.63$), the level of significance represented by the p value obtained for Table 16.2 will be much lower than the p value obtained for Table 16.21.

When Equation 28.74 is employed to compute the effect size index (*ES*) for Tables 16.2 and 16.21, the following values are obtained: $ES = 18.18/200 = .091$ (for Table 16.2) and $ES = 9.1/100 = .091$ (for Table 16.21). The aforementioned effect size values correspond to the square of the values which will be computed for the **phi coefficient** (ϕ) (computed with Equation 16.20/16.21) for each of the tables. Note that the chi-square values computed for the two tables are proportional to the sample sizes — in other words, the chi-square value computed for Table 16.2 is two times the chi-square value computed for Table 16.21, and the sample size employed in Table 16.2 is two times the size of the sample employed in Table 16.21. Yet, in spite of the latter, the identical effect size is computed for both tables. The fact that the two effect sizes are equal illustrates that unlike the computed value of a test statistic and its associated probability value, the effect size is independent of sample size.

Thus, a computed test statistic (e.g., a t, F, χ^2, etc. value) does not in itself provide information regarding the magnitude of a treatment effect. The reason for this is that the value of a test statistic is not only a function of the treatment effect, but is also a function of the size of the sample employed in an experiment. Since the power of a statistical test is directly related to sample size, the larger the sample the more likely a significant value will be obtained for the test statistic if there is an effect of any magnitude present in the underlying populations. Regardless of how small a treatment effect is present, the probability value associated with the computed test statistic will decrease as the size of the sample employed to detect it increases. Thus, it is impossible to determine from a significant test statistic and its associated p value (e.g., .05, .01, .001, etc.) whether the significant result is due to a large, medium, or small treatment effect.

The wisdom of using the conventional hypothesis testing model (which employs the result of a test of statistical significance) is addressed in detail at the end of the discussion of meta-analysis. For some time there has been controversy regarding the wisdom of employing tests of significance, insofar as the results of such tests are a function of power, which, as noted previously, is a function of sample size. The material to be presented on this issue later will describe an alternative hypothesis testing model which some researchers believe should be employed in lieu of the **classical hypothesis testing model** (which is discussed in detail in the **Introduction**). As you will see, regardless of which hypothesis testing model a researcher employs, the key to effective hypothesis testing ultimately boils down to using a representative sample that is large enough to detect any meaningful effect(s) present in the underlying population(s).

At this point a summary of the various indices which are employed to measure **effect size** will be presented. Throughout the discussion to follow, the terms **effect size** and **treatment effect** will be used interchangeably, since all of the measures described below are variously referred to as **measures of effect size**, **measures of magnitude of treatment effect**, **measures of association**, and **correlation coefficients**. All of these measures have been discussed previously in the book in reference to specific tests.

There are essentially two types of effect size indices. One type of index expresses effect size in the form of a correlation coefficient. This type of effect size index (which Rosenthal (1994) and Schulze (2004, p. 20) refer to as representing the r family) is computed at the conclusion of an experiment to indicate the proportion of variability on the dependent variable which can be attributed to the independent variable. Later in the discussion it will be illustrated that the summary value computed for an inferential test statistic (e.g., a t, F, χ^2 value) can be transformed into a correlation coefficient, in order that the latter can be employed as a measure of effect size for a set of data. An example of the latter (which can be found in the first section of this **Addendum**) is the computation of the **point-biserial correlation coefficient** to represent a measure of effect size for a t value.

A second type of effect size index is most commonly employed prior to conducting an experiment, in order to allow a researcher to determine the appropriate sample size to use to identify a hypothesized effect size. The latter type of index (which Rosenthal (1994) and Schulze (2004, p. 20) refer to as representing the d family) expresses effect size in terms of a difference score, which represents the difference (often in standard deviation units) between two underlying population parameters represented by two sample statistics, or the difference between a population parameter represented by a sample statistic and a hypothesized population parameter. Most of the effect size indices of this type were developed by Jacob Cohen, and are described in detail in his classic book on statistical power (Cohen (1977, 1988)). Since within the context of a meta-analysis there are times when a researcher may wish or be required to employ both types of effect size indices, equations are available for converting an effect size index based on a difference score into a correlational effect size index, and vice versa. Cohen (1977, 1988) describes the following effect size indices based on a difference score that are relevant to some of the inferential statistical tests discussed in this book: d, f, q, g, h, and w.[27] It should be noted that although some of the aforementioned indices can be employed to measure effect size in a study that involves two or more experimental conditions, as will be noted later in this section, meta-analysis is generally confined to evaluating hypotheses which contrast only two experimental conditions. A brief description of each of the effect size indices described by Cohen (1977, 1988) follows.

The *d* index The *d* index represents the difference between two means expressed in standard deviation units. The *d* index was previously employed in the computation of the power of the **single-sample *t* test** (where using Equation 2.5, $d = |\mu_1 - \mu|/\sigma$) and the *t* tests **for two independent and dependent samples** (where, using Equations 11.10 and 17.14, $d = |\mu_1 - \mu_2|/\sigma$ and $d = |\mu_1 - \mu_2|/\sigma_D$). Cohen (1977; 1988, Ch. 2) has derived tables which allow a researcher to determine, through use of the *d* index, the appropriate sample size to employ if one wants to test a hypothesis about the difference between two means at a specified level of power. Cohen (1977; 1988, pp. 24–27) has proposed the following (admittedly arbitrary) *d* values as criteria for identifying the magnitude of an effect size: a) A **small effect size** is one that is greater than .2 but not more than .5 standard deviation units; b) A **medium effect size** is one that is greater than .5 but not more than .8 standard deviation units; and c) A **large effect size** is greater than .8 standard deviation units. Grissom and Kim (2005, pp. 59–60) describe computation of a confidence interval for *d*.

Equations 28.75 (Mullen and Rosenthal (1985), Rosenthal (1991, 1993)) and 28.76 (Cohen (1977; 1988, p. 23)) can be employed to convert an *r* value into a *d* value, and vice versa. Cohen (1977; 1988, pp. 23-27) states the *r* value computed with Equation 28.76 is a **point-biserial correlation** (r_{pb}), when $p_0 = p_1$. Equation 28.76 becomes $r = d/\sqrt{d^2 + (1/p_0 p_1)}$ when $p_0 \neq p_1$.[28]

$$d = \sqrt{\frac{4r^2}{1 - r^2}}$$

(Equation 28.75)

$$r = \frac{d}{\sqrt{d^2 + 4}}$$

(Equation 28.76)

In Endnote 15 of the *t* test for two independent samples it was noted that since **Cohen's *d* index** is based on population parameters, sample analogues of the *d* index have been developed by and/or described in Glass (1976), Hedges (1981, 1982), and Hedges and Olkin (1985). It was also noted that Equation 28.77 (which is identical to Equation 11.25) is employed to compute the sample analogue which is commonly referred to as the *g* index. Cohen (2001), Rosenthal *et al.* (2000), and Rosnow *et al.* (2000) note that Equations 28.78–28.80 can respectively be employed to transform a *g* value into a *t* value, a *t* value into a *g* value, and a *g* value into an *r* value. (Equation 28.78 is identical to Equation 11.26 and Equation 28.79 is identical to Equation 11.27.) (The notation *N* represents the total sample size employed in a study while *n* represents the number of subjects per group.) A *d* value can be employed in lieu of a *g* value in Equations 28.77–28.80 (Hunter and Schmidt (2004, p. 278)).

$$g = \frac{|\bar{X}_1 - \bar{X}_2|}{\tilde{s}_p}$$

(Equation 28.77)

$$t = g\sqrt{\frac{n}{2}}$$

(Equation 28.78)

$$g = t\sqrt{\frac{2}{n}} = \frac{2t}{\sqrt{N}} \qquad \textbf{(Equation 28.79)}$$

$$r = \frac{g}{\sqrt{g^2 + 4\left(\dfrac{df_{WG}}{N}\right)}} \qquad \textbf{(Equation 28.80)}$$

Note that when the value $g = 1.24$ obtained for Example 11.1 (in Endnote 15 of the *t* **test for two independent samples**) is substituted in Equation 28.78, the value $t = 1.24\sqrt{5/2} = 1.96$ is computed (which is the absolute value obtained for *t* with Equation 11.1). Conversely, when the absolute value $t = 1.96$ is substituted in Equation 28.79 it yields the value $g = 1.24$ ($g = 1.96\sqrt{2/5} = 1.24$ or $g = (2)(1.96)/\sqrt{10} = 1.24$). The value $r = .57$ is obtained through use of equation 28.80 when g is converted into an *r* value: $r = 1.24/\sqrt{(1.24)^2 + 4(8/10)} = .57$ (where $df_{WG} = N - 2$).

Tatsuoka (1993) notes that Glass (1976) has developed another sample analogue of **Cohen's *d* index** which is designed to serve as a measure of association when a researcher has an experimental group and a control group. Equation 28.81 is employed to compute **Glass's *g* index**.

$$g = \frac{\bar{X}_e - \bar{X}_c}{\tilde{s}_c} \qquad \textbf{(Equation 28.81)}$$

In Equation 28.81, $\bar{X}_e$ and $\bar{X}_c$, respectively, represent the means of the experimental and control groups, and $\tilde{s}_c$ represents the estimated population standard deviation of the control group. Note that Glass (1976) employs the standard deviation of the control group rather than a pooled standard deviation involving both groups in the denominator of his equation. He does this since he believes that if pooled variability is employed in the denominator of the equation, the relevant *g* value for a specific experimental group and a control group will be unduly influenced by the variability of other experimental groups which are not involved in the comparison. A more detailed discussion of **Glass's *g* index** can be found in Tatsuoka (1993) as well as in Grissom and Kim (2005, pp. 53–54). The latter author's note that although the measures of effect size developed by Glass (1976) and Hedges (1981, 1982) tend to have some **positive bias** (i.e., they tend to overestimate the effect size in the underlying population), available adjustment procedures to reduce bias are rarely employed since the degree of positive bias will be minimal except in the case of small sample sizes.

The *f* index The *f* index is a generalization of the *d* index to the case where there are three or more means. The *f* index was previously discussed in Section VI of the **single-factor between-subjects analysis of variance**. Cohen (1977; 1988, Ch. 8) has derived tables that allow a researcher to determine, through use of the *f* index, the appropriate sample size to employ if one wants to test a hypothesis about the difference between three or more means at a specified level of power. Cohen (1977; 1988, pp. 284–288) has proposed the following (admittedly arbitrary) *f* values as criteria for identifying the magnitude of an effect size: a) A **small effect size** is one that is greater than .1 but not more than .25; b) A **medium effect size** is one that is greater than .25 but not more than .4; and c) A **large effect size** is greater than .4.

Hedges (1981) has developed the g' index as an alternative to the f index. As is the case with the latter index, g' is designed to be used as a measure of effect size for the analysis of variance. **Hedges' g'** is computed with Equation 28.82 (which is Equation 28.77 employing different notation).

$$g' = \frac{\bar{X}_{e_j} - \bar{X}_c}{\sqrt{MS_{WG}}}$$

(**Equation 28.82**)

In Equation 28.82, $\bar{X}_{e_j}$ represents the mean of the jth group and $\bar{X}_c$ represents the mean of the control group. $\sqrt{MS_{WG}}$ is the square root of the pooled estimate of within-groups variability employed in the analysis of variance. A more detailed discussion of **Hedges' g' index** can be found in Tatsuoka (1993).

The q index The q index represents the difference between two Pearson product-moment correlation coefficients, where the latter values are expressed through use of **Fisher's z_r transformation**. The equation for the q index is $q = z_1 - z_2$, where z_1 and z_2 are the Fisher transformed z_r values for the two correlation coefficients. Cohen (1977; 1988, Ch. 4) has derived tables that allow a researcher to determine, through use of the q index, the appropriate sample size to employ if one wants to test a hypothesis about the difference between two correlation coefficients at a specified level of power. Cohen (1977; 1988, pp. 113–116) has proposed the following (admittedly arbitrary) q values as criteria for identifying the magnitude of an effect size: a) A **small effect size** is one that is greater than .1 but not more than .3 ; b) A **medium effect size** is one that is greater than .3 but not more than .5; and c) A **large effect size** is greater than .5.

The g index The g index (not to be confused with the g indices discussed earlier in this section) can be employed to compute the power of the **binomial sign test for a single sample (Test 9)**. The g index represents the distance in units of proportion from the value .5. The equation Cohen (1977, 1988) employs for the g index is $g = P - .5$ when H_1 is directional and $g = |P - .5|$ when H_1 is nondirectional (where P represents the hypothesized value of the population proportion stated in the alternative hypothesis — in employing the g index it is assumed that the researcher has stated a specific value in the alternative hypothesis as an alternative to the value that is stipulated in the null hypothesis). Cohen (1977; 1988, Ch. 5) has derived tables that allow a researcher, through use of the g index, to determine the appropriate sample size to employ if one wants to test a hypothesis about the distance of a proportion from the value .5 at a specified level of power. Cohen (1977; 1988, pp. 147–150) has proposed the following (admittedly arbitrary) g values as criteria for identifying the magnitude of an effect size: a) A **small effect size** is one that is greater than .05 but not more than .15; b) A **medium effect size** is one that is greater than .15 but not more than .25; and c) A **large effect size** is greater than .25.

The h index The h index can be employed to compute the power of the **z test for two independent proportions (Test16d)**. The value h is an effect size index reflecting the difference between two population proportions. h is computed through use of the arcsine transformation (discussed in Section VII of the t **test for two independent samples**). The equation for the h **index** is $h = \phi_1 - \phi_2$ (where ϕ_1 and ϕ_2 are the arcsine transformed values for the proportions). Cohen (1977; 1988, Ch. 6) has derived tables that allow a researcher, through use of the h **index**, to determine the appropriate sample size to employ if one wants to test a hypothesis about the difference between two population proportions at a specified level of power. Cohen

(1977; 1988, pp. 184–185) has proposed the following (admittedly arbitrary) *h* values as criteria for identifying the magnitude of an effect size: a) A **small effect size** is one that is greater than .2 but not more than .5; b) A **medium effect size** is one that is greater than .5 but not more than .8; and c) A **large effect size** is greater than .8.

The *w* index The *w* index can be employed to compute the power of the **chi-square goodness-of-fit test (Test 8)** and the **chi-square test for *r* × *c* tables**. The value *w* is an effect size index reflecting the difference between expected and observed frequencies. The equation for the *w* index is $w = \sqrt{\Sigma[(P_{alt} - P_{null})^2/P_{null}]}$. The latter equation indicates the following: a) For each of the cells in the chi-square table, the proportion of cases hypothesized in the null hypothesis is subtracted from the proportion of cases hypothesized in the alternative hypothesis; b) The obtained difference in each cell is squared, and then divided by the proportion hypothesized in the null hypothesis for that cell; c) All of the values obtained for the cells in part b) are summed; and d) *w* represents the square root of the sum obtained in part c). Cohen (1977; 1988, Ch. 7) has derived tables that allow a researcher to determine, through use of the *w* **index**, the appropriate sample size to employ if one wants to test a hypothesis about the difference between observed and expected frequencies in a chi-square table at a specified level of power. Cohen (1977; 1988, pp. 224–226) has proposed the following (admittedly arbitrary) *w* values as criteria for identifying the magnitude of an effect size: a) A **small effect size** is one that is greater than .1 but not more than .3; b) A **medium effect size** is one that is greater than .3 but not more than .5; and c) A **large effect size** is greater than .5.

With the exception of the sample analogue *g* and *g'* indices for the *d* **index**, the effect size indices discussed above are typically employed for power computations prior to conducting an inferential statistical test. Consequently, it is far more common that the effect size indices employed within the framework of meta-analysis are correlational measures (an *r* value) which are based on the empirical data obtained in a study.[29] At this point the use of the **Pearson product-moment correlation coefficient** as a measure of effect size will be discussed.

Pearson *r* as a measure of effect size Although outside of the context of meta-analysis r^2 (the **coefficient of determination** which is discussed in Section V) rather than *r* is more commonly used to represent the measure of effect size, either of the values can be employed for this purpose. In discussing the use of the **Pearson product-moment correlation coefficient** as a measure of effect size, Cohen (1977; 1988, pp. 78–81) has proposed the following (admittedly arbitrary) *r* values as criteria for identifying the magnitude of an effect size: a) A **small effect size** is one that is greater than .1 but not more than .3; b) A **medium effect size** is one that is greater than .3 but not more than .5; and c) A **large effect size** is greater than .5. As previously noted in Section VI, Cohen (1977; 1988, Ch. 3) has derived tables for computing power which allow a researcher to determine the appropriate sample size to employ if one wants to evaluate an alternative hypothesis that designates a specific value for a population correlation (when the null hypothesis is H_0: $\rho = 0$). In addition to **Pearson *r***, Rosenthal (1993) notes that any of the following measures of association (all of which are special cases of the **Pearson product-moment correlation coefficient**) can be used to represent an *r* value within the framework of a meta-analysis: a) The **point-biserial correlation coefficient** (discussed earlier in this **Addendum**), which is employed as a measure of association for the *t* **test for two independent samples** (and can also be employed as a measure of association for the *t* **test for two dependent samples (Test 17)**); b) The **phi coefficient** (φ) (which is discussed in Section VII as well as in Section VI of the **chi-square test for *r* × *c* tables**), which is employed when both variables are dichotomous; c) **Spearman's rank-order correlation coefficient**, which is employed when both variables are in a rank-order format. In Section VI of **Spearman's rank-order correlation**

coefficient, it is demonstrated that the latter correlation coefficient is a special case of the **Pearson product-moment correlation coefficient**. Additional discussion of **Pearson *r*** as a measure of effect size as well as other measures of effect size can be found in Rosenthal *et al.* (2000), Rosnow *et al.* (2000), and Grissom and Kim (2005).

Meta-analytic procedures The meta-analytic procedures to be discussed in this section are those described by Robert Rosenthal (1991, 1993).[30] The reader should take note of the fact that alternative procedures are also employed for meta-analysis. One alternative approach developed by Hunter and Schmidt (Schmidt and Hunter (1977) and Hunter and Schmidt (1990, 2004)) referred to as **psychometric meta-analysis** will be discussed at the end of this section,

Rosenthal (1991,1993) describes the following four types of meta-analytic procedures (as well as additional procedures that will not be covered): a) **Procedures that compare two or more studies with respect to significance level.** In the case of two studies, these procedures determine whether or not the *p* values obtained for two studies are significantly different from one another.   In the case of three or more studies, the meta-analytic procedures determine whether the *p* values for the *k* studies (where *k* represents the number of studies) are homogeneous — i.e., consistent with one another; b) **Procedures that combine the significance levels (i.e., *p* values) of two or more studies, and obtain a combined/pooled estimate of the *p* value for all *k* studies**; c) **Procedures that compare two or more studies with respect to effect size.** In the case of two studies, these procedures determine whether or not the computed values for the effect sizes of the two studies are significantly different from one another. In the case of three or more studies, the meta-analytic procedures determine whether the computed values for the effect sizes of the *k* studies are homogeneous — i.e., consistent with one another; and d) **Procedures that combine the effect size values computed for two or more studies, and obtain a combined/pooled estimate of effect size for all *k* studies.**

It should be noted that when the results of two or more studies are compared or combined within the framework of a meta-analysis, it is assumed that the *k* studies are independent of one another (i.e., represent separate studies employing different subjects). To carry out the procedures to be described in this section, it is required that the test statistic representing the outcome of each of the *k* studies is standardized (i.e., that the result of each study is summarized with the same test statistic). The most common statistics employed for this purpose are values of *z* and *r*. In meta-analytic procedures which compare or combine significance levels, *p* values are converted into *z* values. In procedures that compare or combine effect sizes, the effect size is commonly expressed as an *r* value. At this point I will summarize a number of equations (described in Rosenthal (1985, 1991, 1993)) which allow a researcher to convert the results of an inferential statistical test into an *r* value.

a) Within the framework of conducting a *t* **test for two independent samples** or a *t* **test for two dependent samples**, Equation 28.83 can be employed to transform a *t* value into an *r* value.

$$r = \sqrt{\frac{t^2}{t^2 + df}}$$ **(Equation 28.83)**

The value computed with Equation 28.83 is the square root of the value computed with Equation 28.64 (the equation for computing **eta squared** ($\tilde{\eta}^2$), which is equivalent to the square of the **point-biserial correlation** (r_{pb})). It should also be noted that within the framework of conducting a *t* test, Equation 28.76 was presented earlier for conversion of **Cohen's *d* index** into an *r* value.[31]

b) Within the framework of conducting an **analysis of variance** where the degrees of freedom for the numerator equals 1 (i.e., two groups/conditions), Equation 28.84 can be employed to transform an F value into an r value.

$$r = \sqrt{\frac{F}{F + df_{error}}}$$ **(Equation 28.84)**

c) Within the framework of conducting a **chi-square test for $r \times c$ tables**, where $df = 1$ (i.e., a 2×2 contingency table), Equation 28.85 can be employed to transform a chi-square value into an r value.

$$r = \sqrt{\frac{\chi^2}{n}}$$ **(Equation 28.85)**

Note that Equation 28.85 is the same as Equation 16.21, which is employed to compute the **phi coefficient (ϕ)**.

d) If a researcher wants to transform a z value into an r value, Equation 28.86 can be employed.

$$r = \sqrt{\frac{z^2}{n}}$$ **(Equation 28.86)**

Rosenthal (1991, 1993) notes that within the context of meta-analysis, he is only interested in **single degree of freedom** comparisons — which is the term he uses to refer to studies that compare two groups/conditions with one another. He states that meta-analytic procedures are of little use when $df > 1$, since when an omnibus test (e.g., an analysis of variance, chi-square test, etc.) compares more than two groups, it becomes difficult or impossible to answer the questions addressed by meta-analysis with a high degree of precision. In other words, an omnibus test statistic based on more than two experimental conditions does not identify which of the conditions are significantly different from other another. The use of the term **single degree of freedom comparison** within the context of a two group/condition experiment refers to the following: a) When there are two groups/conditions, the between-groups/between-conditions degrees of freedom for the **analysis of variance** is equal to 1. When $df_{BG/BC} = 1$, the **analysis of variance** and the **t tests for two independent and dependent samples** are equivalent procedures; and b) In the case of the **chi-square test for $r \times c$ tables**, $df = 1$ when two groups are contrasted with one another.

Demonstration of meta-analytic procedures In this section the following four meta-analytic procedures described by Rosenthal (1985, 1991, 1993) will be presented: a) A procedure for comparing k studies with respect to homogeneity of significance level; b) A procedure for obtaining a combined significance level (p value) for k studies; c) A procedure for comparing k studies with respect to homogeneity of effect size; and d) A procedure for obtaining a combined effect size for k studies. Example 28.8 will be employed to demonstrate the aforementioned meta-analytic procedures.

Example 28.8 *Five independent studies (to be identified by the letters* **A, B, C, D,** *and* **E**) *evaluating the same general hypothesis (e.g., patients who receive a specific type of therapy will do better than a no-treatment control group) are conducted over a two year period. All of the* studies employ an independent groups design with an independent variable comprised of two levels. The analysis of the data in each of the studies involved the use of the *t* **test for two independent samples** *to determine if there was a difference between the means of an experimental and control groups (the higher the mean of a group the more effective the treatment). In addition, a **point-biserial correlation** (which will be represented by the notation r) was computed for each study, to determine the magnitude of any effect size that was present. In studies* **A, B, C,** *and* **D** *the mean score of the experimental group was higher than the mean score of the control group. The one-tailed probability values (based on the result of the t-test) and the* **point-biserial correlations** *computed for the studies follow:* **A** *(p =.05; r = .60),* **B** *(p = .01; r = .50),* **C** *(p = .10; r = .20),* **D** *(p = .20; r = .30). The result of study* **E** *was in the opposite direction of the other four studies — i.e., the mean of the control group was higher than the mean of the experimental group. The one-tailed probability value (based on the result of the t-test) and the* **point-biserial correlation** *for study* **E** *follow:* **E** *(p = .09; r = .15). The total number of subjects employed in each of the studies were:* **A**(20); **B** (40); **C**(10); **D** (30); **E** (25). *Compare the p values and effect sizes with respect to homogeneity, and compute a combined/pooled p value and effect size for the five studies.*

At this point it should be stated that all probability values (*p*) used within the framework of the meta-analytical procedures to be described will be **one-tailed**. The reason for this is that we want to be able to designate the direction of the outcome of each study — i.e., whether the mean of the experimental group is larger than the mean of the control group, or vice versa. Thus, the left tail of the distribution will represent one directional outcome and the right tail the other directional outcome. Numerically, the direction of an outcome will be designated in reference to a summary statistic (e.g., an *r* or *z* value) by assigning a plus sign to outcomes in one tail of the sampling distribution, and a minus sign to outcomes in the other tail.

Test 28l: Procedure for comparing *k* studies with respect to significance level The procedure to be described in this section evaluates the following null and alternative hypotheses.

Null hypothesis H_0: The *p* values obtained for the *k* studies are consistent/ homogenous with one another.

Alternative hypothesis H_1: The *p* values obtained for the *k* studies are not consistent/ homogenous with one another.

Equation 28.87 can be employed to evaluate whether *k* (where $k \geq 2$) probability (*p*) values are homogeneous (i.e., consistent with one another). When the test to be described is employed with *k* = 2 studies, it simply evaluates whether there is a significant difference between the two *p* values.

$$\chi^2 = \sum_{j=1}^{k} (z_j - \bar{z}_k)^2 \qquad \text{(Equation 28.87)}$$

Where: $\bar{z}_k$ represents the average *z* value computed for the *k* published studies
        z_j represents the *z* value for the jth study

In order to employ Equation 28.87, it is required that each of the *p* values is converted into its corresponding standard normal deviate (i.e., a *z* value). In order to do the latter, we find the *p* value in **Column 3** of **Table A1** in the **Appendix** which corresponds to the *p* value obtained for a given study (note that the *p* values in **Column 3** represent one-tailed probabilities). The *z* value in the same row (i.e., the value in **Column 1**) of **Table A1** which corresponds to the latter p value is employed as its standard normal deviate. Thus, in the case of *Study A*, which has a p value of .05, the value $z = 1.65$ is employed to represent it, since the probability/proportion in **Column 3** of **Table A1** is .0495 (which is the closest value to .05). In the case of *Study B*, which has a p value of .01, the value $z = 2.33$ is employed to represent it, since the probability/ proportion in **Column 3** of **Table A1** is .0099 (which is the closest value to .01). Employing the same methodology with *Studies C* and *D*, we find that the z values which correspond to the p values .10 and .20 are 1.28 and .84. In the case of *Study E*, the z value that corresponds to the p value .09 is 1.34. The z values for *Studies A, B, C,* and *D* (the outcomes of which are in the same direction) will all be assigned a positive sign. The z value for *Study E* (the outcome of which is in the opposite direction of the other studies) will be assigned a negative sign.[32] Thus, the five z values we will employ in Equation 28.87 are 1.65, 2.33, 1.28, .84, and −1.34.

The following protocol is employed for Equation 28.87: a) Compute the mean of the k z_j values; b) Subtract the mean from each of the k z_j values, and square each difference score; and c) Compute the sum of the k squared difference scores. The resulting value, which is a chi-square value, represents the test statistic, for which the degrees of freedom are $df = k - 1$.

We compute the average of the five z values to be $\bar{z}_k = .95$ (i.e., $[1.65 + 2.33 + 1.28 + .84 + (−1.34)]/5 = .95$). When the latter value along with the five z scores which are computed above are substituted in Equation 28.87, we obtain the value $\chi^2 = 7.75$.

$$\chi^2 = (1.65 - .95)^2 + (2.33 - .95)^2 + (1.28 - .95)^2 + (.84 - .95)^2 + (−1.34 - .95)^2 = 7.75$$

Since there are 5 studies, $df = 5 - 1 = 4$. Employing **Table A4** in the **Appendix**, for $df = 4$, $\chi^2_{.05} = 9.49$ and $\chi^2_{.01} = 13.28$ (the probabilities for these critical values are one-tailed). In order to reject the null hypothesis, the computed chi-square value must be equal to or greater than the tabled critical value at the prespecified level of significance. Since the obtained value $\chi^2 = 7.75$ is less than $\chi^2_{.05} = 9.49$, the null hypothesis is retained. In other words, the data do not indicate that the probability values obtained for the five studies are inconsistent (i.e., not homogeneous) with one another.

Test 28m: The Stouffer procedure for obtaining a combined significance level (p value) for k studies A number of procedures have been developed for obtaining a combined/pooled p value for k independent studies which evaluate the same general hypothesis. These procedures are relevant for obtaining a combined probability for studies that involve a directional hypothesis testing situation where $df = 1$ (i.e., a study in which two groups/ conditions are contrasted with one another). Birnbaum (1954) and Rosenthal (1978, 1991) provide a good overview of the various procedures. The specific test to be described here, which was developed by Stouffer *et al.* (1949), computes a combined p value through use of Equation 28.88. Sources which discuss this test in greater detail are Conover (1999), Mosteller and Bush (1954), Rosenthal (1991, 1993), and Wolf (1986).[33]

$$z = \frac{\sum_{j=1}^{k} z_j}{\sqrt{k}} \qquad \textbf{(Equation 28.88)}$$

Where: z_j represents the z value for the jth study

As is the case with Equation 28.87 (employed for **Test 28l**), Equation 28.88 requires that we convert each of the p values obtained for the k studies into its corresponding standard normal deviate (i.e., a z score). Once again it should be emphasized that one-tailed probabilities are always employed. Since we have already computed the appropriate z values for **Test 28l**, we are ready to employ Equation 28.88. The protocol for the equation requires that we sum the k z_j values, and divide the sum by the square root of k. The resulting z value represents the test statistic, which is evaluated with **Table A1**.

When the z scores for Example 28.8 that correspond to the probability values obtained for the $k = 5$ studies are substituted in Equation 28.88, we obtain $z = 2.13$.

$$z = \frac{1.65 + 2.33 + 1.28 + .84 + (-1.34)}{\sqrt{5}} = \frac{4.76}{\sqrt{5}} = 2.13$$

Employing **Table A1**, we determine that the one-tailed probability associated with $z = 2.13$ is .0166. The latter value represents the combined probability for the five studies. Since $p = .0166$ is less than $p = .05$, the combined probability derived from the test is statistically significant.[34] The combined probability value of .0166 is an overall probability in favor of the outcome of the majority of the studies (in which the experimental group obtained a higher mean than the control group). It is important to note that in Example 28.8 the outcomes of the five studies (reflected by the five p values) appear to be reasonably consistent (although some might consider the contrary outcome of *Study E* to be somewhat problematical). As will be noted later, when Equation 28.88 yields a combined probability value that is significant, yet the outcomes of the studies are not homogeneous, the obtained combined probability must be viewed with great caution.

Everitt (1977, 1992) employs the **Stouffer procedure** as an alternative methodology for conducting a **heterogeneity chi-square analysis** (which is described in Section VI of the **chi-square test for $r \times c$ tables**). A **heterogeneity chi-square analysis** is employed to pool the results of k independent studies (where $k \geq 2$), each of which (through use of the **chi-square test**) evaluates the same hypothesis. To be more specific, the data for each study are summarized in a 2×2 contingency table, and each of the contingency tables is evaluated with the **chi-square test for homogeneity**. In the application of the **Stouffer procedure**, each of the computed chi-square values is converted into a z value by taking the square root of a chi-square value (i.e., $z = \sqrt{\chi^2}$). The latter z values along with the value of k are then substituted in Equation 28.88. The z value obtained with Equation 28.88 is then employed to determine a pooled probability value for the k contingency tables.

The data in **Part A** of Table 16.25 will be employed to illustrate the use of the **Stouffer procedure** for a **heterogeneity chi-square analysis**. The three 2×2 contingency tables in **Part A** of Table 16.25 summarize the results of three independent studies evaluating the frequency of occurrence of kidney diseases in males and females. The chi-square values obtained for the three contingency tables and the z values derived for each of the chi-square values are as follows: **Study 1**: $\chi^2 = 1.69$, $z = \sqrt{\chi^2} = \sqrt{1.69} = 1.3$; **Study 2**: $\chi^2 = 2.88$, $z = \sqrt{\chi^2} = \sqrt{2.88} = 1.70$; **Study 3**: $\chi^2 = 2.63$, $z = \sqrt{\chi^2} = \sqrt{2.63} = 1.62$. The appropriate values are substituted below in Equation 28.88 yielding the value $z = 2.67$. Note that since the patterns of the data are homogeneous for all three studies (i.e., kidney disease is more common among females than it is among males), we assign the same sign (in this case a positive sign) to each z value. If in any of the studies we had observed that males had a higher frequency of kidney disease than females, a negative sign would have been assigned to the z value for that study.

$$z = \frac{1.3 + 1.70 + 1.62}{\sqrt{3}} = \frac{4.62}{\sqrt{3}} = 2.67$$

Employing **Table A1**, we determine that the one-tailed probability associated with $z = 2.67$ is .0038. The latter value represents the combined probability for the three studies. Since $p = .0038$ is less than $p = .05$, the combined probability derived from the test is statistically significant. The combined probability value of .0038 is an overall probability in favor of the outcome of the 3 studies, all of which suggest that kidney disease occurs more frequently in females than in males.

Note that the pooled probability value of .0038 computed for the three studies is lower than the pooled probability value computed earlier in the book under the discussion of **heterogeneity chi-square analysis** for the same problem. Although also statistically significant, the latter probability value was slightly above .01. With regard to the discrepancy between the two probability values, it should be noted that a correction of continuity is not employed in computing the chi-square values for the three contingency tables in **Part A** of Table 16.25. (A correction for continuity is employed in **Part D** of Table 16.25, which summarizes the last step of the **heterogeneity chi-square analysis** described earlier in the book.) If a correction for continuity is employed for the three contingency tables in **Part A** of Table 16.25, the following chi-square values and resulting z values are obtained for the three studies: **Study 1:** $\chi^2 = 1.07$, $z = \sqrt{\chi^2} = \sqrt{1.07} = 1.03$; **Study 2:** $\chi^2 = 2.00$, $z = \sqrt{\chi^2} = \sqrt{2.00} = 1.41$; **Study 3:** $\chi^2 = 1.70$, $z = \sqrt{\chi^2} = \sqrt{1.70} = 1.30$. When the latter three z values are substituted in Equation 28.88, the value $z = 2.16$ is obtained. It turns out that the latter z value is associated with a one-tailed probability of .0154, which is approximately the same probability value which was obtained for the **heterogeneity chi-square analysis** described in Section VI of the **chi-square test for $r \times c$ tables**.

$$ z = \frac{1.03 + 1.41 + 1.30}{\sqrt{3}} = \frac{3.74}{\sqrt{3}} = 2.16 $$

It should be emphasized once again that if through use of the **Stouffer procedure** a combined probability value is statistically significant, yet the outcomes of the individual studies are not homogeneous, the obtained combined probability must be viewed with great caution. Inspection of the data suggests that the latter is not an issue here, since in all three studies kidney disease is observed to occur at a reasonably consistent rate more frequently in females than in males. Nevertheless, a formal analysis of the issue of homogeneity of effect size for the problem under discussion is presented later in this section under the discussion of the **procedure for comparing k studies with respect to effect size (Test 28n)**. In addition, a combined effect size is computed for the three studies under the discussion of the **procedure for obtaining a combined effect size for k studies (Test 28o)**.[35]

The file drawer problem Rosenthal (1979) employs the term **file drawer problem** to refer to the fact there may be additional studies which evaluated the same hypothesis evaluated in a meta-analysis, yet were never submitted for publication or were submitted but rejected. Consequently, if one computes a statistically significant combined probability value for k studies with Equation 28.88, the following question might be asked: If, in fact, the null hypothesis is true, how many additional **null studies** would have to be conducted in order to render the combined probability value nonsignificant? A **null study** is one in which there is no difference between the two groups and, consequently, the values $z = 0$ and $p = .50$ will be obtained. Rosenthal (1979) derived Equation 28.90 from Equation 28.89 to answer the latter question. Equation 28.89 calculates the number of studies averaging null results which must be in the file drawer in order to increase the Type I error rate (i.e., combined p value) so that it equals a specific value (typically 5% or greater).

$$X = \frac{k}{z_\alpha^2}(k\bar{z}_k^2 - z_\alpha^2) \qquad \textbf{(Equation 28.89)}$$

Where: X represents the number of additional studies that are required to render the combined probability nonsignificant

z_α represents the critical one-tailed z value at the required level of statistical significance for the combined probability

$\bar{z}_k$ represents the average z value computed for the k published studies

Rosenthal (1979) notes that if we employ the .05 level of significance, the one-tailed value $z_{.05} = 1.645$ (which for purposes of greater precision is used rather than the usual $z = 1.65$) is employed to represent z_α in Equation 28.89. When $z_\alpha = 1.645$ is substituted in the latter equation, it becomes Equation 28.90.

$$X = \frac{k}{2.706}(k\bar{z}_k^2 - 2.706) \qquad \textbf{(Equation 28.90)}$$

By employing Equation 28.90 one can determine how many additional null studies will be required in order for the combined p value for a hypothesis under study to equal .05. One additional null study above the computed value of X will render the combined probability above .05, and thus the combined probability for the general hypothesis will no longer be statistically significant. In the case of Example 28.8, if we substitute the values $k = 5$ and $\bar{z}_k = .95$ in Equation 28.90, we compute the value $X = 3.34$.

$$X = \frac{5}{2.706}[(5)(.95)^2 - 2.706] = 3.34$$

This result tells us that only four additional null studies evaluating the same general hypothesis are required to be in the **file drawer** to produce a nonsignificant combined probability (since 4 is the next integer number above the obtained value $X = 3.34$). If the results of four such studies are combined with the five studies documented in Example 28.8, the resulting probability computed with Equation 28.88 will be greater than .05. Specifically, if four p values of .50 and their corresponding z values of zero are added to Example 28.8, Equation 28.88 will yield the value $z = 1.59$, which is less than the tabled critical .05 value $z_{.05} = 1.65$. Note that the sum of the z values, which equals 4.76, does not change when $k = 9$ (which represents the original five studies plus the four null studies), since $z = 0$ for each of the four additional studies. Thus: $z = 4.76/\sqrt{9} = 1.59$. Rosenthal (1991, 1993) has addressed the question of how one might go about estimating the number of unpublished studies that remain in the file drawer. He suggests a conservative estimate of the upper limit of the number of unpublished studies might be approximated by the value $5k + 10$ (e.g., if $k = 5$, the minimum estimate for the number of studies in the file drawer will be $(5)(5) + 10 = 35$).

The file drawer problem is most commonly discussed within the context of highly significant meta-analytic research — in other words, in instances where the combined probability for a hypothesis is at a statistically significant level, with the value of p being very low. In such an instance, a skeptic would want to rule out the likelihood the null hypothesis is, in fact, true, and that many of the published studies, in reality, represent Type I errors (i.e., spuriously significant results). Since there is a bias toward publishing (as well as submitting) significant results, it is sometimes suggested that if all the unpublished studies in the file drawer were taken into account, support for many a hypothesis would evaporate. Rosenthal (1991) discusses the latter issue, as well as empirical studies which address the question of how publication bias can influence the results of a meta-analysis. Stokes (2001) provides a provocative critique of the use of Equations 28.89/28.90 as a mechanism for estimating the number of unpublished null studies in the file drawer. Employing the logic underlying **randomization tests** (which are discussed in Section IX (the **Addendum**) of the **Mann–Whitney *U* test**), Stokes (2001) demonstrates that in some instances the value computed for X with Equations 28.89/28.90 may overestimate the number of null studies required in the file drawer in order to yield a nonsignificant meta-analysis. It should also be emphasized that Hunter and Schmidt (2004, pp. 499–509) argue that the method described in this section is not optimal for resolving the **file drawer problem** (which they refer to as **availability bias**), and suggest alternative approaches for estimating the number of unpublished null studies in the file drawer.

At this point in the discussion, it is worth noting it is possible to have a set of k studies and obtain a significant result for **Test 28l**, and also obtain a significant combined probability value for **Test 28m**. As an example, let us assume that the outcomes of the five studies in Example 28.8 were such that three favored the experimental group and two favored the control group (in terms of which group obtained the higher mean). Let us also assume the following: a) The p values for the three studies which favored the experimental group are quite low (e.g., in the .001 range); and b) In the case of the other two studies, the results of both are significant at the .05 level in favor of the control group.

It is likely that if **Test 28l** is employed to evaluate the probabilities for the aforementioned five studies, we will conclude the results of the studies are not homogeneous (i.e., a significant chi-square value will be computed with Equation 28.87). Nevertheless, it is conceivable that when the five probabilities are evaluated with **Test 28m**, a combined p value below .05 (in favor of the experimental group) will be obtained. Under such circumstances the combined probability value would have to be viewed with even greater caution than the combined probability of $p = .0166$, which was obtained through use of Equation 28.88 for Example 28.8. This is the case, since if all five studies are statistically significant, but three are in one direction and two are in the opposite direction, the consistency of the outcomes of the five studies leaves a lot to be desired. The lack of consistency in the probability values can be the result of any of the following: a) Differential effect sizes being present in the k studies; b) Differences in the size of the samples employed (which influence the power of each of the k tests); c) Differences in methodology; d) Errors in instrumentation or recording; e) Faulty data analysis; or f) Some combination of one or more of the aforementioned factors. In the discussion to follow, it will be emphasized that in order for a statistically significant combined probability value to be meaningful, there should be sufficient evidence it reflects the fact that the k studies employed in a meta-analysis yielded relatively homogenous results. The studies should not only exhibit consistency with regard to the direction of their outcome, but more importantly should exhibit consistency with respect to the magnitude of the effect size present in the k studies.

Neither of the two analyses conducted up to this point (i.e., **Tests 28l** and **28m**) provide us with any information regarding effect size for Example 28.8. Before proceeding, it is important to reiterate that the p value obtained for any study is always a direct function of the power of the statistical test, and power is a direct function of the sample size employed in a study. In order to reject the null hypothesis if a small (or even modest medium) effect size is present, it will be required that the researcher employ a large sample size. To illustrate this point, assume that two studies are conducted involving two independent samples which evaluate the same hypothesis. Let us also assume that in the underlying populations a small effect size characterizes the relationship between the independent and dependent variables. One study (A) employs a large sample size, while the other study (B) employs a relatively small sample size. In the case of

Study A, it is very likely we will be able to reject the null hypothesis, and the larger the sample size the lower the *p* value which will be obtained for the study. In the case of *Study B*, we will probably not be able to reject the null hypothesis, since the small sample size will severely compromise the power of the test — i.e., the test's ability to detect a difference between the groups. The computed *p* value for *Study B* will most likely be above .05, and, in fact, may be considerably larger. Yet if we compute the effect size for both studies, we obtain the same value. Let us assume that latter value is computed to be *r* = .15, which by Cohen's (1977, 1988) standards constitutes a small effect size. Let us also assume that additional studies of the same hypothesis consistently obtain approximately the same effect size, but only a few of them — specifically, those which happen to employ a large sample size — yield significant results. Obviously, such an occurrence provides support for Hedges and Olkin's (1995) and Rosenthal's (1991, 1993) contention, that within the framework of meta-analysis, it is more prudent to compare and/or combine effect sizes than it is to compare and/or combine *p* values. The next two procedures to be presented are designed to do just that.

Test 28n: Procedure for comparing *k* studies with respect to effect size   The procedure to be described in this section evaluates the following null and alternative hypotheses.

Null hypothesis H_0: The effect size values obtained for the *k* studies are consistent/homogenous with one another.

Alternative hypothesis H_1: The effect size values obtained for the *k* studies are not consistent/homogenous with one another.

Equation 28.91 (Hedges and Olkin (1985, p. 235) and Rosenthal (1993, p. 530)) can be employed to evaluate whether *k* (where $k \geq 2$) effect size values (as measured by *r*) are homogeneous (i.e., consistent with one another). (Equation 28.24 in Section VI is a different but equivalent version of Equation 28.91.) When the test to be described is employed with $k = 2$ studies, it simply evaluates whether there is a significant difference between the two effect size values.

$$\chi^2 = \sum_{j=1}^{k} (n_j - 3)(z_{r_j} - \bar{z}_{r_k})^2 \hspace{2cm} \textbf{(Equation 28.91)}$$

Where: z_{r_j} represents the Fisher transformed z_r value for the jth study

         $\bar{z}_{r_k}$ represents the average Fisher transformed z_j value computed for the *k* published studies. It is a weighted average, since the sample size of each study is taken into account. The value of $\bar{z}_{r_k}$ is computed with Equation 28.25.

         n_j represents the sample size in the jth study

In order to employ Equation 28.91 (or Equation 28.24), it is required that the *r* value for each study (which, as noted earlier, is used to represent the magnitude of effect size) is converted into a corresponding **Fisher transformed value** (discussed in Section VI). The latter is accomplished through use of either Equation 28.18 or **Table A17**. Employing **Table A17**, the Fisher transformed z_r values for the five studies are as follows.

$$z_{r_A} = .693 \quad z_{r_B} = .549 \quad z_{r_C} = .203 \quad z_{r_D} = .310 \quad z_{r_E} = .151$$

We will assign the Fisher transformed z_r values for *Studies A, B, C*, and *D* a positive sign, since they are all in the direction which favors the experimental group. Since the outcome of

Study E is in the opposite direction, it is assigned a negative sign. Thus, for *Study E* we will use the value $z_{r_E} = -.151$.[36]

Employing Equation 28.25, the average of the five Fisher transformed z_r values is computed to be $\bar{z}_{r_k} = .351$.

$$\bar{z}_{r_k} = \frac{\sum_{j=1}^{k}(n_j - 3)z_{r_j}}{\sum_{j=1}^{k}(n_j - 3)}$$

$$= \frac{(20-3)(.693)+(40-3)(.549)+(10-3)(.203)+(30-3)(.310)+(25-3)(-.151)}{(20-3)+(40-3)+(10-3)+(30-3)+(25-3)}$$

$$= \bar{z}_{r_k} = .351$$

We are now ready to substitute the appropriate values in Equation 28.91. The following protocol is employed in using the latter equation: a) For each of the k studies, subtract the average Fisher transformed z_r value (i.e., $\bar{z}_{r_k} = .351$) from the Fisher transformed z_r value for that study. Square the difference score, and multiply that value by the total sample size less three; and b) The sum of the k values computed in part a) is a chi-square value, which represents the test statistic. The degrees of freedom employed in evaluating the chi-square value are $df = k - 1$. The computations employing Equation 28.91 are shown below, and yield the value $\chi^2 = 9.18$.

$$\chi^2 = (20 - 3)(.693 - .351)^2 + (40 - 3)(.549 - .351)^2 + (10 - 3)(.203 - .351)^2$$

$$+ (30 - 3)(.310 - .351)^2 + (25 - 3)(-.151 - .351)^2 = 9.18$$

Since there are 5 studies, $df = 5 - 1 = 4$. Employing **Table A4**, for $df = 4$, $\chi^2_{.05} = 9.49$ and $\chi^2_{.01} = 13.28$ (the probabilities for these critical values are one-tailed). In order to reject the null hypothesis, the computed chi-square value must be equal to or greater than the tabled critical value at the prespecified level of significance. Since the obtained value $\chi^2 = 9.18$ is less (albeit barely) than $\chi^2_{.05} = 9.49$, the null hypothesis is retained. In other words, the data do not indicate the effect sizes obtained for the five studies are inconsistent with one another. Realistically, however, since the outcome is so close to being significant, most researchers would probably be reluctant to accept that a homogeneous effect size has been demonstrated. Certainly, visual inspection of the r values does not suggest homogeneity.

Earlier in the discussion of meta-analysis (under the discussion of the **Stouffer procedure**) a **heterogeneity chi-square analysis** was conducted on the data contained in the three 2×2 contingency tables in **Part A** of Table 16.25. At this point we will determine whether or not the effect size values for the three studies are homogeneous. The analysis to be described will employ the continuity-corrected chi-square values derived for the three studies (since the latter values are most consistent with the results of the **heterogeneity chi-square analysis** described in Section VI of the **chi-square test for $r \times c$ tables**). Initially we must compute an r value for each of the three studies in **Part A** of Table 16.25. Since the data for each study are summarized in a 2×2 contingency table, the **phi coefficient** is employed to represent the correlation coefficient. Employing Equation 16.21, the **phi coefficient** is computed for each of the studies.

$$\textbf{Study 1}: \phi_1 = \sqrt{\chi^2/n} = \sqrt{1.07/60} = .134$$

$$\textbf{Study 2}: \phi_2 = \sqrt{\chi^2/n} = \sqrt{2.00/50} = .200$$

$$\textbf{Study 3}: \phi_3 = \sqrt{\chi^2/n} = \sqrt{1.70/40} = .206$$

In order to employ Equation 28.91 (or Equation 28.24), the **phi coefficients** (which are employed to represent r values) must be converted into corresponding Fisher transformed values. Through use of **Table A17** the Fisher transformed z_r values for the three studies are as follows. (Interpolation was employed in the latter table when exact values are not listed.)

$$z_{r_1} = .135 \qquad z_{r_2} = .203 \qquad z_{r_3} = .209$$

We will assign a positive sign to the Fisher transformed z_r values for all three of the studies, since in all three instances females have a higher rate of kidney disease than males. Employing Equation 28.25, the average of the three Fisher transformed z_r values is computed to be $\bar{z}_{r_k} = .177$.

$$\bar{z}_{r_k} = \frac{(60-3)(.135)+(50-3)(.203)+(40-3)(.209)}{(60-3)+(50-3)+(40-3)} = .177$$

The appropriate values are substituted in Equation 28.91 below.

$$\chi^2 = (60-3)(.135-.177)^2 + (50-3)(.203-.177)^2 + (40-3)(.209-.177)^2 = .190$$

Since there are 3 studies, $df = 3 - 1 = 2$. Employing **Table A4**, for $df = 2$, $\chi^2_{.05} = 5.99$. Since the obtained value $\chi^2 = .190$ is less than $\chi^2_{.05} = 5.99$, the null hypothesis is retained. In other words, the data do not indicate that the effect sizes obtained for the three studies are inconsistent with one another.

Test 28o: Procedure for obtaining a combined effect size for k studies Rosenthal (1993) employs Equation 28.92 to compute a combined/pooled effect size for k studies.

$$\bar{z}_{r_k} = \frac{\sum\limits_{j=1}^{k} z_{r_j}}{k} \qquad \textbf{(Equation 28.92)}$$

Where: z_{r_j} represents the Fisher transformed z_r value for the jth study

    $\bar{z}_{r_k}$ represents the average Fisher transformed z_r value computed for the k published studies

The following protocol is employed in using the Equation 28.92: a) Obtain the sum of the Fisher transformed z_r values for the k studies, and divide that sum by k; and b) Employing **Table A17**, convert the value computed for $\bar{z}_{r_k}$ into its corresponding r value. The latter value represents the combined effect size for the k studies. Employing part a) of the aforementioned protocol, through use of Equation 28.92 the value $\bar{z}_{r_k} = .321$ is computed. Employing **Table A17** the latter value is converted into $r = .31$, which is the combined/pooled effect size.

$$\bar{z}_{r_k} = \frac{.693 + .549 + .203 + .310 + (-.151)}{5} = \frac{1.604}{5} = .321$$

Rosenthal (1993) notes that, if one wished to weight studies on the basis of their sample size, the value $\bar{z}_{r_k} = .351$ previously computed with Equation 28.25 (for use in Equation 28.91) can be used to represent the average Fisher transformed z_r value. He also states that weighting can be done on the basis of other criteria, such as the relative quality of the studies. Kline (2004, p. 259), Lipsey and Wilson (2001, Ch. 3) and Schulze (2004, p. 36) discuss weighting studies based on the standard error of the relevant measure of effect size — more specifically, the inverse variance weight of the effect size.

The r value in the above example that corresponds to the Fisher transformed value $\bar{z}_{r_k} = .351$ is $r = .338$, which will represent the combined/pooled effect size. Using Cohen's (1977, 1988) criteria (for an r value), the unweighted value $r = .31$ and the weighted value $r = .338$ both fall at the lower bound of the range for a medium effect size. (The effect size criteria for an r value are noted earlier in this section under the general discussion of measures of effect size.)

In the latter part of the discussion of the **procedure for comparing k studies with respect to effect size (Test 28n)**, a homogeneity of effect size analysis was conducted for the data contained in the three 2×2 contingency tables in **Part A** of Table 16.25. At this point we will employ Equation 28.92 to compute a combined effect size for the three studies.

$$\bar{z}_{r_k} = \frac{.135 + .203 + .209}{3} = \frac{.547}{3} = .182$$

Employing **Table A17** the latter value is converted into $r = .180$, which is the combined/pooled effect size. Thus, the combined effect size for the three studies in **Part A** of Table 16.25 is .180. Using Cohen's (1977, 1988) criteria, the latter value falls within the range for a small effect size.

Rosenthal (1993) states that one should view a combined effect size value with extreme caution when there is reason to believe that the k effect sizes employed in determining it are not homogeneous. Certainly, in such a case the computed value for the combined effect size is little more than an average of a group of heterogeneous scores, and not reflective of a consistent effect size across studies. In view of this, in order for the combined effect size value computed with Equation 28.92 to be meaningful, it should have been previously demonstrated that the effect sizes for the k studies are relatively homogeneous. As noted earlier, the latter is questionable in the case of Example 28.8, in spite of the fact that the result obtained with Equation 28.91 did not achieve statistical significance. The final point which should be made is that meta-analysts are also encouraged to compute a confidence interval for a combined effect size. The latter is basically determined by adding to the value computed for the combined effect size the product of the standard error employed in computing the effect size and the relevant tabled critical t value (e.g., $t_{.05} = 1.96$ for a 95% confidence interval).

Alternative meta-analytic procedures Hunter and Schmidt (1990, 2004) note that although Glass (1976) coined the term **meta-analysis**, they independently developed meta-analytic methodologies about the same time (Schmidt and Hunter (1977)). In point of fact, Hunter and Schmidt (1990, 2004) provide a unique perspective with respect to meta-analytic methodology and label their approach **psychometric meta-analysis**, which derives from the fact that it integrates data correction procedures which are based on psychometric theory (i.e., measurement theory developed within the field of psychology). Hunter and Schmidt (1990, 2004) note that their approach to meta-analysis differs from that employed by others such as Glass (1976, 1977) (who focuses on magnitude of effect size) and Rosenthal (1979, 1991, 1993, 1994) (who focuses on cumulation of probability values). Schulze (2004,p. 14), however, states that Hunter and Schmidt's methodology has limited applicability — specifically, it is primarily designed to use meta-analysis for evaluating predictive validity in personnel selection.[37]

Succinctly stated, the goal of **psychometric meta-analysis** is to quantify and correct errors attributable to sampling error and experimental artifacts prior to making a final determination with respect to the consistency of the multiple studies evaluating the same hypothesis. In the preface of their most recent book Hunter and Schmidt (2004) state that **psychometric meta-analysis** provides greater precision in integrating research results, since it allows researchers to determine the degree of variability among different studies which can be attributed to both sampling error and experimental **artifacts** — Hunter and Schmidt (2004, p. 12) note that most researchers underestimate the amount of variability in research results which can be attributed to sampling error. Hunter and Schmidt (2004, p. 33) define an **artifact** as a study imperfection, and note that errors in the results of a study which are attributable to an artifact are human made and not the result of nature. Among the experimental artifacts Hunter and Schmidt (2004, p. 35) identify and document correction procedures for are: a) Measurement error on the independent and dependent variables (e.g., lack of reliability and/or validity on measures employed for the independent and/or dependent variables; the use of invalid measures for the underlying constructs represented by the independent and/or dependent variables; dichotomization of a continuous independent and/or dependent variable; issues affecting range variation on the independent and/or dependent variables); and b) Reporting and transcription errors.

Before continuing the discussion of **psychometric meta-analysis** it should be emphasized that use of any meta-analytic procedure should also take into account all of the factors noted in the previous paragraph with respect to making decisions regarding which studies to include (as well as what if any differential weights should be assigned to individual studies) in a meta-analysis. It cannot be emphasized too strongly that the conclusions to be drawn from a meta-analysis will always be a direct function of the quality of the data employed in the analysis.

Among those issues relevant to meta-analysis which are discussed by Hunter and Schmidt (1990, 2004) is that a common situation encountered by a reviewer attempting to integrate the results of studies which test the same research hypothesis are inconsistencies in the results of multiple studies. Hunter and Schmidt (2004) state that until the introduction of meta-analysis, and most specifically **psychometric meta-analysis** (since in their opinion other meta-analytic procedures often lead to erroneous conclusions), such attempts to synthesize research lacked reliability, and, in the final analysis, were little more than exercises based on the subjective judgement of the reviewer. They note that it was not uncommon for reviewers to explain inconsistencies between studies on the basis of one or more undetected **moderator variables**. (As noted in Section VII, a **moderator variable** is a third variable which influences the zero-order correlation between two variables.) It is their contention that **psychometric meta-analysis** will frequently eliminate the need to search for moderator variables by virtue of its ability to resolve inconsistencies which are due to sampling error and/or the artifacts noted above, and that ultimately by employing **psychometric meta-analysis** (in contrast to alternative meta-analytic procedures) one can arrive at a conclusion indicating high degree of consistency among a set of studies which initially were perceived to be in conflict with one another.

Although a full description of **psychometric meta-analysis** is beyond the scope of this book the methodology can be briefly summarized as follows: a) Calculate and average the relevant descriptive statistic for each of the studies; b) Calculate the variance of the relevant statistic across studies; c) Correct the latter variance by subtracting from it the amount of variability attributable to sampling error; d) Correct the mean and variance for error attributable to other artifacts; and d) Contrast the corrected value obtained for the standard deviation (i.e., the final value computed for the square root of the corrected variance) with the mean in order to assess the magnitude of the potential variation across studies, and if the mean is approximately two standard deviations above zero it is reasonable to conclude that the correlation between the two variables is some value larger than zero (Hunter and Schmidt (2004, p. 64)).[38]

Hunter and Schmidt would argue that since they do not correct for the influence of experimental artifacts, the meta-analytic procedures described in this book (as well as other alternative meta-analytic procedures not discussed in this book, which they critique in Chapter 11 of their 2004 book) are not optimal for synthesizing the results of multiple studies.[39] Additionally, the meta-analytic procedures described in this chapter assume a **fixed-effects model** (as opposed to a **random-effects model**), and **psychometric meta-analysis** always assumes a **random-effects model** (Hunter and Schmidt (2000; 2004, p, 67; pp 201–205; Ch. 9)).[40] To go even further, Hunter and Schmidt (2000; 2004, p. 202) state that "most of the meta-analyses appearing in the *Psychological Bulletin* and other journals, because they are based on fixed-effects models, are potentially inaccurate and should be recomputed using random-effects models." One should not, however, get the impression that **psychometric meta-analysis** is immune to criticism. Among others, Lipsey and Wilson (2001, pp. 108–109) and Schulze (2004, pp. 62–70) have criticized certain aspects of it — for example, the information required for adjusting for artifacts is often not available and/or such adjustments are based on assumptions which may, in fact, be invalid. Schulze (2004, pp. 62–70) also criticizes Hunter and Schmidt for failing to adequately reference material which is the basis for some of the methods they employ.

It should be apparent there is a lack of consensus among methodologists with respect to the most appropriate protocol for conducting meta-analysis. In the final analysis, a researcher should view a combined/pooled probability and/or effect size value computed for a meta-analysis as little more than an approximation. Such a value is subject to change as more data on a hypothesis become available. Certainly, if at a given point in time the available data reflect what is true in the underlying population, test statistics generated in a meta-analysis should be reasonably accurate and not change substantially after additional data become available. Readers who are required to conduct a meta-analysis are advised to consult the various sources cited in this section in order to make the best decision with respect to the appropriate methodology to employ, but regardless of the methodology one selects, the more inconsistent the original data employed in an analysis the more equivocal any conclusions derived from a meta-analysis will be. More detailed discussions of meta-analysis can be found in Hedges and Olkin (1985), Hunter and Schmidt (2004), Kline (2004, Ch. 8), Lipsey and Wilson (2001), Mullen and Rosenthal (1985), Rosenthal (1991, 1993), Schulze (2004), Schulze *et al.* (2003), and Wolf (1986). Hunt (1997) provides a good nonmathematical summary of the history and application of meta-analysis in the scientific community.

Schulze's (2004) book is highly recommended, since he provides a comparative analysis of the various meta-analytic methods which have been developed (a number of which have not been discussed in this book). The latter author discusses the merits and liabilities of different methods based on, among other criteria, a Monte Carlo study (i.e., computer simulation) he conducted. Within the latter context, Schulze (2004, p. 194) notes that since many of the equations employed for meta-analysis are based on large-sample theory (i.e., the use of large samples in individual experiments), it is important to conduct simulation studies to determine the accuracy of estimators employed in meta-analysis when the latter assumption is violated. In the final chapter of his book, Schulze (2004, p. 196) concludes the following: a) The meta-analytic methodology a researcher elects to employ can make a difference with respect to what conclusions one reaches; b) Although some meta-analytic methods are superior to others for some tasks, a single best set of procedures has yet to be established; c) He also notes under which conditions some procedures should be used with caution, as well as those procedures which have been underutilized.

In conclusion, researchers employing meta-analysis should be aware of alternative methodologies, as well as the fact that, in the final analysis, a meta-analysis should never be viewed as a completely objective procedure.

Practical implications of magnitude of effect size value Before closing this section, a comment is in order concerning the relationship between the magnitude of an effect size and its practical implications regarding the relationship that exists between the independent and dependent variable. Rosnow and Rosenthal (1989) provide an interesting example (based on a study by the Steering Committee of the Physicians Health Study Research Group (1988)) involving the use of the **phi coefficient**. These authors illustrate that a low correlation coefficient (in this case the correlation is computed with the **phi coefficient**) need not necessarily indicate that the relationship between two variables is trivial. Table 28.16 summarizes the results of the study, which evaluated the effect of aspirin versus a placebo on heart attacks.

Table 28.16 Summary of Data for Heart Attack Study

		Y variable		Row sums
		Heart attack = 0	No heart attack = 1	
X variable	Aspirin = 0	104	10,933	11,037
	Placebo = 1	189	10,845	11,034
Column sums		293	21,778	22,071

In order to determine whether the result of the study is statistically significant, we evaluate the data with the **chi-square test for $r \times c$ tables**. Employing Equation 16.2, the computed chi-square value for Table 28.16 is $\chi^2 = 25.01$. Since $r = 2$ and $c = 2$, the degrees of freedom are computed to be $df = (2 - 1)(2 - 1) = 1$. Employing **Table A4**, we determine that the tabled critical .05 and .01 chi-square values for $df = 1$ are $\chi^2_{.05} = 3.84$ and $\chi^2_{.01} = 6.63$. Since the computed value $\chi^2 = 25.01$ is greater than both of the aforementioned critical values, the null hypothesis can be rejected at both the .05 and .01 levels. The actual probability value associated with the result is less than .001.

Now let us compute the magnitude of effect size for the study. Both Equations 16.21 and 28.1 (as well as Equation 28.85) can be employed to compute the **phi coefficient** (which as noted earlier in this section can be viewed as an r value within the context of meta-analysis). Employing Equation 16.21, the value $\phi = \sqrt{\chi^2/n} = \sqrt{25.01/22,071} = .034$ is computed. As noted earlier, Equation 28.1 can also be employed to compute the value of **phi** (as is done with the data in Table 28.12). Thus, we designate the independent variable (which is comprised of the following two levels: aspirin versus placebo) as the X variable. The value 0 will be assigned as the X score for any subject who received aspirin, and the value 1 will be assigned as the X score for any subject who received the placebo. The dependent variable, which is whether or not a subject had a heart attack, will be the Y variable. The value 0 will be assigned as the Y score for any subject who had a heart attack, and the value 1 will be assigned as the Y score for any subject who did not have a heart attack. Employing the same protocol that is used to analyze the data in Table 28.12, we compute the following values for Table 28.16: $\Sigma X = \Sigma X^2 = 11,034$; $\Sigma Y = \Sigma Y^2 = 21,778$; $\Sigma XY = 10,845$; $n = 22,071$. When the aforementioned values are substituted in Equation 28.1, the value $r = -.034$ is computed, which is the same absolute value that is computed with Equation 16.21 for the **phi coefficient** (which can only be a positive value).

The square of $\phi = .034$ $r = -.034$) is $\phi^2 = r^2 = .001156$. Earlier in the book it was noted that the latter value, which represents the **coefficient of determination**, is commonly employed as a measure of **effect size** for a 2×2 contingency table. The value $r^2 = .001156$ indicates that .1156% of the variability on the dependent variable can be accounted for on the basis of variability on the independent variable. Since .1156% is such a small value, one might get the impression there is little if any relationship between taking aspirin and having a heart attack. Given the fact that the sample is comprised of 22,071 observations, it is not at all surprising that the probability value .001 is obtained when the **chi-square test for $r \times c$ tables** is employed to analyze the data. A sample size of this magnitude insures that the power of the **chi-square test** will be large enough to identify as statistically significant even a minimal effect size. Yet when Equation 16.27 is employed with the same data to compute the **odds ratio** (discussed in Section VI of the **chi-square test for $r \times c$ tables**), we obtain the value $o = 1.83$: $o = [(10,933)(189)]/[(104)(10,845)] = 1.83$. The value $o = 1.83$ indicates that the odds of a person who received the placebo having a heart attack are 1.83 times larger than the odds of a person who received aspirin having a heart attack. If we employ **relative risk** (also discussed in Section VI of the **chi-square test for $r \times c$ tables**) as a measure, through use of Equation 16.25 we can determine that the **relative risk** of having a heart attack if one received the placebo as opposed to getting an aspirin is $[(189)/(11,034)]/[(104)/(11,037)] = 1.82$. The latter value indicates that someone taking the placebo is 1.82 times more likely to have a heart attack than someone taking aspirin.[41] The values computed for the **odds ratio** and **relative risk** clearly indicate there is a definite advantage to a subject taking aspirin. In point of fact, since the researchers came to the latter conclusion, they terminated the study while it was still in progress — deeming it unethical to deprive the control subjects (i.e., the placebo group) of aspirin.

Test 28p: Binomial effect size display   An alternative way of presenting effect size which (like the **odds ratio**) may make it more apparent if a seemingly small effect size is of practical consequence is a methodology developed by Rosenthal and Rubin (1982) referred to as the **binomial effect size display (BESD)**. BESD is a method for summarizing the results of a study within the format of a 2×2 contingency table. In order to describe the **BESD** we will initially consider a simplified hypothetical version of the aspirin study described above. Specifically, we will assume that two groups comprised of fifty-year-old men were either given an aspirin or a placebo daily, and that ten years later it was determined how many of the subjects had had a heart attack. Let us assume at the conclusion of the study it was determined that the correlation between the treatments and whether or not a subject had a heart attack was .30. Table 28.17 employs the **binomial effect size display** to summarize the results of the study

Table 28.17 Binomial Effect Size Display

		Y variable		Row sums
		Heart attack = 0	No heart attack = 1	
X variable	Aspirin = 0	35	65	100
	Placebo = 1	65	35	100
Column sums		100	100	200

It is important to note that the 2×2 contingency table in Table 28.17 which is employed to represent the **BESD** does not assume that the study involved 200 subjects, with 100 subjects in each of the treatment groups. In point of fact, the study in question in which the correlation between the independent and dependent variables is .30 could have involved fewer than 200 subjects or considerably more than 200 subjects.[42] If the square of the correlation coefficient (i.e., the **coefficient of determination**) is employed as a metric of effect size, the value .30 leads one to conclude that only 9% of the variability with respect to whether or not a subject had a heart attack can be explained on the basis of whether the subject received an aspirin or a placebo. (Note that when data are summarized within the format of a 2 × 2 contingency table, the **phi coefficient** is the correlational measure employed to measure effect size. Thus, $\phi^2 = r^2 = (.30)^2 = .09$, which expressed as a percentage is 9%.) Most researchers would probably consider 9% to be a very low value, thus indicating only a minimal relationship between the variables under study.

It happens to be the case that the values in the cells of a **BESD** can also be expressed as percentages (by simply inserting a percentage sign after each of the cell values). The latter is done in Table 28.18, which allows us to conclude the following: 65% of the subjects who received an aspirin did not have a heart attack, while 35% of the subjects who received an aspirin did have a heart attack. 35% of the subjects who received a placebo did not have a heart attack, while 65% of the subjects who received a placebo did have a heart attack. Thus we can say that 30% more of the subjects who received aspirin did not have a heart attack when compared with subjects who received a placebo. Undoubtedly such a result would suggest to many that it is in a 50-year-old man's best interest to take an aspirin daily as a preventative measure against a heart attack. It would thus appear that the value of 9% derived from the **coefficient of determination** may be misleading with respect to the strength of the relationship between the two variables.

Table 28.18 Binomial Effect Size Display in Percentage Format

		Y variable		
		Heart attack = 0	No heart attack = 1	Row sums
X variable	Aspirin = 0	35%	65%	100%
	Placebo = 1	65%	35%	100%
Column sums		100%	100%	

The cell entries in a **BESD** table will vary as a function of the value of the correlation coefficient. Consequently, the cell entries in the **BESD** depicted in Tables 28.17/28.18 are a direct function of the value .30 obtained for the correlation in the study under discussion. The methodology for converting a correlation coefficient into a **BESD** is as follows: a) Assume that the data are to be summarized in the form of a 2 × 2 contingency table. Regardless of how many subjects actually participate in the experiment, it will be assumed that a total of $n = 200$ subjects are involved in the study. We will also assume that subjects are broken down into two groups (which represent the two levels of the independent variable) consisting of 100 subjects per group. Each of the groups is represented by one of the rows of the 2 × 2 contingency table. Thus the sum for each of the rows of the contingency table will always equal 100. The columns of the contingency table will represent the two levels of the dependent variable on which a subject can be categorized. The sum of each of the columns in the contingency table will also always equal 100; b) Initially record a 50 in each of the four cells of the 2 × 2 contingency table. (Values of 50 in each cell will only be the final values in the **BESD** if the correlation between the independent and dependent variables is zero. This is the case, since when the observed frequency for each cell equals 50, Equation 16.2 yields the value $\chi^2 = 0$, which when substituted in Equation 16.21 yields the value $\phi = 0$); c) If the correlation between the independent and dependent variables is some value other than zero do the following: Employing the absolute value of the correlation coefficient, move the decimal point for the correlation coefficient two places to the right. (Thus, in the case of the **BESD** represented in Table 28.17, the value $\phi = r = .30$ is converted into 30.) Divide the latter value by 2. The result of this division will be designated as V. (In the case of Table 28.17, $V = 30/2 = 15$.) Add the value of V to the 50 in the cell which is in the row representing the group that had the larger proportion of subjects exhibiting a positive treatment response/outcome and the column indicating the positive treatment response/outcome category. (In the case of Table 28.17, this would be the cell representing subjects who received an aspirin and did not have a heart attack.) Also add the value of V to the 50 in the cell which is in the row representing the group that had the smaller proportion of subjects exhibiting a positive treatment response/outcome and the column indicating the negative treatment response/outcome category. (In the case of Table 28.17 this would be the cell representing subjects who received a placebo and had a heart attack.) Since $50 + 15 = 65$, the aforementioned protocol indicates why the value 65 is recorded in the upper right cell (**Aspirin/No heart attack**) and the lower left cell (**Placebo/Heart attack**) of Table 28.17; d) Subtract the value of V from the 50 in the cell which is in the row representing the group that had the larger proportion of subjects exhibiting a positive treatment response/outcome and the column indicating the negative treatment response/outcome category. (In the case of Table 28.17, this would be the cell representing subjects who received an aspirin and had a heart attack.) Also subtract the value of V from the 50 in the cell which is in the row representing the group that had the smaller proportion of subjects exhibiting a positive treatment response/outcome and the column indicating the positive treatment response/outcome category. (In the case of Table 28.17 this would be the cell representing subjects who received a placebo and did not have a heart attack.) Since $50 - 15 = 35$, the aforementioned protocol indicates why the value 35 is recorded in the upper left cell (**Aspirin/Heart attack**) and the lower right cell (**Placebo/No heart attack**) of Table 28.18.

Through use of the protocol noted above, Table 28.19 provides a **BESD** (employing the percentage format used in Table 28.18) for the original aspirin/heart attack study summarized in Table 28.16. Recollect that the correlation $\phi = .034$ was computed for the original study. When the decimal point is moved two places to the right we obtain 3.4, which results in $V = 3.4/2 = 1.7$. Thus 1.7 is added to 50, yielding 51.7 for subjects who received an aspirin and did not have a heart attack as well as for subjects who received a placebo and did have a heart attack. When 1.7 is subtracted from 50 it yields 48.3 for subjects who received an aspirin and had a heart attack as well as for subjects who received a placebo and did not have a heart attack. The **BESD** indicates that 3.4% more of the subjects who received aspirin did not have a heart attack when compared with subjects who received a placebo.

Table 28.19 Binomial Effect Size for Data in Table 28.17

		Y variable		Row sums
		Heart attack = 0	No heart attack = 1	
X variable	Aspirin = 0	48.3%	51.7%	100%
	Placebo = 1	51.7%	48.3%	100%
Column sums		100%	100%	

What can we conclude from the above discussion? As Cohen (1977, 1988) and Rosenthal (1991, 1993) note, there are circumstances when the strength of a relationship between an independent and dependent variable will be greater than the magnitude suggested by the **coefficient of determination**. A small r^2 value does not necessarily mean the relationship between the experimental variables is trivial and is of no practical consequence. In the final analysis, depending upon how one expresses effect size, it is conceivable that a researcher may draw different conclusions regarding the strength of the relationship between the variables under study. Rosenthal *et al.* (2000) provide a more in-depth discussion of the **BESD** and related meta-analytic procedures. It should be noted that a number of authors (most notably, Hsu (2004); also see Grissom and Kim (2005, pp. 89–91)) have been critical of the use of the **BESD** as a measure of effect size. Hsu (2004) states that the value computed for the **BESD** is often larger than the actual differences between two conditions in the real world.

The significance test controversy As noted throughout this book, the **classical hypothesis testing model** employs a null hypothesis and an alternative hypothesis. The null hypothesis is a statement of zero difference, which is commensurate with saying that in the underlying population(s) there is zero effect/correlation present. In the **Introduction** it was noted that during the past 20 years an increasing number of researchers and statisticians have become critical of the classical hypothesis testing model. Among those who have recently addressed this issue are Cohen (1994), Falk and Greenbaum (1995), Gigerenzer (1993), Hagen (1997), Harlow *et al.* (1997), Hunter and Schmidt (2004), Krueger (2001), Kline (2004), Meehl (1978), Morrison and Henkel (1970), Murphy and Myors (1998, 2004), and Serlin and Lapsley (1985, 1993). One alternative methodology for hypothesis testing discussed earlier in the book, which has been endorsed by some critics of the classical model, is the use of the **Bayesian hypothesis testing model** (discussed in Section IX (the **Addendum**) of the **binomial sign test for a single sample**).

The crux of the argument against the classical hypothesis testing model is that, in reality, the null hypothesis is always false. Specifically, various sources note that the null hypothesis is a **point hypothesis**, in that it stipulates a precise value — namely zero — for the difference between the experimental conditions. Thus any difference, no matter how negligible, will provide sufficient grounds for rejecting the null hypothesis. It has been pointed out by numerous researchers that the actual difference between two experimental conditions is probably never exactly equal to zero. Although admittedly a difference may be close to zero, if our measuring instrument is sufficiently sensitive and we carry our measurements out to many decimal places, we will probably never record a difference which is exactly equal to zero. And if the latter is true, it means the null hypothesis will always be false.

If, in fact, the null hypothesis is always false, it logically follows that it is not possible to commit a Type I error (which is rejecting a true null hypothesis). If a Type I error becomes impossible, then the only type of error a researcher need concern herself with is a Type II error (which is not rejecting a false null hypothesis). It was noted earlier in the book, that the likelihood of committing a Type II error is inversely related to the power of a statistical test — i.e., the more powerful the test, the lower the likelihood of committing a Type II error. If a researcher wishes to achieve a specific level of power for a statistical test, prior to conducting the test she must stipulate the magnitude of effect size she is trying to detect. The smaller the effect size, the greater the power that will be required for the test. There are essentially three ways a researcher can increase the power of a statistical test. They are: a) Reduction of error variability; b) Increasing the value of alpha (i.e., *p* value) employed in determining statistical significance; and c) Increasing the size of the sample. If a researcher assumes she is unable to reduce error variability any more than she already has, in order to increase power she must either employ a higher *p* value and/or increase the size of the sample employed in a study. Since current scientific convention does not endorse the use of an alpha level larger than .05, at this point in time it probably is not reasonable to expect that higher alpha levels will be an acceptable mechanism for increasing power. Thus, the most practical and effective way to maximize power is by increasing sample size. How large a sample one employs should be dictated by the magnitude of effect size one is trying to detect (which obviously involves a subjective decision on the part of a researcher). Within the context of the classical hypothesis testing model, it will require a relatively large sample size to declare a result significant if the effect size present is small or even in the low medium range. As Murphy and Myors (1998, 2004) note, the outcome of a test of statistical significance employed within the classical model provides more of a commentary on the power of the statistical test than it does on the strength of the relationship (i.e., effect size) between the variables under study.

In the final analysis, since the null hypothesis is always wrong, Murphy and Myors (1998, 2004) state that a researcher can never design a study which has too much power. They recommend that in reference to the effect size one is trying to detect, whenever possible the power of a test should be at least .50, and ideally .80 or greater. If the power of a study is so low that there is little likelihood of detecting the hypothesized effect size, the study probably is not worth conducting. Consequently, prior to conducting a study a researcher should determine (based on previous research or theoretical conjecture) what she considers to be a meaningful effect size, and employ a large enough sample to insure a reasonable likelihood of detecting it (if, in fact, it is present).

The minimum-effect hypothesis testing model An alternative which has been suggested to the classical hypothesis testing model is the **minimum-effect hypothesis testing model**, which is described in detail by Murphy and Myors (1998, 2004). Based on papers by, among others, Meehl (1978) and Serlin and Lapsley (1985, 1993), the model employs the null hypothesis to stipulate a value below which any effect present in the data would be viewed as trivial, and above which would be meaningful (this same approach is employed in **tests of equivalence**, which are described in reference to a number of tests in the book (e.g., the **t test for two independent samples**)). As an example, if one were comparing the IQ scores of two groups, the null hypothesis might stipulate a difference between 0 and 5 points, while the alternative hypothesis would stipulate a difference greater than five points. In such a case, any difference of five points or less would result in retaining the null hypothesis, since a difference within that range would be considered trivial (i.e., of no practical or theoretical value). A difference of more than five points would lead to rejection of the null hypothesis, since a difference equal to or greater than five points would be considered meaningful. Note that the null hypothesis in the minimum-effect model stipulates a range of values, whereas in the classical hypothesis testing model the null hypothesis stipulates a point (i.e., a specific value).

Murphy and Myors (1998, 2004) discuss stipulating differences within the minimum-effect hypothesis testing model in terms of effect size. They suggest the null hypothesis could stipulate a range of values in which the effects of the treatment account for what is considered to be a negligible/trivial amount of variability. They suggest one might stipulate in the null hypothesis that between 0 and 1% of variance on the dependent variable can be accounted for by variation on the independent variable. The alternative hypothesis would be supported if an effect size of greater than 1% is detected. The value of 1% corresponds to the lower limit of Cohen's (1977, 1988) minimum value for a small effect size noted earlier in this section, when an r value is employed to measure effect size (i.e., $r^2 = (.1)^2 = .01$, which, expressed as a percentage, is 1%). Murphy and Myors (1998, 2004) have developed special tables based on the **noncentral F distribution** for evaluating a minimum-effect null hypothesis.[43] In contrast to the **central F distribution**, which has been used throughout the book for evaluating F values within the context of the classical hypothesis testing model, the noncentral F distribution can be employed to evaluate a minimum-effect null hypothesis. The tables developed by Murphy and Myors (1998, 2004) allow for testing significance at the .05 and .01 levels, and also provide for an analysis of power. In addition to evaluating a minimum-effect null hypothesis which stipulates an effect size of 1% or less as trivial, the tables also evaluate a null hypothesis which stipulates an effect size of 5% or less as trivial (since Murphy and Myors (1998, 2004) believe some researchers might prefer to employ the latter value as the upper limit for a trivial effect). Murphy and Myors (1998) also provide equations for converting various test statistics (e.g., χ^2, R^2 (which represents the square of the **multiple correlation coefficient** which is discussed in the chapter on **multiple**

regression), etc.) into F values, in order that they can be employed with the tables for the noncentral F distribution. It should be noted that for a given set of data, the minimum-effect hypothesis testing model will have lower power than the classical hypothesis testing model. This is the case, since it is easier to reject a null hypothesis stipulating a zero difference than it is to reject a null hypothesis which stipulates a range of values between zero and some number above it. Murphy and Myors (1998, 2004) note that the loss of power is offset by the fact that the minimum-effect model provides the researcher with more meaningful results.

The minimum-effect hypothesis testing model is more compatible with the hypothesis testing philosophies of the major proponents of meta-analysis (e.g., Rosenthal, Hedges, Hunter and Schmidt) than is the classical hypothesis testing model. In both the minimum-effect model and meta-analysis, greater emphasis is placed on effect size than on the level of significance. In the final analysis, when all is said and done, regardless of which hypothesis testing model one employs, the key to effective hypothesis testing ultimately boils down to employing a representative sample which is large enough to detect any meaningful effect(s) present in the underlying population(s). Research that abides by the latter principle will ultimately yield results which are both reliable and meaningful.

References

Aaron, B., Kromrey, J. D., & Ferron, J. (1998, November). **Equating *r*-based and *d*-based effect size indices: Problems with a commonly recommended formula**. Paper presented at the annual meeting of the Florida Educational Research Association, Orlando, FL (ERIC Document Reproduction Service No. ED 433353).

Alleni, M. P. (1997). **Understanding regression analysis**. New York: Plenum Press.

Anastasi, A. (1976). **Psychological testing** (4th ed.). New York: Macmillan.

Anderson, R. L. (1942). Distribution of the serial correlation coefficient. **Annals of Mathematical Statistics**, 13, 1–13.

Antman, E. M., Lau, J., Kupelnick, B., Mosteller, F., & Chalmers, T, C. (1992). A comparison of the results of meta-analysis of randomized control trials and recommendations of clinical experts. **Journal of the American Medical Association**, 268, 240–248.

Banks, J. & Carson, J.S. (1984). **Discrete-event system simulation**. Englewood Cliffs, NJ: Prentice Hall.

Beatty, M. J. (2002). Do we know a vector from a scalar? Why measures of association (not their squares) are appropriate indices of effect. **Human Communication Research**, 28, 605–611.

Bennett, C. A. & Franklin, N. L. (1954). **Statistical analysis in chemistry and the chemical industry**. New York: John Wiley & Sons.

Berenson M. L. & Levine, D.M. (1996). **Basic business statistics: Concepts and applications**. Englewood Cliffs, NJ: Prentice Hall.

Berk, R. A. & Rauma, D. (1983). Capitalizing on nonrandom assignment in treatment: A regression discontinuity evaluation of a crime control program. **Journal of the American Statistical Association**, 78, 21–27.

Birnbaum, A. (1954). Combining independent tests of significance. **Journal of the American Statistical Association**, 49, 554–574.

Black, K. (2001). **Business statistics: Contemporary decision making**. Cincinnati, OH: Southwestern-College Publishing.

Box, G. E. P. & Pierce, D. A. (1970). Distribution of residual autocorrelations in autoregressive integrated moving average time series models. **Journal of the American Statistical Association**, 65, 1509–1526.

Bryant, F. B. & Yarnold, P. R. (1995). Principle-components analysis and exploratory and confirmatory factor analysis. In Grimm, L. G. & Yarnold, P. R. (Eds.) (pp. 100–136). **Reading and understanding multivariate statistics**. Washington, D.C.: American Psychological Association.

Campbell, D. T. (1963). From description to experimentation: Interpreting trends as quasi-experiments. In Harris, C. W. (Ed.), **Problems in measuring change** (pp. 212–243). Madison, WI: University of Wisconsin Press.

Campbell, D. T. & Stanley, J. C. (1963). **Experimental and quasi-experimental designs for research**. Chicago: R& McNally. 9, 836–848.

Canavos, G. C. & Miller, D. M. (1995). **Modern buisiness statistics**. Belmont, CA: Duxbury.

Cattell, R. B. (1966). The scree test for the number of factors. **Multivariate Behavioral Research**, 1, 245–266.

Chalmers, T. C., Berrier, J., Sack, H. S., Levin, H., Reitman, & Nagalingam, R. (1987). Meta-analysis of clinical trials as a scientific discipline. **Statistics in Medicine**, 6, 733–744.

Chou, Y. (1989). **Statistical analysis for business and economics**. New York: Elsevier.

Cohen, B. H. (2001). **Explaining psychological Statistics** (2nd ed.). New York: John Wiley & Sons.

Cochran, W. G. (1954). The combination of estimates from different experiments. **Biometrics**, 10, 101–129.

Cohen, J. (1962). The statistical power of abnormal-social psychological research: A review. **Journal of Abnormal and Social Psychology**, 65, 145–153.

Cohen, J. (1977). **Statistical power analysis for the behavioral sciences**. New York: Academic Press.

Cohen, J. (1988). **Statistical power analysis for the behavioral sciences** (2nd ed.). Hillsdale, NJ: Lawrence Erlbaum Associates, Publishers.

Cohen, J. (1992). A power primer. **Psychological Bulletin**, 112, 155–159.

Cohen, J. (1994). The earth is round ($p < .05$). **American Psychologist**, 49, 997–1003.

Cohen, J. & Cohen, P. (1983). **Applied multiple regression/correlation analysis for the behavioral sciences** (2nd ed.). Hillsdale, NJ: Lawrence Erlbaum Associates, Publishers.

Cohen, R. J., Swerdlick, M. E. & Smith, D. K. (1992). **Psychological testing and assessment**. California: Mayfield Publishing Company.

Conover, W. J. (1999). **Practical nonparametric statistics** (3rd ed.). New York: John Wiley & Sons.

Cook, T. D. & Campbell, D. T. (1979). **Quasi-experimentation: Design and analysis issues for field settings**. Boston : Houghton Mifflin Company.

Cook, R. D. & Weisberg, S. (1982). **Residuals and influence in regression**. New York: Chapman & Hall.

David, F. N. (1938). **Tables of the ordinates and probability integral of the distribution of the correlation coefficient in small samples**. Cambridge: University Press.

Durbin, J. & Watson, G. S. (1950). Testing for serial correlation in least squares regression I. **Biometrika**, 37, 409–438.

Durbin, J. & Watson, G. S. (1951). Testing for serial correlation in least squares regression II. **Biometrika**, 38, 159–178.

Durbin, J. & Watson, G. S. (1971). Testing for serial correlation in least squares regression III. **Biometrika**, 58, 1–19.

Edwards, A. L. (1984). **An introduction to linear regression and correlation** (2nd ed.). New York: W. H. Freeman & Company.

Eron, L. D., Huesman, L. R., Lefkowitz, M. M., & Walder, L. O. (1972). Does television violence cause aggression? **American Psychologist**, 27, 253–263.

Everitt, B. S. (1977). **The analysis of contingency tables**. New York: Chapman & Hall.

Everitt, B. S. (1992). **The analysis of contingency tables** (2nd ed.). New York: Chapman & Hall.

Falk, R. W. & Greenbaum, C. W. (1995). Significance tests die hard. **Theory and Psychology**, 5, 75–98.

Field, A. (2005). **Discovering statistics using SPSS** (2nd ed.). Sage Publications: London.

Field, A. (2005). Is the meta-analysis of correlation coefficients accurate when population correlations vary? **Psychological Methods**, 10, 444–467.

Fisher, R. A. (1921). On the "probable error" of a coefficient of correlation deduced from a small sample. **Metron**, 1, Part 4, 3–32.

Freedman, D. A. (2001) Ecological inference and the ecological fallacy. In Smelser, N. J. & Baltes, P. T. (Eds.). **International Encyclopedia of the Social and Behavioral Sciences**, pp. 4027–4030. Amsterdam: Elsevier.

Gigerenzer, G. (1993). The superego, the ego and the id in statistical reasoning. In Keren, G. & Lewis, C. (Eds.), **A handbook for data analysis in the behavioral sciences: Methodological issues** (pp. 311–339). Hillsdale, NJ: Lawrence Erlbaum Associates, Publishers.

Glass, G. (1976). Primary, secondary and meta-analysis of research. **Educational Researcher**, 5, 3–8.

Glass, G. (1977). Integrating findings: The meta-analysis of research. **Review of Research in Education**, 5, 351–379.

Glass, G., McGaw, B., & Smith, M. L. (1981). **Meta-analysis in Social Research**. Beverly Hills, CA: Sage Publications.

Goodwin, C. J. (2002). **Research methods in psychology: Methods and design**. (3rd ed.). New York: John Wiley & Sons.

Grissom , R. J. & Kim, J. J. (2005). **Effect sizes for research: A broad practical approach**. Mahwah, NJ: Lawrence Erlbaum Associates, Publishers.

Guenther, W. C. (1965). **Concepts of statistical inference**. New York: McGraw–Hill Book Company.

Guilford, J. P. (1965). **Fundamental statistics in psychology and education** (4th ed.). New York: McGraw–Hill Book Company.

Hagen, R. L. (1997). In praise of the null hypothesis statistical test. **American Psychologist**, 52, 15–24.

Harlow, L. L., Mulaik, S. A., & Steiger, J. H. (Eds.) (1997). **What if there were no significance tests?** Mahwah, NJ: Lawrence Erlbaum Associates, Publishers.

Hedges, L. V. (1981). Distribution theory for Glass's estimator of effect size and related estimators. **Journal of Educational Statistics**, 6, 107–128.

Hedges, L. V. (1982). Estimation of effect size from a series of independent experiments. **Psychological Bulletin**, 92, 490–499.

Hedges, L. V. & Olkin, I. (1980). Vote counting methods in research synthesis. **Psychological Bulletin**, 88, 359–369.

Hedges, L. V. & Olkin, I. (1985). **Statistical methods for meta-analysis**. Orlando, FL: Academic Press.

Hildebrand, D. K. (1986). **Statistical thinking for behavioral scientists**. Boston, MA: Duxbury.

Hoagalin C. C. & Welsch, R. (1978). The hat matrix in regression and ANOVA. **The American Statistician**, 32, 17–22.

Hotelling, H. (1940) The selection of variates for use in prediction with some comments on the general problem of nuisance parameters. **Annals of Mathematical Statistics**, 11, 271–283.

Hotelling, H. (1953) New light on the correlation coefficient and its transforms. **Journal of the Royal Statistical Society,** Series B, 15, 193– 232.

Howell, D. C. (2002). **Statistical methods for psychology** (5th ed.). Pacific Grove, CA: Duxbury Press.

Hsu, L. M. (2004). Biases of success rate differences shown in binomial effect size displays. **Psychological Bulletin,** 9, 183–197.

Hudeo-Medina, T. B., Sánchez-Meca, J., & Marín-Martinez, F. (2006). Assessing heterogeneity in meta-analysis: Q statistic or I^2 index? **Psychological Methods,** 11, 193–206.

Hunt, M. (1997). **How science takes stock**. New York: Russell Sage Foundation.

Hunter, J. E. & Schmidt, F. L. (1987). **Error in the meta-analysis of correlations: the mean correlation**. Unpublished manuscript, Department of Psychology, Michigan State University.

Hunter, J. E. & Schmidt, F. L. (1990). **Methods of meta-analysis: Correcting error and bias in research findings** (1st ed.). Newbury Park, CA: Sage Publications.

Hunter, J. E. and Schmidt, F. L. (2000). Fixed effects vs. random effects meta-analysis models: Implications for cumulative knowledge in psychology. **International Journal of Selection and Assessment,** 8, 275–292 Newbury Park, CA: Sage Publications.

Hunter, J. E. & Schmidt, F. L. (2004). **Methods of meta-analysis: Correcting error and bias in research findings** (2nd ed.). Thousand Oaks, CA: Sage Publications.

Hunter, J. E., Schmidt, F. L., & Coggin, T. D. (1996). **Meta-analysis of correlation: Bias in the correlation coefficient and the Fisher z transformation**. Unpublished manuscript, University of Iowa.

Kachigan, S. K. (1986). **Statistical analysis**. New York: Radius Press.

Kaiser, H. F. (1960). The application of electronic computers to factor analysis. **Educational and Psychological Measurement,** 20, 140–151.

Kenny, D. A. (1973). Cross-lagged and synchronous common factors in panel data. In A. S. Goldenberger & O. D. Duncan (Eds.), **Structural equation models in the social sciences**. New York: Seminar Press.

Kenny, D. A. (1975). Cross-lagged panel correlations: A test for spuriousness. **Psychological Bulletin,** 82, 887–903.

Kirk, R. (1995). **Experimental design: Procedures for the behavioral sciences**. Pacific Grove, CA: Brooks/Cole Publishing Company.

Kline, R. B. (2004). **Beyond significance testing: Reforming data analysis methods in behavioral research**. Washington, D. C.: American Psychological Association.

Krueger, J. (2001). Null hypothesis significance testing. **American Psychologist,** 56, 16–26.

Lazarsfeld, P. F. (1947). **The mutual effects of statistical variables**. Unpublished manuscript, Columbia University, Bureau of Applied Social Research.

Lazarsfeld, P. F. (1948). The use of panels in social research. **Proceedings of the American Philosophical Society,** 92, 405–410.

Licht, M. H. (1995). Multiple regression and correlation. In Grimm, L. G. & Yarnold, P. R. (Eds.) (pp. 100–136). **Reading and understanding multivariate statistics**. Washington, D.C.: American Psychological Association.

Lin, C. C. & Mudholkar (1980). A simple test for normality against asymmetric alternatives. **Biometrika,** 67, 455–461.

Lindeman, R. H., Merenda, P. F., & Gold, R. Z. (1980). **Introduction to bivariate and multivariate analysis**. Glenview, IL: Scott, Foresman & Company.

Lipsey, M. W. & Wilson, D. B. (2001). **Practical meta-analysis**. Thousand Oaks, CA: Sage Publications.

Ljung, G. M. & Box, G. E. P. (1978). On a measure of lack of fit in time series models. **Biometrika**, 65, 297–303.

Marascuilo, L. A. & Serlin, R. C. (1988). **Statistical methods for the social and behavioral sciences**. New York: W. H. Freeman & Company.

McNemar, Q. (1969). **Psychological statistics** (4th ed.). New York: John Wiley & Sons.

Meehl, P. (1978). Theoretical risks and tabular asterisks: Sir Karl, Sir Ronald, and the slow progress of soft psychology. **Journal of Consulting and Clinical Psychology**, 46, 806–834.

Mertler, C. A. & Vannatta, R. A. (2005). **Advanced and multivariate statistical methods** (3rd ed.). Los Angeles: Pyrczak Publications.

Montgomery, D. C. & Peck, E. A. (1992). **Introduction to linear regression analysis** (2nd ed.). New York: John Wiley & Sons.

Moore, D. S. & McCabe, G. P. (1993). **Introduction to the practice of statistic** (2nd ed.) New York: W. H. Freeman & Company.

Morrison, D. E. & Henkel, R. E. (Eds.) (1970). **The significance test controversy: A reader**. Chicago: Aldine.

Mosteller, F. (1990) Improving research methodology: An overview. In Sechrest, L., Perrin, E., & Bunker, J. (Eds.), **Research methodology: Strengthening causal interpretation of nonexperimental data** (pp. 221–230). Rockville, MD: U. S. Public Health Service Agency for Health Care Policy & Research.

Mosteller, F. M. & Bush, R. R. (1954). Selected quantitative techniques. In Lindzey, G. (Ed.), **Handbook of social psychology: Volume 1. Theory and method** (pp. 289–334). Cambridge, MA: Addison-Wesley.

Mullen, B. & Rosenthal, R. (1985). **BASIC meta-analysis: Procedures and programs**. Hillsdale, NJ: Lawrence Erlbaum Associates, Publishers.

Murphy, K. R. & Myors, B. (1998). **Statistical power analysis: A simple and general model for traditional and modern hypothesis tests**. Hillsdale, NJ: Lawrence Erlbaum Associates, Publishers.

Murphy, K. R. & Myors, B. (2004). **Statistical power analysis: A simple and general model for traditional and modern hypothesis tests** (2nd ed.) Mahwah, NJ: Lawrence Erlbaum Associates, Publishers.

Myers, J. L. & Well, A. D. (2003). **Research design and statistical analysis** (2nd ed.). Mahwah, NJ: Lawrence Erlbaum Associates, Publishers.

Neale, J. M. & Liebert, R. M. (1980). **Science and behavior: An introduction to methods of research**. Englewood Cliffs, NJ: Prentice-Hall, Inc.

Neter, J., Kutner, M. H., Nachtscheim, C. J. & Wasserman, W.(1996). **Applied linear statistical models** (4th ed.). Boston: WCB McGraw-Hill.

Neter, J., Wasserman, W., & Kutner, M. H. (1983). **Applied linear regression models** (3rd ed.). Homewood, IL: Richard D. Irwin, Inc.

Neter, J., Wasserman, W., & Kutner, M. H. (1990). **Applied linear statistical models** (3rd ed.). Homewood, IL: Richard D. Irwin, Inc.

Ozer, D. J. (1985). Correlation and the coefficient of determination. **Psychological Bulletin**, 97, 307–315.

Palumbo, D. J. (1977). **Statistics in political and social science** (Revised ed.). New York: Columbia University Press.

Pearson, K. (1896). Mathematical contributions to the theory of evolution — III. Regression, heredity and panmixia. **Philosophical Transactions of the Royal Society of London**, Series A 187, 253–318.

Pearson, K. (1900). On the criterion that a given system of deviations from the probable in the case of a correlated system of variables is such that it can reasonably be supposed to have arisen in a random sampling. **Philosophical Magazine**, 5, 157–175.

Pearson, K. (1901). On the correlation of characters not quantitatively measured. **Philosophical Transactions of the Royal Society** (Series A), 195, 1–47.

Pelz, D. C. & Andrews, F. M. (1964). Detecting causal priorities in panel study data. **American Sociological Review**, 29, 836–848.

Ramsey, F. L. & Schafer, D. W. (2002). **The statistical sleuth: Course in methods of data analysis**. Pacific Grove, CA: Duxbury.

Rauma, D. & Berk, R. A. (1987). Remuneration and recidivism: The long-term impact of unemployment compensation on ex-offenders. **Journal of Quantitative Criminology**, 3, 3–27.

Robinson, W. S. (1950). Ecological correlations and the behavior of individuals. **American Sociological Review**, 15, 351–357.

Rogosa, D. (1980). A critique of the cross-lagged correlation. **Psychological Bulletin**, 88, 245–258.

Rogosa, D. (1987). Causal models do not support scientific conclusions: A comment in support of Freedman. **Journal of Educational Statistics**, 12, 185–195.

Rosenthal, R. (1978). Combining results of independent studies. **Psychological Bulletin**, 85, 185–193.

Rosenthal, R. (1979). The file drawer problem and tolerance for null results. **Psychological Bulletin**, 86, 638–641.

Rosenthal, R. (1985). **Basic meta-analysis: Procedures and programs**. Hillsdale, NJ: Lawrence Erlbaum Associates, Publishers.

Rosenthal, R. (1991). **Meta-analytic procedures for social research**. Newbury Park, CA: Sage Publications.

Rosenthal, R. (1993). Cumulating evidence. In Keren, G. & Lewis, C. (Eds.), **A handbook for data analysis in the behavioral sciences: Methodological issues** (pp. 519–559). Hillsdale, NJ: Lawrence Erlbaum Associates, Publishers.

Rosenthal, R. (1994). Parametric measures of effect size. In Cooper, H. M. & Hedges, L. V. (Eds.) **The handbook of research synthesis** (pp. 231–244). New York: Russell Sage Foundation.

Rosenthal, R., Rosnow, R. L., & Rubin, D. B. (2000). **Contrasts and effect sizes in behavioral research: A correlational approach**. Cambridge, UK: Cambridge University Press.

Rosenthal, R. & Rubin, D. B. (1978). Interpersonal expectancy effects: The first 345 studies. **The Behavioral and Brain Sciences**, 3, 377–415.

Rosenthal, R. & Rubin, D. B. (1982). A simple general purpose display of magnitude of experimental effect. **Journal of Educational Psychology**, 74, 166–169.

Rosnow, R. L. & Rosenthal, R. (1989). Statistical procedures and the justification of knowledge in psychological science. **American Psychologist**, 44, 1276–1284.

Rosnow, R. L., Rosenthal, R. & Rubin, D. B. (2000). Contrasts and correlations in effect-size estimation. **Psychological Science**, 11, 446–453.

Rozelle, R. M. & Campbell, D. T. (1969). More plausible rival hypotheses in the cross-lagged panel correlation technique. **Psychological Bulletin**, 71, 74–80.

Ryan, T. P. (1997). **Modern regression methods**. New York: John Wiley & Sons.

Schmidt, F. L. & Hunter, J. E. (1977). Development of a general solution to the problem of validity generalization. **Journal of Applied Psychology**, 62, 529–540.

Schmidt, J. W. & Taylor, R. E. (1970). **Simulation and analysis of industrial systems**. Homewood, IL: Richard D. Irwin, Inc.

Schulze, R. (2004). **Meta-analysis - A comparison of approaches**. Toronto: Hogrefe & Huber.

Schulze, R., Holling, H., & Bohning, D. (Eds.) (2003). **Meta-analysis: New developments and applications in medical and social sciences**. Toronto: Hogrefe & Huber.

Serlin, R. A. & Lapsley, D. K. (1985). Rationality in psychological research: The good-enough principle. **American Psychologist**, 40, 73–83.

Serlin, R. A. & Lapsley, D. K. (1993). Rational appraisal of psychological research and the good-enough principle. In Keren, G. & Lewis, C. (Eds.), **A handbook for data analysis in the behavioral sciences: Methodological issues** (pp. 199–228). Hillsdale, NJ: Lawrence Erlbaum Associates, Publishers.

Shadish, W. R., Cook, T. D. & Campbell, D. T. (2002). **Experimental and quasi-experimental designs for generalized causal inference**. Boston: Houghton Mifflin Company.

Shiffler, R. E. & Adams, A. J. (1995). **Introductory business statistics with computer applications**. Belmont, CA: Duxbury.

Sinich, T. (1996). **Business statistics by example**. Upper Saddle River, NJ: Prentice Hall.

Smith, M. L. & Glass, G. V. (1977) Meta-analysis of psychotherapy outcome studies. **American Psychologist**, 32, 752–760.

Smithson, M. (2003). **Confidence Intervals**. Thousand Oaks, CA: Sage Publications

Spata, A. V. (2003). **Research methods: Science and diversity**. New York: John Wiley & Sons.

Steering Committee of the Physicians' Health Study Research Group (1988). Preliminary report: Findings from the aspirin component of the ongoing physicians' health study. **The New England Journal of Medicine**, 318, 262–264.

Steiger, J. H. (1980) Tests for comparing elements of a correlation matrix. **Psychological Bulletin**, 87, 245–251.

Stokes, D. M. (2001). The shrinking file drawer: On the validity of statistical meta-analyses in parapsychology. **The Skeptical Inquirer**, 25(3), May/June 2001, 22–25.

Stouffer, S. A., Suchman, E. A., DeVinney, L. C., Star, S. A., & Williams, Jr., R. M. (1949). **The American soldier: Adjustment during army life: Volume 1**. Princeton, NJ: Princeton University Press.

Tabachnick, B. G. & Fidell, L. S. (1989). **Using multivariate statistics**. (2nd ed.) New York: Harper Collins Publishers.

Tabachnick, B. G. & Fidell, L. S. (1996). **Using multivariate statistics** (3rd ed.). New York: Harper Collins Publishers.

Tatsuoka, M. (1993). Effect size. In Keren, G. & Lewis, C. (Eds.), **A handbook for data analysis in the behavioral sciences: Methodological issues** (pp. 461–479). Hillsdale, NJ: Lawrence Erlbaum Associates, Publishers.

van Belle, G. (2002). **Statistical rules of thumb**. New York: John Wiley & Sons.

Webster, A. L. (1995). **Applied statistics for business and economics** (2nd ed.). Chicago: Irwin.

White, H. L. (1980). A heteroscedasticity-consistent covariance matrix estimator and a direct test for heteroscedasticity. **Econometrika**, 48, 817–838.

Wilcox, R. R., (2003). **Applying contemporary statistical techniques**. San Diego, CA: Academic Press.

Wilson, J. H. & Keating, B. (1990). **Business forecasting**. Homewood, IL: Irwin.

Wolf, F. M. (1986). **Meta-analysis: Quantitative methods for research synthesis**. Newbury Park, CA: Sage Publications.

Zar, J. H. (1999). **Biostatistical analysis** (4th ed.). Upper Saddle River, NJ: Prentice Hall.

Endnotes

1. It is also possible to designate the Y variable as the predictor variable and the X variable as the criterion variable. The use of the Y variable as the predictor variable is discussed in Section VI.

2. a) It should be noted that when the joint distribution of two variables is bivariate normal, only a linear relationship can exist between the variables. As a result of the latter, whenever the population correlation between two bivariate normally distributed variables equals zero, one can conclude that the variables are statistically independent of one another. Under such conditions the null hypothesis H_0: $\rho = 0$, commonly evaluated for the **Pearson product-moment correlation coefficient**, is equivalent to the null hypothesis that the two variables are independent of one another. On the other hand, it is possible for each of two variables to be normally distributed, yet the joint distribution of the two variables not be bivariate normal. When the latter is true, it is possible to compute the value $r = 0$, and at the same time have two variables which are statistically dependent upon one another. Statistical dependence in such a case will be the result of the fact that the variables are curvilinearly related to one another; b) Further discussion of the bivariate normal distribution (which is a special case of a **multivariate normal distribution**) can be found in Section I of the chapter on **multiple regression**.

3. a) Howell (2002) and Grissom and Kim (2005, pp. 71–72) note that the value of r computed with Equation 28.1 is a biased estimate of the underlying population parameter ρ. More specifically, Grissom and Kim (2005, p. 70) note that r and r_{pb} (the **point-biserial correlation**) which is discussed in Section IX (the **Addendum**) are **negatively biased** — i.e., they tend to underestimate the underlying population correlation. The degree to which the computed value of r is biased is inversely related to the size of the sample employed in computing the correlation coefficient. For this reason, when one employs correlational data within the framework of research, it is always recommended that a reasonably large sample size be employed.

   One way of correcting for bias resulting from a small sample size is to employ Equation 28.93 to compute the value $\tilde{r}$, which represents a relatively unbiased estimate of the population parameter ρ. The value $\tilde{r}$ is referred to as a "shrunken" or "adjusted" estimate of the population correlation. The computation of $\tilde{r}$ (the absolute value of which will always be less than r) is demonstrated for Example 28.1.

   (Equation 28.93)

$$\tilde{r} = \sqrt{1 - \frac{(1 - r^2)(n - 1)}{n - 2}} = \sqrt{1 - \frac{[1 - (.955)^2][5 - 1]}{5 - 2}} = .940$$

   Thus, in the case of Example 28.1, $\tilde{r} = .940$ provides a better estimate of the true population correlation than $r = .955$ (although even if $\tilde{r}$ is employed, $n = 5$ is an absurdly low sample size to employ within the framework of serious research). Since most sources use the computed value of r rather than $\tilde{r}$, the former value will be employed as the estimate of the population correlation throughout the discussion of the **Pearson product-moment correlation coefficient**; b) Equation 28.94 is an alternative form of Equation 28.1.

$$r = \frac{n(\Sigma XY) - (\Sigma X)(\Sigma Y)}{\sqrt{[n(\Sigma X^2) - (\Sigma X)^2][n(\Sigma Y^2) - (\Sigma Y)^2]}}$$

   (Equation 28.94)

4.   An alternative form of Equation 28.3 that is based on the relationship $t^2 = F$ (described
      in Section VII of the **single-factor between-subjects analysis of variance**), which yields
      equivalent results, employs the F distribution. The equation employing the F distribution
      for evaluating the significance of r is noted below.

$$F = \frac{r^2(n - 2)}{1 - r^2}$$

Employing the above equation with Example 28.1, the value $F = 31.10$ is computed.

$$F = \frac{(.955)^2(5 - 3)}{1 - (.955)^2} = 31.10$$

The computed value $F = 31.10$ (which is equivalent to $(t = 5.58)^2$ if rounding off
error is ignored) is evaluated with **Table A10**. The degrees of freedom employed in
evaluating the above equation are $df_{num} = 1$, $df_{den} = n - 2$. Thus, $df_{num} = 1$ and $df_{den} = 3$. It is determined that the tabled critical .05 and .01 two-tailed values are $F_{.05} = 10.13$
and $F_{.01} = 34.12$, and the tabled critical .05 and .01 one-tailed values are $F_{.05} = 5.54$
and $F_{.01} = 20.61$. (The latter values, which are not in **Table A10**, were obtained by
squaring the tabled critical one-tailed values $t_{.05} = 2.35$ and $t_{.01} = 4.54$. A full
discussion of one-tailed F values can be found in Section VI of the **t test for two
independent samples** under the discussion of homogeneity of variance.) The same
guidelines for interpreting a computed t value with Equation 28.3 are employed to
interpret the computed F value, with one exception in reference to the directional
alternative hypothesis H_1: $\rho < 0$. Since the value of F will always be a positive
number, if the directional alternative H_1: $\rho < 0$ is employed, in order to reject the null
hypothesis the value of F must be equal to or greater than the tabled critical one-tailed F
value at the prespecified level of significance. However, the sign of r must be negative.
When the F distribution is employed to evaluate the null hypothesis H_0: $\rho = 0$, it
results in identical conclusions to those reached when Equation 28.3 is employed.
Specifically, the nondirectional alternative hypothesis H_1: $\rho \neq 0$ is supported at the .05
level, since $F = 31.10$ is greater than the tabled critical two-tailed value $F_{.05} = 10.13$.
The directional alternative hypothesis H_1: $\rho > 0$ is supported at both the .05 and .01
levels, since r is a positive number, and $F = 31.10$ is greater than the tabled critical one-
tailed values $F_{.05} = 5.54$ and $F_{.01} = 20.61$.

Marascuilo and Serlin (1988) note that the following equation employing the normal
distribution can also be used with large sample sizes to evaluate the null hypothesis
H_0: $\rho = 0$: $z = r\sqrt{n - 1}$. If applied to Example 28.1, the value $z = (.955)\sqrt{5 - 1}$
$= 1.91$ is computed. The value $z = 1.91$ only supports the directional alternative hypothesis
H_1: $\rho > 0$ at the .05 level, since $z = 1.91$ is greater than the tabled critical one-tailed value
$z_{.05} = 1.65$ in **Table A1** in the **Appendix**. The nondirectional alternative hypothesis
H_1: $\rho \neq 0$ is not supported, since $z = 1.91$ is less than the tabled critical two-tailed
value $z_{.05} = 1.96$. The latter result indicates that when employed with a small sample size,
the normal approximation provides a more conservative test of an alternative hypothesis
than does Equation 28.3.

5.   a) The value $(1 - r^2)$ is often referred to as the **coefficient of nondetermination**, since it represents the proportion of variance that the two variables do not hold in common with one another. Further discussion of the **coefficient of determination** can be found in Section IX (the **Addendum**) in the discussion of **meta-analysis and related topics**; b) Ozer (1985) argues that under certain conditions it is more prudent to employ $|r|$ as a measure of the proportion of variance on one variable which can be accounted for by variability on the other variable. Cohen (1988, p. 533) succinctly summarizes Ozer's (1985) point by noting that when there is reason to believe a causal relationship exists between X and Y, the value r^2 provides an appropriate estimate of the percentage/proportion of variance on Y attributable to X. However, if there is reason to believe that both X and Y are caused by a third variable, the absolute value of r is a more appropriate measure to employ to represent the proportion of shared variance between X and Y. In practice, however, most sources always employ the value of r^2 to represent the percentage/ proportion of variance on one variable attributable to the other variable. Further critique of the use of r^2 as a basis for explaining variability can be found in Beatty (2002), Grissom and Kim (2005, pp. 91–95), and Hunter and Schmidt (2004, pp. 189–191).

6.   As noted earlier, for illustrative purposes, the sample size employed for Example 28.1 is very small. Consequently, the values r and r^2 are, in all likelihood, not accurate estimates of the corresponding underlying population parameters ρ and ρ^2.

7.   An equation that is based on the minimum squared distance of all the points from the line reflects the fact that if the distance of each data point from the line is measured, and the resulting value is squared, the sum of the squared values for the n data points is the lowest possible value that can be obtained for that set of data.

8.   van Belle (2002, pp. 70–74) notes that the range of the values (as well as the spacing of values) employed on the predictor variable is a more critical factor in affecting the precision of a regression analysis than the value of n (i.e., the number of pairs of observations).

9.   The values $s_{Y.X}$ and $s_{X.Y}$ can also be computed with the equations noted below:

$$s_{Y.X} = \sqrt{\frac{SS_Y - \frac{(SP_{XY})^2}{SS_X}}{n - 2}} = \sqrt{\frac{29.2 - \frac{(88.6)^2}{294.8}}{5 - 2}} = .92$$

$$s_{X.Y} = \sqrt{\frac{SS_X - \frac{(SP_{XY})^2}{SS_Y}}{n - 2}} = \sqrt{\frac{294.8 - \frac{(88.6)^2}{29.2}}{5 - 2}} = 2.94$$

10. The reader may find it useful to review the discussion of confidence intervals in Section VI of the **single-sample t test** before reading this section.

11. The term SS_X in Equation 28.14 may also be written in the form $SS_X = (n - 1)\hat{s}_X^2$, and the term SS_Y in Equation 28.15 may also be written in the form $SS_Y = (n - 1)\hat{s}_Y^2$.

12. a) Zar (1999, p. 382) notes that the following equation can be employed to convert a z_r value into an r value: $r = (e^{2z_r} - 1)/(e^{2z_r} + 1)$. Thus, if $z_r = 1.886$ then:

$$r = [(2.71828)^{(2)(1.886)} - 1]/[(2.71828)^{(2)(1.886)} + 1] = .955.$$

b) The equation $r = \tanh z_r$ can also be employed to convert a z_r value into an r value (where $\tanh$ represents the **hyperbolic tangent** of a number). Thus, $\tanh 1.886 = .955$. Scientific calculators generally have keys which allow for quick computation of a $\tanh$ value.

13. Equation 28.20 can also be written in the form: $z = (z_r - z_{p_0})\sqrt{n - 3}$. Thus, $z = (1.886 - 1.099)\sqrt{5 - 3} = 1.11$.

14. The value $n = 5$ employed in Example 28.1 is not used, since the method to be described is recommended when $n \geq 25$. For smaller sample sizes, tables in Cohen (1977, 1988) derived by David (1938) can be employed.

15. van Belle (2002, pp. 31–33; pp. 59–61) notes that Equation 28.95 (which is identical to Equation 2.15) can be employed to estimate the sample size required in order to detect a specific value for a population correlation (assuming a nondirectional analysis is conducted in evaluating the null hypothesis H_0: $\rho = 0$, with $\alpha = .05$). The value of k in the numerator of the latter equation will depend on the desired power of the test, and for the power values .50, .80, .90, .95, and .975 the values to employ for k are respectively 4, 8, 11, 13, and 16. The value d in Equation 28.95 represents the **Fisher transformed value** z_r computed with Equation 28.18. In the latter equation the value specified for the population correlation in the alternative hypothesis (which for purposes of demonstration will be H_1: $\rho = .40$) is employed to represent the value of r.

     Equation 28.95 is employed below to compute the sample size required to detect a population correlation of $\rho = .40$ if the power of the analysis is .80. For the latter power value, we employ $k = 8$ in the numerator of Equation 28.95. The value $d = .4236$ (computed below with Equation 28.18, and which in this instance represents the effect size being detected) is employed in the denominator of the equation. When the appropriate values are substituted in Equation 28.95, the sample size estimate for the power of the test to equal .80 is $n = 44.58$ which rounded off equals 45.

$$z_r = \frac{1}{2} \ln \left[\frac{1 + r}{1 - r} \right] = \frac{1}{2} \ln \left[\frac{1 + .4}{1 - .4} \right] = .4236$$

$$n = \frac{k}{d^2} = \frac{8}{(.4236)^2} = 44.58 \qquad \textbf{(Equation 28.95)}$$

     Note that if $n = 45$ is substituted in Equation 28.21, it yields the value $\varphi = 2.75$ ($\varphi = |0 - .4236| \sqrt{45 - 3} = 2.75$), which in **Table A3-C** corresponds to an approximate power of .80.

16. Equation 28.24 can also be employed to evaluate the hypothesis of whether there is a significant difference between $k = 2$ independent correlations — i.e., the same hypothesis evaluated with Equation 28.22. When $k = 2$, the result obtained with Equation 28.24 will be equivalent to the result obtained with Equation 28.22. Specifically, the square of the

obtained value of z obtained with Equation 28.22 will equal the value of χ^2 obtained with Equation 28.24. Thus, if the data employed in Equation 28.22 are employed in Equation 28.24, the obtained value of chi-square equals $\chi^2 = z^2 = (.878)^2 = .771$.

$$\chi^2 = [(5 - 3)(1.886)^2 + (5 - 3)(1.008)^2] - \frac{[(5 - 3)(1.886) + (5 - 3)(1.008)]^2}{(5 - 3) + (5 - 3)} = .771$$

17. a) When $k = 2$, Equation 28.25 is equivalent to Equation 28.23; b) If the sign of a correlation coefficient is negative, the absolute value or r is employed in obtaining the Fisher transformed z value from **Table A17**. A negative sign is then attached to the z value for its use in Equation 28.25; c) Among others, Hunter and Schmidt (1987; 2004, pp. 82–83) and Hunter, Schmidt, and Coggin (1996) contend that Equation 28.25 tends to produce an estimate of the population correlation which is positively/upwardly biased (i.e., it overestimates the population correlation). They note that although utilization of **Fisher's z transformation** compensates for a slight negative bias (i.e., underestimation of the population correlation) associated with a simple averaging of the correlations, it in fact introduces a small positive bias, which tends to increase as the degree of variation among the correlations employed increases. Hunter and Schmidt (2004, pp. 81–82) suggest the use of Equation 28.97 (see Endnote 37), which is a weighted average, in place of Equation 28.25. (Equation 28.97 yields a weighted average correlation of .855 for the example under discussion: $\bar{r} = [(5)(.955) + (5)(.765) + (5)(.845) / [5 + 5 + 5] = .855$.) Schulze (2004, p. 65), however, contends that Equation 28.97 is negatively biased; d) Schulze (2004, pp. 22–28; p. 193) provides further critique of **Fisher's z transformation** and describes alternative methods for estimating the distribution of a population correlation (most notably one proposed by Hotelling (1953, p. 224)).

18. If homogeneity of variance is assumed for the two samples, a pooled error variance can be computed as follows:

$$s_{Y.X}^2 = \frac{(n_1 - 2)(s_{Y.X_1}^2) + (n_2 - 2)(s_{Y.X_2}^2)}{n_1 + n_2 - 4}$$

$$s_{X.Y}^2 = \frac{(n_1 - 2)(s_{X.Y_1}^2) + (n_2 - 2)(s_{X.Y_2}^2)}{n_1 + n_2 - 4}$$

The computed value $s_{Y.X}^2$ is used in place of both $s_{Y.X_1}^2$ and $s_{Y.X_2}^2$ in Equation 28.31, and the computed value $s_{X.Y}^2$ is used in place of both $s_{X.Y_1}^2$ and $s_{X.Y_2}^2$ in Equation 28.32.

19. Marascuilo and Serlin (1988) describe how the procedure described in this section can be extended to the evaluation of a hypothesis contrasting three or more regression coefficients.

20. The equations for z_{X_i} and z_{Y_i} are analogous to Equation I.38 in the **Introduction** (which employs the population parameters μ and σ in computing a z score).

21. The sum of products within this context is not the same as the sum of products that represents the numerator of Equation 28.1. The product of each subject's z_{X_i} and z_{Y_i} score represents a cross-product for that subject's z scores, and the sum of the products of the n

subjects z scores is the sum of the cross-products of the z scores. Consequently, the correlation coefficient is sometimes defined as the mean of the cross-products of the z scores.

22. a) One way of avoiding the problem of dependent pairs is to form pairs in which no digit is used for more than one pair. In other words, the first two digits in the series represent the X and Y variables for first pair, the third and fourth digits in the series represent the X and Y variables for second pair, and so on. Although use of the latter methodology really does not conform to the definition of autocorrelation, if it is employed one can justify employing the critical values in **Table A16**; b) The use of **Table A16** earlier in this section to evaluate the obtained correlation of $r = .909$ between the scores on the X and Y variables in Table 28.6 was justifiable. In the latter situation it was assumed that the X and Y variables represented two different variables. On the other hand, the data in Table 28.11 (although identical to that in Table 28.6) represents one set of scores in which each score is paired with the score which follows it.

23. A discussion of the derivation of Equation 28.60 can be found in Bennett and Franklin (1954).

24. The reader should take note of the fact that the data for Example 28.6 are fictitious, and, in reality, the result of the analysis in this section may not be consistent with actual studies that have been conducted which evaluate the relationship between intelligence and eye-hand coordination.

25. Although the **phi coefficient** is described in the book as a measure of association for the **chi-square test for $r \times c$ tables** (specifically, for 2×2 tables), it is also employed in psychological testing as a measure of association for 2×2 tables in order to evaluate the consistency of n subjects' responses to two questions. The latter type of analysis is essentially a dependent samples analysis for a 2×2 table, which, in fact, is the general model for which the **McNemar test (Test 20)** is employed.

26. Hunter and Schmidt (1990; 2004, p. 62) note that if, in fact, a population correlation is some value other than zero (i.e., $\rho \neq 0$) and the statistical power of each of the m correlation coefficients computed is less the .50, then the greater the number of studies conducted the greater the likelihood the **vote-counting method** will lead a researcher attempting to synthesize the results of all the studies to conclude that $\rho = 0$.

27. Cohen (1977, 1988) also discusses additional effect size indices that are employed for computing the power of various multivariate procedures.

28. a) The values that Cohen (1977; 1988, pp. 24–27) employs for identifying a small versus medium versus large effect size for the d **index** and other indices to be described in this section were developed in reference to behavioral science research. Although these values can be employed for research in areas other than the behavioral sciences, it is conceivable that practitioners in other disciplines may elect to employ different values which they deem more appropriate for their area of specialization; b) Equation 28.75 can also be employed to convert an **omega squared** value ($\tilde{\omega}^2$) (discussed in Section VI of the t **test for two independent samples**) into a d value. If the values .0099, .0588, and .1379 (which are Cohen's lower limits for **omega squared** for a small, medium, and large effect size) are

employed to represent r^2 in Equation 28.75, they yield the following corresponding d values: .2, .5, and .8; c) Alternative equations for converting a d value into an r value (yielding a slightly different value than that obtained with Equation 28.76) were suggested by Aaron *et al.* (1998) (also found in Schulze (2004, p. 31); d) Further clarification of the relationship between r and Cohen's **d index** can be found in Cohen (1977; 1988, pp. 81–83); e) Schulze (2004, p. 29) notes that Hedges and Olkin (1985, p. 86) first noted that the estimated population variance for a d value is computed with Equation 28.96.

$$\tilde{s}_d = \frac{4 + d^2}{n}$$ **(Equation 28.96)**

29. Although **Cohen's d index** is also used as a measure of effect size within the framework of meta-analysis, Rosenthal (1991, pp. 17–18) argues that the use of an r value is preferable to the use of a d value as a measure of effect size.

30. The author is indebted to Robert Rosenthal for clarifying some of the issues discussed in this section.

31. It is important to emphasize that researchers are often not in agreement with regard to the most appropriate estimate of effect size to employ. Hopefully, if an effect of some magnitude is present which has theoretical or practical implications, regardless of which measure of effect size one employs, a reasonably accurate estimate of the effect size will emerge from an analysis.

32. a) The same test result will be obtained if the z value for *Study E* is assigned a positive sign, and the z values for *Studies A, B, C,* and *D* are assigned negative signs; b) If in a given study the means of the two groups are equal, the p value for that study will equal .50, and the corresponding z value will equal 0.

33. Rosenthal (1991) presents a modified form of Equation 28.88 that allows a researcher to differentially weight the k studies employed in a meta-analysis. A weighting system is employed within this context to reflect the relative quality of each of the studies. The magnitude of the weights (which is assigned by a panel of judges) is supposed to be a direct function of the quality of a study.

34. It is interesting to note that the average z value for five studies is $\bar{z}_k = .95$, and that the latter z value in itself is not statistically significant. It is quite common for Equation 28.88 to yield a significant combined p value when the average z value in itself is not statistically significant.

35. a) A Q-statistic (based on effect sizes expressed in standard deviation units as opposed to correlation coefficients) developed by Cochran (1954) (described in Hedges and Olkin (1985, p. 123) and Huedo-Medina *et al.* (2006)) is commonly employed as an alternative to Equation 28.91 to evaluate the null hypothesis that the population effect sizes for a set of k studies are homogeneous. Computation of the latter Q statistic is based on summing the squared deviation of the effect size estimate for each study from the weighted mean estimate of the overall effect size, and in doing so, weighting the contribution of each study by the inverse of its variance. The larger the value of Q (which is distributed as chi-square with $(k - 1)$ degrees of freedom) the more likely the null hypothesis will be rejected. The limitations of the latter Q statistic (e.g., its low power in detecting heterogeneity of effect

size when a small number of studies is employed in a meta-analysis), which are also associated with Equation 28.91, as well as alternative statistics for assessing homogeneity of effect size are discussed in sources such as Hedges and Olkin (1985) and Huedo-Medina *et al.* (2006); b) In Section VI of the **chi-square test for $r \times c$ tables, Test 16l: The Mantel–Haenszel analysis (Test 16l-a: Test of homogeneity of odds ratios for Mantel–Haenszel analysis, Test 16l-b: Summary odds ratio for Mantel–Haenszel analysis, and Test 16l-c: Mantel-Haenszel test of association)** is described as an alternative (and what most sources consider to be a more reliable) procedure for evaluating and pooling the results obtained for multiple 2×2 contingency tables which evaluate the same hypothesis.

36. The same test result will be obtained if, instead, we assign a negative sign to the Fisher transformed z_r values for *Studies A, B, C,* and *D,* and a positive sign for the Fisher transformed z_r value for *Study E.*

37. a) In the **Introduction** it was noted that Hunter and Schmidt (1990, 2004) are among the methodologists who argue that the **classical hypothesis testing model** should no longer be employed (i.e., that tests of statistical significance should no longer be used for hypothesis testing), and that decision making in individual studies should be based in large part or entirely on the values computed for confidence intervals; b) **Predictive validity** refers to the degree to which performance on a test or some other predictor variable is able to predict the performance of subjects on some measure of behavior. Among other things, Hunter and Schmidt (2004) contend that by taking into account the impact of sampling error, **psychometric meta-analysis** corrects for attenuation (i.e., underestimation) in correlations which are employed for predictive purposes; c) Schulze's (2004) book was published prior to publication of Hunter and Schmidt's (2004) most recent book in which they describe the use of **psychometric meta-analysis** for situations other than evaluating predictive validity for personnel selection.

38. To illustrate a simple example of Hunter and Schmidt's (2004; p. 8; pp. 59–64; p. 81; pp. 88–92) methodology a meta-analysis which corrects for just sampling error (but not any other artifacts) will be demonstrated in reference to Example 28.8. Hunter and Schmidt (2004, p. 81) use the term *bare-bones meta-analysis* for an analysis that only corrects for sampling error. Hunter and Schmidt (2004, p. 81) employ Equation 28.97 to compute the best estimate of a population correlation (which will be designated with the notation $\bar{r}$) based on *m* studies. (As noted in Endnote 17, Hunter and Schmidt (1987; 2004, pp. 82–83) and Hunter, Schmidt, and Coggin (1996) contend that in contrast to Equation 28.25 (presented in Section VI), which they state is positively biased, Equation 28.97 provides the optimal weighted estimate of a population correlation. Yet Schulze (2004, p. 65) states that Equation 28.97 is negatively biased — i.e., it underestimates the population correlation.)
Equation 28.97 is employed below to compute the value $\bar{r} = .314$ for the estimated population correlation.

(Equation 28.97)

$$\bar{r} = \frac{\Sigma(n_j r_j)}{\Sigma n_j} = \frac{(20)(.6) + (40)(.5) + (10)(.2) + (30)(.3) + (25)(-.15)}{20 + 40 + 10 + 30 + 25} = \frac{39.25}{125} = .314$$

The corresponding variance across the m studies (to be designated with the notation $\tilde{s}_r^2$) is computed with Equation 28.99 (Hunter and Schmidt (2004, p. 81)). Hunter and Schmidt (2004, p. 83) note that $\tilde{s}_r^2$ is a confounding of the following two elements: a) Variation on the population correlations (if such variability exists) (the estimated value of the latter will be represented by the notation $\tilde{s}_p^2$); and b) Variation in sample correlations which result from sampling error (the estimated value of the latter will be represented by the notation $\tilde{s}_e^2$). The latter relationship is expressed by Equation 28.98 (Hunter and Schmidt (2004, p. 86–87)), which reflects the fact that the variance of the observed correlations (i.e., sample correlations) is often much larger than the variance of the population correlations, and allows for a test of the hypothesis that $\sigma_p^2 = 0$ (Hunter and Schmidt (2004, p. 86–88)). (The latter analysis is analogous to what is evaluated with **Test 28d (Test for evaluating a hypothesis on whether k independent correlations are homogeneous)** through use of Equation 28.24 and 28.25 in Section VI, as well as being analogous to the meta-analysis conducted with **Tests 28n and 28o (Procedures for comparing k studies with respect to effect size and for obtaining a combined effect size for k studies.)**

$$\tilde{s}_r^2 = \tilde{s}_p^2 + \tilde{s}_e^2 \qquad \textbf{(Equation 28.98)}$$

Employing Equation 28.99, the value $\tilde{s}_r^2 = .0683$ is computed.

$$\tilde{s}_r^2 = \frac{\sum[n_j(r_j - \bar{r})^2]}{\sum n_j} \qquad \textbf{(Equation 28.99)}$$

$$\tilde{s}_r^2 = \frac{(20)(.6-.314)^2 + (40)(.5-.314)^2 + (10)(.2-.314)^2 + (30)(.3-.314)^2 + (25)(-.15-.314)^2}{20 + 40 + 10 + 30 + 25}$$

$$\tilde{s}_r^2 = \frac{8.538}{125} = .0683$$

The sampling error variance ($\tilde{s}_e^2$) is computed with Equation 28.100 (Hunter and Schmidt (2004, pp. 86–87)). Note that in the latter equation, the value n_j represents the number of subjects employed in computing each of the m correlation coefficients when the same sample size is employed for all of the correlations. If, however, the number of subjects in each study is unequal (as is the case in Example 28.8), the value employed for n_j in Equation 28.100 is the average of the m sample sizes (i.e., $n_j = (n_1 + n_2 + \ldots + n_m)/m$. Thus, in the case of Example 28.8, $n_j = (20 + 40 + 10 + 30 + 25)/5 = 25$. Employing Equation 28.100, the value $\tilde{s}_e^2 = .0339$ is computed. (Schulze (2004, pp. 65–67) notes methodologists are not in agreement with respect to the appropriate equation to employ to compute the value of $\tilde{s}_e^2$.)

$$\tilde{s}_e^2 = \frac{(1 - \bar{r}^2)^2}{n_j - 1} + \frac{[1 - (.314)^2]^2}{25 - 1} = .0339 \qquad \textbf{(Equation 28.100)}$$

The estimated variance of the population correlation ($\tilde{s}_p^2$) is computed with Equation 28.101 (which is the algebraic transposition of Equation 28.98). Employing Equation 28.101, the value $\tilde{s}_p^2 = .0344$ is computed. The square root of the latter value represents the estimated population standard deviation after correcting for sampling error. Thus, the value $\tilde{s}_p = .1855$ is computed below. (Hunter and Schmidt (2004, p. 66) note that the value computed for $\tilde{s}_p$ should always be regarded as an overestimate of the true population standard deviation, since procedures for data correction only rectify some artifacts which account for spurious variability across studies.)

$$\tilde{s}_p^2 = \tilde{s}_r^2 - \tilde{s}_e^2 = .0683 - .0339 = .0344 \qquad \textbf{(Equation 28.101)}$$

$$\tilde{s}_p = \sqrt{.0344} = .1855$$

Equation 28.102 (Hunter and Schmidt (2004, pp. 88–92)) allows for the computation of a z value by dividing the estimated population correlation by the estimated population standard deviation.

$$z = \frac{\bar{r}}{\tilde{s}_p} = \frac{.314}{.1855} = 1.69 \qquad \textbf{(Equation 28.102)}$$

If the computed value of z is significant, one can conclude the correlations obtained for the m studies are consistent with the hypothesis there is just one value for ρ (a value other than zero), and that all or most of the variability across the correlation values for the m studies is the result of sampling error. Hunter and Schmidt (2004, p. 65) employ a z value of 2 or greater as a criterion for significance. Since the value $z = 1.69$ is less than 2, it does not allow one to conclude that the correlation between the independent and dependent variables in the population is some value other than zero. If the value computed for z had been 2 or greater, the likelihood of a zero correlation between the two variables would be minimal, and one could conclude that the best estimate for the underlying population correlation is the value computed with Equation 28.97 (i.e., $\bar{r} = .314$). It should be noted that one could argue the value $z = 1.69$ computed above with Equation 28.102 would, in fact, achieve significance if a less stringent criterion is employed (e.g., $z_{.05} = 1.65$). Additionally, if, in fact, as Hunter and Schmidt note, $\tilde{s}_p = .1855$ overestimates the true population standard deviation, the actual z value computed with Equation 28.102 would be larger. It is also worth noting that if the value $r = .338$ computed with Equation 28.25 is employed in Equation 28.102 it yields the value $z = .338/.1855 = 1.82$, which is closer to the Hunter and Schmidt's criterion value of 2.

In contrast to the use of the Hunter and Schmidt methodology, analysis of Example 28.8 through use of **Test 28d** or **Test 28n** (with Equation 28.24 or Equation 28.91) yields the value $\chi^2 = 9.18$; $p < .05$, leading to the conclusion that the effect size in the underlying populations represented by the $m = 5$ is homogeneous. Through use of Equation 28.25 the best estimate of the common population correlation is computed to be $r = .338$ (which is larger than $\bar{r} = .314$ computed above with Equation 28.97).

Readers should take note of the fact that the above example is an oversimplification of the procedures employed in **psychometric meta-analysis**. Those interested in employing the methodology, which is considerably more involved that what has been presented in this section, should consult Hunter and Schmidt (2004).

39. An excellent overview of alternative meta-analytic procedures can be found in Schulze (2004).

40. Although **fixed** and **random-effects models** are discussed in Section VII of the **single-factor between-subjects analysis of variance**, in the case of a meta-analysis a **fixed-effects model** assumes that in the underlying populations for all k studies the identical effect size is present (i.e., homogeneity of effect size). In contrast, a **random-effects model**

(which is appropriate to employ when there is heterogeneity of effect size) does not assume equality of effect size, but instead that the effect sizes for the k studies are distributed normally. Field (2005) notes that inferences derived within the framework of a fixed-effects model only apply to studies employed in the meta-analysis, while inferences derived within the framework of a random-effects model can be generalized to other studies evaluating the same hypothesis that are not included in the meta-analysis. If the null hypothesis of homogeneity of effect size is rejected, within the framework of a random-effects model a mean effect size is computed for the k studies that takes into account both the *between-studies variability* for effect size and the *within-study variability* for each of the k studies (with the latter being the only type of variability considered in computing a mean effect size with Equation 28.92 in the case of a fixed-effects model).

41. As noted under the discussion of the **odds ratio** in Section VI of the **chi-square test for $r \times c$ tables**, the values of the **odds ratio** and the **relative risk** will be very close together when the event in question (in this case a heart attack) has a low probability of occurring. The likelihood of someone in the placebo group having a heart attack is 189/11,034 = .01713, while the likelihood of someone in the aspirin group having a heart attack is 104/11,037 = .00942 (note that .01713/.00942 = 1.82, which is the value of **relative risk**). Thus, the values computed for the **odds ratio** and the **relative risk** are almost identical.

42. It is of interest to note, however, that if the **chi-square test for $r \times c$ tables** is employed to evaluate the data in Table 28.17, Equation 16.2 yields the value $\chi^2 = 18$. Employing **Table A4**, we determine that the computed value $\chi^2 = 18$ is greater than the tabled critical values (for $df = 1$) $\chi^2_{.05} = 3.84$ and $\chi^2_{.01} = 6.63$. Thus, the null hypothesis can be rejected at both the .05 and .01 levels. Substituting $\chi^2 = 18$ in Equation 16.21 yields the value $\phi = \sqrt{\chi^2/n} = \sqrt{18/200} = .30$, which corresponds to the value of the correlation noted for the problem under discussion.

43. The **noncentral F distribution** was alluded to previously in Section VI of the **single-factor between-subjects analysis of variance** under the discussion of **power**. Further clarification of the **noncentrality parameter** underlying the **noncentral F distribution** can be found in Endnote 9 of the **single-sample t test**.

Test 29

Spearman's Rank-Order Correlation Coefficient
(Nonparametric Measure of Association/Correlation Employed with Ordinal Data)

I. Hypothesis Evaluated with Test and Relevant Background Information

Spearman's rank-order correlation coefficient is one of a number of measures of correlation or association discussed in this book. Measures of correlation are not inferential statistical tests, but are, instead, descriptive statistical measures which represent the degree of relationship between two or more variables. Upon computing a measure of correlation, it is common practice to employ one or more inferential statistical tests in order to evaluate one or more hypotheses concerning the correlation coefficient. The hypothesis stated below is the most commonly evaluated hypothesis for **Spearman's rank-order correlation coefficient**.

Hypothesis evaluated with test In the underlying population represented by a sample, is the correlation between subjects' scores on two variables some value other than zero? The latter hypothesis can also be stated in the following form: In the underlying population represented by the sample, is there a significant **monotonic** relationship between the two variables? It is important to note that the nature of the relationship described by **Spearman's rank-order correlation coefficient** is based on an analysis of two sets of ranks.

Relevant background information on test Prior to reading the material in this section the reader should review the general discussion of correlation in Section I of the **Pearson product-moment correlation coefficient (Test 28)**. Developed by Spearman (1904), **Spearman's rank-order correlation coefficient** is a bivariate measure of correlation/association which is employed with rank-order data. The population parameter estimated by the correlation coefficient will be represented by the notation ρ_S (where ρ is the lower case Greek letter **rho**). The sample statistic computed to estimate the value of ρ_S will be represented by the notation r_S. In point of fact, **Spearman's rank-order correlation coefficient** is a special case of the **Pearson product-moment correlation coefficient**, when the latter measure is computed for two sets of ranks. The relationship between **Spearman's rank-order correlation coefficient** and the **Pearson product-moment correlation coefficient** is discussed in Section VI.

As is the case for the **Pearson product-moment correlation coefficient**, **Spearman's rank-order correlation coefficient** can be employed to evaluate data for n subjects, each of whom has contributed a score on two variables (designated as the X and Y variables). Within each of the variables, the n scores are rank-ordered. **Spearman's rank-order correlation coefficient** is also commonly employed to evaluate the degree of agreement between the rankings of $m = 2$ judges for n subjects/objects.

In computing **Spearman's rank-order correlation coefficient**, one of the following is true with regard to the rank-order data that are evaluated: a) The data for both variables are in a rank-order format, since it is the only format for which data are available; b) The original data are in a rank-order format for one variable and in an interval/ratio format for the second variable. In such an instance, data on the second variable are converted to a rank-order format in order that both sets of data represent the same level of measurement; and c) The data for both variables have been transformed into a rank-order format from an interval/ratio format, since the researcher has reason to believe that one or more of the assumptions underlying the **Pearson product-moment correlation coefficient** (which is the analogous parametric correlational procedure employed for interval/ratio data) have been saliently violated. It should be noted that since information is sacrificed when interval/ratio data are transformed into a rank-order format, some researchers may elect to employ the **Pearson product-moment correlation coefficient** rather than **Spearman's rank-order correlation coefficient**, even when there is reason to believe that one or more of the assumptions of the former measure have been violated.

Spearman's rank-order correlation coefficient determines the degree to which a **monotonic** relationship exists between two variables. A monotonic relationship can be described as **monotonic increasing** (which is associated with a positive correlation) or **monotonic decreasing** (which is associated with a negative correlation). A relationship between two variables is monotonic increasing, if an increase in the value of one variable is always accompanied by an increase in the value of the other variable. A relationship between two variables is monotonic decreasing, if an increase in the value of one variable is always accompanied by a decrease in the value of the other variable. Based on the above definitions, a positively sloped straight line represents an example of a monotonic increasing function, while a negatively sloped straight line represents an example of a monotonic decreasing function. In addition to the aforementioned linear functions, curvilinear functions can also be monotonic. For instance, the function $Y = X^2$ depicted in Figure 29.1 represents an example of a monotonic increasing function, since an increase in the X variable always results in an increase in the Y variable. It should be noted that when the interval/ratio scores on two variables are monotonically related to one another, a linear function can be employed to describe the relationship between the rank-orderings of the two variables. This latter fact is demonstrated in Section VI.

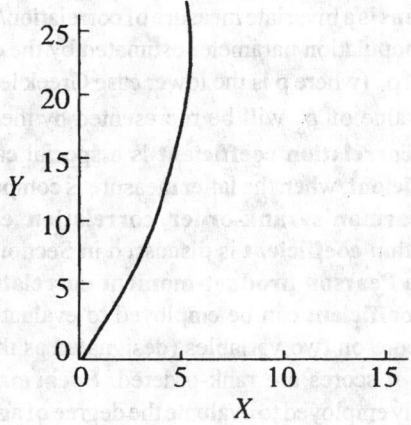

Figure 29.1 Monotonic Increasing Relationship ($Y = X^2$)

The same general guidelines that are described for interpreting the value of the **Pearson product-moment correlation coefficient** can be applied to **Spearman's rank-order correlation coefficient**. Thus, the range of values r_S can assume is defined by the limits -1 to $+1$ (i.e., $-1 \leq r_S \leq +1$). The absolute value of r_S (i.e., $|r_S|$) indicates the strength of the relationship between the two variables. As the absolute value of r_S approaches 1, the strength of the monotonic relationship increases, being the strongest when r_S equals either $+1$ or -1. The closer the absolute value of r_S is to 0, the weaker the monotonic relationship between the two variables, and when $r_S = 0$ no monotonic relationship is present. The sign of r_S indicates the direction of the monotonic relationship (i.e., positive/increasing monotonic versus negative/decreasing monotonic). As is the case for the **Pearson product-moment correlation coefficient**, a positive correlation indicates that an increase (decrease) on one variable is associated with an increase (decrease) on the other variable. A negative correlation indicates that an increase (decrease) on one variable is associated with a decrease (increase) on the other variable.

It is important to note that correlation does not imply causation. Consequently, if there is a strong correlation between two variables (i.e., the absolute value of r_S is close to 1), a researcher is not justified in concluding that one variable causes the other variable. Although it is possible that when a strong correlation exists one variable may, in fact, cause the other variable, the information employed in computing **Spearman's rank-order correlation coefficient** does not allow a researcher to draw such a conclusion. This is the case, since extraneous variables which have not been taken into account by the researcher can be responsible for the observed correlation between the two variables.

II. Example

Example 29.1 is identical to Example 28.1 (which is evaluated with the **Pearson product-moment correlation coefficient**). In evaluating Example 29.1 it will be assumed that the ratio data are rank-ordered, since one or more of the assumptions of the **Pearson product-moment correlation coefficient** have been saliently violated.[1]

Example 29.1 *A psychologist conducts a study employing a sample of five children to determine whether or not there is a statistical relationship between the number of ounces of sugar a ten-year-old child eats per week (which will represent the X variable) and the number of cavities in a child's mouth (which will represent the Y variable). The two scores (ounces of sugar consumed per week and number of cavities) obtained for each of the five children follow:* **Child 1** (20, 7); **Child 2** (0, 0); **Child 3** (1, 2); **Child 4** (12, 5); **Child 5** (3, 3). *Is there a significant correlation between sugar consumption and the number of cavities?*

III. Null versus Alternative Hypotheses

Upon computing **Spearman's rank-order correlation coefficient**, it is common practice to determine whether the obtained absolute value of the correlation coefficient is large enough to allow a researcher to conclude that the underlying population correlation coefficient between the two variables is some value other than zero. Section V describes how the latter hypothesis, which is stated below, can be evaluated through use of tables of critical r_S values or through use of an inferential statistical test which is based on either the t or z distributions.

Null hypothesis H_0: $\rho_S = 0$

(In the underlying population the sample represents, the correlation between the ranks of subjects on Variable X and Variable Y equals 0.)

Alternative hypothesis H_1: $\rho_S \neq 0$

(In the underlying population the sample represents, the correlation between the ranks of subjects on Variable X and Variable Y equals some value other than 0. This is a **nondirectional alternative hypothesis**, and it is evaluated with a **two-tailed test**. Either a significant positive r_S value or a significant negative r_S value will provide support for this alternative hypothesis. In order to be significant, the obtained absolute value of r_S must be equal to or greater than the tabled critical two-tailed r_S value at the prespecified level of significance.)

or

H_1: $\rho_S > 0$

(In the underlying population the sample represents, the correlation between the ranks of subjects on Variable X and Variable Y equals some value greater than 0. This is a **directional alternative hypothesis**, and it is evaluated with a **one-tailed test**. Only a significant positive r_S value will provide support for this alternative hypothesis. In order to be significant (in addition to the requirement of a positive r_S value), the obtained absolute value of r_S must be equal to or greater than the tabled critical one-tailed r_S value at the prespecified level of significance.)

or

H_1: $\rho_S < 0$

(In the underlying population the sample represents, the correlation between the ranks of subjects on Variable X and Variable Y equals some value less than 0. This is a **directional alternative hypothesis**, and it is evaluated with a **one-tailed test**. Only a significant negative r_S value will provide support for this alternative hypothesis. In order to be significant (in addition to the requirement of a negative r_S value), the obtained absolute value of r_S must be equal to or greater than the tabled critical one-tailed r_S value at the prespecified level of significance.)

Note: Only one of the above noted alternative hypotheses is employed. If the alternative hypothesis the researcher selects is supported, the null hypothesis is rejected.[2]

IV. Test Computations

Table 29.1 summarizes the data for Example 29.1. The following should be noted with respect to Table 29.1: a) The number of subjects is $n = 5$. Each subject has an X score and a Y score, and thus there are five X scores and five Y scores; b) The rankings of the five subjects' scores on the X and Y variables are respectively recorded in the columns labeled R_X and R_Y; c) The column labeled $d = R_X - R_Y$ contains a difference score for each subject, which is obtained by subtracting a subject's rank on the Y variable from the subject's rank on the X variable; and d) The column labeled d^2 contains the square of each subject's difference score.

The ranking protocol employed in Table 29.1 is identical to that employed for the **Mann–Whitney U test (Test 12)**. Whereas in the case of the latter test the scores of subjects are ranked

Table 29.1 Summary of Data for Example 29.1

Subject	X	R_X	Y	R_Y	$d = R_X - R_Y$	d^2
1	20	5	7	5	0	0
2	0	1	0	1	0	0
3	1	2	2	2	0	0
4	12	4	5	4	0	0
5	3	3	3	3	0	0
					$\Sigma d = 0$	$\Sigma d^2 = 0$

within each group, in the computation of **Spearman's rho** the scores of the $n = 5$ subjects are ranked within each of the variables. Thus, in Table 29.1 the five subjects' X scores are ranked such that a rank of 1 is assigned to the lowest score on the X variable, a rank of 2 is assigned to the next lowest score on the X variable, and so on until a rank of 5 is assigned to the highest score on the X variable. The identical ranking procedure is employed with respect to the Y scores (i.e., a rank of 1 is assigned to the lowest score on the Y variable, a rank of 2 is assigned to the next lowest score on the Y variable, and so on until a rank of 5 is assigned to the highest score on the Y variable). In the event of tied scores (which do not occur in Example 29.1), as is the case for other rank-order procedures, the average of the ranks involved is assigned to all scores tied for a given rank.

It should be noted that it is permissible to reverse the ranking protocol described above. Specifically, for each variable a rank of 1 can be assigned to the highest score on that variable and a rank of 5 to the lowest score on that variable. Employing this alternative ranking protocol will yield the identical value for r_S as the one yielded by the ranking protocol employed in Table 29.1. It should be emphasized that regardless of which ranking protocol is employed, the same protocol must be employed for both variables. The protocol of assigning the lowest rank to the lowest score and the highest rank to the highest score is employed in Example 29.1, since it allows for easiest interpretation of the results of the study.

In Column 6 of Table 29.1, the sum of the difference scores is computed to be $\Sigma d = 0$. In fact, Σd will always equal zero and if Σd is some value other than zero, it indicates that an error has been made in the rankings and/or computations. In the last column of Table 29.1, the sum of the squared difference scores ($\Sigma d^2 = 0$) is computed. This latter value (which will only equal zero when $r_S = 1$) and the value of n are employed in Equation 29.1, which is the equation for computing **Spearman's rank-order correlation coefficient.**[3]

$$r_S = 1 - \frac{6\Sigma d^2}{n(n^2 - 1)} \qquad \text{(Equation 29.1)}$$

Substituting the appropriate values in Equation 29.1, the value $r_S = 1$ is computed.

$$r_S = 1 - \frac{(6)(0)}{5[(5)^2 - 1]} = 1$$

V. Interpretation of the Test Results

The obtained value $r_S = 1$ is evaluated with **Table A18 (Table of Critical Values for Spearman's Rho)** in the **Appendix**. The critical values in **Table A18** are listed in reference to n.[4] Employing **Table A18**, it can be determined that for $n = 5$ the tabled critical two-tailed r_S

value at the .05 level of significance is $r_{S_{.05}} = 1$. Because of the small sample size, it is not possible to evaluate the nondirectional null hypothesis at the .01 level. The tabled critical one-tailed r_S values at the .05 and .01 levels of significance are $r_{S_{.05}} = .90$ and $r_{S_{.01}} = 1$.

The following guidelines are employed in evaluating the null hypothesis H_0: $\rho_S = 0$.

a) If the nondirectional alternative hypothesis H_1: $\rho_S \neq 0$ is employed, the null hypothesis can be rejected if the obtained absolute value of r_S is equal to or greater than the tabled critical two-tailed value at the prespecified level of significance.

b) If the directional alternative hypothesis H_1: $\rho_S > 0$ is employed, the null hypothesis can be rejected if the sign of r_S is positive, and the value of r_S is equal to or greater than the tabled critical one-tailed value at the prespecified level of significance.

c) If the directional alternative hypothesis H_1: $\rho_S < 0$ is employed, the null hypothesis can be rejected if the sign of r_S is negative, and the absolute value of r_S is equal to or greater than the tabled critical one-tailed value at the prespecified level of significance.

Employing the above guidelines, the nondirectional alternative hypothesis H_1: $\rho_S \neq 0$ is supported at the .05 level, since the computed value $r_S = 1$ is equal to the tabled critical two-tailed value $r_{S_{.05}} = 1$. The directional alternative hypothesis H_1: $\rho_S > 0$ is supported at both the .05 and .01 levels, since the computed value $r_S = 1$ is a positive number that is equal to or greater than the tabled critical one-tailed values $r_{S_{.05}} = .90$ and $r_{S_{.01}} = 1$. The directional alternative hypothesis H_1: $\rho_S < 0$ is not supported, since the computed value $r_S = 1$ is a positive number.

When the **Pearson product-moment correlation coefficient** is employed to evaluate the same set of data (i.e., the ratio scores of subjects are correlated with one another), the nondirectional alternative hypothesis (i.e., H_1: $\rho \neq 0$) is also supported at only the .05 level, and the directional alternative hypothesis (i.e., H_1: $\rho > 0$) is supported at both the .05 and .01 levels. Thus, in this instance, the two correlation coefficients yield comparable results. (However, since **Pearson r** is the more powerful of the two correlational procedures, it is more likely to result in rejection of the null hypothesis at a given level of significance when applied to the same set of data.)

Test 29a: Test of significance for Spearman's rank-order correlation coefficient　　In the event a researcher does not have access to **Table A18**, Equation 29.2, which employs the *t* distribution, provides an alternative way of evaluating the null hypothesis H_0: $\rho_S = 0$. Most sources which recommend Equation 29.2 state that it provides a reasonably good approximation of the underlying sampling distribution when $n > 10$.

$$t = \frac{r_S \sqrt{n - 2}}{\sqrt{1 - r_S^2}} \qquad \text{(Equation 29.2)}$$

The *t* value computed with Equation 29.2 is evaluated with **Table A2 (Table of Student's *t* Distribution)** in the **Appendix**. The degrees of freedom employed are $df = n - 2$. Thus, in the case of Example 29.1, $df = 5 - 2 = 3$. For $df = 3$, the tabled critical two-tailed .05 and .01 values are $t_{.05} = 3.18$ and $t_{.01} = 5.84$, and the tabled critical one-tailed .05 and .01 values are $t_{.05} = 2.35$ and $t_{.01} = 4.54$. Since the sign of the *t* value computed with Equation 29.2 will always be the same as the sign of r_S, the guidelines described earlier in reference to **Table A18** for evaluating an r_S value can also be applied in evaluating the *t* value computed with Equation 29.2 (i.e., substitute *t* in place of r_S in the text of the guidelines for evaluating r_S).

Inspection of Equation 29.2 reveals that if the absolute value of r_S equals 1, the term $\sqrt{1-r^2}$ will equal zero, thus rendering the equation insoluble (i.e., $t = [(1)\sqrt{5-2}]/\sqrt{1-(1)^2} = ?$). Consequently, Equation 29.2 cannot be applied to Example 29.1.

Equation 29.3, which employs the normal distribution, is an alternative equation for evaluating the significance of r_S. When the sample size is large (approximately 200 or greater), Equation 29.3 will yield a result which is equivalent to that obtained with Equation 29.2.[5]

$$z = r_S\sqrt{n-1}$$ (Equation 29.3)

Although the sample size in Example 29.1 is well below the minimum size recommended for Equation 29.3, the appropriate values will be substituted in the latter equation in order to demonstrate its application. Substituting the values $r_S = 1$ and $n = 5$ in Equation 29.3, the value $z = 2.00$ is computed.

$$z = (1)\sqrt{5-1} = 2$$

The computed value $z = 2.00$ is evaluated with **Table A1 (Table of the Normal Distribution)** in the **Appendix**. In the latter table, the tabled critical two-tailed .05 and .01 values are $z_{.05} = 1.96$ and $z_{.01} = 2.58$, and the tabled critical one-tailed .05 and .01 values are $z_{.05} = 1.65$ and $z_{.01} = 2.33$. Since the sign of the z value computed with Equation 29.3 will always be the same as the sign of r_S, the guidelines described earlier in reference to **Table A18** for evaluating an r_S value can also be applied in evaluating the z value computed with Equation 29.3 (i.e., substitute z in place of r_S in the text of the guidelines for evaluating r_S).

Employing the guidelines, the nondirectional alternative hypothesis $H_1: \rho_S \neq 0$ is supported at the .05 level, since the computed value $z = 2.00$ is greater than the tabled critical two-tailed value $z_{.05} = 1.96$. It is not, however, supported at the .01 level, since $z = 2.00$ is less than the tabled critical two-tailed value $z_{.01} = 2.58$.

The directional alternative hypothesis $H_1: \rho_S > 0$ is supported at the .05 level, since the computed value $z = 2.00$ is a positive number that is greater than the tabled critical one-tailed value $z_{.05} = 1.65$. It is not, however, supported at the .01 level, since $z = 2.00$ is less than the tabled critical one-tailed value $z_{.01} = 2.33$.

The directional alternative hypothesis $H_1: \rho < 0$ is not supported, since the computed value $z = 2.00$ is a positive number. In order for the alternative hypothesis $H_1: \rho_S < 0$ to be supported, the computed value of z must be a negative number (as well as the fact that the absolute value of z must be equal to or greater than the tabled critical one-tailed value at the prespecified level of significance). Note that the results obtained through use of Equation 29.3 are reasonably consistent with those that are obtained when **Table A18** is employed.[6]

A summary of the analysis of Example 29.1 follows: It can be concluded there is a significant monotonic increasing/positive relationship between the number of ounces of sugar a ten-year-old child eats and the number of cavities in a child's mouth. This result can be summarized as follows (if it is assumed the nondirectional alternative hypothesis $H_1: \rho_S \neq 0$ is employed): $r_S = 1$, $p < .05$.

VI. Additional Analytical Procedures for Spearman's Rank-Order Correlation Coefficient and/or Related Tests

1. Tie correction for Spearman's rank-order correlation coefficient When one or more ties are present in a set of data, many sources recommend that the r_S value computed with Equation

Table 29.2   Data Employed with Tie Correction Procedure

Subject	X	R_X	Y	R_Y	$d = R_X - R_Y$	d^2
1	0	1.5	1	2	-.5	.25
2	0	1.5	0	1	.5	.25
3	2	3	2	3.5	-.5	.25
4	4	4	2	3.5	.5	.25
5	8	6	8	9.5	-3.5	12.25
6	8	6	8	9.5	-3.5	12.25
7	8	6	3	5	1	1
8	13	8	4	6	2	4
9	16	9.5	6	7	2.5	6.25
10	16	9.5	7	8	1.5	2.25
					$\Sigma d = 0$	$\Sigma d^2 = 39$

29.1 be adjusted. The reason for this is that when ties are present, Equation 29.1 spuriously inflates the absolute value of r_S. In practice, most of the time that ties are present the effect on the value of r_S will be minimal (unless the number of ties is excessive). The tie correction procedure to be demonstrated in this section will employ the data summarized in Table 29.2. Assume that the data are for the same variables evaluated in Example 29.1, except for the fact that a different set of subjects is employed with $n = 10$.

Employing Equation 29.1, it is determined that the value of **Spearman's rho** without employing a tie correction is $r_S = .764$.

$$r_S = 1 - \frac{6(39)}{10[(10)^2 - 1]} = .764$$

The tie correction will now be introduced. In the example under discussion there are $s = 3$ sets of ties involving the ranks of subjects' X scores (Subjects 1 and 2; Subjects 5, 6, and 7; Subjects 9 and 10), and $s = 2$ sets of ties involving the ranks of subjects' Y scores (Subjects 3 and 4; Subjects 5 and 6). Equation 29.8 is employed to compute the tie-corrected **Spearman's rank-order correlation coefficient**, which will be represented by the notation r_{S_c}. Note that the values Σx^2 and Σy^2 in Equation 29.8 are computed with Equations 29.6 and 29.7, and that Equations 29.6 and 29.7 are, respectively, based on the values T_X and T_Y, which are computed with Equations 29.4 and 29.5. In Equation 29.4, $t_{i_{(x)}}$ represents the number of X scores that are tied for a given rank. In Equation 29.5, $t_{i_{(y)}}$ represents the number of Y scores that are tied for a given rank. The notations $\Sigma_{i=1}^s (t_{i_{(x)}}^3 - t_{i_{(x)}})$ and $\Sigma_{i=1}^s (t_{i_{(y)}}^3 - t_{i_{(y)}})$ indicate that the following is done with respect to each of the variables: a) For each set of ties, the number of ties in the set is subtracted from the cube of the number of ties in that set; and b) The sum of all the values computed in part a) is obtained for that variable.

When the data from Table 29.2 are substituted in Equations 29.4–29.8, the tie-corrected value $r_{S_c} = .758$ is computed.

$$T_X = \sum_{i=1}^s (t_{i_{(x)}}^3 - t_{i_{(x)}}) = [(2)^3 - 2] + [(3)^3 - 3] + [(2)^3 - 2] = 36 \quad \textbf{(Equation 29.4)}$$

$$T_Y = \sum_{i=1}^s (t_{i_{(y)}}^3 - t_{i_{(y)}}) = [(2)^3 - 2] + [(2)^3 - 2] = 12 \quad \textbf{(Equation 29.5)}$$

$$\Sigma x^2 = \frac{n^3 - n - T_X}{12} = \frac{(10)^3 - 10 - 36}{12} = 79.5 \quad \textbf{(Equation 29.6)}$$

$$\Sigma y^2 = \frac{n^3 - n - T_Y}{12} = \frac{(10)^3 - 10 - 12}{12} = 81.5 \quad \textbf{(Equation 29.7)}$$

$$r_{S_c} = \frac{\Sigma x^2 + \Sigma y^2 - \Sigma d^2}{2\sqrt{\Sigma x^2 \Sigma y^2}} = \frac{79.5 + 81.5 - 39}{2\sqrt{(79.5)(81.5)}} = .758 \quad \textbf{(Equation 29.8)}$$

Thus, by employing the tie correction, the value of **rho** is reduced from the uncorrected value of $r_S = .764$ to $r_{S_c} = .758$. As noted earlier, the correction is minimal.

2. Spearman's rank-order correlation coefficient as a special case of the Pearson product-moment correlation coefficient Although the procedure described in the previous section for dealing with ties is the one recommended in most sources, in actuality an alternative and, at times, more computationally efficient procedure can be employed. In Section I it is noted that **Spearman's rank-order correlation coefficient** is a special case of the **Pearson product-moment correlation coefficient**. In point of fact, if the **Pearson product-moment correlation coefficient** is computed for the rank-orders in a set of interval/ratio data, the computed r value will be identical to the value computed for r_S with Equation 29.1. This is demonstrated below for Example 29.1, where Equation 28.1 (the equation for computing the **Pearson product-moment correlation coefficient**) is employed to compute the value $r = r_S = 1$. Table 29.3 summarizes the values which are substituted in Equation 28.1. Note that the ranks R_X and R_Y employed in Table 29.1 are used in Table 29.3 to represent the scores on the X and Y variables.[7]

$$r = r_S = \frac{55 - \frac{(15)(15)}{5}}{\sqrt{\left[55 - \frac{(15)^2}{10}\right]\left[55 - \frac{(15)^2}{10}\right]}} = 1$$

Table 29.3 Summary of Data for Example 29.1 for Evaluation with Equation 28.1

Subject	X	X^2	Y	Y^2	XY
1	5	25	5	25	25
2	1	1	1	1	1
3	2	4	2	4	4
4	4	16	4	16	16
5	3	9	3	9	9
	$\Sigma X = 15$	$\Sigma X^2 = 55$	$\Sigma Y = 15$	$\Sigma Y^2 = 55$	$\Sigma XY = 55$

When there are no ties present in the data, Equations 29.1 and 28.1 will always yield the identical value for r_S. However, anytime there is at least one set of ties, the values yielded by the two equations will not be identical. In point of fact, Howell (2002) notes that when ties are present in the data, the r_S value computed with Equation 28.1 will be equivalent to the tie-corrected value r_{S_c} computed with Equation 29.8. When there are no ties present in the data, it is clearly more efficient to employ Equation 29.1 than it is to employ Equation 28.1. However,

Table 29.4   Summary of Data in Table 29.2 for Evaluation with Equation 28.1

Subject	X	X^2	Y	Y^2	XY
1	1.5	2.25	2	4	3
2	1.5	2.25	1	1	1.5
3	3	9	3.5	12.25	10.5
4	4	16	3.5	12.25	14
5	6	36	9.5	90.25	57
6	6	36	9.5	90.25	57
7	6	36	5	25	30
8	8	64	6	36	48
9	9.5	90.25	7	49	66.5
10	9.5	90.25	8	64	76
	$\Sigma X = 55$	$\Sigma X^2 = 382$	$\Sigma Y = 55$	$\Sigma Y^2 = 384$	$\Sigma XY = 363.5$

$$r = \frac{363.5 - \dfrac{(55)(55)}{10}}{\sqrt{\left[382 - \dfrac{(55)^2}{10}\right]\left[384 - \dfrac{(55)^2}{10}\right]}} = .758$$

when ties are present, it can be argued that use of Equation 28.1 is more computationally efficient than Equation 29.8. To demonstrate the equivalency of Equation 28.1 and Equation 29.8, Equation 28.1 is employed below with the rank-orders in Table 29.2. Table 29.4 summarizes the values which are substituted in Equation 28.1. The value $r = .758$ obtained with Equation 28.1 is identical to the value $r_{s_c} = .758$ obtained with Equation 29.8.

3. Regression analysis and Spearman's rank-order correlation coefficient   When **Spearman's rank-order correlation coefficient** is computed for a set of data, a researcher may also want to derive the mathematical function that best allows one to predict a subject's score on one variable through use of the subject's score on the second variable. To do this requires the use of regression analysis which, as noted in Section VI of the **Pearson product-moment correlation coefficient**, is a general term that describes statistical procedures which determine the mathematical function that best describes the relationship between two or more variables. One type of regression analysis which falls within the general category of **nonparametric regression analysis** is referred to as **monotonic regression analysis**. The latter type of analysis is based on the fact that if two variables (which are represented by interval/ratio data) are monotonically related to one another, the rankings on the variables will be linearly related to one another. This can be illustrated in reference to Example 29.1 through use of Figure 29.2. Whereas Figure 28.1 (in Section VI of the **Pearson product-moment correlation coefficient**) represents a scatterplot of the five pairs of ratio scores for Examples 28.1/29.1, Figure 29.2 is a scatterplot of the five pairs of ranks on the two variables. Note that the scatterplot is such that one can draw a positively sloped straight line which passes through all of the data points. The only time all of the data points will fall on the regression line is when the absolute value of the correlation between the two variables equals 1. Although some data points may fall on the line when an imperfect monotonic relationship is present, the others will not. The stronger the monotonic relationship, the closer the proximity of the data points to the line.

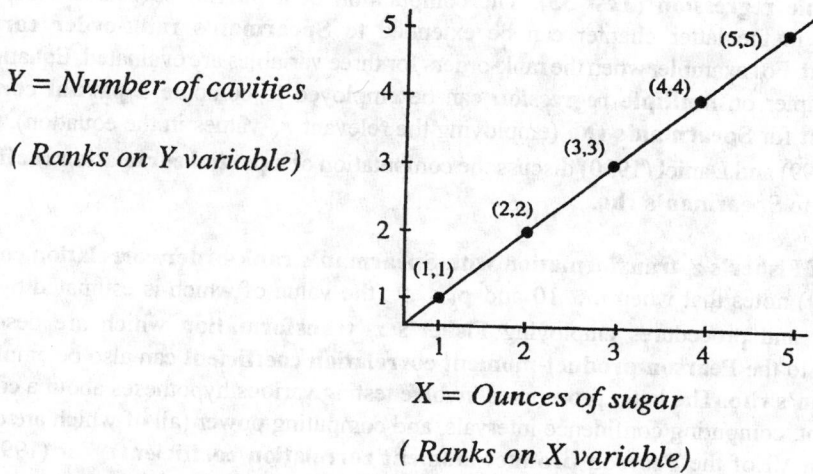

Figure 29.2 Scatterplot of Ranks for Example 29.1

As is noted in Section VI of the **Pearson product-moment correlation coefficient**, the most commonly employed method of regression analysis is the method of least squares (which is a linear regression procedure that derives the straight line which provides the best fit for a set of data). Although visual inspection of Figures 28.1 and 29.2 suggests a strong monotonic increasing relationship between the two variables (i.e., an increase in the number of ounces of sugar consumed is associated with an increase in the number of cavities), it does not allow one to precisely determine whether the function that best describes the relationship is a straight line or a monotonic curve. In order to determine the latter, it is necessary to contrast the predictive accuracy of the method of least squares with some alternative form of regression analysis. Conover (1980, 1999), who provides a bibliography on the general subject of monotonic regression analysis, describes its application in deriving a curve for a set of rank-ordered data. Marascuilo and McSweeney (1977) and Sprent (1989, 1993) also discuss the subject of monotonic regression analysis. In addition to sources on nonparametric statistics that discuss monotonic regression, many books on correlation and regression describe procedures for deriving different types of curvilinear functions. Daniel (1990) discusses a number of different approaches to nonparametric regression analysis, which derive the straight line that best describes the relationship between the interval/ratio scores on the two variables. These latter types of regression analysis (which employ the median instead of the mean as a reference point) are recommended when there is reason to believe that one or more of the assumptions underlying the method of least squares are saliently violated. Among those procedures Daniel (1990) describes are the **Brown-Mood method** (Brown and Mood (1951) and Mood (1950)), and a methodology developed by Theil (1950). Daniel (1990) also provides a comprehensive bibliography on the subject of nonparametric regression analysis.

4. Partial rank correlation When a researcher is evaluating the relationship between three or more variables, it is often useful to compute one or more **partial correlation coefficients**. A **partial correlation** measures the degree of association between two variables after any association one or more additional variables has with the two variables has been removed. For further clarification of **partial correlation**, the reader should refer to Section IV/V of the chapter on **multiple regression (Test 33)**. The computation of a **partial correlation coefficient**, described in the latter chapter can be extended to **Spearman's rank-order correlation coefficient**. For example, when the rank-orders for three variables are evaluated, Equation 33.33 in the chapter on **multiple regression** can be employed to compute a **partial correlation coefficient** for **Spearman's rho** (employing the relevant r_S values in the equation). Conover (1980, 1999) and Daniel (1990) discuss the computation of a **partial correlation coefficient** in reference to **Spearman's rho**.

5. Use of Fisher's z_r transformation with Spearman's rank-order correlation coefficient
Zar (1999) notes that when $n \geq 10$ and $\rho_S < .9$ (the value of which is estimated by r_S), the equations and procedures employing **Fisher's z_r transformation** which are described in reference to the **Pearson product-moment correlation coefficient** can also be employed for **Spearman's rho**. The latter procedures involve testing various hypotheses about a correlation coefficient, computing confidence intervals, and computing power (all of which are described in Section VI of the **Pearson product-moment correlation coefficient**). Zar (1999) notes, however, that when the element $1/(n - 3)$ appears in an equation (the value $\sqrt{1/(n - 3)}$ represents the standard error of **Fisher's z_r**), it should be replaced by the value $1.060/(n - 3)$ when the computations are in reference to **Spearman's rho** (e.g., Equation 28.20 should be in the form $z = (z_r - z_{\rho_0})/\sqrt{1.060/(n - 3)}$ when evaluating the same hypothesis for **Spearman's rho**).

VII. Additional Discussion of Spearman's Rank-Order Correlation Coefficient

1. The relationship between Spearman's rank-order correlation coefficient, Kendall's coefficient of concordance, and the Friedman two-way analysis of variance by ranks
Kendall's coefficient of concordance (Test 31), which is discussed later in the book, is a measure of association which allows a researcher to evaluate the degree of agreement between m sets of ranks on n subjects/objects. In point of fact, **Kendall's coefficient of concordance** is linearly related to **Spearman's rank-order correlation coefficient**.[8] The underlying statistical model upon which **Kendall's coefficient of concordance** is based is identical to the model for the **Friedman two-way analysis of variance by ranks (Test 25)**. As a result of this, the **Friedman two-way analysis of variance by ranks** can be employed to determine whether the value of the **coefficient of concordance** is significant. In point of fact, the **Friedman two-way analysis of variance by ranks** can also be used to determine whether the value of **Spearman's rho** is significant. This will be illustrated with Example 29.2, which represents a type of problem which is commonly evaluated with **Spearman's rank-order correlation coefficient** (as well as **Kendall's coefficient of concordance** when there are more than two sets of ranks). In Example 29.2, $n = 10$ films (i.e., objects/subjects) are rank-ordered by $m = 2$ judges, and a determination is made with respect to the degree of agreement among the rankings of the judges.

Example 29.2 *In order to determine whether or not two critics agree with one another in their evaluation of movies, a newspaper editor asks the two critics to rank-order ten movies (assigning a rank of 1 to the best movie, a rank of 2 to the next best movie, etc.). Table 29.5 summarizes the data for the study. Is there a significant association between the two sets of ranks?*

Table 29.5 Summary of Data for Example 29.2

Movie	Critic 1 R_X	Critic 2 R_Y	$d = R_X - R_Y$	d^2
1	7	10	−3	9
2	1	2	−1	1
3	8	6	2	4
4	10	8	2	4
5	9	7	2	4
6	6	4	2	4
7	5	9	−4	16
8	2.5	3	−.5	.25
9	2.5	1	1.5	2.25
10	4	5	−1	1
			$\Sigma d = 0$	$\Sigma d^2 = 45.5$

Note that in Table 29.5 each of the $n = 10$ rows represents one of the ten movies, instead of representing $n = 10$ subjects (as is the case in Example 29.1). The ranks of Critic 1 are represented in the column labeled R_X, and the ranks of Critic 2 are represented in the column labeled R_Y. Note that Critic 1 places Movies 8 and 9 in a tie for the second best movie. Thus (employing the protocol for tied ranks described in Section IV of the **Mann–Whitney U test**), the two ranks involved (2 and 3) are averaged $((2 + 3)/2 = 2.5)$, and each of the movies is assigned the average rank of 2.5.

Employing Equation 29.1, the value $r_S = .724$ is computed. The tie-corrected value $r_{S_c} = .723$ (for which the calculations are not shown) is almost identical.

$$r = 1 - \frac{(6)(45.5)}{10[(10)^2 - 1]} = .724$$

Employing **Table A18**, it is determined that for $n = 10$, the tabled critical two-tailed .05 and .01 values are $r_{S_{.05}} = .648$ and $r_{S_{.01}} = .794$, and the tabled critical one-tailed .05 and .01 values are $r_{S_{.05}} = .564$ and $r_{S_{.01}} = .745$. Employing the aforementioned critical values, the nondirectional alternative hypothesis $H_1: \rho_S \neq 0$ and the directional alternative hypothesis $H_1: \rho_S > 0$ are supported at the .05 level, since the computed value $r_S = .724$ is greater than the tabled critical two-tailed value $r_{S_{.05}} = .648$ and the tabled critical one-tailed value $r_{S_{.05}} = .564$. The alternative hypotheses are not supported at the .01 level, since $r_S = .724$ is less than the tabled critical two-tailed value $r_{S_{.01}} = .794$ and the tabled critical one-tailed value $r_{S_{.01}} = .745$.

If Equation 29.2 is employed to evaluate the null hypothesis $H_0: \rho_S = 0$, the value $t = 2.97$ is computed.

$$t = \frac{(.724)\sqrt{10 - 2}}{\sqrt{1 - (.724)^2}} = 2.97$$

Employing **Table A2**, it is determined that for $df = 10 - 2 = 8$, the tabled critical two-tailed .05 and .01 values are $t_{.05} = 2.31$ and $t_{.01} = 3.36$, and the tabled critical one-tailed .05 and .01 values are $t_{.05} = 1.86$ and $t_{.01} = 2.90$. Employing the aforementioned critical values, the nondirectional alternative hypothesis H_1: $\rho_S \neq 0$ is supported at the .05 level, since the computed value $t = 2.97$ is greater than the tabled critical two-tailed value $t_{.05} = 2.31$. It is not supported at the .01 level, since $t = 2.97$ is less than $t_{.01} = 3.36$. The directional alternative hypothesis H_1: $\rho_S > 0$ is supported at both the .05 and .01 levels, since the computed value $t = 2.97$ is a positive number (since $r_S = .724$ is a positive number) that is greater than the tabled critical one-tailed values $t_{.05} = 1.86$ and $t_{.01} = 2.90$.

If Equation 29.3 is employed to evaluate the null hypothesis H_0: $\rho_S = 0$, the value $z = 2.17$ is computed.

$$z = (.724)\sqrt{10 - 1} = 2.17$$

Employing **Table A1**, it is determined that the computed value $z = 2.17$ is greater than the tabled critical two-tailed value $z_{.05} = 1.96$ and the tabled critical one-tailed value $z_{.05} = 1.65$, but less than the tabled critical two-tailed value $z_{.01} = 2.58$ and the tabled critical one-tailed value $z_{.01} = 2.33$. Thus, both the nondirectional alternative hypothesis H_1: $\rho_S \neq 0$ and the directional alternative hypothesis H_1: $\rho_S > 0$ are supported at the .05 level, but not at the .01 level. Note that identical conclusions are reached with **Table A18** and Equation 29.3, but the latter conclusions are not identical with those obtained with Equation 29.2 (where the directional alternative hypothesis H_1: $\rho_S > 0$ is also supported at the .01 level). As noted in Section V, the conclusions based on use of **Table A18**, Equation 29.2, and Equation 29.3 will not always be in total agreement.

It is noted earlier in this section that the **Friedman two-way analysis of variance by ranks** can be employed to determine whether the value of **Spearman's rho** is significant. This will now be illustrated in reference to Example 29.2. The data for Example 29.2 are rearranged in Table 29.6 to conform to the test model for the **Friedman two-way analysis of variance by ranks**. Note that the rows and columns employed in Table 29.5 are reversed in Table 29.6. When Table 29.6 is employed within the framework of the **Friedman test** model, the two critics represent $n = 2$ subjects, and the 10 ranks represent $k = 10$ levels of a within-subjects/repeated-measures independent variable.

Table 29.6 Data for Example 29.2 Formatted for Analysis with the Friedman Two-Way Analysis of Variance by Ranks

Movie	1	2	3	4	5	6	7	8	9	10
Critic 1	7	1	8	10	9	6	5	2.5	2.5	4
Critic 2	10	2	6	8	7	4	9	3	1	5
ΣR_j	17	3	14	18	16	10	14	5.5	3.5	9
$(\Sigma R_j)^2$	289	9	196	324	256	100	196	30.25	12.25	81

From the summary information in Table 29.6, the value $\sum_{j=1}^{k}(\Sigma R_j)^2 = 1493.5$ is computed.

$$\sum_{j=1}^{k}(\Sigma R_j)^2 = 289 + 9 + 196 + 324 + 256 + 100 + 196 + 30.25 + 12.25 + 81 = 1493.5$$

Employing the above value, along with the other appropriate values in Equation 25.1 (the equation for the **Friedman two-way analysis of variance by ranks**), the value $\chi_r^2 = 15.46$ is computed.[9]

$$\chi_r^2 = \frac{12}{nk(k + 1)} \sum_{j=1}^{k} (\Sigma R_j)^2 - 3n(k + 1)$$

$$= \left[\frac{12}{(2)(10)(10 + 1)} \right] [1493.5] - (3)(2)(10 + 1) = 15.46$$

The value $\chi_r^2 = 15.46$ is evaluated with **Table A4 (Table of the Chi-Square Distribution)** in the **Appendix**. For $df = k - 1 = 10 - 1 = 9$, the tabled critical two-tailed .05 and .01 values are $\chi_{.05}^2 = 16.92$ and $\chi_{.01}^2 = 21.67$, and the tabled critical one-tailed .05 and .01 values are $\chi_{.05}^2 = 14.68$ and $\chi_{.01}^2 = 19.50$ (the latter value is obtained through interpolation).[10] Employing the aforementioned critical values, the null hypothesis for the **Friedman two-way analysis of variance by ranks** (H_0: $\theta_1 = \theta_2 = \cdots = \theta_{10}$) can be rejected at the .05 level, but only if a one-tailed analysis is conducted (since $\chi_r^2 = 15.46$ is greater than the tabled critical one-tailed value $\chi_{.05}^2 = 14.68$).[11] The result falls short of being significant at the .05 level for a two-tailed analysis, since $\chi_r^2 = 15.46$ is less than the tabled critical two-tailed value $\chi_{.05}^2 = 16.92$. Rejection of the null hypothesis for the **Friedman two-way analysis of variance by ranks** is commensurate with rejection of the null hypothesis H_0: $\rho_S = 0$ for **Spearman's rank-order correlation coefficient**. In actuality, the result derived employing the **Friedman two-way analysis of variance by ranks** is similar, but not identical, to the analysis of **Spearman's rho** with **Table A18**, Equation 29.2, and Equation 29.3 (which, as noted earlier, are not in themselves in total agreement). The slight discrepancy between the results of the **Friedman test** and the more commonly employed methods for assessing the significance of **Spearman's rho** can be attributed to the fact that the test statistics based on the t, normal, and chi-square distributions are large sample approximations, which in the case of Example 29.2 are employed with a small sample size. It was also noted earlier that the values in **Table A18** are approximations of the exact values in the underlying sampling distribution.

2. Power efficiency of Spearman's rank-order correlation coefficient Daniel (1990) and Siegel and Castellan (1988) note that (for large sample sizes) the **asymptotic relative efficiency** (which is discussed in Section VII of the **Wilcoxon signed-ranks test (Test 6)**) of **Spearman's rank-order correlation coefficient** relative to the **Pearson product-moment correlation coefficient** is approximately .91 (when the assumptions underlying the latter test are met).

3. Brief discussion of Kendall's Tau: An alternative measure of association for two sets of ranks Kendall's tau (Test 30) is an alternative measure of association which can be employed to evaluate two sets of ranks. Although **Spearman's rho** and **Kendall's tau** can be employed to measure the degree of association for the same set of data, **Spearman's rho** is the more commonly described of the two measures (primarily because it requires fewer computations). A comparative discussion of **Spearman's rho** and **Kendall's tau** can be found in Section I of the latter test.

4. Weighted rank/top-down correlation There may be occasions when a researcher's primary interest is with respect to the correlation among the most extreme scores in a set of data (i.e., that group of scores which comprise the highest and lowest values for both variables). The latter can be achieved through use of a procedure (developed by Salama and Quade (1981) and Quade and Salama (1992)), which weights scores such that the more extreme a score is the greater its weight in determining the correlation coefficient. The latter procedure, which is referred to as **weighted rank correlation** or **top-down correlation** (Iman and Conover (1985, 1987)), is described in Zar (1999, pp. 398–401).

VIII. Additional Examples Illustrating the Use of Spearman's Rank-Order Correlation Coefficient

If a researcher elects to rank-order the scores of subjects in any of the examples for which the **Pearson product-moment correlation coefficient** is employed, a value can be computed for **Spearman's rank-order correlation coefficient**. Thus, as is the case for Example 28.1, the data for Examples 28.2 and 28.3 can be rank-ordered and evaluated with **Spearman's rho**. Since the rankings for the latter two examples are identical to the rankings for Example 29.1, all three examples yield the identical result. Since **Kendall's tau** and **Spearman's rho** can be employed to evaluate the same data, Example 30.1, as well as the data set presented in Table 30.4, can also be evaluated with **Spearman's rho**.

References

Brown, G. M. & Mood, A. M. (1951). On median tests for linear hypotheses, Jerzy Neyman (ed.), **Proceedings of the Second Berkeley Symposium on Mathematical Statistics and Probability**. Berkeley & Los Angeles: The University of California Press, 159–166.

Conover, W. J. (1980). **Practical nonparametric statistics** (2nd ed.). New York: John Wiley & Sons.

Conover, W. J. (1999). **Practical nonparametric statistics** (3rd ed.). New York: John Wiley & Sons.

Daniel, W. W. (1990). **Applied nonparametric statistics** (2nd ed.). Boston: PWS–Kent Publishing Company.

Edwards, A. L. (1984). **An introduction to linear regression and correlation** (2nd ed.). New York: W. H. Freeman & Company.

Franklin, L. A. (1996). Exact tables for Spearman's rank correlation coefficient for $n = 19$ and $n = 20$. Unpublished paper presented at the joint meetings, Aug 4–8, American Statistical Association, Chicago.

Howell, D. C. (2002). **Statistical methods for psychology** (5th ed.). Pacific Grove, CA.: Duxbury.

Iman, R. L. & Conover, W. J. (1985). A measure of top-down correlation. **Technical Report SAND85-0601**, Sandia National Laboratories, Albuquerque, New Mexico, 44 pp.

Iman, R. L. & Conover, W. J. (1987). A measure of top-down correlation. **Technometrics** 29, 351–357. Correction: **Technometrics**, 1989, 31, 133.

Lindeman, R. H., Merenda, P. F., & Gold, R. Z. (1980). **Introduction to bivariate and multivariate analysis**. Glenview, IL: Scott, Foresman & Company.

Marascuilo, L. A. & McSweeney, M. (1977). **Nonparametric and distribution-free methods for the social sciences**. Monterey, CA: Brooks/Cole Publishing Company.

Mood, A. M. (1950). **Introduction to the theory of statistics**. New York: McGraw–Hill Book Company.

Olds, E. G. (1938). Distribution of sum of squares of rank differences for small numbers of individuals. **Annals of Mathematical Statistics**, 9, 133–148.

Olds, E. G. (1949). The 5% significance levels of sums of squares of rank differences and a correlation. **Annals of Mathematical Statistics**, 20, 117–119.

Quade, D. & Salama, I. (1992). A survey of weighted rank correlation. In Sen, P. K. & Salama, I. (Eds.), **Order statistics and nonparametric theory and appliations** (pp. 213– 224). New York: Elsevier.

Ramsey, P. H. (1989). Critical values for Spearman's rank order correlation. **Journal of Educational Statistics**, 14, 245–253.

Salama, I. & Quade, D. (1981). A nonparametric comparison of two multiple regressions by means of a weighted measure of correlation. **Communic. Statisti. — Theor. Meth.**, A11, 1185–1195.

Siegel, S. & Castellan, N. J., Jr. (1988). **Nonparametric statistics for the behavioral sciences** (2nd ed.). New York: McGraw–Hill Book Company.

Spearman, C. (1904). The proof and measurement of association between two things. **American Journal of Psychology**, 15, 72–101.

Sprent, P. (1989). **Applied nonparametric statistical methods**. London: Chapman & Hall.

Sprent, P. (1993). **Applied nonparametric statistical methods** (2nd ed.). London: Chapman & Hall.

Theil, H. (1950). A rank–invariant method of linear and polynomial regression analysis III. **Nederl. Akad. Wetensch. Proc.**, Series A, 53, 1397–1412.

Zar, J. H. (1972). Significance testing of Spearman rank correlation coefficient. **Journal of the American Statistical Association**, 67, 578–580.

Zar, J. H. (1999). **Biostatistical analysis** (4th ed.). Upper Saddle River, NJ: Prentice Hall.

Endnotes

1. It should be noted that although the scores of subjects in Example 29.1 are ratio data, in most instances when **Spearman's rank-order correlation coefficient** is employed it is more likely the original data for both variables are in a rank-order format. As is noted in Section I, conversion of ratio data to a rank-order format (which is done in Section IV with respect to Example 29.1) is most likely to occur when a researcher has reason to believe that one or more of the assumptions underlying the **Pearson product-moment correlation coefficient** are saliently violated. Example 29.2 in Section VI represents a study involving two variables that are originally in a rank-order format for which **Spearman's rho** is computed.

2. Some sources employ the following statements as the null hypothesis and the nondirectional alternative hypothesis for **Spearman's rank-order correlation coefficient**: **Null hypothesis**: H_0: Variables X and Y are independent of one another; **Nondirectional alternative hypothesis**: H_1: Variables X and Y are not independent of one another.

   It is, in fact, true that if in the underlying population the two variables are independent, the value of ρ_S will equal zero. However, the fact that $\rho_S = 0$, in and of itself does not ensure that the variables are independent of one another. Thus, it is conceivable that in a population in which the correlation between X and Y is $\rho_S = 0$, a nonmonotonic curvilinear function can be employed to describe the relationship between the variables.

3. Daniel (1990) notes that the computed value of r_S is not an unbiased estimate of ρ_S.

4. The reader may find slight discrepancies in the critical values listed for **Spearman's rho** in the tables published in different books. The differences are due to the fact that separate tables derived by Olds (1938, 1949) and Zar (1972), which are not identical, are employed in different sources. Howell (2002) notes that the tabled critical values noted in various sources are approximations and not exact values. Ramsey (1989) and Franklin (1996) have derived critical values which they claim are more accurate than those listed in **Table A18**.

5. The minimum sample size for which Equation 29.3 is recommended varies depending upon which source one consults. Some sources recommend the use of Equation 29.3 for values as low as $n = 25$, whereas others state that n should equal at least 100.

6. The results obtained through use of **Table A18**, Equation 29.2, and Equation 29.3 will not always be in total agreement with one another. In instances where the different methods for evaluating significance do not agree, there will usually not be a major discrepancy among them. In the final analysis, the larger the sample size the more likely it is that the methods will be consistent with one another.

7. The following will always be true when Equation 28.1 is employed in computing **Pearson r** (and r_S) when the rank-orders are employed to represent the scores on the X and Y variables: $\Sigma X = \Sigma Y$ and $\Sigma X^2 = \Sigma Y^2$ (however, the latter will only be true if there are no ties).

8. The relationship between **Spearman's rank-order correlation coefficient** and **Kendall's coefficient of concordance** is discussed in greater detail in Section VII of the latter test. In the latter discussion, it is noted that although, when there are two sets of ranks, the values computed for **Spearman's rho** and **Kendall's coefficient of concordance** will not be identical, one value can be converted into the other through use of Equation 31.7.

9. If the tie correction for the **Friedman two-way analysis of variance by ranks** is employed, the computed value of χ_r^2 will be slightly higher.

10. The tabled critical two-tailed .05 and .01 chi-square values represent the chi-square values at the 95th and 99th percentiles, and the tabled critical one-tailed .05 and .01 chi-square values represent the chi-square values at the 90th and 98th percentiles.

11. In the discussion of the **Friedman two-way analysis of variance by ranks**, it is assumed that a nondirectional analysis is always conducted for the latter test. A directional/one-tailed analysis is used here in order to employ probability values which are comparable to the one-tailed values employed in evaluating **Spearman's rho**. Within the **Friedman test model**, when $k = 10$, the usage of the term one-tailed analysis is really not meaningful. For a clarification of this issue (i.e., conducting a directional analysis when $k > 3$), the reader should read the discussion on the directionality of the **chi-square goodness-of-fit test (Test 8)** in Section VII of the latter test (which can be generalized to the **Friedman test**).

Test 30

Kendall's Tau
(Nonparametric Measure of Association/Correlation Employed with Ordinal Data)

I. Hypothesis Evaluated with Test and Relevant Background Information

Kendall's tau is one of a number of measures of correlation or association discussed in this book. Measures of correlation are not inferential statistical tests, but are, instead, descriptive statistical measures which represent the degree of relationship between two or more variables. Upon computing a measure of correlation, it is common practice to employ one or more inferential statistical tests in order to evaluate one or more hypotheses concerning the correlation coefficient. The hypothesis stated below is the most commonly evaluated hypothesis for **Kendall's tau**.

Hypothesis evaluated with test In the underlying population represented by a sample, is the correlation between subjects' scores on two variables some value other than zero? The latter hypothesis can also be stated in the following form: In the underlying population represented by the sample, is there a significant **monotonic** relationship between the two variables?[1] It is important to note that the nature of the relationship described by **Kendall's tau** is based on an analysis of two sets of ranks.

Relevant background information on test Prior to reading the material in this section the reader should review the general discussion of correlation in Section I of the **Pearson product-moment correlation coefficient (Test 28)**, and the material in Section I of **Spearman's rank-order correlation coefficient (Test 29)** (which also evaluates whether a monotonic relationship exists between two sets of ranks). Developed by Kendall (1938), **tau** is a bivariate measure of correlation/association that is employed with rank-order data. The population parameter estimated by the correlation coefficient will be represented by the notation τ (which is the lower case Greek letter **tau**). The sample statistic computed to estimate the value of τ will be represented by the notation $\tilde{\tau}$. As is the case with **Spearman's rank-order correlation coefficient**, **Kendall's tau** can be employed to evaluate data in which a researcher has scores for n subjects/ objects on two variables (designated as the X and Y variables), both of which are rank-ordered. **Kendall's tau** is also commonly employed to evaluate the degree of agreement between the rankings of $m = 2$ judges for n subjects/objects.

As is the case with **Spearman's rho**, the range of possible values **Kendall's tau** can assume is defined by the limits -1 to $+1$ (i.e., $-1 \le \tilde{\tau} \le +1$). Although **Kendall's tau** and **Spearman's rho** share certain properties in common with one another, they employ a different logic with respect to how they evaluate the degree of association between two variables. **Kendall's tau** measures the degree of agreement between two sets of ranks with respect to the relative ordering of all possible pairs of subjects/objects. One set of ranks represents the ranks on the X variable, and the other set represents the ranks on the Y variable. Specifically, assume data are in the form of the following two pairs of observations expressed in a rank-order format: a) (R_{X_i}, R_{Y_i}) (which, respectively, represent the ranks on Variables X and Y for the ith subject/ object); and b) (R_{X_j}, R_{Y_j}) (which, respectively, represent the ranks on Variables X and Y for the jth subject/object). If the sign/direction of the difference ($R_{X_i} - R_{X_j}$) is the same as the sign/direction of the difference ($R_{Y_i} - R_{Y_j}$), a pair of ranks is said to be **concordant** (i.e., in agreement). If the sign/direction of the difference ($R_{X_i} - R_{X_j}$) is not the same as the sign/direction of the difference ($R_{Y_i} - R_{Y_j}$), a pair of ranks is said to be **discordant** (i.e., disagree). If ($R_{X_i} - R_{X_j}$) and/or ($R_{Y_i} - R_{Y_j}$) result in the value zero, a pair of ranks is neither concordant or discordant (and is conceptualized within the framework of a tie which is discussed in Section VI). **Kendall's tau** is a proportion which represents the difference between the proportion of concordant pairs of ranks less the proportion of discordant pairs of ranks. The computed value of **tau** will equal +1 when there is complete agreement among the rankings (i.e., all of the pairs of ranks are concordant), and will equal −1 when there is complete disagreement among the rankings (i.e., all of the pairs of ranks are discordant).

As a result of the different logic involved in computing **Kendall's tau** and **Spearman's rho**, the two measures have different underlying scales, and, because of this, it is not possible to determine the exact value of one measure if the value of the other measure is known. As a general rule, however, the computed absolute value of $\tilde{\tau}$ will always be less than the computed absolute value of r_S for a set of data, and, as the sample size increases, the ratio $\tilde{\tau}/r_S$ approaches the value .67.[2] Siegel and Castellan (1988) note the following inequality can be employed to describe the relationship between r_S and $\tilde{\tau}$: $-1 \leq (3\tilde{\tau} - 2r_S) \leq 1$.

In spite of the differences between **Kendall's tau** and **Spearman's rho**, the two statistics employ the same amount of information, and, because of this, are equally likely to detect a significant effect in a population. Thus, although for the same set of data different values will be computed for r_S and $\tilde{\tau}$ (unless, as noted in Endnote 2, the correlation between the two variables is +1 or −1), the two measures will essentially result in the same conclusions with respect to whether or not the underlying population correlation equals zero. The comparability of $\tilde{\tau}$ and r_S is discussed in more detail in Section V.

In contrast to **Kendall's tau**, **Spearman's rho** is more commonly discussed in statistics books as a bivariate measure of correlation for ranked data. Two reasons for this are as follows: a) The computations required for computing **tau** are more tedious than those required for computing **rho**; and b) When a sample is derived from a bivariate normal distribution (which is discussed in Section I of the **Pearson product-moment correlation coefficient**), the computed value r_S will generally provide a reasonably good approximation of **Pearson r**, whereas the value of $\tilde{\tau}$ will not. Since r_S provides a good estimate of r, r_S^2 can be employed to represent the **coefficient of determination** (i.e., a measure of the proportion of variability on one variable which can be accounted for by variability on the other variable).[3] One commonly cited advantage of **tau** over **rho** is that $\tilde{\tau}$ is an unbiased estimate of the population parameter τ, whereas the value computed for r_S is not an unbiased estimate of the population parameter ρ_S. Lindeman *et al.* (1980) note another advantage of **tau** is that unlike **rho**, the sampling distribution of **tau** approaches normality very quickly. Because of this, the normal distribution provides a good approximation of the exact sampling distribution of **tau** for small sample sizes. In contrast, a large sample size is required in order to employ the normal distribution to approximate the exact sampling distribution of **rho**.

II. Example

Example 30.1 *Two psychiatrists, Dr. X and Dr. Y, rank-order ten patients with respect to their level of psychological disturbance (assigning a rank of 1 to the least disturbed patient and a rank of 10 to the most disturbed patient). The rankings of the two psychiatrists (along with additional information that allows the value of* **Spearman's rho** *to be computed for the same set of data) are presented in* Table 30.1. *Is there a significant correlation between the rank-orders assigned to the patients by the two doctors?*

III. Null versus Alternative Hypotheses

Upon computing **Kendall's tau**, it is common practice to determine whether the obtained absolute value of the correlation coefficient is large enough to allow a researcher to conclude that the underlying population correlation coefficient between the two variables is some value other than zero. Section V describes how the latter hypothesis, which is stated below, can be evaluated through use of tables of critical $\tilde{\tau}$ values or through use of an inferential statistical test that is based on the normal distribution.

Null hypothesis H_0: $\tau = 0$

(In the underlying population the sample represents, the correlation between the ranks of subjects on Variable X and Variable Y equals 0.)

Alternative hypothesis H_1: $\tau \neq 0$

(In the underlying population the sample represents, the correlation between the ranks of subjects on Variable X and Variable Y equals some value other than 0. This is a **nondirectional alternative hypothesis**, and it is evaluated with a **two-tailed test**. Either a significant positive $\tilde{\tau}$ value or a significant negative $\tilde{\tau}$ value will provide support for this alternative hypothesis. In order to be significant, the obtained absolute value of $\tilde{\tau}$ must be equal to or greater than the tabled critical two-tailed $\tilde{\tau}$ value at the prespecified level of significance.)

or

H_1: $\tau > 0$

(In the underlying population the sample represents, the correlation between the ranks of subjects on Variable X and Variable Y equals some value greater than 0. This is a **directional alternative hypothesis**, and it is evaluated with a **one-tailed test**. Only a significant positive $\tilde{\tau}$ value will provide support for this alternative hypothesis. In order to be significant (in addition to the requirement of a positive $\tilde{\tau}$ value), the obtained absolute value of $\tilde{\tau}$ must be equal to or greater than the tabled critical one-tailed $\tilde{\tau}$ value at the prespecified level of significance.)

or

H_1: $\tau < 0$

(In the underlying population the sample represents, the correlation between the ranks of subjects on Variable X and Variable Y equals some value less than 0. This is a **directional alternative hypothesis**, and it is evaluated with a **one-tailed test**. Only a significant negative $\tilde{\tau}$ value will provide support for this alternative hypothesis. In order to be significant (in addition to the requirement of a negative $\tilde{\tau}$ value), the obtained absolute value of $\tilde{\tau}$ must be equal to or greater than the tabled critical one-tailed $\tilde{\tau}$ value at the prespecified level of significance.)

Note: Only one of the above noted alternative hypotheses is employed. If the alternative hypothesis the researcher selects is supported, the null hypothesis is rejected.[4]

IV. Test Computations

The data for Example 30.1 are summarized in Table 30.1. Although the last two columns of Table 30.1 are not necessary to compute the value of **Kendall's tau**, they are included to allow for the computation of **Spearman's rank-order correlation coefficient** for the same set of data (which is done is Section V).

Table 30.1   Data for Example 30.1

Patient	Rankings of Dr. X R_{X_i}	Rankings of Dr. Y R_{Y_i}	$d_i = R_{X_i} - R_{Y_i}$	d_i^2
1	7	10	-3	9
2	1	2	-1	1
3	8	6	2	4
4	10	8	2	4
5	9	7	2	4
6	6	4	2	4
7	5	9	-4	16
8	3	3	0	0
9	2	1	1	1
10	4	5	-1	1
			$\Sigma d_i = 0$	$\Sigma d_i^2 = 44$

Equation 30.1 is employed to compute the value of **Kendall's tau**

$$\tilde{\tau} = \frac{n_C - n_D}{\left[\dfrac{n(n-1)}{2}\right]}$$ **(Equation 30.1)**

Where: n_C is the number of concordant pairs of ranks
   n_D is the number of discordant pairs of ranks
   $[n(n-1)]/2$ is the total number of possible pairs of ranks

In order to employ Equation 30.1 to compute the value of **Kendall's tau**, it is necessary to determine the number of concordant versus discordant pairs of ranks. In order to do this, the data are recorded in the format employed in Table 30.2.

The first row of Table 30.2 consists of the identification number of each subject (i.e., patient). The order in which subjects are listed is based on their rank-order on the X variable (i.e., R_{X_i}). The latter set of ranks is recorded in the second row of the table. The third row lists each subject's rank-order on the Y variable (i.e., R_{Y_i}). Inspection of Table 30.2 reveals that no ties are present in the data on either the X or the Y variable. The protocol to be described in this section assumes there are no ties. The protocol for handling ties is described in Section VI. The portion of Table 30.2 which lies below the double line consists of cells in which there is either an entry of **C** or **D** (except for the number value to the left of each row). This part of the table provides information with regard to the concordant versus discordant pairs of observations for the two sets of ranks. Specifically, in each of the rows of the table that fall below the double line, the number to the left of a row is the R_{Y_i} value (i.e., rank on the Y variable) of the subject represented by the column in which that value appears. Within each row, the R_{Y_i} value is compared with those R_{Y_j} values that fall in the columns to its right. In any instance where an R_{Y_j} value in a column is larger than the R_{Y_i} value for the row, a **C** is recorded in the cell which is the intersection of that row and column. In any instance where an R_{Y_j} value in a column is smaller than the R_{Y_i} value for the row, a **D** is recorded in the cell that is the intersection of that row and column. The presence of a **C** in a cell indicates a **concordant** pair of observations, since the ordering of the ranks on both the X and Y variables for that pair of observations is in the same direction. The presence of a **D** in a cell indicates a **discordant** pair of observations, since the ordering of the ranks on both the X and Y variables for that pair of observations is in the opposite direction.

Table 30.2   Computational Table for Kendall's Tau

Subject	2	9	8	10	7	6	1	3	5	4	ΣC	ΣD
R_{X_i}	1	2	3	4	5	6	7	8	9	10		
R_{Y_i}	2	1	3	5	9	4	10	6	7	8		
	2	D	C	C	C	C	C	C	C	C	8	1
		1	C	C	C	C	C	C	C	C	8	0
			3	C	C	C	C	C	C	C	7	0
				5	C	D	C	C	C	C	5	1
					9	D	C	D	D	D	1	4
						4	C	C	C	C	4	0
							10	D	D	D	0	3
								6	C	C	2	0
									7	C	1	0
										8	0	0
									$\Sigma\Sigma C = n_C = 36$			$\Sigma\Sigma D = n_D = 9$

The last two columns of Table 30.2 contain the number of concordant (ΣC) versus discordant (ΣD) pairs of observations in each row. The value $\Sigma\Sigma C = n_C = 36$ in the last row of Table 30.2 is the sum of the column labeled ΣC. The value $\Sigma\Sigma C = n_C = 36$ represents the total number of **C** entries in the table, which is the total number of concordant pairs of observations in the data. The value $\Sigma\Sigma D = n_D = 9$ is the sum of the column labeled ΣD. The value $\Sigma\Sigma D = n_D = 9$ represents the total number of **D** entries in the table, which is the total number of discordant pairs of observations in the data (which are also referred to as **inversions**).

Substituting the values $n_C = 36$, $n_D = 9$, and $n = 10$ in Equation 30.1, the value $\tilde{\tau} = .60$ is computed.

$$\tilde{\tau} = \frac{36 - 9}{\left[\dfrac{10(10 - 1)}{2}\right]} = .60$$

The reader should note the following with respect to the sign of $\tilde{\tau}$: a) When $n_C > n_D$, the sign of $\tilde{\tau}$ will be positive; b) When $n_D > n_C$, the sign of $\tilde{\tau}$ will be negative; and c) When $n_C = n_D$, $\tilde{\tau}$ will equal zero.

Some sources employ the notation S to represent the value $(n_C - n_D)$ in the numerator of Equation 30.1. For Example 30.1, $S = 36 - 9 = 27$. If the value S is employed, the value of $\tilde{\tau}$ can be computed with Equation 30.2.

$$\tilde{\tau} = \frac{2S}{n(n-1)} = \frac{(2)(27)}{(10)(10-1)} = .60 \qquad \text{(Equation 30.2)}$$

Equation 30.3 can also be employed to compute $\tilde{\tau}$.

$$\tilde{\tau} = 1 - \frac{4n_D}{n(n-1)} = 1 - \frac{(4)(9)}{10(10-1)} = .60 \qquad \text{(Equation 30.3)}$$

If there are no tied ranks present in the data, Equation 30.3 can be employed to compute **tau** in conjunction with a less tedious method than the one based on use of Table 30.2. The alternative method (which becomes impractical when the sample size is large) involves the use of Figure 30.1

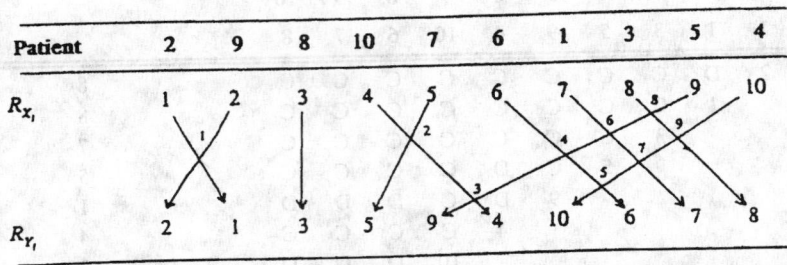

Figure 30.1　Visual Representation of Discordant Pairs of Ranks for Example 30.1

In Figure 30.1 the values of R_{X_i} and R_{Y_i} are recorded as they appear in the second and third rows of Table 30.2. Lines are drawn to connect each of the $n = 10$ corresponding values of R_{X_i} and R_{Y_i}. The total number of intersecting points in the diagram represents the number of discordant pairs of ranks or inversions in the data — i.e., the value of n_D. The value $n_D = 9$ along with the value $n = 10$ is substituted in Equation 30.3 to compute the value $\tilde{\tau} = .60$. Although not required for use in the latter equation, the number of concordant ranks (which, along with n_D, are employed in Equation 30.1) can be computed as follows: $n_C = [n(n-1)/2] - n_D$.

V.　Interpretation of the Test Results

Table 30.3　Exact Tabled Critical Values for $\tilde{\tau}$ and S for $n = 10$

	$\tilde{\tau}_{.05}/S_{.05}$	$\tilde{\tau}_{.01}/S_{.01}$
Two-tailed values	$\tilde{\tau} = .511$	$\tilde{\tau} = .644$
	$S = 23$	$S = 29$
One-tailed values	$\tilde{\tau} = .467$	$\tilde{\tau} = .600$
	$S = 21$	$S = 27$

The obtained value $\tilde{\tau} = .60$ is evaluated with **Table A19 (Table of Critical Values for Kendall's Tau)** in the **Appendix**. Note that **Table A19** lists critical values for both **tau** and S.[5] Table 30.3 lists the tabled critical two-tailed and one-tailed .05 and .01 values for **tau** and S for $n = 10$.

The following guidelines are employed in evaluating the null hypothesis.

a) If the nondirectional alternative hypothesis $H_1: \tau \neq 0$ is employed, the null hypothesis can be rejected if the obtained absolute value of $\tilde{\tau}$ (or S) is equal to or greater than the tabled critical two-tailed value at the prespecified level of significance.

b) If the directional alternative hypothesis $H_1: \tau > 0$ is employed, the null hypothesis can be rejected if the sign of $\tilde{\tau}$ (or S) is positive, and the obtained value of $\tilde{\tau}$ (or S) is equal to or greater than the tabled critical one-tailed value at the prespecified level of significance.

c) If the directional alternative hypothesis $H_1: \tau < 0$ is employed, the null hypothesis can be rejected if the sign of $\tilde{\tau}$ (or S) is negative, and the obtained absolute value of $\tilde{\tau}$ (or S) is equal to or greater than the tabled critical one-tailed value at the prespecified level of significance.

Employing the above guidelines, the nondirectional alternative hypothesis $H_1: \tau \neq 0$ is supported at the .05 level, since the computed value $\tilde{\tau} = .60$ ($S = 27$) is greater than the tabled critical two-tailed value $\tilde{\tau}_{.05} = .511$ ($S_{.05} = 23$). It is not supported at the .01 level, since $\tilde{\tau} = .60$ ($S = 27$) is less than the tabled critical two-tailed value $\tilde{\tau}_{.01} = .644$ ($S_{.01} = 29$).

The directional alternative hypothesis $H_1: \tau > 0$ is supported at the .05 level, since the computed value $\tilde{\tau} = .60$ is a positive number that is greater than the tabled critical one-tailed value $\tilde{\tau}_{.05} = .467$ ($S_{.05} = 21$). It is also supported at the .01 level, since $\tilde{\tau} = .60$ is equal to the tabled critical one-tailed value $\tilde{\tau}_{.01} = .600$ ($S_{.01} = 27$).

The directional alternative hypothesis $H_1: \tau < 0$ is not supported, since the computed value $\tilde{\tau} = .60$ (S) is a positive number.

Test 30a: Test of significance for Kendall's tau When $n > 10$, the normal distribution provides an excellent approximation of the sampling distribution of **tau**. Equation 30.4 is the normal approximation for evaluating the null hypothesis $H_0: \tau = 0$.

$$z = \frac{3\tilde{\tau}\sqrt{n(n-1)}}{\sqrt{2(2n+5)}} \qquad \textbf{(Equation 30.4)}$$

In view of the fact that the sample size $n = 10$ employed in Example 30.1 is just one subject below the minimum value generally recommended for use with Equation 30.4, the normal approximation will still provide a reasonably good approximation of the exact sampling distribution. When the appropriate values from Example 30.1 are substituted in Equation 30.4, the value $z = 2.41$ is computed.

$$z = \frac{(3)(.60)\sqrt{10(10-1)}}{\sqrt{2\,[(2)(10)+5]}} = 2.41$$

Equations 30.5 and 30.6 are alternative equations for computing the value of z which yield the identical result.[6]

$$z = \frac{\tilde{\tau}}{\sqrt{\dfrac{2(2n+5)}{9n(n-1)}}} = \frac{.60}{\sqrt{\dfrac{(2)[(2)(10)+5]}{(9)(10)(10-1)}}} = 2.41 \qquad \textbf{(Equation 30.5)}$$

$$z = \frac{S}{\sqrt{\dfrac{n(n-1)(2n+5)}{18}}} = \frac{27}{\sqrt{\dfrac{(10)(10-1)[(2)(10)+5]}{18}}} = 2.41 \text{ (Equation 30.6)}$$

The computed value $z = 2.41$ is evaluated with **Table A1 (Table of the Normal Distribution)** in the **Appendix**. In the latter table, the tabled critical two-tailed .05 and .01 values are $z_{.05} = 1.96$ and $z_{.01} = 2.58$, and the tabled critical one-tailed .05 and .01 values are $z_{.05} = 1.65$ and $z_{.01} = 2.33$. Since the sign of the z value computed with Equations 30.4–30.6 will always be the same as the sign of $\tilde{\tau}$ (and S), the guidelines which are described earlier in this section for evaluating a $\tilde{\tau}$ (or S) value can also be applied in evaluating the z value computed with Equations 30.4–30.6 (i.e., substitute z in place of $\tilde{\tau}$ (or S) in the text of the guidelines for evaluating $\tilde{\tau}$ (or S)).

Employing the guidelines, the nondirectional alternative hypothesis H_1: $\tau \neq 0$ is supported at the .05 level, since the computed value $z = 2.41$ is greater than the tabled critical two-tailed value $z_{.05} = 1.96$. It is not, however, supported at the .01 level, since $z = 2.41$ is less than the tabled critical two-tailed value $z_{.01} = 2.58$.

The directional alternative hypothesis H_1: $\tau > 0$ is supported at both the .05 and .01 levels, since the computed value $z = 2.41$ is a positive number that is greater than the tabled critical one-tailed values $z_{.05} = 1.65$ and $z_{.01} = 2.33$.

The directional alternative hypothesis H_1: $\tau < 0$ is not supported, since the computed value $z = 2.41$ is a positive number. In order for the alternative hypothesis H_1: $\tau < 0$ to be supported, the computed value of z must be a negative number (as well as the fact that the absolute value of z must be equal to or greater than the tabled critical one-tailed value at the prespecified level of significance).

Note that the results for the normal approximation are identical to those obtained when the exact values of the sampling distribution of **tau** are employed. A summary of the analysis of Example 30.1 follows: It can be concluded there is a significant monotonic increasing/ positive relationship between the rankings of the two judges.[7] The result of the analysis (based on the critical values in Table 30.3 and the normal approximation) can be summarized as follows (if it is assumed the nondirectional alternative hypothesis H_1: $\tau \neq 0$ is employed): $\tilde{\tau} = .60, p < .05$.

It is noted in Section I that if both **Kendall's tau** and **Spearman's rho** are computed for the same set of data, the two measures will result in essentially the same conclusions with respect to whether the value of the underlying population correlation equals zero. In order to demonstrate this, employing the relevant values from Table 30.1 in Equation 29.1, the value $r_S = .733$ is computed for Example 30.1.[8]

$$r_S = 1 - \frac{6\Sigma d^2}{n(n^2-1)} = 1 - \frac{6(44)}{10[(10)^2-1]} = .733$$

Note that the values $\tilde{\tau} = .60$ and $r_S = .733$ computed for Example 30.1 are not identical to one another, and, that as noted in Section I, the absolute value of $\tilde{\tau}$ is less than the absolute value of r_S. It is also the case that the result is consistent with the inequality $-1 \leq (3\tilde{\tau} - 2r_S) \leq 1$ (which is noted in Section I), since $(3)(.60) - (2)(.733) = .334$ (which falls within the range -1 to $+1$).

The computed value $r_S = .733$ is evaluated with **Table A18 (Table of Critical Values for Spearman's Rho)** in the **Appendix**. Employing the latter table, it is determined that for $n = 10$, the tabled critical two-tailed .05 and .01 values are $r_{S_{.05}} = .648$ and $r_{S_{.01}} = .794$, and the tabled critical one-tailed .05 and .01 values are $r_{S_{.05}} = .564$ and $r_{S_{.01}} = .745$. Employing the aforementioned critical values, the nondirectional alternative hypothesis $H_1: \rho_S \neq 0$ and the directional alternative hypothesis $H_1: \rho_S > 0$ are supported at the .05 level, since the computed value $r_S = .733$ is greater than the tabled critical two-tailed value $r_{S_{.05}} = .648$ and the tabled critical one-tailed value $r_{S_{.05}} = .564$. The alternative hypotheses are not supported at the .01 level, since $r_S = .733$ is less than the tabled critical two-tailed value $r_{S_{.01}} = .794$ and the tabled critical one-tailed value $r_{S_{.01}} = .745$. This result is almost identical to that obtained when **Kendall's tau** is employed (although in the analysis for **Kendall's tau**, the directional alternative hypothesis $H_1: \tau > 0$ is supported at the .01 level).

If the computed value $r_S = .733$ is evaluated with Equation 29.2, the value $t = 3.05$ is computed.

$$t = \frac{r_S \sqrt{n-2}}{\sqrt{1 - r_S^2}} = \frac{.733\sqrt{10-2}}{\sqrt{1-(.733)^2}} = 3.05$$

The t value computed with Equation 29.2 is evaluated with **Table A2 (Table of Student's t Distribution)** in the **Appendix**. The degrees of freedom employed are $df = n - 2$. Employing **Table A2**, it is determined that for $df = 10 - 2 = 8$, the tabled critical two-tailed .05 and .01 values are $t_{.05} = 2.31$ and $t_{.01} = 3.36$, and the tabled critical one-tailed .05 and .01 values are $t_{.05} = 1.86$ and $t_{.01} = 2.90$. Employing the aforementioned critical values, the nondirectional alternative hypothesis $H_1: \rho_S \neq 0$ is supported at the .05 level, since the computed value $t = 3.05$ is greater than the tabled critical two-tailed value $t_{.05} = 2.31$. It is not supported at the .01 level, since $t = 3.05$ is less than $t_{.01} = 3.36$. The directional alternative hypothesis $H_1: \rho_S > 0$ is supported at both the .05 and .01 levels, since the computed value $t = 3.05$ is a positive number (since $r_S = .733$ is a positive number) that is greater than the tabled critical one-tailed values $t_{.05} = 1.86$ and $t_{.01} = 2.90$. This result is identical to that obtained when **Kendall's tau** is employed.

The slight discrepancies between the various methods for assessing the significance of $\tilde{\tau}$ and r_S can be attributed to the fact that the values in **Table A18** and the result of Equation 29.2 are approximations of the exact sampling distribution of **Spearman's rho**, as well as the fact that the use of the normal distribution for assessing the significance of **tau** also represents an approximation of an exact sampling distribution. However, for the most part, regardless of whether one elects to compute $\tilde{\tau}$ or r_S as the measure of association for Example 30.1, it will be concluded that the population correlation is some value other than zero, and the latter conclusion will be reached irrespective of whether a nondirectional or directional alternative hypothesis is employed.

VI. Additional Analytical Procedures for Kendall's Tau and/or Related Tests

1. Tie correction for Kendall's tau When one or more ties are present in a set of data, it is necessary to employ a tie correction in order to compute the value of $\tilde{\tau}$. To illustrate how ties

Table 30.4 Computational Table for Kendall's Tau Involving Ties

Subject	2	9	8	10	7	6	1	3	5	4	ΣC	ΣD
R_{X_i}	1.5	1.5	3	4	5	6	7	8	9	10		
R_{Y_i}	2	1	3.5	3.5	9.5	9.5	5	6	7	8		

	2	0*	C	C	C	C	C	C	C	C	8	0
		1	C	C	C	C	C	C	C	C	8	0
			3.5	0	C	C	C	C	C	C	6	0
				3.5	C	C	C	C	C	C	6	0
					9.5	0	D	D	D	D	0	4
						9.5	D	D	D	D	0	4
							5	C	C	C	3	0
								6	C	C	2	0
									7	C	1	0
										8	0	0

$$\Sigma\Sigma C = n_C = 34 \qquad \Sigma\Sigma D = n_D = 8$$

are handled, let us assume that Table 30.4 summarizes the data for Example 30.1. Note that in contrast to Table 30.2, the data in Table 30.4 are characterized by the presence of ties on both the X and Y variables. Specifically, Subjects 2 and 9 are tied for the first ordinal position on the X variable, Subjects 8 and 10 are tied for the third ordinal position on the Y variable, and Subjects 6 and 7 are tied for the ninth ordinal position on the Y variable.

As is the case for Table 30.2, the entries **C** and **D** are employed in the cells of Table 30.4 to indicate concordant versus discordant pairs of ranks. There are, however, three cells in Table 30.4 that involve tied ranks in which the cell entry is **0**. Note that if the R_{Y_i} value for a row is equal to an R_{Y_j} which falls in a column to its right, a **0** is written in the cell that is the intersection of that row and column. Since, however, this protocol only takes into account ties on the Y variable, it will not allow one to identify all of the cells in the table for which **0** is the appropriate entry. In point of fact, a **0** entry should also appear in any cell which involves a pair of tied observations on the X variable, even though the two ranks on the Y variable with which the X variable pair is being contrasted are not tied. In the above example there is just one set of ties on the X variable (the rank of 1.5 for Subjects 2 and 9). Note that the cell identified with an asterisk in the upper left of the table has a **0** entry, even though the rank-order directly to the right of the value $R_{Y_2} = 2$ (which is the rank of Subject 2 on the Y variable) is $R_{Y_9} = 1$ (which is the rank of Subject 9 on the Y variable). If the protocol described for Table 30.2 is employed, since the rank-order $R_{Y_9} = 1$ to the right of $R_{Y_2} = 2$ is less than the latter value, a **D** should be placed in that cell. The reason for employing a **0** in the cell is that if the arrangement of the ranks on the X variable is reversed, with Subject 9 listed first and Subject 2 listed second, the value of R_{Y_i} for that row will be $R_{Y_9} = 1$, and the first rank/R_{Y_j} value that it will be compared with will be $R_{Y_2} = 2$. If the latter arrangement is employed in conjunction with the protocol described for Table 30.2, the appropriate entry for the cell under discussion is a **C**. Thus, whenever a different arrangement of the tied ranks on the X variable will result in a different letter entry for a cell (i.e., **C** versus **D**), that cell is assigned a **0**.

A general protocol for determining whether a cell should be assigned a **0** to represent a tie can be summarized as follows: a) If the R_{Y_i} value at the left of a row is tied with an R_{Y_j} value which falls in a column to its right, a **0** should be placed in the cell that is the intersection of that row and column; and b) If there is a tie between the values of R_{X_i} and R_{X_j} that fall directly above the values of R_{Y_i} and R_{Y_j} being compared, a 0 should be placed in the cell that is the intersection of the row and column where the values R_{Y_i} and R_{Y_j} (as well as R_{X_j}) appear.

It should be noted that when there are no ties present in the data, $n_C + n_D = [n(n-1)]/2$. Thus, in the case of Table 30.2, $(n_C = 36) + (n_D = 9) = [(10)(10-1)]/2 = 45$. When, on the other hand, ties are present in the data, since entries of 0 are not counted as either concordant or discordant pairs, $n_C + n_D \neq [n(n-1)]/2$. The latter can be confirmed by the fact that in Table 30.4, $(n_C = 34) + (n_D = 8) \neq [(10)(10-1)]/2$.

The computation of the value of **Kendall's tau** using the tie correction will now be described. In the example under discussion there is $s = 1$ set of ties involving the ranks of subjects' X scores (Subjects 2 and 9), and $s = 2$ sets of ties involving the ranks of subjects' Y scores (Subjects 8 and 10; Subjects 6 and 7). Equation 30.9 is employed to compute the tie-corrected value of **Kendall's tau**, which will be represented by the notation $\tilde{\tau}_c$. Note that the values ΣT_X and ΣT_Y in Equation 30.9 are computed with Equations 30.7 and 30.8. In Equation 30.7, $t_{i_{(X)}}$ represents the number of X scores that are tied for a given rank. In Equation 30.8, $t_{i_{(Y)}}$ represents the number of Y scores that are tied for a given rank. The notations $\sum_{i=1}^{s}(t_{i_{(X)}}^2 - t_{i_{(X)}})$ and $\sum_{i=1}^{s}(t_{i_{(Y)}}^2 - t_{i_{(Y)}})$ indicate that the following is done with respect to each of the variables: a) For each set of ties, the number of ties in the set is subtracted from the square of the number of ties in that set; and b) The sum of all the values computed in part a) is obtained for that variable.

When the data from Table 30.4 are substituted in Equations 30.7–30.9, the tie-corrected value $\tilde{\tau}_c = .598$ is computed.[9]

$$T_X = \sum_{i=1}^{s}(t_{i_{(X)}}^2 - t_{i_{(X)}}) = [(2)^2 - 2] = 2 \qquad \textbf{(Equation 30.7)}$$

$$T_Y = \sum_{i=1}^{s}(t_{i_{(Y)}}^2 - t_{i_{(Y)}}) = [(2)^2 - 2] + [(2)^2 - 2] = 4 \qquad \textbf{(Equation 30.8)}$$

$$\tilde{\tau}_c = \frac{2(n_C - n_D)}{\sqrt{n(n-1) - T_X}\sqrt{n(n-1) - T_Y}}$$

(Equation 30.9)

$$= \frac{(2)(34 - 8)}{\sqrt{10(10-1) - 2}\sqrt{10(10-1) - 4}} = .598$$

Since $n = 10$, we can employ the critical values in Table 30.3. The nondirectional alternative hypothesis $H_1: \tau \neq 0$ is supported at the .05 level, since the computed value $\tilde{\tau}_c = .598$ is greater than the tabled critical two-tailed value $\tilde{\tau}_{.05} = .511$. It is not supported at the .01 level, since $\tilde{\tau}_c = .598$ is less than the tabled critical two-tailed value $\tilde{\tau}_{.01} = .644$. The directional alternative hypothesis $H_1: \tau > 0$ is supported at the .05 level, since $\tilde{\tau}_c = .598$ is a positive number that is greater than the tabled critical one-tailed value $\tilde{\tau}_{.05} = .467$. It is not supported at the .01 level, since $\tilde{\tau}_c = .598$ is less (albeit barely) than the tabled critical one-tailed value $\tilde{\tau}_{.01} = .600$.

Additional discussion on handling ties can be found in Hollander and Wolfe (1999) who also provide a more in-depth discussion of other issues relating to **Kendall's tau**.

2. Regression analysis and Kendall's tau As noted in the discussion of **Spearman's rank-order correlation coefficient**, regression analysis procedures have been developed for rank-order data. Sources for nonparametric regression analysis (including monotonic regression analysis) are cited in Section VI of the latter test.

3. Partial rank correlation The computation of a **partial correlation coefficient** (described in Section IV/V of the chapter on **multiple regression (Test 33)**, and discussed briefly in Section VI of **Spearman's rank-order correlation coefficient**) can be extended to **Kendall's tau**. Thus, when the rank-orders for three variables are evaluated, Equation 33.33 can be employed to compute one or more **partial correlation coefficients** for **Kendall's tau** (employing the relevant values of $\tilde{\tau}$ in the equation). Conover (1980, 1999), Daniel (1990), Marascuilo and McSweeney (1977), and Siegel and Castellan (1988) discuss the computation of a **partial correlation coefficient** in reference to **Kendall's tau**. It should be noted that the partial rank-order correlation coefficient for **Kendall's tau** employs a different sampling distribution from the one which is employed for evaluating $\tilde{\tau}$. Tables for the appropriate sampling distribution can be found in Daniel (1990) and Siegel and Castellan (1988).

4. Sources for computing a confidence interval for Kendall's tau A procedure (attributed to Noether (1967)) for deriving a confidence interval for **Kendall's tau** is described in Daniel (1990).

VII. Additional Discussion of Kendall's Tau

1. Power efficiency of Kendall's tau Daniel (1990) and Siegel and Castellan (1988) note that (for large sample sizes) the **asymptotic relative efficiency** (which is discussed in Section VII of the **Wilcoxon signed-ranks test (Test 6)**) of **Kendall's tau** relative to the **Pearson product-moment correlation coefficient** is approximately .91 (when the assumptions underlying the latter test are met).

2. Kendall's coefficient of agreement **Kendall's coefficient of agreement** is another measure of association which allows a researcher to evaluate the degree of agreement between m sets of ranks on n subjects/objects. The latter measure, which is described in Siegel and Castellan (1988), is essentially an extension of **Kendall's tau** to more than two sets of ranks. The relationship between **Kendall's tau** and **Kendall's coefficient of agreement** is analogous to the relationship between **Spearman's rho** and **Kendall's coefficient of concordance (Test 31)**.

VIII. Additional Examples Illustrating the Use of Kendall's Tau

Since **Spearman's rho** and **Kendall's tau** can be employed to evaluate the same data, Examples 29.1 and 29.2, as well as the data set presented in Tables 29.2/29.4, can be evaluated with **Kendall's tau**. It is also the case that if a researcher elects to rank-order the scores of subjects in any of the examples for which the **Pearson product-moment correlation coefficient** is employed, a value can be computed for **Kendall's tau**. To illustrate this, Example 28.1 (which is identical to Example 29.1) will be evaluated with **Kendall's tau**. The rank-orders of the scores of subjects on the X and Y variables in Examples 28.1/29.1 are arranged in Figure 30.2. The arrangement of the ranks in Figure 30.2 allows for use of the protocol for determining the number of discordant pairs of ranks which is described in reference to Figure 30.1. Since none of the

Subject	2	3	5	4	1
R_{X_i}	1	2	3	4	5
	↓	↓	↓	↓	↓
R_{Y_i}	1	2	3	4	5

Figure 30.2   Visual Representation of Discordant Pairs of Ranks
for Examples 28.1/29.1

vertical lines intersect, the number of pairs of discordant ranks is $n_D = 0$. Since each subject has the identical rank on both the X and the Y variables, all of the pairs of ranks are concordant. The total number of pairs of ranks is $[(5)(5 - 1)]/2 = 10$, which is also the value of n_C.

Employing the values $n = 5$ and $n_D = 0$ in Equation 30.3, the value $\tilde{\tau} = 1$ is computed. The same value can also be computed with either Equation 30.1 or Equation 30.2, if the values $n_C = 10$ and/or $S = 10$ are employed in the aforementioned equations.

$$\tilde{\tau} = 1 - \frac{(4)(0)}{5(5 - 1)} = 1$$

$\tilde{\tau} = 1$ is identical to the value $r_S = 1$ computed for the same set of data. As noted in Section I, when there is a perfect positive or negative correlation between the variables, identical values are computed for $\tilde{\tau}$ and r_S.

References

Conover, W. J. (1980). **Practical nonparametric statistics** (2nd ed.). New York: John Wiley & Sons.

Conover, W. J. (1999). **Practical nonparametric statistics** (3rd ed.). New York: John Wiley & Sons.

Daniel, W. W. (1990). **Applied nonparametric statistics** (2nd ed.). Boston: PWS–Kent Publishing Company.

Hollander, M. & Wolfe, D. A. (1999). **Nonparametric statistical methods**. New York: John Wiley & Sons.

Howell, D. C. (2002). **Statistical methods for psychology** (5th ed.). Pacific Grove, CA: Duxbury Press.

Kendall, M. G. (1938). A new measure of rank correlation. **Biometrika**, 30, 81–93.

Kendall, M. G. (1952). **The advanced theory of statistics** (Vol. 1). London: Charles Griffin & Co. Ltd.

Kendall, M. G. (1970). **Rank correlation methods** (4th ed.). London: Charles Griffin & Co. Ltd.

Lindeman, R. H., Meranda, P. F., & Gold, R. Z. (1980). **Introduction to bivariate and multivariate analysis**. Glenview, IL: Scott, Foresman & Company.

Marascuilo, L. A. & McSweeney, M. (1977). **Nonparametric and distribution-free methods for the social sciences**. Monterey, CA: Brooks/Cole Publishing Company.

Noether, G. E. (1967). **Elements of nonparametric statistics**. New York: John Wiley & Sons.

Siegel, S. & Castellan, N. J., Jr. (1988). **Nonparametric statistics for the behavioral sciences** (2nd ed.). New York: McGraw–Hill Book Company.

Sprent, P. (1989). **Applied nonparametric statistics**. London: Chapman & Hall.

Sprent, P. (1993). **Applied nonparametric statistics** (2nd ed.). London: Chapman & Hall.

Endnotes

1. A discussion of monotonic relationships can be found in Section I of **Spearman's rank-order correlation coefficient**.

2. The exception to this is that when the computed value of $\tilde{\tau}$ is either +1 or -1, the identical value will be computed for r_S.

3. The **coefficient of determination** is discussed in Section V of the **Pearson product-moment correlation coefficient**.

4. a) Some sources employ the following statements as the null hypothesis and the nondirectional alternative hypothesis for **Kendall's tau**: **Null hypothesis**: H_0: Variables X and Y are independent of one another; **Nondirectional alternative hypothesis**: H_1: Variables X and Y are not independent of one another.

    It is, in fact, true that if in the underlying population the two variables are independent, the value of τ will equal zero. However, the fact that $\tau = 0$, in and of itself does not ensure that the variables are independent of one another. Thus, it is conceivable that in a population in which the correlation between X and Y is $\tau = 0$, a nonmonotonic curvilinear function can be employed to describe the relationship between the variables.

    b) Note that in Example 30.1 the scores of subjects (who are the patients) on the X and Y variables are the respective ranks assigned to the subjects/patients by Dr. X and Dr. Y. Thus, the null hypothesis can also be stated as follows: In the underlying population the sample of subjects/patients represents, the correlation between the rankings of Dr. X and Dr. Y equals 0.

5. If either of the two values $\tilde{\tau}$ or S is known, Equation 30.2 can be employed to compute the other value. Some sources only list critical values for one of the two values $\tilde{\tau}$ or S.

6. The following should be noted with respect to Equations 30.5 and 30.6: a) The denominator of Equation 30.5 is the standard deviation of the sampling distribution of the normal approximation of **tau**; and b) Based on a recommendation by Kendall (1970), Marascuilo and McSweeney (1977) (who employ Equation 30.6) describe the use of a correction for continuity for the normal approximation. In employing the correction for continuity with Equation 30.6, when S is a positive number, the value 1 is subtracted from S, and when S is a negative number, the value 1 is added to S. The correction for continuity (which is not employed by most sources) reduces the absolute value of z, thus resulting in a more conservative test. The rationale for employing a correction for continuity for a normal approximation of a sampling distribution is discussed in Section VI of the **Wilcoxon signed-ranks test**.

7. Howell (2002) notes that the value $\tilde{\tau}$ = .60 indicates that if a pair of subjects is randomly selected, the likelihood the pair will be ranked in the same order is .60 higher than the likelihood they will be ranked in the reverse order.

8. The data for Examples 30.1 and 29.2 are identical, except for the fact that in the latter example there is a tie for the X score in the second ordinal position which involves the X scores in the eighth and ninth rows.

9.   If Equation 30.1 is employed to compute the value of $\tilde{\tau}$ for the data in Table 30.4, the value $\tilde{\tau} = .578$ is computed. Note that the tie correction yields a value for **tau** ($\tilde{\tau}_c = .598$) which is slightly larger than the uncorrected value. As noted in the text, because of the presence of ties, $n_C + n_D \neq [n(n-1)]/2$.

$$\tilde{\tau} = \frac{34 - 8}{\left[\dfrac{10(10-1)}{2}\right]} = .578$$

Test 31

Kendall's Coefficient of Concordance
(Nonparametric Measure of Association/Correlation Employed with Ordinal Data)

I. Hypothesis Evaluated with Test and Relevant Background Information

Kendall's coefficient of concordance is one of a number of measures of correlation or association discussed in this book. Measures of correlation are not inferential statistical tests, but are, instead, descriptive statistical measures which represent the degree of relationship between two or more variables. Upon computing a measure of correlation, it is common practice to employ one or more inferential statistical tests in order to evaluate one or more hypotheses concerning the correlation coefficient. The hypothesis stated below is the most commonly evaluated hypothesis for **Kendall's coefficient of concordance**.

Hypothesis evaluated with test In the underlying population represented by a sample, is the correlation between m sets of ranks some value other than zero? The latter hypothesis can also be stated in the following form: In the underlying population represented by the sample, are m sets of ranks independent of one another?

Relevant background information on test Developed independently by Kendall and Babington–Smith (1939) and Wallis (1939), **Kendall's coefficient of concordance** is a measure of correlation/association that is employed for three or more sets of ranks. Specifically, **Kendall's coefficient of concordance** is a measure which allows a researcher to evaluate the degree of agreement between m sets of ranks for n subjects/objects (which is often referred to as **interjudge reliability**). The population parameter estimated by the correlation coefficient will be represented by the notation W. The sample statistic computed to estimate the value of W will be represented by the notation $\tilde{W}$. The range of possible values within which **Kendall's coefficient of concordance** may fall is $0 \leq \tilde{W} \leq +1$. When there is complete agreement among all m sets of ranks, the value of $\tilde{W}$ will equal 1.[1] When, on the other hand, there is no pattern of agreement among the m sets of ranks, $\tilde{W}$ will equal 0. The value of $\tilde{W}$ cannot be a negative number, since when there are more than two sets of ranks it is not possible to have complete disagreement among all the sets. Because of this, it becomes meaningless to use a negative correlation to describe the degree of association in the data when $m \geq 3$.

It is important to note that **Kendall's coefficient of concordance** is related to both **Spearman's rank-order correlation coefficient (Test 29)** and **Friedman's two-way analysis of variance by ranks (Test 25)**. Specifically: a) The computed value of $\tilde{W}$ for m sets of ranks is linearly related to the average value of **Spearman's rho** which can be computed for all possible pairs of ranks. The relationship between **Kendall's coefficient of concordance** and **Spearman's rank-order correlation coefficient** is discussed in greater detail in Section VII. It should be noted that although **Kendall's coefficient of concordance** can be computed for two sets of ranks, in practice it is not. The latter can be attributed to the fact that in contrast to

Spearman's rho and **Kendall's tau (Test 30)** (which are the measures of association that are employed with two sets of ranks), the value of $\tilde{W}$ cannot be a negative number (which in the case of **Spearman's rho** and **Kendall's tau** indicates the presence of an inverse relationship). Because the measures of association that are employed with two sets of ranks can assume a negative value, $\tilde{W}$ is not directly comparable to them;[2] and b) Although they were developed independently, **Kendall's coefficient of concordance** and **Friedman's two-way analysis of variance by ranks** are based on the same mathematical model. Because of this, for a given set of data the values computed for χ_r^2 (which is the **Friedman test** statistic) and $\tilde{W}$ can be algebraically derived from one another. The relationship between **Kendall's coefficient of concordance** and **Friedman's two-way analysis of variance by ranks** is discussed in Section VII.

As noted earlier, **Kendall's coefficient of concordance** is most commonly employed to assess **interjudge reliability**. In Section VI of the **single-factor within-subjects analysis of variance (Test 24)**, under the discussion of the **intraclass correlation coefficient (Test 24i)**, it was noted that **interjudge reliability** is the degree to which two or more judges are in agreement with one another. It was also noted in the latter discussion that a number of measures of association which can be used to assess interjudge reliability are described in this book. Specifically: a) **Cohen's kappa (Test 16k)** is described in reference to categorical data when two judges are involved; b) **Spearman's rank-order correlation coefficient** and **Kendall's tau** are described in reference to data that are rank-ordered by two judges; c) The **Pearson product-moment correlation coefficient** is described in reference to interval/ratio data when two judges are involved; d) **Kendall's coefficient of concordance** is described in reference to data that are rank-ordered by more than two judges; and e) The **intraclass correlation coefficient** can be employed to assess interjudge reliability in reference to interval/ratio data when more than two judges are involved. In the aforementioned discussion of interjudge reliability, it was noted that unlike all of the other previously noted measures of association which evaluate the **relative differences** between the rankings or ratings of judges, the **intraclass correlation coefficient** evaluates the **absolute differences** between the ratings of judges. Because of the latter, when the ratings of m judges for n subjects/objects are evaluated with the **intraclass correlation coefficient**, the correlation between the ratings will often be lower than the correlation computed for the same set of data when they are rank-ordered and subsequently evaluated with **Kendall's coefficient of concordance**. A more detailed discussion of the difference between the **intraclass correlation coefficient** and **Kendall's coefficient of concordance** can be found in Section VII.

II. Example

Example 31.1 *Six instructors at an art institute rank four students with respect to artistic ability. A rank of 1 is assigned to the student with the highest level of ability and a rank of 4 to the student with the lowest level of ability. The rankings of the six instructors for the four students are summarized in* Table 31.1. *Is there a significant association between the rank-orders assigned to the four students by the six instructors?*

III. Null versus Alternative Hypotheses

Upon computing **Kendall's coefficient of concordance**, it is common practice to determine whether the obtained value of the correlation coefficient is large enough to allow a researcher to conclude that the underlying population correlation coefficient between the m sets of ranks is some value other than zero. Section V describes how the latter hypothesis, which is stated below, can be evaluated through use of tables of critical $\tilde{W}$ values or through use of an inferential statistical test that is based on the chi-square distribution.

Null hypothesis $\qquad\qquad\qquad\qquad H_0: W = 0$

(In the underlying population the sample represents, the correlation between the $m = 6$ sets of ranks equals 0.)

Alternative hypothesis $\qquad\qquad\qquad H_1: W \neq 0$

(In the underlying population the sample represents, the correlation between the $m = 6$ sets of ranks equals some value other than 0. This is equivalent to stating that the $m = 6$ sets of ranks are not independent of one another. When there are more than two sets of ranks, the alternative hypothesis will always be stated **nondirectionally**.[3] In order to be significant, the obtained value of $\tilde{W}$ must be equal to or greater than the tabled critical value of $\tilde{W}$ at the prespecified level of significance.)

IV. Test Computations

The data for Example 31.1 are summarized in Table 31.1. Note that in Table 31.1 there are $m = 6$ instructors, who are represented by the six rows, and $n = 4$ students who are represented by the four columns.

Table 31.1   Data for Example 31.1

Instructor	Student				Totals
	1	2	3	4	
1	3	2	1	4	
2	3	2	1	4	
3	3	2	1	4	
4	4	2	1	3	
5	3	2	1	4	
6	4	1	2	3	
ΣR_j	20	11	7	22	$T = 60$
$(\Sigma R_j)^2$	400	121	49	484	$U = 1054$

The summary values $T = 60$ and $U = 1054$ in Table 31.1 are computed as follows.

$$T = \sum_{j=1}^{n} (\Sigma R_j) = \Sigma R_1 + \Sigma R_2 + \Sigma R_3 + \Sigma R_4 = 20 + 11 + 7 + 22 = 60$$

$$U = \sum_{j=1}^{n} (\Sigma R_j)^2 = (\Sigma R_1)^2 + (\Sigma R_2)^2 + (\Sigma R_3)^2 + (\Sigma R_4)^2$$

$$= (20)^2 + (11)^2 + (7)^2 + (22)^2 = 400 + 121 + 49 + 484 = 1054$$

The **coefficient of concordance** is a ratio of the variance of the sums of the ranks for the subjects (i.e., the variance of the ΣR_j values) divided by the maximum possible value which can be computed for the variance of the sums of the ranks (for the relevant values of m and n). Equation 31.1 summarizes the definition of $\tilde{W}$.

$$\tilde{W} = \frac{\text{Variance of } \Sigma R_j \text{ values}}{\text{Maximum possible variance for } \Sigma R_j} \qquad \textbf{(Equation 31.1)}$$
$$\text{values for relevant values of } m \text{ and } n$$

The variance of the ΣR_j values (which is represented by the notation S) is computed with Equation 31.2.

$$S = \frac{nU - (T)^2}{n} \qquad \textbf{(Equation 31.2)}$$

Substituting the appropriate values from Example 31.1 in Equation 31.2, the value $S = 154$ is computed.

$$S = \frac{(4)(1054) - (60)^2}{4} = 154$$

$\tilde{W}$ is computed with Equation 31.3. The denominator of Equation 31.3 (which for Example 31.1 equals 180) represents the maximum possible value that can be computed for the variance of the sums of the ranks. The only time the value of S will equal the value of the denominator of Equation 31.3 (thus resulting in the value $\tilde{W} = 1$) will be when there is perfect agreement among the m judges with respect to their rankings of the n subjects.

$$\tilde{W} = \frac{S}{\left(\dfrac{m^2 n(n^2 - 1)}{12} \right)} \qquad \textbf{(Equation 31.3)}$$

Substituting the appropriate values in Equation 31.3, the value $\tilde{W} = .856$ is computed.

$$\tilde{W} = \frac{154}{\left(\dfrac{(6)^2(4)[(4)^2 - 1]}{12} \right)} = .856$$

Equation 31.4 is an alternative computationally quicker equation for computing the value of $\tilde{W}$. Equation 31.4, however, does not allow for the direct computation of S. The latter fact is noted, since some of the tables employed to evaluate whether $\tilde{W}$ is significant list critical values for S rather than critical values for $\tilde{W}$.

$$\textbf{(Equation 31.4)}$$

$$\tilde{W} = \frac{12U - 3m^2 n(n + 1)^2}{m^2 n(n^2 - 1)} = \frac{(12)(1054) - (3)(6)^2(4)(4 + 1)^2}{(6)^2(4)[(4)^2 - 1]} = .856$$

The fact the value of $\tilde{W}$ is close to 1 indicates there is a high degree of agreement among the six instructors with respect to how they rank the four students.

V. Interpretation of the Test Results

The obtained value $\tilde{W} = .856$ is evaluated with **Table A20 (Table of Critical Values for Kendall's Coefficient of Concordance)** in the **Appendix**. Note that Table **A20** lists critical values for both $\tilde{W}$ and S. The S values in **Table A20** are extracted from Friedman (1940), and the values of $\tilde{W}$ were computed by substituting the appropriate value of S in Equation 31.3. In order to reject the null hypothesis, the computed value of $\tilde{W}$ (or S) must be equal to or greater than the tabled critical value at the prespecified level of significance. For $m = 6$ and $n = 4$, the tabled critical .05 and .01 values for $\tilde{W}$ (S) in **Table A20** are $\tilde{W}_{.05} = .421$ ($S_{.05} = 75.7$) and $\tilde{W}_{.01} = .553$ ($S_{.01} = 99.5$). Since the computed value $\tilde{W} = .856$ ($S = 154$) is greater than all of the aforementioned critical values, the alternative hypothesis H_1: $W \neq 0$ is supported at both the .05 and .01 levels.

Test 31a: Test of significance for Kendall's coefficient of concordance When exact tables for $\tilde{W}$ (or S) are not available, the chi-square distribution provides a reasonably good approximation of the sampling distribution of $\tilde{W}$. The chi-square approximation of the sampling distribution of $\tilde{W}$ is computed with Equation 31.5. The degrees of freedom employed for Equation 31.5 are $df = n - 1$.

$$\chi^2 = m(n - 1)\tilde{W} \qquad \textbf{(Equation 31.5)}$$

When the appropriate values from Example 31.1 are substituted in Equation 31.5, the value $\chi^2 = 15.41$ is computed.

$$\chi^2 = (6)(4 - 1)(.856) = 15.41$$

The value $\chi^2 = 15.41$ is evaluated with **Table A4 (Table of the Chi-Square Distribution)** in the **Appendix**. In order to reject the null hypothesis, the obtained value of χ^2 must be equal to or greater than the tabled critical value at the prespecified level of significance. For $df = 4 - 1 = 3$, the tabled critical values are $\chi^2_{.05} = 7.81$ and $\chi^2_{.01} = 11.34$ (which are the chi-square values at the 95th and 99th percentiles). Since $\chi^2 = 15.41$ is greater than both of the aforementioned critical values, the alternative hypothesis H_1: $W \neq 0$ is supported at both the .05 and .01 levels.

For small sample sizes, the exact sampling distribution of the **Friedman two-way analysis of variance by ranks** (which, as noted in Section I, is mathematically equivalent to **Kendall's coefficient of concordance**) can be employed to evaluate the significance of $\tilde{W}$. In addition, when the values of m and n are reasonably small, some sources (e.g., Marascuilo and McSweeney (1977) and Siegel and Castellan (1988)) evaluate the significance of $\tilde{W}$ by employing an adjusted chi-square value (discussed in Section VII of the **Friedman two-way analysis of variance by ranks**) which represents an exact value for the underlying sampling distribution. For $m = 6$ and $n = 4$, the adjusted/exact .05 and .01 critical values are $\chi^2_{r_{.05}} = 7.60$ and $\chi^2_{r_{.01}} = 10.00$ (which are reasonably close to the values $\chi^2_{.05} = 7.81$ and $\chi^2_{.01} = 11.34$). Since the computed value $\chi^2 = 15.41$ is greater than both of the aforementioned critical values, the alternative hypothesis H_1: $W \neq 0$ is supported at both the .05 and .01 levels. Thus, regardless of which tables are employed to evaluate the results of Example 31.1, the alternative hypothesis H_1: $W \neq 0$ is supported at both the .05 and .01 levels. Consequently, one can conclude there is a significant association among the six instructors with respect to how they rank the four students.

VI. Additional Analytical Procedures for Kendall's Coefficient of Concordance and/or Related Tests

1. Tie correction for Kendall's coefficient of concordance When ties are present in a set of data, some sources recommend that the value of $\tilde{W}$ computed with Equations 31.3/31.4 be adjusted. Unless there is an excessive number of ties, the difference between the value of $\tilde{W}$ computed with Equations 31.3/31.4 and the value computed with the tie correction will be minimal. The tie correction, which results in a slight increase in the value of $\tilde{W}$, will be illustrated with Example 31.2.

Example 31.2 *Four judges rank four contestants in a beauty contest. The judges are told to assign the most beautiful contestant a rank of 1 and the least beautiful contestant a rank of 4. The rank-orders of the four judges are summarized in* Table 31.2. *Is there a significant association between the rank-orders assigned to the four contestants by the four judges?*

Table 31.2 Data for Example 31.2

Judge	Contestant				Totals
	1	**2**	**3**	**4**	
1	1	3	3	3	
2	1	4	2	3	
3	2	3	1	4	
4	1.5	1.5	3.5	3.5	
ΣR_j	5.5	11.5	9.5	13.5	$T = 40$
$(\Sigma R_j)^2$	30.25	132.25	90.25	182.25	$U = 435$

In Example 31.2, there are $m = 4$ sets of ranks/judges and $n = 4$ subjects/contestants who are ranked. Inspection of Table 31.2 reveals that Judges 1 and 4 employ tied ranks. As is the case with other rank-order tests described in the book, subjects who are tied for a specific rank are assigned the average of the ranks that are involved. Judge 1 assigns a rank of 1 to Contestant 1, and places the other three contestants in a tie for the next ordinal position. Thus, Contestants 2, 3, and 4 are all assigned a rank of 3, which is the average of the three ranks involved (i.e., $(2 + 3 + 4)/3 = 3$). Judge 4 places Contestants 1 and 2 in a tie for the first and second ordinal positions, and Contestants 3 and 4 in a tie for the third and fourth ordinal positions. Thus, the contestants evaluated by Judge 4 are assigned ranks that are the average of those ranks for which they are tied (i.e., $(1 + 2)/2 = 1.5$ and $(3 + 4)/2 = 3.5$).

Equation 31.6 (which is the tie-corrected version of Equation 31.4) is employed to compute the tie-corrected value of **Kendall's coefficient of concordance**, which will be represented by the notation $\tilde{W}_c$.

$$\tilde{W}_c = \frac{12U - 3m^2 n(n + 1)^2}{m^2 n(n^2 - 1) - m \sum_{i=1}^{m} \left[\sum_{a=1}^{s} (t_a^3 - t_a) \right]}$$ **(Equation 31.6)**

The notation $\sum_{i=1}^{m} [\sum_{a=1}^{s} (t_a^3 - t_a)]$ in the denominator of Equation 31.6 indicates the following: a) Within each set of ranks, for each set of ties that is present the number of ties in the set is subtracted from the cube of the number of ties in that set; b) The sum of all the values computed in part a) is obtained for that set of ranks; and c) The sum of the values computed in part b) is computed for the m sets of ranks.

In the case of Example 31.2, Judge 1 has $s = 1$ set of ties involving three contestants. Thus, for Judge 1, $\sum_{a=1}^{s} (t_a^3 - t_a) = [(3)^3 - 3] = 24$. Since Judges 2 and 3 do not employ any ties, the latter two judges will not contribute to the tie correction, and thus the value of $\sum_{a=1}^{s} (t_a^3 - t_a)$

will equal 0 for both of the aforementioned judges. Judge 4 has $s = 2$ sets of ties, each set involving two contestants. Thus, for Judge 4, $\Sigma_{a=1}^{s}(t_a^3 - t_a) = [(2)^3 - 2] + [(2)^3 - 2] = 12$. We can now determine the value $\Sigma_{i=1}^{m}[\Sigma_{a=1}^{s}(t_a^3 - t_a)] = 36$, which is employed in Equation 31.6.

$$\sum_{i=1}^{m}\left[\sum_{a=1}^{s}(t_a^3 - t_a)\right] = 24 + 0 + 0 + 12 = 36$$

When the appropriate values are substituted in Equation 31.6, the tie-corrected value $\tilde{W}_c = .51$ is computed for Example 31.2.[4]

$$\tilde{W}_c = \frac{(12)(435) - (3)(4)^2(4)(4 + 1)^2}{(4)^2(4)[(4)^2 - 1] - (4)(36)} = .51$$

It can be seen below that when Equation 31.4 (which does not employ the tie correction) is employed to compute $\tilde{W}$, the value $\tilde{W} = .44$ is obtained. Note that the latter value is less than $\tilde{W}_c = .51$. The computed correlation $\tilde{W}_c = .51$ (as well as $\tilde{W} = .44$) indicates a moderate degree of association between the four sets of ranks.

$$\tilde{W} = \frac{(12)(435) - (3)(4)^2(4)(4 + 1)^2}{(4)^2(4)[(4)^2 - 1]} = .44$$

Note that in **Table A20** the tabled critical .05 and .01 values for $m = 4$ and $n = 4$ are $\tilde{W}_{.05} = .619$ and $\tilde{W}_{.01} = .768$. Since both $\tilde{W}_c = .51$ and $\tilde{W} = .44$ are less than $\tilde{W}_{.05} = .619$, the null hypothesis H_0: $W = 0$ cannot be rejected.

VII. Additional Discussion of Kendall's Coefficient of Concordance

1. Relationship between Kendall's coefficient of concordance and Spearman's rank-order correlation coefficient The relationship between **Kendall's coefficient of concordance** and **Spearman's rank-order correlation coefficient** is as follows: If for data consisting of m sets of ranks a value for **Spearman's rho** is computed for every possible pair consisting of two sets of ranks (i.e., if $m = 3$, $r_{S_{12}}$, $r_{S_{13}}$, $r_{S_{23}}$), the average of all the r_S values (to be designated $\bar{r}_S$) is a linear function of the value of $\tilde{W}$ computed for the data. Equation 31.7 defines the exact relationship between **Spearman's rho** and $\tilde{W}$ for the same set of data.

$$\bar{r}_S = \frac{m\tilde{W} - 1}{m - 1} \qquad \text{(Equation 31.7)}$$

The above relationship will be demonstrated employing the data in Table 31.3 (which we will assume is a revised set of data for Example 31.1, in which $m = 3$ and $n = 3$).

Substituting the appropriate values in Equation 31.4, the value $\tilde{W} = .111$ is computed. The latter value indicates a weak degree of association between the three sets of ranks.[5]

$$\tilde{W} = \frac{(12)(110) - (3)(3)^2(3)(3 + 1)^2}{(3)^2(3)[(3)^2 - 1]} = .111$$

Table 31.3 Data for Use in Equation 31.7

Instructor	Student 1	Student 2	Student 3	Totals
1	3	1	2	
2	1	2	3	
3	3	2	1	
ΣR_j	7	5	6	$T = 18$
$(\Sigma R_j)^2$	49	25	36	$U = 110$

Substituting $\tilde{W} = .111$ in Equation 31.7, the value $\bar{r}_S = -.333$ is computed.

$$\bar{r}_S = \frac{(3)(.111) - 1}{3 - 1} = -.333$$

We will now confirm that $\bar{r}_S = -.333$. Equation 29.1 is employed to compute the r_S values for the 3 pairs of ranks (i.e., $r_{S_{12}}$ for the ranks of Instructor 1 versus Instructor 2; $r_{S_{13}}$ for the ranks of Instructor 1 versus Instructor 3; and $r_{S_{23}}$ for the ranks of Instructor 2 versus Instructor 3). The resulting values are $r_{S_{12}} = -.5$, $r_{S_{13}} = .5$, and $r_{S_{23}} = -1$. The average of the values of the three pairs of ranks is $\bar{r}_S = [(-.5) + .5 + (-1)]/3 = -.333$, thus confirming the result obtained with Equation 31.7. It should be noted that when Equation 31.7 is employed to compute the value of $\bar{r}_S$, the range of values within which $\bar{r}_S$ can fall is defined by the following limits: $[-1/(m - 1)] \le \bar{r}_S \le +1$. When $m = 3$, as is the case in the example under discussion, the minimum possible value $\bar{r}_S$ can assume is $-1/(3 - 1) = -.5$. Note that even though the sign of $\tilde{W}$ cannot be negative, Equation 31.7 can convert a positive $\tilde{W}$ value into either a positive or negative $\bar{r}_S$ value.

The relationship described by Equation 31.7 can also be demonstrated for any of the examples employed in illustrating **Spearman's rank-order correlation coefficient**, where $m = 2$. To illustrate, in the case of Example 29.2 the value $r_S = .72$ is computed for two sets of ranks. When the relevant values from Example 29.2 (which are summarized in Table 29.6)[6] are substituted in Equation 31.4, the value $\tilde{W} = .86$ is computed. Note that in Example 29.2, $m = 2$ and $n = 10$.

$$\tilde{W} = \frac{(12)(1493.5) - (3)(2)^2(10)(10 + 1)^2}{(2)^2(10)[(10)^2 - 1]} = .86$$

Substituting $\tilde{W} = .86$ in Equation 31.7 yields the value $\bar{r}_S = .72$, which equals $r_S = .72$ computed with Equation 29.1.

$$\bar{r}_S = \frac{(2)(.86) - 1}{2 - 1} = .72$$

Thus, when $m = 2$, the value of $\bar{r}_S$ will equal r_S, since the average of a single value (based on one pair of ranks) is that value.

2. Relationship between Kendall's coefficient of concordance and the Friedman two-way analysis of variance by ranks   In Section I it is noted that **Kendall's coefficient of concordance** and the **Friedman two-way analysis of variance by ranks** are based on the same mathematical model. Equation 31.8 defines the relationship between the computed values of

$\tilde{W}$ and χ_r^2. The chi-square value (χ_r^2) in Equation 31.8 can be employed to represent the test statistic for the **Friedman two-way analysis of variance by ranks** (which is more commonly computed with Equation 25.1). Note that Equation 31.8 is identical to Equation 31.5.

$$\chi_r^2 = m(n - 1)\tilde{W} \qquad \textbf{(Equation 31.8)}$$

Equation 31.9, which is the algebraic transposition of Equation 31.8, provides an alternative way of computing the value $\tilde{W}$.

$$\tilde{W} = \frac{\chi_r^2}{m(n - 1)} \qquad \textbf{(Equation 31.9)}$$

In order to employ Equation 31.9 to compute the value of $\tilde{W}$, it is necessary to evaluate the data for m sets of ranks on n subjects/objects with the **Friedman two-way analysis of variance by ranks**. To illustrate the equivalence of **Kendall's coefficient of concordance** and the **Friedman two-way analysis of variance by ranks**, consider Example 31.3 which employs the same variables employed in Example 25.1 (which is used to illustrate the **Friedman two-way analysis of variance by ranks**). Note that in Example 31.3 there are $m = 6$ judges (who are represented by the six subjects) and $n = 3$ objects (which are represented by the three levels of noise).

Example 31.3 *Six subjects rank three levels of noise (based on the presence or absence of different types of music) with respect to the degree they believe each level of noise will disrupt one's ability to learn a list of nonsense syllables. The subjects are instructed to assign a rank of 1 to the most disruptive level of noise and a rank of 3 to the least disruptive level of noise.* Table 31.4 *summarizes the rankings of the subjects. Is there a significant association between the rank-orders assigned to the three levels of noise by the six subjects?*

Employing Equation 31.4, the value $\tilde{W} = .92$ is computed.[7] The value $\tilde{W} = .92$ indicates a strong degree of association between the six sets of ranks.

$$\tilde{W} = \frac{(12)(498.5) - (3)(6)^2(3)(3 + 1)^2}{(6)^2(3)[(3)^2 - 1]} = .92$$

Table 31.4   Data for Example 31.3

		Type of noise		
Subject	No noise	Classical music	Rock music	Totals
1	3	2	1	
2	3	2	1	
3	3	2	1	
4	3	2	1	
5	3	2	1	
6	3	1.5	1.5	
ΣR_j	18	11.5	6.5	$T = 36$
$(\Sigma R_j)^2$	324	132.25	42.25	$U = 498.5$

It happens to be the case that the configuration of ranks in Example 31.3 is identical to the configuration of ranks employed in Example 25.1. When the **Friedman two-way analysis of**

variance by ranks is employed to evaluate the same six sets of ranks, the value $\chi_r^2 = 11.08$ is computed. The reader should take note of the fact that when the data are evaluated with Equation 25.1 in Section IV of the **Friedman test**, k is employed to represent the number of levels of the independent variable and n is employed to represent the number of subjects. In Table 25.1, the three columns of R_j values represent the $k = 3$ levels of the independent variable, and the six rows represent the $n = 6$ subjects. In the model employed for **Kendall's coefficient of concordance**, the value of n corresponds to the value employed for k in the **Friedman model**, and thus, $n = k = 3$. The value of m in the **Kendall model** corresponds to the value employed for n in the **Friedman model**, and thus, $m = n = 6$. The equations used in this section employ notation which is consistent with the **Kendall model**.

When the value $\chi_r^2 = 11.08$ is substituted in Equation 31.9, the value $\tilde{W} = .92$ is computed.

$$\tilde{W} = \frac{11.08}{(6)(3 - 1)} = .92$$

In the same respect, if $\tilde{W} = .92$ is substituted in Equations 31.8/31.5, it yields the value $\chi_r^2 = 11.08$.[8]

$$\chi_r^2 = (6)(3 - 1)(.92) = 11.08$$

Since the value of $\tilde{W}$ can be computed for the **Friedman test** model, **Kendall's coefficient of concordance** can be employed as a measure of effect size for a within-subjects design (involving data that are rank-ordered) with an independent variable which has three or more levels. The closer the value of $\tilde{W}$ is to 1, the stronger the relationship between the independent and dependent variables. Consequently, the value $\tilde{W} = .92$ computed for Example 25.1 (as well as Example 31.3), indicates there is a strong degree of association between the independent variable (noise) and the dependent variable (the rank-ordering on number of nonsense syllables recalled/disruptive potential of noise).

3. Weighted rank/top-down concordance   Section VII of **Spearman's rank-order correlation coefficient** briefly discusses a method referred to as **weighted/top-down correlation** which can be employed for differentially weighting the most extreme scores in a set of data. Zar (1999, pp. 449–450) describes the extension of this method to **Kendall's coefficient of concordance**. Thus, a **weighted/top-down correlation coefficient** can be employed if a researcher's primary concern is with the degree of agreement for objects/subjects that are ranked the highest by a set of judges. In such a case the correlation coefficient would minimally weight scores of a lower rank.

4. Kendall's coefficient of concordance versus the intraclass correlation coefficient   In Section VI of the **single-factor within-subjects analysis of variance** it was noted that the **intraclass correlation coefficient** can be employed to assess interjudge reliability in reference to interval/ratio data when more than two judges are involved. However, unlike **Kendall's coefficient of concordance** (as well as **Spearman's rank-order correlation coefficient** and **Kendall's tau**), which by virtue of rank-ordering data evaluate the **relative differences** between the ratings of judges, the **intraclass correlation coefficient** evaluates the interval/ratio ratings of the judges, and by virtue of the latter it evaluates the **absolute differences** between the ratings. It was also noted that the **Pearson product-moment correlation coefficient**, which although employed with interval/ratio data, also evaluates the **relative differences** between the ratings of two judges. By virtue of evaluating the absolute differences between the ratings of judges, the **intraclass correlation** computed for a set of data will often be lower than the value computed

Table 31.5 Data for Example 24.1 Formatted to Evaluate Interjudge Reliability

Judge	Subject					
	1	2	3	4	5	6
1	9	10	7	10	7	8
2	7	8	5	8	5	6
3	4	7	3	7	2	6

Table 31.6 Data for Example 24.1 Formatted for Kendall's Coefficient of Concordance

Judge	Subject						Totals
	1	2	3	4	5	6	
1	3	1.5	5.5	1.5	5.5	4	
2	3	1.5	5.5	1.5	5.5	4	
3	4	1.5	5	1.5	6	3	
ΣR_j	10	4.5	16	4.5	17	11	$T = 63$
$(\Sigma R_j)^2$	100	20.25	256	20.25	289	121	$U = 806.5$

with the rank-order methods of correlation and a **Pearson product-moment correlation coefficient**. To illustrate the latter, consider Tables 31.5 and 31.6.

Table 31.5 summarizes the original interval/ratio ratings of the judges employed in the example to illustrate the use of the **intraclass correlation coefficient** earlier in the book. (In the earlier discussion of the data, in Table 24.13 in Section VI of the **single-factor within-subjects analysis of variance**, the letter k was employed to represent the number of judges.) As was noted in the latter discussion, the data are the same as those employed in Example 24.1. When the interval/ratio ratings of the judges were evaluated with the **intraclass correlation**, the value r_{IC} = .38 was obtained. It was noted that since the latter value is far removed from the maximum value of 1, it would most likely be deemed unacceptable as a measure of interjudge reliability. The low value computed for the **intraclass correlation** was attributed to the fact that there are considerable differences between the **mean ratings** of the three judges (specifically, the mean ratings of Judges 1, 2, and 3 are respectively 8.5, 6.5, 4.83).

The ratings in Table 31.5 have been rank-ordered in Table 31.6 in order that the data can be evaluated with **Kendall's coefficient of concordance**. Note that when the data are rank-ordered the mean rankings of the k judges will always be equal to one another (specifically, in this instance the average ranking for each of the $k = 3$ judges is 3.5). By virtue of ranking the data, any absolute differences between the ratings of the judges are rendered irrelevant. As a result of the latter, since the rankings of all three judges in Table 31.6 are extremely similar, the value computed for **Kendall's coefficient of concordance** will be considerably higher than that computed for r_{IC}. Thus, when the appropriate values in Table 31.6 are substituted in Equation 31.4 the value $\tilde{W} = .92$ is computed. (Equation 31.6 yields the tie-corrected value $\tilde{W}_c = .97$.)

$$\tilde{W} = \frac{(12)(806.5) - (3)(3)^2(6)(6 + 1)^2}{(3)^2(6)[(6)^2 - 1]} = .92$$

In **Table A20** the tabled critical .05 and .01 values for $m = 3$ and $n = 6$ are $\tilde{W}_{.05} = .660$ and $\tilde{W}_{.01} = .780$. Since both $\tilde{W} = .92$ and $\tilde{W}_c = .97$ are greater than $\tilde{W}_{.01} = .780$, the null hypothesis H_0: $W = 0$ can be rejected at both the .05 and .01 levels. More importantly an interjudge reliability coefficient of .9 or greater indicates extremely high interjudge agreement.

Since the **Pearson product-moment correlation coefficient** also considers the relative differences between the rankings of the judges, it logically follows that if one were to average the three **Pearson product-moment correlation coefficients** for the ratings of each possible pair of judges (i.e., Judge 1 versus Judge 2, Judge 1 versus Judge 3, and Judge 2 versus Judge 3), the latter average will be much closer to the value of $\tilde{W}$ than it will to r_{IC}. In point of fact, the latter average is computed to be $r = .951$ (through use of Equation 28.25) or $r = .9$ (if the three r values of 1, .85, and .85 obtained in Section VI of the **single-factor within-subjects analysis of variance** under the discussion of the sphericity assumption are averaged).

In closing this discussion, it should be noted that the issue of whether one should employ **Kendall's coefficient of concordance** or the **intraclass correlation coefficient** as a measure of interjudge reliability will depend on whether a researcher is concerned about the absolute differences between the ratings of the m judges. If the latter is true, one should employ the **intraclass correlation coefficient**, whereas if it is not, **Kendall's coefficient of concordance** should be used. In reality, most of the time interjudge reliability is assessed, researchers appear to only be interested in the relative differences between ratings, and because of the latter **Kendall's coefficient of concordance** is employed more frequently than the **intraclass correlation coefficient**. Further discussion of the relationship between **Kendall's coefficient of concordance** the **intraclass correlation coefficient** can be found in Cohen (2001).

VIII. Additional Examples Illustrating the Use of Kendall's Coefficient of Concordance

Examples 31.4 and 31.5 are two additional examples that can be evaluated with **Kendall's coefficient of concordance**. Example 31.4 addresses the same question evaluated by Example 29.2, but in Example 31.4 the values $m = 6$ and $n = 4$ are employed in place of the values $m = 2$ and $n = 10$ employed in Example 29.2. Since Examples 31.4 and 31.5 employ the same data as Example 31.1, they yield the same result.

Example 31.4 *In order to determine whether or not critics agree with one another in their evaluation of movies, a newspaper editor asks six critics to rank four movies (assigning a rank of 1 to the best movie, a rank of 2 to the next best movie, etc.). Table 31.7 summarizes the data for the study. Is there a significant association between the six sets of ranks?*

Example 31.5 *Four members of a track team are ranked by the head coach with respect to their ability on six track and field events. For each event, the coach assigns a rank of 1 to the athlete who is best at the event and a rank of 4 to the athlete who is worst at the event. Table 31.8 summarizes the data for the study. Is there a significant association between the rank-orders assigned to the athletes on the six events?*

Note that in Example 31.5, even though one judge (the coach) is employed, the judge generates six sets of ranks (i.e., six sets of judgements). If there is a significant association between the six sets of ranks/judgements, it indicates that the athletes are perceived to be consistent with respect to performance on the six events.

Table 31.7 Data for Example 31.4

Critic	Movie 1	2	3	4	Totals
1	3	2	1	4	
2	3	2	1	4	
3	3	2	1	4	
4	4	2	1	3	
5	3	2	1	4	
6	4	1	2	3	
ΣR_j	20	11	7	22	$T = 60$
$(\Sigma R_j)^2$	400	121	49	484	$U = 1054$

Table 31.8 Data for Example 31.5

Event	Athlete 1	2	3	4	Totals
Sprint	3	2	1	4	
1500 meters	3	2	1	4	
Pole vault	3	2	1	4	
Long jump	4	2	1	3	
Shot put	3	2	1	4	
400 meters	4	1	2	3	
ΣR_j	20	11	7	22	$T = 60$
$(\Sigma R_j)^2$	400	121	49	484	$U = 1054$

References

Cohen, B. H. (2001). **Explaining psychological Statistics** (2nd ed.). New York: John Wiley & Sons.

Conover, W. J. (1980). **Practical nonparametric statistics** (2nd ed.). New York: John Wiley & Sons.

Conover, W. J. (1999). **Practical nonparametric statistics** (3rd ed.). New York: John Wiley & Sons.

Daniel, W. W. (1990). **Applied nonparametric statistics** (2nd ed.). Boston: PWS–Kent Publishing Company.

Friedman, M. (1940). A comparison of alternative tests of significance for the problem of *m* rankings, **Annals of Mathematical Statistics**, 11, 86–92.

Kendall, M. G. (1970). **Rank correlation methods** (4th ed.). London: Charles Griffin & Co. Ltd.

Kendall, M. G. & Babington–Smith, B. (1939). The problem of *m* rankings. **Annals of Mathematical Statistics**, 10, 275–287.

Lindeman, R. H., Meranda, P. F., & Gold, R. Z. (1980). **Introduction to bivariate and multivariate analysis**. Glenview, IL: Scott, Foresman & Company.

Marascuilo, L. A. & McSweeney, M. (1977). **Nonparametric and distribution-free methods for the social sciences**. Monterey, CA: Brooks/Cole Publishing Company.

Siegel, S. & Castellan, N. J., Jr. (1988). **Nonparametric statistics for the behavioral sciences**
    (2nd ed.). New York: McGraw–Hill Book Company.

Sprent, P. (1989). **Applied nonparametric statistics**. London: Chapman & Hall.

Sprent, P. (1993). **Applied nonparametric statistics** (2nd ed.). London: Chapman & Hall.

Wallis, W. A. (1939). The correlation ratio for ranked data. **Journal of the American**
    Statistical Association, 34, 533–538.

Zar, J. H. (1999). **Biostatistical analysis** (4th ed.). Upper Saddle River, NJ: Prentice Hall.

Endnotes

1. Siegel and Castellan (1988) emphasize the fact that a correlation equal to or close to 1 does not in itself indicate that the rankings are correct. A high correlation only indicates that there is agreement among the m sets of ranks. It is entirely possible that there can be complete agreement among two or more sets of ranks, but that all of the rankings are, in fact, incorrect. In other words, the ranks may not reflect what is actually true with regard to the subjects/objects that are evaluated. Another way of stating the above is that although there may be **interjudge reliability** (i.e., agreement between the rankings of the judges), the latter does not necessarily mean that their rankings are **valid** (i.e., the concept of **validity** refers to whether the rankings are correct).

2. In point of fact, if the values of r_S and $\tilde{W}$ are computed for $m = 2$ sets of ranks, when the computed values for r_S are, respectively, 1, –1, and 0, the computed values of $\tilde{W}$ will, respectively, be 1, 0, and .5. The latter sets of values can be obtained through use of Equation 31.7, which is presented in Section VII.

3. Some sources state that the alternative hypothesis is **directional**, since $\tilde{W}$ can only be a positive value. Related to this is the fact that only the upper tail of the chi-square distribution (which is discussed in Section V) is employed in approximating the exact sampling distribution of $\tilde{W}$. In the final analysis, it becomes academic whether one elects to identify the alternative hypothesis as directional or nondirectional.

4. The tie-corrected version of Equation 31.3 is noted below:

$$\tilde{W} = \frac{S}{\left(\dfrac{m^2 n(n^2 - 1) - m \sum\limits_{i=1}^{m} \left[\sum\limits_{a=1}^{s} (t_a^3 - t_a) \right]}{12} \right)} = \frac{35}{\left(\dfrac{(4)^2(4)[(4)^2 - 1] - (4)(36)}{12} \right)} = .51$$

5. Note that for $m = 3$ and $n = 3$, no tabled critical values are listed in **Table A20**. This is the case, since critical values cannot be computed for values of m and n which fall below specific minimum values. If Equation 31.5 is employed to evaluate $\tilde{W} = .111$, it yields the following result: $\chi^2 = (3)(3 - 1)(.111) = .666$. Since $\chi^2 = .666$ is less than the tabled critical two-tailed value (for $df = 2$) $\chi^2_{05} = 5.99$, the obtained value $\tilde{W} = .111$ is not significant. In point of fact, even if the maximum possible value $\tilde{W} = 1$ is substituted in Equation 31.5, it yields the value $\chi^2 = 6$, which is barely above $\chi^2_{05} = 5.99$. Since the chi-square distribution provides an approximation of the exact sampling distribution, in this instance it would appear that the tabled value $\chi^2_{05} = 5.99$ is a little too high and, in actuality, is associated with a Type I error rate which is slightly above .05.

6. The summary of the data for Example 29.2 in Table 29.6 provides the necessary values required to compute the value of $\tilde{W}$. The latter values are not computed in Table 29.5, which (employing a different format) also summarizes the data for Example 29.2.

7. Although there is one set of ties in the data, the tie correction described in Section VI is not employed for Example 31.3.

8. The exact value $\chi_r^2 = 11.08$ is computed if the value $\tilde{W} = .9236$ (which carries the computation of $\tilde{W}$ to four decimal places) is employed in Equations 31.8/31.5.

Test 32

Goodman and Kruskal's Gamma
(Nonparametric Measure of Association/Correlation Employed with Ordinal Data)

I. Hypothesis Evaluated with Test and Relevant Background Information

Goodman and Kruskal's gamma is one of a number of measures of correlation or association discussed in this book. Measures of correlation are not inferential statistical tests, but are, instead, descriptive statistical measures which represent the degree of relationship between two or more variables. Upon computing a measure of correlation, it is common practice to employ one or more inferential statistical tests in order to evaluate one or more hypotheses concerning the correlation coefficient. The hypothesis stated below is the most commonly evaluated hypothesis for **Goodman and Kruskal's gamma**.

Hypothesis evaluated with test In the underlying population represented by a sample, is the correlation between subjects' scores on two variables some value other than zero?

Relevant background information on test Prior to reading the material in this section the reader should review the general discussion of correlation in Section I of the **Pearson product-moment correlation coefficient (Test 28)**, as well as the material in Section I of **Kendall's tau (Test 30)**. Developed by Goodman and Kruskal (1954, 1959, 1963, 1972), **gamma** is a bivariate measure of correlation/association that is employed with rank-order data which is summarized within the format of an **ordered contingency table**. The population parameter estimated by the correlation coefficient will be represented by the notation γ (which is the lower case Greek letter **gamma**). The sample statistic computed to estimate the value of γ will be represented by the notation G. As is the case with **Spearman's rank-order correlation coefficient (Test 29)** and **Kendall's tau**, **Goodman and Kruskal's gamma** can be employed to evaluate data in which a researcher has scores for n subjects/objects on two variables (designated as the X and Y variables), both of which have been rank-ordered. However, in contrast to **Spearman's rho** and **Kendall's tau**, computation of **gamma** is recommended when there are many ties in a set of data, and thus it becomes more efficient to summarize the data within the format of an ordered $r \times c$ contingency table.

An ordered $r \times c$ contingency table consists of $r \times c$ cells, and is comprised of r rows and c columns.[1] In the model employed for **Goodman and Kruskal's gamma**, each of the rows in the contingency table represents one of the r levels of the X variable, and each of the columns represents one of the c levels of the Y variable (or vice versa). Since the contingency table that is employed to summarize the data is ordered, the categories for both the row and the column variables are arranged sequentially with respect to magnitude/ordinal position. To be more specific, the first row in the table represents the category that is lowest in magnitude on the X variable and the rth row represents the category that is highest in magnitude on the X variable. In the same respect, the first column represents the category that is lowest in magnitude on the

Y variable and the cth column represents the category that is highest in magnitude on the *Y* variable.[2] Recorded within each of the $r \times c$ cells of the contingency table are the number of subjects whose categorization on the *X* and *Y* variables corresponds to the row and column of a specific cell.

The value of **gamma** computed for a set of data represents the difference $p(C) - p(D)$, where: a) $p(C)$ is the probability that the ordering of the scores on the row and column variables for a pair of subjects is concordant (i.e., in agreement); and b) $p(D)$ is the probability that the ordering of the scores on the row and column variables for a pair of subjects is discordant (i.e., disagree).

To illustrate, if a subject is categorized on the lowest level of the row variable and the highest level of the column variable, that subject is concordant with respect to ordering when compared with any other subject who is assigned to a lower category on the row variable than he is on the column variable. On the other hand, that subject is discordant with respect to ordering when compared with another subject who is assigned to a higher category on the row variable than he is on the column variable. For a more thorough discussion of the concepts of concordance and discordance the reader should review Section I of **Kendall's tau**.

The range of possible values within which a computed value of **gamma** may fall is $-1 \leq G \leq +1$. As is the case for **Kendall's tau**, a positive value of *G* indicates that the number of concordant pairs in a set of data is greater than the number of discordant pairs, while a negative value indicates that the number of discordant pairs is greater than the number of concordant pairs. The computed value of *G* will equal 1 when the ordering of scores for all of the pairs of subjects in a set of data is concordant, and will equal -1 when the ordering of scores for all of the pairs of subjects is discordant. When $G = 0$, the number of concordant and discordant pairs of subjects in a set of data is equal.

Since **Goodman and Kruskal's gamma** and **Kendall's tau** both involve evaluating pairs of scores with respect to concordance versus discordance, the two measures of association are related to one another. Marascuilo and McSweeney (1977), who provide a detailed discussion on the nature of the relationship between **gamma** and **tau**, note that if $\tilde{\tau}$ and *G* are computed for the same set of data, as the number of pairs of ties increase, the absolute value computed for *G* will become increasingly larger relative to the absolute value of $\tilde{\tau}$. As a result of the latter, researchers who want to safeguard against obtaining an inflated value for the degree of association between the two variables may prefer to compute $\tilde{\tau}$ for a set of rank-order data, as opposed to computing the value of *G* (although, as noted earlier, **gamma** allows for a more efficient summary of data when there are a large number of ties).

It should be noted that **Yule's *Q* (Test 16i)** (which is one of a number of measures of association that can only be employed to evaluate a 2×2 contingency table) represents a special case of **Goodman and Kruskal's gamma**. Although **gamma** can be employed with a 2×2 contingency table, it is typically employed with ordered contingency tables in which there are at least three levels on either the row or column variable. A more detailed discussion of the relationship between **Yule's *Q*** and **Goodman and Kruskal's gamma** can be found in Section VII.

II. Example

Example 32.1 *A researcher wants to determine whether or not a relationship exists between a person's weight (which will be designated as the X variable) and birth order (which will be designated as the Y variable). Upon determining the weight and birth order of 300 subjects, each subject is categorized with respect to one of three weight categories and one of four birth order categories. Specifically, the following three categories are employed with respect to weight:*

below average, average, above average. *The following four categories are employed with respect to birth order:* **first born, second born, third born, fourth born and all subsequent birth orders.** Table 32.1 *(which is a 3 × 4 ordered contingency table, with r = 3 and c = 4) summarizes the data. Do the data indicate there is a significant association between a person's weight and birth order?*

Table 32.1 Summary of Data for Example 32.1

		Birth order				Row sums
		1st born	2nd born	3rd born	4th born+	
	Below average	70	15	10	5	100
Weight	Average	10	60	20	10	100
	Above average	10	15	35	40	100
Column sums		90	90	65	55	300

III. Null versus Alternative Hypotheses

Upon computing **Goodman and Kruskal's gamma**, it is common practice to determine whether the obtained absolute value of the correlation coefficient is large enough to allow a researcher to conclude that the underlying population correlation coefficient between the two variables is some value other than zero. Section V describes how the latter hypothesis, which is stated below, can be evaluated through use of an inferential statistical test that is based on the normal distribution.

Null hypothesis $H_0: \gamma = 0$

(In the underlying population the sample represents, the correlation between the scores/ categorization of subjects on Variable X and Variable Y equals 0.)

Alternative hypothesis $H_1: \gamma \neq 0$

(In the underlying population the sample represents, the correlation between the scores/ categorization of subjects on Variable X and Variable Y equals some value other than 0. This is a **nondirectional alternative hypothesis**, and it is evaluated with a **two-tailed test**. Either a significant positive G value or a significant negative G value will provide support for this alternative hypothesis. In order to be significant, the obtained absolute value of G must be equal to or greater than the tabled critical two-tailed G value at the prespecified level of significance.)

or

$H_1: \gamma > 0$

(In the underlying population the sample represents, the correlation between the scores/ categorization of subjects on Variable X and Variable Y equals some value greater than 0. This is a **directional alternative hypothesis**, and it is evaluated with a **one-tailed test**. Only a significant positive G value will provide support for this alternative hypothesis. In order to be significant (in addition to the requirement of a positive G value), the obtained absolute value of G must be equal to or greater than the tabled critical one-tailed G value at the prespecified level of significance.)

or

$$H_1: \gamma < 0$$

(In the underlying population the sample represents, the correlation between the scores/
categorization of subjects on Variable X and Variable Y equals some value less than 0. This is
a **directional alternative hypothesis**, and it is evaluated with a **one-tailed test**. Only a
significant negative G value will provide support for this alternative hypothesis. In order to be
significant (in addition to the requirement of a negative G value), the obtained absolute value of
G must be equal to or greater than the tabled critical one-tailed G value at the prespecified level
of significance.)

Note: Only one of the above noted alternative hypotheses is employed. If the alternative
hypothesis the researcher selects is supported, the null hypothesis is rejected.[3]

IV. Test Computations

In order to compute the value of G, it is necessary to determine the number of pairs of subjects
who are concordant with respect to the ordering of their scores on the X and Y variables (which
will be represented by the notation n_C), and the number of pairs of subjects who are discordant
with respect to the ordering of their scores on the X and Y variables (which will be represented
by the notation n_D). Upon computing the values of n_C and n_D, Equation 32.1 is employed to
compute the value of **Goodman and Kruskal's gamma**.

$$G = \frac{n_C - n_D}{n_C + n_D} \qquad \textbf{(Equation 32.1)}$$

The determination of the values of n_C and n_D is based on an analysis of the frequencies
in the ordered contingency table (i.e., Table 32.1). Each of the cells in the table will be identified
by two digits. The first digit will represent the row within which the cell falls, and the second
digit will represent the column within which the cell falls. Thus, Cell$_{ij}$ is the cell in the ith row
and j^{th} column. As an example, since it is in both the first row and first column, the cell in the
upper left-hand corner of Table 32.1 is Cell$_{11}$. The number of subjects within each cell is
identified by the notation n_{ij}. Thus, in the case of Cell$_{11}$, $n_{11} = 70$.

The protocol for determining the values of n_C and n_D will now be described. The following
procedure is employed to determine the value of n_C.

    a) Begin with the cell in the **upper left-hand corner of the table** (i.e., Cell$_{11}$). Determine
the frequency of that cell (which will be referred to as the target cell), and multiply the frequency
by the sum of the frequencies of **all other cells in the table that fall both below it and to the
right of it**. In Table 32.1, the following six cells meet the criteria of being both below and to the
right of Cell$_{11}$: Cell$_{22}$, Cell$_{23}$, Cell$_{24}$, Cell$_{32}$, Cell$_{33}$, Cell$_{34}$. Note that although Cell$_{21}$ and Cell$_{31}$ are
below Cell$_{11}$, they are not to the right of it, and although Cell$_{12}$, Cell$_{13}$, and Cell$_{14}$ fall to the right
of Cell$_{11}$, they do not fall below it. Any subject who falls within a cell that is both below and to
the right of Cell$_{11}$ will form a concordant pair with any subject in Cell$_{11}$. The rationale for this is
as follows: Assume that the values R_{X_i} and R_{Y_i} represent the score/ranking/category of Subject
i on the X and Y variables, and that R_{X_j} and R_{Y_j} represent the score/ranking/category of Subject
j on the X and Y variables. Assume that Subject i is a subject in the target cell, and that Subject

j is a subject in a cell which falls below and to the right of the target cell. We can state that the sign of the difference $(R_{X_i} - R_{X_j})$ will be the same as the sign of the difference $(R_{Y_i} - R_{Y_j})$ when the scores of any subject in the target cell are compared with any subject in a cell that falls below and to the right of the target cell. When for any pair of subjects the signs of the differences $(R_{X_i} - R_{X_j})$ and $(R_{Y_i} - R_{Y_j})$ are identical, that pair of subjects is concordant with respect to their ordering on the two variables.

To illustrate, each of the 70 subjects in $Cell_{11}$ has a rank of 1 on both of the variables, and each of the 60 subjects in $Cell_{22}$ (which is one of the cells below and to the right of $Cell_{11}$) has a rank of 2 on both of the variables. Any pair of subjects that is formed by employing one subject from $Cell_{11}$ and one subject from $Cell_{22}$ will be concordant with respect to their ordering on the two variables, since for each pair the sign of the difference between the ranks on both variables will be negative (i.e., $(R_{X_i} - R_{X_j}) = (1 - 2) = -1$ and $(R_{Y_i} - R_{Y_j}) = (1 - 2) = -1$). If, on the other hand, we compare the ranks on both variables for any subject who is in the target cell with the ranks on both variables for any subject who is in a cell that is not below and to the right of the target cell, $(R_{X_i} - R_{X_j})$ and $(R_{Y_i} - R_{Y_j})$ will have different signs or will equal zero.

The expression which summarizes the product of the frequency of $Cell_{11}$ and the sum of the frequencies of all the cells that fall both below and to the right of it is as follows: $n_{11}(n_{22} + n_{23} + n_{24} + n_{32} + n_{33} + n_{34})$. Substituting the appropriate frequencies from Table 32.1, we obtain $70(60 + 20 + 10 + 15 + 35 + 40) = (70)(180) = 12{,}600$. This latter value will be designated as **Product 1**.

b) The same procedure employed with $Cell_{11}$ is applied to all remaining cells. Moving to the right in Row 1, the procedure is next employed with $Cell_{12}$. **Product 2**, which represents the product for the second target cell, can be summarized by the expression $n_{12}(n_{23} + n_{24} + n_{33} + n_{34})$, since $Cell_{23}$, $Cell_{24}$, $Cell_{33}$, and $Cell_{34}$ are the only cells that fall both below and to the right of $Cell_{12}$. Thus, **Product 2** will equal $15(20 + 10 + 35 + 40) = (15)(105) = 1575$.

c) Upon computing **Product 2**, products for the two remaining cells in Row 1 are computed, after which products are computed for each of the cells in Rows 2 and 3. The computation of the products for all 12 cells in the ordered contingency table is summarized in Table 32.2. Note that since many of the cells have no cell which falls both below and to the right of them, the value that the frequency of these cells will be multiplied by will equal zero, and thus the resulting product will equal zero. The value of n_C is the sum of all the products in Table 32.2. For Example 32.1, $n_C = 20{,}875$.

Table 32.2 Computation of n_C for Example 32.1

$Cell_{11}$:	$70(60 + 20 + 10 + 15 + 35 + 40)$	=	12600	Product 1
$Cell_{12}$:	$15(20 + 10 + 35 + 40)$	=	1575	Product 2
$Cell_{13}$:	$10(10 + 40)$	=	500	Product 3
$Cell_{14}$:	$5(0)$	=	0	Product 4
$Cell_{21}$:	$10(15 + 35 + 40)$	=	900	Product 5
$Cell_{22}$:	$60(35 + 40)$	=	4500	Product 6
$Cell_{23}$:	$20(40)$	=	800	Product 7
$Cell_{24}$:	$10(0)$	=	0	Product 8
$Cell_{31}$:	$10(0)$	=	0	Product 9
$Cell_{32}$:	$15(0)$	=	0	Product 10
$Cell_{33}$:	$35(0)$	=	0	Product 11
$Cell_{34}$:	$40(0)$	=	0	Product 12

$$n_C = \text{Sum of products} = 20875$$

Table 32.3 Computation of n_D for Example 32.1

$Cell_{14}$: $5(10 + 60 + 20 + 10 + 15 + 35)$	=	750	Product 1
$Cell_{13}$: $10(10 + 60 + 10 + 5)$	=	950	Product 2
$Cell_{12}$: $15(10 + 10)$	=	300	Product 3
$Cell_{11}$: $70(0)$	=	0	Product 4
$Cell_{24}$: $10(10 + 15 + 35)$	=	600	Product 5
$Cell_{23}$: $20(10 + 15)$	=	500	Product 6
$Cell_{22}$: $60(10)$	=	600	Product 7
$Cell_{21}$: $10(0)$	=	0	Product 8
$Cell_{34}$: $40(0)$	=	0	Product 9
$Cell_{33}$: $35(0)$	=	0	Product 10
$Cell_{32}$: $15(0)$	=	0	Product 11
$Cell_{31}$: $10(0)$	=	0	Product 12
	n_D = Sum of products =	3700	

Upon computing the value of n_C, the following protocol is employed to compute the value of n_D.

a) Begin with the cell in the **upper right-hand corner of the table** (i.e., $Cell_{14}$). Determine the frequency of that cell, and multiply the frequency by the sum of the frequencies of **all other cells in the table that fall both below it and to the left of it**. In Table 32.1, the following six cells meet the criteria of being both below and to the left of $Cell_{14}$: $Cell_{21}$, $Cell_{22}$, $Cell_{23}$, $Cell_{31}$, $Cell_{32}$, $Cell_{33}$. Note that although $Cell_{24}$ and $Cell_{34}$ are below $Cell_{14}$, they are not to the left of it, and although $Cell_{11}$, $Cell_{12}$, and $Cell_{13}$ fall to the left of $Cell_{14}$, they do not fall below it. Any subject who falls within a cell that is both below and to the left of $Cell_{14}$ will form a discordant pair with any subject in $Cell_{14}$. The general rule that can be stated with respect to discordant pairs is as follows (if we assume that Subject i is a subject in the target cell, and Subject j is a subject in some other cell): The sign of the difference $(R_{X_i} - R_{X_j})$ will be different than the sign of the difference $(R_{Y_i} - R_{Y_j})$ when the scores of any subject in the target cell are compared with those of any subject in a cell that falls below and to the left of the target cell. When for any pair of subjects the signs of the differences $(R_{X_i} - R_{X_j})$ and $(R_{Y_i} - R_{Y_j})$ are different, that pair of subjects is discordant with respect to their ordering on the two variables.

To illustrate, each of the 5 subjects in $Cell_{14}$ has a rank of 1 on weight and a rank of 4 on birth order. Each of the 20 subjects in $Cell_{23}$ (which is one of the cells below and to the left of $Cell_{14}$) has a rank of 2 on weight and a rank of 3 on birth order. Any pair of subjects that is formed by employing one subject from $Cell_{14}$ and one subject from $Cell_{23}$, will be discordant with respect to the ordering of the ranks of the subjects on the two variables, since for each pair the signs of the difference between the ranks on both variables will be different (i.e., $(R_{X_i} - R_{X_j})$ = $(1 - 2) = -1$ and $(R_{Y_i} - R_{Y_j}) = (4 - 3) = +1$). If, on the other hand, we compare the ranks on both variables for any subject who is in the target cell with the ranks on both variables for any subject who is in a cell that is not below and to the left of the target cell, $(R_{X_i} - R_{X_j})$ and $(R_{Y_i} - R_{Y_j})$ will have the same sign or will equal zero.

The expression which summarizes the product of the frequency of $Cell_{14}$ and the sum of the frequencies of all cells that fall both below and to the left of it is as follows: $n_{14}(n_{21} + n_{22} + n_{23} + n_{31} + n_{32} + n_{33})$. Substituting the appropriate frequencies, we obtain $5(10 + 60 + 20 + 10 + 15 + 35) = (5)(150) = 750$. As is the case in determining the number of concordant pairs, we will designate the product for the first cell that is analyzed as **Product 1**.

b) The same procedure employed with $Cell_{14}$ is applied to all remaining cells. Moving to the left in Row 1, the procedure is next employed with $Cell_{13}$. **Product 2**, which represents the product for the second target cell, can be summarized by the expression $n_{13}(n_{21} + n_{22} + n_{31} + n_{32})$, since $Cell_{21}$, $Cell_{22}$, $Cell_{31}$, and $Cell_{32}$ are the only cells that fall both below and to the left of $Cell_{13}$. Thus, **Product 2** will equal $10(10 + 60 + 10 + 15) = (10)(95) = 950$.

c) Upon computing **Product 2**, products for the two remaining cells in Row 1 are computed, after which products are computed for each of the cells in Rows 2 and 3. The computation of the products for all 12 cells in the ordered contingency table is summarized in Table 32.3. Note that since many of the cells have no cell which falls both below and to the left of them, the value that the frequency of such cells will be multiplied by will equal zero, and thus the resulting product will equal zero. The value of n_D will be the sum of all the products in Table 32.3. For Example 32.1, $n_D = 3700$.

Substituting the values $n_C = 20,875$ and $n_D = 3700$ in Equation 32.1, the value $G = .70$ is computed. Note that the value of G is positive, since the number of concordant pairs is greater than the number of discordant pairs.

$$G = \frac{20,875 - 3700}{20,875 + 3700} = .70$$

The value $G = .70$ can also be computed employing the definition of **gamma** presented in Section I. Specifically:

$$G = p(C) - p(D) = \frac{20,875}{24,575} - \frac{3700}{24,575} = .70$$

In the above equation the value 24,575 is the total number of pairs (to be designated n_T), which is the denominator of Equation 32.1 (i.e., $20,875 + 3700 = 24,575$). Thus, $p(C) = n_C/n_T$ and $p(D) = n_D/n_T$.

Since the computed value $G = .70$ is close to 1, it indicates the presence of a strong positive/direct relationship between the two variables. Specifically, it suggests the higher the rank of a subject's weight category, the higher the rank of the subject's birth order category.

V. Interpretation of the Test Results

Test 32a: Test of significance for Goodman and Kruskal's gamma When the sample size is relatively large (which will generally be the case when **gamma** is computed), the computed value of G can be evaluated with Equation 32.2. To be more specific, Equation 32.2 (which employs the normal distribution) is employed to evaluate the null hypothesis $H_0: \gamma = 0$.[4] The sign of the z value computed with Equation 32.2 will be the same as the sign of the value computed for G.

$$z = G \sqrt{\frac{n_C + n_D}{N(1 - G^2)}} \qquad \textbf{(Equation 32.2)}$$

Where: N is the total number of subjects for whom scores are recorded in the ordered contingency table

When the appropriate values from Example 32.1 are substituted in Equation 32.2, the value $z = 8.87$ is computed.

$$z = .70 \sqrt{\frac{20{,}875 + 3700}{300[1 - (.70)^2]}} = 8.87$$

Equation 32.3 is an alternative equation for computing the value of z. The denominator of Equation 32.3 represents the **standard error** of the G statistic (which will be represented by the notation SE_G). In Section VI, SE_G is employed to compute a confidence interval for **gamma**.

(Equation 32.3)

$$z = \frac{G}{SE_G} = \frac{G}{\sqrt{\dfrac{1}{\dfrac{n_C + n_D}{N(1 - G^2)}}}} = \frac{.70}{\sqrt{\dfrac{1}{\dfrac{20{,}875 + 3700}{(300)[1 - (.70)^2]}}}} = \frac{.70}{.0789} = 8.87$$

The computed value $z = 8.87$ is evaluated with **Table A1 (Table of the Normal Distribution)** in the **Appendix**.[5] In the latter table, the tabled critical two-tailed .05 and .01 values are $z_{.05} = 1.96$ and $z_{.01} = 2.58$, and the tabled critical one-tailed .05 and .01 values are $z_{.05} = 1.65$ and $z_{.01} = 2.33$.

The following guidelines are employed in evaluating the null hypothesis.

a) If the nondirectional alternative hypothesis $H_1: \gamma \neq 0$ is employed, the null hypothesis can be rejected if the obtained absolute value of z is equal to or greater than the tabled critical two-tailed value at the prespecified level of significance.

b) If the directional alternative hypothesis $H_1: \gamma > 0$ is employed, the null hypothesis can be rejected if the sign of z is positive, and the value of z is equal to or greater than the tabled critical one-tailed value at the prespecified level of significance.

c) If the directional alternative hypothesis $H_1: \gamma < 0$ is employed, the null hypothesis can be rejected if the sign of z is negative, and the absolute value of z is equal to or greater than the tabled critical one-tailed value at the prespecified level of significance.

Employing the above guidelines, the nondirectional alternative hypothesis $H_1: \gamma \neq 0$ is supported at both the .05 and .01 levels, since the computed value $z = 8.87$ is greater than the tabled critical two-tailed values $z_{.05} = 1.96$ and $z_{.01} = 2.58$. The directional alternative hypothesis $H_1: \gamma > 0$ is supported at both the .05 and .01 levels, since the computed value $z = 8.87$ is a positive number that is greater than the tabled critical one-tailed values $z_{.05} = 1.65$ and $z_{.01} = 2.33$. The directional alternative hypothesis $H_1: \gamma < 0$ is not supported, since the computed value $z = 8.87$ is a positive number.

A summary of the analysis of Example 32.1 follows: It can be concluded there is a significant positive relationship between weight and birth order.

VI. Additional Analytical Procedures for Goodman and Kruskal's Gamma and/or Related Tests

1. The computation of a confidence interval for the value of Goodman and Kruskal's gamma   Equation 32.4 is employed to compute a confidence interval for a computed value of **gamma**.

$$CI_{(1 - \alpha)} = G \pm (z_{\alpha/2})(SE_G) \qquad \textbf{(Equation 32.4)}$$

Where: $z_{\alpha/2}$ represents the tabled critical two-tailed value in the normal distribution below which a proportion (percentage) equal to $[1 - (\alpha/2)]$ of the cases falls. If the proportion (percentage) of the distribution that falls within the confidence interval is subtracted from 1 (100%), it will equal the value of α.

Equation 32.4 will be employed to compute the 95% confidence interval for **gamma**. Along with the tabled critical two-tailed .05 value $z_{.05} = 1.96$, the following values computed for Example 32.1 are substituted in Equation 32.4: $G = .70$ and $SE_G = .0789$ (which is the value computed for the denominator of Equation 32.3, which as noted in Section V represents the **standard error** of G).

$$CI_{.95} = .70 \pm (1.96)(.0789) = .70 \pm .15$$

Subtracting from and adding .15 to .70 yields the values .55 and .85. Thus, the researcher can be 95% confident (or the probability is .95) the interval .55 to .85 contains the true value of **gamma** in the underlying population. Symbolically, this can be written as follows: $.55 \le \gamma \le .85$.

2. Test 32b: Test for evaluating the null hypothesis H_0: $\gamma_1 = \gamma_2$ Marascuilo and McSweeney (1977) note that Equation 32.5 can be employed to determine whether or not there is a significant difference between two independent values of **gamma**. Use of Equation 32.5 assumes that the following conditions have been met: a) The sample size in each of two ordered contingency tables is large enough for evaluation with the normal approximation; b) The values of r and c are identical in the two ordered contingency tables; and c) The same row and column categories are employed in the two ordered contingency tables.

$$z = \frac{G_1 - G_2}{\sqrt{SE_{G_1} + SE_{G_2}}}$$

(**Equation 32.5**)

Where: G_1 and G_2 are the computed values of **gamma** for the two ordered contingency tables, and SE_{G_1} and SE_{G_2} are the computed values of the **standard error** for the two values of **gamma**

To illustrate the use of Equation 32.5, assume that the study described in Example 32.1 is replicated with a different sample comprised of $N = 600$ subjects. The obtained value of **gamma** for the sample is $G = .50$, with $SE_G = .0438$. By employing the values $G_1 = .70$, $SE_{G_1} = .0789$, $G_2 = .50$, and $SE_{G_2} = .0438$ in Equation 32.5, the researcher can evaluate the null hypothesis H_0: $\gamma_1 = \gamma_2$. Substituting the appropriate values in Equation 32.5 yields the value $z = .57$.

$$z = \frac{.7 - .5}{\sqrt{.0789 + .0438}} = .57$$

In the order listed in Section V, the same guidelines described for evaluating the alternative hypotheses H_1: $\gamma \ne 0$, H_1: $\gamma > 0$, and H_1: $\gamma < 0$ are respectively employed for evaluating the alternative hypotheses H_1: $\gamma_1 \ne \gamma_2$, H_1: $\gamma_1 > \gamma_2$, and H_1: $\gamma_1 < \gamma_2$. The nondirectional alternative hypothesis H_1: $\gamma_1 \ne \gamma_2$ is not supported, since the computed value $z = .57$ is less than the tabled critical two-tailed value $z_{.05} = 1.96$. The directional alternative hypothesis H_1: $\gamma_1 > \gamma_2$ is not supported, since the computed value $z = .57$ is less than the tabled critical one-tailed value $z_{.05} = 1.65$. The directional alternative hypothesis H_1: $\gamma_1 < \gamma_2$ is not supported, since the computed value $z = .57$ is a positive number. The fact that the difference $\gamma_1 - \gamma_2 = .7$ $- .5 = .2$ (which is reasonably large) is not significant can be attributed to the fact that both samples have relatively large standard errors.

3. Sources for computing a partial correlation coefficient for Goodman and Kruskal's gamma The computation of a **partial correlation coefficient** (described in Section IV/V of the chapter on **multiple regression (Test 33)**, and discussed briefly in Section VI of **Spearman's rank-order correlation coefficient** and **Kendall's tau**) can be extended to **Goodman and Krukal's gamma**. More specifically, when the rank-orders for three variables are evaluated, a procedure developed by Davis (1967) (described in Marascuilo and McSweeney (1977, pp. 470–471)) can be employed to compute one or more **partial correlation coefficients** for **gamma**.

VII. Additional Discussion of Goodman and Kruskal's Gamma

1. Relationship between Goodman and Kruskal's gamma and Yule's *Q* In Section I it is noted that **Yule's *Q*** is a special case of **Goodman and Kruskal's gamma**. To illustrate this, assume that the four cells in Tables 16.2/16.3 (for which **Yule's *Q*** is computed) represent a 2×2 contingency table in which the cells on both the row and the column variables are ordered. If the procedure described for determining concordant pairs is employed with the data in Tables 16.2/16.3, the only cell that will generate a product other than zero is Cell_{11} (which corresponds to **Cell a** within the framework of the notation used for a 2×2 contingency table). Specifically, the product for Cell_{11}, which will correspond to the value of n_C, is $(n_{11})(n_{12}) = (30)(40) = 1200$. In the same respect, if the procedure described for determining discordant pairs is employed, the only cell that will generate a product other than zero is Cell_{12} (which corresponds to **Cell b** within the framework of the notation used for a 2×2 contingency table). Specifically, the product for Cell_{12}, which will correspond to the value of n_D, is $(n_{12})(n_{21}) = (70)(60) = 4200$. When the values $n_C = 1200$ and $n_D = 4200$ are substituted in Equation 32.1, $G = (1200 - 4200)/(1200 + 4200) = -.56$. Note that this result is identical to that obtained when Equation 16.23 is employed to compute **Yule's *Q*** for the same set of data: $Q = (ad - bc)/(ad + bc) = [(30)(40) - (70)(60)]/[(30)(40) + (70)(60)] = -.56$. It should be noted that unlike **gamma**, which is only employed with ordered contingency tables, **Yule's *Q*** can be employed with both ordered and unordered 2×2 contingency tables.

2. Somers' delta as an alternative measure of association for an ordered contingency table Somers (1962) has developed an alternative measure of association for ordered contingency tables referred to as **delta** (which is represented by the upper case Greek letter Δ). Siegel and Castellan (1988) identify **delta** as an **asymmetric measure of association** (as opposed to a **symmetric measure of association**). An **asymmetric measure of association** is employed when one variable is distinguished in a meaningful way from the other variable (e.g., within the context of the study, one variable is more important than the other, or one variable represents an independent variable and the other a dependent variable). Within this framework, **gamma** is viewed as a symmetric measure of association, since it does not assume a meaningful distinction between the variables within the context noted above. A full discussion of **Somers' delta** can be found in Siegel and Castellan (1988). Grissom and Kim (2005, pp. 211–213; 217) discuss the use of **Somers' delta** as a measure of effect size, as well as citing references which discuss alternative measures that can be employed as measures of effect size for ordered categorical data.

VIII. Additional Examples Illustrating the Use of Goodman and Kruskal's Gamma

Examples 32.2 and 32.3 are two additional examples that can be evaluated with **Goodman and Kruskal's gamma**. Since Examples 32.2 and 32.3 employ the same data as Example 32.1, they yield the same result. Example 32.4 describes the identical study described by Example 32.1, but uses a different configuration of data in order to illustrate the computation of a negative value for **gamma**.

Example 32.2 *A consumer group conducts a survey in order to determine whether or not a relationship exists between customer satisfaction and the price a person pays for an automobile. Each of 300 individuals who has purchased a new vehicle within the past year is classified in one of four categories based on the purchase price of one's automobile. Each subject is also classified in one of three categories with respect to how satisfied he or she is with his or her automobile. The results are summarized in* Table 32.4. *Do the data indicate there is a relationship between the price of an automobile and degree of satisfaction?*

Table 32.4 Summary of Data for Example 32.2

		Purchase price				Row sums
		Under $10,000	$10,000 to $18,000	$18,001 to $30,000	More than $30,000	
Level of satisfaction	Below average	70	15	10	5	100
	Average	10	60	20	10	100
	Above average	10	15	35	40	100
Column sums		90	90	65	55	300

Example 32.3 *A panel of psychiatrists wants to determine whether or not a relationship exists between the number of years a patient is in psychotherapy and the degree of change in a patient's behavior. Each of 300 patients is categorized with respect to one of four time periods during which he or she is in psychotherapy, and one of three categories with respect to the change in behavior he or she has exhibited since initiating therapy. Specifically, the following four categories are employed with respect to psychotherapy duration:* **less than one year, one to two years, more than two years to three years, more than three years.** *The following three categories are employed with respect to changes in behavior:* **deteriorated (−), no change, improved (+).** *Table 32.5 summarizes the data. Do the data indicate there is an association between the amount of time a patient is in psychotherapy and the degree to which he or she changes?*[6]

Table 32.5 Summary of Data for Example 32.3

		Number of years in psychotherapy				Row sums
		Less than one year	One to two years	More than two years to three years	More than three years	
Amount of change	−	70	15	10	5	100
	No change	10	60	20	10	100
	+	10	15	35	40	100
Column sums		90	90	65	55	300

Example 32.4 *A researcher wants to determine whether or not a relationship exists between a person's weight and birth order. Upon determining the weight and birth order of 300 subjects, each subject is categorized with respect to one of three weight categories and one of four birth order categories. Specifically, the following three categories are employed with respect to weight:* **below average, average, above average.** *The following four categories are employed with respect to birth order:* **first born, second born, third born, fourth born and all subsequent birth orders.** *Table 32.6 summarizes the data. Do the data indicate there is a significant association between a person's weight and birth order?*

Table 32.6 Summary of Data for Example 32.4

		Birth order				Row sums
		1st born	2nd born	3rd born	4th born+	
Weight	Below average	5	10	15	70	100
	Average	10	20	60	10	100
	Above average	40	35	15	10	100
Column sums		55	65	90	90	300

Inspection of the data reveals that the cell frequencies in Table 32.6 are the mirror image of those employed in Table 32.1. By virtue of employing the same frequencies in an inverted format, the values of n_C and n_D for Table 32.6 are the reverse of those obtained for Table 32.1. Thus, for Table 32.6, $n_C = 3700$ and $n_D = 20,875$. Consequently, employing Equation 32.1, $G = (3700 - 20,875)/(3700 + 20,875) = -.70$. Because the same configuration of data is employed in an inverted format, the value $G = -.70$ computed for Table 32.6 is the same absolute value computed for Table 32.1. Note that the negative correlation $G = -.70$ indicates that a subject's birth order is inversely related to his weight. Specifically, subjects in a low birth order category are more likely to be above average in weight, while subjects in a high birth order category are more likely to be below average in weight.

References

Daniel, W. W. (1990). **Applied nonparametric statistics** (2nd ed.). Boston: PWS–Kent Publishing Company.

Davis, J. A. (1967). A partial coefficient for Goodman and Kruskal's gamma. **Journal of the American Statistical Association**, 62, 189–193.

Goodman, L. A. & Kruskal, W. H. (1954). Measures of association for cross-classification. **Journal of the American Statistical Association**, 49, 732–764.

Goodman, L. A. & Kruskal, W. H. (1959). Measures of association for cross-classification II: Further discussion and references. **Journal of the American Statistical Association**, 54, 123–163.

Goodman, L. A. & Kruskal, W. H. (1963). Measures of association for cross-classification III: Approximate sample theory. **Journal of the American Statistical Association**, 58, 310–364.

Goodman, L. A. & Kruskal, W. H. (1972). Measures of association for cross-classification IV: Simplification for asymptotic variances. **Journal of the American Statistical Association**, 67, 415–421.

Grissom, R. J. & Kim, J. J. (2005). **Effect sizes for research: A broad practical approach.** Mahwah, NJ: Lawrence Erlbaum Associates Publishers.

Marascuilo, L. A. & McSweeney, M. (1977). **Nonparametric and distribution-free methods for the social sciences**. Monterey, CA: Brooks/Cole Publishing Company.

Ott, R. L, Larson, R., Rexroat, C., & Mendenhall, W. (1992). **Statistics: A tool for the social sciences** (5th ed.). Boston: PWS–Kent Publishing Company.

Siegel, S. & Castellan, N. J., Jr. (1988). **Nonparametric statistics for the behavioral sciences** (2nd ed.). New York: McGraw–Hill Book Company.

Somers, R. H. (1962). A new asymmetric measure of association for ordinal variables. **American Sociological Review**, 27, 799–811.

Endnotes

1. The general model for an $r \times c$ contingency table (which is summarized in Table 16.1) is discussed in Section I of the **chi-square test for $r \times c$ tables (Test 16)**.

2. **Gamma** can also be computed if the ordering is reversed — i.e., within both variables, the first row/column represents the category with the highest magnitude, and the last row/ column represents the category with the lowest magnitude.

3. Some sources employ the following statements as the null hypothesis and the nondirectional alternative hypothesis for **Goodman and Kruskal's gamma: Null hypothesis**: H_0: Variables X and Y are independent of one another; **Nondirectional alternative hypothesis**: H_1: Variables X and Y are not independent of one another.

   It is, in fact, true that if in the underlying population the two variables are independent, the value of γ will equal zero. However, Siegel and Castellan (1988) note that if $\gamma = 0$, the latter does not in and of itself ensure that the two variables are independent of one another (unless the contingency table is a 2×2 table).

4. Equation 32.2 can also be written in the following form:

$$z = (G - \gamma) \sqrt{\frac{n_C + n_D}{N(1 - G^2)}}$$

   In the above equation, γ represents the value of **gamma** stated in the null hypothesis. When the latter value equals zero, the above equation reduces to Equation 32.2. When some value other than zero is stipulated for **gamma** in the null hypothesis, the equation noted above can be employed to evaluate the null hypothesis H_0: $\gamma = \gamma_0$ (where γ_0 represents the value stipulated for the population correlation).

5. Sources which discuss the evaluation of the null hypothesis H_0: $\gamma = 0$ note that the normal approximation computed with Equations 32.2/32.3 tends to be overly conservative. Consequently, the likelihood of committing a Type I error (i.e., rejecting H_0 when it is true) is actually less than the value of alpha employed in the analysis.

6. It could be argued that it might be more appropriate to employ **Somers' delta** (which is briefly discussed in Section VII) rather than **gamma** as a measure of association for Example 32.3. The use of **delta** could be justified, if within the framework of a study the number of years of therapy represents an independent variable and the amount of change represents the

dependent variable. In point of fact, depending upon how one conceptualizes the relationship between the two variables, one could also argue for the use of **delta** as a measure of association for Example 32.1. In the final analysis, it will not always be clear whether it is more appropriate to employ **gamma** or **delta** as a measure of association for an ordered contingency table.

Multivariate
Statistical Analysis

Matrix Algebra and Multivariate Analysis

Test 33: Multiple Regression

Test 34: Hotelling's T^2

Test 35: Multivariate Analysis of Variance

Test 36: Multivariate Analysis of Covariance

Test 37: Discriminant Function Analysis

Test 38: Canonical Correlation

Test 39: Logistic Regression

Test 40: Principal Components Analysis and Factor Analysis

Matrix Algebra and Multivariate Analysis

I. Introductory Comments on Multivariate Statistical Analysis

Grimm and Yarnold (1995, p. 4) note that although researchers are not in complete agreement with respect to the use of the term **multivariate statistics**, it is often employed for procedures which **simultaneously** evaluate **multiple** (i.e., two or more) **independent/predictor variables** and **multiple** (i.e., two or more) **dependent/criterion variables**. Some sources, however, reserve the use of the term **multivariate** to primarily identify procedures which involve two or more dependent variables. It should be noted, however, that **discriminant function analysis (Test 37)** and **logistic regression (Test 39)** (both of which involve one dependent variable), as well as **principal components analysis and factor analysis (Test 40)** (both of which do not make a distinction between independent versus dependent variables) are among the procedures which are commonly identified as multivariate. The analytical procedures to be described in this section all involve more than two variables, and, aside from the latter, are categorized as multivariate on the basis of the fact that they are commonly described in books which deal with multivariate analysis.

Multivariate statistical procedures afford researchers with a number of advantages over **univariate** and **bivariate procedures**. Among others, Harlow (2005, Ch. 1) and Thompson (2000, 285–289) note the following: a) Because multivariate procedures are able to simultaneously assess the role of multiple variables, they represent a more realistic methodology for evaluating real world phenomena and theoretical models (which are typically complex and involve multiple variables). Thompson (2000, p. 286) notes that multivariate analysis can often result in dramatically different conclusions (with respect to both statistical significance and effect size) when contrasted with evaluating a set of data with multiple univariate or bivariate tests, since the latter types of analyses are unable to effectively investigate the simultaneous impact of multiple variables, and are thus ill equipped to detect often complex patterns of relationships between variables. In the chapters to follow, it will be demonstrated that one important characteristic of multivariate procedures is their ability to assign weights to variables which reflect their overall importance in explaining one or more other variables; b) Because they evaluate multiple variables, multivariate procedures can minimize the amount of unexplained variability which results from random error; c) Because it takes into account interrelationships/intercorrelations between variables, multivariate statistics are better able to rule out the role of extraneous variables; d) Multivariate procedures allow a researcher to control the overall Type I error rate, which would otherwise be inflated if multiple bivariate analyses were conducted on the same set of variables; e) Within the context of multivariate analysis, a researcher can conduct an overall analysis involving all of the variables (a macro-analysis) as well as more specific analyses (micro-analyses) assessing the role of the individual variables.

Disadvantages commonly cited in reference to multivariate procedures are: a) The mathematical operations involved in implementing multivariate procedures are considerably more complex than those required for bivariate procedures, and for the most part, multivariate procedures are impractical to conduct without the availability of appropriate computer software; b) Interpreting multivariate data is more complex by virtue of the fact that a researcher is simultaneously examining multiple variables; c) Multivariate analysis requires that certain assumptions (e.g., normality, homogeneity of variance-covariance, linearity, etc.), most of which are also associated with parametric bivariate procedures, not be violated. The assumptions with regard to continuous variables underlying many multivariate procedures are summarized in Section I of the chapter on **multiple regression (Test 33)**[1]; d) Multivariate analysis requires the use of larger sample sizes when contrasted with univariate and bivariate procedures. Harlow (2005, p. 6) notes that some researchers recommend 5 to 10 subjects per variable assuming no assumptions have been violated, and between 20 to 50 subjects per variable when one or more assumptions are violated. Other sources, however, recommend sample sizes for specific multivariate procedures which exceed the aforementioned guidelines.

II. Introduction to Matrix Algebra

Many of the multivariate procedures to be described are most efficiently conducted with data that are organized in a matrix format. In order to facilitate understanding of the material to follow some elementary concepts and operations involved in **matrix algebra**, which is a methodology for performing mathematical operations on a matrix, will be presented in this chapter.

A **matrix** is a array of data comprised of rows and columns that is arranged in a rectangular format. A matrix can provide a convenient way of organizing data which can subsequently be analyzed by implementing mathematical operations such as addition, subtraction, multiplication and division on the information recorded in the matrix. In point of fact, a single value such as the mean of a distribution can be viewed as representing a **scalar matrix** comprised of one row and one column. (The term **scalar algebra** refers to operations performed on a single number.)

An $r \times c$ matrix is comprised of r rows and c columns. The two 3×3 matrices below, summarized by the **bold upper case letters** A and B, will be employed to illustrate some of the terminology and operations employed within the framework of matrix algebra. Note that two examples of **Matrices A and B** are presented, with the first example employing letters to represent the 9 elements in each matrix and the second example employing numbers in place of the letters.

$$\mathbf{A} = \begin{bmatrix} a & b & c \\ d & e & f \\ g & h & i \end{bmatrix} \quad \mathbf{B} = \begin{bmatrix} j & k & l \\ m & n & o \\ p & q & r \end{bmatrix}$$

$$\mathbf{A} = \begin{bmatrix} 2 & -3 & 1 \\ 1 & 1 & -2 \\ 3 & 1 & 5 \end{bmatrix} \quad \mathbf{B} = \begin{bmatrix} 0 & 4 & 2 \\ 2 & 3 & 8 \\ 6 & 2 & -1 \end{bmatrix}$$

The numbers in the above matrices can be viewed as representing data — more specifically, they can be viewed as representing the scores of 3 subjects on 3 variables, with the 3 rows of the matrix representing the three subjects and the 3 columns representing the three variables. Alternatively, the above matrices can be viewed as a **variance-covariance matrix** for 3 variables in which each of the variables is represented by one of the rows as well as by one of the columns. The values in the **main diagonal** of the table (i.e., the diagonal beginning at the upper left of the matrix descending to the lower right) would represent the variance for each of the variables, while the other 6 elements in the table would represent covariances between pairs of variables (although one should take note of the fact that in the case of **Matrix B** it is impossible to have a negative value for a variance). Later in this section **data matrices** and **variance-covariance matrices** will be discussed in greater detail. At this point, however, the terminology and operations employed in matrix algebra will be described.

A **square matrix** is one in which the number of rows is equal to the number of columns (i.e., $r = c$). Whereas both **Matrices A** and **B** are square matrices, **Matrix C** below (which is a 2×3 rectangular matrix) is not square.

$$C = \begin{bmatrix} 2 & -3 & 1 \\ 1 & 1 & -2 \end{bmatrix}$$

As noted above, the **main diagonal** of a matrix is comprised of the elements/values on the diagonal beginning at the upper left and moving down to the lower right. Thus, the elements which comprise the main diagonal in **Matrix A** are $a = 2$, $e = 1$, and $i = 5$. The elements which comprise the main diagonal in **Matrix B** are $j = 0$, $n = 3$, and $r = -1$. In order have a main diagonal a matrix must be square.

A **symmetrical matrix** is one in which the elements above the main diagonal are the mirror image of the elements below the main diagonal. Another way to state that a matrix is **symmetrical** is as follows: If x_{ij} represents the element in the ith row and j^{th} column, then, excluding the main diagonal, $x_{ij} = x_{ji}$ and $i \neq j$ (i.e., excluding the main diagonal of the matrix, the element in the ith row and j^{th} column will be identical to the element in the jth row and i^{th} column). Note that neither **Matrix A** or **B** are symmetrical. On the other hand, **Matrix D** below is a symmetrical matrix.

$$D = \begin{bmatrix} 2 & -3 & 1 \\ -3 & 1 & -2 \\ 1 & -2 & 5 \end{bmatrix}$$

A **vector** (which is summarized by a **bold lower case letter**) is a matrix comprised of a single row or a single column (and is often a single row or column which has been extracted from a larger matrix). A matrix comprised of a single row is referred to as a **row vector**, while a matrix comprised of a single column is referred to as a **column vector**. Note that **Vector a** below represents a **row vector** (which represents the first row of **Matrix A**), while **Vector b** represents a **column vector** (which represents the first column of **Matrix B**).

$$a = \begin{bmatrix} a & b & c \end{bmatrix} \qquad a = \begin{bmatrix} 2 & -3 & 1 \end{bmatrix}$$

$$b = \begin{bmatrix} 0 \\ 2 \\ 6 \end{bmatrix} \qquad b = \begin{bmatrix} j \\ m \\ p \end{bmatrix}$$

As an example to illustrate a vector, assume a researcher summarizes the data for n subjects on k variables in the form of an $n \times k$ **data matrix** comprised of n rows and k columns. The first row will represent a **row vector** containing the k scores of the subject represented by that row. In the same respect, the first column will represent a **column vector** containing the n scores on the variable represented by that column. From the $n \times k$ data matrix a researcher can construct a **mean row vector** which will contain the means of the k variables across the n subjects, or alternatively, a **mean column vector** that contains the means for the n subjects on the k variables.

The **trace** of a matrix is the sum of the values in the **main diagonal** of the matrix. The trace of **Matrix A** is $a + e + i = 2 + 1 + 5 = 8$. The trace of **Matrix B** is $j + n + r = 0 + 3 + -1 = 2$.

In order for two matrices to be **equal** to one another, each of the elements in one matrix must be equal to the corresponding element in the other matrix. **Matrices E and F** below are equal.

$$
E = \begin{bmatrix} 5 & 4 & 1 \\ 11 & -1 & -2 \end{bmatrix}
\qquad
F = \begin{bmatrix} 5 & 4 & 1 \\ 11 & -1 & -2 \end{bmatrix}
$$

The **transpose** of a matrix is obtained by interchanging the rows and the columns of the matrix. The transposes of **Matrices A and B** (respectively labeled **A′** and **B′**) are presented below. Note that the first row of **Matrix A** is the first column of **Matrix A′**, the second row of **Matrix A** is the second column of **Matrix A′**, and the third row of **Matrix A** is the third column of **A′**. In the same respect, the first row of **Matrix B** is the first column of **Matrix B′**, the second row of **Matrix B** is the second column of **Matrix B′**, and the third row of **Matrix B** is the third column of **Matrix B′**.

$$
A' = \begin{bmatrix} 2 & 1 & 3 \\ -3 & 1 & 1 \\ 1 & -2 & 5 \end{bmatrix}
\qquad
B' = \begin{bmatrix} 0 & 2 & 6 \\ 4 & 3 & 2 \\ 2 & 8 & -1 \end{bmatrix}
$$

The **transpose of a row vector** are the values of the row vector aligned in a single column — in other words, the **row vector** is transposed into a **column vector**. Thus **a′**, the 3×1 **column vector** below, is the transpose of the row vector **a** presented earlier. In the same respect, the **transpose of a column vector** are the values of the column vector aligned in a single row — in other words, the **column vector** is transposed into a **row vector**. Thus **b′**, the 1×3 **row vector** below, is the transpose of the column vector **b** presented earlier.

$$
a' = \begin{bmatrix} 2 \\ -3 \\ 1 \end{bmatrix}
$$

$$
b' = \begin{bmatrix} 0 & 2 & 6 \end{bmatrix}
$$

Addition of a constant to a matrix In order to add a constant (t) to a matrix the constant is added to each element in the matrix, as noted below where the constant $t = 4$ is added to **Matrix A**.

$$
A = \begin{bmatrix} a & b & c \\ d & e & f \\ g & h & i \end{bmatrix}
+ t =
\begin{bmatrix} a+t & b+t & c+t \\ d+t & e+t & f+t \\ g+t & h+t & i+t \end{bmatrix}
$$

$$\mathbf{A} = \begin{bmatrix} 2 & -3 & 1 \\ 1 & 1 & -2 \\ 3 & 1 & 5 \end{bmatrix} + 4 = \begin{bmatrix} 6 & 1 & 5 \\ 5 & 5 & 2 \\ 7 & 5 & 9 \end{bmatrix}$$

Subtraction of a constant from a matrix In order to subtract a constant from a matrix the constant (t) is subtracted from each element in the matrix, as noted below where the constant t = 4 is subtracted from **Matrix A**.

$$\mathbf{A} = \begin{bmatrix} a & b & c \\ d & e & f \\ g & h & i \end{bmatrix} - t = \begin{bmatrix} a-t & b-t & c-t \\ d-t & e-t & f-t \\ g-t & h-t & i-t \end{bmatrix}$$

$$\mathbf{A} = \begin{bmatrix} 2 & -3 & 1 \\ 1 & 1 & -2 \\ 3 & 1 & 5 \end{bmatrix} - 4 = \begin{bmatrix} -2 & -7 & -3 \\ -3 & -3 & -6 \\ -1 & -3 & 1 \end{bmatrix}$$

Addition of two matrices In order to add two matrices it is required that the two matrices have the same dimensions — in other words, be equivalent with respect to the number of rows and the number of columns. The operation involved in adding two matrices is to add the corresponding elements in each of the matrices. Addition of matrices is illustrated below with respect to **Matrices A** and **B**.

$$\mathbf{A} + \mathbf{B} = \begin{bmatrix} a & b & c \\ d & e & f \\ g & h & i \end{bmatrix} + \begin{bmatrix} j & k & l \\ m & n & o \\ p & q & r \end{bmatrix} = \begin{bmatrix} a+j & b+k & c+l \\ d+m & e+n & f+o \\ g+p & h+q & i+r \end{bmatrix}$$

$$\mathbf{A} + \mathbf{B} = \begin{bmatrix} 2 & -3 & 1 \\ 1 & 1 & -2 \\ 3 & 1 & 5 \end{bmatrix} + \begin{bmatrix} 0 & 4 & 2 \\ 2 & 3 & 8 \\ 6 & 2 & -1 \end{bmatrix} = \begin{bmatrix} 2 & 1 & 3 \\ 3 & 4 & 6 \\ 9 & 3 & 4 \end{bmatrix}$$

Subtraction of two matrices In order to subtract two matrices it is required that the two matrices have the same dimensions — in other words, be equivalent with respect to the number of rows and the number of columns. The operation involved in subtracting two matrices is to subtract the corresponding elements in the matrices in the order stipulated. Subtraction of matrices is illustrated below with respect to **Matrices A** and **B**.

$$\mathbf{A} - \mathbf{B} = \begin{bmatrix} a & b & c \\ d & e & f \\ g & h & i \end{bmatrix} - \begin{bmatrix} j & k & l \\ m & n & o \\ p & q & r \end{bmatrix} = \begin{bmatrix} a-j & b-k & c-l \\ d-m & e-n & f-o \\ g-p & h-q & i-r \end{bmatrix}$$

$$\mathbf{A} - \mathbf{B} = \begin{bmatrix} 2 & -3 & 1 \\ 1 & 1 & -2 \\ 3 & 1 & 5 \end{bmatrix} - \begin{bmatrix} 0 & 4 & 2 \\ 2 & 3 & 8 \\ 6 & 2 & -1 \end{bmatrix} = \begin{bmatrix} 2 & -7 & -1 \\ -1 & -2 & -10 \\ -3 & -1 & 6 \end{bmatrix}$$

Multiplication of a matrix by a constant   In order to multiply a matrix by a constant (t) each element in the matrix is multiplied by the constant, as noted below where **Matrix A** is multiplied by the constant $t = 4$.

$$\mathbf{A} = \begin{bmatrix} a & b & c \\ d & e & f \\ g & h & i \end{bmatrix} \times t = \begin{bmatrix} a \times t & b \times t & c \times t \\ d \times t & e \times t & f \times t \\ g \times t & h \times t & i \times t \end{bmatrix}$$

$$\mathbf{A} = \begin{bmatrix} 2 & -3 & 1 \\ 1 & 1 & -2 \\ 3 & 1 & 5 \end{bmatrix} \times 4 = \begin{bmatrix} 8 & -12 & 4 \\ 4 & 4 & -8 \\ 12 & 4 & 20 \end{bmatrix}$$

Division of a matrix by a constant   In order to divide a matrix by a constant (t) each element in the matrix is divided by the constant in the order stipulated, as noted below where **Matrix A** is divided by the constant $t = 4$.

$$\mathbf{A} = \begin{bmatrix} a & b & c \\ d & e & f \\ g & h & i \end{bmatrix} \div t = \begin{bmatrix} a \div t & b \div t & c \div t \\ d \div t & e \div t & f \div t \\ g \div t & h \div t & i \div t \end{bmatrix}$$

$$\mathbf{A} = \begin{bmatrix} 2 & -3 & 1 \\ 1 & 1 & -2 \\ 3 & 1 & 5 \end{bmatrix} \div 4 = \begin{bmatrix} .5 & -.75 & .25 \\ .25 & .25 & -.5 \\ .75 & .25 & 1.25 \end{bmatrix}$$

Multiplication of two matrices   Multiplication of two matrices is considerably more complex than any of the mathematical operations introduced up to this point. The following should be noted with respect to multiplication of matrices.

a) If **Matrix A** is multiplied by **Matrix B** then $\mathbf{A} \times \mathbf{B} = \mathbf{AB} = \mathbf{C}$, where **C** represents the resulting product matrix. In order to multiply the latter two matrices it is required that the number of columns in **Matrix A** is equal to the number of rows in **Matrix B**. The dimensions of the resulting product **Matrix C** will be the number of rows in **Matrix A** and the number of columns in **Matrix B**. As an example, **Matrix A**, which is a 3 × 3 matrix, can be multiplied by another 3 × 3 matrix, a 3 × 2 matrix, a 3 × 5 matrix or a 3 × 1 **row vector**, since in all instances the second matrix has 3 rows. **Matrix A**, however, cannot not be multiplied by a 2 × 3 matrix, a 2 × 2 matrix, a 3 × 5 matrix or a 1 × 3 **column vector**, since the number of rows in the latter matrices is some value other than 3. Two matrices which can be multiplied by one another are said to be **conformable**.

b) The order in which the matrices are arranged in the matrix multiplication is important. If the order in which the matrices are arranged in the multiplication is that **Matrix B** is multiplied by **Matrix A**, then $B \times A = BA = D$, where **D** represents the resulting product matrix. In the latter case, it would be required that the number of columns in **Matrix B** is equal to the number of rows in **Matrix A**. The dimensions of the resulting product **Matrix D** will be the number of rows in **Matrix B** and the number of columns in **Matrix A**. Now we will once again consider the following two matrices alluded to above: a 3×3 matrix to be designated **Matrix A** and a 3×2 matrix to be designated Matrix **B**. Note that since Matrix **B** has 2 columns and Matrix **A** has 3 rows, the multiplication $B \times A$ is not possible, since in the order presented the two matrices are not conformable. It should be noted that even if the two multiplications $A \times B = C$ and $B \times A = D$ are conformable, the resulting product matrices will usually not be equivalent (i.e., $AB \neq BA$ or equivalently $C \neq D$). Stevens (2002, p. 61) notes that in mathematical terms, since $AB \neq BA$, we can state that multiplication of matrices is not **commutative**.

In order to illustrate matrix multiplication we will consider the product **Matrix C** resulting from the multiplication $A \times B$. The following rule is employed to derive the elements in the product **Matrix C**: Each of the elements in **Matrix C** is the result of summing the products of the corresponding elements with respect to ordinal position in the row of **Matrix A** (moving from left to right in a given row), which corresponds to the specified row for **Matrix C**, and a column of **Matrix B** (moving from top to bottom in a given column), which corresponds to the specified column for **Matrix C**. The latter rule is illustrated below with respect to three of the elements in **Matrix C**.

The element in **Row 1** and **Column 1** of **Matrix C** (i.e., Element x_{11}) will be obtained as follows: 1) Multiply the first element in **Row 1** of **Matrix A** (a) by the first element in **Column 1** of **Matrix B** (j), thus obtaining the product aj; 2) Multiply the second element in **Row 1** of **Matrix A** (b) by the second element in **Column 1** of **Matrix B** (m), thus obtaining the product bm; 3) Multiply the third element in **Row 1** of **Matrix A** (c) by the third element in **Column 1** of **Matrix B** (p), thus obtaining the product cp; 4) Obtain the sum of the three products (i.e., $aj + bm + cp$), which will represent the element in **Row 1, Column 1** of **Matrix C**.

The element in **Row 2** and **Column 1** of **Matrix C** (i.e., Element x_{21}) will be obtained as follows: 1) Multiply the first element in **Row 2** of **Matrix A** (d) by the first element in **Column 1** of **Matrix B** (j), thus obtaining the product dj; 2) Multiply the second element in **Row 2** of **Matrix A** (e) by the second element in **Column 1** of **Matrix B** (m), thus obtaining the product em; 3) Multiply the third element in **Row 2** of **Matrix A** (f) by the third element in **Column 1** of **Matrix B** (p), thus obtaining the product fp; 4) Obtain the sum of the three products (i.e., $dj + em + fp$), which will represent the element in **Row 2, Column 1** of **Matrix C**.

The element in **Row 1** and **Column 2** of **Matrix C** (i.e., Element x_{12}) will be obtained as follows: 1) Multiply the first element in **Row 1** of **Matrix A** (a) by the first element in **Column 2** of **Matrix B** (k), thus obtaining the product ak; 2) Multiply the second element in **Row 1** of **Matrix A** (b) by the second element in **Column 2** of **Matrix B** (n), thus obtaining the product bn; 3) Multiply the third element in **Row 1** of **Matrix A** (c) by the third element in **Column 2** of **Matrix B** (q), thus obtaining the product cq; 4) Obtain the sum of the three products (i.e., $ak + bn + cq$), which will represent the element in **Row 1, Column 2** of **Matrix C**.

The above protocol is continued for the remaining 6 elements in **Matrix C** obtaining the result below.

$$C = A \times B = \begin{bmatrix} a & b & c \\ d & e & f \\ g & h & i \end{bmatrix} \times \begin{bmatrix} j & k & l \\ m & n & o \\ p & q & r \end{bmatrix}$$

$$\mathbf{C} = \mathbf{A} \times \mathbf{B} = \begin{bmatrix} aj+bm+cp & ak+bn+cq & al+bo+cr \\ dj+em+fp & dk+en+fq & dl+eo+fr \\ gj+hm+ip & gk+hn+iq & gl+ho+ir \end{bmatrix}$$

$$\mathbf{C} = \mathbf{A} \times \mathbf{B} = \begin{bmatrix} 2 & -3 & 1 \\ 1 & 1 & -2 \\ 3 & 1 & 5 \end{bmatrix} \times \begin{bmatrix} 0 & 4 & 2 \\ 2 & 3 & 8 \\ 6 & 2 & -1 \end{bmatrix}$$

$$\mathbf{C} = \begin{bmatrix} (2)(0)+(-3)(2)+(1)(6)=0 & (2)(4)+(-3)(3)+(1)(2)=1 & (2)(2)+(-3)(8)+(1)(-1)=-21 \\ (1)(0)+(1)(2)+(-2)(6)=-10 & (1)(4)+(1)(3)+(-2)(2)=3 & (1)(2)+(1)(8)+(-2)(-1)=12 \\ (3)(0)+(1)(2)+(5)(6)=32 & (3)(4)+(1)(3)+(5)(2)=25 & (3)(2)+(1)(8)+(5)(-1)=9 \end{bmatrix}$$

$$\mathbf{C} = \begin{bmatrix} 0 & 1 & -21 \\ -10 & 3 & 12 \\ 32 & 25 & 9 \end{bmatrix}$$

Determinant of a matrix   The **determinant** of a matrix, which will be represented by the notation $|\mathbf{A}|$, is a single number that provides information about a **square matrix**. The determinant is useful for computing the **inverse** of a matrix which will be described in the next section. Stevens (2002, p. 64) notes that the determinant of a matrix containing variances and covariances for a set of k variables (which is discussed later in this section under the label of a **variance-covariance matrix**) represents a generalized variance for the k variables — in other words, the single value represented by the determinant is an index of variance for the set of variables. Readers should take note of the fact that although the symbolic representation of a determinant may suggest it is an absolute value, it is not, and the value of a determinant can be either positive or negative. However, among others, Harlow (2005, p. 91) and Marascuilo and Levin (1983, p. 180) note that in multivariate analysis a negative value for a determinant (or even a value close to 0) creates major problems for interpretation (since in the case of a **variance-covariance matrix** a measure of variance cannot be negative). Since it is assumed that multivariate computations will be conducted with a computer, rather than presenting the formal rule for deriving a determinant (which is somewhat complex), this section will only present guidelines for deriving the determinant of 2×2 and 3×3 matrices.

The determinant of the 2×2 **Matrix E** below is $|\mathbf{E}| = ad - bc$. Note that the determinant of a 2×2 matrix is obtained by subtracting the product of the **off-diagonal elements** from the product of the **main diagonal elements**. (The **off-diagonal elements** of a matrix refers to all of the elements which are not on the main diagonal of the matrix.)

$$\mathbf{E} = \begin{bmatrix} a & b \\ c & d \end{bmatrix} = \begin{bmatrix} 2 & 4 \\ 1 & 9 \end{bmatrix}$$

The determinant $|\mathbf{E}| = 14$ for **Matrix E** is computed below by substituting the appropriate values from **Matrix E** in $|\mathbf{E}| = ad - bc$.

$$|\mathbf{E}| = \begin{vmatrix} 2 & 4 \\ 1 & 9 \end{vmatrix} = (2)(9) - (4)(1) = 14$$

The determinant of the 3×3 **Matrix A** below is determined by: $|\mathbf{A}| = a(ei - fh) - b(di - fg) + c(dh - eg)$. (The latter can also be written as $|\mathbf{A}| = a(ei - fh) + b(fg - di) + c(dh - eg)$.)

$$\mathbf{A} = \begin{bmatrix} a & b & c \\ d & e & f \\ g & h & i \end{bmatrix} = \begin{bmatrix} 2 & -3 & 1 \\ 1 & 1 & -2 \\ 3 & 1 & 5 \end{bmatrix}$$

The determinant $|\mathbf{A}| = 45$ for **Matrix A** is computed below by substituting the appropriate values from **Matrix A** in $|\mathbf{A}| = a(ei - fh) - b(di - fg) + c(dh - eg)$.

$$|\mathbf{A}| = \begin{vmatrix} 2 & -3 & 1 \\ 1 & 1 & -2 \\ 3 & 1 & 5 \end{vmatrix} = 2[(1)(5) - (-2)(1)] - -3[(1)(5) - (-2)(3)] + 1[(1)(1) - (1)(3)]$$

$$|\mathbf{A}| = (2)(7) - (-3)(11) + (1)(-2) = 14 + 33 - 2 = 45$$

The relevance of determinants to multivariate analysis can be illustrated by the fact that Stevens (2002, p. 64) notes **Wilk's Λ** (where Λ is the upper case Greek letter **lambda**), which is a test statistic for the **multivariate analysis of variance (Test 35)**, is a ratio of the two determinants $|\mathbf{W}|/|\mathbf{T}|$, with $|\mathbf{W}|$ representing a multivariate generalization of a within-subjects sum of squares and $|\mathbf{T}|$ representing a multivariate generalization of a total sum of squares.

Inverse of a matrix   Computing the inverse of a matrix is also referred to as **matrix inversion**. An **inverse** can be computed for a **square matrix** except in the case where the determinant of the matrix equals 0. Stevens (2002, p. 73) notes that if the determinant of a matrix equals 0 the matrix is said to be **singular**, whereas if the determinant of a matrix is some value other than zero the matrix is said to be **nonsingular**.

If **Matrix E** is a square $r \times c$ matrix, then there is an $r \times c$ **Matrix $\mathbf{E}^{-1}$** which represents the inverse of **Matrix E**. The **inverse** of **Matrix E** is derived by dividing the **adjoint** of **Matrix E** (represented by the notation *Adj* (**E**)) by the **determinant** of the **Matrix E**. Thus, $\mathbf{E}^{-1} = Adj$ (**E**)/$|\mathbf{E}|$. The latter relationship is defined below for a 2×2 matrix. Note that to create the adjoint of a 2×2 matrix, the elements on the main diagonal are reversed, while the elements on the off-diagonal, although kept in their original position, are both multiplied by -1.

$$\mathbf{E} = \begin{bmatrix} a & b \\ c & d \end{bmatrix}$$

$$Adj\ (\mathbf{E}) = \begin{bmatrix} d & -b \\ -c & a \end{bmatrix}$$

$$\mathbf{E}^{-1} = \begin{bmatrix} a & b \\ c & d \end{bmatrix}^{-1} = \frac{1}{|E|}\, Adj(E) = \frac{1}{ad - bc}\begin{bmatrix} d & -b \\ -c & a \end{bmatrix}$$

The inverse of the 2×2 **Matrix E** (which in the previous section it was demonstrated had a determinant of $|\mathbf{E}| = 14$) is computed below.

$$\mathbf{E} = \begin{bmatrix} a & b \\ c & d \end{bmatrix} = \begin{bmatrix} 2 & 4 \\ 1 & 9 \end{bmatrix}$$

$$\mathbf{E}^{-1} = \begin{bmatrix} 2 & 4 \\ 1 & 9 \end{bmatrix}^{-1} = \frac{1}{|E|}\, Adj(E) = \frac{1}{(2)(9) - (4)(1)}\begin{bmatrix} 9 & -4 \\ -1 & 2 \end{bmatrix} = \frac{1}{14}\begin{bmatrix} 9 & -4 \\ -1 & 2 \end{bmatrix}$$

At this point, employing the procedure described earlier for multiplying a matrix by a constant (or dividing a matrix by a constant), each of the elements in the **adjoint matrix** $\begin{bmatrix} 9 & -4 \\ -1 & 2 \end{bmatrix}$ is multiplied by 1/14 (or divided by 14), yielding the matrix below which represents $\mathbf{E}^{-1}$, the **inverse** of **Matrix E**.

$$\mathbf{E}^{-1} = \frac{1}{14}\begin{bmatrix} 9 & -4 \\ -1 & 2 \end{bmatrix} = \begin{bmatrix} 9(1/14) & -4(1/14) \\ -1(1/14) & 2(1/14) \end{bmatrix} = \begin{bmatrix} .643 & -.286 \\ -.071 & .143 \end{bmatrix}$$

If the inverse of a matrix is multiplied by the original matrix, the resulting matrix is referred to as an **identity matrix** (represented by the notation **I**). An **identity matrix** is a square matrix in which all of the elements on the main diagonal are 1, and all of the remaining elements are 0. The aforementioned relationship between the original **Matrix E** and its inverse $\mathbf{E}^{-1}$, can be summarized as follows: $\mathbf{E}\mathbf{E}^{-1} = \mathbf{I}$. The latter relationship is demonstrated below.

$$\mathbf{E}\mathbf{E}^{-1} = \mathbf{I} = \begin{bmatrix} 2 & 4 \\ 1 & 9 \end{bmatrix} \times \begin{bmatrix} .643 & -.286 \\ -.071 & .143 \end{bmatrix}$$

$$\mathbf{E}\mathbf{E}^{-1} = \mathbf{I} = \begin{bmatrix} a & b \\ c & d \end{bmatrix} \times \begin{bmatrix} e & f \\ g & h \end{bmatrix} = \begin{bmatrix} ae + bg & af + bh \\ ce + dg & cf + dh \end{bmatrix}$$

$$\mathbf{I} = \mathbf{E}\mathbf{E}^{-1} = \begin{bmatrix} 2 & 4 \\ 1 & 9 \end{bmatrix} \times \begin{bmatrix} .643 & -.286 \\ -.071 & .143 \end{bmatrix} = \begin{bmatrix} (2)(.643) + (4)(-.071) & (2)(-.286) + (4)(.143) \\ (1)(.643) + (9)(-.071) & (1)(-.286) + (9)(.143) \end{bmatrix} = \begin{bmatrix} 1 & 0 \\ 0 & 1 \end{bmatrix}$$

Stevens (2002, p. 70) notes that since there is not a literal mechanism for division of one matrix by another matrix, the matrix analogue of division is computing the inverse of a matrix. To illustrate, in the case of the **single-factor between-subjects analysis of variance (Test 21)** the F ratio is computed by dividing MS_{BG} by MS_{WG}. The latter can also be written as $F = (MS_{BG})(MS_{WG})^{-1}$ (since $(MS_{WG})^{-1}$ represents the reciprocal of MS_{WG}). Stevens (2002, p. 70) notes that in **multivariate analysis of variance** the analogue of the latter ratio is $\mathbf{BW^{-1}}$, where **B** is a matrix that represents a multivariate analogue of the between groups sum of squares, and **W** is a matrix which represents a multivariate analogue of the within groups sum of squares. Multiplication of **Matrix B** by the inverse of **Matrix W** accomplishes the analogue of division required in computing an *F* ratio when between groups variability is divided by within groups variability.

Discussion of applications of matrix algebra in multivariate analysis   Within the framework of multivariate analysis a number of different types of matrices are evaluated. The discussion to follow is based on Harlow (2005, p. 97–100) who describes how the following types of matrices can be employed in multivariate analysis.

   Data matrix   A researcher can begin with a **data matrix** (which will be represented with the notation **X**), which is a rectangular $n \times k$ matrix that summarizes the scores of *n* subjects (with each subject represented by one of the *n* rows of the matrix) on a set of *k* variables (with one variable represented by one of the *k* columns of the matrix). In a **data matrix**, *k* scores are recorded in each of the *n* rows, and within each row the score recorded in given column represents a subject's score on the variable in question. Sometimes, a **data matrix** may be converted into a **standard score matrix**, in which scores on each of the variables are expressed in the form of **standard scores** (i.e., *z* scores, which are obtained through use of Equation I.38 employing $\tilde{s}$ as an estimate of σ). **Matrix X** below represents the general format of a **data matrix**, where X_{ij} represents the score of the ith subject on the jth variable.

$$\mathbf{X} = \begin{bmatrix} X_{11} & X_{12} & X_{13} \\ X_{21} & X_{22} & X_{23} \\ X_{31} & X_{32} & X_{33} \end{bmatrix}$$

   Mean matrix   From a **data matrix** a researcher can derive a rectangular $n \times k$ **mean matrix** (which will be represented with the notation **M**) that contains the mean score on each of the *k* variables (which are represented by the *k* columns in the matrix) for each of the *n* subjects (with each subject represented by one of the *n* rows of the matrix). In a **mean matrix**, within each column the same mean score is recorded for every subject for the variable represented by that column. **Matrix M** below represents the general format of a **mean matrix**, where $\bar{X}_{.k}$ represents the mean score on the kth variable.

$$\mathbf{M} = \begin{bmatrix} \bar{X}_{.1} & \bar{X}_{.2} & \bar{X}_{.3} \\ \bar{X}_{.1} & \bar{X}_{.2} & \bar{X}_{.3} \\ \bar{X}_{.1} & \bar{X}_{.2} & \bar{X}_{.3} \end{bmatrix}$$

   Deviation score matrix   A **deviation score matrix** (which will be represented with the notation **D**) can be derived by subtracting the **mean matrix** from the **data matrix**: Thus $\mathbf{D} = \mathbf{X} - \mathbf{M}$. The resulting **deviation score matrix** is a rectangular $n \times k$ matrix which summarizes the difference scores of the *n* subjects on the *k* variables. Specifically, each of the *k* scores recorded in each of the *n* rows of the **deviation score matrix** represents the difference between the score of the subject represented in that row less the mean of the variable represented by the column in which the score appears. **Matrix D** below represents the general format of a deviation score matrix, where D_{ij} represents the deviation score of the ith subject on the jth variable.

$$
\mathbf{D} = \begin{bmatrix} \bar{X}_{11} & \bar{X}_{12} & \bar{X}_{13} \\ \bar{X}_{21} & \bar{X}_{22} & \bar{X}_{23} \\ \bar{X}_{31} & \bar{X}_{32} & \bar{X}_{33} \end{bmatrix} - \begin{bmatrix} \bar{X}_{.1} & \bar{X}_{.2} & \bar{X}_{.3} \\ \bar{X}_{.1} & \bar{X}_{.2} & \bar{X}_{.3} \\ \bar{X}_{.1} & \bar{X}_{.2} & \bar{X}_{.3} \end{bmatrix} = \begin{bmatrix} X_{11} - \bar{X}_{.1} & X_{12} - \bar{X}_{.2} & X_{13} - \bar{X}_{.3} \\ X_{21} - \bar{X}_{.1} & X_{22} - \bar{X}_{.2} & X_{23} - \bar{X}_{.3} \\ X_{31} - \bar{X}_{.1} & X_{32} - \bar{X}_{.2} & X_{33} - \bar{X}_{.3} \end{bmatrix}
$$

$$
\mathbf{D} = \begin{bmatrix} D_{11} & D_{12} & D_{13} \\ D_{21} & D_{22} & D_{23} \\ D_{31} & D_{32} & D_{33} \end{bmatrix}
$$

Sum of squares and cross-products matrix A **sum of squares and cross-products matrix** (which will be represented with the notation **S**) can be computed from a **deviation score matrix** by multiplying the latter matrix by its **transpose** — in other words, if **D** represents the **deviation score matrix** and **S** represents the **sum of squares and cross-products matrix**, then **S** = **DD′**. A **sum of squares and cross-products matrix** is a square symmetrical matrix which contains the **sums of squares and cross-products** for the k variables, with one of the rows representing each of the variables and one of the columns representing each of the variables. To be more specific, a **sum of squares and cross-products matrix** contains the numerators of the k variances (i.e., $SS_{X_j} = \Sigma X_j^2 - [(\Sigma X_j)^2 / n_j]$) which are represented by elements in the main diagonal, and the numerators of the covariances (i.e., $SP_{X_i Y_j} = \Sigma X_i Y_j - [[(\Sigma X_i)(\Sigma Y_j)]/n]$) for pairs of variables which are represented by the off-diagonal elements. The **trace** of a **sum of squares and sum of cross-products matrix** will be the sum of the sum of squares values for the k variables. **Matrix S** below represents the general format of a **sum of squares and cross-products matrix**, where SS_j represents the sum of squares of the jth variable and SP_{ij} represents the sum of products of the ith and j^{th} variables. Note that $SP_{ij} = SP_{ji}$.

$$
\mathbf{S} = \mathbf{DD'} = \begin{bmatrix} D_{11} & D_{12} & D_{13} \\ D_{21} & D_{22} & D_{23} \\ D_{31} & D_{32} & D_{33} \end{bmatrix} \times \begin{bmatrix} D_{11} & D_{21} & D_{31} \\ D_{12} & D_{22} & D_{32} \\ D_{13} & D_{23} & D_{33} \end{bmatrix}
$$

$$
\mathbf{S} = \begin{bmatrix} SS_1 & SP_{12} & SP_{13} \\ SP_{21} & SS_2 & SP_{23} \\ SP_{31} & SP_{32} & SS_3 \end{bmatrix}
$$

Variance-covariance matrix A **variance-covariance matrix** (which will be represented with the notation **V**) can be derived from a **sum of squares and cross-products matrix** by multiplying the latter matrix by $(1/n-1)$ (which is equivalent to dividing the matrix by $(n-1)$), since the latter will convert a sum of squares into a variance and a sum of cross-products into a covariance. Thus, $\mathbf{V} = \mathbf{S}(1/n-1)$. **A variance-covariance matrix** is a square symmetrical matrix which will contain the **variances** and **covariances** for k variables, with one of the rows representing each of the variables and one of the columns representing each of the variables. More specifically, the elements in the main diagonal of a **variance-covariance matrix** represent the variances of the k variables, and the off-diagonal elements represent the covariances for pairs of the k variables. The **trace** of a **variance-covariance matrix** will be the sum of the variances of the k variables. Stevens (2002, p. 67) notes that the **determinant** of a **variance-covariance matrix** (i.e., in the case of **Matrix E** discussed previously, $|\mathbf{E}| = ad-bc$) can be viewed as a generalized variance for the set of variables. **Matrix V** below represents the general format of a **variance-covariance matrix**, where $\tilde{s}_j^2$ represents the variance of the jth variable and cov_{ij} represents the covariance of the ith and j^{th} variables. Note that $cov_{ij} = cov_{ji}$.

$$\mathbf{V} = \begin{bmatrix} SS_1 & SP_{12} & SP_{13} \\ SP_{21} & SS_2 & SP_{23} \\ SP_{31} & SP_{32} & SS_3 \end{bmatrix} \times \frac{1}{n-1}$$

$$\mathbf{V} = \begin{bmatrix} \tilde{s}_1^2 & cov_{12} & cov_{13} \\ cov_{21} & \tilde{s}_2^2 & cov_{23} \\ cov_{31} & cov_{32} & \tilde{s}_3^2 \end{bmatrix}$$

Correlation matrix A **correlation matrix** (sometimes referred to as an **R-matrix**) can be derived from a **variance-covariance matrix** by dividing each of the elements in the latter matrix by the square root of the product of the respective variances (which is accomplished by multiplying each element in the **variance-covariance matrix** by 1 over the square root of the product of the respective variances). A **correlation matrix** is a square symmetrical matrix which summarizes the intercorrelations between k variables, with one of the rows representing each of the variables and one of the columns representing each of the variables. The main diagonal of a **correlation matrix** contains the correlation of each of the k variables with itself— specifically, $r = 1$ is recorded in the main diagonal for all k variables (except in certain types of **factor analysis** where estimates of common variance (also referred to as communality) or reliability are recorded in the main diagonal). When the value 1 is recorded in the main diagonal for all k variables, the **trace** of a **correlation matrix** will equal k. The off-diagonal elements in a **correlation matrix** represent correlations between pairs of variables. **Matrix R** below represents the general format of a **correlation matrix**, where r_{ij} represents the correlation between the ith and j^{th} variables. Note that $r_{ij} = r_{ji}$.

$$\mathbf{R} = \begin{bmatrix} r_{11} & r_{12} & r_{13} \\ r_{21} & r_{22} & r_{23} \\ r_{31} & r_{32} & r_{33} \end{bmatrix}$$

Diagonal matrix A **diagonal matrix** is a square symmetrical matrix which has zeros as all of the off-diagonal elements and values other than zero as the diagonal elements. An **identity matrix** (referred to earlier in the discussion of the derivation of the inverse of a matrix) is an example of a **diagonal matrix**.

The matrices most commonly encountered in multivariate analysis are those which contain information about variability. Representative of the latter type of matrices are **sum of squares and cross-product matrices** and **variance-covariance matrices**, both of which are evaluated in **multivariate analysis of variance** and **discriminant function analysis**. Linear combinations of variables are often derived in multivariate analysis from information contained in matrices. **Canonical correlation (Test 38)** and **principal components analysis and factor analysis** (as well as in **multivariate analysis of variance** and **discriminant function analysis**) are examples of multivariate procedures which derive linear combinations of variables. Within the context of deriving linear combinations it is common practice to compute **eigenvalues**. **Eigenvalues** are numerical indices that reflect the proportion of the overall variance which can be explained by each of the linear combinations of variables (linear combinations of variables are referred to as **factors** within the context of **factor analysis**), and as a general rule there are as many eigenvalues as there are variables. The sum of the **eigenvalues** derived from a **correlation matrix** will equal k (i.e., the number of variables employed in the analysis).[2]

References

Grimm , L. G. & Yarnold, P. R. (1995). Introduction to multivariate statistics. In Grimm, L. G. & Yarnold, P. R. (Eds.) (pp. 1–18). **Reading and understanding multivariate statistics**. Washington, D.C.: American Psychological Association.

Harlow, L. L. (2005). **The essence of multivariate thinking**. Mahwah, NJ: Lawrence Erlbaum Associates.

Marascuilo, L. A. & Levin, J. R. (1983). **Multivariate statistics in the social science: A researchers guide**. Monterey, CA: Brooks/Cole Publishing Company.

Stevens, J. (2002). **Applied multivariate statistics for the social sciences** (4th ed.). Hillsdale, NJ: Lawrence Erlbaum Associates, Publishers.

Tabachnick, B. G. & Fidell, L. S. (1996). **Using multivariate statistics** (3rd ed.). New York: Harper Collins Publishers.

Tabachnick, B. G. & Fidell, L. S. (2001). **Using multivariate statistics** (4th ed.). Boston: Allyn & Bacon.

Thompson, B. T. (2000). Canonical correlation analysis. In Grimm, L. G. & Yarnold, P. R. (Eds.) (pp. 285–316). **Reading and understanding more multivariate statistics**. Washington, D.C.: American Psychological Association.

Endnotes

1. The use of the term **continuous variable** refers to a variable that can be represented on an interval/ratio scale of measurement.

2. Although the mathematical definition of an **eigenvalue** is complex, Field (2005, pp. 197–198) provides a good illustration of what the latter concept represents. In the latter discussion he notes that an **eigenvalue** is directly related to an **eigenvector**, which is a linear representation of some dimension in a multidimensional space — e.g., the height, length, or width of a three-dimensional space. An **eigenvalue** is simply a numerical expression of the dimension of an **eigenvector**. The nature of the distribution of the variances in a data matrix can be understood by examining the interrelationships between the eigenvalues computed for a matrix.

Test 33

Multiple Regression
(Parametric Test Employed with Interval/Ratio Data)

The use of the term **multiple regression** in this chapter refers to both the computation of a **multiple correlation coefficient** and the derivation of a linear equation which best describes the relationship between two or more of predictor variables (which may represent interval/ratio or categorical data) and an interval/ratio criterion variable. Although some sources define a multivariate procedure as one in which there are two or more dependent variables, **multiple regression**, which involves one criterion/dependent variable, is commonly categorized as a multivariate procedure by virtue of the fact that it involves the use of multiple predictor/ independent variables.[1]

Hypothesis evaluated with test   The primary hypothesis evaluated within the framework of **multiple regression** is whether or not there is a significant association between two or more independent/predictor variables and a dependent/criterion variable.

I. Hypothesis Evaluated with Test and Relevant Background Information

General introduction to multiple regression analysis   Multiple regression (as opposed to **simple, bivariate, univariate,** or **zero-order regression/correlation** — all of which refer to an analysis involving one predictor and one criterion variable) represents the generalization of simple correlation and regression (discussed under the **Pearson product-moment correlation coefficient (Test 28)**) to an analysis that involve more than two predictor variables. Such an analysis is considerably more complex than **bivariate correlation/regression** (which, as noted, involves two variables, the X variable which is generally conceptualized as the predictor variable, and the Y variable which is generally conceptualized as the criterion variable). Because of the laborious computations involved in **multiple regression**, it is all but impractical to conduct without the aid of computer software.

   In contrast to **simple linear regression**, where scores on one predictor variable are employed to predict the scores on a criterion variable, in **multiple regression** a researcher attempts to increase the accuracy of prediction through the use of multiple predictor variables. By employing multiple predictors, one can often account for a greater amount of the variability on the criterion variable. Thus, the major goal of **multiple regression** is to identify a limited number of predictor variables which optimize one's ability to predict scores on a criterion variable. Although the level of measurement for the criterion variable must be interval/ratio data, the predictor variables can be interval/ratio data (which will be the primary focus of this chapter) and/or categorical data — the latter being expressed within the format of a **dummy variable.**[2]

   Multiple regression analysis allows a researcher to compute a correlation coefficient between a criterion variable and k predictor variables (where $k \geq 2$). The value computed for the **multiple correlation coefficient** (which is the analog of the r value computed in simple correlation/regression) is represented by the notation R. The latter value is an estimate of the underlying population **multiple correlation** P (which is the upper case Greek letter **rho**). A computed value of R must fall within the range 0 to +1 (i.e., $0 \leq R \leq +1$). Unlike an r value computed for two variables, the **multiple correlation coefficient** cannot be a negative number.

The closer the value of R is to 1, the stronger the linear relationship between the criterion variable and the k predictor variables, whereas the closer it is to 0, the weaker the linear relationship.

Within the framework of **multiple regression** analysis there are a number of other types of correlation coefficients which can be computed that can further clarify the nature of the relationship between predictor variables and the criterion variable. In evaluating the relationship between a criterion variable and a single predictor variable, it is not uncommon that the correlation is influenced by a third variable. As an example, the relationship between frequency of violent crimes (which will represent the criterion variable Y) and level of stress (which will represent the predictor variable X_1) is undoubtedly influenced by extraneous variables such as social class, which if included in a **multiple regression** analysis would represent a second predictor variable X_2. In some instances, by measuring the influence of a third variable (in this case the second predictor variable of social class), a researcher will be better able to understand the nature of the relationship between the other two variables (violent crime and stress). By allowing the third variable to serve in the role of a meditating variable, it often increases the researcher's ability to predict the scores of subjects on the criterion variable. In other instances, however, the researcher may view the contribution of a third variable as interfering with the study of the relationship between the other two variables. Thus, if a researcher wants to obtain a "purer" measure of the relationship between violent crime and stress, he might want to eliminate the influence of social class from the analysis. In instances where one wants to control for the influence of an extraneous variable, the latter variable is viewed as a **nuisance variable**. Fortunately, correlational procedures have been developed which allow researchers to statistically control for the influence of extraneous variables. Two of these correlations (which are computed within the context of **multiple regression** and are discussed in Section IV/V) are **partial correlation** and **semipartial correlation** (which is also referred to as **part correlation**).

Since in **multiple regression** analysis a researcher will want to derive the simplest possible predictive model, it is unusual to find a model that involves more than five predictor variables. Among the reasons for limiting the number of predictors are the following: a) Selection of predictors should be based on sound theoretical reasoning, and because of the latter, in most instances a researcher will only be able to justify inclusion of a limited number of such variables. It should be emphasized that researchers are discouraged from conducting an exploratory **multiple regression** analysis which initially begins with a large number of predictor variables, with the hope of identifying a limited number of them as significant. Sources are in agreement that the latter strategy inevitably results in Type I errors, as well as identifying unreliable predictors which are idiosyncratic with respect to the sample employed in the study; b) Once a limited number of predictor variables has been identified which explain a relatively large proportion of the variability on the criterion variable, it becomes increasingly unlikely that any new predictors that are identified will result in a significant increase in predictive power; c) Although the researcher wants to identify predictor variables which are highly correlated with the criterion variable, he also wants to make sure that the predictor variables employed account for different proportions of the variability on the criterion variable. In order to accomplish the latter, none of the predictor variables should be highly correlated with one another, since, if the latter is true, the variables will be redundant with respect to the variation on the criterion variable they explain. As a general rule, it is difficult to find a large number of predictors which are highly correlated with a criterion variable, yet not correlated with one another. The term **multicollinearity** (which is also referred to as **collinearity** when just two variables are involved) refers to a situation where two or more predictor variables are highly intercorrelated with one another. When **multicollinearity** (which is discussed in greater detail later in this section as well as in Section IV/V) is present, the reliability of a **multiple regression** analysis may be severely compromised; d) A final reason for limiting the number of predictor variables is that from a time and cost perspective, a model which involves a large number of predictors will entail more time and expense on the part of a researcher.

Within the framework of **multiple regression** the following methodologies can be employed in selecting which variables that comprise a set of k potential predictor variables should be employed to predict a criterion variable.

a) In **standard multiple regression** (also called **direct** or **simultaneous multiple regression**) all k of the predictor variables are included in the analysis simultaneously, including those which may only explain a minimal amount of variability on the criterion variable. This type of regression may be employed when a researcher wants to explore, for theoretical or other reasons, the relationship between a large set of predictor variables and a criterion variable. As noted above, the latter strategy is generally discouraged. **Standard multiple regression** is atypical when compared with the other methods of regression analysis, in that it is more likely to result in a large number of predictor variables.

b) In **hierarchical multiple regression** (also called **sequential multiple regression**) a researcher uses preexisting empirical data/information or some model/theory to determine the order in the which each of the predictor variables is entered into the analysis. The order in which the predictors are entered is dictated by the degree to which they are believed to influence the criterion variable. Predictors are entered into the analysis beginning with the variable thought to be most important, and subsequently entering the remaining variables in descending order of importance. Within the context of **hierarchical multiple regression**, a researcher can determine whether the addition of each new predictor variable increases predictability on the criterion variable — more specifically, one can determine the amount of variance each predictor variable can account for which is significantly above and beyond variance already accounted for by previously entered predictors.

c) In **stepwise multiple regression** (also called **statistical multiple regression**) the order in which the variables are entered into the analysis is based on some statistical rule — most commonly, which predictor variable has the strongest **partial correlation** with the criterion variable, after controlling for the effect of variables already entered into the regression equation. (A **partial correlation coefficient** allows a researcher to measure the degree of association between two variables after any linear association one or more additional variables have with the other two variables has been removed.) Sources (e.g., Field (2005, p. 161), Licht (1995, p. 53), and Harlow (2005, p. 45)) note that since **stepwise multiple regression** depends more on chance variation (e.g., idiosyncrasies associated with a specific sample) than the other **multiple regression** methods, researchers should employ extreme caution in using it (although it may be of value when research is exploratory in nature — i.e., no preexisting empirical information or theory is available to guide a researcher). Three types of **stepwise multiple regression** are **forward selection**, **backward selection**, and **stepwise selection**. In **forward selection** the predictor variable which contributes most to prediction is entered into the analysis. The other predictor variables are then entered into the analysis one by one, based on the degree to which they produce a significant increase in prediction until such time predictor variables no longer make significant contributions. The final regression equation includes only those predictors which made a significant contribution in computing the value of the **coefficient of multiple correlation**. In **backward selection** a regression equation is initially computed employing all of the predictor variables. The predictor contributing least to prediction is dropped from the analysis (i.e., if the variable contributes at all, its contribution is not statistically significant). Employing the remaining predictors, a new regression equation is computed and the same process is repeated. The aforementioned process is repeated until only those predictors remain which make a significant contribution in computing the value of the **coefficient of multiple correlation**. **Stepwise selection** combines both forward and backward selection by employing a stepwise methodology to include predictors which add the most (and significantly) to prediction and exclude predictors which contribute the least (and not significantly) to prediction. More specifically, the procedure is extremely similar to forward selection, except that after the addition of each new predictor to the regression equation, all of the predictors already in the equation are reevaluated to determine whether they should be removed. If an *F* test conducted on the predictor already in the equation which is responsible for the for the smallest increment in R^2 (which is the square of the **multiple correlation coefficient**) is not significant, it is removed from the regression equation.

d) In **all possible subsets multiple regression** (described by Licht (1995, p. 53) all possible permutations and combination of predictor variables are examined, and the combination of variables which yields the largest **squared multiple correlation** is selected for the final model. Licht (1995, p. 53) notes that this methodology is not commonly employed, probably because chance factors can result in identifying a predictive model which is not reflective of what is true with regard to the underlying population.

Upon conducting a **multiple regression** analysis, it is recommended that the resulting regression model be **cross-validated**. By cross-validating a model, a researcher can demonstrate that it generates consistent results, and will thus be of theoretical and/or of practical value in reference to the target population. Among the procedures that can be employed for cross-validation are the following: a) Replicate the results of a **multiple regression** analysis on two subsamples, each representing a different half of the original sample; b) Replicate the results of a **multiple regression** analysis on one or more independent samples that are representative of the target population to which the researcher wishes to apply the regression model; and c) A third method of cross-validation is to initially conduct a **multiple regression** analysis on a sample of subjects. A second sample is then obtained from the same target population, and the **multiple regression** equation derived for the first sample is employed to predict the scores of subjects on the criterion variable in the second sample. The latter set of predicted scores are then correlated with the actual scores of the subjects on the criterion variable in the second sample. The obtained correlation coefficient will reflect the accuracy of prediction when the original regression equation is employed to predict scores on the criterion variable for subjects in the second sample. As Licht (1995, p. 23) notes, this method of cross-validation is not the same as the previously described method. Whereas in the latter method the results of a second **multiple regression** analysis are compared with an initial **multiple regression** analysis, in this type of cross-validation the results of the initial analysis are evaluated by employing them with data obtained from a second sample.

It cannot be emphasized too strongly that prior to conducting a **multiple regression** analysis (or for that matter any multivariate analysis) a researcher should carefully scrutinize his data. Prior screening allows one to do the following: a) Correct any erroneous data entries; b) Identify and develop a strategy for dealing with missing data and/or **outliers**; c) Utilize diagnostic indices of **distance**, **leverage**, and **influence** which are available with most computer software. The latter measures, which are described in Section IV/V, are useful in identifying multivariate observations that exert undue influence on a set of data. More specifically, these indices most often allow the researcher to identify **outliers** on any of the predictor variables and/or the criterion variable, which by virtue of their presence can dramatically impact the values of the **multiple correlation coefficient**, **regression coefficients**, and other statistics that are computed within the framework of a **multiple regression** analysis; d) Assess the data for **multicollinearity**; and e) Determine whether or not any of the assumptions underlying **multiple regression** analysis have been violated. The latter applies to assumptions involving the overall relationship between the variables, as well as to assumptions with respect to the individual variables. A check of the assumptions underlying **multiple regression** can be made both prior to an analysis as well as after deriving a regression model. At this point the following assumptions underlying **multiple regression** (as well as most other multivariate procedures involving continuous variables) will be discussed: 1) Multivariate normality; 2) Homoscedasticity; 3) Linearity; 4) Normality and independence of residuals.

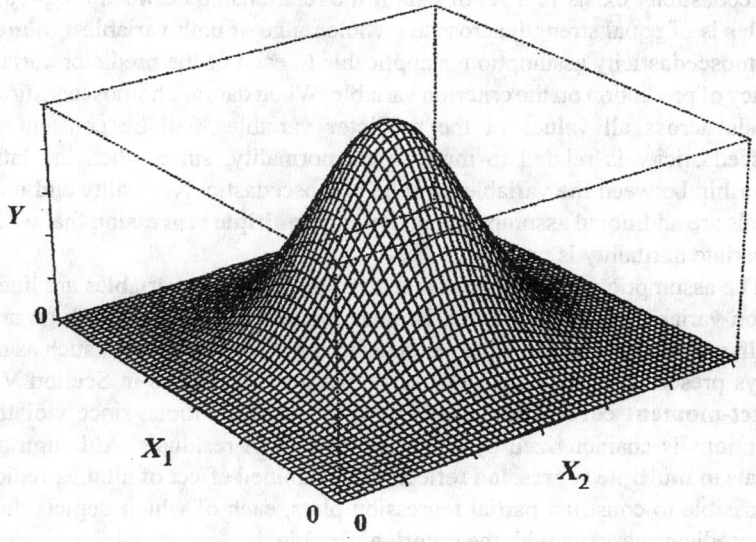

Figure 33.1 Bivariate Normal Distribution

The assumption of **multivariate normality** means that: a) All of the variables are normally distributed. Tabachnick and Fidel (1996, p. 70–71; p. 640) note that the latter is generally assessed through analysis of skewness and kurtosis, and any variable for which either condition is violated should probably be modified through use of a data transformation; b) Multivariate normality also assumes that linear combinations between each pair of variables are normally distributed. In addition to confirming the linear relationship between a pair of variables, the latter is assessed by demonstrating homoscedasticity, as well as by determining that the residuals are distributed normally (most commonly through use of scatterplots).

The simplest case of a multivariate normal distribution is a **bivariate normal distribution** in which there are two predictor variables and a criterion variable. A graph of the latter distribution is a three dimensional image resembling a mountain such as that depicted in Figure 33.1. Multivariate normal distributions involving more than two predictor variables are generally not depicted graphically since they require more than three dimensions.

As previously noted, **multiple regression** (as well as other parametric multivariate analytical procedures) is based on the assumption that the underlying population distribution for any continuous variable is normal. Within the context of assessing normality, computer software can compute descriptive statistics such as skewness and kurtosis for all of the variables. For most computer software, the values computed for both skewness and kurtosis will equal 0 when a distribution is normal. (The reader should refer to the **Introduction** and the **single-sample tests for evaluating population skewness and kurtosis (Tests 4 and 5)** for further clarification of the measurement of skewness and kurtosis, and how the latter values are represented in computer software.) Normality can also be assessed with a test of goodness-of-fit such as those described in the chapters on the **single-sample test for evaluating population kurtosis** and the **Kolmogorov–Smirnov goodness-of-fit test for a single sample (Test 7)**. Harlow (2005, p. 34) notes that violation of the normality assumption can sometimes be resolved through use of a data transformation (most often logarithmic).

A second assumption underlying **multiple regression** is that of homoscedasticity. In Section VII of the **Pearson product-moment correlation coefficient** it was noted that homoscedasticity exists in a set of data if the relationship between the predictor and criterion variables is of equal strength across the whole range of both variables. In **multiple regression** the homoscedasticity assumption is applicable to each of the predictor variable with regard to accuracy of prediction on the criterion variable. When data are homoscedastic the variance of the residuals across all values of the predictor variables will be constant. As noted above, homoscedasticity is related to multivariate normality, since when the latter is present the relationship between the variables will be homoscedastic. Normality and independence of the residuals are additional assumptions underlying **multiple regression** that will also be met when multivariate normality is present.

The assumption of linearity is that all of the predictor variables are linearly related to the criterion variable. This assumption is most commonly evaluated through analysis of residual plots. In point of fact, plots of the residuals in **multiple regression** (such as the type of graphic displays presented in the discussion of regression diagnostics in Section VII of the **Pearson product-moment correlation coefficient**) are valuable tools, since violation of each of the assumptions is characterized by a specific pattern of residuals. Although an examination of residuals in **multiple regression** reflects the combined effect of all the predictor variables, it is also possible to construct partial regression plots, each of which depicts the relationship of a single predictor variable with the criterion variable.

It was noted earlier that the term **multicollinearity** refers to a situation where predictor variables are highly intercorrelated with one another, and that when **multicollinearity** exists the reliability of **multiple regression** analysis may be severely compromised. Licht (1995, p. 45) notes that the presence of **multicollinearity** can result in problems with respect to mathematical solution as well as practical prediction and theoretical interpretation. Although there is no consensus with respect to a specific cutoff value indicative of the presence of **multicollinearity**, among others, Harlow (2005, p. 34) states that if the correlation between two predictor variables is greater than .90, **multicollinearity** is present (as well as the fact that the presence of correlations of .70 or larger may be a red flag suggesting a potential problem). Similarly, Licht (1995, p. 45) views values greater than .80 as an indicator of a potential problem. Correlations between predictors in the range of the latter values strongly suggest that the variables in question may be measuring the same construct (i.e., concept). When **multicollinearity** is present, it is generally suggested that one of the related variables be dropped from the analysis, or else the variables be combined into a single composite variable.

The readers should also take note of these additional points with reference to **multicollinearity**: a) Licht (1995, pp. 45–46) emphasizes the fact that the results of a **multiple regression** analysis will be affected by any degree of correlation between predictor variables. On the basis of the latter it might appear that the ideal situation is to find a set of two or more predictor variables which are highly correlated with the criterion variable, yet are minimally correlated or uncorrelated with one another. Aside from the fact that it is unlikely a researcher will be able to identify such a set of predictors, it is also the case that sometimes inclusion of a variable which is highly correlated with the criterion variable and moderately correlated with one or more predictors may actually result in a superior predictive or theoretical model when contrasted with the model that would result if the variable in question was omitted from the analysis. For the latter reason, Licht (1995, p. 50) states that the ultimate decision with respect to whether or not a specific predictor variable be included in an analysis should be based on a sound theoretical rationale; b) Although Licht (1995, p. 46) notes that one exception to the preference for low **multicollinearity** is inclusion of one or more **suppressor variables**, many researchers view the inclusion of **suppressor variables** as more problematic than beneficial (largely due to the fact that their inclusion can make it more difficult to clearly interpret the results). (A **suppressor variable**, which is discussed in detail in Section IV/V, is a predictor variable which is useful in predicting the criterion variable purely on the basis of the fact that it correlates highly with the other predictor variables.); c) Sources note that the greater the **multicollinearity** in a set of data, the more unstable (i.e., unreliable) the regression coefficients computed for the analysis. An unstable regression coefficient will have an inflated standard error, which will compromise both the power of a test of significance on the coefficient as well as increase the range of values which define the confidence interval for the coefficient.

A term that is related to but not identical to **multicollinearity** is **singularity**, which is when two or more predictor variables are perfectly correlated with one another (Field (2005, p. 641). Tabachnick and Fidell (1996, p. 84–86) (who define **singularity** as being present when the predictor variables are redundant — i.e., one of the variables is a combination of two or more of the other variables) note that both **multicollinearity** and **singularity** can dramatically impact **matrix inversion** (discussed in the previous chapter), which is one of the mathematical operations employed in **multiple regression** analysis. The latter authors state that whereas **multicollinearity** renders matrix inversion unstable, **singularity** makes it impossible.

Since the reliability of a **multiple regression** analysis will be a function of the number of predictors and the sample size, with respect to the latter there is a general consensus that the larger the sample the better. Readers should take note of the following with respect to the sample size employed in **multiple regression**: a) Field (2005, p. 172) notes that the expected value of a **multiple correlation coefficient** (R) for random data is $k/(n - 1)$, and because of the latter a small sample size can yield an R value substantially greater than 0; b) Howell (2002, p. 548) points out that although some researchers (e.g., Marascuilo and Levin (1983)) have adhered to the rule of employing a minimum of 10 observations for each predictor variable, there is no empirical evidence to support the reliability of the latter rule; c) Harlow (2004, p. 46) states that when all of assumptions underlying **multiple regression** are adhered to, the minimum acceptable sample size is $n = 100$, and when one or more of the assumptions have been violated the sample size should be substantially larger; d) Tabachnick and Fidell (1996, p. 132) state the following rules for computing sample size: $n \geq 50 + 8k$ for evaluating a **multiple correlation coefficient**, and $n \geq 104 + k$ for testing individual predictors. If the researcher is interested in using both multiple and individual predictors, the larger of the two sample sizes computed with the two aforementioned rules should be employed; e) Cohen and Cohen (1983) emphasize the importance of taking statistical power into consideration in establishing sample size. With regard to the latter, Field (2005, p. 173) recommends the use of graphs developed by Miles and Shevlin (2001) which plot sample size in relation to power values for different effect sizes. Field (2005, p. 174) notes that use of the latter graphs (assuming an analysis with no more than 20 predictor variables) indicates that sample sizes of 80, 100 to 200, and 600 should be sufficient to respectively detect large, medium and small effect sizes.

II. Examples

Example 33.1 is identical to Example 28.1 (in Section II of the **Pearson product-moment correlation coefficient**), except for the addition of a second predictor variable. The use of the small sample size in Example 33.1 is to facilitate understanding of the computational procedures

Table 33.1   Raw Data for Example 33.2

sub	x1agilit	x2streng	x3intell	yperfor
1	1.00	38.00	24.00	22.00	90.00
2	2.00	32.00	21.00	29.00	62.00
3	3.00	18.00	8.00	34.00	65.00
4	4.00	10.00	16.00	20.00	57.00
5	5.00	37.00	17.00	24.00	72.00
6	6.00	24.00	27.00	15.00	98.00
7	7.00	32.00	23.00	32.00	100.00
8	8.00	36.00	29.00	30.00	106.00
9	9.00	9.00	15.00	16.00	62.00
10	10.00	4.00	19.00	18.00	63.00
11	11.00	39.00	25.00	21.00	90.00
12	12.00	39.00	16.00	22.00	80.00
13	13.00	29.00	21.00	27.00	95.00
14	14.00	19.00	24.00	2.00	79.00
15	15.00	17.00	18.00	25.00	77.00
16	16.00	.00	17.00	48.00	35.00
17	17.00	15.00	18.00	22.00	58.00
18	18.00	40.00	22.00	17.00	84.00
19	19.00	42.00	21.00	39.00	103.00
20	20.00	13.00	13.00	24.00	58.00
21	21.00	49.00	16.00	32.00	112.00
22	22.00	.00	25.00	33.00	57.00
23	23.00	24.00	27.00	44.00	108.00
24	24.00	23.00	28.00	34.00	92.00
25	25.00	26.00	22.00	28.00	83.00

sub	x1agilit	x2streng	x3intell	yperfor
26	26.00	26.00	22.00	21.00	88.00
27	27.00	28.00	22.00	30.00	101.00
28	28.00	23.00	12.00	35.00	60.00
29	29.00	20.00	20.00	20.00	60.00
30	30.00	29.00	10.00	48.00	102.00
31	31.00	36.00	22.00	38.00	100.00
32	32.00	33.00	21.00	30.00	64.00
33	33.00	20.00	10.00	30.00	67.00
34	34.00	11.00	14.00	24.00	60.00
35	35.00	34.00	19.00	23.00	71.00
36	36.00	25.00	27.00	28.00	99.00
37	37.00	30.00	20.00	30.00	90.00
38	38.00	37.00	31.00	48.00	105.00
39	39.00	8.00	14.00	13.00	62.00
40	40.00	5.00	20.00	20.00	65.00
41	41.00	38.00	20.00	32.00	99.00
42	42.00	33.00	15.00	21.00	81.00
43	43.00	28.00	22.00	28.00	96.00
44	44.00	18.00	23.00	11.00	77.00
45	45.00	14.00	16.00	27.00	80.00
46	46.00	1.00	15.00	48.00	35.00
47	47.00	13.00	1.00	23.00	55.00
48	48.00	41.00	20.00	18.00	83.00
49	49.00	44.00	22.00	40.00	104.00
50	50.00	14.00	12.00	28.00	60.00

for a **multiple regression** analysis involving two predictor variables (computations become considerably more complex when additional predictors are involved in an analysis). Example 33.2, which employs a larger sample size (albeit still less than most researchers would find ideal), is presented in order to demonstrate the use of a computer software package (specifically, *SPSS*) in conducting a **multiple regression** analysis. Readers should take note of the fact that the data employed for the Example 33.2 are hypothetical and are not based on an actual study.

Example 33.1 *A nutritionist conducts a study employing a sample of five children to determine whether or not there is a statistical relationship between the number of ounces of sugar and the number of ounces of salt a ten-year-old child eats per week and the number of cavities in the child's mouth. The three scores, ounces of sugar consumed per week (which will represent predictor variable X_1), ounces of salt consumed per week (which will represent predictor variable X_2), and number of cavities (which will represent the criterion variable Y) obtained for each of the five children follow:* **Child 1** (20, 4, 7); **Child 2** (0, 1, 0); **Child 3** (1, 1, 2); **Child 4** (12, 3, 5); **Child 5** (3, 6, 3). *Evaluate what, if any, relationship there is between the two predictor variables and the criterion variable.*

Example 33.2 *A professional football team conducts a study to evaluate whether or not the variables of agility, strength, and intelligence will allow the team to predict how well a player performs during his first year in the league. The researcher conducting the study employs a sample of 50 first year players, each of whom prior to joining the team is administered a test of agility (which will represent predictor variable X_1), strength (which will represent predictor variable X_2), and intelligence (which will represent predictor variable X_3). On the test of agility a subject can obtain a score between 0 and 50, with the higher the score the more agile the subject; on the test of strength a subject can obtain a score between 0 and 35, with the higher the score the stronger the subject; and on the test of intelligence a subject can obtain a score between 0 and 50, with the higher the score the more intelligent the subject. At the conclusion of the season each of the players is rated by the coaching staff (who are not privy to a subject's test scores) on a scale from 0 to 120 with respect to how well he performed during his first year in the league (which will represent the criterion variable Y — with the higher a subject's score the better his performance). The data for the study are presented in* Table 33.1. *Do the data indicate there is a relationship between any of the predictor variables and how well a player performed?*

III. Null versus Alternative Hypotheses

The primary null hypothesis evaluated within the framework of **multiple regression** is that there is no relationship between the predictor/independent variables and the criterion/dependent variable. In order to evaluate the latter hypothesis it is common practice to determine whether the obtained value of the **multiple correlation coefficient** is large enough to conclude that the underlying population **multiple correlation** between the predictor variables and the criterion variable is some value other than zero. The latter is expressed symbolically below in reference to Example 33.1.[3]

Null hypothesis H_0: P = 0

(In the underlying population the sample represents, the **multiple correlation** between the scores of subjects on the predictor variables X_1 and X_2 versus the criterion variable Y equals 0.)

Alternative hypothesis H_1: P ≠ 0

(In the underlying population the sample represents, the **multiple correlation** between the scores of subjects on the predictor variables X_1 and X_2 versus the criterion variable Y does not equal 0.)

An alternative way of stating the null hypothesis is that all of the regression coefficients for the predictor/independent variables are equal to zero. The alternative hypothesis is that at least one of the regression coefficients for the independent/predictor variables is some value other than zero. The latter null and alternative hypotheses are discussed in more detail in the next section.

IV/V. Test Computations and Interpretation of the Test Results

Because of the complexity of the computations involved in **multiple regression** analysis (which generally requires the use of matrix algebra), simple algebraic equations will only be presented for the analysis of Example 33.1 (which involves two predictor variables). The analysis of Example 33.2 (which involves three predictor variables) will focus on the use of computer software (specifically, *SPSS*) in evaluating and interpreting a **multiple regression** analysis. The latter analysis will not describe the matrix algebra operations involved in evaluating Example 33.2. Readers interested in detailed discussions of the mathematics underlying **multiple regression** should consult sources such as Cohen and Cohen (1983), Cohen *et al.* (2003), Harris (2001), Marascuilo and Serlin (1983), Neter *et al.* (1983), Neter *et al.* (1990), Neter *et al.* (1996), and Tabachnick and Fidell (1996, 2001). Field (2005) and Norušis (2004) are excellent sources for those requiring additional information on the use of *SPSS* for **multiple regression** analysis.

A. Test computations and interpretation of results for Example 33.1   In Section I it was noted that prior to conducting a **multiple regression** analysis a researcher should carefully scrutinize the data, and that most computer software can provide relevant information which will alert one to the presence **outliers** and other scores that can exert undue influence on statistics computed for the analysis. Additionally, it was noted that prior screening of data should assess the predictor variables with respect to **multicollinearity**. Since Example 33.1 is primarily employed to illustrate the use of the equations used in **multiple regression analysis** when there are two predictor variables, the aforementioned data screening procedures will be described in detail later in this section within the framework of the discussion of the analysis of Example 33.2. In the latter discussion, Table 33.5 will provide a summary of the **multicollinearity** analysis for Example 33.1.

Computation of the multiple correlation coefficient   When $k = 2$, Equation 33.1 is employed to compute the value of the **multiple correlation coefficient**, which as noted in Section I will be represented by the notation R. More specifically, the notation $R_{Y.X_1X_2}$ can be employed to represent the **multiple correlation coefficient** between the criterion variable Y and the linear combination of the two predictor variables X_1 and X_2.

$$R_{Y.X_1X_2} = \sqrt{\frac{r_{YX_1}^2 + r_{YX_2}^2 - 2r_{YX_1}r_{YX_2}r_{X_1X_2}}{1 - r_{X_1X_2}^2}} . \qquad \textbf{(Equation 33.1)}$$

Prior to computing the values of $R_{Y.X_1X_2}$, the reader should take note of the fact that the following correlation coefficients can be computed for the variables involved in Example 33.1: a) The correlation between the number of ounces of sugar a child eats (which, as noted, represents predictor variable X_1) and the number of cavities (which, as noted, represents criterion variable Y) is $r_{YX_1} = .955$; b) The correlation between the number of ounces of salt a child eats (which, as noted, represents predictor variable X_2) and the number of cavities is $r_{YX_2} = .52$; and c) The correlation between the number of ounces of sugar a child eats and the number of ounces of salt a child eats is $r_{X_1X_2} = .37$.

The three above noted correlations result from the following information: a) The number of ounces of sugar a child eats (X_1) and the number of cavities (Y) are identical to the data recorded in Table 28.1, where $r_{XY} = .955$. Thus, $r_{XY} = r_{YX_1} = .955$; and b) The correlations $r_{YX_2} = .52$ and $r_{X_1X_2} = .37$ are based on the following values computed for salt consumption (X_2) noted for Example 33.1: 4, 1, 1, 3, 6. The mean and estimated population standard deviation for salt consumption are $\bar{X}_2 = 15/5 = 3$ and $\hat{s}_{X_2} = \sqrt{63 - [((15)^2/5)/4]} = \sqrt{18/4} = 2.12$ (with $SS_{X_2} = 18$). Additionally, the sum of products for the X_2Y scores is $SP_{X_2Y} = 12$. The latter value represents the numerator of the covariance between the X_2 and Y scores which is calculated below.

$$SP_{X_2Y} = \Sigma X_2Y - [((\Sigma X_2)(\Sigma Y))/n] = 63 - [((15)(17))/5] = 12$$

Substituting the correlations $r_{YX_1} = .955$, $r_{YX_2} = .52$, and $r_{X_1X_2} = .37$ in Equation 33.1, the **multiple correlation coefficient** $R_{Y.X_1X_2} = .972$ is computed.

$$R_{Y.X_1X_2} = \sqrt{\frac{(.955)^2 + (.52)^2 - 2(.955)(.52)(.37)}{1 - (.37)^2}} = \sqrt{.944} = .972$$

Note that the value $R_{Y.X_1X_2} = .972$ is larger than either value computed when each of the predictor variables is correlated separately with the criterion variable. Of course, the value of $R_{Y.X_1X_2}$ can be only minimally above $r_{YX_1} = .955$, since the maximum value R can attain is 1 (recollect that in Section I it was noted that R must always be a positive value which falls between 0 and 1).

The coefficient of multiple determination R^2, which is the square of the **multiple correlation coefficient**, is referred to as the **coefficient of multiple determination**. The **coefficient of multiple determination** indicates the proportion of variance on the criterion variable which can be accounted for on the basis of variability on the k predictor variables. In our example $R_{Y.X_1X_2}^2 = (.972)^2 = .944$. In point of fact, R^2 is a biased measure of P^2, which is the population parameter it is employed to estimate. The degree to which the computed value of R^2 is a biased estimate of P^2 will be a function of the sample size and the number of predictor variables employed in the analysis. The value of R^2 will be spuriously inflated when the sample size is close in value to the number of predictor variables employed in the analysis.[4] For this reason, sources emphasize that the number of subjects employed in a multiple regression analysis should always be substantially larger than the number of predictor variables. For example (as noted in Section I), Marascuilo and Levin (1983) recommend that the value of n should be at least ten times the value of k.[5]

The information generated in a **multiple regression** analysis is optimized for the sample upon which it is based, and because of the latter is a function of error variance and idiosyncracies associated with that sample. Because of the latter, the value computed for a **multiple correlation** for another sample employing the same variables will not be as high as that computed for the original sample. This latter phenomenon is referred to as **shrinkage**. One way of correcting for the latter, as well as bias resulting from a small sample size, is to employ Equation 33.2 to compute an adjusted value for $\tilde{R}^2$, which is a relatively unbiased estimate of P^2. The value $\tilde{R}^2$ is commonly referred to as a "shrunken" estimate of the **coefficient of multiple determination**.[6]

$$\tilde{R}^2 = 1 - \frac{(1 - R^2)(n - 1)}{n - k - 1}$$ **(Equation 33.2)**

It can be seen below that substituting the values from our example in Equation 33.2 yields the value $\tilde{R}^2_{Y.X_1X_2} = .89$, which is lower than $R^2_{Y.X_1X_2} = .944$ computed with Equation 33.1 ($R^2_{Y.X_1X_2} = .944$ is the value in the radical of Equation 33.1 prior to computing the square root).

$$\tilde{R}^2_{Y.X_1X_2} = 1 - \frac{[1 - (.944)][5 - 1]}{5 - 2 - 1} = .89$$

As is the case for the r^2 value computed in simple correlation, the value computed for R^2 can be employed as a measure of effect size for **multiple correlation**. More specifically, R^2 represents the amount of shared variance between the best linear combination of the predictor variables and the criterion variable. Cohen (1992) has proposed the following (admittedly arbitrary) R^2 values as criteria for identifying the magnitude of an effect size: a) $.02 \leq R^2 < .13$ represents a **small multivariate effect size**; b) $.13 \leq R^2 < .26$ represents a **medium multivariate effect size**; and c) $R^2 \geq .26$ represents a **large multivariate effect size**. Regardless of which equation is employed to compute the **coefficient of multiple determination**, the effect size for Example 33.1 would be categorized as large.

Test of significance for a multiple correlation coefficient Equation 33.3 is employed to evaluate the null hypothesis H_0: $P^2 = 0$. If the latter null hypothesis is supported, the value of P will also equal zero and, consequently, the null hypothesis H_0: P = 0 is also supported.

$$F = \frac{(n - k - 1)R^2}{k(1 - R^2)}$$ **(Equation 33.3)**

The computed value $R^2_{Y.X_1X_2} = .944$, as well as the shrunken estimate $\tilde{R}^2_{Y.X_1X_2} = .89$, is substituted in Equation 33.3 below. If one has to choose which of the two values to employ, researchers would probably consider it more prudent to employ the shrunken estimate $\tilde{R}^2_{Y.X_1X_2} = .89$ (especially when the sample size is small).

$$F = \frac{(5 - 2 - 1)(.944)}{2(1 - .944)} = 16.86$$

$$F = \frac{(5 - 2 - 1)(.89)}{2(1 - .89)} = 8.09$$

The computed F value is evaluated with **Table A10**. In order to reject the null hypothesis, the F value must be equal to or greater than the tabled critical value at the prespecified level of significance. The degrees of freedom employed for the analysis are $df_{num} = k$ and $df_{den} = n - k - 1$. Thus, for our example, $df_{num} = 2$ and $df_{den} = 5 - 2 - 1 = 2$. In **Table A10**, for $df_{num} = 2$ and $df_{den} = 2$, the tabled critical .05 and .01 values are $F_{.05} = 19.00$ and $F_{.01} = 99.00$ (which respectively correspond to the relevant values listed for $F_{.95}$ and $F_{.99}$). Since both of the obtained values $F = 16.86$ and $F = 8.09$ are less than $F_{.05} = 19.00$, regardless of whether R^2 or $\tilde{R}^2$ is computed, the null hypothesis cannot be rejected. Thus, in spite of the fact that the obtained value of R is close to 1, the data still do not allow one to conclude that the population **multiple correlation coefficient** is some value other than zero. The lack of significance for such a large R value can be attributed to the small sample size.

The multiple regression equation[7] A major goal of **multiple regression** analysis is to derive a **multiple regression equation** which utilizes scores on the k predictor variables to predict the scores of subjects on the criterion variable. Equation 33.4 (which is the multivariable analog of Equation 28.4) is the general form of the **multiple regression** equation.

$$Y' = a + b_1 X_1 + b_2 X_2 + \cdots + b_k X_k \qquad \textbf{(Equation 33.4)}$$

Note that the **multiple regression** equation contains a **regression coefficient** (b_i) for each of the predictor variables (X_1, X_2, ..., X_k) and a **regression constant** (a). The latter regression coefficients (i.e., b_i values) are sometimes referred to as **raw** or **unstandardized regression coefficients** in order to distinguish them from **standardized regression coefficients**, which are discussed later in this section. In contrast to the regression line employed in simple linear regression, the **multiple regression** equation describes a **regression plane** which provides the best fit through a set of data points that exists in a **multidimensional space**. The values computed for the regression equation minimize the sum of the squared residuals, which in the case of **multiple regression** are the sum of the squared distances of all the data points from the regression plane.[8]

The reader should take note of the fact that, technically speaking, the regression coefficients represented in Equation 33.4 are in actuality **partial regression coefficients**, since they represent coefficients of regression when the other predictor variable(s) have been partialed out of the analysis (i.e., held constant or removed). The latter is often referred to as statistical control, since it allows a researcher to compute the independent contribution of each predictor variable by controlling for (i.e., eliminating) the effects of any of the other predictor variables. The discussion of **partial** and **semipartial correlation** later in this section will clarify what is meant when a variable is partialed out of an analysis.

Equations 33.5 and 33.6 can be employed to determine the values of the regression coefficients b_1 and b_2, which are the coefficients for predictor variables X_1 and X_2. Each of the regression coefficients indicates the amount of change on the criterion variable Y that will be associated with a one unit change on that predictor variable if the effect of the second predictor variable is held constant (or, to put it another way, is partialed out of the analysis — which is commensurate with being controlled for statistically). In the event there are three or more predictor variables, the value of a regression coefficient for a given predictor variable will indicate the amount of change on the criterion variable Y that will be associated with a one unit change on that predictor variable, if the effects of the other predictor variables are held constant (i.e., partialed out of the analysis).

$$b_1 = \left[\frac{\tilde{s}_Y}{\tilde{s}_{X_1}}\right]\left[\frac{r_{YX_1} - r_{YX_2} r_{X_1 X_2}}{1 - r_{X_1 X_2}^2}\right] \qquad \textbf{(Equation 33.5)}$$

$$b_2 = \left[\frac{\tilde{s}_Y}{\tilde{s}_{X_2}}\right]\left[\frac{r_{YX_2} - r_{YX_1}r_{X_1X_2}}{1 - r_{X_1X_2}^2}\right] \qquad \textbf{(Equation 33.6)}$$

Equation 33.7 is employed to compute the regression constant a (which is analogous to the Y intercept computed in simple linear regression).

$$a = \bar{Y} - b_1\bar{X}_1 - b_2\bar{X}_2 \qquad \textbf{(Equation 33.7)}$$

The **multiple regression** equation will now be computed. From earlier discussion we know the following values: $\bar{X}_1 = 7.2$, $\tilde{s}_{X_1} = 8.58$, $\bar{X}_2 = 3$, $\tilde{s}_{X_2} = 2.12$, $\bar{Y} = 3.4$, $\tilde{s}_Y = 2.70$, $r_{YX_1} = .955$, $r_{YX_2} = .52$, $r_{X_1X_2} = .37$. Substituting the appropriate values in Equations 33.5 –33.7, the **multiple regression** equation is derived below.

$$b_1 = \left[\frac{2.70}{8.58}\right]\left[\frac{.955 - (.52)(.37)}{1 - (.37)^2}\right] = .278$$

$$b_2 = \left[\frac{2.70}{2.12}\right]\left[\frac{.52 - (.955)(.37)}{1 - (.37)^2}\right] = .246$$

$$a = 3.4 - (.278)(7.2) - (.246)(3) = .660$$

$$Y' = .660 + .278X_1 + .246X_2$$

The value $b_1 = .278$ indicates that if the other predictor variable (i.e., salt) is partialed out of the analysis (i.e., statistically controlled for), when the amount of sugar a child ingests is increased by one ounce, the increase in the number of cavities (Y) is .278. In the same respect, the value $b_2 = .246$ indicates that if the other predictor variable (i.e., sugar) is partialed out of the analysis, when the amount of salt a child ingests is increased by one ounce, the increase in the number of cavities (Y) is .246. If there were three predictor variables in Example 33.1 (e.g., let us assume that in addition to sugar and salt, ounces of potassium ingested by the children are added as a third predictor variable) and the value $b_1 = .278$ was computed for the coefficient for the predictor variable X_1 (i.e., sugar), the value $b_1 = .278$ would indicate that if the other two predictor variables (salt and potassium) are partialed out of the analysis, when the amount of sugar a child ingests is increased by one ounce, the increase in the number of cavities (Y) is .278.

To illustrate the application of the **multiple regression** equation, when the appropriate values are substituted in Equation 33.4, a child who consumes 4 ounces of sugar (X_1) and 2 ounces of salt (X_2) per week is predicted to have 2.264 cavities. It should be noted that since it was previously determined the value of the **multiple correlation coefficient** was not statistically significant, in reality, the researcher would not be justified in employing the regression equation derived for the analysis for predictive purposes.

$$Y' = .660 + (.278)(4) + (.246)(2) = 2.264$$

The standard error of multiple estimate　As is the case with simple linear regression, a **standard error of estimate** can be computed which can be employed to determine how accurately the **multiple regression** equation will predict a subject's score on the criterion variable. Employing this error term, which in the case of **multiple regression** is referred to as the **standard error of multiple estimate**, one can compute a confidence interval for a predicted score. The standard error of multiple estimate will be represented by the notation $s_{Y.X_1X_2}$. Equation 33.8 is employed to compute $s_{Y.X_1X_2}$ if R^2 (i.e., the biased estimate of P^2) is used to represent the **coefficient of multiple determination**.[9] If, on the other hand, the unbiased estimate $\tilde{R}^2$ is employed to represent the **coefficient of multiple determination**, Equation 33.9 can be used to compute $s_{Y.X_1X_2}$. The two equations are employed below to compute the value $s_{Y.X_1X_2}$ = .90. (*SPSS* refers to $s_{Y.X_1X_2}$ as the **standard error of the estimate** and yields the value $s_{Y.X_1X_2}$ = .8934.)

(Equation 33.8)

$$s_{Y.X_1X_2} = \tilde{s}_Y \sqrt{\left[\frac{n-1}{n-k-1}\right](1 - R_{Y.X_1X_2}^2)} = (2.70)\sqrt{\left[\frac{5-1}{5-2-1}\right](1 - .944)} = .90$$

(Equation 33.9)

$$s_{Y.X_1X_2} = \tilde{s}_Y \sqrt{1 - \tilde{R}_{Y.X_1X_2}^2} = (2.70)\sqrt{1 - .89} = .90$$

Computation of a confidence interval for Y' Equation 33.10 (which is analogous to Equation 28.16) can be employed to compute a confidence interval for the predicted value Y'. The value $s_{Y.X_1X_2}$ = .90 is employed in Equation 33.10 to compute the 95% confidence interval for the subject who is predicted to have $Y' = 2.264$ cavities. Also employed in the latter equation is the tabled critical two-tailed .05 t value $t_{.05}$ = 4.30 (for $df = n - k - 1 = 5 - 2 - 1 = 2$), which delineates to the 95% confidence interval.

(Equation 33.10)

$$CI_{(1-\alpha)} = Y' \pm (t_{\alpha/2})(s_{Y.X_1X_2}) = 2.264 \pm (4.30)(.90) = 2.264 \pm 3.87$$

This result indicates the researcher can be 95% confident (or the probability is .95) that the interval -1.606 to 6.134 contains the actual number of cavities for the subject (i.e., $-1.606 \le Y \le 6.134$). Since a person cannot have a negative number of cavities, the latter result translates into between 0 and 6.134 cavities.

Evaluation of the relative importance of the predictor variables If the result of a **multiple regression** analysis is significant, a researcher will want to assess the relative importance of the predictor variables in explaining variability on the criterion variable. It should be noted that although the value computed for the **multiple correlation coefficient** is not significant for Example 33.1, within the framework of the discussion of the material in this section, it will be assumed that it is. Intuitively, it might appear that one can evaluate the relative importance of the predictor variables based on the relative magnitude of the regression coefficients. However, since the different predictor variables represent different units of measurement, comparison of the regression coefficients will not allow a researcher to make such an estimate. One approach to solving this problem is to standardize each of the variables so that scores on all of the variables are based on standard normal distributions. As a result of standardizing all of the variables, Equation 33.11 (which is referred to as the **standardized multiple regression equation**) becomes the general form of the **multiple regression** equation.

$$z_{Y'} = \beta_1 z_1 + \beta_2 z_2 + \cdots + \beta_k z_k \qquad \textbf{(Equation 33.11)}$$

In Equation 33.11 the predicted value Y', as well as the scores on the predictor variables X_1 and X_2, are expressed as standard deviation scores (i.e., $z_{Y'}$, z_1, z_2). The standardized equivalent of a regression coefficient, referred to as a **beta weight**, is represented by the notation β_i. The regression constant a is not included in the standardized **multiple regression** equation since it will always equal zero. When there are two predictor variables, Equations 33.12 and 33.13 can be employed to compute the values of β_1 and β_2. Note that each equation is expressed in two equivalent forms, one form employing the regression coefficients and the relevant estimated population standard deviations, and the other form employing the correlations between the three variables.[10]

$$\beta_1 = b_1 \left[\frac{\tilde{s}_{X_1}}{\tilde{s}_Y} \right] = \frac{r_{YX_1} - r_{YX_2} r_{X_1 X_2}}{1 - r_{X_1 X_2}^2} \qquad \text{(Equation 33.12)}$$

$$\beta_2 = b_2 \left[\frac{\tilde{s}_{X_2}}{\tilde{s}_Y} \right] = \frac{r_{YX_2} - r_{YX_1} r_{X_1 X_2}}{1 - r_{X_1 X_2}^2} \qquad \text{(Equation 33.13)}$$

Substituting the appropriate values in Equations 33.12 and 33.13, the values $\beta_1 = .883$ and $\beta_2 = .193$ are computed.

$$\beta_1 = .278 \left[\frac{8.58}{2.70} \right] = \frac{.955 - (.52)(.37)}{[1 - (.37)^2]} = .883$$

$$\beta_2 = .246 \left[\frac{2.12}{2.70} \right] = \frac{.52 - (.955)(.37)}{[1 - (.37)^2]} = .193$$

Thus, the standardized **multiple regression** equation is as follows: $z_{Y'} = .883 z_1 + .193 z_2$. The value $\beta_1 = .883$ indicates that an increase of one standard deviation unit on variable X_1 (i.e., sugar) is associated with an increase of .883 standard deviation units on the criterion variable (if the predictor variable X_2 (i.e., salt) remains at a fixed value — i.e., is partialed out of the analysis). In the same respect the value $\beta_2 = .193$ indicates that an increase of one standard deviation unit on variable X_2 (i.e., salt) is associated with an increase of .193 standard deviation units on the criterion variable (if the predictor variable X_1 (i.e., sugar) remains at a fixed value — i.e., is partialed out of the analysis).

When there are two predictor variables, Equation 33.14 employs the standardized beta weights to provide an alternative method for computing the value $R_{Y.X_1 X_2}^2$. The value derived with Equation 33.14 will be equivalent to the square of the value computed with Equation 33.1.

$$R_{Y.X_1 X_2}^2 = \beta_1 r_{YX_1} + \beta_2 r_{YX_2}$$

(Equation 33.14)

$$R_{Y.X_1 X_2}^2 = (.883)(.955) + (.193)(.52) = .944$$

Some sources note that the absolute values of the beta weights reflect the rank-ordering of the predictor variables with respect to the role they play in accounting for variability on the criterion variable. However, the only time one can actually employ the **standardized regression coefficients** as definitive indicators with respect to the relative importance of the predictor variables is in the unlikely case in which all of the predictor variables are independent of one another.[11] Drawing conclusions with respect to the relative importance of the predictor variables based on the absolute value of their **standardized regression coefficients** becomes more problematical the more highly correlated the predictor variables are with one another. Kachigan (1986) suggests that by dividing the square of a larger beta weight by the square of a smaller beta weight, a researcher can determine the relative influence of two predictor variables on the criterion variable. The problem with the latter approach (as Kachigan (1986) himself notes) is that beta weights do not allow a researcher to separate the joint contribution of two or more predictor variables. The fact that predictor variables are usually correlated with one another (as is the case in the example under discussion) makes it difficult to determine how much variability on the criterion variable can be accounted for by any single predictor variable in and of itself. To put it another way, the **standardized regression coefficients** reflect the **unique** contribution of each predictor variable. However, **joint** contributions not attributed to any specific predictor variable also contribute to the value of R^2, and the latter is not reflected in the values of the **standardized regression coefficients**. Thus, it is conceivable that the contribution of a predictor variable with a substantial joint contribution may be underestimated if one only considers the absolute value of the beta weight for that variable. Because of the latter it is important to report the **zero-order correlation** between each predictor variable and the criterion variable.

Howell (2002, p. 573) notes that although some researchers conducting **stepwise multiple regression** use the rank order in which the predictor variables are included in the regression equation as a criterion for assessing their importance, the latter methodology is inappropriate since it ignores the interrelationship between the predictor variables.[12] In the final analysis, researchers are not in agreement with respect to what, if any, methodology is appropriate for determining the precise amount of variability attributable to each of the predictor variables.

In closing this discussion it should be emphasized that without adequate cross-validation a researcher should not expect regression coefficients (standardized or unstandardized) computed for one sample to necessarily correspond to those computed for another sample derived from the same target population. Thus, in order to employ a set of regression coefficients for predictive purposes, the model in question must be cross-validated.

Evaluating the significance of a regression coefficient In the event it is determined that the **coefficient of multiple correlation** is statistically significant, it is common practice to evaluate whether each of the regression coefficients is statistically significant. Since the **unstandardized/ raw score regression coefficients** and the **standardized regression coefficients** are linear transformations of one another, a statistical test on either set of coefficients will yield the same result. The null hypothesis which is evaluated is that the true value of a regression coefficient in the population equals zero. Thus, H_0: $B_i = 0$ (where B_i, which is the upper case Greek letter **beta**, represents the value of the coefficient for the ith predictor variable in the underlying population). For purpose of discussion it will be assumed that the aforementioned null hypothesis is stated in reference to the unstandardized coefficients. In order to evaluate the null hypothesis H_0: $B_i = 0$, a standard error of estimate must be computed for a coefficient. When there are two predictor variables, the standard error of estimate for an **unstandardized regression coefficient** (represented by the notation s_{b_i}) is computed with Equation 33.15.[13]

$$s_{b_i} = \frac{\tilde{s}_Y}{\tilde{s}_{X_i}} \sqrt{\frac{1 - R_{Y.X_1X_2}^2}{(1 - r_{X_1X_2}^2)(n - k - 1)}}$$ **(Equation 33.15)**

Employing Equation 33.15, the values s_{b_1} = .057 and s_{b_2} = .229 are computed.

$$s_{b_1} = \frac{2.70}{8.58} \sqrt{\frac{1 - .944}{[1 - (.37)^2][5 - 2 - 1]}} = .057$$

$$s_{b_2} = \frac{2.70}{2.12} \sqrt{\frac{1 - .944}{[1 - (.37)^2][5 - 2 - 1]}} = .229$$

Each of the values s_{b_1} = .057 and s_{b_2} = .229 can be substituted in Equation 33.16, which employs the t distribution to evaluate the null hypothesis H_0: B_i = 0.[14] In the analysis to be described it will be assumed that the nondirectional alternative hypothesis H_1: B_i ≠ 0 is evaluated for each regression coefficient. The degrees of freedom employed in evaluating a t value computed with Equation 33.16 are $df = n - k - 1$. Thus, $df = 5 - 2 - 1 = 2$. If the null hypothesis cannot be rejected for a specific coefficient, the researcher can conclude the predictor variable in question will not be of any use in predicting scores on the criterion variable.

$$t_{b_i} = \frac{b_i}{s_{b_i}} \qquad \text{(Equation 33.16)}$$

Employing Equation 33.16, the null hypotheses H_0: B_1 = 0 and H_0: B_2 = 0 are evaluated.[15]

$$t_{b_1} = \frac{.278}{.057} = 4.88 \qquad t_{b_2} = \frac{.246}{.229} = 1.07$$

Employing **Table A2**, for $df = 2$, the tabled critical .05 and .01 t values are $t_{.05}$ = 4.30 and $t_{.01}$ = 9.93.[16] Since the value t_{b_1} = 4.88 is greater than $t_{.05}$ = 4.30, the nondirectional alternative hypothesis H_1: B_1 ≠ 0 is supported at the .05 level. It is not supported at the .01 level, since t_{b_1} = 4.88 is less than $t_{.01}$ = 9.93. Since t_{b_2} = 1.07 is less than $t_{.05}$ = 4.30, the nondirectional alternative hypothesis H_1: B_2 ≠ 0 is not supported. Thus, we can conclude that, whereas predictor variable X_1 (sugar consumption) contributes significantly in predicting variability on the criterion variable (number of cavities), predictor variable X_2 (salt consumption) does not. Consequently, the latter predictor variable can be removed from the analysis.[17]

It should be emphasized that regardless of whether or not the regression coefficient for a predictor variable is significant, if a researcher elects to drop one or more predictor variables from the regression equation, a new regression equation should be computed (which just involves the data for the remaining predictor variables), since by virtue of eliminating one or more predictor variables the original coefficients for the remaining predictor variables will in all likelihood change.

Computation of a confidence interval for a regression coefficient Equation 33.17 can be employed to compute a confidence interval for a regression coefficient.

$$CI_{b_{i(1 - \alpha)}} = b_i \pm (t_{\alpha/2})(s_{b_i}) \qquad \text{(Equation 33.17)}$$

Using Equation 33.17, the 95% confidence intervals for the two regression coefficients are computed. The t value employed in Equation 33.17 is $t_{.05}$ = 4.30, which is the tabled critical two-tailed .05 t value for $df = n - k - 1 = 5 - 2 - 1 = 2$.

$$CI_{.95_{b_1}} = .278 \pm (4.30)(.057) = .278 \pm .245$$

$$CI_{.95_{b_2}} = .246 \pm (4.30)(.229) = .246 \pm .985$$

This result indicates the researcher can be 95% confident (or the probability is .95) that the following intervals contain the true values of the coefficients: $.033 \le B_1 \le .523$ and $-.739 \le B_2 \le 1.231$).

Table 33.2 is a printout from the *SPSS* analysis of Example 28.8 (using the **standard multiple regression** option) documenting relevant information regarding the **multiple regression** equation. As noted in Endnote 7, any discrepancy between the values recorded in Table 33.2 and values computed with the equations in this section are due to rounding off error. The relevant values in the **multiple regression** equation are recorded in Column 2 of Table 33.2 labeled **B**. Specifically, the values of the constant a, and the **unstandardized regression coefficients** b_1 and b_2 are respectively recorded in the rows labeled **(Constant)**, **XSUGAR**, and **ZSALT**. The values of the **standardized regression coefficients** β_1 (for sugar) and β_2 (for salt) are in Column 4 labeled **Beta**. The results of the t tests evaluating the **unstandardized regression coefficients** b_1 and b_2 are in Columns 5 (labeled **t**) and 6 (labeled **Sig.**). In order for a t test to be significant at the .05 level (or .01 level), the probability value listed in Column 6 must be equal to or less than .05 (or .01). The lower and upper bounds of the confidence intervals for the **unstandardized regression coefficients** b_1 and b_2 (computed with Equation 33.17) are recorded in Columns 7 (**Lower Bound**) and 8 (**Upper Bound**). Note than when a regression coefficient is significant, the value 0 will not be included in the confidence interval (which, in fact, is the case for sugar), but will be included when the coefficient is not significant (which is the case for salt).

Table 33.2 *SPSS* Printout of Relevant Information for
Multiple Regression Equation for Example 33.1

Coefficients[a]

Model		Unstandardized Coefficients		Standardized Coefficients	t	Sig.	95% Confidence Interval for B	
		B	Std. Error	Beta			Lower Bound	Upper Bound
1	(Constant)	.650	.763		.852	.484	-2.634	3.935
	XSUGAR	.278	.056	.882	4.950	.038	.036	.519
	ZSALT	.250	.227	.196	1.103	.385	-.726	1.227

Analysis of variance for multiple regression   In the discussion of simple linear regression in the chapter on the **Pearson product-moment correlation coefficient** (specifically, in Section VI under the discussion of **residuals**), it was demonstrated that within the framework of bivariate correlation an **analysis of variance** can be employed to determine how well either of the regression lines fits the data. The latter can also be done with respect to the regression plane within the framework of **multiple regression** analysis. Specifically, an F test can be conducted to simultaneously evaluate the following equivalent null hypotheses: a) H_0: P = 0; b) H_0: $B_1 = B_2 = 0$. Rejection of the latter null hypothesis implies that in the underlying population, at least one of the $k = 2$ regression parameters (B_1, B_2) is some value other than zero. If the latter is true, any predictor variable for which the population parameter B_i (which is estimated by the regression coefficient b_i) is some value other than zero is useful in predicting the value of Y. Within the framework of the analysis of variance, total variability on the Y variable can be expressed by Equation 33.18.

$$SS_Y = SS_{Total} = SS_{reg} + SS_{res} \qquad \textbf{(Equation 33.18)}$$

Equations 33.19–33.28 are employed below to compute the relevant sum of squares, degrees of freedom, and mean square values for the **analysis of variance**. (Rounding off error accounts for the slight discrepancies between equivalent values.) In Equation 33.20 the value $SP_{X_1Y} = 88.6$ was previously computed in Section IV of the **Pearson product-moment correlation coefficient**, while the value $SP_{X_2Y} = 12$ was computed earlier in this section.

(Equation 33.19)

$$SS_{Total} = SS_Y = (n - 1)(s_{\hat{Y}}^2) = (5 - 1)(7.29) = 29.16$$

(Equation 33.20)

$$SS_{reg} = b_1 SP_{X_1Y} + b_2 SP_{X_2Y} = (.278)(88.6) + (.246)(12) = 27.58$$

or

(Equation 33.21)

$$SS_{reg} = (n - 1)(s_{\hat{Y}}^2) R^2 = (5 - 1)(7.29)(.944) = 27.52$$

$$SS_{res} = SS_Y - SS_{reg} = 29.2 - 27.58 = 1.62 \qquad \textbf{(Equation 33.22)}$$

or

(Equation 33.23)

$$SS_{res} = (n - 1)(s_{\hat{Y}}^2)(1 - R^2) = (5 - 1)(7.29)(1 - .944) = 1.63$$

$$df_{Total} = df_Y = n - 1 = 5 - 1 = 4 \qquad \textbf{(Equation 33.24)}$$

$$df_{reg} = k = 2 \qquad \textbf{(Equation 33.25)}$$

$$df_{res} = n - k - 1 = n - 2 - 1 = 5 - 3 = 2 \qquad \textbf{(Equation 33.26)}$$

$$MS_{reg} = \frac{SS_{reg}}{df_{reg}} = \frac{27.58}{2} = 13.79 \qquad \textbf{(Equation 33.27)}$$

$$MS_{res} = \frac{SS_{res}}{df_{res}} = \frac{1.62}{2} = .81 \qquad \textbf{(Equation 33.28)}$$

The F ratio is computed with Equation 33.29. The result of the analysis is summarized in Table 33.3.

$$F = \frac{MS_{reg}}{MS_{res}} = \frac{13.79}{.81} = 17.02 \qquad \textbf{(Equation 33.29)}$$

Table 33.3 Summary Table of Analysis of Variance for Multiple Regression Analysis for Example 33.1

Source of variation	SS	DF	MS	F
Regression	27.58	2	13.79	17.02
Residual	1.62	2	.81	
Total	29.20	4		

The obtained value $F = 17.02$ is evaluated with **Table A10**. The degrees of freedom employed for the analysis are the numerator and denominator degrees of freedom employed in computing the F ratio. Thus, $df_{num} = df_{reg} = 2$ and $df_{den} = df_{res} = 2$. For $df_{num} = 2$ and $df_{den} = 2$, the tabled $F_{.95}$ and $F_{.99}$ values are $F_{.95} = 19.00$ and $F_{.99} = 99.00$. In order to reject the null hypothesis, the obtained F value must be equal to or greater than the tabled critical value at the prespecified level of significance. Since $F = 17.02$ is less than $F_{.95} = 19.00$, the null hypothesis H_0: $P = 0$ cannot be rejected at the .05 level. This result (as well as the F value) is identical to that obtained earlier in this section when Equation 33.3 was employed to evaluate the same null hypothesis (the minimal difference between F values is due to rounding off error). (It should also be noted that when the above analysis of variance is implemented through use of SPSS, the following result is obtained: $F (2, 2) = 17.254$, $p = .055$, which just falls short of significance at the .05 level (since $p = .055$ is greater than .05).)[18]

It is also the case that R^2 will always equal the proportion of total variability (on the Y variable) attributable to the regression sum of squares. Specifically, the value of R^2 can be defined by the relationship described in Equation 33.30. Note that the value $R^2 = .944$ obtained with Equation 33.30 is identical to the square of the value obtained with Equation 33.1 (which equals R).

$$R^2 = \frac{SS_{reg}}{SS_{Total}} = \frac{27.58}{29.2} = .944 \qquad \textbf{(Equation 33.30)}$$

Semipartial and partial correlation   This section will describe the concepts of **semipartial** and **partial correlation**, both of which will allow the reader to better understand the concept of statistical control.

A **semipartial (or part) correlation coefficient** measures the degree of association between two variables, with the linear association of one or more other variables removed from only one of the two variables which are being correlated with one another. In the case of two predictor variables and a criterion variable, a **semipartial correlation coefficient** measures the degree of association between two variables, with the influence of the third variable removed from only one of the two variables that are being correlated with one another. Thus, the **semipartial correlation coefficient** $r_{Y(X_1 \cdot X_2)}$ represents the correlation between Y and X_1 after any linear association that X_2 has with X_1 has been removed. In the same respect, the **semipartial correlation coefficient** $r_{Y(X_2 \cdot X_1)}$ represents the correlation between Y and X_2 after any linear association that X_1 has with X_2 has been removed.

Whereas **multiple correlation** combines variables in order to assess their cumulative effect, **partial correlation** removes the effects of variables in order to determine what effect remains when one or more of the variables have been eliminated. A **partial correlation coefficient** measures the degree of association between two variables, after any linear association one or more additional variables has with the two variables has been removed. In the case of two predictor variables X_1 and X_2 and the criterion variable Y, the **partial correlation coefficient** $r_{YX_1 . X_2}$ represents the correlation between Y and X_1 after any linear association that X_2 has with either Y or X_1 has been removed. It can also be stated that the **partial correlation coefficient** $r_{YX_1 . X_2}$ represents the correlation between Y and X_1 if X_2 is held constant. By computing the **partial correlation coefficient** $r_{YX_1 . X_2}$, one is able to have a "purer" measure of the relationship between a criterion variable Y and the predictor variable X_1. Licht (1995, p. 43) notes that unlike a **semipartial correlation**, which only changes the meaning of a specific predictor variable, a **partial correlation** changes the meaning of both the predictor variable in question and the criterion variable by removing everything that has to do with the other predictors from both the predictor in question and criterion variables involved in the **partial correlation**.

Licht (1995, p. 36 and 42) notes that the bivariate correlation (i.e. the **zero-order correlation**) between a predictor variable and the criterion variable is not a measure which reflects the independent contribution of the predictor variable in explaining the criterion variable, since a **zero-order correlation** ignores rather than eliminates the effects of any other predictor variables included in an analysis. Through use of statistical control, however, a researcher can compute the independent contribution of a predictor variable by controlling for (i.e., eliminating) the effects of any of the other predictor variables. To illustrate the latter in reference to Example 33.1, although the value $r_{YX_1}^2 = (r_{YX_1} = .955)^2 = .912$ (which is the **squared zero-order correlation** between the number of ounces of sugar consumed (X_1) and the criterion variable Y (number of cavities)) is a measure of the variance shared by the latter two variables, it can include variance which is also shared with the other predictor variables employed in the study (i.e., the number of ounces of salt consumed (X_2), which, as previously noted, has a correlation of $r_{X_1 X_2} = .37$ with the predictor variable X_1). Statistical control will allow the researcher to obtain a pure measure of the correlation (and consequently a pure measure of the amount of shared variance) between the predictor variable X_1 (ounces of sugar) and the criterion variable Y (number of cavities). In point of fact, the latter correlation is what is represented by the **semipartial correlation coefficient** $r_{Y(X_1 . X_2)} = .82$ computed later in this section with Equation 33.31, which represents the correlation between Y and X_1 after any linear association that X_2 has with X_1 has been removed. Note that the difference between $R_{Y . X_1 X_2}^2 = .944$ (which is the proportion of shared variance between the criterion variable and the two predictor variables) and the **squared zero-order correlation** ($r_{YX_2}^2 = (.52)^2 = .27$) (which just reflects the shared variance between the number of cavities and salt consumption) is equal to the **squared semipartial correlation** ($r_{Y(X_1 . X_2)}^2 = (.82)^2 = .67$) (which is the proportion of shared variance between the number of cavities and the amount of sugar consumed after any linear association that X_1 (amount of sugar consumed) has with X_2 (amount of salt consumed) has been removed). In other words, $R_{Y . X_1 X_2}^2 = .944 = [(r_{Y(X_1 . X_2)}^2 = .67] + [(r_{YX_2}^2 = .27]$.

In the same respect, the difference between $R_{Y . X_1 X_2}^2 = .944$ and the **squared zero-order correlation** $r_{YX_1}^2 = (r_{YX_1} = .955)^2 = .912$ (which just reflects the shared variance between the number of cavities and sugar consumption) is equal to the **squared semipartial correlation** ($r_{Y(X_2 . X_1)}^2 = (.18)^2 = .032$) computed later in this section with Equation 33.31 (which is the proportion of shared variance between the number of cavities and the amount of salt consumed after any linear association that X_2 (amount of salt consumed) has with X_1 (amount of sugar consumed) has been removed). In other words, $R^2_{Y.X_1X_2} = .944 = [(r^2_{Y(X_2.X_1)} = .032] + [(r^2_{YX_1} = .912]$.

If there were three predictor variables (let us assume that ounces of potassium consumed is added to Example 33.1 as a third predictor variable), the **squared multiple correlation coefficient**, represented by the notation $R^2_{Y.X_1X_2X_3}$, will reflect the proportion of shared variance between the criterion variable and the three predictor variables. The square of the **semipartial correlation** $r_{Y(X_1.X_2X_3)}$ will reflect the proportion of shared variance between the number of cavities and the amount of sugar consumed after any linear association that X_1 (amount of sugar consumed) has with X_2 (amount of salt consumed) and X_3 (amount of potassium consumed) has been removed. Thus, the latter **semipartial correlation** will allow the researcher to obtain a pure measure of the correlation (and consequently a pure measure of the amount of shared variance) between the predictor variable X_1 (ounces of sugar) and the criterion variable Y (number of cavities). The researcher can also compute the **semipartial correlations** $r_{Y(X_2.X_1X_3)}$ and $r_{Y(X_3.X_1X_2)}$ which would allow her to respectively obtain a pure measure of the correlation (and consequently a pure measure of the amount of shared variance) between the number of cavities (Y) and each of the predictor variables X_2 (ounces of salt) and X_3 (ounces of potassium). In the case of three predictor variables, the researcher could initially compute the **zero-order correlation** between X_1 and Y. Upon doing the latter she could compute a **multiple correlation** employing the second predictor variable — in other words, compute $R_{Y.X_1X_2}$. The square of the latter value will be equal to the sum of the **squared zero-order correlation** r_{YX_1} and the **squared semipartial correlation** $r_{Y(X_2.X_1)}$. When the third predictor variable is added to the analysis, the square of the **multiple correlation** $R_{Y.X_1X_2X_3}$ will be equal to the sum of the square of the multiple correlation $R_{Y.X_1X_2}$ and the square of the **semipartial correlation** $r_{Y(X_3.X_1X_2)}$. In other words, as each new predictor variable is added to an analysis, the value of the previously computed **squared multiple correlation** will be incremented by the square of the **semipartial correlation** computed for the new predictor variable.[19]

The relationships discussed in this section can be confirmed when conducting a **hierarchical** or **stepwise multiple regression** analysis in that the computer will print out a **multiple correlation coefficient** and a **semipartial correlation** for each predictor variable which is added or removed (in the case of **backward analysis**) during each step of an analysis. Examination of the **semipartial correlations** (which are labeled **part correlation** by *SPSS*) and the **multiple correlations** computed during the different steps of the analysis should confirm what has been discussed in this section.

Although computer output for a **multiple regression** analysis usually includes **partial correlation** coefficients, they are generally not of as much interest as the **semipartial correlations**. The reason for the latter is that most often a researcher is interested in the relationship between the combination of predictor variables and the criterion variable. Since the **partial correlation** between a predictor variable and the criterion variable eliminates the effect of other predictors variables on both the predictor in question and the criterion variable, a **partial correlation** alters the nature of the relationship being examined by virtue of it partialing out not only the other predictors but also the criterion variable.

Test 33a: Computation of a semipartial correlation coefficient When there are three variables, it is possible to compute the following six **semipartial correlation coefficients**: $r_{Y(X_1.X_2)}$, $r_{Y(X_2.X_1)}$, $r_{X_1(Y.X_2)}$, $r_{X_1(X_2.Y)}$, $r_{X_2(Y.X_1)}$, $r_{X_2(X_1.Y)}$. Equation 33.31 is the general equation for computing a **semipartial correlation coefficient** involving three variables (where A, B, and C

represent the three variables). The notation $r_{A(B.C)}$ represents the correlation between A and B, after any linear relationship that C has with B has been removed.

$$r_{A(B.C)} = \frac{r_{AB} - r_{AC}r_{BC}}{\sqrt{1 - r_{BC}^2}} \qquad \text{(Equation 33.31)}$$

Employing the values computed for Example 33.1 in Equation 33.31, the **semipartial correlation coefficients** $r_{Y(X_1.X_2)} = .82$ and $r_{Y(X_2.X_1)} = .18$ are computed.

$$r_{Y(X_1.X_2)} = \frac{r_{YX_1} - r_{YX_2}r_{X_1X_2}}{\sqrt{1 - r_{X_1X_2}^2}} = \frac{.955 - (.52)(.37)}{\sqrt{1 - (.37)^2}} = .82$$

$$r_{Y(X_2.X_1)} = \frac{r_{YX_2} - r_{YX_1}r_{X_1X_2}}{\sqrt{1 - r_{X_1X_2}^2}} = \frac{.52 - (.955)(.37)}{\sqrt{1 - (.37)^2}} = .18$$

The value $r_{Y(X_1.X_2)} = .82$ indicates the correlation between the number of cavities and sugar consumption, after any linear relationship salt consumption has with sugar consumption has been removed. The value $r_{Y(X_2.X_1)} = .18$ indicates the correlation between the number of cavities and salt consumption, after any linear relationship sugar consumption has with salt consumption has been removed. As noted earlier, the square of a **semipartial correlation coefficient** represents the proportion of variability explained on one of the variables by a second variable, after removing the linear effect of a third variable from the second variable. In the case of $r_{Y(X_1.X_2)} = .82$, $r_{Y(X_1.X_2)}^2 = (.82)^2 = .67$. Thus, 67% of the variability on Y can be accounted for on the basis of X_1 when the linear effect of X_2 is removed from X_1. In the case of $r_{Y(X_2.X_1)} = .18$, $r_{Y(X_2.X_1)}^2 = (.18)^2 = .03$. Thus, only 3% of the variability on Y can be accounted for on the basis of X_2 when the linear effect of X_1 is removed from X_2.

Test of significance for a semipartial correlation coefficient   The null hypothesis $H_0: \rho_{sp} = 0$ can be evaluated with Equation 33.32 (where ρ_{sp} represents the population **semipartial correlation coefficient**).[20]

$$t = \frac{r_{sp}\sqrt{n - v}}{\sqrt{1 - r_{sp}^2}} \qquad \text{(Equation 33.32)}$$

Where: r_{sp} is the **semipartial correlation coefficient**
        v is the total number of variables employed in the analysis

When there are two predictor variables and one criterion variable, the total number of variables employed in the analysis is 3, and thus $\sqrt{n - v} = \sqrt{n - 3}$.[21] The value $n - v = n - 3$ represents the number of degrees of freedom employed for the analysis. Employing Equation 33.32, the null hypothesis is evaluated in reference to the **semipartial correlation coefficients** $r_{Y(X_1.X_2)} = .82$ and $r_{Y(X_2.X_1)} = .18$.

$$t = \frac{(.82)\sqrt{5-3}}{\sqrt{1-(.82)^2}} = 2.03$$

$$t = \frac{(.18)\sqrt{5-3}}{\sqrt{1-(.18)^2}} = .26$$

It will be assumed that the nondirectional alternative hypothesis $H_1: \rho_{sp} \neq 0$ is evaluated. Employing **Table A2**, for $df = 2$ (since $df = 5 - 3 = 2$), the tabled critical two-tailed .05 and .01 values are $t_{.05} = 4.30$ and $t_{.01} = 9.93$. Since both of the obtained t values are less than $t_{.05} = 4.30$, the nondirectional alternative hypotheses $H_1: \rho_{Y(X_1 \cdot X_2)} \neq 0$ and $H_1: \rho_{Y(X_2 \cdot X_1)} \neq 0$ are not supported.

Test 33b: Computation of a partial correlation coefficient When there are three variables, it is possible to compute the following **partial correlation coefficients**: $r_{YX_1 \cdot X_2}$, $r_{YX_2 \cdot X_1}$, $r_{X_1 X_2 \cdot Y}$. Equation 33.33 is the general equation for computing a **partial correlation coefficient** involving three variables (where A, B, and C represent the three variables). The notation $r_{AB \cdot C}$ represents the correlation between A and B, after any linear relationship C has with A and B has been removed.

$$r_{AB \cdot C} = \frac{r_{AB} - r_{AC}r_{BC}}{\sqrt{(1 - r_{AC}^2)(1 - r_{BC}^2)}} \qquad \textbf{(Equation 33.33)}$$

Employing Equation 33.33, the **partial correlation coefficients** $r_{YX_1 \cdot X_2}$ (which is the partial correlation of Y and X_1, with the effect of X_2 removed from both Y and X_1) and $r_{YX_2 \cdot X_1}$ (which is the **partial correlation** of Y and X_2, with the effect of X_1 removed from both Y and X_2) will be computed. When the appropriate correlations from Example 33.1 are substituted in Equation 33.3, the **partial correlation coefficients** $r_{YX_1 \cdot X_2} = .96$ and $r_{YX_2 \cdot X_1} = .60$ are computed.

$$r_{YX_1 \cdot X_2} = \frac{r_{YX_1} - r_{YX_2}r_{X_1 X_2}}{\sqrt{(1 - r_{YX_2}^2)(1 - r_{X_1 X_2}^2)}} = \frac{.955 - (.52)(.37)}{\sqrt{[1 - (.52)^2][1 - (.37)^2]}} = .96$$

$$r_{YX_2 \cdot X_1} = \frac{r_{YX_2} - r_{YX_1}r_{X_1 X_2}}{\sqrt{(1 - r_{YX_1}^2)(1 - r_{X_1 X_2}^2)}} = \frac{.52 - (.955)(.37)}{\sqrt{[1 - (.955)^2][1 - (.37)^2]}} = .60$$

The value $r_{YX_1 \cdot X_2} = .96$ indicates that the correlation between the number of cavities and sugar consumption, after any linear relationship salt consumption has with the number of cavities and sugar consumption has been removed. The value $r_{YX_2 \cdot X_1} = .60$ indicates that the correlation between the number of cavities and salt consumption, after any linear relationship sugar consumption has with the number of cavities and salt consumption has been removed. The square of a **partial correlation coefficient** represents the proportion of variability explained on the criterion variable by one of the predictor variables, after removing any linear effects of the other predictor variable from the other two variables. In the case of $r_{YX_1 \cdot X_2} = .96$, $r_{YX_1 \cdot X_2}^2 = (.96)^2 = .92$. Thus, 92% of the variability on the criterion variable can be accounted for on the basis of the predictor variable X_1, when the linear effects of variable X_2 are removed from the other two variables. In the case of $r_{YX_2 \cdot X_1} = .60$, $r^2_{YX_2 \cdot X_1} = (.60)^2 = .36$. Thus, 36% of the variability on the criterion variable can be accounted for on the basis of the predictor variable X_2, when the linear effects of variable X_1 are removed from the other two variables.

The following should also be noted with respect to a **partial correlation**.

a) Although a **partial correlation** between two variables is generally (but not always) smaller than the **zero-order correlation** (which is the correlation between the two variables before the effect of the third variable has been removed), this is not the case in the example under discussion (since the **partial correlations** $r_{YX_1 \cdot X_2} = .96$ and $r_{YX_2 \cdot X_1} = .60$ are larger than the **zero-order correlations** $r_{YX_1} = .955$ and $r_{YX_2} = .52$).

b) Marascuilo and Serlin (1983) note that although it is theoretically possible for the two values to be equal (in reference to the same variables), a **semipartial correlation coefficient** will have a smaller absolute value than a **partial correlation coefficient**. The latter can be confirmed by the fact that the **partial correlation coefficient** $r_{YX_1 \cdot X_2} = .96$ is larger than the **semipartial correlation coefficient** $r_{Y(X_1 \cdot X_2)} = .82$, and the **partial correlation coefficient** $r_{YX_2 \cdot X_1} = .60$ is larger than the **semipartial correlation coefficient** $r_{Y(X_2 \cdot X_1)} = .18$.

c) When a **partial correlation** is substantially different from the corresponding **zero-order correlation** (especially when the absolute value of the **partial correlation** is substantially larger), it may indicate the presence of a **suppressor variable**. Cohen and Cohen (1983) and Tabachnick and Fidell (1996, p. 165) note that a **suppressor variable** is a predictor variable which is useful in predicting the criterion variable purely on the basis of the fact that it correlates highly with the other predictor variables. A **suppressor variable** (which in itself has little or no correlation with the criterion variable) improves prediction on the criterion variable by suppressing variance that is irrelevant in predicting the latter variable. In a set of three variables, a **suppressor variable** is a predictor variable (X_i) that has a low correlation with the criterion variable (Y), but a high correlation with the other predictor variable (X_j). By virtue of the latter, inclusion of the **suppressor variable** in the analysis may result in a **multiple correlation coefficient** ($R_{Y \cdot X_i X_j}$) that has a larger absolute value (or even a different sign) than the **zero-order correlation** coefficient between the criterion variable and the **suppressor variable** (r_{YX_i}).

Stevens (2002, p. 124) notes that Lord and Novick (1968) state the following two rules of thumb for selecting predictor variables: 1) Select predictor variables which have low intercorrelations with one another, but are highly correlated with the criterion variable; 2) Add additional variables to the regression analysis which are highly correlated with the predictor variables, but have low correlations with the criterion variable.

Note that the second rule of thumb recommends the use of **suppressor variables** to maximize prediction. To illustrate how a **suppressor variable** can increase predictability, Stevens (2002, p. 124) presents an example in which a predictor variable X_1 has a **zero-order correlation** of $r_{YX_1} = .60$ with the criterion variable Y. Also included in the analysis is predictor variable X_2 (which represents a **suppressor variable**) that has a **zero-order correlation** of $r_{YX_2} = 0$ with the criterion variable. The **zero-order correlation** between X_1 and X_2, however, is $r_{X_1 X_2} = .50$. If Equation 33.31 is employed to compute the **semipartial correlation** between X_1 and Y with variable X_2 partialed out, the value computed for the **semipartial correlation** is $r_{Y(X_1 \cdot X_2)} = .693$, which is larger than the **zero-order correlation** of $r_{X_1 Y} = .60$. The latter is the case, since the variance on X_1 which is irrelevant to Y that is related to X_2 is partialed out of the analysis (i.e., suppressed), and the variance which remains on X_1 is more strongly related to Y resulting in the higher value $r_{Y(X_1 \cdot X_2)} = .693$ for the **semipartial correlation coefficient**.

$$r_{Y(X_1.X_2)} = \frac{r_{YX_1} - r_{YX_2}r_{X_1X_2}}{\sqrt{1 - r_{X_1X_2}^2}} = \frac{.60 - (0)(.50)}{\sqrt{1 - (.50)^2}} = .693$$

Although, as illustrated in the above example, **suppressor variables** can be employed to increase predictability, if a researcher is unaware of their presence and the role they are playing in the analysis, such variables can make it extremely difficult to interpret the result of a **multiple regression** analysis.

Test of significance for a partial correlation coefficient The null hypothesis H_0: $\rho_p = 0$ can be evaluated with Equation 33.34 (where ρ_p represents the population **partial correlation coefficient**).[22]

$$t = \frac{r_p\sqrt{n - v}}{\sqrt{1 - r_p^2}}$$

(Equation 33.34)

Where: r_p is the **partial correlation coefficient**
v is the total number of variables employed in the analysis

When there are two predictor variables and one criterion variable, the total number of variables employed in the analysis is 3, and thus $\sqrt{n - v} = \sqrt{n - 3}$. The value $n - v = n - 3$ represents the number of degrees of freedom employed for the analysis.[23] Employing Equation 33.34, the null hypothesis is evaluated in reference to the **partial correlation coefficients** $r_{YX_1.X_2} = .96$ and $r_{YX_2.X_1} = .60$.

$$t = \frac{(.96)\sqrt{5 - 3}}{\sqrt{1 - (.96)^2}} = 4.85 \qquad t = \frac{(.60)\sqrt{5 - 3}}{\sqrt{1 - (.60)^2}} = 1.06$$

It will be assumed that the nondirectional alternative hypothesis H_1: $\rho_p \neq 0$ is evaluated. Employing **Table A2**, for $df = 2$ (since $df = 5 - 3 = 2$), the tabled critical two-tailed .05 and .01 values are $t_{.05} = 4.30$ and $t_{.01} = 9.93$. Since the obtained value $t = 4.85$ is greater than $t_{.05} = 4.30$, the nondirectional alternative hypothesis H_1: $\rho_{YX_1.X_2} \neq 0$ is supported at the .05 level (but not at the .01 level). Since the obtained value $t = 1.06$ is less than $t_{.05} = 4.30$, the nondirectional alternative hypothesis H_1: $\rho_{YX_2.X_1} \neq 0$ is not supported.

When there are two predictor variables, a **partial correlation** can obviously only eliminate the effect of one other predictor variable. This kind of **partial correlation** is often referred to as a **first-order partial correlation**. When there are more than two predictor variables, it is possible to compute **higher-order partial correlations** in which the effects of two or more predictor variables are eliminated. Thus, a **second-order partial correlation** is one in which the effects of two predictor variables are eliminated. A discussion of **higher-order partial correlation** can be found in sources such as Cohen and Cohen (1983), Cohen *et al.* (2003), Hays (1994), and Marascuilo and Levin (1983).

B. Test computations and interpretation of results for Example 33.2 with *SPSS* [24] In Section I it was noted that prior to conducting a **multiple regression** analysis a researcher should carefully scrutinize the data, and that most computer software can provide relevant information that will alert one to the presence of **outliers** or other scores which can exert undue influence on statistics computed for the analysis. Additionally, it was noted that prior screening of data should assess the predictor variables with respect to **multicollinearity**. The aforementioned analyses will be conducted on the data for Example 33.2 prior to conducting the **multiple regression analysis**.

Preliminary screening of the data The table labeled **Descriptive Statistics** in Table 33.4 displays values for a variety of descriptive statistics computed for the three predictor variables (**Agility** displayed in the row **X1AGILIT**, **Strength** displayed in the row **X2STRENG**, and **Intelligence** displayed in the row **X3INTELL**) and the criterion variable (**Performance** displayed in the row **YPERFOR**). Additionally, the intercorrelations between the four variables involved in the analysis (i.e., the three predictors and the criterion variable) are displayed in the table labeled **Correlations**. Note that none of the intercorrelations among the predictor variables are above .332 — the latter suggesting the absence of **multicollinearity**, which will subsequently be confirmed by a more formal analysis.

**Table 33.4 Descriptive Statistics and Intercorrelations Between
Variables for Example 33.2**

Descriptive Statistics

	N	Range	Minimum	Maximum	Mean		Std.	Variance
	Statistic	Statistic	Statistic	Statistic	Statistic	Std. Error	Statistic	Statistic
X1AGILIT	50	49.00	.00	49.00	24.4800	1.7910	12.66401	160.377
X2STRENG	50	30.00	1.00	31.00	19.2400	.8174	5.78019	33.411
X3INTELL	50	46.00	2.00	48.00	27.4400	1.3998	9.89776	97.966
YPERFOR	50	77.00	35.00	112.00	79.0000	2.7524	19.46216	378.776
Valid N (listwise)	50							

	Skewness		Kurtosis	
	Statistic	Std. Error	Statistic	Std. Error
X1AGILIT	-.232	.337	-.837	.662
X2STRENG	-.577	.337	.945	.662
X3INTELL	.275	.337	.357	.662
YPERFOR	-.214	.337	-.803	.662
Valid N (listwise)				

Correlations

		YPERFOR	X1AGILIT	X2STRENG	X3INTELL
Pearson Correlation	YPERFOR	1.000	.746	.547	.163
	X1AGILIT	.746	1.000	.332	.105
	X2STRENG	.547	.332	1.000	.017
	X3INTELL	.163	.105	.017	1.000
Sig. (1-tailed)	YPERFOR	.	.000	.000	.128
	X1AGILIT	.000	.	.009	.233
	X2STRENG	.000	.009	.	.454
	X3INTELL	.128	.233	.454	.

Assessment of variable for skewness and kurtosis As noted in Section I the assumption of multivariate normality requires that all of the variables are normally distributed.[25] With respect to the latter, if a value computed by *SPSS* for either skewness (which corresponds to the value g_1 computed with Equation I.22 in the **Introduction**) or kurtosis (which corresponds to the value g_2 computed with Equation I.30 in the **Introduction**) deviates significantly from zero it may indicate a problem.

Equation 4.9 (i.e., $z = skewness / standard\ error$) is employed to determine whether a value computed for skewness deviates significantly from zero. When the aforementioned equation is applied to each of the variables (using the skewness and standard error values displayed in Table 33.4), all of the resulting z values are less than the tabled critical two-tailed value $z_{.05} = 1.96$. The researcher can thus conclude there is no evidence of skewness for any of the predictor variables or the criterion variable.

Equation 5.10 (i.e., $z = kurtosis / standard\ error$) is employed to determine whether a value computed for kurtosis deviates significantly from zero. When the aforementioned equation is applied to each of the variables (using the kurtosis and standard error values displayed in Table 33.4), all of the resulting z values are less than the tabled critical two-tailed value $z_{.05} = 1.96$. The researcher can thus conclude there is no evidence of kurtosis (i.e., leptokurtosis or platykurtosis) for any of the predictor variables or the criterion variable.

Screening data for influential observations A question a researcher should always address is whether or not one or more observations might have a disproportionate influence over the regression model derived in an analysis. The latter question can be answered through use of indices of **distance, leverage,** and **influence.** Hair *et al.* (1995, p. 122) identify the following three types of influential observations which are directly related to the aforementioned indices: **outliers, leverage points,** and **influentials.**

a) **Outliers** Hair *et al.* (1995, p. 122) note that, in the strictest sense, within the context of regression analysis an outlier is an observation which has a large residual (i.e., an observation for which there is a large discrepancy between the observed and predicted value on the criterion variable). The term **distance** is employed to represent a measure which is useful in identifying potential outliers on the criterion variable. The most commonly employed measure of **distance** are the **residuals.** As noted in Section VII of the **Pearson product-moment correlation coefficient**, residuals are easiest to interpret when they are standardized (i.e., converted to z scores). A **standardized residual** (computed with Equation 28.57) is obtained by dividing an unstandardized residual by the standard deviation of the residuals.

Employing the probabilities for the normal distribution, by chance it would be expected that 5% of the observations in a sample will have a **standardized residual** with an absolute value greater than $z = 1.96$, and that 1% of the observations will have a **standardized residual** with an absolute value greater than $z = 2.58$. If, in fact, the percentage of observations with **standardized residuals** in excess of 1.96 and 2.58 exceed the latter percentages, it would indicate that the derived regression model does not provide a good fit for the data. With respect to identifying **outliers**, any observation with a **standardized residual** with an absolute value greater than 3 (i.e., $|z| > 3$, where $|z| = 3$ is associated with a probability of .0026) can be viewed as a potential outlier (some sources employ $|z| \geq 2$ (which is associated with a probability of .0456)).

SPSS and other software also allow for computation of **Studentized residuals** (sometimes referred to as a **Studentized deleted residual** or **jackknife residuals**). Since a **Studentized residual** can provide a more accurate estimate of error than a **standardized residual**, it is more commonly employed for identifying outliers than the latter type of residual. In order to understand how a **Studentized residual** is computed consider the following. For each observation in a set of data an **adjusted predicted value** on the criterion variable can be computed by obtaining a new model for the data which is based on all of the data except the observation in question. The **adjusted predicted value** will be the value predicted for the latter observation using the model in which it was omitted. For the observation in question, the difference between its **adjusted predicted value** and its observed value will represent what is referred to as a **deleted residual.** A **Studentized residual** (which is based on Student's distribution — i.e., the t distribution) is a standardized version of the **deleted residual** —

specifically, it is the result of dividing a **deleted residual** by the standard deviation computed for the residuals for the model employed to compute the **adjusted predicted value**.[26] The *t* value obtained for a **Studentized residual** can be evaluated with $(n - k - 1)$ degrees of freedom. For further clarification of **Studentized residuals** the reader is referred to Hoagalin and Welsch (1978).

   b) **Leverage points**: A **leverage point** is an observation on a predictor variable which deviates substantially from other observations on that variable, and, by virtue of the latter, inclusion of such an observation in an analysis can dramatically impact the value of the regression coefficient computed for the predictor variable in question. The use of the term **leverage** indicates the degree to which a subject's score on a specific predictor variable deviates from the mean value for that predictor, and consequently **leverage** is considered to be useful in identifying potential outliers on any of the predictor variables. The most commonly employed measure of **leverage** is the **hat element** (or **hat diagonal**) **statistic** which is based on a special matrix referred to as the **hat matrix**. A related statistic is **Mahalanobis distance** which Hair *et al.* (1995, p. 152) note is a measure of the impact of a single observation based on differences between the observation's value and the mean value for all other observations across all the predictor variables. Alternatively, Tabachnick and Fidell (1996, p. 67) define **Mahalanobis distance** as the distance of an observation from the **centroid** of the remaining observations, where the **centroid** is the point created by the means of all the variables. Although the **hat elements** are measured on a different scale than **Mahalanobis distance**, the two measures are related (an equation for converting a value of one measure into the other can be found in Tabachnick and Fidell (1996, p. 68; 2001, pp. 68–69)).

   Sources (e.g., Field (2005, p. 165), Garson (2006), and Stevens (2002, p. 126)) note that a **hat element** can assume a value between 0 and 1, with a value of 0 indicating an observation has no influence on prediction and a value of 1 indicating maximum influence. The mean value of the statistic is $(k + 1)/n$, and if none of the observations exerts disproportionate influence on prediction it would be expected that all of the values computed for the statistic would be close to the average. Although Hoaglin and Welsch (1978) recommended that any observation with a value greater than $[2(k + 1)]/n$ be considered a potential outlier, Stevens (2002, p. 126) recommends the somewhat higher value $[3(k + 1)]/n$ as the limit. Neter *et al.* (1996) states that an observation with a **hat element** value between .2 and .5 indicates moderate leverage, and above .5 indicates very high leverage (and thus should be viewed as a potential outlier).[27]

   Field (2005, p. 165) notes that Barnett and Lewis (1978) have generated tables of critical values for **Mahalanobis distance**, which are a function of the sample size and number of predictor variables. To illustrate, for large sample sizes (e.g., $n > 500$) with $k = 5$, a **Mahalanobis distance** value greater than 25 indicates excessive leverage. For very small samples (e.g., $n < 30$) with $k = 2$, a **Mahalanobis distance** value greater than 11 warrants further scrutiny. Hair *et al.* (1995, p. 156) note that even without tables, observation of a **Mahalanobis distance** value two to three times greater than the next highest value for a set of observations may be indicative of a potential outlier.

   c) **Influentials**: This is a generic term for any observation which has a disproportionate effect on the results of a regression analysis. Although an **influential** will most commonly be an **outlier** and/or **leverage point**, it can also include an observation which does not fall in either of the latter categories. When an observation is said to be of **influence** it implies that the observation affects the position and/or orientation of the regression surface/plane computed for a regression model. Howell (1992, p. 509) notes that although observations which yield high values for either **distance** or **leverage** do not necessarily exert undue **influence** on a regression model, there is the possibility they may. Observations, on the other hand, which are high on both **distance** and **leverage** are likely to be high on **influence**. The most commonly employed measure of **influence** is **Cook's *D*** (Cook and Weisberg (1982)), which Stevens (2002, p. 132) (who provides a more detailed discussion of the measure) notes is an index of the change in the regression coefficients which would result if the observation in question were removed from the analysis. Cook and Weisberg (1982, p. 118) indicated that if an observation yields a value for **Cook's *D*** greater than 1 it may exert undue influence on the data.

The following additional diagnostic indices are available with computer software for identifying influential observations: a) **DFFIT** (Belsey *et al.* (1980)) and its standardized version (**SDFFIT**) are measures of the difference between the predicted value for an observation when the regression model employs the observation versus its value when the model omits the observation. If a model provides a good fit for the data, for a given observation the difference between the latter two predicted values should be minimal. Hair *et al.* (1995, p. 151) note that an observation with a **SDFIT** value greater than $2/\sqrt{k/n}$ be viewed as influential; b) **DFBeta** is the difference between a regression coefficient obtained using all of the observations versus the value of the regression coefficient when a specific observation is omitted from the analysis. Hair *et al.* (1995, p. 151) note that an observation with a **DFBeta** value substantially different from the other observations may possibly be influential; and c) The **covariance ratio (CVR)** is a measure of the effect of an observation on the values computed for all of the regression coefficients. For a given observation, a **CVR** value close to 1 indicates minimal influence. However, if the absolute value of $(CVR - 1)$ is greater than $3k/n$ the observation is influential. For further details on the latter measure the reader should consult Belsey *et al.* (1980), Field (2005, p. 167), and Hair *et al.* (1995, p. 151).[28]

In closing this discussion, the reader should take note of the fact that a researcher should employ sound judgement in employing any of the diagnostic indicators discussed in this section. More specifically, an extreme value on a diagnostic measure does not in itself necessarily justify deleting an observation from a set of data — especially if by doing so it allows the researcher to obtain a regression model that is more compatible with his or her expectations. In the final analysis, deletion of data from an analysis should be supported by sound logic rather than by experimenter convenience and/or bias.

Analysis of Example 33.2 with respect to distance, leverage and influence An analysis of the data for Example 33.2 with *SPSS* did not yield values for any of the $n = 50$ observations which exceeded the acceptable limits for **standardized residuals, Studentized residuals, hat elements, Mahalanobis distance**, and **Cook's *D***. The following observations are in order regarding the latter measures: a) No absolute value computed for a **standardized residual** or a **Studentized residual** exceeded 3. The largest absolute value obtained for a **standardized residual** was 2.281 (Subject 2), and the largest absolute value obtained for a **Studentized residual** was 2.412 (Subject 30). Although the absolute values of the **Studentized residuals** for Subjects 2, 16, 30, 32 were above 2, their inclusion in the analysis did not significantly impact the values computed for the correlation and regression coefficients for the overall analysis; b) The maximum value obtained for a **hat element** (label **lev** in *SPSS*) was .207 (Subject 47) — the latter value being the only **hat element** above .2; c) The largest **Mahalanobis distance** values were 10.152 (Subject 47), 9.097 (Subject 16), and 8.667(Subject 46). All of the latter values were, however, less than the suggested maximum acceptable value 11 noted earlier for small sample sizes; d) The largest values computed for **Cook's distance** were .284 (Subject 16) and .235 (Subject 46), both of which were well below 1.

In the final analysis, there was insufficient evidence to indicate that any of the observations exerted undue influence on the regression model to be computed for the data.

Evaluation of multicollinearity Within the context of assessing **multicollinearity**, most computer software can compute a number of diagnostic indicators. Two such indicators are **tolerance** and **variance inflation factor (VIF)** which are computed for each of the predictor variables. A **tolerance** value, which can range from 0 and 1, is a measure of how nonredundant a predictor is with respect to the other predictor variables included in the regression equation. In other words, it is an index of the amount of variability on a predictor variable which cannot be explained by the other predictor variables. The closer a **tolerance** value is to zero the more redundant a predictor variable is with the other predictors — i.e., the greater the indication of **multicollinearity** for the predictor variable in question. Myers and Well (2003, p. 571) note that if a predictor variable has a **tolerance** of 0 it can be perfectly expressed as a linear combination of all the other predictor variables in the regression equation. The regression coefficient for a variable with a low **tolerance** will be unstable as a result of the fact that it has a high standard error. Typically, if **tolerance** for a variable is below the minimal stipulated value, the variable should be omitted from the regression equation, and thus any equation already computed should be recomputed omitting those variables exhibiting unacceptable levels of **tolerance**.

Hair *et al*. (1995, p. 85) and Myers and Well (2003, p. 571) note that a **tolerance** value is equal to $1 - R_{j.}^2$, where $R_{j.}^2$ is the square of the **multiple correlation** of the jth predictor variable with all of the other predictor variables that are included in the regression equation. Most computer software allows one to set a minimal acceptable **tolerance** value, below which a predictor variable will not be added to the regression equation. Various sources note that the default **tolerance** value for computer software is set between .01 and .0001, but that the latter value can be adjusted by the researcher, and, in fact, it is generally recommended it be elevated to between .1 and .2. More specifically, sources (e.g., Bowerman and O'Connell (1990)) suggest a **tolerance** value less than .1 is an indicator of serious multicollinearity and a value between .1 and .2 (e.g., Menard (1995)) is suggestive of a problem.

Since as Hair *et al*. (1995, p. 127) note, **tolerance** is the amount of variability on a predictor variable which cannot be explained by the other predictor variables, we can determine that if the value for **tolerance** is .0001, then *Tolerance* $= 1 - R_{j.}^2 = .0001$. Consequently, *Tolerance* $= 1 - (.99995)^2 = .0001$ (which equals .01%). The latter indicates only .01% of the variability on the predictor variable in question cannot be explained on the basis of the other predictor variables, and that since $R_{j.}^2 = .99995$ (which is equal to 99.995%), 99.995% of the variability on the predictor variable in question can be explained on the basis of the other predictor variables. On the other hand, if the value of the **tolerance** is .10, then *Tolerance* $= 1 - (.95)^2 = .0975$ (which is 9.75%). The latter indicates that 9.75% (i.e., approximately 10%) of the variability on the predictor variable in question cannot be explained on the basis of the other predictor variables, and that since $R_{j.}^2 = .95$ (which is equal to 95%), 95% of the variability on the predictor variable in question can be explained on the basis of the other predictor variables.

The value noted for **variance inflation factor (VIF)** is the reciprocal of the **tolerance** (i.e., *VIF* $= 1/Tolerance$), and thus **tolerance** and **VIF** are inversely related to one another. Consequently, the larger the value of **VIF** the higher the likelihood a variable exhibits **multicollinearity**. The **tolerance** values .1 and .2 stipulated earlier as respectively indicating or suggesting a **multicollinearity** problem correspond to the **VIF** values 10 (i.e., 1/.1) and 5 (i.e., 1/.2).

Tabachnick and Fidell (1996, p. 136) state that in some cases a researcher may want to determine which predictor variable(s) should be eliminated from an analysis based on logical rather than statistical grounds by considering issues such as the reliability of variables or the cost of measuring variables. Thus, one may elect to eliminate the least reliable variable rather than a variable with very low **tolerance**, and by virtue of the latter the **tolerance** value for the latter variable may be incremented to an appropriate level.

Computer software provides for additional optional **multicollinearity** indicators labeled **eigenvalues, condition index**, and **variance proportions**. In describing the latter three indices,

Myer and Wells (2003, pp. 597–598) note that a **principal components analysis** (described in the chapter on **principal components analysis and factor analysis (Test 40))** is conducted on the variables in order to identify linear combinations of variables which account for substantial amounts of variability in the data. Each of the derived linear combinations is referred to as a **component** or **factor** (commonly labeled **dimension** in a computer printout), and an **eigenvalue** represents the amount of variability in the data accounted for by a specific component. If one or more of the variables is completely redundant (i.e., all of the variability on the variable can be explained by the other variables, thus indicating the presence of perfect **multicollinearity**), one or more of the **eigenvalues** will equal zero. The square root of the ratio of the largest **eigenvalue** to the **eigenvalue** for a specific component is what is represented by the **condition index** of a component (the larger the **condition index** for a component the smaller its **eigenvalue**). The presence of a **condition index** above 30 suggests serious **multicollinearity**, while a **condition index** between 15 and 30 indicates the possibility of **multicollinearity**. **Variance proportions** (also discussed in Hair *et al.* (2004, p. 153)) are the proportion of variance on each variable accounted for by each of the components. If two or more variables have a **variance proportion** of .50 or greater on a component with a high **condition index**, there is a **multicollinearity** problem with respect to those variables.

Table 33.5 *SPSS* **Printout of Collinearity Analysis for Example 33.1**

Coefficients[a]

Model		Correlations			Collinearity Statistics	
		Zero-order	Partial	Part	Tolerance	VIF
1	(Constant)					
	XSUGAR	.955	.962	.819	.863	1.159
	ZSALT	.523	.615	.182	.863	1.159

a. Dependent Variable: YCAV

Collinearity Diagnostics[a]

Model	Dimension	Eigenvalue	Condition Index	Variance Proportions		
				(Constant)	XSUGAR	ZSALT
1	1	2.503	1.000	.04	.06	.03
	2	.345	2.695	.18	.91	.07
	3	.152	4.058	.78	.03	.90

a. Dependent Variable: YCAV

Evaluation of multicollinearity for Example 33.1   Table 33.5 summarizes the *SPSS* collinearity analysis for Example 33.1.[29] The values computed for **Tolerance** and **VIF** are respectively recorded in the rows labeled **XSUGAR** and **ZSALT** in Columns 5 and 6 in the upper part of the table labeled **Coefficients**. Note that both of the **tolerance** values equal .863, which is close to the maximum possible value of 1, and that the latter result is consistent with the absence of **multicollinearity**. (The reader should also take note of the fact that Columns 2, 3, and 4 of the table contain the **zero order correlation** for each of the predictor variables and the criterion variable, the **semi-partial** (labeled **part**) **correlations**, and **partial correlations** (computed earlier with Equations 33.31 and 33.33).) The lower part of Table 33.5 labeled **Collinearity Diagnostics** contains the **eigenvalues, condition indices**, and **variance proportions**. Note that the largest **Condition Index** of 4.058 recorded in Column 3 associated with the smallest **Eigenvalue** .152 (recorded in Column 2 of the row labeled **Dimension 3**) is far removed from the value 15 (which is the minimum value that would suggest the possibility of

multicollinearity). (As noted above, the three dimensions listed in each of the rows of the lower part of the table are linear combinations derived when the data were further evaluated through use of a **principal components analysis**.) In conclusion, the values recorded in Table 33.5 do not indicate there is any evidence of **multicollinearity** in the data.

Table 33.6　*SPSS* Printout of Collinearity Analysis for Example 33.2

Coefficients[a]

Model		Correlations			Collinearity Statistics	
		Zero-order	Partial	Part	Tolerance	VIF
1	(Constant)					
	X1AGILIT	.746	.711	.585	.880	1.137
	X2STRENG	.547	.483	.319	.889	1.124
	X3INTELL	.163	.156	.091	.989	1.012

a. Dependent Variable: YPERFOR

Collinearity Diagnostics[a]

Model	Dimension	Eigenvalue	Condition Index	Variance Proportions			
				(Constant)	X1AGILIT	X2STRENG	X3INTELL
1	1	3.724	1.000	.00	.01	.00	.01
	2	.155	4.900	.02	.78	.00	.20
	3	.089	6.485	.03	.20	.38	.48
	4	.033	10.646	.95	.00	.62	.31

a. Dependent Variable: YPERFOR

Evaluation of multicollinearity for Example 33.2　Table 33.6 summarizes the *SPSS* collinearity analysis for Example 33.2. The values computed for **Tolerance** and **VIF** are respectively recorded in Columns 5 and 6 in the upper part of the table labeled **Coefficients** in the rows labeled **X1AGILT, X2STRENG,** and **X3INTELL.** Note that the three **tolerance** values .880, .889, .989 for the predictor variables are close to the maximum possible value of 1, and that the latter result is consistent with absence of **multicollinearity.** This conclusion is further reinforced by the fact that none of the values computed for a **Condition Index** exceeds 15 (the largest value being 10.646). The researcher can thus conclude there is no evidence of **multicollinearity.**

SPSS multiple regression analysis of Example 33.2　Employing all three predictor variables simultaneously, a **standard multiple regression analysis** (which employs the **Enter** option on the *SPSS* method menu) was conducted. Table 33.7 summarizes the results of the analysis.

The table labeled **Model Summary** in Table 33.7 displays the following information.

a) The reader should take note of the fact that the **1** at the upper left of the table under **Model** (which is also in Tables 33.5 and 33.6) represents that it is possible to compute more than one model for the set of data. The number **1** indicates the information in the table is based on the first (and in this case only) **multiple regression** model computed for the data. In the event a **stepwise** or **hierarchical multiple regression analysis** were conducted, each time a predictor variable was added to or eliminated from the analysis the relevant values which comprised the table would be displayed in a separate row that would be identified with a different model number. For example, if a **forward stepwise analysis** were conducted and the final model contained two of the three predictor variables, there would be information for two models in the **Model Summary** table. The first model (numbered **1**) would represent the analysis involving the predictor variable with the highest **partial correlation** with the criterion variable, while the

Table 33.7 *SPSS* **Printout of Standard Multiple Regression Analysis of Example 33.2**

Model Summary[b]

Model	R	R Square	Adjusted R Square	Std. Error of the Estimate	R Square Change	F Change	df1	df2	Sig. F Change	Durbin-Watson
					Change Statistics					
1	.816[a]	.665	.644	11.61787	.665	30.502	3	46	.000	2.280

a. Predictors: (Constant), X3INTELL, X2STRENG, X1AGILIT
b. Dependent Variable: YPERFOR

ANOVA[b]

Model		Sum of Squares	df	Mean Square	F	Sig.
1	Regression	12351.156	3	4117.052	30.502	.000[a]
	Residual	6208.844	46	134.975		
	Total	18560.000	49			

a. Predictors: (Constant), X3INTELL, X2STRENG, X1AGILIT
b. Dependent Variable: YPERFOR

Coefficients[a]

Model		Unstandardized Coefficients		Standardized Coefficients	t	Sig.	95% Confidence Interval for B	
		B	Std. Error	Beta			Lower Bound	Upper Bound
1	(Constant)	28.645	7.384		3.879	.000	13.782	43.509
	X1AGILIT	.959	.140	.624	6.861	.000	.677	1.240
	X2STRENG	1.140	.304	.338	3.743	.001	.527	1.752
	X3INTELL	.181	.169	.092	1.072	.289	-.159	.520

second model (numbered **2**) would represent the analysis that involved the two predictor variables with the highest **partial correlations** with the criterion variable.

b) The values computed for the **multiple correlation coefficient** ($R_{Y.X_1X_2X_3}$ = .816), the **squared multiple correlation** ($R^2_{Y.X_1X_2X_3}$ = .665), the **adjusted squared multiple correlation** ($\tilde{R}^2_{Y.X_1X_2X_3}$ = .644), and the **standard error of the estimate** (11.61787) are displayed in Columns 2–5. Note that the magnitude of the **squared multiple correlation** represents a large multivariate effect size, since it exceeds Cohen's (1992) criterion of $R^2 \geq .26$ for the latter effect size.

c) With respect to the values in the section of the table labeled **Change Statistics**, one should take note of the following: a) The value .665 in the column labeled **R Square Change** is identical to the value recorded for the **squared multiple correlation**; b) The values in the columns labeled **F Change**, **df1**, **df2**, and **Sig F Change** are identical to the values of **F**, **df**$_{\text{Regression}}$, **df**$_{\text{Residual}}$, and **Sig** in the table below labeled **ANOVA**. It should be noted, however, that when a **stepwise** or **hierarchical multiple regression analysis** is conducted, the values printed under **Change Statistics** will not correspond to the aforementioned values in the **ANOVA** table. The latter is the case since each time a predictor variable is added to or removed from the analysis, values for a new regression model are computed. With respect to the latter, the **Change Statistics** section of the **Coefficients** table will reflect any change (relative to previous values computed) in the **multiple correlation** and F statistic which results from the addition (or elimination) of a variable.

d) The result of the **Durbin–Watson test** (which evaluates independence of the residuals) is displayed in the column labeled **Durbin–Watson**. Employing the guidelines in Table 28.7 (in Section VII of the **Pearson product-moment correlation coefficient**) for interpreting the latter test statistic, the value 2.280 is not significant. More specifically, employing **Table A27 (Table**

of Critical Values for the Durbin–Watson test statistic) in the **Appendix**, it can be determined that for $n = 50$ and $k = 3$, $d_{u_{05}} = 1.67$, and thus $(4 - 1.67) = 2.33$. In order to conclude the residuals are not correlated (i.e., independent), the following condition must be satisfied: $2 < d < (4 - d_U)$. Since the computed value $d = 2.28$ is greater than 2 but less than $(4 - d_U) = 2.33$, the researcher can conclude there is no evidence of violation of the assumption of independence of the residuals.

The table labeled **ANOVA** in Table 33.7 summarizes the result of the **analysis of variance for the regression analysis**. Specifically, the latter table summarizes the analysis evaluating the following null hypotheses: a) H_0: P $= 0$; and b) H_0: $B_1 = B_2 = B_3 = 0$. Rejection of the latter null hypotheses implies that in the underlying population: a) The **multiple correlation** is some value other than zero; and b) At least one of the $k = 3$ regression coefficients (B_1, B_2, B_3) is some value other than zero. Since the probability (i.e., the value .000 in the column labeled **Sig**) associated with value $F = 30.502$ is less than .05, the null hypotheses can be rejected (in fact, the F value is significant beyond the .00049 level, since the latter value rounds off to .000). The researcher can thus conclude the computed value for the **multiple correlation** is statistically significant (i.e., is some value other than zero), and that at least one of the regression coefficients is statistically significant (i.e., is some value other than zero).

The table labeled **Coefficients** in Table 33.7 contains the following information.

a) The values of the constant $a = 28.645$ and the **unstandardized regression coefficients** $b_1 = .959$ (for agility), $b_2 = 1.140$ (for strength), and $b_3 = .181$ (for intelligence) are respectively recorded in Column 2 (labeled **B**) in the rows labeled **(Constant)**, **X1AGILT**, **X2STRENG**, and **X3INTELL**. (The values of the standard errors are recorded in Column 3 labeled **Std. Error**.) The **multiple regression equation** (Equation 33.4) resulting from the latter values is $Y' = 28.645 + .959X_1 + 1.140X_2 + .181X_3$. The value $b_1 = .959$ indicates that if the other predictor variables are partialed out of the analysis, when agility is increased by one point, performance will increase by .959 points. In the same respect, the value $b_2 = 1.140$ indicates that if the other predictor variables are partialed out of the analysis, when strength is increased by one point, performance will increase by 1.140 points. And finally, the value $b_3 = .181$ indicates that if the other predictor variables are partialed out of the analysis, when intelligence is increased by one point, performance will increase by .181 points.

b) The values of the **standardized regression coefficients** $\beta_1 = .624$ (for agility), $\beta_2 = .338$ (for strength), and $\beta_3 = .092$ (for intelligence) are displayed in Column 4 labeled **Beta**. The value $\beta_1 = .624$ indicates that an increase of one standard deviation unit on the predictor variable of agility is associated with an increase of .624 standard deviation units on the criterion variable of performance (if the predictor variables of strength and intelligence are partialed out of the analysis). In the same respect, the value $\beta_2 = .338$ indicates that an increase of one standard deviation unit on the predictor variable of strength is associated with an increase of .338 standard deviation units on the criterion variable of performance (if the predictor variables agility and intelligence are partialed out of the analysis). And finally, the value $\beta_3 = .092$ indicates that an increase of one standard deviation unit on the predictor variable of intelligence is associated with an increase of .092 standard deviation units on the criterion variable of performance (if the predictor variables agility and strength are partialed out of the analysis).

c) The results of the t **test**, which simultaneously evaluate the **unstandardized** and **standardized regression coefficients**, are summarized in Column 5 (labeled **t**) and Column 6 (labeled **Sig.**). In order for a t value to be significant, the probability value listed in Column 6 must be equal to or less than .05. Inspection of the table reveals that the t values for agility ($t = 6.861$, $p = .000$) and strength ($t = 3.743$, $p = .001$) are significant, while the t value for intelligence ($t = 1.072$, $p = .289$) is not. The researcher can thus conclude that the **regression coefficients** computed for agility and strength are some value other than zero. The fact that one cannot conclude the coefficient for intelligence is some value other than zero suggests the latter predictor should be dropped from the regression model.

d) The lower and upper bounds of the **95% confidence intervals** for the **unstandardized regression coefficients** are recorded in Columns 7 (**Lower Bound**) and 8 (**Upper Bound**). Note that when a regression coefficient is significant, the value 0 will not be included in the confidence interval (which is the case for agility and strength), but will be included when the coefficient is not significant (which is the case for intelligence).

e) At this point in the discussion it is instructive to return to the table labeled **Coefficients** in Table 33.6 (which in actuality is the last section printed out by *SPSS* for the **Coefficients** table displayed in Table 33.7). Note that Columns 2, 3, and 4 of the **Coefficients** table in Table 33.6 respectively contain the **zero-order**, **partial**, and **part** (i.e., **semi-partial**) correlations between the criterion variable and the predictor variables. The square of the **semipartial correlation** ($r_{Y(X_1 . X_2 X_3)} = .585)^2 = .342$ reflects the proportion of shared variance between agility and performance after any linear association agility has with strength and intelligence has been removed. Thus, the latter **semipartial correlation** will allow the researcher to obtain a pure measure of the correlation (and consequently a pure measure of the amount of shared variance) between the predictor variable X_1 (agility) and the criterion variable Y (performance). In the same respect, the square of the **semipartial correlation** ($r_{Y(X_2 . X_1 X_3)} = .319)^2 = .102$ reflects the proportion of shared variance between strength and performance after any linear association strength has with agility and intelligence has been removed. And finally, the square of the **semipartial correlation** ($r_{Y(X_3 . X_1 X_2)} = .091)^2 = .008$ reflects the proportion of shared variance between intelligence and performance after any linear association intelligence has with agility and strength has been removed. Because the value $r^2_{Y(X_3 . X_1 X_2)} = .008$ is so close to zero, plus the fact that the regression coefficient for intelligence was nonsignificant, a decision was made to eliminate the predictor variable of intelligence from the **multiple regression model**.

In the case of Example 33.2, the relative magnitude of the **standardized regression coefficients** appears to be a good indicator of the relative importance of predictor variables in explaining variability on the criterion variable. Of the three predictors, it appears that agility (which has the largest **standardized regression coefficient**, $\beta_1 = .624$) is most strongly related to performance followed by strength ($\beta_2 = .338$), and finally by intelligence, which has a coefficient barely above zero ($\beta_3 = .092$). As noted earlier, since the predictor of intelligence does not make a significant contribution, it would seem prudent to drop it from the analysis. The decision to eliminate the latter variable was also predicated on the fact that the author evaluated the data employing a **backward stepwise analysis**. The latter analysis (which began with all three predictors and eliminated any predictor that did not make a significant contribution) resulted in a model which only contained the predictor variables of agility and strength. The latter analysis yielded the **multiple correlation** $R_{Y . X_1 X_2} = .811$, as well as $R^2_{Y . X_1 X_2} = .657$, and $\tilde{R}^2_{Y . X_1 X_2} = .643$. The model with two predictors also yielded the additional values displayed in Table 33.8. Employing the **unstandardized regression coefficients**, the **multiple regression equation** for the new model which just employs the predictors of agility and strength is $Y' = 33.342 + .975 X_1 + 1.133 X_2$. Note that proportional relationship between the **standardized regression coefficients** for agility (.634) and strength (.337) is about the same as in the original model containing all three predictors.

The reader should take note of the fact that if the square of the **semi-partial correlation** for intelligence ($r_{Y(X_3 . X_1 X_2)} = .091)^2 = .008$ (computed when all three predictors are included in

Table 33.8   SPSS Printout of Backward Stepwise Multiple Regression Analysis for Example 33.2

Coefficients[a]

Model		Unstandardized Coefficients		Standardized Coefficients	t	Sig.	95% Confidence Interval for B	
		B	Std. Error	Beta			Lower Bound	Upper Bound
1	(Constant)	33.342	5.954		5.600	.000	21.364	45.320
	X1AGILIT	.975	.139	.634	7.003	.000	.695	1.254
	X2STRENG	1.133	.305	.337	3.717	.001	.520	1.747

the analysis) is added to $R^2_{Y.X_1X_2}$ =.657 (i.e., the multiple correlation when only agility and strength are included in the analysis), it yields the value = .665 computed for the **squared multiple correlation** when all three predictor variables are included in the analysis. The latter one again indicates that inclusion of intelligence in the analysis only minimally increases the amount of explained variability on the criterion variable.

In conclusion, assuming the results obtained for Example 33.2 can be confirmed by cross-validation studies, it would be realistic for the researcher to omit intelligence in the model for predicting performance.[30] The minimal increase in predictability in the sample data resulting from the addition of intelligence is probably the result of chance, and even if the latter is not true, the time and expense required to obtain a measure of intelligence for every subject probably would not justify its inclusion in the predictive model.

VI. Additional Analytical Procedures for Multiple Regression and/or Related Tests

1. **Cross-validation of sample data**   Keeping in mind that the data for Example 33.2 are hypothetical, the only attempt made to cross-validate the results was to conduct separate **multiple regression** analyses on two subsamples, each representing a different half of the original sample. Thus, separate analyses were conducted employing **Subjects 1-25** versus **Subjects 26-50** in Table 33.1. For **Subjects 1-25** the following values were obtained: $R_{Y.X_1X_2X_3}$ = .847; b_1 = .959, β_1 = .663 (t = 5.568, p = 000); b_2 = 1.538, β_2 = .395 (t = 3.313, p = .003); b_3 = .145, β_3 = .071 (t = .613, p = .546). For **Subjects 26-50** the following values were obtained: $R_{Y.X_1X_2X_3}$ = .794; b_1 = .963, β_1 = .581 (t = 3.778, p = 001); b_2 = .921, β_2 = .297 (t = 1.983, p = .061); b_3 = .202, β_3 = .105 (t = .770, p = .450). Although, for the most part, the analyses of the two subsamples yielded results which were similar to those obtained for the full sample, they were not identical. Note that in the case of **Subjects 26-50**, the analysis of the predictor variable of strength (X_2) just falls short of being statistically significant. The point to be made with this particular illustration is that any attempt to cross-validate the results of a **multiple regression** analysis will never yield an identical outcome. Because of this it cannot be emphasized too strongly, that before employing **multiple regression** for predictive purposes it is essential that a researcher obtain sufficient cross-validation data which utilizes large samples that are representative of the target population. Only by doing the latter will one be able to have a reasonable degree of confidence in the reliability of the coefficients employed in a regression equation employed for predictive purposes.

VII. Additional Discussion of Multiple Regression

1. Final comments on multiple regression analysis a) Figure 33.2 (based on Cohen and Cohen (1983)), which is known as a **Venn diagram** (previously discussed in the **Introduction**), provides a visual summary of the proportion of variance represented by **zero-order, multiple, partial**, and **semipartial correlation coefficients** (when there are two predictor variables and a criterion variable). Each of the three circles represents the variance of one of the three variables. Areas of overlap between circles represent shared variance between variables.

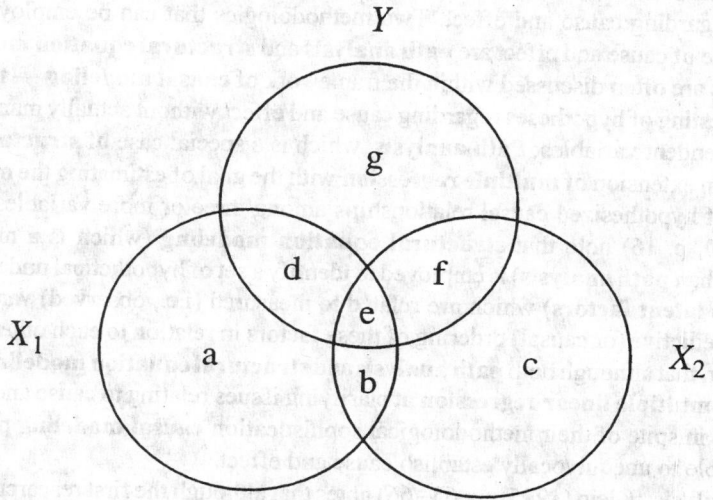

Zero order correlations: $r_{Y.X_1}^2 = d + e$ $\quad r_{Y.X_2}^2 = e + f$ $\quad r_{X_1.X_2}^2 = e + b$

Multiple correlation: $R_{Y.X_1X_2}^2 = d + e + f$

Partial correlations: $r_{YX_1.X_2}^2 = \dfrac{d}{d + g}$ $\quad r_{YX_2.X_1}^2 = \dfrac{f}{f + g}$

Semipartial correlations: $r_{Y(X_1.X_2)}^2 = d$ $\quad r_{Y(X_2.X_1)}^2 = f$

Figure 33.2 Venn Diagram of Variance Components Represented by Squared Correlation Coefficients

b) Although not commonly discussed, within the framework of a **multiple regression** analysis it is possible for two or more predictor variables to interact with one another. For example, consider a situation in which a researcher is interested in the relationship the two predictor variables of agility (X_1) and strength (X_2) have with the criterion variable performance (Y). However, rather than examining whether the relationship between agility and strength with performance is additive and linear (i.e., the general case for linear **multiple regression** described by Equation 33.4), the researcher instead wants to evaluate the hypothesis that although agility is positively related to performance, the degree of its relationship with performance weakens as a subject's score on strength decreases. In such a case, the researcher is hypothesizing an interaction between agility and strength in predicting performance. The appropriate form of the **multiple regression equation** for describing the latter relationship would not be Equation 33.4, but instead would be Equation 33.35 (where the term $b_3 X_1 X_2$ stipulates the interaction multiplied by its coefficient). Readers interested in an in-depth discussion of evaluating interactions in multiple **regression analysis** are referred to Aiken and West (1991).[31]

$$Y' = a + b_1 X_1 + b_2 X_2 + b_3 X_1 X_2 \qquad \text{(Equation 33.35)}$$

2. Causal modeling: Path analysis and structural equation modeling Regression analysis does not provide sufficient control over the variables under study to allow a researcher to draw conclusions regarding cause and effect. Two methodologies that can be employed to further clarify the issue of cause and effect are **path analysis** and **structural equation modeling**. Both methodologies are often discussed within the framework of **causal modeling** — the latter term implying the testing of hypotheses regarding cause and effect without actually manipulating one or more independent variables. **Path analysis**, which is a special case of **structural equation modeling**, is an extension of **multiple regression** with the goal of estimating the magnitude and significance of hypothesized causal relationships among three or more variables. Grimm and Yarnold (2000, p. 16) note that **structural equation modeling** (which is a more powerful methodology than **path analysis**) is employed to identify a set of hypothetical underlying factors (referred to as **latent factors**) which are related to measured (i.e., observed) variables, and to identify the predictive (or causal) ordering of these factors in relation to each other. It should be noted however, that although both **path analysis** and **structural equation modeling** are superior to **simple** and **multiple linear regression** at clarifying issues relating to cause and effect, in the final analysis, in spite of their methodological sophistication **causal modeling** procedures are not actually able to unequivocally establish cause and effect.

Path analysis Klem (1995, pp. 65–66) notes that although the first researcher to describe the use of **path analysis** was the geneticist Sewall Wright (e.g., Wright (1921)), it wasn't until the 1960s that the methodology came to the attention of researchers in other academic disciplines — most notably the social sciences. The latter author states that **path analysis** is an extension of **multiple regression** to situations in which a researcher is: a) Interested in prediction in reference to more than one dependent variable; and b) Particularly concerned with the predictive ordering of the multiple variables. To illustrate the difference between a **regression model** and a **path analysis model**, Klem (1995, p. 65) notes that a model employing statements such as X causes Y or X_1 and X_2 cause Y illustrate a **regression model**, while a **path analysis model** would be illustrated by a statement such as X causes Y and Y causes Z.

In employing **path analysis** a researcher begins with a theoretical model describing the relationship between a set comprised of three or more variables. Such a model is typically summarized by a **path diagram**, which is a visual display resembling a flow chart that depicts the hypothesized causal relationship between the variables. The way in which variables are connected to one another (e.g., straight arrows versus curved arrows) is employed to indicate whether or not pairs of variables are causally related to one another. In **path analysis** two types of variables are employed. **Exogenous variables** are variables which are only employed as predictors (i.e., the hypothesized model makes no attempt to explain them), while **endogenous variables** are dependent measures that are hypothesized to be caused by one or more of the exogenous variables. It is possible, however, in a given **path model** that one or more endogenous variables may also play a role as a causative agent in reference to another endogenous variable. To illustrate the latter, a person's intelligence (which will represent an exogenous variable) may be hypothesized to be responsible for how much money the person earns (which will represent an endogenous variable), which in turn may be responsible the magnitude of a person's self esteem (which also represents an endogenous variable).

Within the framework of **path analysis, path coefficients** (which are essentially **standardized regression coefficients**) are computed that indicate the magnitude of the direct effect of one variable on another, and the latter are employed to help determine whether or not the data are consistent with the hypothesized **path model** – which is the ultimate goal of **path analysis**. If, in fact, the data are consistent with the model, the best the researcher will be able to do is conclude that the model is plausible. She will not, however, be able to conclude it has been proven, since there is always the possibility an alternative model can provide a better fit for a set of data.

Structural equation modeling (commonly identified with the acronym **SEM**) is also referred to as **causal modeling, causal analysis, simultaneous equation modeling,** and **analysis of covariance structures**. Sources note that **structural equation modeling** was first introduced around 1970 in a series of papers written by the Swedish statistician Karl Jöreskog (e.g., Jöreskog (1969)). The distinguishing feature between **path analysis** and **structural equation modeling** is that **path analysis** is employed with observed variables (i.e., measured variables) while **structural equation modeling** can be employed with both observed and **latent variables** (also referred to as **factors**). A **latent variable** (which is essentially the same thing as a **hypothetical construct** as defined in the **Introduction**) is one that cannot be measured directly, but can be represented by obtaining measures on one or more variables ((Hair *et al.* (1995, p. 619); Klem (2000, p. 227)). Whereas **path analysis** involves the use of simple bivariate correlations between variables in a system of equations intended to describe a hypothesized set of relationships among the variables, **structural equation modeling** employs multivariate methods such as **multiple regression** and **factor analysis** to evaluate the relationship between the set of variables. Hair *et al.* (1995, p. 622) note that in **structural equation modeling** a series of separate but interdependent **multiple regression** equations are simultaneously estimated by specifying a specific structural model. The latter authors also note that whereas certain other multivariate procedures (e.g., **multivariate analysis of variance (Test 35)** and **canonical correlation (Test 38)**) allow for only a single relationship between multiple dependent variables and two or more independent variables, **structural equation modeling** can be used to describe multiple relationships between multiple independent variables and multiple dependent variables. Among others, Thompson (2000, p. 276) notes that although **structural equation modeling** can provide researchers with insights regarding causality, in the final analysis, definitive causal evidence can only be extrapolated from appropriately controlled experiments involving manipulated independent variables. Because of the complexity of the computations involved in **structural equation modeling**, the methodology is generally implemented with statistical software packages such as **LISREL, AMOS,** and **EQS**. Selected sources which provide in-depth discussions and/or illustrations of **path analysis** and/or **structural equation modeling** are Byrne (2001), Hair *et al.* (1995, Ch. 11), Klem (1995, 2000), Kline (2004), Loehlin (2004), Mertler and Vannatta (2005), Raykov and Marcoulides (2000), Schumacker and Lomax (2004), Thompson (2000), and Ullman (2001). Freedman (2005) provides an authoritative discussion on the dangers and limitations associated with modeling methodologies, especially when the assumptions underlying a modeling methodology have been violated.

3. Brief note on logistic regression   An alternative to simple and **multiple regression** is **logistic regression (Test 39)**, which is described in detail in a subsequent chapter. As is the case with simple and **multiple regression, logistic regression** regresses one or more independent variables on a dependent variable. In **logistic regression**, however, the independent/predictor variables can be discrete/categorical, continuous, or a combination of both, while the dependent/

criterion variable is categorical (most commonly the dependent variable is a dichotomous measure — although it is possible to have more than two categories). Instead of predicting the score of a subject on the dependent variable, in **logistic regression** what is predicted is the probability a subject will fall in specific category. **Logistic regression**, which is often categorized as a nonparametric procedure, is more flexible than simple and **multiple regression,** since its reliability does not depend on certain restrictive normality assumptions regarding the underlying population distributions for the predictor variables associated with the latter type of analyses.

VIII. Additional Examples Illustrating the Use of Multiple Regression

No additional examples will be presented in this section.

References

Aiken, L. S. & West, S. G. (1991) **Multiple regression: Testing and interpreting interactions**. Newbury Park, CA: Sage Publications.

Alleni, M. P. (1997). **Understanding regression analysis**. New York: Plenum Press.

Barnett, V. & Lewis, T. (1994). **Outliers in statistical data** (3rd ed.). Chichester: John Wiley & Sons.

Belsey, D. A., Kuh, E. & Welsch, R. (1980). **Regression diagnostics: Identifying influential data and sources of collinearity**. New York: John Wiley & Sons.

Berenson M. L. & Levine, D.M. (1996). **Basic business statistics: Concepts and applications**. Englewood Cliffs, NJ: Prentice Hall.

Bernstein, I. H., Garbin, C. P. & Teng, G. K. (1988). **Applied multivariate analysis**. New York: Springer-Verlag.

Bowerman, B. L. & O'Connell, R. T. (1990). **Linear statistical models: An applied approach** (2nd ed.). Belmont, CA: Duxbury.

Browne, M. W. (1975). Predictive validity of a linear regression equation. **British Journal of Mathematical and Statistical Psychology**, 28, 79–87.

Byrne, B. M (2001). **Structural equation modeling with AMOS: Basic concepts, applications and programming**. Mahwah, NJ: Lawrence Erlbaum Associates, Publishers.

Cattin, P. (1980). Note on the estimation of the squared cross-validated multiple correlation of a regression model. **Psychological Bulletin**, 87, 63–65.

Cohen, J. (1992). A power primer. **Psychological Bulletin**, 112, 155–159.

Cohen, J. & Cohen, P. (1983). **Applied multiple regression/correlation analysis for the behavioral sciences** (2nd ed.). Hillsdale, NJ: Lawrence Erlbaum Associates, Publishers.

Cohen, J., Cohen, P., West, S. G., & Aiken, L. S. (2003). **Applied multiple regression/correlation analysis for the behavioral sciences** (3rd ed.). Mahwah, NJ: Lawrence Erlbaum Associates, Publishers.

Cook, R. D. & Weisberg, S. (1982). **Residuals and influence in regression**. New York: Chapman & Hall.

Darlington, R. B. (1990). **Regression and linear models**. New York: McGraw-Hill.

Diekhoff, G. (1992). **Statistics for the social and behavioral sciences: Univariate, bivariate, multivariate**. Dubuque, IA: Wm. C. Brown Publishers.

Field, A. (2005). **Discovering statistics using SPSS** (2nd ed.). Sage Publications: London.

Freedman, D. A. (2005). **Statistical models: Theory and practice**. New York: Cambridge University Press.

Garson, D. G. (2006). Statistics: Topics in multivariate analysis. Website: **http://www2. chas.ncsu.edu/garson/pa765/statnote.htm.**

Grimm, L. G. & Yarnold, P. R. (Eds.) (1995). **Reading and understanding multivariate statistics**. Washington, D.C.: American Psychological Association.

Grimm, L. G. & Yarnold, P. R. (Eds.) (2000). **Reading and understanding more multivariate statistics**. Washington, D.C.: American Psychological Association.

Hair, J. F., Anderson, R. E., Tatham, R. L. & Black, W. C. (1995) **Multivariate data analysis** (4th ed.). Upper Saddle River, N. J.: Prentice Hall.

Hair, J. F., Anderson, R. E., Tatham, R. L. & Black, W. C. (2004) **Multivariate data analysis** (6th ed.). Upper Saddle River, N. J.: Prentice Hall.

Hair, J. F. & Black, W. C. (2000). Cluster analysis. In Grimm, L. G. & Yarnold, P. R. (Eds.) (pp. 147–205). **Reading and understanding more multivariate statistics**. Washington, D.C.: American Psychological Association.

Harlow, L. L. (2005). **The essence of multivariate thinking**. Mahwah, NJ: Lawrence Erlbaum Associates.

Harris, R. J. (2001) **A primer of multivariate statistics** (3rd ed.). Mahwah, NJ: Lawrence Erlbaum Associates.

Hays, W. L. (1994). **Statistics** (5th ed.). Fort Worth: Harcourt Brace College Publishers.

Hoagalin, C. C. & Welsch, R. (1978). The hat matrix in regression and ANOVA. **The American Statistician**, 32, 17–22.

Howell, D. C. (2002). **Statistical methods for psychology** (5th ed.). Pacific Grove, CA: Duxbury Press.

Jöreskog, K. G. (1969). A general approach to confirmatory maximum likelihood factor analysis. **Psychometrika**, 34, 183–220.

Kachigan, S. K. (1986). **Statistical analysis**. New York: Radius Press.

Kachigan, S. K. (1991). **Multivariate statistical analysis** (2nd ed.). New York: Radius Press.

Klem, L. (1995). Path analysis. In Grimm, L. G. & Yarnold, P. R. (Eds.) (pp. 65–97). **Reading and understanding multivariate statistics**. Washington, D.C.: American Psychological Association.

Klem, L. (2000). Structural equation modeling. In Grimm, L. G. & Yarnold, P. R. (Eds.) (pp. 227–260). **Reading and understanding more multivariate statistics**. Washington, D.C.: American Psychological Association.

Kline, R. (2004). **Principles and practices of structural equation modeling** (2nd ed.). New York: Guilford.

Licht, M. H. (1995). Multiple regression and correlation. In Grimm, L. G. & Yarnold, P. R. (Eds.) (pp. 100–136). **Reading and understanding multivariate statistics**. Washington, D.C.: American Psychological Association.

Lindeman, R. H., Merenda, P. F., & Gold, R. Z. (1980). **Introduction to bivariate and multivariate analysis**. Glenview, IL: Scott, Foresman & Company.

Loehlin, J. C. (2004). **Latent variable models: An introduction to factor, path, and structural analysis** (4th ed.). Mahwah, NJ: Lawrence Erlbaum Associates, Publishers.

Lord, R. & Novick, M. (1968). **Statistical theories of mental test scores**. Reading, MA: Addison-Wesley.

Marascuilo, L. A. & Levin, J. R. (1983). **Multivariate statistics in the social science: A researchers guide**. Monterey, CA: Brooks/Cole Publishing Company.

Menard, S. W. (1995) **Applied logistic regression analysis**. Thousand Oaks, CA: Sage Publications.

Mertler, C. A. & Vannatta, R. A. (2005). **Advanced and multivariate statistical methods** (3rd ed.). Los Angeles: Pyrczak Publications.

Miles, J. & Shevlin, M. (2001). **Applying regression and correlation: A guide for students and researchers**. London: Sage Publications.

Montgomery, D. C. & Peck, E. A. (1992). **Introduction to linear regression analysis** (2nd ed.). New York: John Wiley & Sons.

Neter, J., Kutner, M. H., Nachtscheim, C. J. & Wasserman, W.(1996). **Applied linear statistical models** (4th ed.). Boston: WCB McGraw-Hill.

Neter, J., Wasserman, W., & Kutner, M. H. (1983). **Applied linear regression models** (3rd ed.). Homewood, IL: Richard D. Irwin, Inc.

Neter, J., Wasserman, W., & Kutner, M. H. (1990). **Applied linear statistical models** (3rd ed.). Homewood, IL: Richard D. Irwin, Inc.

Norušis, N. J. (2004). **SPSS 13.0 advanced statistical procedures companion**. Upper Saddle River, NJ: Prentice Hall.

Pedhazur, E. J. (1982). **Multiple regression in behavioral research: Explanation and prediction** (2nd ed.). New York: Holt, Rinehart & Winston.

Raykov, T. & Marcoulides, G. A. (2000). **A first course in structural equation modeling**. Mahwah, NJ: Lawrence Erlbaum Associates, Publishers.

Schumacker, R. E. & Lomax, R. G. (2004). **A beginner's guide to structural equation modeling**. Mahwah, NJ: Lawrence Erlbaum Associates, Publishers.

Stevens, J. (1986). **Applied multivariate statistics for the social sciences**. Hillsdale, NJ: Lawrence Erlbaum Associates, Publishers.

Stevens, J. (1996). **Applied multivariate statistics for the social sciences** (3rd ed.). Hillsdale, NJ: Lawrence Erlbaum Associates, Publishers.

Stevens, J. (2002). **Applied multivariate statistics for the social sciences** (4th ed.). Hillsdale, NJ: Lawrence Erlbaum Associates, Publishers.

Tabachnick, B. G. & Fidell, L. S. (1989). **Using multivariate statistics**. (2nd ed.) New York: Harper Collins Publishers.

Tabachnick, B. G. & Fidell, L. S. (1996). **Using multivariate statistics** (3rd ed.). New York: Harper Collins Publishers.

Tabachnick, B. G. & Fidell, L. S. (2001). **Using multivariate statistics** (4th ed.). Boston: Allyn & Bacon.

Tatsuoka, M. M. (1975). **The general linear model: A new trend in analysis of variance**. Champaign, IL: Institute for Personality & Ability Testing.

Thompson, B. (2000). Ten commandments of structural equation modeling. In Grimm, L. G. & Yarnold, P. R. (Eds.) (pp. 261–283). **Reading and understanding more multivariate statistics**. Washington, D.C.: American Psychological Association.

Ullman, J. B. (2001). Structural equation modeling. In Tabachnick, B. G. & Fidell, L. S. (pp. 653–671). **Using multivariate statistics** (4th ed.). Boston: Allyn & Bacon.

Wright, S. (1921). Correlation and causation. **Journal of Agricultural Research**, 20, 557–585.

Endnotes

1. a) Some sources limit the use of the terms independent and dependent variable to the variables employed in a well controlled experimental study — in other words, a **true experiment** (an experiment involving a manipulated independent variable), which is discussed in the **Introduction**. However, many sources, as well as computer software (e.g., *SPSS*), employ the latter terms in reference to correlational research — specifically, they use the terms independent variable and predictor variable interchangeably as well as the terms dependent variable and criterion variable interchangeably; b) Cohen and Cohen (1983), Licht (1995, p. 20), and Tatsuoka (1975) note that in developing the **analysis of**

variance (e.g., the **single-factor between-subjects analysis of variance (Test 21)**) Ronald Fisher originally employed **multiple regression**. Ultimately, however, he used the more familiar computational procedures described for the **analysis of variance** rather than **multiple regression**, because of the complexity of the computations required for the latter type of analysis. The point to be made is that the **analysis of variance** can be conceptualized as a special case of **multiple regression analysis**. Among others, Field (2005, 311–316) provides a good discussion of this subject. The relationship between **multiple regression** and the **analysis of variance** is also discussed in Section VII of the **single-factor between-subjects analysis of variance** and in Endnote 2 in the chapter on **canonical correlation** under the discussion of the **general linear model**—the latter model stating that parametric statistical procedures can be summarized in the form of one or more linear equations, and that in the final analysis a correlational format can be employed to summarize all such procedures (or, to put it another way, all parametric procedures can be conceptualized as special cases of regression analysis).

2. As noted in Section IX (the **Addendum**) of the **Pearson product-moment correlation coefficient**, a **dummy variable** is a dichotomous variable which employs two values (typically 0 and 1) to indicate membership in one of two categories. An excellent discussion on the use of dummy variables in **multiple regression** can be found in Field (2005, pp. 208–210). Endnote 4 in the chapter on **logistic regression** also discusses relevant information regarding the use of dummy variables.

3. The null and alternative hypotheses are sometimes expressed in an alternative format — specifically, employing the **squared population multiple correlation coefficient** (which as noted in Section IV/V represents the **population coefficient of multiple determination**) in lieu of the population **multiple correlation coefficient**. Thus: H_0: $P^2 = 0$ versus H_1: $P^2 \neq 0$.

4. a) It is also the case that the greater the number of predictor variables in a set of data involving a fixed number of subjects, the larger the value of R^2; b) Among others, Tabachnick and Fidell (1996, p. 12) note that **overfitting** may occur when there are too many variables relative to the size of the sample. **Overfitting** is when the information derived from a sample is tailored to the idiosyncrasies of the sample, yet will not provide a good fit for the target population to which the researcher wishes to generalize.

5. This principle has obviously not been adhered to in the example under discussion in order to minimize computations.

6. Tabachnick and Fidell (1996, pp. 164–165) provide equations developed by Browne (1975) and Cattin (1980) for small sample sizes that allow for an even more severe adjustment than that which results from using Equation 33.2. Field (2005, p. 172), employing Equation 33.36 (attributed to Stein), suggests a more conservative adjusted $\tilde{R}^2$ value which can be employed to estimate how well the regression model derived from the data will predict the scores of subjects from an entirely different sample. The fact that the value $\tilde{R}^2 = .5968$ obtained below with Equation 33.36 is substantially lower than the values obtained with Equations 33.1 and 33.2 can be attributed to the small sample size employed in Example 33.1.

(Equation 33.36)

$$\tilde{R}^2 = 1 - \left[\left(\frac{n-1}{n-k-1} \right) \left(\frac{n-2}{n-k-2} \right) \left(\frac{n+1}{n} \right) \right] (1 - R^2)$$

$$\tilde{R}^2 = 1 - \left[\left(\frac{5-1}{5-2-1} \right) \left(\frac{5-2}{5-2-2} \right) \left(\frac{5+1}{5} \right) \right] (1 - .944) = .5968$$

7. The reader should take note of the fact that the values obtained with the remaining equations to be presented in this section are based on computations which were done with the aid of a hand calculator. If the data for Example 33.1 are evaluated with statistical software, one may obtain a slight discrepancy between the values obtained with the equations described in this section and the output generated by a computer. Any differences will be minimal and of no practical consequence, and can be attributed to rounding off error.

8. Harlow (2005, p. 47) notes that the equation $Y' = X' + e$ is a succinct way of summarizing what the **multiple regression equation** represents. The latter equation indicates that the value of Y' is a function of a linear combination of the X variables (X') with the addition of some prediction error (e), which represents variability that is unrelated to any of the predictor variables. Within the latter context, some sources may present Equation 33.4 in the form $Y' = a + b_1 X_1 + b_2 X_2 + \cdots + b_k X_k + e$.

9. The equation noted below is equivalent to Equation 33.8.

$$s_{Y.X_1X_2} = \sqrt{\frac{SS_Y(1 - R^2_{Y.X_1X_2})}{n-k-1}} = \sqrt{\frac{(29.2)(1 - .944)}{5-2-1}} = .90$$

10. a) It should be noted that the regression equation is usually reported in terms of unstandardized coefficients since it allows direct prediction of a Y score from an X score. Use of standardized coefficients would require converting a standardized Y score into a raw score value; b) A **standardized regression coefficient** can easily be converted into an unstandardized coefficient by algebraically transposing the terms in Equations 33.12 and 33.13. In other words, since the generic form of Equations 33.12/33.13 is $\beta_i = b_i \left[\tilde{s}_{X_i} / \tilde{s}_Y \right]$, then through algebraic transposition $b_i = \beta_i \left[\tilde{s}_Y / \tilde{s}_{X_i} \right]$; c) Allen (1997, p. 46) and Marascuilo and Levin (1983, p. 91) note that if there is one predictor variable then the value of the **standardized regression coefficient** is equal to the value of r, since $r = \text{cov}_{XY} / (\tilde{s}_X \tilde{s}_Y) = \text{cov}_{z_X z_Y} / (\tilde{s}_{z_X} \tilde{s}_{z_Y}) = \text{cov}_{z_X z_Y} / (1)(1) = \text{cov}_{z_X z_Y}$, and that another way of computing/defining a **standardized regression coefficient** for the ith predictor variable is $\beta_{X_i} = \text{cov}_{z_{X_i} z_Y} / \tilde{s}_{z_{X_i}}$ $= \text{cov}_{z_{X_i} z_Y} / 1 = \text{cov}_{z_{X_i} z_Y}$; d) Allen (1997, p. 50) notes that it can be mathematically demonstrated that in bivariate correlation the value of a **standardized regression coefficient** (which will be equivalent to the value of r) cannot exceed 1. In point of fact, in **multiple regression**, though in theory **standardized regression coefficients** can range from minus infinity to plus infinity, in practice they range from -1 to +1, and values outside that range indicate a major problem with the regression analysis in question (such as

collinearity, nonnormality, etc. (Field (2005, p. 611), and Harlow (2005, personal communication)).

11. Licht (1995, pp. 46–47) notes that when **multicollinearity** is small, as a general rule, the magnitude of a **standardized regression coefficient** will be proportionally related to the magnitude of the **zero-order correlation** between the predictor variable in question and the criterion variable. However, as **multicollinearity** increases, there is also an increase in the likelihood there will be a discrepancy between the relative magnitude of a **standardized regression coefficient** and the **zero-order correlation** between the predictor in question and the criterion variable.

12. Howell (2002, p. 571–573) also notes some sources argue that an effective measure of importance with respect to the predictor variables, especially when the use of the term *importance* is directly related to predictive power, is the **squared semipartial correlation** (clarified later in this section) between the ith predictor variable and the criterion variable Y (with all of other predictor variables partialed out), which can be computed with Equation 33.37 (where F_{b_i} is the F value computed for the regression coefficient for the ith variable). An F_{b_i} value for a regression coefficient can be obtained by squaring the t_{b_i} value computed with Equation 33.16 (or through use of an equation for computing F_{b_i} in Neter *et al.* (1990, p. 283)).

$$r^2_{Y.(X_i.X_j.......X_k)} = \frac{F_{b_i}(1 - R^2_{Y.(X_i,X_j,X_k)})}{n - p - 1}$$ **(Equation 33.37)**

13. a) Once again the reader should take note of the fact that since, in actuality, the value $R_{Y.X_1X_2} = .972$ was determined not to be significant, it would not be considered appropriate to evaluate the regression coefficients with respect to significance; b)The following should be noted with respect to Equation 33.15: 1) When the sample size is small and/or the number of subjects is not substantially larger than the number of predictor variables, the "shrunken" estimate $\tilde{R}^2$ (computed with Equation 33.2) should be employed in Equation 33.15; 2) When there are more than two predictor variables, the **multiple correlation coefficient** for the k variables is employed in the numerator of the radical of Equation 33,15 in place of $R^2_{Y.X_1X_2}$. The value $r^2_{X_1X_2}$ in the denominator of the radical of Equation 33.15 is replaced by the **squared multiple correlation coefficient** of variable i with all of the remaining predictor variables. Thus, if there are three predictor variables and s_{b_1} is computed, the values employed in the numerator and denominator of the radical are respectively $R^2_{Y.X_1X_2X_3}$ and $R^2_{X_1.X_2X_3}$.

14. Howell (2002, p. 544) cites sources who argue that the t distribution does not provide a precise approximation of the underlying sampling distribution for the standard error of estimate of the coefficients. On the basis of this he states that caution should be employed in interpreting the results of the t test.

15. a) The same results are obtained if the analysis is done employing the **standardized regression coefficients**. This is demonstrated below employing the appropriate equations for the standardized coefficients. The minimal discrepancy between the values t_{β_1} and t_{b_1} is due to rounding off error.

$$s_{\beta_i} = \sqrt{\frac{1 - R_{Y.X_1 X_2}^2}{(1 - r_{X_1 X_2}^2)(n - k - 1)}} = \sqrt{\frac{1 - .944}{[1 - (.37)^2][5 - 2 - 1]}} = .180$$

$$t_{\beta_i} = \frac{\beta_i}{s_{\beta_i}} \qquad t_{\beta_1} = \frac{.883}{.180} = 4.91 \qquad t_{\beta_2} = \frac{.193}{.180} = 1.07$$

b) Harlow (2005, p. 46) notes that in order for regression coefficients to be unbiased, the predictor variables should be reliable measures of the constructs they represent (e.g., have reliability coefficients of at least .70, assuming reliability information is available).

16. Marascuilo and Levin (1983) and Marascuilo and Serlin (1988) recommend that in order to control the Type I error rate, a more conservative t value should be employed when the number of regression coefficients evaluated is greater than one. The latter sources describe the use of the **Bonferroni–Dunn** and **Scheffé procedures** (which are described in reference to multiple comparisons for analysis of variance procedures) in adjusting the t value.

17. a) Since the **zero-order correlation** between sugar and cavities is so high (i.e., $r_{YX_1} = .955$), there is, so to speak, no room left for salt to make a significant contribution.; b) Licht (1995, p. 40) and Neter *et al.* (1990, p. 304) note that when **multicollinearity** is present it is possible for a set of predictor variables to be significantly correlated with the criterion variable, yet all of the individual tests on the regression coefficients yield nonsignificant results. The latter result is also possible (although not likely to occur in practice) when no **multicollinearity** is present.

18. The reader should take note of the fact that the significant result obtained earlier for the regression coefficient $b_1 = .278$ can be attributed to the fact that when sugar is employed as the only predictor variable (as is the case in Example 28.1), the correlation between sugar and the number of cavities is, in fact, statistically significant. Yet, paradoxically, when both sugar and salt are included in the analysis, the **multiple correlation coefficient** is not significant. The latter can be attributed to the fact that the degrees of freedom in the **analysis of variance** involving two predictor variables is $df_{num} = 2$ and $df_{den} = 2$ in contrast to the degrees of freedom for an **analysis of variance** with one predictor variable, where $df_{num} = 1$ and $df_{den} = 3$ (see Table 28.3). Thus, when $df = 1, 3$ (as in Example 28.1), $F_{.05} = 10.13$, yet when $df = 2, 2$ (as in Example 33.1), $F_{.05} = 19.00$. Consequently, it is easier to reject the null hypothesis that the correlation in underlying population equals zero when only one predictor is employed.

19. Licht (1995, p. 37) notes that an alternative way of conceptualizing statistical control can be demonstrated through use of the residuals. Since a residual represents a part of a subject's score which cannot be predicted from the predictor variables, the information contained in a residual will be independent of the predictor variables employed in an analysis. To illustrate how residuals can be employed to demonstrate statistical control let us assume the researcher conducts a **multiple regression** analysis which involves the three predictor variables of sugar (X_1), salt (X_2) and potassium (X_3) consumption along with the criterion variable of the number of cavities (Y). What the researcher does, however, is to

initially conduct a separate analysis in which the sugar consumption scores of subjects are employed as a criterion variable and the salt and potassium consumption scores of subjects are employed as the predictor variables. The researcher then computes residuals for sugar consumption by obtaining the difference between the predicted sugar consumption for each subject and a subject's actual sugar consumption. The latter values (i.e., the residuals) would represent sugar consumption after controlling for the other two predictor variables of salt and potassium. If, in fact, the residual values obtained for sugar consumption are then correlated with the number of cavities, the resulting correlation should correspond to the **semipartial correlation** $r_{Y(X_1 \cdot X_2 X_3)}$ (which when squared will reflect the proportion of shared variance between the number of cavities and the amount of sugar consumed after any linear association that X_1 (amount of sugar consumed) has with X_2 (amount of salt consumed) and X_3 (amount of potassium consumed) has been removed).

20. The computed value r_{sp} can also be evaluated through use of the critical values in **Table A16** (for $df = n - v$). In the latter table, for $df = 2$, the tabled critical two-tailed values are $r_{.05} = .950$ and $r_{.01} = .990$. Since both $r_{Y(X_1 \cdot X_2)} = .82$ and $r_{Y(X_2 \cdot X_1)} = .18$ are less than $r_{.05} = .950$, the nondirectional alternative hypotheses H_1: $\rho_{Y(X_1 \cdot X_2)} \neq 0$ and H_1: $\rho_{Y(X_2 \cdot X_1)} \neq 0$ are not supported.

21. Equation 33.32 becomes identical to Equation 28.3 when $n - v = n - 2$.

22. a) The computed value r_p can be evaluated through use of the critical values in **Table A16** (for $df = n - v$). In the latter table, for $df = 2$, the tabled critical two-tailed values are $r_{.05} = .950$ and $r_{.01} = .990$. Since $r_{YX_1 \cdot X_2} = .96$ is greater than $r_{.05} = .950$, the nondirectional alternative hypothesis H_1: $\rho_{YX_1 \cdot X_2} \neq 0$ is supported at the .05 level (but not at the .01 level). Since $r_{YX_2 \cdot X_1} = .60$ is less than $r_{.05} = .950$, the nondirectional alternative hypothesis H_1: $\rho_{YX_2 \cdot X_1} \neq 0$ is not supported; b) Howell (2002, p. 555) and Licht (1995, p. 43) note that for a given predictor variable, a **partial correlation**, a **semipartial correlation**, and a regression coefficient are different indices of the independent contribution that the predictor has with the criterion variable, and because they all essentially measure the same thing, the test of significance for a regression coefficient will yield a comparable result to that obtained for a test of significance on either the **partial** or **semipartial correlation** obtained for that predictor.

23. Note that when a simple **bivariate/zero-order correlation** is computed, $n - v = n - 2$, and thus Equation 33.34 becomes identical to Equation 28.3 (which is used to evaluate the significance of the **zero-order correlation** coefficient $r_{X_1 X_2}$).

24. The following *SPSS* command sequence was employed in this chapter in conducting a **multiple regression analysis**: a) Click **Analyze**; b) Click **Regression**; c) Click **Linear**; d) Highlight the predictor variables (i.e., **agility, strength**, and **intelligence** in Example 33.2) and move them to the **Independent(s)** window; e) Highlight the criterion variable (i.e., **performance** in Example 33.2) and move it to the **Dependent** window; f) Click **Statistics**, check off desired information (e.g., **Confidence intervals, Descriptives, Part and partial correlations, Collinearity diagnostics**, etc.), and then click **Continue**; g) Click **Plots**, highlight desired graphic displays, and then click **Continue**; h) In the **Methods** window select the type of **multiple regression** analysis to be conducted. The default option of

Enter represents **standard multiple regression** analysis; i) Click **OK** to obtain the output for the analysis.

25. Field (2005, pp.180–181; pp. 202–206) notes that in *SPSS* a regression plot of ***ZRESID** (*y*-axis) versus ***ZPRED** (*x*-axis) can be informative in evaluating the assumptions of linearity, independence of residuals, and homoscedasticity underlying **multiple regression**. In point of fact, the same guidelines described in Section VII of the **Pearson product-moment correlation coefficient** (with reference to Figure 28.8) for evaluating the assumptions underlying simple linear regression can be applied in evaluating the latter plot. More specifically, if none of the aforementioned assumptions are violated the scatterplot should resemble Figure 28.8a — i.e., the points should be randomly distributed above and below the horizontal line corresponding to 0. Field (2005, p. 181) also notes that heteroscedasticity can be detected with a plot of ***SRESID** (*y*-axis) versus ***ZPRED** (*x*-axis). In addition, he discusses other plots available with *SPSS* that can be useful for evaluating the normality of residuals and detection of outliers (Field (2005, pp. 204–206)).

26. When the **Studentized residual** is computed as just described it is sometimes referred to as an **external** or **externally Studentized residual**. When, however, the denominator for computation of the standard deviation includes the omitted observation, the **Studentized residual** is referred to as an **internal** or **internally Studentized residual**.

27. Berenson and Levine (1996, p. 756) note that Hoagalin and Welsch (1978) determined that for **simple linear regression** (i.e., one predictor variable) the **hat element** (h_i) for the ith observation in a sample comprised of *n* scores can be computed with Equation 33.38. Computation of the hat elements for **multiple regression** is more involved and is generally implemented with the aid of computer software.

$$h_i = \frac{1}{n} + \frac{(X_i - \bar{X})^2}{\sum\limits_{i=1}^{n} X_i^2 - n\bar{X}^2} \qquad \textbf{(Equation 33.38)}$$

28. A comprehensive discussion of regression diagnostics can be found in Field (2005, Ch. 5).

29. The **1** at the upper left of the table under **Model** will be clarified in the discussion of Table 33.7.

30. The identical values obtained for the correlation and regression coefficients for the **backward stepwise analysis** in Table 33.8 are also obtained if a **standard multiple regression** analysis is conducted only employing the two predictor variables of agility and strength.

31. Citing Darlington (1990), Aiken and West (1991, pp. 92–93) note that in instances where predictor variables are highly correlated with one another it is difficult to distinguish between a regression equation involving an interaction versus a curvilinear regression function.

Test 34

Hotelling's T^2
(Parametric Test Employed with Interval/Ratio Data)

I. Hypothesis Evaluated with Test and Relevant Background Information

Hypothesis evaluated with test Do two independent samples represent two populations with different values with respect to their **mean vectors** (where a **mean vector** represents a composite score comprised of mean values on two or more dependent variables)?

Relevant background information on test Hotelling's T^2 (Hotelling (1931)), which is a special case of the **multivariate analysis of variance (Test 35)**, is the **multivariate** analog of the **t test for two independent samples (Test 11)**. Hotelling's T^2 can be employed to analyze the data for an experiment that involves a single independent variable comprised of two levels and multiple dependent variables. Just as the t distribution can be employed to evaluate a null hypothesis of no treatment effect in an experiment involving $k = 2$ independent samples and one dependent variable, **Hotelling's T^2** can be employed to evaluate a null hypothesis of no treatment effect in an experiment involving 2 independent samples and two or more dependent variables (where p will represent the number of dependent variables).[1]

To illustrate, let us assume that (as is the case in Example 11.1) the efficacy of an antidepressant drug is assessed by employing two independent groups, with one group receiving the drug and the other group a placebo. However, in the study two scores are obtained on each subject. One score represents a subject's level of depression and a second score represents the subject's level of anxiety. The two scores for each subject represent the two dependent variables of depression and anxiety, and the composite score for a subject (i.e., a value resulting from algebraically combining a subject's depression and anxiety scores into a single measure) is viewed as a measure of emotional instability. Within the framework of **Hotelling's T^2**, a composite mean on emotional instability is computed for each group. The latter composite means are referred to as **mean vectors** or **centroids**. As is the case with the t test for two independent samples, the means (in this case **composite**) for the two groups are then compared with one another.

The assumptions underlying **Hotelling's T^2** are the same as those underlying the t test for two independent samples, except for the fact they are generalized to a **multivariate normal distribution** (which is discussed in Section I in the chapter on **multiple regression (Test 33)**). Specifically, among the assumptions underlying **Hotelling's T^2** are multivariate normality on the dependent variables in each of the $k = 2$ populations and homogeneity of the **variance-covariance matrices** for the dependent variables in each group (which are respectively analogous to the normality and homogeneity of variance assumptions underlying the t test for two independent samples). In addition, the dependent variables should be linearly related to one another, and the observations should be independent and randomly selected from the target population to which one intends to generalize the results.

II. Example

To illustrate the use of **Hotelling's T^2** consider Example 34.1 which is identical to Example 11.1, except for the fact there are two dependent variables represented by two scores for each subject — with one score representing a subject's level of depression and the second score a subject's level of anxiety. The use of two dependent variables in Example 34.1 is based on the researcher's view that, when taken together, a subject's level of depression and anxiety represents a generalized measure of the construct/concept of **emotional instability**. As noted in Section I, within the framework of **Hotelling's** T^2, a composite mean (i.e., **mean vector** or **centroid**) based on both the depression and anxiety scores of subjects will be computed for each group, and the two mean vectors will be compared.

Example 34.1   *In order to assess the impact of a new antidepressant drug on emotional instability, ten patients at a mental health clinic who are diagnosed with both clinical depression and excessive anxiety are randomly assigned to one of two groups. Five patients are assigned to Group 1, which is administered the antidepressant drug for a period of six months. The other five patients are assigned to Group 2, which is administered a placebo during the same six-month period. Assume that prior to introducing the experimental treatments, the experimenter confirmed that the levels of both depression and anxiety in the two groups were equal. After six months elapse all ten subjects are rated by a psychiatrist (who is blind with respect to a subject's experimental condition) on both their level of depression and anxiety (which when taken together are considered to represent a generalized measure of emotional instability). The psychiatrist's depression and anxiety ratings for the five subjects in each group are noted below. Each subject's depression score is listed first followed by his or her anxiety score. The higher the score on both depression and anxiety, the more depressed or anxious a subject (and consequently, the larger the composite mean for a group on the dependent variable the higher the degree of emotional instability):* **Group 1: Subject 1** *(11, 8);* **Subject 2** *(1, 3);* **Subject 3** *(0, 2);* **Subject 4** *(2, 1);* **Subject 5** *(0, 4);* **Group 2: Subject 1** *(11, 9);* **Subject 2** *(11, 6);* **Subject 3** *(5, 7);* **Subject 4** *(8, 6);* **Subject 5** *(4, 9). Do the data indicate that the drug is effective in decreasing emotional instability?*

III. Null versus Alternative Hypotheses

Null hypothesis

The null and alternative hypotheses evaluated with the **Hotelling's** T^2 test statistic are noted below.

$$H_0: \begin{bmatrix} \mu_{11} \\ \mu_{12} \\ . \\ . \\ . \\ \mu_{1p} \end{bmatrix} = \begin{bmatrix} \mu_{21} \\ \mu_{22} \\ . \\ . \\ . \\ \mu_{2p} \end{bmatrix}$$

Where: μ_{ij} represents the mean of the population represented by Group i on dependent variable j

(The null hypothesis states that the vector of the means of the p dependent variables for the population Group 1 represents equals the vector of the means of the p dependent variables for the population Group 2 represents.)

Alternative hypothesis

$$H_1: \begin{bmatrix} \mu_{11} \\ \mu_{12} \\ . \\ . \\ . \\ \mu_{1p} \end{bmatrix} \ne \begin{bmatrix} \mu_{21} \\ \mu_{22} \\ . \\ . \\ . \\ \mu_{2p} \end{bmatrix}$$

(The alternative hypothesis states that the vector of the means of the p dependent variables for the population Group 1 represents does not equal the vector of the means of the p dependent variables for the population Group 2 represents. The above alternative hypothesis is **nondirectional**.[2])

Since Example 34.1 involves two groups and two dependent variables, the null hypothesis and nondirectional alternative hypotheses for the example is noted below.

$$H_0: \begin{bmatrix} \mu_{11} \\ \mu_{12} \end{bmatrix} = \begin{bmatrix} \mu_{21} \\ \mu_{22} \end{bmatrix}$$

$$H_1: \begin{bmatrix} \mu_{11} \\ \mu_{12} \end{bmatrix} \ne \begin{bmatrix} \mu_{21} \\ \mu_{22} \end{bmatrix}$$

IV. Test Computations

The calculation of **Hotelling's T^2** is generally done through use of computer software. Consequently, instead of a step by step description on how to compute the value of **Hotelling's T^2** for Example 34.1, the intent of this section is to provide a succinct summary of the matrix algebra involved in the analysis. Readers interested in reviewing a numerical example documenting the step by step calculations of a **Hotelling's T^2** test statistic are referred to Stevens (2002, Ch. 4).

The two mean column vectors noted below from the sample data are employed to estimate the population mean vectors.

$$\bar{X}_1 = \begin{bmatrix} \bar{X}_{11} \\ \bar{X}_{12} \end{bmatrix} \qquad \bar{X}_2 = \begin{bmatrix} \bar{X}_{21} \\ \bar{X}_{22} \end{bmatrix}$$

The **deviation score vector (d)** noted below can be derived by subtracting **mean vector 2** from **mean vector 1** (i.e., $\mathbf{d} = \bar{X}_1 - \bar{X}_2$).

$$\mathbf{d} = \begin{bmatrix} \bar{X}_{11} - \bar{X}_{21} \\ \bar{X}_{12} - \bar{X}_{22} \end{bmatrix} = \begin{bmatrix} D_{.1} \\ D_{.2} \end{bmatrix}$$

Where: $D_{.j}$ represents the difference score between the two group means on dependent variable j

Equation 34.1 is employed to compute the value of **Hotelling's T^2**. Various sources (e.g., Harris (2001, p. 171), Marascuilo and Levin (1983, p. 282), and Stevens (2002, pp. 176–177)) note that Equation 34.1 can be viewed as an algorithm for comparing between-groups variability (which is represented by the vectors **d** and **d'**) with within-groups variability (which is represented by **Matrix V**, which comprises the **variance-covariance matrix**). Note that inversion of the latter matrix (i.e., $\mathbf{V^{-1}}$) is the matrix algebra analogue of scalar division.

$$T^2 = \frac{n_1 n_2}{n_1 + n_2}\,(\mathbf{d'})\,(\mathbf{V^{-1}})(\mathbf{d}) \qquad \textbf{(Equation 34.1)}$$

Where: **d** represents the **deviation score vector** (which is the column vector)
              d' (which is a row vector) represents the transpose of the **deviation score vector d**
              $\mathbf{V^{-1}}$ represents the inverse of the **variance-covariance matrix**

V. Interpretation of the Test Results

Hotelling (1931) determined that the value computed for T^2 with Equation 34.1 can be converted into an F value through use of Equation 34.2 (Marascuilo and Levin (1983, p. 282), Stevens (2002, p. 180) and Zar (1999, pp. 321–322)).[3]

$$F = \left(\frac{n_1 + n_2 - p - 1}{p(n_1 + n_2 - 2)} \right) T^2 \qquad \textbf{(Equation 34.2)}$$

By algebraic transposition, Equation 34.3 allows for computation of the value of T^2 from the F value computed with Equation 34.2.

$$T^2 = \left(\frac{p(n_1 + n_2 - 2)}{n_1 + n_2 - p - 1} \right) F \qquad \textbf{(Equation 34.3)}$$

As will be noted shortly, analysis of Example 34.1 with the statistical software package *SPSS* yields the value $F = 3.437$, which when substituted in Equation 34.3 yields the value $T^2 = 7.856$.

$$T^2 = \left(\frac{2(5 + 5 - 2)}{5 + 5 - 2 - 1} \right)(3.437) = 7.856$$

The value $F = 3.437$ (which can be obtained with the computer or by using Equations 34.1 and 34.2) can be evaluated with **Table A10 (Table of the F Distribution)** in the **Appendix** employing $df_{num} = p$ and $df_{den} = n_1 + n_2 - p - 1$. In the case of Example 34.1 where $p = 2$ and $n_1 = n_2 = 5$, $df_{num} = p = 2$ and $df_{den} = n_1 + n_2 - p - 1 = 5 + 5 - 2 - 1 = 7$.[4] In order to reject the null hypothesis the computed value of F must be equal to or be greater than the tabled critical value at the prespecified level of significance. In the case of Example 34.1, the tabled critical .05 and .01 values for $df_{num} = 2$ and $df_{den} = 7$ are $F_{.05} = 4.74$ and $F_{.01} = 9.55$.

Hair *et al.* (1995, p. 265) note that a critical T^2 value can be computed with Equation 34.4. In order to reject the null hypothesis the computed value of T^2 must be equal to or be greater than the critical value of $T^2_{critical}$.

$$T^2_{critical} = \left(\frac{p(n_1 + n_2 - 2)}{n_1 + n_2 - p - 1} \right) F_{critical} \qquad \textbf{(Equation 34.4)}$$

Where: $F_{critical}$ = The tabled critical F value in **Table A10** for $df_{num} = p$ and df_{den}
$= n_1 + n_2 - p - 1$

Substituting the value $F_{.05} = 4.74$ in Equation 34.4, the critical value $T^2_{.05} = 10.834$ is computed below. Thus, in order to reject the null hypothesis the computed value of T^2 must be equal to or be greater than 10.834.

$$T^2_{.05} = \left(\frac{2(5 + 5 - 2)}{5 + 5 - 2 - 1} \right) (4.74) = 10.834$$

Since the computed values $F = 3.437$ and $T^2 = 7.856$ are respectively smaller than the tabled critical values $F_{.05} = 4.74$ and $T^2_{.05} = 10.834$, the null hypothesis cannot be rejected at the .05 level. Thus, the researcher cannot conclude there is a significant difference between the two groups with respect to their composite means for emotional instability.

At this point the discussion will focus attention on the use of computer software to evaluate Example 34.1. As noted earlier, **Hotelling's T^2** is a special case of the **multivariate analysis of variance**. In actuality, *SPSS* (and other statistical software packages) does not compute a value for **Hotelling's T^2**, but instead conducts a **multivariate analysis of variance** and computes an F value, which is associated with four test statistics that are computed for the latter type of analysis of variance. The **multivariate analysis of variance** is a generalization of **Hotelling's T^2** to designs with two or more dependent variables involving two or more groups. In point of fact the **multivariate analysis of variance** can be employed to evaluate any experimental design (e.g., dependent samples, factorial, etc.) for which an analysis of variance is appropriate, but which involves multiple dependent variables. (Within the latter context, **Hotelling's T^2** can also be employed to evaluate a dependent samples design with two experimental conditions involving two or more dependent variables.) The discussion in this book, however, will be limited to the use of the **multivariate analysis of variance** in evaluating an independent samples design involving multiple dependent variables.

Table 34.1 presents a partial *SPSS* printout for the analysis of Example 34.1.[5] The reader should attend to the **lower section** of the table (labeled **Group**) which contains information regarding the following four test statistics which are commonly employed for the **multivariate analysis of variance**: **Pillai's trace**, **Wilks' lambda**, **Hotelling's trace**, and **Roy's largest root**.[6] All four of the aforementioned test statistics can be converted into an F value. Note that the value $F = 3.437$ alluded to earlier is recorded in Column 3 in the row for each of the test statistics. Tabachnick and Fidell (1996, p. 400) note that when there are $k = 2$ groups, the same F value will be computed for **Wilks' lambda**, **Pillai's trace**, and **Hotelling's trace** (and **Roy's largest root** also in this instance).

To illustrate the latter conversion of a test statistic into an F value, the conversion will be briefly discussed in reference to **Wilks' lambda**. Marascuilo and Levin (1983, p. 282–283) note that the relationship between the value of **Hotelling's T^2** and **Wilks' lambda** (Λ) is described by Equation 34.5, and that the relationship between **Wilks' lambda** and F is described by Equation 34.6. (The minimal difference between the value $T^2 = 7.856$ computed with Equation 34.3 and $T^2 = 7.842$ computed with Equation 34.5 is due to rounding off error. The latter also accounts for the trivial difference between $F = 3.437$ and $F = 3.431$ computed with Equation 34.6.)

(Equation 34.5)

$$T^2 = \frac{(n_1 + n_2 - 2)(1 - \Lambda)}{\Lambda} = \frac{(5 + 5 - 2)(1 - .505)}{.505} = 7.842$$

(Equation 34.6)

$$F = \left(\frac{n_1 + n_2 - p - 1}{p}\right)\left(\frac{1 - \Lambda}{\Lambda}\right) = \left(\frac{5 + 5 - 2 - 1}{2}\right)\left(\frac{1 - .505}{.505}\right) = 3.431$$

The reader should take note of the following additional information contained in the **lower section** of Table 34.1 (labeled **Group**) for all four test statistics: a) The values listed in Column 4 (**Hypothesis df**) and Column 5 (**Error df**) are respectively the values $df_{num} = 2$ and $df_{den} = 7$ computed earlier for evaluating the F value for Example 34.1; b) The value .091 recorded in Column 6 (**Sig.**) is the probability associated with the result. If the value recorded for **Sig.** is equal to or less than .05 the result of the **multivariate analysis of variance** is significant at the .05 level, and if it is equal to or less than .01 it is significant at the .01 level. Since the recorded value .091 is greater than .05, it indicates the result does not achieve significance at the .05 level. Thus, regardless of which test statistic one employs, in the case of Example 34.1 the researcher cannot conclude there are differences in emotional instability between the drug and placebo groups; and c) The value recorded in Column 7 (**Partial Eta Squared**) is a measure of effect size indicating the proportion of variance on the composite dependent variable (.495) which can be explained on the basis of the independent variable.

Table 34.1   *SPSS* Printout of Hotelling's T^2 Analysis for Example 34.1

Multivariate Tests[c]

Effect		Value	F	Hypothesis df	Error df	Sig.	Partial Eta Squared
Intercept	Pillai's Trace	.887	27.579[b]	2.000	7.000	.000	.887
	Wilks' Lambda	.113	27.579[b]	2.000	7.000	.000	.887
	Hotelling's Trace	7.880	27.579[b]	2.000	7.000	.000	.887
	Roy's Largest Root	7.880	27.579[b]	2.000	7.000	.000	.887
GROUP	Pillai's Trace	.495	3.437[b]	2.000	7.000	.091	.495
	Wilks' Lambda	.505	3.437[b]	2.000	7.000	.091	.495
	Hotelling's Trace	.982	3.437[b]	2.000	7.000	.091	.495
	Roy's Largest Root	.982	3.437[b]	2.000	7.000	.091	.495

a. Computed using alpha = .05

b. Exact statistic

c. Design: Intercept+GROUP

VI. Additional Analytical Procedures for Hotelling's T^2 and/or Related Tests

1. Additional analyses following the test of the omnibus null hypothesis   Following the **multivariate analysis of variance** a researcher can assess the impact of the independent variable on each of the dependent variables, by conducting a separate *t* **test for two independent samples** on each dependent variable. In the case of Example 34.1, when Example 11.1 (which employs the identical depression scores for the 10 subjects as a single dependent variable) was evaluated with the *t* test for two independent samples, the nondirectional alternative hypothesis H_1: $\mu_1 \neq \mu_2$ was not supported, since the obtained absolute value $t = 1.96$ was less than the tabled critical .05 value $t_{.05} = 2.31$. When a *t* test for two independent samples is employed to evaluate the anxiety scores of the 10 subjects, the mean values $\bar{X}_1 = 3.6$ and $\bar{X}_2 = 7.4$ are obtained, which subsequently yield the absolute value $t = 2.74$. The latter value is significant at the .05 level but not at the .01 level, since $t = 2.74$ is greater than $t_{.05} = 2.31$ but less than $t_{.01} = 3.36$ (for $df = 8$). The results of the two *t* tests when viewed within the context of the omnibus test of the composite means summarized in Table 34.1 suggest that although group differences on anxiety were more salient than on depression, when the two measures are combined to represent a generalized measure of emotional instability they do not yield a significant difference between the groups.

The reader should take note of the fact that because of the small sample size employed in Example 34.1 the power of the **multivariate analysis of variance/Hotelling's T^2** analysis was relatively low, thus making it difficult to reject the null hypothesis stated in Section III. In point of fact, the probability value .091 associated with the result of the **multivariate analysis of variance/Hotelling's T^2** analysis is not that far removed from the .05 level of significance, and if a comparable difference between the composite means of the groups had been obtained with a larger sample size the result of the **multivariate analysis of variance/Hotelling's T^2** analysis would have been significant.

In the discussion of the **multivariate analysis of variance** additional analyses which can be conducted following the **Hotelling's T^2** macro-analysis (i.e., the overall analysis of the data summarized in Table 34.1) will be described— specifically, conducting a **single-factor between subjects analysis of covariance (Test 21I)** on each of the dependent variables employing the other dependent variable as a covariate.

2. Test 34a: The single-sample Hotelling's T^2

Various sources (e.g., Harris (2001, pp. 155–165) and Marascuilo and Levin (1983, p. 284)) describe how **Hotelling's T^2** can also be employed to evaluate the null hypothesis noted below with respect to a single sample. In the latter instance the scores of a sample of n subjects on p variables are evaluated in reference to a target population in order to determine whether or not the sample is likely to have come from the target population.

The null and alternative hypotheses evaluated with the **single-sample Hotelling's T^2** test statistic are noted below.

Null hypothesis

$$H_0: \begin{bmatrix} \mu_1 \\ \mu_2 \\ . \\ . \\ . \\ \mu_p \end{bmatrix} = \begin{bmatrix} \mu_{1_0} \\ \mu_{2_0} \\ . \\ . \\ . \\ \mu_{p_0} \end{bmatrix}$$

Where: μ_j represents the mean of the population represented by the sample on variable j, and μ_{j_0} represents the mean of the target population on variable j

(The null hypothesis states that the vector of the means of the p variables for the population the sample represents equals the vector of the means of the p variables for the target population.)

Alternative hypothesis

$$H_1: \begin{bmatrix} \mu_1 \\ \mu_2 \\ . \\ . \\ . \\ \mu_p \end{bmatrix} \neq \begin{bmatrix} \mu_{1_0} \\ \mu_{2_0} \\ . \\ . \\ . \\ \mu_{p_0} \end{bmatrix}$$

(The alternative hypothesis states that the vector of the means of the p variables for the population the sample represents does not equal the vector of the means of the p variables for the target population. The above alternative hypothesis is **nondirectional**.)

Equation 34.7 is employed to compute the value of **Hotelling's T^2** (Harris (2001, p. 159) and Marsascuilo and Levin (1983, p. 284)).

$$T^2 = n\,(\mathbf{d'})\,(\mathbf{V^{-1}})(\mathbf{d}) \qquad \text{(Equation 34.7)}$$

Where: $\mathbf{d}$ represents the **deviation score vector** (reflecting for each variable the difference between the sample mean and the hypothesized target population mean)
$\mathbf{d'}$ represents the transpose of the **deviation score vector d**
$\mathbf{V^{-1}}$ represents the inverse of the **sample variance-covariance matrix**

As is the case when **Hotelling's T^2** is employed to evaluate two independent samples, the T^2 value computed with Equation 34.7 can be converted into an F value through use of Equation 34.8, which is evaluated with $df_{num} = p$ and $df_{den} = n - p$ (Harris (2001, p. 160) and Marascuilo and Levin (1983, p. 284)).

$$F = \left(\frac{n - p}{p(n - 1)} \right) T^2 \qquad \text{(Equation 34.8)}$$

Marascuilo and Levin (1983, p. 284) note that a researcher would not be likely to conduct a study involving the use of the **single-sample Hotelling's T^2** since it would be quite unusual to know the value of a target population mean vector. Although the test statistic will not be computed, Example 34.2 illustrates a study involving a single sample which could be evaluated with **Hotelling's T^2**.

Example 34.2   *In order to assess the impact of a new antidepressant drug on emotional instability, ten patients at a mental health clinic who have taken the medication for six months are rated by a psychiatrist on both their level of depression and anxiety (which when taken together are considered to represent a generalized measure of emotional instability). The psychiatrist's depression and anxiety ratings for the ten subjects are noted below. Each subject's depression score is listed first followed by his or her anxiety score. The higher the score on both depression and anxiety, the more depressed or anxious a subject:* **Subject 1** (11, 8); **Subject 2** (1, 3); **Subject 3** (0, 2); **Subject 4** (2, 1); **Subject 5** (0, 4); **Subject 6** (2, 4); **Subject 7** (1, 4); **Subject 8** (2, 6); **Subject 9** (3, 5); **Subject 10** (3, 3).
    Do the data indicate that the composite depression and anxiety ratings of the patients are consistent with the composite depression and anxiety ratings of a target population of emotionally stable patients (i.e., patients low in emotional instability) in which the mean

depression score is 1 and the mean anxiety score is 3? More specifically, do the data indicate that the mean vector for the 10 patients is consistent with the following normative mean vector for emotionally stable individuals: $\begin{bmatrix} \mu_{1_0} \\ \mu_{2_0} \end{bmatrix} = \begin{bmatrix} 1 \\ 3 \end{bmatrix}$?

The null and alternative hypotheses for Example 34.2 are noted below.

$$H_0: \begin{bmatrix} \mu_1 \\ \mu_2 \end{bmatrix} = \begin{bmatrix} \mu_{1_0} \\ \mu_{2_0} \end{bmatrix}$$

$$H_1: \begin{bmatrix} \mu_1 \\ \mu_2 \end{bmatrix} \neq \begin{bmatrix} \mu_{1_0} \\ \mu_{2_0} \end{bmatrix}$$

3. Test 34b: The use of the single-sample Hotelling's T^2 to evaluate a dependent samples design Prior to reading this section the reader may find it useful to review the discussion of the sphericity assumption in Section VI of the chapter on the **single-factor within-subjects analysis of variance (Test 24)**. In the latter discussion it is noted that many sources state when there is reason to believe the sphericity assumption underlying the **single-factor within-subjects analysis of variance** is violated, the data should be evaluated with a **multivariate analysis of variance**, since sphericity is not an assumption of the latter procedure.[7] This section will describe the multivariate analysis of a dependent samples design involving one independent variable and one dependent variable. Specifically, a multivariate analysis will be employed to evaluate Example 24.1, which previously was evaluated with a **single-factor within-subjects analysis of variance**.

The multivariate analysis of a dependent samples design (also referred to as within-subjects or repeated-measures designs) is implemented through use of the **multivariate analysis of variance**, which in this instance is actually an application of the **single-sample Hotelling's T^2** described above. The null and alternative hypotheses evaluated in the multivariate analysis of a dependent samples design involving one independent variable and one dependent variable (which is transformed into multiple dependent variables) are noted below (Marascuilo and Levin (1983, pp. 374–375), Maxwell and Delaney (2004, Ch. 13), and Meyers and Well (2003, pp. 359–360)).

Null hypothesis

$$H_0: \begin{bmatrix} \mu_{D_1} \\ \mu_{D_2} \\ . \\ . \\ . \\ \mu_{D_{k-1}} \end{bmatrix} = \begin{bmatrix} 0 \\ 0 \\ . \\ . \\ . \\ 0 \end{bmatrix}$$

Where: μ_{D_j} represents the mean difference score for the jth dependent variable

(The null hypothesis states that a vector comprised of $k-1$ mean difference scores equals a vector of mean differences scores in which all $k - 1$ elements equal zero.)

Alternative hypothesis

$$H_1: \begin{bmatrix} \mu_{D_1} \\ \mu_{D_2} \\ . \\ . \\ . \\ \mu_{D_{k-1}} \end{bmatrix} \neq \begin{bmatrix} 0 \\ 0 \\ . \\ . \\ . \\ 0 \end{bmatrix}$$

(The alternative hypothesis states that a vector comprised of $k - 1$ mean difference scores does not equal a vector of mean differences scores in which all $k - 1$ elements equal zero. The alternative hypothesis is **nondirectional**.)

Prior to evaluating the null and alternative hypotheses in reference to Example 24.1, consider the following.

a) Although not applicable to Example 24.1, dependent samples designs involving one independent variable and one dependent variable commonly evaluate subjects over k time intervals, predicting either progressive improvement or progressive deterioration as subjects are sequentially evaluated at Time interval 1 (which corresponds to Condition 1), Time interval 2 (which corresponds to Condition 2), and so on until the final time interval (which corresponds to Time interval/Condition k). In the latter type of study the k time intervals represent k levels of the independent variable. It is also the case that in other types of dependent samples designs the ordering of the k experimental conditions (i.e., Condition 1, Condition 2, .., Condition k) are sequentially related to the magnitude of their predicted impact on the dependent variable. The latter applies to Example 24.1 where Condition 1 corresponds to **no noise**, Condition 2 to **moderate noise**, and Condition 3 to **loud noise**. Note that moving sequentially from Condition 1 to Condition 3, Condition 1 is predicted to have the least impact on the dependent variable (number of nonsense syllables correctly recalled), Condition 2 the second smallest impact, until Condition k (which in the case of Example 24.1 is Condition 3), which is predicted to have the largest impact on the dependent variable. Alternatively, if the ordering of the conditions had been reversed (such that Condition 1 corresponded to **loud noise**, Condition 2 to **moderate noise**, and Condition 1 to **no noise**), Condition 1 would be predicted to have the greatest impact on the dependent variable, Condition 2 the second greatest impact, and Condition 3 the smallest impact. The main point to be made is that in a dependent samples design in which no treatment effect is present, one would expect there would be no difference in the scores of subjects on the dependent variable in adjacent experimental conditions (or, for that matter, in experimental conditions that were not adjacent to one another).

b) Within the context of a single-factor within-subjects design comprised of k experimental conditions, it is possible to compute $k - 1$ **difference scores** for each subject. In order to compute the $k - 1$ difference scores for a subject the following protocol is employed: **Subtract a subject's score in a given condition from the subject's score in the condition which precedes it. Note that since no condition precedes Condition 1, no difference score can be computed by subtracting a score from the score in Condition 1.**

c) After the k scores of each of the n subjects are transformed into $k - 1$ difference scores, the mean of each set of n difference scores is computed (where μ_{D_j} represents the mean of the jth set of difference scores). The $k - 1$ mean difference scores are employed to represent $k - 1$ dependent variables which comprise the elements of the column vector to the left of the equals and unequals signs respectively in the null and alternative hypotheses stipulated above.

The above protocol will now be illustrated in reference to Example 24.1. Since in the latter example there are $k = 3$ experimental conditions, $k - 1 = 2$ difference scores can be computed for each subject. By reformatting the data through computation of difference scores, the data for Example 24.1 will be transformed into a multivariate model comprised of $k - 1 = 2$ dependent variables. As noted above, in order to compute the $k - 1$ difference scores for any subject, the subject's score in a given condition will be subtracted from his or her score in the preceding condition. Following the aforementioned protocol, the first difference score for the ith subject (D_{i1}) is obtained by subtracting the ith subject's score in Condition 1 (X_{i1}) from his or her score in Condition 2 (X_{i2}) — i.e., $D_{i1} = X_{i2} - X_{i1}$. The second difference score for the ith subject (D_{i2}) is obtained by subtracting the ith subject's score in Condition 2 (X_{i2}) from his or her score in Condition 3 (X_{i3}) — i.e., $D_{i2} = X_{i3} - X_{i2}$.[8]

To illustrate the above protocol, we will compute the two difference scores for Subject 1 and Subject 2 in Table 24.1. Subject 1's three scores in Conditions 1, 2, and 3 are respectively 9, 7, and 4. Thus, $D_{11} = X_{12} - X_{11} = 7 - 9 = -2$ and $D_{12} = X_{13} - X_{12} = 4 - 7 = -3$. Subject 2's three scores in Conditions 1, 2, and 3 are respectively 10, 8, and 7. Thus, $D_{21} = X_{22} - X_{21} = 8 - 10 = -2$ and $D_{22} = X_{23} - X_{22} = 7 - 8 = -1$. If a complete analysis was presented, two difference scores would be computed for each of the four remaining subjects. Upon completing the latter, a mean difference score is computed for each set of $n = 6$ difference scores. Specifically, the two mean values computed are $\bar{D}_1 = \Sigma D_{.1} / n$ (i.e., adding the values $D_{11}, D_{21}, D_{31}, D_{41}, D_{51}, D_{61}$ and dividing the sum by $n = 6$) and $\bar{D}_2 = \Sigma D_{.2} / n$ (i.e., adding the values $D_{12}, D_{22}, D_{32}, D_{42}, D_{52}, D_{62}$ and dividing the sum by $n = 6$).[9] If there is no treatment effect, the means of the two difference scores (each of which represents a separate dependent variable) should equal zero.

The null and alternative hypotheses for Example 24.1 are noted below.

$$H_0: \begin{bmatrix} \mu_{D_1} \\ \mu_{D_2} \end{bmatrix} = \begin{bmatrix} 0 \\ 0 \end{bmatrix}$$

$$H_1: \begin{bmatrix} \mu_{D_1} \\ \mu_{D_2} \end{bmatrix} \neq \begin{bmatrix} 0 \\ 0 \end{bmatrix}$$

The column vector **d** below is a **mean difference score vector** comprised of the sample mean difference scores on the two dependent variables. With respect to the latter vector, the null hypothesis predicts that for the sample data $\bar{D}_1 = \bar{D}_2 = 0$.

$$\mathbf{d} = \begin{bmatrix} \bar{D}_1 \\ \bar{D}_2 \end{bmatrix}$$

Equation 34.9 is employed to compute the value of **Hotelling's T^2** (Davis (2002, pp. 49–50), Tabachnick and Fidell (1996, p. 451; 2001, p. 403) and Stevens (2002, p. 500)).[10]

$$T^2 = n\,(\mathbf{d'})\,(\mathbf{V^{-1}})(\mathbf{d}) \qquad \textbf{(Equation 34.9)}$$

Where: **d** represents the **mean difference score vector**

   d' represents the transpose of the **mean difference score vector d**

   V^{-1} represents the inverse of the **variance-covariance matrix**

Table 34.2 presents a partial printout from the statistical software package *SPSS* for the multivariate analysis of Example 24.1.[11] Note that the value $F = 55.00$ is computed for all four of the test statistics (**Pillai's trace, Wilks' lambda, Hotelling's trace,** and **Roy's largest root**), and that the probability recorded in Column 6 for the latter F value is $p = .001$.

Table 34.2   *SPSS* Printout of Multivariate Analysis of Variance for Example 24.1

Multivariate Tests[b]

Effect		Value	F	Hypothesis df	Error df	Sig.	Parti Squ
COND	Pillai's Trace	.917	55.000[a]	1.000	5.000	.001	
	Wilks' Lambda	.083	55.000[a]	1.000	5.000	.001	
	Hotelling's Trace	11.000	55.000[a]	1.000	5.000	.001	
	Roy's Largest Root	11.000	55.000[a]	1.000	5.000	.001	

a. Exact statistic

b.

Design: Intercept
Within Subjects Design: COND

Davis (2002, p. 50), Myers and Well (2003, p. 360) and Stevens (2002, p. 500) note that the value computed for **Hotelling's** T^2 with Equation 34.9 can be converted into an F value through use of Equation 34.10, and thus through algebraic transposition the value $T^2 = 137.5$ can be computed with Equation 34.11.

$$F = \left(\frac{n - k + 1}{(n - 1)(k - 1)} \right) T^2 \qquad \textbf{(Equation 34.10)}$$

$$\textbf{(Equation 34.11)}$$

$$T^2 = \left(\frac{(n - 1)(k - 1)}{n - k + 1} \right) F = \left(\frac{(6 - 1)(3 - 1)}{6 - 3 + 1} \right) 55 = 137.5$$

The F value computed for **Hotelling's** T^2/**multivariate analysis of variance** can be evaluated with **Table A10** employing $df_{num} = k - 1$ and $df_{den} = n - k + 1$, which in the case of Example 24.1 are $df_{num} = 3 - 1 = 2$ and $df_{den} = 6 - 3 + 1 = 4$. In **Table A10** the tabled critical .05 and .01 values for $df_{num} = 2$ and $df_{den} = 4$ are $F_{.05} = 6.94$ and $F_{.01} = 18.00$. Since the computed value $F = 55.00$ is greater than $F_{.01} = 18.00$ the result is significant at the .01 level. As noted in Table 34.2, the exact probability associated with $F(2, 4) = 55.00$ is $p = .001$.

To summarize, the result of the multivariate analysis of Example 24.1 through use of **Hotelling's** T^2/**multivariate analysis of variance** is consistent with that obtained when the data are evaluated with the **single-factor within-subjects analysis of variance** (summarized in Section V of the latter test) in which the result is also significant at the .01 level with $F(2,10) = 41.29, p < .01$. Thus, as was the case when the data for Example 24.1 were evaluated with the **single-factor within-subjects analysis of variance**, the multivariate analysis allows the researcher to conclude there is a significant difference between at least two of the three experimental conditions.

Procedures for conducting contrasts and/or post hoc comparisons for a multivariate analysis of a dependent samples design, as well discussion of the use of the **multivariate analysis of variance** in evaluating more complex within-subjects designs (e.g., factorial) can be found in texts such as Davis (2002), Marascuilo and Levin (1983, pp. 377–381), Maxwell and Delaney (2004, Ch. 13), Stevens (2002, pp. 506–509), and Tabachnick and Fidell (1996, Ch. 10; 2001, Ch. 10).

VII. Additional Discussion of Hotelling's T^2

1. Hotelling's T^2 and Mahalanobis' D^2 statistic   Kachigan (1986, p. 329) and Marascuilo and Levin (1983, p. 292) note that when two independent samples are evaluated with respect to multiple dependent variables either **Hotelling's T^2** or **Mahalanobis' D^2** statistic (which is discussed in Section IV/V in the chapter on **multiple regression**) can be employed to evaluate the data. Marascuilo and Levin (1983, pp. 292–293) note that the corresponding values of **Hotelling's T^2** and **Mahalanobis' D^2** can be computed from one another through use of Equations 34.12 and 34.13.

$$T^2 = \left(\frac{n_1 n_2}{n_1 + n_2} \right) D^2 \qquad \text{(Equation 34.12)}$$

$$D^2 = \left(\frac{1}{n_1} + \frac{1}{n_2} \right) T^2 \qquad \text{(Equation 34.13)}$$

VIII. Additional Example Illustrating the Use of Hotelling's T^2

No additional examples will be presented in this section.

References

Davis, D. S. (2002). **Statistical methods for the analysis of repeated measurements**. New York: Springer.

Hair, J. F., Anderson, R. E., Tatham, R. L. & Black, W. C. (1995). **Multivariate data analysis** (4th ed.). Upper Saddle River, N. J.: Prentice Hall.

Hair, J. F., Anderson, R. E., Tatham, R. L. & Black, W. C. (2004). **Multivariate data analysis** (6th ed.). Upper Saddle River, N. J.: Prentice Hall.

Harris, R. J. (2001) **A primer of multivariate statistics** (3rd ed.). Mahwah, NJ: Lawrence Erlbaum Associates.

Hotelling, H. (1931). The generalization of Student's ratio. **Annals of Mathematical Statistics**, 2, 361–378.

Kachigan, S. K. (1986). **Statistical analysis**. New York: Radius Press.

Marascuilo, L. A. & Levin, J. R. (1983). **Multivariate statistics in the social science: A researchers guide**. Monterey, CA: Brooks/Cole Publishing Company.

Mauchly, J. W. (1940). Significance test for sphericity of a normal *n*-variate distribution. **Annals of Mathematical Statistics**, 11, 204–209.

Maxwell, S. E. & Delaney, H. D. (2004) **Designing experiments and analyzing data: A model comparison perspective** (2nd ed.). Mahwah, NJ: Lawrence Erlbaum Associates.

Myers, J. L. & Well, A. D. (2003). **Research design and statistical analysis** (2nd ed.). Mahwah, NJ: Lawrence Erlbaum Associates, Publishers.

Stevens, J. (2002). **Applied multivariate statistics for the social sciences** (4th ed.). Hillsdale, NJ: Lawrence Erlbaum Associates, Publishers.

Tabachnick, B. G. & Fidell, L. S. (1996). **Using multivariate statistics** (3rd ed.). New York: Harper Collins Publishers.

Tabachnick, B. G. & Fidell, L. S. (2001). **Using multivariate statistics** (4th ed.). Boston: Allyn & Bacon.

Zar, J. H. (1999). **Biostatistical analysis** (4th ed.). Upper Saddle River, NJ: Prentice Hall.

Endnotes

1. As will be noted in Section VI within the framework of the discussion of **Test 34b: The use of the single-sample Hotelling's T^2 to evaluate a dependent samples design**, Hotelling's T^2 can also be employed to evaluate a dependent samples design involving an independent variable with two or more levels and two or more dependent variables.

2. Stevens (2002, p. 193) notes it is not possible to state an alternative hypothesis directionally for a multivariate test. In the case of **Hotelling's T^2** Stevens (2002, p. 175) notes the statement in the null hypothesis that the vectors of the two populations are equal "implies that the groups are equal on all p dependent variables." Consequently, if a significant result is obtained for **Hotelling's T^2**, it can be due to either of the following: a) All of the univariate t values are significant (i.e., t **tests** conducted on each of the p dependent variables yield significant results); b) One or more of the univariate t values is significant; or c) None of the univariate t values is significant, but instead the significant T^2 value is the result of one or more significant linear combinations of the two dependent variables (Stevens (2002, pp. 184–186)).

3. Marascuilo and Levin (1983, p. 282) note that when $p = 1$, $F = T^2$, which represents the univariate case where $F = t^2$ (i.e., the case of the F value computed for the **single-factor between-subjects analysis of variance (Test 21)** and the t value computed for the t **test for two independent samples**).

4. a) The reader should take note of the fact that the latter equations for computing degrees of freedom are only applicable when there are $k = 2$ groups and 2 or more dependent variables. They will not be applicable when there are three or more groups and 2 or more dependent variables; b) A more detailed description of **Pillai's trace**, **Wilks' lambda**, **Hotelling's trace**, and **Roy's largest root** can be found in Section IV of the **multivariate analysis of variance**.

5. The following *SPSS* command sequence was employed in this chapter in conducting a **Hotelling's T^2 analysis**: a) Click **Analyze**; b) Click **General linear model**; c) Click **Multivariate**; d) Highlight the dependent variables (i.e., **anxiety** and **depression** for Example 34.1) and move them to the **Dependent Variables** window; e) Highlight the independent variable (i.e., the variable indicating which group a subject is in — **drug** versus **placebo** in Example 34.1) and move it to the **Fixed Factor(s)** window; f) Click **Options**, check off desired information (e.g., **Descriptive statistics**, **Estimates of effect size**, etc.), and then click **Continue**; g) Click **OK** to obtain the output for the analysis.

6.   Although it will not be necessary to consider the values printed out in the upper section of Table 34.1 labeled **Intercept**, the latter information is derived through use of a **regression model** to evaluate a **multivariate analysis of variance**. As discussed in greater detail in Section VII of the **single-factor between subjects analysis of variance** under the subject of the **general linear model** and in Endnote 2 in the chapter on **canonical correlation (Test 38)**, all parametric statistical procedures can be summarized in the form of one or more linear equations, and in the final analysis a correlational format can be employed to summarize all such procedures (or, to put it another way, all parametric procedures can be conceptualized as special cases of regression analysis). It is through use of the latter type of analysis that the values in the **Intercept** section of Table 34.1 are derived. For further clarification regarding the **multivariate analysis of variance** as a special case of regression analysis the reader should consult Stevens (2002, pp. 188–192).

7.   a) *SPSS* evaluates the sphericity assumption with **Mauchly's test of sphericity** (Mauchly (1940)) (which is displayed in a separate table labeled **Mauchly's Test of Sphericity**). If the probability value for the latter test (which is displayed in a column labeled **Sig.**) is less than .05, the sphericity assumption is violated. Adjusted F values (based on Geisser and Greenhouse (1958) and Huynh and Feldt (1976)) (along with their associated probabilities) are printed out which should be employed when the sphericity assumption is violated; b) Various sources note that when the sphericity assumption underlying the **single-factor within-subjects analysis of variance** is not violated, the latter test provides a more powerful test of an alternative hypothesis than **Hotelling's T^2/multivariate analysis of variance**. However, when with a large sample size the sphericity assumption is not violated, the power of the **Hotelling's T^2/multivariate analysis of variance** is comparable to that of the **single-factor within-subjects analysis of variance**. For further discussion of the relative merits and liabilities associated with **Hotelling's T^2/multivariate analysis of variance** versus the **single-factor within-subjects analysis of variance** in evaluating a repeated-measures design (i.e., dependent samples design), the reader should refer to the discussion of the sphericity assumption in Section VI of the **single-factor within-subjects analysis of variance**.

8.   a) The same result for the analysis will be obtained if the first difference score for the ith subject (D_{i1}) is obtained by subtracting the ith subject's score in Condition 2 (X_{i2}) from his or her score in Condition 1 (X_{i1}) — i.e., $D_{i1} = X_{i1} - X_{i2}$, and the second difference score for the ith subject (D_{i2}) is obtained by subtracting the ith subject's score in Condition 3 (X_{i3}) from his or her score in Condition 2 (X_{i2}) — i.e., $D_{i2} = X_{i2} - X_{i3}$; b) An alternative protocol which will yield the same test result is to subtract each subject's score in a given condition from his or her score in a standard condition. Thus, in the case of Example 24.1, if Condition 1 is designated as a standard condition, the two difference scores for each subject will be $D_{i1} = X_{i1} - X_{i2}$ and $D_{i2} = X_{i1} - X_{i3}$; c) Myers and Well (2003, p. 360) note that in order to conduct a multivariate analysis on a dependent samples design, the value of n must be greater than the value of $k - 1$. The latter, however, is not a requirement for evaluating the data with a **single-factor within-subjects analysis of variance**.

9.   The notation $\bar{D}_j$, which is the mean of the jth difference score, can also be written as $\bar{D}_{.j}$.

10.   Some sources (e.g., Tabachnick and Fidell (1996, Ch. 10; 2001, Ch. 10)) refer to a multivariate analysis of a dependent samples design as **profile analysis**.

11. It should be noted that analysis of Example 24.1 with *SPSS* did not indicate violation of the sphericity assumption. *SPSS* summarizes the latter analysis in a table labeled **Mauchly's test of sphericity**, which in the case of Example 24.1 yielded a probability value above .05.

Test 35
Multivariate Analysis of Variance
(Parametric Test Employed with Interval/Ratio Data)

I. Hypothesis Evaluated with Test and Relevant Background Information

Hypothesis evaluated with test Do k (where $k \geq 2$) independent samples represent k populations with different values with respect to their **mean vectors** (where a **mean vector** represents a composite score comprised of mean values on two or more dependent variables)?

Relevant background information on test In its simplest form the **multivariate analysis of variance** (for which the acronym **MANOVA** is commonly employed) is the multivariate analog of the **single-factor between-subjects analysis of variance** (Test 21). The **multivariate analysis of variance** is a generalization of **Hotelling's T^2** (Test 34) to experimental designs involving more than two groups, and, as noted earlier, **Hotelling's T^2** represents a special case of the **multivariate analysis of variance**. The **multivariate analysis of variance** can be employed to analyze data for an experiment which involves one or more independent variables (each of which is comprised of two or more levels) and multiple dependent variables. With regard to the latter, instead of a single score, each subject produces scores on two or more dependent variables.

To illustrate, let us assume that the efficacy of an antidepressant drug is assessed by employing three independent groups, with one group receiving the drug, a second group receiving a placebo, and the third group not receiving anything. In the study two scores are obtained on each subject. One score represents a subject's level of depression and a second score represents the subject's level of anxiety. The two scores for each subject represent the two dependent variables of depression and anxiety, and the composite score for a subject (i.e., a value resulting from algebraically combining a subject's depression and anxiety scores into a single measure) is viewed as a measure of emotional instability. Within the framework of the **multivariate analysis of variance**, a composite mean on emotional instability is computed for each group. The latter composite means are referred to as **mean vectors** or **centroids**. As is the case with the **single-factor between-subject analysis of variance**, the means (in this case **composite**) for the three groups are then compared with one another.

The following reasons are commonly cited for employing a **multivariate analysis of variance** in lieu of a **univariate analysis of variance** (i.e., an **analysis of variance** involving one dependent variable): a) Most effective treatments will impact subjects in a variety of ways. Consequently, an experimental design which evaluates multiple effects resulting from a treatment will provide a more comprehensive picture of any treatment effect; b) Conducting multiple statistical tests on different dependent variables (through the use of multiple t tests or **univariate analyses of variance**) can inflate the overall Type I error rate associated with a body of data. A **multivariate analysis of variance** can avoid the latter, since it allows a researcher to control the overall/experimentwise Type I error rate; c) A **univariate analysis of variance** cannot take into account whether or not two or more dependent variables are correlated with one another. A **multivariate analysis of variance**, on the other hand, is able to incorporate intercorrelations between dependent variables in computing the test statistic[1]; and d) The effect of one or more dependent variables by themselves may not be strong enough to yield a significant result, yet when employed together within the framework of a design which is evaluated with a **multivariate analysis of variance**, their combined effect may yield a significant result.

The assumptions underlying the **multivariate analysis of variance** are the same as those underlying the **single-factor between-subjects analysis of variance**, except for the fact they are generalized to a **multivariate normal distribution** (which is discussed in Section I in the chapter on **multiple regression (Test 33)**). Specifically, among the assumptions underlying the **multivariate analysis of variance** are multivariate normality on the dependent variables in each of the k populations and homogeneity of the **variance-covariance matrices** for the dependent variable in each group (which are respectively analogous to the normality and homogeneity of variance assumptions underlying the **single-factor between-subjects analysis of variance**). In addition, the dependent variables should be linearly related to one another, and the observations should be independent and randomly selected from the target population to which one intends to generalize the results. Stevens (2002, p. 276–280) provides a detailed discussion of procedures for assessing violation of the assumptions of the **multivariate analysis of variance**.

Mechanisms which can be useful in evaluating the multivariate normality assumption are to check (through use of scatterplots and/or goodness-of-fit tests) whether each of the p dependent variables are normally distributed. Additionally, the researcher can construct for each of the k groups two-dimensional scatterplots for each pair of dependent variables. The latter plots should reveal either no relationship between pairs of variables or a linear relationship, with evidence of one or more curvilinear relationships suggesting violation of the assumption of multivariate normality. Silva and Stam (1995, p. 285) note that Koziol (1986) and Baringhaus, Danschke, and Henze (1989) describe more formalized tests of the multivariate normality assumption. As is the case for a variety of multivariate procedures, the **Mahalanobis distance** and **hat element statistics** (discussed in Section IV/V in the chapter on **multiple regression**) can be computed in order to identify potential outliers which can adversely impact the results of a **multivariate analysis of variance**. As a general rule, the **multivariate analysis of variance** is reasonably robust with respect to assumption violation, especially with a large sample size and an equal number of subjects in each group.

The reader should take note of the fact that although the discussion of the **multivariate analysis of variance** in this chapter will be limited to an independent samples design involving one independent variable, the procedure can be employed to evaluate any experimental design involving two or more dependent variables for which an analysis of variance is appropriate. Consequently, the **multivariate analysis of variance** can be generalized to all varieties of experimental design such as a dependent samples designs, factorial designs, etc.

II. Example

To illustrate the use of the **multivariate analysis of variance**, consider Example 35.1 which involves $k = 3$ independent groups. As is the case with Example 34.1, the two scores for each subject represent the two dependent variables of depression and anxiety, and the composite score for a subject is viewed as a measure of **emotional instability**.

Example 35.1 *In order to assess the impact of a new antidepressant drug on emotional instability, 15 patients at a mental health clinic who are diagnosed with both clinical depression and excessive anxiety are randomly assigned to one of three groups. Five patients are assigned to Group 1, which is administered the antidepressant drug for a period of six months; five patients are assigned to Group 2, which is administered a placebo during the same six-month period; and five patients are assigned to Group 3, which is a no treatment group that is just evaluated at the beginning and end of the study. Assume that prior to introducing the experimental treatments, the experimenter confirmed that the levels of both depression and* anxiety in the three groups were equal. After six months elapse all 15 subjects are rated by a psychiatrist (who is blind with respect to a subject's experimental condition) on both their level of depression and anxiety (which when taken together are considered to represent a generalized measure of emotional instability). The psychiatrist's depression and anxiety ratings for the five subjects in each group are noted below. Each subject's depression score is listed first followed by his or her anxiety score. The higher the score on both depression and anxiety, the more depressed or anxious a subject (and consequently, the larger the composite mean for a group on the dependent variable the higher the degree of emotional instability): **Group 1: Subject 1** (11, 8); **Subject 2** (1, 3); **Subject 3** (0, 2); **Subject 4** (4, 1); **Subject 5** (0, 8); **Group 2: Subject 1** (11, 9); **Subject 2** (11, 6); **Subject 3** (5, 7); **Subject 4** (8, 6); **Subject 5** (4, 9); **Group 3: Subject 1** (12, 11); **Subject 2** (8, 8); **Subject 3** (14, 6); **Subject 4** (8, 10); **Subject 5** (11, 4). *Do the data indicate that the drug is effective in decreasing emotional instability?*

III. Null versus Alternative Hypotheses

The null and alternative hypotheses evaluated with the **multivariate analysis of variance** are noted below.

Null hypothesis

$$H_0: \begin{bmatrix} \mu_{11} \\ \mu_{12} \\ . \\ . \\ . \\ \mu_{1p} \end{bmatrix} = \begin{bmatrix} \mu_{21} \\ \mu_{22} \\ . \\ . \\ . \\ \mu_{2p} \end{bmatrix} = \dots\dots = \begin{bmatrix} \mu_{k1} \\ \mu_{k2} \\ . \\ . \\ . \\ \mu_{kp} \end{bmatrix}$$

Where: μ_{ij} represents the mean of the population represented by Group i on dependent variable j

(The null hypothesis states that the vectors of the means of the p dependent variables for all k populations which are represented by the k samples are equal.[2])

Alternative hypothesis

$$H_1: H_0 \text{ is false}$$

(The alternative hypothesis is **nondirectional**, and implies a difference between at least two of the k mean vectors.)

Since Example 35.1 involves three groups and two dependent variables, the null and alternative hypotheses for the example are noted below.

$$H_0: \begin{bmatrix} \mu_{11} \\ \mu_{12} \end{bmatrix} = \begin{bmatrix} \mu_{21} \\ \mu_{22} \end{bmatrix} = \begin{bmatrix} \mu_{31} \\ \mu_{32} \end{bmatrix}$$

$$H_1: H_0 \text{ is false}$$

IV. Test Computations

As is the case with **Hotelling's T^2**, the following four test statistics (each of which can be converted into an F value) are commonly computed to evaluate the null hypothesis for the **multivariate analysis of variance: Wilks' lambda, Pillai's trace, Hotelling's trace**, and **Roy's largest root**. Since the calculations for the **multivariate analysis of variance** are generally done through use of computer software, the intent of this section is to provide a succinct summary of what each of the four test statistics represents. Readers interested in a detailed description of the matrix algebra operations involved in deriving the test statistics are referred to sources such as Bruning and Kintz (1997, pp. 264–270), Field (2005, Ch. 14), Harris (2001, Ch. 4), Marascuilo and Levin (1983, Ch. 8), Stevens (2002, Ch. 6), and Tabachnick and Fidell (1996, Ch. 9; 2001, Ch. 9). A brief description of **Wilks' lambda, Pillai's trace, Hotelling's trace**, and **Roy's largest root** follows.

    a) **Wilks' lambda** (Wilks, 1932) (also known as **Wilks' likelihood ratio**) is probably the most commonly employed test statistic for the **multivariate analysis of variance**. As noted in the chapter on **matrix algebra**, Wilk's lambda (Λ) is a ratio of two determinants such that $\Lambda = |\mathbf{W}|/|\mathbf{T}|$ (Stevens (2002, p. 64)). In the latter equation, **W** represents a multivariate generalization of a within-subjects sum of squares (in the form of a **within-groups variance-covariance matrix**) and **T** represents a multivariate generalization of a total sum of squares (in the form of a **total variance-covariance matrix**, which is the **between-groups variance-covariance matrix** (to be designated **B**) plus the **within-groups variance-covariance matrix** — i.e., **T = B + W**). **Wilks' lambda** can assume a value between 0 and 1, and the lower the value of Λ the greater the likelihood the null hypothesis is false and should be rejected. The value of Λ reflects the amount of variability in the data that is not explained by the independent variable, and thus the proportion of variability which is explained by the independent variable is equal to $\eta^2 = 1 - \Lambda$ (where η^2 represents the measure of effect size **eta squared**). The value of **Wilks' lambda** can be transformed into an F value or a chi-square value. Stevens (2002, p. 244) notes that **Wilks' lambda** can be expressed as the product of **eigenvalues** of $\mathbf{WT}^{-1}$. Within the context of the **multivariate analysis of variance** an **eigenvalue** is a number which indicates the amount of variability in a matrix. Field (2005, p. 592) notes that the presence of large eigenvalues is indicative of a large effect, which in turn yields a small value for **Wilks' lambda**.[3] Through use of the statistical software package *SPSS*, the value computed for **Wilks' lambda** for Example 35.1 is .451 (which is displayed in Table 35.1 in Section V).

    b) **Pillai's trace** (also referred to as **Pillai's criterion** or the **Pillai–Bartlett trace**) is attributed to Bartlett (1939) and Pillai (1955). The larger the value of **Pillai's trace** the greater the likelihood the null hypothesis is false and should be rejected. Harlow (2005, p. 112) and Stevens (2002, p. 244) note that **Pillai's trace** is the sum of the diagonal elements of the **between-groups variance-covariance matrix B** over the **total variance-covariance matrix T** (i.e., **Pillai's trace** = $[\mathbf{B} + \mathbf{W}]^{-1}\mathbf{B}$, which can also be written as $\mathbf{T}^{-1}\mathbf{B}$). Harlow (2005, p. 113) notes that the value computed for **Pillai's trace** can be interpreted as representing the proportion of variance in the linear combination of the dependent variables that is explained by the independent variable. Employing *SPSS*, the value computed for **Pillai's trace** for Example 35.1 is .561 (which is displayed in Table 35.1 in Section V).

    c) **Hotelling's trace** (also known as the **Hotelling–Lawley trace**) is attributed to Hotelling (1931) and Lawley (1938). Stevens (2002, p. 244) notes that **Hotelling's trace** is the sum of the eigenvalues of $\mathbf{W}^{-1}\mathbf{B}$, and that the latter represents the multivariate analogue of an F ratio contrasting between-groups variability with within-groups variability. Or to put it another way, Harlow (2005, p. 112) notes that **Hotelling's trace** is the sum of the diagonal elements of the matrix formed by the ratio of the **between-groups variance-covariance matrix** over the **within-**

groups variance-covariance matrix (i.e., $W^{-1}B$ noted above, since the sum of the diagonals represents the trace of the matrix, and, as noted in Endnote 3, the latter will always equal the sum of the eigenvalues). Employing *SPSS*, the value computed for **Hotelling's trace** for Example 35.1 is 1.193 (which is displayed in Table 35.1 in Section V).

d) **Roy's largest root** (also referred to as **Roy's maximum root** and **Roy's criterion**) is attributed to Roy (1945, 1953). Harlow (2005, p. 112) and Stevens (2002, p. 244) note that **Roy's largest root** is the largest root (i.e., eigenvalue) from the $W^{-1}B$ matrix, and that the latter represents the multivariate analogue of an F ratio contrasting between-groups variability with within-groups variability. Harlow (2005, p. 112) notes the number represented by **Roy's largest root** is the variance of the largest linear combination in the $W^{-1}B$ matrix. Employing *SPSS*, the value computed for **Roy's largest root** for Example 35.1 is 1.170 (which is displayed in Table 35.1 in Section V).

A number of researchers address the question of which of the four aforementioned multivariate test statistics is best to employ. Field (2005, pp. 593–594) and Stevens (2002, p. 244) essentially conclude that under most circumstances **Wilks' lambda**, **Pillai's trace**, and **Hotelling's trace** are equally robust with respect to violations of the assumptions underlying the **multivariate analysis of variance**, but that **Roys's largest root** should only be employed if the homogeneity of the variance-covariance assumption is not violated. Hair *et al.* (1995, p. 278), Harlow (2005, p. 113), and Tabachnick and Fidell (1996, pp. 400–401) state that when the conditions for a **multivariate analysis of variance** are less than ideal (i.e., unequal sample sizes, heterogeneity of variance-covariance), **Pillai's trace** is the most robust of the above noted four test statistics. In the final analysis, the most frequently recommended test statistics are **Wilks' lambda** and **Pillai's trace**.

As noted in the discussion of **Hotelling's T^2**, Tabachnick and Fidell (1996, p. 400) state that when there are two groups the F values for **Wilks' lambda**, **Pillai's trace**, and **Hotelling's trace** will be identical. When there are three or more groups, although the F values for the latter three measures (as well as **Roy's largest root**) will not be identical, they will usually all be close with respect to whether or not they yield a statistically significant result. When there are inconsistencies between two or more of the measures further studies are recommended to resolve the issue.

V. Interpretation of the Test Results

Table 35.1 presents a partial *SPSS* printout for the analysis of Example 35.1.[4] The reader should attend to the **lower section** of the table (labeled **Group**) which contains information regarding the four test statistics.[5] As noted earlier, all four of the test statistics can be converted into an F value, and consequently F values are recorded in Column 3 in the row designating each of the test statistics.[6] Each of the F values can be interpreted through use of **Table A10 (Table of the F Distribution)** in the **Appendix**, employing the degrees of freedom values listed in Column 4 (**Hypothesis df** which represents df_{num}) and Column 5 (**Error df** which represents df_{den}) for a given row. Column 6 (**Sig.**) contains the probability values associated with the result for each of the test statistics. The value recorded in Column 7 (**Partial Eta Squared**) is a measure of effect size indicating the proportion of variance on the composite dependent variable which can be explained on the basis of the independent variable.

Although computation of the degrees of freedom values is complicated and not necessary to do by hand, it will be described for **Wilks' lambda**, which, as noted in Section IV, is one of the two most commonly employed test statistics for the **multivariate analysis of variance**. A protocol outlined in Marascuilo and Levin (1983, p. 315) and Stevens (2002, p. 216) for computing the values of **Hypothesis df** $= df_{num}$ and **Error df** $= df_{den}$ is described below, and employed to compute the df values for Example 35.1. Marascuilo and Levin (1983, p. 314) note that the value s computed below represents the number of **eigenvalues** generated for the data.

Table 35.1 *SPSS* Printout of Results for Example 35.1:
Omnibus Test Results for Multivariate Analysis of Variance

Multivariate Tests[d]

Effect		Value	F	Hypothesis df	Error df	Sig.	Partial Eta Squared
Intercept	Pillai's Trace	.919	62.153[b]	2.000	11.000	.000	.91
	Wilks' Lambda	.081	62.153[b]	2.000	11.000	.000	.91
	Hotelling's Trace	11.301	62.153[b]	2.000	11.000	.000	.91
	Roy's Largest Root	11.301	62.153[b]	2.000	11.000	.000	.91
GP	Pillai's Trace	.561	2.341	4.000	24.000	.084	.28
	Wilks' Lambda	.451	2.693[b]	4.000	22.000	.058	.32
	Hotelling's Trace	1.193	2.981	4.000	20.000	.044	.37
	Roy's Largest Root	1.170	7.020[c]	2.000	12.000	.010	.53

a. Computed using alpha = .05
b. Exact statistic
c. The statistic is an upper bound on F that yields a lower bound on the significance level.
d. Design: Intercept+GP

$$df_{num} = p(k - 1) = 2(3 - 1) = 4$$

$$df_{den} = ms - \frac{p(k - 1)}{2} + 1 = (11.5)(2) - \frac{2(3 - 1)}{2} + 1 = 22$$

Where:

$$N = n_1 + n_2 + \ldots + n_k = n_1 + n_2 + n_3 = 5 + 5 + 5 = 15$$

$$m = N - 1 - \frac{(p + k)}{2} = 15 - 1 - \frac{(2 + 3)}{2} = 11.5$$

$$s = \sqrt{\frac{p^2(k - 1)^2 - 4}{p^2 + (k - 1)^2 - 5}} = \sqrt{\frac{(2)^2(3 - 1)^2 - 4}{(2)^2 + (3 - 1)^2 - 5}} = 2$$

Inspection of Table 35.1 reveals that the value $\Lambda = .451$ computed for **Wilks' lambda** corresponds to the value $F = 2.693$, which for $df_{num} = 4$ and $df_{den} = 22$ is associated with a probability of $p = .058$. Note that in Table A10 for $df_{num} = 4$ and $df_{den} = 22$, $F_{.05} = 2.82$, which is slightly larger than $F = 2.693$. Since $p = .058$ is greater (albeit minimally) than .05, the null hypothesis cannot be rejected. Note that the p values associated with **Pillai's trace**, **Hotelling's trace**, and **Roy's largest root** are respectively .084, .044, and .010, with the latter two indicating significance at the .05 and .01 levels respectively. Although the p values for all four test statistics are reasonably close to one another, depending upon which one is employed the researcher may retain (in the case of **Pillai's trace** and **Wilks' lambda**) or reject (in the case of **Hotelling's trace** and **Roy's largest root**) the null hypothesis. One might rationalize the failure to achieve significance in the case of **Pillai's trace** and **Wilks' lambda** on the basis of the small sample size employed in the study (which could be viewed as having compromised the power of the analysis). In the final analysis, although the results recorded in Table 35.1 suggest the possibility of differences between two or more of the underlying populations with respect to their mean vectors on the dependent variable of emotional instability (which is a linear combination of the two dependent variables of depression and anxiety), most researchers would probably conclude that further studies are required to clarify the exact nature of the relationship between the independent variable and the composite dependent variable.

The information provided by *SPSS* and other statistical software packages on a **multivariate analysis of variance** goes well beyond what it printed in Table 35.1. Additional information available from an analysis is discussed below.

a) **Descriptive statistics for the groups on the dependent variables**   One can elect to have values printed out for the mean and estimated population standard deviation of each group for each of the dependent variables. The following values were computed for Example 35.1: 1) Depression: $\bar{X}_1 = 3.2$, $\tilde{s}_1 = 4.66$, $\bar{X}_2 = 7.8$, $\tilde{s}_2 = 3.27$, $\bar{X}_3 = 10.6$, $\tilde{s}_3 = 2.61$; 2) Anxiety: $\bar{X}_1 = 4.4$, $\tilde{s}_1 = 3.36$, $\bar{X}_2 = 7.4$, $\tilde{s}_2 = 1.52$, $\bar{X}_3 = 7.8$, $\tilde{s}_3 = 2.86$.

b) **Levene's test of equality of error variance (Test 21g)**   This test evaluates whether or not the homogeneity of variance assumption has been violated with respect to each of the dependent variables. In the case of Example 35.1 the *F* values, degrees of freedom, and probability values obtained for **Levene's test** (for which the test statistic is an *F* value) with respect to depression and anxiety are respectively $F(2, 12) = .681$, $p = .525$ and $F(2, 12) = 3.126$, $p = .081$. Since neither probability value (.525 for depression and .081 for anxiety) is less than .05, the homogeneity of variance assumption is not violated for either variable. In the event the latter assumption is violated for one or more of the dependent variables, it is recommended that a more conservative *F* test be employed (i.e., that a value of alpha below .05 (e.g., .01) should be employed as the criterion in determining significance for the **multivariate analysis of variance**).

c) **Box's test of equality of covariance matrices**   As the name indicates this test evaluates whether there is homogeneity of covariance matrices which is the multivariate generalization of the homogeneity of variance assumption underlying the **single-factor between-subjects analysis of variance**. The test statistic for **Box's test of equality of covariance matrices** is an *F* value. Hair *et al.* (1995, pp. 275–276) and Tabachnick and Fidell (1996, p. 382) note that when the **Box test** is significant with equal sample sizes, robustness of the **multivariate analysis of variance** will not be affected. However, in the case of unequal sample sizes a significant result (especially in a study employing a large number of dependent variables) indicates that unless some adjustment is made, the robustness of the **multivariate analysis of variance** will be compromised. Among the adjustments suggested are data transformation and equalization of sample sizes by judicious deletion of scores. Additionally, the use of **Pillai's trace** as opposed to **Wilks' lambda** is recommended in interpreting the results, since the former measure is seen as more robust under such conditions. Field (2005, p. 593) notes that when the multivariate normality assumption is violated, the likelihood of a Type II error increases for **Box's test** (i.e., the test is more likely to yield a nonsignificant result when, in fact, the covariance matrices are not homogeneous).

In the case of Example 35.1, the probability value (associated with an *F* value) obtained for **Box's test of equality of covariance matrices** was $p = .711$. Since $p = .711$ is greater than .05, the test result is not significant. Consequently, the homogeneity of covariance matrices assumption was not violated.

d) **A single-factor between-subjects analysis of variance conducted on each of the dependent variables**   In order to better assess the relative impact of each of the dependent

Table 35.2   *SPSS* Printout of Results for Example 35.1:
Single-Factor Between-Subjects Analyses of Variance

Tests of Between-Subjects Effects

Source	Dependent Variable	Type III Sum of Squares	df	Mean Square	F	Sig.	Partial Eta Squared
Corrected Model	DEP	139.600[b]	2	69.800	5.342	.022	.471
	ANX	34.533[c]	2	17.267	2.376	.135	.284
Intercept	DEP	777.600	1	777.600	59.510	.000	.832
	ANX	640.267	1	640.267	88.110	.000	.880
GP	DEP	139.600	2	69.800	5.342	.022	.471
	ANX	34.533	2	17.267	2.376	.135	.284
Error	DEP	156.800	12	13.067			
	ANX	87.200	12	7.267			
Total	DEP	1074.000	15				
	ANX	762.000	15				
Corrected Total	DEP	296.400	14				
	ANX	121.733	14				

a. Computed using alpha = .05

b. R Squared = .471 (Adjusted R Squared = .383)

c. R Squared = .284 (Adjusted R Squared = .164)

variables, a **single-factor between-subjects analysis of variance** can be conducted on each of the dependent variables. Table 35.2 displays the result of the latter analysis.

The identical **between-groups information** for each analysis of variance is recorded in both the top section of Table 35.2 labeled **Corrected Model** and the section lower down labeled **GP** (which represents **group** — i.e., the independent variable). The **between-groups information** when **depression** is evaluated as a single dependent variable is in the row labeled **DEP**. The between-groups sum of squares (139.600) for the analysis of variance is in the column labeled **Type III Sum of Squares**, the between-groups degrees of freedom ($df = 2$) is in the column labeled **df**, and the between-groups mean square (69.800) is in the column labeled **Mean Square**. The value $F = 5.342$ is in Column 5, followed by the probability value .022 associated with the result in Column 6 (**Sig.**), which is followed by the value of **Partial Eta Squared** of .471 (which is a measure of effect size) in Column 7. The relevant information for **within-groups variability** is in the section of the table labeled **Error**. The **within-groups information** when **depression** is evaluated as a single dependent variable is in the row labeled **DEP**, where the values recorded for within-groups sum of squares, degrees of freedom, and mean square are respectively 156.800, 12, and 13.067. The **single-factor between-subjects analysis of variance of the depression scores** can be summarized as follows: $F(2, 12) = 5.342$, $p = .022$. Since $p = .022$ is less than .05 level but greater than .01 level, the F value is significant at the .05 level. Thus, there is a significant difference between at least two of the three treatments when only the depression scores are employed as the dependent variable.

In the same respect, the relevant **between-groups information** when **anxiety** is evaluated as a single dependent variable is in the row labeled **ANX** in the sections of the table labeled **Corrected Model** and **GP**. In the row labeled **ANX** the following values are respectively recorded for between-groups sum of squares, degrees of freedom, and mean square: 34.533, 2, and 17.267. Additionally, $F = 2.376$, which is associated with $p = .135$, and a **Partial Eta Squared** value of .284 are respectively recorded in Columns 5, 6, and 7. The relevant information for **within-groups variability** is in the section of the table labeled **Error**, and the **within-groups information** when **anxiety** is evaluated as a single dependent variable is in the row labeled **ANX**, where the values recorded for within-groups sum of squares, degrees of freedom, and mean square are respectively 87.200, 12, and 7.267. The **single-factor between-subjects analysis of variance** of the anxiety scores can be summarized as follows: $F(2, 12) = 2.376, p = .135$. Since $p = .135$ is greater than .05, the F value is not significant. Consequently, there is no significant difference between the three treatments when only the anxiety scores are employed as the dependent variable. The result of the two **single-factor between-subjects analyses** of variance indicate that it is possible to have at least one significant univariate analysis of variance without a significant multivariate result.

e) **Analysis of covariance** Harlow (2005, p. 114) notes that when the overall result for the **multivariate analysis of variance** is statistically significant, a **single-factor between-subjects analysis of covariance** can be employed to further assess the impact of the independent variable on each of the dependent variables. Specifically, each of the dependent variables employed in the **multivariate analysis of variance** is successively employed as a single dependent variable in a **single-factor between-subjects analysis of covariance (Test 21I)** in which the other dependent variable(s) are employed as covariate(s). The latter methodology will allow a researcher to assess the impact of the independent variable on each dependent variable independent of the impact of the independent variable on any of the other dependent variables.

In the case of Example 35.1 a **single-factor between-subjects analysis of covariance** which employs the **depression** scores as the **dependent variable** and the **anxiety** scores as a **covariate** yields the result $F(2, 11) = 3.242, p = .078$. Since $p = .078$ is greater than .05 (albeit barely), the F value is not significant. A **single-factor between-subjects analysis of covariance** which employs the **anxiety** scores as the **dependent variable** and the **depression scores** as a **covariate** yields the result $F(2, 11) = .956, p = .414$. Since $p = .414$ is greater than .05, the F value is not significant.[7]

The results of the two **single factor between-subjects analyses of covariance** and the two **single-factor between-subjects analyses of variance** indicate greater differences between the groups with respect to the dependent variable of depression when contrasted with anxiety. The latter is indicated by the following: a) A significant result was obtained in the analysis of variance in which depression was the dependent variable, while a nonsignificant result was obtained for the analysis of variance in which anxiety was the dependent variable; and b) The analysis of covariance in which depression was the dependent variable, though not significant, was close to being significant at the .05 level, while the analysis of covariance in which anxiety was the dependent variable was far removed from being significant.

f) **Planned and unplanned comparisons** To further isolate the nature of the impact of the independent variable, a researcher can conduct planned or unplanned comparisons in which two or more groups or combination of groups are compared with one another. Stevens (2002, p. 217) notes that when a significant multivariate result is obtained, pairwise multivariate tests can be employed to compare the different groups with respect to mean vectors. Thus, in the case of Example 35.1 the following comparisons can be made between pairs of groups with respect to their mean vectors: Group 1 versus Group 2; Group 1 versus Group 3; Group 2 versus Group 3. The latter comparisons would entail employing **Hotelling's T^2** to compare each set of $k = 2$ groups (as noted, in the case of **Hotelling's T^2** the computer conducts a **multivariate analysis of variance** for $k = 2$). If the researcher wants to maintain an overall Type I error rate of .05, using the **Bonferroni inequality** (discussed in Section VI of the **single-sample between-subjects analysis of variance**), the obtained probability for each comparison would be required to be $.05/3 = .0167$ or lower in order to declare significance. It should be noted that the researcher has the option of employing a larger overall Type I error rate than .05 (which some people might consider too low), and consequently tolerate a per comparison error rate greater than .0167.

With respect to Example 35.1 for each of the above noted comparisons involving pairs of mean vectors, the following values were obtained for **Wilks' lambda**, F (df_{num}, df_{den}), and **sig** (i.e., the p value) (the same values were also obtained for **Pillai's trace**, **Hotelling's trace**, and **Roy's largest root**): **Group 1** versus **Group 2** ($\Lambda = .601$; $F(2, 7) = 2.323$; $p = .168$); **Group 1** versus **Group 3** ($\Lambda = .418$; $F(2, 7) = 4.874$; $p = .047$); **Group 2** versus **Group 3** ($\Lambda = .747$; $F(2, 7) = 1.186$; $p = .360$). Note that none of the probability values is less than .0167, and thus none achieves significance if the **Bonferroni method** is employed to maintain an overall Type I error rate of .05. One should keep in mind, however, that since the **Bonferroni method** is (along with the **Scheffé method**) the most conservative of the comparison procedures (and consequently requires the largest difference between a pair of groups in order to declare significance), some researchers would use a less conservative comparison method (e.g., one based on **Tukey's HSD** methodology (see Stevens (2002, p. 217)) and/or tolerate a higher overall Type I error rate (e.g., .10 or even larger). If, for the moment we assume all of the comparisons were planned, and the usual .05 level of significance is employed for each comparison, the only significant result is for Group 1 versus Group 3, since $p = .047$ is less than .05. To summarize, of the three comparisons involving mean vectors the only one which suggests the likelihood of a significant difference is the comparison between the group which received the medication (Group 1) and the no treatment group (Group 3), with the former group exhibiting a lower level of emotional instability.

In addition to the above comparisons of mean vectors, planned or unplanned comparisons can also be conducted on each of the dependent variables. In the case of Example 35.1 the following three multiple comparison procedures were employed to compare the three groups on the different dependent variables: **Fisher's LSD test (Test 21a)**, **Tukey's HSD test (Test 21d)**, and the **Scheffé test (Test 21e)**. Table 35.3 summarizes the results of the latter comparisons. As indicated in Column 4 (**Sig.**) of the latter table, when each of the dependent variables was considered separately, significant differences were found between pairs of groups for the **dependent variable** of depression (in the upper part of the table where the dependent variable is labeled **DEP** at the extreme left), but only for the Group 1 versus Group 3 comparison. Regardless of which comparison procedure was employed (if, for the moment, one assumes the overall Type I error rate was not set at .05), the difference between Groups 1 and 3 was significant at either the .05 level or .01 level ($p = .007$ for **Fisher's LSD test**, $p = .018$ for **Tukey's HSD test**, and $p = .023$ for **Scheffé test**). In contrast, no significant differences were found between pairs of groups for the **dependent variable** of anxiety (in the lower part of the table where the dependent variable is labeled **ANX** at the extreme left), regardless of which comparison procedure was employed. The only comparison which came close to significance at the .05 level was the comparison of Groups 1 and 3 with **Fisher's LSD test** which yielded the $p = .069$.

The results for the comparisons, when viewed within the context of the other analyses described in this section, suggest that the most dramatic differences in the study were with respect to the **dependent variable** of depression between subjects who received the medication (Group 1) and subjects who received no treatment (Group 3) (with the former group exhibiting a lower level of depression). In the final analysis, the study did not provide evidence for the medication condition (Group 1) being superior to the placebo condition (Group 2). Again, it should be emphasized that the small sample size limited the power of the analysis.

In closing this discussion, the reader should take note of the fact that computer software can provide additional analytical as well as graphical information for a **mutivariate analysis of variance** which has not been included in this section — for example, plots of the data can provide the researcher with valuable information regarding assumption violation, presence of outliers, etc. Those requiring more extensive documentation of such procedures should consult books which specialize in multivariate statistics.

Table 35.3 *SPSS* **Printout of Multiple Comparisons for Each Dependent Variable in Example 35.1**

Multiple Comparisons

Dependent Variable		(I) GP	(J) GP	Mean Difference (I-J)	Std. Error	Sig.
DEP	Tukey HSD	1.00	2.00	-4.6000	2.28619	.152
			3.00	-7.4000*	2.28619	.018
		2.00	1.00	4.6000	2.28619	.152
			3.00	-2.8000	2.28619	.462
		3.00	1.00	7.4000*	2.28619	.018
			2.00	2.8000	2.28619	.462
	Scheffe	1.00	2.00	-4.6000	2.28619	.175
			3.00	-7.4000*	2.28619	.023
		2.00	1.00	4.6000	2.28619	.175
			3.00	-2.8000	2.28619	.493
		3.00	1.00	7.4000*	2.28619	.023
			2.00	2.8000	2.28619	.493
	LSD	1.00	2.00	-4.6000	2.28619	.067
			3.00	-7.4000*	2.28619	.007
		2.00	1.00	4.6000	2.28619	.067
			3.00	-2.8000	2.28619	.244
		3.00	1.00	7.4000*	2.28619	.007
			2.00	2.8000	2.28619	.244

Based on observed means.

Multiple Comparisons

Dependent Variable		(I) GP	(J) GP	Mean Difference (I-J)	Std. Error	Sig.
ANX	Tukey HSD	1.00	2.00	-3.0000	1.70489	.224
			3.00	-3.4000	1.70489	.156
		2.00	1.00	3.0000	1.70489	.224
			3.00	-.4000	1.70489	.970
		3.00	1.00	3.4000	1.70489	.156
			2.00	.4000	1.70489	.970
	Scheffe	1.00	2.00	-3.0000	1.70489	.252
			3.00	-3.4000	1.70489	.180
		2.00	1.00	3.0000	1.70489	.252
			3.00	-.4000	1.70489	.973
		3.00	1.00	3.4000	1.70489	.180
			2.00	.4000	1.70489	.973
	LSD	1.00	2.00	-3.0000	1.70489	.104
			3.00	-3.4000	1.70489	.069
		2.00	1.00	3.0000	1.70489	.104
			3.00	-.4000	1.70489	.818
		3.00	1.00	3.4000	1.70489	.069
			2.00	.4000	1.70489	.818

Based on observed means.

*. The mean difference is significant at the .05 level.

VI. Additional Analytical Procedures for the Multivariate Analysis of Variance and/or Related Tests

No additional procedures will be described in this section.

VII. Additional Discussion of the Multivariate Analysis of Variance

1. Conceptualizing the hypothesis for the multivariate analysis of variance within the context of a linear combination   Harlow (2005, p. 105) states that an alternative way of expressing the hypothesis evaluated with the **multivariate analysis of variance** is as follows: Are there are significant group differences on the best linear combinations of the p dependent variables? Additionally, Harlow (2005, p. 105) and Weinfurt (1995, p. 252) note that a major reason why a **multivariate analysis of variance** can detect group differences on a set of dependent variables is because it transforms the p dependent variables into a composite dependent variable. The aforementioned composite dependent variable is, in fact, the linear combination of the p dependent variables which maximizes the separation between the mean vectors of the k groups, and because of the latter it maximizes the likelihood a researcher will be able to reject the null hypothesis. The linear equation which maximizes the distance between the mean vectors of the k groups is often referred to as a **discriminant function**, which is typically computed when the same set of data are evaluated with **discriminant function analysis (Test 37)** which is the opposite of the **multivariate analysis of variance** (insofar as the roles of the independent and dependent variables are reversed).

Within the context of the **multivariate analysis of variance**, the general form of the linear equation which maximizes differences between the groups is $Y_{composite} = a_1 Y_1 + a_2 Y_2 + ... + a_p Y_p$, where Y_j represents a subject's score on the jth dependent variable and a_j represents a weight reflecting the importance of the jth variable in the equation. The larger the value of a_j the greater the contribution of the jth variable to the composite mean. The weights in the equation (referred to as **discriminant coefficients** or **eigenvector weights**) insure that the distance(s) between the k mean vectors is maximized. Another way of viewing a test statistic for the **multivariate analysis of variance** (e.g., **Wilks' lambda**) is that it evaluates the degree of separation which has been achieved between the mean vectors of the k groups.

In addition to the maximally discriminative linear combination referred to above, it is possible to derive $(m - 1)$ additional orthogonal (i.e., independent) linear combinations (or to put it another way, a total of m possible linear combinations can be derived from a set of data). The value of m will be the smaller of the two values p or $(k - 1)$, and thus if $p = 3$ and $k = 3$, $m = 2$. Each of the linear combinations will have its own unique set of weights and will differ with respect to the degree that it differentiates between the distance between the k mean vectors. Consequently, since $m = 2$ a second linear combination can be derived, yet will not be as discriminative as the linear combination described initially.

2. Multicollinearity and the multivariate analysis of variance   One advantage of employing the **multivariate analysis of variance** is that it is entirely possible there will not be a significant difference between the means of k groups if separate analyses of variance are conducted on each of the individual dependent variables, yet when the dependent variables are considered jointly (assuming there is a sound rationale for combining the multiple dependent variables as a generalized measure of a specific construct) a **multivariate analysis of variance** will yield a significant result. Tabachnick and Fidell (1996, p. 380; p. 383) note that although a researcher may have a rationale for employing two or more dependent variables, the latter variables should not be too highly correlated with one another, and ideally the different dependent variables will be uncorrelated with one another since they each measure a separate facet of the dependent variable. In reality, researchers conducting a **multivariate analysis of variance** typically employ dependent variables which are correlated with one another, but not to an extreme degree. In the case of Example 35.1, the **Pearson product-moment correlation** between the two dependent variables of depression and anxiety was $r = .4075$ (which is a moderate correlation).

When dependent variables are substantially correlated with one another (which constitutes the condition of **multicollinearity**, which is discussed in Sections I and IV/V in the chapter on **multiple regression**), they probably measure the same facet(s) of the independent variable and there may be nothing to gain by including multiple measures of the identical facet in the analysis. In such a case, the overlapping variables should be combined into a single measure or else only include the dependent variable which is believed to have the most impact on the independent variable. Tabachnick and Fidell (1996, p. 383) note, however, if for some reason a researcher feels compelled to include multiple dependent variables which are highly correlated with one another, **factor analysis** (specifically, **principal components analysis (Test 40)**) can be employed to reconform the original dependent variables into a new set of less redundant dependent variables. Tabachnick and Fidell (1996, p. 380; pp. 403–405) and other sources on multivariate analysis describe **Roy–Bargmann stepdown analysis** as an alternative procedure for dealing with the problem of highly correlated dependent variables. The latter procedure, which is analogous to **hierarchical multiple regression**, evaluates the highest priority dependent variable with a **single-factor between-subjects analysis of variance**, and then employs a series of **single-factor between-subjects analysis of covariances** to evaluate the other dependent variables (employing previously evaluated dependent variables as covariates) in order of perceived importance.

VIII. Additional Examples Illustrating the Use of the Multivariate Analysis of Variance

No additional examples will be presented in this section.

References

Baringhaus, L., Danschke, R. & Henze, N. (1989). Recent and classical tests for normality – a comparative study. **Communication in Statistics - Simulation**, 18, 363–379.

Bartlett, M. S. (1939). A note on tests of significance in multivariate analysis. **Proceedings of the Cambridge Philosophical Society**, 35, 180–185.

Bruning, J. L. & Kintz, B. L. (1997). **Computational handbook of statistics** (4th ed.) New York: Addison-Wesley Longman

Cole, D. A., Maxwell, S. E., Arvey, R. & Salas, E. (1994). How the power of the MANOVA can both increase and decrease as a function of the intercorrelations among the dependent variables. **Psychological Bulletin**, 115, 465–474.

Field, A. (2005). **Discovering statistics using SPSS** (2nd ed.). Sage Publications: London.

George, D. & Mallery, P. (2001). **SPSS for Windows: Step by Step** (3rd ed.). Boston: Allyn & Bacon.

Hair, J. F., Anderson, R. E., Tatham, R. L. & Black, W. C. (1995). **Multivariate data analysis** (4th ed.). Upper Saddle River, N. J.: Prentice Hall.

Hair, J. F., Anderson, R. E., Tatham, R. L. & Black, W. C. (2004). **Multivariate data analysis** (6th ed.). Upper Saddle River, N. J.: Prentice Hall.

Harlow, L. L. (2005). **The essence of multivariate thinking**. Mahwah, NJ: Lawrence Erlbaum Associates.

Harris, R. J. (2001) **A primer of multivariate statistics** (3rd ed.). Mahwah, NJ: Lawrence Erlbaum Associates.

Hotelling, H. (1931). The generalization of Student's ratio. **Annals of Mathematical Statistics**, 2, 361–378.

Koziol, J. A. (1986). Assessing multivariate normality: A compendium. **Communication in Statistics – Theory and Methods**, 15, 2763–2783.

Lawley, D. N. (1938). A generalization of Fisher's z test. **Biometrika**, 30, 180–187.

Marascuilo, L. A. & Levin, J. R. (1983). **Multivariate statistics in the social science: A researchers guide**. Monterey, CA: Brooks/Cole Publishing Company.

Maxwell, S. E. & Delaney, H. D. (2004) **Designing experiments and analyzing data: A model comparison perspective** (2nd ed.). Mahwah, NJ: Lawrence Erlbaum Associates.

Mertler, C. A. & Vannatta, R. A. (2005). **Advanced and multivariate statistical methods** (3rd ed.). Los Angeles: Pyrczak Publications.

Pillai, K. C. S. (1955). Some new test criteria in multivariate analysis. **Annals of Mathematical Statistics**, 26, 117–121.

Roy, S. N. (1945). The individual sampling distribution of the maximum, minimum, and any intermediates of the p-statistics on the null-hypothesis. **Sankhyā**, 7, 133–158.

Roy, S. N. (1953). On a heuristic method of test construction and its use in multivariate analysis. **Annals of Mathematical Statistics**, 24, 220–238.

Silva, A. P. D. & Stam, A. (1995). Discriminant analysis. In Grimm, L. G. & Yarnold, P. R. (Eds.) (pp. 277–318). **Reading and understanding multivariate statistics**. Washington, D.C.: American Psychological Association.

Stevens, J. (2002). **Applied multivariate statistics for the social sciences** (4th ed.). Hillsdale, NJ: Lawrence Erlbaum Associates, Publishers.

Tabachnick, B. G. & Fidell, L. S. (1996). **Using multivariate statistics** (3rd ed.). New York: Harper Collins Publishers.

Tabachnick, B. G. & Fidell, L. S. (2001). **Using multivariate statistics** (4th ed.). Boston: Allyn & Bacon.

Weinfurt, K. P. (1995). Multivariate analysis of variance. In Grimm, L. G. & Yarnold, P. R. (Eds.) (pp. 245–276). **Reading and understanding multivariate statistics**. Washington, D.C.: American Psychological Association.

Wilks, S. S. (1932). Certain generalizations in the analysis of variance. **Biometrika**, 24, 471–494.

Endnotes

1. Sources are not in agreement with respect to the relationship between the magnitude of the correlation between the dependent variables and the power of the **multivariate analysis of variance**. In discussing the latter issue, Field (2005, p. 574), citing a study by Cole *et al.* (1994), notes that the power of the **multivariate analysis of variance** appears to depend on a combination of the correlation between the dependent variables and the magnitude of the effect size being measured.

2. The notation μ_k is employed by some sources to summarize the vector of the means of the kth group noted below, which is commonly referred to as the **mean vector** of the kth group.

$$\begin{bmatrix} \mu_{k1} \\ \mu_{k2} \\ \cdot \\ \cdot \\ \mu_{kp} \end{bmatrix}$$

Employing the notation μ_k for the mean vector of the kth group, the null hypothesis can be stated as follows: H_0: $\mu_1 = \mu_2 = \mu_3$.

3. Marascuilo and Levin (1983, pp. 179–181) state that **eigenvalues** (also referred to as **roots**) are most commonly computed for a **variance-covariance matrix**, and Harlow (2005, pp. 93–94) notes that eigenvalues are just a redistribution of the original variances of the variables. Marascuilo and Levin (1983, pp. 179–181) and Stevens (2002, p. 73) define the eigenvalues of **Matrix A** (which is always assumed to be a **square matrix**) as the solution to the determinantal equation $|A - \lambda I| = 0$ (where λ, which is the lower case Greek letter **lambda**, represents the value of an eigenvalue). In the latter equation **Matrix I** is an **identity matrix** with the same number of rows and columns as **Matrix A**. As demonstrated below, in the above determinantal equation an eigenvalue is multiplied by the identity matrix and after the product is subtracted from **Matrix A**, the determinant of the resulting matrix must equal zero. The number of eigenvalues derived for **Matrix A** will equal the number of rows or columns (where $r = c$) in **Matrix A**. To compute the eigenvalues for **Matrix A** the determinantal equation above is solved as noted below, where a, b, c, and d represent the four elements (i.e., four numerical values) that comprise **Matrix A** and λ represents the unknown eigenvalues for which the equation is solved. Since the dimensions of **Matrix A** are $r = c = 2$, the determinantal equation will be reduced to a polynomial equation (in this case a quadratic equation) with two solutions that will represent the two eigenvalues which are designated as λ_1 and λ_2.

The **trace** of **Matrix A** (the sum of all of the values in the diagonal — i.e., $a + d$) will be equal to the sum of its eigenvalues (which in the case of **Matrix A** will be $\lambda_1 + \lambda_2$). One should take note of the fact that the specific values in the diagonal of **Matrix A** (a and d) will not equal the values of each of the eigenvalues (λ_1 and λ_2). Thus, although $a \neq \lambda_1$ and $d \neq \lambda_2$, $a + d = \lambda_1 + \lambda_2$. As a general rule, the values derived for the eigenvalues will not be equal to one another and will all be positive values.

$$A = |A - \lambda I| = \begin{bmatrix} a & b \\ c & d \end{bmatrix} - \lambda \begin{bmatrix} 1 & 0 \\ 0 & 1 \end{bmatrix} = \begin{bmatrix} a & b \\ c & d \end{bmatrix} - \begin{bmatrix} \lambda & 0 \\ 0 & \lambda \end{bmatrix} = \begin{vmatrix} a - \lambda & b \\ c & d - \lambda \end{vmatrix} = 0$$

The equation $\begin{vmatrix} a - \lambda & b \\ c & d - \lambda \end{vmatrix} = 0$ is then solved yielding two values for λ.

4. The following *SPSS* command sequence was employed in this chapter in conducting a **multivariate analysis of variance**: a) Click **Analyze**; b) Click **General linear model**; c) Click **Multivariate**; d) Highlight the dependent variables (i.e., **anxiety** and **depression** in Example 35.1) and move them to the **Dependent Variables** window; e) Highlight the independent variable (i.e., the variable indicating which group a subject is in — **drug** versus **placebo** versus **no treatment** in Example 35.1) and move it to the **Fixed Factor(s)** window; f) Click **Options**, check off desired information (e.g., **Descriptive statistics**,

Estimates of effect size, etc.), and then click **Continue**; g) If you wish to conduct comparisons between pairs of means, click **Post Hoc**, highlight the independent variable in the left window (i.e., the variable indicating which group a subject is in) and move it to the **Post Hoc tests for** window, and then click **Continue**; h) Click **OK** to obtain the output for the analysis.

5.   a) Although it will not be necessary to consider the values printed in the upper section of Table 35.1 labeled **Intercept**, the latter information is derived through use of a **regression model** to evaluate a **multivariate analysis of variance**. As discussed in greater detail in Endnote 2 in the chapter on **canonical correlation (Test 38)**, all parametric statistical procedures can be summarized in the form of one or more linear equations, and in the final analysis a correlational format can be employed to summarize all such procedures (or, to put it another way, all parametric procedures can be conceptualized as special cases of regression analysis). It is through use of the latter type of analysis that the values in the **Intercept** section of Table 35.1 (as well as Table 35.2) are derived; b)Fields (2005) represents the best user friendly reference for those who are interested in employing *SPSS* to conduct statistical analysis. Although nowhere as comprehensive in their discussion of statistical tests as Fields (2005), George and Mallery (2001) is another good reference on *SPSS*. The author also recommends Mertler and Vannatta (2005) for readers interested in using *SPSS* for multivariate analysis.

6.   The relationship between F and Λ as described by Equation 34.6 is only applicable when there are $k = 2$ groups. It will not result in the correct value for F if the total sample size is substituted for $n_1 + n_2$ in the numerator of the latter equation along with the value computed for Λ.

7.   The fact that the analysis of covariance does not provide for a more sensitive test of the null hypothesis for the dependent variable of depression than does a simple analysis of variance on the depression scores is primarily due to the fact that the error variance computed for the analysis of covariance was not substantially different than the error variance computed for the analysis of variance. Additionally, Stevens (2002, pp. 345–347) notes that if the covariate is measured after the treatments (as is the case in this analysis) and it is influenced by the treatments, the change in the covariate may be correlated with the dependent variable, and under the latter circumstances part of the treatment effect will be removed.

Test 36

Multivariate Analysis of Covariance
(Parametric Test Employed with Interval/Ratio Data)

I. Hypothesis Evaluated with Test and Relevant Background Information

Hypothesis evaluated with test   Do k (where $k \geq 2$) independent samples represent k populations with different values with respect to their **mean vectors** (where a **mean vector** represents a composite score comprised of mean values on two or more dependent variables)?

Relevant background information on test   Prior to reading this section the reader should review the discussion of both the **single-factor between-subjects analysis of covariance (Test 21I)** (in Section IX (the **Addendum**) of the **single-factor between-subjects analysis of variance (Test 21)**) and the **multivariate analysis of variance (Test 35)**. In the discussion of the **single-factor between-subjects analysis of covariance** it is noted that the **analysis of covariance** is an analysis of variance procedure which employs a statistical adjustment involving regression analysis to control for the effect of one or more **extraneous variables** known as **covariates** (also referred to as **concomitant variables**) on a dependent variable. By utilizing the correlation between a covariate and the dependent variable, the researcher is able to remove variability on the dependent variable which is attributable to the covariate. The effect of the latter is a reduction in the error variability employed in computing the F ratio, thereby resulting in a more powerful test of the alternative hypothesis. A second potential effect of an analysis of covariance is that by utilizing the correlation between scores on a covariate and the dependent variable, the mean scores of the different groups can be adjusted for any pre-existing differences on the dependent variable which are present prior to the administration of the experimental treatments.

A **multivariate analysis of covariance** (for which the acronym **MANCOVA** is commonly employed) can be viewed as a procedure which combines the **multivariate analysis of variance** with an **analysis of covariance**. In its simplest form (specifically, its use with k independent samples) the **multivariate analysis of covariance** is the multivariate analog of the **single-factor between-subjects analysis of covariance**. Whereas one dependent variable is employed in the latter analysis of covariance, the **multivariate analysis of covariance** employs two or more dependent variables. As is the case in the **multivariate analysis of variance**, the **multivariate analysis of covariance** algebraically converts the mean scores of each group on p dependent variables into a composite group mean (referred to as a **mean vector** or **centroid**), and then evaluates whether or not there are differences among the groups with respect to their composite means. The composite means employed for the **multivariate analysis of covariance** are not the same as those which would be employed for a **multivariate analysis of variance**, since the latter procedure does not take into account the potential impact one or more covariates might have on the composite dependent variable. Tabachnick and Fidell (1996, p. 394) note that by virtue of including one or more covariates in the analysis, the composite means employed for a **multivariate analysis of covariance** represent the adjusted linear combination of the p dependent variables which would be obtained if all subjects had the same score on the covariate(s).

The assumptions underlying the **multivariate analysis of covariance** are the same as those for both the **multivariate analysis of variance** and the **analysis of covariance**. Additionally, a linear relationship is assumed between all pairs of dependent variables, all pairs of covariates, and all pairs involving a dependent variable and a covariate. As is the case for the univariate **analysis of covariance**, the **homogeneity of regression assumption** is still applicable — specifically, that within each of the k groups there is a linear correlation between the dependent variable and the covariate, and that the k group regression lines have the same slope (i.e., are parallel to one another). However, when two or more covariates are involved, the **homogeneity of regression assumption** becomes **homogeneity on regression planes** (for 2 covariates) or **hyperplanes** (for more than two covariates), due to the fact the analysis involves two or more dimensions.

The reader should take note of the fact that although the discussion of the **multivariate analysis of covariance** in this chapter will be limited to an independent samples design involving one independent variable and one covariate, the procedure can be employed to evaluate any experimental design involving two or more dependent variables and one or more covariates. Consequently, the **multivariate analysis of covariance** can be generalized to all varieties of experimental design such as dependent samples designs, factorial designs, etc.

II. Example

To illustrate the use of the **multivariate analysis of covariance**, consider Example 36.1 which is identical to Example 35.1 except for the fact it includes a covariate. The **covariate** is a measure of **social chaos** obtained on each subject (where social chaos is a measure of instability in a subject's home and/or familial environment). As is the case with Example 35.1, the two scores for each subject represent the two dependent variables of depression and anxiety, and the composite score for a subject is viewed as a measure of **emotional instability**. Within the context of the **multivariate analysis of covariance**, the group means for the latter composite dependent variable will be statistically adjusted to account for differences between subjects on the covariate of social chaos.

Example 36.1　*In order to assess the impact of a new antidepressant drug on emotional instability, 15 patients at a mental health clinic who are diagnosed with both clinical depression and excessive anxiety are randomly assigned to one of three groups. Five patients are assigned to Group 1, which is administered the antidepressant drug for a period of six months; five patients are assigned to Group 2, which is administered a placebo during the same six-month period; and five patients are assigned to Group 3, which is a no treatment group that is just evaluated at the beginning and end of the study. Assume that prior to introducing the experimental treatments, the experimenter confirmed that the levels of both depression and anxiety in the three groups were equal. After six months elapse all 15 subjects are rated by a psychiatrist (who is blind with respect to a subject's experimental condition) on both their level of depression and anxiety (which when taken together are considered to represent a generalized measure of emotional instability). Prior to conducting the study the researcher determined that social chaos represents a potentially confounding variable, since research indicates it is moderately correlated with both depression and anxiety. Because of the latter, prior to introducing the experimental treatments a measure of social chaos was obtained for each subject and employed as a covariate in the study.*

The psychiatrist's depression, anxiety, and social chaos ratings for the five subjects in each group are noted below. Each subject's depression score is listed first, followed by his or her anxiety score, followed by his or her social chaos score. The higher the score on depression, anxiety, and social chaos, the greater the degree of depression, anxiety, or social chaos: **Group 1: Subject 1** *(11, 8, 9);* **Subject 2** *(1, 3, 3);* **Subject 3** *(0, 2, 1);* **Subject 4** *(4, 1, 6);* **Subject 5** *(0,*

8, 4); **Group 2: Subject 1** (11, 9, 8); **Subject 2** (11, 6, 6); **Subject 3** (5, 7, 4); **Subject 4** (8, 6, 5); **Subject 5** (4, 9, 4); **Group 3: Subject 1** (12, 11, 6); **Subject 2** (8, 8, 5); **Subject 3** (14, 6, 7); **Subject 4** (8, 10, 8); **Subject 5** (11, 4, 3). *Do the data indicate that the drug is effective in decreasing emotional instability?*

III. Null versus Alternative Hypotheses

The null and alternative hypotheses evaluated with the **multivariate analysis of covariance** are identical to those evaluated with the **multivariate analysis of variance**, except for the fact that the k mean vectors are adjusted for the effect of the covariate. The latter adjustment is indicated by the asterisk (*) above the mean of each group for the specified dependent variable.

Null hypothesis

$$H_0: \begin{bmatrix} \mu_{11}^* \\ \mu_{12}^* \\ \cdot \\ \cdot \\ \mu_{1p}^* \end{bmatrix} = \begin{bmatrix} \mu_{21}^* \\ \mu_{22}^* \\ \cdot \\ \cdot \\ \mu_{2p}^* \end{bmatrix} = \cdots = \begin{bmatrix} \mu_{k1}^* \\ \mu_{k2}^* \\ \cdot \\ \cdot \\ \mu_{kp}^* \end{bmatrix}$$

Where: μ_{ij}^* represents the adjusted mean of the population represented by Group i on dependent variable j

(The null hypothesis states that the adjusted mean vectors on the p dependent variables for the k populations which are represented by the k samples are equal.)

Alternative hypothesis

$$H_1: H_0 \text{ is false}$$

(The alternative hypothesis is **nondirectional**, and implies a difference between at least two of the k adjusted mean vectors.)

Since Example 36.1 involves three groups and two dependent variables, the null and alternative hypotheses for the example are noted below.

$$H_0: \begin{bmatrix} \mu_{11}^* \\ \mu_{12}^* \end{bmatrix} = \begin{bmatrix} \mu_{21}^* \\ \mu_{22}^* \end{bmatrix} = \begin{bmatrix} \mu_{31}^* \\ \mu_{32}^* \end{bmatrix}$$

$$H_1: H_0 \text{ is false}$$

IV. Test Computations

The test statistics for the **multivariate analysis of covariance** are identical to those computed for the **multivariate analysis of variance**, except for the fact that adjusted matrices are employed to compute **Wilks' lambda, Pillai's trace, Hotelling's trace**, and **Roy's largest root**. Sources (e.g., Stevens (2002, p. 354), Mertler and Vannatta (2005, p. 143), and Tabachnick and

Fidell (1996, p. 399); 2001) note that for each of the aforementioned test statistics the sum of squares and cross-products matrices are adjusted for the effects of the covariate(s).

To illustrate, the equation for **Wilks' lambda** is summarized as $\Lambda^* = |W^*|/|T^*|$, where the asterisks denote the use of adjusted matrices. As noted in the discussion of the **multivariate analysis of variance**, in the latter equation, W represents a multivariate generalization of a within-subjects sum of squares (in the form of a **within-groups variance-covariance matrix**) and T represents a multivariate generalization of a total sum of squares (in the form of a **total variance-covariance matrix**, which is the **between-groups variance-covariance matrix** (designated B) plus the **within-groups variance-covariance matrix** — i.e., $T = B + W$).

Further discussion of the test statistics for the **multivariate analysis of covariance** can be found in Section V of the **multivariate analysis of variance** (which, as noted, employ the identical test statistics, except for the fact that the latter are not adjusted for covariate(s).)

V. Interpretation of the Test Results

In the discussion of the **single-factor between-subjects analysis of covariance** it was noted that in order for the latter procedure to provide effective statistical control the following two requirements must be met:

a) There must be a linear relationship between the dependent variable and the covariate(s) (since if they are not linearly related, the adjusted mean values computed for the k experimental conditions will be biased). With respect to the latter, sources generally recommend there be at least a moderate correlation (e.g., $r \geq .30$) between a covariate and the dependent variable.

In the case of Example 36.1, the correlation between subjects' scores on depression and social chaos was $r_{dep \, vs. \, chaos} = .67$, and between anxiety and social chaos was $r_{anx \, vs. \, chaos} = .52$. Both of the correlations are statistically significant. Specifically, employing **Table A16 (Table of Critical Values for Pearson r)** in the **Appendix**, the tabled critical two-tailed .05 and .01 levels of significance values for $df = n - 2 = 13$ are $r_{.05} = .514$ and $r_{.01} = .641$. The computed value $r_{dep \, vs. \, chaos} = .67$ is significant at the .01 level since it exceeds $r_{.01} = .641$, and $r_{anx \, vs. \, chaos} = .52$ is significant at the .05 level since it exceeds $r_{.05} = .514$.

b) A covariate should be independent of (i.e., not be influenced by) the experimental treatments/independent variable. The simplest way to evaluate the latter is to conduct an analysis of variance using the covariate as the dependent variable. Retention of the null hypothesis (i.e., H_0: Equality of the covariate treatment means) indicates independence between the covariate and the independent variable.

Table 36.1 Summary Table of Analysis of Variance on Covariate for Example 36.1

Source of variation	SS	df	MS	F
Between-groups	3.73	2	1.87	.35
Within-groups	63.20	12	5.27	
Total	66.93	14		

Table 36.1 is a summary table of an analysis of variance on the scores of subjects on the covariate of social chaos. Note that the number of degrees of freedom employed are $df_{BG} = k - 1 = 2$ and $df_{WG} = N - k = 12$, for which in **Table A10 (Table of the F Distribution)** in the **Appendix** $F_{.05} = 3.89$ and $F_{.01} = 6.83$. Since the computed value $F = .35$ is less than

$F_{.05}$ = 3.89, the null hypothesis (that the mean social chaos values for the populations represented by the three groups are equal) cannot be rejected. Thus, $F (2, 12) = .35$ ($p > .05$). Based on the latter, we can conclude there is no difference between the groups with respect to their mean scores on the covariate (which are $\overline{X}_1$ = 4.6, $\overline{X}_2$ = 5.4, $\overline{X}_3$ = 5.8), and that the covariate is independent of the experimental treatments.

With the above two requirements met, the analysis at this point for the most part involves conducting the same tests employed in evaluating Example 35.1 with a **multivariate analysis of variance**. The primary information derived with *SPSS* for the analysis of Example 36.1 is summarized in Tables 36.2 and 36.3.[1] Table 36.2, labeled **Multivariate Tests**, contains the results of evaluating the null hypothesis of no difference between the adjusted mean vectors of the three groups (which can be referred to as the omnibus null hypothesis). Our focus will be on the lower section of the table labeled **GP**, where in each row a value is recorded for one of the four test statistics computed for the **multivariate analysis of covariance** (which, as noted above, are the same test statistics computed for the **multivariate analysis of variance**).[2] As is the case in Table 35.1, Columns 2 through 7 contain values for the test statistic (**Value**), the associated *F* value, **Hypothesis df** = **df**$_{num}$, **Error df** = df_{den}, **Sig.**, and **Partial Eta Squared**. The *p* values associated with **Pillai's trace** (p = .035) and **Wilks' lambda** (p = .012) are less than .05, and the *p* values associated with **Hotelling's trace** (p = .005), and **Roy's largest root** (p = .001) are less than .01. Thus, the result of the **mulitvariate analysis of covariance** is statistically significant at either the .05 or .01 level, depending upon which test statistic one employs. Note that all four probability values are lower than the analogous *p* values computed in Table 35.1 (which summarizes the **multivariate analysis of variance** for Example 35.1). It would thus appear that inclusion of the covariate provided for a slightly more powerful test of the alternative hypothesis.

In order to understand the superior power of the **multivariate analysis of covariance** when contrasted with the **multivariate analysis of variance**, recollect that in the discussion of the **single-factor between-subjects analysis of covariance** it was noted that when one or more covariates (which are moderately correlated with the dependent variable) are employed in the latter procedure, it is more likely to yield a larger *F* value (and consequently be associated with a lower probability) than an analysis of variance on the same set of data. The superior power of the analysis of covariance can be attributed to lower error variability and/or larger between-groups variability when contrasted with the analysis of variance. The latter explanation is also applicable on a multivariate level in explaining why a **multivariate analysis of covariance** is often (but not always) more powerful than a **multivariate analysis of variance**.

Table 36.3, labeled **Tests of Between-Subjects Effects**, provides additional information about the analysis of Example 36.1. Specifically, in the section labeled **GP** (which represents **group** — i.e., the independent variable) there are two rows labeled **DEP** (for depression) and **ANX** (for anxiety). The values contained in each row represent for that dependent variable the **between-groups** information derived from a **single-factor between-subjects analysis of covariance** in which social chaos is employed as a covariate. The **within-groups** information for each of the dependent variables is contained in the section below labeled **Error**. Inspection of Table 36.3 reveals that when depression is the only dependent variable and social chaos a covariate, $F (2, 11) = 6.335$, which is significant at the .05 level (since $p = .015$ listed in Column 6 is less than.05). When anxiety is the only dependent variable and social chaos a covariate, $F (2, 11) = 1.808$, which is not significant (since $p = .209$ listed in Column 6 is greater than .05). The results of the two **single-factor between-subjects analyses of covariance** indicate that depression appears to be a more discriminative variable than anxiety in differentiating between

Table 36.2   *SPSS* Printout of Results for Example 36.1:
Omnibus Test Results for Multivariate Analysis of Covariance

Multivariate Tests[c]

Effect		Value	F	Hypothesis df	Error df	Sig.	Partial Eta Squared
Intercept	Pillai's Trace	.385	3.126[a]	2.000	10.000	.088	.385
	Wilks' Lambda	.615	3.126[a]	2.000	10.000	.088	.385
	Hotelling's Trace	.625	3.126[a]	2.000	10.000	.088	.385
	Roy's Largest Root	.625	3.126[a]	2.000	10.000	.088	.385
COVCHAOS	Pillai's Trace	.690	11.127[a]	2.000	10.000	.003	.690
	Wilks' Lambda	.310	11.127[a]	2.000	10.000	.003	.690
	Hotelling's Trace	2.225	11.127[a]	2.000	10.000	.003	.690
	Roy's Largest Root	2.225	11.127[a]	2.000	10.000	.003	.690
GP	Pillai's Trace	.724	3.123	4.000	22.000	.035	.362
	Wilks' Lambda	.291	4.265[a]	4.000	20.000	.012	.460
	Hotelling's Trace	2.380	5.356	4.000	18.000	.005	.543
	Roy's Largest Root	2.358	12.967[b]	2.000	11.000	.001	.702

a. Exact statistic

b. The statistic is an upper bound on F that yields a lower bound on the significance level.

c. Design: Intercept+COVCHAOS+GP

Table 36.3   *SPSS* Printout of Additional Results for Example 36.1

Tests of Between-Subjects Effects

Source	Dependent Variable	Type III Sum of Squares	df	Mean Square	F	Sig.	Partial Eta Squared
Corrected Model	DEP	220.264[a]	3	73.421	10.608	.001	.743
	ANX	54.812[b]	3	18.271	3.003	.077	.450
Intercept	DEP	3.091	1	3.091	.447	.518	.039
	ANX	24.928	1	24.928	4.097	.068	.271
COVCHAOS	DEP	80.664	1	80.664	11.654	.006	.514
	ANX	20.279	1	20.279	3.333	.095	.233
GP	DEP	87.690	2	43.845	6.335	.015	.535
	ANX	21.996	2	10.998	1.808	.209	.247
Error	DEP	76.136	11	6.921			
	ANX	66.921	11	6.084			
Total	DEP	1074.000	15				
	ANX	762.000	15				
Corrected Total	DEP	296.400	14				
	ANX	121.733	14				

a. R Squared = .743 (Adjusted R Squared = .673)

b. R Squared = .450 (Adjusted R Squared = .300)

the groups, and subsequently is more responsible for the significant results in the omnibus analysis of the multivariate dependent variable emotional instability summarized in Table 36.2.

In Section VII of the **multivariate analysis of variance** it was noted that the correlation between the two dependent variables of depression and anxiety was $r = .4075$, indicating a moderate degree of overlap between the two variables. Because of the latter, in addition to the two **single-factor between-subjects analyses of covariance** summarized above, the author conducted the following two additional analyses of covariance: a) A **single-factor between-subjects analysis of covariance** was conducted in which depression was the dependent variable with both social chaos and anxiety categorized as covariates. The latter analysis yielded the value $F(2, 10) = 7.92$, $p = .009$. Thus, when the correlation that both social chaos and anxiety have with depression is partialed out of the analysis, the result is significant at the .01 level; and b) A **single-factor between-subjects analysis of covariance** was conducted in which anxiety was the dependent variable with both social chaos and depression as covariates. The latter analysis yielded the value $F(2, 10) = 2.98$, $p = .097$. Thus, when the correlation that both social chaos and depression have with anxiety is partialed out of the analysis, the result falls short of being significant at the .05 level. The results of the latter two analyses of covariance merely reinforce the earlier conclusion that depression appears to be a more discriminative variable than anxiety in differentiating between the groups, and subsequently is more responsible for the overall significant result with respect to the multivariate dependent variable emotional instability (which is a linear combination of the two dependent variables of depression and anxiety).

As was the case with the **multivariate analysis of variance**, it is also possible to conduct planned or unplanned comparisons in which the adjusted mean values of two or more groups or combination of groups are compared with one another. Although they are not presented here, the latter comparisons reinforced the previous conclusion reached within the framework of the comparisons conducted for the **multivariate analysis of variance**. Specifically: a) That of the three comparisons the only one which strongly suggests the possibility of a significant difference is the comparison between the group which received the medication (Group 1) and the no treatment group (Group 3), with the former group exhibiting a lower level of emotional instability; and b) The most dramatic differences in the study were with respect to the dependent variable of depression between subjects who received the medication (Group 1) and subjects who received no treatment (Group 3) (with the former group exhibiting a lower level of depression). As was the case for the **multivariate analysis of variance**, it should be emphasized that the small sample size limited the power of the analysis, and consequently might have compromised its ability to identify other underlying population differences.

In closing this discussion, the reader should take note of the fact that there is additional analytical as well as graphical information associated with the **multivariate analysis of covariance** which has not been included in this section — for example, computer generated plots of the data can provide the researcher with valuable information regarding assumption violation, presence of outliers, etc. Those requiring more extensive documentation of such procedures should consult books which specialize in multivariate statistics.

VI. Additional Analytical Procedures for the Multivariate Analysis of Covariance and/or Related Tests

No additional procedures will be described in this section.

VII. Additional Discussion of the Multivariate Analysis of Covariance

1. Multiple covariates A researcher may elect to employ more than one covariate in a study. With regard to the latter, it is important that covariates are not highly correlated with one another, since, as Stevens (2002, pp. 345–346) notes, highly correlated covariates will be removing much of the same error variance from the dependent variable (and consequently will not substantially reduce error variability). If, on the other hand, the correlation between two covariates is minimal, they will each remove relatively distinct portions of error variability, and the latter will result in a substantially lower error variance for a **multivariate analysis of covariance**. In addition, Harlow (2005, p 67) notes that when multiple covariates are employed, since each covariate uses up one degree of freedom, the more degrees of freedom used up by the covariates, the less powerful will be the test of the null hypothesis. In such a case the **multivariate analysis of covariance** may be no more powerful (or perhaps even less powerful) than a **multivariate analysis of variance** involving just the independent variable.[3]

VIII. Additional Examples Illustrating the Use of the Multivariate Analysis of Covariance

No additional examples will be presented in this section.

References

Grimm, L. G. & Yarnold, P. R. (Eds.) (1995). **Reading and understanding multivariate statistics**. Washington, D.C.: American Psychological Association.

Harlow, L. L. (2005). **The essence of multivariate thinking**. Mahwah, NJ: Lawrence Erlbaum Associates.

Huitema, B. (1980). **The analysis of covariance and alternatives**. New York: John Wiley & Sons.

Mertler, C. A. & Vannatta, R. A. (2005). **Advanced and multivariate statistical methods** (3rd ed.). Los Angeles: Pyrczak Publications.

Stevens, J. (2002). **Applied multivariate statistics for the social sciences** (4th ed.). Hillsdale, NJ: Lawrence Erlbaum Associates, Publishers.

Tabachnick, B. G. & Fidell, L. S. (1996). **Using multivariate statistics** (3rd ed.). New York: Harper Collins Publishers.

Tabachnick, B. G. & Fidell, L. S. (2001). **Using multivariate statistics** (4th ed.). Boston: Allyn & Bacon.

Weinfurt, K. P. (1995). Multivariate analysis of variance. In Grimm, L. G. & Yarnold, P. R. (Eds.) (pp. 245–276). **Reading and understanding more multivariate statistics**. Washington, D.C.: American Psychological Association.

Endnotes

1.   The following *SPSS* command sequence was employed in this chapter in conducting a **multivariate analysis of covariance**: a) Click **Analyze**; b) Click **General linear model**; c) Click **Multivariate**; d) Highlight the dependent variables (i.e., **anxiety** and **depression** in Example 36.1) and move them to the **Dependent Variables** window; e) Highlight the independent variable (i.e., the variable indicating which group a subject is in — **drug** versus **placebo** versus **no treatment** in Example 36.1) and move it to the **Fixed Factor(s)** window; f) Highlight the covariate (i.e., **social chaos** in Example 36.1) and move it to the **Covariate(s)** window; g) Click **Options**, check off desired information (e.g., **Descriptive statistics, Estimates of effect size**, etc.), and then click **Continue**; h) Click **OK** to obtain the output for the analysis.

2.   Although it will not be necessary to consider the values printed in the sections of Tables 36.2 and 36.2 labeled **Intercept**, the latter information is derived through use of a

regression model to evaluate a **multivariate analysis of covariance**. As discussed in greater detail in Endnote 2 in the chapter on **canonical correlation (Test 38)**, all parametric statistical procedures can be summarized in the form of one or more linear equations, and in the final analysis a correlational format can be employed to summarize all such procedures (or, to put it another way, all parametric procedures can be conceptualized as special cases of regression analysis). It is through use of the latter type of analysis that the values in the **Intercept** section of Tables 36.2 and 36.3 are derived.

3. Stevens (2002, p. 346) notes that when a researcher elects to employ more than one covariate, Huitema (1980, p. 161) recommends limiting the number of covariates such that the ratio [*Number of covariates* + (*Number of groups* – 1)] / *Total sample size* is less than .10.

Test 37

Discriminant Function Analysis
(Parametric Test Employed with Interval/Ratio Data)

I. Hypothesis Evaluated with Test and Relevant Background Information

Discriminant function analysis is not an inferential statistical test, but instead is a methodology for group classification. Specifically, one or more linear equations (referred to as **discriminant functions**) are derived which can be employed to assign a subject to two or more groups based on the subject's scores on a set of predictor variables. It should be noted that although some sources define a multivariate procedure as one in which there are two or more dependent variables, **discriminant function analysis**, which involves one dependent variable (that is evaluated with respect to multiple independent/predictor variables), is generally categorized as a multivariate procedure.

Hypothesis evaluated with test The primary hypothesis evaluated within the framework of **discriminant function analysis** is whether or not a specific discriminant function contributes significantly with respect to differentiating between the k groups.

Relevant background information on test **Discriminant function analysis** is a statistical procedure which derives equations that are designed to predict group membership (which represents the dependent/criterion variable) employing a set of quantitative measures (i.e., continuous variables measured on an interval/ratio scale) to represent the independent/predictor variables. The criterion variable in **discriminant function analysis** is a discrete/qualitative variable which is comprised of two or more categories (e.g., breast cancer survivors versus breast cancer fatalities; religious affiliation, ethnic category, etc.). Silva and Stam (1995, p. 279) note that the k groups/categories employed in **discriminant function analysis** should be **mutually exclusive** (i.e., not overlap one another) and **collectively exhaustive** (i.e., each subject can be definitively classified in one of the k categories), and that the k groups should be defined based on objective criteria. The equations derived in **discriminant function analysis** (which are called **discriminant functions** or **canonical discriminant functions**) are similar to those in **regression analysis**, in that in both procedures the equations are linear combinations of predictors which are correlated with a dependent variable. The major difference between the two procedures is that whereas the criterion variable in **regression analysis** is measured on an interval/ratio scale, the criterion variable in **discriminant function analysis** is categorical.

    Discriminant function analysis addresses the same questions which are evaluated with the **multivariate analysis of variance (Test 35)**. However, in the latter procedure group membership serves as the independent variable, and the multiple quantitative measures represent the dependent variables. If within the context of a **multivariate analysis of variance** a significant difference is found between the groups, it reflects the fact that the dependent variables are reliable predictors of group membership. Whereas in the **multivariate analysis of variance** the focus is on whether k groups differ significantly with respect to differences on p dependent variables (which are expressed within the framework of a composite mean), in **discriminant function analysis** the focus is on whether scores on p predictor variables can be combined in linear combinations which allow a researcher to reliably differentiate between the k groups (and on an individual level, predict in which of k groups a subject is most likely to be a member). If the omnibus test for a **multivariate analysis of variance** is statistically significant, it would indicate there is a significant relationship between at least one linear combination of the predictor variables and the grouping variable — i.e., that at least one discriminant function can be derived from the data.

To illustrate the application of **discriminant function analysis**, assume a researcher wants to categorize people into one of two groups (which will represent the dependent/criterion variable) — those who have had a silent heart attack and those who have not. The categorization with respect to group would be based on equations that employ subjects' scores on the following four independent/predictor variables: a) Cholesterol level; b) Diastolic blood pressure; c) Body fat ratio; and d) Age. The equations derived in the **discriminant function analysis** (i.e., the **discriminant functions**) will be linear combinations of the predictors which are correlated with the criterion variable (i.e., whether or not a subject has had a silent heart attack).

Sources often differentiate between **descriptive discriminant function analysis** and **predictive discriminant function analysis**. The purpose of **descriptive discriminant function analysis** is to identify p predictor variables which optimally discriminate k groups (where $k \geq 2$) from one another. Initially, a researcher identifies what he or she considers potential predictor variables in differentiating the k groups from one another, after which **linear combinations** of the selected p predictor variables are derived, which as noted above are referred to as **discriminant functions**. Within the context of the analysis, the relative importance of each of the p predictor variables is assessed for each the derived discriminant functions.

The purpose of **predictive discriminant function analysis** is to derive a rule which will allow a researcher to employ a person's (or object's) scores on p predictor variables to predict which of k groups the person (or object) is most likely to be a member. The latter rule can be expressed in the form of a set of linear equations referred to as **classification functions**. Two criteria commonly employed for deriving a rule are its ability to **minimize the probability of misclassifying a person (or object)** and/or its ability to **minimize the cost associated with misclassification**. **Predictive discriminant function analysis** is sometimes the last step in evaluating a body of data which was initially evaluated with a **multivariate analysis of variance** (which yielded a significant result) and a **descriptive discriminant function analysis**.

As is the case for **multiple regression (Test 33)**, the following three strategies can be employed for **discriminant function analysis: standard, hierarchical**, and **stepwise**.

In **standard discriminant function analysis** (also called **simultaneous** or **direct discriminant function analysis**) all of the predictor variables are entered into the analysis simultaneously (including those which may only contribute minimally with respect to group separation), and employed in deriving the discriminant functions.

In **hierarchical discriminant function analysis** (also called **sequential discriminant function analysis**) a researcher uses preexisting empirical data/information or some model/theory to determine the order in which each of the predictor variables is entered into the analysis. The order in which the predictor variables are entered is dictated by the degree to which they are believed to influence group separation. Predictors are entered into the analysis beginning with the predictor variable thought to be most important, and subsequently entering the remaining variables in descending order of importance. Within the context of **hierarchical discriminant function analysis**, a researcher can determine whether the addition of each new predictor variable contributes significantly to group separation above and beyond the information available from previously entered predictor variables. Tabachnick and Fidell (1996, p. 529) note that **hierarchical discriminant function analysis** can be useful when a researcher is interested in deriving a smaller subset of predictor variables.

In **stepwise discriminant function analysis** (which will also generally result in a smaller subset of predictor variables) statistical criteria are employed in entering the predictor variables.

As is the case with **multiple regression**, in **stepwise analysis** the order in which the variables are entered into the analysis is based on some statistical rule — most commonly, which predictor variable is most highly correlated with the grouping variable. Three types of **stepwise analysis** are **forward selection, backward selection**, and **stepwise selection**. In **forward selection** predictor variables are entered sequentially based on the magnitude of their correlation with the grouping variable. Thus, the predictor variable that is most highly correlated with grouping variable is entered first, and followed in the statistically designated order by the other predictors until an entered variable no longer contributes significantly toward group separation. The resulting discriminant functions will only include only those predictor variables which significantly contribute toward group separation. In **backward selection** all of the predictor variables are initially evaluated. The predictor variable contributing least to group separation (i.e., the variable with the lowest absolute correlation with the grouping variable) is eliminated from the analysis. The evaluation continues by sequentially eliminating each variable which does not contribute toward group separation, until only those variables that are significantly related to the grouping variable remain. **Stepwise selection** combines forward and backward selection by employing a stepwise methodology to include predictor variables which add the most (and significantly) toward group separation and exclude predictors which contribute the least (and not significantly) toward group separation.

It should be emphasized that in order for the results of a **discriminant function analysis** to be of practical value, a researcher must ultimately demonstrate that the discriminant functions derived in the analysis are applicable not only to the sample from which they were derived, but also to other members of the target population to which one wants to generalize the results. This latter process, which is referred to as **cross-validation**, is discussed in Section V.

The assumptions for **discriminant function analysis** are identical to those underlying the **multivariate analysis of variance** (i.e., multivariate normality (in this case, of the independent variables), homogeneity of the variance-covariance matrices, linear relationships between all predictor variables within each group, and absence of multicollinearity and outliers on the independent variables). As a general rule, **discriminant function analysis** is reasonably robust with respect to assumption violation, especially with a large sample and an equal number of subjects per group. If assumption violation is salient and cannot be corrected through use of data transformation, a researcher can employ **logistic regression (Test 39)**, which is a **nonparametric** analog of **discriminant function analysis**.

With regard to sample size, Stevens (2002, p. 289) cites simulation studies which suggest that unless the ratio of the total number of subjects employed (N) to the number of predictor variables (p) is very large (e.g., 20 to 1), a researcher should use caution in interpreting the results of a **discriminant function analysis** (since the values computed for the **standardized coefficients** and **canonical loadings** (discussed in Sections IV and V) may not be reliable). As is the case in **multiple regression**, in **discriminant function analysis**, **overfitting** (i.e., when the information derived from a sample is tailored to the idiosyncrasies of the sample, yet will not provide a good fit for the target population to which the researcher wishes to generalize) can also occur when the number of predictor variables employed is large relative to sample size, resulting in a loss of predictive accuracy (i.e., **shrinkage**).

II. Examples

Examples 37.1 and 37.2 will be employed to illustrate **discriminant function analysis**.

Example 37.1 *A study is conducted at a mental health clinic to evaluate whether or not the type of treatment psychologically disturbed adolescents receive is related to the following three* *measures which have been demonstrated to be related to psychological adjustment: depression, anxiety, and social chaos (where social chaos is a measure of instability in a subject's home and/or familial environment). The following three comparable groups of patients (randomly assigned to treatment conditions) are employed in the study:* Group 1 *is comprised of 5 subjects who for six months receive family therapy for their psychological problems (i.e., treatment sessions involve the patient as well as members of his or her immediate family);* Group 2 *is comprised of 5 subjects who for a period of six months receive individual psychotherapy;* Group 3 *is comprised of 5 subjects who do not receive any treatment for their psychological problems.*

After six months elapse, each subject is rated by a team of experienced clinicians (who are blind with respect to subjects' group membership) on depression, anxiety, and social chaos. The scores on the three aforementioned dependent variables are noted below. Each subject's depression score is listed first, followed by his or her anxiety score, followed by his or her social chaos score. The higher the a subject's score on any of the variables, the greater the degree of depression, anxiety, or social chaos: **Group 1: Subject 1** (6, 8, 9); **Subject 2** (4, 3, 3); **Subject 3** (0, 2, 8); **Subject 4** (4, 1, 6); **Subject 5** (0, 8, 4); **Group 2: Subject 1** (11, 9, 8); **Subject 2** (11, 6, 6); **Subject 3** (5, 7, 4); **Subject 4** (8, 6, 5); **Subject 5** (4, 9, 4); **Group 3: Subject 1** (12, 11, 6); **Subject 2** (8, 8, 5); **Subject 3** (9, 6, 7); **Subject 4** (8, 10, 8); **Subject 5** (11, 4, 3).

The intent of the study is to answer the following questions: a) Is there a difference between the three groups with respect to depression, anxiety, and social chaos, if the latter three variables are conceptualized as a multivariate dependent variable; b) Can linear combinations of scores on the three measures be identified which significantly differentiate between the three groups, and if so, how accurately can the latter combinations predict group membership?

Example 37.2   *A cardiologist conducts a study to evaluate the relationship between four risk factors (which will be labeled* **Risk factor 1**, **Risk factor 2**, **Risk factor 3**, *and* **Risk factor 4**) *and whether or not a 50 year African-American male will have a heart attack. The cardiologist has access to a database which contains scores on the four risk factors obtained for 100,000 African-American males when they were 40 years old. The cardiologist forms the following four groups: Group 1 is comprised of 20 men (randomly selected from a pool of 2,500 men in the database) who had a fatal heart attack prior to the age of 50; Group 2 is comprised of 20 men (randomly selected from a pool of 5,000 men in the database) who had a serious but nonfatal heart attack prior to the age of 50; Group 3 is comprised of 20 men (randomly selected from a pool of 10,000 men in the database) who had a mild heart attack prior to the age of 50; Group 4 is comprised of 20 men (randomly selected from a pool of 82,500 men in the database) who did not have a heart attack prior to the age of 50. The range of values on which a subject can score on each of the risk factors follows, with the higher a score the more likely a subject is assumed to be at risk for a heart attack.* **Risk factor 1:** 0–20; **Risk factor 2:** 0–30; **Risk factor 3:** 0–40; **Risk factor 4:** 0–50. *The raw data for* Example 37.2 *are presented in* Table 37.4 *in* Section V.

The intent of the study is to answer the following questions: a) Is there a difference between the four groups with respect to the four risk factors, if the risk factors are conceptualized as a multivariate dependent variable; b) Can linear combinations of scores on the four risk factors be identified which significantly differentiate between the four groups, and if so, how accurately can the latter combinations predict group membership?

III. Null versus Alternative Hypotheses

The primary null hypothesis evaluated within the framework of **discriminant function analysis** is assessed for each of the discriminant functions derived in the analysis. The **null hypothesis** evaluated for a discriminant function states that function in question does not contribute significantly with respect to differentiating between the k groups. The **alternative hypothesis** for each discriminant function is that the function in question does contribute significantly with respect to differentiating between the k groups. The analysis only retains those discriminant functions for which the null hypothesis has been rejected.[1]

IV. Test Computations

Prior to reading this section the reader should review the discussion of the **Wilks' lambda statistic** employed to evaluate the omnibus hypothesis for the **multivariate analysis of variance** (which is discussed in Section IV of the latter test), since the latter statistic is also employed as a measure of significance in **discriminant function analysis**. As is the case for the **multivariate analysis of variance**, the matrix algebra operations involved in **discriminant function analysis** focus on the ratio of the **between-groups variance-covariance matrix** versus the **within-groups variance-covariance matrix**. Since the algebraic operations (which are documented in sources such as Huberty (1994), Marascuilo and Levin (1983) and Tabachnick and Fidell (1996, 2001)) are generally implemented through use of computer software, they will not be described here. The intent of the discussion in this section (as well as the next) is to provide the reader with a succinct description of the underlying basis for **discriminant function analysis**, along with an explanation of the summary information generated by computer software.

Within the context of **discriminant function analysis** q discriminant functions are derived — where the value of q is the smaller of the two values p versus $(k - 1)$. Each of the q discriminant functions will be **orthogonal** (i.e., independent) of the others, and will involve a unique weighting of the p predictor variables. The order in which the q discriminant functions are derived will be directly related to how important the unique combination of the p predictor variables in a given discriminant function is with respect to separating the k groups from one another. Thus, the first discriminant function will represent the unique combination of the p predictor variables that is most important with respect to providing discrimination among the k groups, the second discriminant function will be the second most important, and so on.[2]

A **Wilks' lambda** value is computed for each of the q derived discriminant functions. In previous chapters it was noted that **lambda** can be converted into either an F value or a chi-square value (which is what *SPSS* converts **lambda** into in **discriminant function analysis**). If at least one of the q discriminant functions (specifically, the first of the derived functions) contributes significantly toward distinguishing between the k groups, the first value computed for **lambda** (as well as the associated chi-square statistic) will be statistically significant. If the second value computed for **lambda** is statistically significant it would indicate that the second of the derived discriminant functions contributes significantly toward distinguishing between the groups. This assessment process is repeated until a value for **lambda** has been computed for each of the q discriminant functions. Once a **lambda** value is nonsignificant, any subsequent values will also be nonsignificant.[3]

In **discriminant function analysis** q (where $q \geq 1$) linear combinations of the p predictor variables are derived such that the ith linear combination can be described by Equation 37.1. Each of the derived linear combinations is referred to as a **discriminant function**.[4]

$$V_i = a + b_1 X_1 + b_2 X_2 + \dots + b_q X_q \qquad \textbf{(Equation 37.1)}$$

Where: V_i defines the ith **discriminant function**, and the value computed for V_i will represent a subject's score on the ith discriminant function

a represents the **discriminant function constant**

b_j represents the **unstandardized coefficient** (also referred to as an **unstandardized weight** or **eigenvector weight**) for the jth predictor variable

X_j represents a subject's score on the jth predictor variable

It is important to note that for a given discriminant function, the relative absolute magnitude of the **unstandardized discriminant coefficients** in Equation 37.1 (i.e., b_i values) do not indicate the importance of each predictor variable, since the latter coefficients are usually based on different scales of measurement. Because of this, **standardized discriminant coefficients** are computed to determine the relative importance of each of the predictor variables. The larger the absolute value of a standardized discriminant coefficient, the more prominent the role of that predictor variable in defining the discriminant function in question. Although typically calculated by the computer, standardized coefficients can be computed by multiplying the unstandardized coefficients by the within-groups standard deviation of the predictor variables (Silva and Stam (1995, p. 293)). Through use of the standardized coefficients, a **standardized discriminant function** can be derived which has the general form of Equation 37.2 (Tabachnick and Fidell (1996, p. 517)). Note that unlike Equation 37.1, Equation 37.2 does not include a constant.

$$Z_i = d_1 z_1 + d_2 z_2 + ... + d_q z_q \qquad \textbf{(Equation 37.2)}$$

Where: Z_i defines the ith **discriminant function**, and the value computed for Z_i will represent a subject's standardized score on the ith discriminant function

d_j represents the **standardized coefficient** (or **standardized weight**) for the jth predictor variable

z_j represents a subject's standardized score on the jth predictor variable

Within the framework of **discriminant function analysis**, a second set of measures, referred to as **canonical loadings**, are also computed which provide information regarding the relative importance of each of the predictor variables. The latter measure, along with the other values described above will be discussed in greater detail in the next section.

As noted in Section I, the goal of **predictive discriminant function analysis** is to derive a rule which will allow a researcher to employ a person's scores on p predictor variables to predict which group the person is most likely to be a member. One way of assigning a subject whose group membership is unknown to a group is through graphic analysis. In order to do the latter the following must be done: a) The p scores of each subject are substituted in each of the q discriminant functions. As a result of doing the latter a discriminant function score (designated V_i in Equation 37.1) can be computed for each subject on each of the q discriminant functions; b) For each of the k groups, a mean value is computed for each of the q discriminant functions. Thus, the mean of the ith discriminant function in the mth group will be $\bar{V}_{im} = \Sigma V_{im} / n_{im}$; c) The scores on the p predictor variables of the subject to be classified are represented in a vector, and the latter vector is evaluated in reference to k group mean vectors (also referred to as **centroids**). Each group mean vector contains the mean scores on the p predictor variables for one of the k groups. The subject in question is assigned to the group whose mean vector is in closest physical proximity to the vector containing that subject's scores. When there are two discriminant functions the latter distance can be assessed with a two-dimensional scatterplot, but when there are more than two functions distance is assessed on a multidimensional plane (generally through use of the **Mahalanobis D^2** statistic).

Within the context of **predictive discriminant function analysis**, another set of coefficients referred to as **classification function coefficients** can be computed, and the latter coefficients are commonly employed in classifying subjects with respect to group membership. The computation of **classification function coefficients** will be described in the next section.

V. Interpretation of the Test Results

Examples 37.1 and 37.2 will both initially be analyzed with a **multivariate analysis of variance**, after which the data will be evaluated with a **standard discriminant function analysis.**[5] Readers should take note of the fact that in the case of Example 37.1, a small sample size was employed for illustrative purposes, and in reality, if a researcher wished to conduct a reliable analysis a substantially larger sample size would have to be employed.

Analysis of Example 37.1 The data were initially evaluated with a **multivariate analysis of variance**. In conjunction with the analysis, descriptive statistics and intercorrelations between the three predictor variables were computed. The latter information is presented below followed by a summary of the information generated by *SPSS* for the **multivariate analysis of variance** (which is summarized in Table 37.1).

a) **Descriptive statistics for the predictor variables** The following values were computed for each of the groups on the three predictor variables (where $\bar{X}_j$ represents the mean of the jth group on the specified variable): 1) Depression: $\bar{X}_1 = 2.8$, $\tilde{s}_1 = 2.68$, $\bar{X}_2 = 7.8$, $\tilde{s}_2 = 3.27$, $\bar{X}_3 = 9.6$, $\tilde{s}_3 = 1.82$; 2) Anxiety: $\bar{X}_1 = 4.4$, $\tilde{s}_1 = 3.36$, $\bar{X}_2 = 7.4$, $\tilde{s}_2 = 1.52$, $\bar{X}_3 = 7.8$, $\tilde{s}_3 = 2.86$; 3) Social chaos: $\bar{X}_1 = 6.0$, $\tilde{s}_1 = 2.55$, $\bar{X}_2 = 5.4$, $\tilde{s}_2 = 1.67$, $\bar{X}_3 = 5.8$, $\tilde{s}_3 = 1.92$. In addition, the means and standard deviations for all 15 subjects on the three predictor variables were computed to be: $\bar{X}_{dep} = 6.73$, $\tilde{s}_{dep} = 3.86$, $\bar{X}_{anx} = 6.53$, $\tilde{s}_{anx} = 2.95$, $\bar{X}_{chaos} = 5.73$, $\tilde{s}_{chaos} = 1.94$.

b) **Intercorrelations between the predictor variables** The following intercorrelations were computed between subjects' scores on the three predictor variables: $r_{dep\,vs.\,anx} = .39$, $r_{dep\,vs.\,chaos} = .12$, $r_{anx\,vs.\,chaos} = .19$. None of the intercorrelations are statistically significant at the .05 level. Although, in and of itself, the latter is not sufficient to rule out **multicollinearity**, formal assessment of the latter resulted in **tolerance** values of .846, .828, and .962 respectively for the predictor variables of depression, anxiety, and social chaos. Since all of the obtained values are close to the maximum possible value of 1, there is no evidence of **multicollinearity**. (The **tolerance** statistic for **multicollinearity** is discussed in Section IV/V of the chapter on **multiple regression**.)

c) **Levene's test of equality of error variance (Test 21g)** This test evaluates whether or not the homogeneity of variance assumption has been violated with respect to each of the predictor variables. The results of **Levene's test** for Example 37.1 are presented in Table 37.1. Specifically: 1) Depression: $F(2, 12) = 1.452$, $p = .272$; 2) Anxiety: $F(2, 12) = 3.126$, $p = .081$; 3) Social chaos: $F(2, 12) = .638$, $p = .545$. Since none of the probability values is less than .05, the homogeneity of variance assumption is not violated for any of the variables.

d) **Box's test of equality of covariance matrices** The results of **Box's test** for Example 37.1 are presented in Table 37.1. Since the obtained probability value $p = .377$ is greater than .05, the result is not significant. Consequently, the homogeneity of covariance matrices assumption was not violated.

e) **Multivariate analysis of variance** Inspection of the table labeled **Multivariate Tests** (in the section labeled **GP**) in Table 37.1 reveals that the probabilities for the values computed for **Pillai's trace**, **Wilks' lambda**, **Hotelling's trace**, and **Roy's largest root** are respectively .098, .036, .017 and .003. Since the probabilities for **Wilks' lambda** and **Hotelling's trace** are less than .05, they are significant at the .05 level. **Roy's largest root** is significant at the .01

Table 37.1   *SPSS* Printout of Results of Multivariate Analysis of Variance for Example 37.1

Box's Test of Equality of Covariance Matrices[a]

Box's M	20.776
F	1.076
df1	12
df2	697.846
Sig.	.377

Tests the null hypothesis that the observed covariance matrices of the dependent variables are equal across groups.

a. Design: Intercept+GP

Levene's Test of Equality of Error Variances[a]

	F	df1	df2	Sig.
DEP	1.452	2	12	.272
ANX	3.126	2	12	.081
CHAOS	.638	2	12	.545

Tests the null hypothesis that the error variance of the dependent variable is equal across groups.

a. Design: Intercept+GP

Multivariate Tests[c]

Effect		Value	F	Hypothesis df	Error df	Sig.	Partial Eta Squared
Intercept	Pillai's Trace	.948	61.273[a]	3.000	10.000	.000	.948
	Wilks' Lambda	.052	61.273[a]	3.000	10.000	.000	.948
	Hotelling's Trace	18.382	61.273[a]	3.000	10.000	.000	.948
	Roy's Largest Root	18.382	61.273[a]	3.000	10.000	.000	.948
GP	Pillai's Trace	.722	2.073	6.000	22.000	.098	.361
	Wilks' Lambda	.292	2.840[a]	6.000	20.000	.036	.460
	Hotelling's Trace	2.382	3.572	6.000	18.000	.017	.544
	Roy's Largest Root	2.361	8.658[b]	3.000	11.000	.003	.702

a. Exact statistic

b. The statistic is an upper bound on F that yields a lower bound on the significance level.

c. Design: Intercept+GP

Tests of Between-Subjects Effects

Source	Dependent Variable	Type III Sum of Squares	df	Mean Square	F	Sig.	Partial Eta Squared
Corrected Model	DEP	124.133[a]	2	62.067	8.783	.004	.594
	ANX	34.533[b]	2	17.267	2.376	.135	.284
	CHAOS	.933[c]	2	.467	.108	.899	.018
Intercept	DEP	680.067	1	680.067	96.236	.000	.889
	ANX	640.267	1	640.267	88.110	.000	.880
	CHAOS	493.067	1	493.067	113.785	.000	.905
GP	DEP	124.133	2	62.067	8.783	.004	.594
	ANX	34.533	2	17.267	2.376	.135	.284
	CHAOS	.933	2	.467	.108	.899	.018
Error	DEP	84.800	12	7.067			
	ANX	87.200	12	7.267			
	CHAOS	52.000	12	4.333			
Total	DEP	889.000	15				
	ANX	762.000	15				
	CHAOS	546.000	15				
Corrected Total	DEP	208.933	14				
	ANX	121.733	14				
	CHAOS	52.933	14				

a. R Squared = .594 (Adjusted R Squared = .526)

b. R Squared = .284 (Adjusted R Squared = .164)

c. R Squared = .018 (Adjusted R Squared = -.146)

level, since $p = .003$ is less than .01. **Pillai's trace** falls short of significance at the .05 level. The aforementioned results suggest a strong possibility of differences between two or more of the underlying populations with respect to their mean vectors (i.e., centroids). (It should be noted that by employing a small sample size the power of the analysis was compromised, thus making it more difficult to obtain a significant result.)

Comparisons involving pairs of mean vectors employing **Hotelling's T^2 (Test 34)** (i.e., a **multivariate analysis of variance** with $k = 2$) yielded the following values for **Wilks' lambda**: **Group 1** versus **Group 2** ($\Lambda = .333$; $F (3, 6) = 4.006$; $p = .070$); **Group 1** versus **Group 3** ($\Lambda = .244$; $F (3, 6) = 6.198$; $p = .029$); **Group 2** versus **Group 3** ($\Lambda = .852$; $F (3, 6) = .347$; $p = .793$). Note that none of the probability values is less than .0167 (i.e., $.05/3 = .0167$), and thus none achieves significance if the overall Type I error rate is set at .05 (employing the **Bonferroni–Dunn** adjustment). If, for the moment, we assume all of the comparisons were planned and a per comparison .05 level of significance is employed, the only significant result is for Group 1 versus Group 3, since $p = .029$ is less than .05. Thus, of the three comparisons the one which is most indicative of a significant difference is the comparison between Group 1 (Family therapy) versus Group 3 (No therapy), although the comparison of Group 1 versus Group 2 (Individual therapy) approaches significance at the .05 level (when the per comparison significance level is .05).

In order to better assess the relative impact of the predictor variables when considered separately, three **single-factor between-subjects analyses of variance (Test 21)** were conducted. The results of the latter analyses are summarized in the lower part of Table 37.1 labeled **Tests of Between-Subjects Effects**. In the section labeled **GP** (which represents **group**) there are three rows labeled **DEP** (for depression), **ANX** (for anxiety) and **CHAOS** (for social chaos). The values contained in each row represent for that predictor variable the **between-groups** information derived from the **single-factor between-subjects analysis of variance**. The **within-groups** information for each of the predictor variables is contained in the section labeled **Error**. Inspection of the **Tests of Between-Subjects Effects** table reveals that when depression is the only predictor variable, $F (2, 12) = 8.783$, which is significant at the .01 level (since $p = .004$ listed in Column 6 is less than .01). When anxiety is the only predictor variable, $F (2, 12) = 2.376$, which is not significant (since $p = .135$ listed in Column 6 is greater than .05). When social chaos is the only predictor variable, $F (2, 12) = .108$, which is not significant (since $p = .899$ listed in Column 6 is greater than .05). The results of the three **single-factor between-subjects analyses of variance** indicate that without question depression (which is the only variable that yields a significant result) appears to be the most discriminative of the predictor variables in differentiating between the groups, and subsequently is probably most responsible for the significant result obtained for the omnibus **multivariate analysis of variance**.

In conjunction with the above noted **single-factor between-subjects analyses of variance**, comparisons (employing **Fisher's LSD test (Test 21a)**, **Tukey's HSD test (Test 21d)** and the **Scheffé test (Test 21e)**) were also conducted on each of the predictor variables. The only predictor variable for which significant differences were detected was depression. Specifically, the differences between the depression scores of Group 1 versus Group 2 and Group 1 versus Group 3 were significant at the .05 level (for all three comparison procedures).

In conclusion, the analysis of the data with a **multivariate analysis of variance** and related procedures strongly suggests that the scores of Group 1 (Family therapy) were significantly lower (indicating superior psychological adjustment) than the scores of both Groups 2 and 3. In addition, it appears that differences with respect to depression primarily account for the overall significant effect detected with the **multivariate analysis of variance**.

At this point the data will be evaluated with a **discriminant function analysis**. Since the number of **discriminant functions** that can be derived (q) will be the smaller of the two values *p* versus $(k - 1)$, and in Example 37.1, $(k - 1) = (3 - 1) = 2$ and $p = 3$, $q = 2$ discriminant functions will be derived. The results of the **discriminant function analysis** are summarized in Tables 37.2 and 37.3, as well as in Figure 37.1. Initially the information in Table 37.2, which is comprised of six separate tables, will be discussed.

a) **The Eigenvalues table** This table contains information reflecting the degree to which each of the derived discriminant functions contributes meaningfully toward differentiating between the $k = 3$ groups. The magnitude of the eigenvalues 2.361 and .020 respectively recorded in Column 2 (labeled **Eigenvalue**) indicates the relative importance of each function with respect to its ability to differentiate between the groups. The fact that 2.361 is substantially larger than .020 suggests that **Function 1** plays a much more important role in differentiating the groups from one another than **Function 2**.

Each of the values in Column 3 (labeled **% of Variance**) was obtained by dividing the eigenvalue for the function in a row by the sum of all of the eigenvalues. The percentage value recorded for a function represents the proportion of between-groups variability which can be explained by that function. Thus, for **Function 1**, $2.361/(2.361 + .020) = 99.1\%$, and for **Function 2**, $.020/(2.361 + .020) = .9\%$.

The last column labeled **Canonical Correlation** contains the correlation between the function listed in a row and the grouping variable. The value of a canonical correlation can never be negative and will always fall between 0 and 1 (Diekhoff (1992, p. 297)). The square of the value recorded for a canonical correlation is a measure of effect size (typically equated with the **eta squared** (η^2) statistic), which indicates the proportion of variability associated with a function that can be attributed to the grouping variable. The larger the absolute value of the canonical correlation, the stronger the relationship between the function and the grouping variable (i.e., the better able the function is in discriminating between the k groups). In the case of **Function 1**, the square of the canonical correlation .838 is .702, with the latter value indicating that about 70% of the variability associated with the function can be attributed to the grouping variable. In the case of **Function 2**, the square of the canonical correlation .141 is .020, indicating that only 2% of the variability associated with the latter function can be attributed to the grouping variable. The information in the **Eigenvalues** table clearly indicates that of the two functions, only **Function 1** appears to be related to the grouping variable.

It should be noted that when more than one function is derived in **discriminant function analysis** (the only time one function will be derived is when there are $k = 2$ groups), it is common for just one or two of the functions to be significant and thus account for the overwhelming majority of the discriminating power between groups. The amount of discriminating power attributable to the nonsignificant functions will be trivial, and thus those functions will not be employed in classifying subjects. In the next paragraph it will be noted that in the case of Example 37.1 **Function 2** will be determined to be nonsignificant, and thus will not be employed in classifying subjects.

b) **The Wilks' lambda table** The discriminatory power of each of the q discriminant functions is more formally evaluated through use of the **Wilks' lambda** statistic (which is discussed in Section IV of the **multivariate analysis of variance**). The testing of the discriminatory power of the q derived functions is done sequentially beginning with **Function 1** (which, as noted above, is most strongly related to the grouping variable) and continuing with the other functions in descending order of importance. If at any point a nonsignificant result is obtained for a function, the analysis is terminated, since any additional functions derived in the analysis will also yield a nonsignificant result. The value computed for **Wilks' lambda** (which is converted into a chi-square value) reflects the degree to which a function differentiates between the groups after the effects of any previously evaluated functions have been removed.

Table 37.2 *SPSS* Printout of Discriminant Analysis for Example 37.1

Summary of Canonical Discriminant Functions

Eigenvalues

Function	Eigenvalue	% of Variance	Cumulative %	Canonical Correlation
1	2.361[a]	99.1	99.1	.838
2	.020[a]	.9	100.0	.141

a. First 2 canonical discriminant functions were used in the analysis.

Wilks' Lambda

Test of Function(s)	Wilks' Lambda	Chi-square	df	Sig.
1 through 2	.292	13.556	6	.035
2	.980	.221	2	.895

Canonical Discriminant Function Coefficients

	Function 1	Function 2
DEP	.354	.132
ANX	.213	-.211
CHAOS	-.233	.384
(Constant)	-2.439	-1.708

Unstandardized coefficients

Standardized Canonical Discriminant Function Coefficients

	Function 1	Function 2
DEP	.942	.351
ANX	.574	-.569
CHAOS	-.485	.798

Structure Matrix

	Function 1	Function 2
DEP	.785*	.596
ANX	.408*	-.354
CHAOS	-.054	.738*

Pooled within-groups correlations between discriminating variables and standardized canonical discriminant functions

Variables ordered by absolute size of correlation within function.

* Largest absolute correlation between each variable and any discriminant function

Functions at Group Centroids

GP	Function 1	Function 2
1.00	-1.909	.034
2.00	.640	-.170
3.00	1.269	.136

Unstandardized canonical discriminant functions evaluated at group means

Table 37.3   *SPSS* Printout of Classification Analysis for Example 37.1

Classification Results[b,c]

		GP	Predicted Group Membership			Total
			1.00	2.00	3.00	
Original	Count	1.00	5	0	0	5
		2.00	0	3	2	5
		3.00	0	3	2	5
	%	1.00	100.0	.0	.0	100.0
		2.00	.0	60.0	40.0	100.0
		3.00	.0	60.0	40.0	100.0
Cross-validated[a]	Count	1.00	2	3	0	5
		2.00	0	3	2	5
		3.00	0	4	1	5
	%	1.00	40.0	60.0	.0	100.0
		2.00	.0	60.0	40.0	100.0
		3.00	.0	80.0	20.0	100.0

a. Cross validation is done only for those cases in the analysis. In cross validation, each case is classified by the functions derived from all cases other than that case.

b. 66.7% of original grouped cases correctly classified.

c. 40.0% of cross-validated grouped cases correctly classified.

Classification Function Coefficients

	GP		
	1.00	2.00	3.00
DEP	.136	1.012	1.276
ANX	.346	.932	1.001
CHAOS	1.208	.536	.507
(Constant)	-5.676	-9.941	-12.595

Fisher's linear discriminant functions

Note that in Column 1 (labeled **Test of Function(s)**) of the **Wilks' lambda table** in the row labeled **1 through 2**, the value .292 computed for **Wilks' lambda** is statistically significant at the .05 level, since $p = .035$ (in Column 5) is less than .05.[6] This result indicates significant group differentiation using both **Functions 1** and **2**. In the row labeled **2**, the probability $p = .895$ associated with the value $\Lambda = .980$ is not significant (since it exceeds .05). The latter result indicates that **Function 2** does not contribute significantly toward group differentiation, and that the significant result in the first row of the table was almost completely attributable to the role of **Function 1**. The information contained in the **Wilks' lambda** table can also be construed as indicating that one underlying linear function accounts for the significant result obtained for the **multivariate analysis of variance** conducted on Example 37.1.

c) **Canonical discriminant function coefficients, standardized canonical discriminant function coefficients, and structure matrix tables**   Computer software generally lists two measures which allow the researcher to examine the degree of relationship between the predictor variables and each of the discriminant functions. The first of these measures are the **standardized discriminant function coefficients** which are recorded in the table labeled **Standardized Canonical Discriminant Function Coefficients**. As noted in Section IV, the relative absolute magnitude of the **unstandardized discriminant coefficients** (which are displayed in the table labeled **Canonical Discriminant Function Coefficients**) do not indicate the importance of the predictor variables, since, as a general rule, the latter coefficients are computed for predictor variables which are evaluated on different scales of measurement. Because of the this the **standardized discriminant coefficients** are computed to determine the relative importance of each of the predictor variables. The larger the absolute value of a standardized discriminant coefficient, the more prominent the role of that predictor variable in defining the discriminant function in question.[7]

The second measure which allows the researcher to examine the degree of relationship between the predictor variables and each of the discriminant functions is recorded in the table labeled **Structure Matrix**. In the latter table, the values recorded in each column are referred to as **canonical loadings** (or, alternatively, as **canonical variate correlation coefficients**, **discriminant loadings**, or **structure correlations/coefficients**). These values represent the correlation the predictor variable in a given row has with the function represented in each of the columns. Harlow (2005, p. 136) notes that the square of a canonical loading represents the proportion of shared variance between a predictor variable and a function, or to say it another way, the greater the absolute value of a canonical loading the larger the proportion of variability on a function which is accounted for by that variable — and consequently, the greater the role played by the predictor variable in defining that function. The sign of a canonical loading reflects the direction of the relationship for a predictor variable with respect to the function in question.[8] Hair *et al.* (1995, p. 221) and Harlow (2005, p. 136) note that although not formally evaluated with a test of statistical significance, as a general rule, a predictor variable with a canonical loading with an absolute value equal to or greater than .30 is viewed as making a worthwhile contribution to a discriminant function. (Tabachnick and Fidell (1996, p. 540) employ the value $r = .33$, for which the value of r^2 indicates would account for approximately 10% of the variance.) Tabachnick and Fidell (1996, p. 540) state, however, that researchers should employ caution in interpreting a canonical loading (which is a zero-order correlation), since in reality the actual correlation could be lower because the canonical loading does not partial out the correlation of a specific variable with the other predictor variables.

Although the canonical loadings in the **Structure Matrix** table are designed to measure the same thing as the previously noted standardized discriminant function coefficients, the two measures are not always consistent with one another (i.e., they may differ with respect to the relative absolute values computed for the predictor variables). Inconsistencies between the two measures makes it more difficult for a researcher to arrive at a straightforward interpretation of the analysis (i.e., to determine which predictor variables are, in fact, most strongly related to a given function). Stevens (2002, p. 288–289) cites research which suggests that when there are inconsistencies, it is better to employ the canonical loadings, since (especially with smaller samples) the latter measures are more stable than the standardized coefficients (although with very small sample sizes the canonical loadings will also be unstable). Hair *et al.* (1995, p. 206) note that preference for use of canonical loadings can also, in part, be attributed to the fact that although a small standardized coefficient may indicate the predictor variable in question is minimally related to a function, it could also be the result of the fact the variable has been partialed out of the analysis due to a high degree of multicollinearity. To put it another way, the absolute value of a standardized coefficient may underestimate the role of a predictor variable, since it reflects the **unique** contribution of that variable and does not take into account its **joint** contribution with the other predictor variables. (The later criticism is also applicable to standardized coefficients computed for regression analysis.)

At this point in the discussion, the values computed for the **standardized discriminant function coefficients** and **canonical loadings** for Example 37.1 will be examined. The latter values will only be considered for **Function 1**, since the analysis with **Wilks' lambda** indicated that **Function 2** did not contribute significantly toward group separation. Examination of the absolute values of the standardized coefficients computed for **Function 1** reveals that the predictor variable with the highest absolute value is depression (.942), followed by anxiety (.574) and social chaos (.485). The negative sign associated with social chaos indicates that group separation is determined by subtracting the information for the latter variable from the information on depression and anxiety. The ordering of the canonical loadings with respect to absolute values on **Function 1** in the **Structure Matrix** table is depression (.785), anxiety (.408) and social chaos (.054).[9] Note that the value .054 computed for social chaos is below the generally accepted minimal level for a canonical loading of .30. The squares of the canonical loadings $((.785)^2 = .616, (.408)^2 = .166,$ and $(.054)^2 = .003)$ reflect the proportion of variance on each of the predictor variables accounted for by the function. Note that the ordering of the predictor variables from highest to lowest for both the standardized coefficients and canonical loadings is depression, followed by anxiety and social chaos. The latter ordering (along with the squared canonical loading of .616 for depression) strongly indicates that depression plays the most prominent role in defining **Function 1**, followed by anxiety and lastly by social chaos (which does not make a significant contribution).

    d) Functions at group centroids This table contains for each of the discriminant functions the values of the centroids (i.e., composite means) for each of the k groups. Groups which have centroid values with different signs for a specific function are usually discriminated from one another by that function. Thus, the fact the value -1.909 is computed for Group 1's centroid and the centroids of Groups 2 and 3 equal .640 and 1.269 strongly suggests that **Function 1** may differentiate Group 1 from the latter two groups (plus the fact the value -1.909 is more distant from 1.269 than .640 suggests the differentiation is more the result of Group 3's centroid being further removed from Group 1's than Group 2's).

    The same information contained in the **Functions at Group Centroids** table is also provided in Figure 37.1, which is a plot of the group centroids in reference to the two discriminant functions (we will, however, only concern ourselves with **Function 1**, since **Function 2** was determined to be nonsignificant). Note the substantial horizontal distance between Group 1's centroid (represented by the number **1** enclosed within the axes) in relation to the centroids of Groups 2 and 3 (represented by the numbers **2** and **3** enclosed within the axes). In addition, the centroids of Groups 2 and 3 are relatively close to one another. In the graph the value on the abscissa (i.e. X axis) a group centroid falls directly above corresponds to the value recorded for that group on **Function 1** in the **Functions at Group Centroids** table. The value on the ordinate (i.e., Y axis) which is directly below a group centroid on the vertical plane corresponds to the value recorded for that group on **Function 2**. The fact that **Function 2** was not significant is further reinforced by viewing the vertical plane of the graph and noting the minimal vertical distances between the three group centroids. In the final analysis, the information provided by Figure 37.1 and the **Functions at Group Centroids** table indicates that **Function 1** appears to be effective in discriminating whether a subject should be categorized in Group 1 (family therapy) versus Group 2 (individual therapy) or Group 3 (no therapy), but is not good in discriminating whether a subject not categorized in Group1 should be categorized in Group 2 or Group 3.

    e) Discriminant function scores A score can be computed for each of the N subjects employed in a **discriminant function analysis** on each of the q derived discriminant functions. The latter score is computed by substituting a subject's scores on the three predictor variables in Equation 37.1 (Equation 37.2 can also be employed if one elects to compute standardized scores). To illustrate the latter with respect to Example 37.1, we will employ Equation 37.1 to compute a score on **Function 1** for Subject 1 in Group 1. The information required to derive the equation for **Function 1** is in the table of the unstandardized coefficients (i.e., the table labeled **Canonical Discriminant Function Coefficients**). The latter table indicates that in the linear

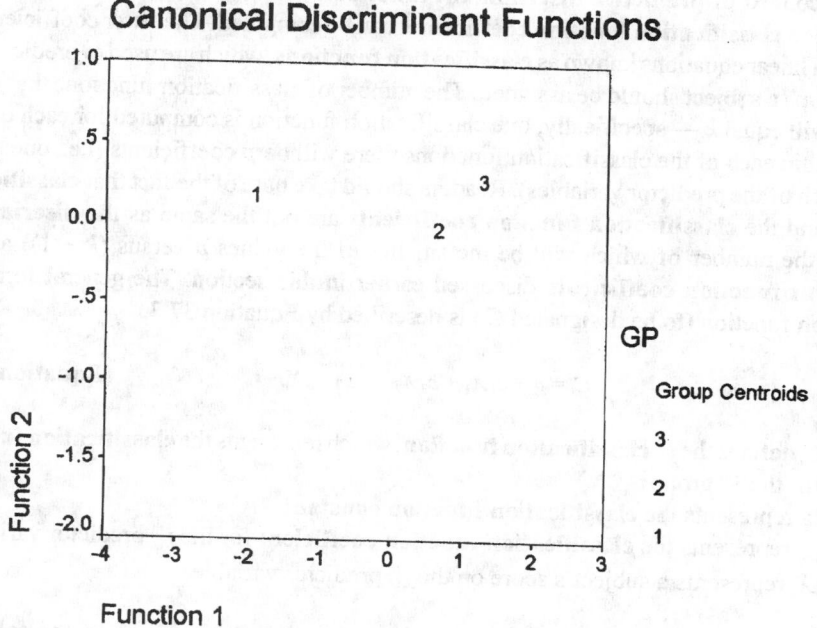

Canonical Discriminant Functions

Figure 37.1 *SPSS* **Graph of Group Centroids on Functions for Example 37.1**

equation for **Function 1**, the constant equals –2.439 and the unstandardized coefficients for depression (which will represent X_1 in Equation 37.1), anxiety (which will represent X_2), and chaos (which will represent X_3) are respectively .354, .213, and –.233. Equation 37.1 for **Function 1** is presented below.

$$V_1 = -2.439 + .354 \, X_1 + .213 \, X_2 - .233 \, X_3$$

When the scores on the predictor variables for Subject 1 in Group 1 are substituted in the above equation the value $V_1 = -2.439 + (.354)(6) + (.213)(8) + (-.233)(9) = -.71$ is computed. Thus, the discriminant function score for Subject 1 in Group 1 on **Function 1** is –.71. This process can be repeated for the remaining 14 subjects. Upon computing a score for each of the 15 subjects on **Function 1**, a mean for each of the $k = 3$ groups comprised of $n = 5$ subjects per group can be computed. In point of fact, the latter means are the group centroid values (–1.909 for Group 1, .640 for Group 2, and 1.269 for Group 3) in the column for **Function 1** in the **Functions of Group Centroid** table.[10] Classification of subjects is predicated on the degree of fit between a subject's scores on the discriminant function(s) determined to be significant and the configuration of group centroids for the latter discriminant function(s). Degree of fit can be assessed mathematically, as well as geometrically (employing two dimensional space when there are two discriminant functions and multidimensional space when there are more than two).

f) **Classification function coefficients** As noted in Section IV, when the intent of **discriminant function analysis** is to use the data for predictive purposes additional analytical procedures are implemented. **Predictive discriminant function analysis** involves the derivation of a classification rule which is most commonly derived through use of the sample data.[11] The accuracy of the classification rule will depend upon the degree to which the sample employed in the analysis is representative of the target population to which the rule is intended to be applied.

Within the context of **predictive discriminant function analysis**, another set of coefficients referred to as a **classification function coefficients** can be computed. The latter coefficients are employed in linear equations known as **classification functions**, which are used to predict which of the k groups a subject should be assigned. The number of classification functions that can be computed will equal k — specifically, one classification function is computed for each of the k groups. Within each of the classification functions there will be p coefficients (i.e., one coefficient for each of the predictor variables). Readers should take note of the fact that **classification functions** and the **classification function coefficients** are not the same as the **discriminant functions** (the number of which will be the smaller of the values p versus $(k - 1)$) and the **discriminant function coefficients** discussed earlier in this section. The general form of a classification function (to be designated C_i) is described by Equation 37.3.

$$C_i = g + c_1 X_1 + c_2 X_2 + ... + c_q X_q \qquad \textbf{(Equation 37.3)}$$

Where: C_i defines the ith **classification function**, which represents the **classification function** for the ith group

g represents the **classification function constant**

c_j represents the **classification function coefficient** for the jth predictor variable.

X_j represents a subject's score on the jth predictor variable

In order to predict membership for a subject whose group is unknown the following protocol is employed: a) k classification scores (i.e., C_i values) are computed for the subject— in other words, one classification score for each of the k groups. Each classification score is computed by substituting a subject's scores on the p predictor variables in Equation 37.3; b) The subject is assigned to the group which yields the largest C_i value.

The classification function coefficients are displayed in Table 37.3 in the lower table labeled **Classification Function Coefficients**. Note that for each group a classification function coefficient is listed for each of the predictor variables along with a classification constant for the group. To illustrate how classification functions are used to classify a subject, Equation 37.3 is employed below to compute the group classification function scores for Subject 1 in Group 1. In the latter equation, the values employed for c_1, c_2, and c_3 respectively represent the classification coefficients recorded in Table 37.3 for depression, anxiety, and social chaos for the group represented in a given column. In each of the equations the values 6, 8, 9, are employed for X_1, X_2, and X_3, since they respectively represent the scores of Subject 1 in Group 1 on depression, anxiety, and social chaos.

$$C_1 = -5.676 + .136 X_1 + .346 X_2 + 1.208 X_3 = -5.676 + (.136)(6) + (.346)(8) + (1.208)(9) = 8.78$$

$$C_2 = -9.941 + 1.012 X_1 + .932 X_2 + .536 X_3 = -9.941 + (1.012)(6) + (.932)(8) + (.536)(9) = 8.411$$

$$C_3 = -12.595 + 1.276 X_1 + 1.001 X_2 + .507 X_3 = -12.595 + (1.276)(6) + (1.001)(8) + (.507)(9) = 7.632$$

Since the largest of the three values computed is $C_1 = 8.78$, the subject would be classified in Group 1. If the same procedure is repeated for the remaining 14 subjects, the resulting classifications will correspond to those displayed in the upper part (labeled **Original**) of the table labeled **Classification Results** in Table 37.3. The table labeled **Classification Results** represents a **classification matrix** (which is sometimes referred to as a **confusion matrix**). The upper part of the latter matrix (labeled **Original**) summarizes the actual group membership of the N subjects in the sample in relation to the group to which each subject was assigned based on the subject's scores on the discriminant functions derived in the analysis. The diagonal of the matrix contains all correct classifications, while the off-diagonal cells contain incorrect classifications.

It was noted in Section I that in order for the results of a **discriminant function analysis** to be of practical value a researcher must ultimately demonstrate the discriminant functions derived in the analysis are applicable not only to the sample from which they were derived, but also to other members of the target population to which one wants to generalize the results, and that this process is referred to as **cross-validation**. Thus, in the case of **predictive discriminant function analysis**, although the percentage of correct classifications may be extremely high for the original sample, the researcher must demonstrate that the q derived discriminant functions can be employed with a comparable degree of accuracy in classifying subjects who were not members of the original sample. A common method employed for cross-validation is to divide the original sample of N subjects in half, and conduct a **discriminant function analysis** on $N/2$ subjects, after which the accuracy of the model is evaluated on the remaining $N/2$ subjects (often referred to as the **holdout sample**).[12] An alternative cross-validation methodology employed in computer analysis is **jackknifing** (which is discussed in Section IX (the **Addendum**) of the **Mann–Whitney U test (Test 12)**). Within the context of **discriminant function analysis** the jackknifing protocol omits the data for each of the subjects one at a time and classifies the latter subject based on the discriminative information generated for the remaining $(N - 1)$ subjects. This procedure generates the lower part of the **Classification Results** matrix (labeled **Cross-validated**).[13] Note that 10 of the 15 subjects are classified correctly in the **Original** part of the **Classification Results** table, but only 6 of the 15 in the **Cross-validated** part of the table (which is barely above chance, since if subjects were randomly assigned to a group the expected number of correct classifications would be $(15)(1/3) = 5$).

Among others, Hair *et al.* (1995, p. 205) note that the discriminatory power of a classification matrix can be evaluated with **Press's Q statistic**, which is computed with Equation 37.4. The latter statistic is evaluated with the chi-square distribution with $df = 1$.

$$Press's\ Q = \frac{[N - (n_c)(k)]^2}{N(k - 1)} \qquad \textbf{(Equation 37.4)}$$

Where: N represents the total sample size
n_c represents the number of correct classifications
k represents the number groups

Hair *et al.* (1995, p. 205) note that **Press's Q statistic** should be used with caution since it is a function of sample size, such that the larger the sample size the lower the classification rate which will be declared significant. Once again it should be emphasized that in order for a **discriminant function analysis** to be employed for predictive purposes, the derived discriminant functions should be based on a large sample which is representative of the target population to which the researcher wishes to generalize. Obviously, in the case of Example 37.1 the use of a small sample would not give one reason to believe that the derived discriminant function could be used reliably with other members of the population from which the sample was drawn.

Naming a discriminant function Within the framework of **discriminant function analysis** a researcher may elect to assign a name to any functions which are determined to be statistically significant. This is done by carefully examining the predictor variables that are highly correlated with the function and assigning a name which captures whatever it is the predictors share in common with one another. For example, if four predictor variables such as depression, anxiety, obsessive thinking, and somatic symptoms all had a high correlation on a function, the researcher might elect to refer to the function as *neuroticism*, since all four of the variables might be viewed as components of the latter trait. Citing Tatsuoka (1973), Stevens (2002, p. 289) states that when the ordering of the predictors on the canonical loadings versus the standardized discriminant coefficients are inconsistent with one another, it is preferable to employ the canonical loadings in interpreting the meaning of the discriminant functions (e.g., using the content of the strongest predictor variables in naming a function), while the standardized coefficients are more useful in determining which of the variables are redundant (specifically, standardized coefficients which are close to zero would be viewed as redundant).

   g) **Stepwise analysis of Example 37.1**   In addition to the **standard discriminant function analysis** described above, a **stepwise analysis** was also conducted. The latter analysis derived only one discriminant function for which depression was the only predictor variable determined to be significantly related to the grouping variable. The equation for the function derived in the stepwise analysis was $V_1 = -2.533 + .376\,X_1$. By employing a subject's score for depression as the value of X_1 in the latter equation, the researcher can compute the score of each subject on the function. Examination of the results of the stepwise analysis revealed that accuracy of classification was about the same when depression was employed as a single predictor variable with the above noted discriminant function when compared with the use of the three predictor variables for **Function 1** derived in the standard analysis. (In the **Classification Results** table for the **stepwise analysis**, 8 of the 15 subjects were classified correctly in both the **Original** and **Cross-validated** parts of the table.) In the final analysis, the overall results for the **stepwise analysis** led to the same conclusion derived from the standard analysis — specifically, that a subject's score on depression appeared to be most clearly related to whether the subject was treated with family therapy (Group 1) as opposed to either individual therapy (Group 2) or no therapy (Group 3).

   A reason commonly cited for employing stepwise analysis is that it can reduce a large number of predictor variables into a smaller subset of predictors. Various sources (e.g., Silva and Stam (1995, p. 288), Stevens (2002, p. 297) and Tabachnick and Fidell (1996, p. 532)), however, note that the subset of predictors may be an artifact of sampling, and not reflect the relative importance of the variables in the underlying population, and thus not be optimal with respect to predictive accuracy. Because of the latter, cross-validation is always recommended for stepwise analysis (although, as noted earlier, cross-validation is always recommended if a researcher intends to generalize one's results beyond a sample, regardless of the method of **discriminant function analysis** employed).[14] Since in stepwise analysis (especially in the case of small sample sizes) the results of the significance tests on the derived discriminant functions tend to be biased in the direction of eliminating predictor variables that may play a significant role, sources recommend researchers employ a higher alpha value (e.g., .10 in lieu of .05) to insure inclusion of all the predictor variables which may be related to group separation. Stevens (2002, p. 297) states stepwise analysis should be used with caution, especially if the ratio N/p is ≤ 5. Huberty (1994, p. 132, p. 227, pp. 261–262) takes the extreme position of arguing against the use of **stepwise discriminant function analysis**, especially when it is employed for predictive purposes. Hair *et al.* (1995, Ch. 4), who provide comprehensive illustrations of stepwise analysis, note it is especially subject to bias if multicollinearity is present with respect to the predictor variables.

   h) **Discriminant function analysis versus factor analysis**   Although the functions derived in **discriminant function analysis** are conceptually similar to the factors derived in **factors analysis** (as well as the fact that the naming of functions is similar to the naming of factors), **discriminant function analysis** and **factor analysis** (which is discussed in the chapter entitled **principal components analysis and factor analysis (Test 40)**) differ mathematically. More specifically, the functions derived in **discriminant function analysis** maximize group differences, whereas the factors in factor analysis maximizes the variance for each of the variables that can be attributed to a given factor.

Hair *et al.* (1995, p. 207) and Stevens (2002, p. 296) note a researcher can elect to **rotate** (typically through use of a **varimax rotation**) those discriminant functions which are statistically significant in order to redistribute the variance, and as a result of the latter facilitate their interpretation (specifically, more clearly delineate the relationship between each function and the groups it is best able to discriminate from one another). The concept of **rotation** is discussed in detail in the chapter on **factor analysis**. At this point all that will be said regarding rotation is that discriminant functions can be represented geometrically as well as mathematically, and rotation is a procedure which involves rotating geometrical axes that serve as reference points for identifying the functions. Among other things, rotation may alter the original ordering of the functions, and thus the first derived function may no longer maximally discriminate between the *k* groups. Although Hair *et al.* (1995, pp. 233–237) provide examples illustrating the use of rotation, Tabachnick and Fidell (1996, p.540) do not recommend that it be used by novice researchers.

Analysis of Example 37.2   Table 37.4 contains the scores of the 80 subjects on the four risk factors which were initially evaluated with a **multivariate analysis of variance**. In conjunction with the latter analysis, descriptive statistics and intercorrelations between the four predictor variables were computed. The latter values are summarized in Table 37.5. The results of the **multivariate analysis of variance** are presented in Table 37.6.

a) **Descriptive statistics for the predictor variables** The means and standard deviations for the four groups on the four risk factors are displayed in the table labeled **Descriptive Statistics** in Table 37.5. Cursory visual inspection of the mean values suggests group differences on all four risk factors.

b) **Intercorrelations between the predictor variables** The intercorrelations between the four risk factors are displayed in the table labeled **Correlations** in Table 37.5. Note that all of the risk factors are positively correlated with one another, and all of the intercorrelations are statistically significant at the .01 level (which indicates that for any pair of risk factors, the researcher can reject the null hypothesis that the underlying population correlation between that pair of risk factors equals zero). Formal assessment of multicollinearity yielded tolerance values of .752, .739, .596, and .628 respectively for Risk factors 1, 2, 3, and 4. Since all of the obtained values are close to the maximum possible value of 1, there is no evidence of multicollinearity.

c) **Levene's test of equality of error variance** The results of **Levene's test** are displayed in the table labeled **Levene's test of Equality of Error Variances** in Table 37.6. Since none of the probability values (in the column labeled **Sig.**) is less than .05, the homogeneity of variance assumption is not violated for any of the predictor variables.

d) **Box's test of equality of covariance matrices** The results of **Box's test** are displayed in the table labeled **Box's test of Equality of Covariance Matrices** in Table 37.6. Since the obtained probability value $p = .659$ (in the row labeled **Sig.**) is greater than .05, the result is not significant. Consequently, the homogeneity of covariance matrices assumption was not violated.

e) **Multivariate analysis of variance** Inspection of the table labeled **Multivariate Tests** (in the section labeled **Group**) in Table 37.6 reveals that the probabilities (in the column labeled **Sig.**) for the values computed for **Pillai's trace**, **Wilks' lambda**, **Hotelling's trace**, and **Roy's largest root** are all .000 (which indicates a probability less than .00049). Since all the probability values are less than .01, they are significant at the .01 level. The aforementioned results indicates differences between two or more of the underlying populations with respect to their mean vectors (i.e., centroids).

Table 37.4 Raw Data for Example 37.2

subject	group	risk1	risk2	risk3	risk4
77	4.00	9.00	13.00	1.00	11.00
78	4.00	5.00	5.00	18.00	14.00
79	4.00	10.00	5.00	2.00	4.00
80	4.00	4.00	14.00	14.00	2.00

subject	group	risk1	risk2	risk3	risk4
39	2.00	12.00	14.00	18.00	14.00
40	3.00	16.00	20.00	19.00	22.00
41	3.00	10.00	14.00	2.00	12.00
42	3.00	11.00	13.00	12.00	30.00
43	3.00	14.00	5.00	22.00	30.00
44	3.00	15.00	14.00	24.00	33.00
45	3.00	13.00	5.00	17.00	27.00
46	3.00	16.00	5.00	18.00	18.00
47	3.00	16.00	15.00	14.00	21.00
48	3.00	10.00	10.00	4.00	2.00
49	3.00	3.00	16.00	14.00	19.00
50	3.00	8.00	18.00	14.00	9.00
51	3.00	12.00	23.00	1.00	18.00
52	3.00	11.00	28.00	10.00	23.00
53	3.00	8.00	8.00	7.00	28.00
54	3.00	11.00	19.00	3.00	24.00
55	3.00	6.00	29.00	11.00	22.00
56	3.00	15.00	19.00	20.00	22.00
57	3.00	19.00	28.00	13.00	19.00
58	3.00	13.00	10.00	7.00	40.00
59	3.00	2.00	10.00	9.00	26.00
60	3.00	17.00	9.00	3.00	28.00
61	4.00	4.00	10.00	13.00	12.00
62	4.00	11.00	9.00	2.00	18.00
63	4.00	12.00	9.00	.00	.00
64	4.00	15.00	14.00	11.00	26.00
65	4.00	9.00	10.00	7.00	1.00
66	4.00	16.00	5.00	17.00	14.00
67	4.00	9.00	6.00	13.00	8.00
68	4.00	6.00	12.00	4.00	8.00
69	4.00	12.00	2.00	7.00	33.00
70	4.00	10.00	13.00	10.00	40.00
71	4.00	12.00	14.00	11.00	3.00
72	4.00	14.00	12.00	12.00	3.00
73	4.00	14.00	9.00	6.00	18.00
74	4.00	13.00	5.00	20.00	6.00
75	4.00	3.00	8.00	21.00	7.00
76	4.00	7.00	3.00	3.00	2.00

subject	group	risk1	risk2	risk3	risk4
1	1.00	12.00	15.00	37.00	44.00
2	1.00	18.00	30.00	34.00	34.00
3	1.00	20.00	23.00	37.00	49.00
4	1.00	19.00	26.00	23.00	41.00
5	1.00	12.00	27.00	14.00	40.00
6	1.00	15.00	24.00	30.00	40.00
7	1.00	12.00	25.00	29.00	15.00
8	1.00	17.00	29.00	34.00	24.00
9	1.00	16.00	23.00	27.00	44.00
10	1.00	15.00	24.00	23.00	39.00
11	1.00	12.00	26.00	16.00	32.00
12	1.00	16.00	10.00	18.00	29.00
13	1.00	17.00	11.00	29.00	48.00
14	1.00	14.00	29.00	35.00	50.00
15	1.00	18.00	23.00	32.00	16.00
16	1.00	19.00	18.00	23.00	29.00
17	1.00	13.00	23.00	33.00	36.00
18	1.00	18.00	22.00	36.00	44.00
19	1.00	16.00	21.00	37.00	40.00
20	1.00	19.00	18.00	34.00	38.00
21	2.00	14.00	12.00	23.00	32.00
22	2.00	15.00	16.00	22.00	34.00
23	2.00	15.00	22.00	16.00	14.00
24	2.00	12.00	24.00	34.00	12.00
25	2.00	13.00	17.00	3.00	16.00
26	2.00	10.00	18.00	33.00	49.00
27	2.00	8.00	20.00	18.00	24.00
28	2.00	20.00	22.00	18.00	39.00
29	2.00	19.00	26.00	19.00	47.00
30	2.00	11.00	14.00	23.00	19.00
31	2.00	16.00	10.00	24.00	47.00
32	2.00	7.00	11.00	20.00	30.00
33	2.00	18.00	21.00	18.00	33.00
34	2.00	15.00	11.00	16.00	25.00
35	2.00	20.00	26.00	14.00	6.00
36	2.00	5.00	20.00	21.00	23.00
37	2.00	13.00	14.00	20.00	16.00
38	2.00	13.00	13.00	11.00	18.00

Table 37.5   Descriptive Statistics and Intercorrelations for Example 37.2

Descriptive Statistics

	GROUP	Mean	Std. Deviation	N
RISK1	1.00	15.9000	2.67346	20
	2.00	13.4000	4.17259	20
	3.00	11.4500	4.45415	20
	4.00	9.7500	3.89162	20
	Total	12.6250	4.43055	80
RISK2	1.00	22.3500	5.54669	20
	2.00	17.6000	5.08248	20
	3.00	14.9000	7.62889	20
	4.00	8.5000	4.22399	20
	Total	15.8375	7.56816	80
RISK3	1.00	29.0500	7.27270	20
	2.00	19.3000	6.78311	20
	3.00	11.2500	6.81233	20
	4.00	9.1000	6.51234	20
	Total	17.1750	10.36276	80
RISK4	1.00	36.7500	10.03087	20
	2.00	27.4000	12.32200	20
	3.00	22.2500	8.30266	20
	4.00	12.1500	11.06096	20
	Total	24.6375	13.68437	80

Correlations

		RISK1	RISK2	RISK3	RISK4
RISK1	Pearson Correlation	1	.378**	.417**	.408**
	Sig. (2-tailed)	.	.001	.000	.000
	N	80	80	80	80
RISK2	Pearson Correlation	.378**	1	.450**	.395**
	Sig. (2-tailed)	.001	.	.000	.000
	N	80	80	80	80
RISK3	Pearson Correlation	.417**	.450**	1	.569**
	Sig. (2-tailed)	.000	.000	.	.000
	N	80	80	80	80
RISK4	Pearson Correlation	.408**	.395**	.569**	1
	Sig. (2-tailed)	.000	.000	.000	.
	N	80	80	80	80

**. Correlation is significant at the 0.01 level (2-tailed).

Comparisons involving pairs of mean vectors employing **Hotelling's T^2** (i.e., a **multivariate analysis of variance** with $k = 2$) yielded the following values for **Wilks' lambda**: **Group 1** versus **Group 2** ($\Lambda = .542$; $F(4, 35) = 7.390$; $p = .000$); **Group 1** versus **Group 3** ($\Lambda = .278$; $F(4, 35) = 22.706$; $p = .000$); **Group 1** versus **Group 4** ($\Lambda = .139$; $F(4, 35) = 54.264$; $p = .000$); **Group 2** versus **Group 3** ($\Lambda = .677$; $F(4, 35) = 4.166$; $p = .007$); **Group 2** versus **Group 4** ($\Lambda = .332$; $F(4, 35) = 17.568$; $p = .000$); **Group 3** versus **Group 4** ($\Lambda = .606$; $F(4, 35) = 5.684$; $p = .001$). Note that all of the probability values are less than .0083 (i.e., .05/6 = .0083), and thus all achieve significance if the overall Type I error rate is set at .05 (employing the **Bonferroni–Dunn** adjustment). The above results indicate significant differences for all six pairs of group centroids.

In order to better assess the relative impact of the predictor variables when considered separately, four **single-factor between-subjects analyses of variance** were conducted. The results of the latter analyses are displayed in the lower part of Table 37.6 labeled **Tests of Equality of Group Means** (which provides a succinct version of the critical information

**Table 37.6   SPSS Printout of Results of Multivariate Analysis
of Variance for Example 37.2**

Box's Test of Equality of Covariance Matrices[a]

Box's M	29.116
F	.877
df1	30
df2	15880.55
Sig.	.659

Tests the null hypothesis that the observed covariance matrices of the dependent variables are equal across grou

a. Design: Intercept+GROUP

Levene's Test of Equality of Error Variances[a]

	F	df1	df2	Sig.
RISK1	1.251	3	76	.297
RISK2	2.184	3	76	.097
RISK3	.386	3	76	.763
RISK4	2.039	3	76	.115

Tests the null hypothesis that the error variance of the dependent variable is equal across groups.

a. Design: Intercept+GROUP

Multivariate Tests[c]

Effect		Value	F	Hypothesis df	Error df	Sig.	Partial Eta Squared
Intercept	Pillai's Trace	.965	508.442[a]	4.000	73.000	.000	.965
	Wilks' Lambda	.035	508.442[a]	4.000	73.000	.000	.965
	Hotelling's Trace	27.860	508.442[a]	4.000	73.000	.000	.965
	Roy's Largest Root	27.860	508.442[a]	4.000	73.000	.000	.965
GROUP	Pillai's Trace	.835	7.229	12.000	225.000	.000	.278
	Wilks' Lambda	.234	11.771	12.000	193.431	.000	.383
	Hotelling's Trace	2.970	17.740	12.000	215.000	.000	.498
	Roy's Largest Root	2.867	53.764[b]	4.000	75.000	.000	.741

a. Exact statistic

b. The statistic is an upper bound on F that yields a lower bound on the significance level.

c. Design: Intercept+GROUP

Tests of Equality of Group Means

	Wilks' Lambda	F	df1	df2	Sig.
RISK1	.730	9.393	3	76	.000
RISK2	.557	20.152	3	76	.000
RISK3	.420	34.923	3	76	.000
RISK4	.572	18.989	3	76	.000

displayed in the **Tests of Between-Subjects Effects** table in Table 37.1 for Example 37.1). The values of F, df_{num}, df_{den}, and probability (i.e., **Sig.**) computed for each of the risk factors are noted in the **Tests of Equality of Group Means** table. Note that all of the analyses of variance were significant at the .01 level, since in all four instances $p = .000$ (which, as previously noted, indicates a probability less than .00049). Thus, it would appear that each of the predictor variables is a contributor with respect to the significant result obtained for the omnibus **multivariate analysis of variance**. If the relative magnitudes of the computed F values are employed as a criterion, Risk factor 3 (which yields the largest F value) is associated with the most salient effect, followed by Risk factors 2, 4, and 1. In point of fact, the identical ordering for the risk factors is obtained later in this section for the standardized coefficients and canonical loadings of the function derived in the **discriminant function analysis** of the data — i.e., Risk factor 3 is the most important variable in defining the derived function followed by Risk factors 2, 4, and 1.

In conjunction with the above noted **single-factor between-subjects analyses of variance**, comparisons (employing **Fisher's LSD test, Tukey's HSD test** and the **Scheffé test**) were also conducted on each of the predictor variables. The following differences between the means of pairs of groups for each of the four risk factors were significant at the .05 level (with all three of the comparison methods unless otherwise indicated): 1) **Risk factor 1**: Group 1 versus Group 3; Group 1 versus Group 4; Group 2 versus Group 4; 2) **Risk factor 2**: Group 1 versus Group 3; Group 1 versus Group 4; Group 2 versus Group 4; Group 3 versus Group 4. The Group 1 versus Group 2 comparison was significant at the .05 level, but only with **Fisher's LSD test**; 3) **Risk factor 3**: Group 1 versus Group 2; Group 1 versus Group 3; Group 1 versus Group 4; Group 2 versus Group 3; Group 2 versus Group 4; 4) **Risk factor 4**: Group 1 versus Group 2 (significant with **Fisher's LSD** and **Tukey's HSD test** but not with the **Scheffé test**); Group 1 versus Group 3; Group 1 versus Group 4; Group 2 versus Group 4; Group 3 versus Group 4.

In conclusion, the analysis of the data with a **multivariate analysis of variance** and the other above noted procedures strongly suggests there are differences between the four group centroids, and that all four predictor variables are probably involved in contributing to those differences.

At this point the data will be evaluated with a **discriminant function analysis**. Specifically, $q = 3$ discriminant functions will be derived, since the latter is the smaller of the two values $p = 4$ versus $(k - 1) = (4 - 1) = 3$. The results of the **discriminant function analysis** are summarized in Tables 37.7 and 37.8, as well as in Figure 37.2. Initially the information in Table 37.7, which is comprised of six separate tables, will be discussed.

a) **The Eigenvalues table** The eigenvalues (along with the percentage of between-groups variability accounted for by the functions) computed for **Functions 1, 2,** and **3** are respectively 2.867 (96.5%), .103 (3.5%), and .000 (0%). The squared canonical correlations (i.e., the square of the value in the last column, which, as noted earlier, indicates the proportion of variability associated with a function that can be attributed to the grouping variable) for **Functions 1, 2,** and **3** are respectively $(.861)^2 = .741$, $(.305)^2 = .093$, and $(.017)^2 = .0003$. The information in the **Eigenvalues** table clearly indicates that **Function 1** is most strongly related to the grouping variable (and most likely statistically significant) with **Functions 2** and **3** making minor contributions (which in all likelihood will not be statistically significant).

b) **The Wilks' lambda table** The value $\Lambda = .234$ displayed in the **Wilks' lambda table** in the row labeled **1 through 3** is statistically significant at the .01 level, since $p = .000$ (i.e., **Sig.**) is less than .01. This result indicates significant group differentiation using all three functions. In the row labeled **2 through 3**, the probability value $p = .290$ associated with the value $\Lambda = .907$ is not significant (since it exceeds .05). The latter result indicates that **Function 2** (as well as **Function 3** for which the value $p = .990$ is computed) does not contribute significantly toward group differentiation, and that the significant result in the first row of the table was almost completely attributable to the role of **Function 1**. Consequently, the latter function will be the only one considered in the analysis from this point on.

c) **Canonical discriminant function coefficients, standardized canonical discriminant function coefficients, and structure matrix tables** As noted in the discussion of Example 37.1, the relative magnitudes of the unstandardized discriminant coefficients (in the table labeled **Canonical Discriminant Function Coefficients**) do not provide useful information regarding the degree of relationship between the predictor variables and each of the discriminant functions, since the four predictor variables are generally evaluated on different scales of measurement.

Table 37.7 *SPSS* **Printout of Discriminant Analysis for Example 37.2**

Eigenvalues

Function	Eigenvalue	% of Variance	Cumulative %	Canonical Correlation
1	2.867[a]	96.5	96.5	.861
2	.103[a]	3.5	100.0	.305
3	.000[a]	.0	100.0	.017

a. First 3 canonical discriminant functions were used in the analysis.

Wilks' Lambda

Test of Function(s)	Wilks' Lambda	Chi-square	df	Sig.
1 through 3	.234	108.795	12	.000
2 through 3	.907	7.351	6	.290
3	1.000	.021	2	.990

Canonical Discriminant Function Coefficients

	Function		
	1	2	3
RISK1	.065	-.022	.172
RISK2	.097	.093	.059
RISK3	.090	-.112	-.016
RISK4	.038	.052	-.063
(Constant)	-4.832	-.544	-1.265

Unstandardized coefficients

Standardized Canonical Discriminant Function Coefficients

	Function		
	1	2	3
RISK1	.252	-.085	.665
RISK2	.557	.534	.338
RISK3	.615	-.767	-.110
RISK4	.399	.546	-.668

Structure Matrix

	Function		
	1	2	3
RISK3	.681	-.688*	-.223
RISK2	.517	.523*	.430
RISK4	.507	.349	-.640*
RISK1	.359	-.030	.606*

Pooled within-groups correlations between discriminating variables and standardized canonical discriminant functions

Variables ordered by absolute size of correlation within function.

*. Largest absolute correlation between each variable and any discriminant function

Functions at Group Centroids

GROUP	Function		
	1	2	3
1.00	2.369	-.170	-.013
2.00	.517	.052	.028
3.00	-.790	.479	-.011
4.00	-2.096	-.360	-.004

Unstandardized canonical discriminant functions evaluated at group means

Table 37.8 *SPSS* **Printout of Classification Analysis for Example 37.2**

Classification Results[b,c]

		GROUP	Predicted Group Membership				Total
			1.00	2.00	3.00	4.00	
Original	Count	1.00	16	4	0	0	20
		2.00	4	12	4	0	20
		3.00	0	4	11	5	20
		4.00	0	0	3	17	20
	%	1.00	80.0	20.0	.0	.0	100.0
		2.00	20.0	60.0	20.0	.0	100.0
		3.00	.0	20.0	55.0	25.0	100.0
		4.00	.0	.0	15.0	85.0	100.0
Cross-validated[a]	Count	1.00	15	5	0	0	20
		2.00	4	10	6	0	20
		3.00	0	7	7	6	20
		4.00	0	0	3	17	20
	%	1.00	75.0	25.0	.0	.0	100.0
		2.00	20.0	50.0	30.0	.0	100.0
		3.00	.0	35.0	35.0	30.0	100.0
		4.00	.0	.0	15.0	85.0	100.0

a. Cross validation is done only for those cases in the analysis. In cross validation, each case is classified by the functions derived from all cases other than that case.

b. 70.0% of original grouped cases correctly classified.

c. 61.3% of cross-validated grouped cases correctly classified.

Classification Function Coefficients

	GROUP			
	1.00	2.00	3.00	4.00
RISK1	.881	.762	.661	.596
RISK2	.700	.543	.454	.250
RISK3	.545	.353	.188	.165
RISK4	.254	.192	.168	.074
(Constant)	-28.779	-17.314	-11.476	-6.554

Fisher's linear discriminant functions

The relative importance of the predictor variables is determined by examining the absolute values computed for the **standardized discriminant coefficients** (in the **Standardized Canonical Discriminant Function Coefficients** table) and the **canonical loadings** (in the **Structure Matrix** table). The latter values will only be considered for **Function 1**, since the analysis with **Wilks' lambda** indicated the other two derived functions did not contribute significantly toward group separation. Examination of the absolute values of the standardized coefficients computed for **Function 1** reveals that the predictor variable with the highest absolute value is Risk factor 3 (.615), followed by Risk factor 2 (.557), Risk factor 4 (.399), and Risk factor 1 (.252). The same ordering of the risk factors with respect to absolute values is observed on the canonical loadings (all of which are larger than the accepted minimal value of .30 for a canonical loading): Risk factor 3 (.681), followed by Risk factor 2 (.517), Risk factor 4 (.507), and Risk factor 1 (.359). The squares of the latter values $((.681)^2 = .464, (.517)^2 = .267, (.507)^2 = .257,$ and $(.359)^2 = .129)$ reflect the proportion of the variance on each of the predictor variables accounted for by the function. The magnitude of the values of the squared canonical loadings indicates that Risk factors 3, 2, and 4 are more important in defining the derived function relative to Risk factor 1.

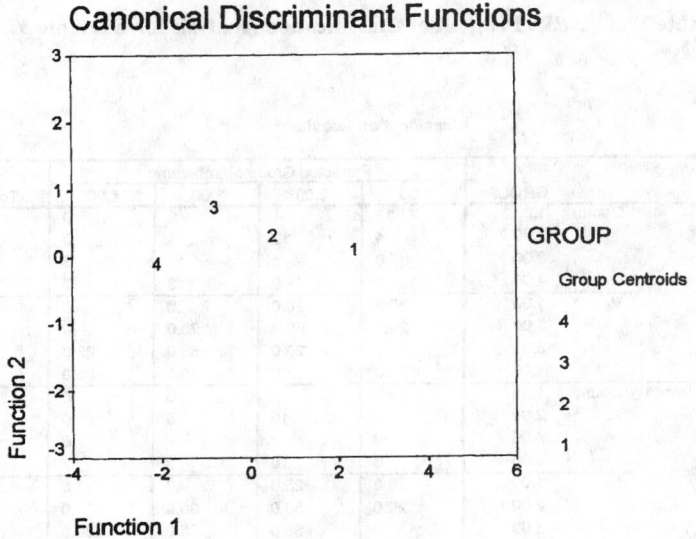

Canonical Discriminant Functions

    d) **Functions at group centroids**   As noted in the discussion of Example 37.1, this table contains for each of the discriminant functions the values of the centroids (e.g., composite means) for each of the $k = 4$ groups. Groups which have centroid values with different signs for a specific function are usually discriminated from one another by that function. Thus, the fact the values 2.369 and .517 are respectively computed for the centroids of Groups 1 and 2 and the centroids of Groups 3 and 4 are respectively −.790 and −2.096, strongly suggests that **Function 1** may differentiate Groups 1 and 2 from Groups 3 and 4. The most extreme distance between the centroids of two groups is that between Groups 1 and 4, whereas the least distance between the centroids of two groups is between Groups 2 and 3.

    The same information contained in the **Functions at Group Centroids** table is also provided in Figure 37.2, where the salient separation on the horizontal plane between Groups 1 and 4 is obvious, as well as the much smaller separation between Groups 2 and 3.

    e) **Classification function coefficients**   Table 37.8 contains the classification matrix (in the table labeled **Classification Results**) and the **classification function coefficients** (in the table with that title). Examination of the classification matrix reveals that 56 of the 80 subjects (i.e., 70%) are classified correctly in the **Original** part of the **Classification Results** table, and 49 of the 80 (i.e., 61.3%) in the **Cross-validated** part of the table. Note that in both parts of the table, the largest number of correct classifications are for Groups 1 and 4.

    Through use of Equation 37.4 the value of **Press's Q statistic** is computed below for the 56 correct classifications in the **Original** part of the **Classification Results** table, and the 49 correct classifications in the **Cross-validation** part of the table. In **Table A4**, for $df = 1$, $\chi^2_{.05} = 3.84$ and $\chi^2_{.01} = 6.63$. Since both $Q = 86.4$ and $Q = 56.07$ are greater than $\chi^2_{.01} = 6.63$, the results for both the **Original** and **Cross-validated** matrices are significant at the .05 level. This indicates that the discriminatory power of both matrices is significant — i.e., the predictions were significantly better than chance.

$$Press\text{'s } Q = \frac{[N - (n_c)(k)]^2}{N(k - 1)} = \frac{[80 - (56)(4)]^2}{80\,(4 - 1)} = 86.4$$

$$Press\text{'s } Q = \frac{[N - (n_c)(k)]^2}{N(k - 1)} = \frac{[80 - (49)(4)]^2}{80\,(4 - 1)} = 56.07$$

f) **Stepwise analysis of Example 37.2**   In addition to the **standard discriminant function analysis** described above, a **stepwise analysis** was also conducted. In the latter analysis, a single significant discriminant function was derived which accounted for 96.3% of the between-groups variability. The latter analysis, however, eliminated Risk factor 1, and thus only employed Risk factors 2, 3, and 4 in classifying subjects. The ordering of the risk factors with respect to their importance in defining the function was different than the ordering for the standard analysis — specifically, Risk factor 2 had the highest standardized coefficient and canonical loading followed by Risk factors 3 and 4. The different ordering of the predictors in the stepwise analysis can be attributed to the fact that the intercorrelation Risk factor 1 had with the other risk factors was partialed out of the analysis.

The classification matrix indicated that 56 of the 80 subjects (i.e., 70%) are classified correctly in the **Original** part of the **Classification Results** table, and 55 of the 80 (i.e., 68.8%) in the **Cross-validated** part of the table. Both of the latter classification frequencies were statistically significant at the .01 level when evaluated with **Press's Q**. The result of the stepwise analysis suggests that comparable (even slightly superior) group differentiation could be obtained through use of a discriminant function comprised of only Risk factors 2, 3, and 4.

g) **The role of posterior versus prior probabilities in classification**   Spicer (2005, p. 143) notes the probabilities displayed in a classification matrix represent **posterior probabilities**. A **posterior probability** is the probability of a subject being classified in a specific group, given the discriminant function score computed for that subject. The latter probability, however, is adjusted for a subject's **prior probability**. A subject's **prior probability** is the chance likelihood of the subject being in a specific group regardless of his or her discriminant function score. In the case of Example 37.1, the researcher would be justified in assuming that since subjects were randomly assigned to one of the three groups, each subject had an equal prior probability (i.e., $p = 1/3$) of being assigned to any of the three groups. In employing *SPSS* a researcher is asked to estimate prior probabilities in a menu window labeled **Discriminant Analysis: Classification** which requires that one of the following two options be selected: **All groups equal** versus **Compute from group size**. Some sources state that when the number of subjects in each group is equal (as is the case in both Example 37.1 and 37.2), the researcher should employ the default setting **All groups equal** in the latter window, while with unequal sample sizes the **Compute from group size** option should be selected. The latter statement, however, requires qualification. Among others, Hair *et al.* (1995, p. 200), Huberty (1994, p. 65, p. 92), and Stevens (2002, p. 310) note the **Compute from group size** option should only be employed if a researcher has reason to believe the sample employed in a study was randomly drawn from the population, and thus the proportion of subjects in each group represents the actual population proportion for that group (and thus the latter proportions would represent the prior probabilities for the k groups).

In the case of Example 37.2, since the same sample size was employed for each group, regardless of whether the researcher employed the **All groups equal** or **Compute from group size** option, the prior probability of a subject being assigned to one of the four groups will be $1/4 = .25$. Note, however, that in actuality the prior probabilities are the proportions in the population which comprise each of the groups — specifically, Group 1: $2,500/100,000 = .025$; Group 2: $5,000/100,000 = .05$: Group 3: $10,000/100,000 = .10$; Group 4: $82,500/100,000 = .825$.

However, it would only be appropriate for the researcher to employ the **Compute from group size** option if the sample sizes reflected the latter prior probabilities. For example, if the study had employed a **stratified sample** (discussed in the **Introduction**) involving 1000 subjects of whom prior to the age of 50, 25 had died from a heart attack, 50 had suffered a serious but nonfatal heart attack, 100 had experienced a mild heart attack, and 825 did not have a heart attack, the use of the **Compute from group size** option would be justified, and the researcher would have reasonable expectation that if the classification matrix for the sample data yielded a high percentage of correct classifications, the use of the discriminant function in classifying other members of the population who were not in the original sample would also result in accurate classification.

The question could be raised that since the number of subjects per group employed in Example 37.2 were not proportional to the actual population proportions, would the latter compromise the researcher's ability to predict with the same degree of accuracy to other members of the target population? The answer to the latter question is "no", since the subjects in each group were randomly selected from four subsets, each of which represented one of the four cardiac relevant categories in the target population. Because of the latter, within the framework of the study each subject had a prior probability of one out of four of being assigned to any of the groups.

VI. Additional Analytical Procedures for Discriminant Function Analysis and/or Related Tests

No additional procedures will be described in this section.

VII. Additional Discussion of Discriminant Function Analysis

No additional material will be discussed in this section.

VIII. Additional Examples Illustrating the Use of Discriminant Function Analysis

No additional examples will be presented in this section.

References

Diekhoff, G. (1992). **Statistics for the social and behavioral sciences: Univariate, bivariate, and multivariate**. Dubuque, IA: Wm. C. Brown Publishers.

Field, A. (2005). **Discovering statistics using SPSS** (2nd ed.). Sage Publications: London.

Grimm, L. G. & Yarnold, P. R. (Eds.) (1995). **Reading and understanding multivariate statistics**. Washington, D.C.: American Psychological Association.

Hair, J. F., Anderson, R. E., Tatham, R. L. & Black, W. C. (1995). **Multivariate data analysis** (4th ed.). Upper Saddle River, N. J.: Prentice Hall.

Hair, J. F., Anderson, R. E., Tatham, R. L. & Black, W. C. (2004). **Multivariate data analysis** (6th ed.). Upper Saddle River, N. J.: Prentice Hall.

Harlow, L. L. (2005). **The essence of multivariate thinking**. Mahwah, NJ: Lawrence Erlbaum Associates.

Huberty, C. J. (1994). **Applied discriminant analysis**. New York: John Wiley & Sons.

Marascuilo, L. A. & Levin, J. R. (1983). **Multivariate statistics in the social science: A researchers guide**. Monterey, CA: Brooks/Cole Publishing Company.

Silva, A. P. D. & Stam, A. (1995). Discriminant analysis. In Grimm, L. G. & Yarnold, P. R. (Eds.) (pp. 277–318). **Reading and understanding multivariate statistics.** Washington, D.C.: American Psychological Association.

Spicer, J. N. (2005). **Making sense of multivariate data analysis.** Thousand Oaks, CA: Sage Publications.

Stevens, J. (2002). **Applied multivariate statistics for the social sciences** (4th ed.). Hillsdale, NJ: Lawrence Erlbaum Associates, Publishers.

Tabachnick, B. G. & Fidell, L. S. (1996). **Using multivariate statistics** (3rd ed.). New York: Harper Collins Publishers.

Tabachnick, B. G. & Fidell, L. S. (2001). **Using multivariate statistics** (4th ed.). Boston: Allyn & Bacon.

Tatsuoka, M. M. (1973). Multivariate analysis in behavioral research. In Kerlinger, F. (Ed.). **Review of research in education.** Itasca, IL: Peacock.

Endnotes

1. Silva and Stam (1995, p. 284) note that the null hypothesis evaluated with the omnibus **multivariate analysis of variance** (i.e., H_0: Equality of the k composite treatment means) is equivalent to evaluating the null hypothesis that none of the discriminant functions makes a statistically significant contribution to group separation.

2. a) On a more technical level, Stevens (2002, p. 287) notes that the first discriminant function will have the largest eigenvalue in the $\mathbf{W^{-1}B}$ matrix (which is the definition of **Roy's largest root**), the second discriminant function will have the second largest eigenvalue in the latter matrix, and so on. Stevens (2002, pp. 286–287) notes that **discriminant function analysis** is often referred to as a **maximization procedure**, since it identifies the linear combinations which maximize the between to within association in the aforementioned matrices; b) Silva and Stam (1995, p. 283) note that in most applications of **discriminant function analysis** a small subset of discriminant functions will explain most of the differences between groups. It is not uncommon for just one or at most two discriminant functions to account for most of the between-groups variability in the data. Silva and Stam (1995, p. 283) also note that since in most cases $(k-1)$ is less than p, the number of predictor variables will usually be greater than the number of discriminant functions.

3. a) The first value of **lambda** value computed in **discriminant function analysis** will always equal the value of **lambda** computed if a **multivariate analysis of variance** is employed to evaluate the same set of data; b) Hair *et al.* (1995, p. 198) and Tabachnick and Fidell (1996, p. 533) note that **Pillai's trace** and **Hotelling's trace** can also be used to evaluate the discriminatory power of all the discriminant functions, but that **Roy's largest root** can only evaluate the discriminatory power of the first discriminant function. The latter sources also state that if a **stepwise discriminant function analysis** is conducted, the **Mahalanobis' D^2 statistic** (which is most useful when there are large number of predictor variables) and a measure referred to as **Rao's V** can also be employed to evaluate the discriminatory power of the functions.

4. Some sources conceptualize **discriminant function analysis** as a form of **canonical correlation (Test 38)**, which in contrast to **discriminant function analysis** evaluates the relationship between **a set of predictor variables** and **a set of criterion variables**.

Canonical correlational analysis extracts two or more linear combinations involving the two sets of variables which are maximally correlated with one another. Each pair of linear combinations that is extracted can be represented by Equation 37.5. Note that the right side of the latter equation is similar (although not identical) to the right side of Equation 37.1. Within the framework of **canonical correlational analysis**, Equation 37.5 is referred to as a **canonical function** (or **canonical root**), and each side of the equation is labeled a **canonical variable** (also referred to as a **canonical variate**). Note that a canonical variable is a linear combination of a set predictor or criterion variables.

(Equation 37.5)

$$b_1 Y_1 + b_2 Y_2 + \dots + b_q Y_q = a_1 X_1 + a_2 X_2 + \dots + a_p X_p$$

Sources that conceptualize **discriminant function analysis** as a form of **canonical correlational analysis** may label Equation 37.1 as representing a **canonical function**, with the right side of the equation (which is a linear combination of the predictor variables) referred to as a **canonical variable**. (Some sources, in fact, may refer to all of Equation 37.1 as a **canonical variable**.) The use of the term **canonical correlation** in Section V to describe the correlation between a discriminant function and the grouping variable reflects the relationship between **discriminant function analysis** and **canonical correlational analysis**.

5. The following *SPSS* command sequence was employed in this chapter in conducting a **discriminant function analysis**: a) Click **Analyze**; b) Click **Classify**; c) Click **Discriminant**; d) Highlight the criterion/grouping variable (i.e., the variable indicating whether or not a subject had a **fatal attack** versus a **serious nonfatal heart attack** versus a **mild heart attack** versus **no heart attack** in Example 37.2) and move it to the **Grouping Variable** window. Click **Define Range**, enter the lowest and highest group identification numbers in the **Minimum** and **Maximum** windows, and then click **Continue**; e) Highlight the predictor variables (i.e., the four **risk factors** in Example 37.2) and move them to the **Independents** window; f) Click **Statistics**, check off desired information (e.g., **Means**, **Univariate ANOVAs**, etc.), and then click **Continue**; g) Click **Classify**, check off desired information (e.g., **Casewise results**, **Summary table**, **Leave-one-out Classification**, etc.), and then click **Continue**; h) Click **Save**, check off desired information (e.g., **Predicted group membership**, **Discriminant scores**, **Probabilities of group membership**), and then click **Continue**; i) Indicate whether you wish to **Enter independents together** or conduct a **Stepwise** analysis; j) Click **OK** to obtain the output for the analysis.

6. a) The value $\Lambda = .292$ is identical to the value computed for **Wilks' lambda** in Table 37.1 for the omnibus **multivariate analysis of variance**. In point of fact, it will always be the case that the value computed in the first row of the **Wilks' lambda** table will equal the value of **lambda** computed for the omnibus **multivariate analysis of variance**; b)The computed chi-square value in the **Wilk's lambda** table is evaluated through use of the critical values in **Table A4**, employing the degrees of freedom value noted in Table 37.2. For the row labeled **1 through 2**, the critical .05 and .01 values employed for $df = 6$ are $\chi^2_{.05} = 12.59$ and $\chi^2_{.01} = 16.81$. Since the computed value $\chi^2 = 13.556$ is greater than $\chi^2_{.05} = 12.59$, the result is significant at the .05 level. Since the value $\chi^2 = .221$ computed for **Function 2** is less than $\chi^2_{.05} = 5.99$ (for $df = 2$), the latter result is not significant.

7. a) In Endnote 10 in the chapter on **multiple regression** it was noted that in **multiple regression** the absolute value of a **standardized regression coefficient** cannot be greater than 1 (and if it exceeds 1 it indicates a major problem with the analysis — e.g., collinearity, nonnormality, etc.). In **discriminant function analysis**, however, the absolute value of a **standardized discriminant function coefficient** can exceed 1 (Lisa Harlow, (personal communication)); b) The author is indebted to Lisa Harlow for clarifying the limiting values which can be assumed by coefficients in **regression analysis** versus **discriminant function analysis**.

8. a) Field (2005, p. 612) and Hair *et al.* (1995, p. 206) state that canonical loadings are analogous to the factor loadings derived in factor analysis (which is discussed later in this chapter); b) The value of a **canonical loading** must fall between −1 and +1, and if a value falls outside of that range it indicates there is a problem with the analysis (Lisa Harlow, personal communication).

9. Stevens (2002, p. 296) notes that if the canonical loading of a predictor variable has a negative sign, the groups that scored highest on that variable would score lower on the function in question.

10. Upon conducting a discriminant function analysis *SPSS* will print out the discriminant function score for each subject on the spreadsheet containing the original data. A good example demonstrating the computation of function scores for subjects can be found in Diekhoff (1992, pp. 294–295).

11. Silva and Stam (1995, p. 296–298) provide a good discussion of alternative methods for deriving classification rules.

12. The proportions of the full sample employed for the two subsamples can be values other than .5. For example, the discriminant functions can be derived using .6 of the subjects and cross-validated using the remaining .4 of the subjects.

13. a) Some sources employ the term *U*-method interchangeably with **jackknifing**. Hair *et al.* (1995, p. 210), however, make a distinction such that the *U*-method focuses on classification accuracy, while **jackknifing** focuses on the stability of the discriminant function coefficients; b) One method for assessing the utility of a classification matrix is the **maximum chance criterion**. The latter method compares the percentage of correct classifications in the matrix with the percentage of subjects which comprised the largest group (in the sample employed to derive the discriminant functions), since if the latter percentage is large the most accurate strategy might be to classify subjects whose group membership is unknown in the largest group; c) Tabachnick and Fidell (1996, p. 520) note that classification procedures for **discriminant function analysis** are extremely sensitive to violation of the homogeneity of variance-covariance matrices assumption. When the latter assumption is violated, subjects are more likely to be classified in the group with the greatest dispersion (i.e., the group with the largest determinant for its within-groups covariance matrix).

14. In this instance cross-validation means employing a new sample in order to determine whether the discriminant function(s) obtained for the original sample can be employed to accurately categorize subjects in the new sample whose group membership is known.

Test 38

Canonical Correlation
(Parametric Test Employed with Interval/Ratio Data)

I. Hypothesis Evaluated with Test and Relevant Background Information

Canonical correlation is not an inferential statistical test, but instead is a statistical procedure which assesses whether or not there is an association between two sets of variables. More specifically, it identifies linear combinations of variables from each set which are maximally correlated with one another.[1]

Hypothesis evaluated with test The primary hypothesis evaluated within the framework of **canonical correlation** is that there is a significant association/correlation between one or more linear combinations of two sets of variables.

Relevant background information on test Thompson (2000, p. 285) notes that although **canonical correlation** was developed in 1935 by Hotelling, it was largely ignored as an analytical procedure until fairly recently when computers could be employed to conduct complex statistical analyses. **Canonical correlation** is an extension of **multiple regression (Test 33)** analysis in that multiple regression analysis evaluates the relationship between a **set of predictor variables** and a **single criterion variable**, while **canonical correlation** evaluates the relationship between a **set of predictor variables** and a **set of criterion variables**. Kachigan (1986) notes that **canonical correlation** is most likely to be useful in situations in which there are multiple predictor variables and a researcher believes that more than one variable is required in order to adequately represent the criterion variable.[2]

As noted above, in **canonical correlation** a researcher has two sets of variables. The first set, comprised of p variables, is variously referred to as X **variables, predictor variables, independent variables, covariates,** or **variables on the left side of the equation**. The second set, comprised of q variables, is variously referred to as Y **variables, criterion variables, dependent variables** or **variables on the right side of the equation**. The goal in **canonical correlation** is to identify pairs of linear combinations involving the two sets of variables which are maximally correlated with one another. The term **canonical variate** (also referred to as a **canonical variable** or **canonical factor**) is employed to identify any linear combination comprised of X (or Y) variables that is correlated with a linear combination of Y (or X) variables. In conducting a **canonical correlation** the set of canonical variates which yields the maximum correlation is initially identified, followed by a second set of canonical variates (uncorrelated with the first) that yields the next highest correlation, and so on until the maximum possible number of sets (all of which are uncorrelated with one another) have been extracted from the data.

A **canonical variate** is expressed as a sum of two or more weighted variables which comprise that set, with the variables weighted with respect to their importance in defining the canonical variate. Each of the canonical variates can be conceptualized as a **latent variable**, in that the linear combination represented by a canonical variate can be viewed as representing some underlying dimension. The latter dimension may sometimes be assigned a name based on the relative importance of the variables which comprise that linear combination.

The linear combination for the set of p X variables can be defined by Equation 38.1. Note that in the latter equation CV_X (which represents the **canonical variate for the X variables**) is the sum of the p variables which comprise the set of X variables, and each of the X variables is optimally weighted such that its weight/coefficient reflects the importance of an X variable in defining CV_X. In Equation 38.1 the value a_i represents the weight assigned to the ith X variable.

$$CV_X = a_1 X_1 + a_2 X_2 + ... + a_p X_p \qquad \text{(Equation 38.1)}$$

In the same respect, Equation 38.2 defines the linear combination for the set of q Y variables. CV_Y (which represents the **canonical variate for the Y variables**) is the sum of the q variables which comprise the set of Y variables, and each of the Y variables is optimally weighted such that its weight/coefficient reflects the importance of a Y variable in defining CV_Y. In Equation 38.2 the value b_j represents the weight assigned to the jth Y variable.

$$CV_Y = b_1 Y_1 + b_2 Y_2 + ... + b_q Y_q \qquad \text{(Equation 38.2)}$$

Any **pair of canonical variates** derived in **canonical correlation** is said to represent a **canonical function** (also referred to as a **canonical correlation, root,** or **characteristic root**). Equation 38.3 is often employed as the general form of the equation for a canonical function. Note that the left side of the latter equation is comprised of an optimally weighted set of criterion variables, while the right side is comprised of an optimally weighted set of predictor variables.

$$\text{(Equation 38.3)}$$
$$b_1 Y_1 + b_2 Y_2 + ... + b_q Y_q = a_1 X_1 + a_2 X_2 + ... + a_p X_p$$

or

$$CV_Y = CV_X$$

A **canonical correlation** will always derive two or more canonical functions, and all of the functions will be independent of (uncorrelated with) one another (i.e., they will be **orthogonal**). The exact number of canonical functions that can be derived will be equal to the smaller of the two values p versus q. Upon deriving the canonical functions, inferential statistical tests are employed to determine which ones are statistically significant (a significant result indicates the correlation between the pair of canonical variates that define a canonical function is some value other than zero). When a significant result is obtained for a **canonical correlation**, the first of the derived canonical functions (i.e., the function which maximizes the correlation between a pair of canonical variates) will in all likelihood be significant (although there are rare instances where it may not be significant). The second derived canonical function (which will always have the second highest correlation between a pair of canonical variates) is then evaluated, and so on until all of the derived canonical functions have been evaluated. It is important to note that the magnitude of the correlation between a pair of canonical variates which define a canonical function will always be lower than the magnitude of the correlation between the canonical variates for any previously derived canonical functions, and only those canonical functions which are determined to be statistically significant are retained in the analysis — in other words, once a canonical function has been determined to be nonsignificant, any subsequently derived functions will also be nonsignificant. It is not uncommon in **canonical correlation** for only the first of the derived canonical functions to be significant, since the latter function will often contain the bulk of information which accounts for the relationship between the two sets of variables.

To illustrate **canonical correlation**, consider the following example. Let us assume a researcher has the following five lifestyle measures on a sample of subjects, which will represent the predictor (X) variables: a) Number of hours of exercise per week; b) Number of grams of fat consumed per week; c) Number of milligrams of caffeine consumed per week; d) Number of grams of sugar consumed per week; and e) Scores on a test assessing daily stress. In addition, the researcher has the following three scores as indices of health, which will represent the criterion (Y) variables: a) Diastolic blood pressure; b) Body fat ratio; and c) A composite blood chemistry index of physical health. **Canonical correlation** can be employed to determine if there are reliable ways in which measures within the two sets of variables are related to one another. Thus, for example, the researcher might find that a canonical variate comprised of two of the predictor variables (e.g., number of milligrams of caffeine consumed per week and number of grams of sugar consumed per week) is highly correlated with a canonical variate comprised of two criterion variables (e.g., diastolic blood pressure and the composite blood chemistry index of physical health). It is important to note that (as is the case with any correlational procedure) **canonical correlation** does not allow one to draw conclusions regarding cause and effect.

As is the case for other multivariate procedures, before generalizing the results of a **canonical correlation** a researcher would be expected to conduct an **external replicability analysis** (which just means replicating the results of an analysis on one or more additional random samples obtained from the same target population) and/or confirm one's results through use of **cross-validation** (as described in Section I of the chapter on **multiple regression**). It should also be noted that given the large number of coefficients that are computed within the framework of **canonical correlation** (most of which are described in Section V), there is an increased likelihood a significant result for any coefficient or test statistic is due to chance (i.e., represents a Type I error) rather than a genuine effect in the underlying population.

In order to be reliable a **canonical correlation** requires a large sample size. Stevens (2002, p. 475) recommends that minimally the ratio of subjects to variables employed be at least 20 to 1 when only one canonical function is interpreted, and at least 42 to 1 when two canonical functions are interpreted. Tabachnick and Fidell (1996, p. 198), on the other hand, state the minimum acceptable sample size will be a function of the reliability of the variables. The latter authors recommend 10 subjects per variable in social science research (where they assume a reliability coefficient of approximately .80 per variable). Fewer or greater subjects per variable would be recommended if a researcher has reason to assume higher or lower variable variability.

As well as the requirement of a large sample size, the reliability of a **canonical correlation** will depend upon certain assumptions having been met, as well as other issues relating to the nature of the data. Specifically:

1) **Multivariate normality** As is the case for most multivariate procedures, multivariate normality (which is discussed in greater detail in Section I of the chapter on **multiple regression**) is an assumption underlying **canonical correlation**. The following conditions must be met to satisfy multivariate normality: a) All of the variables are normally distributed. Tabachnick and Fidell (1996, pp. 70–71; p. 640) state that normality of the variables is generally assessed through analysis of skewness and kurtosis, and any variable for which either condition is violated should probably be modified through use of a data transformation; and b) Multivariate normality requires that linear combinations between each pair of variables should be normally distributed. In addition to confirming the linear relationship between a pair of variables, normality of linear combinations is assessed by demonstrating homoscedasticity, as well as by determining that the residuals are distributed normally (most commonly through use of scatterplots).

2) **Outliers** Prior to conducting an analysis, outliers on any of the variables should be identified. Suspected outliers should be deleted or otherwise dealt with as noted in Section VII of the *t* **test for two independent samples (Test 11)**.

3) **Multicollinearity and singularity** Multicollinearity and/or singularity between variables within sets and across sets can compromise the reliability of a **canonical correlation** (Tabachnick and Fidell (1996, p. 199)). Variables exhibiting the latter should be deleted from the analysis or combined with other variables. (Multicollinearity and singularity are discussed in Sections I and IV/V of the chapter on **multiple regression**.)

II. Example

Example 38.1 will be employed to illustrate the use of **canonical correlation**.[3]

Example 38.1 *A researcher wants to determine whether or not there is a relationship between three physiological measures and three measures of exercise fitness in a population of women who are between 25 and 35 years of age. Data are obtained from sample of 40 women who are randomly selected from an available database that is representative of the target population. The researcher obtains the following three physiological measures for each subject, which will represent the predictor/independent (X) variables: a) Weight; b) Waist size; and c) Resting pulse rate. The following three measures of exercise fitness are employed as the criterion/dependent (Y) variables: a) Number of sit ups; b) Number of pull ups; and c) Number of jumping jacks. The scores of the 40 subjects on the six variables are displayed in Table 38.1. A canonical correlational analysis was conducted in order to determine whether or not the two sets of variables are associated with one another.*

III. Null versus Alternative Hypotheses

The primary **null hypothesis** evaluated within the framework of **canonical correlation** is that there is no significant association/correlation between any of the sets of canonical variates. Within the latter context, a correlation (referred to as a **canonical correlation** which will be represented by the notation r_C) is computed for each of the canonical functions, and the null hypothesis states that in the underlying population represented by the sample, the population canonical correlation for all of the canonical functions equals zero. The **alternative hypothesis** is that at least one of the canonical correlations in the underlying population is some value other than zero. The analysis only retains those canonical functions for which the null hypothesis has been rejected.

IV. Test Computations

Equation 38.4 is the fundamental equation underlying **canonical correlation** (Harlow (2005, p. 188), Tabachnick and Fidell (1996, p. 200), Tacq (2004) and Thompson (2000, p. 291)). The latter equation derives the correlation matrix R_{CC}, which is the product of four correlation submatrices (i.e., the smaller correlation matrices to the right of the equals sign). R_{CC} is sometimes referred to as a **supermatrix** or **quadruple-product matrix** (since it is the result of the multiplication of four submatrices). Subsequent analysis of R_{CC} will yield the test statistics and coefficients that are computed for **canonical correlation** (which will be presented in Section V).

$$R_{CC} = (R_{YY}^{-1})\ (R_{YX})\ (R_{XX}^{-1})\ (R_{XY}) \qquad \text{(Equation 38.4)}$$

Where: R_{YY}^{-1} is the inverse of the $q \times q$ matrix containing the intercorrelations between the q Y variables.

R_{YX} is the matrix containing the intercorrelations between the Y variables and X variables (in which Y represents the row variable and X the column variable).

R_{XX}^{-1} is the inverse of the $p \times p$ matrix containing the intercorrelations between the p X variables.

R_{XY} is the **transpose** of the R_{YX} matrix. (A **transpose** of a matrix interchanges the rows and the columns of the original matrix.)

All of the values employed in the four above noted correlation submatrices are displayed in Table 38.2, which contains the intercorrelations between all six variables employed in Example 38.1. To determine the two variables represented by any of the correlations in Table 38.2, identify the variable at the left of the row and the variable at the top of the column in which a correlation appears. The row and column variables represent the two variables for which the correlation in that cell was computed. Note that of the four above noted submatrices, only R_{XY} and R_{YX} will contain all of the correlations displayed in Table 38.2.

Since the calculations involved in **canonical correlations analysis** are complex and the matrix algebra operations are impractical to implement without the aid of a computer, they will not be demonstrated here. (Readers interested in a detailed description of the mathematical operations involved in **canonical correlation** should consult sources which specialize in multivariate statistics (e.g., Tabachnick and Fidell (1996, Ch. 6; 2001).) Recollect that since the number of canonical functions will equal the smaller of the two values p versus q, in the case of Example 38.1 $p = q = 3$.

V. Interpretation of the Test Results

The intent of this section is to provide the reader with guidelines for interpreting the output generated by computer software (specifically, *SPSS*) for **canonical correlation**.[4]

a) **Descriptive statistics for the variables** The following values were computed for the mean and standard deviation of each of the six variables employed in the analysis: *Weight*: $\bar{X} = 129.65$, $\tilde{s} = 36.09$; *Waist*: $\bar{X} = 26.60$, $\tilde{s} = 4.85$; *Pulse*: $\bar{X} = 68.53$, $\tilde{s} = 7.60$; *Sit ups*: $\bar{X} = 121.45$, $\tilde{s} = 65.42$; *Pull ups*: $\bar{X} = 6.78$, $\tilde{s} = 4.74$; *Jumping jacks*: $\bar{X} = 62.50$, $\tilde{s} = 49.06$.

b) **Intercorrelations between the variables** Examination of Table 38.2 reveals the three predictor variables are significantly correlated with one another, as well as the fact that the three criterion variables are significantly correlated with one another. There was no evidence of collinearity between any of the variables — i.e., all tolerance values and condition indices were within the range of acceptable values.

c) **Canonical correlation** A **canonical correlation** (which will be represented by the notation r_C) represents a correlation coefficient computed for each of the canonical functions. The **square of a canonical correlation** (i.e., r_C^2) is the proportion of variance on one canonical variate (e.g., the set of Y variables) which can be explained by the variance on (or, to put it another way, by its linear relationship with) the other canonical variate (e.g., the set of X variables). A **canonical correlation** is analogous to a **multiple correlation coefficient**, which is discussed in the chapter on **multiple regression**. Whereas the latter type of correlation indicates how accurately scores on a criterion variable can be predicted from the scores on a set of predictor variables, a **canonical correlation** indicates how accurately a **set of dependent canonical variates** can be predicted from a **set of independent canonical variates**. An arbitrary value employed by many sources is that in the case of a large sample size a canonical function may be of interest (i.e., perhaps be useful for predictive purposes) if its canonical correlation is .30 or greater (a larger minimum value would be required for small samples (e.g., $r_C \geq .7$)).

It is important to note that the **square of a canonical correlation** is not the proportion of total variability explained in the original variables, but is only the proportion of variability shared

Table 38.1   Raw Data for Example 38.1

	subject	weight	waist	pulse	pullup	situp	jumpjk
1	1.00	126.00	30.00	60.00	4.00	142.00	55.00
2	2.00	160.00	27.00	62.00	.00	90.00	55.00
3	3.00	128.00	28.00	68.00	10.00	81.00	96.00
4	4.00	97.00	27.00	72.00	10.00	85.00	32.00
5	5.00	124.00	25.00	56.00	11.00	135.00	53.00
6	6.00	99.00	20.00	66.00	2.00	81.00	37.00
7	7.00	146.00	32.00	66.00	6.00	81.00	33.00
8	8.00	112.00	24.00	70.00	4.00	105.00	35.00
9	9.00	111.00	27.00	84.00	13.00	180.00	35.00
10	10.00	89.00	23.00	66.00	15.00	231.00	245.00
11	11.00	114.00	24.00	60.00	15.00	100.00	33.00
12	12.00	101.00	23.00	62.00	11.00	190.00	105.00
13	13.00	89.00	24.00	74.00	12.00	195.00	100.00
14	14.00	182.00	30.00	60.00	.00	30.00	45.00
15	15.00	128.00	26.00	56.00	4.00	50.00	26.00
16	16.00	137.00	27.00	72.00	10.00	190.00	115.00
17	17.00	111.00	29.00	64.00	2.00	40.00	20.00
18	18.00	92.00	20.00	62.00	9.00	210.00	75.00
19	19.00	91.00	27.00	64.00	13.00	205.00	68.00
20	20.00	73.00	23.00	78.00	.00	90.00	38.00
21	21.00	157.00	26.00	77.00	7.00	170.00	105.00
22	22.00	221.00	34.00	69.00	.00	20.00	10.00
23	23.00	134.00	24.00	67.00	6.00	190.00	65.00
24	24.00	111.00	22.00	69.00	10.00	185.00	58.00
25	25.00	93.00	25.00	83.00	.00	70.00	28.00

	subject	weight	waist	pulse	pullup	situp	jumpjk
26	26.00	184.00	29.00	67.00	.00	70.00	45.00
27	27.00	148.00	30.00	73.00	7.00	61.00	86.00
28	28.00	117.00	21.00	77.00	7.00	65.00	22.00
29	29.00	144.00	22.00	61.00	8.00	115.00	43.00
30	30.00	131.00	23.00	89.00	10.00	160.00	25.00
31	31.00	109.00	25.00	71.00	12.00	211.00	215.00
32	32.00	134.00	26.00	65.00	12.00	80.00	23.00
33	33.00	121.00	25.00	67.00	8.00	170.00	95.00
34	34.00	109.00	26.00	79.00	9.00	175.00	90.00
35	35.00	254.00	44.00	65.00	.00	10.00	35.00
36	36.00	148.00	34.00	61.00	1.00	30.00	16.00
37	37.00	157.00	29.00	77.00	7.00	170.00	105.00
38	38.00	131.00	38.00	69.00	.00	20.00	10.00
39	39.00	112.00	24.00	64.00	6.00	190.00	65.00
40	40.00	161.00	21.00	69.00	10.00	185.00	58.00

Table 38.2   *SPSS* Intercorrelation Matrix for Variables Employed in Example 38.1

		WEIGHT	WAIST	PULSE	PULLUP	SITUP	JUMPJK
WEIGHT	Pearson Correlation	1	.672**	-.159	-.448**	-.476**	-.242
	Sig. (2-tailed)	.	.000	.328	.004	.002	.132
	N	40	40	40	40	40	40
WAIST	Pearson Correlation	.672**	1	-.122	-.479**	-.589**	-.248
	Sig. (2-tailed)	.000	.	.455	.002	.000	.124
	N	40	40	40	40	40	40
PULSE	Pearson Correlation	-.159	-.122	1	.085	.179	.046
	Sig. (2-tailed)	.328	.455	.	.601	.270	.780
	N	40	40	40	40	40	40
PULLUP	Pearson Correlation	-.448**	-.479**	.085	1	.707**	.506**
	Sig. (2-tailed)	.004	.002	.601	.	.000	.001
	N	40	40	40	40	40	40
SITUP	Pearson Correlation	-.476**	-.589**	.179	.707**	1	.664**
	Sig. (2-tailed)	.002	.000	.270	.000	.	.000
	N	40	40	40	40	40	40
JUMPJK	Pearson Correlation	-.242	-.248	.046	.506**	.664**	1
	Sig. (2-tailed)	.132	.124	.780	.001	.000	.
	N	40	40	40	40	40	40

** Correlation is significant at the 0.01 level (2-tailed).

by the variables with respect to the canonical function in question. As Wuensch (2005) notes, it is possible to obtain a large squared canonical correlation for a canonical function (i.e., the function is comprised of two linear combinations which are highly correlated with one another), yet the function in question only extracts a minimum amount of the total variability in the original variables. The determination of how much of the total variability in the original variables is accounted for by the extracted canonical functions is described later in this section under the discussion of **redundancy coefficients**.

The canonical correlations computed for the three canonical functions for Example 38.1 are displayed in Table 38.3 in Column 5 (**Canon. Cor.**) of the table labeled **Eigenvalues and Canonical Correlations**. Note that the term **root** is employed in Column 1 to designate the canonical functions, which are numbered 1 through 3. The values .646, .170, and .011 in Column 5 are respectively the canonical correlations computed for *Functions 1, 2,* and *3.*[5] The values .417, .029, and .000 in Column 6 (**Sq. Cor**) are the squared canonical correlations for each of the functions. For a given canonical function, the latter value represents the proportion of variance on one canonical variate (e.g., the set of Y variables) which can be explained by the variance on the other canonical variate (e.g., the set of X variables). It can also be defined as the amount of shared variance between the two canonical variates.

The **eigenvalues** in Column 2 indicate the ratio of explained variance to unexplained variance for each canonical function. For example, if a function has an eigenvalue of 2, it would indicate that the proportion of explained variance for that function is two times the proportion of unexplained variance. An eigenvalue is computed by dividing the squared canonical correlation (i.e., r_C^2 in Column 6) by $(1 - r_C^2)$. Thus, the proportion of explained variance for *Function 1* is $.417/(1 - .417) = .715$.[6] Note that the number of derived eigenvalues corresponds to the number of canonical functions, and the magnitude of eigenvalues decrease with each successive function. The values displayed in Column 3 (**Pct.**) are obtained by dividing the eigenvalue for a function by the sum of all the eigenvalues. Thus, in the case of *Function 1*, $.715/(.715 + .013 + .000) = .715/.745 = 95.98\%$.

Table 38.3   Partial *SPSS* Summary of Analysis of Example 38.1

```
EFFECT .. WITHIN CELLS Regression
Multivariate Tests of Significance (S = 3, M = -1/2, N = 16 )

Test Name        Value  Approx. F Hypoth. DF   Error DF  Sig. of F

Pillais         .44595   2.09526       9.00     108.00      .036
Hotellings      .74484   2.70348       9.00      98.00      .007
Wilks           .56617   2.42534       9.00      82.90      .017
Roys            .41687

- - - - - - - - - - - - - - - - - - - - - - - - - - - - - - - - -

Eigenvalues and Canonical Correlations

Root No.   Eigenvalue       Pct.   Cum. Pct.  Canon Cor.   Sq. Cor

    1          .715       95.980     95.980        .646       .417
    2          .030        4.005     99.985        .170       .029
    3          .000         .015    100.000        .011       .000

- - - - - - - - - - - - - - - - - - - - - - - - - - - - - - - - -

Dimension Reduction Analysis

Roots      Wilks L.        F Hypoth. DF   Error DF  Sig. of F

1 TO 3      .56617   2.42534       9.00      82.90      .017
2 TO 3      .97093    .26007       4.00      70.00      .903
3 TO 3      .99989    .00403       1.00      36.00      .950

- - - - - - - - - - - - - - - - - - - - - - - - - - - - - - - - -
```

d) **Assessing the statistical significance of the derived canonical correlations and the canonical functions**   Multivariate statistical measures discussed in Section V of both Hotelling's T^2 (Test 34) and the **multivariate analysis of variance (Test 35)** are employed to determine whether or not a significant relationship is detected between any pair of canonical variates (i.e., whether any of the canonical functions is significant). In Table 38.3 in the table labeled **EFFECT .. WITHIN CELLS Regression** the results of multivariate tests of significance employing **Pillai's trace**, **Wilks' lambda**, **Hotelling's trace**, and **Roy's largest root** are displayed. As noted in the discussion of **Hotelling's T^2** and the **multivariate analysis of variance**, all four of the aforementioned test statistics (which, as a general rule, yield similar results) can be converted into F values which are recorded in Column 3 (**Approx. F**) (some software employ a chi-square value in lieu of an F value). Each of the F values can be interpreted through use of **Table A10 (Table of the F Distribution)** in the **Appendix**, employing the degrees of freedom values listed in Column 4 (**Hypoth. DF**, which represents df_{num}) and Column 5 (**Error DF**, which represents df_{den}) for a given row. Column 6 (**Sig. of F**) contains the probability associated with the result for each of the test statistics. If for a given test statistic the probability in the Column 6 is greater than .05, the researcher can conclude that in the case of all three derived canonical functions none of the canonical correlations are significant (or, to put it another way, in the underlying population none of the three derived canonical correlations is some value other than zero). If, on the other hand, the probability in Column 6 is equal to or less than .05, the researcher can conclude that in all likelihood (see Endnote 8) the canonical correlation computed for the first canonical function is significant at the .05 level. A significant result, however, does not indicate whether any of the **canonical correlations** for the remaining canonical functions are significant. Note that since the probabilities .036 (**Pillai's trace**), .007 (**Hotelling's trace**), and .017 (**Wilk's lambda**) in Column 6 are all below .05, we can conclude the canonical correlation computed for the **first canonical function** is significant at the .05 level (and in the case of **Hotelling's trace** at the .01 level, since .007 is less than .01).[7]

The question of whether or not more than one canonical function is statistically significant (based on the use of the **Wilk's lambda statistic** (which once again is converted into an F value)) can be answered by inspecting Column 6 (**Sig. Of F**) of the table labeled **Dimension Reduction Analysis** in Table 38.3. If a significant result (which would be indicated by a probability of .05 or less in Column 6) is obtained for the row labeled **1 TO 3**, it would indicate the canonical correlation for the first canonical function is significant (which was previously suggested in the **EFFECTS .. WITHIN CELLS Regression** table).[8] Note that the value .56617 displayed in Column 2 for **Wilk's lambda** is identical to the value computed for **lambda** in Column 2 of the table labeled **EFFECT .. WITHIN CELLS Regression**. A significant result in the row labeled **2 TO 3** would indicate the canonical correlation for the second canonical function is significant, and a significant result in the row labeled **3 TO 3** would indicate that the canonical correlation for the third canonical function is significant. Since the probabilities .903 and .950 recorded in the last column for the second and third canonical functions are well above .05, the canonical correlations for the latter two canonical functions are not significant.

In instances where two or more canonical functions are found to be statistically significant, it would not be unusual that the amount of variability which can be explained by all but the first canonical function is of little or no practical value. In other words, it is important to make a distinction between statistical and practical significance. For example, if the size of a sample employed in a study was very large, it would not be unusual for a canonical correlation below .30 to be statistically significant. However, given the fact that when $r_C < .30$ the amount of variance shared by the two canonical variates will be less than .09 (since $(.30)^2 = .09$), such a result may be of little practical value, and thus the canonical function in question should probably be ignored.

If a nonsignificant result is obtained for all of the canonical correlations, the analysis would be terminated at that point. When, however, one or more of the canonical correlations are found to be significant, further analytical procedures are conducted on the significant function(s). Since in the case of Example 38.1 only *Function 1* was determined to be significant, the discussion from this point on will, for the most part, focus on additional information obtained for the latter function.

e) **Pooled squared canonical correlation** A **pooled squared canonical correlation** is the sum of all the squared canonical correlations (Garson, 2005), which in the case of Example 38.1 is .417 + .029 + .000 = .446. The latter value represents the proportion of variability shared by the variables with respect to the functions derived in the analysis (it does not, however, represent the proportion of total variability explained in the original variables).

f) **Canonical coefficients** A **canonical coefficient** (also referred to as a **canonical function coefficient** or **canonical weight**) can sometimes provide information regarding the relative importance of individual variables with respect to a canonical function. Computer software generally prints out **raw/unstandardized canonical coefficients** and **standardized canonical coefficients**. Canonical coefficients are analogous to the regression coefficients computed in **multiple regression analysis**, and (as is the case with unstandardized/raw regression coefficients, which cannot be employed to evaluate the relative importance of the predictor variables), the **raw canonical coefficients** cannot be employed to evaluate the relative importance of the variables which comprise the canonical variates for a given canonical function. **Standardized canonical coefficients**, on the other hand, can be of some use with respect to the latter. Harlow (2005, p. 184) notes that although a standardized canonical coefficient is analogous to a standardized regression coefficient (i.e., **beta weight**), unlike a standardized regression coefficient a standardized canonical coefficient can have an absolute value greater than 1. The larger the absolute value of a standardized canonical coefficient computed for a specific variable, the more important that variable may be in defining the canonical variate in question (although,

Table 38.4   Canonical Coefficients and Structure Loadings for Example 38.1

Raw canonical coefficients for DEPENDENT variables
        Function No.

Variable	1	2	3
PULLUP	-.051	.255	.147
SITUP	-.016	-.019	-.001
JUMPJK	.008	.013	-.023

- -

Standardized canonical coefficients for DEPENDENT variables
        Function No.

Variable	1	2	3
PULLUP	-.241	1.211	.697
SITUP	-1.044	-1.256	-.094
JUMPJK	.413	.616	-1.117

- -

Correlations between DEPENDENT and canonical variables
        Function No.

Variable	1	2	3
PULLUP	-.770	.635	.065
SITUP	-.940	.010	-.342
JUMPJK	-.401	.395	-.826

- -

Raw canonical coefficients for COVARIATES
        Function No.

COVARIATE	1	2	3
WEIGHT	.006	-.032	.020
WAIST	.167	.149	-.166
PULSE	-.021	-.084	-.101

- -

Standardized canonical coefficients for COVARIATES
        CAN. VAR.

COVARIATE	1	2	3
WEIGHT	.212	-1.138	.711
WAIST	.811	.724	-.803
PULSE	-.159	-.638	-.771

- -

Correlations between COVARIATES and canonical variables
        CAN. VAR.

Covariate	1	2	3
WEIGHT	.782	-.550	.294
WAIST	.972	.037	-.231
PULSE	-.291	-.545	-.786

as will be noted later, there are exceptions to this rule). Standardized canonical coefficients can be employed in Equations 38.1–38.3 to represent the weights in the equations which define the canonical variates.

At this point we will examine the raw and standardized canonical coefficients computed for Example 38.1. We will only consider the coefficients computed for *Function 1*, since it was the only one that was determined to be significant. In Table 38.4 in the table labeled **Raw canonical coefficients for DEPENDENT variables**, the following values are displayed for the **raw coefficients** for the dependent (Y) variables for *Function 1*: *Pull ups* = $-.051$; *Sit ups* = $-.016$; *Jumping jacks* = $.008$. In the table labeled **Raw canonical coefficients for COVARIATES** the following values are displayed for the raw coefficients for the independent (X) variables for *Function 1*: *Weight* = $.006$; *Waist* = $.167$; *Pulse* = $-.021$. As noted above, the values of the raw canonical coefficients do not provide useful information concerning the relative importance of each variable within the **canonical variate** of which it is a member.

In the table labeled **Standardized canonical coefficients for DEPENDENT variables**, the following values are displayed for the **standardized coefficients** for the dependent (Y) variables for *Function 1*: *Pull ups* = $-.241$; *Sit ups* = -1.044; *Jumping jacks* = $.413$. In the table labeled **Standardized canonical coefficients for COVARIATES** the following values are displayed for the standardized coefficients for the independent (X) variables for *Function 1*: *Weight* = $.212$; *Waist* = $.811$; *Pulse* = $-.159$.

If the magnitude of the absolute values of the standardized canonical coefficients is employed as a criterion for evaluating the relative importance of the variables which comprise the two canonical variates that define *Function 1*, the following is suggested:

1) In the case of the canonical variate for the dependent (Y) variables, the most important variable is *Sit ups* which has an absolute value of 1.044, followed by *Jumping jacks* (.413) and *Pull ups* (.241). The fact that the sign for *Jumping jacks* is the opposite of the other two variables suggests there is an inverse relationship between the number of *Jumping jacks* a person is able to do versus the number of *Sit ups* and *Pull ups*. In the next section, however, it will be noted the latter really is not the case.

2) In the case of the canonical variate for the independent (X) variables, the most important variable is *Waist* which has an absolute value of .811, followed by *Weight* (.212) and *Pulse* (.159). The fact that the sign for *Pulse* is the opposite of the other two variables suggests there is an inverse relationship between a person's *Pulse* and his or her *Waist* and *Weight* (which intuitively does make sense).

Thompson (2000, pp. 305–306) states that although variables with a standardized canonical coefficient close to zero may play a minimal role in defining a canonical variate, a variable can also have a standardized canonical coefficient close or equal to zero, because whatever it does contribute to the canonical variate in question it shares with one or more other variables which comprise that variate (and those other variables have arbitrarily been given credit for that portion of variability in the analysis). The latter is analogous to a variable being partialed out of an analysis in **multiple regression** because of a high degree of multicollinearity (Hair *et al.* (1994, p. 336). For this reason, most sources on **canonical correlation** state that **structure coefficients** (discussed later in this section) are more useful for interpreting the relative contribution of each of the variables which comprise a canonical variate.[9] With respect to the latter, sources state that when a variable with a low standardized canonical coefficient, in fact, does contribute substantially in defining a canonical variate, the absolute value of the structure coefficient for that variable will be substantial. Consequently, the only time a researcher can conclude that a variable makes no contribution in defining a canonical variate is when both the standardized canonical coefficient and the structure coefficient for that variable are close to zero.

g) Geometrical representation of a canonical function   For any canonical function it is possible to compute a score for a subject on each of the two canonical variates which comprise that function — keep in mind that a canonical variate can be viewed as representing an underlying dimension, and consequently a subject's score on a canonical variate can be viewed as his or her score on that dimension. For each of the n subjects which comprise a sample a researcher can compute a score on the **canonical variate for the X variables** (which will represent the number of horizontal units associated with a subject's X score if a scatterplot were employed to depict the results visually) and a score on the **canonical variate for the Y variables** (which will represent the number of vertical units associated with a subject's Y score in the scatterplot). The straight line which comes closest to passing through all of the points in the scatterplot will visually represent the canonical correlation (r_C) computed for the canonical function in question. The larger the value of r_C, the closer the n data points will be to the regression line.

The following protocol can be employed to compute the scores of each subject on the two canonical variates which comprise a canonical function: a) For each variable which comprises the **canonical variate for the X variables**, multiply the subject's standardized score on the X variable by the standardized canonical coefficient computed for that variable. (A standard score (see Equation I.38) on a variable is computed by dividing the difference between the mean raw score on the variable and a subject's raw score on the variable by the standard deviation of the raw scores for the variable.); b) The sum of the products of the standardized scores and standardized canonical coefficients will represent the subject's score on the **canonical variate for the X variables**; c) For each variable which comprises the **canonical variate for the Y variables**, multiply the subject's standardized score on the Y variable by the standardized canonical coefficient computed for that variable; d) The sum of the products of the standardized scores and standardized canonical coefficients will represent the subject's score on the **canonical variate for the Y variables**.[10]

h) Structure coefficients   As noted in the discussion of standardized canonical coefficients, **structure coefficients** (which will be represented by the notation r_S) are generally considered to be the most interpretable weights derived in a canonical analysis. A **structure coefficient** (also referred to as a **structure correlation coefficient, canonical loading** or **canonical factor loading**) is a **Pearson product-moment correlation** between a canonical variate and a specific variable in the set which comprises that variate.[11] Whereas a structure coefficient reflects the **overall** correlation a variable has with a canonical variate, a standardized canonical coefficient reflects the **unique** contribution each variable has with a canonical variate — i.e., the information provided by a standardized canonical coefficient is analogous to that provided by a **partial correlation**.

Since they are correlation coefficients, structure coefficients must fall within the range of values -1 to $+1$. For a given variable, the square of a structure coefficient represents the proportion of variability on that variable which can be accounted for by the canonical variate in question. The latter can also be restated by saying that the square of a structure coefficient is the proportion of variance linearly shared between a variable and a canonical variate.

As noted earlier, structure coefficients are considered by most researchers to be more important than standardized canonical coefficients when interpreting the relative importance of specific variables for a canonical variate. Among others, Harlow (2005, p. 184) notes that an absolute value of .30 or greater for a structure coefficient is informally employed to indicate that a variable makes a meaningful contribution in defining a canonical variate. However, as noted earlier, a correlation of .30 only accounts for $(.30)^2 = .09 = 9\%$ of shared variance — certainly not a large amount by anyone's standards. As is the case with any correlation coefficient, the magnitude of a structure coefficient must be viewed in reference to sample size. In the case of a small sample (e.g., $n = 50$), most researchers would require a larger absolute value for a structure coefficient before viewing it as meaningful (e.g., $r_S \geq .70$). A positive sign for a structure coefficient indicates a high score on that variable is associated with a high score on the relevant canonical variate, while a negative sign on a structure coefficient indicates a high score on the variable is associated with a low score on the relevant canonical variate.

At this point we will examine the structure coefficients computed for Example 38.1. Again, we will only consider the coefficients computed for *Function 1*. In Table 38.4 in the table labeled **Correlations between DEPENDENT and canonical variables** the following values are displayed for the structure coefficients for the dependent (Y) variables for *Function 1*: *Pull ups* $= -.770$; *Sit ups* $= -.940$; *Jumping jacks* $= -.401$. In the table labeled **Correlations between COVARIATES and canonical variables** the following values are displayed for the structure coefficients for the independent (X) variables for *Function 1*: *Weight* $= .782$; *Waist* $= .972$; *Pulse* $= -.291$.

If the magnitude of the absolute values of the structure coefficients is employed as a criterion for evaluating the relative importance of the variables which comprise the two **canonical variates** that define *Function 1*, the following is indicated:

1) In the case of the canonical variate for the dependent (Y) variables, the most important variable is *Sit ups* which has an absolute value of .940, followed by *Pull ups* (.770) and *Jumping jacks* (.401). Note that the variables *Pull ups* and *Jumping jacks* have been reversed with respect to ordering of importance when contrasted with the standardized canonical coefficients — more specifically, use of the structure coefficients indicates that *Pull ups* are more important than *Jumping jacks*. The fact that the absolute value of the standardized canonical coefficient for *Pull ups* is small (.241), yet the absolute value for its structure coefficient is large (.782) can be interpreted as the contribution of *Pull ups* to the canonical variate for the dependent variables being partialed out of the analysis when its contribution is assessed on the basis of the magnitude of its standardized canonical coefficient.[12] The reader should also take note of the fact that the absolute value for *Pull ups* is relatively close to that for *Sit ups*, as well as the fact the sign of the structure coefficient is the same for all three variables. Use of the structure coefficients suggests that *Sit ups* and *Pull ups* are more prominent in defining the canonical variate for the dependent variables than *Jumping jacks*. The fact that the latter variable has a structure coefficient with an absolute value above .30 is diminished by the small sample size upon which it is based. Nevertheless, the implication is that all three variables are positively correlated with one another (since their structure coefficients all have the same sign).

2) In the case of the canonical variate for the independent (X) variables, the most important variable is *Waist* which has an absolute value of .972, followed by *Weight* (.782) and *Pulse* ($-.291$). The fact that the sign for *Pulse* is the opposite of the other two variables suggests an inverse relationship between a person's *Pulse* and his or her *Waist* and *Weight*. Although the relative ordering of the values of the structure coefficients is consistent with those obtained for the standardized canonical coefficients, when the structure coefficients are employed *Weight* appears to make a greater contribution in defining the canonical variate for the independent (X) variables. The fact that *Weight* has a small standardized canonical coefficient (.212) yet a large structure coefficient (.782) can be interpreted as *Weight*'s contribution to the canonical variate for the independent variables being partialed out of the analysis when its contribution is assessed on the basis of the magnitude of its standardized canonical coefficient. Use of the structure coefficients suggests that *Waist* and *Weight* are more prominent than *Pulse* in defining the canonical variate for the independent variables, and that the latter variable appears to contribute minimally if at all (given the absolute value of its coefficient, which is below .30, is obtained with a small sample).

To summarize, if the structure coefficients are employed as the criterion in assessing the importance of the variables, the number of *Pull ups* and *Sit ups* a person can do are the most important variables which comprise the canonical variate for the dependent (Y) variables, while a person's *Waist* size and *Weight* are the most important variables which comprise the **canonical variate** for the independent (X) variables. The fact that the sign of the **structure coefficients** for *Pull ups* and *Sit ups* is negative and for *Waist* and *Weight* is positive indicates an inverse relationship between scores on the canonical variates. Simply put, the latter suggests the greater the number of *Pull ups* and *Sit ups* a person can do, the lower the persons's *Waist* size and *Weight*.

It should be noted that it is possible for a variable to have a structure coefficient close or equal to zero and a standardized canonical coefficient with a large absolute value. Thompson (2000, p. 306) and Wuensch (2005) state the latter indicates the role of a **suppressor variable**. As noted in the chapter on **multiple regression**, a **suppressor variable** is a predictor variable which is useful in predicting a criterion variable purely on the basis of the fact that it correlates highly with the other predictor variables. A suppressor variable (which in itself has little or no correlation with the criterion variable) improves prediction on the criterion variable by suppressing variance that is irrelevant in predicting the latter variable.

A final point that should be made is that since both structure coefficients and canonical correlations can be unstable, and especially with small samples, researchers should use caution in generalizing the results of **canonical correlation** without conducting one or more **external replicability analyses** and/or **cross-validation studies** (Hair *et al.* (1994, p. 337) and Stevens (2002, p. 475)).

i) **Canonical adequacy coefficient** The **mean** of all the **squared structure coefficients** for a given canonical variate represents the **canonical adequacy coefficient** for that variate. A **canonical adequacy coefficient** (also referred to as the **canonical variate adequacy coefficient**) indicates on average how adequately a canonical function represents the variability on the variables which comprise a canonical variate. The largest obtainable value for a canonical adequacy coefficient is 1, which would indicate all (i.e., 100%) of the total variance on the variables which comprise a canonical variate is reproduced within a canonical function (Thompson (2000, p. 296).

In the case of Example 38.1 for the canonical variate for the Y variables, the average of the squared structure coefficients is [(*Pull ups* = $-.770)^2$ + (*Sit ups* = $-.940)^2$ + (*Jumping jacks* = $-.401)^2$] / 3 = (.593 + .884 + .161) / 3 = 1.638/3 = .5460, which represents the **canonical adequacy coefficient for the canonical variate for the Y variables**. Thus, on average the canonical function reproduces approximately 54.60% of the variance on the variables which comprise the canonical variate for the Y variables. In the case of the canonical variate for the X variables, the average of the squared structure coefficients is [(*Weight* = $.782)^2$ + (*Waist* = $.972)^2$ + (*Pulse* = $-.291)^2$] / 3 = (.612 + .945 + .085) / 3 = 1.642/3 = .5473, which represents the **canonical adequacy coefficient for the canonical variate for the X variables**. It indicates, on average, the canonical function reproduces approximately 54.73% of the variance on the variables which comprise the canonical variate for the X variables.

j) **Redundancy coefficients** Within the framework of **canonical correlation** the term **redundancy** indicates how much of the **total variance** on the variables which comprise one canonical variate can be predicted from the opposite canonical variate. The greater the value computed for a **redundancy coefficient** (which can be expressed either as a proportion or a percentage) the greater the predictability from the opposite canonical variate. Hair *et al.* (1994, p. 335) depict a redundancy coefficient as analogous to a **squared multiple correlation coefficient** in **multiple regression analysis**.

Although redundancy coefficients for both variates are commonly computed, Hair *et al.* (1994, p 336) note for a given canonical function a researcher is more likely to be interested in the degree to which the variables which comprise the canonical variate for the dependent (Y) variables can be predicted from the canonical variate for the independent (X) variables (since the latter relationship is the most realistic indicator of the predictive ability of a canonical function). Garson (2005) notes that (up to the point of not having multicollinearity) it would be desirable to obtain a high value for a redundancy coefficient, since the latter would indicate the canonical variate for the X variables accounts for a large proportion of the total variability in original dependent variables which comprise the canonical variate for the Y variables.[13] Note that the proportion represented by a redundancy coefficient is not the same as the proportion represented by the squared canonical correlation (which represents for the canonical function in question, the proportion of variability accounted for in the canonical variate for the Y variables by the canonical variate for the X variables).

Stewart and Love (1968) proposed a **redundancy coefficient** which is the product of the **canonical adequacy coefficient** for a **canonical variate** multiplied by the **squared canonical correlation**. Thus, in the case of the **canonical variate for the Y variables**, the **redundancy coefficient** = [**canonical adequacy coefficient** = .5460] × (r_C = .646)^2 = (.5460)(.4173) = .2278. The latter value indicates that 22.78% of the total variability on the variables which comprise the canonical variate for the Y variables can be explained by the canonical variate for the X variables. Note that both the canonical adequacy coefficient and the canonical correlation must be high in order for a redundancy coefficient to be high. In the case of the **canonical variate for the X variables**, the **redundancy coefficient** = [**canonical adequacy coefficient** = .5473] × (r_C = .646)^2 = (.5473)(.4173) = .2284. The latter value indicates that 22.84% of the total variability on the variables which comprise the canonical variate for the X variables can be explained by the canonical variate for the Y variables.

The redundancy coefficients computed with *SPSS* are displayed in Table 38.5. For *Function 1* (which is listed in the row labeled **1**), the value 22.740 in Column 4 (**Pct Var CO**) of the table labeled **Variance in dependent variables explained by canonical variables** represents the redundancy coefficient for the **canonical variate for the Y variables**, which (except for rounding off error) corresponds to the value 22.78% computed above. The value 54.548 in Column 2 (**Pct Var DE**) is the percentage of variance on the original Y variables explained by the canonical variate for the Y variables (Wuensch (2005)). The value 100.000 in the last row of Column 3 (**Cum Pct DE**) indicates that when taken together all three of the derived canonical functions account for 100% of the variance on the original Y variables.

For *Function 1*, the value 22.813 in Column 2 (**Pct Var DE**) of the table labeled **Variance in covariates explained by canonical variables** represents the redundancy coefficient for the **canonical variate for the X variables**, which (except for rounding off error) corresponds to the value 22.84% computed above. The value 54.724 in Column 4 (**Pct Var CO**) is the proportion of variance on the original X variables explained by the canonical variate for the X variables (Wuensch (2005)).[14] The value 100.000 in the last row of Column 5 (**Cum Pct CO**) indicates that when taken together all three of the derived canonical functions account for 100% of the variance on the original X variables.

Another correlational measure which can also be employed to compute redundancy coefficients are **cross-loadings**. A **cross-loading** is the correlation between a variable on one side of a canonical function with the canonical variate on the opposite side of the function. Wuensch (2005) notes that the average of the squared cross-loadings of the variables on one side of a canonical function with the canonical variate on the opposite side of the function will yield the same value for the redundancy coefficient obtained earlier when a canonical adequacy

Table 38.5   *SPSS* Redundancy Analysis for Example 38.1

Variance in dependent variables explained by canonical variables

CAN. VAR.	Pct Var DE	Cum Pct DE	Pct Var CO	Cum Pct CO
1	54.548	54.548	22.740	22.740
2	18.654	73.202	.540	23.280
3	26.798	100.000	.003	23.283

Variance in covariates explained by canonical variables

CAN. VAR.	Pct Var DE	Cum Pct DE	Pct Var CO	Cum Pct CO
1	22.813	22.813	54.724	54.724
2	.580	23.393	20.034	74.758
3	.003	23.396	25.242	100.000

- -

coefficient was multiplied by the squared canonical correlation. Thus in the case of Example 38. 1, the average of the squared correlations (i.e., the squared cross-loadings) between the Y variables and the canonical variate for the X variables will equal .2278, and the average of the squared correlations between the X variables and the canonical variate for the Y variables will equal .2284.[15]

Although Hair *et al.* (1994, pp. 337–338) endorse the use of cross-loadings in interpreting canonical relationships, Harlow (2005, p. 191) states that since the latter measures can be cumbersome to interpret a researcher would be wise to employ multiple regression analysis to clarify the relationships in a canonical analysis.[16] Specifically, the researcher should conduct a number of multiple regression analyses equivalent to the number of dependent variables used in a study, and in each analysis employ one of the criterion variables as the dependent variable with all of the predictors as the independent variables. In point of fact, Hair *et al.* (1994, p. 335) and Stevens (2002, p. 184) note that one can obtain the redundancy coefficient for the canonical variate for the Y variables by computing q multiple correlation coefficients, and in each of the multiple correlations employ the p X variables as the predictor variables and one of the Y variables as the criterion variable. The average of the q squared multiple correlation coefficients will equal the value computed earlier for the redundancy coefficient for the canonical variate for the Y variables. In the same respect, one can compute the redundancy coefficient for the canonical variate for the X variables by computing p multiple correlation coefficients, and in each of the multiple correlations employ the q Y variables as the predictor variables and one of the X variables as the criterion variable. The average of the p squared multiple correlation coefficients will equal the value computed earlier for the redundancy coefficient for the canonical variate for the X variables.

The above described relationship between multiple correlations and redundancy coefficients, draws attention to a problem associated with the latter type of coefficient (first noted by Cramer and Nicewander (1979) and discussed in Stevens (2002, p. 484) and Thompson (1984)). Specifically, a redundancy coefficient is insensitive to the degree that the criterion variables which comprise a canonical variate are intercorrelated with one another. Because of the latter, the same value will be obtained for the redundancy coefficient for the canonical variate for the Y variable regardless of whether the intercorrelations between the q Y variables are high or low. Stevens (2002, p. 484) notes that the latter is undesirable, just as it would be undesirable if the multiple correlation coefficient computed in a multiple regression analysis was unaffected by the degree of intercorrelation between the predictor variables. Because of this insensitivity problem, use of redundancy coefficients is often criticized on the basis of it violating the primary reason for employing a multivariate procedure — specifically that such a procedure will simultaneously take into consideration the impact of all of the variables involved in an analysis. Because of this problem Thompson (1984; 2000, p. 309) generally advocates caution in interpreting such coefficients (also see Stevens (2002, pp. 483–485)). Further discussion of redundancy coefficients can be found in Hair *et al.* (1994, pp. 334–336), Harris (2001, pp. 293–295), Stevens (2002, pp. 483–485), Tabachnick and Fidell (1996, Ch. 6), and Thompson (1984, 2000).

k) **Canonical communality coefficients** A **canonical communality coefficient** is the **sum of the squared structure coefficients** for a given variable across the canonical functions (Garson (2005) and Thompson (2000, p. 297)). This coefficient provides a measure of how much of the variability on a specific variable can be attributed to the derived canonical functions. To illustrate the latter with respect to Example 38.1, the canonical communality coefficient for *Weight* (employing the structure coefficients in the table labeled **Correlations between COVARIATES and canonical variables** in Table 38.4) is: $(.782)^2 + (-.550)^2 + (.294)^2 = 1.00$. The result indicates that the three derived canonical functions account for 100% of the variance on the independent variable of *Weight*. In point of fact, in the case of Example 38.1 a canonical communality coefficient equal to or close to 1 is obtained for the other five variables. However, it will not always be the case that all or most of the communality coefficients are close to 1. Thompson (2000, p. 297) notes that if the canonical communality coefficient computed for a specific variable is substantially lower than that computed for the other variables, by omitting the latter variable from the analysis a researcher might obtain a more parsimonious solution for an canonical analysis. He also states that any variable which has a standardized canonical coefficient close to zero as well as a canonical communality coefficient close to zero contributes nothing to a canonical analysis (Thompson (2000, p. 306)).

l) **Use of multiple regression to determine which independent variables are significant predictors of the dependent variables** As noted earlier, among others, Harlow (2005, p. 191) recommends the use of **multiple regression analysis** to provide further clarification with respect to the relationship between the canonical variate for the X variables and each of the variables which comprise the canonical variate for the Y variables. As previously noted, the latter entails conducting q multiple regression analyses, in which in each analysis the independent variables are all p variables which comprise the canonical variate for the X variables, and the dependent variable is one of the q variables which comprise the canonical variate for the Y variables. The results of the latter analyses are displayed in Tables 38.6–38.8. Each of the tables summarizes an **analysis of variance for a multiple regression analysis** (see Table 33.3) with *Waist*, *Weight*, and *Pulse* as the independent variables, and respectively employing *Sit ups*, *Pull ups*, and *Jumping jacks* as the dependent variable. Note that the result for both *Sit ups* $(F(3, 36) = 6.987, p = .001$, with $R_{Y_1 \cdot X_1 X_2 X_3} = .607)$ and *Pull ups* $(F(3, 36) = 4.188, p = .012$ with $R_{Y_2 \cdot X_1 X_2 X_3} = .509)$ are significant (with the probability value for *Sit ups* being slightly lower than that for *Pull ups*), but the result for *Jumping jacks* $(F(3, 36) = .928, p = .437$ with $R_{Y_3 \cdot X_1 X_2 X_3} = .268)$ is nonsignificant.[17] The results of the analyses of variance are consistent with the conclusions reached when the structure coefficients were employed to evaluate the relative importance of the three Y variables in defining the canonical variate for the dependent variables. Recollect that employing the structure coefficients, it was concluded that *Sit ups* and *Pull ups* are more prominent in defining the canonical variate for the dependent variables than *Jumping jacks*

Table 38.6   Summary Table of Analysis of Variance for
Multiple Regression Analysis (Where: *Y = Sit ups*)

Source of variation	SS	df	MS	F
Regression	61417.29	3	20472.43	6.987
Residual	105478.61	36	2929.96	
Total	166895.90	39		

Table 38.7   Summary Table of Analysis of Variance for Multiple
Regression Analysis (Where: *Y = Pull ups*)

Source of variation	SS	df	MS	F
Regression	226.87	3	75.62	4.188
Residual	650.11	36	18.06	
Total	876.98	39		

Table 38.8   Summary Table of Analysis of Variance for Multiple
Regression Analysis (Where: *Y = Jumping jacks*)

Source of variation	SS	df	MS	F
Regression	6739.93	3	2246.64	.928
Residual	87126.07	36	2420.17	
Total	93866.00	39		

(which appeared to contribute minimally). The latter is concordant with the significant F values (and associated multiple correlation coefficients) computed for *Sit ups* and *Pull ups* and the nonsignificant result obtained for *Jumping jacks*.

In the discussion of redundancy coefficients it was noted that the redundancy coefficient for the canonical variate for the Y variables is equivalent to the average of the squared multiple correlations between the total predictor set and each of the criterion variables. In the case of Example 38.1, $(R_{Y_1 \cdot X_1 X_2 X_3} = .607)^2 = .3684$, $(R_{Y_2 \cdot X_1 X_2 X_3} = .509)^2 = .2591$, and $(R_{Y_3 \cdot X_1 X_2 X_3} = .268)^2 = .0718$. Thus, $(.3684 + .2591 + .0718)/3 = .2331$ which (except for rounding off error) would be equivalent to the value .2278 computed for the redundancy coefficient for the canonical variate for the Y variables.

VI. Additional Analytical Procedures for Canonical Correlation and/or Related Tests

No additional procedures will be described in this section.

VII. Additional Discussion of Canonical Correlation

1. Over the years a major criticism of **canonical correlation** has been that although the methodology can maximize correlation, it often yields results which are difficult if not impossible to clearly interpret (Tabachnick and Fidell (1996, p. 197)). One technique employed by some researchers to facilitate interpretation is **rotating** the canonical variates. (The concept of rotation is discussed in Section V of **Principal Components Analysis and Factor Analysis (Test 40)**.)

Use of rotation in **canonical correlation** is somewhat controversial and is not described in most sources (as well as not an available option with most computer software). Readers interested in employing rotation within the context of **canonical correlation** are referred to Harris (2001, Ch. 5).

2. Although not commonly employed, Thompson (1984) discusses the use of **backward canonical correlation** as a possible methodology for eliminating variables which contribute minimally to a canonical solution. The rationale behind **backward canonical correlation** is similar to that employed for **backward selection** in **multiple regression analysis**.

3. **Canonical correlation** is viewed as an exploratory procedure by many researchers because, as noted above, it often yields results which are difficult to interpret, as well as the fact it is unable to establish cause-effect relationships between the variables. Among others, Harlow (2005, p. 180) notes that because of the latter **structural equation modeling** can be employed as a rigorous follow up procedure to a canonical analysis in order to better clarify the precise nature of the relationships between the variables of interest.

VIII. Additional Examples Illustrating the Use of Canonical Correlation

No additional examples will be presented in this section.

References

Campbell, K. T. & Taylor, D. L. (1996). Canonical correlational analysis as a general linear model: A heuristic lesson for teachers and students. **Journal of Experimental Education**, 64, 157–171.

Cohen, J. (1968). Multiple regression as a general data-analytic system. **Psychological Bulletin**, 70, 426–443.

Cramer, E. & Nicewander, W. A. (1979). Some symmetric, invariant measures of multivariate association. **Psychometrika**, 44, 43–54.

Garson, D. G. (2004). Statistics: Topics in multivariate analysis. Website: **http://www2.chas.ncsu.edu/garson/pa765/statnote.htm**.

Grimm, L. G. & Yarnold, P. R. (Eds.) (2000). **Reading and understanding more multivariate statistics**. Washington, D.C.: American Psychological Association.

Hair, J. F., Anderson, R. E., Tatham, R. L. & Black, W. C. (1995). **Multivariate data analysis** (4th ed.). Upper Saddle River, N. J.: Prentice Hall.

Hair, J. F., Anderson, R. E., Tatham, R. L. & Black, W. C. (2004). **Multivariate data analysis** (6th ed.). Upper Saddle River, N. J.: Prentice Hall.

Harlow, L. L. (2005). **The essence of multivariate thinking**. Mahwah, NJ: Lawrence Erlbaum Associates.

Harris, R. J. (2001) **A primer of multivariate statistics** (3rd ed.). Mahwah, NJ: Lawrence Erlbaum Associates.

Horst, P. (1961). Generalized canonical correlations and their applications to experimental data. **Journal of Clinical Psychology**, 26, 331–347.

Hotelling, H. (1935). The most predictable criterion. **Journal of Experimental Psychology**, 26, 139–142.

Howell, D. C. (2002). **Statistical methods for psychology** (5th ed.). Pacific Grove, CA: Duxbury Press.

Kachigan, S. K. (1986). **Statistical analysis**. New York: Radius Press.

Knapp, T. R. (1978). Canonical correlational analysis: A general parametric significance testing system. **Psychological Bulletin**, 85, 410–416.

Marascuilo, L. A. & Levin, J. R. (1983). **Multivariate statistics in the social science: A researchers guide**. Monterey, CA: Brooks/Cole Publishing Company.

Miller, J. K. (1975). The sampling distribution and a test of significance of the bimultivariate redundancy statistic: A Monte Carlo study. **Multivariate Behavior Research**, 10, 233–244.

Stevens, J. (2002). **Applied multivariate statistics for the social sciences** (4th ed.). Hillsdale, NJ: Lawrence Erlbaum Associates, Publishers.

Stewart, D. K. & Love, W. A. (1968). A general canonical correlation index. **Psychological Bulletin**, 70, 160–163.

Tabachnick, B. G. & Fidell, L. S. (1996). **Using multivariate statistics** (3rd ed.). New York: Harper Collins Publishers.

Tabachnick, B. G. & Fidell, L. S. (2001). **Using multivariate statistics** (4th ed.). Boston: Allyn & Bacon.

Tacq, J. (2004). Canonical correlational analysis. In Lewis-Beck, M. S., Bryman, A. & Liao, T. F. (Eds.) (pp. 83–86). **The Sage encyclopedia of social science research methods (Vol 1)**. Thousand Oaks, CA: Sage Publications.

Thompson, B. T. (1984). **Canonical correlational analysis: Uses and interpretation**. Newbury Park, CA: Sage Publications.

Thompson, B. T. (2000). Canonical correlational analysis. In Grimm, L. G. & Yarnold, P. R. (Eds.) (pp. 285–316). **Reading and understanding more multivariate statistics**. Washington, D.C.: American Psychological Association.

Trochim, W. M. (2005). **Research methods: The concise knowledge base**. Cincinnati: Atomic Dog Publishing Company.

Webster's new collegiate dictionary (1981). Springfield, MA: G. & C. Merriman Co.

Wilks, S. S. (1932). Certain generalizations in the analysis of variance. **Biometrika**, 24, 471–494.

Wuensch, K. (2005). Statistics lessons. **Website: http://core.ecu.edu/psyc/wuenschk/SPSS/SPSS-Mv.htm**.

Endnotes

1. According to Webster's New Collegiate dictionary (1981) the word **canonical** means, "reduced to the simplest or clearest schema possible."

2. a) Because it employs multiple predictor variables and multiple criterion variables **canonical correlation** is sometimes referred to as **multiple-multiple correlation** (Wuensch (2005)); b) Horst (1960) notes that although infrequently employed, **canonical correlation** can be generalized to situations involving more than two variable sets; c) Thompson (2000, p. 297) states that **canonical correlation** represents the most general case of the **general linear model** (discussed in Section VII of the **single-factor between-subjects analysis of variance (Test 21)**), and cites Knapp's (1978, p. 410) contention that essentially all commonly employed parametric tests of statistical significance can be viewed as special cases of **canonical correlation** (also see Campbell and Taylor (1996)). In order to clarify the latter, it is necessary to define the **general linear model** (often identified with the acronym **GLM**), which can be summarized (if for the moment one is not overly concerned with mathematical rigor) with Equation 38.5 (based on Trochim (2005); also see Howell (2002, pp. 604–607).

$$\mathbf{Y} = b_0 + b_i\mathbf{X} + e \qquad \text{(Equation 38.5)}$$

Where: **Y** represents a set of criterion variables
           X represents a set of predictor variables
           b_0 represents the set of intercepts on the axes for the criterion variables
           b_i represents the weight computed for the ith predictor variable
           e represents error variability

Equation 38.5 represents the general form of Equation 28.4 (the equation for simple linear regression) and Equation 33.4 (the equation for multiple regression). The general linear model states that parametric statistical procedures can be summarized in the form of one or more linear equations, and in the final analysis a correlational format can be employed to summarize all such procedures (or, to put it another way, all parametric procedures can be conceptualized as special cases of regression analysis). Illustrative of the latter are the fact that in Section VII of the **Pearson product-moment correlation coefficient** it was demonstrated how regression analysis could be employed to obtain a result equivalent to that obtained for the **single-factor between-subjects analysis of variance**, and in Section IX (the Addendum) of the **Pearson product-moment correlation coefficient** under the discussion of the **point-biserial correlation coefficient (Test 28i)** it was demonstrated how the latter measure could be employed to obtain a result equivalent to that obtained with the *t* **test for two independent samples/single-factor between-subjects analysis of variance.**

3.   The data for Example 38.1 are hypothetical and for demonstration purposes only. The sample size $n = 40$ is smaller than the minimum most researchers would employ for a **canonical correlation**.

4.   In order to conduct a **canonical correlational analysis** with *SPSS* it is necessary to employ the syntax editor. The analysis is conducted within the framework of a **multivariate analysis of variance**. The syntax employed for Example 38.1 is reproduced below.
    manova situps pullups jumpingjacks with weight waist pulse
    /discrim all alpha (1)
    /print=sig (eig. raw dim).

5.   The author is indebted to Karl Wuensch for clarifying what the eigenvalues printed out by *SPSS* represent. Many sources (e.g., Garson (2005), Hair *et al.* (1994, p. 333) and Tabachnick and Fidell (1996, p. 201)) define an eigenvalue as equivalent to the value of the squared canonical correlation (i.e., r_C^2). (Tabachnick and Fidell (1996, p. 201), however, acknowledge this discrepancy in the use of the term **eigenvalue** in a footnote on page 201.)

6.   Some computer software also prints out **adjusted** or **shrunken** values for the canonical correlations which corrects for bias resulting from a small sample size. Adjusted coefficients are discussed in Section IV/V of the chapter on **multiple regression**.

7.   In the case of **Roy's largest root**, *SPSS* only prints out the value of the test statistic.

8.   Tabachnick and Fidell (1996, p. 221) note it is theoretically possible for the first canonical function by itself to not be significant, but rather only to achieve significance when considered in combination with the remaining canonical functions. In such an instance, the probability in the row **1 TO 3** will be greater than .05. The latter authors note there is no acceptable test to evaluate each of the canonical functions separately.

9.   For a contrary position the reader is referred to Harris (2003, Ch. 5), who attaches more importance to the interpretation of standardized canonical coefficients than most researchers.

10. An excellent illustration of the geometrical representation of a canonical function can be found in Thompson (2000, pp. 292–295), who refers to subject's scores on the canonical variates as **synthetic-latent variable scores** (which reflects that fact that such a score is assumed to represent a subject's standing on some underlying dimension).

11. More specifically, a structure coefficient is the correlation between the scores of subjects on a specific canonical variate with the standardized scores of subjects on one of the variables which comprise that variate.

12. Stevens (2002, pp. 476–477) states that if the absolute value of the structure coefficient for a given variable which comprises a canonical variate is high, yet its standardized canonical coefficient is low, such a variable would be viewed as **redundant**. Inspection of Table 38.4 suggests the latter logic could be applied to Y variable of *Pull ups* and the X variable of *Weight*. The concept of **redundancy** is discussed later in this section.

13. Garson (2005) notes that a unique solution for **canonical correlation** cannot be obtained if variables are redundant — i.e., have a perfect or near perfect correlation with one another. A correlation matrix with redundancy is said to be **singular** or **ill-conditioned**, which falls within the general rubric of the multicollinearity problem.

14. a) Although in the case of Example 38.1 the two values computed for the redundancy coefficients (.2284 and .2278) are almost identical, Thompson (1984) and Wuensch (2005) note the redundancy coefficient is an asymmetric index — i.e., the redundancy coefficient computed for one variate will rarely equal that computed for the other; b) Thompson (2000, p. 309) notes the only time a redundancy coefficient will equal 1 (which would be highly unusual) is when a canonical variate accounts for all of the variance on every variable which comprises the variate (which is commensurate with all the squared structure coefficients for the variate being equal to 1) and the squared canonical correlation for the relevant canonical function equals 1; c) Although Miller (1975) developed a test of significance for redundancy coefficients, there are no generally accepted guidelines with respect to when a redundancy coefficient should be viewed as excessive.

15. a) Harlow (2005, p. 184) notes that although it is desirable for variables to have a correlation greater than .30 (once again, in the case of a large sample) with the opposite variate, as a general rule it would be expected that variables will be more highly correlated with their own canonical variate; b) *SPSS* will not print out the cross-loadings for **canonical correlation** unless additional instructions are included in the syntax editor.

16. Hair *et al.* (1994, pp. 334–335) note a statistically significant canonical correlation (even a large value) for a canonical function does not necessarily indicate that such a function is of practical significance. Although the latter authors state that no generally accepted guidelines have been established with respect to a minimum acceptable value for a redundancy coefficient, they attach a lot more importance than most sources to using redundancy coefficients in determining whether or not a canonical function is of practical significance (Hair *et al.* (1994, p. 336)). Although the values .2278 and .2284 computed in

Example 38.1 for the redundancy coefficients might be considered low by some researchers, the information derived from the structure coefficients suggest that *Function 1* may be of practical significance.

17. Wuensch (2005) notes the value of the canonical correlation for the first canonical function will always be equal to or greater than the largest of the multiple correlations. The latter is the case for Example 38.1 where $r_C = .646$ is greater than the largest multiple correlation .607(for *Sit ups*).

Test 39

Logistic Regression
(Nonparametric Test Employed with Interval/Ratio and/or Categorical Data)

I. Hypothesis Evaluated with Test and Relevant Background Information

Logistic regression is not an inferential statistical test, but instead is a methodology for group classification. Employing the relevant predictor variable(s), **logistic regression** derives an equation which provides a probability a subject/observation will be a member of specific category/group.

Although some sources define a multivariate procedure as one in which there are two or more dependent variables, **logistic regression**, which involves one dependent variable (that is evaluated with respect to one or more independent/predictor variables), is commonly discussed within the framework of multivariate procedures. Although some sources employ the term **logistic regression** when the analysis involves only one independent variable and the term **multiple logistic regression** to refer to an analysis involving two or more independent variables, in this chapter the term **logistic regression** will be employed generically, regardless of how many independent variables are employed in an analysis. **Logistic regression** is categorized as a nonparametric procedure, since in contrast to the procedures described in the preceding six chapters, it is not affected by certain restrictive normality assumptions that apply to parametric analyses.

Hypothesis evaluated with test The primary hypothesis evaluated within the framework of **logistic regression** is whether or not one or more predictor variable(s) can significantly predict a subject's category/group on the dependent variable.

Relevant background information on test **Logistic regression** is an alternative to **discriminant function analysis (Test 37)** for predicting group membership. As is the case for **discriminant function analysis**, in **logistic regression** a discrete/qualitative criterion/dependent variable is employed. However, in **logistic regression** the independent/predictor variable(s) can be discrete/categorical, continuous, or a combination of both. Whereas **discriminant function analysis** focuses on correlational weights or percentage of correct classifications, the focus of **logistic regression** is on the odds or likelihood of a specific outcome.

Logistic regression is also an alternative to **standard linear regression** described in the chapters on the **Pearson product-moment correlation coefficient (Test 28)** and **multiple regression (Test 33)**. As is the case with **standard linear regression**, **logistic regression** regresses one or more independent variables on a dependent variable. However, as noted above, in **logistic regression** the independent/predictor variable(s) need not be continuous (i.e., interval/ratio data), and, unlike **standard linear regression**, **logistic regression**, involves a categorical dependent/criterion variable. The most common type of **logistic regression** is **binary logistic regression** in which the dependent variable is comprised of two categories. Instead of predicting the score of a subject on the dependent variable, **logistic regression** derives a probability that a subject will fall in one of the two categories.[1]

Like **discriminant function analysis** and **standard linear regression**, one assumption of **logistic regression** is that the n observations employed in an analysis are independent of one another. However, **logistic regression** is more flexible than the other procedures in that it does not depend on certain restrictive normality assumptions regarding the underlying population distributions for the predictor variables — specifically, the predictor variables do not have to be normally distributed, linearly related, or of equal variance within each group (Tabachnick and Fidell (1996, p. 575)).[2] Unlike **standard linear regression** and **discriminant function analysis**, **logistic regression** is appropriate to employ if the relationship between the independent and dependent variables is nonlinear. In point of fact, in **logistic regression** the relationship between the independent variable(s) and dependent variable is assumed to be nonlinear — specifically, a logarithmic function. Another reason why **logistic regression** is employed with a dichotomous dependent variable is that use of **standard linear regression** with the latter type of dependent variable can result in violation of the homoscedasticity assumption, plus the fact the residuals will not be normally distributed. Additionally, if **standard linear regression** (as opposed to **logistic regression**) is employed to predict the probability of a subject being in a specific group, it can yield a probability less than 0 or greater than 1, both of which are impossible.

Logistic regression is particularly useful when the relationship between the distribution of scores on a dichotomous dependent variable and one or more independent variables is best described by a nonlinear function known as a **logistic function**. The latter type of relationship is most commonly encountered in the field of medicine where researchers often want to predict whether a person will develop a disease, respond to a treatment, or survive beyond a certain point in time. Figure 39.1 depicts a hypothetical **logistic function** which describes the relationship between a ratio level predictor/independent variable (systolic blood pressure of elderly adults) and a categorical criterion/dependent variable (whether or not a person has a stroke). Inspection of Figure 39.1 reveals that among elderly adults (i.e., 70 or older) whose systolic blood pressure is in the range 80 to 180, there is a minimal likelihood of having a stroke, and there is only a minimal increase in the probability of a person having a stroke as one's pressure increases from the lowest value in the range (80) to the highest value (180). In the systolic pressure range 200 to 300 there is a very high likelihood of having a stroke, but once again there is only a minimal increase in the probability of a person having a stroke as one's pressure increases from the lowest value in the latter range (200) to the highest value (300). On the other hand, the probability of a person having a stroke increases precipitously as systolic pressure increases within the range of values 181 to 199. Note that although a straight line provides an imperfect fit for the data, a **logistic function** (which generally resembles the S-shaped curve — commonly referred to as a **sigmoidal function**) depicted in Figure 39.1, provides an excellent fit.[3]

A major difference between **logistic regression** and other procedures discussed in this book is that the dependent variable in **logistic regression** is expressed in the form of what is referred to as a **logit variable** — specifically, the **natural logarithm of the odds** of a subject being a member of a particular category/group. The basis for employing a logarithmic transformation within the context of **logistic regression** is that the latter transformation converts a nonlinear relationship (such as that depicted in Figure 39.1) into one that approximates linearity — more specifically, the natural logarithm of the odds can be employed to represent a linear combination of the independent variables. In addition, the logarithmic transformation allows the researcher to compute a probability (which will always fall between 0 and 1) that a subject is a member of a specific category based on his or her score on the predictor variable(s).

Among others, Tabachnick and Fidell (1996, p. 576) note that **logistic regression** is employed to fit and compare two or more models (which may be employed for predictive or theoretical purposes) that describe the relationship between one or more predictor variables and a dependent variable. The model which typically provides the poorest fit for a set of data is

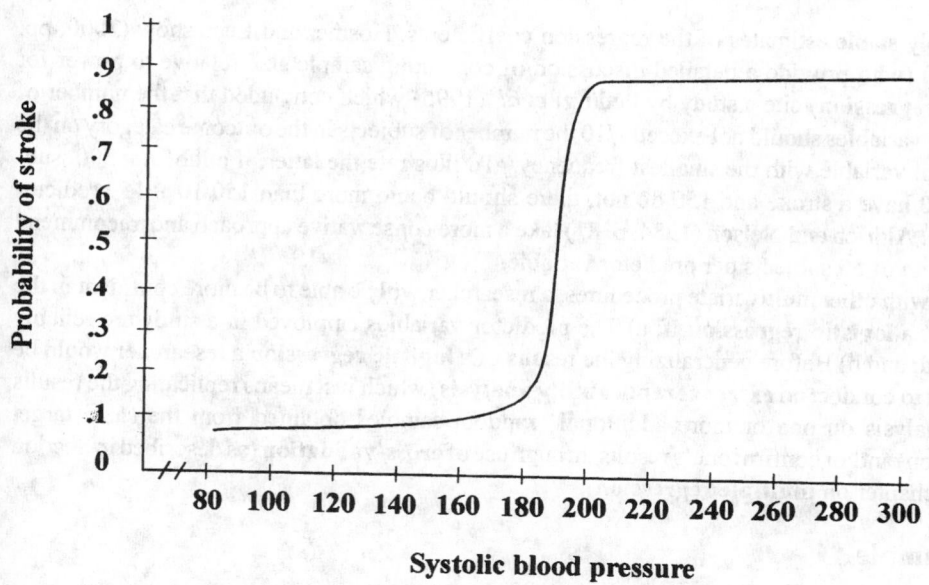

**Figure 39.1   Systolic Blood Pressure versus Probability of Stroke
Among Elderly Adults**

referred to as the **simple model** (as well as the **reduced** or **null model**), and the latter model is characterized by the fact that it does not include any of the predictor variables (and is thus described by an equation which only contains a constant but no predictor coefficients). The most complex, and possibly best fitting, of the models will include the constant, all predictors (with each one having a coefficient), and perhaps interactions between the predictors. The latter model is referred to as the **full** or **saturated model**. However, since all of the predictor variables may not be related to the dependent variable, goodness-of-fit tests can be employed to select the model which provides optimal prediction employing the fewest predictor variables. Within the context of the latter model, a predictive equation is derived that employs the natural logarithm of the odds to represent a linear combination of the predictor variables.

As is the case for **multiple regression** and **discriminant function analysis**, **standard** and **stepwise** analyses can be employed for **logistic regression**. (The latter procedures are reviewed in Section I of **discriminant function analysis**.) Most sources state that **stepwise logistic regression** can be used as an exploratory tool during the early stages of research, but should not be employed for testing theories. For example, Hosmer and Lemeshow (2000, p. 116) state that stepwise methods can be useful in screening a large number of predictor variables with respect to having a potential relationship with a dependent variable. In using stepwise methods, Field (2005, p. 226) endorses the use of **backward selection** over **forward selection**. **Backward stepwise logistic regression** begins with the full or saturated model, and variables are eliminated from the model one by one. The model is reevaluated after each elimination to ensure that it still adequately fits the data. When no more variables can be removed, the remaining model is retained. In this chapter I will focus on the use of **standard logistic regression** as opposed to **stepwise regression**.

Sources (e.g., Wright (1995, p. 221)) note that **logistic regression** requires a larger sample size than **standard linear regression**, since in **logistic regression** too small a sample can result in unreliable regression coefficients characterized by high standard errors. van Belle (2002, p. 87) states that approximately 10 subjects per predictor variable are necessary in order to get reasonably stable estimates of the regression coefficients. Hosmer and Lemeshow (2000, pp. 339–347) (who provide a detailed discussion of computing sample size relative to power for **logistic regression**) cite a study by Peduzzi *et al.* (1996) which concluded that the number of predictor variables should not exceed 1/10 the number of subjects in the outcome category on the dependent variable with the smallest frequency. To illustrate the latter, if out of $n = 500$ subjects, 350 have a stroke and 150 do not, there should be no more than $150/10 = 15$ predictor variables. Aldrich and Nelson (1984, p. 81) take a more conservative approach and recommend a minimum of 50 subjects per predictor variable.

As with other multivariate procedures, a researcher will be able to be more confident in the results of a **logistic regression** if: a) The predictor variables employed in a study are reliably measured; and b) Before generalizing the results of a **logistic regression** a researcher would be expected to conduct an **external replicability analysis** (which just means replicating the results of an analysis on one or more additional random samples obtained from the same target population) and/or confirm one's results through use of **cross-validation** (as described in Section I of the chapter on **multiple regression**).

II. Example

Example 39.1 investigates the use of one continuous and two binary predictor variables to predict a binary dependent variable. Initially, only the continuous predictor variable of systolic blood pressure will be employed in Sections IV and V to illustrate the analytical basis for **logistic regression**. Following that, a **logistic regression** involving all three predictor variables will be conducted.

Example 39.1 *A gerontologist conducted a study to examine what if any relationship systolic blood pressure, smoking, and social support have with respect to whether or not a 70 year old person will have a stroke. Data were obtained from 80 subjects who had no previous history of a stroke. For each of the subjects a score was obtained on: a) Systolic blood pressure (based on an average of 10 readings obtained over a one month period); b) Whether or not the person smoked — with smokers assigned a score of 0 and nonsmokers a score of 1; and c) Social support — with people categorized as not having social support (i.e., no affiliation with a family member(s) and/or caretaker) assigned a score of 0 and people with social support a score of 1. During a two year period subjects were closely monitored, and at the conclusion of the study each subject was categorized on the dichotomous dependent variable of whether or not he or she had a stroke (with subjects who had a stroke assigned a score of 1 and subjects who did not have a stroke a score of 0). Table 39.1 summarizes the data for the study.[4]*

III. Null versus Alternative Hypotheses

The primary null hypothesis evaluated within the framework of **logistic regression** is that there is no relationship between the predictor variable(s) and the categorical dependent variable. The **alternative hypothesis** is that there is a significant relationship between one more of the predictor variables and the categorical dependent variable. An alternative way of stating the null hypothesis is that all of the coefficients for the predictor variables in the logistic function equation are equal to zero. The alternative hypothesis is that at least one of the coefficients for the predictor variables is some value other than zero.

Table 39.1   Raw Data for Example 39.1

	subject	bldpres	smoke	socsup	stroke
1	1.00	116.00	1.00	.00	.00
2	2.00	96.00	1.00	.00	.00
3	3.00	146.00	1.00	1.00	.00
4	4.00	196.00	.00	.00	1.00
5	5.00	111.00	1.00	1.00	.00
6	6.00	103.00	1.00	.00	.00
7	7.00	118.00	1.00	.00	.00
8	8.00	128.00	1.00	1.00	.00
9	9.00	178.00	1.00	1.00	1.00
10	10.00	123.00	.00	.00	1.00
11	11.00	182.00	.00	.00	1.00
12	12.00	144.00	1.00	1.00	.00
13	13.00	145.00	1.00	.00	.00
14	14.00	122.00	1.00	1.00	.00
15	15.00	105.00	1.00	.00	.00
16	16.00	99.00	1.00	1.00	.00
17	17.00	131.00	1.00	.00	.00
18	18.00	126.00	1.00	1.00	.00
19	19.00	129.00	1.00	1.00	.00
20	20.00	189.00	.00	1.00	1.00
21	21.00	178.00	.00	.00	1.00
22	22.00	122.00	1.00	1.00	.00
23	23.00	100.00	1.00	.00	.00
24	24.00	190.00	.00	1.00	1.00
25	25.00	100.00	1.00	1.00	.00
26	26.00	204.00	.00	.00	1.00
27	27.00	143.00	1.00	1.00	.00
28	28.00	162.00	.00	.00	1.00
29	29.00	222.00	.00	1.00	1.00
30	30.00	167.00	.00	.00	.00
31	31.00	119.00	1.00	1.00	.00
32	32.00	98.00	1.00	1.00	.00
33	33.00	120.00	.00	1.00	1.00
34	34.00	115.00	1.00	.00	.00
35	35.00	135.00	1.00	1.00	.00
36	36.00	167.00	1.00	1.00	1.00
37	37.00	165.00	.00	.00	1.00
38	38.00	159.00	.00	1.00	.00

Table 39.1   Raw Data for Example 39.1 (continued)

	subject	bldpres	smoke	socsup	stroke
39	39.00	102.00	1.00	.00	.00
40	40.00	109.00	1.00	1.00	.00
41	41.00	131.00	1.00	.00	.00
42	42.00	83.00	1.00	1.00	.00
43	43.00	101.00	1.00	1.00	.00
44	44.00	194.00	1.00	1.00	1.00
45	45.00	210.00	1.00	1.00	1.00
46	46.00	215.00	1.00	.00	1.00
47	47.00	177.00	.00	.00	1.00
48	48.00	133.00	1.00	1.00	.00
49	49.00	123.00	.00	1.00	1.00
50	50.00	193.00	.00	1.00	1.00
51	51.00	186.00	.00	1.00	1.00
52	52.00	176.00	1.00	.00	1.00
53	53.00	222.00	.00	1.00	1.00
54	54.00	192.00	.00	.00	1.00
55	55.00	136.00	1.00	1.00	.00
56	56.00	156.00	1.00	.00	.00
57	57.00	167.00	1.00	1.00	1.00
58	58.00	159.00	1.00	1.00	.00
59	59.00	120.00	.00	.00	.00
60	60.00	119.00	1.00	1.00	.00
61	61.00	158.00	.00	.00	1.00
62	62.00	234.00	.00	1.00	.00
63	63.00	116.00	1.00	1.00	.00
64	64.00	166.00	.00	1.00	1.00
65	65.00	151.00	1.00	1.00	.00
66	66.00	140.00	.00	.00	1.00
67	67.00	138.00	1.00	1.00	.00
68	68.00	107.00	.00	.00	.00
69	69.00	114.00	1.00	.00	.00
70	70.00	179.00	.00	.00	1.00
71	71.00	99.00	1.00	1.00	.00
72	72.00	188.00	.00	.00	1.00
73	73.00	106.00	1.00	1.00	.00
74	74.00	123.00	1.00	1.00	.00
75	75.00	192.00	.00	.00	1.00
76	76.00	114.00	1.00	1.00	.00

	subject	bldpres	smoke	socsup	stroke
77	77.00	102.00	1.00	1.00	.00
78	78.00	206.00	1.00	.00	1.00
79	79.00	187.00	.00	.00	.00
80	80.00	164.00	1.00	1.00	.00

IV. Test Computations

Like most multivariate procedures, the mathematics involved in **logistic regression** are complex and impractical to implement without the aid of a computer. The intent of the discussion in this section (as well as the next) is to provide the reader with a succinct description of the underlying basis for **logistic regression**, along with an explanation of the summary information generated by computer software.[5] Those interested in a more detailed description of the underlying mathematical basis for **logistic regression** should consult sources such as Aldrich and Nelson (1984) and Hosmer and Lemeshow (1989, 2000).

In Section I it was noted the dependent variable in **logistic regression** is expressed as a **logit variable**, and that a **logit variable** is the **natural logarithim of the odds** of a subject being a member of a specific category.[6] The **natural logarithm of the odds** (written as **log odds** or **ln (odds)**) is also commonly referred to as the **logit**, and the transformation of **odds** to **log odds** is referred to as a **logit transformation**. **Log odds** are on a continuous scale and may assume a range of values between $-\infty$ to $+\infty$. The sign of **log odds** will be positive for odds greater than 1/1 and negative for odds less than 1/1. **Log odds** are undefined when the odds equal 0.

Equation 39.1 (which is identical to Equation 16.26) defines the odds of an event Y occurring — which in the case of Example 39.1 will be the odds a subject will have a stroke. Note that Equation 39.1 indicates the odds of Y are the probability of Y occurring divided by the probability of Y not occurring. Equation 39.2 (which represents a **logit transformation**) computes the natural logarithm of Equation 39.1 — i.e., the natural logarithm of the odds, which as noted earlier is referred to as the **log odds**. The value computed with Equation 39.2 represents a **logit variable**. Note that the value on the extreme right of Equation 39.2 is equivalent to the equation for multiple linear regression (i.e., Equation 33.4). (If there is one independent variable, the right side of Equation 39.2 is simply $a + bX$ (which is equivalent to Equation 28.4).) To put it another way, the logarithm of the odds is represented by a linear combination of the independent variables, which Selvin (1995, p. 365) notes is the same as describing the probability associated with a binary outcome with a logistic function. The latter is illustrated below when Equations 39.3 and 39.4 are derived from Equation 39.2

$$Odds(Y) = \frac{p(Y \ will \ occur)}{p(Y \ will \ not \ occur)} \qquad \textbf{(Equation 39.1)}$$

$$\textbf{(Equation 39.2)}$$
$$log \ odds = ln\left[\frac{p(Y)}{1 - p(Y)}\right] = a + b_1 X_1 + b_2 X_2 \ ... \ + \ ... \ + b_p X_p$$

Equation 39.2 (specifically, $ln[p(Y)/(1 - p(Y))]$) can be employed to derive Equation 39.3, which can also be written in the form of Equation 39.4.[7] Both of the latter equations represent a **logistic function**, and either equation can be used to predict the probability of a subject being a member of a specific category on the dependent variable. As noted below, when there is one predictor/independent variable, the value of the exponent z in Equations 39.3 and 39.4 equals $a + bX$ (which is identical to Equation 28.4), and when there are two or more predictor variables, $z = a + b_1 X_1 + b_2 X_2 + ... + b_p X_p$ (which is identical to Equation 33.4), which represents a linear combination of p predictor variables.[8] The value $p(Y)$ computed with Equations 39.3 and 39.4 represents the probability a subject will be a member of a specified category (typically, the category of primary interest) on the dependent variable. When the value of z in the latter equations is small (e.g., is negative but has a large absolute value), $p(Y)$ will be close to 0; if $z = 0$, $p(Y) = .5$; and if z is large, $p(Y)$ will be close to 1.

$$p(Y) = \frac{e^z}{1 + e^z}$$
    (Equation 39.3)

$$p(Y) = \frac{1}{1 + e^{-z}}$$
    (Equation 39.4)

Where: $e = 2.71828...$ (which is the base of a natural logarithm)

$z = a + bX$ (when there is one predictor variable), and $z = a + b_1 X_1 + b_2 X_2 + ... + b_p X_p$ when there are multiple predictor variables.

Upon transforming a dependent variable into a **logit variable**, **logistic regression** employs a methodology called **maximum likelihood estimation** to estimate the probability of an event occurring. Maximum likelihood estimation is an **iterative** procedure which begins with a guess with respect to the values of the logit coefficients — the latter being the coefficients (i.e. b_1, b_2, etc.) for each of the predictor variables in the exponent z (i.e., in the equation $z = a + b_1 X_1 + b_2 X_2 + ... + b_p X_p$) in Equations 39.3 and 39.4.[9] The use of maximum likelihood estimation in **logistic regression** allows a researcher to compute values for the logit coefficients which maximize the probability of correctly assigning subjects in the sample to the appropriate category on the dependent variable.

Within the framework of **logistic regression** a number of statistical tests are conducted that are employed to determine whether one or more of the predictor variables contributes significantly toward predicting a subject's category on the dependent variable. The following tests (the results of which are printed out by most computer software) are commonly conducted.

a) **Likelihood ratio test** The **likelihood ratio test** (which is sometimes called the **model chi-square test**) is most commonly employed to assess the significance of a **logistic regression** analysis. The test evaluates the difference between the simple model (i.e., the model which omits any of the predictor variables) and one or more alternative models (i.e., models which include one or more of the predictor variables). Prior to describing the **likelihood ratio test**, it is necessary to define some measures upon which the latter test statistic is based.

In **logistic regression** a researcher wants to employ one or more predictor variables to determine the likelihood (i.e., probability) a subject is a member of a specific category on the dependent variable. The natural logarithm of the latter likelihood is referred to as the **log-likelihood** (which will be represented with the notation LL). The value of a **log-likelihood** will fall between 0 and $-\infty$. Except for the rare exception when $p = 1$, the value of a **log-likelihood** will always be a negative number. The computation of a **log-likelihood** through use of maximum likelihood estimation serves as the basis for evaluating the result of a **logistic regression** analysis.

A **log-likelihood** can be employed to compute a **likelihood ratio** (which will be represented by the notation LR). A **likelihood ratio** can be defined by Equation 39.5. (The value computed with Equation 39.5 is sometimes referred to as the **deviance**.)

$$LR = -2LL$$
    (Equation 39.5)

The **likelihood ratio** (which some sources refer to as the **log-likelihood test**) is used as an element in computing the **likelihood ratio test statistic** (which is also called the **likelihood ratio chi-square statistic**). The **likelihood ratio test statistic** (which is interpreted as a chi-square value, although it is most commonly represented with the notation G) evaluates the difference between the **likelihood ratio** for the simple model minus the **likelihood ratio** for an alternative model. To put it another way, the **likelihood ratio test statistic** (which is analogous to the F test statistic computed with Equations 28.49 and 33.3/33.29 for a linear regression analysis (Wright (1995, p. 227 and p. 247)) is used to evaluate whether the coefficients for one or more predictor variables differ from zero. As is the case for an F statistic, a large value for the **likelihood ratio test statistic** indicates there is a high likelihood that one or more of the coefficients for the predictor variables in the underlying population are some value other than zero.

The **likelihood ratio test** is based on Equation 39.6, which contrasts the maximized value of the likelihood function for the simple model (i.e., the model without the predictor variable(s)) with the maximized value of the likelihood function for the full model (i.e., the model which contains the predictor variable(s)). Note that since division using logarithms involves subtracting the value of the denominator from the numerator, Equation 39.6 becomes Equation 39.7.[10]

(Equation 39.6)

$$G = \chi^2 = -2\ln\left[\frac{likelihood\ of\ model\ without\ IV(s)}{likelihood\ of\ model\ with\ IV(s)}\right]$$

(Equation 39.7)

$$G = [-2\ln[likelihood\ of\ model\ without\ IV(s)]] - [-2\ln[likelihood\ of\ model\ with\ IV(s)]] =$$

$$G = \chi^2 = -2[log\ likelihood\ of\ model\ without\ IV(s) - log\ likelihood\ of\ model\ with\ IV(s)]$$

The larger the value computed for $G = \chi^2$, the greater the superiority of the model containing the predictor variable(s) over the model not containing the predictor variable(s). However, in order to reject the null hypothesis, the value of the test statistic must be large enough to achieve statistical significance, and if so, the researcher can conclude that one or more of the predictor variable(s) contributes significantly toward predicting the dependent variable (i.e., category membership). When a stepwise analysis is conducted involving two or more predictor variables, a **likelihood ratio test statistic** can be computed each time a new predictor variable is removed from (in the case of a backward analysis) or added to (in the case of forward analysis) the analysis.

b) **Wald test** This test essentially evaluates the same hypothesis as the **likelihood ratio test**. Specifically, the **Wald test** is employed to determine whether or not each of the coefficients in a model containing one or more predictor variables is significant. Insofar as it does the latter by evaluating the null hypothesis that for a given predictor variable the value of the slope in the underlying population equals 0, the test is analogous to **Test 28f: Test for evaluating the null hypothesis H_0: $\beta = 0$**. The **Wald test** employs Equation 39.8 to compute a z value (where z equals the maximum likelihood estimate of the slope parameter divided by the standard error of the latter). The value of z can be evaluated with **Table A1 (Table of the Normal Distribution)** in the **Appendix** or can be squared, with z^2 interpreted as a chi-square value with one degree of freedom (using **Table A4 (Table of the Chi-Square Distribution)**) (Hosmer and Lemeshow (2000, p. 16)).

$$z = \frac{b_j}{se_{b_j}}$$ **(Equation 39.8)**

Where: b_j represents the coefficient computed for the jth predictor variable
se_{b_j} represents the estimated standard error of b_j

Sources (e.g., Hosmer and Lemeshow (2000, p. 16) and Menard (2002)) note that a study by Hauck and Donner (1977) indicated the power of the **Wald test** may be compromised (i.e. the test may be overly conservative), since the test has a tendency to inflate the standard error of large predictor coefficients (with the latter resulting in too low a chi-square value). Agresti (1996) notes that for small sample sizes, the **likelihood ratio test** yields a more reliable result than the **Wald test**.[11]

c) **The Hosmer and Lemeshow goodness-of-fit test** (also referred to as the **Hosmer and Lemeshow test**) This test, which yields a chi-square value for a test statistic, divides subjects into ten ordered deciles and evaluates the observed frequencies with respect to goodness-of-fit as predicted by the **logistic regression** model. To be more specific, if a model containing one or more predictor variables provides good prediction, the majority of subjects in the first five deciles will be categorized in one of the categories for a dichotomous dependent variable, and the majority of subjects in the latter five deciles will be categorized in the other category of the dependent variable. Since the test evaluates the null hypothesis that the observed data do not differ significantly from the values predicted from the model being evaluated, the researcher, in fact, wants to **retain** the null hypothesis. (The alternative hypothesis is that the observed data do differ significantly from the values predicted from the model being evaluated — the latter implying the model in question is no better than the simple model.) Consequently, a **nonsignificant** result for the test allows a researcher to conclude that the model being evaluated significantly predicts subjects' scores on the dependent variable.

V. Interpretation of the Test Results

Preliminary analysis of the data for Example 39.1 yielded the following information: a) Of the 80 subjects employed in the study, 31 had a stroke while 49 did not; b) Twenty-nine of the 80 subjects were smokers while 51 were nonsmokers; c) Thirty-four of the 80 subjects did not have social support while 46 did have social support; d) There was no evidence of multicollinearity between the predictor variables — i.e., all tolerance values and condition indices were with the range of acceptable values; e) No outliers were detected on the continuous predictor variable of systolic blood pressure. The mean systolic blood pressure for the 80 subjects on the latter variable was $\bar{X} = 146.76$ with $\tilde{s} = 37.38$.

Results for a binary logistic regression analysis with one predictor variable Initially we will consider a **logistic regression** analysis on the data for Example 39.1 using only a single predictor variable. Specifically, in the discussion to follow, only the systolic blood pressure of the 80 subjects was employed to predict whether or not a person will have a stroke. The *SPSS* output for the latter analysis is presented in Table 39.2. (Instruction on the use of *SPSS* in conducting **logistic regression** can be found in George and Mallery (2005), Field (2005), Mertler and Vannatta (2005), and Norušis (2004).)

The section labeled **Block 0: Beginning Block** in Table 39.2 summarizes a preliminary analysis conducted on the **simple model** (i.e., the model associated with the null hypothesis stating the predictor variable does not contribute to group classification). If the latter analysis yields a significant result, it indicates the simple model should be rejected, and that the predictor variable does, in fact, contribute significantly toward predicting categorization on the dependent variable. The reader should note the following with respect to the information displayed under **Block 0: Beginning Block**.

a) Note the value in the last row of Column 2 (labeled **-2 Log likelihood**) in the table labeled **Iteration History**. The latter value $LR = 106.819$ represents the **likelihood ratio** (as

Table 39.2 *SPSS* Output for Example 39.1 When Systolic Blood Pressure Only Predictor Variable

Block 0: Beginning Block

Iteration History[a,b,c]

Iteration		-2 Log likelihood	Coefficients Constant
Step 0	1	106.820	-.450
	2	106.819	-.458
	3	106.819	-.458

a. Constant is included in the model.

b. Initial -2 Log Likelihood: 106.819

c. Estimation terminated at iteration number 3 because parameter estimates changed by less than .001.

Classification Table[a,b]

			Predicted		
			STROKE		Percentage Correct
Observed			healthy	stroke	
Step 0	STROKE	healthy	49	0	100.0
		stroke	31	0	.0
	Overall Percentage				61.3

a. Constant is included in the model.

b. The cut value is .500

Variables in the Equation

		B	S.E.	Wald	df	Sig.	Exp(B)
Step 0	Constant	-.458	.229	3.980	1	.046	.633

Variables not in the Equation

			Score	df	Sig.
Step 0	Variables	BLDPRES	38.964	1	.000
	Overall Statistics		38.964	1	.000

Block 1: Method = Enter

Iteration History[a,b,c,d]

Iteration		-2 Log likelihood	Coefficients	
			Constant	BLDPRES
Step 1	1	64.963	-5.823	.037
	2	60.510	-8.385	.053
	3	60.158	-9.362	.059
	4	60.154	-9.479	.059
	5	60.154	-9.480	.059
	6	60.154	-9.480	.059

a. Method: Enter

b. Constant is included in the model.

c. Initial -2 Log Likelihood: 106.819

d. Estimation terminated at iteration number 6 because parameter estimates changed by less than .001.

Table 39.2 *SPSS* **Output for Example 39.1 When Systolic Blood Pressure Only Predictor Variable (continued)**

Omnibus Tests of Model Coefficients

		Chi-square	df	Sig.
Step 1	Step	46.665	1	.000
	Block	46.665	1	.000
	Model	46.665	1	.000

Model Summary

Step	-2 Log likelihood	Cox & Snell R Square	Nagelkerke R Square
1	60.154[a]	.442	.600

a. Estimation terminated at iteration number 6 because parameter estimates changed by less than .

Hosmer and Lemeshow Test

Step	Chi-square	df	Sig.
1	13.565	8	.094

Contingency Table for Hosmer and Lemeshow Test

		STROKE = healthy		STROKE = stroke		Total
		Observed	Expected	Observed	Expected	
Step 1	1	8	7.801	0	.199	8
	2	8	7.681	0	.319	8
	3	8	7.422	0	.578	8
	4	5	7.199	3	.801	8
	5	8	6.632	0	1.368	8
	6	6	5.283	2	2.717	8
	7	4	3.889	5	5.111	9
	8	1	1.803	7	6.197	8
	9	0	1.025	8	6.975	8
	10	1	.265	6	6.735	7

Classification Table[a]

			Predicted		
			STROKE		Percentage Correct
	Observed		healthy	stroke	
Step 1	STROKE	healthy	45	4	91.8
		stroke	5	26	83.9
	Overall Percentage				88.8

a. The cut value is .500

Variables in the Equation

		B	S.E.	Wald	df	Sig.	Exp(B)	95.0% C.I.for EXP(B)	
								Lower	Upper
Step 1[a]	BLDPRES	.059	.012	23.699	1	.000	1.061	1.036	1.08
	Constant	-9.480	1.915	24.498	1	.000	.000		

a. Variable(s) entered on step 1: BLDPRES.

defined by Equation 39.5) for the simple model. The larger the value of *LR* the poorer the fit of the statistical model. The value *LR* = 106.819 will be compared later with a value for *LR* computed for the model which contains the predictor variable.

b) The 2 × 2 contingency tabled labeled **Classification Table** (where observed frequencies for the dependent variable are represented by the rows and predicted frequencies by the columns) summarizes accuracy of prediction when the simple model is employed. Inspection of the table reveals that all 80 subjects are predicted to not have a stroke using the latter model. Since, in fact, 49 of the subjects did not have a stroke, the simple model predicts with 49/80 = 61.25% accuracy. The distribution of the frequencies in the table is based on the fact that given the **baserate** of 61.25% of subjects not having a stroke, the optimal strategy to employ without any predictor variable is to predict that all 80 subjects will not have a stroke.[12] Using the latter strategy will result in being correct 61.25% of the time. The alternative to this strategy would be to predict that all of the subjects will have a stroke, which would only result in 31/80 = 38.75% accuracy.

c) The table labeled **Variables in the Equation** summarizes the formal analysis of the simple model. A significant result for the latter analysis (which computes a **Wald statistic** recorded in Column 4) indicates the simple model should be rejected — i.e., that a model which does not employ any of the predictor variables does not provide the best possible fit for the data. In order to reject the simple model (and consequently the null hypothesis that the coefficient for the predictor variable in the equation for the logistic regression function is equal to zero), the probability value in Column 6 (labeled **Sig.**) must be equal to or less than .05. Since the value of the **Wald statistic** is 3.980 (which is interpreted as a chi-square value with $df = 1$) with an associated probability of .046, the simple model can be rejected. If the latter result was not significant the simple model would be retained and we would conclude that the predictor variable does not contribute significantly toward prediction. The value **Exp(B)** = .633 in Column 7 represents the odds of having a stroke. More specifically, the value .633 indicates that if the predictor variable is not taken into account, a person is .633 times as likely to have a stroke than to not have a stroke.[13]

d) The table labeled **Variables not in the Equation** summarizes the analysis with respect to whether or not the coefficient for the predictor variable not included in the simple model will significantly affect prediction (or, to put it another way, whether the coefficient for the predictor variable is some value other than zero). The value 38.964 in the last row (**Overall Statistics**) of Column 2 (labeled **Score**), which represents a chi-square value, and its associated probability .000 in the last column (labeled **Sig.**) indicates a significant result, since .000 is less than .05. We can thus conclude the predictor variable of systolic blood pressure makes a significant contribution toward prediction. The value 38.964 in Column 2 (**Score**) in the row for systolic blood pressure (**BLDPRES**) (which is identical to the value noted in the row labeled **Overall Statistics**) represents what is sometimes referred to as **Roa's efficient score statistic**, computed in this instance for the predictor variable systolic blood pressure. A significant result for a predictor variable (i.e., a probability equal to or less than .05 in Column 4 (**Sig.**)) indicates the variable in question may possibly make a significant contribution in prediction (Field (2005, p. 237). The value .000 in Column 4 thus indicates that systolic blood pressure may make a significant contribution in prediction.

The section labeled **Block 1: Method = Enter** in Table 39.2 summarizes the analysis for the model which includes the predictor variable of systolic blood pressure. The reader should note the following with respect to the information displayed in the **Block 1: Method = Enter** section of Table 39.2.

a) In the table labeled **Iteration History** the value *LR* = 60.154 listed in the last row of Column 2 (labeled **−2 Log Likelihood**) represents the **likelihood ratio** when the predictor variable is included in the model. Recollect that the larger the latter value the poorer the fit of the statistical model. In b) below the value $LR = 60.154$ will be contrasted with the value $LR = 106.819$ computed in the **Block 0: Beginning Block** section (for the simple model) of Table 39.2 under **Iteration History**.

b) The table labeled **Omnibus Test of Model Coefficients** summarizes the analysis of the model which includes the predictor variable (and specifically, evaluates the null hypothesis that adding the predictor variable to the analysis does not significantly increase predictability). The null hypothesis is rejected if the probability listed in the last row (labeled **Model**) of Column 4 (labeled **Sig.**) is equal to or less than .05. The value 46.665 in Column 2 (labeled **Chi-square**) is a chi-square value which is evaluated with the degrees of freedom noted in Column 3 (labeled **df**). The degrees of freedom will be equivalent to the number of variables in the model being evaluated (which at this point equals 2, since the model is comprised of the predictor variable and the dependent variable) minus the number of variables in the simple model (which equals 1, since the latter model only involves the dependent variable). (Some sources just state that the degrees of freedom equals the number of predictor variables, which will yield the same value.) Thus $df = 2 - 1 = 1$. The significant probability .000 in Column 4 indicates the model which includes the predictor variable significantly increases predictability when contrasted with the simple model. The reader should note, however, that once again it is important to distinguish between statistical significance and practical significance. In other words, just because the researcher can conclude that a model which includes the predictor variable is significantly better than the simple model, it does not necessarily indicate the superior predictability of the new model is of practical value.

In point of fact, the table labeled **Omnibus Tests of Model Coefficients** summarizes the result of the **likelihood ratio test** described in Section IV. The chi-square value 46.665 in the last row (labeled **Model**) of Column 2 (labeled **Chi-square**) represents the test statistic for the latter test. The value 46.665 was obtained through use of Equation 39.7. Specifically, it was obtained by subtracting the value 60.154 in Column 2 (labeled **−2 Log likelihood**) in the last row of the **Iteration History table** under **Block 1: Method = Enter** from the value 106.819 in Column 2 (labeled **−2 Log likelihood**) in the last row of the **Iteration History table** under **Block 0: Beginning Block**. Note that the values 60.154 and 106.819 each represent a **likelihood ratio** (as defined by Equation 39.5). The poorer the fit of the statistical model being tested, the larger the value of the **likelihood ratio** (i.e., **− 2 Log likelihood**) and the smaller the value of chi-square. Thus if, in fact, the addition of the predictor variable increases predictability, the value computed for the **likelihood ratio** for the model being evaluated should be less than the value $LR = 106.819$ computed previously for the simple model. As noted above, the latter is the case since $LR = 60.154$ is less than $LR = 106.819$, and the difference between the two **likelihood ratios** is significant.

c) The table labeled **Model Summary** contains the **likelihood ratio** and two measures of effect size for the model containing the predictor variable. Column 2 (labeled **−2 Log likelihood**) reproduces the **likelihood ratio** $LR = 60.154$ displayed in the last row of Column 2 in the **Iteration History** table. The values in Column 3 (**Cox & Snell R Square** = .442) and Column 4 (**Nagelkerke R Square** = .600) (Nagelkerke (1991)) represent measures of effect size which are analogous to the squared multiple correlation R^2, computed in **multiple regression** analysis. Note that both of the latter measures suggest a large multivariate effect size, since they exceed Cohen's (1992) criterion of $R^2 \geq .26$ for the latter.[14]

d) The table labeled **Hosmer and Lemeshow Test** contains the result of the **Hosmer and Lemeshow goodness-of-fit test** discussed in Section IV. As noted earlier, a **nonsignificant** result for the **Hosmer and Lemeshow test** allows the researcher to conclude the model being evaluated significantly predicts subjects' scores on the dependent variable. The fact the probability .094 in Column 4 (labeled **Sig.**) is greater than .05 indicates that the computed test statistic $\chi^2 = 13.565$ (displayed in Column 2 labeled **Chi-square**) is not significant. The latter is consistent with previous results which indicated that the model with the predictor variable contributes significantly to prediction.[15] The table labeled **Contingency Table for Hosmer and Lemeshow Test** summarizes the calculations for the latter test which resulted in the nonsignificant chi-square value $\chi^2 = 13.565$.

e) The table labeled **Classification Table** summarizes the accuracy of classification using the model with the predictor variable. Prior to reading the discussion of this table the reader may want to review terminology used in reference to it which is defined in Section IX (the **Addendum**) of the **binomial sign test for a single sample (Test 9)**. The frequencies in the diagonal of the table represent correct classifications which result from employing the model, while the frequencies in the off-diagonal represent incorrect classifications. Inspection of the table reveals the following: a) Twenty-six of the 31 people who have a stroke were correctly classified by the model. The latter subjects represent **true positives** and the percentage 26/31 = 83.9% represents the **sensitivity** of the analysis; b) Forty-five out of the 49 subjects who did not have a stroke were correctly classified. The latter subjects represent **true negatives** and the percentage 45/49 = 91.8% represents the **specificity** of the analysis; c) Four of the 49 subjects who did not have a stroke were incorrectly classified as having a stroke. The latter subjects represent **false positives**; d) Five of the 31 subjects who have a stroke were incorrectly classified as not having a stroke. The latter subjects represent **false negatives**; e) As noted at the bottom of the table the overall percentage of accuracy is 88.8% (i.e., (45 + 26)/80 = 88.75%). The latter percentage is higher than the value 61.3% recorded at the bottom of the table labeled **Classification Table** under **Block 0: Beginning Block** (the latter value being the accuracy of prediction for the simple model only containing the constant). Note that if the classification table for the simple model is employed, although the **specificity** of the analysis is 100% (i.e., 49/49 = 100% of the people who do not have a stroke are correctly classified), the **sensitivity** is 0% (i.e., 0/31 = 0% of the people who have a stroke are correctly classified).[16]

The reader should take note of the following with respect to the **Classification Table**: 1) Garson (2006) states that a **Classification Table** should not be employed to assess the goodness-of-fit of a model, since rather than employing the actual values of the predicted probabilities, the table merely employs the .5 probability as a cutoff in assigning subjects to one of two dichotomous categories. Because of the latter, a poor fitting model for which most of the predicted probabilities (correct or incorrect) are close to .5 can yield the same classification table as a good fitting model for which most of the predicted probabilities are close to 0 and 1; 2) The results in the **Classification Table** should be interpreted with caution, especially when there are an unequal number of subjects in the categories for the dependent variable, since under the latter conditions the accuracy of prediction for the two categories may differ substantially. Wright (1995, p. 230) notes that under such circumstances some researchers prefer to take the average accuracy for the two rows as an overall measure of accuracy. Use of the latter for the simple model yields (100% + 0%)/2 = 50% and (91.8% + 83.9%)/2 = 87.8% for the model with the predictor variable; 3) The ultimate utility of a predictive model will be a function of the degree of relative importance one attaches to correct classification in some or all of the four cells which comprise the **Classification Table**; and 4) Regardless of the values displayed in a **Classification Table**, before applying a model for predictive purposes a researcher should conduct cross-validation studies using the same variables to confirm the reliability of the original analysis.

f) The table labeled **Variables in the Equation** displays the following information for the model containing the predictor variable: 1) Column 2 (labeled **B**) contains the coefficient .059 computed for the predictor variable (i.e., .059 represents the value of b in the equation $z = a + bX$, where z is the exponent in Equations 39.3 and 39.4) and the value -9.480 obtained for the constant. Employing the latter values, the **logistic regression** function for the data is presented below (employing the format of Equation 39.3).

$$p(Y) = \frac{e^{-9.480 + .059X}}{1 + e^{-9.480 + .059X}}$$

2) Column 3 contains the standard errors for the latter measures (.012 and 1.915); 3) Column 4 contains the result of the **Wald test** (described in Section IV) which evaluates the null hypothesis that the value of a predictor coefficient is equal to zero. Rejection of the null hypothesis (indicated by a probability equal to or less than .05 in Column 6 (labeled **Sig.**) indicates a predictor variable contributes significantly to the model. As noted in Section IV, the value of the **Wald statistic** (which is interpreted as a chi-square value with one degree of freedom) is obtained by dividing the value of a predictor's coefficient by its standard error and squaring the result. To illustrate the latter, $(.059/.012)^2 = 24.174$ which is the **Wald statistic** computed for the predictor variable systolic blood pressure. (The minimal discrepancy between 24.174 and 23.699 is due to rounding off error.) Note that the coefficient .059 computed for systolic blood pressure is significant beyond the .01 level, since $p = .000$; 4) The value **Exp(B)** = 1.061 in Column 7 is the **odds ratio** for the predictor variable. The discussion below is intended to clarify what the latter value represents.

Recollect that in the chapter on the **Pearson product-moment correlation coefficient** it was noted in **simple linear regression** a regression coefficient (which represents a slope of a straight line) indicates the amount of change on the criterion variable which will be associated with a one unit change on the predictor variable. The latter interpretation of a regression coefficient cannot, however, be employed for **logistic regression**. When there is one predictor variable in **logistic regression**, the value of b represents the amount of increase (if the sign of b is positive, and the amount of decrease if the sign of b is negative) in the **log odds** for a one-unit increase in the predictor variable. To clarify the latter one must understand the meaning of the **odds ratio** (which will be represented by the notation o).[17] In **logistic regression**, in the case of an interval/ratio predictor variable, the **odds ratio**, which can be computed with Equation 39.9, indicates the change in the odds of membership in the category of primary interest (which in the case of Example 39.1 is **having a stroke**) associated with a one unit change in the predictor variable. As illustrated below, since for Example 39.1 $b = .059$, $o = 1.061$. Thus the odds of someone with a systolic pressure equal to S having a stroke are 1.061 times greater than the odds of someone with a systolic pressure of $(S - 1)$ (i.e., one point less) having a stroke. An alternative way of stating the latter is that the odds in favor of having a stroke increase .061 or 6.1% for each additional point increase in systolic pressure. To illustrate, we can say that the odds of a person with a systolic pressure of 199 having a stroke are 1.061 times greater than odds of a person with a systolic pressure of 198 having a stroke, or alternatively, the odds of a person with a systolic pressure of 199 having a stroke are 6.1% greater than the odds of a person with a systolic pressure of 198 having a stroke.[18]

$$o = e^b = e^{.059} = 1.061 \qquad \text{(Equation 39.9)}$$

If one wishes to compute an **odds ratio** for an increment on the predictor variable for some value (we will designate v) other than one unit, the exponent of Equation 39.9 becomes the regression coefficient multiplied by the relevant increment. The latter is illustrated below where the value $o = 2.726$ is computed when the value $v = 17$ is employed in Equation 39.9. Thus, the odds of a person with a systolic pressure of 215 having a stroke are 2.726 times greater than odds of a person with a systolic pressure of 198 having a stroke (since $215 - 17 = 198$).

$$o = e^{v \times b} = e^{17 \times .059} = e^{1.003} = 2.726 \qquad \textbf{(Equation 39.9)}$$

3) The values 1.036 and 1.087 in Columns 8 and 9 (labeled **95.0% C. I. For EXP(B)**) are the lower and upper limits of the 95% confidence interval for the **odds ratio** of 1.061 in Column 7. The values 1.036 and 1087 are computed by raising e to the power of the lower and upper limits of the 95% confidence interval of the regression coefficient $b = .059$. The confidence interval for the regression coefficient is computed by multiplying its standard error (.012) by $z_{.05} = 1.96$, and adding and subtracting the product from the value of the coefficient. Thus, $CI_{.95} = .059 \pm (.012)(1.96) = .059 \pm .02352$, yielding the values .03548 and .08352. When e is raised to the latter values, the lower and upper limits of 1.036 and 1.087 are obtained for the 95% confidence interval of the **odds ratio** (i.e., $e^{.03548} = 1.036$ and $e^{.08352} = 1.087$). The latter values computed for the confidence interval indicate there is a 95% likelihood the interval 1.036 to 1.087 contains the true value of the **odds ratio** in the underlying population.[19]

Equations 39.3 or 39.4 can be employed to predict the probability a subject will be a member of a particular group if he or she has a specific score on the predictor variable. Wright (1995, p. 222) notes when the value computed for $p(Y)$ is greater than .5, the subject is classified in the category of primary interest, and when $p(Y)$ is less than .5, the subject is classified in the other group. To illustrate, employing the data for Example 39.1, we will determine the probability a subject with a systolic blood pressure of 198 has a stroke. Since we know from Table 39.2 that $a = -9.480$ and $b = .059$, when $X = 198$, $z = -9.480 + .059 (198) = 2.202$. When the latter value is substituted below in Equation 39.3 it yields the value $p(Y) = .900$.[20] If the exact **logistic regression** function computed for Example 39.1 were plotted (in a graph analogous to Figure 39.1), .90 would be the probability value on the Y-axis which corresponded to the systolic blood pressure of 198 on the X-axis.

$$p(Y) = \frac{e^z}{1 + e^z} = \frac{e^{2.202}}{1 + e^{2.202}} = .900$$

Wright (1995, p. 222) notes when there is one predictor variable, by reversing the sign of the constant and dividing the latter value by the predictor coefficient one can compute the .5 probability cutoff point for the predictor variable. Thus, for Example 39.1, *Cutoff value* = 9.480/.059 = 160.68. Consequently, anyone with a systolic pressure above 160.68 would be predicted to have a stroke, and anyone with pressure below that value would be predicted not to have a stroke.

As is the case for **standard linear regression**, residuals can be computed in **logistic regression** by subtracting the predicted score for a subject on the dependent variable (which will be 0 or 1) from the subject's actual score (which will be the probability computed for the subject's score with Equations 39.3 or 39.4). Thus, if the subject who had a systolic blood pressure of 198 in fact had a stroke, the residual for the subject would be $1 - .90 = .10$. On the other hand, if the subject did not have a stroke, the residual would be $0 - .90 = -.90$. Scores on the predictor variable which have a standardized residual (the values of which can be displayed by *SPSS*) with an absolute value greater than $z_{.01} = 2.58$ (some sources employ the more conservative criterion $z_{.05} = 1.96$) should be considered with respect to either fitting the obtained model poorly, exerting undue influence on the model and/or being possible outliers. Field (2005, pp. 245–248) provides a good discussion on the evaluation of residuals in **logistic regression**.

g) **Classification plot** Figure 39.2 is a **classification plot** of the data printed out by *SPSS*. The latter is essentially a histogram of predicted probabilities with respect to whether a subject will have a stroke (recorded along the X-axis) versus the frequency of subjects associated with a given probability (recorded on the Y-axis). Note that the bars of the histogram are represented

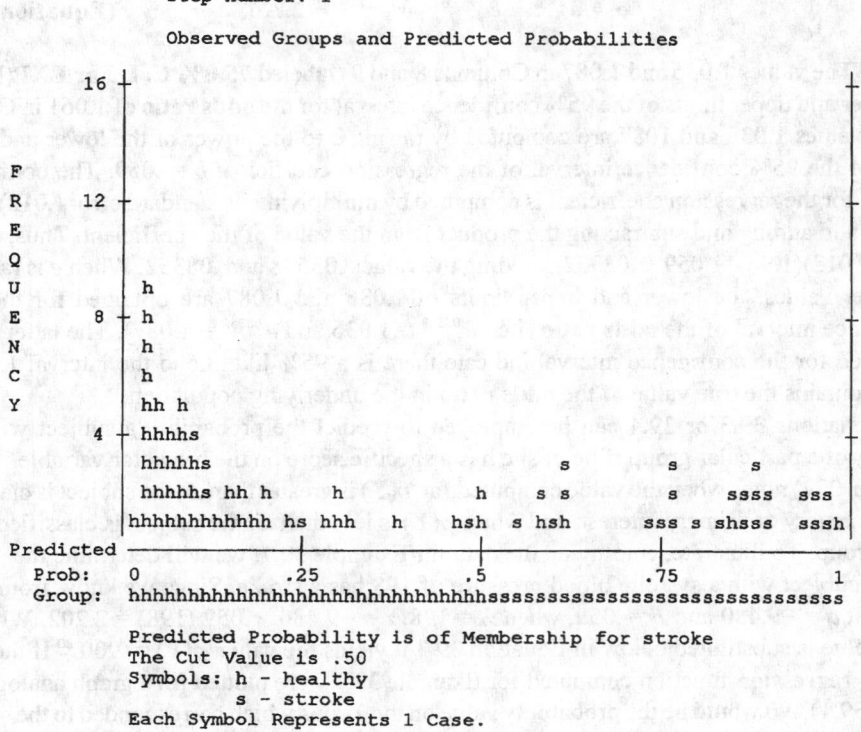

```
                    Step number: 1

                    Observed Groups and Predicted Probabilities

        16 +                                                              +
      F
      R     12 +                                                          +
      E
      Q
      U           h
      E      8 + h                                                        +
      N           h
      C           h
      Y           hh h
         4 +  hhhhs                                                       +
              hhhhhs                            s          s       s
              hhhhhs hh h        h         h    s s        s     ssss  sss
              hhhhhhhhhhhhh hs hhh    h     hsh  s hsh     sss s shsss  sssh
   Predicted -+------------+------------+------------+------------+
     Prob:    0          .25          .5          .75          1
     Group:   hhhhhhhhhhhhhhhhhhhhhhhhhhhhhhhhhhhhhhhssssssssssssssssssssssssssssssssssssss

           Predicted Probability is of Membership for stroke
           The Cut Value is .50
           Symbols: h - healthy
                    s - stroke
           Each Symbol Represents 1 Case.
```

Figure 39.2   Classification Plot for Example 39.1 Employing One Predictor Variable

by one or more of the letters, with the letter **S** indicating subjects who have a stroke and **H** indicating subjects who do not have a stroke. If the model being evaluated provides a good fit for the data, subjects who have a stroke should be clustered on the right side of the graph (i.e., to the right of $p = .5$), and subjects who do not have a stroke should be clustered on the left side of the graph (to the left of $p = .5$), which for the most part appears to be the case in Figure 39.2.

Results for a binary logistic regression analysis with multiple predictor variables A standard **logistic regression** was employed to evaluate the continuous predictor variable of systolic blood pressure and the two dichotomous predictor variables of smoking and social support. The *SPSS* output for the latter analysis is presented in Table 39.3. Once again the section labeled **Block 0: Beginning Block** in the latter table summarizes the preliminary analysis on the **simple model** (which in this instance is the model associated with the null hypothesis which states that none of the three predictor variables contribute to group classification).

a) Note that the first three tables in the **Block 0: Beginning Block** section of Table 39.3 (i.e., the tables labeled **Iteration History, Classification Table**, and **Variables in the Equation**) are identical to those in the **Block 0: Beginning Block** section of Table 39.2 (which summarized the results of the analysis when only one predictor variable was employed). The latter is the case, since without any predictor variables the simple model will be identical for either analysis — regardless of how many predictor variables will subsequently be evaluated. Since, once again, the result 3.980 for the **Wald test** in the table labeled **Variables in the Equation** is significant (with $p = .046 < .05$), the simple model can be rejected. Note that for the analysis under discussion, the value $LR = 106.819$ (i.e., the value of the **likelihood ratio** in the last row of

Table 39.3 *SPSS* Output for Example 39.1 Employing All Three Predictor Variables

Block 0: Beginning Block

Iteration History[a,b,c]

Iteration		-2 Log likelihood	Coefficients Constant
Step 0	1	106.820	-.450
	2	106.819	-.458
	3	106.819	-.458

a. Constant is included in the model.

b. Initial -2 Log Likelihood: 106.819

c. Estimation terminated at iteration number 3 because parameter estimates changed by less than .001.

Classification Table[a,b]

			Predicted		
			STROKE		Percentage Correct
Observed			healthy	stroke	
Step 0	STROKE	healthy	49	0	100.0
		stroke	31	0	.0
	Overall Percentage				61.3

a. Constant is included in the model.

b. The cut value is .500

Variables in the Equation

		B	S.E.	Wald	df	Sig.	Exp(B)
Step 0	Constant	-.458	.229	3.980	1	.046	.633

Variables not in the Equation

			Score	df	Sig.
Step 0	Variables	SOCSUP(1)	3.153	1	.076
		BLDPRES	38.964	1	.000
		SMOKE(1)	31.531	1	.000
	Overall Statistics		46.337	3	.000

Block 1: Method = Enter

Iteration History[a,b,c,d]

Iteration		-2 Log likelihood	Coefficients			
			Constant	SOCSUP(1)	BLDPRES	SMOKE(1)
Step 1	1	56.976	-4.952	.166	.027	1.386
	2	51.022	-7.505	.144	.041	1.891
	3	50.341	-8.697	.081	.048	2.160
	4	50.324	-8.916	.061	.050	2.216
	5	50.324	-8.923	.060	.050	2.218
	6	50.324	-8.923	.060	.050	2.218

a. Method: Enter

b. Constant is included in the model.

c. Initial -2 Log Likelihood: 106.819

d. Estimation terminated at iteration number 6 because parameter estimates changed by less than .001.

Table 39.3 *SPSS* **Output for Example 39.1 Employing All Three Predictor Variables (continued)**

Omnibus Tests of Model Coefficients

		Chi-square	df	Sig.
Step 1	Step	56.495	3	.000
	Block	56.495	3	.000
	Model	56.495	3	.000

Model Summary

Step	-2 Log likelihood	Cox & Snell R Square	Nagelkerke R Square
1	50.324[a]	.506	.687

a. Estimation terminated at iteration number 6 because parameter estimates changed by less than .001.

Hosmer and Lemeshow Test

Step	Chi-square	df	Sig.
1	9.916	8	.271

Contingency Table for Hosmer and Lemeshow Test

		STROKE = healthy		STROKE = stroke		
		Observed	Expected	Observed	Expected	Total
Step 1	1	8	7.865	0	.135	8
	2	8	7.787	0	.213	8
	3	9	8.589	0	.411	9
	4	8	7.374	0	.626	8
	5	8	6.784	0	1.216	8
	6	4	5.474	4	2.526	8
	7	1	3.110	7	4.890	8
	8	1	1.242	7	6.758	8
	9	1	.564	7	7.436	8
	10	1	.213	6	6.787	7

Classification Table[a]

			Predicted		
			STROKE		Percentage
	Observed		healthy	stroke	Correct
Step 1	STROKE	healthy	45	4	91.8
		stroke	7	24	77.4
	Overall Percentage				86.3

a. The cut value is .500

Variables in the Equation

		B	S.E.	Wald	df	Sig.	Exp(B)	95.0% C.I.for EXP(B)	
								Lower	Upper
Step 1	SOCSUP(1)	.060	.791	.006	1	.940	1.062	.225	5.007
	BLDPRES	.050	.012	15.843	1	.000	1.051	1.026	1.077
	SMOKE(1)	2.218	.797	7.746	1	.005	9.188	1.927	43.807
	Constant	-8.923	1.981	20.281	1	.000	.000		

a. Variable(s) entered on step 1: SOCSUP, BLDPRES, SMOKE.

Column 2 (labeled **-2 Log likelihood**) of the **Iteration History** table) will subsequently be compared with a value of *LR* to be computed for the model that contains the three predictor variables.

b) The table labeled **Variables not in the Equation** summarizes the analysis with respect to whether or not the coefficients for the predictor variables not included in the simple model will significantly affect prediction (or, to put it another way, whether the coefficients for one or more of the predictors variables are some value other than zero). As noted in reference to Table 39.2, the value 46.337 in the last row (labeled **Overall Statistics**) of Column 2 (labeled **Score**) represents a chi-square value. The latter value is significant, since its associated probability .000 (in the last column labeled **Sig.**) is less than .05. We can thus conclude that at least one of the predictor variables makes a significant contribution toward prediction. The values 3.153, 38.964, and 31.531 in Column 2 (**Score**) in the rows for social support (**SOCSUP**), systolic blood pressure (**BLDPRES**) and smoking (**SMOKE**) represent **Roa's efficient score statistic** computed for each of the predictor variables. A significant result for any predictor variable (i.e., a probability equal to or less than .05 in Column 4 (**Sig.**)) indicates the variable in question may possibly make a significant contribution in prediction. In point of fact, a significant probability is recorded for systolic blood pressure and smoking (for both of which $p = .000$, which is less than .05), but not for social support ($p = .076$, which is greater than .05).[21]

The section labeled **Block 1: Method = Enter** in Table 39.3 summarizes the analysis for the model which includes all three predictor variables. The reader should note the following with respect to the information displayed in the **Block 1: Method = Enter** section of Table 39.3.

a) In the table labeled **Iteration History** the value $LR = 50.324$ listed in the last row of Column 2 (labeled **-2 Log Likelihood**) represents the **likelihood ratio** when all three predictor variables are included in the model. Recollect that the larger the latter value the poorer the fit of the statistical model. In b) below the value $LR = 50.324$ will be contrasted with the value $LR = 106.819$ computed in the **Block 0: Beginning Block** section of Table 39.3 under **Iteration History**.

b) The table labeled **Omnibus Test of Model Coefficients** summarizes the analysis of the model which includes the three predictor variables (and specifically, evaluates the null hypothesis that adding the predictor variables to the analysis does not significantly increase predictability). The null hypothesis is rejected if the probability listed in the last column (labeled **Sig.**) of the last row (labeled **Model**) is equal to or less than .05. The value 56.495 in Column 2 (labeled **Chi-square**) is a chi-square value which is evaluated with the degrees of freedom noted in Column 3 (labeled **df**). Since, as noted earlier, the degrees of freedom will equal the number of variables in the model being evaluated (which at this point equals 4, since the model is comprised of the three predictor variables and the dependent variable) minus the number of variables in the simple model (which equals 1), $df = 4 - 1 = 3$. The significant probability .000 in the last row of Column 4 (labeled **Sig.**) indicates the model which includes the three predictor variables significantly increases predictability when contrasted with the simple model.

As noted earlier, the table labeled **Omnibus Tests of Model Coefficients** summarizes the result of the **likelihood ratio test** described in Section IV, and the chi-square value 56.495 in the last row (labeled **Model**) of Column 2 (labeled **Chi-square**) represents the test statistic for the latter test. The value 56.495 was obtained by subtracting the value 50.324 in Column 2 (labeled **-2 Log likelihood**) in the last row of the **Iteration History table** under **Block 1: Method = Enter** from the value 106.819 in Column 2 (labeled **-2 Log likelihood**) in the last row of the **Iteration History table** under **Block 0: Beginning Block**. Note that the values 50.324 and 106.819 each represent a **likelihood ratio** (as defined by Equation 39.5). The poorer the fit of the statistical model being tested, the larger the value of the **likelihood ratio** (i.e., **- 2 Log likelihood**) and the smaller the value of chi-square. Thus if, in fact, the addition of the predictor variables increases predictability, the value computed for the **likelihood ratio** for the model being evaluated should be less than the value $LR = 106.819$ computed previously for the simple model. As noted above, the latter is the case since $LR = 50.324$ is less than $LR = 106.819$, and the difference between the two **likelihood ratios** is significant.

c) The table labeled **Model Summary** contains the **likelihood ratio** (in Column 2 (labeled –2 **Log likelihood**), which reproduces the value $LR = 50.324$ displayed in the last row of Column 2 in the **Iteration History** table), and two measures of effect size, **Cox & Snell R Square** = .506 (in Column 3) and **Nagelkerke R Square** = .687 (in Column 4). As noted earlier, the latter two measures are analogous to the squared multiple correlation R^2, computed in **multiple regression** analysis. Note that both of the latter measures suggest a large multivariate effect size, since they exceed Cohen's (1992) criterion of $R^2 \geq .26$ for the latter.

d) The table labeled **Hosmer and Lemeshow Test** contains the result of the **Hosmer and Lemeshow goodness-of-fit test** discussed in Section IV. As noted earlier, a **nonsignificant** result for the **Hosmer and Lemeshow test** allows the researcher to conclude the model being evaluated significantly predicts subjects' scores on the dependent variable. The fact the probability .271 in Column 4 (labeled **Sig.**) is greater than .05 indicates that the computed test statistic $\chi^2 = 9.916$ (displayed in Column 2 labeled **Chi-square**) is not significant. The latter is consistent with previous results which indicated that the model containing the predictor variables contributes significantly toward prediction. The table labeled **Contingency Table for Hosmer and Lemeshow Test** summarizes the calculations for the latter test which resulted in the nonsignificant chi-square value $\chi^2 = 9.916$.

e) The table labeled **Classification Table** summarizes the accuracy of classification using the model employing the three predictor variables. Inspection of the table reveals the following: a) Twenty-four of the 31 people who have a stroke were correctly classified by the model. The latter subjects represent **true positives** and the percentage $24/31 = 77.4\%$ represents the **sensitivity** of the analysis; b) Forty-five out of the 49 subjects who did not have a stroke were correctly classified. The latter subjects represent **true negatives** and the percentage $45/49 = 91.8\%$ represents the **specificity** of the analysis; c) Four of the 49 subjects who did not have a stroke were incorrectly classified as having a stroke. The latter subjects represent **false positives**; d) Seven of the 31 subjects who have a stroke were incorrectly classified as not having a stroke. The latter subjects represent **false negatives**; e) As noted at the bottom of the table the overall percentage of accuracy is 86.3% (i.e., $(45 + 24)/80 = 86.25\%$). The latter percentage is higher than the value 61.3% recorded at the bottom of the table labeled **Classification Table** in the **Block 0: Beginning Block** section of Table 39.3 (the latter value being the accuracy of prediction for the simple model only containing the constant).

Curiously, inspection of the analogous **Classification Table** in Table 39.2 reveals that when systolic blood pressure is employed as a single predictor variable, overall accuracy of classification is slightly higher than overall accuracy of classification when all three predictor variables are employed. To explain the latter, the author conducted additional **logistic regression** analyses on the data which revealed the following: a) If smoking is employed as the only predictor variable, the resulting model is significantly better than the simple model, yet is slightly inferior with respect to predictability to the model just employing the predictor variable of systolic blood pressure (if frequencies in the **Classification tables** derived for both analyses are employed as a criterion for assessing accuracy of predictability); b) If the dichotomous predictor variable of social support is employed as the only predictor variable, the resulting model is not significantly better than the simple model; c) If smoking and systolic blood pressure (but not social support) are employed as the predictor variables, the resulting model is significantly better than the simple model, yet is slightly inferior with respect to predictability to the model just employing the predictor variable of systolic blood pressure (if frequencies in the **Classification**

tables derived for both analyses are employed as a criterion for assessing accuracy of predictability).

Earlier in this section it was noted that sources do not recommend employing a **Classification table** for assessing goodness-of-fit for a model. A more reliable method for contrasting the relative goodness-of-fit of one or more models is comparison of the chi-square values obtained in the last row (labeled **Model**) of the table labeled **Omnibus Tests of Model Coefficients**. With respect to the latter, Howell (2002, p. 592–593) notes that a difference between the chi-square values obtained for two models is itself distributed as a chi-square value with df equal to the difference in df between the two models. Thus, employing the chi-square values in the last row (**Model**) of the table labeled **Omnibus Tests of Model Coefficients** we can compare the value $\chi^2 = 56.495$, $df = 3$ computed for the model containing all three predictor variables with the value $\chi^2 = 46.665$, $df = 1$ computed (in Table 39.2) when systolic blood pressure is employed as the only predictor variable. The latter analysis yields $56.495 - 46.665 = 9.83$, with $df = 3 - 1 = 2$. For $df = 2$ the tabled critical .05 and .01 chi-square values are $\chi^2_{.05} = 5.99$ and $\chi^2_{.01} = 9.21$. Since $\chi^2 = 9.83 > \chi^2_{.01} = 9.21$, we can conclude that the model employing all three predictor variables is significantly more predictive than the model which only employs the predictor of systolic blood pressure.

Since social support did not contribute significantly toward prediction, a separate **logistic regression** was conducted in which just systolic blood pressure and smoking were employed as predictor variables. The latter analysis yielded the value $\chi^2 = 56.489$ (with $df = 2$). Comparing the model containing systolic blood pressure and smoking with the model containing just systolic blood pressure (where $\chi^2 = 46.665$) yielded the following result: $56.489 - 46.665 = 9.824$, with $df = 2 - 1 = 1$. For $df = 1$ the tabled critical .05 and .01 chi-square values are $\chi^2_{.05} = 3.84$ and $\chi^2_{.01} = 6.63$. Since $\chi^2 = 9.824 > \chi^2_{.01} = 6.63$, we can conclude that the model employing systolic blood pressure and smoking is significantly more predictive than the model which only employs the predictor of systolic blood pressure. Additionally, the fact that 9.824 is further removed from $\chi^2_{.01} = 6.63$ (for $df = 1$) than 9.83 is from $\chi^2_{.01} = 9.21$ (for $df = 2$) suggests we might conclude that the model containing systolic blood pressure and smoking is more predictive than the model containing all three predictor variables. However, paradoxically, if we make the latter comparison (i.e., compare the model containing systolic blood pressure and smoking with the model containing all three predictor variables) by comparing chi-square values, we obtain $56.495 - 54.489 = .006$, with $df = 3 - 2 = 1$. The latter result is not significant, since for $df = 1$, $\chi^2 = .006$ is less than $\chi^2_{.05} = 3.84$. Since, however, a separate **logistic regression** employing smoking as the only predictor variable yielded the value $\chi^2 = 32.937$, with $df = 1$, $p < .000$, it would appear that smoking by itself significantly improves predictability when contrasted with the simple model. (As noted earlier, a separate **logistic regression** employing social support as the only predictor variable yielded a nonsignificant result — specifically, $\chi^2 = 3.150$, with $df = 1$, $p < .076$.).

In the final analysis, the absence of a significant difference between the model containing all three predictor variables and the model containing systolic blood pressure and smoking can probably be attributed to the fact that by virtue of including social support in the model involving three predictors, the latter variable diluted the predictive impact of smoking. Consequently, it would seem prudent to replicate the study employing the two predictor variables of systolic blood pressure and smoking. The latter study (which should use a larger sample size) could then be employed to determine the relative accuracy of prediction employing systolic blood pressure alone versus the model containing smoking and systolic blood pressure.

f) The table labeled **Variables in the Equation** displays the following information for the model containing the three predictor variables: 1) Column 2 (labeled **B**) contains the coefficients .050, 2.218, and .060 respectively computed for the predictor variables systolic blood pressure, smoking, and social support (i.e., .the latter coefficients represents the value of b_1, b_2, and b_3, in the equation $z = a + b_1 X_1 + b_2 X_2 + b_3 X_3$, where z is the exponent in Equations 39.3 and 39.4), and the value -8.923 obtained for the constant.[22] Employing the latter values, the **logistic regression** function for the data is presented below (employing the format of Equation 39.3).

$$p(Y) = \frac{e^{-8.923 + .050X_1 + 2.218X_2 + .060X_3}}{1 + e^{-8.923 + .050X_1 + 2.218X_2 + .060X_3}}$$

2) Column 3 contains the standard errors for the coefficients and the constant (.012, 797, .791 and 1.981); 3) Column 4 contains the result of the **Wald test** which evaluates the null hypothesis that the value of a predictor coefficient is equal to zero. As noted earlier, rejection of the null hypothesis (indicated by a probability equal to or less than .05 in Column 6 (labeled **Sig.**)) indicates a predictor variable contributes significantly to the model. Note that the coefficients .050 and 2.218 computed for systolic blood pressure and smoking are significant beyond the .01 level (since $p = .000$ and .005), but the coefficient .060 for social support is not (since $p = .940$); d) The values in Column 7 (**Exp(B)**) represent the **odds ratios** for each of the predictor variables: 1.051 for systolic blood pressure, 9.188 smoking, and 1.062 for social support.

Note that by employing Equation 39.9, we can compute the above noted **odds ratios**: **Systolic blood pressure**: $o = e^b = e^{.050} = 1.051$; **Smoking**: $o = e^{2.218} = 9.188$; **Social support**: $o = e^{.060} = 1.062$. The following statements can be made with respect to the latter values computed for the **odds ratios**: a) For the continuous (i.e., interval/ratio) predictor variable of systolic blood pressure we can say the following: After controlling for the impact of smoking and social support, the odds of someone with a given systolic pressure having a stroke are 1.051 times greater than the odds of someone with a systolic pressure that is one point less having a stroke. Or alternatively, the odds in favor of having a stroke increase .051 or 5.1% for each additional point increase in blood pressure after controlling for the other predictor variables; b) For the dichotomous predictor variable of smoking we can say the following: After controlling for the impact of systolic blood pressure and social support, the odds in favor of having a stroke are 9.188 times larger for a smoker than a nonsmoker; c) For the dichotomous predictor variable of social support we can say the following: After controlling for the impact of systolic blood pressure and smoking, the odds in favor of having a stroke are 1.062 times larger for a person who does not have social support than a person who does have social support. However, since the analysis indicated that social support did not contribute significantly toward prediction, we would not employ the latter variable in our final model (which as noted earlier would only retain the predictor variables of systolic blood pressure and smoking).

3) The values in Columns 8 and 9 (labeled **95.0% C. I. For EXP(B)**) are the lower and upper limits of the 95% confidence interval for the **odds ratios** in Column 7.

Equations 39.3 or 39.4 can be employed to predict the probability that a subject will be a member of a particular group if he or she has specific scores on one or more predictor variables. As noted in the discussion with reference to a single predictor variable, when the value computed for $p(Y)$ is greater than .5, the subject is classified in the category of primary interest, and when the latter value is less than .5, the subject is classified in the other group. To illustrate, we will determine the probability a subject with a blood pressure of 198, who is a smoker (which was represented in the *SPSS* data editor with a score of 0, but must be represented in Equation 39.3 as 1), and who does not have social support (which was represented in the *SPSS* data editor with a score of 0, but must be represented in Equation 39.3 as 1) has a stroke.[23] Since we know from Table 39.3 that $a = -8.923$, $b_1 = .050$, $b_2 = 2.218$, and $b_3 = .060$, when $X_1 = 198$, $X_2 = 1$, and $X_3 = 1$, $z = -8.923 + .050(198) + 2.218(1) + .060(1) = 3.255$. When the latter value is substituted in

Equation 39.3 it yields the value $p(Y) = .963$. Thus, the likelihood of the subject having a stroke is .963. As noted in Endnote 19, a 95% confidence interval should be computed for the probability computed with Equations 39.3 or 39.4. In order to predict that a subject will fall in the category of having a stroke, both the lower and upper limit of the confidence interval should be above .5.

$$p(Y) = \frac{e^z}{1 + e^z} = \frac{e^{3.255}}{1 + e^{3.255}} = .963$$

g) **Classification plot**   Figure 39.3 is a **classification plot** of the data. As is the case for Figure 39.2 (in which only the predictor variable of systolic blood pressure is employed), Figure 39.3 appears to provide a good fit for the data, since most of the subjects who have a stroke are clustered on the right side of the graph (i.e., above $p = .5$) and most of the subjects who do not have a stroke are clustered on the left side of the graph (i.e., below $p = .5$).

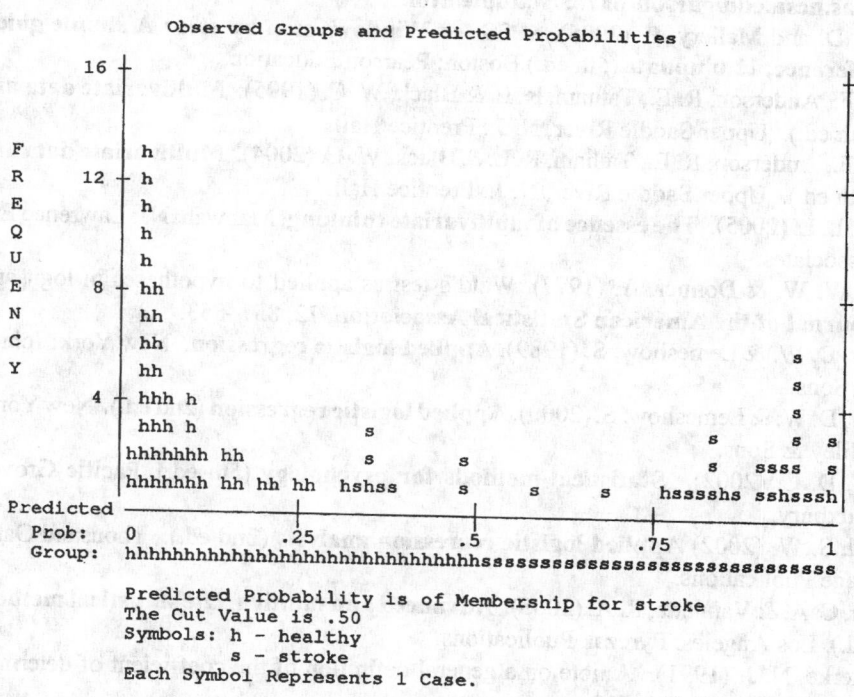

Figure 39.3   Classification Plot for Example 39.1 Employing Three Predictor Variables

VI. Additional Analytical Procedures for Logistic Regression and/or Related Tests

No additional procedures will be described in this section.

VII. Additional Discussion of Logistic Regression

No additional material will be discussed in this section.

VIII. Additional Examples Illustrating the Use of Logistic Regression

No additional examples will be presented in this section.

References

Agresti, A. (1996). **An introduction to categorical data analysis**. New York: John Wiley & Sons.

Aldrich, J. H. & Nelson, F. D (1984). **Linear probability, logit, and probit models**. Beverly Hills, CA: Sage Publications.

Cohen, J. (1992). A power primer. **Psychological Bulletin**, 112, 155–159.

Field, A. (2005). **Discovering statistics using SPSS** (2nd ed.). Sage Publications: London.

Garson, D. G. (2006). Statistics: Topics in multivariate analysis. Website: **http://www2. chas.ncsu.edu/garson/pa765/statnote.htm.**

George, D. and Mallery, P. (2005). **SPSS for Windows step by step: A simple guide and reference, 12.0 update** (5th ed.).Boston: Pearson Education.

Hair, J. F., Anderson, R. E., Tatham, R. L. & Black, W. C. (1995). **Multivariate data analysis** (4th ed.). Upper Saddle River, N. J.: Prentice Hall.

Hair, J. F., Anderson, R. E., Tatham, R. L. & Black, W. C. (2004). **Multivariate data analysis** (6th ed.). Upper Saddle River, N. J.: Prentice Hall.

Harlow, L. L. (2005). **The essence of multivariate thinking**. Mahwah, NJ: Lawrence Erlbaum Associates.

Hauck, W. W. & Donner, A. (1977). Wald's test as applied to hypotheses in logit analysis. **Journal of the American Statistical Association**, 72, 851–853.

Hosmer, D. W. & Lemeshow, S. (1989). **Applied logistic regression**. New York: John Wiley & Sons.

Hosmer, D. W. & Lemeshow, S. (2000). **Applied logistic regression** (2nd ed.). New York: John Wiley & Sons.

Howell, D. C. (2002). **Statistical methods for psychology** (5th ed.). Pacific Grove, CA.: Duxbury.

Menard, S. W. (2002) **Applied logistic regression analysis** (2nd ed.). Thousand Oaks, CA: Sage Publications.

Mertler, C. A. & Vannatta, R. A. (2005). **Advanced and multivariate statistical methods** (3rd ed.). Los Angeles: Pyrczak Publications.

Nagelkerke, N. J. (1991). A note on a general definition of the coefficient of determination. **Biometrika**, 78, 691–692.

Norušis, N. J. (2004). **SPSS 13.0 advanced statistical procedures companion**. Upper Saddle River, NJ: Prentice Hall.

Pagano, M. & Gauvreau, K. (1993). **Principles of biostatistics**. Belmont, CA: Duxbury Press.

Peduzzi, P. N., Concato, J., Kemper, E., Holford, T. R. & Feinstein, A. (1996). A simulation study of the number of events per variable in logistic regression analysis. **Journal of Clinical Epidemiology**, 99, 1373–1379.

Rosner, B. (1995). **Fundamental of biostatistics** (4th ed.). Belmont, CA: Duxbury Press.

Rosner, B. (2000). **Fundamental of biostatistics** (5th ed.). Pacific Grove, CA: Duxbury Press.

Selvin, S. (1995). **Practical biostatistical methods**. Belmont, CA: Duxbury.

Spicer, J. (2005). **Making sense of multivariate data analysis.** Thousand Oaks, CA: Sage Publication.

Stevens, J. (2002). **Applied multivariate statistics for the social sciences** (4th ed.). Hillsdale, NJ: Lawrence Erlbaum Associates, Publishers.

Tabachnick, B. G. & Fidell, L. S. (1996). **Using multivariate statistics** (3rd ed.). New York: Harper Collins Publishers.

Tabachnick, B. G. & Fidell, L. S. (2001). **Using multivariate statistics** (4th ed.). Boston: Allyn & Bacon.

van Belle, G. (2002). **Statistical rules of thumb.** New York: John Wiley &Sons.

Wright, R. E. (1995). Logistic regression. In Grimm, L. G. & Yarnold, P. R. (Eds.) (pp. 245–276). **Reading and understanding multivariate statistics.** Washington, D.C.: American Psychological Association.

Wuensch, K. (2005). Statistics lessons. **Website: http://core.ecu.edu/psyc/wuenschk/ SPSS/SPSS-Mv.htm.**

Endnotes

1. a) The term **multinomial logistic regression** is commonly employed when there are three or more categories on the dependent variable; b) The terms **multinomial** and **polychotomous** (or **polytomous**) are used in identifying a categorical independent or dependent variable which is comprised of more than two categories.

2. a) Harlow (2005, p. 154) and Tabachnick and Fidell (1996, p. 580) note, however, that, as is the case for **standard linear regression**, the reliability of **logistic regression** can also be compromised by multicollinearity — i.e., extremely high intercorrelations between the predictor variables; b) Some sources note that when the dependent variable is dichotomous, **standard linear regression** may work fairly well in predicting a probability when the value of a probability is not extreme — i.e., when the probability falls between .20 and .80 (since a logistic function (e.g., Figure 39.1) is to a large degree linear at its center).

3. a) Hosmer and Lemeshow (1989; 2000, p. 5) demonstrate that if you compute the average score on the dependent variable for each value of the independent variable, the latter relationship yields the same S-shaped function. To illustrate in reference to Figure 39.1, assume that the values **0** and **1** are respectively employed as scores on the dependent variable for subjects who **do not have a stroke** versus subjects who **have a stroke**. Since the overwhelming majority of subjects with a systolic blood pressure under 180 will not have a stroke (with the lower the blood pressure, the more likely someone will not have a stroke), the average score on the dependent variable for each blood pressure value under 180 will be close to 0 (where a subject's score is 0 or 1, and the average score for a given blood pressure value can be viewed as representing the proportion of subjects with that blood pressure). In the same respect, since most of the subjects with a systolic blood pressure over 200 will have a stroke (with the higher the blood pressure, the more likely someone will have a stroke), the average score on the dependent variable for each blood pressure value above 200 will be close to 1. In the case of blood pressure values between 180 and 200 there will be more variability among subjects with respect to whether or not a person will have a stroke, and thus (as is the case in Figure 39.1) the average score for pressures values in the latter range will fall between .25 (for lower pressures in that range) and .75 (for higher pressures in that range); b) Note that a **logistic function,** such as that

depicted in Figure 39.1, closely resembles the plot of a **cumulative frequency distribution** (see Figure I.6 in the **Introduction**), which also conforms to a sigmoidal curve.

4.   a) The data for Example 39.1 (as well as that depicted in Figure 39.1) are hypothetical and not based on the results of an actual study; b) In order to employ computer software for **logistic regression**, it is necessary to employ **dummy coding** for the dichotomous predictor/ independent variables and the dichotomous dependent variable. **Dummy coding** (which is also discussed in Section IX (the **Addendum**) of the **Pearson product-moment correlation coefficient**) assigns the scores 0 and 1 to subjects who are at different levels of a dichotomous predictor variable and different groups representing a dichotomous dependent variable. In employing *SPSS* for **logistic regression** with Example 39.1, the following guidelines were employed in dummy coding dichotomous variables: a) In the case of the dichotomous independent/ predictor variables the following protocol was employed. A code of 0 was assigned to the **category/group of primary interest** (sometimes referred to as the **baseline** or **reference group**), and a code of 1 was assigned to the **other category/group**. In the case of Example 39.1, the categories **smokers** and **not having social support** were identified as the **reference groups** (since the researcher hypothesized that smokers and people who did not have social support would be more likely to have a stroke than nonsmokers and people who did have social support). Consequently, any subject who was a smoker and any subject who did not have social support was assigned a score of 0 on the appropriate predictor variable, and any subject who was nonsmoker and any subject who did have social support was assigned a score on 1 on the appropriate predictor variable; c) In the case of a dichotomous dependent variable, subjects who are in the **category/group of primary interest** are generally assigned a score of 1, and subjects in the **other category/group** are assigned a score of 0. (In medical research it is common for people who are afflicted with a specific condition, such as having a stroke or who die, to be designated as the group of primary interest.) Since for Example 39.1 **stroke** was the category of primary interest on the dependent variable, subjects who had a stroke were assigned a score of 1, and subjects who did not have a stroke were assigned a score of 0. Although intuitively it would seem that if smokers and people who did not have social support are predicted to be more likely to have a stroke, it would be more logical to assign the latter categories of people a score of 1 on the relevant predictor variables, use of the latter coding with *SPSS* yields results which suggest the opposite of what actually occurs. Although reversing the codes employed for categories will yield the same absolute values for some of the relevant statistics, the signs of the latter values may make the analysis more confusing to interpret. Readers, however, should be aware of the fact that the coding protocol for dummy variables resulting in the most straightforward interpretation may differ depending upon the software one employs; d) The use of dummy variable coding can be extended to predictor variables with more than two categories. In the latter case, the predictor variable in question is converted into a **set of dummy variables**. The number of dummy variables will always be one less than the number of categories on a predictor variable. To illustrate, assume that a predictor variable is comprised of the three categories A, B, and C, and that Category A is designated as the reference or baseline category. The following two dummy variables can be derived: **Dummy variable 1**: A = 0, B = 0, C = 1 (which breaks the predictor variable into a dichotomous variable comprised of A & B (members of both of which are assigned a score of 0) versus C (members of which are assigned a score of 1)); **Dummy variable 2**: A = 0, B = 1, C = 0 (which breaks the predictor variable into a dichotomous variable comprised of A & C (members of both of which are assigned a score of 0) versus B (members of which are assigned a score of 1)).

Note that subjects in any category categorized with the reference category are also assigned the score 0, while subjects in the remaining category in the dichotomization are assigned a score of 1. A more detailed description of the latter procedure can be found in Field (2005, pp. 208–210) and Hair *et al.* (1995, pp. 109–110); e) In **multinomial logistic regression** (i.e., the use of a categorical dependent variable involving three or more categories) the category of primary interest should have the highest code number. For example, if three categories were employed with reference to a subject having a heart attack, the following codes might be used: **0**: Did not have a heart attack; **1**: Had a nonfatal heart attack; **2**: Had a fatal heart attack. Since the last category received the highest code number, it would be assumed to represent the category of primary interest; f) *SPSS* refers to the predictor variables as **covariates**.

5. The following *SPSS* command sequence was employed in this chapter in conducting a **logistic regression analysis**: a) Click **Analyze**; b) Click **Regression**; c) Click **Binary Logistic**; d) Highlight the criterion variable (i.e., the variable indicating whether or not a person has a **stroke** in Example 39.1) and move it to the **Dependent** window; e) Highlight the predictor variables (i.e., **blood pressure, smoking** and **social support** in Example 39.1) and move them to the **Covariates** window; f) Click **Categorical** if any of the predictor variables are categorical. Highlight the categorical variables (i.e., **smoking** and **social support** in Example 39.1) and move them to the **Categorical Covariates** window, and then click **Continue**; g) Click **Options**, check off desired information (e.g., **Classification plots, Hosmer-Lemeshow goodness-of-fit**, etc.), and then click **Continue**; h) Click **Save**, check off desired information (e.g. **Predicted values, Influence**, and **Residuals** options), and then click **Continue**; i) In the **Methods** window select the type of **logistic regression** analysis to be conducted. The default option of **Enter** represents **standard logistic regression** in which all of the predictor variables are entered simultaneously; j) Click **OK** to obtain the output for the analysis.

6. a) The reader may want to review the concept of **natural logarithms** which is discussed in Endnote 14 in the **Introduction**, as well as the concept of **odds** which is discussed in Section VI of the **chi-square test for $r \times c$ tables**; b) **Log odds** is sometimes written as *logit* (p); c) Although some sources may refer to **logistic regression** as **logit analysis**, the latter term is more commonly employed to refer to an analysis involving multiple discrete/categorical variables, with one of the variables designated as the dependent variable(s) (Garson (2006), Spicer (2005, p. 204) and Tabachnick and Fidell (1996, p. 281)).

7. $\ln [p(Y)/(1 - p(Y))]$ can also be written as $\ln_e [p(Y)/(1 - p(Y))]$, where e represents the base value of a natural logarithm which is equal to 2.71828..... . If we let z represent the exponent that e must be raised to in order to compute the value $[p(Y)/(1 - p(Y))]$, then $e^z = [p(Y)/(1 - p(Y))]$. Solving the latter equation for $p(Y)$ yields Equation 39.3, which can be algebraically transformed into Equation 39.4.

8. Tabachnick and Fidell (1996, p. 579) note that although not a requirement of **logistic regression**, multivariate normality and linearity among two or more predictor variables may increase the power of an analysis since a linear combination of the predictor variables is employed to form the exponent in Equations 39.3 and 39.4; b) Garson (2006) notes that one assumption underlying **logistic regression** is the **logit** of the predictor variable(s) is linearly related to the dependent variable.

9.   a) An iterative procedure employs **successive approximations** to arrive at an optimal solution for an analysis. In other words, a set of operations (referred to as an **iteration**) is sequentially repeated until the best possible approximation is computed for the value(s) in question. The latter will be illustrated later in Tables 39.2 and 39.3 (in Section V) in a table labeled **Iteration History**. Each row of an **Iteration History** table displays the result for one iteration, with the last row containing the result of the final iteration. The value displayed in the last row will represent the best approximation of the **likelihood ratio** (see Equation 39.5) computed in **logistic regression**; b) Tabachnick and Fidell (1996, p. 579) note the maximum likelihood procedure for estimating regression coefficients will not be able to derive a solution for **logistic regression** when there is perfect separation for a sample of n subjects with respect to category placement — for example, if all subjects in a sample with a systolic blood pressure below a specific value do not have a stroke, while all subjects whose systolic pressure is above that value do have a stroke. The latter authors note the latter situation, which constitutes what is sometimes referred to as **overfitting** the data, is unlikely to occur except perhaps in the case of a very small sample size. As noted in the discussion of **multiple regression**, more generally, **overfitting** is when the information derived from a sample is tailored to the idiosyncrasies of the sample, yet will not provide a good fit for the target population to which a researcher wishes to generalize.

10.   a) The **likelihood ratio test** can be employed to compare any two models with one another. Thus, if there are p predictor variables, the simple model can be contrasted with a model that is comprised of one or more of the p predictors, or alternatively, two models comprised of a different number of predictor variables can be contrasted with one another; b) Some sources employ Equation 39.10 as an alternative way of expressing Equation 39.7.

(Equation 39.10)

$$G = \chi^2 = 2 \,[log \; likelihood \; of \; model \; with \; IV(s) \; - \; log \; likelihood \; of \; model \; without \; IV(s)]$$

11.   Hosmer and Lemeshow (2000, pp. 16–17) also describe the **Scores test** (which is not available with most computer software) as another alternative for evaluating the significance of each of the predictor variables.

12.   As noted in Section IX (the **Addendum**) of the **binomial sign test for a single sample**, the **baserate** of an event is the proportion of times the event occurs within a population (or, in this instance, a sample).

13.   It was previously determined that by just employing the baserates in the **Classification table** (i.e., not taking into account the predictor variable), the probability of having a stroke is $31/80 = .3875$ and the probability of not having a stroke is $49/80 = .6125$. Using the latter values, Equation 39.1 can be employed to compute that the odds of having a stroke are $.3875/.6125 = .633$, which, in fact, is the value of **Exp(B)** in Column 7. Note that the natural logarithm of .633 is $-.458$, which is the value recorded for the constant in Column 2 of the **Variables in the Equation** table. The latter value is derived for the constant since when the **logistic regression** function (as defined by Equations 39.3 and 39.4) omits a predictor variable, the value computed for the exponent z is $z = a = -.458$. Thus, the value $-.458$ represents the **log odds** of having a stroke (i.e., it is equivalent to the value computed with Equation 39.2 if all of the predictor coefficients equal zero). Note that since the odds of having a stroke are less than 1, the **log odds** have a negative sign.

14. a) Garson (2006) and Wuensch (2005) note that since the maximum value **Cox & Snell** R^2 can achieve will generally be less than 1, **Nagelkerke** R^2 is obtained by dividing the value computed for **Cox & Snell** R^2 by the maximum value it can achieve, in order to allow for the possibility an effect size can equal 1; b) Additional measures of effect size for **logistic regression** which may be computed with software other than *SPSS* are **Somer's delta**, **gamma, tau–a, c,** and **McFadden's** ρ^2. **McFadden's** ρ^2 is the most conservative of the latter measures while **c** usually results in the highest value (Harlow (2005, p. 157). The value **McFadden's** $\rho^2 = .437$ is computed below with Equation 39.11.

(Equation 39.11)

$$McFadden's\ \rho^2 = 1 - \left[\frac{LR\ (Model\ with\ predictor\ variable(s))}{LR\ (Simple\ model)} \right] = 1 - \left[\frac{60.154}{106.819} \right] = .437$$

Harlow (2005, p. 157) states that an average of all of the aforementioned measures of effect size can be employed as a summary measure of effect size indicating the proportion of shared variance between one or more of predictor variables and the categorical dependent variable. Cohen's (1992) guidelines for a multivariate effect size (discussed in Section IV/V of the chapter on **multiple regression**) can be employed to assess the effect size (i.e., the average shared variance) for **logistic regression**, employing the values (for a squared correlation) .02, .13, and .26 as respectively small, medium, and large effect sizes.

15. A test statistic for the **Hosmer and Lemeshow goodness-of-fit test** cannot be computed when there is only one predictor variable and the latter is a dichotomous variable (Field (2005, p. 254)).

16. The **chi-square test for** $r \times c$ **tables (Test 16)** can also be employed to evaluate the data in the 2 × 2 **Classification Table**. When it is employed for the table summarizing the predicted model, it yields the significant result χ^2 (1) = 46.43, $p < .01$. The test, however, cannot be employed to evaluate the classification table for the simple model, because the latter model yields expected frequencies of 0 which render the test insoluble.

17. The **odds ratio** is also discussed in a different context in Section VI of the **chi-square test for** $r \times c$ **tables**.

18. a) **Odds ratios** and predictor coefficients for **logistic regression** computed by most software are in reference to the category on the dichotomous dependent variable coded 1 (i.e., the category of primary interest) — which in the case of Example 39.1 are subjects who have a stroke (Tabachnick and Fidell (1996, p. 605)); b) Wright (1995, p. 223–224) notes the reason why the **logistic regression** coefficient is used to determine an **odds ratio** associated with an increment of one or more units on the predictor variable, as opposed to determining the increase in probability associated with the latter increment, is that a change in probability will not only be a function of the predictor coefficient but also of the magnitude of the predictor variable. To be more specific, a one unit increment on the predictor variable at one point on the range of values for the latter variable may be associated with a minimal increase in the probability of a person being a member of the category of primary interest, yet at another point on the range of values for the predictor variable a one unit increment may be associated with a substantial increase in the probability of a person being a member of the category of primary interest. The change in

odds, on the other hand, will be the same for a one unit increase on the predictor variable at any point on the scale of values for the latter variable; c)Wright (1995, p. 223) notes that although the value of *b* (which is a measure of the change in the natural logarithm of the **odds ratio**) is in and of itself difficult to interpret, a positive coefficient indicates that a predicted **odds ratio** will increase as the value of the predictor variable increases, a negative coefficient indicates a predicted **odds ratio** will decrease as the value of the predictor variable increases, and a coefficient of 0 indicates the predicted **odds ratio** will be the same for all values of the predictor variable; d) The steepness and direction of a **logistic regression** curve (e.g., Figure 39.1) will be a function of the value of the coefficient (i.e., *b*). The larger the absolute value of the coefficient, the steeper the curve (i.e., the more perpendicular the central element of the curve will be in reference to the *X*-axis). When the value of the coefficient is positive, the **logistic regression** curve will ascend upwards from left to right as in Figure 39.1 (indicating that membership in the category of primary interest is associated with high values on the predictor variable, and membership in the other group with low values on the predictor variable). A negative *b* value, on the other hand, will yield a mirror image curve which descends from left to right. A coefficient of zero (which yields the same probability of being a member of the category of primary interest for all values of the predictor variable) results in a horizontal line parallel to the *X*-axis. The constant in the equation determines the location of the logistic curve in reference to the *X*-axis, with the curve shifting to the left as the constant increases (assuming the value for the coefficient remains unchanged) (Wright (1995, pp. 224–225)).

19. Wright (1995, p. 228) notes the following with respect to a confidence interval for an **odds ratio**: a) Although not the case for the example under discussion, the confidence interval will often be skewed (with the lower limit being much closer to the sample confidence interval than the upper limit), since the smallest possible value for an **odds ratio** is zero, while there is no boundary for the upper limit; b) If the value 0 is included in a 95% confidence interval, the predictor coefficient will not be significant at the .05 level.

20. a) It is recommended that a 95% confidence interval be computed (or 99% interval if one wants to be more conservative) for the probability obtained with Equations 39.3 or 39.4. In order to predict that a subject will fall in the category of having a stroke, both the lower and upper limit of the confidence interval should be above .5. Since computation of the latter confidence interval involves matrix algebra, it generally requires the use of a computer (Hosmer and Lemeshow (2000, p. 20) and Rosner (1995, p. 535)); b) In the case of two or more predictor variables, the identical procedure is employed, except for the fact that the value of z is obtained through use of the equation $z = a + b_1 X_1 + b_2 X_2 + ... + b_p X_p$, where b_j is the coefficient for each of the k predictor variable and X_j will be the score of the subject in question on each of the predictor variables.

21. Garson (2006) notes **Roa's efficient score statistic** is primarily employed in **forward stepwise logistic regression** as a criterion for determining whether or not a variable should be included in a model.

22. The fact that the absolute value .50 for the coefficient of systolic blood pressure is smaller than the absolute value .059 obtained earlier when systolic blood pressure was the only predictor employed indicates that systolic blood pressure has slightly less of an explanatory role in predicting group membership in the model containing three predictor variables.

23. a) As noted, even though smokers and people without social support were coded as 0 in the
     SPSS data editor, in order to compute the correct probability it is necessary to reverse
     the coding in Equation 39.3; b) Although it was noted earlier that social support should be
     eliminated from the model, it will be employed here to illustrate how to compute a
     probability for a model containing all three predictor variables; c) A separate **logistic
     regression** employing systolic blood pressure and smoking as the predictor variables
     obtained the values $a = -8.915$, $b_1 = .050$ (for systolic blood pressure) and $b_2 = 2.241$ (for
     smoking). When $X_1 = 198$ and $X_2 = 1$, $z = -8.915 + .050(198) + 2.241(1) = 3.226$. When
     the latter value is substituted in Equation 39.3 it yields the value $p(Y) = .962$. Thus, the
     likelihood of the subject having a stroke is .962.

$$p(Y) = \frac{e^z}{1 + e^z} = \frac{e^{3.226}}{1 + e^{3.226}} = .962$$

Test 40
Principal Components Analysis and Factor Analysis
(Parametric Test Employed with Interval/Ratio Data)

I. Hypothesis Evaluated with Test and Relevant Background Information

Principal components analysis and **factor analysis** are not inferential statistical tests, but instead are statistical procedures which attempt to reduce a set of p observable variables into a smaller number of dimensions. Or to put it another way, **principal components analysis** and **factor analysis** are methodologies with the goal of reducing a large number of variables into a limited number of dimensions referred to as **components** or **factors** which are able to account for all or most of the variability in the original set of data.[1]

Hypothesis evaluated with test   The primary hypothesis evaluated within the framework of procedures described in this chapter is the determination with respect to whether or not each of the derived components or factors makes a significant contribution in explaining the total variability in the data.

Relevant background information on test   **Principal component analysis** and **factor analysis** are statistical procedures which are commonly employed in a broad spectrum of academic disciplines in order to eliminate redundancy in a large body of data. To be more specific, these procedures reduce a set of intercorrelated variables into a smaller subset which accounts for all or most of the variability in the original data. The latter subset of variables, which are referred to as **components** or **factors**, are assumed to represent the basic underlying dimensions (often referred to as **latent variables**) which comprise the data. The components or factors derived in the analysis should be minimal in number, yet at the same time should account for all or most of the variability in the original set of data. Most often (although not required), a researcher will want to derive components or factors that are independent of one another.

Most sources note that **principal components analysis** is the preferred methodology when the goal of a researcher is to simplify a large body of data by reducing it to a limited number of components or factors, while **factor analysis** is preferred when a researcher's goal is to identify structure and/or to explore the underlying dimensions for theoretical purposes. In reality, however, many researchers also employ **principal components analysis** as a mechanism for identifying structure.

Since the goal of both **principal components analysis** and **factor analysis** is to identify the basic elements which comprise a body of data, in many respects what they attempt to do is similar to breaking matter down into its basic elements. Just as by virtue of combining the existing chemical elements the chemist is able to account for all varieties of matter, in **principal components analysis** and **factor analysis** it is assumed that when the derived components or factors (which are analogous to the elements) are combined with one another they will allow a researcher to account for all or most of the variability in a set of data.

The reader should be aware of the fact that the terms **factor analysis** and **principal components analysis** (as well as the term **exploratory factor analysis**) are often employed interchangeably. Some sources use the term **factor analysis** as a generic term to refer to a group of procedures that are designed to identify the underlying dimensions which comprise a body of data. The feature which distinguishes the different **factor analytic procedures** from one another is the method they employ in extracting factors. Although in this chapter the terms **factor analysis** and **principal components analysis** will both be conceptualized as **factor analytic procedures**, the main distinction that will be made will be between **principal components analysis** and **principal axis factor analysis**, which is the most commonly employed type of **factor analysis**.[2]

In order to understand the difference between **principal components analysis** and **principal axis factor analysis** it is necessary to examine how the variance is distributed in a set of data comprised of p variables (which it is assumed is summarized in a correlation matrix). The data is comprised of the following three types of variance which when summed account for the **total variance**: a) **Common or shared variance**: This is variance that is shared by two or more of the p variables. Common variance is assumed to be reliable (i.e., stable); b) **Specific or unique variance**: This is variance that is unique to one specific variable. Specific variance is also assumed to be reliable; and c) **Error variance**: This is random variance which is beyond the control of the researcher. Error variance is assumed to be unreliable.

If a researcher elects to extract components or factors based on **all** of the variance in a set of data, the extraction technique of choice is **principal component analysis,** which is both mathematically and conceptually the simplest of the **factor analytic procedures**. Thus, in **principal components analysis** common, specific and error variance associated with each of the observed variables are evaluated, and the derived **components** (also referred to as **factors** or **eigenvectors** in many sources) reflect the totality of variance in the set of data. The **components** maximize explained variability, and the number of derived components will always be equivalent to p — i.e., the original number of observed variables. Each of the components derived in **principal components analysis** represents a linear combination of the variables which to varying degrees can be employed to discriminate between subjects in the sample. More specifically, the first derived principal component will represent a linear combination of variables that maximally discriminates between subjects, the second derived principal component will represent the linear combination of variables which is second best at discriminating between subjects, and so on until p principal components have been derived.[3]

Since all of the information about each of the variables is evaluated in **principal components analysis**, it is inevitable that the derived components will not only reflect common variance among the variables, but also include specific variance which is totally unrelated to the other variables. Thus, an alternative methodology which excludes specific variance from the analysis might be expected to do a better job in describing the relationship between the variables. In **factor analysis**, and most specifically **principal axis factor analysis**, only the variance which each of the variables shares with the other variables (i.e., common variance) is available for analysis. Tabachnick and Fidell (1996, p. 663) note that, in the final analysis, **principal axis factor analysis** evaluates covariance (i.e., common variance), while **principal components analysis** evaluates variance.

The fact that **principal axis factor analysis** excludes specific variance is predicated on the belief that inclusion of such variance will not allow a researcher to derive an optimal picture of the underlying dimensional structure of the data. **Principal axis factor analysis** is more likely to be used for model/theory testing (e.g., that the most parsimonious model for describing the underlying structure of a set of data comprised of 15 observed variables is comprised of three factors), while **principal components analysis** is more likely to be employed within an exploratory context to obtain a general picture of the underlying dimensional structure of a set of data. It should be noted, however, that some sources also describe the use of **principal axis factor analysis** as an exploratory procedure.[4]

Mathematically, the critical difference between **principal components analysis** and **principal axis factor analysis** are the values (which are correlation coefficients) employed in the main diagonal of the correlation matrix (i.e., the matrix which contains the intercorrelations between the p variables). Since **principal components analysis** evaluates all of the variance, the value 1 (which is the maximum possible value, with 0 being the minimum possible value) is placed in each of the cells of the main diagonal. In **principal axis factor analysis** the values typically placed in the main diagonal represent estimates of communality (i.e., estimates of common variance — specifically, the squared multiple correlation of each variable with the other variables (Field (2005, p. 630)). Because of the fact that communalities have to be estimated (and because of the latter to some degree the result of the analysis will be determined by the accuracy of the communality estimates), **principal axis factor analysis** is viewed as more mathematically problematical than **principal components analysis** (which provides a unique mathematical solution).[5] For a detailed discussion of how communality values are estimated the reader is referred to Tabachnick and Fidell (1996, p. 671).

In order for a **factor analytic procedure** (which is the generic term that will be used in this chapter when referring to **principal components analysis**, **principal axis factor analysis**, or other factor analytic methodologies) to be reliable, a researcher should employ a large representative sample from the target population toward which he or she intends to generalize the results. Some sources (e.g. Tabachnick and Fidell (1996, p. 640)) state a minimum of 300 subjects should be used for a **factor analytic procedure** (although an even larger number between 500 and 1000 would be preferable). Bryant and Yarnold (1995, p. 100), on the other hand, note that sample size should be based on a subject-to-variables ratio of at least 5 to 10 (i.e., the number of subjects employed should be at least five to ten times the number of variables involved in the analysis), but that regardless of the value of the subject-to-variables ratio, a minimum of 100 subjects should be employed. In the final analysis, although all sources note the larger the sample size the better, there is a lack of agreement with respect to the acceptable minimum value.

As well as the requirement of a large sample size, the reliability of a **factor analytic procedure** will depend upon certain assumptions having been met, as well as other issues relating to the nature of the data. Specifically:

1) **Multivariate normality** As is the case for most multivariate procedures, multivariate normality is an assumption underlying **factor analytic procedures**. The following conditions must be met to satisfy the latter requirement:

a) Multivariate normality requires that all of the variables are normally distributed. Tabachnick and Fidell (1996, pp. 70–71; p. 640) state that normality of the variables is generally assessed through analysis of skewness and kurtosis, and any variable for which either condition is violated should probably be modified through use of a data transformation. Tabachnick and Fidell (1996, pp. 640) do, however, note that although violation of normality for one or more of the variables may compromise the reliability of a **factor analytic procedure**, the analysis may still yield a worthwhile result.

b) Multivariate normality requires that linear combinations between each pair of variables should be normally distributed. In addition to confirming the linear relationship between a pair of variables, normality of linear combinations is assessed by demonstrating homoscedasticity, as well as by determining that the residuals are distributed normally (most commonly through use of scatterplots).

2) **Outliers** Outlier detection should be implemented for the observations on each variable prior to conducting an analysis. Suspected outliers should be deleted or otherwise dealt with as noted in Section VII of the *t* **test for two independent samples (Test 11)**. On the subject of outliers, Tabachnick and Fidell (1996, p. 642) also note that if any of the p variables has a low

squared multiple correlation with all the other variables and low correlations with the most important of the derived factors or components, such a variable should be considered as an **outlier among variables**, and should thus be ignored in the analysis.

3) **Multicollinearity and singularity** Tabachnick and Fidell (1996, pp. 640–642) note that in **principal components analysis** multicollinearity is not a problem, yet for most forms of **factor analysis** singularity or extreme multicollinearity is a problem, and that variables exhibiting the latter should be deleted from the analysis. (Multicollinearity and singularity are discussed in the chapter on **multiple regression (Test 33)**).

II. Example

Example 40.1 *A researcher wants to determine whether or not the traits measured by six commonly used personality tests can be expressed within the framework of a more limited number of dimensions. The researcher elects to employ principal components analysis, since she believes there is considerable redundancy among the six tests. Eighty people are administered the six tests which measure the following traits (The range of possible values on each test is noted in parentheses.) :* **Test A – Anxiety** (0–20); **Test B – Somatic Complaints** (0–30); **Test C – Guilt** (0–40); **Test D – Friendliness** (0–50); **Test E – Sensation Seeking** (0–10); **Test F – Dominance** (0–100). *The higher a subject's score on any of the six traits, the higher the subject's standing on the trait in question.* Table 40.1 *displays the scores of the 80 subjects on the six variables, and* Table 40.2 *summarizes the intercorrelations between the scores of subjects' on the six tests.*

III. Null versus Alternative Hypotheses

The primary null hypothesis evaluated within the framework of **factor analytic procedures** is evaluated in reference to each of the derived components or factors. For each component or factor the **null hypothesis** is that the component or factor in question does not make a significant contribution in explaining the total variability in the data. The **alternative hypothesis** is that the component or factor in question does make a significant contribution in explaining the total variability in the data.

IV. Test Computations

Since the calculations involved in a **principal components analysis** (as well as a **principal axis factor analysis**) are complex and the matrix algebra operations are impractical to implement without the aid of a computer, they will not be demonstrated here. (Readers interested in a detailed description of the mathematical operations involved in **factor analytic procedures** should consult sources which specialize in multivariate statistics.) This section (as well as Section V) is designed to provide the reader with a succinct summary of the steps involved in a **principal components analysis**, along with guidelines for interpretation. Interpretive guidelines will focus on information available from an *SPSS* analysis.

The following steps are involved in conducting a **principal components analysis** (as well as a **principal axis factor analysis**, with the qualification that in the latter type of analysis different values would be employed for the communalities in the main diagonal of the correlation matrix). In the discussion to follow, the term **factor analysis** will be employed in reference to the operations involved during a **principal components analysis** as well as a **principal axis factor analysis**.

Table 40.1 Raw Data for Example 40.1

	subject	anx	soma	guilt	frnd	sen	dom
1	1.00	12.00	15.00	27.00	32.00	6.00	52.00
2	2.00	18.00	30.00	34.00	34.00	8.00	34.00
3	3.00	20.00	23.00	37.00	14.00	4.00	34.00
4	4.00	19.00	26.00	34.00	12.00	3.00	23.00
5	5.00	12.00	27.00	18.00	10.00	1.00	10.00
6	6.00	15.00	24.00	30.00	49.00	9.00	49.00
7	7.00	12.00	25.00	29.00	24.00	5.00	38.00
8	8.00	17.00	29.00	34.00	39.00	9.00	87.00
9	9.00	16.00	23.00	27.00	47.00	3.00	39.00
10	10.00	15.00	24.00	28.00	19.00	6.00	19.00
11	11.00	18.00	26.00	16.00	47.00	9.00	89.00
12	12.00	6.00	10.00	11.00	40.00	8.00	70.00
13	13.00	11.00	11.00	19.00	30.00	4.00	30.00
14	14.00	20.00	29.00	35.00	33.00	6.00	33.00
15	15.00	18.00	23.00	32.00	25.00	3.00	25.00
16	16.00	19.00	18.00	35.00	23.00	5.00	23.00
17	17.00	13.00	23.00	23.00	16.00	1.00	16.00
18	18.00	18.00	22.00	36.00	32.00	8.00	67.00
19	19.00	16.00	21.00	37.00	34.00	8.00	88.00
20	20.00	19.00	18.00	40.00	14.00	2.00	14.00
21	21.00	14.00	12.00	23.00	12.00	.00	12.00
22	22.00	15.00	16.00	22.00	10.00	4.00	10.00
23	23.00	15.00	22.00	16.00	49.00	9.00	69.00
24	24.00	12.00	24.00	27.00	24.00	8.00	78.00
25	25.00	13.00	17.00	23.00	39.00	10.00	99.00

	subject	anx	soma	guilt	frnd	sen	dom
26	26.00	10.00	19.00	33.00	47.00	10.00	81.00
27	27.00	15.00	20.00	16.00	19.00	6.00	52.00
28	28.00	20.00	22.00	18.00	47.00	8.00	63.00
29	29.00	19.00	26.00	29.00	40.00	7.00	45.00
30	30.00	11.00	14.00	23.00	30.00	4.00	52.00
31	31.00	10.00	15.00	9.00	33.00	8.00	66.00
32	32.00	7.00	11.00	15.00	25.00	5.00	44.00
33	33.00	18.00	21.00	21.00	23.00	5.00	48.00
34	34.00	15.00	11.00	16.00	16.00	4.00	26.00
35	35.00	20.00	26.00	28.00	48.00	9.00	89.00
36	36.00	15.00	20.00	21.00	50.00	8.00	52.00
37	37.00	15.00	14.00	20.00	16.00	2.00	18.00
38	38.00	13.00	13.00	11.00	29.00	8.00	77.00
39	39.00	12.00	14.00	16.00	36.00	7.00	60.00
40	40.00	16.00	20.00	19.00	44.00	3.00	43.00
41	41.00	10.00	14.00	16.00	12.00	4.00	13.00
42	42.00	11.00	13.00	12.00	22.00	5.00	18.00
43	43.00	14.00	15.00	10.00	30.00	7.00	32.00
44	44.00	15.00	14.00	29.00	33.00	8.00	93.00
45	45.00	13.00	9.00	17.00	27.00	6.00	71.00
46	46.00	15.00	10.00	18.00	22.00	6.00	44.00
47	47.00	16.00	15.00	14.00	21.00	4.00	39.00
48	48.00	10.00	17.00	22.00	2.00	1.00	16.00
49	49.00	10.00	16.00	14.00	19.00	5.00	29.00
50	50.00	14.00	18.00	14.00	9.00	.00	7.00

Table 40.1   Raw Data for Example 40.1 (continued)

	subject	anx	soma	guilt	frnd	sen	dom
51	51.00	12.00	23.00	18.00	18.00	5.00	38.00
52	52.00	11.00	18.00	11.00	32.00	7.00	64.00
53	53.00	8.00	13.00	17.00	29.00	7.00	29.00
54	54.00	11.00	19.00	12.00	48.00	8.00	62.00
55	55.00	16.00	29.00	11.00	50.00	7.00	56.00
56	56.00	15.00	19.00	20.00	16.00	5.00	43.00
57	57.00	19.00	28.00	13.00	29.00	7.00	92.00
58	58.00	13.00	10.00	17.00	36.00	7.00	82.00
59	59.00	6.00	10.00	9.00	44.00	10.00	77.00
60	60.00	17.00	9.00	16.00	40.00	7.00	48.00
61	61.00	12.00	10.00	13.00	38.00	9.00	67.00
62	62.00	11.00	14.00	16.00	32.00	6.00	80.00
63	63.00	12.00	9.00	15.00	34.00	9.00	79.00
64	64.00	15.00	14.00	11.00	14.00	4.00	13.00
65	65.00	9.00	10.00	12.00	12.00	5.00	39.00
66	66.00	16.00	13.00	17.00	10.00	3.00	10.00
67	67.00	9.00	6.00	13.00	49.00	10.00	34.00
68	68.00	6.00	12.00	10.00	24.00	6.00	61.00
69	69.00	12.00	12.00	14.00	39.00	7.00	49.00
70	70.00	10.00	13.00	10.00	47.00	7.00	78.00
71	71.00	12.00	14.00	11.00	19.00	4.00	36.00
72	72.00	14.00	12.00	12.00	47.00	10.00	47.00
73	73.00	14.00	9.00	21.00	40.00	6.00	39.00
74	74.00	13.00	11.00	20.00	30.00	4.00	42.00
75	75.00	3.00	8.00	11.00	33.00	7.00	73.00

	subject	anx	soma	guilt	frnd	sen	dom
76	76.00	7.00	3.00	13.00	25.00	7.00	66.00
77	77.00	9.00	13.00	11.00	23.00	8.00	60.00
78	78.00	5.00	4.00	18.00	16.00	4.00	41.00
79	79.00	10.00	5.00	12.00	18.00	6.00	51.00
80	80.00	4.00	7.00	12.00	14.00	3.00	52.00

Table 40.2   Correlation Matrix (Personality Test Intercorrelations)

Test	A	B	C	D	E	F
A	1	.665	.586	.092	-.055	-.105
B	x	1	.560	.136	.006	-.003
C	–	–	1	-.032	-.096	-.095
D	–	–	–	1	.753	.622
E	–	–	–	–	1	.757
F	–	–	–	–	–	1

Step 1 — Accumulating the data: In conducting a **factor analysis** the first thing a researcher must do is to select a set of variables to factor analyze. Which set of variables the researcher selects will depend upon the nature of the problem one is studying. For example, if the researcher wishes to identify the basic traits that can be employed to explain individual differences in personality, she should select a large number of variables which encompass all aspects of human behavior (that are assumed to be a function of one's personality). It is important to note that the specific types (as well as the number) of variables a researcher selects will have a direct impact on the components or factors one will derive from the analysis. The fact that two or more researchers employing a **factor analytic procedure** may reach different conclusions in studying the same subject matter can, among other things, be attributed to the fact that they may have employed different sets of variables for their analyses, and/or used different measures for defining the same variables. In the case of Example 40.1, it is entirely possible that if the traits measured by the six tests were evaluated with alternative measures (e.g., different tests evaluating the same traits, ratings on the traits by a family member, or ratings on specific behaviors viewed as representing the traits), then a **principal components analysis** on the latter data would yield a different result than the one to be described in Section V.

Step 2 — Constructing the correlation matrix: Once scores are obtained for the n subjects (in the case of Example 40.1, $n = 80$) on each of the p variables (in the case of Example 40.1, $p = 6$), correlations are obtained between the scores of subjects on all of the variables. The results of the latter can be summarized in a **correlation matrix** (sometimes referred to as an **R-matrix**), which is represented by Table 40.2.

Each entry in Table 40.2 represents the value of the correlation between two of the six tests. To determine the two tests represented by any of the correlations in the table, identify the letter at the left of the row and the letter at the top of the column in which a specific correlation appears. The row and column letters represent the two tests for which that correlation was computed. Thus, as an example, the correlation between *Test A* and *Test B* is .665, since $r = .665$ appears in Row A and Column B. Note that correlations only appear in the upper half of the table, since the same information would be repeated in the lower half of the table (i.e., as an example, the correlation .665 would also appear in the cell with an x, which is in Row B and Column A). Since a **principal components analysis** will be conducted, the value 1 has been employed in each of the six cells which comprise the main diagonal of the Table 40.2.

Step 3 — Assess the factorability of the correlation matrix: A correlation matrix which is suitable to be factor analyzed should contain several substantial correlations. The latter translates into the presence of some correlations with an absolute value of .30 or greater when the sample size is large. In the case of a smaller sample sizes, larger correlations are necessary. Tabachnick and Fidell (1996, pp. 641) recommend employing **Bartlett's test of sphericity** (Bartlett (1954)) (which evaluates the null hypothesis that all of the correlations in a correlation matrix are zero), but note that the test (which employs the chi-square distribution) should only be used for small sample sizes, since it is overly sensitive to detecting significance with large samples.[6] The latter authors also point out that high bivariate correlations do not guarantee that a correlation matrix can be broken down into meaningful factors.

In the case of Example 40.1, *SPSS* obtained the result $\chi^2 (15) = 219.872$, $p = .000$ (which indicates a probability less than .0005) for **Bartlett's test of sphericity**. Note that in **Table A4 (Table of the Chi-Square Distribution)** in the **Appendix**, for $df = 15$, $\chi^2_{.05} = 25.00$ and $\chi^2_{.01} = 30.58$. Since $\chi^2 = 219.872$ is greater than $\chi^2_{.01} = 30.58$, the null hypothesis can be rejected at the .01 level. The researcher can thus conclude that in the underlying population from which the sample was derived, at least one of the correlations in Table 40.2 is some value other than zero.

The **Kaiser–Meyer–Olkin statistic** is commonly employed to determine whether or not the sample size is suitable for a **factor analysis** (more specifically, whether the size of the sample is adequate to yield reliable (i.e., stable) correlations between the variables). A **Kaiser–Meyer –Olkin statistic** is computed for each variable and the sum of all the latter values is employed to represent an overall **Kaiser–Meyer–Olkin statistic** which can range between 0 and 1. Most sources state that the overall **Kaiser–Meyer–Olkin statistic** should be .6 or greater, and if the latter condition is not met variables should be deleted from the analysis beginning with the variable with the lowest **Kaiser–Meyer–Olkin statistic** until an overall statistic with a value above .6 is achieved. In the case of Example 40.1, where $n = 80$, *SPSS* obtained the value .704 for the **Kaiser–Meyer–Olkin statistic**. Since the latter value is greater than .6, it indicates the correlation matrix is suitable to be factor analyzed.[7]

Although as noted in Section I, Tabachnick and Fidell (1996, pp. 640–642) state that multicollinearity is not a problem for a **principal components analysis**, it can be an issue for a **principal axis factor analysis**. With regard to the latter, Field (2005, pp. 641–642) notes multicollinearity can be assessed by computing the **determinant** of the correlation matrix, and that a value greater than .00001 should be obtained. In the case of Example 40.1 the value .056 was computed for the determinant, which indicates there is no evidence of multicollinearity. Field (2005, p. 642) states if multicollinearity is detected, one or more of the variables which are highly correlated with one another should be deleted from the analysis until an acceptable value is obtained for the determinant of the resulting correlation matrix.

Step 4 — Conducting the factor analysis: Factor analytic procedures involve a complex set of mathematical operations which are conducted on the correlation matrix.[8] The process of factor analyzing such a matrix is often summarized by the term *decomposing the matrix*, since the goal of the analysis is to allow a researcher to determine whether or not the *p* variables can be broken down into a more limited number of dimensions.

The term **factor** derives from mathematics. More specifically, recollect in basic algebra how you were taught to factor an equation such as the one below.

$$(x^2 - y^2) = (x + y)(x - y)$$

By factoring the above equation one has broken it down into two basic elements, which when combined yield the original equation. In the same respect, **factor analysis** of a correlation matrix, such as that depicted in Table 40.2, will determine whether or not a more limited number of basic elements/dimensions can be employed to summarize and explain the information provided by the six tests. At this point it will be assumed that the appropriate mathematical operations associated with a **principal components analysis** have been conducted through use of computer software.[9]

V. Interpretation of the Test Results

The results of a **factor analytic procedure** can be summarized both mathematically (which will be discussed initially) and geometrically (which will be discussed later in this section). Table 40.3 provides a succinct summary of the **principal components analysis** of the data for Example 40.1. In Section I it was noted that when a **principal components analysis** is conducted, the number of derived components (which are referred to as factors) will equal the number of variables. Note, however, that Table 40.3 displays information for just two of the six factors which were derived in the analysis. The reason why information for only two of the factors is displayed is because when a **principal** components analysis is conducted it is not uncommon for the researcher to focus on a subset of factors which accounts for most of the variability in the data. In the case of Example 40.1, it will subsequently be noted that the two factors documented in Table 40.3 account for approximately 80% of the variability in the data, and that the other four factors derived in the analysis only accounted for a minimal amount of variability (which sum to the remaining 20% of the total variability), and because of the latter were discarded from the analysis.[10]

Table 40.3 Summary Table of Factor Analysis of Six
Personality Tests in Example 40.1

| Test | Factor[11] | | Communality |
	I	II	
A (Anx)	-.208	.856	.7760
B (Soma)	-.118	.863	.7587
C (Guilt)	-.267	.782	.6828
D (Frnd)	.838	.295	.7893
E (Sen)	.921	.156	.8726
F (Dom)	.876	.112	.7799
Eigenvalue	2.446	2.214	Sum = 4.660
Percent of Total Variance	40.76%	36.90%	

Note that a set of numerical values are recorded in the columns for *Factor I* and *Factor II*. These values are called **factor loadings**. A factor loading can be interpreted as a correlation coefficient (and thus it will always fall within the range +1 to -1) that tells the researcher how much each of the variables (in this case, each of the tests) correlates with each of the factors. As is also the case with a correlation coefficient, the absolute value of a factor loading indicates the strength of the relationship between that factor and a given variable. The higher the absolute value of a factor loading, the more that variable contributes to that factor or, to put it another way, the higher the factor loading, the purer measure that variable is of that factor. Thus, *Test A* has a loading of -.208 on *Factor I* and a loading of .856 on *Factor II*. This indicates that *Test A* is measuring *Factor II* to a much greater extent than it is measuring *Factor I*. By squaring the factor loading of a test, one can determine how much of the variance on the test can be accounted for by that factor. Thus, only the variance on *Test A* can be accounted for on the basis of *Factor I*, whereas *Factor II* accounts for $(.856)^2 = .7327 = 73.27\%$ of the variance on *Test A*. Taken together, *Factors I* and *II* account for 77.60% (i.e., 4.33% + 73.27%) of the variance on *Test A*. The value 77.60% (which expressed as a proportion = .7760) corresponds to the **communality** of the test (which is listed in the last column of Table 40.3). The concept of communality is discussed in greater detail later in this section.

Since all of the variance on *Test A* (as well as on the other five tests) is not accounted for on the basis of the two factors displayed in Table 40.3, one might ask why only two factors have been employed in Table 40.3 to describe the results of the **principal components analysis**. Or, to put it another way, how does one decide how many factors to derive in a **factor analytic procedure**?

As Kachigan (1986) notes, in interpreting the results of a **factor analytic procedure**, a researcher must weigh **parsimony** against **comprehensiveness**. Thus, although the researcher wishes to account for as much of the variability in the data as possible (comprehensiveness), at the same time she wants to do it in the simplest possible manner (parsimony — i.e., with the fewest number of factors). There is no set rule with respect to how much of the total variance must be accounted for by the factors a researcher derives. (For instance, Stevens (1996) notes that some researchers attempt to account for a minimum of 70% of the variability.) In essence, how many factors one decides to employ will ultimately depend on the purpose for which one intends to use the results of the analysis. Those factors which explain the greatest amount of variability in the data almost always represent what are referred to as **common factors**. Common factors are factors that load on more than one of the variables. Those common factors, which account for a substantial amount of the variability, are designated as **significant factors** in an analysis. *Factors I* and *II* in Table 40.3 represent common factors, since on both of these factors all six tests have loadings above zero, and at least three of the tests have substantial loadings on one of the two factors. In contrast to common factors are **specific factors**, which are factors that load on only one of the variables. Within the framework of an analysis, specific factors generally account for only a small portion of the total variance, and typically do not play an important role in explaining the data. In addition to specific factors, another element that accounts for a small portion of the total variance are **error factors** (also referred to as **error variance**). Error factors represent uncontrolled variability — i.e., such things as poor reliability in measuring the variables, and/or other sources of error in the data which are beyond the control of the researcher.

Returning to our example, the exact amount of the variance which can be accounted for by the two factors represented in Table 40.3 can be obtained by adding up the numbers in the last row of the table (labeled **Percent of Total Variance**). Thus: 40.76% + 36.90% = 77.66%. This tells us that 77.66% of the total variability in the data can be explained by the two factors, and that of that 77.66%, *Factor I* accounts for 40.76% of the variability, while *Factor II* accounts for the remaining 36.90% of the variability. In a **factor analysis** summary table, the first factor listed (identified as *Factor I*) will always account for the greatest amount of variability, followed by *Factor II* which will account for the next largest amount of variability, and so forth. If the results of the analysis were depicted in greater detail (as in Table 40.5), and those additional factors that accounted for minimal variability were included in Table 40.3, 100% of the variability would be accounted for. In any event, a **principal components analysis** (or, for that matter, a **principal axis factor analysis**) based on six initial variables, in which two factors account for 77.66% of the variance, would be considered an excellent compromise between parsimony and comprehensiveness — i.e., one that explains most of the variability through use of a minimum number of factors.

At this point it should be noted that, more often than not, the final factorial structure will be based on **rotated factors**. **Rotation**, which is described later in this section, allows a researcher to derive the simplest possible factorial structure for the data. In the case of Example 40.1 the **unrotated factorial structure** (which is summarized in Table 40.3) and the **rotated factorial structure** (which will subsequently be summarized in Table 40.4) are almost identical.

Eigenvalues and communalities Two other sets of values depicted in Table 40.3 are the **eigenvalues** and **communalities**. An **eigenvalue** is a numerical index that indicates the relative strength of each of the derived factors. On a more technical level, Kachigan (1986) notes that an eigenvalue (also known as a **characteristic** or **latent root**) is the equivalent number of variables a factor represents. As an example, a factor with an eigenvalue of 4 accounts for as much variance in the overall data as one would expect for four variables if the total variability were evenly distributed among all of the variables. Thus the higher the eigenvalue associated with a factor, the larger the role that factor plays in explaining variability in the complete set of data. The value of an eigenvalue can range from any number above zero up to the number of variables being factor analyzed (which in our example is six). An eigenvalue can be computed for each factor by summing the squares of the loadings on each of the variables for that factor. Thus, in the case of *Factor 1*: $(-.208)^2 + (-.118)^2 + (-.267)^2 + (.838)^2 + (.921)^2 + (.876)^2 = 2.446$; in the case of *Factor 2*: $(.856)^2 + (.863)^2 + (.782)^2 + (.295)^2 + (.156)^2 + (.112)^2 = 2.214$.

In order to employ an eigenvalue to determine the relative strength of a factor (in terms of percentage of variability that factor accounts for) one should do the following: a) Divide the value of the eigenvalue by the number of variables employed in the factor analysis; and b) Multiply the result of the division by 100. The resulting value will be the percentage of variability in the data that can be accounted for by that factor. Thus, in our example: For *Factor I*: 2.446 ÷ 6 = .4076, and .4076 × 100 = 40.76%; for *Factor II*: 2.214 ÷ 6 = .3690, and .3690 × 100 = 39.90%. Note that the values 40.76% and 36.90% correspond to the values in the bottom row (labeled **Percent of Total Variance**) of Table 40.3.

A widely accepted criterion for determining the number of factors to retain is the **Kaiser stopping rule** (Kaiser (1960)), which states that only factors with an eigenvalue of 1 or greater should be retained (i.e., factors that at least account for the same amount of variance as one variable would be expected to). As a general rule, factors that have eigenvalues less than 1 will represent specific factors or error factors whose contribution in explaining the overall variability in the data is minimal.[12] In the case of Example 40.1, the eigenvalues computed for *Factors III* through *VI* are all less than 1. Specifically, the eigenvalues (along with the percent of the total variance accounted for by each factor in parentheses) for *Factors III–VI* are: *Factor III*: .478 (7.96%); *Factor IV*: .364 (6.07%); *Factor V*: .303 (5.06%); *Factor VI*: .195 (3.25%). Note that the eigenvalues for all six factors sum to the value $p = 6$: (2.446 + 2.214 + .478 + .364 + .303 + .195) = 6, and the percent of variance accounted for by all six factors sums to 40.76% + 36.90% + 7.96% + 6.07% + 5.06% + 3.25% = 100%. The latter information is summarized in Table 40.5 (which displays part to the *SPSS* output for a **principal components analysis**) in the table labeled **Total Variance Explained**.

An alternative way of evaluating **eigenvalues** is through use of a graphic technique developed by Cattell (1966) referred to as a **scree plot**. In a scree plot a researcher identifies a break in the curve which delineates factors (which are recorded on the *X*-axis) that have substantial eigenvalues (which are recorded on the *Y*-axis) from those factors which have small eigenvalues, and retains only the former category of factors. Figure 40.1 is a **scree plot** for Example 40.1. Note that although six factors are delineated on the abscissa, the eigenvalues for *Factors III* through *VI* are minimal. Note that the plot indicates the eigenvalues for *Factors I* and *II* are respectively 2.446 and 2.214, and that the curve takes a sharp downward turn after *Factor II*.

As noted earlier, the amount of variance on any variable which can be explained by the derived common factors is referred to as **communality** (which is commonly represented by the notation h^2), and the communality of a variable is obtained by squaring the factor loadings of the variable on each of the factors and summing the squared correlations. Communality values will always fall within the range 0 to 1. Since, as noted earlier, the communality for *Test A* is .7660, 76.60% of the variance on *Test A* can be explained by *Factors I* and *II*. Of the 100% total variance on *Test A*, only 23.40% (i.e., 100% − 76.60% = 23.40%) cannot be accounted for by either *Factor I* or *Factor II*. The remaining 23.40% of the variance would be explainable through *Factors III* through *VI*.

The reader should take note of the following additional points with respect to the values depicted in Table 40.3: a) The sum of the communalities is equal to the sum of the eigenvalues. This will always be the case in a **factor analysis** summary table that has the same basic structure as Table 40.3; b) If, in fact, all possible factors in a set of data, including specific factors and error factors, are derived, and the squared factor loadings for each variable are summed, the communality of each variable will equal 1. In such a case, the sum of the communalities of all the variables will always equal the total number of variables (p) employed in the analysis; c) It is also the case that if all possible factors in a set of data are derived, the sum of the eigenvalues of all of the factors will equal the total number of variables (p); d) For any given variable, the

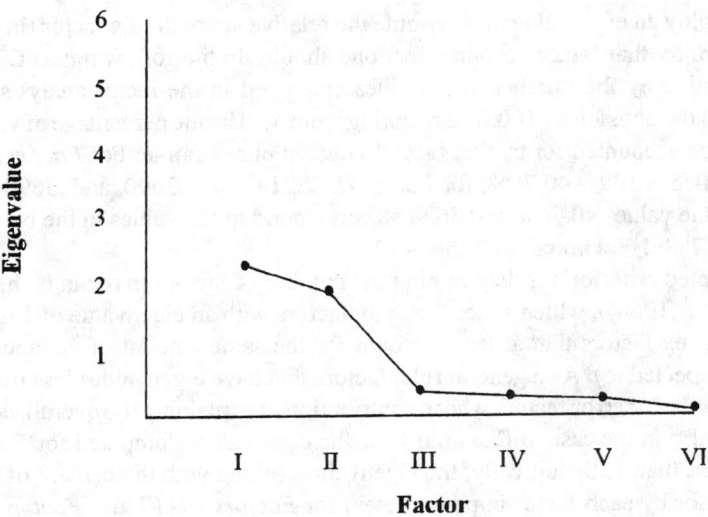

Figure 40.1   Scree Plot for Example 40.1

total variance on that variable can be attributed to the following: 1) The common variance between that variable and all the factors derived in the analysis for which the variable has a factor loading other than zero; 2) Error variance, which is obtained by subtracting the reliability coefficient of the variable from the value 1; and 3) Variance specific to the variable, which is computed by subtracting from the value 1, the total common variance (i.e., communality) of the variable on all common factors and the error variance. The value which remains after this subtraction represents the specific variance unique to that variable. This **specific** or **unique** variance is independent of all other variables. The **uniqueness** of a variable is defined as 1 **minus the communality of the variable** (i.e., *Uniqueness* $= 1 - h^2$). Alternatively **1 minus a variable's uniqueness equals its communality** (i.e., $h^2 = 1 - Uniqueness$). In defining the **uniqueness** of a variable, sources commonly state that the total variance on a variable can be broken down into two components, **common variance** it shares with the other variables (i.e., its **communality**) and **specific** or **unique variance** which can be attributed to that variable (which represents the **uniqueness** of a variable). However, some of the variance which comprises the unique variance for a variable may not be reliable, and this latter type of unreliable variance is viewed as **error variance**. Thus, within the latter context, the **uniqueness** of a variable is comprised of reliably measured unique variance plus unreliable error variance; and e) *SPSS* prints out a separate table labeled **Communalities** for a **principal components analysis**. The latter table is comprised of two columns labeled **Initial** and **Extraction**, as well as p rows for each of the variables. Since **principal components analysis** operates on the assumption that all of the variance is shared, in the row labeled **Initial** it lists a communality value of 1 for each of the variables (which represents the value in the main diagonal of the original correlation matrix — i.e., Table 40.2). The values printed in the rows for each of the six variables in the column labeled **Extracted** are new communality values for each of the variables which reflect the proportion of shared variability associated with each variable, when only the retained principal components are taken into consideration. These communality values for the variables are identical to those in the last column of Table 40.3.

    Factor scores   Within the framework of a **factor analytic procedure** a **factor score** can be computed for each subject on each of the factors. A subject's **factor score** is a composite score based on the relative contribution of all the variables which represent that factor. Computation of factor scores allows a researcher to determine the relative standing of each subject with respect to each of the factors derived in the analysis. On a mathematical level, each factor can be defined by a linear equation in the form of Equation 40.1, and through use of the latter equation a factor score can be computed for any subject. Note that the latter equation requires standardized scores be employed for each of the variables, since in most instances the variables will involve different scales of measurement. To compute a subject's factor score on a specific factor, the subject's standardized score on each variable is multiplied by a **weighting coefficient** computed for each of the variables, after which all of the products are summed. It is important to note that although a factor **weighting coefficient** for a variable is similar to the **factor loading** for that variable, the two measures are not identical. Since computation of factor loadings and weighting coefficients are complex and typically implemented with the use of a computer, they will not be demonstrated here.[13]

$$Y_i = b_1 Z_1 + b_2 Z_2 + \ldots + b_q Z_q \qquad \textbf{(Equation 40.1)}$$

Where: Y_i represents the **factor score** of a subject on the ith **factor**

b_j represents the **weighting coefficient** for the jth predictor variable on the ith **factor**

Z_j represents a subject's standardized score on the jth variable

Since the sign of a factor loading is interpreted the same way as the sign of a correlation coefficient, the sign of a factor loading indicates the direction of the relationship between a subject's score on a variable and his or her score on that factor. Specifically, in the case of *Factor II*, the factor loading of .856 for *Test A* indicates that a subject who obtains a high score on *Test A* will have a high score on *Factor II*, and that a subject who has a low score on *Test A* will have a low score on *Factor II*. Just as positive factor loadings are interpreted as positive correlations, negative factor loadings are interpreted as negative correlations. Thus, the Factor loading of –.208 on *Factor I* for *Test A* indicates that a subject who has a high score on *Test A* will have a somewhat low score on *Factor I*, and a subject who has a low score on *Test A* will have a somewhat high score on *Factor I*.

Significance of factors   Diekhoff (1992), among others, notes the issue of statistical significance in reference to factor analysis addresses the question of whether or not the obtained factor structure for a set of data is a reliable indicator of the factor structure in the underlying population. To be more specific, if it is determined the factor structure is statistically significant it would be expected the same factor structure will be obtained if the same **factor analytic procedure** is conducted on another sample that is drawn from the same target population. Although a number of tests of statistical significance have been developed (all of which are mathematically complex and assume a relatively large sample size) for **factor analytic procedures**, there is a lack of agreement among sources with respect to which test is most appropriate to employ. Some sources, however, use the following guidelines for determining whether or not a specific factor loading is statistically significant: a) For smaller sample sizes, any factor loading with an absolute value of .40 or greater is considered significant; and b) For larger sample sizes, any factor loading with an absolute value of .30 or greater is considered significant. (The reader should note that the amount of explainable variance on a variable attributable to a factor which has a loading of .30 is only $(.30)^2 = .09 = 9\%$ — certainly not a large amount by anyone's standards.) Sources, however, do not agree on the values that constitute a "smaller" versus "larger" sample size.

Based on the results of a study by Cliff and Hamburger (1967) (which indicated high values for the standard errors of factor loadings, as well as the increased likelihood of committing a Type I error as a result of evaluating a large number of loadings within the framework of a **factor analytic procedure**), Stevens (2002, p. 393–394) (also see Hair *et al.* (1995, pp. 384–386)) argues a more conservative approach should be taken with respect to whether or not a factor loading be declared significant. The latter author suggests the criterion for significance for a factor loading (which is interpreted as a **Pearson product-moment correlation coefficient (Test 28)**) should be (for the relevant sample size) two times the tabled critical two-tailed .01 value in **Table A16 (Table of Critical Values for Pearson *r*)** in the **Appendix**. For "smaller" sample sizes the latter criterion will stipulate a critical value substantially above .40. In point of fact, for the sample size of $n = 80$ employed in Example 40.1, the two-tailed tabled critical .01 value in **Table A16** is $r_{.01} = .283$, and two times the latter values is .566. Employing the latter criterion, a factor loading for Example 40.1 would have to have an absolute value equal or greater .566 in order to be declared significant. Interestingly, for an extremely large sample size such as $n = 1000$, the tabled critical two-tailed value is $r_{.01} = .081$, which when doubled equals .162. Thus, in the latter case, in order for a factor loading to be significant its absolute value need only be equal to or greater than .162 (which is obviously less than the more commonly cited value of .30). Once again, as noted throughout the book, the researcher has to make a distinction between statistical versus practical significance. Consequently, a variable which has a statistically significant loading on a specific factor may ultimately prove to have little or no practical value in terms of predictive utility. In the final analysis, the smaller the sample size the more stringent the criterion should be for determining significance.

In the case of Example 40.1, regardless of whether the researcher employs the criterion of .40 for determining significance (since $n = 80$ would be considered by most researchers to be a small sample size), or for that matter .566, Table 40.3 reveals that *Tests D, E,* and *F* have significant loadings on *Factor I*, but not on *Factor II*, while *Tests A, B,* and *C* have significant loadings on *Factor II* but not on *Factor I*. This same information is, in part, suggested in Table 40.2 by virtue of the following: a) *Tests A, B,* and *C* are highly correlated with one another, and b) *Tests D, E,* and *F* are highly correlated with one another. Although the pattern of intercorrelations in Table 40.2 suggests the underlying factorial structure of the data (i.e., that there are two primary factors, with one factor comprised of *Tests A, B,* and *C*, and the other factor comprised of *Tests D, E,* and *F*), such information will not always be obvious through inspection of a correlation matrix. To illustrate this point, if instead of six tests, the researcher had started out with 50 tests, it would be cumbersome (to say the least), and very likely impossible to discern the underlying factorial structure of the data by visual inspection of the correlation matrix. Additionally, high intercorrelations between variables does not insure they will necessarily be significant components of the same factors.

Geometrical representation of the result of a factor analysis Figure 40.2 geometrically represents the result of the **principal components analysis** summarized in Table 40.3. (The names accompanying each of the factors are discussed in the next section.) Note that *Factor I* is represented on the abscissa (*X*-axis) and *Factor II* on the ordinate (*Y*-axis).[14] The fact the axes are perpendicular to one another indicates the two factors are orthogonal (i.e., independent/uncorrelated). The values recorded on the abscissa and ordinate (which range from 0 to ± 1) represent factor loading values (which, as noted, can be viewed as correlation coefficients). The coordinates determining the location of each of the variables on the graph represent a variable's loadings on *Factor I* and *Factor II*. Thus, the coordinates for *Test A* are (−.208, .856) (i.e., −.208 units horizontally and .856 units vertically), and the coordinates for

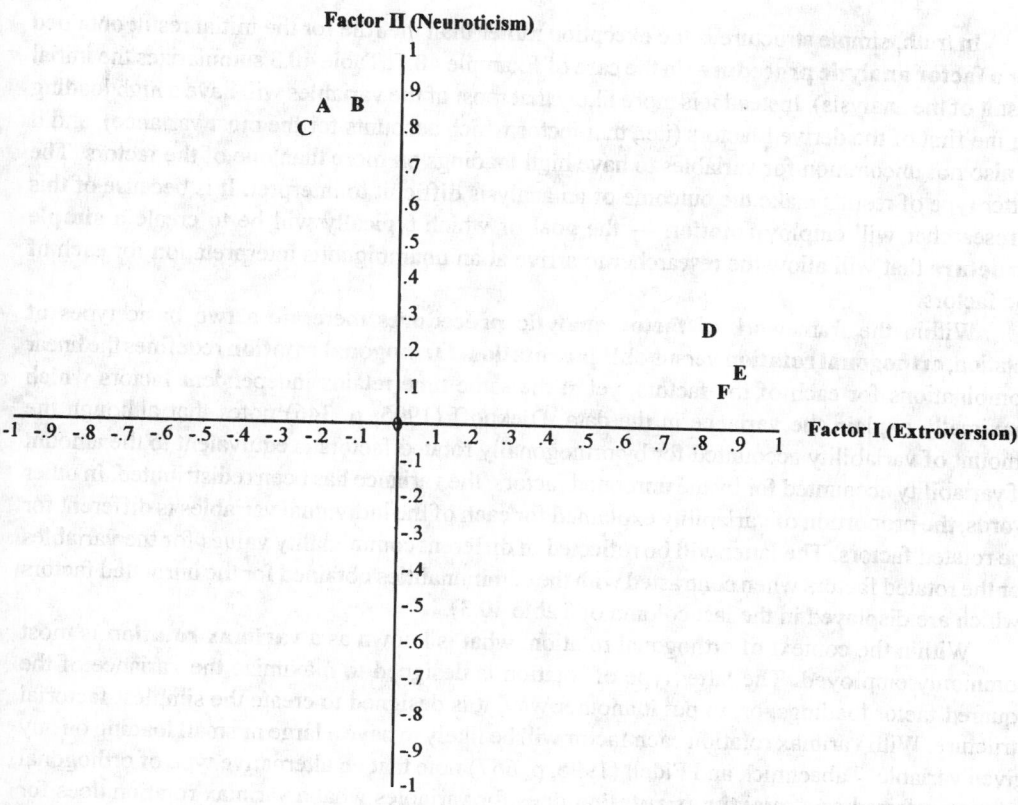

Figure 40.2 Geometrical Representation of Factor Analysis Summarized in Table 40.3

Tests B, C, D, E, and *F* are respectively (−.118, .863), (−.267, .782), (.838, .295), (.921, .156), and (.876, .112).

Rotation In actuality there are a number of different potential solutions which might result from a **factor analytic procedure**. An operation referred to as **rotation** is commonly employed in order to allow a researcher to discriminate more clearly between factors. Since, as noted above, the derived factors can be represented geometrically as well as mathematically, one can conceptualize rotation as a procedure which involves rotating the geometrical axes which serve as reference points for identifying the factors.[15]

When viewed together, the factor loadings in Columns 2 and 3 of Table 40.3 are said to represent the **factor structure**. Factor structure is easiest to interpret when it conforms to what is referred to as **simple structure**. **Simple structure** is present when each variable loads high on only one of the factors and low on the other factors. Ideally, if an analysis involves a large number of variables, the number of derived factors should be limited in number, and on each of the factors there should be several variables which correlate highly with that factor. Diekhoff (1995, p. 34) notes when simple structure is present most of the factor loadings should be either high or low with few loadings of intermediate value. A variable which has a high loading on only one factor (and thus, for the most part, only measures that factor) is called a **factorially simple variable**. On the other hand, a variable which has a high loading on more than one factor is referred to as a **factorially complex variable**.

In truth, simple structure is the exception rather than the rule for the initial result obtained for a **factor analytic procedure** (in the case of Example 40.1, Table 40.3 summarizes the initial result of the analysis). Instead it is more likely that most of the variables will have a high loading on the first of the derived factors (i.e., that factor which accounts for the most variance), and it is also not uncommon for variables to have high loadings on more than one of the factors. The latter type of results make the outcome of an analysis difficult to interpret. It is because of this a researcher will employ **rotation** — the goal of which typically will be to create a **simple structure** that will allow the researcher to arrive at an unambiguous interpretation for each of the factors.

Within the framework of **factor analytic procedures** there are a two basic types of rotation, **orthogonal rotation** versus **oblique rotation**. **Orthogonal rotation** redefines the linear combinations for each of the factors, yet at the same time retains independent factors which maximally explain the variance in the data. Diekhoff (1995, p. 346) notes that although the amount of variability accounted for by orthogonally rotated factors is equivalent to the amount of variability accounted for by the unrotated factors, the variance has been redistributed. In other words, the proportion of variability explained for each of the individual variables is different for the rotated factors. The latter will be reflected in different communality values for the variables for the rotated factors when contrasted with the communalities obtained for the unrotated factors (which are displayed in the last column of Table 40.3).

Within the context of orthogonal rotation, what is known as a **varimax rotation** is most commonly employed. The latter type of rotation is designed to maximize the variance of the squared factor loadings, or, to put it another way, it is designed to create the simplest factorial structure. With varimax rotation each factor will be likely to have a large or small loading on any given variable. Tabachnick and Fidell (1996, p. 667) note that an alternative type of orthogonal rotation referred as a **quartimax rotation** does for variables what a varimax rotation does for factors — specifically, it simplifies variables by increasing the factor loadings within variables across factors. The latter authors note that a quartimax rotation is less popular than a varimax rotation, since most researchers are more interested in deriving simple factors as opposed to simple variables. A third alternative is an **equimax rotation**, which represents a compromise between the varimax and quartimax rotations, and thus simultaneously attempts to simplify both the factors and the variables.

From a geometrical perspective, an orthogonal rotation involves rotating the axes clockwise or counterclockwise about the origin until an optimal fit is achieved between the axes (which represent the factors) and the p variables. Figure 40.3 illustrates a varimax rotation of the two factors derived for Example 40.1. The axes depicted with dotted lines represent the rotated factors, which you will note are in closer proximity to the six variables than the unrotated axes (which are the solid lines axes that are identical to those in Figure 40.2).

Table 40.4 summarizes the distribution of the variability for the two rotated factors. Note that the factor loadings in the latter table are more congruent with simple structure — in other words, when compared with the factor loadings in Table 40.3, each of the variables' factor loadings in Table 40.4 are slightly larger for the factor on which they load the highest and closer to zero for the factor on which they load the lowest. Note that the communalities of the variables for the rotated factors in Table 40.4 are slightly different than their communalities for the unrotated factors in Table 40.3. Although the sum of the communalities (4.660) for the rotated and unrotated factors are identical in both tables, the eigenvalues computed for the rotated factors are slightly different than those computed for the unrotated factors — specifically, the rotated factor eigenvalues in Table 40.4 are *Factor 1*: $(-.014)^2 + (.075)^2 + (-.088)^2 + (.883)^2 + (.932)^2 + (.879)^2 = 2.434$, and *Factor 2*: $(.881)^2 + (.868)^2 + (.822)^2 + (.103)^2 + (-.051)^2 + (-.084)^2 = 2.226$. Although it is the exception rather than the rule, in the final analysis, the rotated and

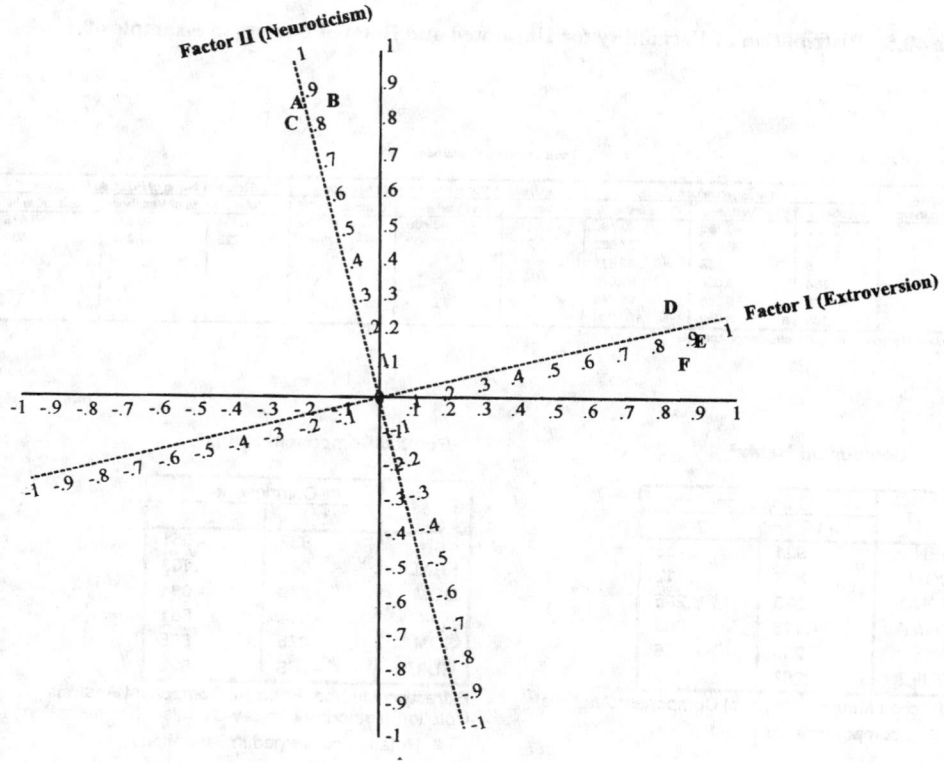

Figure 40.3 Geometrical Representation of Rotated Axes for Example 40.1

Table 40.4 Summary Table of Rotated Factors For Example 40.1

Test	Factor I	Factor II	Communality
A (Anx)	-.014	.881	.7764
B (Soma)	.075	.868	.7590
C (Guilt)	-.088	.822	.6834
D (Frnd)	.883	.103	.7885
E (Sen)	.932	-.051	.8712
F (Dom)	.879	-.084	.7797
Eigenvalue	2.434	2.226	**Sum = 4.660**
Percent of Total Variance	40.57%	37.10%	

unrotated solutions for Example 40.1 are very similar to one another (although the rotated solution is somewhat more congruent with simple structure). It should also be noted that although it is not the case for Example 40.1, it is possible for rotation to alter the ordering of the factors (i.e., *Factor II* instead of *Factor I* might account for the maximum amount of variability).

Note also that in Table 40.5 (which displays part of the *SPSS* output for the **principal components analysis** of Example 40.1), in the table labeled **Total Variance Explained** there are differences (albeit slight) between the two factors in the distribution of variability for the unrotated factors (in the **Extraction Sums of Squared Loadings** section of the table) and rotated factors (in the **Rotation Sums of Squared Loadings** sections of the table). The latter indicates that as a result of rotation, *Factor II* accounts for a slightly larger amount of the overall

Table 40.5 Distribution of Variability for Unrotated and Rotated Factors in Example 40.1

Total Variance Explained

Component	Initial Eigenvalues			Extraction Sums of Squared Loadings			Rotation Sums of Squared Loadings		
	Total	% of Variance	Cumulative %	Total	% of Variance	Cumulative %	Total	% of Variance	Cumulative %
1	2.446	40.760	40.760	2.446	40.760	40.760	2.434	40.573	40.573
2	2.214	36.902	77.662	2.214	36.902	77.662	2.225	37.089	77.662
3	.478	7.959	85.621						
4	.364	6.069	91.690						
5	.303	5.058	96.748						
6	.195	3.252	100.000						

Extraction Method: Principal Component Analysis.

Component Matrix[a]

	Component	
	1	2
SEN	.921	.156
DOM	.876	.112
FRND	.838	.295
SOMA	-.118	.863
ANX	-.208	.856
GUILT	-.267	.782

Extraction Method: Principal Component Analysis.
a. 2 components extracted.

Rotated Component Matrix[a]

	Component	
	1	2
SEN	.932	-.051
FRND	.883	.103
DOM	.879	-.084
ANX	-.014	.881
SOMA	.075	.868
GUILT	-.088	.822

Extraction Method: Principal Component Analysis.
Rotation Method: Varimax with Kaiser Normalization.
a. Rotation converged in 3 iterations.

Component Score Coefficient Matrix

	Component	
	1	2
ANX	.002	.396
SOMA	.039	.391
GUILT	-.029	.369
FRND	.364	.055
SEN	.383	-.014
DOM	.360	-.030

Extraction Method: Principal Component Analysis.
Rotation Method: Varimax with Kaiser Normalization.
Component Scores.

variability in the data and *Factor I* a slightly lower amount of the overall variability (yet the total amount of variability in the data is unchanged). Note that in Table 40.5, the tables labeled **Component Matrix** and **Rotated Component Matrix** contain the same factor loadings for Example 40.1 which are displayed respectively in Tables 40.3 and 40.4.

The table labeled **Component Score Coefficient Matrix** in Table 40.5 contains the coefficients that would be employed for Example 40.1 to compute factor scores for each subject. The latter scores can be displayed in the *SPSS* data editor (i.e., the original spreadsheet in which the data are recorded). Since the coefficients in the **Component Score Coefficient Matrix** are based on the rotated factors, the factor scores for subjects that will be printed in the data editor will be based on the rotated factors as opposed to the unrotated factors.

Oblique rotation, which is less common than orthogonal rotation, also redefines the linear combinations for the factors, but allows the rotated factors to be correlated with one another. Whereas a geometrical representation involving two orthogonal factors (such as Figures 40.2 and 40.3) will position the abscissa and ordinate of the graph perpendicular to one another, in the case of an oblique interpretation the axes will not be perpendicular. Oblique rotation is less frequently employed than orthogonal rotation because, as a general rule, it is more difficult to interpret. Diekhoff (1995, p. 350) notes oblique rotation tends to obscure the factor structure of the variables, since to the degree that oblique factors are correlated with one another all factor loadings will be high. The latter author notes, however, that if there is a sound basis (theoretical or empirical) indicating the factors are, in fact, not independent of one another, a researcher is better off employing an oblique rotation. The most common types of oblique rotations are the **promax** and **direct oblimax** rotations, while **orthoblique**, **dquart**, and **procrustes** are alternative oblique rotations.[16]

Field (2005, p. 636) notes Pedhazur and Schmelkin (1991) suggest employing both an orthogonal and oblique rotation with the framework of an analysis, and if the oblique rotation indicates a minimal correlation between the derived factors then it would probably be prudent to employ the orthogonal rotation. If, on the other hand, the oblique rotation indicates the factors are correlated with one another, the researcher could then employ the latter rotation, but only if she had a sound theoretical basis for expecting correlated factors. Stevens (2002, p. 392) cites a more extreme minority position with respect to rotation which argues that, in reality, factors will probably never be totally independent of one another, and because of the latter a researcher should always try to determine the best oblique rotation rather than an orthogonal one. In terms of resolving the question of which solution to a factor analysis a researcher should ultimately employ, Tabachnick and Fidell (1996, p. 670) make the following statement: "The results of extraction are similar regardless of which method is used when there is a large number of variables with some strong correlations among them, with the same, well-chosen number of factors, and with similar values for communality. Further, differences that are apparent after extraction tend to disappear after rotation." In the final analysis, the conclusions derived from the use of one or more **factor analytic procedures** employed to evaluate a set of variables will in large part depend on the experience and judgement of the researcher.

Naming the factors   At the conclusion of a **factor analytic procedure**, a researcher will generally assign a name to each of the factors. This is done by carefully examining the content of the variables which load high on a given factor. In our example, since *Factor I* is essentially comprised of tests that measure *Friendliness* (*Test D*), *Sensation Seeking* (*Test E*), and *Dominance* (*Test F*), one might elect to label *Factor I* **Extroversion**. This would be based on the premise that the behaviors/traits measured by *Tests D, E,* and *F* would be viewed by many psychologists as underlying components of the more general trait of extroversion. Thus, someone who has high scores on these tests would be likely to be viewed as high on extroversion, while a person with a low score on these tests would be viewed as low on extroversion.

In the same respect, since *Factor II* is essentially comprised of tests that measure *Anxiety* (*Test A*), *Somatic Complaints* (*Test B*), and *Guilt* (*Test C*), one might elect to label *Factor II* **Neuroticism**. The latter would be based on the fact that many mental health professionals would include the behavior/traits measured by *Tests A, B,* and *C* as characteristic of a neurotic individual. Thus, someone who has high scores on these tests would be likely to be viewed as high on neuroticism, while a person with a low score on these tests would be viewed as low on neuroticism.

Of course, one might challenge the above labels. For instance, one might, among other things, prefer to call *Factor I* **Energy Level** and *Factor II* Mental Health. The point to be made is that the naming of factors is based upon a subjective decision made by the researcher. It is conceivable, and not all that uncommon, that a name assigned to one or more factors by a researcher may be challenged by another researcher. Typically, in selecting a name for a factor,

Table 40.6   Factor Loadings for Principal Axis Factor Analysis on Example 40.1

Total Variance Explained

Factor	Initial Eigenvalues			Extraction Sums of Squared Loadings			Rotation Sums of Squared Loadings		
	Total	% of Variance	Cumulative %	Total	% of Variance	Cumulative %	Total	% of Variance	Cumulative
1	2.446	40.760	40.760	2.182	36.366	36.366	2.174	36.241	36.2
2	2.214	36.902	77.662	1.841	30.688	67.054	1.849	30.813	67.0
3	.478	7.959	85.621						
4	.364	6.069	91.690						
5	.303	5.058	96.748						
6	.195	3.252	100.000						

Extraction Method: Principal Axis Factoring.

Factor Matrix[a]

	Factor	
	1	2
SEN	.944	.088
DOM	.795	.037
FRND	.778	.213
ANX	-.138	.826
SOMA	-.054	.803
GUILT	-.178	.678

Extraction Method: Principal Axis Factoring.

a. 2 factors extracted. 13 iterations required.

Rotated Factor Matrix[a]

	Factor	
	1	2
SEN	.947	-.053
FRND	.801	.095
DOM	.792	-.081
ANX	-.014	.837
SOMA	.066	.802
GUILT	-.076	.697

Extraction Method: Principal Axis Factoring.
Rotation Method: Varimax with Kaiser Normalization.

a. Rotation converged in 3 iterations.

the larger the loading of a specific variable on that factor, the greater the role it should play in determining its name.

VI. Additional Analytical Procedures for Principal Components Analysis and Factor Analysis and/or Related Tests

1. Principal axis factor analysis of Example 40.1   A **principal axis factor analysis** was also conducted on the data for Example 40.1. The unrotated and varimax rotated factors loadings for the latter analysis are summarized in Table 40.6. Note that, as was the case for the **principal components analysis**, two factors accounted for most of the variability in the data (recollect that in a **principal axis factor analysis** only shared variability is evaluated). Although the factor loadings for the **principal axis factor analysis** are not identical to those obtained for the **principal components analysis**, the two analyses yield similar results. Once again *Factor I* (which accounts for the largest amount of variability) loads high on *Friendliness* (*Test D*), *Sensation Seeking* (*Test E*), and *Dominance* (*Test F*), while *Factor II* loads high on *Anxiety* (*Test A*), *Somatic Complaints* (*Test B*), and *Guilt* (*Test C*). Note that, when contrasted with the loadings in the unrotated factor table labeled **Factor Matrix**, for both of the factors the loadings displayed in the table labeled **Rotated Factor Matrix** are higher for the variables which load high on a given factor and are lower on the variables which load low on that factor. Although not displayed, different communality values are computed for each of the six variables on the two derived factors (when contrasted with the communality values for the unrotated and rotated factors in Tables 40.3 and 40.4). Note also that (as is the case in Table 40.5) in the table labeled **Total Variance Explained**, there are differences (albeit slight) between the two factors in the distribution of variability for the unrotated factors (in the **Extraction Sums of Squared Loadings** section of the table) and rotated factors (in the **Rotation Sums of Squared Loadings** sections of the table).

VII. Additional Discussion of Principal Components Analysis and Factor Analysis

1. Criticisms of factor analytic procedures   The major criticisms directed toward **factor analytic procedures** revolve around the subjectivity involved in such analyses. As previously noted, the type and degree of rotation employed by a researcher, as well as the names one assigns to the derived factors, are based on subjective decisions. The criticism with respect to rotation is germane to the more general problem, that within the framework of **factor analytic procedures** there are multiple methodologies (e.g., **principal components analysis**, **principal axis factor analysis**, or one of the other alternative procedures alluded to in Endnote 2) from which a researcher can choose, and the different methodologies may not yield identical or even similar results. It is also important to realize that a **factor analytic procedure** will only provide useful information if it is employed with appropriate data. The variables a researcher initially elects to use in an analysis must be carefully thought out in relation to whatever subject matter is being evaluated. If, due to prejudice or ignorance, a researcher ignores certain variables, the analysis will be unable to take such variables into account in describing the factorial structure of whatever it is that is ostensibly being studied. As Kachigan (1986, p. 400) notes, "Factor analysis does not create new information. It merely organizes, summarizes, and quantifies information that is fed into the system." In spite of the above noted criticisms, **factor analytic procedures** are commonly employed by many researchers in multiple academic disciplines. Most researchers who have familiarity and experience with such procedures would agree that, used judiciously, they can be powerful tools for evaluating a large body of data.

2. Cluster analysis   Although there are similarities between **factor analysis** (which some sources refer to as **R-type factor analysis**) and **cluster analysis** (which some sources refer to as **Q-type factor analysis**), the two types of analyses are applied to different types of problems. Whereas **factor analysis** is employed to describe the underlying structure of a set of variables by identifying variables that are highly correlated with one another (with each homogenous set of variables comprising a **factor**), **cluster analysis** is employed to identify groups of objects or subjects who are highly correlated with one another (with each homogenous group comprising a **cluster**). Simply put, the goal of **cluster analysis** (which is also referred to as **taxonomy** or **segmentation analysis** (Garson (2006))) is to group objects or individuals into homogenous clusters such that objects or subjects in a given cluster are more similar to one another than objects or subjects in a different cluster. In **cluster analysis**, cases are assigned numerical values and/or categorized on dimensions that are relevant in reflecting the degree of similarity between them. Although correlational measures can be employed as an index of similarity, it is more common in **cluster analysis** for cases to be represented in a dimensional space, and to employ the distance between them as the criterion for forming clusters.

Although some sources employ the terms **cluster analysis** and **Q-type factor analysis** (also referred to as **inverse factor analysis**) interchangeably, a distinction is often made between them. With regard to the latter, Diekhoff (1992, p. 375) notes that **Q-type factor analysis** was a precursor to modern methods of **cluster analysis**. Detailed discussions of **cluster analysis** can be found in Blashfield and Aldenderfer (1984), Diekhoff (1992, Ch. 17), Everitt *et al.* (2001), Hair *et al.* (1995, Ch. 8), and Kachigan (1986, Ch. 16). **Q-type factor analysis** is described by Thompson (2000).

3. Multidimensional scaling is a methodology which is related to but different than both **factor analysis** and **cluster analysis**. Stalans (1995, p. 138) notes that **multidimensional scaling** (also known as **perceptual mapping**) is a statistical tool designed to detect the hidden structure of similarity judgements. Simply put, **multidimensional scaling** attempts to identify dimensions (i.e., constructs or latent variables) which are responsible for one or more people perceiving similarity between a set of objects. There are two types of multidimensional scaling models: **metric**, which is based on proximities (i.e., perceived distance) that are measured on a interval/ratio scale, and the less restrictive **nonmetric**, which is based on judgements employing rank ordering.

The starting point in **multidimensional scaling** is to construct a matrix of pairwise distances between m objects (imagine a 20×20 table displaying the distance in miles between 20 different cities). An example of such a matrix employed within the context of **multidimensional scaling** could be a 20×20 table summarizing the perceived similarity between 20 automobiles manufactured by different companies. The latter table is referred to as a **distance**, **dissimilarity** or **proximities matrix**. The magnitude of the distance assigned by a subject(s) between any pair of objects will be directly related to the degree of perceived dissimilarity between the objects. Statistical analysis (e.g., use of linear regression and/or other procedures) is then employed to analyze the dissimilarity matrix, culminating in a spatial plot (referred to as a **configuration**) which visually depicts the perceived distance between objects. The researcher must then identify/label the dimensions that can account for perceived similarity of objects. Hair *et al.* (1995, p. 504) note the determination of how many dimensions should be employed to represent the data is based on employing one of the following three criteria: subjective evaluation, scree plots, or an overall index of fit. For pragmatic reasons the dimensions of the projected space are usually kept minimal (i.e., usually two or three dimensions), since with more than three dimensions it becomes impossible to generate a visual map. One index employed in **multidimensional scaling** is referred to as **stress**, which serves as an measure of how well a configuration fits the data. Values of **stress** range from 0 to 1, with the lower the value the better the fit. The final step in **multidimensional scaling** is interpreting the dimensions, which is similar to naming of the factors derived in a **factor analysis**, except in this case what is named are the dimensions a subject(s) employed in classifying objects with respect to similarity. Recommended sources which describe **multidimensional scaling** are Borg and Groenen (2005), Cox and Cox (2001), Hair *et al.* (1995, Ch. 9), Kachigan (1986, Ch. 17), Kruskal and Wish (1978), Norušis (2004), and Stalans (1995).

VIII. Additional Examples Illustrating the Use of Principal Components Analysis and Factor Analysis

Example 40.2 illustrates the use of **principal components analysis** involving $p = 7$ variables which yields a dramatic difference between the rotated and unrotated factor loadings. As noted in Section VI, the latter is not uncommon when data are evaluated with a **factor analytic procedure**.

Example 40.2 *A psychologist administers the following seven tests to 50 children between six and ten years of age in order to measure emotional and/or intellectual maturity. (The higher a subject's score on any of the seven tests, the higher the subject's standing on the ability/behavior in question.)*: **Test A: Vocabulary; Test B: Information; Test C: Verbal Comprehension; Test D: Emotional Maturity; Test E: Object Assembly; Test F: Block Design; Test G: Mazes.**[17] *The psychologist evaluates the data with a principal components analysis in order to determine whether or not it can be decomposed into a small set of dimensions.* Table 40.7 *summarizes the results of the analysis.*

Table 40.7 is comprised of five smaller tables which display information derived from the *SPSS* **principal components analysis** of Example 40.2. Specifically:

The table in Table 40.7 labeled **Correlation Matrix** summarizes the intercorrelations between the seven tests/variables. Cursory inspection of the latter table suggests moderate to high intercorrelations between **Tests A, B,** and **C** as well as between **Tests E, F,** and **G.**

The table labeled **KMO and Bartlett's Test** summarizes the results for the **Kaiser–Meyer–Olkin statistic** and **Bartlett's test of sphericity.** Since the value .568 obtained for the **Kaiser–Meyer–Olkin statistic** is below .6 (albeit barely), it indicates the correlation matrix may be too unreliable to be factor analyzed because of the small sample size. Since, however, Example 40.2 is being employed for illustrative purposes we will proceed with the analysis. The result $\chi^2 (21) = 128.428, p = .000$ (which indicates a probability less than .00049) for **Bartlett's test of sphericity** indicates the researcher can conclude that in the underlying population from which the sample was derived, at least one of the correlations is some value other than zero.

The table labeled **Total Variance Explained** indicates that $p = 7$ components (which corresponds to the number of tests administered) were derived in the analysis. Of the seven components, only the first three were retained since each of them has an eigenvalue greater than 1. As indicated in Column 4 of the table, cumulatively, *Components 1, 2,* and *3* account for 79.616% of the variability in the data. The amount of variance accounted for by each of the components can be computed by dividing the eigenvalue of each component by $p = 7$. Thus, [*Component 1*: 2.504 / 7 = 35.77%] + [*Component 2*: 2.013 / 7 = 28.76%] + [*Component 3* = 1.056 / 7 = 15.09%] = 79.62%

The tables labeled **Component Matrix** and **Rotated Component Matrix** display the factor loadings (which in the case of a **principal components analysis** may also be referred to as **component loadings**) for the seven variables on the three components. Note that in the **Component Matrix** all of the variables with the exception of **Test D: Emotional Maturity** have a loading of .5 or greater on *Component 1*, and that many of the variables have loadings above an absolute value of .5 on more than one of the components. As noted in Section V, it is not uncommon for most of the variables to have a high loading on the first of the derived factors (i.e., that factor which accounts for the most variance), as well as high loadings on more than one of the factors. As also noted earlier, the latter is the major reason why rotation is employed.

Table 40.7 Results of Principal Components Analysis on Example 40.2

Correlation Matrix

		VOCAB	INFORM	VERBCOMP	EMOMAT	OBJASSEM	BLOCKDES	MAZES
Correlation	VOCAB	1.000	.590	.635	.229	-.012	-.041	.074
	INFORM	.590	1.000	.417	.299	.179	.193	.189
	VERBCOMP	.635	.417	1.000	.003	-.027	.069	.276
	EMOMAT	.229	.299	.003	1.000	-.230	.073	-.091
	OBJASSEM	-.012	.179	-.027	-.230	1.000	.579	.684
	BLOCKDES	-.041	.193	.069	.073	.579	1.000	.620
	MAZES	.074	.189	.276	-.091	.684	.620	1.000

KMO and Bartlett's Test

Kaiser-Meyer-Olkin Measure of Sampling Adequacy.		.568
Bartlett's Test of Sphericity	Approx. Chi-Square	128.428
	df	21
	Sig.	.000

Total Variance Explained

	Initial Eigenvalues			Extraction Sums of Squared Loadings			Rotation Sums of Squared Loadings		
Component	Total	% of Variance	Cumulative %	Total	% of Variance	Cumulative %	Total	% of Variance	Cumulative %
1	2.504	35.771	35.771	2.504	35.771	35.771	2.299	32.843	32.843
2	2.013	28.757	64.527	2.013	28.757	64.527	2.079	29.705	62.548
3	1.056	15.089	79.616	1.056	15.089	79.616	1.195	17.069	79.616
4	.577	8.238	87.854						
5	.372	5.308	93.162						
6	.303	4.331	97.493						
7	.175	2.507	100.000						

Extraction Method: Principal Component Analysis.

Component Matrix[a]

	Component		
	1	2	3
MAZES	.784	-.426	-.056
BLOCKDES	.677	-.460	.317
OBJASSEM	.665	-.587	-.030
INFORM	.638	.507	.171
VERBCOMP	.557	.534	-.448
VOCAB	.514	.727	-.167
EMOMAT	.090	.454	.833

Extraction Method: Principal Component Analysis.
a. 3 components extracted.

Rotated Component Matrix[a]

	Component		
	1	2	3
OBJASSEM	.873	-.005	-.162
MAZES	.865	.191	-.117
BLOCKDES	.852	-.027	.206
VOCAB	-.056	.890	.160
VERBCOMP	.056	.875	-.164
INFORM	.221	.690	.410
EMOMAT	-.089	.091	.944

Extraction Method: Principal Component Analysis.
Rotation Method: Varimax with Kaiser Normalization.
a. Rotation converged in 4 iterations.

Inspection of the **Rotated Components Matrix** allows for a much clearer interpretation of the analysis, since it provide a reasonably good approximation of a **simple factorial structure**. Specifically: a) **Test E: Object Assembly, Test F: Block Design**, and **Test G: Mazes** all load high on *Component 1* and low on the other two components; b) **Test A: Vocabulary, Test B: Information**, and **Test C: Verbal Comprehension** all load high on *Component 2* and low on the other two components (the loading of .410 for **Test B: Information** on *Component 3* would probably not be considered to be of significance, given the small sample size); c) **Test D: Emotional Maturity** loads high on Component 3 and low on the other two components.

      As noted in Endnote 17, it might be appropriate to label *Component 1* as representing some form of **nonverbal (i.e., performance) intelligence** and *Component 2* as representing some form of **verbal intelligence**. *Component 3*, which accounts for the least variability, would probably best be labeled **emotional maturity** (i.e. the name of the test which, for the most part, it represents).

References

Bartlett, M. S. (1954). A note on the multiplying factors for various chi-square approximations. **Journal of the Royal Statistical Society**, 16 (Series B), 296–298.

Blashfield, R. K. & Aldenderfer, M. S. (1984). **Cluster analysis** (4th ed.). Beverly Hills, CA: Sage Publications.

Borg, I. & Groenen, P. J. (2005). **Modern multidimensional scaling: Theory and applications** (2nd ed.) New York: Springer.

Bryant, F. B. & Yarnold, P. R. (1995). Principal-components analysis and exploratory and confirmatory factor analysis. In Grimm, L. G. & Yarnold, P. R. (Eds.) (pp. 100–136). **Reading and understanding multivariate statistics**. Washington, D.C.: American Psychological Association.

Cattell, R. B. (1966). The scree test for the number of factors. **Multivariate Behavioral Research**, 1, 245–266.

Cliff, N. & Hamburger, C. D. (1967). The study of sampling errors in factor analysis by means of artificial experiments. **Psychological Bulletin**, 68, 430–445.

Cox, T. F. & Cox, M. A. (2001). **Multidimensional scaling** (2nd ed.) Boca Raton, FL: Chapman & Hall.

Diekhoff, G. (1992). **Statistics for the social and behavioral sciences: Univariate, bivariate, and multivariate**. Dubuque, IA: Wm. C. Brown Publishers.

Field, A. (2005). **Discovering statistics using SPSS** (2nd ed.). Sage Publications: London.

Garson, D. G. (2006). Statistics: Topics in multivariate analysis. Website: **http://www2.chas.ncsu.edu/garson/pa765/statnote.htm.**

George, D. & Mallery, P. (2001). **SPSS for Windows: Step by Step** (3rd ed.). Boston: Allyn & Bacon.

Grimm, L. G. & Yarnold, P. R. (Eds.) (1995). **Reading and understanding multivariate statistics**. Washington, D.C.: American Psychological Association.

Grimm, L. G. & Yarnold, P. R. (Eds.) (2000). **Reading and understanding more multivariate statistics**. Washington, D.C.: American Psychological Association.

Hair, J. F., Anderson, R. E., Tatham, R. L. & Black, W. C. (1995). **Multivariate data analysis** (4th ed.). Upper Saddle River, N. J.: Prentice Hall.

Hair, J. F., Anderson, R. E., Tatham, R. L. & Black, W. C. (2004). **Multivariate data analysis** (6th ed.). Upper Saddle River, N. J.: Prentice Hall.

Hair, J. F. & Black, W. C. (2000). Cluster analysis. In Grimm, L. G. & Yarnold, P. R. (Eds.) (pp. 147–205). **Reading and understanding more multivariate statistics**. Washington, D.C.: American Psychological Association.

Harlow, L. L. (2005). **The essence of multivariate thinking**. Mahwah, NJ: Lawrence Erlbaum Associates.

Harris, R. J. (2001) **A primer of multivariate statistics** (3rd ed.). Mahwah, NJ: Lawrence Erlbaum Associates.

Jolliffe, I. T. (1972). Discarding variables in a principal components analysis, I: artificial data. **Applied Statistics**, 21, 160–163.

Jolliffe, I. T. (1986). **Principal components analysis**. New York: Springer-Verlag.

Kachigan, S. K. (1986). **Statistical analysis**. New York: Radius Press.

Kachigan, S. K. (1991). **Multivariate statistical analysis** (2nd ed.). New York: Radius Press.

Kaiser, H. F. (1960). The application of electronic computers to factor analysis. **Educational and Psychological Measurement**, 20, 140–151.

Kaiser, H. F. (1970). A second generation Little Jiffy. **Psychometrika**, 35, 401–415.

Kaiser, H. F. (1974). An index of factorial simplicity. **Psychometrika**, 39, 31–36.

Kruskal, J. B. & Wish, M. (1978). **Multidimensional scaling**. Beverly Hills, CA: Sage Publication.

Marascuilo, L. A. & Levin, J. R. (1983). **Multivariate statistics in the social science: A researchers guide**. Monterey, CA: Brooks/Cole Publishing Company.

Mauchly, J. W. (1940). Significance test for sphericity of a normal *n*-variate distribution. **Annals of Mathematical Statistics**, 11, 204–209.

Mertler, C. A. & Vannatta, R. A. (2005). **Advanced and multivariate statistical methods** (3rd ed.). Los Angeles: Pyrczak Publications.

Norušis, N. J. (2004). **SPSS 13.0 advanced statistical procedures companion**. Upper Saddle River, NJ: Prentice Hall.

Pedhazur, E. & Schmelkin, L. (1991). **Measurement design and analysis**. Hillsdale, NJ: Lawrence Erlbaum Associates, Publishers.

Spicer, J. N. (2005). **Making sense of multivariate data analysis**. Thousand Oaks, CA: Sage Publications.

Stalans, L. J. (1995). Multidimensionl scaling. In Grimm, L. G. & Yarnold, P. R. (Eds.) (pp. 137–168). **Reading and understanding multivariate statistics**. Washington, D.C.: American Psychological Association.

Stevens, J. (1996). **Applied multivariate statistics for the social sciences** (3rd ed.). Hillsdale, NJ: Lawrence Erlbaum Associates, Publishers.

Stevens, J. (2002). **Applied multivariate statistics for the social sciences** (4th ed.). Hillsdale, NJ: Lawrence Erlbaum Associates, Publishers.

Tabachnick, B. G. & Fidell, L. S. (1996). **Using multivariate statistics** (3rd ed.). New York: Harper Collins Publishers.

Tabachnick, B. G. & Fidell, L. S. (2001). **Using multivariate statistics** (4th ed.). Boston: Allyn & Bacon.

Thompson, B. (2000). Q-technique factor analysis: One variation on the two-mode factor analysis of variables. In Grimm, L. G. & Yarnold, P. R. (Eds.) (pp. 207–226). **Reading and understanding more multivariate statistics**. Washington, D.C.: American Psychological Association.

Endnotes

1. Readers should take note of the fact that the use of the term **factor** within the framework of **factor analysis** is not the same as its usage within the context of the **between-subjects factorial analysis of variance (Test 27)** (as well as within other factorial designs), where the term **factor** is employed to represent each of the independent variables.

2. a) **Principal axis factor analysis** is also referred to as **principal factor analysis, principal axis factoring, common factor analysis,** and **principal components factor**

analysis; b) The term **confirmatory factor analysis** is employed when **factor analysis** is used within the context of hypothesis testing — more specifically, for evaluating goodness-of-fit with respect to the most appropriate factorial model for describing the dimensions underlying a set of observed variables. In implementing **confirmatory factor analysis** (which is discussed in greater detail in Bryant and Yarnold (1995) and Pedhazur and Schmelkin (1991)) a researcher initially specifies one or more factor analytic models to be evaluated (i.e., the researcher stipulates the specific factors which constitute a given model along with relevant factor loadings and other relevant statistical information), and then evaluates the data with respect to goodness-of-fit for a hypothesized model; c) **Image factor extraction, maximum likelihood factor extraction, unweighted and weighted least squares factoring,** and **alpha factoring** are other examples of **factor analytic procedures** (some or all of which are available with different computer software packages). Since a discussion of the latter procedures is beyond the scope of this book, interested readers are referred to sources such as Field (2005, pp. 628–630) and Tabachnick and Fidell (1996, pp. 664–665).

3.   a) Tabachnick and Fidell (1996, p. 684) note that **principal components analysis** is recommended as a first step when a researcher wants to obtain an overall picture of the likely number as well as the nature of the factors which comprise a body of data. Hair *et al.* (1995, p. 376) state that **principal components analysis** is appropriate to employ when the primary intent of a researcher is to optimize prediction or to identify a small number of factors which account for most of the variability in a set of p variables. The latter authors also note that, as a general rule, the use of **principal components analysis** implies (although in reality this is not the case) the researcher has prior knowledge indicating that only a small proportion of the total variance can be accounted for by specific or error variance. If the latter is, in fact, true Hair *et al.* (1995, p. 375) note the derived components will not contain enough specific or error variance to distort the overall factorial structure of the data; b) Stevens (2002, p. 386) notes that **principal components analysis** has some similarity to **discriminant function analysis (Test 37),** in that both methodologies derive uncorrelated linear combinations. More specifically, **principal components analysis** derives uncorrelated linear combinations of the original variables which additively partition the variance for the original set of variables. **Discriminant function analysis,** on the other hand, derives uncorrelated linear combinations which are employed to additively partition the association between the grouping variable and the set of predictor variables.

4.   a) Hair *et al.* (1995, p. 376) note that when the major goal of a researcher is to identify latent dimensions underlying a set of p variables, and that if the researcher is ignorant with respect to the amount of specific and error variance in the overall data, **principal axis factor analysis** is a more appropriate choice; b) Tabachnick and Fidell (1996, p. 663) note that since **principal axis factor analysis** omits specific variance, it is less likely than **principal components analysis** to mirror the relationships between the variables depicted in the original correlation matrix.

5.   a) Bryant and Yarnold (1995, p. 108) note that, less frequently, reliability coefficients between 0 and 1 are employed in the main diagonal for **factor analysis** when the goal of research is to study the theoretical factors which best account for the stable variance on the observed variables; b) There is some evidence that when the number of variables exceeds 30 and the communalities exceed .60, **principal components analysis** and

principal axis factor analysis will yield essentially the same results (Field (2005, p. 631) and Hair *et al.* (1995, p. 376)). The latter, however, is by no means guaranteed, and with a smaller number of variables and lower communalities, the likelihood of differences in results between the two methodologies increases; c) Hair *et al* (1995, p. 376) and Stevens (2002, p. 386) note another criticism of **principal axis factor analysis** is the **factor indeterminancy issue**, which is that unlike **principal components analysis** the former type of analysis does not yield one unique solution. Since more than one solution is possible for a **principal axis factor analysis**, several different factor scores (which are discussed in Section V) can be computed for a subject on each of the derived factors (instead of a single factor score, as in the case of **principal components analysis**).

6.   **Bartletts test of sphericity** (Bartlett (1954)) evaluates a different null hypothesis than **Mauchly's test of sphericity** discussed in Endnotes 15 and 7 respectively of the **single-factor within-subjects analysis of variance (Test 24)** and **Hotelling's T^2 (Test 34)**.

7.   a) In spite of the acceptable value obtained for the **Kaiser–Meyer–Olkin statistic**, most researchers would consider a sample size of 80 to be too small for a factor analysis; b) On a more technical level, Tabachnick and Fidell (1996, p. 642) note the **Kaiser–Meyer–Olkin statistic** (which is attributed to Kaiser (1970, 1974)) is a ratio of the sum of squared correlations to the sum of squared correlations plus the sum of squared partial correlations. The rationale underlying the statistic is that the partial correlations in a correlation matrix should not be large if the sample size is adequate; c) In the *SPSS* **Factor Analysis: Descriptives** menu, the researcher is offered the option of displaying an **Anti-Image Matrix** (which is comprised of an **Anti-Image Correlation Matrix** and an **Anti-Image Covariance Matrix**) which will display a **Kaiser–Meyer–Olkin statistic** for each of the variables. The **Anti-Image Correlation Matrix** provides information regarding the sampling adequacy of each of the variables. Field (2005, p. 642) notes that the rule of thumb with respect to employing the **Anti-Image Correlation matrix** is that all of the values in the main diagonal should be high (i.e., .5 or greater), and all of the values in the off-diagonal cells should be close to zero. The researcher should consider deleting from the analysis any variable with a main diagonal value less than .5. In the case of Example 40.1, acceptable values were obtained for the **Anti-image correlation matrix**.

8.   The *SPSS* **Factor Analysis Extraction** menu offers the researcher the choice of analyzing the **correlation matrix** or the **covariance matrix** (the result of the analysis will often differ depending upon which matrix is selected). Field's (2005, p. 643) notes that the correlation matrix (which is the default option) is the standardized version of the covariance matrix, and is the matrix which should be selected under most circumstances. More specifically, the correlation matrix should be selected when the *p* variables are not all measured on the same scale (which, as in Example 40.1, is usually the case). If, on the other hand, the variables are all measured on the same scale it may be more prudent to employ the covariance matrix for the analysis

9.   The following *SPSS* command sequence was employed in this chapter in conducting a **principal components analysis**: a) Click **Analyze**; b) Click **Data Reduction**; c) Click **Factor**; d) Highlight the variables to be factor analyzed (i.e., **anxiety, somatic complaints, guilt, friendliness, sensation seeking,** and **dominance** in Example 40.1)

and move them to the **Variables** window; e) Click **Descriptives**, check off desired information, and then click **Continue**; f) Click **Extraction** and select the type of analysis to be conducted. Since a **principal components analysis** was conducted, the default option of **Principal components** was employed. (When a **principal axis factor analysis** was employed for the same data, the **Principal axis factoring** option was selected in this window.) After checking off any desired information available in the window, click **Continue**; g) Click **Rotation**, check off the type of rotation desired, and then click **Continue**. For the example under discussion a **Varimax** rotation was selected; h) If factor scores are desired for subjects, click **Scores**, check **Save as variables** (the default option of **Regression** is generally recommended), and then click **Continue**; i) If any data are missing click on **Options** and check the method you elect, and then click **Continue**; j) Click **OK** to obtain the output for the analysis.

10. If a **principal axis factor analysis** is conducted, only the variance which each of the variables shares with the other variables (i.e., common variance) is available for analysis, since the latter analysis excludes the unique variance associated with each of the variables. In point of fact, a **principal axis factor analysis** on the same set of data is described in Section VI which yields two factors that are very similar to the two factors documented in Table 40.3.

11. Although a **principal components analysis** was employed for the Example 40.1, in the interest of accuracy it should be noted that when *SPSS* is employed to conduct the latter type of analysis, the term **Components** is used in the summary table in lieu of **Factors**. (The latter is illustrated in Tables 40.5 as well as in 40.7 (which summarizes the results of Example 40.2) in Section VIII.) When *SPSS* is used for a **principal axis factor analysis**, the term **Factors** is displayed in the summary table. The term **Factors** is used in Table 40.3 and throughout this section, since it is more commonly employed to refer to both the components derived in **principal component analysis**, as well as the factors derived with other **factor analytic procedures**.

12. Jolliffe (1972, 1986) employs a more liberal criterion value of .7 or larger. Although researchers are not in agreement with respect to which criterion should be used (since use of different criteria could result in different conclusions), Kaiser's rule is most commonly employed.

13. a) A more simplistic but less accurate method for computing factor scores would be to employ the unstandardized scores of subjects and factor loadings for each variable in place of the standardized scores and weighting coefficients; b) If requested, *SPSS* will print out the **standardized factor scores** (which can range from approximately -3 to $+3$) in the data editor (i.e., the original spreadsheet in which the data are recorded) for each subject as new variables. Although *SPSS* has three options for computing factor scores, the default method labeled **Regression Method** (which yields the highest correlation between the derived factor scores and their factors — although the former can also correlate with factor scores on other factors) was employed for the analysis of Example 40.1. The other methods are the **Bartlett method** (which derives unbiased factor scores that only correlate with their own factors) and the **Anderson–Rubin method** (which derives standardized factor scores that are uncorrelated with each other even if the factors are correlated — this method is best to employ if one wishes to minimize multicollinearity) (Field (2005, p.628)). Further discussion of computing

factor scores can be found in Diekhoff (1992, Ch. 16), Field (2005, pp. 625–628) and Tabachnick and Fidell (1996, pp. 678–679); c) The *SPSS* output includes a table labeled **Component Score Coefficient Matrix** (which is displayed in Table 40.5) that contains the coefficients which are employed to compute factor scores for each subject (which as noted above can be displayed in the data editor at the end of the analysis). The **component scores** in the latter matrix express the principal components in terms of the variables (i.e., each principal component can be expressed as a linear combination of the original variables, which is what Equation 40.1 represents), while the factor loadings in Table 40.3 express the variables in terms of the principal components.

14.   a) Alternatively, one could elect to have *Factor I* represented on the ordinate and *Factor II* on the abscissa; b) If three factors are derived in a **factor analytical procedure**, three-dimensional space is necessary in order to geometrically represent the result. If there are more than three factors, the appropriate multi-dimensional space is required (yet cannot be depicted visually, since geometrical space beyond three dimensions cannot be represented graphically).

15.   Rotation can only be employed when two or more factors are derived. If the rotation option is employed in *SPSS* when only one factor has been derived, the same result will be obtained as that for the unrotated factor.

16.   a) In oblique rotation a researcher can set a value for a variable (referred to as **delta** in *SPSS*) which establishes the maximum degree to which factors may be correlated with one another. The default value for **delta** is zero, and as its value is decreased the lower the correlation between factors. Higher values increase the degree of correlation between the factors (for further details see Field (2005, p. 637) and Tabachnick and Fidell (1996, p. 668)). Tabachnick and Fidell (1996, p. 668), note, however, that use of an oblique rotation does not guarantee the derived factors will be correlated with one another; b) When an **oblique rotation** is employed, the *SPSS* output will include the following three tables: 1) A table labeled **Pattern Matrix** contains factor loadings for the obliquely rotated factors; 2) A table labeled **Structure Matrix** contains the correlations between the factors and the variables. This will not be the same as the **Pattern Matrix**, since the factors may be correlated with one another; 3) A table labeled **Component Correlation Matrix** contains the correlations between the rotated factors. (When an orthogonal rotation is employed, the *SPSS* output will include a table labeled the **Components Transformation Matrix** which shows the correlation of the factors before and after rotation); c) Oblique factors can themselves be treated as variables, and factor analyzed into lower order factors. A possible situation in which oblique rotation might be employed is within the context of **confirmatory factor analysis** when a specific theoretical model a researcher is trying to confirm factorially is comprised of factors that are conceptualized as being correlated with one another.

17.   With the exception of **Test D: Emotional Maturity**, the other tests are elements of commonly administered intelligence tests. Three of the tests (**Test A: Vocabulary** (which involves defining words); **Test B: Information** (which tests age appropriate knowledge); and **Test C: Verbal Comprehension** (which involves explaining the meaning of sentences such as proverbs)) are often categorized as components of what is referred to as *verbal intelligence*, while the remaining three tests (**Test E: Object Assembly** (which involves assembling puzzles); **Test F: Block Design** (which involves

arranging blocks imprinted with designs to match designs printed on cards); and **Test G: Mazes** (which involves using a pencil to traverse a maze)) are categorized as components of *nonverbal* or *performance intelligence*.

Appendix: Tables
Acknowledgments and Sources for Tables in Appendix

Table A1 Table of the Normal Distribution

Reprinted with permission of CRC Press, Boca Raton, Florida from W. H. Beyer (1968), **CRC Handbook of Tables for Probability and Statistics** (2nd ed.), Table II.1 (The normal probability function and related functions), pp. 127–134.

Table A2 Table of Student's *t* Distribution

Reprinted with permission from Table 12 (Percentage points for the *t* distribution) in E. S. Pearson & H. O. Hartley, eds. (1970), **Biometrika Tables for Statisticians** (3rd ed., Volume I). New York: Cambridge University Press. Reproduced with kind permission of **Biometrika** trustees.

Table A3 Power Curves for Student's *t* Distribution

Reprinted with permission of Pearson Education, Inc. from Table 2.2 (Graphs of the operating characteristics of Student's *t* test) in D. B. Owen (©1962), **Handbook of Statistical Tables**. Reading, MA: Addison–Wesley, pp. 32–35.

Table A4 Table of the Chi-Square Distribution

Reprinted with permission from Table 8 (Percentage points of the χ^2 distribution) in E. S. Pearson & H. O. Hartley, eds. (1970), **Biometrika Tables for Statisticians** (3rd ed., Volume I). New York: Cambridge University Press. Reproduced with kind permission of **Biometrika** trustees.

Table A5 Table of Critical *T* Values for Wilcoxon's Signed Ranks and Matched-Pairs Signed-Ranks Tests

Material from Table II in F. Wilcoxon, S. K. Katti & R. A. Wilcox (1963), **Critical Values and Probability Levels for the Wilcoxon Rank Sum Test and the Wilcoxon Signed Rank Test**. Copyright © 1963, American Cyanamid Company, Lederle Laboratories Division. All rights reserved and reprinted with permission.

Table A6 Table of the Binomial Distribution, Individual Probabilities

Reprinted with permission of CRC Press, Boca Raton, Florida from W. H. Beyer (1968), **CRC Handbook of Tables for Probability and Statistics** (2nd ed.), Table III.1 (Individual terms, binomial distribution), pp. 182–193.

Table A7 Table of the Binomial Distribution, Cumulative Probabilities

Reprinted with permission of CRC Press, Boca Raton, Florida from W. H. Beyer (1968), **CRC Handbook of Tables for Probability and Statistics** (2nd ed.), Table III.2 (Cumulative terms, binomial distribution), pp. 194–205.

Table A8 Table of Critical Values for the Single-Sample Runs Test

Reprinted with permission of Institute of Mathematical Statistics, Hayward, CA from the following: Portions of **Table II** on pp. 83–87 from: F. S. Swed and C. Eisenhart (1943).

Tables for testing randomness of grouping in a sequence of alternatives, **Annals of Mathematical Statistics**, 14, 66–87.

Table A9   Table of the F_{max} Distribution
Reprinted with permission from Table 31 (Percentage points of the ratio s^2_{max}/s^2_{min}) in E. S. Pearson & H. O. Hartley, eds. (1970). **Biometrika Tables for Statisticians** (3rd ed., Volume 1). New York: Cambridge University Press. Reproduced with the kind permission of the **Biometrika** trustees.

Table A10   Table of the F Distribution
Reprinted with permission from Table 18 (Percentage points of the F-distribution (variance ratio)) in E. S. Pearson & H. O. Hartley (eds.) (1970), **Biometrika Tables for Statisticians** (3rd ed., Volume 1). New York: Cambridge University Press. Reproduced with the kind permission of the **Biometrika** trustees. Table reproduced with permission of CRC Press, Boca Raton, Florida from W. H. Beyer (1968), **CRC Handbook of Tables for Probability and Statistics** (2nd ed.), Table VI.1 (Percentage Points, F Distribution), pp. 304–310.

Table A11   Table of Critical Values for Mann-Whitney U Statistic
Reprinted with permission of Indiana University from D. Auble (1953), Extended Tables for the Mann–Whitney Statistic. **Bulletin of the Institute of Educational Research at Indiana University** Vol. 1, No. 2. Table reproduced with permission of CRC Press, Boca Raton, Florida from W. H. Beyer (1968), **CRC Handbook of Tables for Probability and Statistics** (2nd ed.), Table X.4 (Critical Values of U in the Wilcoxon (Mann–Whitney) Two-Sample Statistic), pp. 405–408.

Table A12   Table of Sandler's A Statistic
Reprinted with permission of British Psychological Society and Joseph Sandler from J. Sandler (1955), A test of the significance of difference between the means of correlated measures based on a simplification of Student's t. **British Journal of Psychology**, 46, pp. 225–226.

Table A13   Table of the Studentized Range Statistic
Reprinted with permission from Table 29 (Percentage points of the studentized range) in E. S. Pearson & H. O. Hartley, eds. (1970), **Biometrika Tables for Statisticians** (3rd ed., Volume 1). New York: Cambridge University Press. Reproduced with the kind permission of the **Biometrika** trustees.

Table A14   Table of Dunnett's Modified t Statistic for a Control Group Comparison
Two-tailed values: Reprinted with permission of the Biometric Society, Alexandria, VA from: C. W. Dunnett (1964), New tables for multiple comparisons with a control. **Biometrics**, 20, pp. 482–491.
One-tailed values: Reprinted with permission of the American Statistical Association, Alexandria, VA from: C. W. Dunnett (1955). A multiple comparison procedure for comparing several treatments with a control. **Journal of the American Statistical Association**, 50, pp. 1096–1121.

Table A15   Graphs of the Power Function for the Analysis of Variance
Reprinted with permission of **Biometrika** from E. S. Pearson & H. O. Hartley (1951), Charts of the power function for analysis of variance tests, derived from the non-central F distribution, **Biometrika**, 38, pp. 112–130.

Table A16 Table of Critical Values for Pearson *r*
Reprinted with permission from Table 13 (Percentage points for the distribution of the correlation coefficient, *r*, when $\rho = 0$) in E. S. Pearson & H. O. Hartley, eds. (1970), **Biometrika Tables for Statisticians** (3rd ed., Volume 1). New York: Cambridge University Press. Reproduced with the kind permission of the **Biometrika** trustees.

Table A17 Table of Fisher's z_r Transformation
Reprinted with permission from Table 14 (The *z*-transformation of the correlation coefficient, $z = \tanh^{-1} r$) in E. S. Pearson & H. O. Hartley, eds. (1970), **Biometrika Tables for Statisticians** (3rd ed., Volume 1). New York: Cambridge University Press. Reproduced with the kind permission of the **Biometrika** trustees.

Table A18 Table of Critical Values for Spearman's Rho
Reprinted with permission of the American Statistical Association, Alexandria, VA from: J. H. Zar (1972), Significance testing of the Spearman rank correlation coefficient. **Journal of the American Statistical Association**, 67, pp. 578–580 (Table 1, p. 579).

Table A19 Table of Critical Values for Kendall's Tau
Reprinted with permission of Blackwell Publishers and Statistica Neerlandica, from Table III in L. Kaarsemaker & A. van Wijngaarden (1953), Tables for use in rank correlation. **Statistica Neerlandica,** 7, pp. 41–54 (Copyright: The Netherlands Statistical Society (VVS)).

Table A20 Table of Critical Values for Kendall's Coefficient of Concordance
Reprinted with permission of Institute of Mathematical Statistics, Hayward, CA from: M. Friedman (1940), A comparison of alternative tests of significance for the problem of *m* rankings. **Annals of Mathematical Statistics**, 11, 86–92 (Table III, p. 91).

Table A21 Table of Critical Values for the Kolmogorov-Smirnov Goodness-of-Fit Test for a Single Sample
Reprinted with permission of Institute of Mathematical Statistics, Hayward, CA from: L. H. Miller (1956). Table of percentage points of Kolmogorov statistics. **Journal of the American Statistical Association**, 51, pp. 111–121.

Table A22 Table of Critical Values for the Lilliefors Test for Normality
Reprinted with permission of the American Statistical Association, Alexandria, VA from: H. W. Lilliefors (1967). On the Kolmogorov-Smirnov test for normality with mean and variance unknown. **Journal of the American Statistical Association**, 62, pp. 399–402.

Table A23 Table of Critical Values for the Kolmogorov-Smirnov Test for Two Independent Samples
Reprinted with permission of Institute of Mathematical Statistics, Hayward, CA from: F. J. Massey Jr. (1952). Distribution tables for the deviation between two sample cumulatives. **Annals of Mathematical Statistics**, 23, pp. 435–441.

Table A24 Table of Critical Values for the Jonckheere–Terpstra Test Statistic
Reprinted with permission of the American Statistical Association, Alexandria, VA from: R. E. Odeh (1971), On Jonckheere's *k* sample test against ordered alternatives. **Technometrics**, 13, pp. 912–918 (Tables 1 and 2, pp. 914–917).

Table A25 Table of Critical Values for Page's *L* Statistic
Reprinted with permission of the American Statistical Association, Alexandria, VA from: E. B. Page (1963). Ordered hypotheses for multiple treatments: A significance test for linear ranks. **Journal of the American Statistical Association**, 58, pp. 216–230 (Table 2, pp. 220–223).

Table A26 Table of Extreme Studentized Deviate Outlier Statistic
Reprinted with permission of the American Statistical Association, Alexandria, VA from: B. Rosner (1983). Percentage points for a generalized ESD many-outlier procedure. **Technometrics**, 25(2), 165–172 (Table 3, pp. 168–170).

Table A27 Table of Durbin–Watson Test Statistic
Reprinted with permission of the Biometrika Trustees, Oxford University Press, Oxford UK from: Durbin, J. & Watson, G. S. (1951) Testing for serial correlation in least squares regression (II). **Biometrika**, 38, 159–178. (Tables 4-6, pp. 173–175).

Table A1   Table of the Normal Distribution

z	$p(\mu$ to $z)$	$p(z$ to tail)	ordinate		z	$p(\mu$ to $z)$	$p(z$ to tail)	ordinate
.00	.0000	.5000	.3989		.45	.1736	.3264	.3605
.01	.0040	.4960	.3989		.46	.1772	.3228	.3589
.02	.0080	.4920	.3989		.47	.1808	.3192	.3572
.03	.0120	.4880	.3988		.48	.1844	.3156	.3555
.04	.0160	.4840	.3986		.49	.1879	.3121	.3538
.05	.0199	.4801	.3984		.50	.1915	.3085	.3521
.06	.0239	.4761	.3982		.51	.1950	.3050	.3503
.07	.0279	.4721	.3980		.52	.1985	.3015	.3485
.08	.0319	.4681	.3977		.53	.2019	.2981	.3467
.09	.0359	.4641	.3973		.54	.2054	.2946	.3448
.10	.0398	.4602	.3970		.55	.2088	.2912	.3429
.11	.0438	.4562	.3965		.56	.2123	.2877	.3410
.12	.0478	.4522	.3961		.57	.2157	.2843	.3391
.13	.0517	.4483	.3956		.58	.2190	.2810	.3372
.14	.0557	.4443	.3951		.59	.2224	.2776	.3352
.15	.0596	.4404	.3945		.60	.2257	.2743	.3332
.16	.0636	.4364	.3939		.61	.2291	.2709	.3312
.17	.0675	.4325	.3932		.62	.2324	.2676	.3292
.18	.0714	.4286	.3925		.63	.2357	.2643	.3271
.19	.0753	.4247	.3918		.64	.2389	.2611	.3251
.20	.0793	.4207	.3910		.65	.2422	.2578	.3230
.21	.0832	.4168	.3902		.66	.2454	.2546	.3209
.22	.0871	.4129	.3894		.67	.2486	.2514	.3187
.23	.0901	.4090	.3885		.68	.2517	.2483	.3166
.24	.0948	.4052	.3876		.69	.2549	.2451	.3144
.25	.0987	.4013	.3867		.70	.2580	.2420	.3123
.26	.1026	.3974	.3857		.71	.2611	.2389	.3101
.27	.1064	.3936	.3847		.72	.2642	.2358	.3079
.28	.1103	.3897	.3836		.73	.2673	.2327	.3056
.29	.1141	.3859	.3825		.74	.2704	.2296	.3034
.30	.1179	.3821	.3814		.75	.2734	.2266	.3011
.31	.1217	.3783	.3802		.76	.2764	.2236	.2989
.32	.1255	.3745	.3790		.77	.2794	.2206	.2966
.33	.1293	.3707	.3778		.78	.2823	.2177	.2943
.34	.1331	.3669	.3765		.79	.2852	.2148	.2920
.35	.1368	.3632	.3752		.80	.2881	.2119	.2897
.36	.1406	.3594	.3739		.81	.2910	.2090	.2874
.37	.1443	.3557	.3725		.82	.2939	.2061	.2850
.38	.1480	.3520	.3712		.83	.2967	.2033	.2827
.39	.1517	.3483	.3697		.84	.2995	.2005	.2803
.40	.1554	.3446	.3683		.85	.3023	.1977	.2780
.41	.1591	.3409	.3668		.86	.3051	.1949	.2756
.42	.1628	.3372	.3653		.87	.3078	.1922	.2732
.43	.1664	.3336	.3637		.88	.3106	.1894	.2709
.44	.1700	.3300	.3621		.89	.3133	.1867	.2685

Table A1 Table of the Normal Distribution (continued)

z	p(μ to z)	p(z to tail)	ordinate	z	p(μ to z)	p(z to tail)	ordinate
.90	.3159	.1841	.2661	1.35	.4115	.0885	.1604
.91	.3186	.1814	.2637	1.36	.4131	.0869	.1582
.92	.3212	.1788	.2613	1.37	.4147	.0853	.1561
.93	.3238	.1762	.2589	1.38	.4162	.0838	.1539
.94	.3264	.1736	.2565	1.39	.4177	.0823	.1518
.95	.3289	.1711	.2541	1.40	.4192	.0808	.1497
.96	.3315	.1685	.2516	1.41	.4207	.0793	.1476
.97	.3340	.1660	.2492	1.42	.4222	.0778	.1456
.98	.3365	.1635	.2468	1.43	.4236	.0764	.1435
.99	.3389	.1611	.2444	1.44	.4251	.0749	.1415
1.00	.3413	.1587	.2420	1.45	.4265	.0735	.1394
1.01	.3438	.1562	.2396	1.46	.4279	.0721	.1374
1.02	.3461	.1539	.2371	1.47	.4292	.0708	.1354
1.03	.3485	.1515	.2347	1.48	.4306	.0694	.1334
1.04	.3508	.1492	.2323	1.49	.4319	.0681	.1315
1.05	.3531	.1469	.2299	1.50	.4332	.0668	.1295
1.06	.3554	.1446	.2275	1.51	.4345	.0655	.1276
1.07	.3577	.1423	.2251	1.52	.4357	.0643	.1257
1.08	.3599	.1401	.2227	1.53	.4370	.0630	.1238
1.09	.3621	.1379	.2203	1.54	.4382	.0618	.1219
1.10	.3643	.1357	.2179	1.55	.4394	.0606	.1200
1.11	.3665	.1335	.2155	1.56	.4406	.0594	.1182
1.12	.3686	.1314	.2131	1.57	.4418	.0582	.1163
1.13	.3708	.1292	.2107	1.58	.4429	.0571	.1145
1.14	.3729	.1271	.2083	1.59	.4441	.0559	.1127
1.15	.3749	.1251	.2059	1.60	.4452	.0548	.1109
1.16	.3770	.1230	.2036	1.61	.4463	.0537	.1092
1.17	.3790	.1210	.2012	1.62	.4474	.0526	.1074
1.18	.3810	.1190	.1989	1.63	.4484	.0516	.1057
1.19	.3830	.1170	.1965	1.64	.4495	.0505	.1040
1.20	.3849	.1151	.1942	1.65	.4505	.0495	.1023
1.21	.3869	.1131	.1919	1.66	.4515	.0485	.1006
1.22	.3888	.1112	.1895	1.67	.4525	.0475	.0989
1.23	.3907	.1093	.1872	1.68	.4535	.0465	.0973
1.24	.3925	.1075	.1849	1.69	.4545	.0455	.0957
1.25	.3944	.1056	.1826	1.70	.4554	.0446	.0940
1.26	.3962	.1038	.1804	1.71	.4564	.0436	.0925
1.27	.3980	.1020	.1781	1.72	.4573	.0427	.0909
1.28	.3997	.1003	.1758	1.73	.4582	.0418	.0893
1.29	.4015	.0985	.1736	1.74	.4591	.0409	.0878
1.30	.4032	.0968	.1714	1.75	.4599	.0401	.0863
1.31	.4049	.0951	.1691	1.76	.4608	.0392	.0848
1.32	.4066	.0934	.1669	1.77	.4616	.0384	.0833
1.33	.4082	.0918	.1447	1.78	.4625	.0375	.0818
1.34	.4099	.0901	.1626	1.79	.4633	.0367	.0804

Table A1 Table of the Normal Distribution (continued)

z	p(μ to z)	p(z to tail)	ordinate	z	p(μ to z)	p(z to tail)	ordinate
1.80	.4641	.0359	.0790	2.25	.4878	.0122	.0317
1.81	.4649	.0351	.0775	2.26	.4881	.0119	.0310
1.82	.4656	.0344	.0761	2.27	.4884	.0116	.0303
1.83	.4664	.0336	.0748	2.28	.4887	.0113	.0297
1.84	.4671	.0329	.0734	2.29	.4890	.0110	.0290
1.85	.4678	.0322	.0721	2.30	.4893	.0107	.0283
1.86	.4686	.0314	.0707	2.31	.4896	.0104	.0277
1.87	.4693	.0307	.0694	2.32	.4898	.0102	.0270
1.88	.4699	.0301	.0681	2.33	.4901	.0099	.0264
1.89	.4706	.0294	.0669	2.34	.4904	.0096	.0258
1.90	.4713	.0287	.0656	2.35	.4906	.0094	.0252
1.91	.4719	.0281	.0644	2.36	.4909	.0091	.0246
1.92	.4726	.0274	.0632	2.37	.4911	.0089	.0241
1.93	.4732	.0268	.0620	2.38	.4913	.0087	.0235
1.94	.4738	.0262	.0608	2.39	.4916	.0084	.0229
1.95	.4744	.0256	.0596	2.40	.4918	.0082	.0224
1.96	.4750	.0250	.0584	2.41	.4920	.0080	.0219
1.97	.4756	.0244	.0573	2.42	.4922	.0078	.0213
1.98	.4761	.0239	.0562	2.43	.4925	.0075	.0208
1.99	.4767	.0233	.0551	2.44	.4927	.0073	.0203
2.00	.4772	.0228	.0540	2.45	.4929	.0071	.0198
2.01	.4778	.0222	.0529	2.46	.4931	.0069	.0194
2.02	.4783	.0217	.0519	2.47	.4932	.0068	.0189
2.03	.4788	.0212	.0508	2.48	.4934	.0066	.0184
2.04	.4793	.0207	.0498	2.49	.4936	.0064	.0180
2.05	.4798	.0202	.0488	2.50	.4938	.0062	.0175
2.06	.4803	.0197	.0478	2.51	.4940	.0060	.0171
2.07	.4808	.0192	.0468	2.52	.4941	.0059	.0167
2.08	.4812	.0188	.0459	2.53	.4943	.0057	.0163
2.09	.4817	.0183	.0449	2.54	.4945	.0055	.0158
2.10	.4821	.0179	.0440	2.55	.4946	.0054	.0155
2.11	.4826	.0174	.0431	2.56	.4948	.0052	.0151
2.12	.4830	.0170	.0422	2.57	.4949	.0051	.0147
2.13	.4834	.0166	.0413	2.58	.4951	.0049	.0143
2.14	.4838	.0162	.0404	2.59	.4952	.0048	.0139
2.15	.4842	.0158	.0396	2.60	.4953	.0047	.0136
2.16	.4846	.0154	.0387	2.61	.4955	.0045	.0132
2.17	.4850	.0150	.0379	2.62	.4956	.0044	.0129
2.18	.4854	.0146	.0371	2.63	.4957	.0043	.0126
2.19	.4857	.0143	.0363	2.64	.4959	.0041	.0122
2.20	.4861	.0139	.0355	2.65	.4960	.0040	.0119
2.21	.4864	.0136	.0347	2.66	.4961	.0039	.0116
2.22	.4868	.0132	.0339	2.67	.4962	.0038	.0113
2.23	.4871	.0129	.0332	2.68	.4963	.0037	.0110
2.24	.4875	.0125	.0325	2.69	.4964	.0036	.0107

Table A1   Table of the Normal Distribution (continued)

z	$p(\mu$ to $z)$	$p(z$ to tail$)$	ordinate	z	$p(\mu$ to $z)$	$p(z$ to tail$)$	ordinate
2.70	.4965	.0035	.0104	3.15	.4992	.0008	.0028
2.71	.4966	.0034	.0101	3.16	.4992	.0008	.0027
2.72	.4967	.0033	.0099	3.17	.4992	.0008	.0026
2.73	.4968	.0032	.0096	3.18	.4993	.0007	.0025
2.74	.4969	.0031	.0093	3.19	.4993	.0007	.0025
2.75	.4970	.0030	.0091	3.20	.4993	.0007	.0024
2.76	.4971	.0029	.0088	3.21	.4993	.0007	.0023
2.77	.4972	.0028	.0086	3.22	.4994	.0006	.0022
2.78	.4973	.0027	.0084	3.23	.4994	.0006	.0022
2.79	.4974	.0026	.0081	3.24	.4994	.0006	.0021
2.80	.4974	.0026	.0079	3.25	.4994	.0006	.0020
2.81	.4975	.0025	.0077	3.26	.4994	.0006	.0020
2.82	.4976	.0024	.0075	3.27	.4995	.0005	.0019
2.83	.4977	.0023	.0073	3.28	.4995	.0005	.0018
2.84	.4977	.0023	.0071	3.29	.4995	.0005	.0018
2.85	.4978	.0022	.0069	3.30	.4995	.0005	.0017
2.86	.4979	.0021	.0067	3.31	.4995	.0005	.0017
2.87	.4979	.0021	.0065	3.32	.4995	.0005	.0016
2.88	.4980	.0020	.0063	3.33	.4996	.0004	.0016
2.89	.4981	.0019	.0061	3.34	.4996	.0004	.0015
2.90	.4981	.0019	.0060	3.35	.4996	.0004	.0015
2.91	.4982	.0018	.0058	3.36	.4996	.0004	.0014
2.92	.4982	.0018	.0056	3.37	.4996	.0004	.0014
2.93	.4983	.0017	.0055	3.38	.4996	.0004	.0013
2.94	.4984	.0016	.0053	3.39	.4997	.0003	.0013
2.95	.4984	.0016	.0051	3.40	.4997	.0003	.0012
2.96	.4985	.0015	.0050	3.41	.4997	.0003	.0012
2.97	.4985	.0015	.0048	3.42	.4997	.0003	.0012
2.98	.4986	.0014	.0047	3.43	.4997	.0003	.0011
2.99	.4986	.0014	.0046	3.44	.4997	.0003	.0011
3.00	.4987	.0013	.0044	3.45	.4997	.0003	.0010
3.01	.4987	.0013	.0043	3.46	.4997	.0003	.0010
3.02	.4987	.0013	.0042	3.47	.4997	.0003	.0010
3.03	.4988	.0012	.0040	3.48	.4997	.0003	.0009
3.04	.4988	.0012	.0039	3.49	.4998	.0002	.0009
3.05	.4989	.0011	.0038	3.50	.4998	.0002	.0009
3.06	.4989	.0011	.0037	3.51	.4998	.0002	.0008
3.07	.4989	.0011	.0036	3.52	.4998	.0002	.0008
3.08	.4990	.0010	.0035	3.53	.4998	.0002	.0008
3.09	.4990	.0010	.0034	3.54	.4998	.0002	.0008
3.10	.4990	.0010	.0033	3.55	.4998	.0002	.0007
3.11	.4991	.0009	.0032	3.56	.4998	.0002	.0007
3.12	.4991	.0009	.0031	3.57	.4998	.0002	.0007
3.13	.4991	.0009	.0030	3.58	.4998	.0002	.0007
3.14	.4992	.0008	.0029	3.59	.4998	.0002	.0006

Table A1 Table of the Normal Distribution (continued)

z	p(μ to z)	p(z to tail)	ordinate	z	p(μ to z)	p(z to tail)	ordinate
3.60	.4998	.0002	.0006	3.80	.4999	.0001	.0003
3.61	.4998	.0002	.0006	3.81	.4999	.0001	.0003
3.62	.4999	.0001	.0006	3.82	.4999	.0001	.0003
3.63	.4999	.0001	.0005	3.83	.4999	.0001	.0003
3.64	.4999	.0001	.0005	3.84	.4999	.0001	.0003
3.65	.4999	.0001	.0005	3.85	.4999	.0001	.0002
3.66	.4999	.0001	.0005	3.86	.4999	.0001	.0002
3.67	.4999	.0001	.0005	3.87	.4999	.0001	.0002
3.68	.4999	.0001	.0005	3.88	.4999	.0001	.0002
3.69	.4999	.0001	.0004	3.89	1.0000	.0000	.0002
3.70	.4999	.0001	.0004	3.90	1.0000	.0000	.0002
3.71	.4999	.0001	.0004	3.91	1.0000	.0000	.0002
3.72	.4999	.0001	.0004	3.92	1.0000	.0000	.0002
3.73	.4999	.0001	.0004	3.93	1.0000	.0000	.0002
3.74	.4999	.0001	.0004	3.94	1.0000	.0000	.0002
3.75	.4999	.0001	.0004	3.95	1.0000	.0000	.0002
3.76	.4999	.0001	.0003	3.96	1.0000	.0000	.0002
3.77	.4999	.0001	.0003	3.97	1.0000	.0000	.0002
3.78	.4999	.0001	.0003	3.98	1.0000	.0000	.0001
3.79	.4999	.0001	.0003	3.99	1.0000	.0000	.0001
				4.00	1.0000	.0000	.0001

Table A2   Table of Student's *t* Distribution

Two-tailed	.80	.50	.20	.10	.05	.02	.01	.001
One-tailed	.40	.25	.10	.05	.025	.01	.005	.0005
p	.60	.75	.90	.95	.975	.99	.995	.9995
df								
1	.325	1.000	3.078	6.314	12.706	31.821	63.657	636.619
2	.289	.816	1.886	2.920	4.303	6.965	9.925	31.598
3	.277	.765	1.638	2.353	3.182	4.541	5.841	12.924
4	.271	.741	1.533	2.132	2.776	3.747	4.604	8.610
5	.267	.727	1.476	2.015	2.571	3.365	4.032	6.869
6	.265	.718	1.440	1.943	2.447	3.143	3.707	5.959
7	.263	.711	1.415	1.895	2.365	2.998	3.499	5.408
8	.262	.706	1.397	1.860	2.306	2.896	3.355	5.041
9	.261	.703	1.383	1.833	2.262	2.821	3.250	4.781
10	.260	.700	1.372	1.812	2.228	2.764	3.169	4.587
11	.260	.697	1.363	1.796	2.201	2.718	3.106	4.437
12	.259	.695	1.356	1.782	2.179	2.681	3.055	4.318
13	.259	.694	1.350	1.771	2.160	2.650	3.012	4.221
14	.258	.692	1.345	1.761	2.145	2.624	2.977	4.140
15	.258	.691	1.341	1.753	2.131	2.602	2.947	4.073
16	.258	.690	1.337	1.746	2.120	2.583	2.921	4.015
17	.257	.689	1.333	1.740	2.110	2.567	2.898	3.965
18	.257	.688	1.330	1.734	2.101	2.552	2.878	3.922
19	.257	.688	1.328	1.729	2.093	2.539	2.861	3.883
20	.257	.687	1.325	1.725	2.086	2.528	2.845	3.850
21	.257	.686	1.323	1.721	2.080	2.518	2.831	3.819
22	.256	.686	1.321	1.717	2.074	2.508	2.819	3.792
23	.256	.685	1.319	1.714	2.069	2.500	2.807	3.767
24	.256	.685	1.318	1.711	2.064	2.492	2.797	3.745
25	.256	.684	1.316	1.708	2.060	2.485	2.787	3.725
26	.256	.684	1.315	1.706	2.056	2.479	2.779	3.707
27	.256	.684	1.314	1.703	2.052	2.473	2.771	3.690
28	.256	.683	1.313	1.701	2.048	2.467	2.763	3.674
29	.256	.683	1.311	1.699	2.045	2.462	2.756	3.659
30	.256	.683	1.310	1.697	2.042	2.457	2.750	3.646
40	.255	.681	1.303	1.684	2.021	2.423	2.704	3.551
60	.254	.679	1.296	1.671	2.000	2.390	2.660	3.460
120	.254	.677	1.289	1.658	1.980	2.358	2.617	3.373
∞	.253	.674	1.282	1.645	1.960	2.326	2.576	3.291

Table A3   Power Curves for Student's *t* Distribution

Table A3-A (Two-Tailed .01 and One-Tailed .005 Values)

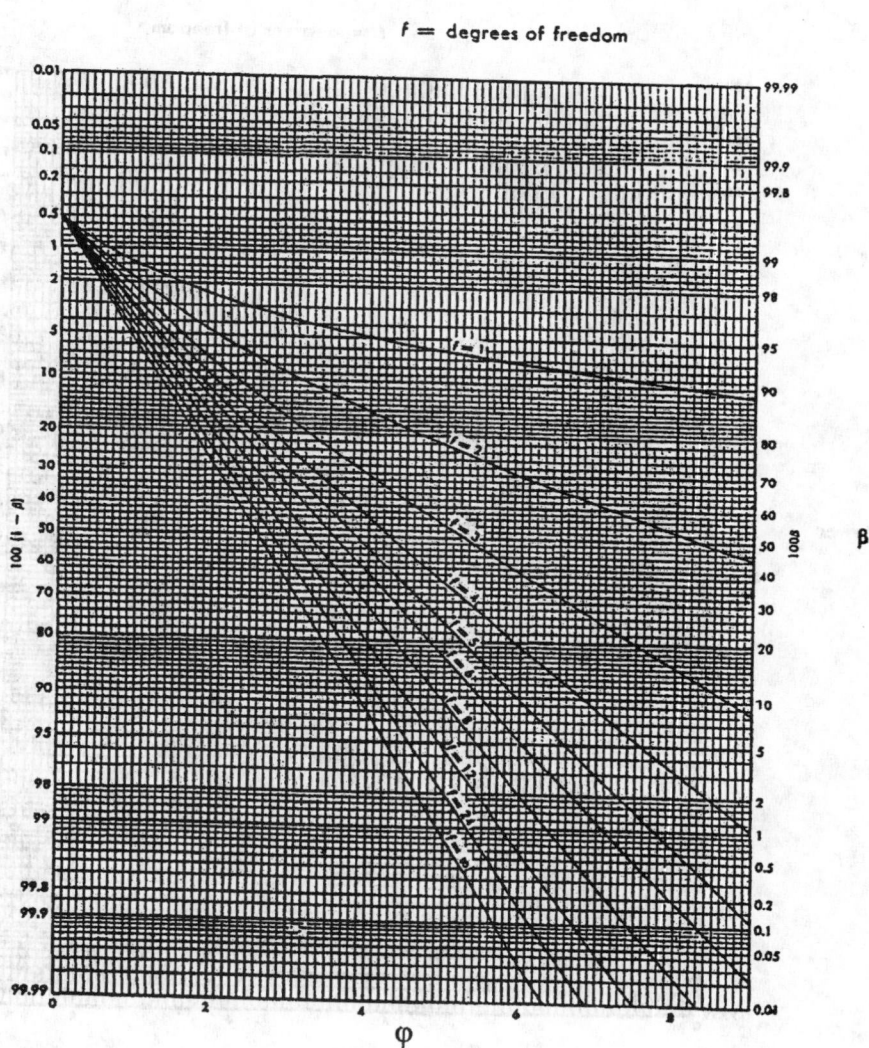

Table A3   Power Curves for Student's *t* Distribution (continued)

Table A3-B (Two-Tailed .02 and One-Tailed .01 Values)

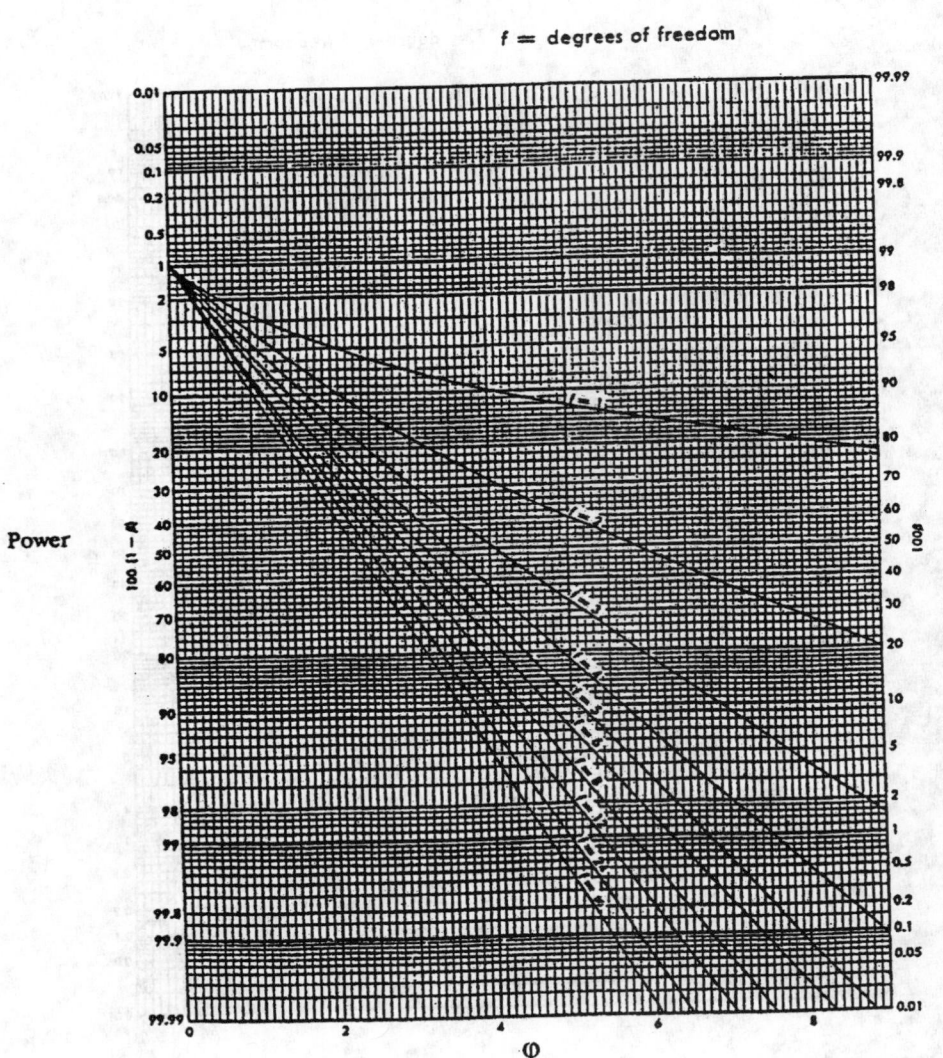

f = degrees of freedom

Power

$100(1-\beta)$

φ

β

100β

Table A3   Power Curves for Student's *t* Distribution (continued)

Table A3-C (Two-Tailed .05 and One-Tailed .025 Values)

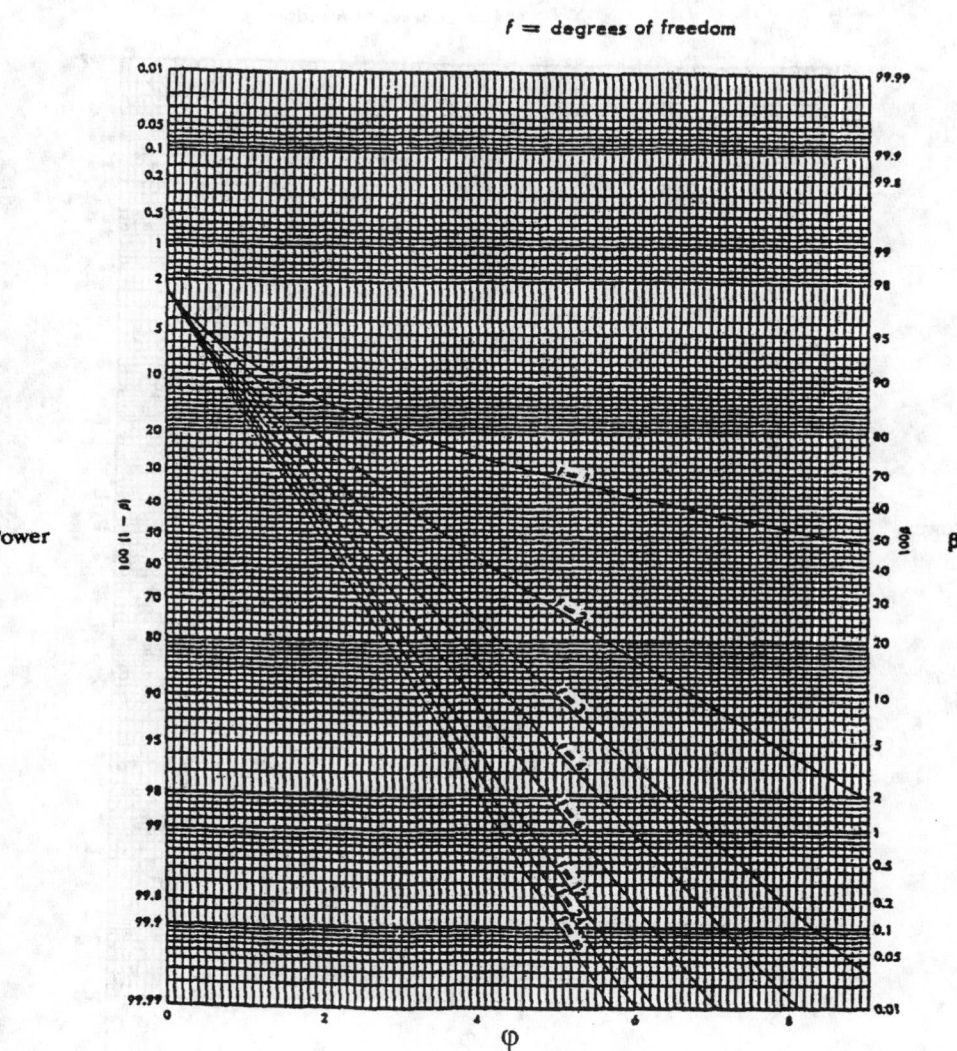

Power

Table A3 Power Curves for Student's *t* Distribution (continued)

Table A3-D (Two-Tailed .10 and One-Tailed .05 Values)

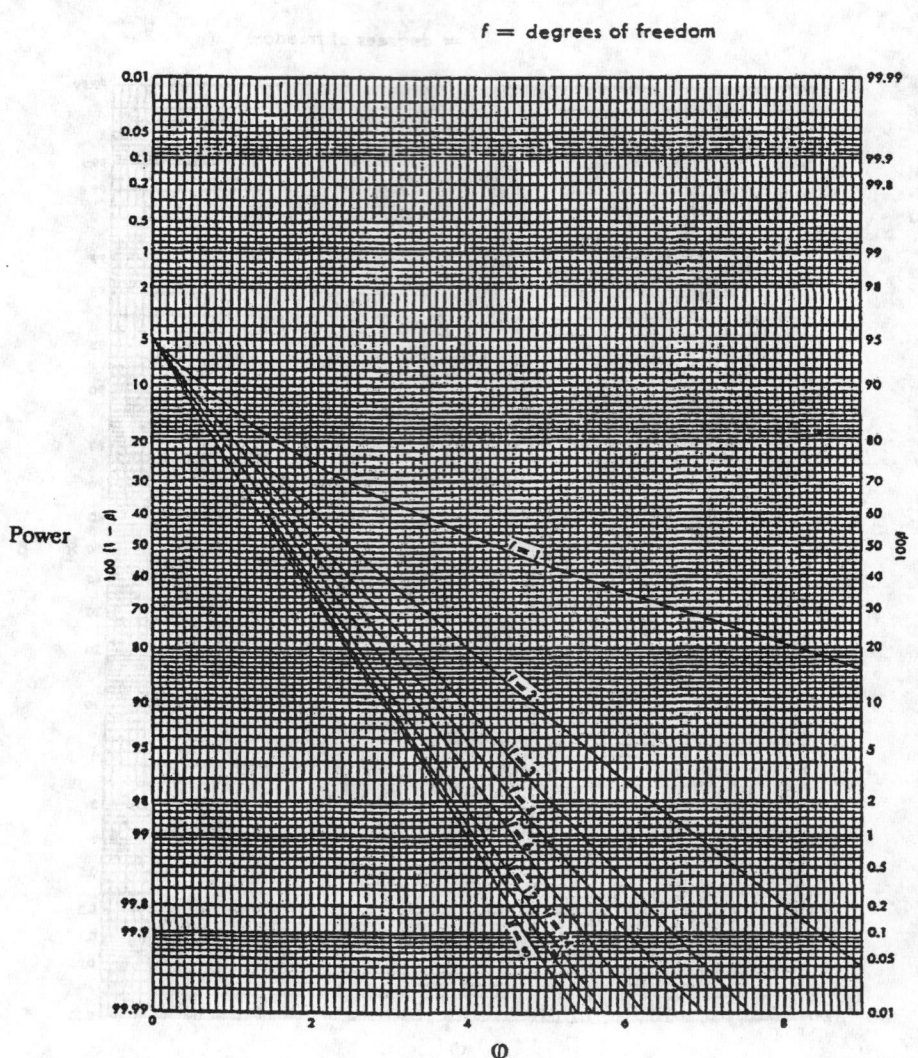

Table A4 Table of the Chi-Square Distribution

p	.005	.010	.025	.050	.100	.900	.950	.975	.990	.995	.999
df											
1	.0393	.0157	.0982	.0393	.0158	2.71	3.84	5.02	6.63	7.88	10.83
2	.0100	.0201	.0506	.103	.211	4.61	5.99	7.38	9.21	10.60	13.82
3	.072	.115	.216	.352	.584	6.25	7.81	9.35	11.34	12.84	16.27
4	.0207	.297	.484	.711	1.064	7.78	9.49	11.14	13.28	14.86	18.47
5	.412	.554	.831	1.145	1.61	9.24	11.07	12.83	15.09	16.75	20.52
6	.676	.872	1.24	1.64	2.20	10.64	12.59	14.45	16.81	18.55	22.46
7	.989	1.24	1.69	2.17	2.83	12.02	14.07	16.01	18.48	20.28	24.32
8	1.34	1.65	2.18	2.73	3.49	13.36	15.51	17.53	20.09	21.96	26.13
9	1.73	2.09	2.70	3.33	4.17	14.68	16.92	19.02	21.67	23.59	27.88
10	2.16	2.56	3.25	3.94	4.87	15.99	18.31	20.48	23.21	25.19	29.59
11	2.60	3.05	3.82	4.57	5.58	17.28	19.68	21.92	24.72	26.76	31.26
12	3.07	3.57	4.40	5.23	6.30	18.55	21.03	23.34	26.22	28.30	32.91
13	3.57	4.11	5.01	5.89	7.04	19.81	22.36	24.74	27.69	29.82	34.53
14	4.07	4.66	5.63	6.57	7.79	21.06	23.68	26.12	29.14	31.32	36.12
15	4.60	5.23	6.26	7.26	8.55	22.31	25.00	27.49	30.58	32.80	37.70
16	5.14	5.81	6.91	7.96	9.31	23.54	26.30	28.85	32.00	34.27	39.25
17	5.70	6.41	7.56	8.67	10.09	24.77	27.59	30.19	33.41	35.72	40.79
18	6.26	7.01	8.23	9.39	10.86	25.99	28.87	31.53	34.81	37.16	42.31
19	6.84	7.63	8.91	10.12	11.65	27.20	30.14	32.85	36.19	38.58	43.82
20	7.43	8.26	8.59	10.85	12.44	28.41	31.41	34.17	37.57	40.00	43.32
21	8.03	8.90	10.28	11.59	13.24	29.62	32.67	35.48	38.93	41.40	46.80
22	8.64	9.54	10.98	12.34	14.04	30.81	33.92	36.78	40.29	42.80	48.27
23	9.26	10.20	11.69	13.09	14.85	32.01	35.17	38.08	41.64	44.18	49.73
24	9.89	10.86	12.40	13.85	15.66	33.20	36.42	39.36	42.98	45.56	51.18
25	10.52	11.52	13.12	14.61	16.47	34.38	37.65	40.65	44.31	46.93	52.62
26	11.16	12.20	13.84	15.38	17.29	35.56	38.89	41.92	45.64	48.29	54.05
27	11.81	12.88	14.57	16.15	18.11	36.74	40.11	43.19	46.96	49.64	55.48
28	12.46	13.56	15.31	16.93	18.94	37.92	41.34	44.46	48.28	50.99	56.89
29	13.21	14.26	16.05	17.71	19.77	39.09	42.56	45.72	49.59	52.34	58.30
30	13.79	14.95	16.79	18.49	20.60	40.26	43.77	46.98	50.89	53.67	59.70
40	20.71	22.16	24.43	26.51	29.05	51.80	55.76	59.34	63.69	66.77	73.40
50	27.99	29.71	32.36	34.76	37.69	63.17	67.50	71.42	76.15	79.49	86.66
60	35.53	37.48	40.48	43.19	46.46	74.40	79.08	83.30	88.38	91.95	99.61
70	43.28	45.44	48.76	51.74	55.33	85.53	90.53	95.02	100.43	104.22	112.32
80	51.17	53.54	57.15	60.39	64.28	96.58	101.88	106.63	112.33	116.32	124.84
90	59.20	61.75	65.65	69.13	73.29	107.56	113.15	118.14	124.12	128.30	137.21
100	67.33	70.06	74.22	77.93	82.36	118.50	124.34	129.56	135.81	140.17	149.45

Table A5   Table of Critical *T* Values for Wilcoxon's Signed-Ranks and Matched-Pairs Signed-Ranks Test

	One-tailed level of significance					One-tailed level of significance			
	.05	.025	.01	.005		.05	.025	.01	.005
	Two-tailed level of significance					Two-tailed level of significance			
n	.10	.05	.02	.01	*n*	.10	.05	.02	.01
5	0	—	—	—	28	130	116	101	91
6	2	0	—	—	29	140	126	110	100
7	3	2	0	—	30	151	137	120	109
8	5	3	1	0	31	163	147	130	118
9	8	5	3	1	32	175	159	140	128
10	10	8	5	3	33	187	170	151	138
11	13	10	7	5	34	200	182	162	148
12	17	13	9	7	35	213	195	173	159
13	21	17	12	9	36	227	208	185	171
14	25	21	15	12	37	241	221	198	182
15	30	25	19	15	38	256	235	211	194
16	35	29	23	19	39	271	249	224	207
17	41	34	27	23	40	286	264	238	220
18	47	40	32	27	41	302	279	252	233
19	53	46	37	32	42	319	294	266	247
20	60	52	43	37	43	336	310	281	261
21	67	58	49	42	44	353	327	296	276
22	75	65	55	48	45	371	343	312	291
23	83	73	62	54	46	389	361	328	307
24	91	81	69	61	47	407	378	345	322
25	100	89	76	68	48	426	396	362	339
26	110	98	84	75	49	446	415	379	355
27	119	107	92	83	50	466	434	397	373

Table A6 Table of the Binomial Distribution, Individual Probabilities

n	x	.05	.10	.15	.20	π .25	.30	.35	.40	.45	.50
1	0	.9500	.9000	.8500	.8000	.7500	.7000	.6500	.6000	.5500	.5000
	1	.0500	.1000	.1500	.2000	.2500	.3000	.3500	.4000	.4500	.5000
2	0	.9025	.8100	.7225	.6400	.5625	.4900	.4225	.3600	.3025	.2500
	1	.0950	.1800	.2550	.3200	.3750	.4200	.4550	.4800	.4950	.5000
	2	.0025	.0100	.0225	.0400	.0625	.0900	.1225	.1600	.2025	.2500
3	0	.8574	.7290	.6141	.5120	.4219	.3430	.2746	.2160	.1664	.1250
	1	.1354	.2430	.3251	.3840	.4219	.4410	.4436	.4320	.4084	.3750
	2	.0071	.0270	.0574	.0960	.1406	.1890	.2389	.2880	.3341	.3750
	3	.0001	.0010	.0034	.0080	.0156	.0270	.0429	.0640	.0911	.1250
4	0	.8145	.6561	.5220	.4096	.3164	.2401	.1785	.1296	.0915	.0625
	1	.1715	.2916	.3685	.4096	.4219	.4116	.3845	.3456	.2995	.2500
	2	.0135	.0486	.0975	.1536	.2109	.2646	.3105	.3456	.3675	.3750
	3	.0005	.0036	.0115	.0256	.0469	.0756	.1115	.1536	.2005	.2500
	4	.0000	.0001	.0005	.0016	.0039	.0081	.0150	.0256	.0410	.0625
5	0	.7738	.5905	.4437	.3277	.2373	.1681	.1160	.0778	.0503	.0312
	1	.2036	.3280	.3915	.4096	.3955	.3602	.3124	.2592	.2059	.1562
	2	.0214	.0729	.1382	.2048	.2637	.3087	.3364	.3456	.3369	.3125
	3	.0011	.0081	.0244	.0512	.0879	.1323	.1811	.2304	.2757	.3125
	4	.0000	.0004	.0022	.0064	.0146	.0284	.0488	.0768	.1128	.1562
	5	.0000	.0000	.0001	.0003	.0010	.0024	.0053	.0102	.0185	.0312
6	0	.7351	.5314	.3771	.2621	.1780	.1176	.0754	.0467	.0277	.0156
	1	.2321	.3543	.3993	.3932	.3560	.3025	.2437	.1866	.1359	.0938
	2	.0305	.0984	.1762	.2458	.2966	.3241	.3280	.3110	.2780	.2344
	3	.0021	.0146	.0415	.0819	.1318	.1852	.2355	.2765	.3032	.3125
	4	.0001	.0012	.0055	.0154	.0330	.0595	.0951	.1382	.1861	.2344
	5	.0000	.0001	.0004	.0015	.0044	.0102	.0205	.0369	.0609	.0938
	6	.0000	.0000	.0000	.0001	.0002	.0007	.0018	.0041	.0083	.0156
7	0	.6983	.4783	.3206	.2097	.1335	.0824	.0490	.0280	.0152	.0078
	1	.2573	.3720	.3960	.3670	.3115	.2471	.1848	.1306	.0872	.0547
	2	.0406	.1240	.2097	.2753	.3115	.3177	.2985	.2613	.2140	.1641
	3	.0036	.0230	.0617	.1147	.1730	.2269	.2679	.2903	.2918	.2734
	4	.0002	.0026	.0109	.0287	.0577	.0972	.1442	.1935	.2388	.2734
	5	.0000	.0002	.0012	.0043	.0115	.0250	.0466	.0774	.1172	.1641
	6	.0000	.0000	.0001	.0004	.0013	.0036	.0084	.0172	.0320	.0547
	7	.0000	.0000	.0000	.0000	.0001	.0002	.0006	.0016	.0037	.0078
8	0	.6634	.4305	.2725	.1678	.1001	.0576	.0319	.0168	.0084	.0039
	1	.2793	.3826	.3847	.3355	.2670	.1977	.1373	.0896	.0548	.0312
	2	.0515	.1488	.2376	.2936	.3115	.2965	.2587	.2090	.1569	.1094
	3	.0054	.0331	.0839	.1468	.2076	.2541	.2786	.2787	.2568	.2188
	4	.0004	.0046	.0185	.0459	.0865	.1361	.1875	.2322	.2627	.2734
	5	.0000	.0004	.0026	.0092	.0231	.0467	.0808	.1239	.1719	.2188
	6	.0000	.0000	.0002	.0011	.0038	.0100	.0217	.0413	.0703	.1094
	7	.0000	.0000	.0000	.0001	.0004	.0012	.0033	.0079	.0164	.0312
	8	.0000	.0000	.0000	.0000	.0000	.0001	.0002	.0007	.0017	.0039

Table A6   Table of the Binomial Distribution, Individual Probabilities (continued)

n	x	.05	.10	.15	.20	.25	.30	.35	.40	.45	.50
9	0	.6302	.3874	.2316	.1342	.0751	.0404	.0207	.0101	.0046	.0020
	1	.2985	.3874	.3679	.3020	.2253	.1556	.1004	.0605	.0339	.0176
	2	.0629	.1722	.2597	.3020	.3003	.2668	.2162	.1612	.1110	.0703
	3	.0077	.0446	.1069	.1762	.2336	.2668	.2716	.2508	.2119	.1641
	4	.0006	.0074	.0283	.0661	.1168	.1715	.2194	.2508	.2600	.2461
	5	.0000	.0008	.0050	.0165	.0389	.0735	.1181	.1672	.2128	.2461
	6	.0000	.0001	.0006	.0028	.0087	.0210	.0424	.0743	.1160	.1641
	7	.0000	.0000	.0000	.0003	.0012	.0039	.0098	.0212	.0407	.0703
	8	.0000	.0000	.0000	.0000	.0001	.0004	.0013	.0035	.0083	.0176
	9	.0000	.0000	.0000	.0000	.0000	.0000	.0001	.0003	.0008	.0020
10	0	.5987	.3487	.1969	.1074	.0563	.0282	.0135	.0060	.0025	.0010
	1	.3151	.3874	.3474	.2684	.1877	.1211	.0725	.0403	.0207	.0098
	2	.0746	.1937	.2759	.3020	.2816	.2335	.1757	.1209	.0763	.0439
	3	.0105	.0574	.1298	.2013	.2503	.2668	.2522	.2150	.1665	.1172
	4	.0010	.0112	.0401	.0881	.1460	.2001	.2377	.2508	.2384	.2051
	5	.0001	.0015	.0085	.0264	.0584	.1029	.1536	.2007	.2340	.2461
	6	.0000	.0001	.0012	.0055	.0162	.0368	.0689	.1115	.1596	.2051
	7	.0000	.0000	.0001	.0008	.0031	.0090	.0212	.0425	.0746	.1172
	8	.0000	.0000	.0000	.0001	.0004	.0014	.0043	.0106	.0229	.0439
	9	.0000	.0000	.0000	.0000	.0000	.0001	.0005	.0016	.0042	.0098
	10	.0000	.0000	.0000	.0000	.0000	.0000	.0000	.0001	.0003	.0010
11	0	.5688	.3138	.1673	.0859	.0422	.0198	.0088	.0036	.0014	.0004
	1	.3293	.3835	.3248	.2362	.1549	.0932	.0518	.0266	.0125	.0055
	2	.0867	.2131	.2866	.2953	.2581	.1998	.1395	.0887	.1259	.0269
	3	.0137	.0710	.1517	.2215	.2581	.2568	.2254	.1774	.1259	.0806
	4	.0014	.0158	.0536	.1107	.1721	.2201	.2428	.2365	.2060	.1611
	5	.0001	.0025	.0132	.0388	.0803	.1321	.1830	.2207	.2360	.2256
	6	.0000	.0003	.0023	.0097	.0268	.0566	.0985	.1471	.1931	.2256
	7	.0000	.0000	.0003	.0017	.0064	.0173	.0379	.0701	.1128	.1611
	8	.0000	.0000	.0000	.0002	.0011	.0037	.0102	.0234	.0462	.0806
	9	.0000	.0000	.0000	.0000	.0001	.0005	.0018	.0052	.0126	.0269
	10	.0000	.0000	.0000	.0000	.0000	.0000	.0002	.0007	.0021	.0054
	11	.0000	.0000	.0000	.0000	.0000	.0000	.0000	.0000	.0002	.0005
12	0	.5404	.2824	.1422	.0687	.0317	.0138	.0057	.0022	.0008	.0002
	1	.3413	.3766	.3012	.2062	.1267	.0712	.0368	.0174	.0075	.0029
	2	.0988	.2301	.2924	.2835	.2323	.1678	.1088	.0639	.0339	.0161
	3	.0173	.0852	.1720	.2362	.2581	.2397	.1954	.1419	.0923	.0537
	4	.0021	.0213	.0683	.1329	.1936	.2311	.2367	.2128	.1700	.1208
	5	.0002	.0038	.0193	.0532	.1032	.1585	.2039	.2270	.2225	.1934
	6	.0000	.0005	.0040	.0155	.0401	.0792	.1281	.1766	.2124	.2256
	7	.0000	.0000	.0006	.0033	.0115	.0291	.0591	.1009	.1489	.1934
	8	.0000	.0000	.0001	.0005	.0024	.0078	.0199	.0420	.0762	.1208
	9	.0000	.0000	.0000	.0001	.0004	.0015	.0048	.0125	.0277	.0537
	10	.0000	.0000	.0000	.0000	.0000	.0002	.0008	.0025	.0068	.0161
	11	.0000	.0000	.0000	.8000	.0000	.0000	.0001	.0003	.0010	.0029
	12	.0000	.0000	.0000	.0000	.0000	.0000	.0000	.0000	.0001	.0002

Table A6 Table of the Binomial Distribution, Individual Probabilities (continued)

n	x	.05	.10	.15	.20	.25	.30	.35	.40	.45	.50
13	0	.5133	.2542	.1209	.0550	.0238	.0097	.0037	.0013	.0004	.0001
	1	.3512	.3672	.2774	.1787	.1029	.0540	.0259	.0113	.0045	.0016
	2	.1109	.2448	.2937	.2680	.2059	.1388	.0836	.0453	.0220	.0095
	3	.0214	.0997	.1900	.2457	.2517	.2181	.1651	.1107	.0660	.0349
	4	.0028	.0277	.0838	.1535	.2097	.2337	.2222	.1845	.1350	.0873
	5	.0003	.0055	.0266	.0691	.1258	.1803	.2154	.2214	.1989	.1571
	6	.0000	.0008	.0063	.0230	.0559	.1030	.1546	.1968	.2169	.2095
	7	.0000	.0001	.0011	.0058	.0186	.0442	.0833	.1312	.1775	.2095
	8	.0000	.0000	.0001	.0011	.0047	.0142	.0336	.0656	.1089	.1571
	9	.0000	.0000	.0000	.0001	.0009	.0034	.0101	.0243	.0495	.0873
	10	.0000	.0000	.0000	.0000	.0001	.0006	.0022	.0065	.0162	.0349
	11	.0000	.0000	.0000	.0000	.0000	.0001	.0003	.0012	.0036	.0095
	12	.0000	.0000	.0000	.0000	.0000	.0000	.0000	.0001	.0005	.0016
	13	.0000	.0000	.0000	.0000	.0000	.0000	.0000	.0000	.0000	.0001
14	0	.4877	.2288	.1028	.0440	.0178	.0068	.0024	.0008	.0002	.0001
	1	.3593	.3559	.2539	.1539	.0832	.0407	.0181	.0073	.0027	.0009
	2	.1229	.2570	.2912	.2501	.1802	.1134	.0634	.0317	.0141	.0056
	3	.0259	.1142	.2056	.2501	.2402	.1943	.1366	.0845	.0462	.0222
	4	.0037	.0349	.0998	.1720	.2202	.2290	.2022	.1549	.1040	.0611
	5	.0004	.0078	.0352	.0860	.1468	.1963	.2178	.2066	.1701	.1222
	6	.0000	.0013	.0093	.0322	.0734	.1262	.1759	.2066	.2088	.1833
	7	.0000	.0002	.0019	.0092	.0280	.0618	.1082	.1574	.1952	.2095
	8	.0000	.0000	.0003	.0020	.0082	.0232	.0510	.0918	.1398	.1833
	9	.0000	.0000	.0000	.0003	.0018	.0066	.0183	.0408	.0762	.1222
	10	.0000	.0000	.0000	.0000	.0003	.0014	.0049	.0136	.0312	.0611
	11	.0000	.0000	.0000	.0000	.0000	.0002	.0010	.0033	.0093	.0222
	12	.0000	.0000	.0000	.0000	.0000	.0000	.0001	.0005	.0019	.0056
	13	.0000	.0000	.0000	.0000	.0000	.0000	.0000	.0001	.0002	.0009
	14	.0000	.0000	.0000	.0000	.0000	.0000	.0000	.0000	.0000	.0001
15	0	.4633	.2059	.0874	.0352	.0134	.0047	.0016	.0005	.0001	.0000
	1	.3658	.3432	.2312	.1319	.0668	.0305	.0126	.0047	.0016	.0005
	2	.1348	.2669	.2856	.2309	.1559	.0916	.0476	.0219	.0090	.0032
	3	.0307	.1285	.2184	.2501	.2252	.1700	.1110	.0634	.0318	.0139
	4	.0049	.0428	.1156	.1876	.2253	.2186	.1792	.1268	.0780	.0417
	5	.0006	.0105	.0449	.1032	.1651	.2061	.2123	.1859	.1404	.0916
	6	.0000	.0019	.0132	.0430	.0917	.1472	.1906	.2066	.1914	.1527
	7	.0000	.0003	.0030	.0138	.0393	.0811	.1319	.1771	.2013	.1964
	8	.0000	.0000	.0005	.0035	.0131	.0348	.0710	.1181	.1647	.1964
	9	.0000	.0000	.0001	.0007	.0034	.0116	.0298	.0612	.1048	.1527
	10	.0000	.0000	.0000	.0001	.0007	.0030	.0096	.0245	.0515	.0916
	11	.0000	.0000	.0000	.0000	.0001	.0006	.0024	.0074	.0191	.0417
	12	.0000	.0000	.0000	.0000	.0000	.0001	.0004	.0016	.0052	.0139
	13	.0000	.0000	.0000	.0000	.0000	.0000	.0001	.0003	.0010	.0032
	14	.0000	.0000	.0000	.0000	.0000	.0000	.0000	.0000	.0001	.0005
	15	.0000	.0000	.0000	.0000	.0000	.0000	.0000	.0000	.0000	.0000

Table A7   Table of the Binomial Distribution, Cumulative Probabilities

n	z	.05	.10	.15	.20	π .25	.30	.35	.40	.45	.50
2	1	.0975	.1900	.2775	.3600	.4375	.5100	.5775	.6400	.6975	.7500
	2	.0025	.0100	.0225	.0400	.0625	.0900	.1225	.1600	.2025	.2500
3	1	.1426	.2710	.3859	.4880	.5781	.6570	.7254	.7840	.8336	.8750
	2	.0072	.0280	.0608	.1040	.1562	.2160	.2818	.3520	.4252	.5000
	3	.0001	.0010	.0034	.0080	.0156	.0270	.0429	.0640	.0911	.1250
4	1	.1855	.3439	.4780	.5904	.6836	.7599	.8215	.8704	.9085	.9375
	2	.0140	.0523	.1095	.1808	.2617	.3483	.4370	.5248	.6090	.6875
	3	.0005	.0037	.0120	.0272	.0508	.0837	.1265	.1792	.2415	.3125
	4	.0000	.0001	.0005	.0016	.0039	.0081	.0150	.0256	.0410	.0625
5	1	.2262	.4095	.5563	.6723	.7627	.8319	.8840	.9222	.9497	.9688
	2	.0226	.0815	.1648	.2627	.3672	.4718	.5716	.6630	.7438	.8125
	3	.0012	.0086	.0266	.0579	.1035	.1631	.2352	.3174	.4069	.5000
	4	.0000	.0005	.0022	.0067	.0156	.0308	.0540	.0870	.1312	.1875
	5	.0000	.0000	.0001	.0003	.0010	.0024	.0053	.0102	.0185	.0312
6	1	.2649	.4686	.6229	.7379	.8220	.8824	.9246	.9533	.9723	.9844
	2	.0328	.1143	.2235	.3447	.4661	.5798	.6809	.7667	.8364	.8906
	3	.0022	.0158	.0473	.0989	.1694	.2557	.3529	.4557	.5585	.6562
	4	.0001	.0013	.0059	.0170	.0376	.0705	.1174	.1792	.2553	.3438
	5	.0000	.0001	.0004	.0016	.0046	.0109	.0223	.0410	.0692	.1094
	6	.0000	.0000	.0000	.0001	.0002	.0007	.0018	.0041	.0083	.0156
7	1	.3017	.5217	.6794	.7903	.8665	.9176	.9510	.9720	.9848	.9922
	2	.0444	.1497	.2834	.4233	.5551	.6706	.7662	.8414	.8976	.9375
	3	.0038	.0257	.0738	.1480	.2436	.3529	.4677	.5801	.6836	.7734
	4	.0002	.0027	.0121	.0333	.0706	.1260	.1998	.2898	.3917	.5000
	5	.0000	.0002	.0012	.0047	.0129	.0288	.0556	.0963	.1529	.2266
	6	.0000	.0000	.0001	.0004	.0013	.0038	.0090	.0188	.0357	.0625
	7	.0000	.0000	.0000	.0000	.0001	.0002	.0006	.0016	.0037	.0078
8	1	.3366	.5695	.7275	.8322	.8999	.9424	.9681	.9832	.9916	.9961
	2	.0572	.1869	.3428	.4967	.6329	.7447	.8309	.8936	.9368	.9648
	3	.0058	.0381	.1052	.2031	.3215	.4482	.5722	.6846	.7799	.8555
	4	.0004	.0050	.0214	.0563	.1138	.1941	.2936	.4059	.5230	.6367
	5	.0000	.0004	.0029	.0104	.0273	.0580	.1061	.1737	.2604	.3633
	6	.0000	.0000	.0002	.0012	.0042	.0113	.0253	.0498	.0885	.1445
	7	.0000	.0000	.0000	.0001	.0004	.0013	.0036	.0085	.0181	.0352
	8	.0000	.0000	.0000	.0000	.0000	.0001	.0002	.0007	.0017	.0039
9	1	.3698	.6126	.7684	.8658	.9249	.9596	.9793	.9899	.9954	.9980
	2	.0712	.2252	.4005	.5638	.6997	.8040	.8789	.9295	.9615	.9805
	3	.0084	.0530	.1409	.2618	.3993	.5372	.6627	.7682	.8505	.9102
	4	.0006	.0083	.0339	.0856	.1657	.2703	.3911	.5174	.6386	.7461
	5	.0000	.0009	.0056	.0196	.0489	.0988	.1717	.2666	.3786	.5000
	6	.0000	.0001	.0006	.0031	.0100	.0253	.0536	.0994	.1658	.2539
	7	.0000	.0000	.0000	.0003	.0013	.0043	.0112	.0250	.0498	.0898
	8	.0000	.0000	.0000	.0000	.0001	.0004	.0014	.0038	.0091	.0195
	9	.0000	.0000	.0000	.0000	.0000	.0000	.0001	.0003	.0008	.0020

Table A7 Table of the Binomial Distribution, Cumulative Probabilities (continued)

n	x	.05	.10	.15	.20	.25	.30	.35	.40	.45	.50
10	1	.4013	.6513	.8031	.8926	.9437	.9718	.9865	.9940	.9975	.9990
	2	.0861	.2639	.4557	.6242	.7560	.8507	.9140	.9536	.9767	.9893
	3	.0115	.0702	.1798	.3222	.4744	.6172	.7384	.8327	.9004	.9453
	4	.0010	.0128	.0500	.1209	.2241	.3504	.4862	.6177	.7340	.8281
	5	.0001	.0016	.0099	.0328	.0781	.1503	.2485	.3669	.4956	.6230
	6	.0000	.0001	.0014	.0064	.0197	.0473	.0949	.1662	.2616	.3770
	7	.0000	.0000	.0001	.0009	.0035	.0106	.0260	.0548	.1020	.1719
	8	.0000	.0000	.0000	.0001	.0004	.0016	.0048	.0123	.0274	.0547
	9	.0000	.0000	.0000	.0000	.0000	.0001	.0005	.0017	.0045	.0107
	10	.0000	.0000	.0000	.0000	.0000	.0000	.0000	.0001	.0003	.0010
11	1	.4312	.6862	.8327	.9141	.9578	.9802	.9912	.9964	.9986	.9995
	2	.1019	.3026	.5078	.6779	.8029	.8870	.9394	.9698	.9861	.9941
	3	.0152	.0896	.2212	.3826	.5448	.6873	.7999	.8811	.9348	.9673
	4	.0016	.0185	.0694	.1611	.2867	.4304	.5744	.7037	.8089	.8867
	5	.0001	.0028	.0159	.0504	.1146	.2103	.3317	.4672	.6029	.7256
	6	.0000	.0003	.0027	.0117	.0343	.0782	.1487	.2465	.3669	.5000
	7	.0000	.0000	.0003	.0020	.0076	.0216	.0501	.0994	.1738	.2744
	8	.0000	.0000	.0000	.0002	.0012	.0043	.0122	.0293	.0610	.1133
	9	.0000	.0000	.0000	.0000	.0001	.0006	.0020	.0059	.0148	.0327
	10	.0000	.0000	.0000	.0000	.0000	.0000	.0002	.0007	.0022	.0059
	11	.0000	.0000	.0000	.0000	.0000	.0000	.0000	.0000	.0002	.0005
12	1	.4596	.7176	.8578	.9313	.9683	.9862	.9943	.9978	.9992	.9998
	2	.1184	.3410	.5565	.7251	.8416	.9150	.9576	.9804	.9917	.9968
	3	.0196	.1109	.2642	.4417	.6093	.7472	.8487	.9166	.9579	.9807
	4	.0022	.0256	.0922	.2054	.3512	.5075	.6533	.7747	.8655	.9270
	5	.0002	.0043	.0239	.0726	.1576	.2763	.4167	.5618	.6956	.8062
	6	.0000	.0005	.0046	.0194	.0544	.1178	.2127	.3348	.4731	.6128
	7	.0000	.0001	.0007	.0039	.0143	.0386	.0846	.1582	.2607	.3872
	8	.0000	.0000	.0001	.0006	.0028	.0095	.0255	.0573	.1117	.1938
	9	.0000	.0000	.0000	.0001	.0004	.0017	.0056	.0153	.0356	.0730
	10	.0000	.0000	.0000	.0000	.0000	.0002	.0008	.0028	.0079	.0193
	11	.0000	.0000	.0000	.0000	.0000	.0000	.0001	.0003	.0011	.0032
	12	.0000	.0000	.0000	.0000	.0000	.0000	.0000	.0000	.0001	.0002
13	1	.4867	.7458	.8791	.9450	.9762	.9903	.9963	.9987	.9996	.9999
	2	.1354	.3787	.6017	.7664	.8733	.9363	.9704	.9874	.9951	.9983
	3	.0245	.1339	.2704	.4983	.6674	.7975	.8868	.9421	.9731	.9888
	4	.0031	.0342	.0967	.2527	.4157	.5794	.7217	.8314	.9071	.9539
	5	.0003	.0065	.0260	.0991	.2060	.3457	.4995	.6470	.7721	.8666
	6	.0000	.0009	.0053	.0300	.0802	.1654	.2841	.4256	.5732	.7095
	7	.0000	.0001	.0013	.0070	.0243	.0624	.1295	.2288	.3563	.5000
	8	.0000	.0000	.0002	.0012	.0056	.0182	.0462	.0977	.1788	.2905
	9	.0000	.0000	.0000	.0002	.0010	.0040	.0126	.0321	.0698	.1334
	10	.0000	.0000	.0000	.0000	.0001	.0007	.0025	.0078	.0203	.0461
	11	.0000	.0000	.0000	.0000	.0000	.0001	.0003	.0013	.0041	.0112
	12	.0000	.0000	.0000	.0000	.0000	.0000	.0000	.0001	.0005	.0017
	13	.0000	.0000	.0000	.0000	.0000	.0000	.0000	.0000	.0000	.0001

The column group header is π.

Table A7   Table of the Binomial Distribution, Cumulative Probabilities (continued)

						π					
n	x	.05	.10	.15	.20	.25	.30	.35	.40	.45	.50
14	1	.5123	.7712	.8972	.9560	.9822	.9932	.9976	.9992	.9998	.9999
	2	.1530	.4154	.6433	.8021	.8990	.9525	.9795	.9919	.9971	.9991
	3	.0301	.1584	.3521	.5519	.7189	.8392	.9161	.9602	.9830	.9935
	4	.0042	.0441	.1465	.3018	.4787	.6448	.7795	.8757	.9368	.9713
	5	.0004	.0092	.0467	.1298	.2585	.4158	.5773	.7207	.8328	.9102
	6	.0000	.0015	.0115	.0439	.1117	.2195	.3595	.5141	.6627	.7880
	7	.0000	.0002	.0022	.0116	.0383	.0933	.1836	.3075	.4539	.6047
	8	.0000	.0000	.0003	.0024	.0103	.0315	.0753	.1501	.2586	.3953
	9	.0000	.0000	.0000	.0004	.0022	.0083	.0243	.0583	.1189	.2120
	10	.0000	.0000	.0000	.0000	.0003	.0017	.0060	.0175	.0426	.0898
	11	.0000	.0000	.0000	.0000	.0000	.0002	.0011	.0039	.0114	.0287
	12	.0000	.0000	.0000	.0000	.0000	.0000	.0001	.0006	.0022	.0065
	13	.0000	.0000	.0000	.0000	.0000	.0000	.0000	.0001	.0003	.0009
	14	.0000	.0000	.0000	.0000	.0000	.0000	.0000	.0000	.0000	.0001
15	1	.5367	.7941	.9126	.9648	.9866	.9953	.9984	.9995	.9999	1.0000
	2	.1710	.4510	.6814	.8329	.9198	.9647	.9858	.9948	.9983	.9995
	3	.0362	.1841	.3958	.6020	.7639	.8732	.9383	.9729	.9893	.9963
	4	.0055	.0556	.1773	.3518	.5387	.7031	.8273	.9095	.9576	.9824
	5	.0006	.0127	.0617	.1642	.3135	.4845	.6481	.7827	.8796	.9408
	6	.0001	.0022	.0168	.0611	.1484	.2784	.4357	.5968	.7392	.8491
	7	.0000	.0003	.0036	.0181	.0566	.1311	.2452	.3902	.5478	.6964
	8	.0000	.0000	.0006	.0042	.0173	.0500	.1132	.2131	.3465	.5000
	9	.0000	.0000	.0001	.0008	.0042	.0152	.0422	.0950	.1818	.3036
	10	.0000	.0000	.0000	.0001	.0008	.0037	.0124	.0338	.0769	.1509
	11	.0000	.0000	.0000	.0000	.0001	.0007	.0028	.0093	.0255	.0592
	12	.0000	.0000	.0000	.0000	.0000	.0001	.0005	.0019	.0063	.0176
	13	.0000	.0000	.0000	.0000	.0000	.0000	.0001	.0003	.0011	.0037
	14	.0000	.0000	.0000	.0000	.0000	.0000	.0000	.0000	.0001	.0005
	15	.0000	.0000	.0000	.0000	.0000	.0000	.0000	.0000	.0000	.0000

Table A8   Table of Critical Values for the Single-Sample Runs Test

Numbers listed are tabled critical two-tailed .05 and one-tailed .025 values.

n_2 \ n_1	2	3	4	5	6	7	8	9	10	11	12	13	14	15	16	17	18	19	20
2											2	2	2	2	2	2	2	2	2
											−	−	−	−	−	−	−	−	−
3					2	2	2	2	2	2	2	2	2	3	3	3	3	3	3
					−	−	−	−	−	−	−	−	−	−	−	−	−	−	−
4				2	2	2	3	3	3	3	3	3	3	3	4	4	4	4	4
				9	9	−	−	−	−	−	−	−	−	−	−	−	−	−	−
5			2	2	3	3	3	3	3	4	4	4	4	4	4	4	5	5	5
			9	10	10	11	11	−	−	−	−	−	−	−	−	−	−	−	−
6		2	2	3	3	3	3	4	4	4	4	5	5	5	5	5	5	6	6
		−	9	10	11	12	12	13	13	13	13	−	−	−	−	−	−	−	−
7		2	2	3	3	3	4	4	5	5	5	5	5	6	6	6	6	6	6
		−	−	11	12	13	13	14	14	14	14	15	15	15	−	−	−	−	−
8		2	3	3	3	4	4	5	5	5	6	6	6	6	6	7	7	7	7
		−	−	11	12	13	14	14	15	15	16	16	16	16	17	17	17	17	17
9		2	3	3	4	4	5	5	5	6	6	6	7	7	7	7	8	8	8
		−	−	−	13	14	14	15	16	16	16	17	17	18	18	18	18	18	18
10		2	3	3	4	5	5	5	6	6	7	7	7	7	8	8	8	8	9
		−	−	−	13	14	15	16	16	17	17	18	18	18	19	19	19	20	20
11		2	3	4	4	5	5	6	6	7	7	7	8	8	8	9	9	9	9
		−	−	−	13	14	15	16	17	17	18	19	19	19	20	20	20	21	21
12	2	2	3	4	4	5	6	6	7	7	7	8	8	8	9	9	9	10	10
	−	−	−	−	13	14	16	16	17	18	19	19	20	20	21	21	21	22	22
13	2	2	3	4	5	5	6	6	7	7	8	8	9	9	9	10	10	10	10
	−	−	−	−	−	15	16	17	18	19	19	20	20	21	21	22	22	23	23
14	2	2	3	4	5	5	6	7	7	8	8	9	9	9	10	10	10	11	11
	−	−	−	−	−	15	16	17	18	19	20	20	21	22	22	23	23	23	24
15	2	3	3	4	5	6	6	7	7	8	8	9	9	10	10	11	11	11	12
	−	−	−	−	−	15	16	18	18	19	20	21	22	22	23	23	24	24	25
16	2	3	4	4	5	6	6	7	8	8	9	9	10	10	11	11	11	12	12
	−	−	−	−	−	−	17	18	19	20	21	21	22	23	23	24	25	25	25
17	2	3	4	4	5	6	7	7	8	9	9	10	10	11	11	11	12	12	13
	−	−	−	−	−	−	17	18	19	20	21	22	23	23	24	25	25	26	26
18	2	3	4	5	5	6	7	8	8	9	9	10	10	11	11	12	12	13	13
	−	−	−	−	−	−	17	18	19	20	21	22	23	24	25	25	26	26	27
19	2	3	4	5	6	6	7	8	8	9	10	10	11	11	12	12	13	13	13
	−	−	−	−	−	−	17	18	20	21	22	23	23	24	25	26	26	27	27
20	2	3	4	5	6	6	7	8	9	9	10	10	11	12	12	13	13	13	14
	−	−	−	−	−	−	17	18	20	21	22	23	24	25	25	26	27	27	28

Table A9   Table of the F_{max} Distribution

The .05 critical values are in lightface type, and the .01 critical values are in **bold** type.

$n-1$ \ k	2	3	4	5	6	7	8	9	10	11	12
2	39	87.5	142	202	266	333	403	475	550	626	704
	199	**448**	**729**	**1036**	**1362**	**1705**	**2063**	**2432**	**2813**	**3204**	**3605**
3	15.4	27.8	39.2	50.7	62	72.9	83.5	93.9	104	114	124
	47.5	**85**	**120**	**151**	**184**	**216***	**249***	**281***	**310***	**337***	**361***
4	9.60	15.5	20.6	25.2	29.5	33.6	37.5	41.4	44.6	48.0	51.4
	23.2	**37**	**49**	**59**	**69**	**79**	**89**	**97**	**106**	**113**	**120**
5	7.15	10.8	13.7	16.3	18.7	20.8	22.9	24.7	26.5	28.2	29.9
	14.9	**22**	**28**	**33**	**38**	**42**	**46**	**50**	**54**	**57**	**60**
6	5.82	8.38	10.4	12.1	13.7	15.0	16.3	17.5	18.6	19.7	20.7
	11.1	**15.5**	**19.1**	**22**	**25**	**27**	**30**	**32**	**34**	**36**	**37**
7	4.99	6.94	8.44	9.70	10.8	11.8	12.7	13.5	14.3	15.1	15.8
	8.89	**12.1**	**14.5**	**16.5**	**18.4**	**20.**	**22**	**23**	**24**	**26**	**27**
8	4.43	6.00	7.18	8.12	9.03	9.78	10.5	11.1	11.7	12.2	12.7
	7.50	**9.9**	**11.7**	**13.2**	**14.5**	**15.8**	**16.9**	**17.9**	**18.9**	**19.8**	**21**
9	4.03	5.34	6.31	7.11	7.80	8.41	8.95	9.45	9.91	10.3	10.7
	6.54	**8.5**	**9.9**	**11.1**	**12.1**	**13.1**	**13.9**	**14.7**	**15.3**	**16.0**	**16.6**
10	3.72	4.85	5.67	6.34	6.92	7.42	7.87	8.28	8.66	9.01	9.34
	5.85	**7.4**	**8.6**	**9.6**	**10.4**	**11.1**	**11.8**	**12.4**	**12.9**	**13.4**	**13.9**
12	3.28	4.16	4.79	5.30	5.72	6.09	6.42	6.72	7.00	7.25	7.48
	4.91	**6.1**	**6.9**	**7.6**	**8.2**	**8.7**	**9.1**	**9.5**	**9.9**	**10.2**	**10.6**
15	2.86	3.54	4.01	4.37	4.68	4.95	5.19	5.40	5.59	5.77	5.93
	4.07	**4.9**	**5.5**	**6.0**	**6.4**	**6.7**	**7.1**	**7.3**	**7.5**	**7.8**	**8.0**
20	2.46	2.95	3.29	3.54	3.76	3.94	4.10	4.24	4.37	4.49	4.59
	3.32	**3.8**	**4.3**	**4.6**	**4.9**	**5.1**	**5.3**	**5.5**	**5.6**	**5.8**	**5.9**
30	2.07	2.40	2.61	2.78	2.91	3.02	3.12	3.21	3.29	3.36	3.39
	2.63	**3.0**	**3.3**	**3.5**	**3.6**	**3.7**	**3.8**	**3.9**	**4.0**	**4.1**	**4.2**
60	1.67	1.85	1.96	2.04	2.11	2.17	2.22	2.26	2.30	2.33	2.36
	1.96	**2.2**	**2.3**	**2.4**	**2.4**	**2.5**	**2.5**	**2.6**	**2.6**	**2.7**	**2.7**

* The third digit in these values is an approximation

Table A10 Table of the F Distribution
$F_{.95}$

df_{den} \ df_{num}	1	2	3	4	5	6	7	8	9	10	12	15	20	30	40	60	120	∞
1	161.4	199.5	215.7	224.6	230.2	234.0	236.8	238.9	240.5	241.9	243.9	245.9	248.0	250.1	251.1	252.2	253.3	254.3
2	18.51	19.0	19.16	19.25	19.30	19.33	19.33	19.37	19.38	19.40	19.41	19.43	19.45	19.46	19.47	19.48	19.49	19.50
3	10.13	9.55	9.28	9.12	9.01	8.94	8.89	8.85	8.81	8.79	8.74	8.70	8.66	8.62	8.59	8.57	8.55	8.53
4	7.71	6.94	6.59	6.39	6.26	6.16	6.09	6.04	6.00	5.96	5.91	5.86	5.80	5.75	5.72	5.69	5.66	5.63
5	6.61	5.79	5.41	5.19	5.05	4.95	4.88	4.82	4.77	4.74	4.68	4.62	4.56	4.50	4.46	4.43	4.40	4.36
6	5.99	5.14	4.76	4.53	4.39	4.28	4.21	4.15	4.10	4.06	4.00	3.94	3.87	3.81	3.77	3.74	3.70	3.67
7	5.59	4.74	4.35	4.12	3.97	3.87	3.79	3.73	3.68	3.64	3.57	3.51	3.44	3.38	3.34	3.30	3.27	3.23
8	5.32	4.46	4.07	3.84	3.69	3.58	3.50	3.44	3.39	3.35	3.28	3.22	3.15	3.08	3.04	3.01	2.97	2.93
9	5.12	4.26	3.86	3.63	3.48	3.37	3.29	3.23	3.18	3.14	3.07	3.01	2.94	2.86	2.83	2.79	2.75	2.71
10	4.96	4.10	3.71	3.48	3.33	3.22	3.14	3.07	3.02	2.98	2.91	2.85	2.77	2.70	2.66	2.62	2.58	2.54
11	4.84	3.98	3.59	3.36	3.20	3.09	3.01	2.95	2.90	2.85	2.79	2.72	2.65	2.57	2.53	2.49	2.45	2.40
12	4.75	3.89	3.49	3.26	3.11	3.00	2.91	2.85	2.80	2.75	2.69	2.62	2.54	2.47	2.43	2.38	2.34	2.30
13	4.67	3.81	3.41	3.18	3.03	2.92	2.83	2.77	2.71	2.67	2.60	2.53	2.46	2.38	2.34	2.30	2.25	2.21
14	4.60	3.74	3.34	3.11	2.96	2.85	2.76	2.70	2.65	2.60	2.53	2.46	2.39	2.31	2.27	2.22	2.18	2.13
15	4.54	3.68	3.29	3.06	2.90	2.79	2.71	2.64	2.59	2.54	2.48	2.40	2.33	2.25	2.20	2.16	2.11	2.07
16	4.49	3.63	3.24	3.01	2.85	2.74	2.66	2.59	2.54	2.49	2.42	2.35	2.28	2.19	2.15	2.11	2.06	2.01
17	4.45	3.59	3.20	2.96	2.81	2.70	2.61	2.55	2.49	2.45	2.38	2.31	2.23	2.15	2.10	2.06	2.01	1.96
18	4.41	3.55	3.16	2.93	2.77	2.66	2.58	2.51	2.46	2.41	2.34	2.27	2.19	2.11	2.06	2.02	1.97	1.92
19	4.38	3.52	3.13	2.90	2.74	2.63	2.54	2.48	2.42	2.38	2.31	2.23	2.16	2.07	2.03	1.98	1.93	1.88
20	4.35	3.49	3.10	2.87	2.71	2.60	2.51	2.45	2.39	2.35	2.28	2.20	2.12	2.04	1.99	1.95	1.90	1.84

Table A10   Table of the *F* Distribution
$F_{.95}$ (continued)

df_num → df_den ↓	1	2	3	4	5	6	7	8	9	10	12	15	20	30	40	60	120	∞
21	4.32	3.47	3.07	2.84	2.68	2.57	2.49	2.42	2.37	2.32	2.25	2.18	2.10	2.01	1.96	1.92	1.87	1.81
22	4.30	3.44	3.05	2.82	2.66	2.55	2.46	2.40	2.34	2.30	2.23	2.15	2.07	1.98	1.94	1.89	1.84	1.78
23	4.28	3.42	3.03	2.80	2.64	2.53	2.44	2.37	2.32	2.27	2.20	2.13	2.05	1.96	1.91	1.86	1.81	1.76
24	4.26	3.40	3.01	2.78	2.62	2.51	2.42	2.36	2.30	2.25	2.18	2.11	2.03	1.94	1.89	1.84	1.79	1.73
25	4.24	3.39	2.99	2.76	2.60	2.49	2.40	2.34	2.28	2.24	2.16	2.09	2.01	1.92	1.87	1.82	1.77	1.71
26	4.23	3.37	2.98	2.74	2.59	2.47	2.39	2.32	2.27	2.22	2.15	2.07	1.99	1.90	1.85	1.80	1.75	1.69
27	4.21	3.35	2.96	2.73	2.57	2.46	2.37	2.31	2.25	2.20	2.13	2.06	1.97	1.88	1.84	1.79	1.73	1.67
28	4.20	3.34	2.95	2.71	2.56	2.45	2.36	2.29	2.24	2.19	2.13	2.04	1.96	1.87	1.82	1.77	1.71	1.65
29	4.18	3.33	2.93	2.70	2.55	2.43	2.35	2.28	2.22	2.18	2.10	2.03	1.94	1.85	1.81	1.75	1.70	1.64
30	4.17	3.32	3.92	2.69	2.53	2.42	2.33	2.27	2.21	2.16	2.09	2.01	1.93	1.84	1.79	1.74	1.68	1.62
40	4.08	3.23	2.84	2.61	2.45	2.34	2.25	2.18	2.12	2.08	2.00	1.92	1.84	1.74	1.69	1.64	1.58	1.51
60	4.00	3.15	2.76	2.53	2.37	2.25	2.17	2.10	2.04	1.99	1.92	1.84	1.75	1.65	1.59	1.53	1.47	1.39
120	3.92	3.07	2.68	2.45	2.29	2.17	2.09	2.02	1.96	1.91	1.83	1.75	1.66	1.55	1.50	1.43	1.35	1.25
∞	3.84	3.00	2.60	2.37	2.21	2.10	2.01	1.94	1.88	1.83	1.75	1.67	1.57	1.46	1.39	1.32	1.22	1.00

Table A10   Table of the F Distribution

$$F_{.975}$$

df_{den} ↓ \ df_{num} →	1	2	3	4	5	6	7	8	9	10	12	15	20	30	40	60	120	∞
1	647.8	799.2	864.2	899.6	921.8	937.1	948.2	956.7	963.3	968.6	976.7	984.9	993.1	1001	1006	1010	1014	1018
2	38.51	39.00	39.17	39.25	39.30	39.33	39.36	39.37	39.39	39.40	39.41	39.43	39.45	39.46	39.47	39.48	39.49	39.50
3	17.44	16.04	15.44	15.10	14.88	14.73	14.62	14.54	14.47	14.42	14.34	14.25	14.17	14.08	14.04	13.99	13.95	13.90
4	12.22	10.65	9.98	9.60	9.36	9.20	9.07	8.98	8.90	8.84	8.75	8.66	8.56	8.46	8.41	8.36	8.31	8.26
5	10.01	8.43	7.76	7.39	7.15	6.98	6.85	6.76	6.68	6.62	6.52	6.43	6.33	6.23	6.18	6.12	6.07	6.02
6	8.81	7.26	6.60	6.23	5.99	5.82	5.70	5.60	5.52	5.46	5.37	5.27	5.17	5.07	5.01	4.96	4.90	4.85
7	8.07	6.54	5.89	5.52	5.29	5.12	4.99	4.90	4.82	4.76	4.67	4.57	4.47	4.36	4.31	4.25	4.20	4.14
8	7.57	6.06	5.42	5.05	4.82	4.65	4.53	4.43	4.36	4.30	4.20	4.10	4.00	3.89	3.84	3.78	3.73	3.67
9	7.21	5.71	5.08	4.72	4.48	4.32	4.20	4.10	4.03	3.96	3.87	3.77	3.67	3.56	3.51	3.45	3.39	3.33
10	6.94	5.46	4.83	4.47	4.24	4.07	3.95	3.85	3.78	3.72	3.62	3.52	3.42	3.31	3.26	3.20	3.14	3.08
11	6.72	5.26	4.63	4.28	4.04	3.88	3.76	3.66	3.59	3.53	3.43	3.33	3.23	3.12	3.06	3.00	2.94	2.88
12	6.55	5.10	4.47	4.12	3.89	3.73	3.61	3.51	3.44	3.37	3.28	3.18	3.07	2.96	2.91	2.85	2.79	2.72
13	6.41	4.97	4.35	4.00	3.77	3.60	3.48	3.39	3.31	3.25	3.15	3.05	2.95	2.84	2.78	2.72	2.66	2.60
14	6.30	4.86	4.24	3.89	3.66	3.50	3.38	3.29	3.21	3.15	3.05	2.95	2.84	2.73	2.67	2.61	2.55	2.49
15	6.20	4.77	4.15	3.80	3.58	3.41	3.29	3.20	3.12	3.06	2.96	2.86	2.76	2.64	2.59	2.52	2.46	2.40
16	6.12	4.69	4.08	3.73	3.50	3.34	3.22	3.12	3.05	2.99	2.89	2.79	2.68	2.57	2.51	2.45	2.48	2.32
17	6.04	4.62	4.01	3.66	3.44	3.28	3.16	3.06	2.98	2.92	2.82	2.72	2.62	2.50	2.44	2.38	2.32	2.25
18	5.98	4.56	3.95	3.61	3.38	3.22	3.10	3.01	2.93	2.87	2.77	2.67	2.56	2.44	2.38	2.32	2.26	2.19
19	5.92	4.51	3.90	3.56	3.33	3.17	3.05	2.96	2.88	2.82	2.72	2.62	2.51	2.39	2.33	2.27	2.20	2.13
20	5.87	4.46	3.86	3.51	3.29	3.13	3.01	2.91	2.84	2.77	2.68	2.57	2.46	2.35	2.29	2.22	2.16	2.09

Table A10　Table of the F Distribution
$F_{.975}$ (continued)

df_num → / df_den ↓	1	2	3	4	5	6	7	8	9	10	12	15	20	30	40	60	120	∞
21	5.83	4.42	3.82	3.48	3.25	3.09	2.97	2.87	2.80	2.73	2.64	2.53	2.42	2.31	2.25	2.18	2.11	2.04
22	5.79	4.38	3.78	3.44	3.22	3.05	2.93	2.84	2.76	2.70	2.60	2.50	2.39	2.27	2.21	2.14	2.08	2.00
23	5.75	4.35	3.75	3.41	3.18	3.03	2.90	2.81	2.73	2.67	2.57	2.47	2.36	2.24	2.18	2.11	2.04	1.97
24	5.72	4.32	3.72	3.38	3.15	2.99	2.87	2.78	2.70	2.64	2.54	2.44	2.33	2.21	2.15	2.08	2.01	1.94
25	5.69	4.29	3.69	3.35	3.13	2.97	2.85	2.75	2.68	2.62	2.51	2.41	2.30	2.18	2.12	2.05	1.98	1.91
26	5.66	4.27	3.67	3.33	3.10	2.94	2.82	2.73	2.65	2.59	2.49	2.39	2.28	2.16	2.09	2.03	1.95	1.88
27	5.63	4.24	3.65	3.31	3.08	2.92	2.80	2.71	2.63	2.57	2.47	2.36	2.25	2.13	2.07	2.00	1.93	1.85
28	5.61	4.22	3.63	3.29	3.06	2.90	2.78	2.69	2.61	2.55	2.45	2.34	2.23	2.11	2.05	1.98	1.91	1.83
29	5.59	4.20	3.61	3.27	3.04	2.88	2.76	2.67	2.59	2.53	2.43	2.32	2.21	2.09	2.03	1.96	1.89	1.81
30	5.57	4.18	3.59	3.25	3.03	2.87	2.75	2.65	2.57	2.51	2.41	2.31	2.20	2.07	2.01	1.94	1.87	1.79
40	5.42	4.05	3.46	3.13	2.90	2.74	2.62	2.53	2.45	2.39	2.29	2.18	2.07	1.94	1.88	1.80	1.72	1.64
60	5.29	3.93	3.34	3.01	2.79	2.63	2.51	2.41	2.33	2.27	2.17	2.06	1.94	1.83	1.74	1.67	1.58	1.48
120	5.15	3.80	3.23	2.89	2.67	2.52	2.39	2.30	2.22	2.16	2.05	1.94	1.82	1.69	1.61	1.53	1.43	1.31
∞	5.02	3.69	3.12	2.79	2.57	2.41	2.29	2.19	2.11	2.05	1.94	1.83	1.71	1.57	1.48	1.39	1.27	1.00

Table A10 Table of the F Distribution
$$F_{.99}$$

$df_{num} \to$ df_{den}	1	2	3	4	5	6	7	8	9	10	12	15	20	30	40	60	120	∞
1	4052	4999	5403	5625	5674	5859	5928	5982	6022	6056	6106	6209	6235	6261	6287	6313	6339	6366
2	98.50	90.00	99.17	99.25	99.30	99.33	99.36	99.37	99.39	99.40	99.42	99.45	99.46	99.47	99.47	99.48	99.49	99.50
3	34.12	30.82	29.46	28.71	28.24	27.91	27.67	27.49	27.35	27.23	27.05	26.69	26.60	26.50	26.41	26.32	26.22	26.13
4	21.20	18.00	16.69	15.98	15.52	15.21	14.98	14.80	14.66	14.55	14.37	14.02	13.93	13.84	13.75	13.65	13.56	13.46
5	16.36	13.27	12.06	11.39	10.97	10.67	10.46	10.29	10.16	10.05	9.89	9.55	9.47	9.38	9.20	9.20	9.11	9.02
6	13.75	10.92	9.78	9.15	8.75	8.47	8.26	8.10	7.98	7.87	7.72	7.40	7.31	7.23	7.14	7.06	6.97	6.88
7	12.25	9.55	8.45	7.85	7.46	7.19	6.99	6.84	6.72	6.62	6.47	6.16	6.07	5.99	5.91	5.82	5.74	5.65
8	11.36	8.65	7.59	7.01	6.63	6.37	6.18	6.03	5.91	5.81	5.67	5.36	5.28	5.20	5.12	5.03	4.95	4.86
9	10.56	8.02	6.99	6.42	6.06	5.80	5.61	5.47	5.35	5.25	5.11	4.81	4.73	4.65	4.57	4.48	4.40	4.31
10	10.04	7.56	6.55	5.99	5.64	5.39	5.20	5.06	4.94	4.85	4.71	4.41	4.33	4.25	4.17	4.08	4.00	3.91
11	9.65	7.21	6.22	5.67	5.32	5.07	4.89	4.74	4.63	4.54	4.40	4.10	4.02	3.94	3.86	3.78	3.69	3.60
12	9.33	6.93	5.95	5.41	5.06	4.82	4.64	4.50	4.39	4.30	4.16	3.86	3.78	3.70	3.62	3.54	3.45	3.36
13	9.07	6.70	5.74	5.21	4.86	4.62	4.44	4.30	4.19	4.10	3.96	3.66	3.59	3.51	3.43	3.34	3.25	3.17
14	8.86	6.51	5.56	5.04	4.69	4.46	4.28	4.14	4.03	3.94	3.80	3.51	3.43	3.35	3.27	3.18	3.09	3.00
15	8.68	6.36	5.42	4.89	4.56	4.32	4.14	4.00	3.89	3.80	3.67	3.37	3.29	3.21	3.13	3.05	2.96	2.87
16	8.53	6.23	5.29	4.77	4.44	4.20	4.03	3.89	3.78	3.69	3.55	3.26	3.18	3.10	3.02	2.93	2.84	2.75
17	8.40	6.11	5.18	4.67	4.34	4.10	3.93	3.79	3.68	3.59	3.46	3.16	3.08	3.00	2.92	2.83	2.75	2.65
18	8.29	6.01	5.09	4.58	4.25	4.01	3.84	3.71	3.60	3.51	3.37	3.08	3.00	2.92	2.84	2.75	2.66	2.57
19	8.18	5.93	5.01	4.50	4.17	3.94	3.77	3.63	3.52	3.43	3.30	3.00	2.92	2.84	2.76	2.67	2.58	2.49
20	8.10	5.85	4.94	4.43	4.10	3.87	3.70	3.56	3.46	3.37	3.23	2.94	2.86	2.78	2.69	2.61	2.52	2.42

Table A10 Table of the F Distribution
$F_{.99}$ (continued)

df_{den} \ df_{num}	1	2	3	4	5	6	7	8	9	10	12	15	20	30	40	60	120	∞
21	8.02	5.78	4.87	4.37	4.04	3.81	3.64	3.51	3.40	3.31	3.17	3.03	2.88	2.72	2.64	2.55	2.46	2.36
22	7.95	5.72	4.82	4.31	3.99	3.76	3.59	3.45	3.35	3.26	3.12	2.98	2.83	2.67	2.58	2.50	2.40	2.31
23	7.88	5.66	4.76	4.26	3.94	3.71	3.54	3.41	3.30	3.21	3.07	2.93	2.78	2.62	2.54	2.45	2.35	2.26
24	7.82	5.61	4.72	4.22	3.90	3.67	3.50	3.36	3.26	3.17	3.03	2.89	2.74	2.58	2.49	2.40	2.31	2.21
25	7.77	5.57	4.68	4.18	3.85	3.63	3.46	3.32	3.22	3.13	2.99	2.85	2.70	2.54	2.45	2.36	2.27	2.17
26	7.72	5.53	4.64	4.14	3.82	3.59	3.42	3.29	3.18	3.09	2.96	2.81	2.66	2.50	2.42	2.33	2.23	2.13
27	7.68	5.49	4.60	4.11	3.78	3.56	3.39	3.26	3.15	3.06	2.93	2.78	2.63	2.47	2.38	2.29	2.20	2.10
28	7.64	5.45	4.57	4.07	3.75	3.53	3.36	3.23	3.12	3.03	2.90	2.75	2.60	2.44	2.35	2.26	2.17	2.06
29	7.60	5.42	4.54	4.04	3.73	3.50	3.33	3.20	3.09	3.00	2.87	2.73	2.57	2.41	2.33	2.23	2.14	2.03
30	7.56	5.39	4.51	4.02	3.70	3.47	3.30	3.17	3.07	2.98	2.84	2.70	2.55	2.39	2.30	2.21	2.11	2.01
40	7.31	5.18	4.31	3.83	3.51	3.29	3.12	2.99	2.89	2.80	2.66	2.52	2.37	2.20	2.11	2.02	1.92	1.80
60	7.08	4.98	4.13	3.65	3.34	3.12	2.95	2.82	2.72	2.63	2.50	2.35	2.20	2.03	1.94	1.84	1.73	1.60
120	6.85	4.79	3.95	3.48	3.17	2.96	2.79	2.66	2.56	2.47	2.34	2.19	2.03	1.86	1.76	1.66	1.53	1.38
∞	6.63	4.61	3.78	3.32	3.02	2.80	2.64	2.51	2.41	2.32	2.18	2.04	1.88	1.70	1.59	1.47	1.32	1.00

Table A10 Table of the F Distribution
$$F_{.995}$$

df_{den} \ df_{num}	1	2	3	4	5	6	7	8	9	10	12	15	20	30	40	60	120	∞
1	16211	20000	21615	22500	23056	23437	23715	23925	24091	24224	24426	24630	24836	24940	25044	25148	25253	25465
2	198.5	199.0	199.2	199.2	199.3	199.3	199.4	199.4	199.4	199.4	199.4	199.4	199.4	199.5	199.5	199.5	199.5	199.5
3	55.55	49.80	47.47	46.19	45.39	44.84	44.43	44.13	43.88	43.69	43.39	43.08	42.78	42.62	42.47	42.31	42.15	41.83
4	31.33	26.28	24.26	23.15	22.46	21.97	21.62	21.35	21.14	20.97	20.70	20.44	20.17	20.03	19.89	19.75	19.61	19.32
5	22.78	18.31	16.53	15.56	14.94	14.51	14.20	13.96	13.77	13.62	13.38	13.15	12.90	12.78	12.66	12.53	12.40	12.14
6	18.63	14.54	12.92	12.03	11.46	11.07	10.79	10.57	10.39	10.25	10.03	9.81	9.59	9.47	9.36	9.24	9.12	8.88
7	16.24	12.40	10.88	10.05	9.52	9.16	8.89	8.68	8.51	8.38	8.18	7.97	7.75	7.65	7.53	7.42	7.31	7.08
8	14.69	11.04	9.60	8.81	8.30	7.95	7.69	7.50	7.34	7.21	7.01	6.81	6.61	6.50	6.40	6.29	6.18	5.95
9	13.61	10.11	8.72	7.96	7.47	7.13	6.88	6.69	6.54	6.42	6.23	6.03	5.83	5.73	5.62	5.52	5.41	5.19
10	12.83	9.43	8.08	7.34	6.87	6.54	6.30	6.12	5.97	5.85	5.66	5.47	5.27	5.17	5.07	4.97	4.86	4.64
11	12.23	8.91	7.60	6.88	6.42	6.10	5.86	5.68	5.54	5.42	5.24	5.05	4.86	4.76	4.66	4.55	4.44	4.23
12	11.75	8.51	7.23	6.52	6.07	5.76	5.52	5.35	5.20	5.09	4.91	4.72	4.53	4.43	4.33	4.23	4.12	3.90
13	11.37	8.19	6.93	6.23	5.79	5.48	5.25	5.08	4.94	4.82	4.64	4.46	4.27	4.17	4.07	3.97	3.87	3.65
14	11.06	7.92	6.68	6.00	5.56	5.26	5.03	4.86	4.72	4.60	4.43	4.25	4.06	3.96	3.86	3.76	3.66	3.44
15	10.80	7.70	6.48	5.80	5.37	5.07	4.85	4.67	4.54	4.42	4.25	4.07	3.88	3.79	3.69	3.58	3.48	3.26
16	10.58	7.51	6.30	5.64	5.21	4.91	4.69	4.52	4.38	4.27	4.10	3.92	3.73	3.64	3.54	3.44	3.33	3.11
17	10.38	7.35	6.16	5.50	5.07	4.78	4.56	4.39	4.25	4.14	3.97	3.79	3.61	3.51	3.41	3.31	3.21	2.98
18	10.23	7.21	6.03	5.37	4.96	4.66	4.44	4.28	4.14	4.03	3.86	3.68	3.50	3.40	3.30	3.20	3.10	2.87
19	10.07	7.09	5.92	5.27	4.85	4.56	4.34	4.18	4.04	3.93	3.76	3.59	3.40	3.31	3.21	3.11	3.00	2.78
20	9.94	6.99	5.82	5.17	4.76	4.47	4.26	4.09	3.96	3.85	3.68	3.50	3.32	3.22	3.12	3.02	2.92	2.69

Table A10　Table of the F Distribution
$F_{.995}$ (continued)

df$_{den}$ ↓ \ df$_{num}$ →	1	2	3	4	5	6	7	8	9	10	12	15	20	30	40	60	120	∞
21	9.83	6.89	5.73	5.09	4.68	4.39	4.18	4.01	3.88	3.77	3.60	3.43	3.24	3.05	2.95	2.84	2.73	2.61
22	9.73	6.81	5.65	5.02	4.61	4.32	4.11	3.94	3.81	3.70	3.54	3.36	3.18	2.98	2.88	2.77	2.66	2.55
23	9.63	6.73	5.58	4.95	4.54	4.26	4.05	3.88	3.75	3.64	3.47	3.30	3.12	2.92	2.82	2.71	2.60	2.48
24	9.55	6.66	5.52	4.89	4.49	4.20	3.99	3.83	3.69	3.59	3.42	3.25	3.06	2.87	2.77	2.66	2.55	2.43
25	9.48	6.60	5.46	4.84	4.43	4.15	3.94	3.78	3.64	3.54	3.37	3.20	3.01	2.82	2.72	2.61	2.50	2.38
26	9.41	6.54	5.41	4.79	4.38	4.10	3.89	3.73	3.60	3.49	3.33	3.15	2.97	2.77	2.67	2.56	2.45	2.33
27	9.34	6.49	5.36	4.74	4.34	4.06	3.85	3.69	3.56	3.45	3.28	3.11	2.93	2.73	2.63	2.52	2.41	2.29
28	9.28	6.44	5.32	4.70	4.30	4.02	3.81	3.65	3.52	3.41	3.25	3.07	2.89	2.69	2.59	2.48	2.37	2.25
29	9.23	6.40	5.28	4.66	4.26	3.98	3.77	3.61	3.48	3.38	3.21	3.04	2.86	2.66	2.56	2.45	2.33	2.21
30	9.18	6.35	5.24	4.62	4.23	3.95	3.74	3.58	3.45	3.34	3.18	3.01	2.82	2.63	2.52	2.42	2.30	2.18
40	8.83	6.07	4.98	4.37	3.99	3.71	3.51	3.35	3.22	3.12	2.95	2.78	2.60	2.40	2.30	2.18	2.06	1.93
60	8.49	5.79	4.73	4.14	3.76	3.49	3.29	3.13	3.01	2.90	2.74	2.57	2.39	2.19	2.08	1.96	1.83	1.69
120	8.18	5.54	4.50	3.92	3.55	3.28	3.09	2.93	2.81	2.71	2.54	2.37	2.19	1.98	1.87	1.75	1.61	1.43
∞	7.88	5.30	4.28	3.72	3.35	3.09	2.90	2.74	2.62	2.52	2.36	2.19	2.00	1.79	1.67	1.53	1.36	1.00

Table A11 Table of Critical Values for Mann–Whitney U Statistic

(Two–Tailed .05 Values)

n_1	1	2	3	4	5	6	7	8	9	10	11	12	13	14	15	16	17	18	19	20
1																				
2								0	0	0	0	1	1	1	1	1	2	2	2	2
3					0	1	1	2	2	3	3	4	4	5	5	6	6	7	7	8
4				0	1	2	3	4	4	5	6	7	8	9	10	11	11	12	13	13
5			0	1	2	3	5	6	7	8	9	11	12	13	14	15	17	18	19	20
6			1	2	3	5	6	8	10	11	13	14	16	17	19	21	22	24	25	27
7			1	3	5	6	8	10	12	14	16	18	20	22	24	26	28	30	32	34
8		0	2	4	6	8	10	13	15	17	19	22	24	26	29	31	34	36	38	41
9		0	2	4	7	10	12	15	17	20	23	26	28	31	34	37	39	42	45	48
10		0	3	5	8	11	14	17	20	23	26	29	33	36	39	42	45	48	52	55
11		0	3	6	9	13	16	19	23	26	30	33	37	40	44	47	51	55	58	62
12		1	4	7	11	14	18	22	26	29	33	37	41	45	49	53	57	61	65	69
13		1	4	8	12	16	20	24	28	33	37	41	45	50	54	59	63	67	72	76
14		1	5	9	13	17	22	26	31	36	40	45	50	55	59	64	67	74	78	83
15		1	5	10	14	19	24	29	34	39	44	49	54	59	64	70	75	80	85	90
16		1	6	11	15	21	26	31	37	42	47	53	59	64	70	75	81	86	92	98
17		2	6	11	17	22	28	34	39	45	51	57	63	67	75	81	87	93	99	105
18		2	7	12	18	24	30	36	42	48	55	61	67	74	80	86	93	99	106	112
19		2	7	13	19	25	32	38	45	52	58	65	72	78	85	92	99	106	113	119
20		2	8	13	20	27	34	41	48	55	62	69	76	83	90	98	105	112	119	127

(One–Tailed .05 Values)

n_2 \ n_1	1	2	3	4	5	6	7	8	9	10	11	12	13	14	15	16	17	18	19	20
1																			0	0
2					0	0	0	1	1	1	1	2	2	2	3	3	3	4	4	4
3			0	0	1	2	2	3	3	4	5	5	6	7	7	8	9	9	10	11
4			0	1	2	3	4	5	6	7	8	9	10	11	12	14	15	16	17	18
5		0	1	2	4	5	6	8	9	11	12	13	15	16	18	19	20	22	23	25
6		0	2	3	5	7	8	10	12	14	16	17	19	21	23	25	26	28	30	32
7		0	2	4	6	8	11	13	15	17	19	21	24	26	28	30	33	35	37	39
8		1	3	5	8	10	13	15	18	20	23	26	28	31	33	36	39	41	44	47
9		1	3	6	9	12	15	18	21	24	27	30	33	36	39	42	45	48	51	54
10		1	4	7	11	14	17	20	24	27	31	34	37	41	44	48	51	55	58	62
11		1	5	8	12	16	19	23	27	31	34	38	42	46	50	54	57	61	65	69
12		2	5	9	13	17	21	26	30	34	38	42	47	51	55	60	64	68	72	77
13		2	6	10	15	19	24	28	33	37	42	47	51	56	61	65	70	75	80	84
14		2	7	11	16	21	26	31	36	41	46	51	56	61	66	71	77	82	87	92
15		3	7	12	18	23	28	33	39	44	50	55	61	66	72	77	83	88	94	100
16		3	8	14	19	25	30	36	42	48	54	60	65	71	77	83	89	95	101	107
17		3	9	15	20	26	33	39	45	51	57	64	70	77	83	89	96	102	109	115
18		4	9	16	22	28	35	41	48	55	61	68	75	82	88	95	102	109	116	123
19	0	4	10	17	23	30	37	44	51	58	65	72	80	87	94	101	109	116	123	130
20	0	4	11	18	25	32	39	47	54	62	69	77	84	92	100	107	115	123	130	138

Table A11   Table of Critical Values for Mann–Whitney U Statistic (continued)

(Two–Tailed .01 Values)

n_2 ＼ n_1	1	2	3	4	5	6	7	8	9	10	11	12	13	14	15	16	17	18	19	20
1																				
2																			0	0
3									0	0	0	1	1	1	2	2	2	2	3	3
4						0	0	1	1	2	2	3	3	4	5	5	6	6	7	8
5					0	1	1	2	3	4	5	6	7	7	8	9	10	11	12	13
6				0	1	2	3	4	5	6	7	9	10	11	12	13	15	16	17	18
7				0	1	3	4	6	7	9	10	12	13	15	16	18	19	21	22	24
8				1	2	4	6	7	9	11	13	15	17	18	20	22	24	26	28	30
9			0	1	3	5	7	9	11	13	16	18	20	22	24	27	29	31	33	36
10			0	2	4	6	9	11	13	16	18	21	24	26	29	31	34	37	39	42
11			0	2	5	7	10	13	16	18	21	24	27	30	33	36	39	42	45	48
12			1	3	6	9	12	15	18	21	24	27	31	34	37	41	44	47	51	54
13			1	3	7	10	13	17	20	24	27	31	34	38	42	45	49	53	56	60
14			1	4	7	11	15	18	22	26	30	34	38	42	46	50	54	58	63	67
15			2	5	8	12	16	20	24	29	33	37	42	46	51	55	60	64	69	73
16			2	5	9	13	18	22	27	31	36	41	45	50	55	60	65	70	74	79
17			2	6	10	15	19	24	29	34	39	44	49	54	60	65	70	75	81	86
18			2	6	11	16	21	26	31	37	42	47	53	58	64	70	75	81	87	92
19		0	3	7	12	17	22	28	33	39	45	51	56	63	69	74	81	87	93	99
20		0	3	8	13	18	24	30	36	42	48	54	60	67	73	79	86	92	99	105

(One–Tailed .01 Values)

n_2 ＼ n_1	1	2	3	4	5	6	7	8	9	10	11	12	13	14	15	16	17	18	19	20
1																				
2													0	0	0	0	0	0	1	1
3							0	0	1	1	1	2	2	2	3	3	4	4	4	5
4					0	1	1	2	3	3	4	5	5	6	7	7	8	9	9	10
5				0	1	2	3	4	5	6	7	8	9	10	11	12	13	14	15	16
6				1	2	3	4	6	7	8	9	11	12	13	15	16	18	19	20	22
7			0	1	3	4	6	7	9	11	12	14	16	17	19	21	23	24	26	28
8			0	2	4	6	7	9	11	13	15	17	20	22	24	26	28	30	32	34
9			1	3	5	7	9	11	14	16	18	21	23	26	28	31	33	36	38	40
10			1	3	6	8	11	13	16	19	22	24	27	30	33	36	38	41	44	47
11			1	4	7	9	12	15	18	22	25	28	31	34	37	41	44	47	50	53
12			2	5	8	11	14	17	21	24	28	31	35	38	42	46	49	53	56	60
13		0	2	5	9	12	16	20	23	27	31	35	39	43	47	51	55	59	63	67
14		0	2	6	10	13	17	22	26	30	34	38	43	47	51	56	60	65	69	73
15		0	3	7	11	15	19	24	28	33	37	42	47	51	56	61	66	70	75	80
16		0	3	7	12	16	21	26	31	36	41	46	51	56	61	66	71	76	82	87
17		0	4	8	13	18	23	28	33	38	44	49	55	60	66	71	77	82	88	93
18		0	4	9	14	19	24	30	36	41	47	53	59	65	70	76	82	88	94	100
19		1	4	9	15	20	26	32	38	44	50	56	63	69	75	82	88	94	101	107
20		1	5	10	16	22	28	34	40	47	53	60	67	73	80	87	93	100	107	114

Table A12 Table of Sandler's *A* Statistic

	One–tailed level of significance				
	.05	.025	.01	.005	.0005
	Two–tailed level of significance				
	.10	.05	.02	.01	.001
df = *n*–1					
1	.5125	.5031	.50049	.50012	.5000012
2	.412	.369	.347	.340	.334
3	.385	.324	.286	.272	.254
4	.376	.304	.257	.238	.211
5	.372	.293	.240	.218	.184
6	.370	.286	.230	.205	.167
7	.369	.281	.222	.196	.155
8	.368	.278	.217	.190	.146
9	.368	.276	.213	.185	.139
10	.368	.274	.210	.181	.134
11	.368	.273	.207	.178	.130
12	.368	.271	.205	.176	.126
13	.368	.270	.204	.174	.124
14	.368	.270	.202	.172	.121
15	.368	.269	.201	.170	.119
16	.368	.268	.200	.169	.117
17	.368	.268	.199	.168	.116
18	.368	.267	.198	.167	.114
19	.368	.267	.197	.166	.113
20	.368	.266	.197	.165	.112
21	.368	.266	.196	.165	.111
22	.368	.266	.196	.164	.110
23	.368	.266	.195	.163	.109
24	.368	.265	.195	.163	.108
25	.368	.265	.194	.162	.108
26	.368	.265	.194	.162	.107
27	.368	.265	.193	.161	.107
28	.368	.265	.193	.161	.106
29	.368	.264	.193	.161	.106
30	.368	.264	.193	.160	.105
40	.368	.263	.191	.158	.102
60	.369	.262	.189	.155	.099
120	.369	.261	.187	.153	.095
∞	.370	.260	.185	.151	.092

Table A13 Table of the Studentized Range Statistic
$q_{.95}$ ($\alpha = .05$)

k	2	3	4	5	6	7	8	9	10
df_{error}									
1	17.97	26.98	32.82	37.08	40.41	43.12	45.40	47.36	49.07
2	6.08	8.33	9.80	10.88	11.74	12.44	13.03	13.54	13.99
3	4.50	5.91	6.82	7.50	8.04	8.48	8.85	9.18	9.46
4	3.93	5.04	5.76	6.29	6.71	7.05	7.35	7.60	7.83
5	3.64	4.60	5.22	5.67	6.03	6.33	6.58	6.80	6.99
6	3.46	4.34	4.90	5.30	5.63	5.90	6.12	6.32	6.49
7	3.34	4.16	4.68	5.06	5.36	5.61	5.82	6.00	6.16
8	3.26	4.04	4.53	4.89	5.17	5.40	5.60	5.77	5.92
9	3.20	3.95	4.41	4.76	5.02	5.24	5.43	5.59	5.74
10	3.15	3.88	4.33	4.65	4.91	5.12	5.30	5.46	5.60
11	3.11	3.82	4.26	4.57	4.82	5.03	5.20	5.35	5.49
12	3.08	3.77	4.20	4.51	4.75	4.95	5.12	5.27	5.39
13	3.06	3.73	4.15	4.45	4.69	4.88	5.05	5.19	5.32
14	3.03	3.70	4.11	4.41	4.64	4.83	4.99	5.13	5.25
15	3.01	3.67	4.08	4.37	4.59	4.78	4.94	5.08	5.20
16	3.00	3.65	4.05	4.33	4.56	4.74	4.90	5.03	5.15
17	2.98	3.63	4.02	4.30	4.52	4.70	4.86	4.99	5.11
18	2.97	3.61	4.00	4.28	4.49	4.67	4.82	4.96	5.07
19	2.96	3.59	3.98	4.25	4.47	4.65	4.79	4.92	5.04
20	2.95	3.58	3.96	4.23	4.45	4.62	4.77	4.90	5.01
24	2.92	3.53	3.90	4.17	4.37	4.54	4.68	4.81	4.92
30	2.89	3.49	3.85	4.10	4.30	4.46	4.60	4.72	4.82
40	2.86	3.44	3.79	4.04	4.23	4.39	4.52	4.63	4.73
60	2.83	3.40	3.74	3.98	4.16	4.31	4.44	4.55	4.65
120	2.80	3.36	3.68	3.92	4.10	4.24	4.36	4.47	4.56
∞	2.77	3.31	3.63	3.86	4.03	4.17	4.29	4.39	4.47

k	11	12	13	14	15	16	17	18	19	20
df_{error}										
1	50.59	51.96	53.20	54.33	55.36	56.32	57.22	58.04	58.83	59.56
2	14.39	14.75	15.08	15.38	15.65	15.91	16.14	16.37	16.57	16.77
3	9.72	9.95	10.15	10.35	10.52	10.69	10.84	10.98	11.11	11.24
4	8.03	8.21	8.37	8.52	8.66	8.79	8.91	9.03	9.13	9.23
5	7.17	7.32	7.47	7.60	7.72	7.83	7.93	8.03	8.12	8.21
6	6.65	6.79	6.92	7.03	7.14	7.24	7.34	7.43	7.51	7.59
7	6.30	6.43	6.55	6.66	6.76	6.85	6.94	7.02	7.10	7.17
8	6.05	6.18	6.29	6.39	6.48	6.57	6.65	6.73	6.80	6.87
9	5.87	5.98	6.09	6.19	6.28	6.36	6.44	6.51	6.58	6.64
10	5.72	5.83	5.93	6.03	6.11	6.19	6.27	6.34	6.40	6.47
11	5.61	5.71	5.81	5.90	5.98	6.06	6.13	6.20	6.27	6.33
12	5.51	5.61	5.71	5.80	5.88	5.95	6.02	6.09	6.15	6.21
13	5.43	5.53	5.63	5.71	5.79	5.86	5.93	5.99	6.05	6.11
14	5.36	5.46	5.55	5.64	5.71	5.79	5.85	5.91	5.97	6.03
15	5.31	5.40	5.49	5.57	5.65	5.72	5.78	5.85	5.90	5.96
16	5.26	5.35	5.44	5.52	5.59	5.66	5.73	5.79	5.84	5.90
17	5.21	5.31	5.39	5.47	5.54	5.61	5.67	5.73	5.79	5.84
18	5.17	5.27	5.35	5.43	5.50	5.57	5.63	5.69	5.74	5.79
19	5.14	5.23	5.31	5.39	5.46	5.53	5.59	5.65	5.70	5.75
20	5.11	5.20	5.28	5.36	5.43	5.49	5.55	5.61	5.66	5.71
24	5.01	5.10	5.18	5.25	5.32	5.38	5.44	5.49	5.55	5.59
30	4.92	5.00	5.08	5.15	5.21	5.27	5.33	5.38	5.43	5.47
40	4.82	4.90	4.98	5.04	5.11	5.16	5.22	5.27	5.31	5.36
60	4.73	4.81	4.88	4.94	5.00	5.06	5.11	5.15	5.20	5.24
120	4.64	4.71	4.78	4.84	4.90	4.95	5.00	5.04	5.09	5.13
∞	4.55	4.62	4.68	4.74	4.80	4.85	4.89	4.93	4.97	5.01

Table A13 Table of the Studentized Range Statistic (continued)
$q_{.99}$ ($\alpha = .01$)

k / df_{error}	2	3	4	5	6	7	8	9	10
1	90.03	135.0	164.3	185.6	202.2	215.8	227.2	237.0	245.6
2	14.04	19.02	22.29	24.72	26.63	28.20	29.53	30.68	31.69
3	8.26	10.62	12.17	13.33	14.24	15.00	15.64	16.20	16.69
4	6.51	8.12	9.17	9.96	10.58	11.10	11.55	11.93	12.27
5	5.70	6.98	7.80	8.42	8.91	9.32	9.67	9.97	10.24
6	5.24	6.33	7.03	7.56	7.97	8.32	8.61	8.87	9.10
7	4.95	5.92	6.54	7.01	7.37	7.68	7.94	8.17	8.37
8	4.75	5.64	6.20	6.62	6.96	7.24	7.47	7.68	7.86
9	4.60	5.43	5.96	6.35	6.66	6.91	7.13	7.33	7.49
10	4.48	5.27	5.77	6.14	6.43	6.67	6.87	7.05	7.21
11	4.39	5.15	5.62	5.97	6.25	6.48	6.67	6.84	6.99
12	4.32	5.05	5.50	5.84	6.10	6.32	6.51	6.67	6.81
13	4.26	4.96	5.40	5.73	5.98	6.19	6.37	6.53	6.67
14	4.21	4.89	5.32	5.63	5.88	6.08	6.26	6.41	6.54
15	4.17	4.84	5.25	5.56	5.80	5.99	6.16	6.31	6.44
16	4.13	4.79	5.19	5.49	5.72	5.92	6.08	6.22	6.35
17	4.10	4.74	5.14	5.43	5.66	5.85	6.01	6.15	6.27
18	4.07	4.70	5.09	5.38	5.60	5.79	5.94	6.08	6.20
19	4.05	4.67	5.05	5.33	5.55	5.73	5.89	6.02	6.14
20	4.02	4.64	5.02	5.29	5.51	5.69	5.84	5.97	6.09
24	3.96	4.55	4.91	5.17	5.37	5.54	5.69	5.81	5.92
30	3.89	4.45	4.80	5.05	5.24	5.40	5.54	5.65	5.76
40	3.82	4.37	4.70	4.93	5.11	5.26	5.39	5.50	5.60
60	3.76	4.28	4.59	4.82	4.99	5.13	5.25	5.36	5.45
120	3.70	4.20	4.50	4.71	4.87	5.01	5.12	5.21	5.30
∞	3.64	4.12	4.40	4.60	4.76	4.88	4.99	5.08	5.16

k / df_{error}	11	12	13	14	15	16	17	18	19	20
1	253.2	260.0	266.2	271.8	277.0	281.8	286.3	290.4	294.3	298.0
2	32.59	33.40	34.13	34.81	35.43	36.00	36.53	37.03	37.50	37.95
3	17.13	17.53	17.89	18.22	18.52	18.81	19.07	19.32	19.55	19.77
4	12.57	12.84	13.09	13.32	13.53	13.73	13.91	14.08	14.24	14.40
5	10.48	10.70	10.89	11.08	11.24	11.40	11.55	11.68	11.81	11.93
6	9.30	9.48	9.65	9.81	9.95	10.08	10.21	10.32	10.43	10.54
7	8.55	8.71	8.86	9.00	9.12	9.24	9.35	9.46	9.55	9.65
8	8.03	8.18	8.31	8.44	8.55	8.66	8.76	8.85	8.94	9.03
9	7.65	7.78	7.91	8.03	8.13	8.23	8.33	8.41	8.49	8.57
10	7.36	7.49	7.60	7.71	7.81	7.91	7.99	8.08	8.15	8.23
11	7.13	7.25	7.36	7.46	7.56	7.65	7.73	7.81	7.88	7.95
12	6.94	7.06	7.17	7.26	7.36	7.44	7.52	7.59	7.66	7.73
13	6.79	6.90	7.01	7.10	7.19	7.27	7.35	7.42	7.48	7.55
14	6.66	6.77	6.87	6.96	7.05	7.13	7.20	7.27	7.33	7.39
15	6.55	6.66	6.76	6.84	6.93	7.00	7.07	7.14	7.20	7.26
16	6.46	6.56	6.66	6.74	6.82	6.90	6.97	7.03	7.09	7.15
17	6.38	6.48	6.57	6.66	6.73	6.81	6.87	6.94	7.00	7.05
18	6.31	6.41	6.50	6.58	6.65	6.73	6.79	6.85	6.91	6.97
19	6.25	6.34	6.43	6.51	6.58	6.65	6.72	6.78	6.84	6.89
20	6.19	6.28	6.37	6.45	6.52	6.59	6.65	6.71	6.77	6.82
24	6.02	6.11	6.19	6.26	6.33	6.39	6.45	6.51	6.56	6.61
30	5.85	5.93	6.01	6.08	6.14	6.20	6.26	6.31	6.36	6.41
40	5.69	5.76	5.83	5.90	5.96	6.02	6.07	6.12	6.16	6.21
60	5.53	5.60	5.67	5.73	5.78	5.84	5.89	5.93	5.97	6.01
120	5.37	5.44	5.50	5.56	5.61	5.66	5.71	5.75	5.79	5.83
∞	5.23	5.29	5.35	5.40	5.45	5.49	5.54	5.57	5.61	5.65

Table A14   Table of Dunnett's Modified *t* Statistic for a Control Group Comparison
Two–Tailed Values

The .05 critical values are in lightface type, and the .01 critical values are in **bold** type.

df_{error}	**2**	**3**	**4**	**5**	**6**	**7**	**8**	**9**	**10**
	k = number of treatment means, including control								
5	2.57	3.03	3.29	3.48	3.62	3.73	3.82	3.90	3.97
	4.03	**4.63**	**4.98**	**5.22**	**5.41**	**5.56**	**5.69**	**5.80**	**5.89**
6	2.45	2.86	3.10	3.26	3.39	3.49	3.57	3.64	3.71
	3.71	**4.21**	**4.51**	**4.71**	**4.87**	**5.00**	**5.10**	**5.20**	**5.28**
7	2.36	2.75	2.97	3.12	3.24	3.33	3.41	3.47	3.53
	3.50	**3.95**	**4.21**	**4.39**	**4.53**	**4.64**	**4.74**	**4.82**	**4.89**
8	2.31	2.67	2.88	3.02	3.13	3.22	3.29	3.35	3.41
	3.36	**3.77**	**4.00**	**4.17**	**4.29**	**4.40**	**4.48**	**4.56**	**4.62**
9	2.26	2.61	2.81	2.95	3.05	3.14	3.20	3.26	3.32
	3.25	**3.63**	**3.85**	**4.01**	**4.12**	**4.22**	**4.30**	**4.37**	**4.43**
10	2.23	2.57	2.76	2.89	2.99	3.07	3.14	3.19	3.24
	3.17	**3.53**	**3.74**	**3.88**	**3.99**	**4.08**	**4.16**	**4.22**	**4.28**
11	2.20	2.53	2.72	2.84	2.94	3.02	3.08	3.14	3.19
	3.11	**3.45**	**3.65**	**3.79**	**3.89**	**3.98**	**4.05**	**4.11**	**4.16**
12	2.18	2.50	2.68	2.81	2.90	2.98	3.04	3.09	3.14
	3.05	**3.39**	**3.58**	**3.71**	**3.81**	**3.89**	**3.96**	**4.02**	**4.07**
13	2.16	2.48	2.65	2.78	2.87	2.94	3.00	3.06	3.10
	3.01	**3.33**	**3.52**	**3.65**	**3.74**	**3.82**	**3.89**	**3.94**	**3.99**
14	2.14	2.46	2.63	2.75	2.84	2.91	2.97	3.02	3.07
	2.98	**3.29**	**3.47**	**3.59**	**3.69**	**3.76**	**3.83**	**3.88**	**3.93**
15	2.13	2.44	2.61	2.73	2.82	2.89	2.95	3.00	3.04
	2.95	**3.25**	**3.43**	**3.55**	**3.64**	**3.71**	**3.78**	**3.83**	**3.88**
16	2.12	2.42	2.59	2.71	2.80	2.87	2.92	2.97	3.02
	2.92	**3.22**	**3.39**	**3.51**	**3.60**	**3.67**	**3.73**	**3.78**	**3.83**
17	2.11	2.41	2.58	2.69	2.78	2.85	2.90	2.95	3.00
	2.90	**3.19**	**3.36**	**3.47**	**3.56**	**3.63**	**3.69**	**3.74**	**3.79**
18	2.10	2.40	2.56	2.68	2.76	2.83	2.89	2.94	2.98
	2.88	**3.17**	**3.33**	**3.44**	**3.53**	**3.60**	**3.66**	**3.71**	**3.75**
19	2.09	2.39	2.55	2.66	2.75	2.81	2.87	2.92	2.96
	2.86	**3.15**	**3.31**	**3.42**	**3.50**	**3.57**	**3.63**	**3.68**	**3.72**
20	2.09	2.38	2.54	2.65	2.73	2.80	2.86	2.90	2.95
	2.85	**3.13**	**3.29**	**3.40**	**3.48**	**3.55**	**3.60**	**3.65**	**3.69**
24	2.06	2.35	2.51	2.61	2.70	2.76	2.81	2.86	2.90
	2.80	**3.07**	**3.22**	**3.32**	**3.40**	**3.47**	**3.52**	**3.57**	**3.61**
30	2.04	2.32	2.47	2.58	2.66	2.72	2.77	2.82	2.86
	2.75	**3.01**	**3.15**	**3.25**	**3.33**	**3.39**	**3.44**	**3.49**	**3.52**
40	2.02	2.29	2.44	2.54	2.62	2.68	2.73	2.77	2.81
	2.70	**2.95**	**3.09**	**3.19**	**3.26**	**3.32**	**3.37**	**3.41**	**3.44**
60	2.00	2.27	2.41	2.51	2.58	2.64	2.69	2.73	2.77
	2.66	**2.90**	**3.03**	**3.12**	**3.19**	**3.25**	**3.29**	**3.33**	**3.37**
120	1.98	2.24	2.38	2.47	2.55	2.60	2.65	2.69	2.73
	2.62	**2.85**	**2.97**	**3.06**	**3.12**	**3.18**	**3.22**	**3.26**	**3.29**
∞	1.96	2.21	2.35	2.44	2.51	2.57	2.61	2.65	2.69
	2.58	**2.79**	**2.92**	**3.00**	**3.06**	**3.11**	**3.15**	**3.19**	**3.22**

Table A14 Table of Dunnett's Modified *t* Statistic for a Control Group Comparison
(continued)

One–Tailed Values

df_{error}	\multicolumn{9}{c}{k = number of treatment means, including control}
	2	3	4	5	6	7	8	9	10
5	2.02	2.44	2.68	2.85	2.98	3.08	3.16	3.24	3.30
	3.37	3.90	4.21	4.43	4.60	4.73	4.85	4.94	5.03
6	1.94	2.34	2.56	2.71	2.83	2.92	3.00	3.07	3.12
	3.14	3.61	3.88	4.07	4.21	4.33	4.43	4.51	4.59
7	1.89	2.27	2.48	2.62	2.73	2.82	2.89	2.95	3.01
	3.00	3.42	3.66	3.83	3.96	4.07	4.15	4.23	4.30
8	1.86	2.22	2.42	2.55	2.66	2.74	2.81	2.87	2.92
	2.90	3.29	3.51	3.67	3.79	3.88	3.96	4.03	4.09
9	1.83	2.18	2.37	2.50	2.20	2.68	2.75	2.81	2.86
	2.82	3.19	3.40	3.55	3.66	3.75	3.82	3.89	3.94
10	1.81	2.15	2.34	2.47	2.56	2.64	2.70	2.76	2.81
	2.76	3.11	3.31	3.45	3.56	3.64	3.71	3.78	3.83
11	1.80	2.13	2.31	2.44	2.53	2.60	2.67	2.72	2.77
	2.72	3.06	3.25	3.38	3.48	3.56	3.63	3.69	3.74
12	1.78	2.11	2.29	2.41	2.50	2.58	2.64	2.69	2.74
	2.68	3.01	3.19	3.32	3.42	3.50	3.56	3.62	3.67
13	1.77	2.09	2.27	2.39	2.48	2.55	2.61	2.66	2.71
	2.65	2.97	3.15	3.27	3.37	3.44	3.51	3.56	3.61
14	1.76	2.08	2.25	2.37	2.46	2.53	2.59	2.64	2.69
	2.62	2.94	3.11	3.23	3.32	3.40	3.46	3.51	3.56
15	1.75	2.07	2.24	2.36	2.44	2.51	2.57	2.62	2.67
	2.60	2.91	3.08	3.20	3.29	3.36	3.42	3.47	3.52
16	1.75	2.06	2.23	2.34	2.43	2.50	2.56	2.61	2.65
	2.58	2.88	3.05	3.17	3.26	3.33	3.39	3.44	3.48
17	1.74	2.05	2.22	2.33	2.42	2.49	2.54	2.59	2.64
	2.57	2.86	3.03	3.14	3.23	3.30	3.36	3.41	3.45
18	1.73	2.04	2.21	2.32	2.41	2.48	2.53	2.58	2.62
	2.55	2.84	3.01	3.12	3.21	3.27	3.33	3.38	3.42
19	1.73	2.03	2.20	2.31	2.40	2.47	2.52	2.57	2.61
	2.54	2.83	2.99	3.10	3.18	3.25	3.31	3.36	3.40
20	1.72	2.03	2.19	2.30	2.39	2.46	2.51	2.56	2.60
	2.53	2.81	2.97	3.08	3.17	3.23	3.29	3.34	3.38
24	1.71	2.01	2.17	2.28	2.36	2.43	2.48	2.53	2.57
	2.49	2.77	2.92	3.03	3.11	3.17	3.22	3.27	3.31
30	1.70	1.99	2.15	2.25	2.33	2.40	2.45	2.50	2.54
	2.46	2.72	2.87	2.97	3.05	3.11	3.16	3.21	3.24
40	1.68	1.97	2.13	2.23	2.31	2.37	2.42	2.47	2.51
	2.42	2.68	2.82	2.92	2.99	3.05	3.10	3.14	3.18
60	1.67	1.95	2.10	2.21	2.28	2.35	2.39	2.44	2.48
	2.39	2.64	2.78	2.87	2.94	3.00	3.04	3.08	3.12
120	1.66	1.93	2.08	2.18	2.26	2.32	2.37	2.41	2.45
	2.36	2.60	2.73	2.82	2.89	2.94	2.99	3.03	3.06
∞	1.64	1.92	2.06	2.16	2.23	2.29	2.34	2.38	2.42
	2.33	2.56	2.68	2.77	2.84	2.89	2.93	2.97	3.00

Table A15   Graphs of the Power Function for the Analysis of Variance
(Fixed–Effects Model)

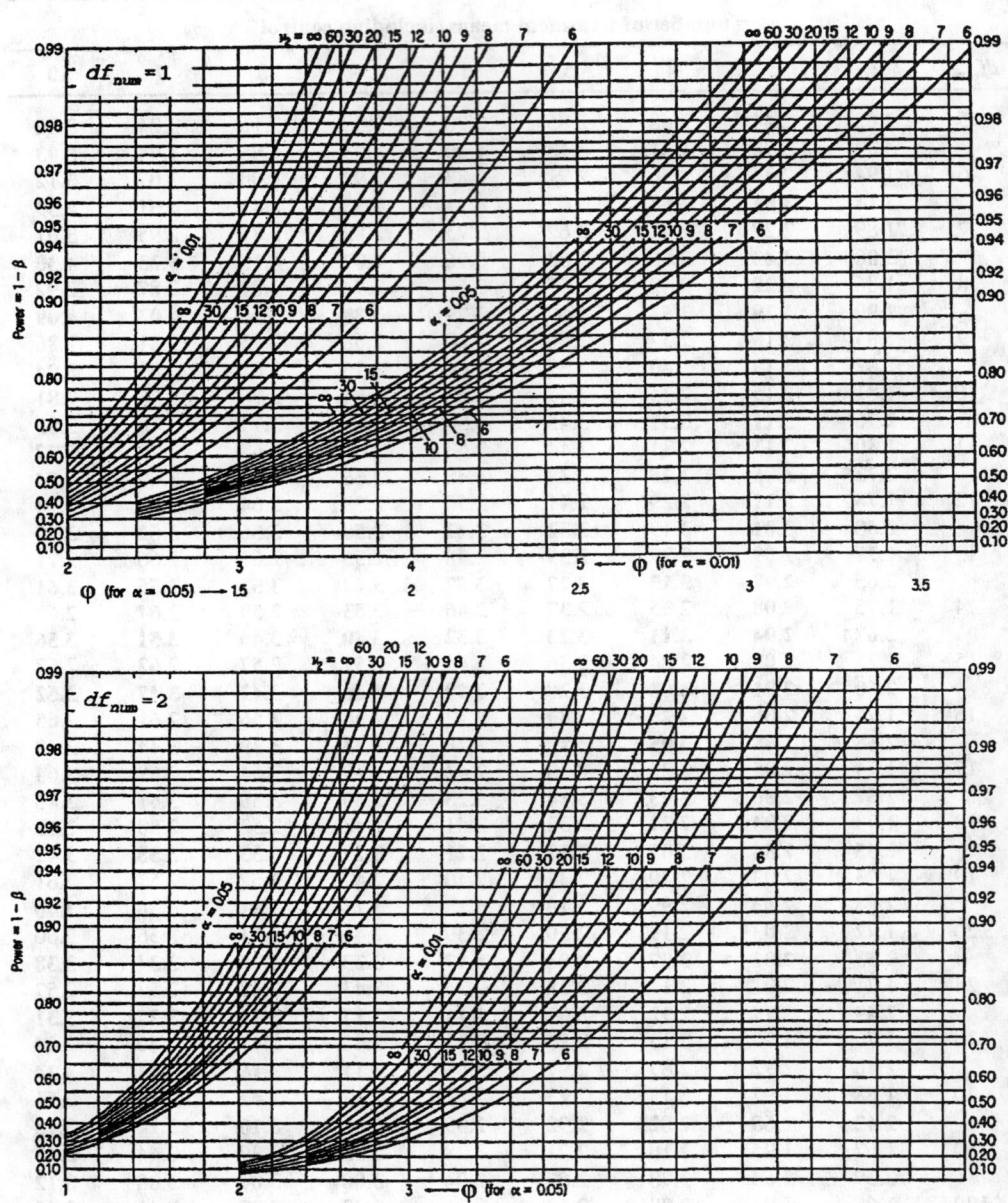

Table A15   Graphs of the Power Function for the Analysis of Variance (continued)

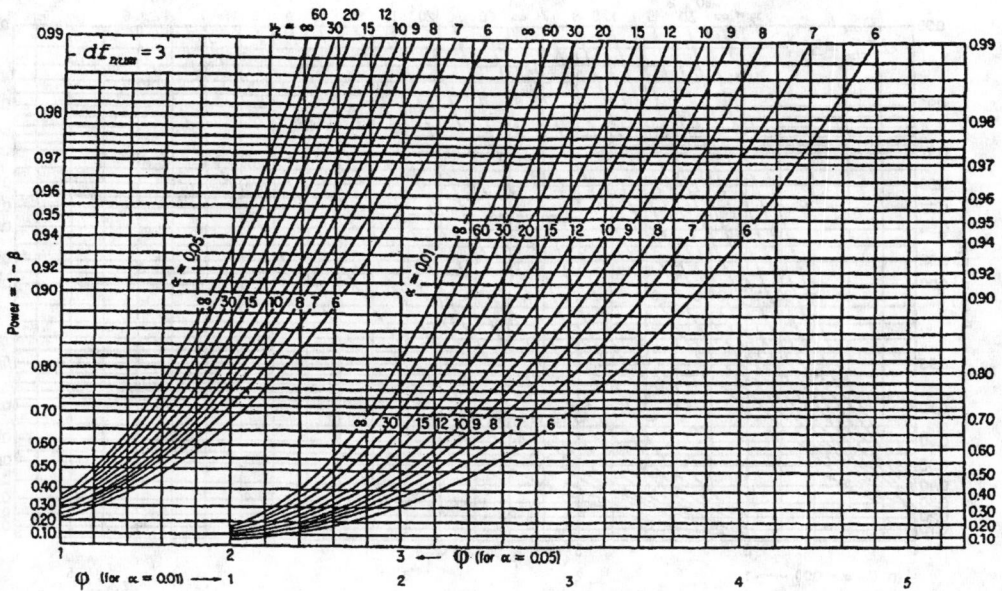

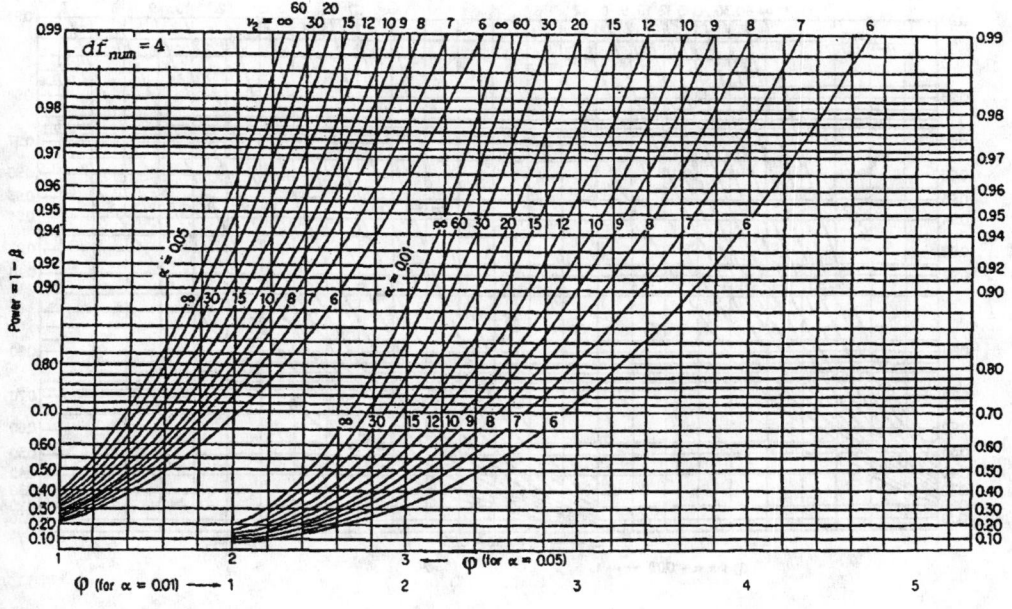

Table A15 Graphs of the Power Function for the Analysis of Variance (continued)

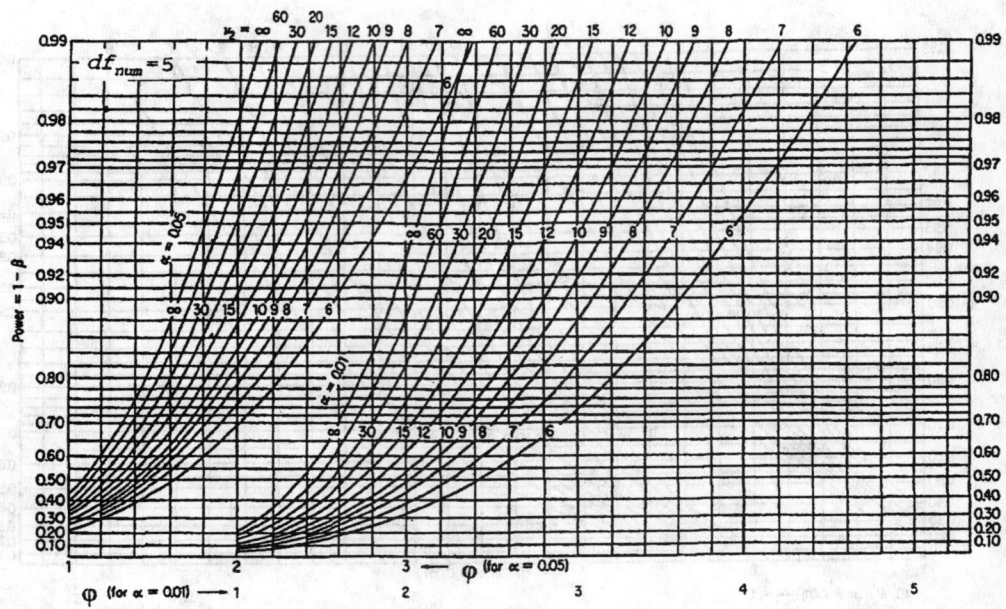

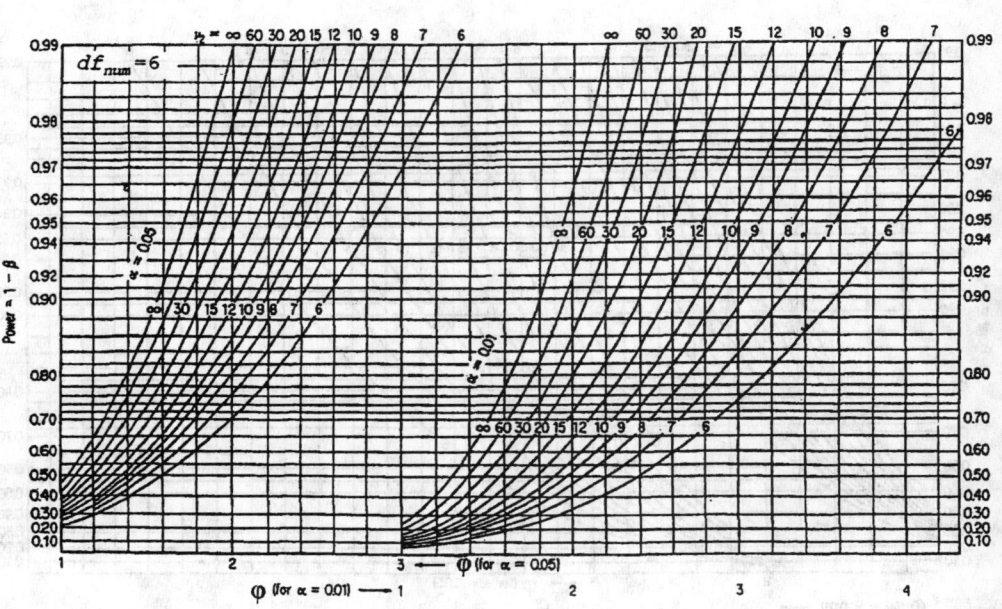

Table A15   Graphs of the Power Function for the Analysis of Variance (continued)

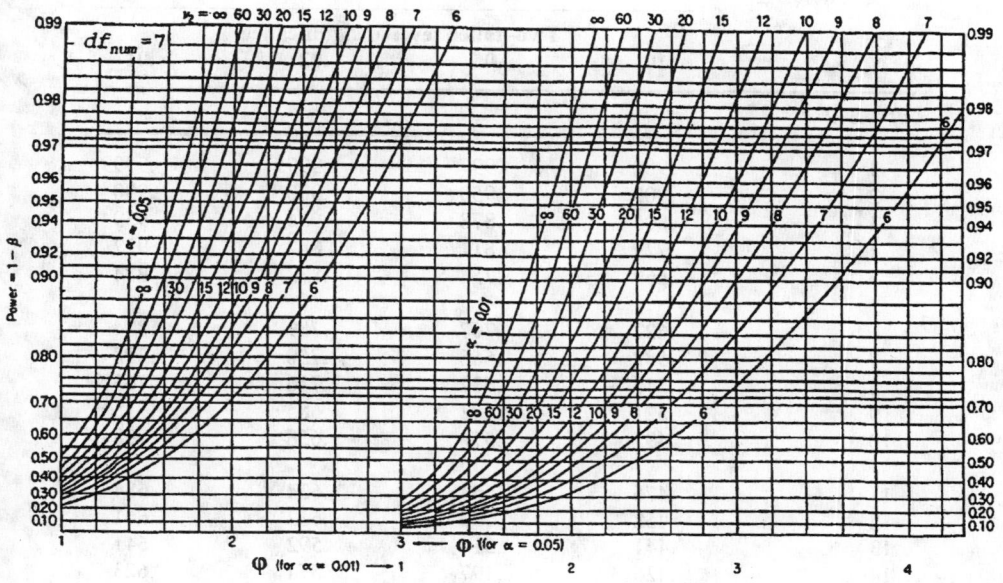

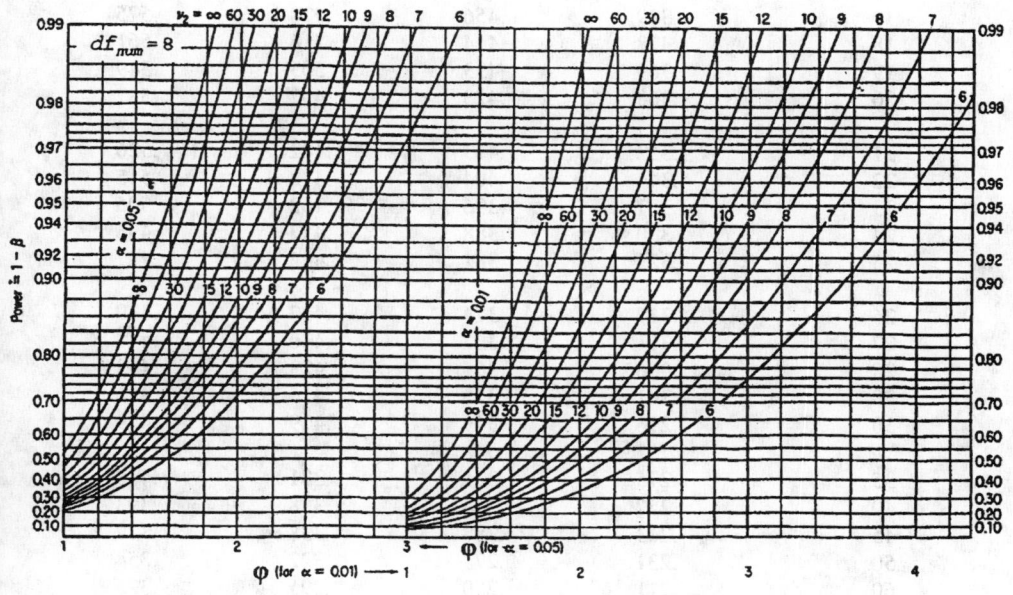

Table A16 Table of Critical Values for Pearson *r*

	One–tailed level of significance			
	.05	.025	.01	.005
	Two–tailed level of significance			
	.10	.05	.02	.01
df = n–2				
1	.988	.997	.9995	.9999
2	.900	.950	.980	.990
3	.805	.878	.934	.959
4	.729	.811	.882	.917
5	.669	.754	.833	.874
6	.622	.707	.789	.834
7	.582	.666	.750	.798
8	.549	.632	.716	.765
9	.521	.602	.685	.735
10	.497	.576	.658	.708
11	.476	.553	.634	.684
12	.458	.532	.612	.661
13	.441	.514	.592	.641
14	.426	.497	.574	.623
15	.412	.482	.558	.606
16	.400	.468	.542	.590
17	.389	.456	.528	.575
18	.378	.444	.516	.561
19	.369	.433	.503	.549
20	.360	.423	.492	.537
21	.352	.413	.482	.526
22	.344	.404	.472	.515
23	.337	.396	.462	.505
24	.330	.388	.453	.496
25	.323	.381	.445	.487
26	.317	.374	.437	.479
27	.311	.367	.430	.471
28	.306	.361	.423	.463
29	.301	.355	.416	.456
30	.296	.349	.409	.449
35	.275	.325	.381	.418
40	.257	.304	.358	.393
45	.243	.288	.338	.372
50	.231	.273	.322	.354
60	.211	.250	.295	.325
70	.195	.232	.274	.302
80	.183	.217	.256	.283
90	.173	.205	.242	.267
100	.164	.195	.230	.254

Table A17 Table of Fisher's z_r Transformation

r	z_r	r	z_r	r	z_r	r	z_r	r	z_r
.000	.000	.200	.203	.400	.424	.600	.693	.800	1.099
.005	.005	.205	.208	.405	.430	.605	.701	.805	1.113
.010	.010	.210	.213	.410	.436	.610	.709	.810	1.127
.015	.015	.215	.218	.415	.442	.615	.717	.815	1.142
.020	.020	.220	.224	.420	.448	.620	.725	.820	1.157
.025	.025	.225	.229	.425	.454	.625	.733	.825	1.172
.030	.030	.230	.234	.430	.460	.630	.741	.830	1.188
.035	.035	.235	.239	.435	.466	.635	.750	.835	1.204
.040	.040	.240	.245	.440	.472	.640	.758	.840	1.221
.045	.045	.245	.250	.445	.478	.645	.767	.845	1.238
.050	.050	.250	.255	.450	.485	.650	.775	.850	1.256
.055	.055	.255	.261	.455	.491	.655	.784	.855	1.274
.060	.060	.260	.266	.460	.497	.660	.793	.860	1.293
.065	.065	.265	.271	.465	.504	.665	.802	.865	1.313
.070	.070	.270	.277	.470	.510	.670	.811	.870	1.333
.075	.075	.275	.282	.475	.517	.675	.820	.875	1.354
.080	.080	.280	.288	.480	.523	.680	.829	.880	1.376
.085	.085	.285	.293	.485	.530	.685	.838	.885	1.398
.090	.090	.290	.299	.490	.536	.690	.848	.890	1.422
.095	.095	.295	.304	.495	.543	.695	.858	.895	1.447
.100	.100	.300	.310	.500	.549	.700	.867	.900	1.472
.105	.105	.305	.315	.505	.556	.705	.877	.905	1.499
.110	.110	.310	.321	.510	.563	.710	.887	.910	1.528
.115	.116	.315	.326	.515	.570	.715	.897	.915	1.557
.120	.121	.320	.332	.520	.576	.720	.908	.920	1.589
.125	.126	.325	.337	.525	.583	.725	.918	.925	1.623
.130	.131	.330	.343	.530	.590	.730	.929	.930	1.658
.135	.136	.335	.348	.535	.597	.735	.940	.935	1.697
.140	.141	.340	.354	.540	.604	.740	.950	.940	1.738
.145	.146	.345	.360	.545	.611	.745	.962	.945	1.783
.150	.151	.350	.365	.550	.618	.750	.973	.950	1.832
.155	.156	.355	.371	.555	.626	.755	.984	.955	1.886
.160	.161	.360	.377	.560	.633	.760	.996	.960	1.946
.165	.167	.365	.383	.565	.640	.765	1.008	.965	2.014
.170	.172	.370	.388	.570	.648	.770	1.020	.970	2.092
.175	.177	.375	.394	.575	.655	.775	1.033	.975	2.185
.180	.182	.380	.400	.580	.662	.780	1.045	.980	2.298
.185	.187	.385	.406	.585	.670	.785	1.058	.985	2.443
.190	.192	.390	.412	.590	.678	.790	1.071	.990	2.647
.195	.198	.395	.418	.595	.685	.795	1.085	.995	2.994

Table A18   Table of Critical Values for Spearman's Rho

	One–tailed level of significance			
	.05	.025	.01	.005
	Two–tailed level of significance			
n	.10	.05	.02	.01
4	1.000	–	–	–
5	.900	1.000	1.000	–
6	.829	.886	.943	1.000
7	.714	.786	.893	.929
8	.643	.738	.833	.881
9	.600	.700	.783	.833
10	.564	.648	.745	.794
11	.536	.618	.709	.755
12	.503	.587	.671	.727
13	.484	.560	.648	.703
14	.464	.538	.622	.675
15	.443	.521	.604	.654
16	.429	.503	.582	.635
17	.414	.485	.566	.615
18	.401	.472	.550	.600
19	.391	.460	.535	.584
20	.380	.447	.520	.570
21	.370	.435	.508	.556
22	.361	.425	.496	.544
23	.353	.415	.486	.532
24	.344	.406	.476	.521
25	.337	.398	.466	.511
26	.331	.390	.457	.501
27	.324	.382	.448	.491
28	.317	.375	.440	.483
29	.312	.368	.433	.475
30	.306	.362	.425	.467
35	.283	.335	.394	.433
40	.264	.313	.368	.405
45	.248	.294	.347	.382
50	.235	.279	.329	.363
60	.214	.255	.300	.331
70	.190	.235	.278	.307
80	.185	.220	.260	.287
90	.174	.207	.245	.271
100	.165	.197	.233	.257

Table A19 Table of Critical Values for Kendall's Tau

Critical values for both τ and S are listed in the table.

| Two–tailed | .01 | | .02 | | .05 | | .10 | | .20 | |
| One tailed | .005 | | .01 | | .025 | | .05 | | .10 | |
n	S	τ	S	τ	S	τ	S	τ	S	τ
4	8	1.000	8	1.000	8	1.000	6	1.000	6	1.000
5	12	1.000	10	1.000	10	1.000	8	.800	8	.800
6	15	1.000	13	.867	13	.867	11	.733	9	.600
7	19	.905	17	.810	15	.714	13	.619	11	.524
8	22	.786	20	.714	18	.643	16	.571	12	.429
9	26	.722	24	.667	20	.556	18	.500	14	.389
10	29	.644	27	.600	23	.511	21	.467	17	.378
11	33	.600	31	.564	27	.491	23	.418	19	.345
12	38	.576	36	.545	30	.455	26	.394	20	.303
13	44	.564	40	.513	34	.436	28	.359	24	.308
14	47	.516	43	.473	37	.407	33	.363	25	.275
15	53	.505	49	.467	41	.390	35	.333	29	.276
16	58	.483	52	.433	46	.383	38	.317	30	.250
17	64	.471	58	.426	50	.368	42	.309	34	.250
18	69	.451	63	.412	53	.346	45	.294	37	.242
19	75	.439	67	.392	57	.333	49	.287	39	.228
20	80	.421	72	.379	62	.326	52	.274	42	.221
21	86	.410	78	.371	66	.314	56	.267	44	.210
22	91	.394	83	.359	71	.307	61	.264	47	.203
23	99	.391	89	.352	75	.296	65	.257	51	.202
24	104	.377	94	.341	80	.290	68	.246	54	.196
25	110	.367	100	.333	86	.287	72	.240	58	.193
26	117	.360	107	.329	91	.280	77	.237	61	.188
27	125	.356	113	.322	95	.271	81	.231	63	.179
28	130	.344	118	.312	100	.265	86	.228	68	.180
29	138	.340	126	.310	106	.261	90	.222	70	.172
30	145	.333	131	.301	111	.255	95	.218	75	.172
31	151	.325	137	.295	117	.252	99	.213	77	.166
32	160	.323	144	.290	122	.246	104	.210	82	.165
33	166	.314	152	.288	128	.242	108	.205	86	.163
34	175	.312	157	.280	133	.237	113	.201	89	.159
35	181	.304	165	.277	139	.234	117	.197	93	.156
36	190	.302	172	.273	146	.232	122	.194	96	.152
37	198	.297	178	.267	152	.228	128	.192	100	.150
38	205	.292	185	.263	157	.223	133	.189	105	.149
39	213	.287	193	.260	163	.220	139	.188	109	.147
40	222	.285	200	.256	170	.218	144	.185	112	.144

Table A20   Table of Critical Values for Kendall's Coefficient of Concordance

Critical values for both S and $\tilde{W}$ are listed. The values of $\tilde{W}$ in the table were computed by substituting the tabled S values in Equation 31.3.

					n					
m	3		4		5		6		7	
	S	$\tilde{W}$	S	$\tilde{W}$	S	$\tilde{W}$	S	$\tilde{W}$	S	$\tilde{W}$
							Values at .05 level of significance			
3					64.4	.716	103.9	.660	157.3	.624
4			49.5	.619	88.4	.552	143.3	.512	217.0	.484
5			62.6	.501	112.3	.449	182.4	.417	276.2	.395
6			75.7	.421	136.1	.378	221.4	.351	335.2	.333
8	48.1	.376	101.7	.318	183.7	.287	299.0	.267	453.1	.253
10	60.0	.300	127.8	.256	231.2	.231	376.7	.215	571.0	.204
15	89.8	.200	192.9	.171	349.8	.155	570.5	.145	864.9	.137
20	119.7	.150	258.0	.129	468.5	.117	764.4	.109	1158.7	.103
							Values at .01 level of significance			
3					75.6	.840	122.8	.780	185.6	.737
4			61.4	.768	109.3	.683	176.2	.629	265.0	.592
5			80.5	.644	142.8	.571	229.4	.524	343.8	.491
6			99.5	.553	176.1	.489	282.4	.448	422.6	.419
8	66.8	.522	137.4	.429	242.7	.379	388.3	.347	579.9	.324
10	85.1	.425	175.3	.351	309.1	.309	494.0	.282	737.0	.263
15	131.0	.291	269.8	.240	475.2	.211	758.2	.193	1129.5	.179
20	177.0	.221	364.2	.182	641.2	.160	1022.2	.146	1521.9	.136

Additional values for $n = 3$

	At .05 level		At .01 level	
m	S	$\tilde{W}$	S	$\tilde{W}$
9	54.0	.333	75.9	.469
12	71.9	.250	103.5	.359
14	83.8	.214	121.9	.311
16	95.8	.187	140.2	.274
18	107.7	.166	158.6	.245

Table A21 **Table of Critical Values for the Kolmogorov–Smirnov Goodness-of-Fit Test for a Single Sample**

One-tailed	.10	.05	.025	.01	.005
Two-tailed	.20	.10	.050	.02	.010
$n = 1$	.900	.950	.975	.990	.995
2	.684	.776	.842	.900	.929
3	.565	.636	.708	.785	.829
4	.493	.565	.624	.689	.734
5	.447	.509	.563	.627	.669
6	.410	.468	.519	.577	.617
7	.381	.436	.483	.538	.576
8	.358	.410	.454	.507	.542
9	.339	.387	.430	.480	.513
10	.323	.369	.409	.457	.489
11	.308	.352	.391	.437	.468
12	.296	.338	.375	.419	.449
13	.285	.325	.361	.404	.432
14	.275	.314	.349	.390	.418
15	.266	.304	.338	.377	.404
16	.258	.295	.327	.366	.392
17	.250	.286	.318	.355	.381
18	.244	.279	.309	.346	.371
19	.237	.271	.301	.337	.361
20	.232	.265	.294	.329	.352
21	.226	.259	.287	.321	.344
22	.221	.253	.281	.314	.337
23	.216	.247	.275	.307	.330
24	.212	.242	.269	.301	.323
25	.208	.238	.264	.295	.317
26	.204	.233	.259	.290	.311
27	.200	.229	.254	.284	.305
28	.197	.225	.250	.279	.300
29	.193	.221	.246	.275	.295
30	.190	.218	.242	.270	.290
31	.187	.214	.238	.266	.285
32	.184	.211	.234	.262	.281
33	.182	.208	.231	.258	.277
34	.179	.205	.227	.254	.273
35	.177	.202	.224	.251	.269
36	.174	.199	.221	.247	.265
37	.172	.196	.218	.244	.262
38	.170	.194	.215	.241	.258
39	.168	.191	.213	.238	.255
40	.165	.189	.210	.235	.252
$n > 40$	$1.07/\sqrt{n}$	$1.22/\sqrt{n}$	$1.36/\sqrt{n}$	$1.52/\sqrt{n}$	$1.63/\sqrt{n}$

Table A22   Table of Critical Values for the Lilliefors Test for Normality

One-tailed Two-tailed	.20 .40	.15 .30	.10 .20	.05 .10	.01 .02
n = 4	.300	.319	.352	.381	.417
5	.285	.299	.315	.337	.405
6	.265	.277	.294	.319	.364
7	.247	.258	.276	.300	.348
8	.233	.244	.261	.285	.331
9	.223	.233	.249	.271	.311
10	.215	.224	.239	.258	.294
11	.206	.217	.230	.249	.284
12	.199	.212	.223	.242	.275
13	.190	.202	.214	.234	.268
14	.183	.194	.207	.227	.261
15	.177	.187	.201	.220	.257
16	.173	.182	.195	.213	.250
17	.169	.177	.189	.206	.245
18	.166	.173	.184	.200	.239
19	.163	.169	.179	.195	.235
20	.160	.166	.174	.190	.231
25	.142	.147	.158	.173	.200
30	.131	.136	.144	.161	.187
n > 30	$.736/\sqrt{n}$	$.768/\sqrt{n}$	$.805/\sqrt{n}$	$.886/\sqrt{n}$	$1.031/\sqrt{n}$

Table A23 Table of Critical Values for the Kolmogorov–Smirnov Test for Two Independent Samples

One-tailed		.10	.05	.025	.01	.005
Two-tailed		.20	.10	.05	.02	.01
n_1	n_2					
3	3	.667	.667			
3	4	.750	.750			
3	5	.667	.800	.800		
3	6	.667	.667	.833		
3	7	.667	.714	.857	.857	
3	8	.625	.750	.750	.875	
3	9	.667	.667	.778	.889	.889
3	10	.600	.700	.800	.900	.900
3	12	.583	.667	.750	.833	.917
4	4	.750	.750	.750		
4	5	.600	.750	.800	.800	
4	6	.583	.667	.750	.833	.833
4	7	.607	.714	.750	.857	.857
4	8	.625	.625	.750	.875	.875
4	9	.556	.667	.750	.778	.889
4	10	.550	.650	.700	.800	.800
4	12	.583	.667	.667	.750	.833
4	16	.563	.625	.688	.750	.812
5	5	.600	.600	.800	.800	.800
5	6	.600	.667	.667	.833	.833
5	7	.571	.657	.714	.829	.857
5	8	.550	.625	.675	.800	.800
5	9	.556	.600	.689	.778	.800
5	10	.500	.600	.700	.700	.800
5	15	.533	.600	.667	.733	.733
5	20	.500	.550	.600	.700	.750
6	6	.500	.667	.667	.833	.833
6	7	.548	.571	.690	.714	.833
6	8	.500	.583	.667	.750	.750
6	9	.500	.556	.667	.722	.778
6	10	.500	.567	.633	.700	.733
6	12	.500	.583	.583	.667	.750
6	18	.444	.556	.611	.667	.722
6	24	.458	.500	.583	.625	.667
7	7	.571	.571	.714	.714	.714
7	8	.482	.589	.625	.732	.750
7	9	.492	.556	.635	.714	.746
7	10	.471	.557	.614	.700	.714
7	14	.429	.500	.571	.643	.714
7	28	.429	.464	.536	.607	.643
8	8	.500	.500	.625	.625	.750
8	9	.444	.542	.625	.667	.750
8	10	.475	.525	.575	.675	.700
8	12	.458	.500	.583	.625	.667
8	16	.438	.500	.563	.625	.625
8	32	.406	.438	.500	.563	.594

Table A23   Table of Critical Values for the Kolmogorov–Smirnov Test for Two Independent Samples (continued)

One-tailed		.10	.05	.025	.01	.005
Two-tailed		.20	.10	.05	.02	.01
n_1	n_2					
9	9	.444	.556	.556	.667	.667
9	10	.467	.500	.578	.667	.689
9	12	.444	.500	.556	.611	.667
9	15	.422	.489	.533	.600	.644
9	18	.389	.444	.500	.556	.611
9	36	.361	.417	.472	.528	.556
10	10	.400	.500	.600	.600	.700
10	15	.400	.467	.500	.567	.633
10	20	.400	.450	.500	.550	.600
10	40	.350	.400	.450	.500	.576
11	11	.454	.454	.545	.636	.636
12	12	.417	.417	.500	.583	.583
12	15	.383	.450	.500	.550	.583
12	16	.375	.438	.479	.542	.583
12	18	.361	.417	.472	.528	.556
12	20	.367	.417	.467	.517	.567
13	13	.385	.462	.462	.538	.615
14	14	.357	.429	.500	.500	.571
15	15	.333	.400	.467	.467	.533
16	16	.375	.375	.438	.500	.563
17	17	.353	.412	.412	.471	.529
18	18	.333	.389	.444	.500	.500
19	19	.316	.368	.421	.473	.473
20	20	.300	.350	.400	.450	.500
21	21	.286	.333	.381	.429	.476
22	22	.318	.364	.364	.454	.454
23	23	.304	.348	.391	.435	.435
24	24	.292	.333	.375	.417	.458
25	25	.280	.320	.360	.400	.440
For all other sample sizes		$1.07K$	$1.22K$	$1.36K$	$1.52K$	$1.63K$

$$\text{Where: } K = \sqrt{\frac{n_1 + n_2}{n_1 n_2}}$$

Table A24 Table of Critical Values for the Jonckheere–Terpstra Test Statistic

One-tailed critical J values listed are listed in relation to the alpha values noted in the table. The J value at a given level of significance is also the critical two-tailed value at a probability that is twice the value indicated for *alpha* (e.g., the value recorded for *alpha* = .05 represents the tabled critical one-tailed .05 value, as well as the tabled critical two-tailed .10 value).

Sample sizes			alpha .05	.025	.01	.005	Sample sizes			alpha .05	.025	.01	.005
2	2	2	11	12	–	–	3	6	6	52	55	58	60
2	2	3	14	15	16	16	3	6	7	58	61	66	67
2	2	4	17	18	19	20	3	6	8	64	68	71	74
2	2	5	20	21	22	23	3	7	7	65	68	72	74
2	2	6	23	24	26	27	3	7	8	71	75	79	82
2	2	7	26	28	29	30	3	8	8	79	82	87	90
2	2	8	29	31	33	34	4	4	4	36	38	40	42
2	3	3	18	19	20	21	4	4	5	42	44	46	48
2	3	4	21	23	24	25	4	4	6	47	49	52	54
2	3	5	25	26	28	29	4	4	7	52	55	58	60
2	3	6	28	30	32	33	4	4	8	59	61	64	66
2	3	7	30	34	36	37	4	5	5	48	50	53	55
2	3	8	36	38	40	41	4	5	6	54	56	59	62
2	4	4	26	27	29	30	4	5	7	59	63	66	68
2	4	5	30	31	33	34	4	5	8	65	69	72	75
2	4	6	34	36	38	39	4	6	6	60	63	67	69
2	4	7	38	40	43	44	4	6	7	66	70	74	76
2	4	8	42	45	47	49	4	6	8	73	77	81	84
2	5	5	35	36	39	40	4	7	7	74	77	82	84
2	5	6	39	41	44	45	4	7	8	81	85	89	92
2	5	7	44	47	49	51	4	8	8	88	93	98	101
2	5	8	49	52	54	56	5	5	5	54	57	60	62
2	6	6	45	47	50	52	5	5	6	61	64	67	70
2	6	7	50	53	56	58	5	5	7	67	71	72	77
2	6	8	55	58	62	64	5	5	8	74	77	81	84
2	7	7	56	59	62	65	5	6	6	68	71	75	77
2	7	8	62	65	69	71	5	6	7	75	79	83	85
2	8	8	69	72	76	79	5	6	8	82	86	90	93
3	3	3	22	23	25	25	5	7	7	82	86	91	94
3	3	4	26	28	29	30	5	7	8	90	94	99	103
3	3	5	30	32	34	35	5	8	8	98	103	108	112
3	3	6	34	36	39	40	6	6	6	75	79	83	86
3	3	7	39	41	43	45	6	6	7	83	87	92	95
3	3	8	43	45	48	49	6	6	8	91	95	100	103
3	4	4	31	33	35	36	6	7	7	91	96	101	104
3	4	5	36	38	40	41	6	7	8	100	104	110	113
3	4	6	40	43	45	47	6	8	8	108	113	119	123
3	4	7	45	48	50	52	7	7	7	100	105	110	114
3	4	8	50	53	56	57	7	7	8	109	114	120	123
3	5	5	41	43	46	47	7	8	8	118	124	130	134
3	5	6	46	49	52	54	8	8	8	128	134	140	145
3	5	7	52	55	58	60							
3	5	8	57	60	64	66							

Table A24   Table of Critical Values for the Jonckheere–Terpstra Test Statistic
(continued)

Sample sizes	alpha			
	.05	.025	.01	.005
2 2 2 2	19	21	22	23
2 2 2 2 2	30	32	33	35
2 2 2 2 2 2	43	45	47	49
3 3 3 3	40	42	44	45
3 3 3 3 3	62	65	69	71
3 3 3 3 3 3	90	94	98	101
4 4 4 4	67	70	73	76
4 4 4 4 4	106	110	116	119
4 4 4 4 4 4	154	160	167	171
5 5 5 5	100	105	110	114
5 5 5 5 5	160	167	174	179
5 5 5 5 5 5	234	242	252	258
6 6 6 6	141	147	154	158
6 6 6 6 6	226	235	244	251
6 6 6 6 6 6	330	342	354	363

Table A25 Table of Critical Values for the Page Test Statistic

One-tailed critical L values are listed in relation to the alpha values noted in the table. The L value at a given level of significance is also the critical two-tailed value at a probability that is twice the value indicated for *alpha* (e.g., the value recorded for *alpha* = .05 represents the tabled critical one-tailed .05 value, as well as the tabled critical two-tailed .10 value).

| | $k = 3$ | | | | $k = 4$ | | | | $k = 5$ | | |
| | alpha | | | | alpha | | | | alpha | | |
n	.05	.01	.001		.05	.01	.001		.05	.01	.001
2	28				58	60			103	106	109
3	41	42			84	87	89		150	155	160
4	54	55	56		111	114	117		197	204	210
5	66	68	70		137	141	145		244	251	259
6	79	81	83		163	167	172		291	299	307
7	91	93	96		189	193	198		338	346	355
8	104	106	109		214	220	225		384	393	403
9	116	119	121		240	246	252		431	441	451
10	128	131	134		266	272	278		477	487	499
11	141	144	147		293	298	305		523	534	546
12	153	156	160		317	324	331		570	581	593
13	165	169	172								
14	178	181	185								
15	190	194	197								
16	202	206	210								
17	215	218	223								
18	227	231	235								
19	239	243	248								
20	251	256	260								

Table A25 Table of Critical Values for the Page Test Statistic
(continued)

| | k = 6 | | | | k = 7 | | | | k = 8 | | |
| | alpha | | | | alpha | | | | alpha | | |
n	.05	.01	.001		.05	.01	.001		.05	.01	.001
2	166	173	178		252	261	269		362	376	388
3	244	252	260		370	382	394		532	549	567
4	321	331	341		487	501	516		701	722	743
5	397	409	420		603	620	637		869	893	917
6	474	486	499		719	737	757		1037	1063	1090
7	550	563	577		835	855	876		1204	1232	1262
8	625	640	655		950	972	994		1371	1401	1433
9	701	717	733		1065	1088	1113		1537	1569	1603
10	777	793	811		1180	1205	1230		1703	1736	1773
11	852	869	888		1295	1321	1348		1868	1905	1943
12	928	946	965		1410	1437	1465		2035	2072	2112

| | k = 9 | | | | k = 10 | | |
| | alpha | | | | alpha | | |
n	.05	.01	.001		.05	.01	.001
2	500	520	544		670	696	726
3	736	761	790		987	1019	1056
4	971	999	1032		1301	1339	1382
5	1204	1236	1273		1614	1656	1704
6	1436	1472	1512		1927	1972	2025
7	1668	1706	1750		2238	2288	2344
8	1900	1940	1987		2549	2602	2662
9	2131	2174	2223		2859	2915	2980
10	2361	2407	2459		3169	3228	3296
11	2592	2639	2694		3478	3541	3612
12	2822	2872	2929		3788	3852	3927

Table A26 Table of Extreme Studentized Deviate Outlier Statistic

	alpha			alpha			alpha	
n	.05	.01	*n*	.05	.01	*n*	.05	.01
5	1.72	1.76	25	2.82	3.14	45	3.09	3.44
6	1.89	1.97	26	2.84	3.16	46	3.09	3.45
7	2.02	2.14	27	2.86	3.18	47	3.10	3.46
8	2.13	2.28	28	2.88	3.20	48	3.11	3.46
9	2.21	2.39	29	2.89	3.22	49	3.12	3.47
10	2.29	2.48	30	2.91	3.24	50	3.13	3.48
11	2.36	2.56	31	2.92	3.25	60	3.20	3.56
12	2.41	2.64	32	2.94	3.27	70	3.26	3.62
13	2.46	2.70	33	2.95	3.29	80	3.31	3.67
14	2.51	2.75	34	2.97	3.30	90	3.35	3.72
15	2.55	2.81	35	2.98	3.32	100	3.38	3.75
16	2.59	2.85	36	2.99	3.33	150	3.52	3.89
17	2.62	2.90	37	3.00	3.34	200	3.61	3.98
18	2.65	2.93	38	3.01	3.36	250	3.67	4.04
19	2.68	2.97	39	3.03	3.37	300	3.72	4.09
20	2.71	3.00	40	3.04	3.38	350	3.77	4.14
21	2.73	3.03	41	3.05	3.39	400	3.80	4.17
22	2.76	3.06	42	3.06	3.40	450	3.84	4.20
23	2.78	3.08	43	3.07	3.41	500	3.86	4.23
24	2.80	3.11	44	3.08	3.43			

Table A27 Table of Durbin–Watson Test Statistic

One-tailed alpha = .05 values for d_L and d_U

	$k=1$		$k=2$		$k=3$		$k=4$		$k=5$	
n	d_L	d_U	d_L	d_U	d_L	d_U	d_L	d_U	d_L	d_U
15	1.08	1.36	0.95	1.54	0.82	1.75	0.69	1.97	0.56	2.21
16	1.10	1.37	0.98	1.54	0.86	1.73	0.74	1.93	0.62	2.15
17	1.13	1.38	1.02	1.54	0.90	1,71	0.78	1.90	0.67	2.10
18	1.16	1.39	1.05	1.53	0.93	1.69	0.82	1.87	0.71	2.06
19	1.18	1.40	1.08	1.53	0,97	1.68	0.86	1.85	0.75	2.02
20	1.20	1.41	1.10	1.54	1.00	1.68	0.90	1.83	0.79	1.99
21	1.22	1.42	1.13	1.54	1.03	1.67	0.93	1.81	0.83	1.96
22	1.24	1.43	1.15	1.54	1.05	1.66	0.96	1.80	0.86	1.94
23	1.26	1.44	1.17	1.54	1.08	1.66	0.99	1.79	0.90	1.92
24	1.27	1.45	1.19	1.55	1.10	1.66	1.01	1.78	0.93	1.90
25	1.29	1.45	1.21	1.55	1.12	1.66	1.04	1.77	0.95	1.89
26	1.30	1.46	1.22	1.55	1.14	1.65	1.06	1.76	0.98	1.88
27	1.32	1.47	1.24	1.56	1.16	1.65	1.08	1.76	1.01	1.86
28	1.33	1.48	1.26	1.56	1.18	1.65	1.10	1.75	1.03	1.85
29	1.34	1.48	1.27	1.56	1.20	1.65	1.12	1.74	1.05	1.84
30	1.35	1.49	1.28	1.57	1.21	1.65	1.14	1.74	1.07	1.83
31	1.36	1.50	1.30	1.57	1.23	1.65	1.16	1.74	1.09	1.83
32	1.37	1.50	1.31	1.57	1.24	1.65	1.18	1.73	1.11	1.82
33	1.38	1.51	1.32	1.58	1.26	1.65	1.19	1.73	1.13	1.81
34	1.39	1.51	1.33	1.58	1.27	1.65	1.21	1.73	1.15	1.81
35	1.40	1.52	1.34	1.58	1.28	1.65	1.22	1.73	1.16	1.80
36	1.41	1.52	1.35	1.59	1.29	1.65	1.24	1.73	1.18	1.80
37	1.42	1.53	1.36	1.59	1.31	1.66	1.25	1.72	1.19	1.80
38	1.43	1.54	1.37	1.59	1.32	1.66	1.26	1.72	1.21	1.79
39	1.43	1.54	1.38	1.60	1.33	1.66	1.27	1.72	1.22	1.79
40	1.44	1.54	1.39	1.60	1.34	1.66	1.29	1.72	1.23	1.79
45	1.48	1.57	1.43	1.62	1.38	1.67	1.34	1.72	1.29	1.78
50	1.50	1.59	1.46	1.63	1.42	1.67	1.38	1.72	1.34	1.77
55	1.53	1.60	1.49	1.64	1.45	1.68	1.41	1.72	1.38	1.77
60	1.55	1.62	1.51	1.65	1.48	1.69	1.44	1.73	1.41	1.77
65	1.57	1.63	1.54	1.66	1.50	1.70	1.47	1.73	1.44	1.77
70	1.58	1.64	1.55	1.67	1.52	1.70	1.49	1.74	1.46	1.77
75	1.60	1.65	1.57	1.68	1.54	1.71	1.51	1.74	1.49	1.77
80	1.61	1.66	1.59	1.69	1.56	1.72	1.53	1.74	1.51	1.77
85	1.62	1.67	1.60	1.70	1.57	1.72	1.55	1.75	1.52	1.77
90	1.63	1.68	1.61	1.70	1.59	1.73	1.57	1.75	1.54	1.78
95	1.64	1.69	1.62	1.71	1.60	1.73	1.58	1.75	1.56	1.78
100	1.65	1.69	1.63	1.72	1.61	1.74	1.59	1.76	1.57	1.78

Table A27 Table of Durbin–Watson Test Statistic (continued)

One-tailed alpha = .01 values for d_L and d_U

n	$k=1$		$k=2$		$k=3$		$k=4$		$k=5$	
	d_L	d_U	d_L	d_U	d_L	d_U	d_L	d_U	d_L	d_U
15	0.81	1.07	0.70	1.25	0.59	1.46	0.49	1.70	0.39	1.96
16	0.84	1.09	0.74	1.25	0.63	1.44	0.53	1.66	0.44	1.90
17	0.87	1.10	0.77	1.25	0.67	1.43	0.57	1.63	0.48	1.85
18	0.90	1.12	0.80	1.26	0.71	1.42	0.61	1.60	0.52	1.80
19	0.93	1.13	0,83	1.26	0.74	1.41	0.65	1.58	0.56	1.77
20	0.95	1.15	0.86	1.27	0.77	1.41	0.68	1.57	0.60	1.74
21	0.97	1.16	0.89	1.27	0.80	1.41	0.72	1.55	0.63	1.71
22	1.00	1.17	0.91	1.28	0.83	1.40	0.75	1.54	0.66	1.69
23	1.02	1.19	0.94	1.29	0.86	1.40	0.77	1.53	0.70	1.67
24	1.04	1.20	0.96	1.30	0.88	1.41	0.80	1.53	0.72	1.66
25	1.05	1.21	0.98	1.30	0.90	1.41	0.83	1.52	0.75	1.65
26	1.07	1.22	1.00	1.31	0.93	1.41	0.85	1.52	0.78	1.64
27	1.09	1.23	1.02	1.32	0.95	1.41	0.88	1.51	0.81	1.63
28	1.10	1.24	1.04	1.32	0.97	1.41	0.90	1.51	0.83	1.62
29	1.12	1.25	1.05	1.33	0.99	1.42	0.92	1.51	0.85	1.61
30	1.13	1.26	1.07	1.34	1.01	1.42	0.94	1.51	0.88	1.61
31	1.15	1.27	1.08	1.34	1.02	1.42	0.96	1.51	0.90	1.60
32	1.16	1.28	1.10	1.35	1.04	1.43	0.98	1.51	0.92	1.60
33	1.17	1.29	1.11	1.36	1.05	1.43	1.00	1.51	0.94	1.59
34	1.18	1.30	1.13	1.36	1.07	1.43	1.01	1.51	.095	1.59
35	1.19	1.31	1.14	1.37	1.08	1.44	1.03	1.51	0.97	1.59
36	1.21	1.32	1.15	1.38	1.10	1.44	1.04	1.51	0.99	1.59
37	1.22	1.32	1.16	1.38	1.11	1.45	1.06	1.51	1.00	1.59
38	1.23	1.33	1.18	1.39	1.12	1.45	1.07	1.52	1.02	1.58
39	1.24	1.34	1.19	1.39	1.14	1.45	1.09	1.52	1.03	1.58
40	1.25	1.34	1.20	1.40	1.15	1.46	1.10	1.52	1.05	1.58
45	1.29	1.38	1.24	1.42	1.20	1.48	1.16	1.53	1.11	1.58
50	1.32	1.40	1.28	1.45	1.24	1.49	1.20	1.54	1.16	1.59
55	1.36	1.43	1.32	1.47	1.28	1.51	1.25	1.55	1.21	1.59
60	1.38	1.45	1.35	1.48	1.32	1.52	1.28	1.56	1.25	1.60
65	1.41	1.47	1.38	1.50	1.35	1.53	1.31	1.57	1.28	1.61
70	1.43	1.49	1.40	1.52	1.37	1.55	1.34	1.58	1.31	1.61
75	1.45	1.50	1.42	1.53	1.39	1.56	1.37	1.59	1.34	1.62
80	1.47	1.52	1.44	1.54	1.42	1.57	1.39	1.60	1.36	1.62
85	1.48	1.53	1.46	1.55	1.43	1.58	1.41	1.60	1.39	1.63
90	1.50	1.54	1.47	1.56	1.45	1.59	1.43	1.61	1.41	1.64
95	1.51	1.55	1.49	1.57	1.47	1.60	1.45	1.62	1.42	1.64
100	1.52	1.56	1.50	1.58	1.48	1.60	1.46	1.63	1.44	1.65

Index